T0336507

ORNITHOLOGY

ORNITHOLOGY

Foundation, Analysis, and Application

EDITED BY

Michael L. Morrison
DEPARTMENT OF WILDLIFE AND FISHERIES SCIENCES
TEXAS A&M UNIVERSITY

Amanda D. Rodewald
CORNELL LAB OF ORNITHOLOGY AND DEPARTMENT OF NATURAL RESOURCES
CORNELL UNIVERSITY

Gary Voelker
DEPARTMENT OF WILDLIFE AND FISHERIES SCIENCES
TEXAS A&M UNIVERSITY

Melanie R. Colón
NATURAL RESOURCES INSTITUTE
TEXAS A&M UNIVERSITY

Jonathan F. Prather
DEPARTMENT OF ZOOLOGY AND PHYSIOLOGY
UNIVERSITY OF WYOMING

JOHNS HOPKINS UNIVERSITY PRESS   BALTIMORE

Johns Hopkins University Press
2715 North Charles Street
Baltimore, Maryland 21218-4363
www.press.jhu.edu

Library of Congress Cataloging-in-Publication Data

Names: Morrison, Michael L., editor.
Title: Ornithology : foundation, analysis, and application / edited by Michael L. Morrison, Amanda D. Rodewald, Gary Voelker, Melanie R. Colón, Jonathan F. Prather
Description: Baltimore : Johns Hopkins University Press, 2018. | Includes bibliographical references and index.
Identifiers: LCCN 2017019811|
   ISBN 9781421424712 (hardcover : alk. paper) | ISBN 9781421424729 (electronic) | ISBN 1421424711 (hardcover : alk. paper) | ISBN 142142472X (electronic)
Subjects: LCSH: Ornithology. | Birds.
Classification: LCC QL673 .O76 2018 | DDC 598—dc23
LC record available at https://lccn.loc.gov/2017019811

A catalog record for this book is available from the British Library.

Special discounts are available for bulk purchases of this book. For more information, please contact Special Sales at 410-516-6936 or specialsales@press.jhu.edu.

Johns Hopkins University Press uses environmentally friendly book materials, including recycled text paper that is composed of at least 30 percent post-consumer waste, whenever possible.

Contents

Contributors

EDITORS

Michael L. Morrison
Texas A&M University, College Station

Amanda D. Rodewald
Cornell University

Gary Voelker
Texas A&M University, College Station

Melanie R. Colón
Texas A&M University, College Station

Jonathan F. Prather
University of Wyoming

Editorial Assistant
Elliott Foxley
Texas A&M University, College Station

AUTHORS

Peter Arcese
University of British Columbia, Canada

George E. Bentley
University of California, Berkeley

Lori A. Blanc
Virginia Polytechnic Institute

William M. Block
Rocky Mountain Research Station,
US Forest Service

Alice Boyle
Kansas State University

Leonard A. Brennan
Texas A&M University, Kingsville

Luke K. Butler
The College of New Jersey

Zac Cheviron
University of Montana

Luis M. Chiappe
Natural History Museum of
Los Angeles County

Caren B. Cooper
North Carolina State University
North Carolina Museum of Natural
Sciences

Robert J. Cooper
University of Georgia

Jamie M. Cornelius
Eastern Michigan University

Carlos Martinez Del Rio
University of Wyoming

John Dumbacher
California Academy of Sciences

Shannon Farrell
State University of New York
College of Environmental Science
and Forestry

Maureen Flannery
California Academy of Sciences

Geoffrey Geupel
Point Blue Conservation Science

Patricia Adair Gowaty
University of California, Los Angeles

Thomas P. Hahn
University of California, Davis

Ashley M. Heers
American Museum of Natural History

Fritz Hertel
California State University, Northridge

Geoffrey E. Hill
Auburn University

Matthew Johnson
Humboldt State University

Lukas F. Keller
University of Zurich, Switzerland

Dylan C. Kesler
Institute for Bird Populations

Pablo Sabat Kirkwood
Universidad de Chile, Chile

John Klicka
University of Washington

Christopher A. Lepczyk
Auburn University

Ashley M. Long
Texas A&M University, College Station

Scott R. Loss
Oklahoma State University

Graham R. Martin
University of Birmingham, UK

John M. Marzluff
University of Washington

Susan B. McRae
East Carolina University

Timothy J. O'Connell
Oklahoma State University

Jen C. Owen
Michigan State University

Marco Pavia
University of Torino, Italy

Jeffrey Podos
University of Massachusetts, Amherst

Lars Pomara
Southern Research Station,
US Forest Service

Marco Restani
St. Cloud State University

Alejandro Rico-Guevara
University of California, Berkeley

Vanya G. Rohwer
Cornell University Museum
of Vertebrates

Matthias Starck
Ludwig-Maximilians Universität
Munchën, Germany

Michael W. Strohbach
Technische Universität Braunschweig,
Germany

S. Mažeika P. Sullivan
Ohio State University

Diego Sustaita
California State University, San Marcos

Kerri T. Vierling
University of Idaho

Margaret A. Voss
Syracuse University

Jeff R. Walters
Virginia Polytechnic Institute

Paige S. Warren
University of Massachusetts, Amherst

Elisabeth B. Webb
US Geological Survey
Missouri Cooperative Fish and Wildlife
Research Unit

Michael S. Webster
Cornell University

Eric M. Wood
California State University, Los Angeles

Robert M. Zink
University of Nebraska

Benjamin Zuckerberg
University of Wisconsin–Madison

Preface

When Vincent Burke, then executive editor at Johns Hopkins University Press, first approached the chief editor, Michael Morrison, to lead the production of an ornithology textbook, Morrison knew he had to assemble a diverse team to accomplish such a task. This team of editors spanned a wide range of expertise and career stages, ranging from a finishing doctoral student through established scientists. The editors then spent considerable time developing a detailed outline of chapters and selecting leading experts in each subject area who would provide the latest information on research and applications in ornithology.

Our diverse team of chapter authors includes scientists from six different countries, all of whom frequently engage with students, teach or have taught ornithology, and have active research programs. In the course of creating each chapter, authors submitted detailed outlines and then drafted manuscripts that were reviewed by the volume editors. At each stage, we emphasized the breadth of topics, the depth of coverage among topics, and the suitability of the writing for the intended readership. Independent peer reviewers also reviewed each chapter and commented on depth, accuracy, and style. Although we strove, at a broad level, for consistency in writing styles, we also recognized that chapters are most compelling when authors can speak in their own voices. The result is a text in which the voice of each author is preserved, yet the content is still easily approachable for students and experts alike.

Although the textbook is intended for undergraduates and graduate students, it is also written to be useful for professionals who are experts in one or a few subject areas (e.g., ecology and behavior) and want to strengthen their understanding of others (e.g., anatomy and physiology). Each chapter includes both historical perspective and contemporary knowledge so that readers can appreciate how the understanding of each subject area has developed over time. We emphasize the way in which the diversity of research led by ornithologists reflects not only innovation in methodological approaches and tools but also maturation of our ideas and the guiding paradigms that shape how we frame questions and interpret data.

Our intent is that each chapter will help the reader come away with an awareness of the relevant foundational ideas and emerging topics in that subject area. As such, each chapter provides a window into the dynamic field of ornithology that can serve as a starting point from which topics can be more deeply explored. We assembled the chapters into seven parts to help any reader, student, or instructor move through the material in an organized manner.

Knowing that many readers will be in early stages in their academic and professional careers, we specifically included a chapter that introduces readers to the ornithological profession, highlighting examples from different career tracks. We further incorporate into each chapter boxes that highlight and personalize the contributions of many individual ornithologists. These boxes are not intended to portray the complete history of ornithology, but rather to provide examples of major contributors to ornithological subdisciplines so students can become acquainted with some of the people responsible for the discoveries cited in the book. To facilitate understanding and recall, each chapter ends with a summary of the material covered, a discussion of potential management and conservation applications, and several suggested study questions.

The editors of this textbook are indebted to numerous colleagues, friends, and family members for ensuring that this project came to a successful completion. Most important are the nearly 60 chapter authors who spent countless hours writing their chapters, obtaining and securing permission for high-quality images and other materials, and maintaining positive spirits as they were being continually pestered by us.

The editors specifically acknowledge the guidance provided by our former editor, Vincent Burke. As someone who once taught ornithology, Dr. Burke combined years of experience working in the publishing field with years of experience as a field biologist. Tiffany Gasbarrini took over the editing duties from Burke and saw the project to completion; her guidance was extremely valuable. Meagan Szekely and Sahara Clement, editorial assistants at JHUP, were extremely knowledgeable and provided detailed guidance on a continuing basis.

Elliott Foxley, a graduate student at Texas A&M University, served as our editorial assistant and worked directly with chapter authors and JHUP to ensure that all formatting requirements were met. Ms. Foxley did a substantial amount of work on this project and was largely responsible for pulling the package together for publication.

We also thank Carlos Martinez Del Rio for helping us during the early phases of this project by assisting with development of the initial outlines. We thank Joyce VanDeWater for producing high-quality images for many chapters. We thank Kimberly Giambattisto, production editor, Westchester Publishing Services, for her diligent efforts at insuring the copyediting and production process ran smoothly.

Here we list additional acknowledgments provided by authors for individual chapters: *Chapter 2:* Stephanie Abramowicz produced the illustrations, and Tyler Hayden assisted with editorial duties. *Chapter 5:* The authors thank Daniel Blackburn for discussion of chapter content. *Chapter 6:* A heartfelt thank-you to Susan Ellis-Felege for constructive comments on the chapter. *Chapter 8:* The authors thank John C. Wingfield and Stephan J. Schoech for comments on a draft of the chapter. The authors also are grateful to Julia L. Coombs for expert preparation of the figures. *Chapter 9:* Sievert Rohwer provided helpful comments and discussions on content and organization of the chapter, and Philina English provided helpful comments on early sections of the chapter. *Chapter 10:* Many thanks to Ken Dial, Bret Tobalske, Tom Martin, Brandon Jackson, John Hutchinson, and Mark Norell—for an introduction to the world of birds and bird flight. *Chapter 13:* The author is grateful to his students and colleagues with whom he has had many conversations about the content of the chapter. This work is part of the Broader Impacts of an NSF CAREER Award to the author (NSF IOS 1453084). *Chapter 15:* The authors thank Thomas W. Sherry for his very thoughtful and constructive review. They extend special thanks to Dave Furseth for his vast contribution of exquisite images, as well as to Glenn Conlan, Pat Gaines, Ron Dudley, Earl Orf, Thomas Smith, and Kerri Farley for their generous photographic contributions. Thanks also to Bruno Ens, Carole A. Bonga Tomlinson, Ella Cole, Angelique van

der Leeuw, and Jorge E. Schondube-Friedewold for their permission to reproduce their figures. Finally, the authors thank Margaret A. Rubega for inspiring the chapter. *Chapter 16:* The author thanks Lee Drickamer, Mark Friedman, Geoff Hill, and David Kinneer for the use of original photographs; Steve Hubbell, Gordon Orians, James Briskie and Steve Emlen for allowing reproduction of their original work; and Patty Brennan, Dee Boesma, Malcolm Gordon, Brant Faircloth, and Gordon Orians for writing content and box text. *Chapter 17:* This chapter was written while the author was visiting the Museum of Vertebrate Zoology, University of California, Berkeley. She thanks Michael Nachman, Eileen Lacey, Rauri Bowie, Steven Beissinger, and the staff of the museum for hosting her. Social Behavior Lab Group members provided helpful suggestions during the planning stages. Contributions of photographs and artwork from Nick Athanas, Nick Davies, Janis Dickinson, Walt Koenig, Ben Hatchwell, Christie Riehl, Mark Stanback, and Warwick Tarboton were particularly appreciated. *Chapter 18:* The authors thank Dr. Tim Bean and the numerous graduate students associated with Humboldt State University's graduate Wildlife Habitat Ecology class for stimulating discussion of the topics in this chapter. *Chapter 19:* The author thanks students of the Kansas State University spring 2016 Ornithology class for discussing the chapter. Also thanks to Suzanne Replogle for administrative assistance and for financial support from KSU (contribution no. 17-124-B from the Kansas Agricultural Experiment Station). *Chapter 20:* The authors are grateful for support from the FRBC Chair in Conservation Biology and Department of Forest and Conservation Science at the University of British Columbia, and from the Department of Evolutionary Biology and Environmental Studies, University of Zurich. *Chapter 21:* Lenny Brennan was supported by the C. C. Winn Endowed Chair in the Richard M. Kleberg, Jr. Center for Quail Research. *Chapter 23:* Thanks to Elizabeth Throckmorton for her invaluable assistance reviewing and editing final drafts of this chapter; Zak Pohlen, who gave permission to use photos; Rebecca Tegtmeyer for her work on graphic designs; and Dana Hawley and Carter Atkinson for their time and contributions to the chapter. *Chapter 24:* The authors are grateful for the assistance of Molly H. Polk and Kenneth R. Young in preparing the Pastoruri glacier images. While working on this chapter, Ben Zuckerberg received support from NSF EF-1340632 and NASA NNX14AC36G. *Chapter 25:* The author thanks the James W. Ridgeway Professorship at the University of Washington for support. *Chapter 26:* Thanks to R. Dwayne Elmore, Joelle Gehring, Daniel Klem Jr., and Peter G. Saenger for contributing boxes to this chapter and to the thousands of researchers and citizens who have collected field data to build an understanding of human-caused bird mortality. Scott R. Loss and Timothy J. O'Connell were funded during writing by the Oklahoma Agricultural Experiment Station through NIFA/USDA Hatch and McIntire-Stennis funds, respectively. *Chapter 27:* The Institute for Bird Populations sponsored Dylan C. Kesler's contributions to this text. Elisabeth B. Webb was supported by the Missouri Cooperative Fish and Wildlife Research Unit, which is jointly sponsored by the Missouri Department of Conservation, the University of Missouri, the US Fish and Wildlife Service, the US Geological Survey, and the Wildlife Management Institute. The Harold H. Bailey fund and Department of Biological Sciences at Virginia Polytechnic Institute and State University sponsored Jeffrey R. Walters's contributions to this text. *Chapter 28:* Multiple undergraduate and graduate students provided valuable feedback on this chapter, and the authors very much appreciate the generosity of the many people who provided photos or shared graphics. *Chapter 29:* Material covering Phoenix, Arizona, is based on work supported by the National Science Foundation under grant no. BCS-1026865. Michael W. Strohbach was supported by the EU FP7 collaborative project GREEN SURGE (FP7-ENV.2013.6.2-5-603567) and the MWK-Verbundvorhaben "METAPOLIS—eine interdisziplinäre Plattform für eine nachhaltige Entwicklung der Stadt-Land-Beziehungen in Niedersachsen" (Förderkennz. ZN3121). *Chapter 31:* The authors thank the California Academy of Sciences, the Makray Foundation, and Point Blue Conservation Science for financial support.

REVIEWERS

We sincerely thank the many individuals who critically and constructively reviewed drafts of the chapters: Jim Belthoff, Bill Block, Rauri C. K. Bowie, Kevin Burns, Chris Butler, Adriana Chaparro, Chris Clark, Robert J. Cooper, Charles Deeming, Elizabeth Derryberry, Susan Ellis-Felege, Vicki Garcia, Tom Gardali, Erick Greene, Ben Hatchwell, Jeffrey Hepinstall-Cymerman, Mark Herse, Richard Hutto, Danny Ingold, Nina Karnovsky, John Klicka, Eileen Lacey, Hayley Christine Lanier, Brenda Larison, Dave Manuwal, Ron Mumme, Troy Murphy, Charlie Nilon, Brian Olsen, Diana Outlaw, Ken Petren, Anna Pidgeon, Steve Portugal, J. Powell, Peter Pyle, Jim Reynolds, Dina Roberts, Sievert Rohwer, Steve Schoech, Michael Schummer, Jim Sedinger, Thomas W. Sherry, Elsie Shogren, Brian Smith, Nathan Smith, Robert J. Smith, Garth Spellman, Greg Spicer, Mark A. Stanback, James R. Stewart, Jeff Stratford, Allan Strong, David Swanson, Sarah Turner, Laura Wilkinson, Emily Williams, John Wingfield, Blair Wolf, Jared Wolfe, Stefan Woltmann, and Doug Wood.

PART I

THE BIG PICTURE

CHAPTER 1

What Makes a Bird?

Amanda D. Rodewald, Michael L. Morrison, Melanie R. Colón,
Gary Voelker, and Jonathan F. Prather

In the grand diversity of life, birds have long held a special place in the interest and imagination of humans. Part of that interest is because birds provide important resources—being directly consumed as food or providing materials such as feathers for bedding and winter clothing, or bones for making tools. Additionally, birds may be useful guides to other resources. The Greater Honeyguide (*Indicator indicator*), for example, has been used by generations of people in sub-Saharan Africa to find bee colonies. Birds can also be useful indicators of environmental quality. Historically, coal miners brought caged canaries into mines to alert them of dangerous air conditions—the proverbial "canary in a coal mine"—and, in the more recent past, birds like the Peregrine Falcon (*Falco peregrinus*; fig. 1.1) alerted us to the harmful effects of the once commonly used insecticide DDT.

While we clearly derive many benefits from birds, our interest in them transcends the practical and is heightened by their conspicuous and ubiquitous presence in the world around us. Birds have been widely depicted in our culture for thousands of years in a variety of forms, including song, dance, poetry, paintings, and fashion. Our artistic, cultural, and recreational connections to birds remain strong today. The millions of people engaging in bird-watching each year provide evidence of this connection.

Given the visibility and prominence of birds, almost everyone is familiar with them and can distinguish them from other groups of animals. Even so, many people would be hard-pressed to explain the defining attributes of birds. What makes a bird a bird?

CLASSIFICATION OF BIRDS

All life on Earth can be classified, or organized, into a series of hierarchical groups based on shared characteristics. This classification is called "taxonomy," and its father was Carl

Linnaeus, a Swedish biologist who developed the Linnaean classification system for naming organisms with two names that reflect those groupings (e.g., humans are *Homo sapiens*). More recently, Carl Woese and colleagues have used molecular tools to add additional resolution to taxonomic classification. Their work resulted in the addition of the classification of domain to divide cellular life into three groups, largely reflecting differences between prokaryotes and eukaryotes. Domains called Eubacteria and Archaebacteria (sometimes called Bacteria and Archaea, respectively) contain unicellular organisms with prokaryotic cells that lack nuclei. In contrast, the domain Eukaryota contains eukaryotic organisms with cells that are nucleated. Only protists (kingdom Pro-

Figure 1.1. The Peregrine Falcon (*Falco peregrinus*) was one of several species whose populations plummeted because of eggshell thinning associated with the pesticide DDT. Following the banning of the pesticide and intensive conservation efforts, its populations have rebounded. *Photo © Ian Davies / Macaulay Library at the Cornell Lab of Ornithology.*

tista), which includes the famed *Paramecium* and amoebas, have both unicellular and multicellular eukaryotes.

Among members of the domain Eukayota, the highest level of classification is the kingdom, of which there are five or six, depending on how one classifies bacteria (fig. 1.2). The remaining kingdoms are composed of multicellular eukaryotic organisms. Species in kingdom Plantae can make their own food (i.e., they are autotrophic), whereas those in kingdoms Fungi and Animalia must consume other organisms to survive (i.e., they are heterotrophic). Like the other members of kingdom Animalia, birds are multicellular and heterotrophic organisms that have eukaryotic cells.

Within kingdom Animalia, birds belong to phylum Chordata, which means that at some point in their development they have a notochord, a dorsal and hollow nerve cord, pharyngeal slits, and a post-anal tail. Birds also belong to the subphylum Vertebrata because they have vertebrae surrounding the nerve cord, a brain enclosed by a skull, an endoskeleton, a well-developed coelom that contains the viscera, a closed circulatory system with a ventral heart, a complete digestive system, and excretion via kidneys. At the class level of organization, animals are classified into several familiar groups—birds, mammals, reptiles, amphibians, jawless fish, cartilaginous fish, bony fish, and so on (fig. 1.3).

At the time of writing, there are 82 families and 28 orders within class Aves, and those groups comprise roughly 9,800

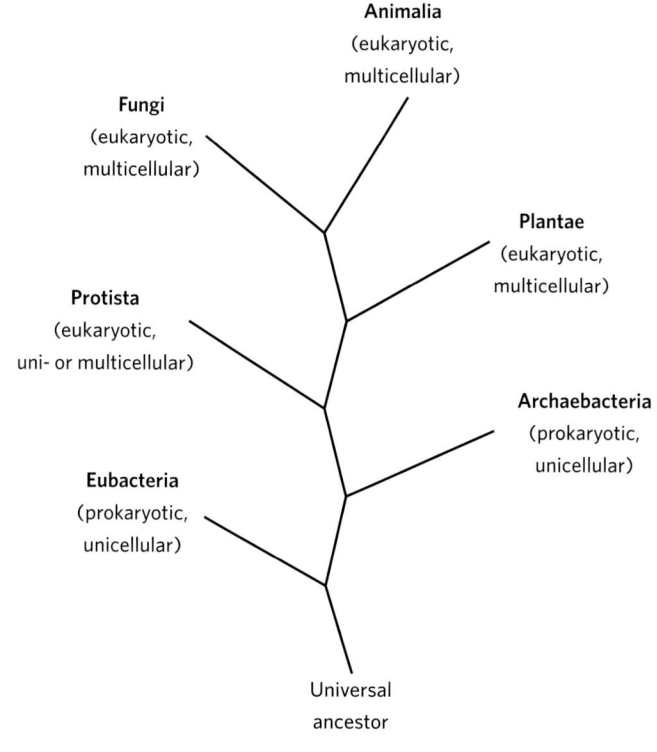

Figure 1.2. Kingdoms of life.

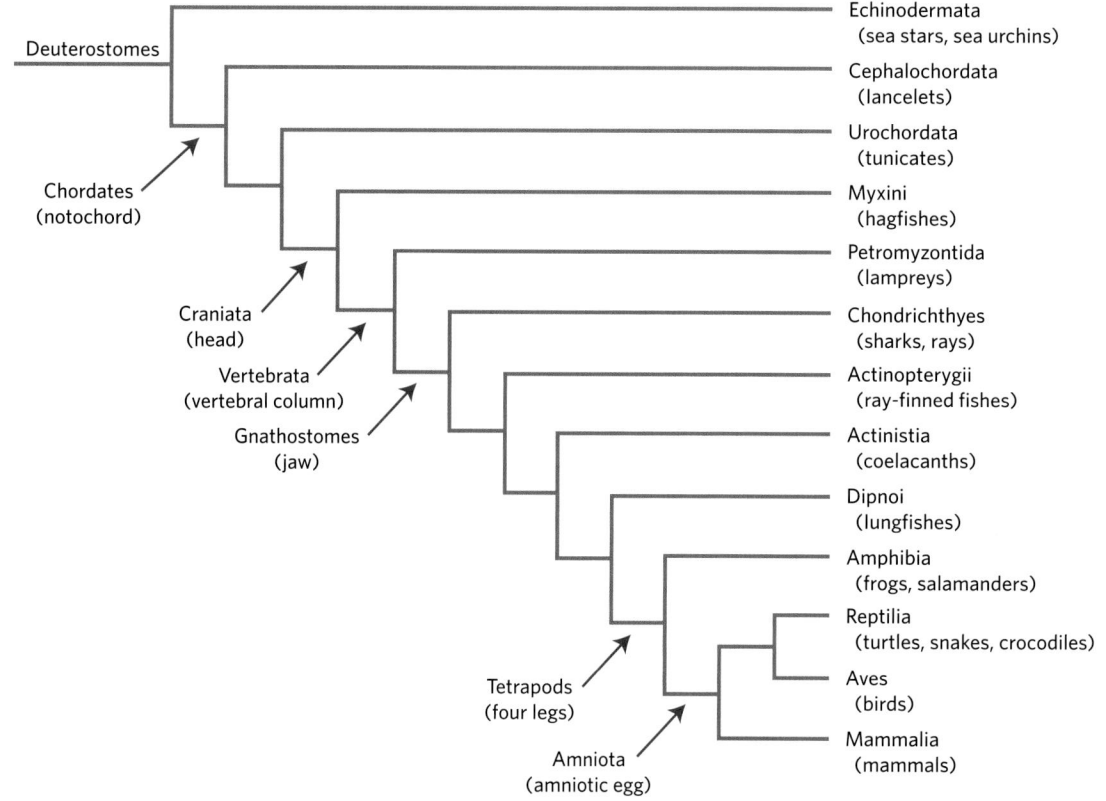

Figure 1.3. Phylogeny of chordates. Image modified from "Chordates" by OpenStax College, OpenStax-CNX, licensed under a Creative Commons Attribution 4.0 License. https://cnx.org/contents/VB2yhrAh@7/Chordates.

Figure 1.4. Taxonomic classification of the American Robin (*Turdus migratorius*), a common species in North America. *Photo © Ian Davies / Macaulay Library at the Cornell Lab of Ornithology.*

species of birds, each with a unique classification (fig. 1.4). You might wonder why we specify "at the time of writing" or mention *"roughly* 9,800 species." The reason is that taxonomy is a living field, and classifications change as we learn more about birds, improve our molecular techniques, and even discover new species. Species concepts can also impact how scientists assess how many species of birds there are. For example, proponents of the biological species concept tend to recognize 9,800 species, whereas advocates of the phylogenetic species concept would recognize many more species (perhaps even twice as many). You will read more about these concepts and their role in species recognition later in the book.

TRAITS SHARED WITH OTHER ANIMAL GROUPS

Have you ever heard someone refer to birds as living dinosaurs? Usually what the person means is that birds descended from reptilian ancestors during the Mesozoic (65 to 250 million years ago) and, later, from dinosaurs. That evolutionary history is why modern birds share a number of traits with reptiles, including:

Single occipital condyle. The occipital condyle is where the skull connects to the vertebral column. Both birds and reptiles have only one, whereas mammals and amphibians have two.

Single middle ear bone. Unlike mammals, which have three bones in the middle ear, birds and reptiles have only one bone in the middle ear.

Fused jawbones. The jaws of birds and reptiles comprise five fused bones. Mammals, in contrast, have a single jawbone.

Nucleated red blood cells. In humans and other groups, red blood cells lack nuclei. In contrast, red blood cells in birds and reptiles contain a nucleus.

Egg structure. Reptiles and birds, along with monotremes (i.e., three species of echidnas and the duckbill platypus [*Ornithorhynchus anatinus*]), lay eggs out of water and thus share a similar egg structure that requires a protective shell. They also share the same membranes: the amnion, which protects the embryo in a fluid-filled sac; the yolk sac, which contains food for the embryo; the chorion, which lines the inner surface of the shell and helps with gas exchange; and the allantois, which stores metabolic waste.

Egg tooth. Hatchling birds and reptiles have an egg tooth, or small hook on the beak, that helps the young break through the eggshell when hatching.

Uncinate processes of the ribs. Most species of birds and reptiles have extensions of bone from each rib that help strengthen the rib cage (fig. 1.5). These uncinate processes can also improve the effectiveness of muscles involved in respiration.

Even though birds are more closely related to reptiles, they share the following two key traits with mammals:

Endothermic. Birds and mammals can regulate and maintain warm body temperatures. Although some species of sharks and other fishes (e.g., tuna) are endotherms, birds and mammals represent the only classes of animals that are universally endothermic.

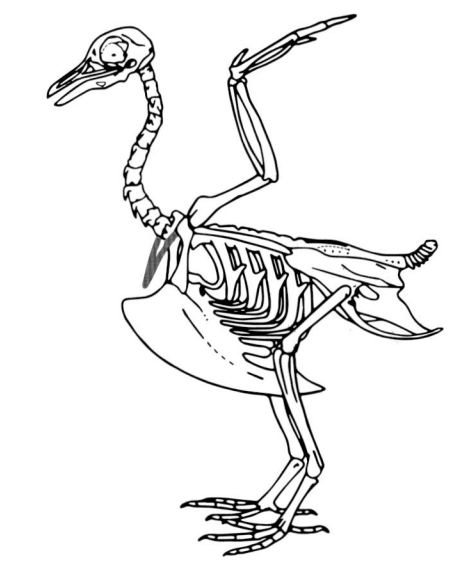

Figure 1.5. The furcula is highlighted in red. Uncinate processes are evident along the posterior margin of each rib and are a trait shared by modern birds and reptiles. *Wikimedia Commons.*

Four-chambered heart. The hearts of birds and mammals have four chambers—two atria and two ventricles—that separate oxygenated and deoxygenated blood, allowing for a very efficient "double circulatory system." Despite the similar anatomical design, the hearts of birds are proportionately larger than those of mammals, which helps birds meet the high metabolic demands of flight.

WHAT MAKES BIRDS UNIQUE?

All birds share certain characteristics that distinguish them from other groups of animals. Surprisingly, some of the traits that we most closely associate with birds—wings and laying eggs—are not unique to them. After all, flying mammals (bats) and insects also have wings, and eggs are laid by a wide variety of animals, including insects, fish, reptiles, amphibians, and monotremes. So how are birds distinct?

If we consider the traits that are unique to birds, most are adaptations for flight. Birds have many adaptations for flight that reduce weight while increasing strength and power.

Pneumatic bones. Certain flightless birds, including penguins and ostriches, have solid bones. Birds that fly, however, have some bones that are pneumatic, which are lighter than solid bones because they lack marrow and have, instead, air-filled canals strengthened by internal struts.

Fused and reduced bones. Birds have many fused and reduced bones, which decrease weight while providing the strength and stability required for flight. For example, the thoracic, pelvic, and tail regions of the spine are fused to maintain body posture during flight. Likewise, the bones of the hand are fused and reduced to support the flight feathers. An important skeletal modification is the fusion of the clavicles into a single bone called the furcula, or wishbone (fig. 1.5). The furcula prevents the chest from collapsing on the downward stroke of the wings.

Keel. Birds have two powerful flight muscles that are responsible for the flapping of the wing—the pectoralis and supracoracoideus muscles. These muscles are so well-developed in birds that they require a greater area of attachment than is typically found on the breastbone, or sternum, of an animal that does not fly. The keel is a blade-shaped projection of the sternum where the large flight muscles attach.

Toothless and horny beak. Teeth may not seem to weigh much, but every little bit counts for a flying bird. Early birds had teeth, but modern birds have toothless jaws covered with a horny sheath that are much lighter than jawbones with teeth. Beaks show an impressive range of shapes and sizes, each adapted to the ecology of the bird, typically related to what it eats (fig. 1.6).

Reduced urogenital system. The urogenital system of birds is highly modified to reduce weight. Birds lack a urinary bladder and urethra and, hence, can avoid carrying water. Avian kidneys contribute to water conservation by producing uric acid that is excreted as a semisolid paste. Most species also lack a phallus and instead rely on a single opening, called a cloaca, for reproductive and excretory functions. Gonads are either absent, as with most females, which have only a single ovary, or recess greatly during nonbreeding seasons.

Respiratory efficiency—air sacs. To meet the oxygen demands for flight and to fuel their high metabolism, the respiratory system of birds is much more efficient than that of other animals. Unlike mammals, birds do not have a diaphragm to help move air through the respiratory system. Instead, birds have air sacs that increase the amount of air that can be inhaled and also function like bellows, moving air so that gas exchange occurs continuously.

Feathers. Although you might think of feathers as adaptations for flight, they actually serve a wide variety of purposes, including insulation, courtship display, camouflage, and communication (fig. 1.7). Among living animals, feathers are unique to birds, though the fossil record shows some of the dinosaurs from which birds evolved also had feather-like structures. Scientists long thought that feathers were simply modified scales, but more recent research has suggested that feathers are actually novel epidermal structures (i.e., not developed directly from reptilian-type scales).

Hallux. Birds have a unique foot structure with a modified big toe called the "hallux." In most birds, the hallux is oriented backward, but foot structure in birds is highly variable and related to the behavior and ecology of a species.

Syrinx. The vocal organ of birds is called the syrinx, which is located at the base of the bird's trachea. Unlike a mammal's larynx, the syrinx lacks vocal cords and instead acts as a resonating chamber of elastic and vibrating membranes. In further contrast to the mammalian vocal system, which contains only one set of vocal cords, the avian syrinx contains multiple sets of vibrating structures, enabling birds to produce two different sounds at the same time. Birds can vary those sounds by changing air pressure and the tension of the syringeal muscles, and those vocal performances have long fascinated professional and amateur naturalists.

Highly developed vision. Vision is more developed in birds than in most other animals. Birds not only can see

Figure 1.6. The impressive diversity in avian bills reflects the varied ways species use resources. Pictured are *A*, Northern Hawk-Owl (*Surnia ulula*); *B*, Great Egret (*Ardea alba*); *C*, Greater Flamingo (*Phoenicopterus roseus*); and *D*, Black-faced Spoonbill (*Platalea minor*). *Photo © Ian Davies / Macaulay Library at the Cornell Lab of Ornithology.*

objects at great distance, but they also see color. Interestingly, the eyes are among the few organs that are proportionately larger in birds than in most other animals. The size and placement of the eyes can tell a lot about the ecology of a species. Predatory birds that pursue prey usually have eyes toward the front of the head, whereas many prey species have eyes on the sides or top of the head to better detect predators.

You will learn more about these and other characteristics of birds as you move through the chapters of this book.

ORNITHOLOGY: THE STUDY OF BIRDS

Ornithology, the topic of this textbook, is a branch of science that focuses on the study of birds. The word "ornithology" comes from the ancient Greek words ὄρνις *ornis* (bird) and λόγος *logos* (explanation). Ornithology has old roots, and some of the earliest scientific accounts of birds date back to 1500 BC, where the Vedas, a Sanskrit text from ancient India,

included careful observations about distribution, life history, and ecology of birds. Despite birds being a subject of study for thousands of years, ornithology did not emerge as a professional discipline until the eighteenth century.

Since then, ornithology has played an important role in the development of ideas that are foundational to the biological and ecological sciences today. The ideas of Charles Darwin and his articulation of the theory of natural selection were strongly shaped by his study of Galapagos finches. Our understanding of density-dependence in population regulation was influenced by David Lack's studies of birds, and theories about competition and competitive exclusion were influenced by Robert MacArthur's study of boreal forest warblers. Research by Robert MacArthur and E. O. Wilson provided the foundation for island biogeography theory and, subsequently, the field of landscape ecology. Even the establishment of the field of ethology (or behavior) was heavily influenced by pioneering work on imprinting behavior and instinct in birds by Konrad Lorenz and Niko Tinbergen, respectively.

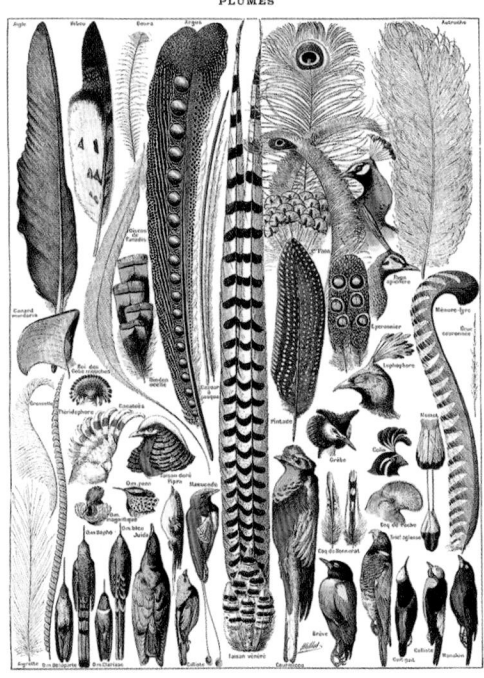

Figure 1.7. Feathers serve a wide variety of functions including flight, insulation, camouflage, sound production, and visual display. *Illustration by Adolphe Millot/PD 1923.*

Although a focus on birds might be viewed by some as narrow, ornithology includes a broad range of topics, and most scientists specialize in only one or a few of them. Areas of specialization include molecular genetics, taxonomy, physiology, communication, behavior, evolution, population demography, disease ecology, and many more. Traditionally

aligned with natural sciences, modern ornithology now includes the social sciences and topics such as the economic impacts of bird-feeding, social psychology of bird-watching, cultural relevance of birds, and birds as learning tools in educational programs. For some, the decision to study birds may reflect personal curiosity and fascination with birds. For others, the focus might be more practical, because birds provide a tractable system for addressing questions and problems that cut across many systems. Indeed, birds can offer an effective lens through which we can view and understand the world around us.

Throughout this textbook, in the boxes included in each chapter, you will discover the diverse interests and professional affiliations of ornithologists. Chapter 30 provides a more in-depth view of the various career paths that are available in ornithology and the ways you might best position yourself to encounter professional opportunities. Even if you are not interested in becoming an ornithologist, you can still contribute in meaningful ways to the discipline. One thing that distinguishes ornithology from many other sciences is the important role played by amateurs and citizen scientists, who have made remarkable contributions to advancing and refining our knowledge of birds and their habitats. You also will find descriptions of the ways that amateurs can contribute in chapter 30.

We hope that as you read this textbook and reflect on the questions presented in each chapter, you will develop a better appreciation of how diverse ornithology is in terms of topics and contributors, as well as the ways that the field has contributed to our understanding of basic and applied sciences.

CHAPTER 2

Origin and Early Evolution

Luis M. Chiappe

Birds in flight look very different from the stereotypical dinosaurs of the Mesozoic era, the geologic time contained between 251 and 66 million years ago. Yet, fossil discoveries from around the world have revealed a cohort of intermediate forms blurring this apparent divide, and studies on this paleontological bonanza have demonstrated that birds and the extinct dinosaurs that lived many millions of years ago actually share many features in common.

Birds have a very deep evolutionary past, one that was recognized long ago with the discovery of the first specimen of *Archaeopteryx lithographica* in the nineteenth century (Wellnhofer 2009) (fig. 2.1). This oldest and most primitive bird lived in what is today southern Germany at the end of the Jurassic period, some 150 million years ago. Many other groups of younger and more advanced birds evolved, thrived, and became extinct during the subsequent Cretaceous period, their remains unearthed worldwide and from rocks dating approximately between 131 and 66 million years ago (Chiappe 2007, O'Connor et al. 2011, Zhou 2014). Modern birds, those belonging to lineages represented by more than 10,000 species alive today, also trace their origin to the Mesozoic era; their earliest recognized fossils date back some 80 to 75 million years ago (Mayr 2014).

The reptilian appearance of the skeleton of *Archaeopteryx*, together with a host of studies ranging from embryological to molecular, have consolidated the notion that birds evolved from within animals traditionally classified as reptiles (Gauthier et al. 1988, Hedges 1994, Gill 2007). In evolutionary terms, birds are a highly specialized group of feathered reptiles, just as bats are a highly specialized group of winged mammals. While the long evolutionary history of birds and their reptilian ancestry has been long recognized and universally accepted by the scientific community, details of the group's evolutionary origin—the specific relationship of birds to other groups of reptiles—remained contentious for most of the

nineteenth and twentieth centuries (Feduccia 1999, Chiappe 2007, Pickrell 2014).

Indeed, the earliest ideas about the evolutionary descent of birds date back to the 1860s (Padian and Chiappe 1998). Since then, many different groups of reptiles—from duck-billed dinosaurs to the extinct flying reptiles of the Mesozoic

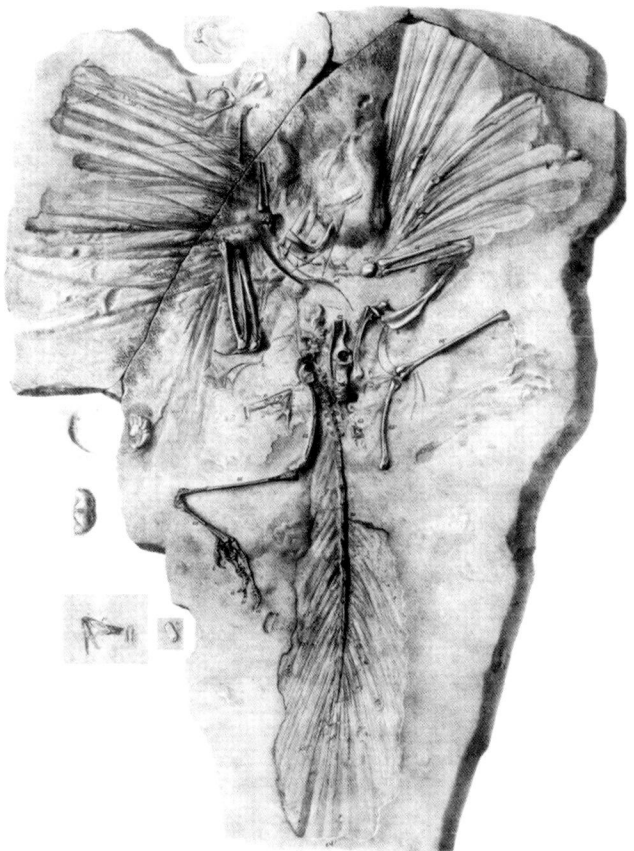

Figure 2.1. The London specimen of *Archaeopteryx lithographica* from the Late Jurassic of southern Germany. This first skeletal specimen of the oldest known bird was unearthed from a limestone quarry in Bavaria in 1861. *From Owen 1863.*

era (pterosaurs) and from crocodiles to other groups of reptiles that originated more than 200 million years ago in the Triassic period—have been identified as the closest relatives of *Archaeopteryx* and living birds (Chiappe 2007, 2009). Theropod dinosaurs, the group that includes famous dinosaurs such as *Tyrannosaurus rex* and *Velociraptor*, have been among the earliest proposed candidates for the ancestry of birds, a hypothesis that has gone through periods of contention and widespread popularity since its inception (Chiappe 2007, Wellnhofer 2009). Today, however, the hypothesis that birds evolved from these dinosaurs receives almost universal support, largely because of the enormous body of evidence from disparate fields of science accumulated over the last five decades (Padian and Chiappe 1998, Chiappe 2007, Xu et al. 2014, Brussatte et al. 2015). These studies have cemented the notion that birds originated, presumably in the Jurassic (some 190 to 160 million years ago), from animals traditionally classified as theropod dinosaurs. Within these dinosaur groups, birds are overwhelmingly interpreted as belonging to the maniraptorans and, particularly, to the more birdlike paravians (fig. 2.2), a group that includes a variety of small and feathered species, some even capable of flight (Xu et al. 2003, Han et al. 2014). Since birds are evolutionarily nested within theropod dinosaurs, they are regarded as living dinosaurs, namely "avian dinosaurs"; for clarity, paleontologists and evolutionary biologists refer to those dinosaurs that are not birds as "non-avian" dinosaurs.

This chapter summarizes the diverse lines of evidence that converge in support of the hypothesis that birds are living theropod dinosaurs and how feathers and flapping flight—

two of the most distinct attributes of birds—might have evolved during the long evolutionary history of the group. It also reviews the many different groups of extinct birds that thrived during the Mesozoic era, fossils of which bridge the gap between the extinct non-avian dinosaurs and their present-day relatives, recounting the saga of how living birds became the remarkable animals they are.

THE DINOSAURIAN ANCESTRY OF BIRDS

The notion that the ancestry of birds can be traced back to the theropod dinosaurs that lived during the Mesozoic era is not new. Several researchers argued in favor of this relationship soon after the publication of Charles Darwin's *Origin of Species*, but it was Thomas Henry Huxley who, nearly 150 years ago, provided the most compelling evidence for the dinosaurian ancestry of birds. As early as 1868, Huxley had amassed a list of more than 30 features of the hindlimbs and pelvis that were shared between non-avian theropods such as the Jurassic *Megalosaurus* and *Compsognathus*, and present-day birds (Huxley 1868). In spite of this evidence, and the discovery of birdlike dinosaurs with even greater similarity to modern birds in subsequent years—most remarkably the skeletons of small-bodied theropods such as the now-familiar *Velociraptor* and *Oviraptor* (Osborn 1924)—the hypothesis of a dinosaurian origin of birds fell into disgrace for much of the twentieth century.

The main reason for such a change in popularity is likely rooted in the evolutionary framework of the time. Just as Huxley was amassing morphological evidence in support of a dinosaurian origin of birds, other researchers were claiming that the similarities between birds and non-avian theropods were instead the result of evolutionary convergence (i.e., similar traits that evolved independently among these organisms and in response to the similar functional requirements of bipedalism). These claims were strengthened by what was viewed as a major weakness of the theropod hypothesis of bird origins: the fact that none of the fossils of non-avian theropods known at the turn of the nineteenth century had clavicles. Such an absence was viewed as a critical flaw of this hypothesis because it resonated with an evolutionary principle popular at that time—Dollo's law of irreversibility—which stated that structures lost during the course of evolution could not re-evolve. Therefore, if the dinosaurs assumed to be the ancestors of birds had lost their clavicles, birds (whose clavicles form the notorious wishbone) could not be the descendants of those animals. In other words, dinosaurs had evolved into a dead end; they were too specialized to be avian ancestors. The 1913 discovery of the Middle Triassic *Euparkeria* in the Karoo Desert of South Africa (Broom 1913), a small and morphologically

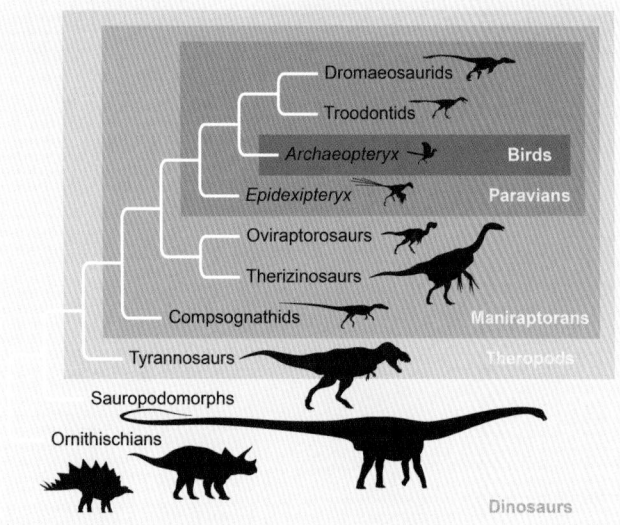

Figure 2.2. Simplified cladogram of dinosaur evolutionary relationships. Birds are nested within paravians, maniraptorans, theropods, and dinosaurs.

generalized archosauromorph reptile (and a distant relative of dinosaurs), provided a candidate for the ancestry of birds that suited the evolutionary thinking of the time. Not surprisingly, *Euparkeria* was quickly adopted as the evolutionary paradigm for the origin of birds, and this view was widely popularized by an influential early twentieth-century book, *The Origin of Birds* (Heilmann 1926), by Danish artist and amateur paleontologist Gerhard Heilmann.

It took decades for the arguments about the evolutionary convergence between non-avian theropods and birds, and the ideas that the latter had evolved from a Triassic stock of primitive archosauromorphs, to be reconsidered. This task fell primarily on the shoulders of a dinosaur paleontologist from Yale University, John Ostrom, who, in the 1960s and 1970s, conducted detailed comparative studies between *Archaeopteryx* and a series of paravian theropods such as the dromaeosaurid *Deinonychus antirrhopus* (Ostrom 1969, 1975, 1976). These studies led to the resurgence in support of the dinosaurian origin of birds that has grown ever since with more and more discoveries. Today, this hypothesis argues that birds evolved from within a group of generally small-bodied theropods known as maniraptorans and that their closest relatives include paravian dinosaurs such as the sickle-clawed dromaeosaurids (*Deinonychus, Velociraptor mongoliensis*, and their kin; Turner et al. 2007) and the lightly built *Troodon formosus, Mei long*, and relatives (Makovicky and Norell 2004); most current studies argue that birds are the sister-group of a clade (Deinonychosauria) composed of Dromaeosauridae and Troodontidae (e.g., Turner et al. 2007, Xu et al. 2014). Nonetheless, the various similarities between living birds and their theropod predecessors can be recognized in many other dinosaur groups, including those that lived in the Triassic and that represent some of the earliest branches of the theropod tree (e.g., *Tawa hallae*; Nesbitt et al. 2009, Brussatte et al. 2015) (fig. 2.3). From an osteological perspective, these similarities include a suite of very general features (bipedal posture with vertically oriented hindlimbs,

three fully developed toes facing forward, S-shaped neck, among others) and many more specific attributes including vertebral pneumatization by air sacs (presumably developing a connection with the lungs similar to that of modern birds), unique structures in the vertebral column and rib cage, elongated forelimbs with wrist bones allowing swivel-like movements of the hand, and similar structures in the pelvis and hindlimbs, as well as numerous other characteristics distributed throughout the entire skeleton (Gauthier 1986, Brusatte 2013). Indeed, many skeletal features previously thought to be exclusively avian—such as wishbones, laterally facing wing-pits, and large breastbones—are today known among different lineages of non-avian maniraptoran dinosaurs.

Although museums around the world emphasize the colossal dimensions reached by the extinct dinosaurs of the Mesozoic, such a superlative approach to the evolutionary history of these animals overlooks the fact that many non-avian dinosaurs had body sizes that fall within the range of living birds (Xu et al. 2003, Turner et al. 2007, Butler et al. 2010). In fact, a number of theropod groups—tyrannosaurs (Rauhut 2003), oviraptorosaurs (Xu et al. 2002), and others—had rather small early relatives. Particularly small are many early members of the two best-known groups of non-avian maniraptorans, dromaeosaurids (Norell et al. 2004) and troodontids (Makovicky et al. 2004), which document the extreme size reduction experienced at the base of the paravian clade of maniraptorans (the clade that includes troodontids, dromaeosaurids, and birds) (Turner et al. 2007) (fig. 2.4). Recent studies (Lee et al. 2014a) have shown how such a trend towards miniaturization started at least 50 million years prior to the divergence of *Archaeopteryx* and paedomorphosis (an heterochronic process in which the adults of the derived species retain traits of the juveniles of the ancestral species) has been claimed to be the underlying developmental mechanism for such a notable trend in size reduction (Bhullar et al. 2012).

In the last few decades, numerous discoveries from around the world and novel studies of fossil material have

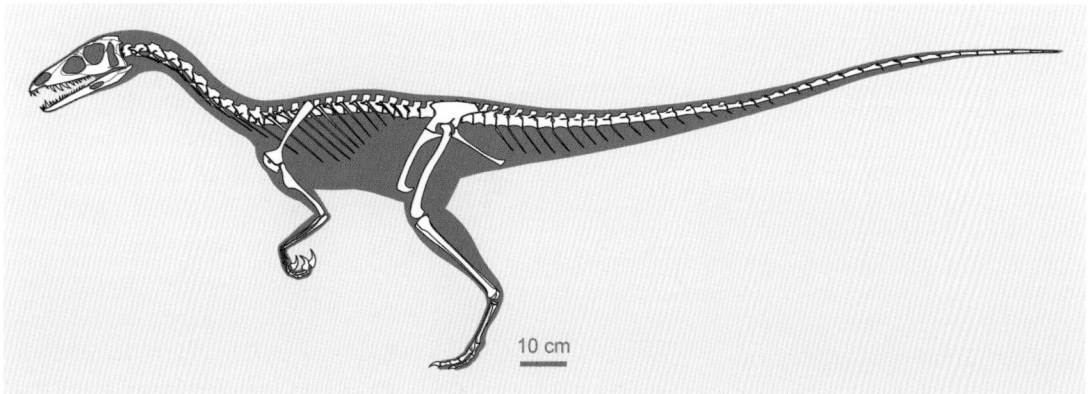

10 cm

Figure 2.3. Skeletal reconstruction of the Late Triassic theropod dinosaur *Tawa hallae* of New Mexico. *From Nesbitt et al. 2009; image courtesy of S. Nesbitt.*

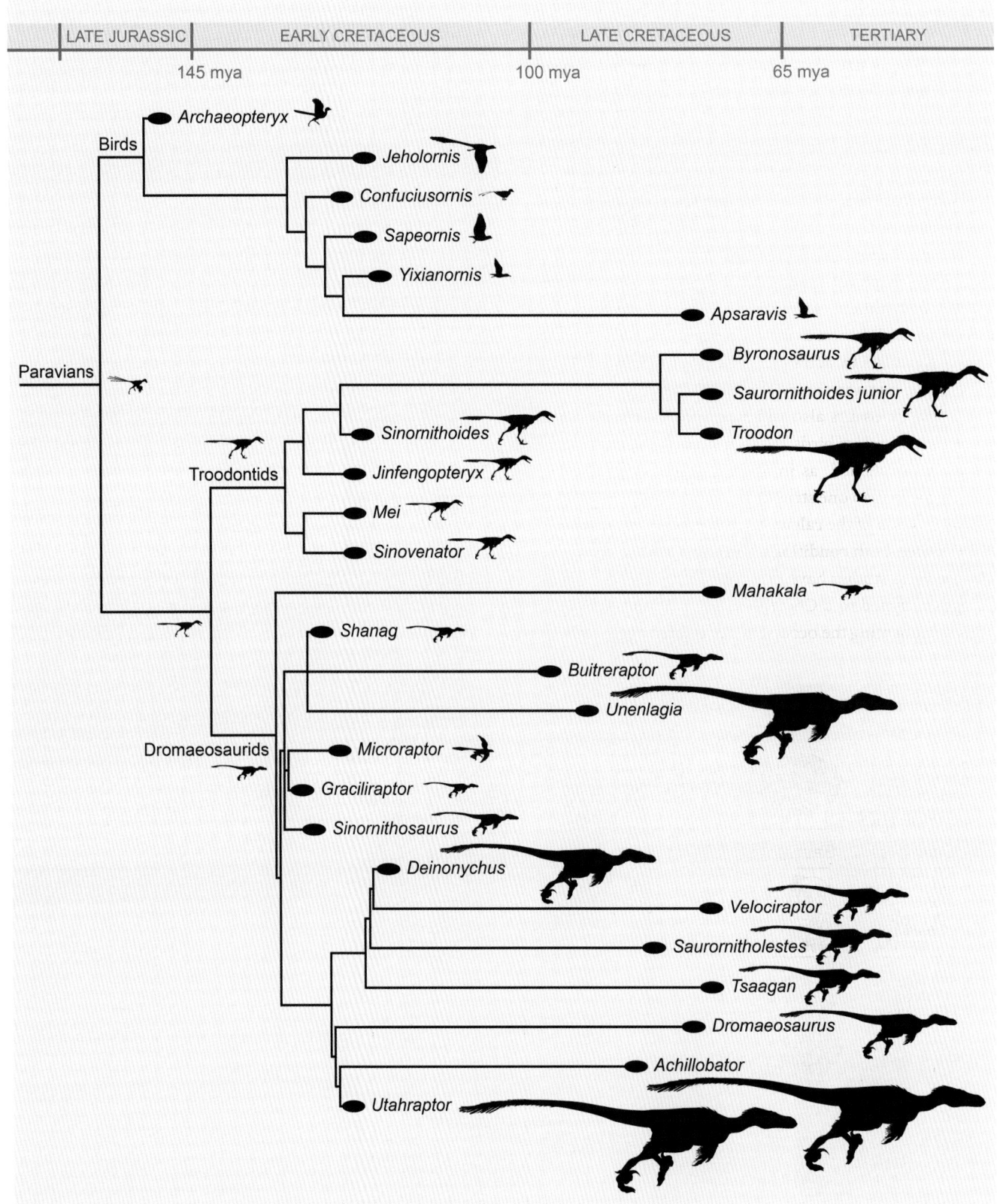

Figure 2.4. Cladogram showing the diminutive sizes that characterize the most basal lineages of paravians. New discoveries and recent phylogenetic studies document that many lineages of non-avian maniraptoran theropods underwent a prolonged process of miniaturization that was particularly marked among paravians. *Modified from Turner et al. 2007.*

revealed a wealth of similarities between non-avian theropods and living birds that goes beyond the skeletal evidence and small body size. Fossilized eggs of theropod dinosaurs—some containing the remains of rare embryos—have shown that the shape of the eggs of non-avian maniraptorans and the microstructure of their shell were in many respects alike to those of present-day birds. Some of these dinosaurs (e.g., oviraptorids, troodontids) laid eggs of various degrees of asymmetry, with one pole narrower than the other and approaching the lopsided shape of a typical avian egg (fig. 2.5). While egg shape varies significantly among amniotes, departure from symmetry around the equatorial axis is predominantly limited to non-avian theropods and birds (Deeming and Ruta 2014). In addition, calculations between the volume of the egg and the body mass of the egg-layer indicate that these dinosaurs also laid relatively larger egg volumes, which are typical of birds. Furthermore, the eggshell of these dinosaurs—calcified, as in all birds—was characterized by having more than one structural layer (denoted by a differential disposition of the calcite crystals) and by being pierced by fewer pores, both conditions common to avian eggs. Recent geochemical studies have even detected traces of coloration in eggs assigned to a Chinese Late Cretaceous oviraptorid, thus documenting the occurrence of colored eggs—otherwise

known from birds, among amniotes—in non-avian maniraptorans (Wiemann et al. 2015).

Rarely, the fossil record has provided us with a glimpse of the behaviors of the extinct maniraptorans. An oviraptorid specimen containing a pair of shelled eggs inside its pelvic canal has provided support for interpretations arguing that these dinosaurs laid their eggs in a sequential pattern as opposed to en masse. Such interpretations have been long supported on the basis of the spatial arrangement of eggs within clutches of non-avian maniraptorans, which show that the eggs were arranged in pairs, as opposed to typical reptilian clutches (turtles, crocodiles, and other dinosaurs) in which they lack a particular spatial arrangement. Collectively, all this evidence indicates that non-avian maniraptorans laid their eggs at discrete intervals, perhaps taking up to several days to lay an entire clutch. Skeletons of troodontids and oviraptorids have been found in close association with, and often literally on top of, their egg clutches (fig. 2.6). In troodontids, an adult skeleton was found directly on top of the vertically arranged eggs (Varricchio et al. 1997). In oviraptorids, several adult skeletons show the hindlimbs tucked against the belly of the animal and placed within an open space at the center of a clutch formed by dozens of pair-arranged eggs; the long forelimbs of the animal are often preserved embracing the

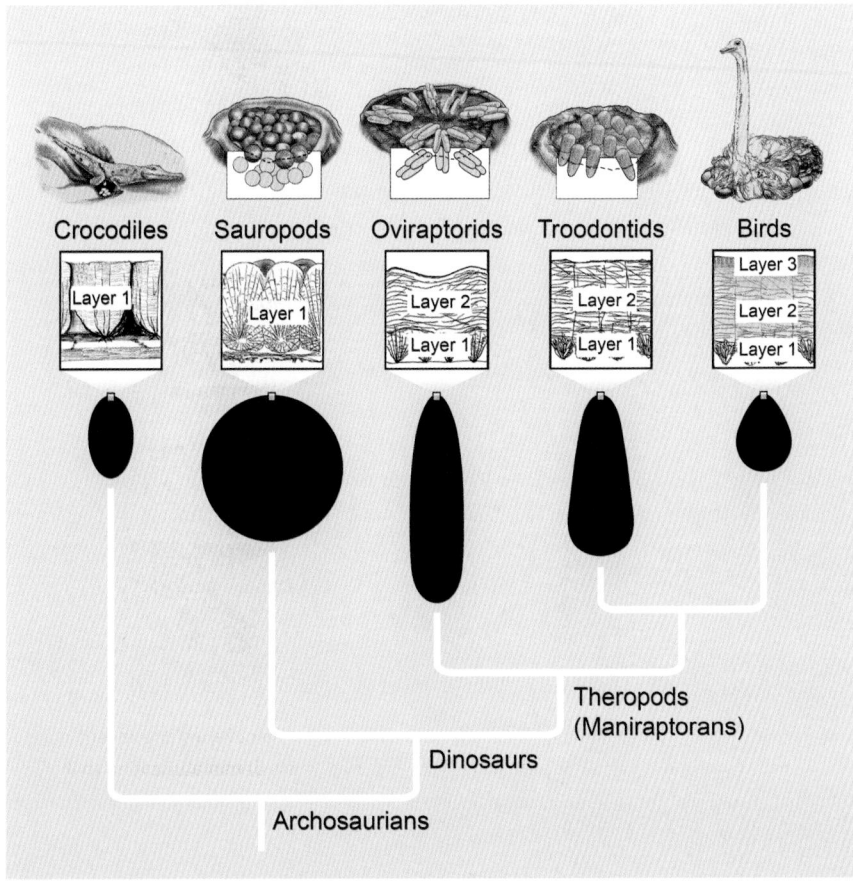

Figure 2.5. Characteristics of the eggs (e.g., presence of at least two distinct crystalline layers in the eggshell, asymmetric egg shape) of several non-avian maniraptorans (e.g., oviraptorids, troodontids) support the inclusion of birds within these theropod dinosaurs. *Modified from Chiappe 2004.*

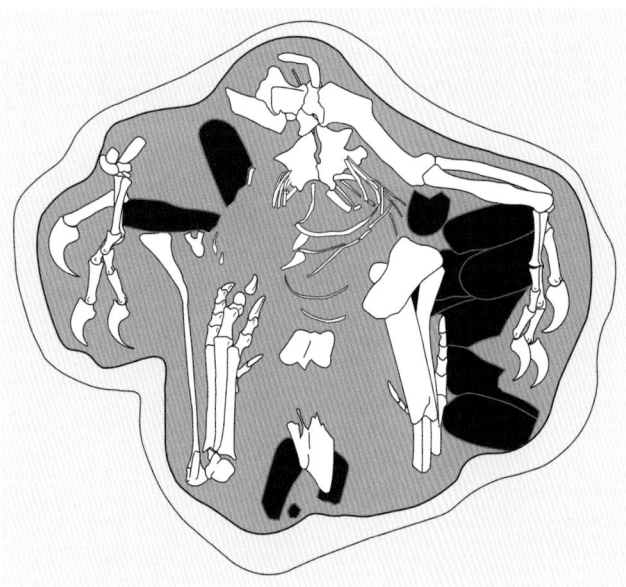

Figure 2.6. Discoveries of adult oviraptorid skeletons on top of clutches demonstrate that, as in the case of birds, these maniraptorans brooded their eggs. *Modified from Clark et al. 1999.*

periphery of the egg clutch. Such evidence indicates that several species within these dinosaur groups had nesting behaviors resembling those of brooding birds. These fossils suggest that, regardless of the specific purpose (shelter, incubation, or other), avian brooding behaviors had already evolved among non-avian maniraptorans (Norell et al. 1995, Clark et al. 1999).

Further evidence documents additional behavioral similarities with birds. The skeletons of a few small troodontids from the Early Cretaceous of China (e.g., *M. long, Sinornithoides youngi*) are preserved with the hindlimbs flexed beneath the belly, the neck folded backward, and the skull tucked beneath a forelimb and against the ribcage (Xu and Norell 2004, Gao et al. 2012a). The arrangement of the skeletons together with characteristics of the sediments entombing them suggests that these dinosaurs died while sleeping. Such a skeletal arrangement hints at resting poses that are highly reminiscent of the stereotypical sleeping postures of many present-day birds. Additionally, a range of fossil traces from Cretaceous sites in Colorado has shed light on other aspects of the behavioral repertoire of non-avian theropods. These exceptional sites show series of bilobate depressions with a middle ridge, and in some instances with a tridactyl theropod footprint inside, that are difficult to associate with any behavior other than non-nesting courtships displays (Lockley et al. 2016). Such traces suggest that some non-avian dinosaur species had stereotypical courtship behaviors akin to those observed among some groups of living birds (e.g., scraping, arena display, lekking). Altogether, these snapshots

of the behavior of non-avian theropods give us rare glimpses into the conducts of these extinct animals and extraordinary evidence in support of the dinosaurian ancestry of birds.

Further similarities are revealed by the microstructure of the bone tissues preserved in many fossil maniraptorans. While typical characteristics of bone histology (particularly the degree of vascularity and the rate of growth or deposition) are partially the result of the function performed by bone, studies have shown that they also carry a strong phylogenetic signal (Montes et al. 2010). In the last few decades, the study of these tissues has clearly demonstrated that although dinosaurs have been traditionally characterized as slow-growing behemoths, these animals had growth rates comparable to those of many living birds, even if they grew over multiple years (Chinsamy 2005). Furthermore, specialized bone tissues such as the medullary bone that female birds deposit inside the medullary cavity of their long bones during breeding season (a calcium reservoir recycled during eggshell production; see Dacke et al. 1993), have been documented among non-avian theropod dinosaurs such as *Tyrannosaurus rex* (Schweitzer et al. 2005). Once again, these discoveries highlight how non-avian dinosaurs shared key aspects of their reproductive physiology with modern birds.

Additional similarities between birds and the dinosaurs frequently identified as their closest relatives are now being hinted by provocative biochemical, immunological, and genomic investigations. Amino-acid sequences of putative proteins—particularly collagen, bone's most abundant protein—found in soft tissues associated with bones of *T. rex* have highlighted an evolutionary closeness between this dinosaur and birds (Schweitzer et al. 2005, Asara et al. 2007). More recently, a study of eight bone samples from different types of dinosaurs (Bertazzo et al. 2015) observed structures similar to the calcified collagen fibers found in modern bone and other structures resembling the red blood cells of birds. The elemental analysis of the latter structures resulted in a composition that was remarkably similar to that of emu blood. While these studies remain controversial (see Dalton 2008, Service 2015), mounting evidence is suggesting that complex biomolecules can survive under unusual conditions for millions of years (Schweitzer 2011, Bertazzo et al. 2015). If correctly interpreted, the studies of the putative collagen of *T. rex* and other non-avian dinosaurs provide additional evidence in support of the dinosaurian origin of birds. Equally tantalizing are other studies that, by correlating the sizes of bone cells with those of genomes (the entire genetic material of an organism), have revealed that non-avian theropods had the small genomic dimensions characteristic of modern birds (Organ et al. 2007, 2008). These studies used a well-established correlation between the dimensions of bone cells

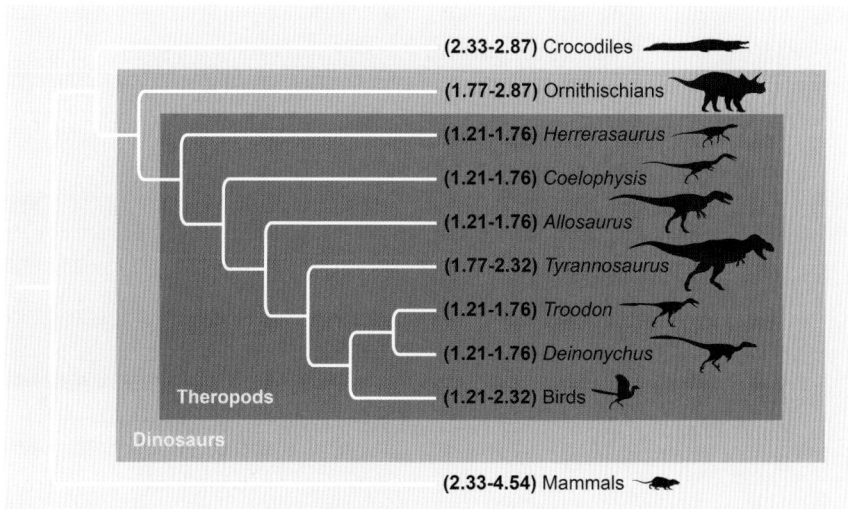

Figure 2.7. Inferences about the sizes of the genomes of non-avian theropods (in picograms, 10^{-12}g) indicate that these dinosaurs evolved the small genome sizes of birds more than 230 million years ago. Numbers in parentheses indicate the range of genome sizes estimated for different lineages of non-avian dinosaurs as well as those known for birds, crocodiles, and mammals. *Modified from Organ et al. 2007.*

and the sizes of the genomes of modern organisms and made estimates of genome sizes in extinct animals based on the sizes of the osteocytes lacunae (the spaces originally occupied by bone cells) visible in thin sections of fossilized bones. Most of the non-avian theropod bones examined pointed at genome sizes significantly smaller than those estimated for other Mesozoic dinosaurs; the genome sizes of these non-avian theropods fell within the range of living birds, which have genomes that are about half the size of those of crocodiles and mammals (fig. 2.7). Recent studies (Zhang et al. 2014) have identified several genetic processes (deletion of repetitive DNA portions, entire gene loss, and others) as the causal factor for the characteristic small size of the genome of modern birds. It is likely that these same processes contributed to the reduction of the genome sizes in non-avian theropods. These processes may have been acting on the genomes of other volant groups of vertebrates, as Organ and Shedlock (2009) also found relatively small genome sizes in pterosaurs and bats, thus providing fresh evidence pointing at a correlation between a decrease in genome size and the evolution of flight in vertebrates.

In spite of the many different lines of evidence in support of the theropod origin of birds, nothing has consolidated the universal acceptance of this hypothesis as much as the discovery of a cohort of non-avian dinosaurs with feathers. After two decades of paleontological discoveries from around the world, fossils of more than 20 species of non-avian theropod dinosaurs provide direct evidence of integumentary coverings interpreted as feathers, and a number of other fossils reveal circumstantial evidence (e.g., quill knobs) for the presence of these structures in additional species. Known primarily from the Late Jurassic and Early Cretaceous of northeastern China (Norell and Xu 2005, Long and Schouten 2008, Xu

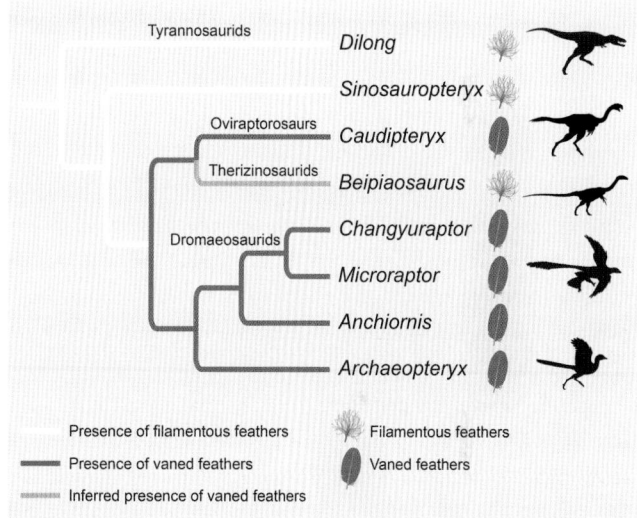

Figure 2.8. The broad phylogenetic distribution of different lineages of non-avian theropods carrying either protofeathers or vaned feathers of modern appearance suggests that the bodies of most theropod dinosaurs were covered by plumage. *Modified from Chiappe 2007.*

et al. 2014, 2015) but also from Late Jurassic and Late Cretaceous sites in Europe (Rauhut et al. 2012) and North America (Zelenitsky et al. 2012, Van Der Reest et al. 2016), respectively, these species are phylogenetically distributed over a large portion of the theropod tree (fig. 2.8). The characteristics of the plumage also differ greatly among these species. In some, the plumage was made exclusively of simple or loosely branched filaments, which are referred to as protofeathers. These structures are known for examples of basal tetanurans (e.g., *Sciurumimus albersdoerferi*; Rauhut et al. 2012), early tyrannosaurus (from the two-meter-long *Dilong paradoxus* to the nine-meter-long *Yutyrannus huali*), and a variety of compsognathids (e.g., *Sinosauropteryx prima*, *Sinocalliopteryx gigas*; Ji et al. 2007). These protofeathers range significantly in size (e.g., 2 cm in *Dilong* but 20 cm in *Yutyrannus*) and their length also varies according to their position within the body. Evidence of protofeathers has also been found among herbivorous and omnivorous theropod dinosaurs such as in the small-headed and typically toothless ornithomimosaurs and the therizinosaurs, the last of which had extended necks, small heads, long arms with powerful claws, and stout feet. Fossils of the therizinosaur *Beipiaosaurus inexpectus* (Xu et al. 1999) show that these animals also carried a small number of broader and longer (10 to 15 centimeters) single filaments projecting from the head, neck, and tail; these structures have been interpreted as specialized protofeathers that may have played a role in the sexual displays of these dinosaurs (Xu et al. 2009).

In species closer to the origin of birds, the plumage consisted of both protofeathers and additional structures with the characteristics typical of a modern feather. Abundant examples of the latter feathers are known among the oviraptorosaurs (Ji et al. 1998, Zhou and Wang 2000), a group of herbivorous/omnivorous dinosaurs that are more closely related to birds than any of the previously mentioned feathered dinosaurs. One of these dinosaurs, the turkey-sized *Caudipteryx zoui*, had forelimb and tail feathers with multiple thin filaments branching off either side of a central shaft in a parallel arrangement; these are essentially feathers of modern aspect in which barbs form two large and flattened vanes on either side of a rachis. At least 14 symmetrically vaned feathers, some reaching up to 20 cm in length, projected from the distal half of the forelimbs of *Caudipteryx*, and a series of equally vaned, long feathers formed a fan at the end of its relatively short tail. The distalmost feathers of the forelimb were attached to the middle digit of its three-digit hand, the same digit that supports the primaries of today's birds. Fossils of this dinosaur also show that the longest feathers of the forelimb were sided by shorter feathers toward both the tip and the base of the wing, an arrangement typical of modern avian wings.

Many fossils of Chinese Early Cretaceous dromaeosaurids, particularly the pheasant-sized *Microraptor zhaoianus* (Xu et al. 2003, Turner et al. 2012), show how the plumage of the non-avian theropods closer to the origin of birds evolved a further degree of feather differentiation (fig. 2.9). In addition to having two- to five-cm-long simple protofeathers, the elongated forelimbs of these paravian dinosaurs had long feathers with vanes of disproportionate widths; as in their modern counterparts, the leading vanes of these asymmetric feathers were narrower than the trailing vanes of the feathers. *Microraptor* also carried a fan of long, vaned feathers at the end of the tail, although these had symmetrical vanes. The

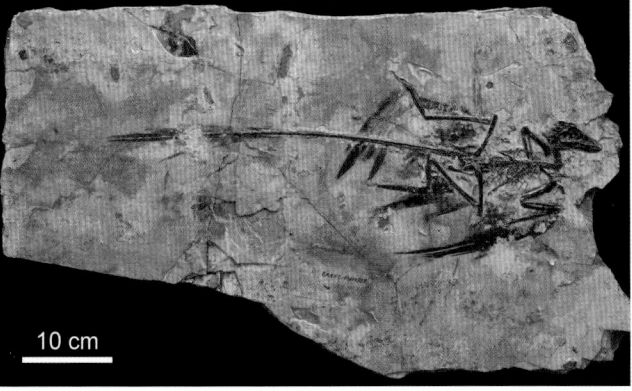

Figure 2.9. The Early Cretaceous non-avian maniraptoran *Microraptor zhaianus*.

distalmost pair of tail feathers stretched more than 22 cm in *Microraptor*, close to a third of the length of the entire animal; in its heavier close relative, *Changyuraptor yangi*, they reached more than 30 cm in length (Han et al. 2014). The forelimbs of these Chinese dinosaurs became transformed in long and slender wings of essentially modern design, which were formed by asymmetrically vaned feathers. As is typical of the wings of modern birds, the primaries of these theropods were longer but fewer in number than the secondaries. Similarly, the farthest primaries were aligned with the direction of the digits, while the remaining flight feathers were angled with respect to the skeletal forelimb. Additionally, and most surprisingly, these dromaeosaurids had a set of long, vaned feathers (some reaching 20 cm in length) projecting from the lower leg (i.e., shin and foot) (Xu et al. 2003).

A similarly differentiated plumage is also known in several troodontids. Well-known examples of these paravians include the Middle to Late Jurassic *Anchiornis huxleyi* (Hu et al. 2009) and the younger, Early Cretaceous, *Jinfengopteryx elegans* (Ji et al. 2005). Both carried abundant protofeathers but fossils of the former show a set of long, vaned feathers projecting from its hindlimbs (including feathers covering its toes), and the only known specimen of the latter displays a frond-like tail formed by long and symmetrically vaned feathers that become increasingly longer toward the tip.

Despite that, evidence of plumage is restricted to a fraction of the non-avian dinosaurs known to date. The fact that these fossils span over a large portion of the phylogeny of theropods and belong to species exhibiting a large diversity of sizes, appearances, and lifestyles suggests that feathers of different kinds covered the bodies of most theropod dinosaurs (fig. 2.8). Just as modern-looking feathers evolved among the dinosaurian predecessors of birds, so did a variety of molecular, morphological, physiological, and behavioral traits that were traditionally assumed to have evolved at the onset of, or after, the rise of birds. Altogether, the discovery that these traits first evolved among non-avian theropods provides compelling evidence in support of the hypothesis that argues that birds are indeed living dinosaurs.

THE ORIGIN OF FEATHERS

Feathers are a universal and fundamental characteristic of birds. These lightweight and yet strong keratin structures have evolved over millions of years into many types, differentiated by variations in their organization, size, and the characteristics of their shafts, barbs, and barbules (fig. 2.10). Discoveries of feathered non-avian dinosaurs have demonstrated that the origin of these quintessential avian structures pre-dated the rise of birds, yet the original function of these singular integumentary appendages remains controversial (Dececchi et al. 2016).

The first direct evidence of feathers in non-avian dinosaurs came with the discovery, in 1996, of a complete skeleton of the compsognathid theropod *Sinosauropteryx prima* from the Early Cretaceous of northeastern China (Chen et al. 1998, Ji et al. 1998). This 60-cm-long skeleton was surrounded by a halo of dark, filament-like structures, which varied in length substantially across the body. The fact that the bones of the skeleton were variably separated from the bases of these structures, a pattern that clearly reflected the thickness of the connective tissue along different regions of the body, indicated that these were external structures. Furthermore, the structures were hollowed and subdivided into thinner filaments that branched off from their bases, conditions seen among modern feathers (Currie and Chen 2001).

Since the discovery of the first fossil of *Sinosauropteryx*, numerous other feathered non-avian theropods have provided evidence of the evolution of different feather types among the predecessors of birds. The discovery of superficially similar structures in small ornithischian dinosaurs such as psittacosaurs (Mayr et al. 2002) and heterodontosaurs (Zheng et al. 2009) has advanced the possibility that protofeathers might have first evolved outside theropods and, perhaps, even prior to the origin of all dinosaurs. The fact that the earliest phylogenetic divergences among the known non-avian feathered theropods correspond to animals with bodies that were covered by filament-like protofeathers, structures with obvious minimal (if any) aerodynamic capacity, has also provided evidence that the original function of the feather was not for the purpose of flight. Nonetheless, the presence of vaned feathers, with possible aerodynamic functions, anchored to the extremities of the limbs and tails of non-avian maniraptorans (e.g., *Caudipteryx*, *Microraptor*, *Changyuraptor*, *Anchiornis*, among others) suggests that feathers may have

Figure 2.10. Main types of modern feathers (down, body, and flight feathers) showing details of their hierarchical branching order (rachis, barb, and barbule).

quickly been co-opted into incipient airfoils capable of generating aerodynamic forces (Burgers and Chiappe 1999, Chiappe 2007, Heers et al. 2014). The firmness and cohesiveness shown by the vanes of the pennaceous feathers of dromaeosaurids and oviraptorosaurs indicates that a Velcro-like system of interlocking barbules had already evolved among these dinosaurs.

Classic interpretations about the origin of feathers were influenced by the now-contested hypothesis that these structures evolved from reptilian scales that gradually became long, frayed, and flexible (Prum and Brush 2002). Advances in our understanding of feather development, however, have documented fundamental differences between scales and feathers. These studies, combined with the evidence revealed by the spectacular discoveries of feathered non-avian dinosaurs, led to the formulation of a now broadly accepted model for the origin and subsequent evolution of feathers (Prum 1999, Prum and Brush 2002). This model argues that modern feathers evolved as the result of a series of evolutionary transitions at the feather follicle that produced the variety of feathers observed today and in the fossil record (fig. 2.11). The earliest stage of this evolutionary transition, Stage I, consists of an undifferentiated tubular collar (the follicle's innermost epidermal layer) generating a simple,

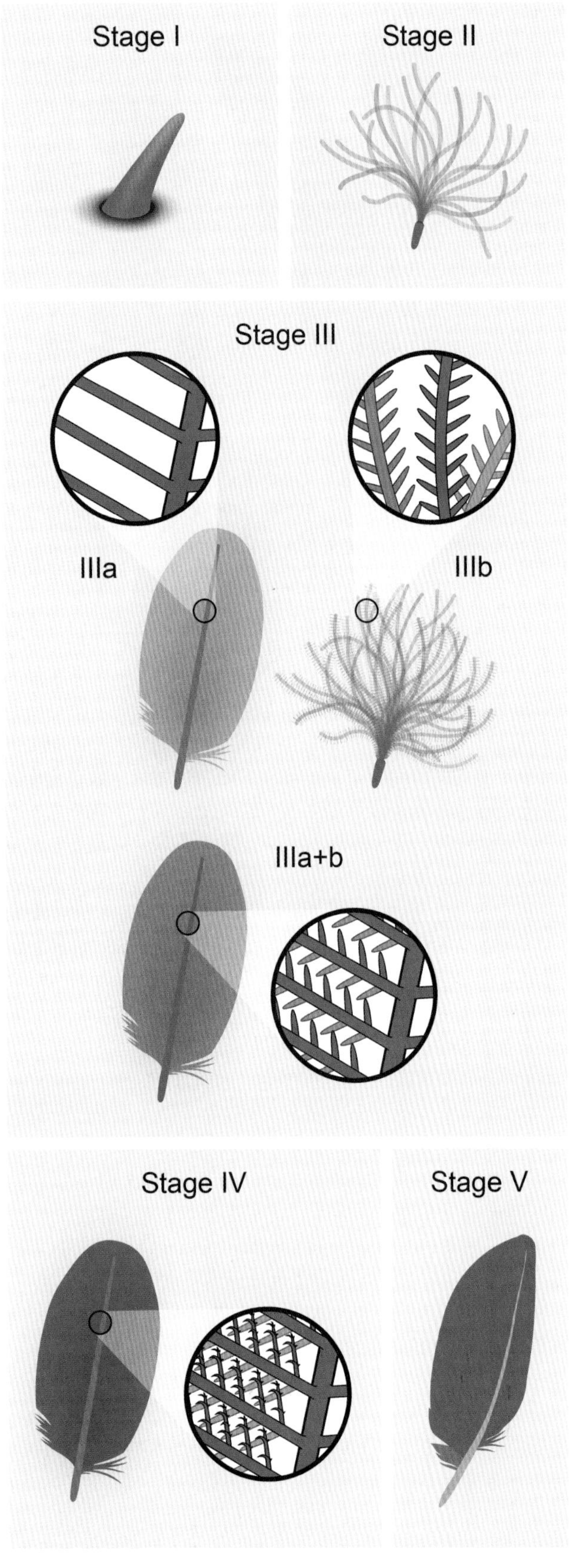

Stage I

Stage II

Stage III

IIIa

IIIb

IIIa+b

Stage IV

Stage V

Figure 2.11. Prum's model of feather evolution: Stage I, an undifferentiated tubular collar generates a simple, hollow cylinder; Stage II, collar differentiation into barb ridges produces a feather with a tuft of unbranched barbs and a calamus; additional collar differentiation leads to either Stage IIIa (vaned feather with simple barbs fused to a rachis) or Stage IIIb (many barbs with barbules attached to a basal calamus); Stage IIIa+b results in a feather with barbs and simple barbules; Stage IV produces a feather with differentiated proximal and distal (hooked) barbules; Stage V leads to an asymmetric flight feather. *Modified from Prum 1999.*

hollow cylinder. During Stage II, the collar differentiates barb ridges, which result in a feather with a tuft of unbranched barbs and a calamus. Further differentiation of the follicle's collar could have led to two alternative stages: Stage IIIa, which leads to the development of the vaned feather with simple barbs fused to a central rachis, and Stage IIIb, which yields a feather with numerous branched barbs (i.e., with barbules) attached to a basal calamus. Regardless of the specific evolutionary path taken (either Stage IIIa or IIIb), the Stage IIIa+b is envisioned as resulting in a feather that is vaned and that has barbs that bear simple barbules. The Stage IV of this transition would have produced the differentiation of proximal and distal barbules and, therefore, closed vanes (i.e., the distal barbules developed hooks capable of interlocking with the grooved proximal barbules of the adjacent barb). Finally, further modifications of the follicle's collar, in Stage V, would have led to the evolution of an asymmetric flight feather. In general terms, the evolutionary sequence proposed by this model corresponds well to the phylogenetic distribution of feathers in non-avian dinosaurs, namely the earliest phylogenetic divergences (e.g., *Sciurumimus albersdoerferi*, *Sinosauropteryx prima*, *Dilong paradoxus*) have structures that are morphologically similar to those of the model's Stage I (i.e., simple protofeathers) and taxa closer to the origin of birds (e.g., paravians such as *Microraptor zhaoianus*) have more complex structures (i.e., feathers with asymmetric vanes) that correspond to Stage V (fig. 2.11). As more discoveries of the plumage of non-avian theropods and their avian descendants come to light, it is becoming evident (not surprisingly) that the evolution of feathers took a more complex path than previously envisioned. Several reports of extinct feather types (Perrichot et al. 2008, O'Connor et al. 2012a) do not precisely fit Prum's proposed model; the documented developmental pathways of modern feathers may not be able to explain the development of the variety of feathers structures seen among non-avian theropods and basal birds.

While we have gained a significant understanding of the evolution of feathers through important paleontological and developmental evidence, the original function of the earliest feathers (or their structural precursors) remains controversial. Over the years, many different hypotheses for the original function of feathers have been put forward. Nowadays, however, explanations for the primal function of these structures are centered on two of their modern basic functions: insulating the body from the environmental conditions and providing the basis for a wide range of behavioral and sexual displays. The former hypothesis argues that feathers evolved as non-avian dinosaurs developed higher metabolic rates. Although there are clear challenges to determining with precision the thermal physiology of extinct animals, mounting evidence indicates that non-avian dinosaurs—big and small—had body temperatures that were higher than those typical of living reptiles (Eagle et al. 2015). The bone histology of these animals, for example, has consistently shown that they grew at rates that clearly exceeded those typical of turtles, lizards, and crocodiles (Varricchio 1993, Chinsamy 2005, Erickson 2005, 2014), thus suggesting higher metabolic rates than in these reptiles. Recent geochemical studies have added fresh evidence in support of this conclusion (Eagle et al. 2011, 2015). These advances have estimated that the body temperatures of non-avian dinosaurs, though variable among groups, were more elevated than those of modern ectothermic reptiles, albeit not as high as those of living endothermic birds; for this reason, non-avian dinosaurs are thought to have been metabolically "mesotherms" (Grady et al. 2014, Eagle et al. 2015). New developments have also demonstrated that many lineages of theropod dinosaurs went through a significant trend of body size reduction during the Jurassic (Lee et al. 2014a); because small-sized animals lose body heat more easily than larger ones, such a reduction in body size would have created selective pressures for the evolution of insulating structures. Therefore, if non-avian dinosaurs were transitioning toward the metabolic rates common of living birds, it is reasonable to conceive an adaptive scenario in which feathers evolved to insulate the bodies of animals (particularly those that had small sizes) that have crossed a thermal threshold; the most basic protofeathers—filament-like structures covering the bodies of the earliest known feathered dinosaurs—seem to be well-suited for this purpose, just like hair and down shield the body of modern mammals and birds, respectively.

Birds are renowned for their ornamental plumages and colorfulness—it is thus not surprising that some researchers have argued that feathers first evolved for the purpose of visual display. This hypothesis claims that ornamental structures and complex color patterns evolved to enhance mating competitiveness, a process that results from sexual selection. Feathers interpreted as ornamental (i.e., lacking other apparent function) have been found in association with the skeletons of a variety of non-avian theropods. These include long ribbon-like feathers attached to the tail of the paravian scansoriopterygid *Epidexipteryx hui* (Zhang et al. 2008a) and the broad filaments sparsely distributed among the protofeathers of the therizinosaur *B. inexpectus* (Xu et al. 2009). Feather coloration among these dinosaurs has been inferred from preserved microscopic structures interpreted as melanosomes, cellular organelles that in living animals store melanin (a pigment responsible for black, brown, and reddish coloration). Melanosome shape varies according to the type of melanin, and the differential concentration of melanosome types has been used to approximate the color of the feathers of some non-avian dinosaurs (as well as early birds) (Li et al. 2010, Zhang et al. 2010). While such an approach has been criticized by those who argue that shape alone is not enough to differentiate between melanosomes and similarly shaped bacteria (Lindgren et al. 2015, Schweitzer et al. 2015), which do not carry melanin and are known to mediate in the preservation of feathers and other soft structures, definitive examples of these structures have been confirmed by both their microscopic appearance and their chemical fingerprinting (Lindgren et al. 2015). The discovery of ornamental feathers together with proxies for coloration among some of the most basal feathered non-avian theropods has encouraged the notion that feathers might have evolved within the context of display, a major role feathers play in modern birds.

In sum, the avian feather is the result of many millions of years of evolutionary experimentation in multiple lineages of non-avian theropods and their early descendants. Some controversial evidence (i.e., integumentary coverings in non-theropod dinosaurs and pterosaurs) suggests that the origin of these quintessential avian structures may be traceable to animals that are far removed from the ancestry of birds. A well-accepted model for the evolution of modern feathers—including coverts, flight feathers, and others—has been devised, but the primal function of the earliest feathers remains controversial. Solid evidence, however, rules out flight as the driving force underlying the origin of feathers. In fact, the fossil record combined with our understanding of the evolutionary relationships of birds and their dinosaurian predecessors clearly indicates that the aerodynamic functions characteristic of modern flight feathers are the result of exaptation, an evolutionary process by which physical traits develop a new function different from the one they originally served (in the case of feathers, insulation or sexual display). The available evidence thus indicates that avian flight developed among animals that had already evolved an array of feather types, including those that would allow them to generate the forces necessary to take off.

THE ORIGIN OF AVIAN FLIGHT

The evolution of flight in birds, and the ecologic context of the animals that first acquired structures critical for this means of locomotion, continue to be at the center of a century-old debate that at times has been intertwined (inappropriately) with disputes surrounding the ancestry of birds. Today, numerous fossils have documented that critical flight-related characteristics—long and foldable wings, relatively small sizes, and others—arose prior to the rise of birds, as some non-avian maniraptorans decreased in body size, augmented the size of feathered forelimbs, and evolved specific structures that enabled a suite of movements akin to those of modern birds during flight.

For most of the history of this debate, the problem of the origin of bird flight has been polarized between two ostensibly incompatible arguments. On the one hand, those supporting the "arboreal theory" have argued that avian flight evolved through incremental stages in tree-climbing animals that leaped between branches, evolved gliding capabilities, and subsequently developed flapping flight. On the other hand, followers of the "cursorial theory" have claimed that the land-dwelling ancestors of birds never developed arboreal habits but instead evolved the ability to generate aerodynamic forces by flapping their feathered forelimbs in a fashion similar to the flight stroke of present-day birds; this view argues that the evolution of such an ability helped the land-dwelling ancestors of birds leap among fallen trees, climb rocks and other topographic heights, and increase their running speed, and in time, it allowed them to become airborne.

Numerous details about how the maniraptoran ancestors of birds became airborne remain unanswered, but some fundamental aspects of this evolutionary process have become clear. Fossil evidence shows that key characteristics enabling motions similar to the avian flight stroke—skeletal features that allow the distal forelimb to swivel around the wrist, to sweep the entire forelimb up and down, and to fold it in birdlike manner—were already present in a number of non-avian maniraptoran theropods. Well-preserved fossils also document that the glenoid facet of the scapulocoracoid (the shoulder socket), which faces laterally in birds and allows the wings to move up and down with significant amplitude, had already evolved among these dinosaurs. These discoveries show that regardless of the functional context in which these non-avian maniraptorans evolved such characteristics of the forelimb and shoulder—for climbing trees, grasping their prey, or other purposes—these animals had evolved the ability to perform movements similar to those of the flight stroke of living birds and, thus, to generate thrust and lift

using their forelimbs (Burgers and Chiappe 1999). Fossil discoveries have also shown that feathers suitable for forming an airfoil—feathers with a strong shaft and cohesive vanes made up of tightly closed barbs—evolved among non-avian maniraptorans that were incapable of flying. For example, the vaned feathers of the oviraptorosaur *Caudipteryx* formed wings that were too small in relation to its body to produce the lift necessary for taking off. The anatomy of *Caudipteryx* also points at a ground-dwelling existence. In fact, a thorough analysis of the skeletal characteristics of both tree-climbing and terrestrial vertebrates demonstrated that many other non-avian maniraptorans (including the paravian *Microraptor*) had morphologies and skeletal proportions similar to those of terrestrial mammals and ground-based birds (Dececchi and Larsson 2011). This and other studies have provided compelling evidence indicating that the dinosaurian forerunners of birds lived primarily on the ground. The feathered forelimbs of these dinosaurs, however, could have generated a certain degree of aerodynamic forces (Burgers and Chiappe 1999; Dececchi et al. 2016) (fig. 2.12). Numerous land birds—including pheasants, partridges, turkeys, and megapodes—use the thrust and lift produced by their wings to ascend all sorts of objects (Dial 2003, Dial et al. 2006, 2008). Hence, fossil discoveries of the last few decades and new analyses are congruent with the main tenet of the "cursorial theory": these findings demonstrate that the flight stroke first evolved in the ground-dwelling relatives of birds (fig. 2.12).

Fossil discoveries also suggest that not all non-avian maniraptorans were flightless. In the last two decades, a series of winged non-avian paravians from northeastern China during the Late Jurassic and Early Cretaceous has provided evidence indicating that the origin of flight along the avian line very likely preceded the rise of birds. The troodontid

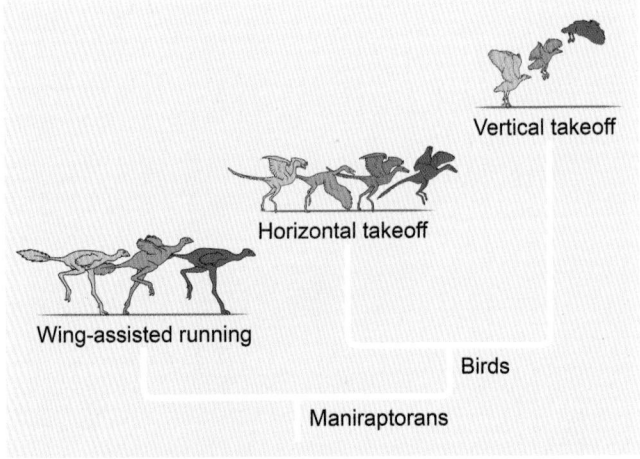

Figure 2.12. Flight may have evolved as a by-product of aerodynamic forces produced by ground-dwelling dinosaurs. *From Chiappe 2007.*

Anchiornis, the dromaeosaurids *Microraptor* and *Changyuraptor*, and a number of other similarly small dinosaurs show how, in addition to their well-developed wings, these animals possessed a set of long, vaned feathers attached to the lower legs that formed extensive "hindwings" (Xu et al. 2003, Hu et al. 2009, Han et al. 2014) (fig. 2.9). The morphology of the shoulder, thorax, and wings of these dinosaurs strongly suggest that they were capable of becoming airborne; in spite of this, precisely how (e.g., gliding, flapping) and how well they flew remains controversial (Dyke et al. 2013, Han et al. 2014; Dececchi et al. 2016). Likewise, the function of their "hindwings" has also not been entirely worked out. The soundest interpretations claim that these feathers trailed behind the lower leg and that they played a role in flight control (Han et al. 2014).

The discovery of non-avian paravians with well-developed "hindwings" highlights once again the degree of evolutionary experimentation that took place around the origin of birds. Nonetheless, the fact that the phylogenetic positions of the few known examples of flighted paravians remain unstable (i.e., various studies resolving them along different branches of the basal paravian tree) makes it difficult to determine whether their four-winged body plan (and inferred flight mode) was ancestral to the fore-winged body plan of birds or whether it was a specialization of some non-avian paravian lineages. Some researchers, who believe it to be a body plan ancestral to birds, have argued that the "hindwings" of basal paravians became gradually reduced during early avian evolution (Zhang and Zhou 2004, Longrich 2006, Zheng et al. 2013). These researchers ascribe the aerodynamic functions inferred for the large paravian "hindwings" to the plumage covering portions of the middle leg (i.e., the shin) of *Archaeopteryx* (Longrich 2006) and other basal birds (Zhang and Zhou 2004, Zheng et al. 2013). However, the fact that only short feathers extended down the legs of basal birds points at these being examples of "feather trousers," common among groups of modern birds (e.g., Accipitridae, Falconidae) and with negligible (if any) aerodynamic functions.

Further evolutionary experimentation related to the development of flight capabilities among dinosaurs close to the origin of birds is illustrated by the bizarre scansoriopterygids, a group of diminutive non-avian paravians from China during the Jurassic (Czerkas and Yuan 2002, Zhang et al. 2002, 2008a, Xu et al. 2015). The recent discovery of the peculiar *Yi qi* (Xu et al. 2015) revealed that these animals had a unique configuration of the forelimb in which a rod-like bone projected from the wrist and was apparently connected to a skinfold. Such a finding suggested that while feathered (their bodies covered by filament-like structures frayed distally), the skeletal forelimb of the scansoriopterygids sup-

ported a membranous airfoil similar to those of present-day flying squirrels and bats (and the extinct pterosaurs) but unlike the feathered airfoil of other non-avian maniraptorans and birds. Overall, *Yi* and its relatives look much less bird-like than either the troodontids or the dromaeosaurids, yet phylogenetically they appear to be a maniraptoran divergence close to the avian clade. Although much remains to be learned about these highly specialized avian forerunners, the unusual configuration of the forelimb of *Y. qi* points at potential aerodynamic specializations unlike those of any other dinosaur, thus adding to the puzzle of how the flight of birds first evolved.

In sum, discoveries in the last few decades suggest that flight evolved as a byproduct of functions performed by a suite of structures that originated among flightless, land-based dinosaurs—aerial behaviors among non-avian dinosaurs may have also evolved multiple times independently (Brusatte 2017). These findings have also shown how some lineages of non-avian dinosaurs evolved a variety of structures (e.g., "hindwings," membranous airfoils) that conferred to them the ability to fly in a very different fashion than their modern descendants. As with many attributes previously thought to be exclusive to birds, these recent fossil discoveries have demonstrated that flight (as a locomotory adaptation and with specifics notwithstanding) was also an innovation passed to their descendants by the small dinosaurian ancestors of birds.

EVOLUTIONARY AND ECOLOGICAL DIVERSITY AMONG EARLY BIRDS

Although discoveries of Mesozoic birds in the nineteenth century—the Late Jurassic *Archaeopteryx lithographica* and the Late Cretaceous *Hesperornis regalis* and *Ichthyornis dispar*—provided evidence for a very ancient origin of the group (Chiappe 1995, 2007, Feduccia 1999, Wellnhofer 2009), these fossils were limited to a handful of taxa that differed greatly anatomically and were largely restricted to nearshore and marine environments. Our poor understanding of the morphological disparity and ecological diversity of early birds remained mostly unchanged during much of the twentieth century. Such a scant fossil record led many researchers to believe that bird diversity had remained low during the Mesozoic, overshadowed by the spectacular diversity of the large dinosaurs that dominated the land ecosystems of this era. However, some key discoveries in the last few decades of the twentieth century (e.g., Elzanowski 1974, Brodkorb 1976, Walker 1981) started to document the presence of another important group of early birds—the Enantiornithes. These initial enantiornithine discoveries consisted of incomplete fossils, which in some cases were misinterpreted as belonging to modern groups (e.g., paleognaths, piciforms, and coraciiforms). Fossils of the early forerunners of modern birds—stem ornithuromorphs—were also discovered in Cretaceous sediments during the last decades of the twentieth century (e.g., Chiappe 1991, Forster et al. 1996). Consisting of rather incomplete fossils, these too were identified as belonging to modern clades in some instances (e.g., Alvarenga and Bonaparte 1992). Despite the limited (and sometimes misleading) information revealed by the studies of these fossils, their discovery helped establish the idea that the Mesozoic diversity of birds was indeed much greater than previously envisioned (Martin 1983, Chiappe 1991). This notion was to be confirmed by many subsequent discoveries, primarily from Asia.

Since the early 1990s, excavations at various Early Cretaceous sites in northeastern China (mostly in western Liaoning Province, but also in northeastern Hebei Province and the southeastern corner of the Inner Mongolia Autonomous Region) have revealed a wealth of exquisitely preserved fossils of early birds in association with an extraordinarily diverse fossil assemblage known as the Jehol Biota (Chang et al. 2003, Zhou 2014, Chiappe and Meng 2016). This explosion of fossil discoveries—together with a surge of findings of Cretaceous birds from around the world (e.g., Chiappe and Witmer 2002, Bell et al. 2010, Souza Carvalho et al. 2015)—has filled many of the anatomical, phylogenetic, and temporal gaps that existed previously (Chiappe and Dyke 2007, O'Connor et al. 2011), since the first discoveries in the 1800s (fig. 2.13). The complete skeletons of these Chinese birds, often surrounded by their plumage, have enlightened our view of the ecological diversity of early birds, refined our understanding of their evolutionary relationships, and provided information about the evolution of key traits related to the life history, physiology, ecology, and locomotion of the group like no other previous evidence.

The new discoveries have illustrated that the vast diversity of living birds is in fact a remnant of an archaic evolutionary radiation that can be traced back to *Archaeopteryx*. This crow-sized animal still stands alone as the only record of birds in the Jurassic, and it is broadly considered the most

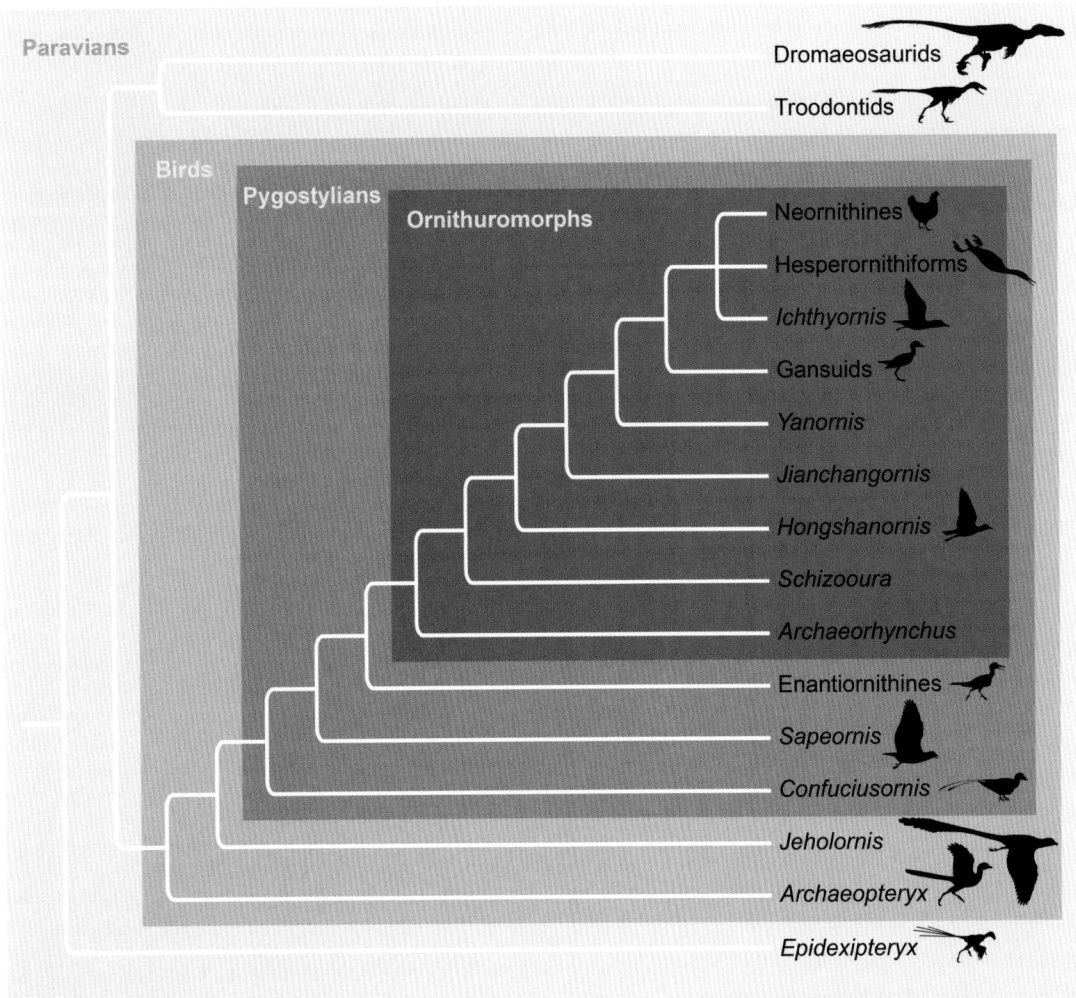

Figure 2.13. Cladogram depicting the phylogenetic relationships of the main lineages of pre-modern birds. *Modified from O'Connor et al. 2011.*

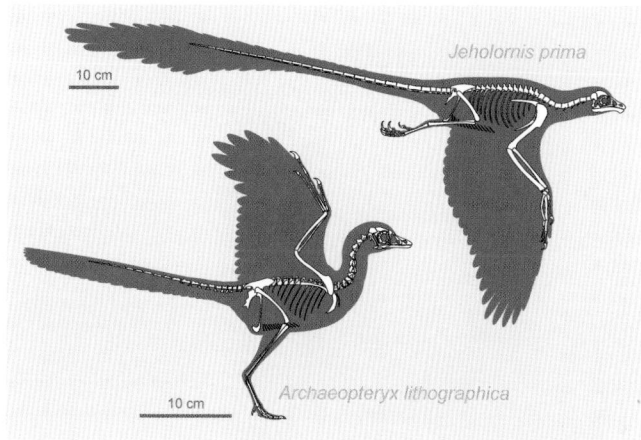

Figure 2.14. Skeletal reconstructions of *Archaeopteryx lithographica* and *Jeholornis prima,* two basal lineages of long bony-tailed birds.

primitive member of the group (Chiappe 1995, Wellnhofer 2009, Brussatte et al. 2015). The 150-million-year-old *Archaeopteryx* had toothed jaws, clawed wings, and a long bony tail (fig. 2.14), and few features that set it apart from its paravian predecessors, to the point that some phylogenetic hypotheses have placed it within Deinonychosauria (Dromaeosauridae + Troodontidae) (e.g., Xu et al. 2011, Godefroit et al. 2013), albeit not convincingly. Computer tomography of the braincase of one fossil (*Archaeopteryx*'s London specimen) reveals a brain that is comparable in relative size and architecture to those of its non-avian maniraptoran predecessors but substantially larger and more complex than those of more primitive dinosaurs. Details of the inner ear also indicate that *Archaeopteryx* was able to move in a three-dimensional space in ways comparable to present-day birds (Dominguez Alonso et al. 2004). These studies have shown that *Archaeopteryx* (and its non-avian maniraptoran relatives) had evolved the degree of neurological complexity required for flight (Balanoff et al. 2013). Additionally, *Archaeopteryx* had proportionally larger wings (and therefore, lower wing-loading) when compared with non-avian maniraptorans of similar size. Nonetheless, the body plan of the earliest bird suggests an aerodynamic performance similar to that of landfowl, rails, and other heavy fliers. In fact, new estimations of the wing-loading and wingbeat frequency of extinct birds based on the statistical relationship between these aerodynamic parameters and the proportions of the wing bones in modern birds suggest that *Archaeopteryx* was most likely incapable of any prolonged flapping flight.

New discoveries from the Early Cretaceous Jehol Biota have revealed that other primitive avians with long bony tails preceded the evolution of birds with an abbreviated bony tail, one composed of fewer vertebrae ending in a pygostyle (the structure that supports the rectricial bulbs that anchor the flight feathers of the tail). Several taxa have been named based on fossils unearthed from 120- to 125-million-year-old deposits of northeastern China (e.g., *Jeholornis prima, J. palmapenis, J. curvipes, Shenzhouraptor sinensis, Jixiangornis orientalis*) (fig. 2.15). Nonetheless, the similarity of these fossils suggests they are all members of the Jeholornithidae (most likely just *Jeholornis*), a clade that was named in honor of the Jehol Biota. With a size similar to that of a turkey, the best-known species of this clade, *Jeholornis prima* (Zhou and Zhang 2002a, 2003a) had a relatively tall skull with small teeth on the maxillae and at the tip of the dentaries, and wings ending in three clawed fingers. The overall morphology of the skeleton of *Jeholornis* was very primitive with respect to that of modern birds; nonetheless, it displayed several advanced characteristics—particularly in the shoulder girdle and forelimb—when compared with *Archaeopteryx* (Zhou and Zhang 2003a)

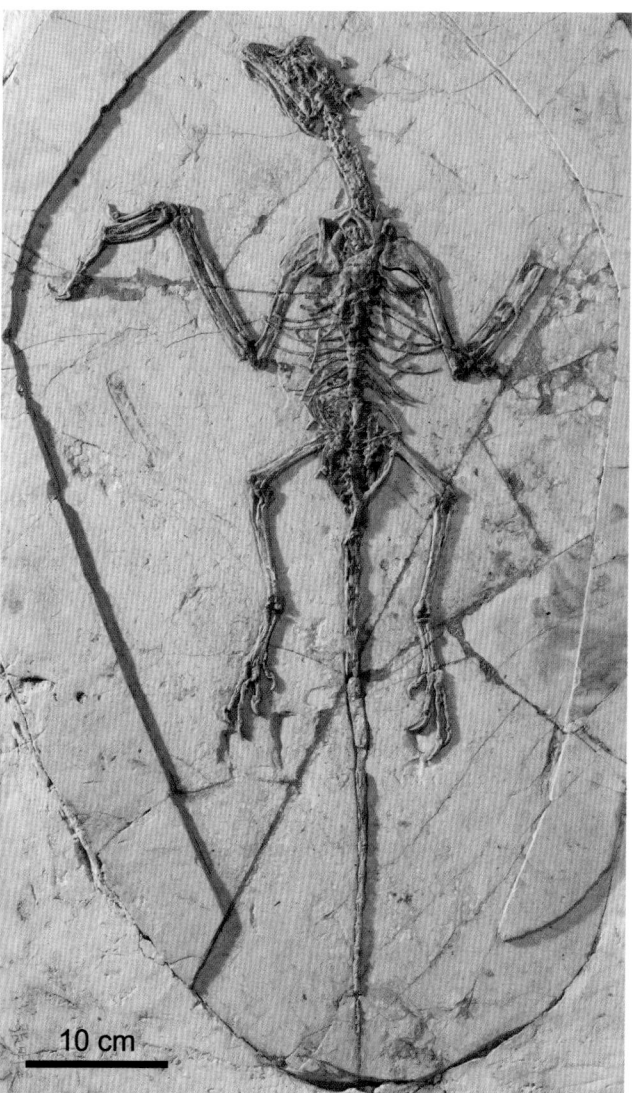

Figure 2.15. The long, bony-tailed Early Cretaceous *Jeholornis prima* from the Jehol Biota of northeastern China.

(fig. 2.14). Despite these morphological advances, the tail of *Je-holornis* was proportionally longer (with a greater number of vertebrae) than that of *Archaeopteryx* (Zhou and Zhang 2003a). The tail feathers were also differently organized: unlike the broad, frond-like feathered tail of *Archaeopteryx*, the feathered tail of the jeholornithids was narrower, and it ended in either a terminal fan (e.g., *J. prima*) or in a palm-like arrangement of feathers (e.g., *J. palmapenis*) (O'Connor et al. 2012b). Such differences in the size of the caudal airfoil indicate that the tail of *Jeholornis* was able to generate less lift than that of *Archaeopteryx*, which in turn points to the wings of *Jeholornis* playing a more important role in the generation of aerodynamic forces. Characteristics of their plumage, their large wings, and details of their "flight-ready" brains indicate that *Archaeopteryx* and *Jeholornis* had a degree of aerial proficiency. Comparisons with a variety of living vertebrates that lived on either trees or the ground (Dececchi and Larsson 2011) suggest that *Archaeopteryx* and *Jeholornis* were predominantly land dwellers. All things considered, these animals might have been able to fly for short distances, requiring either a take-off run or an elevated launch to become airborne (Burgers and Chiappe 1999) (fig. 2.12).

A large ensemble of more advanced birds with abbreviated bony tails is also known from the Jehol Biota (fig. 2.13). The varying morphology of the skulls, teeth, wings, and feet of these fossils indicates that even at this early phase of their evolutionary history, birds had expanded into a wide range of ecological niches. Important evolutionary innovations evident in the wings, shoulders, and tails of these 131- to 120-million-year-old fossils also indicate that birds evolved sophisticated flying capabilities soon after *Archaeopteryx*; such aerodynamic improvements most likely influenced the ecological diversification witnessed by the Jehol avifauna.

Paramount among these flight-correlated transformations was the abbreviation of the tail and the development of a pygostyle (a platelike, compound bone at the end of the tail that in modern birds supports the fatty rectricial bulbs also known as the pope's nose; see Gill 2007). The reduction of the bony tail moved the center of gravity forward, decoupling the function that the tail had in terrestrial locomotion and transforming its aerodynamic role (Gatesy and Dial 1996)—namely, the reduction of the tail led to new ways of walking and flying. Such a skeletal abbreviation most likely took place in concert with developments in the forelimbs that boosted the wings' role in flight control as the latter became responsible for the subtle adjustments (in the wing planform, angle of attack, and others) that are critical for aerial proficiency. Evidence for this is hinted by the large size and shape of wings of the most primitive short-tailed birds—confuciusornithids and sapeornithids—that are recorded in

5 cm

Sapeornis chaoyangensis

5 cm

Confuciusornis sanctus

Figure 2.16. Skeletal reconstructions of the most basal pygostylian lineages: *Confuciusornis sanctus* and *Sapeornis chaoyangensis*.

the Jehol Biota (fig. 2.16). Unfortunately, little is known about the evolutionary pattern by which the avian tail was reduced. One fossil from the Jehol Biota, the 125-million-year-old *Zhongornis haoae* (Gao et al. 2008), suggests that the reduction in the number of caudal vertebrae pre-dated the evolution of the pygostyle. New developmental studies have the promise of clarifying aspects of this important transformation (Rashid et al. 2014); nonetheless, additional fossil evidence is needed to elucidate more precisely when and how the abbreviation of the tail (and the evolution of the pygostyle) took place during the early evolution of birds.

The 125-million-year-old *Confuciusornis sanctus* (Chiappe et al. 1999)—known from hundreds of specimens from the Jehol Biota—is possibly the most primitive known pygostylian (the group containing all birds with a pygostyle, including living ones) (fig. 2.17). A handful of other confuciusornithids (e.g., *Changchengornis hengdaoziensis*, *Eoconfuciusornis zhengi*) are also known from the Jehol Biota (Ji et al. 1999, Zhang et al. 2008b). These jay- to crow-sized birds had strong forelimbs ending in long hands with enormous claws, elongated wings of high aspect ratio, and stout jaws lacking teeth (Chiappe et al. 1999, Chiappe 2007) (fig. 2.16); in fact, the confuciusornithids are the earliest examples of fully beaked birds. The

10 cm

Figure 2.17. Confuciusornis sanctus, male (with ornamental tail feathers) and female (without ornamental tail feathers).

less abundant and phylogenetically more derived *Sapeornis chaoyangensis* (Zhou and Zhang 2002b, 2003b, Gao et al. 2012b), also from the Jehol Biota, had long wings as well; these, however, were much broader than the tapering wings of the confuciusornithids. In fact, a recent study combining computational modeling and morphofunctional analyses suggests that *Sapeornis* would have been able to use its broad wings for soaring on continental thermals (Serrano and Chiappe 2017). Several species of sapeornithids have been named (e.g., *Didactylornis jii, Sapeornis angustis, Omnivoropteryx sinousaorum, Shenshiornis primita*), although all these are likely synonyms of *Sapeornis chaoyangensis* (Gao et al. 2012b, Pu et al. 2013). Reaching a size similar to that of a large gull (e.g., *Larus occi-*

dentalis), *Sapeornis* was larger than *Confuciusornis*. Its tall skull had a few spade-shaped teeth; the forelimbs were long and carried large claws on only two fingers (inner and middle), and the feet had powerfully grasping toes (Zhou and Zhang 2003b, Gao et al. 2012b, Pu et al. 2013, Chiappe and Meng 2016) (fig. 2.16). The wings of both *Sapeornis* and *Confuciusornis* also showed an additional skinfold, the propatagium, which connected the wrist with the shoulder (Chiappe et al. 1999, Chiappe and Meng 2016) (fig. 2.16). This delta-shaped skinfold, either absent or poorly developed in both *Archaeopteryx* and *Jeholornis*, would have increased the surface area and, hence, the lift-generating capabilities, of the wing. While the wing bones of these early pygostylians show many primitive anatomical details when compared with those of modern birds, the proportionally larger size of their wings and their reduced bony tail suggest that the flight performance of these birds was superior to that of their long bony-tailed predecessors (Serrano et al. 2017, Serrano and Chiappe 2017).

The hundreds of known fossils of *Confuciusornis sanctus* include some that are almost half the size of others (based on linear measurements of long bones), yet all of them belong to fledged individuals that are anatomically indistinguishable from one another. Such an extensive growth series indicates that, unlike most living birds, *C. sanctus* (and presumably its close relatives) grew over several years before becoming full-grown (Chiappe et al. 2008, Marugan-Lobon et al. 2011). The same growth pattern can be inferred for *Sapeornis chaoyangensis*, for which a similar growth series is known, albeit one in which the size of the specimens falls within a narrower range (Gao et al. 2012b, Pu et al. 2013). These life history inferences are also supported by histological studies of their bone tissues, which reveal annual growth rings (lines of arrested growth), indicating that *Confuciusornis* and *Sapeornis* grew over multiple years.

The extraordinary sample of *C. sanctus* also shows that some specimens carried a pair of long ornamental feathers projecting from the tail, while others had unornamented tails (fig. 2.17). The discovery of medullary bone (a female-only type of bone tissue) in a specimen lacking these feathers (Chinsamy et al. 2013) has provided support to the interpretation of this differential plumage as evidence of sexual dimorphism, in which ornamented specimens are regarded as males (Chiappe et al. 1999, Marugan-Lobon et al. 2011). Furthermore, the fact that these ornamental tail feathers (a secondary sexual trait that could be used as a proxy for the onset of sexual maturity) are observed in specimens of *C. sanctus* that are significantly smaller than the largest known specimens of this species indicates that, unlike modern birds, these ancient birds begun reproducing well before they reached

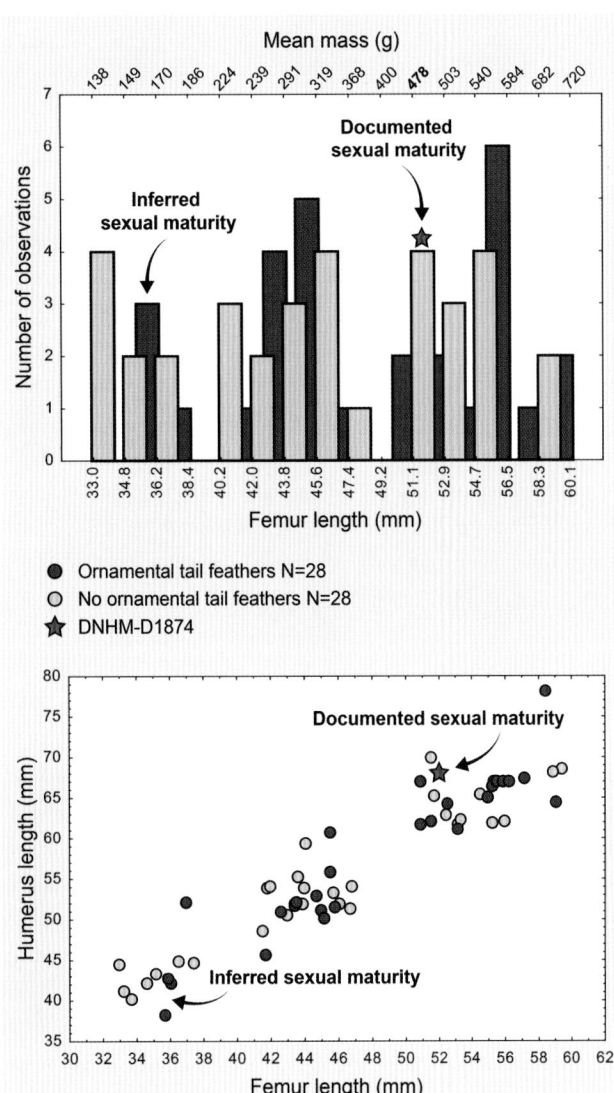

Figure 2.18. Distribution of specimen sizes within a growth series of *Confuciusornis sanctus*. The star points at a female for which the gender was confirmed histologically by the discovery of medullary bone. The onset of sexual maturity is inferred by the presence of ornamental feathers in individuals interpreted as males. The available evidence suggests that early birds such as *Confuciusornis sanctus* reached sexual maturity prior to reaching somatic maturity. *From Chinsamy et al. 2013.*

full-grown size (Chinsamy et al. 2013), a reproductive pattern inherited from their dinosaurian forerunners (Lee and Werning 2008, Makovicky and Zanno 2011) (fig. 2.18).

A great variety of more modern-looking birds are also recorded in the Jehol Biota. Among these fossils are the earliest records of enantiornithines worldwide (e.g., *Protopteryx fengningensis*, *Eopengornis martini*) (Zhang and Zhou 2000, Wang et al. 2014), which are found in 131-million-year-old layers representing the earliest phases of the Jehol Biota. The enantiornithines represent the largest radiation of premodern birds (Chiappe and Walker 2002). As mentioned above, these

birds were not recognized as a distinct clade until the early 1980s (Walker 1981); the more than 70 enantiornithine species named in the last 35 years are a testimony to the remarkable burst of recent discoveries of Mesozoic birds (fig. 2.19).

Enantiornithine fossils are indeed extremely abundant in the fine sediments of Jehol Biota. These rocks alone have yielded the remains of an embryo (Zhou and Zhang 2004), numerous hatchlings (e.g., Chiappe et al. 2007, Zheng et al. 2012), and hundreds of subadult and adult individuals (including the holotypes of more than 30 named species; see Chiappe and Meng 2016). Yet, fossils of these birds are known from Cretaceous deposits in every continent except Antarctica (Chiappe and Walker 2002) and tantalizing discoveries in 100-million-year-old amber are revealing exceptional details of their anatomy (Xing et al. 2016, 2017). The abundant fossil record of this diverse clade shows that during their long evolutionary history—approximately from 131 to 66 million years ago—the enantiornithines evolved many different lifestyles and ecological specializations, inhabiting a variety of environments (from deserts to temperate forests and from marshes to oceanic littorals).

The skulls of most enantiornithines were toothed (upper and lower jaws), although the teeth differ greatly in morphology, size, and position within the jaws (O'Connor and Chiappe

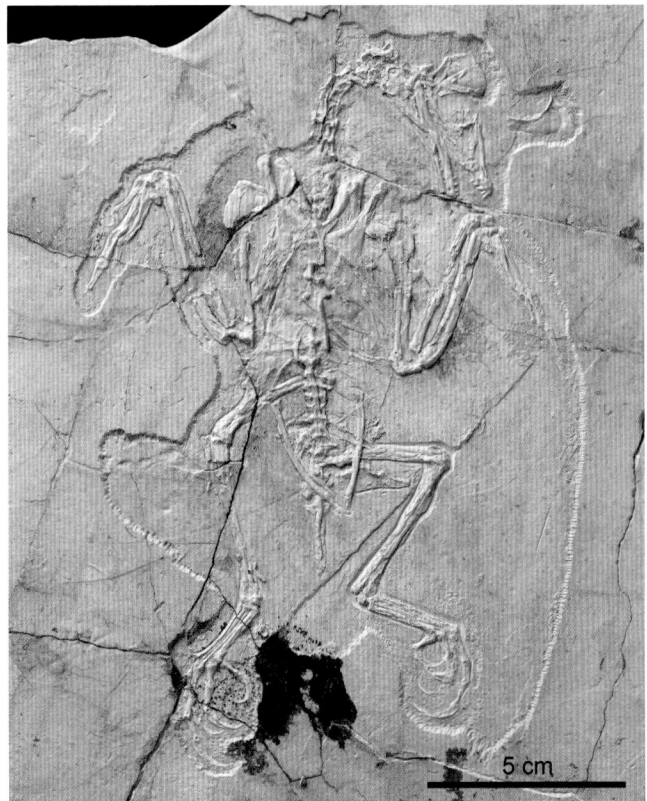

Figure 2.19. The Early Cretaceous enantiornithine *Zhouornis hani* from the Jehol Biota.

2011) (fig. 2.20). The skeletal anatomy of their wings evolved proportions approaching those of living birds, but the fore-limbs retained small claws on two of the three fingers (inner and middle) (Chiappe and Walker 2002) (fig. 2.21). The feet of most enantiornithines unmistakably show grasping capabilities; their perching design is consistent with the general interpretation of these birds as tree-dwellers (fig. 2.19). Some taxa, however, show different pedal morphologies indicating that wading and swimming lifestyles also evolved among these birds (Chiappe 1993). A variety of feeding specializations (e.g., insectivory, durophagy, and piscivory, among others) are also interpreted on the basis of the morphology of their jaws and teeth (O'Connor and Chiappe 2011) (fig. 2.20). Some small Jehol enantiornithines (e.g., *Longirostravis hani, Shan-weiniao cooperorum, Rapaxavis pani*) with long and delicate snouts carrying tiny peg-like teeth at their tips (Hou et al. 2004, Morschauser et al. 2009, O'Connor et al. 2009) have

Figure 2.21. Skeletal reconstruction of the Early Cretaceous enantiornithine *Longipteryx chaoyangensis* from the Jehol Biota.

been consistently interpreted as mud probers that foraged along lakeshores (Chiappe and Meng 2016). Coeval sites in South Korea reveal a diversity of probing traces (mixed with avian webbed footprints) documenting that different types of birds had evolved probing feeding behaviors by the Early Cretaceous (Falk et al. 2014). Gut contents and other associated evidence of diet are exceedingly rare among enantiornithines. Nonetheless, the available evidence suggests that some of these birds had highly specialized feeding behaviors (Sanz et al. 1996, Wang et al. 2016a), including sap eating (Dalla Vecchia and Chiappe 2002).

Enantiornithines also evolved important innovations in their plumage and skeletal flight apparatus. Their wings carried flight feathers (primaries and secondaries) similar in number and arrangement to those of today's flying birds. Their shoulder bones developed features that would have imparted more power to the muscles most responsible for the wing stroke. Rare fossils show evidence of an alula (a tuft of small feathers attached to the inner finger), which forms a leading-edge slot that enhances low-speed flight when deployed (critical during their takeoff and landing) (Sanz et al. 1996); they have also shown that these birds had evolved a modern-looking network of ligaments and muscles, which in living birds controls the movements of the flight feathers (Navalon et al. 2015). The feathered tail of enantiornithines varied significantly (O'Connor et al. 2016), although most fossils from the Jehol Biota had a wedge-like tail composed of down-like feathers and a pair of long ornamental feathers (fig. 2.22). Such tails clearly lacked any aerodynamic function. Nonetheless, in two taxa (*Shanweiniao cooperorum* and *Chiappeavis magnapremaxillo*) the feathered tail included several pairs of shafted feathers capable of generating lift and playing a more significant aerodynamic role (O'Connor et al.

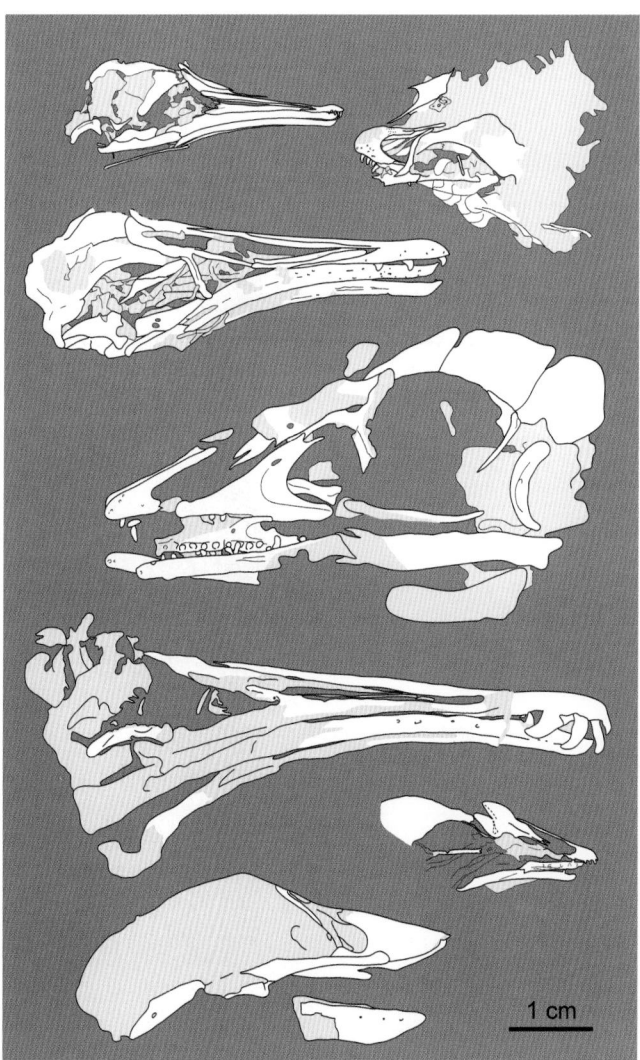

Figure 2.20. Different types of skull morphologies among enantiornithines. *From O'Connor and Chiappe 2011.*

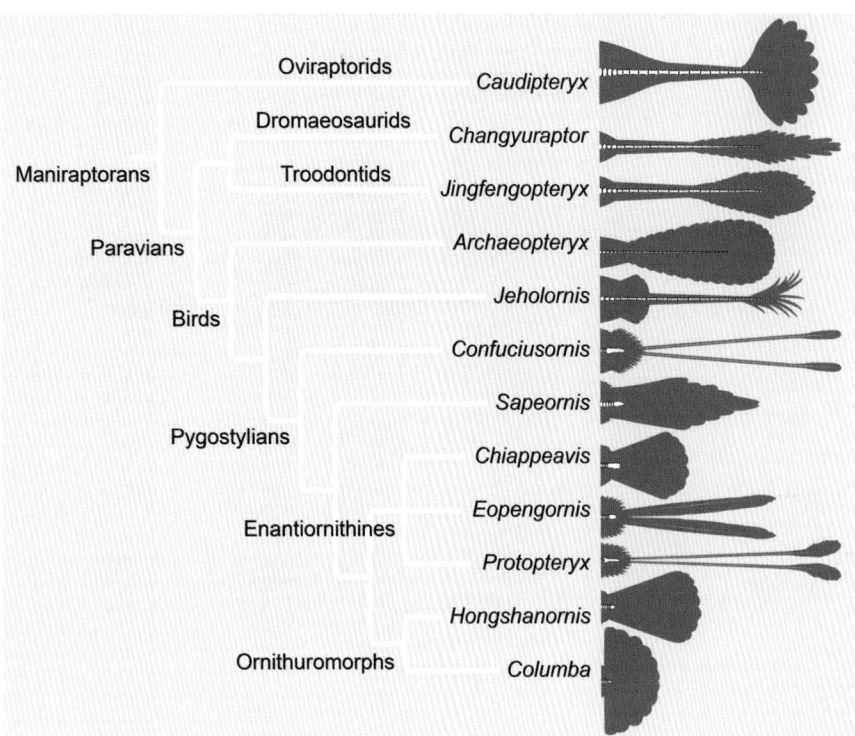

Figure 2.22. Different morphologies of the caudal plumage of Mesozoic birds. *Modified from O'Connor et al. 2016.*

2009, 2016). The presence of these aerodynamic tails in at least two enantiornithines suggests that these birds may have evolved a version of rectricial bulbs, the fatty structure anchored by the pygostyle that in modern birds plays a key role in controlling the fanning of the feathered tail. However, soft tissues visible around the pygostyle of some exceptionally preserved fossils do not reveal the presence of well-formed rectricial bulbs, suggesting that these were not developed in many enantiornithines. In sum, the development of a more compact distal segment of the forelimb carrying modern-looking skinfolds and an alula, the modern proportions of wing bones, the important modifications of their shoulder girdle (related to the size and functioning of the flight muscles), and the presence in some taxa of an aerodynamic tail are some of the most noticeable transformations shown by these birds when compared with the most primitive short-tailed birds (e.g., *Confuciusornis*, *Sapeornis*). Furthermore, while their sizes ranged significantly during their long evolutionary history (from that of a small songbird to that of a vulture), many Jehol enantiornithines were substantially smaller than the most primitive short-tailed birds (e.g., *Confuciusornis*, *Sapeornis*). The miniaturization experienced by the earliest enantiornithines has a well-understood functional significance: flight performance correlates positively with size reduction. Furthermore, smaller birds—typically with comparatively smaller wing-loading—are more maneuverable, and their flight is energetically more efficient. Estimates for the wing-loadings of enantiornithines are significantly

lower than for those of either *Sapeornis* or *Confuciusornis*, evidence that once again underscores the superior flight competence of the former. In all, the significant transformations of the skeleton, plumage, and dimensions of these birds suggest that, even at the onset of their evolutionary history, enantiornithines were able to fly in ways similar to those seen among small living birds. It is most likely that the evolution of these enhanced flying capabilities played a key role in the evolutionary success of the enantiornithines, which by about 120 million years ago seem to have risen to dominance.

A wealth of other enantiornithine fossils shows key evidence for understanding several aspects of their biology. An embryo and multiple hatchlings from the Jehol Biota (Zhou and Zhang 2004, Chiappe et al. 2007), a neonate contained in mid-Cretaceous amber (Xing et al. 2017), and histological evidence from other fossils, indicate that most enantiornithines had precocial hatching strategies in which their chicks were nidifugous, able to find food by themselves, and capable of flying soon after they hatched. The exceptional preservation of the Jehol Biota has also revealed some cases in which ovarian follicles have been preserved; these remarkable fossils tell us that enantiornithines had reproductive physiologies involving hierarchical ovulation (O'Connor et al. 2013). Some researchers (e.g., Zheng et al. 2013, Wang et al. 2016b) have argued that these fossils also show evidence of a reduced right ovary (an adaptation typical of most living birds; Gill 2007). Nonetheless, the preservation of the known specimens with these structures makes this interpretation problematic (it is

questionable whether ovaries from either side can be indi-
vidualized in laterally preserved specimens). The different
examples of these birds with ornamental tail feathers (e.g.,
Zhang and Zhou 2000, Souza Carvalho et al. 2015) and their
differing morphology (Wang et al. 2014) suggest the presence
of sexually dimorphic species and mating systems in which
ornamentation plumage played an important role.

Rocks from the Early Cretaceous also record a number of
transitional fossils that heralded the evolution of the closest
relatives of modern birds (fig. 2.13). In some respects these
primitive ornithuromorphs resemble the enantiornithines,
but their skeletons show for the first time clear trademarks
of their living counterparts. The majority of these primitive
ornithuromorphs were lightly built flying birds, whose sizes
tended to be larger than those of the contemporaneous enantiornithines. Like the latter, both their skeletons and plumage
showed clear evidence of enhanced aerodynamic capabilities.
It is within these birds that we witness the origin of the extremely fast rates of body maturation characteristic of modern birds, which reach their full body size within a year after
hatching.

Many studies have supported a sister-group relationship between the enantiornithines and the ornithuromorphs (e.g., Chiappe 1996, 2002, Gao et al. 2008, O'Connor
and Zhou 2012); the latter clade includes all living birds and a
number of Mesozoic lineages that are more closely related
to modern avians than to the enantiornithines (fig. 2.13). Ornithuromorphs and enantiornithines split from a shared ancestor that most likely had already evolved key traits with

advanced aerodynamic functions. Like the enantiornithines,
the earliest ornithuromorphs—known from throughout the
Jehol Biota—had wing bones of modern proportions (those
from the forearm were the longest and those from the hand
formed a compact unit, a design best for thrust generation).
The morphology of the shoulder girdle and the size of the
keeled breastbone indicate that even the earliest ornithuromorphs (*Archaeornithura meemannae*, *Archaeorhynchus spathula*,
and *Xinghaiornis lini*, among other Jehol species) (Zhou and
Zhang 2006, Zhou et al. 2013, Wang et al. 2013, 2015) had strong
flight muscles capable of elevating and depressing their wings
with large amplitude. They also had evolved adaptations for
landing, as shown by the modernization of the skeletal anatomy of the pelvis and hindlimb. Fossils with well-preserved
plumage also show that these archaic birds had large wings
(with moderate aspect ratios and an alula) and aerodynamic
tails capable of fanning their feathers (Chiappe et al. 2014)
(fig. 2.22). Their short pygostyle most likely supported a version of the rectricial bulbs, which muscles controlled the fanning of the rectrices (main caudal feathers), thus adjusting it
to the requirements of flight. In all, the anatomy of the Jehol
ornithuromorphs, which include the oldest known members
of this main avian clade, indicates that these birds evolved
aerodynamic surfaces and flight modes comparable to those
of many modern birds more than 130 million years ago (Chiappe et al. 2014) (fig. 2.23).

Fossils of Jehol ornithuromorphs document several stages
of the evolution of the group that includes all living birds;
phylogenetically, their different lineages constitute successive

Figure 2.23. The
Early Cretaceous
*Hongshangornis
longicresta*, one of
the basal lineages
of ornithuromorphs
from the Jehol
Biota.

5 cm

Figure 2.24. Skeletal reconstruction of *Gansus zheni*. This taxon is phylogenetically more closely related to modern birds than any other Early Cretaceous bird from the Jehol Biota.

outgroups to the neornithine clade. On the one hand, the most basal ornithuromorph—the 125-million-year-old Jehol *Archaeorhynchus spathula* (Zhou and Zhang 2006)—possesses morphologies that are highly reminiscent of contemporaneous enantiornithines from the Jehol Biota, thus indicating a proximity to the split between ornithuromorphs and enantiornithines. On the other hand, the anatomically modern *Gansus zheni* (Liu et al. 2014) and *Iteravis huchzermeyeri* (Zhou et al. 2014)—possibly two synonymous species—represent a very close outgroup of the neornithine clade (fig. 2.24). Fossils of the Jehol ornithuromorphs reveal that, early in their evolutionary history, these birds differed ecologically from their coeval enantiornithines. In contrast to the small sizes and perching adaptations typical of the latter, the most basal lineages of ornithuromorphs show larger sizes and anatomical structures that are better adapted for either a semiaquatic or a terrestrial existence (Zhou et al. 2012, 2013, Chiappe et al. 2014).

Several other lineages of stem ornithuromorphs are known from younger Cretaceous rocks (e.g., Chiappe 1996, Norell and Clarke 2001, Bell et al. 2010). One of these groups, the hesperornithiforms, includes a large radiation of medium-to-large toothed birds found in sediments deposited at the bottom of shallow seas and near shore environments in North America and Asia (Marsh 1880, Bell 2013). These animals, with flight capabilities from reduced to none and foot-propelled swimming specializations, are often compared to living diving birds such as grebes and loons (indeed, these three lineages have evolved a number of convergent morphologies in their

hindlimbs and pelves). Albeit entirely restricted to the aquatic realm, these birds had a rich evolutionary history represented by more than 20 species that, collectively, span from about 113 million years ago to the latest Cretaceous mass extinction, 65.5 million years ago (Bell and Chiappe 2015). During this long history, the hesperornithiforms transitioned from birds with limited diving capabilities and, at best, rudimentary flight performance to flightless animals with highly advanced diving specializations. Throughout their evolution, these birds also experienced multiple independent events of body size escalation that were uncorrelated with the stepwise acquisition of diving adaptations (Bell and Chiappe 2015). These advanced diving capabilities are best illustrated by the skeletal anatomy of *Hesperornis regalis* (Marsh 1880) (fig. 2.25), a species that reached the length of an Emperor Penguin (*Aptenodytes forsteri*). Despite this, and the fact that hesperornithiforms had teeth in the maxillae and dentaries, and highly unusual cranial features (Elzanowski 1991), the overall morphology of their skeletons was in many respects modern-looking to the point that some studies have identified these birds as members of modern lineages (e.g., Cracraft 1982). Current studies, however, overwhelmingly support

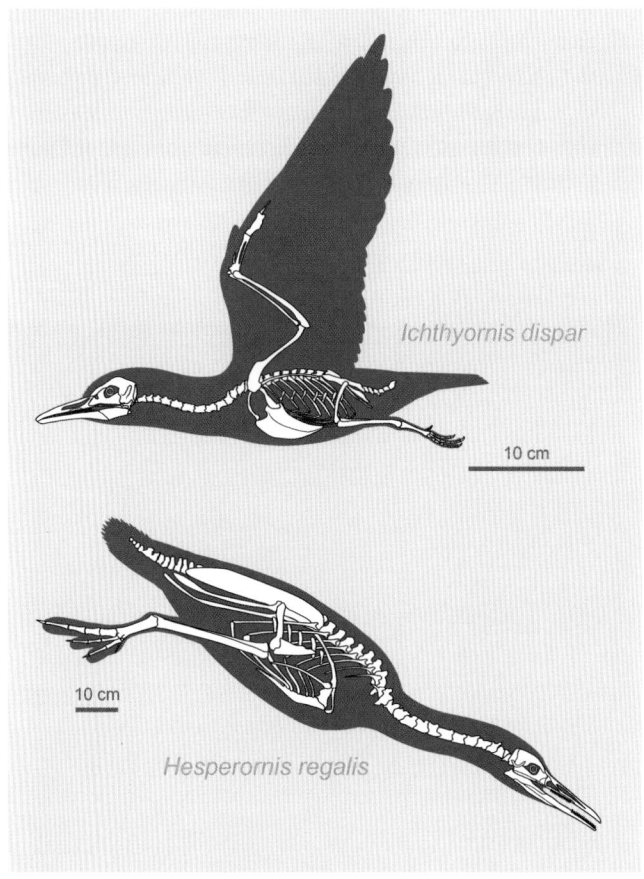

Figure 2.25. Skeletal reconstructions of the Late Cretaceous *Hesperornis regalis* and *Ichthyornis dispar*.

their phylogenetic placement as a close outgroup of the modern bird clade (neornithines) (e.g., O'Connor and Zhou 2012, Liu et al. 2014, Bell and Chiappe 2015). Other close outgroups of neornithines include *Ichthyornis dispar* (fig. 2.25), which in most anatomical details represents a step closer to modern birds than the hesperornithiforms. Despite the advanced degree of modernization of the skeleton, this small flying bird retained teeth in the maxillae and dentaries. Abundant remains of *Ichthyornis* have been found in many of the same sites that contain the remains of *Hesperornis regalis* (Clarke 2004), with which it shared the nearshore and littoral environments of the Western Interior Seaway, a shallow sea that flooded much of North America during the Late Cretaceous.

In addition to the large diversity of premodern birds that lived during the Mesozoic, early representatives of present-day lineages have also been unearthed in rocks from the last part of the Cretaceous period (Hope 2002, Kurochkin et al. 2002, Clarke et al. 2005, Ksepka and Clarke 2015). The divergence of early neornithine lineages in pre-Cenozoic times has been inferred from many molecular phylogenies, which have consistently concluded that modern birds first differentiated more than 100 million years ago, in the Early Cretaceous (e.g., Cooper and Penny 1997, Lee et al. 2014b, Jarvis et al. 2014). While documenting the existence of some early divergences prior to the end of the Mesozoic, even the oldest Late Cretaceous neornithines are many millions of years younger than the origin of the group as predicted by molecular studies (Mayr 2014, Ksepka and Clarke 2015). Such discrepancy has been somewhat mitigated by the discovery of the Jehol gansuids (e.g., *Gansus zheni* and *Iteravis huchzermeyeri*) (fig. 2.26), which are remarkably modern looking and yet lived 120 million years ago. While Early Cretaceous fossils that can be confidently identified as belonging to modern lineages are yet to be found, the existence of close relatives of neornithines at such an early phase of avian evolution eases the discrepancy between the molecular and paleontological inferences in what refers to the origin of modern birds (Liu et al. 2014).

The factors that led to the survival of the earliest neornithine divergences through the end-Cretaceous mass extinction remain unknown, but the fossil record shows that most major clades of modern birds (corresponding to traditional neornithine orders) had already originated 50 to 55 million years ago (Mayr 2009, Ksepka and Clarke 2015). Molecular inferences indicate that the last descendants of the dinosaurian radiation that started more than 230 million years ago (Brusatte et al. 2010) underwent a large diversification event at the beginning of the Paleogene (early in the Cenozoic era) (Prum et al. 2015). This explosive radiation of neornithines in the aftermath of the Cretaceous-Paleogene extinction es-

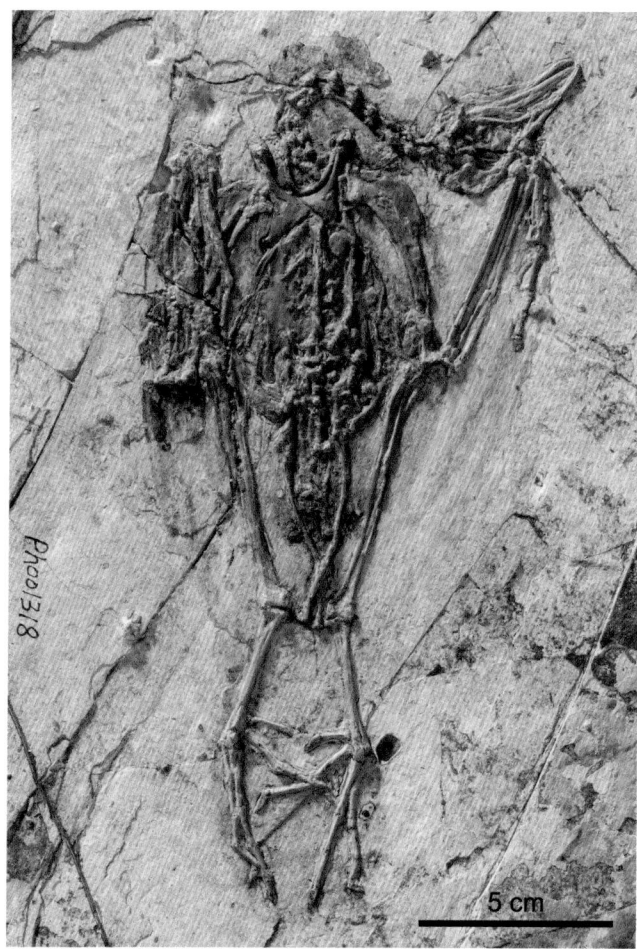

Figure 2.26. The 120-million-year-old *Gansus zheni* is the anatomically most-modern bird in the Jehol Biota.

tablished the foundations for the morphological disparity and ecological diversity of today's birds, recapitulating the evolution of lifestyles that characterized many archaic lineages of birds that had thrived millions of years earlier, during the Mesozoic era.

BIRDS AS LIVING DINOSAURS

In the last few decades, our understanding of the origin and subsequent diversification of birds has advanced at an unparalleled pace. A wealth of new evidence now shows that many of the features previously considered avian trademarks first evolved within theropod dinosaurs. The strength of the hypothesis that birds evolved within the maniraptoran theropods is demonstrated by the convergent results of a diversity of studies within a multitude of scientific disciplines. Today, the theropod origin of birds is supported by disparate evidence ranging from skeletal anatomy to molecular data. This evolutionary conclusion indicates that modern birds are but a branch of the tree of dinosaurs.

Modern feathers—already present among the closest dinosaurian relatives of birds—evolved as the result of stepwise transformations at the follicle level, which led to increased feather complexity. While the primary function of the feather remains controversial, compelling evidence indicates that simple filamentous feathers covered the bodies of many non-avian theropods and that these structures did not originate for the purpose of flight. Feathers evolved aerodynamic morphologies (shaft and cohesive vanes) and functions as some non-volant theropod dinosaurs developed airfoils with the capability of generating flight forces. Thus, flight could have evolved as a by-product of the aerodynamic forces generated by the feathered wings of non-avian maniraptorans and as a result of changes in wing-loading (the result of an evolutionary trend toward smaller body sizes and larger wing surfaces). A diversity of non-avian dinosaur lineages close to the origin of birds (*Microraptor, Yi*, and others) experimented with different flight modalities—perhaps evolving independently—based on structures that lack analogues among living birds; theories on the flight performance of these animals remain contentious. Over the subsequent course of evolution, the refinement of flight involved numerous transformations, including not only modifications in the skeleton and plumage but also changes in physiology, musculature, and sensory perception. While the specifics of their flight modes and performance are far from being fully understood, evidence from a number of premodern lineages suggests that birds reached flight capabilities similar to those of their living relatives very early in their evolutionary history.

A wealth of new discoveries of Mesozoic avian lineages—from rocks greatly separated in time and distributed across the globe—has documented an enormous morphological and ecological diversity among the early predecessors of modern birds. These discoveries have helped us understand the incremental refinement of the aerodynamic tool kit of birds and the origin of key traits related to their life history and physiology. Fossils from the Mesozoic have also documented the earliest divergences of modern birds, providing evidence for the deep origin of the group. The transformations experienced by the numerous lineages of birds that diverged, evolved, and became extinct during the Mesozoic led to the development of the modern avian body plan and set the evolutionary stage for the main diversification of present-day lineages in post-Cretaceous times.

In all, a spectacular array of fossils discovered worldwide has consolidated the notion that birds have a deep evolutionary history rooted among the magnificent dinosaurs of the Mesozoic; these discoveries have documented the stepwise nature of one of the most fascinating evolutionary transitions in the history of life, one connecting animals akin to *Tyrannosaurus rex* with the diversity of today's birds.

KEY POINTS

- Birds are living dinosaurs. A wealth of evidence, from diverse morphological traits to genome size, support the hypothesis that birds are evolutionarily nested within theropod dinosaurs, with dromaeosaurids and troodontids as their closest relatives.
- Many of the typical characteristics of birds, including feathers, brooding, and flight, had their earliest manifestations among birds' dinosaurian forerunners. Many of the unique adaptations of the avian body—rapid growth, large egg volume, and endothermy, among others—evolved in a stepwise fashion over millions of years.
- Avian flight evolved from small-sized, ground-dwelling theropod dinosaurs capable of generating aerodynamic forces with feathered wings. While flight refinement during avian evolution appears to have followed a stepwise fashion, different flight modes were attained early in the history of birds.
- Numerous lineages of early, premodern birds diverged, radiated, and became extinct during the Mesozoic era. Fossils of these archaic groups provide evidence of a wide range of ecological differentiation.
- Modern birds originated in the Cretaceous, more than 80 million years ago. While the Mesozoic diversity of these birds remained low and limited to a few basal lineages, these birds experienced an extraordinary radiation in the aftermath of the Cretaceous-Paleogene extinction.
- The fossil record of Mesozoic birds, and of their dinosaurian predecessors, provides a deep-time, historical context for conservation efforts on the only living descendants of the magnificent dinosaurs that dominated the world's terrestrial vertebrate faunas across more than 150 million years.

References

Alvarenga, H. M. F., and J. F. Bonaparte (1992). A new flightless land bird from the Cretaceous of Patagonia. *In* Papers in avian paleontology, honoring Pierce Brodkorb, K. E. Campbell, Editor. Natural History Museum of Los Angeles County, Science Series 36:51–64.

Asara, J. M., M. H. Schweitzer, L. M. Freimark, M. Phillips, and L. C. Cantley (2007). Protein sequences from mastodon and *Tyrannosaurus rex* revealed by mass spectrometry. Science 316 (5822): 280–285.

Balanoff, A., G. S. Bever, T. B. Rowe, and M. A. Norell (2013). Evolutionary origins of the avian brain. Nature 501:93–96.

Bell, A. (2013). Evolution and ecology of Mesozoic birds: A case study of the derived Hesperornithiformes and the use of morphometric data in quantifying avian paleoecology. PhD diss., University of Southern California, Los Angeles.

Bell, A., and L. M. Chiappe (2015). A species-level phylogeny of the Cretaceous Hesperornithiformes (Aves: Ornithuromorpha): Implications for body size evolution amongst the earliest diving birds. Journal of Systematic Palaeontology. doi:10.1080/147 72019.2015.1036141.

Bell, A. K., L. M. Chiappe, G. M. Erickson, S. Suzuki, M. Watabe, R. Barsbold, and K. Tsogtbaatar (2010). Description and ecologic analysis of *Hollanda luceria*, a Late Cretaceous bird from the Gobi Desert (Mongolia). Cretaceous Research 31:16–26.

Benson, R. B. J., N. E. Campione, M. T. Carrano, P. D. Mannion, C. Sullivan, et al. (2014). Rates of dinosaur body mass evolution indicate 170 million years of sustained ecological innovation on the avian stem lineage. PLoS Biology. doi:10.1371/journal.pbio.1001853.

Bertazzo, S., S. C. R. Maidment, C. Kallepitis, S. Fearn, M. M. Stevens, and H.-N. Xie (2015). Fibres and cellular structures preserved in 75-million-year-old dinosaur specimens. Nature Communications. doi:10.1038/ncomms8352.

Bhullar, B., J. Marugán-Lobón, F. B. Racimo, T. Rowe, M. Norell, and A. Abzhanov (2012). Birds have paedomorphic dinosaur skulls. Nature 7406:223–226.

Brodkorb, P. (1976). Discovery of a Cretaceous bird, apparently ancestral to the orders Coraciiformes and Piciformes (Aves: Carinatae). Smithsonian Contributions to Paleobiology 27:67–73.

Broom, R. (1913). On the South African pseudosuchian *Euparkeria* and allied genera. Proceedings of the Zoological Society 83:619–633.

Brusatte, S. L. (2013). The phylogeny of basal coelurosaurian theropods (Archosauria: Dinosauria) and patterns of morphological evolution during the dinosaur-bird transition. PhD diss., Columbia University, New York.

Brusatte, S. L. (2017). A Mesozoic aviary. Science 355:792–794.

Brusatte, S. L., S. J. Nesbitt, R. B. Irmis, R. J. Butler, M. J. Benton, and M. A. Norell (2010). The origin and early radiation of dinosaurs. Earth-Science Reviews 101:68–100.

Brusatte, S. L., J. K. O'Connor, and E. D. Jarvis (2015). The origin and diversification of birds. Current Biology 25:R888–R898.

Burgers, P., and L. M. Chiappe (1999). The wing of *Archaeopteryx* as a primary thrust generator. Nature 399:60–62.

Butler, R. J., P. M. Galton, L. B. Porro, L. M. Chiappe, D. M. Henderson, and G. M. Erickson (2010). Lower limits of ornithischian dinosaur body size inferred from a new Upper Jurassic heterodontosaurid from North America. Proceedings of the Royal Society B 277 (1680): 375–381.

Carroll, R. L. (1988). Vertebrate Paleontology and Evolution. W. H. Freeman, New York.

Carvalho, I. de S., F. E. Novas, F. L. Agnolín, M. P. Isasi, F. I. Freitas, and J. A. Andrade (2015). A Mesozoic bird from Gondwana preserving feathers. Nature Communications 6 (7141): 1–5.

Chang, M.-M., P.-J. Chen, Y.-Q. Wang, Y. Wang, and D.-S. Miao, Editors (2003). The Jehol fossils: The emergence of feathered

dinosaurs, beaked birds and flowering plants. Shanghai Scientific & Technical Publishers.

Chen, P., Z. Dong, and S. Zhen (1998). An exceptionally well-preserved theropod dinosaur from the Yixian Formation of China. Nature 391:147–152.

Chiappe, L. M. (1991). Cretaceous birds of Latin-America. Cretaceous Research 12:55–63.

Chiappe, L. M. (1993). Enantiornithine (Aves) tarsometatarsi from the Cretaceous Lecho Formation of northwestern Argentina. American Museum Novitates 3083:1–27.

Chiappe, L. M. (1995). The first 85 million years of avian evolution. Nature 378:349–355.

Chiappe, L. M. (1996). Late Cretaceous birds of southern South America: Anatomy and systematics of enantiornithines and *Patagopteryx deferrariisi*. In Contributions of Southern South America to Vertebrate Paleontology, G. Arratia, Editor. Münchner Geowissenschaftliche Abhandlungen (A) 30:203–244.

Chiappe, L. M. (2002). Basal bird phylogeny: Problems and solutions. In Mesozoic birds: Above the heads of dinosaurs, L. M. Chiappe and L. M. Witmer, Editors. University of California Press, Berkeley, pp. 448–472.

Chiappe, L. M. (2004). The closest relatives of birds. Ornitologia Tropical 15:1–16.

Chiappe, L. M. (2007). Glorified dinosaurs: The origin and evolution of birds. John Wiley, New York.

Chiappe, L. M. (2009). Downsized dinosaurs: The evolutionary transition to modern birds. Evolution: Education and Outreach 2:248–256.

Chiappe, L. M., S. Ji, Q. Ji, and M. A. Norell (1999). Anatomy and systematics of the Confuciusornithidae (Theropoda: Aves) from the Late Mesozoic of northeastern China. Bulletin of the American Museum of Natural History 242:1–89.

Chiappe, L. M., and C. A. Walker (2002). Skeletal morphology and systematics of the Cretaceous Euenantiornithes (Ornithothoraces: Enantiornithes). In Mesozoic birds: Above the heads of dinosaurs, L. M. Chiappe and L. M. Witmer, Editors. University of California Press, Berkeley, pp. 240–267.

Chiappe, L. M., and L. M. Witmer (2002). Mesozoic birds: Above the heads of dinosaurs. University of California Press, Berkeley.

Chiappe, L. M., and G. Dyke (2007). The beginnings of birds: Recent discoveries, ongoing arguments and new directions. In Major transitions in vertebrate evolution (Series: Life of the past), J. S. Anderson and H. Sues, Editors. Columbia University Press, New York, pp. 303–336.

Chiappe, L. M., S. Ji, and Q. Ji (2007). Juvenile birds from the Early Cretaceous of China: Implications for enantiornithine ontogeny. American Museum Novitates 3594:1–46.

Chiappe, L. M., J. Marugán-Lobón, S. Ji, and Z. Zhou (2008). Life history of a basal bird: Morphometrics of the Early Cretaceous *Confuciusornis*. Biology Letters 4:719–723.

Chiappe, L. M., B. Zhao, J. O'Connor, C. Gao, X. Wang, M. Habib, J. Marugan-Lobon, Q. Meng, and X. Cheng (2014). A new specimen of the Early Cretaceous bird *Hongshanornis longicresta*: Insights into the aerodynamics and diet of a basal ornithuromorph. PeerJ 2. doi:10.7717/peerj.234.

Chiappe, L. M., and J. Meng (2016). Birds of Stone. Johns Hopkins University Press, Baltimore, MD.

Chinsamy, A. (2002). *In* Mesozoic birds: Above the heads of dinosaurs, L. M. Chiappe and L. M. Witmer, Editors. University of California Press, Berkeley, pp. 421–431.

Chinsamy, A. (2005). The microstructure of dinosaur bone. Johns Hopkins University Press, Baltimore, MD.

Chinsamy, A., L. M. Chiappe, J. Marugán-Lobón, C. Gao, and F. Zhang (2013). Gender identification of the Mesozoic bird *Confuciusornis sanctus*. Nature Communications 4. doi:10.1038/ncomms2377.

Clark, J. M., M. A. Norell, and L. M. Chiappe (1999). An oviraptorid skeleton from the Late Cretaceous of Ukhaa Tolgod, Mongolia, preserved in an avian-like brooding position over an oviraptorid nest. American Museum Novitates 3265:1–36.

Clarke, J. A. (2004). Morphology, phylogenetic taxonomy, and systematics of *Ichthyornis* and *Apatornis* (Avialae: Ornithurae). Bulletin of the American Museum of Natural History Number 286:1–179.

Clarke, J. A., C. P. Tambussi, J. I. Noriega, G. M. Erickson, and R. A. Ketchum (2005). Definitive fossil evidence for the extant avian radiation in the Cretaceous. Nature 433:305–308.

Clarke, J. A., Z. H. Zhou, and F. C. Zhang (2006). Insight into the evolution of avian flight from a new clade of Early Cretaceous ornithurines from China and the morphology of *Yixianornis grabaui*. Journal of Anatomy 208:287–308.

Cooper, A., and D. Penny (1997). Mass survival of birds across the Cretaceous-Tertiary boundary: Molecular evidence. Science 275 (5303): 1109–1113.

Cracraft, J. (1982). Phylogenetic relationships and monophyly of loons, grebes, and hesperornithiform birds, with comments on the early history of birds. Systematic Zoology 31:35–56.

Currie, P., and P.-J. Chen (2001). Anatomy of *Sinosauropteryx prima* from Liaoning, northeastern China. Canadian Journal of Earth Sciences 38 (12): 1705–1727.

Czerkas, S. A., and C. Yuan (2002). An arboreal maniraptoran from northeast China. *In* Feathered dinosaurs and the origin of flight, S. A. Czerkas, Editor. The Dinosaur Museum Journal 1. The Dinosaur Museum, Blanding, UT, pp. 63–95.

Dacke, C. G., S. Arkle, D. J. Cook, I. M. Wormstone, S. Jones, M. Zaidi, and A. Bascal (1993). Medullary bone and avian calcium regulation. Journal of Experimental Biology 184:63–88.

Dalla Vecchia, F. M., and L. M. Chiappe (2002). First avian skeleton from the Mesozoic of northern Gondwana. Journal of Vertebrate Paleontology 22:856–860.

Dalton, R. (2008). Fresh doubts over *T. rex* chicken link. Nature 454:1035.

Dececchi, T. A., and H. C. E. Larsson (2011). Assessing arboreal adaptations of bird antecedents: Testing the ecological setting of the origin of the avian flight stroke. PLoS ONE 6 (8). doi:10.1371/journal.pone.0022292.

Dececchi, T. A., H. C. E. Larsson, and M. B. Habib (2016). The wings before the birds: An evaluation of flapping-based locomotory hypotheses in bird antecedents. PeerJ. doi:10.7717/peerj.2159.

Deeming, D. C., and M. Ruta (2014). Egg shape changes at the theropod–bird transition, and a morphometric study of amniote eggs. Royal Society Open Science 1:140311.

Dial, K. P. (2003). Wing-assisted incline running and the evolution of flight. Science 299:402–404.

Dial, K. P., R. J. Randall, and T. R. Dial (2006). What use is half a wing in the ecology and evolution of birds? BioScience 56 (5): 437–445.

Dial, K. P., B. E. Jackson, and P. Segre (2008). A fundamental avian wing-stroke provides a new perspective on the evolution of flight. Nature 451:985–989.

Dominguez Alonso, P., A. C. Milner, R. A. Ketcham, M. J. Cookson, and T. B. Rowe (2004). The avian nature of the brain and inner ear of *Archaeopteryx*. Nature 430:666–669.

Dyke, G. J., and M. Van Tuinen (2004). The evolutionary radiation of modern birds (Neornithes): Reconciling molecules, morphology, and the fossil record. Zoological Journal of the Linnean Society 141:153–177.

Dyke, G., R. de Kat, C. Palmer, J. van der Kindere, D. Naish, and B. Ganapathisubramani (2013). Aerodynamic performance of the feathered dinosaur *Microraptor* and the evolution of feathered flight. Nature Communications 4:2489. doi:10.1038/ncomms3489.

Eagle, R. A., T. Tütken, T. S. Martin, A. K. Tripati, H. C. Fricke, M. Connely, R. L. Cifelli, and J. M. Eiler (2011). Dinosaur body temperatures determined from isotopic (13C 18O) ordering in fossil biominerals. Science 333:443–445.

Eagle, R. A., M. Enriquez, G. Grellet-Tinner, A. Pérez-Huerta, D. Hu, T. Tütken . . . , L. M. Chiappe, et al. (2015). Isotopic ordering in eggshells reflects body temperatures and suggests differing thermophysiology in two Cretaceous dinosaurs. Nature Communications 6:8296. doi:10.1038/ncomms9296.

Elzanowski, A. (1974). Preliminary note on the palaeognathous bird from the Upper Cretaceous of Mongolia. Acta Palaeontologica Polonica 30:103–109.

Elzanowski, A. (1991). New observations on the skull of *Hesperornis* with reconstructions of the bony palate and otic region. Postilla 207:1–20.

Erickson, G. M. (2005). Assessing dinosaur growth patterns: A microscopic revolution. Trends in Ecology and Evolution 20 (12): 677–684.

Erickson, G. M. (2014). On dinosaur growth. Annual Review of Earth and Planetary Sciences 42:675–697.

Erickson, G. M., K. Curry-Rogers, and S. A. Yerby (2001). Dinosaurian growth patterns and rapid avian growth rates. Nature 412:429–433.

Falk, A., J. Lim, and S. Hasiotis (2014). A behavioral analysis of fossil bird tracks from the Haman Formation (Republic of Korea) shows a nearly modern avian ecosystem. Vertebrata PalAsiatica 52:129–152.

Feduccia, A. (1999). The origin and evolution of birds. 2nd ed. Yale University Press, New Haven, CT.

Forster, C. A., L. M. Chiappe, D. W. Krause, and S. D. Sampson (1996). The first Cretaceous bird from Madagascar. Nature 382:532–533.

Gao, C., L. M. Chiappe, Q. Meng, J. K. O'Connor, X. Wang, X. Cheng, and J. Liu (2008). A new basal lineage of Early Cretaceous birds from China and its implications on the evolution of the avian tail. Palaeontology 51:775–791.

Gao, C., L. M. Chiappe, F. Zhang, D. Pomeroy, C. Shen, A. Chinsamy, and M. O. Walsh (2012a). A subadult specimen of the Early Cretaceous bird Sapeornis chaoyangensis and a taxonomic reassessment of sapeornithids. Journal of Vertebrate Paleontology 32 (5): 1103–1112.

Gao, C., E. M. Morschhauser, D. J. Varricchio, J. Liu, and B. Zhao (2012b). A second soundly sleeping dragon: New anatomical details of the Chinese Troodontid Mei long with implications for phylogeny and taphonomy. PLoS ONE 7 (9): e45203. doi:10.1371/journal.pone.0045203.

Gatesy, S. M., and K. P. Dial (1996). Locomotor modules and the evolution of avian flight. Evolution 50:331–340.

Gauthier, J. A. (1986). Saurischian monophyly and the origin of birds. In The origin of birds and the evolution of flight, K. Padian, Editor. California Academy of Science, Berkeley, pp. 1–55.

Gauthier, J., A. G. Kluge, and T. Rowe (1988). Amniote phylogeny and the importance of fossils. Cladistics 4:105–209.

Gill, F. (2007). Ornithology. 3rd ed. W. H. Freeman, New York.

Godefroit, P., A. Cau, D.-Y. Hu, F. Escuillié, W. Wu, and G. Dyke (2013). A Jurassic avialan dinosaur from China resolves the early phylogenetic history of birds. Nature 498:359–362.

Grady, J. M., B. J. Enquist, E. Dettweiler-Robinson, N. A. Wright, and F. A. Smith (2014). Evidence for mesothermy in dinosaurs. Science 344:1268–1272.

Grellet-Tinner, G., L. M. Chiappe, M. Norell, and D. Bottjer (2006). Dinosaur eggs and nesting behaviors: Their paleobiological inferences. Palaeogeography, Palaeoclimatology, and Paleoecology 232:294–321.

Han, G., L. M. Chiappe, S. Ji, M. Habib, A. Turner, A. Chinsamy, X. Liu, and L. Han (2014). A new raptorial dinosaur with exceptionally long feathering provides insights into dromaeosaurid flight performance. Nature Communications 5:4382. doi:10.1038/ncomms5382.

Hedges, S. B. (1994). Molecular evidence for the origin of birds. Proceedings of the National Academy of Sciences 91:2621–2624.

Heers, A., K. Dial, and B. Tobalske (2014). From baby birds to feathered dinosaurs: Incipient wings and the evolution of flight. Paleobiology 40 (3): 459–476.

Heilmann, Gerhard (1926). The origin of birds. Witherby, London.

Hope, S. (2002). The Mesozoic record of Neornithes. In Mesozoic birds: Above the heads of dinosaurs, L. M. Chiappe and L. M. Witmer, Editors. University of California Press, Berkeley, pp. 339–388.

Hou, L., L. M. Chiappe, F. Zhang, and C. Chuong (2004). New Early Cretaceous fossil from China documents a novel trophic specialization for Mesozoic birds. Naturwissenschaften 91 (1): 22–25.

Hu, D., L. Hou, L. Zhang, and X. Xu (2009). A pre-Archaeopteryx troodontid Theropod from China with long feathers on the metatarsus. Nature 461:640–643.

Hu, H., J. K. O'Connor, and Z. Zhou (2015). A new species of pengornithidae (Aves: Enantiornithes) from the Lower Cretaceous of China suggests a specialized scansorial habitat previously unknown in early birds. PLoS ONE 10 (6): e0126791.

Huxley, T. H. (1868). On the animals which are most nearly intermediate between the birds and the reptiles. Annals and Magazine of Natural History 2:66–75.

Jarvis, E. D., S. Mirarab, A. J. Aberer, B. Li, P. Houde, C. Li, S. Y. W. Ho, et al. (2014). Whole-genome analyses resolve early branches in the tree of life of modern birds. Science 346:1320–1331.

Ji, Q., P. Currie, M. A. Norell, and S.-A. Ji (1998). Two feathered dinosaurs from northeastern China. Nature 393:753–761.

Ji, Q., L. M. Chiappe, and S. Ji (1999). A new Late Mesozoic confuciusornithid bird from China. Journal of Vertebrate Paleontology 19 (1): 1–7.

Ji, Q., S. Ji, J. Lü, H. You, W. Chen, Y. Liu, and Y. Liu (2005). First avialian bird from China: Jinfengopteryx elegans gen et sp nov. Geological Bulletin of China 24:197–210.

Ji, S., Q. Ji, J. Lü, and C. Yuan (2007). A new giant compsognathid dinosaur with long filamentous integuments from Lower Cretaceous of northeastern China. Acta Geologica Sinica (English edition) 81:8–15.

Jin, F., F.-C. Zhang, Z.-H. Li, J.-Y. Zhang, C. Li, and Z. Zhou (2008). On the horizon of Protopteryx and the early vertebrate fossil assemblages of the Jehol Biota. Chinese Science Bulletin 53 (18): 2820–2827.

Ksepka, D., and J. Clarke (2015). Phylogenetically vetted and stratigraphically constrained fossil calibrations within Aves. Palaeontologia Electronica. Art. no. 18.1.3FC:1–25.

Kurochkin, E., G. Dyke, and A. Karhu (2002). A new presbyornithid bird (Aves, Anseriformes) from the Late Cretaceous of southern Mongolia. American Museum Novitates 3386:1–11.

Lee, A. H., and S. Werning (2008). Sexual maturity in growing dinosaurs does not fit reptilian growth models. Proceedings of the National Academy of Sciences 105 (2): 582–587. doi:10.1073/pnas.0708903105.

Lee, M. S. Y., A. Cau, D. Naish, and G. J. Dyke (2014a). Sustained miniaturization and anatomical innovation in the dinosaurian ancestors of birds. Science 345:562–566.

Lee, M. S. Y., A. Cau, D. Naish, and G. J. Dyke (2014b). Morphological clocks in palaeontology, and a mid-Cretaceous origin of crown Aves. Systematic Biology 63 (3): 442–449.

Li, Q., K. Ga., J. Vinther, M. Shawkey, J. Clarke, L. D'Alba, Q. Meng, D. Briggs, and R. Prum (2010). Plumage color patterns of an extinct dinosaur. Science 327:1369–1372.

Lindgren, J., A. Moyer, M. Schweitzer, P. Sjövall, P. Uvdal, D. Nilsson, J. Heimdal, et al. (2015). Interpreting melanin-based coloration through deep time: A critical review. Proceedings of the Royal Society B 282:20150614.

Liu, D., L. M. Chiappe, Y. Zhang, A. Bell, Q. Meng, Q. Ji, and X. Wang (2014). An advanced, new long-legged bird from the Early Cretaceous of the Jehol Group (northeastern China): Insights into the temporal divergence of modern birds. Zootaxa 3884 (3): 253–266.

Lockley, M., R. McCrea, L. Buckley, J. Lim, N. Matthews, B. Breithaupt, K. Houck, et al. (2016). Theropod courtship: Large scale physical evidence of display arenas and avian-like scrape ceremony behaviour by Cretaceous dinosaurs. Scientific Reports 6:18952. doi:10.1038/srep18952.

Long, J. A., and P. Schouten (2008). Feathered dinosaurs: The origin of birds. Oxford University Press, Oxford.

Longrich, N. (2006). Structure and function of hindlimb feathers in *Archaeopteryx lithographica*. Paleobiology 32:417–431.

Makovicky, P. J., and M. A. Norell (2004). Troodontidae. *In* The Dinosauria, 2nd ed., D. B. Weishampel, P. Dodson, and H. Osmólska, Editors. University of California Press, Berkeley, pp. 184–195.

Makovicky, P. J., and L. E. Zanno (2011). Theropod diversity and the refinement of avian characteristics. *In* Living dinosaurs: The evolutionary history of modern birds, G. Dyke and G. Kaiser, Editors. John Wiley, Hoboken, NJ, pp. 9–29.

Marsh, O. C. (1880). Odontornithes, a monograph on the extinct toothed birds of North America. Government Printing Office, Washington, DC.

Martin, L. D. (1983). The origin and early radiation of birds. *In* Perspectives in ornithology, A. H. Brush and G. A. Clark Jr., Editors. Cambridge University Press, New York, pp. 291–338.

Marugán-Lobón, J., L. M. Chiappe, S. Ji, Z. Zhou, C. Gao, D. Hu, and M. Qinjing (2011). Quantitative patterns of morphological variation in the appendicular skeleton of the Early Cretaceous bird *Confuciusornis*. Journal of Systematic Palaeontology 9 (1): 91–101.

Mayr, G. (2009). Paleogene fossil birds. Springer-Verlag, Berlin.

Mayr, G. (2014). The origin of crown group birds: Molecules and fossils. Palaeontology 57:231–242.

Mayr, G., D. S. Peters, G. Plowdowski, and O. Vogel (2002). Bristle-like integumentary structures at the tail of the horned dinosaur *Psittacosaurus*. Naturwissenschaften 89:361–365.

Mayr, G., B. Pohl, and D. S. Peters (2005). A well-preserved *Archaeopteryx* specimen with theropod features. Science 310:1483–1486.

Montes, L., J. Castanet, and J. Cubo (2010). Relationship between bone growth rate and bone tissue organization in amniotes: First test of Amprino's rule in a phylogenetic context. Animal Biology 60:25–41.

Morschhauser, E. M., D. J. Varricchio, C.-H. Gao, J.-Y. Liu, X.-R. Wang, X.-D. Cheng, and Q.-J. Meng (2009). Anatomy of the Early Cretaceous bird *Rapaxavis pani*, a new species from Liaoning Province, China. Journal of Vertebrate Paleontology 29:545–554.

Navalon, G., J. Marugán-Lobón, L. M. Chiappe, J. L. Sanz, and A. Buscalioni (2015). Soft-tissue and dermal arrangement in the wing of an Early Cretaceous bird: Implications for the evolution of avian flight. Scientific Reports 5:14864. doi:10.1038/srep14864.

Nesbitt, S. J., N. D. Smith, R. B. Irmis, A. H. Turner, A. Downs, and M. A. Norell (2009). A complete skeleton of a Late Triassic saurischian and the early evolution of dinosaurs. Science 326 (5959): 1530–1533.

Norell, M. A., J. M. Clark, L. M. Chiappe, and D. Dashzeveg (1995). A nesting dinosaur. Nature 378:774–776.

Norell, M. A., and J. A. Clarke (2001). Fossil that fills a critical gap in avian evolution. Nature 409:181–184.

Norell, M. A., and P. J. Makovicky (2004). Dromaeosauridae. *In* The Dinosauria, 2nd ed., D. B. Weishampel, P. Dodson, and H. Osmólska, Editors. University of California Press, Berkeley, pp. 196–209.

Norell, M. A., and X. Xu (2005). Feathered dinosaurs. Annual Review of Earth and Planetary Sciences 33:277–299.

O'Connor, J. K., X.-R. Wang, L. M. Chiappe, C.-H. Gao, Q.-J. Meng, X.-D. Cheng, and J.-Y. Liu (2009). Phylogenetic support for a specialized clade of Cretaceous enantiornithine pre-modern birds: Avian divergences in the Mesozoic birds with information from a new species. Journal of Vertebrate Paleontology 29:188–204.

O'Connor, J. K., and L. M. Chiappe (2011). A revision of enantiornithine (Aves: Ornithothoraces) skull morphology. Journal of Systematic Palaeontology 9 (1): 135–157.

O'Connor, J. K., L. M. Chiappe, and A. Bell (2011). Pre-modern birds: Avian divergences in the Mesozoic. *In* Living dinosaurs: The evolutionary history of modern birds, D. G. Dyke and G. Kaiser, Editors. John Wiley, Hoboken, NJ, pp. 39–114.

O'Connor, J. K., L. M. Chiappe, C. Chuong, D. Bottjer, and H. You (2012a). Homology and potential cellular and molecular mechanisms for the development of unique feather morphologies in early birds. Geosciences 2:157–177. doi:10.3390/geosciences2030157.

O'Connor, J. K., C. Sun, X. Xu, X. Wang, and Z. Zhou (2012b). A new species of *Jeholornis* with complete caudal integument. Historical Biology 24 (1): 29–41.

O'Connor, J. K., and Z. Zhou (2012). A redescription of *Chaoyangia beishanensis* (Aves) and a comprehensive phylogeny of Mesozoic birds. Journal of Systematic Palaeontology. doi:10.1080/14772019.2012.690455.

O'Connor, J. K., X. Zheng, X. Wang, Y. Wang, and Z. Zhou (2013). Ovarian follicles shed new light on dinosaur reproduction during the transition towards birds. National Science Review 1: 1–3. doi:10.1093/nsr/nwt012.

O'Connor, J. K., X. Wang, X. Zheng, X. Zhang, and Z. Zhou (2016). An enantiornithine with a fan-shaped tail, and the evolution of the rectricial complex in early birds. Current Biology 26:114–119.

Organ, C. L., A. M. Shedlock, M. Pagel, and S. V. Edwards (2007). Origin of avian genome size and structure in non-avian dinosaurs. Nature 446:180–184.

Organ, C. L., M. H. Schweitzer, W. Zheng, L. M. Freimark, L. C. Cantley, and J. M. Asara (2008). Molecular phylogenetics of mastodon and *Tyrannosaurus rex*. Science 320:499.

Organ, C., and A. Shedlock (2009). Palaeogenomics of pterosaurs and the evolution of small genome size in flying vertebrates. Biology Letters 5 (1): 47–50.

Osborn, H. F. (1924). Three new Theropoda, *Protoceratops* zone, central Mongolia. American Museum Novitates 144:1–12.

Ostrom, J. H. (1969). Osteology of *Deinonychus antirrhopus*, an unusual theropod from the lower Cretaceous of Montana. Peabody Museum of Natural History Bulletin 30:1–165.

Ostrom, J. H. (1975). The origin of birds. Annual Review of Earth and Planetary Sciences 3:55–77.

Ostrom, J. H. (1976). *Archaeopteryx* and the origin of birds. Biological Journal of the Linnean Society 8:91–182.

Owen, R. (1863). On the *Archeopteryx* of Von Meyer, with a description of the fossil remains of a long-tailed species, from the Lithographic Stone of Solenhofen. Philosophical Transactions of the Royal Society London 153:33–47.

Padian, K., and L. M. Chiappe (1998). The origin and early evolution of birds. Biological Reviews 73 (1): 1–42.

Perrichot, V., L. Marion, D. Néraudeau, R. Vullo, and P. Tafforeau (2008). The early evolution of feathers: Fossil evidence from Cretaceous amber of France. Proceedings of the Royal Society B 275:1197–1202.

Pickrell, J. (2014). Flying dinosaurs: How fearsome reptiles became birds. Columbia University Press, New York.

Prum, R. O. (1999). Development and evolutionary origin of feathers. Journal of Experimental Zoology (Molecular and Developmental Evolution) 285:291–306.

Prum, R. O., and A. H. Brush (2002). The evolutionary origin and diversification of feathers. Quarterly Review of Biology 77:261–295.

Prum, R., J. Berv, A. Dornburg, D. Field, J. Townsend, E. Moriarty Lemmon, and A. Lemmon (2015). A comprehensive phylogeny of birds (Aves) using targeted next-generation DNA sequencing. Nature 526:569–573.

Pu, H. Y., H. L. Chang, J. C. Lü, Y. Wu, L. Xu, J.-M. Zhang, and S. Jia (2013). A new juvenile specimen of *Sapeornis* (Pygostylia: Aves) from the Lower Cretaceous of Northeast China and allometric scaling of this basal bird. Paleontological Research 17:27–38.

Rashid, D. J., S. Chapman, L. Larsson, C. L. Organ, A. Bebin, C. Merzdorf, R. Bradley, and J. R. Horner (2014). From dinosaurs to birds: A tail of evolution. EvoDevo 5:25. doi: 10.1186/2041 9139 5 25.

Rauhut, O. W. M. (2003). A tyrannosauroid dinosaur from the Upper Jurassic of Portugal. Palaeontology 46 (5): 903–910.

Rauhut, O., C. Foth, H. Tischlinger, and M. A. Norell (2012). Exceptionally preserved juvenile megalosauroid theropod dinosaur with filamentous integument from the Late Jurassic of Germany. Proceedings of the National Academy of Sciences 109 (29): 11746–11751.

Sanz, J. L., L. M. Chiappe, B. P. Pérez-Moreno, A. D. Buscalioni, J. J. Moratalla, F. Ortega, and F. J. Poyato-Ariza (1996). An Early Cretaceous bird from Spain and its implications for the evolution of avian flight. Nature 382:442–445.

Sato, T., Y. Cheng, X. Wu, D. K. Zelenitsky, and Y. Hsiao (2005). A pair of shelled eggs inside a female dinosaur. Science 308:375.

Schweitzer, M. H. (2011). Soft tissue preservation in terrestrial Mesozoic vertebrates. Annual Review of Earth and Planetary Sciences 39:187–216.

Schweitzer, M. H., J. L. Wittmeyer, and J. R. Horner (2005). Gender-specific reproductive tissue in ratites and *Tyrannosaurus rex*. Science 308:1456–1460.

Schweitzer, M. H., W. Zheng, C. L. Organ, R. Avci, Z. Suo, L. M. Freimark, V. S. Lebleu, et al. (2009). Biomolecular characterization and protein sequences of the Campanian hadrosaur *B. canadensis*. Science 324:626–631.

Schweitzer, M., J. Lindgren, and A. Moyen (2015). Melanosomes and ancient coloration re-examined: A response to Vinther 2015. Bioessays 37:1174–1183.

Serrano, F. J., P. Palmqvist, L. M. Chiappe, and J. L. Sanz (2017). Inferring flight parameters of Mesozoic avians through multivariate analyses of forelimb elements in their living relatives. Paleobiology 43:144–169.

Serrano, F. J., and L. M. Chiappe (2017). Aerodynamic modelling of a Cretaceous bird reveals thermal soaring capabilities during early avian evolution. Journal of the Royal Society Interface. http://dx.doi.org/10.1098/rsif.2017.0182.

Service, R. F. (2015). Signs of ancient cells and proteins found in dinosaur fossils. Science. doi:10.1126/science.aac6808.

Souza de Carvalho, I., F. E. Novas, F. L. Agnolín, M. P. Isasi, F. I. Freitas, and J. A. Andrade (2015). A Mesozoic bird from Gondwana preserving feathers. Nature Communications 6:7141. doi:10.1038/ncomms8141.

Turner, A. H., D. Pol, J. A. Clarke, G. M. Erickson, and M. A. Norell (2007). A basal dromaeosaurid and size evolution preceding avian flight. Science 317:1378–1381.

Turner, A., P. Makovicky, and M. Norell (2012). A review of dromaeosaurids systematics and paravian phylogeny. Bulletin of the American Museum of Natural History 371:1–206.

Van Der Reest, A. J., A. P. Wolfe, and P. Currie (2016). A densely feathered ornithomimid (Dinosauria: Theropoda) from the Upper Cretaceous Dinosaur Park Formation, Alberta, Canada. Cretaceous Research 58:108–117.

Varricchio, D. (1993). Bone microstructure of the Upper Cretaceous theropod dinosaur *Troodon formosus*. Journal of Vertebrate Paleontology 1 (13): 99–104.

Varricchio, D. J., F. Jackson, J. Borkowski, and J. R. Horner (1997). Nest and egg clutches of the dinosaur *Troodon formosus* and the evolution of avian reproductive traits. Nature 385:247–250.

Varricchio, D. V., and F. Jackson (2004). Two eggs sunny-side up: Reproductive physiology in the dinosaur *Troodon formosus*. In Feathered dragons: Studies on the transition from dinosaurs to birds, P. J. Currie, E. B. Koppelhus, M. A. Shugar, and J. L. Wright, Editors. Indiana University Press, Bloomington, pp. 215–233.

Varricchio, D. J., A. M. Balanoff, and M. A. Norell (2015). Reidentification of avian embryonic remains from the Cretaceous of Mongolia. PLoS ONE 10 (6): e0128458.

Walker, C. A. (1981). New subclass of birds from the Cretaceous of South America. Nature 292:51–53.

Wang, X., L. M. Chiappe, F. F. Teng, and Q. Ji (2013). *Xinghaiornis lini* from the Early Cretaceous of Liaoning: An example of evolutionary mosaic in early birds. Acta Geologica Sinica 87 (3): 686–689.

Wang, X., J. K. O'Connor, X. Zheng, M. Wang, H. Hu, and Z. Zhou (2014). Insights into the evolution of rachis dominated tail feathers from a new basal enantiornithine (Aves: Ornithothoraces). Biological Journal of the Linnean Society 113:806–819.

Wang, M., X. Zheng, J. K. O'Connor, G. T. Lloyd, X. Wang, Y. Wang, X. Zhang, and Z. Zhou (2015). The oldest record of Ornithuromorpha from the early Cretaceous of China. Nature Communications 6:6987. doi:10.1038/ncomms7987.

Wang, M., Z. Zhou, and C. Sullivan (2016a). A fish-eating enantiornithine bird from the Early Cretaceous of China provides evidence of modern avian digestive features. Current Biology 26:1170–1176.

Wang, Y., M. Wang, J. K. O'Connor, X. Wang, X. Zheng, and X. Zhang (2016b). A new Jehol enantiornithine bird with three-dimensional preservation and ovarian follicles. Journal of Vertebrate Paleontology. doi:10.1080/02724634.2015.1054496.

Wellnhofer, P. (2009). *Archaeopteryx*: The icon of evolution. F. Pfeil, Munich, Germany.

Wiemann, J., T. Yang, P. Sander, M. Schneider, M. Engeser, S. Kath-Schorr, C. Müller, and M. Sander (2015). The blue-green eggs of dinosaurs: How fossil metabolites provide insights into the evolution of bird reproduction. PeerJ. https://peerj.com/preprints/1080v1.pdf.

Xing, L. (2016). Mummified precocial bird wings in mid-Cretaceous Burmese amber. Nature Communications 7:12089. doi:10.1038/ncomms12089.

Xing, L., J. K. O'Connor, R. C. McKellar, L. M. Chiappe, K. Tseng, G. Li, and M. Bai (2017). Mid-Cretaceous enantiornithine (Aves) hatchling preserved in Burmese amber with unusual plumage. Gondwana Research 49:264–277.

Xu, X., Z.-L. Tang, and X.-L. Wang (1999). A therizinosauroid dinosaur with integumentary structures from China. Nature 399 (6734): 350–354.

Xu, X., Y.-N. Cheng, X.-L. Wang, and C.-H. Chang (2002). An unusual oviraptorosaurian dinosaur from China. Nature 419:291–293.

Xu, X., Z. Zhou, X. Wang, X. Kuang, and X. Du (2003). Four winged dinosaurs from China. Nature 421:335–340.

Xu, X., and M. A. Norell (2004). A new troodontid dinosaur from China with avian-like sleeping posture. Nature 431:838–841.

Xu, X., X. Zheng, and H. You (2009). A new feather type in a nonavian theropod and the early evolution of feathers. Proceedings of the National Academy of Sciences 106 (3): 832–834.

Xu, X., H. You, K. Du, and F. Han (2011). An *Archaeopteryx*-like theropod from China and the origin of Avialae. Nature 475:465–470.

Xu, X., Z. Zhou, R. Dudley, S. Mackem, C. Chuong, G. Erickson, and D. J. Varricchio (2014). An integrative approach to understanding bird origins. Science. doi:10.1126/science.1253293.

Xu, X., X. Zheng, C. Sullivan, X. Wang, X. Lida, Y. Wang, X. Zhang, J. K. O'Connor, F. Zhang, and Y. Pan (2015). A bizarre Jurassic maniraptoran theropod with preserved evidence of membranous wings. Nature. doi:10.1038/nature14423.

Zelenitsky, D. (2006). Reproductive traits of non-avian theropods. Journal of Paleontological Society of Korea 22:209–216.

Zelenitsky, D., F. Therrien, G. Erickson, C. DeBuhr, Y. Kobayashi, D. Eberth, and F. Hadfield (2012). Feathered non-avian dinosaurs from North America provide insight into wing origins. Science 338:510–514.

Zhang, F., and Z. Zhou (2000). A primitive enantiornithine bird and the origin of feathers. Science 290:1955–1959.

Zhang, F., Z. Zhou, X. Xu, and X. Wang (2002). A juvenile coelurosaurian theropod from China indicates arboreal habits. Naturwissenschaften 89 (9): 394–398.

Zhang, F., and Z. Zhou (2004). Leg feathers in an Early Cretaceous bird. Nature 431:925.

Zhang, F., Z. Zhou, X. Xu, X. Wang, and C. Sullivan (2008a). A bizarre Jurassic maniraptoran from China with elongate ribbon-like feathers. Nature 455:1105–1108.

Zhang, F., Z. H. Zhou, and M. J. Benton (2008b). A primitive confuciusornithid bird from China and its implications for early avian flight. Science in China, Series D—Earth Sciences 51:625–639.

Zhang, F., S. L. Kearns, P. J. Orr, M. J. Benton, Z. Zhou, D. Johnson, X. Xu, and X. Wang (2010). Fossilized melanosomes and the colour of Cretaceous dinosaurs and birds. Nature 463:1075–1078.

Zhang, G., C. Li, Q. Li, B. Li, D. M. Larkin, C. Lee, J. F. Storz, et al. (2014). Comparative genomics reveals insights into avian genome evolution and adaptation. Science 346:1311–1320.

Zheng, X., H. You, X. Xu, and Z. Dong (2009). An Early Cretaceous heterodontosaurid dinosaur with filamentous integumentary structures. Nature 458 (7236): 333–336.

Zheng, X., X. Wang, J. K. O'Connor, and Z. Zhou (2012). Insight into the early evolution of the avian sternum from juvenile enantiornithines. Nature Communications 3. doi:10.1038/ncomms2104.

Zheng, X., Z. Zhou, X. Wang, F. Zhang, X. Zhang, Y. Wang, G. Wei, S. Wang, and X. Xu (2013). Hind wings in basal birds and the evolution of leg feathers. Science 339:1309–1312.

Zhou, Z. (2014). The Jehol Biota, an Early Cretaceous terrestrial Lagerstätten: New discoveries and implications. National Science Review 1:543–559.

Zhou, Z., and X. Wang (2000). A new species of *Caudipteryx* from the Yixian Formation of Liaoning, northeastern China. Vertebrata PalAsiatica 38 (2): 113–130.

Zhou, Z. H., and F. Zhang (2002a). A long-tailed, seed-eating bird from the Early Cretaceous of China. Nature 418:405–409.

Zhou, Z., and F. Zhang (2002b). Largest bird from the Early Cretaceous and its implications for the earliest avian ecological diversification. Naturwissenschaften 89:34–38.

Zhou, Z., and F. Zhang (2003a). *Jeholornis* compared to *Archaeopteryx*, with a new understanding of the earliest avian evolution. Naturwissenschaften 90:220–225.

Zhou, Z., and F. Zhang (2003b). Anatomy of the primitive bird *Sapeornis chaoyangensis* from the Early Cretaceous of Liaoning, China. Canadian Journal of Earth Sciences 40:731–747.

Zhou, Z., and F. A. Zhang (2004). Precocial avian embryo from the Lower Cretaceous of China. Science 306:653.

Zhou, Z. H., and F. C. Zhang (2006). A beaked basal ornithurine bird (Aves, Ornithurae) from the Lower Cretaceous of China. Zoologica Scripta 35:363–373.

Zhou, S., Z. H. Zhou, and J. K. O'Connor (2012). A new toothless ornithurine bird (*Schizooura lii* gen. et sp. nov.) from the Lower Cretaceous of China. Vertebrata Palasiatica 50 (1): 9–24.

Zhou, S., Z. H. Zhou, and J. K. O'Connor (2013). Anatomy of the basal ornithuromorph bird *Archaeorhynchus spathula* from the Early Cretaceous of Liaoning, China. Journal of Vertebrate Paleontology 33 (1): 141–152.

Zhou, S., J. K. O'Connor, and M. Wang (2014). A new species from an ornithuromorph dominated locality of the Jehol Group. Chinese Science Bulletin 59 (36): 5366–5378.

CHAPTER 3

Species Concepts
and Speciation Analysis

Robert M. Zink and John Klicka

What is a species?

Very few ideas impinge on more biological research than the concept of the species. Probably all biological studies reference their study organisms by using a Linnaean binomial, consisting of its Latin rendition of genus and species. Yet, how one defines a species remains a controversial topic. Indeed, more than two dozen differing "species concepts" have been proposed (for a summary, see Mayden 1997, 2002). In table 3.1, we show a sampling of these, revealing the debate over how to define species that has been ongoing since before Darwin (1859) remarked "No one definition has as yet satisfied all naturalists; yet every naturalist knows vaguely what he means when he speaks of a species." In some cases, the way in which a species has been defined is taxon-specific. Some properties of plants, such as their tendency to hybridize with each other, have resulted in botanists defining species in different ways than do entomologists working on beetles. Paleontologists have the special problem of trying to decide from morphology alone whether fragments of fossils from thousands of years apart in the sedimentary record represent the same species. Scientists studying microbes have their own set of problems—for example, when the influenza virus mutates each year, requiring us to get a new vaccination, has it become a new species?

Among competing species concepts, at least for sexually reproducing vertebrates such as birds, we think that the underlying debate involves the difference between concepts that emphasize *a process*, such as the evolution of reproductive isolation mechanisms, and concepts that emphasize *the pattern* of evolutionary history, as evidenced by the evolution of shared diagnostic traits. Differences between the two approaches result in biologists constructing different species taxonomies from the same data, depending on whether process or pattern is the primary point of emphasis. The biological species concept (BSC) and phylogenetic species concept

(PSC) emphasize process and pattern, respectively, and we focus our discussion on the application of these two species concepts in ornithology.

BIOLOGICAL SPECIES CONCEPT

Historically, ornithologists have contributed significantly to the development and application of the BSC. The eminent evolutionary biologist and ornithologist Ernst Mayr (see the box on page 41) was responsible for popularizing the BSC, although it was originally formulated by the geneticist Theodosius G. Dobzhansky (1937; also known for his famous statement [1973] that "Nothing in biology makes sense except in the light of evolution"). An often-quoted definition of the BSC is "Species are groups of actually or potentially interbreeding natural populations, which are reproductively isolated from other such groups" (Mayr 1942).

The BSC has been the prevalent species concept in ornithology, as well as in many other disciplines, especially the vertebrates. In a nutshell, it considers two recognizably different (or diagnosable) groups of individuals as representing two species if individuals in each group are unable to reproduce successfully with each other. If the ranges of the two groups are allopatric, they are recognized as biological species if they are presumed to be incapable of reproducing. Thus, to be considered species, the two groups must have acquired reproductive isolating mechanisms. These may act to prevent individuals from mating directly or, if they do mate, to prevent hybrids from developing or reproducing. The net effect is the cessation of gene flow from one group to another, which allows the descendant species to evolve independently.

Advocates of a biological species concept emphasize the intrinsic value of studying reproductive isolating mechanisms. For one thing, species sometimes become sympatric with other species of the same genus, and if not reproductively

Table 3.1. **Some species definitions.**

"A species is a lineage (or a closely related set of lineages) which occupies an adaptive zone minimally different from any other lineage in its range and which evolves separately from all lineages outside its range." (Van Valen 1976: 233)

"Species are the smallest groups that are consistently and persistently distinct, and distinguishable by ordinary means." (Cronquist 1978: 15)

"Species are simply the smallest detected samples of self-perpetuating organisms that have unique sets of characters." (Nelson and Platnick 1981: 12)

"Each species is an internally similar part of a phylogenetic tree." (Willis 1981: 84)

"A species is a group of animals or plants all of which are similar enough in form to be considered as minor variations of the same organism. Members of the group normally interbreed and reproduce their own kind over considerable periods of time." (Trueman 1979: 764)

"We can, therefore, regard as a species that most inclusive population of individual biparental organisms which share a common fertilization system." (Paterson 1985: 25)

"At the outset I confess a disbelief in species, as that word is commonly understood to refer to the basic taxonomic unit or to the taxonomic unit of evolution. . . . There seem to be no basic taxonomic units and no particular taxonomic unit of evolution, . . . and as Agassiz said in 1859 'species do not exist in nature in a different way from the higher groups.' " (Nelson 1989)

"[Consider] species as the most inclusive group of organisms having the potential for genetic and/or demographic exchangeability." (Templeton 1989)

"A species is a lineage of ancestral descendant populations which maintains its identity from other such lineages and which has its own evolutionary tendencies and historical fate." (Wiley 1978)

"No one definition has as yet satisfied all naturalists; yet every naturalist knows vaguely what he means when he speaks of a species." (Darwin 1859)

isolated, then species diversity would be lost through introgression. Indeed, the evolution of reproductive isolation can be considered a "point of no return" in the process of evolutionary divergence, and is a reason for the popularity of the BSC. It seems that the majority of reproductive isolating mechanisms, or at least those best known, function during the mate choice period, as opposed to those that act after mating. In birds, most species have ritualized behaviors, diagnostic male songs, and differently colored patches of plumage, which can be variously displayed as signals to the same or opposite sex

(chapter 16). These appear to function in the context of species recognition and as premating barriers to hybridization between different species.

Plumage Cues as Premating Isolation Mechanisms

At least some potentially important plumage traits that could function as isolating mechanisms are in the ultraviolet spectrum and thus not visible to humans (Eaton and Lanyon 2003, Loyau et al. 2007). However, most species of bird can be identified using plumage cues visible to humans; indeed, generations of workers at avian museums have used degree of plumage differences to infer how distantly related two taxa might be, and whether or not the two taxa could potentially interbreed. Birds are visual creatures and clearly use differences in plumage color or pattern to help identify conspecifics. These same plumage cues may also be used to identify sex, age, and desirability as a potential mate. For example, plumage manipulation studies have shown that facial pattern is important for Golden-winged Warblers (*Vermivora chrysoptera*; fig. 3.1), wherein males that had their dark throat patch and contrasting eye mask artificially lightened both lost their territories *and* failed to obtain mates (Leichty and Grier 2006). In this experiment, it is unclear whether females are responding directly to the loss of an important diagnostic trait (facial pattern) or whether experimental males are abandoned because they are no longer able to hold territories. In Red-winged Blackbirds (*Agelaius phoeniceus*), experiments in which the red epaulettes were enlarged, or painted black, had effects on male-male interactions (Røskraft and Rohwer 1987) and experimentally dulling or reddening epaulets affects female choice (Yasukawa et al. 2009). In the Northern Flicker (*Colaptes auratus*), Noble (1936) found that gluing a black mustache (malar stripe) on to a female caused her mate to attack her mercilessly, apparently convinced that his plumage-doctored mate was a rival male. Conversely, female flickers apparently choose mates irrespective of the color of the male's malar stripe in the Great Plains hybrid zone (see below).

Stereotyped Behaviors as Premating Isolating Mechanisms

Stereotyped behaviors likely play an important yet poorly understood role in species recognition/premating isolation. For example, in an effort to impress potential mates, male Sharp-tailed Grouse (*Tympanuchus phasianellus*; fig. 3.2) and Greater Prairie Chicken (*Tympanuchus cupido*) perform species-specific vocalizations along with "dances" that include inflating colored air sacs, head bobbing and bowing, and foot shuffling. Hybrids of these two forms perform an ineffective intermediate display that leads to less frequent mating attempts (Evans 1966, Sparling 1983). Similarly, hybrid crosses of Costa's Hummingbird (*Calypte costae*) and Anna's

ERNST MAYR

Ernst Mayr is one of the greatest evolutionary biologists of the twentieth century. He attained his PhD in Berlin (under Erwin Stressman) at the age of 21. Like most young naturalists and ornithologists of his time, he soon embarked on an expedition of discovery. Between 1928 and 1930, he explored the Solomon Islands and New Guinea, collecting the skins of more than 3,000 bird specimens (while eating the flesh of many; it is said that Ernst Mayr ate more birds-of-paradise than any other modern ornithologist). The experiences and insights obtained during these years in the South Pacific were to stimulate Mayr's thinking about biology and the development of species for decades to come. Over the course of his career, he would name 26 new species of birds (and 38 new species of orchids), and 410 new subspecies, more than any other taxonomist. Throughout his lifetime, he published over 300 papers that discussed and described the geographic variation and distribution of birds.

In the 1930s, Mayr's interests began to diversify beyond taxonomy into evolutionary biology. Mayr, along with Theodosius Dobzhansky, George Gaylord Simpson, and others, became architect of the "modern synthesis" of the 1930s and 1940s, a paradigm that integrated Mendel's theory of heredity with Darwin's theory of evolution and natural selection. Mayr's main (and lasting) contribution to this endeavor was his first book, *Systematics and the Origin of Species*, published in 1942. Arguably his most important book, this volume combined insights and methods from paleontology, population genetics, systematics, and natural history, thus providing a unified modern evolutionary theory.

Ernst Mayr is perhaps best known as the framer of the "biological species concept." His experiences and knowledge obtained in the South Pacific also led him to develop concepts about the geographic mechanisms of speciation. He concluded that most species originate in allopatry, an idea that is widely accepted today. Mayr however, favored a special case of allopatric speciation called "peripatric" speciation or "founder-effect" speciation. He noted that on many New Guinea offshore islands, individuals differed from individuals in the mainland population. He reasoned that this differentiation could result from a small number of individuals founding the island population. By bringing only a subset of all the genes of the main population (causing a genetic bottleneck), genetic drift (random fixation) and natural selection (caused by a different set of selection pressures on these islands) would not only promote the formation of new species but would also do so rapidly.

Ernst Mayr lived a long and productive life, passing away in 2005 at the age of 100. His scientific legacy includes two dozen single-authored or edited books and more than 600 scientific publications. For his scientific achievements, he was given the Coues Award in ornithological research in 1977 (with J. Delacour).

Figure 3.1. Golden-winged Warbler. Differently patterned patches of plumage play a role in mate choice.

Figure 3.2. Sharp-tailed Grouse. This species is renowned for the male's spectacular mating dances.

Hummingbird (*Calypte anna*) are less effective in courtship because they are unable to perform correctly the circular courtship flight displays characteristic of each of these species (Wells et al. 1978). Thus, even if plumage and vocalizations pass muster with females, a poor performance in the display category could be a deal-breaker.

Song as an Isolating Mechanism

Remsen (2005: 406) wrote, "One can question whether plumage differences in the absence of song differences can promote reproductive isolation," thereby focusing attention on the perhaps primary role of vocalizations in isolating species behaviorally and in allowing taxonomists to judge species limits. Most bird species have diagnostic vocalizations and male song functions both to advertise territory ownership and to attract mates. In theory, to prevent hybridization and a loss of fitness, members of one species should not be attracted to the songs of another species. To test whether different songs play a role in reproductive isolation, a common research technique has been to play a recording of a male from one area to males in another area (Kroodsma 1986) and measure their response. A lack of response could indicate that the geographically separated groups have evolved into separate biological species and that if they were to come together, vocal differences might prevent hybridization. However, males might be responding to an ancestral vocal component that simply elicits territorial aggression. Determining the significance of female response to vocalizations of males from different areas, or even from different species, requires a number of assumptions (Ballentine et al. 2004). Thus whether results of playback experiments show unambiguously that different biological species are involved is unresolved.

As an example, Lanyon (1957) performed some elegant vocal playback experiments with Eastern Meadowlarks (*Sturnella magna*; fig. 3.3) and Western Meadowlarks (*S. neglecta*). He measured the response of males to recordings of conspecifics in areas where only one species occurred, and also in areas where both occur. He found that in areas with a single species, recordings of the other species elicited relatively little response. But in areas where both occurred, males responded equally to recordings of either species. Thus, at least males recognize individuals from the other species and respond aggressively to their vocalizations, presumably because the similar morphologies result in ecological competition. Interestingly, Lanyon (1979) also detected hybrid sterility (a postmating isolating mechanism) among interspecific crosses of these same species in an aviary, thereby placing a premium on females to choose males with the correct vocalization. In this case, the qualitatively different male songs are indicative of two species.

The structures of vocalizations can be used as taxonomic characters (see chapter 13), particularly in bird species for which vocalizations are innate, such as suboscine songbirds. Isler et al. (2005) suggested a protocol for determining from Neotropical antbird (suboscine) vocalizations whether allopatric populations represented more than one biological species. They suggested that at least three vocal character differences were necessary to assure that isolated antbird populations had achieved biological species status. The generality of this finding is not clear, as playback experiments cannot fully replicate natural conditions.

Sorting Out the Roles of Plumage, Song, and Behavior

Complicating the assessment of the role of vocalizations in mate choice is the observation that most premating displays include behavioral, plumage, and vocal components. Isolating the effects of each is difficult. In some cases, the effects of one component seem clear. Great-crested Flycatchers (*Myiarchus crinitus*; fig. 3.4) attacked stuffed mounts of conspecifics—Eastern Kingbird (*Tyrannus tyrannus*), Yellow-bellied Flycatcher

Figure 3.3. Eastern Meadowlark. This species was used by Lanyon (1957) in playback experiments to explore the role of vocalizations as reproductive isolating mechanisms.

Figure 3.4. Great-crested Flycatcher. This species was used in experiments to test whether vocalizations or plumages were important in species recognition.

(*Empidonax flaviventris*), Yellow-browed Tyrant (*Satrapa icterophrys*), Baltimore Oriole (*Icterus galbula*), and Red-eyed Vireo (*Vireo olivaceus*)—as long as each mount was placed near a speaker playing Great-crested Flycatcher vocalizations (Lanyon 1963). In this case then, vocalizations appear to take precedence over physical appearance. Without such experimentation, however, it is often difficult to discern which particular cues are meant for which particular class of recipients (males, females, conspecifics, heterospecifics). For example, the extravagant display plumages of male birds-of-paradise are well known, but for many species, a stereotyped "dance" and distinctive and complex vocalizations are simultaneously a part of a comprehensive elaborate visual display. Precisely which elements of such a display function to assure proper mate choice (premating isolating mechanism) are unknown for many groups of birds and require extensive experimentation (Tarof et al. 2005, Dunn et al. 2008).

Critiques of the Biological Species Concept

Advocates of alternative species concepts have pointed out several problems with the implementation of the BSC. First, speciation occurs in allopatry, and the degree of ecological differentiation that would allow two descendant groups to coexist (i.e., exhibit sympatry) does not evolve for long periods of time following speciation (Peterson et al. 1999, Zink 2012). Thus, because most newly evolved groups are allopatric, it is not possible to determine whether or not they could hybridize under natural conditions. As discussed above, to circumvent this issue, ornithologists examine characteristics such as songs and plumage patterns of allopatric groups to see whether they possess the sorts of differences that may function in sympatry to prevent hybridization in related groups of birds.

It is often thought that hybrid zones give birds a chance to "tell us" whether they are different species, and therefore hy-

brid zones are often considered a "proving grounds" for species status. Birds are renowned for their ability to hybridize despite relatively deep evolutionary divergences (Prager and Wilson 1975). That is, hybrids are known not just between close relatives but also between species rather distant on the avian evolutionary tree. For example, it is probably the case that every species of New World quail has hybridized with one or more of its congeners in the wild (Johnsgard 1973). However, these hybridizations are relatively infrequent and tend to occur when one sex (usually females) of a species is present in small numbers, but another species is common in the same area. Thus, this facultative (and infrequent) hybridization does not lead to the breakdown of species barriers. Thus, taxonomic interpretation of hybridization has to be tempered by whether there are just a few hybrids or a zone in which hybridization is frequent and without apparent reproductive penalty (e.g., sterile offspring). The relevant question is, how much hybridization is too much for two taxa to be considered separate species?

All species concepts encounter problems in certain situations. For example, the Audubon's Warbler (*Setophaga coronata auduboni*) and Myrtle Warbler (*S. c. coronata*) were once considered separate species (figs. 3.5A, B), but it was discovered that in Pine Pass in British Columbia and Jasper Park, Alberta (a combined area of less than 1/10,000 of the range of the two forms), there are a few hybrids, and hence the two well-differentiated groups were reclassified by the American Ornithologists' Union (AOU 1998) North American Classification Committee as a single species, the Yellow-rumped Warbler (the AOU has now merged with the Cooper Ornithological Society under the new name the American Ornithological Society, AOS). New data from genomics (Toews et al. 2016a) suggest that there are fixed differences in various gene regions between the two taxa and that each should be returned to species status; the fixed differences indicate species status under the PSC, and lack of introgression beyond the narrow hybrid zone, hence reproductive incapability, under the BSC. In contrast, Golden-winged Warblers (*Vermivora chrysoptera*) and Blue-winged Warblers (*Vermivora cyanoptera*) hybridize extensively across a broad geographic front but are recognized as separate species. A recent study (Toews et al. 2016b) documented that the two species share 99.97 percent of their genomes, with just a few islands of genes controlling diagnostic plumage differences. Thus there is considerable latitude in how the BSC is applied with respect to the extent of hybridization.

One might think that a large (wide) hybrid zone would be clear evidence that reproductive isolation had not reached completion and that the two hybridizing groups should be classified as the same species. However, Mayr (1970) noted

Figure 3.5. Two species that have figured in speciation research. A, Audubon's Warbler; and B, Myrtle Warbler.

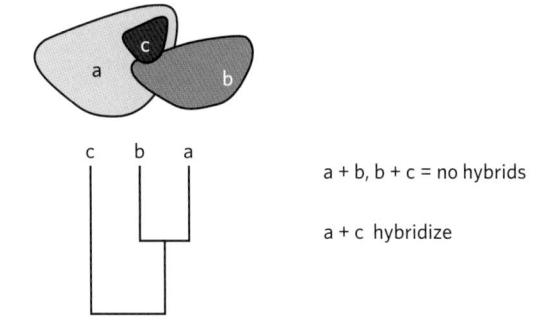

a + b, b + c = no hybrids

a + c hybridize

Figure 3.6. Hypothetical example to show that reproductive isolation need not evolve concomitantly with lineage divergence, resulting in different taxonomic classifications depending on whether mate choice or evolutionary history is given primary emphasis.

that if the hybrid zone is stable, it implies a barrier to free gene exchange, and the two groups should be classified as separate species. Unfortunately, subsequent studies have shown that it can take thousands of years to determine whether a hybrid zone is stable. Furthermore, if a "density trough" (an area of low density that impedes gene flow) exists between the two taxa, as Barrowclough et al. (2005) suggest exists between the ranges of the Northern (*Strix occidentalis caurina*) and California (*S. o. occidentalis*) Spotted owls, gene flow can be much reduced, diminishing the potential significance of a hybrid zone. With regard to the *Setophaga* warblers noted above, one could note that the almost nonexistent spread of phenotypic traits away from the hybrid zone despite 10,000 years of contact shows a temporally stable reproductive barrier, irrespective of occasional hybridization.

Finally, it is known that reproductive isolation does not evolve at a set (single) rate as groups become diagnosably dis-

tinct. For example, imagine three sympatric diagnostic groups for which the phylogenetic pattern is known (fig. 3.6). Then suppose that the pairwise occurrence of reproductive isolation is known and it does not match the phylogenetic pattern. Under the BSC, one would combine groups a+c to the exclusion of b, which would misrepresent evolutionary history. The species a+c has no value in biological studies because, in the parlance of systematics, it is "paraphyletic." Are there actual examples of this situation? Indeed, there are many, as in the case of the North American Flicker (*Colaptes* species) complex discussed below.

The BSC continues to be used in the most recent treatments of avian taxonomy, including the AOU *Check-List of North American Birds* (1998; see below). The list of species and their phylogenetic arrangement in this publication is used by private, state, and federal agencies in compiling lists of species for states and for regional and national parks. The US Fish and Wildlife Service also uses it when listing species or subspecies under the Endangered Species Act (ESA).

PHYLOGENETIC SPECIES CONCEPT

A prominent ornithologist, Joel Cracraft (1983; see the box on page 47), introduced the phylogenetic species concept to ornithology. He wrote that a phylogenetic species "is the smallest diagnosable cluster of individual organisms within which there is a parental pattern of ancestry and descent." In theory, a phylogenetic species is a geographically coherent group of individuals that share one or more (nonconflicting) heritable diagnostic traits. These traits can be either phenotypic or genotypic in nature. It is assumed that these shared traits evolved in isolation and that they represent evidence of an independent evolutionary history, which Cracraft and his supporters argue is the main requirement for delimiting species. Perhaps unfortunately, the "phylogenetic" in phylogenetic species does not necessarily mean that they are discovered by phylogenetic methods (although they may be). Instead, they are the smallest irreducible clusters of individuals that serve as the basic unit in phylogenetic, biogeographic, or comparative studies. We note that several slightly varied versions of the phylogenetic species concept have subsequently been proposed (summarized in de Queiroz 1998, Wheeler and Meier 2000). For all of these versions, the contrast with the BSC is immediately apparent in that there is no stipulation that knowledge of reproductive isolation is required to assign species limits.

There have been relatively few attempts to apply alternative species concepts to classifying the birds of the world. Cracraft (1992) studied morphological variation within and among the 40 to 42 biological species of birds-of-paradise and concluded there were approximately 90 phylogenetic species, or about twice as many species as previously thought. Zink (2004) found that many temperate-zone biological species contained approximately two distinct groups based on mitochondrial DNA (mtDNA), in agreement with Cracraft's (1992) assessment of increase in diversity of species of birds-of-paradise. Peterson and Navarro-Sigüenza (1999) analyzed patterns of species diversity in the birds of Mexico. Using the BSC, there are 100 endemic species, whereas using a phylogenetic species concept approach, there were 249 species. When mapping the distribution of these species, it was found that the PSC provided a better picture of the geographic distribution of avian diversity and one more appropriate for conservation (also see Soltis and Gitzendanner 1999, Agapow et al. 2004). Thus, for birds, there appears to be approximately twice as many phylogenetic species than there are biological species (Barrowclough et al. 2016).

Critiques of the Phylogenetic Species Concept

Critics of the PSC (e.g., Remsen 2005) have questioned what is meant by "diagnosable." Are not all individuals diagnosable, except identical twins, and would not that mean that all individuals are, in effect, species? Furthermore, if one considers a geographically bounded population, is the criterion that 100 percent of individuals exhibit the diagnostic state? Most advocates of the PSC consider species to be geographically coherent populations that share a diagnostic trait or traits that do not conflict with other traits, but nonetheless can show variation in other traits that could allow identification of every individual (which is true for all species concepts). Typically, these populations or groups of populations are the smallest geographic groupings that cannot be further subdivided. That does not mean, however, that individuals or family groups qualify as phylogenetic species (see section

below on numbers of species). It might be asked whether "trivial" traits could be used to diagnose a population and whether they might be evolutionarily transient, reducing the biological significance of the species. Some authors (e.g., Edwards et al. 2005) have argued that the often-used mtDNA is a "single" character because it reflects a single gene tree; however, the many diagnostic characters in a mtDNA gene tree often support previous taxonomic limits or are consistent with geographic barriers (McKay and Zink 2010). In fact, any single character with a genetic basis indicates that the individuals that share it also share an independent evolutionary history; hence, it is the pattern and not the level of divergence that reveals species limits. If diagnostic characters conflict, then one must reconcile this disagreement, as is done at higher levels in constructing consensus phylogenetic trees when there is character conflict.

The notion that some phylogenetic species can retain the primitive ability to hybridize (and do so) is emphasized by Remsen (2005) as a failing of the PSC because it presumably deemphasizes the role of gene flow cessation in the evolutionary process. This overlooks the fact that species limits should reflect evolutionary or phylogenetic history, whereas simply pooling groups into the same species because some level of hybridization occurs can lead to nonhistorical species (fig. 3.6).

It is possible that current or future hybridization will result in the decay of diagnostic traits, but this can take a very long time, up to hundreds of thousands of generations (Zink and McKitrick 1995). Alternatively, in some species the process can be rapid. For example, the Snow Goose and Blue Goose (fig. 3.7) were once considered distinct species, *Chen hyperborea* and *C. caerulescens*, respectively, and they had largely allopatric breeding ranges in arctic North America, the Snow Goose in the west and the Blue Goose in the east. They were considered biological species and would be considered phylogenetic species as well (the color phases have a simple genetic basis). Owing to favorable conditions on the wintering grounds that resulted in mixing of the two color phases, and the tendency for birds to pair on the wintering grounds with males following females to the female's natal area, a renewal of gene flow has subsequently resulted in extensive mixing of blue and white morphs across the arctic breeding range. Thus, there is currently little geographic segregation of different color phases that was once considered evidence for two species. Today the two color phases are considered part of the same biological species. Given the known history, it is also an example of two phylogenetic species that have fused owing to hybridization. The current distribution of the two historical phylogenetic species therefore obscures their distinctive recent evolutionary histories. However,

Figure 3.7. Blue Goose (shown here) is the sister taxon of the Snow Goose, and they were once considered separate species because of their plumage differences. It is now known that the plumage differences have a relatively simply basis; the two forms hybridize extensively, and are considered the same biological species. They are an example of phylogenetic species that are losing their identities owing to dispersal into each other's range.

knowledge of their prior historical independence would inform biogeographers reconstructing the history of biotas.

In practice, all "biological species" meet the criteria required to be recognized as "phylogenetic species" but not all phylogenetic species are recognizable as biological species. The latter frequently exhibit geographic variation in size, shape, or plumage characteristics, and this variation has been used by taxonomists to describe subspecies. Subspecies represent a taxonomic rank below the level of biological species and can be defined as "geographically defined aggregates of local populations which differ taxonomically from other such subdivisions of a species" (Mayr 1970). Under the BSC, subspecies continue to be considered a legitimate taxonomic category. Some (e.g., Remsen 2005) suggest that diagnostic taxa that hybridize with other such taxa should be recognized as subspecies. In contrast, the subspecies category has no role under the PSC because such diagnosable units *are* species.

Subspecies have historically been an important category in ornithology, with over 20,000 having been named by taxonomists. How are subspecies identified? In practice, methods (and rigor) have varied widely, with some taxonomists naming subspecies based on only a couple of characters scored from a handful of specimens from scattered localities, whereas others are described after carefully examining hundreds of specimens from all parts of the distribution. Typically, differences in size and coloration form the basis for subspecies descriptions, with the number of characters ranging

from one to a few. Although subspecies can be listed under the ESA, molecular methods often show them to not be distinct evolutionary entities (Zink 2015). A somewhat more rigorous approach to describing subspecies invokes the "75 percent rule" (Amadon 1949), in which subspecies status is warranted if 75 percent of individuals in a population differ from all (or from >99 percent of) members of an overlapping population. Although the 75 percent rule provides a quantitative measure of interpopulation differences, there is disagreement about the 75 percent threshold and the number of characters that should be required when comparing populations. For example, recent applications of the rule advocate a 95 percent threshold as the standard for diagnosability (Patten and Unitt 2002). Although the 75 percent rule has a reasonably long history in ornithology, its application has been erratic (Remsen 2010). The rule was popularized after the majority of subspecies had already been named, and it is certain that many currently recognized subspecies would not qualify as valid taxa under this rule, particularly those delimited by mensural data (Remsen 2010).

The Flickers: Hybridization and Plumage Evolution Lead to Different Species Limits

When early naturalists examined specimens from throughout the range of the Northern Flicker (*Colaptes auratus*), a common North American woodpecker that spends considerable time on the ground, they discovered that in the east (*C. a. auratus*), males had a black "moustache" mark and both sexes had yellow shafts to their flight feathers, whereas in the west (*C. a. cafer*), the male's moustache was red as were the feather shafts. Thus, taxonomists could easily sort specimens into two groups, and the moustache and feather shaft color was in effect a bar code as to whether the birds were from east or west of the Great Plains.

From the perspective of the PSC, the two types of flickers are distinct species, because their inherited diagnostic traits indicate that each has been evolving independently. Furthermore, it has been discovered that a third flicker form, the Gilded Flicker (red moustache but with yellow feather shafts), is more closely related to the red-shafted form than the red-shafted form is to the yellow-shafted form (Moore et al. 1991). Thus, there are three phylogenetic species in the complex, the Yellow-shafted Flicker, Red-shafted Flicker, and Gilded Flicker (fig. 3.8). From the perspective of the biological species concept, however, the question is whether these diagnosable groups can hybridize. In the case of the Red- and Yellow-shafted flickers, there is a relatively wide hybrid zone in the central Great Plains. Within this zone, male moustaches range from mostly red with a few black feathers to mostly black with a few red feathers, as well as to just about every

JOEL CRACRAFT

A hallmark of great scientists is that they keep focused on important conceptual issues in their field and integrate as needed the latest technologies throughout their careers. This description epitomizes the career of Dr. Joel Cracraft.

Dr. Cracraft received a BS in zoology from the University of Oklahoma, an MS in zoology from Louisiana State University (Baton Rouge), and a PhD in biology from Columbia University. He was a postdoctoral fellow at the American Museum of Natural History. From 1970 to 1992, he was a professor at the University of Illinois, Chicago. Since 1992, he has been a curator at the American Museum of Natural History in New York City, serving as chairman of the Ornithology Department during his tenure at the American Museum.

Early in his career, Cracraft investigated the evolutionary history of birds from morphological and anatomical perspectives. He also contributed to the literature on avian paleontology, describing first records of the Wild Turkey (*Meleagris gallopavo*) and the Whooping Crane (*Grus americana*) from the fossil record. A related work explored the functional morphology of the hindlimb of the domestic pigeon (*Columba livia*). He published the results of many other empirical studies, which reported on his investigations of evolutionary relationships within a variety of bird groups.

In the early 1970s, Cracraft began over a decade of contributions to discussions of a developing field of systematics termed "cladistics," or phylogenetic systematics. In 1972, for example, he published "The relationships of the higher taxa of birds: Problems in phylogenetic reasoning" in *The Condor*. Papers on continental drift and biogeography also started to appear. In 1974, he published a landmark study in *The Ibis*, "Phylogeny and evolution of the ratite birds," which set the stage for a wealth of subsequent studies. His 1981 paper in *The Auk*, "Toward a phylogenetic classification of the recent birds of the world," has been heavily cited by subsequent authors. His 1980 book *Phylogenetic Patterns and the Evolutionary Process: Method and Theory in Comparative Biology*, coauthored with N. Eldredge, was a must-read for workers in systematics and remains his most heavily cited publication.

In a landmark paper in 1983, Dr. Cracraft turned his attention to the topic of species concept, and introduced into systematics an early iteration of the "phylogenetic species concept." His views would catch hold and give rise to a strong reaction against the prevailing biological species concept, so heavily promoted by Ernst Mayr. His paper has been cited over 1,500 times as of this writing. In 1989, he published an important position paper on speciation called "Speciation and its ontology: The empirical consequences of alternative species concepts for understanding patterns and processes of differentiation," in which he pointed out that the study of speciation depended entirely on the concept of species used by the author. He is one of the few people who have rigorously applied the phylogenetic species concept to a group of birds: in 1992, he analyzed the birds-of-paradise and concluded that species diversity under the biological species concept was underestimated by 50 percent. Cracraft's contribution to the debate on species concepts is second to none, and continues to the present time.

Cracraft also energetically challenged those that have portrayed creationism as science. He has written many influential papers on the morphological and biogeographic relationships of various groups of birds, often blending in paleontological knowledge. His approach to incorporating molecular data into his research changed in the early to mid-1990s, when several of his students and postdocs began using mitochondrial DNA sequence data to address phylogenetic relationships. His work in molecular systematics continues to the present day, and in 2015 he published, with Santiago Claramunt, "A new time tree reveals Earth history's imprint on the evolution of modern birds" (Claramunt and Cracraft 2015), which integrated molecular and morphological systematics, and provided a novel view of the biogeography of birds.

One also cannot underestimate Cracraft's service to the field of systematics and biodiversity. He published many position statements on the importance of systematics and led the Systematics Agenda 2000 effort, which helped promote the value of systematics at a national and international level. His influence on his students, postdocs, and the scientific community at large is immeasurable.

For his work in avian phylogeny, paleontology, systematic philosophy, and species concepts and speciation analysis, Cracraft was awarded the Elliott Coues Medal by the American Ornithologists' Union in 1993. Dr. Cracraft epitomizes an outstanding scientist by his attention to significant scientific questions and his use of the technologies best suited to answering those questions.

Figure 3.8. Gilded Flicker. One of three closely related taxa (with Yellow-shafted Flicker and Red-shafted Flicker) that figure prominently in discussions of species limits. The Gilded Flicker is the sister taxon to the Red-shafted Flicker, but the Red-shafted Flicker hybridizes extensively with the Yellow-shafted Flicker.

intermediate condition. Thus, because there does not appear to be any reproductive isolating barriers, the two flickers are considered the same biological species by the AOU.

The case of the flickers highlights the differences between the biological species concept and the phylogenetic species concept. Not only are there differing numbers of species, three for the PSC and two for the BSC, but also the BSC classification creates a species, the Northern Flicker, that is not evolutionarily coherent because the two components (red-shafted, yellow-shafted) are not each other's nearest living relatives. In other words, as a consequence of focusing on the *process* of reproductive isolation, the BSC taxonomy does not reflect the true underlying historical *pattern* of evolution.

SPECIES CONCEPTS IN OPERATION IN ORNITHOLOGY

Most species concepts have several features in common, at least in practice. First, a taxonomist attempts to examine all of the possible specimens of a potentially new species, usually adding specimens of closely related species for comparison. Then, they gather evidence from the specimens that might include the coloration of body regions, plumage patterns, or estimates of size and shape derived from measurements of the wing, tail, tarsus, and bill. Modern assessments include comparisons of DNA sequences and large-scale genomic comparisons (Ellegren 2013, Harvey and Brumfield 2015). To illustrate how ornithologists decide species limits, we discuss three examples, towhees, grouse, and gnatcatchers.

Evolution of Brown Towhees, Species Limits, and the Problem of Classifying Allopatric Populations

The "brown towhee complex" was once classified as three species, the Brown Towhee (*Melozone fusca*), the Abert's Towhee (*M. aberti*), and the White-throated Towhee (*M. albicollis*). The Brown Towhee was known to consist of two clusters of subspecies, a western group called *crissalis* and an eastern group called *fusca* (fig. 3.9). The two diagnosable groups were considered part of the same species by the AOU because, although their ranges do not overlap, it was presumed that the morphological differences between them were too slight to induce birds to mate assortatively if they ever were to meet during the breeding season. That is, there was no direct "test of sympatry" in which one could observe under natural conditions whether there was premating or postmating reproductive isolation. Several molecular studies (e.g., DaCosta et al. 2009, Klicka et al. 2014) have subsequently shown conclusively that *crissalis* and *aberti* are sister species, rather than *crissalis* and *fusca*. The AOU Classification Committee responded by dividing the Brown Towhee into two species, the California Towhee (*M. crissalis*) and the Canyon Towhee (*M. fusca*).

However, this decision was made in the absence of information about actual interbreeding (fig. 3.9; see matrix on lower left). A fundamental difference between the PSC and BSC would result if *crissalis* and *fusca* were found to be interbreeding. Advocates of a BSC would unite them in the same species (fig. 3.9; lower right), whereas under the PSC the four species have had independent evolutionary histories and the species classification (fig. 3.9; upper right) would correctly represent this history irrespective of patterns of reproductive compatibility. Advocates of a PSC would not unite

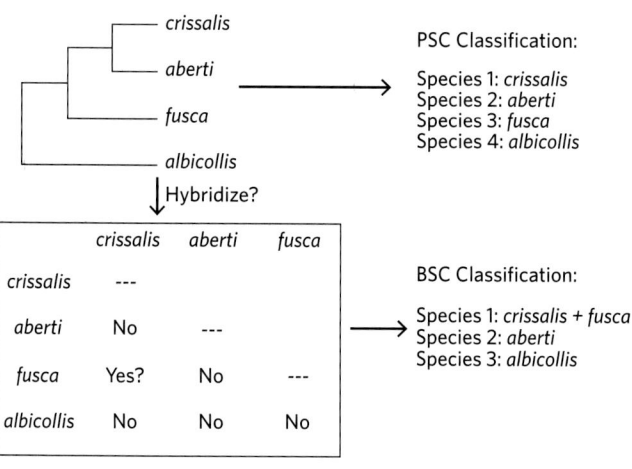

Figure 3.9. Evolutionary history of the brown towhee complex (genus *Melozone*).

fusca plus *crissalis* into one species because this species would misrepresent known evolutionary history (fig. 3.9; upper left) and would not qualify as a unit in studies of biogeography, ecology, phylogeny, conservation, or behavioral ecology. Importantly, this is not a semantic difference over simply how many species there might be, but rather which diagnostic groups are combined into particular species and how it is done, based on history (PSC) or reproductive isolation (BSC).

Molecular Study Results in the (Taxonomic) Resurrection of a Grouse Species

In western North America exist populations of a relatively large and tame grouse that presents a complicated taxonomic history. Taxonomists struggled with the question of whether this grouse, often known as the Blue Grouse, represented one or two species (fig. 3.10A). Originally a single species was described that nonetheless contained considerable morphological variation across its wide geographic distribution. The eastern and western groups differed in the coloration of the bare skin on the male apteria, presence or absence of a tail band, number of tail feathers, and the coloration of the downy young (Zwickel et al. 1991, Zwickel 1992). In addition, there are differences in mating calls and general courtship behavior. In 1931 the species was split into two by the AOU, but reunited by the AOU in 1944. This decision, however, was based not on comprehensive analyses, but only on the observation that there were a few known hybrids between the two forms. Several authors (e.g., Grinnell and Miller 1944) questioned the basis for a single species, given that hybridization was rare and that substantial morphological and behavioral differences existed.

This species then, was ripe for reanalysis using modern taxonomic methods. George Barrowclough from the American Museum of Natural History and his colleagues investigated geographic patterns of genetic variation using mtDNA. They analyzed specimens from many places throughout the range and found two evolutionarily distinct groups. A couple of individuals appeared to be hybrids, but their low frequency does not merit subsuming the considerable genetic divergence, and the morphological and behavioral differences observed among groups, into a single species. Interestingly, in one of the groups, there were two additional distinct mtDNA lineages that were previously unknown and did not reflect subspecies limits, but the researchers suggested more study was warranted before considering their status as species. As a result of Barrowclough et al.'s (2004) study, the AOU has (re)elevated the eastern and western forms of "Blue Grouse" to species status, and they are now known as the Dusky Grouse and Sooty Grouse (fig. 3.10B).

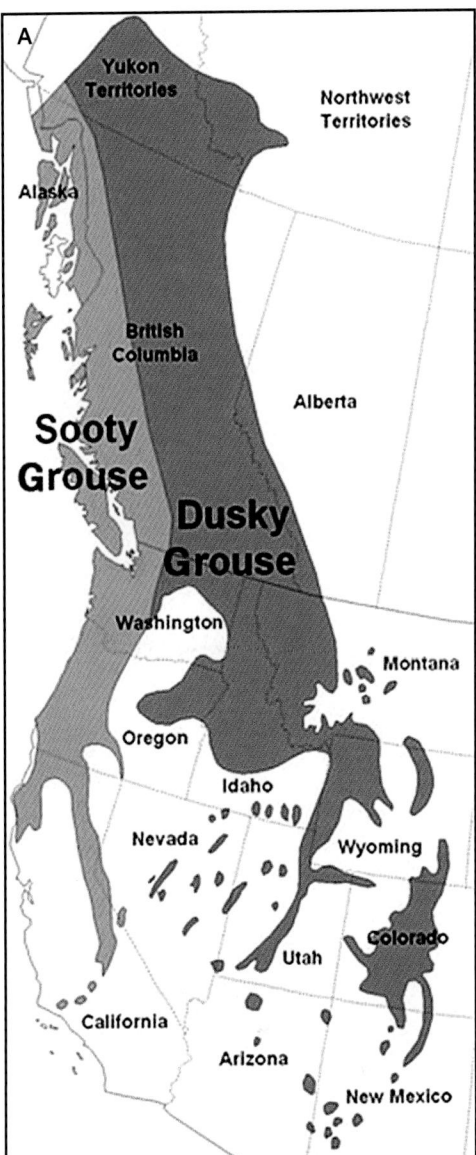

Figure 3.10. A, Distribution of Sooty Grouse and Dusky Grouse, once considered parts of the same species ("Blue Grouse"); *B,* Sooty Grouse.

Morphological and Behavior Reassessment Leads to the "Splitting" of a Widespread Gnatcatcher Species

The nonmigratory Black-tailed Gnatcatcher (*Polioptila melanura*; fig. 3.11A) was once considered a widespread species in the arid areas of the southwestern United States, northern Mexico, and the Baja California Peninsula (fig. 3.11B). Considerable taxonomic confusion preceded the work of

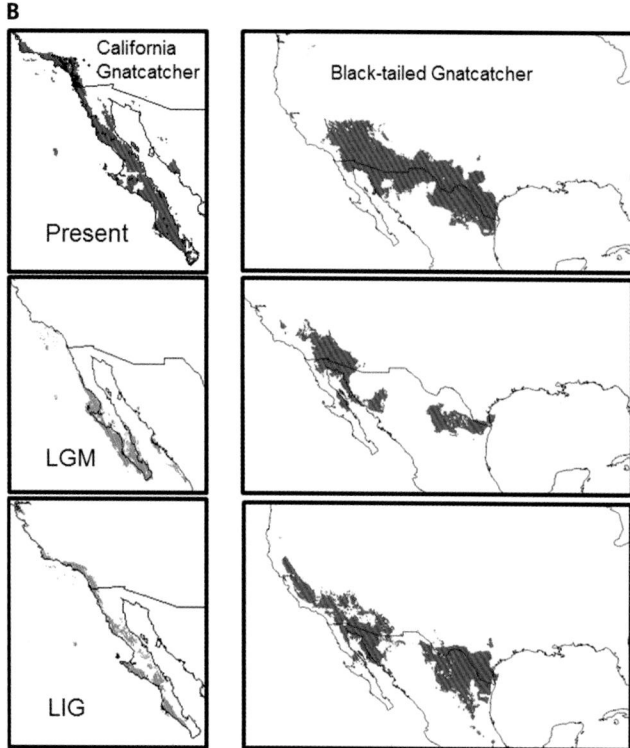

Jonathan Atwood (1988), who studied birds that had been collected throughout the range and spent considerable time in the field recording vocalizations. Of interest were two groups of gnatcatchers: those found from southern California south through the Baja Peninsula to its southernmost tip (Cabo San Lucas), and those in the Sonoran and Chihuahuan Deserts to the east. In the past, these had been variously referred to as members of the same species, or as different species, depending on the taxonomist. From 851 museum scientific study skins, Atwood measured 19 external characters that captured a variety of aspects of size and shape of the wing, bill, and feet, as well as using a spectrophotometer to measure coloration of 12 body regions (the same type of machine that your paint store uses to match paint color). From recorded vocalizations, he used a sonograph to represent visually the frequency and tempo of sounds given by males. Atwood's results were unequivocal. Using a principal components analysis, which accounts for correlations among variables and plots individuals or populations in multivariate "morphospace" along uncorrelated dimensions (called "principal components"), Atwood showed that the two groups exhibit considerable and nonoverlapping morphological differences (fig. 3.12A). To establish the relative degree of dif-

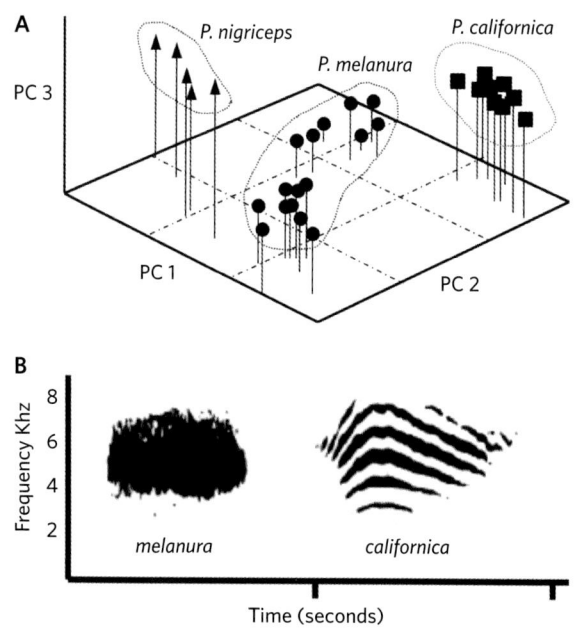

Figure 3.11. A, Black-tailed Gnatcatcher; *B*, Distribution of two sister species, Black-tailed Gnatcatcher and California Gnatcatcher, during the present, the Last Glacial Maximum (21,000 ybp), and the Last Interglacial (120,000 ybp), estimated by ecological niche modeling (Zink 2012). These two species were formerly considered the same species, but their plumage, song, and mtDNA differences clearly indicate that they are separate species.

Figure 3.12. A, Results of principal components analysis of morphological characters showing the distinctiveness of California Gnatcatcher, Black-tailed Gnatcatcher, and Black-capped Gnatcatcher (*from Atwood 1991*); *B*, Diagnostic vocal differences between Black-tailed Gnatcatcher (*Polioptila melanura*) and California Gnatcatcher (*P. californica*).

ference, he added a third species, *P. nigriceps*, and it can be seen that all three are morphologically distinguishable (e.g., they occupy three distinct regions of the morphospace). He also was able to show diagnostic vocal differences (fig. 3.12B) that, when used in playback experiments, showed that birds tended to respond to playbacks of their own species but not of the other. Thus, Atwood's study revealed that, not only were the two gnatcatcher groups phylogenetic species but, given the lack of hybridization and their response to vocalizations, they were probably biological species as well. It was later determined that the two species were also very distinct in their mtDNA (Zink and Blackwell 1998). Today there is no doubt that two species are represented.

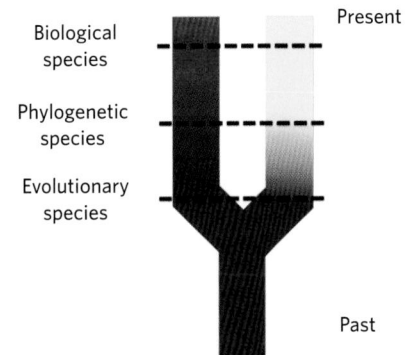

Figure 3.13. Diagrammatic representation of lineage divergence and approximate points at which different species concepts recognize taxa as species.

PUTTING DIFFERENT SPECIES CONCEPTS IN A TEMPORAL PERSPECTIVE

Avise and Walker (1998) noted that speciation is not a "point event," as phylogenetic trees appear to suggest by depicting the instantaneous splitting of one lineage into two descendant lineages. Instead, speciation can be a protracted process, taking perhaps thousands of years, owing to the long time it takes for gene pools to diverge and the fact that some geographic isolating barriers form over time rather than all at once. However, from the initial splitting of the ancestral species, the two diverging lineages are evolving independently of one another, assuming there is no gene flow in either direction. At this stage, each taxon would represent an evolutionary species (under the evolutionary species concept [ESC] sensu Simpson 1961). Wiley (1978; see table 3.1 and fig. 3.13) defines an evolutionary species in this way: "A species is a lineage of ancestral descendant populations which maintains its identity from other such lineages and which has its own evolutionary tendencies and historical fate." Thus, under the ESC, new (incipient) species exist when population isolation is initiated and gene flow ceases. This may be a protracted process, or it could occur rapidly via a long-distance dispersal event. In either case, species defined under the ESC are the first to be recognized when viewed from a temporal perspective (see fig. 3.13). Unfortunately, the ESC is not operational in practice because it does not explicitly identify any criteria by which a species might be identified (the PSC and BSC do provide criteria—fixed diagnostic characters and reproductive isolation, respectively). Given more time (fig. 3.13), genetic drift, natural selection, or sexual selection will yield diagnostic traits that, under the PSC, allow members of each lineage to be identified as distinct phylogenetic species. In theory, it will take an even longer period of time for randomly evolved diagnostic traits to function as reproductive isolating mechanisms. When this does occur,

the diverging lineages (already diagnosable as phylogenetic species) would then be recognized under the BSC as biological species. Thus, different species concepts can be thought of as representing different stages along the continuum of lineage divergence (Harrison 1998). However, this diagram should not be interpreted to mean that biological species are always "older" than phylogenetic species. It is possible to have very old phylogenetic species that share a common ancestor farther in the past than two biological sister species.

Do the BSC and PSC Lead to the Same Number of Species of Birds of the World?

In North America there are about 3.3 subspecies per bird species. If each of these were evolutionarily distinct and diagnosable, then one could simply elevate each subspecies to the phylogenetic species level and increase species diversity by a factor of three. However, it is not that simple. In some species, none of the subspecies have been shown to be distinct using modern molecular methods. For example, the Common Yellowthroat (*Geothlypis trichas*), a widespread North American warbler, was found to consist of just two phylogenetic species, among the 12 subspecies that were analyzed (Escalante et al. 2009). When Ernst Mayr set about classifying the birds of the world using the BSC, there were approximately 20,000 avian species. When judgments were made as to the actual or potential reproductive isolation of many species, over half were reclassified as subspecies, reducing the number of species by around 50 percent. Many viewed this as a welcome house cleaning, whereas others viewed it as burying taxonomically important diversity at the subspecies level. A recent study suggested that the number of phylogenetic species of birds of the world is in the range of 15,000 to 20,000 (Zink 1996, Barrowclough et al. 2016). Thus, the PSC would result in an approximate doubling of the number of species of birds of the world.

How Do Authors of Field Guides Know How Many Species of Birds Exist in North America?

The North American Classification Committee (NACC), an official committee of the AOU (now AOS), produces "Checklist of North and Middle American Birds" (http://www.american ornithology.org/content/checklist-north-and-middle -american-birds), which is the official source, according to the committee, on the taxonomy of birds found in North and Middle America. In short, the NACC members decide what is and what is not a species via a democratic process—they vote. Historically, their deliberations have been private and the resulting checklists much-anticipated, as ornithologists waited to see what subspecies or geographic groupings were made full species and which were demoted to subspecies. Whether you are writing a field guide, making a county checklist of birds, doing an ecological or evolutionary study, or considering a species as threatened or endangered for conservation, the authoritative list of birds in North America is the AOU checklist. Today, committee deliberations can be followed online (http://checklist.aou.org/). One can even see what proposals have been submitted for committee consideration, the resulting discussion, and the final vote. Elsewhere in the world, other ornithological organizations produce lists that are used by writers of field guides, bird-watchers, and ecotourists. Several world bird lists have been compiled recently, including *The Clements Checklist of Birds of the World* (Clements 2007; see Clements et al. 2015 for regularly updated online version), *The Howard and Moore Complete Checklist of the Birds of the World* (Dickinson and Remsen 2013, non-passerines; Dickinson and Christidis 2014, passerines), and the *BirdLife International Illustrated Checklist of the Birds of the World* (del Hoyo et al. 2014, 2016). Because each differs slightly in their primary goals and taxonomic philosophy, they differ from one another in their accounting of the number of bird species, with each also differing from the official AOU NACC list. So, the number of bird species that occur in a given area depends on which checklist you consult.

THE FUTURE OF THE "SPECIES CONCEPTS" DEBATES

Most biologists (but not all) believe that "species" are real biological entities and the fundamental units of evolution. As such, species are routinely used as the units of analysis in biogeography, ecology, macroevolution, and conservation biology (Sites and Marshall 2004). As we have outlined above, biologists have long struggled (debated) to agree on exactly how these important units should be defined. In the modern era, Mayr's (1942) introduction of the BSC influenced the thinking of evolutionary biologists in most fields of study, and his views on species and the speciation process became the prevailing views among most ornithologists. With the advance of molecular phylogenetics in the early 1980s, the PSC was introduced (Cracraft 1983), providing a different way of thinking about species and rekindling a vigorous debate over species concepts. In this time of increasing habitat loss and increasing extinction rates, the need to identify, delineate, and name taxa is especially acute. As alternatives to the PSC and BSC, several novel methods have recently been proposed for identifying species in a statistical framework. Collectively these are referred to as "species delimitation" methods (e.g., Sites and Marshall 2004).

Species delimitation methods are based on the fundamental notion that species should represent independently evolving evolutionary lineages (Wiley 1978; see Mayden 1997, de Queiroz 1998). A goal of these new methods is to bridge the historically independent disciplines of phylogenetics and population genetics, with the goal of identifying the point at which population level processes begin to produce phylogenetic patterns (Carstens et al. 2013). Most take advantage of the relative ease with which large amounts of genomic data can now be generated. Typically, coalescence theory is applied to such data, allowing for the modeling of population-level processes such as genetic drift, migration, and population expansion. Among the approaches developed so far (see Fujita et al. 2012, Carstens et al. 2013, Rannala 2015), perhaps the most promising are those model-based methods that employ Bayesian species delimitation in a multispecies coalescent (MSC) framework (e.g., Yang and Rannala 2010). This method calculates the probability distribution of many individual gene trees under a coalescent model and then uses this distribution to identify all distinct evolutionary lineages to assess which species hypothesis (e.g., number of species) is best supported by the evidence. Species identified in this way are in theory reproductively isolated, as required by the BSC, and diagnosable, as required by the PSC, in both cases by virtue of fixed genetic differences. However, this species delimitation method cannot infer whether or not two reproductively isolated species might interbreed in the future. Several recent papers have explored differences between coalescent-based and more traditional methods of species delimitation (McKay et al. 2013, Hung et al. 2014). The development of species delimitation methods is a relatively new field within systematics and an active area of research.

What impact will new methods of identifying species have on the ongoing "species concept" debate? Biologists have argued about how to define a species for more than 300 years, and a consensus has not been reached. Advocates of both the BSC and PSC have well-established criteria for identifying species and deeply entrenched points of view. In addition it seems likely that biological classification will continue to use a Linnaean, hierarchical taxonomic system. As more and more of the genome is charted, and exact differences between species (of whatever ilk) are documented, perhaps a new consensus will be forthcoming. If history is a guide, we suggest that the species question will not be resolved in the near future.

WHAT IS SPECIATION?

Hundreds of millions of species have existed on earth, although over 99 percent of these are now extinct. Because new species arise from a single ancestor, it follows that the process of speciation is very common. This suggests we should know a great deal about the evolutionary process of speciation, the generator of biodiversity. As noted above, speciation is a protracted process, and it is controversial to decide where in the process of lineage divergence two diverging populations might be, assuming they actually complete the process of becoming new species. Nonetheless, two major aspects typify the process of speciation, and we refer to those as the geography of speciation and the biology of speciation. The first aspect is relatively noncontroversial, but whether the second aspect is controversial depends in part on one's choice of species concept (fig. 3.13). In other words, species define speciation.

THE GEOGRAPHY OF SPECIATION AND THE IMPORTANCE OF ALLOPATRY

A little gene flow goes a long way. This axiom of population genetics highlights the importance of allopatry. If populations continue to exchange breeding individuals, the process of divergence will be prevented or seriously retarded. Thus, most biologists infer that speciation occurs primarily in geographic isolation (allopatry). Perhaps the most compelling factor in support of the importance of allopatry is the observation that most sister species are allopatric. From this observation, we infer that the "geography of speciation" was allopatric. Many allopatric sister species occupy somewhat different habitats, suggesting that local adaptation is a mechanism that prevents them from becoming sympatric. Alternatively, it is possible that insufficient ecological divergence has taken place to allow coexistence, and although such nascent species are different in other respects, competition keeps them geographically isolated. The famous ecologist G. E. Hutchinson (1959) noted that most sympatric species differ by at least a factor of 1.1 to 1.4 in morphological structures associated with food gathering (a figure often called Hutchinson's ratio). Thus, it might be that, until allopatric populations reach this level of divergence, coexistence (sympatry) is not likely.

For example, using the gnatcatcher example discussed above, we can examine the recent history of their distributions using ecological niche modeling (Peterson et al. 1999). In brief, one takes a series of locations where a species breeds at present and builds a model using various climatic factors related to temperature and humidity. By building the model with fewer than the full number of localities, one can then subsequently check how well the excluded points are predicted. The model for gnatcatchers performed very well, predicting over 90 percent of the locality points excluded from model construction. Next, one finds where the areas with climate suitable for today's populations existed at the Last Glacial Maximum (21,000 years before present, ybp) and the Last Interglacial (120,000 ybp). This is made possible because the same climatic variables measured in the present have also been estimated worldwide for these two time periods (there are other periods estimated as well, including the future). The exercise gives an estimate of where the species lived at these two times in the past, assuming their climatic niche preferences have remained stable. In the case of the gnatcatchers (fig. 3.11B), one can see that the two species were allopatric at both prior time periods, showing the long-term maintenance of allopatry (Zink 2012). In theory, the two species are morphologically too similar to invade each other's ranges, given that the differences between their bills do not exceed a Hutchinsonian ratio. In addition, it is possible they are adapted to differing habitats.

Contrary to the widely held assertion that allopatry is required for avian speciation to occur, some evidence suggests that species have formed in sympatry. For example, Fjeldså (1983) suggested that a flightless grebe evolved from within the range of a volant species. It is also possible that when brood parasites switch hosts, their young will be tied to the new host and in effect speciate without geographic isolation (Price 2008, Sorenson et al. 2003). These examples, however compelling, are in a small minority relative to the overall prevalence of allopatric speciation events (Phillimore et al. 2008, Fitzpatrick et al. 2009). We note Dobzhansky's famous statement: "Sympatric speciation is like measles; everyone gets it, and we all get over it."

Another mode of speciation envisions populations being isolated along an ecological gradient (ecological speciation). There are several variants of the model, but in general, one assumes that gene flow between groups living in different areas can be curtailed by environmental differences, not discrete geographic barriers (such as rivers or mountain ranges). Rundle and Nosil (2005) envision ecological speciation as a situation in which "barriers to gene flow evolve between populations as a result of ecologically-based divergent selection." This mode of speciation has received considerable

Figure 3.14. Red Crossbill. This species exhibits many different "call types" that might be independently evolving taxa.

recent attention (e.g., Schluter 2009). However, in ornithology, very few concrete examples exist. The two main examples in Price's (2008) review are both special cases. The Red Crossbill (*Loxia curvirostra*; fig. 3.14) consists of nine nomadic populations that have different call types, and at least some of these have specialized on different food sources; whether there are two or more species is not clear at present (Parchman et al. 2016). The other example of potential ecological effects on speciation, the Galapagos finches, relies on naturally selected shifts of bill sizes within populations as a result of temporally varying seed size availability (Grant and Grant 2011), but this does not explain whether the species of finches arose this way. McKay and Zink (2015) proposed that finches on the Galapagos move up and down transient adaptive peaks, a process they termed Sisyphean evolution. Hence, although there is much recent interest in ecological speciation, little evidence exists for it in birds, and it would therefore be a fruitful area for greater research.

THE BIOLOGY OF SPECIATION, OR HOW DOES DIVERGENCE IN ALLOPATRY RESULT IN NEW SPECIES, AND IS IT MYSTERIOUS?

The origin of species is only slightly less mysterious now than it was 150 years ago when Darwin published his famous book.
—Via (2009:9939)

We believe that the state of knowledge concerning speciation is a little less mysterious than Via suggests. Lineages diverge in allopatry as a result of at least three mechanisms, including natural selection, sexual selection, and random factors such as genetic drift, and perhaps combinations of all three work at once. A role for natural selection is suggested by the relatively large number of species whose plumage matches

predictions of Gloger's rule, posited by Constantin Gloger (1833), which states that the degree of plumage pigmentation is related to the relative humidity of the environment. As an example, Song Sparrows (*Melospiza melodia*; figs. 3.15A, 3.15B) from the Aleutian Islands and nearby mainland are very darkly plumaged (e.g., heavily pigmented), which allows them to match the environmental—mainly vegetation—background. In contrast, Song Sparrows from southern California near the Salton Sea are very light colored, adaptively matching the lighter background of their more arid environment. Burtt and Ichida (2004) suggested an alternative explanation for Gloger's observation. They concluded that more darkly pigmented feathers may provide greater protection from feather-degrading bacteria that are more common in humid environments. Either explanation suggests that natural selection acts on plumage traits, yielding phenotypes that are better adapted to a given environment.

Another ecological "rule" stems from the German biologist Carl Bergmann (1847), who noted that for warm-blooded organisms, body size was larger in colder environments. The presumed causal mechanism is that when animals get larger, their surface area increases as a square whereas their volume increases as a cube. Hence, larger animals have relatively smaller surface areas from which to lose metabolic heat from a relatively larger body core. The relationship holds in large part for many birds, especially nonmigratory species (Meiri and Dayan 2003). The usual result is a geographical gradient or cline in size for widespread species, where those populations living in colder climates commonly have larger bodies than their conspecific relatives living in warmer climates. A related ecogeographic rule, Allen's rule (Allen 1877), posits that birds living in colder regions generally have shorter bills, legs, and wings than their nearest relatives in warmer climates, to reduce thermoregulatory costs through heat loss

Figure 3.15. Song Sparrow. *A,* Typical eastern bird; *B,* Sooty plumaged bird from Alaska. This species is highly morphologically variable throughout its North American range and is divided into many subspecies. It is unclear how many, if any, are evolutionarily distinct, or whether clinal variation connects all subspecies.

via the extremities. A recent comparative study (Symonds and Tattersall 2010) of 214 bird species found strong support for Allen's rule. Across all species, there were strongly significant relationships between bill length and both latitude and temperature, with species in colder regions having significantly shorter bills. Others have noted that migratory species tend to have longer wings, which promote efficient long-distance flights, and the fact that many species in north temperate regions tend to be migratory might also in part explain geographic patterns in wing length. Thus, adaptation to differing environments yields morphological differences among allopatric populations.

Many sister species differ qualitatively in plumage pattern and song. Although these characteristics can differ between allopatric conspecific populations, they are more typically the kinds of diagnostic traits that distinguish species. For example, the three North American species of bluebirds (Eastern, *Sialia sialis*; Western, *S. mexicana*; Mountain, *S. currucoides*)

differ conspicuously in plumage, and this sort of qualitative morphological difference is relatively rare in the geographic variability that occurs within most continentally distributed bird species (it is more common on islands). In another example, the sister species Rose-breasted Grosbeak (*Pheucticus ludovicianus*) and Black-headed Grosbeak (*P. melanocephalus*) are strikingly different in their plumage patterns. In contrast, a few species, like the flycatchers in the genus *Empidonax*, can be reliably identified only by their vocalizations (and then only the males). Other closely related species, such as the Eastern Meadowlark and Western Meadowlark, have relatively subtle morphological differences, but differ in habitat use and vocalizations. Thus, although geographic isolation is a usual prerequisite for speciation, it is not certain that natural selection is always the driving force behind the acquisition of the diagnostic features that distinguish sister species. That is, differences between sister species are usually not simple extensions of geographic variation in size, shape, and overall coloration.

Given that many closely related species differ in song and plumage pattern, one potential mechanism of divergence is sexual selection. A possible example concerns two warblers, the Audubon's Warbler and the Myrtle Warbler discussed previously. Principal differences include a yellow throat and more black in the plumage in the Audubon's Warbler, whereas the Myrtle Warbler has a white throat and less black (see figs. 3.5A, 3.5B). Each is widespread over a varied environmental gradient spanning more than 1,000 km. It is difficult to imagine that throat color is a "local adaptation" that functions in attracting prey or providing camouflage across such a huge range, because environments are rarely homogeneous across wide areas. However, one could imagine a mutation arising that conferred a change in throat color, and that if it were preferred by females, the trait could spread quickly and over a large environmentally heterogeneous area because it did not jeopardize local adaptability. Shutler and Weatherhead (1990) referred to the diagnostic plumage patterns of closely related warblers as "targets" of sexual selection. Thus sexual selection might be very important in some groups, and can easily apply to vocalizations as well (Badyaev et al. 2002). It has been observed that bird species are less differentiated genetically than species in other groups, which might mean that the speciation process in birds is relatively faster (Johns and Avise 1998), which is predicted by the sexual selection process.

As we noted above, what one thinks about speciation depends on one's species concept (Wiens 2004; fig. 3.13). Following the BSC, speciation is completed when character changes that delimit two or more groups also result in assortative mating. Of course there is no logical reason for characters that would specifically influence mate choice relative to the sister lineage to change in allopatry, given that the two diverging forms do not interact. If character divergence between allopatric populations occurs randomly, or as an incidental by-product of divergence (Mayr 1970), it stands to reason that many of the initial characters that evolve to become diagnostic characters in allopatric populations might not influence mate choice. Bush (1982) noted that reproductive isolation is a consequence, not a cause, of speciation. Therefore, because phylogenetic species are diagnosable groups that will often not be reproductively isolated, they will likely be recognized as species earlier in the process of lineage divergence, as opposed to biological species, which require that traits also function as reproductive isolating mechanisms (e.g., fig. 3.13).

Another recent suggestion is that speciation is driven by mitochondrial-nuclear incompatibilities (Gershoni et al. 2009, Wolff et al. 2014, Hill 2015). The physiology of most organisms is powered by energy produced in the mitochondrion by the electron transport chain (ETS). The circular DNA molecule contained in the mitochondrion codes for genes that function in the ETS in concert with genes encoded in the nucleus. It is thought that mtDNA sequences evolve relatively rapidly, with some changes requiring concomitant changes in nuclear loci. If in allopatry this mito-nuclear co-evolution occurs, it could result in reproductive barriers (e.g., isolating mechanisms) between allopatric populations owing to an incompatibility that would put hybrids at a physiological disadvantage (Cheviron and Brumfield 2009). Whether this will prove to be an important driver of speciation in birds is yet to be determined.

Perhaps one of the more important points to remember in attempting to deduce changes that caused speciation, by whatever means, is that at the minimum, sister species must be the framework for comparison. That is, one documents differences between sister species and attempts to deduce which are related to speciation and which evolved after speciation occurred. If two sister species are relatively old, one cannot easily determine which changes occurred before, during, or after the speciation process. Thus, in speciation research, attention should focus on determining where in the process of divergence particular features might have arisen and why.

What about the mystery in speciation? Under both the BSC and PSC, speciation almost certainly has an allopatric component, which is not mysterious at all. Under the PSC, speciation is completed when two descendant lineages acquire diagnostic characters, and the logical processes in play are natural selection, sexual selection, and genetic drift. The time it takes for these processes to operate depends on the effective population size of each population and the extent of gene flow between them. Hence, there is nothing particu-

larly mysterious about speciation under the PSC. Under the BSC, genetic changes must occur that lead to premating or postmating reproductive isolation mechanisms. It would seem likely that there are many ways these changes could be accomplished, but none are particularly mysterious. Although some authors have posited a "speciation gene," we consider this an unlikely outcome. Thus, it is possible that speciation might be considered mysterious under the BSC because there is not a general pathway to reproductive isolation. Lack of generality, however, should not mean that the process is mysterious. Instead, we think that speciation is actually relatively well understood.

KEY POINTS

- The biological species concept has long been the dominant species concept used in ornithology. Under the BSC, if two diagnosable groups of individuals are unable to reproduce successfully with each other, they represent two different species.
- The phylogenetic species concept was introduced to ornithology in 1983 and in a short time has gained many supporters. A phylogenetic species is a geographically coherent group of individuals that share one or more heritable, diagnostic traits. Such traits provide evidence of an independent evolutionary history.
- The PSC and the BSC may best be viewed as occupying different stages along a speciation continuum; i.e., speciation is a "process," during which the evolution of diagnostic traits (PSC) typically occurs before the evolution of complete reproductive isolation (BSC) is achieved.
- If the BSC were to be replaced by the PSC globally, the number of recognized extant bird species would approximately double.
- New molecular methods will potentially change the way species are recognized.
- Although sympatric speciation and ecological speciation have been suggested as modes of speciation in some bird groups, most ornithologists agree that speciation usually occurs in geographic isolation (allopatric speciation). This view is supported by the observation that most sister species have allopatric distributions.
- Birds presumably use signals that act as premating isolating mechanisms. However, many birds possess an array of signals (plumages, behaviors, vocalizations) with which to communicate. Without experimentation, it is difficult to know which particular signal is in use at a given time, what is the actual intent of the signal, and who is the intended recipient of the signal.

KEY MANAGEMENT AND CONSERVATION IMPLICATIONS

- A phylogenetic species is a diagnostic clusters of individuals, typically geographically circumscribed, with an independent evolutionary history.
- Phylogenetic species are the units identified and used in ecological, behavioral, and evolutionary studies.
- Many phylogenetic species could hybridize with other phylogenetic species and therefore are not recognized as biological species.
- Biological species are equivalent to one or more phylogenetic species.
- There are about twice as many phylogenetic species as biological species.
- The PSC is more appropriate for use when management and conservation implications are concerned.

DISCUSSION QUESTIONS

1. Describe a situation in which use of the PSC would be more appropriate than the BSC.
2. Describe a situation in which use of the BSC would be preferred.
3. In discussing the relative strengths of the BSC and PSC, one major difference noted was that the former emphasizes "process" and the latter emphasizes "pattern." What is meant by this?
4. In the example given regarding speciation in aridland gnatcatchers, which evidence provided would cause the *californica* and *melanura* groups to be recognized as phylogenetic species? Which evidence suggested that they were biological species as well?
5. Several "biogeographic" (Gloger's, Bergmann's, Allen's) rules were described. Summarize each and provide examples from the bird world. What is the relevance of these in speciation?
6. Can sexual selection on phenotypic or vocal traits lead to speciation? How? Does sexual selection on such traits necessarily lead to speciation? Why or why not?
7. Male Prairie Chickens perform a choreographed dance on the "booming grounds" during which they simultaneously display a fanned tail and a brightly colored, inflated air sac on either side of their neck. This is accompanied by a variety of vocalizations, including a "booming" sound. How could one reliably determine which (if any) of these traits are used as reproductive isolating mechanisms, and which might be intended for other purposes (e.g., male-to-male competition)?

References

Agapow, P. M., O. R. P. Bininda-Emonds, K. A. Crandall, J. L. Gittleman, G. M. Mace, J. C. Marshall, and A. Purvis (2004). The impact of species concept on biodiversity studies. Quarterly Review of Biology 79:161–179.

Allen, J. A. (1877). The influence of physical conditions in the genesis of species. Radical Review 1:108–140.

Amadon, D. (1949). The seventy-five per cent rule for subspecies. The Condor 51:250–258.

American Ornithologists' Union (1998). Check-list of North American birds. 7th ed. American Ornithologists' Union, Washington, DC.

Atwood, J. L. (1988). Speciation and geographic variation in Black-tailed Gnatcatchers. Ornithological Monographs 42.

Atwood, J. L. (1991). Species limits and geographic patterns of morphological variation in California Gnatcatchers (*Polioptila californica*). Bulletin of the Southern California Academy of Sciences 90:118–133.

Avise, J. C., and D. Walker (1998). Pleistocene phylogeographic effects on avian populations and the speciation process. Proceedings of the Royal Society of London B 265:457–463.

Badyaev, A. V., G. E. Hill, and B. V. Weckworth (2002). Species divergence in sexually selected traits: Increase in song elaboration is related to decrease in plumage ornamentation in finches. Evolution 56:412–419.

Ballentine, B., J. Hyman, and S. Nowicki (2004). Vocal performance influences female response to male bird song: An experimental test. Behavioral Ecology 15:163–168.

Barrowclough, G. F., J. G. Groth, L. A. Mertz, and R. J. Gutiérrez (2004). Phylogeographic structure, gene flow and species status in Blue Grouse (*Dendragapus obscurus*). Molecular Ecology 13:1911–1922.

Barrowclough, G. F., J. G. Groth, L. A. Mertz, and R. J. Gutiérrez (2005). Genetic structure, introgression, and a narrow hybrid zone between Northern and California Spotted Owls (*Strix occidentalis*). Molecular Ecology 14:1109–1120.

Barrowclough, G. F., J. Cracraft, J. Klicka, and R. M. Zink (2016). How many kinds of birds are there and why does it matter? PLoS ONE 11 (11): e0166307. doi:10.1371/journal.pone.0166307.

Bergmann, C. (1847). Ueber die verhältnisse der wärmeokönomie der thiere zu ihrer grösse. Gottinger studien 3:595–708.

Burtt, E. H., and J. M. Ichida (2004). Gloger's rule, feather-degrading bacteria, and color variation among song sparrows. The Condor 106:681–686.

Bush, G. L. (1982). What do we really know about speciation? *In* Perspectives on evolution, R. Milkman, Editor. Sinauer, Sunderland, MA, pp. 119–128.

Carstens, B. C., T. A. Pelletier, N. M. Reid, and J. D. Satler (2013). How to fail at species delimitation. Molecular Ecology 22:4369–4383.

Cheviron, Z. A., and R. T. Brumfield (2009). Migration-selection balance and local adaptation of mitochondrial haplotypes in rufous-collared sparrows (*Zonotrichia capensis*) along an elevational gradient. Evolution 63:1593–1605.

Claramunt, S., and J. Cracraft (2015). A new time tree reveals Earth history's imprint on the evolution of modern birds. Science Advances 1 (11): e1501005.

Clements, J. F. (2007). The Clements checklist of birds of the world. Cornell University Press, New York.

Clements, J. F., T. S. Schulenberg, M. J. Iliff, D. Roberson, T. A. Fredericks, B. L. Sullivan, and C. L. Wood (2015). The eBird/Clements checklist of birds of the world: v2015. http://www.birds.cornell.edu/clementschecklist/download/.

Cracraft, J. (1983). Species concepts and speciation analysis. Current Ornithology 1:159–187.

Cracraft, J. (1992). The species of the birds-of-paradise (Paradisaeidae): Applying the phylogenetic species concept to a complex pattern of diversification. Cladistics 8:1–43.

Cronquist, A. (1978). Once again, what is a species? In Biosystematics in agriculture, J. A. Romberger, Editor. John Wiley & Sons, New York, pp. 3–20.

DaCosta, J., G. M. Spellman, P. Escalante, and J. Klicka (2009). A molecular systematic revision of two historically problematic songbird clades: Aimophila and Pipilo. Journal of Avian Biology 40:206–216.

Darwin, C. (1859). On the origin of species. Or the preservation of favoured races in the struggle for life. John Murray, London.

Del Hoyo, J., N. J. Collar, D. A. Christie, A. Elliot, and L. D. C. Fishpool (2014). Handbook of the birds of the world/BirdLife International illustrated checklist of the birds of the world, Volume 1: Non-passerines. Lynx Edicions, Barcelona, Spain.

Del Hoyo, J., N. J. Collar, D. A. Christie, A. Elliot, L. D. C. Fishpool, P. Boesman, and G. M. Kirwan (2016). Handbook of the birds of the world/BirdLife International illustrated checklist of the birds of the world, Volume 2: Passerines. Lynx Edicions, Barcelona, Spain

De Queiroz, K. (1998). The general lineage concept of species, species criteria, and the process of speciation. In Endless forms: Species and speciation, D. J. Howard and S. H. Berlocher, Editors. Oxford University Press, New York, pp. 57–75.

Dickinson, E. C., and J. V. Remsen Jr., Editors (2013). The Howard and Moore complete checklist of the birds of the world. 4th ed. Volume 1: Non-passerines. Aves Press, Eastbourne, UK.

Dickinson, E. C., and L. Christidis, Editors (2014). The Howard and Moore complete checklist of the birds of the world. 4th ed. Volume 2: Passerines. Aves Press, Eastbourne, UK.

Dobzhansky, T. G. (1937). Genetics and the origin of species. Columbia University Press, New York.

Dobzhansky, T. G. (1973). Nothing in biology makes sense except in the light of evolution. American Biology Teacher 35:125–129.

Dunn, P. O., L. A. Whittingham, C. R. Freeman-Gallant, and J. DeCoste (2008). Geographic variation in the function of ornaments in the common yellowthroat Geothlypis trichas. Journal of Avian Biology 39:66–72.

Eaton, M. D., and S. M. Lanyon (2003). The ubiquity of avian ultraviolet plumage reflectance. Proceedings of the Royal Society of London B: Biological Sciences 270:1721–1726.

Edwards, S. V., S. B. Kingan, J. D. Calkins, C. N. Balakrishnan, W. B. Jennings, W. J. Swanson, and M. D. Sorenson (2005). Speciation in birds: Genes, geography and sexual selection. Proceedings of the National Academy of Sciences USA 102:6550–6557.

Ellegren, H. (2013). The evolutionary genomics of birds. Annual Review of Ecology and Systematics 44:239–259.

Escalante, P., L. Márquez-Valdelamar, P. de La Torre, J. P. Laclette, and J. Klicka (2009). Evolutionary history of a prominent North American warbler clade: The Oporornis–Geothlypis complex. Molecular Phylogenetics and Evolution 53:668–678.

Evans, K. (1966). Observations on a hybrid between the Sharp-tailed Grouse and the Greater Prairie Chicken. The Auk 83:128–129.

Fitzpatrick, B. M., J. A. Fordyce, and S. Gavrilets (2009). Pattern, process and geographic modes of speciation. Journal of Evolutionary Biology 22:2342–2347.

Fjeldså, J. (1983). Ecological character displacement and character release in grebes Podicipedidae. Ibis 125:463–481.

Fujita, M. K., A. D. Leaché, F. T. Burbrink, J. A. Mcguire, and C. Moritz (2012). Coalescent-based species delimitation in an integrative taxonomy. Trends in Ecology and Evolution 27:480–488.

Gershoni, M., A. R. Templeton, and D. Mishmar (2009). Mitochondrial bioenergetics as a major motive force of speciation. Bioessays 31:642–650.

Gloger, C. L. (1833). Das Abändern der Vögel durch Einfluss des Klimas August Schulz, Breslau, Germany.

Grant, P. R., and B. R. Grant (2011). How and why species multiply: The radiation of Darwin's finches. Princeton University Press, Princeton, NJ.

Grinnell, J., and A. H. Miller (1944). The distribution of the birds of California. Pacific Coast Avifauna 27:1–608.

Harrison, R. G. (1998). Linking evolutionary pattern and process. In Endless forms: Species and speciation, D. J. Howard and S. H. Berlocher, Editors. Oxford University Press, New York.

Harvey, M. G., and R. T. Brumfield (2015). Genomic variation in a widespread Neotropical bird (Xenops minutus) reveals divergence, population expansion, and gene flow. Molecular Phylogenetics and Evolution 83:305–316.

Hill, G. E. (2015). Mitonuclear ecology. Molecular Biology and Evolution 32:1917–1927.

Hung, C. M., H. Y. Hung, C. F. Yeh, Y. Q. Fu, D. Chen, F. Lei, C. T. Yao, et al. (2014). Species delimitation in the Chinese bamboo partridge Bambusicola thoracica (Phasianidae; Aves). Zoologica Scripta 43:562–575.

Hutchinson, G. E. (1959). Homage to Santa Rosalia or why are there so many kinds of animals? American Naturalist 93:145–159.

Isler, M. L., P. R. Isler, and R. T. Brumfield (2005). Clinal variation in vocalizations of an antbird (Thamnophilidae) and implications for defining species limits. The Auk 122:433–444.

Johns, G. C., and J. C. Avise (1998). A comparative summary of genetic distances in the vertebrates from the mitochondrial cytochrome b gene. Molecular Biology and Evolution 15:1481–1490.

Johnsgard, P. A. (1973). Grouse and quails of North America. University of Nebraska Press, Lincoln.

Klicka, J., F. K. Barker, K. J. Burns, S. M. Lanyon, I. J. Lovette, J. A. Chaves, and R. W. Bryson Jr. (2014). A comprehensive multilocus assessment of sparrow (Aves: Passerellidae) relationships. Molecular Phylogenetics and Evolution 77:177–182.

Kroodsma, D. E. (1986). Design of song playback experiments. The Auk 103:640–642.

Lanyon, W. E. (1957). The comparative biology of the meadowlarks (*Sturnella*) in Wisconsin. Publications of the Nuttall Ornithological Club no. 1, Cambridge, MA.

Lanyon, W. E. (1963). Experiments on species discrimination in *Myiarchus* flycatchers. American Museum Novitates 2126:1–16.

Lanyon, W. E. (1979). Hybrid sterility in meadowlarks. Nature 279:557–558.

Leichty, E. R., and J. W. Grier (2006). Importance of facial pattern to sexual selection in Golden-winged Warbler (*Vermivora chrysoptera*). The Auk 123:962–966.

Loyau, A., D. Gomez, B. Moureau, M. Théry, N. S. Hart, M. Saint Jalme, A. T. D. Bennett, and G. Sorci (2007). Iridescent structurally based coloration of eyespots correlates with mating success in the peacock. Behavioral Ecology 18:1123–1131.

Mayden, R. L. (1997). A hierarchy of species concepts: The denouement in the saga of the species problem. *In* Species: The units of biodiversity, M. F. Claridge, H. A. Dawah, and M. R. Wilson, Editors. Chapman and Hall, London.

Mayden, R. L. (2002). On biological species, species concepts and individuation in the natural world. Fish and Fisheries 3:171–196.

Mayr, E. (1942). Systematics and the origin of species. Columbia University Press, New York.

Mayr, E. (1970). Populations, species, and evolution: An abridgment of animal species and evolution. Harvard University Press, Cambridge, MA.

McKay, B. D., and R. M. Zink (2010). The causes of mitochondrial DNA gene tree paraphyly in birds. Molecular Phylogenetics and Evolution 54:647–650.

McKay, B. D., H. L. Mays, Y. Wu, H. Li, C. T. Yao, I. Nishiumi, and F. Zou (2013). An empirical comparison of character-based and coalescent-based approaches to species delimitation in a young avian complex. Molecular Ecology 22:4943–4957.

McKay, B. D., and R. M. Zink (2015). Sisyphean evolution in Darwin's finches. Biological Reviews 90:689–698.

Meiri, S., and T. Dayan (2003). On the validity of Bergmann's rule. Journal of Biogeography 30:331–351.

Moore, W. S., J. H. Graham, and J. T. Price (1991). Mitochondrial DNA variation in the Northern Flicker (*Colaptes auratus*, Aves). Molecular Biology and Evolution 8:327–344.

Nelson, G. (1989). Species and taxa: systematics and evolution. *In* Speciation and its consequences, D. Otte and J. A. Endler, Editors. Sinauer Assoc. Inc., Sunderland, MA, pp. 60–84.

Nelson, G., and N. Platnick (1981). Systematics and biogeography. Columbia University Press, New York.

Noble, G. K. (1936). Courtship and sexual selection of the flicker (*Colaptes auratus luteus*). The Auk 53:269–282.

Parchman, T. L., C. A. Buerkle, V. Soria-Carrasco, and C. W. Benkman (2016). Genome divergence and diversification within a geographic mosaic of coevolution. Molecular Ecology 25:5705–5718.

Paterson, H. E. H. (1985). The recognition concept of species. Transvaal Museum Monograph 4:21–29.

Patten, M. A., and P. Unitt (2002). Diagnosability versus mean differences of Sage Sparrow subspecies. The Auk 119:26–35.

Peterson, A. T., and A. G. Navarro-Sigüenza (1999). Alternate species concepts as bases for determining priority conservation areas. Conservation Biology 13:427–431.

Peterson, A. T., J. Soberón, and V. Sánchez-Cordero (1999). Conservatism of ecological niches in evolutionary time. Science 285:1265–1267.

Phillimore, A. B., C. D. L. Orme, G. H. Thomas, T. M. Blackburn, P. M. Bennett, K. J. Gaston, and I. P. F. Owens (2008). Sympatric speciation in birds is rare: Insights from range data and simulations. American Naturalist 171:646–657.

Podos, J. (2001). Correlated evolution of morphology and vocal signal structure in Darwin's finches. Nature 409:185–188.

Prager, E. M., and A. C. Wilson (1975). Slow evolutionary loss of the potential for interspecific hybridization in birds: A manifestation of slow regulatory evolution. Proceedings of the National Academy of Sciences 72:200–204.

Price, T. (2008). Speciation in birds. Roberts and Co., Greenwood Village, CO.

Rannala, B. (2015). The art and science of species delimitation. Current Zoology 61:846–853.

Remsen Jr., J. V. (2005). Pattern, process, and rigor meet classification. The Auk 122:403–413.

Remsen Jr., J. V. (2010). Subspecies as a meaningful taxonomic rank in avian classification. Ornithological Monographs 67:62–78.

Røskaft, E., and S. Rohwer (1987). An experimental study of the function of the red epaulettes and the black body colour of male Red-winged Blackbirds. Animal Behaviour 35:1070–1077.

Rundle, H. D., and P. Nosil (2005). Ecological speciation. Ecology Letters 8:336–352.

Schluter, D. (2009). Evidence for ecological speciation and its alternative. Science 323:737–741.

Shutler, D., and P. J. Weatherhead (1990). Targets of sexual selection: Song and plumage of wood warblers. Evolution 44:1967–1977.

Simpson, G. G. (1961). Principles of animal taxonomy. Columbia University Press, New York.

Sites, J. W., and J. C. Marshall (2004). Operational criteria for delimiting species. Annual Review of Ecology, Evolution, and Systematics 35:199–227.

Soltis, P. S., and M. A. Gitzendanner (1999). Molecular systematics and the conservation of rare species. Conservation Biology 13:471–483.

Sorenson, M. D., K. M. Sefc, and R. B. Payne (2003). Speciation by host switch in brood parasitic indigobirds. Nature 424:928–931.

Sparling, D. W. (1983). Quantitative analysis of Prairie Grouse vocalizations. The Condor 85:30–42.

Symonds, M. R. E., and G. J. Tattersall (2010). Geographical variation in bill size across bird species provides evidence for Allen's rule. American Naturalist 176:188–197.

Tarof, S. A., P. O. Dunn, and L. A. Whittingham (2005). Dual functions of a melanin-based ornament in the common yellowthroat. Proceedings of the Royal Society of London B: Biological Sciences 272:1121–1127.

Templeton, A. R. (1989). The meaning of species and speciation: A genetic perspective. *In* Speciation and its consequences, D. Otte and J. A. Endler, Editors. Sinauer Assoc. Inc., Sunderland, MA, pp. 3–27.

Toews, D. P. L., A. Brelsford, C. Grossen, B. Mila, and D. E. Irwin (2016a). Genomic variation across the yellow-rumped warbler species complex. The Auk 133:698–717.

Toews, D. P. L., S. A. Taylor, R. Vallender, A. Brelsford, B. G. Butcher, P. W. Messer, and I. J. Lovette (2016b). Plumage genes and little else distinguish the genomes of hybridizing warblers. Current Biology. http://dx.doi.org/10.1016/j.cub.2016.06.034.

Trueman, E. R. (1979). Species concept. *In* The encyclopedia of paleontology, R. W. Fairbridge and D. Jablonski, Editors. Dowden, Hutchinson & Ross, Stroudsburg, PA, pp. 764–767.

Valen, L. van (1976). Ecological species, multispecies, and oaks. Taxon 25:233–239.

Via, S. (2009). Natural selection in action during speciation. Proceedings of the National Academy of Sciences 106 (Supplement 1): 9939–9946.

Waples, R. S. (1991) Pacific salmon, *Oncorhynchus* spp., and the definition of "species" under the endangered species act. Marine Fisheries Review 53:11–22

Wells, S., R. A. Bradley, and L. F. Baptista (1978). Hybridization in *Calypte* hummingbirds. The Auk 95:537–549.

Wheeler, Q. D., and R. Meier (2000). Species concepts and phylogenetic theory: A debate. Columbia University Press, New York.

Wiens, J. J. (2004). What is speciation and how should we study it? American Naturalist 163:914–923.

Wiens, J. J. (2007). Species delimitation: New approaches for discovering diversity. Systematic Biology 56:875–878.

Wiley, E. O. (1978). The evolutionary species concept reconsidered. Systematic Zoology 27:17–26.

Willis, E. O. (1981). Is a species an interbreeding unit, or an internally similar part of a phylogenetic tree? Systematic Zoology 30:84–85.

Wolff, J. N., E. D. Ladoukakis, J. A. Enríquez, and D. K. Dowling (2014). Mitonuclear interactions: Evolutionary consequences over multiple biological scales. Philosophical Transactions of the Royal Society of London B 369:20130443. http://dx.doi.org/10.1098/rstb.2013.0443.

Yang, Z., and B. Rannala (2010). Bayesian species delimitation using multilocus sequence data. Proceedings of the National Academy of Sciences USA 107:9264–9269.

Yasukawa, K., L. K. Butler, and D. A. Enstrom (2009). Intersexual and intrasexual consequences of epaulet colour in male Red-winged Blackbirds: An experimental approach. Animal Behaviour 77:531–540.

Zink, R. M. (1996). Bird species diversity. Nature 381:566.

Zink, R. M. (2004). The role of subspecies in obscuring biological diversity and misleading conservation policy. Proceedings of the Royal Society of London B: Biological Sciences 271:561–564.

Zink, R. M. (2012). The geography of speciation: Case studies from birds. Evolution: Education and Outreach 5:541–546.

Zink, R. M. (2015). Genetics, morphology and ecological niche modeling do not support the subspecies status of the Southwestern Willow Flycatcher (*Empidonax traillii extimus*). The Condor 117:6–86.

Zink, R. M., and M. C. McKitrick (1995). The debate over species concepts and its implications for ornithology. The Auk 112:701–719.

Zink, R. M., and R. C. Blackwell (1998). Molecular systematics and biogeography of aridland gnatcatchers (genus *Polioptila*) and evidence supporting species status of the California Gnatcatcher (*Polioptila californica*). Molecular Phylogenetics and Evolution 9:26–32.

Zwickel, F. C. (1992). Blue Grouse (*Dendragapus obscurus*). *In* The birds of North America 15, A. Poole, P. Stettenheim, and F. B. Gill, Editors. Academy of Natural Sciences, Philadelphia, PA, and American Ornithologists' Union, Washington, DC.

Zwickel, F. C., M. A. Degner, D. T. McKinnon, and D. A. Boag (1991). Sexual and subspecific variation in the numbers of rectrices of Blue Grouse. Canadian Journal of Zoology 69:134–140.

CHAPTER 4

Evolutionary and Ecological Perspectives on Avian Distributions

Robert M. Zink and Gary Voelker

If you are going on a trip anywhere in the world, pack your binoculars and bring a field guide, because you will see birds. A few species have a worldwide distribution (excluding Antarctica); these are often referred to as "cosmopolitan birds" (e.g., Osprey, *Pandion haliaetus*; Peregrine Falcon, *Falco peregrinus*; and Barn Owl, *Otus asio*; fig. 4.1). Although you typically will see few of the same species in your travels to different places, you will recognize similar types of birds. For example, if you are familiar with the local soaring raptors in Sweden, you will find species that fill this niche on most of your travels. If you are from Japan and are familiar with the Great Spotted Woodpecker (*Dendrocopos major*; fig. 4.2), you will see similarly sized woodpeckers elsewhere that fill a parallel niche. In this chapter, we will consider the world's bird species diversity from ecological and evolutionary per-

spectives, and we will explore radiations that are remarkable in their own right.

DISTRIBUTIONS AND PHYLOGENY: UNRAVELING BIOGEOGRAPHIC HISTORY

In any one place, the community of species includes old and new species, and probably species that have emigrated from other places. Deciphering how these communities arose has been an active line of ornithological inquiry since at least Darwin's time, and we call this endeavor "biogeography" (see the box on page 62). Although biogeography has had a long and often contentious history, the use of phylogenies (trees of relatedness) has added rigor to our assessments of biogeographic history.

Figure 4.1. Barn Owl

Figure 4.2. Great Spotted Woodpecker

JON FJELDSÅ

Many biologists have made significant contributions to our knowledge of avian distributions, spanning centuries and including names such as Darwin, Wallace, and Audubon. Naturally, most very early studies of distributions involved basic documentation of species and their ranges. More recently, studies have included assessments of how those species are related to one another, how they diversified, and how they achieved their modern distributions.

One ornithologist who has contributed a wide variety of distribution-related studies is Jon Fjeldså. Jon was born in 1942 in Bodø, Norway, which lies just north of the Arctic Circle. Jon's early life included interesting pastimes: he and his friends explored old German bunkers and searched for explosives left over from World War II. In 1948, Jon's father bought a farm in the mountains, and Jon set to exploring the lakes, forests, and mountain habitats, leading to a lifelong interest in nature.

Jon's father was a self-taught taxidermist, a skill he shared with Jon, adding another dimension to Jon's interests. Jon essentially had his own museum filled with stuffed birds, fishes in formalin, and insects pinned in drawers. By the age of nine, Jon was determined to be the director of a natural history museum (something he later wished he'd avoided). Following high school, Jon studied geology, chemistry, and zoology at the University of Bergen, graduating in 1970. His fieldwork and studies focused on the ecology of the Horned Grebe. Jon defended his doctor of science dissertation in Copenhagen in 1975, having been employed at the university since 1971 as curator of birds at the Zoological Museum.

Jon continued his work with grebes for years thereafter and expanded it to include a focus on community assembly, with fieldwork in Europe, Australia, and the Andes. While in the Andes, Jon realized that research in the region was constrained by the lack of a field guide; with Niels Krabbe, he started work on such a field guide in 1982. The result was the *Birds of the High Andes*, published in 1990 by Apollo Booksellers.

Since that time, Jon has focused largely on passerines and has been involved in extensive fieldwork on all continents. Importantly, Jon targeted much of his fieldwork on little-known areas of those continents (such as the montane regions of Tanzania), thereby expanding our knowledge of distributions. Jon's interests in broader conceptual topics has grown to include a range of questions that seek to explain global-scale patterns of distributions, diversification, and variation in species richness; his methodology has drawn from ecology, biogeography, and phylogenetics in the pursuit of these questions.

Additionally, Jon's work has taken a path through conservation with a goal of ensuring that the discovery/identification of biodiversity hotspots would be followed up with practical conservation initiatives to ensure the maintenance of that diversity. Jon was heavily involved in the transition of the International Council of Bird Preservation into the modern BirdLife International, serving on the first board of directors of the latter. BirdLife is the official IUCN Red List authority for birds, coordinating the categorization and documentation of all species, of which they recognize one in eight as being threatened with extinction. Working with BirdLife, Jon has had a tremendous influence on our knowledge of bird distributions and conversation priorities.

Although "formally" retired from his position at the Zoological Museum (now the Natural History Museum of Denmark) since he turned 70, Jon now occupies an externally funded research position at the Center for Macroecology, Evolution and Climate at that same institution. In this new role, Jon continues to pursue his interests in exploring and explaining avian distributions. Overall, Jon's life work to date has had major impacts on our knowledge of birds and their distributions, reflected in his exceptional record of 450 publications, ranging from short notes to books to over 300 peer-reviewed publications.

A simple hypothetical model (fig. 4.3) will illustrate how we use phylogenies to understand the history of species in a region. We start with an island with a single widespread species (species A). The island is split apart by a geological process (e.g., drifting tectonic plates) and the now isolated populations on each island evolve into new species (B and C). This process is referred to as vicariance, where a once-continuous distribution is split by a fragmenting process (e.g., plate tectonics, river formation, mountain or land-bridge uplift), leading to the cessation of gene flow between, and subsequent evolution of, independent populations in each of the areas separated by the barrier—i.e., speciation (see chapter 3). As part of this process, we assume that the ancestral species (here, A) becomes extinct, because evolution is proceeding in both new species B and C. Next, on one of our islands, species C becomes species D and E owing to an isolating barrier to gene flow (e.g., a new river). Next, species D is divided by another vicariant event (e.g., a new mountain range arising) and becomes species F and G (E is unchanged by this new vicariant event). Now assume that some individuals from species G colonize the other island, which is termed "dispersal," which can lead to either range expansion (if gene flow can still occur between our islands) or two new species (if gene flow ceases; i.e., G would be an ancestral extinction and we would have two new species, one on each island). Assume here that the dispersal event by G is a range expansion, with continued gene flow between islands. Now we assume that species B experiences vicariance and undergoes speciation to become species H and I, and subsequently, members of species I colonize the other island via dispersal with the subsequent loss of gene flow between islands. This is our second dispersal outcome, and here the new colonist population undergoes evolutionary change to become species J. Thus, we have six species on our two islands, G, H, and I on one island, and E, F, G, and J on the other island, with one species (G) shared between islands.

Given this history, we can reconstruct a phylogenetic tree (fig. 4.3) that would result from this sequence of vicariant and dispersal events. Typically we do not have the step-by-step history, but by translating the phylogenetic pattern, we can reconstruct what happened. One would code each species as to its geographic area, and reconstruct cases of both dispersal and vicariance; in one case the dispersal led to speciation, whereas in the other it did not. Although all of the speciation events attributable to vicariance are apparent, it is not clear from our tree where there were potential barriers that species managed to traverse (and thereby avoid isolation and speciation). That is, some species can cross areas of environmental change whereas others might not.

An example of this sort of analysis applied to a real situation comes from a study by Barker et al. (2015). They produced one of the most complete phylogenetic hypotheses of a large group done to date for any lineage: 832 species of blackbirds, cardinals, sparrows, tanagers, warblers and "allies," which was remarkable in having included more than 95 percent of all species known or suspected to be in these groups. Barker et al. concluded that most of the species present in North America diversified there, as opposed to having colonized North America from different continents. An exception was the tanagers, which underwent diversification mostly in South America and subsequently colonized North America. Of interest, Barker et al. were able to exclude areas with high species diversity, such as the Caribbean and other islands, as potential centers of increased speciation. This re-

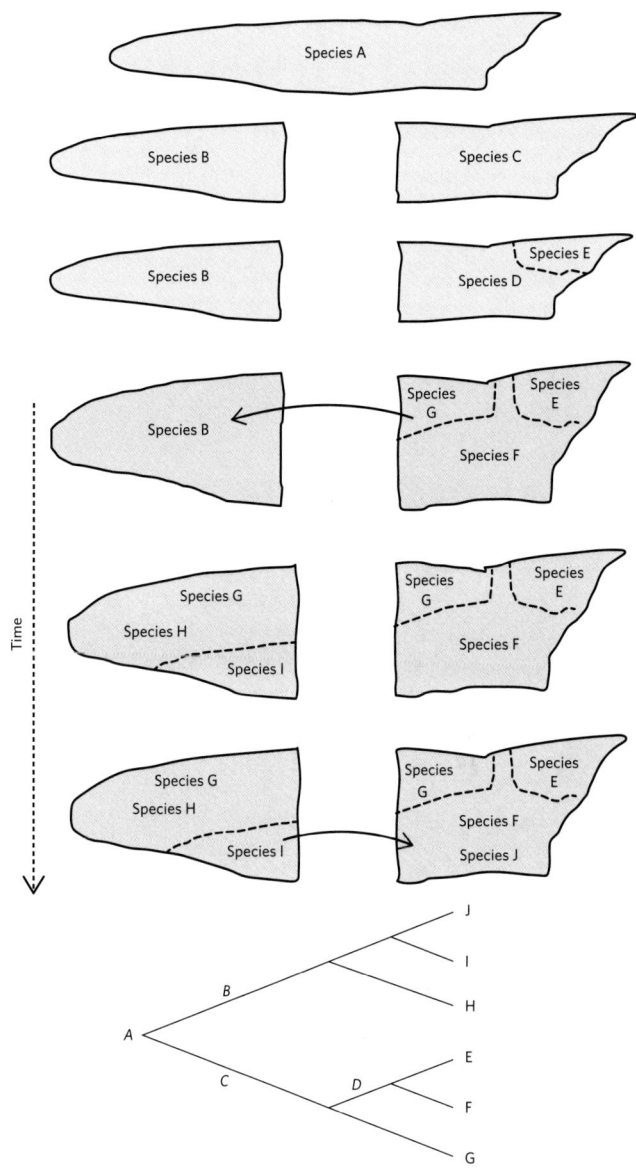

Figure 4.3. Evolution on islands and phylogenies.

sult was in line with other studies (e.g., Voelker 1999, Voelker et al. 2009, 2010), suggesting that regions with high levels of endemicity often result from multiple colonizations rather than from speciation within that area. Such a conclusion is important because, historically, biogeographers would often assume that the "center of origin" for a given lineage was the area with the highest species diversity. Linking phylogenetics with biogeography has been a major stride forward in our knowledge of how distribution patterns evolve(d).

The ratites provide another empirical example of how phylogenies and biogeographic history are reciprocally illuminating, although recent evidence suggests the picture is not as straightforward as was once believed. Extant ratites include kiwis, emus, tinamous (fig. 4.4), rheas, ostriches, and cassowaries. Whereas these species might not seem to be closely related, molecular data indicate that they are. Furthermore, ratites share a distinct palate structure (chapter 6), and their feathers are not like those of other birds in that they lack barbules (chapter 9) and are almost fluffy. Ratites are large, except for the kiwis, and flightless, except for the tinamous. Distributionally, they are almost entirely restricted to southern hemisphere landmasses: the kiwi to New Zealand, the ostrich to Africa, the rhea to South America, the emu to Australia, and the cassowary to Australia and New Guinea; although the tinamous are found largely in South America, some species occur in Central America and Mexico.

The geographic distribution of ratites has fascinated ornithologists and evolutionary biologists since they were recognized as being closely related. Given that most cannot fly, how did these large flightless birds come to be distributed on most of the southern landmasses? The possibility that some absurdly lucky individual(s) survived an ocean crossing on a floating tree trunk defies statistical imagination (this possibility, called "rafting," was often invoked in older biogeographic literature). Thus, there are at least two scenarios to explain ratite distributions. First, the ancestral ratite was widespread on Gondwanaland, and when that landmass separated to form modern continents and islands, each continent/island carried with it a population of the ancestral lineage, and in the subsequent absence of gene flow, each new continent/island lineage evolved the amazing differences that now exist among modern ratites. Second, flying ancestors of modern ratites made ocean crossings after the fragmentation of Gondwanaland, and secondarily became flightless. This scenario is viable, because a volant (flighted) ratite ancestry is revealed by their possession of highly reduced wing skeletal elements of the same shape and design as those found in modern flying birds (see chapter 6). To distinguish between these two scenarios requires a phylogeny.

Reconstructing ratite evolution requires considering several extinct lineages, including moas and elephant birds. Overhunting by Polynesians on arrival in New Zealand led to the relatively abrupt extinction, 600 years ago, of the entire moa radiation. Subfossil remains of moas reveal the existence of at least nine species, which had existed for millions of years. They ranged in size from 12 to 250 kg, and the tallest was 2 meters and able to forage 3.6 meters above the ground. A recent study of moa (fig. 4.5) DNA obtained from bone fragments (Allentoft et al. 2014) showed that moa species retained high levels of genetic variability at the time of their extinction, suggesting that humans, and not climate or environmental changes, caused their demise. That is, if

Figure 4.4. *A*, North Island Brown Kiwi; *B*, Emu; *C*, Tinamous

Figure 4.5. Moa. Richard Owen, who became director of London's Museum of Natural History, was the first to recognize that a bone fragment he was shown in 1839 came from a large bird. When later sent collections of bird bones, he managed to reconstruct moa skeletons. In this photograph, published in 1879, he stands next to the largest of all moa, *Dinornis maximus* (now *D. novaezealandiae*), while holding the first bone fragment he had examined 40 years earlier. *From Owen 1879.*

the moas were on an extinction trajectory, one would expect decreases in genetic variability owing to decreasing population sizes.

Another large extinct ratite lineage consists of seven species of elephant birds, which once lived on Madagascar. These birds are known from historical accounts, often only observations of sailors, although eggs and skeletons remain in museum collections. It is thought that the species were driven to extinction by humans in the seventeenth or eighteen century. These species were also extremely large, at least one possibly reaching 500 kg. Their large size and vulnerability, coupled with their food value, led to a recipe for their demise.

To understand the evolution and biogeography of ratites requires a phylogeny that includes extinct forms. As for any phylogenetic analysis, missing species or lineages could lead to erroneous reconstructions of phylogeny and subsequent

erroneous hypotheses regarding biogeography or character evolution. For example, the missing species could provide a novel character condition or be from an area not inhabited by other species, and would therefore render reconstructions either incomplete or incorrect.

Many authors have provided phylogenetic hypotheses for the ratites, with relatively little consensus, other than concluding that the tinamous, the only volant members of the clade, were the sister to all other ratites. Such a relationship would mean that flightlessness evolved just once, early in ratite history. Early attempts to understand ratite evolution (e.g., Cracraft 1974) relied on morphological characters, whereas more recent attempts (e.g., Smith et al. 2012b, Baker et al. 2014) used molecular characters. Mitchell et al. (2014) analyzed complete mitochondrial genomes (ca. 16,000 base pairs), and for the first time included high-quality DNA data from the extinct elephant bird genera (*Aepyornis* and

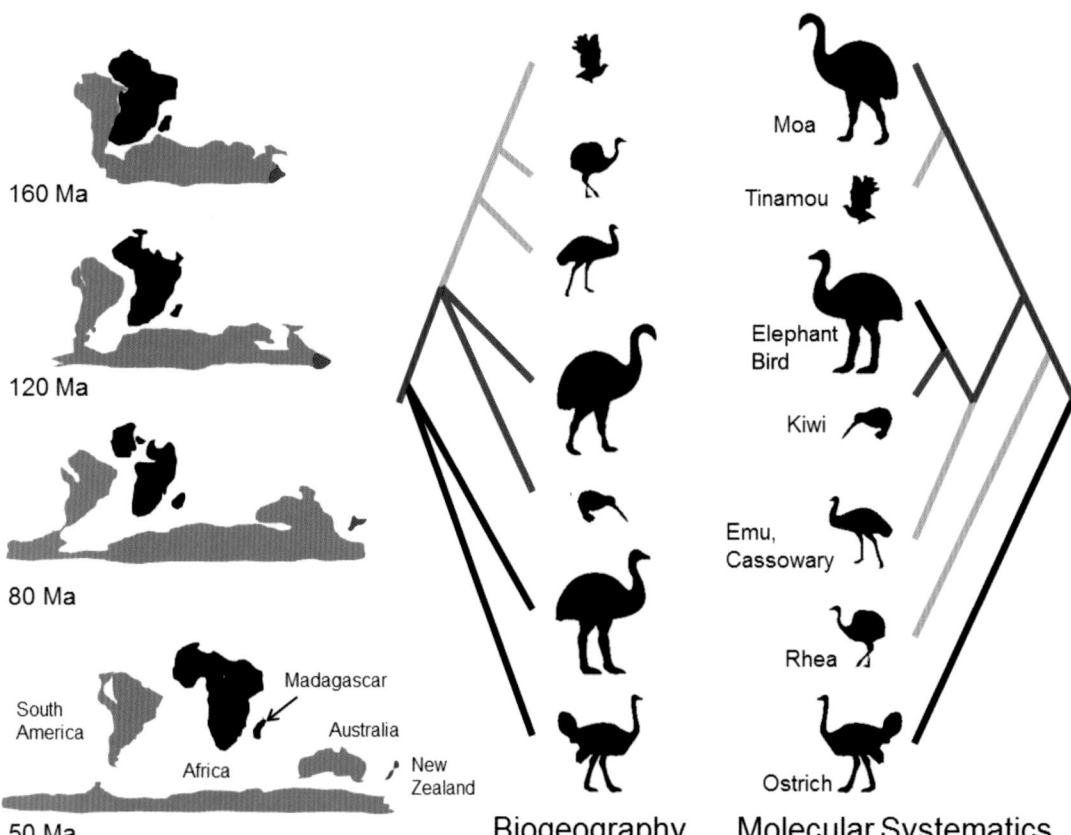

160 Ma

120 Ma

80 Ma

South
America   Madagascar

Africa   Australia

New
Zealand

50 Ma

Biogeography   Molecular Systematics

Moa

Tinamou

Elephant
Bird

Kiwi

Emu,
Cassowary

Rhea

Ostrich

Mullerornis) as well as the moas. Their analysis provides both a fascinating glimpse of the evolutionary history of ratites, especially the evolution of flightlessness, and a lesson in methods of inferring patterns of historical biogeography.

Figure 4.6 shows the geological history of the continental landmasses on which ratites exist today. If one assumes that ratites have been unable to fly since their origin, then the temporal sequence of the breakup of continents ought to provide a vicariance-based hypothesis for how they are related. As shown in the biogeographic hypothesis, one would assume that moas and kiwis (New Zealand) were each other's nearest relatives, as would also be the case for elephant birds and ostriches (Madagascar and Africa, respectively). Although seemingly plausible on biogeographic grounds, a close relationship between kiwis and moas has been difficult to accept given their differences in morphology and size, although molecular phylogenies (Claramunt and Cracraft 2015) show that their common ancestor existed long ago, providing ample time for major morphological shifts. Lastly, this hypothesis suggests that flight evolved secondarily in tinamous (having been lost initially in all other ratites), perhaps concomitant with their radiation in Central and South America. This vicariance scenario for ratite evolution has been assumed to be the most likely hypothesis by many previous authors.

In contrast, Mitchell et al.'s (2014) phylogeny yields novel ideas about both the evolution of biogeographic distributions and the evolution of flight. Their molecular systematic results (fig. 4.6) provided support for a growing consensus that the ostrich is the sister lineage to all other ratites (e.g., Baker et al. 2014), and that the volant tinamous and the flightless moas are sister lineages. Two surprises were the sister-group relationship between the elephant birds from Madagascar and the kiwis from New Zealand, sharing a common ancestor roughly 50 million years ago, and the sister-group relationship between tinamous and moas. These results provide an almost complete transformation of the pre-molecular understanding of ratite evolution and strongly contradict the hypothesis that the sequence of continental breakup explains ratite relationships. Instead, the new results support dispersal by the major ratite lineages *after* modern continents separated from Gondwanaland. This suggests the major lineages were capable of strong flight and lost that ability after arriving on new landmasses. Indeed, Mitchell et al. (2014) suggested that ratite flightlessness evolved as many as six times. Although these conclusions are not easily inferred from the molecular phylogeny, they are based on assumptions that, for the elephant birds and kiwis to be sister lineages, there had to have been a flighted ancestor, with subsequent loss of flight. The authors also suggest that lineages independently gained large size (gigantism) by transitioning into diurnal herbivores made possible by the extinction of dinosaurs, which were either competitors or predators. This example nicely shows the value of phylogenies in biogeographic reconstructions.

THE GLOBAL PATTERN OF AVIAN DIVERSITY

Species diversity varies spatially across the world, with the highest breeding diversity in South America and Africa, and lower values toward the poles (fig. 4.7). In general, more bird species are found in tropical regions with higher vegetation diversity than in temperate realms and deserts with correspondingly lower plant species richness. Another broad generalization is that as taxonomic groups become less and less inclusive, they are less likely to be widespread. From our previous example, the Paleognathae (ratites) is spread across all southern hemisphere landmasses, whereas one of its constituents, the tinamous, is confined to Mesoamerica and South America, with tinamou genera and species having successively smaller ranges. This sort of pattern is typical for most organisms.

Describing the numerical distribution of species requires specifying a geographic unit of a given size. In grid cells of 2° × 2° (fig. 4.8), it can be seen that both North America and South America show considerable geographic heterogeneity in numbers of species (Hawkins et al. 2003). In a similar analysis, Rahbek and Graves (2001) plotted species diversity in 1° blocks, and found the highest diversity in Andean Ecuador (with a peak of 854 species), southeastern Peru (782), and southern Bolivia (698 species). If you consider larger and larger blocks, the heterogeneity washes out somewhat,

showing the importance of geographic scale in describing species diversity.

Adding to distributional complexity are montane or elevationally complex regions. For example, the Amazon basin of South America is species-dense (fig. 4.8). At the western margin of the basin (largely outlined in red for high species density) the thin band of yellow shows the medium species-density region of the Andes. This pattern of high diversity montane regions is easier to see in Africa, where most red is associated with the montane Albertine Rift and Eastern Arc Mountain systems (fig. 4.7). And in Asia, a yellow band indicates the Himalayas, with the eastern margins of that system shifting to higher-density color indicators. Why do these areas house such high levels of diversity? Janzen (1967) suggested that this was because "mountain passes are higher in the tropics," and Huey (1978) modified this by suggesting that "mountains might be "higher" in the tropics." In a nutshell, they suggested that differing physiological tolerances of species will lead to high species packing along elevational gradients. Indeed, as you move up montane slopes in tropical regions (e.g., the Andes), you will encounter very different ecological communities (think of this as primarily sharp elevational stratification in floral assemblages), and as a function of this, the broader ecological community (think of this as the associated faunal assemblages) in each floral assemblage will often comprise very different species of birds. In contrast, montane ecological communities in the temperate zone (e.g., the Rocky Mountains) show far less elevational stratification in habitat, and as such, bird species tend to have much broader distributions along temperate mountain slopes.

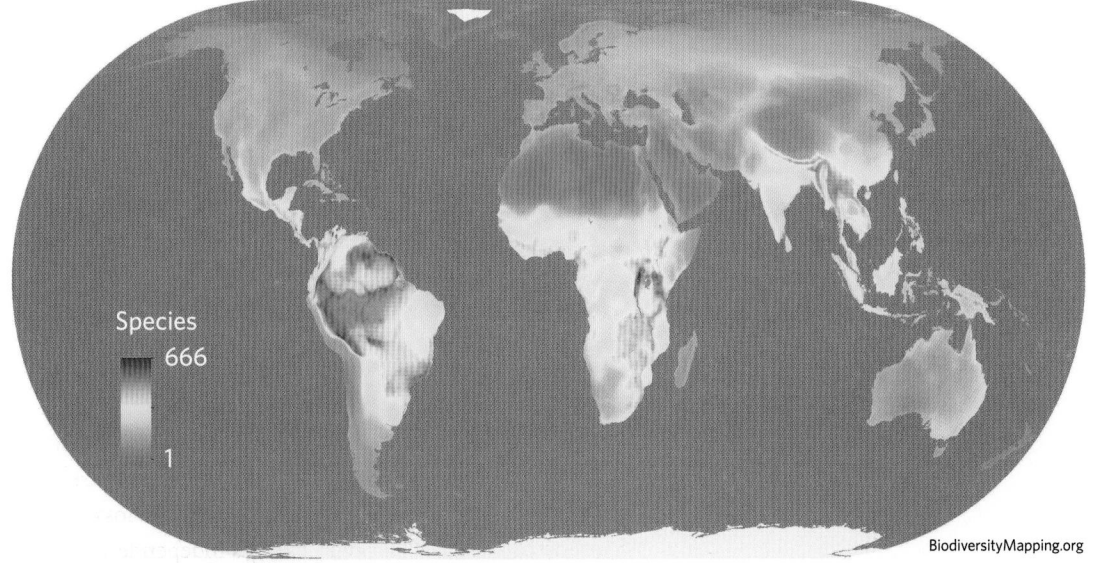

Bird Diversity

Species
666

1

BiodiversityMapping.org

Figure 4.7. Bird species diversity across the world.

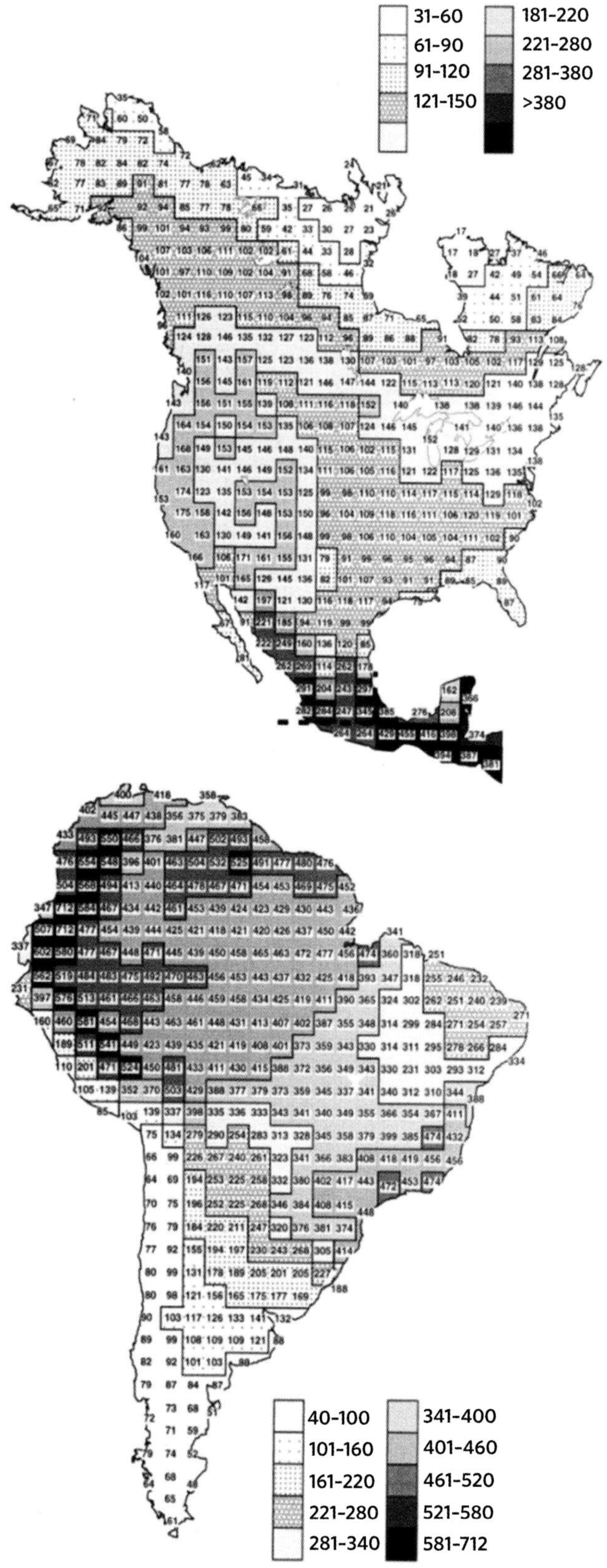

THE DISTRIBUTION OF AVIAN DIVERSITY

As we outlined above, drifting continents provided an evolutionary stage for lineages to diversify. In some cases, diversifying lineages remained geographically localized, whereas other lineages produced species that were good at dispersing into new areas, across the widening water barriers that separated the nascent continents. These biogeographic processes and events led to the distribution of major lineages we see today (see the box on page 69).

One way of summarizing bird distributions is to consider the major biogeographical realms that were proposed by Philip Lutley Sclater and Alfred Russel Wallace in the 1870s (fig. 4.9) and later modified (Holt et al. 2012). The Earth's landmasses can be divided into six realms, the Palearctic (Europe, Asia, north Africa), Afrotropical, Nearctic, Neotropical, Australian, Oriental (Indo-Malayan), plus Oceania. These realms were based on species distributions, and therefore represent areas where diversity accumulated more-or-less independently of other areas. Although these regions are somewhat arbitrary and subject to debate, they broadly reflect past history of regional trends in the distribution of species groups. In some reviews, New Zealand is separated out. Antarctica is often not mentioned, as its role in avian evolution was as a past stepping stone (during Gondwanaland breakup), although there are species like Emperor Penguins (*Aptenodytes forsteri*) that breed there today.

Afrotropical Realm

Africa contains approximately 2,500 species from more than 100 families. The countries with the highest species counts are the Democratic Republic of the Congo (1,140), Tanzania (1,140), and Kenya (1,080); a checklist can be found at http://www.africanbirdclub.org/resources/checklist/intro.

Africa's high bird diversity results from its great latitudinal span, large topographically variable area, and high habitat diversity, with tropical and temperate forests and deserts, extensive savannas, high mountains, and long coastlines. Africa has many endemic and highly specialized species. The ostrich, discussed above, is a prominent member of open savannahs, where males incubate large clutches laid by several females. The Hamerkop (*Scopus umbretta*; fig. 4.10A) is a heronlike bird with a head, crest, and long bill shaped like a hammer, hence its name. The Shoebill (*Balaeniceps rex*; fig. 4.10B) has a large—almost comically large—bill, and looks superficially like a stork. One of the best-known African birds is the Secretarybird (*Sagittarius serpentarius*), a long-legged raptorial bird that inhabits open grasslands and savannas. It is most famous for its habit of stomping on lizards and snakes with its feet to render them unconscious or sufficiently dazed so that they can be swallowed. The mousebirds are confined to

Figure 4.8. Distribution of bird species diversity in North America and South America in 2 × 2 blocks, showing high diversity in neotropical regions, and marked variation in the number of species from area to area. *From Hawkins et al. 2003.*

PHILIP LUTLEY SCLATER (1829–1913)

It is sobering to many biologists to realize that few of us will have our works still used a century after we published them. Such is not the case for Philip Lutley Sclater. Sclater grew up in England, favoring birds as a youth. From a family with considerable means, he was educated by some of the best naturalists of his time. Slater studied ornithology under H. E. Strickland, a foremost ornithologist. Like most youths of his class at the time, he traveled abroad, with one of his major trips being to Lake Superior and the St. Croix River in the upper Midwest of the United States. He traveled later to Philadelphia, where he met several major figures in North American ornithology, including Spencer Baird, John Cassin, and Joseph Leidy (the first two in this list have birds named after them). His interest in ornithology was kindled and started to burn.

Although his passion was birds, he returned to England to practice law. However, he kept his hand in natural history by attending meetings of the Zoological Society of London. It is also fair to say that his interest was stimulated by the collection of bird skins he made as an undergraduate and added to later in life. In this day and age, less and less emphasis is placed on the growth and use of scientific collections, which is a major oversight. Sclater's collection of scientific specimens reached 9,000 by the end of his career, and they were transferred to the British Museum (now in Tring [Hertfordshire]). Specimens, such as those that inspired Sclater, continue to be the gist from which novel ideas stem.

At least six species or subspecies of birds were named after Sclater, including the Dusky-fronted Parrotlet (*Forpus modestus sclateri*), the Erect-crested Penguin (*Eudyptes sclateri*), Sclater's Monal (*Lophophorus sclateri*), Ecuadorian Cacique (*Cacicus sclateri*), Bay-vented Cotinga (*Doliornis sclateri*), and the Mexican Chickadee (*Poecile sclateri*). These scientific names alone assure that Sclater will be remembered. But he is also remembered for his scientific contributions.

From Sclater's knowledge of birds gleaned from studying museum specimens, he recognized the major geographic patterns in the biogeographic distribution of birds. He published his findings in a 1858 paper titled "On the General Geographical Distribution of the Members of the Class Aves" in the *Journal of the Proceedings of the Linnaean Society: Zoology 2:130–145*. Unlike writers of his time who took arbitrary regions and figured out how similar they were in terms of plants and animals, Sclater looked at the biota first and let the concordant distributions of plants and animals tell him what the principal regions were. And his thinking, although pre-evolutionary, was very insightful when he commented, "Two or more of these geographical divisions may have much closer relations to each other than to any third." Thus, he recognized an evolutionary signal in the distribution of the world's birds.

Sclater ended up dividing the world's birds into the six major divisions (Palaearctic, Aethiopian, Indian, Australasian, Nearctic, and Neotropical) that were adopted by Wallace (see text). Although most modern writers credit Wallace for these regions, it was Sclater who laid the groundwork for our understanding of the major biogeographic patterns of birds.

Sclater contributed more to ornithology and science in general by his professional activities. He was a president of the Biological Section of the British Association for the Advancement of Science. He was secretary of the Zoological Society of London from 1860 to 1902, a run of 43 years! Sclater provided a scientific description of the Okapi (the sister species of the giraffe), although he himself had never seen one. He published several books, including *Exotic Ornithology and Nomenclator Avium*, both co-authored with O. Salvin. Importantly, Sclater founded the journal *The Ibis*, one of the most important ornithological journals in the world.

Philip Sclater was a contemporary of Charles Darwin and Alfred Russel Wallace. It is a testament to his intellectual contributions that his work in 1858 is still a foundation for our thinking about the distributional patterns of the world's birds.

---

sub-Saharan Africa and include six species of slender, grayish birds that are gregarious and often form conspicuous flocks. Many a visitor, especially to South African game parks, may need to take a second glance at the Southern Ground Hornbill (*Bucorvus leadbeateri*; fig. 4.11) before confirming that it is a bird. A radiation of 30 species called wattle-eyes or batises occur in open forests or bush, where they hunt by flycatching or taking prey from the ground. The "bald crows," named for the featherless but brightly colored heads, occur in rain forests of central and western Africa. The two species of sugarbirds, which obtain nectar with long tongues, are confined to Africa, and resemble long-tailed, non-iridescent sunbirds. Finch-like birds, called indigobirds and whydahs, have exceptionally long forked tails and are obligate brood para-

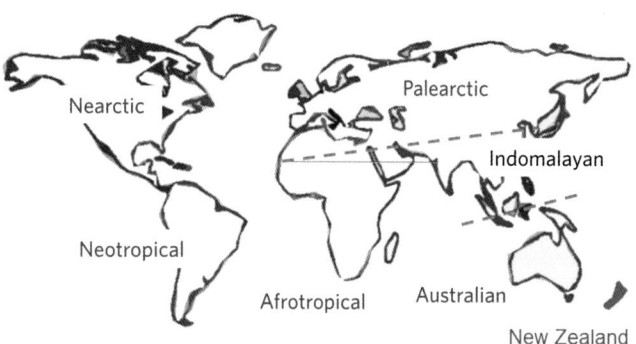

Figure 4.9. Wallace regions

Figure 4.10. A, Hamerkop; B, Shoebill

Figure 4.11. Southern Ground Hornbill

colored, often sporting plumages with notable patches of iridescence. Because of their similar foraging ecology and iridescent plumage, sunbirds and hummingbirds are considered one of the premier examples of evolutionary convergence.

Aquatic habitats support a diversity of grebes, ducks, geese, herons, darters, cormorants, storks, flamingos, jacanas, and a painted snipe. The African Openbill (*Anastomus lamelligerus*) is a stork in which the bill, as the name suggests, does not close when the bill tips occlude. Openbills regurgitate water on their nestlings to keep them cool. Other aquatic birds include rails, coots, and gallinules. The African Finfoot (*Podica senegalensis*; fig. 4.13) is a member of a small group that also includes the Sungrebe (*Heliornis fulica*) of Central and South America. Like many other continents, Africa is home to a large number of kingfisher (fig. 4.14) species that can be seen along rivers and shallow lakes, although some are dry forest–adapted species that prey on insects and lizards

Open habitats in Africa are home to a diversity of game birds, including francolins and partridges, many of which are extensively hunted. Guineafowl are common in many areas. Buttonquail, small drab birds that run amid ground vegetation are also widespread. A diversity of cranes inhabits open areas. The bustards, large and mostly terrestrial birds, are well represented in Africa. Africa has several species of sandgrouse, birds known for nesting far from water and carrying water back to nest sites in their specialized downy breast feathers. The stone-curlews, or "thick-knees" occur along with pratincoles and coursers, plover-like shorebirds that are often found in relatively dry areas. A large group of Old World orioles (not closely related to New World orioles; fig. 4.15) and shrikes can be found, along with many members of the widely distributed silver-eye group.

sites. Finches, buntings, and grosbeaks are also found in open country or woodlands.

One of the most conspicuous Old World radiations includes the sunbirds (fig. 4.12A, 4.12B), of which approximately 80 species are found in Africa. These ecological equivalents of New World hummingbirds forage on nectar and are brightly

Figure 4.13. African Finfoot

Figure 4.12. A, Loten's Sunbird; *B,* Malachite Sunbird

Figure 4.14. Mangrove Kingfisher

Shorebirds include most of the major northern hemisphere breeders that winter along the coasts, including oystercatchers, avocets, stilts, and a variety of sandpipers and typical plovers. One of the three living species of crab plover (*Dromas ardeola*) breeds along the northeast coast. A number of familiar birds including gannets, albatross, petrels, storm petrels, loons, tropicbirds, penguins, gulls, terns, a skimmer, skuas, and jaegers, exist offshore and along the coasts.

Africa is home to a large array of diurnal raptors, including typical soaring hawks, bird catchers (accipiters), falcons, and large eagles. In addition, Africa is home to a large diversity of vultures, which partition carrion according to how long it has "aged." One specialized species, the bearded vulture, or Lammergeier (*Gypaetus barbatus*; fig. 4.16), feeds almost exclusively on bone marrow that it obtains by dropping bones onto rocks to break them open. Large and small owls are also common.

Terrestrial savannahs, grasslands, and open woodlands abound with interesting species. African forests are home to many species of pigeons and doves, and about 20 species of parrots. Turacos include plantain-eaters and the "go-away"

Figure 4.15. African Oriole

birds, named for the vocalization they give when near hunters trying to hide from prey or stealthy birders trying to glimpse a shy bird. Crepuscular periods are often accompanied by sounds of nightjars, of which Africa has approximately 25 species. Bee-eaters and rollers are common in open habitats, and the former is well known for cooperative breeding. Three species of the pan-tropically distributed trogons are found in Africa. The iconic African Hoopoe (*Upupa africana*; fig. 4.17A) and the related woodhoopoes, including the scimitarbills with their long decurved bills, inhabit open woodlands. A large number of larks, wagtails, and pipits inhabit grassland areas. Bush-shrikes and helmet shrikes use open areas. Drongos (e.g., Fork-tailed Drongo, *Dicrurus adsimilis*; fig. 4.17B) and some other woodland species mimic the alarm calls of other birds and even meerkats, to drive them away from a captured prey item, which the drongos then take. Crows and relatives are found in Africa, including the Piapiac (*Ptilostomus afer*; fig. 4.17C), a species restricted to central western Africa. Many sparrows and almost a hundred species of weavers can be found; it is the large communal nests of the weavers (the largest nest of any bird) that often reveal their presence. The estrildid finches, including the waxbills (named because the bill color resembles red sealing

wax), are common, small, often colonial seedeaters. Aerial environments are home to swifts and swallows, including both African breeders and migrants from Europe.

In some parts of the world, starlings are introduced pests (you are likely familiar with the European Starling), but up to 50 species of native, often colorful, starlings (fig. 4.18A) reside in Africa. The oxpeckers are seen most everywhere there are large ungulates; oxpeckers (fig. 4.18B) hitchhike on them and remove engorged ticks and pick at wounds, keeping them from healing and thereby providing renewable blood meals.

Forests are inhabited by a variety of woodpeckers, honeyguides, barbets, broadbills, the two African pittas, bulbuls, and cuckoo-shrikes. Many species of thrushes occur in forested habitats, including both breeders and migrants from Eurasia. Cisticolas are small birds named for their woven-basket nests, but also sometimes called fantails because of their habit of tail-flicking; all but two of the 45 described species occur in Africa. A variety of Old World warblers occur in Africa, including sylviid, cettid, locustellid, phylloscopid, and acrocephalid warblers familiar to Eurasian birders. Some African species go by additional common names including crombecs (fig. 4.19A) and longbills. Many Old World flycatchers (fig. 4.19B) are found in Africa, as are the bluethroat (*Luscinia svecica*), robin-chats, akalats, redstarts (fig. 4.19C), chats, stonechats, and wheatears. A number of monarch flycatchers, also found in southeastern Asia and Australia, forage on insects. Babblers, a large Old World group of mostly terrestrial species, are conspicuous members of Africa's avifauna. Other species include a parrotbill, the Long-tailed Tit (*Aegithalos caudatus*), and many small tits (fig. 4.19D), nuthatches, creepers, and penduline tits.

How did some of this diversity evolve? In some cases, there is clear evidence for colonizations from the Palearctic and Australasian regions, with subsequent intra-African radiations by those colonizing lineages; the aforementioned pipits and wagtails (fig. 4.20) are examples of this (Voelker 1999, 2002). Closely linked to these colonizations are climatic shifts that affected the geographic extent of the Afrotropical forests. These forests have alternately been extended to the Kenyan/Tanzanian coast (from five to three million years ago, mya), or retracted to the west (prior to five mya, and since three mya to the present). Arid-adapted lineages, such as pipits, colonized southern Africa during forest retraction periods and subsequently became isolated there during forest expansion (a vicariant event for arid-adapted species), after which further pipit speciation occurred. Alternatively, forest-adapted species like thrushes were able to colonize Africa during periods of forest expansion, and were isolated in montane forest refugia as the forest retracted westward (e.g., Voelker and

Figure 4.16. Lammergeier

Figure 4.17. A, African Hoopoe; *B*, Fork-tailed Drongo; *C*, Piapiac

Figure 4.18. A, Superb Starling; *B*, Red-billed Oxpecker

Outlaw 2008). These same forest dynamics have worked to drive diversity in largely African lineages such as forest robins and sunbirds, many of which have rather tiny ranges on east African mountains, as compared with others that have widespread distributions in the extensive lowland tropical forests (Bowie et al. 2004, Voelker et al. 2010). And, there is

evidence that elevational gradients have provided additional opportunities for speciation on African montane areas (Jetz and Rahbek 2002). There is also growing evidence that there are deeply divergent genetic lineages within Afrotropical species that are monochromatic across their central-west African ranges (Marks 2010, Huntley and Voelker 2016). This genetic diversity may well indicate cryptic speciation, and in some cases research on such taxa has resulted in the description of new species (Fjeldså et al. 2006, Voelker et al. 2010)—a rather uncommon occurrence for Africa in recent decades.

Evidence of other factors driving (or perhaps just maintaining) avian diversity in Africa are less clear, but there is some evidence that riverine barriers have played a role. A study based on species sampled from opposite banks of the

Figure 4.19. A, Long-billed Crombec; B, Black and Orange Flycatcher; C, Black Redstart; D, Great Tit

Congo River found substantial genetic differentiation in four of those species (Voelker et al. 2013). The species showing differentiation across the river were all understory specialists, and as such are less prone to major movements than are mid-level or canopy dwelling species. Elsewhere in Africa, diversification can be related to climate and habitat heterogeneity. This is the case for white-eyes, several species of which have extensive hybrid zones in South Africa, which correspond to habitat transitions (Oatley et al. 2012). Habitat heterogeneity also seems to have been important in driving speciation in African lineages of *Muscicapa* flycatchers and their allies, where sister lineages occupy often adjacent yet very different habitat types (Voelker et al. 2016).

Neotropical Realm

Extending from Mexico to the tip of South America, this bird-rich region contains approximately 3,500 species from over 70 families. The countries with the highest species counts are Colombia (1,837), Ecuador (1,619), and Peru (1,774); a checklist can be found at http://www.museum.lsu.edu

Figure 4.20. African Wagtail

/~Remsen/SACCCountryLists.htm. The region has the greatest endemism of any of the world's regions, being home to two orders, 20 families, 686 genera, and over 3,100 species (Newton 2003) found only in the Neotropics. Although it shares only 7 percent of breeding species with the Nearctic, many migrants from the north spend nonbreeding periods in this region.

South America's diversity comes largely from its immense tropical area that is thought to have been stable for long geological periods, allowing plant and insect species to undergo great species diversification and bird species to finely partition the resulting rich niche-space. In addition, major barriers exist, including the Amazon River and its tributaries, and the Andes. As in Africa, open savannahs are home to large ratites, in this case two species of rheas. The tinamous are found in both forested and open habitats. Many of the Neotropical specialties are residents of tropical forests. One of these is the Hoatzin (*Opisthocomus hoazin*; fig. 4.21), an extremely old lineage with no close relatives. Hoatzins live in swamps and riparian forests and are unusual among birds because they eat plants, which they are able to digest with the help of microbes in a multichambered enlarged crop. The Hoatzin is considered a "primitive" bird because the young have a wing claw, which they use to clamber among branches; this may represent an intermediate stage between modern birds and dinosaur ancestors. Another Neotropical species with a unique life history is the Oilbird (*Steatornis caripensis*; fig. 4.22), a colonial, nocturnal frugivore that is one of only two birds known to use echolocation to navigate to their nests in caves (Cave Swiftlet is the other). Oilbirds are so named because chicks were once captured and boiled down to make oil. The Lyre-tailed Nightjar (*Uropsalis lyra*), a patchily distributed species from the Andes, sports tail streamers that extend more than twice the length of the body.

Many of the families typical of the Neotropics have undergone extensive differentiation, including the ovenbirds and woodcreepers (280 species); antbirds (225 species);

Figure 4.22. Oilbird

tyrant flycatchers, cotingas, and manakins (ca. 550 species); toucans (55 species); tanagers and finches (340 species); guans (50 species); tinamous (45 species); and ground antbirds (ca. 60 species).

One of the most impressive radiations has occurred in the hummingbirds (fig. 4.23A, B, C), with close to 350 species, which are all restricted to the New World (and the vast majority to the Neotropics). Known for their iridescent plumages, often extreme sexual dimorphism, and remarkable variation in bill size and shape, hummingbirds have become favorites of the public and scientists alike. The coevolutionary relationships between hummingbird bills and the flowers they pollinate are well known. Thus, some think that the extraordinary diversity of this group is a result of South America's many millions of years of isolation; one might also consider the transoceanic colonization ability of hummingbirds relatively low.

Not all specialized groups include a large number of species. For example, the trumpeters (genus *Psophia*) consist of three species of chicken-sized birds found in humid forests of the Amazon and Guiana Shield. They are named for the cackling or trumpeting threat calls of males.

Some groups not confined to the Neotropics are also diverse there, such as the doves and pigeons (50 species), which range from large forest-dwelling species to small open-country birds. Over 80 woodpeckers occur, including the smaller-bodied species called piculets. Over 50 species of toucans and 150 species of parrots, parrotlets, and macaws can be found in South America. Different taxonomists carve up the "flycatchers" in different ways, but under old definitions there are over 400 species, including tyrants, pewees, monjitas, "flycatchers," kingbirds, sirystes, attilas, tyrannulets, doraditos, elaenias, mourners, spadebills, becards, and schiffornis.

A substantial number of waterbirds occur in South America. The high elevation torrent ducks, the Coscoroba Swan

Figure 4.21. Hoatzin

Figure 4.23. A, Speckled Hummingbird; *B*, Crowned Woodnymph; *C*, Ruby-throated Hummingbird

(*Coscoroba coscoroba*), the sheldgeese, and the steamer ducks are examples. In addition there are a large number of dabbling ducks, as well as herons, spoonbills, ibises, night herons, egrets, bitterns, and rails. About 10 species of grebes are found here, including the flightless Titicaca Grebe (*Rollandia microptera*; fig. 4.24), an endangered species found in

the Lake Titicaca basin in the high Altiplano of southern Peru and western Bolivia. Tops on many bird-watchers' lists are the Sungrebe and Sunbittern (*Eurypyga helias*), the latter of which has spots that are visible on the extended wings. These appear to be "eye spots" and might function to startle predators. Seabirds such as albatrosses, storm petrels, petrels, and shearwaters, as well as penguins, boobies, gulls, terns, noddys, skuas, oystercatchers, and cormorants can be found in near- and offshore waters. Beaches and tidal flats support large concentrations of plovers, avocets, and sandpipers (including many migrants from North America).

Neotropical forests include an impressive diversity, and one can see manakins, cotingas, tyrant flycatchers, wrens, antwrens, antbirds (fig. 4.25A), gnateaters, woodpeckers, tinamous, woodcreepers, antpittas, jacamars, puffbirds (fig. 4.25B), antshrikes, miners, ant-thrushes, hummingbirds, ovenbirds, vireos, wood warblers (including many migrants from the Nearctic), and euphonias. In some Neotropical areas, species diversity reaches the highest levels of anywhere on the planet (fig. 4.7). Favorites of bird-watchers are the trogons and especially the fruit-eating, often iridescent, quetzals, including the exceptionally long-tailed Resplendent Quetzal (*Pharomachrus mocinno*) in Central America. Impressive to a visitor is the habit of antbirds following, as the name suggests, ant swarms. As the ants swarm across the forest floor, insects and other prey are dislodged and eaten by the antbirds, some of which are obligate ant followers. It was once thought that the birds ate the ants too, but it is now known that a small antbird can be killed by as few as four stings from a fire ant. Wood quail are short-tailed, plumpish, and short-billed birds that are usually exceedingly difficult to see, although some bird-watching centers have habituated some species to feeders, allowing visitors to see them.

Figure 4.24. Titicaca Grebe

Figure 4.25. A, Ocellated Antbird; *B*, Pied Puffbird; *C*, Harpy Eagle; *D*, Barbet

Raptorial birds are well represented and include soaring raptors, swift forest-dwelling bird hawks, kites, harriers, and large eagles, including the Harpy Eagle (*Harpia harpyja*; fig. 4.25C), which hunts monkeys and sloths. The carrion niche is filled by six vultures, including the Andean Condor, although there are far fewer species than found in Africa. Over 40 species of owls have radiated into most habitats, from forests to open country. Open habitats include some representatives of the same groups that inhabit forests, including the caprimulgids (over 30 species), tapaculos, ovenbirds, corvids, pipits, sparrows, icterids, tanagers, grosbeaks, finches, and flowerpiercers.

How did some of this diversity evolve? We have already discussed elevation as a probable factor for African

diversity—it is also a factor in the Neotropics but perhaps more so, given that the Andes run north to south across the entirety of South America, and that Central America is extensively mountainous. The extent and complexity of the Andes has allowed for much broader assessments of elevational and latitudinal diversification than are possible in most other regions. For example, recent work on *Metallura* hummingbirds (Benham et al. 2015) has found isolation across major Andean topographic barriers, as well as clades that are either tree line specialists (very high elevation) or habitat generalists— both vicariance and dispersal have been important at various times in the evolution of this diversity. And, the topography of the Andes is such that it includes smaller montane regions or valleys that are highly isolated or disconnected from other such subregions. As a function of this isolation, there are generally several new bird species described each year as researchers visit these isolated valleys. A recent example is a new species of barbet (fig. 4.25D) from an isolated range in Peru (Seeholzer et al. 2012).

Ultimately, elevation and habitat stratification affect diversification across all regions. The Neotropics have a rich history of speciation studies, some of which reveals underlying mechanisms that facilitated diversification. For example, Galen et al. (2015) found genetic mutations in Andean house wrens (*Troglodytes aedon*) related to physiological differentiation in blood oxygen carrying capacity and were able to relate this difference to populations found at different elevations. Similar results were found in Andean hummingbirds (Projecto-Garcia et al. 2013), where there are clear elevational transitions in hemoglobin function. These studies have gone beyond phylogenetic-based analyses of pattern and timing of diversifications to delve into causal physiological mechanisms.

Neotropical diversity has also been affected by interchanges with the Nearctic. A recent phylogeny-based example by Barker et al. (2015) has documented extensive interchange between North and South America. This study included over 95 percent of all species of blackbirds, cardinals, sparrows, tanagers, warblers, and their allies, and was able to show interchanges between continents both before but mostly after the closure of the Isthmus of Panama. Indeed the resulting uplift of the isthmus had an incredible influence on avian interchange between regions, with species inhabiting a variety of habitats and elevations utilizing this new dispersal corridor, although the overall pattern suggests that most invasions were from the Nearctic into the Neotropics (Smith and Klicka 2010). Further, the dispersal of lineages was related to expansion and contraction of humid and dry habitats, such that lineages differentially occupying those habitats dispersed and speciated at different times (Smith et al. 2012a). We see similar isthmus-related effects in Mexico (the northern extent of the Neotropics), where montane bird species diversified several times across the low-lying habitats of the Isthmus of Tehuantepec (Barber and Klicka 2010). A rather unique dispersal event occurred from Africa, establishing *Turdus* thrushes (such as the American Robin, *Turdus migratorius*) in the Neotropics (Voelker et al. 2009), where they subsequently diversified extensively in Central America, the Caribbean, and especially South America.

And as we discussed for Africa, there are studies of Neotropical birds that document lowland diversification related to habitat. Examples include understory specialists such as *Sclerurus* leaftossers (fig. 4.26A; d'Horta et al. 2013), and Amazonian birds like *Xiphorhynchus* woodcreepers (fig. 4.26B)

Figure 4.26. A, Rufous-breasted Leaftosser; B, Spotted Woodcreeper

that are associated with seasonally flooded forests (Aleixo 2006), while other members of the genus are not. Rivers have also played a role in driving diversification in the Amazonian basin, at multiple taxonomic levels, as shown for example by studies on *Psophia* trumpeters (Ribas et al. 2012) and the Blue-crowned Manakin (*Lepidothrix coronata*; Cheviron et al. 2005).

Nearctic Realm

Although a large landmass, much of the Nearctic is in the north temperate zone and relatively few bird species, around 730, breed here. There are no endemic bird families, but about 20 percent of genera and 50 percent of species are found nowhere else (in the breeding season, as many migrate to Central and South America). Possibly because of the relatively few species, or the existence of a large number of professional and amateur ornithologists, the avifauna of this region is probably the best known. Regular breeding surveys (BBS) and Christmas Bird Counts (see chapter 30) provide annual updates, and most states have a bird atlas project and state ornithological journal. The official species list is kept by the American Ornithological Society (aou.org). Given that North America and Eurasia were once connected, many species co-occurring in both places today are termed Holarctic. Within North America, there is a distinct difference between the birds found in the west and east, and in the early days of field identification, field guides were produced for each region.

Different biogeographers include different parts of Mexico and Central America within the Nearctic. Obviously, by including Mexico, one obtains a higher species count because many tropical species range northward into Mexico, but very few range into the southernmost United States. For example, the Elegant Trogon (*Trogon elegans*; fig. 4.27A) reaches southern Arizona, and the Green Jay (*Cyanocorax yncas*) reaches south Texas; these are the only places these species occur north of Mexico. Compared with the Neotropics or Africa, there are few endemic species; one is the Wrentit (*Chamaea fasciata*; fig. 4.27B), the only babbler in the New World. A small radiation of nightjars can be found, including the Common Poorwill (*Phalaenoptilus nuttallii*), one of the largest-bodied birds that regularly enters a state of torpor that can last for weeks to months in colder winter periods.

Approximately 75 waterfowl species occur in the Nearctic, including ducks, geese, swans, and mergansers. Several species of loons breed in northern lakes; some are Holarctic. A diversity of sandpipers and other shorebirds breed in the arctic reaches, although relatively few songbirds range that far north. Herons and egrets are found around most bodies of water, as are two species of kingfisher. Rails, coots, and bitterns inhabit marshes, along with wrens, blackbirds, and

Figure 4.27. A, Elegant Trogon; *B*, Wrentit

some sparrows. Gulls and terns are both found in the interior of the continent and along coastlines.

Both permanent residents and migrants inhabit Nearctic forests. Residents include woodpeckers, grouse, chickadees, nuthatches, some corvids, and some owls. A large number of groups are migratory, especially those in the northern reaches, including warblers (fig. 4.28), vireos, tanagers, orioles, thrushes, and flycatchers. Once the world's most common bird, numbering from three to five billion individuals, the forest-breeding Passenger Pigeon (*Ectopistes migratorius*), went extinct over 100 years ago as a result of market hunting and habitat alteration.

Perhaps the best-studied and most well-known group of forest birds is the New World wood warblers. This group includes small-bodied colorful species that are often highly sexually dimorphic in breeding plumage and that migrate long distances. They are examples of sexual selection owing to the high degree of plumage evolution relative to size and shape. Mostly tree-dwellers, some wood warblers are found on the ground. Robert MacArthur (1958) made his famous observation

Figure 4.28. Blackburnian Warbler

Figure 4.29. California Condor

on ecological niche partitioning by studying these birds and finding that, although several species used the same trees, each species used a different part of the tree.

Open areas are inhabited by larks and pipits, thrashers (especially in desert areas), bluebirds, doves, sparrows, some flycatchers, two species of shrikes, large soaring raptors and eagles, some owls, cranes, and lekking prairie grouse such as the Sharp-tailed Grouse (*Tympanuchus phasianellus*). Although there was once a large radiation of vulturine birds that fed on the carcasses of the now-extinct Pleistocene megafauna (species like mammoths, rhinos, camels, and glyptodonts), two vultures and the endangered California Condor (*Gymnogyps californianus*; fig. 4.29) are all that remain. The condor is the symbol of the former Cooper Ornithological Society and its journal *The Condor* (http://americanornithology.org/content /condor-and-condor-ornithological-applications). Although most cuckoos are forest or woodland dwellers, the Greater Roadrunner (*Geococcyx californianus*) has adaptations that enable it to live in extreme desert environments. The species is a public favorite that spawned a children's cartoon character.

Offshore and in coastal wetlands a number of species can be found, including skimmers, oystercatchers, skuas, and a diversity of alcids (e.g., murres, murrelets, guillemots, auklets, puffins). An extinct member of the last group, the Great Auk, serves as a symbol of conservation and as the namesake of the former American Ornithologists' Union primary publication, *The Auk* (http://americanornithology.org/content /auk-and-auk-ornithological-advances).

Upon the connection of North America and South America via the Panamanian land bridge, approximately three million years ago, there ensued the Great American Interchange (Smith and Klicka 2010). This biogeographic event resulted in the commingling of faunas from North and South America. The hummingbirds that occur in North America likely arose from South American species that moved north-

ward and diverged from their common ancestors. Most hummingbird diversity in North America is in Mexico and the western United States. Only the migratory Ruby-throated Hummingbird (*Archilochus colubris*) commonly resides in the eastern half of the continent.

How did some of this diversity evolve? Speciation in the Nearctic has long been associated with glaciation. At glacial maxima, the last one occurring 21,000 years ago (termed the Last Glacial Maximum, LGM), species were displaced south of the ice and were thought to have been isolated in geographically disjunct areas of suitable habitat called refugia. In refugia, it was thought that many current North American species evolved during the LGM from their most recent common ancestor, facilitated by small population size. Many sister species have east–west distributions, and this pattern was thought to reflect evolution in disjunct refugia. Klicka and Zink (1997) examined genetic distances between specific species pairs thought to have evolved this way and, by employing a molecular clock, found that most of these speciation events predated the LGM. They did not discount the role of glaciation in speciation; rather they concluded that glacially induced refugial speciation was a protracted process that spanned many glacial cycles, some even predating the Pleistocene.

Palearctic Realm

Europe and Asia are considered different regions, although there are no major geographic barriers between the two realms; together these regions constitute the Palearctic. The Palearctic is the largest landmass of the areas considered here; it spans 13 time zones. Its 46 million km² are bounded by the Atlantic, Arctic, and Pacific Oceans, to the west, north, and east, respectively. The southern boundary is less clear, and most suggest it extends to the southern limit of the Sahara Desert in Africa, because the birds of the northern and central Sahara are mostly Palearctic lineages, occurring mostly at oases. To the east the approximate dividing line stretches through the Middle East to south of Japan.

The Palearctic hosts roughly 940 species, relatively few, given its large area. No families, and just 9 percent of genera and 47 percent of species are restricted to this region, the latter a result of the relative lack of isolating barriers and largely similar habitats across vast expanses of Europe and Russia. Although the distance from Russia to Alaska is exceptionally short, fewer than 10 species have "invaded" the Nearctic from the Palearctic across the Aleutian Islands, and only three have made the reverse colonization (Newton 2003). On the Atlantic side, few species have colonized from either continent, with the Northern Wheatear (*Oenanthe oenanthe*; fig. 4.30A) colonizing northeastern Canada from Europe. As Newton (2003) points out, this is surprising, given the large number of individuals of different species that each year are blown from one continent to another by storms during migrations. Perhaps this reveals how difficult it is for some species to become established far from their native ranges, through either lack of suitable habitat or competition from native species. A biogeographic oddity is the "Winter Wren" (*Troglodytes troglodytes*; fig. 4.30B), which has a Holarctic distribution and as such is the only wren to occur in the Old World. The winter wren was once considered one species but has since been divided into three species following molecular studies (Drovetski et al. 2004), and two of these species occur in North America.

European Palearctic Region

A specialty of the Palearctic is the coniferous-forest dwelling Capercaillie (*Tetrao urogallus*; fig. 4.31), the largest grouse in the world. Male Capercaillies are twice the body weight of females, and they vigorously defend their leks against other males; displaying males will even attack a person that ventures too close to the lek site. Several species of sandgrouse occur in the southern reaches. The Eurasian Wryneck (*Jynx torquilla*; fig. 4.32A), a woodpecker, is the only wryneck in Europe (it extends to Japan). It can turn its head almost 180 degrees; the generic name is from a hissing sound it makes, supposedly putting a "jinx" on whomever disturbed it.

Palearctic forests have a blend of species similar to the Nearctic, with sedentary species including owls, woodpeckers, tits, kinglets, creepers, Old World flycatchers (ecological equivalents of New World flycatchers), corvids, grouse, pigeons and doves, and cuckoos. The Common Cuckoo (*Cuculus canorus*) is well known for its parasitism of reed warblers and the ongoing arms race between egg markings of host and parasite. The Palearctic has an interesting set of corvids, including jays and magpies, as well as birds that have taken to

Figure 4.30. A, Northern Wheatear; B, Winter Wren

cities, including the Rook (*Corvus frugilegus*; fig. 4.32B), Western Jackdaw (*C. monedula*), and Hooded Crow (*C. cornix*).

Aquatic habitats are home to many swans, mergansers, ducks and geese, pelicans, cormorants, and grebes. Several species of herons, bitterns, and one flamingo are found. A few rails and coots occupy marshes. Northern boreal reaches are home to many sandpipers, and woodcock and snipe inhabit wet areas to the south. Loons, gulls, terns, plovers, stilts, and oystercatchers inhabit lakes and shoreline. As in the Nearctic, there is a single species of dipper. Offshore one can find typical ocean birds including alcids, petrels, storm petrels, shearwaters, albatrosses (especially in the south), gulls, and terns.

Open grassland and woodlands are home to bustards, cranes, and storks, some of the latter of which nest on

Figure 4.31. Capercaillie

Figure 4.32. A, Eurasian Wryneck; B, Rook

are found in both forests and open country. Several species of shrikes occur, including one (*Lanius excubitor*) that is Holarctic. Thrushes occur in forests and open country; North American ornithologists are sometimes confused when they see a Eurasian Blackbird (*Turdus merula*; fig. 4.33A) and realize it is a robin to them; similarly, the European Robin (*Erithacus rubecula*) is an Old World flycatcher relative. The Common Nightingale (*Luscinia megarhynchos*) is a famous songster that often sings at night, as the name implies (to the chagrin of sleepy people). This species was once thought to be a thrush but is now thought to be closer to the Old World flycatchers. A diversity of larks and pipits inhabit open grasslands and semideserts.

A well-studied group is the Old World warblers. They are generally not as brightly colored as the New World wood warblers, and considerable speciation has occurred, with over 35 species in Europe alone. Studies of some of these species, such as the Blackcap (*Sylvia atricapilla*), have provided definitive evidence of the role of a magnetic compass in orientation and the genetic basis of migratory duration and directionality. The Dunnock (*Prunella modularis*; fig. 4.33B) has become famous because, despite its drab appearance, the species exhibits nearly every mating system known. Sparrows, finches, and buntings are common in woodlands, forests, and open areas.

Although not limited to the Palearctic, a favorite of ornithologists and the public alike is the Hoopoe (*Upupa epops*), a bird with a clown-like crest. A famous ornithologist and World War II fighter pilot ace, Harrison "Bud" Tordoff, named his P-51 Mustang after this bird.

Asian Palearctic and Indomalayan Regions

This part of the Palearctic region includes Russia east of the Ural Mountains, Turkey, Sri Lanka, Japan, Taiwan, the Philippines, Malaysia, and the part of Indonesia that does not belong with the Australian region. It is called the "Indo-Malayan" region by Newton (2003) and the "Sino-Japanese/Oriental" by Holt et al. (2012). Large regions of rain forest used to cover this region. It is famous for its southern border, historically called Wallace's line, after Alfred Russel Wallace, who recognized that Australian and Oriental faunas met in Southeast Asia (Wallace 1863). Subsequent authors have proposed revisions (e.g., Holt et al. 2012) often related to different vertebrate or invertebrate lineages.

The region is the third richest in bird diversity, with approximately 1,700 species, although just over 400 are found only on islands. Many of the same groups occur in the European Palearctic, as the habitat across much of the Palearctic is rather similar. In addition, many groups are shared with Africa, which reflects a shared history between India and Af-

rooftops in small towns; however, because the nests can weigh 450 kg, they are monitored and removed when threatening roofs and chimneys. Unlike in the Nearctic, here one can find thick-knees, coursers, and some large noisy lapwings. Several eagles and other large hawks, falcons, and accipiters

Figure 4.33. A, Eurasian Blackbird; *B,* Dunnock (also called Hedge Sparrow)

rica before India broke away and drifted north, eventually crashing into Asia, as well as aforementioned instances of dispersal into Africa. In general, many groups exhibit more species per group, such as the herons (30 species), storks (14), owls (80), cuckoos (60), cuckoo-shrikes (40), nightjars (20), swifts (30), barbets (25), broadbills (10), bulbuls (60), pittas (30), woodpeckers (60), kingfishers (35), larks (35), swallows (25), babblers (225), wagtails (25), thrushes (70), monarch fly-catchers (14), whistlers (12), cisticolas (20), Old World warblers (165), Old World flycatchers (including wheatears, bushchats, stonechats, redstarts, robins) (175), sunbirds (45), flowerpeckers (35), white-eyes (25), jays (50), starlings (40), waxbills (30), finches (80), and buntings (45). Many other groups are represented by fewer species. Although most honeyguides occur in Africa, two species occur in Asia.

A large number of hawks, kites, and Old World vultures can be found, including the Philippine Eagle (*Pithecophaga jefferyi*), known for eating tropical monkeys. A few species of mound-builders, or megapodes, occur in this region. Megapodes moderate the temperature of their eggs by adding or removing soil to the nest mound. The treeswifts (fig. 4.34),

a group of four species, have plumage that differs from other swifts, and they often have crests. A large radiation of hornbills (30 species) exists in Asia and the Oriental regions. Male hornbills (fig. 4.35) wall females into nest cavities during incubation using a mixture of mud and feces. A group of 10 tropical frogmouths (*Batrachostomus*) might represent an endemic radiation in Asia. The 10–15 species of leafbirds represent an endemic family of bulbul-like birds with mostly green plumage (hence their name); these birds drop many feathers when handled, which might be an adaptation to escaping predation by snakes. The Rail-babbler (*Eupetes macrocercus*), a shy secretive bird found on the forest floor, has puzzled taxonomists, because it does not appear to be closely related to other extant species. The four species of ioras (*Aegithina*) found in India and Southeast Asia might be one of only three endemic families in the Indomalayan region.

Forests contain most of the groups found in the European Palearctic. Of note is the large number of pigeons and doves, nearly 100 species, in the Asian region, including the ground dwelling "bleeding-hearts," so named because of the touch of vivid red color in the center of their otherwise white breasts. Over 50 species of cuckoos occur here, including some Palearctic species. In open country are a large number of gallinaceous birds, including some Palearctic grouse and ptarmigan, snowcock, partridges, francolins, and Old World quail, as well as the large radiation of over 20 pheasants, peafowl, guineafowl, and buttonquail. The same aquatic groups as found in the European Palearctic occur in this region. An interesting species is the Ibisbill (*Ibidorhyncha struthersii*; fig. 4.36), a wader, which is so distinctive that some place it in its own family.

How did some of this diversity evolve? Much of the diversification across the Palearctic is caused by glacial cycles and isolation of lineages into refugia. Major glacial refugia were known to exist in southern Europe (e.g., Iberia and Italy), the Caucasus Mountains, southeastern China, and the Russian

Figure 4.34. Whiskered Treeswift

Figure 4.35. Oriental Pied Hornbill

Figure 4.36. Ibisbill

Far East. There is broad support for diverse lineages having been repeatedly isolated in western, southeastern, and far eastern refugia since the Miocene (e.g., Voelker 2010), and that the Caucasus Mountains harbor a genetically unique avifauna (e.g. Hung et al. 2012, 2013). In the species-rich Himalayas (fig. 4.7), there is evidence that these mountains have served as a source area for several lineages. Both *Phoenicurus* redstarts and *Prunella* accentors originated in that region, and subsequently diversified across Eurasia (Drovetski et al. 2013, Voelker et al. 2015). In the latter genus, diversification was also linked to habitat, where some species were alpine-associated and others occupied shrub habitat.

Elsewhere, Robin et al. (2015) have documented deep divergences within 10 of 23 species distributed on different mountain types in the Western Ghats of India. And Pons et al. (2011) showed that the Green Woodpecker (*Picus viridis*) comprises three lineages distributed in North Africa (a part of the Palearctic for avian distributions), Iberia, and across Europe. Although these were intraspecific comparisons, the results collectively indicate how isolated mountains in (or subregions of) the Palearctic can be potential drivers of diversity. Given sufficient time in isolation, we would expect increased diversification of such lineages.

In Indomalaya, many studies have focused on the role of sea level change on and around the Sunda Shelf. These climate-driven changes in sea level have alternately connected many Sunda Shelf islands (e.g., from Sumatra and Java through Borneo to the Philippine Island of Palawan) during low sea stands and isolated these islands during high sea stands (as we have today). As a result, both vicariance and dis-

persal may have played roles in diversification in the region. For example, vicariance likely drove early speciation within *Ficedula* flycatchers, but more recent speciation events are clearly driven by dispersal (Outlaw and Voelker 2008). With respect to the Philippine Islands, Oliveros and Moyle (2010) were able to show that eight independent colonizations of those islands were performed by Philippine bulbuls. These colonizations occurred either via Palawan (a Sunda Shelf island) or via colonization of the oceanic Philippine Islands. Interestingly each clade that invaded, invaded via one route or the other; none did both.

Australasian Realm

Different authors attribute different areas to this region. Newton (2003) includes New Zealand, but we have separated it out in our discussion. The main landmasses in this region then are New Guinea and Australia, along with some islands, including Tasmania. Relatively small in landmass, the extent of the region is about nine million km², or 20 percent of the entire Palearctic. Much of the land is tropical, although Australia is the driest of all continents, with frequent droughts lasting several years. As a result, Australian habitats are fire-prone, and many bird species can be described as nomadic and might not breed for several years in a row (Newton 2003).

The Australasian region as a whole has about 1,600 breeding landbirds, with 925 on New Guinea, Australia, and Tasmania together, and about 540 on Australia and 580 on New Guinea; relatively few species are shared between Australia and New Guinea. There are around 18 endemic families, and 1,415 endemic landbird species, second only to the Neotropical realm. Australia has been isolated from other major land areas for over 50 million years. The long isolation of Australia is no doubt a major reason for its diversity. In addition, there is little migration of Australian birds to Asia and vice versa. Parrots, pigeons, doves, honeyeaters, and kingfishers are well represented (Newton 2003). The approximately 90 species of honeyeaters have diversified to fill most habitats, from arid regions to rain forest to high mountains. A number of groups may have arisen from ancestors shared from Gondwanaland before Australia drifted away. These probably include scrubbirds, some corvids, the Magpie-lark (*Grallina cyanoleuca*; fig. 4.37), Torrent-lark (*G. bruijni*), White-winged Chough (*Corcorax melanorhamphos*), and Apostlebird (*Struthidae cinerea*). The latter species was named for its habitat of traveling in groups of about 12 individuals, presumably like biblical apostles.

Certainly one of the most interesting groups are the bowerbirds (fig. 4.38A), with 10 species found only in New Guinea, eight in Australia, and two in both regions. The bowerbirds rival in many ways the birds-of-paradise in gaudiness (fig. 4.38B), only it is a result of colorfully decorated display arenas (not nests), termed bowers, that males decorate, often with colorful objects, to attract females. Hence, the bower's accoutrements substitute for a brightly colored bird, although some bowerbirds are quite colorful and have less elaborate bowers. Another well-known Australian bird is the lyrebird, of which there are two species. The Superb Lyrebird (*Menura novaehollandiae*) is famous for its ability to mimic sounds in its environment, including other birds as well as human sounds such as car alarms, camera shutters, and chainsaws (https://www.youtube.com/watch?v=VjE0Kdfos4Y). The Superb Fairywren (*Malurus cyaneus*; fig. 4.38C) is a favorite of avian ethologists because of its mating behavior, which involves being socially monogamous and sexually promiscuous.

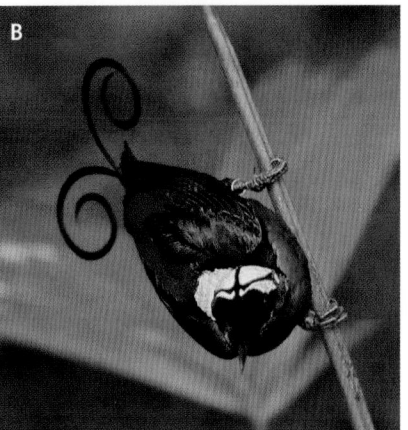

Figure 4.38.
A, Satin Bowerbird; B, Raggiana Bird-of-paradise; C, Superb Fairywren

Figure 4.37. Magpie-lark

What is remarkable has been the discovery from molecular systematics in the past few decades that the Australian avifauna is filled with dramatic examples of convergence. Apparently an early member of the crows made it to Australia, and subsequent speciation resulted in a new lineage that filled many niches occupied by other birds in other biogeographic regions. Without competitors in place, convergent evolution yielded honeyeaters that resemble African sunbirds, sittellas (fig. 4.39A) that resemble Holarctic nuthatches (fig. 4.39B), Australasian babblers that resemble Himalayan scimitar babblers, Australian robins that resemble chats from other parts of the Old World, and Australian scrub robins that resemble thrushes. That is, given the absence of these very different lineages in Australia, the early corvid ancestor radiated into

Figure 4.39. A, Varied Sittella; B, European Nuthatch

niches typically occupied by them. Hence, if one expected that an Australian robin would be a member of the continentally derived Old World chats that made it to Australia, one would be wrong. All of these convergent Australasian lineages trace to a common Australian ancestor that excludes the lineages on which they are convergent. The avifauna of the Australasian region also includes species that are in fact related to groups found elsewhere and arrived in the area via dispersal at different times; as a result, some are barely differentiated, others more so (such as sylviid warblers).

Newton (2003) points out that the number of African bird groups progressively diminishes as one approaches Australia, which suggests more recent colonization rather than the remnants of a Gondwanaland avifauna. It has also been noted that very few Australian genera have breached Wallace's line to the north, and none have made it to Africa. Newton (2003) suggests that this indicates an asymmetry in the dispersal or colonization ability of the birds of the different regions. Interestingly, Old World vultures, pheasants, skimmers, sandgrouse, trogons, rollers, woodpeckers, broadbills, finches, and buntings have not colonized Australasia from the Palearctic.

How did this diversity evolve? Naturally, the same factors that we've discussed in other regions have also influenced avian diversification in Australasia. For example, a recent study by Kearns et al. (2014) was able to show that ranges of Australian Pied Butcherbird (*Cracticus nigrogularis*; fig. 4.40A) in Australia and New Guinea were greatly affected by climate shifts during the last glacial maximum. An interesting result is that the group shows high levels of genetic introgression across species, but most relevant to diversity is that Australia has experienced similar shifts in habitat that have facilitated diversification of lineages. Other studies have focused on Australia's Wet Tropics, and its northern Black Mountain Barrier, to name a few (see Joseph and Omland 2009 for some examples).

Also of interest are the biotic interactions between Australia and New Guinea (and other islands). The Collared Kingfisher (*Todiramphus chloris*) has no fewer than 50 subspecies distributed throughout Australasia (and as far west as the Red Sea). Perhaps not surprisingly with that many races, a detailed phylogenetic assessment of this species and its relatives found that many other *Todiramphus* species are embedded within *T. chloris* (Anderson et al. 2015). As such, many *"chloris"* races must be elevated to species status—in other words, taxonomy was masking species-level diversity. This same study also provided a detailed biogeographic assessment of the group, and as such provides an excellent example of how island biodiversity can be assembled. A similar example is exemplified

Figure 4.40. A, Australian Pied Butcherbird; B, Golden Whistler

sal event from Australia across the Indian Ocean to Africa resulted in the large radiation of the Passerida. This second result was particularly surprising, as a connection to Africa had previously been assumed to have involved movements through southern Asia.

ISLANDS AS A DRIVING FORCE IN BIRD DIVERSITY

A well-known ornithologist, George F. Barrowclough, at the American Museum of Natural History, once remarked that "visionaries need islands" (1988: 522). His comment was based on the fact that islands often accentuate evolutionary processes, owing to several factors. Loss of flight is a relatively common outcome for birds that manage to reach distant islands. Species also often change greatly in size when isolated on islands. An island's isolation and (generally) low species diversity can mean that, once it colonizes, a species can evolve—often radically—such that the clues to its ancestry fades. The Kagu (*Rhynochetos jubatus*; fig. 4.41), from New Caledonia, is a ground-dwelling bird that looks like a small heron and is essentially flightless. The Kagu posed a difficult taxonomic riddle, and it was suggested to be related to herons, but Claramunt and Cracraft (2015) have confirmed that it is a distant relative to the sunbitterns.

Adaptive radiations have long been of interest to evolutionary biologists, and birds are known for several adaptive radiations. Examples include the Darwin's finches, the Hawaiian honeycreepers, the Madagascan vangas, and the birds-of-paradise. At least two ways exist to recognize an adaptive radiation, and both are ends of a continuum. First, an adaptive radiation can be a group of species that evolve

by the Golden Whistler (*Pachycephala pectoralis/melanura*; fig. 4.40B) species complex. This species is the most geographically variable bird species in the world. As with the Collared Kingfisher, complex Australasian biogeographic patterns are evident from the phylogeny (Anderson et al. 2015).

Importantly, the Australasian region has been a source area from which old lineages have colonized other areas. A recent phylogeny-based biogeographic assessment by Jønsson and Fjeldså (2006) found that "crown" Corvida dispersed from Australia to Southeast Asia, and that a single disper-

Figure 4.41. Kagu

an innovation and radiate rapidly, often filling unexploited niches. Second, an adaptive radiation is recognized if a group possessing an innovation is far more speciose than its sister lineage. It is difficult to find examples of the second type of radiation because phylogenies must be extremely imbalanced for the difference in species diversity to be statistically significant (Guyer and Slowinski 1993). However, the passerine birds as a whole probably qualify as this type of adaptive radiation owing to the extremely large number of species (about half of all bird species) relative to their sister group (Raikow 1986). Below we consider three famous avian radiations that qualify as adaptive under the first definition; interested readers can explore Jønsson et al. (2012) for details on the Madagascan vangas.

Darwin's Finches

The finches of the Galápagos, often referred to as Darwin's finches, consist of a handful of species that vary remarkably in beak size and ecological specialization, but in little else. Indeed, even the number of species is in doubt (McKay and Zink 2014). Often considered a prime example of an adaptive radiation, these finches qualify as such only in that considerable morphological diversity apparently evolved in a short time. The Woodpecker Finch (*Camarhynchus pallidus*; fig. 4.42) has evolved tool use, in which individuals use cactus spines to fish beetle larvae out of wood. This seems a clear case of convergence to a way of life normally filled by woodpeckers, which are not found on the Galápagos. The alternative definition of adaptive radiation is that one group, such as Darwin's finches, is far more speciose than their sister lineage, and this does not hold for the finches of the Galápagos.

Nonetheless, Darwin's finches are one of the most famous examples of natural selection in the world (Grant and Grant 2008). In response to shifts in seed size, the distribution of bill sizes shifted accordingly, in just a couple of years. However, this shift did not involve the origin of new bill sizes, but rather the shifting of the frequency of ones already in existence. It is clear that this morphological shift follows resource peaks, which are ephemeral over time, and therefore revisits the same adaptive peaks probably multiple times as environmental conditions shift.

Figure 4.42. Woodpecker Finch

Hawaiian Honeycreepers

It was unfortunate that Charles Darwin's voyage on the HMS *Beagle* did not bring him to the Hawaiian Islands as well as the Galápagos. He would have found one of the most dramatic adaptive radiations in all of vertebrates, the Hawaiian honeycreepers, with a large number of species showing extreme divergence in bill sizes and coloration. But, Darwin might not have grasped the novelty of what he was seeing. Illustrating this fact is a quote by James (2004: 208): "Unlike the case of the relatively uniform Galápagos finches, the Hawaiian finches were not widely recognized as a related group of species until about a century after their discovery." In other words, they are so diverse, early taxonomists assumed they must be an assemblage consisting of many unrelated lineages that independently colonized the islands.

The Hawaiian Islands are the most remote of any major island group, and therefore speciation has occurred without multiple colonizations from continents. The 19 extant species of Hawaiian honeycreepers possess bills that range from seed-eating to nectar-feeding forms. By colonizing an archipelago with many available niches, the honeycreepers evolved bills that are convergent on different groups from continental avifaunas, as well as some that are known nowhere else, such as the ʻAkiapōlāʻau (*Hemignathus munroi*), a species in which the mandible is short and the maxilla curves over the mandible and downward. Unfortunately, probably close to 40 additional species of honeycreepers did not survive contact by Polynesians, who upon colonization of the islands introduced predators and diseases and extensively altered habitat for agriculture (James and Olson 1991).

Molecular data (Lerner et al. 2011) show that the Holarctic *Carpodacus* finches (fig. 4.43A) are the sister-group to the Hawaiian honeycreepers (fig. 4.43B). Lerner et al. (2011) suggest that the evolution of distinctive extant honeycreeper morphologies occurred between 5.8 and 2.4 million years ago, which overlaps with the emergence of Oʻahu, but is prior to the emergence of the Maui Nui island complex (fig. 4.43C).

It would seem likely that the environmental heterogeneity of the Hawaiian Islands and the conveyor-like emergence and northwestward drift of the islands provided a unique evolutionary setting. Like other remote islands, the lack of predators, at least pre-dating human colonization, allowed for high species diversity and relatively tame, hence vulnerable, species.

The birds of Hawaiʻi are not limited to the honeycreepers. Another 58 native, non-passerine species have been identified (not to mention numerous introduced exotics). Unfortunately, like the honeycreepers, about 40 of these native non-passerines are extinct.

A

B

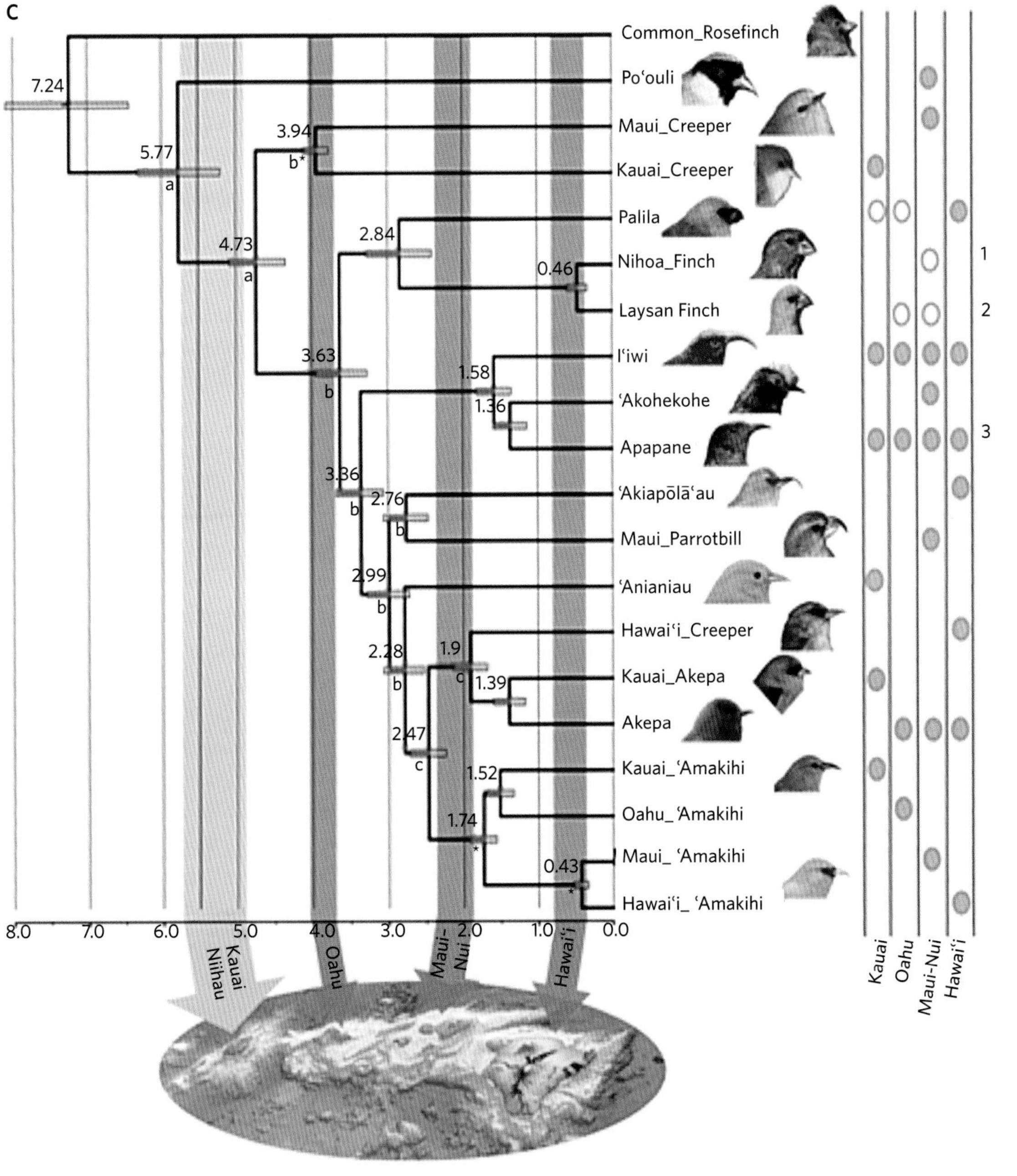

C

Common_Rosefinch				
Po'ouli			Maui-Nui	
Maui_Creeper			Maui-Nui	
Kauai_Creeper	Kauai			
Palila	Kauai	Oahu		Hawai'i
Nihoa_Finch		Oahu		
Laysan_Finch		Oahu		
I'iwi	Kauai	Oahu	Maui-Nui	Hawai'i
'Akohekohe			Maui-Nui	
Apapane	Kauai	Oahu	Maui-Nui	Hawai'i
'Akiapōlā'au				
Maui_Parrotbill				
'Anianiau	Kauai			
Hawai'i_Creeper				Hawai'i
Kauai_Akepa	Kauai			
Akepa			Maui-Nui	Hawai'i
Kauai_'Amakihi	Kauai			
Oahu_'Amakihi		Oahu		
Maui_'Amakihi			Maui-Nui	
Hawai'i_'Amakihi				Hawai'i

7.24

5.77
a

4.73
a

3.94
b*

2.84

0.46

3.63
b

1.58

1.36

3.36
b

2.76
b

2.99
b

2.28
b

1.9
c

1.39

2.47
c

1.52

1.74
*

0.43
*

1

2

3

8.0   7.0   6.0   5.0   4.0   3.0   2.0   1.0   0.0

Kauai–
Niihau

Oahu

Maui–
Nui

Hawai'i

Kauai   Oahu   Maui-Nui   Hawai'i

Figure 4.43.
A, House Finch;
B, I'iwi; *C*, Phylogeny
of Hawaiian Honey-
creepers, showing
general bill sizes
and island distribu-
tions. The map at
the bottom shows
that the basal (old-
est) lineages are
found on the oldest
islands. *From Lerner*
et al. 2011.

Many of the extinct birds of Hawai'i are known because of the work of Helen James and Storrs Olson at the Smithsonian Institution. They point out that many of the extinct forms they discover and describe are not fossils in the typical sense. Instead, the remains are from birds that expired only hundreds of years ago. Although they are primarily paleontologists, James and Olson provided an extremely insightful ecological observation, best read in their own words: "Had not *Homo sapiens* arrived in these islands some 16 centuries ago, these birds would still be alive today—skin, feathers, songs, enzymes and all. The poor remnants of fauna and flora that still persist in the Hawaiian Islands evolved alongside and interacted with a diverse array of browsing, seed-eating, frugivorous, insectivorous, malacophagous, and raptorial birds whose former existence had never been suspected. Consequently, the distribution, adaptations, and history of the existing biota of the islands can no longer be sensibly interpreted without consideration of the fossil record" (Olson and James 1991). This is a poignant reminder of the influence that man has had on ecosystems, especially on islands, and that birds alive today likely evolved in an ecological setting that involved a very different set of species. And unfortunately, these effects of human colonization can be seen from many islands across the Pacific (Steadman 2006).

Birds-of-Paradise

It is perhaps even more disappointing that Darwin did not find himself in New Guinea watching birds-of-paradise. Viewing the high density of amazingly colorful males and their bizarre courtships, all in a small area, who could imagine what ideas and hypotheses his fertile mind might have generated? If his studies of Galápagos finches eventually revolutionized evolutionary biology, the ecological, morphological, and behavioral diversity of the birds-of-paradise might have more quickly led him to his revelations. Between 40 and 90 species exist, depending on what species concept is used (Cracraft 1992). Many species have extremely small ranges and are poor dispersers. Most species have extreme sexual dimorphism, and some species described in the past were probably hybrids.

Irrespective of numbers, nowhere on Earth are there more gaudy-colored males that engage in spectacular displays than the birds-of-paradise. Watching video of just about any species makes one wonder if it is an animation. But, in their remote island homes, these birds have not only radiated into an adaptive radiation but also attained an incredible degree of species-specific uniqueness, even more so than the Hawaiian honeycreepers.

How could this complexity evolve? Several interacting factors seem relevant, mostly related to isolation. There are very few predators, hence, a gaudy male displaying and calling with the utmost conspicuousness does not risk being easy prey. However, the fact that females are generally drab-colored suggests that there might be predators that could find and depredate nests. The low level of dispersal means that populations can differentiate over small geographic scales, without gene flow from neighboring areas arresting or halting speciation. There are also many small islands, and many sedentary birds rarely cross water barriers.

The Cornell Laboratory of Ornithology has produced an excellent web page including information on most aspects of the biology of birds-of-paradise.

New Zealand

New Zealand has been isolated from Australia for over 80 million years, and unlike Australia, it is smaller, farther south, colder, and has been glaciated. It has a few native bats, but lacks native land mammals and has only a few reptiles. Newton (2003) considers New Zealand to have three groups of birds: an ancient group, relatively more recent colonists from Australia, and species from the north. There is an abundance of seabirds, with over 80 species. About 65 species of landbirds survive today, and New Zealand has many tales of extinction and species on the brink thereof. What is impressive about the birds of New Zealand is not just the endemic species, but how different in appearance many of them are from relatives. Some of the diversity of New Zealand's birds is likely a result of multiple colonization events (Trewick and Gibb 2010), owing to the fact that few of the endemic groups are monophyletic.

The most impressive part of the New Zealand avifauna is the "ancient" group. The 9 or 10 species of extinct moas (fig. 4.5) included smaller-bodied species and a couple that reached 12 feet in height. The opportunity for the moa radiation to have occurred suggests a long period of isolation and an ecological base that allowed specialization and subsequent sympatry. The wattlebirds, not to be confused with unrelated birds of the same common name in Australia, are a distinctive group including the extinct Huia (*Heteralocha carunculatus*), Kōkako (*Callaeas cinereus*; fig. 4.44A), and the Saddleback (*Philesturnus carunculatus*; fig. 4.44B), endemic birds with no near relatives, as is the Stitchbird (*Notiomystis cincta*; fig. 4.44C), another distinct endemic (confined to the North Island). The Fernbird (*Megalurus punctatus*) is a ground-dweller that was once common and now is quite rare.

New Zealand is home to several species of kiwis, flightless birds that possess so many unique adaptations that their time in isolation can only be assumed to have been very long: characteristics include being nocturnal and having poor eyesight, very good hearing, and nostrils located in the tip of the bill. Two very distinctive parrots are the Kea (*Nestor*

Figure 4.44.
A, Kōkako;
B, Saddleback;
C, Stitchbird

Figure 4.45. A, Kakapo; B, Rifleman

rail, the Weka (*Gallirallus australis*), represent other highly divergent species. The New Zealand Wren, or rifleman (*Acanthisitta chloris*; fig. 4.45B), has long puzzled taxonomists, as it is not a wren, except in outward appearance, and might be the sister species to all other passerine birds, and hence the oldest songbird (Barker et al. 2004). The list of New Zealand oddities could be much longer, but the point that the long period of isolation has led to extensive evolutionary change in many species should be clear.

KEY POINTS

- Birds are found on all continents and most islands.
- Avian biogeography is the study of the distribution of birds in space and time.
- Both dispersal and vicariance have influenced the geographic distribution of birds.
- The Neotropics have the greatest bird species diversity.
- Islands have produced adaptive radiations of Hawaiian honeycreepers, birds-of-paradise, and Galápagos finches.
- Species composition in any given area is a function of speciation and immigration, and changes over time.

KEY MANAGEMENT AND CONSERVATION IMPLICATIONS

- Islands reveal the vulnerability of small populations to human pressures.
- Taxonomy can often mask underlying complexity and diversity, particularly in widespread and morphologically variable species.

DISCUSSION QUESTIONS

1. Does the avifauna of New Zealand qualify as an adaptive radiation?
2. What is the principal distinction between vicariance and dispersal?
3. Why would the birds of Hawai'i necessarily be strong evidence for dispersal?

notabilis) and the Kaka (*N. meridionalis*). A flightless parrot, the Kakapo (*Strigops habroptilus*; fig. 4.45A), is another highly divergent form that conservation biologists are working diligently to save from extinction. A large flightless coot relative, the Takahē (*Porphyrio hochstetteri*), and the flightless

4. What are some reasons that might explain why the Neotropics has the greatest species diversity?

5. Why might the evolution of ratites be explained by both vicariance and dispersal?

6. How might a museum collection of bird specimens incite the imagination of scientists?

References

Aleixo, A. (2006). Historical diversification of floodplain forest specialist species in the Amazon: A case study with two species of the avian genus *Xiphorhynchus* (Aves: Dendrocolaptidae). Biological Journal of the Linnean Society 89:383–395.

Allentoft, M. E., R. Heller, C. L. Oskam, E. D. Lorenzen, M. L. Hale, M. T. P. Gilbert, C. Jacomb, R. N. Holdaway, and M. Bunce (2014). Extinct New Zealand megafauna were not in decline before human colonization. Proceedings of the National Academy of Sciences USA 111:4922–4927.

Andersen, M. J., Á. S. Nyári, I. Mason, L. Joseph, J. P. Dumbacher, C. E. Filardi, and R. G. Moyle (2014). Molecular systematics of the world's most polytypic bird: The *Pachycephala pectoralis/melanura* (Aves: Pachycephalidae) species complex. Zoological Journal of the Linnean Society 170:566–588.

Andersen, M. J., H. T. Shult, A. Cibois, J.-C. Thibault, C. E. Filardi, and R. G. Moyle (2015). Rapid diversification and secondary sympatry in Australo-Pacific kingfishers (Aves: Alcedinidae: *Todiramphus*). Royal Society Open Science 2:140375.

Baker, A. J., O. Haddrath, J. D. McPherson, and A. Cloutier (2014). Genomic support for a moa–tinamou clade and adaptive morphological convergence in flightless ratites. Molecular Biology and Evolution 31:1686–1696.

Barber, B. R., and J. Klicka (2010). Two pulses of diversification across the Isthmus of Tehuantepec in a montane Mexican bird fauna. Proceedings of the Royal Society of London B 277:2675–2681.

Barker, F. K., A. Cibois, P. Schikler, J. Feinstein, and J. Cracraft (2004). Phylogeny and diversification of the largest avian radiation. Proceedings of the National Academy of Sciences USA 101:11040–11045.

Barker, F. K., K. J. Burns, J. Klicka, S. M. Lanyon, and I. J. Lovette (2015). New insights into New World biogeography: An integrated view from the phylogeny of blackbirds, cardinals, sparrows, tanagers, warblers, and allies. The Auk 132:333–348.

Barrowclough, G. F. (1988). Review of "Ecology and evolution of Darwin's finches." The Condor 90:522–523.

Benham, P. M., A. M. Cuervo, J. A. McGuire, and C. C. Witt (2015). Biogeography of the Andean metaltail hummingbirds: contrasting evolutionary histories of tree line and habitat-generalist clades. Journal of Biogeography 42:763–777.

Bowie, R. C. K., J. Fjeldså, S. J. Hackett, and T. M. Crowe (2004). Systematics and biogeography of Double-Collared Sunbirds from the Eastern Arc Mountains, Tanzania. The Auk 121:660–681.

Cheviron, Z. A., S. J. Hackett, and A. P. Capparella (2005). Complex evolutionary history of a Neotropical lowland forest bird (*Lepidothrix coronata*) and its implications for historical hypotheses of the origin of Neotropical avian diversity. Molecular Phylogenetics and Evolution 46:338–357.

Claramunt, S., and J. Cracraft (2015). A new time tree reveals Earth history's imprint on the evolution of modern birds. Science Advances 1 (11): e1501005.

Cracraft, J. (1974). Phylogeny and evolution of the ratite birds. Ibis 116:494–521.

Cracraft, J. (1992). The species of the birds-of-paradise (Paradisaeidae): Applying the phylogenetic species concept to a complex pattern of diversification. Cladistics 8:1–43.

d'Horta, F. M., A. M. Cuervo, C. C. Ribas, R. T. Brumfield, and C. Y. Miyaki (2013). Phylogeny and comparative phylogeography of *Sclerurus* (Aves: Furnariidae) reveal constant and cryptic diversification in an old radiation of rain forest understorey specialists. Journal of Biogeography 40:37–49.

Drovetski, S. V., R. M. Zink, S. Rohwer, I. V. Fadeev, Y. V. Nesterov, I. Karagodin, and Y. A. Red'kin (2004). Cryptic vicariant speciation in the Holarctic Winter Wren (*Troglodytes troglodytes*). Proceedings of the Royal Society of London B 271:545–551.

Drovetski, S. V., G. Semenov, S. S. Drovetskaya, I. V. Fadeev, Y. A. Red'kin, and G. Voelker (2013). Geographic mode of speciation in a mountain specialist avian family endemic to the Palearctic. Ecology and Evolution 3:1518–1528.

Fjeldså, J., R. C. K. Bowie, and J. Kiure (2006). The Forest Batis, *Batis mixta*, is two species: description of a new, narrowly distributed Batis species in the Eastern Arc biodiversity hotspot. Journal of Ornithology 147:578–590.

Galen, S. C., C. Natarajan, H. Moriyama, R. E. Weber, A. Fago, P. M. Benham, A. N. Chavez, Z. A. Cheviron, J. F. Storz, and C. C. Witt (2015). Contribution of a mutational hot spot to hemoglobin adaptation in high-altitude Andean house wrens. Proceedings of the National Academy of Sciences 112:13958–13963.

Grant, P. R., and B. R. Grant (2008). How and why species multiply: The radiation of Darwin's finches. Princeton University Press, Princeton, NJ.

Guyer, C., and J. B. Slowinski (1993). Adaptive radiation and the topology of large phylogenies. Evolution 47:253–263.

Hawkins, B. A., E. E. Porter, and J. A. Felizola Diniz-Filho (2003). Productivity and history as predictors of the latitudinal diversity gradient of terrestrial birds. Ecology 84:1608–1623.

Holt, B. G., J.-P. Lessard, M. K. Borregaard, S. A. Fritz, M. B. Araújo, D. Dimitrov, P.-H. Fabre, C. H. Graham, G. R. Graves, K. A. Jønsson, D. Nogués-Bravo, Z. Wang, R. J. Whittaker, J. Fjeldså, and C. Rahbek (2012). An update of Wallace's zoogeographic regions of the World. Science 339:74–78.

Holt, B. G., J. P. Lessard, M. K. Borregaard, S. A. Fritz, M. B. Araújo, D. Dimitrov, P. H. Fabre, et al. (2013). An update of Wallace's zoogeographic regions of the world. Science 339:74–78.

Huey, R. B. (1978). Latitudinal pattern of between-altitude faunal similarity: Mountains might be "higher" in the tropics. American Naturalist 112:225–229.

Hung, C. M., S. V. Drovetski, and R. M. Zink (2012). Multilocus coalescence analyses support a mtDNA-based phylogeographic history for widespread Palearctic passerine bird, *Sitta europaea*. Evolution 66:2850–2864.

Hung, C. M., S. V. Drovetski, and R. M. Zink (2013). Recent allopatric divergence and niche evolution in a widespread Palearctic bird, the common rosefinch (*Carpodacus erythrinus*). Molecular Phylogenetics and Evolution 66:103–111.

Huntley, J. W., and G. Voelker (2016). Cryptic diversity in Afrotropical forests: The systematics and biogeography of the avian genus *Bleda*. Molecular Phylogenetics and Evolution 99:297–308.

James, H. F. (2004). The osteology and phylogeny of the Hawaiian finch radiation (Fringillidae: Drepanidini), including extinct taxa. Zoological Journal of the Linnean Society 141:207–255.

James, H. F., and S. L. Olson (1991). Descriptions of thirty-two new species of Hawaiian birds. Part II. Passeriformes. Ornithological Monographs 46:1–88.

Janzen, D. H. (1967). Why mountain passes are higher in the tropics. American Naturalist 101:233–249.

Jarvis, E. D., S. Mirarab, A. J. Aberer, B. Li, P. Houde, Li, C., . . . and A. Suh (2014). Whole-genome analyses resolve early branches in the tree of life of modern birds. Science 346:1320–1331.

Jetz, W., and C. Rahbek (2002). Geographic range size and determinants of avian species richness. Science 297:1548–1551.

Jønsson, K. A., and J. Fjeldså (2006). Determining biogeographic patterns of dispersal and diversification in oscine passerine birds in Australia, Southeast Asia and Africa. Journal of Biogeography 33:1155–1165.

Jønsson, K. A., P. H. Fabre, S. A. Fritz, R. S. Etienne, R. E. Ricklefs, T. B. Jørgensen, J. Fjeldså, C. Rahbek, P. G. Ericson, F. Woog, and E. Pasquet (2012). Ecological and evolutionary determinants for the adaptive radiation of the Madagascan vangas. Proceedings of the National Academy of Sciences 109:6620–6625.

Joseph, L., and K. E. Omland (2009). Phylogeography: Its development and impact in Australo-Papuan ornithology with special reference to paraphyly in Australian birds. Emu 109:1–23.

Kearns, A. M., L. Joseph, and L. G. Cook (2010). The impact of Pleistocene changes of climate and landscape on Australian birds: A test using the Pied Butcherbird (*Cracticus nigrogularis*). Emu 110:285–295.

Kearns, A. M., L. Joseph, A. Toon, and L. G. Cook (2014). Australia's arid-adapted butcherbirds experienced range expansions during Pleistocene glacial Maxima. Nature Communications 5:3994.

Klicka, J., and R. M. Zink (1997). The importance of recent ice ages in speciation: A failed paradigm. Science 277:1666–1669.

Lerner, H. R. L., M. Meyer, H. F. James, M. Hofreiter, and R. C. Fleischer (2011). Multilocus resolution of phylogeny and timescale in the extant adaptive radiation of Hawaiian honeycreepers. Current Biology 21:1838–1844.

MacArthur, R. H. (1958). Population ecology of some warblers of northeastern coniferous forests. Ecology 39:599–619.

Marks, B. D. (2010). Are lowland rainforests really evolutionary museums? Phylogeography of the green hylia (*Hylia prasina*) in the Afrotropics. Molecular Phylogenetics and Evolution 55:178–184.

Mayr, E. (1946). History of the North American bird fauna. Wilson Bulletin 58:1–68.

McKay, B. D., and R. M. Zink (2014). Sisyphean evolution in Darwin's finches. Biological Reviews 90:689–698.

Mitchell, K. J., B. Llamas, J. Soubrier, N. J. Rawlence, T. H. Worthy, J. Wood, M. S. Y. Lee, and A. Cooper (2014). Ancient DNA reveals elephant birds and kiwi are sister taxa and clarifies ratite bird evolution. Science 344:898–900.

Newton, I. (2003). Speciation and biogeography of birds. Academic Press.

Oatley, G., G. Voelker, T. M. Crowe, and R. C. K. Bowie (2012). A multi-locus phylogeny reveals a complex pattern of diversification related to climate and habitat heterogeneity in Southern African White-eyes. Molecular Phylogenetics and Evolution 64:633–644.

Oliveros, C. H., and R. G. Moyle (2010). Origin and diversification of Philippine bulbuls. Molecular Phylogenetics and Evolution 54:822–832.

Olson, S. L., and H. F. James (1991). Descriptions of 32 new species of Hawaiian birds. Part I. Non-Passeriformes. Ornithological Monographs 45:1–88.

Outlaw, D. C., and G. Voelker (2008). Pliocene climatic change in insular Southeast Asia as an engine of diversification in *Ficedula* flycatchers. Journal of Biogeography 35:739–752.

Owen, Richard (1879). Memoirs on the extinct wingless birds of New Zealand. Volume 2. John van Voorst, London, plate XCVII.

Pons, J.-M., G. Olioso, C. Cruaud, and J. Fuchs (2011). Phylogeography of the Eurasian green woodpecker (*Picus viridis*). Journal of Biogeography 38:311–325.

Projecto-Garcia, J., C. Natarajan, H. Moriyama, R. E. Weber, A. Fago, Z. A. Cheviron, and J. F. Storz (2013). Repeated elevational transitions in hemoglobin function during the evolution of Andean hummingbirds. Proceedings of the National Academy of Sciences USA 110:20669–20674.

Prum, R. O., J. S. Berv, A. Dornburg, D. J. Field, J. P. Townsend, E. M. Lemmon, and A. R. Lemmon (2015). A comprehensive phylogeny of birds (Aves) using targeted next-generation DNA sequencing. Nature 526:569–577.

Rahbek, C., and G. R. Graves (2001). Multiscale assessment of patterns of avian species richness. Proceedings of the National Academy of Sciences USA 98:4534–4539.

Raikow, R. J. (1986). Why are there so many kinds of passerine birds? Systematic Zoology 1986:255–259.

Ribas, C. C., A. Aleixo, A. C. R. Nogueira, C. Y. Miyaki, and J. Cracraft (2012). A palaeobiogeographic model for biotic diversification within Amazonia over the past three million years. Proceedings of the Royal Society B 279:681–689.

Robin, V. V., C. K. Vishnudas, P. Gupta, and U. Ramakrishnan (2015). Deep and wide valleys drive nested phylogeographic patterns across a montane bird community. Proceedings of the Royal Society of London B 282:20150861.

Sclater, P. L. (1858). On the general geographical distribution of the members of the class Aves. Journal of the Proceedings of the Linnean Society of London Zoology 2:130–136.

Seeholzer, G. F., B. M. Winger, M. G. Harvey, D. Caceres A., and J. D. Weckstein (2012). A new species of barbet (Capitonidae: *Capito*) from the Cerros del Sira, Ucayali, Peru. The Auk 129:551–559.

Smith, B. T., and J. Klicka (2010). The profound influence of the Late Pliocene Panamanian uplift on the exchange, diversification, and distribution of New World birds. Ecography 33:333–342.

Smith, B. T., A. Amei, and J. Klicka (2012a). Evaluating the role of contracting and expanding rainforest in initiating cycles of speciation across the Isthmus of Panama. Proceedings of the Royal Society of London B 279:3520–3526.

Smith, J. V., E. L. Braun, and R. T. Kimball (2012b). Ratite non-monophyly: Independent evidence from 40 novel loci. Systematic Biology 62:35–49.

Smith, J. V., E. L. Braun, and R. T. Kimball (2013). Ratite non-monophyly: Independent evidence from 40 novel loci. Systematic Biology 62:35–49.

Spellman, G. M., B. Riddle, and J. Klicka (2007). Phylogeography of the mountain chickadee (*Poecile gambeli*): Diversification, introgression, and expansion in response to Quaternary climate change. Molecular Ecology 16:1055–1068.

Steadman, D. (2006). Extinction and biogeography in tropical Pacific birds. University of Chicago Press, Chicago.

Trewick, S. A., and G. C. Gibb (2010). Vicars, tramps and assembly of the New Zealand avifauna: A review of molecular phylogenetic evidence. Ibis 152:226–253.

Voelker, G. (1999). Dispersal, vicariance and clocks: Historical biogeography and speciation in a cosmopolitan passerine genus (Anthus: Motacillidae). Evolution 53:1536–1552.

Voelker, G. (2002). Systematics and historical biogeography of wagtails (Aves: *Motacilla*): Dispersal versus vicariance revisited. Condor 104:725–739.

Voelker, G. (2010). Repeated vicariance of Eurasian songbird lineages since the late Miocene. Journal of Biogeography 37:1251–1261.

Voelker, G., and R. K. Outlaw (2008). Establishing a perimeter position: Thrush speciation around the Indian Ocean Basin. Journal of Evolutionary Biology 21:1779–1788.

Voelker, G., S. Rohwer, D. C. Outlaw, and R. C. K. Bowie (2009). Repeated trans-Atlantic dispersal catalyzed a global songbird radiation. Global Ecology Biogeography 18:41–49.

Voelker, G., R. K. Outlaw, and R. C. K. Bowie (2010). Pliocene forest dynamics as a primary driver of African bird speciation. Global Ecology and Biogeography 19:111–121.

Voelker, G., B. D. Marks, C. Kahindo, U. A'genonga, F. Bapeamoni, L. E. Duffie, J. W. Huntley, E. Mulotwa, S. A. Rosenbaum, and J. E. Light (2013). River barriers and cryptic biodiversity in an evolutionary museum. Ecology and Evolution 3:536–545. doi:10.1002/ece3.482.

Voelker, G., G. Semenov, I. V. Fadeev, A. Blick†, and S. V. Drovetski (2015). The biogeographic history of *Phoenicurus* redstarts reveals an allopatric mode of speciation and an out-of-Himalayas colonization pattern. Systematics and Biodiversity 13:296–305.

Voelker, G., J. W. Huntley, J. V. Penalba, and R. C. K. Bowie (2016). Resolving taxonomic uncertainty and historical biogeographic patterns in *Muscicapa* flycatchers and their allies. Molecular Phylogenetics and Evolution 94:618–625.

Wallace, A. R. (1863). On the physical geography of the Malay Archipelago. Proceedings of the Royal Geographical Society 7:205–212.

PART II

BUILDING THE BIRD

CHAPTER 5

From Fertilization to Independence

Matthias Starck

From fertilization to independence is a most fascinating period in the life of a bird, because it represents the time when all (genetic and nongenetic) information contained in the single-celled egg is translated into the development of a bird. In this chapter we discuss how evolution has created a diversity of developmental pathways among birds and which components of avian development are constant and unchangeable and which respond to the various selection forces. Avian development can be separated into roughly two phases: embryogenesis (from fertilization to hatching) and the postnatal period of growth (from hatching to adult). The structural and functional features of the egg determine, directly and indirectly, both phases. Because the egg contains all the energy, nutrients, and water for embryonic development, it takes a pivotal role not only in development but also in avian life histories. With this in mind, the first part of this chapter focuses on the avian egg and analyzes its structural and functional properties and how they relate to the evolution of avian diversity. The second part of this chapter presents a somewhat arbitrary selection of topics to introduce (1) key events in avian embryogenesis, (2) how embryogenesis creates different hatchlings (ranging from feathered, actively running, and independently feeding chicken hatchlings to naked, helpless, altricial songbird hatchlings), and (3) how this diversity may have evolved.

Finally, the postnatal growth period, from hatching to adult, will be analyzed for selection, internal constraint, and phylogenetic relatedness. Again, the selection of examples is to a certain degree arbitrary, but it might help understanding that evolution does not follow law-like rules, and that each species represents a unique evolutionary compromise among selection, internal constraint, and phylogenetic relatedness.

EGG

All birds lay eggs; there are no viviparous birds. The shape of the typical avian egg is symmetrical about the long axis, with one end being more pointed than the other (usually referred to as pointed and blunt ends of the egg; figs. 5.1, 5.2). The avian egg is a **cleidoic** egg (from Greek κλειστοσ = closed), in which a large amount yolk and albumen are enclosed by a protective layer of shell membranes and a calcified eggshell (fig. 5.1). Only water vapor, respiratory gases, and heat are exchanged with the environment. This cleidoic nature of the egg determines many aspects of avian embryonic and posthatching development, as well as many life history parameters of adult birds; therefore, it is important to understand the structural and quantitative design of the egg and how it may change in response to selection. Yolk, albumen, and eggshells are the main components of the egg, but they are deposits of female glands; i.e., they are not cellular. The actual cytoplasmatic compartment, containing the haploid nucleus, is a very small region floating on top of the yolk.

Egg Size and Quality

Size and quality depend on the female's condition. Therefore, egg metrics can vary widely within and among individuals of a species. Because this variation ultimately reflects variation in energy content of the eggs, it plays a critical role in avian life history. The adaptive value of intra-clutch egg size variation, the relationship between egg size and offspring quality, and the effects of egg size on hatching asynchrony are central themes of avian life history research.

Egg size (fresh weight) of extant birds ranges from 0.3 gram for the Vervain Hummingbird (*Mellisuga minima*) to about 1,500 grams for the Common Ostrich (*Struthio camelus*). Semi-fossil eggs of the extinct Giant Elephant Bird

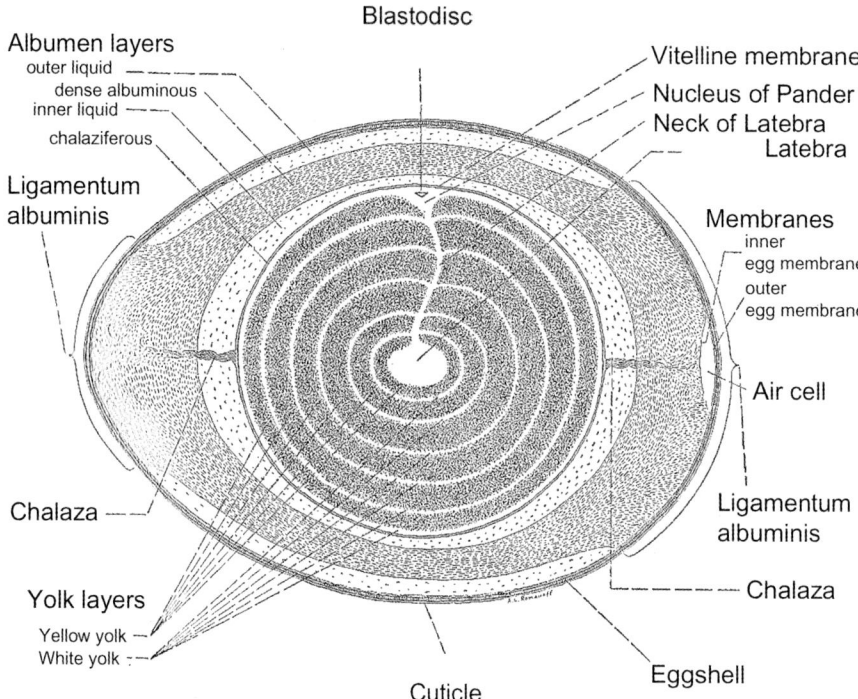

Figure 5.1. Schematic drawing of the avian egg, showing all relevant morphological features. *From Romanoff and Romanoff 1949.*

(*Aepyornis maximus*) have been estimated to have reached up to 12,000 grams (Schönwetter and Meise 1971). Besides weight, other relevant metrics of the avian egg are egg volume, surface area, length and width, and their relationship to each other. Because of the geometry of the avian egg, relevant parameters like egg volume, weight, and surface can be estimated from linear measures of the egg (important for work with museum specimens; Schönwetter 1961–1992, Hoyt 1976, 1979, Smart 1991, Troscianko 2014).

Egg Coloration and Patterning

Eggshell coloration is remarkably diverse among birds (fig. 5.2). This diversity stimulates questions about the evolutionary history and functional importance of different egg coloration. Most explanations are based on human vision capabilities and simple color scoring, but not on avian vision, which clearly differs from mammalian (human) in the spectral perception (see chapter 12). Kilner (2006) summarizes that the most plausible general explanation for egg pigmentation is that white eggs without pattern are the ancestral condition for birds (i.e., inherited from their non-avian sauropsid ancestors) and that egg colors and color patterning evolved in birds primarily for the cryptic appearance of the eggs. She showed that egg coloration and patterning occurred independently among clades of birds. Phil Cassey and coworkers (e.g., Cassey et al. 2010) attempted to provide a quantification of the variation in eggshell colors, relate it to the real visual percep-

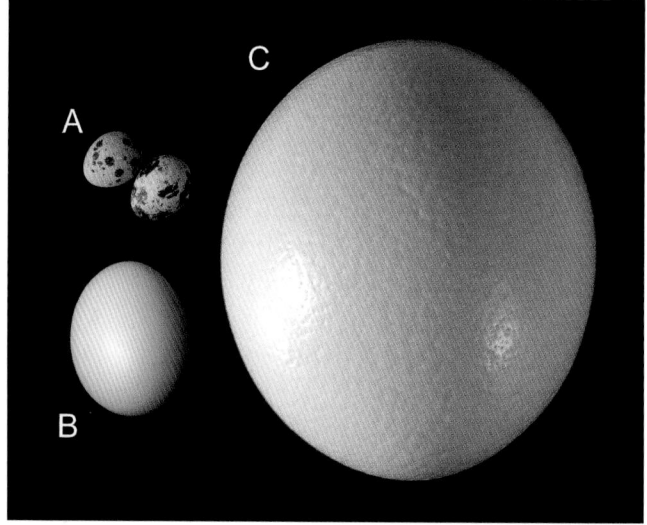

Figure 5.2. Illustration of color range and color pattern of bird eggs. *Image available from https://en.wikipedia.org/wiki/Egg_(food)#/media /File:Vogeleier.jpg - CC-BY-SA Rainer Zenz 2005.*

tion of birds, and place it in an explicit phylogenetic context. In addition to the above-mentioned selection factors, they recognize UV protection of the developing embryo and brood parasitism as important factors that affect egg coloration and patterning. In particular, brood parasitism appears to result in an arms race between hosts and parasites, a topic that has recently moved into the center of interest of evolutionary ornithologists.

Eggshell

The avian eggshell provides protection and stability but also constrains gas exchange for the developing embryo. How is the eggshell constructed to allow for these contrasting functional demands? The eggshell is a composite of an organic matrix and inorganic material, mainly calcium carbonate (i.e., biomineralized). The eggshell is three-layered, with (1) an **outer cuticle** (ca. 10 μm, organic layer, proteoglycans), (2) a **biomineralized layer** (ca. 300 μm), and (3) the **outer** (ca. 50 μm) and **inner** (ca. 20 μm) **shell membranes** (collagen fibers; figs. 5.1, 5.3). The outer cuticle is a thin organic layer that may carry pigments of egg coloration and has antimicrobial properties. The biomineralized layer is highly structured (fig. 5.3). Its basic unit is the **eisospherite**, an inorganic crystallization center that is inserted in the outer shell membrane. Calciferous cones grow radially away from the eisospherite, forming the mammillary tips. Outside the **mammillary layer**, the calcium crystals form a columnar layer (**palisade layer**), which constitutes the thickest layer of the biomineralized eggshell. Outside the column layer is an external zone of irregular arrangement with no crystallized structure (**vertical matrix layer**). Pore canals penetrate the biomineralized eggshell between the palisades (figs. 5.3, 5.4). The pore canals are functionally important, because they provide the conduits for gas exchange for the developing embryo. Simple pores may be straight tubes or funnel-shaped with the large opening to the outside, but more complex branching patterns have been described too. The structure of eggshell pores is generally associated with egg size—larger eggs generally have more complex pore structures.

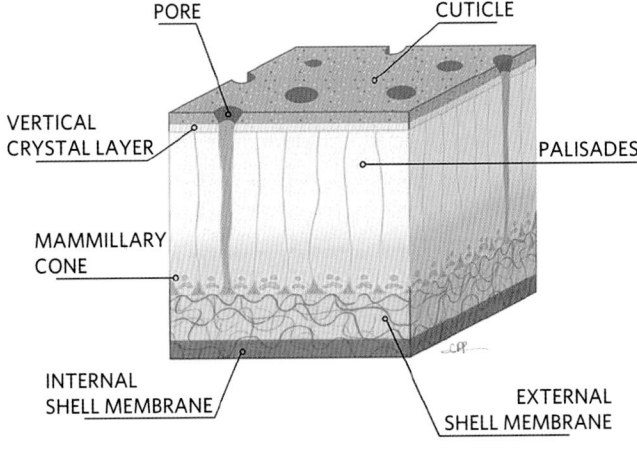

Figure 5.3. Schematic rendition of the fine structure of the chicken eggshell. The major morphological features are discussed in the text. *From Hincke et al. 2012.*

Eggshell Porosity, Water Vapor Conductance, and Incubation Time

The eggshell protects the developing bird, but it also presents a diffusion barrier for respiratory gases. Therefore, eggshell pores are vital structures for gas exchange between the developing embryo and the environment. Eggshell porosity differs between regions of the egg. For example, the number of pores per mm² and pore size of chicken eggs is highest at the equator of the egg, followed by the blunt end, and lowest at the pointed end. The differences are highly significant (Riley et al. 2014) and can most certainly be related to regions of respiratory function of the developing embryo (fig. 5.4). Together, eggshell thickness and eggshell porosity are contrasting determinants of an egg's capacity for respiratory gas exchange and egg water loss. With increasing eggshell thickness, an egg's capacity for gas exchange (diffusion capacity) decreases; but, gas exchange increases with increasing number of pores per area. A series of studies by Rahn, Paganelli, and Ar (and others) between 1970 and 1990 established a general relationship between egg size, water loss, respiratory capacity, and incubation time. On a gross scale, they showed that gas exchange through the eggshell is determined by eggshell porosity (i.e., the number of pores per eggshell area), eggshell thickness, and partial pressure difference for the gases. They suggested that the eggshell capacity for respiratory gas exchange is matched to the maximum oxygen demand of the embryo, but that ultimately the respiratory requirements of the growing embryo limit the length of the incubation period, and consequently, the embryo has to hatch when its demand for oxygen exceeds the eggshell capacity for gas exchange.

Another fundamental functional trade-off to respiratory gas exchange through the eggshell is water vapor conductance and invasion of pathogens. Because of the large pores, the biomineralized layer does not provide any reasonable barrier against water vapor loss, or against bacterial or viral invasion of the egg. However, while the biomineralized eggshell is pervasive for microorganisms, the outer cuticle and the eggshell membranes seem to provide not only some "mechanical" protection against bacterial invasion of the egg but also some biochemical defense mechanism (Berrang et al. 1999). Recent studies reported also antimicrobial proteins in the eggshell cuticle (Rose-Martel et al. 2012). The outer and inner eggshell membranes are a network of fibers that act as filters for invading pathogens, blocking their entry into the egg white. Integrated into the membrane fibers are a number of active antimicrobial proteins, which augment protection against invading microbes (Wellmann-Labadie et al. 2008).

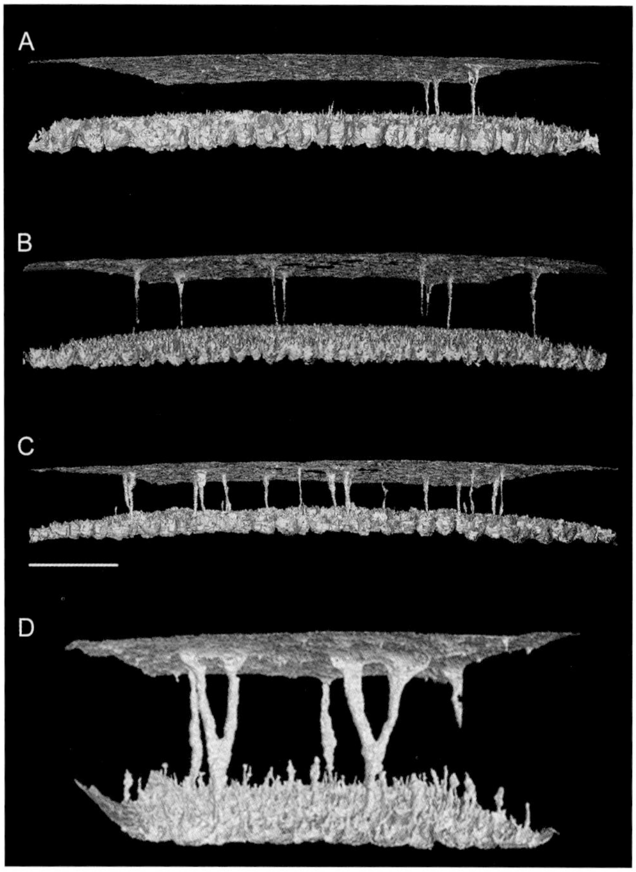

Figure 5.4. Three-dimensional X-ray µCT visualization of shell pore canals of the avian eggshell. The pore canals are shown as solid structures while the solid eggshell is not shown (inverse image). The three-dimensional images show the quantity, shape, and arrangement of pores in the *A,* sharp; *B,* equator; *C,* blunt regions of chicken eggshell. Note that pores vary in the size, shape, and completeness with which they traverse the whole shell. Scale bar = 400 µm. *D, Rhea Americana,* three-dimensional X-ray µCT visualization of the shell of equatorial region of an egg. Note the branching pore canals. *A–C* from Riley et al. 2014, https://www.ncbi.nlm.nih.gov/pubmed/24875292, PMC open access; *all images courtesy of M. R. Luck, J. Snow, A. Riley, C. Sturrock, and S. Mooney, University of Nottingham.*

Eggshell as Calcium Depot and Embryo-Induced Eggshell Thinning

The mammillary layer of the biomineralized eggshell is an important source of calcium for the developing embryo. Once the embryo has developed the extraembryonic membranes and the chorioallantois (CA) is formed (see below), a specialized type of cell in the CA membrane (villus-cavity cells) supposedly secretes hydrogen ions, which are necessary, along with carbonic anhydrase, for mobilizing Ca^{++} from the insoluble calcium carbonate of the mammillary tips of the eggshell (Coleman and Terepka 1972, Gabrielli 2004). The mobilization of calcium from the eggshell results in an erosion of the mammillary layer (embryo-induced eggshell thinning)

and, consequently, embryo-induced morphological changes of the eggshell structure. The amount of erosion differs topographically between the blunt pole, equator, and sharp pole of an egg. On average, the thickness of the eggshell is reduced by 6.4 percent. Given the fact that the biomineralized eggshell serves as a calcium depot for the developing bird, it is not surprising to find the degree of embryo-induced eggshell thinning related to the degree of ossification of the skeleton at hatching (Bond et al. 1988, Blom and Lilja 2004, Österström and Lilja 2012); however, the postulated relationship has wide prediction limits and should be considered with caution (Orlowski and Halupka 2015).

Variation in eggshell thickness is caused by additive genetic variance, seasonal variation, nutrition, physiological condition of the female (including aging), environmental factors, and the embryo. Consequently, there is considerable variation of eggshell thickness, not only between species but also within species and even individuals. Environmental pollutants (e.g., halogenated hydrocarbons, such as polychlorinated biphenyls (PCBs), DDE and DDT, and dioxins and dibenzofurans) have been recognized as increasingly important external factors contributing to eggshell thinning and, as a consequence of reduced eggshell thickness, to potentially increased mortality, and eggshell thickness has been used as a bio-indicator for environmental pollution.

Yolk

Yolk is the energy and nutrient resource for the developing embryo. Consequently, it takes a central role in the reproductive biology of birds. Yolk size, quality, and composition are affected by genetic factors, the physiological condition of the female bird, and environmental factors (through effects on the female). When an egg is laid, the yolk is a ball centered in the middle of the egg. It is surrounded by a thin **vitelline membrane** (fig. 5.1). The yolk contents of eggs range between 15 and 70 percent of the fresh egg weight, depending on egg size, length of incubation period, and mode of posthatching development (Ricklefs 1977, Sotherland and Rahn 1987). In species with long incubation periods such as the Kiwi (*Apteryx australis*), the yolk content of the egg ranges between 60 and 70 percent.

About 50 percent of the yolk is water, 16 percent is protein, 32 percent is fat, and about 2 percent is ash (Romanoff and Romanoff 1949). The yolk also contains vitamins (A, B-complex, C, E), minerals (calcium, magnesium, potassium, sodium, phosphorus), hormones, antibodies, and carotenoids, providing essential components for the developing embryo. Recent proteomic analysis reported 119 different proteins from (unincubated) chicken egg yolk (Mann and Mann 2008). Through variation of any of these components, the viability of the developing embryo and the hatchling are affected.

The yolk ball is not homogeneous. Instead it has a unique structure that provides vital functions for the developing embryo. The center of the yolk ball contains a liquid, whitish yolk called the **latebra**. The latebra extends with a long "neck" from the center of the yolk to its surface, where it forms the **nucleus of Pander** (figs. 5.1, 5.5) right under the blastodisc (the area where the embryo develops). The latebra, its neck, and the nucleus of Pander are of lower density than the remaining yellow yolk, so that the blastodisc (the cytoplasmatic region of the egg or, later, the early developing embryo) always returns to an upper position in the egg.

Many studies exist investigating the effects of the environment and physiology of the female bird on yolk/egg composition. Indeed, the energy content of an egg may affect hatchling size, rate of postnatal growth, nestling hierarchy, or survival probability (mortality). The amount of maternal testosterone deposited in the yolk may affect dominance behavior of growing and even adult birds, and environmental factors such as temperature and humidity, but also pollution, affect egg quality and development of embryos and, thus, life history of birds.

Albumen

Albumen (i.e., egg white) consists of 88 percent water, 10 percent protein, >0.1 percent lipids, and ca. 1 percent carbohydrates. Albumen is an important source of water and protein for the developing embryo. Proteomic analyses have revealed 87 different proteins in the albumen (Mann 2007). Primary functions of egg white proteins are regulation of cell development and proliferation, cell-to-cell interaction, development of the hematological system, and cell migration (D'Alessandro et al. 2010). Lysozyme-C plays a major role in antimicrobial functions of the albumen (Deeming 2002).

The albumen surrounds the yolk in the egg and forms the **chalazae,** i.e., proteinaceous cords that extend from the vitelline membrane into the albumen, where the fibers interlace with similar fibers in the albumen (figs. 5.1, 5.5).

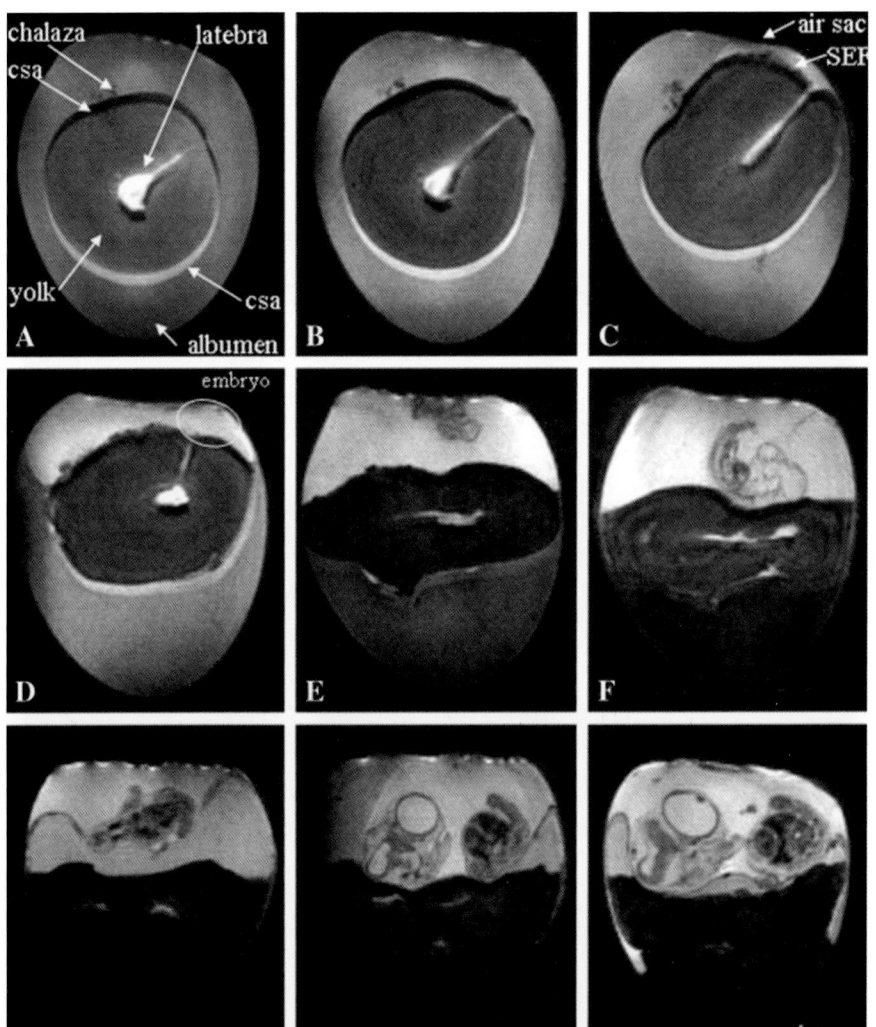

Figure 5.5. Magnetic resonance image of the quail egg during successive stages of development. The images are acquired in a longitudinal mid-plane of the egg. The pointed pole of the egg is to the bottom of the image (note that the orientation is not natural): *A,* Day 0; *B,* Day 1; *C,* Day 2; *D,* Day 3; *E,* Day 4; *F,* Day 5; *G,* Day 6; *H,* Day 7; *I,* Day 8. csa = chemical shift artifact. *From Duce et al. 2011. Retrieved from https://www .ncbi.nlm.nih.gov/pmc/articles/PMC3006493 /figure/f0005/. Open access under CC BY 3.0 license.*

The chalazae suspend the yolk in the center of the egg, and when the egg is turned, the yolk rotates until the blastodisc is uppermost (see above).

The Egg during Incubation
Yolk and Albumen

The described distribution and shape of the yolk and the albumen refer to the unincubated egg (i.e., in a nonfunctional condition). When incubation begins, they undergo considerable changes, turning yolk and albumen into functional components for the developing embryo. The most obvious change is water transfer from the albumen to the yolk, resulting in a rapid increase of the yolk volume, followed by a separation into two compartments, a liquefied yolk and a depot-yolk. The compartments of the egg stratify, with the liquefied yolk as the top level, storage yolk in the middle, and albumen in the lower level (fig. 5.5). These processes are continuous, and the amount of liquefied yolk increases during development until, successively, the entire yolk is liquefied. During this process, the developing embryo becomes deeper and deeper embedded in the yolk, like on a pillow. The proteins contained in the yolk (and albumen) also undergo quantitative changes during development. Recent studies show (not surprisingly) that the relative abundance of proteins change during incubation, some increasing in their relative abundance, others decreasing (Réhault-Godbert et al. 2014). Certainly, these structural and proteomic changes indicate functional changes and activation of the egg contents, but details of how, for example, nutrients become available to the developing embryo and how regulatory proteins affect development need to be clarified.

Turning the Egg

Avian eggs need to be turned during the first week of incubation. Extensive studies on commercial breeding of chickens show that hatchability of eggs is highest when the eggs are turned at least five times a day. However, eggs vary in their rates of egg turning; i.e., eggs of precocial species are turned about once an hour, whereas eggs of altricial birds are turned five to 10 times an hour. The positive effect of turning disappears during the second week of incubation in chickens, and turning shortly before hatching might even be detrimental for hatchability. Egg turning somehow affects the growth of the extraembryonic membranes. The reason for high mortality of unturned eggs is that the extraembryonic membranes adhere to the inner shell membrane, which obstructs early embryogenesis (Deeming 1991). The latebra, nucleus of Pander, and chalazae provide the specific avian features (see the box on page 102) that allow rotations of the yolk so that the developing embryo always moves to the

EVOLUTIONARY HISTORY OF THE AVIAN EGG
The Ancestral Sauropsids Egg

Non-avian sauropsids also lay eggs. However, their morphological features and physiological functions differ from those found in birds. The avian egg represents a mosaic of plesiomorphic and derived (autapomorphic avian) traits. Therefore, it is important to look at the evolutionary history of the avian egg as derived from amniote eggs to understand which features of the egg are typical for birds (numbers in the text refer to the characters in the cladogram).

The ancestral amniote egg contained a large albumen/jelly layer (1). Phylogenetically, it occurred already in tetrapods (or before) and therefore must be considered a plesiomorphic feature for amniotes and certainly for sauropsids eggs. A large yolk ball (megalecithal egg (2) is also a feature that can be found already among amniotes, if not tetrapods; (3) probably in all amniotes, the calcium derives from the yolk.

The eggs of sauropsids possess a more or less rigid, composite eggshell, consisting of an outer calcareous (biomineralized) layer and an inner organic layer. Among extant sauropsids, the calcareous layer is thick in turtles and archosaurs (crocodiles, dinosaurs,† and birds), providing a stable shape of the egg (4). It is built on mammillae as basic units. The inner, organic layer of the eggshell is typically formed by a single layer of collagenous fibers (5).

Within sauropsids, turtle eggs represent this plesiomorphic condition, but squamate and archosaur eggs show autapomorphic features characterizing each clade. In squamates (i.e., tuatara, lizards, and snakes), the biomineralized eggshell is only a thin calcium crust on the organic shell membrane, rendering it soft and flexible (6). The soft and flexible eggshell is an autapomorphic feature of this clade and must be related to the uptake of water from the environment (7); i.e., a few days after deposition, squamate eggs become fully turgescent. In geckos, a clade of the squamates, a solid, calcareous eggshell has independently re-evolved. Strong indication for re-evolution of a solid eggshell comes from phylogenetic analyses and the observation that the structure of the biomineralized eggshell of geckos is different from the ancestral condition. In squamates, the originally thick jelly coat (character 1 in the cladogram) is reduced in thickness (8).

The evolution of the eggs in the clade of archosaurs has resulted in features that clearly differ from the plesiomorphic condition and from squamates. The eggshell membrane is now two-layered (inner and outer eggshell membrane, **9a**) allowing the formation of an air space at the blunt pole of the egg **(9b)**. Both features are clearly related and therefore are coded as one character **(9a,b)**. In archosaurs, the biomineralized eggshell is the primary source of calcium for the developing embryo (derived condition **10**), while in all other sauropsids the calcium derives from the yolk (plesiomorphic condition **3**). In archosaurs, the eggs are also now truly cleidoic, i.e., they do not absorb water or other materials (except oxygen) from the environment **(11)**.

Finally, avian eggs are characterized by distinct features such as an external organic cuticle on the eggshell **(12)** and a fundamental reorganization of the albumen/jelly layer **(13)** and the yolk **(14a, b)**. The albumen forms collagenous strings (the chalazae), fixing the yolk in the center of the egg **(13)**. Chalazae are an avian autapomorphic feature. Also, the yolk shows autapomorphic features like the latebra **(14a)** and the nucleus of Pander **(14b)**. Functionally, they integrate with the chalazae and allow free rotation of the yolk so that the developing embryo always moves to the upper side of the egg where brooding is most effective. Egg turning **(15)** occurs only in birds. Chalazae **(13)**, latebra **(14a)**, nucleus of Pander **(14b)**, and egg turning **(15)** are correlated characters, i.e., only together do they provide a certain physiological function and therefore must have evolved together. In summary, only characters 12–15 are truly avian; all other features that also occur in the avian egg have evolved earlier in amniote outgroups to birds. The avian egg represents a mosaic of plesiomorphic and autapomorphic characters.

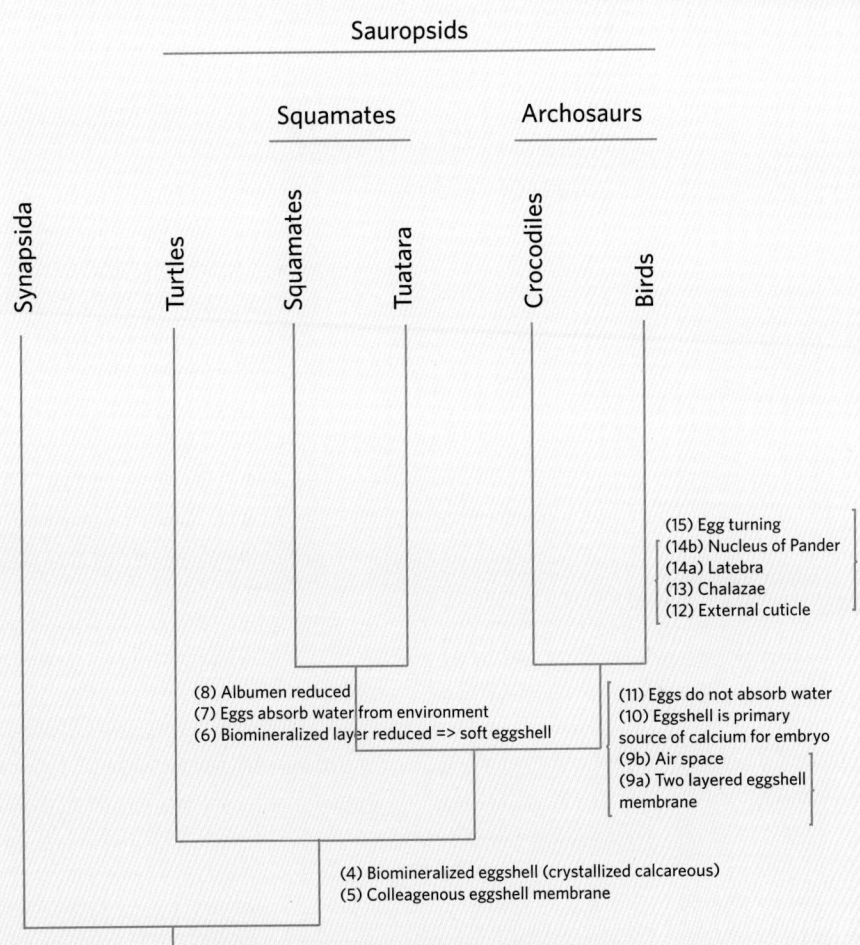

Phylogeny of amniotes indicating the evolutionary transformation of an ancestral amniote egg into derived eggs of sauropsids, including birds. Note that squamate and archosaur evolution takes different pathways of forming the egg (see text for details of character explanation).

region of the oxygen supply (highest eggshell porosity, largest pores, thinnest eggshell).

EMBRYONIC DEVELOPMENT AND INCUBATION PERIOD

Avian embryonic development covers the period from fertilization of the egg through hatching of the young bird. The length of the embryonic period of birds ranges between 10 days for some small songbirds and cuckoos and up to 90 days for some albatrosses (Diomedeidae). A functional framework for embryonic development, including length of the incubation period, is set by morphological features of the egg; these are a very large yolk ball, containing all energy and nutrients for the developing embryo, and the biomineralized eggshells limiting the exchange with the environment to respiratory gases (oxygen, carbon dioxide, water vapor) and temperature. It follows that the length of the incubation period is set by a trade-off between the embryonic demand for oxygen and the capacity of the eggshell for respiratory gases exchange. Of course selective factors, the evolutionary history, and phylogenetic relationships among bird species may also determine the length of the incubation period.

Indeed, Deeming et al. (2006) presented a phylogenetically informed comparative analysis of the relationship between initial egg mass and incubation time on 1,525 species, showing that incubation time is strongly dependent on phylogenetic relationship (i.e., species within a clade tend to have more similar incubation periods). The slope and intercept of a relationship between initial egg mass and incubation period are primarily dependent on phylogenetic relationship and only to a lesser degree dependent on egg mass or developmental mode.

Besides morphology, physiology, and phylogenetic relatedness, life history parameters are important factors affecting the length of incubation period (Lack 1968). In the theoretical framework of life history evolution, natural selection should favor short incubation periods. Nest predation, energetic costs of embryonic development, evaporative water loss of the egg, and parental cost to incubation all select for short incubation periods. Observed long incubation periods present a certain paradox, because they increase the cumulative risk of time-dependent mortality to young without providing a clear benefit. Long incubation time also increases the energetic costs of embryogenesis, because the maintenance metabolism of the embryo needs to be sustained for a longer period and results in higher evaporative water loss (Ricklefs 1993). Indeed, few studies have found evidence for a fitness advantage to long incubation time. Ricklefs (1993) showed a correlative relationship between prolonged incubation period

and delayed senescence of adult birds. He suggested therefore that longer incubation period may allow building of organs and tissues in a qualitatively better way (e.g., improved immune system), so that they ensure a longer lifespan. Indeed, Martin (2002) showed that parents of (some) species with low adult mortality accept increased risk of mortality to their young from longer incubation if this allows reduced risk of mortality to themselves. In addition, Martin et al. (2007) analyzed the relationship between nest predation, nest attentiveness, egg temperature, and incubation period across a broad diversity of species and geographic areas. Supported by nest-switching experiments, they showed that, within a phylogenetically defined group, selection would favor long incubation periods if the fitness benefits of prolonged incubation exceed the disadvantages. Figure 5.6 summarizes in a nonquantitative manner the factors that affect the length of the incubation period.

Conflict and dissent in the explanations of incubation period in the context of life history evolution raises interesting questions stimulating further analysis and critical thinking. The work mentioned above is important because it shows that even "simple" parameters like length of incubation period evolve in a complex hierarchical system of constraints and selection, and it appears to be differently modified in different species of birds. As we analyze more data, it becomes more evident that biology does not follow general rules (as implicitly assumed in many comparative and correlative studies) but rather that the life history of each species (even closely related species) is the result of its unique evolutionary history (del Rio 2008). Even if all factors summarized in fig. 5.6 apply to all birds, they may differ quantitatively, so that one factor may be stronger in one species than in another. Consequently, the right and wrong in life history explanations becomes confounded by quantitative divergence.

Embryonic Development

Research examining avian embryonic development requires detailed "timing" so that sequential events during development can be compared or experimentally manipulated. Because physical time and developmental time are not equal, researchers have been seeking normal stages of development (i.e., stages of development that are defined by embryonic events/structures and can clearly be distinguished from one another). Using normal stages of development, one can compare development independent of time and detect patterns of heterochrony in development (i.e., differences in absolute or relative time in the development of organ systems/tissues/cells when comparing two or more species). Normal stages of development of the domestic chicken, defined on the appearance of morphological characters during development, were first presented by Hamburger and Hamilton (1951). Their

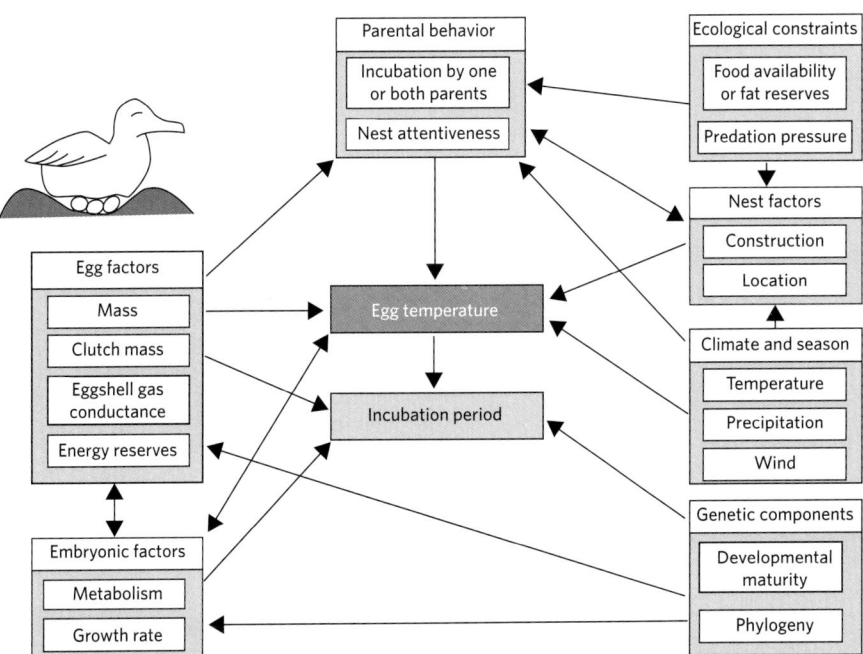

Figure 5.6. Factors affecting egg temperature and incubation period in the bird-nest incubation unit (i.e., all potential factors are visualized with the same weight, but in fact their impacts on incubation periods differ). Phylogenetic relatedness is a strong factor affecting incubation time. In contrast to earlier studies, initial egg mass becomes insignificant when phylogenetic relatedness is considered a factor in the analyses. *From Deeming et al. 2006.*

system has become a standard for avian developmental biology. This is particularly useful because little heterochrony is observed in avian embryonic development, so that Hamburger-Hamilton stages (HH stages) can be applied across species (Starck 1989, 1993). The HH stages have been complemented and detailed for the very early stages by Eyal-Giladi and Kochav (1975) for the chicken. The following summary (see the box on page 108) follows the classic normal HH stages of development for the chicken. It is important to note that embryonic stages are defined by distinct morphologies but not by physical time. Therefore, embryonic stages differ in duration. Generally, early embryonic stages, during formation of the basic body organization and period of rapid cell divisions, are short compared with later embryonic stages.

Embryonic Stages for Developmental Biology

After fertilization, the zygote undergoes a series of mitotic divisions (i.e., cleavage divisions). Because of the very large yolk, cleavage extends only through the cytoplasmatic region (i.e., the **blastodisc**) of the egg, resulting in a small disc of cells (**discoidal cleavage**). This cell layer is the **blastoderm** (fig. 5.7A), which is four to six cell layers thick; the covering layer is the **epiblast** and the deeper layers are the **primary hypoblast** (syn. endophyll cells, polyinvaginated cells). The cells of the primary hypoblast later give rise to the primordial germ cells and, probably, components of the yolk stalk. During the cleavage stages, a small **cleavage cavity** forms between the blastoderm and the yolk so that the blastoderm is in contact with the yolk only at its edges. The central part of the blastoderm, where it covers the cleavage cavity, is lighter

(**area pellucida**), while the part residing on the yolk ball is darker (**area opaca**). The distinction is important, because the actual embryo will develop from the area pellucida, while the area opaca forms the extraembryonic material from which the cellular yolk sac, amnion, and chorion develop. The development to area pellucida and area opaca usually occurs in the oviduct of the bird, so a two layered embryo is the stage when the egg is laid (HH stage 2, 3; fig. 5.7B). At the same time, the posterior border of the area pellucida (**Koller's sickle**) is an important region of primary cell migration. From here, cells migrate under the upper layer of cells, forming the **endoblast**. The endoblast cells form a layer of cells covering the underside of the blastodisc and, later, start growing over the yolk sac. The endoblast has important inductive and regulatory functions in the early bird embryo, in particular with respect to endoderm formation (positioning of the primitive streak), head formation, and forming vital vascular connections. The cells of the endoblast give rise to extraembryonic material (i.e., the yolk sac endoderm), but there is also evidence that they contribute to the formation of the embryonic gut (Stern and Downs 2012).

The endoblast induces the formation of the **primitive streak** by condensation of cells in the upper layer of cells, the epiblast. The primitive streak forms in the midline of the blastodisc, beginning at the posterior and slowly extending anteriorly. It determines lateralization of the embryo and, from here, cells migrate below the epiblast, forming the endoderm (**gastrulation**) while the epiblast now becomes the **ectoderm**.

In the fully formed primitive streak stage (HH4), cells emigrate from the primitive streak and form a cell layer

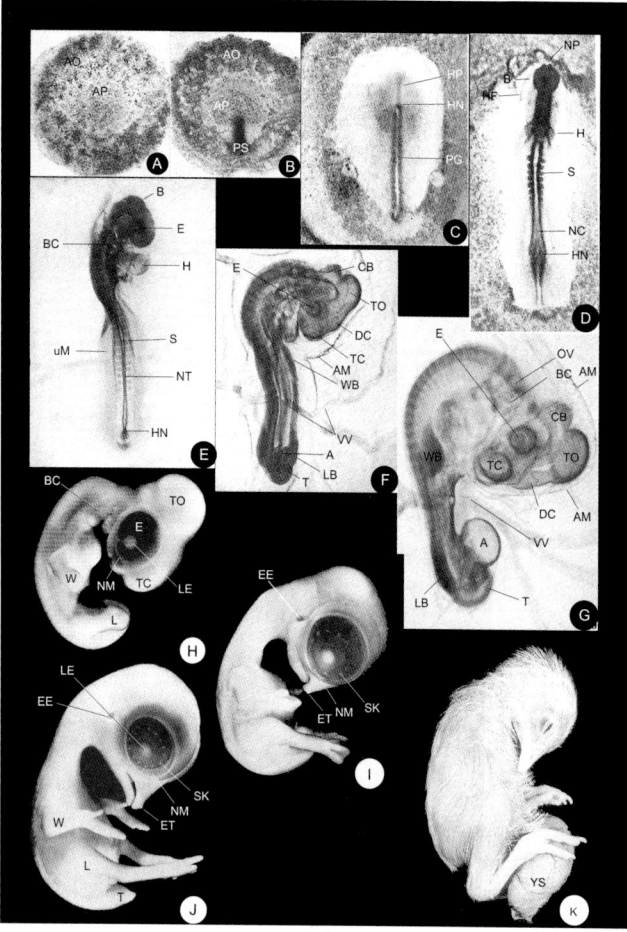

Figure 5.7. Selected stages of chicken normal development (stages according to Hamburger and Hamilton): *A*, HH1—blastodisc with blastoderm; *B*, HH3—primitive streak; *C*, HH5—head process and Hensen's node; *D*, HH9—seven pairs of somites, primary optic vesicles present, paired primordia of the heart begin to fuse; *E*, HH15—lateral body folds extend to the level of somites 15-17, 24-27 somites, limb primordia faint condensations of mesoderm, amnion extends to the level of somites 11-13; *F*, HH18—limb buds enlarged, somites 30-36, amnion closed, allantois short, thick-walled pocket; *G*, HH20—wing buds symmetrical, leg buds slightly asymmetrical, somites 40-43 pairs, visceral arches clearly recognizable, allantois vesicle of variable size; *H*, HH28—limbs second and third toe longer than others, three digits and four toes distinct, beak recognizable; *I*, HH33—three distinct rows of feather germs visible on tail, 13 scleral papillae; *J*, HH35—fingers and toes free, beak lengthened, egg tooth recognizable, nictitating membrane reaches the outer scleral papillae, feather germs distinct and elongate over the body; *K*, HH45—chick almost ready to hatch, yolk sac half enclosed in the body. Abbreviations: A, allantois; AM, amnion; AO, area opaca; AP, area pellucida; B, rudiment of brain; BC, branchial cleft; CB, cerebellum; DC, rudiment of the diencephalon; E, eye; EE, external ear opening; ET, egg tooth; G, primitive grove; H, rudiment of the heart (paired); HF, head fold; HN, Hensen's node; HP, head process; L, leg; LE, lens; LB, leg bud; NC, notochord; NM, nictitating membrane; NP, neuropore; NT, neural tube; OV, otic vesicle; PS, primitive streak; S, somite; SK, scleral papilla; T, tail; TC, rudiment of the telencephalon (forebrain); TO, rudiment of tectum opticum (visual center of the brain); uM, unsegmented mesoderm; VV, vitelline vein (yolk sac vein); W, wing; WB, wing bud; YS, yolk sac. *From Hamburger and Hamilton 1951; images not to scale.*

between ectoderm and endoderm, i.e., the mesoderm. This stage, again, is of central importance for the development of the embryo, because it forms the **neural plate**, **neural crest**, **notochord**, **somites**, **lateral plate**, and **intermediate mesoderm** as fundamental embryonic compartments that give rise to all later organs and tissues. The primitive streak is an important transitional embryonic structure. The midline of the primitive streak deepens, forming the **primitive groove** from which cells emigrate to form mesoderm. The anterior end of the primitive streak differentiates into **Hensen's node** (syn. primitive node, primitive knot, chordoneural hinge) from which the notochord forms. The anterior end of the notochord is called the head process. Hensen's node moves from its anterior position caudalward, thereby giving rise to the trunk notochord. While Hensen's node regresses from anterior to posterior, the neuronal folds emerge anterior to the node, forming the neuronal tube (HH5, HH6; fig. 5.7C). With just a little time lag, the mesoderm condensates left and right of the notochord, forming distinct packages of cells, i.e., the somites (HH7–HH14, defined by the number of pairs of somites formed; fig. 5.7D).

In parallel to these developments, the rudiment of the central nervous system is regionalized and the first rudiments of the eyes, ears, gut, liver, and heart appear. In stage HH13, a fold of the extraembryonic ectoderm covers the head, thus indicating the beginning development of the amnion and chorion (below).

The following stages, HH14–HH18 (e.g., fig. 5.7E) are characterized primarily by a continued development of somites (in chickens there are 36), the first appearance of wing and leg buds, and the continued regionalization and differentiation of the central nervous system, visceral organs, sensory organs, and facial development. During these early developmental stages, the anterior end of the embryo turns so that the head comes to lie on the left side (fig. 5.7E, F). The amniotic folds now enclose the embryo into two membranes: the inner membrane is the amnion (the cavity formed by this membrane is the amniotic cavity) and the outer membrane is the chorion, with the cavity being the extraembryonic coelom. In HH18, the **allantois** becomes visible as a small, thick-walled bud growing from the rudiment of the embryonic hindgut (fig. 5.7G). Later during development, the allantois, together with the chorion, forms the extraembryonic respiratory organ.

During stages HH19–HH26 (fig. 5.7G), the limb buds grow and differentiate so that simple wings and legs can be recognized but no digits or toes are formed, yet. The development of the visceral organs, in particular the morphological changes of the transitory visceral arches (from which jaws, pharyngeal tissues, and larynx form), are used to define developmental stages. In stage HH26, the production of erythrocytes begins (in the extraembryonic yolk sac membrane), first movements of the trunk can be observed, the heart has four chambers, and the first nerves from the eye reach the optic lobe in the developing midbrain.

During stages HH27–HH35 (fig. 5.7H, I, J) the wings and legs differentiate, forming digits and toes. First ossifications occur in the long bones of wing (humerus, radius, and ulna) and leg (femur, tibia, fibula, and tarsus). The integument forms the first feather papillae on the hips. Scleral papillae are visible on the eye, but the eyelids and nictitating membrane do not yet overgrow fully the eyeball.

Stages HH36–HH39 are mainly based on differentiations of the integument. The eyelids expand over the eyeball. Depending on developmental mode, the development of the feathers and other cornified structures of the integument differentiate (see below). The ossification of the skeleton progresses so that all elements of the postcranial and cranial skeleton show ossification centers. Some interspecific variation and minor heterochrony in the sequence of ossification can be observed in the comparison of species following different developmental modes.

Stages HH40–HH44 (fig. 5.7K) are mainly based on quantitative development of the length of the beak and the third toe. Those stages are species- if not breed-specific and cannot simply be transferred to other species. Generally, the physiology of the embryo is slowly transferred to functioning outside the egg; during these stages the egg is pipped (first internally, later also externally), the chorioallantois ceases functioning and the chorioallantoic circulation is reduced, and the lungs become functional as respiratory organs. The yolk sac residues are drawn into the body cavity and the embryo finally hatches.

Development and Function of the Extraembryonic Membranes

The amnion, chorion, allantois, and the yolk sac membrane are the four extraembryonic membranes that characterize amniotes. They enclose the developing embryo in the egg and provide vital transitory functions for the developing embryo (fig. 5.8). Although the amnion, chorion, and allantois are amniote characters, their morphological and functional integration into avian development has taken its own evolutionary pathway (compared with other sauropsids or mammals). The extraembryonic membranes establish fluid-filled compartments around the embryo. The primary functions of the extraembryonic membranes are protection, gas exchange, nutrient absorption, storage of excreta, and erythropoiesis (i.e., production of blood cells).

Amnion

The amnion develops during the early stages of embryonic development, from the extraembryonic ectoderm and mesoderm. The amniotic folds rise in front of the embryo and to its sides (fig. 5.7E); within a few days they extend over the entire embryo and ultimately enclose it with two membranes, the inner being the amnion the outer being the chorion. Thus, the embryo resides in the amniotic cavity, a fluid-filled

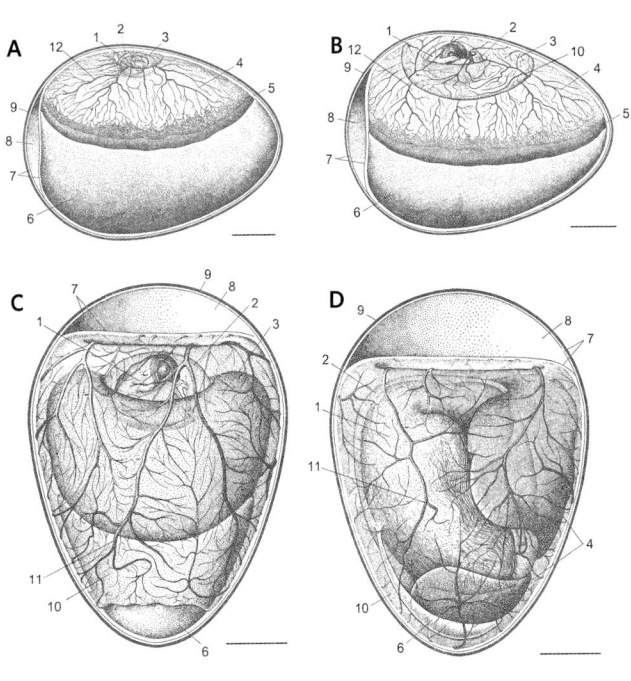

Figure 5.8. Extraembryonic membranes of the developing chicken. *A*, After 4 days of incubation, the yolk sac membrane has grown halfway over the yolk and develops a dense vascular network; the chorioallantois is just a very small rudiment; *B*, After 6 days of incubation, the chorioallantois has increased in size and major chorioallantoic vessels are recognizable; *C*, After 10 days of incubation, the chorioallantois has reached maximum size, extending over the entire inner surface of the eggshell. The chorioallantois functions now as the respiratory exchange organ of the embryo. The yolk sac membrane fully encloses the yolk sac; *D*, After 18 days of incubation, no further morphological changes of the chorioallantois. The embryo is a few days before hatching. Key: 1, embryo; 2, amnion; 3, allantois; 4, yolk sac membrane; 5, extraembryonic endoderm; 6, albumen; 7, eggshell membrane; 8, air space; 9, biomineralized eggshell; 10, chorioallantois membrane; 11, blood vessel in the chorioallantois membrane; 12, blood vessel in the yolk sac membrane. *Redrawn from Raginosa 1961.* Scale bar = 10 mm.

RELEVANCE OF EMBRYOLOGICAL STUDIES

The study of avian embryonic development has been fundamental to understanding the developmental biology of vertebrates. Early descriptive embryological work provided the fundamentals of our knowledge on vertebrate embryonic development. These early descriptive studies were important because they documented in detail development over time, thus allowing us to compare the embryonic development of different species by stages. Today, the description of the embryonic normal stages of chicken development by Hamburger and Hamilton (1951) ranks among the most cited scientific papers (more than 12,000 citations). A detailed account and histological atlas of the development of the chicken, including normal stages of development from cleavage through hatching, was presented by Bellairs and Osmond (2005). Today, tables of normal development exist for 30 bird species (see below), all recognizing between 42 and 46 developmental stages, thus indicating rather uniform patterns of embryonic development largely independent of developmental mode.

The easy availability, accessibility, and relatively large size of chicken eggs and embryos made them a logical choice for embryological research. The fenestration technique (i.e., opening the eggshell and replacing it with a glass coverslip, so that the development of the embryo can be observed in vivo) has illustrated the considerable advantage of the chicken over other amniote species and offered access to observational as well as experimental studies. Later, the development of the chicken-quail chimera technique (e.g., Le Douarin 1973) provided an ingenious approach to studying the cellular component of cell proliferation, migration, and differentiation. Embryonic chicken tissue can be easily transferred to quail embryos (of the same stage) and vice versa. The grafted tissue develops normally in the host organism but can be recognized by differences in the cell nucleus; i.e., the nuclei of chicken cells look different from those of quail, so that, despite their close relationship, cells in chicken and quail tissue can always be recognized. Le Douarin's studies using the chicken-quail chimera system provided important new insight into understanding new crest development and the embryogenesis of the vertebrate head.

Today, tables of normal stages are important tools for studying the time course of gene expression pattern in normal and manipulated embryos. Gene constructs can be readily introduced into chicken embryos using retroviral methods or electroporation. These techniques have been widely used to overexpress genes at particular times and at specific locations within the developing chicken embryo. This highlights the possibility of using the chick embryo as a high-throughput tool for testing vertebrate gene function (reviewed by Brown et al. 2003). With the complete sequencing of the chicken genome in 2004 (Hillier et al. 2004), this species now serves mainly as an economic laboratory model for the study of gene expression and function, gene organization and regulation, gene families and pathways, and, more recently, genomic and proteomic functions.

List of tables of normal embryonic stages available for birds

Species	Stages	Incubation period	Reference
Dromaius novaehollandiae	43	50–56	Nagai et al. (2011)
Gallus gallus f. dom.		21	Keibel and Abraham (1900)
	46	21	Hamburger and Hamilton (1951)
			Rempel and Eastlick (1957)
	46	21	Yamasaki and Tonosaki (1988)
	45	20–21	Klepáček (1991)
only early development	10	<20hrs	Eyal-Giladi and Kochav (1976)
Gallus gallus		20	Künzel (1963)
Meleagris gallopavo f. dom.	46	28	Mun and Kosin (1960)
only early development	>11	>12hrs	Gupta and Bakst (1993)
only early development	>11	>12hrs	Bakst et al. (1997)
Phasianus colchicus		23–24	Fant (1957)
		23	Labisky and Opsahl (1958)
Coturnix c. japonica		16	Padgett and Ivey (1960)
only early development	1–15	<96 hours	Sato et al. (1971)

Species	Stages	Incubation period	Reference
	42	16–18	Starck (1989)
	45	16–18	Klepáček (1991)
only early development	1–19	<72 hours	Sellier et al. (2006)
	46	16.5	Ainsworth et al. (2010)
Colinus virginianus		22	Hanson (1954)
	41	23	Hendrickx and Hanzlik (1965)
Branta canadensis		27	Cooper and Batt (1972)
Anas platyrhynchos	46	26–28	Koecke (1958)
Not to hatching		<25	Hanson (1954)
Not to hatching		27	Caldwell and Snart (1974)
Aix sponsa		<28	Hanson (1954)
		30	Burcke et al. (1978)
Anas domestica	descriptive		Kaltofen (1971)
only early development	>33 (=HH20)	>72h	Dupuy et al. (2002)
Cairina moschata f. dom.	42	35	Starck (1989)
Falco sparverius	42	30	Scheibe (1991)
		27	Bird et al. (1984)
Falco tinnunculus	42	28	Scheibe (1991)
Microhierax caerulescens	42	35	Scheibe (1991)
Milvus migrans govinda		30	Desai and Malhotra (1980)
Vanellus vanellus	>33	to mid development	Grosser and Tandler (1909)
		25	Graul (1907)
Zenaida macroura		14	Hanson and Kossack (1957)
Columba livia f. dom.	42	16	Starck (1989)
	43	17	Olea and Sandoval (2016)
Agapornis roseicollis	descriptive		Mebes (1984)
Melopsittacus undulatus			Abraham (1901)
	42	18	Starck (1989)
Myiopsitta monachus	41	24	Carril and Tambussi (2015)
Turnix suscitator	42	14	Starck (1989)
Pygoscelis adeliae	45	>36	Herbert (1967, 1969)
Aptenodytes forsteri	38 +		Glenister (1954)
Corvus corone	42/43	17	Schurakov (1985)
Sturnus vulgaris	42/43	12	Schurakov (1985)
Agelaius phoeniceus	descriptive	to mid development	Daniel (1957)
Turdus philomelos	42/43	12	Schurakov (1985)
Turdus migratorius	42		Schurakov (1975)
Passer domesticus	42/43	11	Schurakov (1985)
Lonchura striata	46	17–18	Yamasaki and Tomosaki (1988)
Lonchura oryzivora	42	18	Starck (1989)

space that completely surrounds the embryo, isolating and protecting it from all other structures in the egg (fig. 5.8A, B). The (mechanical) isolation of the embryo is important because during later periods of development it needs to move freely and must not adhere to any extraembryonic structure. Because the amniotic folds develop from extraembryonic ectoderm underlain by mesoderm, the amniotic cavity is internally covered by ectoderm and externally by mesoderm. The amniotic cavity remains a small cavity enclosing the embryo; its mesoderm is the origin of smooth muscle cells, providing contractibility of the amnion; however, despite its mesodermal contributions, it does not develop vascularization. Once closed, the amnion performs slow contractions, thus gently moving the embryo before it can move by itself.

Chorion and Chorioallantois

The chorion develops together with the amnion, but continues to grow, and extends as the outer membrane across almost the entire inner eggshell membrane (fig. 5.8C, D). The epithelium of the chorion is ectodermal in origin, and it is internally accompanied by a thin layer of mesoderm. However, the chorion becomes functional as the embryonic respiratory organ when it fuses with the allantois. The allantois is a saclike structure growing from the hindgut of the embryo. From the beginning of its development, it has a rich vascularization (fig. 5.7F, G). The inner surface of the allantois is endodermal in origin; its outer surface is mesodermal and vascularized. Because of its position, it grows into the lumen of the extraembryonic coelom, which, after several days, is completely filled by the allantoic sac. The membranes of the chorion and the allantois fuse and adhere to the inner shell membrane, where they differentiate into a functional gas exchange organ, the **chorioallantois**.

By day 10 of embryonic development, the chorioallantoic membrane covers the entire inner surface of the egg, thus utilizing all available area for gas exchange. However, the respiratory diffusion barrier is thick, from outside to inside consisting of biomineralized eggshell, eggshell membranes, chorionic epithelium, chorionic mesoderm, and capillary wall of the blood allantoic vessel. Once the inner eggshell area is completely occupied by the chorioallantoic membrane, the respiratory exchange capacity can be enhanced only by changes of its microstructure. Indeed, during the next days of development, a considerable reorganization of the microstructure of the chorioallantoic membrane increases the respiratory gas exchange capacity: first the capillarization of the chorioallantoic membrane intensifies (i.e., more capillaries per are unit), and second, capillaries move between the cells of the original chorion ectoderm and later even into the eggshell membrane,

thus decreasing the thickness of the diffusion barrier. Although the development of the extraembryonic membranes has been studied only in chickens (Fitze-Gschwind 1973), these processes appear to be quite universal among birds.

Yolk Sac

The yolk sac develops from the extraembryonic endoderm and mesoderm. It is the site of early formation of blood cells (**erythropoiesis**), formation of blood vessels, and nutrient absorption (fig. 5.8). The yolk sac membrane is specialized to function as an extraembryonic absorptive organ, but it is neither morphologically nor functionally part of the embryonic gut; also, at least during later developmental stages, there is no open connection between the lumen of the embryonic gut and the yolk sac. Yolk absorption is largely by phagocytotic activity of the cells of the yolk sac epithelium (Raginosa 1961, Lambson 1970, Noble and Cocchi 1990). Lipase activity and the presence of bile in the yolk sac membrane suggests that phagocytosed lipids are hydrolyzed in the membrane before they are transported to the embryo (Yadgary et al. 2013). Also, the yolk sac membrane expresses genes that are related to carbohydrate uptake, protein, and amino acid digestion (Yadgary et al. 2011).

During embryogenesis, the yolk sac endoderm forms large villi that increase the absorptive surface and reach deep into the yolk ball. The histology of the absorptive epithelium is unique and structurally specialized for phagocytotic absorption of yolk. Very large epithelial cells are filled with lipid droplets. The epithelium of phagocytic cells may be several layers thick (fig. 5.9). Below the yolk sac epithelium one finds an extensive layer of hematopoietic tissue; deep in this tissue lie the yolk sac capillaries. Thus, yolk that was phagocytosed by the surface cells is transported through several layers of cells before it comes to the capillaries that finally transport the nutrients to the embryo. The yolk sac is a large extraembryonic structure that is not completely absorbed during embryogenesis; about 20–30 percent of the yolk remains as residue at hatching. Immediately before hatching, the yolk residues are drawn into the body cavity. It has often been claimed that the residual nutrients function as a source of energy for the hatchling during the first few days of life (Noy and Sklan, 2001). However, feeding experiments show that the yolk residues are absorbed faster when hatchlings are well fed, while food-restricted hatchlings take longer to absorb the nutrients. These experiments suggest that absorption of yolk residues requires energy, rather than serving as energy reserves. A simple explanation for yolk residues is that yolk is deposited in the egg in excess so that normal development (including some delay in hatching) is not limited by the amount of food.

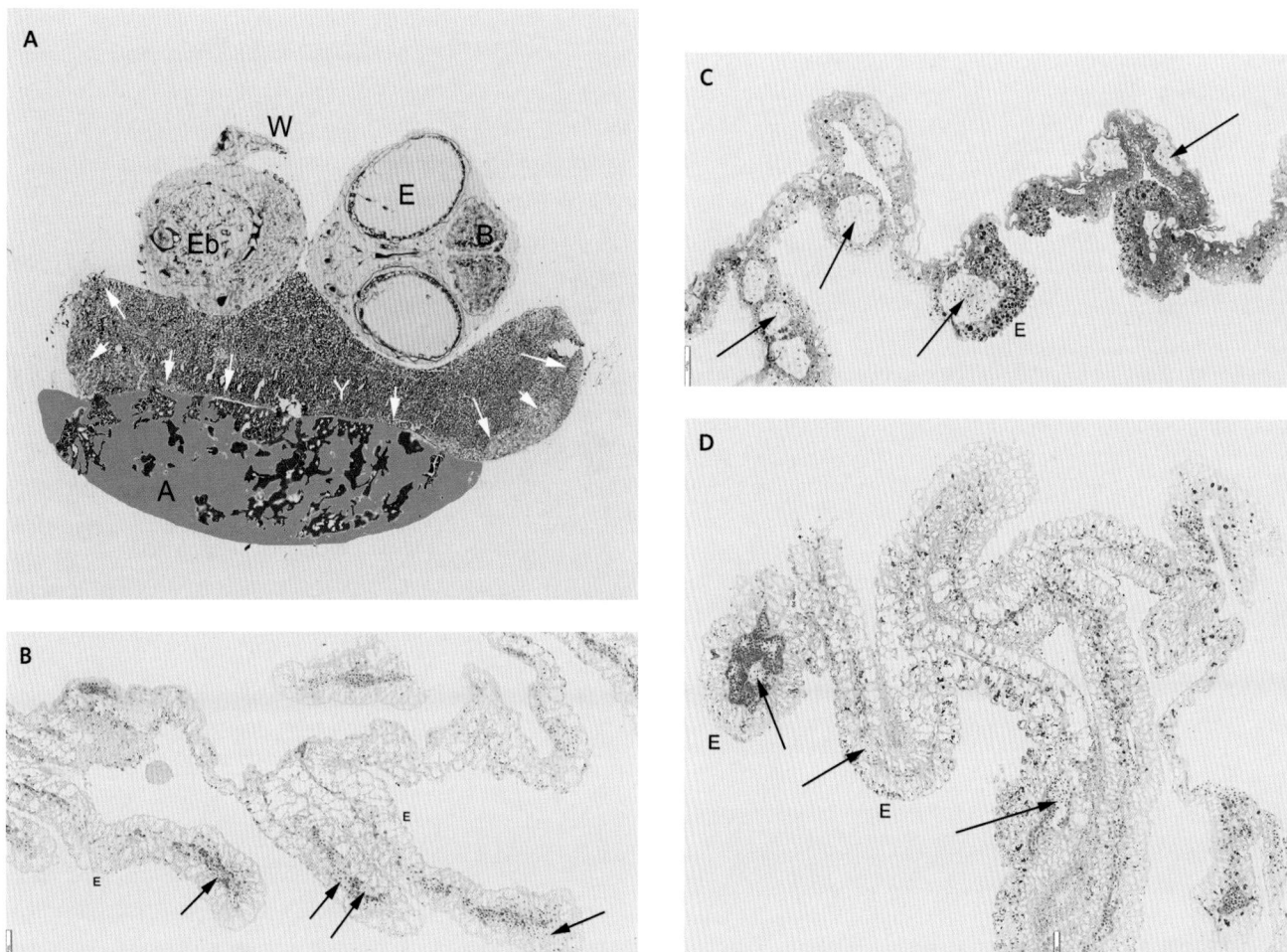

Figure 5.9. The yolk sac endoderm is an extraembryonic digestive organ that is specialized to absorb yolk. It shows several structural peculiarities in its cellular components. *A*, Quail embryo, histological section through the entire egg, with embryo, yolk, and albumen. At this stage of development, the embryo lies on the yolk like on a pillow. White arrows show villi of the yolk sac membrane (red) reaching into the yolk (black; see also micrographs in B–D); *B*, Micrograph of the yolk sac membrane of a Japanese quail embryo after eight days of incubation (stage HH27). The yolk sac endoderm is single layered and forms villi that reach into the yolk; blood vessels (arrows) are present in the yolk mesoderm and surrounded by erythropoietic tissue (dark stain); *C*, Micrograph of the yolk sac membrane of a Japanese quail embryo after nine days of incubation (stage HH29). The yolk sac endoderm becomes multilayered, and erythropoietic tissue forms a thick layer around the blood vessels; *D*, Micrograph of the yolk sac membrane of a Japanese quail embryo after 11 days of incubation (stage HH33, HH34). The yolk sac endoderm is multilayered, and cells are characterized by large vesicles containing phagocytosed yolk (purple granules in cells). Abbreviations: A, albumen; B, brain; E, yolk sac endoderm; Eb, embryo; Ey, eye of embryo; W, rudiment of wing.

Hatching

Hatching is the transition from embryonic life to emergence from the egg. In birds, this process takes several days and involves physiological transitions from transitory organs to final functioning organ systems. This is most obvious in the transition from embryonic respiration (i.e., chorioallantois) to lung respiration. During embryogenesis, the developing lungs are filled with liquid, and it takes several days to absorb the liquid from the respiratory structures of the lung and fill it with air. The transition to lung respiration is ini-

tiated when the embryo pierces the inner eggshell with its beak, pecking into the air cell (internal pipping). The air in the air space is used by the bird after internal pipping to fill the lungs and air sacs that are required for avian breathing—this is one reason why the perinatal period between internal pipping and hatching is long. Later the embryo breaks through the biomineralized eggshell (external pipping) and the tip of the beak reaches outside the egg. Another important process preparing the embryo for living outside the egg is the withdrawal of the yolk sac into the body cavity. During embryogenesis, the yolk sac with the yolk sac membrane and parts of

Figure 5.10. Drawing of the head of a developing chicken at 12 days of incubation. The egg tooth (gray, arrow) is an epithelial thickening on the upper beak. It will be lost shortly after hatching. See also images on fig. 5.11, where some of the hatchlings still bear the whitish egg tooth on the tips of their beaks. *Modified after Linde-Medina and Newman 2014.*

the small intestine lay outside the embryo. Just before hatching they are all incorporated into the belly and the body wall closes over it (umbilicus).

Finally, the embryo needs to emerge from the egg. During the actual hatching process, the embryo uses its beak with a little cornified hook (the egg tooth; fig. 5.10) to cut the eggshell at the blunt pole of the egg. The egg tooth is a transitory cornified hook on the bill (not a real tooth), and it is shed within one or two days after hatching (Starck 1989). A specialized muscle in the neck of the hatching bird allows for powerful movements of the head so that the eggshell can be broken. This **musculus complexus** (syn. m. semispinalis capitis) is a muscle extending from the vertebral column to the dorsal surface of the head. Its action is to raise the head and bill. In chickens, the muscle reaches its maximum size at 20 or 21 days of incubation (i.e., just before hatching), when it is loaded with glycogen as energy storage. The muscle also is a transitory structure and gradually disappears after hatching (John et al. 1987). During the actual hatching process, the chick rotates in the egg and the eggshell is broken through a series of consecutive pip-holes that join together. Once the eggshell has been opened, the young bird pushes itself out of the egg using its hind legs.

POSTNATAL DEVELOPMENT

The Hatchling Bird: The Altricial-Precocial Spectrum

At hatching, young birds differ in their degree of differentiation, whether morphological, physiological, or behavioral.

Early researchers described hatchlings as **nidifugous** (fleeing away from the nest) and **nidicolous** (inhabiting the nest; Oken 1816) or as **precocial** (early maturing) and **altricial** (requiring a nurse; Sundeval 1836). Both pairs of terms are still in use to characterize differences in bird hatchlings: altricial and precocial refer to the developmental stage, while nidicolous and nidifugous refer to nest attendance. The distinction between altricial and precocial has not changed to the present, but a number of additional categories have been introduced to differentiate more precisely the hatchling condition, now recognizing seven categories, reaching from superprecocial through semiprecocial and semialtricial to altricial (Portmann 1935, 1938, Nice 1962, Ricklefs 1983, O'Connor 1984, Starck 1993, Starck and Ricklefs 1998a, Botelho and Faunes 2015; table 5.1).

Precocial hatchlings leave the nest relatively soon after hatching and follow their parents. Usually they are covered in a downy feather coat, their sense organs are well developed, they walk (and/or swim), and feed on their own. Their thermoregulatory capacities are not fully developed, and they require occasional warming by their parents. There are different degrees of independence, in particular with respect to interaction with parental birds, independent feeding, and searching for food. Accordingly, different types of precocial birds are recognized, from superprecocial to precocial-3 (table 5.1 and figs. 5.11A–H; see the box on page 108). Postnatal growth of precocial chicks is relatively slow, and they need several weeks to reach adult size.

Semiprecocial hatchlings show all morphological characters of precocial hatchlings. Many semiprecocial species are capable of movement from hatching but are incapable of catching their food, thus they closely attend their parents for some period after hatching and are fed by their parents. During the period of nest attendance, they leave the nest only when endangered by predators.

Semialtricial hatchlings also have a downy feather coat but their eyes are closed at hatching and not fully functional. They attend the nest for an extended period until the first juvenile feathers are grown and they attain flight capabilities.

Altricial hatchlings are naked at hatching, their sense organs (eyes and ears) are closed, and they fully depend on parental support for an extended nestling period. Altricial hatchlings usually grow quickly and may become independent within a few weeks after hatching (table 5.1; fig. 5.11). The observed diversity of bird hatchlings raises interesting evolutionary questions, in particular with respect to the relatively uniform patterns of development and lack of obvious heterochrony. How could such a morphological and physiological diversity evolve? What are the selective factors and fitness advantages of being altricial or precocial?

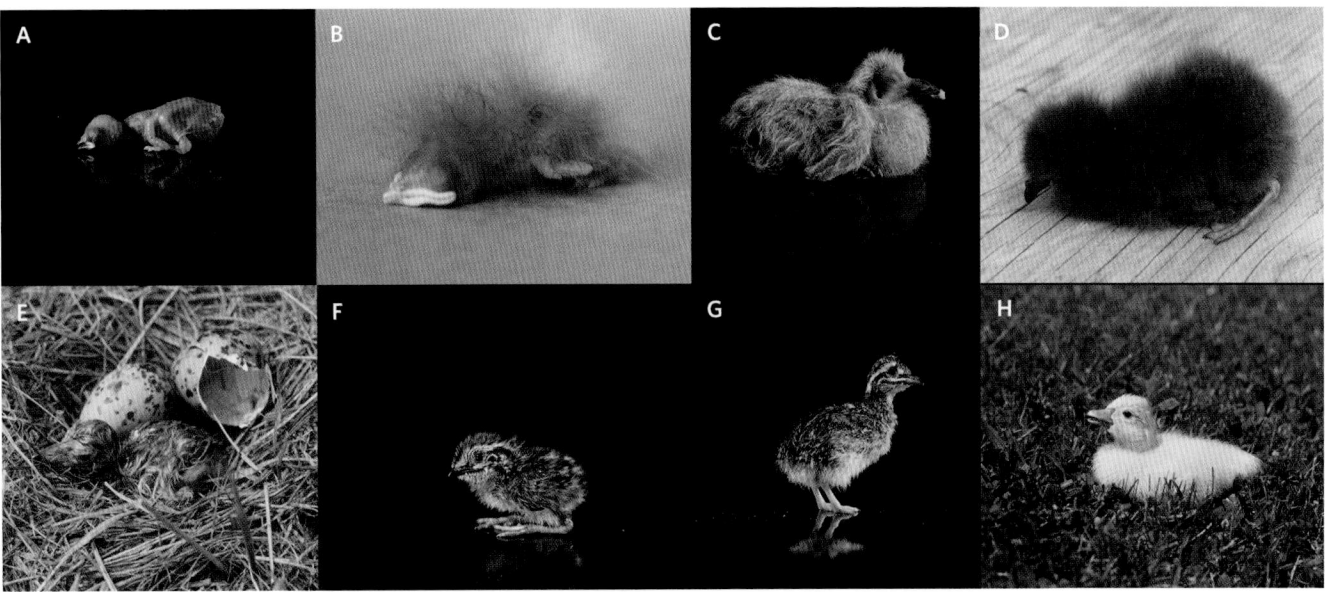

Figure 5.11. Hatchlings of precocial and altricial bird species; for details of their descriptions, see table 5.1. *A*, Reis Finch (*Lonchura oryzivora*), altricial 2; *B*, European Starling (*Sturnus vulgaris*), altricial 2; *C*, Rock Pigeon (*Columba livia*), altricial 1; *D*, Storm Petrel (*Oceanodroma leucorhoa*), semialtricial; *E*, Forster's Tern (*Sterna forsteri*), semiprecocial; *F*, Buttonquail (*Turnix suscitator*), precocial 3; *G*, Crested Tinamou (*Eudromia elegans*), precocial 2; *H*, Domestic Duck (*Cairina moschata*), precocial 1. See the box on page 114 for a description of superprecocial megapodes.

Table 5.1. **Characterization of eight different developmental modes in birds.**

Nice (1962)	Starck (1993)	1	2	3	4	5	6	7	8	9	10	11	Example
Precocial 1	Superprecocial												Megapodiidae
Precocial 2	Precocial 1												Anatidae, many charadriiform birds
Precocial 3	Precocial 2												Rheidae, Numididae, Phasianidae
Precocial 4	Precocial 3												Cracidae, Turnicidae, Gruidae
Semiprecocial	Semiprecocial												Alcidae, Laridae, Stercorariidae
Semialtricial 1	Semialtricial												Accipitridae, Ciconiidae
Semialtricial 2	Altricial 1												Columbidae, Phaethontidae
Altricial 2	Altricial 2												Passeriformes, Psittaciformes

Modified from Starck and Ricklefs 1998a.
1—No parent-chick interaction
2—Contour feathers at hatching
3—Downy hatching plumage
4—Motor activity
5—Locomotor activity
6—Follow parents
7—Search food and feed alone
8—Young fed by parents
9—Stay in nest
10—Eyes closed at hatching
11—Without external feathers

Phylogenetic analysis showed that precocial hatchlings represent the ancestral condition for birds, and evolutionary diversification among birds resulted in multiple and parallel origins of semiprecocial, semialtricial, and altricial hatchlings (Starck and Ricklefs 1998a, Botelho and Faunes 2015; fig. 5.12). While phylogenetic analyses suggest that the various modes of development evolved multiple times and independently, the associated morphological and physiological character transitions are similar among all groups of birds. Thus, the characteristic features associated with modes of development appear to change along a single axis of variation, defined by the functional maturity of tissue at hatching (Starck and Ricklefs 1998b). This observation implies that patterns of development are highly integrated and can change only in a concerted manner. It suggests that tissues and organs of young birds can be either mature (tissues and organs being differentiated and functional) or grow (high rates of cell proliferation). If, indeed, the trade-off between functional maturity and growth-rate constrains evolutionary variation, it is not surprising that multiple and independent evolutionary events always resulted in the same developmental "syndrome."

MEGAPODES

Megapodes form a small clade of galliform birds (7 genera, 22 species) living in the Indo-Australian region and on some Pacific islands. These birds have attracted the interest of researchers since the early sixteenth century because of their unusual breeding behavior. Megapodes do not incubate their eggs like other birds. Instead they bury their eggs in the ground and employ physical heat generated by bacterial decomposition of organic materials (e.g., *Alectura lathami*, *Leipoa ocellata*, *Megapodius* spp.), geothermic activity (e.g., *Megapodius* spp., *Macrocephalon maleo*), or solar heating of sandy soils (e.g., *Eulipoa wallacei*, *Megapodius* spp., *Macrocephalon maleo*). Some species do not attend their eggs at all after they have been laid in the ground. Others, mainly those employing piles of decomposing leaves to produce incubation heat, evolved complex nest guarding behavior and carefully regulate the temperature of the incubation pile (see figure). For example, the Malleefowl (*Leipoa ocellata*), spends several months per year compiling, activating, and maintaining an incubation mound. When eggs are buried in the mount, both parents carefully guard the mount, adding and removing material to control temperature and humidity. Certainly, the Malleefowl shows the most complex incubation mounts and most sophisticated breeding behavior among megapodes (Frith 1962, Jones et al. 1995). Because the eggs are buried in the ground they cannot be turned. Buried in the ground, the eggs are particularly susceptible to bacterial infections; e.g., the Australian Brushturkeys (*Alectura lathami*) bury their eggs in piles of decomposing leaves. Probably as an adaptation preventing bacterial infections, their eggs have an additional layer of calcium phosphate nanospheres that covers the external surface, providing a hydrophobic surface coat that decreases bacterial attachment and is most likely the major component preventing trans-shell bacterial penetration (D'Alba et al. 2014).

Besides burying eggs in thermal ground, megapodes show some outstanding features in their reproductive biology. They lay very large eggs with a high yolk content (50–70 percent of egg weight) compared with 32–49 percent in other Galliformes (Sotherland and Rahn 1987) or 14–35 percent in altricial birds (Sotherland and Rahn 1987). Incubation periods are long, ranging from 47 to 52 days in *Alectura lathami*, from 42 to 63 days in *Megapodius eremita*, and from 49 to 101 days in *Eulipoa wallacei* (Jones and Birks 1992) compared with 21 days in domestic chickens. Unique among birds, megapodes hatch without an egg tooth (Clark 1960, 1964) by cracking the eggshell with the legs. However, they develop an egg tooth during embryogenesis, but they shed it when still in the egg, approximately at the time

other galliforms would hatch (around three weeks). Because they lose the egg tooth in the egg, they can neither perforate the egg nor use an egg tooth for opening the eggshell. When they hatch they break the eggshell with their feet and kick themselves out of the egg. They use legs and wings to dig themselves out of the incubation mound. Megapode hatching is derived and modified because of their prolonged incubation period.

Megapode hatchlings are **superprecocial** (Nice 1962, Skutch 1976, O'Connor 1984, Starck 1993, Starck and Ricklefs 1998); they start an independent life immediately after emergence from the incubation mound (see figure). Social interactions between parents and their offspring have not been observed (Frith 1962, Immelmann and Böhner 1984, Jones et al. 1995). Hatchlings are relatively large (about 6–13 percent of adult body

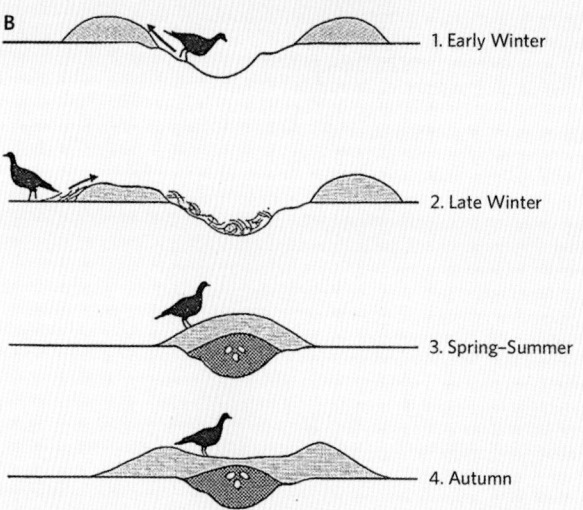

A, Hatchling of Brush Turkey (*Alectura lathami*), after emergence from the breeding mound. Note the well-developed wing feathers that enable the hatchling to fly on the day of emergence, *photo by D. Starck*; B, Schematic representation of construction and structure of the breeding mound of the Malleefowl (*Leipoa ocellata*). *Illustration from Rowley 1974 in Jones et al. 1995.*

mass), and their wings wear well-developed flight feathers, which enable the chicks to perform short bursts of flight from the moment of emerging from the ground. Superficially, their breeding behavior is similar to that of non-avian archosaurs (e.g., crocodiles), but phylogenetic analysis shows that it originated from galliform precocity and must be considered a phylogenetically derived trait (Clark 1960, 1964, Brom and Dekker 1992, Jones et al. 1995).

The spectacular breeding biology of megapodes raises the question how this breeding behavior and superprecocial hatchlings could evolve from galliform biology. Megapode breeding behavior evolved in regions where eutherian mammals were lacking as predators. Marsupials and reptilian sauropsids do not possess a fine sense of smell like eutherian mammals and do not detect the megapode eggs in the ground. Eutherian mammals do, however, and after being introduced on many of the Indomalayan Islands and Australia, they became a serious threat for megapodes. The superprecocial hatchlings of megapodes evolved by simple prolongation of the incubation period, compared with other galliform birds. Megapodes have large eggs with large energy reserves for the embryo, which enables chicks to remain in the egg for an extended incubation period and hatch in more developed condition. Indeed, megapode chicks hatch in a developmental stage equivalent to a two-to-three-week-old turkey chick. Comparative developmental studies have shown that there is a clear trend of a heterochronic delay of hatching in galliforms, with the megapodes representing an extreme. Indeed, megapodes follow exactly the same developmental trajectory as other galliform birds, but hatch later. They have shifted part of the posthatching development into the egg and hatch at a later moment (Starck and Sutter 2000).

Postnatal Organ Development

During the past 20 years, the study of postnatal development of avian organ systems has gained little attention because the focus of biological research has shifted to more cellular or molecular questions and methods. However, these new methods contributed little to the understanding of evolu-

tionary diversification of avian ontogenies and how avian life histories integrate with developmental modes. This is regrettable, because new and refined methods provide a better mechanistic understanding of many developmental processes, and we have better-resolved phylogenies, but when it comes to systemic physiology or morphology, we are referred to data that have been collected 20–30 years ago. Studies in meat production, however, have gathered detailed data on some economically relevant species; i.e., chicken, turkey, and Muscovy Duck (Cairina moschata). These studies provide important information about the effect of artificial selection on patterns of growth and muscle development, but they are focused on economically relevant aspects and do not answer evolutionary questions.

Skeleton

Irrespective of developmental maturity, birds hatch with most bones present, independent of their mode of development, size, or incubation period. Indeed, the time patterns of ossification (i.e., the embryonic stage during which bones are formed) are relatively consistent among birds, and little heterochrony has been described (e.g., Starck 1993, 1998, Maxwell 2008a, 2008b). However, the quantitative design of skeletons (i.e., the amount of cartilage retained and the degree of ossification attained at hatching) differ remarkably among species and developmental mode. Because cartilage is a relatively fast-growing tissue, and because bone grows rather slowly, it has been suggested that the degree of cartilage at hatching correlates with the postnatal growth rate (Starck 1996, 1998). Indeed, the degree of cartilage in long bones correlates with the rate of postnatal growth; however, the correlation is nonlinear, with an asymptotic curve (i.e., an increasing proportion of cartilage correlates with increasing rates of postnatal growth for part of the relationship, but from a certain degree of cartilage on, the growth rate does not increase any further), indicating that other constraints on postnatal growth become effective. Also, and importantly, the degree of ossification does not correlate with mode of development. There are (slow-growing) altricial species with a high degree of ossification and fast-growing precocial species with a low degree of ossification.

This distinction is important because paleontologists have used the degree of ossification as predictor of mode of development in fossil birds and even non-avian dinosaurs (e.g., Geist and Jones 1996, Weishampel and Horner 1994, Paul 1994). While it is notoriously difficult to determine the degree of ossification in a fossil, the degree of ossification in living birds does not allow us to deduce mode of development—so it is highly improbable that it works for non-avian dinosaurs. While the contention about altricial dinosaur hatchlings cannot be solved on the existing material basis of dinosaur hatchlings, the findings of dinosaur nests, and even incubating dinosaurs on their nests (Horner 2000), suggests that mode of development and attention of parents to the young was variable among archosaurs, ranging from protection from predators to possible parental feeding of nest-bound hatchlings. Thus, some reproductive behaviors, once thought exclusive to Aves, arose first in non-avian dinosaurs.

The microstructure of bone has been used extensively to deduce rates of growth. The basic assumptions are that different histological types of bone are laid down at different rates, thus reflecting different overall growth rates of individuals. Much of this is based on typological assumptions and spurious comparisons. Experimental testing showed that the measurable rates of bone deposition in just one type of bone (e.g., fibrolamellar bone) overlaps with the full range of bone deposition rates associated with other types of bone (Starck and Chinsamy 2002, Margerie et al. 2002, 2004). These results do not support the idea that similar types of bone tissue could be used to deduce postnatal growth rates.

Bone Growth and Functional Maturity of the Skeleton

Postnatal development of bone is not about only patterns and degree of ossification. What really counts for birds is attaining biomechanical strengths, so that bone can be loaded by weight and finally used. During growth and also during adult stages, bone is dynamically reorganized and adjusted to the actual biomechanical requirements. Bone is plastic tissue that is laid down not only according to the genetic blueprint of a species but also in response to environmental and physiological conditions. However, little research has been done on the relationship of postnatal growth and attainment of functional, biomechanical strength in birds.

Carrier and Leon (1990) studied skeletal growth and function in the California Gull (*Larus californicus*). This is a semiprecocial species that starts walking soon after hatching but does not use its wings until it leaves the nest about 45 days after hatching. This imposes an interesting contrast on the development of legs and wings; legs are functional while growing (precocial), while wings become functional only after growth has ceased (altricial). Interestingly, growth of both, legs, and wings follows a typical asymptotic exponential curve, with the wings showing higher growth rate compared with the legs. If one compares, however, growth with breaking load (i.e., the force that breaks the bone), the pattern differs strikingly. The breaking load of the hindlimb bones increases in a gradual and continuous manner that parallels

the growth curve. The bones of the wings, however, remain weak and break easily throughout most of the growth period. Only when the gulls reach maximum body mass and begin to exercise their wings does the breaking load of the wing bones increase rapidly. By the time the young birds begin to fly, the breaking load of their wing bones has approached that of the adults.

The case of the California Gull shows that growth and functionality of the wing bones can be dissociated. A soft bone with low loading resistance but high growth rate is deposited as a "placeholder" during development until shortly before the birds start using their wings. Only then, the skeletal elements are remodeled to become mechanically resistant. In this example, the contrast between wing and leg development may be explained by the growth rate–functional maturity trade-off; we find legs that are functionally mature but grow slowly (i.e., precocial legs) and wings that are functionally immature but grow rapidly (altricial). A potential corollary of that observation is that birds with long wings should fledge later than birds with shorter wings, because they need more time to attain functionality in their wings. A comparative study by Carrier and Auriemma (1992) presented partial support for this idea by correlating wing bone length with time to fledging. However, it was also evident that different patterns of development could be observed in different clades of semiprecocial and precocial birds. Charadriiform (shorebirds) and procellariiform birds (seabirds) appeared to follow the leg-wing disparity while galliform birds (chicken and relatives) showed a different pattern in which wings became functional well before final length was reached. Thus there is clearly divergent evolution in how postnatal growth of the skeleton diversifies. In a follow-up study, Dial and Carrier (2012) demonstrated a similar developmental disparity between wings and legs in ducks. The fact that bone may grow differently depending on the mechanical load has important implications for the understanding of skeletal growth in general but also, potentially, for aspects of bone healing and regeneration.

The patterns of postnatal skeletal growth have been studied on a comparative perspective also in galliform bird species (Starck and Sutter 2000). Galliform birds are interesting because they show a broad range of precocial developmental pathways, from relatively fast-growing quail to superprecocial megapodes that are independent of their parents and can fly on the day of hatching. This variety is the result of two different ontogenetic trajectories that evolved in galliforms. Some species, like quail, hatch at lower functional maturity of their skeletal tissue but grow comparatively faster and reach adult size more rapidly than other galliform species. Other species like megapodes are more fully functional (can fly at

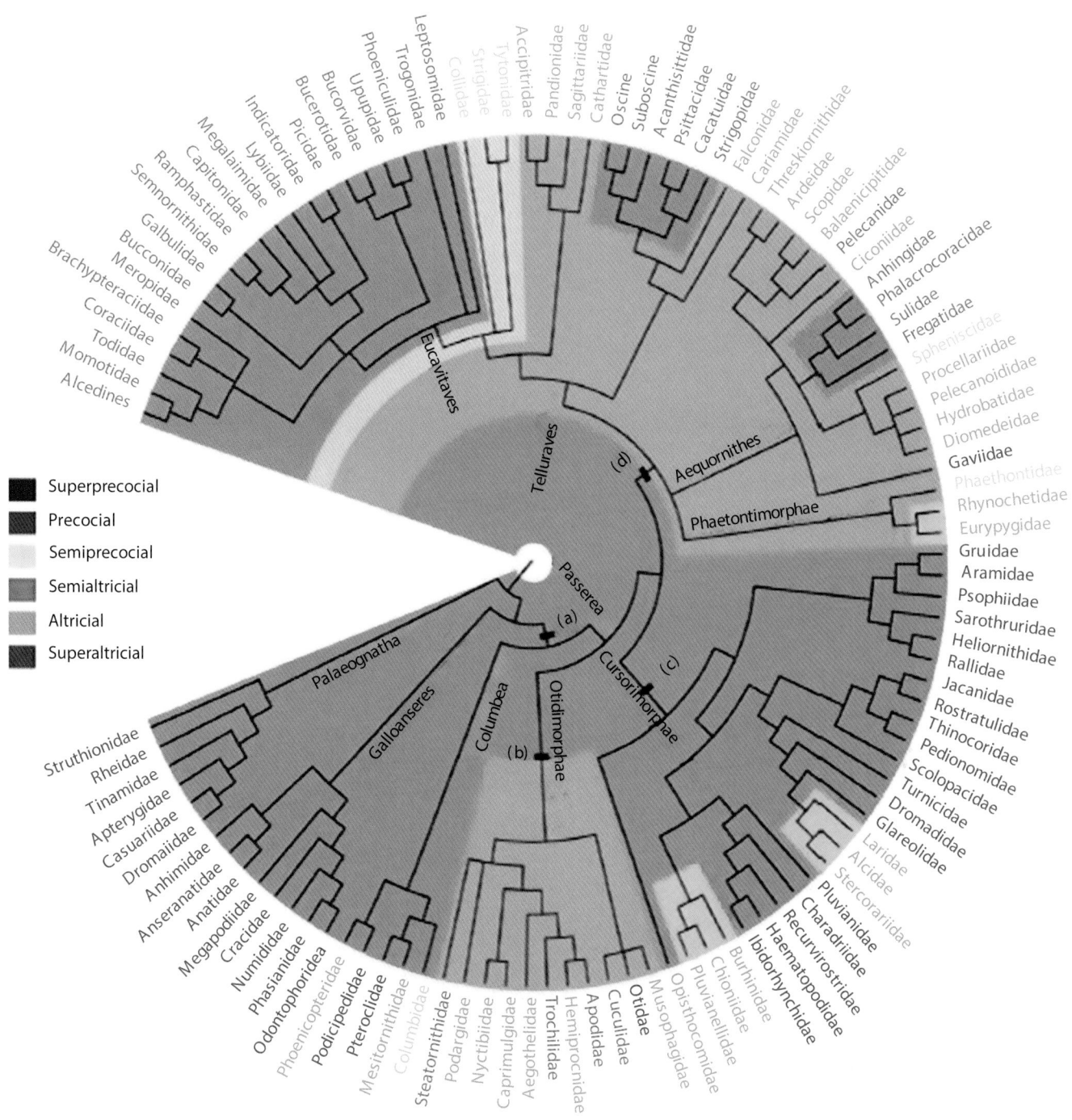

Figure 5.12. Phylogenetic tree depicting all extant bird families labeled by mode of development. *a–d* indicate important relationships (see text) robustly supported by whole-genome phylogenetic analysis, Jarvis et al. 2014. Developmental modes follow Nice 1962 and Starck 1993; *from Botehlo and Faunes 2015.*

hatching) but grow more slowly. Obviously, even within the clearly defined clade of galliform birds, some species are more altricial than others. Among other species, a more precocial mode of development is apparent as a result of hatching heterochrony. These species do not differ in their developmental pathway or the relationship between functional maturity and growth rate; instead, hatching is delayed. Some species (e.g.,

megapodes) delay hatching (up to three weeks) compared with other galliform species. Because of this delay they are much more developed when hatching, but, in fact, they are fully comparable to three-week-old chicks of other galliform species. While we have embarrassingly few comparative data on skeletal growth in birds, we recognize at least three different pathways of evolutionary diversification: (1) changes of the developmental trajectory (e.g., becoming more altricial), (2) decoupling of the development of different modules (e.g., wings and legs), and (3) heterochrony. Of course, various combinations of these can be expected.

Artificial Selection and Bone Growth

Postnatal development of the skeleton is also of interest in the poultry production industries. With high growth–selected lines reaching two to three times the body weight of their unselected domestic counterparts, their skeletons have to carry a much higher weight than the traditional breeds. Williams et al. (2004a, 2004b) showed that skeletal development in selected strains showed many similarities to control strains and produced bone of the correct dimensions to support the greater weights attained by the selected birds. However, the quality of bone in the selected strain was relatively poor in terms of porosity and mineral content and is likely to have a lower effective breaking strength. Rapid bone deposition was associated with decreased mineralization, increased porosity, and altered biomechanical properties (Williams et al. 2004a, 2004b). Artificial selection appears to be limited by the functional maturity–growth rate trade-off. While the poultry industry has shown that it is possible to select for very fast postnatal growth, the functional maturity, i.e., the bones' ability to support the weight, is reduced, resulting in increased incidences of leg disorders.

Muscle

Skeletal (striated) muscle has key functions in locomotion and thermogenesis of neonatal birds. The functional maturity of muscle tissue (i.e., the capacity of the tissue to perform adult function) is high in precocial and low in altricial chicks. Precocial hatchlings exhibit high tissue function and are relatively independent of environmental temperature but grow slowly. In contrast, altricial hatchlings have high growth rates but are thermally unstable and have low functional capacities of muscle. These generalized statements are supported by studies of ultrastructure of muscle (Eppley and Russel 1995), measures of dry matter fraction (as a measure of functional maturity of muscle; Dietz and Ricklefs 1997), cell proliferation (Ricklefs and Weremiuk 1977; as a measure of muscle growth capacity), enzyme activity (Ricklefs et al. 1994, Shea et al. 2007), electrolyte concentration in muscle (Choi and Ricklefs 1997), and performance (Dial and Carrier 2012). With respect to chemical, physiological, and ultrastructural properties of muscle, precocial bird chicks are essentially miniature adults with highly developed functional capacities. The slow growth of these species is consistent with a basic incompatibility between functional maturity and tissue growth. Altricial birds, in contrast, grow rapidly as nestlings, but their tissues reach functional maturation only at the end of the growth period.

Ducks present an interesting model in which this trade-off can be studied during development within individuals. Postnatal development of ducks is characterized by precocial development of the hindlimbs but altricial development of the wings. These divergent patterns of growth correlate with the development of performance: chicks run and swim immediately after hatching, but it takes two months until they start flying (Dial and Carrier 2012). Growth of leg muscles starts immediately after hatching, but growth rates are relatively slow. In contrast, growth of wing muscles begins very late, but then it follows a steep growth trajectory, reaching adult size quickly (Gille and Salomon 1998). Again, the example demonstrates that pattern of development can be decoupled in certain clades of birds. This is in contrast to young galliform birds that show equally precocial development in legs and wings. Actually, comparisons of different species of galliforms suggest a trend toward early flight capability, with the megapodes as an extreme (see above).

Artificial Selection and Muscle Growth

The poultry industry has a strong interest in economy and efficiency of meat production. Because chicken, turkey, guineafowl, and duck are all relatively slow-growing species, artificial selection is directed toward fast growth and high yield of industrially valuable meat; altricial birds, e.g., doves, play no significant role in industrial meat production. Artificial selection for high four-week body weight of Japanese Quail (*Coturnix japonica*) has resulted in more than doubling of body weight after about 100 generations, but selection limits were not reached in these lines (Noordwijk and Marks 1998). Much of this gain in adult body mass is attributed to an increased pectoralis muscle and the intestinal system (Lilja et al. 1984). While interested in improving economic efficiency and yield of meat production, the poultry industry has involuntarily provided an interesting model for testing the growth rate–functional maturity trade-off. One would expect that selection for high growth rate would result in a lower functional maturity of muscle compared with unselected control lines. Indeed, Choi et al. (2014) described that the expression of myogenic regulatory factors is prolonged in heavy-weight quail lines compared with control lines, resulting in delayed muscle maturation and thus permitting hypertrophic growth of the muscles in heavy-weight lines. Similarly, Choi et al. (2013) showed that DNA content of muscle fibers is heterochronically delayed in quail lines selected for heavy weight. Effects of selection can be observed even during early embryonic stages. In quail, selected for fast growth, the expression of myosin heavy chain is significantly heterochronically delayed compared with control lines, resulting in a lower functional maturity of muscle at hatching (Lilja et al. 2001). All studies show that selection for fast growth and large size obviously resulted in delayed maturity of tissues at hatching, thus supporting the hypothesis of a growth rate–functional maturity trade-off.

Although no mechanistic explanation can be given yet, these findings coincide with increasing concerns of the poultry production industry about bird lines selected for fast growth showing problematic issues with meat quality (pale, soft, and exudative meat; immaturity of intramuscular connective tissue). These two examples (bone and muscle development) of how artificial selection disrupts developmental pathways, resulting in increased pathologies, call for more explicit and thoughtful discussion of ethical aspects in animal mass production.

Feathers

In all birds, first feather germs develop during mid-embryogenesis. At hatching, however, some hatchlings are feathered while others are naked. The first feathers to grow differ from juvenile and adult feathers by structure and function; the final plumage is reached only during several juvenile molts. To understand the differences among species, we need to look at feather development from several different perspectives: (1) morphological development of feathers, (2) heterochrony in feather development, (3) selection on presence/absence of feathers at hatching, and (4) feather generations and development of final plumage. Because of the relative morphological simplicity of feather development during early cellular differentiation, its study has become a successful model system for studying mesenchymal-epithelial interactions and analyzing the establishment of anterior-posterior and proximal-distal axes of structures (e.g., Chuong 1998).

Morphological Feather Development

First rudiments of feathers are seen as little epidermal humps in stage HH28 (see above) of embryonic development. These visually detectable humps are preceded by molecular tract formation, i.e., lines of embryonic skin that become determined to form feathers. The feather tracts are later periodically patterned so that they are divided into feather buds (humps) and interbud regions (see chapters 6 and 9). These epidermal humps are induced by a condensation of mesenchyme under the epiderm. This mesenchymal condensation is the actual inductive tissue for feather development, causing cell proliferation in the epiderm so that it becomes multilayered. Epidermal humps occur in precisely those topographic positions of the skin that later represent feather tracts; thus the final topographic position of a feather on the skin of a bird is deter-mined from the very beginning of its development. Because of continued cell proliferation, the humps grow into papillae that reach beyond the surface of the skin and consist of a multilayered epiderm and a core of condensed mesenchyme containing the first capillaries. The contribution of two germinal epithelia (ectoderm = epiderm; mesoderm = mesenchyme) is an important feature for the many functions of a differentiated feather.

Not visible using standard methods but detectable using molecular probes, an anterior-posterior axis is determined in each feather germ. The further development of the feather is characterized by elongation of the papilla into a filiform feather rudiment and synchronous invagination of the feather germ into a pocket of the skin. From a molecular developmental perspective, a different set of regulator genes now becomes active, determining the proximal-distal axis of the feather germ. Thus, when the feather germ elongates, the anterior-posterior axis and proximal-distal axis are already out and determine the fundamental symmetry of the feather anlage. During the short bud stage, epithelial cell proliferation localizes mainly to the distal compartment, and as growth progresses, the proliferation zone shifts proximally. While the feather rudiment continues to grow, it becomes incorporated into a skin pocket that reaches deeper and deeper into the mesoderm of the skin. The pocket will become the later feather follicle. Thus the feather rudiment always consists of a mesenchymal compartment and an epithelial compartment. The mesenchymal compartment provides supply structures: e.g., capillaries and attachment; e.g., feather muscles and dense connective tissue fixing the feather follicle into the dermis. All cornified structures of the feather derive from the epithelial compartment. Later the feather stem cells reside in a specialized region of the feather sheath, the collar.

The development of a feather creates a complex adult feather structure, e.g., contour feather (see chapter 6), from a simple cellular cone. The principle of shaping a branching keratinized feather is always the same, and different types of (adult) feathers (e.g., from bristle to contour feather) develop because of quantitative difference in cell proliferation, number of cells, heterochrony, and programmed cell death (Maderson et al. 2009, Chen et al. 2015). Feather growth and differentiation occur in the proximal follicle zone, i.e., the collar or ramogenic zone, where the stem cells reside. Morphogenesis begins by infolding of the multilayered epidermal epithelium, forming the barb ridges. Each barb ridge consists of centrally aligned axial plates and centripetally oriented barbule plates. The basal layer flanking each barb ridge becomes the marginal plate epithelia. Barb ridges at the anterior end of the feather fuse to form the rachis or feather shaft. The posterior end of the feather develops the barb generative zone. As barbule plates keratinize, the marginal plate and axial cells undergo controlled cell death (apoptosis). Expression of feather keratins is indeed specific for the barbule plates (Haake et al. 1984); thus, only cells of the barb ridges keratinize and retain intercellular connections, while axial cells and cells of the marginal plate undergo apoptosis and lose the connections. Because of the differentiated keratinization and apoptosis of cells, a keratinized feather develops from the cellular cone of the feather filament like a sculptor carves a sculpture from a block of wood or stone. More to the distal end of the developing feather, the feather sheath and the pulp epithelium also undergo apoptosis, finally releasing the feather branches to open. The detailed geometry of cells in the barb ridges (e.g., Lucas and Stettenheim 1972, Maderson et al. 2009), the pattern of apoptosis, and, of course, regulatory gene networks (Chen et al. 2015) differ among feather types, feather generation, and among species of birds. However, the fundamental principle of feather development is the same for all.

Many birds hatch with a feather coat, the neoptile feathers. Although neoptile feathers are usually downy, different clades of birds show diverging morphologies of their neoptile feathers. Some altricial bird species have completely reduced the hatching feather coat, hatch naked, and develop a neoptile only later during the nestling period. Reducing the hatchling feathers is a derived condition that evolved independently in several clades of birds (e.g., Portmann 1938, Starck 1993). As far as we know today, hatching naked evolves from quantitative changes in development but not from heterochrony; i.e., those species with naked young have miniaturized feathers that are hidden in epidermal pockets of the follicles, but they are fully developed little feathers. So the time patterns of development are unchanged but the size of the feathers is reduced, and they grow only after hatching. An adaptive explanation for the loss of feathers in altricial and largely heterothermic hatchlings that are brooded by their parents is that removing the feather coat removes an insulating layer (Starck 1993). Brooding by parental birds is much more effective without the insulating neoptile layer. Vice versa, cooling down during periods of parental desertion may be faster and help save the energy of the young nestling.

Gastrointestinal Tract

The gastrointestinal (GI) tract is the place of food digestion and absorption. It has to be functional in hatchling birds, independent of mode of development, because they all start feeding the day of hatching. Because of its required functionality at hatching, the gastrointestinal tract escapes the functional maturity–growth rate trade-off. This is possible because, in the GI tract, areas of tissue growth and areas of tissue function (e.g., nutrient absorption) are separated, so that the GI tract can grow and function at the same time (see chapter 6). However, the GI tract is an energetically expensive organ to maintain, because the rate of cell renewal is high (e.g., cells are replaced every two to three days). Therefore, one would predict that the capacity of the GI tract to process and absorb food would always be matched to the amount of food available, because any overcapacity incurs energetic costs that reduce energy allocation to growth.

The relative size of the gastrointestinal tract (the weight of the GI tract in relation to hatchling body mass) is much larger in altricial species than it is in precocial species (Portmann and Vischer 1942, Lilja 1982). This correlates with the high postnatal growth rates of altricial and semialtricial birds, i.e., with an over-proportionally large intestine they can process and absorb large quantities of food/energy, thereby sustaining high growth rates. In contrast, precocial hatchlings have proportionally smaller GI tracts, grow on average at a slower rate, and allocate absorbed energy not only to growth but also for activity, maintenance, and thermoregulation. Based on these overall observations, it was hypothesized that the size of the gastrointestinal tract might represent an internal constraint on growth of precocial chicks (Ricklefs 1969a); i.e., they cannot grow faster because they cannot absorb more energy, independent of how much food they find. In this context, it was assumed that the GI tract of altricial hatchlings were matched to maximize the energy input of the growing nestlings, again, independent of the parental capacity of food provisioning. Alternatively, and considering that the GI tract is an energetically expensive organ (see above), it was suggested that size and functional capacities of the GI tract were optimized to feeding conditions, i.e., flexibly adjusted to the actual food availability (e.g., Ricklefs 1983, Konarzewski et al. 1989, 1990). Optimization would allow for adjusting the size and functional capacity of the GI tract to the actual need and would avoid overcapacity, i.e., the GI-tract being larger than needed and thus wasting energy (for maintenance) that could otherwise be used for growth.

The question of whether the GI tract is optimized or maximized to support energy uptake can be tested in feeding experiments: overfeeding versus food-limitations. If the GI tract were set to maximized size to sustain growth, then artificially overfeeding of food limitations to nestlings should have no effect on its size or uptake capacity for nutrients. If however, the GI tract were optimized to match food supply, then changes in food supply should result in correlated changes of gut size and uptake capacity. Results of feeding experiments are heterogeneous; some, mainly on altricial passerine birds, support the idea that the GI tract is maximized for energy uptake (Konarzewski et al. 1989, Konarzew-

ski and Starck 2000). Other experiments show that precocial chicks were more flexible and could adjust gut size and
growth rates to changing conditions (Schew and Ricklefs
1998). However, measuring gut weight, gut length, or absorptive surface provides only a rough measure of the true
ability of the GI tract for absorbing nutrients (energy). What
determines the nutrient absorption capacity is the activity
of enzymes and membrane-bound nutrient transporter systems together with the morphological measures. Only few
data from a limited number of species are presently available, and they reveal more variability and diversity than expected. It appears that the activities of carbohydrases but not
proteases are matched to diet composition in passerine birds
(Brzęk et al. 2009, Kohl et al. 2011). Again, a possible explanation for the heterogeneous results is that adaptation of bird
species does not follow generalized law-like rules in evolutionary physiology, but rather each species evolves under a
unique set of environmental and genetic conditions resulting
in inimitable adaptations.

The question of whether and how nestling birds adjust
their GI tracts to fluctuations in food quality and quantity is
not only important with respect to their immediate response
to the actual conditions but also determines their potential
to compensate for periods of limited food supply or poor
food quality. In particular precocial birds (mainly tested in
galliform species) show a remarkable capability to adjust to
periods of food limitation. Chickens can suspend growth
for several days if not weeks if kept on a maintenance diet
(i.e., enough food to maintain metabolic functions, but not
enough to grow). If refed, they may compensate for periods
of delay by increased growth rates (catch-up growth) and
quickly return on their normal growth trajectory (Schew
and Ricklefs 1989). Other species, in particular aerial insectivores, may interrupt growth during periods of bad weather
and food shortage and resume growth when refed.

In some species, the food of the parent birds may be inappropriate for feeding their young. In particular, seeds, grains,
or fruits do not provide sufficient nutrients for fast growth.
Therefore, many granivorous or frugivorous bird species
feed to their young different, protein-rich foods that support
growth, such as insects. Only later during their postnatal
life do these young birds switch to food items of adult birds.
The postnatal diet switch has been studied only in some
granivorous passerines (Brzęk et al. 2009, Kohl et al. 2011)
and showed flexible adaptations of enzyme activity to food
composition. Diet switches during development incur physiological and morphological changes of the GI tract, because
the functional requirements differ between different foods.
Today, it is unresolved whether the adaptive responses of the
GI tract to food quality and quantity are genetically hard-

wired or triggered by environmental effects. In some groups
of birds, parents process food enzymatically before feeding
it to their young. For example parrots predigest seeds fed
to their chicks, and Procellariiformes (albatrosses and kin)
produce stomach oil (an oily concentrate of the food they
collect during long foraging trips away from the nest). Only
two taxa of birds evolved parentally secreted nutrition for
their chicks. Pigeons and doves produce a fat and protein-
rich crop milk. The crop milk is produced by both parents
as holocrine secretion (i.e., whole cells are secreted into the
crop) of the crop epithelium. The crop-milk is regurgitated
during feeding and serves as exclusive food during the first
days of hatchling growth. In flamingos, the epithelium of
the esophagus proliferates and produces a lipid-rich secretion containing blood and white blood cells that is fed to the
young (Lang 1963).

POSTNATAL GROWTH

One theme is common in avian postnatal growth, and that
is that all birds reach determinate adult body mass within
one reproductive season. This is remarkable when compared
with other sauropsids, in which indeterminate growth pre-
vails; i.e., individuals grow during most of their life, though
with decreasing growth rates (e.g., crocodiles and alligators).
But still, there is remarkable variation in avian postnatal
growth, and birds of the same body mass may vary by 10
times in their rates of postnatal growth.

Some bird species reach asymptotic body size within 10
days after hatching, others require three months or longer
and still remain in the nest for an extended period after
reaching asymptotic body mass. The realized growth rate is
a compromise between contrasting selective factors favoring
longer and shorter growth periods (Lack 1968, Ricklefs 1969b,
1979b, Starck and Ricklefs 1998b), internal constraints, and
phylogenetic relatedness. For example, (1) time-dependent
mortality and sibling competition (for food) may be factors
selecting for fast postnatal growth, but limited energy supply may select for slow growth; (2) internal constraints like
the growth rate–functional maturity trade-off, or the gut's
capacity to process food, may impose limits to the maximum
possible growth rate; and finally, (3) closely related species in
a clade tend to grow at similar rates.

Measuring the rate of postnatal growth is not trivial,
because it is affected by size and body proportions. Therefore,
postnatal growth is typically described by sigmoidal curves
fitted to empirical data. Deriving parameters of growth curve
estimates is necessary to obtain size-independent measures
of the dynamics of growth and the shape of the growth trajectory that can then be compared among species. Because

NESTLING OBESITY

Postnatal increase in body mass follows a sigmoidal growth curve. In many altricial and semialtricial bird species, however, postnatal body mass exceeds adult mass, resulting in a body mass maximum followed by a period of weight recession. Peak postnatal body mass occurs, e.g., in many songbirds and albatrosses, shearwaters, and petrels (Procellariiformes). An extreme is found in the nestlings of the South American Oilbird (*Steatornis caripensis*, Steatornithidae), which reach peak body mass of 145 percent adult body mass (415 g). The young of oilbirds get so fat that indigenous inhabitants used them as a source of oil; history tells that early voyagers killed the nestlings, pulled a candlewick through their bodies, and used them as candles. At peak weight they cannot fly because their weight exceeds by far the maximum possible wing load. Weight recession takes a period of 30–35 days, during which the nestlings reduce body mass to adult weight.

Peak postnatal weight and hatching obesity are in striking contrast to the observation that food limitation is a major selective force on postnatal growth. Several hypotheses aim at explaining the evolutionary history and selective advantages of hatching obesity: (1) large fat stores may ensure the survival of chicks during extended fasting periods; (2) they may buffer against stochastic fluctuations in food supply; (3) large fat deposits fuel chicks during the initial critical period away from the nest site; (4) chicks deposit lipid early in the nestling period when their energy requirements are comparatively low, and use these stores to subsidize higher metabolic costs later in development; and (5) they are a side product of a diet rich in energy but poor in nitrogen, i.e., chicks assimilate more energy than necessary to get the required nutrients and therefore have to dump excess energy in adipose tissue (energy sink hypothesis; Ricklefs 1979a). Evidence from different clades of birds has been documented supporting all hypotheses, and it is most probable that post-

Antarctic Fulmar (*Fulmarus glacialoides*) parent feeding chick. *Image courtesy of Colin Miskelly, New Zealand Birds Online, www.nzbirdsonline.org.nz.*

natal peak weight and nestling obesity may have different evolutionary causes in different clades of birds.

In songbirds and many other altricial hatchlings, postnatal peak weight may indeed maintain chicks through fasting periods or buffer against stochastic fluctuations in food supply (Lack 1968). In other clades of birds (e.g., Procellariiformes) a likely explanation for large fat deposits may be to fuel chicks during the initial critical period away from the nest site while they learn to forage effectively (Phillips and Hamer 1999). Many procellariform birds feed stomach oil to their young. This is an extremely energy-rich, nutrient-poor food concentrate parent birds produce in their stomach during long foraging trips. Considering the low nutrient content of stomach oil, the energy sink hypothesis provides a valuable explanation. Oilbirds, again, are special. They feed their young on a plain fruit diet, which also is rich in energy but poor in nitrogen and amino acids. Here again, the energy sink hypothesis provides a useful explanation for nestling obesity.

---

of the size differences, absolute values of size increments would not be helpful for any comparison among species. A derived growth parameter such as the growth rate constant (*K*), however, is a valuable parameter because it is directly proportional to the rate at which size approaches the asymptotic value.

The quantitative analysis of growth rates of birds was initialized by Ricklefs (e.g., Ricklefs 1967, 1968, 1973). A comprehensive review of growth rate data (body mass) consider-

ing 557 species was presented by Starck and Ricklefs (1998c). Body mass as a measure of size is a simple but probably not sufficiently precise measure of growth. In many altricial and semialtricial bird species, body mass increases to values exceeding adult body mass. The postnatal maximum in body mass can reach up to 140 percent of adult weight (see the box on page 122). Therefore, skeletal measures of growth are potentially better descriptors of growth. Tarsus length (correctly tarsometatarsus length; see chapter 6) is a commonly used measure to describe growth of nestlings. It a sufficiently good indicator of overall growth, but multivariate descriptions, e.g., several body measures, might be even better and more precise. In reality, however, they are often more time-consuming and difficult to take.

Selection on Growth Rates
Mortality Rates

The growth period is a particularly vulnerable time for birds. Precocial and semiprecocial chicks may run away from predators, but they are comparatively slow, inexperienced, and cannot fly. Nestlings of altricial and semialtricial birds cannot even run away and, confined to their nest, need to be guarded by their parents, otherwise they are openly presented to potential predators (depending on nesting sites, of course). It has therefore been hypothesized, that the rates of postnatal growth are selected for by postnatal **mortality rates** (e.g., Williams 1966, Lack 1968, Ricklefs 1969b). Overall mortality during the period of postnatal growth may be caused by different factors (e.g., predation, food limitation, weather condition, brood parasites, adult death), and each of these may be of different importance in different species (Ricklefs 1969a, 1979b). A number of studies have tested the relationship between postnatal mortality and growth rate, but the results are contradictory. Indeed, the postulated relationship might be difficult to test at all because of the notorious inaccuracy of mortality data, differences among clades of birds, and the many other factors that affect postnatal growth. Ricklefs et al. (1998) analyzed by far the largest data set of postnatal growth rate with respect to average daily nestling mortality with mode of development and sibling competition as covariates. This analysis showed that body mass, phylogenetic relationship, and mode of development are certainly the strongest factors in the evolutionary diversification of growth rates, but sibling competition and average daily mortality could not be excluded convincingly. Comparative analyses of life history parameters traditionally assume that consistent causal relationships exist for the tested parameters across all taxa in an analysis. Refinement of methods by considering an increasing number of covariates and phylogenetic relatedness of species in an analysis do not solve the conceptual weakness that evolution does not obeys laws (as defined by biologists), and each clade and each species of birds has its own evolutionary history with selection and constraints acting on it. Contrasting results as summarized above may not necessarily indicate that one of the analyses is incorrect; rather, they show that variation in life history parameters is shaped by a multitude of hierarchically structured factors that need to be analyzed for each individual species.

Sibling Competition

Sibling competition has also been discussed as a selective factor on growth rate in altricial and semialtricial birds when parental birds distribute food according to offspring size (Ricklefs 1993). Royle et al. (1999) tested the hypothesis that sibling competition for resources is a major selection pressure on the evolution of maximal growth rates in birds in a large comparative analysis of altricial and semialtricial birds with multiple nestmates. While their analysis generally supports the idea that sibling competition for parentally provided food resources has a positive effect on growth rate, they also showed that other factors of kin-related selection may affect growth rate too. The evolutionary consequences of sibling competition may be reduced if parent birds would lay eggs across long intervals, thus creating hatching asynchrony among their offspring (Ricklefs 1993). With increased age difference and thus differing food requirements among chicks of a brood, parent birds may dampen the maximum food requirements of their brood, relax food provisioning, and thus increase survival probability in an energy-limited environment.

Energy Allocation to Growth

Parental food provisioning to the growing chicks is certainly an important factor that may select for slow growth rates. However, a chick's ability to process the food and absorb the energy may impose limits on how much energy can be absorbed and finally allocated to growth. Indeed, part of the difference in growth rates between altricial and precocial birds could be explained by the fact that precocial chicks spend energy not only on growth and maintenance but also on thermoregulation and locomotion. Thus altricial chicks of the same size and energy assimilation as a precocial chick could allocate a much larger proportion of the assimilated energy to growth. Of course, the efficiency of food conversion into metabolizable energy and the efficiency of biosynthesis may play important but difficult-to-test roles. Thus the amount of energy allocated to growth may be determined by (1) the amount of food available for the chick (external limitation), (2) the ability of the gut to digest and absorb the food (structural constraint), and (3) the efficiency of converting metabolizable energy into growth (internal limitation). While food provisioning to the chick can be studied relatively easily, the two other levels of constraint are rather difficult to grasp. Conventionally, gut size (dry mass) is used as a measure of the gut's ability to digest and absorb food, but this is only a crude approximation. Absorptive surface and enzyme activity, as well as density and kinetics of trans-membrane transporters, are more direct measures of the functional capacity of the gut.

KEY POINTS

- The avian egg is a semi-closed container that holds all energy, nutrients, and water for the developing embryo; respiratory gas exchange with the environment is constrained by the eggshell. Variation in energy content of the egg and its capacity for respiratory gas exchange ultimately affect many avian life history parameters. Differences between species, but also between broods within species and even differences between chicks of a brood, can be explained by variation of quantitative parameters of the egg.

- The eggshell and the developing extraembryonic membranes provide an adapted environment for the developing bird embryo. These functions are partially conflicting; thus the eggshell and eggshell membranes present a compromise between the contrasting functional demands of mechanical protection, gas transfer (through pores), and protection against bacterial infection. Extraembryonic membranes are transitory embryonic structures that maintain the embryo through the incubation periods; they provide gas exchange, nutrient uptake, and storage of excreta.

- Time patterns of avian embryogenesis show relatively little heterochrony. Embryogenesis of birds (mainly chickens and quail) is an important high throughput study system for molecular developmental studies.

- The length of incubation period is determined by a hierarchical system of internal constraints, selective factors, and phylogenetic relationship. Nest predation, energetic costs of embryonic development, evaporative water loss from the egg, and parental cost to incubation all select for short incubation periods. Selection for long incubation time is difficult to prove and can be shown convincingly only when the fitness benefits of prolonged incubation exceed the disadvantages.

- The diversity of avian hatchlings, i.e., the altricial-precocial spectrum, evolved along a trade-off between functional maturity and growth rate. Among birds, precocial hatchlings are ancestral, and altricial species (and the many other modes of development in between) evolved multiple times independently.

- Variation in the rate of postnatal growth among birds is limited by internal constraints like the trade-off between growth rate and functional maturity, but it is also maintained by external selective factors. Average daily mortality rates of chicks, food availability to hatchlings, and sibling competition are discussed as the key factors that select growth rates.

- Postnatal organ development is largely determined by a growth rate–functional maturity trade-off; i.e., organs and tissues can either grow or function. This trade-off results in common syndromes; thus, altricial bird hatchlings all appear very similar, despite having evolved many independent times.

- Artificial selection on posthatching growth rates has resulted in an enormous increase in the rate of growth and increase in body mass of commercially interesting species. Even after many years of artificial selection, the genetic variance has not been exhausted. Exemplified by bone development and muscle development, artificial selection for fast growth can disrupt this trade-off, but only at the cost of a serious decline in meat quality and the mechanical resistance of bone.

- The rate of postnatal growth is determined by natural selection, internal constraints, and phylogenetic relatedness. Life history theory has failed to provide generalized explanations because evolution does not obey law-like rules; rather each species is a unique result of its evolutionary history and needs to be analyzed individually.

KEY MANAGEMENT AND CONSERVATION IMPLICATIONS

- The egg has a central role in the evolution of avian life histories. Because of the strong effects of the environment on the physiology of egg production, egg quality is potentially affected by environmental disturbance. Reduced egg quality may result in increased mortality and a decline of a species.

- Egg quality may be used to monitor environmental (including stress) disturbance.

- Genetic, nongenetic, maternal, and environmental factors contribute to egg quality. They are of central importance in economic settings.

- Environmental disturbance may affect parental food supply to chicks and thus affect postnatal growth.

- Selection for fast growth (e.g., in meat production) has probably disturbed the basic trade-off between functional maturity and growth rate. Thus fast growth is achieved on the cost of tissue quality.

DISCUSSION QUESTIONS

1. Compare an avian egg with a sauropsid egg. What are the autapomorphic avian characters?
2. What are the structures and key functions of the avian eggshell?
3. Why do eggs need to be turned?

4. Explain why we need to know normal stages of development; who uses them?

5. What are the functions of the extraembryonic membranes in bird embryos?

6. How does a bird hatch?

7. What is a central constraint on the evolution of modes of development in birds?

8. Which selective factors have been recognized to affect avian postnatal growth?

9. Discuss the ethical topics arising with artificial selection on "high performance" lines in domestic birds.

References

Bellairs, R., and M. Osmond (2005). The atlas of chick development. 2nd ed. Elsevier Academic Press, San Diego.

Berrang, M. E., N. A. Cox, J. F. Frank, and R. J. Buhr (1999). Bacterial penetration of the eggshell and shell membranes of the chicken hatching egg: A review. Applied Poultry Research 8:499–504.

Blom, J., and C. Lilja (2004). A comparative study of growth, skeletal development and eggshell composition in some species of birds. Journal of Zoology London 262:361–369.

Bond, G. M., R. G. Board, and V. D. Scott (1988). A comparative study of changes in the structure of avian eggshells during incubation. Zoological Journal of the Linnean Society 92:105–113.

Botelho, J. F., and M. Faunes (2015). The evolution of developmental modes in the new avian phylogenetic tree. Evolution and Development 17:221–223.

Brom, T. G., and R. W. R. J. Dekker (1992). Current studies on megapode phylogeny. Zoologische Verhandelingen 278.

Brown, W., S. J. Hubbard, C. A. Tickle, and S. W. Wilson (2003). The chicken as a model for large-scale analysis of vertebrate gene function. Nature Reviews Genetics 4:87–98.

Brzęk, P., K. Kohl, E. Caviedes-Vidal, and W. H. Karasov (2009). Developmental adjustments of house sparrow (Passer domesticus) nestlings to diet composition. Journal of Experimental Biology 212:1284–1293.

Carrier, D., and L. R. Leon (1990). Skeletal growth and function in the California gull (Larus californicus). Journal of Zoology 222:375–389.

Carrier, D. R., and J. Auriemma (1992). A developmental constraint on the fledging time of birds. Biological Journal of the Linnean Society 47:61–77.

Cassey, P., S. J. Portugal, G. Maurer, J. G. Ewen, R. L. Boulton, M. E. Hauber, and T. M. Blackburn (2010). Variability in avian eggshell colour: A comparative study of museum eggshells. PLoS ONE 5:e12054.

Chen, C. F., J. Foley, P. C. Tang, A. Li, T. X. Jiang, P. Wu, and C. M. Chuong (2015). Development, regeneration, and evolution of feathers. Annual Review of Animal Biosciences 3:169–195.

Choi, I. H., and R. E. Ricklefs (1997). Changes in protein and electrolyte concentrations in the pectoral and leg muscles during avian development. The Auk 114:688–694.

Choi, Y. M., D. Sarah, S. Shin, M. P. Wick, B. C. Kim, and K. Lee (2013). Comparative growth performance in different Japanese quail lines: The effect of muscle DNA content and fiber morphology. Poultry Science 92:1870–1877.

Choi, Y. M., Y. Suh, J. Ahn, and K. Lee (2014). Muscle hypertrophy in heavy weight Japanese quail line: Delayed muscle maturation and continued muscle growth with prolonged upregulation of myogenic regulatory factors. Poultry Science 93:2271–2277.

Chuong, C. M., Editor (1998). Molecular basis of epithelial appendage morphogenesis. R.G. Landes, Austin, TX.

Clark, G. A. (1960). Notes on the embryology and evolution of the megapodes (Aves: Galliformes). Postilla Yale Peabody Museum of Natural History 45:1–7.

Clark, G. A. (1964). Ontogeny and evolution in the megapodes (Aves: Galliformes). Postilla Yale Peabody Museum of Natural History 78:1–37.

Coleman, J. R., and A. R. Terepka (1972). Fine structural changes associated with the onset of calcium, sodium and water transport by the chick chorioallantoic membrane. Journal of Membrane Biology 7:111–127.

D'Alba, L., D. N. Jones, H. T. Badawy, C. M. Eliason, and M. D. Shawkey (2014). Antimicrobial properties of a nanostructured eggshell from a compost-nesting bird. Journal of Experimental Biology 217 (7): 1116–1121.

D'Alessandro, A. P., G. Righetti, E. Fasoli, and L. Zolla (2010). The egg white and yolk interactomes as gleaned from extensive proteomic data. Journal of Proteomics 73:1028–1042.

Deeming, D. C. (1991). Reasons for the dichotomy in egg turning in birds and reptiles. In Egg incubation: Its effects on embryonic development in birds and reptiles, C. D. Deeming and M. W. Ferguson, Editors. Cambridge University Press, Cambridge, UK.

Deeming, D. C. (2002). Functional characteristics of eggs. Oxford Ornithology Series 13:28–42.

Deeming, D. C., G. F. Birchard, R. Crafer, and P. E. Eady (2006). Egg mass and incubation period allometry in birds and reptiles: Effects of phylogeny. Journal of Zoology 270:209–218.

Dekker, R. W. R. J., and T. G. Brom (1992). Megapode phylogeny and the interpretation of incubation strategies. Zoologische Verhandelingen (Leiden) 278:19–31.

Del Rio, C. M. (2008). Metabolic theory or metabolic models? Trends in Ecology and Evolution 23:256–260.

Dial, T. R., and D. R. Carrier (2012). Precocial hindlimbs and altricial forelimbs: Partitioning ontogenetic strategies in Mallard Ducks (Anas platyrhynchos). Journal of Experimental Biology 215:3703–3710.

Dietz, M. W., and R. E. Ricklefs (1997). Growth rate and maturation of skeletal muscles over a size range of galliform birds. Physiological and Biochemical Zoology 70:502–510.

Duce, S., F. Morrison, M. Welten, G. Baggott, and C. Tickle (2011). Micro-magnetic resonance imaging study of live quail embryos during embryonic development. Magnetic Resonance Imaging 29 (1): 132–139.

Eppley, Z. A., and B. Russell (1995). Perinatal changes in avian muscle: Implications from ultrastructure for the development of endothermy. Journal of Morphology 225:357–367.

Eyal-Giladi, H., and S. Kochav (1975). From cleavage to primitive streak formation: A complementary normal table and a new look at the first stages of the development of the chick. Volume 1. General morphology. Developmental Biology 49:321–337.

Fitze-Gschwind, V. (1973). Zur Entwicklung der Chorioallantois-membran des Hühnchens. Advances in Anatomy, Embryology and Cell Biology 47:1–51.

Frith, H. J. (1962). The Mallee Fowl. Angus and Robertson, Sydney.

Gabrielli, M. G. (2004). Carbonic anhydrases in chick extra-embryonic structures: A role for CA in bicarbonate reabsorption through the chorioallantoic membrane. Journal of Enzyme Inhibition and Medicinal Chemistry 19:283–286.

Geist, N. R., and T. D. Jones (1996). Juvenile skeletal structure and the reproductive habits of dinosaurs. Science 272:712.

Gille, U., and F. V. Salomon (1998). Muscle growth in wild and domestic ducks. British Poultry Science 29:500–505.

Haake, A. R., G. König, and R. H. Sawyer (1984). Avian feather development: Relationships between morphogenesis and kera-tinization. Developmental Biology 106:406–413.

Hamburger, V., and H. L. Hamilton (1951). A series of normal stages in the development of the chick embryo. Journal of Morphology 88:49–92.

Hillier, L. W., W. Miller, E. Birney, W. Warren, R. C. Hardison, C. P. Ponting, . . . and J. B. Dodgson (2004). Sequence and comparative analysis of the chicken genome provide unique perspectives on vertebrate evolution. Nature 432:695–716.

Hincke, M. T., Y. Nys, K. Mann, A. B. Rodriguez-Navarro, and M. D. McKee (2012). The eggshell: structure, composition and mineralization. Frontiers in Bioscience 17:1266–1280.

Horner, J. R. (2000). Dinosaur reproduction and parenting. Annual Review of Earth and Planetary Sciences 28:19–45.

Hoyt, D. F. (1976). The effect of shape on the surface-volume relationships of birds' eggs. The Condor 78:343–349.

Hoyt, D. F. (1979). Practical methods of estimating volume and fresh weight of bird eggs. The Auk 96:73–77.

Immelmann, K., and J. Böhner (1984). Beobachtungen am Thermometerhuhn (Leipoa ocellata) in Australien. Journal für Ornithologie 125:141–155.

Jarvis, E. D., S. Mirab, A. J. Aberer, B. Li, P. Houde, C. Li, S. Y. Hou, B. C. Faircloth, B. Nabholz, and J. T. Howard (2014). Whole-genome analyses resolve early branches in the tree of life of modern birds. Science 346:1320–1331.

John, T. M., J. C. George, and E. T. Moran Jr. (1987). Pre- and post-hatch ultrastructural and metabolic changes in the hatching muscle of turkey embryos from antibiotic and glucose treated eggs. Cytobios 49:197–210.

Jones, D., and S. Birks (1992). Megapodes: recent ideas on origins, adaptations and reproduction. Trends in Ecology and Evolution 7 (3): 88–91.

Jones, D. N., R. W. R. J. Dekker, and C. S. Roselaar (1995). The Megapodes (Megapodiidae). In Bird families of the world, vol. 3, C. M. Perrins, W. J. Bock, and J. Kikkawa, Editors. Oxford University Press, Oxford.

Kilner, R. M. (2006). The evolution of egg colour and patterning in birds. Biological Reviews 81:383–406.

Kohl, K. D., P. Brzęk, E. Caviedes-Vidal, and W. H. Karasov (2011). Pancreatic and intestinal carbohydrases are matched to dietary starch level in wild passerine birds. Physiological and Biochemical Zoology 84:195–203.

Konarzewski, M., J. Kozłowski, and M. Ziółko (1989). Optimal allocation of energy to growth of the alimentary tract in birds. Functional Ecology 3:589–596.

Konarzewski, M., C. Lilja, J. Kozłowski, and B. Lewończuk (1990). On the optimal growth of the alimentary tract in avian post-embryonic development. Journal of Zoology 222:89–101.

Konarzewski, M., and J. M. Starck (2000). Effects of food shortage and oversupply on energy utilization, histology, and function of the gut in nestling song thrushes (Turdus philomelos). Physiological and Biochemical Zoology 73:416–427.

Lack, D. L. (1968). Ecological adaptations for breeding in birds. Methuen, London.

Lambson, R. O. (1970). An electron microscopic study of the ento-dermal cells of the yolk sac of the chick during incubation and after hatching. American Journal of Anatomy 129:1–19.

Lang, E. M. (1963). Flamingoes raise their young on a liquid containing blood. Experientia 19:532–533.

Le Douarin, N. (1973). A biological cell labeling technique and its use in experimental embryology. Developmental Biology 30:217–222.

Lilja, C. (1982). A comparative study of postnatal growth and organ development in some species of birds. Growth 47:317–339.

Lilja, C., I. Sperber, and H. L. Marks (1984). Postnatal growth and organ development in Japanese quail selected for high growth rate. Growth 49:51–62.

Lilja, C., J. Blom, and H. L. Marks (2001). A comparative study of embryonic development of Japanese quail selected for different patterns of postnatal growth. Zoology 104:115–122.

Linde-Medina, M., and S. A. Newman (2014). Limb, tooth, beak: Three modes of development and evolutionary innovation of form. Journal of Biosciences 39 (2): 211–223.

Lucas, A. M., and P. R. Stettenheim (1972). Avian anatomy. Integument. Agriculture Handbook 362, United States Department of Agriculture, Superintendent of Documents, Washington, DC.

Maderson, P. F., W. J. Hillenius, U. Hiller, and C. C. Dove (2009). Towards a comprehensive model of feather regeneration. Journal of Morphology 270:1166–1208.

Mann, K. (2007). The chicken egg white proteome. Proteomics 7:3558–3568.

Mann, K., and M. Mann (2008). The chicken egg yolk plasma and granule proteomes. Proteomics 8:187–191.

Margerie, E. de, J. Cubo, and J. Castanet (2002). Bone typology and growth rate: Testing and quantifying "Amprino's rule" in the mallard (Anas platyrhynchos). Comptes Rendus Biologies 325:221–230.

Margerie, E. de, J. P. Robin, D. Verrier, J. Cubo, R. Groscolas, and J. Castanet (2004). Assessing a relationship between bone microstructure and growth rate: A fluorescent labelling study in the King Penguin chick (Aptenodytes patagonicus). Journal of Experimental Biology 207:869–879.

Martin, T. E. (2002). A new view of avian life-history evolution tested on an incubation paradox. Proceedings of the Royal Society of London B 269:309–316.

Martin, T. E., S. K. Auer, R. D. Bassar, A. M. Niklison, and P. Lloyd (2007). Geographic variation in avian incubation periods and parental influences on embryonic temperature. Evolution 61:2558–2569.

Maxwell, E. E. (2008a). Comparative embryonic development of the skeleton of the domestic turkey (*Meleagris gallopavo*) and other galliform birds. Zoology 111:242–257.

Maxwell, E. E. (2008b). Ossification sequence of the avian order Anseriformes, with comparison to other precocial birds. Journal of Morphology 269:1095–1113.

Nice, M. M. (1962). Development of behavior in precocial birds. Transactions of the Linnean Society NY 8:1–211.

Noble, R. C., and M. Cocchi (1990). Lipid metabolism and the neonatal chicken. Progress in Lipid Research 29:107–140.

Noordwijk, A. van, and H. L. Marks (1998). Genetic aspects of growth. *In* Avian growth and development: Evolution within the altricial-precocial spectrum, J. M. Starck and R. E. Ricklefs, Editors. Oxford University Press, New York, pp. 305–323.

Noy, Y., and D. Sklan (2001). Yolk and exogenous feed utilization in the posthatch chick. Poultry Science 80:1490–1495.

O'Connor, R. J. (1984). The growth and development of birds. Wiley, Chichester, UK.

Oken, L. (1816). Lehrbuch der Zoologie-mit vierzig Kupfertafeln. Zweite Abtheilung, Fleischthiere. August Schmid, Jena.

Orlowski, G., and L. L. Halupka (2015). Embryonic eggshell thickness erosion: A literature survey re-assessing embryo-induced eggshell thinning in birds. Environmental Pollution 205:218–224.

Österström, O., and C. Lilja (2012). Evolution of avian eggshell structure. Journal of Morphology 273:241–247.

Paul, G. S. (1994). Dinosaur reproduction in the fast lane: Implications for size, success, and extinction. *In* Dinosaur eggs and babies, K. Carpenter, K. F. Hirsch, and J. R. Horner, Editors. Cambridge University Press, Cambridge, MA, pp. 244–255.

Phillips, R. A., and K. C. Hamer (1999). Lipid reserves, fasting capability and the evolution of nestling obesity in procellariiform seabirds. Proceedings of the Royal Society of London B: Biological Sciences 266 (1426): 1329–1334.

Portmann, A. (1935). Die Ontogenese der Vögel als Evolutionsproblem. Acta Biotheor 1A:59–90.

Portmann, A. (1938). Die Ontogenese der Vögel als Evolutionsproblem II. Zahl der Jungen, Tragzeit und Ausbildungsgrad der Jungen bei der Geburt. Bio-Morphosis 1:109–126.

Portmann, A., and L. Vischer (1942). Über das Verhältnis von Sinnesorganen, Stoffwechselorganen und Bewegungsapparat in der Körpermasse der Vögel. Revue Suisse De Zoologie 49: 277–282.

Raginosa, M. N. (1961). Embryonic development of the chicken. Academy of Science of the USSR.

Réhault-Godbert, S., K. Mann, M. Bourin, A. Brionne, and Y. Nys (2014). Effect of embryonic development on the chicken egg yolk plasma proteome after 12 days of incubation. Journal of Agricultural and Food Chemistry 62:2531–2540.

Ricklefs, R. E. (1967). A graphical method of fitting equations to growth curves. Ecology 48:978–983.

Ricklefs, R. E. (1968). Patterns of growth in birds. Ibis 110:419–451.

Ricklefs, R. E. (1969a). Preliminary models for growth rates in altricial birds. Ecology 50:1031–1039.

Ricklefs, R. E. (1969b). An analysis of nestling mortality in birds. Smithsonian Contributions to Zoology 9:1–48.

Ricklefs, R. E. (1973). Patterns of growth in birds II. Growth rate and mode of development. Ibis 115:177–201.

Ricklefs, R. E. (1977). Composition of eggs of several bird species. The Auk 94:350–356.

Ricklefs, R. E. (1979a). Adaptation, constraint, and compromise in avian postnatal development. Biological Reviews 54:269–290.

Ricklefs, R. E. (1979b). Patterns of growth in birds. V. A comparative study of development in the starling, common tern, and Japanese quail. The Auk 96:10–30.

Ricklefs, R. E. (1983). Avian postnatal development. *In* Avian biology, vol. 7, D. S. Farner, J. R. King, and K. C. Parkes, Editors. Academic Press, New York, pp. 1–83.

Ricklefs, R. E. (1993). Sibling competition, hatching asynchrony, incubation period, and lifespan in altricial birds. Current Ornithology 11:199–276.

Ricklefs, R. E. (2012). Avian energetics, ecology and evolution. *In* Avian energetics and nutritional ecology, Cynthia Carey, Editor. Chapman and Hall, New York, pp. 1–30.

Ricklefs, R. E., and S. Weremiuk (1977). Dynamics of muscle growth in the starling and Japanese quail: A preliminary study. Comparative Biochemistry and Physiology Part A: Physiology 56:419–423.

Ricklefs, R. E., A. R. Place, and D. J. Anderson (1987). An experimental investigation of the influence of diet quality on growth in Leach's storm-petrel. American Naturalist 130:300–305.

Ricklefs, R. E., R. E. Shea, and I. H. Choi (1994). Inverse relationship between functional maturity and exponential growth rate of avian skeletal muscle: A constraint on evolutionary response. Evolution 48:1080–1088.

Ricklefs, R. E., J. M. Starck, and M. Konarzewski (1998). Internal constraints on growth in birds. *In* Avian growth and development: Evolution within the altricial-precocial spectrum, J. M. Starck and R. E. Ricklefs, Editors. Oxford University Press, New York, pp. 266–287.

Riley, A., C. J. Sturrock, S. J. Mooney, and M. R. Luck (2014). Quantification of eggshell microstructure using X-ray micro computed tomography. British Poultry Science 55:311–320.

Romanoff, A. L., and A. J. Romanoff (1949). The avian egg. John Wiley and Sons, New York.

Rose-Martel, M., D. Jungwen, and M. T. Hincke (2012). Proteomic analysis provides new insight into the chicken eggshell cuticle. Journal of Proteomics 75:2697–2706.

Rose-Martel, M., and M. T. Hincke (2013). Eggshell as a source of novel bioactive molecules. Journal of Food Science and Engineering 3:219–225.

Royle, N. J., I. R. Hartley, I. P. Owens, and G. A. Parker (1999). Sibling competition and the evolution of growth rates in birds. Proceedings of the Royal Society of London B: Biological Sciences 266:923–932.

Schew, W. A., and R. E. Ricklefs (1998). Developmental plasticity. *In* Avian growth and development: Evolution within the altricial-precocial spectrum, J. M. Starck and R. E. Ricklefs, Editors. Oxford University Press, New York, pp. 288–304.

Schönwetter, M., and W. Meise (1960–1992). Handbuch der Oologie. Vols. 1–4, Akademie Verlag, Berlin.

Shea, R. E., J. M. Olson, and R. E. Ricklefs (2007). Growth rate, protein accumulation, and catabolic enzyme activity of skeletal muscles of galliform birds. Physiological and Biochemical Zoology 80:306–316.

Skutch, Alexander F. (1976). Parent birds and their young. Austin University of Texas Press, Austin.

Smart, I. H. M. (1991). Egg shape in birds. In Egg incubation: Its effects on embryonic development in birds and reptiles, C. D. Deeming and M. W. Ferguson, Editors. Cambridge University Press, Cambridge, UK, pp. 101–116.

Sotherland, P. R., and H. Rahn (1987). On the composition of bird eggs. The Condor 89:48–65.

Starck, J. M. (1989). Zeitmuster der Ontogenesen bei nestflüchtenden und nesthockenden Vögeln. Courier Forschungs-Institut Senckenberg 114:1–318.

Starck, J. M. (1993). Evolution of avian ontogenies. Current Ornithology 10:275–366.

Starck, J. M. (1996). Comparative morphology and cytokinetics of skeletal growth in hatchlings of altricial and precocial birds. Zoologischer Anzeiger 235:53–75.

Starck, J. M. (1998). Structural variants and invariants in avian development. In Avian growth and development: Evolution within the altricial-precocial spectrum, J. M. Starck and R. E. Ricklefs, Editors. Oxford University Press, New York, pp. 59–88.

Starck, J. M., and R. E. Ricklefs (1998a). Patterns of development: The altricial-precocial spectrum. In Avian growth and development: Evolution within the altricial-precocial spectrum, J. M. Starck and R. E. Ricklefs, Editors. Oxford University Press, New York, pp. 3–30.

Starck, J. M., and R. E. Ricklefs (1998b). Variation, constraint and phylogeny: Comparative analysis of variation in growth. In Avian growth and development: Evolution within the altricial-precocial spectrum, J. M. Starck and R. E. Ricklefs, Editors. Oxford University Press, New York, pp. 247–265.

Starck, J. M., and R. E. Ricklefs (1998c). Avian growth rate data set. In Avian growth and development: Evolution within the altricial-precocial spectrum, J. M. Starck and R. E. Ricklefs, Editors. Oxford University Press, New York, pp. 381–423.

Starck, J. M., and E. Sutter (2000). Patterns of growth and heterochrony in moundbuilders (Megapodiidae) and fowl (Phasianidae). Journal of Avian Biology 31:527–547.

Starck, J. M., and A. Chinsamy (2002). Bone microstructure and developmental plasticity in birds and other dinosaurs. Journal of Morphology 254:232–246.

Stern, C. D., and K. M. Downs (2012). The hypoblast (visceral endoderm): An evo-devo perspective. Development 139:1059–1069.

Sundeval, C. E. (1836). Ornithologiskt System. Kungliga Svenska Vetenskapsakademiens Handlingar, pp. 43–130.

Troscianko, J. (2014). A simple tool for calculating egg shape, volume and surface area from digital images. Ibis 156:874–878.

Weishampel, D. B., and J. R. Horner (1994). Life history syndromes, heterochrony, and the evolution of the Dinosauria. In Dinosaur eggs and babies, K. Carpenter, K. F. Hirsch, and J. R. Horner, Editors. Cambridge University Press, Cambridge, MA, pp. 229–243.

Wellmann-Labadie, O., J. Picman, and M. T. Hincke (2008). Antimicrobial activity of the anseriform outer eggshell and cuticle. Comparative Biochemistry and Physiology B 149:640–649.

Williams, B., S. Solomon, D. Waddington, B. Thorp, and C. Farquharson (2004a). Skeletal development in the meat-type chicken. British Poultry Sciences 41:141–149.

Williams, B., D. Waddington, D. H. Murray, and C. Farquharson (2004b). Bone strength during growth: Influence of growth rate on cortical porosity and mineralization. Calcified Tissue International 74:236–245.

Williams, G. C. (1966). Natural selection, the costs of reproduction, and the refinement of Lack's principle. American Naturalist 100:687–690.

Yadgary, L., R. Yair, and Z. Uni (2011). The chick embryo yolk sac membrane expresses nutrient transporter and digestive enzyme genes. Poultry Science 90:410–416.

Yadgary, L., O. Kedar, O. Adepeju, and Z. Uni (2013). Changes in yolk sac membrane absorptive area and fat digestion during chick embryonic development. Poultry Science 92:1634–1640.

CHAPTER 6

Anatomy

Margaret A. Voss and Marco Pavia

Birds are capable of sustaining high levels of physical activity across a wide range of temperatures because they are **endotherms**. This ability to maintain stable internal body temperatures (i.e., **thermoregulate**) makes flight—which is a novel but energetically costly mode of locomotion—possible. The biomechanical requirements of flight began to modify the vertebrate body plan when the earliest birds took to the air. This is apparent when we compare the basic body plan of birds with the diversity of anatomical structures and forms of locomotion seen in mammals; the laws of aerodynamics clearly demand a high degree of anatomical uniformity. All extant bird species share a common set of physical characteristics that are recognized as uniquely avian, even though some modern birds no longer rely on flight to get around. The most ornamental and beautifully distinctive feature of birds is the **feather**, the defining characteristic of the **class Aves**. Additional adaptations to flight include a relatively small body size, high metabolism, lack of heavy structures like teeth, extensive fusion and reduction of bones to reduce body mass, hollow bones, and specialized forelimbs we recognize as wings (Proctor and Lynch 1993). Birds are derived from **tetrapod** ancestors (**theropods**, see chapter 2) that walked on land and are therefore constrained to a four-appendage body plan. Wings, however, are of limited use for walking. Birds are thus limited to a bipedal mode of locomotion on the ground, using only their hindlimbs for walking and hopping.

Adaptation to flight required modifications in other anatomical systems as well, such as a uniquely efficient respiratory system that makes use of **pneumatic air sacs** within hollow bones and a lung ventilation mechanism (see chapter 7) that works within the structural reinforcement necessary to stabilize the rib cage during flight. Although certain bird bones can be quite dense regionally, to resist compression and torsion in flight (Dumont 2010), hollow spaces lower the specific gravity of the avian body overall. The body of a feathered duck reportedly has a specific gravity of 0.6, while that of our own species is approximately 1.0 (Welty and Baptista 1988). The avian skull has been lightened through the fusion and reduction of bones to the point that it cannot sustain the pressures generated by chewing and grinding. A lightweight **keratin**-covered bill is far more practical than heavy mineralized teeth for finding and manipulating food and for grooming feathers. The lack of teeth, however, requires major modification to the vertebrate digestive system. The digestive system must have spaces (e.g., a **crop**) to temporarily store and soften food acquired "on the wing." Even in the absence of teeth, food must be broken up and reduced to small pieces (**masticated**) for proper digestion. This function is fulfilled by the muscular stomach of birds, a novel structure called the **ventriculus**, or **gizzard**. Soft structures such as fluid-filled bladders are also heavy and impractical in flight, so the avian **urogenital system** has been modified to produce a relatively lightweight, pasty **uric acid** waste product (**guano**) that is excreted through a **cloaca**, a common exit for both excretory and reproductive products.

The avian reproductive system has also been modified for flight and to meet the energetic demands of long-distance migration. The **gonads** of both male and female birds are often so small they are almost undetectable for most of the year. As the breeding season approaches, gonadal mass may increase by an order of magnitude in preparation for reproduction. Birds do not give birth to live young. They encapsulate their developing offspring in a calcareous, **cleidoic** egg that requires some form of external incubation. Incubation occurs either through contact with a vascularized parental **brood patch** (e.g., in passerine species) or from environmental heat sources, such as within an insulated mound of composting vegetation (e.g., the Australasian megapodes).

COMMON ANATOMICAL TERMS

There are many pairs of anatomical terms that describe the opposite directions or opposing actions of the vertebrate body. In birds, the long axis of the body is usually along a horizontal plane (fig. 6.1), so many of these terms become synonymous with each other (e.g., posterior and caudal).

Dorsal *Ventral*	Imagine a line running along the horizontal plane of a bird's body from its beak to its tail. The dorsal surface would be the upper region (the bird's back) and the ventral surface the lower region (the bird's chest and abdomen).
Anterior *Posterior*	Imagine a line running along a vertical plane from a bird's shoulder to its chest. The front half of the bird's body, including the head, is anterior. The back half of the body, including the tail, is posterior. The words rostral, cranial, and cephalic are occasionally used to refer to the anterior portion of the body. Caudal is often used to refer to the back half of the bird's body.
Cephalic *Caudal*	These terms are synonymous with anterior and posterior in birds. Cephalic means toward the head and caudal toward the tail.
Lateral *Medial*	Imagine a line that splits the bird's body into right and left halves. Medial structures are located closer to the midline of the body. Lateral structures are farther from the midline of the body.
Proximal *Distal*	These terms describe the location of a structure with reference to the midline of the body (see above) or the point of attachment. Proximal structures are closer to the midline, or point of attachment, and distal structures are farther from the point of attachment.
Extension *Flexion*	Extension is the movement of a body part or appendage away from midline of the body. Flexion is the movement of a body part or appendage toward midline.
Pronation *Supination*	Pronation is the rotation of an appendage forward and ventrally. In birds, the leading edge of the wing is pronated, or rotated downward, during the downstroke of flapping flight. Supination is the rotation of an appendage backward and dorsally. In the upstroke of flight, the wing is supinated and the leading edge is rotated upward.

The only avian anatomical system that does not show marked reduction in size and weight is the nervous system. Bird brains are anything but small. Compared to its body mass, the brain of a bird is larger than that of other vertebrates of comparable size (e.g., reptiles, small mammals). This is also an adaptation for flight. Large eyes relative to body size and fine coordination of muscles are necessary to achieve the aerial acrobatics and swift maneuvers observed in the many bird species that feed on the wing. Aerial feeding, and flight in general, requires a large amount of information to be processed simultaneously. To meet this demand, the avian brain is equipped with relatively large **cerebral**, **cerebellar**, and **optic lobes** when compared with body mass. As this chapter and the next will demonstrate, birds are truly remarkable and efficient flying "machines," with unique anatomical and physiological adaptations to meet the energetic and biomechanical demands of life on the wing.

INTEGUMENT

Skin, or more properly, the **integument**, serves as a physical boundary to separate the tightly controlled internal chemical environment of animals from the harsh (from the cellular perspective) and variable external environment. The integument must integrate this protective role with many other functions crucial to the success of any living organism. As such, it has a variety of sensory structures that inform the central nervous system about what the animal physically contacts (e.g., cavity, nest, egg, other organisms) and what the surrounding environment is like (e.g., temperature, wind, precipitation). The basic structure of the avian integument is similar to all other vertebrates, although it is notably thinner than the skin of mammals in the areas normally covered with feathers. Most feathers are distributed across the integument in well-defined tracts called **pterylae** (fig. 6.1). The thin, pliable skin in these regions is loosely attached to the

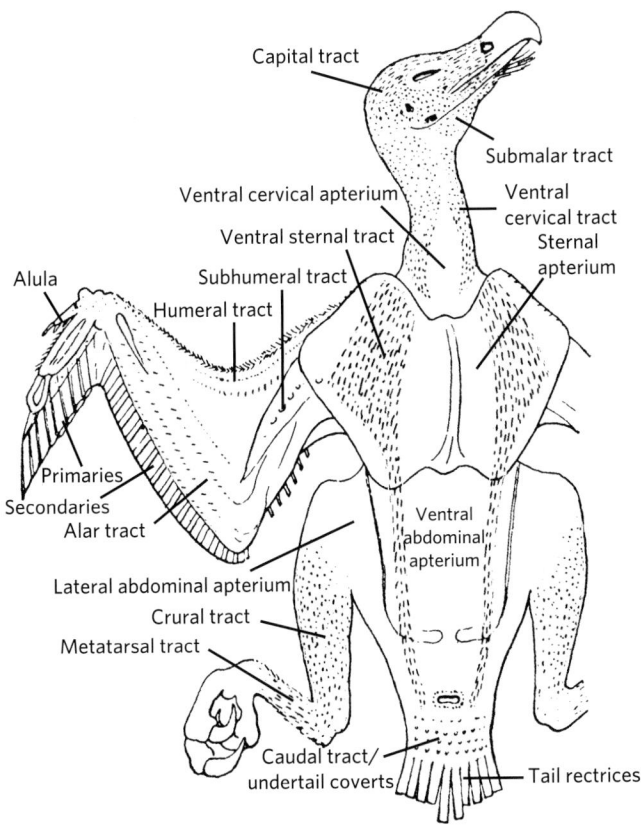

Capital tract

Submalar tract

Ventral cervical apterium

Ventral cervical tract

Ventral sternal tract

Sternal apterium

Subhumeral tract

Alula

Humeral tract

Primaries

Secondaries

Alar tract

Ventral abdominal apterium

Lateral abdominal apterium

Crural tract

Metatarsal tract

Caudal tract/ undertail coverts

Tail rectrices

Figure 6.1. The feathers covering the body of the bird and those used for flight are collectively called **contour feathers** *(see chapter 9 for additional detail). These feathers streamline the bird's body and contribute to its aerodynamic form. Contour feathers are usually distributed in well-defined tracts called* **pterylae.** *The bare, unfeathered areas between the tracts are called* **apteria.** *The distribution of the pterylae can vary greatly among species and is characteristic within taxonomic groups of birds. In some flightless birds, such as penguins, distinct pterylae may be absent and the feathers distributed evenly across the skin. This figure is a drawing of a juvenile bearded griffin (Gypaetus barbatus), modified from Thomson 1916, with reference to Proctor and Lynch 1993.*

underlying muscle to accommodate the freedom of movement necessary for flight. The skin is more firmly attached in the areas that cover the skull, the wing tips, and the skeleton of the beak and feet, where it is unfeathered and often highly specialized (Stettenheim 2000).

Bird skin, like that of all vertebrates, consists of two layers: a thin **epidermis** that covers a deeper **dermis.** The thicker dermal layer contains blood vessels, fat deposits, free nerve endings, nerves, neuroreceptors, and smooth muscles that move the feathers (Stettenheim 2000). The epidermis is further divided into two main layers: a superficial **stratum corneum** and a deeper **stratum germinativum** (fig. 6.2). The avian stratum corneum is similar in structure and function

to that of mammals and reptiles. It consists of flattened cells (**keratinocytes**) that contain a fibrous protein called **keratin** that helps create a barrier to prevent excessive evaporative water loss. The epidermis of a bird also contains unique cells called **sebokeratinocytes** that secrete sebum-like lipids (Menon and Menon 2000). Sebokeratinocytes play an important role in thermoregulation. Mammals, the other group of endothermic vertebrates, use sweat glands for thermoregulation by evaporative cooling; however, sweat glands are of limited use to a bird. Sweat would saturate down feathers to produce a lumpy, wet mat that would inhibit evaporative cooling of the skin (Welty and Baptista 1988). On the other hand, the complete lack of any evaporative cooling mechanism would also be problematic given the high internal body temperature of birds, their insulating layer of plumage, and their ability to generate a lot of heat during flight. The sebokeratinocytes solve this problem by responding adaptively to a range of thermal conditions. When there is a need to conserve water, the cells secrete a lipid barrier to prevent water loss. When evaporative cooling is required to alleviate high body temperatures, the lipid secretion is inhibited (Menon and Menon 2000).

Glands

Although birds lack sweat glands, other types of integumental glands are present. Most vertebrates have some form of oral or salivary gland. Since these glands are important to the digestive process, their presence or absence in birds is discussed with digestion (below). At the opposite end of the digestive tract, many bird species have **vent** glands that open into the **cloaca** near its opening. The location of these glands varies widely, but all secrete a mixture of mucopolysaccharides. The function of these glands is not entirely understood. They are thought to play an important role in reproduction, and there is some evidence that they are important in internal fertilization in at least one species, the Japanese Quail (*Coturnix coturnix*) (Quay 1967). Birds also have **ceruminous,** or **ear glands,** that open along the floor of the outer ear canal (Menon 1984). The function of the wax secreted from the ceruminous glands is also not entirely clear, although there is some evidence that it may trap particles to keep the ear canal clear (Stettenheim 2000).

The avian **uropygial** or preen gland is better understood. This is usually a bilobed structure (figs. 6.3A, B), although it can take on a wide range of species-specific sizes and shapes (Salibian and Montalti 2009). The gland has been found in the developing embryos of all bird species examined so far, although it regresses to become a vestigial structure in the adults of certain species. Adults of species in at least nine families of birds lack an uropygial gland entirely. In other

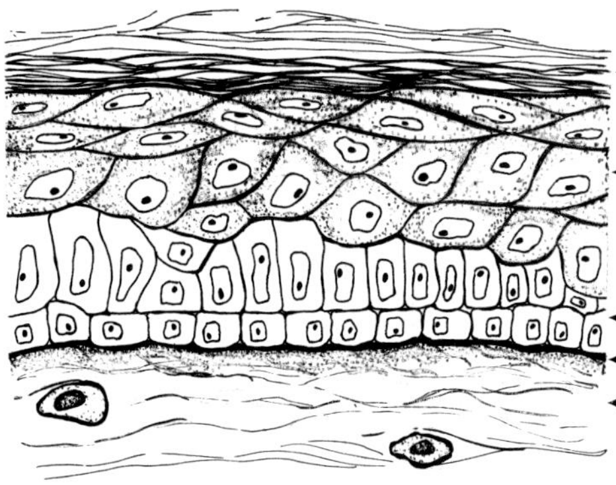

Figure 6.2. Diagram of avian skin in cross section. The epidermis has two main layers, the super-ficial **stratum corneum** and a deeper **stratum germinativum**. The stratum corneum consists of flattened, keratinized cells. The outer layer of transitional cells of the stratum germinativum is granular **sebokeratinocytes** that gradually fill with extracellular lipids over time. Together, the two epidermal layers provide a tough barrier that regulates evaporative water loss. When there is a need to conserve water, the sebokeratinocytes secrete a lipid barrier to prevent water loss. When evaporative cooling is required to alleviate high body temperatures, the lipid secretion is inhibited. *Modified from original drawing in A. M. Lucas and P. R. Stettenheim 1972.*

taxonomic groups, the glands may be small and nonfunctional or entirely absent. This includes many flightless species such as ostriches (Struthionidae), rheas (Rheidae), cassowaries (Casuariidae), emu (Dromaiidae), kiwis (Apterygidae), weak fliers such as the mesites (Mesitornithidae) and bustards (Otididae), and parrots (psittacines) (Welty and Baptista 1988). Species in which the uropygial gland is nonfunctional or absent usually produce a powder down or rely on dust baths for feather maintenance.

Recent phylogenetic analyses have been unable to reliably correlate the presence or absence of the uropygial gland with factors such as distribution, climate, or ecology (Salibian and Montalti 2009), although, historically, preen gland secretions have been assumed to be important waterproofing agents in waterfowl (Vincze et al. 2013). It is more likely that the secretions have a broad range of functions that differ among species. Preen gland secretions are chemically complex, often consisting of more than 100 different stable wax esters (Haribal et al. 2005). Preen-wax compositions can vary with season (Piersma et al. 1999, Reneerkens et al. 2008), and some compounds make better water repellents than others (Sweeney et al. 2004). The interaction between preen gland secretion and the microflora of the integument is also an emerging area of interest (Reneerkens et al. 2008) and might be of importance in avian chemical ecology. For example, Hagelin and Jones (2007) have raised the intriguing possibility that avian glandular secretions might be used for chemical communication within bird species (see also chapter 14).

Keratin and Integumentary Derivatives

As we saw in the preceding section, **keratins** are a class of long-lasting fibrous structural proteins produced by the keratinocytes of the epidermal stratum corneum (fig. 6.2). As keratin builds up in these cells, changes occur in cell shape and function. With time, the normal cycle ceases and the keratinocyte loses its nucleus and dies. When the cells are fully **cornified**, they have been converted to sheets of flattened keratin-filled squamous epithelia. Cornified epithelial sheets are the major structural component of epidermal derivatives such as **feathers**, **beaks**, **talons**, and **scales** (Stettenheim 1972). The outermost layer of cornified cells and some of the keratinized epidermal structures (e.g., feathers) are shed at regular intervals, such as during periods of **molt** (see chapter 9).

Feathers

Some mammals may share the modified forelimbs required for flight, but birds alone bear the unique and colorful integumentary derivatives we call **feathers**. Feathers are a product of a cell signaling interaction between the dermis and the epidermis of the integument. With cues from the dermis, the epidermis begins to invaginate and push down into the dermis to form the **feather follicle** during embryonic development (chapters 5 and 9). The fully formed feather that we see above the skin is purely epidermal in origin, although the internal pulp, blood vessels, and **papilla** that support feather growth are all dermal structures

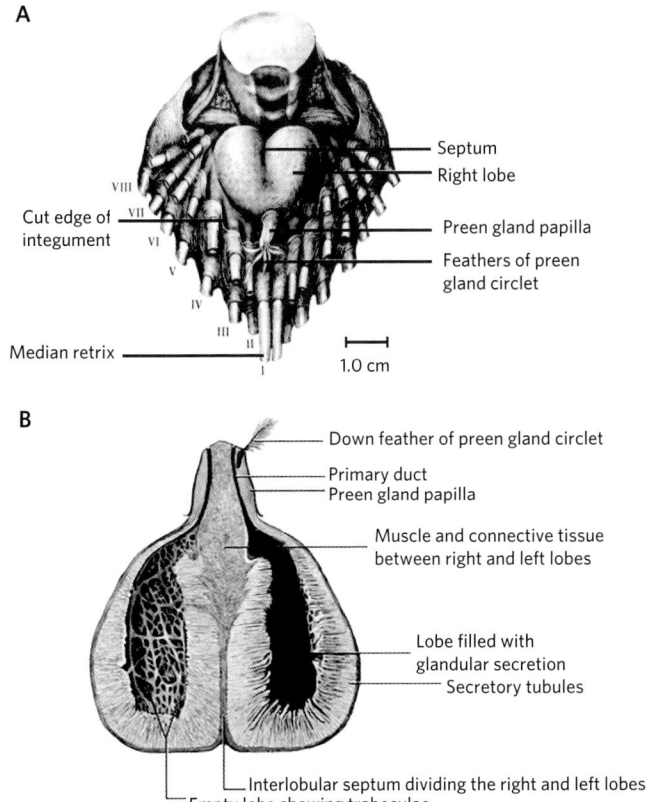

A

Septum
Right lobe
Cut edge of integument
Preen gland papilla
Feathers of preen gland circlet
Median retrix
1.0 cm

B

Down feather of preen gland circlet
Primary duct
Preen gland papilla
Muscle and connective tissue between right and left lobes
Lobe filled with glandular secretion
Secretory tubules
Interlobular septum dividing the right and left lobes
Empty lobe showing trabeculae

Figure 6.3. A, Dorsal view of the dissected **uropygial** (preen) gland of a Single Comb White Leghorn Chicken (*Gallus gallus domesticus*); *B,* Cross section of the dissected uropygial gland illustrating the uropygial papilla and ducts. The empty right lobe shows the internal structure of the gland, including trabeculae. *Modified from the original illustrations in Lucas and Stettenheim 1972.*

(fig. 6.4). The dermal layer determines what type of feather the epidermal structure will become (see types of feathers, chapter 9). The feather arises from its dermal base and will continue its growth until it reaches a predetermined length, at which time its dermal cells cease the normal cell cycle, become dormant, and are eventually reabsorbed. The epidermal cells have become almost fully keratinized and filled with β-**keratin** at this point. In the mature feather, the only evidence of the once-living dermal pulp and blood vessels is a hole called the **inferior umbilicus**, at the bottom of the **feather shaft** in a region called the **calamus** (figs. 6.4 and 6.5). In addition to the embedded calamus, the feather shaft has a region outside the follicle called the **rachis**. At the top of the calamus, before the rachis begins, is a second small opening called the **superior umbilicus**. This is where blood vessels that enter through the inferior umbilicus (fig. 6.4) exit the feather shaft and return through the follicle to the dermis (fig. 6.5). Above the level of the follicle,

each feather has two parts: the **vane**, divided into two parts by the rachis, and a small **afterfeather** or **aftershaft**, which is attached at the superior umbilicus. All feather types, with the exception of **filoplumes** (see chapter 9), are attached to the follicle by tiny bundles of smooth muscle that link adjacent follicles. This allows the feathers in a single tract (**pteryla**, fig. 6.1) to be raised, lowered, or rotated together (Stettenheim 2000).

Compared with other vertebrate epidermal derivatives (i.e., hair, scales, etc.), feathers are very complex and have a highly specialized architecture. An interlocking mesh of **barbs** and **barbules** form the bulk of the **vane**. The microscopic structure and the level of organization of the barbs and barbules (i.e., how tightly the mesh locks together) differs according to the function of the feather. This gives rise to many specialized types of feathers (see chapters 7, 9, and 10). Regardless of specialization, all feathers are replaced as they become worn and damaged over time. This is a process called molt (chapter 9), and it may occur gradually in some species (e.g., the wear molt of starlings and house sparrows) or at a discrete point in time in others (e.g., penguins, grebes, pelicans, auks, and migratory waterfowl). In some species a partial molt just prior to the reproductive season will produce showy courtship plumage (chapter 11). During molt, a new feather will develop within the existing feather follicle, eventually pushing the old feather out.

At this point you probably recognize the importance of feathers to both birds and ornithologists. We characterize birds taxonomically by the presence of feathers, and we use field marks to distinguish between species from afar (chapter 31). Feathers contribute to distinctive silhouettes and are often used by birds themselves to identify potential mates during courtship (chapter 11). Feathers also serve as structural specializations that facilitate the physiology of thermoregulation (chapter 7) and the physics of flight (chapter 10). Recently, feathers have become important for noninvasive genetic tracking of bird populations. Given their diverse functions and unique roles in avian behavior and ecology, feathers are discussed more fully in chapters that follow (chapters 7, 9, 10, and 11).

Beaks

Beaks take on a diversity of shapes and sizes governed by the selective pressures of diverse foraging habitats and behaviors (fig. 6.22; also see chapters 15 and 18). There is a common structure and pattern of development, however, that gives rise to this diversity. Every beak consists of a dermal core (i.e., the **premaxilla** and **maxilla/dentary**

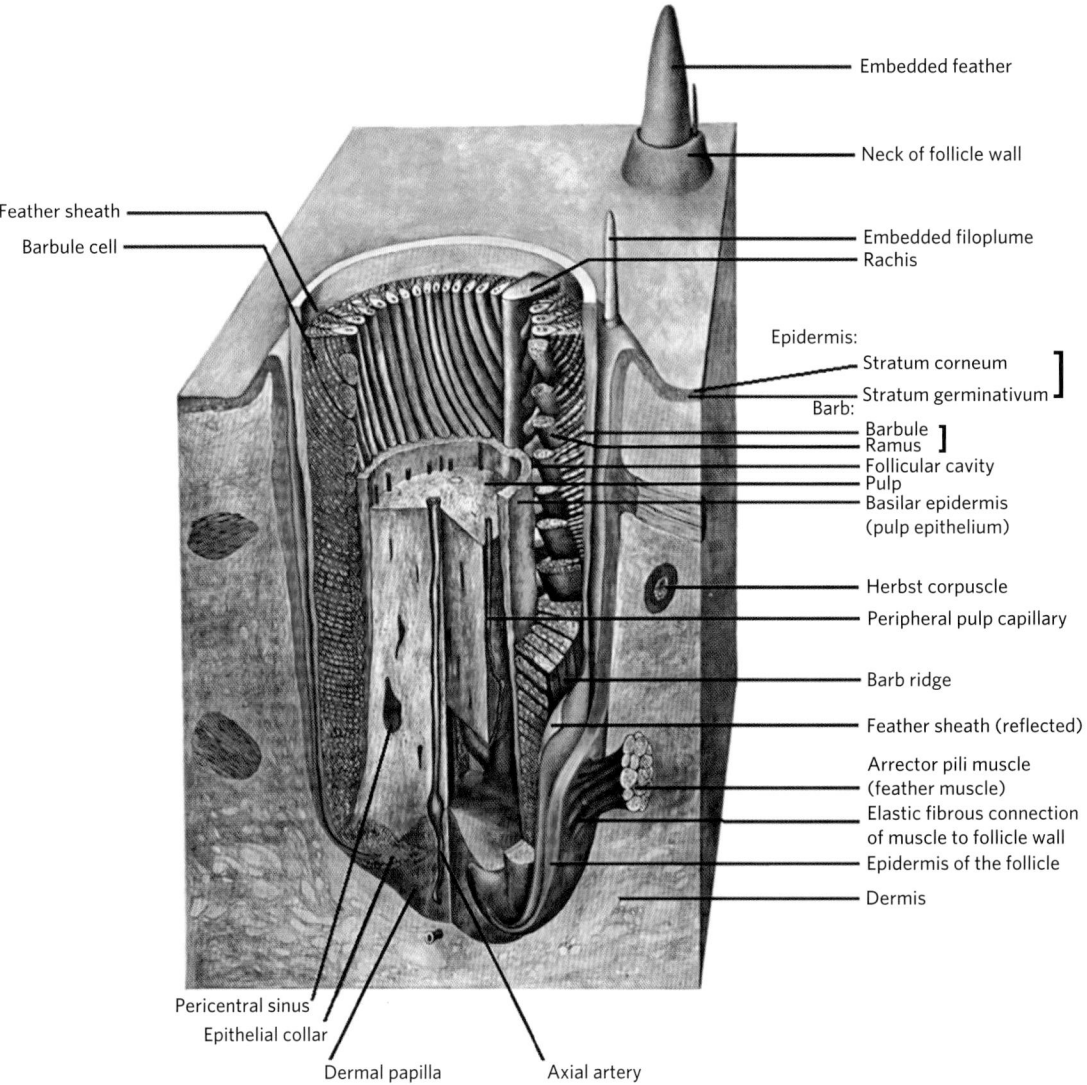

Feather sheath
Barbule cell

Embedded feather
Neck of follicle wall
Embedded filoplume
Rachis
Epidermis:
Stratum corneum ⎤
Stratum germinativum ⎦
Barb:
Barbule ⎤
Ramus ⎦
Follicular cavity
Pulp
Basilar epidermis
(pulp epithelium)
Herbst corpuscle
Peripheral pulp capillary
Barb ridge
Feather sheath (reflected)
Arrector pili muscle
(feather muscle)
Elastic fibrous connection
of muscle to follicle wall
Epidermis of the follicle
Dermis

Pericentral sinus
Epithelial collar
Dermal papilla   Axial artery

Figure 6.4. Diagram of a longitudinal section of a growing **contour feather**. Although the diagram is oriented vertically, note that feathers would actually grow at an angle from the skin. *Figure from Lucas and Stettenheim 1972.*

bones) sheathed in a compact layer of epidermal origin. The epidermal layer has the same basic structure previously discussed and illustrated in figure 6.2. The stratum corneum makes up the outermost surface of the beak, a thin hard layer of β-keratin called the **rhamphotheca** (fig. 6.6). The keratin of the rhamphotheca develops from layers of cells that originate in the stratum germinativum and migrate upward to the stratum corneum in a delamination process (Campbell and Lack 1985). The word delamination literally means the "separation of sheets." The rhamphotheca grows continuously and may vary in color by sex and between seasons. The rhamphotheca may also be shed yearly in some species (e.g., some penguins; male American White Pelicans, *Pelecanus erythrorhynchos*; and puffins; Stettenheim 1972).

Many avian species are **altricial**, meaning they are born in a relatively underdeveloped state and require an extended

period of parental care before they become self-sufficient. In altricial species, the beak and the claws of the embryos are not fully developed at hatch and are ineffective in helping the neonate break free of the egg. In these species, one or more **egg teeth** are present. The egg tooth of most species is a conical structure of hard keratin that will be lost as the beak further develops posthatch (Lucas and Stettenheim 1972).

Scales and Claws

Feathers do not extend all the way down the legs to the **digits** (toes) in most birds, although certain grouse, owls, and domestic fowl are notable and often showy exceptions to this rule. Most birds have **scales**, another epidermal-derived keratin structure, that cover the legs and feet. Scales usually increase in prominence distally toward the toes and on the anterior surface of the legs (Stettenheim 2000) (refer to the

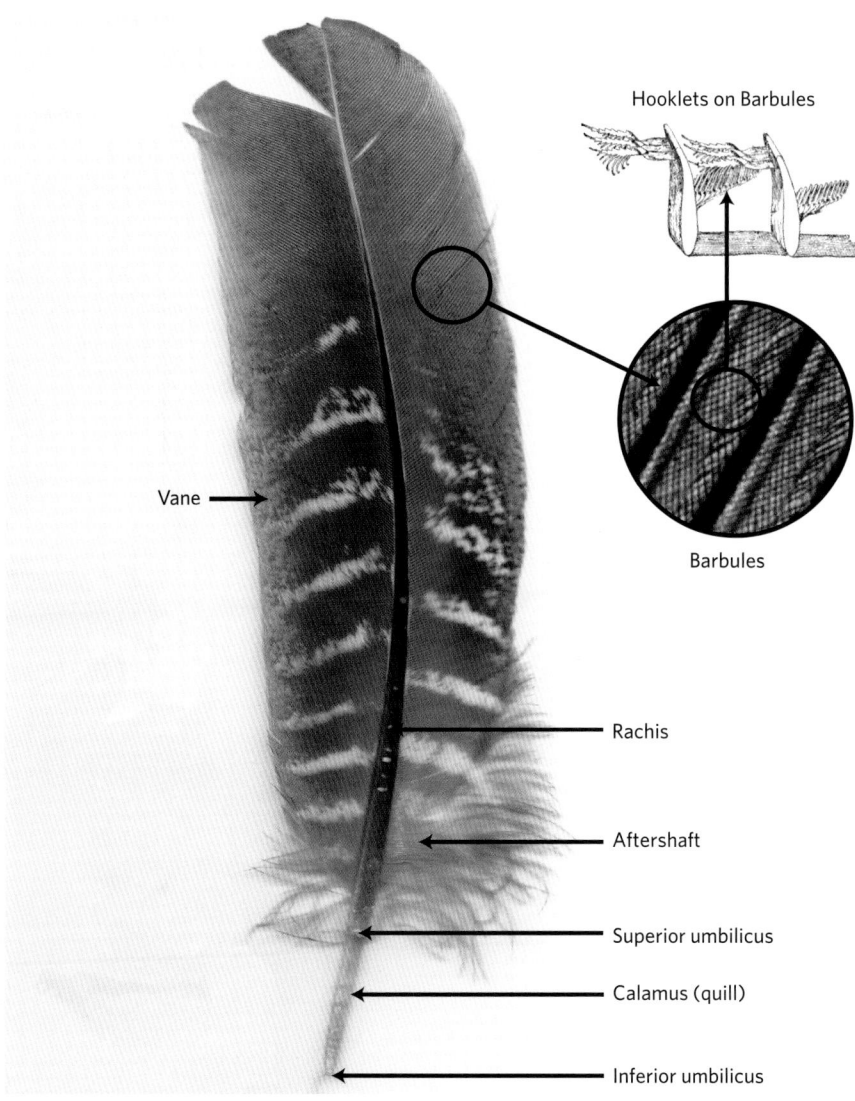

Hooklets on Barbules

Vane

Barbules

Rachis

Aftershaft

Superior umbilicus

Calamus (quill)

Inferior umbilicus

Figure 6.5. Anatomy of a **primary flight feather** (Green Pheasant, *Phasianus versicolor*). The anatomical structure of the primary flight feather is uniquely adapted to its role in generating lift during flight (see chapter 10). Note that the **vane** of a primary feather is composed of **barbs** that interlock by way of a network of **barbules** and **hooklets**. *Photos modified from the originals by Alpsdake via Wikimedia Commons, https:// commons.wikimedia.org/wiki/File:Phasianus _versicolor_(flight_feather).jpg; https:// commons.wikimedia.org/wiki/File:Feather01 .jpg; illustration from Bahr 1907.*

box on page 130 for anatomical terms). Avian scales differ in many important anatomical and developmental ways from the scales of fish and reptiles—which is an extensive topic of discussion for another book (Dhouailly 2009). Bird scales follow a pattern of development similar to that of feathers, dependent on a biochemical interaction between a dermal papilla and the overlying epidermis. The structures produced by this interaction are flat, rounded, or occasionally conical raised areas of highly keratinized epidermis (Stettenheim 2000; fig. 6.7). Thin and flexible sheets of α-keratin fold inward between each of these structures, similar to the scales of some reptiles (e.g., snakes). Like most other vertebrate epidermal structures, avian scales build up and then gradually wear away, or shed, by a process called **functional epithelial extinction**. Shedding depends on the processes of keratinization and cornification, as previously described. This ensures

that the leg is always protected by a tough, flexible, waterproof covering.

The size, shape, pattern, and organization of scales reflect a bird's habitat (fig. 6.8). Avian scales can be classified by their distinct morphology. **Scutes** are the largest scales; these are usually found on the anterior surface of the leg (the **metatarsus**) and on the dorsal (upper) surface of the digits (toes) (see the box on page 130 for anatomical terms). **Scutella** occur in the same general region, but are not quite as large as scutes (see fig. 6.7). Scutes and scutella are mostly composed of tough, resistant, and rigid β-keratin. **Reticula** (from the Latin word for "net") are smaller structures that form a mesh-like continuous network of flexible soft α-keratin fibers located on the lateral and medial surfaces (the sides) of the foot (Stettenheim 2000, Dhouailly 2009). The scutellation patterns found on different species of bird legs are often used for taxonomic

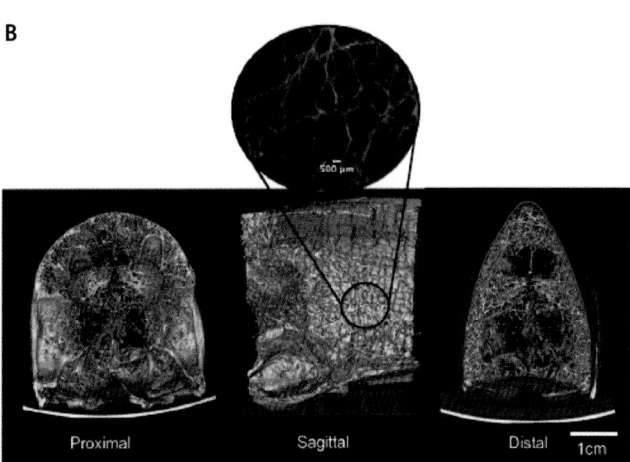

Figure 6.6. A, The **rhamphotheca**; B, distal and proximal cross-sectional views of the dermal core of the **maxilla** (upper beak) of the Toco Toucan (*Ramphastos toco*). The outer shell, or rhamphotheca, of the toucan beak is composed of β-keratin and encases the bony, interior foam of the dermal bones. *Images courtesy of Marc Meyers; from Yasuaki et al. 2009, https://www.ncbi.nlm.nih.gov/pubmed/19699818.*

classification and can be important field marks for species identification in the wild (Proctor and Lynch 1993).

Claws and **nails** (a form of flattened claw) are specialized scales that protect the toes. Claws vary in length, cur-

vature, and how pointed the tips are. The shape and sharpness depend on where the bird lives and how it catches its food or protects itself (fig. 6.8). One of the more extreme examples of this is the adult cassowary, which has just three toes. The inner (medial) toe of each foot bears a large pointed claw on its tip (Stettenheim 2000). When combined with the large bird's powerful thigh muscles, this is a formidable weapon used to kick a potential attacker. One species of extant bird has wing claws as well as toe claws: Hoatzin (*Opisthocomus hoazin*) chicks have functional claws on two digits of the forearm (the wing) that are used for climbing around branches. These wing claws are shed as the juveniles approach maturity (between 70 and 100 days; Thomas 1996).

The claws of most bird species experience gradual wear and continual replacement. Certain grouse and some passerines, however, shed their claws annually each summer during the annual molt. In the case of the grouse, the previous year's growth of tough, horny epidermal keratin over the toes loosens above the soft regenerating vascular matrix of the dermis. Similar to a new feather pushing out the old and worn, the developing nail will eventually shed the old. At this point the keratinization process is incomplete, and the short new nail is soft, pink, and still vascular. Exposure to the air (specifically oxygen) rearranges the sulfur bonds between the amino acids on the keratin fibers, causing the new nail to darken, harden, and become a functional tough rigid nail. It turns out that the molt of claws is useful not only to the grouse but also to the ornithologist. A shallow groove or indentation is left behind where the old nail falls off. This indicates to the ornithologist or wildlife biologist that the claws have been shed at least once and that the bird must therefore be more than twelve months old.

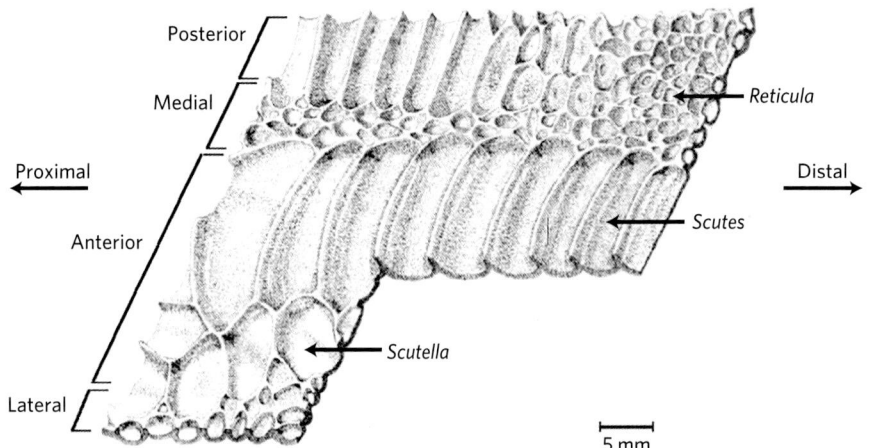

Figure 6.7. Scales of the metatarsus of a domestic chicken. The view is of the inner surface of the stratum corneum that has been peeled from the stratum germinativum. The anatomical orientation of the dissected skin, which was removed from the entire metatarsus, is noted by arrows and brackets (posterior surface at top left, lateral surface at bottom left). The **scutes** are the largest scales on the anterior surface of the leg. **Scutella** occur in the same general region, but are not quite as large as scutes. The **reticula** are the small scales often observed on the medial and posterior surface of the leg. *Figure from Lucas and Stettenheim 1972.*

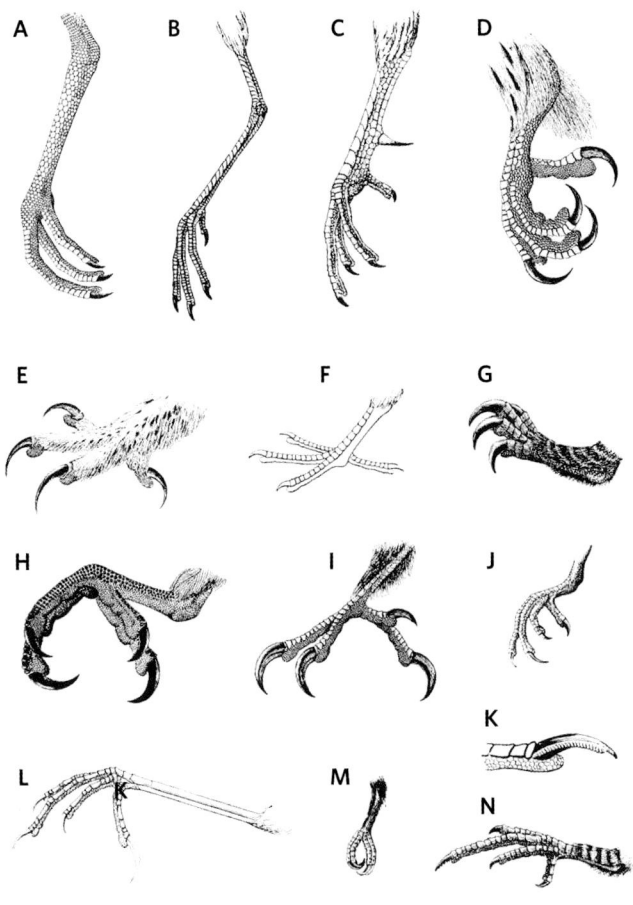

Figure 6.8. The ecology of a given species places a strong selective pressure on the integumentary derivatives of the foot, including scales, scutes, claws, nails, and toe configurations. What do the following examples tell you about the habitats each bird might occupy? Each illustration is of a generalized species: *A*, oystercatcher; *B*, heron; *C*, pheasant; *D*, falcon; *E*, owl; *F*, dove; *G*, swift; *H*, parrot; *I*, woodpecker; *J*, kingfisher; *K*, the middle toe of a heron; *L*, thrush (passerines in general); *M*, trogon; *N*, nightjar. *From Reichenow 1913.*

SKELETAL SYSTEM

Like that of all vertebrates, the avian skeleton gives the body rigidity, anchors muscles and tendons, supports movement, and protects vital organs. The bird skeleton, however, has been extensively modified from the basic vertebrate "blueprint" by many unique adaptations to support flight. Evolution, however, is not a perfect process. There are constraints on the extent of changes that can be made to the basic vertebrate design. Despite the ability to fly, birds are still essentially bipedal terrestrial endotherms. Only a few species live almost completely in the air, such as swifts (Apodidae) are able to do, or even spend the majority of their lifetime flying, like the albatross (Procellariiformes, family Diomede-

idae). Most species do not even participate in aerial feeding. Birds are to a great extent terrestrial—they forage and feed on the ground or while perched. In fact, most birds mainly fly to escape predators, to migrate, or to reach distant foraging sites. The dependence on the forelimbs for flight and on the hindlimbs for walking or perching has left birds with few options for carrying and manipulating food. In many ways, as a result, the bird's neck has become its arm and the beak its hand.

There are approximately 100 bones in the bird skeleton, divided into discrete anatomical regions, as described in subsequent sections (see also the box on page 130 for anatomical terms). The organization of the skeleton is remarkably similar among the various bird groups, despite very different lifestyles (e.g., flight vs. flightless species, terrestrial perching species vs. diving birds). The observed variations reflect differences in evolution and behavior, with the result that some differences between species are subtle and others quite obvious.

Bone Development and Organization

The basic organization of avian bone differs somewhat from that of mammals. The process of **ossification**, the transformation of cartilage into bone, is the same in birds as other vertebrates. The pattern of ossification, however, differs. During development, avian long bones grow from the shaft to the end, instead of growing outward from a center of ossification between the **epiphysis** and the **diaphysis**, as seen in mammals. Small **Haversian canals** predominate, as opposed to the clearly defined **osteons** commonly seen in mammalian long bones (Castrogiovanni et al. 2011). Other osteological peculiarities of birds include a light and thin **cortex**, or outer layer of bone; extensive bone **pneumatization** with **trabeculae** (i.e., hollow bones reinforced by a mesh of small rods of bone); the absence of teeth in all living species; and extremely high bone mineral content. In addition, the bones of the forelimb and hindlimb are very different from each other. In most mammals, both the fore- and hindlimbs are used for terrestrial locomotion (e.g., running, hopping). In birds, the forelimbs have been extensively modified by the demands of flight (aerial locomotion), while the hindlimbs retain the functions of walking, running, and hopping (terrestrial locomotion). The forelimbs have been completely modified to support **primary flight feathers** with **digits** that are no longer useful for gripping or other functions. The hindlimbs retain their original function of terrestrial locomotion but have been reorganized through the fusion of several bones for weight reduction and the lengthening of the toes to accommodate stability in bipedal locomotion. The bones of the skull have been reduced and fused to lighten the head. The neck is long and the vertebral column extremely flexible in the **cervical** regional. The **thoracic** and **lumbar** vertebrae are highly modified and have fused into two separate rigid units. **Caudal** vertebrae have been reduced in number, with about half fused into a single broad bone called the **pygostyle** that supports tail feathers.

With the exception of some diving species (e.g., grebes, auklets, puffins, loons, and penguins), most birds exhibit extensive **pneumatization** of bones. Pneumatized bones have extensive hollow spaces that are internally reinforced by a network of bony rods called **trabeculae** (from the Latin word for beam), creating a lightweight and strong structure that efficiently diffuses stress. Trabeculae are normally concentrated near points of articulation. As a result, **pneumatic foramen**, or "holes," are found primarily in the proximal and distal regions of a given bone. The calcification of bone tends to follow lines of mechanical stress, resulting in a crisscross pattern of calcified trabeculae that divert stress and strain to locations where it can be more easily absorbed and transferred. The function is similar to a "shock absorber" in the strut of an automobile. The pneumatic foramen mentioned above should not be confused with the small **nutrient foramina** located in the middle of the shaft of long bones. These smaller foramen accommodate blood vessels that support bone remodeling in response to mechanical stress. The degree of osseous pneumatization observed in a given bird species will vary with lifestyle. Large-bodied species that spend most of their life on the wing (i.e., soaring, gliding) have more pneumatized bones than smaller, more sedentary species (Welty and Baptista 1988).

It is important to note that the pneumatization of the avian body is not strictly confined to the bones. The avian respiratory system has nine hollow **air sacs** that originate in the lungs and extend into the pneumatic cavities of some bones (e.g., cervical vertebrae, thoracic vertebrae, clavicle, humerus, and synsacrum) and body cavities. For example, the **clavicular air** sac surrounds the heart, while the **abdominal air sac** comes in contact with the kidneys, reproductive organs, small intestine, and large intestine. These air sacs reduce body density while increasing the lung ventilation efficiency (avian respiration is discussed in more detail in chapter 7).

The unique hollow bones of birds also play an important role in calcium regulation and egg production. The pneumatization of some bones, especially the more proximal long bones, provides space for so-called **medullary bone**. Medullary bone is an unusual, highly mineralized, nonstructural form of bone with a "woven" morphology that may completely fill bone marrow spaces (Dacke et al. 1993) (fig. 6.9). This bone has no mechanical function; it is simply a temporary

storage site for calcium. It is found primarily in the less pneumatized long bones of reproductive female birds (e.g., ribs, femur, tibiotarsus, and ulna), where it is used to buffer blood calcium levels and maintain a supply of calcium for eggshell formation. Medullary bone is under hormonal control and accompanies the maturation of **ovarian follicles**. It accumulates rapidly during the pre-laying period. Although it is present to some degree throughout the laying period, the amount of medullary bone fluctuates as calcium is removed

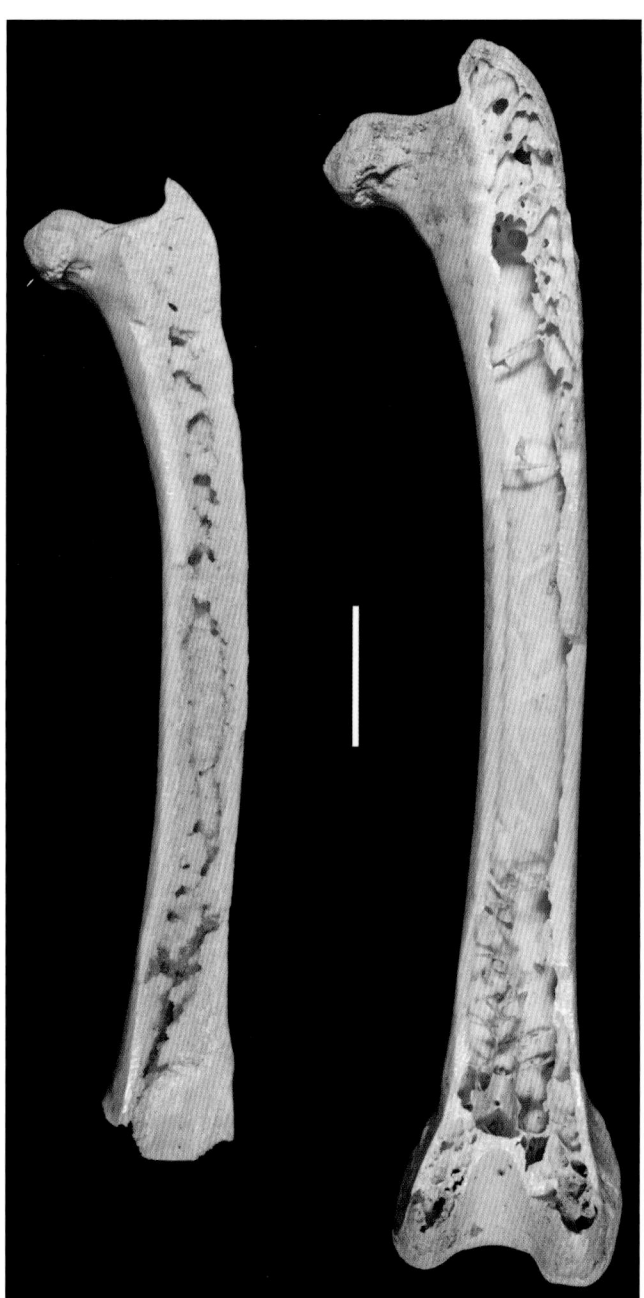

Figure 6.9. Femora of the Domestic Chicken (*Gallus gallus*) filled with **medullary bone** (*left*) in comparison with normal structure (*right*). Scale bar = 1 cm.

for eggshell production. Once egg production has ended for the season, medullary bone will regress and be resorbed over a period of one to three weeks (Rick 1975). The seasonality of medullary bone and its rapid turnover can been used by paleontologists, archeologists, and forensic ecologists to understand seasonality in paleontological and archeological digs and crime scenes.

The Skull

The main functions of the vertebrate skull are to protect the brain and support the feeding apparatus (i.e., jaws). Bird skulls, however, have lost much of their protective capacity. The bird skull has thin boned walls and is hollow with large **orbits** separated only by a thin **mesethmoid** bone called the **interorbital septum**. In some species, the eyes are so large that the interorbital septum is incomplete and there is a **foramina**, a window through the septum that links the two orbits. Within each orbit, a ring of bony plates called the **sclerotic ring** supports the huge avian eyeball. This structure is found only in birds and reptiles and is evidence of a shared dinosaur ancestry. The avian brain and eyes are very large, as they are extremely important for navigation and foraging on the wing (see the nervous system and sense organs, discussed below); accordingly, soft tissue, not bone, will make up the heaviest part of the skull. To offset the weight of eyes and brain, most birds have extensive pneumatization of the skull. Most bird species have one air sac that extends from the **nasal cavity** to the maxilla, and another that originates in the **tympanic cavity** and extends up through the intraorbital septum and into the roof and base of the **cranium**. The extent of the **cranial air sacs** depends on the ecology of the species. Flightless birds, like the kiwi, may have little to no pneumatization, while some woodpeckers have unusual pneumatized **frontal bones** that assist in **cranial kinesis** (described below; Welty and Baptista 1988).

Birds are derived from small theropod dinosaurs whose skulls had numerous bones and articulations. It is no longer possible to recognize these bones individually in modern birds, as they are seamlessly fused with their neighbors. This is useful to the ornithologist, as the approximate age of a living bird can be determined to some extent by the degree of ossification of the **frontal** bones. This is a field technique called **skulling** (skull ossification), which is often used by bird banders to separate the immature birds from the mature birds during fall migration (see chapter 31). At maturity, the fusion of skull bones is complete; only the fused **cranium**, the **maxilla**, **quadrate** bones, **pterygoids**, and the lower **mandible** remain as independent structures. The adult cranial surface is normally smooth and rounded, although crests and specialized ridges to support exceptionally heavy

beaks are found in some groups of birds. A fascinating example of crest structure is the frontal bones of the cassowary, which vault upward into a large, keratin-covered **casque** filled with fragile bony trabeculae (fig. 6.10) (Naish and Perron 2016). The cassowary casque was once assumed to act as a protective helmet, but is now thought to be important in visual displays for mate choice (Naish and Perron 2016) or possibly used for low frequency acoustic communication (Mack and Jones 2003).

Since birds use their beaks to grasp and manipulate food, it should not be surprising that their skulls often have unusual adaptions for foraging behavior specific to taxonomic groups. For example, some species (e.g., swifts, hummingbirds, cranes, and shorebirds) have a joint (the **nasofrontal hinge**) between the maxilla (the upper jaw) and the fused frontal bones of the skull that allows the upper beak (**rhinotheca**) to move independently. This movement, known as **rhynchokinesis** (from the Greek words *rynchos* for beak and *kinesis* for movement), is unique to birds among living vertebrates. A much rarer form of cranial kinesis is found only in some shorebirds and the American Woodcock (*Scolopax minor*). This movement, known as **distal rhynchokinesis**, allows a bird to probe in sand or earth, and then efficiently grasp and extract its prey with just the tip of the beak. Another unusual foraging adaptation is seen in variations in the size and shape of the **hyoid apparatus**. The hyoid apparatus is a

Figure 6.10. Skull of Australian Cassowary (*Casuarius casuarius*). *From Flower 1871.*

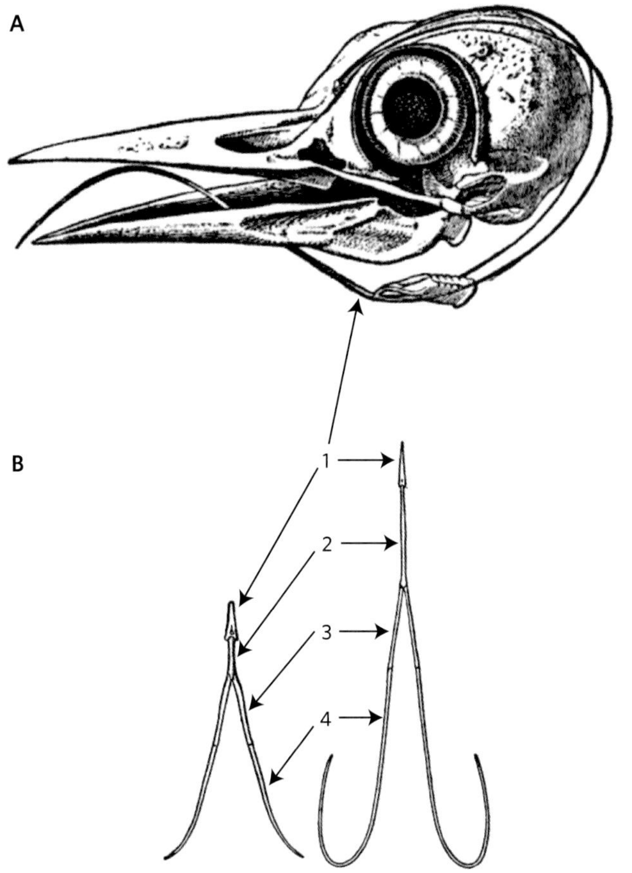

A

B

Figure 6.11. A, Skull of a Golden-fronted Woodpecker (Picidae: *Melanerpes aurifrons*) showing the position of the bones that support the tongue in the skull. The bones of the vertebrate tongue are collectively known as the **hyoid apparatus**. Most birds have small tongues with little musculature, so the hyoid apparatus is fairly short. Woodpeckers, however, have exceptionally long tongues that can probe for insects in the cracks and crevices of tree bark; B, The hyoid apparatus of a Sapsucker (genus *Sphyrapicus*, *left*) and Golden-fronted Woodpecker (*right*). The basal bone of the hyoid apparatus that connects to the tongue is called the (1) **paraglossale**. The paraglossale connects to the (2) **basihyal** bone just in front of the trachea. At this junction, two horns of bone arise and fork to the right and left sides of the skull. The horns consist of (3) the **ceratobranchial** bone and (4) the **epibranchial** bone. In the Picidae, the bone and cartilage of the horns of the hyoid are extremely long, extending up and around the back of the skull. The longer the hyoid apparatus, the further the bird can extend its tongue. The bony horns of the hyoid are covered by muscles that insert near the base of the upper mandible. When these muscles contract, the long slender hyoid bones are pulled down over the skull and forward until the tongue extends from the mouth. When the muscles relax, the tongue is retracted as the bones move back and around the skull within the sheaths of muscle. *Modified from the drawings of Eckstrom 1901.*

series of more or less ossified bones that support the vertebrate tongue. The hyoid bones of most birds are fairly short. However, in species that use their tongues to forage (e.g., woodpeckers, hummingbirds, sunbirds), the hyoid bones are long, to support the exceptionally long tongue. In woodpeckers, the **ceratobranchial** and **epibranchial** horns of the hyoid apparatus wrap around the back of the skull and extend to the base of the upper mandible (fig. 6.11). The tongue covers the anterior part of the hyoid apparatus near the base of the lower jaw. The long hyoid horns support the additional musculature required to extend the tongue and probe into crevices of trees for insects. Whether long or short, however, it should be noted that the hyoid bones of all birds connect to the skull by ligaments attached to the lower jaw (the **mandible**) and at the base of the skull (Bock 1999). The lower jaw of birds, the mandible, is a fusion of bones. The first four bones proximal to the articulation with the maxilla (the **periarticular**, **surangular**, **angular**, and **splenial**) are completely fused, with no obvious demarcation to identify individual bones. The distal end of the mandible is the **dentary** bone, which is tooth-bearing in mammals. Modern birds have neither teeth nor true "jaws," and are unable to chew, yet their mandibles still perform a variety of functions that range from the capture and transport of food to the care of body and feathers (i.e., preening).

The Vertebral Column

The shape and function of the vertebrae vary between regions of the vertebral column. **Cervical vertebrae** support the neck, the **thoracic vertebrae** bear ribs, **sacral vertebrae** are completely fused for stability during flight, and the **caudal vertebrae** are reduced in number and somewhat fused to support the tail. The total number of vertebrae in the avian spinal column varies between taxonomic groups. A small passerine such as a sparrow may have as few as 39 vertebrae, while a long-necked, larger species such as a swan might have as many as 63 (Welty and Baptista 1988, Bellairs and Jenkins 1960). In general, long-necked birds have higher numbers of cervical vertebrae (Hyman 1992). By comparison, mammals have fewer vertebrae overall (approximately 33 in humans), and with the exception of the tail, there is less variability between species in the number of vertebrae found in each anatomical region. For example, mammals have seven cervical vertebrae, with only a few notable exceptions (Hyman 1992). The extreme variation in the number of cervical vertebrae of birds reflects the ecological importance of the avian neck. While the avian neck is variable in morphology and extremely flexible, the rest of the backbone is rigid, with little structural variability between taxonomic groups. All birds exhibit a high degree of fusion and rigidity in the thoracic, sacral, and caudal regions of the spinal column. This is necessary to accommodate bipedal locomotion, buffer the stresses and strains of muscle contraction during flight, and act as a shock absorber during landing and hopping. The overall pattern of fusion and amount of spinal rigidity may, however, be slightly modified according to the ecology of a given species.

We begin our detailed description of the bird backbone with the neck. The cervical vertebrae are connected by a complex series of more than 200 muscles and tendons and vary in number from eight to 25 bones, depending on species. Some cervical vertebrae may bear ribs, but these do not articulate with the **sternum**, as described below. Because the neck is normally folded and completely hidden by the plumage, it often appears to be much shorter than its true length. The first cervical vertebra, the **atlas**, has a unique shape that cradles the single **occipital condyle** of the skull and facilitates the turning and twisting of the head. Most birds can turn their heads nearly 180° in either direction (Hyman 1992). The degree of flexibility in the spinal cord is due in part to the shape of the **centrum**, or body, of the vertebrae. The centrum of the cervical vertebrae in birds is a saddle-shaped (**heterocoelous**) articulation that permits extensive lateral and vertical flexion without stretching or twisting the nerve cord. This makes the bird's neck extremely flexible and strong when extended. The neck is necessarily more than a mere link between a bird's head and body, as it must assist in many of the activities normally devoted to the forelimb, such as nest building and capturing food. A good example of this is found in the herons (Ardeidae). In this family, the sixth cervical vertebra is modified to fold the neck into a distinctive S-shape that allows the birds to quickly strike distant prey with their harpoon-shaped bills.

The **thoracic vertebrae** bear **ribs** and will vary in number from three to 10 bones. Depending on the species, three to five of the thoracic vertebrae are usually fused together into a single bone. Two or three free thoracic vertebrae will be found behind these fused bones. The free thoracic vertebrae form a flexible joint with the **synsacrum** in the lumbar region. The synsacrum is a rigid fused shield of bone that consists of all of the **lumbar vertebrae** (variable in number, with three at minimum) fused to a series of **sacral vertebrae**, the pelvic bones, and six of the **caudal vertebrae** (the tail). The total number of vertebrae fused within the synsacrum can vary from 10 to 23 bones. The position of the joint between the thoracic vertebrae and the synsacrum will also vary according to the number of free (unfused) thoracic vertebrae present. The fused and composite nature of the synsacrum results in a lightweight, inflexible framework of trusses, beams, and ridges to support the legs while walking on land, yet stabilize the body against muscle contraction during flight. Posterior to the synsacrum are about 12 **caudal vertebrae**, six of which are fused into the **pygostyle**. The pygostyle is a broad triangular bone that supports the tail feathers. The pygostyle is often very large in species that use the tail as a support and counterbalance for the body (e.g., woodpeckers). The extensive fusion of the vertebral column means that birds are able to move their backbone only in three distinct areas: the highly flexible neck (the cervical vertebrae), the joint between the fused thoracic vertebra and the synsacrum, and the joint between the free caudal vertebrae and the pygostyle.

The vertebral column is connected to the **sternum** by ossified **thoracic** and **sternal ribs**. Together, these bones constitute the **rib cage**. The vertebral ribs extend down from the thoracic vertebrae to articulate with sternal ribs that extend up from the ventral side of the body. The number of fully articulated thoracic and sternal ribs that completely connect vertebrae to sternum varies by species, from three in some pigeons to nine in swans (Welty and Baptista 1988, Bellairs and Jenkins 1960). The thoracic ribs are reinforced by **uncinate processes** that extend backward (caudally) from the posterior margin of each rib. The uncinate processes are unique structures, found only in birds and a few extant reptiles. In birds, these processes brace and strengthen the rib cage against the flexion and torsion of the flight muscles. In diving species such as loons and guillemots, the uncinate processes are well developed and very prominent, extending across two consecutive ribs to further reinforce the rib cage against the pressures experienced during deep dives (Welty and Baptista 1988).

The Sternum

The **sternum** of most bird species has been modified into a large ventral projection called a **keel** or **carina**, which acts as an anchor for massive wing muscles. The keel is absent in ratites (ostriches and allies) and reduced in secondarily flightless birds (for additional discussion, see the box on page 148). As such, the sternum has historically been classified as one of two types: **ratite** or **carinate**. The ratite sternum of flightless birds lacks a distinctive keel. The carinate sternum of flying birds, on the other hand, has a greatly enlarged keel to accommodate the equally large pectoralis muscles. Outside the ratites, most bird species, including those that rarely fly, have a carinate sternum. Even flightless penguins have a keel to support the pectoral muscles for "flying" through water. Hovering species such as hummingbirds have the largest keels of all birds with respect to body size.

The Pectoral Girdle

The **pectoral** and **pelvic girdles** of terrestrial vertebrates evolved as appendicular suspension systems to support locomotion on land. It should therefore be no surprise that the biomechanical demands of flight extensively remodeled the **pectoral girdle**; it is, after all, the structure that connects the wing to the body. The reorganization, fusion, and pneumatization of bones are all readily apparent in this region of the skeleton.

The avian pectoral girdle consists of the sternum and the right and left **coracoids**, **scapulae**, and **clavicles**. The two large clavicles are often fused into a symmetrical V-shaped **furcula**, the so-called wishbone, which attaches to the sternum. The furcula acts as a spring, and therefore an energy recovery mechanism, during flapping flight. As the bird pushes its wings down (the **downstroke**) the furcula bends under pressure from the flight muscles. At the end of the downstroke, the furcula springs back to assist in the **upstroke** of the wings (see chapter 10 for more information on flight). The structure of the furcula can vary greatly between species according to body size, wingbeat frequency, and mode of flight (fig. 6.12, Goslow et al. 1990). The scapula, coracoid, and clavicle on each side of the body come together to form the **triosseal canal** or foramen, a tiny opening that is possibly the most peculiar skeletal trait of birds. The triosseal canal holds the secret to a perplexing biomechanical problem. When you look at the large pectoralis muscle covering the sternum and inserting on the **humerus**, it is easy to understand how the bird pulls its wing down. However, there does not appear to be a comparable large antagonist muscle on its back to pull the wing up. The mystery is solved when you consider that the muscle underneath the pectoralis (the **supracoracoideus**) attaches to the upper, dorsal surface of the humerus by way of a long tendon that passes through the triosseal canal and up and over the bird's "shoulder." So the triosseal canal forms a pulley for the tendon of the supracoracoideus to rides over. When the supracoracoideus contracts, the wing is pulled upward even though the muscle is positioned on the ventral surface of the body (refer to the box on page 130 for anatomical terms). This adaptation allows the bird to maximize the muscle mass that surrounds the center of gravity, thus reducing the energetic costs for flight. This mechanism is also described in the section on muscles below and in chapter 10.

The Forelimb: The Wing

Despite major differences in external appearance, the bird wing is homologous with the human arm. The forelimbs of all vertebrates consist of a **humerus**, **radius**, **ulna**, **carpal**

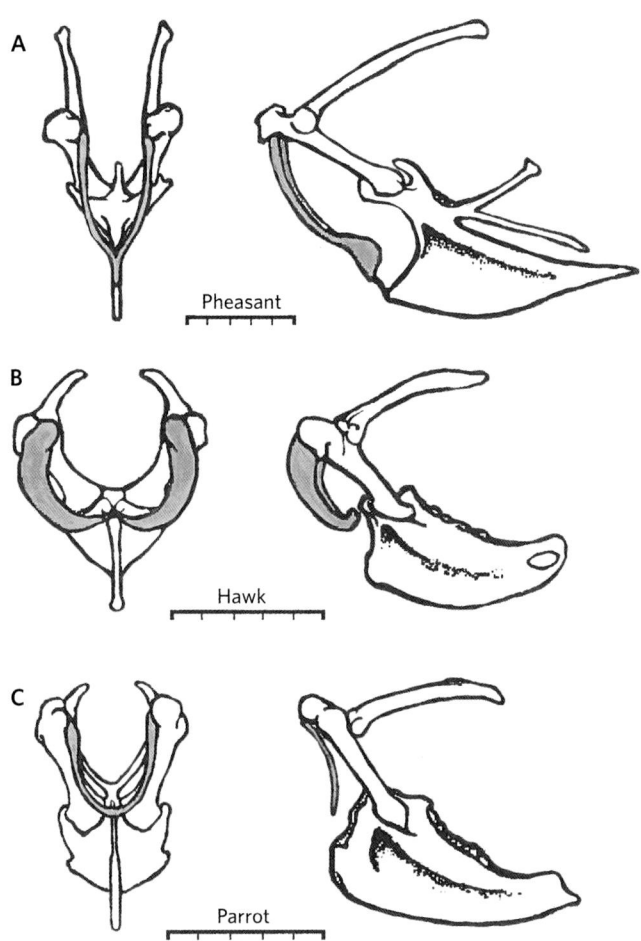

Pheasant

Hawk

Parrot

Figure 6.12. The structure of the **furcula** (highlighted in gray) can vary greatly between species depending on body size, wingbeat frequency, and flight mode. Shown are *A*, Common Pheasant (*Phasianus colchicus*); *B*, Common Buzzard (*Buteo buteo*); *C*, Scarlet Macaw (*Ara macao*). *From Goslow et al. © 1990, American Institute of Biological Sciences, http://scholarworks.umt.edu/biosci_pubs/92; used with permission, courtesy of K. P. Dial.*

bones, and **digits** (fig. 6.13). Beyond that generic description, however, the forelimbs of birds diverge from those of other vertebrates. Bone remodels itself according to the forces placed on it, and the avian humerus must resist the forces of both major flight muscles. The avian humerus is accordingly thick and strong, with two large crests that accommodate the attachment of each of the powerful antagonistic flight muscles (i.e., the pectoralis and supracoracoideus muscles) near the shoulder joint. Recall the pneumatization of the humerus discussed above. An extensive network of internal trabecular struts necessarily reinforce and strengthen this bone. The humerus also has a large hole near its proximal end to connect the intraosseous air spaces of the forelimb to the **interclavicular air sacs**. The **condyle** of the humerus articulates with the **glenoid fossa**, a cup-shaped depression formed by the junction of the **scapula** and the **coracoid** bones. The large articulating surface of the glenoid fossa allows the resting wing to fold neatly against the body. The lateral orientation of the articulation between the glenoid fossa and the humerus facilitates a wide range of movement in this joint; this allows the wing to change shape during the various flight phases (chapter 10).

As in all vertebrates, the avian forearm consists of a radius and a larger ulna. The ulna is normally similar in size to the humerus, with a distinctive triangular cross section and **papillae ulnare**, or **quill knobs**, for the attachment of the **secondary flight feathers**. The carpal, metacarpal, and digits of the bird are also highly modified for the attachment of flight feathers. The carpal bones in the wrist have been reduced to two small bones, the **radiale** and **ulnare**, that articulate with the highly fused bones of the avian **manus**. The remaining carpal bones fuse during development with the first three metacarpals to form a wedge-like structure, the **carpometa-**

carpus, that supports the first six **primary flight feathers** (fig. 6.14). The fusion of the metacarpals is completed shortly after hatching. Only three distinct **digits** remain in the avian hand. The fist digit, along with its fused **phalanges**, supports the feathers (three to five in number) of the **alula**. The alula acts as a sort of "brake" during landing, when flying at slow speeds, and at high angles of attack. A bird moves its alula slightly upward and forward to form a small slot on the wing's leading edge that allows air to pass through. This slows the speed of forward motion while maintaining some lift and allows the bird to decrease its momentum without crashing. The second digit and its phalanges form the tip of the wing. The phalanges of this digit support the remaining four **primary flight feathers** (fig. 6.14). The third digit bears a single flight feather on one small phalanx that projects from just below the second digit at its junction with the carpometacarpus. It is worth noting that the humerus is the only wing bone whose shape varies greatly according to

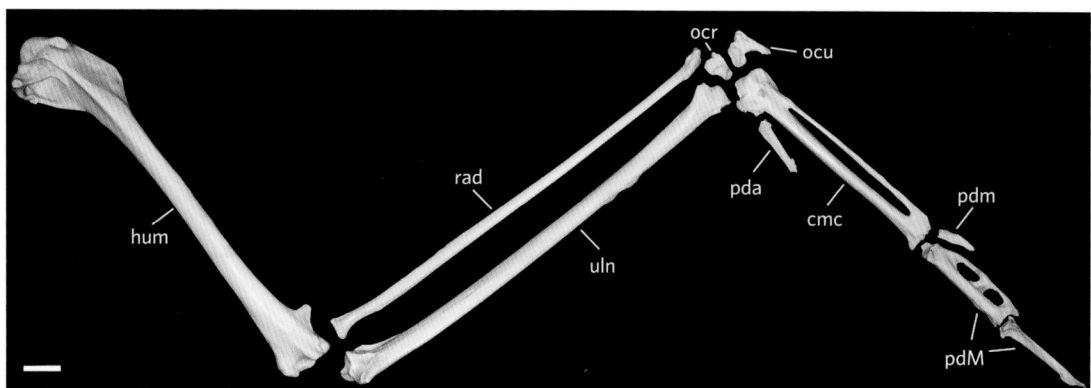

Figure 6.13. The forelimbs of all vertebrates consist of a **humerus, radius,** and **ulna,** as well as **carpal** bones and **digits.** The wing of the Lesser Black-backed Gull (*Larus fuscus*) is shown to illustrate the position and orientation of the forelimb bones in the bird: hum = humerus; rad = radius; uln = ulna; ocr = os carpi radiale; ocu = os carpi ulnare; pda = phalanx digit alulae; cmc = carpometacarpus; pdm = phalanx digit minoris; pdM = phalanges digit majoris. Scale bar = 1 cm.

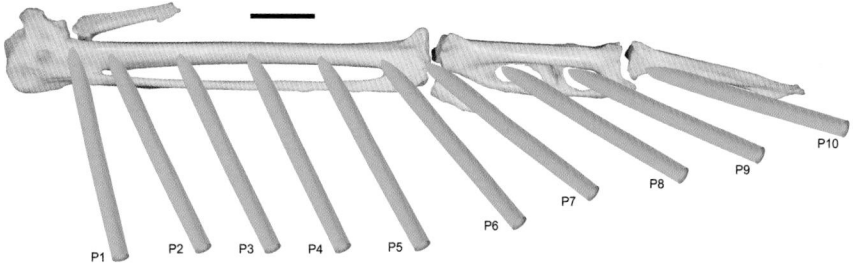

Figure 6.14. During development the fusion of some carpal bones with the first three metacarpals forms a wedge-like structure, the **carpometacarpus**, that supports the first six primary feathers. The carpometacarpus and wing phalanges of the Lesser Black-backed Gull (*Larus fuscus*) are shown, along with the points of insertion for the ten primary feathers (P1–P10). Scale bars = 1 cm; feathers not to scale.

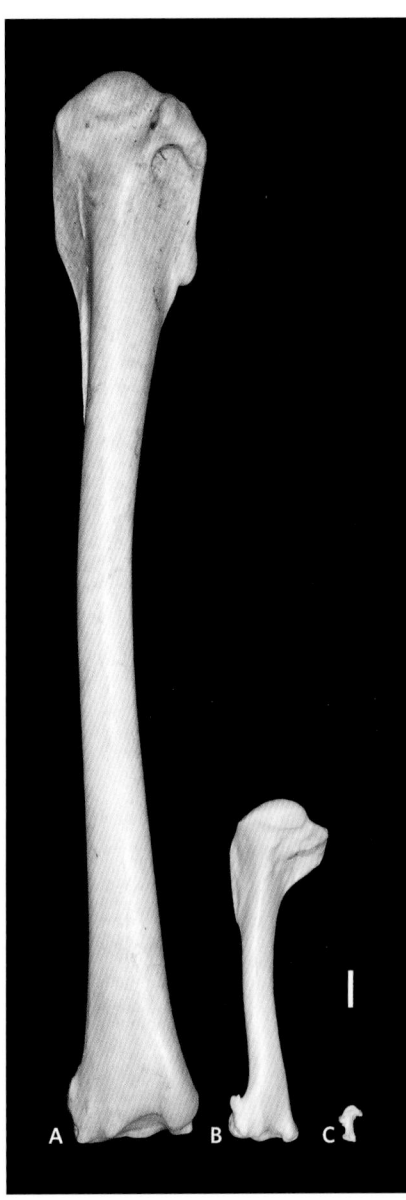

Figure 6.15. Extreme variability in the avian **humerus**: A, Brown Pelican (*Pelecanus occidentalis*); B, Common Raven (*Corvus corax*); C, Little Swift (*Apus affinis*). Scale bar = 1 cm.

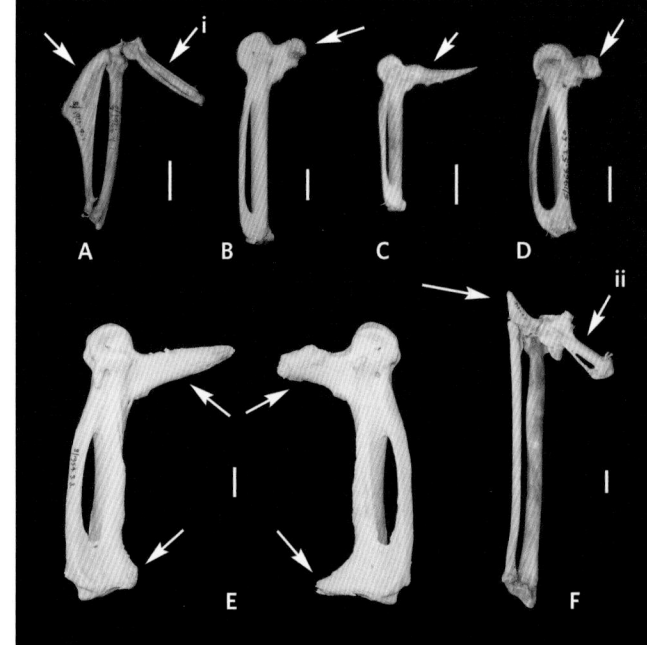

Figure 6.16. Selected examples of carpal weapons in other aves, all in ventral aspect, left side, except for F: A, Actophilornis africana NHMUK S/1964.91 has a blade-like extension of the radius (arrowed); B, Tachyeres brachypterus NHMUK S/2004.7.1 has a carpal knob (arrowed); C, Vanellus novaehollandiae NHMUK S/1966.50.8 has a carpal spike (arrowed); D, Goura cristata NHMUK S/1966.52.60 has a carpal knob (arrowed); E, Chauna chavaria NHMUK S/1954.3.3 has carpal spike or knobs on either end of the carpometacarpus (arrowed); F, Plectropterus gambensis (NHMUK 1898.5.9.5) has a carpal spike (arrowed). Note that in A and F, the carpometacarpus (indicated by i and ii) does not form a carpal weapon. Scale bar = 10mm. *From Hume and Steel 2013 with permission of the authors.*

mode of flight (Kaiser 2010). In soaring birds like the pelican, the humerus is extremely long, while in birds that depend on flapping flight, such as swifts and hummingbirds, the humerus is shorter and stout (fig. 6.15). The carpometacarpus of some bird species in various families (Anhimidae, Anatidae, Cracidae, Burhinidae, Charadriidae, Chionidae, and Columbidae) shows spurs or knobs projecting from the processus extensorius, which have been interpreted as weapons used during intra- and interspecific fighting (Hume and Steel 2013) (fig. 6.16).

The Pelvic Girdle and Hindlimb

Birds inherited their erect, bipedal, **digitigrade** (walking on toes) mode of walking on land from their theropod ancestors

(Grossi et al. 2014). This is important, as the structural shift of bipedalism freed the forearms for new functions; without it, the forearms of birds would very likely have remained legs instead of becoming wings (Hyman 1992). The pelvic girdles of bipedal dinosaurs and modern birds are similar in many aspects. The entire structure consists of three bones: the **ilium**, **ischium**, and **pubis**. The ilia of bipedal dinosaurs and birds are elongated and attached to several sacral vertebrae to support the efficient transfer of the entire body mass onto the hindlimbs. The pubis is likewise elongated into a rod-like structure in both groups of vertebrates. The pelvis of birds, however, was modified from that of its reptilian ancestors as the center of gravity moved far forward to accommodate flight. The ilium, ischium, and pubis of reptiles past and present are individual bones that radiate out from the **acetabulum** (the hip socket) like blades of a fan. These bones are separate at hatch in birds, but by maturity they have fused to the point of being indistinguishable from each other. The pelvis of modern birds has also lengthened considerably, with the long thin ischium and pubis extending caudally toward the tail. Both bones are attached to the ilium anteriorly and extend back almost parallel to each other and to the backbone. There is no **symphysis** (joint at the midline of the body) to join the distal ends of the right and left pubis. Instead, there is an open space in the abdomen between each blade of the pubis that accommodates egg-laying. In some species the width of this gap is wider in females (i.e., it is sexually dimorphic, as seen in some parrot species, for example). Mode of locomotion dictates the width of the gap in other species; diving birds tend to have a narrower pelvis, while those that climb or run have wider pelves. There are some exceptions to this open abdomen morphology; ostrich have a **pubic symphysis**, and rhea have an **ischial symphysis** (Welty and Baptista 1988). In most bird species, the ilium is completely fused to the vertebral column to become part of the synsacrum, although there are taxonomic exceptions in which this fusion is incomplete (e.g., Podicipediformes, Gaviiformes, Charadriiformes).

The **femur** of the leg is connected to the body at the acetabulum of the pelvis. The acetabulum of birds differs from mammals in that it is perforated at its center, the point at which the ilium, ischium, and pubis meet. The bones of the leg consist of the **femur**, **patella**, **tibiotarsus**, **fibula**, **tarsometatarsus**, and a variable number of **pedal phalanges**. The femur is similar to that of mammals; its function in both taxa is to transmit the weight of the body to the distal leg bones and, eventually, to the ground. The femur is completely covered by muscle, skin, and feathers in a living bird. Below the femur, two long bones form the visible part of the leg, the **tibiotarsus** and the **tarsometatarsus**. The lower leg bones have been extensively fused and modified to support the force of landing and takeoff. The tibiotarsus is a fusion of the **tibia** and the proximal row of **tarsal bones** normally found in terrestrial vertebrates. The tibiotarsus articulates proximally with the femur at the knee. This proximal articulation of the tibiotarsus is normally flattened, with the exception of some groups of diving birds, which have an elongated **cnemial crest**. This crest is a ridge at the front of the tibiotarsus that facilitates the insertion of the main extensor muscle of the thigh. It is extraordinarily large and well developed in loons (Gaviiformes, fig. 6.17A), allowing the birds to use powerful extensions of their legs to navigate underwater. The distal articulation of the tibiotarsus (i.e., the ankle) is characterized by two **condyles**. An ossified tunnel, called the **supratendinal bridge**, can be just found above the condyles, covering the tendon of the long digital extensor muscle. This bridge is present in almost all bird groups, with a few exceptions (owls and parrots, fig. 6.17B).

The tibiotarsus articulates with the tarsometatarsus at the ankle. The tarsometatarsus represents a fusion of the distal tarsal bones and three metatarsal (II, III, and IV). The fusion of the foot bones is complete a few weeks after fledging. The plantar side of the proximal end of the tarsometatarsus is characterized by the **hypotarsus**, a complex structure that guides the tendons of the flexor muscles of the toes. This structure varies widely among bird groups, from simple crests to very complex systems of grooves, ridges, and channels (Mayr 2016). The distal end of the tarsometatarsus is characterized by three **trochleae**, which articulate with digits II–IV. In the living birds, only the ostrich has two trochleae (for the second and third digits), while woodpeckers (Piciformes) have four trochleae. A few groups (parrots and cuckoos, among others) show an accessory trochlea on metatarsi IV.

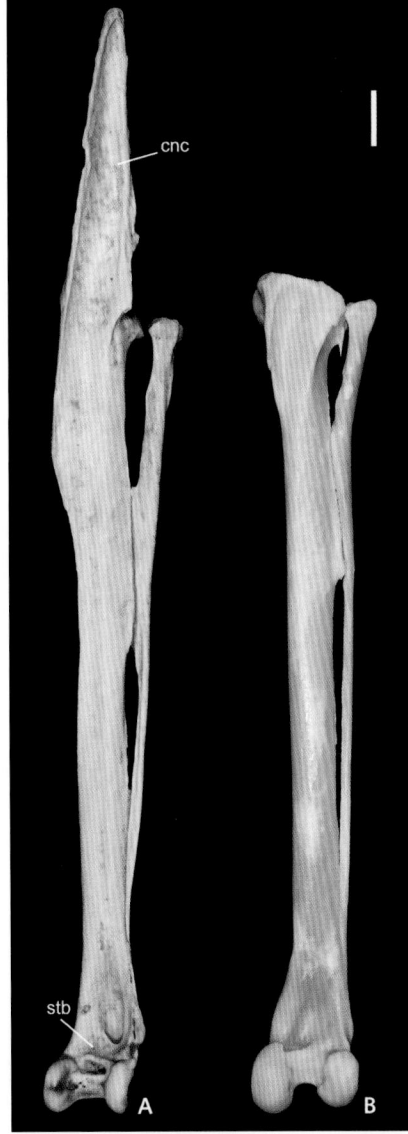

Figure 6.17. A, The **tibiotarsus** of the Common Loon (*Gavia immer*) with the extremely elongated **cnemial crest** (cnc) and a well-developed **supratendinal bridge** (stb); B, tibiotarsus of the Eurasian Eagle-Owl (*Bubo bubo*), without the suprantendinal bridge. Scale bar = 1cm.

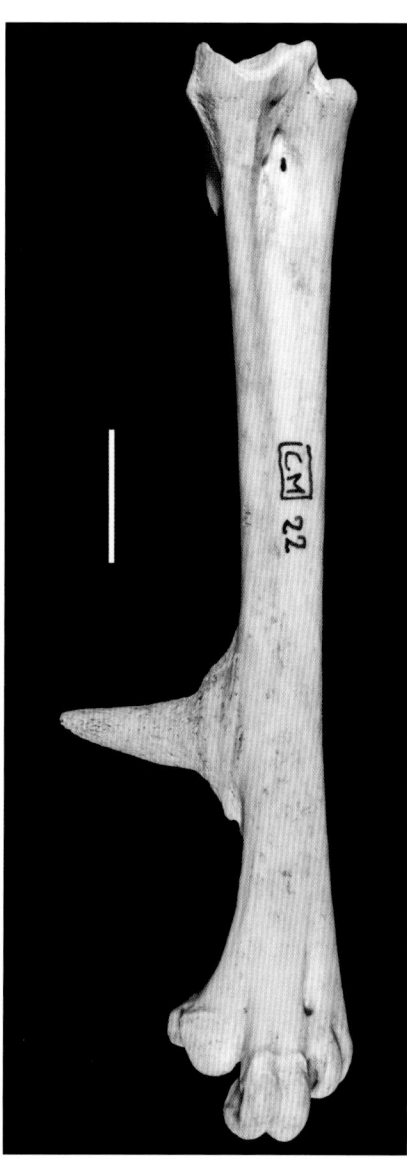

Figure 6.18. The tarsometatarsus of the Domestic Chicken (*Gallus gallus*) with a well-developed **spur**. Scale bar = 1 cm.

The tarsometatarsus of some species of Phasianidae (domestic chicken, pheasants, partridges, and francolins) have a more or less developed **spur**, which can be double in some cases (fig. 6.18). The **toes** are the most variable part of the hindlimb, as they can greatly vary in number and position, as well as in shape and size, up to a maximum of nine combinations (fig. 6.19). The scales and claws, previously discussed with the integument, are usually more modified for special ecological functions than the underlying bones.

THE MUSCULAR SYSTEM

The combination of flight and bipedal locomotion on land requires a major reorganization of the tetrapod body plan, and the musculature of modern birds has certainly come a long way from that of their theropod ancestors. In this chapter,

we focus only on the modifications to the muscular system required for flight. Additional information is covered in chapter 7 (physiology) and chapter 10 (flight and locomotion). In general, the muscles controlling both wings and legs have moved closer to the center of gravity on the ventral side of the body to create a compact, streamlined shape that is properly balanced for efficient flight. The **pectoralis major** and its antagonist, the **supracoracoideus**, are excellent examples of adaptations to flight. The pectoralis major originates on the keel and furcula and attaches to the ventral side of the humerus. This muscle is responsible for powering the downstroke during flight (i.e., it depresses the wings). The supracoracoideus muscle, which lies between the pectoralis and sternum (fig. 6.20), also originates on the keel, but attaches to the dorsal surface of the humerus by way of a tendon that travels through the triosseal canal. The triosseal canal acts as a pulley that allows the supracoracoideus tendon to lift the wing up, even though the contracting supracoracoideus is found on the ventral surface of the body (fig. 6.20). The pectoralis muscles are more prominent in birds than in tetrapod mammals, accounting for as much as 40 percent of the body weight in strong fliers (Hartman 1961). At the same time, the fused synsacrum reduces the need for extensive musculature to stabilize the back, so the dorsal muscles of birds are generally decreased in size.

The Musculature of the Forelimb

The musculature of the avian forelimb (the wing) has been modified into thin tough bands of extensors and flexors attached to bone by wiry lightweight tendons. The general pattern of the forelimb musculature is as follows. The muscles of the humerus attach the wing to the body. The antagonistic contractions of the flight muscles (described above) are transferred to the rest of the wing through the humerus. Muscles that cross the anterior surface of the humerus to insert on the radius (e.g., the **biceps brachii**) flex the forearm of the wing. Muscles that cross the posterior surface of the humerus to insert on the ulna (e.g., the **triceps brachii**) extend the forearm. Note that this description groups muscles by their location (anterior or posterior) with respect to the humerus. It does not provide detailed information about muscular origins and insertions, which is best left to a graphical anatomical atlas or lab manual. The musculature of the forearm follows a similar pattern. Muscles anterior to the forearm extend the manus and flight feathers. Muscles posterior to the forearm flex these structures. The **expansor secondarium**, a muscle unique to the forearm of birds, extends the secondary flight feathers. Together, the forearm muscles work to **supinate** (i.e., twist upward) and **pronate** (i.e., turn outward) both the manus and outer wing (see the box on

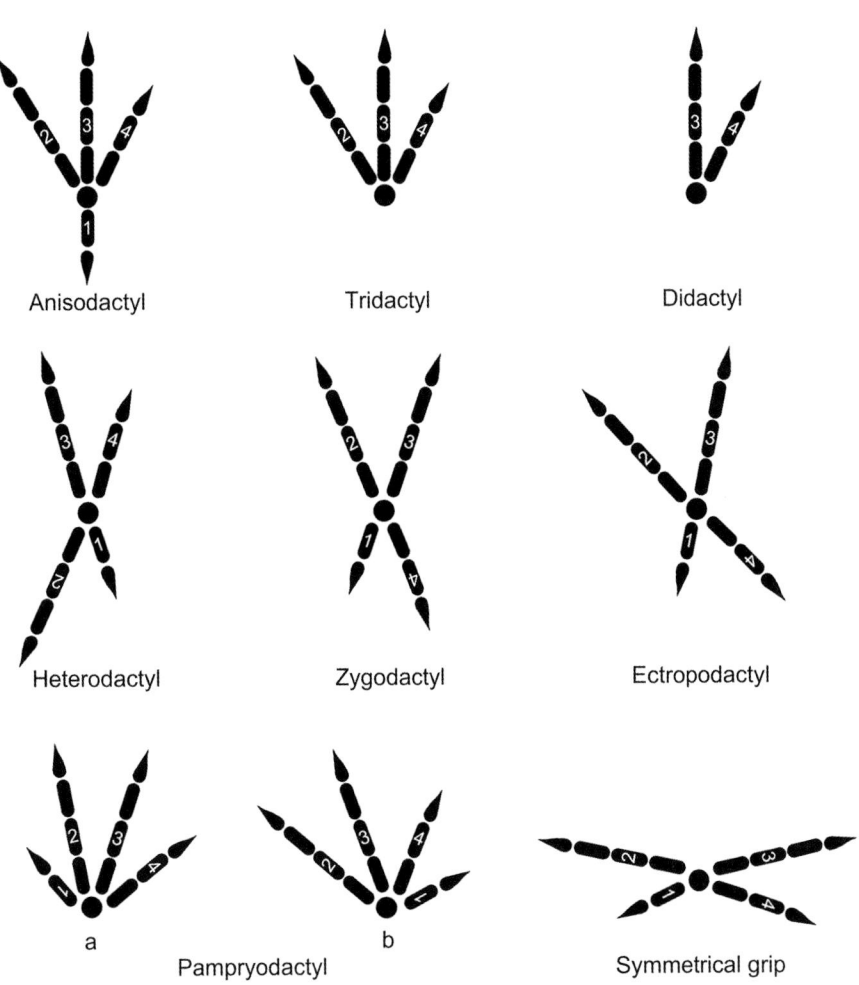

Anisodactyl

Tridactyl

Didactyl

Heterodactyl

Zygodactyl

Ectropodactyl

a
b
Pampryodactyl

Symmetrical grip

Figure 6.19. Patterns of toe placement in birds. **Anisodactyl** is the basic perching arrangement found in most Passerines; **tridactyl** is for walking and running (waders, bustards, and sandgrouses); **didactyl** is present only in ostrich, extremely adapted for running; **heterodactyl** is a perching arrangement present only in trogons; **zygodactyl** is for perching in woodpeckers and allies; **ectropodactyl** is a wide **anisodactyl** pattern used for grasping by raptors. A **pamprodactyl** arrangement is used for gripping surfaces, such as can be seen in some swifts, mousebirds, and some parrots (a). Another version of pamprodactyl toe arrangement (b) had digit 1 in the opposite position; this was only present in the extinct Ivory-billed Woodpecker (*Campephilus principalis*). Finally, some species of swifts exhibit a **symmetrical** grip, as seen in the last diagram.

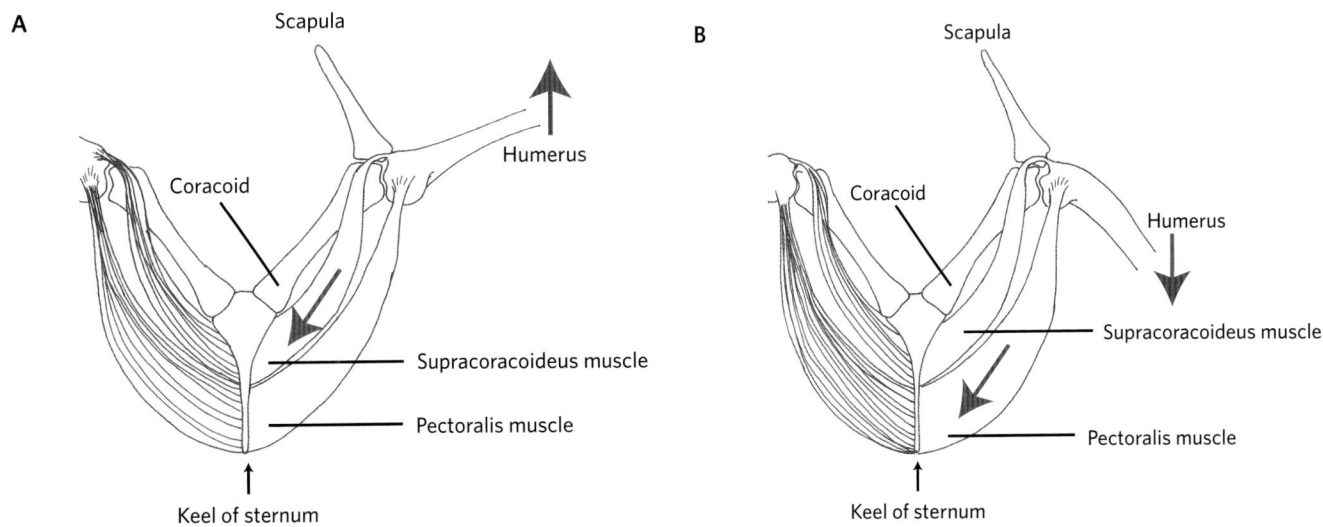

Figure 6.20. The **triosseal canal** acts as a pulley that allows the **supracoracoideus tendon** to lift the wing up, even though the contracting **supracoracoideus muscle** is found on the ventral surface of the body, near the keel. *A*, The red arrow shows the direction of muscle contraction. When the supracoracoideus muscle contracts toward the keel, the wing moves upward; *B*, When the antagonistic **pectoralis muscle** contracts, the wing is pulled into a downstroke. *Drawing modified from Hinchliffe and Johnson 1980.*

SKELETAL MODIFICATIONS IN SECONDARILY FLIGHTLESS BIRDS

Flight is the most important common characteristic of all the Neognathous birds, with the exception of penguins. At the same time, not all birds fly. In fact, secondarily flightless bird species are found in many groups, including grebes, pigeons, parrots, waterfowl, cormorants, auks, rails, and passerines. Most flight-less bird species have evolved in insular, predator-free domains. As we have established in this chapter, the evolution of flight has allowed birds, which are essentially bipedal terrestrial animals, to move fast and far to reach food, to migrate, and to escape from predators. The more or less remote islands of the Atlantic, Indian, and Pacific Oceans, and of the Mediterranean Sea, host a great number of terrestrial birds that became flightless soon after island colonization. This raises the question of why did these species become flightless in a short period of time if the ability to fly is of such great advantage to birds? The answer must be found in the trade-off between the function of flight and the cost of maintaining the ability to fly, which is very high. In small, predatory-free areas, such as the smaller oceanic islands, terrestrial feeding species such as rails, pigeons, and waterfowl may rapidly lose energetically expensive flight capabilities to maximize scarce food resources. It is difficult to assess the time span over which these changes occur, as each will vary, but in all cases some hundreds of years are likely required.

In terms of skeletal morphology, the first bone affected by the loss of flight is the sternum and its keel, which become small and reduced in depth as the pectoral muscles weaken. Disuse of muscle results in atrophy, and muscular atrophy in turn reduces the stimulation to build bone. The limited calcium and protein available in a resource-scarce ecosystem would be better put to use for other functions, such as egg production and reinforc-ing the leg bones. For similar reasons, the wing bones would be rapidly reduced in prominence and length. These modifications would first be seen in reductions in the radius and ulna, then the humerus, and lastly the carpometacarpus. At the same time, the hindlimb bones would become stronger and, sometimes, longer.

The skeleton of the Dodo (*Raphus cucullatus*). Note the vestigial fore-limbs and the powerful hindlimbs.

In cases where the endemic avian community survived for a long time, some species underwent major changes across the entire skeleton; eventually they became very different from their flying ancestors. This is the case of the world-famous Dodo (*Raphus cucullatus*) (see figure above) or the Hawaiian Moa-Nalos (*Thambetochen*) (Olson and James 1991). In both of these taxa, but also in other taxa as well, the forelimbs became somewhat vestigial while the hindlimbs became very powerful.

Flightlessness is not the only modification seen in bird species endemic to isolated islands. In fact, the birds of prey on islands, both diurnal and nocturnal, often exhibit great allometric variation or increased levels of speciation, the latter in accor-

page 130 for anatomical terms). These motions are necessary to control powered flight (see the box on page 150). Small slips of muscles that insert on the manus assist in this twisting motion and as well extend the primary flight feathers. This enables precise muscular control of wing position in the airstream, which facilitates turning and gliding (Proctor and Lynch 1993; see also chapter 10).

The Musculature of the Hindlimb

You will recall from the description of the avian vertebral column and the pelvis that the synsacrum is a large compos-ite bone that provides structural stability and support for the unusual bimodal locomotion pattern of birds. We will now consider its role as a point of attachment for the musculature

dance with prey diversity and availability. This is the case, for example, of the Strigiformes of the Caribbean, where at least 13 species have been described (Louchart 2005), or of the Gargano paleo-archipelago, where at least six species have been recorded (Ballmann 1976) (see figure below, B–F). Unfortunately, all very specialized taxa are extremely sensitive to minimal changes in their environment. Flightless endemic island species are particularly ill-equipped to develop countermeasures to the introduction of terrestrial predators after human colonization. As a matter of fact, most island endemics have become extinct in historical times or are currently listed as critically endangered. The scientific community is now undertaking great efforts to save those that remain from extinction.

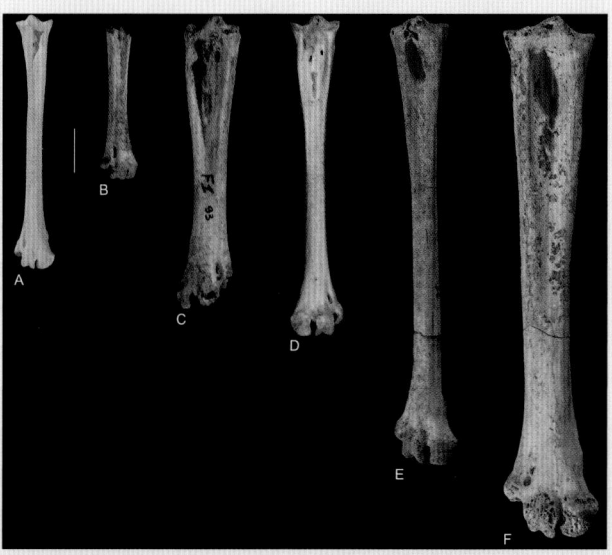

Diversity in the tarsometatarsus of fossil Strigiformes in the Late Miocene of Gargano paleo-archipelago (Italy), compared with the recent Barn Owl *Tyto alba* (A in the figure). Note the differences in length and shape of the tarsometatarsus in the various taxa, in response to the feeding adaptations.

of the hindlimb muscles. The avian leg has 38 muscles that primarily support forward and backward movement of the legs while restricting lateral movement or rotation. These muscles support walking, running, hopping, leaping into the air, landing, and grasping (e.g., perching, manipulating food, etc.). The bulk of the leg musculature surrounds the femur in the thigh, thus concentrating the muscle mass close

to the body's center of gravity. The thigh muscles are similar to those found in other vertebrate hindlimbs. You will recognize a series of hip flexors (e.g., the **sartorius** and **iliotibialis**, also called the **gluteus maximus**), knee extensors (e.g., the **sartorius, iliotibialis, quadriceps**), knee flexors (e.g., **gastrocnemius**), and thigh extensors (e.g., the **adductor longus**, the **semitendinosus** and **semimembranosus**) (refer to the box on page 130 for anatomical terms). Few muscles are located in the shank (the **tibiotarsus**) beneath the thigh; as a result some thigh muscles control the lower leg and foot through long, strong tendons. A small number of **tibiotarsal muscles** and a few additional thin threadlike muscles along the tarsometatarsus also connect to the toes by way of long slender tendons. Beyond this very general description, the avian hindlimb has undergone many modifications to accommodate diverse ecological roles. For example, swifts live their entire lives primarily in the air. The family name (Apodidae) literally means "without feet." The Chimney Swift (*Chaetura pelagica*) has short legs that are incapable of perching and tiny feet that are adapted to clinging to vertical surfaces. At the other end of the spectrum are most passerine songbirds (e.g., sparrows, wrens, warblers), in which the tendons of the **flexor digitorum longus** and the **flexor hallucis** muscles form a locking mechanism that results in a **perching reflex**. The flexor tendons run down the back of the tibiotarsus, across the bottom of the foot, and insert on the plantar (bottom) surface of the digits. When a bird bends its ankle (the tibiotarsal joint) to perch, the tendons are pulled across the sole of the foot and automatically close the toes around the perch. While passerine feet may automatically flex, the African Ostrich (*Struthio camelus*) has a leg-stabilizing mechanism that allows the ankle joint to passively extend (Schaller et al. 2009). This passive extension is an energy conservation mechanism that contributes to the exceptional speed and running endurance of these large, heavy, flightless birds.

THE DIGESTIVE SYSTEM

The organization of the avian digestive system reflects a compromise to simultaneously meet several ecological needs. As discussed in above, heavy structures like teeth and grinding jaws are inconsistent with the reduction in weight required to accommodate flight. Birds must swallow their food whole. Unlike snakes, however, a bird cannot crawl under a rock and wait for gastric enzymes and hydrochloric acid to take their toll on a meal. Natural selection dictates the need for a lightweight alternative for breaking food into small pieces and keeping digestion efficient. This is no small constraint, as the process of digestion must be extremely efficient to

BRET W. TOBALSKE, ASSOCIATE PROFESSOR, DIRECTOR OF FIELD RESEARCH STATION, UNIVERSITY OF MONTANA

Dr. Bret Tobalske is the director of the University of Montana Field Research Station and associate professor of biology. With over 20 years of bird-related research experience, Dr. Tobalske is one of the world's foremost experts on the biomechanics of bird flight. His contributions to the study of bird flight, comparative biomechanics, and functional morphology are broad and varied, with publications on topics ranging from the scaling of muscle physiology to hummingbird aerodynamics. He earned his PhD at the University of Montana studying the effects of speed and body mass on the intermittent flight of birds. His work then took him to a fellowship at Harvard and professorships at Allegheny College and University of Portland. In 2008, Dr. Tobalske returned to the University of Montana as an associate professor and director of the Field Research Station. His research utilizes a variety of techniques, including high-speed video, flow visualization (particle image velocimetry), and muscle electromyography to describe the aerodynamic and biomechanical complexities of animal flight. His work has been featured in *National Geographic*, the *New York Times*, on the Discovery Channel and the BBC, and in numerous peer-reviewed journals including *Nature* and the *Journal of Experimental Biology*.

cess (crack, peel, or open) food. The specifics of avian foraging behavior are covered in more detail in chapter 15, and we have already covered the fundamental anatomy of the beak in this chapter. Here we focus on the important structures within the **oral cavity** that work with the beak to catch, process, manipulate, and swallow food. Like the beak, the anatomy of the avian mouth is often specialized for handling different types of food. For example, the oral cavity can stretch into a flexible large pouch (e.g., pelicans) to temporarily hold or store food. Birds lack a **soft palate** in their oral cavity, and the **hard palate** is connected to the nasal passage above by a slit-like opening called the **choanal opening**. The choanal opening and the lack of a soft palate prevent the use of negative pressure to create suction within the oral cavity. Birds must therefore use their lower mandible to scoop water into the mouth and allow it to flow back into the throat by gravity. Oral glands are usually present in the mouths of bird species that consume dry food (e.g., seeds) or live in arid environments (e.g., ostrich), but may be absent or poorly devel-

support the avian high basal metabolic rate. Since the products of digestion must cross an anatomical boundary to be used as substrates for metabolic pathways, a discussion of digestive anatomy often crosses into the area of physiology (chapter 7). We concern ourselves here with the basic organization (fig. 6.21) and some examples of specialized adaptations, while leaving the physiology, or anatomy in action, to the next chapter.

The Beak, Oral Cavity, and Tongue

Before you can digest food, you must first acquire it—and the beak is a handy multipurpose tool for this function. Bird beaks have undergone an amazing degree of specialization to accommodate diverse foraging techniques and food types (fig. 6.22). The beak can be used to find (probe, feel, or sense), uncover (dig, peck, and bore), grasp, kill (if need be), and pro-

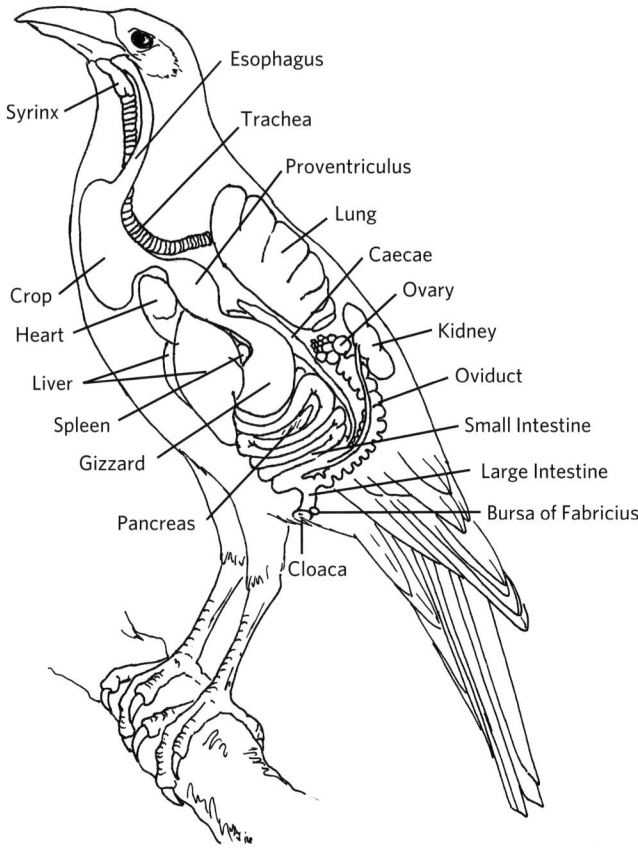

Figure 6.21. Overview of the avian digestive system. Figure is not to scale and stylized to show general placement of the internal organs with reference to each other.

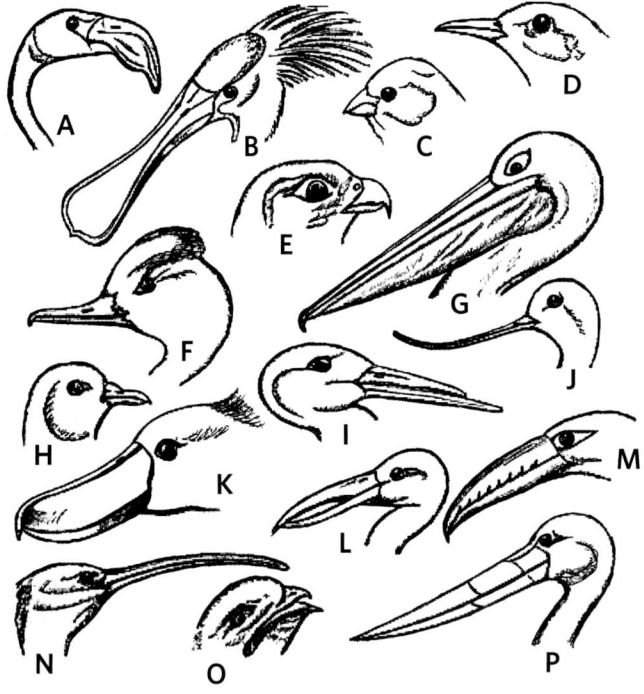

Figure 6.22. Birds use their beaks for a wide range of purposes—everything from feeding to nest building, preening, and manipulating objects in their environment. The following examples focus on adaptations for feeding and are just a representation of the amazing diversity of bird beaks. *A*, The filtering beak of the Greater Flamingo (*Phoenicopterus roseus*); *B*, The beak of the Royal Spoonbill (*Platalea regia*) is used to swish water and mud from side to side to filter insects and crustaceans from mud and silt. The tip of the beak is used like a pair of tongs to grasp its prey; *C*, The small conical beaks of sparrows are adapted to picking and crushing seeds; *D*, The long sharp beaks of thrushes are ideal for probing and capturing insects; *E*, The sharp pointed beak of the falcon (*Falco* spp.) is equipped with a small triangular shaped **tomial** tooth on its outer edge strong enough to sever the ligaments between the cervical vertebrae of their prey; *F*, The sharp and narrow beak of the merganser has sawlike ridges to help trap small fish and amphibians; *G*, The pouch-like beak of pelicans is used to scoop up fish; *H*, The beak of the dove is adapted to feeding on a variety of seeds, fruits, and plants; *I*, The unusual uneven beak of the skimmer is used to plow through the water as the bird skims low over its surface. If the bird encounters a fish, its beak snaps shut around its prey; *J*, The long, slender, up-curved beak of an avocet (*Recurvirostra* spp.) is used to swish water from side to side while foraging for aquatic insects in marshes; *K*, The Shoebill (*Balaeniceps rex*) stalks its prey like a stork, but its large sharp-edged beak with a huge gape allows it to take much larger fish than other wading birds; *L*, The beaks of openbill storks (*Anastomus* spp.) have a distinctive gap between the upper and lower mandibles that develops as the birds mature, with the lower mandible often twisted to one side. The mandibles have rough fringed edges that are thought to assist in grasping slippery snail shells; *M*, The beak of the toucan (*Ramphastos* spp.) is a multipurpose tool that can peel fruit, their primary food, as well as capture small prey. The toucan beak also houses a long flat tongue that assists in catching insects and small frogs and reptiles; *N*, The slim curved

oped in fish-eating species. In some birds, such as the swifts and some swallows, the salivary glands are large and well developed. In these birds, the size of the salivary glands is under hormonal control. The glands often double in size during the reproductive season to produce large amounts of a sticky salivary secretion that is used as a form of cement during nest building. Other bird species use sticky salivary secretions on their tongue to capture prey (e.g., the Grey Jay, *Perisoreus canadensis*, and the European Green Woodpecker, *Picus viridis*).

Bird tongues are almost as diverse as bird beaks when it comes to shapes and sizes. They can be practically nonexistent in some species (e.g., the cormorants and ostriches), and are usually very short in the majority of other species (e.g., most passerines). However, some bird tongues are exceptionally long (e.g., sunbirds, hummingbirds, flickers, woodpeckers). Some bird tongues may be feathered or fringed at the tip (e.g., flowerpeckers, honeyeaters, sugarbirds), while others will have a variety of **spines** (penguins), **barbs** (woodpeckers), **barbules**, **bristles** (ducks, lories), **papillae**, or other adaptations that support unique feeding modalities. All bird tongues are located on the floor of the oral cavity within the lower mandible of the beak. Most birds have a narrow triangular, pointed tongue that lacks intrinsic musculature, although the muscular tongue of parrots is an exception to this rule. The epithelium of the avian tongue tends to be tough and partially keratinized. In some species (e.g., ducks, geese, swan, parrots, penguins), a hard keratinized structure called a "**lingual nail**" forms at the tip of the tongue (Erdoğan and Iwasaki 2014). The lingual nail of some species is spoon-shaped, possibly useful in manipulating or cracking seeds (Jackowiak et al. 2011). Papillae of various sizes and shapes cover the tongues of many bird species. In fish-eating species, the papillae are usually sharp, rigid cones that point backward to hold their slippery prey. Penguins, for example, catch fish using large, densely packed spinelike papillae that completely cover the tongue and point back toward the throat. In raptors, a V-shaped row of conical papillae, known as a **papillary crest**, point to the back of the tongue to assist in guiding food into the **esophagus** and preventing regurgitation (Erdoğan and Iwasaki 2014). The

beak of the ibis is specialized for probing in mud and shallow water; *O*, The short wide beak of the swifts and nighthawks opens into a large gape to efficiently scoop up insects in flight; *P*, The Saddle-billed Stork (*Ephippiorhynchus senegalensis*) uses its magnificent red beak to strike and grab frogs, fish, small birds, and reptiles. *Modified from original illustration, Koerperteile der Voegel, in 4th edition of Meyers Konversations-Lexikon (1885), https://archive.org/stream/bub_gb_wpAGAQAAIAAJ /bub_gb_wpAGAQAAIAAJ_djvu.txt.*

tongues of flickers and woodpeckers also have barbs and pa-pillae that face toward the back of the oral cavity. When the barbed tip of the tongue skewers an insect, it retracts the food item into the mouth and guides it to the back of the oral cavity to be swallowed (fig. 6.23). Tactile sensory cor-puscles provide the tongues of many species (e.g., finches, woodpeckers, geese) with a well-developed sense of touch. In addition to tactile receptors, birds have thermoreceptors, primarily sensitive to cold, concentrated on their tongues and beaks (Beason 2003). And although birds tend to eat "on the fly" and swallow their food whole, they do have a lim-ited sense of taste. The taste buds of birds are distributed not only in the less-keratinized epithelium at the back of the tongue but also frequently on the roof or floor of the oral cavity (Iwasaki 2002).

The tongue is attached to a boney structure in the neck common to all vertebrates, the **hyoid apparatus** (previously described). The structure of the hyoid helps define the shape of the tongue and support its movement. In swans, parrots, and most other birds, the "horns" of the hyoid extend backward from the tongue on either side of the larynx. Hummingbirds, flickers, and woodpeckers have unique hyoid structures that wrap backward and over the top of the skull (see fig 6.11). The elongated hyoid structure allows these birds to extend their tongues far outside the oral cavity and into crevices to hunt for insects (flickers, woodpeckers), or deep into flowers to pump nectar out (hummingbirds). Hummingbird tongues are unique in their function as micropumps that transport fluid (Rico-Guevara et al. 2015). A common misconception has been that hummingbirds use capillary action to "suck" fluid into the oral cavity. In reality, a hummingbird's tongue changes shape rapidly while flicking in and out of a nectar pool to quickly and efficiently pump liquid to its mouth. The mechanism depends on the unique shape of the tongue and its biomechanical properties. A hummingbird compresses its forked tongue as it dips into nectar. When the tongue comes in contact with the fluid, the tongue reshapes to pull fluid into the grooves in the distal part of the tongue (see the box on page 130). The tongue then bends near the oral cavity to draw the nectar into the mouth.

The Esophagus and Its Derivatives

Once food has been swallowed, it travels from the oral cav-ity to the stomach through a muscular tube lined with **mu-cous glands**. This is the **esophagus**. In birds, the size of the esophagus corresponds to the size of the food it transports; this may range from insects in some species to whole fish or small mammals in others (Welty and Baptista 1988). Al-though some humans have been rumored to swallow whole

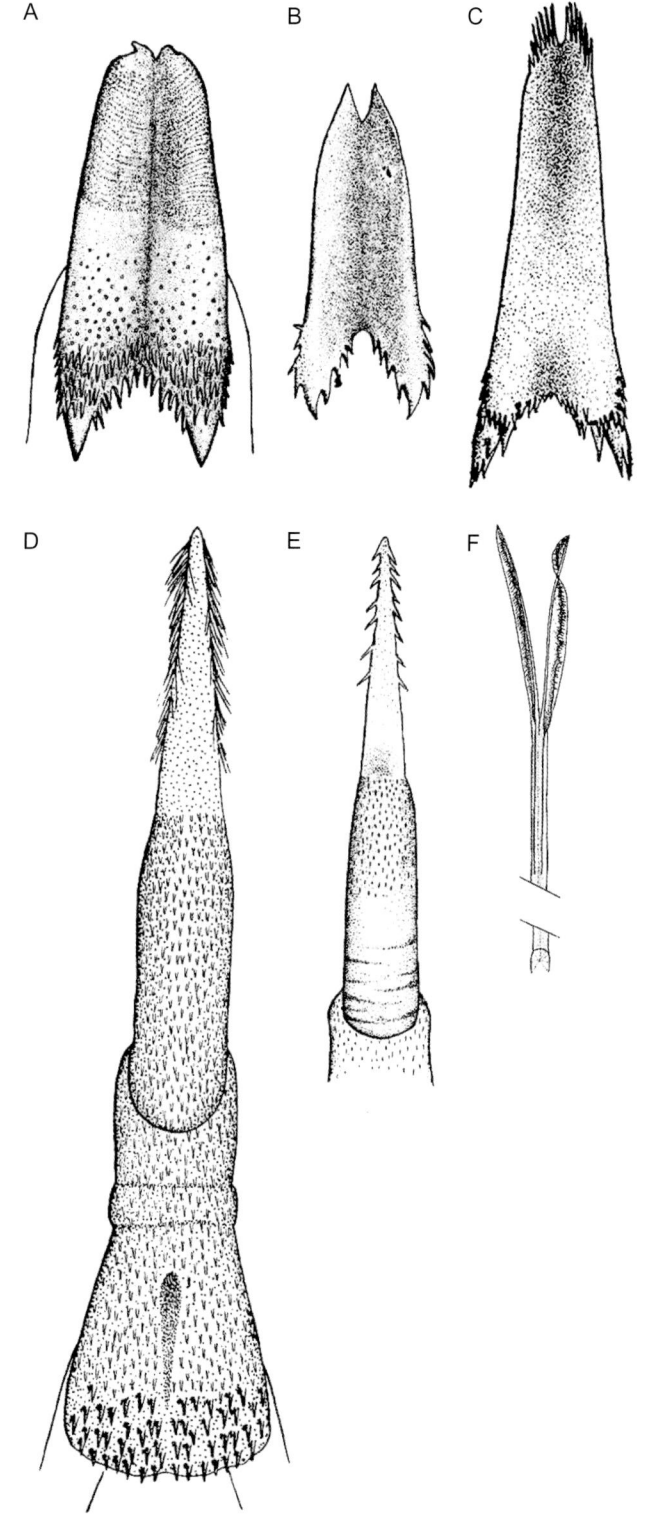

Figure 6.23. Diversity of avian tongue structures; note that figure is not drawn to scale. *A*, Northern Mockingbird (*Mimus polyglottos*); *B*, Say's Phoebe (*Sayornis saya*); *C*, American Kestrel (*Falco sparverius*); *D*, Anna's Hummingbird (*Calypte anna*); *E*, Nuttall's Woodpecker (*Picoides nuttallii*); *F*, Acorn Woodpecker (*Melanerpes formicivorus*). *Drawing modified from L. L. Gardner 1925.*

fish as a trite party trick, many bird species make this their way of life. The extreme distensible nature of the esophagus allows seabirds, unlike humans, to show no visible discomfort if a fish remains lodged there for extended periods of time. Some seed-eating species such as finches (e.g., Common Redpoll, *Acanthis flammea*), on the other hand, have specialized small out-pocketings (**diverticula**) of the esophagus that can be used for temporary food storage. The stored food is later regurgitated to be swallowed or possibly fed to nestlings (Fisher and Dater 1961). Specializations aside, the esophagus of most birds empties into the **crop**, a larger out-pocketing of the gut used for food storage. This structure is diverse in its morphology, as its shape depends on its species-specific use. In most species, the crop is primarily a temporary holding area for food waiting to be processed by the **stomach**. In larger species with more storage capacity, this holding time can often be a day or more. The general consensus is that the crop is particularly well developed in bird species that must feed on the move to avoid predation, such as game birds and doves. The fight-or-flight responses of the vertebrate nervous system oppose efficient digestion, so food could be held until a flushed bird reached safety. Some birds species use crop fluids to feed their developing young (e.g., doves, pigeons, flamingos, and penguins), although the sources of the secretions differ between species. Pigeons and doves secrete a milky protein-rich liquid from fluid-filled cells in a special pocket of the crop in response to reproductive hormones (Beams and Meyer 1931). This is used to feed developing nestlings. Flamingos and penguins use glands in the esophagus for a similar purpose, although the fluid has a higher fat composition than that of pigeon crop-milk. In all cases, the production of "milk" is in response to a period of limited fat and protein availability for the diet of the developing chicks, albeit for very different ecological reasons.

The Visceral Organs

The avian stomach is divided into two distinct regions. The first region is called the **proventriculus**, or glandular stomach. This structure is analogous to the mammalian monogastric stomach, the type found in humans. It is an expansion of the soft-walled tube of the digestive tract and heavily lined with tubular mucosal glands. Like the monogastric mammalian stomach, the gastric mucosa secretes hydrochloric acid and acid-activated digestive enzymes such as pepsinogen. This glandular stomach is best developed in fish-eating birds and raptors that swallow their prey whole. The ingestion of whole bones requires a low stomach pH (0.7–2.3) and strong digestive enzymes. Some species use the oily glandular secretions of the proventriculus for nondigestive func-

tions such as food for their young (e.g., petrels), as a weapon against competitors (e.g., fulmars), or in self-defense (e.g., eiders). One of the most unusual examples of this is found in the Eurasian Roller (*Coracias garrulus*). When a juvenile of this species is threatened, it vomits a powerfully pungent orange oil on itself that repels predators and sends a chemical cue that calls the parents to help (Parejo et al. 2012).

The second region of the stomach, or muscular stomach, is also known as the **gizzard**. This part of the digestive system is lined with rough convoluted grooves and ridges. It breaks food particles into smaller pieces, thereby increasing the surface area available to peptic enzymes and optimizing digestive enzyme function. Some birds will ingest grit and small rocks to enhance the grinding process. In other birds, the gizzard is so muscular it is capable of crushing the bones of prey. In many species (e.g., raptors such as owls, fish-eating species such as cormorants), fur, feather, scales, and bones cannot pass the gizzard. These birds will periodically egest (regurgitate) a pellet or **bolus** that can be useful for the ornithologist interested in dietary patterns, and for the paleontologist, as this is the main way to collect small mammal remains in caves (Andrews 1990). In some species, changes in stomach structure and function accommodate seasonal changes in diet. Digestion and absorption times and mechanisms may vary further depending on the structure of the food ingested and its macronutrient composition (e.g., fat, carbohydrate, protein).

The avian gut includes several associated organs (e.g., **liver**, **pancreas**, **kidney**) that begin as out-pocketings along the **alimentary canal** during embryonic development. As animal form and function are always tightly linked, the anatomical specializations associated with these structures become increasingly diverse in adult birds, based on dietary patterns and ecology. Given the close association between the anatomy and physiology of the remaining avian internal organs, they are covered in detail in the next chapter. However, this chapter briefly explores one unique internal structure that differs from all other vertebrates, a modification of the trachea called the **syrinx**.

The Trachea and Syrinx

The **trachea** conducts air from the mouth to the lungs. As in mammals, it consists of a tough fibrous tube reinforced by a series of C-shaped rings of cartilage that often ossify and turn to bone in adult birds. The trachea branches into a right and left bronchus just above the sternum. The **syrinx**, a structure unique to birds, is located at this bronchial junction. It is analogous to the larynx in mammals. The word syrinx means "double flute," a perfect description of the system of muscles, membranes, and rings that wrap around the bronchi to produce birdsong. The structure of the syrinx varies between avian families and determines the characteristics of the song produced. In some birds, the sound produced by the syrinx is modified by the tongue, oral cavity, and beak. Parrots in particular may use their tongue to modify the amplitude and frequencies of sounds produced by the syrinx to mimic human speech (Beckers et al. 2004). Many groups of birds have no syrinx at all, like vultures and ostriches, which produce only grunts or hissing noises. There are three main structural variations of the syrinx: the bronchial, tracheal, and tracheobronchial forms (Suthers 2004). The tracheal syrinx common to parrots and the tracheobronchial form found in songbirds are illustrated in fig 6.24. Unlike the mammalian **larynx**, the syrinx has no vocal cords. Sound is produced by the vibration of a bone (the **pessulus**) at the base of the syrinx between the **bronchi** and **tympaniform** membranes in the wall of the structure. The pessulus and the tympaniform membranes vibrate and resonate with each other as air rushes up from the lungs (Greenwalt 1968). The syringeal muscles constrict on each side, changing the tension of the membranes and modulating the sound (Larsen and Goller 2002). In songbirds, each bronchus has its own muscles, tympanic membranes, and innervation. This allows modulation of the airflow through each bronchus independently and can produce flutelike double notes, such as those of the Wood Thrush, *Hylocichla mustelina* (Greenwalt 1968, Suthers 2004).

Beyond the trachea, the anatomy of the avian respiratory system has undergone extensive modification. The lungs of birds are quite small given the oxygen demands of flight and the high metabolic rate of birds in general. The lungs are not as elastic as those of mammals and are attached to the ribs and thoracic vertebrae. The bronchi are also organized in a manner unique from other vertebrates. After the trachea divides into the left and right bronchus, each bronchus is highly modified into a series of interconnecting bronchial tubes that together form a structure called the **mesobronchus**. The organization of the mesobronchus permits **unidirectional ventilation** of the avian lung, a mechanism discussed in detail in the next chapter.

THE CARDIOVASCULAR SYSTEM

It is clear by now that flight is energetically demanding and that birds must maintain a high metabolic rate to meet that demand. Birds are **endothermic homeotherms**, so they must maintain high, stable internal temperatures. Mammals are the only other group of truly homeothermic vertebrates. The need to regulate internal body temperatures as well as main-

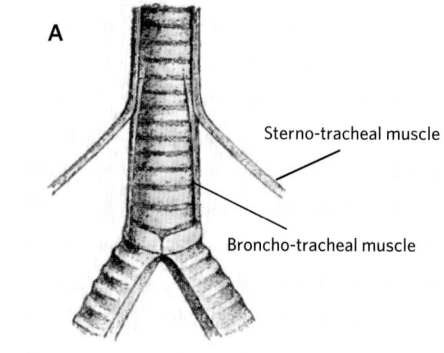

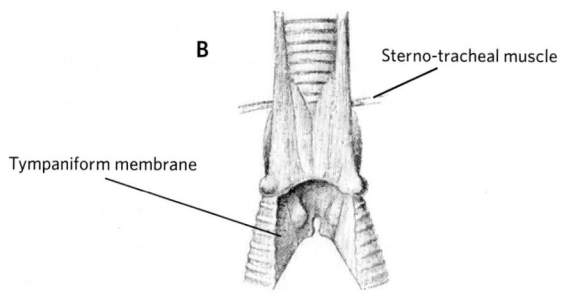

Figure 6.24. Two types of syrinx. *A*, The **tracheal syrinx** characteristic of the suboscine Noisy Pitta (*Pitta versicolor*); *B*, The **tracheobronchial syrinx** of the oscine Rook (*Corvus frugilegus*). *From R. Bowdler Sharpe 1891; public domain via the Internet Archive, https://archive.org/details /catalogueofspeci00shar.*

tain stable internal chemistry is a strong selection pressure. The structure of the avian circulatory system is therefore similar to that of mammals. In general, the endotherm circulatory system must deliver the chemical products of digestion (glucose, lipids, and amino acids), oxygen from the lungs, and hormones, as well as regulate blood pH, body temperature, mineral metabolism, and the function of the immune system. Metabolic wastes and carbon dioxide must also be cleared from the body by way of the circulatory system. Our survey of the avian cardiovascular system begins with the heart.

The Heart

Like mammals, birds have a four-chambered heart that consists of two **atria** and two **ventricles**. The high metabolic rates and body temperatures of **homeotherms** require high concentrations of oxygen. The four-chambered heart design completely separates the **pulmonary circulation** from the **systemic**, or general body, circulation. This prevents the mixing of oxygenated **arterial blood** with deoxygenated **venous blood**, as is seen in the single ventricle of the reptilian three-chambered heart. The dual circuit design also allows blood to circulate through the capillary beds of the lungs at substantially lower pressures than that of the systemic circulation that supplies the rest of the body. Lower pulmonary blood pressure enhances gas exchange efficiency while protecting the delicate capillary beds of the lungs. The avian **aorta** originates from the left ventricle but differs from mammals in that it arches to the right of the body. Although this arrangement seems to resemble that of crocodiles, birds lack the left systemic arch seen in crocodilian reptiles. In an arrangement unique to birds, two large **brachiocephalic vessels** branch off the avian aorta near its base. These vessels are extremely large in diameter and under great pressure as they lead to the arteries that supply blood to the massive pectoral muscles responsible for flight. Since the avian left ventricle must generate high pressure to pump blood throughout the entire body, in contrast to the right ventricle, which pumps blood only to the lungs, the walls of the left ventricle are much thicker and more muscular than those of the right ventricle. The structure of the avian **superior vena cava** also resembles that of reptiles, with two **precava veins** joining to form a **sinus venosus** just before entering the right atrium. The structure of the **inferior vena cava**, however, is similar to that of mammals (Proctor and Lynch 1993).

Relative to body mass, bird hearts tend to be larger than mammalian hearts. In spite of this, their heart rates are lower than those of mammals of comparable size. This is an adaptation to meet the high metabolic demands of flight (Bishop 1997). The combination of relatively low heart rate with efficient ventricular refill and high arterial blood pressure allows birds to have a higher **cardiac output** and a higher volume of blood oxygen than similar-sized mammals (Grubb 1983). Cardiac output (the volume of blood pumped per minute) is a function of both heart rate and stroke volume (the volume of blood pumped from the left ventricle with each beat). **Stroke volume** increases with increased efficiency of ventricular emptying during each contraction. Cardiac output can therefore be used as a measure of mechanical efficiency of the vertebrate heart, and by this metric, bird hearts are very efficient hydraulic pumps. However, form follows function in the natural world, and variations in habitat and life history put selection pressure on the size and function of the avian heart beyond the mechanical requirements for flight. The size of the avian heart scales with altitude (low temperatures and low oxygen levels require larger hearts to meet metabolic demand), latitude (colder environments, greater metabolic demand), migratory patterns (metabolic demand of long distance flight; Welty and Baptista 1988), and body size (small birds have larger hearts than bigger birds relative to body mass; Seymour and Blaylock 2000, Welty

and Baptista 1988). Hummingbirds are an excellent example of how heart size scales according to the energetics of flight. Hovering flight is extremely energetically costly; as a result, hummingbirds have the largest heart relative to body size of all birds (Suarez 1992).

Although bird heart rates tend to be lower than comparably sized mammals, within the class Aves heart rate generally increases with decreased body size (Machida and Aohagi 2001). For example, the Wild Turkey (*Meleagris gallopavo*) has a resting heart rate of about 93 beats per minute at rest, while the resting heart rate of an American Robin (*Turdus migratorius*) will be about 570 beats per minute (Welty and Baptista 1988). Some small passerines can attain heart rates greater than 700 beats per minute. General trends in heart rate must be adjusted to meet a wide range of environmentally induced metabolic demands, including those of environmental temperature, physiological state (e.g., stress, aggression, fear), physical exertion, and modes of locomotion (e.g., running, flapping flight, soaring flight). The heart rate of some species during flight may be as much as four times higher than at rest. The heart rate of the Rock Pigeon (*Columba livia*), for example, has been measured at about 170 beats per minute at rest, but can be as high as 670 beats per minute while flying (Bishop 1997). The heart rate of the Herring Gull (*Larus argentatus*) has been reported to increase from about 130 beats per minute at rest to 625 beats per minute during sustained flapping flight. Soaring flight, however, produced heart rates similar to those at rest (Kanwisher et al. 1978). Diving birds can experience substantial decreases in heart rate and may even exhibit a classic "diving response," in which heart rate is suppressed below baseline. The South Georgian Shag (*Phalacrocorax georgianus*) dives deep for extended periods of time. Bevan et al. (1997) monitored heart rates of shag while resting (104.0 ± 13.1 beats per minute), flying (309.5 ± 18.0 beats per minute), and diving (103.7 ± 13.7 beats per minute). Although the average diving heart rate was not statistically different from that at rest, the minimum heart rate recorded during a dive was 64.8 ± 5.8 beats per minute. In general, heart rate decreased with increasing dive depth and duration. This is an example of how the basic anatomical design of the avian heart is flexible enough to function as an efficient hydraulic pump to support flight while still accommodating some variations in lifestyle, such as foraging at depth.

The Circulatory System

The remainder of the avian vascular system is similar to mammals, with only a few modifications for flight. In general, the blood pressure of a bird will be somewhat higher that of a comparably sized mammal, due in part to artery walls that are stiffer than those of mammals. The increased resistance of stiff arterial walls improves peripheral blood flow under pressure during exertion (flight), but it also increases the susceptibility of birds to atherosclerosis (Clarkson et al. 1965). Although the avian heart and vascular system are models of efficiency, finely tuned to the energetic demands of flight, they also operate at the limit of structural integrity. Aortic ruptures, aneurysms, cardiac failure, and atherosclerosis have long been recognized as major health problems in Domestic Turkey (*Meleagris gallopavo*), Domestic Chickens (*Gallus gallus domesticus*), Rock Pigeons (*Columba livia domestica*), and in pet birds such as parrots. The combination of artificially high-fat diets, weight gain in captivity, and stress all contribute to these disorders. However, stress-induced heart failure or aortic rupture is also a possibility the wildlife biologist must consider when working with small wild birds in the field.

The one notable modification to the avian vascular system is a **renal portal system** not found in mammals. The renal portal system is a specialized low-pressure arterial network that carries blood from the hindlimbs directly to the kidneys. This is thought to be an efficient way to excrete uric acid, the semisolid nitrogenous waste produced by birds. The avian renal portal system is similar to the reptilian structure by the same name, although it differs in that it contains valves to bypass renal portal circulation during the exertion of flight (Akester 1967). The avian renal portal system is extremely effective at redirecting blood from the peripheral vascular system to offset drops in systemic blood pressure (Raidal and Raidal 2006), making birds somewhat resistant to shock due to blood loss.

THE LYMPHATIC SYSTEM AND ASSOCIATED ORGANS

Compared with that of reptiles and mammals, the **lymph system** of birds is not particularly well developed. Lymph "hearts" are apparent in the embryo, but these disappear over the course of development, and there are no prominent lymph nodes in the adults of most species. Some waterfowl are the exception to this rule, with lymph nodes found in the neck and lumbar regions (King and McLelland 1984). The avian **spleen** is a small oval structure adjacent to the liver; it is not particularly important for blood storage, although it does experience changes in size and activity with season (John 1994). In spite of the differences noted, there are some structural similarities between the lymph system of birds and mammals. Patches of lymph tissue can be found near the **intestines** and **mesenteries**, and in the **thymus gland**. Lymph tissue is also found within a pouch (a **bursa**) on the wall of the

cloaca called the **bursa of Fabricius** (fig. 6.21). This glandular structure is unique to birds and is apparent only in immature birds. The bursa of Fabricius changes in structure and function with age, eventually regressing to a firm nodule in adults (Ciriaco et al. 2003). The avian thymus, similar in function to the mammalian thymus gland, also regresses with age. Both structures, when active in young birds, are responsible for B-cell formation and acquired immunity. Birds also have a **cecal tonsil** closely associated with the gastrointestinal tract that contains T-cells, B-cells, and plasma cells, and **Peyer's patches** can be found along the intestine just above the ileocecal junction of the small and large intestine.

THE UROGENITAL SYSTEM

Vertebrate excretory and reproductive structures share common embryonic origins and developmental pathways, are located near each other within the abdominal cavity, and are often physically linked through the use of the same tubules and ducts. Collectively, these structures are called the **urogenital system**. In the mature adult bird, both reproductive and excretory products exit the body through a common anatomical structure called the **cloaca**, a term aptly derived from the Latin word for "sewer." The avian urogenital system has several other features that are novel and function to reduce overall body mass. One example of this is the "use it or lose it" seasonality seen in the reproductive organs. Outside the reproductive season, male and female **gonads** are significantly reduced in both size and weight. Most of the year, the internal male **testes** are almost impossible to see in dissection or with modern laparoscopic surgical devices. This seasonal pattern of regression and dormancy outside the reproductive season effectively reduces body mass to accommodate other important activities, such as migration. As the reproductive season (spring) approaches, and day length increases, photic stimulation of the endocrine system triggers increases in size and weight of both male and female gonads, often by several orders of magnitude (Wingfield and Farner 1993).

The Female Reproductive System

Most birds have only a single **ovary**, on the left side of the body, although two ovaries may be present in some species of raptors (Guioli et al. 2014). Functional left and right ovaries have been observed in the Long-eared Owl (*Asio otus*), Common Buzzard (*Buteo buteo*), European Sparrowhawk (*Accipiter nisus*), and Northern Goshawk (*Accipiter gentilis*) (Rodler et al. 2015). In all species, the avian ovary is a cluster of developing **primary oocytes**. Although the ovary is fully formed at hatch, it will remain small until the female bird reaches sexual maturity. The mature ovary increases 10–50 times its dormant size when the appropriate seasonal cues for reproduction are present. Similar to mammals, no new **ova** form after embryonic development is complete. The maximum number of eggs a bird can lay is therefore determined at hatch. The transition from a small and bare ovum to what we recognize as bird's fully shelled egg begins with release from the ovary into the mouth (**ostium**) of the **oviduct**. Each ovum will increase by more than an order of magnitude in size and weight by the time it has passed through the oviduct. **Yolk**, **albumen** (egg white), and a hard calcareous shell (derived from calcium; see chapter 5) will be sequentially added during this journey (fig. 6.25).

The **oviduct** itself is a long tubular structure divided into functional regions: the **ostium** (or infundibulum), **magnum**, **isthmus**, **uterus**, and **vagina**. A maturing ovum is first released from the ovary into the ostium. The next four regions of the oviduct are not visibly distinct and can only be distinguished by their unique functions and histology. As the ovum passes through the magnum region, it receives a layer of albumen from **glandular cells** (fig. 6.25). Although albumen is mostly water, it contains about 50 percent of the protein available in the egg (in the form of albumins and globins) to support embryo development. The next region, called the isthmus, is not present in all species; it is most notably lacking in parrots. In other species, the ovum receives a coating of flexible keratin proteins as it passes through the isthmus. This layer forms the base of what will eventually become the familiar external calcareous eggshell. The hard calcium carbonate outer layer is added by the **shell gland** in the uterus, the region where the egg will spend most of its time in the oviduct. The pigments (primarily **protoporphyrin** and **biliverdin**) that often give eggs species-specific colors and patterns are secreted by **pigment glands**, also found in the uterus. Also located in the uterus of many species, near the beginning of the vagina, are deep **sperm storage glands** (Shugart 1988). The sperm of birds can remain viable at body temperature for extended periods of time. When a female bird ovulates, she may release stored sperm into the oviduct, which will migrate to the ostium to fertilize the newly released ovum. The next, final, section of the oviduct, the vagina, is simply a passageway that has no role in egg formation. A newly formed egg will be held in this region until the female is ready to **oviposit** (lay). The entire process from ovulation to oviposition generally takes about 24 hours in most species but may be as long as two to three days in some (Haywood 2013). As the egg exits the oviduct, it passes into the cloaca. The cloaca is a common passageway that receives the excretory products from the colon and the ureters, as well as the reproductive products (eggs through the oviduct, and sperm through the **ductus**

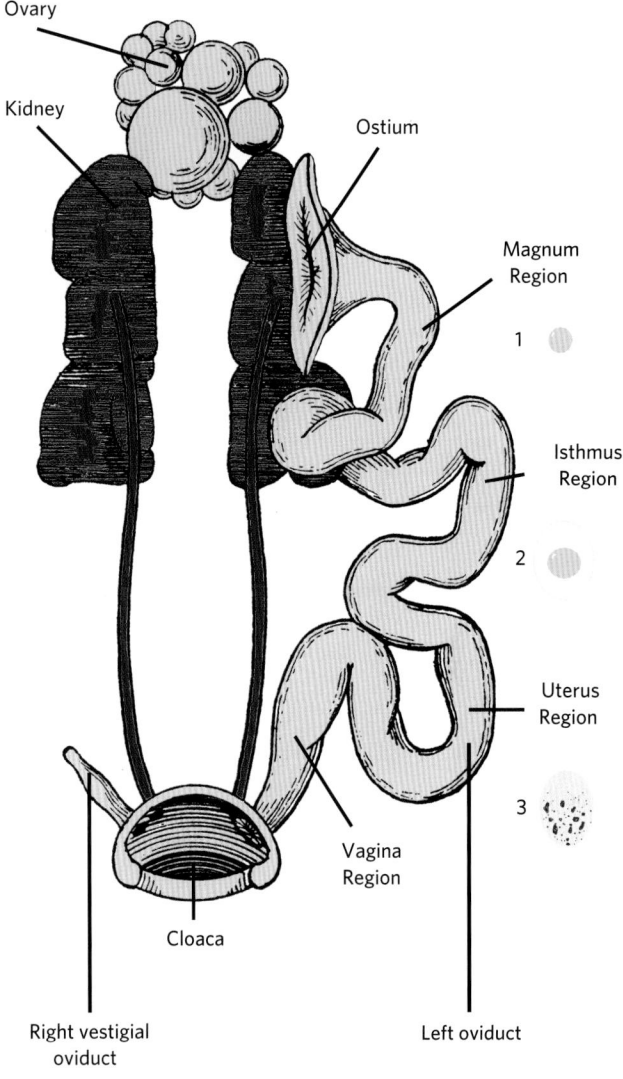

Figure 6.25. Female reproductive system and stages of egg formation. A mature ovum is released into the **ostium** of the **oviduct**. As the ovum passes through the **magnum** region, it receives a layer of albumen from **glandular cells** (1). In most species, the ovum also receives a coating of flexible keratin proteins as it passes through the **isthmus** (2). The hard calcium carbonate outer layer, along with any pigmentation, is added by the **shell gland** in the **uterus**. *Modified from Coues 1884.*

deferens in males). The contents of the cloaca pass out of the body through a horizontal opening called the **vent**.

The Male Reproductive System

The reproductive structures of male birds are entirely internal. Although males of most bird species lack a true penis, or **intromittent organ**, some may have glandular erectile tissue located on the ventral surface of the cloaca. Ducks, storks, flamingos, ostriches, emus, rheas, and tinamous, as well as certain species of parrots and passerines, have some form of intromittent organ (Briskie and Montgomerie 1997), although the structures vary greatly by species. Most of the year, the avian **testes** are small, bean-shaped structures located ventral to the kidneys. Like the female gonad, the male gonad is only prominent during reproductive season. When a male is in breeding condition, testes mass (corrected for body mass) is considered a reliable measure of sperm competition (Birkhead and Møller 1998). Males with larger testes can produce larger quantities of sperm; this increases the probability of a male fertilizing a female's eggs (see chapter 16). Sperm are carried from the testes through a duct called the **vas deferens** to the cloaca. The width of the vas deferens increases as it approaches its junction with the cloaca. This region of the vas deferens is a temporary storage area for sperm called the **seminal vesicle**; it is significantly reduced in size outside of the breeding season. The seminal vesical is similar in function to the mammalian scrotum. Its position at the far end of the male reproductive tract, away from the abdomen, allows developing sperm to be stored at a reduced temperature to increase viability.

The Kidney

In both sexes, the **kidney** is a trilobed structure that rests in pockets on the dorsal wall of the abdominal cavity (**retroperitoneal**) just beneath the synsacrum. Each kidney is drained by a **ureter** that originates between the anterior and middle lobes. The functional unit of the vertebrate kidney is the **nephron**, a microscopic filtration structure. Tens of thousands of nephrons are housed in the lobes of each kidney. Nephrons regulate the concentration of water and soluble substances, such as sodium chloride, by filtering the blood. This is the process of **osmoregulation**. Regions of the nephron reabsorb water and solutes (e.g., sodium, glucose) and excrete the remaining filtrate, including any nitrogenous compounds, as waste. Birds are the only vertebrates other than mammals that can produce urine that is more concentrated than their blood (i.e., hyperosmotic). Mammalian nephrons use a countercurrent exchange system driven by Na^+/K^+ pumps, located in the ascending side of the long **loop of Henle**, to reabsorb water into **peritubular capillary beds**. Only about one-quarter of the nephrons in a bird are equipped with loops of Henle, and these are short with limited surface area for the sodium exchange necessary to drive water reabsorption. This nephron morphology is more similar to that of reptiles than mammals. In the bird, dilute urine accumulates in the **urodeum** or middle chamber of the cloaca and flows up into the **coprodeum**, or top chamber, where it is concentrated through the reabsorption of water and electrolytes (e.g., sodium and chloride). This results in a relatively lightweight colloidal suspension of **uric acid** (Gold-

stein and Skadhauge 1998), the nitrogenous waste produced by birds and reptiles. The semisolid nature of uric acid is very efficient with respect to the amount of water required for its excretion (Braun 1999). Birds use about 1 ml of water to excrete 370 ml of uric acid. In contrast, a mammal requires 20 times the amount of water to excrete the same amount of nitrogen as a bird. This is useful for both osmoregulation and for the reduced-weight requirements of flight. Furthermore, the nitrogenous waste produced by mammals (urea), must be suspended in water and held in temporary storage in a **urinary bladder**. The dehydrated urates produced by birds do not require a bladder for storage, and most species lack this structure altogether. The only exceptions are flightless birds such as ostriches and rheas. In these birds, an expansion of the proctodeum forms what appears to be a small bladder (Duke et al. 1995). In all birds, the white uric acid paste is ultimately excreted through the cloaca, along with darker fecal wastes from the digestive system. In seabirds, this waste product is referred to as **guano**.

Extrarenal Osmoregulation: Salt Glands

Like all terrestrial vertebrates, birds must maintain a balanced hydration state while living in a relatively arid environment. The problem is greatest for marine bird species that forage in saltwater and eat foods with high salt concentrations. Seawater contains three times the amount of sodium found in avian blood. The avian kidney filters salt from the blood, but it would need an extremely long loop of Henle to produce urine as concentrated as seawater. Given the structure of the avian kidney, it would take approximately one and one half quarts of freshwater for a bird to process, concentrate, and flush the salt accumulated from drinking one quart of seawater (Proctor and Lynch 1993). In short, seawater contains more salt than the avian kidney can filter. To compensate, marine birds must rely on multiple organs and mechanisms of osmoregulation. Water can be produced as a by-product of metabolism, and water can reabsorbed by the kidneys, the hindgut (cloaca, rectum, and cecae), and **nasal salt (supraorbital) glands** (Braun 1999).

Salt glands have been observed in ten avian orders (Struthioniformes, Anseriformes, Phoenicopteriformes, Gruiformes, Charadriiformes, some Gaviiformes, Sphenisciformes, Procellariiformes, Pelecaniformes, Falconiformes, and one species of Galliforme, *Ammoperdix heyi*) (Maclean 2013). Although studies of avian salt glands have focused primarily on marine species, functional nasal glands are also found in terrestrial birds that consume little water (Sabat 2000). Dehydration and/or the ingestion of many different types of osmolytes stimulate salt gland secretion. Evidence suggests birds have **osmoreceptors** in the heart and major blood vessels that monitor the **tonicity**, or effective osmotic gradient, of their body fluids (Hanwell et al. 1972). **Osmolytes** are osmotically active molecules that include electrolytes (e.g., NaCl, MgCl) and common dietary components such as sugars (sucrose, glucose, mannitol) and proteins. For example, the protein-rich diets of Roadrunners (*Geococcyx californianus*) and Savanah Hawks (*Buteogallus meridionalis*) induce the production of hypertonic salt gland secretions to minimize water losses. Dehydration can also stimulate salt glands. The salt glands of desert birds, such as the Sand Partridge (*Ammoperdix heyi*) and the Ostrich (*Struthio camelus*), respond to elevated body temperature to offset potential dehydration (Maclean 2013).

Marine bird species have large, paired nasal salt glands located above the eyes that connect with the nasal cavity through a duct. The microstructure of a salt gland is similar to the nephron of the vertebrate kidney. Each gland consists of lobes that contain secretory tubules connected by a central canal. Like the nephron, the secretory tubules are surrounded by capillary beds that allow countercurrent exchange of NaCl and water. Water is reabsorbed into the bloodstream, and salt is discharged into a central canal that joins with the central canals of other lobes to form ducts. The ducts eventually discharge the accumulated hypertonic fluid through the nares (Sturkie 1976). The glands are normally inactive. However, when a bird experiences an osmotic stress, such as increased salt load in the diet, the glands increase in size. The salt glands are incredibly efficient, producing a secretion several times as concentrated as the maximum urine concentration (Gohary et al. 2013), yet using less water than consumed. This is, in effect, a way of generating free water to maintain osmotic equilibrium and balanced hydration (Hughes 2003).

THE NERVOUS SYSTEM AND SENSE ORGANS

The anatomy of the vertebrate nervous system is divided into two functional categories, the **central nervous system (CNS)** and the **peripheral nervous system (PNS)**. The central nervous system includes the brain, retina, optic nerve, olfactory nerves, and spinal cord. The peripheral nervous system consists of the remaining cranial and spinal nerves (both sensory and motor). The central nervous system receives information (stimuli) from the external environment (e.g., chemical, magnetic, pressure, tactile, light, sound, temperature cues, etc.) and from within the body (e.g., hormonal, chemical, and cell signaling information). These signals are received through a diversity of sensory receptors and are then passed to the CNS by a network of sensory neurons. The sensory receptors and sensory neurons are part of the PNS. The CNS processes and coordinates the received sensory information and then relays the appropriate response back through PNS motor neurons to effectors (e.g., muscles, glands).

The integration of social and environmental information to coordinate the correct behavioral response is a complex undertaking. As such, and contrary to popular belief, bird brains are definitely not small, at least with respect to body size. The complexity of birdsong, in terms of the acoustic patterns produced and context-specific information encoded, is but one example of a unique avian cognitive function (chapters 13 and 14). The ways in which different taxa use their brains influences anatomical structure and organization. The size, volume, and shape of structures within the brain will reflect the behavioral repertoire, cognitive capacity, and evolutionary history of each species (Kawabe 2013). For example, the ability to process and integrate social and environmental information to complete a range of complex tasks has been demonstrated in crows, ravens, magpies, and parrots. All of these birds display large, well-developed cerebral hemispheres. A famous and oft-cited example of this is the ability of some African grey parrots to use human words and labels to express numerical and relational concepts (e.g., addition, the "zero-like" concept of "none"), in a manner similar to very young children (Pepperberg 2006). Complex cognitive abilities such as these were once thought to be beyond the scope of avian intelligence. This is an idea that is now challenged by our better understanding of the architecture of the avian brain (Jarvis et al. 2005).

The Central Nervous System (CNS)
The Brain

The organization of the avian brain differs significantly from that of mammals (see chapter 13). Cues to these differences are apparent when observing the surface of the avian **cerebrum**, which is smooth and lacks the characteristic folds of mammalian brains (fig. 6.26). The bird brain is similar in appearance to a reptile brain in this respect. Unlike reptiles, birds and mammals both have relatively large cerebral hemispheres in relation to body size, reflecting superior cognitive abilities compared with the other vertebrate classes. The neural mechanisms of cognition are strikingly different between birds and mammals, however. The complex folding of the mammalian **cerebral cortex**, the upper and outermost layer of neuronal tissue in the vertebrate brain, reflects its importance in cognition. Beneath the cortex, the cerebral hemispheres of all vertebrates contain layers of gray and white matter called **pallia**. The smooth avian cerebrum is unique in that it houses a particularly large and well-developed pallium that processes information related to memory and learning in a manner similar to the sensory and motor cortices of mammals (Butler and Cotterill 2006).

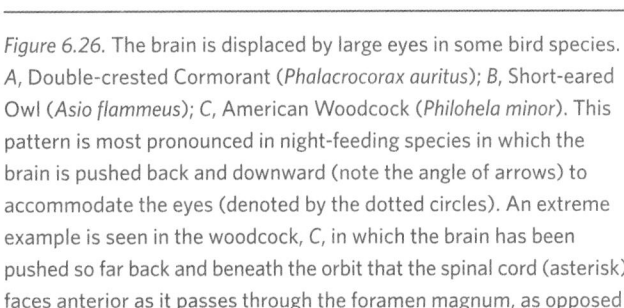

Figure 6.26. The brain is displaced by large eyes in some bird species. A, Double-crested Cormorant (*Phalacrocorax auritus*); B, Short-eared Owl (*Asio flammeus*); C, American Woodcock (*Philohela minor*). This pattern is most pronounced in night-feeding species in which the brain is pushed back and downward (note the angle of arrows) to accommodate the eyes (denoted by the dotted circles). An extreme example is seen in the woodcock, *C*, in which the brain has been pushed so far back and beneath the orbit that the spinal cord (asterisk) faces anterior as it passes through the foramen magnum, as opposed to the posterior position found in most vertebrates. *Modified from original drawings in Cobb 1960.*

In general, the vertebrate **cerebellum** is responsible for coordinating balance, muscular coordination, and complex patterns of movement. Compared with mammals, birds have a large cerebellum with respect to the size of their cerebrum. This highlights the importance of the complex motor control required for flight. Birds also have large **optic lobes** that are very complex in structure. In mammals, visual information is coordinated in the cerebral cortex. In birds, the optic lobes are well developed to quickly coordinate visual information during rapid movement, a crucial aspect of life on the wing. The speed of flight, coupled with the need to detect prey or predator quickly, has placed a strong selection pressure on avian eye size. Raptor eyes, for example, tend to be 1.4 times larger than those of other bird species of similar body mass. The depth of the raptor eye is also longer, presumably to facilitate the high level of visual acuity required for prey detection. In owl species, the eyes are larger by a factor of 2.2. This anatomical modification accommodates a wider pupil to improve night vision (Brooke et al. 1999).

The importance of vision drives other structural and organizational changes in the avian brain. The overall ori-

entation of the brain within the skull varies greatly among different groups of birds and usually reflects species-specific feeding modalities (e.g., using binocular vision to forage as opposed to tactile information). Corvid species (crows) that frequently probe structures with their beaks for food have relatively longer bills, with orbits oriented to the side, in contrast to species that frequently peck (Kulemeyer et al. 2009). The placement of the orbits is important, as this constrains the way the brain is positioned within the skull (Duijm 1951). Bird eyes, and therefore orbits, are proportionally larger than mammalian eyes (Wylie et al. 2015). As such, in some species there is little room for the brain to sit between or beneath the eyes, so it is pushed into a vertical position near the back of the skull (Kaiser 2010). Woodcocks are an extreme example of this morphology. The brain is pushed so far to the back of the skull that the cerebellum is almost upside down. In bird species with long narrow skulls (e.g., cormorants) the brain is positioned horizontally, in a manner similar to terrestrial tetrapods (fig. 6.26).

Photoreceptors: The Avian Eye

The unique organization and orientation of the brain underscores the importance of vision to birds. A keen, well-developed visual system is necessary to precisely maneuver at high speeds, to find and acquire food, to identify and escape from predators, and to identify potential mates. Birds also engage in complex and colorful courtship displays that may include wavelengths of light outside the visible range of the human eye, such as ultraviolet (UV) light, which can encode information about sex and physical condition for use in mate choice (chapters 11 and 16). Although the avian eye is similar in its overall anatomy to the mammalian eye, it also retains vestiges of its reptilian history (e.g., a pecten projecting from the retina, a bony sclerotic ring within the orbit, fig. 6.27), and several features that make it entirely unique from other classes of vertebrates.

The avian eye is divided into two compartments, the anterior and posterior chambers, divided by a **sclerotic ring** of bones (fig. 6.27). These bones, or **ossicles**, are found only in fish, reptiles, and birds. The sclerotic ring of birds is thought to aid the **ciliary muscles** of the eye in **accommodation**, or focusing, of the **lens**. The avian lens is a transparent and somewhat spherical structure with a tough outer layer and a series of softer flexible inner layers. The flexibility of the lens is supported by a watery fluid (the **aqueous humor**) contained within the anterior chamber. There is substantial variation in the shape of the lens among bird species. In some species, the anterior surface of the lens is flattened and the posterior surface is convex (e.g., parrots, storm petrels, and hoopoes). In other species, the lens is strongly convex on both the anterior

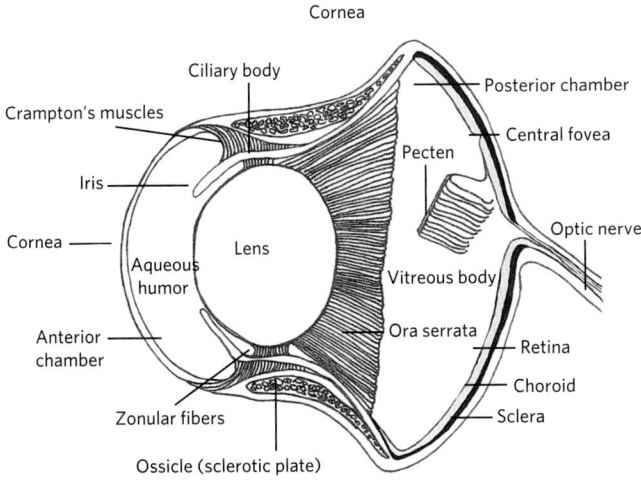

Figure 6.27. A horizontal section of the eye of an owl, showing the major anatomical features. The avian eye is divided into two compartments, the anterior and posterior chambers, that are divided by a **sclerotic ring** of bones. Although the overall structure is similar to that of the human eye, the bony reinforcement of ossicles is not found in mammals. Not surprisingly, bird eyes focus in a manner that differs slightly from that of other vertebrates. This is yet another adaption to life on the wing. In all bird species, the zonular fibers associated with **Brücke's muscles** act on the sclerotic ring to change the shape of the lens and focus light on the retina. This is achieved by increasing the pressure in the posterior compartment of the eye, causing the lens to bulge through the pupil. Some species (e.g., diving birds, raptors) also have a ring of ciliary muscles (**Crampton's muscles**) that change the curvature of the cornea to increase the power of its focus (Levy and Sivak 1980). This allows many species to achieve a faster and greater range of accommodation than is commonly seen in mammals. *From Welty and Baptista 1988 and Proctor and Lynch 1993.*

and posterior surfaces (e.g., ducks, owls, and nightjars). In raptors and passerines, the posterior curvature is more strongly convex than the anterior curvature (Welty and Baptista 1988). This variation in shape is likely a functional response to the ecology of each species, as it often corresponds to behavioral patterns. For example, diurnal species tend to have a flatter anterior lens surface than is observed in nocturnal and aquatic species (Jones et al. 2007). Birds also have the unique ability to change the curvature of their **corneas** through the contraction of a specialized set of muscles (**Crampton's muscles**) to achieve a greater range of accommodation when focusing. The mammalian eye can change the shape of only the lens for this purpose. This ability to rapidly focus using both lens and cornea is extremely well developed in some species of diving birds (e.g., mergansers).

Light passes through the cornea, the aqueous humor, the iris, and the lens to focus on the retina in the **posterior chamber** of the eye. The **iris** is a colored, muscular diaphragm that controls the amount of light entering the eye. Iris color varies greatly to include shades of brown, red, yellow, green, blue, and white, and may play a role in species recognition. The posterior chamber contains the **vitreous humor**, a clear jelly-like substance that maintains the shape of the eye. The posterior chamber is protected by a tough white collagen layer (the **sclera**) that permits muscle attachment for movement of the eye. Beneath the sclera is a dark and heavily pigmented **choroid** layer that contains the thin-walled blood vessels that supply the retina. The **retina** is the avascular innermost layer of the eye, held in place by the pressure of the vitreous humor. In some nocturnal bird species, a vascular layer of tissue behind the retina (the **tapetum lucidum**) reflects visible light. This reflection increases the light available to the photoreceptors. It is the source of the "eye-shine" you would see if you were to shine a flash light on an owl at night (Welty and Baptista 1988).

Visual acuity and resolution are dependent on both the density and type of **photoreceptors** in the retina. In general, the photoreceptors of the avian retina are densely packed compared with those of mammals. The retina of the house sparrow averages 400,000 photoreceptors and that of the common buzzard 100,000,000 photoreceptors per mm², respectively (Sinclair 1985). The human retina, in comparison, averages a mere 275,500 receptors per mm² at its densest concentration in the **fovea** (Jonas et al. 1992). The word "fovea" comes from the Latin word for "small pit." It is a concave depression in the vertebrate retina that contains the greatest density of photoreceptors and is the area of greatest forward visual acuity. Most bird species have at least one well-defined fovea within a central area of photoreceptors. In some species, such as the domestic fowl and the Black-Footed Penguin (*Spheniscus demersus*), the central area is present, but a clearly defined fovea is not (**afoveal**). Diurnal species, birds that fly swiftly, and those that feed on the wing often have two foveas (**bifoveal**), a central fovea and a second, temporal fovea that assists in both lateral and binocular vision (e.g., swifts, hummingbirds, terns, raptors, parrots, swallows). The temporal fovea is particularly important in improving a bird's perception of distance and speed of prey. Some bird species, such as owls, Andean Condors (*Vultur gryphus*), and American Black Vultures (*Coragyps atratus*) are **monofoveal**; they have only one fovea. Owls have only temporal foveae, whereas condors and Black Vultures have only a nasal fovea. The monofoveal condition in owls may be due to the frontal position of their eyes, which allows for a greater binocular field while maximizing the light received from both eyes in dim light (Jones et al. 2007). Within the central area of the retina, birds may have up to five types of visual pigments housed in seven different types of photoreceptors (rods, double cones, and five types of single cone). The ratio of photoreceptor types differs by species, behavior, and ecology (e.g., diurnal vs. nocturnal, diving vs. feeding on the wing). **Rods** are simple photoreceptors that respond to differences in light and dark, and are thus important in image formation at night. The visual pigments contained within **cones** are sensitive to light in the extreme ultraviolet, ultraviolet, blue, green, and red wavelengths. Light must also pass through colored **oil droplets** in the distal end of each cone before reaching the **photopigments**, forming a sort of low-pass color filter. The oil droplets vary in carotenoid concentration and can be classified into six types based on color (clear, transparent, pale yellow, orange, red, or green; Stavenga and Wilts 2014). Each droplet filters a specific wavelength of light. This means that the response of each cone cell depends on both the spectral sensitivity of its photopigment and the filter characteristics of its oil droplet (Hart 2001). As a result, birds have the ability to discriminate between more wavelengths of light than human trichromatic vision can possibly perceive.

The avian retina is avascular, meaning it is not directly perfused by blood vessels; direct vascularization would likely interfere with the path of light to the retina. The lack of vascularization is potentially problematic, as birds have large eyes, a relatively thick retina, and high metabolic rates. Two adaptations help address the issue of oxygen supply to the retina. First, birds have retained a reptilian structure called the **pecten**, which is attached to the retina near the optic nerve. The pecten is a highly vascularized structure that oscillates when the eyes move, allowing oxygen and nutrients to diffuse through the vitreous humor to the retina (Kiama et al. 1998). The shape of the pecten varies among nocturnal and diurnal birds. In most birds, the pecten has many accordion-style pleats, or folds. Nocturnal birds tend to have smaller pectens with fewer folds than diurnal birds. Some birds, such as the kiwi, have no folds at all. Body size and activity may also be a factor in the size and number of folds found in the pecten. Small birds, with higher mass-specific metabolic rates, seem to have larger pectens with more folds. Within the raptors, for example, the Red-tailed Hawk (*Buteo jamaicensis*) has a very large pecten with 17 to 18 folds; the pecten of the Golden Eagle (*Aquila chrysaetos*) is 4 to 8 mm high with 10 folds; the Barred Owl (*Strix varia*) has smaller pecten, with 8 to 10 folds; and the large Great Horned Owl (*Bubo virginianus*) has an even smaller pecten with only seven to eight folds (Jones et al. 2007). In addition to the pecten, birds have a unique eye-specific respiratory protein, **Globin E** (GbE), which helps maintain an adequate supply of oxygen to the retina (Blank et al. 2011).

Magnetoreception

Birds sense magnetic fields and are able to use this information for navigation (see the box on page 163). Birds appear to have at least two independent **magnetoreception** systems located in different regions of the body. One is a form of "chemical" magnetoreception mediated by photopigments embedded in the retina. This is a blue-light-dependent magnetic compass (Beason 2005, Kishkinev and Chernetsov 2015) that operates through wavelength-specific (550–630 nm) photosensitive molecules called cryptochromes. Four types of cryptochrome proteins (CRY1a, CRY1b, CRY2, and CRY4) have been identified in the retinas of two species of migrating passerines, European Robin (*Erithacus rubecula*) and Garden Warbler (*Sylvia borin*; Mouritsen et al. 2004).

The second magnetoreception system enables a bird to determine the appropriate geographical direction of migration from its current location. This is an iron-based (biogenic magnetite, Fe_3O_4, and maghemite, Fe_2O_3) system located in the upper beak and innervated by the ophthalmic branch of the trigeminal nerve (Beason and Semm 1987). It is hypothesized that small (<100 nm) biogenic iron particles might function like magnetic needles turning in the presence of magnetic field lines (Kishkinev and Chernetsov 2015). Although the biophysical mechanism is not fully understood at this time, substantial experimental evidence demonstrates the sensitivity of this system to small changes in the intensity of the magnetic field (Beason 2005). Recent data suggest this iron-based magnetoreceptor is a common feature of birds; its presence has been documented across a wide range of avian species, including homing pigeons and garden warblers (Falkenberg et al. 2010).

Olfaction and Chemoreception

Historically, it was thought that most birds had a poorly developed sense of smell. This assumption was based largely on early electrophysiological studies and studies of the anatomy of the olfactory lobe, which is relatively small in most bird species. Recent genomic evidence now suggests that a high proportion of avian olfactory receptor genes are functional and that the importance of the sense of smell for birds has probably been greatly underestimated in the past (Steiger et al. 2008). **Chemoreception** in birds is mediated through the olfactory receptors, primarily located in the epithelium of the third olfactory (nasal) chamber near the **choanae** (internal nares, Wenzel 1971) and a few **taste buds** located at the rear of the tongue, on the palate, and on the floor of the pharynx. Birds appear to have an acute sense of taste, although their taste buds are slightly different in structure (Kudo et al. 2008) and far fewer in number than those of mammals (Kudo et al.

2010). The position of the olfactory receptors close to the choanae that open at the back of the oral cavity most likely allow birds to "smell" food that is held in the mouth. Impulses from the olfactory receptors are carried to the brain by way of the olfactory nerve. Researchers have begun to put historic misconceptions to rest by demonstrating that many bird species rely on olfactory cues to detect food (e.g., kiwis, albatrosses, petrels, shearwaters, turkey vultures), nest material with biocidal properties (Roper 1999), conspecifics (e.g., petrels, shearwaters, prions; the Crested Auklet, *Aethia cristatella*; Hagelin and Jones 2007), and alarm odors (e.g., Eurasian Hoopoes, *Upupa epops*; Green Wood Hoopoes, *Phoeniculus purpureus*; Hagelin and Jones 2007). Birds lack the specialized **vomeronasal** structure used by other vertebrates to detect pheromones, yet they have retained the **terminal nerve** associated with pheromonal control of reproductive behavior (Hagelin and Jones 2007). This suggests there is much to be done before we fully understand the complexity of avian olfaction and its role in the ecology and life history of birds.

The Peripheral Nervous System (PNS)
Mechanoreceptors

All vertebrates are equipped with a variety of sensory structures that respond to mechanical stimuli. The ability to perceive pressure (e.g., barometric pressure, touch), vibration, pain, temperature, and muscle tension are as important to birds as to any other organism. Given the common anatomical history of vertebrates, birds necessarily have free nerve endings and mechanoreceptors that are similar in structure to those of other animals. However, the dermal derivatives

ROBERT C. BEASON, RADAR ORNITHOLOGIST

Bob Beason, now retired, was a professor and head of biology at the University of Louisiana at Monroe, a distinguished professor of biology and biophysics at SUNY Geneseo, a project leader for the USDA-APHIS Wildlife Services, and, most recently, a radar ornithologist. Bob did his doctoral work at Clemson University, where he used radar to study waterbird migration in the southwestern United States. Bob's research career has focused on many aspects of avian sensory perception (magnetic and vision), migration, orientation, and navigation. In the 1980s, Bob and Peter Semm of Goethe University Frankfurt in Germany conducted a series of experiments to test whether the avian trigeminal nerve might be involved in magnetoreception. The three branches of the trigeminal nerve (the ophthalmic, maxillary, and mandibular) are well placed to quickly convey sensory information from the face and beak to the brain. When Bob and Peter began their work, it was already known that one branch of the trigeminal nerve was important for magnetoreception in fish, so that seemed to be good place to begin looking in birds as well. Bobolinks (*Dolichonyx oryzivorus*) are migratory birds that were suspected to use magnetic fields to navigate. Bob and Peter exposed Bobolinks to a set of coils that could produce both vertical and horizontal magnetic fields. They then recorded the electrical activity of individual sensory neurons in the trigeminal nerve while exposing the birds to different magnetic fields. They found that many of these trigeminal sensory neurons, especially in the ophthalmic nerve

branch, responded to the magnetic fields by either increasing or decreasing their electrical activity. Horizontal and vertical magnetic fields elicited different responses from different sensory neurons. Also, some sensory neurons responded to increases in magnetic strength and others responded to decreases in magnetic strength. The pattern of neuron activity could be a way that the nervous system could communicate the direction, and change in direction, of the magnetic field to the brain during navigation. In order to test whether the ophthalmic nerve branch carries magnetic information to the brain, Bob and Peter tested another group of Bobolinks that were preparing for migration. For each bird, they first tested what direction it preferred to go (for a control). Then, they magnetized the birds such that, if their beak were iron, the tip of it would attract the south end of a compass, anticipating that this process would send a confusing magnetic signal to the birds' brains. Then they tested the birds' preferred flying directions again. As expected, the birds got confused and went the wrong way. Finally, Bob and Peter numbed the ophthalmic nerve by putting a drop of lidocaine on it and tested the birds' preferred directions again. They found that although magnetizing the birds made them go the wrong direction, when their ophthalmic nerve was numbed, they ignored this incorrect magnetic information and went the right way again. Clearly, the ophthalmic nerve was sending magnetic information to the brain.

of birds (e.g., feathers, a keratinized bill) necessitate some structural modifications and variations in receptor distribution. Feathers, for example, cover much of the skin and substantially reduce the need for widespread distribution of tactile sensors across the body. At the same time, many birds rely on sensitive tactile reception in the heavily keratinized bill to locate prey.

A network of free nerve endings can be found in the dermis in regions of the body that lack feathers. These simple sensory receptors receive environmental information that is ultimately perceived as temperature (hot and cold) or pain (nociception). In birds, cold receptors are more common that warm receptors (Necker 2000). Structurally complex tactile and pressure receptors are also found in many regions of the avian dermis. The **Herbst corpuscle** is the most widely distributed tactile receptor in birds; this type of receptor is highly sensitive to vibration and acceleration. The Herbst corpuscle is similar in structure to the mammalian Pacinian corpuscle (see fig. 6.4). It is a nerve fiber encapsulated in a layered, onion-like receptor consisting of a series of membrane lamellae (sheets) separated by a fluid-filled space. The lamellae can "slip" past each other, thus allowing the receptor to respond quickly to even the slightest touch while retaining the ability to adapt to constant stimuli over time. Herbst corpuscles are found in the deep dermal layer of the legs (e.g., cranes, herons), wing tips and joints, the bills of many birds (e.g., waterfowl, shorebirds), and the tongues of woodpeckers (Portmann 1961). Herbst corpuscles are also associated with the filoplumes, rictal, and facial bristles in many species of birds. The Herbst corpuscles found in the bills of ducks and geese and in the legs of cranes and herons are used to detect vibrations in water. Those found in the wings detect vibrations produced by air currents (Dorward 1970, Brown and Fedde 1993).

Merkel cells are also pressure and vibration receptors often associated with feathers. Merkel receptors are clear oval cells in contact with enlarged terminal branches of myelinated nerve fibers. In birds, these receptors are found in the dermal layer of the beak, tongue, toes, legs, and skin (both feathered and featherless; Halata et al. 2003). Together, Herbst and Merkel receptors are thought to be important in detecting feather position during flight. **Grandry corpuscles** are modified Merkel cells that are found only in the upper bill and tongue of waterfowl. Although they are tactile mechanoreceptors, Grandry corpuscles do not necessarily sense vibrations (Leitner and Roumy 1974). The corpuscles are located in the dermis, just below the epidermal layer; each contains a discoid network of nerve fibrils sandwiched between small groups of two to five large oval or flat cells. In the mucosa of the tongue, small Grandry corpuscles consisting of only two or three cells are found just below the epithelium (Halata et al. 2003). Pain sensors, or **nociceptors**, are also located in the skin and mouth (Necker 2000).

Hearing and Equilibrium: The Avian Ear

Birds do not have **pinnae**, which are the external structures we identify as "ears" on mammals. The external auditory structures of the avian ear are inconspicuous and camouflaged by **auricular feathers**. Beyond this notable difference, the overall structure and function of the avian ear is similar to that of mammals. The ear is divided into three compartments, the external, middle, and inner ear (fig. 6.28). The **external ear**, or **acoustic meatus**, is a tube that conducts sound waves from the surrounding air inward to the ear drum, or **tympanum**, at its base. The opening to this relatively short auditory canal is behind and slightly below the eye (Kühne and Lewis 1985). In ostriches, vultures, and most galliform species, the acoustic meatus is featherless and completely exposed. In most other bird species, feathers surround the external ear for protection during flight. In many birds, these **auricular feathers** are attached to a muscular ridge, the **operculum**, that can be raised to help direct sound waves into the acoustic meatus, or closed to protect the auditory canal when diving. The concentric rings of feathers around the eyes of owls, which extend back and over the external ear, are a familiar example of auricular feathers. The auricular feathers of owls exhibit a unique placement pattern that improves the ability to locate the source of sounds when hunting at night. Owls have one ear positioned slightly lower than the other. This orientation slightly offsets the time at which a given sound arrives at each tympanum. The phase delay allows the bird to use a precise form of two-point discrimination to accurately pinpoint the source of sound. The auricular feathers surrounding the lower ear point downward to direct sounds into auditory meatus. The feathers around the other, slightly higher, ear funnel sound upward. This asymmetrical feather arrangement further enhances the time differential effect and improves the two-point sound discrimination system (Knudsen 1981).

The external ear ends at the **tympanic membrane** or ear drum. This membrane separates the external ear from the **middle ear**. Like mammals, the middle ear is an air-filled cavity that extends from the tympanic membrane to the **oval window** of the fluid-filled **inner ear**. Birds, like reptiles and amphibians, have a single rod-like middle ear **ossicle** called the **columella**. This differs from mammals, which have three middle ear ossicles. Sound waves bounce through the acoustic meatus to vibrate the tympanic membrane. The vibrations of

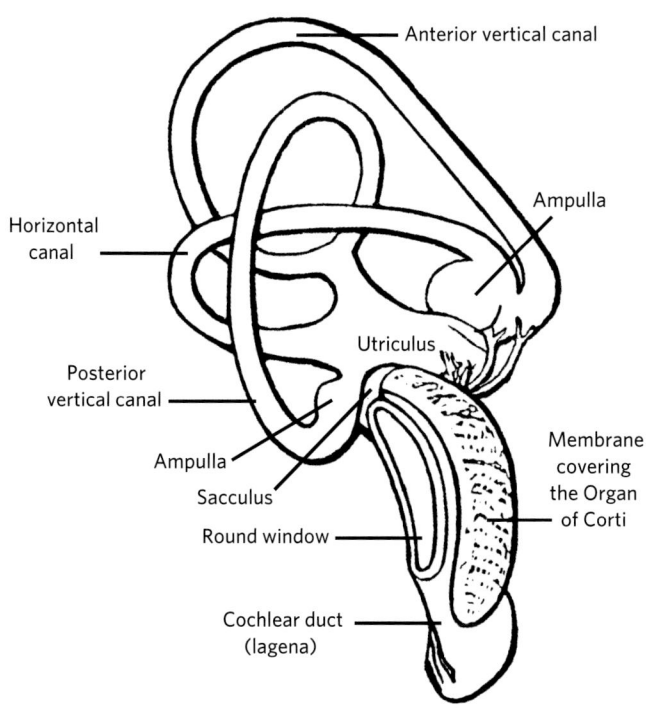

Anterior vertical canal

Ampulla

Horizontal
canal

Utriculus

Posterior
vertical canal

Membrane
covering
the Organ
of Corti

Ampulla

Sacculus

Round window

Cochlear duct
(lagena)

Figure 6.28. The ear is divided into three compartments, the external, middle, and inner ear. The external ear ends at the tympanic membrane or ear drum. The middle ear extends from the tympanic membrane to the oval window of the fluid-filled inner ear. Birds have a single middle ear ossicle called the columella that transfers the vibrations of the tympanum to the membranous oval window of the inner ear. The membranous labyrinth of the inner ear in the White-tailed Sea-eagle (*Haliaeetus albicilla*) is shown here. The three **semicircular canals** have **ampullae** at their base that each contain a gelatinous **cupula** that responds to angular and rotational acceleration (dynamic equilibrium and spatial orientation). Beneath the semicircular canals are the **utriculus** and **sacculus**, each containing a gelatinous macula topped with **otoliths**. Otoliths function as gravity detectors to tell the bird which way is up (static equilibrium) and relay information about linear acceleration. The final compartment, the **lagena**, contains the cochlear duct that houses the **Organ of Corti**, or auditory apparatus. The oval window, not shown, is the point of entry to the inner ear for sound waves amplified by the columella in the middle ear. The round window allows the pressure generated by the columella on the fluid-filled cochlea to exit the inner ear back into the middle ear chamber. *Redrawn and modified from Coues 1884.*

the tympanum are then transferred to the membranous **oval window** of the inner ear by the columella. Movement of the columella is controlled by a single **columellar muscle**, which changes the tension on the tympanic membrane and the position of the columella to dampen vibrations and protect the inner ear. The difference in surface area between the large tympanum and the smaller oval window amplifies the sounds waves transmitted from the air to the fluid in the inner ear. The

middle ear connects to the oropharynx and to several sinuses and cavities in the skull through a series of tubes (**pharyngotympanic tubes**). These tubes equalize pressures in a manner similar to mammalian eustachian tubes. The middle ear also connects with cavities in the mandible (Kühne and Lewis 1985).

The **inner ear** plays a role in two very different sensory functions: hearing and the maintenance of equilibrium. The inner ear is separated from the middle ear by the oval window of the cochlea. In all vertebrates, the cochlea houses the auditory receptor. The avian cochlea is quite different from the coiled structure found in mammals—it is a straight and slightly flattened bony tube. The length of the cochlea varies among bird species, but is shorter than that of mammals in general. The bony cochlea contains a membranous **cochlear duct**, which is a section of a larger **membranous labyrinth**. The membranous labyrinth is a delicate network of tubes encased within a protective bony labyrinth and filled with a viscous fluid called **endolymph**. The membranous labyrinth floats in a plasma-like fluid called **perilymph** that cushions and separates it from the bony labyrinth. The membranous labyrinth contains two saclike regions, the utriculus and sacculus. The upper region, the **utriculus**, connects to three **semicircular canals**: the anterior, the posterior, and the horizontal canal. The lower region, the **sacculus**, connects to the endolymph-filled cochlear duct, where the vertebrate auditory receptor, the **organ of Corti**, is located. The organ of Corti rests on a **basilar membrane** that contains sensitive hair cells. The top of each hair cell is embedded in a **tectorial membrane** that floats within the endolymph (Smith 1985). When the endolymph within the cochlear duct moves in response to the vibration of the columella against the oval window, the hair cells rock back and forth to generate action potentials. The resulting impulses travel from the hair cells through the auditory nerve and to the brain. Vibrations of different frequencies stimulate hair cells in different regions along the basilar membrane from its base to apex. These impulses are interpreted by the brain as sounds of different pitch. Most songbirds have good discrimination of pitch, with an optimum range of one to five kHz. For reference, this is a slightly narrower band than the optimal range of hearing in the human ear (Dooling 1982, Smith 1985). Within bird species, the length of cochlea and the number of hair cells housed in the basilar membrane govern the range and sensitivity of hearing (Walsh et al. 2009). Barn Owls (*Tyto alba*) have longer cochlea than most species and can detect sounds up to about 12 kHz (Dooling 1982). Even though songbirds have shorter and smaller cochlea than owls, they still have more sensitive hearing at higher frequencies than non-songbirds, whose ears are more attuned to low frequencies (Dooling 1980).

The inner ear is also responsible for the maintenance of equilibrium (but also see the box on page 167). Each semicircular canal of membranous labyrinth houses a mechanoreceptor that provides information to the brain about position and movement of the head in three-dimensional space. Collectively, the three semicircular canals and their mechanoreceptors are referred to as the **vestibular organ**; this is the balance organ of the inner ear. Two of the semicircular canals are oriented vertically and one is positioned horizontally. The end of each semicircular canal forms a bulbous structure called an **ampulla**. The three **ampullae**, the **utriculus**, the **sacculus**, and the tip of the cochlea (the **lagena**) all contain hair cells whose tips are embedded within a gelatinous membrane called a **macula**. In the utriculus and sacculus, calcium carbonate crystals called **otoliths** ("ear rocks") rest on top of the macula. When the head is in an upright position, the mass of the otoliths is pulled downward by gravity. This bends the hair cells, sending impulses to the brain that are interpreted as "my head is above my feet." When the head is tipped forward or lowered (a positional change), the weight of the otoliths shifts forward as well, sending impulses to the brain about the position of the head in reference to the limbs. In the ampullae at the base of the fluid-filled semicircular canals, the hair cells respond to displacement of a gelatinous cap (a **cupula**) that floats above the sensory cells in the endolymph. When the head rotates, the endolymph moves through the semicircular canal, pushing the cupula and stimulating the embedded hair cells. The pattern of otolith and cupula stimulation transmitted to the brain through the vestibulocochlear nerve allows a bird to sense gravity (which way is up), as well as linear and rotational acceleration. Birds have more hair cells per mm^2 of macula than mammals (Smith 1985), reflecting the ecological importance of detailed sensory information about gravitational pull and acceleration.

Echolocation

Echolocation is the use of reflected vocalizations to navigate. It is an uncommon vertebrate trait, and it is usually only associated with a few well-known mammalian groups (e.g., shrews, bats, dolphins). The only other vertebrates currently known to echolocate are found in two groups of birds: the Caprimulgiformes (Oilbirds, Steatornithidae: *Steatornis caripensis*) and the Apodiformes (several species of swiftlets in the family Apodidae: *Aerodramus* spp. and *Collocalia troglodytes*; Brinkløv et al. 2013). These birds use the **syrinx** to produce echolocation "clicks" at frequencies between 2 to 15 kHz, well within the range of human hearing. This differs from bats, which use more precise echolocation signals within the ultrasound spectrum. As best we understand, birds use subsy-

ringeal pressure to produce an echolocation "click" as they exhale. Two antagonistic muscle pairs are involved. First, the extrinsic sternotrachealis muscles contract to fold the external tympaniform membranes into the syrinx **lumen** (i.e., opening) and bring them closer to the internal tympaniform membranes. Air leaving the trachea causes the tympaniform membranes to vibrate and produce clicking sounds (Brinkløv et al. 2013). Oilbirds contract a single pair of intrinsic syringeal muscles (broncholateralis; Suthers and Hector 1985) to control the duration of their clicks (between 1 and 50 ms) within a frequency spectrum of 1.5–10 kHz (Brinkløv et al. 2013). Oilbird clicks appear to reach a peak signal energy between 1.5 to 2.5 kHz, which is consistent with the most sensitive region of this bird's range of hearing (Konishi and Knudsen 1979). Swiftlets lack the intrinsic muscles found in the Oilbird and therefore terminate their clicks by contracting the extrinsic tracheolateralis muscles (Suthers and Hector 1985, Thomassen 2005). Most species of echolocating swiftlet produce single clicks as well as double clicks. The double click may be produced by briefly blocking the airflow exiting the syrinx just as the external and internal tympaniform membranes touch. Single clicks are thought to occur when the tympaniform membranes are pulled together just before the rush of air leaving the trachea generates enough pressure to start the vibration of the membranes (Suthers and Hector 1985, Brinkløv et al. 2013). Whether single or double, the sharp clicks produced in this manner bounce off objects in the bird's flight path, creating echoes. The echoed sound returns to the birds' ears at levels of loudness and intensity that inform the bird of the size, shape, and location of the objects that surround it. Larger structures, for example, will deflect more sound waves and created louder echoes. Oilbirds and swiftlets both use this relatively simple yet effective click and echo system to navigate in the total darkness of the caves they roost in.

KEY POINTS

- Avian anatomy has been shaped by the demands of two separate forms of locomotion: flight in the air and bipedalism on the ground.
- Feathers are the anatomical feature that characterizes all species in the class Aves.
- All integumentary derivatives of birds (e.g., feathers, beaks, claws) are highly adapted to species-specific foraging ecology and habitat.
- The biomechanical demands of flight require the pneumatization and fusion of bones to provide rigid, stable structures that withstand powerful muscular contractions while reducing body mass.

THE GLYCOGEN BODY OF THE SYNSACRUM

It is clear by now that birds are first and foremost terrestrial organisms that took to the skies secondarily at some point in their evolutionary history. But long before there were birds, the earliest terrestrial tetrapods would have ventured onto land carrying the vestibular organs of their aquatic ancestors in their ears. Maintaining equilibrium under the conditions of neutral buoyancy in water requires the ability to sense and stabilize the body in three dimensions (roll, pitch, and yaw). The three semicircular canals of our inner ear, incredibly useful inheritances from our fishy predecessors, respond to changes in head position in these three dimensions. We also inherited the macula and its otoliths, which are useful for detecting gravity and linear acceleration. But life on land also requires the ability to look around and watch your back. If you have never noticed before, fish do not have necks. A neck allows the head to move around independent from the body, so the position of the head does not necessarily accurately reflect the position of the hindlimbs. Most terrestrial bipeds do just fine with minor size and shape modifications to the vestibular apparatus of the inner ear. For example, our heads are positioned over our hindlimbs, so the vestibular organ accurately informs us of where our head is with respect to our feet. But birds, terrestrial bipeds of a different feather, have a unique problem—they use two different kinds of locomotion: they fly in the air with their modified forelimbs and they walk on the ground with their hindlimbs. Their body is held in a position nearly horizontal to the ground, with a head on a flexible neck located far to the front of the center of gravity. A bird's hindlimbs, however, are positioned far to the rear of its center of gravity. Bipedal dinosaurs compensated for a similar problem by using large tails as a counterbalance to adjust the center of gravity; but heavy tails are not conducive to flight. The bipedal bird, with its far-forward center of gravity and altered hindlimb gait, needs additional sensory information to maintain its sense of equilibrium (and sense of dignity). The logical place to put a second equilibrium sensor in a bird would be somewhere in the abdomen (Necker et al 2006). Ornithologists had long wondered about the function of the large glycogen body often observed in the lumbosacral region of the spine beneath the synsacrum. This structure appears to be an enlarged portion of the neural tube that sits behind the hip joint (the acetabulum; Kaiser 2010). Recent investigations into the function of the glycogen body has led to the conclusion that its accessory lobes, which look very similar to the semicircular canals of the inner ear, play a sensory role in maintaining equilibrium and control of hindlimbs during terrestrial locomotion (Necker 2006).

- The presence of winged forelimbs has altered the center of gravity, modified pelvic and hindlimb structure, and changed the way birds use their feet to interact with their environment (e.g., perching, grasping, scratching, preening, prey manipulation, egg rolling).
- The overall reduction of the urogenital system and the seasonal regression of reproductive structures are adaptations to flight that have had major implications for avian life history.
- The avian brain and eyes are enlarged and enhanced to adequately process visual information during rapid movement.

KEY MANAGEMENT AND CONSERVATION IMPLICATIONS

- Keratinized integumentary derivatives such as feathers are useful storehouses of avian DNA that can be used in noninvasive studies for the management and conservation of birds. For example, archived feathers can provide genetic information to identify the migration patterns of bird species (https://www.allaboutbirds.org/capturing-migration-in-a-strand-of-dna-feathers-reveal-a-birds-origins/). Genetic tracking techniques using material extracted from keratinized structures can be used to address a wide range of conservation questions, including tracking the effects of human disturbance and climate change on migration patterns and following population changes in declining or endangered species.

DISCUSSION QUESTIONS

1. In terms of anatomy, what makes a bird a bird?
2. How are birds similar in structure to reptiles? How are they similar to mammals? Is there a pattern to these

anatomical similarities? Why? What explains the differences you observe?

3. Birds are primarily terrestrial organisms. How does the anatomy of a bird balance the need to both walk and hop on land with the ability to fly?

4. Birds lack teeth, yet many species consume foods that require physical processing before chemical digestion (e.g., whole seeds and nuts, whole fish with bones). Explain the ways this can be accomplished and provide specific examples of the structures involved.

5. The structure of bird beaks, feet, and claws vary greatly across species. If you had only study skins and museum reference material to work with, what could you infer about the ecology of a species from these structures?

6. Describe the structure of the syrinx and briefly explain how it works. Why is this structure so important in relation to the complexity of birdsong and vocalization?

7. The functional unit in the kidney of both birds and mammals is the nephron, yet they excrete different nitrogenous waste compounds. Explain why this is so. How do birds produce concentrated guano? Why are some bird species able to drink saltwater while mammals normally cannot?

8. Every anatomical system in the bird has been modified in some way by flight. How many of these adaptations can you describe and explain?

References

Akester, A. R. (1967). Renal portal shunts in the kidney of the domestic fowl. Journal of Anatomy 101 (3): 569.

Andrews, P. (1990). Owls, caves, and fossils. Natural History Museum Publications, London.

Bahr, P. H. (1907). On the "Bleating" or "Drumming" of the Snipe (*Gallinago cœlestis*). Proceedings of the Zoological Society of London 77:12–35.

Ballmann, P. (1976). Fossile Vögel aus dem Neogen der Halbinsel Gargano (Italien), zweiter Teil. Scripta Geologica 38:1–59.

Beams, H. W., and R. K. Meyer (1931). The formation of pigeon "milk." Physiological Zoology 4 (3): 486–500.

Beason, R. C. (2003). Through a bird's eye-exploring avian sensory perception. *In* 2003 Bird Strike Committee USA/Canada, 5th Joint Annual Meeting, August 2003, Toronto, Ontario.

Beason, R. C. (2005). Mechanisms of magnetic orientation in birds. Integrative and Comparative Biology 45 (3): 565–573.

Beason, R. C., and P. Semm (1987). Magnetic responses of the trigeminal nerve system of the Bobolink (*Dolichonyx oryzivorus*). Neuroscience Letters 80 (2): 229–234.

Beckers, G. J., B. S. Nelson, and R. A. Suthers (2004). Vocal-tract filtering by lingual articulation in a parrot. Current Biology 14 (17): 1592–1597.

Bellairs, A. D. A., and C. R. Jenkins (1960). The skeleton of birds. Biology and Comparative Physiology of Birds 1:241–300.

Bevan, R. M., I. L. Boyd, P. J. Butler, K. Reid, A. J. Woakes, and J. P. Croxall (1997). Heart rates and abdominal temperatures of free-ranging South Georgian Shags, *Phalacrocorax georgianus*. Journal of Experimental Biology 200 (4): 661–675.

Birkhead, T. R., and A. P. Møller (1998). Sperm competition and sexual selection. Academic Press, London.

Bishop, C. M. (1997). Heart mass and the maximum cardiac output of birds and mammals: Implications for estimating the maximum aerobic power input of flying animals. Philosophical Transactions of the Royal Society of London B: Biological Sciences 352 (1352): 447–456.

Blank, M., L. Kiger, A. Thielebein, F. Gerlach, T. Hankeln, M. C. Marden, and T. Burmester (2011). Oxygen supply from the bird's eye perspective: Globin E is a respiratory protein in the chicken retina. Journal of Biological Chemistry 286 (30): 26507–26515.

Bock, W. J. (1999). Functional and evolutionary morphology of woodpeckers. Ostrich 70 (1): 23–31.

Braun, E. J. (1999). Integration of organ systems in avian osmoregulation. Journal of Experimental Zoology 283 (7): 702–707.

Brinkløv, S., M. B. Fenton, and J. M. Ratcliffe (2013). Echolocation in oilbirds and swiftlets. Frontiers in Physiology 4 (123): 188–197. doi:10.3389/fphys.2013.00123.

Briskie, J., and R. Montgomerie (1997). Sexual selection and the intromittent organ of birds. Journal of Avian Biology 28 (1): 73–86. http://www.jstor.org/stable/3677097. doi:10.2307/3677097.

Brooke, M. D. L., S. Hanley, and S. B. Laughlin (1999). The scaling of eye size with body mass in birds. Proceedings of the Royal Society of London B: Biological Sciences 266 (1417): 405–412.

Brown, R. E., and M. R. Fedde (1993). Airflow sensors in the avian wing. Journal of Experimental Biology 179 (1): 13–30.

Butler, A. B., and R. M. Cotterill (2006). Mammalian and avian neuroanatomy and the question of consciousness in birds. The Biological Bulletin 211 (2): 106–127.

Campbell, B., and E. Lack (1985). A dictionary of birds. Poyser, Calton, UK.

Castrogiovanni, P., R. Imbesi, M. Fisichella, and V. Mazzone (2011). Osteonic organization of limb bones in mammals, including humans, and birds: A preliminary study. Italian Journal of Anatomy and Embryology 116 (1): 30–37.

Ciriaco, E., P. P. Píñera, B. Díaz-Esnal, and R. Laurà (2003). Age-related changes in the avian primary lymphoid organs (thymus and bursa of Fabricius). Microscopy Research and Technique 62 (6): 482–487.

Clarkson, T. B., C. C. Middleton, R. W. Prichard, and H. B. Lofland (1965). Naturally-occurring atherosclerosis in birds. Annals of the New York Academy of Sciences 127 (1): 685–693.

Cobb, S. (1960). Observations on the comparative anatomy of the avian brain. Perspectives in Biology and Medicine 3 (3): 383–408.

Coues, E. (1884). Coues' key to North American birds. Estes and Lauriat, Boston, MA.

Dacke, C. G., S. Arkle, D. J. Cook, I. M. Wormstone, S. Jones, M. Zaidi, and Z. A. Bascal (1993). Medullary bone and avian calcium regulation. Journal of Experimental Biology 184 (1): 63–88.

Dhouailly, D. (2009). A new scenario for the evolutionary origin of hair, feather, and avian scales. Journal of Anatomy 214:587–606. doi:10.1111/j.1469 7580.2008.01041.x.

Dooling, R. J. (1980). Behavior and psychophysics of hearing in birds. *In* Comparative studies of hearing in vertebrates, A. N. Popper and R. R. Fay, Editors. Springer-Verlag, New York, pp. 261–288.

Dooling, R. J. (1982). Auditory perception in birds. *In* Acoustic communication in birds, vol. 1, D. Kroodsma, E. H. Miller, and H. Ouellet, Editors. Academic Press, New York, pp. 96–130.

Dorward, P. K. (1970). Response patterns of cutaneous mechano-receptors in the domestic duck. Comparative Biochemistry and Physiology 35 (3): 729–735.

Duijm, M. (1951). On the head posture in birds and its relation to some anatomical features: I-II. Proceedings of the Koninklijke Nederlandse Akademie van Wetenschappen, Series C 54:260–271.

Duke, G. E., A. A. Degen, and J. K. Reynhout (1995). Movement of urine in the lower colon and cloaca of ostriches. The Condor 97:165–173.

Dumont, E. R. (2010). Bone density and the lightweight skeletons of birds. Proceedings of the Royal Society of London B 277:2193–2198.

Eckstrom, F. H. (1901). The woodpeckers. Houghton, Mifflin, Riverside Press, Cambridge. The Project Gutenberg EBook of the woodpeckers. http://gutenberg.readingroo.ms/3/5/0/6/35062/35062h/35062h.htm.

Erdoğan, S., and S. Iwasaki (2014). Function-related morphological characteristics and specialized structures of the avian tongue. Annals of Anatomy-Anatomischer Anzeiger 196 (2): 75–87.

Falkenberg, G., G. Fleissner, K. Schuchardt, M. Kuehbacher, P. Thalau, H. Mouritsen, D. Heyers, G. Wellenreuther, and G. Fleissner (2010). Avian magnetoreception: Elaborate iron mineral containing dendrites in the upper beak seem to be a common feature of birds. PLoS ONE 5 (2): e9231.

Fisher, H. I., and E. E. Dater (1961). Esophageal diverticula in the Redpoll, *Acanthis flammea*. The Auk 78:528–531.

Flower, W. H. (1871). On the skeleton of the Australian cassowary. Proceedings of the Zoological Society 1871:33.

Gardner, L. L. (1925). The adaptive modifications and the taxonomic value of the tongue in birds. US Government Printing Office, Washington, DC. http://www.jstor.org/stable/4157845.

Gohary, Z. M. E., F. I. E. Sayad, H. A. Hassan, A. Mohammed, and M. Hamoda (2013). The functional alterations of the avian salt gland subsequent to osmotic stress. Egyptian Journal of Hospital Medicine 51:346–360. http://go.galegroup.com.libezproxy2.syr.edu/ps/i.do?id=GALE%7CA409698766&v=2.1&u=nysl_ce_syr&it=r&p=HRCA&sw=w&asid=8f80106781db7c50051f9cf4a8111616.

Goldstein, D. L., and E. Skadhauge (1998). Renal and extrarenal regulation of body fluid composition. *In* Sturkie's avian physiology, Academic Press, London.

Goslow, G. E. J., K. P. Dial, and F. A. Jenkins (1990). Bird flight: Insights and complications: New techniques show that more than the wing participates in flying. Bioscience 40:108–115.

Greenewalt, C. H. (1968). Bird song: Acoustics and physiology. Smithsonian Institution, Washington, DC.

Grossi, B., J. Iriarte-Díaz, O. Larach, M. Canals, and R. A. Vásquez (2014). Walking like dinosaurs: Chickens with artificial tails provide clues about non-avian theropod locomotion. PLoS ONE 9 (2): e88458.

Grubb, B. R. (1983). Allometric relations of cardiovascular function in birds. American Journal of Physiology-Heart and Circulatory Physiology 245 (4): H567–H572.

Guioli, S., S. Nandi, D. Zhao, J. Burgess-Shannon, R. Lovell-Badge, and M. Clinton (2014). Gonadal asymmetry and sex determination in birds. Sexual Development 8 (5): 227–242.

Hagelin, J. C., and I. L. Jones (2007). Bird odors and other chemical substances: A defense mechanism or overlooked mode of intraspecific communication? The Auk 124 (3): 741–761. doi: http://dx.doi.org/10.1642/0004 8038(2007)124[741:BOAOCS]2.0.CO;2.

Halata, Z., M. Grim, and K. I. Bauman (2003). Friedrich Sigmund Merkel and his "Merkel cell," morphology, development, and physiology: Review and new results. Anatomical Record Part A: Discoveries in Molecular, Cellular, and Evolutionary Biology 271 (1): 225–239.

Hanson, H. C. (1962). Characters of age, sex and sexual maturity in Canada geese. Biological Notes 049.

Hanwell, A., J. L. Linzell, and M. Peaker (1972). Nature and location of the receptors for salt-gland secretion in the goose. Journal of Physiology 226 (2): 453–472.

Haribal, M., A. A. Dhondt, D. Rosane, and E. Rodríguez (2005). Chemistry of preen gland secretions of passerines: Different pathways to same goal? Why? Chemoecology 15 (4): 251–260.

Hart, N. S. (2001). The visual ecology of avian photoreceptors. Progress in Retinal and Eye Research 20 (5): 675–703. doi:10.1016/S1350 9462(01)00009 X.

Hartman, F. A. (1961). Locomotor mechanisms of birds. Smithsonian Miscellaneous Collections 143:1–91.

Haywood, S. (2013). Origin of evolutionary change in avian clutch size. Biological Reviews 88 (4): 895–911. doi:10.1111/brv.12035.

Hinchliffe, J. R., and D. R. Johnson (1980). The development of the vertebrate limb: An approach through experiment, genetics, and evolution. Oxford University Press, Oxford.

Hughes, M. R. (2003). Regulation of salt gland, gut and kidney interactions. Comparative Biochemistry and Physiology—Part A: Molecular and Integrative Physiology 136:507–524.

Hume, J. P., and L. Steel (2013). Fight club: A unique weapon in the wing of the solitaire, *Pezophaps solitaria* (Aves: Columbidae), an extinct flightless bird from Rodrigues, Mascarene Islands. Biological Journal of the Linnean Society 110:32–44.

Hyman, L. H. (1992). Hyman's comparative vertebrate anatomy. University of Chicago Press.

Iwasaki, S. (2002). Evolution of the structure and function of the vertebrate tongue. Journal of Anatomy 201:1–13.

Jackowiak, H., K. Skieresz-Szewczyk, S. Godynicki, S. Iwasaki, and W. Meyer (2011). Functional morphology of the tongue in the Domestic Goose (*Anser anser f. domestica*). Anatomical Record 294 (9): 1574–1584.

Jarvis, E. D., O. Güntürkün, L. Bruce, A. Csillag, H. Karten, W. Kuenzel, L. Medina, et al. (2005). Avian brains and a new understanding of vertebrate brain evolution. Nature Reviews Neuroscience 6 (2): 151–159.

John, J. L. (1994). The avian spleen: A neglected organ. Quarterly Review of Biology 69:327–351.

Jonas, J. B., U. Schneider, and G. O. Naumann (1992). Count and density of human retinal photoreceptors. Graefe's Archive for Clinical and Experimental Ophthalmology 230 (6): 505–510.

Jones, M. P., K. E. Pierce, and D. Ward (2007). Avian vision: A review of form and function with special consideration to birds of prey. Journal of Exotic Pet Medicine 16 (2): 69–87. doi:10.1053/j.jepm.2007.03.012.

Kaiser, G. W. (2010). The inner bird: Anatomy and evolution. UBC Press, Vancouver.

Kanwisher, J. W., T. C. Williams, J. M. Teal, and K. O. Lawson (1978). Radiotelemetry of heart rates from free-ranging gulls. The Auk 95:288–293.

Kawabe, S., T. Shimokawa, H. Miki, S. Matsuda, and H. Endo (2013). Variation in avian brain shape: Relationship with size and orbital shape. Journal of Anatomy 223 (5): 495–508. doi:10.1111/joa.12109.

Kiama, S. G., J. N. Maina, J. Bhattacharjee, K. D. Weyrauch, and P. Gehr (1998). A scanning electron microscope study of the luminal surface specializations in the blood vessels of the pecten oculi in a diurnal bird, the Black Kite (Milvus migrans). Annals of Anatomy 180 (5): 455–460. doi:10.1016/S0940 9602(98)801088.

King, A. S., and J. McLelland (1984). Lymphatic system. In Birds: Their structure and function. Baillere Tindall, London, pp. 229–236.

Kishkinev, D. A., and N. S. Chernetsov (2015). Magnetoreception systems in birds: A review of current research. Biology Bulletin Reviews 5 (1): 46–62. doi:10.1134/S2079086415010041.

Knudsen, E. I. (1981). The hearing of the barn owl. Scientific American 245 (6): 112–125.

Konishi, M., and E. I. Knudsen (1979). The Oilbird: Hearing and echolocation. Science 204:425–427. doi:10.1126/science.441731.

Kudo, K. I., S. Nishimura, and S. Tabata (2008). Distribution of taste buds in layer-type chickens: Scanning electron microscopic observations. Animal Science Journal 79:680–685.

Kudo, K. I., J. I. Shiraishi, S. Nishimura, T. Bungo, and S. Tabata (2010). The number of taste buds is related to bitter taste sensitivity in layer and broiler chickens. Animal Science Journal 81:240–244.

Kühne, R., and B. Lewis (1985). External and middle ears. In Form and function in birds, vol. 3, A. King and J. McLelland, Editors. Academic Press, New York, pp. 227–271.

Kulemeyer, C., K. Asbahr, P. Gunz, S. Frahnert, and F. Bairlein (2009). Functional morphology and integration of corvid skulls: A 3D geometric morphometric approach. Frontiers in Zoology 6 (1): 2. doi:10.1186/1742 9994 6 2.

Larsen, O. N., and F. Goller (2002). Direct observation of syringeal muscle function in songbirds and a parrot. Journal of Experimental Biology 205 (Pt 1): 25–35. PMID 11818409.

Leitner, L.-M., and M. Roumy (1974). Thermosensitive units in the tongue and in the skin of the duck's bill. Pflügers Archiv 346 (2): 151–155.

Levy, B., and J. G. Sivak (1980). Mechanisms of accommodation in the bird eye. Journal of Comparative Physiology 137: 267–272.

Louchart, A. (2005). Integrating the fossil record in the study of insular body size evolution: Example of owls (Aves, Strigiformes). In International Symposium "Insular Vertebrate Evolution: the Paleontological Approach." Monografies de la Societat d'Historia Natural de le Balears 12:155–174.

Lucas, A. M. (1979). Integumentum commune. In Handbook of Avian Anatomy: Nomina Anatomica Avium. Academic Press, London, pp. 19–51.

Lucas, A. M., and P. R. Stettenheim (1972). Avian Anatomy: Integument. Vol. 1. Agricultural Handbook 362, USDA.

Machida, N., and Y. Aohagi (2001). Electrocardiography, heart rates, and heart weights of free-living birds. Journal of Zoo and Wildlife Medicine 32 (1): 47–54.

Mack, A. L., and J. Jones (2003). Low-frequency vocalizations by cassowaries (Casuarius spp.). The Auk 120 (4): 1062–1068.

Maclean, G. L. (2013). Ecophysiology of desert birds. Springer Science and Business Media.

Mayr, G. (2016). Variations in the hypotarsus morphology of birds and their evolutionary significance. Acta Zoologica 97 (2): 196–210.

Menon, G. K. (1984). Glandular functions of avian integument: An overview. Journal of the Yamashina Institute for Ornithology 16:1–12

Menon, G. K., and J. Menon (2000). Avian epidermal lipids: Functional considerations and relationship to feathering. American Zoologist 40:540–552.

Meyer, H. J. (1885). Koerperteile der Voegel. Meyers Konversations-Lexikon. 4th ed. Leipzig, Verlag des Bibliographischen Instituts, Germany. https://archive.org/stream/bub_gb_wpAGAQAAIAAJ/bub_gb_wpAGAQAAIAAJ_djvu.txt.

Mouritsen, H., U. Janssen-Bienhold, M. Liedvogel, G. Feenders, J. Stalleicken, P. Dirks, and R. Weiler (2004). Cryptochromes and neuronal-activity markers colocalize in the retina of migratory birds during magnetic orientation. Proceedings of the National Academy of Sciences USA 101 (39): 14294–14299.

Naish, D., and R. Perron (2016). Structure and function of the cassowary's casque and its implications for cassowary history, biology and evolution. Historical Biology 28 (4): 507–518.

Necker, R. (2000). The somatosensory system. In Sturkie's avian physiology, 5th ed., G. C. Whittow, Editor, pp. 57–69.

Necker, R. (2006). Specializations in the lumbosacral vertebral canal and spinal cord of birds: Evidence of a function as a sense organ which is involved in the control of walking. Journal of Comparative Physiology A 192 (5): 439–448.

Olson, S. L., and H. F. James (1991). Descriptions of thirty-two new species of birds from the Hawaiian Islands: Part I. Non-Passeriformes. Ornithological Monographs 45:1–88.

Owen, R. (1872). On the Dodo (Part II). Notes on the articulated skeleton of the Dodo (Didus ineptus, Linn.) in the British

Museum. Transactions of the Zoological Society of London 7:513–525.

Parejo, D., L. Amo, J. Rodríguez, and J. M. Avilés (2012). Rollers smell the fear of nestlings. Biology Letters 8 (4): 502–504. doi:10.1098/rsbl.2012.0124.

Pepperberg, I. M. (2006). Grey parrot (*Psittacus erithacus*) numerical abilities: Addition and further experiments on a zero-like concept. Journal of Comparative Psychology 120 (1): 1.

Piersma, T., M. Dekker, and J. S. Sinninghe Damsté (1999). An avian equivalent of make up? Ecology Letters 2 (4): 201–203.

Portmann, A. (1961). Part I. Sensory organs: Skin, taste and olfaction. Biology and Comparative Physiology of Birds 2:37–48.

Proctor, N. S., and P. J. Lynch (1993). Manual of ornithology: Avian structure and function. Yale University Press, Princeton, NJ.

Quay, W. B. (1967). Comparative survey of the anal glands of birds. The Auk 84 (3): 379–389. doi:10.2307/4083087.

Raidal, S. R., and S. L. Raidal (2006). Comparative renal physiology of exotic species. Veterinary Clinics of North America: Exotic Animal Practice 9 (1): 13–31.

Reichenow, A. (1913). Die Vogel: Handbuch der Systematischen Ornithologie. Vol. 1. http://www.archive.org/stream/dievgelhandbuc01reic#page/24/mode/1up.

Reneerkens, J., M. A. Versteegh, A. M. Schneider, T. Piersma, and E. H. Burtt Jr. (2008). Seasonally changing preen-wax composition: Red Knots'(*Calidris canutus*) flexible defense against feather-degrading bacteria. The Auk 125 (2): 285–290.

Rick, A. M. (1975). Bird medullary bone: A seasonal dating technique for faunal analysis. Bulletin of the Canadian Archaeological Association 7:183–190.

Rico-Guevara, A., T. H. Fan, and M. A. Rubega (2015). Hummingbird tongues are elastic micropumps. Proceedings of the Royal Society of London B 282:1–8.

Rodler, D., K. Stein, and R. Korbel (2015). Observations on the right ovary of birds of prey: A histological and immunohistochemical study. Anatomia, Histologia, Embryologia 44 (3): 68–177.

Roper, T. J. (1999). Olfaction in birds. Advances in the Study of Behavior 28:247–332.

Sabat, P. (2000). Birds in marine and saline environments: Living in dry habitats. Revista Chilena de Historia Natural 73 (2000): 401–410.

Salibian, A., and D. Montalti (2009). Physiological and biochemical aspects of the avian uropygial gland. Brazilian Journal of Biology 69 (2): 437–446.

Schaller, N. U., B. Herkner, R. Villa, and P. Aerts (2009). The intertarsal joint of the ostrich (*Struthio camelus*): Anatomical examination and function of passive structures in locomotion. Journal of Anatomy 214 (6): 830–847.

Seymour, R. S., and A. J. Blaylock (2000). The principle of Laplace and scaling of ventricular wall stress and blood pressure in mammals and birds. Physiological and Biochemical Zoology 73 (4): 389–405.

Sharpe, R. Bowdler (1891). Catalogue of the specimens illustrating the osteology of vertebrated animals, recent and extinct, contained in the Museum of the Royal College of Surgeons of England (part 3), London. Public domain via Internet Archive. https://archive.org/details/catalogueofspeci00shar.

Shugart, G. W. (1988). Uterovaginal sperm-storage glands in sixteen species with comments on morphological differences. The Auk 105 (2): 379–384.

Sinclair, S. (1985). How animals see: Other visions of our world. Croom Helm, London.

Smith, C. A. (1985). Inner ear. In Form and function in birds, vol. 3, A. King and J. McLelland, Editors. Academic Press, New York, pp. 273–310.

Stavenga, D. G., and B. D. Wilts (2014). Oil droplets of bird eyes: Microlenses acting as spectral filters. Philosophical Transactions of the Royal Society of London B: Biological Sciences 369 (1636): 20130041–20130041. doi:10.1098/rstb.2013.0041.

Steiger, S. S., A. E. Fidler, M. Valcu, and B. Kempenaers (2008). Avian olfactory receptor gene repertoires: Evidence for a well-developed sense of smell in birds? Proceedings of the Royal Society of London B: Biological Sciences 275 (1649): 2309–2317.

Stettenheim, P. (1972). The integument of birds. In Avian biology, vol. 2, D. S. Farner and J. R. King, Editors. Academic Press, New York, pp. 1–63.

Stettenheim, P. R. (2000). The integumentary morphology of modern birds: An overview. American Zoologist 40:461–477.

Sturkie, P. D. (1976). Kidneys, extrarenal salt excretion, and urine. In Avian physiology. Springer Berlin Heidelberg, pp. 263–285.

Suarez, R. K. (1992). Hummingbird flight: Sustaining the highest mass-specific metabolic rates among vertebrates. Experientia 48 (6): 565–70. doi:10.1007/bf01920240. PMID 1612136.

Suthers, R. A. (2004). How birds sing and why it matters. In Nature's music: The science of birdsong, P. Marler and H. Slabbekoorn, Editors. Elsevier, New York, pp. 272–295.

Suthers, R. A., and D. H. Hector (1985). The physiology of vocalization by the echolocating Oilbird, *Steatornis caripensis*. Journal of Comparative Physiology A 156:243–266. doi:10.1007/BF00610867.

Sweeney, R. I., I. J. Lovette, and E. L. Harvey (2004). Evolutionary variation in feather waxes of passerine birds. The Auk 121 (2): 435–445.

Thomas, B. T. (1996). Family Opisthocomidae (Hoatzin). In Handbook of the birds of the world, vol. 3, J. del Hoyo, A. Elliott, and J. Sargatal, Editors. Lynx Edicions, Barcelona, pp. 24–32.

Thomassen, H. A. (2005). Swift as sound: Design and evolution of the echolocation system in swiftlets (Apodidae: Collocaliini). PhD diss., Leiden University.

Thomson, J. A. (1916). Outlines of zoology. 5th ed. D. Appleton, New York. http://dx.doi.org/10.5962/bhl.title.54196.

Tickle, P. G., A. R. Ennos, L. E. Lennox, S. F. Perry, and J. R. Codd (2007). Functional significance of the uncinate processes in birds. Journal of Experimental Biology 210:3955–3961. doi:10.1242/jeb.008953.

Vincze, O., C. I. Vágási, I. Kovács, I. Galván, and P. L. Pap (2013). Sources of variation in uropygial gland size in European birds. Biological Journal of the Linnean Society 110 (3): 543–563.

Walsh, S. A., P. M. Barrett, A. C. Milner, G. Manley, and L. M. Witmer (2009). Inner ear anatomy is a proxy for deducing auditory capability and behaviour in reptiles and birds. Proceedings of the Royal Society B: Biological Sciences 276 (1660): 1355–1360.

Welty, J. C., and L. Baptista (1988). The life of birds. Saunders College, New York.

Wenzel, B. M. (1971). Olfaction in birds. *In* Olfaction. Springer, Berlin Heidelberg, pp. 432–448.

Wingfield, J. C., and D. S. Farner (1993). Endocrinology of reproduction in wild species. Avian Biology 9 (163): 327.

Wylie, D. R., C. Gutiérrez-Ibáñez, and A. N. Iwaniuk (2015). Integrating brain, behavior, and phylogeny to understand the evolution of sensory systems in birds. Frontiers in Neuroscience 9:281. http://doi.org/10.3389/fnins.2015.00281.

Yasuaki, S., S. G. Bodde, and M. A. Meyers (2009). Toucan and hornbill beaks: A comparative study. Acta Biomaterialia 6 (2): 331–343. doi:10.1016/j.actbio.2009.08.026. PMID 19699818.

CHAPTER 7

Physiology

Carlos Martinez Del Rio, Pablo Sabat Kirkwood, and Zac Cheviron

This chapter is about the physiological characteristics that allow birds to live in the places that they do and take advantage of the many resources that they use; it is about how birds work. Birds can be found from sea level to flying over the world's highest mountains, and they can feed on foods that range from nectar to toxic leaves. Natural selection has shaped all aspects of their physiology, from the pigments in their blood to the structure of their kidneys. However, rather than just describing the functions of the molecules and organs of birds, we focus on how these mechanisms can be used to explain the features of birds that allow them to live in different habitats (and sometimes prevent them from using others), perform amazing athletic feats, and feed on a wide variety of foods. We adopt both an ecological and a broadly comparative approach because it is only by placing the characteristics of birds in these contexts that one can fully appreciate their remarkable physiological diversity. We also take an integrative perspective. This means that we consider case studies of how birds work from the properties of the molecules within their cells and the genes that code for them, all the way up to how they function as fully integrated feathered organisms. Physiology has two major complementary branches. One branch deals with the structures and mechanisms that make up animals' bodies and their performances. The other branch of physiology is about how the nervous system, together with a system of hormone-producing glands, regulates the function of these structures and mechanisms (see chapter 8). This chapter emphasizes the former branch. It focuses on the design of structures and systems that allow birds to cope with environmental and ecological challenges.

WHY BODY SIZE MATTERS

George Bartholomew was one of the greatest comparative physiologists of the twentieth century. He stated that "the most important attribute of an animal, both physiologically and ecologically, is its size" (Bartholomew 1981). Body size is a major determinant of how an animal functions and of the role that it plays in ecological communities. In a subsequent section, we will argue that you can make coarse (but useful) predictions of many features in birds and other animals if you know how big they are. This is because many features of animals depend on body size in a predictable way. The metric that we use as an estimate of an animal's size is its body mass. Using very simple equations, we can estimate many of an animal's characteristics just by knowing the animal's body mass. The dependence of animal functions and traits on body mass requires that we take an animal's mass into account when we make comparisons. For example, we will see that hummingbirds tend to do things at a rapid pace. Hummingbirds have incredibly high heart and breathing rates. Per unit of body mass they use energy very rapidly, and they must feed frequently and abundantly. Is this because they feed on nectar, a high-octane food, and fly by hovering, which is a demanding way to move? Or is it because they are tiny?

Birds range in body mass from the diminutive Bee Hummingbird (*Mellisuga helenae*), which weighs slightly over two grams (less than one-tenth of an ounce), to the Ostrich (*Struthio camelus*), the largest living bird species, which can weigh over 100 kg (220.5 lb). If we consider extinct species, the range of avian body sizes gets even larger. The flightless Elephant Bird (*Aepyornis maximus*) from Madagascar and the gigantic carnivorous *Dromornis* from Australia could weigh half a ton (500 kg, or over 1,000 lb) and reach over three meters in height (9.8 ft; fig. 7.1). One of the largest flying birds to ever live, *Argentavis magnificens,* weighed around 80 kg (176.4 lbs). These vulture-like birds had wingspans of as much as seven meters (23 ft) (Palmqvist and Vizcaíno 2003). *Argentavis magnificens* was at least five times heavier than the California Condor (*Gymnogyps californianus*), which weighs up to 14 kg

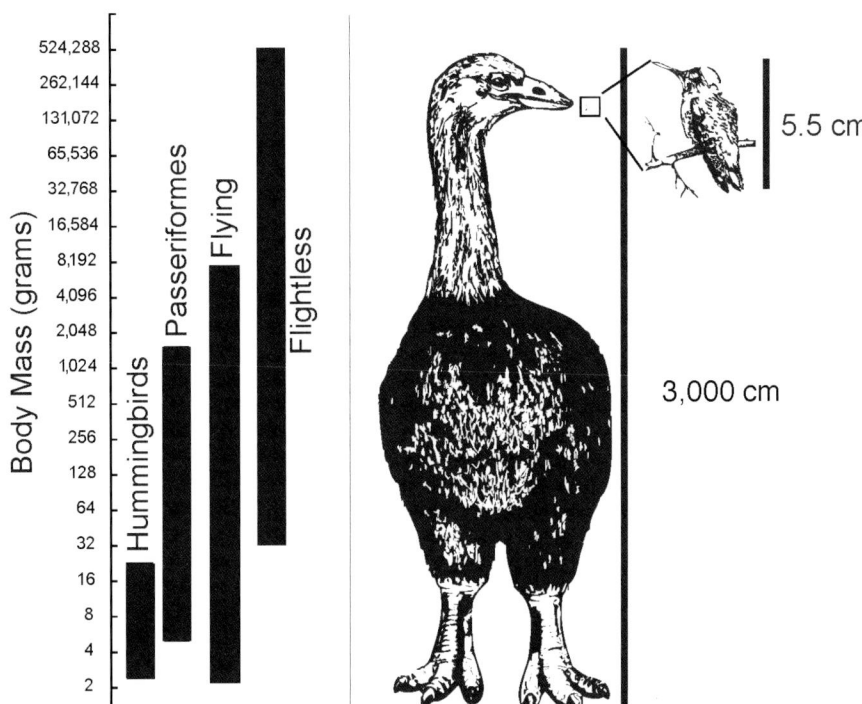

(31 lb) and can have a wingspan of up to three meters (10 ft). Kori Bustards (*Ardeotis kori*) can also be very large (up to 15 kg or 33 lb) and compete with California Condors as the largest extant flying birds. Thus, there is about a 250,000-fold difference in body mass between the Bee Hummingbird and the largest known birds; the masses of birds span five orders of magnitude (fig. 7.1). This large spread in body masses has profound physiological and ecological consequences. Table 7.1 shows a few of the physiological and ecological traits that are dependent on and can be predicted by body mass. The study of the relationship between body mass and biological function is called allometry, and the equations that relate body mass with form and function are called allometric equations.

BODY SIZE AND METABOLIC RATE

Knowing how much energy a bird spends per unit time is useful. We will call this rate of energy use **metabolic rate**. A bird's metabolic rate determines how much the bird should eat each day, and hence the size of the area that it must travel around to find food (the bird's home range), or the size of the area that it must defend (its territory). Consequently, if we know the abundance of resources, we might be able to predict, from an estimate of metabolic rates in individual birds, the number of birds that an area could support (Nee et al. 1991, table 7.1). Physiologically, we expect that the rate at which birds breathe air in and out of their lungs and transport blood to their tissues (and hence the rate at which their

hearts pump blood) should be also dependent on metabolic rate. Many bird characteristics ranging from the physiological to the ecological depend on metabolic rate, and metabolic rate depends on body size (table 7.1).

The dependence of all these properties on body size is, not surprisingly, because larger birds use more energy than smaller ones. Field metabolic rate, basal metabolic rate, and summit cold-induced metabolism (these three measures are defined below) all increase with body mass. However, the relationship between the rate at which birds use energy and their size is not one of proportionality. You cannot just take the metabolic rate of a small bird and multiply it by a constant factor to get the metabolic rate of a larger bird. Elephant Birds did not have metabolic rates that were 250,000 times larger than that of Bee Hummingbirds; the rate was much lower. The relationship between metabolic rate and body mass is a power function of this form:

Rate of energy use $= a * (\text{body mass})^b$

where a is a constant and b is less than one (typically a value around 2/3 or 3/4; Hudson et al. 2013). Recall from your mathematics courses that if the value of b in a power function is less than 1 but larger than 0, then the relation between the dependent and the independent variable is an increasing, but decelerating, function. Therefore, standardizing metabolic rate by body mass (this is called mass-specific metabolic rate) produces a negative relationship between

Table 7.1. **Several characteristics of birds that depend predictably on body mass.**

Type of trait	Equation	Reference
Morphological		
Skeletal mass (kg)	$Y = 0.06X^{1.1}$	Calder (1983)
Heart mass (kg)	$Y = 0.009X^{0.91}$	Calder (1983)
Kidney mass (kg)	$Y = 0.009X^{0.91}$	Calder (1983)
Physiological		
Respiratory rate (breaths/min)	$Y = 18.6^{-0.33}$	Calder (1968)
Heart hate (beats/min)	$Y = 155.8X^{-0.23}$	Calder (1968)
Glomerular filtration rate (ml/hour)	$Y = 141.1X^{0.73}$	Bakken et al. (2004)
Basal metabolic rate (J/second)	$Y = 3.5X^{0.67}$	McKechnie and Wolf (2004)
Maximum lifespan potential of flying birds (years)	$Y = 30.2X^{0.25}$	Healy et al. (2014)
Maximum lifespan potential of non-flying birds (years)	$Y = 41.6X^{0.13}$	Healy et al. (2014)
Ecological		
Population density (individuals/km²)	$Y = 24.7X^{-0.604}$	Silva et al. (1997)
Raptor home range (ha)	$Y = 3664X^{1.34}$	Ottaviani et al. (2006)

Note: We emphasize that the predictions from these scaling equations are approximations that are often very inaccurate. The characteristic in question (Y) depends on body mass (X), but also on a variety of other factors. For example, population density depends not only on body mass but also on diet, the productivity of the habitat inhabited by the species in question, and the taxonomic affinities of a group.

mass-specific metabolic rate and body mass ($b-1 < 0$). Consequently, small birds have higher mass-specific metabolic rates than large birds. Hummingbirds have extraordinarily high mass-specific metabolic rates, but these are simply the result of their tiny size (fig. 7.2). Recall also that we can make power functions look linear if we plot them on log-log axes:

$$\text{Log(rate of energy use)} = \text{Log}(a) + b * \text{Log(body mass)}.$$

Figure 7.2 shows the relationship between the **field metabolic rate** of birds and their body mass in a log-log plot. Field metabolic rate is the rate at which birds use energy during the day and under natural conditions. The slope of this relationship (or the exponent *b*) is about 0.7 (Hudson et al. 2013). If this rate is standardized by 24 hours, this rate is called daily energy expenditure. The two primary methods for estimating field metabolic rate are described in some detail in the box on page 176.

Metabolic rate is measured in a variety of units and at a variety of time scales. You will find that metabolic rate is expressed in kilocalories per unit time, kilojoules per unit time, watts (joules/second), and as the rate of oxygen consumption or carbon dioxide production (in units of volume per unit time). The box on page 176 explains how to transform among different units of measurement. We often use the rate at which birds exchange respiratory gases with the atmosphere to estimate metabolic rate; this method is called **respirometry** (see the box on page 176). Our knowledge of the biochemical pro-

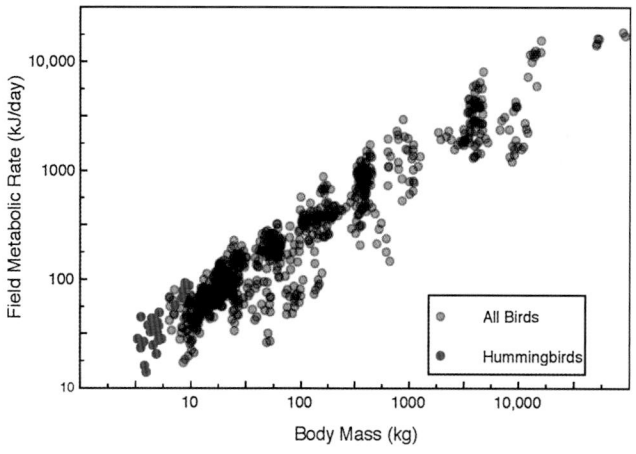

Figure 7.2. The relationship between field metabolic rate and body mass is roughly linear on a log-log scale. Each point is a measurement on an individual bird. The figure represents 895 individuals in 90 species and 15 orders. Note that measurements for hummingbirds fall roughly where they would be expected. Because the slope of the line is lower than 1 (it is about 2/3), the mass-specific field metabolic rate of small birds is higher than that of larger birds. *After Hudson et al. 2013.*

cess of cellular respiration allows us to estimate metabolic rate because oxygen is consumed and carbon dioxide produced and exhaled in rather specific proportions during respiration.

Birds spend energy to produce heat and maintain a relatively constant body temperature that is independent of that

METHODS OF MEASURING METABOLIC RATE

Metabolic rate is the rate at which chemical energy is converted to heat and external work. Energy is expressed in calories (cal) and joules (J; 1J = 4.184 cal), and metabolic rates are expressed as energy units per units of time, usually in cal/s or watts (W). When an animal oxidizes glucose or any other chemical foodstuff, a fixed proportional relation is obtained between the heat produced and the amount of O_2 used. The same is true for CO_2 production. Although the volumes of O_2 consumed per gram of each of the main foodstuff types that are oxidized (carbohydrates, lipids, and proteins) are very different, the amount of heat produced by a liter of oxygen consumed is quite similar (4.8 Kcal per liter O_2 on average). Thus an indirect measure of metabolic rate is the rate of O_2 consumption. The rate of heat produced per liter of CO_2 produced is very dependent on the type of food oxidized; therefore, researchers prefer to measure O_2 consumption as an indirect measure of metabolic rate.

Two methods are commonly used for estimating rates of O_2 consumption: open and closed system respirometry. In the open system, a continuous flow of air flows past the animal placed inside a chamber, and the difference in O_2 concentration between incurrent and excurrent air is measured. Knowing the airflow rate allows the rate of oxygen uptake by the animal to be calculated. In the closed system, the organism is placed in a sealed chamber, which leads to a decrease of O_2 concentration due to respiration. The average rate of oxygen consumption can be computed by using the difference in gas concentration at the start and end of a period of measurements, the chamber volume, and the time of measurement. The advantage of these methods lies in the fact that metabolic rates can be estimated in controlled conditions.

[18]Oxygen and [2]Hydrogen are artificially enriched in the body water pool and their subsequent elimination is measured. This occurs at different rates for the two isotopes because hydrogen is primarily eliminated by the outflow of body water, while oxygen is eliminated both by water loss and from expiration of respiratory CO_2. The difference between the elimination rates of the two isotopes therefore provides an estimation of CO_2 production in an individual over a given time. The use of the doubly labeled water method is useful for estimating the average metabolic rate (field metabolic rate) in free-living birds. When animals spend energy in the body, carbon dioxide (CO_2) and water are produced; this technique measures the production of carbon dioxide over relatively long periods of time (days or weeks). The technique requires the administration (usually by injection) of water in which both the hydrogen and oxygen have been substituted partially or completely, for tracking purposes (i.e., labeled), with an uncommon stable isotope of these elements: deuterium (^{2}H) and oxygen (^{18}O). The method is based on the principle that, after a loading dose of $^2H_2{}^{18}O$, [18]O is eliminated as CO_2 and water, while deuterium is removed from the body as water. The rate of CO_2 production can be estimated from the difference between the two isotope elimination rates and the total amount of water in the bird's body (see figure). Knowing the production rate of CO_2, total oxygen consumption (metabolic rate) can be estimated from the respiratory quotient, defined by the ratio of CO_2 produced to O_2 consumed while food is being metabolized. This ratio can be measured by respirometry and usually exhibits values of from 0.7, when fat is being oxidized, to 1.0 when carbohydrates are being used; and in a mixed diet is usually about 0.8 (Frayn 1983). One can find out the "fuels" that birds use to power their metabolism by measuring the rate at which they consume oxygen and produce carbon dioxide. More recently, measurement of the heart rates of free-living birds have also been used to estimate field metabolic rate (Steiger et al. 2009). This approach is called heart rate telemetry, and it offers some key advantages over the doubly labeled water method. Most importantly, it allows for "real-time" measures of changes in heart rate (and thus metabolic rate) that are associated with specific activities or behaviors. To use heart rate telemetry to estimate field metabolic rate, however, changes in heart rate must be related to changes in O_2 consumption or CO_2 production. This is accomplished using respirometry on birds that are equipped with heart rate monitors, effectively calibrating the heart rate monitors so that they can be used as a proxy to measure respiratory gas exchange.

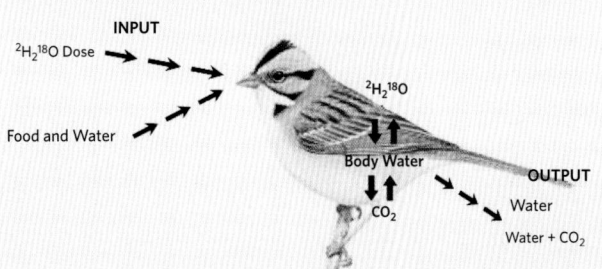

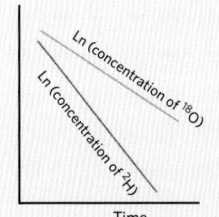

Field metabolic rate is commonly measured by injecting a bird with a dose of water labeled with Deuterium and [18]O (the heavy stable isotopes of hydrogen and oxygen) and measuring the bird's water volume and the rate at which the concentration of the isotopes decreases with time.

[18]O elimination rate (in water and CO_2) - [2]H elimination rate (in water) = CO_2 production

of their environments. These features make them homeo-thermic endotherms. In birds, resting body temperature is lower in the larger Paleognathae (cassowaries and ostriches, mean body temperature about 38.5°C) than in other birds (average body temperature about 42°C; Clarke and Rothery 2008). In 1950, Per Scholander and Larry Irving, two adventurous, pioneering physiologists, proposed that we could study endotherms as if they were furnaces (Scholander et al. 1950). They suggested that in order to maintain a constant body temperature, an animal has to spend as much energy as it loses—much as the furnace that warms your house. The amount of energy lost to the environment, in turn, should be proportional to the difference in temperature between the animal's body and that of the environment. Thus, if you make a plot of an animal's metabolic rate against environmental temperature, you should get a straight line with a negative slope that projects into the x-axis at the value of the animal's body temperature (fig. 7.3). As temperature increases, rather than reaching a value of 0, metabolic rate becomes relatively stable. This minimal metabolic rate is called the **basal metabolic rate** if the animal is at rest and post-absorptive (meaning that it is not digesting a meal). The range of temperatures at which metabolic rate is constant is called the **thermoneutral zone**. The slope of the relationship between metabolic rate and ambient temperature is called **thermal conductance**.

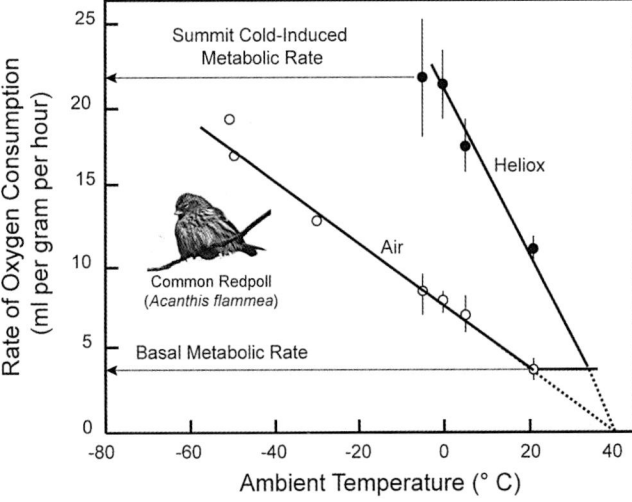

Figure 7.3. Scholander and Irving's model of thermoregulation in birds exposed to air and heliox. In order to maintain constant body temperature, heat output must match heat losses, which depend on the gradient between the bird's body temperature and air temperatures. Heat loss is steeper when insulation is low or when birds are exposed to a medium such as heliox that conducts heat efficiently. The relationship between energy used in thermoregulation and air temperature is flat for temperatures above a temperature called the **lower critical temperature**. *After Rosenmann and Morrison 1974.*

The slope is shallow if the bird is well insulated and steep if it is poorly insulated and loses a lot of heat (fig. 7.3).

Basal metabolic rate estimates the minimal amount of energy that a bird needs to maintain its body temperature when it is resting and when ambient temperature is within the thermoneutral zone. How much can birds increase their metabolism in response to cold temperatures? Chilean physiologist Mario Rosenmann took advantage of the observation that a mixture of 80 percent helium and 20 percent oxygen (called heliox) conducts heat four times more rapidly than air (Rosenmann and Morrison 1974). If you place a bird in heliox, the gas increases heat loss. When they placed a Common Redpoll (*Acanthis flammea*) in heliox and measured its rate of oxygen consumption, they found that at 20°C it lost as much heat, and hence consumed as much oxygen per unit time, as if it was at −20°C. At only 0°C, the redpoll reached its maximal possible metabolic rate. This rate is called **summit** (or peak) **cold-induced metabolism**. To force the bird to reach this peak metabolic rate in air, Rosenmann would have had to lower ambient temperature to about −75°C. Many similar experiments have been conducted to find out the maximal rate at which birds can spend energy to keep body temperature in the cold.

Measurements of basal metabolic rate and summit metabolism are accomplished using respirometry to measure the rate at which an individual consumes oxygen or produces CO_2 at different environmental temperatures. This gives a standardized way to compare these values across species. Hundreds of measurements of the metabolic rates of birds have been made using respirometry, and we have learned a lot from them. But measuring the metabolic rate of a bird under these standardized conditions can be an oversimplification, because in nature there are lots of factors that influence the rate at which birds exchange heat with the environment (Wolf et al. 2000). Birds exchange heat with the sun and sky, with the ground, the air, and, in aquatic birds, with water. The rate at which these exchanges take place is affected by the speed of the wind or water. Figure 7.4 describes these mechanisms in detail. We have emphasized heat loss as a determinant of energy use. Other factors are also as important. All forms of locomotion are costly (see chapter 10); incubation, growth, and molt are also major determinants of how much energy is used by birds and are addressed in the corresponding chapters.

AVIAN ENERGETICS

How Do Birds Generate Heat?

The generation of heat is called thermogenesis. Because metabolic processes are inefficient, one of the end products of

Figure 7.4. Birds exchange heat with the environment in many ways. When an object is in direct contact with another object, there is net transfer of the kinetic energy of the warmer object to the cooler one by conduction. The Flightless Cormorant (*Phalacrocorax harrisi*) is exchanging heat with the lava by conduction. Depending on the lava's temperature, net heat flow will flow into the bird's body or from the body into lava. Heat transfer is enhanced by a moving fluid, such as wind or water. This mode of heat transfer is called convection. Birds also emit and receive electromagnetic waves, with net movement of heat energy from high to low temperatures. The cormorant is emitting heat to the sky and receiving heat from the sun as it basks. Finally, because it takes energy to evaporate water, birds lose heat to the environment when they evaporate water. The cormorant evaporates water, and therefore cools down by panting. *Photo of birds on land by Nathan Bell; photo of diving bird by Greg Estes. With permission.*

their action is heat. Flight muscles (the *pectoralis* and *supracoracoideus*, chapter 6) represent a large fraction of a bird's body mass (from 15 to 25 percent) and are the most important heat generators (Swanson 2010). In some birds, leg muscles also contribute to thermogenesis. In mammals, non-shivering thermogenesis can be an important contributor to an individual's heat budget. In these animals, proteins found in mitochondria uncouple the electron gradient from the production of adenosine triphosphate (ATP) and therefore generate heat rather than store it in chemical bonds. These uncoupling proteins (or UCPs) are found in a type of fatty tissue called brown fat. The occurrence of non-shivering thermogenesis in birds is controversial, because birds do not have brown fat. However, several studies show that when Muscovy Ducks (*Cairina moschata*) and Chickens (*Gallus gallus*) are exposed to cold temperatures for several weeks, they express an uncoupling protein in their muscles and may experience non-shivering thermogenesis. Perplexingly, this protein does not seem to uncouple electron transport from ADP phosphorylation. No one knows exactly how (or if) this protein contributes to thermogenesis (Bicudo et al. 2001), making this an area of exciting new research.

Birds in the Cold

Tiny birds such as chickadees can survive the winter, even in places with very cold winters. Birds survive cold temperatures by increasing heat production to counteract heat loss and by conserving heat. David Swanson and Karen Olmstead (1999) captured Black-capped Chickadees (*Poecile atricapillus*), Dark-eyed Juncos (*Junco hyemalis*), and American Tree Sparrows (*Spizella arborea*) during several winters in South Dakota. They immediately brought them to the laboratory and measured their basal and summit metabolic rates. They found that basal and summit metabolic rates were higher in birds caught in colder than in warmer winters. They also found that a good predictor of how high these rates were was how cold it was the week before the birds were caught. Recent studies of juncos in the lab have revealed that cold exposure leads to dramatic increases in summit metabolic rate and the expression of genes that encode key metabolic enzymes (Stager et al. 2015a). These changes were more dramatic than those induced by changes in photoperiod, an environmental cue that signals the onset of winter. These results suggest that increases in the capacity to generate heat in juncos is a direct response to cold exposure, rather than an anticipatory response to winter onset. Chickadees also appear to use cold temperature as a cue to up-regulate their capacity to generate heat (Swanson and Olmstead 1999). Moreover, in comparisons among species that occur in a variety of environments, mean daily winter temperatures strongly predict mass-specific summit metabolic rates among species, and minimum winter temperature predicts variation in metabolic scope (the difference between mass-specific basal and summit metabolic rates) (Stager et al. 2015b). Taken together, these results demonstrate that the ability to up-regulate thermogenic capacity in response to cold temperatures strongly influences the geographic distributions of birds.

Small birds in winter not only increase their ability to generate heat, they also reduce heat losses. Agnes Lewden and her collaborators (2014) measured the body temperature of Black-capped Chickadees in the winter and found that on colder days chickadees become hypothermic. The normal temperature of chickadees is 41°C, whereas that of some individuals on really cold days (−15°C) was as low as 35.5°C. During the night, Black-capped Chickadees find an old nest or a hollow stump to roost and lower their body temperature even more than during the day, to about 30°C. By lowering their body temperature on cold days and nights, chickadees reduce the gradient with the environment and hence decrease the rate of heat loss. The increase in metabolic rate and body temperature that birds can undergo when exposed to natural cold temperatures is an example of both flexibility in physiological characteristics (what is called **phenotypic flexibility**; box on page 179) and of **acclimatization**. The term acclimatization refers to changes in the expression of a characteristic that result from exposure to changes in environmental conditions in free-living birds. In contrast, the changes of bird characteristics due to changing conditions in the laboratory is called **acclimation**.

Emperor Penguins (*Aptenodytes forsteri*) are the only birds that breed in the Antarctic winter. Males and females court for about six weeks. Females then lay a single egg and pass it

PHENOTYPIC PLASTICITY IN BIRDS

The phenotype of an organism (i.e., the set of an individual's observable traits) is the result of the interaction between genes and the environment. The phenomenon of genotypes that lead to different phenotypes under different environmental conditions is called **phenotypic plasticity**. Phenotypic plasticity can take many forms: birds can differ in phenotype because they developed in different environments. This is called **developmental plasticity** and is typically irreversible. For example, when Peter Boag (1987) fed Zebra Finch (*Taeniopygia guttata*) nestlings on diets with high and low protein content, the chicks fed low-protein diets were smaller than those fed high-protein diets and remained small throughout their lives. Birds can also change their phenotypes reversibly and temporarily in response to changing environments or ecological demands such as migration and reproduction. This chapter has many examples of this type of **phenotypic flexibility**. The laboratory studies of physiological acclimation and the field studies of acclimatization that bird physiologists do all the time are examples of research into the phenotypic flexibility of birds. Finally, many bird species change their morphology and physiological characteristics

seasonally. Rock Ptarmigan (*Lagopus muta*) are cryptic most of the year. In summer, their plumage is mottled dark brown, black, and grayish—a good camouflage for the tundra's summer vegetation. In winter, it is spotless white and matches the snow-covered ground. The ptarmigan's changing colors are an example of **life-cycle staging**, in which phenotype depends on season and/or the life stage of the individual. The following table, extracted from Piersma and Drent (2003), summarizes the three types of phenotypic plasticity that we often find in birds.

Plasticity Category	Is phenotypic change reversible?	Does it occur within a single individual?	Is it seasonally cyclic?
Developmental Plasticity	No	No	No
Phenotypic flexibility	Yes	Yes	No
Life-cycle staging	Yes	Yes	Yes

to a male who cradles it in a skinfold between his legs called a brood pouch. Then the female departs in a long march to the ocean to feed. Males incubate the eggs for over two months. Males fast from the beginning of courting to the time when females return to feed chicks after they have hatched. The success of males while courting and incubating the egg depends on how good they are at using their energy reserves efficiently. To save energy, Emperor Penguins aggregate in huge groups packed close together. Huddling in groups has two major advantages: (1) it reduces cold-exposed surface area and (2) it increases the temperature inside of the group. Huddling penguins have a metabolic rate that is about half of what they would have if they were alone (Gilbert et al. 2008), and they burn a lot less body reserves. Penguins constantly shuffle from the cold-exposed edge of the group to the inside, forming wavelike movements within the huddle (Gerum et al. 2013). The result is that all penguins in the group share energy savings fairly. Huddling has been studied in 25 bird species in nine families (Gilbert et al. 2010), but it is likely more widespread. Wojciechowski et al. (2011) found that migrating Blackcaps (*Sylvia atricapilla*) huddle when they stop to rest and feed at stopover sites. Huddling reduces metabolic rate and therefore facilitates the accumulation of fat reserves at stopover sites. Huddling is common in small birds that must cope with very cold winters, such as chickadees and Golden-crowned Kinglets (*Regulus satrapa*), and may be more common in small, migratory birds than previously thought.

Maintaining a high body temperature by the production of metabolic heat is very costly. As we have seen, birds reduce these costs by behavioral and physiological means: they choose microenvironments that reduce heat loss, they huddle, and they sometimes lower their body temperature. Birds can lower their body temperature to save energy in response to increased energy demands caused by cold temperatures or reduced availability of food, or to be able to gain more reserves from food even if it is abundant. This lowering of body temperature is not an uncontrolled process, which is why physiologists call the phenomenon **facultative hypothermia**. Facultative hypothermia has been observed in birds ranging in size from tiny hummingbirds (Booted Racket-tails, *Ocreatus underwoodii*, 2.7 g) to massive Eurasian Griffons (*Gyps fulvus*, 6,500 g). McKechnie and Lovegrove (2002) found reports of facultative hypothermia in 96 bird species in 29 families and 11 orders. The phenomenon is clearly widespread. Lowering of body temperature comes in two forms: rest-phase hypothermia and torpor. Birds in rest-phase hypothermia remain alert and are capable of flight. In contrast, birds in torpor are lethargic, often completely immobile, and unable to respond to external stimuli.

Facultative hypothermia appears to be very common in birds, and torpor has been reported in eight orders (Coraciiformes, Coliiformes, Apodiformes, Trochiliformes, Strigiformes, Caprimulgiformes, Columbiformes, and Passeriformes; Ruf and Geiser 2015). Because they are small, hummingbirds have very high mass-specific metabolic rates, and therefore they must feed on prodigious amounts of nectar. Because they find flowers by sight, they cannot feed at night. If hummingbirds maintain their high rates of metabolism throughout the night, they use their fat reserves very quickly. This is especially true if the night is cold. Furthermore, migrant hummingbirds can accumulate fat faster if they reduce the rate at which they use energy to thermoregulate at night. To save energy, hummingbirds often enter into torpor. When they enter into torpor, their body temperature drops precipitously, and with it their metabolic rate (Bicudo 1996). The metabolic rate of a torpid hummingbird can be 20 times lower than that of an active one (fig. 7.5). Hummingbirds can maintain a low metabolic rate for several hours. To arouse from torpor, hummingbirds increase their metabolic rate by shivering vigorously. Their body temperature increases by about 1°C per minute (Bicudo 1996). The rate at which temperature increases during arousal is faster in smaller than in larger birds (this is yet another example of a trait that depends predictably on body size). Hence, it is not very surprising that deep hypothermia and torpor are more common in small than in large birds (Ruf and Geiser 2015). The eight-gram Andean Hillstar (*Oreotrochilus estella*) can drop its body temperature to about 7°C (Carpenter 1974), whereas the Australian Tawny Frogmouth (*Podargus strigoides*, 500 g), one of the largest birds capable of entering torpor, drops it to slightly less than 30°C (Körtner

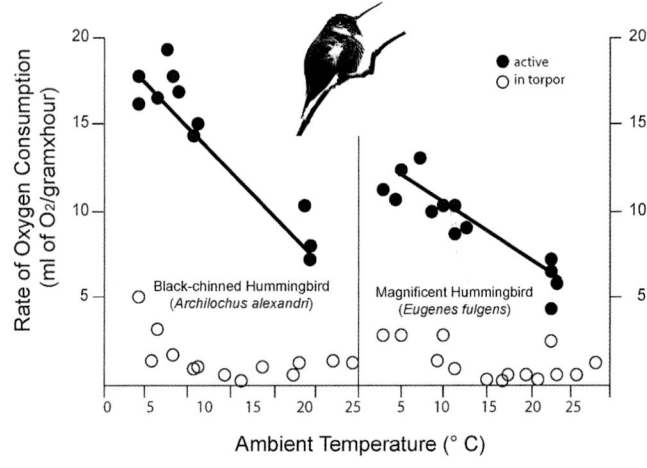

Figure 7.5. When in torpor, hummingbirds lower their body temperature and therefore the gradient between body and environmental temperature. Their metabolic rate drops and they save energy. *Redrawn from Hainsworth and Wolf 1978.*

et al. 2001). Hibernation is a form of torpor that lasts longer than 24 hours (Ruf and Geiser 2015). Many mammals hibernate, but as far as we know only one bird species, the Common Poorwill (*Phalaenoptilus nuttallii*), does. The Native America Hopi call Common Poorwills Hölchoco, "the sleeping ones." They likely knew that these birds were capable not only of entering torpor, but of staying inactive and seemingly torpid for over 10 days in the winter (Woods and Brigham 2004).

Birds in a Warming World

In 1932, a heat wave hit south central Australia. Temperatures reached over 49°C (over 120°F) and thousands of Budgerigars (*Melopsittacus undulatus*) and Zebra Finches (*Taeniopygia guttata*) died (McKechnie and Wolf 2010). The frequency of heat waves in the arid and semiarid areas of the world is increasing as the planet's atmosphere warms due to human emission of greenhouse gases. How will birds cope with these extreme events? Will we witness bird die-offs resulting from heat waves more frequently? Birds rely on both finding suitable cooler microclimates to avoid the hottest part of the day and on the evaporation of body water to cool themselves. Several species of storks and New World vultures excrete on their legs, and the evaporation of the liquid cools them down (Kahl 1963). When ambient temperature is higher than body temperature, birds lose water through their skin (**cutaneous evaporation**) and through the respiratory tract by panting. Some bird species supplement panting by vibrating the throat muscles in a behavior called gular fluttering. As ambient temperature increases above body temperature, the rate at which birds evaporate water (evaporative water loss) increases linearly (fig. 7.6). When temperatures become too high, though, birds lose the capacity to thermoregulate, become too hot (**hyperthermic**), and die. Andrew McKechnie and Blair Wolf (2010) combined data on body mass and evaporation rates and predicted ambient temperatures in two desert localities in Australia (Birdsville) and the United States (Yuma) to estimate bird survival during heat waves. Their results are sobering. They predicted reduced survival times and presumably increased frequency of bird die-offs caused by heat waves in the near future. Because the rate of evaporative water loss with temperature is steeper in smaller birds, they predicted stronger effects on smaller than on larger birds. Climate change, therefore, might have an impact on bird community composition in hot deserts (McKechnie and Wolf 2010).

RESPIRATION

Pathway of Oxygen

To maintain their active lifestyles and high body temperatures, birds need to transport oxygen efficiently from the at-

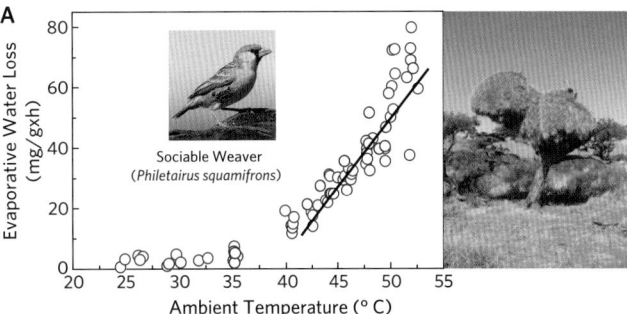

Figure 7.6. If ambient temperature (Ta) becomes too hot, the rate at which birds lose water by evaporation to regulate their body temperature increases linearly. *After Whitfield et al. 2015.* The photographs depict a Sociable Weaver and the massive nests of this species. *Photo of Sociable Weaver by Johan Grobbelar. Photo of nest by Robert Thomson. With permission.*

mosphere to respiring cells inside their bodies (fig. 7.7). The transport of oxygen from the lungs, through the blood, and on to working tissues is accomplished by the coordinated efforts of the respiratory and circulatory systems. Birds have evolved remarkably efficient lungs, powerful hearts, and extensive capillary networks to accomplish this vital, life-sustaining process. Perhaps the most distinctive feature of

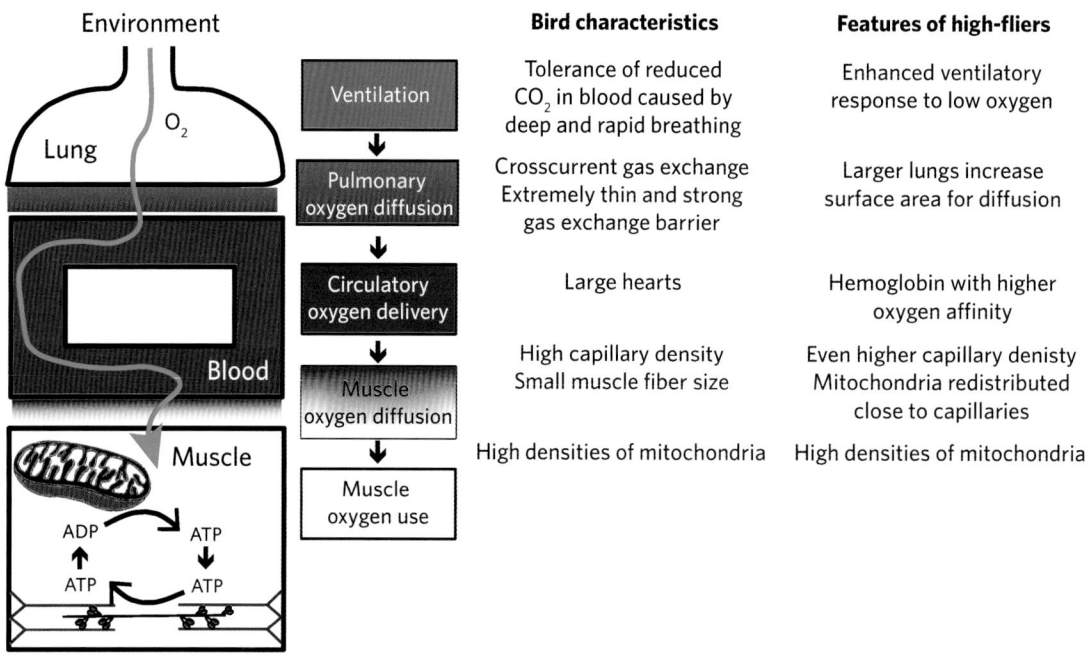

Figure 7.7. Increased aerobic performance under hypoxia requires modifications of several integrated steps in the oxygen transport cascade: these include breathing, diffusion of oxygen at the lung into the bloodstream, bloodstream circulation to transport oxygen throughout the body, diffusion of oxygen from the bloodstream into the cells of respiring tissues (e.g., muscle cells), and, finally, transport of oxygen to mitochondria to generate ATP via oxidative phosphorylation. All birds may be preadapted for increased aerobic performance under hypoxia due to modifications of each of the integrated steps in the oxygen cascade. The features that are common to all birds and enhance aerobic performance are accentuated in birds that fly at very high altitudes, such as Bar-headed Geese. *Modified from Scott 2011.*

the avian respiratory system is related to the structure and function of the lungs. In contrast to mammalian lungs, air flows in a single direction across the gas exchange surfaces of the avian lung. This unidirectional flow of air is accomplished by structural modifications of the lung itself, and by the presence of air sacs that act as baffles to store and move air through the respiratory system (fig. 7.8). A single breath of air moves through the respiratory system in a four-step process that includes two inhalation and exhalation cycles (fig. 7.8). Birds lack a diaphragm, the muscle that mammals use to create the changes in pressure within the thoracic cavity that inflate and deflate the lungs. Instead, birds breathe by muscular contractions that lower and raise the sternum and expand and contract the rib cage. The movement of the sternum and rib cage allows the air sacs to inflate (inhalation) and deflate (exhalation), forcing air to move through the intricate series of air sacs and across the parabronchioles within the lung.

With the first inhalation, the sternum lowers and the rib cage expands, allowing a breath of air to travel down the trachea and into the posterior air sacs. Upon the first exhalation, muscular contractions again raise the sternum and compress the posterior air sac, which forces the fresh, oxygen-rich air into the lungs through a series of finely subdivided parabronchioles. Radiating from the parabronchioles are air capillaries, the surfaces where gas exchange occurs. A dense network of capillaries within the lung crosses the parabronchioles perpendicularly, and this orientation has important consequences for the efficiency of oxygen and carbon dioxide exchange (see below). With the second inhalation, the oxygen depleted, carbon dioxide–rich air moves from the lungs into the anterior air sacs, where it is held until the second exhalation. During this final exhalation, the air leaves the anterior air sac, returns to the trachea, and is expelled. This four-step process ensures that nearly all of the air within the lung is turned over with each cycle—nearly all of the air within the lung is expelled, and it prevents mixing of fresh incoming air with expelled oxygen-depleted air. Birds avoid the problem that mammals have of having a dead space of stale air that stays in the lungs when breathing in and out. The bird's complex breathing pattern is enhanced during powered flight, which is an exceptionally aerobically demanding form of locomotion (chapter 10). During flight, contractions of the flight muscles enhance the movement of the sternum, rib cage, and furculum (or wishbone) to pump air more forcefully through the respiratory system (Jenkins

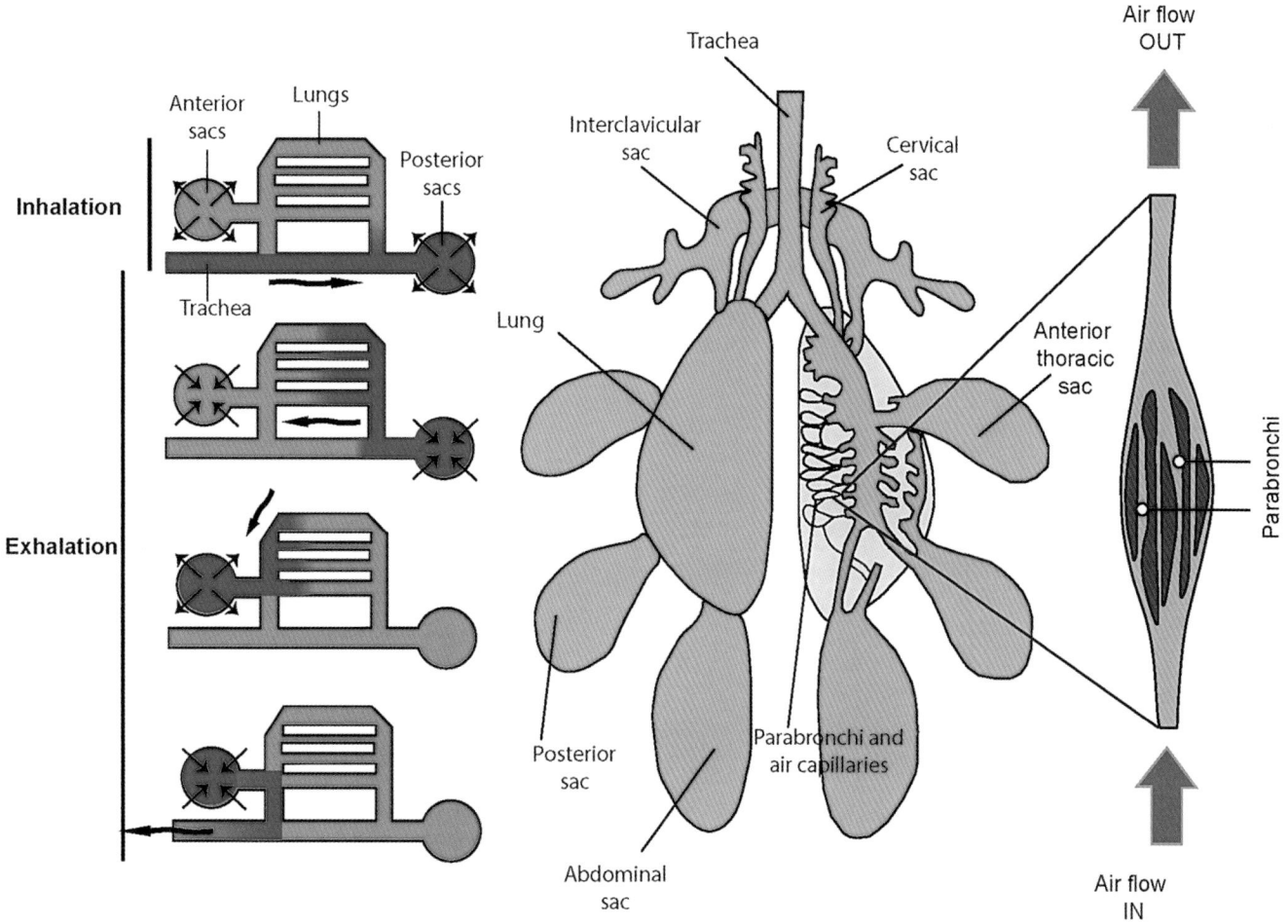

Figure 7.8. The avian respiratory system. *Left*, Air movement is unidirectional through the avian respiratory system and a single breath of air requires two full respiratory cycles to move through the system. In the first inhalation, air moves into posterior air sacs through the trachea. Upon exhalation, air is forced from the posterior air sac across the lung, completing the first respiratory cycle. In the second inhalation, air move from the lungs into the anterior air sacs, and a new second breath enters the system. Upon the second exhalation, the first breath is forced out from the anterior air sacs and out the trachea, completing the second respiratory cycle, while the second breath is moved into the lungs. This cycle then repeats in a living, breathing bird. *Center*, Major components of the avian respiratory system. *Right*, Detailed view of air movement through primary and secondary bronchi within the lung, and the orientation of the parabronchi.

et al. 1988). Thus, the action of flying allows birds to breathe more deeply just when they need it the most.

In addition to the air sacs that allow for unidirectional air flow, another key innovation of the avian respiratory system lies in the structural properties of the parabronchioles and the orientation of the capillaries that service them. Air enters the lung from the posterior air sac through a pair of primary bronchi, which are further subdivided into parabronchioles, rigid air tubes that are intertwined with capillaries. Because the parabronchioles run perpendicular to the capillaries in the lung, they create a crosscurrent exchange system that is exceptionally efficient for oxygen and carbon dioxide exchange. As air flows through the parabronchioles, oxygen passively diffuses along its concentration gradient into the capillaries

carrying the deoxygenated blood from the body. When blood flows through capillaries that are perpendicular to the flow of air through the parabronchioles, the orientation creates a crosscurrent exchange system that allows birds to obtain higher oxygen concentrations in their blood than that of the expired air. This mechanism is in stark contrast to that in mammalian lung, where air flows into tiny sacs inside of the lung (called alveoli) that are enmeshed in capillaries. Because the mammalian lung lacks the avian crosscurrent exchanger, the oxygen concentration in mammalian blood can be *at best* equal to that in the expired air. In addition, the gas exchange surfaces of birds are extremely thin, which reduces the distance over which oxygen must diffuse to enter the bloodstream. However, despite their thinness, the gas exchange surfaces of birds are mechanically stronger than those of mammals, which allows them to withstand greater increases in pulmonary pressure and blood flow (West 2009). Together these anatomical modifications allow birds to extract more oxygen from the air they breathe than do mammals.

Once oxygen enters the bloodstream, the vast majority is bound to hemoglobin within red blood cells, as it is in other vertebrates. Birds differ from other terrestrial vertebrates in their comparably large and powerful hearts, which pump this hemoglobin-bound oxygen throughout the body. Like mammals, birds have four-chambered hearts, but birds have larger hearts compared to similarly sized mammals. Birds also have slower resting heart rates than mammals (Grubb 1983). However, their increased size enables avian hearts to pump greater volumes of blood per stroke than those of a similarly sized mammal (Grubb 1983). These large heart masses and stroke volumes allow bird hearts to move more blood per unit time with fewer heart contractions than a mammal of similar size, suggesting that birds have a very high capacity for circulatory oxygen transport. Birds also have a very high capacity for oxygen diffusion from the blood into working tissues. Compared with the locomotory muscles of mammals, the flight muscles of birds have a much higher surface area of capillaries to muscle fiber ratio (Matheiu-Costello 1991). Bird brains and hearts also appear to have greater capillary densities than those of mammals (Faraci 1991). Because capillaries supply oxygen to muscles and other working tissues, these increases in capillary density should provide birds with a greater surface area for oxygen diffusion into metabolically active tissues.

In summary, modifications of nearly every step of the oxygen transport cascade allow birds to meet the demands of their active lifestyles and have transformed them into some of the most impressive aerobic athletes in the animal world. These modifications have also allowed them to exploit oxygen-poor environments during critical periods of their lives. During their annual migrations, some birds fly higher than any other terrestrial animal, and many species breed and forage at high-elevation habitats around the world. In the next section, we consider how the respiratory and circulatory systems of these amazing birds are modified to allow them to exploit these inhospitable environments.

Birds at High Altitude

Among the most remarkable aspects of avian physiological performance is that many birds utilize the most metabolically demanding form of animal locomotion—powered flight—under severely oxygen-poor high-altitude conditions. The partial pressure of oxygen drops rapidly with altitude. Although the percentage of oxygen in air is about the same at all elevations (21 percent in dry air), at high elevation there are fewer oxygen molecules than at sea level. For example, the partial pressure of oxygen in the air on the summit of Mount Everest is only 35 percent of that at sea level, which means that a single breath of air contains only about 35 percent of the oxygen that is available in a breath at sea level. Bar-headed Geese (Anser indicus) famously migrate over the highest peaks in the Himalayas while elite mountaineers struggle to climb the same peaks with the help of supplemental oxygen. Black-breasted Hillstar hummingbirds (Oreotrochilus melanogaster) hover in front of flowers on the Andean Altiplano at elevations above 4,000 m. In 1973, a soaring Ruppell's Griffon (Gyps rueppellii) collided with a commercial jetliner soaring at an altitude of 11,278 m (37,000 ft) (Laybourne 1974). Exposure to this altitude could induce unconsciousness and death within minutes of exposure in humans. The ability of these birds to perform such amazing physiological feats stems, in part, from the efficient cardiorespiratory systems that are common to all birds. Birds in general are preadapted for high performance in hypoxic environments, but extreme high flyers also exhibit a number of specific cardiorespiratory adaptations that underlie their remarkable abilities. High-altitude athletic performance has been best studied in Bar-headed Geese (Scott et al. 2015), so we use them as a case study in evolutionary adaptation for high-altitude performance.

In early reports, mountaineers described observing Bar-headed Geese flying over the highest peaks in the Himalayas during their spring migration to India (Swan 1970). Ever since these early reports, news of these remarkable flights have captivated avian physiologists, and several decades of research have documented the numerous physiological attributes that give Bar-headed Geese their abilities. First among these is an efficient breathing pattern: Bar-headed Geese take fewer but deeper breaths than lowland geese when exposed to low oxygen conditions, and this breathing pattern increases their total ventilation nearly twofold over their lowland counterparts (Scott 2011). These alterations of breathing pattern are enhanced by a dramatic increase in surface area of the lungs; the lungs of Bar-headed Geese are more than 25 percent larger than those of the lowland geese of similar body size (Scott 2011). In addition, Bar-headed Geese are also capable of greater circulatory oxygen delivery under hypoxia. This enhancement is driven in large part by changes in the functional properties of hemoglobin and physiological changes that enhance oxygen supply to the heart under hypoxia. The hemoglobin of adult Bar-headed Geese has much greater binding affinity for oxygen than that of lowland waterfowl, and this enhancement of binding affinity should help increase blood oxygen saturation under severe hypoxia. Changes in temperature also have profound effects on the ability of hemoglobin to bind oxygen. The hemoglobin of Bar-headed Geese is remarkably sensitive to changes in temperature. It has high affinity for oxygen in the lung's capillaries, where it takes up the scanty oxygen from the cold air, and low affinity when it delivers it to the hot flying muscles. These changes in hemoglobin function may allow Bar-headed Geese to extract more oxygen from the cold, hypoxic air at high altitude, while also allowing effective unloading of oxygen to warm muscles, to power their long-distance migratory flights.

Bar-headed Geese also possess cardiac adaptations to ensure that, once oxygen is bound to hemoglobin, it can be transported to the cells that need it. Their hearts have nearly 40 percent greater capillary density than those of geese that fly at lower altitudes, and this enhancement helps maintain adequate oxygen supplies to support high cardiac output during high-altitude flight (Scott 2011). Finally, Bar-headed Geese have also altered the structural properties and metabolic profiles of their flight muscles (Scott 2011). The flight muscles of Bar-headed Geese have greater capillary densities than lowland waterfowl, but the differences in muscle go even deeper. Within muscle fibers of the flight muscles, the mitochondria tend to be located closer to the cell membrane, which reduces the distance that oxygen must diffuse to reach these cellular power plants. Finally, the functional properties of a key metabolic enzyme that is expressed in the mitochondria of flight muscles, cytochrome c oxidase, are altered in Bar-headed Geese (Scott 2011). Although the physiological significance of this alteration is still being determined, it may play a role in reducing oxidative stress and cellular damage by reducing the production of damaging reactive oxygen species (ROS) during high-altitude flight (Scott 2011). In short, the amazing athletic performance of Bar-headed Geese is thanks to multiple complementary physiological traits that alter nearly every aspect of the oxygen-transport cascade (fig. 7.9). No other extreme high-altitude flyer has been as intensively studied as the Bar-headed Goose, but a number of these modifications have been documented in other high-altitude specialists. Modifications of hemoglobin function seem to be common among high-altitude bird species (see below), and Gray-crowned Rosy Finches (*Leucosticte tephrocotis*), which breed above the tree line in the mountain ranges of western North America, also have very high capillary densities in their leg and flight muscles compared with lowland finches (Mathieu-Costello 1998). Determining whether adaptation to life at high altitude generally involves the type of wholesale alteration of the oxygen transport cascade that is seen in Bar-headed Geese is an area of active research.

A recurring theme in avian adaptation to high-altitude flight and life in high-elevation habitats is the repeated adaptive modifications of hemoglobin function in high-altitude

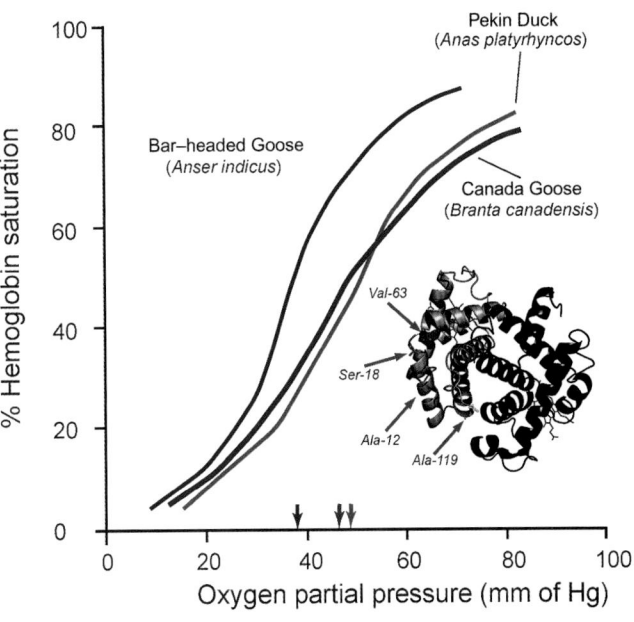

Figure 7.9. Hemoglobin oxygen dissociation curves for three species of waterfowl. The curve of the Bar-headed Goose is shifted to left, indicating that its hemoglobin has a higher affinity for oxygen than those of lowland species (Canada Goose and Pekin Duck). Hemoglobin oxygen affinity is often measured as a P_{50} value, the partial pressure of oxygen at which 50 percent of the hemoglobin in a sample is bound to oxygen. The P_{50} values for each species are shown as arrows in the x-axis and represent the partial pressures of oxygen at which 50 percent of the binding sites of hemoglobin are occupied by oxygen. The low P_{50} value and left-shifted curve for Bar-headed Goose indicates that its hemoglobin is better able to saturate at low partial pressures of oxygen, which is beneficial at high altitude where the partial pressure of oxygen is reduced. The inset depicts the four amino acid substitutions that distinguish the α- subunits of Bar-headed Goose hemoglobin from those of the lowland species (Natarajan et al. 2015). The increased hemoglobin affinity of Bar-headed Geese is due, at least in part, to the effects of these mutations. *Modified from Scott 2011.*

specialists. In addition to Bar-headed Geese, a number of other high-altitude species also possess hemoglobins with very high binding affinities for oxygen. An increase in hemoglobin oxygen affinity is beneficial at very high elevations because it helps ensure adequate oxygen loading at the lungs. To understand why this is the case, consider the relationship illustrated in figure 7.9, where it can easily be seen that at high oxygen partial pressures, the vast majority of hemoglobin is bound to oxygen, but the opposite is true at very low partial pressures of oxygen. The relationship between hemoglobin saturation and oxygen partial pressure is often referred to as the hemoglobin-oxygen dissociation curve, and this curve describes many of the functional properties of hemoglobin variants that are relevant for adapting to low-oxygen conditions. The oxygen affinity of a particular hemoglobin variant is often measured as a P_{50} value, which is simply the partial pressure of oxygen at which 50 percent of the hemoglobin in blood is bound to oxygen. Thus, when comparing two hemoglobin variants, the one with the lowest P_{50} value will have the highest oxygen affinity, because it is able to achieve 50 percent saturation at lower oxygen partial pressures. In fact, a variant with a low P_{50} values often has a higher percent saturation at most physiologically relevant oxygen partial pressures; its entire oxygen dissociation curve is shifted to the left. Because the primary challenge of high altitude is the reduced partial pressure of oxygen, a hemoglobin variant with a high oxygen affinity and a left-shifted oxygen dissociation curve would be beneficial to high-altitude specialists, because it would allow for greater hemoglobin oxygen saturation in face of the reduced oxygen partial pressure that occurs at high elevation. Consistent with this expectation, an increase in hemoglobin oxygen affinity is perhaps the most common evolutionary response to life at high altitude.

This pattern of repeated convergent evolution is perhaps best illustrated in Andean hummingbirds. Hummingbird diversity peaks in the Andes Mountains of South America, where several species occur at very high elevations, which is somewhat surprising, given that hummingbirds have exceedingly high oxygen demands. Upward of 150 species of hummingbirds are known to occur in the Andes. This impressive radiation is associated with several independent colonizations of elevations above 3,000 m, and subsequent recolonizations of lowland habitats by highland ancestors (McGuire et al. 2014, Projecto-Garcia et al. 2013). These elevational shifts are associated with repeated evolution of increased hemoglobin oxygen affinity in highland species. Remarkably, however, colonization of lowland habitats by highland ancestors is associated with evolved *reductions* in hemoglobin oxygen affinity (Projecto-Garcia et al. 2013). Why should reduced oxygen affinity evolve at low elevation? Again, examination of the oxygen dissociation curve in figure 7.9 holds the answer. Because a leftward shift in the oxygen dissociation curve means that oxygen will remain bound to hemoglobin at low partial pressures of oxygen, increasing hemoglobin's affinity for oxygen may reduce its ability to unload oxygen at the peripheral tissues. In working tissues, the partial pressure of oxygen is reduced through the consumption of oxygen in respiring cells. All else being equal, a hemoglobin variant with high oxygen affinity will necessarily unload less oxygen for a given reduction in the partial pressure of oxygen. Thus, at high elevations the challenge lies in simply ensuring that hemoglobin is able to saturate with oxygen at the lungs. Issues of oxygen delivery at the peripheral tissues are secondary if adequate arterial oxygen saturation cannot be maintained. Conversely, at low

elevations, pulmonary oxygen loading is not a challenge, so the adaptive premium shifts toward enhancing hemoglobin's ability to unload oxygen at working tissues. Remarkably, these adaptive shifts in hemoglobin structure and function have been linked to repeated genetic changes at just two amino acid residues in the beta chains of adult hemoglobin (Projecto-Garcia et al. 2013). Whether this remarkable level of convergent evolution is common in other high-altitude birds remains to be seen. A similar study of Andean ducks has revealed similar patterns of convergent evolution in hemoglobin function at high altitude, but in these cases, the genetic basis for the convergent trait is not conserved. Different mutations tended to underlie the enhancement of hemoglobin oxygen affinity in independent highland lineages (Natarajan et al. 2015). In other cases, colonization of high elevations is not associated with modification of hemoglobin function at all (Cheviron et al. 2014), suggesting that modification of hemoglobin oxygen affinity is not necessarily a requirement for avian life at high elevation.

DIGESTION AND NUTRITION
The Gut and the Digestive Process

Birds feed on a wide variety of food types. These range from nectar and pollen to other animals. Chapter 15 illustrates the diversity of beak morphologies associated with these diets. Because different food types contain widely different compositions of macronutrients (lipids, proteins, and carbohydrates), and because each of these macronutrients must be assimilated by different processes, the digestive system of birds is also highly variable in morphology and physiological function. To appreciate the diversity in avian digestive systems (fig. 7.10), it is useful to review briefly how vertebrates assimilate different food types. We use the word **assimilation** to mean the chemical breakdown of complex molecules and the transport of the smaller molecules that result from these enzymatic processes across the walls of the gut. The box on page 188 shows the mechanisms that birds use to assimilate macronutrients.

The gut is a (relatively) simple organ in which each compartment is specialized to perform different functions (box on page 188). The esophagus delivers food to the crop, an extensible sac where food can be stored temporarily and released into the stomach. Because birds do not have teeth, they swallow food mostly whole into the stomach. The bird's stomach has two parts. A glandular section called the proventriculus secretes hydrochloric acid and the enzyme pepsin. The function of this section is to initiate the digestion of protein. The proventriculus is followed by an organ called the gizzard. The gizzard is a muscular mill lined with

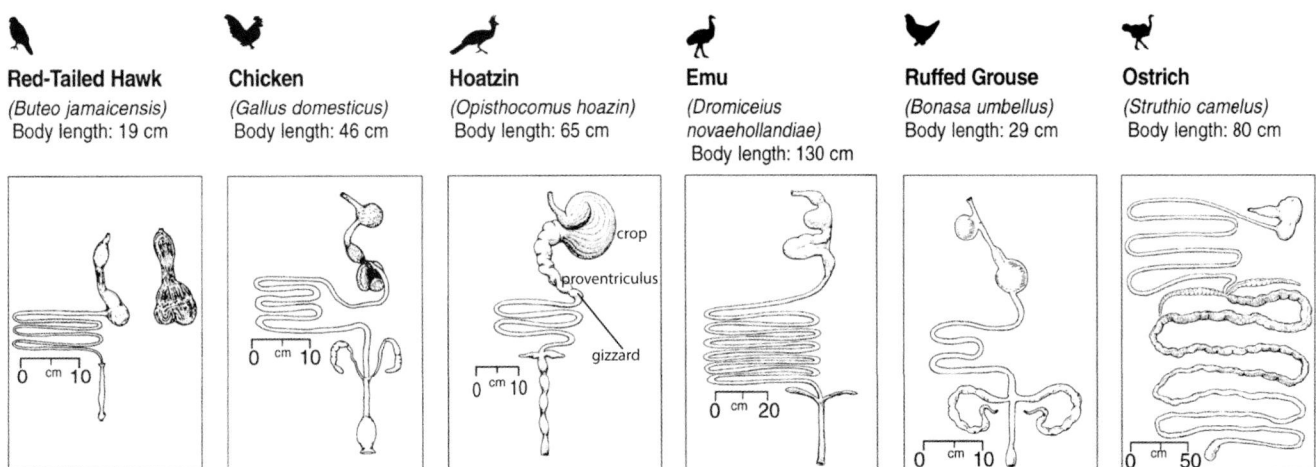

Red-Tailed Hawk
(*Buteo jamaicensis*)
Body length: 19 cm

Chicken
(*Gallus domesticus*)
Body length: 46 cm

Hoatzin
(*Opisthocomus hoazin*)
Body length: 65 cm

crop

proventriculus

gizzard

Emu
(*Dromiceius novaehollandiae*)
Body length: 130 cm

Ruffed Grouse
(*Bonasa umbellus*)
Body length: 29 cm

Ostrich
(*Struthio camelus*)
Body length: 80 cm

Figure 7.10. The morphology of the digestive system in birds varies among groups. For example, some bird groups lack cecae (hummingbirds and swifts, woodpeckers, and kingfishers). Others have very small ones (e.g., passerines, falcons, and storks). Some groups have large cecae (e.g., ostriches and tinamous, pheasants, quail, and grouse). The relative proportion of each segment of the gastrointestinal tract differs between species depending on diet. *After Stevens and Hume 1998. With permission from the American Physiological Society.*

a horny (keratinous) material (called koilin) that protects its inner walls and helps grind up food. The gizzard of birds appears to be as effective in reducing the size of food particles as the teeth of mammals (Fritz et al. 2011). The gizzard often contains small stones (called grit or gastroliths) that the birds ingest and that aid in the mechanical grinding of foods. Ducks and geese (Anseriformes), grain-eating birds such as quail (Galliformes), and many dove species (Columbidae) can confuse lead shotgun pellets and granular pesticides for grit and become poisoned (Gionfrido and Best 1996).

The gizzard varies in size and strength. It is large and very strong in granivores and herbivores and weaker in insectivores and carnivores (Piersma et al. 1993). Red Knots (*Calidris canutus*) feed on mollusks and use their muscular gizzards to crack the shell. When Theunis Piersma and his collaborators fed captive Red Knots on soft food pellets, their gizzards became smaller and they were reluctant to feed on hard-shelled molluscs (Piersma et al. 1993). Wild Red Knots have larger gizzards and feed on hard-shelled mollusks readily. This is an example of how a digestive feature of birds can shape feeding preferences. In another set of experiments, Matthias Starck (1999) fed Japanese Quail (*Coturnix japonica*) on diets with increased fiber content and measured the size of their gizzards using ultrasonography (much as physicians use ultrasound to evaluate a human fetus' growth). When they were fed on high-fiber diets, the birds ate more, and their gizzards became larger. When Starck shifted birds to low-fiber diets, their gizzards became smaller. The gizzard shows reversible changes in response to diet: it shows phenotypic flexibility (box on page 179).

Many bird species use the gizzard as a filter that retains undigested parts of prey such as bones, feathers, and insect exoskeletons. The gizzard molds the indigestible materials in a pellet that the bird regurgitates. Owls (Strigidae), hawks (Accipitridae), falcons (Falconidae), and bee-eaters (Meropidae) regurgitate pellets daily. Terns (Sternidae), herons (Ardeidae), and gulls (Laridae) regurgitate them less regularly (Errington 1930). Another interesting use of the gizzard is found in Phainopeplas (*Phainopepla nitens*), which feed on the berries of mistletoes and use their reduced gizzard to separate the berries' exocarps (peels) from the seed and pulp. Unlike seed-eating birds that grind up and digest seeds, Phainopeplas defecate long strings of sticky intact seeds that alternate with neatly folded packets of peels (Walsberg 1975; fig. 7.11). The seeds attach to branches, germinate, and start a new mistletoe infection.

Food exits the gizzard by the pylorus (box on page 188) and enters the small intestine. The small intestine is the site where most nutrients are digested and absorbed. The pancreas secretes its enzymes into the small intestine, and the cells that line the walls of the small intestine have digestive enzymes on their membranes. They also have transporters that carry small molecules from the lumen of the intestine into the bird's body. Enzymes secreted into the lumen of the intestine and those found at the surface of intestinal cells act in series. For example, if a bird eats a starchy seed, the enzyme amylase secreted by the pancreas breaks starch's long chain of glucose into molecules made of two or three glucoses. These smaller molecules are then broken down into glucose by enzymes bound to the membrane of intestinal

HOW BIRDS ASSIMILATE NUTRIENTS

The assimilation of nutrients in food can be divided into three steps: First, food must be processed mechanically to increase the surface area in relation to volume of particles. Second, large molecules must be broken down into smaller molecules by enzymes secreted into the lumen of the digestive tract by a variety of cells in the stomach and the pancreas. These smaller molecules must then be reduced in size further by enzymes bound to the membrane of intestinal cells. The third and last step is the absorption of these molecules across the wall of the gut. For example, if a bird eats an insect, it first grinds it in the gizzard. The pepsin secreted by the proventriculus begins breaking the insect's protein into smaller peptides. Digestion continues in the intestine as a result of protein-digesting enzymes secreted by the pancreas. The resulting smaller peptides are broken up further by enzymes bound to the membrane of intestinal cells.

The resulting amino acids and very small peptides are then transported into these cells. The exception to this general description of assimilation as enzymatic digestion followed by nutrient absorption are lipids and complex carbohydrates such as cellulose. Lipids are emulsified in the small intestine by bile salts secreted by the liver, digested by pancreatic enzymes called lipases, and the resulting molecules (single fatty acids and monoacylglycerides) diffuse into the cells of the intestine. Complex carbohydrates can be difficult to assimilate and are sometimes assimilated with the aid of microorganisms, and we discuss them in a later section. Undigested materials end up in the cloaca, where they are mixed with excreta. The gastrointestinal tract illustrated here in the figure is a schematic representation of that of a woodpecker. Woodpeckers lack cecae.

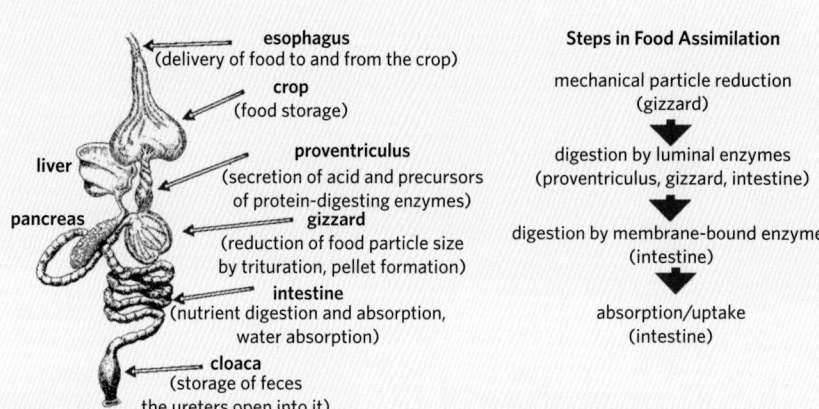

Birds assimilate the nutrients in food with a series of stepwise processes that take place in the gastrointestinal tract. First food must be physically processed to reduce the size of particles, then nutrients are broken by digestive anzymes, and finally, the smallest chemical components of nutrients are absorbed across the walls of the gut.

cells. Finally, intestinal cells transport glucose. As you would expect, birds that eat a lot of starchy seeds, such as House Finches (*Carpodacus mexicanus*) and House Sparrows (*Passer domesticus*) have higher levels of pancreatic and intestinal enzymes that break up starch than birds with low intakes of starchy foods, such as American Robins (*Turdus migratorius*) and Barn Swallows (*Hirundo rustica*; Kohl et al. 2011). Curiously, when Kevin Kohl and his collaborators (2011) measured pancreatic and intestinal enzymes that digest protein in these birds, they found no correlation with protein intake. They hypothesized that even animals with low protein intakes must maintain relatively high levels of protein-digesting enzymes to absorb adequate amounts.

Specialized nectar-feeding birds such as hummingbirds (Trochilidae) have very high levels of the intestinal and membrane-bound enzyme that breaks up sucrose in nectar into glucose and fructose (Schondube and Martinez del Rio 2004). These birds feed on nectar with very high sucrose content. Hummingbirds love to visit feeders that contain only a solution of table sugar (almost pure sucrose) and water. In contrast, starlings (Sturnidae), thrushes (Turdidae), and mockingbirds (Mimidae) seem to lack the ability to digest sucrose and get sick when they feed on sucrose-containing food (Martinez del Rio 1990). However, European Starlings (*Sturnus vulgaris*) and American Robins (*Turdus migratorius*) are notorious pests of plantations of blueberries and grapes. They can feed on these sweet fruits because their pulp contains glucose and fructose rather than sucrose.

Like the gizzard, the small intestine has remarkable phenotypic flexibility in both size and function. It grows and

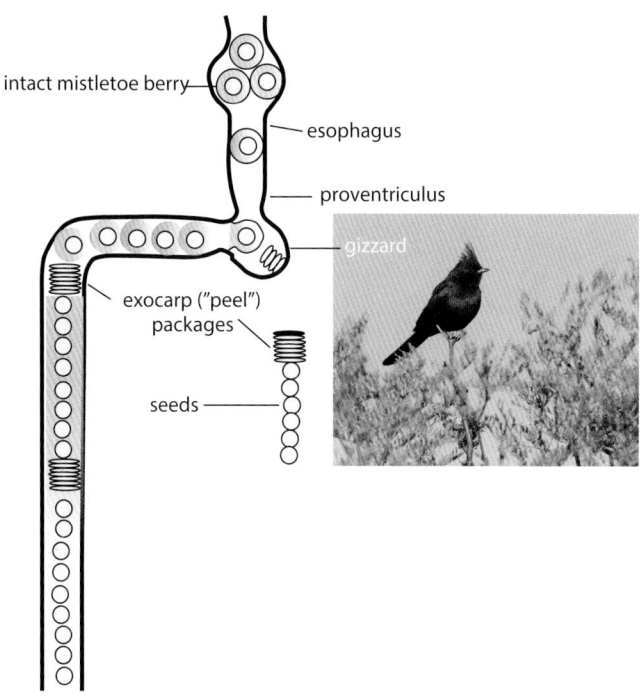

Figure 7.11. Phainopeplas (*Phainopepla nitens*) are fruit-eating birds that specialize in feeding on mistletoe berries. They use their gizzard to separate the seeds from the exocarps in order to assimilate the nutrients in the sticky pulp that surrounds the seeds. They defecate long strings of seeds that alternate with packages of peels/skins (technically called exocarps) neatly folded by the gizzard. You can sometimes find mounds of defecated mistletoe seeds under the perches favored by Phainopeplas. *After Walsberg 1975. Photo by Dave Kutilek used with permission.*

shrinks in response to environmental stimuli, and the enzymes in it change with diet (McWhorter et al. 2009). When birds are exposed to cold temperature, they must eat more to satisfy the increased energy costs of staying warm, and their gastrointestinal tracts grow. Scott McWilliams and his collaborators (1999) acclimated Cedar Waxwings (*Bombycilla cedrorum*) to either −21°C or 21°C. Cold-acclimated birds had to use more energy to thermoregulate and ate over 2.5 times more food. Their intestines were 22 percent heavier. Although the time that food stayed in the gut of cold-acclimated birds was shorter, the efficiency with which they extracted nutrients from it was only slightly (1.5 percent) lower than that of birds at 20°C (McWilliams et al. 1999). White-throated Sparrows acclimated to 21°C can increase their food intake by about 45 percent when temperature drops rapidly to −21°C. However this increase in intake in response to a quick change in energy demands is not sufficient to satisfy their energy costs, and they lose weight. If they are given enough time (50 days) to acclimate to −21°C, their guts increase in size, and they can then process enough

food to maintain constant body weight (McWilliams and Karasov 2014). The bird gut has some spare capacity that can be harnessed immediately, but if energy demands increase too much, this spare capacity is not sufficient and the birds lose weight. Given time, birds can increase the gut's capacity and process more food, but clearly the gut can only increase in size up to a point (fig. 7.12). White-throated Sparrows increased their food intake by about 126 percent over their intake at 21°C when acclimated to really cold temperatures (−21°C), but could not increase it more.

Migratory birds also have phenotypically flexible guts. In preparation for migration, birds increase food intake (they become **hyperphagic**) so that they can store sufficient fuel reserves for the flight. The gut sometimes increases in size during these pre-migratory hyperphagic episodes (McWilliams and Karasov 2001). Migratory songbirds arrive to stopover sites to recover body condition and accumulate energy for the next leg of their journey. Often, these birds feed at low rates, and hence recover body mass slowly during the first days after arrival, and then much more rapidly. Birds in the laboratory also have this two-step recovery after they have been fasted and then fed as much as they want to eat. Why don't they eat as much immediately after they arrive to stopover sites? Most birds fast during migratory flights;

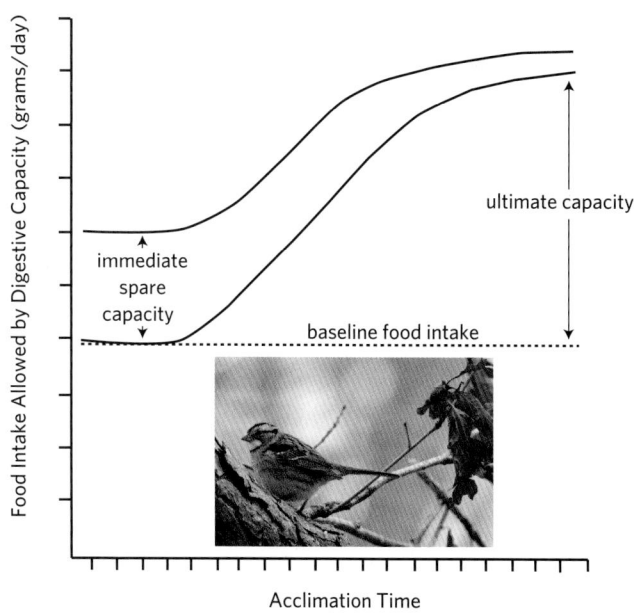

Figure 7.12. Schematic representation of the effect of the capacity to digest and absorb food (**digestive capacity**) and time of acclimation to conditions that demand higher intake such as increased energy losses in the cold. Birds appear to have modest instantaneous capacity to increase intake at first, but increase this capacity with exposure to the new conditions and to increased food intake. *After McWhorter et al. 2009. Photo by Shawn Billerman, used with permission.*

WILLIAM H. (BILL) KARASOV

Bill Karasov is a physiological ecologist that spends much of his time finding out how birds acquire resources such as food and water and how they cope with toxins and extreme environments.

As a junior at the University of Minnesota, Bill Karasov took a semester off to work at a kibbutz in Israel. Ironically, he was assigned to work at the chicken house. Little did he know that observing and doing research on birds would occupy a lot of his time during his future as a researcher. In the free time left from tending the chicken flock, he read Michael Evenari and colleagues' (1982) book on the ecology of the Negev desert, and he became fascinated by the questions that physiological ecologists ask: How is it that birds can survive the scorching temperatures of tropical deserts or the frigid ones of the tundra? How is it that they can meet their nutritional requirements on diets as different as grass and nectar? Although his doctoral research at the University of California Los Angeles

(UCLA) was on the adaptations of desert ground squirrels and lizards, he had been a bird-watcher since college and was soon hooked on birds as research organisms while working as a postdoctoral researcher in Jared Diamond's lab at UCLA. In his stint as a postdoc, Bill studied the digestive system of birds as well as that of members of all vertebrate classes. His work illustrates the great benefits of a broad comparative perspective, and of an integrative approach to research. Bill's work runs the gamut from working with molecules and cells in the laboratory to capturing and studying animals in the field. Because Bill's work is comparative and integrative, it is of interest to a broad swath of biologists. He has published over 200 papers, and a book, and has made remarkable discoveries (some of which are described in this chapter). As important as his scientific contributions are his efforts to train and mentor future generations of comparative biologists. Bill has trained many comparative physiologists over his career, both formally, as trainees in his laboratory (24 doctoral students, 16 masters students, 17 postdocs, and myriad undergraduates), and informally as one of the wise gurus of ecological physiology. When he was a graduate student, one of us (CMR) received a life-changing phone call from Bill. Bill invited Carlos to spend time in his laboratory in Madison, Wisconsin, to pursue common interests. He invited him and funded his research, although they had never met. Scientists can be competitive; we care about priority in discovery and of receiving credit for our findings. Bill Karasov's generosity as a mentor and adviser is unusual and, therefore, legendary.

exceptions are some swifts and swallows that feed on the wing. The guts of birds often decrease in size during long fasts and long migratory flights (see the box on page 190). Scott McWilliams and Bill Karasov (2001) suggest that, because birds lose gut tissue and therefore digestive function during migratory flights, they must rebuild the gut before they can take advantage of the food at stopover sites.

Herbivory in Birds

Herbivory is rare among birds, even though it is very common in mammals. We use the word herbivory to refer to the eating of plant parts (leaves, buds, and twigs). We exclude nectar, fruit, and seeds from this definition. Plant parts are difficult to process because cell walls contain materials such as cel-

lulose and lignin that are difficult to digest (nutritionists call these substances **refractory**). Although cellulose is one of the most abundant substances in terrestrial ecosystems, no vertebrate is able to produce the enzymes necessary to digest it. Herbivores, including birds, follow two strategies: they either eat and process large amounts of plant materials and assimilate only the cell contents (they skim the digestible "cream"), or they enlist the help of microbes capable of breaking up and fermenting cell wall materials. Small herbivorous birds such as the South American plantcutters (genus *Phytotoma*, Cotingidae), follow the former strategy. They use their sturdy serrated bills to "masticate" young leaves and buds. Plantcutters have short robust intestines with very high digestive enzyme levels (Meynard et al. 1999), but they have to be very selective and eat primarily nutrient-rich young plant parts. Another example of a bird that relies on cell contents of plant parts is the Kakapo (*Strigops habroptilus*). The Kakapo is found in New Zealand and is the largest parrot. Although they were once widespread and common, they are now endangered. Kakapos are nocturnal, flightless, and herbivorous. They pull foliage through their beaks with their feet while squeezing the nutritive juices out with their tongues. They leave small crescent-shaped fibrous bundles attached to the plant. The underside of the Kakapo's thick tongue has a hard and keratinized band that facilitates mashing foliage (Kirk et al. 1993).

Relying on microbes to assimilate the refractory materials in plant cell walls requires that the gut has a compartment that houses large microbial populations. The Hoatzin (*Opisthocomus hoazin*) is the only example of a bird species with a crop that has evolved into a fermentation vat (Grajal 1995; see the Hoatzin's enlarged crop in fig. 7.10). Hoatzins are the only known **foregut fermenters** among birds. Like cows, their fermentation vat is anterior to the stomach. Filipa Godoy-Vitorino and her collaborators (2012) characterized the microbial communities of Hoatzin crops (called **microbiomes**) and found a very diverse community of bacteria (many of which were completely new to science) with similarities between them and those of cows. Having the fermentation chamber before the stomach allows Hoatzins to digest the bacteria that they are associated with and benefit from bacterial protein. Many other herbivorous species house fermenting bacteria in paired structures called cecae. Cecae are sacs that open at the junction between the small and the large intestine. Herbivorous birds that use the cecae to house microbes are **hindgut fermenters**. Hummingbirds and swifts (Apodiformes) and parrots (Psittaciformes) lack cecae; passerines (Passeriformes) have tiny ones with unknown function; grouse (Tetraonidae), waterfowl (Anatidae), and ostriches (Struthionidae) have large ones. Herbivorous ducks and geese have larger gizzards and cecae than omnivorous and carnivorous species (Barnes and Thomas 1987). A. Starker Leopold (1953) measured the cecae of herbivorous and granivorous Galliformes and found that herbivorous grouse (Tetraonidae) had much larger cecae than seed-eating quail (Odontophoridae), pheasants (Phasianidae), and turkeys (*Meleagris gallopavo*). Avian cecae have diverse microbiomes that seem to help birds assimilate materials in cell walls (Waite and Taylor 2014), but birds in general appear not to be as good as mammals at digesting this abundant but hard-to-digest material (McWhorter et al. 2009). The explanation for the paucity of herbivorous birds with the capacity to assimilate refractory plant materials is likely that birds fly and cannot afford to carry a large chamber to house a big microbiome. The cecae of herbivorous birds might play a role in their nitrogen economy. The ureters are the conduits that deliver urine to the rectum. The uric acid in urine can then be refluxed up to the cecae and used by bacteria to synthesize protein that can then potentially be used by birds. The problem with this scenario is that bacterial protein must be digested. You might recall that protein digestion takes place first in the stomach and then in the small intestine. Many mammals solve this problem by eating their own feces. Although there are anecdotal observations of birds eating their own feces, the significance of these sparse observations is very unclear. The function of the cecae and of avian microbiomes remains an area where we still have much to learn.

OSMOREGULATION: COPING WITH SALT AND WATER

Bird Kidneys

Birds must maintain water balance. If they live in hot and dry environments, they must reduce losses. If they inhabit oceanic environments without freshwater, they must get rid of the salt in seawater and their food. In addition birds can lose astounding amounts of water through their respiratory system when they are engaged in active flight. The converse challenge is that birds that feed on nectar can sometimes ingest prodigious amounts of water and then must eliminate it. Birds meet these challenges with kidneys and other organs such as salt glands, and with the help of the lower end of the digestive system. This section is about how birds in different environments cope with water and salts. It begins with kidneys. The two bird kidneys are embedded in recesses on each side of the ventral side of the synsacrum. The ureters carry urine to the urodeum of the cloaca—there is no urinary bladder as in mammals (fig. 7.13). Unlike mammals, which void feces and urine through different orifices, birds void it into the cloaca, where it mixes with undigested material from the gastrointestinal tract. Therefore birds do not defecate, they excrete, and the appropriate way to refer to bird poop is excreta. Bird kidneys filter and reabsorb water and substances such as electrolytes and small molecules. They also secrete toxins. Kidney size depends on need and therefore on the habitats occupied by birds. Birds that live in marine environments must cope with salt and drink large amounts of water; so, their kidneys are larger (over 1 percent of body mass) than those of birds that live on land (under 1 percent of body mass; Hughes 1970). In contrast, birds that live in the desert have very small kidneys. The functional units of kidneys are called nephrons. These tiny structures filter, reabsorb, and secrete substances. Some birds have nephrons that lack a loop of Henle, located in the external layer of the kidney (the renal cortex). Because these nephrons resemble those of reptiles, they are called reptilian. Birds also have nephrons with a loop

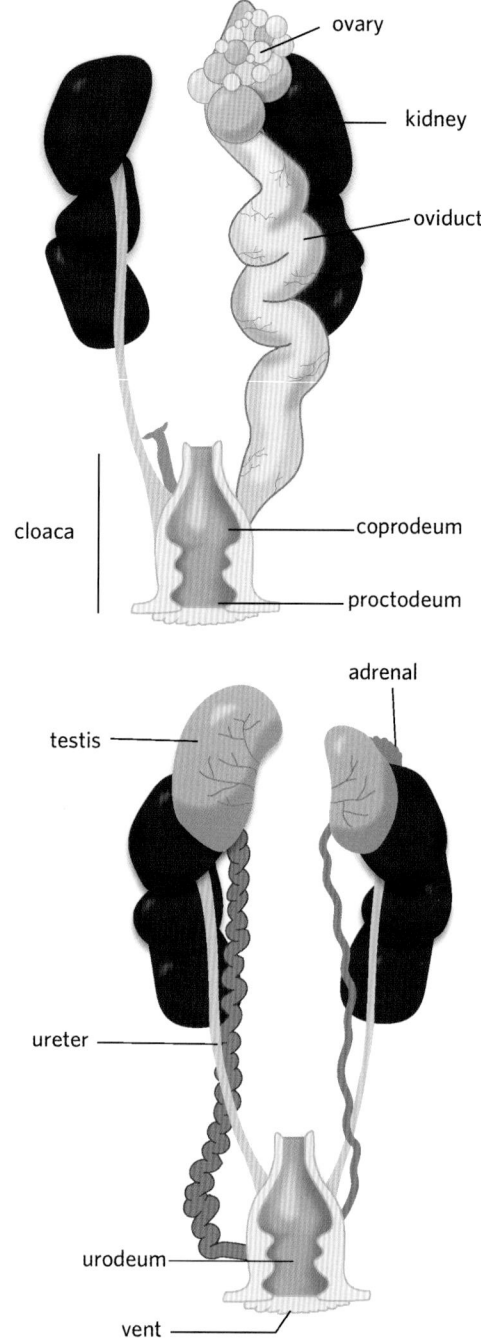

collecting ducts. Bird kidneys are composed of a multitude of lobules, each of which has a bundle of medulla surrounded by connective tissue called the renal cone and associated cortical tissue (fig. 7.14).

Ecological Correlates of Kidney Structure and Function

Only two groups of vertebrates are capable of making urine that is more concentrated than blood plasma: mammals and birds. We recommend Poulson's (1965) and Goldstein and Skadhauge's (2000) descriptions of the mechanisms that birds use to concentrate urine. It is sufficient to say that mammalian nephrons are responsible for the concentration of urine and that having more of them is associated with the production of more concentrated urine. Although birds can produce urine that is up to two to three times more concentrated than plasma, birds are not nearly as good urine concentrators as many mammals. Desert Hopping Mice (*Leggadina hermanburgensis*), for example, are champion urine concentrators and can produce urine that is 27 times more concentrated than plasma. There are no birds that can even come close. The

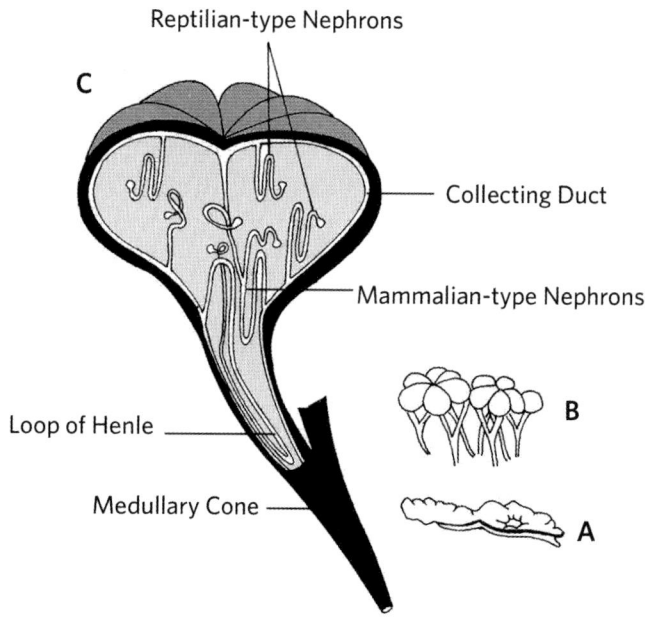

Figure 7.13. A schematic of the urogenital organs of birds. The kidneys are lodged within the synsacrum. The urine that they produce joins the rectum by the ureters at the urodeum. Note that the digestive, urinary, and reproductive systems converge in the cloaca.

of Henle, located in the kidney's interior. Because, in addition to birds, mammals are the only vertebrates with nephrons with loops of Henle, these nephrons are called mammalian. In addition to these two extreme nephron types, birds have an intermediate type, in which nephrons empty their filtered (and modified by reabsorption and secretion) contents into

Figure 7.14. Renal structure in bird kidneys. Avian kidneys have two types of nephrons, mammalian-type and reptilian-type (this is a slight simplification, as these nephron types are extremes in a continuum). Only mammalian-type nephrons have the ability to generate a significant osmotic gradient to produce a urine more concentrated than other body fluids, but the functional link between both nephron types through the collecting duct inside of the medullary cone determines that the ability to concentrate the urine is less efficient than in mammals. The whole kidney is shown in *A*, and two successive enlargements (*B* and *C*) are shown, each showing greater detail than the previous one. *Modified from Braun 1982.*

presence of longer cones (i.e., cones with nephrons with longer loops of Henle) in some birds enables them to retain more water and produce more concentrated urine. Therefore, the ability to produce concentrated urine (maximum urine concentrating ability), and cope with aridity and diets with high salt loads, is correlated with the relative size of the kidney, with the fraction of the kidney that is made of medullary tissue, and with the number of medullary cones (Goldstein and Braun 1989).

Giovanni Casotti and Eldon Braun (2000) compared the kidney morphology of Savannah Sparrows (*Passerculus sandwichensis*) that live in dry salt marshes without freshwater with those of House Sparrows (*Passer domesticus*, Passeridae) and Song Sparrows (*Melospiza melodia*, Emberizidae) that live in semiarid and mesic environments, respectively. They found that Savannah Sparrows were capable of producing more concentrated urine than the other two species and had kidneys with more mammalian-type nephrons and larger medullas than the other two species. This pattern seems to be repeated in other bird groups: Casotti and Richardson (1992) found that honeyeaters (Meliphagidae) from arid areas have kidneys with a higher percentage of tissue made of renal medulla than those from humid environments. McNabb (1969) found that desert Gambel's Quail (*Callipepla gambelii*) have kidneys with a larger proportion of renal medulla than Northern Bobwhite (*Colinus virginianus*) and California Quail (*Callipepla californica*), which occupy humid and mesic areas. Birds, such as hummingbirds, that must process large volumes of nectar and hence produce copious dilute urine, have kidneys in which 99 percent of nephrons are of the reptilian kind with almost no medullary tissue (Casotti et al. 1998). These birds are unable to produce concentrated urine and may face a difficult challenge. Like other birds, they absorb almost all the water they ingest, and the renal system has to eliminate it. When they are feeding, hummingbirds are diuretic. However, when they are not feeding, their high metabolic rates (and hence high rates of respiration) cause them to lose a lot of water by evaporation. Hummingbirds must get rid of a lot of water when they are feeding but conserve it when they are not. Unlike other birds, they cannot save water by making concentrated urine. Bradley Bakken and his collaborators (2004) found that Broad-tailed Hummingbirds (*Selasphorus platycercus*) solve this quandary by decreasing how much water gets reabsorbed by the kidney when they are feeding and by ceasing to filter water in the kidneys when they are not. Ceasing filtration in the kidneys in humans is called acute renal failure, which is a serious medical condition. Hummingbirds cope with it just fine.

Unlike seabirds (see following section), songbirds (Passeriformes) lack functional salt glands. Because birds are not very good at concentrating urine, the renal traits of this very large group of birds might limit their ability to use marine environments that impose a heavy salt load. Pablo Sabat and his collaborators (2006) studied the renal structure and the ability to concentrate urine in five species of ovenbirds (Furnariidae) in the genus *Cinclodes*. Because carbon molecules in food from marine sources have a higher content of ^{13}C, the heavy isotope of carbon, they used the content of this isotope to estimate how much individuals of each species fed on marine invertebrates relative to terrestrial ones (Sabat and Martinez del Rio 2002). They assumed that birds feeding on marine sources, with more ^{13}C in their tissues, would face higher loads of salt in their food. They found that birds with more ^{13}C in their tissues, and thus more marine invertebrates in their diets, had larger kidneys with a higher fraction of the kidney made of medulla and with a higher number of medullary cones per unit of kidney mass than closely related birds that fed on terrestrial invertebrates (fig. 7.15). One of the species, the Chilean Seaside Cinclodes (*Cinclodes nigrofumosus*), fed almost exclusively on seafood plucked from intertidal pools. Chilean Seaside Cinclodes are the most marine of all passerines. They have huge kidneys and are unusual among passerines in their capacity to deal with salt by producing concentrated urine. This ability allows them to live by the ocean in the Atacama Desert, one of the world's driest deserts, while feeding on marine invertebrates.

At this point you might wonder if kidneys are as phenotypically flexible and plastic as guts are. Several studies have shown that the composition of diet can affect kidney structure and function in birds (Ward et al. 1975a, 1975b, Singer 2003). This is because kidneys are not only used to deal with water and salt, but also to get rid of the end products of the breakdown of protein. Unlike mammals, which excrete primarily urea, birds excrete primarily uric acid. They also secrete urea and ammonia (Tsahar et al. 2005), but in most species uric acid is the primary form of nitrogenated excretion product in urine. Unlike urea and ammonia, which are soluble, uric acid is very water insoluble. It is secreted by nephrons in the cortex of the kidney. The urine of birds has uric acid both in solution and as precipitate (the precipitated uric acid is the white material that you find in the bird excreta splattered in the windshield of cars). The uric acid that precipitates in the cloaca of birds is in the form of tiny spheres (Braun 1982). These microspheres form a stable colloidal suspension in the liquid phase of urine. Thus, birds can conserve water by excreting a semisolid paste that has little water (Braun 1982). The microspheres are not only osmotically inactive, they also contain ions (such as Na^+) within them. Birds, as we have seen, have limited ability to produce concentrated urine. They can use these spheres to dispose

Figure 7.15. In the genus *Cinclodes* (Furnariidae), the relative size of the kidney, the number of medullary cones, and the ability to produce concentrated urine increase with the use of marine habitats as measured by the abundance of ^{13}C in tissues. The Chilean Seaside Cinclodes inhabits the intertidal areas on Central and Northern Chile, feeds on marine invertebrates primarily, and has the kidneys to cope with this very salty diet and a very dry environment. Like all passerines, this species lacks functional salt glands. *Photos (from top) by Pablo Sabat Kirkwood, Bob Lewis (https://www.flickr.com/photos/boblewis/), and Raúl Demoli. With permission.*

of ions. About 50 percent of the sodium excreted by European Starlings (*Sturnus vulgaris*) can be excreted associated with uric acid (Braun 1982). Birds that eat more protein must excrete more uric acid. Consequently, when birds are acclimated to diets with high protein content, the size of their kidneys increases, and the morphology of nephrons changes so that birds on high protein diets have kidneys with higher fractions of medullary tissue (Goldstein et al. 2001, Sabat et al. 2004). To evaluate whether the changes in kidney form and function observed in short-term experiments were also found in the renal characteristics of birds across species, Gonzalo Barceló and his collaborators (2012) measured the kidneys of 16 passerine species. They found the opposite of what they expected. The mass of the medullary portion of the kidney and the length of medullary cones were negatively correlated (rather than positively correlated, as expected) with the percentage of invertebrates in the birds' diet. They hypothesized that the renal medulla of grain-

eating birds is large because there is very little water in seeds. They argued that although getting rid of nitrogenated waste products is important for the design of the kidney, across species this need might be overridden by the need to conserve water.

Salt Glands

Shearwaters (Procellariidae), albatrosses (Diomedeidae), and petrels (Procellariidae and Hydrobatidae) spend most of their lives at sea. They feed offshore on marine organisms and drink seawater. They come to land only to breed. Their kidneys, like those of most birds, are not very good at making concentrated urine and getting rid of salt. They can drink seawater and eat marine salty prey because they have salt glands. Salt glands are located in depressions in or above the eye orbits and empty their briny secretion into the nasal cavity (fig. 7.16) where it drips out or birds shake it off. The fossil skulls of two birds from the Cretaceous, *Ichthyornis* (Ichtyor-

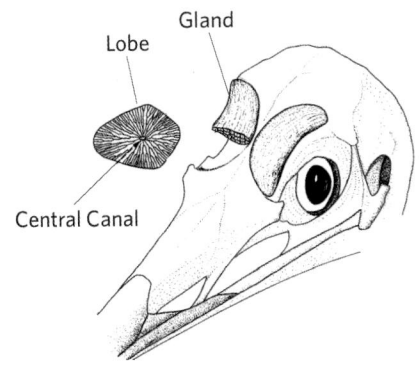

Figure 7.16. Many birds have salt glands, also called supraorbital glands, that can produce a concentrated salt solution. These glands are an extrarenal mechanism that allows many species to inhabit marine environments. *After Schmidt-Nielsen et al. 1958.*

niothidae) and *Hesperornis* (Hesperornithidae), show clear impressions of similarly positioned salt glands. These birds were most likely marine. Although the existence of salt glands in birds has been known for a very long time, their role in salt-riddance was discovered by Knuth Schmidt-Nielsen and his collaborators (1958) after they gave seawater to a Double-crested Cormorant (*Phalacrocorax auritus*) and, after a very short time, found drops of salty brine in the bird's nares.

Seabirds can drink seawater and eat seafood because the secretions of the salt gland can be more concentrated than seawater. Nevertheless, to get rid of salt, birds must drink a lot of water. Birds with salt glands have water fluxes (i.e., rates of water intake and loss) that are about twice as high as those of birds of the same size that lack them. The kidneys of birds that experience large salt loads must be able to get rid of absorbed water. Maryanne Hughes (1970) has shown that, for birds of similar sizes, birds with large salt glands have large kidneys as well. She found that marine birds that have larger kidneys and salt glands are capable of filtering more water. Jorge Gutierrez and his collaborators (2012) measured the size of salt glands in 29 species of shorebirds in the order Charadriiformes. They found that, for a given body mass, marine-dwelling species had larger salt glands than birds that live in terrestrial wetlands. They also found that Dunlins (*Calidris alpina*) living in marine environments had larger salt glands than those living inland in freshwater habitats. The salt gland is phenotypically flexible. Franklin Gulls (*Larus pipixcan*) breed in freshwater marshes in North America but spend the winter in the west coast of South America. They have larger salt glands in the winter than in the summer. When Mallards (*Anas platyrhynchos*) are acclimated to salty water, their salt glands quadruple in mass due to an increase in cell numbers and size. The cells in the glands increase the number of mitochondria and proteins that transport sodium in them (Hildebrandt 2001).

The Digestive System Aids in Osmoregulation

The last segment of the avian digestive system is the cloaca. The cloaca has three compartments. The first one, closest to the colon, is called the coprodeum; the second one is the urodeum; and the third one is the proctodeum, which ends in an opening called the vent (fig. 7.13). The ureters deliver urine into the urodeum. Before voiding urine, some birds can modify its composition and volume by refluxing it into the hindgut (the coprodeum, colon, and cecae). The hindgut of birds reabsorbs both salts and water and relies on bacteria to use uric acid. The Emu (*Dromaius novaehollandiae*), for example, produces dilute urine but has a big rectum with large folds capable of reabsorbing both salt and water. The capacity of the lower gut to save water allows Emus to live in semidesert areas in spite of having kidneys with relatively poor concentrating capacity. The salt that flows into the gut is reabsorbed, and Emus get rid of it using salt glands. The ability to recover salt depends on intake. When domestic chickens are fed on low-salt (NaCl) diets, their coprodeum and colon absorb sodium at a very high rate, presumably by increasing both the number of cells in the tissue and the number of sodium transporters in the membranes of the epithelia of these cells (Goldstein and Skadhauge 2000). When birds are fed on high salt diets, the recovery of salt is reduced. The hindgut differs in physiology and morphology among species. These differences are sometimes accompanied by differences in osmoregulation. For example, subspecies of Savannah Sparrows (*Passerculus sandwichensis*) inhabit salt marshes or upland environments. The most distal section of the large intestine (the rectum) of the salt-marsh subspecies (*P. s. beldingi*) has smoother apical surfaces and shorter microvilli than subspecies inhabiting upland areas with available freshwater (*P. s. anthinus*, *P. s. nevadensis*, and *P. s. brooksi*; Goldstein et al. 1990). These structural differences of the large intestine match the differences between chickens acclimated to either high or low intake of salts (Clauss et al. 1988).

The retrograde flow of urine into the large intestine represents a potential physiological complication for birds without salt glands. If urine is very concentrated, water may be drained from the plasma to the urine in the cloaca, counteracting the work of the kidney. Birds such as the House Sparrow (*Passer domesticus*) are capable of regulating the flow from the cloaca to the ileum when hydrated, but the flow is inhibited when birds are dehydrated and produce concentrated (hypertonic) urine (Goldstein and Braun 1988). The complementarity of the digestive and urinary systems in birds poses many interesting questions: Are many birds without salt glands capable of regulating the flow from the cloaca to the intestine? How many species can regulate the water permeability of the walls of the cloaca? These questions remain unanswered. They are some of the many questions that await answers by the bird physiologists of the future.

KEY POINTS

- Body size is a determinant of a bird's characteristics. It can be used to predict many traits, ranging from how much energy a bird uses to how dense the populations of a species are.
- Birds cope with cold temperatures by increasing heat production and by minimizing heat loss. They cope with hot temperatures by choosing cool microenvironments and by losing heat by evaporation of water.
- Many bird species are able to reduce their body temperatures to reduce the rate of heat loss in the cold, and many species are even capable of entering torpor. Only a single species, as far as we know, is capable of hibernation.
- Birds are very good at delivering oxygen to their tissues and thereby fueling their expensive mode of locomotion. Birds that live at the highest places on earth have very efficient lungs, hemoglobin with high affinity for oxygen, and extraordinary circulatory systems that include large hearts and dense capillary beds.
- The anatomy and physiology of bird digestive systems varies with diet. The characteristics of the gastrointestinal tract of birds are often phenotypically plastic and vary with energetic demands and diet.
- Bird kidneys have limited capacity to concentrate urine, at least compared with those of mammals. Marine bird species rely on supraorbital salt glands to get rid of the salt that they ingest with seawater and prey. The lower gastrointestinal tract of birds complements the function of the kidneys in the regulation of water and salt balance.

KEY MANAGEMENT AND CONSERVATION IMPLICATIONS

- Managers can use simple allometric equations to make predictions about avian traits of management and conservation interest.
- Habitat characteristics determine rates of heat gain and loss in birds. Thus, changes in bird habitat can have profound consequences for the rates at which animals spend energy.
- A warming earth can lead to changes in the composition of bird communities, especially in hot tropical and subtropical deserts.
- Although birds have remarkable phenotypic plasticity, this plasticity has limits, and thus rapid environmental

changes can have deleterious effects on the performance of individual birds and, through these, on their populations.

DISCUSSION QUESTIONS

1. How is it that body size can be used to predict things as disparate as the rate at which kidneys filter urine and how long birds live?
2. Birds are much better at coping with high elevations than mammals. Why?
3. Can physiological traits (i.e., the capacity to cope with high and low temperatures, to get rid of salt in food and water, and to feed on certain types of food) shape the broad ecological characteristics of bird species, including their use of certain habitats and their geographical distribution?
4. How is it that gut microbiomes shape and are shaped by the ecological characteristics of a bird species?
5. Individuals of a species can differ from each other in physiological characteristics. Also, species can differ from each other. Discuss the factors that influence the "expression" of a physiological characteristic in an individual and in a species.

References

Bakken, B. H., T. J. McWhorter, E. Tsahar, and C. Martínez del Rio (2004). Hummingbirds arrest their kidneys at night: Diel variation in glomerular filtration rate in *Selasphorus platycercus*. Journal of Experimental Biology 207:4383–4391.

Barceló, G., J. Salinas, and P. Sabat (2012). Body mass, phylogeny and diet composition affects kidney morphology in passerine birds. Journal of Morphology 273:842–849.

Barnes, G. G., and V. G. Thomas (1987). Digestive organ morphology, diet, and guild structure of North American Anatidae. Canadian Journal of Zoology 65:1812–1817.

Bartholomew, G. (1981). A matter of size: An examination of endothermy in insects and terrestrial vertebrates. *In* Insect thermoregulation, B. Heinrich, Editor. Wiley, New York, pp. 45–78.

Bicudo, J. E. P. W. (1996). Physiological correlates of daily torpor in hummingbirds. *In* Animals and temperature: Phenotypic and evolutionary adaptation I, A. Johnston and A. F. Bennett, Editors. Cambridge University Press, Cambridge, UK, pp. 293–311.

Bicudo, J. E. P. W., C. R. Vianna, and J. G. Chaui-Berlinck (2001). Thermogenesis in birds. Bioscience Reports 21:181–188.

Boag, P. T. (1987). Effects of nestlings' diet on growth and adult size of Zebra Finches (*Poephila guttata*). The Auk 104:155–166.

Braun, E. J. (1982). Renal function. Comparative Biochemistry and Physiology 71A:511–517.

Calder, W. A. (1968). Respiratory and heart rates of birds at rest. The Condor 70:358–365.

Calder, W. A. (1983). Size, function, and life history. Harvard University Press, Cambridge, MA.

Carpenter, F. L. (1974). Torpor in an Andean hummingbird: Its ecological significance. Science 183:545–547.

Casotti, G., and K. Richardson (1992). A stereological analysis of kidney structure of honeyeater birds (Meliphagidae) inhabiting either arid or wet environments. Journal of Anatomy 180:281–288.

Casotti, G., C. A. Beuchat, and E. J. Braun (1998). Morphology of the kidney in a nectarivorous bird, the Anna's hummingbirds Calypte anna. Journal of Zoology 244:175–184.

Casotti, G., and E. J. Braun (2000). Renal anatomy in sparrows from different environments. Journal of Morphology 243:283–291.

Cheviron, Z. A., C. Natarajan, J. Projecto-Garcia, D. K. Eddy, J. Jones, M. D. Carling, C. C. Witt, et al. (2014). Integrating evolutionary and functional tests of adaptive hypotheses: A case study of altitudinal differentiation in hemoglobin function in an Andean sparrow, Zonotrichia capensis. Molecular Biology and Evolution 31:2948–2962.

Clarke, A., and P. Rothery (2008). Scaling of body temperature in mammals and birds. Functional Ecology 22:58–67.

Clauss, W., V. Dantzler, and E. Skadhauge (1988). A low-salt diet facilitates Cl secretions in hen lower intestine. Journal of Membrane Biology 102:83–96.

Errington, P. L. (1930). The pellet analysis method of raptor food study habits. The Condor 32:292–296.

Evenari, M., L. Shanan, and N. Tadmor (1982). The challenge of a desert. Harvard University Press, Cambridge, MA.

Faraci, F. M. (1991). Adaptations to hypoxia in birds: How to fly high. Annual Review of Physiology 53:59–70.

Frayn, K. N. (1983). Calculation of substrate oxidation rates in vivo from gaseous exchange. Journal of Applied Physiology 55:628–634.

Fritz, J., E. Kienzle, J. Hummel, O. Wings, W. J. Streich, and M. Clauss (2011). Gizzards vs. teeth: It's a tie: Food processing efficiency in herbivorous birds and mammals and implications for dinosaur feeding strategies. Paleobiology 37:577–586.

Gerum, R. C., B. Fabry, C. Metzner, M. Baulieu, A. Ancel, and D. P. Zitterbart (2013). The origin of traveling waves in an emperor penguin huddle. New Journal of Physics 15. doi:10.108 8/1367 2630/15/12/125022.

Gilbert, C., S. Blanc, Y. Le Maho, and A. Ancel (2008). Energy saving processes in huddling emperor penguins: From experiments to theory. Journal of Experimental Biology 211:1–8.

Gilbert, C., D. McCafferty, Y. Le Maho, J. Martrette, S. Giroud, S. Blanc, and A. Ancel (2010). One for all and all for one: The energetic benefits of huddling in endotherms. Biological Reviews 85:545–569.

Gionfrido, J. P., and L. B. Best (1996). Grit-use in North American birds: The influence of diet, body size, and gender. Wilson Bulletin 108:685–696.

Godoy-Vitorino, F., K. C. Goldfarb, U. Karaoz, S. Leal, M. A. Garcia-Amado, P. Hugenholtz, S. G. Tringe, E. L. Brodie, and M. G. Dominguez-Bello (2012). Comparative analyses of foregut and hindgut bacterial communities in Hoatzins and cows. ISME Journal 6:531–541.

Goldstein, D. L., and E. J. Braun (1988). Contributions of the kidneys and intestines to water conservation, and plasma levels of antidiuretic hormone, during dehydration in House Sparrows (Passer domesticus). Journal of Comparative Physiology B 158:353–361.

Goldstein, D. L., and E. J. Braun (1989). Structure and concentrating ability in the avian kidney. American Journal of Physiology 256:R501–R509.

Goldstein, D. L., J. B. Williams, and E. J. Braun (1990). Osmoregulation in the field by saltmarsh Savannah sparrows Passerculus sandwichensis beldingi. Physiological Zoology 63:669–682.

Goldstein, D. L., and E. Skadhauge (2000). Renal and extrarenal regulation of body fluid composition. In Sturkie's avian physiology, G. C. Whittow, Editor. Academic Press, San Diego, pp. 265–297.

Goldstein, D. L., L. Guntle, and C. Flaugher (2001). Renal response to dietary protein in the house sparrow Passer domesticus. Physiological and Biochemical Zoology 74:461–467.

Grajal, A. (1995). Structure and function of the digestive tract of the hoatzin (Opisthocomus hoazin): A folivorous bird with foregut fermentation. The Auk 112:20–28.

Grubb, B. R. (1983). Allometric relations of cardiovascular function in birds. American Journal of Physiology 245:H567–H572.

Gutiérrez, J. S., M. W. Dietz, J. A. Masero, R. E Gill Jr., A. Dekinga, P. F. Battley, J. M. Sánchez-Guzmán, and T. Piersma (2012). Functional ecology of salt glands in shorebirds: Flexible responses to variable environmental conditions. Functional Ecology 26:236–244.

Hainsworth, F. R., and L. L. Wolf (1978). Regulation of metabolism during torpor in "temperate" zone hummingbirds. The Auk 95:197–199.

Healy, K., T. Guillerme, S. Finlay, A. Kane, S. B. A. Kelly, D. McLean, D. Kelly, I. Donohue, A. L. Jackson, and N. Cooper (2014). Ecology and mode-of-life explain lifespan variation in birds and mammals. Proceedings of the Royal Society of London B 281. doi:10.1098/rspb.2014.0298.

Hildebrandt, J. P. (2001). Coping with excess salt: Adaptive functions of extrarenal osmoregulatory organs in vertebrates. Zoology 104:209–220.

Hudson, L. N., N. J. B. Isaac, and D. C. Reuman (2013). The relationship between body mass and field metabolic rate among individual birds and mammals. Journal of Animal Ecology 82:1009–1020.

Hughes, M. R. (1970). Relative kidney size in nonpasserine birds with functional salt glands. The Condor 72:164–168.

Jenkins, F. A., K. P. Dial, and G. E. Goslow (1988). A cineradiographic analysis of bird flight: The wishbone in starlings is a spring. Science 241:1495–1498.

Kahl, M. P. (1963). Thermoregulation in the Wood Stork, with special reference to the role of the legs. Physiological Zoology 36:141–151.

Kirk, E. J., R. G. Powlesland, and S. G. Cork (1993). Anatomy of the mandibles, tongue, and alimentary tract of Kakapo with some comparative information from Kea and Kaka. Notornis 40:55–63.

Kohl, K. D., P. Brzek, E. Caviedes-Vidal, and W. H. Karasov (2011). Pancreatic and intestinal carbohydrases are matched to dietary starch level in wild passerine birds. Physiological and Biochemical Zoology 84:195–203.

Körtner, G., R. M. Brigham, and F. Geiser (2001). Torpor in free-ranging tawny frogmouths (*Podargus strigoides*). Physiological and Biochemical Zoology 74:789–797.

Laybourne, R. (1974). Collision between a vulture and an aircraft at an altitude of 37000 feet. Wilson Bulletin 86:461–462.

Leopold, A. S. (1953). Intestinal morphology of gallinaceous birds in relation to food habits. Journal of Wildlife Management 17:197–203.

Lewden, A., M. Petit, M. Malbergue, S. Orio, and F. Vezina (2014). Evidence of facultative daytime hypothermia in a small passerine wintering at northern latitudes. Ibis 156:321–329.

Martinez del Rio, C. (1990). Dietary, phylogenetic, and ecological correlate of intestinal sucrase and maltase activity in birds. Physiological Zoology 63:987–1011.

Mathieu-Costello, O. (1991). Morphometric analysis of capillary geometry in pigeon pectoralis muscle. American Journal of Anatomy 191:74–84.

McGuire, J. A., C. C. Witt, J. V. Remsen Jr., A. Corl, D. L. Rabosky, D. L. Altshuler, and R. Dudley (2014). Molecular phylogenetics and the diversification of hummingbirds (Apodiformes: Trochilidae). Current Biology 24:910–916.

McKechnie, A. E., and B. G. Lovegrove (2002). Avian facultative hypothermic responses: A review. The Condor 104:705–724.

McKechnie, A. E., and B. O. Wolf (2004). The allometry of metabolic rate: Good predictions need good data. Physiological and Biochemical Zoology 77:502–521.

McKechnie, A. E., and B. O. Wolf (2010). Climate change increases the likelihood of catastrophic avian mortality events during extreme heat waves. Biology Letters 6:253–256.

McNabb, F. M. A. (1969). A comparative study of water balance in three species of quail II: Utilization of saline drinking solutions. Comparative Biochemistry and Physiology 28:1059–1074.

McWhorter, T. J., E. Caviedes-Vidal, and W. H. Karasov (2009). The integration of digestion and osmoregulation in the avian gut. Biological Reviews 84:533–565.

McWilliams, S., E. Caviedes-Vidal, and W. H. Karasov (1999). Digestive adjustments in Cedar Waxwings to high feeding rate. Journal of Experimental Biology 283:394–407.

McWilliams, S. R., and W. H. Karasov (2001). Phenotypic flexibility in digestive systems structure and function in migratory birds and its ecological significance. Comparative Biochemistry and Physiology A 128:577–591.

McWilliams, S. E., and W. H. Karasov (2014). Spare capacity and phenotypic flexibility of a migratory bird: Defining the limits of animal design. Proceedings of the Royal Society of London B 281:20140308.

Meynard, C., M. V. Lopez-Calleja, F. Bozinoic, and P. Sabat (1999). Digestive enzymes of a small avian gerbivore, the rufous-tailed plantcutter. The Condor 101:904–907.

Natarajan, C., J. Projecto-Garcia, H. Moriyama, R. E. Weber, V. Munoz-Fuentes, A. J. Green, C. Kopuchian, et al. (2015). Convergent evolution of hemoglobin function in high-altitude Andean waterfowl involves limited parallelism at the molecular sequence level. PLoS Genetics 11:e1005681.

Nee, S., A. F. Read, J. J. D. Greenwood, and P. H. Harvey (1991). The relationship between abundance and body size in British birds. Nature 351:312–313.

Ottaviani, D., S. C. Cairns, M. Oliverio, and L. Boitani (2006). Body mass as a predictive variable of home-range size among Italian mammals and birds. Journal of Zoology 269:317–330.

Palmqvist, P., and S. F. Vizcaíno (2003). Ecological and reproductive constraints of body size in the gigantic *Argentavis magnificens* (Aves: Teratornithidae) from the Miocene of Argentina. Ameghiniana 40:379–385.

Piersma, T., A. Koolhas, and A. Dekinga (1993). Interaction between stomach structure and diet choice in shorebirds. The Auk 110:552–564.

Piersma, T., and J. Drent (2003). Phenotypic flexibility and the evolution of organismal design. Trends in Ecology and Evolution 18:228–233.

Poulson, T. L. (1965). Countercurrent multipliers in avian kidneys. Science 148:389–391.

Projecto-Garcia, J., C. Natarajan, H. Moriyama, R. E. Weber, A. Fago, Z. A. Cheviron, R. Dudley, J. A. McGuire, C. C. Witt, and J. F. Storz (2013). Repeated elevational transitions in hemoglobin function during the evolution of Andean hummingbirds. Proceedings of the National Academy of Sciences USA 110:20669–20674.

Rosenmann, M., and P. Morrison (1974). Maximum oxygen consumption and heat loss facilitation in small homeotherms by He-O2. American Journal of Physiology 226:490–495.

Ruf, T., and F. Geiser (2015). Daily torpor and hibernation in birds and mammals. Biological Reviews 90:891–926.

Sabat, P., and C. Martínez del Rio (2002). Inter- and intraspecific variation in the use of marine food resources by three *Cinclodes* (Furnariidae, Aves) species: Carbon isotopes and osmoregulatory physiology. Zoology 105:247–256.

Sabat, P., E. Sepúlveda-Kattan, and K. Maldonado (2004). Physiological and biochemical responses to dietary protein in the omnivore passerine *Zonotrichia capensis* (Emberizidae). Comparative Biochemistry and Physiology A 137:391–396.

Sabat P., K. Maldonado, M. Canals, and C. Martínez Del Río (2006). Osmoregulation and adaptive radiation in the ovenbird genus *Cinclodes* (Passeriformes: Furnariidae). Functional Ecology 20:799–805.

Schmidt-Nielsen, K., C. B. Jörgensen, and H. Osaki (1958). Extra-renal salt excretion in birds. American Journal of Physiology 193:101–107.

Scholander, P. F., R. Hock, V. Walters, F. Johnson, and L. Irving (1950). Heat regulation in some arctic and tropical mammals and birds. Biological Bulletin 99:237–258.

Schondube, J. E., and C. Martinez del Rio (2004). Sugar and protein digestion in flowerpiercers and hummingbirds: A comparative test of adaptive convergence. Journal of Comparative Physiology B 174:263–273.

Scott, G. R. (2011). Elevated performance: The unique physiology of birds that fly at high altitudes. Journal of Experimental Biology 214:2455–2462.

Scott, G. R., L. A. Hawkes, P. B. Frappell, P. J. Butler, C. M. Bishop, and W. K. Milsom (2015). How Bar-headed Geese fly over the Himalayas. Physiology 30:107–115.

Silva, M., J. H. Brown, and J. H. Downing (1997). Differences in population density and energy use between birds and mammals: A macroecological approach. Journal of Animal Ecology 66:327–340.

Singer, M. A. (2003). Do mammals, birds, reptiles and fish have similar nitrogen conserving systems? Comparative Biochemistry and Physiology B 134:543–558.

Stager, M., D. L. Swanson, and Z. A. Cheviron (2015a). Regulatory mechanisms of seasonal metabolic flexibility in the Dark-eyed Junco (*Junco hyemalis*). Journal of Experimental Biology 218:767–777.

Stager, M., H. P. Pollock, P. M. Benham, N. D. Sly, J. D. Brawn, and Z. A. Cheviron (2015b). Disentangling environmental drivers of metabolic flexibility in birds: The importance of temperature extremes vs. temperature variability. Ecography 39:787–795.

Starck, J. M. (1999). Phenotypic flexibility of the avian gizzard: Rapid, reversible, and repeated changes of organ size in response to changes in dietary fiber content. Journal of Experimental Biology 202:3171–3179.

Steiger, S. S., J. P. Kelly, W. W. Cochran, and M. Wikelski (2009). Low metabolism and inactive lifestyle of a tropical rain forest bird investigated via heart-rate telemetry. Physiological and Biochemical Zoology 82:580–589.

Stevens, C. E., and I. D. Hume (1998). Contribution of microbes in vertebrate gastrointestinal tract to production and conservation of nutrients. Physiological Reviews 78:393–427.

Swan, L. W. (1970). Goose of the Himalayas. Natural History 79:68–75.

Swanson, D. L. (2010). Seasonal metabolic variation in birds: Functional and mechanistic correlates. Current Ornithology 17:75–129.

Swanson, D. L., and K. L. Olmstead (1999). Evidence for a proximate influence of winter temperature on metabolism in passerine birds. Physiological and Biochemical Zoology 72:566–575.

Tsahar, E., C. Martínez Del Río, and A. Zeev (2005). Can birds be ammoniotelic? Nitrogen balance and excretion in two frugivores. Journal of Experimental Biology 208:1025–1034.

Waite, D. W., and M. W. Taylor (2014). Characterizing the avian gut microbiota: Membership, driving influences, and potential function. Frontiers in Microbiology 5. doi:10.3389/fmicb.2014.00223.

Walsberg, G. E. (1975). Digestive adaptations of *Phainopepla nitens* associated with the eating of mistletoe berries. The Condor 77:169–174.

Ward, J. M., R. A. McNabb, and F. M. A. McNabb (1975a). Effects of changes in dietary protein and water availability on urinary nitrogen compounds in rooster, *Gallus domesticus* I: Urine flow and excretion of uric-acid and ammonia. Comparative Biochemistry and Physiology A 51:165–169.

Ward, J. M., R. A. McNabb, and F. M. A. McNabb (1975b). The effects of changes in dietary protein and water availability on urinary nitrogen compounds in the rooster, *Gallus domesticus* II: Diurnal patterns in urine flowrates, and urinary uric acid and ammonia concentrations. Comparative Biochemistry and Physiology A 51:171–174.

West, J. B. (2009). Comparative physiology of the pulmonary blood-gas barrier: The unique avian solution. American Journal of Physiology: Regulatory, Integrative and Comparative Physiology 297:R1625–R1634.

Whitfield, M. C., B. Smit, A. E. McKechnie, and B. O. Wolf (2015). Avian thermoregulation in the heat: Scaling of heat tolerance and evaporative cooling capacity in three southern African arid-zone passerines. Journal of Experimental Biology 218:1705–1714.

Wojciechowski, M. S., M. Jefimow, and B. Pinshow (2011). Heterothermy, and the energetic consequences of huddling in small migrating passerine birds. Integrative and Comparative Biology. doi:10.1093/icb/icr055.

Wolf, B. O., K. M. Wooden, and G. E. Walsberg (2000). Effects of complex radiative and convective environments on the thermal biology of the White-crowned Sparrow (*Zonotrichia leucophrys gambelii*). Journal of Experimental Biology 203:803–811.

Woods, C. P., and R. M. Brigham (2004). The avian enigma: "Hibernation" by Common Poorwills (*Phalaenoptilus nuttallii*). *In* Life in the cold: Evolution, mechanisms, and applications, B. M. Barnes and H. V. Carey, Editors. Biological Papers of the University of Alaska 27:231–240.

CHAPTER 8

Endocrinology

Thomas P. Hahn, Jamie M. Cornelius, and George E. Bentley

It is early April in central California, and you are standing near a thicket of chaparral shrubs, marveling at the din of singing coming from within. Dozens of Gambel's White-crowned Sparrows (*Zonotrichia leucophrys gambelii*) and Golden-crowned Sparrows (*Zonotrichia atricapilla*)—which have been in the area since the previous September—are gathered here, feeding determinedly on the ground around the edges of the thicket and singing from the cover. Two weeks later you happen by the same thicket and stop to enjoy the serenade, but . . . not a sound. They've gone. To find them now, you would need to head north, perhaps catching up with them along their migration route or at their breeding grounds in western Canada and Alaska. How did they know when to leave? How did they prepare for the trip? How did they manage to arrive at their breeding ground physiologically ready to breed?

A couple of weeks later, you head to the high Sierra Nevada to begin your seasonal studies of Mountain White-crowned Sparrows (*Z. l. oriantha*). They have just returned from their wintering grounds in Mexico and are preparing to breed in the subalpine meadows along the Sierra crest. It has been a year of heavy winter snow in the high Sierra. You snowshoe in to Tioga Pass Meadow, at 3,000 m elevation, on a sunny May morning, and over the course of the day you observe males testing territory boundaries in willows and small pines peeking above the snow. You capture several sparrows in seed-baited traps; some wear colored leg bands from a previous year's research, and others are new birds that you band for the first time. Later that afternoon, clouds loom in the west, and by evening it is snowing and blowing, covering the birds' food and presenting thermoregulatory challenges to small birds and humans alike. You hightail it down into the Mono Basin, 1,000 m lower in elevation, to escape the winter-like conditions and ponder from your sleeping bag that night how the little birds you left at the crest will fare in the storm.

In the morning, the storm has passed, but there in your campsite, out in the shrub-steppe of the Mono Basin, is a flock of Mountain White-crowned Sparrows, feeding quietly. Several are color-banded birds that you saw or trapped at Tioga Pass the previous morning. How, you ask yourself, do these birds decide whether and when to leave the high country for this low-elevation refuge from the storm? And how will they decide when to return to the Sierra crest to reclaim their breeding territories?

Later in the season, about half the snow has melted from Tioga Pass, exposing nesting sites in the willows and small pines. Male Mountain White-crowned Sparrows sing from conspicuous perches and are attentive to their mates, who are building nests, laying eggs, and beginning incubation. The males are not interested in building nests or incubating; the females are, for the most part, not interested in singing. In the nests you are monitoring, one egg appears each day, for up to five eggs—a huge investment in energy and nutrients by the female, including a massive need for calcium as each eggshell is formed. What, you wonder, determines the distinct behaviors and physiological capabilities of the different sexes of these birds? What causes the males to produce those highly stereotyped songs? And how do the females coordinate the massive investment in eggs?

The endocrine system is the key coordinator of these and many other fascinating activities that are critical to the survival and reproductive success of birds. The avian endocrine system resembles that of other vertebrates in terms of anatomy, mechanisms of action, and functional roles in regulation of growth, development, physiology, morphology, and behavior. The endocrine system works with the central and autonomic nervous systems to coordinate the activities of the bird, including both internal processes and responses to the external environment. There is no clear-cut functional separation between the nervous and the endocrine systems;

the neuroendocrine hypothalamus forms a direct link between neural processing and hormonal regulation, and the adrenal medulla, which secretes catecholamines (mostly epinephrine) into the general circulation is a neuroendocrine extension of the sympathetic branch of the autonomic nervous system. However, in broad terms, whereas the nervous system tends to deal with issues requiring rapid or very specific responses, the neuroendocrine/endocrine system tends to orchestrate broadscale, long-term, homeostatic, and allostatic (maintaining stability through changing, see below) processes. Because of the important roles hormones play in the regulation of such a wide variety of processes, neuroendocrine/endocrine physiology touches on every aspect of birds' lives. Further, endocrine mechanisms can either facilitate or constrain adjustment to particular environmental circumstances, including those related to human-induced rapid environmental change, and may influence the evolutionary paths of adaptation available to populations. Some compounds in the environment also have direct "endocrine disrupting" effects on the physiology and behavior of birds. This chapter examines endocrine function in the context of how birds regulate both stasis (stability) and change in physiology, morphology, and behavior in the naturally changing world they occupy, and discusses implications for conservation in modern times. To understand and appreciate how the endocrine system serves the needs of birds in their natural environments, it is first necessary to summarize general principles of endocrine function and to place the endocrine system into the context of general physiological processes such as homeostasis.

PRINCIPLES AND DETAILS OF THE AVIAN ENDOCRINE SYSTEM

Cues, Signals, and Control Processes

Environmental and Physiological Cues

All living things exist in environments, the conditions of which affect their ability to function. Environmental conditions are seldom constant; organisms are faced with change in time and space. The ability to adjust appropriately to these changes—in some cases to respond directly to the changes as they occur, and in others to anticipate them—is critical to survival and reproductive success. Further, the process of coping with the environment involves challenges to maintenance of the internal conditions of organisms. In most cases, organisms have been selected to possess mechanisms that resist dramatic deviations from particular "set points" in internal conditions such as body temperature, osmotic status, and energy balance. This **homeostatic regulation** (cf. Cannon 1932) of internal conditions depends on the organism sensing

and responding to **cues** from both the external and internal environments (fig. 8.1). Cues are attributes of the environment (external or internal) that provide information about relevant conditions, and that organisms detect and respond to with changes in physiology, morphology, and behavior. Examples of external environmental cues include ambient temperature, food supply, and photoperiod. Examples of internal environmental cues include body temperature, levels of metabolic fuels in the circulation, and osmolality of the blood or extracellular fluid.

Environmental and Physiological Signals

Information can also be transferred to or within the organism via **signals** (fig. 8.1). Signals differ from cues in that there is a **sender** and a **receiver**, and natural and sexual selection can affect both, leading to the evolution of a **signaling system**. The physical environment does not "send" signals, but external environmental stimuli do come from con- and heterospecific individuals, and these stimuli may constitute signals. For instance, conspecific individuals (e.g., a male songbird) may produce an acoustic stimulus (e.g., song). The nature of both the stimulus and the response of another individual (e.g., a female) may affect the fitness of both the sender and receiver, and thus selection may lead to the evolution of changes in both the stimulus and the response to it. Likewise, within the organism, stimuli originating in one part of the body can carry information to other parts of the body, and both the nature of the stimulus and the nature of the response influence the functional effectiveness (fitness) of the organism as a whole. Sending, receiving, interpreting, and orchestrating responses to signals are fundamental roles of nervous (see also chapter 14) and endocrine systems. Further, these systems are well-suited both to coordinating changes in internal function via signals and to transducing both cues and signals from the external environment into internal signals. Interestingly, both nervous and endocrine systems share many common signaling pathways at the cellular and subcellular levels, involving both electrical and chemical messages. A major difference between the two systems is the form taken by long-distance signals: in the nervous system, signals are propagated electrically over long distances, with only the transmission across the synapse typically being chemical (see chapter 13), whereas in the endocrine system, long-distance transmission is entirely chemical. Another difference is that whereas the nervous system achieves target specificity largely through point-to-point electrical and electrochemical connections, specificity within the endocrine system is achieved primarily through the distribution of receptors specifically tuned to particular signaling molecules.

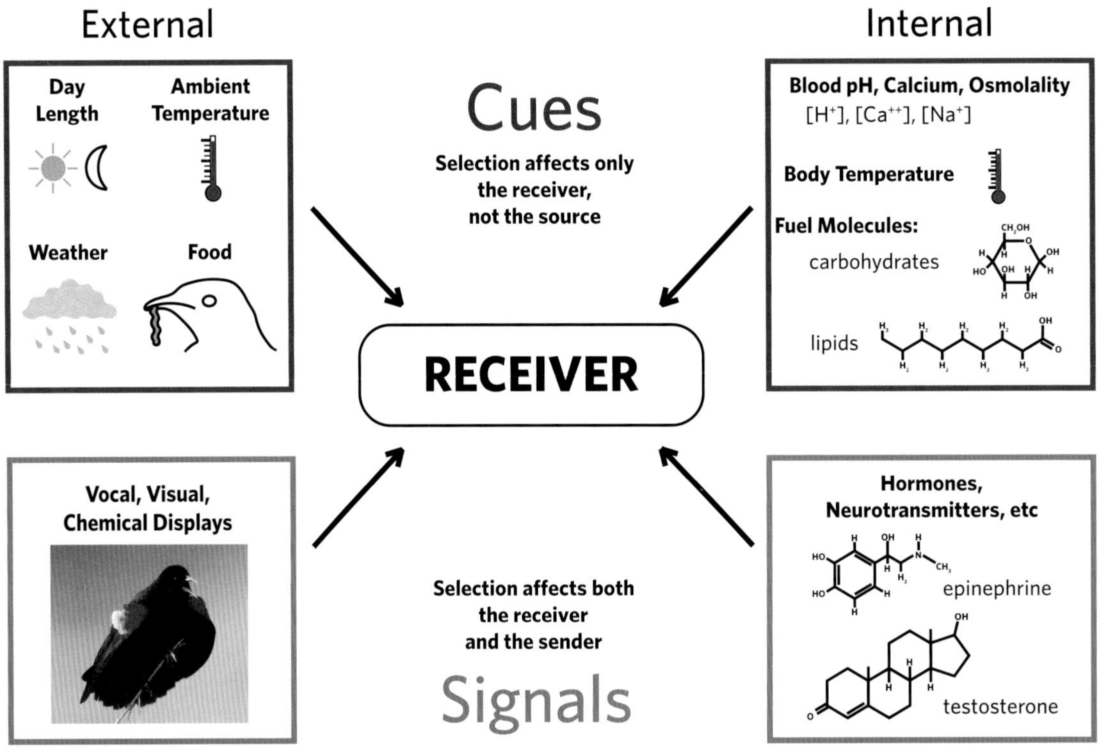

External

Day Length Ambient Temperature

Weather Food

Cues

Selection affects only
the receiver,
not the source

Internal

Blood pH, Calcium, Osmolality
[H⁺], [Ca⁺⁺], [Na⁺]

Body Temperature

Fuel Molecules:

carbohydrates

lipids

RECEIVER

Vocal, Visual,
Chemical Displays

Selection affects both
the receiver
and the sender

Signals

Hormones,
Neurotransmitters, etc

epinephrine

testosterone

Figure 8.1. Information passes from the environment to individuals (*left*), and within individuals (*right*), via cues (top, red boxes) and signals (bottom, green boxes). There is always a "receiver" for both cues and signals. This receiver may be an individual bird, or an organ, tissue, cell, or even molecule within a bird. Cues are effectively "public information" that is present for reasons unrelated to the receiver. The receiver may possess adaptations allowing it to detect and respond to the cue, but the source of the cue is not responsive to selection. Signals are components of evolved communication systems, in which there is a sender and a receiver, both of which are under natural and/or sexual selection. For inter-individual signaling (e.g., birds responding to songs of conspecifics), selection affects both signaler and receiver. For intra-individual signaling (e.g., hormone-receptor systems), the functional consequences to the individual (individual fitness) are affected by both the signal and the response, and both can be modified in populations through selection.

Modes of Regulation, Integration, and Mechanisms of Action

Endocrine regulation involves several components, including top-down and feedback pathways; transport in the blood, which may or may not involve transport molecules to which the hormones are bound; and, finally, actions at target cells involving receptors, transporters, metabolizing enzymes, and so forth. Each of these will be touched on below.

Top-Down Regulation Endocrine regulation takes two closely interconnected forms: top-down regulation and feedback regulation (fig. 8.2). Top-down regulation occurs when a particular endocrine organ or tissue initiates an **endocrine cascade**, sending a signaling molecule to target tissues that respond with either further endocrine responses or changes in other aspects of physiological function, or both. For instance, the neuroendocrine hypothalamus secretes releasing and inhibiting hormones to exert top-down endocrine control over the anterior pituitary, which in turn secretes various peptide hormones that exert top-down control over numerous other tissues' secretion of other hormones (e.g., thyroid hormones, gonadal steroid hormones; see below) and other processes (e.g., production of gametes, changes in membrane permeability, etc.).

Feedback Regulation: Positive and Negative Feedback regulation occurs when a product "downstream" in an endocrine cascade influences the activity of one or more of the higher level (top-down) controlling steps in the pathway (fig. 8.2). Feedback most commonly is negative, leading to stabilization of hormone secretion around some targeted level (a homeostatic process, see below), but in a few notable cases may be positive, leading to destabilization (usually temporary). A good example of negative feedback is the homeostatic stabilization of blood calcium levels via balancing of top-down control and negative feedback, leading to changes in calcium release from and deposition to stores, via changes in calcitonin, parathyroid hormone, and vitamin D3. A good example of positive feedback is the temporary switch from negative to positive feedback of sex steroid hormones (especially progesterone in birds) on the hypothalamo-pituitary (GnRH-gonadotropin) system to produce a preovulatory surge of LH (rather than the decline that negative feedback would have induced), and thence ovulation.

Integration: Interactions among Hormones and Hormone Pathways Hormones along particular top-down/feedback pathways do not function in isolation. Activity along one pathway may be affected by hormones of another pathway. For instance, activity at different levels along the hypothalamo-pituitary-gonad pathway (see below) may be modulated by inputs from adrenal steroids. Cross-talk between pathways

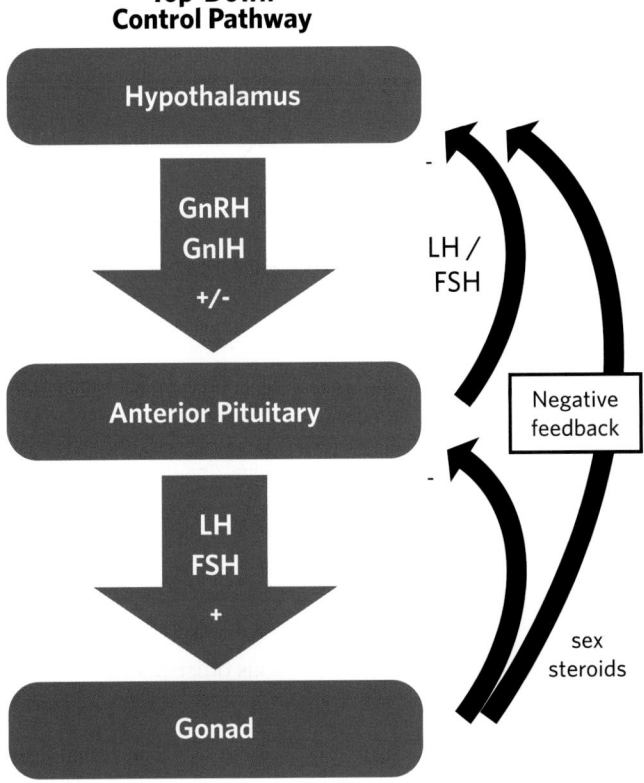

Figure 8.2. Top-down and feedback signaling. External and internal cues and signals impinge on the hypothalamus, changing top-down output of releasing and inhibiting hormones such as gonadotropin-releasing hormone, GnRH, and gonadotropin-inhibitory hormone, GnIH, which regulate a downstream cascade, including pituitary output of gonadotropins (such as luteinizing hormone, LH, and follicle-stimulating hormone, FSH), that controls gamete production and hormone output (mainly sex steroids) of gonads. Negative feedback serves to reduce drive from top-down control points, leading to stabilization of circulating levels of hormones in the cascade. Feedback can occasionally be positive, leading to temporary destabilization of hormone levels. For instance, rising LH during an ovulatory cycle leads to increasing progesterone (note: different than in mammals, in which estradiol responds to LH in this way), which further enhances LH secretion, leading to ovulation.

contributes to the coordinated integration of multiple complex processes (see below).

Actions at the Target Cell Hormones can regulate processes in a variety of ways that differ mechanistically and in their time courses. These can be grouped broadly into two categories: (1) modifications of the activity of "existing machinery," such as extant ion channels, enzymes, and compounds standing ready for secretion; and (2) production of new proteins (hormones, enzymes, channels, receptors, etc.) via gene transcription, message translation, and posttranslational processing (fig. 8.3). The specific effects hormones have at target tissues are dictated by which, if any, receptors are present, and in some cases by which intracellular signal transduction systems they are connected to. The effects hormones have can also be affected by changes in blood-borne carrier proteins such as binding globulins, by enzymatic conversions to different metabolites at target tissues, and by deactivation or clearance. A key attribute of endocrine signaling pathways is **amplification**, whereby a low-amplitude signal (perhaps just a few molecules of the hormone) leads to a greatly enhanced secondary or tertiary signal, and a large physiological response (fig. 8.3).

Components of the Avian Endocrine System
Chemical Messaging

Categories of Messages: Endocrine to Intracrine The diverse array of chemicals that acts as physiological messengers can influence their targets at any distance from their point of origin (fig. 8.4). Chemicals secreted into the blood and reaching distant targets qualify as **neuroendocrine** secretions if they are produced and secreted by neurons, and as **endocrine** secretions if they are produced by any other type of cell. **Paracrine** secretions affect other cells in the immediate vicinity of the source cell, **autocrine** secretions affect the same individual cell that produces and releases the chemical, and **intracrine** factors affect the cell in which they are produced through intracellular mechanisms without ever leaving the cell. Specific examples of each of these forms of chemical signaling are indicated in the figure legend for figure 8.4. All these forms of chemical communication share common mechanisms of action, in that the response at the target cell requires receptors that recognize and bind the chemical messenger molecule. Particular signaling molecules may act through two or more of these types of mechanisms. For instance, norepinephrine can be a neurotransmitter (paracrine) if released neuron-to-neuron, or a hormone (endocrine) if released from the adrenal medulla into the general circulation. Likewise, estradiol can act as an intracrine signal if produced by conversion from testosterone within a target cell, or as an endocrine signal if released into the blood from a cell that produces and releases it.

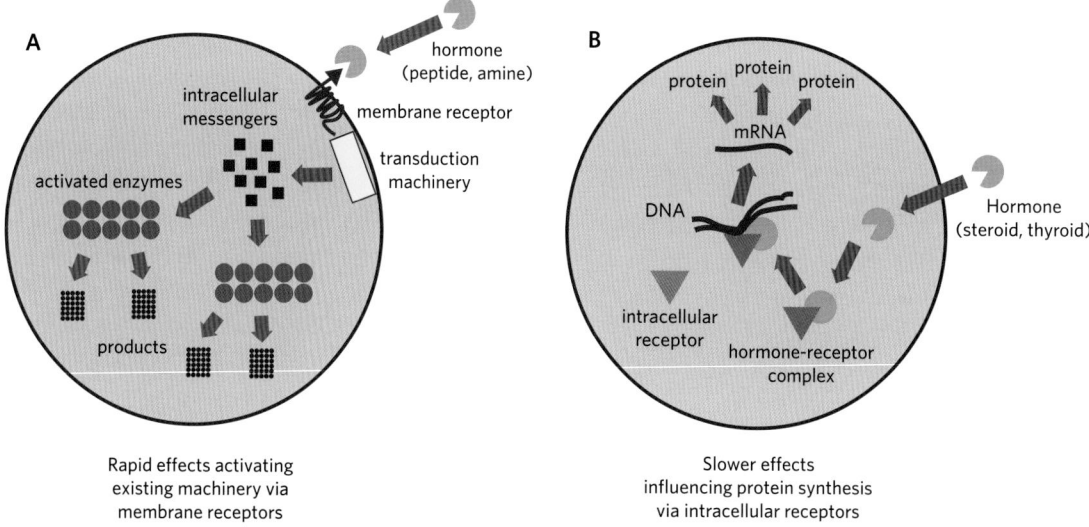

Rapid effects activating
existing machinery via
membrane receptors

Slower effects
influencing protein synthesis
via intracellular receptors

Figure 8.3. Schematic and simplified illustration of two main mechanisms of action of hormones, including amplification. A, Rapid effects of peptide or amine hormone activation of existing machinery via a membrane receptor and an intracellular messenger cascade. These lipid-insoluble hormones cannot easily cross the cell plasma membrane themselves, so their signal enters the cell via the hormone binding to a membrane receptor, and this hormone-receptor complex activates existing transduction machinery associated with the inner surface of the membrane. This activated transduction machinery amplifies the signal by producing or releasing numerous intracellular messenger molecules (e.g., cAMP, IP3, Ca++, *black squares*), which then activate an existing set of enzyme molecules (often via phosphorylation, *green circles*), which then further amplify the signal by each enzyme catalyzing production of numerous product molecules (*small dark blue boxes*).

Red arrows identify steps where amplification of the signal occurs. This process occurs rapidly, because no gene transcription or message translation is required; existing machinery is simply activated; B, Relatively slow effects of steroid or thyroid hormone activating gene transcription via an intracellular receptor. These lipid-soluble hormones cross the membrane into the cytoplasm, where they can interact with the intracellular receptors. Amplification (*red arrows*) occurs via each hormone-receptor complex interacting with the DNA to cause transcription of multiple mRNA copies, which are then translated into multiple copies of the gene product. Note that steroid hormones are now known also to interact with membrane receptors in some circumstances, and to have rapid actions via existing machinery through this mechanism. However, influencing gene transcription is still considered the primary mechanism of action for steroids and thyroid hormones.

Classes of Chemical Messengers: Structures and Mechanisms of Action Hormones can be grouped broadly into two categories based on how water or lipid soluble (hydrophilic or lipophilic) they are, and mechanisms of action. Relatively hydrophilic hormones such as peptides and amines readily reside in the aqueous solution of the blood and the extracellular fluid, and typically activate membrane-bound receptors and thereby influence activity of intracellular messenger cascades within target cells. That is, they activate existing machinery (see fig. 8.3). Lipophilic hormones such as steroids and thyroid hormones primarily pass through cell membranes to activate intracellular receptors, creating hormone-receptor complexes inside the cell that act as transcription factors and induce mRNA production and protein synthesis. However, recent evidence indicates that steroids may sometimes induce their effects relatively rapidly via membrane receptors, and this less familiar pathway may be more widespread than was formerly realized. For example, corticosterone has long-term effects on lipid and protein metabolism that enhance

ready energy availability through classic "genomic" mechanisms of action, but can also have very rapid effects on behavior (within minutes of an increase of the hormone in the blood) that probably involve a membrane receptor pathway (Breuner et al. 1998, Breuner and Wingfield 2000). The **affinity** of receptors for hormones (the strength of binding between hormone and receptor) plays an important role in the physiological effects of hormones. When affinity is high, a hormone may saturate the available receptor binding sites while hormone levels are very low, leading to strong cellular and physiological responses at very low hormone levels. In contrast, when affinity is lower, hormone levels may need to become quite high before substantial cellular and physiological effects occur. Variation in affinity between receptors and hormones, neurotransmitters, and other chemicals is important both to natural physiological responses to hormones and to effects of **endocrine disrupting chemicals** (discussed further below; see also the box on page 216).

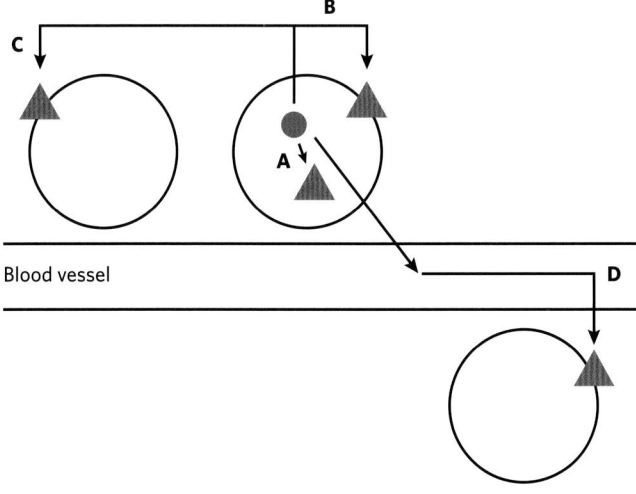

Figure 8.4. Chemical signaling can occur over different spatial scales. *A*, Intracrine signaling occurs when a chemical messenger (*blue dot*) produced within a cell interacts with a receptor (*orange triangle*) within that same cell. An example would be production of estradiol from testosterone within a cell, and the estradiol then interacting with an estrogen receptor within that same cell; *B*, Autocrine signaling occurs when a chemical messenger produced within a cell leaves the cell but then interacts with a receptor on that same cell. This type of signaling is common in immune cells; *C*, Paracrine signaling occurs when a chemical messenger produced within a cell diffuses locally to interact with receptors on or in an adjacent cell, without being transported in the blood. An example would be when testosterone produced within Leydig cells of the testis interacts with Sertoli cells nearby to influence spermatogenesis without ever entering the blood stream. Note that chemical neurotransmission, when chemical messengers from the presynaptic cell diffuse across the tiny space of the synaptic cleft to interact with receptors on the postsynaptic cell membrane, is a specialized form of paracrine signaling; *D*, Classical endocrine and neuroendocrine signaling occur when the chemical messenger from a cell passes out of the cell, through the extracellular space into the blood stream, is transported to a distant site, where it leaves the bloodstream and interacts with a receptor on or in a target cell there. A neuroendocrine example is GnRH produced in hypothalamic cells being released into the hypothalamo-pituitary-portal blood vessels and then interacting with receptors on gonadotroph cells in the anterior pituitary. An endocrine example is LH produced in the anterior pituitary being released into the general circulation and interacting with receptors on Leydig cells in the testis or granulosa cells in the ovary.

Gross Anatomy of the Avian Endocrine System

Virtually every tissue in the vertebrate body produces and responds to chemical signaling molecules; the classical view that there are a few specific "endocrine glands" that produce all known hormones clearly no longer holds true (Schmidt et al. 2008; see the box on page 206). Figure 8.5 shows schematically the whole-body distribution of various endocrine

tissues, and table 8.1 provides an abbreviated list of major hormones and their sources, their chemical types, and major targets and effects. Note that many hormones have numerous targets and many effects. The list of included compounds is by no means exhaustive. In recent years, the number of different molecules known or suspected to play roles in chemical signaling has been increasing dramatically. Readers interested in a more comprehensive discussion of the array of compounds acting as chemical messengers in birds should refer to Scanes (2015a).

It is important to keep in mind that for many hormones, little or nothing is known about their distribution or functions in any birds other than domestic fowl. It is wise not to assume that the information from chickens applies across different groups of birds (Scanes 2015a), but the information from studies of domestic fowl can be used as a starting point for exploring endocrine regulation mechanisms in other groups. Some hormones, such as leptin—known for many years and studied extensively in mammals—have remained enigmatic in birds until very recently.

A central component of all vertebrate endocrine systems is the **hypothalamo-pituitary unit** (fig. 8.6). Top-down signals from the hypothalamus influence many aspects of behavior, physiology, and morphology. They integrate cues and signals from the external environment, from the internal environment via feedback loops, and from endogenous clocks (see below). The neurosecretions from the hypothalamus fall into two main groups: (1) those being released directly into the general circulation at the **posterior pituitary**, and (2) those being released at the **median eminence** into the

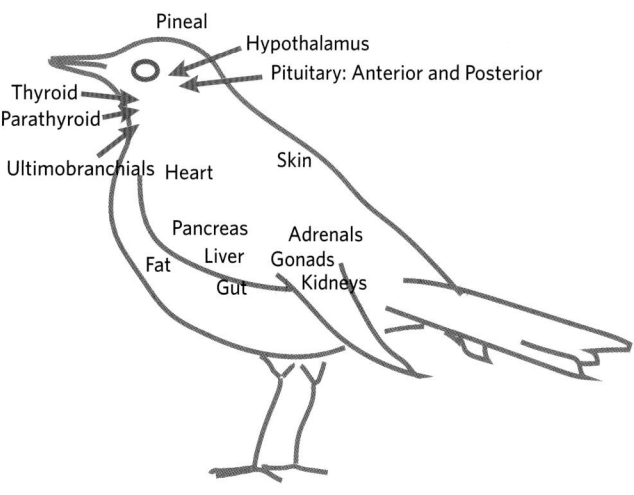

Figure 8.5. A schematic illustration of sites in the avian body acting as sources of hormones. The primary endocrine organs are distributed throughout the body. Virtually all tissues of the bird produce and respond to one or more hormones.

SOURCES OF HORMONES

Historically, particular hormones were generally considered to come from one or perhaps a couple of sources among a suite of a few "endocrine glands" or "ductless glands." Over the years, it became evident that at least some signaling molecules could be produced in multiple locations, such as in both the gut and the brain. It has become increasingly clear that many hormones have multiple sites of origin, and that these production sites are by no means restricted to a few endocrine glands. The recent evidence for production of "neurosteroids" within target tissues in the brain is particularly striking. The aromatase enzyme that catalyzes the conversion of testosterone (T) to estradiol (E2) has long been known to be present in multiple brain regions. The possibility that E2 can be produced from other non-testosterone androgens such as DHEA (dehydroepiandrosterone, an androgen produced in the adrenal gland) if other enzymes are also present, or even that sex steroids might be produced "de novo" (from the original precursor cholesterol) within the brain, now has some support as well. There is likewise evidence for local production and paracrine functions of other classically "endocrine" factors, such as TSH production within the hypothalamus to regulate enzyme activities in glial cells and thereby local supplies of thyroid hormones. In general, it is prudent to think of the endocrine system as representing a complex collection of sources of and targets for signaling molecules, with many of the molecules having multiple sources and diverse functions at multiple target tissues.

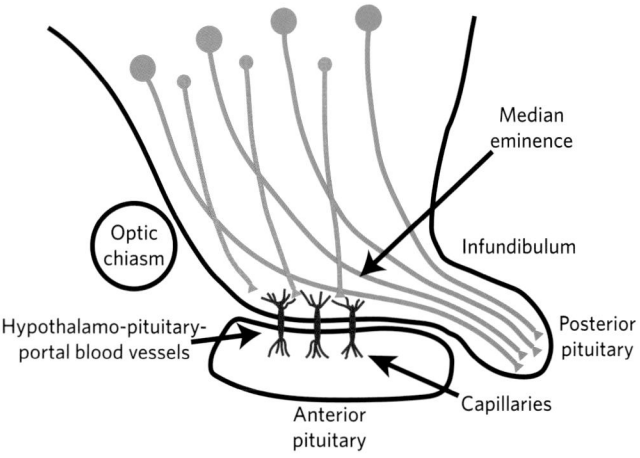

Figure 8.6. The hypothalamo-pituitary unit of birds. Various peptides produced in the hypothalamus are transported down axons to terminals either in the posterior pituitary (a neural component of the pituitary), from which they are released directly into the general circulation, or to the median eminence, from which they enter the hypothalamo-pituitary-portal blood vessels for private transport directly to the anterior pituitary. Compounds released at the posterior pituitary are primarily the peptides arginine vasotocin (AVT) and mesotocin (MT) in birds. A range of releasing and inhibiting hormones are released at the median eminence.

localized **hypothalamo-pituitary-portal vessels** that connect directly to the **anterior pituitary** without being diluted into the systemic blood supply. The neurohormones secreted at the posterior pituitary are primarily the two nonapeptides (nona = nine; these have nine amino acids) arginine vasotocin and mesotocin, which are produced in cell bodies in the hypothalamus and then transported down axons to release terminals at the posterior pituitary. The compounds released at the median eminence are a suite of releasing and inhibiting hormones that target specific groups of cells within the anterior pituitary and regulate the production and release of the systemic hormones produced by those cells. The pituitary cells include those that produce growth hormone (also known as somatotropin, the somatotrophs), prolactin (the lactotrophs), luteinizing hormone (LH) or follicle-stimulating

hormone (FSH, the gonadotrophs), adrenocorticotropic hormone (ACTH, the corticotrophs, which cleave ACTH from proopiomelanocortin, POMC—the precursor for several other peptide hormones), or TSH (the thyrotrophs). Growth hormone and prolactin induce responses, including secretion of a variety of other signaling molecules such as various growth factors (e.g., insulin-like growth factors), at a range of sites throughout the body. For the other hormones, the combination of the hypothalamus with these specific groups of anterior pituitary cells and the peripheral targets of the anterior pituitary hormones form distinct "axes," known as the **hypothalamo-pituitary-gonad (HPG) axis, hypothalamo-pituitary-adrenal (HPA) axis, and hypothalamo-pituitary-thyroid (HPT) axis**.

ENDOCRINE REGULATION AND COORDINATION OF PHYSIOLOGY, MORPHOLOGY, AND BEHAVIOR

General Issues: Stasis and Change

Birds must maintain control of behavior, physiology, and morphology throughout their lives in the face of variable environmental conditions and life stages. At any given time, this may involve maintenance of a relatively constant status quo (homeostasis), modest to major changes in anatomy and physiology (growth and development of young birds, transi-

Table 8.1. Avian hormones listed by source, molecular type, and targets.

HORMONE	ABBREV	TYPE	Target	Effects
Hypothalamic Hormones				
Gonadotropin-releasing Hormone	GnRH	Neuropeptide	Pituitary gonadotroph cells	Increase LH and FSH production and release
Gonadotropin-inhibitory Hormone	GnIH	Neuropeptide	Pituitary gonadotroph cells	Reduce LH and FSH production and release
Corticotropin Releasing Hormone	CRH	Neuropeptide	Pituitary corticotroph cells	Increase ACTH production and release
Thyrotropin Releasing Hormone	TRH	Neuropeptide	Pituitary thyrotroph cells	Increase TSH production and release
Growth Hormone Releasing Hormone	GHRH	Neuropeptide	Pituitary somatotroph cells	Increase GH production and release
Somatostatin	SS	Neuropeptide	Pituitary somatotroph cells	Decrease GH production and release
Arginine Vasotocin	AVT	Neuropeptide	Kidney, smooth muscle, various	Social behavior, water balance
Mesotocin	MT	Neuropeptide	Brain, smooth muscle, kidney	Affiliative behavior (esp. females), song learning, water balance
Anterior Pituitary Hormones				
Luteinizing Hormone	LH	Glycoprotein	Granulosa cells, Leydig cells	Increase T production, ovulation, spermiation
Follicle-stimulating Hormone	FSH	Glycoprotein	Theca cells, Sertoli cells	Follicular development, spermatogenesis
Thyroid Stimulating Hormone	TSH	Glycoprotein	Thyroid gland	T3/T4 production and release
Growth Hormone	GH / STH	Polypeptide	Many	Many
Prolactin	PRL	Polypeptide	Many	Many
Adrenocorticotropic Hormone	ACTH	Peptide	Adrenal cortex	Increase glucocorticosteroid production and release
Gonadal Hormones				
Testosterone	T	Sex Steroid (androgen)	Testis, muscle, brain; many	Sperm maturation, behavior, etc
Estradiol	E2	Sex Steroid (estrogen)	Ovary, brain, liver; many	Oviduct development, vitellogenesis
Progesterone	PROG	Sex Steroid (progestin)	Brain, ovary, various	Aggressive behavior, courtship, sex ratio of offspring
Adrenal Hormones				
Corticosterone	CORT	Glucocorticoid	Many	Many
Aldosterone	ALDO	Mineralocorticoid	Gut, many	Sodium and glucose balance, many
Dehydroepiandrosterone (an important precursor hormone)	DHEA	Sex steroid (androgen)	Brain, many	Territorial behavior, many
Epinephrine	EPI	Catecholamine	Liver, many	Glycogenolysis, many
Norepinephrine	NOREP	Catecholamine	Brain, many	Auditory processing, many
Thyroid Hormones				
Thyroxine	T4	Thyroid hormone	Many	Many
Tri-iodo thyronine	T3	Thyroid hormone	Many	Many

HORMONE	ABBREV	TYPE	Target	Effects
Pineal				
Melatonin	MEL	Indoleamine	Suprachiasmatic nucleus, many other	Inhibit catecholaminergic tone in circadian system, many other
Pancreas				
Insulin	INS	Peptide	Many	Reduce blood glucose, amino acid uptake by cells
Glucagon	GLU	Peptide	Liver and other	Increase blood glucose, release FFAs
Parathyroid Gland				
Parathyroid Hormone	PTH	Peptide	Bone, kidney	Increase blood calcium and phosphate
Ultimobranchials				
Calcitonin		Peptide	Bone, kidney	Lower blood calcium and phosphate
Skin				
Vitamin D3		Steroid	Liver, kidney, intestine, bone, shell gland	Increased bone deposition, shell mineralization
Diverse locations				
Growth factors	IGF, NGF, etc	Peptides	Many	Many
Angiotensin II (made in blood, converted from liver angiotensinogen)	AII	Peptide	Kidney, salt gland	Osmoregulation
Atrial Natriuretic Factor (Heart)	ANF	Peptide	Heart, kidney, brain, salt gland	Water and salt balance
Leptin	Lep	Peptide	Under study	Under study

tions among life cycle stages in adults), or coping with unpredictable, potentially life-threatening challenges (inclement weather, predators).

From an endocrine perspective, homeostatic regulation relies heavily on feedback loops that help prevent dramatic deviations from relatively stable set points. However, birds living in natural environments may seldom be in a truly steady, "homeostasis only" state. Early in life they are of course growing and developing extremely rapidly. Although internal physiological state (e.g., temperature and osmotic conditions) must be maintained during this period, food intake and allocation of nutritional and energetic resources are far from steady state. Gene-environment interactions involving endocrine top-down and feedback control orchestrate this process. Hormones also play critical roles in directing paths of differentiation (sexual differentiation and other permanent modifications that will affect functionality later in life) during growth and development.

Even once birds reach maturity, their lives are subdivided into life-cycle stages with dramatically different demands and phenotypic characteristics. Processes such as reproduction, plumage molt, and migration have the potential to impose extremely different nutritional, energetic, and temporal scheduling demands. Because most birds live in temporally and spatially variable environments, the endocrine system has evolved to coordinate **allostatic regulation**—a homeostasis-like regulation that takes into account demands imposed by a changing environment and particular life-cycle stages. Whereas homeostasis is entirely reactive (e.g., to feedback loops based fundamentally on physiological cues), allostasis is fundamentally proactive and dependent on top-down signals originating with endogenous clocks, stimuli from environmental cues, and genetic programs. It can be useful to think of allostatic regulation of coping with environmental variation as akin to recurring periods of growth and development combined with regression of particular tissues and physiological capabilities throughout the lives of the animals. For example, an adult male bird cycles annually through stages in which it grows its testes and parts of its brain in the spring in anticipation of breeding, experiences many testosterone-regulated changes in physiology and behavior, and then cycles away from these reproductive behaviors and physiology as the testes shrink and testosterone levels decline again.

Below, we will discuss the role of the endocrine system in homeostasis, growth and development, and allostasis using specific examples important to the lives of birds occupying natural environments.

Endocrine Regulation of Homeostasis (Physiological and Behavioral)

Energy Homeostasis

Energy balance is critical to the survival and reproductive success of all organisms. The neuroendocrine/endocrine system plays fundamental roles in the coordination of all aspects of energy homeostasis, including regulating (1) food intake, digestion, transport and storage of fuel molecules; (2) pathways of metabolic processing of different fuel types; (3) overall metabolic rate; and (4) the timing and nature of activity versus rest. Much of what is known concerning hormones and energy homeostasis comes from studies of chickens in the food industry. While these data can form the basis for hypotheses concerning other birds, it is important to remember that these mechanisms may vary (and even do among chicken strains bred selectively for meat versus egg production). Many hormones play roles in each of the processes listed above. Here we focus on just a few of them, to highlight the complexity of control over energy balance in birds.

Food Intake and Body Mass Regulation   Hormones influence appetite and body mass in birds (Denbow and Cline 2015). The avian hypothalamus is the main starting point of top-down control of food intake, and performs the integrative processes involved in regulating overall energy balance and, therefore, body mass. The hypothalamus and brain stem can detect how much and which types of fuel molecules are present in the blood and respond by influencing appetite through neurohormone release. Some neurons in the hypothalamus release neuropeptide Y (NPY) to increase food intake, called **orexia**, and stimulate anabolic (building) processes that result in increased body mass. Other neurons in the hypothalamus release melanocortins (POMC derivatives—POMC, or pro-opio-melanocortin, is a pro-hormone that can be enzymatically processed into a variety of different active hormones, including ACTH) to decrease food intake, called **anorexia**, and stimulate catabolic (breakdown) processes that result in decreased body mass. Other hormones produced by energy-

regulating organs such as the pancreas, adipocytes, adrenals, and GI tract (see below) may all influence food intake and overall energy balance through effects at either the hypothalamic or brain-stem levels. For example, experimental food deprivation (i.e., fasting) causes decreased insulin (from the pancreas) and increased glucocorticosteroids (gluco=increase glucose availability, cortico=made in adrenal cortex, steroid= steroid hormone) and glucagon, which stimulates NPY release in the hypothalamus to increase food intake. The bird will eat at a faster rate and for a longer period compared with a normal, unfasted state. These feedback loops are particularly important during the intensive fasting and feeding cycles that migrant birds experience as they cycle between long bouts of flight and periods of refueling (reviewed in Cornelius et al. 2013).

Metabolism of Carbohydrates, Lipids, Proteins: Storage versus Mobilization   Two important metabolic hormones critical to energy balance in mammals are also found in birds: insulin and glucagon. These hormones have generally similar roles for glucose storage versus mobilization in birds: insulin is released when glucose levels are high (i.e., in fed birds) and stimulates glucose storage, whereas glucagon is released when glucose levels are low (i.e., in unfed birds) and stimulates glucose release. However, birds differ rather dramatically from mammals in that they tend to have much higher plasma glucose levels, and they are less responsive to the hypoglycemic (plasma glucose lowering) effects of insulin (Dupont et al. 2015). Plasma glucagon levels also tend to be higher in birds, perhaps underlying the high plasma glucose levels typical of birds (Dupont et al. 2015). Glucagon promotes conversion of stored fuels into glucose (i.e., gluconeogenesis) and is the primary lipolytic hormone in birds (i.e., stimulates the liver and adipocytes to convert lipids into glucose and free fatty acids for delivery to working cells). Numerous other chemical messengers besides insulin and glucagon contribute to the regulation of carbohydrate and lipid metabolism in birds, including thyroid hormones, catecholamines (e.g., epinephrine, norepinephrine), growth hormone, and glucocorticoids—to name a few. As mentioned above, these hormones influence both top-down processes in the hypothalamus related to energy balance and food intake, as well as metabolism of fuels in energy-storing organs (e.g., adipocytes, liver).

Osmotic Balance

The first unicellular organisms all the way up to relatively complex vertebrates evolved in an aqueous environment. One consequence of this is that water, a dynamic matrix of H_2O molecules interacting with each other, provides the chemical infrastructure within which physiology (i.e., billions of chemical interactions) has evolved. The chemical reactions that sustain life depend on this water infrastructure. Water accounts for more than 60 percent of body mass in most organisms, and birds are no exception. Being an air breather and an often-terrestrial animal, the challenge for birds usually lies in water acquisition and retention so that their cells—and the working molecules within them—can exist happily in an aqueous environment. Dehydration is a serious risk for birds because cells and whole organ systems could lose function as water content and blood pressure drop. Over-hydration is similarly problematic and manifests itself in high blood pressure. Water, however, is only part of the picture. Salts and other osmolytes (i.e., dissolved particles in the body fluids) must be carefully regulated to maintain an optimal chemical environment to support physiology. Disruption of osmotic balance can lead to loss of cellular functions. Several organ systems coordinate responses to such disruptions to achieve homeostasis of water and salt levels in a process collectively known as osmoregulation.

Different species of bird face different types of osmoregulatory challenges. Nectarivorous hummingbirds, for example, consume large amounts of aqueous solutions and must excrete excess water or risk high blood pressure. Desert-dwelling birds, on the other hand, face the challenge of conserving the little water they gain through their diet, because of limited drinking-water sources. Seabirds face the simultaneous challenge of excreting excess salt gained through their marine diet and the potentially large volume of water that is often consumed in the process of eating salty food or drinking seawater. These challenges are met through the interacting contributions of various organ systems (e.g., kidneys, intestinal tract, and, in some species, salt glands)—many of which are coordinated hormonally.

Bird kidneys function in much the same way as mammalian kidneys (i.e., to reabsorb water and needed solutes from filtered blood plasma), but are not as effective as mammalian kidneys at producing a highly hyperosmotic solution relative to the blood plasma (Braun 2015). In other words, birds make only a weakly concentrated urine in the kidney. Another difference between birds and mammals is that the avian kidney empties into the GI tract rather than into a bladder. The GI tract is responsible for further reabsorption of water and solutes, depending on homeostatic needs, and both the kidney and GI tract respond to osmoregulatory hormones. An important structure involved in the water balance of some species, primarily marine birds, is the salt gland. Salt glands are capable of excreting a highly concentrated salt solution, allowing seabirds to achieve a net water gain by drinking seawater. The salt gland is not under direct hormonal control like the kidney and parts of the GI tract, but osmoregulatory hormones can apparently influence salt gland activity.

When blood becomes hypertonic (i.e., too salty) or blood pressure falls (indicating there is not enough water in the body), the hypothalamus responds by releasing arginine vasotocin (AVT) via the posterior pituitary. AVT is an antidiuretic hormone that influences the expression of proteins called aquaporins, which function as pores for water. When more aquaporin proteins are inserted into the membrane of the collecting ducts of kidney nephrons, more water leaves the filtrate (urine) and is reabsorbed by the body. This can help dilute hyperosmotic blood and create a more hyperosmotic urine for excretion to regain homeostatic balance. Conversely, if salt levels drop in the blood (or blood pressure falls), osmoregulatory sensors in the kidney produce an enzyme called renin, which catalyzes the production of angiotensin II. Angiotensin II has coordinating effects on many different organs to increase blood pressure, and one of these targets is the adrenal gland. The adrenal releases the mineralocorticoid hormone aldosterone in response, which acts on the collecting duct of the nephron and on the intestine to increase expression of a protein called ENaC (epithelial sodium channel). This protein essentially allows movement of salt from the hypertonic filtrate in the nephron back toward the extracellular fluid and blood, thus increasing sodium reabsorption. This not only serves to retain salts but also encourages water to follow (if aquaporin is present in the membrane) and is thereby important in the response to low blood pressure. Angiotensin II also inhibits salt gland activity, thereby slowing salt excretion.

Calcium Balance

Calcium is vital in birds because it plays a role in the intracellular and extracellular environments, is crucial for normal neuron function and muscle activity, and is a major component of the skeleton and eggshell. Calcium is acquired in the diet, but the amount present in the fluid compartments of the body (extracellular and intracellular fluids) must be tightly regulated. Calcium levels that are too low or too high can cause overactive nerve and muscle cells or sluggish reflexes and lethargy, respectively. The bone, intestine, and kidney are major players in the storage, release, and reuptake of calcium, and they work together to regulate calcium levels in the extracellular fluids. Hormones communicate among these organs and the parathyroid and ultimobranchial glands, which are endocrine glands that can monitor and respond to changes in blood calcium levels. The parathyroid gland increases production and secretion of parathyroid hormone when blood calcium levels are too low, and the ultimobranchial gland increases production and secretion of calcitonin when calcium levels are too high. These two hormones work in opposing directions on target cells to change calcium levels. Parathyroid hormone stimulates three main pathways for increasing blood calcium levels: (1) bone reabsorption (i.e., breakdown of bone to increase calcium availability in the blood), (2) increased reabsorption of calcium by the kidney (i.e., to conserve calcium from excretion), and (3) increased calcium absorption in the intestine, facilitated by conversion of vitamin D3 to its active form (i.e., increased dietary acquisition). Calcitonin has inhibitory effects on each of these pathways and can, therefore, reduce blood calcium levels when they become too high. (Note that vitamin D is technically not a vitamin, since it is synthesized when the skin is exposed to sunlight and is not required in the diet except when exposure to sunshine is insufficient. It acts in a fashion similar to thyroid hormones and steroid hormones—and indeed is a form of steroid—via intracellular receptors that regulate gene transcription.)

Female birds develop a transient form of bone called medullary bone in the hollow spaces of the structural bones just prior to egg formation during reproduction. This provides a place to store large amounts of calcium that can be liberated quickly and incorporated into growing shells at the appropriate time of development. These temporary storehouses of calcium are necessary in reproducing females because each egg can require up to 10 percent of the normal body calcium stores, and the very high instantaneous demands for calcium cannot be met from dietary sources alone. Dinosaurs (from which birds evolved) appeared to develop similar bone prior to egg laying (Schweitzer et al. 2016). The upset of normal calcium regulation can have drastic consequences for eggshell formation and reproductive fitness in birds. This became well-known around the world after the widely used pesticide DDT was found to cause shell thinning and subsequent population declines in raptors and seabirds. In healthy birds, ovaries of reproductive females release large amounts of estradiol and moderate amounts of testosterone into the blood. Elevated levels of these two hormones are necessary to stimulate medullary bone formation. Breeding males typically have high circulating testosterone levels, but only have the enzymatic capacity to convert the testosterone to estrogens in very specific tissues (see below). Consequently, males have low circulating estradiol levels and do not form medullary bone (although they can do it if given estrogen experimentally). The release of calcium from medullary bone stores is controlled by opposing influences of parathyroid hormone and calcitonin. The increased reabsorption of medullary bone only occurs when an egg is in the shell gland of the oviduct. Parathyroid hormone stimulates bone reabsorption from medullary bone; however, calcitonin inhibits this process until the developing egg enters the shell gland. These two hormones thus ensure that increased blood calcium levels will be available during the critical step of shell formation but that the bone will not be unnecessarily degraded when calcium is not in high demand.

Endocrine Regulation of Growth, Development, and Sexual Differentiation

Growth and Development

Growth of avian embryos and chicks is under broad endocrine control, primarily by circulating peptide growth factors such as insulin-like growth factor-1 (IGF-1). These growth factors are generally under the control of pituitary growth hormone (GH, also known as somatotropic hormone, or STH), although the effects of removal of pituitary growth hormone on growth varies among species. For instance, growth rate of young chickens whose pituitary glands have been removed is dramatically reduced (by about 50 percent), and this is at least partially reversed if they are given growth hormone replacement. However, no similar effects are observed in young turkeys; hypophysectomy (removal of the pituitary) has at most a modest effect on daily weight gain, and growth hormone replacement has little reversing effect (see Scanes 2015b for review).

Thyroid hormones are also critical to normal growth, though the effects appear to be indirect or permissive (i.e., necessary for the process, but not directly stimulatory, and not sufficient alone). Extremely low levels of thyroid hormones following thyroidectomy or exposure to compounds that prevent production of thyroid hormones lead to reduced growth rates. Interestingly, so do extremely high levels, because this greatly increases metabolic rate and shifts the animal away from anabolism (metabolic pathways focused on building) and toward catabolism (pathways focused on breakdown). Manipulation of circulating thyroid hormones within the normal physiological range has little effect on growth rates, however (reviewed in McNabb and Darras 2015).

Thyroid hormones are also important in regulating differentiation during development. Throughout development, levels of enzymes involved in production of thyroid hormones (the deiodinases, or Dio enzymes, especially Dio2, which catalyzes conversion of T4, or thyroxine, to the more metabolically active form T3, or tri-iodothyronine), expression of thyroid hormone receptors, thyroid hormone transporter molecules, and local levels of thyroid hormones themselves fluctuate in region- and tissue-specific patterns that appear to be important to differentiation (e.g., Van Herck et al. 2012). Many of these effects involve interactions between thyroid hormones and other hormones. For instance, the thyroid hormones and glucocorticoids interact to induce differentiation of glucose transport mechanisms in the developing avian gut (see McNabb et al. 2006). Further, thyroid hormones interact with growth hormone to influence skeletal muscle maturation, and with IGF-1 to facilitate cartilage and skeletal maturation (see McNabb et al. 2006).

Thyroid hormones, specifically T3, play a particularly fascinating role in the process of filial imprinting, an important component of behavioral development in birds. Filial imprinting is a specialized form of learning whereby very young birds (it is generally studied in precocial birds such as fowl chicks, ducklings, or goslings) develop a strong, irreversible attachment to the mother. Imprinting is only possible during an extremely limited **critical period** of development immediately after hatching. Experimental studies have shown that these young birds can be imprinted on a variety of objects, including some rather bizarre but experimentally convenient ones. One-day-old leghorn fowl chicks exposed to yel-

low LEGO pieces readily imprint on them, showing a strong preference for them over otherwise identical red LEGO pieces when tested an hour later (fig. 8.7). The development of this preference is blocked if Dio2 blockers, which prevent the conversion of T4 to T3, are administered to the chicks 30 minutes prior to the training period of exposure to the yellow LEGO piece. Giving these blocker-treated chicks supplemental T3 allows them to imprint normally, but T4 is ineffective (fig. 8.7). Thus, thyroid hormones are critical to this developmental phenomenon, and ability to convert T4 (the primary form released by the thyroid gland into the circulation) to T3 via Dio2 appears to be essential (Yamaguchi et al. 2012).

Thyroid hormones also play roles in reversible "programming" phenomena during adulthood. For example, they are necessary early in the process of photostimulation for subse-

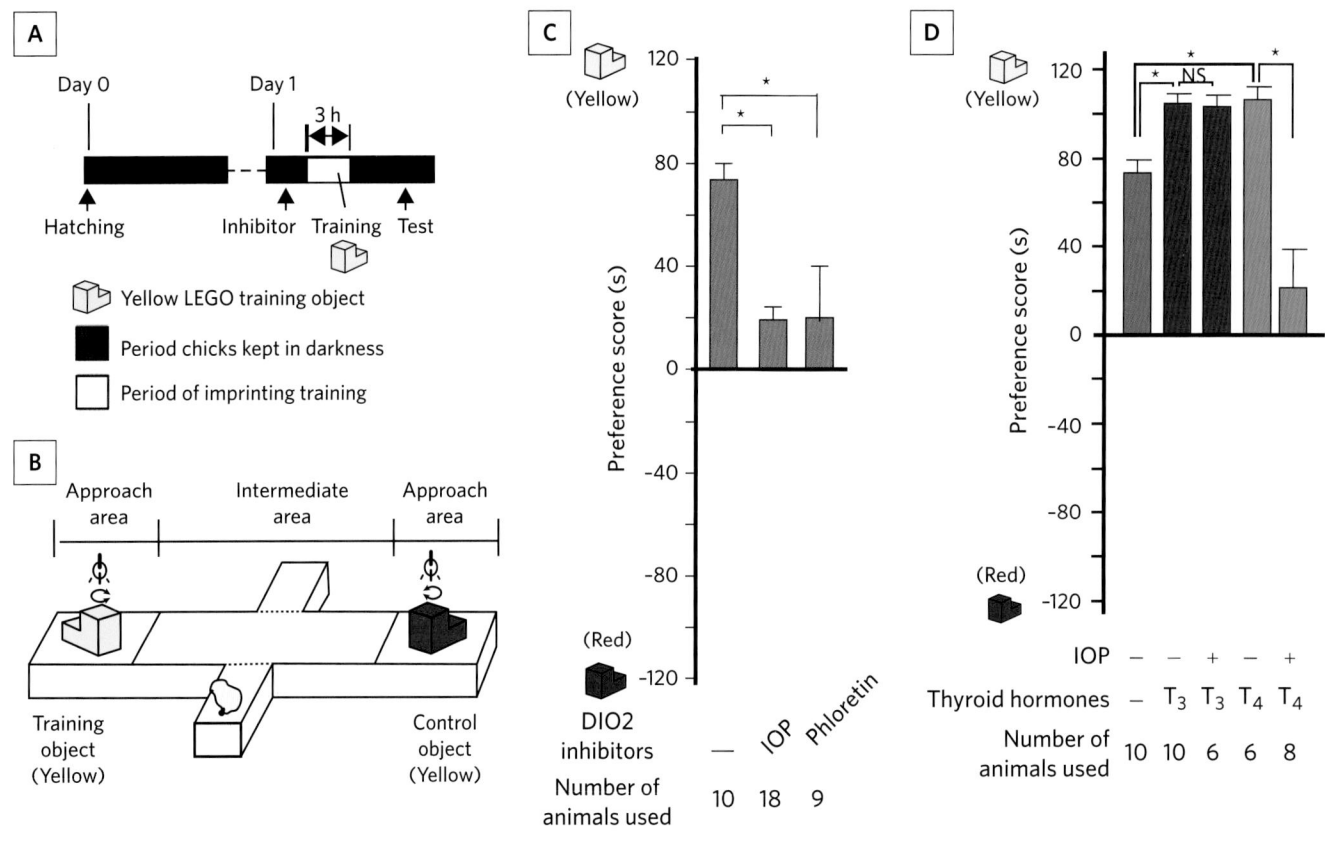

Figure 8.7. Importance of T3 to filial imprinting in chicks of domestic fowl. *A,* Illustration of experimental schedule. Leghorn chicks were exposed at day 1 after hatching to a yellow LEGO object during a "training" period, and then tested for a preference for that yellow LEGO object compared with an identically shaped but previously never encountered red LEGO object one hour later. Some individuals were given inhibitors of the DIO2 enzyme that converts T4 to T3 prior to the LEGO exposure "training" period. Others (controls) were not; *B,* Testing occurred in an apparatus that gave the chick the opportunity to approach either the yellow or the red LEGO object. Time spent in each arm of the apparatus was measured to determine level

of interest in each LEGO object; *C,* All groups (controls and animals receiving two different DIO2 inhibitors, IOP and phloretin) tended to spend more time in proximity to the yellow LEGO object (i.e., positive preference scores, *blue bars*), but the preference score was significantly greater in the controls than in either group receiving a DIO2 inhibitor; *D,* Replacement therapy with T3 (*red bar, third from left*) but not T4 (*green, rightmost bar*) in chicks treated with the DIO2 inhibitor IOP reinstates ability to imprint on the yellow LEGO object. Thus, the ability to convert T4 to T3 is critical to the imprinting process. *From Yamaguchi et al. 2012, https://www.nature.com/articles /ncomms2088.*

	Organizational Effects	Activational Effects
Timing:	Early life / critical period	Any time
Persistence:	Permanent / irreversible once they occur	Reversible
Hormone dependence:	Hormone only required during critical period	Hormone influential for persistence
Nature of effects:	Modify anatomy dramatically	Primarily affect functionality of existing circuits, muscles, etc.
Examples:	Sexual differentiation of Song System	Amount and stereotypy of song

Figure 8.8. Defining characteristics of organizational and activational effects of hormones. The concepts can be traced back to the seminal paper by Phoenix et al. (1959) on the role of androgens in sexual differentiation of guinea pigs.

quent completion of the sequence of events of gonad maturation followed by photorefractoriness and plumage molt (see below), although there is disagreement among researchers as to whether these effects are direct (e.g., Reinert and Wilson 1996) or permissive (e.g., Bentley et al. 1997).

Sexual Differentiation: A Role for Sex Steroids

It has been known since the 1950s that sex steroids have dramatic effects on sexual differentiation. Pioneering work on guinea pigs in the laboratory of William C. Young (Phoenix et al. 1959) documented effects of prenatal exposure to testosterone on the sexual differentiation of morphology and behavior. This work further made an important distinction between different kinds of effects of steroid hormones on development and expression of behavior—a distinction that has been fundamental to how researchers think about hormonal effects for over 50 years. Specifically, Young and colleagues distinguished between **organizational effects** of hormones, which occurred early in life, generally during a critical period of sensitivity, and had major, irreversible effects on morphology and eventual behavior, and **activational effects**, which were not restricted to a critical period of development and involved less dramatic and generally reversible effects of the hormones on behavior and morphology (fig. 8.8).

Application of the organization-activation paradigm to hormonal effects on development of brain and behavior of birds has been quite fruitful. Many species of birds display dramatic behavioral differences between males and females. Extensive work in Japanese Quail shows that estradiol produced by the developing ovary is responsible for **demasculinizing** embryonic females so that they will show female-typical copulatory behavior as adults and not show male-typical copulatory behavior (e.g., mounting, neck grabbing, and cloacal contact movements; see fig. 8.9). If female embryos are treated with aromatase inhibitors (i.e., agents that block functioning of the enzyme that catalyzes conversion of T [testosterone] to E2 [estradiol]), those genetic females display male-typical copulatory behavior (e.g., mounting) as adults. In addition, if male embryos are treated with estradiol, they are demasculinized and fail to display normal male-typical copulatory behavior as adults (reviewed in Balthazart et al. 2009; fig. 8.9). Interestingly, in addition to these orga-

nizational effects of estrogens that, in this case, lead to demasculinization of brain and behavior, normal adult male copulatory behavior is also dependent on activational effects of estradiol. Specifically, testosterone secreted by adult male testes is locally converted to estradiol via effects of aromatase within a hypothalamic area called the medial preoptic nucleus (POM). This is an example of what may seem a somewhat confusing role for "female hormones" (estrogens, such as estradiol) in the regulation of male behavior. It may help to keep in mind that testosterone is primarily a precursor hormone that is readily converted either to estrogens, via aromatase, or to other androgens such as 5α-DHT, via another enzyme called 5α-reductase. Further, estrogens are actually

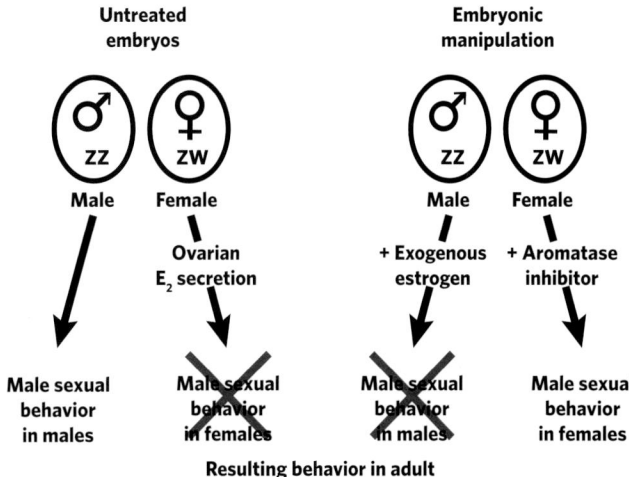

Figure 8.9. Role of estradiol in the sexual differentiation of behavior in Japanese Quail. In untreated embryos of normal female quail (*left*), estradiol from the ovary of young females demasculinizes behavior. These females show no male-typical sexual behaviors (mounting, neck biting, cloacal contact movements) as adults. If embryos of very young females are treated with an aromatase inhibitor (which blocks enzymatic conversion of testosterone to estradiol), behavior is not demasculinized, and these females display male-typical behaviors as adults (*right*). In unmanipulated male embryos, little or no estradiol is produced during development so behavior remains masculinized (*left*). If young male embryos are given estradiol supplementation, their behavior is demasculinized and at maturity they fail to display male-typical sexual behaviors (*right*). *Modified from Balthazart et al. 2009.*

responsible for many male-typical traits. For instance, in rodents, **estradiol** is essential to the organizational **masculinization** of both structure and function of multiple brain regions involved in sexually differentiated behavior (e.g., aggression, copulatory behavior) and physiology (e.g., loss of the ability of the GnRH "pulse generator" to display a positive feedback response to estradiol, as is required for the preovulatory LH surge in adult females). It is best to try to discard preconceptions that "estrogens are female hormones" and "androgens are male hormones," since the biochemistry of these two classes of sex steroids is so closely linked, and since both are present in and play important roles in development and expression of traits in both sexes.

The steroid-dependent organization and activation of song in passerines differs significantly from that of copulatory behavior in quail. In this case, early (around the time of hatching) exposure to **estrogens** masculinizes neural circuitry involved in song production, and **androgens**, not their estrogenic metabolites, are involved in activation of that circuitry later in life. Experiments investigating this have made use of the fact that certain androgens, such as 5α-DHT, are "nonaromatizable"—i.e., cannot be converted to estradiol. Early exposure to DHT is ineffective at organizing the ability to sing, whereas adult exposure to either T or DHT is effective at enhancing singing behavior (e.g., Gurney and Konishi 1980).

Interestingly, unlike in the case of endocrine regulation of development and expression of copulatory behavior in quail, sexual differentiation of brain and behavior of songbirds cannot be explained by endocrine control alone. It has long been known that early exposure to estradiol masculinizes the brains and subsequent behavior of female Zebra Finches (i.e., they do respond to T administered in adulthood by singing, whereas control females do not). But it is not possible to completely masculinize the song control brain nuclei (see chapter 13) of females by exposure to estrogens. Rather, there appears to be a strong sex-specific genetic contribution to differentiation of the song system and of singing behavior. To study this issue, Arthur Arnold and colleagues took advantage of a serendipitous and rather bizarre opportunity presented by the appearance of a "gynandromorphic" Zebra Finch that appeared in their laboratory finch colony (Agate et al. 2003; fig. 8.10). On the right side of its body, this individual possessed male plumage, a testis, and gene activity patterns typical of the homogametic (ZZ) sex of male birds. That is, it was "male" on its right side. On its left side, it possessed female plumage, an ovary, and gene activity patterns typical of the heterogametic (ZW) sex of female birds; it was "female" on its left side. The song control nuclei of this individual were exceptionally strongly **lateralized**. Specifically, the HVC and RA song control nuclei (see chapter 13) were

much larger on the right (male) side than they were on the left (female) side. Song nuclei are generally lateralized to some degree; this just means that the nucleus on one half of the brain is a bit larger than the same nucleus in the other half. But the right HVC and RA of the gynandromorph were so much larger than the left that the difference was outside the normal range of differences observed in normal male Zebra Finches. Since both the left and right sides of this bird's brain must have been exposed to similar circulating sex steroid levels during development, it follows that a difference in the responsiveness of the left and right sides to those steroids must exist, otherwise there would not have been such a huge difference in the size of the left and the right song nuclei. This argues for a strong gene-hormone interaction determining the development of the song control nuclei, rather than it being "purely genetic" or "purely hormonal."

Although birds have been an important source of critical insight into the effects of hormones on developmental neural plasticity, their role in revealing adult plasticity of the nervous system, and specifically the hormone dependence of that plasticity, has been even more substantial. Plasticity of the brains of adult songbirds was called to everyone's attention by Fernando Nottebohm when he demonstrated that adult female canaries treated with testosterone showed dramatic increases in the size of several song control nuclei (Nottebohm 1980), and adult males showed seasonal fluctuations in the size of those nuclei (Nottebohm 1981). Nottebohm originally suspected that the increases in volume were related primarily to growth of dendritic segments and synapses, which facilitated development and expression of new motor patterns associated with increases in singing behavior in both sexes. It has subsequently proved true that changes in the volumes of these song nuclei do relate to changes in complexity of neural branching patterns, as well as size and density of neurons, and—most notably and to many researchers, most excitingly—neuron number. That is, the HVC of the adult male canary brain actually incorporates new neurons annually (see Brenowitz 2004, for a review). It is now known that there is dramatic sex steroid–dependent seasonal brain plasticity in both quail and songbirds (Balthazart et al. 2010). For example, in male quail, the medial preoptic nuclei (POM, referred to above) change in volume seasonally in concert with changing gonad size and steroid-producing activity, and administering T to castrated males causes rapid and dramatic increases in POM size. These changes are associated with changes in neural spacing and dendritic branching. In the HVC of male canaries, seasonal increases in volume are the result of incorporation of new neurons as well as increases in spacing and branching, and these effects are mediated by T acting through androgen receptors in the song system (fig. 8.11).

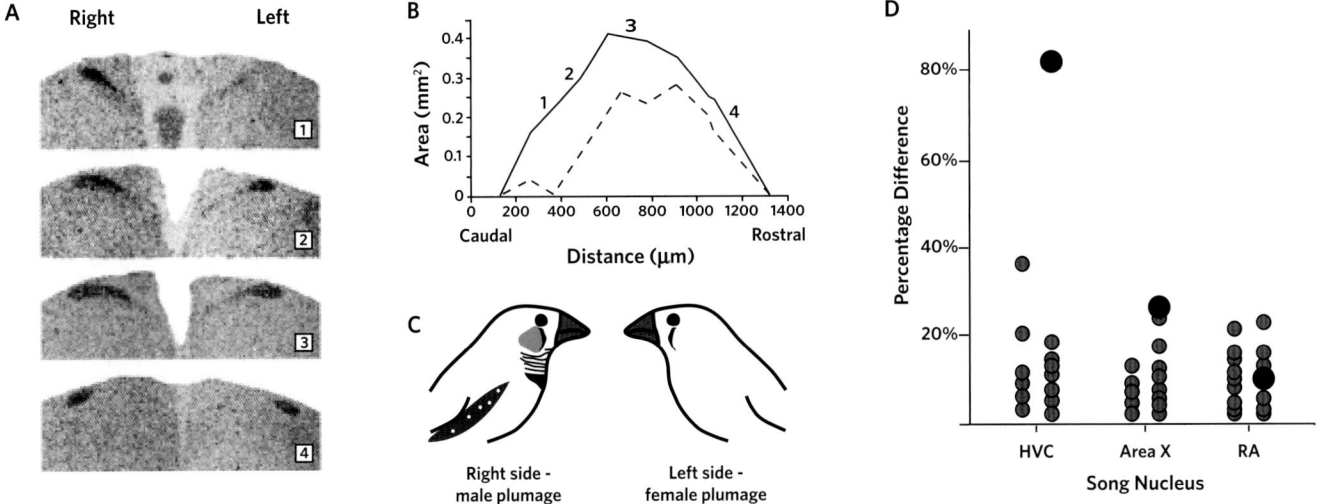

Figure 8.10. An illustration of the interplay between genetic and hormonal factors in the sexual differentiation of the song system of songbirds, using the unusual case of a gynandromorphic (possessing both male and female features) Zebra Finch. A, Typically, song control brain nuclei such as HVC (shown) are "lateralized," such that the nucleus on one half of the brain is somewhat larger than its counterpart on the other half of the brain. In this example, the right HVC is larger (note: it is shown on the left, as if the bird is facing the viewer). Each brain section, 1–4, is taken at a different point moving from the back (1) to the front (4) of the brain; B, A graph showing the "volume reconstruction" for the left (dotted line) and right (solid line) HVCs of that same male Zebra Finch from which the sections in part A were taken. The area under each curve is proportional to the nucleus volume; the right HVC is much larger than the left; C, In the gynandromorphic individual, the right side of the bird displayed male typical plumage (orange cheek plus zebra barring on chest and brown with white spotted feathers on flank), whereas the left side displayed female typical plumage (all gray). The right side of the bird also possessed a testis, and expressed greater amounts of Z-linked genes (of which male birds have two

copies, compared with only a single copy in the genetically ZW females). The left side possessed an ovary and expressed W-linked genes that would only be present in females (ZW), not males (ZZ). All this supports the interpretation that the gynandromorph indeed possessed "male tissues" on its right, and "female tissues" on its left; D, Figure showing degree of lateralization of three song nuclei: HVC (left), Area X (middle), and RA (right) of the gynandromorph (*black circles*) compared with a set of normal males (*red circles*). Each dot represents the lateralization for a single bird. Dots in the left column for a particular brain nucleus represent individuals whose left nucleus was larger than the right. Dots in the right column for a particular brain nucleus represent individuals whose right nucleus was larger. The right HVC of the gynandromorph was enormous (nearly twice as large) compared with the left, a degree of lateralization that was far outside the range for normal male finches. Although Area X was not as dramatically lateralized as HVC, the degree of difference was still the most extreme observed in the sample of birds. RA was no more lateralized in the gynandromorph than in other males. *Modified from Agate et al. 2003, copyright (2003) National Academy of Sciences, U.S.A.*

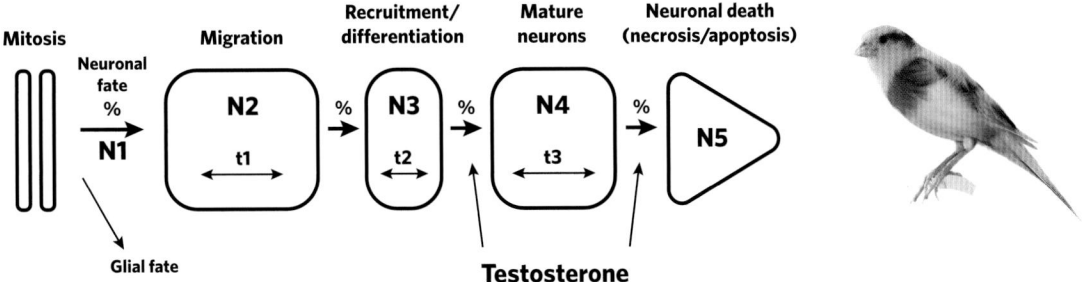

Figure 8.11. Steroid-dependent neural plasticity in POM (medial preoptic nucleus) and HVC (song control nucleus) of canaries. T is known

to influence the differentiation to mature neurons, and the transition to neuronal death. See text for details. *From Balthazart et al. 2010.*

ENDOCRINE DISRUPTORS

It has been known since the 1970s that plant chemicals could influence endocrine function of birds. Specifically, phytoestrogens—secondary plant compounds that interact with estrogen receptors of birds and other animals—were found to disrupt reproductive function of California Quail (*Callipepla californica*) (Leopold 1977, Leopold et al. 1976). A wide variety of compounds in animals' external environments are now known to influence endocrine pathways. Although many of these compounds are of natural origin (e.g., phytoestrogens), a plethora of other compounds with endocrine-disrupting effects are released into the environment by humans. All of these so-called endocrine disrupting chemicals (EDCs), both natural and artificial, can affect endocrine signaling, thereby influencing anatomy, physiology, and behavior, both during development and in adults (Rochester and Millam 2009, Ottinger et al. 2013). The specific chemicals involved are quite variable. For instance, many phytoestrogens belong to one of three general categories of compounds—phenolics, terpenoids, or saponins—that can interact with estrogen receptors. Others, such as lignans, are not inherently estrogenic but can be converted to estrogenic compounds by animals' gut flora. Human-produced EDCs include many compounds used as pesticides, fungicides, or herbicides, most famously the pesticide DDT, which caused the near-extinction of a number of species of predatory birds by causing extreme eggshell thinning. Although DDT (and its metabolite, DDE) may have some effects through non-endocrine mechanisms, some problems it creates probably stem from endocrine disruption of female reproductive tract development. Other human-produced EDCs include poly-brominated diphenyl ethers (PBDEs) used as flame retardants. These are just a few of the known or suspected EDCs; the list of chemicals known to have endocrine-disrupting effects is long and will surely only grow as more information is obtained. Mechanisms of action of EDCs are also variable, including activating or inhibiting estrogen or androgen receptor function, as well as having effects via indirect pathways that do not involve binding to the hormone receptors. Concerns related to effects of these compounds are particularly high for birds living in parts of the world where restrictions are minimal or nonexistent on use of pesticides. Because of the phenomenon of amplification (see figure 8.3), very low exposures to such compounds—or exposure to compounds with relatively low affinities for steroid receptors if the exposure level is high—may affect phenotypic traits, particularly if exposure occurs during critical periods of development. Further, some EDCs are particularly prone to bio-accumulation as they rise up the food chain. Overall, EDCs present both interesting examples of effects of the natural environment on birds (e.g., phytoestrogen effects on reproductive function) and serious threats (e.g., pesticide disruption of endocrine signaling) to many populations of birds.

Maternal Effects: Effects of Yolk Steroids on Development

The above studies demonstrate effects of sex steroid exposure during development on sexual differentiation of gonad morphology, brain, and behavior, and use experimental manipulations of the hormone environment in the egg to investigate these processes. Until relatively recently, the steroid hormones influencing development were thought to originate entirely through direct production by the developing embryo or nestling. In 1993, Hubert Schwabl published a paper demonstrating that yolks of freshly laid canary and Zebra Finch eggs contained T. Further, the amount of T varied between species, among clutches, and by laying order within clutches, with eggs laid later generally containing more yolk T than eggs laid earlier. These yolk T levels were unrelated to the sex of the embryo. This study also showed a tantalizing positive correlation between yolk T levels and the eventual dominance rank of the individuals that hatched from those eggs (Schwabl 1993). Subsequent work has demonstrated that yolk testosterone can organize a variety of important traits, including musculature (e.g., Lipar et al. 2000), plumage characters, and behavior (e.g., Strasser and Schwabl 2004).

The role of maternally derived T in development of avian embryos has important implications for populations of birds in natural contexts. More rapid development is generally favored in environments where predation on eggs and nestlings is high. Comparing across several species of songbirds, yolk androgen levels turn out to be positively correlated with both daily predation rate and with rate of development (fig. 8.12). Specifically, species experiencing higher egg and nestling predation rates display higher yolk T and DHT, embryo period and nestling period are shorter in species with higher yolk T and DHT, and overall growth rate is higher in species with higher yolk T and DHT (Schwabl et al. 2007).

	A4	T	DHT
Daily predation	Ø	+	+
Embryonic period	Ø	−	−
Nestling period	Ø	−	−
Growth rate	Ø	+	+

Figure 8.12. Daily predation rate (measured empirically in field studies of wild songbirds) is positively correlated (green "plus" signs) to yolk levels of the strong androgens T and DHT, but unrelated to yolk levels of the precursor androgen androstenedione (A4, *top row*). Nestling growth rates are also positively correlated with yolk T and DHT (*bottom row*). Both embryonic period and nestling period are negatively correlated with yolk T and DHT; that is, young birds develop more rapidly the higher yolk T and DHT levels are (*middle rows*). Together, these data support the interpretation that selection for faster development has favored females incorporating higher amounts of androgens—which enhance development rate—in the yolks of their eggs. *Based on Schwabl et al. 2007.*

Endocrine Basis of Allostatic Coordination

Changing Resources, Changing Demand, Changing Allocation

The challenges associated with maintenance of internal conditions (homeostasis) vary over the life of a bird as functions of (1) environmental conditions (e.g., temperature or the amount of food available) and (2) investments the bird makes in different processes (such as growth/development, reproduction, plumage molt, or migration). The process of **allostasis**—which is generally defined as the maintenance of homeostasis through change (McEwen and Wingfield 2003, Wingfield and McEwen 2010)—encompasses all of these functions. Allostatic regulation is the means by which an individual assesses whether a set of circumstances represent a major challenge, requiring that emergency measures be taken, or can be handled with relatively little effort. This will depend on the availability of resources required as well as the nature and magnitude of the demands being experienced. Some demands, such as thermoregulatory demands, are the result of physical environmental conditions. Although the individual cannot control the environmental conditions, it can alter exposure to them through behavioral choices (migration, habitat selection) or the consequences of them through physiological options (e.g., use of torpor to reduce thermoregulatory energy requirements). Other demands are

the direct consequence of "decisions" the individual makes regarding commitment to different costly activities, such as breeding, migrating, or molting of the plumage. This is one of the most fundamental ways that allostasis differs both conceptually and mechanistically from classical homeostasis: Allostasis is inherently proactive, involving preparation that the animal makes for anticipated changes in demands, whereas homeostasis is entirely reactive, responding to the circumstances imposed, primarily via feedback loops.

There are two main ways that allostatic regulation helps birds cope adaptively with varying resources and demands. The first is to prepare for predictable changes in the environment by responding to predictive environmental cues (e.g., changes in day length) and/or through endogenous rhythms. This allows individuals to make appropriate physiological and behavioral preparations for anticipated challenging conditions, such as elevating capacity for metabolic heat production or migrating to more favorable areas. It also permits them to restrict investment in exceptionally demanding processes (e.g., breeding) to periods of the year when (1) resources are likely to be sufficient for successful completion, and (2) demands of competing processes are acceptably low. Finally, it permits them to reduce conflicts between demanding processes through scheduling so that important processes are not in direct competition for a resource base that may be insufficient to meet simultaneous demands. For example, a bird may breed first and, only upon completion of breeding, begin to replace feathers in a plumage molt.

The second way that allostatic regulation helps birds deal adaptively with varying conditions is, perhaps paradoxically, to be prepared for the unexpected and to modulate response to unpredictable challenges appropriately. Virtually all environments display some component of unpredictable conditions. Erratic changes in weather, encounters with predators, and conflicts with conspecifics all can commence with little warning and potentially threaten the well-being of individuals. When the challenges imposed threaten to exceed the capacity of an individual to cope, an emergency life history stage (ELHS) can be initiated (Wingfield 2003), characterized by reallocation of resources (energy, nutrients, time) to the processes and activities most critical to immediate survival.

All of these processes are under endocrine influence. For instance, seasonal investments in reproduction, plumage molt, and migration have a neuroendocrine basis through the HPG axis. Likewise, the entire physiological and behavioral response to unpredictable events is fundamentally a neuroendocrine process involving the HPA axis. The following sections will illustrate the endocrine regulation of all of these examples.

Cyclic Adjustments: Daily and Annual Rhythms

Most environments on earth display cyclic changes in conditions. These cycles vary in their durations, or **periods**. The environmental cycles most relevant to birds are daily, annual, and (for some) tidal. Day-night cycles of light, temperature, relative humidity, accessibility of food/prey, and risk of predators are experienced by all bird species. Annual (i.e., seasonal) changes in **photoperiod** (the number of hours of light per 24 hours), temperature, precipitation, and food supply are also typical of the lives of most birds. Survival and reproductive success depend on birds making appropriate adjustments of behavior, physiology, and morphology to these changing conditions. One way that this could be achieved would be for individuals simply to respond to environmental cues as direct **drivers** of changes in physiology and behavior. For instance, when bright light occurs shortly after dawn, a diurnal (day-active) individual could switch from inactivity to activity in direct response to experiencing the light cue. A problem with this type of mechanism is that many processes require preparations that take time, so changes from, say, a sleep state to an active state, or from wintering reproductive inactivity to spring breeding, simply cannot be made instantly. Consequently, to be prepared for changes in the environment, preparations must be initiated well before the actual change in environmental state occurs. It is now known that the vast majority of organisms possess internal timing mechanisms called **endogenous biological rhythms** that depend on molecular/physiological clocks and allow individuals to anticipate changes in conditions and be ready to respond appropriately. These rhythms allow preparation for impending changes in the environment, and help avoid mistiming of demanding processes in response to unusual circumstances (such as a warm spell in winter). The endocrine system is crucial to the generation and functioning of these endogenous rhythms.

Diel Cycles and Circadian Rhythms Conditions that change regularly on the 24-hour-period day-night cycle are called **diel cycles**, and the endogenous biological rhythms that have evolved to track these are called **circadian rhythms** (*circa* = about, *dia* = day). The animals' cyclic changes in physiology and behavior are fundamentally endogenous (internally generated) and do not require external daylight cues. This can be revealed experimentally by placing individuals in constant dim light for an extended period of time. Rhythmic cycles of activity (e.g., wake versus sleep) will **persist** in these conditions for many repetitions of the cycle despite there being no environmental cues indicating day versus night. However, the period of the rhythmic activity expressed in constant dim light will not exactly coincide with a 24-hour day and may be

slightly shorter (e.g., a 23-hour "day") or slightly longer (e.g., a 25-hour "day"). This is called **free running**, and suggests that the mechanism responsible for the endogenous rhythm needs information from the environment to remain synchronized with a 24-hour cycle (fig. 8.13). Environmental cues, such as the timing of dawn and dusk, can **entrain** endogenous rhythms (i.e., set the clock), which allows the bird's endogenous rhythm to remain synchronized to the environment. Thus, the rhythm is not **driven** by the external cues: it is generated endogenously, and **entrained** by environmental cues.

The pineal gland and its primary hormone product, melatonin, play key roles in the maintenance of these endogenous behavioral rhythms. The avian pineal produces melatonin cyclically, with elevated production at night (and during the **subjective night** if the bird is free-running in constant experimental conditions) (see the box on page 220 for definitions of terminology). This rhythmic production of melatonin is based on a roughly 24-hour oscillation in the production of the enzymes necessary for conversion of the amino acid tryptophan to the hormone melatonin, and this oscillation depends on an oscillating feedback loop in the transcription of a set of so-called clock genes within pineal and retina cells. Removal of the pineal eliminates behavioral rhythms, and cyclic administration of melatonin by injection reinstates them (fig. 8.14). In the live bird, circadian rhythmicity is controlled by interactive neural and endocrine signaling between the pineal and the suprachiasmatic nuclei of the hypothalamus, although species variation in how disruptive the removal of the pineal is to rhythmicity complicates our ability to make broad generalizations (see Cassone and Kumar 2015).

Annual Cycles: Life History Cycles The vast majority of birds occupy environments that also vary predictably on longer time scales, such as seasonal changes in temperature, weather, food supply, and photoperiod. Birds display endogenous annual rhythms (i.e., circannual rhythms) that can be synchronized to the natural world by environmental cues. Synchronizing cues include some that can be used to predict long-term changes in environment patterns (e.g., the seasonal change in photoperiod can predict seasonally consistent environmental patterns such as a north temperate winter versus summer or a tropical wet versus dry season) and other cues that provide more immediate, short-term information about the environment (such as local food supply, temperature, or social interactions with mates and other conspecifics). Species of birds vary in how they use these cues to organize their lives and can vary quite substantially in the complexity of their annual schedules. In the temperate zone, for example, birds often subdivide the year into a relatively fixed sequence of life history, or life cycle, stages (fig. 8.15; Jacobs and Wingfield 2000, Wingfield 2008). Subdivision of the year into multiple distinct stages, includ-

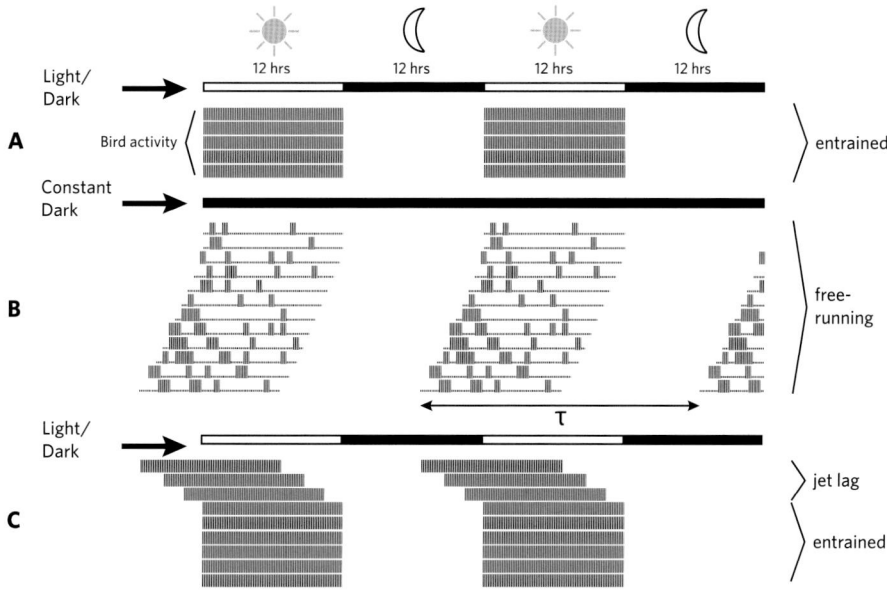

Figure 8.13. Illustration of the key characteristics of persistence, free-running, and entrainment of circadian rhythms of perch hopping in a Zebra Finch. Chronobiologists (see the box on page 220) typically display results on figures consisting of rows and columns, with hatch marks depicting time intervals where activity occurs. Time is shown on the x-axis, and usually two 24-hour days are shown side by side (e.g., day 1 at left, day 2 immediately to its right), and then they repeat the day shown in the right column of a row as the left column on the next row below. The duplication of the right-hand day as the left-hand day on the next row is purely a presentation method to make it easier to visualize whether active and inactive periods are entrained (remain perfectly aligned vertically) or free-running (drift to left or right); it is just easier to see the pattern when each row is 48 hours rather than only 24 hours. Thus, row one depicts perch-hopping activity on days 1 and 2, row two depicts activity on days 2 and 3, and so forth. In this figure, six days of activity are shown (days 1 and 2 on the first row, and days 5 and 6 on the last row of the top panel). In this figure, the bird is first exposed to a normal light-dark cycle of 12L:12D for a total of six days (top panel, designated "entrained" at right). The active period of this diurnal bird corresponds to the 12-hour light periods, and remains entrained; activity consistently is restricted only to the light period. The bird is then exposed to a period (13 days shown) of constant near darkness (middle). Perch-hopping activity is reduced overall, but remains limited to approximately 12-hour intervals that gradually shift in phase. Specifically, this bird becomes active and then ceases activity a bit earlier each day than it would have if provided an entraining light-dark cycle. This is why the active period shifts gradually to the left; the bird is free-running on a period determined by its own endogenous circadian clock. When the bird is returned to a 12L:12D light-dark cycle (bottom) identical in phase to the original L:D cycle, it takes a few days to re-entrain—i.e., there is a "jet lag" period—but once entrained it again remains synchronized to the light-dark cycle. Adapted from Cassone and Kumar 2015.

ing molt, migration, and reproduction, allows birds to focus resources toward a given subset of processes at any particular time of year. This has the advantage of increasing the range of environmental conditions that can be tolerated but has the disadvantage of reducing temporal flexibility in when they might perform each of those processes (e.g., breeding can only occur within a set window of time) (fig. 8.16). For example, long-distance migrant taxa such as Gambel's White-crowned Sparrows schedule their annual cycle largely using predictive changes in photoperiod. They have wide environmental tolerances across their lives but are temporally constrained to migrate, molt, and breed during narrow windows of opportunity. In contrast, species like the Zebra Finch rely more heavily on local, short-term environmental cues to determine level of investment in a particular process. Zebra Finches in at least

some parts of their range remain in a kind of constant "chimera" stage, overlapping reproduction with a very slow plumage molt to take advantage of unpredictable rain-dependent pulses of resources. They may be constrained in the range of environmental conditions they can tolerate as a result, but are capable of very rapid changes to take advantage of sudden or unpredictable resource fluctuations. These two extreme examples demonstrate the wide variety of annual schedules and responses to environmental cues that exist in birds.

The precise timing of how birds progress through a sequence of life cycle stages is regulated by responses to proximate environmental cues, including the predictive cue of photoperiod and immediate cues of temperature, food supply, and social factors. When photoperiod increases above a basic threshold level in late winter or early spring, the testes

CONCEPTS AND TERMINOLOGY OF ENDOGENOUS BIOLOGICAL RHYTHMS

Chronobiologists—researchers interested in how biological clocks work—use a set of standard terminology to discuss key concepts regarding endogenous rhythms and how they are studied. One key attribute of all biological rhythms is **persistence**. For example, when an animal is isolated from external stimuli (e.g., by placing the animal on **constant conditions** of continuous very dim light), cycles of activity-inactivity and many other processes **persist** indefinitely (for many repetitions of the cycle). However, when **free-running** in isolation from environmental stimuli, these endogenous rhythms drift out of phase with the real world because the **period** of the endogenous rhythm (the duration of one turn of the cycle) deviates a bit from the period of the environmental cycle it has evolved to track. In the case of **circadian rhythms**, endogenous periods typically vary up to an hour or so from exactly 24 hours, and for **circannual rhythms**, periods vary somewhat from exactly 365 days. Thus, although these rhythms are endogenous and persist in the absence of environmental cues, they must regularly be reset, or **entrained**, in order to remain synchronized to the real world. The process

of entrainment is akin to the setting of a clock, with specific environmental stimuli acting as **_zeitgebers_** ("time givers") that influence the **phase** of the rhythm. When an individual animal is free-running under constant conditions, particular time points in its cycle are referred to as **subjective** times, because it does not have access to information telling it when actual events (such as sunrise or sunset) are occurring in the external environment. For instance, the times when the animal's endogenous clock is expecting dawn and dusk to occur are called **subjective dawn** and **subjective dusk**, and the intervals when the clock expects it to be light and dark are called **subjective day** and **subjective night**. Typically, the time of day that a free-running animal's system "thinks" it is at any given moment is referred to as its **circadian time** or **clock time**, and is expressed relative to subjective dawn. For instance, circadian hour 12 would be 12 hours after subjective dawn. For an individual under natural conditions, with access to actual dawn and dusk cues, circadian time is the same as "real time," because the circadian clock will be entrained to actual time in the external environment.

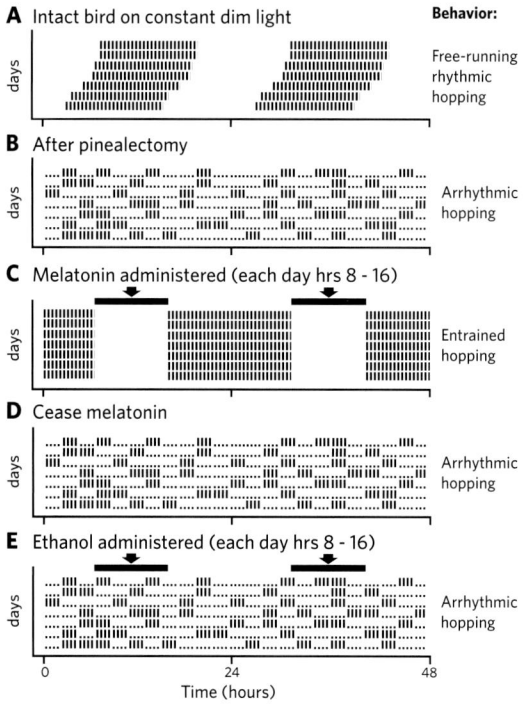

A Intact bird on constant dim light

Behavior:

Free-running rhythmic hopping

B After pinealectomy

Arrhythmic hopping

C Melatonin administered (each day hrs 8 - 16)

Entrained hopping

D Cease melatonin

Arrhythmic hopping

E Ethanol administered (each day hrs 8 - 16)

Arrhythmic hopping

Time (hours)

Figure 8.14. Effect of melatonin on activity patterns of House Sparrows held on constant dim light. *A*, Birds that possess their pineal gland free-run on constant dim light but *B*, become arrhythmic (no specific phases of activity versus inactivity) after the pineal is removed. *C*, If melatonin is administered to the bird in a cyclic fashion, the period of activity follows the cessation of melatonin administration on a regular cycle that mimics an entrained rhythm. *D*, If melatonin administration ceases, arrhythmic activity resumes. *E*, If ethanol alone (the "vehicle" in which melatonin was previously administered) is administered on the same cycle as melatonin had been administered, there is no effect on the arrhythmic activity. *Modified from Gwinner et al. 1997.*

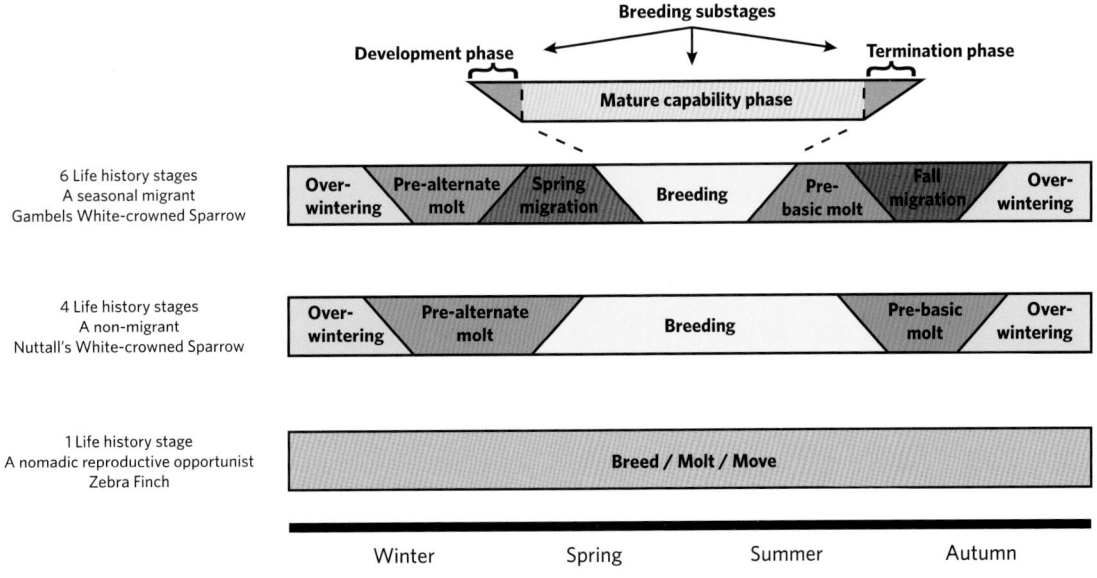

Figure 8.15. Examples of how different bird species can subdivide the year in different ways into separate life cycle, or life history, stages. The complex cycle shown on top allows the bird greater capacity to cope with varying environmental conditions, but imposes temporal constraints. The simpler life cycle shown on the bottom confers great temporal flexibility but may restrict the range of tolerable environmental conditions. Each stage is also subdivided into a development phase, mature capability phase, and termination phase (*inset at top*). *Concept based on Wingfield 2008.*

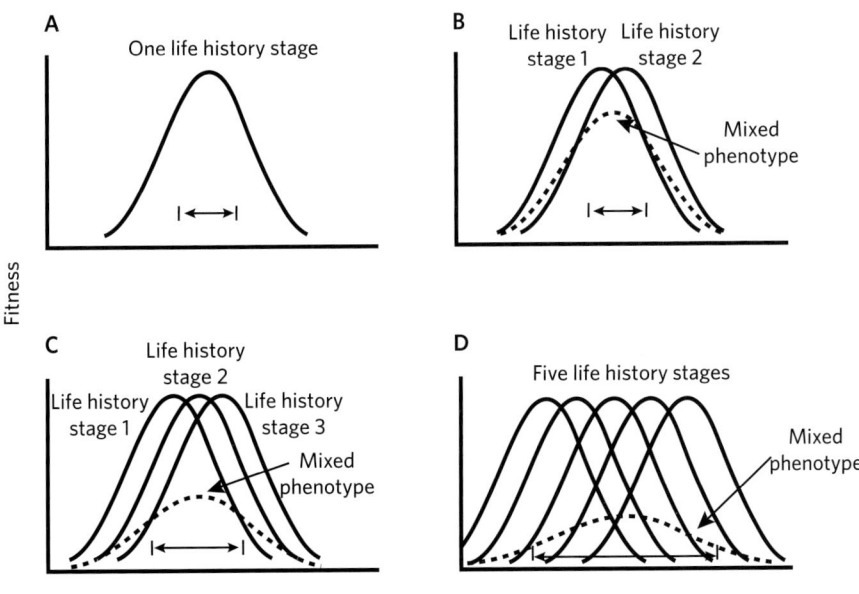

Figure 8.16. Relationship between fitness (y-axis), range of environmental conditions experienced (x-axis), and number of life history stages for several hypothetical situations. By exploiting multiple separate life history stages (solid curves), overall fitness is maximized compared with a hypothetical mixed phenotype exploiting fewer or only a single life history stage combining components of the others (dotted curves). *From Wingfield 2008.*

and the ovaries begin to develop via a neuroendocrine cascade (e.g., Dawson 2002; see below). This allows individuals to respond rapidly to more immediate cues of temperature, food, or social cues when the breeding season arrives. Reproductive development is followed eventually by spontaneous collapse of the gonads in most species, even if day length is experimentally prevented from declining (fig. 8.17). In other words, the entire reproductive axis is inactive because the system is no longer responsive to the top-down stimulus of long days. Individuals that are insensitive to the stimulatory effects of long days are said to be **photorefractory**. As long as the refractory state persists, individuals will remain nonreproductive. Many, though not all, species require a period of exposure to short days before the refractory state will

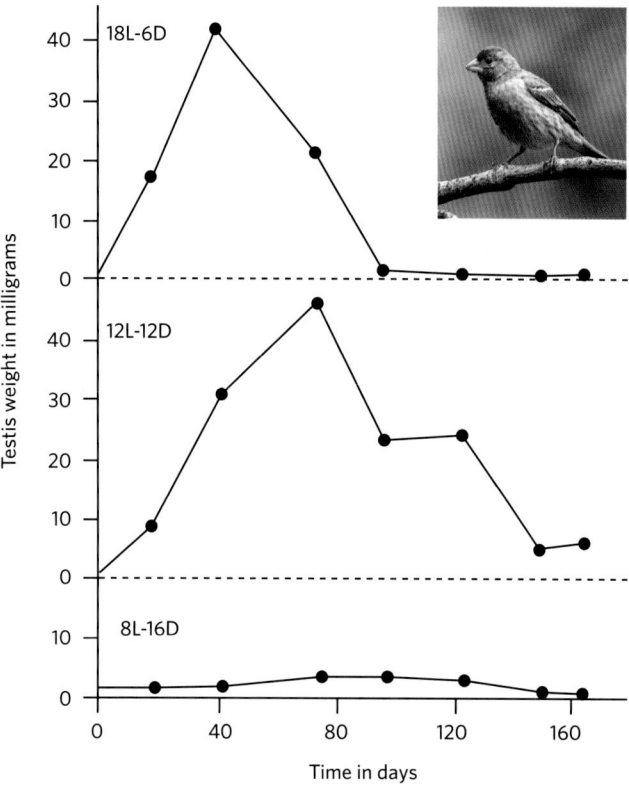

Figure 8.17. Illustration of reproductive photorefractoriness in House Finches. Males moved from short days (e.g., 8 hours light, 16 hours dark) directly onto long days of either 16L:8D (*top*) or 12L:12D (*middle*) rapidly grow the gonads, but then spontaneously collapse them again (i.e., become reproductively photorefractory) even with no further change in photoperiod. Both growth of the gonads and onset of the refractory state occur more quickly on the longer photoperiod. Birds held on short days (8L:16D, *bottom*) throughout the study do not grow the gonads. *Redrawn and modified from Hamner 1966.*

dissipate. This resetting of the system to being photosensitive again normally occurs during autumn. Note, however, that at this point the gonads remain inactive because by then the photoperiod is too short to be stimulatory. Once photoperiod rises again above the photostimulation threshold, in late winter or early spring, the reproductive neuroendocrine cascade stimulates rapid development of the gonads, and the cycle repeats itself. This entire "three stage" sequence, including photosensitivity, photostimulation, and photorefractoriness, is illustrated in figure 8.18 (Hahn et al. 2008).

The neuroendocrine system is fundamental to this process, and neuroendocrine processing of photoperiod cues plays a central role. Long days have been known to cause changes in reproductive and migratory physiology and behavior since the 1920s. Since then, a great deal has been learned about the neuroendocrine mechanisms underlying avian photoperiod-

ism, especially in Japanese Quail. The current hypothesis (see Ikegami and Yoshimura 2012, Cassone and Yoshimura 2015, Hahn et al. 2015) is that deep brain photoreceptors in the hypothalamus communicate with thyrotroph cells nearby in the pars tuberalis, and these then communicate in a paracrine fashion with another set of cells, called ependymal cells, in the median eminence and lining the third ventricle. Under long days, the increased production of TSH in these thyrotroph cells acts in a paracrine fashion to up-regulate activity of the deiodinase-2 (Dio2) enzyme within the ependymal cells, and this leads to locally enhanced conversion of T4 to the more biologically active T3. Through mechanisms that are not yet fully understood, GnRH release is enhanced under long days, leading to the endocrine cascade that produces gonad growth and enhanced production of gametes and sex steroids. It is not currently known how widespread this mechanism is in birds, although some evidence from other species studied under natural conditions, such as European Starlings, suggests that at least some details may differ (Bentley et al. 2013). Further research is necessary on a greater variety of bird species and in natural environments, rather than in the laboratory.

Birds' reproductive response to long days is based on a circadian cycle of changing sensitivity to light. In short, there is an **inducible phase** that occurs during the latter half of the subjective day. The total duration of the light period is not important, but rather the birds' reproductive system responds to the fact that light coincides with the interval between roughly circadian hour 12 and circadian hour 20 (see the box on page 220 for terminology regarding circadian rhythms). Consequently, a short day in winter (e.g., the eight-hour day of Seattle, Washington, near the winter solstice) does not provide light during the inducible phase, and thus, the birds remain reproductively quiescent. These same birds will, however, show increased LH secretion and growth of the gonads if the light cycle coincides with the inducible phase (e.g., a 13-hour day of late March, where the last hour of daylight overlaps with the inducible phase) (fig. 8.19; Hamner 1963, Farner 1964, Follett et al. 1974). Some of the fascinating history of how the circadian basis of avian photoperiodism was demonstrated, as well as specific details of the experimental tests and interpretation of figure 8.19 can be found in the box on Brian Follett on page 228.

Although photoperiod is certainly a key proximate environmental cue regulating avian reproductive cycles, and indeed serving to synchronize entire annual schedules, other environmental cues are also extremely important. Temperature, food supply, and social factors all can serve to accelerate

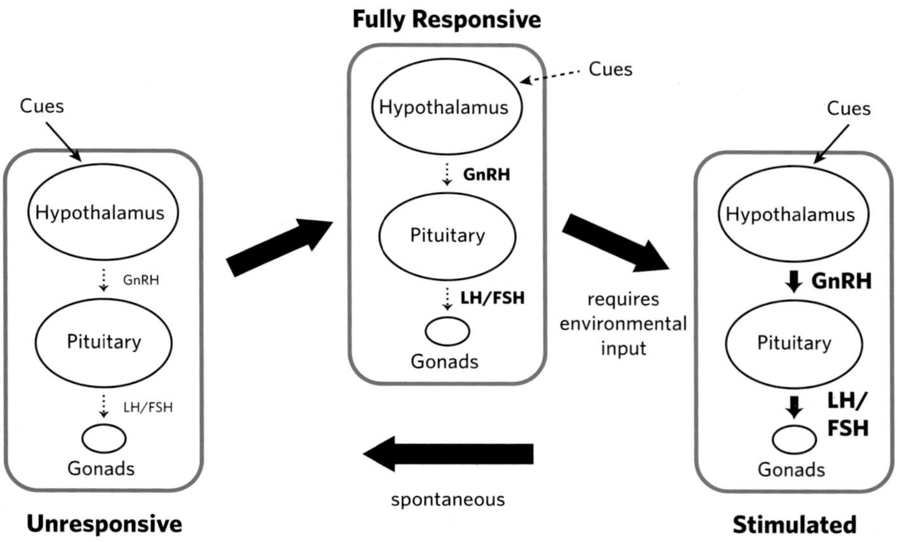

Figure 8.18. The "three stage" seasonal cycle of changing photosensitivity, reproductive condition, and neuroendocrine status in a typical photoperiodic seasonal breeder such as a White-crowned Sparrow, Dark-eyed Junco, or European Starling. Photosensitive individuals (*top*) are "fully responsive" to environmental cues such as photoperiod, food, and social factors, but those cues are not yet stimulatory (dotted arrows from cues to hypothalamus) so the hypothalamus, pituitary, and gonads are relatively inactive (dotted arrows inside blue box). Photostimulated individuals (*right*) are also fully responsive, and since cues are actually stimulatory (solid arrow from cues to hypothalamus), there is GnRH release from the hypothalamus, and gonadotropin (LH, FSH) release from the pituitary (solid thick arrows inside blue box) that fully stimulate the gonads. The transition from "fully responsive" to "stimulated" requires environmental input (a change in environmental cues from nonstimulatory to stimulatory). When the birds become absolutely photorefractory, there is a spontaneous change (no further environmental input required) of the system to an "unresponsive" state (*left*); even though environmental cues (photoperiod, food, social factors) may remain highly stimulatory (solid arrow from cues to hypothalamus), the hypothalamus becomes unresponsive, and no neuroendocrine or endocrine response occurs (dotted arrows inside blue box). The transition back to the "fully responsive" state occurs spontaneously in some species, but generally requires, or is at least facilitated by, a period of exposure to short days to dissipate the refractory state. *After Hahn et al. 2008.*

or delay both the initial development of the reproductive system and the onset of photorefractoriness (Silverin et al. 2008; fig. 8.20). The importance of such non-photic environmental cues to species' responses to changes in climate is discussed below.

Coping with Unpredictable Changes in the Environment

Deep ocean thermal vents are probably the only example of an environment in which there are essentially no unpredictable changes. All other environments, and all environments that a bird might inhabit, can generate unpredictable fluctuations in resources, weather conditions, social factors, predation risk, immune challenge, or risk of injury. Birds must cope with these changes by altering behavior or physiology, but the degree to which they do so is dependent on the nature of the unpredictable change, the bird's current life cycle stage, and its own condition or health status. For example, a bird that is invested in raising young may have a more

limited range of responses available (reduced "perturbation resistance potential"; Wingfield et al. 2011a, 2011b) than a bird that is not raising young. Further, a reproductive individual generally has higher baseline energy costs and may, therefore, be more at risk of becoming stressed by an unpredictable environmental change. We often use the term "stress" in our daily lives to indicate a feeling of anxiety or lack of control, but the word has a much more technical and, unfortunately, ambiguous meaning in the world of avian research. The stress response typically refers to the process of a bird responding to an environmental change in an adaptive way (i.e., a way that prevents or reduces negative impacts on the bird). Thus, the "stress response" is better thought of as an "antistress response." However, if the adaptive responses to a stressor fail to restore the bird to a healthy norm, then those same adaptive mechanisms can become problematic in and of themselves, leading to a state of chronic stress. The use of the word "stress" has many different connotations and meanings

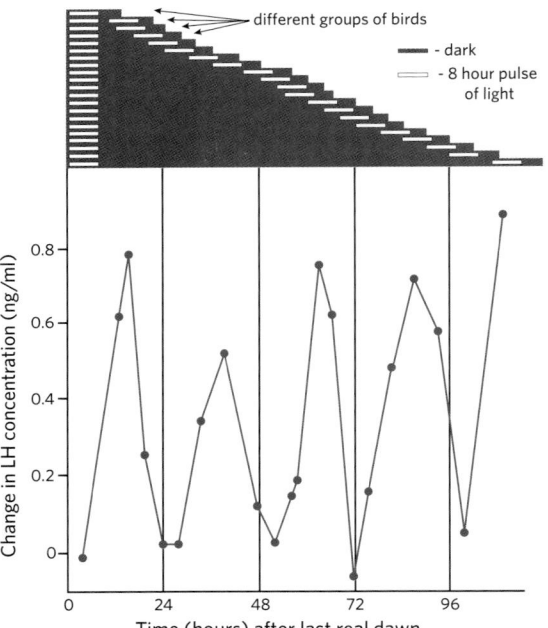

Gambel's
White-crowned Sparrow

Figure 8.19. Illustration of an experiment definitively demonstrating that White-crowned Sparrows display a circadian cycle of reproductive responsiveness to light, rather than measuring the total duration of light to determine whether days are long or short. Multiple separate groups of birds (rows in top part of figure) were all initially held on normal short-day photocycles (8L:16D; light periods depicted by pale bars in each row). They were then allowed to free-run on constant darkness for different periods of time, up to about 96 hours after the end of the last normal short day before being exposed to eight-hour "test pulses" of light falling at different circadian times. Circulating LH levels were measured during the last short day and again near the end of the eight-hour test pulse of light, and the average change in LH secretion in each group was plotted (bottom portion of figure). Circulating LH rose dramatically in groups exposed to a test pulse of light that spanned all or part of the latter half of the circadian day (after circadian hour 12), but was unchanged in groups where the test pulse fell during the first half of the circadian day.

and can be confusing when not understood in context. Researchers have recently developed several models for thinking about responses to environmental change that eliminate this confusing terminology and incorporate more nuanced ways of thinking about environmental change, physiological responses, and interactions between the two.

The Allostatic Model: It's All about Energy! Many different environmental changes (i.e., perturbations) boil down to having an energy cost for the bird. Storms, cold weather, and reduced food availability have obvious energy costs for birds that are trying to maintain a homeostatic body temperature and meet the energy costs of maintaining optimal (or sufficient) tissue function. Other perturbations may create more complex demands. For example, the introduction of a new male on a breeding territory induces established males to increase singing rates and aggressive interactions. These not only take energy to perform but also take time away from energy assimilating behaviors like foraging. The additional energy cost of a perturbation to a bird, relative to the costs it was already having to meet in a given life cycle stage, determine whether or not the bird will have to enter an emer-gency life history stage (ELHS, i.e., abandon other pursuits such as breeding or migration in favor of behaviors and physiology that enhance immediate survival). The allostatic model (fig. 8.21) describes these different contributing factors or contexts and theoretically predicts when a bird will enter an ELHS or enter a state of pathology if the ELHS fails to restore the system to a homeostatic norm. In short, if energy costs of a particular life history stage become greater than the energy available in the environment, then the individual will experience allostatic overload and change its behavior and physiology to try to restore allostatic load to a level that better matches the environmental conditions.

Allostatic Mediators and Systems of Classification The body must have the means to coordinate behavioral and cellular allostatic responses to environmental or endogenous change. Hormones, such as glucocorticoids, catecholamines, cytokines, and so on play a central part in this and are called **allostatic mediators**. These mediators fluctuate in response to many of the energy variables and environmental conditions described in figure 8.21. A growing body of research focuses on fluctuations of the steroid hormone corticoste-

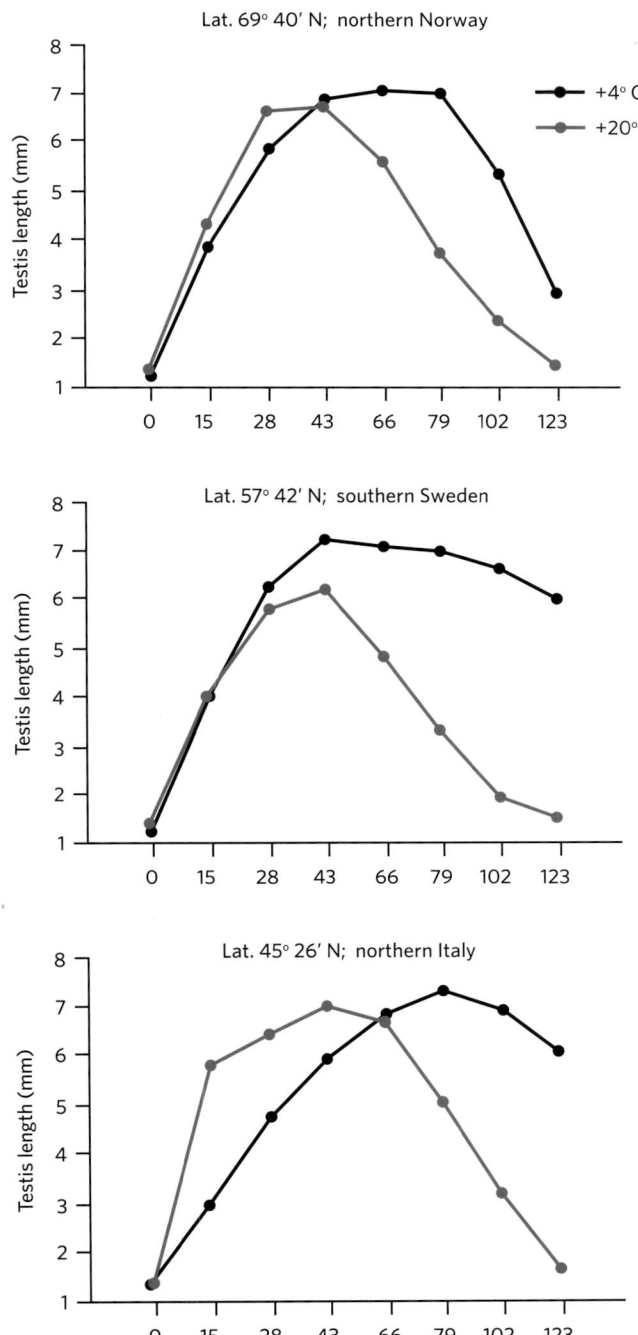

Figure 8.20. Effects of non-photic cues (ambient temperature) on the avian reproductive system in Great Tits from different latitudes. Long-day induced gonad development was similar at high and low ambient temperatures in the two populations from relatively high latitudes, but onset of photorefractoriness occurred sooner in birds held at higher temperatures than in birds held at lower temperatures. In birds from the lowest latitude population, both gonadal development and onset of refractoriness were delayed in birds held on low temperatures compared with those held on high temperatures. *Redrawn from Silverin et al. 2008.*

rone (CORT) in birds. CORT is considered a central player in the coordination of behavioral and physiological responses to increased allostatic load. Allostatic mediators like CORT offer an opportunity to estimate allostatic load in wild populations. This general concept has been around for many decades and was pioneered by John Wingfield (box on page 230). Another system for classifying CORT levels that also complements the allostatic energy model is the reactive scope concept described by Romero and colleagues (2009), which has the capacity to predict homeostatic failure and allows for incorporation of wear and tear that chronic stress can impose on birds.

Responses to Environmental Change: Coping with the Unpredictable Birds are often found to tolerate severe conditions that would seriously challenge a human's ability to survive. Some birds, for example, can survive for many months without a water source other than the seeds or vegetation they consume, and others can survive extreme low temperatures of high latitudes or elevations. The caveat, of course, is that these extreme conditions are predictable (i.e., repetitive and consistent seasonal or diel phenomena) to which the species has evolved relevant coping mechanisms. A cold winter night in the Arctic would not, therefore, be likely to cause allostatic overload so long as the temperature was not significantly outside the normal range experienced in that season and adequate food resources were available. The evolved physiological mechanisms for enhancing thermoregulatory abilities in the winter obviate activation of an emergency life history stage. The same temperature, however, might induce a strong reactive response and allostatic overload if experienced in the summer when conditions are normally more benign and the bird is in an entirely different physiological state. The real challenge in maintaining homeostasis and meeting energy demand is in response to *unpredictable* changes in the environment, for which the bird may not have the time to prepare or possess the evolved mechanisms to easily cope.

Unpredictable changes in the environment can be categorized based on how strongly they impose demands on the bird. Very short-term perturbations, such as a single predator attack or a rare aggressive interaction, may induce a catecholamine response (i.e., fight-or-flight autonomic nervous system response), and can cause at least temporary increases in CORT (e.g., Jones et al. 2016), but probably do not lead to a prolonged increase in CORT or energy demand. These are called **indirect labile perturbation factors**, and birds typically resume normal activities within seconds to hours of having experienced one. Perturbations that last slightly longer (minutes to days) and cause allostatic overload leading to a true ELHS are called **direct labile perturbation factors**. The ELHS involves suppression of nonsurvival

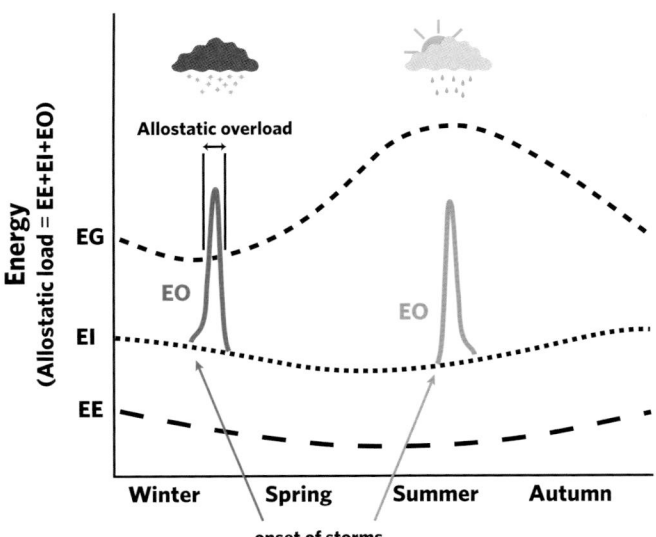

Figure 8.21. Illustration of the allostatic model for how birds cope with varying environmental conditions (supplies and demands) and commitments to different life cycle stages, including unpredictable challenges potentially requiring an emergency life history stage. The existence energy (EE, bottom trace) is highest in winter, when environmental conditions are most thermally challenging. The ideal energy (EI), which includes additional demands imposed by reproduction, migration, and so on, is elevated above EE at all times of year, but still lower in summer when conditions are relatively benign. Energy to be gained from the environment (EG) increases dramatically in spring and summer, when food supply increases. When the energy demand placed on the bird increases owing to a perturbation such as a storm (EO), the effect depends on several things, including what the demand level was before the perturbation and the resource availability. Two storms that increase the energy demand by equal amounts (*left*, red, EO trace, and *right*, orange, EO trace) may nevertheless differ dramatically in how challenging they are to the bird. Because of initially high EI and low EG, the winter storm rapidly puts the bird into allostatic overload. In contrast, because of the initially low EI and very high EG, a similar storm in summer does not even approach forcing the bird into allostatic overload. The bird would be expected to mount a stress response to the winter storm, but not to the summer storm shown in the figure. *Modified from Blas 2015.*

behaviors (e.g., territorial or courtship behavior) and adoption of a coping strategy, which can vary from initiating a set of survival tactics (e.g., seeking shelter or increased foraging behavior) to leaving an area altogether in the event that survival tactics are not sufficient to cope with conditions. Activation of the ELHS is induced by allostatic mediators such as CORT, which have a wide variety of effects on target tissues, including (1) increasing protein catabolism, gluconeogenesis, and triglyceride catabolism to provide cells with ready access to energy for meeting increased metabolic demands;

(2) altering activity levels to promote either exploration of the environment for additional resources or seeking shelter; (3) suppression of nonsurvival behaviors such as territoriality and reproductive behavior; and (4) increase of night restfulness to save energy. These effects mediated by CORT are adaptive responses that normally bring a bird from a state of allostatic overload back into a homeostatic norm.

White-crowned Sparrows provide an example of how this system of allostatic regulation is hypothesized to work (fig. 8.22). A very early study in avian field endocrinology showed the response of Puget Sound White-crowned Sparrows (*Zonotrichia leucophrys pugetensis*) to nasty weather in May during the early breeding season. Low temperatures, high winds, and rain were associated with greatly elevated circulating CORT in these birds while they weathered the storm. Although subcutaneous fat deposits were reduced during this storm, indicating that the conditions were taxing to the birds, circulating LH and T remained unchanged, suggesting that the elevated CORT levels likely were redirecting behavior away from reproduction while leaving reproductive physiology intact so as to facilitate resumption of breeding activities immediately upon return of better weather. A very similar pattern to this occurred in a study of eastern Song Sparrows (*Melospiza melodia melodia*) in the mid–Hudson Valley of New York State. Foul weather in spring caused birds to abandon territories, form flocks, and spend time foraging rather than engaging in reproductive behavior. These changes were associated with increased plasma CORT levels, but reproductive physiology again was relatively unaffected (Wingfield 1985a, 1985b). These studies first demonstrated a correlation between increased CORT and behavioral changes during allostatic overload, which has since been tested through experimental manipulation of hormone levels. Mountain White-crowned Sparrows (*Zonotrichia leucophrys oriantha*) occupy much more extreme, high-elevation environments than either Puget Sound White-crowned Sparrows or eastern Song Sparrows, and they display dramatic movements to low elevations during severe spring snowstorms if food becomes limited (fig. 8.22; Hahn et al. 2004). The decision of how to respond to adverse weather is hypothesized to depend on their ability to meet energy demands via adjustments in behavior. The thermoregulatory demands imposed by snow storms, in combination with snow covering up available food, is thought to cause modest increases in CORT that enhance local foraging activity. If this behavioral change successfully meets demand then the birds remain on or near their territory. If not, then CORT continues to rise and the birds shift into an ELHS where they may abandon the breeding territory and escape to low-

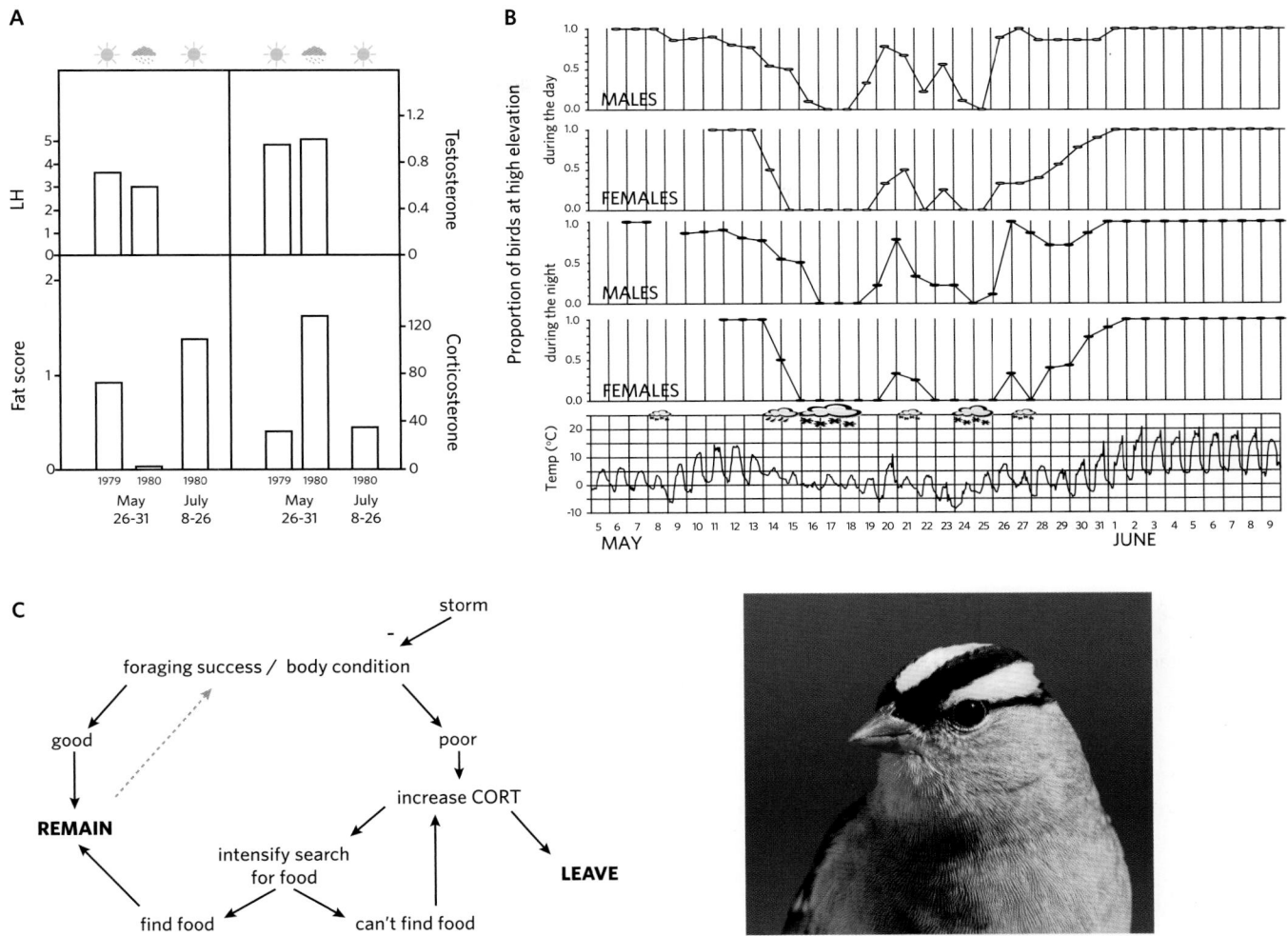

Figure 8.22. Potential role of CORT in management of allostatic load in White-crowned Sparrows. *A*, Effects of a severe spring storm in the Puget Sound lowlands on circulating levels of CORT, subcutaneous fat deposits, and reproductive hormones in free-living Puget Sound White-crowned Sparrows; *B*, Mountain White-crowned Sparrows display dramatic movements to low elevations (about 1,000 meters elevation drop) east of the Sierra Nevada crest when severe spring snowstorms present high thermoregulatory costs combined with limited access to food. They return to the high elevation breeding area when conditions improve; *C*, The decision of Mountain White-crowned Sparrows how to respond to adverse weather in spring is hypothesized to depend on their ability to meet energy demands via adjustments in behavior. Increasing thermoregulatory demands combined with reduced food accessibility initially leads to modest increases in CORT and increased local foraging activity. If this successfully meets demands, the birds remain on or near their territory. If not, and if further increases in local foraging still fail, then CORT is expected to eventually rise sufficiently to shift the bird into an ELHS involving abandonment of the breeding territory and escape to low-elevation refugia within a few kilometers of, and about 1,000 meters below, the breeding site. *A, from Wingfield et al. 1983; B, from Hahn et al. 2004; C, from Breuner and Hahn 2003.*

elevation refugia (Breuner and Hahn 2003). Experimental studies using CORT implants support this hypothesis. Specifically, CORT implants cause the sparrows to increase the size of their foraging areas but not to abandon territories during good weather when food remains accessible. However, CORT-implanted sparrows spend much more time at low elevations than do control individuals during severe storms that limit availability of food (Breuner and Hahn 2003).

In the above examples, the flexible adjustments employed by the birds typically alleviate the taxing conditions. That is, the ELHS successfully facilitates survival, and after the storm, the birds return to their territories in good condition and resume reproductive activities. If, however, these mechanisms fail to accommodate the challenge, or if the perturbation becomes more permanent, the bird may enter a state of homeostatic overload, or chronic, pathological stress. At this

BRIAN K. FOLLETT: REPRODUCTIVE NEUROENDOCRINOLOGY AND MECHANISMS OF AVIAN PHOTOPERIODISM

Sir Brian K. Follett is a central figure in the field of vertebrate endocrinology, particularly regarding the mechanisms of avian photoperiodism. He received his PhD from the University of Bristol, where he eventually returned as head of the Zoology Department after stretches at the University of Leeds and at the University of Wales, Bangor. When he began studying avian photoperiodism in the 1960s, it was not even clear whether birds possessed two separate gonadotropic hormones or a single gonadotropin with both LH-like and FSH-like effects. Further, methods for measuring the content of gonadotropins in avian tissues were restricted to various bioassays. These suffered from relatively poor sensitivity as well as a requirement for very large numbers of animals. In 1971, Follett and his student Colin Scanes and colleague Frank Cunningham succeeded in developing a highly sensitive radioimmunoassay for chicken luteinizing hormone (LH). This technical advance required collection of hundreds of chicken pituitary glands from which to purify a sufficient quantity of the hormone to be used for development of a specific antibody to chicken LH, as well as for use in generating standard curves as part of the assay.

Radioimmunoassays are a form of "competitive binding assay," involving competition of the "cold" hormone (either the naturally occurring hormone in a sample of interest or a known quantity of the "standard" purified hormone added to a standard curve) with "hot" (radioactively labeled) hormone for binding sites on a high-affinity antibody specific to the hormone of interest. After a period of equilibration with the antibody in the assay tubes, the "bound" fraction (the antibody, plus any hormone to which antibody is bound) is separated from the "free" fraction (any of the hormone, whether "cold" or "hot," not bound to antibody) and the amount of radioactivity in one of the fractions is counted on an appropriate machine (e.g., a gamma counter, for an assay employing a gamma-emitting isotope such as Iodine-125, as was typically used for these particular assays). The amount of radioactivity in the "bound" fraction would be high if there had been little "cold" hormone present to compete for antibody binding sites, but low if large quantities of "cold" hormone had occupied the available binding sites. Using a standard curve generated by competing a range of known amounts of "cold" hormone with a fixed quantity of "hot" hormone, the amount of "cold" hormone in blood samples or extracts of tissues could be inferred from the amount of radioactivity measured in the "bound" fraction.

Follett and colleagues' development of this specific radioimmunoassay for avian LH was a game-changer when it came to the options available in experimental studies of reproductive function, and particularly for exploring mechanisms of response to environmental cues such as photoperiod. With this assay in hand, the researchers were able to select for a strain of Japanese Quail (*Coturnix japonica*) with very strong photoperiodic responsiveness; individuals held on short photoperiods (e.g., 8L:16D) all consistently showed extremely robust increases in circulating LH, measured using the new radioimmunoassay, after exposure to just a single long day (e.g., 16L:8D). The fact that these birds showed such a strong response, and that the response could be measured using a small blood sample, made it possible to study mechanisms of avian photoperiodism in detail. Further, the assay (or variants of it) proved effective for measuring LH in songbirds in addition to galliforms such as chickens and quail. The convenience and broad applicability of this one hormone assay provided a reliable, relatively nonintrusive way of "interrogating" individual birds regarding their photoperiodic status—whether they were photosensitive or photorefractory. Further, it provided an important window into the physiological level at which regulation was occurring. For instance, the fact that gonadectomy (removal of the gonads) of a photorefractory European Starling (*Sturnus vulgaris*) leads to no change in circulating LH, whereas the same procedure performed on a photosensitive bird held on short days leads to an immediate dramatic rise in LH, even without any increase in photoperiod, confirms that the refractory state is not a consequence of enhanced sensitivity to negative feedback on the brain or pituitary by gonadal sex steroids (i.e., it is a "central" or "brain-mediated" phenomenon). However, the small amounts of sex steroids produced after photorefractoriness dissipates can play a role in keeping the reproductive system inactive until lengthening days greatly increase the "hypothalamic drive" on the pituitary via an increase in secretion of GnRH.

One of the greatest benefits of this sensitive LH assay was that it allowed repeated, nonterminal sampling of individual small birds over short time scales, so that *changes* in circulating

LH within individuals exposed to different treatments (not just *average* differences among treatment groups) could be evaluated. This fact is responsible for Follett being able to conduct his absolutely classic experimental demonstration of the circadian basis of day length measurement by birds. Up until the 1960s, there had been a debate as to whether birds measured day length using an "hourglass" type mechanism or a circadian mechanism. The hourglass mechanism was posited to tally up the entire duration of a light period, and treat it (physiologically) as "long" only if it exceeded some total duration threshold (e.g., at least 12 hours of continuous light). In contrast, the circadian mechanism was posited to determine *when* it was light on the background of an *oscillating rhythm of sensitivity to light*. Under this hypothesis, light periods longer than 12 hours led to reproductive development not because the total duration of light was so long, but because it was light during a portion of the animal's circadian rhythm that was sensitive to light, and this phase began a number of hours (e.g., about 12 hours) after dawn, or subjective dawn (see the box on page 220 for biological rhythm terminology). William Hamner had demonstrated in the mid-1960s that House Finches (*Haemorhous mexicanus*) grew their gonads when held on "resonance photocycles" on which short light pulses (e.g., six hours at a time) coincided with the latter half of birds' "circadian days," whereas identical duration light pulses failed to induce gonadal development if they remained in phase with normal dawn time. This finding was consistent with a circadian, rather than an hourglass, mechanism of photoperiodic time measurement. However, resonance experiments suffer from an extremely thorny entrainment problem. Specifically, the very light pulse used to "interrogate" the bird as to whether its reproductive system "thinks" the day is long or short also would re-entrain the timing of the hypothesized "photoinducible phases" of the circadian rhythm that followed. Since Hamner could use only gonad growth as his metric for a reproductive response, and since detecting an effect on the gonads required prolonged exposure to "long days" (many days of treatment, because the gonads take time to change size), his resonance experiments suffered from an intractable entrainment problem.

Employing the advantages of the sensitive LH assay, Follett was able to design an experiment that solved in one stroke this entrainment problem and confirmed unequivocally that birds use a circadian system of photoperiod measurement. While visiting Donald Farner's laboratory in Seattle, Follett kept groups of Gambel's White-crowned Sparrows (*Zonotrichia leucophrys gambelii*) on constant short days of 8L:16D. On the first day of the experiment, he collected small blood samples from all the birds during the morning of their last normal "short" day. After the end of this day, all birds went into a period of constant darkness, after which they were exposed to a "test" pulse of light that was also "short"—only eight hours in length (a duration that would normally never induce reproductive activation in these birds). For different treatment groups, the test pulses began anywhere from 10 to 108 hours after the last normal dawn. A second small blood sample was then taken from each bird 7–16 hours after the beginning of the test light pulse, i.e., after the bird had had time to experience this light pulse. After performing the assay for LH on all these blood samples, the "change" in circulating LH between blood sample one (during the last short day) and blood sample two (after exposure to the eight-hour duration test light pulse) was calculated for each individual bird, and the results plotted according to the time of the test light pulse after the last normal dawn (see fig. 8.19). Sparrows that experienced their test pulse during their circadian "morning" (i.e., for which the test pulse began in phase with "expected dawn") did not show any change in circulating LH; they treated the eight-hour test pulse as another short day. Those for which the test pulse illuminated all or part of the latter half of the circadian day showed dramatic increases in circulating LH; they treated the eight-hour test pulse as a long day. Since no further data needed to be collected from each bird after the second blood sample was taken immediately following the test light pulse, any entraining effect the test pulse had on the timing of future photoinducible phases was irrelevant to this experiment; the entrainment problem had been solved.

Follett's work on avian photoperiodism extended to (1) studies of interspecies variation in photoperiodic mechanisms, particularly differences between "absolute" and "relative" photorefractoriness; (2) the role of thyroid hormones in programming photoinduction and photorefractoriness; (3) localization of the anatomical sites involved in processing and transducing light cues; and (4) understanding the extent to which photoperiodic responses reflected light acting as a "driver" of the annual cycle as opposed to an "entraining stimulus" for an endogenous circannual rhythm. His relentless focus on mechanisms at the level of neuroendocrine and endocrine physiology has had a huge impact on the fields of photoperiodism, avian endocrinology, and organism-environment interactions more generally.

JOHN C. WINGFIELD: PIONEER IN FIELD AND ENVIRONMENTAL ENDOCRINOLOGY OF FREE-LIVING BIRDS

John Wingfield followed a circuitous route to his stature as the father of modern avian field and environmental endocrinology. He had originally hoped to conduct his graduate work with David Lack at the Edward Grey Institute of Field Ornithology, Oxford University. When he was not accepted to pursue that dream, he instead joined the laboratory of Andrew Grimm at University College of North Wales, where he studied the endocrinology of a flatfish called the plaice, focusing in particular on the fish's steroid hormone biochemistry and physiology. Upon completion of his doctoral work, he moved in the mid-1970s to the laboratory of Donald Farner in the Department of Zoology at the University of Washington in Seattle. Farner, long interested in how birds processed environmental cues (especially photoperiod) to regulate their annual cycle of reproduction, molt, and migration, humored Wingfield's ambitious plan to attempt to study the endocrinology of these birds "in the field," though initially Farner was a bit skeptical that the undertaking would be successful, because of the huge technical challenges involved and because so many uncontrolled factors were potentially affecting free-living birds' endocrine physiology. Wingfield initiated studies of the field endocrinology of Puget Sound White-crowned Sparrows (*Zonotrichia leucophrys pugetensis*) at Farner's land on Camano Island, Washington, as well as of Farner's beloved Gambel's White-crowned Sparrows (*Z. l. gambelii*) near Fairbanks, Alaska. When Farner saw the data, with hormone levels plotted according to life cycle stage, he knew he had been wise to take a chance on supporting Wingfield's field studies. The papers Wingfield and Farner published in the mid- to late 1970s on the temporal patterns of circulating sex steroid hormones and glucocorticosteroids of these birds were the first of their kind and constitute the founding publications in the now-massive research area known as "field endocrinology."

The small size of white-crowned sparrows presented a daunting challenge to someone interested in measuring changes in their circulating hormone levels without disrupting the birds' survival or reproductive success: they could tolerate donation of only tiny blood samples (about 300 microliters maximum). As a result, hormone assays needed to be extremely sensitive. The tiny blood volume available created the further conundrum that, even with very sensitive assays, virtually the entire plasma sample from a particular blood collection would be needed just to measure levels of a single hormone. Wingfield used his expertise with steroid biochemistry and endocrine physiology to develop extremely sensitive radioimmunoassays for multiple steroid hormones. Further, he employed the technique of column chromatography to separate steroids of different polarities out of each individual plasma sample so that he could measure as many as five different individual steroids from a single tiny plasma aliquot. This approach had the added benefit of allowing specific quantification of extremely similar compounds such as testosterone (T) and 5-α-dihydrotestosterone (DHT) that could not otherwise be distinguished reliably because available antibodies cross-reacted strongly with both. The methods developed by Wingfield while in Farner's lab formed the foundation for years of further field and laboratory work on the endocrinology of small birds, as well as of numerous non-avian animals. Despite the recent advent of other assay methods, such as enzyme-linked assays, many of the advantages of Wingfield's original column chromatography followed by radioimmunoassay remain valid today, and researchers interested in measuring specific androgens (e.g., T and DHT) still employ these methods developed back in the 1970s in Seattle.

Wingfield's successful application of endocrinology techniques to free-living small songbirds laid the groundwork for his own subsequent work, now spanning 40 years, on the environmental and behavioral endocrinology of a wide variety of birds. He and his students and postdocs have pioneered the use of experimental modification of endocrine function

point the perturbation is called long-term and can be classified as a "modifying factor." Modifying factors are perturbations lasting from months to years and can severely disrupt the normal life cycle of the bird by continuously imposing high-energy costs. Examples of modifying factors include permanent destruction or damage to a habitat, shifting resources caused by climate change, and injury or illness. The allostatic overload induced by such factors results in chronically high CORT secretions, which can lead to maladaptive consequences for the bird. Reproductive behavior and physiology is suppressed—a consequence that is not in and of itself a problem for the adult bird, but immunosuppression can increase susceptibility to infection or disease, and growth suppression can stunt normal development in juveniles. The

(via hormone and blocker implants) to manipulate behavioral and physiological traits, thereby testing hypotheses regarding hormone-behavior interactions, mechanisms of endocrine regulation, and adaptive significance of variation in life histories. In addition, they have laid out conceptually broad frameworks for understanding such diverse topics as (1) the effects of social factors and other cues on androgen physiology, reproduction, and behavior under natural conditions (e.g., the Challenge Hypothesis); (2) the mechanisms by which animals partition their lives into separate life history stages (e.g., reproduction, plumage molt, migration, in birds); (3) the mechanisms by which different kinds of environmental cues (e.g., photoperiod, food, social factors) are processed and integrated at the levels of brain and peripheral physiology; (4) the ecological and evolutionary importance, and mechanistic regulation, of adaptive responses to unpredictable environmental challenges (i.e., the emergency life history stage); (5) the application of the broad concepts of allostasis, allostatic load, and allostatic overload to birds' and other animals' natural lives; and (6) how mechanisms of endocrine regulation and response to the environment affect how species will be affected by human-induced alteration of the environment (e.g., climate change). These conceptual frameworks have been used by numerous researchers as bases for developing and testing mechanistic and evolutionary hypotheses in a wide variety of both avian and non-avian study systems.

The importance of Wingfield's visionary work in field endocrinology is difficult to exaggerate. Following the original papers on white-crowned sparrows in the 1970s, which established nonterminal methods for sampling endocrine function of small free-living animals, there was a massive blossoming of interest in applying endocrine techniques to study the behavioral ecology and environmental physiology of free-living animals. The influence of his work on our understanding of the biology of birds and other animals has been huge and lasting.

energy-mobilizing effects of CORT at the cellular level can also cause severe muscle wasting, which can result in reduced capacity to fly. Birds experiencing these symptoms are clearly in dire straits, and even if the modifying factor causing the imbalance is removed, they may not fully recover.

Beyond the Hormone: Other Key Players in Stress Physiology   Glucocorticoid (e.g., CORT) secretion is one very important and highly plastic (i.e., responsive) element of the stress response system in birds and other vertebrates. The levels of CORT in the blood plasma reflect upstream control mechanisms involving HPA axis (see above) hormones such as CRH and ACTH. The actions of CORT are also influenced by factors such as the presence of corticosteroid-binding globulins (CBGs) in the blood, and changing sensitivity of target cells to CORT. CBGs are proteins produced in the liver that bind CORT in the blood and thereby influence how much CORT the aqueous blood plasma can hold, and potentially affect the availability of CORT to cellular receptors at targets. It is not entirely clear whether CORT must be "free" (not bound to CBG) to gain access to receptors, or if CBGs somehow facilitate the translocation of CORT into cells to reach intracellular receptors. In any case, the quantity of CBGs in the blood is potentially important to the effects of CORT in allostatic coordination. Likewise, the nature and number of CORT receptors will affect tissue responses to circulating CORT. CORT receptors are located largely inside the cell (i.e., intracellular receptors), but faster-acting protein-coupled receptors can also be present on the cell membrane. There are multiple types of intracellular CORT receptors: the mineralocorticoid receptor (MR) has a high affinity for CORT and therefore is active at relatively low CORT levels, and the glucocorticoid receptor (GR) has a lower affinity for CORT and therefore does not become activated until CORT levels reach relatively high levels. Thus, MR is generally considered to regulate basal responses to CORT, while GR is generally considered to regulate stress-induced responses to CORT. Both are probably involved in relatively complex ways in maintaining allostatic equilibrium, and research into receptor-mediated mechanisms is on the rise.

THE AVIAN ENDOCRINE SYSTEM IN AN ECOLOGICAL AND EVOLUTIONARY CONTEXT

Endocrine System as Regulator of Life History Trade-Offs

One of the most important fitness-enhancing processes of coordination in birds' lives is balancing life history trade-offs between survival and reproduction. In general, processes and activities that enhance immediate and long-term survival are temporally segregated from those enhancing immediate reproductive success, and the endocrine system plays an integral part in this coordination.

An excellent example of a life history trade-off coordinated by the endocrine system is the timing of the transition from breeding to plumage molt in many temperate-zone birds. Reproduction is energetically and nutritionally demanding, as well as time consuming, and is generally restricted to times when resources are abundant and conditions are relatively benign. However, molt also is an energetically and nutritionally demanding process that requires a high level of resources and is most easily completed when conditions are relatively benign. Consequently, times that are good for reproduction are also among the best times to molt. Obviously, breeding represents an investment in the "reproductive success" component of fitness. Molt actually contributes to both reproductive success and survival, because it can affect both sexually selected appearance traits and naturally selected traits such as feather durability, insulation quality, flight surface quality, and camouflage. In most temperate-zone species, plumage is completely replaced once per year during an annual prebasic molt (see chapter 9), and although this molt does affect appearance, potentially even of the future "breeding plumage" (often acquired through wear during the months following molt), it is arguably its effect on survival through refreshing of insulation and flight surfaces that is most important to birds' fitness. Rather than breed and molt simultaneously, most species breed while conditions are still improving, e.g., in spring and early summer, and defer the prebasic molt to just after breeding (although specific patterns vary, and some species do not molt until after migration). Endocrine regulation of the timing of plumage molt is not well understood; however, one endocrine-molt interaction that is critically important is that elevated sex steroid hormones typical of breeding inhibit molt (fig. 8.23). As a result, individuals prolonging reproductive activity will begin molt later, and molt more slowly once begun, thereby reducing energetic and nutritional conflicts between breeding and molt. A negative consequence of delaying and slowing molt when breeding is prolonged is that a more rapid molt may be required, leading to reduced feather quality (see below).

Another example of hormonally mediated coordination of a life history trade-off is the role played by CORT in redirecting time and energy toward survival and away from reproduction when birds enter an emergency life history stage (ELHS, see above).

Effects of Rapid Environmental Change
Endocrine-Based Annual Schedule Mismatches to the Environment

One of the most conspicuous effects of human-induced rapid environmental change is the disruption of long-standing correlations between environmental conditions pertinent to the survival and reproductive success of birds (ultimate factors) and cues (proximate factors; Baker 1938) that avian populations have evolved to use to appropriately time adjustments in behavior, physiology, and morphology. For instance, changing climate can lead to overall average advancement of the date of peak food supply critical for successful reproduction. The consequence of this change for avian populations will depend on the nature of the cue-processing mechanisms used to time life cycle events. In cases where the neuroendocrine regulation of changing reproductive condition depends primarily on a response to photoperiod, with relatively little plasticity for adjusting scheduling details using non-photic cues, correction of the emerging mismatch between reproductive schedule and timing of maximum food supply will require an evolutionary change in the population. If the adaptive modification of the timing mechanisms proceeds more slowly than the rate of the change in relationship between date and peak food supply, then the mismatch will gradually increase (fig. 8.24).

There already is evidence of such mismatches in some free-living populations. For instance, in a Dutch population of Great Tits (*Parus major*) that has been studied since the 1970s, reproductive success is dependent on how well pairs of birds match the time when they have nestlings to the seasonal peak in abundance of caterpillars that feed on the forest trees. Historically, the Great Tit population has done a very good job of anticipating when the peak caterpillar abundance would be—the birds need to lay eggs two to three weeks before peak caterpillar abundance is reached, so that peak nestling demand will coincide with the maximum abundance of food. As spring temperatures have increased in recent years, the date of peak caterpillar availability has advanced. Apparently because these birds rely heavily on photoperiod as the primary cue timing reproduction, and perhaps because of details of how they respond to temperature cues to fine-tune that schedule, the population is not tracking the changing appearance of caterpillars with appropriate changes in nesting timing (Visser et al. 1998; fig. 8.25). As a result, there is an increasing "selection differential" caused by the increasing discrepancy between ideal timing of onset of nesting and actual timing of onset of nesting in the population. Populations like this one that lack the temporal flexibility to adjust to the changing climate, and that do not respond to selection with adaptations of their timing mechanisms, are potentially at risk of declines or even extinction in the face of climate change.

Populations that rely more heavily on the fine-tuning effects of non-photic cues to adjust annual schedules are generally expected to be better at coping with rapid environmental change. Even if a population is photoperiodic, a strong response to accelerating or slowing effects of stimulatory or inhibitory non-photic cues would allow closer tracking of the environment. For example, the effects of temperature on photoperiodically regulated reproductive schedules (see fig. 8.20) could help bring the annual schedule into closer syn-

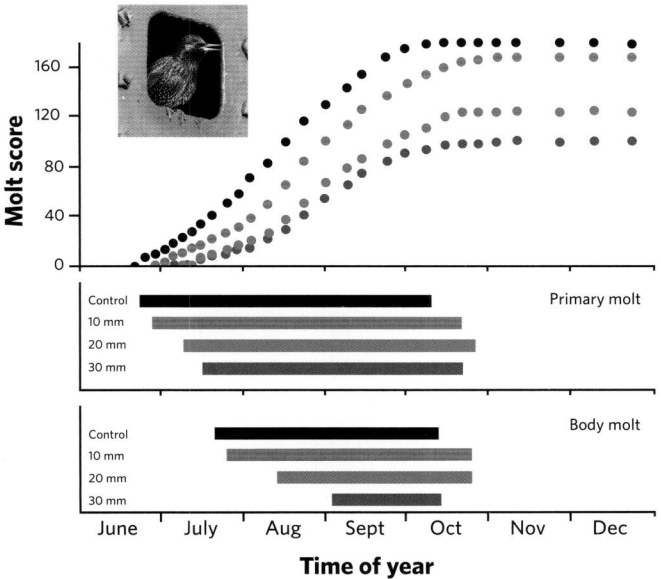

Figure 8.23. Effects of testosterone implants on timing and rate of plumage molt in European Starlings. Persistently elevated T levels such as would occur during prolonged reproduction lead to delay and sometimes to truncation of molt before it is complete. Black symbols and bars = controls (empty implants). Red symbols and bars = 10 mm T implants. Green symbols and bars = 20 mm T implants. Blue symbols and bars = 30 mm T implants. Note that carryover effects would be on timing of preparation for migration as well as on potential for surviving winter if plumage quality is compromised. *Modified from Schleussner et al. 1985.*

chrony with changing food supply in the environment. However, it is also possible that too-strong responses to cues other than photoperiod, such as temperature, could lead to inappropriately timed breeding, if the population responds to erratic fluctuations in temperature that can sometimes produce unseasonably high temperatures long before it is appropriate to initiate nesting. Clearly, further work is needed on the relationships between the mechanisms by which birds process environmental cues, and on the fitness and population-level consequences.

Carryover Effects

Responsiveness to non-photic cues is not necessarily a panacea for maintaining annual schedules that are well synchronized to a changing environment. Even if the response to a non-photic cue improves the timing of one event (e.g., onset of egg-laying) relative to a critical ultimate factor (e.g., date of peak food supply), such an adjustment can produce undesirable sequelae that reverberate through the annual schedule. These effects are termed **carryover effects**, because they carry over from one life cycle stage to others that fol-

low. For example, high ambient temperatures may accelerate reproductive development early in the season, and this may be appropriate and adaptive in terms of short-term enhancement of reproductive success within that season. However, if onset of photorefractoriness is also accelerated, this could lead to inappropriately timed molt or migration. Most work has been done on carryover effects between events occurring during overwintering or spring migration and breeding (e.g., Norris and Marra 2007), but carryover effects can occur among any stages of the annual cycle, and may even carry through for an entire year (Brazeal 2014).

Hormonal mechanisms underlie many carryover effects. For instance, prebasic plumage molt is highly sensitive to inhibitory effects of elevated sex steroid levels (e.g., Schleussner et al. 1985, Hahn et al. 1992; see fig. 8.23). This sensitivity may have evolved as a mechanism to avoid intractable overlap or other conflict (resource commitment) between two demanding processes: reproduction and molt. However, in cases where reproduction is prolonged by a response to non-photic cues, leading to prolonged elevation of sex steroid levels, the timing of onset of plumage molt can be delayed, either throwing off the entire sequence of following stages (if molt duration remains unchanged) or leading to production of inferior feathers (if molt is accelerated to make up for lost time; Dawson et al. 2000). Either of these carryover effects has the potential to be deleterious in the context of a changing climate where non-photic cues push the system outside the normal bounds for which it had been optimized by selection under different conditions.

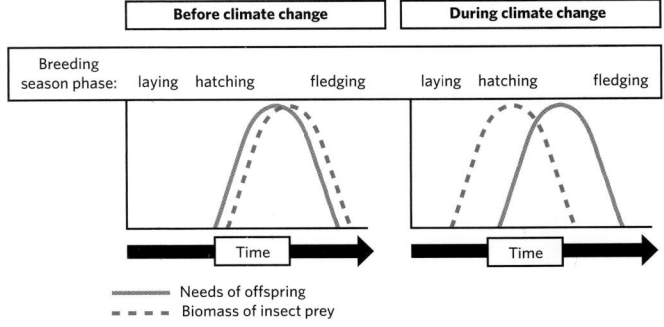

Figure 8.24. Diagram illustrating how a timing mismatch can be caused by rapid environmental change. Prior to the environmental change, timing of nesting is such that peak demand by the offspring (gray trace) corresponds very closely with the timing of peak resource availability (dotted red trace). With climate change, early spring warming causes insect development to occur much earlier. If timing of breeding does not also advance, then peak offspring needs no longer match peak resource availability. *Based on Visser et al. 2004.*

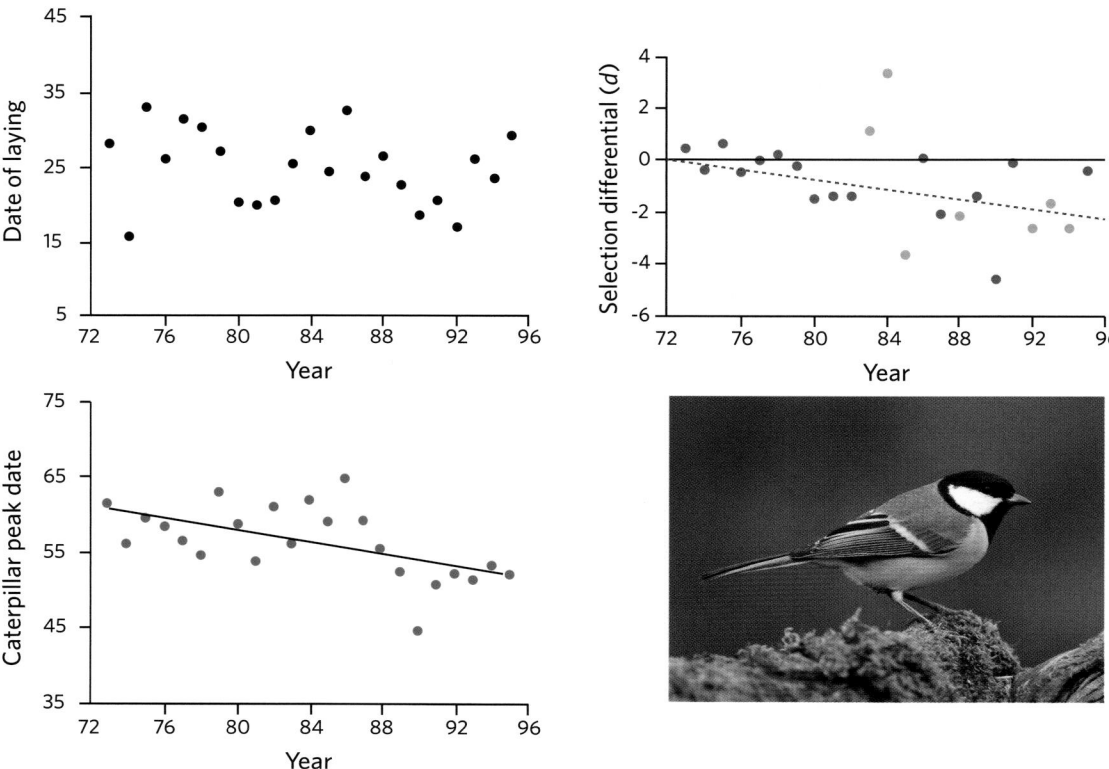

Figure 8.25. Timing mismatch between availability of caterpillars critical to feeding nestlings and the egg-laying date of Great Tits. Date of laying in the population fluctuates between years, but has not been advancing to earlier dates over time (*top left panel*), whereas peak availability of caterpillars has been advancing to earlier and earlier dates as climate changes (*bottom left panel*). This leads to a "selection differential" (dotted line, *top right panel*), which can be thought of as the strength of selection imposed by the environ- ment, and which is increasing in magnitude the further the popula- tion mean laying date gets from the "ideal" laying date dictated by the timing of peak caterpillar abundance. Specifically, pairs that lay earlier produce more recruits to the population than pairs that lay later. Nevertheless, this population is not keeping up with the chang- ing climate, probably because of a strong dependence of the timing of the reproductive schedule on long-term cues such as photoperiod. *From Visser et al. 1998.*

Endocrine Mechanisms under Natural Selection: Case Studies

As the environments birds experience change over time, either through processes like climate change or because of range expansions, existing behavioral and physiological plasticity may be sufficient to cope with the new conditions. However, selection may also lead to changes in behavior, physiology, or morphology that are better suited—adaptively special- ized—to the new conditions. A basic question in the biology of organisms generally, and of birds in particular, is, to what extent do the specific details of physiology and behavior rep- resent adaptive specializations to the particular environmen- tal challenges faced by particular species of birds? This ques- tion has been studied from a variety of specific perspectives.

This section addresses two examples: the endocrine mecha- nisms underlying variation in annual schedules, and the de- tails of testosterone physiology and its regulation.

Interspecies Variation in Neuroendocrine Mechanisms of Photoperiodism

Recent research on the endocrine mechanisms underlying variation in reproductive cycles of birds has addressed the is- sue of adaptive specialization of annual scheduling using a comparative approach. The vast majority of birds that have been studied in any detail display annual cycles of changing reproductive physiology and behavior that track changing environmental conditions. This includes temperate zone and high-latitude species that occupy extremely seasonal environments as well as many tropical and subtropical species that live in less temporally variable environments. As described above, the mechanisms underlying annual cycles of changing reproductive condition entail a combination of endogenous rhythms and responses to environmental cues. The expression of a "reproductive season" requires that the birds initiate reproductive development, reach a state of full reproductive competence and maintain it for some period of time, and then terminate reproduction. The process of termination of reproductive competence is particularly interesting, because most species of birds studied do not simply wait for stimulatory environmental cues to wane as the season progresses. Rather, they spontaneously enter a state of reproductive refractoriness to the stimulatory effects of long days, leading to quiescence of the reproductive neuroendocrine system, collapse of reproductive competence, and the transition into the annual plumage molt. The state of photorefractoriness is of particular interest here because, while it clearly is adaptive, in that it helps guarantee that molt will be completed while the food supply is still good, it also imposes a profound if temporary constraint on the ability to reacquire reproductive competence, thereby imposing strict limits on any individuals that display it.

A comparative examination of the phylogenetic distribution of photorefractoriness in birds has revealed that most species studied display what is termed **absolute photorefractoriness**. That is, they spontaneously terminate reproductive competence without the need to be exposed to any decline in photoperiod (Hahn and MacDougall-Shackleton 2008). Refractoriness is essentially universal in seasonal breeders, and is only known to be absent in highly flexible and opportunistic taxa. For instance, within the cardueline finches (canaries, goldfinches, rosefinches, crossbills, etc.), only the crossbills (temporal reproductive opportunists) fail to display absolute photorefractoriness (fig. 8.26). The fact that crossbills evolved recently, and all sister-groups and groups that branched off more basally on the cardueline phylogenetic tree display absolute refractoriness, suggests that a loss of photorefractoriness is an adaptation facilitating temporal reproductive flexibility in crossbills. Crossbills are adapted to exploiting an unpredictable food source—the seeds of coniferous trees—and are capable of nesting in most months of the year (apart from autumn, when they terminate reproduction to complete plumage molt; Hahn 1998, Cornelius et al. 2011). This pattern appears to hold when considering all birds for which data are available regarding presence or absence of absolute photorefractoriness; only the handful of species that display exceptional temporal reproductive flexibility fail to display absolute photorefractoriness (Hahn and MacDougall-Shackleton 2008). In the context of

annual progression through a series of life cycle stages, absence of a refractory period is expected in species displaying annual schedules that confer greater temporal flexibility (see Wingfield 2008).

At the neuroendocrine level, absolute photorefractoriness is closely correlated with dramatic reductions in the presence of GnRH—the primary top-down neuroendocrine regulator of reproductive function—in the septal and infundibular regions of the brain (Ball and Hahn 1997, MacDougall-Shackleton et al. 2009). As for refractoriness itself, this down-regulation appears to be absent only in extremely temporally flexible species such as crossbills and Zebra Finches. Figure 8.27 illustrates the differences in the GnRH system of three closely related cardueline finches: Common Redpolls, Pine Siskins and White-winged Crossbills. Only the temporally extremely flexible crossbills fail to become refractory and down-regulate the amount of GnRH present after prolonged exposure to very long days (Pereyra et al. 2005).

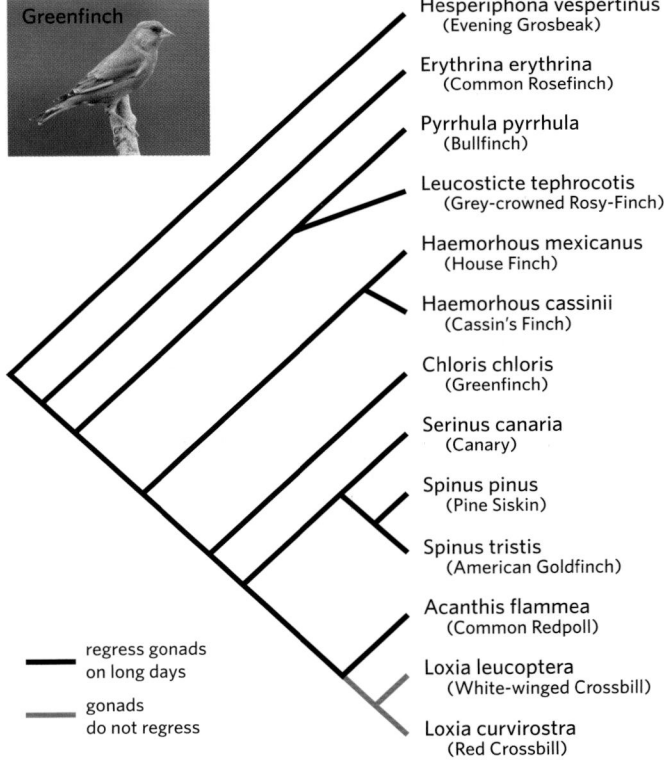

Figure 8.26. Distribution of absolute photorefractoriness in cardueline finches. Only two species of crossbills—by far the most temporally flexible breeders of all of the finches tested—fail to display it, indicating that loss of refractoriness represents a derived adaptation facilitating temporal reproductive flexibility. *Adapted and updated from Hahn and MacDougall-Shackleton 2008, and Hahn, unpublished.*

SD (5L:19D)   LD (20L:4D)

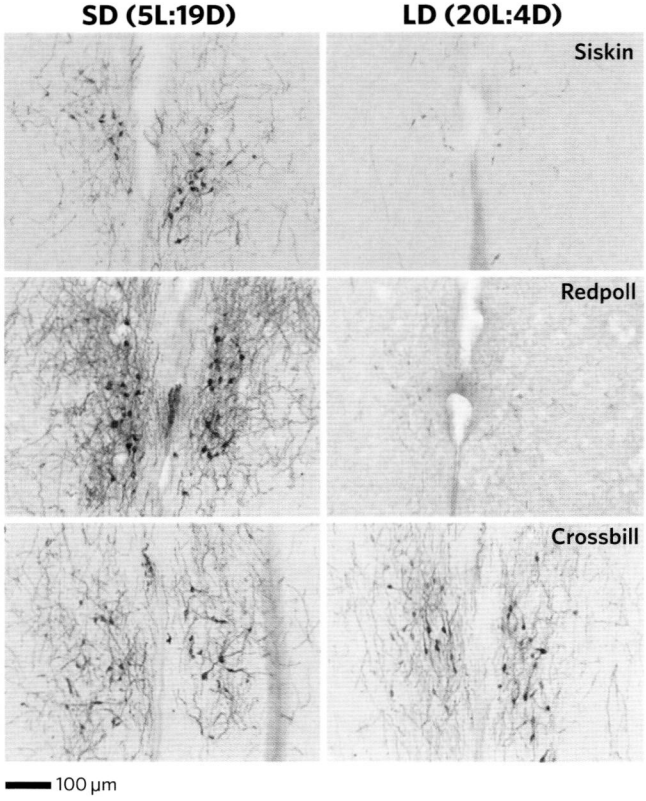

Siskin

Redpoll

Crossbill

■■ 100 µm

Figure 8.27. GnRH immunoreactivity in the septo-infundibular region of the brain is dramatically down-related following prolonged exposure to long photoperiod in two seasonal species (Common Redpoll, Pine Siskin) but not in the opportunistic breeder (White-winged Crossbill). Micrographs on the left are from birds held on constant very short days, and those on the right are from birds placed on very long days for five months. *From Pereyra et al. 2005.*

Testosterone, Behavior, and Morphology of Juncos

Species with wide geographic ranges, and those colonizing new habitats, provide excellent opportunities to explore how natural selection may have modified relationships between endocrine mechanisms and the adaptive physiological and behavioral processes they regulate. A long-term study of Dark-eyed Juncos (*Junco hyemalis*) has been particularly fruitful in revealing how adaptations to different environments relate to the hormonal regulation of behavior and morphology. Ellen Ketterson and her students, postdocs, and colleagues have been studying juncos across their wide geographic range in North America, from the mountains of Virginia through the Rocky Mountains to the west coast. They have focused on a number of behavioral and morphological traits related to testosterone levels and to fitness, including the pattern of white in the tail feathers (used in behavioral displays), aggressiveness in response to simu-

lated territorial intrusions (experimental tests of territorial behavior using song playback), parental behavior, and the timing and duration of breeding. One of the topics they are most interested in is how the relationship between testosterone levels and these behavioral and morphological traits may affect (facilitate, or potentially interfere with) the response of populations to a changing environment. As noted above, this could relate to something like climate change, but it also is relevant to the process of populations establishing themselves in new locations where environmental conditions may be quite different from those typical of where the source population originated. Dark-eyed Juncos commonly breed in the mountains of the west, and spend the winter in lowland areas. Back in the 1980s, apparently some of the juncos that typically wintered around San Diego, California, ceased to migrate back to their normal breeding areas at nearby Mount Laguna and instead began breeding on the UC San Diego campus. The more benign climate, with a much longer potential breeding season, experienced by the birds that colonized the UCSD campus potentially selects for males to place greater emphasis on parental care and less emphasis on sexual and aggressive behaviors and morphological traits compared with the source population on Mount Laguna. How has the relationship between circulating testosterone and these behaviors and morphological traits affected the ability of the population that has colonized the UCSD campus to adapt to the novel conditions experienced there?

In general, male juncos on the UCSD campus show a substantially lengthened period of elevated testosterone, with a lower overall population average than the mountain population. This is consistent with the lowland population's longer breeding season (females laid multiple clutches of eggs, whereas the mountain females typically laid only a single clutch), and a generally "slower pace of life" there. Within both populations, individuals displaying the highest parental care tended to be those with the lowest testosterone, and individuals displaying the greatest aggression tended to be those with the highest testosterone. Likewise, the extent of the black hood and white in the tail feathers (plumage characters important in social interactions) were positively correlated with testosterone in both populations. However, overall, males in the lowland population showed reduced expression of the plumage characters, reduced aggressive responsiveness to simulated territorial intrusion (STI), and increased paternal care compared with the Mount Laguna population (fig. 8.28). Interestingly, when these birds were kept under "common garden" conditions (brought into captivity as juveniles and held for several months under identical conditions), the population level differences in the plumage

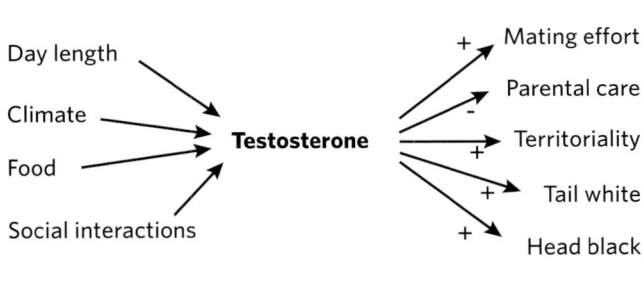

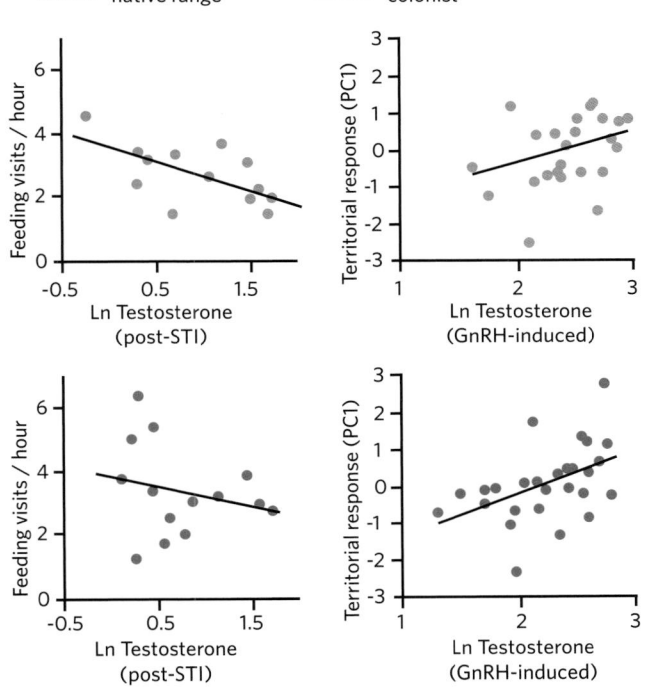

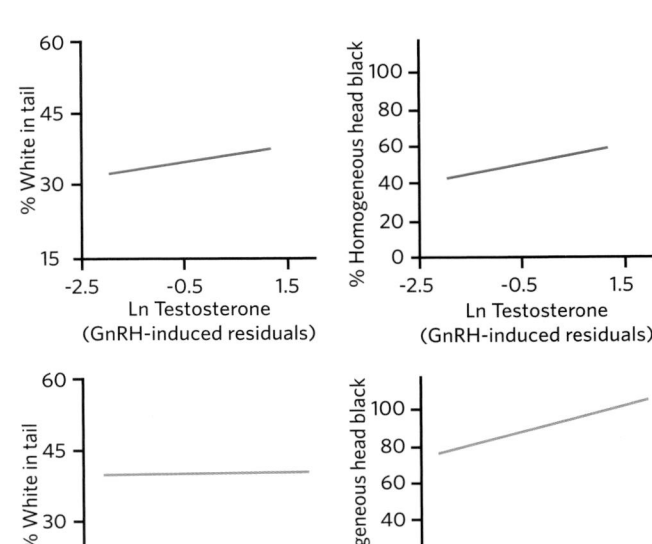

Figure 8.28. Relationships between testosterone and behavioral and morphological phenotypic traits in two populations of juncos. Testosterone of juncos is responsive to various environmental stimuli, including day length, weather, food supply, and social interactions with conspecifics (*top left panel*). In two populations, one breeding in the mountains of southern California (native range, pale blue dots) and an- other that has become resident on the UC San Diego campus since the 1980s (colonists, red dots), circulating testosterone levels correlated negatively with nestling feeding rate and positively with aggressive responsiveness to simulated territorial intrusions (*bottom left panel*). Testosterone levels also correlated positively with two plumage characters in both populations (*bottom right panel*). *From Atwell et al. 2014.*

characters persisted when the birds molted, but the differences in circulating testosterone levels disappeared, suggesting an evolved change in the plumage, but simple plasticity of T secretion in response to the different environments (Atwell et al. 2014). This is a really interesting result; the fact that the relationship between testosterone and these behavioral traits is the same both within and between the two populations (i.e., positively related to amount of aggression and socially relevant plumage characters, negatively related to

paternal care) means that testosterone dependence of these traits can actually facilitate adaptation to the novel environment, rather than conspiring against it.

KEY POINTS

- The endocrine system is involved in the regulation of virtually all behavioral, physiological, and morphological processes in birds.

- Every tissue in the bird produces and is influenced by hormones.
- Hormones are very important in homeostatic regulation of internal conditions of energy balance, calcium balance, and osmotic balance.
- Hormones are also extremely important in allostatic regulation, such as occurs when resources change, when environmental demands change, and when commitments to different demanding processes change.
- Hormones play key roles in managing life history trade-offs, such as between reproduction and plumage molt.

KEY MANAGEMENT AND CONSERVATION IMPLICATIONS

- Because of the exquisite sensitivity of hormone receptors to signaling molecules, it is relatively easy for the endocrine system to be "fooled" by chemicals that mimic natural hormones, or by chemicals that affect endocrine signaling cascades in other ways. Consequently, endocrine disrupting chemicals (EDCs) have the potential to affect many aspects of birds' lives. These chemicals can be of natural (e.g., plant secondary compounds) or human origin, and include many agricultural and industrial chemicals released widely into the environment by human activities.
- The endocrine system plays fundamental roles in the mechanisms by which human-induced environmental change, such as climate change, affects bird populations. For example, changes in the correlations between environmental conditions (temperature, food) and the cues birds use to track the environment (e.g., photoperiod) can lead to timing mismatches between birds' life cycles and seasonal changes in environmental conditions. Likewise, increases in frequency of extreme environmental events can lead to more frequent occurrence of circumstances causing allostatic overload. The nature of endocrine mechanisms thus can influence the way the effects of these environmental factors affect bird populations.
- Habitat modification, climate change, and extreme environmental events can also affect bird populations by creating conditions leading to chronic stress within individuals in populations. The negative effects of chronic exposure to endocrine mediators of the stress response (e.g., glucocorticoids) have the potential to compromise health and stability of populations if conditions consistently exceed those that populations have evolved to cope with.
- It is possible to use endocrine metrics, such as CORT in blood, feces, or feathers, as indicators of how well populations are faring in the face of habitat modifica-tions, changing climate, environmental pollutants, and so forth.

DISCUSSION QUESTIONS

1. A common feature of urban, suburban, and even many rural environments is the presence of light pollution at night. In fact, some species of birds (e.g., House Sparrows) commonly nest within light fixtures that provide bright light all night long. Use what you know about how photoperiod affects avian reproduction to discuss potential effects of light pollution on birds' annual cycles of reproduction, plumage molt, and migration.

2. The effects of endocrine disrupting chemicals (EDCs) are frequently demonstrated in their interactions with hormone receptors. Discuss how **amplification** and **affinity**—two critical features of endocrine signaling cascades initiated by hormone-receptor binding are likely to affect the nature and magnitude of the effects of EDCs.

3. Two of the main functions of the HPA axis are to help individual birds avoid disruptions to homeostasis and to restore homeostatic control quickly after a disruption occurs. Discuss how the HPA axis can achieve these functions, and yet can become destructive if homeostasis is not quickly restored (i.e., if chronic stress ensues).

4. Discuss how homeostatic endocrine systems allow birds to occupy extreme environments, such as deserts, open ocean, high elevation/latitude, and the like.

5. The concepts of homeostasis and allostasis are broadly important in avian physiology. Use the example of a population of birds that lives in a seasonally variable environment, breeding during spring and early summer, molting in late summer, and remaining reproductively quiescent in autumn and winter, to discuss both concepts as they relate to the fitness of birds living in natural environments.

6. Endogenous biological rhythms are conspicuous and widespread features of the physiology and behavior of birds. Discuss the fitness benefits of biological rhythms, with specific reference to circadian rhythms and circannual rhythms.

7. Historically, there have been two hypotheses for how birds determine whether days are long or short: the hourglass hypothesis and the circadian hypothesis. The hourglass hypothesis states that day length is measured as the total duration of the light period, whereas the circadian hypothesis states that it is not the duration of light but when, during a circadian cycle of sensitivity to light, it is light. Discuss how the development of a sensitive

assay for luteinizing hormone (LH) allowed performance of a definitive experiment confirming the validity of the circadian hypothesis (hint: see the box on Brian K. Follett on page 228).

References

Agate, R. J., W. Grisham, J. Wade, S. Mann, J. Wingfield, C. Schanen, A. Palotie, and A. P. Arnold (2003). Neural, not gonadal, origin of brain sex differences in a gynandromorphic finch. Proceedings of the National Academy of Sciences USA 100:4873–4878.

Atwell, J. W., G. C. Cardoso, D. J. Whittaker, T. D. Price, and E. D. Ketterson (2014). Hormonal, behavioral, and life-history traits exhibit correlated shifts in relation to population establishment in a novel environment. American Naturalist 184:E147–E160.

Ball, G. F., and T. P. Hahn (1997). GnRH neuronal systems in birds and their relation to the control of seasonal reproduction. In GnRH neurons: Gene to behavior, I. S. Parhar and Y. Sakuma, Editors. Brain Shuppan, Tokyo, pp. 325–342.

Balthazart, J., C. A. Cornil, T. D. Charlier, M. Taziaux, and G. F. Ball (2009). Estradiol, a key endocrine signal in the sexual differentiation and activation of reproductive behavior in quail. Journal of Experimental Zoology 311A:323–345.

Balthazart, J., T. D. Charlier, J. M. Barker, T. Yamamura, and G. F. Ball (2010). Sex steroid–induced neuroplasticity and behavioral activation in birds. European Journal of Neuroscience 32:2116–2132.

Baker, J. R. (1938). The evolution of breeding seasons. In Evolution: Essays on aspects of evolutionary biology, G. B. De Beer, Editor. Clarendon Press, Oxford, pp. 161–177.

Bentley, G. E., A. R. Goldsmith, A. Dawson, L. Glennie, R. T. Talbot, and P. J. Sharp (1997). Photorefractoriness in European Starlings (Sturnus vulgaris) is not dependent upon the long-day-induced rise in plasma thyroxine. General and Comparative Endocrinology 107:428–438.

Bentley, G. E., S. Tucker, H. Chou, M. Hau, and N. Perfito (2013). Testicular growth and regression are not correlated with Dio2 expression in a wild male songbird, Sturnus vulgaris, exposed to natural changes in photoperiod. Endocrinology 154:1813–1819.

Blas, J. (2015). Stress in birds. In Sturkie's avian physiology, 6th ed., C. G. Scanes, Editor. Elsevier, London, pp. 769–810.

Braun, E. J. (2015). Osmoregulatory systems of birds. In Sturkie's avian physiology, 6th ed., C. G. Scanes, Editor. Elsevier, London, pp. 285–300.

Brazeal, K. R. (2014). Proximate factors influencing timing and overlap of the breeding-molt transition in cardueline finches. PhD diss., University of California, Davis.

Brenowitz, E. A. (2004). Plasticity of the adult avian song control system. Annals of the New York Academy of Sciences 1016:560–585.

Breuner, C. W., A. L. Greenberg, and J. C. Wingfield (1998). Non-invasive corticosterone treatment rapidly increases activity in Gambel's White-crowned Sparrow (Zonotrichia leucophrys gambelii). General and Comparative Endocrinology 111:386–394.

Breuner, C. W., and J. C. Wingfield (2000). Rapid behavioral response to corticosterone varies with photoperiod and dose. Hormones and Behavior 37:23–30.

Breuner, C. W., and T. P. Hahn (2003). Integrating stress physiology, environmental change, and behavior in free-living sparrows. Hormones and Behavior 43:115–123.

Cannon, W. B. (1932). The wisdom of the body. W. W. Norton, New York.

Cassone, V. M., and T. Yoshimura (2015). Circannual cycles and photoperiodism. In Sturkie's avian physiology, 6th ed., C. G. Scanes, Editor. Elsevier, London, pp. 829–845.

Cassone, V. M., and V. Kumar (2015). Circadian rhythms. In Sturkie's avian physiology, 6th ed., C. G. Scanes, Editor. Elsevier, London, pp. 811–827.

Cornelius, J. M., N. Perfito, R. Zann, C. W. Breuner, and T. P. Hahn (2011). Physiological trade-offs in self-maintenance: Plumage molt and stress physiology in birds. Journal of Experimental Biology 214:2768–2777.

Cornelius, J. M., T. Boswell, S. Jenni-Eiermann, C. W. Breuner, and M. Ramenofsky (2013). Contributions of endocrinology to the migration life history of birds. General and Comparative Endocrinology 190:47–60.

Dawson, A. (2002). Photoperiodic control of the annual cycle in birds and comparison with mammals. Ardea 90:355–367.

Dawson, A., S. A. Hinsley, P. N. Ferns, R. H. C. Bonser, and L. Eccleston (2000). Rate of moult affects feather quality: A mechanism linking current reproductive effort to future survival. Proceedings of the Royal Society of London B 267:2093–2098.

Denbow, M. D., and M. A. Cline (2015). Food intake regulation. In Sturkie's avian physiology, 6th ed., C. G. Scanes, Editor. Elsevier, London, pp. 469–485.

Dupont, J., N. Rideau, and J. Simon (2015). Endocrine pancreas. In Sturkie's avian physiology, 6th ed., C. G. Scanes, Editor. Elsevier, London, pp. 613–631.

Farner, D. S. (1964). Time measurement in vertebrate photoperiodism. American Naturalist 95:375–386.

Follett, B. K., P. W. Mattocks Jr., and D. S. Farner (1974). Circadian function in the photoperiodic induction of gonadotropin secretion in the White-crowned Sparrow, Zonotrichia leucophrys gambelii. Proceedings of the National Academy of Sciences USA 71:1666–1669.

Gurney, M. E., and M. Konishi (1980). Hormone-induced sexual differentiation of brain and behavior in Zebra Finches. Science 208:1380–1383.

Gwinner, E., M. Hau, and S. Heigl (1997). Melatonin: Generation and modulation of avian circadian rhythms. Brain Research Bulletin 44:439–444.

Hahn, T. P. (1998). Reproductive seasonality in an opportunistic breeder, the red crossbill, Loxia curvirostra. Ecology 79:2365–2375.

Hahn, T. P., J. Swingle, J. C. Wingfield, and M. Ramenofsky (1992). Adjustments of the prebasic molt schedule in birds. Ornis Scandinavica 23:314–321.

Hahn, T. P., K. W. Sockman, C. W. Breuner, and M. L. Morton (2004). Facultative altitudinal movements by Mountain White-crowned Sparrows (*Zonotrichia leucophrys oriantha*) in the Sierra Nevada. The Auk 121:1269–1281.

Hahn, T. P., and S. A. MacDougall-Shackleton (2008). Adaptive specialization, conditional plasticity and phylogenetic history in the reproductive cue response systems of birds. Philosophical Transactions of the Royal Society B 363:267–286.

Hahn, T. P., J. M. Cornelius, K. B. Sewall, T. R. Kelsey, M. Hau, and N. Perfito (2008). Environmental regulation of annual schedules in opportunistically-breeding songbirds: Adaptive specializations or variations on a theme of White-crowned Sparrow? General and Comparative Endocrinology 157:217–226.

Hahn, T. P., K. R. Brazeal, E. M. Schultz, H. E. Chmura, J. M. Cornelius, H. E. Watts, and S. A. MacDougall-Shackleton (2015). Annual schedules. *In* Sturkie's avian physiology, 6th ed., C. G. Scanes, Editor. Elsevier, London, pp. 847–867.

Hamner, W. M. (1963). Diurnal rhythm and photoperiodism in testicular recrudescence of the House Finch. Science 142:1294–1295.

Hamner, W. M. (1966). Photoperiodic control of annual testicular cycle in House Finch *Carpodacus mexicanus*. General and Comparative Endocrinology 7:224–233.

Ikegami, K., and T. Yoshimura (2012). Circadian clocks and the measurement of daylength in seasonal reproduction. Molecular and Cellular Endocrinology 349:76–81.

Jacobs, J. D., and J. C. Wingfield (2000). Endocrine control of life-cycle stages: A constraint on response to the environment? The Condor 102:35–51.

Jones, B. C., A. D. Smith, S. E. Bebus, and S. J. Schoech (2016). Two seconds is all it takes: European Starlings (*Sturnus vulgaris*) increase levels of circulating glucocorticoids after witnessing a brief raptor attack. Hormones and Behavior 78:72–78.

Leopold, A. S. (1977). The California Quail. University of California Press, Berkeley.

Leopold, A. S., M. Erwin, J. Oh, and B. Browning (1976). Phytoestrogens—adverse effects on reproduction in California Quail. Science 191:98–100.

Lipar, J. L., and E. D. Ketterson (2000). Maternally derived yolk testosterone enhances the development of the hatching muscle in the Red-winged Blackbird, *Agelaius phoeniceus*. Proceedings of the Royal Society of London B 267:2005–2010.

MacDougall-Shackleton, S. A., T. J. Stevenson, H. E. Watts, M. E. Pereyra, and T. P. Hahn (2009). The evolution of photoperiod response systems and seasonal GnRH plasticity in birds. Integrative and Comparative Biology 49:580–589.

McEwen, B. S., and J. C. Wingfield (2003). The concept of allostasis in biology and biomedicine. Hormones and Behavior 43:2–15.

McNabb, F. M. A., M. J. Hooper, E. E. Smith, S. McMurry, and A. B. Gentles (2006). Perchlorate effects in birds. *In* Perchlorate ecotoxicology, R. Kendall and P. N. Smith, Editors. Society for Environmental Toxicology and Chemistry Press, Pensacola, FL, pp. 99–126.

McNabb, F. M. A., and V. M. Darras (2015). Thyroids. *In* Sturkie's avian physiology, 6th ed., C. G. Scanes, Editor. Elsevier, London, pp. 535–547.

Norris, D. R., and P. P. Marra (2007). Seasonal interactions, habitat quality, and population dynamics in migratory birds. The Condor 109:535–547.

Nottebohm, F. (1980). Testosterone triggers growth of brain vocal control nuclei in adult female Canaries. Brain Research 189:429–436.

Nottebohm, F. (1981). A brain for all seasons: Cyclical anatomical changes in song control nuclei of the canary brain. Science 214:1368–1370.

Ottinger, M. A., T. Carro, M. Bohannon, L. Baltos, A. M. Marcell, M. McKernan, K. M. Dean, E. Lavoie, and M. Abdelnabi (2013). Assessing effects of environmental chemicals on neuroendocrine systems: Potential mechanisms and functional outcomes. General and Comparative Endocrinology 190:194–202.

Pereyra, M. E., S. M. Sharbaugh, and T. P. Hahn (2005). Interspecific variation in photo-induced GnRH plasticity among nomadic cardueline finches. Brain, Behavior and Evolution 66:35–49.

Phoenix, C. H., R. W. Goy, A. A. Gerall, and W. C. Young (1959). Organizing action of prenatally administered testosterone propionate on the tissues mediating mating behavior in the female guinea pig. Endocrinology 65:369–382.

Reinert, B. D., and F. E. Wilson (1996). Thyroid dysfunction and thyroxine-dependent programming of photoinduced ovarian growth in American Tree Sparrows (*Spizella arborea*). General and Comparative Endocrinology 103:71–81.

Rochester, J. R., and J. R. Millam (2009). Phytoestrogens and avian reproduction: Exploring the evolution and function of phytoestrogens and possible role of plant compounds in the breeding ecology of wild birds. Comparative Biochemistry and Physiology A 154:279–288.

Romero, L. M., M. J. Dickens, and N. E. Cyr (2009). The reactive scope model—A new model integrating homeostasis, allostasis and stress. Hormones and Behavior 55:375–389.

Scanes, C. G. (2015a). Avian endocrine system. *In* Sturkie's avian physiology, 6th ed., C. G. Scanes, Editor. Elsevier, London, pp. 489–496.

Scanes, C. G. (2015b). Pituitary gland. *In* Sturkie's avian physiology, 6th ed., C. G. Scanes, Editor. Elsevier, London, pp. 497–533.

Schleussner, G., J. P. Dittami, and E. Gwinner (1985). Testosterone implants affect molt in male European Starlings, *Sturnus vulgaris*. Physiological Zoology 58:597–604.

Schmidt, K. L., D. S. Pradhan, A. H. Shah, T. D. Charlier, E. H. Chin, and K. K. Soma (2008). Neurosteroids, immunosteroids, and the Balkanization of endocrinology. General and Comparative Endocrinology 157:266–274.

Schwabl, H. (1993). Yolk is a source of maternal testosterone for developing birds. Proceedings of the National Academy of Sciences USA 90:11439–11441.

Schwabl, H., M. G. Palacios, and T. E. Martin (2007). Selection for rapid embryo development correlates with embryo exposure to maternal androgens among passerine birds. American Naturalist 170:196–206.

Schweitzer, M. H., W. Zheng, L. Zanno, S. Werning, and T. Sugiyama (2016). Chemistry supports the identification of gender-
specific reproductive tissue in *Tyrannosaurus rex*. Scientific
Reports 6:23099. http://doi.org/10.1038/srep23099.

Silverin, B., J. Wingfield, K. A. Stokkan, R. Massa, A. Jarvinen, N. A.
Andersson, M. Lambrechts, A. Sorace, and D. Blomqvist (2008).
Ambient temperature effects on photoinduced gonadal cycles
and hormonal secretion patterns in Great Tits from three different breeding latitudes. Hormones and Behavior 54:60–68.

Strasser, R., and H. Schwabl (2004). Yolk testosterone organizes
behavior and male plumage coloration in House Sparrows
(*Passer domesticus*). Behavioral Ecology and Sociobiology
56:491–497.

Van Herck, S. L., S. Geysens, J. Delbaere, P. Tylzanowski, and
D. M. Varras (2012). Expression profile and thyroid hormone
responsiveness of transporters and deiodinases in early embryonic chicken brain development. Molecular and Cellular
Endocrinology 349:289–297.

Visser, M. E., A. J. van Noordwijk, J. M. Tinbergen, and C. M.
Lessels (1998). Warmer springs lead to mistimed reproduction
in Great Tits (*Parus major*). Proceedings of the Royal Society of
London B 265:1867–1870.

Visser, M. E., C. Both, and M. M. Lambrechts (2004). Global climate change leads to mistimed avian reproduction. Advances
in Ecological Research 35:89–110.

Wingfield, J. C. (1985a). Influences of weather on reproductive
function in male Song Sparrows, *Melospiza melodia*. Journal of
Zoology 205:525–544.

Wingfield, J. C. (1985b). Influences of weather on reproductive
function in female Song Sparrows, *Melospiza melodia*. Journal
of Zoology 205:545–558.

Wingfield, J. C. (2003). Control of behavioural strategies for capri-
cious environments. Animal Behaviour 66:807–816.

Wingfield, J. C. (2008). Organization of vertebrate annual cycles:
Implications for control mechanisms. Philosophical Transactions of the Royal Society B 363:425–441.

Wingfield, J. C., M. C. Moore, and D. S. Farner (1983). Endocrine
responses to inclement weather in naturally breeding populations of White-crowned Sparrows (*Zonotrichia leucophrys
pugetensis*). The Auk 100:56–62.

Wingfield, J. C., and B. S. McEwen (2010). What is in a name?
Integrating homeostasis, allostasis and stress. Hormones and
Behavior 57:105–111.

Wingfield, J. C., J. P. Kelley, and F. Angelier (2011a). What are extreme environmental conditions and how do organisms cope
with them? Current Zoology 57:363–374.

Wingfield, J. C., J. P. Kelley, F. Angelier, O. Chastel, F. Lei,
S. E. Lynn, B. Miner, J. E. Davis, D. Li, and G. Wang (2011b).
Organism-environment interactions in a changing world: A
mechanistic approach. Journal of Ornithology 152 (Suppl.):
S279–S288.

Yamaguchi, S., N. Aoki, T. Kitajima, E. Iikubo, S. Katagiri, T.
Matsushima, and K. J. Homma (2012). Thyroid hormone
determines the start of the sensitive period of imprinting and
primes later learning. Nature Communications. doi:10.1038/
ncomms2088.

CHAPTER 9

Feathers and Molt

Luke K. Butler and Vanya G. Rohwer

As we learned in chapter 2, feathers are one of the defining traits of modern birds. But beyond being a unique anatomical feature, feathers perform a wide range of functions that allow birds to do the things they do. In fact, you would probably have a hard time thinking of an analogous structure, anywhere in the animal world, that contributes so much to a broad range of basic organismal functions: locomotion, thermoregulation, protection, and sensation. And, in many avian lineages, feathers have evolved to serve purposes other than supporting life's basic functions. The long train feathers of male peafowl are a spectacular example of how a key evolutionary innovation, like the feather, can evolve to serve purposes much different from its original purpose. The train feathers are not actually tail feathers, but heavily modified body feathers that cover the base of the tail. When displaying to a potential mate, a peacock raises and spreads his train to produce an impressive visual display. When the female's interest declines, the male rattles the train to produce a sound that helps recapture her attention (Yorzinski et al. 2013). Research on peahen mate choice has shown that females prefer to mate with males that possess trains with more eyespots (Petrie et al. 1991). Thus in peafowl, as in many other species, feathers play a role in nearly every aspect of life and have evolved into a huge variety of forms.

The importance of feathers in the lives of birds has made feathers the subject of much research. However, much remains unknown about how feathers work and how birds renew their feathers through the process of **molt,** making the modern study of feathers an area of exciting new discoveries.

We start this chapter by examining the plumage as a whole and the distinctive structural features of individual feathers. We then explore four major questions that ornithologists continue to ask about feathers and plumage:

- How are feathers produced?
- What do feathers do, and how do they do it?
- How are feathers maintained and replaced?
- How do birds change their plumage to fit different seasons and life stages?

Because feathers are so important to birds, we next examine how molt is related to broad patterns in avian life histories. We end by highlighting new techniques that use feathers to gather information that would be difficult or impossible to obtain from any other source.

PLUMAGE IS A DYNAMIC SYSTEM OF MANY INTERACTING PARTS

A bird's **plumage** is the total of all of its feathers. At the broadest level of biological organization, the plumage is one of the major parts of a bird's integumentary system, which also includes the body's other protective outer tissues: the skin, claws, and tough tissues that cover the feet (podotheca) and bill (rhamphotheca). From this perspective, we see that a **feather** is an appendage of a bird's integumentary system.

Each feather is produced and held in place by a **feather follicle,** a structure unique to birds that is derived from and embedded within the skin. Each feather follicle can be thought of as a separate "mini-organ" composed of specialized layers of cells that manage the production, loss, and renewal of feathers throughout a bird's life.

Extensive networks of smooth muscle lying just below the skin allow birds to raise (ptiloerect) and lower (ptilocompress) many feathers on the body and rotate the long feathers of the wing and tail. Many feathers also transmit information to sensory neurons that terminate within the follicle. These features allow the plumage to be much more than a passive surface that simply shields the body from the environment or generates lift when pulled through air. Instead, the plumage is an adjustable, dynamic system of thousands of interacting parts, working with other major body systems to maintain homeostasis and facilitate complex behaviors such as flight, prey capture, and social signaling. In fact, the extent to which birds can adjust the shape and volume of their plumage has made it challenging for researchers to study the role of the plumage in flight and thermoregulation, because it is difficult to experimentally manipulate feathers in the ways that a bird's neuromuscular system manipulates them.

Although feathers and mammalian hair share some functions and are each produced by follicles in the skin, feathers and hair have separate evolutionary origins in ancestral vertebrates. Thus, plumage and pelage (a mammal's coat of fur) are evolutionarily analogous (*not* homologous) coverings for a homeothermic vertebrate body. But feathers are unique for their huge diversity of forms and functions—a much greater diversity than is found in mammalian hair.

Most of a bird's plumage is **contour feathers**, which cover the head, neck, torso, and legs (body feathers) and also form the flight surfaces of the wing and tail (flight feathers; fig. 9.1A). Contour feathers are what you see when you observe a bird in the field. The plumage surface is continuous over most of the body in adults of most birds, so it may be surprising to learn that if you parted the feathers of most birds you would expose large patches of bare skin. Follicles are found only in distinct tracts called **pterylae** ('ter-ə-lē), which may occur over a very small area of the skin compared with the corresponding area of the plumage surface (fig. 9.1B). This is possible partly because body feathers are typically much wider at the tip than the base, and partly because feathers spread and curve apart from each other, covering the patches of skin that have no follicles (the **apterylae**). One of the largest apterylae in many songbirds occurs on the belly, making it possible to observe abdominal

fat directly through the skin—a simple and widely used field technique for indexing a bird's energy stores.

Individual contour feathers project from the skin along an anterior-to-posterior axis. Anterior feathers layer like shingles on top of neighboring posterior feathers, so the plumage may be several feathers deep over some parts of the body. The body plumage fills space around the neck and torso, creating an outer surface that is much larger and much more streamlined than the body of a naked bird (fig. 9.1B).

By plucking and counting all of the feathers on dead specimens, ornithologists have found that the size and number of feathers varies greatly among species and among different parts of the body. Across species, the total number of contour feathers that make up the plumage is positively correlated with body mass, with larger species possessing tens of thousands of feathers (table 9.1). Interestingly, small-bodied birds have relatively more feathers *per unit of body mass* than large birds (fig. 9.2), meaning that a given area of the body surface has more feathers in a small bird than in a large bird. This may be an adaptation for thermoregulation in small-bodied birds, along with many other adaptations (Hutt and Ball 1938). Contour feathers are shortest and most numerous on the face and head, and much longer and less abundant on the belly and rump. In the Tundra Swan (*Cygnus columbianus*), 80 percent of the feathers are on the head and neck alone (Ammann 1937). Similarly, species with exceptionally long inner wings, such as albatrosses, have many more inner wing feathers (>30) than most species (typically 9–12).

An individual feather may seem extremely light to the touch, but, when combined in the thousands, their masses account for approximately 3 to 10 percent of a bird's total body mass (Wetmore 1936). In the Bald Eagle (*Haliaeetus leucocephalus*), plumage mass is twice that of the skeleton

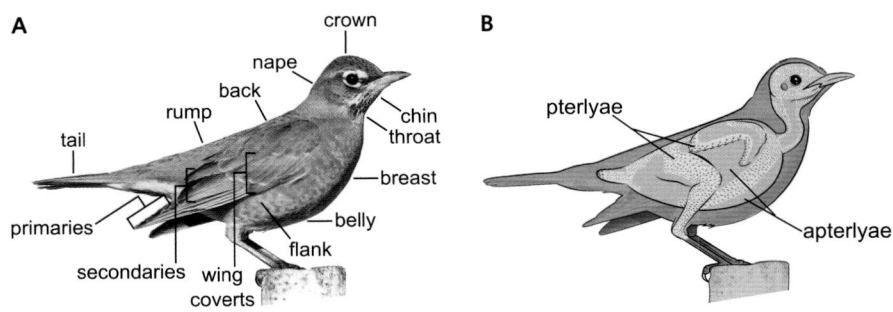

Figure 9.1. Contour feathers form the plumage surface. *A,* Major body feather regions and flight feather tracts visible when viewing a bird in the field; *B,* Although the contour feathers form a nearly continuous plumage surface, the feather follicles are actually organized in tracts (pterylae) that only cover about half of the skin's surface and are interspersed with large patches of bare skin (apterylae). American Robin (*Turdus migratorius*), *photo by Geoffrey Hill. Illustration by Cecilia Johnson.*

Table 9.1. **Number of contour feathers in six groups of birds.**

Taxon	Number of contour feathers (approximate)
Hummingbird	1000[1]
Songbird	1000–3000[1]
Dove	2500–4300[1]
Gull	6500[2]
Duck	7000–11,000[3]
Swan	25,000[4]

1 Wetmore 1936, Markus 1963
2 McGregor 1902
3 Knappen 1932, Hopps 2002
4 Ammann 1937

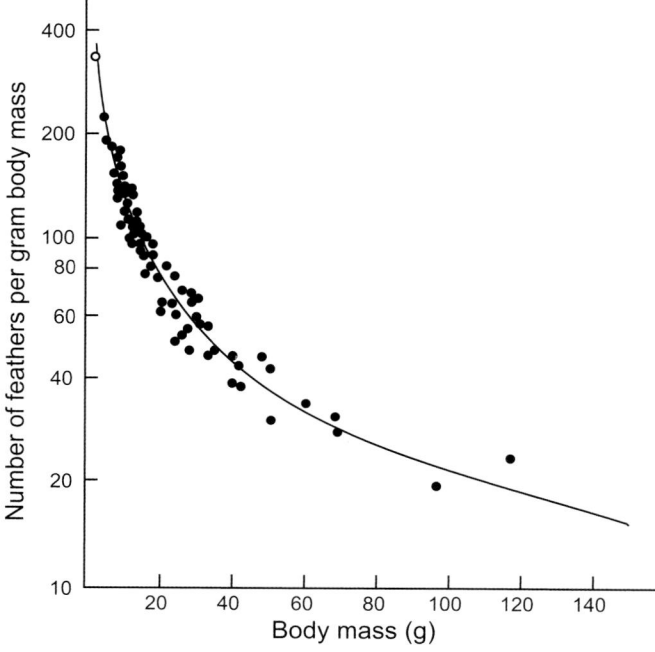

Figure 9.2. Number of body feathers in relation to body size. Although bigger bird species have more feathers (see table 9.1), smaller species have more feathers per gram of body mass. Filled circles = passerines (one circle per species). Open circle = Ruby-throated Hummingbird (*Archilochus colubris*). Adapted from Hutt and Ball 1938.

(Brodkorb 1955). The vast majority of a bird's total plumage mass comes from the body feathers, not the long feathers of the wing and tail.

STRUCTURE-FUNCTION RELATIONSHIPS WITHIN FEATHERS

So far we have discussed plumage as a part of the integumentary system, without saying much about feathers themselves. In this section we identify the distinctive structural

properties of feathers and how they vary according to their functions.

A typical contour feather probably has the structure you envision when you think of what a feather looks like: a central **shaft** with thin projections that emanate from opposite sides, forming the feather **vane** (fig. 9.3). The shaft has two parts: a short, barbless, hollow section, called the **calamus**, which inserts into the skin, and a thinner, tapering, pith-filled section, called the **rachis**. **Barbs** emanate from the rachis, forming the feather vane. Barbs have two parts: a central **ramus** (plural, rami), which fuses, at its base, to the rachis, and numerous **barbules**, which emanate along most or all of the ramus. Many feathers also have an **afterfeather**, which attaches to the skin side of the calamus and has a similar structure to a feather.

The texture of the distal part of a body feather, which helps form the plumage surface, often has a relatively smooth, sheetlike appearance (called "pennaceous"). The part of the feather that is not exposed to the plumage surface typically has a fluffy or shaggy appearance (plumulaceous). This difference is primarily because of differences in barbules. In the plumulaceous part of a contour feather, the barbules are relatively long, and they emanate from the rami in many directions, providing "depth" to the feather vane. In the pennaceous part of the feather, barbules are relatively short, and they emanate from the rami primarily in opposite directions, and in the same plane as the barbs. The very planar, fabric-like structure of the vanes of wing and tail feathers is further caused by interactions between barbules. Hook-shaped projections called **barbicels** emanate from the underside of distal barbules (those projecting toward the tip of the feather), anchoring each distal barbule onto its proximal neighbor (fig. 9.3).

Down feathers appear similar to the plumulaceous part of contour feathers because both have flexible, wispy barbules that emanate from the ramus in multiple directions and do not interlock or form a smooth plane. The rachis is also very short or absent in down feathers, and the overall number of barbs is much lower than in a contour feather. In bristles and plumes the rachises are mostly barb-free—bristles have barbs at the base and plumes have barbs at the tip (fig. 9.3).

All feathers are made of several varieties of beta-keratin, filamentous proteins held together by hydrogen bonds, forming beta sheets (a beta sheet is a wavy secondary protein structure you may have learned about in your introductory biology class). Keratins are a large family of proteins found in many vertebrate integumentary tissues (e.g., skin, hair, and horns); however, beta-keratins are found only in birds and reptiles. Feather beta-keratin is distinct for its strength, flexibility, and low molecular weight compared with other avian

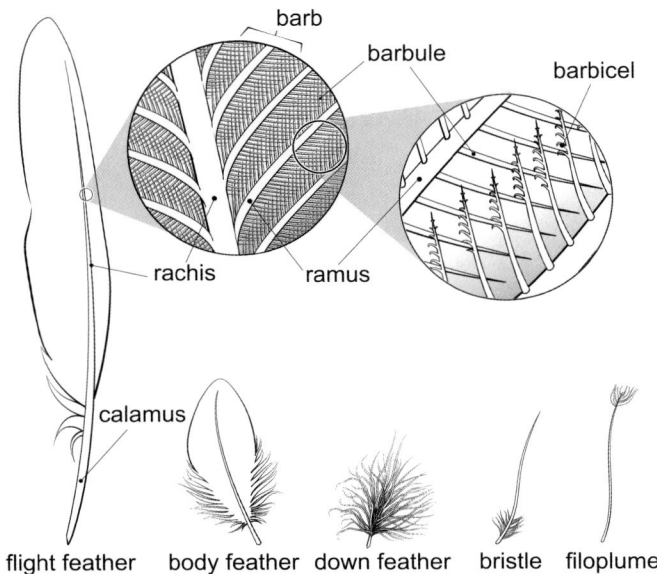

Figure 9.3. Major types and parts of feathers. The plumage comprises a variety of feather types, defined according to their structure or function. The smooth, sheetlike character of typical flight and body feathers is a result of relatively stiff barbs that are "zipped" together by barbules from proximal barbs hooking onto the neighboring barbules of distal barbs via barbicels. *Illustration by Cecilia Johnson.*

keratins, because of differences in the primary structure (amino acid sequence) of the keratins in different tissues. Having a low molecular weight means that feather keratins are relatively light compared with how strong they are.

Because the evolution of feathers is so closely tied to the evolution of modern birds (see chapter 2), understanding the molecular structure of beta-keratin is a topic of great interest. By comparing the DNA sequences of beta-keratin genes in birds and reptiles, Matthew Greenwold and Roger Sawyer (2011) revealed that the beta-keratins found in modern bird feathers likely evolved millions of years after the evolution of the first feathers in theropod dinosaurs. Therefore, the earliest feathers may have had different biophysical properties than the feathers of birds today.

HOW ARE FEATHERS PRODUCED?

In this section we summarize the process of feather production and highlight important relationships between the way feathers are produced and other aspects of feather biology. Additional details of feather development are provided in chapter 5.

Every feather is produced by the same basic structure of the skin: a feather follicle (see fig. 6.4). To create a new feather, stem cells in the follicle divide to create new cells, called **ke-**

ratinocytes, that eventually fill themselves with keratin, die, and fuse together to form the feather parts. Feathers are produced tip-first. As a feather grows, keratinocytes are added proximally to the tips of developing barbs, creating a spiral pattern of barb elongation that results in the bases of the medial and lateral barbs converging on the middorsal line of the follicle. There, the bases of the rami and cells from a few middorsal barbs fuse together via keratinous fibers to form the rachis. Similarly, the cells that form barbules fuse at their proximal ends to the ramus. New barbs arise in the ventral region of the follicle, and the process repeats until all of the barbs of the feather have been produced. As the feather nears completion, new keratinocytes stop forming barbs, instead forming a complete ring and eventually forming the calamus. A given follicle will typically produce a new feather many times within a bird's life (discussed below). Nerves, muscles, and capillaries connect the follicle to the body's other systems and make the plumage an adjustable and renewable part of a bird's integument.

Several important features of feathers are the result of the way feathers are produced. First, feathers are produced from tip to base: the cells that will form the distalmost barbs and the distal tip of the rachis arise first, and growth proceeds proximally—calamus last. In a long wing feather, the tip of the feather is at least a week older than the base of the feather. Second, the point where new barbs form on the ventral side of the follicle will determine the relative lengths of the barbs on either side of the rachis, and, ultimately, whether the feather is symmetrical or asymmetrical. The overall shape of the vane is determined by differences in the lengths of barbs and the angle at which they emanate from the rachis. Third, the transition from the pennaceous to the plumulaceous part of a contour feather is determined by differences in the ways cells differentiate in pennaceous and plumulaceous barbs. Similarly, differences in cellular processes that occur within the follicle account for differences between feathers in different parts of the body, between the sexes, or among species, and for differences in the structure and color of the feathers created at different times in a bird's life. Currently, the genetics underlying the incredible functional variability of an individual feather follicle is poorly understood but of great interest among biologists.

WHAT DO FEATHERS DO, AND HOW DO THEY DO IT?

In this section we examine the roles of the plumage and relationships between the structural properties of feathers and the functions that feathers perform. It is important to remember that feathers serve a variety of functions, and so the structural properties of a given feather or plumage represent the result of evolutionary trade-offs among several competing demands. An underlying theme of this section is that every feather function involves some kind of interaction with the external environment. To understand these interactions, we often need to turn to physics.

Modulating Heat Transfer

Birds often experience air temperatures that are outside of the thermoneutral zone, requiring them to expend a great deal of their daily energy budget on thermoregulation (chapter 7). Body plumage helps reduce the costs of thermoregulation by reducing the rate of heat transfer between a bird's body and the surrounding environment. Heat can move toward or away from a bird's skin surface in just two ways—between the feathers (via radiation or convection) or within the feathers (via conduction)—but numerous factors affect the rate of heat transfer between a bird and its environment, and understanding the roles of plumage in thermoregulation is an area of ongoing research.

Radiative heat transfer occurs when heat waves are emitted or absorbed by an object (e.g., when a bird is warmed by sunlight). Considering the many feather elements that separate a bird's skin from the surrounding environment, it might not be surprising that radiation accounts for only about 5–10 percent of overall heat transfer through the plumage (Walsberg 1988, McCafferty et al. 1997).

The remaining 90–95 percent of heat transfer through the plumage is attributed to approximately equal proportions of conduction and convection. **Conduction** occurs when heat is transferred from the skin directly to matter that the skin is in contact with—in most cases, air molecules or feather elements. Interestingly, keratin is a better heat conductor than air, so a bird's plumage actually provides less insulation than an equal depth of still air. Fluffing the plumage increases the volume of air surrounding the body, reducing heat loss greatly. However, the air beneath the plumage is never completely still—it moves by **convection**, carrying heat with it. The numerous long barbules of the plumulaceous part of the contour feathers create many barriers to airflow, thus reducing heat loss caused by convection. So even though feather elements conduct heat faster than air, heat transfer via convection (and radiation) would be much higher if a bird's plumage lacked its complex, subsurface matrix.

Understanding heat transfer through the plumage is a big challenge. Birds can rapidly change the depth of their plumage via ptiloerection, and they can adjust their exposure to solar radiation and wind. And of course, birds vary widely in plumage color. To isolate the relative influence of these variables, Glenn Walsberg and colleagues conducted an experi-

ment in which they manipulated plumage depth and wind speed while measuring solar heat gain in the skin of birds with white or black plumage (fig. 9.4; Walsberg et al. 1978). In low winds, white plumage transferred less solar heat to the skin than black plumage. But, under high winds, erected black plumage transferred less heat to the skin than white plumage because black feathers absorb more solar radiation at the plumage surface, where the wind then carries the heat away. These data may help explain why dark-colored birds are unexpectedly common in deserts, which are often both sunny and windy.

The number, distribution, and structure of body feathers play important roles in reducing heat loss. For example the smallest diving birds, dippers (*Cinclus* spp.), possess an exceptionally high number of body feathers (3,000–4,000) for their body size (~60 g), and their feathers cover the body completely (no apterylae). Dipper down is very finely textured and is found on the head and extensively on the wings, unlike other passerines (Davenport et al. 2009). Penguins also lack apterylae, and they have a high density of down feath-

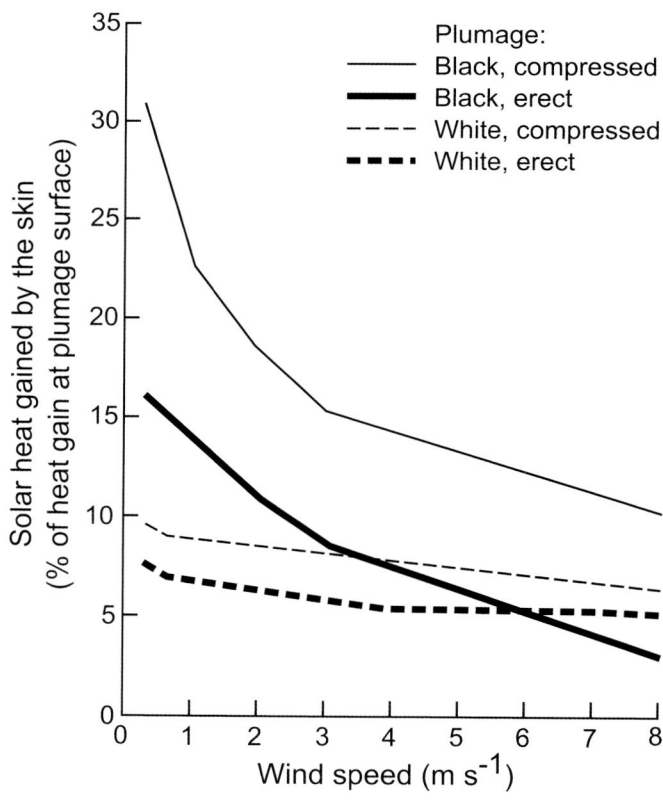

Figure 9.4. Effect of plumage on solar heat gain by the skin. The plumage plays a major role in thermoregulation, partly by blocking most of the sun's rays from reaching the skin. What do these data say about the role of plumage color in thermoregulation? *Adapted from Wolf and Walsberg 2000.*

ers. As penguins fast during the breeding season their bodies shrink, increasing feather density and helping counteract the loss of their insulative fat layer (Davenport et al. 2009). Other cold-weather specialists, such as the snowfinches (*Montifringilla* spp.), have loosely textured contour feathers (Lei et al. 2002) that increase the penetration depth of solar radiation.

One of the fastest ways to lose heat is by getting wet, and feathers help prevent water from reaching the skin. In terrestrial birds, the compact structure of the pennaceous ends of most contour feathers causes light rain and water transferred from vegetation to bead on the plumage surface. Heavy rain may penetrate the plumage, so birds often seek shelter during rainstorms. How diving birds stay dry is still an area of active research, and the solution may vary among species (Pap et al. 2015). Using a fluid mechanics approach, Arie Rijke and his colleagues have shown that in some waterbirds, such as gulls and ducks, special spacing of the barbs and barbules may make feathers water-resistant by trapping tiny air bubbles in the plumage surface (Rijke and Jesser 2011).

In some instances, having wet feathers is highly advantageous. Working in the Kalahari Desert, Tom Cade and Gordon Maclean showed through careful field observation that young sandgrouse (*Pterocles* spp.) use their parents' wet plumage as a water source (Cade and Maclean 1967). Breeding males fly to distant watering holes, soak their breast feathers, and return to their young—who strip the water from his plumage with their bills.

Locomotion

Feathers exhibit a wide array of adaptations for locomotion, primarily flight. Elongated wing feathers are an obvious adaptation for creating an airfoil (see chapter 10). The lengths of individual wing feathers determine overall wing shape and vary among species depending on evolutionary history and dominant flight mode. Tail feathers are used for steering and braking, and they help reduce body drag caused by turbulence by promoting attached airflow on the downstream (posterior) end of a bird's body. Woodpeckers use their exceptionally stiff tail feathers as a prop when climbing tree trunks. Flight feathers are especially short in flightless birds, but Common Ostriches (*Struthio camelus*) extend their wings to facilitate turning and braking.

Flight feather vanes are asymmetrical, with the rachis positioned toward the leading edge of the vane. This forward placement puts the rachis near the point of greatest aerodynamic force, preventing the feather from twisting. The calamus is enlarged and the rachis is stiffened in proportion to aerodynamic bending forces. By comparing feather structure among a wide diversity of species, Peter Pap and colleagues found that greater flight demands are correlated

with broader rachises and higher barb densities in the primaries (Pap et al. 2015). Ligaments attach the flight feathers firmly to bones of the hand (primaries), forearm (secondaries), or tail (central two rectrices). The plumulaceous segment of flight feathers is short or absent.

Body plumage reduces drag caused by turbulence by filling spaces around bulky body parts and creating a streamlined shape with about 20 percent less surface area than the area of the skin (fig. 9.1B; Walsberg and King 1978). Individual body feathers may also help reduce body drag. The curvature and elasticity of body feathers provides a flexible surface that can dampen turbulence. In many species, the barbs that form the plumage surface curve toward the rachis, so that the tips exposed to the airstream are oriented longitudinally. This creates a ribbed plumage surface that may reduce skin friction in a mechanism similar to shark scales. Finally, similar to flight feathers, greater flight demands are correlated with greater barb densities in body feathers (Butler et al. 2008).

Mechanosensory

The skin is the major sensory organ of the peripheral nervous system, and much of the tactile information that a bird gains from the environment is transferred through the feathers to nerves that terminate near feather follicles. These nerves have a wide distribution in the body and wings, allowing birds to sense feather position, aerodynamic forces during flight, and objects in the environment. Some feathers appear to be specially adapted for tactile sensing.

Filoplumes have a very thin rachis that is barbless except at the tip (fig. 9.3). One or more filoplumes are positioned near each body and flight feather. Neurologist Reinhold Necker conducted a detailed anatomical study of the nerves associated with filoplumes in the wing plumage of the Rock Dove (*Columba livia*), finding that lateral movements of the contour feathers caused the filoplumes to move, triggering signals from the skin to the central nervous system and providing information about feather position (Necker 1985). Filoplumes usually terminate below the surface of the contour plumage, but they extend beyond the plumage surface on the napes of some species, suggesting that they have an unknown aerodynamic function.

Bristles have a thick rachis that is barbless except at the base (fig. 9.3). Unlike filoplumes, bristles function independently of other feathers. Bristles are found on the face and head of most bird species, and like filoplumes, bristles are closely associated with pressure-sensitive nerves in the skin. Bristles probably have different functions in different species. Some species of birds have regularly spaced bristles on the eyelids that function as eyelashes, but there have been no systematic studies of avian eyelashes. Nocturnal and flycatching species have especially long facial bristles, suggesting roles in sensing prey or other objects, or protecting the eyes in a manner similar to eyelashes, but these hypotheses need more testing.

Elongated contour feathers of the face and head may also serve sensory functions (fig. 9.5). Experimentally flattening the head plumes of Whiskered (*Aethia pygmaea*) and Crested Auklets (*A. cristatella*) causes them to bump their heads when navigating a narrow path in the dark—an important skill in these crevice-nesting seabirds (Seneviratne and Jones 2010). In several owl species, the facial feathers form a concave disc that directs sound waves toward the ears, making it easier to detect noises created by prey.

Social Signaling

Because of the wide variety of lighting conditions and background substrates found on earth, the niches available for producing visual signals with plumage are immense. Feathers have major roles in social signaling, which is the topic of chapter 11. In this section we focus on the physical bases of signal production by plumages and individual feathers.

Feathers produce visual signals in a variety of ways, but all depend on two processes: reflecting or absorbing photons of light in a way that is perceived by a potential receiver of the signal. Individual feathers exhibit extreme specializations in size, shape, structure, and color for creating conspicuous visual displays. Simply being larger means that more light waves can be reflected or absorbed, creating a larger signal, and indeed, elongation is a widespread feature of feathers used for signaling. Within the birds-of-paradise (Paradisaeidae) alone, male courtship dances involve elongated feathers of the face, crown, nape, breast, flank, and tail, depending on species (fig. 9.6). Broad, paddle-shaped tips on elongated flight feathers ("rackets") have evolved independently in birds-of-paradise and several other lineages. Head feathers take a wide variety of shapes, including rackets, commas, wires, and hairs.

The type of light that feathers reflect or absorb depends on the color pigments deposited in them and the microstructure of the feather itself. Feather colors are typically attributable to pigments deposited within the keratin fibers by special cells that migrate into the papilla during feather production. Because of their chemical structures, different types of pigment molecules absorb some wavelengths of light and reflect others, giving each pigment type a characteristic color (hue). Variation in the concentration and combination of pigments can produce fine differences in hue and brightness among feathers. The two most common types of pig-

Figure 9.5. Mechanosensory adaptations in head feathers. *A*, Crested (*top*) and Whiskered (*bottom*) Auklets use their head plumes to navigate the dark crevices used for nesting. *Photo by Ian Jones. B*, The facial disc of Barn Owls directs sound waves toward the ears, improving detection of prey while hunting at night. *Photo by Geoffrey Hill.*

Figure 9.6. Ballerina display by male Wahnes's Parotia (*Parotia wahnesi*). Spectacular structural feathers have evolved in several feather tracts of male Wahnes's Parotias. In a display that females observe from above, males wobble their heads and flutter their elongated breast feathers in an attempt to attract a mate. *Photo by Tim Laman.*

ment are melanins (responsible for browns, grays, and black) and carotenoids (responsible for reds, yellows, oranges, and UV—visible to birds, but not humans). The less common porphyrins, psittacofulvins, and metal oxides also produce reds, yellows, and browns.

Alternatively, feather colors can be produced by physical properties of the cortex and medulla (inner pith) of the barbs. Photons of light that pass through the outer cortex reach an inner, sponge-like matrix of keratin walls and air spaces that precisely scatter the light waves to return only a subset of the incident wavelengths, thus appearing to the eye as a particular color (fig. 9.7). Structural coloration is typically responsible for feathers that appear green, blue, or iridescent to the eye. The iridescent green on the trains of peacocks is caused by structural coloration.

The train-shaking behavior of peacocks described at the beginning of this chapter shows that feathers can also be used to make audial signals. Like conspicuous plumage colors, sounds produced by feathers are typically employed by males as part of a display during the breeding season. Breeding male American Woodcocks (*Scolopax minor*) and Broad-tailed Hummingbirds (*Selasphorus platycercus*) use outer primaries with reduced vanes to create whistles that accompany acrobatic flights. Club-winged Manakins (*Machaeropterus deliciosus*) produce an intense beep-like tone by rapidly vibrating their inner secondaries, similar to the way male crickets chirp by rubbing their wings together (Bostwick et al. 2010).

Feathers even play indirect and direct roles in chemical signaling. Until recently, it was generally believed that olfactory systems in birds were weak or absent. However, birds do indeed have functional olfactory systems, and odorous species are widely distributed among taxonomic families. Furthermore, dozens of volatile lipid compounds are secreted by skin cells or the uropygial gland and transferred to the plumage, providing the basis for a chemical signal that may contain information about genetic compatibility or immune system quality. The tangerine smell of Crested Auklets peaks during the summer breeding season, and courting pairs perform a "ruff sniff" display, burying their bills in their partner's nape feathers, perhaps gaining information about their partner via chemicals on the feathers or skin. Indeed, chemical mixture on the feathers of Antarctic Prions (*Pachyptila desolata*) is consistent over time within in an individual but differs among individuals, suggesting that individuals possess an "odor signature." Given a choice, prions preferentially approach the scent of their partner over the scent of another conspecific (Bonadonna and Nevitt 2004).

Deception and Threats

Birds also use feathers to hide information about themselves, usually to either avoid predation or succeed as a predator. Visual crypsis is used widely, especially in relation to concealing a nest or its contents. The evolution of sexual dichromatism may often be driven by selection on nesting females to have cryptic plumages, rather than selection on males to be bright and flashy. Common Nighthawks (*Chordeiles minor*) use two forms of visual deception to protect their young. The nest is placed on the ground, but the adult's heavily mottled dorsal plumage makes it very difficult to detect the nest by

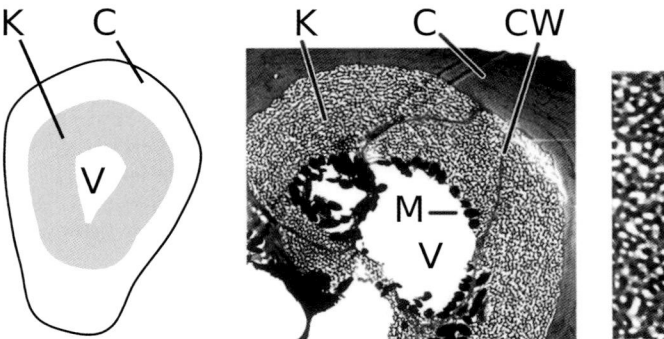
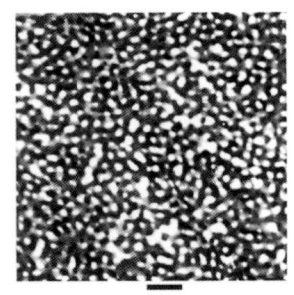

Figure 9.7. Feather microstructure produces UV coloration. *Left*, Basic cross-sectional structure of a feather barb ramus that produces UV coloration: a central vacuole (V) is surrounded by an inner layer of spongy keratin (K) and an outer cortex (C) of solid keratin. *Center*, Cross section of a UV-colored barb from a Blue Whistling Thrush (*Myophonus caeruleus*), showing three spongy keratin cells (distinguished by cell walls, CW) and melanin granules (M) surrounding the central vacuole. *Right*, Because of precise, regular spacing of air pockets (white) and keratin (black/gray) within the spongy keratin layer, only a subset of incident wavelengths are reflected by the spongy keratin (scale bar = 500 nm). When sunlight passes through the outer cortex, some wavelengths (typically blue) are reflected by the spongy keratin, while other wavelengths pass through the matrix and are absorbed by the underlying melanin layer, creating a brilliant color without the use of colorful pigments. *Center and right panels from Prum et al. 2003.*

sight. If a predator does get too close, the female scurries away, extending and dragging her wings awkwardly in a "broken wing" display that suggests she is an easy meal. Once the predator is lured away from the nest, the uninjured female flies to safety.

Owls are visually cryptic while roosting in trees during the day, which may reduce mobbing by other birds. Owls also exhibit "auditory crypsis" during attack flights via specializations of the wing feathers. The leading edge of the wing is serrated (barb tips are separated and curved dorsally), and the vanes and trailing edges of all wing feathers are finely fringed. Together these specializations are thought to reduce noise from air passing over the wing that may alert a potential prey animal or interfere with the owl's own hearing. In the opposite way, Slate-throated Redstarts (*Myioborus miniatus*) flash white patches on their tail feathers to flush insects into the air, increasing foraging success (Mumme et al. 2006). A bird that uses feather-based noises to increase foraging performance is yet to be discovered.

Feathers make poor armor, but they may help prey species avoid being eaten in some cases. In response to a predator attack, many species drop large numbers of feathers from the back, rump, and tail (fright molt), confusing the predator and allowing the prey to escape (Lindström and Nilsson 1988). The skin and feathers of pitohuis contain a rapidly acting neurotoxin that causes a numbing, tingling sensation in persons handling them (Dumbacher et al. 2004). Thus a predator with a pitohui in its grasp may be discouraged from continuing with its meal.

FEATHER DEGRADATION AND LOSS

Over time, feathers weaken and break, just like your fingernails and hair. Keratin fibers are weakened by UV radiation, and normal physical interactions with each other and the environment cause bits of barb and rachis to break off over time. During flight, air pushing against the feathers causes them to flex, vibrate, and abrade one another, and feathers also break from impacts with vegetation, the ground, and airborne particles. A host of microorganisms degrade feathers by chewing on, burrowing into, or chemically degrading feathers, accelerating feather wear caused by the abiotic environment.

Feather wear caused by routine use typically starts with the tips of the distal barbs, progresses proximally, and eventually affects the rachis (fig. 9.8). In Stonechats (*Saxicola rubicola*), wing tips shorten by 0.2–0.5 percent per month (fig. 9.9; Flinks and Salewski 2012). In open grasslands, UV radiation and contact with siliceous plants like grasses and sedges cause feathers to degrade rapidly (Stresemann and Stresemann 1966). In some cases—after fights, struggles with prey, or flights to escape predators—wing and tail feathers can break or fall out, creating asymmetry in the flight surfaces and amplifying the effects of normal wear and tear and reducing a bird's ability to escape from predators or catch prey (Swaddle et al. 1996).

Feathers host a diversity of lice, mites, bacteria, and fungi, some of which consume feather keratin. These organisms interact with each other and their avian hosts in a complex micro-ecosystem that we are only beginning to understand. Chewing lice inhabit the feathers (suborder Ischnocera) and skin (suborder Amblycera) of a wide variety of birds, and up to 95 percent of individuals may be infested with lice in some species (Silva et al. 2014). Some mites (superorder Acariformes) live on the surface of feathers, and others burrow into the calamus, feeding on the pith and possibly weakening the feather, though evidence for this is lacking. Bacteria that secrete keratinolytic enzymes are also widespread in birds. Working in Ohio, Jed Burtt (see chapter 30) and Jann Ichida rubbed sterile swabs on the back, belly, and tail of nearly 1,600 free-living birds from 83 species. By culturing the swabs in the lab and then incubating the cultures with chicken feathers, Burtt and Ichida (1999) found that the plumages of about 10 percent of free-living birds, and approximately half of all species, are infected with feather-degrading bacteria. Infection rates were highest in aquatic and ground-foraging species and lowest in aerial-foraging species, suggesting that birds exposed to wetter habitats provide a more hospitable plumage for microbiota.

Although we know that some ectoparasites degrade and consume feather parts, determining the costs of natural levels of parasite-induced feather degradation on the fitness of free-living hosts is difficult. Numerous interspecific interactions occur within the plumage, complicating the relationship between a host and any one of its parasites. For example, feather lice depend on a symbiotic bacterium of the genus

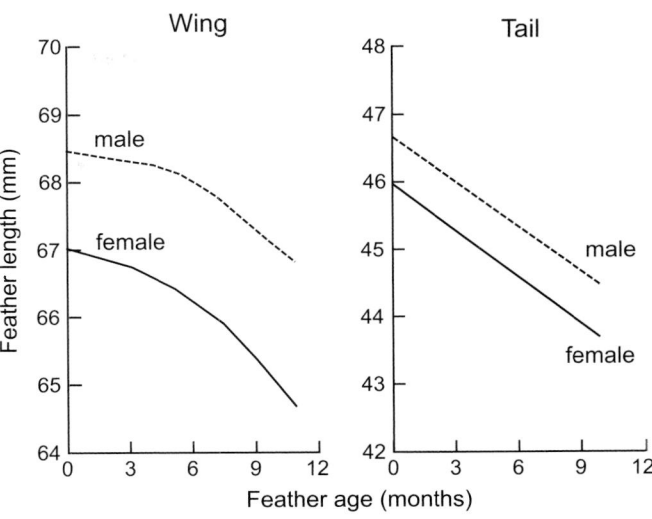

Figure 9.9. Feather wear in the wing and tail. Change in feather length over time in adult European Stonechats (*Saxicola rubicola*) caused by feather wear. Feather age = 0 months in October, shortly after molt is complete. Stonechats inhabit open scrub and grasslands, so their feathers experience degradation from high levels of UV radiation as well as from contact with siliceous vegetation. Stonechats feed by short flights to capture insect prey, so worn wing feathers probably carry considerable aerodynamic cost. *Adapted from Flinks and Salewski 2012.*

Rickettsia to digest keratin, creating a three-species interaction (Marshall 1981). In Red-billed Choughs (*Pyrrhocorax pyrrhocorax*), mite load is positively correlated with host body condition, suggesting that mite infestation is beneficial to chough health, perhaps because the mites consume or compete with parasitic fungi or other detrimental microorganisms (Blanco et al. 1997). Bird hosts may also use chemical and behavioral defenses to maintain ectoparasites at levels generally inconsequential to plumage function. Bacteria may inhibit each other by secreting antibiotics. Indeed, thus far, direct evidence for costly effects of lice, mites, and bacteria on feather quality under natural conditions is limited (Proctor and Owens 2000, Vágási 2014). Much of the cost of ectoparasites may come in the cost of feather care, which is described in the next section.

Some feather degradation is adaptive. In many species, the dull tips of body feathers break off to expose bright plumage badges just in time for the breeding season, allowing birds to change from cryptic to bright without producing new feathers. Through carefully timed preening and dust-bathing, male House Sparrows (*Passer domesticus*) abrade the light-gray tips of their throat feathers to expose a dark black badge that signals their social dominance rank, with higher-ranking males exposing their badges earlier in the breeding season (Møller and Erritzøe 1992). Similarly, the racket-shaped

Figure 9.8. Wear of a wing feather over time. *A*, New wing feathers gradually wear down and lose function; *B*, Early signs of wear include broken barbs and slight fading caused by pigment breakdown; *C*, Heavily worn feathers exhibit distinct fading and the loss of many barbs and even part of the rachis. *Illustration by Cecilia Johnson.*

tail feathers of motmots (Momotidae) result from the loss of especially weak barbs midway along the rachis.

ADAPTATIONS TO MAINTAIN PLUMAGE FUNCTION

Feathers are dead structures that are cut off from the blood supply, so after they degrade, they cannot be repaired by the follicle. However, feather wear can be delayed or reduced by producing feathers that resist wear by their intrinsic chemical properties, and by caring for the feathers so that their intrinsic properties are preserved. If a feather is lost accidentally, it may be replaced, especially if it is a flight feather.

Adding pigments to feathers increases wear resistance. Melanin forms granules within the keratin matrix, increasing the hardness of the material and making it more difficult to crack and break apart (Bonser 1995). Melanins also dissipate the energy of UV photons into small amounts of heat. To test for effects of melanization on feather wear, Jed Burtt applied a stream of airborne silicaceous particles to warbler (Parulidae) tail feathers of varying colors. Rates of degradation were much lower in melanistic than non-melanistic feathers (fig. 9.10). Melanin, and a class of pigments unique to parrots called psittacofulvins, also make feathers more resistant to bacterial degradation (Goldstein et al. 2004, Gunderson et al. 2008).

Birds care for their feathers with a diversity of behaviors—they spend hours preening, bathing, scratching, and vigor-

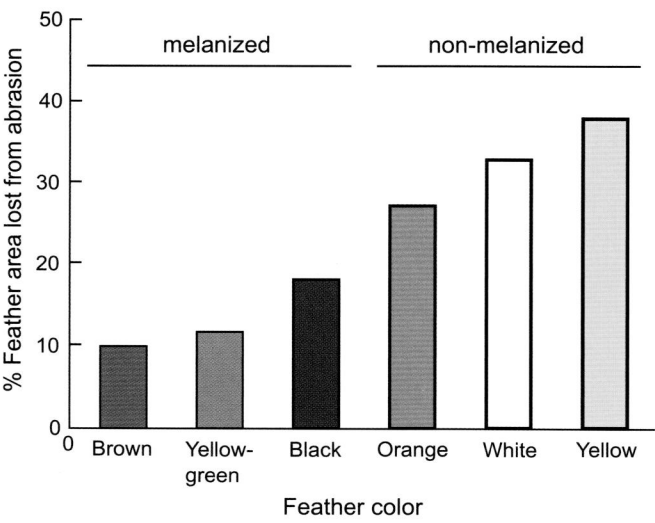

Figure 9.10. Wear rates of feathers of different colors. Pigments incorporated into feathers during feather production affect how quickly a feather breaks down over time. This may partly explain color patterns in some bird species, such as the black wing tips in many gulls. *Adapted from Burtt 1986.*

ously shaking themselves. Time spent maintaining the plumage takes away from time available for foraging and other important activities. When Peter Arcese provided free-living Song Sparrows (*Melospiza melodia*) with extra food, they spent much more time preening than their unfed conspecifics, suggesting a trade-off between benefits of preening and other time-consuming activities (Arcese 1989).

Birds preen their plumage to reposition misaligned feathers, zip neighboring barbules together, remove parasites, and distribute an oily secretion from the uropygial gland, which is located on the rump at the base of the tail (fig. 9.11). Uropygial oil is thought to help maintain the plumage by increasing the water repellency of feathers and by preventing microorganisms from degrading feathers. Interestingly, the hydrophobic properties typically associated with oil are not what increase water repellency. Rather, uropygial oils are thought to keep the fine structures of the barbs and barbules supple, and these microstructures maintain water repellency of feathers. Preening oil is also poisonous to feather-feeding lice (Moyer et al. 2003) and keratinolytic bacteria (Shawkey et al. 2003). Thus, the largest uropygial glands (relative to body size) are found in highly aquatic species or species inhabiting riparian habitats (Montalti and Salibián 2000, Galván et al. 2008), suggesting that the oil's primary functions are to aid in water repellency and combat feather degrading organisms associated with these environments.

Birds may bathe in water, snow, dust, smoke, and sunlight, and each kind of bathing may serve multiple functions. Wetting removes surface particles and may serve to relax the keratin, making barbs easier to reposition (Healy and Thomas 1973). Dust applied to the feathers absorbs oil and moisture, returning loft to greasy plumulaceous barbs (Healy and Thomas 1973), and injuring or killing ectoparasites, or removing their food sources (Clayton et al. 2010). Snow bathing probably functions similarly to dust bathing, because birds tend to use powdery snow, which contacts the feathers only briefly and without melting (Hendricks 2009). The benefits of smoke bathing are unknown, but smoke's acidity and warmth may inhibit parasites.

Using the sun's rays to bathe the plumage is different from basking: sunbathing is for plumage maintenance, and basking is for warming the body. Sunbathers extend the flight surfaces, but basking birds do not. Temperate birds tend to sunbathe during the hottest months, during the hottest part of the day, and in still air conditions (Blem and Blem 1992), suggesting the benefit comes from increasing the temperature of the plumage, perhaps to dry it out or inhibit ectoparasites.

"Anting"—allowing ants to crawl through the plumage—is one of the most puzzling bird behaviors. The benefits of

Figure 9.11. Bird applying preen oil. A Marbled Godwit (*Limosa fedoa*) reaches back with its bill to draw preen oil from its uropygial gland, which sits at the base of the tail. In this photo the tail is directed toward the camera, and the bird has raised its uppertail coverts to make the uropygial gland more accessible. *Photo by Geoffrey Hill.*

this behavior are unknown, but may be related to combating ectoparasites and feather-degrading microorganisms. Birds may also capture ants in the bill and aggressively rub them into the plumage. Ants produce formic acid as a defensive chemical, and while it seems plausible that rubbing ants into the feathers transfers the formic acid and helps reduce ectoparasite load, the concentration of formic acid in ant bodies is too low to inhibit microbial growth (Revis and Waller 2004).

FEATHER RENEWAL (MOLT)

Despite birds' adaptations for preventing feather degradation, all feathers eventually wear out, so birds undergo periodic molts to replace old feathers and regain plumage quality. Some molts renew all of the plumage (complete molt) and some molts only renew subsets of the body and flight feathers depending on species, the time of year, and the sex and age of the individual. Through molt birds can also change plumage color and ornamentation to suit different stages within the annual cycle, such as between nonbreeding and breeding seasons.

Feather Loss and Replacement

New feathers are produced as described above and in chapter 5. Growth of a new feather causes the old feather to release from the follicle and fall out. Species vary in the sequence and rate of feather replacement within feather tracts, but replacement patterns remain undescribed for the vast majority of

birds. In most described species, wing feather molt starts with one or more feathers near the middle of the wing and proceeds in waves of replacement that move distally and proximally, with typically three or four feathers molting at the same time within a wave (see the boxes on pages 254 and 255). Some birds create multiple waves within a flight feather tract, and some birds even molt all of the flight feathers simultaneously, becoming flightless. Body molt starts on the torso and proceeds toward the head and tail, with dozens or hundreds of feathers growing simultaneously. How information about feather color and structure is transmitted across rounds of replacement within an individual's lifetime is unknown, but follicles transplanted from one bird to another continue to produce the donor's feather type, meaning that cells within the follicle retain the information needed to produce each type of feather within an individual's lifetime (Lin et al. 2013).

Control and Regulation of Molt

Like other life history stages, molt is controlled by the neuroendocrine system, and timed in response to internal and external cues to optimize scheduling relative to other demanding activities within the annual cycle. Knowledge of the physiological mechanisms that regulate molt comes primarily from studies of domesticated and temperate species, and much of what we know comes from elegant experiments on captive birds by Alistair Dawson (box on page 256) and colleagues. Seasonal peaks in prolactin and thyroxine are correlated with the onset of molt (Dawson 2006), whereas high levels of gonadal (Nolan et al. 1992) and adrenal (Romero et al. 2005) hormones inhibit molt. As the breeding season approaches in spring, prolactin levels increase in response to longer days and warmer temperatures (Dawson and Sharp 2010), peak during incubation and brooding, and drop sharply after nesting (Sharp et al. 1998). In captive European Starlings, onset of molt is delayed by high prolactin (Dawson 2006) and closely follows the postbreeding drop in prolactin (Crossin et al. 2012). Prolactin also causes a postbreeding drop in testosterone, a gonadal hormone that delays the onset of molt (Nolan et al. 1992). Thus, prolactin appears to be involved in the shift from reproduction to molt in seasonally breeding birds, though changes in prolactin and the onset of molt may simply both be caused by another, unknown factor (Dawson 2006). During molt the vast majority of birds also downregulate corticosterone (Romero 2002), the major stress-related hormone in birds (see chapter 8), perhaps because of its negative effects on protein synthesis and feather quality (DesRochers et al. 2009). Shortening days hasten the end of molt, and late-molting birds molt faster than early molters (Dawson et al. 2000), apparently as an adaptation to finish molt before summer ends.

MOLT TABLES AND QUANTIFYING MOLT RULES

Quantifying patterns of feathers replacement in birds is important for understanding avian life histories and the evolution of molt strategies. Unfortunately, reliable summaries of molt patterns are rarely reported, especially for large birds. Indeed, molts of many large species that had multiple series of growing feathers were originally described as "chaotic." Part of the challenge associated with quantifying the sequence of feather replacement arises from making inferences about molt sequences using "snap shots" (e.g., single molting specimens, netted birds, or photographs) of individuals in molt.

One method employed for describing the sequence and pattern of feather replacement using individuals is a molt table. Molt tables summarize molt patterns using a sample of molting individuals, which can then be used to infer the number of molt series that proceed through the wing, and the boundaries between these series.

Here we present a brief summary of how to create molt tables (but we recommend the following papers for more detailed explanations and considerations when creating and interpreting these tables: Rohwer 2008, Rohwer and Wang 2010, Rohwer and Broms 2013). The first step is scoring feathers as either old (0), new (1), or growing; growing feathers are typically scored as fractions of total feather length, so a growing feather with score 0.7 would be 70 percent of its full length. These assignments are easy for species that replace all their feathers annually but can be challenging for species that carry flight feathers for multiple years, because feather groups or parts of feathers that are exposed to ultraviolet light or excessive wear (tertials, outer primaries, and wing tips) degrade at different rates, making them appear older than they are. Importantly, for species that carry individual flight feathers for multiple years, feathers must be aged as best as possible (e.g., 1 = newly replaced during the most recent molt, 2 = a feather that is over one molt cycle old, 3 = a feather that is over two molt cycles old). Once all molting feathers have been scored, these data can be distilled into tables to infer what groups of feathers constitute a molt series and in what direction the molt proceeds.

To identify where a molt series begins and ends, and the direction in which it proceeds, we need to correctly assign nodal feathers, terminal feathers, and directionality. We highlight three molting Yellow-billed Cuckoo (*Coccyzus americanus*) wings scored by Rohwer and Broms (2013) to illustrate how to infer the beginning and end of a molt series and the direction in which it proceeds (see figure). Nodal feathers are those that start a molt series and can be identified as growing feathers that are (1) surrounded by old feathers (as in P6 and P9 in *B*), (2) flanked by an old feather on one side and a growing feather that is shorter than the focal feather on the other side (similar to that seen in P1 of *B*), or (3) flanked by old or growing feathers on both sides, both of which are shorter than the focal feather. Terminal feathers are those that mark the end of a molt series and are identified in a similar way to nodal feathers, except that the feathers surrounding terminal feathers must meet the opposite criteria than those surrounding nodal feathers: for example, terminal feathers are surrounded by new feathers (as in P5 and P8 in *C*). Assessing directionality of a molt series is best done by comparing adjacent pairs of growing feathers. If the proximal feather is longer than the distal feather, we would score directionality as distal between this pair. Using pairs of only growing feathers can limit our ability to identify directionality, especially in tropical species that rarely lose adjacent feathers in quick succession. Thus, directionality can also be scored when growing feathers are flanked by newly replaced feathers and old feathers, although this has the potential to assign directionality of one molt series onto a separate series, as old feathers may not be recognized as terminal feathers.

Primary molts in cuckoos were historically thought to be a single molt series, but, using molt tables, Rohwer and Broms (2013) showed that Yellow-billed Cuckoos have broken up their primaries into three molt series. In these examples, without summarizing data across a series of museum specimens, molt appears chaotic and random. However, summarizing molt data from just 12 individuals clearly shows that Yellow-billed Cuckoos have broken up their primary molt into three independent series. Molt tables illustrate and succinctly summarize the pattern of primary replacement in cuckoos while providing the necessary data and sample

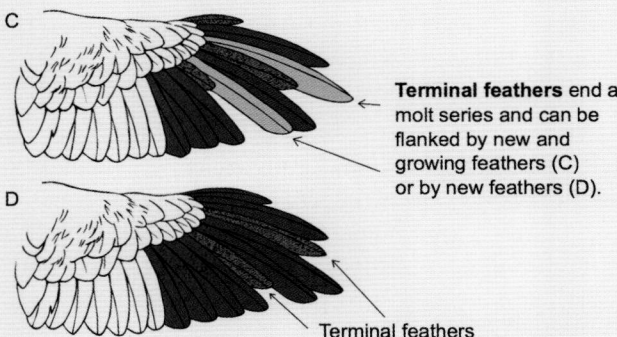

Feather age: ■ New ▨ Growing ▨ Old

A P10

Primaries are numbered sequentially from the inner most feather outward.

Secondaries | P1 | Primaries

B

Nodal feathers initiate molt series and are flanked by old feathers. On this wing, P1, 6, and 9 are nodal.

Direction that molt proceeds in a wing is determined by multiple adjacent growing feathers. The longest growing feather was lost first, so, in this series, molt proceeds outward.

C

Terminal feathers end a molt series and can be flanked by new and growing feathers (C) or by new feathers (D).

D

Terminal feathers

Yellow-billed Cuckoo wings in various stages of molt. These "snap-shots" of molt can be used to infer the number of feather replacement series that proceed through a wing and which feathers initiate or terminate a series. Additionally, comparing adjacent feathers allows you to infer the direction in which feathers are replaced. *Cuckoo wings illustrated by Liz Clayton Fuller.*

sizes used to infer the sequence of feather replacement, adding a level of transparency and rigor to quantifying molt. Ecologically, dividing the primaries into three different molt series enables cuckoos to replace several primaries at once, while minimizing both the time required to molt and the performances costs to flight by spacing out growing feathers, which reduces large gaps in the wing.

Costs of Molt

Producing a new set of feathers takes energy and time, and the requirements for molt are substantial. Åke Lindström and colleagues measured the metabolic rate of Bluethroats (*Luscinia svecica*) before, during, and after molt, finding that daily energy use *doubled* during peak molt (fig. 9.12; Lindström et al. 1993). In adult White-crowned Sparrows (*Zonotrichia leucophrys*), the postbreeding molt of the wing, tail, and body feathers is roughly equivalent to a total turnover in body protein (Murphy and King 1992), and juveniles replacing just 25 percent of their body feathers increased their daily food intake by 17 percent on average (Bonier et al. 2007). While the energetic costs of molt appear high for small birds, these costs drop and appear to remain low for birds 100 grams and larger. Why do energetic costs of molt vary so much with body size?

Feathers grow continuously day and night (Murphy and King 1986), and the energetic costs of feather synthesis are especially severe at night when birds are not feeding and protein for growing feathers must be mobilized from body reserves. Once the blood supply of protein is exhausted, birds must break down body protein to replenish the blood stores for feather growth while fasting (Groscolas and Cherel 1992, Jenni and Winkler 1994). Large birds have ample protein reserves, and thus feather synthesis at night minimally strains their blood resources for protein. By contrast, small birds must break down a much larger percentage of their body protein every night to meet the demands of feather generation, which presumably makes the energetic cost of molting high for small birds because they must rebuild these reserves by day.

The energetic costs of molt mean that most birds must set aside a substantial amount of time each year for plumage renewal. For example, among passerines at northern latitudes, the annual postbreeding molt may take up to 100 days in starlings (Rothery et al. 2001) and swallows (Yuri and Rohwer 1997). Molt duration tends to be even longer in the tropics, often lasting over 150 days in passerines, including up to 300 days in the White-plumed Antbird (*Pithys albifrons*; Johnson et al. 2012). Diversity in time spent molting is discussed further in the next section.

MOLT AS A DRIVER OF AVIAN LIFE HISTORY DIVERSITY

In this section we illustrate how the energetic and temporal costs of molt have influenced the evolution of avian life histories. Much of this section draws from work by Sievert Rohwer and colleagues (box on page 257), who helped transform studies of molt biology from descriptive works that

ALISTAIR DAWSON
By David DesRochers

Alistair Dawson's research, which spans nearly four decades, has provided key insights into the physiology and phenology of avian molt. In his over 100 published papers, his controlled, lab-based approach to research provides us with a critical counterpart to the many field-based studies exploring molt physiology. His research has gained for us a deep understanding of how and why birds time molt and the consequences of those decisions.

Dawson's early research in avian endocrinology, focusing on European Starlings, has established a foundation for many contemporary studies into avian hormonal physiology. While much of his research is lab-based and focuses on individual bird health and survival, he also has explored the physiological consequences at larger biological scales such as the population and community level and the ways those relationships change over time. In fact, his publication exploring long-term effects of the timing of molt in songbirds is considered central to modern molt research.

Dawson continues to expand our understanding of the seasonal physiology of birds and has begun to explore how changing climate affects the physiology and survival of birds. As an impressive extension of his research into seasonality of biological systems, Dawson began exploring the consequences of seasonal disruption in human populations as well, demonstrating the broad application of his research approach. His storied research in avian physiology is regarded as central to avian endocrinology, as evidenced by many of his key publications being cited hundreds of times by his colleagues. His long career investigating molt physiology also earned him the opportunity to author the molt chapter in the most recent edition of *Sturkie's Avian Physiology*.

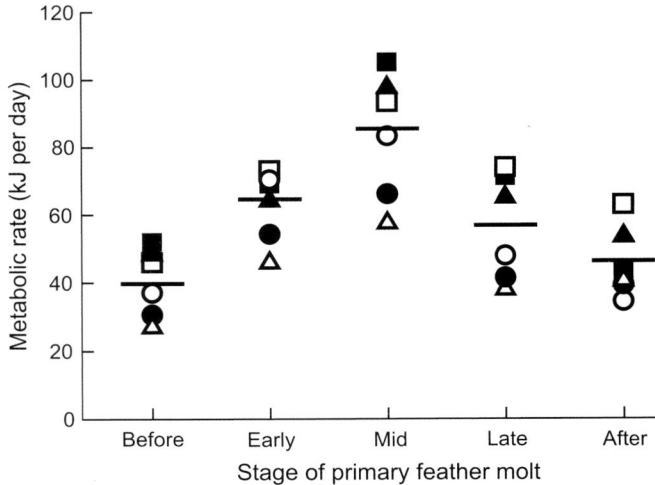

Figure 9.12. Metabolic costs of molt. Daily energy use doubles in male Bluethroats (*Luscinia svecica*) during mid-molt. Molt of the primary feathers lasts about 54 days in this species, representing an extended period of high caloric demands in these 20-gram songbirds. Each symbol represents one individual at different stages of molt. Horizontal lines = means of all individual metabolic rates within each stage of molt. *Adapted from Lindstrom et al. 1993.*

were "as exciting as watching paint dry" to big-picture comparative works that shed new light on life history evolution in birds. We focus on species that breed seasonally at northern latitudes because molt has been studied most extensively in those taxa. We also focus on molt of the primary flight feathers because the costs of primary molt can most clearly be linked to the life history trade-offs that shape the evolution of molt strategies, and because the time required to replace the primaries encompasses the majority of time required to

replace all other feathers (wing, tail, and body). The box on page 258 explains how ornithologists estimate the time required to molt, a critical life history variable that influences many aspects of the annual cycle, as discussed in this section.

Body Size, Time Constraints, and Molt Strategy

Time constraints associated with molting affect small and large birds in different ways. Small birds can typically replace all their flight feathers during a single episode of molt, suggesting time constraints do not prevent a complete molt. By contrast, most large birds do not replace all their flight feathers during a single episode of molt and must carry individual feathers for two and sometimes three years before they are replaced. This suggests that time constraints to complete the molt are much more severe for large birds, but why?

Comparing feather growth rates and the total length of the flight feathers across a diversity of body sizes reveals striking contrasts—the rate at which individual feathers grow shows little variation across body size, while the summed length of the flight feathers increases dramatically with body size (Rohwer et al. 2009; box on page 263). Thus, large birds have a longer summed length of flight feathers to replace during molt, but the rate at which their feathers grow does not compensate for this increase in feather length, presumably because developmental constraints limit the rate at which feathers can grow (Rohwer et al. 2009).

SIEVERT ROHWER
By Chris Filardi and Catherine Filardi

At its heart, field ornithology is an endeavor that combines the intellectual with the practical. Through far-reaching and stalwart adherence to this balance between the visionary and the mundane in studying the lives of birds, Sievert Rohwer has become one of the most innovative and influential forces in ornithology. Rooted in direct, lifelong observations of the natural world, Sievert has combined laboratory, museum, and field techniques to investigate big theoretical questions that have advanced the fields of evolution and ecology well beyond birds.

Sievert pioneered difficult, elegant field studies on plumage variation and social signaling. His early work dyeing the throats of wintering Harris's Sparrows stimulated numerous studies on the social function of color variation in birds and other animals. Later, with Eivin Røskaft, Sievert dyed breeding male Yellow-headed Blackbirds, work that ultimately led to his "arbitrary badge hypothesis," providing a mechanism by which novel color patches may serve aggressive competition, or other signaling function, and account for rapid color divergence among geographically disjunct populations.

Interest in plumage signaling led to studies of molt and the implications of replacing feathers for fitness. Carefully prepared museum specimens provided critical data for these studies, resulting in influential work on molt-migration trade-offs, hybridization, and speciation. His commitment to developing contemporary specimen collections at the Burke Museum resulted in the world's most expansive spread-wing collection, along with heavily used collections of data-rich traditional specimens and tissue (DNA) samples.

Sievert's legacy extends beyond his theoretical and museum-based accomplishments and is a product of his focused and generous investment in a broad spectrum of student naturalists, young scientists, and curators. Sievert instills in those around him a purposeful concentration, in rigorous yet inherently joyful pursuit of natural history as a foundation for not only natural science, but an inspired life.

How might time constraints affect the evolution of molt strategies in birds? Because increasing the rate at which feathers grow has little effect on molt durations, birds primarily reduce the duration of their molt by growing more feathers at once. But this strategy has serious costs. Growing multiple adjacent feathers at once can generate large gaps in the wing that reduce flight performance. Thus, selection favoring molt patterns that distribute missing feathers across the wing should be strong and drive patterns of feather replacement across body sizes.

Birds have three kinds of strategies to replace their flight feathers—simple, complex, and simultaneous—and these strategies vary with body size. Species that have simple molts replace all their primaries annually in a single wave of feather regeneration. Accelerating these molts results in large gaps in the wing caused by missing adjacent feathers. Costs to flight as a result of these gaps should be strongest in large, heavy-bodied birds because the energy required for powered flight and the proportional loss of wing surface area (and subsequent increase in wing loading) is high relative to small birds (Tucker 1991, Swaddle and Witter 1997, Hedenström and Sunada 1999). Examining body sizes of species with simple molts shows that this molt strategy is most common among small-bodied birds (fig. 9.13).

In complex molts, two or more waves of feather regeneration proceed through the primaries simultaneously (see Rohwer 1999 for details). Complex molts allow replacement of multiple feathers at once while distributing growing feathers throughout the wing. This reduces both the time spent molting and the aerodynamic costs of replacing multiple feathers simultaneously and should be most common in larger birds that face strong time constraints to complete molt. Body sizes for species that undergo complex molts are larger than those of species with simple molts and approach the upper size limits for flying birds (fig. 9.13). However, because complex molts often do not result in the replacement of all flight feathers, species undergoing complex molts accumulate overworn feathers and likely suffer costs of asymmetric molts (Swaddle et al. 1996, Brommer et al. 2003). These costs should disfavor this strategy for species with ecologies that permit replacing all feathers during a single episode of molt.

Simultaneous molts occur when all flight feathers are dropped at once, rendering birds flightless until primary growth is nearly complete. These molts reduce the time required to replace the flight feathers to the time needed to grow the longest primary. If a species' ecology permits a period of flightlessness, simultaneous molts reduce trade-offs in time allocation between molting and other activities (e.g., breeding) in the annual cycle. Comparing body size distributions

MEASURING MOLT DURATION

Estimates of molt duration are essential for comparative studies of life histories and for understanding time budgets in the annual cycle. Most studies measure the duration of the primary molt as an index of the total time spent molting because the duration it take to replace the primaries encompasses the majority of time spent replacing all other feathers. Additionally, the primary feathers are the most accessible feathers on traditional museum specimens, allowing researchers to estimate molt duration using natural history collections. There are three general methods for estimating molt duration: recapturing molting individuals (Newton 1966, Bancroft and Woolfenden 1982), comparing the progression of molt from single observations of molting individuals through time (Pimm 1976, Underhill and Zucchini 1988, Rothery and Newton 2002), and calculating feather growth rates and the intervals between feather loss (Rohwer and Broms 2012). We very briefly discuss each approach but recommend reading the key papers associated with each method for more details.

Recapture methods are best for marked individuals that can be captured multiple times during their molt. This method is logistically impossible when individuals are difficult to capture, or impossible to recapture, such as species that molt and migrate simultaneously (like swallows). The timing of recapture can influence the estimate of molt duration, especially in species that initiate molt by dropping several flight feathers in quick succession then replace outer feathers more slowly. Thus, for this method, maximizing the time interval between recaptures of molting individuals provides the most realistic estimates of molt duration.

Duration estimates made using regression and maximum likelihood techniques are the most common. These methods examine molt scores from individuals examined just once during the molt for the entire molting process. Individuals receive a molt score that describes their relative progress with the molt, and then molt scores of many individuals are compared across time to generate estimates of molt duration. These methods assume that all individuals have an equal probability of being sampled, molting individuals come from the same population with similar molt schedules, molt is seasonal, and molt initiates at P1 and proceeds as a single wave to the outermost primary. Of course, sometimes these assumptions cannot be met. When molt initiation and termination dates overlap with each other within populations, both methods underestimate duration. Thus for tropical species that typically have long, low-intensity molts, these methods will likely provide unrealistic duration estimates. Overall, regression estimates (Pimm 1976) are often (heavily) criticized because samples of molting individuals are unevenly distributed with fewer individuals in early and late stages of the molt, which tends to overestimate molt duration compared with maximum likelihood techniques (Underhill and Zucchini 1988). However, the differences observed between these methods for a single species is often only a matter of days, which, compared with the variation observed across species, is trivial. Thus, despite the statistical drawbacks of regression techniques, their estimates of molt duration are generally realistic and very close to estimates made using maximum likelihood techniques.

When recapture methods are impractical and molt data do not fit the assumptions of regression and maximum likelihood techniques, using feather loss intervals provides an alternative method to estimate molt duration (Rohwer and Broms 2012). This method requires three pieces of data: (1) the length of each primary feather when fully grown, (2) estimates of feather growth rates, and (3) the number of days between successive feather losses (estimated from feather growth rates and the difference in feather length between adjacent pairs of growing feathers). With these data, the number of days between successive feather losses for each successive feather pair is multiplied by the summed length of the first focal feather then divided by the feather growth rate. This process is repeated for each primary in the wing, then summed across the primaries to estimate molt duration (see Rohwer and Broms 2012 for more details). Estimating molt durations using feather loss intervals will be most valuable for tropical species with prolonged molting periods or for species with asynchronous molts.

Ultimately, the best estimates of molt duration will likely come from using a combination of these approaches, and the appropriate technique for estimating duration will depend on the species' ecology.

of species that molt their flight feathers simultaneously shows that these species crowd the upper size limits of flying birds even more strongly than species with complex molts (fig. 9.13). In order to achieve larger body sizes, birds have evolved either complex or simultaneous molt strategies to accommodate sufficiently regular replacement of their flight feathers.

Molt-Breeding Trade-Offs

Molt and breeding tend to be segregated in the annual cycle because both events increase energy demands, and impaired flight performance caused by missing flight feathers is thought to lower reproductive success (Foster 1974, Johnson et al. 2012). For example, in Black-footed Albatrosses (*Phoebastria nigripes*), successful breeding requires over 260 days, leaving little time in the annual cycle to molt. Accordingly, adults molt only three to five primary feathers in most years, which results in the accumulation of worn flight feathers (Langston and Rohwer 1996). Adults that have accumulated heavily worn feathers (~three years old) are more likely to skip breeding than adults with fresher feathers, and those that do breed are 60 percent less likely to successfully fledge young relative to pairs with freshly replaced feathers (Rohwer et al. 2011). Thus, individuals may breed successfully for one to three years, then take a one-year hiatus from breeding, presumably to replace an accumulation of overworn feathers (Langston and Rohwer 1996).

Molt-breeding trade-offs are likely most pronounced in large, long-lived birds, in part because in long-lived animals selection for adult survival should be stronger relative to selection for short-term reproductive gains (Stearns 1992). Because large birds cannot effectively reduce the duration of molt by increasing feather growth rates (Rohwer et al. 2009), their only other option is to grow more feathers simultaneously (fig. 9.14). For long-lived birds the cost of growing many feathers simultaneously likely reduces adult survival, so selection apparently favors forgoing breeding attempts in order to replace accumulated worn feathers in the wing, rather than increasing the number of feathers replaced at once. By contrast, in smaller, short-lived birds, selection for high reproductive investment should be stronger than selection for high adult survival. In these species, selection should favor short-term reproductive investment at the costs of future feather quality (Nilsson and Svensson 1996) or increased risk of mortality by growing more feathers simultaneously (Jenni and Winkler 1994, Hall and Fransson 2000).

Molt in Relation to Migration

Migration requires time and energy, and thus migration and molt rarely occur simultaneously in migratory species. Breeding in north temperate regions fixes the timing of spring migration and reproduction for migratory passerines;

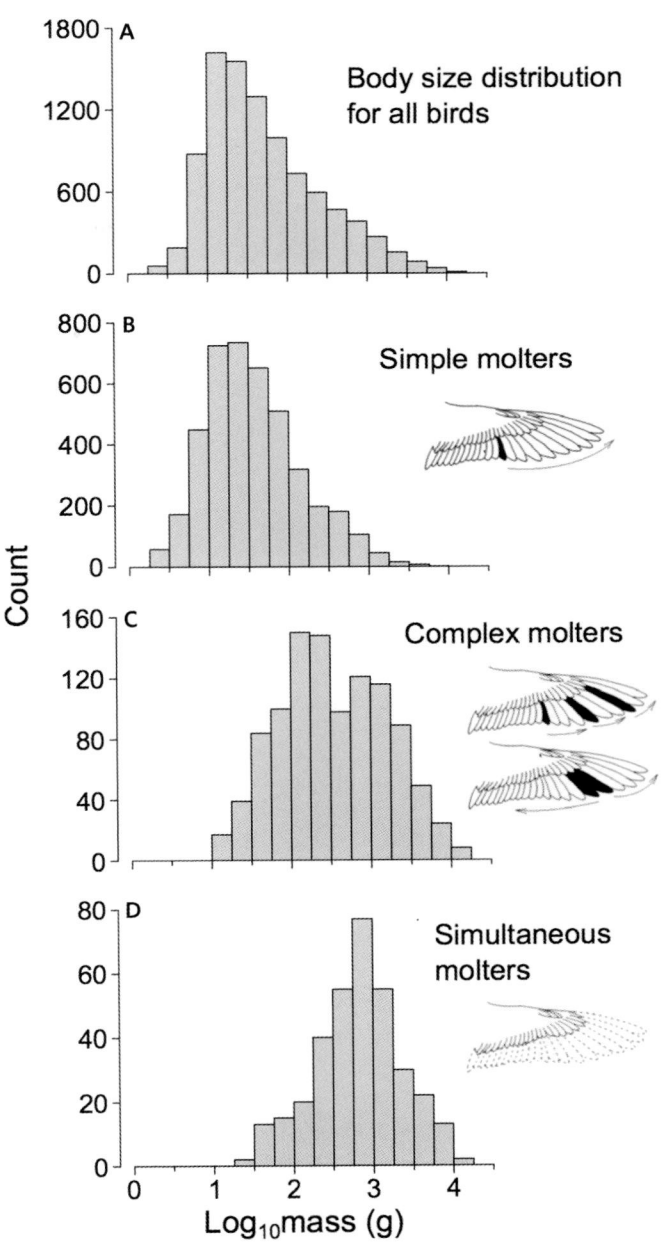

Figure 9.13. Body size distributions vary with molt strategies. Histogram *A* shows the body size distribution for all birds, and the histograms below show size distributions for species with *B*, simple molts; *C*, complex molts; *D*, simultaneous molts. Complex and simultaneous molts are most common in larger species, likely because these patterns of feather replacement reduce the time needed to complete the molt. *Adapted from Rohwer et al. 2009.*

this leaves flexibility only in scheduling the postbreeding molt relative to the fall migration (fig. 9.15). The scheduling of these events shows remarkable evolutionary flexibility among migratory passerines of North America and Europe, and may act as a species isolating mechanism (Svensson and Hedenström 1999, Rohwer et al. 2005).

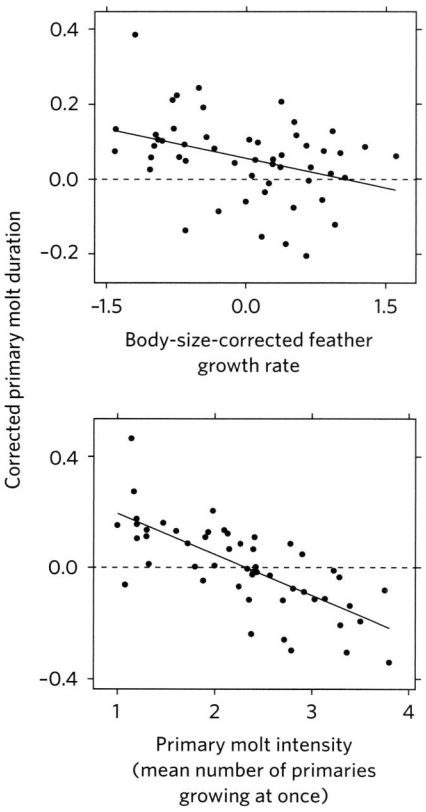

Figure 9.14. Reducing the time required to molt. Birds could reduce the time spent molting either by increasing the rate they grow individual feathers or by growing more feathers at once. This figure compares these mechanisms by regressing primary molt durations that have been corrected for body size (because larger birds take longer to molt than smaller birds) on feather growth rate (again corrected for body size, as larger birds grow their feathers faster than smaller birds) and the number of feathers grown at once. Comparing these two mechanisms suggests that growing more feathers at once is the most effective strategy for reducing the time required to molt. All points in top and bottom figures are matched so that data on molt duration, feather growth rate, and the number of feathers grown at once come from the same species. *Adapted from Rohwer and Rohwer 2013.*

Contrasts in environmental conditions that can support the energetic demands of molt appear to drive the evolution of molt schedules in migratory birds. Among small migratory passerines that breed in eastern or western North America, the vast majority of eastern species undergo their postbreeding molt on the breeding grounds. By contrast, only about half of western species remain on the breeding ground to molt, while the other half leave the breeding range and move to northwestern Mexico to molt (Rohwer et al. 2005). This contrast between eastern and western species is thought to be driven by geographic patterns of precipitation and food

availability. Throughout much of eastern North America, late-summer rains maintain high levels of primary productivity and food availability for species to molt on the breeding grounds. In the west, late-summer droughts reduce primary productivity and food availability, especially in low-elevation habitats. To find food resources needed for molting, western species either move upslope to wetter, more productive habitats (Greenberg et al. 1974, Rohwer et al. 2008), or they migrate south to the region of the Mexican monsoon, where summer rains transform the habitat from leafless thorn-scrub into a lush, food-rich forest (fig. 9.16). For these monsoon molt-migrants, summer aridity is thought to "push" birds away from the breeding range, while the flush of productivity brought by the monsoon rains is thought to "pull" birds south to molt (Rohwer et al. 2005).

Molt schedules that differ between closely related species have the potential to create discrete, species-level differences that impose fitness costs to hybrids, making molt a potential species-isolating mechanism. Some Neotropical species pairs with contrasting molt schedules, such as Rose-breasted (*Pheucticus ludovicianus*) and Black-headed Grosbeaks (*P. melanocephalus*), have narrow hybrid zones suggesting strong selection against hybrids. Differences in the scheduling of molt relative to migration likely have a heritable basis (Berthold and Querner 1982, Helm and Gwinner 1999), and hybrids may inherit suboptimal molt schedules relative to their parents. Hybrids or back-crossed individuals could be selected against if they inherit combinations of genes that (1) mismatch the timing of molt relative to migration for the region in which they breed, (2) increase or decrease the number of molts in the annual cycle, or (3) overlap molt and migration (fig. 9.15). Among 12 well-studied contact zones of closely related eastern and western species, all have strong contrasts in either their migratory orientation or the scheduling of their molt or both (Rohwer and Irwin 2011). These patterns suggest that both orientation and molt scheduling divides play an important role in selecting against hybrids and thus favor assortative mating among closely related species.

Molt Constraints on Color Change

Many birds undergo molts that dramatically change the color of the body plumage between the breeding and nonbreeding seasons. Perhaps the most readily visible and frequently studied contrast in seasonal plumage variation is among male Neotropical migrants. Nearly all these species undergo complete molts of both body and flight feathers after the breeding season, and many undergo a second spring molt of only the body plumage before the start of the breeding season (Froehlich et al. 2005). Why might birds change their appearance for different seasons?

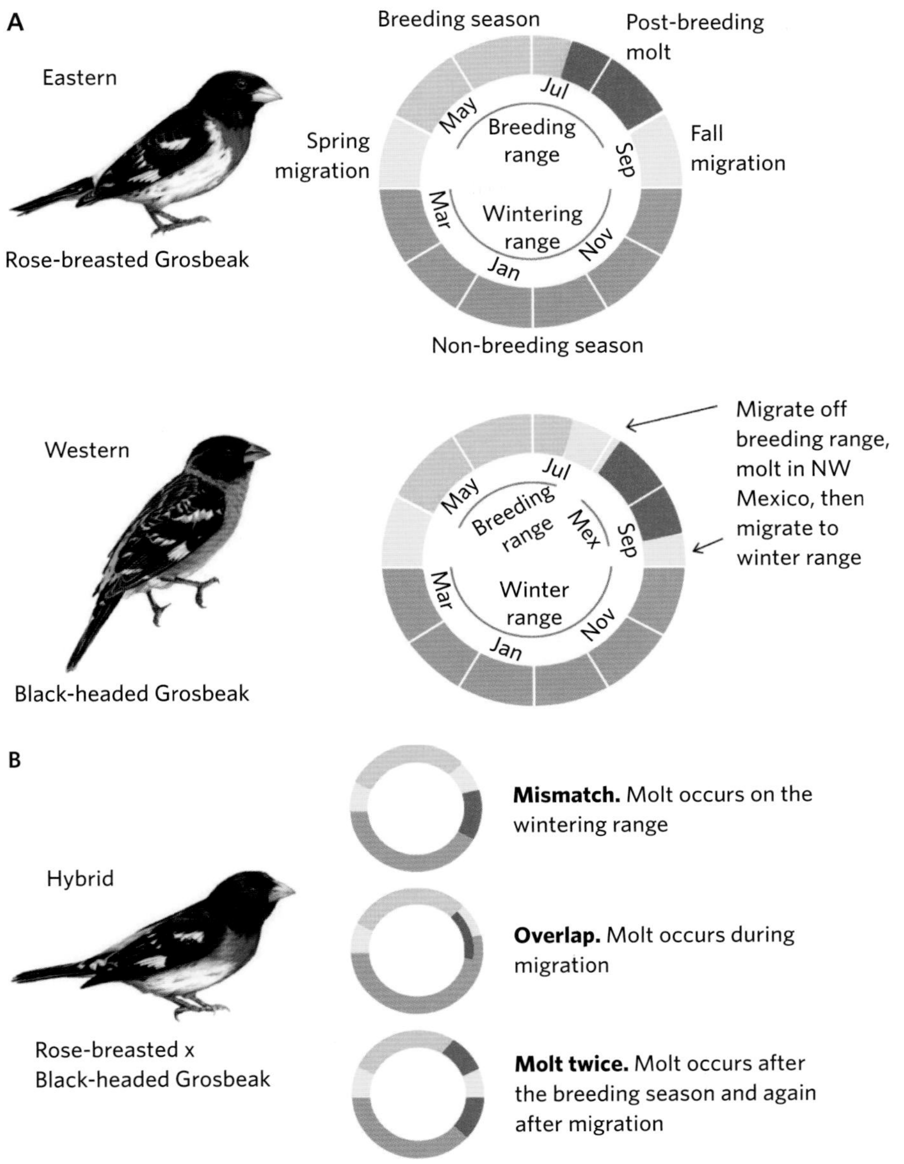

Figure 9.15. Contrast in molt scheduling. A, Differences in the timing and location of the postbreeding molt between Rose-breasted (eastern) and Black-headed (western) Grosbeaks, closely related passerines that breed in North America. Most eastern passerines molt on the breeding range prior to migrating south, while many western passerines migrate to the American Southwest and northwestern Mexico for their postbreeding molt, then continue their migration south to their wintering range; B, Molt and migration are likely genetically controlled traits (each controlled by independent genes), thus hybrids between eastern and western species may inherit combinations of these genes that cause them to (i) molt in the wrong location, (ii) molt during migration, or (iii) undergo two complete molts during the annual cycle. Suboptimal scheduling of the postbreeding molt likely functions as a strong selective mechanism against hybrids, favoring assortative mating between closely related species. Grosbeaks illustrated by Chloe Ohmori.

In species with seasonally distinct plumages, breeding plumage is generally brighter than nonbreeding plumage, and this difference is most extreme among males. While sexual selection is thought to explain the evolution of bright breeding plumage in many species (see chapter 11), hypotheses for bright or dull plumage in the nonbreeding season remain less well understood. Adaptive explanations for seasonal plumages changes include crypsis (Hamilton and Barth 1962), social signaling (Rohwer 1975, Lyon and Montgomerie 1986), and signaling high vigilance and agility (the "unprofitable prey hypothesis"; Baker and Parker 1979).

Two observations of plumage cycles in males of many passerines suggest that the energetic requirements of molt may constrain the production of bright spring plumages on the wintering grounds. First, males of some species do not have a spring molt and must maintain a bright winter plumage even though they do not need to attract a mate during winter (Froehlich et al. 2005). While these bright plumages are hypothesized to be costly because of increased predation, data for this hypothesis are lacking (but see Berggren et al. 2004). Second, first-year males in many sexually dichromatic species of Neotropical migrants delay plumage maturity, but not sexual maturity, until after their first breeding season, a phenomenon called **delayed plumage maturation** (Lyon and Montgomerie 1986, Hawkins et al. 2012). First-year males of species with delayed plumage maturation have mostly dull plumage during the breeding season and often look similar to females.

Figure 9.16A and B. Molt in the Mexican monsoon region. Late-summer monsoon rains dramatically transform the deciduous thorn forest of northwestern Mexico into a verdant, food-rich region where many migrants come to molt after breeding in western North America. *Photos by Vanya Rohwer.*

Both bright winter plumage of adult males and delayed plumage maturation of first-year males may be explained in part by molt constraints. If costs of winter molt are high, adult males may be forced to carry conspicuous plumage on the wintering grounds, and first-year males may be prevented from producing a conspicuous adult-like plumage (Rohwer and Butcher 1988). Thus, these plumages for adult males and first-year males may not be adaptations to the winter and summer range, respectively, but may simply reflect the costs of winter molts. Unfortunately, no studies have directly tested the molt constraints hypothesis with food supplementation experiments (Hawkins et al. 2012). Alternatively, bright winter plumages in adults and dull breeding-season plumages in first-year males could be adaptations that reduce aggressive encounters (Hawkins et al. 2012).

Costs associated with producing conspicuous breeding plumage suggest that plumage coloration may be an honest signal of individual quality. However, few studies have linked reproductive benefits with the timing and extent of the spring molt. In cooperatively breeding Superb Fairy-wrens (*Malurus cyaneus*), the earliest males to molt into conspicuous breeding plumage have a higher probability of siring extra-group young (Mulder and Magrath 1994, Dunn and Cockburn 1999). Early molting males complete molt in winter when resources are most limiting. Once males have completed their molt, they immediately start displaying to females of other cooperatively breeding groups (Dunn and Cockburn 1999). Extra-pair breeding is never forced, and females appear to make breeding decisions based on how early in the season a male begins displaying to her. Thus, the timing of molt in this system is likely an honest signal of individual quality, as the costs of winter molts should keep the frequency of cheaters low.

Feather Quality as the "Currency" of Molt

Thus far in this section we have examined how the temporal and energetic costs of molt influence the evolution of molt strategies in large birds and the scheduling of molt in small migratory passerines. Regardless of body size, however, shorter molt durations would allow more time for breeding and reduce the time spent paying the immediate costs of growing feathers (e.g., increased energy demands and impaired flight). So, why don't birds molt more quickly?

While rapid molts would appear beneficial, a growing body of evidence suggests that molting too quickly compromises the quality of feathers grown. Captive European Starlings (*Sturnus vulgaris*) with experimentally hurried molts grew shorter, lighter flight feathers with thinner and more flexible rachises, and their feathers degraded more rapidly compared with starlings with unhurried molts (Dawson et al. 2000). Populations of Grey Plovers (*Pluvialis squatarola*) with the least available time to molt show the greatest loss in wing feather length between new and old feathers, suggesting that populations with hurried molts grow lower-quality feathers compared with those with unhurried molts (Serra 2001). And, Eurasian Blue Tits (*Cyanistes caeruleus*) experimentally forced to nest late in the season—reducing the time available to molt—had lower winter survival, increased thermoregulatory costs, and lower reproductive success during the following breeding season (Nilsson and Svensson 1996).

These studies illustrate a trade-off between time spent molting and feather quality: individuals that molt rapidly produce low-quality feathers. These observations suggest

ALLOMETRIES OF MOLT

Allometric relationships describe how traits of organisms scale with body size and are powerful tools for comparative studies. But, unless you are familiar with interpreting these plots, they can often subdue differences across traits or body sizes that are in reality quite dramatic. This is because the scaling of these plots is logarithmic, which disproportionately contracts very large values and expands very small values. For example, the variation in the summed length of the primary flight feathers (prior to $\log_{10}$ transformation) across body size presented in Rohwer et al. 2009 ranges from 454 mm to 4869 mm, while the variation in feather growth rate varies from 1.7 mm to 11 mm, across the same range of body sizes. Variation between primary feather length and primary feather growth rate differ by several orders of magnitude. When these variables are log transformed and presented as allometric relationships, their differences are not nearly as obvious because the coefficient describing this variation becomes an exponent. Thus, in the allometric relationships presented by Rohwer et al. (2009), the key variable for understanding how feather length and feather growth rate scale with body size is the coefficient (the slope in allometric relationships) for each variable. Feather growth rate increases with body mass (M) at a rate of $M^{0.171}$, while feather length increases with body mass at a rate of $M^{0.313}$, nearly twice as fast as feather growth rates. While these differences in slopes do not appear dramatic, they are. Because these slopes are exponential, a difference of nearly two results in a difference of nearly two orders of magnitude (e.g., $10^2 = 100$; $10^4 = 10000$).

To illustrate the seemingly minor differences in how feather growth rate and the summed length of the primary flight feathers scale with body size, we replot these variables using untransformed data. This comparison illustrates the relative variation observed between these variables prior to $\log_{10}$ transformation. In the plot generated using untransformed data, the slopes of the two lines vary dramatically and illustrate why large birds face ever-increasing time demands to complete their molt. Feather growth rates simply

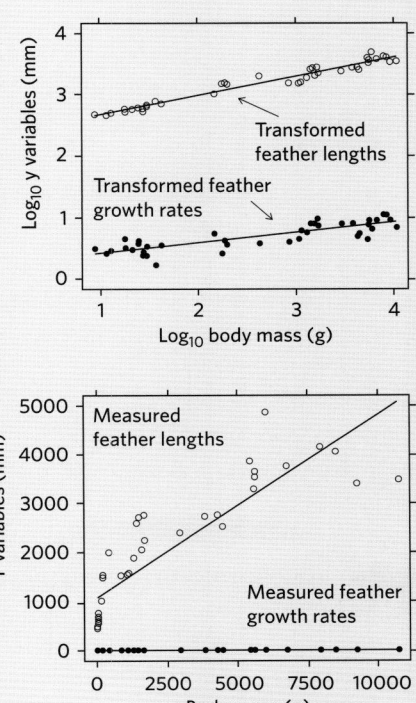

Comparisons between feather growth rate and the summed length of the primary flight feathers. Open circles are the summed length of the primary feathers for 43 species and filled circles are the average feather growth rates for the same species; measures of feather length and feather growth rate are paired for each species. The *top* plot shows allometric relationships between these two variables when data are $\log_{10}$ transformed, while the *bottom* plot shows the same data prior to transformation. The variation in both feather length and growth rate across body sizes is readily visible using untransformed data, but more difficult to recognize using $\log_{10}$ transformed data. *Adapted from Rohwer et al. 2009.*

do not increase with body size as quickly as the summed length of the flight feathers, making molt one of the most time-consuming events in the annual cycle of large birds that fly while molting.

that factors that force individuals to molt quickly cause a loss in feather quality, which can carry forward to affect survival and future reproductive success. Thus, feather quality appears to be the currency that selection acts on to reduce year-to-year variation in trade-offs among important life history events.

APPLICATION: LEARNING ABOUT BIRDS THROUGH THEIR FEATHERS

Imagine you wanted to learn something fundamental about robins: What do they eat? You might observe a robin eating mulberries, but then the bird flies away. What does it eat in the next woodlot, out of sight? And what does it eat when it flies south for the winter? Have robins' diets changed over time? Do their diets include environmental pollutants? These questions, and many others regarding the lives of birds, are difficult to answer because they span broad scales of space and time. However, feathers contain chemical signatures that provide a durable source of information about a bird's past. During molt, various elements and compounds in the blood are incorporated into the keratin matrix and persist there until the feather is replaced. Today, feather chemistry is one of the most powerful tools for learning about birds (Bortolotti 2010).

Isotopes of common elements found in biological molecules, such as hydrogen, carbon, and nitrogen, vary in their frequencies according to geography (fig. 9.17) and food-web dynamics. **Stable isotope analysis** takes advantage of these variations in order to identify a bird's location before or during the time the feather was produced, and to gather information about its diet (see Hobson 2005 for a review). For example the ratios of hydrogen isotopes in the feathers of migratory songbirds captured on their wintering grounds in Central America revealed the birds' breeding (and molting) grounds thousands of kilometers to the north (Hobson and Wassenaar 1997). Before stable isotope analysis, biologists had to rely on marking a bird on the breeding grounds and capturing it again on the wintering grounds—an extremely unlikely event across so much space and time.

Feather isotopes also help answer questions like those we asked about robins at the beginning of this section. For example, ratios of nitrogen and carbon isotopes in the body feathers of two tropical petrel species revealed that one species consumed mainly fish, while the other consumed crustaceans and squid, revealing a dietary niche separation that explains why the two species can molt sympatrically. This discovery would be very difficult to make by direct observation, as both species forage primarily at night on the open sea (Rayner et al. 2016). Feather isotopes are so stable that they can be sampled from museum specimens collected decades ago and compared with modern samples. Differences in isotope ratios between historical and modern feathers revealed that Northern Fulmars (*Fulmarus glacialis*) made a major dietary shift during the 1900s as the whaling industry declined (Thompson et al. 1995).

Birds also acquire potentially toxic pollutants from the environment that become incorporated into feathers. Heavy metals, such as mercury, enter food webs through water and soil and accumulate at the higher trophic levels that birds occupy. During molt, mercury stored in the tissues is mobilized and deposited in the feathers, so feather mercury levels provide an index of cumulative exposure between molts (Furness et al. 1986). Recently, feathers are being used to study

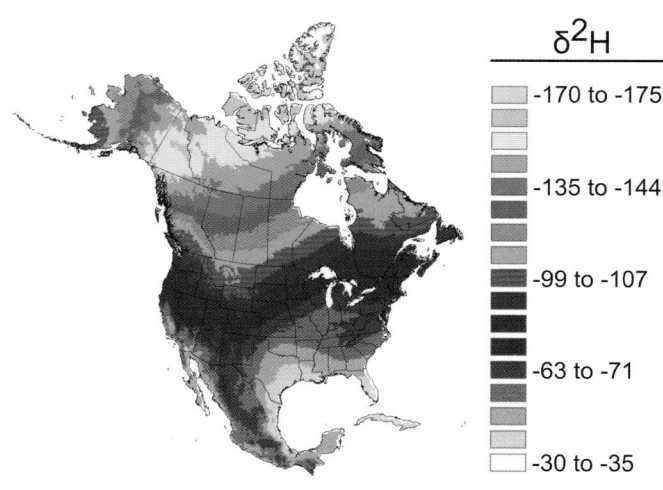

Figure 9.17. Geographic variation in stable isotope ratios. North America exhibits distinct geographical variation in the naturally occurring ratio (δ^2H) between the hydrogen isotopes protium to deuterium found in the environment. Local isotope ratios persist in the food chain and are incorporated into the feathers of birds during feather production, providing a potential way to identify a bird's natal or molting grounds. *Adapted from Hobson et al. 2015.*

birds' exposure to organic pollutants, such as those found in pesticides (Jaspers et al. 2007).

Groundbreaking work by Gary Bortolotti has recently advanced feathers as a source of information about endocrine physiology. Endogenous steroid hormones are deposited in feathers and can be assayed using techniques similar to those used to assay steroids in the blood plasma (Bortolotti et al. 2008). In contrast to heavy metals, feather hormones derive from hormones circulating in the blood during feather production. Thus the total amount of a hormone deposited in a long wing feather reflects plasma concentrations over several days. This tool is especially useful for studying stress physiology, because a hormone sample can be collected from a dropped feather without causing stress to the bird. Feather corticosterone is so stable that it can even be assayed in historic museum specimens (Kennedy et al. 2013).

KEY POINTS

- A bird's plumage is integrated with the muscular and nervous systems to support nearly every aspect of life.
- Differences in the number, distribution, shape, and structural features of the feathers that make up a plumage reflect adaptations to diverse plumage functions among birds.
- Feathers are produced by coordinated activation of stem cells within the feather follicles. Once produced, feath-

Figure 9.18. Conservation of molt locations. Hundreds of thousands of Eared Grebes congregate in predictable molting sites, like this one at Great Salt Lake in Utah. Changes to habitats in these areas, where so many individuals of a single species come to molt, could have harmful impacts on these species. *Photo by Caitlin Dean.*

ers are cut off from blood supply and can only degrade over time, thus birds have evolved a variety of ways to prolong the useful lifetime of the plumage.

- Birds molt periodically to renew their plumage, but doing so takes substantial amounts of time and energy, and molt of the wing and tail reduces flight efficiency. To fit molt into the annual cycle, birds have evolved diverse strategies that reflect species diversity in ecological and behavioral traits.
- Large birds face strong time constraints to complete the molt and, probably because of these constraints, have evolved complex and simultaneous molt strategies. Time constraints of molting in large birds are so severe that many fail to replace all their flight feathers in an episode of molt, forcing them to trade-off breeding opportunities with replacing accumulated worn feathers.
- Feather quality is crucial for birds. High-quality feathers are long lasting, perform well in flight, and provide

thermal insulation against cold temperatures. Rapid molts compromise feather quality. This makes feather quality the likely currency by which selection operates on variation in life history trade-offs. Investing heavily in reproduction leaves less time for molting, and this trade-off is expressed in poor-quality feathers that must be carried until the following molt.

KEY MANAGEMENT AND CONSERVATION IMPLICATIONS

- Several species of birds congregate in extraordinary numbers at predictable molting sites (Jehl 1990). For example, hundreds of thousands of Eared Grebes (*Podiceps nigricollis*) flock to Mono Lake and Great Salt Lake in California and Utah, respectively, for their fall molts (fig. 9.18; Storer and Jehl 1985). Similarly, over 90 percent of the population of Common Shelducks (*Tadorna tadorna*; over one year of age) from northwestern Europe flock to the Wadden

Sea to molt (Jehl 1990). In these cases, molting areas attract the bulk of the population of certain species, and any changes to these areas that reduce the probability of successfully molting could have population-level consequences for these species. However, our understanding of how habitat changes in molting areas affect species remains poor (Leu and Thompson 2002). For example, northwestern Mexico, where many species of western Neotropical migrants molt, has seen dramatic reductions in the extent of lowland coastal thorn forest, as this habitat is converted to agricultural fields (Rohwer et al. 2015). As agricultural fields become more abundant, so too do irrigation canals that dam large arroyos and capture pulses of rainwater in the lowlands, creating more mesic areas. While this habitat conversion undoubtedly harms species that rely on the original thorn forest habitat, it likely benefits species that used open fields or mesic, riparian zones for their molt (Rohwer et al. 2015). Similar changes may be responsible for declines of many European migrants that winter in sub-Saharan Africa (Vickery et al. 2014) as these species have experienced stronger declines relative to resident and short-distance migrants, suggesting that habitat changes on the winter range, where these species molt, is responsible for population declines (Sanderson et al. 2006).

- In general, we need to know more about breeding and nonbreeding linkages to be able to assess whether molt might be a constraint in the annual cycle (Sillett and Holmes 2002, Holmes 2007). Research on American Redstarts (*Setophaga ruticilla*) has revealed striking effects of winter habitat use that carry over and affect departure dates for spring migration, territory settlement patterns across the breeding range, and reproductive success (Marra et al. 1998, Marra and Holmes 2001, Studds et al. 2008). Loss of molting areas or lower-quality molting habitat will likely produce carryover effects that erode population stability. Overall, we know little about breeding and molting habitat congruencies (whether a species that breeds in one habitat type uses similar habitat for molting), what causes mortality during the molt, and how species will respond to habitat changes or disturbances in their molt locations (Leu and Thompson 2002).

DISCUSSION QUESTIONS

1. Imagine you are trying to understand the function of the body and flight feathers in fossil birds, such as *Archaeopteryx*. What kinds of information would you want to know about fossil bird feathers, and what could the information tell you about the functions of feathers in prehistoric birds?

2. Imagine a bird living in a bitterly cold, windy, and sunny climate. Which combination of these three plumage traits would be most thermodynamically-advantageous: (1) color: black or white; (2) plumage surface texture: loose (open matrix) or tight (sheetlike); and (3) plumage depth: shallow or deep? Explain your reasoning based on the way heat flows through plumage.

3. Based on figure 9.9, is feather wear greater in male or female stonechats? Explain your reasoning based on the information provided in the figure. Next, generate one or more hypotheses for the sex difference in feather wear rates, and describe the observations or data you would need in order to test each hypothesis.

4. What data would you need to test the idea that some molt strategies are easier to evolve than others? How would addressing this question improve our understanding of molt biology across species?

5. How might feather quality vary among simple, complex, and simultaneous molt strategies? How could you test your ideas?

6. Generate multiple hypotheses for why birds have different plumages during the breeding and nonbreeding season. How could you test among your hypotheses?

7. What evidence suggests that increasing the rate at which feathers grow reduces feather quality, and do you believe this evidence? What data would make this hypothesis more robust?

8. Generate several questions related to avian physiology, ecology, and conservation that could be addressed by sampling and analyzing feather structure and feather chemistry.

References

Ammann, G. A. (1937). Number of contour feathers of Cygnus and Xanthocephalus. The Auk 54:201–202.

Arcese, P. (1989). Intrasexual competition and the mating system in primarily monogamous birds: The case of the Song Sparrow. Animal Behaviour 38:96–111.

Baker, R. R., and G. A. Parker (1979). The evolution of bird coloration. Philosophical Transactions of the Royal Society B 287:63–130.

Bancroft, G. T., and G. E. Woolfenden (1982). The molt of Scrub Jays and Blue Jays in Florida. Ornithological Monographs 29:1–51.

Berggren, A., D. P. Armstrong, and R. M. Lewis (2004). Delayed plumage maturation increases overwinter survival in North Island Robins. Proceedings of the Royal Society Series B 271:2123–2130.

Berthold, P., and U. Querner (1982). Genetic basis of moult, wing length, and body weight in a migratory bird species, *Sylvia atricapilla*. Experientia 38:801–802.

Blanco, G., J. L. Tella, and J. Potti (1997). Feather mites on group-living Red-billed Choughs: A non-parasitic interaction? Journal of Avian Biology 28:197–206.

Blem, C. R., and L. B. Blem (1992). Some observations of sunbathing in swallows. Journal of Field Ornithology 63:53–56.

Bonadonna, F., and G. A. Nevitt (2004). Partner-specific odor recognition in an Antarctic seabird. Science 306:835.

Bonier, F., P. Martin, J. Jensen, L. K. Butler, M. Ramenofsky, and J. C. Wingfield (2007). Post-juvenal molt and pre-migratory fattening of an Arctic-breeding bird, *Zonotrichia leucophrys gambelii*: Costs, constraints, and tradeoffs. Ecology 88:2729–2735.

Bonser, R. H. C. (1995). Melanin and the abrasion resistance of feathers. The Condor 97:590–591.

Bortolotti, G. R. (2010). Flaws and pitfalls in the chemical analysis of feathers: Bad news-good news for avian chemoecology and toxicology. Ecological Applications 20:1766–1774.

Bortolotti, G. R., T. A. Marchant, J. Blas, and T. German (2008). Corticosterone in feathers is a long-term, integrated measure of avian stress physiology. Functional Ecology 22:494–500.

Bostwick, K. S., D. O. Elias, A. Mason, and F. Montealgre-Z (2010). Resonating feathers produce courtship song. Proceedings of the Royal Society Series B 277:835–841.

Brodkorb, P. (1955). Number of feathers and weights of various systems in a Bald Eagle. The Wilson Bulletin 67:142.

Brommer, J. E., O. Pihlajamäki, H. Kolunen, and H. Pietiäinen (2003). Life-history consequences of partial-moult asymmetry. Journal of Animal Ecology 72:1057–1063.

Burtt, E. H. (1986). An analysis of physical, physiological, and optical aspects of avian coloration with emphasis on wood-warblers. Ornithological Monographs 38.

Burtt Jr., E. H., and J. M. Ichida (1999). Occurrence of feather-degrading bacilli in the plumage of birds. The Auk 116:364–372.

Butler, L. K., S. Rohwer, and M. G. Speidel (2008). Quantifying structural variation in contour feathers to address functional variation and life history trade-offs. Journal of Avian Biology 39:629–639.

Cade, T. J., and G. Maclean (1967). Transport of water by adult sandgrouse to their young. The Condor 69:323–343.

Clayton, D. H., J. A. Koop, C. W. Harbison, B. R. Moyer, and S. E. Bush (2010). How birds combat ectoparasites. Open Ornithology Journal 3:41–71.

Crossin, G. T., A. Dawson, R. A. Phillips, P. N. Trathan, K. B. Gorman, S. Adlard, and T. D. Williams (2012). Seasonal patterns of prolactin and corticosterone secretion in an Antarctic seabird that moults during reproduction. General and Comparative Endocrinology 175:74–81.

Davenport, J., J. O'Halloran, F. Hannah, O. McLaughlin, and P. Smiddy (2009). Comparison of plumages of White-throated Dipper *Cinclus cinclus* and Blackbird *Turdus merula*. Waterbirds 32:169–178.

Dawson, A. (2006). Control of molt in birds: Association with prolactin and gonadal regression in starlings. General and Comparative Endocrinology 147:314–322.

Dawson, A., S. A. Hinsley, P. N. Ferns, R. H. C. Bonser, and L. Eccleston (2000). Rate of moult affects feather quality: A mechanism linking current reproductive effort to future survival. Proceedings of the Royal Society Series B 267:2093–2098.

Dawson, A., and P. J. Sharp (2010). Seasonal changes in concentrations of plasma LH and prolactin associated with the advance in the development of photorefractoriness and molt by high temperature in the starling. General and Comparative Endocrinology 167:122–127.

DesRochers, D. W., J. M. Reed, J. Awerman, J. A. Kluge, J. Wilkinson, L. I. van Griethuijsen, and L. M. Romero (2009). Exogenous and endogenous corticosterone alter feather quality. Comparative Biochemistry and Physiology Part A: Molecular and Integrative Physiology 152:46–52.

Dumbacher, J. P., A. Wako, S. R. Derrickson, A. Samuelson, T. F. Spande, and J. W. Daly (2004). Melyrid beetles (Choresine): A putative source for the batrachotoxin alkaloids found in poison-dart frogs and toxic passerine birds. Proceedings of the National Academy of Sciences USA 101:15857–15860.

Dunn, P. O., and A. Cockburn (1999). Extrapair mate choice and honest signaling in cooperatively breeding Superb Fairy-wrens. Evolution 53:938–946.

Flinks, H., and V. Salewski (2012). Quantifying the effect of feather abrasion on wing and tail lengths measurements. Journal of Ornithology 153:1053–1065.

Foster, M. S. (1974). A model to explain molt-breeding overlap and clutch size in some tropical birds. Evolution 28:182–190.

Froehlich, D. R., S. Rohwer, and B. J. Stutchbury (2005). Spring molt constraints versus winter territoriality: Is conspicuous winter coloration maladaptive? *In* Birds of two worlds, R. Greenberg and P. R. Marra, Editors. Johns Hopkins University Press, Baltimore, MD, pp. 321–335.

Furness, R. W., S. J. Muirhead, and M. Woodburn (1986). Using bird feathers to measure mercury in the environment: Relationships between mercury content and molt. Marine Pollution Bulletin 17:27–30.

Galván, I., E. Barba, R. Piculo, J. L. Cantó, V. Cortés, J. S. Monrós, F. Atiénzar, and H. Proctor (2008). Feather mites and birds: An interaction mediated by uropygial gland size? Journal of Evolutionary Biology 21:133–144.

Goldstein, G., K. R. Flory, B. A. Browne, S. Majid, J. M. Ichida, and E. H. Burtt Jr. (2004). Bacterial degradation of black and white feathers. The Auk 121:656–659.

Greenberg, R., T. Keeler-Wolf, and V. Keeler-Wolf (1974). Wood warbler populations in the Yolla Bolly mountains of California. Western Birds 5:81–90.

Greenwold, M. J., and R. H. Sawyer (2011). Linking the molecular evolution of avian beta (β) keratins to the evolution of feathers. Journal of Experimental Zoology Part B: Molecular and Developmental Evolution 316:609–616.

Groscolas, R., and Y. Cherel (1992). How to molt while fasting in the cold: The metabolic and hormonal adaptations of Emperor and King Penguins. Ornis Scandinavica 23:328–334.

Gunderson, A. R., A. M. Frame, J. P. Swaddle, and M. H. Forsyth (2008). Resistance of melanized feathers to bacterial degradation: Is it really so black and white? Journal of Avian Biology 39:539–545.

Hall, K. S. S., and T. Fransson (2000). Lesser Whitethroats under time-constraint moult more rapidly and grow shorter wing feathers. Journal of Avian Biology 31:583–587.

Hamilton, T. H., and R. H. Barth Jr. (1962). The biological significance of season change in male plumage appearance in some New World migratory bird species. American Naturalist 96:129–144.

Hawkins, G. L., G. E. Hill, and A. Mercadante (2012). Delayed plumage maturation and delayed reproductive investment in birds. Biological Reviews 87:257–274.

Healy, W. M., and J. W. Thomas (1973). Effects of dusting on plumage of Japanese Quail. The Wilson Bulletin 85:442–448.

Hedenström, A., and S. Sunada (1999). On the aerodynamics of moult gaps in birds. Journal of Experimental Biology 202:67–76.

Helm, B., and E. Gwinner (1999). Timing of postjuvenal molt in African (*Saxicola torquata axillaris*) and European (*Saxicola torquata rubicola*) Stonechats: Effects of genetic and environmental factors. The Auk 116:589–603.

Hendricks, P. (2009). Snow bathing by house finches: A review of this behavior by North American birds. Wilson Journal of Ornithology 121:834–838.

Hobson, K. A. (2005). Using stable isotopes to trace long-distance dispersal in birds and other taxa. Diversity and Distributions 11:157–164.

Hobson, K. A., and L. I. Wassenaar (1997). Linking breeding and wintering grounds of neotropical migrant songbirds using stable hydrogen isotopic analysis of feathers. Oecologia 109:142–148.

Hobson, K. A., S. L. Van Wilgenburg, E. H. Dunn, D. J. T. Hussell, P. D. Taylor, and D. M. Collister (2015). Predicting origins of passerines migrating through Canadian migration monitoring stations using stable-hydrogen isotope analyses of feathers: A new tool for bird conservation. Avian Conservation and Ecology 10:3.

Holmes, R. T. (2007). Understanding population change in migratory songbirds: Long-term and experimental studies of Neotropical migrants in breeding and wintering areas. Ibis 149 (Suppl. 2): 2–13.

Hopps, E. C. (2002). Information on waterfowl feather characteristics. Transactions of the Illinois State Academy of Science 95:229–237.

Hutt, F. B., and L. Ball (1938). Number of feathers and body size in passerine birds. The Auk 55:651–657.

Jaspers, V. L. B., S. Voorspoels, A. Covaci, G. Lepoint, and M. Eens (2007). Evaluation of the usefulness of bird feathers as a non-destructive biomonitoring tool for organic pollutants: A comparative and meta-analytical approach. Environment International 33:328–337.

Jehl Jr., J. R. (1990). Aspects of the molt migration. *In* Bird migration, E. Gwinner, Editor. Springer, Berlin, pp. 102–113.

Jenni, L., and R. Winkler (1994). Moult and aging of European Passerines. Academic Press, London.

Johnson, E. I., P. C. Stouffer, and R. O. Bierregaard Jr. (2012). The phenology of molting, breeding and their overlap in central Amazonian birds. Journal of Avian Biology 43:141–154.

Kennedy, E. A., C. R. Lattin, L. M. Romero, and D. C. Dearborn (2013). Feather coloration in museum specimens is related to feather corticosterone. Behavioral Ecology and Sociobiology 67:341–348.

Knappen, P. (1932). Number of feathers on a duck. The Auk 49:461.

Langston, N. E., and S. Rohwer (1996). Molt-breeding tradeoffs in albatrosses: Life history implications for big birds. Oikos 76:498–510.

Lei, F.-M., Y.-H. Qu, Y.-L. Gan, A. Gebauer, and M. Kaiser (2002). The feather microstructure of passerine sparrows in China. Journal für Ornithologie 143:205–212.

Leu, M., and C. W. Thompson (2002). The potential importance of migratory stopover sites as flight feather molt staging areas: A review for neotropical migrants. Biological Conservation 106:45–56.

Lin, S.-J., R. B. Wideliz, Z. Yue, A. Li, X. Wu, T.-X. Jiang, P. Wu., and C.-M. Chuong (2013). Feather regeneration as a model for organogenesis. Development, Growth and Differentiation 55:139–148.

Lindström, Å., and J.-Å. Nilsson (1988). Birds doing it the octopus way: Fright moulting and distraction of predators. Ornis Scandinavica 19:165–166.

Lindström, Å., G. H. Visser, and S. Daan (1993). The energetic cost of feather synthesis is proportional to basal metabolic rate. Physiological Zoology 66:490–510.

Lyon, B. E., and R. D. Montgomerie (1986). Delayed plumage maturation in passerine birds: Reliable signaling by subordinate males? Evolution 40:605–615.

Markus, M. B. (1963). The number of feathers in the Laughing Dove *Streptopelia senegalensis* (Linnaeus). Ostrich 34:92–94.

Marra, P. P., K. A. Hobson, and R. T. Holmes (1998). Linking winter and summer events in a migratory bird by using stable-carbon isotopes. Science 282:1884–1886.

Marra, P. P., and R. T. Holmes (2001). Consequences of dominance-mediated habitat segregation in American Redstarts during the nonbreeding season. The Auk 118:92–104.

Marshall, A. G. (1981). The ecology of ectoparasitic insects. Academic Press, London.

McCafferty, D. J., J. B. Moncrieff, and I. R. Taylor (1997). The effect of wind speed and wetting on thermal resistance of the barn owl (*Tyto alba*) II: Coat resistance. Journal of Thermal Biology 22:265–273.

McGregor, R. C. (1902). The number of feathers in a bird skin. The Condor 4:17.

Møller, A. P., and J. Erritzøe (1992). Acquisition of breeding coloration depends on badge size in male house sparrows *Passer domesticus*. Behavioral Ecology and Sociobiology 31:271–277.

Montalti, D., and A. Salibián (2000). Uropygial gland size and avian habitat. Ornitologia Neotropical 11:297–306.

Moyer, B., A. N. Rock, and D. H. Clayton (2003). Experimental test of the importance of preen oil in Rock Doves (*Columba livia*). The Auk 120:490–496.

Mulder, R. A., and M. J. L. Magrath (1994). Timing of prenuptial molt as a sexually selected indicator of male quality in Superb Fairy-wrens (*Malurus cyaneus*). Behavioral Ecology 5:393–400.

Mumme, R. L., M. L. Galatowitsch, P. G. Jabłoński, T. M. Stawarczyk, and J. P. Cygan (2006). Evolutionary significance of geographic variation in a plumage-based foraging adaptation:

An experimental test in the Slate-throated Redstart (*Myioborus miniatus*). Evolution 60:1086–1097.

Murphy, M. E., and J. R. King (1986). Diurnal constancy of feather growth rates in White-crowned Sparrows exposed to various photoperiods and feeding schedules during the postnuptial molt. Canadian Journal of Zoology 64:1292–1294.

Murphy, M. E., and J. R. King (1992). Energy and nutrient use during moult by White-crowned Sparrows *Zonotrichia leucophrys gambelii*. Ornis Scandinavica 23:304–313.

Necker, R. (1985). Observations on the function of a slowly-adapting mechanoreceptor associated with filoplumes in the feathered skin of pigeons. Journal of Comparative Physiology A 156:391–394.

Newton, I. (1966). The moult of the Bullfinch *Pyrrhula pyrrhula*. Ibis 108:41–67.

Nilsson, J., and E. Svensson (1996). The cost of reproduction: A new link between current reproductive effort and future reproductive success. Proceedings of the Royal Society of London, Series B 263:711–714.

Nolan, V., E. D. Ketterson, C. Ziegenfus, D. P. Cullen, and C. R. Chandler (1992). Testosterone and avian life histories: Effects of experimentally elevated testosterone on prebasic molt and survival in male Dark-eyed Juncos. The Condor 94:364–370.

Pap, P. L., G. Osváth, K. Sándor, O. Vincze, L. Bărbos, A. Marton, R. L. Nudds, and C. I. Vágási (2015). Interspecific variation in the structural properties of flight feathers in birds indicates adaptation to flight requirements and habitat. Functional Ecology 29:746–757.

Petrie, M., H. Tim, and S. Carolyn (1991). Peahens prefer peacocks with elaborate trains. Animal Behaviour 41:323–331.

Pimm, S. (1976). Estimation of duration of bird molt. The Condor 78:550–550.

Proctor, H., and I. Owens (2000). Mites and birds: Diversity, parasitism and coevolution. Trends in Ecology and Evolution 15:358–364.

Prum, R. O., S. Andersson, and R. H. Torres (2003). Coherent scattering of ultraviolet light by avian feather barbs. The Auk 120:163–170.

Rayner, M. J., N. Carlile, D. Priddel, V. Bretagnolle, M. G. R. Miller, R. A. Phillips, L. Ranjard, S. J. Bury, and L. G. Torres (2016). Niche partitioning by three Pterodroma petrel species during non-breeding in the equatorial Pacific Ocean. Marine Ecology Progress Series 549:217–229.

Revis, H. C., and D. A. Waller (2004). Bactericidal and fungicidal activity of ant chemicals on feather parasites: An evaluation of anting behavior as a method of self-medication in songbirds. The Auk 121:1262–1268.

Rijke, A. M., and W. A. Jesser (2011). The water penetration and repellency of feathers revisited. The Condor 113:245–254.

Rohwer, S. (1975). The social significance of avian winter plumage variability. Evolution 29:593–610.

Rohwer, S. (1999). Time constraints and moult-breeding tradeoffs in large birds. Proceedings of the International Ornithological Congress 22:568–581.

Rohwer, S. (2008). A primer on summarizing molt data for flight feathers. The Condor 110:799–806.

Rohwer, S., and G. S. Butcher (1988). Winter versus summer explanations of delayed plumage maturation in temperate passerine birds. American Naturalist 131:556–572.

Rohwer, S., L. K. Butler, and D. Froehlich (2005). Ecology and demography of east-west differences in molt scheduling of Neotropical migrant passerines. *In* Birds of two worlds, R. Greenberg and P. R. Marra, Editors. Johns Hopkins University Press, Baltimore, MD, pp. 87–105.

Rohwer, V. G., S. Rohwer, and J. H. Barry (2008). Molt scheduling of western Neotropical migrants and up-slope movements of Cassin's Vireo. The Condor 110:365–370.

Rohwer, S., R. E. Ricklefs, V. G. Rohwer, and M. M. Copple (2009). Allometry of the duration of flight feather molt in birds. PLoS Biology 7:e1000132.

Rohwer, S., and L. K. Wang (2010). A quantitative analysis of flight feather replacement in the Moustached Tree Swift *Hemiprocne mystacea*, a tropical aerial forager. PLoS ONE 5:e11586.

Rohwer, S., and D. E. Irwin (2011). Molt, orientation, and avian speciation. The Auk 128:419–425.

Rohwer, S., A. Viggiano, and J. M. Marzluff (2011). Reciprocal tradeoffs between molt and breeding in albatrosses. The Condor 113:61–70.

Rohwer, S., and K. Broms (2012). Use of feather loss intervals to estimate molt duration and to sample feather vein at equal time intervals through the primary replacement. The Auk 129:653–659.

Rohwer, S., and K. Broms (2013). Replacement rules for the flight feathers of Yellow-billed Cuckoos (*Coccyzus americanus*) and Common Cuckoos (*Cuculus canorus*). The Auk 130:599–608.

Rohwer, V. G., and S. Rohwer (2013). How do birds adjust the time required to replace their flight feathers? The Auk 130:699–707.

Rohwer, S., E. Grason, and A. G. Navarro-Siguenza (2015). Irrigation and avifaunal change in coastal Northwest Mexico: Has irrigated habitat attracted threatened migratory species? PeerJ 3:e1187. doi:10.7717/peerj.1187.

Romero, L. M. (2002). Seasonal changes in plasma glucocorticoid concentrations in free-living vertebrates. General and Comparative Endocrinology 128:1–24.

Romero, L. M., D. Strochlic, and J. C. Wingfield (2005). Corticosterone inhibits feather growth: Potential mechanism explaining seasonal down regulation of corticosterone during molt. Comparative Biochemistry and Physiology Part A: Molecular and Integrative Physiology 142:65–73.

Rothery, P., I. Wyllie, I. Newton, A. Dawson, and D. Osborn (2001). The timing and duration of molt in adult Starlings *Sturnus vulgaris* in east-central England. Ibis 143:435–441.

Rothery, P., and I. Newton (2002). A simple method for estimating timing and duration of avian primary moult using field data. Ibis 144:526–528.

Sanderson, F. J., P. F. Donald, D. J. Pain, I. J. Burnfield, and F. P. J. van Bommel (2006). Long-term population declines in Afro-Palearctic migrant birds. Biological Conservation 131:93–105.

Seneviratne, S. S., and I. L. Jones (2010). Origin and maintenance of mechanosensory feather ornaments. Animal Behaviour 79:637–644.

Serra, L. (2001). Duration of primary moult affects primary quality in Grey Plovers (*Pluvialis squatarola*). Journal of Avian Biology 32:377–380.

Sharp, P. J., A. Dawson, and R. W. Lea (1998). Control of luteinizing hormone and prolactin secretion in birds. Comparative Biochemistry and Physiology Part C: Pharmacology, Toxicology and Endocrinology 119:275–282.

Shawkey, M. D., S. R. Pillai, and G. E. Hill (2003). Chemical warfare? Effects of uropygial oil on feather-degrading bacteria. Journal of Avian Biology 34:345–349.

Sillett, T. S., and R. T. Holmes (2002). Variation in survivorship of a migratory songbird throughout its annual cycle. Journal of Animal Ecology 71:296–308.

Silva, H. M., M. P. Valim, and R. A. Gama (2014). Community of chewing lice (Phthiraptera: Amblycera and Ischnocera) parasites of resident birds at the archipelago of São Pedro and São Paulo in northeast Brazil. Journal of Medical Entomology 51:941–947.

Stearns, S. C. (1992). The evolution of life histories. Oxford University Press, Oxford.

Storer, R. W., and J. R. Jehl Jr. (1985). Moult patterns and moult migration in the Black-Necked Grebe *Podiceps nigricollis*. Ornis Scandinavica 16:253–260.

Stresemann, E., and V. Stresemann (1966). Die Mauser der Vögel. Journal für Ornithologie 107:1–447.

Studds, C. E., T. K. Kyser, and P. P. Marra (2008). Natal dispersal driven by environmental conditions interacting across the annual cycle of a migratory songbird. Proceedings of the National Academy of Sciences USA 105:2929–2933.

Svensson, E., and A. Hedenström (1999). A phylogenetic analysis of the evolution of moult strategies in Western Palearctic warblers (Aves: Sylviidae). Biological Journal of the Linnean Society 67:263–276.

Swaddle, J. P., M. S. Witter, I. C. Cuthill, A. Budden, and P. McCowen (1996). Plumage condition affects flight performance in Common Starlings: Implications for developmental homeostasis, abrasion and moult. Journal of Avian Biology 27:103–111.

Swaddle, J. P., and M. S. Witter (1997). The effects of molt on the flight performance, body mass, and behavior of European Starlings (*Sturnus vulgaris*): An experimental approach. Canadian Journal of Zoology 75:1135–1146.

Thompson, D. R., R. W. Furness, and S. A. Lewis (1995). Diets and long-term changes in δ15N and δ13C values in Northern Fulmars *Fulmarus glacialis* from two northeast Atlantic colonies. Marine Ecology Progress Series 125:3–11.

Tucker, V. A. (1991). The effect of molting on the gliding performance of a Harris' Hawk (*Parabuteo unicinctus*). The Auk 10:108–113.

Underhill, L. G., and W. Zucchini (1988). A model for avian primary moult. Ibis 130:358–372.

Vágási, C. I. (2014). The origin of feather holes: A word of caution. Journal of Avian Biology 45:431–436.

Vickery, J. A., S. R. Ewing, K. W. Smith, D. J. Pain, F. Bairlein, J. Škorpilová, and R. D. Gregory (2014). The decline of Afro-Palaearctic migrants and an assessment of potential causes. Ibis 156:1–22.

Walsberg, G. E. (1988). Heat flow through avian plumages: The relative importance of conduction, convection, and radiation. Journal of Thermal Biology 13:89–92.

Walsberg, G. E., and J. R. King (1978). The relationship of the external surface area of birds to skin surface area and body mass. Journal of Experimental Biology 76:185–189.

Walsberg, G. E., G. S. Campbell, and J. R. King (1978). Animal coat color and radiative heat gain: A re-evaluation. Journal of Comparative Physiology 126:211–222.

Wetmore, A. (1936). The number of contour feathers in passeriform and related birds. The Auk 53:159–169.

Wolf, B. O., and G. E. Walsberg (2000). The role of the plumage in heat transfer processes of birds. American Zoologist 40:575–584.

Yorzinski, J. L., G. L. Patricelli, J. S. Babcock, J. M. Pearson, and M. L. Platt (2013). Through their eyes: Selective attention in peahens during courtship. Journal of Experimental Biology 216:3035–3046.

Yuri, T., and S. Rohwer (1997). Molt and migration in the Northern Rough-winged Swallow. The Auk 114:249–262.

PART III

MOVEMENT, PERCEPTION, AND COMMUNICATION

CHAPTER 10

Flight and Locomotion

Ashley M. Heers

The avian body plan is intimately connected to locomotion, particularly flight. Many organisms engage in aerial locomotion: flying squirrels (Sciuridae), flying frogs (Hylidae, Rhacophoridae), and flying snakes (*Chrysopelea*), as well as insects, bats, and birds. However, unlike "flying" squirrels and other gliders—which must fall through the air or take advantage of air currents to generate aerodynamic forces—birds, bats, and insects are experts in true or **powered flight**, using their muscles to flap wings and generate aerodynamic forces. Powered flight is the most physically demanding form of locomotion (Alexander 2002), and in a number of ways, the bird body plan is built around these demands. All flying birds have certain features in common, such as large feathered wings and strong muscles and skeletons. But species also differ with respect to the details of these features (such as wing shape or relative proportion of wings versus legs) because there are many different styles of flight, and because birds do more than just fly. For example, many birds spend large amounts of time foraging on foot, and although this is not surprising to anyone who has watched pigeons or had a pet chicken, among vertebrates, birds' ability to move effectively both on the ground and in the air is rather unique. Some birds can even swim, using their wings (forelimbs), their legs (hindlimbs), or both. This versatility (fig. 10.1) contributes to incredible diversity in locomotor styles, styles which in turn are related to life history strategies, predator avoidance tactics, habitat preferences, foraging strategies, and migratory patterns (Dial 2003a, Heers and Dial 2015). When we think about avian anatomy, it is therefore important to consider both forelimbs and hindlimbs, and different modes of locomotion. The tail is also important, contributing to weight support, drag reduction, and maneuvering during flight, as well as display. In this chapter, however, we focus on the main drivers of avian locomotion: wings and legs.

WINGS VERSUS LEGS: LOCOMOTION AND THE AVIAN BODY PLAN

Why are birds so effective at navigating both aerial and terrestrial environments? Bats—the only other living group of vertebrate fliers—are rarely seen on the ground, and when they are, their movements are generally awkward (with a few exceptions, such as vampire bats, Phyllostomidae; and New Zealand short-tailed bats, *Mystacina*). What sets birds apart?

Among vertebrates, birds are unique in having distinct wing and leg locomotor modules (Gatesy and Dial 1996) (fig. 10.2). A **locomotor module** is a functionally integrated anatomical unit used for locomotion, and many animals have only one. For example, early tetrapods and extant salamanders and lizards swim, walk, or run by moving their trunks, limbs, and tail in concert, and thus have a single locomotor module. Bats also have a single locomotor module because they have incorporated all four limbs into the flight apparatus: part of the wing membrane actually attaches to the legs and, in some cases, the tail. This arrangement works well for flight but makes terrestrial locomotion somewhat challenging (https://www.youtube.com/watch?v=glekFyXHOd8). In contrast, birds are more compartmentalized—they have distinct wing and leg locomotor modules, and can locomote effectively both on the ground and in the air. A specialized wing module for aerial activities and leg module for terrestrial activities presumably helps facilitate aerial locomotion without compromising terrestrial locomotion and vice versa. In other words, locomotor modularity allows birds to exploit a wide range of habitats (terrestrial versus aerial or even aquatic) using a wide range of locomotor styles (more leg-dependent versus more wing-dependent).

One consequence of avian modularity is that scientists have traditionally studied wings and legs independent of one

Figure 10.1. The versatility of avian locomotion. Birds are the only group of living vertebrates that can locomote effectively A, on the ground, B, in the air, and C, underwater. Here, an Atlantic Puffin (*Fratercula arctica*). A, B in the public domain; C, from Richard Shucksmith, with permission.

another. That is, one group of scientists has focused on how wings work during flight, and another group has focused on how legs work during walking or running, with rather little overlap. Though it is intuitive to think about wings and legs as separate entities, as we tour the world of avian locomotion it is actually extremely important to consider both wings and legs, regardless of whether the topic is aerial or terrestrial locomotion. Wings and legs are not truly independent of one another because (1) birds often use their wings and legs cooperatively, and (2) there are probably trade-offs between wings and legs—how birds allocate resources to their wings versus their legs seems to influence their locomotor behavior and performance.

Cooperative Use of Wings and Legs

When we think about birds using their wings, we usually think of flight. However, a growing body of work has shown that birds often engage their wings and legs cooperatively (simultaneously or in coordination) (https://www.youtube .com/watch?v=3USAC-Ky25s). For example, birds in many species recruit their wings and legs cooperatively to flap-run up steep obstacles (wing-assisted incline running, WAIR) (Dial 2003b). Similarly, waterfowl "steam" across water surfaces by using their wings like oars and their feet like paddles (Livezey and Humphrey 1983, Dial and Carrier 2012). Birds that jump to take off are also using their wings and legs cooperatively. During jumping takeoffs, initial flight velocity is proportional to the ground reaction force produced by the legs during takeoff and the aerodynamic force produced by the wings after liftoff; in fact, most of the velocity for take-off comes from the legs (Earls 2000, Tobalske et al. 2004). Such wing-leg cooperation is common and widespread, occurring in all major avian groups (Dial 2011) (https://www.youtube.com /watch?v=k94EDd8aKng). Even though adult birds in most species are flight-capable, behaviors like WAIR are less power-demanding than flight (Jackson et al. 2011), and many flight-capable animals elect to flap-run, rather than fly, up to a nest.

Whereas wing-leg cooperation is facultative (optional) for most adults, it is crucially important to very young birds that have left their nests but are not yet able to fly, and to adult birds that have difficulty becoming airborne (Dial et al. 2015). When threatened by a predator, most adult birds can simply leap into the air and fly to safety, because they are morphologically equipped (large wings and muscles) to deal with aerial environments. In contrast, juvenile birds have very small, rudimentary wings and muscles (Dial 2003b, Dial et al. 2006, Heers and Dial 2012, 2015) and must supplement their underdeveloped wings with their legs until their wings can fully support body weight during flight. Until flight ca-

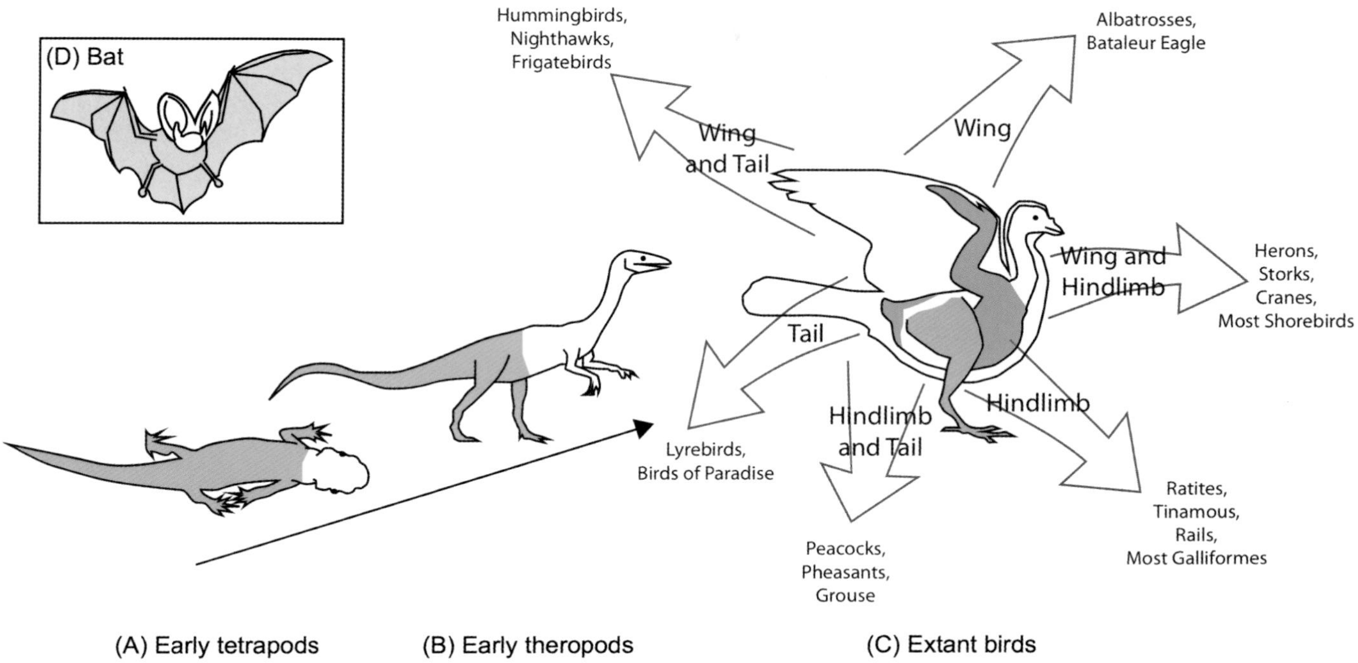

Figure 10.2. Locomotor modules. The evolutionary history of birds reveals increasing compartmentalization of the locomotor apparatus, from A, a single locomotor module incorporating the trunk, limbs, and tail (early tetrapods); to B, a single locomotor module in which the forelimbs are freed from locomotor duties (theropod dinosaurs); to C, multiple locomotor modules in which the wings and tail are used for flight (aerial or aquatic), and the legs are used for terrestrial (or aquatic) activities (birds). Bats, D—the only other living vertebrates capable of powered flight—have a single locomotor apparatus and are much less versatile in locomotion than birds. *Modified, with permission, from Gatesy and Dial 1996; Heers 2013, copyright eLS © 2013 John Wiley & Sons, Ltd. All rights reserved. www.els.net.*

pacity is acquired, many immature birds thus avoid predators by flap-running (WAIR) up obstacles such as trees and boulders, or flap-rowing across water (steaming) (Dial et al. 2015). Adult birds that have difficulty becoming airborne are similarly dependent on wing-leg cooperation. Swans (*Cygnus*), loons (*Gavia*) (Norberg and Norberg 1971), albatrosses (Procellariiformes), and other large-bodied birds with low power-to-mass ratios (which we discuss later) must flap their wings while running to gain speed for takeoff (https://www.youtube.com/watch?v=pcnUPJoh15c). Shearwaters' wings, though smaller than the wings of their albatross relatives, are still so long that they make takeoff challenging. Consequently, shearwaters often flap-run up trees, then "fall" into the air for takeoff (BBC Life of Birds: https://www.youtube.com/watch?v=Cjmtt_B_i4A&t=9s). Even adult birds that do not fly use their wings and legs cooperatively and can be quite dependent on this usage. For example, flightless or weakly flying steamer ducks (*Tachyeres*) elude terrestrial predators not by taking to the air, but by steaming across the water (Livezey and Humphrey 1983). In short, birds flap their wings for much more than flight, and behaviors involving the cooperative use of wings and legs are not only common but also

potentially crucial to large-bodied or weakly flying adult birds and to juvenile birds with developing wings (Dial et al. 2015).

Trade-Offs between Wings and Legs

Behaviors like wing-assisted incline running and steaming demonstrate that wing and leg locomotor modules are not used as independently as traditionally thought. Functional boundaries between wings and legs are further blurred by trade-offs. A **trade-off** occurs when improvements in one feature or function occur at a cost to another feature or function. In birds, trade-offs between wings and legs likely influence both ontogenetic (developmental) and evolutionary trajectories in locomotor style.

Evolutionary Timescales

Birds are highly diverse in terms of locomotor style and limb anatomy, and we discuss this diversity in the next two sections. But in general, across species, birds with higher wing investment tend to have lower leg investment: wing and leg muscle mass (expressed as a percentage of body weight) are negatively correlated (Heers and Dial 2015) (fig. 10.3). This trade-off between wings and legs likely occurs (1) because

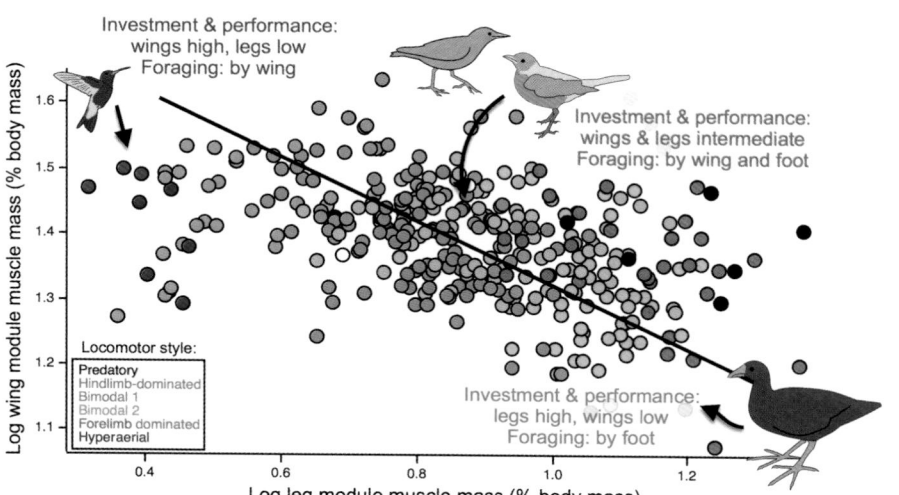

Figure 10.3. Trade-offs between wings and legs: evolutionary timescales. Across species, birds with higher wing investment have lower leg investment, and vice versa: wing versus leg investment, performance, and behavior are negatively correlated. *Plot modified, with permission, from Heers and Dial 2015, Heers 2016.*

resources allocated to wings may reduce those available for legs (and vice versa) if food supplies or time is limited, and/or (2) because legs must be carried as baggage by the wings during flight, whereas wings must be carried as baggage by the legs while running or jumping. In other words, to achieve high wing performance, a bird should have large wings and wing muscles and small legs and leg muscles; to achieve high leg performance, a bird should have larger legs and smaller wings. For example, Rock Doves (*Columba livia*) invest more muscle mass in their wings and less in their legs than similarly sized Chukar Partridges (*Alectoris chukar*), and they have greater flight performance, but worse jumping performance, than chukars (Heers and Dial 2015). Birds with higher wing investment also tend to use more forelimb-dominated foraging behaviors that require high wing use (e.g., screening, fly-catching), whereas birds with higher leg investment tend to use hindlimb-dominated foraging behaviors requiring high leg use (rarely fly or fly only briefly). Birds with intermediate levels of wing and leg investment rely on foraging behaviors that involve moderate use of both wings and legs (e.g., hopping and flying among tree branches to glean insects). In short, birds display a wide array of locomotor styles (more wing-reliant to more leg-reliant), and this diversity is probably at least partially dictated by trade-offs among wing and leg investment, performance, and behavior (fig. 10.3).

Ontogenetic Timescales

Birds are also highly diverse with respect to the timing of wing and leg development (fig. 10.4) (Heers and Dial 2015). Birds that rely on their legs early in development (e.g., Mallard Ducks, *Anas platyrhynchos*) have high leg investment and performance as juveniles, but relatively low wing investment and performance compared to adult counterparts or other

juveniles. In contrast, birds that rely on their wings early in development (e.g., Indian Peafowl, *Pavo cristatus*) exhibit high wing investment and performance as juveniles, but low leg investment and performance compared to adult counterparts or other juveniles (contrast wings of Mallard ducklings, http://jeb.biologists.org/content/jexbio/suppl/2012/10/07/215.21.3703.DC1/Movie3.mov, with wings of peachick in https://www.youtube.com/watch?v=k94EDd8aKng). Birds like Chukar Partridges take a middle-of-the-road path, emphasizing neither wings nor legs at any point during development (https://www.youtube.com/watch?v=3USAC-Ky25s). These different strategies are likely related to whether juveniles rely more on their wings (forested areas) or more their legs (flat areas, including areas near water) to avoid predators and reach refuges (Heers and Dial 2015).

In summary, wings and legs do not function completely independently of one another. As we look at how birds fly, run, and swim, it is important to remember that wings and legs are often used cooperatively, and that trade-offs between wings and legs likely influence avian development and evolution. Understanding these interactions is a key component of understanding variation in bird locomotion. With that in mind, we now explore what birds are best known for: flight.

WINGS AND AERIAL LOCOMOTION

Requirements for Aerial Locomotion: The Anatomy of Flight

Whether an insect, a bat, or a bird, powered flight has several basic requirements: an airfoil or wing, an engine, a flight stroke, coordination and control, relatively small body size (with some exceptions), and fuel. We discuss the biomechanics

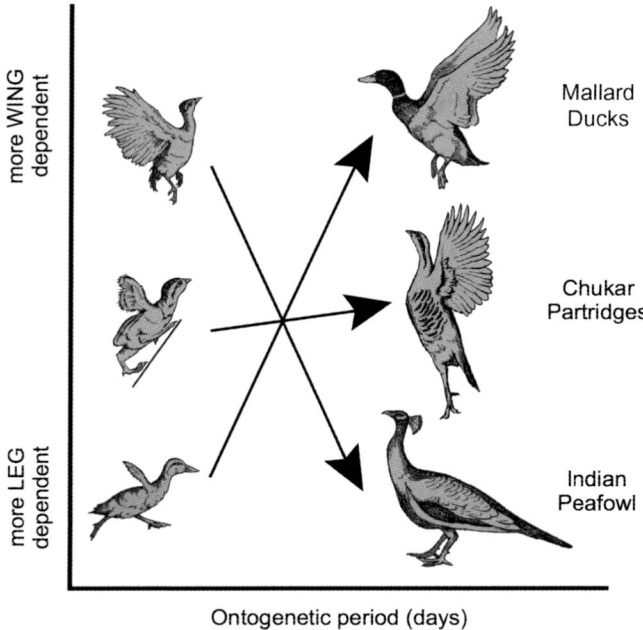

Figure 10.4. Trade-offs between wings and legs: developmental time-scales. During development, wing versus leg investment, performance, and behavior are negatively correlated. In Mallard Ducks, wings increase at the expense of legs, whereas in Indian Peafowl, legs increase at the expense of wings. Chukar Partridges take a middle-of-the-road strategy, emphasizing wings and legs approximately equally. *Drawings by Bob Petty, plot modified, with permission, after Heers and Dial 2015, Heers 2016.*

behind each requirement in the next section, but we begin from an anatomical perspective: How are birds morphologically equipped to meet the basic requirements for flight?

The Airfoil: Wings

Birds, bats, and the extinct pterosaurs arrived at different evolutionary solutions for creating an airfoil, or wing (fig. 10.5 inset). In birds, the wing is formed by a series of **pennaceous** or vaned feathers (chapter 9) that anchor to the forelimb skeleton and overlap to form a continuous surface, which generates aerodynamic force that is transmitted to the environment for weight support and propulsion (fig. 10.5A). Wing size and shape are highly diverse across species and vary throughout a wingbeat, but in general, flight-capable birds must have relatively large wings in order to generate enough aerodynamic force for flight.

Feather shape and microstructure also show variation, particularly between different locations along the wing. In flying birds, **primary feathers**—those attaching to manus or "hand" bones—are generally asymmetrical in shape, meaning that the vanes on either side of the rachis (central shaft) are different widths. **Secondary feathers**—those attaching to the ulna or "forearm"—are more symmetrical in shape (vanes ~ equal width). These shape differences (fig. 10.5) are thought to be a consequence of differences in feather orientation with respect to oncoming airflow (Norberg 1985a). Primary feathers are oriented more parallel to the leading edge or front of the wing, and are stabilized against airflow (blue arrows in fig. 10.5) by positioning the rachis close to the front of the wing (→ asymmetry). This also allows the feathers to twist during the wing stroke (fig. 10.6). Secondary feathers are oriented more perpendicularly to oncoming airflow and probably do not require such stabilization. Regardless of the degree of symmetry, both primary and secondary feathers have curved and relatively stiff rachises that help brace the feathers against flight-related forces. Overlap between adjacent feathers and interlocking **barbicels** or "hooklets" within a feather (fig. 10.5) also provide support, and reduce permeability (Muller and Patone 1998, Heers et al. 2011, Dial et al. 2012) by preventing air from flowing through the wing rather than over it. Collectively, these features constitute the avian airfoil: a large wing composed of strong, curved feathers that become more asymmetrical toward the tip of the wing and that allow the wing to be morphed into different shapes while also helping to stabilize the airfoil against oncoming airflow (Norberg 1985a), prevent excessive deformation (Nudds and Dyke 2010), and reduce airfoil permeability (Muller and Patone 1998, Heers et al. 2011, Dial et al. 2012).

The Engine: Flight Muscles

The avian equivalent of an engine is a specialized set of flight muscles, which provide power and help orient the wing. In terms of power output, flight is the most physically challenging form of locomotion (Alexander 2002), and birds have extremely large wing muscles (~12–43 percent body mass; Hartman 1961) that help meet such demands. Most of the power for flight comes from large muscles that are located proximally, close to the shoulder (fig. 10.5). Smaller, more distal muscles that cross the elbow and/or wrist joints are more important for fine-tuning wing movements (Dial 1992a, 1992b, Biewener 2011). This proximal-distal distribution of mass is important for flight because concentrating muscle mass close to the body wall reduces wing inertia and allows birds to flap their wings rapidly without fatiguing (note that cursorial, or running, animals like horses oscillate their limbs rapidly and also concentrate muscle mass proximally).

Two proximal muscles are particularly important. The **pectoralis muscle** is anchored to a large, bony **keel** on the sternum (keels are present only in birds and pterosaurs; fig. 10.5), inserts on the *ventral* (lower) surface of the deltopectoral crest of the humerus (upper arm bone), and provides

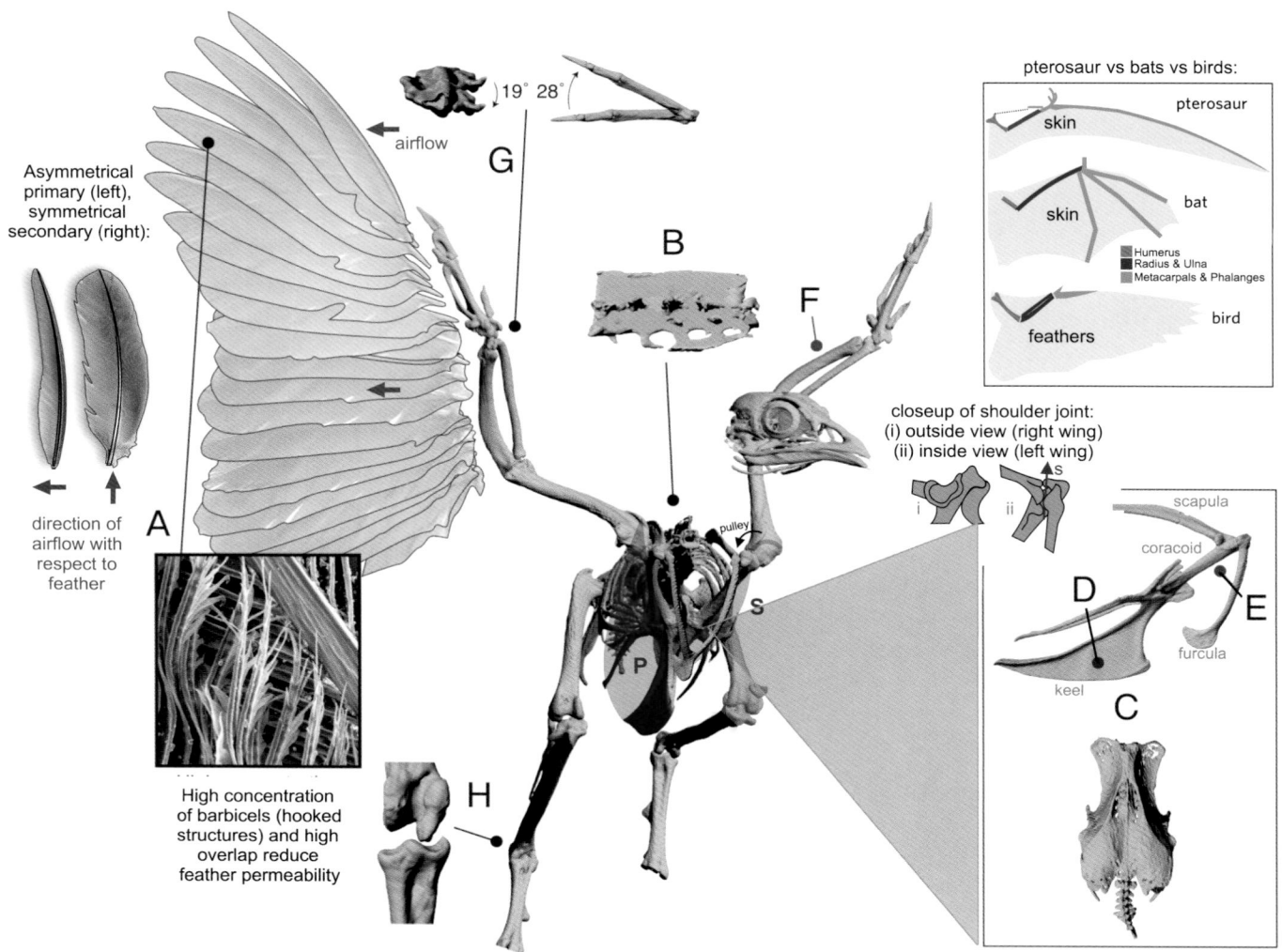

Figure 10.5. The avian body plan: meeting basic requirements for flight. Flight-capable adult birds have many morphological features that help meet aerial challenges: *A*, Large, morphable wings with strong, overlapping, and relatively impermeable feathers generate aerodynamic forces; *B, C*, Fused thoracic and sacral vertebrae probably stabilize the body against flight-related forces (notarium), and possibly act as a shock absorber during landing (synsacrum). With respect to the limbs, a robust shoulder girdle (*D*, sternum with large keel; *E*, strut-like and well-articulated coracoids) and forelimbs (*F*, bowed ulna) allow for the attachment and contraction of powerful flight muscles (*P*, pectoralis, and *S*, supracoracoideus), while the acrocoracoid and triosseal canal (close-up of shoulder joint, ii) allow the supracoracoideus to act as a pulley, contributing to humeral elevation and rotation; *G, H*, Distally, reduced/fused skeletal elements and channelized limb joints probably reduce mass and permit rapid and efficient limb oscillation, coordinate elbow and wrist movements, keep a planar wing orientation during the downstroke, and increase stride effectiveness by restricting ankle movements to a single plane of motion. These features are key components of the avian body plan and a classic example of anatomical specialization. *B, D, E, G* (right) in side view; *C* in top view; *G* (left) in view of fingertips. *Image and legend modified, with permission, after Heers et al. 2016.*

most of the power to pull the wing down during downstroke. The **supracoracoideus muscle** also anchors to the keel, but near the shoulder joint it passes through a specialized structure known as the **triosseal canal** (a hole bounded by the scapula, furcula, and coracoid bones) and inserts on the *dorsal* (upper) surface of the deltopectoral crest. This dorsal attachment allows the supracoracoideus to function like a pulley and elevate and rotate the humerus during upstroke (Poore et al. 1997). Together, the pectoralis and supracoracoideus are the main power generators of the avian flight stroke. As we discuss in the next section, the great capacity for power output by these two muscles is facilitated not only by their large size, but also by muscle microstructure and activation patterns.

The Flight Stroke: Wing Flapping

Engineers designing an airplane do not worry about flight strokes, because in airplanes, wings and engines are decou-

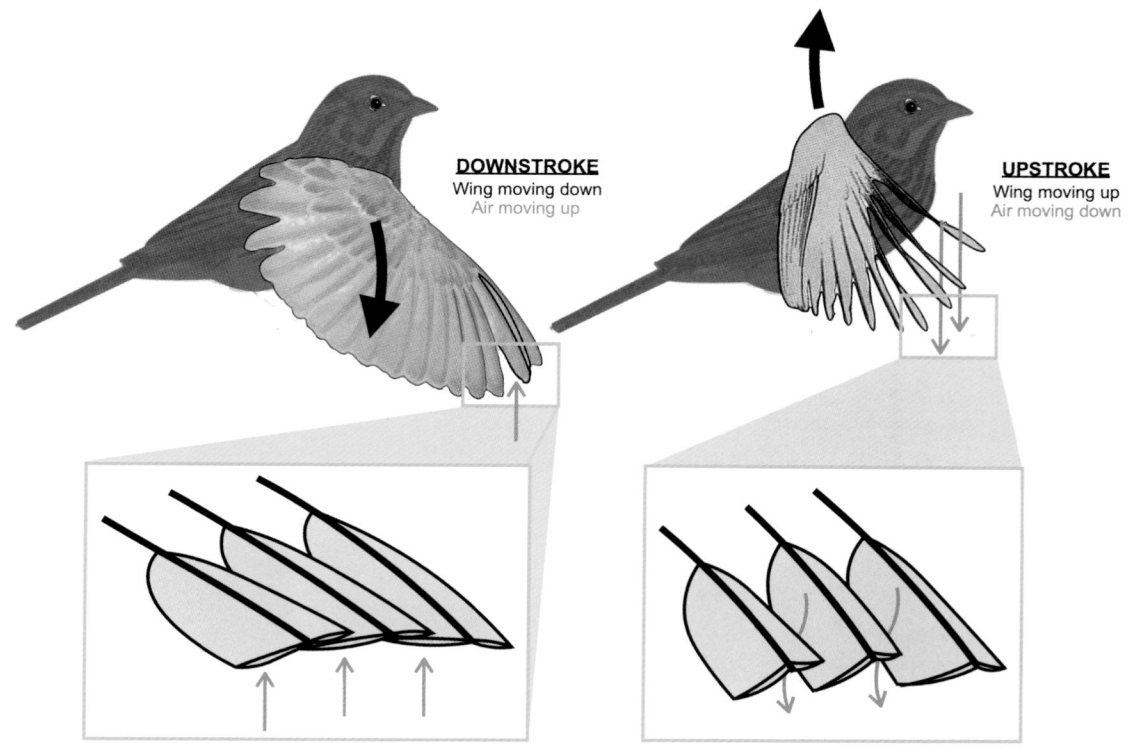

DOWNSTROKE
Wing moving down
Air moving up

UPSTROKE
Wing moving up
Air moving down

Figure 10.6. Feather twist during the wing stroke. During the down-stroke, air pulls and pushes upward on the wing and feathers, causing the primary feathers to pitch downward about their rachises and "slam" against each other. This "closes" the wing and forces air to flow around it—a prerequisite for lift production. During the upstroke, air pushes downward on the wing and feathers, causing the primary feathers to pitch upward about their rachises. This "opens" the wing like a Venetian blind and allows air to pass *through*, thereby reducing drag. *Image by A. Heers.*

pled: engines propel the aircraft forward and provide thrust, while air flowing over the fixed (non-flapping) wings generates a vertical lift force for weight support. In birds, however, the engine and airfoil form a functional unit: pectoral muscles contract (engine) to flap the wings (airfoil), and the resulting motion of the wings through the air generates both weight-supporting and propulsive aerodynamic forces. This wing motion is called the **flight stroke**. Though the flight stroke can be extremely fast (up to 50 or more cycles per second; Greenewalt 1960) and highly complex, it is facilitated by several specialized features of the bony wing joints. Starting at the root of the wing, the **shoulder joint** is formed by a saddle-shaped depression on the coracoid and scapula (fig. 10.5 inset), which faces dorsolaterally (up and sideways) and articulates with the rounded head of the humerus (Jenkins 1993). This arrangement permits a great range of motion, allowing birds to flap their wings through nearly 180° while simultaneously rotating the wing to many different angles. In contrast, the **elbow and wrist joints** are much more constrained, partially due to the fusion of many carpal (wrist) and metacarpal (hand) bones and partially due to interlock-ing ridges and grooves between adjacent bones that permit flexion, extension, and wing folding but limit other types of motion. Collectively, these features of the elbow and wrist probably (1) make the wing easier to move (less mass = less inertia) and control (fewer degrees of possible motion = less neural input), (2) help prevent wing joints from bending or twisting into less aerodynamically effective orientations, and (3) help coordinate elbow and wrist movements (Vazquez 1992). Compared with nonflying animals, the avian forelimb is thus highly specialized for flapping flight.

Coordination and Control

Maneuvering at high flight speeds requires great coordination and control. Birds have large brains with excellent vision and balance (chapter 12), and we briefly discuss how these features relate to flight in the next section.

Small Body Size

Across species, the largest birds that can fly (swans, *Cygnus*; Kori Bustard, *Ardeotis kori*; Andean Condor, *Vultur gryphus*; Wandering Albatross, *Diomedea exulans*; and Wild Turkey,

Meleagris gallopavo) converge on a similar weight: ~13–15 kg (Dial et al. 2008a). Although some extinct birds were much larger (such as Haast's Eagle, *Harpagornis*, and *Argentavis* and *Pelagornis*), these fliers probably relied on soaring and thus may have flown only under certain conditions. What limits body size in flying organisms? Scaling! Bird anatomy typically scales with geometric similarity (Greenewalt 1975), meaning that structures maintain the same shape as body mass increases. Thus, across species, mass increases faster than surface area, and surface area increases faster than length (fig. 10.7). Consequently, flight performance declines with increasing body mass (Dial et al. 2008a, Jackson and Dial 2011). According to the **force-limiting hypothesis** (Marden 1987, 1994), flight performance in large birds is limited by a drop in aerodynamic force production, or the ability to push air downward and backward in order to overcome gravity and air resistance. This would occur because wing surface area (and thus force production) increases more slowly than body mass. The **power-limiting hypothesis** (Hill 1950, Pennycuick 1975) suggests that flight performance declines with size because mass-specific muscle power (rate of work or energy production by muscles) declines with size. This might occur because larger birds have longer wings and flap more slowly (wingbeat frequency inversely proportional to wing length), generating less power for their body size.

There is some support for both hypotheses, but at this point it is not clear which mechanism is more important. Either way, flight performance clearly decreases as body size increases: larger birds are less maneuverable, cannot ascend through the air as quickly, and have more difficulty taking off—they generally must run, leap from heights, or face into the wind (Dial et al. 2008a). Thus, powered, flapping fliers are generally rather small. In addition, birds possess a number of anatomical features that may help reduce weight:

- Pneumatized ("hollow") and lightweight bones supported by internal trabeculae "struts" (although as a percentage of body weight, bird skeletons are as heavy as the skeletons of terrestrial mammals [Prange et al. 1979, Dumont 2010]—so the main advantage for flight is probably an increase in bone strength without an increase in mass because, for the same amount of material, a hollow bone is stronger than a solid bone [Vogel 2013]).
- Reduction or fusion of many bones, particularly distally (digits, carpals, tarsals).
- Reduction of tail (although the tail began shortening before the evolutionary origin of flight; chapter 2).
- Loss of teeth (although bats retain teeth and several extinct non-flying dinosaurs lost them).

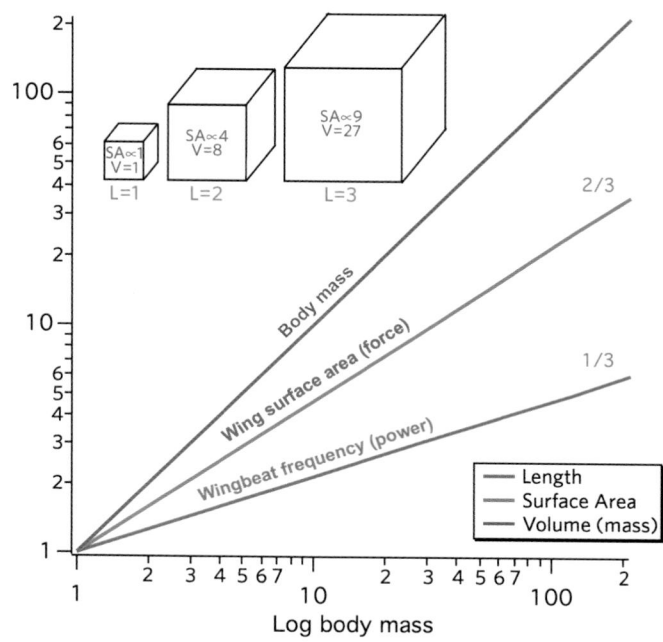

Figure 10.7. Scaling: limits to body size in flying animals. For anatomical features that maintain shape while increasing size (geometric similarity), mass ($\approx$ volume, V; purple) increases faster than surface area (SA; green) and length (L; red):

- mass = length^3 → length = mass$^{1/3}$ → length = 1/3 slope on log-log plot
- length^2 = (mass$^{1/3}$)^2 = area → area = mass$^{2/3}$ → area = 2/3 slope on log-log plot

These scaling relationships cause flight performance to decline with size. According to the power-limiting hypothesis, mass-specific muscle power declines with size because wing length (red line) increases more slowly than body mass (purple line): larger birds have longer wings and flap more slowly (wingbeat frequency inversely proportional to wing length), generating less power for their body size. According to the force-limiting hypothesis, aerodynamic force production declines with size because wing surface area (green line) increases more slowly than body mass. *Image by A. Heers.*

- Reduction of organs: no urinary bladder, reproductive organs that shrink when not in use, single ovary in most species.

Some of these potentially weight-reducing features are more pronounced in heavier birds or more aerial birds, and less pronounced in flightless birds or aquatic diving birds that must fight against buoyancy. For example, soaring pelicans have more pneumatic bones than diving relatives (Gutzwiller et al. 2013). Similarly, nonessential organs are reduced in size prior to migration (reviewed in McWilliams and Karasov 2005). In short, body weight is a major determinant of flight performance and influences all aspects of avian ecology.

Fuel

Compared with running or swimming, flight is extremely metabolically demanding (Alexander 2002). Standing on the ground or floating in the water requires relatively little effort, but a bird cannot simply float in still air. Air is far less dense than water, and unlike the ground, it moves when pushed, so flying vertebrates must constantly struggle to support their body weight. Birds meet the metabolic demands of flight through highly efficient **cardiovascular** and **respiratory** systems. For example, the avian heart is proportionally large, powerful, and beats very rapidly (hovering hummingbirds can achieve heart rates exceeding 1,000 beats per minute) (Lasiewski 1964). Similarly, one-way airflow and continuous, crosscurrent gas exchange allow for highly efficient respiration. These cardiovascular and respiratory features are discussed in chapter 7, but keep them in mind because without oxygen and fuel, flight is not possible.

In summary, flight is the most physically demanding form of locomotion (Alexander 2002), and flight-capable adult birds have a number of anatomical **adaptations** or **exaptations** (originally served another function, later co-opted for flight) that help meet these demands (fig. 10.5):

- Large, morphable wings with strong, overlapping, and relatively impermeable feathers
- Large and powerful pectoral muscles concentrated near the body wall, with robust skeletal areas for attachment
  - Pectoralis muscle (keel ↔ ventral surface of humerus) powers downstroke
  - Supracoracoideus muscle (keel ↔ dorsal surface of humerus, via triosseal canal) powers upstroke
- Elevated shoulder joint with great range of motion, but relatively channelized elbow and wrist joints
  - Fusion of bones in manus (hand) reduces weight and inertia
  - Interlocking ridges and grooves between adjacent bones help channelize elbow and wrist movement for a stereotypic flight stroke
- Large brains with excellent vision and balance
- Low body weight, mainly because of small size but facilitated by reduction of various bones and organs
- Highly efficient cardiovascular and respiratory systems

Many of these features appeared before the evolutionary origin of flight—in the theropod ancestors of birds (chapter 2)—indicating that they initially served some function other than flight. How bird flight evolved is an exciting and active area of research, with many competing hypotheses and new fossils and new data being contributed all the time

THE EVOLUTION OF AVIAN FLIGHT

Traditionally, there were two competing sets of hypotheses to explain how flight evolved (reviewed in Dial 2011, Heers 2013). The **cursorial** or **ground up** hypothesis posited that flight evolved in terrestrial, running theropods that flapped their protowings to run faster and/or jump higher. In contrast, the **arboreal** or **trees down** hypothesis suggested that flight evolved in tree-dwelling animals that initially used their protowings to glide between trees. These hypotheses are not mutually exclusive. For example, many birds spend significant amounts of time in both terrestrial and arboreal habitats (e.g., robins, pigeons, galliforms that forage on the ground but roost in trees). Similarly, developing birds that cannot fly yet often flap their wings to run up steep obstacles, and then flap their wings to slow their descent back down from those obstacles (these behaviors together make up the **ontogenetic transitional wing hypothesis**, which incorporates elements of both the ground up and trees down hypotheses) (Dial 2003, Dial et al. 2008). Finally, most birds spend time both flapping and gliding or soaring. This indicates that avian ancestors need not have been restricted to one style of locomotion, and indeed, extinct theropods show morphological variation suggestive of different locomotor styles. In short, living birds move from "the ground up" and from "the trees down" on a regular basis, both by flapping and gliding, demonstrating that the origin of avian flight cannot be viewed from a black-and-white (up or down, gliding or flapping) perspective.

(reviewed in Heers and Dial 2012). The study of locomotion in living birds—especially developing birds—is an integral component of this research on flight origins, because only among living birds can we observe and test how anatomical features function during flight (box on page 281). With that in mind, how does the avian flight apparatus work?

How Wings Function: The Biomechanics of Flight

The Airfoil and Aerodynamic Force Production

Flight is essentially a balancing act between three types of force: birds must (1) generate enough aerodynamic force with their wings to (2) support their body weight (counteract gravity) and (3) propel themselves forward (counteract drag from the body) (fig. 10.8).

Body Weight Caused by Earth's Gravity We all have an intuitive sense of what gravity is, but for flying animals,

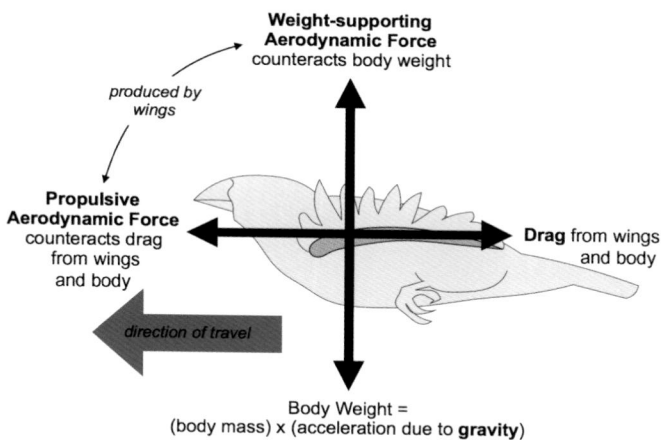

Figure 10.8. Flight is a balancing act. Forces indicated by black arrows. *Image of bird* (Zebra Finch, *Taeniopygia guttata*) *modified after Tobalske et al. 1999.*

gravity is a particular challenge because air is far less dense than living organisms, and it moves out of the way when you push on it. A flying bird is essentially a falling bird that maintains altitude by producing enough aerodynamic force to counteract the force of its own body weight. Just as you are able to stand on the ground because the force of your body weight pulling down is balanced by the force of the ground

pushing back up, a bird is able to fly because the downward force of its body weight is balanced by an upward, weight-supporting aerodynamic force.

Drag   Though perhaps less familiar than body weight, **drag** is also an important force during flight. When you ride a bike and stop pedaling, you slow down because friction between your tires and the road opposes your forward motion. Drag works the same way—it is an aerodynamic (or hydrodynamic) force that acts parallel to the direction of movement and mainly opposes motion through the air. Aerodynamic drag comes in three problematic flavors:

- *Pressure or form drag* (fig. 10.9A) occurs when air flowing around a moving object separates and produces a low-pressure wake behind the object. Low pressure inside a vacuum cleaner generates suction; similarly, low pressure behind an object pulls backward on the object and opposes its forward motion. Pressure drag is very low for **streamlined** shapes, such as the torpedo shape of a bird's body or the tapering cross section of a bird's wing. A bird's tail can also help reduce pressure drag at faster flight speeds by acting as a "splitter plate" that helps prevent air from separating from the body in the first place (Maybury and Rayner 2001).

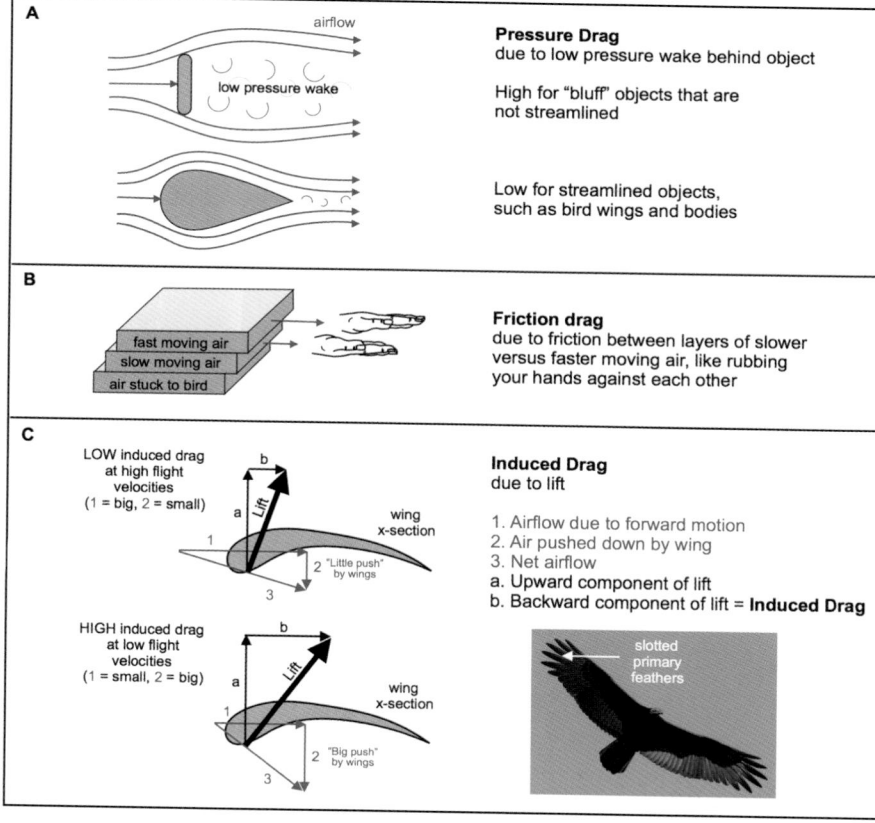

Figure 10.9. Types of drag. Birds must overcome three types of drag. Streamlined bodies, smooth feathers, and long wings or slotted primary feathers help reduce *A*, pressure drag; *B*, friction drag; and *C*, induced drag, respectively. *Image of vulture in public domain, https://pixabay.com/en /turkey-vulture-bird-wildlife-nature-1107362/.*

- *Friction drag* (fig. 10.9B) is caused when air molecules stick to the skin or feathers of a flying animal and pull on (slow down) nearby air molecules ("no-slip condition"). This friction, or shear, between adjacent air molecules opposes forward motion, and is higher for objects with rough surfaces (imagine rubbing your hands together wearing gloves—rough, high friction drag—versus wearing lotion—smooth, low friction drag). Friction drag increases with increasing flight velocity ($\propto v^2$; the faster you rub your hands, the more friction—and heat—they produce).

- *Induced drag* (fig. 10.9C) is incurred by lift production. Lift, as we discuss next, is an aerodynamic force that acts perpendicular to the direction of movement and airflow (and thus perpendicular to drag). Wings generate lift by pushing air downward, just as a fan pushes air toward your face. This downward deflection tilts the net lift vector backward, and the backward component of that lift is induced drag. Induced drag is lowest for birds with long, high aspect ratio wings, and it is likely reduced by slotted primary feathers (both features of soaring birds). Induced drag also decreases with increasing flight velocity ($\propto 1/v^2$). For example, a hovering hummingbird with no forward velocity (low v) supports its body weight by giving a big push to a small amount of air ($F=ma$; aerodynamic *force* = [low air *mass*, because the bird is staying in place and not moving through a high volume of air] × [high air *acceleration*]). This "big push" results in high induced drag. In contrast, when a bird is flying forward (higher v) and moving through a larger volume of air, it can give a smaller push to that larger amount of air (aerodynamic *force* = [high air *mass*] × [low air *acceleration*]), which induces less drag.

A flying bird therefore deals with three types of drag that collectively must be overcome by a propulsive aerodynamic force, just as a weight-supporting aerodynamic force must counteract gravity (fig. 10.8).

Lift: An Aerodynamic Force That Counteracts Body Weight and Drag   Birds produce both weight-supporting and propulsive forces by moving their wings through the air, either passively, as in gliding or soaring (non-flapping/non-powered flight), or actively, by flapping. When **gliding**, birds generate airflow around their wings by falling through the air with their wings outstretched and constantly *losing* altitude—until they land or start flapping. Some birds glide as they come in for a landing and some birds intersperse gliding with flapping, but no bird exclusively glides. Similarly, no bird exclusively soars, though some species come pretty close (e.g., albatrosses, Procellariiformes). When birds **soar**, they gener-

ate airflow around their wings by taking advantage of wind gradients or by positioning themselves in currents of rising air and constantly *gaining* altitude. However, soaring is constrained to habitats with thermals or wind gradients and only a few clades of birds regularly take advantage of it. Thus, for most birds, sustained flight requires active, muscle-powered **flapping**, because flapping is the only way to completely counteract gravity and drag. How do wing movements create weight-supporting and propulsive aerodynamic forces? In other words, how do birds generate lift?

Like drag, **lift** is an aerodynamic (or hydrodynamic) force created when a solid object moves through a fluid medium such as air (or when a fluid moves past a solid object). Whereas drag acts parallel to the direction of movement and is caused by friction with the air or a turbulent wake, lift acts perpendicular to airflow and results from a low-pressure zone forming above the wing. You have probably heard this described as the "Bernoulli effect," but there are actually several different ways to define and calculate lift, and each is useful for different types of analyses (box on pages 284–285). Why so many models? Visualizing airflow and measuring aerodynamic forces is a very challenging problem, and no model is perfect. However, recently developed techniques (e.g., particle image velocimetry [PIV], propeller force-plate models, robotics, computational fluid dynamics [CFD], aerodynamic force platforms) are vastly improving our understanding of flight aerodynamics (box on pages 284–285).

Where is lift produced? Birds generate some lift with their body and tail (Tobalske et al. 2009), but most lift comes from the wings. In airplanes with fixed (non-flapping) wings, lift is directed relatively vertically (fig. 10.11). However, for flapping birds, wing-generated lift is actually directed vertically and slightly forward. The vertical component of lift counteracts the bird's body weight, and the forward component (analogous to thrust produced by airplane engines) counteracts drag produced by the body. Note that both airplane wings and bird wings generate drag as well as lift. In the case of the airplane, wing drag opposes forward motion, but in the case of the bird, a component of wing drag points upward and helps support body weight (fig. 10.11). So drag is not always "bad," and for some behaviors (e.g., vertical flight, hovering), drag is quite useful.

Variation in Lift and Drag   Unlike the effects of gravity, aerodynamic forces—lift and drag—are not constant quantities. The magnitudes of lift (L) and drag (D) depend on several factors:

$$L = \tfrac{1}{2}\rho v^2 S C_L \tag{1}$$
$$D = \tfrac{1}{2}\rho v^2 S C_D \tag{2}$$

HOW DO BIRDS PRODUCE LIFT?

There are several different ways to think about lift.

Bernoulli's Principle. This principle states that the sum of dynamic and static pressure is constant, such that an increase in dynamic pressure (increase in velocity) is accompanied by a decrease in static pressure. In cross section, bird wings are **cambered**, meaning that the dorsal (upper) and ventral (lower) surfaces of the wing are convex. This camber accelerates air (or water) over the top of the wing (increases dynamic pressure) and reduces static pressure, thereby generating lift (https://www.youtube.com/watch?v=UqBmdZ-BNig). Why?

- When the bird moves forward, air flowing over the top of the wing is "constricted" by wing camber, just as if it were passing through the narrow portion of a Venturi tube (https://www.youtube.com/watch?v=Qz1g6kqvUG8). In contrast, air flowing beneath the wing has more space to flow through, as though it were passing through a wider portion of a Venturi tube. Consequently, streamlines are compressed above the wing and expanded below the wing, resulting in a relatively low static pressure above the wing and a relatively high static pressure below the wing. This difference in pressure creates the upward lift force, perpendicular to the direction of airflow. Birds can enhance lift production by pitching the leading edge of their wing upward to achieve a positive angle of attack (+ AOA), which essentially "compresses" the air above the wing even more. But at some point airflow will "detach" from the wing and become turbulent. This results in **stall**, or a drop in lift. The **alula**—a thumb-like group of feathers

(fig. 10.10)—likely delays stall, by keeping air attached to the wing at higher angles of attack.

- Lift can also be thought of as a result of the **Coanda effect**, or the tendency of a fluid to be attracted to a nearby surface (https://www.youtube.com/watch?v=AvLwqRCbGKY). The Coanda effect dictates that fluid will be bent around a cambered wing; this acceleration increases dynamic pressure and decreases static pressure, generating lift.

Scientists can measure differences in pressure using pressure transducers, but this is very challenging with live animals.

Newton's Laws of Motion, and the No-Slip Condition. As air flows over the top of the cambered wing, a very thin layer "sticks" to the wing and doesn't move; this "**no-slip condition**" is a consequence of the air's viscosity or stickiness.

- Just above the thin layer, air molecules begin to move, and velocity continues to increase farther away from the wing until free stream speed is reached. This velocity gradient marks the **boundary layer**, a region characterized by intense shearing of air molecules as they move past one another.

- Because air molecules are viscous, they tend to stick to one another, and thus stick to the wing indirectly. Consequently, air streamlines follow the surface of the wing and are deflected downward both above and below the bird's cambered wing. This downward deflection of air is known as **downwash**.

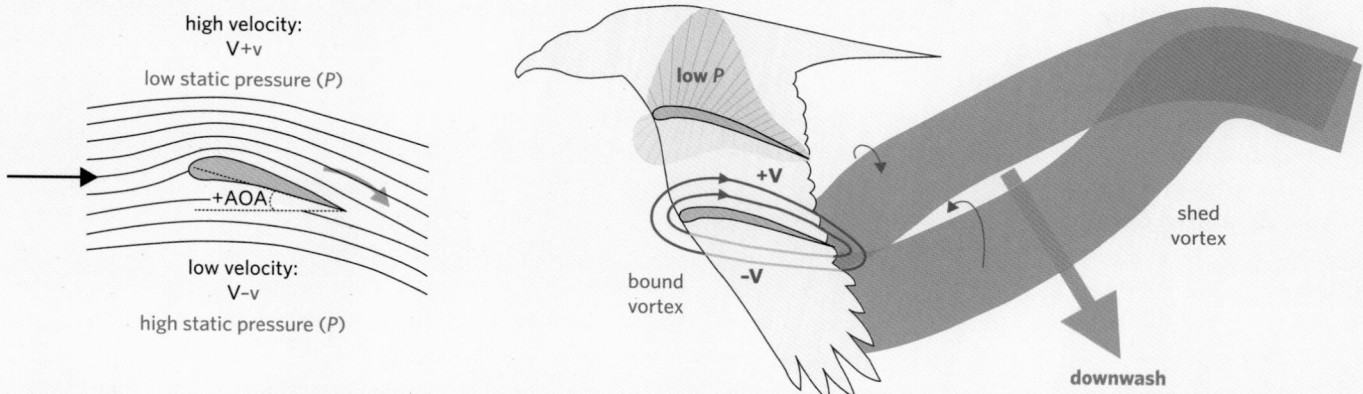

Different ways to calculate lift production: Bernoulli's Principle (differences in pressure, blue), Newton's Laws of Motion (velocity of downwash, brown), or Circulation (strength of vortices, purple). *Image by A. Heers.*

- Newton's **second law of motion** states that any change in motion of a mass is caused by a force acting on that mass. Since air molecules in the downwash were initially moving horizontally (in the oncoming airflow), a downward-directed force must be present to cause the change in direction and produce the downwash; in this case, the force is the attraction between air molecules, or viscosity.

- In accordance with Newton's **third law of motion**—"for every action there is an equal and opposite reaction"—the downward force, or action, is accompanied by an upward force, or reaction, of equal magnitude. This upward reaction can be thought of as pulling up on the wing and causing lift.

Scientists can measure downwash velocity using lasers (particle image velocimetry) to visualize downwash behind the wing. This is the realm of momentum jet theory, although it tells you more about what is happening behind the wing than on the wing itself.

Circulation and the Bound Vortex. A horizontally oriented cylinder rotating clockwise about its longitudinal axis causes nearby air molecules to rotate clockwise as well, because of the no-slip condition and the viscosity of the air. If a left-to-right translational airflow—a breeze, for example—is superimposed on the clockwise rotation of air about the cylinder,

- The two airflows will combine constructively above the cylinder, because they are moving in the same direction, and the net velocity of the air will be amplified (V+v).

- The two airflows will combine destructively beneath the cylinder, because they are moving in opposite directions, and the net velocity of the air will be decreased (V-v).

- A bird's wing can be thought of as generating a similar pattern of *relative* velocities, but because of camber, rather than rotation, of the wing. Streamlines constrict above the wing and move relatively fast (+v); streamlines expand beneath the wing and move relatively slow (-v).

This *relative pattern* of airflow around the wing is known as the **bound vortex**.

- Explained differently, a wing moving to the left (or air flowing to the right over a stationary wing) is *analogous* to superimposing a left-to-right translational airflow with a clockwise circulation of air around the wing (the bound vortex). In reality, air molecules do not actually circulate around the wing: all streamlines flow from the front of the wing to the back. The bound vortex is merely a representation of *relative* air movement that is useful for calculating aerodynamic forces.

Scientists can measure circulation by using lasers (particle image velocimetry), to visualize **vortices** that are shed behind the wing. This is the realm of the Kutta-Joukowski theorem, which, like momentum jet theory, does not tell you what is happening directly on the wing.

Once the magnitude of total aerodynamic force production is known, using one of the methods above, scientists can compare the aerodynamic effectiveness of different wings using **blade-element theory**, which relates the production of lift (L) and drag (D) to air density (ρ) and velocity (v), wing surface area (S), and wing morphology (C_L, C_D):

- $L = \frac{1}{2}\rho v^2 S C_L$
- $D = \frac{1}{2}\rho v^2 S C_D$

Bear in mind that Bernoulli's principle, Newton's laws of motion, and the circulation of bound vortices are all indirect methods for *calculating* aerodynamic force production, based on measurements of airflow around and below or behind a wing. Various alternative methods exist for directly measuring aerodynamic forces produced by dried or model wings in wind tunnels (to mimic gliding flight) or propeller-like apparatuses (to mimic slow flight), while techniques such as computational fluid dynamics rely on computers to model aerodynamic force production. Only very recently have scientists been able to directly measure the aerodynamic forces produced by flying birds (http://lentinklab.stanford.edu/), so this is an exciting time to study bird flight!

Figure 10.10. The alula. The alula, a group of thumb-like feathers, helps delay stall by keeping airflow attached to the wing. *Image of owl in public domain, https://pixabay.com /en/owl-bird-fly-nature-wildlife-922216/.*

ANDERS HEDENSTRÖM

Anders Hedenström grew up in a small town on the east coast of south Sweden, where he explored the resident bird life together with his father from an early age. His interest in bird migration grew during summer holidays while working as ringer (bander) at Ottenby Bird Observatory, on the island Öland in the Baltic Sea, where he met others with a similar interest. Via a detour through engineering school, he ended up studying biology at Lund University, followed by a PhD dissertation titled "Ecology of Avian Flight," under Thomas Alerstam's guidance. During this period, a few field trips to Africa fulfilled Hedenström's desire to study winter ecology of Palearctic songbird migrants, including molt strategies. Research for his PhD combined aerodynamic and optimality modeling to derive flight and migration strategies in birds. After a postdoc at Cambridge University, where his focus was bumblebee flight, Hedenström returned to Lund to set up his research program around the new low-turbulence wind tunnel. There he developed new techniques, in particular applying particle image velocimetry (PIV) to visualize and characterize the wake vortices of unrestrained birds, which in turn reflect the magnitude and time history of aerodynamic forces. This technique has since been applied to several bird, bat, and insect species. In parallel to the more mechanistic wind tunnel studies of animal flight, Hedenström also measured flight performance in wild birds with an "ornithodolite"—a portable instrument for recording winds and flight tracks of birds. The rapid technological development, making use of micro data loggers, geolocation, and accelerometry, now allows Hedenström to study flight activity during whole annual migrations in several bird species and theories developed early on can now be evaluated using relevant data.

where ρ is air density, v is velocity (speed at which bird is flying and flapping), S is wing surface area, and C_L and C_D are the coefficients of lift and drag, respectively. For behaviors like hovering, where flapping velocity (angular velocity, Ω) is much greater than the speed at which the bird is flying (translational velocity; 0 in the case of hovering), the equations for lift and drag can be rewritten as:

$$\text{Vertical Force (proportional to } L) = \tfrac{1}{2}\rho\Omega^2 S_2 C_V \qquad (3)$$
$$\text{Horizontal Torque (proportional to } D) = \tfrac{1}{2}\rho\Omega^2 S_3 C_H, \qquad (4)$$

where C_V and C_H are analogous to the coefficients of lift and drag, respectively, and S_2 and S_3 are the second and third moments of area, respectively. S_2 and S_3 account for the fact that during flapping flight, the tip of the wing (analogous to your hand, if you flap like a bird) will contribute more aerodynamic force than the base or root of the wing (analogous to your shoulder), partially because the wingtip is moving faster and partially because it is farther from the body and has a greater "lever arm," which increases the effectiveness of the applied force (Archimedes: "Give me a lever long enough . . . and I shall move the world").

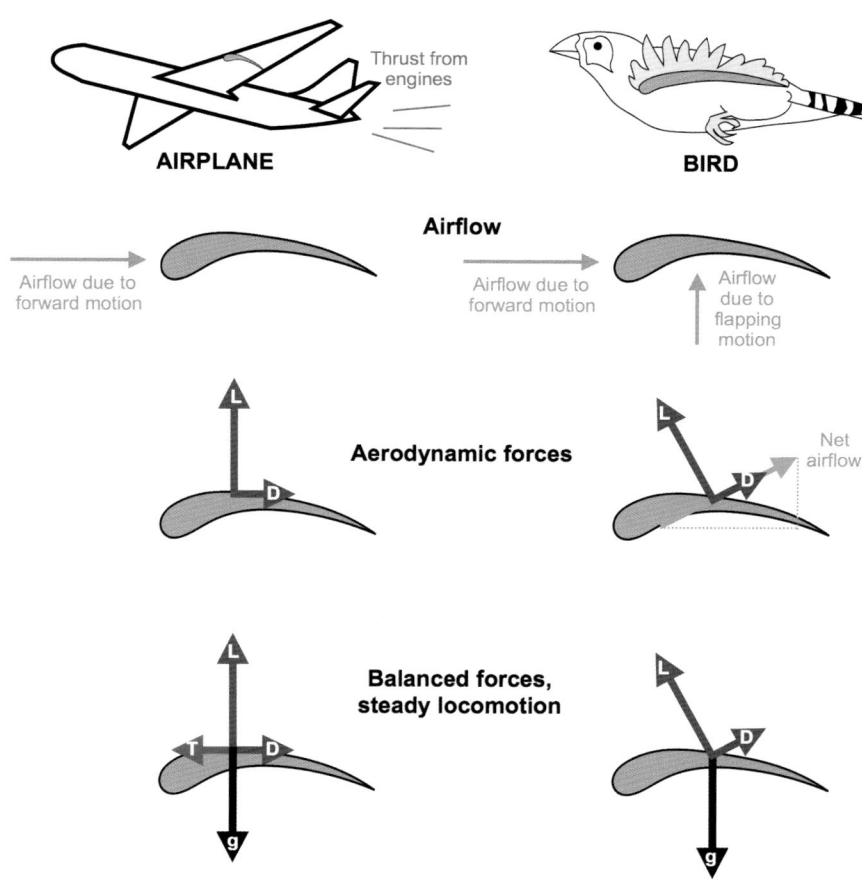

Figure 10.11. Aerodynamic forces produced by flapping wings. In airplanes, wings and engines are decoupled: engines provide thrust (*T*) to overcome drag (*D*; parallel to airflow), and wings generate lift (*L*; perpendicular to airflow) to counteract body weight (*g*). In flapping birds (Zebra Finch, *Taeniopygia guttata*; modified after Tobalske et al. 1999), net airflow (vector sum of all airflow) over the wings is oblique to the direction of travel, such that lift is directed both upward and forward. This lift force both overcomes drag and counteracts body weight.

As equations 1 and 2 or 3 and 4 show, both lift and drag increase with increasing **air density** (relevant at different altitudes), **velocity**, and wing **surface area**. What is not immediately apparent from equations 1–4 is that lift and drag are not equally distributed along the wing, for two reasons (fig. 10.12). First, different sections of the wing have different feather lengths and therefore different surface areas (*S*). Second, in a flapping bird, the tip of the wing moves much faster than the base of the wing (higher *v*), such that primary feathers produce more lift than secondary feathers but also have greater induced drag (air flows around the wingtips from the high-pressure region beneath a wing to the low-pressure region above a wing and gets shed as "tip vortices," which reduce the pressure differential that creates lift). Because of these differences in force production at different points in the wing, scientists integrate forces or coefficients across the wing to obtain a net value.

Lift and drag are also affected by wing morphology (**aspect ratio**, **camber**, feather microstructure) and orientation (**angle of attack**) (fig. 10.12). Differences in aerodynamic performance caused by differences in wing properties are reflected in the final parts of equations 1 and 2, the **coefficients of lift and drag** (C_L and C_D). Comparing coefficients of differ-

ent wings is thus a quantitative way to compare aerodynamic performance of different wing designs and orientations. Different species have different wing designs and individuals constantly morph their wing shape during flight to modulate lift and drag production. Understanding this wing morphing/deformation (https://www.youtube.com/watch?v=2CFckjfP -1E) and the modulation of aerodynamic forces across species is an active area of research (box on page 288).

In short, birds remain airborne by flapping their wings to generate lift: a continuously modulated aerodynamic force that acts perpendicular to the direction of motion and that can be decomposed into a vertical component counteracting body weight, and a forward-directed component (thrust) counteracting drag. Lift and drag production depend on air density, flight/flapping velocity, wing surface area, and wing morphology/orientation, which vary across behaviors, across species, and throughout the wing stroke—birds constantly morph their wings to modulate the magnitude and direction of these forces. With that in mind, we now consider how birds use their muscles to flap their wings and generate the aerodynamic forces necessary for flight.

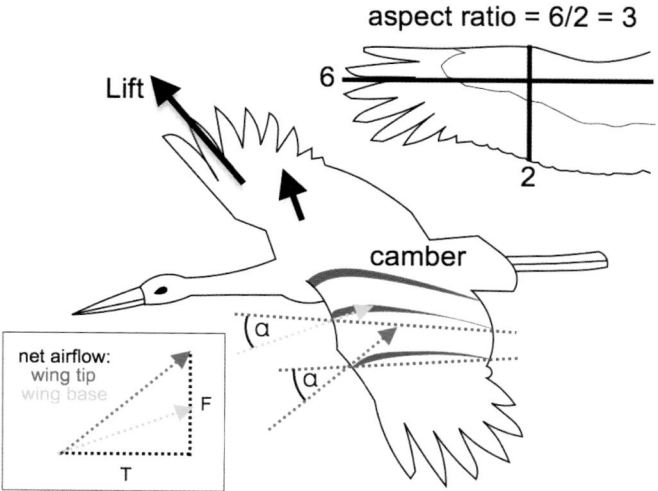

Figure 10.12. Distribution of lift along a flapping wing. In a flapping bird, the magnitude and orientation of aerodynamic forces differ along the wing, because the tip of the wing moves faster than the base of the wing. More lift is produced toward the wing tip, since the magnitude of lift production is proportional to velocity (v^2). Lift produced near the wing tip is also directed more anteriorly (forward), since the vector sum of airflow due to translational (T) and flapping (F) velocity is inclined more vertically. To compensate for this change in airflow and avoid stalling due to a high angle of attack (α), birds often pronate (twist) the tips of their wings forward. *Image by A. Heers.*

The Engine: How Do Muscles Power Flight?

A general theme of the avian flight apparatus is that large, proximal muscles provide power to pump the wing up and down, while small, distal muscles help position the wing and fine-tune its movements (proximal = closer to body wall; distal = closer to wing tip) (Dial 1992a, Biewener 2011; box on page 289). **Power** is defined as the rate of doing **work**, and it is equivalent to the amount of energy produced or consumed as a force displaces an object over a given unit of time:

$$\text{Power} = \frac{\text{work}}{\text{time}} = \frac{\text{force} \times \text{displacement}}{\text{time}} = \frac{F \times d}{t} \qquad (5)$$

All forms of active vertebrate locomotion require muscles to pull on bones and rotate or move those bones around joints. However, not all locomotion requires high muscle power. For example, during running, muscles produce force (F) but much of the length change (d) occurs passively in adjacent tendons. Stretch and recoil of these springlike tendons reduces muscle power requirements during running, by (1) allowing muscles to operate **isometrically** (contract without shortening—better for producing high force), and (2) cyclically storing and releasing energy, a bit like a rubber band (Roberts et al. 1997, Biewener 2011). In contrast, energy

UNSTEADY AIRFLOW AND FLAPPING FLIGHT

So far we have been discussing flight under the assumption that fluid flows over the wing in a **steady** or **quasi-steady** fashion in which kinematics and forces do not change over time—like a fixed-wing airplane flying at constant speed. However, flapping locomotion is anything but steady: wings constantly change size, shape, and orientation, and also reverse direction at the beginning and end of the downstroke. These abrupt changes result in unsteady flow conditions over the wing, and we now know that both insects and small birds can take advantage of **unsteady** effects to augment aerodynamic force production (Dickinson et al. 2000, Biewener 2003, Videler et al. 2004, Warrick et al. 2005):

- **Leading edge vortices** and **delayed stall**: air spinning at the leading (front) edge of the wing creates a low-pressure zone that augments lift and helps keep airflow "attached" at high angles of attack.
- **Rotational lift**: as a wing rotates and reverses direction at the end of downstroke or upstroke, the "spinning" motion of the rotation combines *constructively* with airflow on the upper surface of the wing, and *destructively* with airflow on the lower surface of the wing, such that air flows even faster over the top of the wing and even slower over the bottom. This increases the pressure differential across the wing and augments lift—just like putting backspin on a golf ball.
- **Clap-and-peel**: this "wing-clapping" motion at the upstroke-downstroke transition speeds up the initiation of circulation (Crandell and Tobalske 2015).
- **Wake recapture**: an additional mechanism observed in hovering insects but not yet in birds, whereby a wing generates additional lift by moving back through its own wake.

In short, flapping flight is complex and not truly steady, and we must always consider how unsteady mechanisms contribute to flight.

recovery by tendons does not seem to play as prevalent a role in bird flight (although it is probably still important: see Tobalske and Biewener 2008, Konow et al. 2015). Flight requires high power output by muscles, because flight muscles must produce high force (to pull down or up on wing bones), change substantially in length (to sweep the wing through

KEN DIAL

Ken Dial's father was an aeronautical engineer, and consequently Ken grew up right next to the Los Angeles International Airport, watching airplanes land every two minutes for most of his childhood. Dial was introduced to bird ecology as an undergraduate and to functional vertebrate morphology as a graduate student, and aerodynamics and bird flight merged when he was a postdoctoral fellow at Harvard University. Dial went on to found and direct the University of Montana's Flight Laboratory and Field Research Station at Fort Missoula for more than two decades. During the first half of his academic career, Dial focused on the functional aspects of how adult birds fly (e.g., neuromuscular control, skeletal kinematics, mechanical power development, and wing morphing), and was pivotal in developing new techniques for understanding muscle and skeletal function during flight. For the past 15 years, Dial has dedicated his research activities to understanding the development/ontogeny of bird flight, highlighting strategies among species with contrasting life histories. This work led to a previously unappreciated phenomenon that Dial has termed "wing-assisted incline running," or WAIR, which he and his students suggest is relevant to both the survivorship of extant birds and to the evolution of flight among extinct theropod dinosaurs (ontogenetic transitional wing hypothesis). More recently, Dial has become interested in how birds that seemingly abandon flight (semi-flightless and flightless species) employ their wings and legs cooperatively in order to negotiate their environments, as occurs in species that live in habitats with relaxed predation pressure.

its large range of motion about the shoulder joint), and do both of these things rapidly (high F, high d, low $t \rightarrow$ high power output; for more information on the relationship between muscle power, muscle force, and locomotion, see Biewener 1998, Biewener and Roberts 2000). Flight is thus the most power-demanding mode of locomotion (Alexander 2002), and flight-capable birds have very large and powerful flight muscles—potentially more powerful than the muscles of any other vertebrate (Askew and Ellerby 2007, Jackson and Dial 2011, Biewener 2011, Jackson et al. 2011).

Important Muscles   In birds, the main source of power for pulling the wing down (downstroke) is the large **pectoralis** muscle (fig. 10.13A). The pectoralis actually begins contracting *before* the beginning of downstroke, while the wing is still

being elevated and the pectoralis is being *stretched*. This **eccentric contraction** (fig. 10.13B) helps slow the wing's upward path, and initiate the downstroke. Once the downstroke begins, the pectoralis is contracting and *shortening* to pull on the underside of the humerus and depress the wing. This **concentric contraction** continues through the first part of the downstroke, at which point the pectoralis stops contracting and the wing begins to slow down.

Just as the pectoralis muscle is the main source of power for downstroke, the **supracoracoideus** muscle is the main source of power for upstroke. Whereas the downstroke is always active (driven by muscles), wing elevation during the upstroke may be achieved somewhat passively (without muscle input) at high flight velocities, because upward-pulling aerodynamic forces are sufficient to raise the wing without help from muscles (Tobalske et al. 2003a). The supracoracoideus muscle is more important at slow to moderate flight speeds, contracting to pull on the dorsal (upper) surface of the humerus and elevate and rotate the wing. Like the pectoralis muscle, the supracoracoideus begins contracting before the beginning of upstroke (eccentric contraction), and continues to contract concentrically through the first part of upstroke.

Variation in Power Requirements   Different modes of flight have different power requirements. Figure 10.13C shows that flight at very slow and very high speeds requires more power than flight at moderate speeds. Recall that flying birds must overcome both gravity (body weight) and aerodynamic drag. While body weight is invariant, friction drag increases with increasing velocity (v^2), and induced drag decreases with increasing velocity ($1/v^2$). For a given bird, the muscle power required to fly therefore should be proportional to the U-shaped sum of friction and induced drag (pressure drag is always low) (Pennycuick 1968). Many species do show U-shaped power curves, though not all birds can fly at the same speed, and variation in the shape of the curve exists—probably caused by differences in morphology, flapping movements (kinematics), and general flight style (Tobalske et al. 2003b, Askew and Ellerby 2007). Hovering (zero velocity) is particularly expensive, and only hummingbirds (Apodiformes)—with their small body size and highly specialized flight stroke—can hover for extended periods of time (Tobalske 2010).

Power required for flight also varies with the angle of movement (fig. 10.13D). Ascending flight is more challenging than level flight, which is more challenging than descending flight (and WAIR requires less power than all of these) (Jackson et al. 2011). Ascending flight and vertical takeoffs are particularly power-demanding because a bird must gain altitude in addition to supporting its body weight (think of

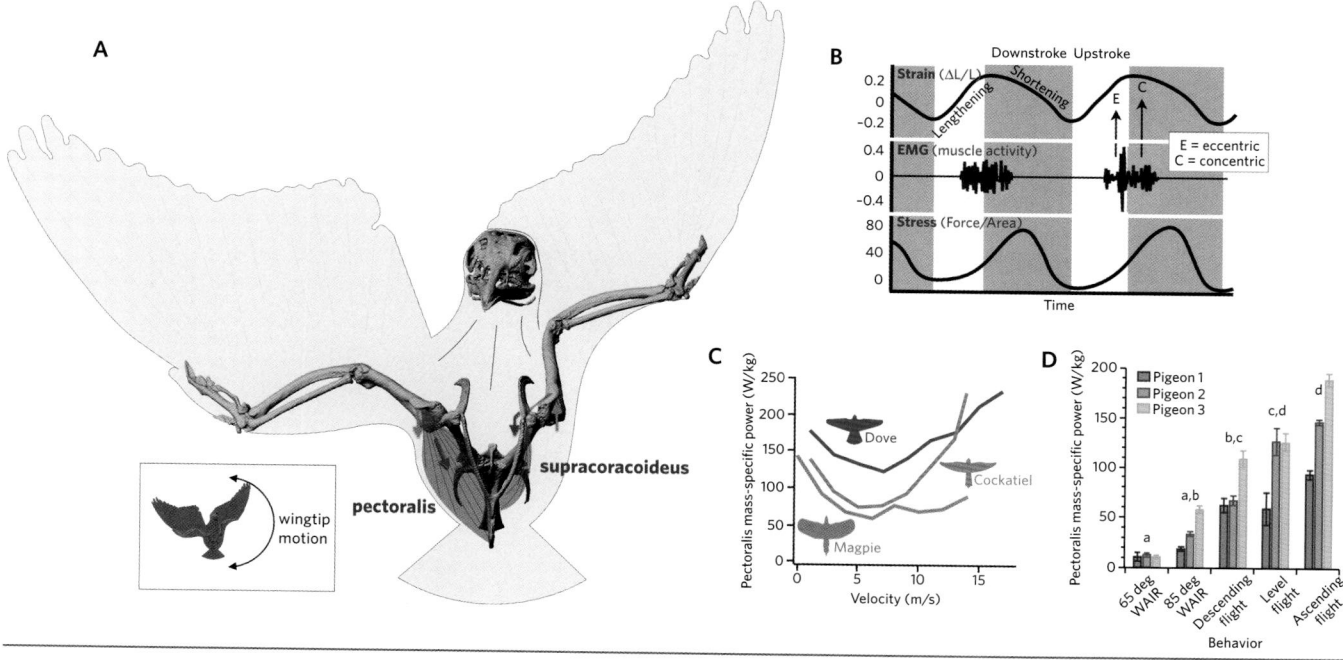

Figure 10.13. Muscle power. *A*, In birds, the downstroke is largely powered by the pectoralis muscle (*image by A. Heers*), which contracts both while lengthening (eccentric contraction) and shortening (concentric contraction) to pull on the deltopectoral crest of the humerus and depress the wing; *B*, This pattern of active lengthening followed by active shortening (*redrawn after Jackson et al. 2011*) indicates that the pectoralis muscle decelerates the wing at the end of upstroke, then reaccelerates it at the beginning of downstroke. Active lengthen-ing also allows for greater force production, because muscles produce more force while lengthening than while shortening. Power output of the pectoralis varies with both *C*, flight speed (*redrawn after Tobalske et al. 2003b*) and *D*, flight mode (*from Jackson et al. 2011, with permission*). Though not shown in *B*, the supracoracoideus muscle also actively lengthens, to decelerate the wing at the end of downstroke. Both muscles have long muscle fibers (red lines), which allow them to lengthen and shorten extensively.

walking on a horizontal surface versus climbing stairs). Consequently, larger birds with less excess muscle power (less power available beyond the power required for flight) tend to have lower performance than smaller birds with greater excess power (Dial et al. 2008a), while birds specialized for short-burst takeoffs—such as the chicken-like galliforms—help meet the exceptional demands of takeoff flight with specialized muscle fiber types (Scanes 2015). Descending flight, on the other hand, is "easier" because a bird does not need to support all of its body weight—it merely needs to slow its descent through the air. Even juvenile birds that cannot fly levelly can flap to slow a descent through the air (Dial et al. 2008b). How do birds modulate power output to fly at these different angles and achieve different flight speeds?

Meeting Power Requirements   Birds seem to modulate power and aerodynamic output in a number of ways. At the kinematic level (joint movement, next section), birds can adjust the amplitude of their wingbeats or the orientation of their wings, or adopt different wing gaits to modulate how much aerodynamic force they produce. At the muscular level, birds can adjust power output through increases in muscle stress, muscle strain, and contractile velocity (Jackson et al. 2011,

Tobalske and Dial 1994, Tobalske 1995, Tobalske et al. 1999, Tobalske 2001):

- *Muscle stress* is the amount of force produced by a given cross-sectional area of muscle. How much force a muscle produces depends on the contractile velocity and length of the muscle, as well as how many muscle cells are contracting: the greater the number of active, contracting cells, the greater the force and the greater the stress. Birds can increase muscle power by recruiting more muscle cells and increasing muscle force (*F* in equation 5) and stress.
- *Muscle strain* is a metric of how much a muscle lengthens or shortens relative to its "relaxed" length. A muscle can rotate or move a bone more by shortening more (*d* in equation 5).
- *Contractile velocity* is how fast a muscle contracts (proportional to $1/t$ in equation 5). All else being equal, the faster the muscle contraction, the greater the muscle power. To adjust the speed of muscle contraction, some birds recruit different muscle **fiber types** that are specialized for contracting at different speeds (table 10.1, simplified

Table 10.1. **Muscle fiber types.**

Fiber Type	Contraction Speed	Metabolism	Color
Slow-twitch oxidative (SO, type I, or red fiber)	Slow: sustained locomotion, like walking or running	Oxidative (aerobic): slow to fatigue	"red meat," like chicken thighs or the legs of a Thanksgiving turkey (high concentration of mitochondria)
Fast-twitch oxidative/glycolytic (FOG, type IIA, or intermediate fiber)	Moderately fast	Oxidative (aerobic) and glycolytic (anaerobic)	Intermediate, pink
Fast-twitch glycolytic (FG, type IIB, or white fiber)	Very fast: explosive locomotion, like the brief bursts of flight in wild galliforms	Glycolytic (anaerobic): quick to fatigue	"white meat," like chicken breast or the breast of a Thanksgiving turkey (low concentration of mitochondria)

from Scanes 2015). Other birds (particularly small birds, like Zebra Finches, *Taeniopygia guttata*, ~10–15 grams) have only one muscle fiber type (not enough muscle mass for multiple fiber types) and are probably limited in how much they can vary muscle contractile velocity. These birds may instead adjust mean power output by interspersing flapping phases (high power) with gliding or bounding phases (low power) (Tobalske et al. 1999).

In short, birds modulate power output in various ways to achieve different flight angles and speeds, though all modes of flight require high power output compared to terrestrial locomotion. Most of the power required for flight is produced by the large, proximally located pectoralis and supracoracoideus muscles, which contract both eccentrically and concentrically to pull down (pectoralis) or up (supracoracoideus) on the deltopectoral crest of the humerus.

The Flight Stroke: How Do Birds Move Their Wings during Flight?

Kinematics, or the translation and rotation of bones about joints, is an important component of locomotor performance. When you walk or run, you place your foot on the ground to support your body weight and to push yourself forward. This stance phase or power stroke is followed by a swing phase or recovery stroke, where you swing your leg forward to reposition it for the next stance. For flying animals, the equivalent of the stance phase (power stroke) is the **downstroke**: the portion of flight stroke when the wing is pulled downward and lift is produced for weight support and propulsion. The downstroke is followed by the **upstroke**, or the recovery phase of flight when the wing is pulled upward and repositioned for the start of downstroke. Downstroke and upstroke kinematics are unique and complex, but with their

specialized joint morphologies, birds are well equipped to achieve these movements.

Downstroke   In general, during the downstroke birds pull their wings downward (depression) and either backward (retraction; wing-assisted incline running, ascending flight) or forward (protraction; hovering, slow forward flight) at the shoulder joint, while extending their elbows and wrists and rotating (pronating or supinating; all joints) their wings to achieve an appropriate angle of attack (fig. 10.12) (Tobalske 2007, Baier et al. 2013, Heers et al. 2016; S3 Video in Heers et al. 2016, http://journals.plos.org/plosone/article?id=10.1371/journal.pone.0153446#sec030). Downstroke kinematics are fairly similar across species, though many species have great capacity for adjusting kinematics to maneuver and meet aerial challenges. For example, like a helicopter, birds can fly at different angles by adjusting the angle at which they pump their wings up and down (stroke plane angle) (Tobalske et al. 2007). Some angles of flight are more challenging than others, and kinematically, birds meet such challenges in a variety of ways: (1) increasing wing stroke amplitude to increase the area swept out by the wings during downstroke (Altshuler et al. 2010); (2) flapping faster (Chai and Millard 1997); (3) orienting the wings at a high angle of attack (Jackson 2009), presumably to increase total aerodynamic force production (lift + drag; recall that drag can help support body weight during takeoff); and/or (4) tucking their wings in during the upstroke, potentially to reduce drag and inertia (Hedrick et al. 2004) when the wings are oscillating quickly. Surprisingly, how birds **maneuver** around obstacles is not well understood. Evidence so far suggests that birds maneuver by reorienting their bodies and by modulating muscle activity (e.g., pectoralis, biceps) and skeletal kinematics (flapping velocity, stroke amplitude, and/or feathering-angle) to produce different amounts and/or directions of lift on their left versus right wings (Tobalske 2007, Biewener 2011, Ros et al. 2011). Even so, compared to the upstroke, downstroke kinematics are relatively conservative across species.

Upstroke   In contrast to downstroke, upstroke kinematics are much more variable, across species and flight speeds. Flight gaits are thus distinguished by differences in upstroke (fig. 10.14) (Tobalske 2000):

- In a **vortex-ring gait**, lift is produced only or mainly during downstroke; during the upstroke, birds tuck their wings in to reduce drag and inertia and either (1) rotate the wing tip backward (tip-reversal upstroke; https://www.youtube.com/watch?v=YhjvUKCQSZc), or (2) arrange their feathers like a venetian blind for air to pass through (feathered upstroke; https://www.youtube.com/watch?v=uAxOTWC57PA; 46 sec. into video). The tip-reversal upstroke may produce some lift, whereas the feathered upstroke mainly reduces drag.
- In a **continuous-vortex gait**, birds keep their wings partially extended during the upstroke so that lift is produced continuously, during both the downstroke and the upstroke (although most lift is produced dur-

ing downstroke) (https://www.youtube.com/watch?v=dACQDs4Pevs).

Which gait a bird uses depends both on flight speed and wing length or shape. Birds with long or high aspect ratio wings (e.g., Rock Doves, *Columba livia*; Budgerigars, *Melopsittacus undulatus*; American Kestrels, *Falco sparverius*; see fig. 10.12) tend to use a vortex-ring gait at slow (tip-reversal upstroke) and intermediate (feathered upstroke) speeds, and a continuous-vortex gait at high speeds. In contrast, birds with short or low aspect ratio wings (e.g., Black-billed Magpies, *Pica hudsonia*) do not adjust wing gait for different flight speeds—most use a vortex-ring gait with a feathered upstroke at every speed and vary aerodynamic output in other ways (note that galliforms and owls are an exception: they have low aspect ratio wings but use tip-reversal upstrokes). Irrespective of wing shape, all birds tend to use a vortex-ring gait to accelerate and a continuous-vortex gait to decelerate. These gait transitions match aerodynamic theory in many respects (for example, continuous lift production is not efficient at slow speeds and should only be used at high speeds). However, the functional relationship between wing shape

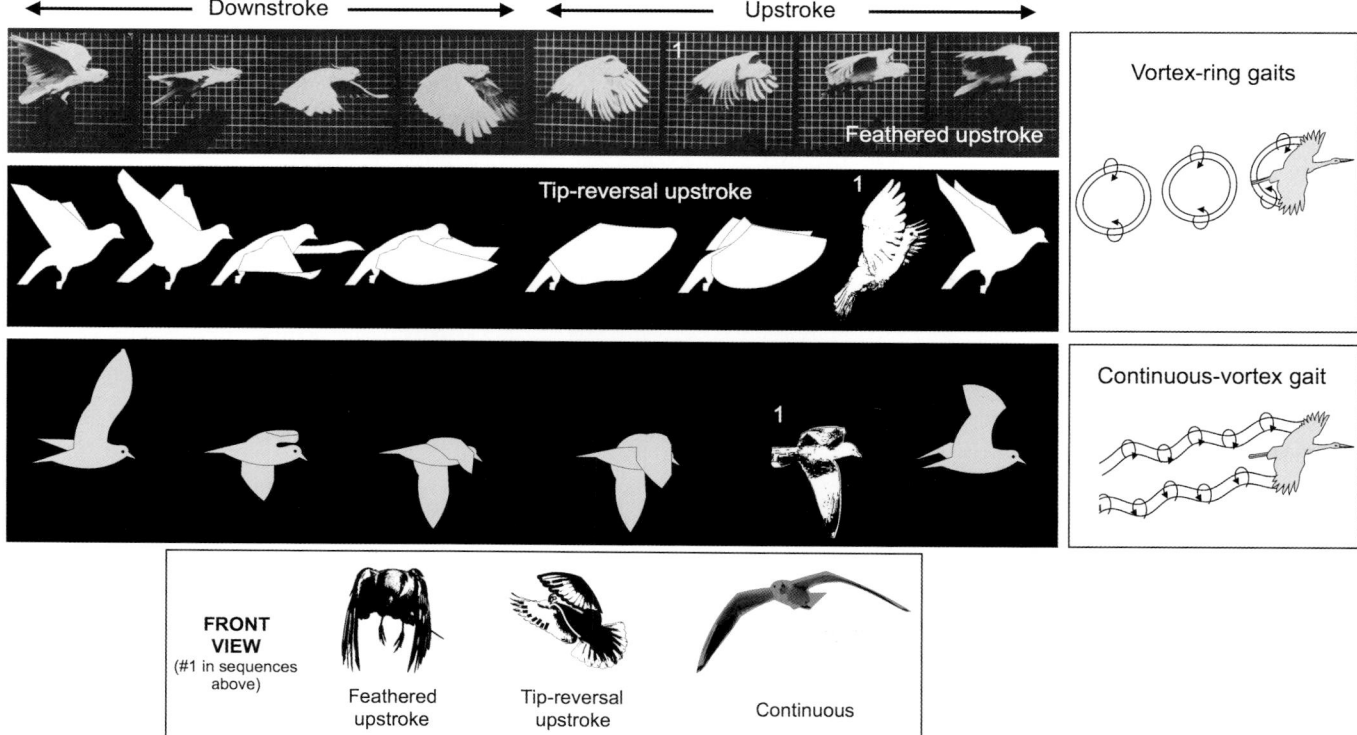

Figure 10.14. Flapping kinematics and upstroke gaits. Flight gaits are distinguished by upstroke kinematics. During the upstroke of a vortex-ring gait, birds either arrange their feathers like a venetian blind for air to pass through (feathered upstroke) or rotate the wing tip backward (tip-reversal upstroke). During the upstroke of a continuous-vortex gait, birds keep their wings partially extended. *Drawings of wing gaits (right) based on Hedrick et al. 2002. Images of cockatoo by Eadweard Muybridge, in public domain, https://www.nls.uk/about-us/publications/discover/2009/muybridge. Other drawings by A. Heers.*

and wing gait, and the aerodynamics of upstroke, are not well understood and are an active area of research.

Overall, flapping kinematics are unique and complex, and birds have specialized joint morphologies that help achieve these movements. The downstroke or power stroke is more conservative across species, involving wing depression, protraction or retraction, and rotation, whereas the upstroke or recovery stroke shows more variation and defines different flapping gaits (vortex-ring versus continuous-vortex). Within this general framework, birds can adjust their stroke plane angle, as well as the timing and/or magnitude of joint movements, to maneuver and meet aerial challenges.

Coordination and Control: Flight and the Nervous System

If you have ever watched a hummingbird zipping around between flowers or swallows darting around each other and chasing insects, you have witnessed some incredibly acrobatic behavior. Remarkably, we know relatively little about the biomechanics or neural control of aerial maneuvering. On a very basic level, birds have large brains, with excellent vision (large optic lobes, rapid motion processing) and balance (large cerebellum, well developed vestibulo-ocular and vestibulocollic reflexes). Both traits are necessary for moving in complex, three-dimensional environments at high speed (Husband and Shimizu 2001) and are discussed in chapter 12. Birds have also evolved the ability to maintain head stability in the presence of rapid body oscillations (Warrick et al. 2002). Why is this important? During flapping flight, a bird's body moves up and down with each wingbeat. These movements occur several times per second (~10 times per second, an average wingbeat frequency) and become exaggerated during maneuvers. Such rapid oscillations could impair vision and static equilibrium by adding extraneous visual flow and continuously accelerating the semicircular canals. Birds bypass this potential problem with "blurred bodies and steady heads": birds isolate their heads from body accelerations by activating the muscles of their head and long neck (also see Pete et al. 2015) to stabilize the head (vestibulocollic reflex) and eyes (vestibulo-ocular reflex) independent of body movement (https://www.youtube.com/watch?v=bLIU18s4cy8). Though the avian brain and nervous system are clearly "flight-ready" and equipped to process a rapid flux of motion-related information, future studies that monitor brain activity during flight will clarify the inner workings of the "control-center" for avian flight.

Putting this all together and summarizing how wings work:

- Birds remain airborne by flapping their wings to generate lift, an aerodynamic force that acts perpendicular to airflow and that can be decomposed into two components: a vertical component counteracting body weight, and a forward-directed component (thrust) counteracting drag. Lift and drag production vary across behaviors, across species, and throughout the wing stroke, and birds constantly morph their wings to modulate the magnitude and direction of these forces.

- Birds adjust power output with flight angle and speed, though all modes of flight require high power output compared with swimming or terrestrial locomotion. Most of this power is produced by the large, proximally located pectoralis and supracoracoideus muscles, which contract eccentrically and then concentrically to slow the wing and then pull down (pectoralis) or up (supracoracoideus) on the deltopectoral crest of the humerus.

- Flapping kinematics are complex. Compared with the upstroke or recovery stroke, which varies both across species and with flight speed and defines different flapping gaits (vortex-ring versus continuous-vortex), the downstroke or power stroke is relatively conservative or standardized, involving wing depression, protraction or retraction, and rotation. Within this general framework, birds can adjust their stroke plane angle, as well as the timing and/or magnitude of joint movements, to maneuver and meet aerial challenges.

- With their large brains and excellent vision and balance, birds are well equipped to process a rapid flux of motion-related information.

Different Styles of Flight: Flapping, Soaring, Gliding

So far we have discussed flapping flight, because all birds rely on flapping flight, to some extent, to remain airborne. However, birds flap their wings to different degrees, and some birds are specialized for flapping very little. These differences in flight style are reflected in differences in wing morphology (fig. 10.15) and habitat preferences.

Continuous Flapping

Most birds flap their wings during takeoff and landing, and many birds, particularly species with average or high **wing loadings** (high body weight for a given wing surface area), flap their wings continuously during flight. Wading birds (e.g., flamingos, Phoenicopteridae; herons, Ardeidae), waterfowl (ducks and geese, Anseriformes), swimmers/divers (e.g., loons, Gaviidae; grebes, Podicipediformes), galliforms (chicken-like birds such as pheasants and quails), tinamous (Tinamiformes), hummingbirds, and many passerines (e.g., American Crow, *Corvus brachyrhynchos*; American Robin, *Turdus migratorius*; other songbirds) all use continuous flapping

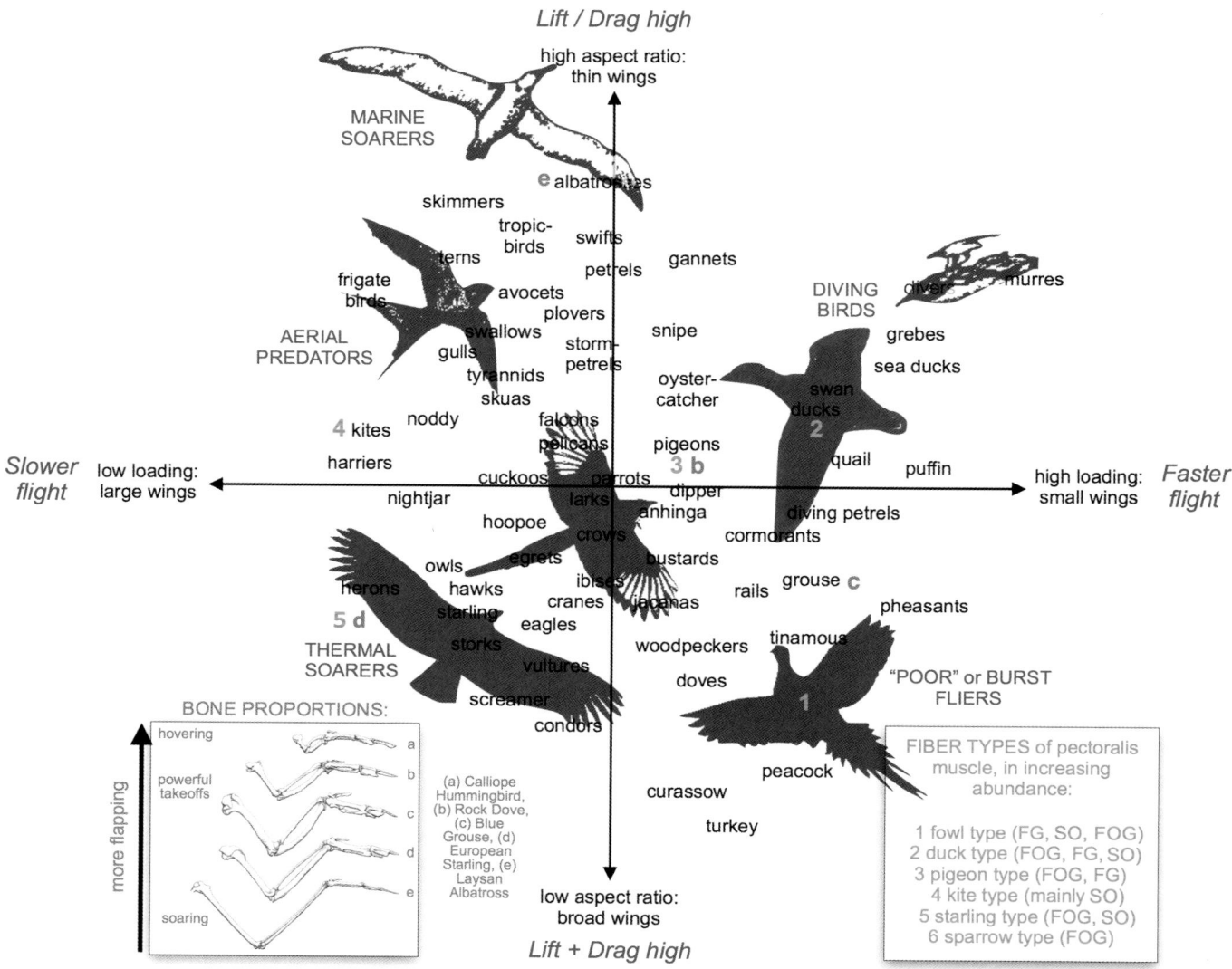

Figure 10.15. Flight style and wing morphology. Different styles of flight are associated with different skeletal proportions, muscle properties, and wing shapes. Birds that flap their wings continuously (*a-c, left inset*) have proportionally larger hand skeletons than birds that spend time flap-gliding (*d*) or soaring (*e*). Birds specialized for rapid takeoffs (*1–3, right inset*) have more FG (fast glycolytic, or anaerobic) muscle fibers in their main flight muscles than birds that flap more slowly (FOG, fast oxidative glycolytic fibers, or SO, slow oxidative fibers) (*4-6, right inset*). Finally, birds with higher wing loadings (greater body weight per unit wing area, main plot) tend to fly faster than birds with lower wing loadings, while birds with higher aspect ratio wings (long and/or thin) may be specialized for "efficient" flight (lift-to-drag ratio = high) and birds with lower aspect ratio wings (short and/or broad) may be specialized for "efficacious" flight (lift + drag = high). *Plot modified after Rayner 1988, wing skeletons reproduced, with permission, from Dial 1992b.*

flight (Bruderer et al. 2010) but differ with respect to flight speed, maneuverability, and distance.

Passerines are generally good fliers and often forage in cluttered habitats requiring slower flight and/or high maneuverability (https://www.youtube.com/watch?v=YlqQBEZvHJA). In contrast, waterfowl and swimmers/divers fly at high speed with less maneuverability and are generally found in more open habitats conducive to this flight style (e.g., lakes) (https://www.youtube.com/watch?v=fbUeq52Cmto). In spite of these differences, many passerines, waterfowl, and wading birds can fly for long distances and even migrate (chapter 19).

Galliforms and tinamous are specialized for **short-burst** flight (https://www.youtube.com/watch?v=Oi01KLGB0TQ). They leap into the air when startled and ascend quickly, but also fatigue quickly and do not remain airborne for long. Most short-burst fliers do not migrate.

Hummingbirds are highly specialized for **hovering**: continuous flapping without translational velocity (https://www.

youtube.com/watch?v=mJ4n_PeyK_E). Some species, such as the Pied Kingfisher (*Ceryle rudis*), can hover for short periods of time. Other species, such as kestrels (Falconidae) or terns (Sternidae), "windhover" by facing into the wind and flapping their wings while staying in one place to hunt (https://www.youtube.com/watch?v=qyScZ4SxDFA). But only hummingbirds can sustain hovering (Tobalske 2010)—they have specialized wing joints and insect-like flight strokes that allow them to produce lift during the downstroke *and* upstroke and achieve this very difficult form of flight.

Finally, many shorebirds rely on continuous flapping flight for **long-distance migration** (chapter 19). A Bar-tailed Godwit (*Limosa lapponica*) was tracked flying nonstop from Alaska to New Zealand (7,000+ miles over eight to nine days) without ever landing.

Intermittent Flight

Not all species flap continuously: many birds alternate periods of flapping with non-flapping periods of gliding (wings extended) or bounding (wings completely or partially tucked against body). This type of flight is known as **intermittent flight**. According to the body-lift hypothesis, intermittent flight saves energy, because birds can rest their pectoral muscles while still producing some aerodynamic force with their body, tail, and wings (if extended). According to the fixed-gear hypotheses, intermittent flight allows small birds with a fixed wing gait and/or muscle contractile velocity to modulate average power or aerodynamic output (Tobalske and Dial 1994, Tobalske 1995, Tobalske et al. 1999, Tobalske 2001). **Flap-bounding** (https://www.youtube.com/watch?v=HS4T7pNzxdY) is common in relatively small birds (e.g., some passerines, and woodpeckers) and tends to occur at higher flight speeds, whereas **flap-gliding** is common in small to moderate sized birds (e.g., swifts, Apodiformes: https://www.youtube.com/watch?v=Zby2ERIVVxE; starlings, Passeriformes; falcons, Falconiformes) and tends to occur at slow to moderate flight speeds (fig. 10.16). Many highly ma-

neuverable aerial hunters (such as swallows, Hirundinidae; fig. 10.15) are flap-gliders.

Gliding and Soaring

Gliding (reviewed in Norberg 1985b) is a form of passive, non-flapping flight used by some birds to land (watch a duck the next time you are outside) and by some birds during intermittent flight. As during active, flapping flight, air flowing over the wings and body produces aerodynamic forces (lift and drag). In the case of gliding, however, airflow is produced passively as the bird "falls" through the air, and the amount of vertically directed force is never enough to completely overcome body weight and sustain altitude. **Sink rate**, the rate at which vertical height is lost, determines how long a gliding animal can remain aloft without flapping and depends on wing loading and lift-to-drag (L:D) ratio. **Glide angle**, or angle of descent, depends only on the L:D ratio and determines how far an animal can travel in a given amount of time. Birds with low wing loadings and high ratios of L:D have the slowest rate of sinking and the shallowest glide angle. Both of these metrics of gliding performance vary with speed. Although several non-avian groups ("flying" fish, Exocoetidae; frogs, Rhacophoridae, Hylidae; lizards, *Draco*; snakes, *Chrysopelea*; squirrels, Petauristinae) are quite proficient at gliding (angle of descent < 45°) or parachuting (angle of descent > 45°), there are actually no birds that *only* glide. Birds always intersperse gliding with flapping, although some birds take advantage of air currents and can flap very little. These are the soaring birds.

Soaring is essentially a form of gliding that involves sinking (1) more slowly than air is rising (thermal and ridge soaring) or (2) more slowly than air currents cause a bird to rise (dynamic soaring). During **thermal soaring** (fig. 10.17A), birds fly in spiraling circles to maintain contact with updrafts of warm air and gain altitude without flapping (https://www.youtube.com/watch?v=7KwsGWVXCPM). After reaching a certain altitude or losing a thermal, the bird glides to an adjacent thermal to start the process over again. This style of flight allows birds to traverse

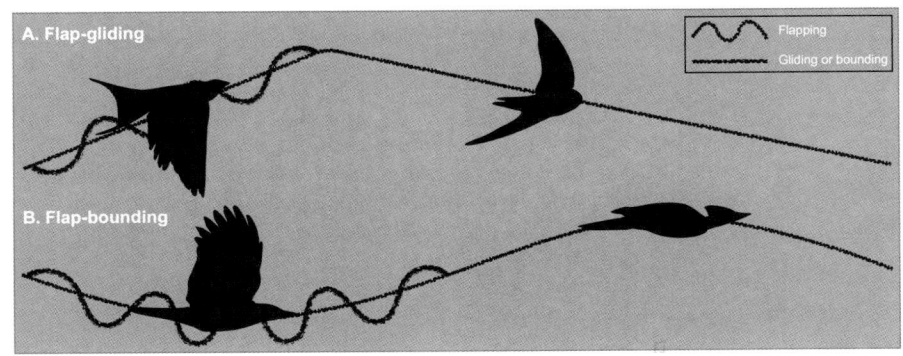

A. Flap-gliding

B. Flap-bounding

Flapping
Gliding or bounding

Figure 10.16. Intermittent flight. *A*, Flap-gliding in a swallow; *B*, Flap-bounding in a woodpecker. *Modified after http://www.biology.leeds.ac.uk/staff/jmvr/Flight/PWV/index1.htm.*

great distances without expending much energy, and is very common in vultures and hawks (Cathartidae, Accipitridae)—birds that have relatively low wing loadings, broad wings, and slotted wing tips (figs. 10.9, 10.15). Similar updrafts of rising air can be found near ridges or along waves, and many birds take advantage of such updrafts for **ridge soaring** (fig. 10.17B; https://www.youtube.com/watch?v=vkVVnx70FgI). A slightly different process occurs during dynamic soaring. **Dynamic soaring** (fig. 10.17C; https://www.youtube.com/watch?v=uMX2wCJga8g) is the specialty of albatrosses and their relatives (Procellariiformes)—birds with long, high aspect ratio wings, low to average wing loadings, and locking shoulder joints (Pennycuick 1982). During dynamic soaring, birds extract energy from wind gradients over the open ocean. Air blowing close to the surface of the water is slowed down by friction with the water, such that wind speed increases with increasing distance from the water surface. Recall from the box on pages 284–285 that lift is produced when air flowing above the wing moves faster than air flowing beneath it. The same thing happens when an albatross flies into a wind gradient: air above the wing blows slightly faster than air below the wing, lift is produced, and the bird gains altitude until the wind gradient is not substantial enough to support it. At that point, the bird glides back to the surface of the ocean and repeats the process. Thermal soaring, ridge soaring, and dynamic soaring are thus all ways to flap less and save energy (see formation flying in chapter 19). However, not all environments are conducive to soaring. Ridge soaring is restricted to areas with cliffs or waves, thermal soaring occurs mainly in open areas where uneven heating of the ground produces columns of rising air, and dynamic soaring is restricted mainly to the windy open ocean.

By now it should be apparent that flight and the avian body plan are intimately related. Powered, flapping flight is a unique and challenging form of locomotion, with a number of requirements that birds meet with specialized forelimb apparatuses and physiologies. Given the stringent requirements for flight, birds are perhaps surprisingly diverse with respect to wing shape, musculoskeletal morphology, upstroke kinematics, and the degree of flapping. As we discuss next, part of this variation may be related to variation in terrestrial or aquatic locomotion. For the more we learn about birds, the more we realize that the story of bird flight would not be complete without a discussion of bird legs.

ON FOOT: LEG-BASED TERRESTRIAL LOCOMOTION

Compared with bats, birds spend a large amount of time—in fact, most of their time—on foot. Whether on the ground

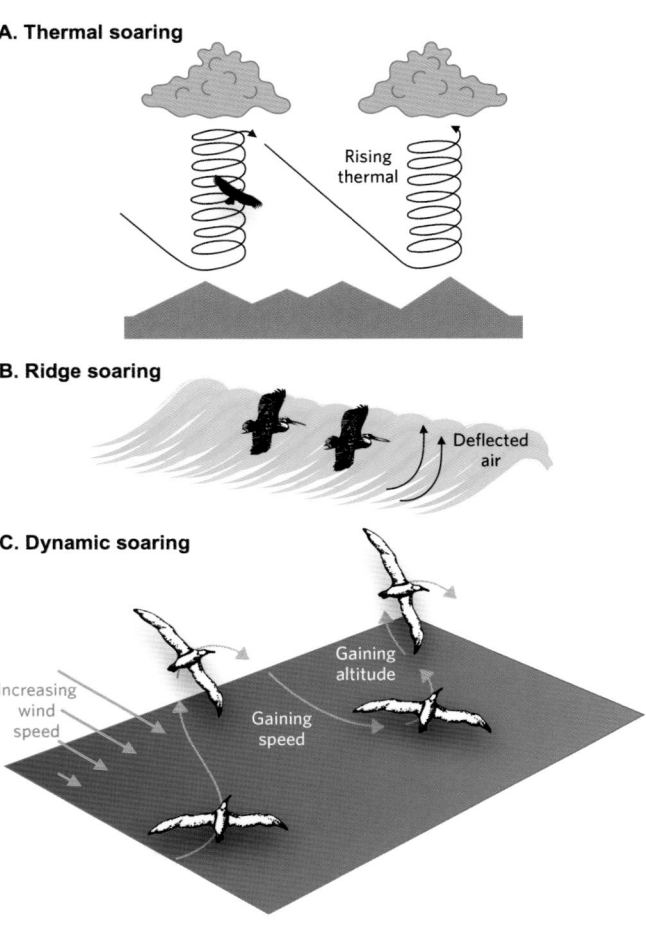

A. Thermal soaring

Rising thermal

B. Ridge soaring

Deflected air

C. Dynamic soaring

Gaining altitude

Increasing wind speed

Gaining speed

Figure 10.17. Soaring. *A*, Vulture taking advantage of rising thermals; *B*, Pelicans taking advantage of "ridge lift" over waves; *C*, Albatross taking advantage of wind gradients over open ocean. *Modified after Sachs 2005.*

or in trees and shrubs, birds are **bipedal**: they locomote on two legs. Among living animals, the only other group with this kind of locomotion is humans. Some animals will stand or run or hop on their two legs under certain conditions ("facultative bipeds" such as kangaroos, basilisk lizards, and jerboas), but most terrestrial vertebrates use all four of their limbs for locomotion (quadrupeds), and only humans and birds walk and run exclusively on two legs. The locomotion of birds is therefore unique both in the air and on the ground.

How Legs Function: The Anatomy and Biomechanics of Terrestrial Locomotion

As humans we are intimately familiar with bipedal locomotion. We walk at slow speeds and run at high speeds, using our legs to support our body weight and propel ourselves

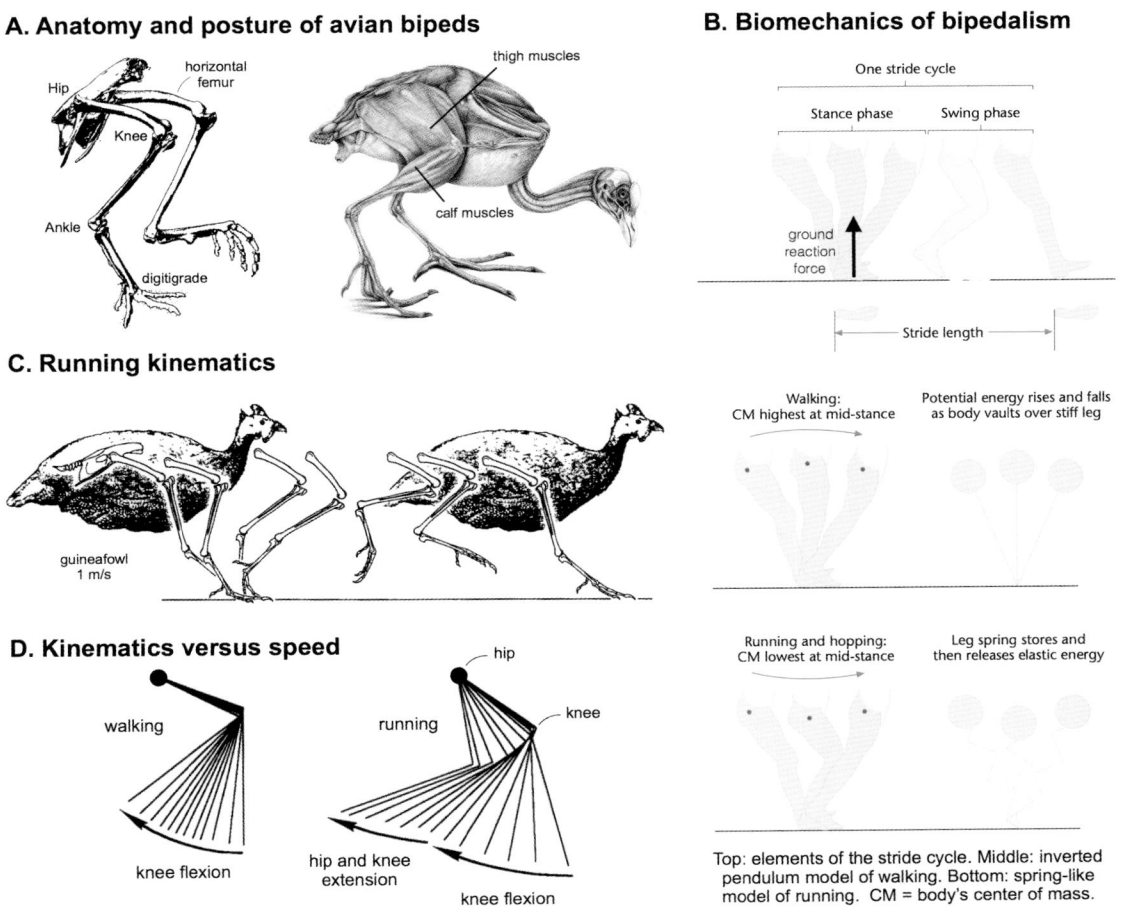

A. Anatomy and posture of avian bipeds

horizontal femur

Hip

Knee

Ankle

digitigrade

thigh muscles

calf muscles

B. Biomechanics of bipedalism

One stride cycle

Stance phase   Swing phase

ground reaction force

Stride length

Walking: CM highest at mid-stance

Potential energy rises and falls as body vaults over stiff leg

Running and hopping: CM lowest at mid-stance

Leg spring stores and then releases elastic energy

C. Running kinematics

guineafowl 1 m/s

D. Kinematics versus speed

hip

walking

knee flexion

running

knee

hip and knee extension

knee flexion

Top: elements of the stride cycle. Middle: inverted pendulum model of walking. Bottom: spring-like model of running. CM = body's center of mass.

Figure 10.18. Terrestrial locomotion. *A (left),* Leg skeleton of Chukar Partridge (*Alectoris chukar*) in walking posture, *image by A. Heers;* (*right*) leg muscles of Common Coot (*Fulica atra*), *reproduced, with permission, from van Grouw 2013. B,* Biomechanics of walking versus running. *Reproduced, with permission, from Hutchinson and Gatesy 2001, copyright © 2001 John Wiley & Sons, Ltd. All rights reserved. C,*

Helmeted Guineafowl (*Numida meleagris*) running at 1m/s, illustrating movement at knee, ankle, and tarsometatarsal-phalangeal joints. *Reproduced, with permission, from Gatesy 1999. D,* To gain speed, birds extend their leg joints more at the end of stance. *Modified, with permission, from Gatesy 1999.*

forward. Birds are similar in many respects, but different in others. Unlike humans, birds stand on their toes and have a **digitigrade** posture (fig. 10.18A). Birds also hold their femurs (thigh bones) in a roughly horizontal position and keep their legs relatively crouched (although this becomes less pronounced in large birds), and most movement occurs at the knee—rather than at the hip as in humans. In spite of these differences, both humans and birds have **erect** posture (legs held beneath body, rather than sprawled sideways), and the biomechanical underpinnings of human and avian bipedal locomotion are thus comparable (details below summarized from Gatesy and Biewener 1991, Roberts et al. 1997, Gatesy 1999, Reilly 2000, Hutchinson and Gatesy 2000, 2001, Rob-

erts and Scales 2004, Hancock et al. 2007, Nyakatura et al. 2012).

During **stance phase** (equivalent to downstroke; fig. 10.18B), a foot is planted on the ground, and the ground produces an equal and opposite **ground reaction force** that passes roughly through the bird's center of mass and provides weight support and deceleration or propulsion. Most propulsive movement occurs at the knee and tarsometatarsal-phalangeal joints (joints between foot and toes), while twisting (long-axis rotation) and outward-inward (abduction-adduction) movements of the femur and tibiotarsus (shin bone) determine lateral placement of the foot (fig. 10.18C). As in humans, knee extension is driven by the avian equivalent of "quadriceps" muscles (femorotibialis and iliotibialis muscles), while knee flexion is driven by the "hamstring" muscles (flexor cruris and iliofibularis muscles). Movement at the ankle and toes is controlled by the gastrocnemius and other calf or shin muscles, which merge to become the Achilles tendon as they cross the ankle or form long tendons to the toes (digital flexors and extensors).

Stance phase is followed by **swing phase** (equivalent to upstroke; fig. 10.18B). The knee and ankle joints, and sometimes the hip joint, flex and then extend to swing the leg forward and reposition it for the next stance (fig. 10.18C). Together, the stance and swing phase sum to one **stride**. During walking or running, the left and right legs move asynchronously—one leg is in stance while the other is in swing. This alternating limb motion is commonly called "striding," and many birds, like humans, are "striders" (e.g., wading birds, galliforms, and roadrunners, Cuculiformes). In contrast, hopping involves the synchronous movement of both limbs, and is a common mode of terrestrial locomotion in passerines and woodpeckers. What distinguishes these different modes of locomotion? Walking and running may superficially seem to be different speeds of a similar gait, but the biomechanical underpinnings of walking, running, and hopping are actually quite different (Hutchinson and Gatesy 2001).

Terrestrial Gaits

Walking is defined as a pendulum-like gait (fig. 10.18B) where each foot is in stance phase for more than 50 percent of a stride. One foot is always on the ground, and sometimes both are. This type of movement is often modeled as an *inverted pendulum* because the stance leg is relatively stiff and the body rotates over the stance foot, a bit like an upside-down pendulum or pole vaulter. The body's center of mass is highest at mid-stance, then arcs down and forward over the stance foot and converts potential energy (height) into kinetic energy (velocity). This increase in kinetic energy at the end of stance is reconverted back to potential energy to propel the body up and over the other leg as it begins stance. Such a continuous exchange of kinetic and potential energy likely conserves energy, by reducing the amount of work that muscles would otherwise do raising and accelerating the body's center of mass.

In contrast, both hopping and running are defined as *springlike* gaits (fig. 10.18B). **Hopping** (https://www.youtube .com/watch?v=UPiFJIzMAXY) always involves an aerial phase where both feet are off the ground, whereas running often, but not always, involves an aerial phase. In humans, for example, an aerial phase (https://www.youtube.com /watch?v=4fjC1Oim0UQ) is so characteristic of running that running was traditionally defined by the presence of this aerial phase. In birds, however, the transition from walking to running is not clear-cut: as speed increases, birds transition from a walking gait, to a **grounded running** or compliant gait with no aerial phase (watch bird on right in, https://www.youtube.com/watch?v=ZcODEGxnHTI), to a **running** gait with an aerial phase (https://www.youtube .com/watch?v=MNyfCWxUfso) (Gatesy and Biewener 1991). Nevertheless, grounded running, running with an aerial phase, and hopping are all springlike gaits. During the first part of stance, the leg is compressed like the spring of a pogo stick, and strain energy is stored in the tendons and elastic components of leg muscles. Unlike a walking gait, in a running gait the body reaches its lowest height at mid-stance. During the second half of stance, the tendons recoil and release energy to propel the body upward and forward. Just as the exchange of kinetic and potential energy conserves energy in walking gaits, this stretch and elastic recoil of tendons helps conserve energy in running and hopping gaits. Proximal thigh muscles power knee movement (some muscle work), while calf muscles generate isometric force (F in equation 5) to control ankle-crossing tendons that store and release strain energy as they change length (d in equation 5; no or very little muscle work) (Roberts et al. 1997, Biewener 2011) (fig. 10.18A). This arrangement is very economical compared with flight, where the pectoralis and supracoracoideus muscles have high power outputs.

Speed

How do birds transition from walking to running and gain speed? Just as aerial challenges can be met through increases in stroke amplitude and/or wingbeat frequency, birds increase walking or running speed through increases in stride length and stride frequency. Stride length is increased mainly by extending the leg more at the end of stance, at the hip, knee, and ankle joints (fig. 10.18D). Increases in stride frequency are achieved mainly by reducing the amount of time that the foot is in contact with the ground and thus decreasing stance duration, which eventually results in an aerial phase (Gatesy and Biewener 1991). Running faster requires some increase in muscle work because increased joint movements at higher speeds result from greater muscle shortening (d in equation 5), rather than greater muscle force (F in equation 5) (Roberts et al. 1997). Muscle shortening similarly increases during accelerations and incline running (Roberts et al. 1997, Roberts and Scales 2004). But even at top speed, the force-producing calf muscles and springlike ankle tendons allow most birds to locomote on foot economically, compared with flight.

Different Styles of Leg-Based Locomotion

Birds are highly diverse with respect to how and how much they use their legs for locomotion, and at this time there are not any widely used schemes for classifying such diversity. However, the manner in which birds use their legs is often closely tied to foraging style and wing use. Here, we will use different foraging strategies to help describe different styles of leg-based locomotion (Remsen and Robinson 1990, Dial 2003a, Habib and Ruff 2008, Heers and Dial 2015). The categories are listed in order of increasing leg use and decreasing wing use—recall that birds that rely more on their legs for foraging tend to rely less on their wings, and vice versa.

Hyperaerial: "no" leg use. Hyperaerial birds feed and locomote almost exclusively by wing, either by "screening" (flapping and gliding continuously to hunt and feed on insects while airborne, like swallows, Hirundinidae; swifts, Apodiformes; and nightjars, Caprimulgidae), or by hovering, like the nectar-feeding hummingbirds. These birds are excellent, specialized fliers with relatively weak (non-muscular) legs that they use mainly for *perching* or *sitting*, and tend to be extremely awkward on foot and sometimes barely capable of walking (https://www.youtube.com/watch?v=pJPGTvzIXRI).

Forelimb-dominated: little leg use. Like hyperaerial birds, forelimb-dominated birds fly to acquire food and are very wing-dependent, with relatively non-muscular legs. However, unlike hyperaerial birds, forelimb-dominated birds use their legs for more than just perching, either to locomote (*hopping* or *climbing* through branches, *walking* on ground) or to dig tunnels for nesting. This category includes birds such as flycatchers (Passeriformes) and kingfishers (Coraciiformes), which forage by "sallying"/"hawking"/"flycatching": flying from a lookout perch to acquire a food item, then returning to the perch. It also includes birds that hunt fish from the air, like terns (Sternidae) and skimmers (Rynchopidae).

Bimodal: moderate leg use. Most birds forage using both their wings and their legs without heavily emphasizing one module over the other. Some bimodal birds rely a little more on their wings, while others rely a little more on their legs. More wing-reliant bimodal birds include animals that forage in trees or undergrowth by *hopping*, *climbing*, or *walking* along branches and frequently flying between branches to glean food (e.g., passerines such as thrushes and many sparrows; woodpeckers, Piciformes; many parrots, Psittaciformes), as well as birds that fly to and from daily foraging areas in large flocks, such as starlings (Passeriformes) and some shorebirds and wading birds. More leg-reliant bimodal birds are birds that forage mainly on foot by *hopping, climbing, skipping, walking, wading,* or *waddling,* and using their wings less frequently to reach adjacent trees or patches (e.g., corvids such as crows; wading birds like herons and egrets; anthrushes, Formicariidae; and antbirds, Thamnophilidae).

Hindlimb-dominated: extensive leg use. Galliforms, tinamous, roadrunners, and rails (Rallidae) locomote and feed almost exclusively by leg. These birds have long and/or powerful hindlimbs and are often excellent runners. They are flight-capable but generally rarely fly, usually only to avoid predators or reach roosts.

Flightless: exclusive leg use. Ostriches, rheas, emus, cassowaries, and kiwis (all ratites), as well as the Flightless Cormorant (Phalacrocoracidae), Kakapo (a parrot), flightless grebes, and a number of rails (Gruiformes) and waterfowl, have completely lost the capacity for flight and rely exclusively on their legs to feed and locomote (penguins, though incapable of aerial flight, use their flippers to "fly" underwater and are considered below). These birds, which are often, but not always, associated with predator-free islands (box on page 300) tend to have reduced wings and wing muscles and often have extremely large and powerful hindlimbs (https://www.youtube.com/watch?v=U2cCV1AP9xc). Ostriches, for example, have the most muscular legs of any terrestrial animal (Hutchinson et al. 2011), and cassowaries can deliver fatal kicks with their legs.

Predatory hindlimbs. Birds that use their feet for killing or holding prey—such as raptors, owls, and vultures—span nearly all of the above categories. In general, birds that handle larger prey have more leg muscle, and birds that handle smaller prey have less leg muscle.

These broad categories of foraging style and wing versus leg use are correlated with a number of morphological features in the hindlimbs, including muscle mass, claw and toe morphology, and bone proportions (Ostrom 1974, Gatesy and Middleton 1997, Hopson 2001, Birn-Jeffery et al. 2012, Heers and Dial 2015). Birds that are more leg-reliant tend to have greater leg muscle mass and less wing muscle mass than birds that are more wing-reliant. In predatory, perching, and climbing birds that use their feet for grasping, the second-to-last phalanx (toe bone) is longer than the toe bone more proximal to it, the claws are curved and sharp, and the hallux (first toe) is long and low. In striding birds such as galliforms, rails, and ratites, the phalanges decrease in length toward the tip of the toe, the claws are less curved and relatively blunt, and the hallux is short and high. With respect to limb proportions (length of femur versus tibiotarsus versus tarsometatarsus), most birds have intermediate proportions except for wading birds and shorebirds, the chicken-like galliforms, flightless ratites, and foot-propelled divers (see below), which all have uniquely proportioned limbs presumably associated with wading, running, and diving styles of locomotion. Although there is substantial overlap between general categories of locomotion, foraging locomotion is thus broadly related to morphology as well as ecology: birds that are more leg-reliant often spend more time on the ground or in water, while birds that are more-wing reliant often spend more time in trees or near cliffs.

Brief Review of the Fossil Record

We have now seen that birds are adept both on the ground and in the air, with a high diversity of locomotor styles and morphologies ranging from more wing-reliant to more leg-reliant. Bipedalism is an extremely important component of this diversity, particularly when viewed through the fossil record. As discussed in chapter 2, the bipedal posture of theropod dinosaurs freed their forelimbs from terrestrial locomotion, and likely facilitated the evolution of flight in paravian theropods and early birds. Once flight was acquired, variation in wing morphology and flight style likely facilitated the evolution of hindlimb diversity. In short, wing- and leg-based locomotion are intimately connected in living birds, and have been throughout their evolutionary history.

AQUATIC LOCOMOTION

So far we have discussed wing- and leg-based locomotion in the context of aerial and terrestrial environments. Many birds also use their wings and/or legs for swimming. In fact, swimming is common in a number of avian groups, such as waterfowl, grebes, loons, charadriiforms (e.g., shorebirds, gulls, auks), cormorants, pelicans (Pelecaniformes), dippers (Passeriformes), and penguins (Sphenisciformes). Distinguishing features (Hinic-Frlog and Motani 2010, Elliott et al. 2013) of these groups often include:

- webbed feet (paddles)
- narrow (drag-reducing) foot and toe bones
- posteriorly positioned feet, due to a posteriorly positioned hip joint and short, posteriorly curved femur (better propulsion underwater)

FLIGHTLESS BIRDS

Though we often associate birds with flight, flightless birds are actually rather common, occurring in many different avian groups. Some birds are **seasonally flightless**, temporarily losing the ability to fly when they molt their feathers (e.g., eclipse plumage of waterfowl) or lay their eggs (e.g., cavity-nesting hornbills, Bucerotidae) (Roots 2006). Other birds are permanently or **secondarily flightless** (e.g., ratites such as ostriches and emus; penguins, Sphenisciformes; and more than 40 other extant species from more than 10 families, along with several extinct groups).

Why become permanently flightless? Energy conservation—particularly in habitats with limited resources—might be a major incentive to give up flight capacity, because flight is a very energetically costly activity. Indeed, flightless birds have reduced pectoral muscles and lower metabolic rates than flight-capable relatives (McNab 1994), and these energy savings are probably advantageous under a very specific set of conditions in which the benefits of flight would not outweigh the costs:

- *Lack of predators, or at least extremely reduced rates of predation.* Many birds fly to escape predators, but a lack of predators could reduce this selective pressure.
- *Relatively mild climates.* Many birds migrate to avoid harsh weather; flightless birds cannot traverse great distances and must be able to survive in one area year-round.
- *Year-round supply of food.* Many birds also migrate to avoid seasonal declines in food availability, so flightlessness requires a continuous supply of food.

These conditions are most often met in remote islands, and with the exception of ratites (which evolved large body size and fast running for predator avoidance) and aquatic groups like penguins (which are not restricted to terrestrial habitats), flightless birds are most common in such areas (see *Key Management and Conservation Implications*).

- small and sturdy wings (less surface area = easier to flap underwater)
- less mobile (more rigid) wing joints, flattened wing bones (penguins only)
- muscles optimized for low wingbeat frequencies (penguins only; flap more slowly underwater)

- wide, heavy bodies (aid floatation, reduce buoyancy while diving)
- oily feathers (waterproofing)

Some of these features interfere with aerial flight (heavy bodies, small and rigid wings, low wingbeat frequencies) or terrestrial locomotion (posteriorly positioned feet) (Raikow et al. 1988, Elliott et al. 2013). Consequently, many specialized diving birds, such as penguins, are secondarily flightless and/ or somewhat awkward on land (https://www.youtube.com /watch?v=COA_vgQjR6s&t=10s; https://www.youtube.com /watch?v=Kq-tGVPKMTY). Though a number of swimming birds are able to balance the demands of aerial flight, terrestrial locomotion, and swimming, there is a cost: better swimmers often have trouble walking or taking off, and often are not very maneuverable once airborne (e.g., loons), whereas better walkers and aerial fliers are not specialized divers (e.g., dippers, some ducks). Different species therefore use different styles of swimming, depending on their evolutionary history and requirements for terrestrial or aerial locomotion.

Surface Swimming: Paddling, Steaming, and Taxiing

Swimming at the surface of the water may seem simple—ducks, seagulls, pelicans, and many other birds float with ease. But surface swimming involves its own set of challenges, because of the costs of making waves on the surface of the water. When an animal swims at the air-water interface, it must constantly impart momentum to the water to push water out of the way as it moves forward, creating surface waves (Aigeldinger and Fish 1995). Surface waves limit the speed at which an animal can travel because the animal essentially becomes trapped in a trough of water behind the very wave it is generating. Many animals therefore avoid swimming at the surface of the water, but birds do not—they use four different styles of surface swimming, depending on speed and body size.

At slow speeds, surface swimming is simply **paddling**: floating on the surface of the water and pushing the feet rapidly backward through the water for propulsion. Propulsion is drag-based at the beginning of the power stroke, meaning that, as the feet push backward on the water parallel to the direction of travel, the water "pushes back" and propels the bird forward, similar to the wheel of a paddle boat. Later in the power stroke, some lift-production occurs as well, as the foot translates toward the body midline and becomes oriented like a delta-wing (fig. 10.19A) (Johansson and Norberg 2003).

To achieve higher speeds, birds lift their bodies above the surface of the water and **hydroplane**. If the bird is small or buoyant enough, sufficient lift can be generated solely by foot paddling (Aigeldinger and Fish 1995). Larger birds rely on

steaming (fig. 10.19B; third and fourth bird in http://jeb.biol ogists.org/content/jexbio/suppl/2012/10/07/215.21.3703.DC1 /Movie2.mov), which is similar to hydroplaning in that the feet are still used to paddle, but additional drag-based propulsion is provided by the wings dipping in and out of the water like oars, allowing the bird to elevate its body and hydroplane or skim the surface of the water at higher speeds (Livezey and Humphrey 1983, Dial and Carrier 2012). **Paddle-assisted flying/skittering/water taxiing** (fig. 10.19C; https://www .youtube.com/watch?v=sRIXOEuTmcE) occurs when birds run across the surface of the water while flapping their wings through the air to generate aerodynamic forces. Taxiing differs from paddling and steaming in two respects. First, the wings do not contact the water, and generate aerodynamic, rather than hydrodynamic, forces. Second, the recovery stroke of the feet occurs above the surface of the water, rather than below. Consequently, weight support (no longer provided by buoyancy) and propulsion occur not only from the feet paddling through the water but also from the feet slapping the surface of the water, like a Basilisk Lizard (Glasheen and McMahon 1996, Clifton et al. 2015).

Though not as well studied as flight, these surface swimming behaviors are quite common: most swimming birds paddle and steam, whereas taxiing is mainly—but not exclusively—used by larger birds to gain speed for takeoff (https://www.youtube .com/watch?v=pcnUPJoh15c).

Underwater Swimming: Wing- and Foot-Propelled Diving

Underwater swimming is used by many birds to avoid terrestrial or aerial predators and by more specialized diving birds to forage for food. Recall that aerial flight requires birds to generate enough aerodynamic force to (1) support their body weight (counteract gravity) and (2) propel themselves forward (counteract drag). Though water and air are both fluids governed by fluid dynamics (aerodynamic forces in air, analogous to hydrodynamic forces in water), water is ~850 times denser than air, making drag far more problematic and buoyancy more important than gravity. Aquatic locomotion thus differs from aerial locomotion in several important respects. Underwater swimming requires birds to generate enough hydrodynamic force to (1) counteract buoyancy, which may be positive or negative depending on species and water depth; (2) provide forward propulsion (counteract drag, which is substantial); and (3) overcome the inertial resistance of water (water moves *with* oscillating wings or feet and "adds mass"). These demands are met in different ways in different groups of birds.

Wing-propelled divers (fig. 10.19D; https://www.you tube.com/watch?v=86nt3yCkbQ4), such as puffins (Alcidae) and penguins, swim underwater by flapping their wings to

A. Paddling

i

ii

Body motion
Foot motion
Net foot motion

Body moves
forward

Drag

Foot moves
toward body
midline

Lift

Drag

Foot moves
backward faster
than body moves
forward

Drag: antiparallel to net foot motion (foot motion ‖ to body motion)
Lift: perpendicular to net foot motion (foot motion ⊥ to body motion)

B. Steaming

Drag from
feet and
wings

C. Taxiing

Aerodynamic
forces

Foot slap then
paddle (drag)

D. Wing-propelled diving

Downstroke

Hydrodynamic forces produced by wings

Upstroke

E. Foot-propelled diving

Hydrodynamic forces produced by legs

foot
motion

Grebes, loons(?):
lift-based propulsion

Cormorants, ducks:
drag-based propulsion

foot
motion

Figure 10.19. Different types of surface and underwater swimming. *A*, Paddling duckling, showing how foot motion generates hydrodynamic force. *Duckling photo in public domain, http://i.imgur.com/f4Cv602 .jpg; diagrams based on Johansson and Norberg 2003. B*, Steamer Duck (*Tachyeres*) steaming. *Copyright © Arthur Grosset, www.arthurgrosset.com, with permission.* *C*, Bufflehead (*Bucephala*) taxiing. *Copyright © Sandra Calderbank, scalderphotography.com, with permission. D*, Murre (*Uria*) using wings to swim underwater. *Modified after Lovvorn 2001. E*, Great Crested Grebe (*Podiceps cristatus*) and Double-crested Cormorant (*Phalacrocorax auritus*) using feet to swim underwater. *Modified after Johansson and Lindhe Norberg 2001.* Note that the magnitudes and directions of aerodynamic and hydrodynamic forces (lift and/or drag; blue arrows) are approximations only; lift and drag vary throughout the stroke cycle and across species, and change with water depth. Dashed blue arrow in *D* indicates that, in wing-propelled divers, force production is highly variable during the upstroke.

generate hydrodynamic force to counteract buoyancy and provide propulsion. Buoyancy is most problematic at shallow water depths (Lovvorn 2001), and in the case of buoyant, shallow divers, the downstroke must actually produce a downward directed force to overcome the buoyancy of the animal. Otherwise, underwater flapping kinematics are fairly similar to aerial flapping, except that movements are more pulsed, and in alcids the wings are kept partly folded during the downstroke because water is much denser than air and hydrodynamic forces are much larger (see equations 1 and 2) (https://www.youtube.com/watch?v=bWP5SU7vhWE). The greater density of water also means that the aquatic up-

stroke, which is often achieved passively during aerial flight, requires more effort. Consequently, the supracoracoideus muscle (figs. 10.5, 10.13) of wing-propelled diving birds is proportionally quite large, compared with birds that do not flap their wings underwater.

Foot-propelled divers (fig. 10.19E), like grebes, loons, cormorants, and diving ducks, rely on their feet for underwater propulsion. Ducks and cormorants oscillate their feet in a vertical plane parallel to the direction of travel (https:// www.youtube.com/watch?v=K7aUroa762o), like a paddle wheel, and probably rely mainly on drag-based propulsion. Loons and grebes oscillate their feet in a horizontal plane (https://www.youtube.com/watch?v=LfP2rqDQovI), like the oar of a kayak, and probably generate lift as well as drag (Johansson and Lindhe Norberg 2001, Hinic-Frlog and Motani 2010). Some birds, such as eiders and scoters, use both their wings and their legs to swim underwater.

In summary, swimming is a common mode of locomotion among birds. Different species rely on aquatic environments to different extents, and the most specialized swimmers have given up the ability to fly or walk well. Neither surface nor underwater swimming are as well studied as aerial or terrestrial locomotion, and future studies can explore the competing constraints and advantages of swimming versus other forms of locomotion in birds that divide their time between air, land, and water.

KEY POINTS

- Birds are common occupants of many environments and, among vertebrates, are uniquely capable of navigating aerial, terrestrial, and aquatic media. Such versatility is likely a product of **wing-leg modularity**, which contributes to incredible diversity in locomotor styles that in turn are related to life history strategies, predator avoidance tactics, habitat preferences, foraging strategies, and migratory patterns (fig. 10.20). When we think about avian anatomy, locomotion, and ecology, it is therefore crucial to consider not only wings and aerial environments, but also legs and terrestrial and aquatic environments. Some birds are more wing-reliant and some are more leg-reliant, but in no bird are wings and legs completely independent of one another: wings and legs are often used **cooperatively**, and **trade-offs** between wings and legs likely influence avian development and evolution.

- Aerial Locomotion. Powered, flapping flight is a unique and extremely power-demanding form of locomotion, with several requirements: an airfoil or wing, an engine, a flight stroke, coordination and control, relatively small body size, and fuel. Birds meet these six requirements by having (1) **large, feathered wings** that generate aerodynamic forces to counteract body weight and drag; (2) **large proximal muscles** that provide power to pump the wing up and down and smaller distal muscles that help position the wing and fine-tune its movements; (3) a **robust forelimb skeleton** with specialized joints allowing for complex but adjustable flapping kinematics; (4) excellent vision and balance, coupled with **large brains** capable of processing a rapid flux of motion-related information; (5) **low body weight**, mainly because of small body size but also facilitated by weight-reducing anatomical features; and (6) highly **efficient cardiovascular and respiratory systems** that help fuel the high metabolic demands of flight. Given flight's stringent requirements, birds are nevertheless surprisingly diverse with respect

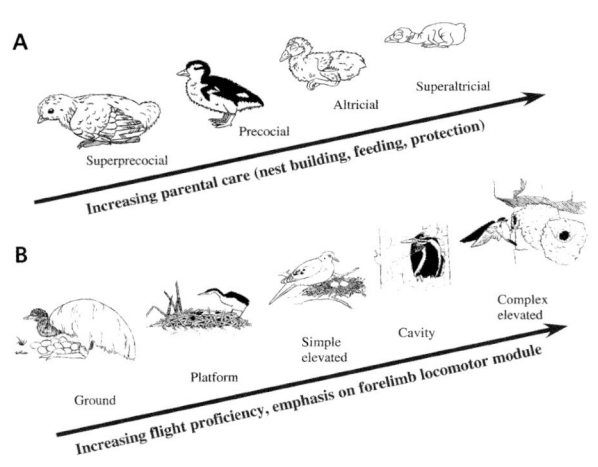

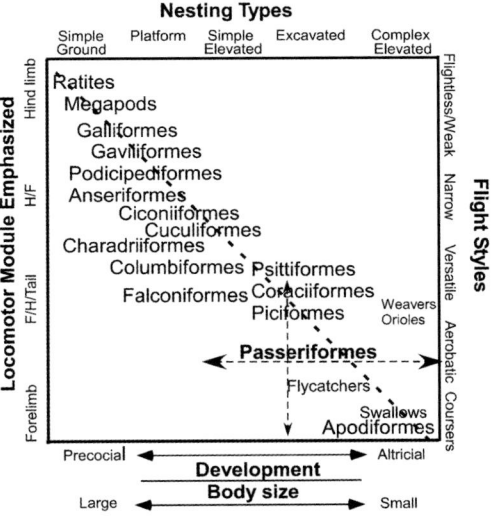

Figure 10.20. Locomotion and other aspects of avian biology. Avian locomotion is related to body size, life history strategy, foraging style, habitat preferences, and migratory capacity. Smaller birds tend to emphasize their wing module more and are better fliers, which may require altricial development and high levels of parental care. *Images reproduced, with permission, from Dial 2003a.*

to wing shape, musculoskeletal morphology, upstroke kinematics, and the degree of flapping. At least part of this variation may be related to variation in terrestrial or aquatic locomotion.

- Terrestrial Locomotion. Whether escaping from predators, foraging, or taking off for flight, leg-based locomotion is an important component of avian locomotion. Like humans, birds are **obligate bipeds** with an erect (non-sprawling) stance. Though the biomechanics of avian and human bipedalism are comparable, birds differ from humans in standing on their toes (**digitigrade**), keeping a relatively crouched (flexed-knee) posture, and having most joint movement occur at the knee rather than at the hip as in humans.
- Aquatic Locomotion. Surface and underwater swimming are common modes of locomotion among birds. Different species rely on aquatic environments to different extents, and the most specialized swimmers have given up the ability to fly or walk well.

KEY MANAGEMENT AND CONSERVATION IMPLICATIONS

- Across avian species, the relative emphasis of wings versus legs broadly correlates with a number of characteristics that have important implications for management and conservation (Dial 2003a, Heers and Dial 2015), including:
  - *Life history.* Birds that are more dependent on their wings for locomotion tend to have altricial development with high levels of parental care, so that chicks are fed and protected until their wings develop (e.g., passerine songbirds).
  - *Habitat preferences.* Birds that rely heavily on their wings for locomotion can be awkward on the ground, and often nest in shrubs/trees (e.g., hummingbird) or on remote, predator-free island habitats (e.g., albatross, Procellariiformes).
  - *Foraging strategies.* Birds specialized for hunting in the air tend to have reduced legs (e.g., swifts, Apodiformes, meaning "no feet")—presumably to reduce their "leg baggage" and improve flight performance.
  - *Migratory patterns.* Migratory birds tend to be more "wingy" and less "leggy" than nonmigratory relatives.
- Though birds are often associated with flight, in many species, adults have **secondarily reduced wings** and fly poorly, rarely, or not at all. Similarly, at **hatching**, nearly all birds are flight-incapable and entirely dependent on their legs for locomotion. These two groups

are among the most vulnerable, yet least studied, birds. For example, the relatively recent introduction of exotic predators—including humans—to many islands has taken a major toll on flightless birds. At least 50 species have been driven extinct since 1600 (Roots 2006), and this is a major and ongoing area of research in conservation biology (e.g., Wright et al. 2016).

DISCUSSION QUESTIONS

1. What is a locomotor module, and how does locomotor modularity contribute to avian diversity?
2. What anatomical features help birds meet the requirements for powered flight? How would you assess whether an extinct animal could fly or not? (Hint: see the box on page 281.)
3. What factors influence how much lift and drag a bird produces during flight?
4. What are the two main "flight" muscles in a bird? What do they do, and why are they so big?
5. Some types of flight are more challenging than others (e.g., ascending flight > level > descending). What are some of the ways that birds modulate power and aerodynamic output to achieve these different types of flight?
6. Describe the different styles of flight (flapping versus gliding versus . . .), and the groups of birds that use them.
7. Not all birds fly—some are seasonally or secondarily flightless. Why do you think we use the term "secondarily" flightless? What conditions might favor the loss of flight?
8. What are some of the similarities and differences in the bipedal locomotion of humans and birds?
9. What are some of the challenges of aquatic locomotion, and how do they differ from the challenges of aerial locomotion? How have different species adapted to these conflicting demands?

References

Aigeldinger, T., and F. Fish (1995). Hydroplaning by ducklings: Overcoming limitations to swimming at the water surface. Journal of Experimental Biology 198:1567–1574.

Alexander, R. M. (2002). The merits and implications of travel by swimming, flight and running for animals of different sizes. Integrative and Comparative Biology 42:1060–1064. doi:10.1093/icb/42.5.1060.

Altshuler, D. L., R. Dudley, S. M. Heredia, and J. A. McGuire (2010). Allometry of hummingbird lifting performance. Journal of Experimental Biology 213:725–734. doi:10.1242/jeb.037002.

Askew, G., and D. Ellerby (2007). The mechanical power requirements of avian flight. Biology Letters 3:445–448. doi:10.1098/rsbl.2007.0182.

Baier, D. B., S. M. Gatesy, and K. P. Dial (2013). Three-dimensional, high-resolution skeletal kinematics of the avian wing and shoulder during ascending flapping flight and uphill flap-running. PLoS ONE 8:e63982.

Biewener, A. A. (1998). Muscle function in vivo: A comparison of muscles used for elastic energy savings versus muscles used to generate mechanical power 1. American Zoologist 38:703–717. doi:10.1093/icb/38.4.703.

Biewener, A. A. (2003). Animal locomotion. Oxford University Press, USA.

Biewener, A. A. (2011). Muscle function in avian flight: Achieving power and control. Philosophical Transactions of the Royal Society of London B: Biological Sciences 366:1496–1506. doi:10.1098/rstb.2010.0353.

Biewener, A. A., and T. J. Roberts (2000). Muscle and tendon contributions to force, work, and elastic energy savings: A comparative perspective. Exercise and Sports Sciences Reviews 28:99–107.

Birn-Jeffery, A. V., C. E. Miller, D. Naish, E. J. Rayfield, and D. W. E. Hone (2012). Pedal claw curvature in birds, lizards and Mesozoic dinosaurs: Complicated categories and compensating for mass-specific and phylogenetic control. PLoS ONE 7:e50555.

Bruderer, B., D. Peter, A. Boldt, and F. Liechti (2010). Wing-beat characteristics of birds recorded with tracking radar and cine camera. Ibis 152:272–291. doi:10.1111/j.1474919X.2010.01014.x.

Chai, P., and D. Millard (1997). Flight and size constraints: Hovering performance of large hummingbirds under maximal loading. Journal of Experimental Biology 200:2757–2763.

Clifton, G. T., T. L. Hedrick, and A. A. Biewener (2015). Western and Clark's grebes use novel strategies for running on water. Journal of Experimental Biology 218:1235–1243. doi:10.1242/jeb.118745.

Crandell, K. E., and B. W. Tobalske (2015). Kinematics and aerodynamics of avian upstrokes during slow flight. Journal of Experimental Biology 218:2518–2527. doi:10.1242/jeb.116228.

Dial, K. P. (1992a). Activity patterns of the wing muscles of the Pigeon (Columba livia) during different modes of flight. Journal of Experimental Zoology 262:357–373.

Dial, K. P. (1992b). Avian forelimb muscles and nonsteady flight: Can birds fly without using the muscle in their wings? The Auk 109:874–885.

Dial, K. P. (2003a). Evolution of avian locomotion: Correlates of flight style, locomotor modules, nesting biology, body size, development, and the origin of flapping flight. The Auk 120:941–952. doi:10.1642/0004 8038(2003)120[0941:EOALCO]2.0.CO;2.

Dial, K. P. (2003b). Wing-assisted incline running and the evolution of flight. Science 299:402–404. doi:10.1126/science.1078237.

Dial, K. P. (2011). From extant to extinct: Empirical studies of transitional forms and allometric correlates delimit boundaries of functional capacity. Journal of Vertebrate Paleontology 31:99.

Dial, K. P., R. J. Randall, and T. R. Dial (2006). What use is half a wing in the ecology and evolution of birds? BioScience 56:437–445. doi:10.1641/0006 3568(2006)056[0437:WUIHAW]2.0.CO;2.

Dial, K. P., E. Greene, and D. J. Irschick (2008a). Allometry of behavior. Trends in Ecology and Evolution 23:394–401. doi:10.1016/j.tree.2008.03.005.

Dial, K. P., B. E. Jackson, and P. Segre (2008b). A fundamental avian wing-stroke provides a new perspective on the evolution of flight. Nature 451:985–989. doi:10.1038/nature06517.

Dial, T. R., and D. R. Carrier (2012). Precocial hindlimbs and altricial forelimbs: Partitioning ontogenetic strategies in Mallard Ducks (Anas platyrhynchos). Journal of Experimental Biology 215:3703–3710. doi:10.1242/jeb.057380.

Dial, T. R., A. M. Heers, and B. W. Tobalske (2012). Ontogeny of aerodynamics in mallards: Comparative performance and developmental implications. Journal of Experimental Biology 215:3693–3702. doi:10.1242/jeb.062018.

Dial, K. P., A. M. Heers, and T. R. Dial (2015). Ontogenetic and evolutionary transformations: The ecological significance of rudimentary structures. In Great transformations in vertebrate evolution, K. P. Dial, N. Shubin, and E. L. Brainerd, Editors. University of Chicago Press, Chicago, pp. 283–301.

Dickinson, M. H., C. T. Farley, R. J. Full, M. A. R. Koehl, R. Kram, and S. Lehman (2000). How animals move: An integrative view. Science 288:100–106. doi:10.1126/science.288.5463.100.

Dumont, E. R. (2010). Bone density and the lightweight skeletons of birds. Proceedings of the Royal Society B: Biological Sciences 277:2193–2198. doi:10.1098/rspb.2010.0117.

Earls, K. D. (2000). Kinematics and mechanics of ground take-off in the starling Sturnis vulgaris and the quail Coturnix coturnix. Journal of Experimental Biology 203:725–739.

Elliott, K. H., R. E. Ricklefs, A. J. Gaston, S. A. Hatch, J. R. Speakman, and G. K. Davoren (2013). High flight costs, but low dive costs, in auks support the biomechanical hypothesis for flightlessness in penguins. Proceedings of the National Academy of Sciences 110:9380–9384. doi:10.1073/pnas.1304838110.

Gatesy, S. M. (1999). Guineafowl hind limb function. I: Cineradiographic analysis and speed effects. Journal of Morphology 240:115–125. doi:10.1002/(SICI)1097 4687(199905)240:2<115::AID-JMOR3>3.0.CO;2 Y.

Gatesy, S. M., and A. A. Biewener (1991). Bipedal locomotion: Effects of speed, size and limb posture in birds and humans. Journal of Zoology 224:127–147. doi:10.1111/j.1469 7998.1991.tb04794.x.

Gatesy, S. M., and K. P. Dial (1996). Locomotor modules and the evolution of avian flight. Evolution 50:331–340.

Gatesy, S. M., and K. M. Middleton (1997). Bipedalism, flight, and the evolution of theropod locomotor diversity. Journal of Vertebrate Paleontology 17:308–329.

Glasheen, J. W., and T. A. McMahon (1996). A hydrodynamic model of locomotion in the Basilisk Lizard. Nature 380:340–342.

Greenewalt, C. H. (1960). The wings of insects and birds as mechanical oscillators. Proceedings of the American Philosophical Society 104:605–611.

Greenewalt, C. H. (1975). The flight of birds: The significant dimensions, their departure from the requirements for dimensional similarity, and the effect on flight aerodynamics of that departure. Transactions of the American Philosophical Society 65:1–67.

Gutzwiller, S. C., A. Su, and P. M. O'Connor (2013). Postcranial pneumaticity and bone structure in two clades of neognath birds. Anatomical Record 296:867–876. doi:10.1002/ar.22691.

Habib, M. B., and C. B. Ruff (2008). The effects of locomotion on the structural characteristics of avian limb bones. Zoological Journal of the Linnean Society 153:601–624.

Hancock, J. A., N. J. Stevens, and A. R. Biknevicius (2007). Whole-body mechanics and kinematics of terrestrial locomotion in the Elegant-crested Tinamou *Eudromia elegans*. Ibis 149:605–614. doi:10.1111/j.1474 919X.2007.00688.x.

Hartman, F. A. (1961). Locomotor mechanisms of birds. Smithsonian Miscellaneous Collections 143:1–91.

Hedrick, T. L., B. W. Tobalske, and A. A. Biewener (2002). Estimates of circulation and gait change based on a three-dimensional kinematic analysis of flight in cockatiels (*Nymphicus hollandicus*) and ringed turtle-doves (*Streptopelia risoria*). Journal of Experimental Biology 205:1389–1409.

Hedrick, T. L., J. R. Usherwood, and A. A. Biewener (2004). Wing inertia and whole-body acceleration: An analysis of instantaneous aerodynamic force production in cockatiels (*Nymphicus hollandicus*) flying across a range of speeds. Journal of Experimental Biology 207:1689–1702. doi:10.1242/jeb.00933.

Heers, A. M. (2013). Evolution of avian flight. In eLS. John Wiley. doi:10.1002/9780470015902.a0024965.

Heers, A. M. (2016). New perspectives on the ontogeny and evolution of avian locomotion. Integrative and Comparative Biology 56:428–444. doi:10.1093/icb/icw065.

Heers, A. M., and K. P. Dial (2012). From extant to extinct: Locomotor ontogeny and the evolution of avian flight. Trends in Ecology and Evolution 27:296–305. doi:10.1016/j.tree.2011.12.003.

Heers, A. M., and K. P. Dial (2015). Wings versus legs in the avian bauplan: Development and evolution of alternative locomotor strategies. Evolution 69:305–320. doi:10.1111/evo.12576.

Heers, A. M., D. B. Baier, B. E. Jackson, and K. P. Dial (2016). Flapping before flight: High resolution, three-dimensional skeletal kinematics of wings and legs during avian development. PLoS ONE 11:e0153446. doi:10.1371/journal.pone.0153446.

Hill, A. V. (1950). The dimensions of animals and their muscular dynamics. Science Progress 38:209–230.

Hinic-Frlog, S., and R. Motani (2010). Relationship between osteology and aquatic locomotion in birds: Determining modes of locomotion in extinct Ornithurae. Journal of Evolutionary Biology 23:372–385.

Hopson, J. A. (2001). Ecomorphology of avian and nonavian theropod phalangeal proportions: Implications for the arboreal versus terrestrial origin of bird flight. *In* New perspectives on the origin and early evolution of birds: Proceedings of the International Symposium in Honor of John H. Ostrom, J. Gauthier and L. F. Gall, Editors. Peabody Museum of Natural History, Yale University, New Haven, CT.

Husband, S., and T. Shimizu (2001). Evolution of the avian visual system. *In* Avian visual cognition, R. G. Cook, Editor. Comparative Cognition Press, Massachusetts. http://www.pigeon.psy.tufts.edu/avc/husband/.

Hutchinson, J. R., and S. M. Gatesy (2000). Adductors, abductors, and the evolution of archosaur locomotion. Paleobiology 26:734–751. doi:10.1666/0094 8373(2000)026<0734:AAATEO>2.0.CO;2.

Hutchinson, J. R., and S. M. Gatesy (2001). Bipedalism. In eLS. John Wiley. doi:10.1038/npg.els.0001869.

Hutchinson, J. R., K. T. Bates, J. Molnar, V. Allen, and P. J. Makovicky (2011). A computational analysis of limb and body dimensions in *Tyrannosaurus rex* with implications for locomotion, ontogeny, and growth. PLoS ONE 6:e26037. doi:10.1371/journal.pone.0026037.

Jackson, B. E. (2009). The allometry of bird flight performance. University of Montana Press, Missoula.

Jackson, B. E., and K. P. Dial (2011). Scaling of mechanical power output during burst escape flight in the Corvidae. Journal of Experimental Biology 214:452–461. doi:10.1242/jeb.046789.

Jackson, B. E., B. W. Tobalske, and K. P. Dial (2011). The broad range of contractile behaviour of the avian pectoralis: Functional and evolutionary implications. Journal of Experimental Biology 214:2354–2361. doi:10.1242/jeb.052829.

Jenkins, F. A. (1993). The evolution of the avian shoulder joint. American Journal of Science 293:253–267. doi:10.2475/ajs.293.A.253.

Johansson, L. C., and U. M. Lindhe Norberg (2001). Lift-based paddling in diving grebe. Journal of Experimental Biology 204:1687–1696.

Johansson, L. C., and R. A. Norberg (2003). Delta-wing function of webbed feet gives hydrodynamic lift for swimming propulsion in birds. Nature 424:65–68. doi:10.1038/nature01695.

Konow, N., J. A. Cheney, T. J. Roberts, J. R. S. Waldman, and S. M. Swartz (2015). Spring or string: Does tendon elastic action influence wing muscle mechanics in bat flight? Proceedings of the Royal Society of London B: Biological Sciences 282:20151832.

Lasiewski, R. C. (1964). Body temperatures, heart and breathing rate, and evaporative water loss in hummingbirds. Physiological Zoology 37:212–223.

Livezey, B. C., and P. S. Humphrey (1983). Mechanics of steaming in steamer-ducks. The Auk 100:485–488.

Lovvorn, J. R. (2001). Upstroke thrust, drag effects, and stroke-glide cycles in wing-propelled swimming by birds. American Zoologist 41:154–165. doi:10.1668/0003 1569(2001)041[0154:UTDEAS]2.0.CO;2.

Marden, J. H. (1987). Maximum lift production during take-off in flying animals. Journal of Experimental Biology 130:235–258.

Marden, J. H. (1994). From damselflies to pterosaurs: How burst and sustainable flight performance scale with size. American Journal of Physiology: Regulatory, Integrative and Comparative Physiology 266:R1077–R1084.

Maybury, W. J., and J. M. Rayner (2001). The avian tail reduces body parasite drag by controlling flow separation and vortex shedding. Proceedings of the Royal Society B: Biological Sciences 268:1405–1410. doi:10.1098/rspb.2001.1635.

McNab, B. K. (1994). Energy conservation and the evolution of flightlessness in birds. The American Naturalist 144:628–642.

McWilliams, S. R., and W. H. Karasov (2005). Migration takes guts: Digestive physiology of migratory birds and its ecological significance. *In* Birds of two worlds: The ecology and evolution of migration, R. Greenberg and P. P. Marra, Editors. Johns Hopkins University Press, Baltimore, MD, pp. 67–78.

Muller, W., and G. Patone (1998). Air transmissivity of feathers. Journal of Experimental Biology 201:2591–2599.

Norberg, R. A. (1985a). Function of vane asymmetry and shaft curvature in bird flight feathers: Inferences on flight ability of Archaeopteryx. *In* The beginnings of birds, J. H. Ostrom, M. K. Hecht, G. Viohl, and P. Wellnhofer, Editors. Jura Museum, Eichstatt, pp. 303–318.

Norberg, U. M. (1985b). Flying, gliding, and soaring. *In* Functional vertebrate morphology, M. Hildebrand, D. M. Bramble, K. F. Liem, and D. B. Wake, Editors. Harvard University Press, Cambridge, MA, pp. 129–158.

Norberg, R. Å., and U. M. Norberg (1971). Take-off, landing, and flight speed during fishing flights of *Gavia stellata* (Pont.). Ornis Scandinavica 2:55–67.

Nudds, R. L., and G. J. Dyke (2010). Narrow primary feather rachises in Confuciusornis and Archaeopteryx suggest poor flight ability. Science 328:887–889. doi:10.1126/science.1188895.

Nyakatura, J. A., E. Andrada, N. Grimm, H. Weise, and M. S. Fischer (2012). Kinematics and center of mass mechanics during terrestrial locomotion in Northern Lapwings (*Vanellus vanellus*, Charadriiformes). Journal of Experimental Zoology Part A: Ecological Genetics and Physiology 317:580–594. doi:10.1002/jez.1750.

Ostrom, J. H. (1974). Archaeopteryx and the origin of flight. Quarterly Review of Biology 49:27–47.

Pennycuick, C. J. (1968). Power requirements for horizontal flight in the pigeon *Columba livia*. Journal of Experimental Biology 49:527–555.

Pennycuick, C. J. (1975). Mechanics of flight. Journal of Avian Biology 5:1–75.

Pennycuick, C. J. (1982). The flight of petrels and albatrosses (Procellariiformes), observed in South Georgia and its vicinity. Philosophical Transactions of the Royal Society of London. Series B: Biological Sciences 300:75–106.

Pete, A. E., D. Kress, M. A. Dimitrov, and D. Lentink (2015). The role of passive avian head stabilization in flapping flight. Journal of the Royal Society Interface 12:20150508. http://doi.org/10.1098/rsif.2015.0508.

Poore, S. O., A. Sanchez-Haiman, and G. E. Goslow (1997). Wing upstroke and the evolution of flapping flight. Nature 387:799–802.

Prange, H. D., J. F. Anderson, and H. Rahn (1979). Scaling of skeletal mass to body mass in birds and mammals. American Naturalist 113:103–122.

Raikow, R. J., Lesley Bicanovsky, and A. H. Bledsoe (1988). Forelimb joint mobility and the evolution of wing-propelled diving in birds. The Auk 105:446–451.

Rayner, J. M. V. (1988). Form and function in avian flight. Current Ornithology 5:1–66.

Reilly, S. M. (2000). Locomotion in the quail (*Coturnix japonica*): The kinematics of walking and increasing speed. Journal of Morphology 243:173–185. doi:10.1002/(SICI)1097 4687(200002)243:2<173::AID-JMOR6>3.0.CO;2 E.

Remsen, J. V., and S. K. Robinson (1990). A classification scheme for foraging behavior of birds in terrestrial habitats. Studies in Avian Biology 13:144–160.

Roberts, T. J., R. L. Marsh, P. G. Weyand, and C. R. Taylor (1997). Muscular force in running turkeys: The economy of minimizing work. Science 275:1113–1115. doi:10.1126/science.275.5303.1113.

Roberts, T. J., and J. A. Scales (2004). Adjusting muscle function to demand: Joint work during acceleration in wild turkeys. Journal of Experimental Biology 207:4165–4174. doi:10.1242/jeb.01253.

Roots, C. (2006). Flightless birds. Greenwood Press, CT.

Ros, Ivo G., Lori C. Bassman, Marc A. Badger, Alyssa N. Pierson, and Andrew A. Biewener (2011). Pigeons steer like helicopters and generate down- and upstroke lift during low speed turns. *PNAS* 108 (50): 19990–19995.

Sachs, G. (2005). Minimum shear wind strength required for dynamic soaring of albatrosses. Ibis 147:1–10.

Scanes, C. C., Editor. (2015). Sturkie's avian physiology. 6th ed. Elsevier, San Diego.

Tobalske, B. (1995). Neuromuscular control and kinematics of intermittent flight in the European starling (*Sturnus vulgaris*). Journal of Experimental Biology 198:1259–1273.

Tobalske, B. W. (2000). Biomechanics and physiology of gait selection in flying birds. Physiological and Biochemical Zoology 73:736–750.

Tobalske, B. W. (2001). Morphology, velocity, and intermittent flight in birds. American Zoologist 41:177–187. doi:10.1668/000 3 1569(2001)041[0177:MVAIFI]2.0.CO;2.

Tobalske, B. W. (2007). Biomechanics of bird flight. Journal of Experimental Biology 210:3135–3146. doi:10.1242/jeb.000273.

Tobalske, B. W. (2010). Hovering and intermittent flight in birds. Bioinspiration and Biomimetics 5:45004.

Tobalske, B., and K. Dial (1994). Neuromuscular control and kinematics of intermittent flight in Budgerigars (*Melopsittacus undulatus*). Journal of Experimental Biology 187:1–18.

Tobalske, B. W., and A. A. Biewener (2008). Contractile properties of the pigeon supracoracoideus during different modes of flight. Journal of Experimental Biology 211:170–179.

Tobalske, B. W., W. L. Peacock, and K. P. Dial (1999). Kinematics of flap-bounding flight in the Zebra Finch over a wide range of speeds. Journal of Experimental Biology 202:1725–1739.

Tobalske, B. W., T. L. Hedrick, and A. A. Biewener (2003a). Wing kinematics of avian flight across speeds. Journal of Avian Biology 34:177–184. doi:10.1034/j.1600 048X.2003.03006.x.

Tobalske, B. W., T. L. Hedrick, K. P. Dial, and A. A. Biewener (2003b). Comparative power curves in bird flight. Nature 421:363–366. doi:10.1038/nature01284.

Tobalske, B. W., D. L. Altshuler, and D. R. Powers (2004). Take-off mechanics in hummingbirds (Trochilidae). Journal of Experimental Biology 207:1345–1352. doi:10.1242/jeb.00889.

Tobalske, B. W., D. R. Warrick, C. J. Clark, D. R. Powers, T. L. Hedrick, G. A. Hyder, and A. A. Biewener (2007). Three-dimensional kinematics of hummingbird flight. Journal of Experimental Biology 210:2368–2382. doi:10.1242/jeb.005686.

Tobalske, B. W., J. W. D. Hearn, and D. R. Warrick (2009). Aerodynamics of intermittent bounds in flying birds. Experimental Fluids 46:963–973.

van Grouw, K. (2013). The unfeathered bird. Princeton University Press, Princeton, NJ.

Vazquez, R. J. (1992). Functional osteology of the avian wrist and the evolution of flapping flight. Journal of Morphology 211:259–268.

Videler, J. J., E. J. Stamhuis, and G. D. E. Povel (2004). Leading-Edge Vortex Lifts Swifts. Science 306:1960–1962. doi:10.1126/science.1104682.

Vogel, S. (2013). Comparative biomechanics: Life's physical world. 2nd ed. Princeton University Press, Princeton, NJ.

Warrick, D. R., M. W. Bundle, and K. P. Dial (2002). Bird maneuvering flight: Blurred bodies, clear heads. Integrative and Comparative Biology 42:141–148.

Warrick, D. R., B. W. Tobalske, and D. R. Powers (2005). Aerodynamics of the hovering hummingbird. Nature 435:1094–1097. doi:10.1038/nature03647.

Wright, N. A., D. W. Steadman, and C. C. Witt (2016). Predictable evolution toward flightlessness in volant island birds. Proceedings of the National Academy of Sciences 113:4765–4770. doi:10.1073/pnas.1522931113.

CHAPTER 11

Color and Ornamentation

Geoffrey E. Hill

Among vertebrates, birds have the greatest percentage of diurnal, surface-dwelling species, and in their brilliantly lit world, birds shine. They are the most fantastically colored, boldly patterned, and highly ornamented taxon of animals. Much of the research on the mechanisms of animal coloration was conducted before World War II (Fox and Vevers 1960), but recent years have seen a rebirth in interest in the mechanisms responsible for animal coloration, and especially bird coloration, with great advances being made in explanations of the mechanisms of structural coloration as well as carotenoid and melanin pigmentation (Hill and McGraw 2006). With the discovery of the gene for MC1R, the enzyme that is a primary controller of melanin coloration (Mundy 2005b), and the gene for ketolase, the enzyme that controls production of red coloration in many birds (Lopes et al. 2016, Mundy et al. 2016), ornithology entered the era of color genomics. We can anticipate major advances in understanding of the genetics of coloration in the near future (Hubbard et al. 2010, Wilkinson et al. 2015) (see the box on page 312).

Feathers, bills, and the skin are all composed of keratin, and without special mechanisms for coloration, all of these parts have a pale off-white coloration. Birds use a combination of **pigments** and **microstructural mechanisms** to produce a magnificent range of hues and color patterns in these otherwise drab keratin structures. Pigments are molecules that can be added to skin, irises, or feathers to alter their coloration, just as pigments can be added to clothing to produce a desired coloration. Structural coloration, in contrast, results from the interaction of light and microscopic structures. For instance, a blue feather gets its coloration because the microstructures of the feather reflect the blue portion of daylight while red, green, and other portions of the daylight are absorbed by microstructural feather components.

The distinction between the production of coloration from pigments versus microstructures is fundamental (Hill 2010). Pigment-based coloration can exist, at least theoretically, separately from the rest of a feather. Colored pigments can be extracted with a solvent, leaving behind a depigmented feather and a colorful solution of dissolved pigment (Hill 2010:60). Structural coloration, in contrast, is a property of the feather itself; structural coloration can no more be extracted from an intact feather than can its length or mass. These distinctions are heuristically valuable in categorizing color types, but they also have actual practical value when ornithologists try to understand the physiology and function of color displays.

FEATHER PIGMENTATION

Melanin

Many of the bold and striking patterns of feathers, including spots, bars, and stripes, are created by melanin pigmentation (fig. 11.1). **Melanin** also gives rise to subtle gray, brown, and reddish earth tones in feathers and is the primary pigment used by birds to create coloration for camouflage (fig. 11.2). Melanin is the most widespread class of pigmentation across kingdom Animalia, and it is also the most common type of pigmentation in birds (Solano 2014). The external coloration of nearly all birds involves at least some melanin pigmentation, and melanin pigmentation is extensive on the surface of most birds. Think of some of the common birds one would encounter in Central Park in New York—Rock Pigeons (*Columba livia*), European Starlings (*Sturnus vulgaris*), and American Robins (*Turdus migratorius*). All of these species have extensive dark brown and black melanin coloration. Birds use melanin to color feathers, eyes, and bare parts such as legs, unfeathered skin, bills, air sacs, and spurs.

Figure 11.1. Melanin pigments can produce sharp and intricate patterns within feathers. *Photo by Geoffrey E. Hill.*

Figure 11.2. The melanin pigmentation of the dorsal plumage of this Sharp-tailed Grouse is deposited in a blurry pattern lacking sharp edges, which breaks up the shape of the bird and makes it hard to spot in dry grasslands where it dwells. The inset photo is an individual grouse feather showing the intricate but blurry color pattern. *Photo by Geoffrey E. Hill.*

There are two classes of melanin. **Eumelanin** produces black coloration, and by depositing melanin granules in feathers at different densities, birds can achieve any shade of gray (McGraw 2006a) (fig. 11.3). Crisp and detailed patterns within feathers are produced with eumelanin (fig. 11.1), and eumelanin commonly produces characteristic black patches such as the facemasks of many shrike species, the midnight black plumage of Common Ravens (*Corvus corax*), and the chest bibs of House Sparrows (*Passer domesticus*). **Phaeomelanin** produces earth tone coloration ranging from yellow to rust to chocolate to red, depending on the chemical characteristics of the phaeomelanin (fig. 11.3). The reds produced by some phaeomelanins are perceptibly different than the reds produced by other red pigments (described below) in the same way that

the color of "red" human hair differs from red human blood. Phaeomelanin can produce intricate patterns within feathers, but they tend to lack sharp boundaries (fig. 11.2). Rather than bars and spots, the patterns formed by phaeomelanins in feathers are reminiscent of the patterns produced for camouflaged hunter gear (fig. 11.2). Indeed, the function of most phaeomelanin patterning within the feathers of most species of birds is likely camouflage, but camouflage is not a universal function of phaeomelanin coloration. Some bold and striking color displays, like the chestnut breast of a Mallard (*Anas platyrhynchos*), the throat patch of a Red-throated Loon (*Gavia stellata*), and the orange breast of a Red Knot (*Calidris canutus*) function as colorful sexual displays.

Even though we characterize black pigmentation as resulting from eumelanin and earth tones as resulting from

Figure 11.3A, B. Phaeomelanin and eumelanin pigmentation produces the earth tones of plumage. *A,* The velvet black body plumage of a male Surf Scoter; *B,* The head feathering of a male American Robin result from eumelanin. The rusty breast coloration of the robin, in contrast, results primarily from phaeomelanin. The gray back of the robin also results from eumelanin, but deposited at lower concentration. *Photo by Geoffrey E. Hill.*

phaeomelanin, phaeomelanin is generally deposited with eumelanin. It is the relative ratio of phaeomelanin to eumelanin that determines the black to rusty to yellowish coloration that is produced (McGraw et al. 2005b). A higher ratio of eumelanin produces darker coloration (e.g., dark chocolate coloration), while a higher ratio of phaeomelanin produces brighter coloration (e.g., bright rusty orange).

Both eumelanin and phaeomelanins are composed of numerous covalently linked indole subunits, so both eumelanin and phaeomelanin exist in the tissues of birds as large and complex polymers (McGraw 2006a) (fig. 11.4). The complexity of the molecular structure of melanin makes biochemical quantification challenging (Ito et al. 2000), and has made it difficult to answer basic questions, such as, what is the pigmentary basis for differences in the darkness or hue of melanin-based plumage coloration? As a result, the relative contributions of phaeomelanin versus eumelanin for the colored parts of birds was, until recently, unknown (McGraw 2006a). Even with recent advances in quantification of melanin pigmentation, the melanin composition of colored structures is known for only a handful of bird species (McGraw 2006a).

Like all vertebrates, birds synthesize melanin from amino acid precursors in specialized cells called melanocytes at the site of deposition in integumentary structures (McGraw 2006a). Eumelanins are synthesized from the amino acid tyrosine via the catalytic action of tyrosinase. Phaeomelanins require both tyrosine and the sulfur containing amino acid cysteine for synthesis, but after an initial conversion of tyrosine, phaeomelanin synthesis is not dependent on tyrosi-nase (Wakamatsu and Ito 2002). Melanocytes synthesize melanin in organelles known as melanosomes, which are transported to the ends of dendritic processes so they can be transferred to the protein-synthesizing cells, keratinocytes (Prum 1999). Melanosomes containing eumelanin tend to be rod-shaped while those containing phaeomelanin tend to be oval. Melanosomes are incorporated into growing feathers in a very precise arrangement so feathers achieve their species-typical color and pattern (Prum and Williamson 2002) (fig. 11.1).

Carotenoids

Where melanins produce bold patterns, **carotenoids** create brilliant yellow, orange, and red coloration in feathers and bare parts (McGraw 2006a). Carotenoids are the second most abundant and widespread class of pigments that birds use for external coloration (McGraw 2006b). They are unique among avian pigments in that they cannot be synthesized within the bodies of birds (Brockmann and Völker 1934); they have to be ingested to be available as colorants. Fruit and some marine crustaceans have abundant carotenoids, seeds and insects have modest amounts, the flesh of vertebrates tends to have few carotenoids (Goodwin 1984); the coloration of the birds that subsist on these various food classes are colored accordingly. Vertebrate predators tend to have little carotenoid coloration; seed and insect eaters with modest carotenoid intake have modest carotenoid coloration; and fruit-eaters and species that eat marine invertebrates tend to have extensive carotenoid coloration (Olson and Owens 2005) (11.5). Carotenoid coloration tends

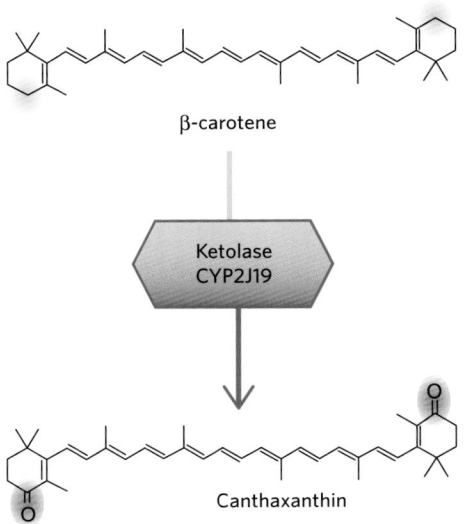

Figure 11.4. The backbone structures of eumelanin and phaeomelanins (*left*). These basic backbone structures are repeated many times in creating the large polymers that melanin comprises. Also shown are the structures of a yellow carotenoid pigment (β-carotene) that is common in the diets of birds (*top right*) and that can be biochemically converted into the red carotenoid pigment, canthaxanthin (*bottom right*), that is common in the feathers of red birds. This biotransformation is catalyzed by a ketolase (CYP2J19). This conversion entails the oxidation of the end ring at the site highlighted in yellow, to produce a ketone, highlighted in red.

FEATHER COLORATION OF EXTINCT ANIMALS

At a subcellular level within animal tissues, melanin pigments reside in membrane-bound compartments called melanosomes. The size and shape of these organelles is specific to the type of melanin pigments that they contain. This observation enables scientists to deduce the coloration of an animal solely by evaluating the ultrastructure of its melanosomes. But when would such an assessment be meaningful? Why not just look at the coloration of the animal? Deduction of coloration from the microanatomy of feathers becomes a key tool when trying to reconstruct the coloration of extinct animals from fossils (Vinther et al. 2008). This technique was first applied not to a fossil bird, but to a well-preserved fossil of a feathered bird-like dinosaur that died more than 150 million years ago (Li et al. 2010). The microstructure of feathers was preserved in this fossilized animal, and by sampling melanosome across the body and deducing coloration from melanosome shape, scientists were able to reconstruct its color (Li et al. 2010). This feathered dinosaur had a bold black-and-white pattern on its wing and leg feathers, with a rufous crown (see figure).

In a second example of ancestral color reconstruction, scientists looked at iridescent coloration of feathers that results from characteristic layers of melanosomes and that can also be observed and measured in fossilized microstructures. Through careful analysis of well-preserved feathers from Eocene fossil beds deposited about 40 million years

The coloration of the 155-million-year-old feathered dinosaur *Anchiornis huxleyi* was reconstructed by examining the size and shape of fossilized melanosomes.

ago, scientists were able to show that some of these ancient birds had iridescent feather coloration (Vinther et al. 2009). This technique was also applied to the feathers of a theropod dinosaur from the Late Jurassic, about 150 million years ago. The researchers found compelling evidence that the feathers of this dinosaur also had iridescent coloration (Li et al. 2012).

Just a few years ago, it was thought to be impossible to reconstruct the coloration of long-extinct animals, but an understanding of the mechanisms of color production, with many insights coming from studies of birds, is enabling scientists to color the past.

to be concentrated toward the tips of individual feathers, but carotenoids are never deposited in detailed patterns within a feather.

Carotenoid coloration of bare parts is widespread in birds, but is lacking in some avian orders (e.g., Caprimulgiformes, Strigiformes) (Olson and Owens 2005). Carotenoid coloration of feathers, in contrast, is uncommon in birds, occurring in less than 25 percent of avian orders (Olson and Owens 2005). In those groups in which carotenoid-based plumage coloration does occur, it often creates signature coloration, such as in flamingos, trogons, and barbets. In both bare parts and feathers, yellow carotenoid coloration is more common than red coloration (Olson and Owens 2005). Whenever there are differences between males and females or adult and juvenile birds in red and yellow coloration, it is invariably the male/adult with red and the female/juvenile with yellow coloration (Hill 1996). Among birds with carotenoid-based

feather coloration, the extent of pigmentation ranges from a small spot involving only a few feathers, as on the head of Ruby-crowned Kinglets (*Regulus calendula*), to nearly every feather on the body, as in male Summer Tanagers (*Piranga rubra*) (fig. 11.6).

Carotenoids are lipid-like molecules built around 40 carbon atoms, typically with two, six-carbon end rings connected by a carbon chain (Brush 1981) (fig. 11.4). The placement of double bonds and the addition of hydroxyl and ketone groups to the end rings determines the characteristic hue of carotenoids (Bauernfiend 1981). About two dozen different carotenoids pigments are used as colorants by birds (McGraw 2006b). Because carotenoids are large and generally nonpolar, they accumulate in fats (creating the yellow coloration of the fats of birds), and they have to be transported through blood (which is mostly water) in lipoproteins (McGraw 2006b).

Figure 11.5. Examples of how carotenoid content of diet shapes carotenoid display. *A*, Species such as American Flamingos (*Phoenicopterus ruber*) that eat food containing abundant carotenoids such a marine invertebrates or fruit tend to have extensive carotenoid coloration; *B*, Species such as Red-winged Blackbirds (Agelaius phoeniceus) that eat insects or seeds tend to have modest carotenoid displays; *C*, Species such as Cooper's Hawks (*Accipiter cooperii*) that eat the flesh of vertebrates tend to have little or no carotenoid coloration. *All photos by Geoffrey E. Hill.*

Birds cannot synthesize the carotenoids that they use as colorants, but some species can biochemically modify ingested carotenoids in ways that change their coloration (Brush 1990). Many bird species use the carotenoids derived from food, such as lutein, without biochemically altering them, and the deposition of unmodified pigments produces the yellow feather coloration, such as the breast coloration of the Yellow-breasted Chat (*Icteria virens*) (fig. 11.7A) and the Great Tit (*Parus major*) (Mays Jr. et al. 2004, Isaksson 2009). A few marine birds such as flamingos ingest red pigments in animal food and deposit dietary red pigments such as astaxanthin (Fox and McBeth 1970), but most birds with red carotenoid coloration obtain red pigments by biochemically modifying ingested yellow pigments (figs. 11.4 and

Figure 11.6. The extent of red and yellow carotenoid pigmentation varies markedly among species. *A,* Ruby-crowned Kinglets (*Regulus calendula*) have just a few red feathers on their crowns, whereas *B,* male Summer Tanagers (*Piranga rubra*) have red coloration on every contour feather of their body. *All photos by Geoffrey E. Hill.*

11.7). Across different bird species, there are two different types of enzyme-mediated transformations of carotenoids: dehydrogenation and ketolation. In dehydrogenation, a hydroxyl functional group on one or both end rings of a dietary carotenoid is converted to a carbonyl functional group (McGraw 2006b). Dehydrogenation transforms a yellow dietary pigment such as lutein to a yellow feather pigment such as canary xanthophyll A, as in the feathers of American Gold-

finches (*Spinus tristis*) (McGraw and Hill 2001) (fig. 11.7B). The purpose of such yellow-to-yellow conversion may have to do with the molecular stability or antioxidant function of the modified pigments (Higginson et al. 2016). In the process of ketolation, a precursor carotenoid is oxidized to produce a ketone at the 4 or 4′ carbon position on one or both end rings and then to transform yellow dietary pigments to red pigments, as in the plumage of House Finches (*Haemorhous mexicanus*) (Inouye et al. 2001, Lopes et al. 2016) (fig. 11.4). Because red pigments produced in this manner have a ketone group on the fourth carbon position, they are sometimes called "C4-keto-carotenoids." C4-keto-carotenoids are a primary source of red carotenoid coloration in birds (Hill and Johnson 2012).

Psittacofulvins

Melanin and carotenoids account for most of the coloration on the surfaces of most species of birds, but across class Aves, a few groups of birds have evolved unique mechanisms for coloration. Parrots have their own unique feather pigments called **psittacofulvins**, which produce red, orange, and yellow coloration (McGraw and Nogare 2005) (fig. 11.8). Even though psittacofulvins generate a range of hues that is similar to the range produced by carotenoid pigments, psittacofulvins are biochemically very distinct from carotenoid pigments (McGraw 2006c). Birds can synthesize their own psittacofulvin pigments, although the pathways involved in psittacofulvin synthesis are entirely unknown. Another curious feature of psittacofulvins is that they are not deposited in legs, bills, or other bare parts—only in feathers. Despite a concerted search, no carotenoid pigments were isolated from the feathers of any species of parrot (McGraw 2006c). Parrots with both red feathers and red bills deposit psittacofulvins in feathers and carotenoids in bills.

Porphyrins

Porphyrins are perhaps the most abundant pigments in nature (fig. 11.9). This large class of molecules includes chlorophyll, accounting for the green coloration that envelops the planet, as well has heme, the pigment at the center of hemoglobin that makes blood red. In birds, however, **natural porphyrins** (uroporphyrin, metalloporphyrin, and bilins) are incorporated into feathers in at least a few avian orders including owls (Strigiformes), nightjars (Caprimulgiformes), and bustards (Gruiformes) (McGraw 2006c). Their contribution to the appearance of birds in ambient light, however, is generally considered to be negligible. For instance, they have been isolated from the feathers of owls and bustards,

Figure 11.7. Examples of the types of carotenoids that birds use to create plumage coloration. *A*, Yellow-breasted Chats deposit dietary lutein unmodified into ventral plumage to achieve yellow coloration; *B*, Male American Goldfinches modify yellow dietary carotenoids such as lutein into yellow canary xanthophylls to achieve yellow feather coloration; *C*, Male Northern Cardinals modify yellow dietary carotenoids such as zeaxanthin into red 4-keto-carotenoids such as astaxanthin to achieve red feather coloration. *All photos by Geoffrey E. Hill.*

but certainly most of the brown coloration of owls and bustards comes from melanin pigments (With 1978, Roulin and Dijkstra 2003). However, in Black-winged Kites (*Elanus caeruleus*), a species in which adults are black and white, the reddish-brown coloration of the breast feathers of juveniles is produced entirely by natural porphyrin pigments that biochemically degrade over a couple of months (Negro et al. 2009). Thus, at least for juvenile kites, natural porphyrin pigments produce a clearly discernable brown coloration. Overall, natural porphyrins in feathers have received little study and are not considered to be important sources of plumage coloration.

Turacos (Musophagiformes) make up a family of birds endemic to sub-Saharan Africa, with their own unique red and green pigments that belong to a class of porphyrins (fig. 11.9). The red pigment of turacos is called **turacin**, while the green

pigment is called **turacoverdin** (McGraw 2006c). Both pigments are constructed around a copper molecule. The colors produced by these copper-based porphyrin pigments are brilliant: deep scarlet red for turacin and emerald green for turacoverdin (fig. 11.9). Even though turacin and turacoverdin are restricted to one order of African birds, and therefore are of marginal importance to the understanding of coloration across Aves, they hold an honored place in the history of bird color research. Turacin was the first avian pigment to be extracted and biochemically characterized (Church 1869). Thus, there was a time when much more was known about the biochemistry of turacin and turacoverdin than melanin or carotenoids. Turacin and turacoverdin were the focus of early biochemical analysis because, unlike carotenoids or melanins, they are weakly bound to feather keratin and easily leached from feathers. As a matter of fact, turacin and turacoverdin are so water soluble that it is claimed that turacos have to protect their colored feathers from rain or their red and green pigments will be washed from plumage. Pigments being flushed out of feathers by rain may be an exaggeration, but dipping a red feather from a turaco into mildly

Figure 11.8. The red and yellow coloration of the plumage of parrots, such as *A*, the red of feathers of the Crimson Rosella (*Platycercus elegans*) and *B*, the yellow feathers of the Blue-and-Yellow Macaw (*Ara ararauna*), results exclusively from psittacofulvin pigments, which only parrots produce. Blue coloration results from feather microstructures. *All photos by Geoffrey E. Hill.*

basic solution causes the majority of red pigments to leach into solution (Hill 2010). Because the fruit diet of turacos has a low copper content, accruing sufficient copper for pigment production may be a major challenge for turacos (Hood and Hill 2009, McGraw 2006c).

EGGSHELL COLORATION

The shells of eggs have a wide range of colors and patterns, from uniform white to metallic turquoise to lightly speckled to massively blotched. The background coloration of eggs is generally produced by a group of porphyrin pigments known as **biliverdin**, which produces blue to green coloration in eggs (McGraw 2006c) (fig. 11.10A). The reddish-brown spots on the surface of the eggs of many species of birds were long thought to be from heme in dried blood, but are now known to be produced by **natural porphyrins** (McGraw 2006c) (fig. 11.10B). The coloration of eggs is generally tied to risk of predation. Eggs laid in cavities or other hidden locations tend to be unmarked and brightly colored, whereas eggs laid in more exposed conditions tend to be blotched and drab (Dale 2006). There are many exceptions to this general pattern, and researchers are investigating the idea that the coloration of an egg is a signal of the body condition of the laying female (Moreno and Osorno 2003, Siefferman et al. 2006).

STRUCTURAL COLORATION

All avian coloration results from the differential emission, absorption, and reflection of ambient light. In the case of pigments, individual molecules interact with light such that the structure of the pigment molecule determines the coloration that is produced. **Structural coloration**, in contrast, results from the interaction of ambient light not with specific colored pigment molecules but with the fine structures of the surface

Figure 11.9. Turacos (order: Musophagiformes), an order of bird found only in sub-Saharan Africa, have two unique feather pigments: turacin, which produces brilliant red coloration, and turacoverdin, which produces bright green coloration. Red turacin pigmentation is always restricted to flight feathers, as in this Knysna Turaco (*Tauraco corythaix*). *Photo by Geoffrey E. Hill.*

Figure 11.10. The blue coloration of eggs, like *A*, the eggs of an Eastern Bluebird, *Sialia sialis*, derive their coloration from biliverdin, while the blotches and spots on eggs, like *B*, the eggs of Killdeer, *Charadrius vociferus*, are produced by natural porphyrins. *All photos by Geoffrey E. Hill.*

tissues of feathers, bare parts, or eyes. The structures that interact with light to produce structural coloration are called **microstructures** or nanostructures, because they are organized at the scale of nanometers. The color produced by microstructures is a function of the size and three-dimensional arrangement of microstructural elements (Prum 2006). Any hue—from red to ultraviolet—can be produced via structural coloration, but short-wavelength coloration from ultraviolet to blue is most common.

Iridescent Coloration

The adjective "iridescent" has its root in the Latin *irido-*, meaning rainbow, and **iridescent coloration** is characterized by the ephemeral and ever-changing colored landscape that it creates on feathers. In a species like Ocellated Turkey (*Meleagris ocellata*), a single feather can appear blue, green, or copper-colored depending on the viewing angle and lighting

(fig. 11.11B). Indeed, most plumage patches with the potential for iridescent structural coloration appear black until the viewing angle and lighting are just right, and then vivid coloration bursts from the feather. Many of the most fantastic color displays of birds, including the gorgets of hummingbirds, the eyespots of peacocks and other pheasants, and the saddle feathers of Nicobar Pigeons (*Caloenas nicobarica*), result from iridescent coloration (fig. 11.12). While iridescent coloration is common in feathers, it never occurs in soft parts.

Figure 11.11. Iridescent coloration of feathers changes with the angle of viewing. *A*, Common Grackles typically appear purple when their structural coloration is visible, but with different viewing angles and lighting conditions they can also appear blue or bronzed; *B*, The dorsal feathers of an Oscillated Turkey typically appear blue with a bronze tip, but the blue can shift to green and bronze to copper depending on angle and lighting. *All photos by Geoffrey E. Hill.*

Iridescent coloration is produced by layers of material with different refractive indices (fig. 11.13). The most familiar manifestation of structural coloration is the variably violet/blue/green coloration that appears on a puddle covered by a film of oil on black asphalt. The air, oil, and water have different refractive indices with sharp boundaries between. As light passes from air to oil and from oil to water, some photons are reflected off the boundaries of layers and some penetrate between layers. Light that makes it all the way through the oil and water is absorbed by the black asphalt. Thus light reflected back to the observer will have traveled different distances through substances with different refractive indices. As light waves reflecting off of the air/oil and oil/water interfaces meet on their way back to the observer, some waves will be shifted out of phase and cancel each other out. Other wavelengths (in this case violet, blue, or green) will be in phase and amplified. The net result is a subtraction of some colors from ambient light and an enhancement of others, and hence the appearance of coloration (Doucet and Meadows 2009) (fig. 11.13).

The iridescent coloration of feathers is created in much the same way, but instead of layers of oil and water, bird feathers use keratin and melanosomes as their refractive layers (Prum 2006). The characteristic hue of iridescent feathers is a product of the refractive indices of air, keratin, and melanosomes and the spacing between these structural layers (Shawkey et al. 2006). Although many birds achieve iridescent plumage coloration with single layers of material, just like oil on water, bird feathers are not restricted to a single layer of refractive material. Some birds, like hummingbirds and peacocks, have many stacked layers that seem to reinforce the effects of one layer and lead to more brilliant and stable coloration (Greenewalt et al. 1960, Doucet and Meadows 2009).

Non-Iridescent Coloration

In contrast to iridescent coloration, non-iridescent coloration tends to have a stable hue regardless of the angle of viewing (Prum 2006) (fig. 11.14). Many of the most familiar color displays of birds come from non-iridescent structural coloration, including the blue feathers of Blue Jays (*Cyanocitta cristata*), and the lilac breasts of Gouldian Finches (*Erythrura gouldiae*). As with iridescent coloration, non-iridescent structural coloration results from the interaction of structural elements with ambient light such that some wavelengths are subtracted and some

Figure 11.12. Some of the most brilliant and beautiful color displays of birds, such as *A*, the gorget of the Lucifer Hummingbird; *B*, the eyespots of Indian Peafowl; *C*, crown and tail of the Green-tailed Sunbird result from iridescence. *All photos by Geoffrey E. Hill.*

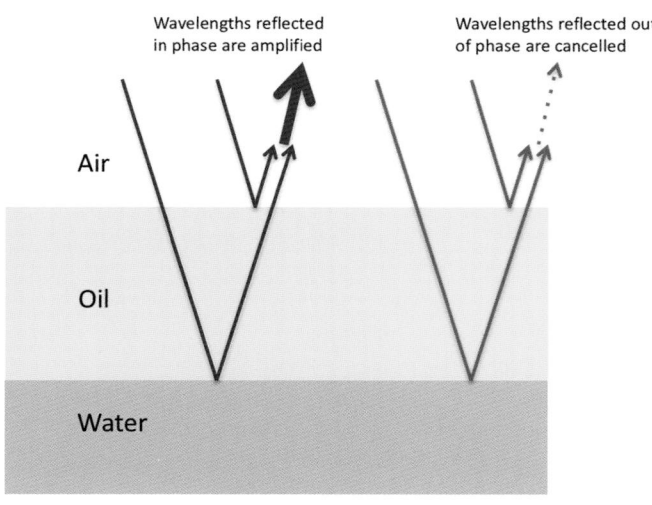

Wavelengths reflected in phase are amplified

Wavelengths reflected out of phase are cancelled

Air

Oil

Water

Figure 11.13. Iridescent coloration is created by layers of substances with different refractive indices, such as oil floating on water. Light is reflected from each interface, and when light reflected from different interfaces meets, it can either be in phase or out of phase. Light that is out of phase is cancelled and lost. Light that is in phase is amplified and becomes the color of the material.

are enhanced (Prum 2006). Instead of two-dimensional stacking of layers of material, however, non-iridescent coloration in birds results from a three-dimensional matrix of interacting particles (see the box on page 320). These particles are constructed of a specific size so that they reflect a small range of wavelengths back to the observer (Shawkey et al. 2003, 2005). Wavelengths that are not of the proper size to interact with the structural elements pass through the scattering matrix and are absorbed by a basal layer of melanin (Shawkey and Hill 2006). The result is coloration that ranges from a faint wash to brilliant and striking.

Pigments Combined with Structural Coloration

Among the most common plumage coloration of birds is green, but paradoxically, green pigments are rare in birds (McGraw 2006c). Most green coloration in the feathers of birds results either from green structural coloration or from a combination of blue structural coloration and yellow pigments (Hill 2010). Blue structural coloration in combination with yellow carotenoid coloration is the source for green coloration in hundreds of species of passerine birds, as well as bee-eaters, jacamars, trogons, todies, pigeons, and toucanets, among others (fig. 11.15). The enigmatic green coloration of parrots is the product of structural coloration and yellow psittacofulvins. Green structural coloration is the other major source of the green coloration in birds and produces such familiar green plumages as the

head of a drake Mallard and the body coloration of many hummingbirds.

COSMETIC COLORATION

Cosmetic coloration describes coloration that is produced through external rather than internal processes (Delhey et al. 2007, Montgomerie 2006). The addition of cosmetic coloration to plumage can be passive, as when the feathers of Sandhill Cranes (*Grus canadensis*) are stained reddish by the iron-rich wetlands in which they feed. Alternatively, cosmetic coloration can be applied actively, such as when Bearded Vultures (*Gypaetus barbatus*) bathe in mud to dye their feathers rusty red (Negro et al. 1999). A few species of birds such as the Great Hornbills (*Buceros bicornis*) secrete pigments, which may contain carotenoids, and rub them into feathers and bare parts to achieve yellow coloration (Montgomerie 2006) (fig. 11.16).

Perhaps the most fantastic form of cosmetic coloration used by birds is the collection of colored objects displayed by male bowerbirds (family Ptilonorhynchidae). In Satin Bowerbirds (*Ptilonorhynchus violaceus*), all copulations occur in stick bowers built by males, and the bower owners decorate their structures with blue objects (Borgia 1985). These blue objects appear to accentuate the blue plumage of males and to help stimulate females to copulate, because males with more blue objects obtain more matings (Borgia 1985). In White-winged Fairywrens (*Malurus leucopterus*), males hold a colorful flower to accentuate their appearance (Rathburn and Montgomerie 2003).

SOURCES OF COLOR VARIATION

It is most common to think of variation in the pattern of avian coloration as existing between species. Indeed, except for drab and cryptic species, taxonomists define most bird species by coloration and color pattern (Remsen Jr. et al. 2010). However, a single pattern of coloration rarely defines the individuals within a species of birds. Different sexes and age classes may have very different coloration. Carl Linnaeus, the father of modern taxonomy and more of a botanist than a zoologist, famously classified male and female Mallards as different species (fig. 11.17A). Two hundred years later, skilled avian taxonomist made the same mistake when they described male and female Williamson Sapsuckers (*Sphyrapicus thyroideus*) as different species based on study skins (Ridgway 1914). Such mistakes are understandable, because there are greater differences in plumage coloration between the males and females of Mallards and Williamson's

Figure 11.14. A male Blue Grosbeak (Passerina caerulea) is resplendent in its cobalt blue feathers. The hue of Blue Grosbeak plumage is much more stable than that of a Common Grackle (fig. 11.11A) because it is non-iridescent, resulting from incoherent scattering of light by the spongy layers of barbs in contour feathers. Photo by Geoffrey E. Hill.

THE PHYSICS OF STRUCTURAL COLORATION

The discovery of the mechanisms that produce non-iridescent coloration in feathers was a major breakthrough in the research on animal coloration. Since the early twentieth century, blue feather coloration was thought to result from incoherent (essentially random) scattering of photons (known as Rayleigh scattering) (Fox 1976). Rayleigh scattering is the process that produces a blue sky and blue coloration of smoke and ice. In the 1970s, Dyke published data on structural coloration of blue feathers, showing that it violated assumptions of Rayleigh scattering, and he proposed new mechanisms for the production of structural coloration (Dyck 1971). However, Dyck lacked the analytical tools to fully explain how non-iridescent coloration is produced in bird structures. Around the turn of the twenty-first century, such tools became available. Using new scanning techniques and a mathematical modeling approach known as Fourier analysis, ornithologists studying the non-iridescent blue and ultraviolet coloration of feathers and bare parts showed such coloration results from coherent scattering of light (Prum et al. 1999, Prum and Dyck 2003). A major ongoing mystery is how genes can code for different structural coloration. The three-dimensional keratin structures that are deposited in semi-ordered three-dimensional patterns and that produce coherent scattering and coloration precipitate out of solution during feather development (Prum et al. 2009). Self-assembly of the microstructures that create structural coloration does not seem to have a place for fine genetic control. Paradoxically, thousands of species of birds have unique, species-typical hues of structurally colored feathers, so genetic control of coloration must occur. Genes may code cellular conditions that dictate assembly of structures that create different color displays, thus explaining the basis for species-typical differences in structural coloration. This is an area of active research.

Sapsuckers than between the breeding males of many sister species, and variation in coloration within a species is not limited to male/female differences. Individuals in some species can have different color morphs with strikingly different appearance, and individuals in different populations of the same species can also be distinctly colored. The appearance of a single individual can also change greatly as progresses through developmental stages and as it transitions between seasons.

Figure 11.15. Green plumage coloration is most commonly created with a combination of yellow pigments and blue structural coloration. In this Emerald Toucanet (*Aulacorhynchus prasinus*) yellow carotenoids combine with blue structural coloration to create a vivid green display. Note that in patches of feathers with fewer carotenoids, such as behind the bill, the plumage shifts to blue. *Photo by Geoffrey E. Hill.*

Figure 11.16. Great Hornbills (*Buceros bicornis*) grow wing feathers lacking yellow coloration. The bright yellow visible in the wing stripe of this bird was rubbed into feathers as a component of uropygial oil. The biochemical basis of this coloration is unknown. *Photo by Geoffrey E. Hill.*

Variation within a Species

Sexual dimorphism describes a difference between males and females in external form or structure. **Sexual dichromatism** is the form of sexual dimorphism concerned with external coloration. **Sexual monochromatism** describes the condition of males and females being indistinguishable in

coloration. The coloration of males and females slides along two axes: degree of similarity/difference between the sexes and overall brightness of coloration (Badyaev and Hill 2003). Thus, a bird species can be sexually monochromatic and drab, sexually monochromatic and bright, sexually dichromatic with bright males and drab females, or sexually dichromatic with drab males and bright females. In about half of all birds—including such familiar species as the Canada Goose and the Blue Jay—males and females look alike to a human observer. In about half of these monochromatic species, both males and females are drably colored (fig. 11.17B). The opposite pattern, with both males and females brightly colored, is especially common in tropical songbirds, including Neotropical migrants that fly to the tropics every winter (Dale et al. 2015). Among sexually dichromatic species, the

Figure 11.17. Even closely related species like the Mallard and Mottled Duck can vary dramatically in sexual dichromatism. *A*, The male and female Mallard have very different feather color and pattern, with the male bright and colorful while the female is drab; *B*, In contrast, both male and female Mottled Ducks are drab, and the sexes are indistinguishable to humans by color or pattern. *All photos by Geoffrey E. Hill.*

Figure 11.18. In a few species of birds, such as the Wilson's Phalaropes (*Phalaropus tricolor*), shown here, females are brightly colored while males are drab. In these species, females are also more aggressive than males and males assume more parental duties. *Photo by Geoffrey E. Hill.*

great majority of species have brightly colored males and drably colored females (for example Northern Cardinals and American Goldfinches), and the differences between males and females vary from barely perceptible differences to the striking differences of Mallards (fig. 11.17A) and Williamson Sapsuckers. In a very few species, such as the Wilson's Phalarope (fig. 11.18), females are bright and males are drab, and such reversed sexual dichromatism is invariably associated with a polyandrous breeding system (Ligon 1999).

Males and females are identical to a human observer in about half of all bird species, but an avian observer would classify a much higher proportion of birds as dichromatic. But how do we know what a bird sees? For a few bird species, physiologists have been able to measure the range of sensitivity for the cone cells in an avian retina and from those data calculate the boundaries of color perception for birds (see chapter 12). Not only do birds perceive UV light and a fourth color dimension with UV-sensitive cones, but they also distinguish between human-visible colors much better than humans. By measuring the feather coloration of males and females using a reflectance spectrometer and calculating the color differences that would be perceived by an avian eye, ornithologists estimated that most birds that a human would classify as monochromatic would be sexually dichromatic to a bird (Eaton and Lanyon 2003, Eaton 2005).

Why birds show such tremendous variation across species in plumage brightness and sexual dichromatism is an open question in ornithology, with many hypotheses advanced (Badyaev and Hill 2003, Dale 2006). Such broad-scale evolutionary questions are tackled with a technique called the "com-

parative method," whereby biologists look for associations between evolutionary change in the coloration and change in other factors affecting the bird species, such as habitat or social environment (Garamszegi 2014).

Variation between Morphs

In a few hundred species of birds, individuals within a population display different **color morphs**, which are genetically determined color types that are not related to age, sex, or population (Fowlie and Kruger 2003, Galeotti et al. 2003). For instance, Reddish Egrets (*Egretta rufescens*) are either blue and reddish in coloration or they are stark white. The color morph of an individual is determined by one or a very few alternative forms of a gene, called alleles. Humans are familiar with color morphs for eye coloration in people. Eye color is controlled by only a few alleles in humans that are inherited in Mendelian fashion such that full sibs can show a different eye color type. In the same way, full sibling Reddish Egrets can be either dark or white. Color morphs are common in herons, hawks, owls, jaegers, and geese, but are rare in most orders of birds including Passeriformes (Galeotti et al. 2003) (fig. 11.19). The function of color morphs remains an unsolved puzzle in ornithology.

Variation within an Individual

By the time most birds are independent of their parents, they have grown a **juvenal plumage** that they will retain from a few weeks to over a year. Note that the first non-downy plumage of birds is called "juvenal"; the adjective "juvenile"

Figure 11.19. In a few species of birds, discrete plumage variants called morphs exist within a single population. In this flock of five Snow Geese (*Chen caerulescens*), four birds are white morph and one bird is blue morph. Such polymorphism is independent of age or sex. *Photo by Geoffrey E. Hill.*

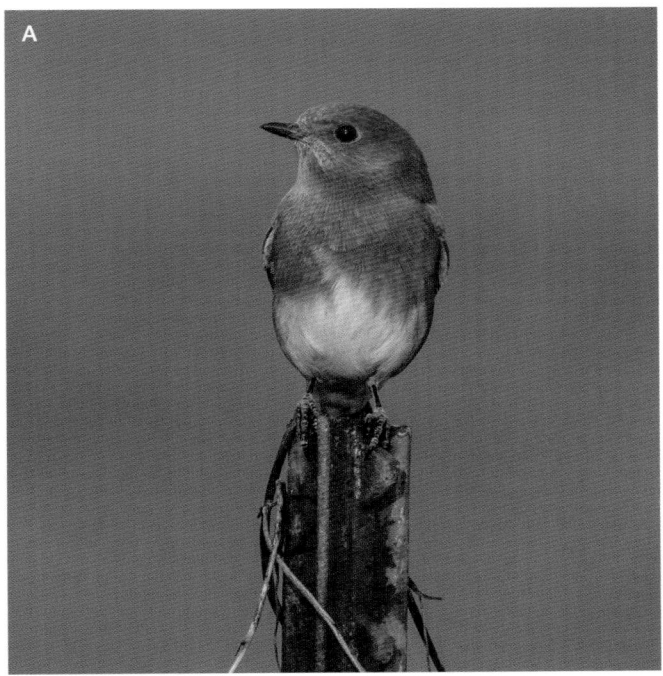

Figure 11.20. Definitive plumage is a plumage (often the terminal plumage in a sequence) worn by fully adult individuals such as *A*, the male Eastern Bluebird (*Sialia sialis*), and *C*, the male Orchard Oriole, *Icterus spurius*. Many species have a juvenal plumage, which is distinct from definitive plumage, as illustrated by *B*, the young Eastern Bluebird in drab spotty plumage. *D*, Some birds take two or more years to achieve definitive plumage and, even after becoming sexually mature, have a subadult plumage that is distinct from definitive plumage, as illustrated by the yearling male Orchard Oriole. *All photos by Geoffrey E. Hill.*

refers to an age class that is not sexually mature. In a few species, juvenal plumage is identical to adult "definitive" plumages. For most species, however, juvenal plumage is the most cryptic of plumages and distinct from later adult plumages. Juvenal plumages are characterized by streaks, spots, and bars that make the juveniles hard to see (Ligon and Hill 2013) (fig. 11.20B).

Most birds breed in their first potential breeding season and have a rapid transition from drab juvenal plumage to definitive adult plumage. In some species, however, individuals—typically males—continue to have a distinctive "subadult" plumage even after they are sexually mature. This pattern of feather development is called **delayed plumage maturation** and has been the object of study, particularly in songbird species in North America and Europe (Hawkins et al. 2012). Most birds with delayed plumage maturation simply incorporate elements of juvenal plumage pattern into their otherwise adult male appearance, for instance the brown instead of black wing and tail of subadult male Rose-breasted Grosbeaks (*Pheucticus ludovicianus*), but some species, such as the Orchard Oriole (*Icterus spurius*), have distinct subadult plumage that is unlike females, juveniles, or males in definitive plumage (Hawkins et al. 2012) (fig. 11.20D).

A few large birds go through several cycles of subadult plumages before reaching definitive plumage and breeding (Hawkins et al. 2012). For instance, most large gulls (Laridae) go through either three- or four-year cycles before reaching a definitive plumage, with each year class having a distinctive appearance. The huge range of within-species color variation that such complex developmental patterns create is frustrating to novice birders trying to master gull identification. First-year Herring Gulls (*Larus argentatus*) and Lesser Black-backed Gulls (*Larus fuscus*), for instance, look much more alike than do first- and fourth-year Herring Gulls.

COLOR GENETICS

It has long been known that there is a strong genetic basis for patterns of coloration in birds. For centuries, aviculturists have taken advantage of the influence of genes on color expression to select on plumage and bare part coloration, creating a huge range of color varieties of common cage birds like Zebra Finches (*Taeniopygia guttata*), Canaries (*Serinus canaria*), and Budgerigars (*Melopsittacus undulates*) (Birkhead 2003) (fig. 11.21). Very recently, with breakthroughs that allow rapid and cheap DNA sequencing, the specific genes that control the coloration of birds have begun to be identified (Hubbard et al. 2010).

Genes exert an influence over the coloration and color patterns of birds in both a qualitative and quantitative manner. **Single-gene effects** describe gross genetic control of major patterns by discrete sets of specific genes (Buckley 1987). Even though such patterns can be complex, they tend to be controlled by one or a small set of genes, such that an entire plumage pattern, or at least major elements of a plumage pattern, is turned on or turned off with a single controlling allele. A good example is the color morphs in Reddish Egrets described in the plumage "Variation between Morphs" section above. Whether feathers are reddish and blue or snowy white is determined by alleles at a single locus. In cage birds, new color varieties are most often created by subtraction of elements from the coloration of a wild-type plumage (Hill

Figure 11.21. In the wild, Budgerigars (*Melopsittacus undulatus*) are green, like the bird at the center of the photo. Common domestic varieties include blue, in which the pathway for yellow psittacofulvin coloration has been lost; yellow, in which the blue structural coloration has been lost; and white, in which both yellow and blue color mechanisms have been lost. *Photo by Geoffrey E. Hill.*

2010). For instance, wild Budgerigars are green, but domestic Budgerigars commonly come in blue, yellow, and white (fig. 11.21). These domestic color forms make perfect sense when one considers that the green of wild budgies is produced through a combination of blue structural coloration and yellow psittacofulvin coloration. Blue budgies carry a genetic mutation that causes the loss of yellow psittacofulvin. Yellow budgies carry a genetic mutation that causes the loss of blue structural coloration. White budgies carry both mutations. In a similar way, yellow "lipochrome canaries," which have unbroken bright yellow coloration, have a genetic mutation that causes the loss of melanin pigmentation in feathers, which in wild canaries adds stripes and dark wings to canary plumage. More rarely, color varieties in domestic birds exist because of the addition of novel color traits, such as Zebra Finches with entirely black ventral plumage. When birds hybridize, such as when a Blue-winged Warbler (*Vermivora cyanoptera*) produces young with a Golden-winged Warbler (*Vermivora chrysoptera*), the offspring invariably have a color pattern that is intermediate between the two parental types, likely reflecting new combinations of pattern-controlling genes.

How a set of genes that code for elements in a complex plumage color pattern become linked so that they are inherited as one trait is not well understood. Interestingly, complex plumage patterns can be turned off and turned on over evolutionary time, apparently without numerous intermediate steps leading to the complex pattern (Omland and Hofmann

2006). Such a pattern of losses and gains in complex color pattern was illustrated most clearly in New World Orioles, family Icteridae, in which both a "Baltimore Oriole" plumage pattern (with a black hood) and an "Altamira Oriole" plumage pattern (with a black face and bib) have been gained and lost repeatedly over millions of years (Omland and Lanyon 2000). One speculation is that the gene complexes that create these color displays can remain viable when turned off, so that they can be gained or lost by mutations to regulatory elements.

The other type of genetic control of coloration involves **quantitative traits**. Unlike traits that are controlled by one or a small set of genes and are inherited in a Mendelian fashion, quantitative traits are controlled by many genes as well as by the environment (Merilä and Sheldon 2001). Quantitative traits are inherited in a fashion that is called non-Mendelian. Offspring tend to resemble their genetic parents, but the effects of genes can be weak. A good example of a familiar quantitative trait is height in humans. Tall parents tend to have tall children, but the exact height of a child cannot be known from the height of the parents. The degree to which variation in a quantitative trait is controlled by the effects of genes is called its **heritability** (Merilä and Sheldon 2001).

Where species-typical plumage pattern tends to be controlled by single-gene effects, the execution of the coloration that makes up the pattern is typically controlled by many genes and is therefore quantitative. For instance, all male House Sparrows inherit, through single-gene effects, a species-typical color pattern with white median wing coverts, rusty nape, gray crown, black bill, and a black throat and chest. However, the extent of the black ventral patch varies greatly among individual (fig. 11.22A, B), and this variation in melanin pigmentation is a quantitative trait. There is a tendency for sons to resemble sires (Griffith et al. 2006), but the environment in which feathers are grown also has a large effect on the size of the black patch. The heritability of most quantitative color traits in birds is relatively low—usually less than 10 percent—but such measures are hard to obtain (Merilä and Sheldon 2001).

Maternal and Nest Environment Effects

The quantitative traits of birds are shaped by the environment as well as by genes (fig. 11.22). Environmental influences on plumage coloration start with the effects of the environment provided by parents and, particularly, the mother (Fitze et al. 2003). These maternal effects begin in the egg and continue until independence from parental care. Such maternal effects come in many forms. Females put ca-

Figure 11.22. The extent of black melanin pigmentation in the ventral plumage of male House Sparrows (*Passer domesticus*) varies greatly. Whether a male has *A*, a large patch or *B*, a small patch is determined partly by complex genetic interactions and partly by the male's environment. *All photos by Geoffrey E. Hill.*

rotenoids in eggs, which seem to directly affect the leg and bill coloration of offspring in Zebra Finches (McGraw et al. 2005a). The carotenoid content of diet also affects the mouth coloration of nestling, and mouth coloration, in turn, affects how much chicks are fed (Saino et al. 2000). Females also put immune elements and hormones in eggs, and both of these factors can affect the expression of quantitative coloration

(Strasser and Schwabl 2004). How the nutrients allocated to eggs shapes the phenotypes of adult birds is an area of active research. The amount and quantity of food brought to the nest can also affect the coloration of the resulting offspring. Nestling that receive fewer resources during growth and development produce less colorful feathers (Horak et al. 2000, Siefferman and Hill 2007).

Environmental Regulation of Coloration

Many features of a bird, such as the length and shape of the bill or skeletal size, are created through one developmental sequence early in the bird's life and then are fixed for the remainder of life. The coloration of both plumage and bare parts, however, have the potential to change through the life of the bird as feathers are replaced one or two times per year and as the integument is continuously renewed. Each time a plumage or bare part color display develops, the quality of its execution is subject to the effects of the environment. A variety of environmental factors have been shown to exert significant effects on the color displays of different species of birds (Hill 2006). A range of pathogens, including especially coccidia (protozoan gut parasites), have long been known to affect the skin coloration of chickens (known as pale bird syndrome in the poultry industry) (Brawner et al. 2000), and studies with wild birds have shown that pathogens ranging from viruses to lice can diminish carotenoid, structural, and perhaps melanin coloration (Hill 2006). Other environmental factors including nutrition, exposure to heavy metals, and the stress of being put in a cage can also have substantial effects on coloration (Hill 2006). The level of circulating stress hormone (Buchanan 2000), degree of oxidative stress to which an individual is exposed (Galván and Alonso-Alvarez 2008), and the social environment prior to feather growth (McGraw et al. 2003) can affect the size of black eumelanin patches (fig. 11.22A).

There has been some controversy regarding whether melanin pigmentation is less affected by environmental factors such as pathogens than carotenoid coloration, and whether structural coloration has any potential for condition dependency (Badyaev and Hill 2000, Griffith et al. 2006, Prum 2006). Several studies have now shown that structural coloration is affected by environmental factors such as parasites and nutrition, and that both melanin and carotenoid coloration can be condition-dependent (Hill 2006, Meunier et al. 2011). Another controversy concerns the mechanisms that link individual condition and performance to carotenoid coloration. The **resource trade-off hypothesis** proposes that carotenoids are rare resources that serve vital roles as antioxidants and immune enhancers as well as colorants, such

that only birds that accrue large stores of carotenoids can both maintain health and have a full color display (Lozano 1994, von Schantz et al. 1999, McGraw et al. 2010). Alternatively, it was proposed that carotenoid coloration is an **honest signal** because production of carotenoid pigmentation is intimately linked to core cellular processes (Hill 2011, 2014). By this **shared pathway hypothesis**, females assess male ornamentation in mate choice to gain information about core biochemical pathways that affect offspring fitness, quality, and condition. What information is encoded in the coloration of feathers and how feather coloration is assessed in aggressive contests and in female mate choice is the subject of ongoing research.

HORMONAL CONTROL OF COLORATION

Both genes and environmental factors can exert an influence on coloration through hormones. Hormones control patterns of dichromatism in birds, but the mechanisms of action vary greatly among different orders of birds. In Struthioniformes, Galliformes, and Anseriformes—the basal lineages of birds—male coloration is the default state (Kimball 2006). Experiments with individuals from these species can dramatically demonstrate this effect: if the gonads of either males or females are removed before molt, a bright male plumage is produced. The female hormone estrogen is the regulatory element in these species, and males will grow hen plumage if they are given estrogen treatments (Owens and Short 1995). In shorebirds and at least some songbirds, drab hen plumage is the default state, and male hormones induce bright coloration (Kimball 2006). In these taxa, genetic females grow male plumage if treated with testosterone, and males lose their male coloration if castrated. In the Red Bishop (*Euplectes orix*), and presumably other songbirds, luteinizing hormone, a gonadotropin hormone, induces males to undergo a prealternate molt that is not found in females, and through this prealternate molt males attain a bright orange plumage not found in females (Witschi 1961). In each of these cases, genes program the pattern of hormone release that control sex differences in plumage. The mechanisms of sexual dichromatism in most orders of birds are unstudied.

Hormone levels can also change in response to the environment and in turn affect bird coloration. High levels of the stress hormone corticosterone are associated with reduced coloration, while high levels of the male hormone testosterone are associated with enhanced coloration (Bókony et al. 2008, Roulin et al. 2008). In terms of plumage coloration, perhaps the most influential environment for most

Figure 11.23. Common Canaries (*Serinus canaria*) typically have yellow feathers. To create a red canary, breeders crossed yellow canaries with a species with red feathers—Red Siskin (*Spinus cucullata*)—and captured the "red factor" from canaries. Recent genomic research revealed that the "red factor" was both a gene for converting yellow carotenoids to red and a gene for depositing red pigments into plumage. *Photo by Rebecca Koch.*

birds is the social environment. Many species of birds honestly signal their social status through plumage coloration, especially patches of black coloration (Tibbetts and Safran 2009). Individuals in these species presumably use hormonal feedback from social interactions to produce an ornament of the correct exaggeration, and this was shown experimentally in House Sparrows (McGraw et al. 2003). Stressors like parasites, poor diet, and toxins might also create effects on feather coloration, partly through their effects on hormones (Buchanan 2000).

THE GENOMICS OF BIRD COLORATION

In recent years, with the availability of powerful genetic tools, there have been breakthroughs in the discovery of specific genes that control the expression of coloration. The age of color genomics began with the MC1R gene, first discovered to control pelage coloration in mice (Mountjoy et al. 1992) and subsequently shown to control dark/light polymorphic melanin variation in a wide range of birds from Passeriformes to Charadriiformes (Mundy 2005). The discovery of the MC1R gene was followed by the discovery of the ketolase gene, the "redness gene" in birds that codes for the enzyme that enables yellow dietary pigments to be converted to red feather and soft-part pigments (Lopes et al. 2016, Mundy et al. 2016). It was the ketolase gene that was transferred from Red Siskins (*Carduelis cucullata*) to Common Canaries through hybridization to create the first red canary (Birkhead 2003, Lopes et al. 2016) (fig. 11.23). MC1R and the ketolase gene are just the beginning of what is sure to be a new age of discovery regarding the genes that underlie the colors and patterns of birds.

PHYLOGENETIC OPPORTUNITIES AND CONSTRAINTS

The colors of birds arise as a response to natural and sexual selection, so evolutionary patterns are shaped more by function than by mechanism of production (see chapters 3 and 9). Nevertheless, natural and sexual selection act only on existing phenotypes, so a mechanism for color production must exist in a population or such a mechanism must appear through mutation for it to be the object of selection. Consequently, availability of mechanisms for coloration has shaped or constrained color evolution in birds. The lack of a mechanistic pathway to an evolutionary endpoint is called a phylogenetic constraint.

Melanin pigments are ubiquitous in organisms from bacteria to birds, and all lineages of birds have the mechanisms available to have dark coloration of any external surface (Solano 2014). Similarly, carotenoid coloration of integumentary structures is widespread among vertebrates. It is reasonable, therefore, to assume that the ancestor of birds had carotenoid-based bare-part coloration and that all extant lineages inherited basic mechanisms for bare-part carotenoid coloration (Hill 2010). The capacity to convert yellow dietary carotenoids to red ketolated carotenoids for red color displays of soft parts also appears to be primitive to birds (Lopes et al. 2016, Mundy et al. 2016), but the evolution of carotenoid ketolation remains to be studied in detail. Carotenoid coloration of feathers, in contrast to its use in bare parts, appears to require special adaptations that evolved in birds only after the major lineage split between Galliformes/Anseriformes and all other Neognathae; with very few exceptions, Galliformes and Anseriformes lack carotenoid feather coloration (Hill 2010). The lack of carotenoid feather coloration in Galliformes and Anseriformes is likely caused by lack of mechanisms for such pigmentation.

Two major classes of pigments are each unique to a single order of birds. The copper-based pigments turacin and turacoverdin provide mechanisms for red and green feather coloration that are unique to species in Musophagiformes, and the highly restricted distribution of these pigments seems very likely to be an outcome of phylogenetic constraint: the mechanisms for production of turacin and turacoverdin simply has never evolved in any other lineage of birds (Hood and Hill 2009). How and why these copper-based pigments evolved in Musophagiformes remains a mystery. Similarly, psittacofulvins are restricted to Psittaciformes, and it is interesting to speculate that red and yellow psittacofulvin pigmentation is a form of dishonest signaling. All red and yellow feather coloration in parrots results from psittacofulvins; carotenoids, the alternative source of red and yellow coloration, are never incorporated into parrot feathers (McGraw and Nogare 2004). At the same time, the red bill coloration of many parrots results from carotenoid pigments, and psittacofulvins are never used as bill pigments (McGraw and Nogare 2004). The potential significance of this pattern lies in the fact that carotenoid color displays are widely recognized as signals that are challenging to produce and that reflect individual condition, while psittacofulvin coloration seems disconnected from individual condition (Berg and Bennett 2010). It seems that whenever possible, psittacofulvin pigmentation is prioritized over carotenoid coloration, perhaps to circumvent the restriction of honest signaling.

KEY POINTS

- Avian coloration seems boundless in diversity and complexity, but the colors of birds actually result from combinations of a limited set of pigment types and structural mechanisms. The two fundamentally distinct sources of coloration—pigments and microstructures—are both widespread across Aves, and both produce a wide spectrum of color displays.
- Genes typically control gross-scale, species-level differences in color patterns, but much of the variation among individuals within a species emerges from the effects of the environment.
- Major advances in understanding the chemistry and physics of pigments and microstructures has been achieved in the last few decades, but the tools for robust

investigation of the genetic basis of coloration have only recently become available.
- Birds derive their coloration either from pigments or from structural elements.
- Most pigmentary coloration in birds is derived from either melanin or carotenoids.
- Structural coloration can be either iridescent, such that hue changes with viewing angle, or non-iridescent, such that hue is constant regardless of viewing angle.
- Genes determine the overall pattern of coloration, but the environment can have a large effect on how coloration is executed.
- The genes responsible for avian coloration are being deduced using new genomic tools.

DISCUSSION QUESTIONS

1. How could one determine whether a patch of coloration is the result of pigments or microstructures?
2. How do birds use carotenoids to color their feathers and bare parts? What is the source of most red carotenoid pigmentation?
3. How did paleontologists use an understanding of the mechanisms of feather coloration to recreate the coloration of extinct animals?
4. What is a color morph?
5. What is the basis for the coloration of eggshells?
6. Which orders of birds have their own unique pigmentation systems, and what pigments are produced?

References

Badyaev, A. V., and G. E. Hill (2000). Evolution of sexual dichromatism: Contribution of carotenoid-versus melanin-based plumage coloration. Biological Journal of the Linnean Society 69:153–172.

Badyaev, A. V., and G. E. Hill (2003). Avian sexual dichromatism in relation to phylogeny and ecology. Annual Review of Ecology and Systematics 34:27–49.

Bauernfiend, J. C. (1981). Carotenoids as colorants and vitamin A precursors. Academic Press, London.

Berg, M. L., and A. T. Bennett (2010). The evolution of plumage colouration in parrots: A review. Emu 110:10–20.

Birkhead, T. (2003). A brand-new bird: How two amateur scientists created the first genetically engineered animal. Basic Books, New York.

Bókony, V., L. Z. Garamszegi, K. Hirschenhauser, and A. Liker (2008). Testosterone and melanin-based black plumage coloration: A comparative study. Behavioral Ecology and Sociobiology 62:1229–1238.

Borgia, G. (1985). Bower quality, number of decorations and mating success of male satin bowerbirds (Ptilonorhynchus violaceus): An experimental analysis. Animal Behaviour 33:266–271.

Brawner, W. R., III, G. E. Hill, and C. A. Sundermann (2000). Effects of coccidial and mycoplasmal infections on carotenoid-based plumage pigmentation in male House Finches. The Auk 117:952–963.

Brockmann, H., and O. Völker. 1934. Der gelbe Federfarbstoff des Kanarienvogels (Serinus canaria canaria [L.]) und das Vorkommen von Carotinoiden bei Vögeln. Hoppe-Seyler's Zeitschrift für physiologische Chemie 224:193–215.

Brush, A. H. (1981). Carotenoids in wild and captive birds. In Carotenoids as colorants and vitamin A precursors, J. C. Bauernfeind, Editor. Academic Press, London, pp. 539–562.

Brush, A. H. (1990). Metabolism of carotenoid pigments in birds. Federation of American Societies for Experimental Biology Journal 4:2969–2977.

Buchanan, K. L. (2000). Stress and the evolution of condition-dependent signals. Trends in Ecology and Evolution 15:156–160.

Church, A. H. (1869). Researches on turacin, an animal pigment containing copper. Philosophical Transactions of the Royal Society of London B: Biological Sciences 159:627–636.

Cooke, F., and P. A. Buckley (1987). Avian genetics: A population and ecological approach. Academic Press, London.

Dale, J. (2006). Intraspecific variation in bird colors. In Bird coloration, vol. 2, Function and evolution, G. E. Hill and K. J. McGraw, Editors. Harvard University Press, Cambridge, MA, pp. 36–86.

Dale, J., C. J. Dey, K. Delhey, B. Kempenaers, and M. Valcu (2015). The effects of life history and sexual selection on male and female plumage colouration. Nature. doi:10.1038/nature15509.

Delhey, K. A. Peters, and B. Kempenaers (2007). Cosmetic coloration in birds: occurrence, function, and evolution. American Naturalist 169:145–158.

Doucet, S. M., and M. G. Meadows (2009). Iridescence: A functional perspective. Journal of the Royal Society Interface 6:S115–S132.

Dyck, J. (1971). Structure and colour-production of the blue barbs of Agapornis roseicollis and Cotinga maynana. Zeitschrift für Zellforschung 115:17–29.

Eaton, M. D. (2005). Human vision fails to distinguish widespread sexual dichromatism among sexually "monochromatic" birds. Proceedings of the National Academy of Sciences 102:10942–10946.

Eaton, M. D., and S. M. Lanyon (2003). The ubiquity of avian ultraviolet plumage reflectance. Proceedings of the Royal Society of London B: Biological Sciences 270:1721–1726.

Fitze, P. S., M. Kolliker, and H. Richner (2003). Effects of common origin and common environment on nestling plumage coloration in the Great Tit (Parus major). Evolution 57:144–150.

Fowlie, M. K., and O. Kruger (2003). The evolution of plumage polymorphism in birds of prey and owls: The apostatic selection hypothesis revisited. Journal of Evolutionary Biology 16:577–583.

Fox, D. L. (1976). Animal biochromes and structural colours. University of California Press, Berkeley.

Fox, H. M., and G. Vevers (1960). The nature of animal colors. Macmillan, New York.

Fox, D. L., and J. W. McBeth (1970). Some dietary and blood carotenoid levels in flamingos. Comparative Biochemical Physiology 34:707–713.

Galeotti, P., D. Rubolini, P. O. Dunn, and M. Fasola (2003). Colour polymorphism in birds: Causes and functions. Journal of Evolutionary Biology 16:635–646.

Galván, I., and C. Alonso-Alvarez (2008). An intracellular antioxidant determines the expression of a melanin-based signal in a bird. PLoS ONE 3:e3335–e3335.

Garamszegi, L. Z. (2014). Modern phylogenetic comparative methods and their application in evolutionary biology. Springer-Verlag Berlin Heidelberg.

Goodwin, T. W. (1984). The biochemistry of carotenoids, vol. 2, Animals. 2nd ed. Chapman and Hall, New York.

Greenewalt, C. H., W. Brandt, and D. D. Friel (1960). The iridescent colors of hummingbird feathers. Journal of the Optical Society of America 50:1005–1013.

Griffith, S. C., T. H. Parker, and V. A. Olson (2006). Melanin-versus carotenoid-based sexual signals: Is the difference really so black and red? Animal Behaviour 71:749–763.

Hawkins, G. L., G. E. Hill, and A. Mercadante (2012). Delayed plumage maturation and delayed reproductive investment in birds. Biological Reviews 87:257–274. doi:10.1111/j.1469 185X.2011.00193.x.

Higginson, D. M., V. Belloni, S. N. Davis, E. S. Morrison, J. E. Andrews, and A. V. Badyaev (2016). Evolution of long-term coloration trends with biochemically unstable ingredients. Proceedings of the Royal Society B 283:20160403. http://dx.doi.org/10.1098/rspb.2016.0403.

Hill, G. E. (1996). Redness as a measure of the production cost of ornamental coloration. Ethology Ecology and Evolution 8:157–175.

Hill, G. E. (2006). Environmental regulation of ornamental coloration. In Bird coloration, vol. 1, Mechanisms and measurements, G. E. Hill and K. J. McGraw, Editors. Harvard University Press, Cambridge, MA.

Hill, G. E. (2010). National Geographic bird coloration. National Geographic Society, Washington, DC.

Hill, G. E. (2011). Condition-dependent traits as signals of the functionality of vital cellular processes. Ecology Letters 14:625–634.

Hill, G. E. (2014). Cellular respiration: The nexus of stress, condition, and ornamentation. Integrative and Comparative Biology 54:645–657.

Hill, G. E., and K. J. McGraw, Editors (2006). Bird coloration. Volume 1. Mechanisms and measurements. Harvard University Press, Cambridge, MA.

Hill, G. E., and J. D. Johnson (2012). The vitamin A-redox hypothesis: A biochemical basis for honest signaling via carotenoid pigmentation. American Naturalist 180:E127–E150. doi:10.1086/667861.

Hood, W. R., and G. E. Hill (2009). The mystery of turacin and turacoverdin: Why do Turacos have unique feather pigments? International Turaco Society Magazine 32:20–27.

Horak, P., H. Vellau, I. Ots, and A. P. Møller (2000). Growth conditions affect carotenoid-based plumage coloration of great tit nestlings. Naturwissenschaften 87:460–464.

Hubbard, J. K., J. A. C. Uy, M. E. Hauber, H. E. Hoekstra, and R. J. Safran (2010). Vertebrate pigmentation: From underlying genes to adaptive function. Trends in Genetics 26:231–239.

Inouye, C. Y., G. E. Hill, R. Montgomerie, and R. D. Stradi (2001). Carotenoid pigments in male House Finch plumage in relation to age, subspecies, and ornamental coloration. The Auk 118:900–915.

Isaksson, C. (2009). The chemical pathway of carotenoids: From plants to birds. Ardea 97:125–128.

Ito, S., K. Wakamatsu, and H. Ozeki (2000). Chemical analysis of melanins and its application to the study of the regulation of melanogenesis. Pigment Cell Research 13:103–109.

Kimball, R. T. (2006). Hormonal control of coloration. *In* Bird coloration, vol. 1, Mechanisms and measurements, G. E. Hill and K. J. McGraw, Editors. Harvard University Press, Cambridge, MA.

Li, Q., K.-Q. Gao, J. Vinther, M. D. Shawkey, J. A. Clarke, L. D'Alba, Q. Meng, D. D. Briggs, and R. O. Prum (2010). Plumage color patterns of an extinct dinosaur. Science 327 (5971): 1369–1372.

Li, Q., K.-Q. Gao, Q. Meng, J. A. Clarke, M. D. Shawkey, L. D'Alba, R. Pei, M. Ellison, M. A. Norell, and J. Vinther (2012). Reconstruction of Microraptor and the evolution of iridescent plumage. Science 335:1215–1219.

Ligon, J. D. (1999). The evolution of avian breeding systems. Oxford University Press, Oxford.

Ligon, R. A., and G. E. Hill (2013). Is the juvenal plumage of altricial songbirds an honest signal of age? Evidence from a comparative study of thrushes (Passeriformes: Turdidae). Journal of Zoological Systematics and Evolutionary Research 51:64–71. doi:10.1111/j.1439 0469.2012.00668.x.

Lopes, R. J., J. D. Johnson, M. B. Toomey, M. S. Ferreira, P. M. Araujo, J. Melo-Ferreira, L. Andersson, G. E. Hill, J. C. Corbo, and M. Carneiro (2016). Genetic basis for red coloration in birds. Current Biology 26:1427–1434.

Lozano, G. A. (1994). Carotenoids, parasites, and sexual selection. Oikos 70:309–311.

Mays Jr., H. L., K. J. McGraw, G. Ritchison, S. Cooper, V. Rush, and R. S. Parker (2004). Sexual dichromatism in the yellow-breasted chat *Icteria virens*: Spectrophotometric analysis and biochemical basis. Journal of Avian Biology 35:125–134.

McGraw, K. J. (2006a). Mechanics of melanin coloration. *In* Bird coloration, vol. 1, Measurements and mechanisms, G. E. Hill and K. J. McGraw, Editors. Harvard University Press, Cambridge, MA.

McGraw, K. J. (2006b). Mechanics of carotenoid coloration. *In* Bird voloration, vol. 1, Measurements and mechanisms, G. E. Hill and K. J. McGraw, Editors. Harvard University Press, Cambridge, MA.

McGraw, K. J. (2006c). Mechanics of uncommon colors: Pterins, porphyrins, and psittacofulvins. *In* Bird coloration, vol. 1,

Measurements and mechanisms, G. E. Hill and K. J. McGraw, Editors. Harvard University Press, Cambridge, MA.

McGraw, K. J., J. Dale, and E. A. Mackillop (2003). Social environment during molt and the expression of melanin-based plumage pigmentation in male House Sparrows (*Passer domesticus*). Behavioral Ecology and Sociobiology 53:116–122.

McGraw, K. J., and M. C. Nogare (2004). Carotenoid pigments and the selectivity of psittacofulvin-based coloration systems in parrots. Comparative Biochemistry and Physiology B: Biochemistry and Molecular Biology 138:229–233. doi:10.1016/j .cbpc.2004.03.011.

McGraw, K. J., and M. C. Nogare (2005). Distribution of unique red feather pigments in parrots. Biology Letters 1:38–43.

McGraw, K., E. Adkins-Regan, and R. Parker (2005a). Maternally derived carotenoid pigments affect offspring survival, sex ratio, and sexual attractiveness in a colorful songbird. Naturwissenschaften 92:375–380.

McGraw, K., R. Safran, and K. Wakamatsu (2005b). How feather colour reflects its melanin content. Functional Ecology 19:816–821.

McGraw, K. J., A. A. Cohen, D. Constantini, and P. Horak (2010). The ecological significance of antioxidants and oxidative stress: A marriage of functional and mechanistic approaches. Functional Ecology 24:947–949.

Merilä, J., and B. C. Sheldon (2001). Avian quantitative genetics. *In* Current ornithology. Springer, pp. 179–255.

Meunier, J., S. F. Pinto, R. Burri, and A. Roulin (2011). Eumelanin-based coloration and fitness parameters in birds: A meta-analysis. Behavioral Ecology and Sociobiology 65:559–567.

Montgomerie, R. M. (2006). Cosmetic and adventitious colors. *In* Bird coloration, vol. 1, Measurements and mechanisms, G. E. Hill and K. J. McGraw, Editors. Oxford University Press, New York, pp. 399–427.

Moreno, J., and J. L. Osorno (2003). Avian egg colour and sexual selection: Does eggshell pigmentation reflect female condition and genetic quality? Ecology Letters 6:803–806.

Mountjoy, K. G., L. S. Robbins, M. T. Mortrud, and R. D. Cone (1992). The cloning of a family of genes that encode the melanocortin receptors. Science 257:1248–1251.

Mundy, N. I. (2005). A window on the genetics of evolution: MC1R and plumage colouration in birds. Proceedings of the Royal Society of London B: Biological Sciences 272:1633–1640. doi:10.1098/rspb.2005.3107.

Mundy, N. I., J. Stapley, C. Bennison, R. Tucker, H. Twyman, K.-W. Kim, T. Burke, T. R. Birkhead, S. Andersson, and J. Slate (2016). Red carotenoid coloration in the Zebra Finch is controlled by a cytochrome P450 gene cluster. Current Biology. doi:10.1016/j.cub.2016.04.047.

Negro, J. J., A. Margalida, F. Hiraldo, and R. Heredia (1999). The function of the cosmetic coloration of bearded vultures: When art imitates life. Animal Behaviour 58:F14–F17.

Negro, J. J., G. R. Bortolotti, R. Mateo, and I. M. García (2009). Porphyrins and pheomelanins contribute to the reddish juvenal plumage of black-shouldered kites. Comparative Biochemistry and Physiology B: Biochemistry and Molecular Biology 153:296–299.

Olson, V., and I. Owens (2005). Interspecific variation in the use of carotenoid-based coloration in birds: Diet, life history and phylogeny. Journal of Evolutionary Biology 18:1534–1546.

Omland, K. E., and S. M. Lanyon (2000). Reconstructing plumage evolution in orioles (*Icterus*): Repeated convergence and reversal in patterns. Evolution 54:2119–2133.

Omland, K. E., and C. M. Hofmann (2006). Adding color to the past: Ancestral state reconstruction of bird coloration. *In* Bird coloration, vol. 2, Function and evolution, G. E. Hill and K. J. McGraw, Editors. Harvard University Press, Cambridge, MA, pp. 417–454.

Owens, I. P. F., and R. V. Short (1995). Hormonal basis of sexual dimorphism in birds: Implications for new theories of sexual selection. Trends in Ecology and Evolution 10:44–47.

Prum, R. O. (1999). Development and evolutionary origin of feathers. Journal of Experimental Zoology 285:291–306.

Prum, R. O. (2006). Mechanisms of structural color production. *In* Bird coloration, vol. 1, Measurements and mechanisms, G. E. Hill and K. J. McGraw, Editors. Harvard University Press, Cambridge, MA.

Prum, R. O., R. H. Torres, S. Williamson, and J. Dyck (1999). Coherent light scattering by blue feather barbs. Nature 396:28–29.

Prum, R. O., and S. Williamson (2002). Reaction-diffusion models of within feather pigmentation patterning. Proceedings of the Royal Society of London B 269:781–792.

Prum, R. O., and J. Dyck (2003). A hierarchical model of plumage: Morphology, development, and evolution. Journal of Experimental Zoology 298B:73–90.

Prum, R. O., E. R. Dufresne, T. Quinn, and K. Waters (2009). Development of colour-producing β-keratin nanostructures in avian feather barbs. Journal of the Royal Society Interface 6:S253–S265.

Prum, R. O., A. M. LaFountain, J. Berro, M. C. Stoddard, and H. A. Frank (2012). Molecular diversity, metabolic transformation, and evolution of carotenoid feather pigments in cotingas (Aves: Cotingidae). Journal of Comparative Physiology B 182:1095–1116.

Rathbun, M. K., and R. Montgomerie (2003). Breeding biology and social structure of White-winged Fairy-wrens (*Malurus leucopterus*): Comparison between island and mainland subspecies having different plumage phenotypes. Emu 103:295–306.

Remsen Jr., J., K. Winker, and S. Haig (2010). Subspecies as a meaningful taxonomic rank in avian classification. Ornithological Monographs 67:62–78.

Ridgway, R. (1914). The birds of North and Middle America 6. US Government Printing Office, Washington, DC.

Roulin, A., and C. Dijkstra (2003). Genetic and environmental components of variation in eumelanin and phaeomelanin sex-traits in the barn owl. Heredity 90:359–364.

Roulin, A., B. Almasi, A. Rossi-Pedruzzi, A.-L. Ducrest, K. Wakamatsu, I. Miksik, J. D. Blount, S. Jenni-Eiermann, and L. Jenni (2008). Corticosterone mediates the condition-dependent component of melanin-based coloration. Animal Behaviour 75:1351–1358.

Saino, N., S. Calza, R. Martinelli, F. De Bernardi, P. Ninni, and A. P. Møller (2000). Better red than dead: Carotenoid-based mouth coloration reveals infection in barn swallow nestlings. Proceedings of the Royal Society of London 267:57–61.

Shawkey, M. D., A. M. Estes, L. Siefferman, and G. E. Hill (2005). The anatomical basis of sexual dichromatism in non-iridescent ultraviolet-blue structural coloration of feathers. Biological Journal of the Linnean Society 84:259–271. doi:10.1111/j.1095 83 12.2005.00428.x.

Shawkey, M. D., and G. E. Hill (2006). Significance of a basal melanin layer to production of non-iridescent structural plumage color: Evidence from an amelanotic Steller's Jay (*Cyanocitta stelleri*). Journal of Experimental Biology 209:1245–1250.

Shawkey, M. D., M. E. Hauber, L. K. Estep, and G. E. Hill (2006). Evolutionary transitions and mechanisms of matte and iridescent plumage coloration in grackles and allies (Icteridae). Journal of the Royal Society Interface 3:777–786.

Siefferman, L., K. J. Navara, and G. E. Hill (2006). Egg coloration is correlated with female condition in eastern bluebirds (*Sialia sialis*). Behavioral Ecology and Sociobiology 59:651–656. doi:10.1007/s00265-005 0092 x.

Siefferman, L., and G. E. Hill (2007). The effect of rearing environment on blue structural coloration of eastern bluebirds (*Sialia sialis*). Behavioral Ecology and Sociobiology 61:1839–1846. doi:10.1007/s00265 007 0416 0.

Solano, F. (2014). Melanins: Skin pigments and much more—types, structural models, biological functions, and formation routes. New Journal of Science. http://dx.doi.org/10.1155/2014/498276.

Strasser, R., and H. Schwabl (2004). Yolk testosterone organizes behavior and male plumage coloration in House Sparrows (*Passer domesticus*). Behavioral Ecology and Sociobiology 56:491–497.

Tibbetts, E., and R. Safran (2009). Co-evolution of plumage characteristics and winter sociality in New and Old World sparrows. Journal of Evolutionary Biology 22:2376–2386.

Vinther, J., D. E. Briggs, R. O. Prum, and V. Saranathan (2008). The colour of fossil feathers. Biology Letters 4:522–525.

Vinther, J., D. E. Briggs, J. Clarke, G. Mayr, and R. O. Prum (2009). Structural coloration in a fossil feather. Biology Letters. doi:10.1098/rsbl20090524.

von Schantz, T., S. Bensch, M. Grahn, D. Hasselquist, and H. Wittzell (1999). Good genes, oxidative stress and condition-dependent sexual signals. Proceedings of the Royal Society of London B 266:1–12.

Wakamatsu, K., and S. Ito (2002). Advanced chemical methods in melanin determination. Pigment Cell Research 15:174–183.

Weidensaul, C. S., B. A. Colvin, D. F. Brinker, and J. S. Huy (2011). Use of ultraviolet light as an aid in age classification of owls. Wilson Journal of Ornithology 123:373–377.

Wilkinson, G. S., F. Breden, J. E. Mank, M. G. Ritchie, A. D. Higginson, J. Radwan, J. Jaquiery, W. Salzburger, E. Arriero, and S. Barribeau (2015). The locus of sexual selection: Moving sexual selection studies into the post-genomics era. Journal of Evolutionary Biology 28:739–755.

With, T. K. (1978). On porphyrins in feathers of owls and bustards. International Journal of Biochemistry 137:597–598.

Witschi, E. (1961). Sex and secondary sexual characters. *In* Biology and comparative physiology of birds, A. J. Marshall, Editor. Academic Press, New York, pp. 115–168.

CHAPTER 12

The Senses

Graham R. Martin

INTRODUCTION: SENSES AND SENSORY ECOLOGY

Understanding what a bird is able to sense is crucial to understanding its behavior. Senses are the only means by which any animal gains information about its environment. The information gained can be very diverse. It includes some seemingly basic information, such as ambient temperature, the direction of gravity, and properties of the earth's magnetic field; however, sensory information also includes highly sophisticated information. For example, birds can detect the presence of chemical compounds in the air or in water, specific information about the chemical and mechanical properties of objects in contact with parts of the body, and very accurate information about the positions and movements of objects both close to and at distance from the body. Information available in the environment changes continually, so the senses can rarely relax in their task of information extraction. Although gaining information is essential, it is also costly in terms of brain metabolism.

Senses are, however, highly selective. Even within a specific sense, such as vision, hearing, or olfaction, only certain types of information are detected and processed by the brain. Crucially, this information differs from one species to another. This is because what an animal senses has been subject to long and continuous processes of natural selection. Sensory information, and its associated mechanisms, has been selected to meet the challenges of particular tasks just as much as other aspects of an animal's biology. Avian senses are finely tuned to detect certain types of information or to give priority to the detection of particular information. This tuning is just as subtle as the more obvious evolutionary tuning of the more readily observed structural features of birds, such as wings, feet, and bills. We readily understand these

structures in terms of how they facilitate the exploitation of the resources of particular habitats. But to fully understand a bird's behavior, we need to first understand its sensory abilities, since they provide the information that allows its specialized wings, feet, and bills to be used.

The Senses: Limits, Differential Sensitivity, and Trade-Offs

The senses have been traditionally viewed as discrete channels of information, each providing a different type of "window on the world," and they have typically been investigated by isolating one particular sense or one particular aspect of a sense. The manipulation of just one or two stimulus parameters has been used to determine the limits of sensory performance. However, this knowledge has greatest value when the sensory capacities are viewed in a comparative context (i.e., compared with what is known of the sensory capacities of other species) so that researchers can begin to identify broadly conserved features of sensory biology and behavior (box on page 334).

Thresholds

Limits of sensory performance are typically identified by measuring **absolute thresholds**, which are the minimum amount of a stimulus that can be detected. In the case of sound, this will be the smallest amount of air disturbance that the hearing system can detect. In vision, it is the lowest number of photons per unit time that can be detected. In the chemical senses (taste and olfaction), it is the smallest number of molecules of a particular substance that can be detected. In mechanical senses (e.g., tactile sensitivity of the bill), it is the smallest displacement of the detecting surface; and in magnetoreception, it is the minimum strength of a magnetic field that can be detected.

COLLISIONS WITH HUMAN ARTIFACTS: A BIRD'S PERSPECTIVE

The world is full of human artifacts, and many pose problems for birds. Collisions with fixed obstacles including buildings, fences, powerlines, wind turbines, glass windows, and fishing nets are sufficiently frequent that they may pose threats to populations of birds of many species and in many parts of the world. An apparently puzzling aspect of many of these collisions is that they occur with objects that to humans appear to be large and conspicuous, and they frequently occur in daylight when visibility is good. If birds are visually guided while flying, why should they fail to see what lies ahead of them when conditions are clear?

First, collisions could result from birds "looking but failing to see." This is a class of explanation used for certain types of accidents involving human driving. It is a perceptual failure, a misinterpretation by the brain of the information provided by the eyes. In birds, certain human artifacts may present properties that do not match the kinds of stimuli that the eyes and brains of particular species have evolved to detect and interpret (Martin 2011).

A second type of collision may results from birds simply not looking ahead. This has been shown to be the case in birds of prey (notably vultures and eagles) and bustards. In flight, these birds may spend a lot of time looking down, searching for prey or suitable foraging patches, and when doing so they do not have any visual coverage of the path directly ahead (Martin and Shaw 2010).

The third type of collision problem can arise from human structures that are deliberately designed to be inconspicuous. This applies particularly to the use of fishing nets, which are designed to be not seen by fish. Unfortunately, the spatial resolution of birds may be very low underwater, especially if birds are foraging for fish at depth or at night when light levels are low. Therefore nets and even the structures associated with underwater turbines may be undetectable until a bird is at very close range (Martin and Crawford 2015).

All three types of collision occur primarily because of fundamental limitations and specializations of the visual systems of birds. In that light, it is clear that it is necessary to look at those obstacles "through birds' eyes." It is important to recognize that making the obstacles more conspicuous to humans will not necessarily render them more conspicuous to birds.

In some situations, it may be better to manage the space around the obstacles so that birds do not fly or forage in their vicinity. For example, by using habitat manipulation to encourage birds to land before they reach a power line. There is no single or easy solution to the problem of bird collisions with human artifacts. Mitigation measures need to be tuned to the sensory world and the ecology of specific bird species, as well as to the specific environments in which collisions occur.

Some common collision hazards. Wind turbines, power lines, fences, glass panes, oil platforms, and gill nets. *Photos by Graham R. Martin.*

Relative Sensitivity within a Sense

Each type of stimulus that an animal can detect may vary in a number of dimensions. For example, the vibrations of air molecules that we detect as sounds can have different frequencies or wavelengths (which humans describe as sounds of different pitch), and light also varies in frequency/wavelength (which we describe as light of different colors). In each individual there are likely to be different detection thresholds for each frequency of sound or light. Therefore, a complete characterization of avian vision or hearing requires knowledge of thresholds across a wide range of frequencies. For light this is presented as a **spectral sensitivity function**; for sounds this is presented as an **audiogram**.

Integration, Costs, and Trade-offs

It is not possible to detect everything that is going on in the environment. Humans are all too keen to think that the world that we know through our senses tells us how the world actually is. However, each species lives in a different world, as defined by its senses. In this way, species may share the same environment, but inhabit different worlds within it. No one of these worlds is more important or special than another; this is something that was first recognized more than two thousand years ago (box on page 335). All such worlds have been shaped by natural selection to extract information through different sensory capacities for the conduct of the life of each species.

An important constraint on how animals detect their environments is the metabolic costs of operating different sensory systems. Vision is particularly costly. Not only are eyes demanding of support and protection in the skull but also their actual running costs are high. There is a rapid and constant turnover of materials, and large amounts of neural processing are necessary to extract information from

EPICURUS AND SEXTUS EMPIRICUS

Illustrations depicting the philosophers Sextus Empiricus and Epicurus. *Creative Commons, Wikipedia.*

The Greek philosopher Epicurus (341–270 BCE) recognized that the senses, especially vision, not only place constraints on our overall understanding of the world but also shape how our personal worlds change from moment to moment. Epicurus argued that human reality is always changing and uncertain, but he did so from a human perspective. The Roman philosopher Sextus Empiricus (c. 160–210 CE) extended the insights of Epicurus to the world of nonhuman animals (Bailey 2002). In so doing, his arguments led to the philosophical foundations of Skepticism, the application of which underpins today's scientific world view (Popkin 2003).

Skepticism is based on a comparative approach to the senses and recognition that the sensory systems of animals differ one from another. From this, it is concluded that humans cannot ever be sure about the status of the world. This is because what is known about the world is very different depending on what information the different senses provide, even about the same set of objects. Different senses provide different information, and different animals have different senses; so where does reality lie? From these observations, Skepticism leads to the argument that it is not possible for humans to either affirm or deny any belief because, at a fundamental level, we cannot trust our senses to provide a foundation for certain truth about the world.

While the early skeptics could assert that there were differences between the senses of different animals, they based this assertion on the application of reason. They could not quantify these differences, nor could they say how or why differences came about. Today we have a range of techniques to investigate and quantify differences between the same senses in different animals.

In addition, an evolutionary framework now provides a way of understanding why and how differences have come about. The skeptical approach of Sextus Empiricus has been reinforced, not diminished, by the more modern understanding of how and why animals differ in their senses.

KIWI AND NOCTURNALITY: TACTILE CUES, OLFACTION, AND THE REGRESSION OF VISION

Nocturnality is rare among birds but is common among mammals. Even among flightless birds nocturnality is rare. One group of birds often compared with mammals because of their nocturnal foraging and their inability to fly is the five species of kiwi of New Zealand.

Kiwi were the first birds in which foraging guided solely by olfactory cues was demonstrated (Wenzel 1968). Their nostrils open at the tip of their long decurved bill, and the birds can be heard sniffing as they sort through leaf litter while foraging on the forest floor. Recent descriptions of the anatomy of the olfactory system have also attested to a prime role for olfaction in kiwi (Corfield et al. 2014). However, olfaction is not the only cue the birds use. Recent experimental and anatomical investigations show that kiwi also gather information from an organ at the tip of their bill that detects tactile information about the presence of prey as the bill is used to turn over and probe into substrata (Cunningham et al. 2007, Cunningham et al. 2009). Kiwi also have conspicuous whisker-like feathers that grow from the bases of their bills. Although the specific role of these specialized feathers has not been investigated, they may also provide tactile cues from somatic receptors around the feather bases.

Unlike owls—in which the eyes have evolved to become very large (axial length of 28 mm) in order to gather available light and produce a large and bright retinal image—kiwi eyes are very small (axial length of 7 mm) and can provide neither the sensitivity nor the resolution equivalent to those of owls at low light levels. Possessing very large and sensitive eyes would not seem to be a problem for a flightless bird, yet evolution has favored smaller rather than larger eyes in kiwi. Correlated with these small eyes is the region of kiwi's brains associated with the processing of visual information. It is small both absolutely and in proportion to the total brain volume (Martin

North Island Brown Kiwi *Apteryx mantelli*. Photo by Graham R. Martin.

et al. 2007b). This seems to suggest that, as the olfactory and somatic senses became more specialized for the extraction of information from the environment, there has been regressive evolution of vision. Without the need to guide flight or capture prey, it seems that the extraction of information from the bird's environment using vision has become of little importance in kiwi (Cunningham and Castro 2011).

It seems unlikely that this reliance on tactile, olfactory, and acoustic information for the guidance of nocturnal foraging and mobility could have evolved had there been mammalian predators in the kiwi's environment. The kiwi's relatively unusual suite of sensory information, perhaps unique among birds, has resulted in them now being particularly vulnerable to predation by the mammals that have been introduced by humans into their habitats. Kiwi can now thrive only on predator-free islands or in places where predator numbers are strictly controlled (Wilson 2004).

visual input. Eye size is a fundamental factor in both visual resolution (the amount of detail that can be extracted from a scene) and sensitivity (the minimum amount of light necessary for the extraction of information). As a general rule the larger the eye, the higher the sensitivity and resolution. The eyes of most birds are small, but there are also plenty of species with large eyes: for example, owls, albatrosses, raptors, and penguins. Interestingly, the eyes of relatively large kiwi species (Apterygidae) are similar in

size to those of small passerines. Because of kiwi's nocturnal habits, we might predict that large eyes capable of high visual sensitivity would be a considerable advantage, but this is not the case. The answer to this apparent paradox lies in the fact that in kiwi, many tasks are conducted with information derived from other senses, most notably olfaction, hearing, and tactile cues (box on page 336). This saves the high metabolic costs of building and maintaining large eyes, and this is a prime example of how the costs of losing visual information can be offset by information from other senses.

There may be complementarity, or trade-offs, between the information received from different sensory systems; however there are also significant trade-offs *within* a sensory system. Even within a particular sensory system, it is simply not possible to collect all the information that is potentially available. This is most clearly seen in the trade-off between visual resolution and sensitivity. At the very limit, both resolution and sensitivity are determined by the quantal nature of light and by noise within the nervous system elements that detect and transmit that information. This trade-off is evident most dramatically in the fact that resolution always decreases as higher sensitivity is gained. It is simply not possible to achieve high resolution at low light levels *and* vice versa. An eye that has evolved to achieve high sensitivity (which means that information can be reliably detected at low light levels), is unable to detect fine spatial information at these low light levels, but neither can a highly sensitive eye readily achieve high resolution when there is a lot of ambient light. Life is full of compromises, and that is certainly true both within and between different sensory systems.

A Unique Property of Vision

The trade-off between resolution and sensitivity highlights a uniquely important aspect of vision. Vision differs from all other senses in one crucial respect: the information that can be extracted from the visual environment varies markedly with the amount of the stimulus that is available. Across a wide range of sound levels, or concentrations of chemical compounds, or physical components of a touch stimulus, the sensory systems are able to extract more or less the same information. In vision this is not the case, because the information that is available changes profoundly as light levels change. This makes it very difficult for an observer to determine what visual information is available to a bird at any one moment. We may have knowledge of what a bird can detect at one light level, but that will not be the same at another light level. These differences are not trivial. At any one location, naturally occurring light levels may change over a range of at least a million-fold (10^6) between maximum sunlight and moonlight. On moonless nights the range is extended by a further 100-fold, and if we take into account how the presence of clouds and tree canopies can alter light intensity, then the total range of light levels in which a visual system could operate varies by a factor of 10^{11}. Over such ranges, the information potentially available through vision changes very markedly.

VISION

Vision is the primary means by which the majority of bird species extract information from their environment (box on page 338). Light has the advantage of direct and effectively instantaneous travel between object and receiver, thus allowing information to be extracted in real time. Vision is therefore the ideal source of information to inform the control of movement or the detection of objects and surfaces in a bird's environment. Recent synthesis of bird vision and behavior has suggested that "a bird is a bill guided by an eye" (Martin 2014). It seems that vision has evolved primarily to control foraging, particularly the accurate positioning of the bill, the timing of its opening. The control of locomotion is achieved within these requirements for accurate bill control (box on page 339).

Spatial Vision and Color Vision

We cannot talk about what an eye can do without immediately breaking vision down into a number of capacities that can be studied more or less independently. The crucial property of all eye types is that they are able to determine the position of a light source relative to the animal (Land and Nilsson 2012). That is, they have the capacity of "spatial vision." Since the first eyes evolved, their subsequent evolution has been in essence a series of refinements of spatial vision. These have resulted in the increasingly accurate determination of the positions of objects within a wide field of view about the animal, and over an increasingly wider range of light levels. It should be noted that color vision, which is often thought of as something rather different, is in fact an elaboration of spatial vision. Color vision allows the extraction of finer spatial detail by detecting differences in the wavelengths of light, not just differences in the intensities of the light.

The Fundamental Trade-Off: Sensitivity Versus Spatial Resolution

There is an important and fundamental trade-off in vision between the capacities of **sensitivity** (the ability to detect light) and **resolution** (the ability to resolve spatial detail using that light). Detailed analysis has shown that high resolution is not possible at low light levels (Land and Nilsson 2012). An eye that has evolved to detect ever-lower levels of light inevitably experiences a decrease in its ability to resolve spatial detail at those lower light levels. The inverse of this is also true; an eye that is highly sensitive to low light levels is unable to employ the same mechanisms to detect fine detail at high light levels. This is in fact a property of any vision system, including man-made ones such as photographic and video cameras (Land and Nilsson 2012).

GORDON LYNN WALLS (1905-1962) AND ANDRÉ ROCHON-DUVIGNEAUD (1863-1952)

Photographs of André Rochon-Duvignead and Gordon Llyn Walls. *Creative Commons, Wikipedia.*

The first landmark publication on comparative vision was Gordon Lynn Wall's book *The Vertebrate Eye and Its Adaptive Radiation* (Walls 1942). This book set a standard of breadth and depth in a single volume on senses that has perhaps not been surpassed. Walls ranged widely and brought together all that was then known about vertebrate eyes, their structure, physiology, and evolution, and much of what he described still stands. His volume contained 785 pages, with an index taking up 60 of them. Given the growth in knowledge about how eyes work and what they can do, a modern publication covering similar ground would have to be multivolume. In fact, a six-volume series, extend-ing to some 4,200 pages, called *The Senses* and focusing primarily on humans, was published in 2008 (Basbaum et al. 2008).

André Rochon-Duvigneaud published *Les Yeux et la Vision de Vertébrés* in 1943, one year after Walls's book. At the time, Rochon-Duvigneaud was 80 years old. He had retired from clinical practice as an ophthalmologist and devoted his retirement to comparative studies of the structure of the eyes in a wide range of vertebrate animals. Like Walls's book, Rochon-Duvigneaud's book was framed around a broad comparative approach. Also like Walls's, the book was encyclopedic and ran to over 700 pages. Rochon-Duvigneaud's book was never published in English, and it did not become as widely known as Walls's. These books complemented each other, and a joint publication covering all of the material would have been a marvelous compilation. The authors were of different generations. Walls graduated with his first degree in mechanical engineering at Tufts University in Boston, Massachusetts, in 1926, the same year that Rochon-Duvigneaud retired from his practice in clinical medicine at the Hotel-Dieu in Paris. They must have been working on their major works on comparative vision in parallel: one author at the start of his career at Wayne State University, Detroit, the other on his second career in retirement in Paris.

The importance of this is that, while it may be possible at low light levels for an eye to gain information about the presence of an object in a scene, the mechanism to extract that information cannot detect that object's presence with a high degree of certainty or discern fine detail of its structure. In fact, the more sensitive the analyzing system becomes, the greater the uncertainty. No matter how sensitive eyes are, they inevitably will reach a limit on the details that they can detect (see the box on page 340).

The clearest manifestations of this trade-off between sensitivity and resolution are studies that show how acuity (resolution) decreases with light level. Such decreases in resolution have been demonstrated in humans, owls, and doves (fig. 12.1) and in Great Cormorants (*Phalacrocorax carbo*) underwater (fig. 12.2) over a wide range of naturally occurring light levels.

In nocturnally active animals, eyes are typically large. This can be interpreted as an adaptation to trap a high number of photons from the scene in order to produce a relatively bright image on the retina (Land and Nilsson 2012). An absolutely large entrance aperture (pupil) is especially important for the detection of point sources of light within a scene. Eyes of large size are found in nocturnally active birds including owls and Oilbirds, which fly and forage regularly at low (crepuscular and nocturnal) light levels in terrestrial environments (see the box on page 340). Although these species probably have eyes that are among the most sensitive found in birds (Martin 1986, Martin et al. 2004), their vision is not sufficient to guide all foraging behavior: hearing is used in different ways by owls and by Oilbirds to complement their vision, and Oilbirds also employ olfaction to aid the location of their foods (fruits). This suggests that although vision in these species is close to the theoretical limit of sensitivity, resolution at low light levels is not sufficient to guide key aspects of their nocturnal behavior, and other sensory information must come into play.

The box on page 341 describes how woodland owls can achieve their nocturnal activity only as a result of complex interactions between vision, hearing, and behavioral adaptations (Martin 1986). Similarly, other bird species that forage regularly at night, such as some waterfowl and shorebirds, achieve this through complementarity between information received through vision and that obtained through touch sensitivity in the bill (Martin 2012). Diving birds that forage at depths where light levels are equivalent to nighttime at the surface may exploit fish prey that reveal their presence through light emitted from photophores on their body surface (King Penguins, *Aptenodytes patagonicus*) (Martin 1999), or they take relatively large but well-camouflaged fish (e.g., Great Cormorants) using a technique that involves the birds forcing the fish to reveal their presence by triggering an es-

cape response, and the escaping prey is detected and caught at short range only (box on page 344). In all these low light level situations, the vision of these birds hits a very real limit to its spatial resolution, and only larger objects can be detected. Thus vision can provide only partial information about these birds' environments; this may be compensated for by information from other senses, the exploitation of specific prey types, and/or specific foraging techniques.

Measures of Spatial Resolution
Acuity

Acuity is a measure of spatial resolution. It is typically measured with high-contrast stimuli. For example, black objects (which reflect little light across the spectrum), positioned against a white surface (which reflects that same light but very strongly) are very easily detected, as evidenced by this text, for example. Acuity is what has been measured in figures 12.1 and 12.2; it is quantified according to how close

A BIRD IS A BILL GUIDED BY AN EYE

Is there a simple way of encapsulating the sensory world of birds? For many years birds were envisaged as "a wing guided by an eye" (Rochon-Duvigneaud 1943). This neatly combined what appeared to be the two key elements of most birds: flight and its apparent dependence on visual information. However, the senses of birds are varied and complex. It appears that the primary task for which they provide information is foraging, not flight. Foraging includes detecting food items, verifying their palatability, and handling and ingesting them.

In most birds the primary task of vision is for seizing food objects in the bill. This is not a trivial task, since it requires

An American Robin *Turdus migratorius* foraging on a lawn. *Photo courtesy of Geoffrey E. Hill.*

getting the bill to the accurate location and opening the bill to seize an item, and all these tasks must be done at the right time.

Foraging is an exacting task and is subject to ever-present natural selection (Martin 2014). Natural selection of vision is, however, also likely to be driven by the demands of predator detection, and there are many examples where the demands for bill control and predator detection appear to be opposed. Only in species that can forage without the need for accurate control of bill position—for example in the filter-feeding ducks—can panoramic vision provide constant surveillance for predators. In most species, visual coverage for predator detection is maximized but it cannot be comprehensive. In these circumstances "vigilance" behaviors have evolved to overcome the lack of comprehensive vision.

Thus, two tasks seem to be the prime drivers of vision in birds: positioning and timing the bill opening with respect to a target, and detecting predators. Other behaviors, including flight, are controlled by information that can be extracted from the environment within the key constraints imposed by bill positioning and predator detection. Thus characterizing a bird as "a bill guided by an eye" (Martin 2014) helps clarify the essential functions and evolutionary drivers of the information gathered by bird sensory systems.

together lines can be in black-and-white striped patterns and still be resolved when presented from a standard viewing distance (fig. 12.1 shows an example of a grating pattern used to determine resolution). Although such high-contrast stimuli rarely occur in nature, acuity gives a useful indication of the "best" spatial performance at a given light level, and it permits between-species comparisons of visual performance at different light levels. The best performance always occurs under the high light levels that occur during daytime. Even so, the highest acuity of birds varies markedly across species, ranging from the highly acute vision of eagles and falcons to the relatively poor resolution of owls and small passerines (table 12.1).

It is important to consider the function of high resolution in birds. The highest resolution may not aid detection of very small objects when they are close, but it is probably particularly important for the detection of objects at a great distance. This seems to be the case in large raptors. For these birds, food items are relatively rare and widely scattered across the environment, and it can be argued that it is the need to detect objects at great distances that has led to the evolution of vision capable of high spatial resolution. The twofold higher acuity of an eagle compared with the highest acuity in young humans means that these eagles could detect a given object at twice the distance, and at an even greater distance when compared with the eye of an older human. This performance in an eagle is about eight times better than a Rock Dove (*Columba livia*). Because their ecology does not require the detection of small objects at large distances, doves would gain no advantage in being able to match the performance of an eagle or even of a human (the most acute humans have acuity about four times better than that of doves). Thus it would seem that, for doves, which feed on small items detected on the ground, eagle-eyed acuity would have no

OILBIRDS AND NOCTURNALITY: VISION, ECHOLOCATION, AND OLFACTION

A fascinating example of how a unique suite of senses are used to guide behavior in very low light conditions is provided by Oilbirds (*Steatornis caripensis*). Not only are Oilbirds strictly nocturnal, they also nest and roost deep inside caves where no light penetrates.

As in the case of kiwi and owls, Oilbird behavior is made possible by the integration of information from a suite of senses. Like owls, their eyes are relatively large and produce perhaps the brightest retinal image found in bird eyes. This bright image is analyzed by a retina that is packed with rod receptors. In some portions of the retina, those receptors appear to be tiered on top of each other, a feature also found in some deepsea fish, in which it provides high sensitivity but low resolution. It appears that Oilbird eyes are at the limits of sensitivity that is possible in the eye of a bird (Martin et al. 2004). However, low resolution means that vision can be used to guide foraging and locomotion only with respect to relatively large structures. Oilbirds overcome this by flying at low speed over the woodland canopy at night, where they use the sense of smell to locate ripe fruits in the upper branches of trees. In caves, the birds use hearing (both active and passive SONAR) to determine the positions of cave walls; to locate nest sites, mates, and young; and to avoid colliding with one another. Slow flight speed and hovering help avoid dangerous encounters. Together with slow locomotion, the integrated use of information from low-resolution but sensitive vision, low-resolution echolocation, and olfaction allows these birds to exploit fruits within the canopies of tropical forests and to exploit roosting and nest sites within caves, while remaining relatively safe from predation under the cover of darkness.

Oilbird *Steatornis caripensis*. Photo by Graham R. Martin.

OWLS AND NOCTURNALITY: VISION, HEARING, AND TERRITORIALITY

Activity, especially flight and foraging, at low light levels poses specific challenges for birds. Owls are the most nocturnal group of birds, but not even all owl species are strictly nocturnal; many species are active in twilight and some are diurnal. Among the most strictly nocturnal species are owls of the genus *Strix*, and the sensory capacities of Tawny Owls (*Strix aluco*) have been studied in some detail (Martin 1990). In order to be active at night, especially under a woodland canopy, *Strix* owls require highly sensitive vision. This alone, however, is not sufficient to account for their nocturnal behavior.

Tawny owl eyes are large and protrude well outside the protection of the skull. The structure of their eyes produces a retinal image with a maximum brightness approximately 2.5 times higher than in human eyes. Owls' threshold of detectable light is also about 2.5 times better than that of human eyes, indicating that it is the optical system, rather than the retina, that gives owls the advantage.

Light levels experienced on a forest floor at night may be well below even an owl's threshold for vision. Therefore exceptional vision cannot be the sole answer to nocturnality in owls. It is well established that hunting owls also rely on hearing, in particular accurate sound localization. Unlike all other birds so far investigated, owls can pinpoint a sound source with high accuracy, both in direction and range, over a distance of a few meters.

Sound cues produced by a prey animal as it moves through leaf litter are sufficient for an owl to capture prey in total darkness. The owl waits quietly on a low perch for a prey item to come within close range, then the bird drops, spreading its talons, and captures the prey without ever needing to see it. This accurate sound localization is made possible by the elaborate outer ear structures that are a unique characteristic of owls.

Impressive as this is, it is clear that the birds cannot do this acoustic location and capture of prey without first developing considerable familiarity with the situation in which they operate. Thus it appears that there is a considerable cognitive component to prey capture. Cognition is involved not only in learning the ways in which sounds are degraded with distance, as in typical sound ranging, but birds must also gain knowledge of the positions of major branches, which can be seen on the darkest nights only in silhouette against the sky.

Gaining this information is not easy, and the birds have to spend extended time in known situations so that these minimal sound and visual cues can be interpreted correctly. It seems likely that an owl will have opportunity to gather this information when moonlight raises light levels above those of absolute visual threshold. At those times, finer details of the structure of the specific woodland habitat in which the bird lives can be determined. Thus each bird has to be resident in the same situation for an extended period; this is borne out in Tawny Owls by their extreme territorial behavior. Tawny owls remain on the same territory throughout their lives, which in northern Europe is typically 5 or more years but can be over 20 years. During this time the owls go nowhere. Such extreme residence is probably necessary to allow the birds to gather sufficient information so that they are able to interpret the minimal visual and hearing information necessary to guide their foraging and movements. But, equally important, once the birds have learned about their territory, they cannot go elsewhere, since they could not operate successfully in a new situation.

A consequence of this is that, because owls stay in one place all their lives, they cannot be selective about their diet; they are prepared to take anything, including earthworms, frogs, beetles, small rodents—in fact anything that will cause a sound as it moves through litter.

Thus it is only by a combination of hearing and vision, plus behavioral adaptations, that owls can be strictly nocturnal.

Tawny Owl *Strix aluco. Photo by Graham R. Martin.*

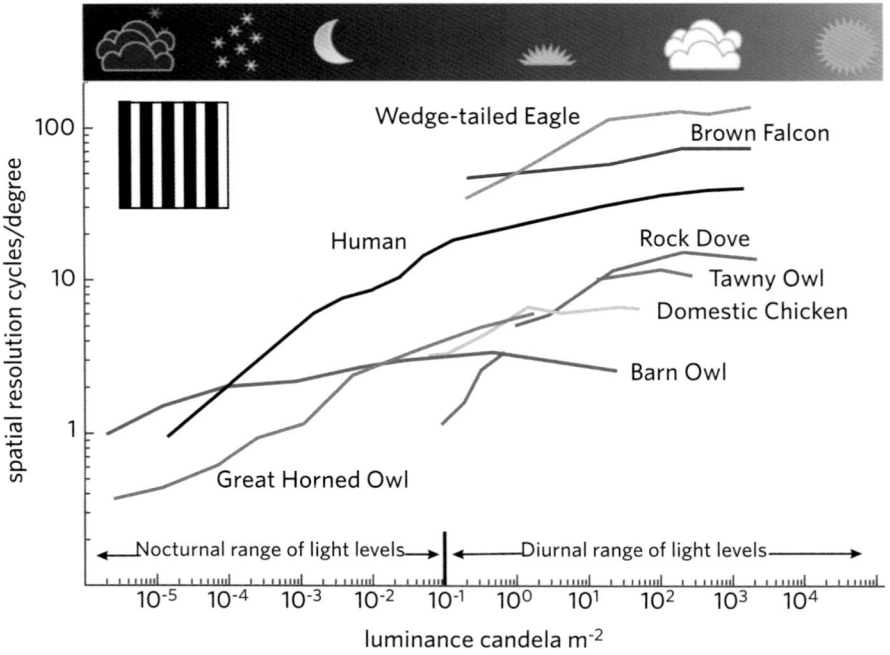

Figure 12.1. How light levels influence the ability to resolve spatial detail. In all species, resolution was measured as the highest frequency (cycles/degree) of a grating pattern of equally spaced black and white stripes (see example, *top left*) that could be resolved at different light levels. The variation is considerable among the species, but they all show a decrease in resolution as light levels fall, although the rate of decrease in acuity also differs among species. The daytime (diurnal) and nighttime (nocturnal) ranges of natural light levels are indicated, and the symbols above give an idea of the natural light sources. Wedge-tailed Eagle (*Aquila audax*) (Reymond 1985); Brown Falcon (*Falco berigora*) (Reymond 1987); Rock Dove (*Columba livia*) (Blough 1971, Hodos and Leibowitz 1977); Tawny Owl (*Strix aluco*) (Martin and Gordon 1974); Domestic Chicken (*Gallus gallus domesticus*) (Gover et al. 2009); Barn Owl (*Tyto alba*) (Orlowski et al. 2012); Great Horned Owl (*Bubo virginianus*) (Fite 1973); Human (Shlaer 1937, Pirenne et al. 1957). *Based on an original drawing, courtesy of Mindaugas Mitkus (Vision Group, Lund University).*

selective advantage. Consequently, they have evolved much lower spatial resolution in their vision.

The high resolution of raptors is not without costs. As light levels fall acuity drops rather rapidly, and there is evidence that this occurs in American Kestrels (*Falco sparverius*) (Hirsch 1982) as well as doves (fig. 12.1). In owls, on the other hand, such high acuity is not achieved at high light levels, but the decline of resolution with light levels is probably less precipitous than in the diurnal raptors. Furthermore, the vision of larger eyes may be more prone to disruption when the image of the sun falls on the retina. This has led to the evolution of various "sunshade" structures, including prominent "brows" above the eyes and relatively reduced fields of view, in large-eyed bird species compared with smaller-eyed and less visually acute bird species (Martin and Katzir 2000).

Contrast Sensitivity

Natural stimuli are rarely black and white. Most real-world tasks require the detection of gray targets (i.e., targets that reflect some light across a range of wavelengths) against gray backgrounds. A way of describing the ability to detect such stimuli is to measure **contrast sensitivity**.

This approach to investigating the effect of contrast on spatial resolution has been to measure contrast sensitivity functions (Ghim and Hodos 2006). In effect, these functions determine the minimum amount of contrast that can be detected for stripes of different widths when they are presented as gratings of different spatial frequencies. From this it is possible to determine how wide stripes have to be for the smallest difference in contrast to be visible. Studies have suggested that birds have surprisingly low contrast sensitivity compared with mammals, including humans. That is, at a given stripe width, contrast has to be higher in birds than in mammals before it can be detected (Ghim and Hodos 2006, Harmening et al. 2009, Lind and Kelber 2011, Lind et al. 2012). In all vertebrate species tested to date, contrast sensitivity declines as stripe widths either increase or decrease away from the widths where contrast sensitivity is highest. Thus, there

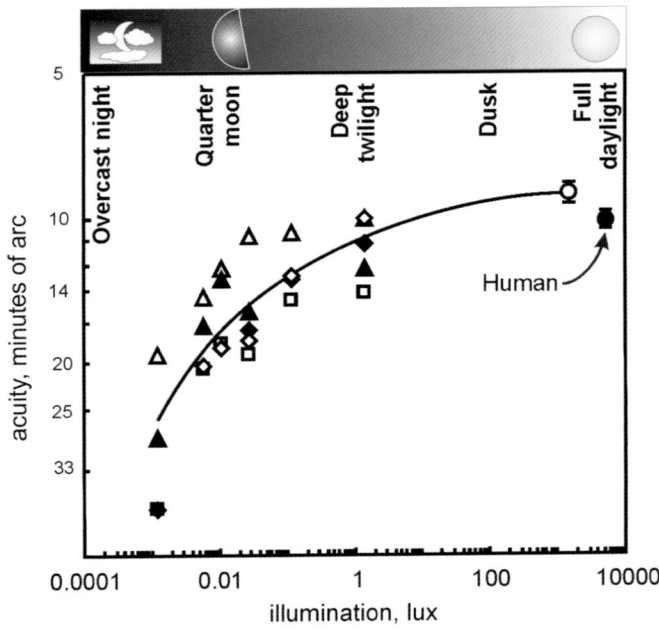

Table 12.1. Spatial resolution and acuity in a sample of bird species measured at high (daytime) light levels, indicating the best performance recorded for each species. The resolution of humans is for young people. Eagle (Reymond 1985); kestrel (Hirsch 1982); dove (Hodos et al. 1976); owl (Fite 1973); sparrow (Dolan and Fernandez-Juricic 2010); human (Land and Nilsson 2012).

Species	Spatial resolution Cycles/degree	Acuity Minutes of arc
Wedge-tailed Eagle (*Aquila audax*)	142	0.21
America Kestrel (*Falco sparverius*)	40	0.75
Rock Dove (*Columba livia*)	18	1.7
Great Horned Owl (*Bubo virginianus*)	7.5	4
House Sparrow (*Passer domesticus*)	4.8	6.3
Human	72	0.4

Figure 12.2. Visual acuity in Great Cormorants underwater at a range of light levels. Acuity was measured in five birds. At high light levels, the birds showed very similar acuity of about 10 minutes of arc, but as light levels decreased, acuity differed among the birds. For example, the visual acuity of the bird indicated by the open triangle symbol was always higher than that of the bird indicated by the open square symbol. The line is the best fit through all of the data points and shows clearly how acuity decreases with falling light levels. The data point for humans is the acuity of European children underwater without the aid of face masks; the best performance of cormorants is very similar. However, underwater acuity twice as good as cormorants (five minutes of arc) has been found in children of the Moken people of Southeast Asia who dive to collect food from the seafloor without the use of visual aids (Gislen et al. 2003). *Illustrations by Craig White and Graham R. Martin, published in White et al. 2007.*

is a middle range of stripe widths where small amounts of contrast differences are relatively easily detected. Most importantly, these studies demonstrate that when stimuli are of low contrast, the optimal size at which they are most likely to be detected is relatively large; both very broad and very narrow striped patterns require a high contrast for them to be detected (fig. 12.3).

An indication of how spatial resolution is affected by both light levels and the degrees of contrast in a target is indicated in the study of underwater vision in Great Cormorants (fig. 12.4). It was possible to combine resolution and contrast data to model how natural-type targets (fishes of a fixed size but of different contrasts with the background), viewed from different distances, would appear to the birds (fig. 12.5). This modelling shows that target contrast has a profound effect on the ability of the birds to detect a naturalistic target, even

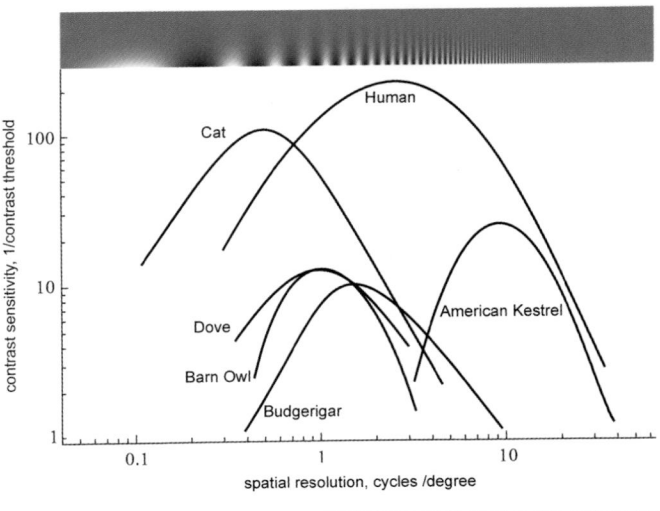

Figure 12.3. Contrast sensitivity functions of four species of birds, cats, and humans. Spatial resolution is measured as described in figure 12.1 using gratings, but in this case the contrast between the stripes is altered at each frquency to determine the minimum contrast at which stripes of each width can be detected. The insert at the top of the illustration provides an indication of how spatial frequency is varied. The illustration is courtesy of Mindaugas Mitkus, Vision Group, Lund University, and assembled from Dove (*Columba livia*), Hodos et al. 2002; Barn Owl (*Tyto alba*), Harmening et al. 2009; Budgerigar (*Melopsittacus undulatus*), Lind et al. 2012; American Kestrel (*Falco sparverius*), Hirsch 1982; cat, Bisti and Maffei 1974; human, Berkley 1976.

THE CHALLENGES OF FORAGING UNDERWATER: CORMORANTS

Many birds are amphibious. They conduct most of their activities in air but seek food underwater. Birds that forage in this way include many species of ducks, auks, penguins, petrels, cormorants and gannets, grebes and divers. There is no evidence that birds can use hearing or olfaction underwater. Some ducks probably use tactile sensitivity in their bills to find and then detach sessile prey from rocks. However, most amphibious birds use vision to catch fish or other prey. Such foraging presents a number of sensory challenges. First, water is often turbid and is usually less transparent than air. Second, light levels fall dramatically with depth. And third, seeing well both above and below the water surface presents a unique sensory challenge. This arises because an eye that is well focused in air will lose that focus upon immersion because the cornea will no longer function as a lens.

Great Cormorants (*Phalacrocorax carbo*) are notable for their skill as underwater foragers and their high foraging efficiency (Gremillet et al. 2004) and their ability to catch prey even in highly turbid waters (Gremillet et al. 2012) and at night (Gremillet et al. 2005). Behavioral measures of acuity in Great Cormorants have shown that their underwater vision is in fact surprisingly poor and that resolution falls steeply with decreasing light levels (Martin et al. 2008) (fig. 12.2). In short, cormorants are unable to detect fine detail even in clear waters; cormorant vision is no better than the vision of a young human underwater without a face mask.

Rather than detecting fish at a distance and then hunting them down, cormorants rely on disturbing fish that are hidden. Cormorants are extremely maneuverable underwater, and video of foraging cormorants reveals that they poke their bills into holes and substrata. They force prey to make an escape, and they use rapid neck extension and precise timing of bill opening to catch the prey as it escapes (Gómez-

Great Cormorant *Phalacrocorax carbo*. Photo by Graham R. Martin.

Laich et al. 2015, Watanuki et al. 2008). Because of their poor visual resolution and their exploitation of often turbid and low light level conditions, cormorants must in effect be lunging at an escaping blur, rather than at a clearly seen fish (White et al. 2007). There is, of course, a high probability that such an escaping blur is suitable prey, but the birds bring the fish to the surface—often to the annoyance of anglers watching from the shore—perhaps to both verify what has been caught and reposition it in the bill for swallowing. Their binocular visual field allows them to examine what they are holding (Martin et al. 2008).

Cormorants provide a good illustration of the fact that even highly efficient predatory behaviors do not necessarily require senses of high performance for their completion. In this case, prey within a particular size range, and showing particular behaviors, are exploited by the combination of relatively low level visual information with specific prey capture behaviors.

at short range. The vision of these underwater predators is nowhere near equal to that of aerial predators; at its best it is about 50 times lower than an eagle in bright daylight.

Color Vision and Sensitivity in the Spectrum

We take it for granted that the environment is full of color. This is not, however, the case. Color is a property of the nervous system, not the environment. This truth was first captured in the phrase "the rays are not colored," stated by Isaac Newton in 1704 (Newton 2010). We cannot know for certain whether birds see the world in "color." All that can be determined with certainty is the range of wavelengths of light to which a bird is sensitive (the avian visible spectrum), and the ways a bird can subdivide that spectrum into discrete sections. We cannot actually know what colors birds see.

It is clear that many birds detect light over a wider range of wavelengths than the range to which humans are sensitive. In other words, birds have a broader visible spectrum. Also, it seems that birds are able to discern more colors within

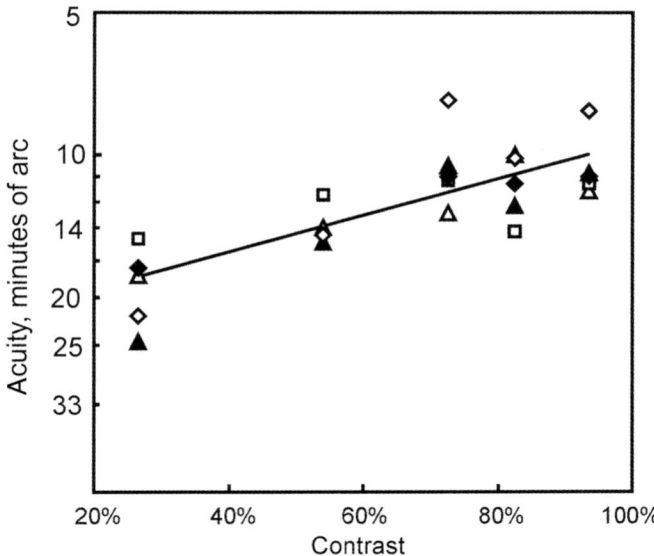

Figure 12.4. How the contrast within a target influences the ability to resolve detail. This is indicated by how acuity falls as the contrast of the target changes from strong black-and-white contrasts (100 percent) to low contrast when the target contains only different shades of gray (20–40 percent contrast). This figure shows that the acuity of Great Cormorants underwater almost halves, from a best resolution of about 10 minutes of arc when contrast is high to 20 minutes of arc when contrast is low, even though the ambient light level has not changed. *Illustrations by Craig White and Graham R. Martin, published in White et al. 2007.*

their spectrum (i.e., they can probably make finer color discriminations). Some birds can detect light in the ultraviolet (UV) part of the spectrum, light to which human vision is insensitive. It is important to note, however, that not all birds see in the UV. In fact, bird species with true UV vision are found only in the gulls (Laridae, Charadriiformes), ostriches (Struthioniformes), parrots (Psittaciformes), and the oscine passerines (Passeriformes), but excluding the Corvidae (Martin and Osorio 2008). Other species may have visual sensitivity that extends into the violet-near UV spectrum, but they lack a specific UV receptor in their eyes and cannot be considered truly UV sensitive.

The prime function of color vision is that it allows the extraction of spatial detail by using differences in the wavelengths of the light that is reflected from structures in the environment. For example, two patches of plumage could reflect equal numbers of photons; thus there would be no differences in their visibility based just on the number of photons. However, photons from each of those patches may be from different parts of the spectrum. A bird with color vision could readily detect differences between the patches, whereas one without color vision could not.

It is not possible to know what kinds of objects and tasks were first detected by the color vision systems of ancestral birds, although today it is possible to identify examples of objects for which detection is enhanced by color vision. These may include objects used in display behaviors (Endler et al. 2014), plumage patterns (Bennett et al. 1997), and particular types of fruits (Burkhardt 1982). Simply seeing more differences between colors is unlikely to have been the driver of natural selection; it is the spatial information about objects that probably drives selection in the direction of detecting fine differences within the spectrum.

The most detailed knowledge available on the detection of differences in the spectrum is in Rock Doves (Wright 1979, Hodos 1993). However, it is possible to say something about color vision, and more generally about breadth of the spectrum that birds can see, from knowledge of the visual pigments found in the cone photoreceptors of bird retinas. Cone-shaped light receptors in the retina (elaborated in the following section) is where analysis of color differences in the image begins. Based on knowledge of those receptors, it seems safe to assume that all birds have color vision. Even the nocturnally active owls seem to have some color vision, although it is not as sophisticated as that of other bird species (Martin 1974, Bowmaker and Martin 1978). Genome analysis, but not the determination of visual pigments directly, suggests that color vision may be absent from kiwi species (Le Duc et al. 2015), and this may be the only group of birds that is not capable of some kind of color discrimination.

Photoreceptors and Visual Pigments

The photoreceptors of all vertebrate retinas are of two functional types: rods and cones. Cones are classified primarily according to the wavelength of light to which their photopigments are most sensitive (fig. 12.6). Photopigments trap photons from the image that is projected onto them by the eye's optical system. Once a photon is trapped, a cascade of chemical reactions is initiated, which eventually results in the conduction of activity into the nervous system. In human eyes, cone receptors containing three types of photopigments (Land and Nilsson 2012) provide the basis for color vision. In birds, there are four types of cone photopigments, and avian eyes also contain rod receptors. The important functional difference between rods and cones is that rods function primarily at low light levels (twilight and below), while cones function at higher light levels (the levels experienced in twilight and daytime). Both types of receptors function within a range of light levels, but during lower twilight and night time, spatial resolution is purely a property of the rods. Because those rods are of one type only, color vision is not possible under those light conditions. The dual different functions of the rod

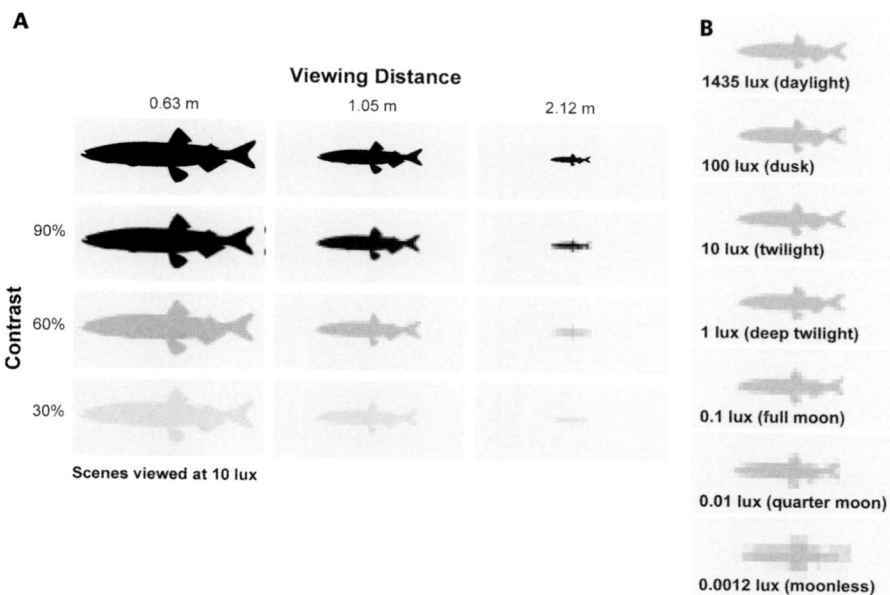

Figure 12.5. Simulations of a cormorant's-eye view of fish underwater under different conditions of viewing. These views are based on the kind of acuity data shown in figures 12.3 and 12.4. In *A*, the light level is held constant at the lower end of the daylight range—how a fish of the same size and shape (a silhouette of a 10-cm-long Capelin, *Mallotus villosus*, which are commonly taken by Great Cormorants) will appear at three different viewing distances and under four different levels of contrast with the background light. Apart from the high contrast fish viewed at close range, all target fish will appear quite indistinct. In *B*, the fish has a realistic mid-range contrast of 60 percent and is viewed at a distance of 1m under a range of naturally occurring light levels from daylight to low moonlight; again the target fish appears as an indistinct blur for many naturally occurring conditions. *Illustrations by Craig White and Graham R. Martin, published in White et al. 2007.*

and cone receptors enable the vertebrate eye to function over a very wide range of naturally occurring light levels.

Spectrophotometric measurements of the sensitivity maxima (λ_{max}) of the photopigments found in bird retinas indicate that these photopigments fall into five classes (four in the cone receptors and one in the rod receptors). Four photopigment classes show a high degree of similarity across a wide range of species. This suggests that color vision arose early in bird ancestry and its properties have been highly conserved. For example, the visual pigments found in the eyes of a species of pelagic seabird (Wedge-tailed Shearwaters, *Puffinus pacificus*, Procellariiformes; Hart 2004) are very similar to those found in a phylogenetically distant species that is terrestrial and lives in open forest habitats (Indian Peafowl, *Pavo cristatus*, Galliformes; Hart 2002, Hart and Hunt 2007). This suggests that color vision in birds has general, all-purpose properties that are not tuned to specific tasks performed by different species.

The five types of visual pigments of birds are labeled and defined as follows:

RH1: rhodopsin type 1, with a sensitivity maximum at
  about 500 nm (λ_{max} 500 nm), found in the rod receptors

RH2: rhodopsin type 2 (λ_{max} 505 nm), found in cone
  receptors
SWS2: short-wave type 2 (λ_{max} 470 nm), found in cone
  receptors
LWS: long wave (λ_{max} 565 nm), found in cone receptors
SWS1: short wave sensitive type 1 is also found in cone
  receptors but is differentiated into two types, and these
  types are found in different species. Pigments with λ_{max}
  at 365 nm are referred to as ultraviolet sensitive, UVS;
  those in which λ_{max} is at 410 nm are referred to as violet
  sensitive, VS (Wilkie et al. 2000).
Possession of the UVS pigment gives the oscine passerines,
  gulls, ostriches, and parrots their visual sensitivity into
  the ultraviolet part of the spectrum.

To date only Humboldt Penguins (*Spheniscus humboldti*) have provided a likely exception to the general uniformity of avian photopigments. This species has been reported to have cone pigments with λ_{max} 403 nm, 450 nm, and 543 nm (Bowmaker and Martin 1985), and it is suggested that the penguin's LWS pigment is shifted to a shorter wavelength maximum (565 nm → 543 nm) and the RH2 pigment is absent.

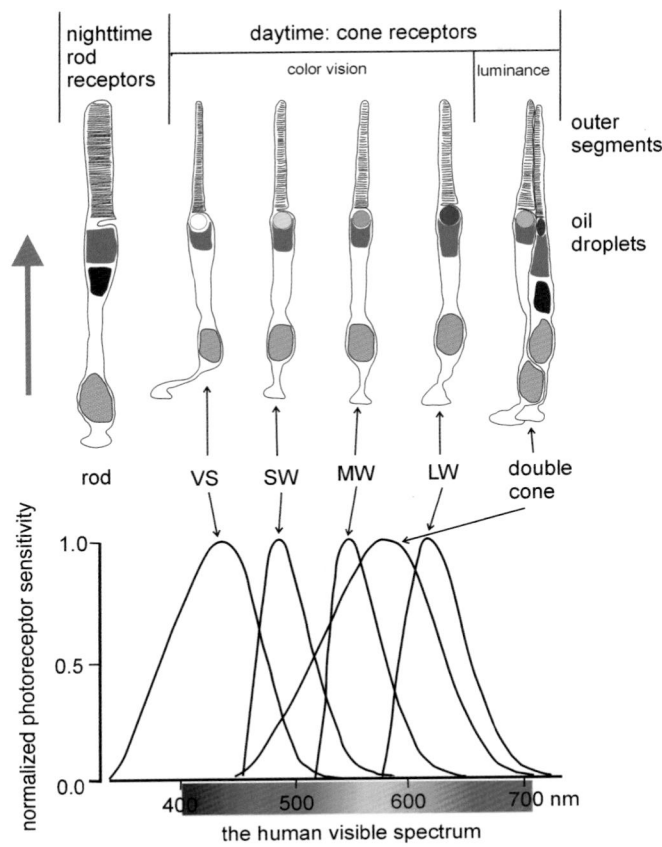

nighttime rod receptors

daytime: cone receptors

color vision

luminance

outer segments

oil droplets

rod VS SW MW LW double cone

normalized photoreceptor sensitivity

1.0

0.5

0.0

400   500   600   700 nm

the human visible spectrum

Figure 12.6. The photoreceptors of birds. The four types of single cones, the double cones, and the rods are shown (*top*). The outer segments contain the photopigments, and the oil droplets are depicted in the characteristic colors as they appear to the human eye. The large blue arrow indicates the direction that light travels to make up the retinal image at the level of the outer segments. In effect the photoreceptors face away from the direction of light travel; the drawings make clear that the oil droplets filter the light that reaches the outer segments and thus modify the part of the light spectrum that each receptor type is selectively sensitive to. The labels designating the main types of cones (VS, SW, MW, and LW) are explained in the text. The resulting spectral sensitivity of each cone type is shown (*below*), along with the colors of the human spectrum and wavelength of light. Note that the sensitivity of each single cone type covers a relatively narrow portion of the spectrum, but they overlap in their spectral sensitivity; it is these single cone photoreceptors that underpin the color vision system. Double cones have a much broader sensitivity; they are probably not involved in color vision but provide a luminance signal at higher daytime light levels. *Illustration drawn from original illustrations courtesy of Peter Olsson, Vision Group, University of Lund, and Daniel Osorio, University of Sussex.*

In addition to the five types of photopigments, the retinas of birds have three morphologically distinct types of photoreceptors: rods, single cones, and double cones (Cserhati et al. 1989) (fig. 12.6). Double cones are widespread in vertebrates (Walls 1942, Bowmaker and Loew 2008) but absent from mammals. In birds, the double cones always contain the LWS (λ_{max} 565 nm) pigment (fig. 12.6).

This relative simplicity of receptor types is, however, complicated by the presence in each cone receptor of an oil droplet, which sits in the proximal part of the outer-segment (Bowmaker 1977, Cserhati et al. 1989, Hart 2001) (fig. 12.6). Such oil droplets are absent from most mammals but are common to both reptiles and birds. The droplets contain carotenoid pigments that are derived from the bird's diet. These carotenoid pigments give the droplet a bright color (reds, greens, and yellows) and are not photosensitive. They serve an important filtering function, sharpening the spectral tuning of the LWS, MWS, and SWS cones (Hart and Vorobyev 2005), and the peak sensitivities of the four receptor types are spaced approximately equally apart within the visible spectrum (fig. 12.6). Such spectral sharpening does not occur in the double cones, but the oil droplets in these receptors do block UV light from reaching the photopigment. The UV/VS cone oil droplet is transparent, which means that UV light can reach the photopigments that they contain. This spectral sharpening is thought to enhance color discrimination within the spectrum, i.e., allowing the visible spectrum to be more minutely subdivided.

The retinas of birds are more complex at the level of the receptors than those of mammals, including humans. There is good evidence that bird color vision should be viewed as based on a system of four receptor types (tetrachromatic vision), as opposed to the human color vision, based on three receptor types (trichromatic vision). A range of investigative techniques (Osorio et al. 1999, Jones and Osorio 2004) have indicated that double cones are not part of the color vision mechanism (fig. 12.6) and may constitute a separate channel that signals luminance at higher light levels. Differences in sensitivity and color vision between birds and mammals apply only at high (daytime) light levels, when vision is mediated by the cone receptors. At lower (twilight and nighttime) light levels, only the rod receptors are functional. Since rod receptors have very similar characteristics across all birds and mammals, vision at low light levels, with respect to sensitivity in the spectrum and the absence of color vision, is likely to be very similar across all bird species, in humans and other mammals, and indeed all vertebrates.

The Three Key Sources of Variation in Bird Eyes

Bird eyes are structurally and conceptually simple (Land 1981). Like cameras, they are composed of two functional units, an **image producing system** and an **image analyzing system** (fig. 12.7). Importantly, however, these functional units can vary independently of each other. Added to this is the simple fact that, although there are always two eyes in an animal's head, they can be placed in different positions in the skull. Different eye positions alter the region of space from which visual information can be retrieved at any one instant. These differences have important implications for how vision can influence behavior.

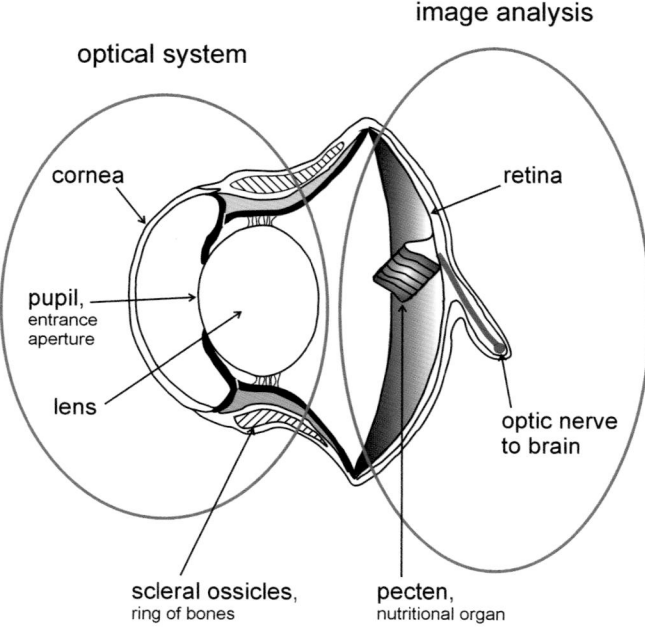

Figure 12.7. The two main functional units of a vertebrate eye depicted in a cross-sectional diagram through the eye of an owl. The optical system has just two main components: the cornea and the lens. These produce an image focused within the layers of the retina, where image analysis begins. Although these two functional units are clearly linked anatomically by virtue of being part of a single organ, each component is relatively independent and hence has been subject to different selection pressures. In birds, the integrity and rigidity of the eye is achieved by a ring of bone plates, the scleral ossicles. These ensure firm anchorage for the muscles and suspensory ligaments that can change the shape of both the lens and the cornea. Characteristics of the image are also modified by the diameter of the pupil, which can change both the brightness and quality of the image. Nutrition to bird eyes is achieved through the pecten, which is a highly pigmented and pleated structure with a large surface area; it is not sensitive to light and constitutes a blind area within the image analyzing system of each eye. Ganglion cells in the retina are the first main mechanism of image analysis; they communicate to the brain via the optic nerve. *Original drawing by Graham R. Martin.*

Variation in the Optical Systems of Bird Eyes

The optical system, which produces an image of the world surrounding an organism, can vary with respect to four key parameters: the maximum brightness of the image (how much light is captured to make the image), the physical size of the image (how large is the area over which the image is spread), the angular size of the image (how wide a field of view is captured in the image), and finally, the quality of the image (how precise is light from a point in the world brought to a focus in the image). Variation in each one of these parameters can alter the amount of information that an eye can potentially collect and convey. Furthermore, differences in these parameters can be quite subtle and allow fine-tuning of performance to match the perceptual demands of different environments and tasks. The fact that these parameters can, within limits, vary independently, means that each eye can provide a unique suite of information for analysis by the image analyzing system.

The optical system has two main elements (fig. 12.7). The **cornea** is the curved surface at the front of the eye. In eyes that operate primarily in air, the cornea is essentially a boundary between air and the fluid-filled chamber of the eye. The radius of curvature of the cornea is the key to its image-forming properties. A more highly curved surface produces a smaller image than a more shallowly curved one.

The **lens**, suspended in the fluids that fill the chambers of the eye, is also relatively simple and derives its primary optical function by virtue of its two convex surfaces. The refractive index of the lens is greater than that of the surrounding fluid, enabling it to bend light that passes through it. The lens surfaces can vary in how curved they are, whether the two surfaces have the same or different curvatures, and whether the curvature is uniform or becomes flatter or more curved toward the equator of the lens. All these properties have subtle influences on the size of the image and how far behind the lens it is focused. Therefore, they also have a significant effect on the quality of the image (i.e., how sharply it is focused and over how much of the image the best focus is maintained).

The overall optical function of the lens is primarily concerned with making relatively fine adjustments to the focus of the image already formed by the cornea, and correcting for image distortions produced by the relatively simple optics of the cornea. In birds, the curvature of both the cornea and the lens surfaces can be altered as part of the mechanism of accommodation, which is the adjustment of the focus of the eye so that objects at different distances are well focused (Schaeffel and Howland 1987, Glasser et al. 1994, Glasser et al. 1995).

The relative amount of focusing contributed by the lens and the cornea can differ quite significantly, even in eyes of the same overall size. An example of this is given in the box on page 349 comparing the eyes of Rock Doves and Manx Shearwaters (*Puffinus puffinus*).

It is important to note that when a bird goes underwater the cornea no longer has an optical function. This is because it no longer separates two surfaces of different refractive index (normally air and the aqueous humour in the front cavity of the eye). This loss of corneal refractive power on entering water poses particular problems for vision in amphibious birds since, presumably, their eyes have to operate effectively in both air and underwater. Interestingly in that regard, the eyes of Great Cormorants may have the same refractive power both in air and in water (Katzir and Howland 2003). There is conflicting evidence as to how this might be achieved. It has been suggested that in some amphibious eyes the refractive power of the lens could be dramatically increased by greatly increasing its surface curvatures (Sivak 1978). An alternative strategy was suggested by the observation that some penguins have corneas that are relatively flat and hence of low refractive power, so the loss of focusing power upon immersion is not that great (Sivak and Millodot 1977). Flatter corneas may in part result from the eyes of some of these amphibious birds being very large. Because of scaling effects, larger eyes must have flatter corneas regardless of whether the bird is amphibious (Martin 1998).

COMPARING THE EYES OF DOVES AND SHEARWATERS

The optical structure of vertebrate eyes can vary in subtle ways that can be related to differences in the visual task faced by different species. This is exemplified by a comparison of the eyes of Rock Doves and Manx Shearwaters (*Puffinus puffinus*). These two species have different evolutionary origins and are placed in different avian orders: Columbiformes and Procellariiformes, respectively. They also differ markedly in their behavior and ecology. The natural habitats of Rock Doves are open and rocky habitats, with the birds preferring to nest and roost on inland cliffs and sea cliffs. Doves have also become naturalized in towns and cities where they use buildings as substitutes for these natural habitat types. Doves are strictly diurnal, going to roost as soon as light levels start to fall. Manx Shearwaters, as their name suggests, are birds of the open oceans. They range very widely and come ashore only to breed in burrows situated on cliff tops. They are able to walk only with a shuffling gait and when on land are vulnerable to attack by predators. To avoid attacks, they come ashore and visit nest burrows mainly under the cover of darkness. Thus, in natural situations it is possible to find Manx Shearwaters nesting along the tops of cliffs while Rock Doves are on the rock faces below. To see the doves, bird-watchers must visit during the day, but to see shearwaters they must visit at night.

A glance at the eyes of these birds would suggest that they are very similar. When looking at the birds in the hand, it is difficult to see any difference in their eyes, apart from the fact that the iris in a dove is usually brightly colored, whereas in shearwaters the iris is dark brown.

Careful analysis of the optical structure of the eyes of these two species, however, reveals quite a different story (Martin and Brooke 1991). The corneas and lenses, and their relative positions to each other, differ so that although both produce a focused image on the retina, in the shearwaters the image is smaller but brighter compared with that of doves. In fact shearwater eyes have an image 1.5 times brighter than that of dove eyes, a difference that is likely to be functionally significant and can be related to the nocturnal behavior of shearwaters and to the reluctance of doves to fly at night.

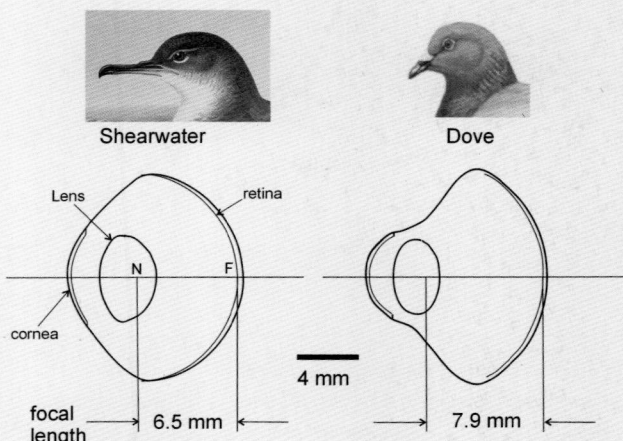

Comparison of eye optical structures in Rock Dove *Columba livia* and Manx Shearwater *Puffinus puffinus*. *Illustration by Graham R. Martin.*

A final very important property of simple eyes is that they can vary greatly in size. Size matters for three important reasons: it inevitably determines the size of the image, it determines what might be the upper limit on the quality of the image, and it determines how much light can be gathered and hence the brightness of the image. Clearly these are very important ways the eye may have changed through the process of natural selection in response to different perceptual challenges associated with life in different habitats and in the conduct of different tasks (see the box on page 349).

Given the possible variations in the two optical components of simple eyes, it can be seen immediately that there is much scope for changing the overall image-forming properties of an eye by virtue of small changes in the absolute size, curvatures, internal structure, and relative positions of these two optical components. Just how different bird eyes can be just with respect to size and optical components can be seen in the comparison of schematic eyes for Common Starlings (*Sturnus vulgaris*) and an owl (*Strix aluco*) (fig. 12.8).

Variation in the Image Analyzing Systems of Bird Eyes

As described above, the image analyzing system starts at the retina (fig. 12.7), and the primary structure within the retina is the layer that contains the photoreceptors. If a photon is absorbed by a photoreceptor, a neural signal is initiated and transmitted to other layers of nervous tissue in the retina, and the retinal ganglion cells send that information to the brain.

Each retina contains many millions of rods and cone receptors. Although there seems to be little variation in the suites of photopigments found in the retinas of different bird species, there can be very marked differences in the relative abundance of receptor types and in their distributions within the retinas of different species (Hart et al. 1998, Hart et al. 2000, Hart 2001, Hart 2004). The result of this is twofold and

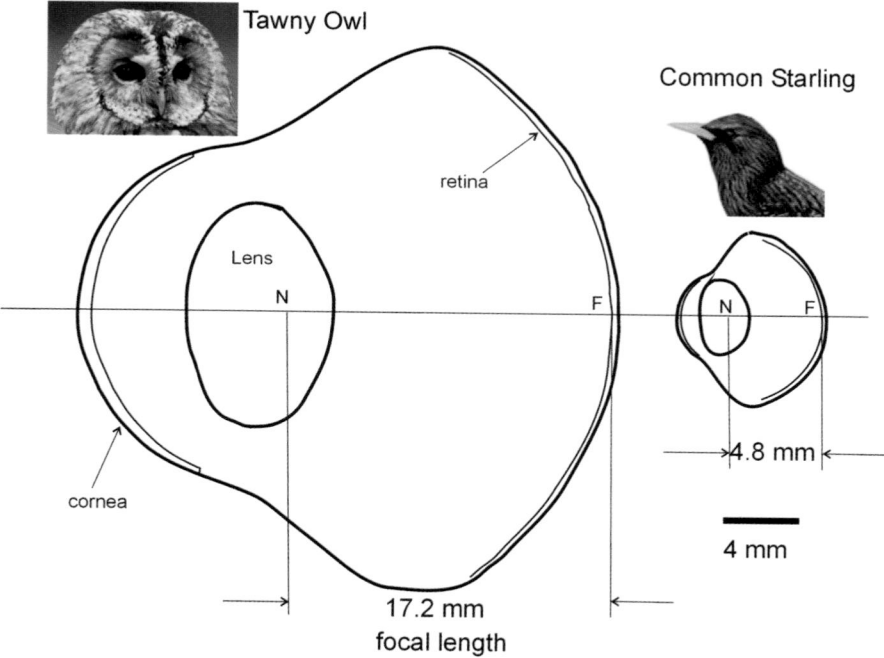

Figure 12.8. Variation in the optical structures of birds. These sectional drawings through the eyes of Tawny Owls and Common Starlings are based on detailed analyses of their overall size, shape, and optical mechanisms. The owl eye embodies the characteristics of an eye that has evolved to operate at lower light levels. It captures a large amount of light and produces a relatively bright image, achieved by virtue of its absolute size and relatively large pupil. The starling eye operates best at high daytime light levels and contributes to these birds' need to roost at low nighttime light levels, while the owl usually becomes active at that time. Differences in the image-analyzing systems (retinal structure) of these two species also contribute further to their abilities to be active at different times of the daily cycle. The principal feature of the owl's eye is its large size, which adds considerable weight and metabolic demands compared with the starling's eye. *Original drawings by Graham R. Martin.*

of great importance. First, in any eye, the image projected onto the retina is not analyzed in a uniform way across the whole of its area. Second, there can be very marked differences in these patterns of analysis between eyes of different species. In short, an image of the same scene in the eye of one bird species can be subject to different analysis in the eyes of another species.

These differences mean that perceptual challenges posed by different environments and different tasks can be met by both variation in the ways that the world is imaged and in how those images are analyzed.

Spatial Variation of Image Analysis   Two major ways have been identified in which variation of image analysis occurs in bird retinas.

The first is that very marked variations in the distributions of cone receptor types can occur across the retina. These variations can be striking, and it seems likely that they result in differences in color vision and spectral sensitivity in different parts of the visual field of a single eye. A striking example is in Rock Doves, where a large area is dominated by receptors containing red oil droplets (that is, they are dominated by LW-sensitive cones, fig. 12.6). This area looks downward within the visual field, while receptors containing yellow oil droplets (MW-sensitive cones, fig. 12.6) predominate in areas that look laterally and upward. These different areas are obvious even to the naked eye in a transilluminated excised retina; there is a very clear boundary between them (fig. 12.9). The visual ecology and function of this striking regional specialization within dove eyes is not understood, although it has been suggested that the yellow field in some way enhances the contrast of objects seen against the blue of the sky (Lythgoe 1979). Although not so striking as the case of dove retinas, other work has also revealed systematic differences in the distributions of cone receptor types

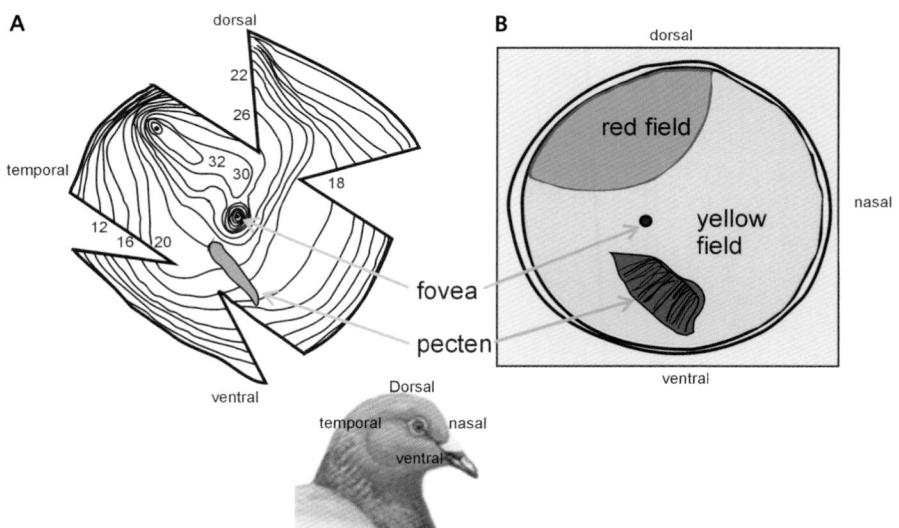

Figure 12.9. Complexity within the image analyzing system of a bird's eye. This example draws on studies of the retina of Rock Doves. The bird's head is shown in typical orientation; the diagrams show two ways of capturing diversity within the retina. *A*, A topographical map of the density of retinal ganglion cells. It should be imagined that the retina, which is a curved surface (fig. 12.7) has been taken from the back of the eye and flattened out. The numbers refer to the number of cells × 1000 per mm². Around the periphery of the retina, cell density is relatively low (about 8 × 1000 per mm²) but there is a steady increase toward the center of the retina, with two small areas with very high cell densities. Density is highest close to the very center of the retina, where a fovea is found; this area projects out laterally from the bird toward the observer (based on the depiction of the bird's head). A second area of high density looks forward and downward. *B*, A diagram of the view of a retina in the same orientation as in *A*; it has been transilluminated to show the relative densities of retinal oil droplets, and hence the relative densities of receptor types. The red field is dominated by red oil droplets and indicates that LW-type photoreceptors dominate in this region (see fig. 12.6 and text for a description of these types of photoreceptors); this region of the retina looks more forward and downward. The remainder of the retina is dominated by photoreceptors that contain yellow oil droplets (MW-type receptors) and look out laterally from the bird. The fovea lies in the yellow field. The other area of very high photoreceptor density is in the red field. *Drawing by Graham R. Martin based on original illustrations of Galifret 1968 and Bingelli and Paule 1969.*

in the eyes of other birds. These tend to show a gradient in the relative abundance of different receptor types across the retina. These suggest that marked interspecific differences exist in color vision and spectral sensitivity, and that these differences are correlated with specific perceptual challenges posed by life in different environments and the conduct of different tasks.

The second type of variation between species and within individual retinas lies in the relative density of receptor and ganglion cells. These differences in the density of cells may be overlaid on patterns of cone receptor types discussed above (fig. 12.9). Very distinct patterns in ganglion cell density are discerned with light microscopy (Galifret 1968, Meyer 1977, Fernandez-Juricic et al. 2011, Lisney et al. 2013, Mitkus et al. 2014, Coimbra et al. 2015). They show that significantly different patterns of photoreceptor and ganglion cell concentrations can occur (fig. 12.10). These patterns may be roughly circular, with a very high concentration at the center and a gradual decline of cell density away from the center. In some species more than one region of high concentration can occur, while in others high densities of receptors may occur in linear bands that are typically arranged so that they project approximately horizontally when the bird's head is held in its usual resting posture. The usual interpretation of these pat-

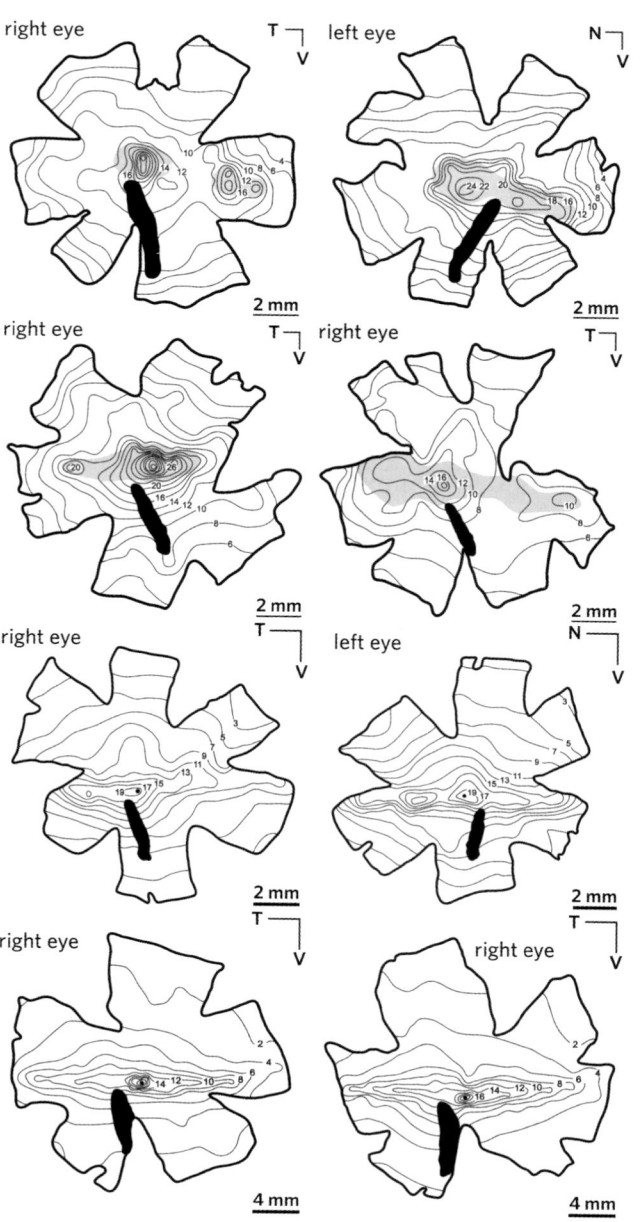

Budgerigar
right and left eyes of
the same individual

Bourke's Parrot
right eyes of two
different birds

Leach's Storm Petrel
right and left eyes of
the same individual

Northern Fulmar
right eyes of two
different birds

Figure 12.10. Patterns of retinal ganglion cells in birds. Each diagram shows the contours of equal ganglion cell densities ($\times 1000$ cells per mm²) on a flattened retina; the pecten (indicated by the black shading) creates a blind area in the eye and is a prominent landmark. The two examples from a Budgerigar show the left and right eyes from the same individual. They show a very high concentration of ganglion cells in the center of the retina, which indicate well-demarked regions of high acuity that look out laterally from the eye, approximately in the center of the field of view. A second concentration in the retina toward the bill looks slightly backward. However, there are also marked differences between the two eyes. In Bourke's Parrots, the two diagrams are for the same eyes from different birds and show marked differences between the individuals. Both have a high concentration of cells in the center of the retina, but in one bird a second concentration looks forward, while in the other the overall density is much lower and the second area of concentration looks backward in the field of view. In the two seabird species, clear linear bands of high ganglion cell concentrations are found running across the retina and project laterally toward the direction of the horizon. However, as in the parrots, there are differences between the left and right eyes of the same individual and differences between individuals. *Budgerigars and Bourke's Parrot, Mitkus et al. 2014; Fulmar and Storm Petrel, Mitkus et al. 2015; courtesy of Mindaugas Mitkus. Photographs, Budgerigar, Michael Cole; Bourke's Parrot, Daniela Parra; Leach's Storm Petrel, Peter R. Flood; Northern Fulmar, Steve Garvie.*

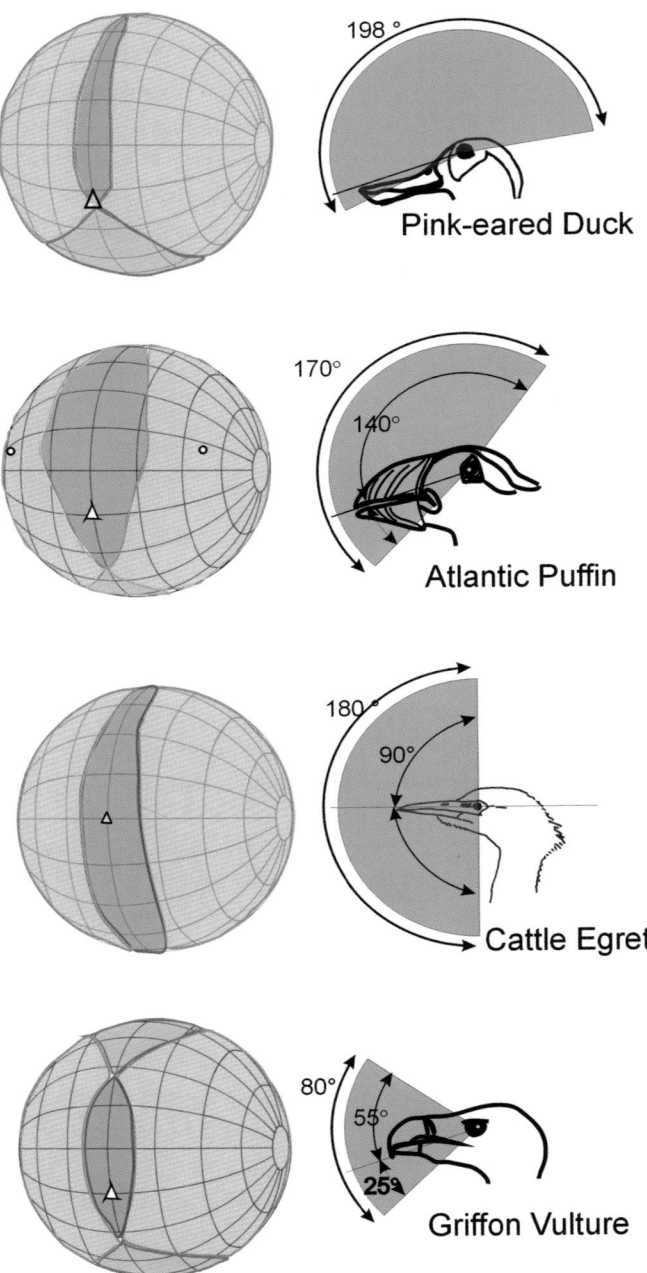

terns is that areas of high receptor concentration are regions of heightened visual acuity, perhaps associated with viewing toward the horizon or looking out laterally or forward. Areas of very high density may be associated with a fovea. This is a small region where the outer layers of the retinal cells are displaced to form a small pit or depression in the surface of the retina.

These two ways in which receptor cell distributions can vary within a retina attest to the fact that birds of two different species, placed in exactly the same position at the same time, are very likely to retrieve different information from the world about them.

Variation in the Visual Fields of Birds

Differences in optical structure affect not only the size and brightness of the image but also determine the extent of the world that is imaged. It is relatively easy to understand why this is important, since it determines the extent of a bird's visual world from which it can gain information. In birds, the field of view of a single eye can be as narrow as 124° (owls) and as wide as 180° (ducks and shorebirds). However, when two eyes are combined, the situation becomes more complicated, and variations in the birds' total visual field are considerably more diverse (Martin 2007). Thus, in a horizontal plane, the total width of the visual field in owls is about 200°, while in some shorebirds and ducks the field provides a total 360° panorama. Such large differences have arisen because the positions of eyes in the skull have evolved to take up a wide range of configurations. The axes of the two eyes may diverge by a relatively small amount (e.g., in owls eyes diverge by about 55°), or they can be almost diametrically opposed (ducks and shorebirds). Thus, the fields of the two eyes can be combined in many ways to provide a wide range in different total fields of view and in the amount by which the two eyes overlap and provide binocular coverage of some part of the total field (fig. 12.11).

Humans are unusual animals in having two eyes placed on the front of the skull. The result of that configuration is that much of what one eye sees is also seen by the other eye.

Figure 12.11. Diversity in the visual fields of birds. The visual fields of four species are shown, with two diagrams for each species. *Left,* The visual fields as projected onto the surface of a globe that surrounds the bird's head. This is a perspective view of the head, with the bill projecting in the directions indicated by the white triangles. The green sector is the region of binocular overlap; the blue is the blind sector; and the orange sector shows the area covered by each eye alone, to the left and right of the head. Clear differences are apparent in the position, width, and vertical extent of the regions of binocular overlap and in the extent of blind regions about the head. *Right,* The width and vertical extent of the binocular regions in the median sagittal plane of the head (the plane that divides the head in two vertically). In Pink-eared Ducks, the binocular area extends from the bill through more than 180°, hence there is no blind region about the head. In Puffins, the binocular region is broader and oriented upward, but not sufficiently to allow the birds to see fully behind their heads. In Cattle Egrets, the binocular region is also long and narrow, but is oriented so that the birds can see directly below them in the frontal field; there is a small blind area behind the head. In Vultures, the region of binocular overlap is very small in width and in vertical extent, and there is a large blind area above and behind the head. *Drawings by Graham R. Martin based on originals in Katzir and Martin 1994, Martin et al. 2007a, Martin et al. 2012, Martin and Wanless 2015.*

In fact, total visual coverage of the visual world in humans is about 200° in the horizontal plane, about 120° of which is seen simultaneously by both eyes. This arrangement is different from that found in any bird (Martin 2007). In the majority of species, the eyes are placed so laterally in the skull that each eye sees something different from what is seen by the other eye. Frontal eye placement in humans leaves us with the subjective experience that the world lies in front of us and that we move into it. For the majority of birds, the world surrounds them and they flow through it. Birds are able to extract information from a scene receding behind them while simultaneously gaining information from the world into which they are traveling.

At one extreme are the visual fields of some ducks and some shorebirds, which can see the whole of the hemisphere above and around their head (Martin 2007, Martin 2014). In fact, the only place from which they cannot retrieve information is the part obscured by their own bodies. The utility of such broad visual coverage for the detection of conspecifics, predators, or food items is clear.

That differences in visual field configuration can occur between closely related species suggests that visual fields can be finely tuned to subtle differences in the perceptual challenges of different foraging tasks. For example, there are significant differences in the visual fields of different species of ibises in the same family (Threskiornithidae) that depend on whether the birds probe their bills into soft substrata or take items from dry surfaces (Martin and Portugal 2011). There are differences in the visual fields of ducks within the same genus depending on whether they are selective grazers or filter feeders (Guillemain et al. 2002), and differences among sandpipers (Scolopacidae) depending on the extent to which they rely on tactile cues from the bill as opposed to visually guided foraging (Piersma et al. 1998, Martin and Piersma 2009).

These examples demonstrate that the vision of different bird species within the same family, and even within the same genus, have been subtly tuned by natural selection to the perceptual challenges of particular foraging tasks. This subtle tuning of vision should be considered as important as the fine-tuning of bill structures as a means of meeting the physical challenges posed by the exploitation of different foods in these same species (Martin 2014).

Vision: Key Points

- Vision is the main means by which birds gain information about their environment.
- The information that vision can provide changes dramatically with the amount of light available. In this, vision is unlike the other senses, which are able to extract more

or less the same information regardless of the amount of stimulus that is around.

- Vision is subject to many fundamental constraints and trade-offs that limit the information that can be extracted from moment to moment.
- Within the basic structures of the image-producing mechanisms and the image-analyzing mechanisms there is a great deal of flexibility, which has been sufficient to allow natural selection to shape visual capacities such that information can be extracted and used to guide the execution of different tasks in different environments.
- The vision of birds is tuned to the challenges of carrying out specific tasks, especially those associated with foraging, under a variety of different light environments.
- The visual worlds of birds differ one from another, down to the level of species.
- Vision alone cannot be the sole source of information for birds in the conduct of their daily tasks throughout their annual cycles.
- A suite of other senses provide other information, which sometimes complements that provided by vision but often is additional to the information that vision can provide.

HEARING

Hearing provides a unique suite of information not available through other senses. Auditory information allows animals to respond rapidly and in a spatially oriented manner to signals produced by sources that are remote and cannot be seen. A degree of coevolution between sound reception and sound production has enabled sounds to become a means of communication between animals, with the properties of vocalizations and other sounds matching the hearing abilities of the intended targets of the communication. Acoustic communication has led to the rich repertoires of sounds deliberately produced by an individual to send information to others, such as songs and calls. These deliberately produced sounds have become elaborated so that the receiver may extract information about the signaler, including its species, sex, readiness to reproduce, position within a social hierarchy, and so on. In some species sound production has also evolved to maximize sound transmission in specific ecological situations, thus increasing the probability of successful communication.

Sounds are well suited for these functions. Although they are transmitted more slowly than light, at short ranges their transmission is effectively instantaneous. They are transmitted in air more rapidly than are odors, but unlike odors they do not linger, so information is received and communication occurs in real time. Sound also has the advantage that it can transmit around opaque objects. As in the case of vision, the information extracted from the world of sound is selective.

The specialized evolution of hearing systems has allowed vertebrates to respond to vibrations in the air, underwater, or in the substrata on which they sit (Fritzsch 1992, Lewis et al. 2006). Evidence suggests that birds, however, are sensitive only to sounds transmitted through air.

Sounds and Hearing

All physical movements of objects cause the medium that surrounds them (air or water) to oscillate. These oscillations propagate away from the object as compression waves; it is these oscillations that hearing organs detect. Properties of the oscillations are correlated with properties of the object that caused the disturbance. For example, the nature of the oscillations can depend on whether the object has hard or soft surfaces, or how fast or how far it moves. Furthermore, the interactions of those oscillations with properties of the environment can be used to determine the direction and distance of the sound source from the observer.

The amplitudes of the oscillations of air molecules that are detected as "sound" are small, even for very loud sounds. Typically, sounds in the normal hearing range are caused by air molecules moving with amplitudes smaller than the wavelength of light. This means that the detection systems, the ears, must be sensitive to these minute movements of air molecules. Close to the threshold of hearing in air, which in humans is approximated by the faint sound of rustling leaves (quantified as having a sound pressure level of between 0 and 10 dB), the displacement of air molecules detected at the ear is about the diameter of a hydrogen atom ($\approx 10^{-10}$ m). For comparison, visible light has wavelengths between 400–700 nm, 4×10^{-7} m. Even sounds that we perceive as very loud (e.g., a car horn sounding five meters away, with a sound pressure level of about 100 dB), are produced by air molecule movements of approximately only 1 μm (10^{-6} m). Molecules of air can be set in vibration over a very wide range of frequencies, and those oscillatory movements can be detected by animals. The frequency range of hearing among vertebrates can extend from approximately 1 Hz (1 cycle per second) to 100 kHz (100,000 cycles per second); however, no one ear can detect this full range of frequencies. The ears of different species of animals are capable of detecting only certain ranges of those frequencies.

Organization of the Hearing System in Birds

Hearing, or the detection of acoustic vibrations, is achieved in birds, reptiles, and mammals by an ear of three main func-

tional parts (Gridi-Papp and Narins 2008): the outer, middle, and inner ears.

The movements of air molecules are detected at the interface between the outer ear and the middle ear. In birds, the outer ear is usually just a short tube protected by the skull that leads into the head. Outer ear structures akin to the conspicuous outer ears (pinnae) of mammals are found only in owls (Strigiformes). The detecting surface is the ear drum (tympanic membrane). This very thin and taught membrane separates the outer ear tube from the middle ear, which is also air-filled. The thinness of the membrane allows it to follow airborne pressure fluctuations (average molecule movements) from instant to instant.

Vertebrate ears are unable to respond directly to airborne vibrations since, as a legacy of the evolution of vertebrates from aquatic ancestors, vibrations are detected in the inner ear by receptors surrounded by fluid. The vibrations of the ear drum are transformed into vibrations of the fluid surrounding the receptors.

The middle ear functions to amplify the vibrations of the ear drum so that they are sufficient to make the fluid of the inner ear vibrate. In birds, reptiles, and amphibians, this amplification is achieved by a single bone, the columella. It is believed that the three-bone system of mammals is a key reason why their hearing sensitivity can extend to sounds that are of much higher frequency than can be detected by animals that have a columella mechanism (Gridi-Papp and Narins 2008).

The vibration receptors of the inner ears are hair cells in vertebrates. These occur in bundles and are fixed rigidly at one end to a stiff membrane. Their other ends are embedded in a flexible membrane that distorts as vibrations are transmitted through the fluid that bathes them. These distortions cause the hair cells to bend, and this bending triggers nerve impulses that are sent to the brain. Both within and between bird species, these hair cells vary in many ways, such as number, size, shape, and the orientation of the hair cell bundles. These differences, as well as the transmission properties of the middle ear, are primarily responsible for differences in the information that different species are able to detect from the acoustic vibrations of the air (Fay 1992).

Hearing Sensitivity

The principal tool for describing and comparing the hearing of different animals is an audiogram, which is a graphic representation of the range of frequencies that an animal can hear and the minimum sound levels necessary for detecting sounds at selected frequencies. By using training techniques, a bird can be made to indicate whether it can detect sounds of specific frequencies and amplitudes. Audiograms can also be constructed using physiological techniques (e.g., recording from the auditory nerve, which runs from the ear to the brain), which on the whole support data obtained using behavioral audiograms.

A broad comparison of audiograms across the five major classes of vertebrates shows significant variations. These attest to both phylogenetic constraints (differences within and between major vertebrate classes) and ecological constraints, especially those associated with hearing systems that function in either air or water (Fay 1992). Among birds, however, comparisons of average audiograms reveal considerable homogeneity across species (fig. 12.12). The audiograms of all vertebrates are U-shaped, and among birds they indicate highest sensitivity to sounds in the frequency range 1–4 kHz. There is a rapid decline in sensitivity toward higher frequencies, with a cutoff in hearing at about 8 kHz, and a similar rapid decline in sensitivity at low frequencies, with a cutoff at about 300 Hz.

Behavioral and observational data suggest that some birds can detect infra-sounds. These are sounds with frequencies below 20 Hz, which are inaudible to humans. Such

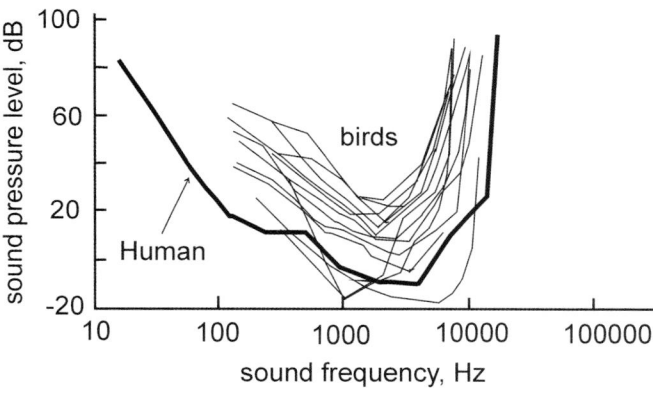

Figure 12.12. The hearing of birds depicted by audiograms. These show the threshold sound pressure level as a function of sound frequency. The lower the sound pressure level that can be heard, the higher the sensitivity. Audiograms for a number of different bird species are shown together alongside the standard audiogram for young humans. The typical U-shape of vertebrate audiograms is clearly evident; there is rapid loss in hearing sensitivity at higher frequencies and a less steep loss at lower frequencies. It is clear that the hearing of birds sits within the overall envelope of hearing of young humans, suggesting that whatever a bird can hear a human will also be able to hear, and that humans can hear a wider range of frequencies than birds. Highest sensitivity in both humans and birds is in the region of sounds with a frequency of about 4000 Hz. *Redrawn by Graham R. Martin, from a diagram of the audiograms of birds from Fay 1992; the typical audiogram of humans is redrawn from Heffner and Heffner 1998.*

low-frequency sensitivity was first reported in Rock Doves, in which it was shown that they could detect sound with a frequency as low as 1 Hz (Kreithen and Quine 1979). This is supported by circumstantial evidence for infra-sound detection in the same species when performing homing flights (Hagstrum 2000).

The function of such low-frequency sound detection has been linked to navigational mechanisms in Rock Doves, but it may also serve to warn of advancing intense weather systems. This has been suggested to have occurred in some Golden-winged Warblers (*Vermivora chrysoptera*) that left their breeding grounds ahead of a tornado event (Streby et al. 2015). Low pressure weather systems emit low-frequency sounds at high intensity; these sounds transmit over distances of hundreds of kilometers and have been used in human weather forecasting using special detector arrays. Infra-sound detection has also been reported in Domestic Chickens (*Gallus gallus domesticus*) and in Helmeted guineafowl (*Numida meleagris*) (Theurich et al. 1984, Hill et al. 2014), but the function is unknown. Across bird species, it is not clear whether the detection of any of these infra-sounds represents instances of hearing. It could be that such low-frequency sounds, at the high intensities at which they have an effect, are detected by somatic receptors, perhaps by the groups of somatic receptors (see below) that provide information about internal conditions within the limbs and the cardiovascular respiratory systems. Thus birds may "feel" these sounds rather than hear them.

The hearing of birds is significantly different from that of mammals. Hearing in many species of mammals, including rodents and bats, extends to high frequencies, typically up to 100 kHz, while in other mammal species, hearing extends into the infra-sound range (Heffner and Heffner 1982). The hearing range of young humans (alas, it commonly lessens with age) is generally considered to extend between 20 Hz and 20 kHz, while a similar rule of thumb for birds is 300 Hz to 8 kHz (Gridi-Papp and Narins 2008). Thus, young humans can certainly hear all the sounds that a bird can hear, as well as frequencies both above and below the birds' range. In terms of absolute sensitivity, birds are generally about 10 dB less sensitive than young humans. However, there is variation between individuals, and some birds will have sensitivity equal to that of some humans.

Locating Sounds

Locating the source of a sound is as important as detecting it. This is true whether the sound is used as a cue in foraging, as a means of communication, or for predator detection. Sound location comprises two components, however—

direction and distance—and these are determined by different mechanisms.

Sound Direction

The primary challenge in determining the direction of a sound source is that it requires two ears. This is unlike eyes, in which the direction of a light source is determined by where its image falls on the retina, and therefore the location of a light source can be achieved by a single eye. Each ear can signal only the presence of a sound source. Accurate location of the direction of a sound source requires simultaneous information from both ears. Differences in the timing of signals received at each ear from the same sound source are used by the brain to compute the direction from which the sound came.

Sound localization mechanisms exploit two major sources of information: differences in the intensity of sound arriving at each ear, and differences in the timing of that arrival (Klump 2000). These differences arise because the ears' entrances are separated by the width of the head. The sound is slightly attenuated when it arrives at the ear farthest from the source, and it will also take longer to arrive at that ear. For most birds, however, the head is so small that differences in the signals arriving at each ear are also small, both in intensity and time of arrival. Intensity differences are dependent on the frequency of the sound, such that between-ear intensity differences in general increase with frequency (Klump and Larsen 1992). Alternatively stated, the "acoustic shadow" of the head is greater for high-frequency sounds than for low-frequency sounds. In Common Starlings, these intensity differences vary between 2 and 8 dB (for frequencies between 1 and 8 kHz), and differences in time of arrival are about 100 microseconds.

Sound Direction Accuracy in Songbirds   These small between-ear differences in birds have the effect of greatly reducing the accuracy with which the positions of sounds can be determined compared with mammals. This is borne out by a number of studies in songbirds. For example, in Great Tits (*Parus major*), depending on frequency of the sound source, location accuracy varies between 20° and 26° (Klump et al. 1986); in Zebra Finches (*Taeniopygia guttata*) between 71° and 180°, Budgerigars (*Melopsittacus undulatus*) 25° and 69°, and Atlantic Canaries (*Serinus canaries*) between 49° and 71° (Park and Dooling 1991); and Common Starlings between 19° and 27° (Feinkohl and Klump 2013). In all these species, sound localization accuracy was studied in a similar way, involving the detection of the direction of a single sound source. However, another approach has been to determine how far apart two simultaneously presented sound sources have to be for a bird to be able to detect them as sepa-

rate. Eastern Towhees (*Pipilo erythrophthalmus*) could detect two sound sources only 7° apart (Nelson and Suthers 2004).

These are not very impressive sound location performances, and if translated to physical distances, they are perhaps even more telling. For example, a 20° localization accuracy means that a sound 20 m away from a bird could be anywhere in an arc 7 m wide, while a sound at 50 m distant could be anywhere within an arc 17 m wide. Among songbirds, it appears that the best accuracy of sound localization is about 20°, and many birds have a much worse performance than this. However, it should be remembered that these birds are highly mobile and can respond to important sounds, such as the songs and calls of other birds, by using short flights to home in on a sound until visual contact with the singer is possible.

Sound Direction Accuracy in Owls   One group of birds achieves sound localization of much higher accuracy than songbirds. Owls (Strigidae) and Barn Owls (Tytonidae) have been shown to be capable of capturing prey using sound cues alone (Payne 1971), and this is based on accurate sound localization (Knudsen and Konishi 1979, Bala et al. 2003). Barn Owls can locate the directions of sounds to an accuracy of 3°–4° if the sound is broadband noise and is sustained for about one second. This degree of accuracy is comparable to that achieved by humans.

The key to this accuracy lies not in the ears themselves but in the fact that owls have elaborate outer ear structures: they are the only group of birds to have ear structures at the entrance to the outer ear canals (Norberg 1968, Norberg 1978). These structures take the form of flaps of skin placed both before and behind the ear openings, to which specialized small hard feathers are attached (fig. 12.13). The positions of these flaps of skin are under muscular control, and changes in their position can result in dramatic alterations in the head shapes of owls. The presence of the outer ears also accounts for the apparently broad head shape of these birds. Despite their skulls having basically the same general shape as that of other birds, owls' heads appear much broader because of the outer ears and the feathers attached to them. Moreover, in a number of owl species, these outer ear structures are asymmetric, being both slightly higher and larger on one side compared with the other (Norberg 1978). Analysis of how these outer ear structures distort sounds that arrive at the ear drums has led to a detailed understanding of how they enhance the accuracy of an owl's ability to determine the direction of a nearby sound. This particular development of auditory localization in owls seems to be intimately linked to the nocturnal hunting habit of these birds, in which hearing is used to complement vision in the location of prey (box on page 341).

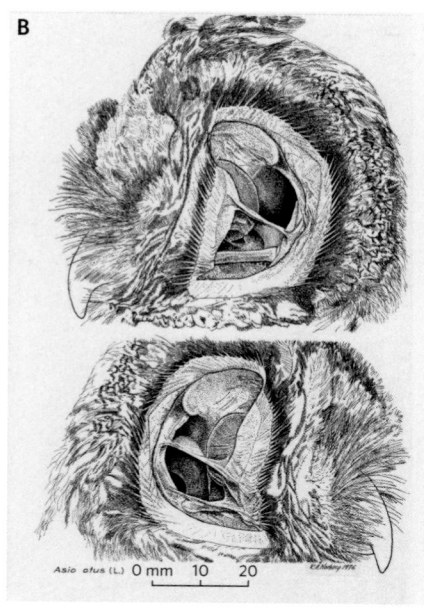

Figure 12.13. The outer ears of owls. *A*, A Barn Owl in which the feathers of the facial disk have been removed to show the flaps of skin positioned in front of the ear openings, which are positioned just behind the eyes. Note that these structures are of different size and in different positions on each side of the head; *B*, Looking into the ear openings of a Long-eared Owl, the feathers at the edge of the facial disk have been parted to reveal the large entrances to the ear canals. The edges of the ear openings are large flaps of skin, which are under muscular control; the flap to the front is the structure equivalent to the skin flaps shown in the Barn Owl. Note that the ear opening in the Long-eared Owl are also of different size and shape on the two sides of the head, and that the side of the eye can be seen when looking into the ears. *Original drawings by Rolf Åke Norberg from Norberg 1977.*

Sound Distance (Sound Ranging)

The ability of birds to determine the distance to a sound source has been measured in passerines using situations that evoke territorial behavior in response to the playback of a song of the same species. For example, a male territory holder may approach the sound source in an attempt to localize its presumed rival, and the distance covered and its direction of travel are used as a measure of the bird's ranging ability (Nelson and Stoddard 1998, Naguib et al. 2000, Holland et al. 2001, Morton et al. 2006).

Manipulation of the playback sound allows researchers to identify cues that the bird might use to estimate the distance of the sound source. It is important to note that the sounds used in such studies are not novel sounds but are species-specific calls and short phrases of song. These studies show that simple reduction in the overall amplitude of a familiar sound is a cue for distance assessment (Naguib 1997, Nelson 2000). However, amplitude is not thought to be a very reliable cue to distance because naturally produced sounds can vary in amplitude, and movements of the singer's head will also alter the amplitude of the sound received by the listener (Nelson 2000). It has been suggested that the listener's knowledge

of the signal's original frequency spectrum and its intensity is required when using attenuation as a distance cue (especially for high frequencies) (Naguib and Wiley 2001). Such a cue may thus be especially useful for signals that are familiar to the receiver, and this is particularly the case for songs and calls used in the interactions between territorial neighbors.

Field studies have demonstrated that familiarity with a specific song type will affect a male's ability to discriminate between degraded and undegraded playback songs, and the ability to assess the distance of a sound source (McGregor et al. 1983, Naguib 1998, Morton et al. 2006). However, other field studies did not find enhanced ranging ability for familiar song types (Wiley and Godard 1996), and even the distance to unfamiliar sounds can be effectively determined in certain situations (Naguib 1997).

The idea that a bird needs to be familiar with the characteristics of a sound source to be able to determine its distance is also supported by studies of the ability of owls to capture prey in total darkness using sound cues emitted as the prey moves. Observations clearly show that the birds do have an estimate of the distance to the sound. However, to achieve this, an owl has to have a high degree of familiarity with the experimental setup. Thus the birds have to learn their way around the experimental situation and get used to capturing prey from fixed perches at low light levels for a couple of weeks before they will readily do this in darkness (Payne 1971; box on page 341).

In summary, studies provide evidence for improved sound ranging with the familiarity of the sounds, usually species-specific songs and calls. A further complication is that birds may not make fine distance discriminations, and their responses to territorial songs may be categorical, basically "near" or "far." Common Chaffinches (*Fringilla coelebs*), for example (Naguib et al. 2000), showed a categorical response to playback of degraded songs corresponding to transmission distances of between 0 and 120 m, indicating that the birds distinguished "near" (0, 20, and 40 m) from "far" (80 and 120 m). In the context of territorial defense, this might be all that is required. It may be sufficient to simply categorize the song of a nearby male as either a potential threat when "near" and possibly inside its territory or no threat when the rival is "far" and outside its territory. In another study, it was found that Great Tits may be able to categorize sounds as coming from several different distances (Pohl et al. 2015), but nevertheless their response may still be categorical and thus express a rather crude determination of distance to a sound source.

Echolocation (Active SONAR)

A specialized use of hearing is echolocation, or active SOund NAvigation Ranging. As in the cases of location and ranging described above, active SONAR depends on knowledge of the sound source and how it is modified by the environment. In this case, however, the familiar sound is produced by the bird itself. Thus SONAR involves the detection of echoes from vocalizations emitted by a bird. In SONAR the ears do not carry out any specialized analysis of the sounds received. SONAR is primarily a function of the brain.

Echolocation is well understood in bats and cetaceans (Johnson 1986), in which it functions to provide finely detailed information about the nature, positions, and movements of objects. The extraction of fine spatial detail by these species is achieved through the use of high-frequency sounds that are well above the auditory range of humans and birds (Pye 1979, 1985). Because the hearing of birds is within a much narrower and lower range of frequencies that those high-frequency sounds, bird species that employ echolocation have only a rather coarse ability to determine the size and the position of objects.

Only two groups of birds are known to use echolocation, and they have in common the use of deep caves or very dimly lit locations as nesting and roosting sites. Echolocation is used primarily to orient the birds within these locations, and

it does not seem to be used outside the caves. Oilbirds (Konishi and Knudsen 1979) and Cave Swiftlets (Apodiformes, Apodidae) (Medway and Pye 1977) are the only birds in which active SONAR has been demonstrated (fig. 12.14). Experiments designed to determine the smallest objects that swiftlets can detect in free flight (birds flying in a space in which wooden rods or wires of known size were arrayed) have produced a range of results. The smallest targets reported to have been detected had a diameter of 4–10 mm, although the size may be as high as 20 mm diameter in White-rumped Swiftlets (*Aerodramus spodiopygius*) (Smythe and Roberts 1983). These performances are not impressive compared with what bats can achieve in comparable tests, where wires as small as 0.1 mm diameter can be detected. Nothing is known about the distance, shape, or volume of the space around the birds where targets can be detected.

The minimum sizes of objects that Oilbirds can detect using active SONAR are even larger than those detected by swiftlets (box on page 340). When plastic discs of various diameter were hung in the passageway leading out of a nesting cave, "all birds hit 5 and 10 cm diameter discs as if nothing had existed in their paths. The first signs of avoidance appeared when 20 cm discs were presented and all birds avoided 40 cm discs" (Konishi and Knudsen 1979). This

Figure 12.14. An Oilbird. Oilbirds, large members of the order Caprimulgiformes, are probably the most nocturnal of all birds. They roost by day in the deep or complete darkness of caves and leave the caves at dusk and return before dawn. They are one of the few bird species able to use active SONAR (echolocation). Their eyes are among the most sensitive bird eyes yet described, with a large entrance pupil and a retina dominated by rod photoreceptors. Around the bill, their many rictal bristles are touch-sensitive, and the birds also use their sense of smell to locate ripe fruits in the canopy of tropical rain forests. From all perspectives, Oilbirds are extraordinary, and the combination of their sensory abilities allows them to complete a unique life cycle that includes nesting on cave ledges in complete darkness. *Photograph by Graham R. Martin.*

performance is surprisingly poor, but it clearly indicates that active SONAR can be used to guide the birds with respect to large objects such as the walls of the cave or other birds. Oilbirds also have wings that provide a lot of lift during flight, giving them the ability to fly slowly yet stably and the ability to hover for short periods. Therefore, collisions that do occur within caves are typically not at high speeds. Oilbirds seem to live highly predictable lives within the darkness of their roosting and nest caves, apparently being attached to the same nest ledge throughout their relatively long lives (Snow 1961). Thus, there is plenty of time for learning about how sounds are degraded within that space, which can be used as a source of spatial information using both active and passive SONAR.

Hearing: Key Points

- Acoustic signals in the form of songs and calls provide rich sources of information that mediate the social behavior of birds (chapter 14).
- The hearing of birds is based on a relatively narrow range of sound frequencies and a relatively poor ability to locate the direction and distance of sound sources.
- In owls specific information can be extracted from sounds to mediate the detection and capture of prey (box on page 341).
- In a few bird species active SONAR underpins the use of dark caves as safe roosting and nesting sites (box on page 340).
- Some birds may be able to employ low-frequency sounds as a means of predicting weather patterns but it is not clear that this is achieved through hearing.

SOMATIC SENSITIVITY (TOUCH)

The general function of somatic sensitivity is to provide information about the environment at, or in contact with, the body surface. Somatic sensitivities may also provide information about the physical conditions of the internal environment of an animal (Gottschaldt 1985). In the section on vision, we discussed an interplay between information derived from vision and information derived from "touch" sensitivity in the bill as a means of guiding foraging behavior in certain bird species. Like vision, "touch" is a multifaceted sense. What we experience as touch is the amalgam of a broad set of information, and each component has separate sensitivities and thresholds. These components can evolve independently of each other, and components can vary in their importance between species. Furthermore, the different touch receptors are distributed across the surface of the body at different densities. The specific stimulus dimensions of touch are de-

tected mainly through sensory structures in the skin covering the body. Stimulus dimensions include pressure, skin stretch, vibration, noxious stimuli, and temperature. Each of these dimensions involves different receptor types, and all of those types are found in birds. Touch receptors are not usually concentrated within a sense organ, although some birds have a specialized clustering of touch receptors in their bill known as "bill-tip organs." Some specializations of touch can be seen as finely tuned to specific tasks, most notably foraging, but touch also has important roles in the guidance of more general tasks.

Somatic Sensitivities: Touch and Other Senses

Receptors in the skin detect mechanical, thermal, and noxious stimuli coming from the outside world. Other somatic receptors provide information about internal conditions within the limbs, the cardiovascular systems, and the respiratory systems. In birds, a multitude of somatic receptors and several classifications have been proposed based on the sensory modalities they respond to, such as mechanical, thermal, chemical, or noxious stimuli (Gottschaldt 1985). Receptors in the bills of birds have received most recent and detailed studies about their form and function (Piersma et al. 1998, Cunningham et al. 2007, Cunningham 2010, Cunningham et al. 2010). These are particularly interesting since they have been shown to be sufficient for prey detection in certain species.

Mechanoreception

Mechanoreceptors respond to the movements of an object or fluid with which they are in direct contact. These movements can have different properties, and different types of mechanoreceptors are capable of responding to these properties separately. Acceleration (vibration) and the amplitude of mechanical stimuli are the key properties that are detected, and it seems that they are signaled to the brain separately.

Herbst Corpuscles Herbst corpuscles detect the acceleration component in vibrations. These receptors can respond directly to a vibrating stimulus in a one-to-one manner for each cycle of a vibration at frequencies between 100 and 1000 Hz. They detect the force of the object impacting the surface of the body, but they do not detect either the amplitude or the velocity of the vibrating stimulus (Gottschaldt 1985). The structures of these receptors vary markedly, depending on their location on the body. Also, there is not just one type of receptor: instead, they are a class of receptors. Herbst corpuscles are the most widely distributed mechanoreceptors in a bird's body, occurring everywhere in the skin; along the large leg and wing bones; in tendons, muscles, and joint capsules; and near large blood vessels. In feathered skin, Herbst corpuscles are located primarily at the base of the feather follicles; these have been interpreted as vibration detectors enabling the birds to detect vibrations of feathers. Among Herbst corpuscles located in different parts of the bird body, the most is known about those that are located in the bill, particularly in ducks.

The distribution and number of Herbst corpuscles in the bills of birds varies greatly, and their number and locations appear to be related to the manner in which the bill is used as a tactile exploratory device during foraging (Cunningham et al. 2007). For example, in shorebirds (Charadriiforms) that use their bills for probing in soft mud or sandy substrata, the tip of the bone of the upper jaw (the premaxillar) may bear large numbers of small cavities (pits), which are packed with Herbst corpuscles. Similar groups of these pits are found in the very distal portions of the bills of ducks and geese, and in kiwi, ibises, and parrots. In songbirds that feed primarily on grain, especially the finches (Fringillidae), Herbst corpuscles and other types of mechanical receptors are located at exactly those places in the bill that are mechanically involved in seed-opening. In the long tongues of woodpeckers (Piciformes), which are used for extracting invertebrates from narrow holes, Herbst corpuscles are also numerous.

Grandry Corpuscles Grandry corpuscles are found particularly in feathered skin. Grandry corpuscles have also been identified in the skin (dermis) of the soft part of the bill covering (rhamphotheca) of geese. Since they cannot follow vibrations, they are thought to be primarily sensitive to tangential movements that cause crumpling of the skin (Gottschaldt 1985).

Thermosensitive Receptors There seems to be good general evidence that the skin of birds is sensitive to temperature; however, the existence of specific thermoreceptive structures has not been demonstrated unequivocally. There appear to be what are termed "thermosensitive units," which respond to both cold and warmth, with input that responds to both the amplitude and rapidity of a temperature change. These may be widely distributed among all bird species, but the existence and mode of action of specific units is not well established.

Cutaneous Nociceptors The somatosensory units called cutaneous nociceptors are probably very important for the survival of an animal; however, little is known about them in birds (Gottschaldt 1985). Nociceptors respond to invasion of the body (e.g., cutting or pinching), and their responses probably elicit changes in the condition of the body, such as an increase in blood pressure, heart rate, or respiratory frequency.

Bill-Tip Organs

One of the most interesting aspects of somatic sensitivity in birds is the concentration of groups of receptors in the bills of certain species. These specialized structures are called "bill-tip organs." They occur only in certain taxa and are readily interpreted as having a specific role in foraging or bill exploratory behavior (Cunningham et al. 2010, Demery et al. 2011). Bill-tip organs turn the bill into a "tactile exploratory organ." They have been described by Gottschaldt (1985) as an avian equivalent of the sinus hair system in moles (Talpidae) (Eimer's organ; Catania and Remple 2004). As such, bill-tip organs allow birds that possess them to detect and probably interpret objects without the need of vision. In that way they can be seen as complementing vision in the foraging of particular species.

Bill-tip organs were first described in the bills of parrots (Goujon 1896). They do not seem to have any specialized receptor types, but they contain high concentrations of known somatosensory receptors, mainly Herbst corpuscles, positioned at or close to the bill tip. The link between bill-tip organs and foraging is clear. If the bill is used to search for, catch, select, and manipulate food, as in parrots, waterfowl, and shorebirds, a well-developed bill-tip organ is present. The structure of bill-tip organs has been most thoroughly studied in geese and ducks (Gottschaldt 1985) although recent research has revealed details of the structure and mode of operation of bill-tip organs in shorebirds (Piersma et al. 1998), kiwi (Cunningham et al. 2007), and ibises (Cunningham et al. 2010).

Bill-Tip Organs in Waterfowl

In geese and ducks, the outer surface of the upper bill and the outer and inner surfaces of the lower bill are covered by a horny plate shaped somewhat like a fingernail. Hidden underneath the hard outer horn of the surface of the nail are many "touch papillae," which protrude from the deeper part of the skin around the rim of the bill, with the tips of the receptors reaching the surface. The number, size, and shape of individual touch papillae varies between species, suggesting a fine-tuning of the structure of the bill-tip organ to meet the sensory demands of detecting various types of food items using touch cues alone. Neighboring touch papillae are mechanically isolated one from another, and it is argued that this should allow fine spatial tactile discrimination by the bill-tip organ (Gottschaldt 1985). In support of that idea is experimental evidence that Mallards (*Anas platyrhynchos*) can distinguish between real and model peas buried in soft substrata using only cues derived from the bill-tip organ (Zweers and Wouterlood 1973).

Bill-Tip Organs in Parrots

In parrots, rather than forming continuous rows around the tip of the bill, touch receptors occur in groups of discrete

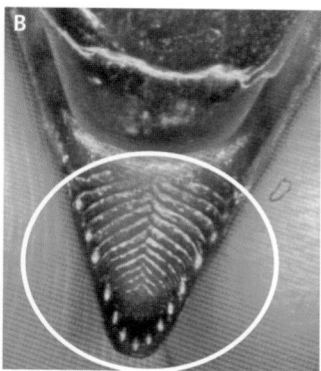

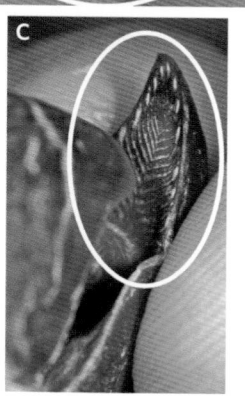

Figure 12.15. Bill-tip organs in parrots. *A,* The bills of parrots are unique in having clusters of touch receptors arranged in discrete groups inside the upper bill in the curved tip section. This tip is used prominently in the manipulation and examination of objects and is also used as a third limb when climbing. Eye position and visual fields do not allow parrots to see what they are grasping in their bill tip. *B, C,* The inside of the bill tip. The clusters of touch receptors can be seen to be symmetrically arranged just inside the edge of the bill. There is no bone in this section of the bill and the touch receptors are embedded in the horny keratin. The patterns of lines inside the bill are probably structures that help grip objects and are not part of the bill-tip sensory organ. *Photographs by Graham R. Martin.*

bundles well separated from each other in a symmetrical pattern along the edges of the maxilla and mandible (fig. 12.15). In Senegal Parrots (*Poicephalus senegalus*), the bill-tip organ in the maxilla consists of seven pairs of pit clusters containing mechanoreceptors spaced out along the edges of the inside of the bill and with a single pit cluster at the bill tip (Demery et al. 2011). The capabilities of such arrangements of touch receptors are not known, although it seems likely that, because of their spaced distribution, they are unlikely to provide the kinds of fine-grained spatial information that is thought to be available through the bill-tip organs of waterfowl. In parrots, bill tip organs may function in the manipulation of objects being held in the bill tip and also in the positioning of the bill when it is used as a "third limb" in climbing (Demery et al. 2011).

Bill-Tip Organs in Shorebirds, Kiwi, and Ibises

A distinctly different type of bill-tip organ occurs in shorebirds (Scolopacidae), kiwi (Apterygidae), and ibises (Threskiornithidae) (fig. 12.16). These birds probe with their long thin bills into soft substrata, which may be covered in murky water, in search of invertebrate prey (see the box on page 369). This type of bill-tip organ consists of clusters of pits within the bone around the tips of the maxilla and mandible. The pits contain bundles of Herbst and Grandry corpuscles, but the receptors are not exposed directly to external stim-

uli. Instead, they lie beneath the soft pliable skin that covers the bill. Using these receptors, birds can also detect pressure patterns produced by hard-shelled sessile prey as the bill is thrust into the substratum. This capability is known as "remote touch" because birds are capable of detecting objects within a cylindrical volume around the bill tip. The bill tip does not have to make direct contact with a prey item in order to extract information about its presence in the mud because of back pressure produced by the object as the bill is thrust in its vicinity (Piersma et al. 1998).

This type of bill-tip organ occurs in distinct avian lineages. They are found in species in the two living superorders of birds: the Paleognathae (kiwi) and the Neognathae (Scolopacidae). Because of the deep taxonomic divide between these bird families, the occurrence of similar bill-tip organs in them has been interpreted as an example of convergent evolution. That is, while Herbst and Grandry corpuscles have deep evolutionary origins, their concentrations into similar-looking bill-tip organs have separate evolutionary origins (Cunningham et al. 2007).

In ibises, there is indication of a subtle tuning of bill-tip organ structure in relation to differences in foraging tasks (Cunningham et al. 2010). Two parameters of bill-tip organ structure—the length along the bill that sensory pits extend from the tip and the total number of sensory pits in the bill—show a clear relationship with the type of habitat in which

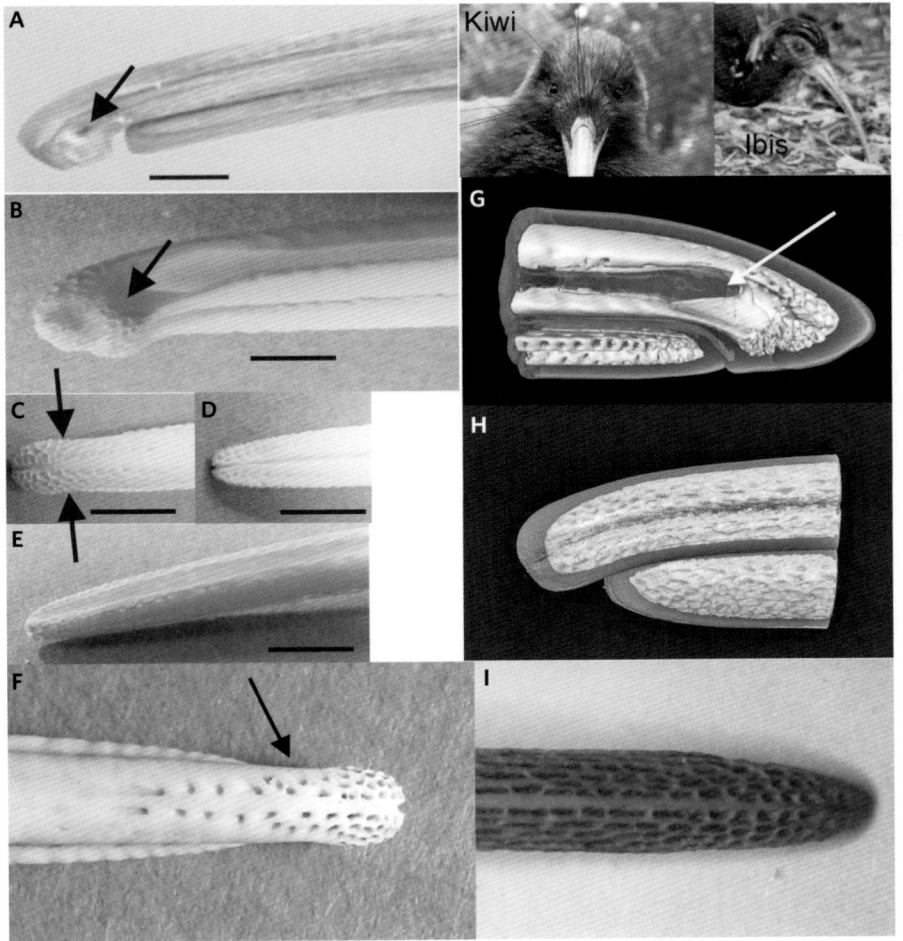

Figure 12.16. Bill-tip organs in kiwi and ibises. *A–G,* Different views of the bill tips in Brown Kiwi. *A,* The intact bill with the keratin sheath in place. *B–F,* The structure of the bone beneath, revealing the numerous pits where clusters of touch receptors are housed. *B, C,* and *F,* The upper bill; *D* and *E,* The lower bill. *G,* A 3-D construction from a μCT scan. In all these photos, the arrows point to the opening of the nostrils. In kiwi, uniquely among birds, the nostrils open laterally just behind the bill tip. *H* and *I,* The bill tip in a Madagascar Crested Ibis (*Lophotibis cristata*), showing the same kind of honeycomb of pits in the bone as in the kiwi, in which clusters of tactile receptors are housed. The distant relationship between kiwi and ibis species suggest that these similar touch-sensitive systems are the result of evolutionary convergence, not common ancestry. *Photos A–E from Martin et al. 2007b; F–I courtesy of Susan Jane Cunningham and Jeremy Corfield (Cunningham 2010, Cunningham et al. 2010, Cunningham, Corfield et al. 2013).*

the birds probe their bills. The more aquatic the habitat, the greater the number of sensory pits and the farther up the bill that they extend. Thus, more extensive bill-tip organs are found in Glossy Ibises (*Plegadis falcinellus*), whereas less extensive organs are found in Buff-necked Ibises (*Theristicus caudatus*). Members of the former species forage almost exclusively in standing water, while members of the latter forage on dry land. A functional interpretation of these differences is that, for the dryland foragers, prey items must occur only near the bill tip, whereas an ibis foraging in water and softer substrata may encounter prey at a wide range of depths.

Somatic Sensitivity: Key Points

- The information that somatosensory receptors provide is about objects at the body surface.
- Somatosensory receptors also provide information for monitoring internal body conditions. This information is vital to all birds.
- The presence of bill-tip organs in certain species shows that the somatosensory system can take on specialized

functions that have become refined for the detection of particular types of prey.

- In parrots, it seems that the bill-tip organs can also provide a vital source of specialized information that guides the use of the bill as a third limb, thus facilitating these birds' ability to climb both upward and downward in structurally complex situations.

OLFACTION

For many years olfaction in birds was considered something rather special, found within relatively few species in which it served specific functions in the guidance of foraging. Now it is possible for authors to write about the excellent sense of smell of birds, its wide distribution across taxa, and its importance as a source of many different types of information (Caro et al. 2015, Corfield et al. 2015). Such information is as crucial to individual bird survival and reproductive success as the information derived from vision or hearing.

Organization of the Olfactory System

The basic structure and function of the olfactory system in birds is similar to that of all vertebrates. Air potentially contains a huge range of chemical compounds, and the presence of certain compounds is detected and signaled to the brain. In the majority of birds, the paired nostrils are positioned toward the base and top of the bill. Exceptions to this are found in kiwi, where the nostrils are positioned at the bill tip, and some plunge-diving birds (Suliformes, gannets and cormorants) in which the nostrils open inside the mouth, such that with the mouth closed the nostrils are effectively sealed.

Inhaled air passes from the nostrils successively into three chambers (conchae) that serve to filter, warm, moisten, and, finally, to sample the chemical composition of the air. The mucosa lining the third concha is innervated by dense neural fibers (olfactory receptors) that detect the presence of spe-

cific compounds. From these receptors, neurons connect to the brain at the olfactory bulbs, which are positioned at the lower front of the brain close to the nostrils.

Olfactory bulbs in birds vary markedly in both their shapes and sizes. They are usually two distinct, paired structures, but they are fused in some species (fig. 12.17). A lot of attention has been paid to the olfactory bulbs in birds, mainly because they are a readily identified brain structure with dimensions that can be measured and compared (Corfield et al. 2015).

The organization of olfactory bulbs of birds shows the same laminar structure across all species that have been studied, suggesting that olfactory bulbs arose early in the evolution of birds' ancestors and their basic features have been highly conserved. The absolute size of olfactory bulbs of birds has been shown to vary markedly across species. The smallest reported is 0.06 mm³ in Spotted Pardalotes (*Pardalotus punctatus*, Passeriformes), and the largest is 217.63 mm³ in

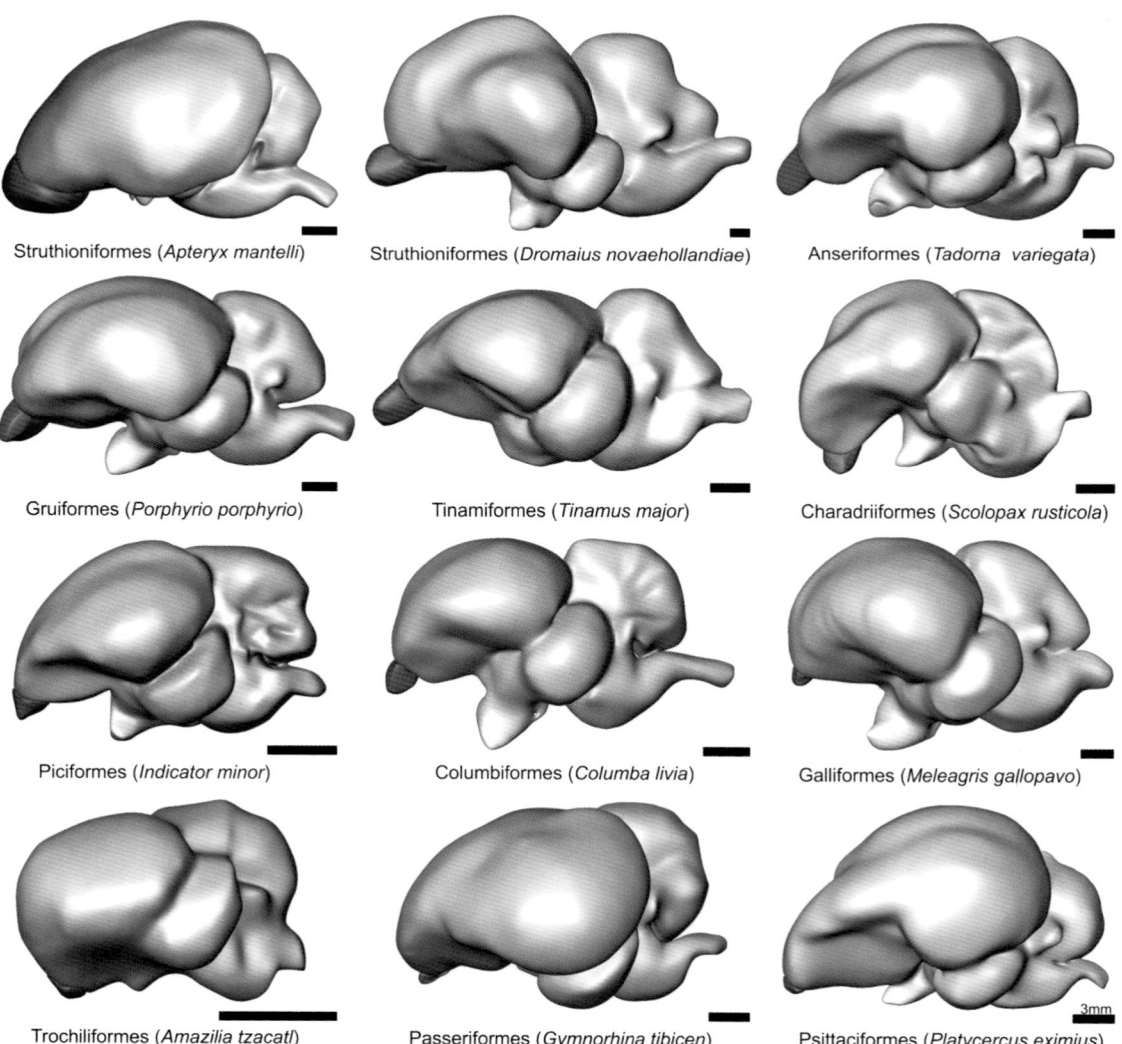

Struthioniformes (*Apteryx mantelli*)

Struthioniformes (*Dromaius novaehollandiae*)

Anseriformes (*Tadorna variegata*)

Gruiformes (*Porphyrio porphyrio*)

Tinamiformes (*Tinamus major*)

Charadriiformes (*Scolopax rusticola*)

Piciformes (*Indicator minor*)

Columbiformes (*Columba livia*)

Galliformes (*Meleagris gallopavo*)

Trochiliformes (*Amazilia tzacatl*)

Passeriformes (*Gymnorhina tibicen*)

Psittaciformes (*Platycercus eximius*)

Figure 12.17. Size and shape of the olfactory bulbs in birds. Lateral views of 3-D models of the brains of 12 bird species, with the olfactory bulbs highlighted in blue. The sequence runs from largest to smallest olfactory bulbs recorded to date. Scale bar = 3 mm. *Drawings courtesy of Jeremy Corfield, adapted from Corfield et al. 2015.*

Emus (*Dromaius novaehollandiae*, Casuariiformes), spanning a nearly 4,000-fold difference in size (Corfield et al. 2015). Clearly, different amounts of space and material are devoted to olfaction in different species, but how this size difference relates to olfactory abilities is not understood.

The Importance of Olfaction and Olfactory Bulb Size

Focusing on those species that have relatively large olfactory bulbs, behavioral studies first showed that olfaction is used for food location in kiwi (Wenzel 1968) and New World vultures (Cathartidae) (Houston 1984, Houston 1986, Graves 1992). More recently, the specific role of olfaction in food location in petrels (Procellariidae) has also been demonstrated (Nevitt 2000, Cunningham et al. 2003, Nevitt 2008). Studies have now also shown that olfaction can play a key role in behaviors other than foraging. These include the detection of predators by some passerines (Amo et al. 2008, Leclaire et al. 2009, Amo et al. 2011), recognition of various odors in chicken chicks (Jones and Roper 1997, Bertin et al. 2010), and navigation in a range of species (Gagliardo 2013), including Rock Doves, Common Swifts (*Apus apus*), Common Starlings, Gray Catbirds (*Dumetella carolinensis*), and Cory's Shearwaters (*Calonectris borealis*).

The range of bird species in which behavioral evidence has demonstrated a specific function of the sense of smell includes those with both relatively large and small olfactory bulbs. This demonstrates that although a greater proportion of the brain is devoted to the analysis of smell in some birds, the presence of a small olfactory bulb does not mean that smell is nonfunctional or of little importance. Recent analysis of olfactory bulb size (Corfield et al. 2015) concluded that the relation between absolute size and proportional size of olfactory bulbs and brain size is complex. Comparative analysis showed that bulb size scaled allometrically with brain size, but that there are also close links between olfactory bulb size and both the ecology and phylogeny of species. This comparative anatomical study also supported the conclusion that olfaction is an important sense in all avian species.

Olfactory Information and Foraging for Specific Items

The two best-studied examples of the use of specific olfactory cues in foraging come from kiwi (Cunningham et al. 2009) and New World vultures (Cathartidae) (Houston 1986, Graves 1992). The phylogeny, behavior, and ecology of these birds are quite different, yet it has been possible to show that olfaction plays a key role in the location of specific food items important in their diets: soil invertebrates (kiwi) and decomposing carcasses (vultures). In both species, these cues are important in finding food sources that cannot be readily detected through information obtained using other senses. This is true in kiwi species because items are buried in the soil (Cunningham et al. 2009), and in the vultures, smell becomes particularly valuable when carcasses are hidden from sight under vegetation (Houston 1986).

Olfactory Information and the Detection of Foraging Locations

Olfactory cues may be used by many birds to locate profitable foraging locations. One particularly important chemical compound is dimethyl sulfide (DMS). High concentrations usually indicate locations in the open oceans where foraging for prey by pelagic seabirds would be profitable. This compound is produced by marine phytoplankton when it is grazed by zooplankton. Therefore when it occurs at higher concentrations, DMS is associated with areas of the ocean with high primary productivity. In turn, these are areas where the prey often sought by seabirds is also likely to be found in greater abundance. The presence of high concentrations of DMS is indeed detected and exploited by various petrel species (Procellariiformes), and large concentrations of these birds may be drawn upwind to DMS concentrations. Furthermore, these birds exhibit particular zigzag flight patterns that seem to be aimed at detecting such odor plumes (Nevitt 2000, 2008) (fig. 12.18). It has also been suggested that these intense sources of DMS are exploited as a means of olfactory navigation, since they may act as temporarily fixed beacons in the seascape (Nevitt and Bonadonna 2005).

A similar use of odor to locate a profitable foraging location has also been demonstrated in Great Tits, a small passerine of mainly woodland habitats. These birds may be attracted to concentrations of odors produced by tree foliage when trees are being browsed by the caterpillars, the main food source of breeding birds. The chemicals that are released provide a signal indicating that a concentration of caterpillars is present in an area or on a particular tree. Therefore, abundance of those chemicals also signals a profitable feeding opportunity. This provides an explanation of how small passerines may be able to detect concentrations of caterpillars that infest trees, particularly during the period when they are in need of abundant food for the provisioning of nestlings (Amo et al. 2013).

Body Odors and Semiochemicals

The chemical environment contains a rich variety of odors emitted by animals. Many of these have been found to be "semiochemicals," chemical substances produced by an organism that carry a message for the purpose of communication between individuals in a species. Semiochemicals have long been established in mammals and insects. It has only

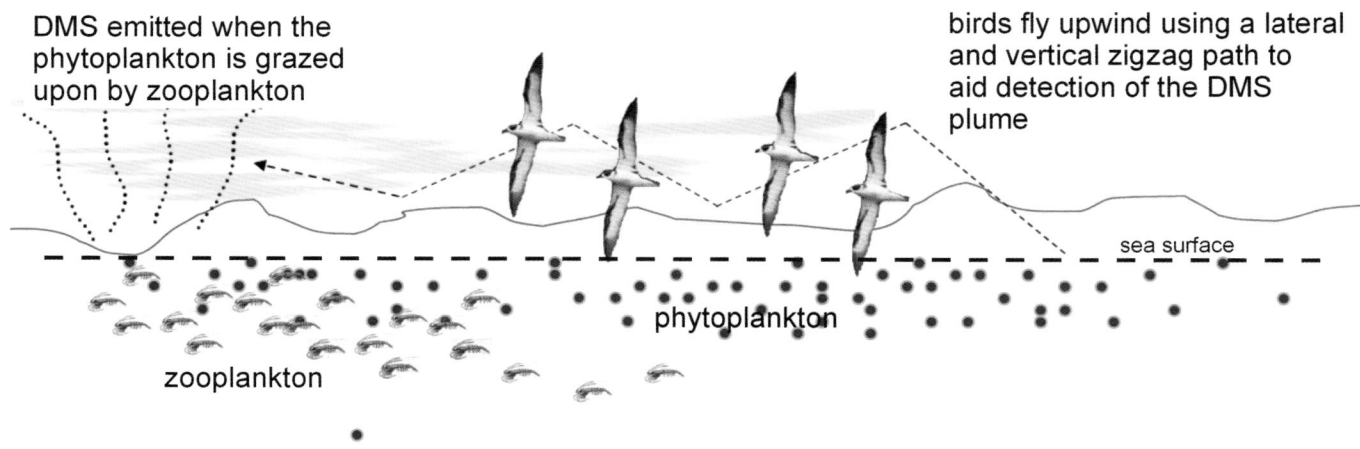

Figure 12.18. Seabirds exploiting an odor landscape to detect prey concentrations in the open sea. Concentrations of dimethyl sulfide (DMS) indicate efficient foraging opportunities. Antarctic Prions (*Pachyptila desolata*) are an example of a species that flies in zigzag paths across the open ocean to detect an odor plume and then fly upwind to the source of the odor concentration. *Illustration courtesy of Gabrielle Nevitt.*

recently become clear that odors produced by birds could also function as semiochemicals.

Odor-Based Recognition of Species

In birds, odors produced by the body mostly originate from secretions of the uropygial gland. This is the only sebaceous gland in birds, and it is situated low down on the back, above the tail. The gland produces waxy fluids (preen oils) that are spread on feathers using the bill during preening. Thus any odors produced by that gland may be spread over much of a bird's outer surfaces.

The secretions of uropygial glands vary markedly between species and evidence suggests that these differences are detected by birds. For example, there is circumstantial evidence that odor information is used for species recognition in certain petrel species that nest underground in mixed species colonies and emerge only at night, when visual recognition may be difficult (Bonadonna and Mardon 2010). Behavioral evidence of species recognition has also been found in Dark-eyed Juncos (*Junco hyemalis*) (Whittaker et al. 2013), Waxwings (*Bombycilla* spp.) (Zhang et al. 2013), and Budgerigars (*Melopsittacus undulates*) (Zhang et al. 2010), and sex recognition has been found in Spotless Starlings (*Sturnus unicolor*) (Amo et al. 2012a). Some authors have speculated that odor signals may even differentiate between subspecies, as in the possible case of the Atlantic and the Mediterranean Cory's Shearwaters (*Calonectris borealis*) (Caro et al. 2015). (Note that these subspecies have been raised to species status—*C. borealis*, Cory's Shearwater; *C. edwardsii* Cape Verde Shearwater—by some authorities; Sangster et al. 2012.) It is not yet known, however, whether these shearwaters exploit these sources of information. Similarly, body odors strongly differ between

closely related and similar-looking species of passerines such as Zebra Finches (*Taeniopygia guttata*) and Diamond Firetails (*Stagonopleura guttata*), which have overlapping distributions (Caro et al. 2015). Odor signals could differentiate between these populations, but to date it remains unclear whether birds actually exploit these sources of information.

Odor-Based Recognition of Individuals

Evidence that birds can recognize individuals based on odor cues alone has come from the chicks of European Storm Petrels (*Hydrobates pelagicus*) (De Leon et al. 2003), adult Antarctic Prions (*Pachyptila desolata*) (Bonadonna and Nevitt 2004), Wilson's Storm Petrels (*Oceanites oceanicus*) (Jouventin et al. 2007), and Blue Petrels (*Halobaena caerulea*) (Mardon and Bonadonna 2009).

An interesting source of evidence that birds could potentially discriminate between individuals based on odors has come from experiments with laboratory mice (*Mus musculus*), which showed that mice could differentiate between the odors of individual chickens. This leaves open the question of whether the birds may also detect and exploit those differences, but it does show that chickens have individual odor signatures (Karlsson et al. 2010).

Odor-Based Recognition of Individual Quality and Mate Choice

Olfactory signals of individual quality have been clearly demonstrated to be detected in Crested Auklets (*Aethia cristatella*) (Hagelin 2007) (fig. 12.19). These seabirds exhibit a seasonally released scent, described by humans as "citrus-like." Production of this odor is associated with a display behavior called a "ruff-sniff," which involves a bird rubbing its face in

Crested Auklet

House Finch

Figure 12.19. The use of semiochemicals in birds. Semiochemicals are substances produced by an organism that carries a message for the purpose of communication between individuals in a species. Personal odors may function in the mate choice behavior of Crested Auklets (*Aethia cristatella*) and in male-male competition in House Finches (*Carpodacus mexicanus*). *Photograph of Crested Auklet, Public Domain, www.fhwa.dot.gov/byways/photos/58710; photograph of House Finch by Geoffrey E. Hill.*

the scented nape of a displaying partner. This odor is not associated with secretions from the uropygial gland. Instead, it comes from feathers in the interscapular region high up on the back. This secretion is thought to function as a chemical repellent of parasites (Douglas 2006, 2013), and it has been suggested (Caro et al. 2015) that exchange of these repellents through ruff-sniffing could also function in the exchange of information regarding the health status of the individuals. In that way these odors could also be contributing to mate choice.

In House Finches (*Carpodacus mexicanus*; fig. 12.19) individual-specific odors may be important in male-male competition. Evidence for this comes from experiments that show that males avoided the scent of other males that were in higher body condition than themselves (Amo et al. 2012b).

Uropygial gland secretions may play an important role in the control of copulation in chickens. Males prefer to mate with females whose uropygial glands are intact as opposed to females that had had the uropygial gland surgically removed. Furthermore, this differential preference is not evident in the behavior of males that had had their olfactory bulbs removed. This strongly suggests that the behavior of the males was based on the detection of odor cues (Hirao et al. 2009). However, copulation involves multifaceted behaviors, and it remains unclear exactly what is being detected through olfaction. It could be the species, the sex, or the reproductive status of the individual, or indeed all of these.

Odors and Nests

The role of olfactory cues during nest-building has been demonstrated in a number of bird species, especially the role of odor cues in the selection of specific nest materials. Some species of tits (Paridae) and starlings (Sturnidae) are well known

for incorporating aromatic plants in their nests, and these plant items may be added during the breeding attempt, not just in the initial building. This aromatic plant material may function to repel ectoparasites and ultimately enhance nestling growth rates. Starlings use odor cues to select suitable nest material (Gwinner 2013), and in tits the parents appear to use odor cues to determine when to replenish the nest with fresh aromatic herbs (Petit et al. 2002, Deeming and Reynolds 2015).

Olfactory recognition of nest sites also occurs in Leach's Storm Petrels (*Oceanodroma leucorhoa*). Like other petrels, these birds breed in dense colonies, which they enter only at night. To humans, these nest burrows have a strong musky odor, and choice experiments with individual birds show they are able to identify their own burrow by smell alone (Benvenuti et al. 1993, Bonadonna and Bretagnolle 2002).

Olfaction: Key Points

- There is clear evidence that olfactory information is detected by birds and can have high importance in the control of certain behaviors.
- Odors can play a role in foraging for specific items, and in the location of profitable foraging areas.
- Olfaction can play a role in social behavior and reproduction.
- Evidence in support of the importance of odors comes from a small sample of the total diversity of birds; because of this, we may be only at the beginning of a full understanding of the role of odors in the lives of birds.

TASTE

In the case of olfaction, certain chemicals in the air are taken into the olfactory cavities and detected. In taste, specific chemical properties are detected from objects that have been taken into the mouth (Berkhoudt 1985, Roura et al. 2013). Taste receptors occur in specific types, and they signal the presence of a relatively small number of chemical compounds. There are marked differences between species in the absolute and relative numbers of different receptors and in their distributions within the mouth. This strongly suggests that natural selection has shaped both the detection of different tastes and the ways that the chemical compositions of specific objects are sampled in different species.

It is clear that taste is vitally important to birds. It evolved primarily to identify nutritious foods through the detection of relevant compounds: carbohydrates, amino acids, lipids, salts, calcium, and toxic compounds. The distribution of taste receptors in the mouth seems to follow ingestion pathways, allowing items taken into the mouth to be chemically screened. Much of what is known about taste in birds is from chickens and ducks. This is because understanding taste preferences in these birds is critical to understanding how to maximize the efficiency of food intake in these commercially exploited species. Another applied aspect of taste research in birds has been in determining which tastes birds find aversive. This information has been utilized with a view toward developing bird repellents (Roura et al. 2013).

Taste Buds

As in all vertebrates, taste signals in birds are generated in taste buds. These barrel or flask-shaped groups of cells are embedded within the skin of the tongue or other structures in the mouth. Those sensory cells are in continuous contact with the oral cavity through a pore, and they sample the chemical composition of objects or liquids resting against the pore entrance. It is not clear how many different types of taste buds are present in birds, but at least three have been classified, primarily according to their shape (Berkhoudt 1985). Each bud contains variable numbers of sensory cells that extend through the pore and detect the presence of particular chemical stimuli. The distribution of taste buds in birds was first described in Mallards in 1906 (Bath 1906) and has subsequently been refined to show that taste buds in this species occur not only at the base of the tongue but also at the bill tip, just behind the bill-tip organ, and in the floor of the mouth, just behind the bill tip. This distribution shows that in Mallards taste discrimination can take place when an object is held at the bill tip. This configuration indicates that an object does not have to be brought farther into the mouth for its chemical composition to be sampled. This correlates with studies of Mallards of the way in which food items are selected and held prior to ingestion (Zweers and Wouterlood 1973). In all birds investigated, taste receptors also occur on the tongue, particularly toward the back, but they do not occur widely over the dorsal surface, as is typical in mammals. Clearly, much sampling of chemical composition occurs when objects and fluids are well inside the mouth, not just at the bill tip.

Taste Genes and Taste Receptors

Much has been learned in recent years about the presence of taste receptors in birds. Genetic analysis has identified "taste genes" in a wide range of vertebrates, and they have also been found in a range of bird species. These genes seem to be highly conserved across vertebrate classes, attesting to the critical role that taste has played in the survival and adaptation of species (Shi and Zhang 2006), and genomic evidence indicates that taste systems are related to dietary regimes among vertebrates in general (Jiang et al. 2012). This work has also led to the identification of taste receptors, as opposed to the taste buds, in which the receptors are housed (Hoon et al. 1999, Matsunami et al. 2000). This work has shown that in mammals, taste receptors occur not only in the mouth but also at other locations involved in nutrient absorption and tissue metabolism, such as the gastrointestinal tract, liver, adipose deposits, and hypothalamus. However, because of the high concentration of taste receptors in the mouth and its obvious role in ingesting food, the mouth should be considered one of the main sensory organs of the body.

Relative Numbers of Taste Receptors

In the past it was argued that birds have an inferior taste acuity compared with mammals. Mainly because of a lack of mastication (chewing), it was thought that chemicals are not so readily exposed in the mouth, and that was thought to explain why taste buds occurred in relatively low numbers in birds (Berkhoudt 1985). However, when the number of taste buds is considered in relation to the size of food that is taken into the mouth in one bite, the number of taste receptors in chickens is not lower compared with mammals. This suggests that the capacity to taste food in birds and mammals is likely to be similar (Roura et al. 2013).

Taste Categories

As elaborated in the following sections, birds probably detect some or all of the "classical" taste categories, but they may also have other chemical sensitivities (Roura et al. 2013).

Sweet

The response of birds to sweet compounds, i.e., chemicals labeled as sweet by humans, varies between species. When faced with a choice, omnivorous and granivorous birds (e.g., chickens) do not respond positively to sugars, while nec-

DIETARY SWITCHING AND CHANGES IN PERCEPTUAL CHALLENGES

Many species of birds change their diet within the annual cycle, and this is usually associated with dramatic changes in anatomy and physiology (Piersma and van Gils 2011). Dietary change can also be associated with changes in foraging behavior, however, and this may bring about a change in the perceptual challenges associated with detecting different prey. An example of how these changes are reflected in the visual system has recently come to light.

Red Knots (*Calidris canutus*) are shorebirds of the Scolopacidae family. Their foraging behavior has been well studied and shown to use touch information from their bill-tip organ for the detection of invertebrates buried in muddy substrata. Studies of this behavior led to the idea of "remote touch" (Piersma et al. 1998). Comparison with other bird species of the same family, such as woodcocks, whose foraging is also guided by cues from a bill-tip organ, predicted that knots would also have comprehensive vision about the head, of the kind found in woodcocks. This is not, however, the case. In knots, the eyes are more forward in the skull than in woodcocks, there is a blind area behind the head, and the birds show conspicuous vigilance behaviors, regularly breaking off from foraging to scan between bouts of probing. The reason for this is that Red Knots feed using blind probing guided by remote touch only outside the breeding season, while woodcocks forage in this way all year round.

During breeding, knots exploit a different type of habitat: high-latitude tundra. In these habitats they may probe occasionally, but they are particularly dependent on capturing large numbers of small surface and flying insects, which are abundant. This prey cannot be detected by touch sensitivity in the bill and must be detected and seized using visual cues to guide accurate location of their bill and accurate timing of its opening. More forward eye position ensures that bill position can be controlled accurately, but this is also associated with noncomprehensive vision and increased vulnerability to predator attack throughout the annual cycle (Martin and Piersma 2009).

Thus, knots switch between visually and nonvisually guided foraging during the annual cycle; it is the more visually demanding foraging of the summer months that has driven the evolution of their visual fields.

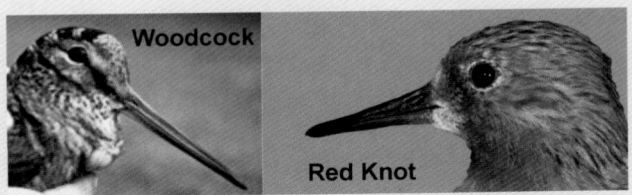

Eurasian Woodcock *Scolopax rusticola* and Red Knot *Calidris canutus. Photos by Graham R. Martin.*

tivorous and frugivorous bird species do respond positively (e.g., Cockatiels, *Nymphicus hollandicus*; and hummingbirds, Trochilidae) (Baldwin et al. 2014). This is evidence of a clear ecology for this particular taste (Kare and Mason 1986, Klasing 1998, Matson et al. 2001). Some bird species (e.g., Common Starlings) show a preference for fructose and glucose; this may be related to their preference for fruits at particular times of the year (Schuler 1983).

Umami

In humans, the umami taste is generally identified as the taste of monosodium glutamate or glutamic acid, and is interpreted as a taste that functions in the detection of proteins or amino acids. In choice experiments, Domestic Chicken chicks showed a preference for a diet containing amino acids. The actual substances detected by umami or amino acid receptors are unclear and will require a good deal of further investigation, but they include 1-alanine, which is detected by both European Starlings (*Sturnus vulgaris*) and Red-winged Blackbirds (*Agelaius phoeniceus*) (Espaillat and Mason 1990, Werner et al. 2008). In Domestic Chickens, umami receptors are thought to be involved in the coordination of post-ingestive and metabolic events, since they occur in hypothalamus, liver, and abdominal fat tissues. Umami receptors in Domestic Chickens may be capable of detecting a wide range of amino acids and may drive protein-specific appetites (Roura et al. 2013).

Bitter

Bitter tastes are usually associated with naturally occurring toxic compounds of plant or animal origin, especially insects. Bitter-tasting substances such as tannins and phenylpropanoids seem to be aversive to birds (e.g., European Starlings and Red-winged Blackbirds), and there have been attempts to exploit this in the development of bird repellents. Chickens and cockatiels find quinine (a substance known particularly for its bitter taste in humans) aversive. Quinine has also been shown to be sufficiently unpleasant to Domestic Chickens that they learn to avoid it after just a few exposures (Skelhorn and Rowe 2005).

Calcium

Calcium requirements in the diets of birds are particularly high during egg-laying. Dietary preferences of many bird species for calcium-rich items such as shells, bones, and calcareous grit have been well documented (Reynolds and Perrins 2010). Domestic Chickens are among the bird species that have a well-defined appetite for calcium during skeletal development and in adult females because of the high demands for calcium to enable the formation of eggs over extended periods (Wood-Gush and Kare 1966). Just how important taste is in the intake of calcium is unclear. It could be that calcium intake is regulated by post-ingestive effects. Recent work suggests that "calcium-like" is a distinct taste (Tordoff et al. 2008), but additional conclusive evidence is still required.

Salt

The taste of salt (sodium chloride, NaCl) in birds is well defined and can result in two different behavioral responses depending on the concentration of NaCl that is presented. High concentrations elicit aversion, but low concentrations usually produce an attraction response, particularly after sodium depletion (Kare and Mason 1986). This suggests a fine discrimination of concentration that is continuously calibrated against the internal state of the birds. High concentrations of salt (>2 percent solution) are toxic to birds, except those that have salt glands (Bradley 2009), but all birds must at times ingest low concentrations of NaCl to ensure good health (Mason and Clark 2000).

Sour

Sourness is related to the acidity of food. The presence of acids is often caused by bacterial fermentation, and in mammals it is a source of tastes that typically evoke a rejection response. However, as in the case of salt, a graded response to acid/sour is evident, depending on the degree of acidity (pH) of the stimulus. For example, Domestic Chickens are tolerant of medium acidic or alkaline solutions (either side of the pH value of 7), but Domestic Chickens avoid extreme pH values (Fuerst and Kare 1962).

The actual taste-sensitive cells for both salt and sour have yet to be identified in birds, although they have been identified in mammals, and the orthologous genes (i.e., inherited

from a shared ancestor) for these receptors have been shown to be present in the Domestic Chicken genome (Roura et al. 2013).

Fat

Evidence suggests that in mammals a fat-sensing taste receptor exists and that it plays an important role in regulating the calorific content of the diet (Cartoni et al. 2010). Little is known, however, regarding the receptor and its function, and to date no avian homolog gene for this receptor has been found in the Domestic Chicken genome database.

Taste: Key Points

- The taste sensory system in birds is complex and can be viewed as a group of nutrient sensors that have evolved to evaluate the nutritional quality and content of foods.
- Arrays of different receptors types have evolved, and their frequencies and distributions within the mouths of birds are complex.
- These arrays of taste receptors constitute a sense organ that may be as equally important and complex as any of the other sense organs found in the heads of birds.
- Taste systems seem to show distinct relationships with the ecology of the species with respect to receptor types and their positioning within the mouth and possibly in other tissues.
- The array of taste receptors in any one species may be closely linked to the diet and the means of ingestion.
- Taste has a role as equally vital in the foraging of birds as other aspects of food acquisition.

MAGNETORECEPTION

Experiments indicating that magnetic fields can control behavior in animals were first published in the 1960s. Since then, a large body of work has demonstrated the central importance of information about the Earth's magnetic field in many animal taxa, especially birds (Wiltschko and Wiltschko 2006). The ability to detect the geomagnetic field probably evolved because it is omnipresent and can provide a source of information to guide navigation and orientation. The geomagnetic field can theoretically be used to provide information on both direction and position. Direction can be derived from the horizontal component of the geomagnetic vector; this is the information that humans have learned to detect with the invention of the technical compass. Position information can be derived from the total intensity and/or inclination of the magnetic field vector, which exhibits gradients between the Earth's magnetic poles and the magnetic equator. Although humans have technology to detect this property of the magnetic field, it has not been commonly employed in human navigation devices (Skiles 1985).

Animals That Detect the Geomagnetic Field

A wide range of animals have been shown to detect one or both types of geomagnetic information, and the uses to which they are able to put this information are becoming well understood (Wiltschko and Wiltschko 1995). Animals that can use an internal magnetic compass include members of phyla Mollusca and Arthropoda, plus all major vertebrate groups including cartilaginous and bony fish, amphibians, reptiles, mammals, and birds. Magnetic cues are known to be used for orientation at different spatial scales: for example, within a home range by birds, for the orientation of buildings by bees and termites, and in the extended migrations of eels (Anguilliformes), salmon (Salmonidae), marine turtles (Chelonioidea), and birds. Magnetoreception is likely to be a widespread ability among birds; it has been demonstrated in a broad range of avian orders including pigeons and doves (Columbiformes), gamebirds (Galliformes), and passerines. Despite this phylogenetic diversity, the actual number of species in which it has been investigated is still very small.

Magnetic Compass Mechanisms

Two types of magnetic compass mechanisms have been described, with fundamental differences in functional characteristics (Wiltschko and Wiltschko 2006). The first type, a polarity compass, works in a similar way to a technical compass, using polarity of the magnetic field to distinguish between magnetic "north" and "south." The second type, an inclination compass, is the type of compass used by birds. It relies upon the axial course of the geomagnetic field lines, with directional information obtained by interpreting the inclination of the field lines with respect to up and down (gravity). This mechanism distinguishes between "poleward," where the field lines run downward, and "equatorward," where they run upward, but it does not distinguish between north and south poles.

Detection of the Geomagnetic Field

Although the presence of a compass mechanism is well established in birds, the way birds detect the Earth's magnetic field and how information is transmitted from those receptors to the nervous system is still unclear. It seems to be established that there are two detector mechanisms in birds: the **magnetite model** and the **radical pair model**. The magnetite model proposes a primary process involving tiny crystals of permanently magnetic material somewhere in

the animal's body, while the radical pair model proposes a so-called chemical compass based on "singlet-triplet transitions" in photopigments in the retina. Thus magnetic information is detected in the eye.

The Magnetite Model of Magnetic Field Detection

Magnetite crystals have been found to function in magnetic field orientation in bacteria (Blakemore 1975), and they have also been found in birds. For example, there are descriptions of deposits of iron oxide (probably magnetite) that lie in sheaths of tissues around the olfactory nerve and bulb and between the eyes, and also in bristles that project into the nasal cavity in Bobolinks (*Dolichonyx oryzivorus*) (Beason and Nichols 1984) (fig. 12.20). Also, very small magnetite crystals have been found in the skin of the upper bill in Rock Doves (Fleissner et al. 2003), and there is evidence from Bobolinks that the nerve fibers that carry information from these regions are responsive to changes in earth-strength magnetic fields (Semm and Beason 1990). This anatomical and physiological evidence has been supported by various behavioral experiments in which birds were placed in high-intensity magnetic fields that overwhelmed the low-strength geomagnetic field. This treatment disrupted the orientation behavior of the birds, but natural behavior was reinstated when the high-strength magnetic field was removed (Wiltschko and Wiltschko 2006).

The Radical Pair Model

Evidence in support of the radical pair model was first found through demonstrations that magnetoreception is light-dependent. Species in which this has been found include

Figure 12.20. Magnetoreception in birds. European Robins have been shown to employ the light-dependent mechanism based on specialized photopigments in the retina of their right eye. Bobolinks have been shown to have another mechanism based on deposits of iron oxide (probably magnetite) in sheaths of tissues around the olfactory nerve and bulb and between the eyes. Each species may have both mechanisms, with photopigments used for the detection of directional information and magnetite-based mechanism used to detect magnetic field intensity. *Photographs: European Robin by Graham R. Martin; Bobolink, JanetandPhil https://www.flickr.com/photos/dharma_for_one.*

European Robins (*Erithacus rubecula*), Australian Silver Eyes (*Zosterops lateralis*), and Garden Warblers (*Sylvia borin*), all passerine species that have migratory populations (fig. 12.20). The radical pair model proposes that magnetic fields are detected by specialized photopigments in the retina, with the most likely candidates being cryptochromes that are known to occur in birds' eyes (these molecules are not the same as the photopigments that underlie vision). This magnetoreception is not only light-dependent, but also depends on the wavelength of light and on the eye that is used. Evidence suggests that radical pair–based detectors are found only in the right eye of these birds (Wiltschko et al. 2002, Wiltschko et al. 2003, Stapput et al. 2010). Furthermore, experiments with Rock Doves and Domestic Chickens have shown the presence of a magnetic compass that works under blue, turquoise, and green light (up to a wavelength of 565 nm), but these birds are normally disoriented under yellow light (light with wavelengths longer than 582 nm) (Freire et al. 2008, Wiltschko et al. 2010).

Magnetoreception: Key Points

- Magnetoreception in birds is complex and has generated some controversy.
- There is now a strong body of evidence that birds, along with many other animals, are able to detect the Earth's magnetic field.
- Birds are able to employ information derived from the Earth's magnetic field to determine migratory orientation at the time of departure and possibly to guide themselves between locations.
- Magnetoreception can play a part in the control of bird movements at much smaller scales, perhaps helping birds find their way back to sites within a home range or territory. As such, magnetoreception is a sense that may be employed on a daily basis, not just at key times of year such as on migration.
- Many fundamental aspects of magnetoreception and its role in bird behaviors remain to be determined, it does seem clear, however, that more than one type of magnetic information is used by birds, and that at least two types of detector mechanisms may be involved.
- Photopigments (cryptochromes) in the right eye are probably used for the detection of directional information.
- Magnetite-based receptors in the upper bill probably extract positional information by recording differences in magnetic intensity.
- It would seem that birds have a compass in their right eye and a magnetometer in their bill, but a great deal remains to be learned about this sense and its uses (Wiltschko and Wiltschko 2006).

The Senses: Key Points

- Birds gather information about their environment using vision, hearing, somatic sensitivity (touch), olfaction, taste, and magnetoreception.
- Vision is the primary means by which birds gain information from their environments.
- Sensory capacities vary markedly between bird species.
- The senses of each species are finely tuned to the perceptual challenges of particular tasks and environments.
- The primary task that has driven the evolution of bird senses is the gathering of information for the guidance of foraging.
- The gathering of information for the guidance of foraging is traded off against the demands for detecting predators.

KEY MANAGEMENT AND CONSERVATION IMPLICATIONS

- Because of differences in sensory capacities, birds live in a very different world from that of humans. This requires that conservation and management issues be analyzed from the perspective of birds, not of humans.
- Collisions with human artifacts such as wind turbines, power lines, buildings, fences, and fishing nets occur primarily because of constraints and limitations on the vision of birds, which must be taken into account when designing mitigation measures.
- Collision-prone flying birds may not always be looking where they are heading. Distracting birds away from dangerous flight paths may be better than trying to make obstacles more conspicuous.
- Nocturnally active birds rely on minimal sensory cues. The interpretation of these cues usually necessitates familiarity with a specific environment, usually built up through prolonged residence. Hence habitat stability is critical to conserving nocturnal species.
- Commercial fishing nets are a hazard to all birds that forage underwater because of the low visual resolution of birds when underwater. Mitigation measures need to either make nets more conspicuous or distract birds away from net locations.

DISCUSSION QUESTIONS

1. What is a bird's-eye view?
2. How important is the sense of smell in birds?
3. Why are songbirds unable to locate sounds accurately?
4. What are bill-tip organs used for?

5. "A bird is a bill guided by an eye." Does this capture the essence of a bird?

6. Why is nocturnality rare among birds?

7. What can be done to prevent birds flying into obstacles?

References

Amo, L., I. Galvan, G. Tomas, and J. J. Sanz (2008). Predator odour recognition and avoidance in a songbird. Functional Ecology 22:289–293.

Amo, L., S. P. Caro, and M. E. Visser (2011). Sleeping birds do not respond to predator odour. PLoS ONE 6. doi:10.1371/journal.pone.0027576.

Amo, L., J. M. Aviles, D. Parejo, A. Pena, J. Rodriguez, and G. Tomas (2012a). Sex recognition by odour and variation in the uropygial gland secretion in starlings. Journal of Animal Ecology 81:605–613.

Amo, L., I. Lopez-Rull, I. Pagan, and C. M. Garcia (2012b). Male quality and conspecific scent preferences in the House Finch, *Carpodacus mexicanus*. Animal Behaviour 84:1483–1489.

Amo, L., J. J. Jansen, N. M. van Dam, M. Dicke, and M. E. Visser (2013). Birds exploit herbivore-induced plant volatiles to locate herbivorous prey. Ecology Letters 16:1348–1355.

Bailey, A. (2002). Sextus Empiricus and Pyrrhonean scepticism. Oxford University Press, Oxford.

Bala, A. D. S., M. W. Spitzer, and T. T. Takahashi (2003). Prediction of auditory spatial acuity from neural images of the owl's auditory space map. Nature 424:771–774.

Baldwin, M. W., Y. Toda, T. Nakagita, M. J. O'Connell, K. C. Klasing, T. Misaka, S. V. Edwards, and S. D. Liberles (2014). Evolution of sweet taste perception in hummingbirds by transformation of the ancestral umami receptor. Science 345:929–933.

Basbaum, A. I., A. Kaneko, T. Shimizu, and G. Westheimer (2008). The senses: A comprehensive reference. Vols. 1 and 2. Elsevier, Amsterdam.

Bath, W. (1906). Die Geschmacksorgane der Vogel und Krokodile. Archiv fur Biontologie 1:5–47.

Beason, R. C., and J. E. Nichols (1984). Magnetic orientation and magnetically sensitive material in a transequatorial migratory bird. Nature 309:151–153.

Bennett, A. T. D., I. C. Cuthill, J. C. Partridge, and K. Lunau (1997). Ultraviolet plumage colors predict mate preferences in starlings. Proceedings of the National Academy of Sciences USA 94:8618–8621.

Benvenuti, S., P. Ioale, and B. Massa (1993). Olfactory experiments on Cory Shearwater (*Calonectris diomedea*): The effect of intranasal zinc-sulfate treatment on short-range homing behavior. Bollettino di zoologia 60:207–210.

Berkhoudt, H. (1985). Structure and function of avian taste receptors. *In* Form and function in birds, A. S. King and J. McLelland, Editors. Academic Press, London, pp. 463–496.

Berkley, M. (1976). Cat visual psychophysics: Neural correlated and comparison with man. *In* Progress in psychobiology, J. Sprague and A. Epstein, Editors. Academic Press, New York, pp. 63–119.

Bertin, A., L. Calandreau, C. Arnould, R. Nowak, F. Levy, V. Noirot, I. Bouvarel, and C. Leterrier (2010). In ovo olfactory experience influences post-hatch feeding behaviour in young chickens. Ethology 116:1027–1037.

Bingelli, R. L., and W. J. Paule (1969). The pigeon retina: Quantitative aspects of the optic nerve and ganglion cell layer. Journal of Comparative Neurology 137:1–18.

Bisti, S., and L. Maffei (1974). Behavioural contrast sensitivity of the cat in various visual meridians. Journal of Physiology 241:201–210.

Blakemore, R. (1975). Magnetotactic bacteria. Science 190:377–379.

Blough, D. S. (1971). The visual acuity of the pigeon for distant targets. Journal of the Experimental Analysis of Behavior 15:57–68.

Bonadonna, F., and V. Bretagnolle (2002). Smelling home: A good solution for burrow-finding in nocturnal petrels? Journal of Experimental Biology 205:2519–2523.

Bonadonna, F., and G. A. Nevitt (2004). Partner-specific odor recognition in an Antarctic seabird. Science 306:835.

Bonadonna, F., and J. Mardon (2010). One house two families: Petrel squatters get a sniff of low-cost breeding opportunities. Ethology 116:176–182.

Bowmaker, J. K. (1977). The visual pigments, oil droplets and spectral sensitivity of the pigeon. Vision Research 17:1129–1138.

Bowmaker, J. K., and G. R. Martin (1978). Visual pigments and colour vision in a nocturnal bird, *Strix aluco* (Tawny Owl). Vision Research 18:1125–1130.

Bowmaker, J. K., and G. R. Martin (1985). Visual pigments and oil droplets in the penguin, *Spheniscus humboldti*. Journal of Comparative Physiology A: Sensory Neural and Behavioral Physiology 156:71–77.

Bowmaker, J. K., and E. Loew (2008). Vision in fish. *In* The senses: A comprehensive reference, vol. 1, Vision, A. I. Basbaum, A. Kaneko, G. M. Shepherd, and G. Westheimer, Editors. Elsevier, Amsterdam, pp. 54–76.

Bradley, T. J. (2009). Animal osmoregulation. Oxford University Press, Oxford.

Burkhardt, D. (1982). Birds, berries and UV: A note on some consequences of UV vision in birds. Naturwissenschaften 69:153–157.

Caro, S. P., J. Balthazart, and F. Bonadonna (2015). The perfume of reproduction in birds: Chemosignaling in avian social life. Hormones and Behavior 68:25–42.

Cartoni, C., K. Yasumatsu, T. Ohkuri, N. Shigemura, R. Yoshida, N. Godinot, J. le Coutre, Y. Ninomiya, and S. Damak (2010). Taste preference for fatty acids is mediated by GPR40 and GPR120. Journal of Neuroscience 30:8376–8382.

Catania, K. C., and F. E. Remple (2004). Tactile foveation in the star-nosed mole. Brain Behavior and Evolution 63:1–12.

Coimbra, J. P., S. P. Collin, and N. S. Hart (2015). Variations in retinal photoreceptor topography and the organization of the rod-free zone reflect behavioral diversity in Australian passerines. Journal of Comparative Neurology 523:1073–1094.

Corfield, J. R., H. L. Eisthen, A. N. Iwaniuk, and S. Parsons (2014). Anatomical specializations for enhanced olfactory sensitivity in kiwi, *Apteryx mantelli*. Brain Behavior and Evolution 84:214–216.

Corfield, J., K. Price, A. N. Iwaniuk, C. Gutiérrez-Ibáñez, T. Birkhead, and D. R. Wylie (2015). Diversity in olfactory bulb size in birds reflects allometry, ecology, and phylogeny. Frontiers in Neuroanatomy. doi:10.3389/fnana.2015.00102.

Cserhati, P., A. Szel, and P. Rohlich (1989). Four cone types characterized by anti-visual pigment antibodies in the pigeon retina. Investigative Ophthalmology and Visual Science 30:74–81.

Cunningham, S. J. (2010). Remote touch prey-detection by Madagascar Crested Ibises *Lophotibis cristat urschi*. Journal of Avian Biology 41:350–353.

Cunningham, G. B., R. W. Van Buskirk, F. Bonadonna, H. Weimerskirch, and G. A. Nevitt (2003). A comparison of the olfactory abilities of three species of procellariiform chicks. Journal of Experimental Biology 206:1615–1620.

Cunningham, S. J., I. Castro, and M. Alley (2007). A new prey-detection mechanism for kiwi (*Apteryx* spp.) suggests convergent evolution between paleognathous and neognathous birds. Journal of Anatomy 211:493–502.

Cunningham, S. J., I. Castro, and M. A. Potter (2009). The relative importance of olfaction and remote touch in prey detection by North Island Brown Kiwis. Animal Behaviour 78:899–905.

Cunningham, S. J., M. R. Alley, I. Castro, M. A. Potter, and M. J. Pyne (2010). Bill morphology of ibises suggests a remote-tactile sensory system for prey detection. The Auk 127:308–316.

Cunningham, S. J., and I. Castro (2011). The secret life of wild Brown Kiwi: Studying behaviour of a cryptic species by direct observation. New Zealand Journal of Ecology 35:209–219.

Cunningham, S. J., J. R. Corfield, A. N. Iwaniuk, I. Castro, M. R. Alley, T. Birkhead, and S. Parsons (2013). The anatomy of the bill tip of kiwis and associated somatosensory regions of the brain: Comparisons with shorebirds. PLoS ONE 8 (11): e80036. doi:10.1371/journal.pone.0080036.

Deeming, D. C., and S. J. Reynolds (2015). Nests, eggs and incubation: New ideas about avian reproduction. Oxford University Press, Oxford.

De Leon, A., E. Minguez, and B. Belliure (2003). Self-odour recognition in European storm-petrel chicks. Behaviour 140:925–933.

Demery, Z. P., J. Chappell, and G. R. Martin (2011). Vision, touch and object manipulation in Senegal parrots *Poicephalus senegalus*. Proceedings of the Royal Society B 278:3687–3693.

Dolan, T., and E. Fernandez-Juricic (2010). Retinal ganglion cell topography of five species of ground-foraging birds. Brain Behavior and Evolution 75:111–121.

Douglas, H. D. (2006). Measurement of chemical emissions in crested auklets (*Aethia cristatella*). Journal of Chemical Ecology 32:2559–2567.

Douglas, H. D. (2013). Colonial seabird's paralytic perfume slows lice down: An opportunity for parasite-mediated selection? International Journal for Parasitology 43:399–407.

Endler, J. A., J. Gaburro, and L. A. Kelley (2014). Visual effects in Great Bowerbird sexual displays and their implications for signal design. Proceedings of the Royal Society Biological Sciences B 281. doi:10.1098/rspb.2014.0864.

Espaillat, J. E., and J. R. Mason (1990). Differences in taste preference between Red-winged Blackbirds and European Starlings. Wilson Bulletin 102:292–299.

Fay, R. R. (1992). The evolutionary biology of hearing. *In* Structure and function in sound discrimination among vertebrates, D. B. Webster, R. R. Fay, and A. N. Popper, Editors. Springer-Verlag, Berlin, pp. 229–264.

Feinkohl, A., and G. M. Klump (2013). Azimuthal sound localization in the European starling (*Sturnus vulgaris*): II. Psychophysical results. Journal of Comparative Physiology A: Neuroethology Sensory Neural and Behavioral Physiology 199:127–138.

Fernandez-Juricic, E., B. A. Moore, M. Doppler, J. Freeman, B. F. Blackwell, S. L. Lima, and T. L. DeVault (2011). Testing the terrain hypothesis: Canada Geese see the world laterally and obliquely. Brain Behavior and Evolution 77:147–158.

Fite, K. V. (1973). Anatomical and behavioral correlates of visual acuity in the Great Horned Owl. Vision Research 13:219–230.

Fleissner, G., E. Holtkamp-Rotzler, M. Hanzlik, M. Winklhofer, G. Fleissner, N. Petersen, and W. Wiltschko (2003). Ultrastructural analysis of a putative magnetoreceptor in the beak of homing pigeons. Journal of Comparative Neurology 458:350–360.

Freire, R., U. Munro, L. J. Rogers, S. Sagasser, R. Wiltschko, and W. Wiltschko (2008). Different responses in two strains of chickens (*Gallus gallus*) in a magnetic orientation test. Animal Cognition 11:547–552.

Fritzsch, B. (1992). The water-to-land transition evolution of the tetrapod basilar papilla, middle ear and auditory nuclei. *In* The evolutionary biology of hearing, D. B. Webster, R. R. Fay, and A. N. Popper, Editors. Springer, New York, pp. 351–376.

Fuerst, F. F., and M. R. Kare (1962). The influence of pH on fluid tolerance and preferences. Poultry Science 41:71–77.

Gagliardo, A. (2013). Forty years of olfactory navigation in birds. Journal of Experimental Biology 216:2165–2171.

Galifret, Y. (1968). Les diverse aires fonctionelles de la retine du pigeon. Zeitschrift fur Zellforschung und Mikroskopische Anatomie 86:535–545.

Ghim, M. M., and W. Hodos (2006). Spatial contrast sensitivity of birds. Journal of Comparative Physiology A 192:523–534.

Gislen, A., M. Dacke, R. H. H. Kroger, M. Abrahamsson, D.-E. Nilsson, and E. Warrant (2003). Superior underwater vision in a human population of sea gypsies. Current Biology 13:833–836.

Glasser, A., D. Troilo, and H. C. Howland (1994). The mechanism of corneal accommodation in chicks. Vision Research 34:1549–1566.

Glasser, A., C. J. Murphy, D. Troilo, and H. C. Howland (1995). The mechanism of lenticular accommodation in chicks. Vision Research 35:1525–1540.

Gómez-Laich, A., K. Yoda, C. Zavalaga, and F. Quintana (2015). Selfies of Imperial Cormorants (*Phalacrocorax atriceps*): What is happening underwater? PLoS ONE 10 (9): e0136980. doi:10.1371/journal.pone.0136980.

Gottschaldt, K. M. (1985). Structure and function of avian somatosensory receptors. *In* Form and function in birds, vol. 3, A. S. King and J. McLelland, Editors. Academic Press, London, pp. 375–461.

Goujon, D. E. (1896). Sur un appareil de corpuscules tactiles situe dans le bec des perroquets. Journal de L'anatomie et de la Physiologie Normale et Pathologique de L'homme 6:449–455.

Gover, N., J. R. Jarvis, S. M. Abeyesinghe, and C. M. Wathes (2009). Stimulus luminance and the spatial acuity of domestic fowl (Gallus g. domesticus). Vision Research 49:2747–2753.

Graves, G. R. (1992). Greater Yellow-headed Vulture (Cathartes melambrotus) locates good by olfaction. Raptor Research 26:38–39.

Gremillet, D., G. Kuntz, F. Delbart, M. Mellet, and A. Kato (2004). Linking the foraging performance of a marine predator to local prey abundance. Functional Ecology 18:793–801.

Gremillet, D., G. Kuntz, C. Gilbert, A. J. Woakes, P. J. Butler, and Y. Le Maho (2005). Cormorants dive through the polar night. Biology Letters 1:469–471.

Gremillet, D., T. Nazirides, H. Nikolaou, and A. J. Crivelli (2012). Fish are not safe from great cormorants in turbid water. Aquatic Biology 15:187–194.

Gridi-Papp, M., and P. M. Narins (2008). Sensory ecology of hearing. In The senses: A comprehensive reference, vol. 3, Audition, A. I. Basbaum, A. Kaneko, G. M. Shepherd, and G. Westheimer, Editors. Elsevier, Amsterdam, pp. 62–74.

Guillemain, M., G. R. Martin, and H. Fritz (2002). Feeding methods, visual fields and vigilance in dabbling ducks (Anatidae). Functional Ecology 16:522–529.

Gwinner, H. (2013). Male European Starlings use odorous herbs as nest material to attract females and benefit nestlings. In Chemical signals in vertebrates, M. L. East and M. Dehnhard, Editors. Springer, New York, pp. 353–362.

Hagelin, J. C. (2007). The citrus-like scent of crested auklets: Reviewing the evidence for an avian olfactory ornament. Journal of Ornithology 148:S195–S201.

Hagstrum, J. T. (2000). Infrasound and the avian navigational map. Journal of Experimental Biology 203:1103–1111.

Harmening, W. M., P. Nikolay, J. Orlowski, and H. Wagner (2009). Spatial contrast sensitivity and grating acuity of barn owls. Journal of Vision 9. doi:10.1167/9.7.13.

Hart, N. S. (2001). Variations in cone photoreceptor abundance and the visual ecology of birds. Journal of Comparative Physiology A: Sensory, Neural, and Behavioral Physiology 187:685–697.

Hart, N. S. (2002). Vision in the peafowl (Aves: Pavo cristatus). Journal of Experimental Biology 205:3925–3935.

Hart, N. S. (2004). Microspectrophotometry of visual pigments and oil droplets in a marine bird, the wedge-tailed shearwater Puffinus pacificus: Topographic variations in photoreceptor spectral characteristics. Journal of Experimental Biology 207:1229–1240.

Hart, N. S., J. C. Partridge, and I. C. Cuthill (1998). Visual pigments, oil droplets and cone photoreceptor distribution in the European Starling (Sturnus Vulgaris). Journal of Experimental Biology 201:1433–1446.

Hart, N. S., J. C. Partridge, I. C. Cuthill, and A. T. D. Bennett (2000). Visual pigments, oil droplets, ocular media and cone photoreceptor distribution in two species of Passerine bird: The Blue Tit (Parus caeruleus L.) and the Blackbird (Turdus merula L.). Journal of Comparative Physiology A: Sensory Neural and Behavioral Physiology 186:375–387.

Hart, N. S., and M. Vorobyev (2005). Modelling oil droplet absorption spectra and spectral sensitivities of bird cone photoreceptors. Journal of Comparative Physiology A 191:381–392.

Hart, N. S., and D. Hunt (2007). Avian visual pigments: Characteristics, spectral tuning, and evolution. American Naturalist 169:S7–S26.

Heffner, H. E., and R. S. Heffner (1998). Hearing. In Comparative psychology: A handbook, G. Greenberg and H. M. Haraway, Editors. Garland, New York, pp. 290–303.

Heffner, R. S., and H. E. Heffner (1982). Hearing in the elephant (Elephas maximus): Absolute sensitivity, frequency discrimination, and sound localization. Journal of Comparative and Physiological Psychology 96:926–944.

Hill, E. M., G. Koay, R. S. Heffner, and H. E. Heffner (2014). Audiogram of the chicken (Gallus gallus domesticus) from 2 Hz to 9 kHz. Journal of Comparative Physiology A: Neuroethology Sensory Neural and Behavioral Physiology 200:863–870.

Hirao, A., M. Aoyama, and S. Sugita (2009). The role of uropygial gland on sexual behavior in Domestic Chicken Gallus gallus domesticus. Behavioural Processes 80:115–120.

Hirsch, J. (1982). Falcon visual sensitivity to grating contrast. Nature 300:57–58.

Hodos, W. (1993). The visual capabilities of birds. In Avian vision, brain and behavior, H. P. Ziegler and H. J. Bischof, Editors. MIT Press, Cambridge, MA, pp. 63–76.

Hodos, W., R. W. Leibowitz, and J. C. Bonbright (1976). Near-field visual acuity of pigeons: Effects of head location and stimulus luminance. Journal of the Experimental Analysis of Behavior 25:129–141.

Hodos, W., and R. W. Leibowitz (1977). Near-field acuity of pigeons: Effect of scotopic adaptation and wavelength. Vision Research 17:463–467.

Hodos, W., M. Ghim, A. Potocki, J. Fields, and T. Storm (2002). Contrast sensitivity in pigeons: A comparison of behavioral and pattern ERG methods. Documenta Ophthalmologica 104:107–118.

Holland, J., T. Dabelsteen, C. P. Bjørn, and S. B. Pedersen (2001). The location of ranging cues in wren song: Evidence from calibrated interactive playback experiments. Behaviour 138:189–206.

Hoon, M. A., E. Adler, J. Lindemeier, J. F. Battey, N. J. P. Ryba, and C. S. Zuker (1999). Putative mammalian taste receptors: A class of taste-specific GPCRs with distinct topographic selectivity. Cell 96:541–551.

Houston, D. C. (1984). Does the King Vulture Sarcoramphus papa use a sense of smell to locate food? Ibis 126:67–69.

Houston, D. C. (1986). Scavenging efficiency of Turkey Vultures in tropical forest. The Condor 88:318–323.

Jiang, P. H., J. Josue, X. Li, D. Glaser, W. H. Li, J. G. Brand, R. F. Margolskee, D. R. Reed, and G. K. Beauchamp (2012). Reply to Zhao and Zhang: Loss of taste receptor function in mammals is directly related to feeding specializations. Proceedings of the National Academy of Sciences USA 109:E1465.

Johnson, C. S. (1986). Dolphin audition and echolocation capacities. In Dolphin cognition and behaviour: A comparative approach, R. Schusterman, J. Thomas, and F. Wood, Editors. Lawrence Erlbaum, Hillsdale, NJ, pp. 43–56.

Jones, C. D., and D. Osorio (2004). Discrimination of oriented visual textures by poultry chicks. Vision Research 44:83–89.

Jones, R. B., and T. J. Roper (1997). Olfaction in the domestic fowl: A critical review. Physiology and Behavior 62:1009–1018.

Jouventin, P., V. Mouret, and F. Bonadonna (2007). Wilson's Storm Petrels *Oceanites oceanicus* recognise the olfactory signature of their mate. Ethology 113:1228–1232.

Kare, M. R., and J. R. Mason (1986). The chemical senses in birds. *In* Avian physiology, 4th ed., P. D. Sturkie, Editor. Springer-Verlag, New York, pp. 59–67.

Karlsson, A. C., P. Jensen, M. Elgland, K. Laur, T. Fyrner, P. Konradsson, and M. Laska (2010). Red junglefowl have individual body odors. Journal of Experimental Biology 213:1619–1624.

Katzir, G., and G. R. Martin (1994). Visual fields in herons (Ardeidae): Panoramic vision beneath the bill. Naturwissenschaften 81:182–184.

Katzir, G., and H. C. Howland (2003). Corneal power and underwater accommodation in great cormorants (*Phalacrocorax carbo sinensis*). Journal of Experimental Biology 206:833–841.

Klasing, K. C. (1998). Anatomy and physiology of the digestive system. *In* Comparative avian nutrition, K. C. Klasing, Editor. CAB International, Oxford, pp. 16–19.

Klump, G. (2000). Sound localization in birds. *In* Comparative hearing: Birds and reptiles, R. J. Dooling, R. R. Fay, and A. N. Popper, Editors. Springer, New York, pp. 249–307.

Klump, G., W. Windt, and E. Curio (1986). The Great Tit's (*Parus major*) auditory resolution in azimuth. Journal of Comparative Physiology A 158:383–390.

Klump, G. M., and O. N. Larsen (1992). Azimuthal sound localization in the European Starling (*Sturnus vulgaris*): I. Physical binaural cues. Journal of Comparative Physiology A 170:243–251.

Knudsen, E. I., and M. Konishi (1979). Mechanisms of sound localization in the Barn Owl (*Tyto alba*). Journal of Comparative Physiology A 133:13–21.

Konishi, M., and E. I. Knudsen (1979). The Oilbird: Hearing and echolocation. Science 204:425–427.

Kreithen, M. L., and D. B. Quine (1979). Infrasound detection by the homing pigeon: Behavioral audiogram. Journal of Comparative Physiology 129:1–4.

Land, M. F. (1981). Optics and vision in invertebrates. *In* Handbook of sensory physiology, vol. 7/6B, H. Autrum, Editor. Springer-Verlag, Berlin, pp. 471–592.

Land, M. F., and D.-E. Nilsson (2012). Animal eyes. 2nd ed. Oxford University Press, Oxford.

Leclaire, S., H. Mulard, R. H. Wagner, S. A. Hatch, and E. Danchin (2009). Can Kittiwakes smell? Experimental evidence in a Larid species. Ibis 151:584–587.

Le Duc, D., G. Renaud, A. Krishnan, and M. S. Almen (2015). Kiwi genome provides insights into evolution of a nocturnal lifestyle. Genome Biology 16:147. doi:10.1186/s13059 015 0711 4.

Lewis, E. R., P. M. Narins, J. U. M. Jarvis, G. Bronner, and M. Mason (2006). Preliminary evidence for the use of microseismic cues for navigation by the Namib golden mole. Journal of the Acoustical Society of America 119:1260–1268.

Lind, O., and A. Kelber (2011). The spatial tuning of achromatic and chromatic vision in budgerigars. Journal of Vision 11. doi:10.1167/11.7.2.

Lind, O., T. Sunesson, M. Mitkus, and A. Kelber (2012). Luminance-dependence of spatial vision in budgerigars (*Melopsittacus undulatus*) and Bourke's parrots (*Neopsephotus bourkii*). Journal of Comparative Physiology A: Neuroethology Sensory Neural and Behavioral Physiology 198:69–77.

Lisney, T. J., K. Stecyk, J. Kolominsky, G. R. Graves, D. R. Wylie, and A. N. Iwaniuk (2013). Comparison of eye morphology and retinal topography in two species of New World vultures (Aves: Cathartidae). Anatomical Record 296:1954–1970.

Lythgoe, J. N. (1979). The ecology of vision. Clarendon Press, Oxford.

Mardon, J., and F. Bonadonna (2009). Atypical homing or self-odour avoidance? Blue petrels (*Halobaena caerulea*) are attracted to their mate's odour but avoid their own. Behavioral Ecology and Sociobiology 63:537–542.

Martin, G. R. (1974). Colour vision in the Tawny Owl *Strix aluco*. Journal of Comparative and Physiological Psychology 86:133–141.

Martin, G. R. (1986). Sensory capacities and the nocturnal habit of owls (Strigiformes). Ibis 128:266–277.

Martin, G. R. (1990). Birds by night. T. and A. D. Poyser, London.

Martin, G. R. (1998). Eye structure and amphibious foraging in albatrosses. Proceedings of the Royal Society of London B: Biological Sciences 265:1–7.

Martin, G. R. (1999). Eye structure and foraging in King Penguins *Aptenodytes patagonicus*. Ibis 141:444–450.

Martin, G. R. (2007). Visual fields and their functions in birds. Journal of Ornithology 148 (Suppl. 2): 547–562.

Martin, G. R. (2011). Understanding bird collisions with man-made objects: A sensory ecology approach. Ibis 153:239–254.

Martin, G. R. (2012). Through birds' eyes: Insights into avian sensory ecology. Journal of Ornithology 153 (Suppl. 1): S23–S48.

Martin, G. R. (2014). The subtlety of simple eyes: The tuning of visual fields to perceptual challenges in birds. Philosophical Transactions of the Royal Society B: Biological Sciences 369. doi.org/10.1098/rstb.2013.0040.

Martin, G. R., and I. E. Gordon (1974). Visual acuity in the Tawny Owl (*Strix aluco*). Vision Research 14:1393–1397.

Martin, G. R., and M. D. L. Brooke (1991). The eye of a procellariiform seabird, the Manx Shearwater, *Puffinus puffinus*: Visual fields and optical structure. Brain, Behavior and Evolution 37:65–78.

Martin, G. R., and G. Katzir (2000). Sun shades and eye size in birds. Brain, Behavior and Evolution 56:340–344.

Martin, G. R., L. M. Rojas, Y. Ramirez, and R. McNeil (2004). The eyes of Oilbirds (*Steatornis caripensis*): Pushing at the limits of sensitivity. Naturwissenschaften 91:26–29.

Martin, G. R., N. Jarrett, and M. Williams (2007a). Visual fields in Blue Ducks and Pink-eared Ducks: Visual and tactile foraging. Ibis 149:112–120.

Martin, G. R., K. J. Wilson, M. J. Wild, S. Parsons, M. F. Kubke, and J. Corfield (2007b). Kiwi forego vision in the guidance of their nocturnal activities. PLoS ONE 2 (2): e198. doi:10.1371 /journal.pone.0000198.

Martin, G. R., and D. Osorio (2008). Vision in birds. *In* The senses: A comprehensive reference, vol. 1, Vision, A. I. Basbaum, A. Kaneko, G. M. Shepherd, and G. Westheimer, Editors. Elsevier, Amsterdam, pp. 25–52.

Martin, G. R., C. R. White, and P. J. Butler (2008). Vision and the foraging technique of Great Cormorants *Phalacrocorax carbo*: Pursuit or flush-foraging? Ibis 150:39–48.

Martin, G. R., and T. Piersma (2009). Vision and touch in relation to foraging and predator detection: Insightful contrasts between a plover and a sandpiper. Proceedings of the Royal Society of London B: Biological Sciences 276:437–445.

Martin, G. R., and J. M. Shaw (2010). Bird collisions with power lines: Failing to see the way ahead? Biological Conservation 143:2695–2702.

Martin, G. R., and S. J. Portugal (2011). Differences in foraging ecology determine variation in visual field in ibises and spoonbills (Threskiornithidae). Ibis 153:662–671.

Martin, G. R., S. J. Portugal, and C. P. Murn (2012). Visual fields, foraging and collision vulnerability in *Gyps* vultures. Ibis 154:626–631.

Martin, G. R., and S. Wanless (2015). The visual fields of Common Guillemots *Uria aalge* and Atlantic Puffins *Fratercula arctica*: Foraging, vigilance and collision vulnerability. Ibis 157:798–807.

Martin, G. R., and R. Crawford (2015). Reducing bycatch in gillnets: A sensory ecology perspective. Global Ecology and Conservation 3:28–50.

Mason, J. R., and L. Clark (2000). The chemical senses in birds. *In* Sturkie's avian physiology, 5th ed., G. C. Whittow, Editor. Academic Press, San Diego, pp. 39–56.

Matson, K. D., J. R. Millam, and K. C. Klasing (2001). Thresholds for sweet, salt, and sour taste stimuli in cockatiels (*Nymphicus hollandicus*). Zoo Biology 20:1–13.

Matsunami, H., J. P. Montmayeur, and L. B. Buck (2000). A family of candidate taste receptors in human and mouse. Nature 404:601–602.

McGregor, P. K., J. R. Krebs, and L. M. Ratcliffe (1983). The reaction of Great Tits (*Parus major*) to playback of degraded and undegraded songs: The effect of familiarity with the stimulus song type. The Auk 100:898–906.

Medway, L., and D. Pye (1977). Echolocation and systematic of swiftlets. *In* Evolutionary Ecology, B. Stonehouse and C. M. Perrins, Editors, Macmillan, London, pp. 225–238.

Meyer, D. B. (1977). The avian eye and its adaptations. *In* Handbook of sensory physiology, vol. 7/5, F. Crescitelli, Editor. Springer-Verlag, Berlin, pp. 549–611.

Mitkus, M., S. Chaib, O. Lind, and A. Kelber (2014). Retinal ganglion cell topography and spatial resolution of two parrot species: Budgerigar *Melopsittacus undulatus* and Bourke's parrot *Neopsephotus bourkii*. Journal of Comparative Physiology A 200:371–384.

Mitkus, M., G. A. Nevitt, J. Danielsen, and A. Kelber (2015). Spatial resolution and optical sensitivity of a DMS-responder and a non-responder: Leach's Storm-petrel and Northern Fulmar. *In*

Spatial vision in birds: Anatomical investigation of spatial resolving power, PhD diss., M. Mitkus, Editor. Lund University, Lund, pp. 75–88.

Morton, E. S., J. Howlett, N. C. Kopysh, and I. Chiver (2006). Song ranging by incubating male Blue-headed Vireos: The importance of song representation in repertoires and implications for song delivery patterns and local/foreign dialect discrimination. Journal of Field Ornithology 77:291–301.

Naguib, M. (1997). Use of song amplitude for ranging in Carolina Wrens, *Thryothorus ludovicianus*. Ethology 103:723–731.

Naguib, M. (1998). Perception of degradation in acoustic signals and its implications for ranging. Behavioral Ecology and Sociobiology 42:139–142.

Naguib, M., G. Klump, E. Hillmann, B. Griebmann, and T. Teige (2000). Assessment of auditory distance in a territorial songbird: Accurate feat or rule of thumb? Animal Behaviour 59:715–721.

Naguib, M., and R. H. Wiley (2001). Estimating the distance to a source of sound: Mechanisms and adaptations for long-range communication. Animal Behaviour 62:825–837.

Nelson, B. S. (2000). Avian dependence on sound-pressure level as an auditory distance cue. Animal Behaviour 59:57–67.

Nelson, B. S., and P. K. Stoddard (1998). Accuracy of auditory distance and azimuth perception by a passerine bird in natural habitat. Animal Behaviour 56:467–477.

Nelson, B. S., and R. A. Suthers (2004). Sound localization in a small passerine bird: Discrimination of azimuth as a function of head orientation and sound frequency. Journal of Experimental Biology 207:4121–4133.

Nevitt, G. A. (2000). Olfactory foraging by Antarctic procellariiform seabirds: Life at high Reynolds numbers. Biology Bulletin (Woods Hole) 198:245–253.

Nevitt, G. A. (2008). Sensory ecology on the high seas: The odor world of the procellariiform seabirds. Journal of Experimental Biology 211:1706–1713.

Nevitt, G. A., and F. Bonadonna (2005). Sensitivity to dimethyl sulphide suggests a mechanism for olfactory navigation by seabirds. Biology Letters 1:303–305.

Newton, I. (2010). Opticks; Or, a treatise of the reflections, refractions, inflections, and colours of light. Project Gutenberg. http://www.gutenberg.org/ebooks/33504.

Norberg, R. A. (1968). Physical factors in directional hearing in *Aegolius funereus* (Strigiformes), with special reference to the significance of the asymmetry of the external ears. Arkive Zoology 20:181–204.

Norberg, R. A. (1977). Occurrence and independent evolution of bilateral ear asymmetry in owls and implications on owl taxonomy. Philosophical Transactions of the Royal Society of London B: Biological Sciences 280:376–408.

Norberg, R. A. (1978). Skull asymmetry, ear structure and function and auditory localization in Tengmalm's Owl, *Aegolius funereus*. Philosophical Transactions of the Royal Society of London B: Biological Sciences 282B:325–410.

Orlowski, J., W. Harmening, and H. Wagner (2012). Night vision in Barn Owls: Visual acuity and contrast sensitivity under dark adaptation. Journal of Vision 12:1–8. doi:10.1167/12.13.4.

Osorio, D., M. Vorobyev, and C. D. Jones (1999). Colour vision of domestic chicks. Journal of Experimental Biology 202:2951–2959.

Park, T. J., and R. J. Dooling (1991). Sound localization in small birds: Absolute localization in azimuth. Journal of Comparative Psychology 105:125–133.

Payne, R. S. (1971). Acoustic location of prey by Barn Owls. Journal of Experimental Biology 54:535–573.

Petit, C., M. Hossaert-McKey, P. Perret, J. Blondel, and M. M. Lambrechts (2002). Blue Tits use selected plants and olfaction to maintain an aromatic environment for nestlings. Ecology Letters 5:585–589.

Piersma, T., R. van Aelst, K. Kurk, H. Berkhoudt, and L. R. M. Maas (1998). A new pressure sensory mechanisms for prey detection in birds: The use of principles of seabed dynamics? Proceedings of the Royal Society of London B: Biological Sciences 265:1377–1383.

Piersma, T., and J. A. van Gils (2011). The flexible phenotype. Oxford University Press, Oxford.

Pirenne, M. H., F. H. C. Marriott, and E. F. O'Doherty (1957). Individual differences in night vision efficiency. Medical Research Council (GB) Special Report Series 294.

Pohl, N. U., G. Klump, and U. Langemann (2015). Effects of signal features and background noise on distance cue discrimination by a songbird. Journal of Experimental Biology 218:1006–1015.

Popkin, R. (2003). The history of scepticism: From Savonarola to Bayle. Oxford University Press, Oxford.

Pye, J. D. (1979). Why ultrasound? Endeavour 3:57–62.

Pye, J. D. (1985). Echolocation. In A dictionary of birds, B. Campbell and E. Lack, Editors. Poyser, Calton, UK, pp. 165–166.

Reymond, L. (1985). Spatial visual acuity of the eagle Aquila audax: A behavioural, optical and anatomical investigation. Vision Research 25:1477–1491.

Reymond, L. (1987). Spatial visual acuity of the falcon, Falco berigora: A behavioural, optical and anatomical investigation. Vision Research 27:1859–1974.

Reynolds, S. J., and C. M. Perrins (2010). Dietary calcium availability and reproduction in birds. In Current ornithology, vol. 17, C. F. Thompson, Editor. Springer, New York, pp. 31–74.

Rochon-Duvigneaud, A. (1943). Les yeux et la vision des vertébrés. Masson, Paris.

Roura, E., M. W. Baldwin, and K. C. Klasing (2013). The avian taste system: Potential implications in poultry nutrition. Animal Feed Science and Technology 180:1–9.

Sangster, G., J. M. Collinson, P.-A. Crochet, A. G. Knox, D. T. Parkin, and S. C. Votier (2012). Taxonomic recommendations for British birds: Eighth report. Ibis 154:874–883.

Schaeffel, F., and H. C. Howland (1987). Corneal accommodation in chick and pigeon. Journal of Comparative Physiology A 160:375–384.

Schuler, W. (1983). Responses to sugars and their behavioral mechanisms in the Starling (Sturnus vulgaris L.). Behavioral Ecology and Sociobiology 13:243–251.

Semm, P., and R. C. Beason (1990). Responses to small magnetic variations by the trigeminal system of the Bobolink. Brain Research Bulletin 25:735–740.

Shi, P., and J. Z. Zhang (2006). Contrasting modes of evolution between vertebrate sweet/umami receptor genes and bitter receptor genes. Molecular Biology and Evolution 23:292–300.

Shlaer, S. (1937). The relation between visual acuity and illumination. Journal of General Physiology 21:165–188.

Sivak, J. G. (1978). A survey of vertebrate strategies for vision in air and water. In Sensory ecology: Review and perspectives, M. A. Ali, Editor. Plenum Press, New York, pp. 503–520.

Sivak, J. G., and M. Millodot (1977). Optical performance of the penguin eye in air and water. Journal of Comparative Physiology A: Sensory, Neural, and Behavioral Physiology 119:241–247.

Skelhorn, J., and C. Rowe (2005). Frequency-dependent taste-rejection by avian predation may select for defence chemical polymorphisms in aposematic prey. Biology Letters 1:500–503.

Skiles, D. D. (1985). The geomagnetic field: Its nature, history and biological relevance. In Magnetite biomineralization and magnetoreception in organisms, J. L. Kirschvink, D. S. Jones, and B. J. Mac Fadden, Editors. Plenum Press, New York, pp. 43–102.

Smythe, D. M., and J. R. Roberts (1983). The sensitivity of echolocation by Grey Swiftlets Aerodramus spodiopygius. Ibis 125:339–345.

Snow, D. W. (1961). The natural history of the Oilbird, Steatornis caripensis, in Trinidad: 1. General behaviour and breeding habits. Zoologica 46:27–48.

Stapput, K., O. Gunturkun, K. P. Hoffmann, R. Wiltschko, and W. Wiltschko (2010). Magnetoreception of directional information in birds requires nondegraded vision. Current Biology 20:1259–1262.

Streby, H. M., G. R. Kramer, S. M. Peterson, J. A. Lehman, D. A. Buehler, and D. E. Andersen (2015). Tornadic storm avoidance behavior in breeding songbirds. Current Biology 25:98–102.

Theurich, M., G. Langner, and H. Scheich (1984). Infrasound responses in the midbrain of the guinea fowl. Neuroscience Letters 49:81–86.

Tordoff, M. G., H. G. Shao, L. K. Alarcon, R. F. Margolskee, B. Mosinger, A. A. Bachmanov, D. R. Reed, and S. McCaughey (2008). Involvement of T1R3 in calcium-magnesium taste. Physiological Genomics 34:338–348.

Walls, G. L. (1942). The vertebrate eye and its adaptive radiation. Cranbrook Institute of Science, Michigan.

Watanuki, Y., F. Daunt, A. Takahashi, M. Newei, S. Wanless, K. Sat, and N. Miyazaki (2008). Microhabitat use and prey capture of a bottom-feeding top predator, the European Shag, shown by camera loggers. Marine Ecology Progress Series 356:283–293.

Wenzel, B. (1968). Olfactory prowess of the kiwi. Nature 220:1133–1134.

Werner, S. J., B. A. Kimball, and F. D. Provenza (2008). Food color, flavor, and conditioned avoidance among Red-winged Blackbirds. Physiology and Behavior 93:110–117.

White, C. R., N. Day, P. J. Butler, and G. R. Martin (2007). Vision and foraging in cormorants: More like herons than hawks? PLoS ONE 2 (7): e639. doi:10.1371/journal.pone.0000639.

Whittaker, D. J., N. M. Gerlach, H. A. Soini, M. V. Novotny, and E. D. Ketterson (2013). Bird odour predicts reproductive success. Animal Behaviour 86:697–703.

Wiley, R. H., and R. Godard (1996). Ranging of conspecific songs by Kentucky Warblers and its implications for interactions of territorial males. Behaviour 133:81–102.

Wilkie, S. E., P. R. Robinson, T. W. Cronin, S. Poopalasun-daram, J. K. Bowmaker, and D. M. Hunt (2000). Spectral tuning of avian violet- and ultraviolet-sensitive visual pigments. Biochemistry 39:7895–7901.

Wilson, K.-J. (2004). Flight of the Huia: Ecology and conservation of New Zealand's frogs, reptiles, birds and mammals. Canterbury University Press, Christchurch, NZ.

Wiltschko, R., and W. Wiltschko (1995). Magnetic orientation in animals. Springer Verlag Berlin Heidelberg.

Wiltschko, W., J. Traudt, O. Gunturkun, H. Prior, and R. Wiltschko (2002). Lateralization of magnetic compass orientation in a migratory bird. Nature 419:467–470.

Wiltschko, W., U. Munro, H. Ford, and R. Wiltschko (2003). Lateralisation of magnetic compass orientation in silvereyes, Zosterops lateralis. Australian Journal of Zoology 51:597–602.

Wiltschko, R., and W. Wiltschko (2006). Magnetoreception. BioEssays 28:157–168.

Wiltschko, R., K. Stapput, P. Thalau, and W. Wiltschko (2010). Directional orientation of birds by the magnetic field under different light conditions. Journal of the Royal Society Interface 7:S163–S177.

Wood-Gush, D. G. M., and M. R. Kare (1966). The behaviour of calcium deficient chicken. British Poultry Science 7:285–290.

Wright, A. A. (1979). Color-vision psychophysics. In Neural mechanisms of behavior in the pigeon, A. M. Granda and J. H. Maxwell, Editors. Plenum Press, New York.

Zhang, J. X., W. Wei, J. H. Zhang, and W. H. Yang (2010). Uropygial gland-secreted alkanols contribute to olfactory sex signals in Budgerigars. Chemical Senses 35:375–382.

Zhang, Y. H., Y. F. Du, and J. X. Zhang (2013). Uropygial gland volatiles facilitate species recognition between two sympatric sibling bird species. Behavioral Ecology 24:1271–1278.

Zweers, G. A., and F. G. Wouterlood (1973). Functional anatomy of the feeding apparatus of the Mallard (Anas platyrhynchos). Proceedings of the 3rd European Anatomical Congress, Manchester 1:88–89.

CHAPTER 13

Song and the Brain

Jonathan F. Prather

WHAT IS A SONGBIRD?

Among all species of birds, over half belong to the order Passeriformes—the perching birds. Commonly called passerines, these birds are characterized by a toe arrangement that facilitates perching (Raikow and Bledsoe 2000). Within the passerines, the term **songbird** refers to species that possess the intriguing ability to learn songs by imitating the sounds they hear performed by other members of their species (**oscine birds**, suborder Passeri). The ability to learn the sounds that an animal uses in vocal communication is called **vocal learning**, and it is noteworthy because it is also a central feature of how we learn the sounds we use in speech, but it is otherwise quite rare across animals (Bolhuis and Everaert 2013). By studying how the structure and function of the songbird brain enables vocal learning through imitation, songbird neurobiologists are revealing not only how songbirds perceive and perform their songs but also how those insights may teach us about how we communicate through speech.

Among songbirds, thousands of species acquire their songs by imitating the sounds they hear performed by other members of their species, and they use those songs to indicate identity, defend territories, and attract mates. Other species of birds that do not rely on auditory experience to develop their songs are classified as **suboscines** and are not considered songbirds. Although the songs of suboscine species can have complex structure, their development has little or no relation to auditory experience (for further consideration of the possible role of learning in some suboscine species, see Kroodsma et al. 2013, Liu et al. 2013, Touchton et al. 2014). The impact of learning is most clearly evident in the songs of oscine songbirds, and those species are the focus of this chapter.

The ability to learn by transforming auditory experience into vocal performance is associated with a network of brain structures that is collectively called the **song system**. The presence of such a network is another defining characteristic of songbird species (reviewed in Mooney et al. 2008). In most songbird species, males sing but females do not. That behavioral difference is also accompanied by a striking sexual dimorphism in the size of the song system structures, with those structures being large and obvious in males but much smaller or absent in females (Ball 2016). If elements of the song system are damaged, deficits emerge in the male's ability to sing. There is a very close relation between the properties of specific sites in the brain and specific aspects of vocal learning and performance. This chapter focuses primarily on male songbirds to provide you with an understanding of song behavior, the role of auditory experience in song learning and maintenance, and the structure and function of the brain system that enables songbirds to learn those complex imitative behaviors.

WHAT IS A SONG?

Some early researchers suggested that birdsong should be viewed as a form of primitive art that was beautiful from a bird's point of view as well as from our human perspective. Darwin (1874) even commented that "on the whole, birds appear to be the most aesthetic of animals, excepting of course man, and they have nearly the same taste for the beautiful as we have." While many people find birdsongs aesthetically pleasing, we now appreciate that song is not simply aesthetic from a bird's point of view. Song plays a very important role in songbird communication, contributing to a host of functions including territorial defense, individual recognition, and mate attraction (Mooney et al. 2008).

Birdsongs are typically elaborate vocalizations that are longer and more complex than calls (box on page 391). For example, the **calls** performed by Canaries (*Serinus canaria*) last only a few tenths of a second, but **songs** can last 30 seconds or more. Across all songbird species, songs are composed of brief periods of sound that are separated by short intervals of silence. Those sounds, called **notes**, can have complex frequency characteristics, and researchers struggled for many years to characterize their content. Drawings in field notebooks, creative mnemonics, and even attempts to place song notes on a musical staff were helpful in summarizing the properties of individual notes, but detailed quantification remained out of reach for many years. The advent of spectrographic technology enabled researchers to draw a picture of how frequency and intensity changed with time over the course of an individual note and throughout the duration of entire songs. By creating those pictures, called **spectrograms**, for the songs performed by different individuals and different species, it quickly became apparent that songs span an enormous diversity of spectral and temporal characteristics. For example, some notes consist of a range of harmonically related frequencies, such as the notes performed by Zebra Finches (*Taeniopygia guttata*) and Bengalese Finches (*Lonchura striata domestica*), whereas other notes are very tonal, such as the notes in Canary or Swamp Sparrow (*Melospiza georgiana*) songs (fig. 13.1). There is also considerable diversity in the way that notes are sequenced to form songs. For example, Zebra Finch songs consist of a repeated multi-note motif, Swamp Sparrow songs consist of a trilled syllable, and Song Sparrow (*Melospiza melodia*) songs consist of a series of different multi-note phrases. Finally, species also vary in the number of songs in their vocal repertoire. Zebra Finches sing only one song type, Swamp Sparrows sing three to five different song types, wrens can sing approximately 20 different songs, and Mockingbirds (*Mimus polyglottos*) and Brown Thrashers (*Toxostoma rufum*) can have repertoires consisting of tens or hundreds of different songs (Boughey and Thompson 1981, Derrickson 1987; box on page 399).

Despite all of the song diversity expressed across the thousands of songbird species, the songs performed by males in each species share specific traits that characterize the song of that species. Those **species-typical traits** are most evident in the patterns and sequences in which sounds are produced to compose songs (the song syntax). For example, the song performed by one Zebra Finch may contain different notes than the song of another Zebra Finch, but both males will express only one song type, and those songs will both consist of a multi-note motif that is repeated several times in each song performance. Similarly, two different Swamp Sparrows may have different song repertoires, but each of their songs will consist of a multi-note syllable that is repeated many times to compose the song. Therefore, despite the fact that songs of different species span such a wide variety of song properties, those properties are generally conserved among songs performed by members of the same species. Interestingly, the high-resolution analysis that was made possible by spectrograms also revealed that the spectral content of individual notes is commonly less complex in birds that were raised in the absence of hearing song of their own species. In order for birds to achieve species-typical complexity, the bird must hear song of its own species during a sensitive period of juvenile development (Konishi 1965, reviewed in Mooney et al. 2008). This and other observations led to an appreciation of the critical role that learning plays in the acquisition of song behavior.

SONG BEHAVIOR IS SHAPED BY AUDITORY EXPERIENCE

Juvenile Song Learning

In the first few weeks after they hatch, young songbirds hear the songs that males of their species are performing on nearby breeding territories (fig. 13.2). Birds are much more likely to memorize and incorporate into their adult repertoires songs heard during that window of development than songs that they hear earlier or later in life (the process of song learning is reviewed in Mooney et al. 2008). For that reason, that especially impressionable period of a juvenile bird's development is referred to as the **sensitive period** of song development (box on page 404). During that time, the young bird relies on auditory experience to memorize song models that will eventually become the basis of its own vocal performances. The bird's adult repertoire will typically contain many copied elements and some improvised elements, so that the song takes on all of its species-typical quality but also is sufficiently individual-specific to serve as an indicator of individual identity. This time when the juvenile bird hears song but has not yet begun its own attempts to sing is called the **sensory phase** of song learning.

A second phase of song development emerges when the young bird begins its vocal attempts to imitate those memorized songs. That phase is called the **sensorimotor phase** of song learning. Initially, the bird performs strings of sounds that bear little or no resemblance to the songs that it heard or the songs that will eventually compose its adult repertoire. Those highly variable performances are called **subsongs**. As development progresses over the course of thousands of trial-and-error rehearsals, bits of sound start to become recognizable versions of sounds that the young bird heard or that will eventually become part of its adult repertoire (Gardner et al. 2005). Those performances in which recognizable song elements begin to emerge but are still variable in the way they are performed are called **plastic songs**. Plastic songs

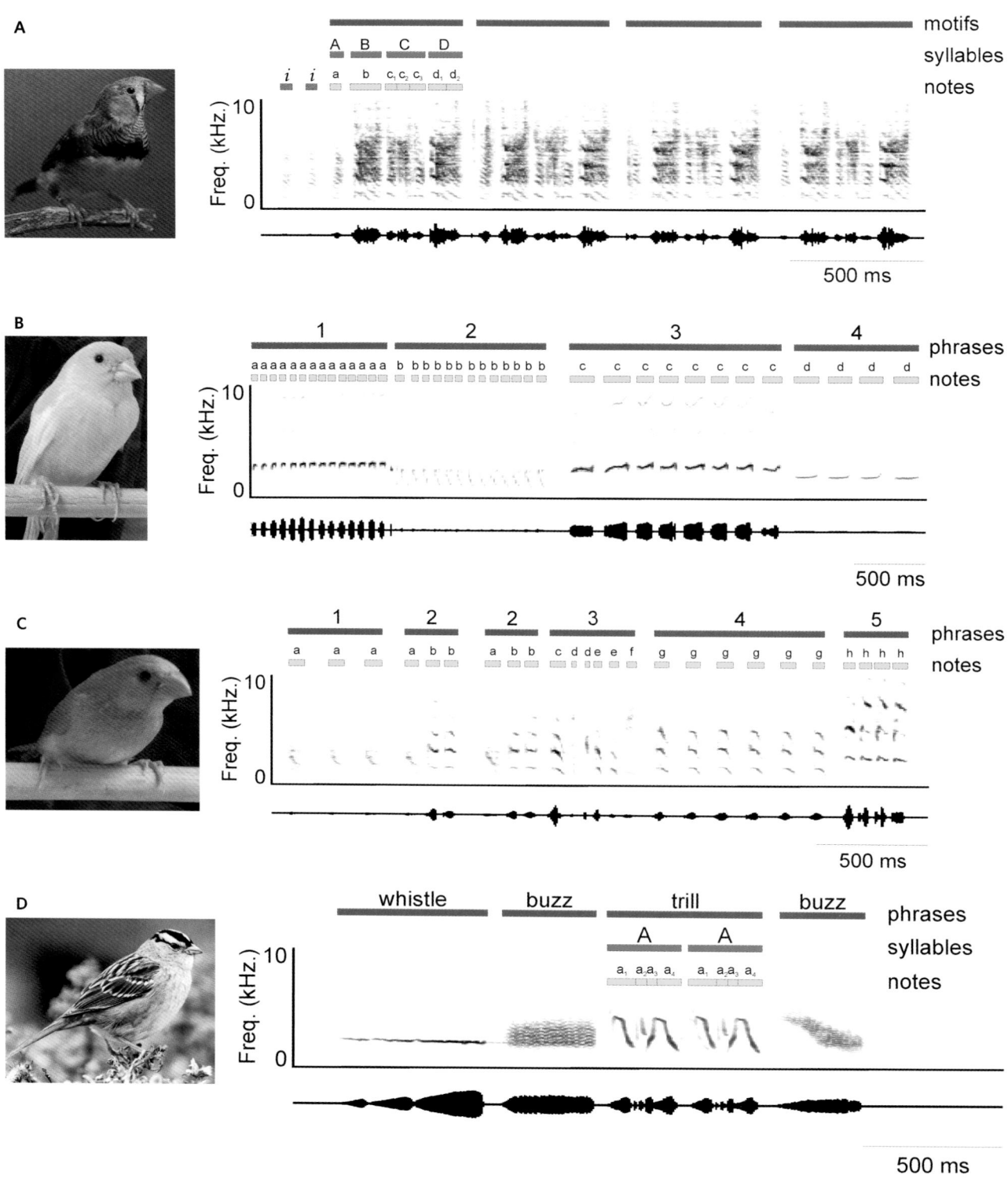

Figure 13.1. Four species of commonly studied songbirds and their adult songs depicted by a spectrogram (*top*) and an oscillogram (*bottom*). Colored boxes above each spectrogram indicate the components of song for each species. In each song, notes are sometimes performed together as syllables, which are then performed in specific sequences known as motifs, phrases, or songs. Songs consist of a wide variety of note types and syntaxes. *A*, Zebra Finch songs begin with a series of introductory notes (denoted by *i*) followed by a motif that is repeated to form the song; *B*, Canary song consists of a series of phrases, and each phrase consists of a single note that is trilled or a multi-note syllable (not shown); *C*, Bengalese Finch song also consists of phrases in which the note sequence can vary across songs performed by the same male; *D*, White-crowned Sparrow song begins with a whistle followed by other phrases referred to as buzzes or trills. *Adapted from figure 1 in Mooney et al. 2008. Zebra Finch image courtesy of Daniel D. Baleckaitis. Canary and Bengalese Finch images courtesy of Jonathan F. Prather. White-crowned sparrow image courtesy of Vladimir Pravosudov.*

also span many trial-and-error performances, and the similarity of plastic songs to memorized models or elements generally increases over the course of juvenile song learning (Marler and Peters 1981, 1982b). Deviations from species-typical spectral features and syntax are common at the start of plastic song, but those deviations become less common as the plastic song phase progresses. Eventually, the bird begins producing a song that has taken on all of its species-typical complexity and syntax. That song is performed in nearly the same way each time the bird sings it, and in species that sing a repertoire of different song types, that maturation typically occurs at approximately the same time for all song types (Marler and Peters 1982b). The advent of highly stereotyped song performances is referred to as the time of **song crystallization** and is typically the point that the bird is said to have reached adulthood. Apart from a brief period of plasticity at the start of subsequent breeding seasons or the plasticity that is evident in the rare species that continue to acquire new songs throughout adulthood (e.g., mockingbirds, lyrebirds), the song properties that are evident at the time of crystallization will define the bird's song repertoire for the rest of its adult life.

Juvenile songbirds perform an amazing feat of learning and memory. In seasonal species that breed each spring and summer, the sensitive period and the sensorimotor period of song learning can be separated by approximately nine months. The young bird maintains its memorized song models for that entire time, then recalls those models and goes on to perform strikingly accurate adult imitations of those models (Marler and Peters 1982a). Therefore, those memories were held without corruption for many months. The mental capabilities of some species, such as Song Sparrows and Nightingales (*Luscinia megarhynchos*), are especially impressive, as they can learn very complex songs even if they have heard the song model only a handful of times (Hultsch and Todt 1992, Peters et al. 1992). Studies in the laboratory can shed additional light on this process, as they can provide a condition in which juvenile birds are raised in isolation from all other singers. Those birds still go on to produce strikingly accurate imitations, revealing that the quality of mental representation and the clarity with which those song models is recalled are sufficient for accurate imitation even in the absence of any ongoing instruction. Laboratory experiments have also revealed that such impressive feats of memorization and recall are also possible in species that live colonially and breed opportunistically, such that young birds naturally pass through both the sensory and the sensorimotor phases within just a few months and in the presence of ongoing singing by other males of their species (Funabiki and Konishi 2003). Therefore, it appears that impressive feats

of learning and memory are not a specialization of any particular songbird species but instead are a general feature of how the songbird brain enables vocal learning.

In addition to listening to other members of their species to memorize their songs, juvenile songbirds must also listen to themselves and rely on **auditory feedback** to refine their developing song performances. The progression of song development from subsong to crystallization typically spans many thousands of song performances. If juvenile birds are deprived of auditory feedback by either deafness or constant presentation of very loud noise, then they fail to mature at the same rate as their counterparts that experience natural auditory feedback (Konishi 1965, Funabiki and Konishi 2003). Songs performed by feedback-deprived juveniles take on gross features of species-typical vocalization, such as the syntax in which notes are produced, but they typically fail to achieve the high-resolution spectral complexity that characterizes songs of normally developing members of their species (Konishi 1965) (fig. 13.3). Therefore, auditory input is essential for guiding all aspects of song memorization and song learning, but the importance of auditory feedback doesn't end there. Auditory feedback also plays an important role in preservation of crystallized song throughout a songbird's adult life.

Adult Song Maintenance

Songbirds rely on auditory feedback not only to refine their juvenile songs but also to maintain that high-quality vocal behavior throughout adult life. The essential role of auditory feedback becomes clear when it is lost or otherwise altered. For example, Zebra Finches that experience adult-onset deafness undergo changes in their song performance (Nordeen and Nordeen 1992, Lombardino and Nottebohm 2000). Those changes are not immediate, indicating that neither the nervous system nor the muscles that control the vocal organ is immediately compromised in the ability to produce song. That delay in the effects of deafening also suggests that the bird has some form of "motor memory" that is not dependent on moment to moment auditory feedback. Instead, changes emerge over the course of days and progress for weeks. Complex syllables that are typical of normal adult song gradually deteriorate and are eventually replaced by a smaller and simpler song repertoire consisting of poorly modulated sounds comparable to the song notes performed by birds that had been deafened very early in life and therefore had never had access to auditory feedback. The emergence of changes in the neuromuscular program only after prolonged mismatches between motor commands and the associated auditory feedback is thought to be advantageous because it prevents the nervous system from

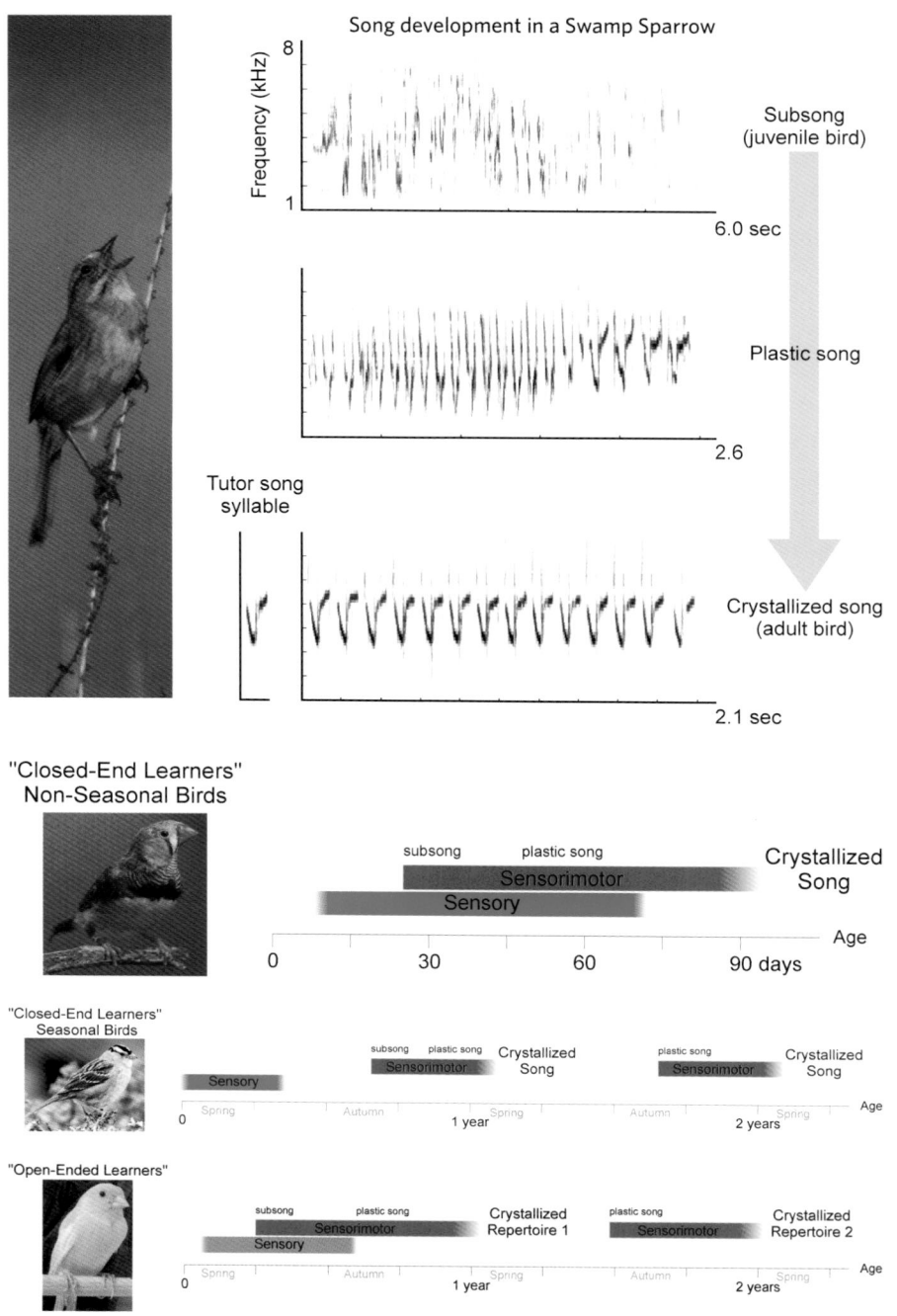

Figure 13.2. Songbirds rely on auditory experience to learn their songs during juvenile development and maintain them during adulthood. Song development for this juvenile Swamp Sparrow (*top* half of this figure) progresses from a series of unstructured sounds (subsong) to the stereotyped pattern of adult song in their species (crystallized song). Plastic song is characterized by the emergence of the acoustic features typical of adult song but with residual juvenile variability in the structure and sequence of the sounds. Swamp sparrow image courtesy of Rob Lachlan. *Adapted from figure 4 in Mooney et al. 2008.* During sensory acquisition (blue boxes in the three panels at the *bottom* of this figure), a juvenile bird memorizes one or more song models. During sensorimotor learning (red box) the juvenile uses auditory feedback to match its performance to those memorized models. These phases can be overlapping, as in the Zebra Finch, or occur many months apart, as in the White-crowned Sparrow. Song crystallization marks adulthood and the end of sensorimotor learning. Closed-end learners retained the same crystallized song repertoire throughout adulthood. In contrast, the song repertoires of open-ended learners, such as Canaries, can change from year to year. *Zebra Finch image courtesy of Daniel D. Baleckaitis. White-crowned Sparrow image courtesy of Vladimir Pravosudov. Canary image courtesy of Jonathan F. Prather. Adapted from figure 2 in Mooney et al. 2008.*

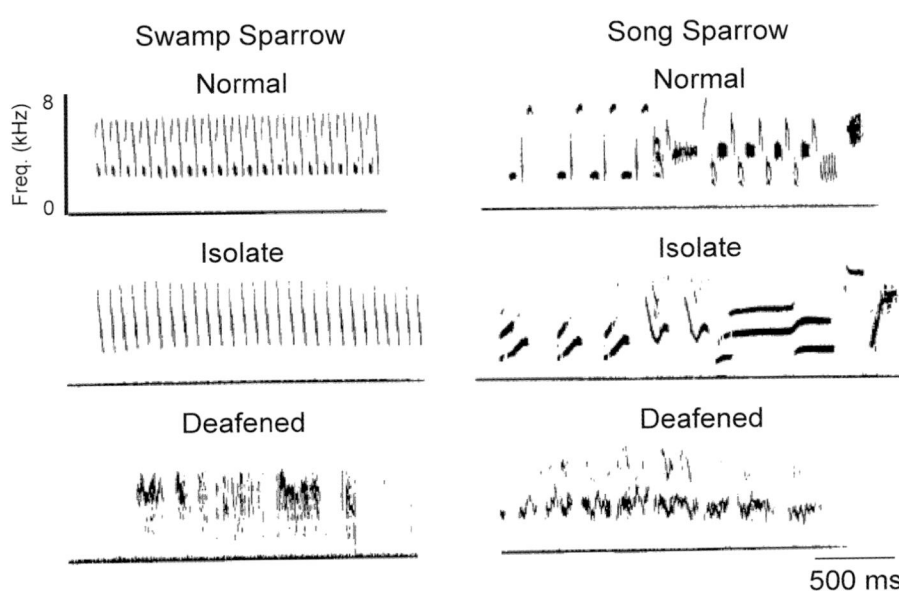

Figure 13.3. Development and maintenance of species-typical song structure requires auditory experience of a conspecific song model and singing-related auditory feedback. Songs of Swamp Sparrows (*left*) and Song Sparrows (*right*) display rudimentary species-typical features but lack their typical complexity in songs recorded from birds raised in isolation from other birds' songs (*middle*). Songs recorded from birds that were deafened before sensorimotor learning (*bottom*) lack even the rudimentary features of isolate song, making clear the importance of auditory feedback and sensorimotor learning. *Adapted from figure 3 in Mooney et al. 2008, originally adapted from figure 3 in Marler and Doupe 2000, copyright 2000, National Academy of Sciences, U.S.A.*

being modified by brief and nearly unavoidable mismatches such as those introduced by ambient noise or the songs of nearby birds.

In further support of the idea that a sensorimotor mismatch plays a key role in causing song variability, Anthony Leonardo and Mark Konishi (1999) (box on page 388) found that even transient mismatches were sufficient to induce changes in song performance. In their elegant experiment, they used song playback to induce a mismatch between what Zebra Finches sang and what they heard. When the bird sang, the computer detected that sound, recorded it, and immediately played it back to the bird so that the signal that arrived at the bird's ears consisted of two things: (1) natural feedback associated with the bird's own vocalization, plus (2) playback of another copy of that song that was played at approximately the same loudness but was slightly out of phase with the natural feedback. Leonardo and Konishi observed a pattern of gradual song deterioration very much like the changes that occur following adult-onset deafness. Over the course of weeks, the songs of birds that sang in the presence of distorted auditory feedback became highly distorted. The truly elegant aspect of their experiment arose from the fact that they could turn off the distorted auditory feedback and observe what happened as the birds recovered. One might expect that if a sensorimotor mismatch induces stereotyped Zebra Finch songs to become variable, then turning off that mismatch might cause song to become stereotyped again. That is what they observed, but the most informative aspect lay in the nature of the song after it became stereotyped (fig. 13.4). Instead of becoming locked in the distorted state that was present when the distorted auditory feedback was removed, the song went back to the original song that the bird had been singing at the start of the study. In other words, the song didn't just become stereotyped in some warped state; it went back to its original state. This restoration of the learned behavior made it clear that the memory of the song model persisted throughout the period of distorted vocal performance. That, in turn, suggests that the song memory is stored independently of the program associated with song performance. Those and other observations provide an especially interesting context in which to explore how birds memorize, recall, learn, and refine their vocal behaviors.

Interestingly, song changes following adult-onset deafness emerge at different rates in different species. For example, the songs of Zebra Finches change over the course of several weeks, and the songs of Bengalese Finches can change even more rapidly (Nordeen and Nordeen 1992, Woolley and Rubel 1997). In contrast, the songs of White-crowned Sparrows change at a much slower pace, if at all. Songs of those birds can remain unchanged even after more than a year of deafness (Konishi 1965). These findings indicate that auditory feedback generally plays an important role in the maintenance of adult song, but the degree to which the song motor program is dependent on auditory feedback varies across species. These data highlight the value of a **comparative approach** to take advantage of the wide variety of songbird species to explore how the songbird brain enables the imitative behavior that lies at the heart of vocal learning (Brenowitz and Zakon 2015).

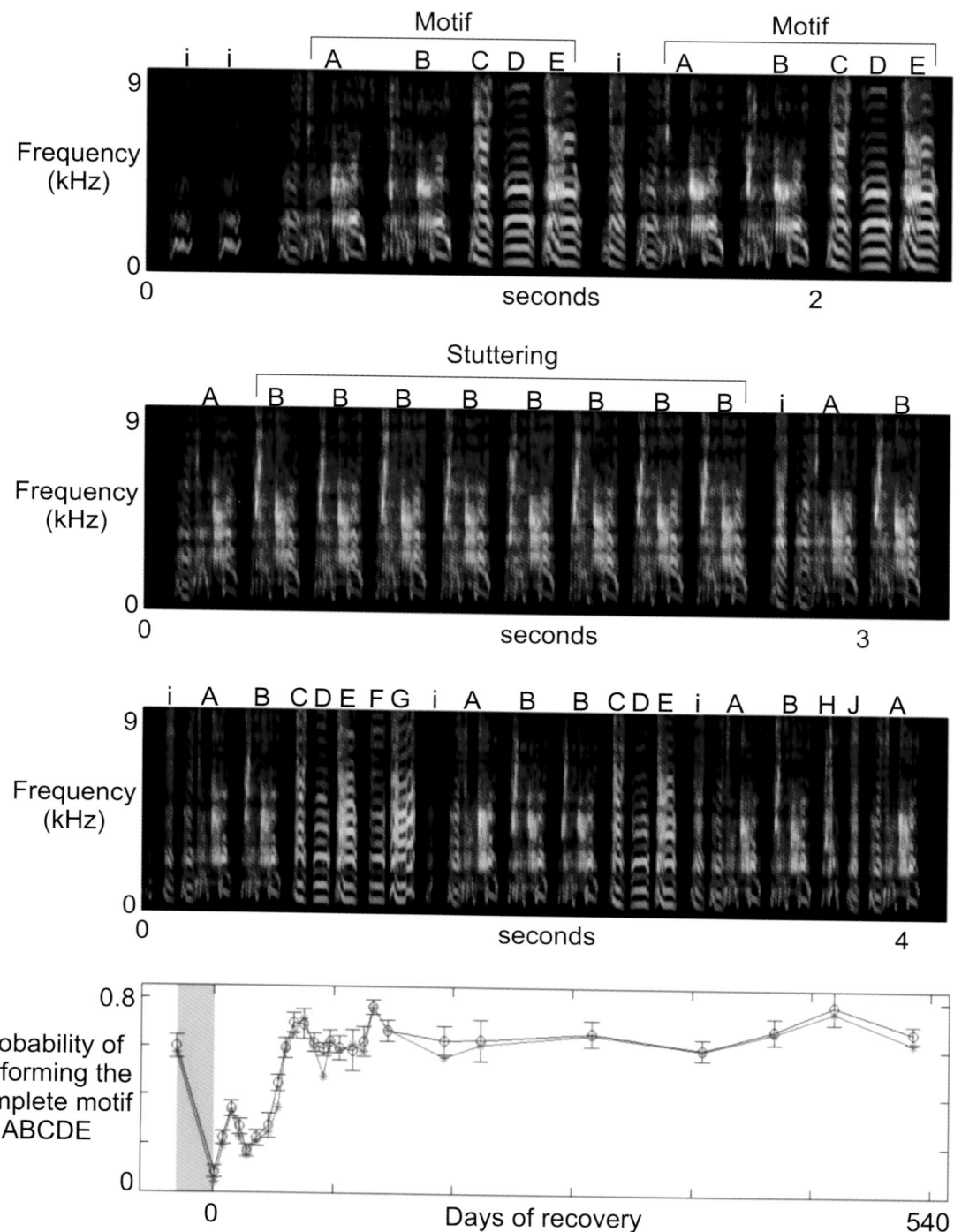

Figure 13.4. Alteration of auditory feedback induces changes to crystallized adult song. In normal, crystallized adult Zebra Finch song (*top*), each syllable has a stable and well-defined structure, and the note sequence is consistent across motifs (ABCDE; i indicates introductory notes). Alterations in auditory feedback can induce changes in song structure, including repeated notes reminiscent of stuttering (*second from top*), or changes in the motif structure (*third from top*), including either addition or removal of notes from the motif. *Adapted from figure 2 in Leonardo and Konishi 1999, reprinted* *by permission from Macmillan Publishers Ltd, copyright 1999.* Following removal of distorted auditory feedback, song structure again becomes crystallized (*bottom*). Importantly, song structure does not simply become fixed in its altered state. Instead, it returns to the original, pre-distortion song structure (in this case, motif sequence ABCDE), indicating that song memory is stored independently of song performance. *Adapted from figure 4 in Leonardo and Konishi 1999, reprinted by permission from Macmillan Publishers Ltd, copyright 1999.*

OVERVIEW OF AVIAN COGNITION: THE STRUCTURE OF THE SONGBIRD BRAIN

The avian brain is capable of some very complex behaviors. That may be surprising because of its relatively compact size compared with human brains (fig. 13.5), but don't be misled by the idea that bigger brains are necessarily better or that smaller brains are necessarily worse. Birds have primate-like numbers of neurons in the most evolutionarily recent portion of the brain (the forebrain; Olkowicz et al. 2016), and small brains can do some very impressive things. For example, many species of birds bury or otherwise hide their food reserves in a behavior called caching, and they are capable of accurately recalling the locations of those items over a span of weeks or months (Krebs et al. 1996). Beyond simply the impressive scale of that task, Scrub Jays (*Aphelocoma coerulescens*) also display context-dependent strategies regarding the manner in which they hide those resources. Nicky Clayton and her colleagues have shown that in the presence of other members of their species, some jays will implement strategies to minimize the likelihood that their precious cache will be stolen by onlookers (Emery and Clayton 2001). When a bird that had cached food in the presence of an onlooker was allowed to retrieve the cached items in private, some birds recovered the caches and relocated them to a new site. Interestingly, the birds that relocated food to a new site were those that had prior experience stealing another bird's cache, whereas jays that had not previously stolen another bird's food did not relocate their own caches. Those data suggest that jays are sufficiently clever to alter their strategies as a function of their own prior social experience.

In another example of impressive avian cognition, crows exhibit behavioral signs of analytical reasoning. When presented with symbolic representations of a specific pattern, crows exhibit the high-level cognitive ability to recognize objects as the same as or different than one another (Smirnova et al. 2015). Crows are also renowned for their ability to craft and use tools. When a crow is presented with an especially difficult task of retrieving food from a small space, it is capable of using nearby materials to craft a tool to help it complete the task (Weir et al. 2002, Bluff et al. 2010). An old fable by Aesop describes a crow that was smart enough to add small rocks to a pitcher of water in order to raise the water high enough that the bird could have a drink. A recent effort to test the plausibility of that fable confirmed that crows are, in fact, capable of such cleverness. Crows added stones into a tube to displace water in order to receive a reward. Furthermore, the birds added stones to tubes of water but not to tubes containing sand, and the birds added solid items that would sink but not hollow items that would float (Jelbert et al. 2014). This striking result suggests that crows understand the cause and effect of their actions. In fact, Jelbert and colleagues (2014) even speculated that the crows' understanding of causal relations rivals that of a 5- to 7-year-old child. Together, these data make it clear that although avian brains may be rather small, especially when compared with the size of your brain, birds are nonetheless very intelligent creatures. The ability of songbirds to memorize complex behaviors, hold them in memory for many months, then recall and learn to imitate them with high fidelity is another prominent example of **avian cognition**. Over the last 40 years, the community of songbird neurobiologists has made great strides in understanding how the brain enables that impressive mental feat.

When researchers first considered the structure of the songbird brain, they named various areas according to how they thought those areas corresponded to related structures in the mammalian brain. The nomenclature that emerged from those efforts suggested that large portions of the songbird brain were devoted to instinctive behaviors, whereas few brain sites were thought to contribute to learned behaviors such as song (fig. 13.6A). For example, many aspects of mammalian cognition are associated with activity in a multilayered structure called the **neocortex** that resides along the outermost margin of the mammalian brain (the green "rind" of the human brain in the images on the right in fig. 13.6). The mammalian neocortex (commonly referred to as simply "the cortex") is what most people think of when they hear the word "brain," and it is implicated in many very high-level thinking processes, but no such layered structure is present in the songbird brain. Therefore, large portions of the songbird brain were named according to what we now realize was the incorrect idea that the songbird brain was composed almost entirely of structures corresponding to mammalian structures that were hierarchically lower than the neocortex (Jarvis et al. 2005). As detailed in subsequent sections of this chapter, the validity of that idea was called into question when it was found that damage to portions of the songbird brain that were thought to underlie instinctive behaviors resulted in a loss of the ability to perform song or to learn by imitating others. For many years, the nomenclature reflected an antiquated view of the songbird brain while new data continued to reveal increasingly detailed knowledge of the role of specific brain sites in song learning and performance.

An important advance in our understanding of the songbird brain occurred at the start of this century, when a consortium of neuroanatomists convened to reconsider the terminology used to describe songbird brain areas (Reiner

MARK KONISHI

Masakazu "Mark" Konishi received his doctorate in the laboratory of Peter Marler in 1963. He has been a professor and head of his own lab at CalTech since 1975, and he and his students have had an immense impact on the field of avian neuroethology. His experiments have been seminal in our understanding of how young birds memorize a tutor song and then rely on auditory feedback to develop and maintain their imitation of that model, and his contributions were recognized in his election into the United States National Academy of Sciences in 1985. In addition to the insights that have emerged from his experiments, he has also trained many students and postdoctoral researchers who have also gone on to become leaders in the field. Together with his first postdoctoral fellow, Eric Knudsen, and others, Mark also discovered auditory neurons that respond only when a source of sound is in a particular location in the space surrounding the bird. Building on those observations in Barn Owls, they eventually realized that the brain contains a map of the auditory world such that owls can localize the source of a sound in both the vertical and the horizontal axes, enabling them to hunt even in complete darkness. Their finding was remarkable because it was the first demonstration of a neural map of space that did not include parts of the animal's body. In other words, they found a mechanism through which the brain encodes location in the space through which the organism moves and interacts. Eric Knudsen and his students have pursued this observation in much greater detail, and they continue to make new discoveries about how animals determine sound location and integrate that information with bodily position and movement. In recognition of his own considerable body of work, Eric Knudsen was also elected to the National Academy of Sciences in 2002.

A   Songbird

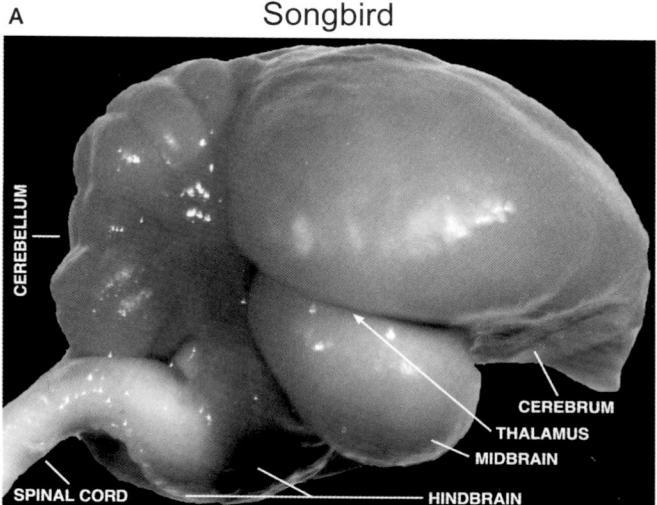

B   Human

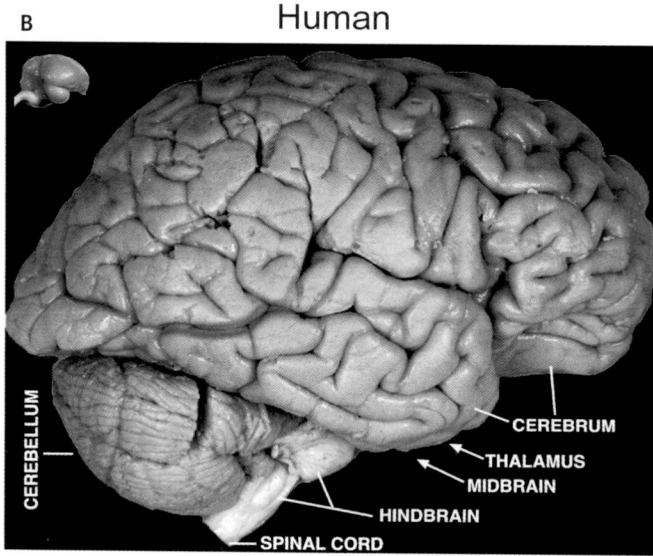

Figure 13.5. Comparison of a songbird brain (*top*) and the human brain (*bottom*). Despite their differences in size and gross appearance (avian brain is to scale in the inset at the top left of the image on the right), both brains contain homologous structures, including the cerebrum, cerebellum, thalamus, midbrain, hindbrain, and an interface with the spinal cord. *Adapted from figure 1 in Jarvis 2009.*

et al. 2004). Those researchers evaluated new evidence regarding the types of cells that are present, the patterns of gene expression, and the connectivity of sites in songbird and mammalian brains. Their analysis revealed a correspondence between mammalian cortex and large portions of the songbird brain (fig. 13.6B). In fact, their data supported the idea proposed by Harvey Karten and his colleagues that specific layers of the mammalian neocortex may have corresponding structures in the songbird brain. They realized

that a layered arrangement and a network of distributed brain regions are both capable of giving rise to very complex aspects of cognition. Both types of neural architecture enable learning, assigning meaning to sensory experiences and using that information to enable an organism to make decisions and enact specific behavioral outcomes in response to information detected through sensory experiences. Each of those cognitive processes is at work in communication through song. Therefore, songbirds provide a behaviorally rich, anatomically defined, and experimentally tractable system in which to explore the neural mechanisms that enable those complex behaviors.

SPECIALIZATIONS ASSOCIATED WITH SONG LEARNING AND PERFORMANCE: PERIPHERAL STRUCTURES

The vocal organ that songbirds use to produce their songs is called the **syrinx**, and it is a specialization of songbirds that is not found in any other animal group (fig. 13.7). The avian syrinx was named for the musical instrument of the same name (sometimes called a pan flute) that is played by blowing air across resonant tubes. For many years it remained unclear whether the avian syrinx produces sound using the same sort of whistle-like mechanism, or whether it produces sound using vibratory structures that are functionally similar to human vocal cords. Experiments by Franz Goller, Roderick Suthers, and their colleagues resolved that question by placing small endoscopes into the throats of singing birds and visualizing the properties of the syrinx in action (reviewed in Suthers 2004). They found that the sound is produced by vi-

bration of labial folds in the interior of the syrinx. They also confirmed previous findings that sound is produced during expiration and that birds take rapid "minibreaths" during the brief intervals between song notes. Their results also confirmed the idea that the syrinx contains two independently controlled sources of sound. Astute listeners had long noted that the sounds produced by the syrinx could consist of frequencies that were not harmonically related and thus could not be attributed to a single source. This gave rise to the "two-voice theory" of syringeal sound production, and the high-resolution insight provided by spectrograms and endoscopy enabled researchers to confirm that idea.

The production of sound from multiple sources is possible because the syrinx consists of two sets of vibrating structures that can be controlled independently of each other. The syrinx is composed of a set of cartilaginous rings that reside at the interface between the caudal end of the trachea and the rostral end of the bronchi that lead to the lungs (Suthers 2004) (fig. 13.7). The two bronchi fuse to join the trachea, and it is at that point of fusion that the two sound-producing structures reside. In the upper portion of the bronchi, the medial portion of the cartilaginous rings is absent. In its

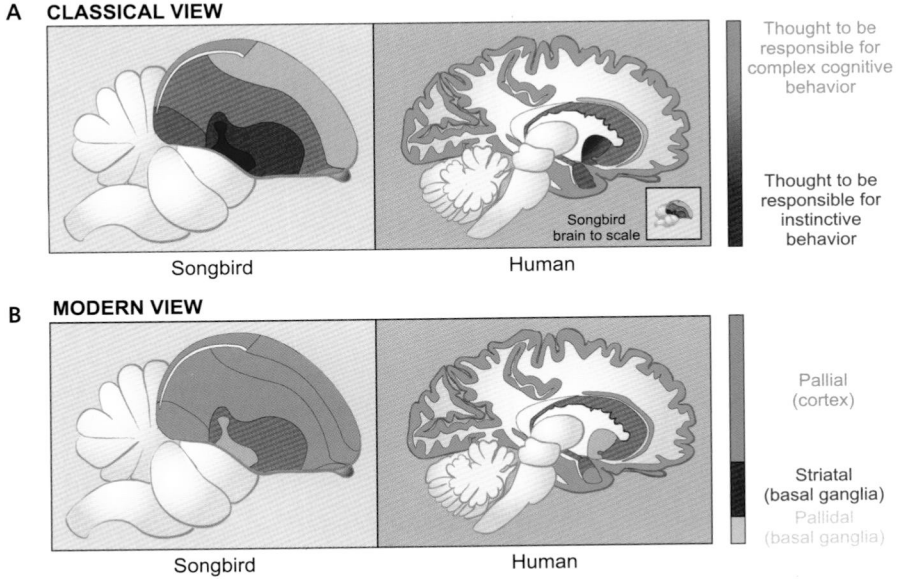

Figure 13.6. Recent studies have revealed far greater similarity between avian and mammalian brains than was previously appreciated. A, In the classic view, the avian brain was thought to be dominated by relatively primitive structures (pink, striatum), with relatively few relatively advanced structures (green, pallium). That was in stark contrast to the dominance of pallial structures (green) in the human forebrain; B, In the modern view, it is now clear that the avian brain is dominated by pallial structures (such as the cortex in the human brain, green) with some striatal (such as the portions of the basal ganglia, purple) and pallidal (such as the globus pallidus, blue) tissue, in proportions much more reminiscent of human and other mammalian brains. This similarity indicates that findings in studies of the avian brain may provide insight into the relation between neural structure and cognitive function in not only birds but also ourselves. *Brain images courtesy of Zina Deretsky, National Science Foundation.*

place are thin folds of tissue called the medial and lateral labia. The recordings performed by Goller and Suthers confirmed that sound is produced by vibration of these syringeal labia, and the vibratory folds on each side of the syrinx can be controlled independently. For example, air flow can be restricted on one side of the syrinx while sound is produced using the other side of the syrinx, and that arrangement can be reversed or otherwise modified very rapidly (Suthers 2004).

Additional studies revealed that the two sides of the syrinx have subtly different diameters and thus different resonant frequencies, and some species exploit that asymmetry in their song performances. Songs of some species contain frequency modulations that sweep through a large range of frequencies. The higher-frequency portion of the sweep is performed using the smaller side of the syrinx, and the lower-frequency portion is performed using the larger side. The transition between the two occurs almost seamlessly, and it is difficult to detect by ear or even on a spectrogram. Experiments also revealed that the frequencies that are produced by syringeal vibration are filtered by not only the properties of the syrinx but also the resonant properties of the entire **vocal tract** (Nowicki 1987, Riede et al. 2006). Even the movements of the beak are synchronized with song performance. The beak acts to elongate the vocal tract, favoring the resonance of lower frequencies, when is it nearly closed. Similarly, opening the beak shortens the tract, favoring the resonance of higher frequencies. As expected from those differences in resonant properties, birds

coordinate their beak movements with the properties of their song production (Podos et al. 1995). Together, these insights reveal that song production is a multifaceted performance requiring precise coordination of a wide range of contributing elements.

The properties of syringeal sound production are under exquisite muscular control. On the exterior surface of the syrinx are a set of syringeal muscles that control aspects of sound production such as the tension of the labial folds and whether air is flowing or occluded on one side or the other. Another finding that emerged from the experiments by Goller and Suthers was that those functions are associated with different sets of muscles. The muscles on the dorsal side of the syrinx control air flow and thus the presence or absence of sound, whereas the muscles on the ventral side control the tension of the labial folds and thus the fundamental frequency of the song (Suthers 2004). Those muscles are more developed in oscines than in their suboscine relatives, and this enhanced control of the vocal organ is thought to be another specialization of vocal-learning species. The spectral and temporal complexity that characterizes the songs of each species reflects the precise coordination of activity in each of these six bilateral pairs of syringeal muscles. That degree of muscular coordination emerges from an equally precise coordination of activity in the motor neurons that control those muscles. Those motor neurons reside in the brainstem, and they are controlled by a network of brain sites that collectively reveal central specializations for song learning and performance.

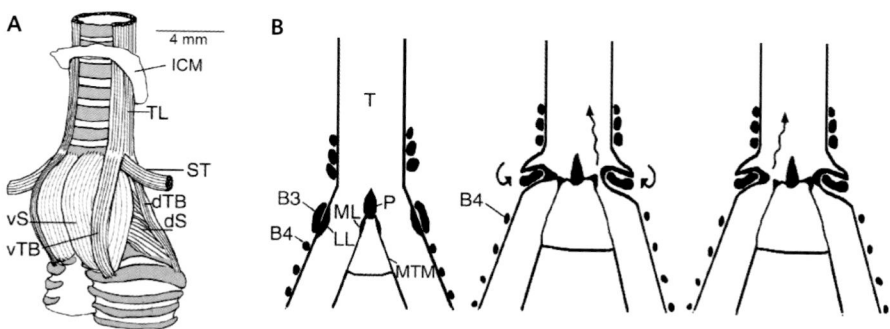

Figure 13.7. The songbird syrinx is a two-sided structure that resides at the junction between the trachea and bronchi. *A*, A ventrolateral external view of the syrinx illustrates the complex syringeal musculature (ICM = membrane of the interclavicular air sac; TL = tracheolateralis muscle; ST = sternotrachealis muscles; dTB = dorsal tracheobrachialis muscle [vTB is ventral portion]; dS = dorsal syringealis muscle [vS is ventral portion]. *Adapted from figure 1 in Suthers and Zollinger 2004.* *B*, During vocalization, the syrinx moves rostrally (*left*). Contraction of the dS and dTB muscles rotates the bronchial cartilages and moves the labia into the airstream (curved arrows), where they are set into vibration to produce sound (wavy arrows). Phonation may occur through either the right side (*middle*) or the left side (*right*) or through both sides simultaneously (not shown). (B3 and B4 = third and fourth bronchial cartilages; ML = medial labium; LL = lateral labium; P = pessulus; MTM = medial tympaniform membrane). *Adapted from figure 1 in Suthers and Zollinger 2004.*

CALLS

The seminal work of Fernando Nottebohm and colleagues revealed a close association between song performance and the function of specific regions of the songbird brain. That insight led to intense focus on the function of the "song system" (fig. 13.8) as a neural basis of learned vocal communication. The behaviors that songbirds use in vocal communication, however, include more than just song. The term "calls" is used to describe the collection of non-song vocalizations that serve a wide range of communication functions. Calls are brief, species-typical vocalizations, and they are produced by both males and females. Given the prevalence and importance of calls in songbird communication, it is somewhat surprising that calls and the underlying neural mechanisms of call production and perception have received relatively little attention. Initial behavioral studies revealed that the sex differences that are evident in the calls of male and female Zebra Finches are dependent on learning in males. Closer inspection in Zebra Finches and other species revealed that call production is dependent on the same forebrain circuits that Nottebohm and others showed to be necessary for song production. Specifically, lesion studies revealed changes in the acoustic structure of calls following removal of function in the vocal motor cortex (nucleus RA, fig. 13.8). The fact that calls are influenced by activity in the "song system" suggests that future studies may lead us to consider that network of neural structures as a more general "vocal learning system." Together with studies of song behavior, future studies of calls can also provide new insight into the neural mechanisms of auditory perception, vocal production, and their interaction in service of vocal learning.

SPECIALIZATIONS ASSOCIATED WITH SONG LEARNING AND PERFORMANCE: CENTRAL STRUCTURES

In a seminal paper, Fernando Nottebohm and his colleagues (1976) (box on page 392) were the first to point to discrete areas of the songbird brain as being involved in specific aspects of song behavior. Anatomical studies revealed discrete collections of cells, called **brain nuclei**, present in brains of Canaries. Those nuclei were distributed throughout the

telencephalon and the brainstem, and they were present in roughly symmetrical locations in each of the two hemispheres (fig. 13.8). In a foreshadowing of the neuroanatomical distinctions between oscines and suboscines, some of those sites that were present in canaries (oscines) had not been noted in an earlier atlas of the pigeon brain (suboscines). Nottebohm and his team used a strategy of altering the function of neural tissue in very small, very focal sites (placing **focal lesions**) either unilaterally or bilaterally to affect each of these discrete nuclei. They used histological and behavioral analyses to determine the interconnectivity of those nuclei and their respective roles in song production. The team found that the nuclei were linked together by specific pathways, and they noted that the "strength and specificity of their connections suggest a marked degree of localization for song control."

Nottebohm and colleagues found that lesions placed throughout large regions of the cerebral hemispheres had no effect on song, but lesions in one discrete nucleus resulted in

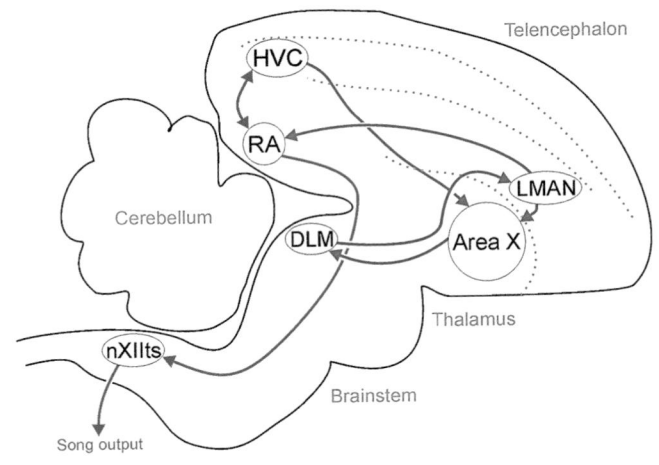

Figure 13.8. The song system is an interconnected network of sites in the songbird brain that are specialized for song learning, performance, and perception. Neurons in HVC give rise to two prominent pathways: one that is essential for song performance (red arrows, SMP described in the text) and another that is important in song learning and perception (blue arrows, AFP described in the text). Those pathways converge on the vocal motor cortex (RA) to influence the activity of brainstem motor neurons (nXIIts) that innervate muscles of the syrinx to control song performance. Diagram depicts a parasagittal section of the songbird brain (HVC = high vocal center; RA = robust nucleus of the arcopallium; nXIIts = tracheosyringeal portion of the hypoglossal nerve; DLM = dorsolateral medial nucleus of the thalamus; Area X = songbird stratum; LMAN = lateral magnocellular nucleus of the anterior nidopallium). *Adapted from figure 2 in Prather 2013.*

FERNANDO NOTTEBOHM

Like Mark Konishi, Fernando Nottebohm also received his doctorate in the laboratory of Peter Marler. Fernando was the younger of the two students, receiving his degree in 1966. Reflecting the interests of the Marler lab at that time, his dissertation was also devoted to understanding the role of sensory feedback in the development of avian vocalizations. He immediately received an invitation to join the faculty of Rockefeller University, and he has remained there throughout his impressive career. His experiments in 1976 were the first demonstration of specific sites in the songbird brain that were specialized for song learning and performance. Those pioneering observations led the way to mapping the circuitry of the "song system" in the songbird brain (fig. 13.8), and they have given rise to an enormous diversity of inquiries into how the brain enables birds to learn from experience. In addition to that contribution, Fernando is perhaps best known for his realization that new neurons are born from neuronal stem cells even in the adult vertebrate brain. He also described the choreography through which those cells migrate to their final destination and become incorporated into functional circuits. The insight that the brain produces new neurons has led the way to new approaches not only for understanding learning and behavior but also for treating pathologies of brain injury and neurodegeneration. Fernando has won many awards in his long career, including election into the United States National Academy of Sciences in 1988. Many of those prizes have been shared with Mark Konishi and Peter Marler, appropriately recognizing the achievements of each of those pioneers in avian neuroethology.

complete elimination of song (Nottebohm and Arnold 1976). That site was later named **HVC** (originally named for its location and now simply described using its abbreviated name; Reiner et al. 2004), and HVC is now recognized to be analogous to portions of the mammalian neocortex. Nottebohm and his team found that lesions in HVC had no effect on the bird's movement or health, but they caused severe deterioration of song performance. Unilateral lesions resulted in a loss of syntactical structure (the sequence in which sounds are produced), and bilateral lesions resulted in complete elimination of song. In further support for the idea that HVC was specialized for song behavior, bilateral lesions of HVC

eliminated song production but not the motivation to sing or the performance of other behaviors associated with singing. For example, a male with experimentally altered function in HVC of both hemispheres (bilaterally lesioned) adopted the upright posture of singing, inflated its throat, pivoted on its perch, and made pumping motions with its head. The beak remained closed and no sound was produced, but other behaviors associated with singing were robustly expressed. When that bird was given testosterone and brought into contact with a female bird, the male again adopted the posture of singing, pulsed its throat rapidly, and opened its beak, but still no song was produced. Those observations led them to conclude that motivation for singing was likely a multifaceted process but that HVC is responsible for production of the complex sounds that characterize song performance.

With the insight that the function of HVC was critical for song performance, they proceeded to investigate its connectivity with other structures. They found that in each cerebral hemisphere, HVC sends axonal connections to two additional nuclei in other parts of the same hemisphere (fig. 13.8). The first of those sites is the robust nucleus of the arcopallium (**nucleus RA**), which resides in the posterior portion of the hemisphere. As in the case of lesions in HVC, lesions in RA also caused profound deterioration of song (Nottebohm and Arnold 1976). Damage to RA resulted in a loss of some individual notes, much smaller frequency bandwidths in the remaining notes, and instability of song structure. Those results indicated that both HVC and RA must be intact in order for proper performance of the bird's adult song, leading to the idea that HVC and RA are part of a direct vocal control pathway.

The second site to which HVC projected was a large nucleus in the anterior portion of the hemisphere. In contrast to the stark and immediate effects of damage to RA, lesions in that area had no detectable effect on adult song. The fact that it received such clear projections from HVC yet apparently played little or no role in adult song performance led the authors to name that enigmatic site **Area X**, and that label has persisted even as researchers have learned much more about its function. The results that emerged from lesions in HVC and the sites to which it projects revealed a strong correlation between song performance and activity in specific brain sites, confirming Nottebohm's idea that specific nuclei in the songbird brain exert control over specific aspects of song production. Those insights were the first steps toward the realization that the songbird brain contains a **song system** comprising a network of brain sites that are specialized for different aspects of song learning and performance.

A NETWORK OF NEURAL CIRCUITS SPECIALIZED FOR SONG LEARNING AND DEVELOPMENT

Building on the work of Nottebohm and colleagues, songbird neurobiologists identified two pathways that emerge from HVC and that eventually converge onto RA (reviewed in Mooney et al. 2008). The first of these is a direct projection in which cell bodies residing in HVC send their axons to make synaptic connections onto cells in RA. RA is the motor cortical area associated with vocal production. Through cells that project from HVC to RA (**HVC_{RA} cells**), HVC influences vocal production. Because damage to either HVC or RA results in a loss of the ability to perform adult song, this circuit is called the **song motor pathway (SMP)**.

Recent studies have provided greater insight into the functional contributions of HVC and RA. Michael Long and Michale Fee (2008) developed a device that relied on a thermoelectric phenomenon called the Peltier effect to selectively cool either HVC or RA in singing birds. When they cooled HVC by several degrees, they observed that the song tempo was dramatically slower, but there was little or no effect on other features such as note pitch or the sequence in which the notes were performed (fig. 13.9). When they used the same technique to selectively cool RA, they did not observe the same effect. This led them to posit that HVC is acting as a sort of metronome for song timing, and that HVC and

RA operate in a manner somewhat like a music box. HVC appears to control song timing, and downstream cells in RA and brainstem sites that send projections to the syrinx determine other features such as the pitch of individual notes or the volume of song production.

The second pathway that emerges from HVC projects indirectly to RA via three intermediate sites. The origin of that pathway is the projection from cell bodies that reside in HVC and make synaptic connections onto cells in Area X. Area X is the input to the avian basal ganglia, commonly called the striatum, and it is homologous to the striatum in mammalian brains (Doupe et al. 2005). Thus, through cells that project from HVC to Area X (**HVC_X cells**), HVC can also influence the activity of cells in the basal ganglia. From Area X, cells project to the thalamus (the dorsolateral portion of the medial thalamus, DLM) and from there back to a forebrain nucleus (the lateral magnocellular nucleus of the anterior nidopallium, LMAN) containing cells that send their connections to RA, completing this indirect projection from HVC to RA (fig. 13.8). In further similarity to the mammalian brain, the arrangement of a cortico-striatal-thalamo-cortical loop is evident not only in this network in the songbird brain but also in the neural architecture of the human brain and brains of other mammals. Because of the location of many of the structures in this indirect circuit from HVC onto RA, this projection is called the **anterior forebrain pathway (AFP)**.

Figure 13.9. Changes in the temperature of HVC affect the pace of song performance. In experiments that used a Peltier device to change the temperature of only HVC, heating resulted in an accelerated song performance (red, *top*), whereas cooling slowed song performance (blue, *bottom*). Song is illustrated as spectrograms in which colors represent greater sound intensity. Slowing of song performance was evident on all timescales, as apparent from the similarity between the maximally slowed performance (+38.2 percent song duration) and digital stretching of a control song performance (spectrogram surrounded by yellow box). *Adapted from figure 1 in Long and Fee 2008, reprinted by permission from Macmillan Publishers Ltd, copyright 2008.*

Experiments have also revealed the functional contributions of HVC and its projection into the AFP. In contrast to the role of the SMP in song production, the AFP is not required for song but is closely associated with song learning and plasticity. For example, lesions placed in Area X of adult birds have little or no effect on adult song, a puzzling lack of effect that gave rise to the name Area X, but lesions in Area X of juvenile birds prevent them from learning from the sounds they hear produced by others (Bottjer et al. 1984). Importantly, this effect of Area X emerges not because of a failure to be able to sing, but rather because of a failure to be able to engage in song plasticity. Area X has also been implicated in not only song performance but also song perception. For example, lesions to either HVC or Area X can result in reduced ability to discriminate different songs in behavioral tests of song perception (Scharff et al. 1998, Gentner et al. 2000). Therefore, the input to the AFP has been implicated in song plasticity and perception.

Recordings from neurons in the AFP as birds are engaged in singing reveal a correlation between variability of song behavior and variability of timing in the activity of individual neurons (Kao and Brainard 2006). Specifically, those recordings indicate that activity in the AFP plays an important role in song plasticity (fig. 13.10). Additional behavioral studies have also revealed that the structures of the AFP play important roles in song learning, perception, and plasticity. These findings regarding the AFP and the SMP make it clear that the two pathways that collectively compose the song system are each responsible for different aspects of song learning and performance. Through the AFP and the SMP, the influence of activity in HVC converges onto RA. Activity in RA is then able to influence song production through projections from cell bodies of RA neurons to brainstem structures that control respiration and the function of the syrinx.

SEXUAL DIMORPHISM IN THE SONG SYSTEM

The network through which activity in HVC and elsewhere can influence song performance, collectively called the song system, is present in adult males but not in females. Those structures are initially present in very young birds of both sexes, but that pattern changes during the course of development. The structures are present in equal measure in males and females up until a few weeks after hatching (Konishi and Akutagawa 1985, 1988). Therefore, the neural circuitry underlying song learning and production is initially present in females, but the properties of that circuit follow very different trajectories as males and females continue to develop. In males, the structures of the song system continue to grow and eventually enable the male to learn and perform his

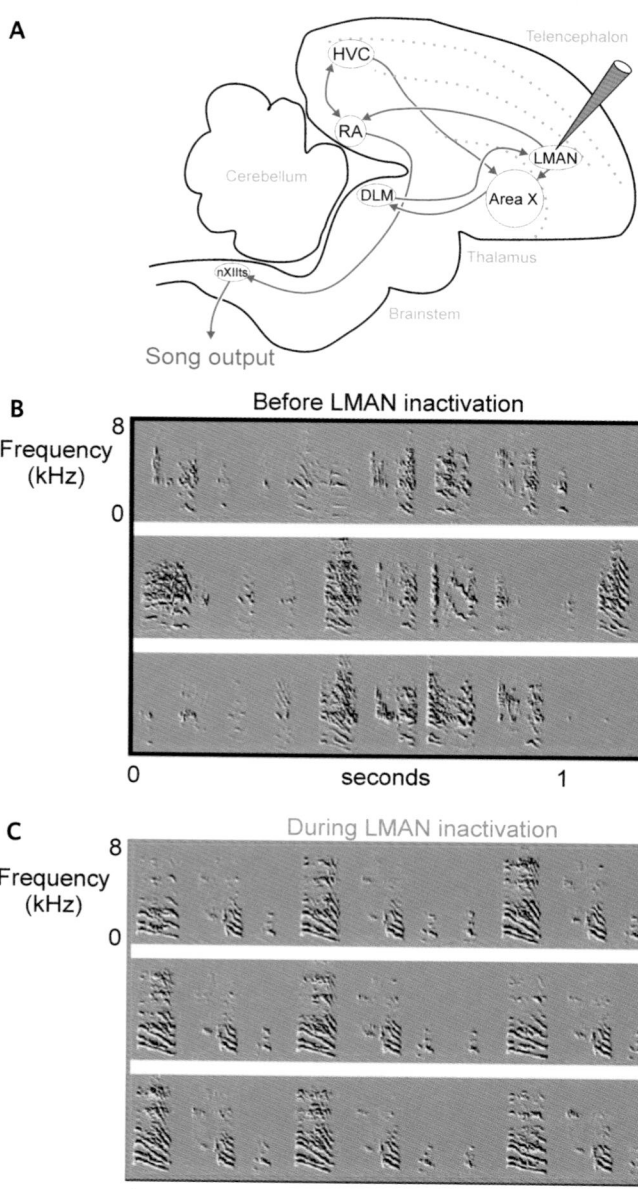

Figure 13.10. *A*, Pharmacological inactivation of neurons in one part of the AFP (LMAN, inactivation indicated by an orange pipette) made the variable song of a juvenile Zebra Finches highly stereotyped; *B*, Before LMAN inactivation, song was variable in spectral content and temporal sequence (spectrograms indicate three performances by the same juvenile Zebra Finch); *C*, During LMAN inactivation, song performances by the same individual as in panel *B* became highly stereotyped in both spectral and temporal properties. Therefore, the AFP appears to play an important role in regulating the variability of song performance. *Adapted from figure 1 in Olveczky et al. 2005. Image available from http://journals.plos.org/plosbiology/article?id=10.1371 /journal.pbio.0030153, Open Access under Creative Commons Attribution License.*

songs. In naturally developing females, however, nearly all of those structures atrophy to become much smaller or undetectable versions of their counterparts in the male brain. That divergence between male and female brains occurs because of hormonal differences that emerge during juvenile development (Wade and Arnold 2004).

In support of a role for hormonal control, hormone receptors are expressed throughout the song system (Ball et al. 2004), and females that are given estradiol (a hormone that plays important roles in development and regulation of sex characteristics) will experience masculinization of their brain and behavior. Those females will go on to develop a song system and be able to sing (Wade and Arnold 2004). Those observations suggested that the song system may develop because of elevated estradiol levels and that males may experience elevated levels of estradiol through metabolism of testosterone into estradiol. Although testosterone and estradiol undoubtedly play a role in songbird development, subsequent studies have revealed a more complicated role for those hormones. For example, some results have supported the rather counterintuitive idea that estradiol normally initiates several events in the masculinization of the male song system; however, other attempts to block masculinization of the male song system by pharmacologically blocking the effects of estradiol have revealed little or no effect (Wade and Arnold 2004). More recent findings have pointed to a role for estradiol that is synthesized in the brain itself. In a striking bit of evidence, cultures of neurons taken from the brains of 25-day-old birds release estrogen into the culture medium, and cultures taken from males produce more estradiol than cultures taken from females (Wade and Arnold 2004). Furthermore, the estradiol produced by the male brain has important functional consequences, as coculturing female slices with male slices affects female cells in a manner similar to that observed after administration of estradiol. Therefore, hormones play an important role in the development of the song system, and their effects may arise from production in a number of different places.

Hormones also play a role in not only the differentiation but also the continued development of the song system. Males that are given testosterone during the plastic phase of their development, when song is still quite variable, will quickly begin singing songs with adultlike stereotyped properties. In the course of that hormone-induced change in behavior, the properties of the synaptic projections from LMAN to RA also take on a prematurely adultlike pattern (White et al. 1999). Together these data reveal a strong hormonal influence on sexual dimorphism of the song system, and they further highlight the close relation between properties of the song system and properties of vocal learning.

SENSORY AND MOTOR ACTIVITY IN THE SONG SYSTEM

In its essence, song is a sensory-guided motor behavior. Over the course of hundreds or thousands of feedback-dependent trial-and-error rehearsals, a young bird eventually becomes proficient in its imitation of the sounds that it heard performed by others. Throughout that process, sensory and motor-related information are closely linked to enable vocal learning. Because young birds that are deafened or raised in the absence of male song fail to develop species-typical songs, it is clear that sensory input somehow shapes motor performance during the course of song learning. A long-standing question in the field of birdsong neurobiology is: where are sensory and motor information integrated in the service of sensory-guided vocal learning? Answers to that question will have a very far-reaching impact throughout the field of behavioral neuroscience, and studies over the course of several decades have begun to provide insight into how sensory and motor-related activity are represented and integrated in the songbird brain.

Building on the observation that HVC was essential for song performance, Larry Katz and Mark Gurney (1981) found that HVC neurons also respond to auditory stimuli. They characterized those auditory responses using the relatively nonspecific stimuli of noise bursts and pure tones, and the activity of those cells was sampled in the anesthetized state, but those data nonetheless showed that a nucleus that is essential to performance of a learned vocal behavior is also responsive to auditory stimuli. James McCasland and Mark Konishi (1981) extended that observation by studying the activity of HVC neurons in awake birds as they were engaged in singing or as they listened to playback of their songs played through speaker. Those data revealed that HVC neurons are active in both states, revealing HVC as a site where both sensory and motor activity were present in the behaviorally relevant awake state. However, those data left open the question of whether sensory and motor activity were *both* expressed by individual neurons in HVC. For example, it could be the case that some cells were active only when the bird sang, and another entirely distinct population was active only when the bird heard song. If that were the case, then that would suggest that sensorimotor integration occurs elsewhere in the brain. Alternatively, individual neurons could be active in both states, generating motor-related activity when the bird sings and auditory activity when the bird hears song. If that were the case, then that group of cells would be a very attractive candidate to serve as a link between sensory input and motor output. Investigating that possibility required songbird neurobiologists to record neural activity of individual neurons, and investigators turned first to the activity of neurons in HVC.

In early studies using anesthetized birds, researchers found that individual HVC neurons express robust auditory responses (e.g., Margoliash 1986, Theunissen and Doupe 1998). The anesthetized preparation was used for ethical reasons, but those higher-resolution recordings still left open the question of how those cells would perform in the behaviorally relevant awake state. In an effort led by Michale Fee and colleagues, songbird neurobiologists have developed miniaturized recording electrodes and devices to house and move them (Fee and Leonardo 2001, Otchy and Olveczky 2012). Those technological advances enabled investigators to sample the activity of song system neurons as birds go about their daily activities of singing, eating, drinking, grooming, and even flying short distances. Importantly, those recording devices are placed under surgical anesthesia then used to record cells after the bird reawakens. Because the brain contains no sensory receptors, that neural activity can be sampled in the absence of any pain or discomfort for the bird. That newfound ability to sample neural activity during the naturally relevant behavioral state of waking and moving freely has yielded important new insights. For example, HVC_X cells that give rise to the AFP pathway implicated in learning and plasticity have robust auditory responses in awake birds. In contrast, HVC_{RA} cells that innervate the pathway implicated in song production have no auditory responses and are almost entirely silent when the bird is not vocalizing (fig. 13.11). If we are seeking to understand how sensory and motor-related activity may be linked within a specific set of neurons, then the absence of auditory responses in HVC_{RA} cells in the behaviorally relevant state indicates that those cells cannot fulfill that role. That led researchers to turn their attention to HVC_X cells.

HVC_X cells are responsive to auditory stimuli in the awake and freely behaving state, and they are some of the most selectively responsive cells ever described. Recordings from individual HVC_X neurons in a variety of species including canaries, sparrows, and finches, have revealed that those cells are selectively responsive to specific song elements (Prather 2013). Each HVC_X neuron expresses temporally precise responses to one or a few song elements, while other portions of the bird's repertoire, random sounds with which it is familiar, or novel sounds have little or no effect on the activity of those cells (fig. 13.12). Across the population of HVC_X cells in each hemisphere, there appears to be thousands of neurons that represent each portion of the bird's song repertoire, resulting in robust representation of each song portion of the vocal repertoire across the complete population of HVC_X

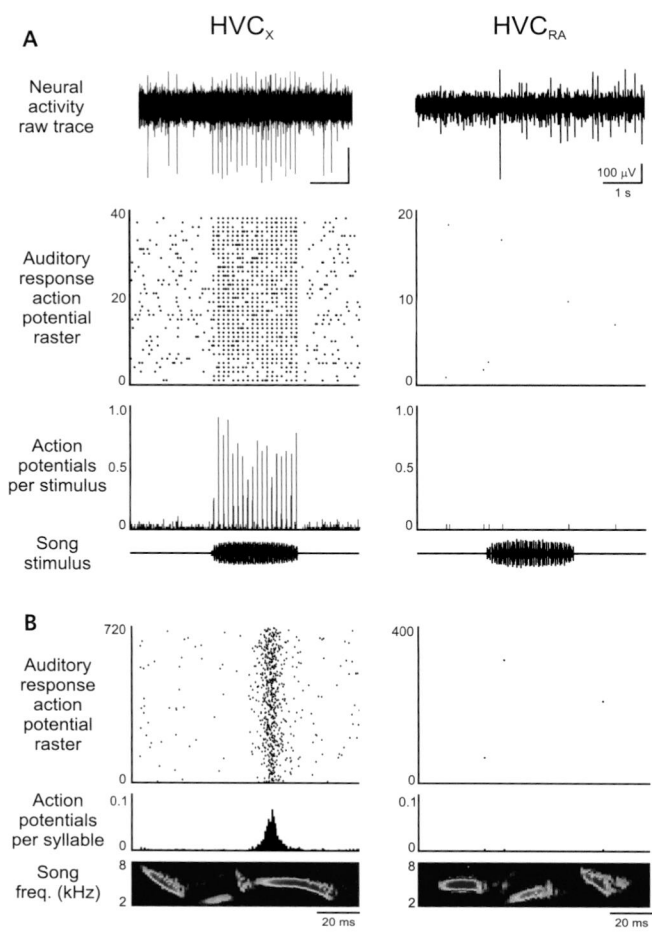

Figure 13.11. In recordings from awake Swamp Sparrows, auditory responses are evident in HVC neurons that project into the AFP pathway involved in learning and plasticity (HVC_X) but not in cells that project into the SMP pathway involved in song performance (HVC_{RA}). *A*, Playback of song through a speaker (*bottom* = oscillogram) resulted in robust response of HVC_X cells (*top* = raw recording, *second row* = raster of responses of this cell to 40 presentations of the song stimulus, *third row* = histogram of auditory responses) but not HVC_{RA} cells; *B*, Auditory responses of HVC_X neurons are phasic and temporally precise, and they occurred at a precise phase in each syllable in the song stimulus. *Adapted from figure 2 in Prather et al. 2009.*

neurons in each bird (Prather et al. 2008, Fujimoto et al. 2011). It is thought that the collective responses of the entire population of HVC_X cells in each bird represent the complete set of vocal elements that compose the bird's entire adult vocal repertoire. Importantly, individual HVC_X neurons respond to specific vocal elements regardless of whether those sounds are performed by the bird himself or by other members of his species (Prather et al. 2008) (fig. 13.13). Furthermore, experiments combining laboratory recordings from HVC_X neurons in Swamp Sparrows and field studies of song perception

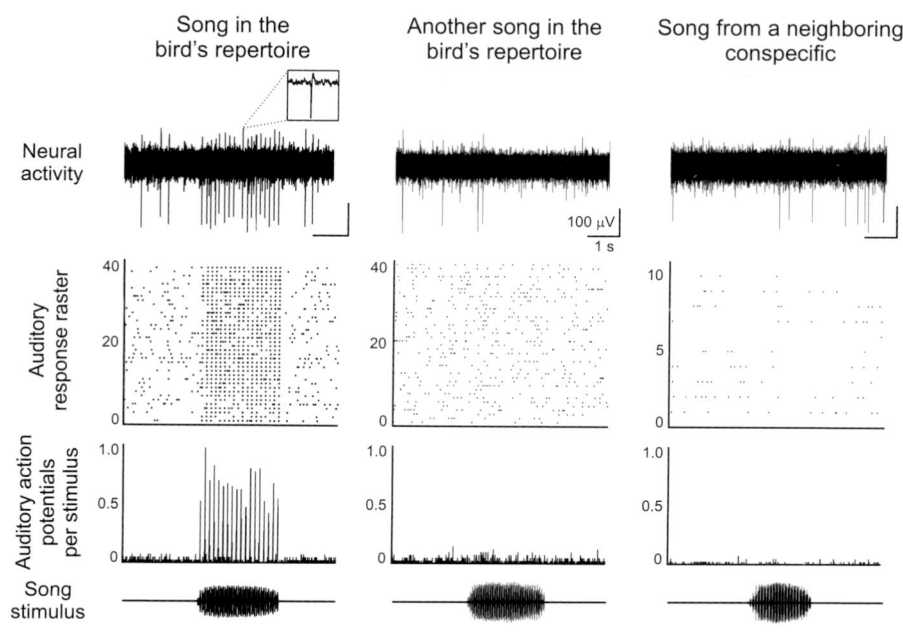

Song in the bird's repertoire

Another song in the bird's repertoire

Song from a neighboring conspecific

Neural activity

100 μV

1 s

Auditory response raster

40

20

0

40

20

0

10

5

0

Auditory action potentials per stimulus

1.0

0.5

0

1.0

0.5

0

1.0

0.5

0

Song stimulus

Figure 13.12. HVC$_X$ neurons respond selectively to one song type in the bird's repertoire. This neuron from an awake Swamp Sparrow responded very strongly to one song type in the bird's adult repertoire (*left*) but not to another song type in its repertoire (*middle*) or song from another bird living in its vicinity (*right*) (*top* = raw recording, second row = raster of 40 presentations of the song stimulus, *third row* = histogram of auditory responses, *bottom* = oscillogram of the song stimulus). Different cells are responsive to different song types (Prather et al. 2008). Therefore, the activity of individual HVC$_X$ cells is highly selective for specific sounds. *Adapted from figure 1 in Prather et al. 2008.*

in the same species revealed that the activity of HVC$_X$ neurons is closely related to the animal's perception of the song stimulus (fig. 13.14). It remains unknown where auditory perception first emerges in the brain, but auditory responses of HVC$_X$ neurons are a neural representation of not just auditory processing but also the more cognitively advanced step of perception. Therefore, auditory responses are selective for specific vocal elements and extend to perception of both self-generated sounds and sounds produced by others. Together, these data suggest that this system is well suited to encode the sensory portion of the sensory-to-motor transformation that is thought to underlie imitative song learning.

NEURAL LINKS BETWEEN PERCEPTION AND PERFORMANCE: A POSSIBLE KEY TO IMITATIVE LEARNING

The precise auditory representation of specific sounds that are generated by self or others makes HVC$_X$ neurons well suited to forge a link between perception of a specific vocalization and self-generated performance of that same sound. In that light, the essential next question is whether HVC$_X$ cells also express singing-related activity. Recordings from awake and freely behaving birds have revealed that HVC$_X$ neurons are active both when the bird hears song and when it performs song.

Just as individual HVC$_X$ neurons express temporally precise activity (brief transients in the voltage of individual neurons, commonly referred to as action potentials) in association with a specific song note or note sequence, those same cells are also active in association with those same sounds when the bird sings them (Prather et al. 2008, Fujimoto et al. 2011) (fig. 13.15). Importantly, activity during singing appears to be related to the vocal performance of those sounds as opposed to auditory experience as the bird hears himself sing (Kozhevnikov and Fee 2007, Prather et al. 2008). In support of that idea, playback of the relevant portion of song evokes robust auditory responses in an HVC$_X$ cell. If another copy of that song portion is simultaneously played slightly out of phase with the first copy, then the cell no longer responds (Prather et al. 2008). Thus, auditory responses are eliminated by distortion of what the bird hears. That insight enabled researchers to identify the nature of the action potentials that occur when the bird sings. The experimenters waited until the bird sang and then used rapid, song-triggered playback to initiate simultaneous playback of another copy of the same song that the bird was singing. In that context, the singing-related action potentials did not disappear, and their timing relative to the salient song element was unchanged from when the bird sang without distorted feedback (Prather et al. 2008). Together, these data revealed that distorted auditory input eliminates auditory action potentials but has no effect on singing-related activity. This led the investigators to conclude that activity of HVC$_X$ neurons during singing is a vocal-related signal. Curiously, that activity is a motor-related signal in a pathway that is not necessary for vocal motor output. Results from other systems and species have also revealed motor-related activity in pathways that are not

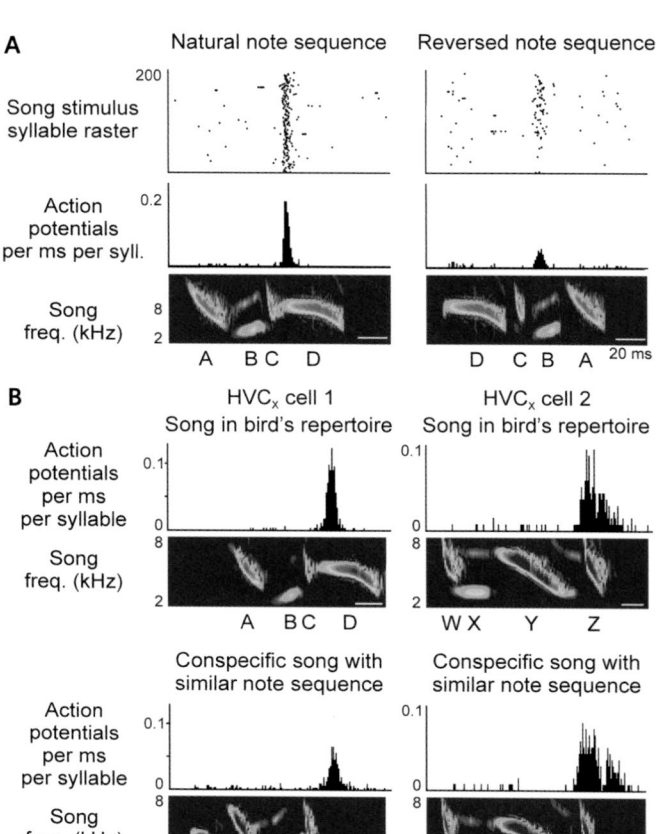

Figure 13.13. HVC_X neurons respond to specific note sequences regardless of whether they are performed by the male himself or by others. *A,* In this recording from an HVC_X cell in an awake Swamp Sparrow, the cell responded robustly to the natural note sequence in its own song (*left*) but much less when the note sequence was destroyed by playing in the notes in the reverse order (*right*) (*top* = raster of auditory responses to each syllable in the trilled song, *middle* = histogram of auditory responses, *bottom* = spectrogram of the song stimulus); *B,* These two HVC_X neurons (*left* = cell 1, *right* = cell 2) responded to specific note sequences in the bird's own song (*top* pair) and similar note sequences in the song of another Swamp Sparrow (*bottom* pair). Therefore, the activity of HVCX cells is a neural representation of specific sounds, regardless of whether those sounds are produced by self or by other members of the bird's species. *Adapted from figure 5 in Prather et al. 2008.*

necessary for behavioral output. That activity is called a **corollary discharge,** and it is thought to facilitate the task of disambiguating sensory input that results from an animal's own actions versus input that arises from external sources (Bell 1989, Crapse and Sommer 2008).

Individual HVC_X cells express a corollary discharge associated with performance of specific song elements. Ad-

ditional studies have revealed that corollary discharge may be present not only in HVC but also in other areas of the auditory system (Keller and Hahnloser 2009). Individual neurons that express both auditory and vocal activity that are each closely related to the same song element are well suited to serve as a site where sensory experience and motor performance are colocalized in the brain. By forging that link, those cells could help to translate sensory experience into vocal performance and thus facilitate the process of imitative learning. It is a very intriguing idea that those cells may guide juvenile learning and adult song maintenance. The advent of optical techniques to rapidly and selectively activate or inactivate specific pathways in the songbird brain while also sampling neural activity and behavior opens the door to testing that idea (e.g., Roberts et al. 2012). The coming decades hold the promise of many more insights into the neural mechanisms through which cells in the song system and elsewhere enable the learning, perception, and performance of the sounds that birds use in vocal communication.

BEYOND THE SONG SYSTEM: ADDITIONAL STRUCTURES THAT CONTRIBUTE TO LEARNING AND PERCEPTION

The song system undoubtedly plays a central role in song learning and performance, but recent studies have shed new light on additional structures that are also important in vocal learning. Cortical areas in the auditory system also play a key role in imitative song learning (fig. 13.16). Sarah London and David Clayton (2008) performed an elegant experiment to identify that contribution. They presented juvenile Zebra Finches with tutor song under different experimental conditions and measured the amount of imitative learning that emerged in the songs of those birds after they reached adulthood. Some birds were presented the tutor song without any disruption of neural activity. Those birds learned to imitate the tutor song quite well, revealing that the experimental paradigm was adequate to enable high-quality learning. Other birds were presented the same tutor song but they received that instruction while also receiving bilateral infusion of a drug into CM and NCM, sites in the avian brain that are the functional homologs of mammalian auditory cortical areas. The drug that those birds received interfered with the function of extracellular signal-related kinase (ERK), a molecule that regulates gene expression and that had been previously implicated in formation of memories in other systems. Birds that received song tutoring only while that drug was present failed to create accurate copies

SONG REPERTOIRES

There are vast differences among species in terms of the number of songs that an individual bird can perform. For example, adult males of some species such as Zebra Finches or Chipping Sparrows sing only one type of song. The machinery of their song systems is associated with performance of a greater number of sounds during juvenile song development, but those behaviors are pruned away during maturation, and the function of the song system is associated with just one type of song in adulthood. Other species such as Swamp Sparrows sing a handful of songs. Those songs are sufficiently distinct that we can begin to investigate how the brain learns and stores a diversity of sounds used in vocal communication, but that number is still small enough to be experimentally tractable. Species such as Chaffinches, Canaries, and Carolina Wrens have repertoires of dozens of song types, and others such as Nightingales and thrashers can have enormous repertoires of hundreds of different learned vocalizations. Somewhat surprisingly, components of the song system are not much larger in birds that sing large repertoires than they are in species that sing a single song type. An important future goal will be to understand how the brain encodes the learning, storage, and performance of distinct vocal behaviors. It is suspected that song repertoires are beneficial because they may facilitate territorial defense, and such a diversity of behavior provides females with considerable insight into the male's quality as a potential mate.

---

of the tutor song model. Thus, normal activity of CM and NCM appears to be necessary for imitative song learning. In further support of that idea, other birds also received both ERK infusion and tutor song exposure, but the drug was not presented at the same time as the tutor song. Those birds learned to produce accurate song copies. Importantly, that failure to imitate was not caused by a failure to hear the tutor song, as additional tests revealed auditory responses in CM and NCM even in the presence of ERK, and those birds were not impaired in behavioral tests of song discrimination. Together, these data indicate that CM and NCM play an important role in memorization and recall of the tutor song model and are therefore essential for imitative song learning.

CM and NCM are present in both male and female songbirds, leading some researchers to investigate whether those sites may also play a role in song perception in female birds (fig. 13.16). Interestingly, activity in those sites has been implicated in female recognition of individual males by their vocalizations (Menardy et al. 2012) and female evaluation of the quality of song as an indicator of the quality of prospective mates (MacDougall-Shackleton et al. 1998). Therefore, CM and NCM may play important roles in song perception, and that may underlie not only song memorization in juvenile males but also vocal communication in adults of both sexes. CM and NCM reside outside of the canonical song system, but they are interconnected with the song system and with each other (Bolhuis and Gahr 2006, Bauer et al. 2008). As songbird neurobiologists expand our understanding of how the brain enables imitative learning, the notion of the song system will likely also be expanded to acknowledge the important functional contributions of these and other brain sites.

BRAIN AREAS THAT UNDERLIE SONG LEARNING ARE NOT A BLANK SLATE

Songbirds face an interesting challenge. They hatch into a world filled with songs of many species. With such acoustically diverse experience, how do they "know" to memorize and imitate only the songs of their own species? The answer is far from understood, but genetics and other innate influences appear to play important roles in directing song properties and the choice of which songs to imitate (Marler 2004). Behavioral studies indicate that the function of brain pathways devoted to song learning and performance is somehow biased toward learning and performing songs of the bird's own species. For example, some features, such as the trilled syntax of Swamp Sparrow songs, emerge reliably even in the absence of hearing other members of the bird's species and even in the presence of song models from other species performing different syntax (Marler and Peters 1977, Searcy et al. 1985, Marler 1990). In addition, birds engaged in song learning are especially receptive to imitating songs of their own species. Very young birds will imitate songs of another species if those foreign songs are the only models that the developing bird ever hears, but developing males will quickly revert to song of their own species if they are later exposed to songs of their own species. For example, juvenile Zebra Finches that are exposed to tutor song models consisting entirely of songs performed by Bengalese Finches will imitate those foreign songs. If those Zebra Finches are later exposed to tutor song models that include songs of their

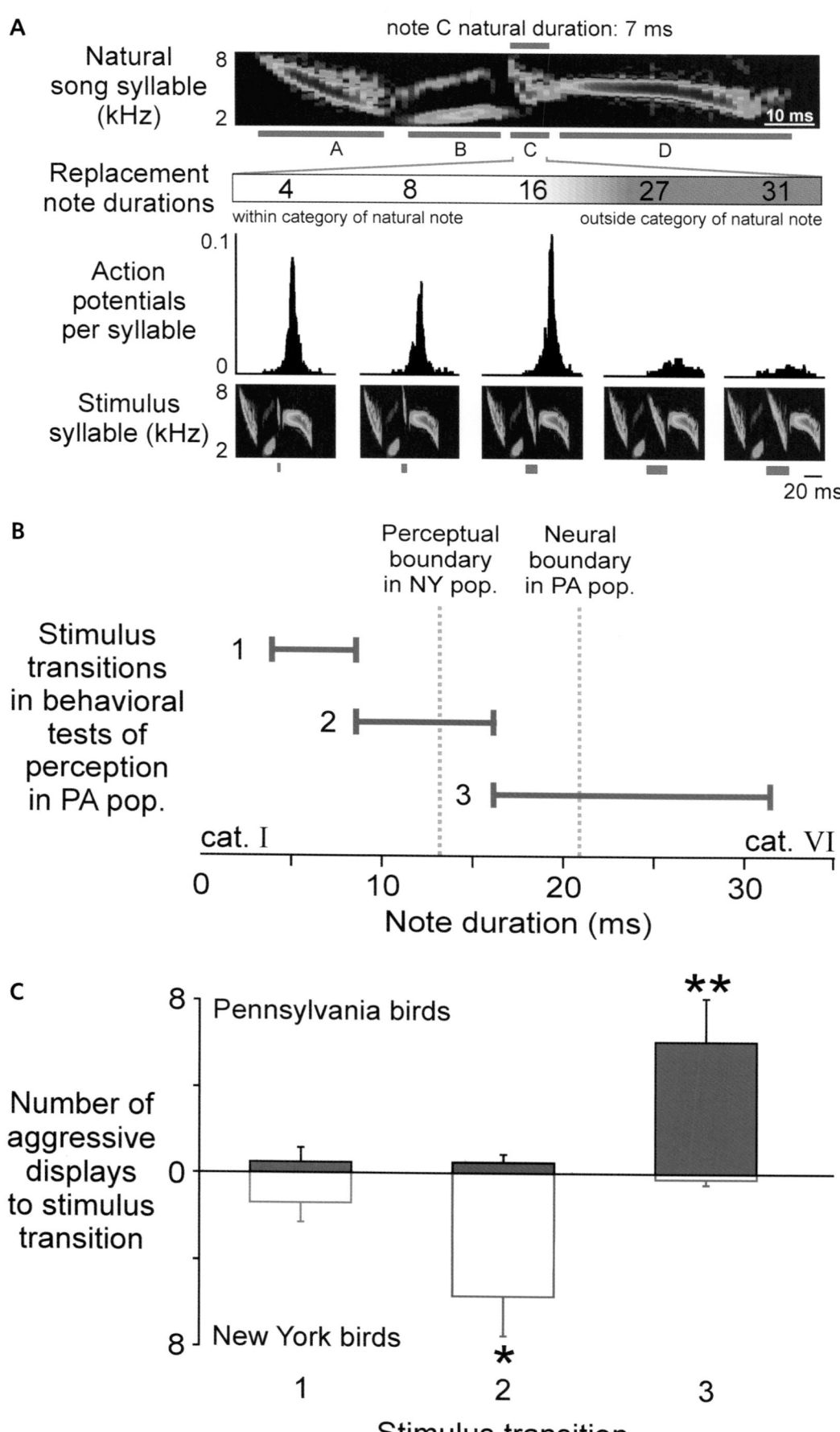

A

Natural song syllable (kHz)

note C natural duration: 7 ms

8
2

10 ms

A   B   C   D

Replacement note durations

4   8   16   27   31

within category of natural note   outside category of natural note

Action potentials per syllable

0.1

0

Stimulus syllable (kHz)

8
2

20 ms

B

Perceptual boundary in NY pop.   Neural boundary in PA pop.

Stimulus transitions in behavioral tests of perception in PA pop.

1

2

3

cat. I   cat. VI

0   10   20   30

Note duration (ms)

C

8   Pennsylvania birds

Number of aggressive displays to stimulus transition

0

8   New York birds

**   *

1   2   3

Stimulus transition

own species, then they will switch to imitating those Zebra Finch songs. That pattern of switching will not happen if the young birds are exposed to those models in the opposite order. Once a young male Zebra Finch has been exposed to songs of his own species, switching the tutor song models to Bengalese Finch songs will have little or no effect on the songs those Zebra Finches produce (Sugiyama and Mooney 2004). Many authors have speculated about the motor and neural constraints that may shape this apparent preference to learn and perform species-typical songs (Marler 1997), but it remains an open question how that functional bias emerges from the anatomical and genetic composition of the associated neuromuscular control sites.

WHAT ADVANTAGES ARE ASSOCIATED WITH THE ABILITY TO LEARN SONGS?

The ability to learn to imitate the sounds performed by others provides direct and indirect benefits. For example, the behavioral flexibility associated with song learning enables males to adapt to local conditions on their breeding grounds. Young Chipping Sparrows (*Spizella passerina*) sing a repertoire comprising a handful of crude songlike vocalizations called subsongs. When the juvenile bird arrives on what will eventually be his territory in his first breeding season, he gradually eliminates subsongs from his repertoire until over the course of just a few days he has eliminated all of his types of subsongs except for one. The one type that remains is typically one that is similar to the song performed by another neighboring male Chipping Sparrow. In just a few days

the new resident morphs that subsong into a mature song that is very much like the song of his neighbor (Liu and Nottebohm 2007). This is an efficient means of ensuring that the first-year male takes on song features that are not only characteristic of his species but also of the breeding population into which he has settled.

The degree to which a male is capable of vocal learning and performance is correlated with various aspects of male fitness. As discussed earlier in this chapter, song performance is intimately tied to neural structure and function, which are shaped by the conditions that a bird experiences during development. Because those same conditions and the consequent brain structure also affect cognition, song and cognitive ability may be linked. Experimental tests have generally supported the idea that high-quality song performance indicates high-quality cognition. For example, higher-quality songs, as defined by features such as longer duration and greater song complexity, have been linked to better performance on foraging and spatial learning tasks (Farrell et al. 2012). In addition to providing insight into the male's cognitive ability, song performance is also correlated with physical status and other aspects of behavior. For example, high-quality songs have been linked to better body condition (Lampe and Espmark 2003), a more robust immune system (Pfaff et al. 2007), greater parental effort (Bartsch et al. 2015), and greater probability of fledglings surviving to independence (Reid et al. 2005). Therefore, song ability appears to be predictive of a diverse suite of physical and cognitive traits. As male songbirds perform their songs to advertise their quality and attract mates, nearby females may eavesdrop

Figure 13.14. (opposite) HVC$_X$ neurons express categorical auditory responses that predict the bird's categorical perception of song stimuli. A, HVC$_X$ cells respond to each syllable in the trills in Swamp Sparrow songs (e.g., fig. 13.11). One of those syllables ("natural song syllable," four notes indicated by letters and gray boxes) is shown in the spectrogram at the top. Note C in that syllable was replaced with one of five different notes from songs of other birds. Those replacement notes had similar spectral characteristics (a frequency downsweep) but different durations (4, 8, 16, 27, 31 ms). Those five syllables, each containing a different replacement note (*bottom* = syllable spectrograms) were trilled at the same rate as the natural song to form five new song stimuli. When those new songs were played to the bird, auditory responses of this cell (histograms) and others from the same bird (not shown) were strong among one group of notes (4, 8, 16 ms) but weak among another group (27, 31 ms). *Adapted from figure 3 in Prather et al. 2009. B,* Those same song stimuli were used in behavioral tests of categorical perception. They were presented in pairs wherein the bird was habituated to one song, then the stimulus was switched to the other song. Switches between stimuli contained transitions between note durations that crossed no

putative boundary of categorical perception (4–8 ms), crossed the perceptual boundary for a New York population that was unfamiliar to the birds used in these neural recordings (8–16 ms), or crossed the putative boundary for Pennsylvania birds used in these neural recordings but not for New York birds (16–32 ms). C, The number of aggressive displays following the stimulus transition was measured as an index of the degree to which birds perceived the stimuli as different; no response would indicate that the birds detected no difference between the stimuli. Behavioral tests revealed that Pennsylvania Swamp Sparrows (*top*, filled bars) perceived strong differences when the transition in note duration spanned the Pennsylvania neural boundary detected in HVC$_X$ neurons (stimulus set 3), but perceived little or no difference when the transition spanned the perceptual boundary detected in New York birds (stimulus set 2) or spanned no putative boundary (stimulus set 1). Comparison of the behavioral data obtained from Pennsylvania birds (*top*, filled bars, mean ± s.e.) and from New York birds (*bottom*, open bars, mean ± s.e.) make clear the population-specific differences in categorical perception of note duration. New York data adapted from Nelson and Marler 1989. *Adapted from figure 4 in Prather et al. 2009.*

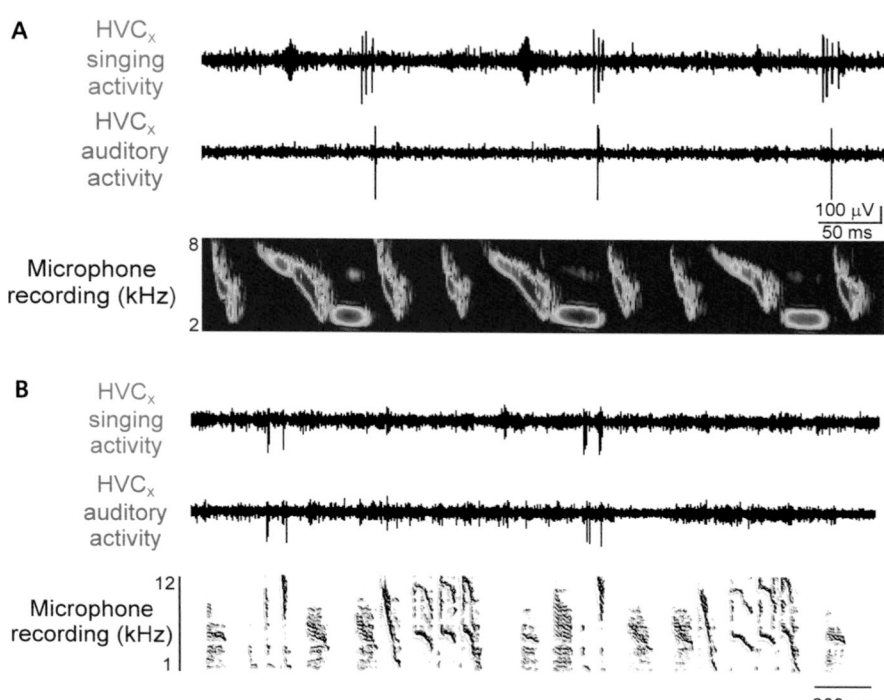

Figure 13.15. HVC$_X$ neurons exhibit a precise sensorimotor correspondence. *A,* In this recording from an HVC$_X$ cell in an awake Swamp Sparrow, the timing of singing-related activity (evident as spikes of voltage called action potentials in the *top row,* red) and auditory activity in the same cell (*middle row,* green) was quite similar in the two states. This correspondence between activity when the bird sang and when it heard the song played as an auditory stimulus was evident across HVC$_X$ cells (*bottom* = spectrogram of the song performance [singing] that was then played back to the bird as an auditory stimulus [auditory]). *Adapted from figure 3 in Prather et al. 2008. B,* The precise sensorimotor correspondence evident in Swamp Sparrows was also evident in the activity of HVC$_X$ neurons in Bengalese Finches, as shown in the activity of this cell when the bird sang (*top row,* red) and when it heard playback of the same song as an auditory stimulus (*middle row,* green) (*bottom* = spectrogram of the song performance that was recorded as the bird sang then played back to the bird as an auditory stimulus). Together, these data reveal that a precise sensorimotor correspondence is a broadly conserved feature among vocal learning species. *Adapted from supplemental figure 3 in Prather et al. 2008.*

and obtain information about the possible benefits associated with choosing a specific male as their mate. Thus, quality of song learning and the traits with which it is associated may confer benefits on not only the male himself but also his mate and their offspring.

IMPLICATIONS FOR UNDERSTANDING HOW HUMANS COMMUNICATE USING SPEECH

The ability of an organism to acquire the sounds used in vocal communication by imitating the sounds that it hears performed by others is rare across animals. We are exceptionally good at it, as evident in the diversity of sounds that we use to communicate with one another through speech. Songbirds, parrots, whales, and a few other groups of animals are similarly skilled, but most animals lack the ability to imitate what they hear (Jarvis 2004, Bolhuis and Everaert 2013). For example, a dog can learn to perform a specific action when he hears you say the word "sit," but even the cleverest of dogs will never be able to learn to say the word "sit." Humans and songbirds are special in that we can each learn to reproduce what we hear performed by others. Just as songbirds learn their songs by listening to models performed by others and then relying on auditory feedback to develop and maintain their imitations of those models, humans go through a strikingly parallel pattern of development as we acquire the sounds we use in speech (Doupe and Kuhl 1999). As a very young child, you learned the sounds you use in speech by imitating what you heard performed by more advanced speakers around you, most likely your parents. You were likely very good at it. In fact, if your parents and the people in your community have a regional accent, such as the gentle lilt of speakers from the American South, then you likely imitated them so well that you also have that accent. Birds are similarly skilled in the degree of their imitation. They imitate so well that regional song dialects emerge among different breeding populations of the same species (reviewed in Podos and Warren 2007). Those dialectical patterns persist across generations because young birds imitate down to the level of the relatively subtle features that define regional dialectical differences, and females prefer to mate with males that perform songs that are characteristic of the female's home group (Anderson 2009).

Because of these and other parallels between how we acquire the sounds used in speech and how birds acquire their songs, understanding the neural basis of how birds sing can provide valuable insight into how we communicate through speech (Doupe and Kuhl 1999). Just as HVC neurons are active in association with both performance and perception of the sounds that compose songs, specific neurons in the human brain have also been implicated in performance and perception of the sounds we use in speech (Hickok 2010). For example, injury to a specific region in the anterior lateral portion of the left cerebral hemisphere (Broca's area) can result in difficulty producing speech. Injury at another region in the posterior lateral portion of the left cerebral hemisphere (Wernicke's area) can result in difficulty processing speech. As neurologists study those areas more thoroughly, the distinctions between neurons that are active in association with production or perception are becoming blurred, and it appears that both areas are involved in both processes with perhaps some functional bias toward speech production or perception (reviewed in Bolhuis and Everaert 2013). Just as injury to those sites in the left hemisphere of the human brain results in impairment in the perception and performance of speech, injury to specific sites in the songbird brain results in impairment of perception and performance of song. The songbird brain also appears to be lateralized such that the left hemisphere is more dominant than the right hemisphere in song control, and that lateralization emerges through auditory-vocal learning as the bird becomes more proficient in his song performance (Chirathivat et al. 2015). Together with our growing understanding of how specific sites in the songbird brain are related to corresponding sites in the mammalian brain, insights into how the songbird brain enables songbirds to learn and perform the songs they use in vocal communication will also shed new light on how our brains enable us to communicate through speech. These sorts of discoveries will help us to understand ourselves more

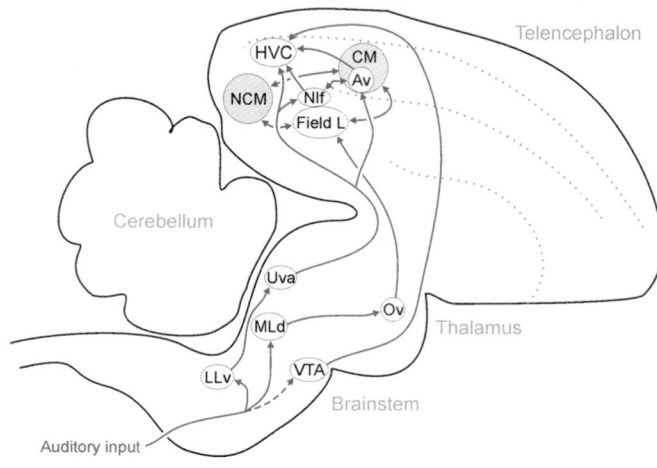

From the ear to HVC - ascending auditory pathways

Figure 13.16. The song system resides in a network of other structures that have also been found to play important roles in imitative song learning. Prominent among these other structures are CM (caudal mesopallium, shaded) and NCM (caudomedial nidopallium, shaded). The collection of song-related brain sites is therefore much more intricate than was thought in the early studies of songbird neurobiology. An important goal of future studies will be to understand how these many structures interact to enable learning, perception, and performance of the sounds used in vocal communication (HVC = high vocal center; AV = nucleus avalanche; NIF = nucleus interfacialis; Field L = primary auditory cortex; Uva = nucleus uvaeformis; OV = nucleus ovoidalis; MLd = midbrain nucleus mesencephalicus lateralis pars dorsalis; VTA = ventral tegmental area; LLv = ventral nucleus of the lateral lemniscus). *Adapted from figure 2 in Prather 2013.*

fully, and insights into how song pathologies emerge following brain injury may also guide new thinking about how we can treat speech pathologies such as stuttering and unintentional rearrangement of the sequences of sound that define words and sentences.

KEY POINTS

- Songs are long, complex vocalizations that young birds learn by listening to the sounds they hear produced by other members of their species.
- Young birds memorize song models then recall them during later development and use those memories to guide their refinement of a high-quality imitation of that model.
- Vocal performance is intimately related to auditory input. Young birds must listen to others in order to memorize their song models, developing birds must listen to themselves to refine their imitative performance, and

PETER MARLER

Peter Marler (1928–2014) was a giant in the history of song-bird neurobiology and animal communication in general. He began his career by earning a PhD in botany at the University of London. Remarkably, he followed that feat with another PhD in zoology from the University of Cambridge just two years later. His main interest was the vocal behaviors of birds, both their songs and their calls. He was one of the first researchers to use a spectrograph to visualize the structure of avian vocalizations, and he used that technology to explore the calls of Chaffinches. Upon moving to the United States in 1957, he turned his attention to White-crowned Sparrows, which he found especially interesting because of their regional song dialects. Marler's studies revealed that not only is song learned, but young birds also somehow innately prefer to learn songs of their own species over songs of other species in the same area. He also found that song learning occurs during a sensitive period of juvenile develop-ment. Together with his demonstration that song acquisition preceded song production, these findings were the first steps toward realization of the striking parallels between how songbirds learn their songs and how humans acquire the sounds we use in speech. His work was strengthened by his integration of field studies and laboratory experiments, and he was also keenly aware of the power of a comparative approach to enable researchers to discern broadly conserved features of learned behavior. Marler's legacy is evident not only in his results but also in the people that he trained. Students and postdoctoral researchers that emerged from the Marler laboratory have gone on to lead research groups of their own, and they continue to have an enormous impact on our understanding of songbird vocal communication (e.g., Ball, Dooling, Harding, Konishi, Kroodsma, Nottebohm, Nowicki, Peters, Searcy, Wiley) and other communication behaviors (e.g., Hopkins, Seyfarth, Cheney).

adult birds must listen to themselves to maintain their song quality throughout life.

- The songbird brain contains a network of highly specialized structures dedicated to song learning, performance, and perception. Properties of cells in that system suggest they may be sites where auditory experience and vocal performance are linked in the brain, making them well suited to facilitate imitative learning.

- Songbirds acquire and maintain their songs through a pattern of development that is strikingly similar to how we acquire the sounds we use in speech. Research into the neural basis of song learning and performance may also reveal important new insights into how humans communicate using speech.

KEY MANAGEMENT AND CONSERVATION IMPLICATIONS

- Songbirds rely on auditory experience to learn and maintain their songs. Increasingly, researchers are finding that songs of urban species are changing in response to the anthropogenic din of noise in big cities (Narango and Rodewald 2016). Noise presents a challenge to effective transmission of vocal signals from sender to receiver. To overcome that challenge, and thus to aide in making their songs more easily detectable in the background of noise, birds have shifted the properties of their songs. Specifically, they have shifted their performance so that more energy is present at frequencies outside of the range of typical urban noise (outside of the frequency range from about one to two kHz and into the band from about 4 to 9 kHz, reviewed in Patricelli and Blickley 2006). Unfortunately, this adaptation has not been completely successful from the birds' point of view (e.g., Luther et al. 2015, Moiron et al. 2015), and there are concerns that song adaptations that occur in response to ambient noise may make those signals less effective in defending territory and attracting mates. This topic has received considerable attention recently, and it will continue to be important to understand the impacts of noise on song and the survival and reproduction of the affected species.

DISCUSSION QUESTIONS

1. Where and how do you think the song model is stored?

2. How do you think the song model is recalled and linked to performance to guide imitation? Through what mechanisms do you think the brain may accomplish those tasks?

3. Song performance is thought to be an "honest signal" of the quality of the associated singer. Honest signals are thought to be reliable because they come at a cost. What do you think are the costs associated with learning and singing songs?

4. Neurobiologists often make claims that a certain area of the brain plays an important role in a certain behavior. How do you think they assign a specific function to a specific brain site? What data set would you need in order to be convinced that a specific brain region was associated with a specific behavior?

5. Birds in many species form very high-quality imitations of the song models they hear performed by other members of their species. How do you think that listeners can identify the singer of the original model versus the singer of the copy? If two birds perform nearly identical songs, how could a listener tell those birds apart? How might your answer differ for species in which males sing only one song type versus species in which males sing a large repertoire?

6. Species such as Northern Mockingbirds and Brown Thrashers can learn to imitate the songs of a very wide range of species, with individual males able to imitate tens or hundreds of the songs and other sounds that they hear performed in their environment. What do you think are the costs and benefits of the ability to imitate such a wide range of songs and other sounds?

References

Anderson, R. C. (2009). Operant conditioning and copulation solicitation display assays reveal a stable preference for local song by female Swamp Sparrows *Melospiza georgiana*. Behavioral Ecology and Sociobiology 64:215–223.

Ball, G. F. (2016). Species variation in the degree of sex differences in brain and behaviour related to birdsong: Adaptations and constraints. Philosophical Transactions of the Royal Society B: Biological Sciences 371. doi:10.1098/rstb.2015.0117.

Ball, G. F., C. J. Auger, D. J. Bernard, T. D. Charlier, J. J. Sartor, L. V. Riters, and J. Balthazart (2004). Seasonal plasticity in the song control system: Multiple brain sites of steroid hormone action and the importance of variation in song behavior. Annals of the New York Academy of Science 1016:586–610.

Bartsch, C., M. Weiss, and S. Kipper (2015). Multiple song features are related to paternal effort in common nightingales. BMC Evolutionary Biology 15. doi:10.1186/s12862-015-0390-5.

Bauer, E. E., M. J. Coleman, T. F. Roberts, A. Roy, J. F. Prather, and R. Mooney (2008). A synaptic basis for auditory-vocal integration in the songbird. Journal of Neuroscience 28:1509–1522.

Bell, C. C. (1989). Sensory coding and corollary discharge effects in mormyrid electric fish. Journal of Experimental Biology 146:229–253.

Bluff, L. A., J. Troscianko, A. A. S. Weir, A. Kacelnik, and C. Rutz (2010). Tool use by wild New Caledonian Crows *Corvus moneduloides* at natural foraging sites. Proceedings of the Royal Society B: Biological Sciences 277:1377–1385.

Bolhuis, J. J., and M. Gahr (2006). Neural mechanisms of birdsong memory. Nature Reviews Neuroscience 7:347–357.

Bolhuis, J. J., and M. Everaert, Editors (2013). Birdsong, speech and language: Exploring the evolution of mind and brain. MIT Press, Cambridge, MA.

Bottjer, S. W., E. A. Miesner, and A. P. Arnold (1984). Forebrain lesions disrupt development but not maintenance of song in passerine birds. Science 224:901–903.

Boughey, M. J., and N. S. Thompson (1981). Song variety in the Brown Thrasher (*Toxostoma rufum*). Zeitschrift Fur Tierpsychologie—Journal of Comparative Ethology 56:47–58.

Brenowitz, E. A., and H. H. Zakon (2015). Emerging from the bottleneck: Benefits of the comparative approach to modern neuroscience. Trends in Neurosciences 38:273–278.

Chirathivat, N., S. C. Raja, and S. M. Gobes (2015). Hemispheric dominance underlying the neural substrate for learned vocalizations develops with experience. Scientific Reports 5:11359.

Crapse, T. B., and M. A. Sommer (2008). Corollary discharge across the animal kingdom. Nature Reviews Neuroscience 9:587–600.

Darwin, C. (1874). The descent of man, and selection in relation to sex. John Murray, London.

Derrickson, K. C. (1987). Yearly and situational changes in the estimate of repertoire size in Northern Mockingbirds (*Mimus polyglottos*). The Auk 104:198–207.

Doupe, A. J., and P. K. Kuhl (1999). Birdsong and human speech: Common themes and mechanisms. Annual Review Neuroscience 22:567–631.

Doupe, A. J., D. J. Perkel, A. Reiner, and E. A. Stern (2005). Birdbrains could teach basal ganglia research a new song. Trends in Neuroscience 28:353–363.

Emery, N., and N. Clayton (2001). Effects of experience and social context on prospective caching strategies in scrub jays. Nature 414:443–446.

Farrell, T. M., K. Weaver, Y. S. An, and S. A. MacDougall-Shackleton (2012). Song bout length is indicative of spatial learning in European Starlings. Behavioral Ecology 23:101–111.

Fee, M. S., and A. Leonardo (2001). Miniature motorized microdrive and commutator system for chronic neural recording in small animals. Journal of Neuroscience Methods 112:83–94.

Fujimoto, H., T. Hasegawa, and D. Watanabe (2011). Neural coding of syntactic structure in learned vocalizations in the songbird. Journal of Neuroscience 31:10023–10033.

Funabiki, Y., and M. Konishi (2003). Long memory in song learning by Zebra Finches. Journal of Neuroscience 23:6928–6935.

Gardner, T. J., F. Naef, and F. Nottebohm (2005). Freedom and rules: The acquisition and reprogramming of a bird's learned song. Science 308:1046–1049.

Gentner, T. Q., S. H. Hulse, G. E. Bentley, and G. F. Ball (2000). Individual vocal recognition and the effect of partial lesions to HVC on discrimination, learning, and categorization of conspecific song in adult songbirds. Journal of Neurobiology 42:117–133.

Hickok, G. (2010). The role of mirror neurons in speech and language processing. Brain Language 112:1–2.

Hultsch, H., and D. Todt (1992). The serial order effect in the song acquisition of birds: Relevance of exposure frequency to song models. Animal Behaviour 44:590–592.

Jarvis, E. D. (2004). Learned birdsong and the neurobiology of human language. Behavioral Neurobiology of Birdsong 1016:749–777.

Jarvis, E. D. (2009). Bird brain: Evolution. In Encyclopedia of neuroscience, vol. 2, L. E. Squire, Editor. Oxford: Academic Press, pp. 209–215.

Jarvis, E., O. Gunturkun, L. Bruce, A. Csillag, H. Karten, W. Kuenzel, L. Medina, et al. (2005). Avian brains and a new understanding of vertebrate brain evolution. Nature Reviews Neuroscience 6:151–159.

Jelbert, S. A., A. H. Taylor, L. G. Cheke, N. S. Clayton, and R. D. Gray (2014). Using the Aesop's Fable paradigm to investigate causal understanding of water displacement by New Caledonian Crows. PLoS ONE 9. https://doi.org/10.1371/journal.pone.0092895.

Kao, M. H., and M. S. Brainard (2006). Lesions of an avian basal ganglia circuit prevent context-dependent changes to song variability. Journal of Neurophysiology 96:1441–1455.

Katz, L. C., and M. E. Gurney (1981). Auditory responses in the Zebra Finch's motor system for song. Brain Research 221:192–197.

Keller, G. B., and R. H. Hahnloser (2009). Neural processing of auditory feedback during vocal practice in a songbird. Nature 457:187–190.

Konishi, M. (1965). The role of auditory feedback in the control of vocalization in the White-crowned Sparrow. Zeitschrift Fur Tierpsychologie 22:770–783.

Konishi, M., and E. Akutagawa (1985). Neuronal growth, atrophy and death in a sexually dimorphic song nucleus in the Zebra Finch brain. Nature 315:145–147.

Konishi, M., and E. Akutagawa (1988). A critical period for estrogen action on neurons of the song control system in the Zebra Finch. Proceedings of the National Academy of Sciences USA 85:7006–7007.

Kozhevnikov, A. A., and M. S. Fee (2007). Singing-related activity of identified HVC neurons in the Zebra Finch. Journal of Neurophysiology 97:4271–4283.

Krebs, J. R., N. S. Clayton, S. D. Healy, D. A. Cristol, S. N. Patel, and A. R. Jolliffe (1996). The ecology of the avian brain: Food-storing memory and the hippocampus. Ibis 138:34–46.

Kroodsma, D., D. Hamilton, J. E. Sanchez, B. E. Byers, H. Fandino-Marino, D. W. Stemple, J. M. Trainer, and G. V. N. Powell (2013). Behavioral evidence for song learning in the suboscine bellbirds (Procnias spp.: Cotingidae). Wilson Journal of Ornithology 125:1–14.

Lampe, H. M., and Y. O. Espmark (2003). Mate choice in Pied Flycatchers Ficedula hypoleuca: Can females use song to find high-quality males and territories? Ibis 145:E24–E33.

Leonardo, A., and M. Konishi (1999). Decrystallization of adult birdsong by perturbation of auditory feedback. Nature 399:466–470.

Liu, W. C., and F. Nottebohm (2007). A learning program that ensures prompt and versatile vocal imitation. Proceedings of the National Academy of Sciences USA 104:20398–20403.

Liu, W. C., K. Wada, E. D. Jarvis, and F. Nottebohm (2013). Rudimentary substrates for vocal learning in a suboscine. Nature Communications 4. doi:10.1038/ncomms3082.

Lombardino, A. J., and F. Nottebohm (2000). Age at deafening affects the stability of learned song in adult male Zebra Finches. Journal of Neuroscience 20:5054–5064.

London, S. E., and D. F. Clayton (2008). Functional identification of sensory mechanisms required for developmental song learning. Nature Neuroscience 11:579–586.

Long, M. A., and M. S. Fee (2008). Using temperature to analyse temporal dynamics in the songbird motor pathway. Nature 456:189–194.

Luther, D. A., J. Phillips, and E. P. Derryberry (2015). Not so sexy in the city: Urban birds adjust songs to noise but compromise vocal performance. Behavioral Ecology 27 (1): 332–340. doi:https://doi.org/10.1093/beheco/arv162.

MacDougall-Shackleton, S., S. Hulse, and G. Ball (1998). Neural bases of song preferences in female Zebra Finches (Taeniopygia guttata). Neuroreport 9:3047–3052.

Margoliash, D. (1986). Preference for autogenous song by auditory neurons in a song system nucleus of the White-crowned Sparrow. Journal of Neuroscience 6:1643–1661.

Marler, P. (1990). Innate learning preferences: Signals for communication. Developmental Psychobiology 23:557–568.

Marler, P. (1997). Three models of song learning: Evidence from behavior. Journal of Neurobiology 33:501–516.

Marler, P. (2004). Innateness and the instinct to learn. Anais Da Academia Brasileira De Ciencias 76:189–200.

Marler, P., and S. Peters (1977). Selective vocal learning in a sparrow. Science 198:519–521.

Marler, P., and S. Peters (1981). Sparrows learn adult song and more from memory. Science 213:780–782.

Marler, P., and S. Peters (1982a). Long-term storage of learned birdsongs prior to production. Animal Behaviour 30: 479–482.

Marler, P., and S. Peters (1982b). Structural-changes in song ontogeny in the Swamp Sparrow Melospiza georgiana. The Auk 99:446–458.

Marler, P., and A. J. Doupe (2000). Singing in the brain. Proceedings of the National Academy of Sciences USA 97:2965–2967.

McCasland, J. S., and M. Konishi (1981). Interaction between auditory and motor activities in an avian song control nucleus. Proceedings of the National Academy of Sciences USA 78:7815–7819.

Menardy, F., K. Touiki, G. Dutrieux, B. Bozon, C. Vignal, N. Mathevon, and C. Del Negro (2012). Social experience affects neuronal responses to male calls in adult female Zebra Finches. European Journal of Neuroscience 35:1322–1336.

Moiron, M., C. Gonzalez-Lagos, H. Slabbekoorn, and D. Sol (2015). Singing in the city: High song frequencies are no guarantee for urban success in birds. Behavioral Ecology 26:843–850.

Mooney, R., J. F. Prather, and T. Roberts (2008). Neurophysiology of birdsong learning. In Learning and memory: A comprehensive reference, vol. 3, Memory systems, H. Eichenbaum, Editor. Elsevier, Oxford, pp. 441–474.

Narango, D. L., and A. D. Rodewald (2016). Urban-associated drivers of song variation along a rural-urban gradient. Behavioral Ecology 27:608–616.

Nelson, D. A., and P. Marler (1989). Categorical perception of a natural stimulus continuum: birdsong. Science 244 (4907): 976–978.

Nordeen, K. W., and E. J. Nordeen (1992). Auditory feedback is necessary for the maintenance of stereotyped song in adult Zebra Finches. Behavioral and Neural Biology 57: 58–66.

Nottebohm, F., and A. P. Arnold (1976). Sexual dimorphism in vocal control areas of the songbird brain. Science 194:211–213.

Nowicki, S. (1987). Vocal-tract resonances in oscine bird sound production: Evidence from birdsongs in a helium atmosphere. Nature 325:53–55.

Olkowicz, S., M. Kocourek, R. Lučana, M. Porteš, W. Fitch, S. Herculano-Houzel, and P. Němec (2016). Birds have primate-like numbers of neurons in the forebrain. Proceedings of the National Academy of Sciences USA 113:7255–7260.

Olveczky, B. P., A. S. Andalman, and M. S. Fee (2005). Vocal experimentation in the juvenile songbird requires a basal ganglia circuit. PLoS Biol. 3 (5): e153.

Otchy, T. M., and B. P. Olveczky (2012). Design and assembly of an ultra-light motorized microdrive for chronic neural recordings in small animals. Journal of Visualized Experiments 69:4314. doi:10.3791/4314.

Patricelli, G. L., and J. L. Blickley (2006). Avian communication in urban noise: Causes and consequences of vocal adjustment. The Auk 123:639–649.

Peters, S., P. Marler, and S. Nowicki (1992). Song Sparrows learn from limited exposure to song models. The Condor 94:1016–1019.

Pfaff, J. A., L. Zanette, S. A. MacDougall-Shackleton, and E. A. MacDougall-Shackleton (2007). Song repertoire size varies with HVC volume and is indicative of male quality in Song Sparrows (Melospiza melodia). Proceedings of the Royal Society B: Biological Sciences 274:2035–2040.

Podos, J., J. K. Sherer, S. Peters, and S. Nowicki (1995). Ontogeny of vocal-tract movements during song production in Song Sparrows. Animal Behaviour 50:1287–1296.

Podos, J., and P. S. Warren, Editors (2007). The evolution of geographic variation in birdsong. Elsevier, London.

Prather, J. F. (2013). Auditory signal processing in communication: Perception and performance of vocal sounds. Hearing Research 305:144–155.

Prather, J. F., S. Peters, S. Nowicki, and R. Mooney (2008). Precise auditory-vocal mirroring in neurons for learned vocal communication. Nature 451:305–310.

Prather, J. F., S. Nowicki, R. C. Anderson, S. Peters, and R. Mooney (2009). Neural correlates of categorical perception in learned vocal communication. Nature Neuroscience 12 (2): 221–228.

Raikow, R. J., and A. H. Bledsoe (2000). Phylogeny and evolution of the passerine birds. BioScience 50:487–499.

Reid, J. M., P. Arcese, A. L. E. V. Cassidy, S. M. Hiebert, J. N. M. Smith, P. K. Stoddard, A. B. Marr, and L. F. Keller (2005). Fitness correlates of song repertoire size in free-living Song Sparrows (Melospiza melodia). American Naturalist 165:299–310.

Reiner, A., D. J. Perkel, L. L. Bruce, A. B. Butler, A. Csillag, W. Kuenzel, L. Medina, et al. (2004). The avian brain nomenclature forum: Terminology for a new century in comparative neuroanatomy. Journal of Comparative Neurology 473:E1–E6.

Riede, T., R. A. Suthers, N. H. Fletcher, and W. E. Blevins (2006). Songbirds tune their vocal tract to the fundamental frequency of their song. Proceedings of the National Academy of Sciences USA 103:5543–5548.

Roberts, T. F., S. M. Gobes, M. Murugan, B. P. Olveczky, and R. Mooney (2012). Motor circuits are required to encode a sensory model for imitative learning. Nature Neuroscience 15:1454–1459.

Scharff, C., F. Nottebohm, and J. Cynx (1998). Conspecific and heterospecific song discrimination in male Zebra Finches with lesions in the anterior forebrain pathway. Journal of Neurobiology 36:81–90.

Searcy, W. A., P. Marler, and S. S. Peters (1985). Songs of isolation-reared sparrows function in communication, but are significantly less effective than learned songs. Behavioral Ecology and Sociobiology 17:223–229.

Smirnova, A., Z. Zorina, T. Obozova, and E. Wasserman (2015). Crows spontaneously exhibit analogical reasoning. Current Biology 25:256–260.

Sugiyama, Y. Y., and R. Mooney (2004). Sequential learning from multiple tutors and serial retuning of neurons in a brain area important to birdsong learning. Journal of Neurophysiology 92:2771–2788.

Suthers, R. A. (2004). How birds sing and why it matters. In Nature's music: The science of birdsong. Elsevier, New York, pp. 272–295.

Suthers, R. A., and S. A. Zollinger (2004). Producing song: The vocal apparatus. In Behavioral neurobiology of birdsong. Annals of the New York Academy of Sciences, pp. 19–129.

Theunissen, F. E., and A. J. Doupe (1998). Temporal and spectral sensitivity of complex auditory neurons in the nucleus HVC of male Zebra Finches. Journal of Neuroscience 18:3786–3802.

Touchton, J. M., N. Seddon, and J. A. Tobias (2014). Captive rearing experiments confirm song development without learning in a tracheophone suboscine bird. PLoS ONE 9. https://doi.org/10.1371/journal.pone.0095746.

Wade, J., and A. P. Arnold (2004). Sexual differentiation of the Zebra Finch song system. Annals of the New York Academy of Science 1016:540–559.

Weir, A. A. S., J. Chappell, and A. Kacelnik (2002). Shaping of hooks in New Caledonian Crows. Science 297:981.

White, S. A., F. S. Livingston, and R. Mooney (1999). Androgens modulate NMDA receptor-mediated EPSCs in the Zebra Finch song system. Journal of Neurophysiology 82:2221–2234.

Woolley, S. M., and E. W. Rubel (1997). Bengalese Finches *Lonchura striata domestica* depend upon auditory feedback for the maintenance of adult song. Journal of Neuroscience 17:6380–6390.

CHAPTER 14

Acoustic Communication

Michael S. Webster and Jeffrey Podos

Most bird species are highly social, and nearly all facets of avian sociality are enabled by signals that allow individuals to coordinate their actions with precision and according to context. Indeed, birds communicate extensively while dealing with most of life's major challenges. They communicate while foraging, migrating, settling on mating grounds, dealing with predators, fighting, courting, and rearing young (chapters 15–19).

Although birds use a number of modalities to communicate with each other, including visual (chapter 11) and possibly even olfactory signals (chapter 12), it is in the acoustic realm that they truly stand out. Birds are vocal creatures, and they make extensive use of acoustic signals. These acoustic signals include vocalizations, which are songs and calls produced with the vocal apparatus, as well as mechanical "sonations" produced with various body parts. Acoustic signals are used to attract mates, repel rivals, coordinate group movements, warn others of impending danger, and ask for food. Chapter 13 covers the neural underpinnings of song production and learning. In this chapter we address the mechanisms and function of acoustic communication in birds: How do birds make acoustic signals, why do they make them, and how have these signals evolved?

Some physical attributes of sound make acoustic signals particularly effective for communication (Bradbury and Vehrencamp 2011). Acoustic signals carry long distances, even across cluttered (e.g., forested) landscapes, allowing animals to maintain contact even when out of visual sight lines or at night. Acoustic signals can be adjusted rapidly, allowing animals high precision and flexibility in response to changing environmental and social circumstances, and acoustic signals can pack substantial information into short periods of time. Not surprisingly, acoustic signals are a hallmark feature of many bird groups, especially the songbirds, which make up over half of all avian species (chapters 3 and 4).

The acoustic signals of birds have been the focus of intense research for decades, and we have learned a great deal about how and why birds produce sounds. Yet there is so much more to learn: for most species we don't even know all of the sounds that they make, let alone the context in which they make those sounds. Even common and well-studied species such as chickadees and robins may make well over 20 different types of sounds, but for the most part we don't know what these various sounds "mean." At the same time, new technologies and analytical methods are allowing us to study these signals in new ways and at scales previously unimaginable. Accordingly, these are exciting times for research on bird communication.

Our chapter begins with a brief look at definitions. What is communication, what are signals, and what are the different types of acoustic signals in birds? We next consider mechanisms of acoustic signal production and transmission in birds. How do animals produce sound, and how are acoustic signals affected by the environment through which they propagate? From there we consider the various functions of acoustic signals: What do signaling animals gain, and why should listeners pay attention to the signals of others? The final part of our chapter turns to questions about acoustic signal evolution. Here we delve into topics that have been subject to lively recent debate and scrutiny: topics including signaling honesty, reliability, and deception.

WHAT IS ACOUSTIC COMMUNICATION?

"Communication" is one of those things that is easy to discuss but challenging to define. **Communication** happens when one individual (the **signaler**) sends information (via a **signal**) through the environment to another individual (the **receiver**, typically of the same species), who then responds in some way (fig. 14.1). For this exchange of information to

Figure 14.1. Communication occurs when one individual (the signaler, S) transmits a signal through the environment to a second individual (the receiver, R), causing that receiver to modify its behavior to the benefit of both. Here a nestling (S) produces begging signals, both visual and vocal, that increase the likelihood that the parental bird (R) will feed it. *Photo of Black-throated Blue Warbler* (Setophaga caerulescens) *nest by Steve Maslowski, USFWS, image available from https://commons .wikimedia.org/wiki/File:Dendroica_caerulescens1.jpg.*

represent true communication, it is necessary that, on average, the receiver's response increase the fitness of both signaler and the receiver (Searcy and Nowicki 2005). Why? First, if a particular class of signal does not typically increase the signaler's fitness, then individuals who do not produce the signal will be favored, and signaling will not spread through the population. Second, if the receiver's response does not typically increase its own fitness, then receivers that respond to the signal will be selected against, and receivers that ignore the signal will be favored. In both cases, signalers would gain no fitness advantages from signaling, and the signal would disappear over time.

The effects on fitness help us distinguish signals from **cues**, which can be defined as morphological traits or behavioral activities that send unintended, potentially detrimental information about an animal. Consider a mouse rustling leaf litter as it moves across the forest floor. The sound of rustling provides would-be predators, such as owls, information about the location of the mouse, which the owl can use to capture and consume the mouse (Payne 1971). In this case the acoustic information given off by the mouse is a cue but not a signal, as the information clearly benefits the owl but not the mouse. In fact, selection should favor mice that move through the environment as quietly as possible, to reduce—not increase—the likelihood of being detected by owls and other predators.

Contrast this with the case of a young raven that discovers a large food source and gives a call to attract other ravens to it. In this case, the responding ravens benefit from getting

food, whereas the calling raven has to share the food with the other birds. Nonetheless, the calling raven would benefit overall if the attracted ravens help defend the food from other species or even other ravens (Heinrich 1988, 1989). So long as the fitness benefits outweigh the costs—such that the calling bird gets more food than it would have otherwise—then selection will favor calling. In this case the call is a true signal, in the sense that both the signaler and the receiver(s) benefit.

As a general rule, for a signal to evolve and/or be maintained in a population, both the signaler and the receiver must benefit from that signal (Searcy and Nowicki 2005). This does not imply, though, that both parties will benefit every time the signal is given—they just need to benefit on average. For example, as we discuss below, nestling birds often give begging calls to obtain food from their parents. Nestlings obviously benefit from being fed. Likewise, parents benefit from feeding their own young, for the obvious reason that it maintains the health of their reproductive investment. However, begging calls might also put the chicks at risk. Sometimes, unintended receivers may pay attention to the signal, which is called **eavesdropping**. In particular, predators may eavesdrop on nestling begging calls to locate the nest and obtain a meal of their own, clearly at a large cost to both nestlings and their parents. Similarly, nestlings sometimes use begging calls to trick adult birds into providing food even when it is against the adults' self-interests. A prime example of this is when the nestling is not the parents' own offspring, but rather a nestling of a brood parasitic species such as a cowbird or cuckoo (fig. 14.2). In this case the parent is feeding a nestling that is not related to it at all, and for that matter is not even the same species. This would be considered a case of **deception** or **manipulation**. Indeed, brood parasites have evolved an array of mimetic begging signals, including vocalizations, to obtain food deceptively from their parental hosts (Feeney et al. 2014). So why would a nestling produce begging calls if those calls attract predators, and why would a parental bird respond to begging calls if it risks wasting time and energy feeding a non-offspring? The answer is that selection will favor giving begging calls, and also responding to those calls, so long as the benefits of doing so outweigh the costs *on average*; most of the time a begging nestling will not attract a predator, and most of the time the parental bird will be feeding its own offspring.

When it comes to vocal communication in birds, the signals given are traditionally divided into two classes. The distinction between these two classes is based partly on acoustic structure and partly on social context (Marler 2004). Bird **songs** are vocal signals that tend to be relatively long and complex (fig. 14.3A) and are often associated with breeding or territorial defense (see below). In contrast, bird **calls** tend

Figure 14.2. An adult Eurasian Reed Warbler (*Acrocephalus scirpaceus*) feeds a nestling Common Cuckoo (*Cuculus canorus*). The cuckoo nestling, which hatched from an egg laid parasitically in the warbler's nest, uses vocal and other signals to induce feeding by the adult. Note that visual cues are available that clearly show the cuckoo is a different species, yet the warbler responds to vocal signals and feeds the parasitic nestling. *Photo by Per Harald Olsen, image available from https://commons.wikimedia.org/wiki/File:Reed_warbler_cuckoo.jpg.*

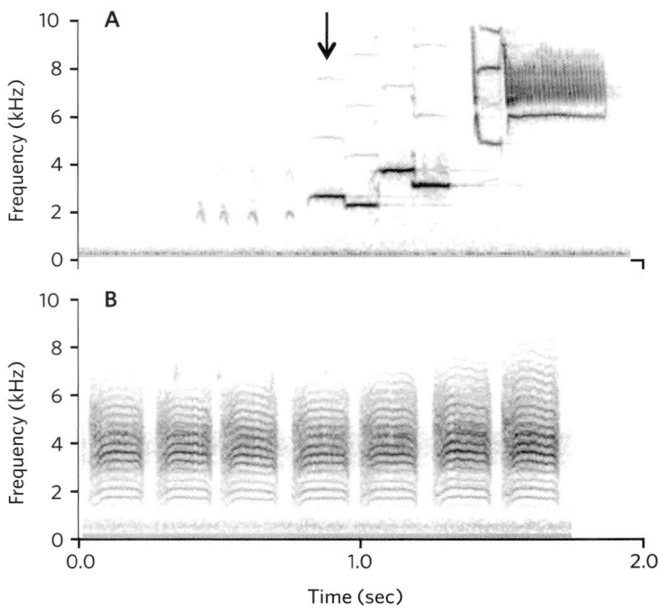

Figure 14.3. A spectrogram is a plot, read left to right, of an acoustic signal's frequency (or pitch) versus time. The darkness of the markings indicates the signal's amplitude at a given frequency and point in time. Spectrograms thus illustrate the distribution of acoustic energy (or loudness) across different frequencies as an acoustic signal progresses. A sound's fundamental frequency, typically its lowest frequency, carries most of the sound's acoustic energy, and very often can be attributed to rates of modulation of the sound source. Individual song notes also often carry some energy in higher-frequency multiples of the fundamental frequency, known as harmonics, which show as lighter bands on the spectrogram. The example spectrograms here illustrate these points with two common types of avian acoustic signals. *A*, The song of a Wood Thrush (*Hylocichla mustelina*) recorded by Arthur Allen in 1951 (*Macaulay Library, Cornell Lab of Ornithology #11316*), shows that songs are often long and acoustically complex. One of the early introductory notes, marked with an arrow, is a tonal whistle of about 2.4 kHz, lasts about 0.1 second, and is followed by a slightly lower-pitched whistle and then two more higher-pitched whistles of about the same duration. All of these introductory notes show low-amplitude harmonics. The song ends with a number of frequency-modulated notes given in rapid succession. *B*, Seven repetitions of the "dee" call of the Black-capped Chickadee (*Poecile atricapillus*) (*Macaulay Library, Cornell Lab of Ornithology #202239, recorded by Jay McGowan*). In contrast to the Wood Thrush song, this call is a repetition of a single note, though that note has a relatively broad bandwidth (large range of frequencies from about 1 to 6 kHz).

to be relatively short and simple (fig. 14.3B) and are associated with a variety of contexts, including begging for food, drawing attention to predators, and maintaining social-group cohesion. In reality, the distinction between songs and calls is often quite fuzzy, as the function of a vocalization may not be known until it has been studied carefully. Additionally, some mating/defense songs can be relatively short in duration compared with alarms and other calls. Another partial distinction between songs and calls is that songs tend to be learned through imitation in some large clades of birds (e.g., oscine passerines, parrots, and hummingbirds), whereas calls are often thought to be innate and not learned. However, this is another fuzzy distinction, as aspects of song are almost certainly innate in some bird groups (including many suboscine passerines), and recent studies suggest that learning can influence aspects of calls in at least some species (see chapter 13).

SIGNAL PRODUCTION

The mechanisms that birds use to produce acoustic signals are varied and complex, but in recent years considerable progress has been made in our understanding of these mechanisms. Much of this work has focused on production of **vocal signals**, which are signals produced with a specialized, bird-specific apparatus that involves the lungs, syrinx, and trachea. We focus most of this section on production of vocal signals. Recently, though, there has been increasing appreciation for the fact that some birds produce acoustic signals using other parts of their bodies. We begin our discussion with these nonvocal acoustic signals.

Sonations

Many animals, particularly in some groups of fish and insects (Bradbury and Vehrencamp 2011), produce acoustic signals by rubbing, scraping, or clicking various body parts together (think of the stridulating songs that male crickets produce by rubbing their wings together). Collectively, such signals are referred to as **mechanical sounds** or **sonations**. The production of these signals has received relatively little attention in birds, but it is clear that many birds do produce sonations, and in fact, the behavior may be much more widespread than previously appreciated.

One excellent example of sonation in birds comes from the Club-winged Manakin (*Machaeropterus deliciosus*). In this species, males produce loud, multiparted calls to attract mates: the call begins with one or more "ticks," followed by a "ting" linked to rapid stridulatory movements of highly modified secondary feathers in the wings. In his book on sexual selection, Charles Darwin (1871) included a figure illustrating this species' unusual secondary flight feathers, but was unable to offer an explanation for how these feathers might contribute to communication (fig. 14.4). Bostwick and Prum (2005) filled this

gap in knowledge over a century later, using high-speed videography to document birds' wing and feather movements during sonation. They found that when producing tings, Club-winged Manakins raise their wings behind their back and knock them together rapidly, about 106 times per second (faster than a hummingbird beats its wings during flight). With each wingbeat, a stiff "pick" on the tip of the modified fifth secondary rubs over a series of seven low ridges on the modified sixth secondary feather, producing a rapid series of seven clicks on the downstroke followed by another seven clicks on the rebound. The mechanism is analogous to that of a zydeco washboard: rubbing causes the wing feathers to produce a resonating tone just below 1.5 kHz, which is roughly equivalent to the frequency of rubs (106 Hz wing knocks × 14 clicks per knock; Bostwick and Prum 2005). Morphological adaptations that allow highly specialized sonations in this species extend beyond feather structure. Incredibly, and in contrast to the general rule of weight reduction in birds, the ulna bone (to which the secondary feathers are attached) in males of this species is solid (rather than hollow, like most bird limb bones), highly mineralized, and triple the volume of the same bones in birds of similar size (Bostwick et al. 2012). It is likely that this modification has evolved to enhance the sound produced by the modified secondaries, for example by decreasing the amount of vibrational energy lost to the bone itself (Bostwick et al. 2012).

The sonation signals and supporting morphology of the Club-winged Manakin are highly developed and extreme, but simpler versions may be common in some bird clades. For example, wing snaps and other percussive behaviors appear to be common components of male displays in many

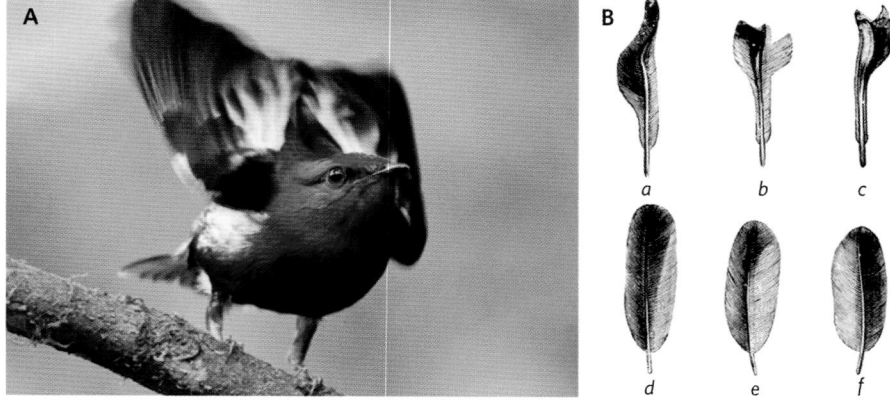

Figure 14.4. A, During courtship, the Club-winged Manakin raises its wings behind its back and knocks them together rapidly—over 100 knocks/second—to produce a tonal signal of about 1.5 kHz. The tone is caused by structures on the secondary wing feathers rubbing against each other (stridulation). *Photo by Dušan Brinkhuizen/sapayoa.com. B*, An illustration, from Darwin (1871), showing the highly modified secondaries of the male Club-winged Manakin (*top row*) compared with those same feathers of the female (*bottom row*).

manakin species (Bostwick and Prum 2003). Similarly, in hummingbirds, males produce acoustic courtship signals in flight, as air moves across modified tail and wing feathers, causing them to vibrate ("aeroelastic flutter") and produce a high-pitched whine (Clark 2008, Clark and Feo 2008, Clark et al. 2011; see fig. 14.5). Sonation courtship signals also are produced by the American Woodcock (*Scolopax minor*) and snipe (species in the genus *Gallinago*), using modified wing feathers (Sheldon 1967). Likewise, many pigeons and doves have highly modified primaries and secondaries that produce sounds during some types of flight (Niese and Tobalske 2016). How common sonation signals are in other groups of birds remains to be seen. It is notable, though, that the ability to flutter appears to be an inherent feature of flight feathers, and modifications of these feathers to produce sounds have accordingly evolved numerous different times in birds (Clark and Prum 2015). This suggests that many sonation signals likely originate as adventitious flutter sounds that are subsequently modified into true signals via evolutionary modifications of the feathers that enhance the sounds produced.

Vocal Signals

Birds in all avian orders use vocal signals of some sort. Many of these are truly impressive to human ears, even those avian vocalizations that are relatively simple. As noted by our colleague Erick Greene, calling Sandhill Cranes are emphatic reminders of birds' dinosaur roots. What we find remarkable is that even fairly simple sounds, such as the honking of geese, can also contain tremendous richness when one looks closely. In one example, a wide range of species including geese, swans, cranes, and curassow have evolved calls that seem to exaggerate these birds' apparent body size—that is, which make the birds seem larger than they really are (Fitch and Kelley 2000). This has been accomplished through the evolution of greatly lengthened trachea, sometimes up to 20 times the standard length for a bird of a given size, subsequently coiled within the sternum or thorax (Fitch 1999). A lengthened trachea apparently leads to a specific change in the acoustic structure of calls that normally indicates a larger-bodied caller (Fitch 1999). Another compelling example of the complexity of seemingly simple vocalizations concerns the calls of penguins, which are given in both mating and familial contexts. While these calls tend to sound both simple and uniform across birds, they actually contain precise variation in their acoustic properties that helps penguins recognize each other on an individual basis. These recognition abilities are a critical glue in penguin social behavior (e.g., Jouventin et al. 1999).

Yet even more impressive are the vocalizations of songbirds, which are especially renowned for their intricacy,

complexity, and musical quality. Some songs comprise short-duration notes produced in rapid-fire sequence and sweeping across octaves, features that would challenge the speed and range of any human musical instrument. Some birds sing large repertoires of notes and songs, with an interspersion of vocal themes and variations that can evoke a session of improvisational jazz. Many songs are highly pure-tonal or whistle-like in quality, with acoustic energy concentrated at a narrow range of frequencies (fig. 14.6). Birds' vocal versatility can be attributed not just to an exquisite neural control system (chapter 13) but also to highly specialized vocal anatomy and physiology.

The principal sound-producing organ in birds is the **syrinx**, situated at the juncture of the bronchi and trachea (Greenewalt 1968; see chapter 13). The syrinx is unique to birds, although mammals and other vertebrates vocalize using an analogous though somewhat less complex structure, the larynx. In birds, membranes and other tissues of the syrinx are stimulated to vibrate pneumatically, as air is pushed outward from the lungs. A notable feature of the syrinx is that its left and right sides can be activated independently via the left and right bronchi/lungs and a set of partially independent bronchial valves. This setup enables birds to produce two distinct sounds in synchrony (an "internal duet"), as well as sounds with unusual degrees of acoustical complexity (Nowicki and Marler 1988, Suthers et al. 2012).

The ways that birds use the syrinx to vocalize has been worked out in recent years through some nicely designed work in the laboratory (box on page 418). Timing patterns in song production—that is, sequences of notes and silent gaps in between—result directly from patterns of breathing. Studies on the physiology of singing birds have revealed that birds typically (but not always) vocalize only while breathing out, and use silent intervals between notes to breathe in (Suthers et al. 1994, Goller and Larsen 1997). Such "minibreaths," with their cycles of breathing in and out, help singing birds maintain a sufficient volume of air in their lungs and air sacs for respiration, which in turn allows for impressively long-duration songs, such as the extended rolling trills typical of canaries (Hartley and Suthers 1989). A major exception to this pattern involves what is referred to as "pulsatile" respiration, in which birds only breathe out, and the respiratory system's tidal volume thus depletes as a song progresses. Pulsatile respiration has been associated with the production of very fast trills or buzzes, both of which have very brief gaps between notes; these gaps are so brief that birds apparently do not have sufficient time to breathe in during note intervals. The volume of air a bird can hold in its lungs and air sacs may thus set limits on the overall duration of fast songs or buzzes.

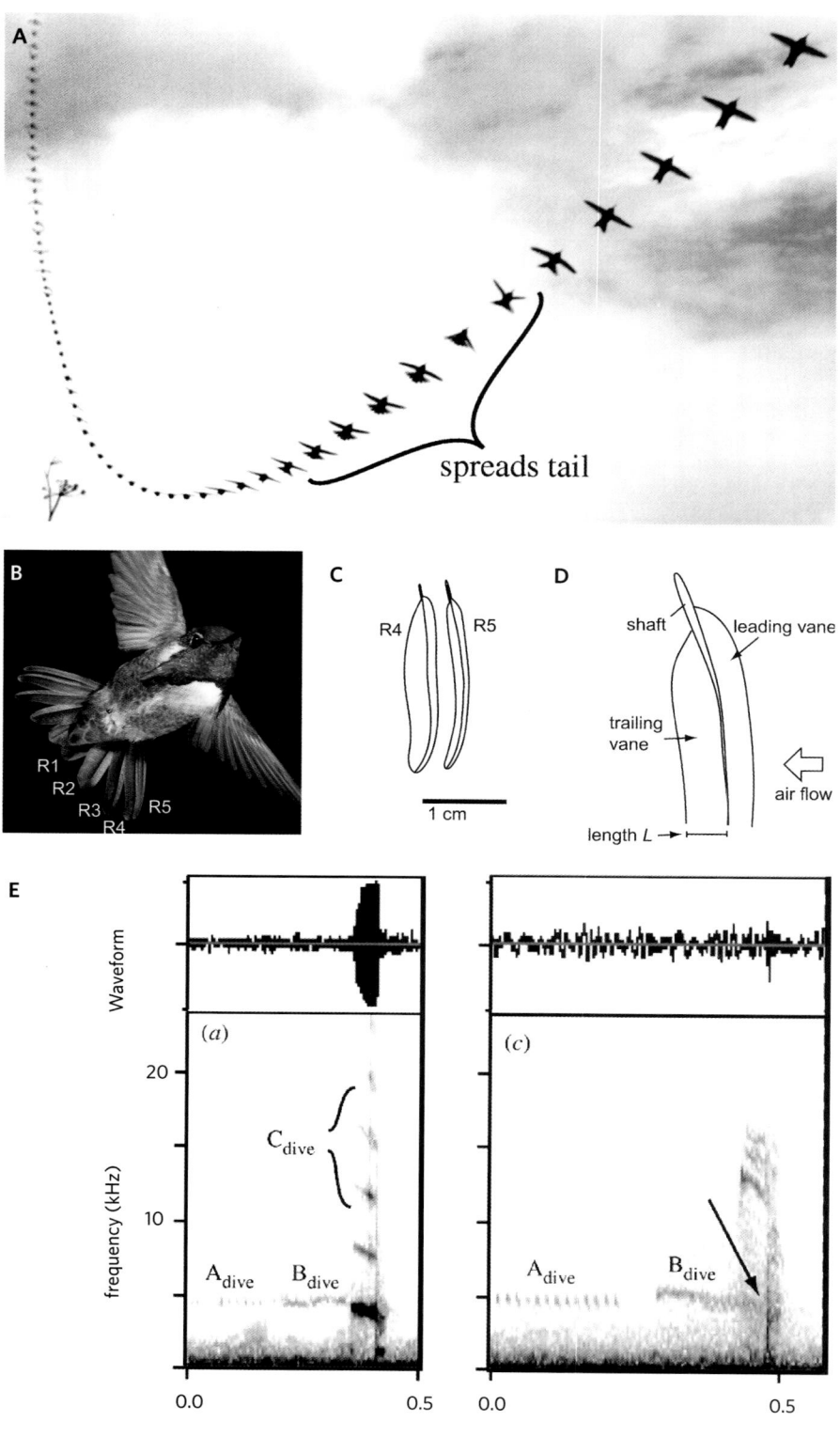

Figure 14.5. Male bee hummingbirds "sing" with their tails. A, Composite photo showing courtship dive of male Anna's Hummingbird (*Calypte anna*). At the bottom of the dive, which is near where the female is perched, the male spreads his tail briefly and simultaneously produces a loud sound with a fundamental frequency of about 4 kHz. The composite image was created using high-speed video, with consecutive images 0.01s apart; B, Male Costa's Hummingbird (*Calypte costae*) with tail fully spread. The individual tail feathers (rectrices) are numbered (R1–R5). During a courtship dive, the male partially spreads its tail to expose the individual tail feathers; C, The R4 and R5 tail feathers from a male Costa's Hummingbird; D, During the dive or fast flight with the tail spread, air flows past the R5 feather, causing the trailing vane to vibrate rapidly and produce the dive sound; E, Waveforms (sound amplitude versus time, *top*) and spectrograms (sound frequency versus time, *bottom*) showing sound produced by a male Anna's Hummingbird during the courtship dive. The sound produced at bottom of the dive, when tail is spread, is labelled "C_{dive}." When the male is unmanipulated (*left panel*), the C_{dive} note is present, but when the R5 feather is removed (*right panel*), the C_{dive} note disappears and the diving male produces only a "whoosh" sound (arrow). *Figures modified with permission from (a, e) Clark and Feo (2008) and (b–d) Feo and Clark (2010).*

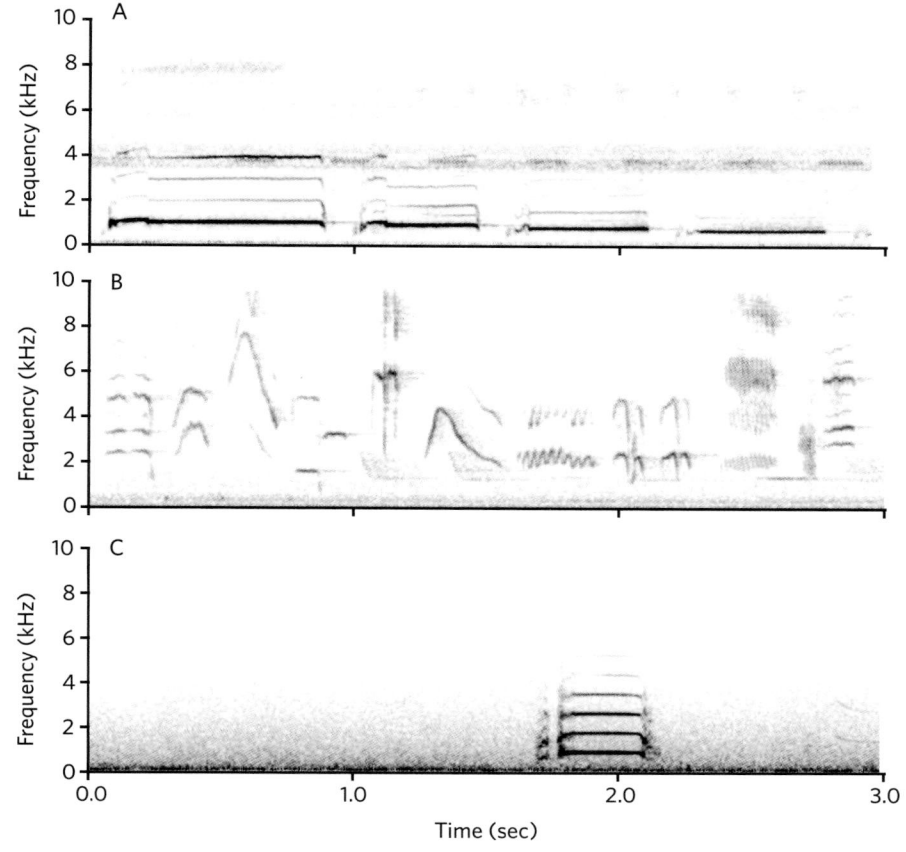

Figure 14.6. A, The song of the Common Potoo (Nyctibius griseus, *Macaulay Library #59964, recorded by Paul Schwartz*), a non-oscine passerine, is a relatively simple series of long whistled notes descending from about 1kHz; *B,* In contrast, the song on New Zealand's endemic Tui (Prosthemadera novaeseelandiae, *Macaulay Library #136001, recorded by Matthew Medler*) is a complex and rapid series of short notes that vary dramatically in pitch; *C,* Some "songs" are extremely simple, like the single note vocalization of the Whooping Crane (Grus americana, *Macaulay Library #2741, recorded by Arthur Allen*). This example illustrates well the often fuzzy distinction between songs and calls. Notes in bird vocalizations can vary in their harmonic structure, from being mostly pure-tonal (Common Potoos and Tuis) to showing substantial harmonic overtones (Whooping Crane).

Supporting this hypothesis is a study of mockingbirds tutored with songs of canary trills that were both rapid and of extended duration (Zollinger and Suthers 2004). These trained mockingbirds were able to match the rapid speed of the canary trills, but only for very short durations. Presumably the mockingbirds couldn't sing the canary song models using minibreaths and instead relied on pulsatile respiration, because their comparatively large body and syrinx size constrained their ability to maintain sufficient tidal volume (in contrast to the smaller-bodied canary). As a result, the mockingbirds ended up introducing long silent gaps between trill segments, presumably in order to catch their breath.

The properties of song also can be influenced by the vocal tract, which includes the trachea, glottis, tongue, and beak (fig. 14.7). The avian vocal tract, along with the surrounding oropharyngeal cavity, can be compared roughly to the tube of a brass or woodwind instrument; both serve as acoustic resonance chambers, and as such enable the production of sounds with a pure-tonal or whistle-like quality (Nowicki and Marler 1988). Thus, the overall configuration of the birds' vocal apparatus—with the sound source located deep in the body and with sounds then passing through a long resonance tube—helps explain why birds are among nature's most

musical animals. One fascinating aspect of the vocal tract is that a singing bird often rapidly adjusts its volume, for example by closing and opening the beak or by expanding or contracting the oropharyngeal cavity (Westneat et al. 1993, Riede et al. 2006). These rapid adjustments appear to have a specific mechanical function: when singing high-frequency notes, birds are better able to dampen harmonic overtones with a relatively low-volume resonance chamber; and, similarly, harmonic overtones over low-frequency notes are best dampened with a high-volume resonance chamber (Nowicki and Marler 1988). Indeed, as songbirds sweep up and down the frequency register, one can observe birds making corresponding adjustments to their vocal tracts, most obviously by opening and closing their beaks (fig. 14.8).

The intricate, multifaceted nature of the vocal apparatus, and its complex neural control system (chapter 13), suggests that songs may be shaped in their expression, development, and even in their evolution by **performance constraints**, that is, by physical limits on birds' ability to sing (Lambrechts 1996, Podos and Nowicki 2004). This idea is supported by various lines of experimental and descriptive work, including the study of mockingbirds mentioned above. A particular focus for studies of performance constraints has been songs

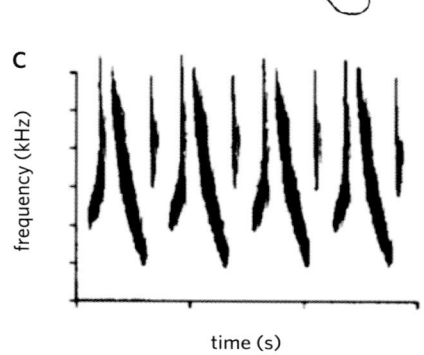

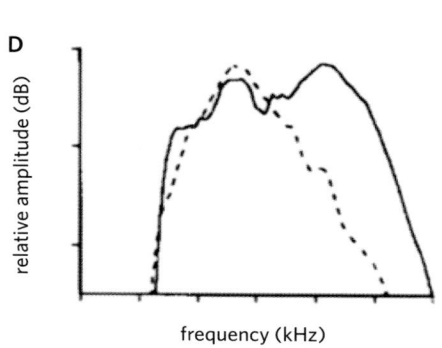

Figure 14.7. A, The sound source (syrinx, shaded gray) is activated to generate sound by air pressure from the lungs. Sound travels through the vocal tract, which includes the trachea and beak. The vocal tract serves as a resonance chamber that filters harmonic overtones and enables pure-tonal, whistle-like song; B, A male Swamp Sparrow, a species in which the resonance function of the vocal tract has been tested experimentally; C, A song segment from a Swamp Sparrow singing normally; D, The amplitude spectrum of the song segment shown in panel C produced normally (dashed line) and then with the beak clamped to a fixed gape (solid line). Immobilization of the beak diminished birds' ability to modulate vocal tract resonances, resulting in a shift of acoustic energy to higher-frequency harmonic overtones. *Figures reproduced with permission from (A) Fitch and Hauser (2003), (B) photo credit J. Podos, (C and D) Hoese et al. (2000).*

with trilled structure, in which individual notes or groups of notes ("syllables") are repeated in quick succession. Maximum rates of note repetition ("trill rates") can be set by limits on the speed of breathing or vocal tract adjustments. Moreover, when birds perform songs with especially fast trill rates, the acoustic structure of individual syllables tends to be relatively simple, for example by having narrower bandwidth (i.e., the range of frequencies spanned, from lowest to highest, within a song segment of interest). This is presumably a concession that allows those syllables to be repeated rapidly (fig. 14.9).

As described above, birds use changes in beak gape as an integral part of song production. But beaks are of course also used for feeding. Thus, selection for precise feeding adaptations could affect how the beak can perform in song production. As a case in point, consider the Darwin's finches (fig. 14.10), which include a radiation of ground finches specialized to feed on seeds of varying hardness. Larger beaks (and associated jaw morphology), specialized for force application, are also less versatile in speed of beak gape changes, because of a trade-off between force and velocity (Herrel et al. 2009). Because of this trade-off, birds with large beaks should be less able to produce high-performance trills. Indeed beak size correlates tightly with vocal performance in the Darwin's finch clade, both between and within species,

as predicted (Podos 2001, Huber and Podos 2006). Similar relationships between beaks and trill structure have been identified in other bird clades (Seddon 2005, Derryberry et al. 2012), although the relationship is not universal (Slabbekoorn and Smith 2000). Beaks are less likely to constrain trills in species with songs that are low-performance and that thus do not push the boundaries of what birds can produce with a given vocal morphology.

TRANSMISSION THROUGH THE ENVIRONMENT

In addition to being shaped by the vocal apparatus, the sounds produced by a bird are also affected by the environment around it. Consider a bird singing near a forest stream. Receivers may not be able to perceive that song well because it is obscured by the sounds of the stream—particularly if the song is relatively quiet. This is the most obvious environmental effect on signals: **masking** of the signal by environmental noise. In this way, environmental noise effectively reduces the **active space** of the signal (that is, the distance over which the signal can be distinguished from background noise). In noisy environments, the active space of a signal should be smaller than that of the same signal in a quiet en-

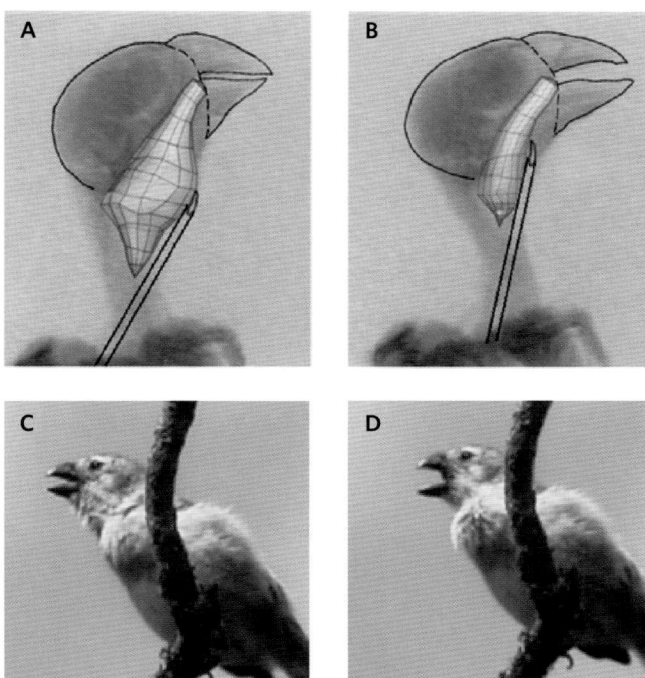

Figure 14.8. Illustrations of vocal tract modulations in singing birds. Panels *A* and *B* show X-ray images of a cardinal singing a low frequency note (~1.5 kHz, *panel a*) and a high frequency note (~5 kHz, *panel b*). Superimposed on the X-ray images are three-dimensional reconstructions of the oropharyngeal-esophageal cavity (OEC). Across various species studied, note frequencies tend to correlate positively with the volume of the vocal tract. Larger vocal tract volumes, enabled by expansion of the OEC and reductions in beak gape (*A*), provide a more effective resonance filter for lower-frequency sounds. Similarly, resonance filtering of higher-frequency notes is more effective at lower vocal tract volumes, achieved by smaller OEC volumes and wider beak gapes (*B*). Panels *C* and *D* illustrate postural changes in a singing Darwin's tree finch, for a low- and a high-frequency note, respectively. Vocal tract modulations can sometimes be inferred from standard video still images. *Illustrations reprinted with permission from (A, B) Riede et al. 2006, copyright (2006) National Academy of Sciences, U.S.A.; and (C, D) Podos et al. (2004).*

lustrate these points. Working at a site in the Brazilian Amazon rainforest, Luther (2009) recorded and analyzed the songs of 82 bird species and showed that the songs occupy an unusually diverse range of acoustic space. Moreover, greater degrees of divergence were identified for species groups occupying the same forest substrata, or singing at the same time of day. Both lines of evidence support the hypothesis that birds have evolved to use songs in times and places that help them avoid interference (Luther 2009). Today, much of the background noise that birds must deal with is anthropogenic, which may have direct negative impacts on birds as well as indirect effects on the ways in which they can communicate with each other (box on page 420).

There are other subtle but important ways by which the environment can alter sounds that birds produce. For example, objects in the environment reflect sounds off of their surfaces, scattering those sounds in various directions. Because of the physics of sound (see Wiley 1991, Bradbury and Vehrencamp 2011), small objects like leaves and twigs tend to scatter short wavelengths (i.e., high-pitched sounds) much more than they scatter long wavelengths (i.e., low-pitched sounds). As a result, the low-frequency elements of song typically

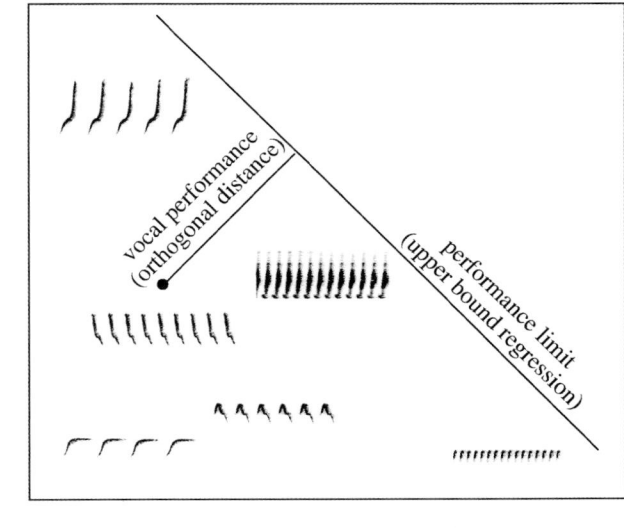

Figure 14.9. Trade-off between maximal values of trill rate and frequency bandwidth, illustrated for a sample of trilled songs of banded wrens (*reprinted with permission from Illes et al. 2006*). Similar trade-offs have been shown for a wide range of trilling species and are thought to arise from performance constraints on song production. Performance limits for trills can be estimated using an upper-bound regression, and the performance level of individual songs can be estimated by calculating the orthogonal distance from the song's data point to the upper bound regression.

vironment (Klump 1996, Lohr et al. 2003). Masking noise can include both abiotic sounds (e.g., water or wind noise) and sounds produced by other noisy animals (Brumm and Slabbekoorn 2005, Brumm and Naguib 2009). In tropical forests, for instance, cicadas are a common source of noise that can interfere with bird vocal communication (Hart et al. 2015). In response, both via behavioral accommodation and evolutionary adaptation, birds in noisy environments can adjust their vocalizations to be louder, to avoid temporal overlap with masking sources, or to use areas of the available "acoustic space" not filled by competing sounds (Nelson and Marler 1990). Recent studies by David Luther (2008, 2009) nicely il-

RODERICK SUTHERS

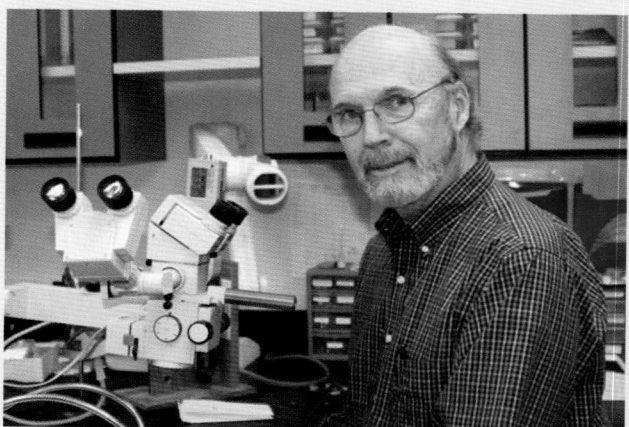

Roderick Suthers. *Photo by Indiana University School of Medicine.*

Rod Suthers began his career in Don Griffin's bat echolocation lab, where he already demonstrated a talent for designing innovative and novel experiments. When Suthers started his own lab, he continued studying bats, developing the novel use of tiny heated thermistor beads to measure tracheal airflow during vocalization.

Suthers then turned his attentions toward rare examples of echolocation in birds. It was during his investigation into the production of sonar clicks in Oilbirds (*Steatornis caripensis*), that Suthers first used a dual-thermistor technique that would lead to a major change in our understanding of song-

bird vocal dynamics. Oilbirds have a two-sided syrinx, with a vibratory structure in each bronchus. To determine the role that the two sides played in click production, Suthers carefully implanted one tiny bead into each bronchus, deep in the thorax of the bird, and recorded airflow through each side.

The work on Oilbirds led Suthers to explore the two-sided syrinx of songbirds. The bipartite anatomy of this apparatus, together with the complex songs produced, had led early researchers to hypothesize that songbirds must have "two voices." Suthers used his dual-thermistor technique to test this hypothesis. The result was his landmark discovery—songbirds can produce two independent notes simultaneously. Suthers went on to investigate vocal production in more detail, measuring air sac pressure during phonation and using EMG electrodes to tease out the role of each syringeal and respiratory muscle in song production.

Suthers and his group have since provided insight into the diversity of vocal behavior among birds, the development of motor patterns, and the biomechanics of voice. Much of what we know about vocalizations in birds, from functional morphology to upper vocal tract filtering to the articulation of the tongue, is thanks to the pioneering work of Suthers and his lab.

By: *Sue Anne Zollinger, Communication and Social Behaviour Group, Max Planck Institute for Ornithology, Seewiesen, Germany (zollinger@orn.mpg.de)*

---

travel through the environment unimpeded, able to reach receivers at long distances, whereas high-frequency elements tend to be scattered and travel shorter distances. Thus, cluttered environments with lots of small objects, like forests, differentially scatter sounds according to wavelength, leading to **frequency distortion** of the signal. More open and uncluttered habitats, like deserts and meadows, have fewer objects in them, and so sounds can travel farther and with less frequency distortion. This is not to say that open habitats are always ideal for vocal communication. Wind speed is often higher in open habitats, and high winds tend to degrade songs or song components through **amplitude distortion** (e.g., the disproportionate drop-out of certain signal components versus others). Finally, sounds echoing off of objects in the environment (i.e., **reverberations**) can lead to "slurring" of the sound, because the sound of the signal arrives at the receiver's ear via multiple pathways—directly from the receiver and indirectly via echoes off other objects. This can

make it difficult for receivers to differentiate the end of one note and the start of the next.

The combination of these and other habitat effects on sound has led to the "acoustic adaptation hypothesis" (Morton 1975), which proposes that song structure should evolve differently in different habitats, in all cases to maximize transmission efficacy (reviewed by Brumm and Naguib 2009). This hypothesis predicts that birds living in cluttered habitats, like forests, should evolve songs that are low-pitched, relatively slow, and structurally simple (i.e., with few quick changes in pitch) relative to birds that live in more open habitats, like deserts and meadows. Indeed, some studies have found this to be the case (e.g., Seddon 2005), even when comparing populations of the same species in different habitats (fig. 14.11). It is worth noting, however, that a number of studies have also failed to find this pattern, and overall the role of habitat structure in shaping birds' acoustic signals may be only minor (Boncoraglio and Saino 2007).

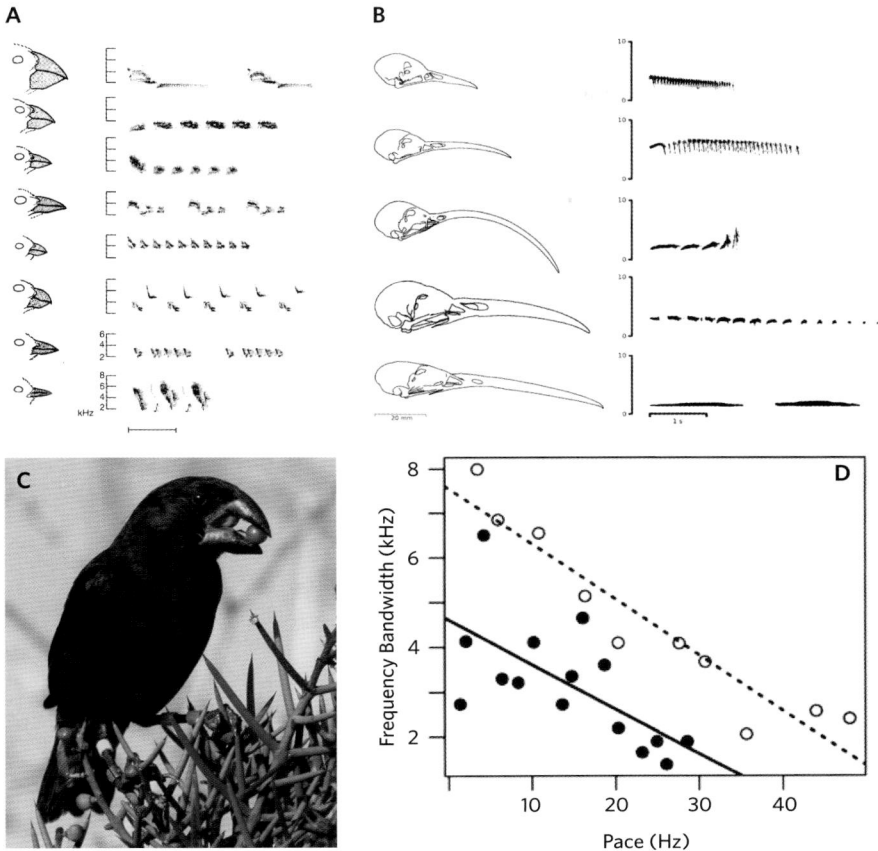

Figure 14.10. Bill morphology correlates with song characteristics in *A*, Darwin's finches (family Emberizidae) and *B*, Neotropical woodcreepers (subfamily Dendrocolaptinae). Head and skull tracings show the bill profiles, and the spectrogram (sound frequency versus time) for each species shows its typical song. In both of these passerine groups, species with larger beaks produced songs that are slower paced and have narrower frequency bandwidths. This suggests that selective factors that affect beak morphology, such as foraging ecology, have an indirect effect on the type of song produced; *C*, A male Medium Ground Finch (*Geospiza fortis*) uses its bill for manipulating food as well as for singing; *D*, The frequency bandwidth of a song (i.e., the difference in frequency between the highest- and lowest-pitched elements of the song) are negatively correlated with the "pace" of a song (i.e., number of syllable repetitions per unit time) for woodcreeper songs (closed circles and solid line) and emberizid songs (open circles and dashed line). To produce this graph, songs were grouped into 2-Hz increments (0–2 Hz, 2–4 Hz . . . 28–30 Hz) bins, and each point shows the frequency bandwidth of the song with the maximum bandwidth in that bin. The lines show a linear regression using these maximum values for each group. The negative relationship indicates that it is difficult for a bird to produce a trilled song that is rapid in pace and also has a broad frequency bandwidth. This suggests that higher-quality individuals within a species, for example those in better condition, should be able to produce songs closer to the upper bound of this trade-off (i.e., the regression line). Species shown in panel *A* are (top to bottom) *Geospiza magnirostris*, *G. fortis*, *G. fuliginosa*, *G. scandens*, *Camarhynchus parvulus*, *C. psittacula*, *Cactospiza pallida*, *Certhidea olivacea*, and the scale bar shows 0.5 seconds. For panel *B*, the species shown are (top to bottom) *Certhiasomus stictolaemus*, *Lepidocolaptes albolineatus*, *Campylorhamphus trochilirostris*, *Xiphocolaptes promeropirhynchus*, and *Nasica longirostris*, and the scale bar shows 1.0 second. *Figures reproduced with permission from (A) Podos (2001) and (B, D) Derryberry et al. (2012). Photo in (C) by Andrew Hendry.*

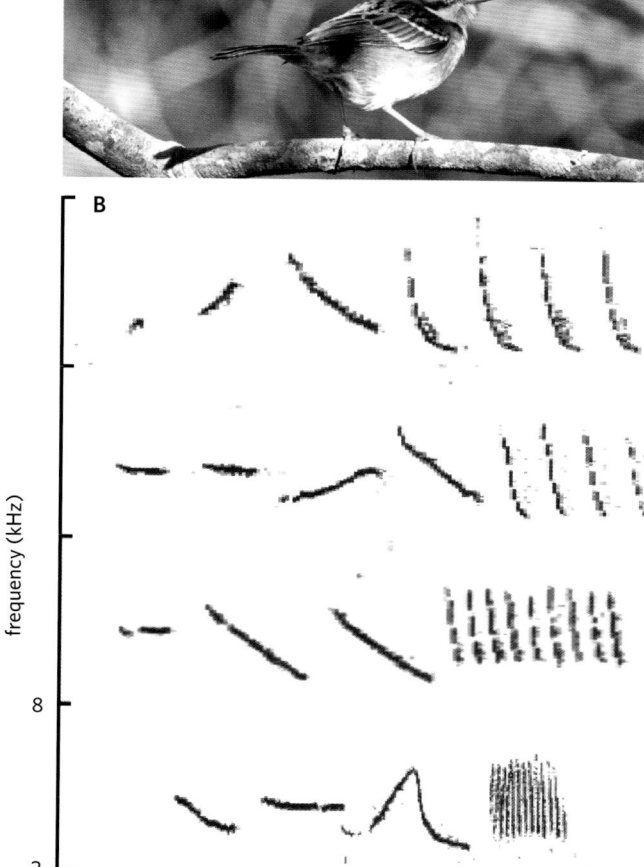

Figure 14.11. Effects of environmental acoustics on the structure of bird songs, illustrated for *A*, male Rufous-collared Sparrows (*Zonotrichia capensis*), panel *A photo credit J. Podos*. Panel *B, Four representative songs reprinted with permission from Handford 1988. B*, The top two were recorded from birds in woodland habitats, and the lower two from birds in open habitats. Songs in woodlands tend to emphasize slow note repetitions and low frequencies, adaptations to minimize the negative effects in cluttered habitats of reverberation and high-frequency attenuation. By contrast, songs in open habitats tend to feature buzzes (very rapidly repeated notes seen at the end of the two examples here), which can transmit reliably in windy conditions typical of open habitats.

THE INFORMATION CONTENT OF SIGNALS

As discussed above, we expect signals to evolve only if, on average, they enhance the fitness of both the signaler and the receiver. The benefits of communication to both signalers and receivers hinge directly on the signals' **information content**—that is, what is revealed about signalers in their signals, and what listeners can learn about signalers by pay-

ing attention to those signals (Seyfarth et al. 2010). Here we discuss calls and songs separately, because their functions are largely thought to be distinct. We reserve our overview of a related topic—signal reliability and honesty—for the final section of this chapter.

Calls

Most bird species produce a wide variety of calls. In some cases the information contained in a call is quite apparent from the context in which it is produced. Consider again the example of a young passerine bird in the nest. Most of the time the nestling is silent, but when a parent bird comes to the nest, the young bird quickly lifts its head, opens its mouth, and produces **begging calls**. Begging calls usually

BIRDS AND ANTHROPOGENIC NOISE

Noise in many modern environments can be attributed to human activities. Anthropogenic noise sources tend to concentrate at the low end of the frequency spectrum (think of the rumble of cars passing on a nearby highway), and thus potentially mask lower-pitched bird vocalizations. We might expect birds living near areas with many humans, such as cities, to be affected by this low-pitched anthropogenic noise much more than birds living in areas with few humans, such as more rural and undisturbed habitats. Several recent studies have shown that birds living in urbanized areas tend to sing higher-pitched songs than birds living in more rural areas, an effect seen even within species (Slabbekoorn and Peet 2003, Brumm 2004, Dowling et al. 2012). Bird species with low-frequency vocal signals might even be reluctant to settle and breed in areas with high traffic noise (Goodwin and Shriver 2011). It is unclear, though, whether these kinds of patterns represent direct behavioral responses by birds living in noisy environments, evolutionary responses to selection in noisy environments, or both. Regardless, recent research shows that birds tend to avoid areas with traffic noise (McClure et al. 2013), and those that do not avoid them suffer from reduced body condition (Ware et al. 2015). Urban areas also tend to be more brightly lit at night than rural areas, and studies have also found that artificial lighting can disrupt daily patterns of singing: birds in urbanized areas tend to sing more at night, and sometimes sing through the night (Miller 2006, da Silva et al. 2015), which in turn can affect various aspects of breeding (Kempenaers et al. 2010).

elicit a quick response from the parent, who stuffs food into the nestling's open gape. Begging calls are also given frequently by fledglings who have recently left the nest and also sometimes by incubating females soliciting food from their mates (Tobias and Seddon 2002, Ellis et al. 2009). Focusing on nestlings, if the function of begging calls is to solicit food, has selection shaped them to convey information that is maximally precise? Toward this end, a begging call should convey not just the hungry status of the nestling but also more subtle information about the nestling's hunger level and/or condition. This does indeed appear to be the case in the Barn Swallow (*Hirundo rustica*) and other species where it has been studied: hungrier nestlings tend to raise their heads higher, beg more loudly, and beg more continuously than do nestlings that recently have been fed (e.g., Lotem 1998).

Nestling begging calls may also be shaped by other factors, including selection to minimize detection by nest predators. Indeed, nestlings of species in vulnerable sites appear to beg less vigorously or often than do nestlings of species that are relatively safe from predators (Haskell 1994). Selection to avoid predation can shape not just how often a nestling begs, but also the call's acoustic characteristics. For example, the begging calls of ground-nesting birds, which are generally highly susceptible to ground predators, tend to be higher in pitch than those of tree-nesting birds, presumably because higher-pitched calls are harder to localize (Briskie et al. 1999). To test this presumption, Haskell (1999) used a clever experiment in which he recorded the begging calls of a tree-nesting species, the Black-throated Blue Warbler (*Setophaga caerulescens*). As with many other tree-nesting species, begging calls of the Black-throated Blue Warbler are lower-pitched than those of ground-nesting birds. Haskell (1999) then played these low-pitched begging calls through tiny speakers within artificial nests placed either in trees or on the ground (the latter were nests that mimicked those of the Ovenbird, *Seiurus aurocapilla*, a ground-nesting warbler). The lower-pitched begging calls of the tree-nesting species attracted predators at higher rates for nests on the ground compared with nests in trees. This result supports the hypothesis that nestling begging calls in ground-nesting species are indeed shaped in pitch by natural selection to increase feeding rates by parents but also to minimize the risk of attracting unwanted attention.

Another commonly used class of calls is the **alarm call**, used to alert conspecifics to potential predatory threats. The numerous types of alarm calls can be categorized by acoustic structure and/or context. Many small passerines give **seet calls** when they detect an aerial predator, such as an accipiter, flying nearby. The typical seet call given by many songbirds is a narrow-bandwidth, high-pitched (about 8–10 kHz)

vocalization with a gradual increase and then decrease in amplitude (Marler 1955). Both of these frequency and amplitude properties make seet calls very difficult to locate, particularly by large raptors who have relatively poor hearing within seet calls' frequency range (Marler 2004). Thus, small birds can use seet calls to alert others to impending danger from above, while at the same time not drastically increasing their own chances of being taken by a predator. The structure of seet calls is strikingly similar across a broad swath of songbird species (fig. 14.12), and some small mammals also emit seet-like calls in response to aerial predators (Greene and Meagher 1998), suggesting evolutionary convergence on this relatively safe but effective call structure.

A second main type of alarm in passerines is the **mobbing call**, usually given in response to detection of terrestrial predators or perched raptors. The structure of a typical mobbing call is quite different from that of a seet call. Mobbing calls typically have an abrupt start and finish, are loud, and are very broad in pitch (i.e., wide-frequency bandwidth, often with harmonics; fig. 14.12). All these characteristics make the call relatively easy to localize. As suggested by its name, mobbing calls elicit a very different sort of behavior from receivers: whereas a bird's typical response to a seet call is to freeze or immediately take cover, birds that hear a mobbing call often fly toward the source of the call, joining in with their own mobbing calls, and attacking (or "mobbing") the predator to drive it away. The difference in response to these two types of calls can be explained by relative level of threat: a flying raptor is an immediate and highly mobile threat to be avoided, whereas a perched raptor or terrestrial nest predator is a future and somewhat less agile threat.

Remarkably, it appears that mobbing calls in some species can specify not just the presence but also the relative threat level of a perched raptor or terrestrial predator. Recent work on Black-capped Chickadees (*Poecile atricapillus*), who give their eponymous "chicka-dee-dee-dee" call to warn of perched and terrestrial predators, has shown that these birds tack more "dee" notes onto the ends of their calls for smaller predators, as these pose a great threat to the chickadees; and, correspondingly, chickadees that hear calls with more "dee" notes (see fig. 14.3) are more likely to mob the predator (Templeton et al. 2005). Follow-up work indicates that the ability to encode information about threat levels is common in parids (Soard and Ritchison 2009, Courter and Ritchison 2010, Sieving et al. 2010), with threat levels encoded not just by graded signals (e.g., numbers of "dee" notes) but also through use of distinct call structures (Suzuki 2014). Moreover, the ability to encode threat levels in alarm calls has now been documented in a number of other bird groups (e.g., Naguib et al. 1999, Leavesley and Magrath 2005, Griesser 2009).

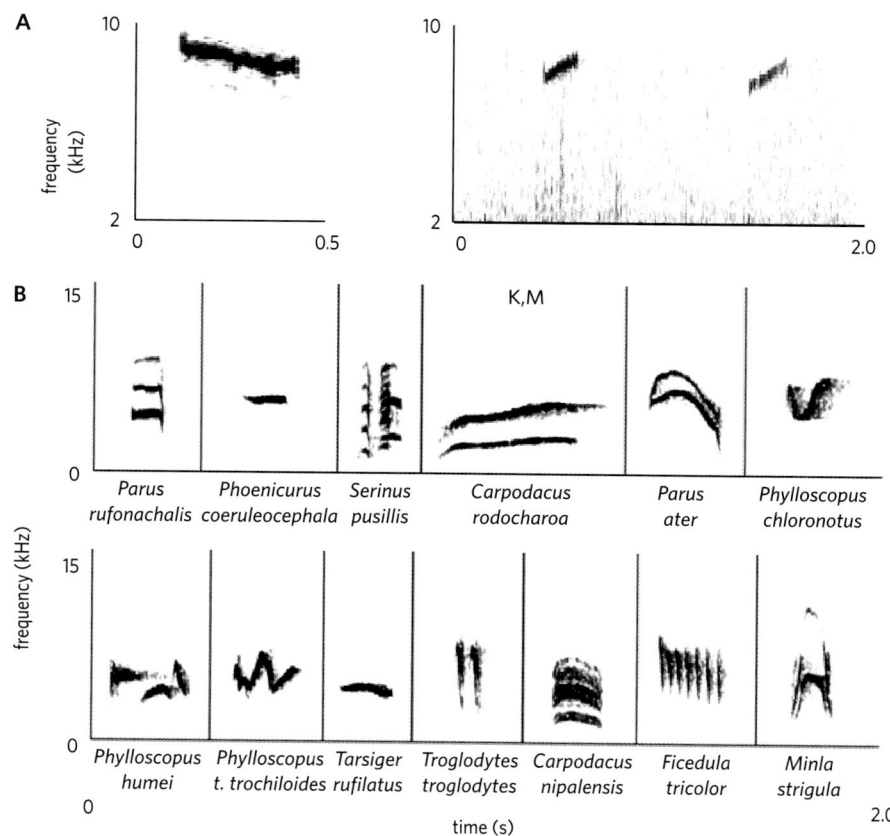

Figure 14.12. *A*, Seet calls. *Reprinted with permission from Jones and Hill 2001 (left) and Krama et al. 2008 (right).* Seet calls tend to be uniform in their structure, featuring narrow-band high-frequency tones with limited frequency or amplitude modulations. These features make seet calls difficult to detect and localize. *B*, Examples of mobbing calls. *Reprinted with permission from Wheatcroft and Price 2013.* Mobbing calls tend to be structurally more complex than seet calls, and also show high diversity in structure across species.

Overall, bird calls remain a fascinating yet largely understudied area, with a rich variety of call structures and contexts (Marler 2004). Additional call categories include **contact calls** that maintain group or mating-pair cohesion, **flight initiation calls** that synchronize group movements, **scolding calls** given in agonistic interactions, and **migratory flight calls** that are given largely at night during migration. As with begging and alarm calls, these classes of calls appear to be produced in specific contexts with clear fitness benefits for both signalers and receivers.

Songs

Songs generally appear to serve two important functions in avian communication, both associated with breeding: territory defense and mate attraction. Before delving into these functions, it is useful to note that song in many bird species, especially in the temperate zone, is produced either exclusively or primarily by males (but see below). As first noted by Darwin (1871), male-exclusive traits are often best explained as products of sexual selection, evolving to help males compete for access to females (Darwin 1871, Searcy and Andersson 1986). The reason that males typically compete for access to females, and not vice versa, is that females typically invest more time and energy into breeding than do males, for example through egg production and incubation of the young (see chapter 16). As such, females tend to be choosy about mating partners, which fosters the evolution of male competition and associated ornaments, armaments, and sexual displays (Andersson 1994).

The importance of song in territory defense and mate attraction is supported by ample observational data (Searcy and Andersson 1986, Searcy and Yasukawa 1996, Catchpole and Slater 2008). Birds of many species sing most often during their breeding seasons, and tend not to sing during overwintering or migration periods. Within breeding seasons, males sing most often during days most critical to their breeding success, such as when first establishing territories or when breeding partners are fertile (Pinxten and Eens 1998). Moreover, correlations between singing behavior and breeding success have been documented for many species, with breeding success correlating with how often a bird sings ("singing rate" or "vocal output"; Alatalo et al. 1990, Nystrom 1997) and also with specific song properties such as repertoire size and vocal performance (fig. 14.13).

We have gained significant insights into song function through experimental approaches, particularly those employ-

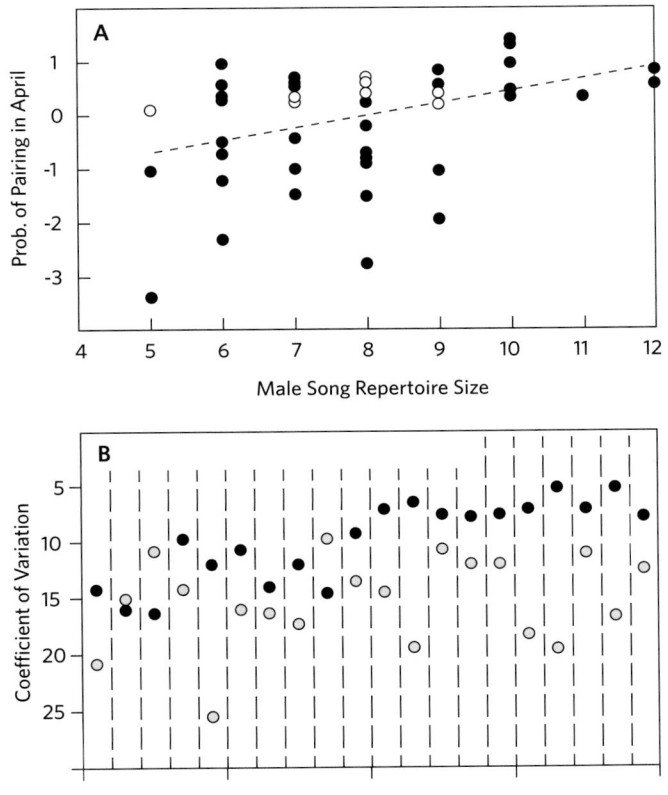

Figure 14.13. A, For first-year breeding Song Sparrows, the probability of obtaining a mate early in the season is correlated with the number of songs in the male's repertoire. The graph shows the probability of pairing in April (corrected for year-to-year variation in territory size and population sex ratio) versus song repertoire size. Pairing early in the season is strongly related to total reproductive success. *Figure modified with permission from Reid et al. 2004.* B, Male Chestnut-sided Warblers with better song performance are more likely to sire extra-pair young. Figure shows song performance (in this case, consistency of the peak frequency of song during dawn chorus) for pairs of males shown in each column. For each pair, one male (shown with black dot) sired extra-pair young in the nest of the other (shown with gray dot). The male fertilizing the extra-pair nestling had better song performance in 17 out of 20 cases, which is significantly different from chance, and other measures of song performance show similar patterns. *Figure modified with permission from Byers 2007.*

ing "song playback" methods (box on page 424). The vast majority of song playbacks in field conditions have been presented to territorial males, with songs simulating territorial intrusion. Males typically respond to territorial intrusion by approaching the playback speaker, singing, and performing aggressive displays. The vigor of these responses can be regarded as reflecting the relative threat and salience of the songs presented. Males have been shown to respond more vigor-

ously to songs of their own species and populations (Ratcliffe and Grant 1985), to sequences of songs presented at higher rates (Rivera-Cáceres et al. 2011), and to songs with higher performance levels (such as faster trill rates; Illes et al. 2006, Moseley et al. 2013).

Playback experiments also have been used to examine the role of song in mate attraction. Female responses to playbacks can be subtle and difficult to score under field conditions (Searcy 1992, Searcy and Yasukawa 1996). Some researchers have assessed female preferences by presenting pairs of songs at two sides of a cage or aviary and asking if females prefer to perch closer to one speaker over the other. Other experiments have shown that females of some species will give "copulation solicitation" displays in response to playback of some songs even if no male is present (Searcy 1992), thus verifying that females use song to assess and help choose mates. Accordingly, detailed song playback experiments have allowed researchers to hone in on the acoustic features that guide mate choice. Playback studies with females have revealed female preferences for songs of birds' own species and population (Baker et al. 1987), for large song repertoires (Wasserman and Cigliano 1991), and for songs that are more challenging to produce (see below; Ballentine et al. 2004).

Female Songs and Duets

Much prior research on the function and evolution of birdsong—including most of the studies we mentioned above—has focused on temperate species in which males sing but females rarely do so. Accordingly, much of that research has focused on the role of male song in defending territories against other males, and in attracting females. However, it is known that females also sing in at least some species, particularly (but not exclusively) in the tropics (e.g., Odom et al. 2015). In fact, detailed phylogenetic studies suggest that all modern songbirds descended from an ancestral species in which both sexes probably sang (Odom et al. 2014). This finding raises basic unanswered questions about song evolution and function in both sexes (Price 2015). For example, if females do not typically defend territories or compete for mates, why should they sing at all? And, if ancestral females used to sing, what led to the loss of song in temperate zone species?

A relevant piece of information in beginning to address both questions is that, in many tropical songbird species, females make significant contributions to territory defense, or at least compete for other important limited resources such as nest sites (Stutchbury and Morton 2001). This stands in contrast to temperate species, in which typically males

PLAYBACK EXPERIMENTS

A powerful experimental approach in the study of bird vocalizations and their functions involves the use of "playbacks" (Searcy and Andersson 1986). In playback studies, previously recorded vocalizations are presented through loudspeakers and the responses of focal birds observed. This approach allows researchers to isolate the information content of the vocalizations themselves, independent of other cues that listeners might glean through interactions with actual signalers (McGregor 1992, Falls 1992, Searcy et al. 2006). A majority of published playback studies have focused on songs and their functions in male-male interactions. Such studies are normally conducted during the breeding season, when males are territorial. Songs played through a speaker on a male's territory thus simulate a territorial intruder, and the strength of the territory-holder's response provides an index of the salience of that particular stimulus. Song playback studies to females are more often conducted in laboratory settings, and female song preferences are gauged by their tendency to associate with certain songs over others, or by the likelihood that females will respond to song playback with copulation solicitation displays (Searcy and Yasukawa 1996). Song playback experiments have helped researchers interpret the function and thus the evolutionary basis for vocal divergence at numerous structural levels, including across species, across geographic location (within species),

within populations (Catchpole and Slater 2008), and across time (Derryberry 2007). One of the most interesting findings from recent playback studies is that females are seen to be more discriminating than males in their assessment of song variations, a sex difference consistent with sexual selection theory (Searcy et al. 2002, Danner et al. 2011).

A Chipping Sparrow responding to a song playback. *Photo by J. Podos.*

defend territories but females do not (Catchpole and Slater 2008). Moreover, a phylogenetic analysis suggests that females likely sang in ancestral species, and subsequently lost that tendency, especially as they diverged in the temperate zone (Price et al. 2009). The extent to which song plays a role in female territory defense in tropical species has not been well studied, although the possibility seems likely (e.g., Illes and Yunes-Jimena 2009). This is especially so given that another class of sexual signal, bright female plumage coloration, has evolved in numerous tropical species in which females defend territories or otherwise compete with each other for resources (Heinsohn et al. 2005, LeBas 2006). In short, the function of female song requires further study, but likely functions in social competition among females for territories, nest sites, or other important resources.

Most of the research on female song has focused on species in which females and males join their songs together

in vocal **duets**. In some species duets are very imprecise, such that each bird sings its own song more or less at the same time as its partner's song without much coordination (e.g., Dowling and Webster 2013). More often, though, duets between males and females are timed with impressive precision, for example with each partner producing its own portion of a complex coordinated duet that is referred to as "antiphonal" (Mennill and Vehrencamp 2005, Mann et al. 2009, Fortune et al. 2011, Templeton et al. 2013; see fig. 14.14). Some research has looked into the function of antiphonal duets, and emerging data suggest they most often serve as a coordinated form of territory defense by both mating partners (Hall 2009, Dahlin and Benedict 2014). The general idea is that a coordinated song by two territory defenders presents a louder and more aggressive signal to intruders, and the degree of coordination may also provide information about the pair's ability to defend the territory. Duets between mated

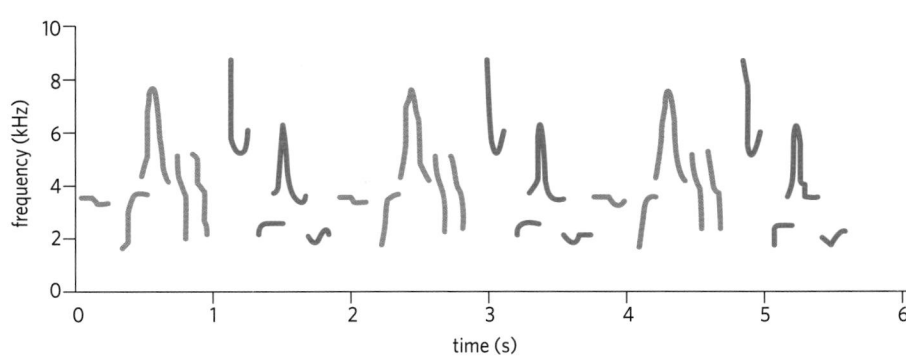

Figure 14.14. In the Happy Wren (*Pheugopedius felix*), males and females sing precisely timed antiphonal duets in which the male and female song phrases are interspersed with each other. In this song spectrogram, the phrases sung by the male are shown in blue, and those sung by the female in red. *Spectrogram and photograph* © *Chris Templeton.*

birds may perform other functions as well (reviewed in Hall 2009, Dahlin and Benedict 2014). In some cases duets appear to enhance the strength of the pair-bond and/or coordinate breeding behaviors (e.g., Hall and Magrath 2007), or to serve as a form of vocal mate-guarding, with the song of each partner signaling to others that the mate is already paired (e.g., Marshall-Ball et al. 2006, Mennill 2006, Rogers et al. 2007). Whether strengthening the pair-bond or guarding the mate, the strength of duetting behavior can affect patterns of parentage (Baldassarre et al. 2016). Overall, the reasons why females sing at all, and why they often join their songs together with males in duets, are certainly complex and deserving of more research. Nevertheless, it seems that the loss of song in female temperate zone species is likely a by-product of changing breeding strategies, such as transitions from joint to male-only territory defense.

Information Content in Songs

The kinds of results described above, from both observational and experimental realms, bring us back to the point raised at the start of this section. Presumably birds glean specific kinds of information from song, and use that information to guide their subsequent responses. But what is that information? Most generally, song provides information about species' identity, with selection favoring distinct songs that help birds identify appropriate mating partners, and thus avoid mating with the wrong species (Grant and Grant 1997). At finer scales, songs can also potentially carry information about a singer's geographic origin. In a large number of species, songs show substantial geographic variation, such that birds from different areas sound different from one another. These

vocal **dialects** (fig. 14.15) likely arise from "cultural drift" caused by song learning errors and other stochastic changes within different populations, much as different human populations develop regional dialects or accents over time (Podos and Warren 2007). It is likely that conspecifics can detect and use this information. For example, females may use dialects to help identify prospective mates from their own subspecies or even local population. Mating within-population may be favored by selection because it can help ensure that offspring are well-adapted to local conditions (Baker et al. 1987, MacDougall-Shackleton and MacDougall-Shackleton 2001). A large number of playback experiments have shown that birds typically respond most strongly to songs from their own population, less strongly to songs of conspecifics from other populations, and least strongly to songs of heterospecifics (fig. 14.16). At an even finer scale, songs or calls could potentially also reveal information about group membership or individual identity (Molles and Vehrencamp 2001, Lovell and Lein 2004, Keen et al. 2013), or even about genetic relatedness among individuals within a population. If some aspects of song are genetically inherited, or if young birds learn songs directly from their parents, then the degree of song similarity between two individuals could provide information regarding the degree of genetic relatedness between them. Some studies have indeed found that closely related individuals have more similar vocalizations than do more distantly related individuals (Greig et al. 2012). Whether receivers use this information, for example females to avoid close inbreeding, is unknown.

A recent body of research suggests that song can also reveal information about a singer's quality beyond species,

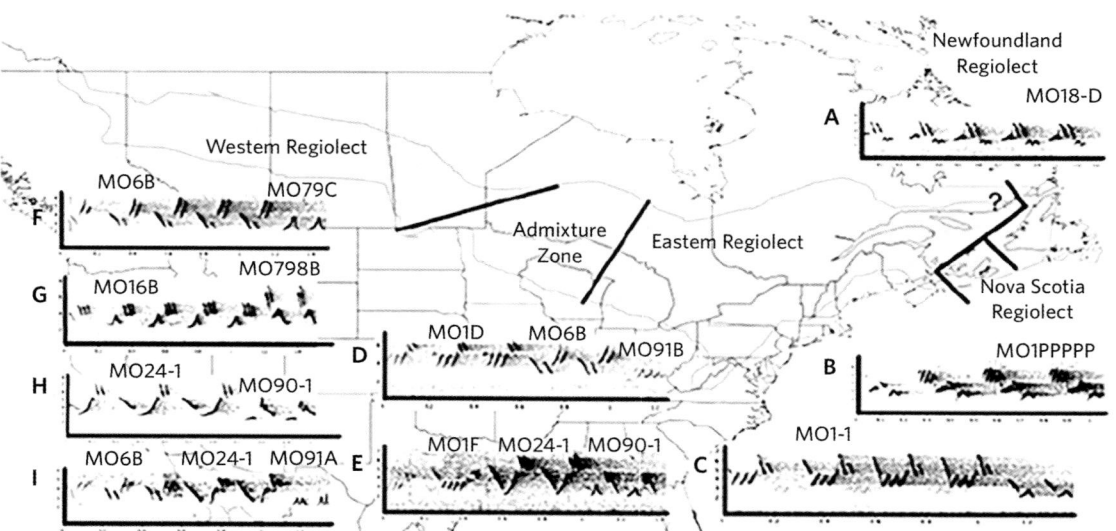

Figure 14.15. Vocal dialects, here shown for Mourning Warblers, *Oporornis philadelphia. Reprinted with permission from Pitocchelli 2011.* Dialects are songs with common structure within a specified range; dialects spanning large geographic areas are sometimes referred to as "regiolects."

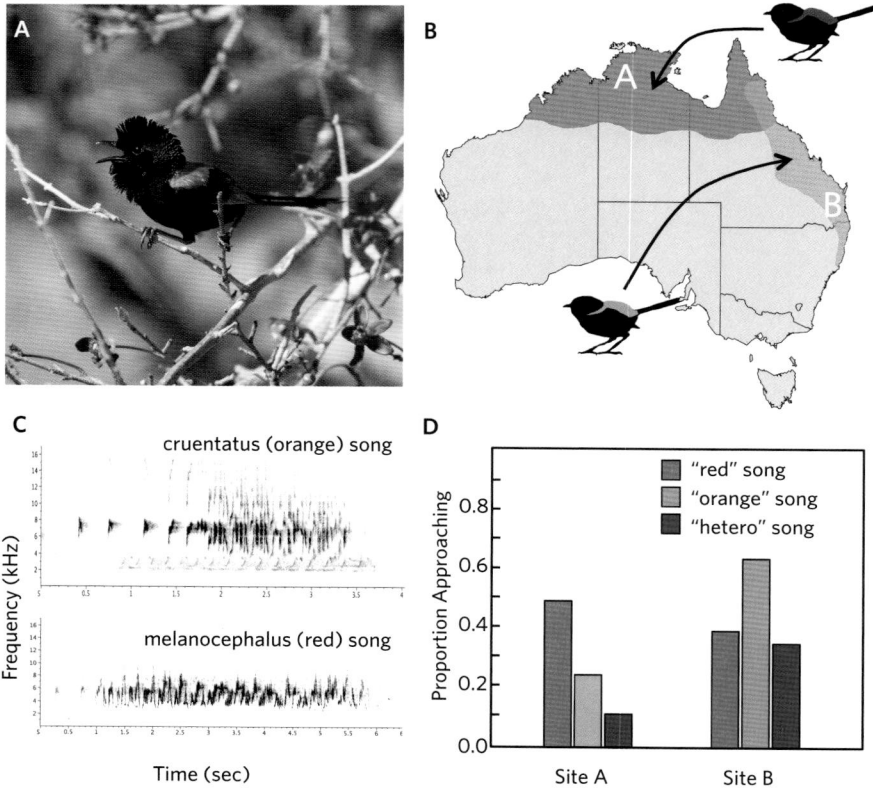

Figure 14.16. A, Male Red-backed Fairywrens (*Malurus melanocephalus*) produce songs to defend territories; B, There are two subspecies, with males of one subspecies (*M. m. cruentatus*, found in northern Australia) having red back plumage and males of the other subspecies (*M. m. melanocephalus*, found in eastern Australia) having orange back plumage. Letters on the map show locations of playback experiments; C, Spectrograms show that the songs of the two subspecies are different; D, At Site A (where males are red), males responded more strongly to playbacks of local ("red") songs than they did to nonlocal ("orange") and heterospecific ("hetero") songs, and comparable results were obtained at Site B (where males are orange). Heterospecific playbacks were recordings of the closely related White-winged Fairywren (*Malurus leucopterus*). *Figures modified with permission from Greig et al. 2015. Photo by Joe Welklin, fairywren cartoon by Emma Greig.*

population, or family identity. Information about singer quality can be used by females to select a high-quality mate, and can be used by males to avoid confrontations with a high-quality rival. Here it is useful to turn again to sexual selection theory, which suggests that animals' responses to elaborate mating signals are guided by two factors: females' intrinsic preferences for complex or intricate mating signals (a result of a process of runaway sexual selection or sensory biases), and the value of song as an indicator of male quality (Andersson 1994). Both processes lead to largely convergent predictions about which song features should be most effective. Recent research on quality hypotheses suggests that song, especially in species that learn their songs through vocal imitation (chapter 13), can serve as an indicator of both a male's **current quality** and his **developmental history**. Current quality refers to the males' status during a given breed-

SUSAN PETERS

Susan Peters. *Photo by Steve Nowicki.*

As a young girl, Susan Peters loved to build Heathkit FM radios. She also grew up making music on a number of instruments, including the piano, guitar, hammer dulcimer, and concertina. Susan brings her natural tinkering and musical abilities to her study of birdsong, which has resulted in a number of important contributions to what we know about the development of learned songs.

Susan began her groundbreaking work on birdsong in 1974, when she joined Peter Marler at the Rockefeller University Field Research Center. In Marler's lab she undertook intensive birdsong studies in the lab with Song Sparrows (*Melospiza melodia*) and Swamp Sparrows (*M. georgiana*) that laid the foundation for what we know about species-specific "instincts to learn." Her research also provided the first detailed information on the sequence of song learning, which became a model for future work in this area.

On moving to Duke University, Susan continued to produce groundbreaking research on song in collaboration with Stephen Nowicki and William Searcy. One of their most important recent contributions is the early nutritional stress hypothesis, which postulates that a male's learned song contains honest information on his early development. This hypothesis stimulated a new area of research on the role of development in sexual selection and the evolution of learned song.

Much of our foundational knowledge about the development of learned song came from painstaking experiments carried out by Susan. On top of that, another significant contribution by Susan to the world of ornithology is her unending support and mentorship to generations of undergraduate, graduate, and postdoctoral students of bird behavior.

By: *Elizabeth Derryberry, Department of Ecology & Evolutionary Biology, Tulane University, New Orleans, LA 70118 (ederrybe@ tulane.edu).*

---

ing season. A male on a high-quality territory, for instance, might have easier access to food and be able to spend time singing that otherwise would have gone to foraging (e.g., Alatalo et al. 1990). As such, singing rate would indicate the high quality of that male's territory, as well as his ability to secure that territory against rival males in the first place.

Song features might also provide information about a bird's developmental history because song is learned and crystallized during the early stages of a bird's life (chapter 13, box on page 427), when the bird is also subject to a variety of developmental stresses. Thus, a male who is able to learn to imitate a tutor song with accuracy—particularly a song that requires high vocal performance—reveals that his neural and vocal mechanisms developed well during his early months of life, which may in turn reflect both his genetic quality and the efficacy of the care he received from parents. These factors may then correlate with that male's future

potential as a sire and parent, making song features indicative of males' developmental history valuable to females prospecting for mates (Nowicki et al. 1998, Spencer and MacDougall-Shackleton 2011). New evidence suggests that females' responses to song also are influenced by their own developmental experiences (e.g., Schmidt et al. 2013).

HONEST SIGNALING?

The above discussion on the function of vocal signals brings us back to an important and challenging question: Do birds provide honest (or reliable) information in their signals, and if so why? Consider again the case of a bird giving an alarm call to conspecifics when detecting an approaching raptor. In this case the interests of the signaler and the receiver often are more or less aligned: the receiver benefits from hearing the warning so that it can take cover, and the signaler benefits because it helps save the life of the receiver, who is likely a mate or close relative. In such cases we expect signals to contain honest information.

But what if the fitness interests of the signaler and receiver do not fully align? What if, for instance, signalers are sometimes favored to give false (dishonest) information to "manipulate" the behavior of receivers? One well-studied example involves alarm calls given by African Fork-tailed Drongos (*Dicrurus adsimilis*) (Flower et al. 2014; see also Munn 1986, Goodale and Kotagama 2006). This species gives a number of calls that closely mimic the alarm calls of other bird species (such as Southern Pied Babblers, *Turdoides bicolor*) and even mammals (such as Meerkats, *Suricata suricatta*). Usually drongos give mimicked alarm calls when they detect a predator, and receivers respond appropriately by fleeing for cover. At times, though, particularly when a babbler or Meerkat has a large food item, drongos give alarm calls when no predator is present. These false alarm calls cause the receiver to drop its food and flee for cover, allowing the drongo to swoop in and claim the food for itself. So here the false alarm benefits the signaler (the drongo) but not the receiver (the babbler that loses its food).

But why would a receiver respond to a signal containing false information, particularly when that response comes at a fitness cost? The answer, just as in our earlier discussion of deceptive mimicry by brood parasites, is that the receiver probably benefits on average by responding to the mimicked call—that is, most of the time the alarm call needs to be honest or else receivers will stop responding. Indeed, in the previous example, babblers reduce their fleeing response if drongos "cry wolf" too many times, and drongos counter this by varying the alarm calls that they give (Flower et al. 2014). Accordingly, false alarm calls should be rare relative to honest signals of approaching danger.

Other cases of potentially honest signaling are more difficult to understand because the fitness interests of signalers and receivers not only do not overlap but in fact stand in direct opposition (see Searcy and Nowicki 2005). Think about a male bird trying to court a female or intimidate a rival male with his song. In this case, it would seem that it is always in the best interests of the male to advertise that he is a high-quality male in good condition, even if he is not. What prevents low-quality males from deceptively exaggerating their own quality in these situations? Theoretical work has shown that honest signaling can be maintained in these systems if the signals are costly to give and, more specifically, if the signals are more costly for low-quality individuals than they are for high-quality individuals (box on page 429). If this is the case, then the optimal level of signaling will be lower for low-quality individuals than it will be for high-quality individuals.

If costs are important to maintaining honest signaling, then what is the nature of those costs? This question has proven difficult to answer, particularly for vocal signals like birdsong, and considerable research has delved into the topic. One possibility is that these costs are "receiver-independent," meaning that they apply regardless of whether or how other animals respond to the signals being produced. For example, singing behavior may be energetically costly, as loud singing would appear to burn a lot of energy, such that low-quality individuals may be unable to sing loudly or for prolonged periods. However research indicates that singing birds expend little energy above baseline rates (Oberweger and Goller 2001). Alternatively, acoustic signals might carry important costs if the physiological mechanisms underlying song production are expensive to develop or maintain. In many temperate-zone birds, increasing day length increases androgen (testosterone) levels in males, which activates singing behavior (see chapters 8 and 13). It turns out that testosterone has a number of collateral negative effects on the physiology of a bird, including increased energy consumption and a suppressed immune system in at least some birds (e.g., Peters 2000, Buchanan et al. 2001, Garamszegi et al. 2004). These costs potentially may be important to maintaining honest signaling (Folstad and Karter 1992, Hilgarth et al. 1997). However, if testosterone is so costly, it is not clear why birds would not evolve to produce song with lower testosterone levels, for example by increasing sensitivity to lower T levels. Moreover, it is not clear that the negative consequences of high testosterone levels are more costly to low-quality males than they are to high-quality males. As such, it remains unclear whether these sorts of receiver-independent costs are important in maintaining honesty in bird songs.

Alternatively, signal reliability may be maintained by "receiver-dependent" costs, which occur when producing signals places a signaler at greater risk of retaliation from conspecific rivals. To illustrate, Song Sparrows (*Melospiza melodia*) that match song types of their neighbors with relatively high frequency are subject to more aggressive responses from neighboring territorial males, and subsequently greater risk of aggressive escalation and its attendant costs, such as risk of injury in a fight (Vehrencamp 2001). Social costs such as these are sometimes defined as "vulnerability" costs, because the signaling animal is more vulnerable to aggressive retaliation than it would be otherwise.

Another possibility is that honest/reliable signaling is maintained by performance constraints (defined earlier in our chapter) and corresponding variation among signalers in their capacities to overcome those constraints. In this case the constraints maintaining honesty are not costs per se, but rather uncheatable indicators of signaler quality (Maynard Smith and Harper 2003). Such signals are sometimes referred to as "index" signals. High structural consistency in song production by Chestnut-sided Warblers (*Setophaga pensylvanica*), for instance, may only be achieved by singers with the highest quality or condition (Byers 2007). The link between signaling reliability and performance constraints is not completely independent of the idea of costs. This is because singers who are able to sing high-performance songs as adults were likely better able, during development, to direct metabolic resources toward the development of brain structures that enable accurate learning and song model representation (Nowicki and Searcy 2005).

COSTS AND HONEST SIGNALING

The fitness interests of a signaler and receiver often do not overlap, for example when a male bird courts a female, or sings to deter a rival from his territory, or aggressively intimidates another bird while competing for food. In cases like these we might expect the signaler to give dishonest signals; that is to "bluff" and exaggerate its own quality so that it gets the mate, the territory, or the food. And yet, for the most part, such signals appear to honestly advertise individual quality. Why would this be? An important insight came from the work of Johnstone (1997; see also Grafen 1990), who modeled the benefits and costs of signal production against signaling intensity (see figure). Johnstone reasoned that the benefits of signaling would increase with the intensity of signaling; for example, males who sang or otherwise displayed at higher rates would attract more mates or be better able to defend their territory than males who sang at low rates. Johnstone further reasoned that these benefits would be independent of male quality: males would benefit from increased singing regardless of whether they were in good or poor condition, for example. Accordingly, in the graphical model there is a single benefits curve that applies to all signalers. In contrast, the costs of signaling also increase with increasing signaling intensity, but Johnstone modeled these costs as depending on the condition of the signaler. This would happen, for example, if males in poor condition suffered more from the physiological costs of high singing rates, or were less able to avoid predators attracted to the song, compared with males in good condition. Thus there are two cost lines in the graphical model, one for high-quality males (in good condition) and one for low-quality males (in poor condition). The cost line increases more rapidly (steeper slope) for low-quality males because they suffer more than high-quality males for any given level of signal expression. The optimal level of signaling for any male is the level that maximizes the net benefit (i.e., the benefit minus the costs). As can be seen from the figure, because the costs of signaling are dependent on signaler quality, the optimal level of signaling is lower for a low-quality individual than for a high-quality individual. This general insight helps us understand why signals are typically honest and, accordingly, much of the research on honest signaling has focused on the costs of signaling and the relationship of those costs to individual quality.

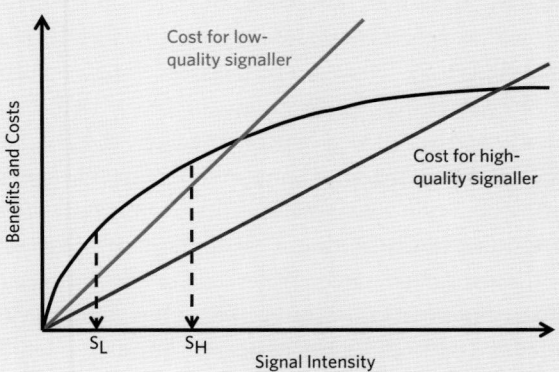

The benefits and costs of signaling plotted against signal intensity. As signal intensity increases, the benefits of signaling (black line) increase. However, the costs of signaling also increase with increasing signaling intensity. The optimal level of signaling is the level that maximizes net benefit, or the difference between the benefits and costs of signaling. If signaling costs increase more rapidly for low quality signalers (red line) than for high quality individuals (blue line), then the optimal level of signaling will be higher for high-quality individuals (S_H) than for low quality individuals (S_L). *Figure modified, with permission, from Johnstone (1997).*

COMMUNICATION NETWORKS

We end this chapter with a brief discussion of an exciting new direction in the study of birdsong. Most research on animal communication has kept things simple, by focusing on a single individual signaler and a single individual receiver. Yet, a more realistic picture of communication acknowledges that most habitats are filled by numerous individuals of multiple species (birds and other taxa), many of whom can be giving and receiving signals at any given time. Real-world communication thus actually occurs in a "network" of signalers and receivers. Thus when a bird gives a signal, some of the other birds hearing it may be the intended receivers that the signaler is trying to reach, whereas others may be unintended receivers that are eavesdropping. For example, when males sing to defend their territories from each other, females in the area could presumably eavesdrop on these vocal interactions to gain information about the relative dominance and quality of different males (i.e., to identify winners and losers).

Mennill et al. (2002) tested this hypothesis by simulating acoustic interactions among Black-capped Chickadees (*Poecile atricapillus*) and recording their songs. By altering the timing of the playback in response to the live bird, Mennill and colleagues were able to simulate interactions in which the focal (live) male won the altercation, and other interactions in which the focal male lost. The researchers then went on to examine, using genetic markers, the parentage of the offspring produced by the mates of these focal males. Surprisingly, they found that females mated to males that lost the simulated vocal battles were much more likely to have sought out extra-pair mates (i.e., extra-pair males), having shunned (to some extent) their own mates. This clever study strongly supports the hypothesis that female chickadees eavesdrop on vocal skirmishes between males, and use the information gleaned to guide mating decisions (Mennill et al. 2002).

Complex communication networks can also extend to multiple species. Alarm calls provide an excellent and well-studied example. When a small bird gives an alarm signaling the approach of a predator, it is intended to warn conspecifics in its same group. However, individuals of other small bird species may also respond, because the approaching predator is likely a threat to them as well. Playback experiments have verified that birds do indeed pay attention to the seet calls of other species and respond appropriately by seeking cover (Magrath et al. 2007, 2015a, Templeton and Greene 2007, Hetrick and Sieving 2012), and that even some mammals have converged on producing and using similar calls to signal danger (Greene and Meagher 1998). Birds also seem to discriminate somewhat across the alarm calls given by other species, responding most strongly to the most reliable signalers and less strongly to species that give alarm calls that are less reliable (Magrath et al. 2009). Interestingly, young birds appear to learn to associate heterospecific calls, and even novel sounds, that signal approaching danger (Haff and Magrath 2013, Magrath et al. 2015b).

How the various intended and unintended receivers shape the signal itself will depend on how they respond to it and on corresponding costs and benefits for the signaler. In some cases, the other species responding to an acoustic signal may actually be the intended receivers of the signal, as in mobbing calls where the signaler benefits by having a large number of other birds respond, regardless of species. In other cases, though, species responding to a signal are not the intended receivers, but instead are eavesdropping. Sometimes the consequences of responses by the unintended receivers on signaler fitness will be minimal. Consider, for example, a bird that gives a seet call to an approaching raptor; other birds may respond to this signal and take cover, but that likely is not costly to the signaler so long as the intended receivers (kin or mate) also take cover and are not taken by the raptor. In fact, there may be subtle benefits to the signaler of warning other species about approaching danger. For example, individuals alerted by a seet call typically become vigilant and produce their own seet calls. In this way the original signaler can benefit by having access to updated information about the location and movement of a dangerous predator (see Bower and Clark 2005).

Sometimes, though, responses by eavesdroppers will carry a fitness cost to the signaler. The most obvious case occurs when a predator finds its dinner by eavesdropping on the courtship signals of a bird trying to attract a mate. This has been well documented for a number of non-avian species (e.g., Halfwerk et al. 2014), and might also occur in birds (Greig and Pruett-Jones 2009, Schmidt and Belinsky 2013). But eavesdropping can have negative fitness consequences for a signaling bird even if it does not end up being somebody's dinner. A good example is the study of chickadees cited above (Mennill et al. 2002), which suggests that a male signaling to a rival male risks having other conspecifics, such as his own mate, eavesdrop on the interaction to assess the male's dominance rank relative to others. Overall, a new picture of communication is beginning to emerge, in which multiple species interact and potentially benefit from signals, and the distinction between signalers and receivers becomes muddied.

KEY POINTS

- Birds produce acoustic signals using a variety of mechanisms that include sonations (nonvocal sounds produced, for example, by modified flight feathers or the bill) as well as vocalizations. Vocalizations are produced by the syrinx, which in many birds has two parts, allowing for highly complex vocal signals.

- An acoustic signal passing through the environment is affected by physical attributes of that environment, leading to frequency and amplitude distortion, reverberations, and masking of the signal. These factors affect the sound the receiver actually hears, as well as the active space of the signal.

- Calls are generally simple vocalizations that can serve a variety of purposes, such as begging for food or alerting others about predators. The acoustic structure of calls is shaped by a variety of selective factors, generally increasing the probability that the signals are detected by intended receivers but not by unintended receivers.

- Songs tend to be longer and more complex vocalizations that generally serve for mate attraction and territorial defense. These signals can reveal considerable information about the signaler, including species identity, population of origin (through dialects), kinship, and individual quality.

- The costs that maintain honesty of avian vocal signals can be receiver-independent (e.g., the physiological costs of signal production) or receiver-dependent (e.g., retaliation by the receiver).

- In natural systems, communication happens in complicated networks composed of multiple individuals from multiple species. These various individuals can respond to signals in various ways, and those responses will have varying costs/benefits for the signalers themselves. Researchers are just now beginning to explore the ways that communication signals evolve in response to these complex networks.

KEY MANAGEMENT AND CONSERVATION IMPLICATIONS

- The study of bird vocalizations is providing unique access into the question of how birds, as individuals or populations, might respond to increasing habitat urbanization and other anthropogenic pressures. For example, birds of some species are seen to respond to urbanized habitats by shifting the structure of their songs to avoid masking (box on page 420). Yet other species might be unable to shift away from urban noise—either because of limited intrinsic plasticity, or greater severity of masking—in which case the only solution to maintain effective vocal communication is to abandon noisy habitats altogether (e.g., Goodwin and Shriver 2011). Future work should address the extent to which different species of birds can adapt to noisier, human-dominated landscapes, for example by adjusting their vocal behavior, versus the extent to which anthropogenic noise degrades the habitat for them.

- On a more positive note, the vocal diversity of songbirds can be used as a tool to help researchers assess and monitor populations, and also species diversity, quickly and comprehensively. As noted earlier, most songbird species sing species-specific vocalizations, and trained researchers can thus sample habitats quickly by collecting recordings with automated recording units, or simply by listening (Dawson and Efford 2009, Bardelia et al. 2010, Wimmer et al. 2013). It also might be possible to attract birds to specific habitats through song playback, for example to aid resettlement of newly restored habitats. This works because birds of some species prefer to settle alongside others of the same species, much as some species of duck are attracted to decoys. This method has proven successful in establishing new breeding colonies of threatened marine birds (Kress 1983, Podolsky and Kress 1992), and might also be suitable for other species (e.g., Rodgers 1992, Betts et al. 2008).

DISCUSSION QUESTIONS

1. In true communication, a signaler produces a signal that elicits a response from a receiver, such that both the signaler and the receiver benefit on average. But sometimes receivers can be manipulated by deceitful signals, and unintended receivers can eavesdrop on signals to their own benefit. What are examples of manipulative signals and eavesdropping, and how can such signals persist in a population?

2. The structures that many birds use to produce sonations are the flight feathers (e.g., wing primaries or tail feathers). What is it about these feathers that might predispose them to evolve into mechanisms for sound production?

3. What are some of the ways that human-influenced landscapes can affect the ways that birds signal and/or the effectiveness of those signals? In terms of acoustic communication, how might birds respond over time to landscapes that have been heavily affected by humans?

4. Mobbing calls and seets are both vocalizations that birds give in response to predators, but they differ from each other in acoustic structure: mobbing calls tend to be loud signals with abrupt on/off and a broad frequency bandwidth, whereas seet calls tend to be quiet and high-pitched with gradual on/off. Why the difference?

5. What sorts of information can songs (or calls) reveal about the signaler, and how?

6. Why does honest signaling require that signals carry some sort of cost to produce? Are some types of costs more likely to maintain honesty than others?

References

Alatalo, R. V., C. Glynn, and A. Lundberg (1990). Singing rate and female attraction in the pied flycatcher: An experiment. Animal Behaviour 39:601–603.

Andersson, M. B. (1994). Sexual selection. Princeton University Press, Princeton, NJ.

Baker, M. C., T. K. Bjerke, H. U. Lampe, and Y. O. Espmark (1987). Sexual response of female yellowhammers to differences in regional song dialects and repertoire sizes. Animal Behaviour 35:395–401.

Baldassarre, D. T., E. I. Greig, and M. S. Webster (2016). The couple that sings together stays together: Duetting, but not male aggression, is associated with extra-pair paternity in Red-backed Fairy-wrens. Biology Letters 12:20151025.

Ballentine, B., J. Hyman, and S. Nowicki (2004). Vocal performance influences female response to male birdsong: An experimental test. Behavioral Ecology 15:163–168.

Bardelia, R., D. Wolff, F. Kurth, M. Koche, K.-H. Tauchertf, and K.-H. Frommolt (2010). Detecting bird sounds in a complex acoustic environment and application to bioacoustic monitoring. Pattern Recognition Letters 31:1524–1534.

Betts, M. G., A. S. Hadley, N. L. Rodenhouse, and J. J. Nocera (2008). Social information trumps vegetation structure in breeding site selection by a migrant songbird. Proceedings of the Royal Society of London B 275:2257–2263.

Boncoraglio, G., and N. Saino (2007). Habitat structure and the evolution of bird song: A meta-analysis of evidence for the acoustic adaptation hypothesis. Functional Ecology 21:134–142.

Bostwick, K. S., and R. O. Prum (2003). High-speed video analysis of wing-snapping in two manakin clades (Pipridae: Aves). Journal of Experimental Biology 206:3693–3706.

Bostwick, K. S., and R. O. Prum (2005). Courting bird sings with stridulating wing feathers. Science 309:736–736.

Bostwick, K. S., M. L. Riccio, and J. M. Humphries (2012). Massive, solidified bone in the wing of a volant courting bird. Biology Letters 8:760–763.

Bower, J. L., and C. W. Clark (2005). A field test of the accuracy of a passive acoustic location system. Bioacoustics 15:1–14.

Bradbury, J. W., and S. L. Vehrencamp (2011). Principles of animal communication. 2nd ed. Sinauer Associates, Sunderland, MA.

Briskie, J. V., P. R. Martin, and T. E. Martin (1999). Nest predation and the evolution of nestling begging calls. Proceedings of the Royal Society of London B 266:2153–2159.

Brumm, H. (2004). The impact of environmental noise on song amplitude in a territorial bird. Journal of Animal Ecology 73:434–440.

Brumm, H., and H. Slabbekoorn (2005). Acoustic communication in noise. Advances in the Study of Behavior 35:151–209.

Brumm, H., and M. Naguib (2009). Environmental acoustics and the evolution of bird song. Advances in the Study of Behavior 40:1–33.

Buchanan, K. L., M. R. Evans, A. R. Goldsmith, D. M. Bryant, and L. V. Rowe (2001). Testosterone influences basal metabolic rate in male House Sparrows: A new cost of dominance signalling? Proceedings of the Royal Society of London B 268:1337–1344.

Byers, B. E. (2007). Extrapair paternity in Chestnut-sided Warblers is correlated with consistent vocal performance. Behavioral Ecology 18:130–136.

Catchpole, C. K., and P. J. B. Slater (2008). Bird song: Biological themes and variations. 2nd ed. Cambridge University Press, Cambridge, UK.

Clark, C. J. (2008). Fluttering wing feathers produce the flight sounds of male streamertail hummingbirds. Biology Letters 4:341–344.

Clark, C. J., and T. J. Feo (2008). The Anna's Hummingbird chirps with its tail: A new mechanism of sonation in birds. Proceedings of the Royal Society of London B 275:955–962.

Clark, C. J., D. Elias, and R. O. Prum (2011). Aeroelastic flutter produces hummingbird feather songs. Science 333:1430–1433.

Clark, C. J., and R. O. Prum (2015). Aeroelastic flutter of feathers, flight and the evolution of non-vocal communication in birds. Journal of Experimental Biology 218:3520–3527.

Courter, J. R., and G. Ritchison (2010). Alarm calls of Tufted Titmice convey information about predator size and threat. Behavioral Ecology 21:936–942.

Dahlin, C. R., and L. Benedict (2014). Angry birds need not apply: A perspective on the flexible form and multifunctionality of avian vocal duets. Ethology 120:1–10.

Danner, J. E., R. M. Danner, F. Bonier, P. R. Martin, T. W. Small, and I. T. Moore (2011). Female, but not male, tropical sparrows respond more strongly to the local song dialect: Implications for population divergence. American Naturalist 178:53–63.

Darwin, C. (1871). The descent of man, and selection in relation to sex. John Murray, London.

da Silva, A., M. Valcu, and B. Kempenaers (2015). Light pollution alters the phenology of dawn and dusk singing in common European songbirds. Proceedings of the Royal Society of London B 370:20140126.

Dawson, D. K., and M. G. Efford (2009). Bird population density estimated from acoustic signals. Journal of Applied Ecology 46:1201–1209.

Derryberry, E. P. (2007). Evolution of bird song affects signal efficacy: An experimental test using historical and current signals. Evolution 61:1938–1945.

Derryberry, E. P., N. Seddon, S. Claramunt, J. A. Tobias, A. Baker, A. Aleixo, and R. T. Brumfield (2012). Correlated evolution of beak morphology and song in the Neotropical woodcreeper radiation. Evolution 66:2784–2797.

Dowling, J. L., D. A. Luther, and P. P. Marra (2012). Comparative effects of urban development and anthropogenic noise on bird songs. Behavioral Ecology 23:201–209.

Dowling, J., and M. S. Webster (2013). The form and function of duets and choruses in Red-backed Fairy-wrens. Emu: Austral Ornithology 113:282–293.

Ellis, J. M. S., T. A. Langen, and E. C. Berg (2009). Signaling for food and sex? Begging by reproductive female White-throated Magpie-Jays. Animal Behaviour 78:615–623.

Falls, J. B. (1992). Playback: A historical perspective. *In* Playback and studies of animal communication, P. K. McGregor, Editor. Plenum Press, New York, pp. 11–33.

Feeney, W. E., J. A. Welbergen, and N. E. Langmore (2014). Advances in the study of coevolution between avian brood parasites and their hosts. Annual Review of Ecology, Evolution, and Systematics 45:227–246.

Feo, T. J., and C. J. Clark (2010). The displays and mechanical sounds of the Black-chinned Hummingbird (Trochilidae: *Archilochus alexandri*). The Auk 127:787–796.

Fitch, W. T. (1999). Acoustic exaggeration of size in birds via tracheal elongation: Comparative and theoretical analyses. Journal of Zoology 248:31–48.

Fitch, W. T., and J. P. Kelley (2000). Perception of vocal tract resonances by Whooping Cranes *Grus americana*. Ethology 106:559–574.

Fitch, W. T., and M. D. Hauser (2003). Unpacking "honesty": Vertebrate vocal production and the evolution of acoustic signals. *In* Acoustic communication, A. Simmons, R. R. Fay, and A. N. Popper, Editors. Springer, New York, pp. 65–136.

Flower, T. P., M. Gribble, and A. R. Ridley (2014). Deception by flexible alarm mimicry in an African bird. Science 344:513–516.

Folstad, I., and A. J. Karter (1992). Parasites, bright males, and the immunocompetence handicap. American Naturalist 159:603–622.

Fortune, E. S., C. Rodríguez, D. Li, Gr. F. Ball, and M. J. Coleman (2011). Neural mechanisms for the coordination of duet singing in wrens. Science 334:666–670.

Garamszegi, L. Z., A. P. Møller, J. Török, G. Michl, P. Péczely, and M. Richard (2004). Immune challenge mediates vocal communication in a passerine bird: An experiment. Behavioral Ecology 15:148–157.

Goller, F., and O. N. Larsen (1997). In situ biomechanics of the syrinx and sound generation in pigeons. Journal of Experimental Biology 200:2165–2176.

Goodale, E., and S. W. Kotagama (2006). Context-dependent vocal mimicry in a passerine bird. Proceedings of the Royal Society of London B 273:875–880.

Goodwin, S. E., and W. G. Shriver (2011). Effects of traffic noise on occupancy patterns of forest birds. Conservation Biology 25:406–411.

Grafen, A. (1990). Biological signals as handicaps. Journal of Theoretical Biology 144:517–546.

Grant, P. R., and B. R. Grant (1997). Genetics and the origin of bird species. Proceedings of the National Academy of Sciences USA 94 (15): 7768–7775.

Greene, E., and T. Meagher (1998). Red Squirrels, *Tamiasciurus hudsonicus*, produce predator-class specific alarm calls. Animal Behaviour 55:511–518.

Greenewalt, C. (1968). Bird song: Acoustics and physiology. Smithsonian Institution Press, Washington, DC.

Greig, E. I., and S. Pruett-Jones (2009). A predator-elicited song in the Splendid Fairy-wren: Warning signal or intraspecific display? Animal Behaviour 78:45–52.

Greig, E. I., B. N. Taft, and S. Pruett-Jones (2012). Sons learn songs from their social fathers in a cooperatively-breeding bird. Proceedings of the Royal Society of London B 279:3154–3160.

Greig, E. I., D. T. Baldassarre, and M. S. Webster (2015). Differential rates of phenotypic introgression are associated with male responses to multimodal signals. Evolution 69:2602–2612.

Griesser, M. (2009). Mobbing calls signal predator category in a kin group-living bird species. Proceedings of the Royal Society of London B 276:2887–2892.

Haff, T. M., and R. D. Magrath (2013). Eavesdropping on the neighbours: Fledglings learn to respond to heterospecific alarm. Animal Behaviour 85:411–418.

Halfwerk, W., P. L. Jones, R. C. Taylor, M. J. Ryan, and R. A. Page (2014). Risky ripples allow bats and frogs to eavesdrop on a multisensory sexual display. Science 343:413–416.

Hall, M. L. (2009). A review of vocal duetting in birds. Advances in the Study of Behavior 40:67–121.

Hall, M. L., and R. D. Magrath (2007). Temporal coordination signals coalition quality. Current Biology 17:R406–R407.

Handford, P. (1988). Trill rate dialects in the Rufous-collared Sparrow, *Zonotrichia capensis*, in northwestern Argentina. Canadian Journal of Zoology 66:2658–2670.

Hart, P. J., R. Hall, W. Ray, A. Beck, and J. Zook (2015). Cicadas impact bird communication in a noisy tropical rainforest. Behavioral Ecology 26:839–842.

Hartley, R. S., and R. A. Suthers (1989). Air-flow and pressure during canary song: Direct evidence for mini-breaths. Journal of Comparative Physiology A 165:15–26.

Haskell, D. G. (1994). Experimental evidence that nestling begging behaviour incurs a cost due to nest predation. Proceedings of the Royal Society of London B 257:161–164.

Haskell, D. G. (1999). The effect of predation on begging-call evolution in nestling wood warblers. Animal Behaviour 57:893–901.

Heinrich, B. (1988). Winter foraging at carcasses by three sympatric corvids, with emphasis on recruitment by the raven, *Corvus corax*. Behavioral Ecology and Sociobiology 23:141–156.

Heinrich, B. (1989). Ravens in winter. Summit Books, New York.

Heinsohn, R., S. Legge, and J. A. Endler (2005). Extreme reversed sexual dichromatism in a bird without sex role reversal. Science 309:617–619.

Herrel, A., J. Podos, B. Vanhooydonck, and A. P. Hendry (2009). Force-velocity trade-off in Darwin's finch jaw function: A biomechanical basis for ecological speciation? Functional Ecology 23:119–125.

Hetrick, S. A., and K. E. Sieving (2012). Antipredator calls of Tufted Titmice and interspecific transfer of encoded threat information. Behavioral Ecology 23:83–92.

Hilgarth, N., M. Ramenofsky, and J. Wingfield (1997). Testosterone and sexual selection. Behavioral Ecology 8:108–112.

Hoese, W. J., J. Podos, N. C. Boetticher, and S. Nowicki (2000). Vocal tract function in birdsong production: Experimental manipulation of beak movements. Journal of Experimental Biology 203:1845–1855.

Huber, S. K., and J. Podos (2006). Beak morphology and song features covary in a population of Darwin's finches (*Geospiza fortis*). Biological Journal of the Linnean Society 88:489–498.

Illes, A. E., M. L. Hall, and S. L. Vehrencamp (2006). Vocal performance influences male receiver response in the Banded Wren. Proceedings of the Royal Society of London B 273:1907–1912.

Illes, A. E., and L. Yunes-Jimenez (2009). A female songbird outsings male conspecifics during simulated territorial intrusions. Proceedings of the Royal Society of London B 276:981–986.

Johnstone, R. A. (1997). The evolution of animal signals. *In* Behavioural ecology: An evolutionary approach, 4th ed., J. R. Krebs and N. B. Davies, Editors. Blackwell, Oxford, pp. 155–178.

Jones, K. J., and W. L. Hill (2001). Auditory perception of hawks and owls for passerine alarm calls. Ethology 107:717–726.

Jouventin, P., T. Aubin, and T. Lengagne (1999). Finding a parent in a King Penguin colony: The acoustic system of individual recognition. Animal Behaviour 57:1175–1183.

Keen, S. C., C. D. Meliza, and D. R. Rubenstein (2013). Flight calls signal group and individual identity but not kinship in a cooperatively breeding bird. Behavioral Ecology 24:1279–1285.

Kempenaers, B., P. Borgström, P. Loës, E. Schlicht, and M. Valcu (2010). Artificial night lighting affects dawn song, extra-pair siring success, and lay date in songbirds. Current Biology 20:1735–1739.

Klump, G. M. (1996). Bird communication in the noisy world. *In* Ecology and evolution of acoustic communication in birds, D. E. Kroodsma and E. H. Miller, Editors. Cornell University Press, Ithaca, NY, pp. 321–338.

Krama, T., I. Krams, and K. Igaune (2008). Effects of cover on loud trill-call and soft seet-call use in the Crested Tit *Parus cristatus*. Ethology 114:656–661.

Kress, S. W. (1983). The use of decoys, sound recordings, and gull control for re-establishing a tern colony in Maine. Colonial Waterbirds 6:185–196.

Lambrechts, M. M. (1996). Organization of birdsong and constraints on performance. *In* Ecology and evolution of acoustic communication in birds, D. E. Kroodsma and E. H. Miller, Editors. Cornell University Press, Ithaca, NY, pp. 305–320.

Leavesley, A. J., and R. D. Magrath (2005). Communicating about danger: Urgency alarm calling in a bird. Animal Behaviour 70:365–373.

LeBas, N. R. (2006). Female finery is not for males. Trends in Ecology and Evolution 21:170–173.

Lohr, B., T. F. Wright, and R. J. Dooling (2003). Detection and discrimination of natural calls in masking noise by birds: Estimating the active space of a signal. Animal Behaviour 65:763–777.

Lotem, A. (1998). Differences in begging behaviour between Barn Swallow, *Hirundo rustica*, nestlings. Animal Behaviour 55:809–818.

Lovell, S. F., and M. R. Lein (2004). Neighbor-stranger discrimination by song in a suboscine bird, the Alder Flycatcher, *Empidonax alnorum*. Behavioral Ecology 15:799–804

Luther, D. A. (2008). Signaller:receiver coordination and the timing of communication in Amazonian birds. Biology Letters 4:651–654.

Luther, D. A. (2009). The influence of the acoustic community on songs of birds in a Neotropical rain forest. Behavioral Ecology 20:864–871.

MacDougall-Shackleton, E. A., and S. A. MacDougall-Shackleton (2001). Cultural and genetic evolution in Mountain White-crowned Sparrows: Song dialects are associated with population structure. Evolution 55:2568–2575.

Magrath, R. D., B. J. Pither, and J. L.Gardner (2007). A mutual understanding? Interspecific responses by birds to each other's aerial alarm calls. Behavioral Ecology 18:944–951.

Magrath, R. D., B. J. Pither, and J. L.Gardner (2009). An avian eavesdropping network: Alarm signal reliability and heterospecific response. Behavioral Ecology 20:745–752.

Magrath, R. D., T. M. Haff, P. M. Fallow, and A. N. Radford (2015a). Eavesdropping on heterospecific alarm calls: From mechanisms to consequences. Biological Reviews 90:560–586.

Magrath, R. D., T. M. Haff, J. R. McLachlan, and B. Igic (2015b). Wild birds learn to eavesdrop on heterospecific alarm calls. Current Biology 25:2047–2050.

Mann, N. I., K. A. Dingess, K. F. Barker, J. A. Graves, and P. J. B. Slater (2009). A comparative study of song form and duetting in Neotropical *Thryothorus* wrens. Behaviour 146:1–43.

Marler, P. (1955). Characteristics of some animal calls. Nature 176:6–8.

Marler, P. (2004). Bird calls: A cornucopia for communication. *In* Nature's music: The science of birdsong, P. Marler and H. Slabbekoorn, Editors. Elsevier, Amsterdam, pp. 132–177.

Marshall-Ball, L., N. Mann, and P. Slater (2006). Multiple functions to duet singing: Hidden conflicts and apparent cooperation. Animal Behaviour 71:823–831.

Maynard Smith, J., and D. Harper (2003). Animal signals. Oxford University Press, Oxford.

McClure, C. J. W., H. E. Ware, J. Carlisle, G. Kaltenecker, and J. R. Barber (2013). An experimental investigation into the effects of traffic noise on distributions of birds: Avoiding the phantom road. Proceedings of the Royal Society of London B 280:20132290.

McGregor, P. K. (1992). Playback and studies of animal communication. Plenum Press, New York.

Mennill, D. J. (2006). Aggressive responses of male and female Rufous-and-White Wrens to stereo duet playback. Animal Behaviour 71:219–226.

Mennill, D. J., L. M. Ratcliffe, and P. T. Boag (2002). Female eavesdropping on male song contests in songbirds. Science 296:873.

Mennill, D. J., and S. L. Vehrencamp (2005). Sex differences in singing and duetting behavior of Neotropical Rufous-and-White Wrens (*Thryothorus rufalbus*). The Auk 122:175–186.

Miller, M. W. (2006) Apparent effects of light pollution on singing behavior of American Robins. The Condor 108:130–139.

Molles, L. E., and S. L. Vehrencamp (2001). Neighbour recognition by resident males in the Banded Wren, *Thryothorus pleurostictus*, a tropical songbird with high song type sharing. Animal Behaviour 61:119–127.

Morton, E. S. (1975). Ecological sources of selection on avian sounds. American Naturalist 109:17–34.

Moseley, D. L., D. C. Lahti, and J. Podos (2013). Responses to song playback vary with the vocal performance of both signal senders and receivers. Proceedings of the Royal Society of London B 280:20131401.

Munn, C. A. (1986). Birds that "cry wolf." Nature 391:143–145.

Naguib, M., R. Mundry, R. Ostreiher, H. Hultsch, L. Chrader, and D. Todt (1999). Cooperatively breeding Arabian babblers call differently when mobbing in different predator-induced situations. Behavioral Ecology 10:636–640.

Nelson, D. A., and P. Marler (1990). The perception of birdsong and an ecological concept of signal space. *In* Comparative perception, complex signals, W. C. Stebbins and M. A. Berkley, Editors. John Wiley, New York, pp. 443–478.

Niese, R. L., and B. W. Tobalske (2016). Specialized primary feathers produce tonal sounds during flight in rock pigeons. Journal of Experimental Biology 219:2173–2181.

Nowicki, S., and P. Marler (1988). How do birds sing? Music Perception 5:391–426.

Nowicki, S., S. Peters, and J. Podos (1998). Song learning, early nutrition and sexual selection in songbirds. American Zoologist 38:179–190.

Nowicki, S., and W. A. Searcy (2005). Song and mate choice in birds: How the development of behavior helps us understand function. The Auk 122:1–14.

Nystrom, K. G. K. (1997). Food density, song rate, and body condition in territory-establishing Willow Warblers (*Phylloscopus trochilus*). Canadian Journal of Zoology 75:47–58.

Oberweger, K., and F. Goller (2001). The metabolic cost of birdsong production. Journal of Experimental Biology 204:3379–3388.

Odom, K. J., M. L. Hall, K. Riebel, K. E. Omland, and N. E. Langmore (2014). Female song is widespread and ancestral in songbirds. Nature Communications 5:3379.

Odom, K. J., K. E. Omland, and J. J. Price (2015). Differentiating the evolution of female song and male-female duets in the New World blackbirds: Can tropical natural history traits explain duets? Evolution 69:839–847.

Payne, R. S. (1971). Acoustic location of prey by Barn Owls (*Tyto alba*). Journal of Experimental Biology 54:535–573.

Peters, A. (2000). Testosterone treatment is immunosuppressive in Superb Fairy-wrens, yet free-living males with high testosterone are more immunocompetent. Proceedings of the Royal Society of London B 267:883–889.

Pinxten, R., and M. Eens (1998). Male starlings sing most in the late morning, following egg-laying: A strategy to protect their paternity? Behaviour 35:1197–1211.

Pitocchelli, J. (2011). Macrogeographic variation in the song of the Mourning Warbler (*Oporornis philadelphia*). Canadian Journal of Zoology 89:1027–1040.

Podolsky, R., and S. W. Kress (1992). Attraction of the endangered Dark-rumped Petrel to recorded vocalizations in the Galapagos Islands. The Condor 94:448–453.

Podos, J. (2001). Correlated evolution of morphology and vocal signal structure in Darwin's finches. Nature 409:185–188.

Podos, J., and S. Nowicki (2004). Performance limits on birdsong. *In* Nature's music: The science of birdsong, P. Marler and H. Slabbekoorn, Editors. Elsevier, Amsterdam, pp. 318–342.

Podos, J., A. Southall, and M. R. Rossi-Santos (2004). Vocal mechanics in Darwin's finches: Correlation of break gape and song frequency. Journal of Experimental Biology 207:607–619.

Podos, J., and P. S. Warren (2007). The evolution of geographic variation in birdsong. Advances in the Study of Behavior 37:403–458.

Price, J. J. (2015). Rethinking our assumptions about the evolution of bird song and other sexually dimorphic signals. Frontiers in Ecology and Evolution 3:40–45.

Price, J. J., S. M. Lanyon, and K. E. Omland (2009). Losses of female song with changes from tropical to temperate breeding in the New World blackbirds. Proceedings of the Royal Society of London B 276:1971–1980.

Ratcliffe, L. M., and P. R. Grant (1985). Species recognition in Darwin finches (*Geospiza*, Gould) 3: Male-responses to playback of different song types, dialects and heterospecific songs. Animal Behaviour 33:290–307.

Reid, J. M., P. Arcese, A. L. E. V. Cassidy, S. M. Hiebert, J. N. M. Smith, P. K. Stoddard, A. B. Marr, and L. F. Keller (2004). Song repertoire size predicts initial mating success in male Song Sparrows, *Melospiza melodia*. Animal Behaviour 68:1055–1063.

Riede, T., R. A. Suthers, N. H. Fletcher, and W. E. Blevins (2006). Songbirds tune their vocal tract to the fundamental frequency of their song. Proceedings of the National Academy of Sciences USA 103:5543–5548.

Rivera-Cáceres, K., C. Macías Garcia, E. Quirós-Guerrero, and A. A. Ríos-Chelén (2011). An interactive playback experiment shows song bout size discrimination in the suboscine Vermilion Flycatcher (*Pyrocephalus rubinus*). Ethology 117: 1120–1127.

Rodgers, R. D. (1992). A technique for establishing Sharp-Tailed Grouse in unoccupied range. Wildlife Society Bulletin 20:101–106.

Rogers, A. C., N. E. Langmore, and R. A. Mulder (2007). Function of pair duets in the eastern whipbird: Cooperative defense or sexual conflict? Behavioral Ecology 18:182–188.

Schmidt, K. L., E. S. McCallum, E. A. MacDougall-Shackleton, and S. A. MacDougall-Shackleton (2013). Early-life stress affects the behavioural and neural response of female Song Sparrows to conspecific song. Animal Behaviour 85:825–837.

Schmidt, K. A., and K. L. Belinsky (2013). Voices in the dark: Predation risk by owls influences dusk singing in a diurnal passerine. Behavioral Ecology and Sociobiology 67:1837–1843.

Searcy, W. A. (1992). Measuring responses of female birds to male song. *In* Playback and studies of animal communication, P. K. McGregor, Editor. Plenum Press, New York, pp. 175–189.

Searcy, W. A., and M. Andersson (1986). Sexual selection and the evolution of song. Annual Review of Ecology and Systematics 17:507–533.

Searcy, W. A., and K. Yasukawa (1996). Song and female choice. *In* Ecology and evolution of acoustic communication in birds, D. E. Kroodsma and E. H. Miller, Editors. Cornell University Press, Ithaca, NY, pp. 454–473.

Searcy, W. A., S. Nowicki, M. Hughes, and S. Peters (2002). Geographic song discrimination in relation to dispersal distances in Song Sparrows. American Naturalist 159:221–230.

Searcy, W. A., and S. Nowicki (2005). The evolution of animal communication: Reliability and deception in signaling systems. Princeton University Press, Princeton, NJ.

Searcy, W. A., R. C. Anderson, and S. Nowicki (2006). Bird song as a signal of aggressive intent. Behavioral Ecology and Sociobiology 60:234–241.

Seddon, N. (2005). Ecological adaptation and species recognition drives vocal evolution in Neotropical suboscine birds. Evolution 59:200–215.

Seyfarth, R. M., D. L. Cheney, T. Bergman, J. Fischer, K. Zuberbuhler, and K. Hammerschmidt (2010). The central importance of information in studies of animal communication. Animal Behaviour 80:3–8.

Sheldon, W. G. (1967). The book of the American Woodcock. University of Massachusetts Press, Amherst.

Sieving, K. E., S. A. Hetrick, and M. L. Avery (2010). The versatility of graded acoustic measures in classification of predation threats by the Tufted Titmouse Baeolophus bicolor: Exploring a mixed framework for threat communication. Oikos 119:264–276.

Slabbekoorn, H., and T. B. Smith (2000). Does bill size polymorphism affect courtship song characteristics in the African finch Pyrenestes ostrinus? Biological Journal of the Linnean Society 71:737–753.

Slabbekoorn, H., and M. Peet (2003). Birds sing at a higher pitch in urban noise. Nature 424:267.

Soard, C. M., and G. Ritchison (2009). "Chick-a-dee" calls: Carolina chickadees convey information about degree of threat posed by avian predators. Animal Behaviour 78:1447–1453.

Spencer, K. A., and S. A. MacDougall-Shackleton (2011). Indicators of development as sexually selected traits: The developmental stress hypothesis in context. Behavioral Ecology 22:1–9.

Stutchbury, B. J. M., and E. S. Morton (2001). Behavioral ecology of tropical birds. Academic Press, San Diego, CA.

Suthers, R. A., F. Goller, and R. S. Hartley (1994). Motor dynamics of song production by mimic thrushes. Journal of Neurobiology 25:917–936.

Suthers, R. A., E. Vallet, and M. Kreutzer (2012). Bilateral coordination and the motor basis of female preference for sexual signals in canary song. Journal of Experimental Biology 215:2950–2959.

Suzuki, T. N. (2014). Communication about predator type by a bird using discrete, graded and combinatorial variation in alarm calls. Animal Behaviour 87:59–65.

Templeton, C. N., E. Greene, and K. Davis (2005). Allometry of alarm calls: Black-capped Chickadees encode information about predator size. Science 308:1934–1937.

Templeton, C. N., and E. Greene (2007). Nuthatches eavesdrop on variations in heterospecific chickadee mobbing alarm calls. Proceedings of the National Academy of Sciences USA 104:5479–5482.

Templeton, C. N., N. I. Mann, A. A. Ríos-Chelén, E. Quiros-Guerreroc, C. M. Garciac, and P. J. B. Slater (2013). An experimental study of duet integration in the Happy Wren, Pheugopedius felix. Animal Behaviour 86:821–827.

Tobias, J. A., and N. Seddon (2002). Female begging in European Robins: Do neighbors eavesdrop for extrapair copulations? Behavioral Ecology 13:637–642.

Vehrencamp, S. L. (2001). Is song-type matching a conventional signal of aggressive intentions? Proceedings of the Royal Society of London B 268:1637–1642.

Ware, H. E., C. J. W. McClure, J. D. Carlisle, and J. R. Barber (2015). A phantom road experiment reveals traffic noise is an invisible source of habitat degradation. Proceedings of the National Academy of Sciences USA 12:12105–12109.

Wasserman, F. E., and J. A. Cigliano (1991). Song output and stimulation of the female in White-throated Sparrows. Behavioral Ecology and Sociobiology 29:55–59.

Westneat, M. W., J. H. Long, W. Hoese, and S. Nowicki (1993). Kinematics of birdsong: Functional correlation of cranial movements and acoustic features in sparrows. Journal of Experimental Biology 182:147–171.

Wheatcroft, D., and T. D. Price (2013). Learning and signal copying facilitate communication among bird species. Proceedings of the Royal Society of London B 280:1–7.

Wiley, R. H. (1991). Associations of song properties with habitats for territorial oscine birds of eastern North America. American Naturalist 138:973–993.

Wimmer, J., M. Towsey, P. Roe, and I. Williamson (2013). Sampling environmental acoustic recordings to determine bird species richness. Ecological Applications 23:1419–1428.

Zollinger, S. A., and R. A. Suthers (2004). Motor mechanisms of a vocal mimic: Implications for birdsong production. Proceedings of the Royal Society of London B 271:483–491.

PART IV

HOW BIRDS LIVE

CHAPTER 15

Foraging Behavior

Diego Sustaita, Alejandro Rico-Guevara, and Fritz Hertel

FORAGING BY LAND, SEA, AND AIR

Surely by now you have generated a sense of awe and appreciation for the amazing phylogenetic and functional avian diversity. Birds have evolved to fill just about every ecological niche (chapters 18, 22), from the North Pole to the South Pole. Each of these niches imposes a different set of challenges and opportunities for foraging and feeding. Birds are adapted to feed on the ground, in trees, in marine and freshwater environments, in caves, in midair, in the light of day, and in the darkness of night (fig. 15.1). Birds eat virtually everything, from seeds to baby seals. Some even eat wax, and yet others consume blood. Foraging behavior may have phylogenetic or ecological bases. In some lineages, species share a common feeding mode (e.g., most galliforms are primarily granivorous and feed on the ground), whereas in other lineages species have diversified in foraging and feeding behavior, presumably in response to different environmental conditions (e.g., passeriforms have exploited the full gamut of terrestrial, arboreal, aquatic, and aerial modalities).

Behavior essentially describes everything an animal does—all of the movements and sensations that mediate their relationships with their external environments (e.g., physical, biotic, and social; Alexander 1975). Accordingly, the study of avian foraging behavior integrates, among others, the fields of ornithology, neurology, psychology, anatomy, physiology, ecology, and evolution and, as such, constitutes a topic for multiple books, let alone a single chapter. Accordingly, we provide a general overview of several of these facets as they relate to avian foraging and feeding behavior. How birds find food; determine what, where, and when to eat; and how they go about the process of eating are dictated by innate and learned, as well as stereotyped and novel, behavioral patterns (Bels and Baussart 2006). Although anatomical (chapter 6) and physiological (chapter 7) mechanisms

ultimately determine birds' abilities to obtain, handle, and assimilate food, proximate behavioral processes intervene to bring consumers to their food sources. For instance, before a Large Ground Finch (*Geospiza magnirostris*) can assimilate the nutrients of a caltrop (*Tribulus cistoides*) seed, it first has to find it, get its body to it, use its evolutionarily tuned beak to crush the mericarp that surrounds it, and transport the seed from the tip of its beak to its oropharynx and esophagus for digestion (Grant 1986). These processes have been described in detail for only a handful of avian taxa (Rubega 2000).

This chapter focuses on the various ways in which birds find, obtain, and ingest food. Zweers et al. (1994) described three main processes in food consumption: foraging—searching for food using the whole body; food acquisition—using the head to access the food and move it into the oropharynx (includes ingestion and swallowing); and digestion. We follow their distinction between "foraging" and "food acquisition" (hereafter, "feeding"); digestion was treated in chapter 7. Note, however, that these are different components along a continuum of behavior, and different sources partition the events differently. However, the act of locating and accessing food typically involves different structures and movements from those involved in the act of consuming it. Accordingly, this chapter begins with a breakdown of the general mechanisms of foraging and feeding that apply to most groups of birds. We continue to explore some of the dietary diversity among taxa, and specifically how different foraging and feeding behaviors are associated with different kinds of foods. We then discuss the energetic and ecological implications of foraging and feeding behavior to address how birds allocate time and energy to these tasks, what kinds of biotic and abiotic factors influence their behavior, and how their behavior, in turn, affects their environment (i.e., other individuals, species, and habitats). Finally, we discuss how foraging behavior develops within

Figure 15.1. Birds employ a wide array of foraging tactics to feed on a great variety of food sources. *Top row* (from left to right): juvenile Pied-billed Grebe (*Podilymbus podiceps*), Costa's Hummingbird (*Calypte costae*), California Scrub-Jay (*Aphelocoma californica*). *Middle row*: Brown Creeper (*Certhia americana*), Black Skimmer (*Rynchops niger*), American Kestrel (*Falco sparverius*). *Bottom row*: Black Oystercatcher (*Haematopus bachmani*), Nashville Warbler (*Oreothlypis ruficapilla*), Short-billed Dowitcher (*Limnodromus griseus*), and Least Sandpiper (*Calidris minutilla*). *Photos by Dave Furseth, www.davesbirds.com.*

individuals (i.e., ontogenetically), and among individuals (e.g., cultural transmission, group foraging). We conclude the chapter with a synopsis of contemporary methods for studying foraging and feeding behavior, and discuss the various ways in which information gleaned from these kinds of studies can address wildlife management and conservation issues.

GENERAL MECHANISMS OF FORAGING AND FEEDING

Foraging encompasses neurosensory mechanisms (e.g., vision, olfaction, memory, etc.; chapter 12) for selecting suitable habitats and detecting and locating food—often at a distance—as well as locomotor mechanisms involved in getting to it (e.g., wing shape, flight; chapter 10). The subjects of object recognition and search image formation (e.g., Pietrewicz and Kamil 1979) are treated much more extensively in the neuroethology and cognitive science literatures. For finding food, birds (as well as other predatory animals) have been traditionally dichotomized into "sit-and-wait predators"— sedentary predators that wait for prey to come within striking distance (e.g., flycatchers) and "active searchers"—those that move throughout the environment to locate and attack prey (e.g., titmice, warblers) (McLaughlin 1989). These strategies have profound implications for foraging theory (treated below), but also practical implications for the probability of encountering prey and energy expenditure during foraging. This dichotomization, however, disguises the reality that there is a gradation of foraging modes within each classification. Indeed, Fitzpatrick (1980) identified 10 foraging modes among New World tyrannid flycatchers, from foliage gleaning (picking prey items off leaf surfaces), to aerial sallying (long periods of stationary perching, punctuated by short flight capture attempts), to ground-related foraging (prey attacks that are initiated, and end, on the ground) (fig. 15.2). Beak, wing, and leg morphology are broadly correlated with these foraging modes (Fitzpatrick 1985) given their direct implications for prey prehension, flight, and terrestrial locomotor capabilities, respectively. "Foliage gleaners" tend to have broad bills, short, rounded wings, and long legs, whereas "aerial hawkers" tend to have long, pointed wings, short legs, and triangular-shaped bills (Fitzpatrick 1985). Robinson and Holmes (1982) identified six categories of searching behavior in a taxonomically diverse group of foliage-gleaning insectivores (table 15.1). Birds of prey have been dichotomized similarly into "sit-and-wait" and "active searcher" classifications, but many species employ mixed strategies. As in flycatchers, these are often correlated with morphological characteristics, such as wing loading (chapter 10); active searchers tend to have lower wing loads (Jaksić and Carothers 1985).

Table 15.1. Searching tactics of various forest-dwelling insectivorous birds and the species, among the 11 studied by Robinson and Holmes (1982) at Hubbard Brook, NH, USA, which performed them with the greatest frequencies.

Behavior	Definition	Representative species
Open-perch searchers	Birds perch in open areas and scan vegetation and air space for active prey, such as flies and bees	Least Flycatcher (*Empidonax minimus*), Scarlet Tanager (*Piranga olivacea*)
Variable-distance searchers	Birds hop and fly about, searching substrates near and far, mostly for lepidopteran larvae	vireos (*Vireo* spp.)
Flush-chasers	Birds move rapidly to flush homopteran prey and actively pursue them	American Redstart (*Setophaga ruticilla*)
Near-surface searchers	Birds make frequent movements of short distances to search nearby substrates for small, cryptic prey	*Dendroica* spp. warblers
Substrate-restricted searchers	Birds seek specific substrates, such as dead, damaged, or curled leaves	Black-capped Chickadee (*Parus atricapillus*)

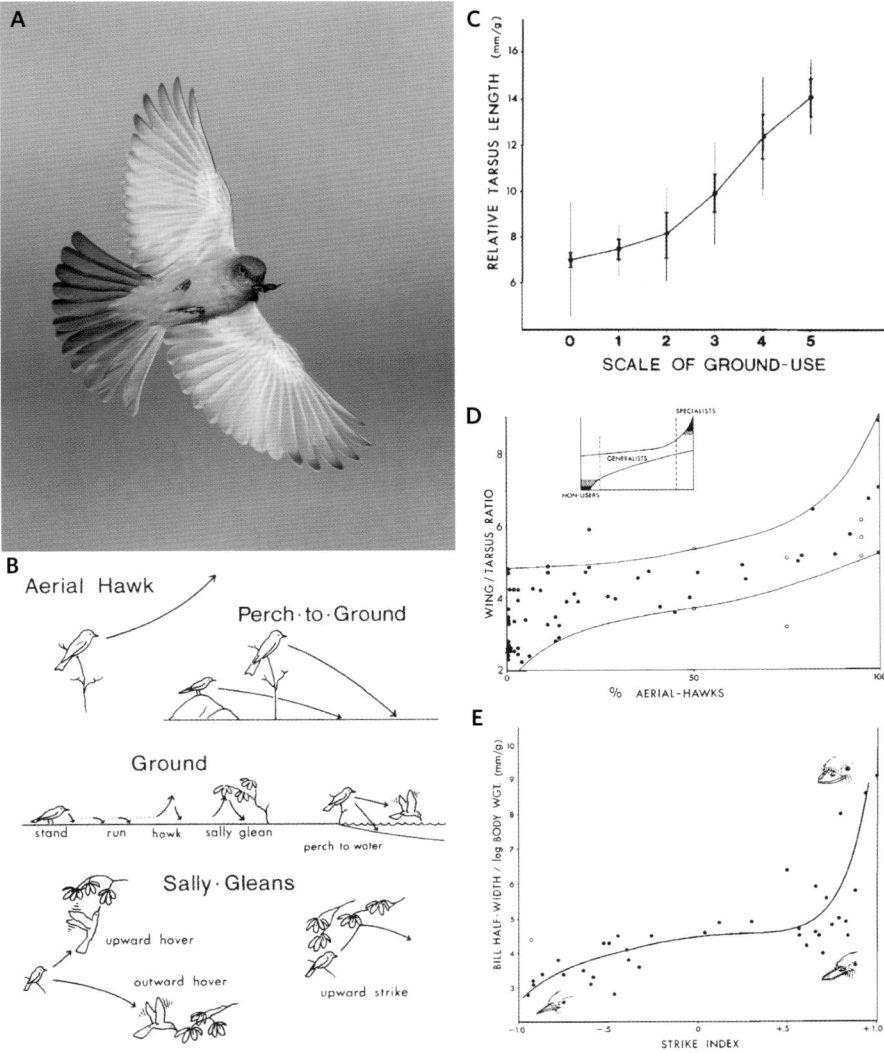

Figure 15.2. A, Representative tyrannid flycatcher, Say's Phoebe (*Sayornis saya*), capturing an insect in mid-flight (*photo by Glenn Conlan*); B, Representative foraging modes of tyrant flycatchers: "sallying" refers to a behavior typical of flycatchers, in which brief, short flights at prey items (e.g., arthropods) are interspersed among relatively longer periods of sedentary perching. *Figure modified from Fitzpatrick 1980, reproduced with permission from* The Condor. Morphology tends to be associated with foraging mode: C, tarsus length is longer in ground foragers; D, relative wing length is longer in those that spend more time aerial-hawking (i.e., taking insects on the wing); E, bill width is greater in those that tend to strike prey along leaf surfaces from below. *From Fitzpatrick 1985, reproduced with permission from* Ornithological Monographs.

Nevertheless, search tactics and foraging patterns can vary within species between years (Robinson and Holmes 1982), and even a single individual could be both a sit-and-wait and an active searcher opportunistically, depending on the circumstances. For instance, hummingbirds will often engage in flycatching from their favorite perches while waiting for flowers to refill (Stiles 1995, Rico-Guevara 2008).

The Feeding Cycle

The "gape cycle" (opening and closing the jaws during feeding), involves ingesting the food, moving it through the mouth, mechanically breaking it down, and finally swallowing (Schwenk 2000a). This entire process has been studied more extensively in lepidosaurs (snakes and lizards) and mammals, but also describes the movement of the jaws and hyolingual apparatus (tongue and hyoid bone; chapter 6) observed in birds (van den Heuvel 1992, Bels and Baussart 2006). Upon "grasping" the food or prey item with the bill (ingestion), many birds must first reduce it into smaller, or otherwise more manageable, bits before it can be transported intraorally. For instance, most raptors take relatively large prey and pluck out fur and feathers to bite off chunks of flesh from their quarry. Sparrows and finches husk seeds from their protective shells. Whereas in mammals such "processing" is performed by the teeth, in birds this is performed along the tomia (cutting margins of the bill) and bill tips of keratinous rhamphotheca (chapter 6) through repetitive

"mandibulation"—or dorsoventral movements of the upper and lower bills to effect "chewing" behavior. Note that birds do not necessarily chew in the way that mammals do; mammalian "mastication" implies a certain degree of cyclic rhythmicity that is driven by novel modulatory sensorimotor mechanisms (Ross et al. 2007). Instead, birds exhibit two general types of intraoral transport: cranioinertial ("catch-and-throw") and lingual ("slide-and-glue") mechanisms (Zweers 1982). In the former, the food item is released from the bill tips as the head is jerked upward and backward, then recaptured more caudally in the bill as the head is moved downward and forward, as the jaws are closed. The latter occurs when the food adheres to the tongue tip as the tongue is pressed against it, and the food is carried caudally upon tongue retraction (fig. 15.3). Cranioinertial transport is characteristic of both paleognaths (Paleognathae: ratites) and neognaths (Neognathae: all other lineages). Most neognaths also perform lingual transport, but tend toward cranioinertial transport with larger food items (Zweers 1982). Although their tongues play a role in swallowing, the reduced tongue sizes of paleognaths preclude them from using lingual transport to move food from the bill tip to the pharynx (Tomlinson 2000).

The jaw and hyolingual movements of the gape cycle during filter-feeding (e.g., in ducks) are not fundamentally different from that of "pecking," or jaw prehension (Zweers and Berkhoudt 1991). As the jaws open, water is drawn into the oral cavity via a "suction pump" mechanism formed by

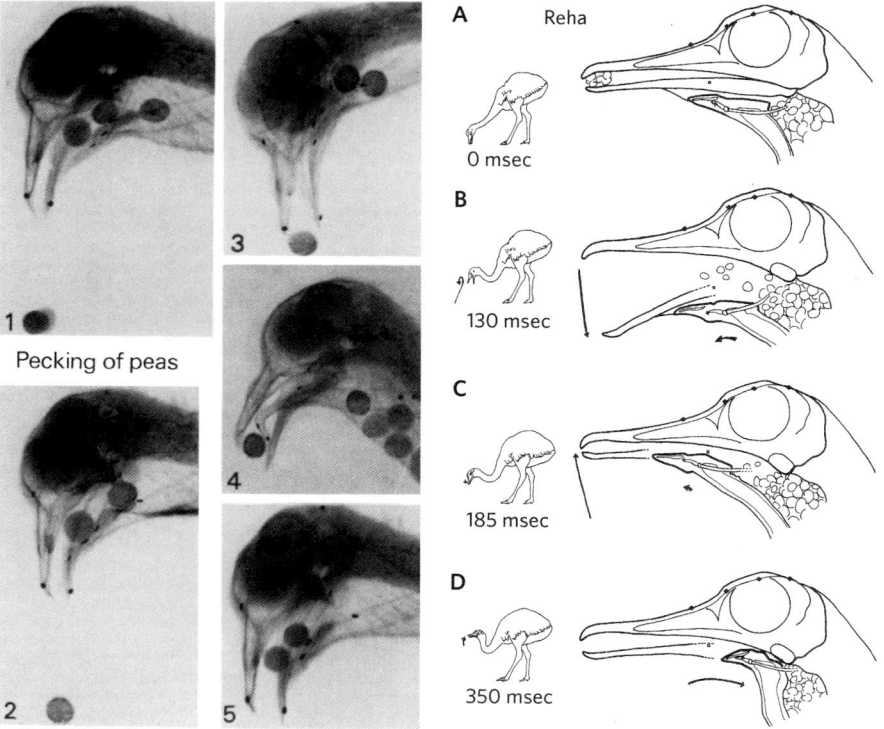

Figure 15.3. *Left,* radiographic sequence showing jaw prehension (1–3) and lingual transport (4, 5) in a pigeon (*Columbia livia*) pecking a pea. Lead markers track the movements of the bill tips and tongue. *From Zweers 1982, reproduced from* Behaviour, *with permission. Right,* illustrated sequence from a cineradiographic film showing cranio-inertial transport and deglutition in a Rhea (*Rhea americana*). *From Tomlinson 2000, reproduced from* Feeding: Form, Function, and Evolution in Tetrapod Vertebrates, *Academic Press, with permission.*

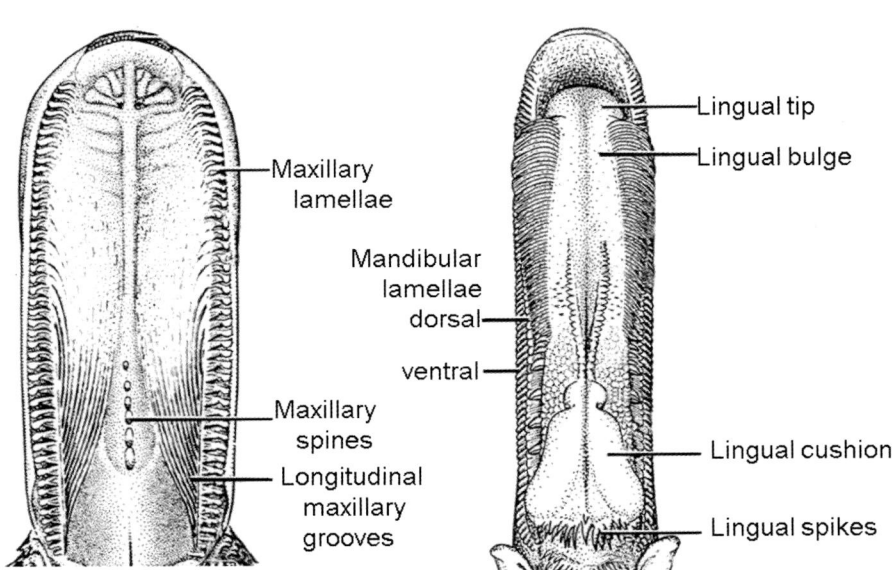

Figure 15.4. Key anatomical features of *left*, the upper bill and *right*, lower bill related to filter-feeding in ducks. *Figure modified from van der Leeuw et al. 2003, reproduced from Animal Biology, with permission.*

the lingual bulges (lateral, longitudinal thickenings of the tongue) and the lingual cushion (a smooth, convex structure located caudally) of the tongue, and pumped out (mostly by movement of the lingual bulges) through the sieve formed by the lamellae along the lateral margins of the upper and lower bill (fig. 15.4). Food particles are brushed from the dorsal lamellae into the longitudinal maxillary grooves, where lingual spikes then sweep the particles caudally during tongue retraction for transport to the esophagus (Zweers et al. 1977). By altering their gape, ducks can adapt the sieve to the size of the food particles; visual, gustatory, and tactile senses are all used to estimate particle size and to regulate their gape during feeding (Dubbeldam 1984). The rhythmic repetition of the gape cycle is thought to be controlled by "central pattern generators" in the brain stem (Tomlinson 2000, Bels and Baussart 2006) and modulated by exteroceptive (mechanoreceptors, such as Herbst and Gandry corpuscles, located in the beak and tongue) and proprioceptive (stretch receptors located within the jaw muscles) systems that monitor the location of food and the position of the feeding apparatus, respectively (Dubbeldam 1984, Zweers and Berkhoudt 1991).

Drinking is a slightly different process from feeding but still depends on a high degree of coordination between the jaws and hyolingual apparatus (Kooloos and Zweers 1989). Water is typically transported from the beak tips to the pharynx in a single gape cycle, although in ducks, two steps are required to navigate around the lingual bulges and cushion (Kooloos and Zweers 1991). Two general types of drinking behavior have been described in birds: "tip-up" and "tip-down" or "suction" drinking (Hiedweiller and Zweers 1989). During tip-up drinking, the beak is immersed in water, the liquid is "scooped" into the lower bill by moving the head forward, and the load of water is then transported to the pharynx and esophagus by the actions of gravity and retraction of the tongue and larynx (Hiedweiller and Zweers 1989, Gussekloo and Bout 2005a). Suction drinking, by contrast, occurs with the head and bill tipped downward and relies on pumping mechanisms of the tongue and pharynx. Tip-up drinking is thought to represent the ancestral condition (typical of paleognaths and basal neognaths), whereas tip-down drinking may represent a derived condition; one that, among estrildid finches, may have been exapted from selection on anatomical elements for rapid seed-eating behavior in predator-rich environments (Hiedweiller and Zweers 1989).

Anatomical Aspects

Cranial kinesis—the mobility of the quadrate bones, portions of the upper bill, and lower jaws relative to the brain case (fig. 15.5)—is an important facet of avian feeding behavior (Bock 1964, Bühler 1981, Zusi 1984). Various bones, ligaments, and muscles of the skull (chapter 6) interact to elevate and depress the upper bill about the craniofacial hinge (prokinesis) or elevate and depress portions of the bill about bending zones located more distally (rhynchokinesis). Prokinesis and some mobility of the quadrates are present in virtually all birds (Bout and Zweers 2001). Rhynchokinesis, in one form or another, is relegated to comparatively fewer groups and is particularly well-developed in shorebirds (Zusi 1984, Estrella and Masero 2007). Beyond this, relationships between types of cranial kinesis and specific feeding behaviors are unclear (Bout and Zweers 2001, Bels and Baussart 2006). Kinetic motions of the upper jaw contribute substantively to gape changes during the feeding cycle (van den Heuvel 1992, Gussekloo and Bout 2005b). Furthermore, prokinesis may

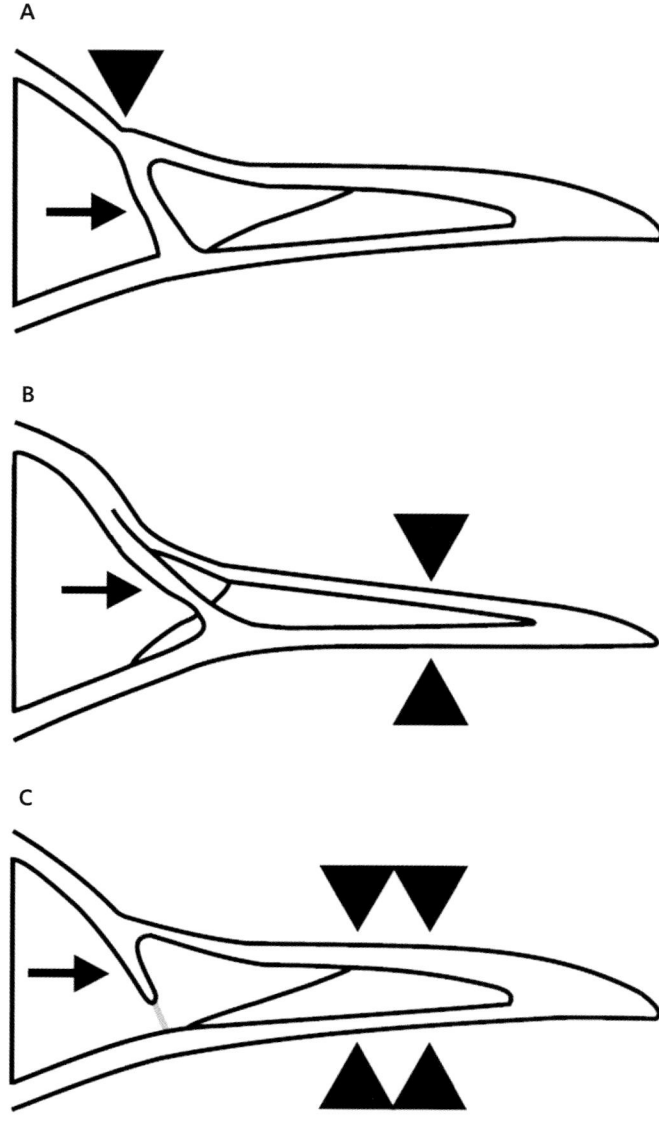

Figure 15.5. Various forms of avian cranial kinesis. *A*, Prokinesis refers to the ability of the upper beak to elevate/depress about the craniofacial hinge (black triangle), and is characteristic of most birds; *B*, Distal rhynchokinesis refers to elevation/depression along bending zones (black triangles) located more rostrally along the upper beak, and is characteristic of shorebirds; *C*, Central rhynchokinesis refers to elevation/depression of the upper beak along bending zones (black triangles) located roughly halfway along the length of the beak, and is characteristic of paleognaths, such as Ostriches (*Struthio camelus*) and Rheas (*Rhea americana*). Arrows indicate the nasal bar. *Figure from Gussekloo and Bout 2005b, reproduced from* The Journal of Experimental Biology, *with permission.*

enhance jaw speed, bite force, and bill-tip dexterity (Bock 1964, Bramble and Wake 1985, Hull 1991, Bout and Zweers 2001, Meekangvan et al. 2006), affording birds greater versatility for manipulating food and other items with their beaks (Liem et al. 2001).

Perhaps it is for obvious reasons that studies of avian feeding mechanics tend to focus on the skull and beak. Many birds, however, rely extensively on their feet for seizing, handling, and processing food (fig. 15.6). Clark (1973) suggested that use of the feet for feeding affords birds a greater range of possible food items that cannot be tackled by the bill alone. Raptors (e.g., eagles, hawks, falcons, and owls)—among the most notable foot-users—make initial contact with prey using their feet by striking or grasping them with their talons (Goslow 1971). The enlarged, sharp claws possessed by these birds on their toes play important roles for latching onto and immobilizing prey for subsequent dispatchment and feeding (Einoder and Richardson 2007, Fowler et al. 2009). In most raptors (e.g., hawks, eagles, and owls), gripping force plays a more important role for inducing death via thoracic compression (Goslow 1971, Csermely et al. 1998, Ward et al. 2002), whereas in others (e.g., falcons) the beak is the primary killing implement (Cade 1982, Hertel 1995). Some raptors, such as owls and Ospreys (*Pandion haliaetus*), can even manipulate the arrangement of their toes, from an anisodactylous to a zygodactylous configuration (chapter 6), to enhance their grasping capabilities (Ward et al. 2002, Einoder and Richardson 2007; Supplementary Materials S1). Although raptors present dramatic examples of foot use for handling prey, using the feet in the context of feeding is broadly distributed phylogenetically (Sustaita et al. 2013). Other notable foot users are parrots (Psittacidae) and mousebirds (Coliidae), which possess unique modifications of the intrinsic foot musculature that facilitate a wide range of toe motions and enhance digital dexterity for grasping, holding, and manipulating food items (Berman and Raikow 1982, Berman 1984; fig. 15.6). Many other birds, such as chickadees and titmice (Paridae), use their feet to support food items to manipulate them with their beak (fig. 15.6; Supplementary Materials S2).

Behavioral Aspects

In addition to these intrinsic anatomical and behavioral mechanisms to foraging and feeding, birds often extend their use of the environment with certain behaviors. By taking advantage of objects around them, they can enhance their efficiencies for obtaining, and maintaining, food resources. The former is accomplished through the use and manufacture of tools, and the latter is accomplished through the use of food caches—often hidden storage sites or structures. Both of these types of behavior are cognitively demanding, and tend

Figure 15.6. In addition to their beaks, birds use their feet in a variety of ways to handle prey items during feeding. *Top,* Bald Eagle (*Haliaeetus leucocephalus*). *Photo by Ron Dudley. Bottom left,* Carolina Chickadee (*Poecile carolinensis*). *Photo by Kerri Farley/New River Nature. Bottom right,* Umbrella Cockatoo (*Cacatua alba*). *Photo by D. Sustaita.*

to be associated with intelligence (Emery and Clayton 2005; chapter 12).

Tool use constitutes a series of actions with an external object to procure food. According to Alcock (1972), "tool-using involves the manipulation of an inanimate object, not internally manufactured, with the effect of improving the animal's efficiency in altering the position or form of some separate object." Tool use in birds spans from relatively simple to exceedingly complex. On one end of the spectrum is the rock-tossing behavior that Egyptian Vultures (*Neophron percnopterus*) perform to crack large eggs (Alcock 1970; Supplementary Materials S3). At the other end of the spectrum are New Caledonian Crows (*Corvus moneduloides*), which not only use tools to extract beetle larvae (and other arthropods) from holes that could not otherwise be accessed (Bluff et al.

2010) but also fashion them from leaf stems and twigs into shapes and sizes fit for specific tasks (Hunt 1996, Weir et al. 2002, Bluff et al. 2010; fig. 15.7). Between these endpoints exist several other avian examples of tool use. Among the better-documented cases are some species of Darwin's finches, most notably the Woodpecker Finch (*Cactospiza pallida*), which use twigs and cactus spines to wedge prey from crevices (Alcock 1972). New Caledonian Crows have been observed to make, use, and keep hooked implements in both field and lab settings (fig. 15.7). Their reliance on tools is so profound that their bills might have been adapted to handle the implements better without obstructing their view during object manipulation (Troscianko et al. 2012, Matsui et al. 2016). The use of these stick tools has important ecological consequences for the crows, because a substantial amount of their dietary protein and lipids is derived from the wood-boring beetle larvae they obtain using them (Rutz et al. 2010). This kind of technical capability is exceedingly rare among non-primates, and bears hallmarks of tool manufacture known only from early human tool-using cultures, such as standardization; discrete, distinctly shaped, tool types; and the use of hooks (Hunt 1996). What is even more exceptional is the fact that this occurs with little prior experience with the setting and materials (Weir et al. 2002; fig. 15.7). This kind of research has further-reaching implications for understanding the ecological and neural preconditions for the evolution of complex cognition in animals (Weir et al. 2002).

As we saw in chapter 10, birds are somewhat limited in their load-bearing capacities as a result of constraints on body size for flight. Although several species have evolved crops that allow the temporary storage of more food than can be digested at any one time (chapter 6), not all birds have one. Even then, the crop is somewhat limited in its capacity. Some species, however, have evolved ways to help hold on to "extra" food resources through caching or hoarding food items in the soil, clumps of vegetation, and/or cracks and crevices in rocks and tree trunks (Vander Wall and Balda 1981). If the Corvidae (crows, ravens, jays, and nutcrackers) are well known for their tendencies to hide food items (e.g., seeds, acorns, nuts), they are renowned for their exquisite spatial memories for retrieving them later (Kamil and Balda 1990, Supplementary Materials S4, S5). Studies with California Scrub-Jays (*Aphelocoma californica*; fig. 15.1) have demonstrated that they possess an "episodic-like" memory that allows them to remember *what* type of food they cached (e.g., perishable versus nonperishable), *where* they cached it, and *when* it was cached (Emery and Clayton 2005). Acorn Woodpeckers (*Melanerpes formicivorus*) persist in extensive family groups, and during autumn they stock large granaries—tree trunks peppered with acorns, on which they feed in addition

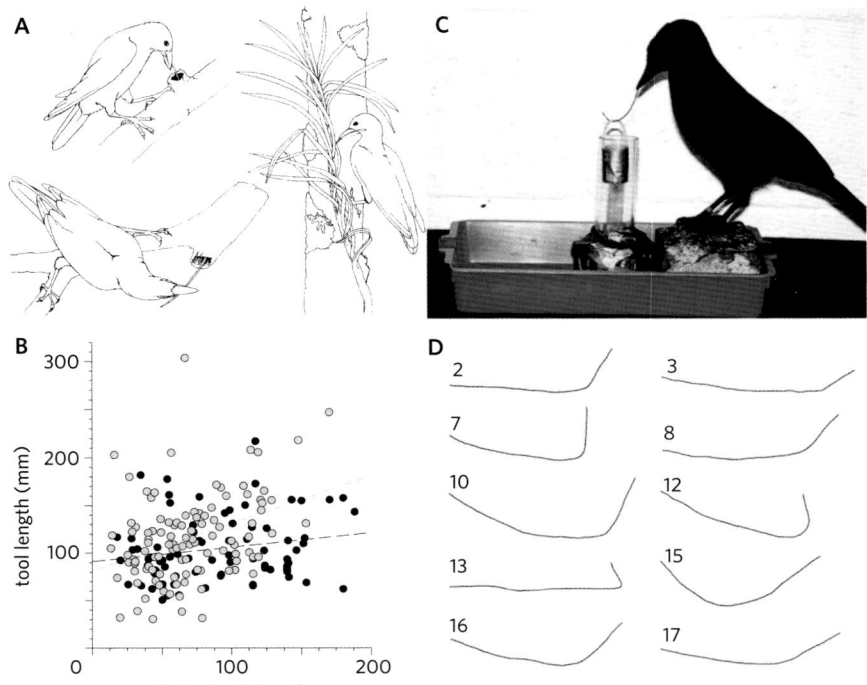

Figure 15.7. A, Illustrations of hooked-tool foraging in New Caledonian Crows (*Corvus moneduloides*). *Illustration from Hunt 1996, adapted with permission from* Nature. B, Relationship between the lengths of twig (gray dots) and leaf-stem (black dots) tools used by New Caledonian Crows (*C. moneduloides*) and the depths of the holes into which they were inserted, showing that crows match tools to holes during natural foraging. *Figure from Bluff et al. 2010, reproduced from* Proceedings of the Royal Society B, *with permission*. C, Experimental apparatus that required a captive New Caledonian Crow—(with little prior experience)—to manufacture and use D, hooked implements to access a piece of food. *Panels C and D from Weir et al. 2002, reproduced with permission from BERG, Oxford, and* Science.

to insects and sap—that they continuously replenish and defend (MacRoberts and MacRoberts 1976; fig. 15.8).

Caching is not limited to granivorous birds. Falcons (*Falco* spp.) and some accipitrid hawks and eagles are known to cache prey as well (Olsen 1995). Many species often take prey larger than their own bodies, which they cannot possibly consume in one sitting, and will often drag the remains of a kill under a log, rock, or into the bushes and return to it at a later time. Caching has the adaptive benefits of minimizing food wastage, buffering against temporary food shortages, and optimizing foraging efficiency by allowing hunting effort to be decoupled from the act of feeding and more strategically allocated during times of greatest prey availability (Olsen 1995). Some Australian butcherbirds (*Cracticus* spp.) and Southern Boubous (*Laniarius ferrugineus*) will actually suspend their insect and small vertebrate prey in trees and bushes, by wedging them in the forks of branches (Yosef and Pinshow 2005). Shrikes (*Lanius* spp.) also do this, but then also take it a step further by skewering their arthropod and vertebrate prey on sharp objects (e.g., thorns, spines, barbed-wire fencing; fig. 15.8). This "impaling" behavior extends beyond the purpose of surplus food storage by functioning in prey dismemberment, mate attraction, demarcation of territories, and even to "age" chemically defended prey (e.g., lubber grasshoppers) to allow them to detoxify (Yosef and Pinshow 2005).

DIVERSITY OF AVIAN DIETS

Birds consume an exceptionally wide range of food items, which is reflected in the great diversity of bill shapes (Rubega 2000). Beaks also play a role in sound communication, thermoregulation, and sensory functions, but their most obvious function is in feeding. The association between beak shape and food is a time-honored tradition, most notably and thoroughly illustrated in Darwin's finches (fig. 15.9). However, Bowman (1961) recognized that the bills do not act by themselves, but rather are powered by jaw muscles, which also vary in size according to the demands imposed on the jaws by different types of food. Furthermore, differences in beak shape may be driven by their abilities to withstand the forces they incur, depending on what and how they eat. This sets the stage for an important facet of adaptation—that the feeding apparatus is not necessarily "tuned" to the food source per se, but rather to the mechanical task of consuming it (Schwenk 2000b, Schwenk and Rubega 2005).

Birds consume food items that range dramatically in their material properties (fig. 15.10) and in regard to behavior, for animal prey. For instance, the relatively greater toughness of prey skin might select for sharp beaks and relatively more complex head and body movements for tearing, whereas the relatively greater stiffness of bone and seeds selects for robust beaks and jaws to generate the requisite crushing forces. Thus, multiple morphological and behavioral solutions have

Figure 15.8. Examples of avian food caches from throughout California, USA. *Left*, Caches of Loggerhead Shrikes (*Lanius ludovicianus*; also fig. 15.14D), harboring different arthropod and vertebrate prey types, and natural and anthropogenic impaling implements. *Photo by Diego Sustaita. Right*, An Acorn Woodpecker (*Melanerpes formicivorus*) granary. *Photo by Dave Furseth.*

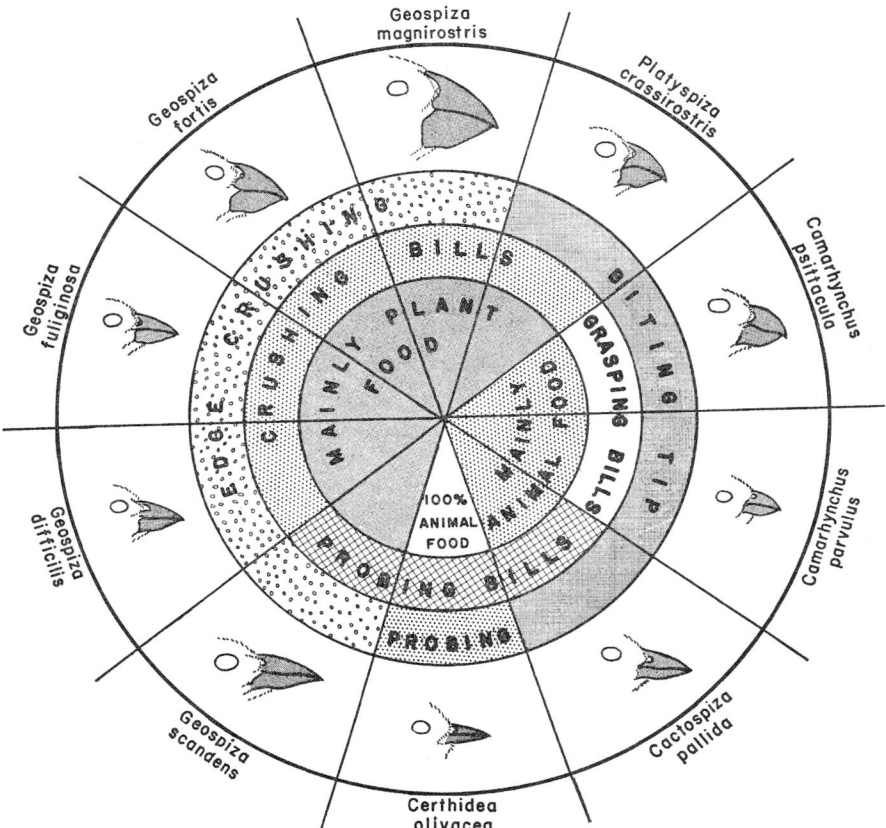

Figure 15.9. In a classic study of adaption in Darwin's Finches, Robert Bowman investigated relationships among bill shape, feeding behavior, and diet among species, based on the forces that the jaws and beak experience during feeding. *Figures from Bowman 1961, reproduced from University of California Publications in Zoology (public domain).*

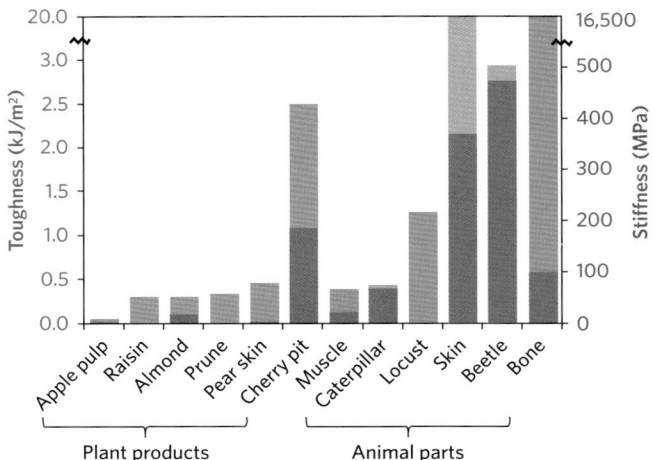

Figure 15.10. Material properties of various types of plant products and animal parts, representative of food items handled and consumed by birds, arranged in order of increasing toughness for each plant and animal sources. "Toughness" expresses a material's resistance to crack propagation and is used to describe the properties of relatively "soft" and compliant materials, whereas "stiffness" (i.e., modulus of elasticity) refers to the resistance to deformation within the plastic range and is used to describe the properties of relatively "hard" and brittle materials (Strait and Vincent 1998). *Graph drawn from mean values for selected plant products reported by Williams et al. 2005, and for animal parts reported by Strait and Vincent 1998.*

evolved to meet these mechanical and locomotor demands. As we will see below, some of these demands require more specialized morphologies and/or foraging behaviors, whereas others are more generalized. For instance, some food sources possess qualities of both stiffness and toughness. In fact, more often than not, birds will consume a variety of food items, despite their apparent adaptation for a particular type of food, particularly when resources are unpredictable.

Below we outline some of the main types of foods that birds consume: plants (including their products), invertebrates, and vertebrates, each of which imposes different challenges. Note that it is beyond the scope of this text to outline the diet of every species—there are great resources available for this information, such as del Hoyo et al. (2014) and Wilman et al. (2014)—so we discuss examples as they relate to these major dietary categories. It is also worth repeating that many species transcend these categories and feed opportunistically on a variety of food sources. Thus, the traditional "X-ivore" classifications (carnivores, frugivores, etc.) may not necessarily be universally applicable for categorizing avian diets. Therefore we describe bird diets in the context of the foraging and feeding modes with which they are associated. "Bulk" foods is a fairly comprehensive category that includes

terrestrial plants and animals, or parts thereof, which birds must pick, capture, pluck, husk, and/or otherwise process in order to ingest. Because diets in this category derive from vastly different plant and/or animal sources, the main thing they have in common is their distinction from diets associated with other feeding modes. "Liquid" foods refer to diets that consist of nectar and other plant and animal exudates. "Aquatic and marine" foods refer to prey that are obtained almost exclusively in freshwater or marine environments and are associated with particular sensory, behavioral, and morphological attributes of consumers.

Bulk Foods

Fruit

Fruit constitutes an attractive resource for birds, because it is sessile and easily accessible. This does not mean that frugivory is not without its challenges; fruiting phenology, spatial distributions, and crop sizes can be quite variable (Moermond and Denslow 1985). Nutritionally, fruit tends to be relatively low in protein, variable in lipids, and high in carbohydrates and water (Moermond and Denslow 1985). Although the low fiber composition does reduce requirement for mechanical breakdown of cell contents, the seed masses within the bulky dilute pulp require some handling (Moermond and Denslow 1985). Nevertheless, eating fruit is probably among the more generalized forms of feeding. Birds such as robins and thrushes (Turdidae) and tanagers (Thraupidae) are characterized by bills that do not appear particularly specialized (Moermond and Denslow 1985). As a result, beaks of various shapes and sizes may be associated with frugivory. In fact, it is not so much the shape and size of the bill that is important for some frugivores, but rather the size of the gape (oral aperture) that limits the sizes of fruits consumed. This is particularly true of species that do not tend to process the fruit, but rather swallow it whole. Wheelwright (1985) found that the average and maximum fruit diameters consumed by various frugivorous species were positively correlated with gape width, such that birds with larger gapes tended to consume larger fruits. However, the minimum diameter of fruits consumed was not correlated with gape size, reflecting the reality that large-gaped birds also feed on small fruits (fig. 15.11). Nevertheless, even large-gaped birds do not necessarily forage randomly, but are size-selective. Theoretically, intermediate-sized fruit provides the greatest payoff: fruits that are too large require more handling, and fruits too small limit detection (Wheelwright 1985).

In addition to gape-size determinants of frugivory, there is considerable variability in feeding behavior. Birds generally "capture" fruit either in flight or from a perch; although most species use both techniques depending on the circumstances,

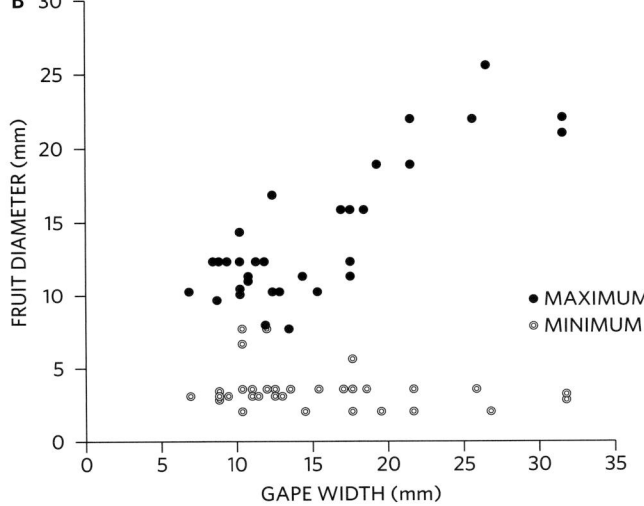

Figure 15.11. A, The Cedar Waxwing (*Bombycilla cedrorum*), a predominantly frugivorous North American species, consuming a berry. *Photo by Glenn Conlan. B*, Maximum and minimum diameters of fruits consumed by frugivorous birds characterized by different gape (oral aperture) widths. Whereas gape size limits the maximum size of fruits that can be swallowed, minimum-sized fruits are available to birds with small and large gape sizes alike. *Figure from Wheelwright 1985, reproduced from* Ecology, *with permission.*

one typically predominates (Moermond and Denslow 1985). In some cases these distinct behaviors have a morphological signature: wide, flat bills are associated with birds that take fruit on the wing; stronger-billed birds tend to bite chunks from the fruit, and weaker-billed birds tend to swallow them whole (Moermond and Denslow 1985). In other cases, behavior is somewhat divorced from morphology. Foster (1987) identified several types of fruit-eaters among the Neotropical species she studied; some "pluck and swallow," others "cut or mash," and still others "push and bite" into fruits. These forms of processing were relatively independent of bill di-

mensions but varied in their efficiencies. The fact that some of these birds could subsist on fruit despite their relative inefficiencies suggests that feeding efficiency in frugivores may not be limited by handling and/or processing time, potentially because of an enhanced digestive efficiency of a fruit diet. The fruit-eating behavior of toucans (Ramphastidae) and hornbills (Bucerotidae) underscores this decoupling between beak morphology and fruit. These birds use a form of cranioinertial feeding dubbed the "ballistic transport mechanism," in recognition of the ballistic curve of the motion path of the food item as it is tossed from the bill tips and recaptured in the oropharynx (Baussart et al. 2009, Baussart and Bels 2011). This mechanism may represent a specialization of birds with large (ostensibly cumbersome) beaks shaped by various selective pressures to feed on relatively large fruits. Yet other birds, such as mousebirds, bypass potential bill and gape size limitations by recruiting their feet to manipulate fruit, which allows them to feed on larger and more formidable fruits (Symes and Downs 2001).

Frugivory in birds has important implications for the coevolution of plants and birds, as do insectivory and nectarivory, described below. Wheelwright's (1985) work (above) suggests that plants that produce larger fruits effectively specialize on larger avian dispersers of their seeds, which could be advantageous (greater disperser fidelity and reliability) or disadvantageous (narrow pool of potential seed dispersers). Avian frugivory has long been hypothesized to drive the evolution of fruit color—particularly red and black—although with mixed support (Willson and Whelan 1990). Evidence in support of birds acting as selective agents on fruit color (according to experiments with Australian Silvereyes, Zosteropidae) suggests that preferences are based on hue (the spectral composition of reflected light), not brightness (luminance) (Puckey et al. 1996).

Grains, Seeds, and Nuts

Granivory is a catchall term for birds that eat relatively brittle foods like seeds and grains. The foraging aspects of this classification are tremendously variable because they span everything from ground-based searching and pecking to accessing seeds and nuts from more complex, arboreal environments. The feeding aspects are similarly variable because they encompass behaviors from grasping and gulping down tiny particles to cracking enormous nutshells to get at the edible interiors. The most basic form of granivory is probably that represented by galliforms (e.g., fowl, pheasants, and quail), which employ primarily pecking and cranioinertial feeding mechanisms described above.

Cracking and husking seeds requires not only force but also maxillary/mandibular dexterity to position the seed properly within the beak (van der Meij and Bout 2006). Indeed,

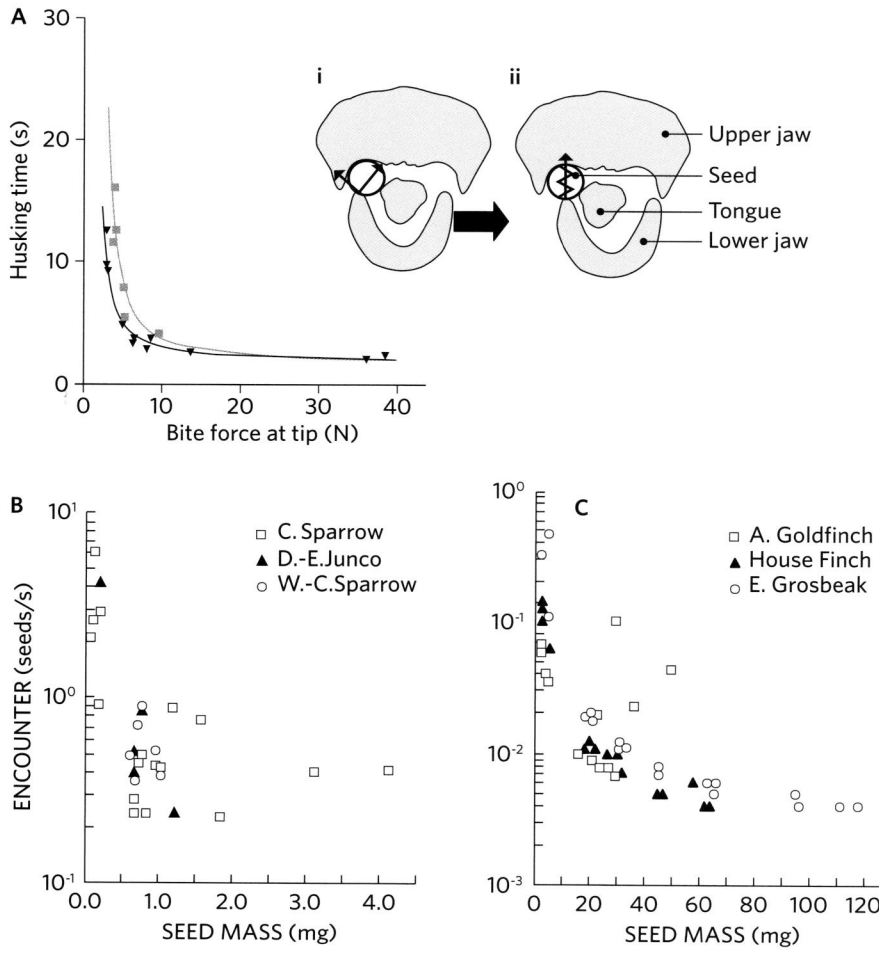

Figure 15.12. A, When seed hardness is just below maximum bite force capacity of estrildid (squares) and fringillid (triangles) finches, greater bite forces serve to decrease the amount of time it takes to husk the seed. Equally important to bite force is the manner in which the seed is positioned along the tomia of the upper and lower bills (inset, showing schematic cross sections of jaws and tongue), which is accomplished by lateral translation of the mandible (i→ii) during mandibulation. *Figures from van der Meij and Bout 2006, reproduced from* The Journal of Experimental Biology, *with permission.* The rates at which B, sparrows, and C, finches encounter seeds during foraging vary with seed size but, overall, are substantially lower in finches (note the difference in scales on the vertical axes). These lower encounter rates for finches suggest that, as a result of their greater handling efficiencies with large seeds, they can subsist on comparatively fewer, more sparsely distributed spatially and temporally, seed resources. *Figure from Benkman and Pulliam 1988, reproduced from* Ecology, *with permission.*

maximum bite force capability sets an upper limit to the size and hardness of the seeds these birds can consume. Larger, stronger-billed birds are able to husk larger, harder seeds faster than are smaller, weaker-billed birds. But efficient processing does not all come down to force, as evidenced by the fact that the reverse is also true—smaller, weaker-billed birds are faster at husking small seeds than are larger, stronger-billed birds (Abbott et al. 1975, Soobramoney and Perrin 2007). The larger and harder the seed, the greater the number of mandibulations (and time) it takes to husk it (van der Meij et al. 2004). However, when seed hardness is just below maximum bite force capability, greater bite forces serve to decrease husking time, but ultimately seeds must be positioned properly along the edges of the upper and lower bill tomia to be cracked effectively (fig. 15.12A).

In fact, bill shape and feeding behavior are often adapted to the specific demands of preferred seed types. For instance, in a comparison among similarly sized sparrows and finches, Benkman and Pulliam (1988) found that, because of their greater efficiency in handling larger and more diverse types of seeds, finches could persist in areas with much lower seed densities where encounter rates were low (fig. 15.12B, C). In addition, finches possess seed-storing structures (e.g., esophageal diverticula and buccal pouches) that delay satiation and bridge gaps in foraging when resources are sparsely distributed. Sparrows, on the other hand, required encounter rates one to two orders of magnitude greater than did finches to meet their daily energetic requirements because of their smaller seed-size selection and relatively lower seed-handling efficiencies (Benkman and Pulliam 1988). Similarly, when crossbills—conifer specialists, whose "crossed" upper and lower bill tips are adapted to prying apart cone scales to extract seeds (Benkman 1987; box on page 454)—were fed other types of seeds, they required two to three times the encounter rates and seed abundances than those required by other seed-eating species to meet their daily energy demands (Benkman 1988).

Larger seeds and nuts with hard or fibrous shells present additional challenges, which parrots, for instance, have answered by evolving a unique jaw mechanism that facilitates the movements and forces necessary to crack even the strongest nuts (Homberger 2003). For example, unlike most other taxa, psittaciforms possess musculoskeletal features that enable transverse motions of the mandible, which are essential for properly positioning and cracking seeds between a corner of the transverse edge of the lower bill and upper bill tip in calyptorhynchid cockatoos (Homberger 2003). Many other species possess a notch in the palatal surface of the upper bill rhamphotheca that forms a shelf against which seeds and nuts are anchored as they are manipulated with the mandible. During these manipulations, the bills and tongue are used to assess the physical attributes of the shell to find the weakest points and, hence, the most effective places to apply bite forces (Homberger 2003).

Plants and Leaves

The consumption of plant matter (herbivory) is among the rarer forms of subsistence in birds. This is in part because the long digestive tracts required to process and assimilate cellulose impose added weight costs, and gut retention times cannot keep pace with the generally high metabolic demands of birds (Downs et al. 2000). Despite the fact that only 2 percent of extant species are characterized by an herbivorous diet (Olsen 2015), herbivory has evolved independently several times and occurs in ratites, Anatidae, Phasianidae, Coliidae, a few species of passerines and Psittacidae, and most notably, in the Hoatzin (*Ophisthocomus hoazin*) (Downs et al. 2000, Olsen 2015).

Among the Anatidae, with their relatively deep, short beaks and terrestrial habits, geese are the most herbivorous. Herbivorous anatids deal with the low digestibility of plant matter by ramping up food intake rates, which in turn decreases gut retention time (Olsen 2015). This "high-throughput" strategy requires that they graze continuously for several hours, in contrast to other herbivores with high retention times, such as Hoatzins and Ostriches (*Struthio camelus*), which rely more on gut fermentation (Grajal et al. 1989, Olsen 2015). Geese possess musculoskeletal and oropharyngeal features of the jaws, beak, and neck that enhance their abilities to grab and maintain a beak-hold on grass (van der Leeuw et al. 2003). As a trade-off, the anatomical (e.g., maxillary spines; fig. 15.4) and behavioral (over-tongue transport) features that make geese (and certain ducks) good at herbivory (fig. 15.13A) also limit pump capacity and make them poor filter-feeders relative to most anatid ducks (van der Leeuw et al. 2003).

Folivory (leaf-eating) requires specialized gut structures and microflora for breaking down cellulose and releasing fatty acids (Downs et al. 2000). The Hoatzin (fig. 15.13B) is rare among avian herbivores, both in terms of the extent to which it depends on leaves for subsistence and in its capacity for foregut fermentation (Grajal et al. 1989). The cooptation of the enlarged crop and esophagus into a fermentation chamber did not come without costs, however, as this modification to their anatomy has contributed to their poor flight capabilities. In South American plantcutters (Cotingidae:

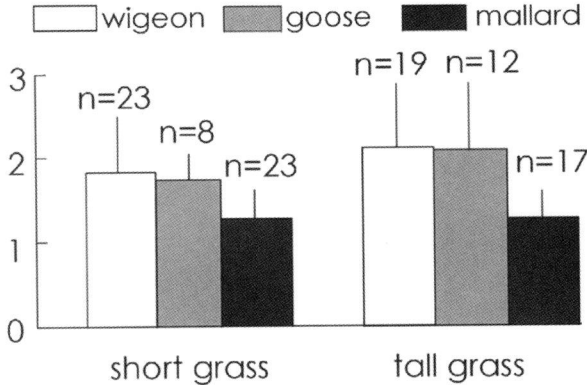

A grass intake rate (g/min.kg)

Figure 15.13. A, Intake rates of ducks and geese during grazing. Geese, such as the White-fronted Goose (*Anser albifrons*) and facultatively herbivorous ducks, such as the Eurasian Wigeon (*Anas penelope*), possess anatomical modifications of the beak and jaws that facilitate higher grass intake rates than primarily filter-feeding ducks, such as the Mallard (*Anas platyrhynchos*). *Figure from van der Leeuw et al. 2003, reproduced from* Animal Biology, *with permission.* B, The Hoatzin (*Opisthocomus hoazin*), a folivorous foregut fermenter, consuming a leaf. *Photo by Pat Gaines.*

Phytotoma spp.)—the only truly folivorous passeriforms—an esophagus with many folds and abundant mucus glands, a roughened muscular stomach inner wall, small intestinal ceca, and bill serrations thought to be adaptive for macerating vegetation reflect their morphological commitment to a leaf-eating habit (López-Calleja and Bozinovic 1999). Some birds feed on leaves more facultatively and thus are less hampered by the demands of a more obligate folivorous lifestyle. For example, the primarily frugivorous mousebirds (Downs et al.

Figure 15.14.
Consumers of
terrestrial animal
prey vary tremen-
dously in form and
function. *A*, Insect-
gleaning Wilson's
Warbler (*Cardellina
pusilla*) with small
coleopteran;
B, Red-shouldered
Hawk (*Buteo linea-
tus*) with avian prey
remains; *C*, Aerial-
foraging Tree
Swallow (*Tachy-
cineta bicolor*)
hot on a fly's tail;
D, Arthropod- and
vertebrate-eating
Loggerhead Shrike
(*Lanius ludovicia-
nus*) casting a pellet
of indigestible prey
parts; *E*, Nocturnal
hunter, Short-eared
Owl (*Asio flam-
meus*), carrying off
a rodent. *Photos
A–C by Dave Furs-
eth; photos D–E by
Ron Dudley.*

2000) and Grayish Saltators (*Saltator coerulescens*) (Rodríguez-Ferraro et al. 2007) will eat more leaves during periods of low fruit availability. Some birds exhibit further behavioral modifications to subsist on herbivory. For instance, despite their small body sizes and the high metabolic demands of endothermy and flight, mousebirds are able to persist on a low-quality folivorous diet by reducing energetic demands associated with thermoregulation and flight by huddling together when roosting and by gliding more, respectively.

Animal Prey

Feeding on animals presents a whole different set of challenges to consumers. In addition to the diverse material properties of animal tissues (fig. 15.10), animals present moving targets. As we saw above, different suites of locomotor (e.g., flight) adaptations accompany different foraging modes, and different suites of feeding (e.g., bill shape) adaptations accompany different feeding modes. Consumers of arthropods (insects and arachnids) and vertebrates (mammals, lepidosaurs, amphibians, and other birds) feed (often variably) on the ground, in trees, and in the air (fig. 15.14). Thus, the various types of terrestrial animals that birds consume select for a wide range of consumer phenotypes. Nevertheless, as in other cases, most species consume a variety of prey types, even if they focus on particular types during particular times of the year, such as during breeding or migration.

Consumers of arthropods are extraordinarily numerous and diverse, particularly in the Neotropics, and range in form, function, and behavior from hummingbirds (box on page 458) to ratites. These consumers contribute substantially to biodiversity, are essential components of food webs, and perform a variety of ecosystem functions (Razeng and Watson 2015). Insects, particularly of the orders Lepidoptera (butterflies and moths), Coleoptera (beetles), and Orthoptera (e.g., grasshoppers), appear to be heavily consumed by birds. One study of Australian ground-foraging insectivores found that these preferred groups, including spiders (Araneae), were high in crude fat and/or protein, suggesting that nutritional quality is a key determinant of prey selection (Razeng and Watson 2015). Among tropical foliage-gleaning species, large, soft-bodied arthropods (mostly orthopterans) are important during breeding, particularly for nestling diets, whereas temperate zone species rely more on small arthropods (Greenberg 1981).

Given the phylogenetic and functional diversity of avian insectivores, there does not appear to be any particular "syndrome" (or suite of morphological and functional characteristics) that is specific to feeding on arthropods. Birds have infiltrated virtually every niche in which arthropods occur. Some insectivorous groups are extremely specialized, such as

the sustained flight-feeding swifts (Apodidae), which spend most of their lives on the wing, and woodpeckers (Picidae), whose crania and pecking kinematics are adapted to minimize impacts to the brain (Gibson 2006, Wang et al. 2011) when drilling into trees to access wood-boring insects and grubs. Within the Passeriformes alone, insectivores range from aerial-foraging swallows (Hirundinidae; fig. 15.14C), to bark-gleaning nuthatches (Sittidae), treecreepers (Certhiidae; fig. 15.1), and woodcreepers (Dendrocolaptidae), to South American ground-foraging, ant-eating antthrushes (Formicariidae). Nevertheless, there are some morphological correlates within certain taxa and communities. Gape size limits noncrushable arthropod prey size, in much the way that we saw in frugivorous species. However, many birds overcome such constraints by crushing and beating prey items in ways that make them easier to swallow (Sherry and McDade 1982). Naturally, the larger the prey item, the greater the handling time required, and this has energetic implications. Nevertheless, benefits exceed costs at the range of prey sizes taken, and perhaps the ability to detect, pursue, swallow, and digest prey may be more important for determining prey choice (Sherry and McDade 1982).

In general, however, selection seems to act more on locomotor attributes involved with the terrestrial, arboreal, or aerial pursuit of prey. Selection also acts on the wings, tails, legs, and feet, as these are extremely important in determining what prey are encountered and captured by these birds (fig. 15.2). Avian insectivores vary less qualitatively but more quantitatively in their feeding behavior (Sherry et al. 2016). For instance, Robinson and Holmes's (1982) study of 11 forest-dwelling insectivores reported that almost all of the species (consisting of a flycatcher, a chickadee, a tanager, a grosbeak, vireos, and warblers) performed several different kinds of prey-attack maneuvers to obtain prey (table 15.2), albeit with different frequencies. The most commonly performed maneuver among species was hovering, followed by gleaning. Robinson and Holmes (1982) emphasized the role of variation in searching behavior (table 15.1), as distinct from prey-attack behavior. These searching tactics set the stage for how insectivores ultimately capture their prey, and thus are critical to the process. Furthermore, searching behavior itself can present a mechanism for resource partitioning. For example, the Least Flycatcher (*Empidonax minimus*) and American Redstart (*Setophaga ruticilla*) are very similar in their use of microhabitats and bill morphology, but differ in search behavior, and the differences in their diets reflect differences in their search tactics. Conversely, species with different morphologies often use the same general searching mode and take similar types of prey, such as the Rose-breasted Grosbeak and vireos, revealing that morphology does not necessarily predeter-

CRAIG W. BENKMAN: CROSSBILLS AND COEVOLUTION

Professor Craig W. Benkman, of the University of Wyoming, has made significant contributions to the fields of ornithology, ecology, and evolution. His research on interactions between crossbills (*Loxia* spp.) and conifers (figure below) has generated key insights into the process of coevolution (whereby species reciprocally select for adaptive changes in response to one another). The Cassia Crossbills have exerted selection favoring thicker distal cone scales, which in turn has favored the evolution of the deep beak

characterizing this reproductively and genetically isolated crossbill, providing a classic example of a diversifying coevolution. Much of Benkman's work has focused on the relative roles of multiple selective agents (crossbills *and* tree squirrels, for instance) on the evolution of conifer cone structure (e.g., Benkman 1999, Benkman 2010), and how variation in the occurrence of tree squirrels alters the coevolutionary pathway and can even promote speciation in crossbills (Smith and Benkman 2007). Benkman, his students, and his colleagues have investigated *Loxia* spp.–conifer interactions throughout North America and abroad. One of the most impressive aspects of his research program is the breadth of methodologies he and his collaborators have utilized, involving not only measurements of crossbill abundance and distribution, morphology, behavior, and genetics but also extensive measurements of pinecone phenotypic traits.

With great foresight into the implications of feeding behavior for crossbill and conifer coevolution, Benkman's earlier work focused on the particulars of crossbill beak morphology and feeding performance (e.g., Benkman 1993). Through extensive bill measurements and experimentation,

Left, Red Crossbill (*Loxia curvirostra*) and, *right*, a lodgepole pinecone (*Pinus contorta*) from one of Benkman's South Hills, Idaho, study sites (not pictured to scale). The crossed bill tips are positioned in a manner that facilitates prying apart the large scales that themselves are adapted to deter animals from accessing the seeds they contain, formulating a classic example of coevolution. *Photos courtesy of C. W. Benkman.*

mine foraging behavior or diet. The most important factors affecting search behavior are vegetation structure—which influences how birds move about the canopy and see prey—and characteristics of the prey themselves, such as their relative abundances, sizes, and locomotor capabilities.

Consumers of vertebrate prey are comparatively less numerous and diverse, as is generally expected for any carnivore feeding at higher trophic levels. The advantage of eating meat is that it is relatively easily digested and mostly assimilated; the nutrient composition (relatively high in protein and secondarily

Benkman identified structural aspects of the beak and palate that account for variation in the crossbill's ability to pry pinecone scales apart to feed on the seeds they protect. One of Benkman's greatest contributions to avian biology was in his demonstration of how subtle variations in these features could have profound effects on food intake rates. Thus, a crossbill's feeding rate quite accurately reflects its level of morphological adaptation, and slight reductions in feeding rates can have substantial fitness consequences (figure below).

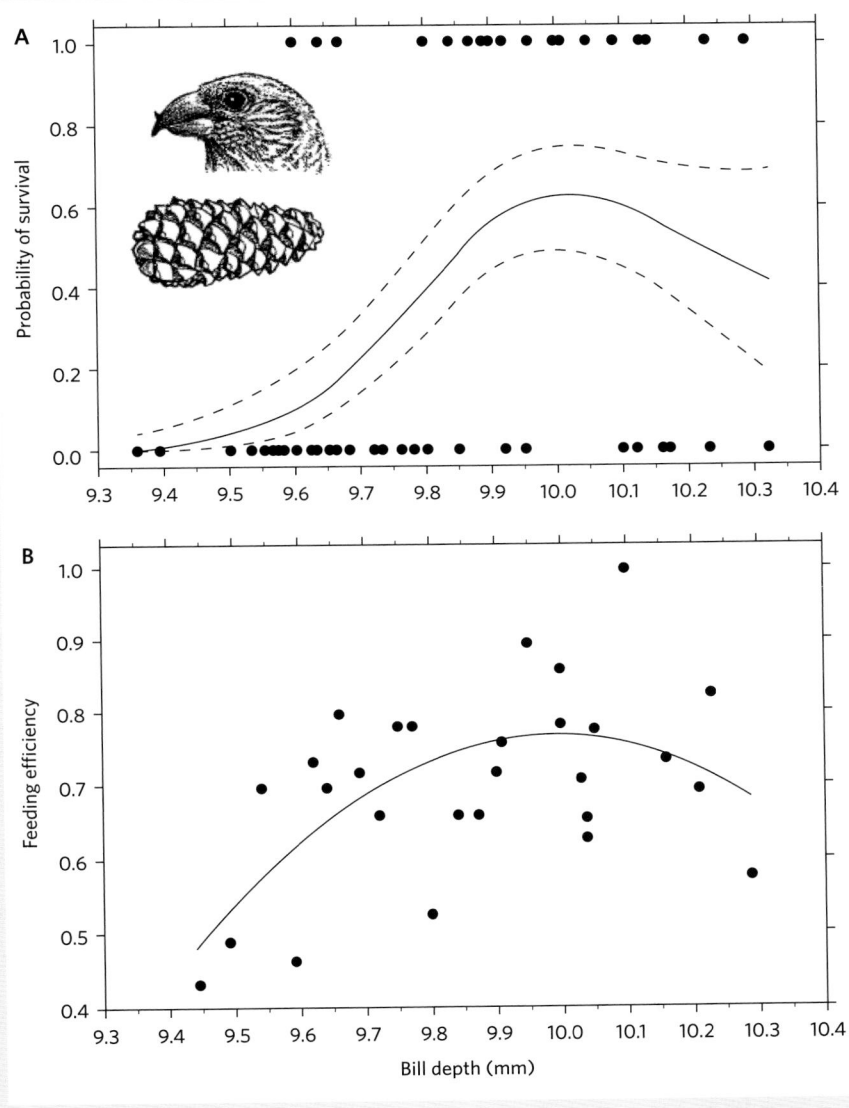

What difference does a millimeter make? When it comes to the depth of the bill, a lot apparently. Panel *B* shows that feeding efficiency on South Hills lodgepole pine cones is highest in crossbills having a bill depth of 10.0 mm, which matches the bill depth of *A*, crossbills that experience the greatest probability of survival. *Figure from Benkman 2003, reproduced with permission from* Evolution.

in fat) is relatively constant, and closely matches that required by birds (Houston and Duke 2007). The major disadvantage is that vertebrate prey are often large (relative to the predator's body size) and mobile, and typically must be hunted down, captured, and reduced into smaller bits prior to ingestion.

Furthermore, vertebrate tissue comes with lots of indigestible parts (hair, feathers, bones, etc.) that must be cast out later in the form of pellets (Houston and Duke 2007; fig. 15.14D).

Meat-eating birds vary in form and function in much the same way that insectivorous birds do; that is, primarily by

Table 15.2. **Prey-attack maneuvers of various forest-dwelling insectivorous birds and the seven species, among the 11 studied by Robinson and Holmes (1982) at Hubbard Brook, NH, USA, that performed them with the greatest frequencies.**

Behavior	Definition	Representative species
Gleaning	Bird takes stationary prey while standing or hopping	Black-capped Chickadee (*Parus atricapillus*), Blackburnian Warbler (*Dendroica fusca*)
Hovering	Bird picks prey from substrate via upward-directed flights	Least Flycatcher (*Empidonax minimus*), Philadelphia Vireo (*Vireo philadelphicus*)
Hanging	Bird flies to leaf or twig and hangs by feet to pluck prey from substrate or uncurl a leaf to access them	Black-capped Chickadee
Flush-chase	Bird flushes prey by fanning tail and flicking wings and chases it in flight, typically downward	American Redstart (*Setophaga ruticilla*)
Hawking	Flying prey are pursued by bird on the wing	Scarlet Tanager (*Piranga olivacea*)

foraging mode—e.g., sit-and-wait ambushing versus active searching and pursuit—and secondarily by prey type—e.g., birds, mammals, and lepidosaurs. As we saw with regard to prey-attacking behavior in forest-dwelling insectivores, the diets of meat-eating birds tend to vary more quantitatively than they do qualitatively, owing to the reality that most are very opportunistic. Many raptors will feed on just about any living, and often dead, animal of the right size, and preferences for avian, mammalian, amphibian, lepidosaurian, and even arthropod prey vary in proportion within and among species (e.g., Johnsgard 1990). Nevertheless, some groups are more strongly associated with certain prey types, such as New (Cathartidae) and Old (Accipitridae) World vultures that scavenge primarily from carcasses, and falcons (Falconidae) that tend to take avian prey on the wing. Others yet, such as the Short-toed Snake Eagle (*Circaetus gallicus*) feed primarily on snakes (Ferguson-Lees and Christie 2001). The Lammergeier (*Gypaetus barbatus*) feeds on bones and marrow—the only bird known to be able to digest bones (Houston and Copsey 1994)—often by carrying bones into the air and dropping them from great heights to break them into smaller, swallowable bits (Ferguson-Lees and Christie 2001).

Carnivorous foraging and feeding is associated with clear morphological and neurosensory attributes. Hooked beaks are common to virtually all species that regularly consume meat in one form or another, from passeriform shrikes (*Lanius* spp.) to strigiform owls (fig. 15.14D, E) to accipitriform vultures; however, beak shapes vary somewhat according to prey regularly consumed, body and skull size (Hertel 1995; see also the box on page 467), and shared ancestry (Bright et al. 2016). Among falconids (falcons and caracaras), accipitrids (hawks and eagles), and strigiforms that actively hunt, capture, and kill prey, large, curved, sharp claws (talons) are

the norm (fig. 15.6). In addition to these common morphological features, meat-eating birds enjoy heightened senses (chapter 12). Detecting prey from afar requires a high degree of visual acuity (i.e., the ability to distinguish two points from a distance), which in raptors is ~2.5 times greater than humans (Olsen 1995). Nocturnal hunting, such as in owls, also relies on enhanced visual capabilities (afforded by relatively large, tube-shaped eyes; Proctor and Lynch 1993) but also depends on extremely sensitive hearing capabilities. The bilaterally asymmetric ears of Barn Owls (*Tyto alba*) afford them the ability to pinpoint the location of their prey in complete darkness (Gill 2007), upon which they pounce with ~15 body weights of force (Usherwood et al. 2014). Turkey Vultures (*Cathartes aura*) are known to possess an extremely well-developed sense of smell, which helps them locate carcasses under the cover of the forest canopy (Stager 1964, Houston 1986). When these morphological and neurosensory mechanisms fail to help a bird acquire prey, they turn to an additional behavioral mechanism—kleptoparasitism (i.e., piracy). Many species of raptorial birds, such as Bald Eagles (*Haliaeetus leucocephalus*; fig. 15.6), regularly steal prey from the clutches of successful con- or heterospecific hunters, by harassing them until they let go or by raiding their caches (Olsen 1995).

Liquid Foods

Unlike bulk foods, liquids—such as nectar—require specialized endeavors that select for a relatively narrow set of characteristics. Nevertheless, opportunistic floral visitation is widespread in birds, because nectar is a highly energetic resource. Nectar is assimilated with >97 percent efficiency, making nectarivory an attractive mode of sustenance for birds (Martínez del Río et al. 1992). However, only some birds have evolved the specialized set of characteristics that make up the "syndrome of anthophily" that allows them to efficiently exploit floral nectar (Stiles 1981). Typical features of this syndrome include a slender, often long and/or curved bill and an extensible tongue that is split, with tips that are grooved, fringed, and/or capable of rolling into tubes (fig. 15.15; see also the box on page 469). Although bills and tongues were very important in the classifications of the earliest nectarivorous birds (Gill 1971), less attention was given to these characters beyond the 1980s. Considering the most recent phylogenies of nectarivores, it is remarkable that specialized nectar-feeding species have evolved independently in at least 18 families (and up to three times within a single family, e.g., Thraupidae), with a grand total of 23 independent developments within the class as a whole (Rico-Guevara, unpublished data). Each independent development of nectarivory may have been preceded by a different ancestral condition (e.g., frugivory or insectivory), which may have provided a different set of raw materials for, as well as different constraints on, how nectarivory evolved.

In addition, some groups engage in both ancestral and nectarivorous forms of feeding (e.g., the genus *Diglossa*, of the primarily frugivorous tanagers, Thraupidae). Consequently, nectarivory is achieved by similar, but not identical, structural changes (Rand 1967).

The most diverse and specialized families of nectar-feeding birds are the passerine sunbirds (Nectariniidae) and honeyeaters (Meliphagidae), as well as the non-passerine hummingbirds (Apodiformes: Trochilidae) (Stiles 1981). Hummingbirds are the most successful nectarivorous vertebrates in terms of species richness, dependence on nectar, and morphological diversity of the feeding apparatus (Stiles 1981). Hummingbirds evolved to feed on flowers well enough to make their living on small volumes of nectar, and their subsistence on nectar has completely modified their lifestyle. Consequently, some clades have achieved nectar-feeding specializations that have fostered coevolutionary mutualistic systems. Much the same way that plants use frugivorous

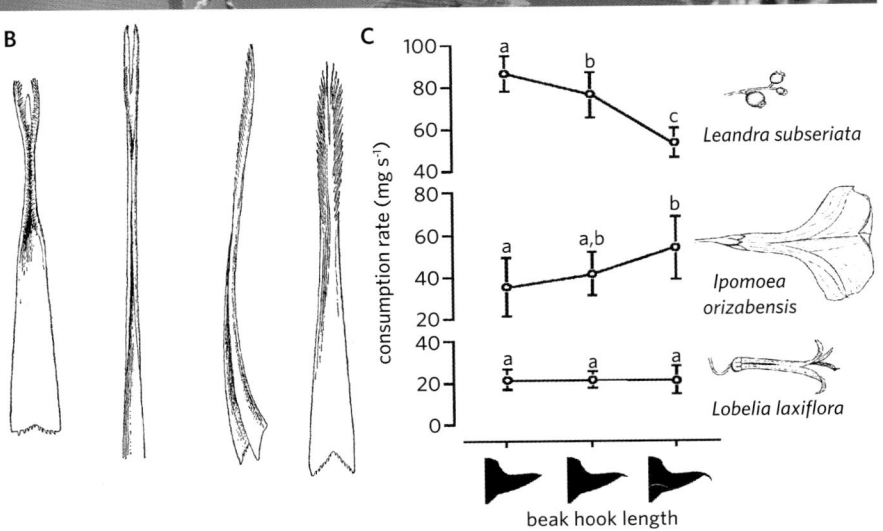

Figure 15.15. Various forms of nectarivory. *A,* Female Anna's Hummingbird (*Calypte anna*). *Photo by Pat Gaines. B,* Illustrations of tongues representative of various species of nectar-lapping songbirds, showing the tubed, bifurcating, and brushed tips that function to collect nectar in various ways. *Figure from Lucas 1894, reproduced from* The Auk, *with permission. C,* Nectar-robbing Flowerpiercers (*Diglossa* spp.) use the hooked tips to pierce the base of *Ipomoea orizabensis* corollas to access the nectary; nectar-robbing feeding performance diminishes when the bill tip is experimentally shortened, but fruit-eating (*Leandra subseriata*) performance is enhanced. The *Lobelia laxiflora* flowers require no special handling, and thus feeding performance remains constant across treatments. *Figure from Schondube and Martinez del Rio 2003, reproduced from* Proceedings of the Royal Society B, *with permission.*

birds as seed dispersers (above), they also use nectarivorous birds as pollination vectors (Stiles 1981). And much the same way that fruit color is ostensibly driven by avian preference, so too is flower coloration of predominantly bird-pollinated flowers, in comparison with insect-pollinated plants (Stiles 1976). Other birds have evolved to take advantage of the floral nectar offer, but by precluding pollination. These "nectar robbers" access the nectaries of the flowers directly, for instance, by piercing holes in the base of the flower without contacting the reproductive parts (stigma and stamen) (Schondube and Martínez del Río 2003) (fig. 15.15C).

Despite the advantages of morphologically and behaviorally committing to an exclusively nectarivorous diet, it does have some drawbacks. Nectar is virtually entirely composed of sugars (sucrose, glucose, and/or fructose) with a very small percentage of protein (and amino acids in general), and trace amounts of vitamins and minerals (Gartrell 2000). Therefore, nectarivorous birds must supplement their sugary diets on a daily basis, particularly during times of greater demand, such as during breeding (Stiles 1995, Gartrell 2000). In hummingbirds, this is typically accomplished by feeding on arthropods, mostly insects (Stiles 1995; box on page 458). Yanega and Rubega (2004) demonstrated how hummingbirds overcome the potential limitations of a long, narrow, tubelike bill honed by natural selection for nectar extraction. When hunting insects, hummingbirds open their beaks and essentially fly and catch them in their mouth. To facilitate capture, the distal half of the mandible flexes downward relative to the rest of the mandible. This intramandibular flexion has the effect of bowing the rami of the mandible outward (laterally), thereby widening the gap substantially. In addition, floral nectar resources are distributed patchily in the environment and are depleted rapidly. Therefore, nectarivores continuously track resources as they replenish themselves (known as "traplining"), or defend clumps of resources from competitors (Feinsinger 1976), both of which impose significant energetic costs.

Several other species feed on a variety of plant (sugar) exudates, such as sap and manna—a fluid that oozes from damaged plants and later crystallizes in white clusters on the foliage (Paton 1980, Case and Edworthy 2016). These include sapsuckers, members of the woodpecker family (Picidae) whose drilling behavior and long, bristled tongue serve equally well for boring into trees and lapping up sap as they do for accessing insects within the bark (Danforth 1938). Passeriform Hawaiian honeycreepers (Fringillidae) and psittaciform lorikeets also feed variably on sap in addition to their predominantly nectar diets (fig. 15.15). Many species native to Australia feed extensively on manna, such as honeyeaters (Meliphagidae) and pardalotes (Pardalotidae). Manna constitutes

F. GARY STILES: ECOMORPHOLOGY OF HUMMINGBIRDS

F. Gary Stiles. *Photo by Elena Stiles.*

For over 50 years, Professor Gary Stiles, now at the National University of Colombia (since 1990), has contributed vastly to Neotropical ecology, evolution, and conservation. He has authored over 150 peer-reviewed papers, five books, and has been the recipient of numerous accolades, including the Coues and Eisenmann Medals from the American Ornithologists' Union and the Linnaean Society of New York, respectively, and the Parker/Gentry Award from the Field Museum. After publishing the *Birds of Costa Rica* field guide, he moved to Colombia, home of the largest number of bird species (~1,900), to pursue his true passion: hummingbirds. He is a pioneer in the study of the interplay among morphology, behavior, and ecology across this family of roughly 340 recognized species. As a member of the South American Classification Committee of birds, he has both described new species and had species named after him (e.g., *Scytalopus stilesi*). He has spent his time in the Neotropics fostering an awareness and appreciation for birds, not only through his numerous students, but also by leading local and national avian scientific associations (e.g., as editor of *Ornitología Colombiana*, Colombia's first international-caliber ornithological journal), and leading research and conservation expeditions. Professor Stiles has studied many hummingbird assemblages and their annual cycles, including the phenology of the plants from which they feed (Stiles 1975), in order to elucidate coevolutionary relationships between plants and birds (Stiles 1981). Through long-term studies, Stiles has shown how compensatory phenological responses to unusual rainfall patterns structure hummingbird-plant communities in

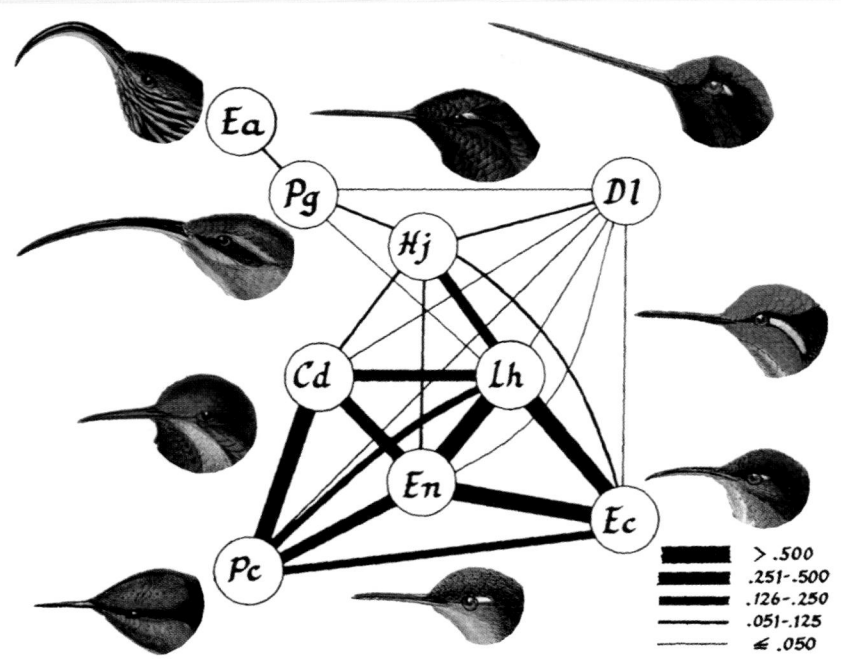

Pairwise relationships among species in terms of their floral sources in a subtropical forest. Species connected by thicker lines experience greater overlap in the flowers they visit. *Bottom*, The five species with short to medium bills form a nuclear community that shares resources. *Top*, The four species with long and/or curved bills are specialized to exploit different resources, consisting of flowers with longer and curved corollas. *Figure from Stiles 1985, modified from Ornithological Monographs, with permission. Drawings by John Gould; images courtesy of Biodiversity Heritage Library, http://www.biodiversitylibrary.org.*

> .500
.251–.500
.126–.250
.051–.125
≤ .050

tropical forests (Stiles 1977, 1980), formulating an important reference for investigating the effects of climate change in the tropics.

Professor Stiles has performed some of the most comprehensive studies of hummingbird foraging to date (figure above), including field and experimental observations on floral preferences (Stiles 1976); on the efficiency of nectar extraction, energetics, and aerodynamics (Wolf et al. 1972, 1976, Stiles et al. 2005, Stiles 2008), and on the underappreciated food sources that make up about 50 percent of their diet: arthropods (figure on the left). Professor Stiles's investigations in hummingbird foraging biology have led to other novel avenues of research, such as courtship display behavior (Stiles 1982), which has inspired the field of animal aeroacoustics (e.g., Clark et al. 2011). Furthermore, his seminal work on lekking in the Long-billed Hermit (Stiles and Wolf 1979) set the stage for understanding the role of male-male combat in the evolution of bill morphology (e.g., Rico-Guevara and Araya-Salas 2015).

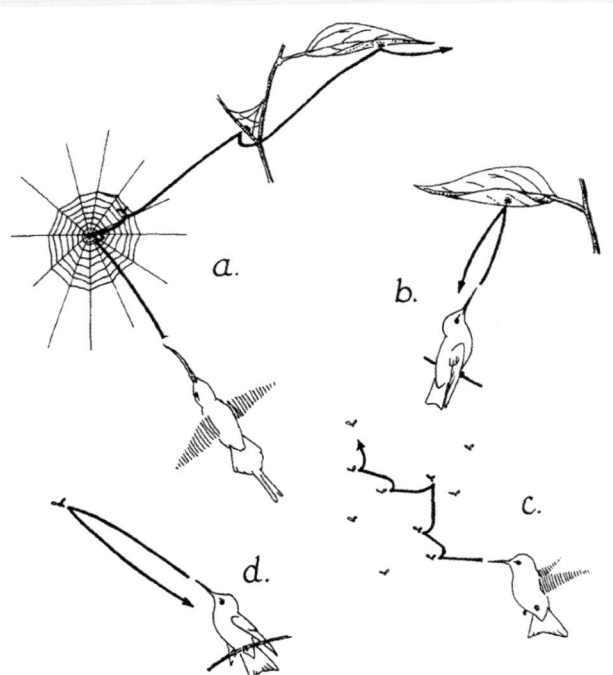

Arthropod foraging tactics of hummingbirds: *A*, hover-gleaning; *B*, sally-gleaning; *C*, hover-hawking; *D*, sally-hawking. Territorial hummingbirds preferentially hunt aerial prey by *D*, flycatching from a perch, whereas *A*, trapliners prefer to catch spiders and other prey from the substrate. *Figure from Stiles 1995, reproduced from* The Condor, *with permission.*

an important seasonal resource for Forty-spotted Pardalotes (*Pardalotus quadragintus*) and constitutes 75–84 percent of the nestling diet (Case and Edworthy 2016). This species has been observed to "mine" manna resources by puncturing stems of *Eucalyptus viminalis* with their conspicuously notched bill tips in a way that stimulates manna production over the course of several days (Case and Edworthy 2016).

There are also other exudates that form essential components of bird diets. These include those formed by plant products, such as honey and wax, as well as those formed from insects, such as honeydew (sugary secretions of nymphal insects, such as aphids, coccids, and psyllids), and lerp (the protective covering of psyllids, mostly carbohydrate; Paton 1980). Honeyguides (Indicatoridae) feed on both honey and beeswax. Wax has high lipid content, and thus provides a great source of energy for those capable of digesting it (Downs et al. 2002). Honeyeaters and pardalotes will feed opportunistically on honeydew and lerp to supplement their nectar and manna diets. Some hummingbirds will also consume fruit fluids by opening small holes on the rind and drinking the sugar-rich liquids (Ruschi 2014) or simply licking the exposed flesh of already open fruits (Wendelken and Martin 1988). Other birds feed on vertebrate fluids like sweat, saliva, tears, and blood (reviewed by Bishop and Bishop 2014). Sharp-beaked Ground Finches (*Geospiza difficilis*) are known to pick at the feather quills of resting boobies (Sulidae) and consume the blood (Bowman and Billeb 1965; Supplementary Materials S6). Oxpeckers (*Buphagus* spp.) have been reported to perch at open wounds, reopen others, and even make new ones to feed on the blood of several African mammals (Bishop and Bishop 2014). Sanguinivory (blood-feeding) is thought to provide a protein-rich supplement when other resources become unpredictable or limiting (Bowman and Billeb 1965). It is not common among birds and could be opportunistic, as in the case of Black Crakes (*Amaurornis flavirostris*) drinking blood from hippos' fight wounds (Bosque 2009). In Sharp-beaked Ground Finches and oxpeckers, however, sanguinivory was likely exapted from the birds' ectoparasite-removal behavior (Bowman and Billeb 1965, Nunn et al. 2011).

Aquatic and Marine Foods

Water covers about 75 percent of the earth's surface, with the oceans covering about two-thirds of that area. Although most animals consumed by birds are captured on land and air, many species take their prey from marine and freshwater environments. About 10 percent of all bird species are strictly marine ("seabirds") in that they extract their food almost exclusively from the marine realm. These include penguins (Spheniscidae), boobies and gannets (Sulidae), tropicbirds

(Phaethontidae), frigatebirds (Fregatidae), petrels and shearwaters (Procellariidae), albatrosses (Diomedeidae), storm petrels (Hydrobatidae), diving petrels (Pelecanoididae), and auks and puffins (Alcidae). Many species (generally termed "waterbirds") use the oceans and/or inland water bodies (e.g., bays, estuaries, lakes) for only part of the year (e.g., as wintering areas) or for part of their resources and survival. These include, but are not limited to, some species of pelicans, loons, grebes, ducks, herons, egrets, rails, gulls, and shorebirds. Naturally, much behavioral diversity accompanies this phylogenetic diversity (fig. 15.16, table 15.3), and the ensuing paragraphs of this subsection are divided according to the various ways in which aquatic and marine foods are consumed.

Feeding on small aquatic prey poses a certain challenge in that birds have to secure enough prey and feed efficiently without swallowing undesirably large amounts of water. Hence, the common denominator for birds hunting in water is their ability to separate the captured animal from the water. Avian diets in aquatic environments range from microbes to large fish; many calidrid sandpipers (small migratory shorebirds) feed on biofilm, a layer of microscopic animals, algae, and bacteria found on the surface of mudflats (reviewed by Jardine et al. 2015). Biofilm feeding constitutes half of their daily energy budgets and total diet in many stopover locations (Jardine et al. 2015). The biofilm intake process has been related to modifications in tongue shape (Elner et al. 2005) and proposed to have similarities with nectar-feeding mechanisms (Elner et al. 2005). To consume prey larger than microorganisms, birds use filter-feeding to ingest the small invertebrates that are suspended near the surface of moderately calm waters. Filter-feeders exhibit specialized behaviors and morphological adaptations (e.g., fig. 15.4). Several kinds of birds have evolved to filter-feed convergently: ducks and flamingos (Zweers et al. 1995, Gurd 2007) are among the best-known, but shearwaters and prions (seabirds) also have been reported to filter-feed (Klages and Cooper 1992, Lovvorn et al. 2001). The specific mechanics vary among the independent evolutions of this behavior, even within each group. For instance, some ducks can control the distance between their upper and lower bill to modify the lamellar distance (sieving mechanism), whereas others modify the vortices in the flow (inertial impact deposition mechanism), thus selecting the size of their desired food (Kooloos et al. 1989; but see Gurd 2007).

Filter-feeding is efficient for relatively small food particles and prey only; therefore birds employ a different strategy to pick up larger invertebrates from the water. Many scolopacids (long-billed shorebirds) visually track prey that swim near the surface, and when within striking range, the birds swiftly submerge their bills to seize shrimp, insect larvae, and other

Table 15.3. **Typical foraging tactics of freshwater and marine species, and the representative groups with which they are associated (adapted from Ashmole 1971 and del Hoyo et al. 1992).**

Behavior	Definition	Representative groups
Foot-propelled (pursuit) diving	Use feet for propulsion during underwater dives	loons (Gaviidae); grebes (Podicipedidae; fig. 15.16); anhingas (Anhingidae); cormorants (Phalacrocoracidae); "diving" ducks (Anatidae: Aythyinae, Merginae)
Wing-propelled (pursuit) diving	Use wings for propulsion during underwater dives	penguins (Spheniscidae); auks, puffins, murres (Alcidae); diving petrels (Pelecanoididae)
Deep-plunging	Plunge deep into the water from the air to seize prey	Tropicbirds (Phaethontidae), boobies and gannets (Sulidae; fig. 15.16)
Surface-plunging	Make shallow plunges into the water from the air to seize prey	terns (Laridae); pelicans (Pelecanidae); Osprey (Pandionidae); kingfishers (Alcedinidae)
Surface-feeding	Feed on surface prey and access it from the air or by sitting on the water	gulls, skimmers (Laridae; fig. 15.16); frigatebirds (Fregatidae); storm petrels (Hydrobatidae), albatrosses (Diomedeidae), petrels, shearwaters, and fulmars (Procellariidae); phalaropes (Scolopacidae)
Kleptoparasitic	Aerially harass other hunters and pirate/steal their prey	skuas, jaegers (Laridae); frigatebirds
Dabbling	Feed by tipping head and bill into water from the surface	"dabbling" ducks (Anatidae: Anatinae)
Nearshore ambushing/wading	Stand motionless or wade slowly in shallow water; seize prey that approaches within striking distance	Herons and egrets (Ardeidae), storks (Ciconiidae), rails (Rallidae)
Nearshore probing	Probe nearshore substrates and shallow water for prey	Shorebirds (Scolopacidae, Charadriidae), rails

fast swimmers. Once the prey is captured at the bill tip, the birds withdraw their bills from the water, slightly opening them to leave a drop of water near the tips, within which the invertebrate is trapped (Rubega and Obst 1993). To move the prey toward the bill base where it can be swallowed, these birds do not use their tongues (as in fig. 15.3), nor do they employ suction, as in filter-feeding or drinking. Rather, all they do is separate their jaws farther. Because of the adhesion between water and bill keratin, combined with the tendency of a drop to minimize its exposed surface, the droplet moves quickly until it reaches the bill base (fig. 15.17A–D). Finally, the bird closes its beak to expel the water and consume the prey item (Rubega and Obst 1993, Rubega 1997). Amazingly, the complete process, dubbed "surface tension transport," takes less than a quarter of a second (Rubega 1997; fig. 15.17E).

To access invertebrates—such as aquatic insects, small crustaceans, and molluscs—from the shores of inland and marine water bodies, other shorebirds (Scolopacidae and Charadriidae) probe beneath the surface with their bills

(fig. 15.16). These species possess an array of bill shapes and sizes that allows them to probe for different types of prey, located at different depths, within the substratum (Goss-Custard 1975; figs. 15.1, 15.16). These differences in bill morphology and foraging behavior—such as the types and rates of substrate pecking and probing actions they perform—provide mechanisms for resource partitioning, particularly during winter when resources are limiting (Baker and Baker 1973, Bocher et al. 2014). Shorebirds rely on distal rhynchokinesis (fig. 15.5) to grasp prey with the bill tips when submerged within the substratum and when capturing prey suspended in the water column (Estrella and Masero 2007). Some species, such as *Calidris* spp., possess arrays of mechanoreceptors (Herbst corpuscles) located in sensory pits under the keratin layer throughout the bill that facilitate tactile detection of prey (Nebel et al. 2005). Although shorebirds are typically associated with aquatic/marine habitats, some species of sandpipers (e.g., Upland Sandpipers, *Bartramia longicauda*) and plovers (e.g., lapwings, Vanellinae) inhabit grassland habitats far from water.

Figure 15.16. Selected waterbird foraging tactics. *Top row,* Torpedo-like Blue-footed Booby (*Sula nebouxii*) deep plunge-diving. *Photos by Pat Gaines. Middle row,* Black Skimmers (*Rynchops niger*) surface-feeding. *Photo by Dave Furseth. Bottom row, left,* A foot-propelled diver, Horned Grebe (*Podiceps auritus*), after a successful dive; *center,* A nearshore wader, Tricolored Heron (*Egretta tricolor*), after a successful strike; and, *right,* A nearshore surface prober, Marbled Godwit (*Limosa fedoa*), probing the substrate (*inset*). *Photos by Dave Furseth. See also Supplementary Materials 15.S1, 15.S8, and 15.S9.*

Molluscivory (feeding on bivalves and gastropods) is a more specialized endeavor, because it often requires force and finesse more than a good reach of the bill. In addition to probing the surface for worms and other foods, Oystercatchers (Haematopodidae; fig. 15.1, 15.18) adopt two main techniques for feeding on bivalves: "stabbers" use their pointed or chiseled bill tips to wedge between and pry apart the gaping shells of surface-bound or superficially buried bivalves to extract the soft bodies within, whereas "hammerers" use their blunt bill tips to pound cracks into the dorsal or ventral surface of the shell, then snip the adductor muscle that keeps the shells closed so they can access the meat within (Norton-

Griffiths 1967, Goss-Custard and Sutherland 1984, van de Pol et al. 2009; fig. 15.18, Supplementary Materials S7). These bill morphologies are necessarily discrete, but there is much variation within these categories among individuals (e.g., between sexes), populations, and species (Swennen et al. 1983, van de Pol et al. 2009). Hook-billed Kites (*Chondrohierax uncinatus*) and Snail Kites (*Rostrhamus sociabilis*), raptors that tend to hunt over water bodies and marshes, possess specialized, sickle-shaped bills for accessing and extracting snail bodies from deep within the coils of their shells (Smith and Temple 1982; fig. 15.18). For some species, it is less the morphology of the bill that matters but more that of the stomach; Red

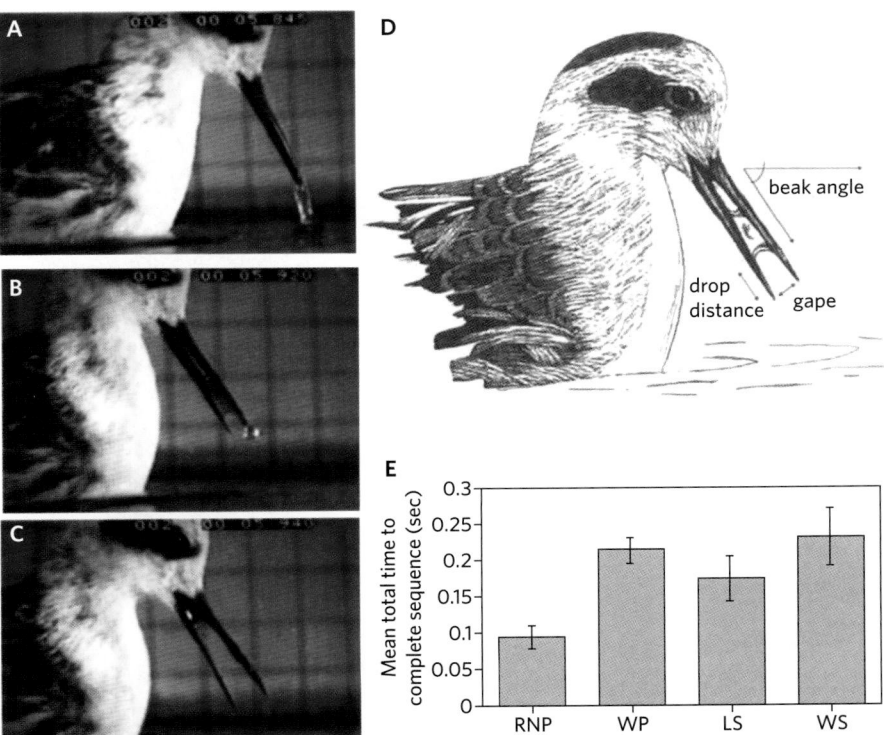

Figure 15.17. Surface tension transport (STT) as exhibited by phalaropes. *A–C*, High-speed video sequence of a Red-necked Phalarope (*Phalaropus lobatus*) drawing in a prey-containing water droplet along the upper and lower bill tomia. *Figure modified from Rubega and Obst 1993, reproduced with permission from* The Auk. *D*, Illustration showing the relevant morphological and kinematic attributes of STT: the angle of bill aperture, gape distance, and the distance traveled by the droplet. *E*, The time required to complete a given prey transport sequence (e.g., A–C) varies within less than 1/4 of a second across phalarope species; Red-necked Phalaropes (RNP) are particularly adept at STT feeding, given their durations of around 1/10 of a second. *Figures modified (D) and reproduced (E) from Rubega 1997, with permission from* Ibis.

Knots (*Calidris canutus*) ingest small gastropods and bivalves whole and crush them in their muscular gizzards (Piersma et al. 1993). For yet other species, it is more about behavior; gulls (Laridae) are known to drop hard-shelled molluscs (e.g., whelks, clams, and scallops) and crustaceans onto hard substrates from heights of 4–15 meters to crack them (Barash et al. 1975, Ingolfsson and Estrella 1978; fig. 15.18).

Piscivory (fish-eating) is a special case of carnivory because of the challenges involved in locating and accessing fish underwater, especially gaining purchase on slippery prey. This is particularly so for non-swimming accipitriform raptors (e.g., Ospreys and sea eagles, *Haliaeetus* spp.), kingfishers (Alcedinidae), and ardeids. These predators tend to be quite specialized consumers of fish, often to the exclusion of other prey types. Piscivorous raptors tend to have large, subequally sized, and strongly curved claws that are presumed to be better apt for handling slippery fish (Fowler et al. 2009). As we have seen in previous sections, foraging tactics vary among groups. In aquatic and marine birds, however, these seem to cut more along taxonomic than along ecological lines. With the exception of terns and some alcids, most seabirds have serrated or hooked beaks that facilitate handling slippery prey. A classic description of behavioral and morphological diversity of foraging adaptations in seabirds (Ashmole 1971) described modes of prey acquisition and associated differences in beak morphologies, which, in turn, formulated a ba-

sis for prey resource partitioning. Some groups tend to feed from the surface, whereas others plunge from great heights to deep depths to chase down fish (table 15.3; fig. 15.16). These are general trends, and some birds use different tactics depending on the circumstances; for instance, Australasian Gannets (*Morus serrator*) have been observed to plunge-dive when dolphins drive schools of fish near the surface, but pursuit-dive to reach deeper, moving fish (Machovsky-Capuska et al. 2011). Nevertheless, there is some degree of dietary structure within communities. Moreno et al. (2016) identified four feeding guilds in an Antarctic procellariform community, consisting of species that feed mainly on crustaceans (diving petrels, petrels, and storm petrels); large fish and squid (Wandering Albatross, *Diomedea exulans*); a mix of small fish, squid, and crustaceans (albatrosses); and carrion (giant petrels). Feeding often occurs in the presence of con- and heterospecifics in large aggregations, and this has implications for prey capture success. In a study of Cape Gannets (*Morus capensis*; Supplementary Materials S8), Thiebault et al. (2016) found that the combined effect of multiple predators attacking simultaneously served to disorganize and disrupt school cohesiveness, resulting in a net benefit to each consumer, up to a certain point of diminishing returns, when too many predators caused confusion among them.

As a result of the often patchy distribution of resources in the ocean, the basic foraging strategy for seabirds is to be

Figure 15.18. There is more than one way to skin a mollusk. *Top,* Snail Kites (*Rostrhamus sociabilis*) use their sickle-shaped bills to extract gastropod bodies from their spiraled shells. *Photo by Earl Orf. Lower left,* Black Oystercatchers (*Haematopus bachmani*) "hammer" shells or "stab" between the gaping valves with their beaks to access bivalve bodies. *Photo by Dave Furseth. Lower right,* Herring Gulls (*Larus argentatus*) drop clams onto hard surfaces to break their shells. *Photo by Thomas Smith.* Whereas some birds possess morphological and behavioral specializations for particular prey types—such as the oystercatcher and kite shown here—others, with ostensibly more generalized bills and feeding habits, such as gulls, may compensate behaviorally.

opportunistic. Some foraging techniques of seabirds were summarized by Ballance and Pitman (1999), which include feeding nocturnally, foraging around large objects or pack ice, feeding at fronts (boundaries between two distinct water masses), feeding in flocks, associating with subsurface predators such as tuna and dolphin, and/or covering large areas—(one Wandering Albatross was tracked 15,000 km on a 33-day feeding bout). Different feeding modalities are also associated with different environments; temperate and polar communities are associated with surface feeding, filtering, and pursuit diving, whereas tropical environments characterized by low seasonality and productivity are associated with surface feeders that forage in multispecies flocks, mostly on flying fish (Moreno et al. 2016). Extrinsic factors that contribute to foraging strategies are areas of productivity, available breeding sites or islands, and substrates on the sites (e.g., ledges, crevices, burrows); intrinsic factors include population density and energetics such as the cost of transport (Cairns 1992).

FORAGING AND FEEDING ECONOMY

"Optimal foraging theory" is one of the most studied principles in ecology (Marquet et al. 2014), and the study of avian foraging was instrumental in its development. In fact, one of its proposers, R. H. MacArthur (MacArthur and Pianka 1966), was a renowned ornithologist (chapter 22). This theory states that natural selection will maximize fitness via the retention of foraging strategies that optimize energy intake and minimize energy expenditures, i.e., maximize net energy gain per unit time (Pyke 1984). This theory has been supported in a variety of birds, for instance in oystercatchers (Meire and Ervynck 1986), Zebra Finches (*Taeniopygia guttata*; Lemon 1991), and diving ducks (Anatidae; Halsey et al. 2003). This idea has direct implications for understanding many aspects of foraging decisions, such as whether or not animals specialize on certain foods, when and how they should move to another patch, what size food they should eat, and which routes they should take when they move among patches (reviewed by Pyke et al. 1977, Griffiths 1980).

In terms of optimality, there are two fundamental foraging strategies that an animal could undertake: (1) stop foraging after acquiring a given amount of energy, thereby allotting more time to other activities—"time minimizer"; or (2) continue foraging throughout the time available, thereby increasing the amount of resources obtained—"energy maximizer" (Schoener 1971, Hixon 1982). These ideas have been tested in wild birds and other animals, and although it is usually complicated to assign definite labels, it has been shown that these two strategies can be employed by different sexes (Schoener 1971), migrant versus resident birds (Hixon and Carpenter 1988), and predatory versus grazing species (Hixon 1982). Another concept emerging from optimal foraging theory is the "marginal value theorem" (Charnov 1976). Because the resource depletes as a bird forages in a given patch, the individual must decide when to leave that patch. The decision would be influenced by the slope of diminishing returns, distance to the nearest patch, rate of travel between patches, and potential patch quality (fig. 15.19).

In most cases, birds forage with a priori information about the location of their food resources (Viswanathan et al. 1999).

For instance, they have some idea of their prey's microhabitat preferences or the position of flowering or fructifying trees and so on. However, some birds have to search with limited or no information on the location of resources across vast, seemingly homogeneous landscapes. For instance, it has been shown that, for pelagic seabirds (e.g., albatrosses), taking multiple short trips interspersed with long relocations—"Lévy flights"—is a more efficient search strategy than taking random search flight forays, as in Brownian motion (Humphries et al. 2012).

Support for optimal foraging theory is mixed among avian study systems (reviewed by Sih and Christensen 2001). Factors such as risk of adult and nest predation, prey defensive strategies, and interference competition can cause animals to deviate from optimal foraging (Rands et al. 2000). Currently the main research avenues in this field are on learning (Teichmann et al. 2014), spatial memory (Milesi and Marone 2015), predatory foraging (Nadjafzadeh et al. 2016a), social foraging decisions (Marshall et al. 2015), and pollination (Devaux et al. 2014).

FORAGING AND FEEDING ECOLOGY

Birds interact with their environment largely through the acts of foraging and feeding; what they eat has direct and indirect effects on their fitness, abundance, and distribution. The spatial and temporal distribution of resources plays a key role in how individuals, populations, and species are able to exist, and coexist, in the environment. In addition, other factors, such as competition and predation, often determine the *availability* of resources, regardless of their abundance in nature.

Competition for food (and other) resources is a major structuring force in all vertebrate communities. MacArthur's (1958) seminal work on resource partitioning among five coniferous forest warbler species provided a test of the Volterra-Gause competitive exclusion principle of limiting similarity; species can coexist by using different resources, by using the same resources in different places, or by using the same resources at different times—all of which are important for warblers. Among MacArthur's warblers, differences in their preferred foraging locations in the tree canopy (fig. 15.20), as well as differences in the way they foraged (e.g., long flights between trees, hawking, or hovering), resulted in differences in the relative proportions of arboreal arthropod taxa consumed. Smith et al. (1998) found that even within one of these species, the Black-throated Green Warbler (*Dendroica virens*), individuals directed their foraging maneuvers differently (i.e., some toward the needles, and others toward the branches of conifers) between shoreline and inland habitats. Stiles (1985) found that a staggered sequence of blooming of five principal

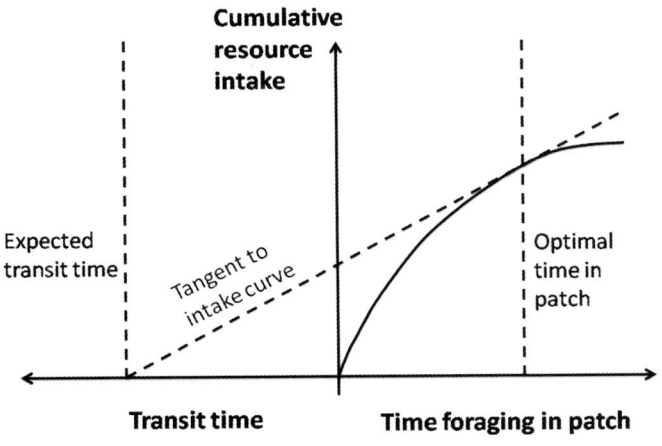

Figure 15.19. The marginal value theorem describes how animals should allocate their time across habitat patches, based on energy gained from foraging/feeding in a patch (i.e., calories per unit time), and the simultaneous depletion of resources as a result. One important factor is transit time to get from patch to patch; the longer it takes, the later the animal should leave the previous patch in order to maximize its energy intake. This relationship between transit time and time in patch is represented in the graph by the tangent-to-intake curve; the shorter the transit time, the sooner the tangent will encounter the resource intake curve. As a result, short expected transit times predict short optimal foraging times in a given patch, and vice versa. *Image available from Wikimedia Commons, https://commons.wikimedia.org/wiki/File: Marginalvaluetheorem.jpg.*

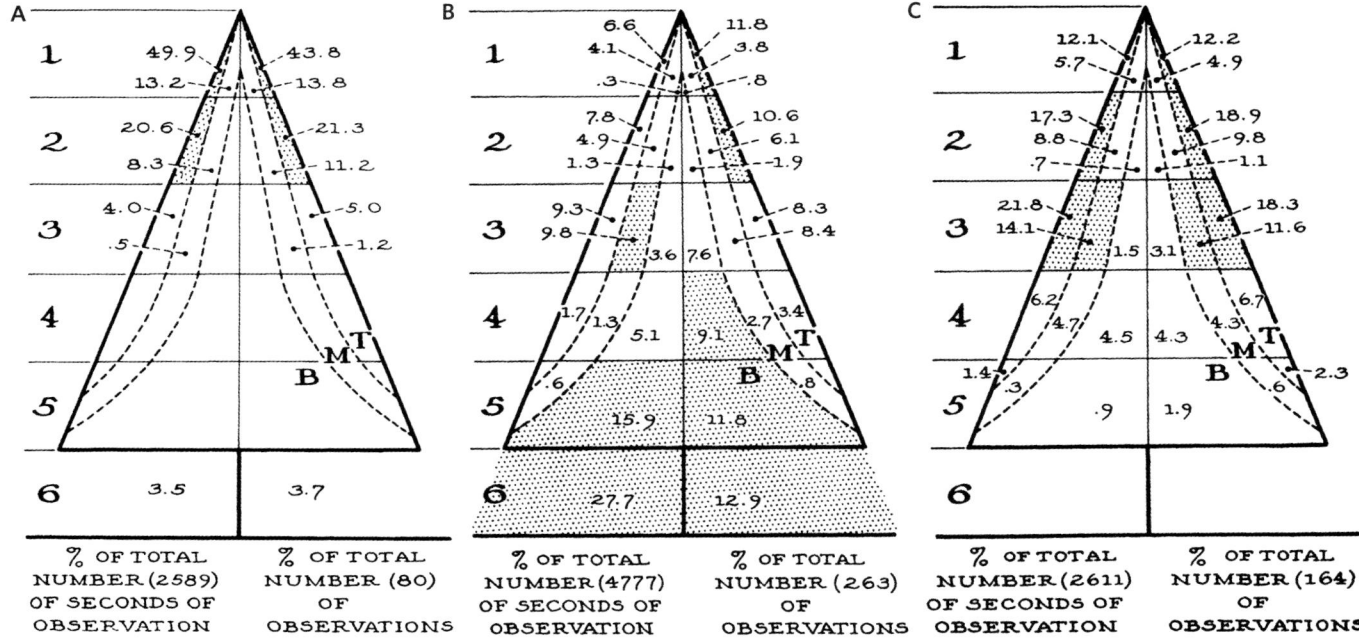

Figure 15.20. A classic example of resource partitioning in the field of ecology emanated from MacArthur's (1958) study of five coexisting coniferous forest warbler species. Shown here are differences among three of them: *A*, Cape May Warbler (*Setophaga tigrina*); *B*, Myrtle Warbler (*Dendroica coronata coronata*); and *C*, Black-throated Green Warbler (*Dendroica virens*), in the distributions of their foraging effort in the tree canopy, expressed in terms of the percentages of time and numbers of observations of individuals feeding in each zone. The shaded regions show the areas of greatest concentration of activity. *Reproduced from* Ecology, *with permission.*

food plants drove differences in flower visitation patterns, which in turn facilitated partitioning in a Caribbean-slope Costa Rican community of 22 hummingbird species. Vestiges of prehistoric feeding-related competition and community structure can even be observed in fossils (box on page 469).

Birds directly affect the abundance and distribution of their prey by eating them. The resultant prey population fluctuations, in turn, affect avian predator populations. Predator-prey interactions are generally characterized by food handling behavior, particularly "handling time"—the time it takes to catch and consume a prey item (Gotelli 2001). Handling time tends to increase with prey size (e.g., Salt and Willard 1971, Craig 1978, Grosch 2003), and can limit predation rates, particularly at higher prey densities (Jeschke et al. 2002). Holling's (1959) classic "disc equation" describes the nature of the relationship between prey density and a predator's intake rate—the "functional response." There are three types of curves that relate prey consumption to prey density: type I is linear, type II is hyperbolic (concave downward), and type III is sigmoid (S-shaped), usually indicating a switch from one type of prey to another (Gotelli 2001) (fig. 15.21). For most vertebrate predators, the curve of this relationship is type II: as prey density increases, so does the number of prey taken per predator, until the rate of increase begins to

slow to a plateau. This leveling-off (asymptote) reflects the predator's inability to continue to consume prey at the same rate at higher prey densities as they can at lower densities, either (or both) because handling prey takes time away from consuming it or/and the predator becomes progressively satiated (Holling 1959, Jeschke et al. 2002). Density-dependence is taken as evidence that a predator population can regulate the prey population; only type III curves really show this effect at low prey densities (Korpimäki and Norrdahl 1991).

The functional response, and the related phenomenon of "numerical response"—the relationship between predator and prey population densities—has probably been studied most extensively in shorebirds and raptors. A comprehensive review and analysis of shorebird food intake rate and density data indicated that the functional response of shorebirds was described by a type II curve (Goss-Custard et al. 2006). However, it seems that in shorebirds it is not handling time that sets the asymptote, but potential time costs associated with prey detection. Studies of raptors have sought the extent to which predators regulate prey populations, based on how they respond to changes in prey densities (Redpath and Thirgood 1999). In the owls and kestrels studied by Korpimäki and Norrdahl (1991), the combined functional response to vole (*Microtus* spp.) densities was linear (type I), suggesting

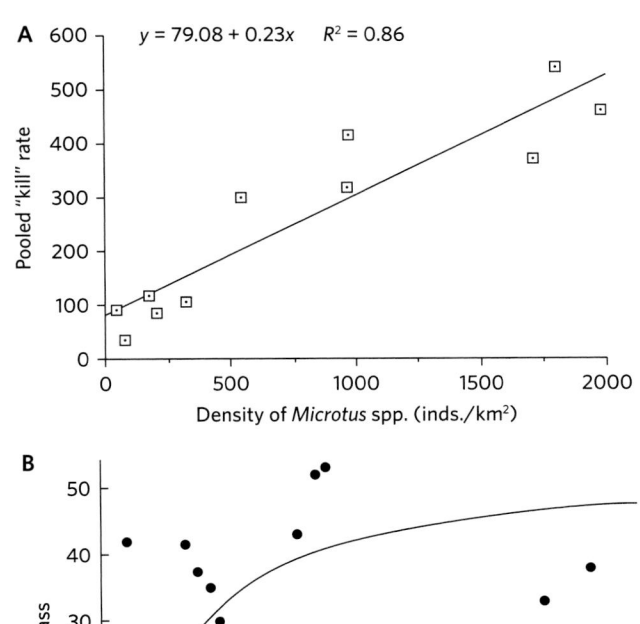

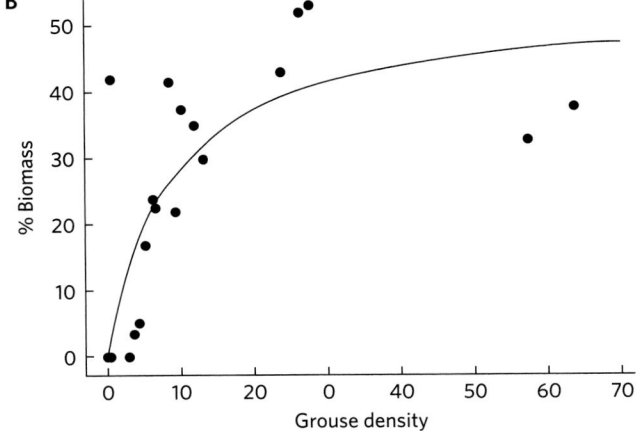

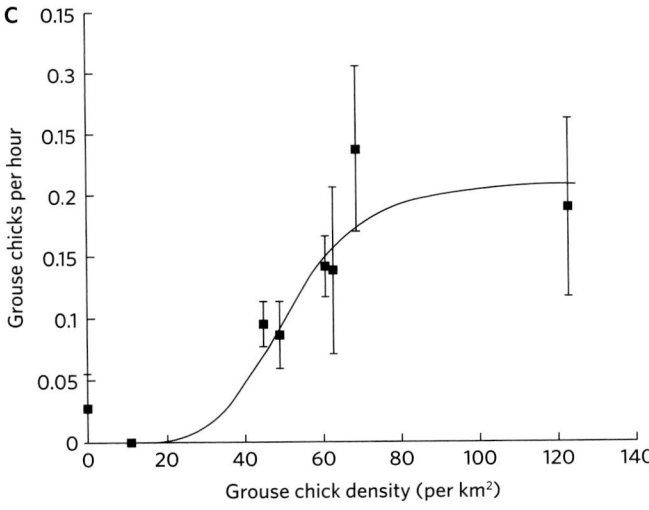

Figure 15.21. Examples of functional responses (per capita intake rates as a function of prey density) of various raptors. *A*, Type I combined response of Common Kestrels (*Falco tinnunculus*), Short-eared Owls (*Asio flammeus*), and Long-eared Owls (*Asio otus*) to vole (*Microtus* spp.) densities. *From Korpimäki and Norrdahl 1991; figure reproduced from* Ecology, *with permission. B*, Type II response (expressed in terms of the percentage of grouse biomass that made up the total prey) of Peregrine Falcons (*Falco peregrinus*); *C*, Type III response of Hen Harriers (*Circus cyaneus*), to Red Grouse (*Lagopus lagopus scotica*)

that these raptors took voles in proportion to their availability, independent of density (fig. 15.21A). Different predators may respond differently to the same type of prey; the Peregrine Falcons (*Falco peregrinus*) and Hen Harriers (*Circus cyaneus*) studied by Redpath and Thirgood (1999) showed type II and type III functional responses, respectively, to Red Grouse (*Lagopus lagopus scotica*) densities (fig. 15.21B, C). These results indicated that harriers, but not peregrines, had the ability to stabilize grouse populations. Conversely, the same predator may respond differently to different types of prey; Hellström et al. (2014) found that the functional response of Rough-legged Buzzards (*Buteo lagopus*) to lemmings (type II) was different from that to voles (type III).

Risk of Predation

Natural selection operates through differential fitness, which has two main components: survival and reproduction. To survive it is essential to eat (well enough to outcompete and outreproduce others) and to avoid being eaten (or killed). Animals can afford failing to reproduce and/or to eat optimally because they may have opportunities to make up for those failures in the future; however, failure to avoid a predator means no further opportunities (Lima and Dill 1990). Therefore while foraging a bird could easily switch from being a predator to becoming prey. Experiments with Great Tits (*Parus major*) have shown that when the birds are starved they spend more time looking down and handling food items, and that as the birds approach satiation they spend more time looking up in vigilance (Krebs 1980). A hungry bird is willing to pay a high predation risk in order to decrease the starvation risk, whereas a less hungry one is willing to pay only the feeding cost while attending more to predators; there exists a trade-off between vigilance and feeding (McCleery 1978).

The basic components of predation risk (Holling 1959, Lima and Dill 1990) are (1) predator-prey encounter rate, (2) mortality in a specific encounter, and (3) time spent vulnerable to an encounter. Birds can reduce their predator encounter rate by avoiding being exposed to predators, for example by following foraging routes and foraging where predators cannot easily detect them (McCabe and Olsen 2015). Birds can reduce mortality by detecting the predator before being detected, escaping, and/or dissuading it from hunting (e.g., mobbing; Altmann 1956). Finally, birds can minimize the amount of time they are vulnerable by gathering the food as

adult and chick densities, respectively. *From Redpath and Thirgood 1999; figures reproduced with permission from* Journal of Animal Ecology. Of these, only *C*, Hen Harriers, exhibited an effect of prey population regulation.

SPATIOTEMPORAL CONSISTENCY IN FEEDING ECOLOGY OF VULTURES

One of the few studies of foraging ecology in paleocommunities in birds is described for vultures. Hertel (1994) described how the vulture community at the Pleistocene Rancho La Brea tar pits (located in what is now the city of Los Angeles, California) was structured similar to modern Old and New World localities. These localities harbor vulture species that exhibit similar ranges of body sizes and that employ similar varieties of feeding behaviors—"gulpers," "rippers," and "scrappers"—as indicated by differences in their skull and beak shapes (figure below).

However, within each locality, the three distinct feeding types are not associated with any particular body size. At

Rancho La Brea, unlike the modern localities, both Old (Accipitridae) and New (Cathartidae) World vulture fossils are present. Nevertheless, these two groups are more similar than would be expected by chance, despite their phylogenetic differences (figure below). Note that no assemblage possesses more than two species of the same body size *and* feeding behavior, and that body sizes at all locations fall into three distinct size categories. These results reveal the importance of body size in structuring consumer communities, and emphasize the role of competition (covered in greater detail in chapter 22) in maintaining vulture guilds through time.

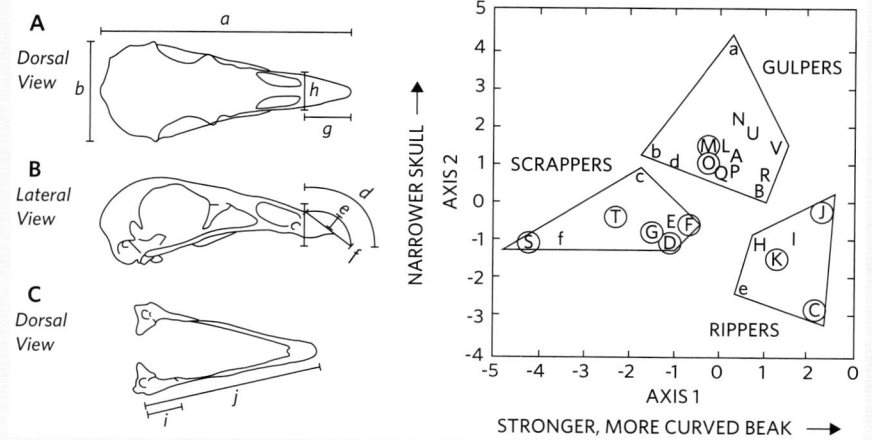

How do you know what a fossil ate? *Left,* Hertel (1994) took a series of skull and beak measurements from a variety of extinct and extant vulture species and, *right,* found that they fell into three groups, based largely on skull width and beak curvature. Based on the feeding modes observed in living members of each group (rippers, gulpers, and scrappers), we can understand the general feeding modes of extinct members by association, from just their skull and beak shapes. *Figures from Hertel 1994, reproduced from* Ecology, *with permission.*

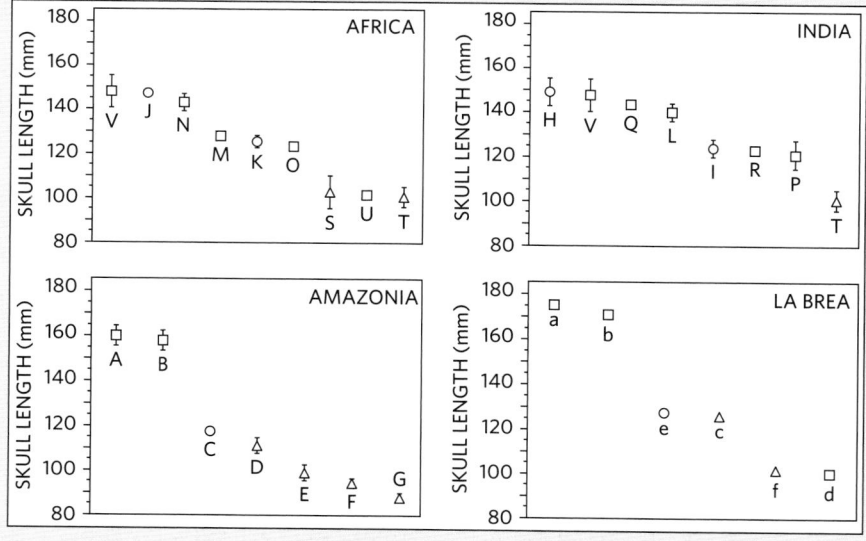

Body size and community structure. Distributions of body size based on overall skull length across four vulture assemblages. Note that each feeding behavior type (circles are "rippers," squares are "gulpers," and triangles are "scrappers") can occur at roughly any body size. *Figure from Hertel 1994, reproduced from* Ecology, *with permission.*

NECTAR FEEDING MECHANICS AND THE ECONOMY OF FORAGING IN HUMMINGBIRDS

Hummingbirds evolved from insectivorous aerial hunting ancestors into tiny, long-billed, hovering machines. Their main caloric source is floral nectar, and everything in their lives is connected to their ability to drink it efficiently. Hovering is convenient for accessing delicate flowers located along thin branches, but it is also the most expensive form of locomotion. For this and other reasons, hummingbirds are under strong selective pressure to collect nectar as fast as they can, and the limits of their nectar extraction capabilities will in turn determine their preferred flowers and foraging strategies (figure below).

For almost two centuries (e.g., from Martin 1833 to Kim et al. 2012) scientists modeled how fast hummingbirds would extract nectar assuming that the tongues behaved as tiny capillary tubes. Only in recent years, aided by the advent of macro high-speed videography, could we study in detail what the birds actually do. This work has uncovered mechanisms that not only are completely different from capillarity but that also work five times faster (figure below). These nectar extraction and transport rates ultimately affect the whole foraging cycle (figure above). As a result, these mechanisms have profound implications for our current understanding of energetic expenditure in nectarivorous birds, on which decades of foraging theories have been based.

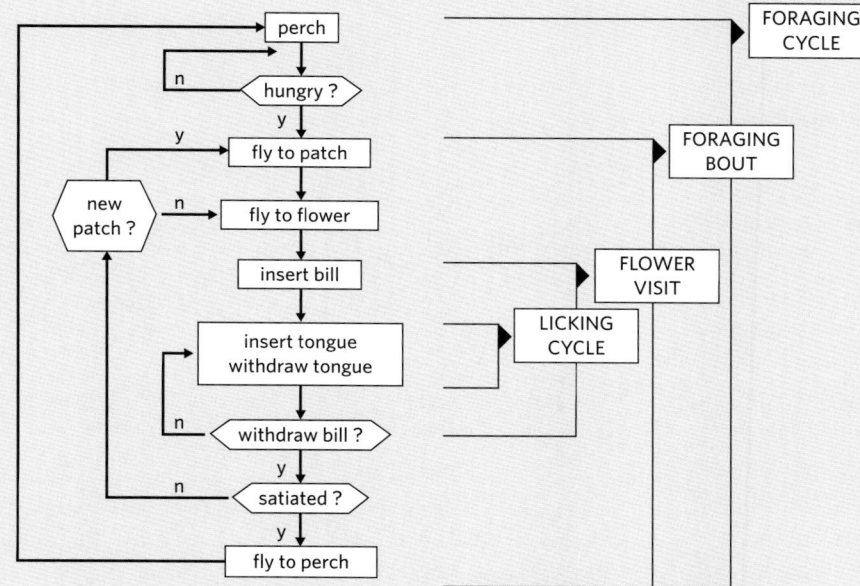

Temporal scales of hummingbird foraging. This figure shows the series of decisions that nectarivores make at each step of the foraging and feeding process. At the center, however, lies the cycling of the tongue inside the nectar chamber. Exactly how fast the tongue can collect the nectar will determine licking and liquid extraction rates, time spent at each flower, and so on, creating a cascade of events that will directly affect the foraging economy of nectarivory. We have only recently started to understand specifically how hummingbirds drink nectar (figure 2). *Figure from Gass and Roberts 1992, reproduced with permission from* The American Naturalist.

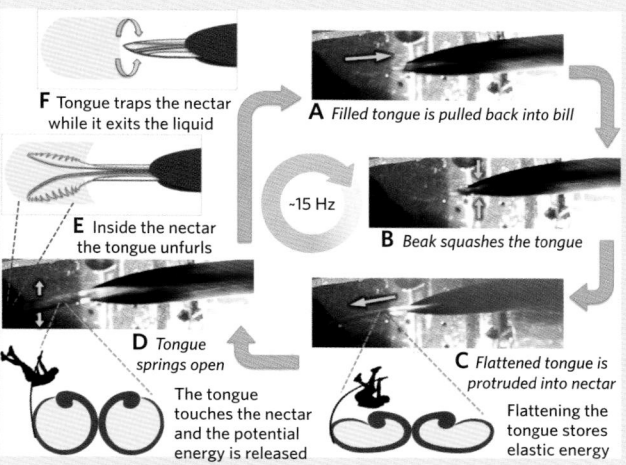

F Tongue traps the nectar while it exits the liquid

E Inside the nectar the tongue unfurls

D *Tongue springs open*
The tongue touches the nectar and the potential energy is released

A *Filled tongue is pulled back into bill*

~15 Hz

B *Beak squashes the tongue*

C *Flattened tongue is protruded into nectar*
Flattening the tongue stores elastic energy

The "nectar trapping" and "elastic micropump" mechanisms. *A*, The tongue cycle starts with the tongue inside the beak. *B*, The tongue is flattened by the bill tips, thereby *C*, loading it like a spring upon protrusion. *D*, When the tongue enters the nectar, the spring action of the tongue walls acts to "pump" nectar up through the tongue while recovering its cylindrical shape. While this is happening in the shaft of the tongue, *E*, the lamellae of the bifurcated tips immersed in nectar unfurl, and *F*, when the tongue is retruded, each lamella furls back at the surface, "trapping" drops of nectar. The horizontal yellow arrows show the anterior-posterior movement of the tongue, and the vertical yellow arrows show changes in the dorsoventral shape of the tongue (matching schematics of tongue cross sections, below). This cycle is repeated about 15 times a second. *Modified from Rico-Guevara and Rubega 2011 and Rico-Guevara et al. 2015, with permission from* Proceedings of the Royal Society B.

THE ROLE OF FEEDING IN POPULATION AND COMMUNITY ECOLOGY
By Thomas W. Sherry

Thomas W. Sherry. *Photo by R. David Sherry.*

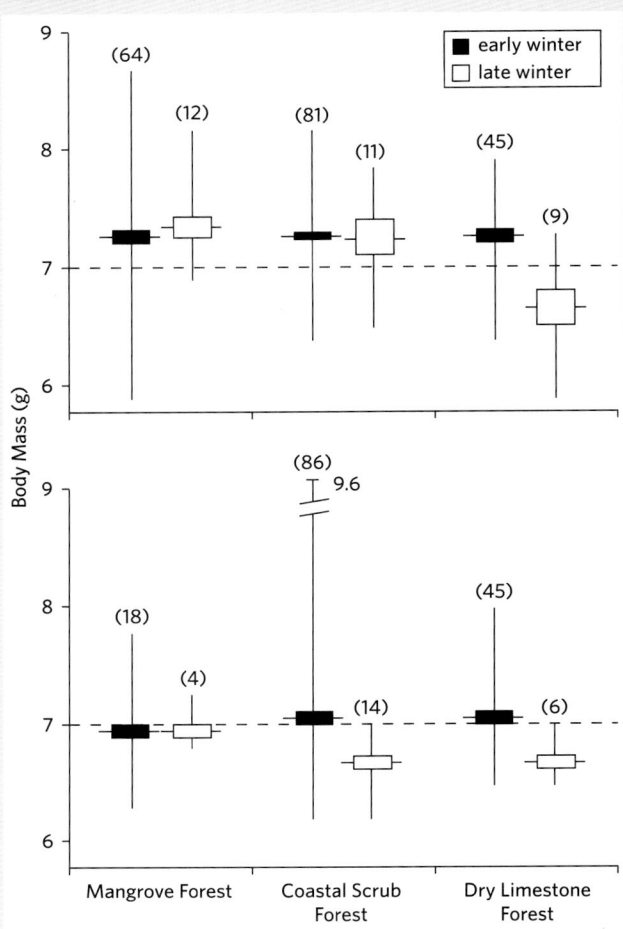

Body mass of, *top,* male, and *bottom,* female American Redstarts (*Setophaga ruticilla*) across habitats and time. Body mass declines, particularly in females, between early (black) and late (white) winter in the most drought-stressed habitats. This decline may reflect depletion of lipid reserves, which may be linked to the low rates of persistence reported for those habitats. *Figure from Sherry and Holmes 1996, reproduced with permission from* Ecology.

For almost 40 years, Professor Thomas W. Sherry, at Tulane University since 1989, has contributed significantly to ornithology, ecology, and conservation. Feeding behavior, resource partitioning, and diet specialization, particularly in resident tropical birds (Sherry 1984, 1990), have provided research foci throughout his career, culminating in a book chapter, "Foods and Foraging," in the third edition of *Handbook of Bird Biology* (Sherry 2016). His diet research links feeding ecology to interspecific competition, exemplified by one of the earliest studies documenting individual foraging specialization in the broadniched Cocos Island Darwin's Finch (*Pinaroloxias inornata*), released from competition by its isolation on a remote tropical island (Werner and Sherry 1987).

Food availability and food competition are central to research Sherry began in Jamaica with Richard T. Holmes, in 1986, on winter- and year-round population limitation of migratory birds (e.g., Sherry and Holmes 1995; figure right). These studies link winter food abundance to body condition and timing of spring migration in the American Redstart (*Setophaga ruticilla*; Johnson et al. 2006, Cooper et al. 2015). Both body condition and spring migration timing carry over

fast as they can, storing it in their crops or caching it (above), and processing it later under protective cover (Pravosudov and Lucas 2001). Studies have also shown that storing energy in the form of fat could allow birds to operate without having to sacrifice predator avoidance, but too much fat could increase predation risk by impairing their flight capabilities (reviewed by Bonter et al. 2013). Other studies have shown that birds minimize predation risk by foraging at lower-risk times, or when the weather conditions hinder detection, or

by selecting microhabitats where predators are less common (reviewed by Lima and Dill 1990, Pravosudov and Lucas 2001, McCabe and Olsen 2015).

SOCIAL ASPECTS OF FORAGING AND FEEDING

Birds often join forces to forage and feed in numbers. European Starlings (*Sturnus vulgaris*) are typically seen foraging for grubs, seeds, and insects en masse. American Robins (*Turdus*

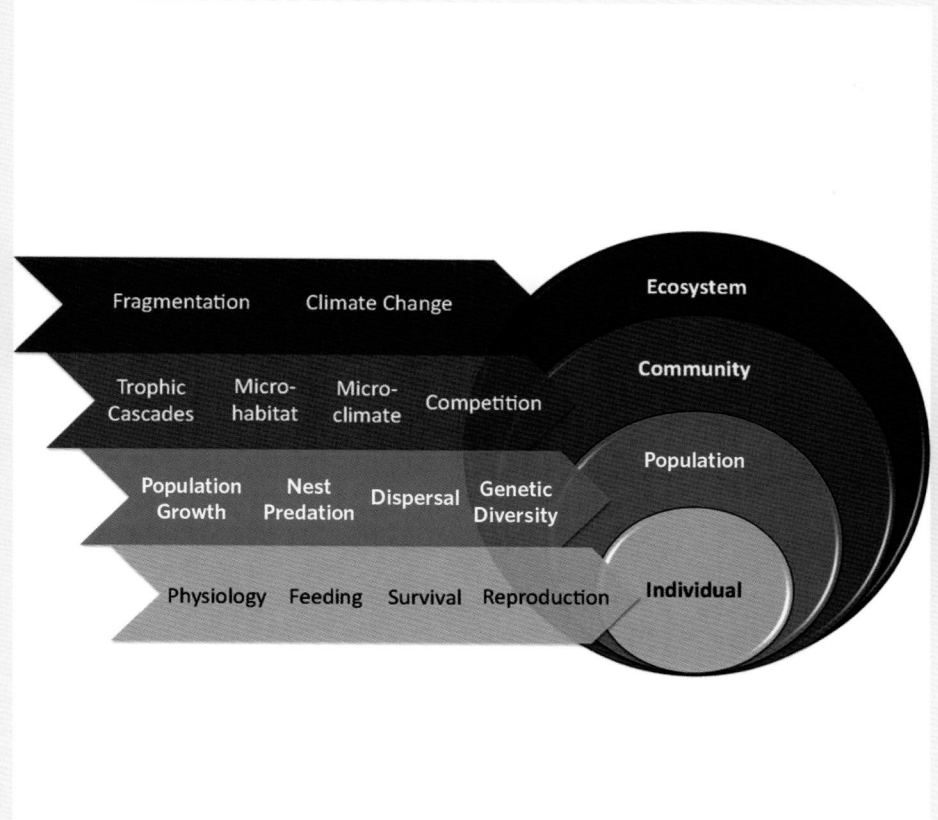

Multiple factors, and at multiple levels of biological organization, causing resident tropical forest interior bird populations to decline in response to global change phenomena. *Figure from Visco et al. 2015, reproduced from* Biological Conservation *with permission.*

in migratory birds to influence reproductive success at higher latitudes (Norris et al. 2004), although nest predators and spring weather also limit migratory bird populations (Sherry et al. 2015). Observed diets and available prey provided the first support of the "breeding currency hypothesis" (sensu Greenberg 1995), involving food limitation of subtropical species that facilitates their winter coexistence with abundant migrant birds (Johnson et al. 2005) and implicates competition for limited winter food among the wintering migrant species themselves (Sherry et al. 2016).

Besides helping us understand how populations are limited and regulated (Rodenhouse et al. 1997, Marra et al. 2015), Sherry's research has identified multiple mechanisms of avian vulnerability to global change phenomena. This research, as well as Sherry's research on interspecific competition, have established important new research needs in community ecology and conservation and helped synthesize our understanding of year-round limiting factors in birds.

migratorius) are often seen in variable numbers at a time searching the ground for the same food. This type of group effort in foraging has distinct advantages and disadvantages, mostly related to the balance between having a greater number of eyes looking for food and potential dangers (advantage), and having to split resources among a greater number of participants (disadvantage). The intricacies of social systems are treated in greater depth in chapter 17; here we focus on some of the causes and consequences of foraging

in groups. There are a few different degrees of group foraging. Relatively less coordinated forms, such as foraging by local enhancement, social facilitation, and flocks (Ellis et al. 1993, Krause and Ruxton 2002, Jackson et al. 2008), typically consist of conspecifics that feed communally. In other cases, birds that may or may not generally flock together actively cooperate specifically to forage more effectively and/or efficiently (Ellis et al. 1993). Members of these foraging associations may or may not be related to one another, and social foraging theory often grades into other theoretical constructs, such as optimal foraging, game theory, reciprocal altruism, and kin selection (Giraldeau and Caraco 2000).

Foraging by local enhancement (whereby foragers are attracted to join foraging groups) and social facilitation (in which information, in this case about the location of food, is provided by conspecifics; Ellis et al. 1993, Jackson et al. 2008) constitute important mechanisms of cooperative foraging. In Griffon Vultures (*Gyps fulvus*), conspecifics act as important cues for finding carcasses, thereby increasing the efficiency with which vultures locate prey (Jackson et al. 2008). A sort of chain reaction ensues, such that as the number of foraging vultures increases, the probability of finding prey increases exponentially (Jackson et al. 2008). However, as the number of vultures becomes larger, the system becomes saturated and the probability of finding prey levels off. Seabirds often feed in large single- or multispecies aggregations, but do not necessarily search for prey in large numbers (Shealer 2002). Instead, they often rely on social facilitation to locate prey, but when they find it, they do not necessarily share resources (Ellis et al. 1993, Shealer 2002). In some cases these foraging associations cross class lines; one study documented an association between Aplomado Falcons (*Falco femoralis*) and Maned Wolves (*Chrysocyon brachyurus*) in Brazil, whereby the falcons would routinely take tinamous (Tinamidae) that the wolves missed and flushed (Silveira et al. 1997).

Birds may not only become exposed to new foods and/ or food sources by associating with other birds but may also learn new foraging techniques by observing them. Behaviors developed this way can spread rapidly within and among populations. Probably the most famous example is the milk bottle cap opening behavior (usually puncturing the cover with their beak) exhibited by tits (*Parus* spp.; predominantly the Blue Tit, *Cyanistes caeruleus*), which spread dramatically throughout Great Britain over a 20-year period (Fisher and Hinde 1949). Although the mechanism of learning (e.g., the role of imitation; Sherry and Galef 1984), and the model describing the pattern of advancement of the behavior throughout the population (e.g., single versus multiple centers of proliferation; Lefebvre 1995) have been debated, this, and other studies of congeneric Black-capped Chickadees (*Parus atricapillus*; Sherry and Galef 1984), illustrate important points about the development of foraging skills in a social context. Birds are capable of learning by observing other experienced birds, or simply by being drawn to a novel food source (e.g., milk bottles opened by others) via local enhancement, and then learning the requisite behaviors (e.g., opening sealed bottles) on their own by trial and error (Sherry and Galef 1984).

Above we saw how birds may converge in their behavior for a common, shared interest. Coordinating foraging effort,

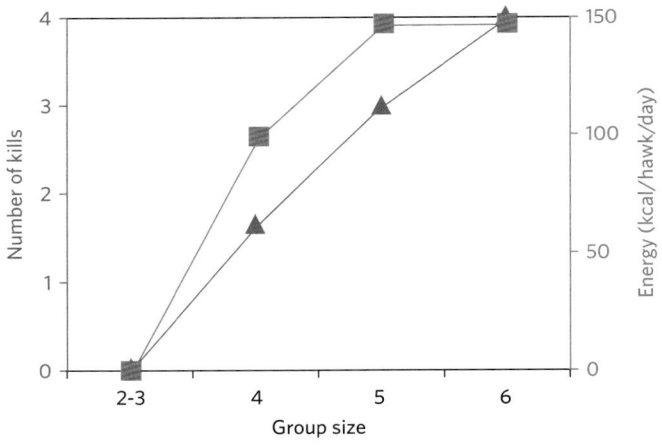

Figure 15.22. The number of rabbits captured (blue triangles) and energy derived (red squares) as a function of group size of Harris's Hawk (*Parabuteo unicinctus*) hunting parties. The larger the group size, the greater the capture success. Despite the larger number of mouths to feed, there was a per capita energetic advantage to hunting in larger groups; in fact, it was only at the larger group sizes that the hawks met their daily caloric needs (estimated at 147.8 kcal/hawk/ day). *Graph redrawn from figure 2 of Bednarz 1988.*

however, is quite a different behavior. Some raptor species, such as bird-eating falcons, Golden Eagles (*Aquila chrysaetos*), and Harris's Hawks (*Parabuteo unicinctus*) have been known to search and hunt cooperatively—that is, participate in capturing relatively large and elusive prey, then share the bounty (Bednarz 1988, Ellis et al. 1993). The best-documented case of cooperative hunting is in Harris's Hawks that occupy the southwestern US deserts, in which parties of two to six individuals were observed "collaborating" to attack large rabbits (Bednarz 1988). The hawks would typically converge upon a rabbit from different directions. When the rabbit took cover, the majority of the members would surround the location, and one or two would enter the cover to flush it out. Group size was positively correlated with capture success, and each individual benefited more—energetically—with increased group size (fig. 15.22).

Mixed-species flock foraging is particularly interesting as it pertains to cooperation, because it entails associations of two or more bird species that move and forage together, and maintain this interaction for long periods (Harrison and Whitehouse 2011). Mixed flocks have been observed in diverse ecosystems and are common in bird communities of tropical forests (Powell 1985). In Neotropical communities differences are found between general patterns of montane versus lowland flocks in terms of species composition, food resources, territoriality, vertical distribution, and frequency of participation (Marín-Gómez and Arbeláez-Cortés 2015).

There are many theories as to why this flocking behavior exists (e.g., Sridhar et al. 2009), but we discuss here only the two main hypotheses about the advantages that have influenced the evolution of mixed flocking (sensu Powell 1985): (1) birds forage in mixed flocks because it increases foraging efficiency; and (2) birds perform this behavior because it decreases their vulnerability to predators. These two main hypotheses are not mutually exclusive, as we describe below.

(1) *Increasing foraging efficiency.* Moving in a large group that covers a variety of substrates and hunting strategies ensures an increased spectrum of prey detection and spatial memory by pooling the capacities of the participant species. Group foraging also ensures learning and exploitation of new techniques and niches for foraging, increases the probability of capturing prey flushed by other individuals, and affords defense of larger territories (Munn and Terborgh 1979, Powell 1985, Harrison and Whitehouse 2011). Larger mixed flocks move faster through the environment than do smaller ones because as inter-individual spacing is diminished, each individual must move farther, which may ultimately afford greater overall efficiency by reducing duplication of effort (Morse 1970).

(2) *Improvement of the detection capacity and "power" against predators* (Powell 1985). As an advantage of any aggregation in nature, moving in a mixed flock may allow the participants to reduce the time spent in sentinel duties, and therefore increase the time available for other activities, e.g., foraging (reviewed by Sridhar et al. 2009). Flocking in general confuses predators by the sensory overload of many moving targets and enables a dilution effect (Sridhar et al. 2009). Additionally, a mixed flock can mob a predator, taking advantage of its strength in numbers.

Social dominance and interference play an important role for the foraging success of individuals in many taxa. These facets of behavioral ecology also represent important parameters for models of foraging optimality, such as the "ideal free distribution" (Fretwell and Lucas 1970), which basically describes how individuals should sort them themselves across habitat patches that vary in quality (Tregenza et al. 1996, Smith et al. 2001). An individual's place in the dominance hierarchy (i.e., "pecking order") largely determines which kinds of, and how much, food becomes available to it. The role of dominance in the context of feeding behavior of birds has probably been studied most thoroughly in tits (*Parus* spp.; box on page 474) and oystercatchers. Based on studies of individually marked oystercatchers, food intake rate (mg of flesh consumed per five-minute period) decreased at higher bird densities, but not for the most dominant individuals (Ens and Goss-Custard 1984). This was due in

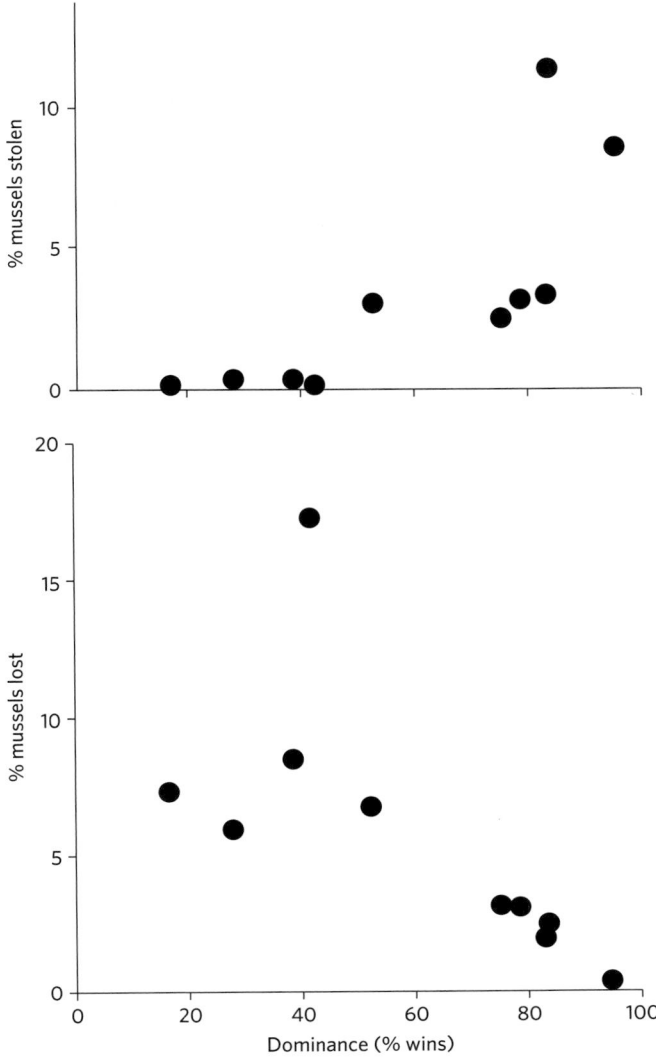

Figure 15.23. Among the color-banded Eurasian Oystercatchers (*Haematopus ostralegus*) studied by Ens and Goss-Custard (1984), the more dominant individuals had the highest food intake rates, in part because, *top*, they stole more mussels from and, *bottom*, lost fewer mussels to, other individuals. *Figure reproduced from the* Journal of Animal Ecology, *with permission.*

part to kleptoparasitism by dominants (fig. 15.23), but also because of reduced capture rates of subdominants and increased time spent on aggression (Ens and Goss-Custard 1984). Furthermore, prey intake rates were generally higher in hammerers, but hammerers were more susceptible to interference than were stabbers, possibly because they were less aggressive and/or were attacked more frequently, possibly because they took longer to open mussels (Goss-Custard and Durell 1988).

PLASTICITY IN FORAGING BEHAVIOR: DO BIRDS HAVE PERSONALITY?

By Katherine A. Shaw

Ask any researcher who has spent numerous hours observing the behavior of their study animal and they will likely tell you individuals exhibit consistent differences in their exploratory behavior (or boldness, sociability, activity, aggression, etc.). In humans, such consistent differences in behavior within and across situations or contexts are described as "personality." However, mainly out of fear of anthropomorphizing animal behavior, such inter-individual differences in behavioral responses have long been interpreted as consequences of inaccurate measurement and often neglected as a source of biologically meaningful variation (Sih et al. 2004, Groothuis and Carere 2005). But consistent individual differences in behavior within and across situations or contexts (personality) are extremely common in species other than humans, including invertebrates, fish, mammals, and birds. Evidence suggests animal personality traits can directly affect dispersal, foraging, and antipredator responses, potentially resulting in differential survival and reproductive success (fitness) among individuals in a population (e.g., Dingemanse et al. 2004, Duckworth 2008, Patrick and Weimerskirch 2014). Today, a growing field of behavioral research aims to describe the various axes of animal personality and the causes and consequences of this inter-individual variation within animal populations (Briffa and Weiss 2010, Carere and Maestripieri 2013).

Behavioral variation among individuals is likely to be continuous and complex, with individuals varying in whole suites of behavioral traits and exhibiting various levels of behavioral plasticity (Sih et al. 2004, Dingemanse et al. 2010). This type of plasticity in birds has previously been described as "ecological plasticity"—an individual's ability to respond to food, competition, and novel resources (Greenberg 1990, Moreno et al. 2001). Competitive ability and resource exploitation are major determinants of fitness in many species, and personality traits can have direct effects on these. Cole and Quinn (2012) examined how competitive ability in Great Tits, measured by the amount of time with sole access to a feeder, relates to exploration behavior (measured by the number

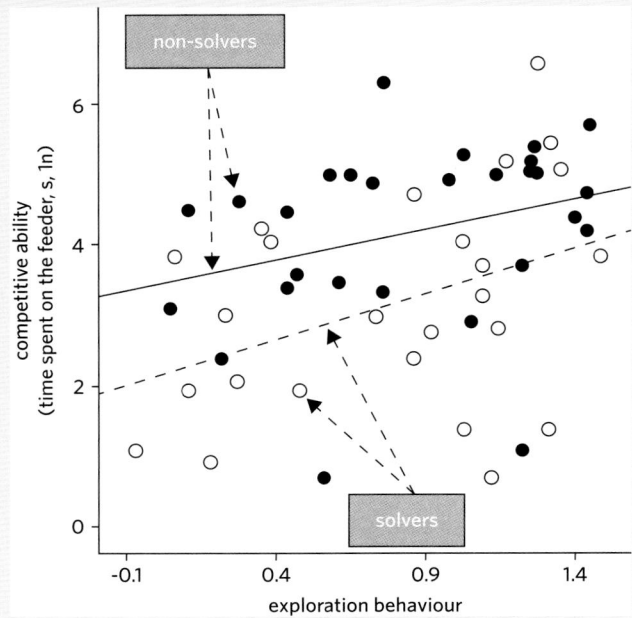

In Great Tits (*Parus major*), differences in competitive ability among individual males were explained by variation in two personality traits: exploration behavior, a measure of proactivity, and problem-solving ability. *Figure from Cole and Quinn 2012, reproduced with permission from* Proceedings of the Royal Society B.

of movements made in a novel environment), a proxy for proactivity that is correlated with boldness, aggressiveness, and risk-taking under captive conditions (Verbeek et al. 1994, 1996, van Oers et al. 2004). They also assayed problem-solving performance, another personality trait, based on the birds' abilities to open a contraption to access food. Individual males that were better competitors tended to be more proactive. Interestingly, however, poor competitors were better at problem-solving (see figure). One potential explanation for this is that subordinates are forced to use innovative foraging strategies (necessity driving innovation) to rely on food sources other than the one subject to intense competition.

ONTOGENY OF FORAGING AND FEEDING BEHAVIOR

Birds are not necessarily born proficient foragers and feeders, but are often made. Juveniles ("young birds who have not reached sexual maturity," Wunderle 1991) tend to perform poorly compared with adults in virtually all aspects of foraging, from prey recognition to capture and handling. For example, adult Ospreys showed preferences for more profitable littoral habitats, whereas juveniles did not, presumably because of their inabilities to recognize differences in fish abundance (Edwards 1989). With regard to prey capture, juveniles tended toward smaller fish size classes and avoided larger ones—likely because of their inabilities to physically handle larger fish (see Wunderle 1991 for numerous other examples across a range of taxa). Foraging proficiency improves over time, and naturally the time course in which birds develop adult capabilities varies considerably across species and prey types, from days (e.g., wheatears; Wunderle 1991) to years (e.g., oystercatchers; Marchetti and Price 1989). This discrepancy between adults and juveniles has important implications because juvenile birds experience relatively higher levels of mortality (Wunderle 1991). Reduced foraging efficiency can force birds to spend more time foraging, potentially at the expense of other vital functions, and has been linked to deferred breeding and reduced survival, both of which are components of fitness (Marchetti and Price 1989).

In general, juveniles tend to drop food items more frequently and exhibit longer handling times (Wunderle 1991). Other age-related differences in foraging performance are typically quantified by "foraging success" (the number of prey items captured per capture attempt), "foraging rate" (the number of prey captured per unit time), and/or by the time interval between the capture of one prey item and the next (Wunderle 1991). There are also qualitative differences, such as in the types and size classes of food consumed. For example, one study reported that juvenile Northern Mockingbirds (*Mimus polyglottos*) often pursued relatively less-challenging fruit items and small arthropods than did their primarily larger-arthropod-eating adult counterparts (Breitwisch et al. 1987). Walton (1979) reported that juvenile Meadow Pipits (*Anthus pratensis*) ate more relatively slow-moving coleopterans and hymenopterans, whereas adults were better equipped to handle faster-moving dipterans.

Based on a comprehensive review of the topic, Marchetti and Price (1989) offered some compelling explanations for the observed differences between adults and juveniles. One possibility is that because juveniles are still undergoing skeletal growth and neurological and functional maturation, they are constrained in their abilities. The beaks of juveniles tend to be shorter, shallower, and less rigid than those of adults, and in many species of finches, juveniles are relegated to smaller and softer seeds (Marchetti and Price 1989). In European Nuthatches (*Sitta europaea*) and European Starlings, maturation of bill length was at least partially implicated in the observed improvement in handling skills of young birds (Wunderle 1991). Prey capture skills develop from more rudimentary pecking motions that progressively become more elaborate. Wunderle (1991) cited examples of Reed Warblers (*Acrocephalus scirpaceus*) and Northern Wheatears (*Oenanthe oenanthe*) progressing from "stand-and-catch" techniques to aerial techniques such as flycatching and hawking.

Another prevailing hypothesis is that foraging techniques may require learning, for instance by way of trial and error. Eurasian Oystercatchers (*Haematopus ostralegus*) learn and develop the mussel-handling technique ("stabbing" or "hammering") of their parents, but will also use other techniques at times (Norton-Griffiths 1967, Goss-Custard and Sutherland 1984). Despite an ostensible innate component to their predatory behavior, young raptors often follow (and continue to be fed by) their parents for extended periods of time, and likely gain valuable experience before they are left to their own devices (Johnsgard 1990; fig. 15.24). Parents may participate in their learning process, for instance, by dropping dead or live prey for them to pursue on the wing (Marchetti and Price 1989, Johnsgard 1990). Learning can compound the effects of maturation, because it can continue to contribute to enhancing proficiency even after birds stop developing (Wunderle 1991).

Other ideas posit that adults and juveniles are subject to other physiological or external constrains that force them

Figure 15.24. Female Peregrine Falcon (*Falco peregrinus*) dropping an avian prey item for her fledglings. Fledglings remain dependent on the parents for one to two months after fledging, and this behavior is thought to provide a form of practice that helps hone their hunting skills. *Photo by Glenn Conlan.*

down different foraging pathways. For instance, adults and juveniles might have different nutritional requirements, and therefore might be required to feed on different prey, irrespective of differences in their abilities (Marchetti and Price 1989). Among the examples cited by Marchetti and Price (1989) are studies of juvenile American Black Ducks (*Anas rubripes*) that consumed more protein than their older counterparts, and adult Northwestern Crows (*Corvus caurinus*) that foraged for larger clams than did juveniles. Alternatively, other external factors may force juveniles and adults to feed differently, such as predator avoidance and social dominance (Marchetti and Price 1989). Wunderle (1991) cited studies of Glaucous-winged Gulls (*Larus glaucescens*) and Eurasian Oystercatchers displacing juveniles from preferred habitats and forcing them into suboptimal ones.

Juveniles often compensate for their inabilities in other ways, such as by spending more time foraging, foraging in alternative habitats, targeting different types or size classes of prey, or stealing food from other birds—essentially by finding a temporary work-around that requires less skill (Wunderle 1991). For example, because juvenile Laughing Gulls (*Leucophaeus atricilla*) were less successful than adults at locating food and plunge-diving for it, they were often observed congregating at garbage dumps and snatching offal in the air (Wunderle 1991). Juveniles of other gull species and Brown Pelicans (*Pelecanus occidentalis*) have been known to join actively feeding flocks to compensate for their inabilities to select profitable resource patches (Wunderle 1991).

CONTEMPORARY APPROACHES FOR STUDYING FORAGING AND FEEDING BEHAVIOR

Methods of Behavioral Observation

A full treatise of animal behavior methodology is beyond the scope of this chapter; here we provide a general overview of the nuts and bolts of avian foraging and feeding studies. These studies not only entail what, when, and how birds perform relevant behaviors but also typically involve some form of quantification of their resources and environment. A comprehensive reference that explains and exemplifies many (and much more) of these techniques is Morrison et al. (1990). These methodologies cover a great range of sophistication, technological requirements, and expertise, depending on the complexity of the behavior of interest, and/or that of the bird's resources (e.g., arthropods, fruit, or fish) and environment (e.g., desert, forest, or ocean).

Within the purview of "foraging behavior" research is the astounding number of dietary studies that have been published for a great diversity of avian taxa. The more traditional meth-

odologies involve direct visual observation and/or video recording of prey use and/or delivery to chicks at the nest, as well as various forms of analyses of stomach contents, regurgitated pellets, droppings, and prey remains at nests (e.g., Rosenberg and Cooper 1990, Marti et al. 2007). Prey are often grouped taxonomically or ecologically (Cooper et al. 1990), and diversity indices that take into account the types and relative abundances of prey consumed are commonly employed to quantify dietary diversity and breadth within and among individuals, populations, or species (box on page 477). Each dietary analysis potentially imposes a different bias. For example, regurgitated pellet analysis is mostly representative of prey characterized by hard, indigestible parts (e.g., bone, hair, feathers, and chitin), whereas other more fully digested "soft" prey consumed may not be represented in the sample. Similarly, analysis on fecal samples may be biased by differential digestibility and rates of passage across prey types (Rosenberg and Cooper 1990).

The fundamentals of quantifying foraging and feeding behavior are derived from general ethological methods commonly employed in behavioral ecological studies (e.g., Altmann 1974, Lehner 1996). Deploying these techniques requires little more than a decent set of binoculars, pencil and notepad, stopwatch, and perhaps some patience (e.g., MacArthur 1958). The typical forms of measurement are qualitative, i.e., the types of behavior performed (see Remsen and Robinson 1990 for a comprehensive classification scheme) or the sequence in which they are performed, and quantitative, e.g., the duration of a particular behavior or the frequencies with which different types of behaviors occur. However, considering logistical constraints, it is impossible to record every single behavior of interest, performed by every single bird in the study area of interest, every single time. For this reason, standardized regimes that allow for measurements to be taken objectively and constantly for the sample units of interest (e.g., individuals, groups, or blocks of time) are necessary. Lehner (1996; see also Altmann 1974) summarized the major types of sampling regimes: "focal"—in which several behaviors are recorded for one focal individual, pair, or group over a set observation period; "all-animals"—in which a few sporadically performed behaviors are opportunistically recorded for a large number of individuals; and "ad libitum"—in which unplanned, opportunistic field observations on one or more individuals are recorded. Data may be recorded continuously to provide a complete account of all occurrences or sequences of behaviors of interest for (typically) a focal individual, pair, or group. Alternatively, data may be recorded at set time points, for example, to determine whether or not a behavior occurs during a sampling interval. "Instantaneous sampling" involves scoring the occur-

QUANTIFYING DIETS: DIET DIVERSITY USING SHANNON INDEX

Determining what, and how much of it, a bird eats is an important facet of avian ecology, evolution, conservation, and management. The proportional composition of prey in the diet is often calculated according to the number of different types of prey consumed relative to the total number of prey items; or according to the mass of each prey item (measured directly, or estimated by the average mass of the prey type multiplied by the number of such prey items of that type) relative to total biomass consumed (Rosenberg and Cooper 1990, Marti et al. 2007); or other metrics (e.g., stable isotope ratios) that represent the contributions of various types of prey to the diet (Inger and Bearhop 2008).

In addition, various metrics have been devised to encapsulate dietary diversity or niche breadth within or among individuals, populations, species, and communities. These metrics are derived mostly from diversity indices, which are more commonly used to characterize species diversity and community structure (Stiling 1999). By taking into account not just the number of different types of things, but also their relative abundances, these indices shed light on the ways in which resources are distributed; whether a diet is dominated by one or few prey types, or if all prey types are represented more evenly. A diet has a high diversity—and represents a broad food niche—if many prey types occur in roughly equal numbers. Conversely, a low-diversity diet consists of few prey types, or many prey types with different abundances (e.g., the sample is dominated by a few prey types; see table). Here we provide a prelude to one of the most popular approaches for measuring dietary diversity, the Shannon index (H'; Shannon and Weaver 1949):

$$H' = -\sum p_i \ln p_i$$

where p_i is the proportion of each prey type in the sample. The Shannon index is affected by both richness (i.e., the number of prey types) and evenness (i.e., how uniformly the various prey types are represented in the sample); a larger number of prey types and a more even distribution will increase H' (Stiling 1999, Marti et al. 2007). A more direct measure of the evenness component can be obtained from:

$$H'/H_{max} = H'/\ln S,$$

where H_{max} is the value of diversity when all species are equally abundant (i.e., $H_{max} = \ln S$), and S is the total number of prey types (after Stiling 1999). Diversity measures assume that all resources are equally available (Marti et al. 2007). Sample size is an important consideration; Marti et al. (2007) suggested that samples of 100 or more prey individuals give a reasonable approximation of a raptor's diet.

Prey of Barn Owls (*Tyto alba*) in Utah, adapted from Marti et al. (2007), and sample calculation of the Shannon index (*H'*).

Prey taxon	No. of individuals	p_i	$p_i \ln p_i$	H'
Shrews	4223	0.038	−0.124	0.252
Bats	15	1.4E-04	−0.001	
Rabbits	3	2.7E-05	0.000	
Rodents	104807	0.944	−0.054	
Weasel	1	9.E-06	0.000	
Birds	1953	0.018	−0.071	
Insects	14	1.3E-04	−0.001	
Sum total	111016	1.000	−0.252	

rence of a behavior (yes/no or 1/0) at predetermined points in time (e.g., every 15 minutes during a three-hour period; Green et al. 1999). "Scan sampling" refers to instantaneous sampling of several individuals simultaneously, by scanning the study area at predetermined time points.

Several things must be considered when deploying these techniques to ensure that data are statistically valid and that broader inferences are sound. Ideally, the individuals, pairs, or groups under study should be representative of all individuals in the population(s) of interest, and should be selected at random, or sampled in random order. The nature of the data is

an important consideration for downstream statistical analysis, and careful consideration must be given to what kinds of data are more appropriate for addressing the research question (Noon and Block 1990). Quantitative data may be continuous (measurements that can take any value between two fixed points, such as bill length) or discrete (measurements that can only take only fixed values, such as the number of acorns cached in a tree) (Sokal and Rohlf 1995), and studies involving one or the other might be designed and analyzed differently. Another critical consideration (for all types of studies, really) is sample size, both in terms of the number

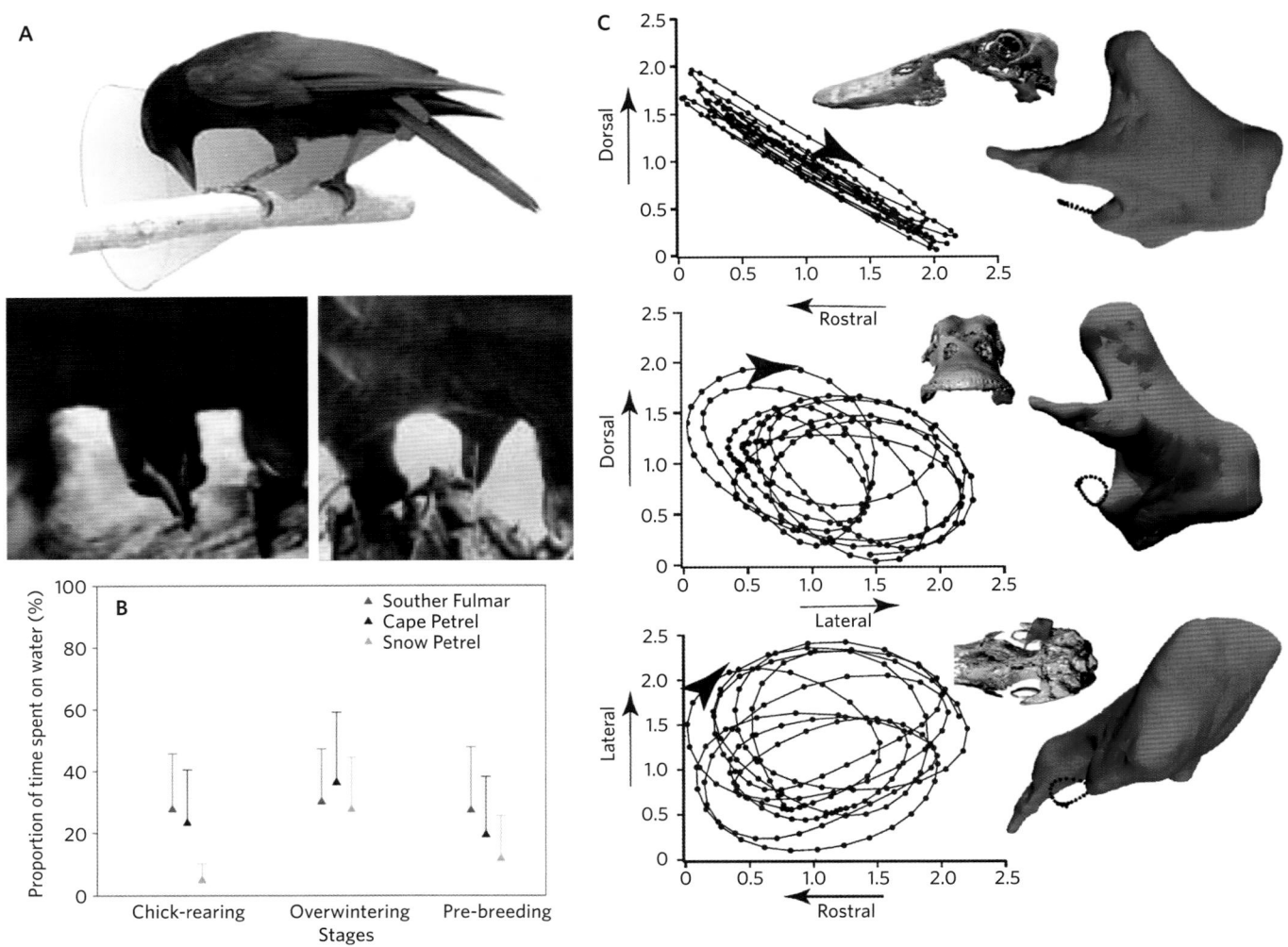

Figure 15.25. Advances in the study of avian foraging and feeding behavior. *A*, Bird-borne video camera mounted to a New Caledonian Crow (*Corvus moneduloides*), showing the camera (red) and its viewshed (blue cone), and captured images from video footage of, *left*, a bird processing a fruit and, *right*, using a tool. *Figure modified from Rutz et al. 2007, reproduced with permission from* Science. *B*, Data generated from geolocator-immersion loggers, showing differences in activity patterns among fulmarine petrels (Procellariidae) during different stages of breeding. *Figure from Delord et al. 2016, reproduced with permission from* Ibis. *C*, Motion paths (in mm) of the quadrate bone (red) of a Mallard (*Anas platyrhynchos*) during feeding for, *top*, lateral, *middle*, anterior, *bottom*, and ventral planar views from an X-ray Reconstruction of Moving Morphology study, showing the complex 3-D motions that are critical for avian cranial kinesis. *Figure from Dawson et al. 2011, reproduced with permission from* The Journal of Experimental Biology. *See also Supplemental Material 15.S10.*

of "states" (discrete behaviors performed by an individual or group) or "events" (changes in states; Lehner 1996) that are recorded for a given individual, and in terms of the total number of individuals under study. For example, measures of central tendency (e.g., arithmetic mean) based only the first sighting of an individual performing a behavior, versus those compiled from sequential observations of a single individual, can bias the results for some aspects of behavior (e.g., Morrison 1984, Brennan and Morrison 1990, Recher and Gebksi 1990). Although sequential observations provide the benefit

of capturing a wider range of activities, they suffer from lack of independence (because of serially correlated observations), and ultimately require greater numbers of observations and individuals (Morrison 1984, Noon and Block 1990).

Recent Technological Advances

The study of avian foraging and feeding has come a long way, from analyzing diets to studying the component acts of feeding behavior. Advances in stable isotope (e.g., carbon and nitrogen) analysis (Hobson and Wassenaar 1999, Inger and

Bearhop 2008) have allowed researchers to quantitatively characterize feeding niches (e.g., Martínez del Rio et al. 2009, Moreno et al. 2016), and infer prey proportions (e.g., Inger et al. 2006, Rutz et al. 2010), from relatively easily sampled consumer and prey tissues (e.g., feathers, keratin, blood, and muscle). Because isotopic ratios reflect assimilated foods and are incorporated into tissues during their growth, one of their greatest advantages is that they provide "a time-integrated average of numerous feeding events" (Hobson 2005), rather than the snapshots in space and time that other methods allow. The transformative technology of next-generation sequencing, which has revolutionized genomics, has permeated into foraging studies (Pompanon et al. 2012), by allowing researchers to take advantage of vast quantities of DNA data to facilitate the identification of prey from fecal samples (e.g., Gerwing et al. 2016). In addition to advances in understanding avian diets, we have also taken great strides in understanding the physical aspects of consuming different types of food. For example, since Bowman's (1961; fig. 15.9) elegant diagrammatic analyses of Darwin's finch beak and jaw biomechanics, researchers have measured their actual bite forces with force transducers (e.g., Herrel et al. 2005) and used finite element modeling to simulate how stresses and strains are propagated throughout the beak and skull (e.g., Soons et al. 2010).

Part of the problem of studying avian feeding behavior (and any behavior, really) is that it happens at a distance, it often happens very fast, and much of the process is concealed inside the beak. Continuous advances in data-logging instrumentation and video-recording technology have steadily been solving these problems. With GPS tracking devices as light as 8 grams—once restricted to only the largest seabirds, such as albatrosses (Weimerskirch et al. 2002)—studying at-sea foraging patterns of ever-smaller seabirds is increasingly possible, both above (Kotzerka et al. 2010) and below (Ryan et al. 2004) the surface of the water. The use of temperature-depth accelerometers has revealed differences in the energetic costs between wing-propelled and foot-propelled (e.g., table 15.3) foraging dives of murres and cormorants, respectively (Elliot et al. 2013). Delord et al. (2016) used 1.5–1.9 g miniaturized saltwater immersion geolocators, which detect saltwater immersion every three seconds and provide estimations of longitude and latitude based on daylight level intensity measured every 60 seconds, to determine differences in foraging movements among Antarctic fulmarine petrels (fig. 15.25B).

Videography is an indispensable tool for behavioral ecology. The miniaturization of camera components has made bird-borne video cameras a reality, and their use offers researchers unprecedented access to aspects of foraging and feeding behavior (Rutz et al. 2007, Troscianko and Rutz 2015; fig. 15.25A). Although high-speed cinematography has been employed for about half a century to capture the intricacies of avian foraging and feeding events (e.g., Goslow 1971), video technology has become much more accessible and streamlined in the past 10 years—to the point that people have high-speed video functions on their cell phones nowadays. The superb frame rates and image resolution of these cameras have shed insights into processes that happen in a fraction of a second, like in the flick of a hummingbird's tongue (Rico-Guevara and Rubega 2011, Rico-Guevara et al. 2015; box on page 469). To understand the mechanics of the movements that we see externally, radiography has been used for decades (e.g., Zweers et al. 1977, Zweers 1982, Kooloos and Zweers 1989). More recently, however, X-ray reconstruction of moving morphology (XROMM), a methodology combining biplanar X-ray videos with high-resolution CT scans to visualize the actual movements of bones (Brainerd et al. 2010, Gatesy et al. 2010), has revolutionized the study of animal movement. Dawson et al. (2011) used XROMM to measure the movements of the quadrate bone, which links the mandible to the cranium and is a key component of cranial kinesis (fig. 15.5), during feeding in Mallards (*Anas platyrhynchos*) (fig. 15.25C; Supplementary Materials S10).

KEY POINTS

- Birds inhabit every environment and as a result, they feed on all forms of organic material and show a variety of behaviors and adaptations for acquiring and processing different food types. Two general modes of locating prey are sit-and-wait and active searching, which represent endpoints on a spectrum used both among and within individuals/species. Locating food involves sensory and locomotor abilities that were discussed in other chapters, but ingesting food is manifested by diverse movements of some skull bones and the beak, termed cranial kinesis. The tongue is also important for manipulating and swallowing food, as well as in filtering food from an aquatic medium. In some groups, such as raptors, the feet play an integral role in subduing, grasping, and manipulating prey prior to ingestion.

- A variety of behaviors are involved with prey acquisition, including tool use. This behavior ranges from simple rock throwing by some vultures to break large eggs, to amazing abilities to fashion twigs and leaves into hooks by New Caledonian Crows to extract beetle larvae from holes. Members of the corvid family (crows, jays, nutcrackers) excel at caching by storing thousands of seeds in autumn and retrieving most of them during times of little food resources. Some raptors hide portions of their prey for later use, shrikes often skewer their prey onto thorns, and some woodpeckers drill seeds into tree bark (granaries) for later use.

- Birds consume a wide range of food items, which is reflected not only in the great diversity of bill shapes but also by the musculature that powers the jaws. Different types of food present different challenges to consumers, and the reality is that many birds often feed on a variety of food types, often defying simple "X-ivore" designations. As a result in this chapter we grouped foods and feeding types very broadly, first by gross physical qualities, such as "bulk" (e.g., plants and animals), "liquid" (e.g., nectar), and "aquatic and marine" (e.g., marine and freshwater invertebrates and vertebrates) foods, then by subtypes within each. Bulk food consumers include frugivores (fruit), granivores (seeds and nuts), herbivores (grass), folivores (leaves), insectivores (arthropods), and carnivores (animals). Liquid consumers comprise nectarivores (nectar), sanguinivores (blood), and consumers of saps, manna, honey, and waxes. Aquatic and marine consumers include diets ranging from small invertebrates to piscivory (eating fish of all sizes and ages). Filter-feeding is common, but other tactics include probing in soft substrates for molluscs, suction, plunging, diving, and underwater pursuit using wings or legs for propulsion. Different modes of feeding are often associated with different characteristics. For example, the diversity of insectivorous birds is manifested in the locomotor adaptations for performing particular foraging maneuvers (e.g., aerial pursuit, gleaning from a substrate, climbing trees, and flycatching) to catch their prey, rather than in beak morphology. Avian folivores are rare, given that leaves are low-quality food as a result of the cellulose surrounding plant cells. Cellulose requires greater energy and longer guts (greater mass) to process, which is not adaptive for the high energetic demands of avian flight.
- Birds have featured prominently in studies of optimal foraging theory, which briefly states that animals should maximize energy intake and minimize energy expenditure to maximize fitness. As resources deplete, birds must make decisions on how long to forage in a patch/area and when to move to another patch, considering distance to the next patch, travel rate between patches, patch quality, etc.
- Competition for limited resources affects fitness, therefore coexistence among species or individuals within a species selects for some form of ecological segregation (e.g., in space or time). Predator-prey interactions have been studied to understand population fluctuations and how each affects the other (i.e., whether one regulates the other). Predation pressure and availability of resources affects how often and how far from cover an individual will forage compared with the time being vigilant for predators.
- Social aspects play a role in foraging such that group size influences costs and benefits. Advantages for larger groups include more eyes to find more food, greater vigilance for predators, greater per capita food intake, learning from others, etc., whereas disadvantages include diminished resources, predator attraction, increased competition for mates, etc. Proposed advantages of mixed-species flocks, common in tropical forests, include greater foraging efficiency and greater power against predators (detection, mobbing, and confusion).
- Young birds become better foragers with age and are typically less efficient than adults at foraging. This is one reason juvenile mortality is typically higher than in adults. Less prey recognition and encounter, lower capture success, increased handling time, and different nutritional requirements all contribute to lower juvenile foraging proficiency. Adults often supplant juveniles when foraging together, and juveniles also need time to learn from adults on what to eat and how to eat it.
- Studies of avian foraging and feeding behavior still depend on traditional ethological methods. However, novel applications of existing technologies, such as DNA and stable isotope analyses, have only recently contributed to our diet analysis tool set. Applications of novel technologies, such as tiny GPS trackers, temperature and depth loggers, accelerometers, and cameras, can now document foraging activities far from the eyes of observers. High-speed videography and XROMM have opened the door to understanding aspects of feeding mechanics that we could only guess at until now.

KEY MANAGEMENT AND CONSERVATION IMPLICATIONS

Understanding foraging and feeding behavior can have profound implications for conservation and management. Land use patterns and other anthropogenic impacts on the environment can directly and indirectly affect foraging resources. Knowing not only what, but also how, birds eat is critical for predicting the extent to which individuals, populations, and species can adapt to changes in their habitats, and ultimately for establishing management and conservation plans for sensitive, threatened, and endangered species. Below we outline six conservation and management issues in which foraging and feeding behavior are directly relevant:

- *Climate change.* Global shifting of weather patterns will affect availability and quality of resources worldwide, and thereby impact foraging and population dynamics (Walther et al. 2002). Many migratory North American birds skip the harsh winter by traveling to more amenable Neotropical habitats. However, because of climate change, tropical seasonal forests are becoming drier (Meir and Pennington 2011), which causes arthropod food to decline (Williams and Middleton 2008). This brings immediate and direct impacts on bird survival and fitness caused by decreased foraging success in their wintering grounds (Cooper et al. 2015). Understanding the foraging patterns of resident and migratory birds is crucial to direct conservation efforts that encompass both reproductive and nonbreeding seasons (Sherry and Holmes 1996, Johnson et al. 2005, 2006, Visco et al. 2015; box on page 470).

- *Invasive species.* Every habitat in the world is changing because of direct or indirect human intervention. Understanding how avian food preferences influence the spread of invasive plants and animals is important for designing effective management strategies. For example, in some cases both native and invasive frugivorous birds prefer invasive fruits to similar native ones (LaFleur et al. 2007), which may impact the survival of some native plants that rely on avian seed dispersal (Sallabanks 1992).

- *Declining bird populations.* Durell (2000) reviewed the degrees of foraging specialization to different habitats and foods among shorebirds, and found that birds often differentiate, (1) between sexes in the types of foods they consume, (2) between age classes in their preferred habitats, and (3) among social dominance classes in feeding patch quality. Identifying these foraging differences within species is vital for channeling conservation efforts toward the most vulnerable demographic sectors of the population (e.g., Durell 2003).

- *Human-wildlife conflicts.* With the rampant growth of the global human population, more and more terrestrial and aquatic systems are being transformed into food production areas for human consumption. The proper management of human-bird conflicts in agroecosystems requires evaluating the main causes of yield reduction and understanding avian food preferences (reviewed by Fox et al. 2016). Similar issues are faced by commercial fisheries in which herons and other avian predators compete for resources with humans. For example, Glahn et al. (2000) showed that although herons consumed large amounts of fish in farmed catfish ponds, they mostly captured naturally occurring sunfish and diseased catfish in relatively shallow areas.

Another source of human-wildlife conflict arises from the fear of predators taking domestic animals. Nevertheless, there is a dearth of studies evaluating the actual propensity of domesticated animals to fall victim to wild birds. In the few studies in which this has been examined, only <10 percent of the diet of raptors living in the vicinity of humans comprised domesticated prey (McPherson et al. 2016). However, a 30-year study has shown that populations of eight African vulture species are in severe decline, spiraling toward extinction (Ogada et al. 2016). One of the leading causes—accounting for 61 percent of recorded deaths—is the intentional and unintentional poisoning of animal carcasses. Long-term studies of the federally endangered California Condor (*Gymnogyps californianus*) have shown that exposure to lead poisoning (from lead-based ammunition—currently banned in California since 2008—associated with big-game and non-game hunting) varies by age, as older individuals gain independence from food subsidies and increase their home ranges (Kelly et al. 2014). Even though vultures are particularly vulnerable to poisoning because they scavenge communally (Kelly et al. 2014), carcasses are important seasonal food sources for other raptors, such as eagles, which makes them also susceptible to lead poisoning (e.g., Nadjafzadeh et al. 2016b).

- *Reintroduction programs.* Humans have driven many species to extinction and pushed many others to the brink. To mitigate this, captive-reared animals are sometimes reintroduced to augment wild populations. However, the chances of survival of these reintroduced individuals are often low compared to those of their wild conspecifics (Fischer and Lindenmayer 2000). Identifying key aspects of foraging behavior of a threatened species is the first step for replicating a naturalistic diet and simulating natural search strategies in captivity. Whiteside et al. (2015) showed that supplementing commercial chick crumb with mealworms, seeds, and fruits led to improved post-release survival, as individuals with supplemented diets were more efficient at finding, handling, and digesting food in the wild.

- *Pollution.* Pollution comes in a large variety of forms such as water, air, and soil contamination; noise and light pollution; and nonbiodegradable residues (e.g., plastic). Plastic and other floating anthropogenic debris is commonly mistaken for food by seabirds, affecting growth and body condition and leading to starvation of adults and, more often, their fledglings (reviewed by Avery-Gomm et al. 2013, Lavers and Bond 2016). Understanding the reasons why these aquatic birds mistake such debris for food (e.g., Lavers and Bond 2016) could help for designing conservation strategies and establishing cleanup priorities. Surfactants, from simple household detergents to untreated industrial residues, drastically reduce the surface tension at the water-air interface, where some birds (such as phalaropes; fig. 15.17) feed exclusively, and on only certain types of prey (Rubega and Inouye 1994). Thus, affecting their feeding mechanics is just another one of the myriad of problems that catastrophic oil spills cause birds (Eppley and Rubega 1990).

DISCUSSION QUESTIONS

1. Describe (with examples) how morphology relates to foraging and/or feeding behavior, in terms of (a) the beak, (b) wings, and (c) legs and feet. Integrating information from chapter 10, discuss the relative importance of flight to granivorous, carnivorous, insectivorous, and frugivorous feeding.

2. Of those presented in this chapter, describe (with examples) two anatomical and two behavioral attributes that are important in avian feeding repertoires.

3. Explain (with examples) some of the challenges faced by consumers of bulk, liquid, and aquatic/ marine foods, and how they overcome them.

4. Of those presented in this chapter, describe two searching behaviors and two prey-attacking or feeding behaviors (with examples of representative taxa) characteristic of each of the following: (a) terrestrial invertebrate-consumers, (b) vertebrate consumers, (c) aquatic invertebrate consumers, and (d) plant consumers. For each, describe a unique challenge posed by their diet, and how birds overcome them.

5. Describe (with examples) the concept of optimal foraging, in the contexts of time-minimization, energy-maximization, and the marginal-value theorem. Do birds really forage optimally?

6. What role does foraging and feeding behavior play in competition? Provide an example.

7. Briefly describe how you would design a field study to address one of the following research questions (your choice): (a) "What does species X eat?" (b) "Does the foraging behavior of insectivorous species X differ between coastal and inland habitats?" (c) "How does group size affect the foraging efficiency of species X?" In your description, be sure to include the following: What are the qualitative or quantitative variables to be measured? What are the statistical units of replication (i.e., the individual entities, or data points)? How will you sample from the population? How will you summarize that data to compare the results?

8. Using the example in the box on page 477), compute Shannon's index (H') with the same total number of prey items, but assuming that they are equally represented in the Barn Owl's diet. Does the value of diversity change, and if so, why? Now compute Evenness for both the actual and the hypothetical scenarios. What do these values reflect? How might sample sizes affect these values?

9. What are some advantages and disadvantages of foraging in a group? What kinds of environmental conditions might select for group foraging?

10. Explain (with examples) how foraging performance differs between adult and juvenile birds. Be sure to include specifically in which aspects of foraging/ feeding behavior (search, capture, handling, ingesting) juveniles might be limited, and specifically how so.

11. Explain how technology has revolutionized the study of avian foraging and feeding behavior. Given the current state and rate of digital technological advancements, what do you think will come next?

12. Explain how action on your part can ameliorate at least two of the conservation threats facing birds described above.

SUPPLEMENTARY MATERIALS

S1: Osprey (*Pandion haliaetus*) fishing: http://www.arkive.org /osprey/pandion-haliaetus/video-00.html.

S2: Blue Tit (*Cyanistes caeruleus*) feeding with aide of feet: http://www.arkive.org/blue-tit/parus-caeruleus/video -08.html.

S3: Egyptian Vulture (*Neophron percnopterus*) using a rock to crack an egg: http://www.arkive.org/egyptian-vulture /neophron-percnopterus/video-00.html.

S4: Corvid caching: http://www.pbslearningmedia.org/resou rce/nvbg-sci-corvidcaching/corvid-caching/.

S5: Common Raven (*Corvus corax*) caching seabird egg: http://www.arkive.org/raven/corvus-corax/video-08a.

S6: Sharp-beaked Ground-Finch (*Geospiza difficilis*) sanguinivory: http://www.arkive.org/sharp-beaked-ground-finch /geospiza-difficilis/video-08a.html.

S7: Eurasian Oystercatcher (*Haematopus ostralegus*) feeding: "Stabbing": http://www.arkive.org/oystercatcher/haema topus-ostralegus/video-01.html; "Hammering": http:// www.arkive.org/common-mussel/mytilus-edulis/video -16.html.

S8: Cape Gannets (*Morus capensis*) feeding at sea: http://www .arkive.org/cape-gannet/morus-capensis/video-08.html.

S9: Black Skimmers (*Rynchops niger*) surface feeding: http:// www.arkive.org/black-skimmer/rynchops-niger/video -08.html.

S10: Biplanar fluoroscopy video: http://jeb.biologists.org/high wire/filestream/1123832/field_highwire_adjunct_files/2 /Movie1.mov; and corresponding XROMM animation: http:// jeb.biologists.org/highwire/filestream/1123832/field _highwire_adjunct_files/3/Movie2.mov of a Mallard (*Anas platyrhynchos*) feeding (from Dawson et al. 2011).

References

Abbott, I., L. K. Abbott, and P. R. Grant (1975). Seed selection and handling ability of four species of Darwin's finches. The Condor 77:332–335.

Alcock, J. (1970). The origin of tool-using by Egyptian Vultures *Neophron percnopterus*. Ibis 112:452.

Alcock, J. (1972). The evolution of the use of tools by feeding animals. Evolution 26:464–473.

Alexander, R. D. (1975). The search for a general theory of behavior. Behavioral Science 20:77–100.

Altmann, J. (1974). Observational study of behavior: Sampling methods. Behaviour 49:227–267.

Altmann, S. A. (1956). Avian mobbing behavior and predator recognition. The Condor 58:241–253.

Ashmole, N. P. (1971). Seabird ecology and marine environments. *In* Avian biology, vol. 1, D. Farner and J. King, Editors. Academic Press, New York, pp. 223–286.

Avery-Gomm, S., J. F. Provencher, K. H. Morgan, and D. F. Bertram (2013). Plastic ingestion in marine-associated bird species from the eastern North Pacific. Marine Pollution Bulletin 72:257–259.

Baker, M. C., and A. E. M. Baker (1973). Niche relationships among six species of shorebirds on their wintering and breeding ranges. Ecological Monographs 43:193–212.

Ballance, L. T., and R. L. Pitman (1999). Foraging ecology of tropical seabirds. *In* Proceedings of the 22nd International Ornithological Congress, N. Adams and R. Slotow, Editors. BirdLife Durban, South Africa, pp. 2057–2207.

Barash, D. P., P. Donovan, and R. Myrick (1975). Clam dropping behavior of the Glaucous-winged Gull (*Larus glaucescens*). Wilson Bulletin 87:60–64.

Baussart, S., L. Korsoun, P. A. Libourel, and V. Bels (2009). Ballistic food transport in toucans. Journal of Experimental Zoology 311:465–474.

Baussart, S., and V. Bels (2011). Tropical hornbills (*Aceros cassidix*, *Aceros undulatus*, and *Buceros hydrocorax*) use ballistic transport to feed with their large beaks. Journal of Experimental Zoology 315:72–83.

Bednarz, J. C. (1988). Cooperative Hunting in Harris Hawks (*Parabuteo unicinctus*). Science 239:1525–1527.

Bels, V., and S. Baussart (2006). Feeding behaviour and mechanisms in domestic birds. *In* Feeding in domestic vertebrates: From structure to behaviour, V. Bels, Editor. CABI Publishing, Cambridge, MA, pp. 33–49.

Benkman, C. (1987). Crossbill foraging behavior, bill structure, and patterns of food profitability. Wilson Bulletin 99:351–368.

Benkman, C. W. (1988). Seed handling efficiency, bill structure, and the cost of specialization for crossbills. The Auk 105:715–719.

Benkman, C. W. (1993). Adaptation to single resources and the evolution of crossbill (*Loxia*) diversity. Ecological Monographs 63:305–325.

Benkman, C. W. (1999). The selection mosaic and diversifying coevolution between crossbills and lodgepole pine. American Naturalist 153:S75–S91.

Benkman, C. W. (2003). Divergent selection drives the adaptive radiation of crossbills. Evolution 57:1176–1181.

Benkman, C. W. (2010). Diversifying coevolution between crossbills and conifers. Evolution: Education and Outreach 3:47–53.

Benkman, C. W., and H. R. Pulliam (1988). The comparative feeding rates of North American sparrows and finches. Ecology 69:1195–1199

Berman, S. L. (1984). The hindlimb musculature of the white-fronted Amazon (*Amazonia albifrons*, Psittaciformes). The Auk 101:74–92.

Berman, S. L., and R. J. Raikow (1982). The hindlimb musculature of the mousebirds (Coliiformes). The Auk 99:41–57.

Bishop, A. L., and R. P. Bishop (2014). Resistance of wild African ungulates to foraging by red-billed oxpeckers (*Buphagus erythrorhynchus*): Evidence that this behaviour modulates a potentially parasitic interaction. African Journal of Ecology 52:103–110.

Bluff, L. A., J. Troscianko, A. A. S. Weir, A. Kacelnik, and C. Rutz (2010). Tool use by wild New Caledonian Crows *Corvus moneduloides* at natural foraging sites. Proceedings of the Royal Society B: 277:1377–1385.

Bocher, P., F. Robin, J. Kojadinovic, P. Delaporte, P. Rousseau, C. Dupuy, and P. Bustamante (2014). Trophic resource partitioning within a shorebird community feeding on intertidal mudflat habitats. Journal of Sea Research 92:115–124.

Bock, W. J. (1964). Kinetics of the avian skull. Journal of Morphology 114:1–42.

Bonter, D. N., B. Zuckerberg, C. W. Sedgwick, and W. M. Hochachka (2013). Daily foraging patterns in free-living birds: Exploring the predation-starvation trade-off. Proceedings of the Royal Society B: Biological Sciences 280. doi:10.1098/rspb.2012.3087.

Bosque, C. (2009). Opportunistic blood-drinking by Black Crake *Amaurornis flavirostris*. Ostrich 80:65.

Bout, R. G., and G. A. Zweers (2001). The role of cranial kinesis in birds. Comparative Biochemistry and Physiology A 131:197–205.

Bowman, R. I. (1961). Morphological differentiation and adaptation in the Galápagos Finches. *In* University of California Publications in Zoology, vol. 58, W. B. Quay, A. S. Leopold, P. Marler, and L. E. Rosenberg, Editors. University of California Press, Los Angeles, pp. 1–302.

Bowman, R. I., and S. L. Billeb (1965). Blood-eating in a Galápagos finch. Living Bird 4:29–44.

Brainerd, E. L., D. B. Baier, S. M. Gatesy, T. L. Hedrick, K. A. Metzger, S. L. Gilbert, and J. J. Crisco (2010). X-ray reconstruction of moving morphology (XROMM): Precision, accuracy and applications in comparative biomechanics research. Journal of Experimental Zoology 313:262–279.

Bramble, D. M., and D. B. Wake (1985). Feeding mechanisms in lower tetrapods. *In* Functional vertebrate morphology, M. Hildebrand, D. M. Bramble, K. F. Liem, and D. B. Wake, Editors. Harvard University Press, Cambridge, MA, pp. 230–261.

Breitwisch, R., M. Diaz, and R. Lee (1987). Foraging efficiencies and techniques of juvenile and adult northern mockingbirds (*Mimus polyglottos*). Behaviour 101:225–235.

Brennan, L. A., and M. L. Morrison (1990). Influence of sample size on interpretations of foraging patterns by Chestnut-backed Chickadees. *In* Avian foraging: Theory, methodology, and applications, vol. 13, Joseph R. Jehl, Editor. Studies in Avian Biology, Cooper Ornithological Society, Lawrence, KS, pp. 187–192.

Briffa, M., and A. Weiss (2010). Animal personality. Current Biology 20:R912–R914.

Bright, J. A., J. Marugan-Lobon, S. N. Cobb, and E. J. Rayfield (2016). The shapes of bird beaks are highly controlled by nondietary factors. Proceedings of the National Academy of Sciences USA 113:5352–5357.

Bühler, P. (1981). Functional anatomy of the avian jaw apparatus. *In* Form and function in birds, vol. 2, A. S. King and J. McLelland, Editors. Academic Press, London, pp. 2439–2468.

Cade, T. J. (1982). The falcons of the world. Cornell University Press, Ithaca, NY, p. 188.

Cairns, D. K. (1992). Population regulation of seabird colonies. *In* Current ornithology, vol. 9, D. Power, Editor. Plenum Press, New York, pp. 37–61.

Carere, C., and D. Maestripieri (2013). Animal personalities: Who cares and why? *In* Animal personalities: Behavior, physiology, and evolution, C. Carere and D. Maestripieri, Editors. University of Chicago Press, Chicago, pp. 1–9.

Case, S. B., and A. B. Edworthy (2016). First report of "mining" as a feeding behaviour among Australian manna-feeding birds. Ibis 158:407–415.

Charnov, E. L. (1976). Optimal foraging, the marginal value theorem. Theoretical Population Biology 9:129–136.

Clark, C. J., D. O. Elias, and R. O. Prum (2011). Aeroelastic flutter produces hummingbird feather songs. Science 333:1430–1433.

Clark, G. A. J. (1973). Holding food with the feet in passerines. Bird Banding 44:91–99.

Cole, E. F., and J. L. Quinn (2012). Personality and problem-solving performance explain competitive ability in the wild. Proceedings of the Royal Society B: Biological Sciences 279:1168–1175.

Cooper, N., T. W. Sherry, and P. P. Marra (2015). Experimental reduction of winter food availability decreases body condition and delays spring migration in a long-distance migratory bird. Ecology 96:1933–1942.

Cooper, R. J., P. J. Martinat, and R. C. Whitmore (1990). Dietary similarity among insectivorous birds: Influence of taxonomic versus ecological categorization of prey. *In* Avian foraging: Theory, methodology, and applications, vol. 13, Joseph R. Jehl, Editor. Studies in Avian Biology, Cooper Ornithological Society, Lawrence, KS, pp. 104–109.

Craig, R. B. (1978). An analysis of the predatory behavior of the loggerhead shrike. The Auk 95:221–234.

Csermely, D., L. Bertè, and R. Camoni (1998). Prey killing by Eurasian Kestrels: The role of the foot and the significance of bill and talons. Journal of Avian Biology 29:10–16.

Danforth, C. G. (1938). Some feeding habits of the Red-Breasted Sapsucker. The Condor 40:219–224.

Dawson, M. M., K. A. Metzger, D. B. Baier, and E. L. Brainerd (2011). Kinematics of the quadrate bone during feeding in Mallard Ducks. Journal of Experimental Biology 214:2036–2046.

del Hoyo, J., A. Elliot, and J. Sargatal (1992). Handbook of the birds of the world, vol. 1. Lynx Edicions, Barcelona.

del Hoyo, J., A. Elliott, J. Sargatal, D. A. Christie, and E. de Juana (2014). Handbook of the birds of the world alive. Lynx Edicions, Barcelona. http://www.hbw.com/.

Delord, K., P. Pinet, D. Pinaud, C. Barbraud, S. De Grissac, A. Lewden, Y. Cherel, and H. Weimerskirch (2016). Species-specific foraging strategies and segregation mechanisms of sympatric Antarctic fulmarine petrels throughout the annual cycle. Ibis 158:569–586.

Devaux, C., C. Lepers, and E. Porcher (2014). Constraints imposed by pollinator behaviour on the ecology and evolution of plant mating systems. Journal of Evolutionary Biology 27:1413–1430.

Dingemanse, N. J., C. Both, P. J. Drent, and J. M. Tinbergen (2004). Fitness consequences of avian personalities in a fluctuating environment. Proceedings of the Royal Society B: Biological Sciences 271:847–852.

Dingemanse, N. J., A. J. N. Kazem, D. Reale, and J. Wright (2010). Behavioural reaction norms: Animal personality meets individual plasticity. Trends in Ecology and Evolution 25:81–89.

Downs, C., J. O. Wirminghaus, and M. J. Lawes (2000). Anatomical and nutritional adaptations of the Speckled Mousebird (*Colius striatus*). The Auk 117:791–794.

Downs, C. T., R. J. van Dyk, and P. Iji (2002). Wax digestion by the Lesser Honeyguide *Indicator minor*. Comparative Biochemistry and Physiology A: Molecular and Integrative Physiology 133:125–134.

Dubbeldam, J. L. (1984). Brainstem mechanisms for feeding in birds: Interaction or plasticity. Brain, Behavior and Evolution 25:85–98.

Duckworth, R. A. (2008). Adaptive dispersal strategies and the colonization of a novel environment. American Naturalist 172:S4–S17.

Durell, S. E. A. l. V. d. (2000). Individual feeding specialisation in shorebirds: Population consequences and conservation implications. Biological Reviews 75:503–518.

Durell, S. E. A. l. V. d. (2003). The implications for conservation of age- and sex-related feeding specialisations in shorebirds. Wader Study Group Bulletin 100:35–39.

Edwards, T. C. (1989). The ontogeny of diet selection in fledgling Ospreys. Ecology 70:881–896.

Einoder, L. D., and A. M. M. Richardson (2007). Aspects of the hindlimb morphology of some Australian birds of prey: A comparative and quantitative study. The Auk 124:773–788.

Elliott, K. H., R. E. Ricklefs, A. J. Gaston, S. A. Hatch, J. R. Speakman, and G. K. Davoren (2013). High flight costs, but low dive costs, in auks support the biomechanical hypothesis for flightlessness in penguins. Proceedings of the National Academy of Sciences USA 110:9380–9384.

Ellis, D. H., J. C. Bednarz, D. G. Smith, and S. P. Flemming (1993). Social foraging classes in raptorial birds. Bioscience 43:14–20.

Elner, R. W., P. G. Beninger, D. L. Jackson, and T. M. Potter (2005). Evidence of a new feeding mode in Western Sandpiper (*Calidris mauri*) and Dunlin (*Calidris alpina*) based on bill and tongue morphology and ultrastructure. Marine Biology 146:1223–1234.

Emery, N. J., and N. S. Clayton (2005). Evolution of the avian brain and intelligence. Current Biology 15:R946–R950.

Ens, B. J., and J. D. Goss-Custard (1984). Interference among Oystercatchers, *Haematopus ostralegus*, feeding on mussels, *Mytilus edulis*, on the Exe estuary. Journal of Animal Ecology 53:217–231.

Eppley, Z. A., and M. A. Rubega (1990). Indirect effects of an oil spill: Reproductive failure in a population of south Polar skuas following the "Bahia Paraiso" oil spill in Antarctica. Marine Ecology Progress Series 67:1–6.

Estrella, S. M., and J. A. Masero (2007). The use of distal rhynchokinesis by birds feeding in water. Journal of Experimental Biology 210:3757–3762.

Feinsinger, P. (1976). Organization of a tropical guild of nectarivorous birds. Ecological Monographs 46:257–291.

Ferguson-Lees, J., and D. A. Christie (2001). Raptors of the world. Helm Identification Guides, London.

Fischer, J., and D. B. Lindenmayer (2000). An assessment of the published results of animal relocations. Biological Conservation 96:1–11.

Fisher, J., and R. A. Hinde (1949). The opening of milk bottles by birds. British Birds 42:347–357.

Fitzpatrick, J. W. (1980). Foraging behavior of Neotropical tyrant flycatchers. The Condor 82:43–57.

Fitzpatrick, J. W. (1985). Form, foraging behavior, and adaptive radiation in the Tyrannidae. Ornithological Monographs 36:447–470.

Foster, M. S. (1987). Feeding methods and efficiencies of selected frugivorous birds. The Condor 89:566–580.

Fowler, D. W., E. A. Freedman, and J. B. Scannella (2009). Predatory functional morphology in raptors: Interdigital variation in talon size is related to prey restraint and immobilisation technique. PLoS ONE 4:e7999.

Fox, A. D., J. Elmberg, I. M. Tombre, and R. Hessel (2016). Agriculture and herbivorous waterfowl: A review of the scientific basis for improved management. Biological Reviews 92 (2): 854–877. doi:10.1111/brv.12258.

Fretwell, S. D., and H. L. Lucas Jr. (1970). On territorial behavior and other factors influencing habitat distribution in birds. I: Theoretical development. Acta Biotheoretica 19:16–36.

Gartrell, B. D. (2000). The nutritional, morphologic, and physiologic bases of nectarivory in Australian birds. Journal of Avian Medicine and Surgery 14:85–94.

Gass, C. L., and W. M. Roberts (1992). The problem of temporal scale in optimization: Three contrasting views of hummingbird visits to flowers. American Naturalist 140:829–853.

Gatesy, S. M., D. B. Baier, F. A. Jenkins, and K. P. Dial (2010). Scientific rotoscoping: A morphology-based method of 3-D motion analysis and visualization. Journal of Experimental Zoology 313:244–261.

Gerwing, T. G., J. H. Kim, D. J. Hamilton, M. A. Barbeau, and J. A. Addison (2016). Diet reconstruction using next-generation sequencing increases the known ecosystem usage by a shorebird. The Auk 133:168–177.

Gibson, L. J. (2006). Woodpecker pecking: How woodpeckers avoid brain injury. Journal of Zoology 270:462–465.

Gill, F. B. (1971). Tongue structure of the sunbird Hypogramma hypogrammica. The Condor 73 (4): 485–486.

Gill, F. (2007). Ornithology. 3rd ed. W. H. Freeman, New York.

Giraldeau, L.-A., and T. Caraco (2000). Social foraging theory. Princeton University Press, Princeton, NJ.

Glahn, J. F., B. Dorr, and M. E. Tobin (2000). Captive Great Blue Heron predation on farmed channel catfish fingerlings. North American Journal of Aquaculture 62:149–156.

Goslow, G. E., Jr. (1971). The attack and strike of some North American raptors. The Auk 88:815–827.

Goss-Custard, J. D. (1975). Beach feast. Birds (September/October): 23–26.

Goss-Custard, J. D., and W. J. Sutherland (1984). Feeding specializations in oystercatchers Haematopus ostralegus. Animal Behaviour 32:200–301.

Goss-Custard, J. D., and S. E. A. L. V. Dit Durell (1988). The effect of dominance and feeding method on the intake rates of oystercatchers, *Haematopus ostralegus,* feeding on mussels. Journal of Animal Ecology 57:827–844.

Goss-Custard, J. D., A. D. West, M. G. Yates, R. W. Caldow, R. A. Stillman, L. Bardsley, J. Castilla, et al. (2006). Intake rates and the functional response in shorebirds (Charadriiformes) eating macro-invertebrates. Biological Reviews 81:501–529.

Gotelli, N. J. (2001). A primer of ecology. 3rd ed. Sinaur, Sunderland, MA.

Grajal, A., S. D. Strahl, R. Parra, M. G. Domínguez, and A. Neher (1989). Foregut fermentation in the Hoatzin, a Neotropical avian folivore. Science 245:1131–1134.

Grant, P. R. (1986). Ecology and evolution of Darwin's finches. Princeton University Press, Princeton, NJ.

Green, A. J., A. D. Fox, B. Hughes, and G. M. Hilton (1999). Time-activity budgets and site selection of White-headed Ducks *Oxyura leucocephala* at Burdur Lake, Turkey in late winter. Bird Study 46:62–73.

Greenberg, R. (1981). Dissimilar bill shapes in New World tropical versus temperate forest foliage-gleaning birds. Oecologia 49:143–147.

Greenberg, R. (1990). Ecological plasticity, neophobia, and resource use in birds. *In* Avian foraging: Theory, methodology, and applications. Studies in Avian Biology, no. 13, J. R. Jehl Jr., Editor. Cooper Ornithological Society, Lawrence, KS, pp. 431–437.

Greenberg, R. (1995). Insectivorous migratory birds in tropical ecosystems: The breeding currency hypothesis. Journal of Avian Biology 26:260–264.

Griffiths, D. (1980). Foraging costs and relative prey size. American Naturalist 116:743–752.

Groothuis, T. G. G., and C. Carere (2005). Avian personalities: Characterization and epigenesis. Neuroscience and Biobehavioural Reviews 29:137–150.

Grosch, K. (2003). Hybridization between two insectivorous bird species and the effect on prey-handling efficiency. Evolutionary Ecology 17:1–17.

Gurd, D. B. (2007). Predicting resource partitioning and community organization of filter-feeding dabbling ducks from functional morphology. American Naturalist 169:335–343.

Gussekloo, S. W. S., and R. G. Bout (2005a). The kinematics of feeding and drinking in palaeognathous birds in relation to cranial morphology. Journal of Experimental Biology 208:3395–3407.

Gussekloo, S. W. S., and R. G. Bout (2005b). Cranial kinesis in paleognathous birds. Journal of Experimental Biology 208:3409–3419.

Halsey, L., A. Woakes, and P. Butler (2003). Testing optimal foraging models for air-breathing divers. Animal Behaviour 65:641–653.

Harrison, N. M., and M. J. Whitehouse (2011). Mixed-species flocks: An example of niche construction? Animal Behaviour 81:675–682.

Heidweiller, J., and G. A. Zweers (1989). Drinking mechanisms in the Zebra Finch and the Bengalese Finch. The Condor 92:1–28.

Hellström, P., J. Nyström, and A. Angerbjörn (2014). Functional responses of the rough-legged buzzard in a multi-prey system. Oecologia 174:1241–1254.

Herrel, A., J. Podos, S. K. Huber, and A. P. Hendry (2005). Bite performance and morphology in a population of Darwin's finches: Implications for the evolution of beak shape. Functional Ecology 19:43–48.

Hertel, F. (1994). Diversity in body size and feeding morphology within past and present vulture assemblages. Ecology 75:1074–1084.

Hertel, F. (1995). Ecomorphological indicators of feeding behavior in recent and fossil raptors. The Auk 112:890–903.

Hixon, M. A. (1982). Energy maximizers and time minimizers: Theory and reality. American Naturalist 119:596–599.

Hixon, M. A., and F. L. Carpenter (1988). Distinguishing energy maximizers from time minimizers: A comparative study of two hummingbird species. Integrative and Comparative Biology 28:913–925.

Hobson, K. A. (2005). Stable isotopes and the determination of avian migratory connectivity and seasonal interactions. The Auk 122:1037–1048.

Hobson, K. A., and L. I. Wassenaar (1999). Stable isotope ecology: An introduction. Oecologia 120:312–313.

Holling, C. S. (1959). The components of predation as revealed by a study of small-mammal predation of the European pine sawfly. Canadian Entomologist 91:293–320.

Homberger, D. G. (2003). The comparative biomechanics of a prey-predator relationship: The adaptive morphologies of the feeding apparatus of Australian Black-Cockatoos and their foods as a basis for the reconstruction of the evolutionary history of the Psittaciformes. In Vertebrate biomechanics and evolution, V. L. Bels, J.-P. Gasc, and A. Casinos, Editors. Bios Scientific, Oxford, pp. 203–228.

Houston, D. C. (1986). Scavenging efficiency of Turkey Vultures in tropical forest. The Condor 88:318–323.

Houston, D. C., and J. Copsey (1994). Bone digestion and intestinal morphology of the Bearded Vulture. Journal of Raptor Research 28:73–78.

Houston, D. C., and G. E. Duke (2007). Physiology. In Raptor research and management techniques, D. M. Bird, K. L. Bildstein, D. R. Barber, and A. Zimmerman, Editors. Hancock House, Blaine, WA, pp. 267–277.

Hull, C. (1991). A comparison of the morphology of the feeding apparatus in the Peregrine Falcon, Falco peregrinus, and the Brown Falcon, F. berigora (Falconiformes). Australian Journal of Zoology 39:67–76.

Humphries, N. E., H. Weimerskirch, N. Queiroz, E. J. Southall, and D. W. Sims (2012). Foraging success of biological Levy flights recorded in situ. Proceedings of the National Academy of Sciences USA 109:7169–7174.

Hunt, G. R. (1996). Manufacture and use of hook-tools by New Caledonian Crows. Nature 379:249–251.

Inger, R., G. D. Ruxton, J. Newton, K. Colhoun, J. A. Robinson, A. L. Jackson, and S. Bearhop (2006). Temporal and intrapopulation variation in prey choice of wintering geese determined by stable isotope analysis. Journal of Animal Ecology 75:1190–1200.

Inger, R., and S. Bearhop (2008). Applications of stable isotope analyses to avian ecology. Ibis 150:447–461.

Ingolfsson, A., and B. T. Estrella (1978). The development of shell-cracking behavior in Herring Gulls. The Auk 95:577–579.

Jackson, A. L., G. D. Ruxton, and D. C. Houston (2008). The effect of social facilitation on foraging success in vultures: A modelling study. Biology Letters 4:311–313.

Jaksić, F. M., and J. H. Carothers (1985). Ecological, morphological, and bioenergetic correlates of hunting mode in hawks and owls. Ornis Scandinavica 16:165–172.

Jardine, C. B., A. L. Bond, P. J. Davidson, R. W. Butler, and T. Kuwae (2015). Biofilm consumption and variable diet composition of Western Sandpipers (Calidris mauri) during migratory stopover. PLoS ONE 10:e0124164.

Jeschke, J. M., M. Kopp, and R. Tollrian (2002). Predator functional responses: Discriminating between handling and digesting prey. Ecological Monographs 72:95–112.

Johnsgard, P. A. (1990). Hawks, eagles, and falcons of North America. Smithsonian Institution Press, Washington, DC.

Johnson, M. D., T. W. Sherry, A. M. Strong, and A. Medori (2005). Migrants in Neotropical bird communities: An assessment of the breeding currency hypothesis. Journal of Animal Ecology 74:333–341.

Johnson, M. D., T. W. Sherry, R. T. Holmes, and P. P. Marra (2006). Assessing habitat quality for a wintering songbird in natural and agricultural areas. Conservation Biology 20:1433–1444.

Kamil, A. C., and R. P. Balda (1990). The psychology of learning and motivation: Advances in research and theory. Academic Press, San Diego, CA.

Kelly, T. R., J. Grantham, D. George, A. Welch, J. Brandt, L. J. Burnett, K. J. Sorenson, et al. (2014). Spatiotemporal patterns and risk factors for lead exposure in endangered California condors during 15 years of reintroduction. Conservation Biology 28:1721–1730.

Kim, W., F. Peaudecerf, M. W. Baldwin, and J. W. Bush (2012). The hummingbird's tongue: A self-assembling capillary syphon. Proceedings of the Royal Society B: Biological Sciences 279:4990–4996.

Klages, N. T., and J. Cooper (1992). Bill morphology and diet of a filter-feeding seabird: The broad-billed prion *Pachyptila vittata* at South Atlantic Gough Island. Journal of Zoology 227:385–396.

Kooloos, J. G. M., and G. A. Zweers (1989). Mechanics of drinking in the Mallard (*Anas platyrhynchos*, Anatidae). Journal of Morphology 199:327–347.

Kooloos, J. G. M., A. R. Kraaijeveld, G. E. J. Langenbach, and G. A. Zweers (1989). Comparative mechanics of filterfeeding in *Anas platyrhynchos*, *Anas clypeata* and *Aythya fuligula* (Aves, Anseriformes). Zoomorphology 108:269–290.

Kooloos, J. G. M., and G. A. Zweers (1991). Integration of pecking, filter feeding and drinking mechanisms in waterfowl. Acta Biotheoretica 39:107–140.

Korpimäki, E., and K. Norrdahl (1991). Numerical and functional-responses of kestrels, Short-Eared Owls, and Long-Eared Owls to vole densities. Ecology 72:814–826.

Kotzerka, J., S. Garthe, and S. A. Hatch (2010). GPS tracking devices reveal foraging strategies of Black-legged Kittiwakes. Journal of Ornithology 151:459–467.

Krause, J., and G. D. Ruxton (2002). Living in groups. Oxford University Press, Oxford.

Krebs, J. R. (1980). Optimal foraging, predation risk and territory defence. Ardea 68:83–90.

LaFleur, N. E., M. A. Rubega, and C. S. Elphick (2007). Invasive fruits, novel foods, and choice: An investigation of European Starling and American Robin frugivory. Wilson Journal of Ornithology 119:429–438.

Lavers, J. L., and A. L. Bond (2016). Selectivity of flesh-footed shearwaters for plastic colour: Evidence for differential provisioning in adults and fledglings. Marine Environmental Research 113:1–6.

Lefebvre, L. (1995). The opening of milk bottles by birds: Evidence for accelerating learning rates, but against the wave-of-advance model of cultural transmission. Behavioural Processes 34:43–53.

Lehner, P. N. (1996). Handbook of ethological methods. 2nd ed. Cambridge University Press, Cambridge, UK.

Lemon, W. C. (1991). Fitness consequences of foraging behaviour in the Zebra Finch. Nature 352:153–155.

Liem, K. F., W. E. Bemis, W. F. J. Walker, and L. Grande (2001). Functional anatomy of the vertebrates: An evolutionary perspective. 3rd ed. Brooks/Cole-Thomson Learning, Belmont, MA.

Lima, S. L., and L. M. Dill (1990). Behavioral decisions made under the risk of predation: A review and prospectus. Canadian Journal of Zoology 68:619–640.

López-Calleja, M. V., and F. Bozinovic (1999). Feeding behavior and assimilation efficiency of the Rufous-tailed Plantcutter: A small avian herbivore. The Condor 101:705–710.

Lovvorn, J. R., C. L. Baduini, and G. L. Hunt (2001). Modeling underwater visual and filter feeding by planktivorous shearwaters in unusual sea conditions. Ecology 82:2342–2356.

Lucas, F. A. (1894). The tongue of the Cape May Warbler. The Auk 11:141–144.

MacArthur, R. H. (1958). Population ecology of some warblers of northeastern coniferous forests. Ecology 39:599–619.

MacArthur, R. H., and E. R. Pianka (1966). On optimal use of a patchy environment. American Naturalist 100:603–609.

Machovsky Capuska, G. E., R. L. Vaughn, B. Würsig, G. Katzir, and D. Raubenheimer (2011). Dive strategies and foraging effort in the Australasian gannet *Morus serrator* revealed by underwater videography. Marine Ecology Progress Series 442:255–261.

MacRoberts, M. H., and B. R. MacRoberts (1976). Social organization and behavior of the Acorn Woodpecker in central coastal California. Ornithological Monographs 21:1–115.

Marchetti, K., and T. Price (1989). Differences in the foraging of juvenile and adult birds: The importance of developmental constraints. Biological Reviews 64:51–70.

Marín-Gómez, O. H., and E. Arbeláez-Cortés (2015). Variation on species composition and richness in mixed bird flocks along an altitudinal gradient in the Central Andes of Colombia. Studies on Neotropical Fauna and Environment 50:113–129.

Marquet, P. A., A. P. Allen, J. H. Brown, J. A. Dunne, B. J. Enquist, J. F. Gillooly, P. A. Gowaty, et al. (2014). On theory in ecology. Bioscience 64:701–710.

Marra, P. P., C. E. Studds, S. Wilson, T. S. Sillett, T. W. Sherry, and R. T. Holmes. (2015). Non-breeding season habitat quality mediates the strength of density-dependence for a migratory bird. Proceedings of the Royal Society B: Biological Sciences 282:20150624.

Marshall, H. H., A. J. Carter, A. Ashford, J. M. Rowcliffe, and G. Cowlishaw (2015). Social effects on foraging behavior and success depend on local environmental conditions. Ecology and Evolution 5:475–492.

Marti, C. D., M. Bechard, and F. M. Jaksic (2007). Food habits. *In* Raptor research and management techniques, D. M. Bird, K. L. Bildstein, D. R. Barber, and A. Zimmerman, Editors. Hancock House, Blaine, WA, pp. 129–151.

Martin, W. C. L. (1833). The naturalist's library: A general history of humming-birds or the Trochilidae. H. G. Bohn, London.

Martínez del Río, C., H. G. Baker, and I. Baker (1992). Ecological and evolutionary implications of digestive processes: Bird preferences and the sugar constituents of floral nectar and fruit pulp. Experientia 48:544–551.

Martínez del Rio, C., N. Wolf, S. A. Carleton, and L. Z. Gannes (2009). Isotopic ecology ten years after a call for more laboratory experiments. Biological Reviews 84:91–111.

Matsui, H., G. R. Hunt, K. Oberhofer, N. Ogihara, K. J. McGowan, K. Mithraratne, T. Yamasaki, R. D. Gray, and E. Izawa (2016). Adaptive bill morphology for enhanced tool manipulation in New Caledonian Crows. Scientific Reports 6:22776.

McCabe, J. D., and B. J. Olsen (2015). Tradeoffs between predation risk and fruit resources shape habitat use of landbirds during autumn migration. The Auk 132:903–913.

McCleery, R. H. (1978). Optimal behaviour sequences and decision making. *In* Behavioural ecology: An evolutionary approach, J. R. Krebs and N. B. Davies, Editors. Wiley-Blackwell, Oxford, pp. 377–410.

McLaughlin, R. L. (1989). Search modes of birds and lizards: Evidence for alternative movement patterns. American Naturalist 133:654–670.

McPherson, S. C., M. Brown, and C. T. Downs (2016). Diet of the Crowned Eagle (*Stephanoaetus coronatus*) in an urban landscape: Potential for human-wildlife conflict? Urban Ecosystems 19:383–396.

Meekangvan, P., A. A. Barhorst, T. D. Burton, S. Chatterjee, and L. Schovanec (2006). Nonlinear dynamical model and response of avian cranial kinesis. Journal of Theoretical Biology 240:32–47.

Meir, P., and R. T. Pennington (2011). Climatic change and seasonally dry tropical forests. *In* Seasonally dry tropical forests: Ecology and conservation, R. Dirzo, H. S. Young, H. A. Mooney, and G. Ceballos, Editors. Island Press, Washington, DC, pp. 279–299.

Meire, P. M., and A. Ervynck (1986). Are Oystercatchers (*Haematopus ostralegus*) selecting the most profitable mussels (*Mytilus edulis*)? Animal Behaviour 34:1427–1435.

Milesi, F. A., and L. Marone (2015). Exploration and exploitation of foraging patches by desert sparrows: Environmental indicators and local evaluation of spatially correlated costs and benefits. Journal of Avian Biology 46:225–235.

Moermond, T. C., and J. S. Denslow (1985). Neotropical avian frugivores: Patterns of behavior, morphology, and nutrition, with consequences for fruit selection. Ornithological Monographs 36:865–897.

Moreno, E., M. Barluenga, and A. Barbosa (2001). Ecological plasticity by morphological design reduces the cost of subordination: Influence on species distribution. Oecologia 128:603–607.

Moreno, R., G. Stowasser, R. A. McGill, S. Bearhop, and R. A. Phillips (2016). Assessing the structure and temporal dynamics of seabird communities: The challenge of capturing marine ecosystem complexity. Journal of Animal Ecology 85:199–212.

Morrison, M. L. (1984). Influence of sample-size and sampling design on analysis of avian foraging behavior. The Condor 86:146–150.

Morrison, M. L., C. J. Ralph, J. Verner, and J. Jehl (1990). Avian foraging: Theory, methodology, and applications. Studies in Avian Biology, no. 13, J. R. Jehl Jr., Editor. Cooper Ornithological Society, Lawrence, KS.

Morse, D. H. (1970). Ecological aspects of some mixed-species foraging flocks of birds. Ecological Monographs 40:119–168.

Munn, C. A., and J. W. Terborgh (1979). Multi-species territoriality in Neotropical foraging flocks. The Condor 81:338–347.

Nadjafzadeh, M., H. Heribert, and O. Krone (2016a). Sit-and-wait for large prey: Foraging strategy and prey choice of White-tailed Eagles. Journal of Ornithology 157:165–178.

Nadjafzadeh, M., C. C. Voigt, and O. Krone (2016b). Spatial, seasonal and individual variation in the diet of White-tailed Eagles *Haliaeetus albicilla* assessed using stable isotope ratios. Ibis 158:1–15.

Nebel, S., D. L. Jackson, and R. W. Elner (2005). Functional association of bill morphology and foraging behaviour in calidrid sandpipers. Animal Biology 55:235–243.

Noon, B. R., and W. M. Block (1990). Analytical considerations for study design. *In* Avian foraging: Theory, methodology, and applications. Studies in Avian Biology, no. 13, J. R. Jehl Jr., Editor. Cooper Ornithological Society, Lawrence, KS, pp. 126–133.

Norris, D. R., P. P. Marra, T. K. Kyser, T. W. Sherry, and L. M. Ratcliffe (2004). Tropical winter habitat limits reproductive success on the temperate breeding grounds in a migratory bird. Proceedings of the Royal Society B: Biological Sciences 271:59–64.

Norton-Griffiths, M. (1967). Some ecological aspects of the feeding behavior of the Oystercatcher (*Haematopus ostralegus*) on the edible mussel *Mytilus edulis*. Ibis 109:412–424.

Nunn, C. L., V. O. Ezenwa, C. A. Arnold, and W. D. Koenig (2011). Mutualism or parasitism? Using a phylogenetic approach to characterize the oxpecker-ungulate relationship. Evolution 65:1297–1304.

Ogada, D., P. Shaw, R. L. Beyers, R. Buij, C. Murn, J. M. Thiollay, C. M. Beale, et al. (2016). Another continental vulture crisis: Africa's vultures collapsing toward extinction. Conservation Letters 9:89–97.

Olsen, A. M. (2015). Exceptional avian herbivores: Multiple transitions toward herbivory in the bird order Anseriformes and its correlation with body mass. Ecology and Evolution 5:5016–5032.

Olsen, P. (1995). Australian birds of prey. Johns Hopkins University Press, Baltimore, MD.

Paton, D. C. (1980). The importance of manna, honeydew and lerp in the diets of honeyeaters. Emu 80:213–226.

Patrick, S. C., and H. Weimerskirch (2014). Personality, foraging and fitness consequences in a long lived seabird. PLoS ONE 9:e87269.

Piersma, T., A. Koolhaas, and A. Dekinga (1993). Interactions between stomach structure and diet choice in shorebirds. The Auk 110:552–564.

Pietrewicz, A. T., and A. C. Kamil (1979). Search image formation in the Blue Jay (*Cyanocitta cristata*). Science 204:1332–1333.

Pompanon, F., B. E. Deagle, W. O. C. Symondson, D. S. Brown, S. N. Jarman, and P. Taberlet (2012). Who is eating what: Diet assessment using next generation sequencing. Molecular Ecology 21:1931–1950.

Powell, G. V. N. (1985). Sociobiology and adaptive significance of interspecific foraging flocks in the Neotropics. Ornithological Monographs 36:713–732.

Pravosudov, V. V., and J. R. Lucas (2001). Daily patterns of energy storage in food-caching birds under variable daily predation risk: A dynamic state variable model. Behavioral Ecology and Sociobiology 50:239–250.

Proctor, N. S., and P. J. Lynch (1993). Manual of ornithology: Avian structure and function. Yale University Press, New Haven, CT.

Puckey, H. L., A. Lill, and D. J. O'Dowd (1996). Fruit color choices of captive Silvereyes (*Zosterops lateralis*). The Condor 98:780–790.

Pyke, G. H. (1984). Optimal foraging theory: A critical review. Annual Review of Ecology and Systematics 15:523–575.

Pyke, G. H., H. R. Puliam, and E. L. Charnov (1977). Optimal foraging: A selective review of theory and tests. Quarterly Review of Biology 52:137–154.

Rand, A. L. (1967). The flower-adapted tongue of a Timaliinae bird and its implications. Fieldiana Zoology 51:53–61.

Rands, S. A., A. I. Houston, and C. E. Gasson (2000). Prey processing in central place foragers. Journal of Theoretical Biology 202:161–174.

Razeng, E., and D. M. Watson (2015). Nutritional composition of the preferred prey of insectivorous birds: Popularity reflects quality. Journal of Avian Biology 46:89–96.

Recher, H. F., and V. Gebski (1990). Analysis of the foraging ecology of eucalypt forest birds: Sequential versus single-point observations. In Avian foraging: Theory, methodology, and applications. Studies in Avian Biology, no. 13, J. R. Jehl Jr., Editor. Cooper Ornithological Society, Lawrence, KS, pp. 174–180.

Redpath, S. M., and S. J. Thirgood (1999). Numerical and functional responses in generalist predators: Hen harriers and peregrines on Scottish grouse moors. Journal of Animal Ecology 68:879–892.

Remsen, J. V. J., and S. K. Robinson (1990). A classification scheme for foraging behavior of birds in terrestrial habitats. In Avian foraging: Theory, methodology, and applications. Studies in Avian Biology, no. 13, J. R. Jehl Jr., Editor. Cooper Ornithological Society, Lawrence, KS, pp. 144–160.

Rico-Guevara, A. (2008). Morphology and arthropod foraging by high Andean hummingbirds. Ornitología Colombiana 7:43–58.

Rico-Guevara, A., and M. A. Rubega (2011). The hummingbird tongue is a fluid trap, not a capillary tube. Proceedings of the National Academy of Sciences USA 108:9356–9360.

Rico-Guevara, A., and M. Araya-Salas (2015). Bills as daggers? A test for sexually dimorphic weapons in a lekking hummingbird. Behavioral Ecology 26:21–29.

Rico-Guevara, A., T. H. Fan, and M. A. Rubega (2015). Hummingbird tongues are elastic micropumps. Proceedings of the Royal Society B: Biological Sciences 282:20151014.

Robinson, S. K., and R. T. Holmes (1982). Foraging behavior of forest birds: The relationships among search tactics, diet, and habitat structure. Ecology 63:1918–1931.

Rodenhouse, N. L., T. W. Sherry, and R. T. Holmes (1997). Site-dependent regulation of population size: A new synthesis. Ecology 78:2025–2042.

Rodríguez-Ferraro, A., M. A. García-Amado, and C. Bosque (2007). Diet, food preferences, and digestive efficiency of the Grayish Saltator, a partly folivorous passerine. The Condor 109:824–840.

Rosenberg, K. V., and R. J. Cooper (1990). Approaches to avian diet analysis. In Avian foraging: Theory, methodology, and applications. Studies in Avian Biology, no. 13, J. R. Jehl Jr., Editor. Cooper Ornithological Society, Lawrence, KS, pp. 80–90.

Ross, C. F., A. Eckhardt, A. Herrel, W. L. Hylander, K. A. Metzger, V. Schaerlaeken, R. L. Washington, and S. H. Williams (2007). Modulation of intra-oral processing in mammals and lepidosaurs. Integrative and Comparative Biology 47:118–136.

Rubega, M. A. (1997). Surface tension prey transport in shorebirds: How widespread is it? Ibis 139:488–493.

Rubega, M. (2000). Feeding in birds: Approaches and opportunities. In Feeding: Form, function, and evolution in tetrapod vertebrates, K. Schwenk, Editor. Academic Press, San Diego, CA, pp. 395–407.

Rubega, M. A., and B. S. Obst (1993). Surface-tension feeding in phalaropes: Discovery of a novel feeding mechanism. The Auk 110:169–178.

Rubega, M., and C. Inouye (1994). Prey switching in Red-necked Phalaropes Phalaropus lobatus: Feeding limitations, the functional response and water management at Mono Lake, California, USA. Biological Conservation 70:205–210.

Ruschi, P. A. (2014). Frugivory by the hummingbird Chlorostilbon notatus (Apodiformes: Trochilidae) in the Brazilian Amazon. Boletim do Museu de Biologia Mello Leitão 35:43–47.

Rutz, C., L. A. Bluff, A. A. Weir, and A. Kacelnik (2007). Video cameras on wild birds. Science 318:765.

Rutz, C., L. A. Bluff, N. Reed, J. Troscianko, J. Newton, R. Inger, A. Kacelnik, and S. Bearhop (2010). The ecological significance of tool use in New Caledonian Crows. Science 329:1523–1526.

Ryan, P. G., S. L. Petersen, G. Peters, and D. Gremillet (2004). GPS tracking a marine predator: The effects of precision, resolution and sampling rate on foraging tracks of African Penguins. Marine Biology 145:215–223.

Sallabanks, R. (1992). Fruit fate, frugivory, and fruit characteristics: A study of the hawthorn, Crataegus monogyna (Rosaceae). Oecologia 91:296–304.

Salt, G. W., and D. E. Willard (1971). The hunting behavior and success of Forster's Tern. Ecology 52:989–998.

Schoener, T. W. (1971). Theory of feeding strategies. Annual Review of Ecology and Systematics 2:369–404.

Schondube, J. E., and C. Martínez del Rio (2003). The flowerpiercers' hook: An experimental test of an evolutionary trade-off. Proceedings of the Royal Society B: Biological Sciences 270:195–198.

Schwenk, K. (2000a). An introduction to tetrapod feeding. In Feeding: Form, function, and evolution in tetrapod vertebrates, K. Schwenk, Editor. Academic Press, San Diego, pp. 21–61.

Schwenk, K. (2000b). Feeding in Lepidosaurs. In Feeding: Form, function, and evolution in tetrapod vertebrates, K. Schwenk, Editor. Academic Press, San Diego, pp. 175–291.

Schwenk, K., and M. A. Rubega (2005). Diversity of vertebrate feeding systems. In Physiological and ecological adaptations to feeding in vertebrates, J. M. Starck and T. Wang, Editors. Science Publishers, Enfield, CT, pp. 1–41.

Shannon, C. E., and W. Weaver (1949). The mathematical theory of communication. University of Illinois Press, Urbana.

Shealer, D. A. (2002). Foraging behavior and food of seabirds. In Biology of marine birds, E. A. Schreiber and J. Burger, Editors. CRC Press, Boca Raton, FL, pp. 137–177.

Sherry, D. F., and B. G. J. Galef (1984). Cultural transmission without imitation: Milk bottle opening by birds. Animal Behaviour 32:937–938.

Sherry, T. W. (1984). Comparative dietary ecology of sympatric, insectivorous Neotropical flycatchers (Tyrannidae). Ecological Monographs 54:313–338.

Sherry, T. W. (1990). When are birds dietarily specialized? Distinguishing ecological from evolutionary approaches. In Avian foraging: Theory, methodology, and applications. Studies in Avian Biology, no. 13, J. R. Jehl Jr., Editor. Cooper Ornithological Society, Lawrence, KS, pp. 337–352.

Sherry, T. W. (2016). Foods and foraging. *In* The Cornell handbook of bird biology. John Wiley, Hoboken, NJ.

Sherry, T. W., and L. A. McDade (1982). Prey selection and handling in two Neotropical hover gleaning birds. Ecology 63:1016–1028.

Sherry, T. W., and R. T. Holmes (1995). Summer versus winter limitation of populations: Issues and the evidence. *In* Ecology and management of Neotropical migratory birds, T. M. Martin and D. Finch, Editors. Oxford University Press, New York, pp. 85–120.

Sherry, T. W., and R. T. Holmes (1996). Winter habitat limitation in Neotropical Nearctic migrant birds: Implications for population dynamics and conservation. Ecology 77:36–48.

Sherry, T. W., D. S. Wilson, C. S. Hunter, and R. T. Holmes (2015). Impacts of nest predators and weather on reproductive success and population limitation in a long-distance migratory bird. Journal of Avian Biology 46:559–569.

Sherry, T. W., M. D. Johnson, K. Williams, J. Kaban, C. McAvoy, A. Medori, S. Rainey, and S. Xu (2016). Dietary opportunism, resource partitioning, and consumption of coffee-berry borers by five migratory wood warblers (Parulidae) wintering in Jamaican shade coffee plantations. Journal of Field Ornithology 87:273–292.

Sih, A., and B. Christensen (2001). Optimal diet theory: When does it work, and when and why does it fail? Animal Behaviour 61:379–390.

Sih, A., A. M. Bell, and J. C. Johnson (2004). Behavioral syndromes: An ecological and evolutionary overview. Trends in Ecology and Evolution 19:372–378.

Silveira, L., A. T. A. Jacomo, F. H. G. Rodrigues, and P. G. Crawshaw (1997). Hunting association between the Aplomado Falcon (*Falco femoralis*) and the Maned Wolf (*Chrysocyon brachyurus*) in Emas National Park, central Brazil. The Condor 99:201–202.

Smith, J. W., and C. W. Benkman (2007). A coevolutionary arms race causes ecological speciation in crossbills. American Naturalist 169:455–465.

Smith, R., M. Hamas, M. Dallman, and D. Ewert (1998). Spatial variation in foraging of the Black-throated Green Warbler along the shoreline of northern Lake Huron. The Condor 100:474–484.

Smith, R. D., G. D. Ruxton, and W. Cresswell (2001). Dominance and feeding interference in small groups of blackbirds. Behavioral Ecology 12:475–481.

Smith, T. B., and S. A. Temple (1982). Feeding habits and bill polymorphism in Hook-billed Kites. The Auk 99:197–207.

Sokal, R. R., and F. J. Rohlf (1995). Biometry. 3rd ed. W. H. Freeman, New York.

Soobramoney, S., and M. R. Perrin (2007). The effect of bill structure on seed selection and handling ability of five species of granivorous birds. Emu 107:169–176.

Soons, J., A. Herrel, A. Genbrugge, P. Aerts, J. Podos, D. Adriaens, Y. de Witte, P. Jacobs, and J. Dirckx (2010). Mechanical stress, fracture risk and beak evolution in Darwin's ground finches (Geospiza). Philosophical Transactions of the Royal Society B: Biological Sciences 365:1093–1098.

Sridhar, H., G. Beauchamp, and K. Shanker (2009). Why do birds participate in mixed-species foraging flocks? A large-scale synthesis. Animal Behaviour 78:337–347.

Stager, K. E. (1964). The role of olfaction in food location by the Turkey Vulture. Los Angeles County Museum Contributions in Science 81:1–63.

Stiles, F. G. (1975). Ecology, flowering phenology, and hummingbird pollination of some Costa Rican *Heliconia* species. Ecology 56:285–301.

Stiles, F. G. (1976). Taste preferences, color preferences, and flower choice in hummingbirds. The Condor 78:10–26.

Stiles, F. G. (1977). Coadapted competitors: The flowering seasons of hummingbird-pollinated plants in a tropical forest. Science 198:1177–1178.

Stiles, F. G. (1980). The annual cycle in a tropical wet forest hummingbird community. Ibis 122:322–343.

Stiles, F. G. (1981). Geographical aspects of bird-flower coevolution, with particular reference to Central-America. Annals of the Missouri Botanical Garden 68:323–351.

Stiles, F. G. (1982). Aggressive and courtship displays of the male Anna's Hummingbird. The Condor 84:208–225.

Stiles, F. G. (1985). Seasonal patterns and coevolution in the hummingbird-flower community of a Costa Rican subtropical forest. Ornithological Monographs 36:757–787.

Stiles, F. G. (1995). Behavioral, ecological and morphological correlates of foraging for arthropods by the hummingbirds of a tropical wet forest. The Condor 97:853–878.

Stiles, F. G. (2008). Ecomorphology and phylogeny of hummingbirds: Divergence and convergence to high elevations. Ornitologia Neotropical 19:511–519.

Stiles, F. G., and L. L. Wolf (1979). Ecology and evolution of lek mating behavior in the Long-tailed Hermit hummingbird. Ornithological Monographs 27:1–78.

Stiles, F. G., D. L. Altshuler, and R. Dudley (2005). Wing morphology and flight behavior of some North American hummingbird species. The Auk 122:872–886.

Stiling, P. (1999). Ecology: Theories and application. 3rd ed. Prentice Hall, Upper Saddle River, NJ.

Strait, S. G., and J. F. V. Vincent (1998). Primate faunivores: Physical properties of prey items. International Journal of Primatology 19:867–878.

Sustaita, D., E. Pouydebat, A. Manzano, V. Abdala, F. Hertel, and A. Herrel (2013). Getting a grip on tetrapod grasping: Form, function, and evolution. Biological Reviews 88:380–405.

Swennen, C., L. L. M. de Bruijn, P. Duiven, M. F. Leopold, and E. C. L. Marteijn (1983). Differences in bill form of the Oystercatcher *Haematopus ostralegus*: A dynamic adaptation to specific foraging techniques. Netherlands Journal of Sea Research 17:57–83.

Symes, C. T., and C. T. Downs (2001). Feeding and energy intake in two avian frugivores, the Black-eyed Bulbul *Pycnonotus barbartus* (Passeriformes: Pycnonotidae) and Speckled Mousebird *Colius striatus* (Passeriformes: Coliidae). Durban Museum Novitates 26:20–24.

Teichmann, J., M. Broom, and E. Alonso (2014). The application of temporal difference learning in optimal diet models. Journal of Theoretical Biology 340:11–16.

Thiebault, A., M. Semeria, C. Lett, and Y. Tremblay (2016). How to capture fish in a school? Effect of successive predator attacks on seabird feeding success. Journal of Animal Ecology 85:157–167.

Tomlinson, C. A. B. (2000). Feeding in paleognathous birds. In Feeding: Form, function, and evolution in tetrapod vertebrates, K. Schwenk, Editor. Academic Press, San Diego, CA, pp. 359–394.

Tregenza, T., G. A. Parker, and D. J. Thompson (1996). Interference and the ideal free distribution: Models and tests. Behavioral Ecology 7:379–386.

Troscianko, J., A. M. von Bayern, J. Chappell, C. Rutz, and G. R. Martin (2012). Extreme binocular vision and a straight bill facilitate tool use in New Caledonian Crows. Nature Communications 3:1110.

Troscianko, J., and C. Rutz (2015). Activity profiles and hook-tool use of New Caledonian Crows recorded by bird-borne video cameras. Biology Letters 11:20150777.

Usherwood, J. R., E. L. Sparkes, and R. Weller (2014). Leap and strike kinetics of an acoustically "hunting" barn owl (Tyto alba). Journal of Experimental Biology 217:3002–3005.

van den Heuvel, W. F. (1992). Kinetics of the skull in the chicken (Gallus gallus domesticus). Netherlands Journal of Zoology 42:561–582.

van de Pol, M., B. J. Ens, K. Oosterbeek, L. Brouwer, S. Verhulst, J. M. Tinbergen, A. L. Rutten, and M. d. Jong (2009). Oystercatchers' bill shapes as a proxy for diet specialization: More differentiation than meets the eye. Ardea 97:335–347.

van der Leeuw, A. H. J., K. Kurk, P. C. Snelderwaard, R. G. Bout, and H. Berkhoudt (2003). Conflicting demands on the trophic system of Anseriformes and their evolutionary implications. Animal Biology 53:259–301.

van der Meij, M. A. A., M. Griekspoor, and R. G. Bout (2004). The effect of seed hardness on husking time in finches. Animal Biology 52:195–205.

van der Meij, M. A. A., and R. G. Bout (2006). Seed husking time and maximal bite force in finches. Journal of Experimental Biology 209:3329–3335.

Vander Wall, S. B., and R. P. Balda (1981). Ecology and evolution of food-storage behavior in conifer-seed-caching corvids. Zeitschrift fur Tierpsychologie 56:217–242.

van Oers, K., P. J. Drent, P. d. Goede, and A. J. v. Noordwijk (2004). Realized heritability and repeatability of risk taking behaviour in relation to avian personalities. Proceedings of the Royal Society B: Biological Sciences 274:65–73.

Verbeek, M. E. M., P. J. Drent, and P. R. Wiepkema (1994). Consistent individual differences in early exploratory behavior of male Great Tits. Animal Behaviour 48:1113–1121.

Verbeek, M. E. M., A. Boon, and P. J. Drent (1996). Exploration, aggressive behaviour and dominance in pair-wise confrontations of juvenile male Great Tits. Behaviour 133:946–963.

Visco, D. M., N. L. Michel, A. W. Boyle, B. J. Sigel, S. Woltmann, and T. W. Sherry (2015). Patterns and causes of understory bird declines from human-disturbed tropical forest landscapes: A case study from Central America. Biological Conservation 191:117–129.

Viswanathan, G. M., S. V. Buldyrev, S. Havlin, M. G. E. Da Luz, E. P. Raposo, and H. E. Stanley (1999). Optimizing the success of random searches. Nature 401:911–914.

Walther, G. R., E. Post, P. Convey, A. Menzel, C. Parmesan, T. J. Beebee, J. M. Fromentin, O. Hoegh-Guldberg, and F. Bairlein (2002). Ecological responses to recent climate change. Nature 416:389–395.

Walton, K. C. (1979). Diet of meadow pipits Anthus pratensis on mountain grassland in Snowdonia. The Ibis 121:325–329

Wang, L. Z., J. T. M. Cheung, F. Pu, D. Y. Li, M. Zhang, and Y. B. Fan (2011). Why do woodpeckers resist head impact injury: A biomechanical investigation. PLoS ONE 6 (10): e26490. doi:10.1371/journal.pone.0026490.

Ward, A. B., P. D. Weigl, and R. M. Conroy (2002). Functional morphology of raptor hindlimbs: Implications for resource partitioning. The Auk 119:1052–1063.

Weimerskirch, H., F. Bonadonna, F. Bailleul, G. Mabille, G. Dell'Omo, and H. P. Lipp (2002). GPS tracking of foraging albatrosses. Science 295:1259.

Weir, A. A. S., J. Chappell, and A. Kacelnik (2002). Shaping of hooks in New Caledonian Crows. Science 297:981.

Wendelken, P. W., and R. F. Martin (1988). Avian consumption of the fruit of the cacti Stenocereus eichlamii and Pilosocereus maxonii in Guatemala. American Midland Naturalist 19:235–243.

Werner, T. K., and T. W. Sherry (1987). Behavioral feeding specialization in Pinaroloxias inornata, the "Darwin's Finch" of Cocos Island, Costa Rica. Proceedings of the National Academy of Sciences USA 84:5506–5510.

Wheelwright, N. T. (1985). Fruit size, gape width, and the diets of fruit-eating birds. Ecology 66:808–818.

Whiteside, M. A., R. Sage, and J. R. Madden (2015). Diet complexity in early life affects survival in released pheasants by altering foraging efficiency, food choice, handling skills and gut morphology. Journal of Animal Ecology 84:1480–1489.

Williams, S. E., and J. Middleton (2008). Climatic seasonality, resource bottlenecks, and abundance of rainforest birds: Implications for global climate change. Diversity and Distributions 14:69–77.

Williams, S. H., B. W. Wright, V. D. Truong, C. R. Daubert, and C. J. Vinyard (2005). Mechanical properties of foods used in experimental studies of primate masticatory function. American Journal of Primatology 67:329–346.

Willson, M. F., and C. J. Whelan (1990). The evolution of fruit color in fleshy-fruited plants. American Naturalist 136:790–809.

Wilman, H., J. Belmaker, J. Simpson, C. d. l. Rosa, M. M. Rivadeneira, and W. Jetz (2014). EltonTraits 1.0: Species-level foraging attributes of the world's birds and mammals. Ecology 95:2027.

Wolf, L. L., F. G. Stiles, and F. R. Hainsworth (1972). Energetics of foraging: Rate and efficiency of nectar extraction by hummingbirds. Science 176:1351–1352.

Wolf, L. L., F. G. Stiles, and F. R. Hainsworth (1976). Ecological organization of a tropical, highland hummingbird community. Journal of Animal Ecology 45:349–379.

Wunderle, J. M. (1991). Age-specific foraging proficiency in birds. *In* Current ornithology, vol. 8, D. M. Power, Editor. Plenum Press, New York, pp. 273–324.

Yanega, G., and M. A. Rubega (2004). Hummingbird jaw bends to aid insect capture. Nature 428:615.

Yosef, R., and B. Pinshow (2005). Impaling in true shrikes (Laniidae): A behavioral and ontogenetic perspective. Behavioural Processes 69:363–367.

Zusi, R. L. (1984). A functional and evolutionary analysis of rhynchokinesis in birds. Smithsonian Contributions to Zoology 395:1–40.

Zweers, G. A. (1982). Pecking in the pigeon (*Columba livia* L.). Behaviour 81:173–230.

Zweers, G. A., A. F. C. Gerritsen, and P. J. van Kranenburg-Voogd (1977). Mechanics of feeding of the Mallard (*Anas platyrhynchos* L., Aves, Anseriformes). Contributions in Vertebrate Evolution 3:1–109.

Zweers, G. A., and H. Berkhoudt (1991). Recognition of food in pecking, probing, and filter feeding birds. *In* The avian feeding system, vol. 1, W. J. Bock and P. Buhler, Editors. New Zealand Ornithological Congress Trust Board, Wellington, NZ, pp. 897–902.

Zweers, G. A., H. Berkhoudt, and J. C. V. Berge (1994). Behavioral mechanisms of avian feeding. Advances in Comparative Environmental Physiology 18:241–279.

Zweers, G. A., F. de Jong, H. Berkhoudt, and J. C. Vanden Berge (1995). Filter feeding in flamingos (*Phoenicopterus ruber*). The Condor 97:297–324.

CHAPTER 16

Reproductive Behavior and Mating Systems

Patricia Adair Gowaty

This chapter is about the socio-ecology of reproduction. Socio-ecology is the study of how ecology and social dynamics of individuals in populations affect what individuals do, including their pairing, mating, and reproduction (Lack 1968). Socio-ecology is the scientific perspective that answers the evolutionary "why" questions (also called ultimate questions about "the functions" of behavior). The principles of socio-ecology are often theoretical, sometimes expressed in terms of formal quantitative mathematics but most often in terms of qualitative statements about the selective forces affecting this or that behavior. The basic principles of reproductive behavior are intuitive: reproduction in sexual species means that females and males must mate, i.e., copulate, which depends irreducibly on two things: reproductively mature individuals must be alive and they must encounter one another in order to share gametes. Being alive and meeting potential mates are two essential things dependent on the social and ecological circumstances that individuals experience. This chapter provides examples of breeding adaptations, discusses socio-ecological theory, and alerts you to remaining questions and new perspectives and suggests where neophyte ornithologists may make their marks—perhaps studying such curious animals as pariah chickens (Lawler 2014) or researching never-before-imagined questions. Therefore, we urge you to think for yourself about "the next questions." This chapter also indicates how studies of birds have applications far beyond conservation, management, and pleasure: birds are often model species for testing hypotheses that originate in our curiosity about the selective pressures acting on social behavior in other socially monogamous animals—like humans. To understand the material in this chapter, it will be useful for students to consider elementary ideas of "scientific ways of knowing"; thus, this chapter—like most of the chapters in this book—encourages you to learn about the nature of scientific inferences.

Birds reproduce in all of the world's environments (fig. 16.1), so it is no surprise that avian breeding biology, which is most often socially monogamous and associated with the collaborative interactions between presumptive genetic parents (Black 1996), nevertheless can be extraordinarily variable in ways that sometimes defy easy description. Understanding avian reproductive behavior requires knowledge of the timing of events in annual cycles (fig. 16.2). What birds do in the non-breeding season and where they breed are important determinants of the timing of events in their annual cycles. Knowing about the constraints and opportunities—where and when—of food getting—"foraging"—is fundamental to understanding reproductive behavior. One must also know about social interactions: within-sex rivalries over territories, nesting sites, and mates (competitive and cooperative), as well as between-sex interactions of "getting together," copulating, and collaborating (or not) on parenting essentials such as nest-building, incubating eggs, and feeding nestlings and fledglings.

Reproductive behavior is a cascade of sequential motor acts (i.e., behavior), including pairing, copulation, and caring for offspring: individuals have to encounter potential mates, and once they do, they must accept or reject those potential mates, and be accepted themselves, before copulation can happen. It sounds oh-so-simple, and sometimes it is, because sometimes mating is random and just happens. Other times, an individual's rejecting or accepting a potential mate for copulation is a matter of assessment, a cognitive process: Does he like her? Does she like him? Sometimes assessment is instantaneous, taking no time at all. Other times individuals must evaluate the complex physical contortions ("dances") of potential mates (think, birds-of-paradise) or songs and calls of opposite sex suitors. Displays take time, even when displays are energetically demanding as in, e.g., Golden-collared Manakins (*Manacus vitellinus*). Assessing

Figure 16.1. Birds breed in a variety of habitats. *Photos courtesy of Geoff Hill and Mark Friedman.*

or sorting between others' traits that provide information or advertisements, which may or may not be "honest," depends on within-sex variation in displays as well as variation in individual assessments. In some species, acceptance of a territorial partner is contingent on the quality of territory "owners," who advertise the quality of essential or superior resources that they defend. Precopulatory displays are often immediate invitations to mate (fig. 16.3).

Other "courtship" displays are stereotypical, complex interactions—"dances"—involving both a female and a male, occurring after individuals have accepted each other as mates, and serve to coordinate neurological and hormonal

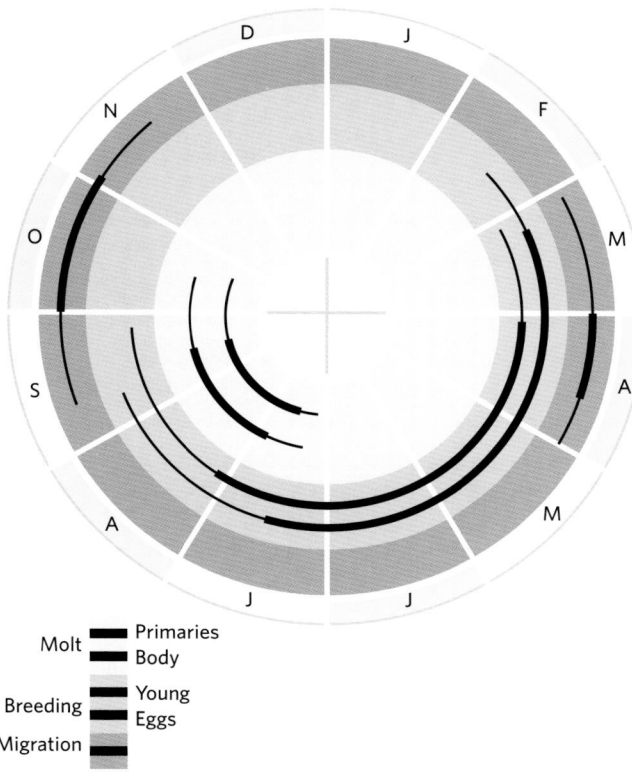

Molt — **Primaries**
     Body

Breeding — Young
        Eggs

Migration — ██

* Timing pertains to North American populations

Eastern Bluebird
Sialia sialis | Order PASSERIFORMES - Family TURDIDAE

Figure 16.2. The annual cycle of Eastern Bluebirds (*Sialia sialis*) shows the timing of molt, migration, and breeding during a single annual cycle. *From Gowaty and Plissner 2015, http://bna.birds.cornell.edu/bna /species/381.*

Figure 16.3. Sharp-tailed Grouse (*Tympanuchus phasianellus*) display on leks to entice females to copulate. *Photograph courtesy of Geoff Hill.*

Figure 16.4. Post-pairing, pre-reproductive display of Gentoo Penguins (*Pygoscelis papua*). *Photos courtesy of Lee C. Drickamer.*

readiness for the collaborative and cooperative aspects of mutual parental care (fig. 16.4).

The act of mating—copulation and the transfer of sperm from males to females—occurs in 97 percent of bird species with a "cloacal kiss" (Fitch and Shugart 1984). In other species, such as ducks, in which males have intromittent organs, the males insert their penises into the cloacae ("vaginas") of females before sperm transfer. Also important to successful reproduction are finding nesting places, building nests, laying eggs, incubating eggs, guarding eggs and nestlings, collaboratively brooding or caring for altricial (nidicolous) nestlings or watching over precocial (nidifugous) young that leave nests as soon as they hatch. Once altricial young fledge, in most passerines and in many other species, both female and male parents have extended parental duties protecting

ON THE ASSUMPTIONS OF HYPOTHESES OF NATURAL AND SEXUAL SELECTION

Natural selection is a process proposed by Charles Darwin (CD) and Alfred Russell Wallace (ARW) (1858) in a joint paper to account for evolutionary change, known then as "transmutation." Their paper described the slow process by which one species evolves into another. The next year, Darwin published his big book *On the Origin of Species by Means of Natural Selection* (Darwin 1859), an idea he mulled over for more than 30 years before publishing. *On the Origin* is a long argument in support of the hypothesis that natural selection accounts for transformation of one species into another: organic change through time. Darwin and Wallace argued that natural selection is a cause of evolution and adaptation. Adaptations are the morphological, physiological, and behavioral traits that evolved because those individuals with those traits survived and reproduced better than others in the environments they all experienced.

A modern view consistent with CD's and ARW's idea of natural selection is that selection is like arithmetic or an algorithm (Dennett 1995). Just as $1+1=2$, or just as "if this is so, and this is so, and this is so, then this new thing is so," natural selection is so, if the following things are true: if, in a population, (1) individuals vary in heritable traits; and (2) if ecological or social "forces" exist (i.e., if "selection pressures" exist) and favor some individuals over others because of their heritable traits; and (3) if favored individuals survive better (and produce more descendants) than other individuals, then one can deduce that natural selection has occurred or is operating. Evolution by natural selection is a fact, but it is also a testable hypothetical-deductive hypothesis of morphological, behavioral, or physiological change between one generation and the next (coded by genes, epigenetics, behavior-culture,

and, in the case of humans and maybe other organisms, symbolic communication). To test the hypothesis of selection of any sort, one must make assumptions about three things: (1) the level of selection or the units of selection (individuals, kin groups, populations, groups, etc.) having variation in heritable traits; (2) the posited selection mechanisms—the social or ecological forces that sort among the units in the particular level of selection; and (3) the component of fitness—i.e., the survival or reproductive success component—that is enhanced or worsened by the possession of the trait variants in the units of the level of selection.

"Sexual selection" (Darwin 1871) is a type of selection in which (1) the units of selection are within a sex, i.e., either among males or among females in a population because of heritable trait variation among males (for male-male intrasexual selection) or among females (for female-female intrasexual selection). (2) The mechanisms are behavioral or physiological interactions between individuals in the population. Mechanisms of selection include opposite-sex mate choice or same-sex rivalry over access to mates or to mate quality. (3) The component(s) of fitness that mate choice or same-sex rivalry affect include the number or quality of offspring as well as the survival probability of rivals or potential mates. It is often overlooked that sexual selection is a matter of selection on both sexes: mate choice and same-sex rivalries over the number (usually in males) or the quality (usually in females) of mates sorts among same-sex rivals and affects the number of offspring (usually in males) or the quality of offspring (usually in females). Much research remains to be done, especially on the fitness payouts for mate choices in both sexes.

---

and teaching young to forage independently and to avoid predators. In contrast to many other egg-laying vertebrates, parenting after egg-laying is crucial in birds and is intimately associated with mating systems (Ar and Yomtov 1978). Birds are famous for opposite-sex collaborations over the care of offspring: profound collaboration of avian parents is made possible by the fact that, unlike mammal males, avian males can perform—just as easily as females—most parental duties (Black 1996). Each reproductive task is accomplished in relation to ecology; together they constitute "ecological adaptations for breeding" within the conceptual frame of the evolutionary origins and maintenance of individual adaptations to socio-ecological conditions. The box on page 496 describes

the essentials of natural and sexual selection hypotheses that organize research on adaptation.

SETTLING: ADVERTISEMENTS FOR MATES AND TERRITORIAL DISPLAYS

In species in which males display on leks, the places where copulation occurs are distinct from where nesting occurs; the pre-mating antics of males in lekking species are spectacular sights (fig. 16.3) in which males sometimes display in groups at a particular site or on individually defended bit of an "arena." In contrast are the more common avian territories in which mating and nesting both occur. These more common territories are defended areas in which individuals of either sex dissuade others (either of the same sex and species or individuals of other species that compete for the same space and the resources in those spaces) from entering by singing, display of "war paint" plumage, or by threat of aggression or outright fights. Residents may display territorial defenses to conspecifics and heterospecifics.

The functions of territorial defense vary with timing in the nesting season. For example, territorial defense by males when females are fertile may differ dramatically from territorial defense before females recruit or when adults are caring for eggs and nestlings. Some displays of territorial owners function to attract potential mates, as if the spatial area is a part of the extended phenotype (Dawkins 1999) of the territorial "owner" indicating some aspect of "mate quality." Other displays appear to be aggressive defense of territory, or sometimes, collaborative group displays. In north temperate habitats, most territorial signaling occurs in the spring, whether for attraction of mates or repulsion of rivals. Some territoriality is accompanied by outright aggressive interactions (fig. 16.5, 16.6) that often vary in intensity (see below). In most well-studied species in which males are first to arrive and fight with other males over a territory, females may aggressively interact with each other too; because males and females in most bird species are sexually monomorphic (look alike), female-female aggression may be commonly overlooked. One of the most interesting and enduring studies

of aggression in birds is Nobel (1936), an excellent article for neophyte ornithologists to read.

Study of the functions (or "adaptive significance") of territorial behavior has a long history in modern ornithology. Elliot Howard (1920) argued for two functions of territoriality: (1) to facilitate pair bonding, and (2) to guarantee a food supply for the young. Other authors have cast more complex classifications of types of territories. Margaret Morse Nice (see the box on page 581 and the box on page 702)—one of the first women ornithologists of the modern era—developed a framework of territorial functions. Nice's (1941) framework illustrated that defended areas often have multiple functions that may vary by species and by habitat, and also by whether males or females or individuals of both sexes are territorial, showing agonistic behavior toward conspecifics or fighting with same-sex conspecifics. Nice based her classification on three criteria: the functions of a territory, and its extent in space and in time. She discussed variation among different species, indicating that defended real estate was for (A) mating, for nesting, providing a feeding ground for the young; (B) mating and nesting, but not as a feeding ground; or (C) as a mating station only. Defended areas could be restricted to narrow surroundings of a nest for either colonial species or solitary nesters. Nice also differentiated the breeding territories of A, B, and C, from winter territories and roosting territories (places were individuals may bivouac while foraging or to sleep during nonbreeding seasons). Robert Hinde

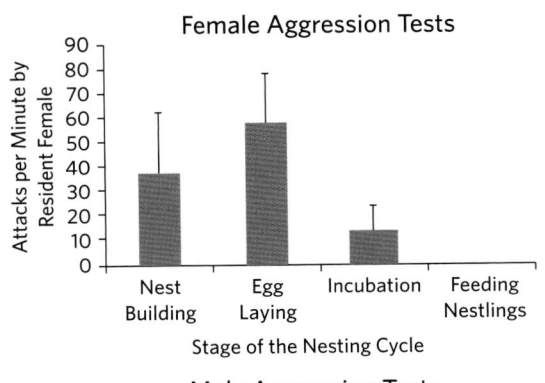

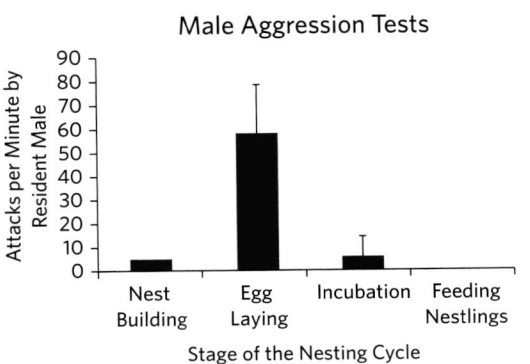

Figure 16.5. Experimental manipulation of aggression in Eastern Bluebirds (*Sialia sialis*) was induced using stuffed skins. The graphs show relative aggression during different phases of nesting cycles. *Photos and data courtesy of the author.*

Figure 16.6. Naturally occurring early nesting season male-male aggression of Eastern Bluebirds (*Sialia sialis*) in Fredericksburg, VA, March 1, 2016. A female looks on. *Photo courtesy of David Kinneer.*

(1956), a pioneer in the study of animal behavior, said the classification schemes should be used only to aid discussion, because "the diversity of nature can never be fitted into a system of pigeon-holes" (342), a claim you might want to challenge in class discussions.

Brown's (1964) response to the pigeonhole problem of the function of territory was to explain the diversity of avian territorial behavior with a general theory. Brown's theory is still useful in ornithology but also in attempts to understand territorial imperatives in humans. Brown's territory theory captured two fundamental determinants of territoriality: the economic defendability of the resource and the aggressiveness of the rivals. Rivals are not always aggressive, because aggressive rivalry depends on the environmental inducers of the urge to compete, and the level of defensive behavior may be an indicator of the value of the resource to one or both competitors. The adaptive advantage of the competitive urge depends in turn on whether the essential resources for reproduction are limited in space and time, as well as on the number of or population density of rivals.

In modern studies, formal attention to "territories" is in decline, yet better and more complete answers to why and how birds contend (or do not contend) over access to space will no doubt produce surprises for dedicated field ornithologists. One could recast even the simplicity and elegance of Brown's theory in terms of the effects of territorial acquisition and mechanisms of acquisition on "owner's" survival and the effects of defense on owners and their rivals—of both sexes. The two most fundamental parameters predicting lifetime mating success—risks to survival and effects on encounter probabilities with potential mates—have yet to be

systematically evaluated in relation to the function of territories in any bird species (box on page 517). In addition, stochastic (chance) models of territorial settlement could provide strongly inferential alternative hypotheses to the hypotheses of adaptation. However, students should not overlook that Brown's model was a first-principles hypothetical-deductive model (box on page 504) that went beyond description and made predictions, and it is still being used productively to make other testable predictions.

Theory of bird territoriality has reached into studies of human societies, modern politics, and even warfare, providing guideposts to further understanding of human behavior and modern human genocides. In addition, future applications in ornithology will include a better understanding of the adaptive value of territories in particular species, aiding in the development of wildlife management schemes.

REPRODUCTIVE DECISIONS: CHOOSY AND INDISCRIMINATE BEHAVIOR

Classical ideas about adaptive mate choice start with Darwin (1859, 1871) and Wallace (1889), the co-discovers of evolution by natural selection, who had lively, late-life arguments about whether sexual selection or natural selection resulted in the evolution of sexual dimorphism in birds. Darwin thought that females' mate choices could be a powerful selection pressure favoring the evolution of bizarre or fancy male traits, while Wallace thought that cryptic females were favored because of natural selection exerted by their nest predators (box on page 499). (These two different paths to sexual dimorphism still offer interesting opportunities for new research. A class discussion could sharpen your understanding of these two different selective pathways.)

R. A. Fisher (1930), one of the architects of "the neosynthesis of selection and genetics," later said that female mate choice could be nonadaptive and result simply from of a type of sensory exploitation of preexisting sensory biases (discussed in chapter 14). Sensory biases could result in a runaway selection process that would ultimately be selected against when exaggerated male traits decreased bearers' likelihood of survival, but would be selectively favored if the males with bizarre or exaggerated traits fathered more offspring. Indeed, most discussions of sexual selection are about the evolution of traits in males. Fisher's runaway version is a vision of how female choices sometimes favor deadly beauty. Fisher also posited that female mate choice under most conditions would likely *enhance the fitness of choosing females* (box on page 500), about which much remains to be discovered. For example, in species with males lacking bizarre or fancy traits, the existence and extent of female choice remains relatively

TESTING HYPOTHESES FOR SEX DIFFERENCES IN CHOOSY AND INDISCRIMINATE MATING BEHAVIOR

Darwin argued that sexual selection (see the box on page 496) via female choice could account for the fancy and bizarre male plumage traits in many species of birds as well as sexual dichromatism, in general. Alfred Russel Wallace argued that natural selection favored cryptic females—the sex that is most often the vulnerable "sitting duck" during bouts of incubation. These two ideas, one about the selective force of mate choice and one about the selective force of predation, each may explain the evolution of sex differences in avian plumages. Darwin's sexual selection idea is the one most often tested, but Wallace's natural selection hypothesis is equally testable. What is of interest in a discussion of behavioral evolution is that the commonness of pretty male birds has focused attention on *females'* choices of mates, and left the question of whether—even in species in which females are dull plumaged—*males* also choose mates. That is a mostly unasked, and therefore unanswered, but interesting question. And, it is also reasonable from many perspectives to ask whether males, fancy or not, have mate preferences, and whether their preferences, like female preferences (see the box on page 500) have associated fitness rewards.

R. A. Fisher (1930) argued that female choice could be a runaway process in which male traits exploited females' sensory biases, resulting in enhanced numbers of mates for fancy males compared with less fancy. This idea requires no fitness reward for females. But others hypothesized that such females would benefit by having sons as sexy as their fathers. Fisher also argued, however, that females might enjoy other benefits of mate choice, something he considered in need of testing. The Hamilton and Zuk hypothesis (1982) is an interesting post hoc addition to Darwin's sexual selection hypothesis. These authors argued that females preferred the fanciest males because of their "good genes" for disease resistance, an idea with two important predictions: (1) fancier males of a species are healthier than less fancy males; and (2) offspring of fancier males, who inherit good genes that resist local diseases, are healthier than the offspring of less fancy males. Many studies support the first prediction about the health of fancy males, but only a handful of studies support the second prediction, suggesting that there is more to discover. If mate preferences are adaptive, investigators should be able to identify fitness rewards for choosers. (Wallace also argued long ago that the brilliant plumages of some sexually monochromatic species were an indicator of the health of individuals, which is a signal that informs individuals, independent of their sexes, of conspecifics' health status.)

Under the assumption that the healthiest offspring are those with highly diverse immune systems, it is reasonable to predict that individuals of both sexes prefer as mates those dissimilar to themselves at immune-coding loci, and that individuals able to mate with those they prefer will produce the healthiest offspring (see the box on page 500 for an avian example).

unknown, just as the existence of male choice of females is not yet commonly evaluated. One of the main reasons that few investigators look for males making mating discriminations among females is that there are few socially polyandrous species of birds. In addition, many investigators assume that the sex that contributes the most parental investment will be the choosiest. The classical theories of sex differences predict that females' typically greater investment in either gametes (Parker et al. 1972) or post-gametic parental care (Trivers 1972) means that females have a greater potential energetic investment than males from any copulation, implying that selection "should act" to make females "choosy." In contrast, the supposed lesser male investments in gametes and parental care "should not exert" such selection, leaving males more indiscriminate in mate choice than females.

FITNESS CONSEQUENCES OF MATING PREFERENCES

Believers in these classical ideas overlook that mate choice can evolve in the absence of sex differences in parental investment or in gamete size asymmetry. In addition, no one can answer a basic related question: Do all individuals *assess* alternative potential mates before *acting* as if "choosy" or "indiscriminate"? There are good reasons to doubt that the parental investment arguments are good signposts to who is choosy and who indiscriminate. For example, tests of mating preferences in birds are often one-sided, focused on the presumed choosy females and their preferences for the most ornamented male, and most investigations have in the past looked for fitness payouts of mating with most- versus less-ornamented males, with little or no attention on the fitness

FITNESS CONSEQUENCES OF MATING PREFERENCES IN MALLARD FEMALES

Mallard ducks pair on the wintering grounds, sometimes months before females are capable of producing eggs. Therefore, to study females' preferences for males, it is a good bet to do experiments in the winter, when presumably the hormonal cascades associated with choosing a mate are operating. When studying mate preferences, it is also important to reduce or eliminate the opportunities for sexual conflict, aggression toward females, or forced copulation, all of which happen to Mallard females on the wintering grounds and could result in females appearing to "like" a male when, in fact, she is being coerced into mating. In such situations it is not uncommon to observe almost all females displaying toward the same dominant male, indicating "consensus mate choice," which may be a mechanism by which females can avoid some aspects of coercion. However, evidence of female mate choice under conditions in which male coercion was controlled revealed another adaptive benefit of female Mallard mate preferences. Investigators (Bluhm and Gowaty 2004a, Bluhm and Gowaty 2004b) in Manitoba, Canada, built an indoor pond and performed mate choice trials in which a subject female was able to indicate, by the amount of time she spent with each, which of three males she "liked best." The males were in separate cages on the far end of the experimental indoor pond; males could not see their rivals, but could hear them. Perhaps surprisingly, none of the females that saw the same set of three males ever preferred the same male, indicating that female mate preferences were "self-referential": there was no "best" male that all females preferred. After winter preference tests, in the spring when females were fertile, the investigators randomly assigned females to breed with the male she liked best or the male she liked least. The investigators then counted the number of eggs females laid, noting which eggs were fertile and which not. Once eggs hatched, the chicks were individually marked and were put into group cages—crèches—until they achieved adult size, when they were put into large flight cages. The results showed that, compared with mothers experimentally paired to males they did not like, mothers that mated with males they liked had healthier offspring: more of the eggs they laid survived to hatch, and more of their hatchlings survived to the age when they were released to the wild. These results are consistent with the idea that mate choice predicts the health of offspring. The conclusion was that female Mallards, like flies, mice, cockroaches, and fish (Gowaty et al. 2007), have fitness-enhancing mate preferences. Given that the results also showed that females who were in forced pairs with males they did not like had fewer offspring with poorer health than the females with males they liked implies a large fitness cost for females coerced into reproduction with males they do not like. Such sexual coercion of mating in wild-living Mallards is a selection pressure lowering female reproductive success so that selection favors female resistance to coercion.

Some of you are probably wondering what "female resistance" looks like. That is a good question. Do females have ways to avoid male coercive sexuality? Do they fly away? Do they forage in groups? Do they hide? Can you think of other female resistance tactics? Do females have ways to kill the sperm of males who force-copulate them? One hypothesis (Gowaty and Buschhaus 1998) is that cells in the lining of the cloacae of female mallards secrete hydrochloric acid or some other chemical that could denature the sperm of a male that force-copulated her. So far there is no direct evidence that female mallards can kill sperm using chemical warfare, but the mechanics of female copulatory behavior are complex, and females are able to eject sperm: males have large, corkscrew-shaped penises, the morphology of which is matched by the baroque complexity of Mallard females' cloacae. Dr. Patricia Brennan (box on page 513) has characterized not only the mechanics of copulation—no simple "in and out"—but a dynamic explosion of ballooning tissues, and she has also described for the first time the many complementary twists and turns in the uteri of female Mallards (Brennan et al. 2007, Brennan et al. 2010, Brennan and Prum 2015). Reading Dr. Brennan's research papers would be an excellent class project.

payouts for the choosy females of liking this or that male independent of traits (that human investigators find attractive or interesting). In other words, most previous studies have not separated out the cognitive process of chooser's perceptions and assessment of potential mates from the motor acts of mating. Investigators usually design their studies to explain advantages for ornamented males rather than advantages to female choosers of preferring ornamented males (if they do). In addition, very few mate choice studies consider alternative hypotheses explaining reproductive decision-making.

For example, an alternative theory says that all individuals, independent of their sex/gender, assess potential mates for the fitness that would be conferred before either accepting or rejecting a copulation with a potential mate (Hubbell and Johnson 1987). Similar theories (Gowaty and Hubbell 2009) also predict that the fanciest males are often likely to say "no" to female suitors, and yet most investigators, even with intimate knowledge of their study species, are unable to describe what it looks like when males say "no" to females' solicitations. One might claim that this is only because most males are universally indiscriminate, but it could also be because the idea of choosy males is theoretically new, so that observers unaware of the possibility never even look for indicators of male choosiness, and so, never see it. Only recently have investigators started to examine experimentally why choosers sometimes change from being choosy to indiscriminate or from being indiscriminate to choosy, but the experimental evidence available now (Ah-King and Gowaty 2016) indicates that individual birds (and other organisms) are not genetically fixed to be choosy or indiscriminate, but are flexible in their reproductive decision-making, depending on their demographic circumstances (meaning their likelihoods of encountering potential mates, or their own likelihood of surviving until tomorrow). Such demographic opportunities and constraints are often dynamic, changing even from moment to moment.

It is often notoriously difficult to get captive animals to breed, and captive breeding schemes of threatened and endangered taxa regularly fail: algorithms that compute the genetic dissimilarity between potential partners often guide managers' decisions of who should breed with whom. This makes sense under the important and reasonable assumption that genetically dissimilar parents produce the healthiest offspring, because having more diverse immune systems means that individuals have greater capacity for building antibodies to hundreds of thousands of antigens, i.e., these offspring are healthier than others. Thus, why even these clever and theoretically robust breeding schemes so often fail is a mystery. One hypothesis is that individuals—both female and male—have mate preferences that develop in most vertebrates during crucial periods *before* individuals are sexually competent: critical periods in which individuals may learn who they are relative to others. The periods during which predicted crucial learning occurs are often periods during which captive individuals are underexposed to conspecifics compared with their wild relatives. Thus, this explanation for failed breeding of captive animals is that they simply do not have the experiences with conspecifics that would facilitate crucial learning about "who one is relative to others." In other words, captive-born individuals may not get the developmental exposures that provide information mediating adaptive breeding decisions. Another hypothesis explaining breeding failures in captivity is that mate preferences in *both* females and males are important to successful reproduction. In most cases, even when managers put genetically dissimilar individuals together to breed, the individuals often do not copulate, so managers assume that it was because the female did not like the male, i.e., they seldom suspect that perhaps it was the male who did not like the female: some managers still believe that only females choose mates. Theory and data, however, suggest that mate choice occurs in both females and males, independent of their sex/gender and their relative parental investment. If these assumptions are correct, captive breeding schemes might productively devise opportunities for males, not just females, to make choices among potential mates. Obviously, questions about captive breeding are ripe for further hypotheses explaining breeding failures, ripe for strong inference testing of alternative hypotheses using clever experimental tests—in other words, better understanding of breeding failures in captivity would produce excellent thesis topics for ornithology students seeking advanced degrees.

SOCIAL AND GENETIC MATING SYSTEMS

"Mating systems" characterize the associations between females and males during breeding seasons. Mating systems include monogamy, polygyny, polyandry, and polygynandry (fig. 16.7). Though the figure 16.7 categories seem comprehensive and simple, what one sees with binoculars is not the whole story. Social mating systems revealed through binoculars tell us "who hangs out with whom," but without other information, it is usually difficult to know who reproduces with whom. In "social mating systems," like monogamy, what we see is often a female and a male behaviorally interacting on a territory, often nearby or at a nest with eggs and nestlings or fledglings that "parents"—the caregivers—collaboratively defend and tend. But mating systems are not just the patterns of "who hangs out with whom" but also patterns of copulation and sharing of gametes. For example, in socially monogamous mating systems, the individuals of both sexes in a pair may copulate and produce offspring with an extraterritorial partner or partners (fig. 16.8). Genetic mating patterns are "cryptic" in the sense that to observe them, investigators must collect tissue samples from individuals (adults and nestlings), characterize DNA variation with various molecular methods, and use quantitative analytical methods of exclusion or assignment of parentage (box on page 505).

Genetic parentage studies that use DNA methods in birds readily reveal when females mate with more than one male,

Mating Patterns

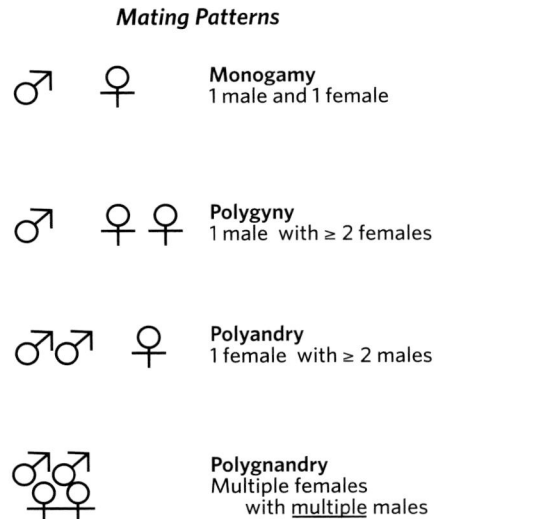

Figure 16.7. Mating patterns may be social or genetic. Ornithologists observe social mating systems with binoculars (observations in nature). Ornithologists observe genetic mating patterns using genetic parentage studies (Gowaty 1985). *Courtesy of the author.*

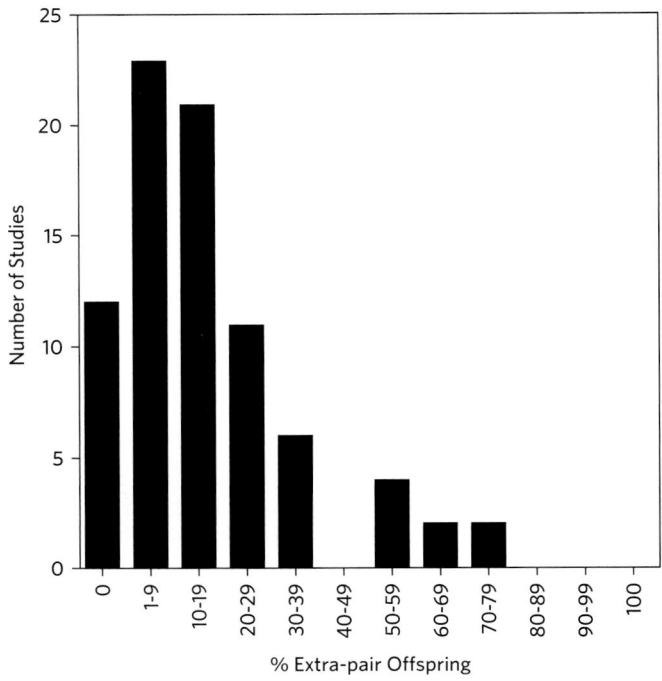

Figure 16.8. The histogram summarizes data from table 1 in Gowaty (1996b) with 78 studies on 61 species of birds with evidence of genetic parentage. Only 12 of the tabled species had no evidence of extra-pair paternity, and half of these were notable because of the very small sample sizes of individuals and broods. Extra-pair paternity is common: the estimates of the frequency of extra-pair young varied between fewer than 10 percent of sampled chicks to almost 80 percent. *Courtesy of the author.*

because females usually lay all their eggs in one nest, so that figuring out who is the mother of each nestling is often straightforward, and it is often easy to figure out that there are multiple sires in a nest, even when there is no clear way to identify exactly who the extra sires are. In contrast to the usual ease of counting females' number of mates, counting males' number of mates is often much more difficult, because progeny from males that multiply mate will be more widely distributed than is the case for females; i.e., field ornithologists would have to sample every nest over a very wide area to find the offspring of every male who mates outside his social pair bond. Because of this sampling bias associated with parental sex, we currently know more about genetic polyandry (female multiple mating) than about genetic polygyny (male multiple mating). Obviously sometimes the patterns of social associations are more complex than the categories we use for binning or categorizing pairs: e.g., just because some species appear to be strictly socially monogamous does not rule out that both females and males can mate and produce offspring with others besides their territorial partner, so that the pattern of the social mating system does not imply the genetic mating system. In fact, in socially monogamous species, strict genetic monogamy is relatively rare (a remarkable exception is described in the box on page 518): in the majority of socially monogamous species in which investigators have completed "genetic parentage studies," females often mate and produce offspring with males besides their social

mate, something called "extra-pair copulation" (EPC), which can produce "extra-pair offspring" (EPO), via "extra-pair paternity" (EPP) (fig. 16.7). "Extra-pair maternity" can also occur when a nonresident female enters another's nest and lays an egg or eggs. Such "conspecific nest parasitism" may most commonly happen when females lay eggs in neighbor's nests (Gowaty and Bridges 1991b). Sometimes "parasitic females" are in cahoots with resident males with whom they have EPCs, and these females subsequently lay eggs in the nest of the male. This type of parasitism is called "quasi-parasitism," in which cases extra-pair maternity is analogous to extra-pair paternity. Asymmetries in parental relatedness to offspring have incited theoretical ideas about the origins of infanticide (Rohwer 1986, Veiga 1990, Moreno 2012), differential male and female parental care, and same-sex aggression. You may be able to imagine such effects, and this topic would be a great group discussion topic; furthermore, you might gain some practice in identifying primary sources for these topics using Web of Science. Besides all the variation we've so far discussed, some of which seems relatively intuitive and easy to explain, some mating systems are so complex that they are almost impossible to efficiently describe, much less "explain." The box on page 506 describes the curious mating system of Northern Bobwhite (*Colinus virginianus*), which Brant Faircloth (box on page 508) first described thoroughly using modern methods. The bobwhite mating system is one that fails to fit the usual categories of mating systems, despite the fact that observational and genetic studies on them are among the most thorough studies ever done on an avian mating system.

HYPOTHESES OF THE EVOLUTION OF MATING SYSTEMS

Gordon Orians's (1969) (box on page 509) polygyny threshold hypothesis (PTH) (fig. 16.9) influenced almost all of the modern ecological theories of mating systems. The PTH describes when it would be beneficial for a female to settle on the territory of an already-mated male compared with an unmated male. The idea assumes that the occurrence of social polygyny is dependent on a female mating decision. In other words, a female who chooses to settle on the territory of an already-mated male makes that male socially polygynous; she is the "king maker." The y-axis of the PTH model is meant to be about the "fitness" of the "king maker," the female making the decision to join or not join an already-mated male. When looking at the graph (fig. 16.9), remember that "fitness" is a measure of an individual's reproductive success relative to everyone else's, so imagine if the axis were quantitative, rather than qualitative, the numbers on the axis would vary between 0 and 1. The x-axis is "environmental quality," a qualitative catchall term, which might in nature indicate variation in food availability and quality, risk of predation, quality

of nesting substrates, etc. The two curves on the graph show a theoretical association between "environmental quality" and fitness for females: as environmental quality increases, the relative reproductive success of the female in question increases. The fact that the lower curve represents the fitness of a female that settles with an already-mated male, while the upper curve represents female fitness if she settles with an unmated male embodies the assumption that females always benefit from the help of a male. The exact shapes of the curve do not matter to the PTH, but their relationships to each other do. The vertical dashed line "1" indicates the difference in relative reproductive success for females settled in territories of the same environmental quality with an already-mated

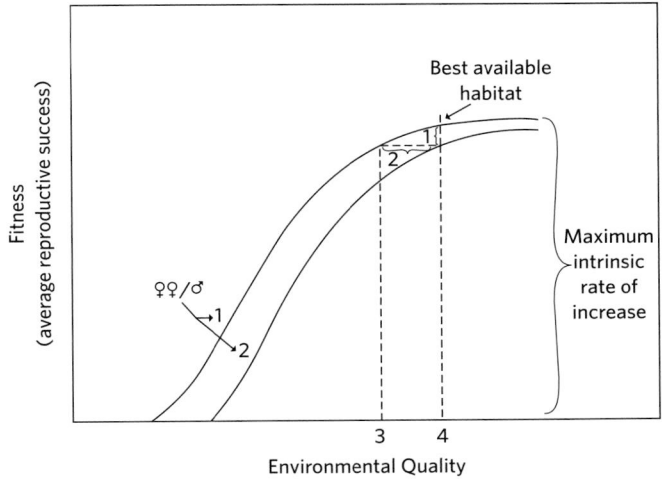

Figure 16.9. A graph describing the polygyny threshold hypothesis (PTH). Given the axes, the curves embed the assumptions of the PTH, and from the relationships of the curves, one can deduce many of the predictions of the PTH (see text for discussion). *From Orians 1969, used with permission of the author.*

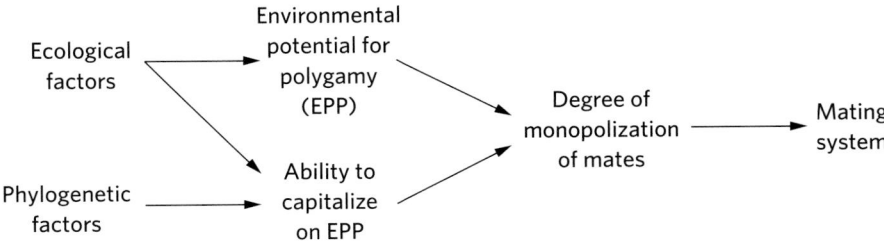

Figure 16.10. A general scheme describing the socio-ecology of mating systems. the envioronmental potential for polygamy hypothesis (Emlen and Oring 1977) assumed that both ecological circumstances (habitat, for example) and evolved intrinsic differences between the sexes in a given species (called "phylogenetic factors") determined the ability of one sex of individuals in a given species to monopolize opposite sex mates. This idea was one that arose before the widespread discovery of cryptic mating—such as extra-pair paternity—in which both females and males had multiple sires or dams in their progenies. *Figure reproduced courtesy of Stephen Emlen.*

A PRIMER OF SCIENTIFIC WAYS OF KNOWING

When we say someone has a PhD in ornithology, it implies that they know how to generate knowledge about birds, i.e., that they practice scientific ways of knowing. Much of ornithology continues to be descriptive biology, a branch of natural history, one of the most venerable of sciences. Ornithology, of course, is about the study of wild-living or fossil birds or birds in a laboratory. Ornithology today is a very rich natural history discipline that also interrogates the biology of birds from both proximate (how) and ultimate (why) perspectives. Chapter 16 is about the "why" questions of adaptive significance of avian behavior related to breeding and reproduction in ecological settings. The answers are often posited as hypotheses and then tested with observations or with field or laboratory experiments. In this chapter we discuss many types of hypotheses, and among your jobs as students is one of starting to understand the form and the power of different kinds of hypotheses.

Different "kinds" of hypotheses provide different standards or quality of inferences and demand different ways of testing. Some of you have already noticed that the authors of this book frequently mention that a hypothesis or its predictions have been rejected or confirmed. Scientists value rejection of assumptions of hypotheses because rejection of its assumptions removes a hypothesis from contention, so that we feel we're getting somewhere, now able to go forward empirically to the next question, often with a new hypothesis. In numbing contrast, confirmation of a prediction of a hypothesis provides weaker inferential power, and often gets scientists no further than they were when they started. For example, do other hypotheses make the same prediction? When the prediction of a hypothesis is confirmed, it means it is still in the running to be rejected, i.e., there is more work to do. Given that scientists wish to be "efficient," learning as much reliable information as possible in the shortest amount of time, tests that are capable of only confirmation are often dismissed as "only confirmatory evidence." Whereas, if one completes tests capable of rejecting a particular hypothesis and the hypothesis continues to stand, one has greater confidence that it is true.

Inductive hypotheses have the form, "something happened before and that predicts that it will happen again." Such correlational hypotheses are confirmatory over and over again until they are not, and they are incapable of informing questions about causation, because "correlation is not causation." Inductive hypotheses are often extremely difficult to reject. The classic example is the hypothesis that "all swans are white," which comes from previous observations of swans, all of which were white. Rejection of this inductive hypothesis requires just one observation of a black swan, which may occur only rarely, making empiricists who depend on induction wait a long time for the resolution of any doubts they have about the white swan hypothesis. Inductive hypotheses are not "efficient."

One form of deductive hypotheses is known as "hypothetico-deductive" or "H-D," indicating that the assumptions (the "first principles" of the hypothesis), taken together, form the hypothesis. Natural and sexual selection (see the box on page 496) are hypothetico-deductive hypotheses, as are the polygyny threshold hypothesis, the female constraint hypothesis, the switch point theorem, and many other hypotheses in this book. Empiricists can test an H-D hypothesis by testing either assumptions or predictions. The best thing to do to test an H-D hypothesis is to test its assumptions: if its assumptions are not met by nature or a particular test system, the H-D hypothesis does not match the system, and you can put the hypothesis aside as an explanation for the system you are investigating. If one tests and rejects a prediction of the H-D hypothesis, one might imagine that something is left out of the H-D hypothesis and that it needs ad hoc additions to it to fully match the nature of your system. H-D hypotheses are efficient means of interrogating nature.

Strong inference developed because simultaneously testing two hypotheses is a more efficient form of science than testing one hypothesis at a time (Platt 1964). Simultaneously testing two alternative hypotheses has an added benefit, because it often decreases a type of bias associated with "investigator pride" about "their idea," which could cloud their ability to evaluate the hypothesis. On the other hand, when investigators test two alternative hypotheses at the same time, they are likely always going to have a "publishable" result and so not be overcommitted to "their" idea. Most of you are familiar with the shorthand of statistical hypothesis tests in which investigators compare a single hypothesis against a null hypothesis. For instance you have probably seen hypotheses indicated this way: H_0: $a = b$ versus H_1: $a > b$. Here H_0 is the "null hypothesis" and H_1 describes the prediction of interest. In contrast, when one tests simultaneously two alternative hypotheses, there are three possibilities: As before, H_0: $a = b$ is the null, but there are two other alternative hypotheses, namely H_1: $a > b$ versus H_2: $a < b$. In strong inference testing, one increases the possibility of getting an answer with a single well-formulated test, simultaneously capable of rejecting one alternative while finding consistency with another.

EXCLUSION AND ASSIGNMENT OF GENETIC PARENTAGE STUDIES IN BIRDS

Mating patterns include social and genetic mating systems. In social mating patterns, such as monogamy, what we see is often a female and male socially interacting on a territory, often nearby or at a nest with eggs and nestlings and, later, fledglings that they collaboratively defend and tend. But mating patterns are not just the patterns of "who hangs out with whom," but also patterns of copulation and sharing of gametes. For example, in socially monogamous mating systems, the individuals of both sexes in a pair may copulate and produce offspring with an extraterritorial partner. Genetic mating patterns are "cryptic" in the sense that, to observe them, investigators must collect tissue samples from individuals (adults and nestlings), characterize DNA variation, and use analytical methods of exclusion or assignment of parentage. The cryptic nature of genetic parentage means that social monogamy is very often genetic polyandry, which is easier to "see" in genetic studies of parentage than is genetic polygyny. Genetic polygyny may also be common, but it is much harder to observe than genetic polyandry, partly because the evidence that a female has multiply mated can usually be found in the brood of nestlings for which she cares, but for males, similar evidence of multiple mates will be distributed in the nests of multiple females. It is just statistically and practically easier to find the evidence that socially monogamous females are **genetically polyandrous** than to find the evidence that socially monogamous males are **genetically polygynous**.

Parentage exclusion studies can produce the most powerful inferences about genetic parentage. In exclusion studies, one ideally knows who all the potential parents are, which is often possible in captive studies. If an investigator has access to tissue samples from which DNA can be extracted from all potential parents, and if there is enough between-individual variation in the DNA or protein markers, alleles in nestlings that have no correspondence with alleles in particular adults exclude those particular adults from parentage.

Parentage assignment studies are today a more common type of genetic parentage study. These use genetic algorithms to evaluate the probability (likelihood) that a particular adult is a parent of a given offspring. These assignment studies are vulnerable to mis-assignment whenever some of the potential parents are unsampled or unknown. There are two ways to estimate the probability of assignments, one based on the distribution of alleles in all sampled offspring and the other based on the distribution of alleles in sampled potential parents. Offspring alleles are a good representation of the parental alleles investigators should look for. Depending on alleles in sampled adults is less reliable than depending on the distributions of alleles in the offspring. Why? Because in field populations, investigators sample only a finite part of the entire population of potential parents. Given that birds fly, there is a good chance, in all studies of wild populations, that some of the potential parents go unsampled, because they are outside the study areas of the investigators, or investigators fail to catch and sample all adults. This being the case, using the distribution of alleles in sampled adults is relatively less reliable than using the distributions of offspring alleles.

versus an unmated male. The horizontal dashed line "2" indicates the "polygyny threshold," which is the difference in environmental quality that makes up for or compensates for the predicted loss in a female's reproductive success from sharing a male's parental care with another female.

The PTH inspired a great deal of research (e.g., Slagsvold and Lifjeld 1994). Today, the consensus is that in many avian systems, some of the explicit and silent assumptions of the PTH are not met, making it inappropriate for empiricists to test the predictions of the PTH in particular systems *in which the assumptions fail* (see the box on page 504). The point of our discussion then—particularly for neophyte ornithologists— is about the utility of models in scientific thought. Good models focus our attention, send us to the field, and demand

stringent testing of assumptions and, only then, evaluation of predictions. Many models of social behavior that ornithologists have previously crafted were based on correlations familiar to observers from their field observations, i.e., the hypotheses were "inductive" (e.g., "I saw such before, so I predict I'll see it again"). The difficulty with such reasoning is obvious: "correlation is not causation," leading most scientists to prefer deductive to inductive logic.

The PTH is a deductive hypothesis, and not just a historical graphical relic: it is a classic idea, well worth your efforts to understand, because its assumptions constitute the hypothesis, providing the guideposts (the assumptions) to evaluating its utility in particular systems (box on page 504). Because of its assumptions, the PTH stimulated research that

THE CURIOUS MATING SYSTEM OF NORTHERN BOBWHITE (*COLINUS VIRGNIANUS*)
By Brant C. Faircloth

Northern Bobwhite Quail are small (140–200 g), "weakly flying," ground-dwelling Galliformes, whose reproductive behavior contrasts with typical familiar passerines. Rather than having relatively small clutches, altricial chicks, and post-hatching parental care, Bobwhite lay multiple clutches per year with relatively large numbers of eggs per clutch (12–14 eggs); the chicks are precocial, and the parents provide little or virtually no posthatching parental care. Both males and females can incubate, but usually not collaboratively; rather, males will incubate separate clutches from those that females incubate. The precocial young are nidifugous, able to forage on their own just after hatching. Unrelated chicks often join foraging groups after leaving nests (Faircloth et al. 2005), perhaps an evolved defense against predation. The social mating system is very hard to infer with binoculars, the main observation tool for most field ornithologists, but observations of radio-telemetered Bobwhite indicated that many males had opportunities to mate with many females and vice versa: there seems to be no defensive behavior against parentage uncertainty, and weak or absent mate-guarding by either males or females. Evidence from telemetry observations of interacting individuals and genetic parentage studies proved that, within a single population, individual Bobwhite were variable within a season (Faircloth 2008). Both females and males were flexibly monogamous or polygamous—meaning that males sometimes were polygynous and females sometimes were polyandrous. Individual mating behavior was not fixed, because individuals flexibly expressed variable reproductive behavior throughout the breeding season. Rarely are individuals of passerine species so individually variable in their behavior, even between years

and breeding seasons: Bobwhite seem even more flexible than Dunnocks, *Prunella modularis*) (Davies 1992) and vary their behavior within a single breeding season.

The assumptions of typical socio-ecological theories explaining avian mating systems such as the polygyny threshold hypothesis (fig. 16.7), its derivatives (fig. 16.10; box on page 509 and box on page 510), the environmental potential for polygamy hypothesis (fig. 16.9) or the hypotheses explaining the origins of monogamy (Wittenberger and Tilson 1980) do not fit with what we know about Bobwhite mating behavior. For example, males do not defend nesting territories on which females nest, nor do they provide or control access to essential resources to females; female success in laying and incubating eggs does not necessarily depend on collaborative parental care; there is no obvious mate-guarding behavior, nor do females seem to guard their clutches from conspecific nest parasitism. In addition, female Bobwhite seldom (never?) lay subsequent clutches in the same place. Bobwhite might be much more likely to experience predation than individuals in typical territorial passerine species that are much better aviators than Bobwhite and so able to exploit safer places for their nests—like high branches among leafy boughs or holes in trees. A significant constraint on individuals' survival probabilities suggests that the predictions of mating theory (box on page 513) that depend on stochastic demography may at last provide an explanation for the flexible patterns of social and genetic mating systems of Bobwhite. A question that remains in Bobwhite biology is whether adult females and males have mating preferences that enhance offspring viability (box on page 499), similar to what happens in Mallard females (box on page 500).

produced novel observations propelling forward the science of mating systems, not just in birds but also in humans. Yes, sometimes observations were inconsistent with the PTH, but impassioned naturalists turned its failures to their advantage. It became the substrate for ad hoc modifications, i.e., adjustments to the underlying assumptions to better match the real-world biology of specific species: in other words, the PTH was the starting point for many other conceptual ideas about mating systems, including (but not limited to) the environmental potential for polygamy hypothesis (Emlen and

Oring 1977; fig. 6.10), the polyandry threshold hypothesis (fig. 16.11), the female constraint hypothesis (box on page 516), and the switch point theorem (box on page 517). Like the PTH, all are conceptual statements of how ecology and social interactions, including sexual selection and sexual conflict, interact to produce the diversity of social behavior that field ornithologists have described over the last 100 years. The environmental potential for polygamy hypothesis had as a key assumption that conditions of the environment and the ability of one sex or the other to "capitalize"

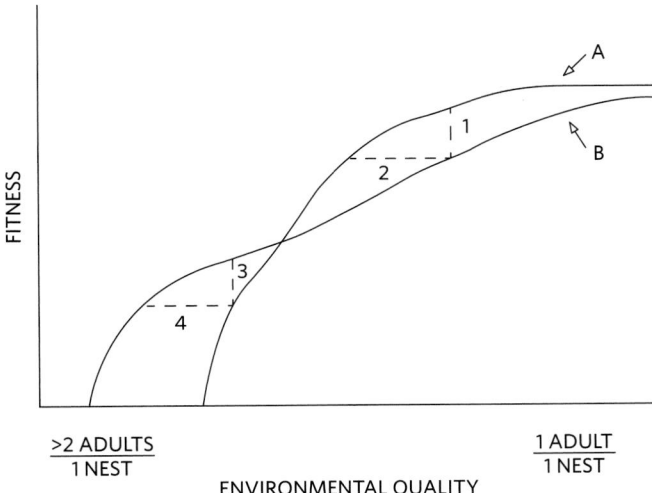

Figure 16.11. A graph shows relative fitness variation associated with different types of social mating systems (Gowaty 1981b). Curve *A* is the expected mean fitness of an individual of either sex mated monogamously; curve *B* is the expected mean relative fitness of an individual mated simultenously to two mates on the same territory. Distance 1 indicates the fitness difference between those individuals choosing to mate monogamously versus polygamously in the same environment. Distance 2 is the difference in habitat quality, called "the polgamy threshold" (either polyandry or polygyny), which is the smallest difference in habitat quality suffcient to make polygamous matings advantageous relative to monogamy for the individual gaining the least advantage in the mating system. Distance 3 is the fitness difference between monogamously mated versus polygamously mated individuals. Disance 4 is the habitat quality distance, known as the "monogamy threshold," making monogamy more advantageous than collaborative nesting, with or without relatives. Distance 1 (dotted line) is the difference in fitness of individuals choosing to mate monogamously and individuals choosing to mate polygamously in the same environment. Keep in mind that graphical models assuming that the social mating system adequately represents fitness were made obsolete with the discovery of extra-pair paternity and extra pair maternity in many passserine species. *Reproduced with permission of the author.*

opposite sex individuals determined the "degree of monopolization of mates."

To date there is no consensus hypothesis of mating systems, so that ornithologists and other ecologists interested in efficient science (box on page 504) suspect that the fundamental first principles of mating systems have not yet been appropriately characterized. It seems as though we are in need of quantitative models/hypotheses to replace our love affairs with qualitative models and to guide us to ever-stronger inferential empirical tests. And, so, the modern question facing the upcoming generation of ornithologists is, what are the first principles (fundamentals, axioms) of avian mating systems? With some creative insight, one of you may

be the discoverer of "the universal theory of avian mating systems."

Socially monogamous birds are proxies for understanding the dynamics of social monogamy in other species, including humans. The models/hypotheses and theories that ornithologists have cast are useful in comparative biology and have inspired observers to make detailed observations of the behavioral ecology of humans (Borgerhoff Mulder and Schacht 2012).

WITHIN-SEX AGGRESSION

Females usually fight with females and males fight with males (Noble 1936). Commonly, the intensity of aggressive defense against an interloper varies with the stage of the nesting cycle (Gowaty 1981a). Experimental studies that use controlled stimuli to induce responses have allowed strong inferences (box on page 504) about the functional or adaptive significance of territorial aggression (Slagsvold 1993, Sandell 1998, Karvonen et al. 2000). The patterns of aggression show that males fight males and females fight females in Eastern Bluebirds (*Sialia sialis*) (Gowaty 1981a) and that their responses varied during different phases of the nest cycle, in patterns that either rejected or were consistent with hypothesized functions of aggression during nesting cycles. Males fight males when females are fertile; females fight females when their nests are vulnerable to takeover, or when the risk of conspecific nest parasitism is high. Selection logic says that aggression is often dangerous for the aggressor, because even slight scratches can make the aggressor sick or disabled in the long run. Thus, evolutionary scientists expect that individuals that pick fights or that are induced to defend their "interests"—territories or their mates or their offspring—are fighting over something worth the risk of injury or death. Experiments during the breeding season with Eastern Bluebirds showed that males were most aggressive to other males when their social mates were fertile and copulating, a result consistent with the hypothesis of male defense of their genetic paternity—in effect increasing the likelihood that they care for only their own offspring, not their neighbors' or rivals' offspring. This hypothesis also assumes that caring for the offspring of a rival is energetically expensive as well as likely to decrease a male's fitness. Similarly, female bluebirds were most reliably aggressive to other females when they were laying eggs, suggesting that female bluebirds were defending their nests from the threat of conspecific nest parasitism and thereby defending their genetic maternity. When these experiments were being carried out, no one knew that female bluebirds lay eggs in each other's nests. In fact, fierce female-female aggression was a surprise, but the costly

BRANT C. FAIRCLOTH

Brant C. Faircloth.

Although my given name suggests otherwise, my interest in studying birds began late. When I was a new grad student at the University of Georgia, taking a vertebrate natural history class, Brian Chapman introduced me to the variety of forms, plumage patterns, life histories, behaviors, and songs of birds. The nomenclature of birds was frightening at first, but also fascinating Here was a group of organisms—delicate, strong, beautiful, and ugly—that are the only remaining lineage of theropod dinosaurs . . . and we are surrounded by them! Of course, in college I had dabbled in other interesting organismal groups, primarily mammals and squamates of the Southeast, but my focus kept returning to these feathered creatures—I could not shake my interest in understanding how birds evolved from their common ancestor. What were the factors that might be responsible for generating the incredible diversity of avian species? That question still fascinates me.

After college I did bird surveys in beautiful longleaf pine habitats, and fell in love with the incredible forests of the southern coastal plain. I managed different habitats for birds, and I banded and resighted Red-cockaded Woodpeckers (*Picoides borealis*) in South Carolina, Georgia, and Florida. RCWs fueled my interest in the socio-ecology of avian reproduction and mating systems. My questions about the origins of avian diversity led me to work with John Carroll and William Palmer on Northern Bobwhites (*Colinus virginianus*)—perhaps the most well-loved and convivial bird of the Southeast. Bobwhites are interesting for many reasons, particularly because we understand so little of their mating behavior.

After completing my PhD and my postdoctoral research at UCLA, where I had enormous freedom to pursue my interests, the variation that I had observed in bobwhites began to gnaw at me, and I began to look at the broader picture of avian phylogenies. Now, a significant portion of my time is devoted to understanding the evolutionary processes that have shaped the genetic variation that underlies phenotypic diversity within and among different bird species.

aggression suggested the protection of maternity hypothesis, which is analogous to the protection of paternity hypothesis described above (fig. 16.12). Female-female aggression tipped the investigators off to the possibility of conspecific nest parasitism in eastern bluebirds.

Very few studies of female-female aggression in birds exist. Needed is more attention to the subtleties of female-female aggressive competition. More hypotheses about the origins and causes of female-female aggressive competition will stimulate more studies. One of the reasons so few studies of female-female aggression exist is that in most passerines, sexual dimorphism and sexual dichromatism are slight or absent (at least to human eyes). In other words, it was difficult for investigators to make inferences about who fights; in the past, observers just assumed that when they saw a fight it was between males. Obviously, if it is hard to tell females

Figure 16.12. Naturally occuring aggression in a fight between female Eastern Bluebirds (*Sialia sialis*) contending over a woodpecker cavity in early fall, when bluebirds often search for available cavities. *Photo courtesy of David Kinneer.*

GORDON ORIANS

Gordon Orians.

I began bird-watching when I was seven year old. When I was about thirteen, I made a momentous discovery—people were actually paid to watch birds! I decided then to pursue a zoology major at a university and to become a professional bird-watcher. I never deviated from that trajectory.

For many years my research, using American blackbirds as my study species, focused on the main challenges that all organisms confront—choosing a suitable habitat, selecting foraging strategies, deciding which prey to pursue and consume, and learning how to avoid becoming food

for other organisms, with whom to mate, and how much to invest in offspring. I chose blackbirds primarily because their social systems are so varied. An early paper in which I developed the "polygyny threshold model" is probably the most important one I published. It has a conciseness and clarity that reflect the great amount of time I devoted to it; it stimulated a substantial amount of research by others.

Because our ancestors made the same set of decisions as my avian subjects, my investigations of avian behavioral ecology segued easily into an interest in human behavioral ecology, my current research emphasis. This research is based on the assumption that evolution predisposed humans, as well as other animal species, to learn easily and quickly, and to preferentially retain, those associations or responses that fostered survival under various environmental situations. We now select habitats and food under circumstances our ancestors would not recognize. Predators, for example, pose little danger to most of us, but a rapidly growing body of research shows that our preferences retain what I call "ghosts" of decisions made in the distant past.

Ecologists typically try to motivate people to act as environmental stewards by emphasizing the goods and services ecosystems provide, but these valid concerns rarely stimulate a level of involvement sufficient to motivate action. As we design policies and institutions to achieve conservation goals, we may improve our success if we make use of our evolutionarily based psychological and physiological responses to nature (Orians 2014).

from males, it would be easy to make the mistake, especially for a holdover from another era, someone who did not believe that females could or would be aggressive.

BETWEEN-SEX CONFLICT: MATE-GUARDING AND SEXUAL POWER DYNAMICS

In the last third of the twentieth century, sociobiologists theorized that socially monogamous male birds would seek extra-pair copulations with already-mated females, which would produce sexual conflict between mated pairs (box on page 510). The ideas predicted that extra-pair mating would result primarily from attempts by males to copulate with already-mated females. A follow-on prediction was that the risk of extra-pair paternity selectively favored males

that would "guard their mates" from the attention of other males. Later thinkers hypothesized that it was as likely for females to seek and accept extra-pair copulations as males, and these visionaries predicted that mate-guarding was something males did to keep their fertile mates from leaving their home territories to seek or accept extra-pair copulations. Operationally, in either case, one inferred the intensity of mate-guarding through documenting the variation in the distance a male was from his fertile mate, the duration of males' attendance to females, and the number of times that a male followed his fertile social partner. First studies of mate-guarding contrasted the variation in the behavioral traits of time males spent near or following their mate during different stages of nesting cycles, and such studies confirmed that males sometimes behave as though they are attempt-

EVOLUTIONARY DYNAMICS OF SEXUAL CONFLICT: INTRASEXUAL SELECTION IN TWO SEXES

In the context of socio-ecology, sexual conflict is a dual process of intrasexual selection on both males and females, but it is also a process whereby individuals of either sex can manipulate the behavior or physiology of the other sex. What this means is that an **agonistic** or **coercive interaction** between males and females can produce selection acting both among males (intrasexual selection among males within a population) and among females (intrasexual selection among females in that same population). As such, it is hard sometimes to figure out who "wins" during coercive interactions. For example, an agonistic interaction—say, a fight—between a male and a female might leave one of the combatants with an injury that may affect their fitness (survival or reproductive success relative to other members of their sex), but even if there is a decrement in reproductive success or survival of the injured individual, it says nothing about the fitness (remember "fitness" is relative to others in the population) of the winner of the fight relative to their same-sex rivals. So, a question one might ask is whether a male who hurts his mate or potential mate enhances his own fitness relative to other males, who are his rivals, who do not hurt their mates or potential mates: not an easy evolutionary question to answer.

In terms of dynamics of individuals, sexual conflict (by definition coercive behavior between opposite-sex individuals) is sexual selection in two sexes. Consider a female making a choice among males: the male who mates with her may enhance his number of mates relative to his rivals, but the female may have fitness rewards relative to her rivals if she produces healthier offspring than females who use other criteria or who are not as good at finding the right male or a series of males with whom they would produce "healthiest offspring." Even mate choice between individuals is an evolutionary dynamic of within-sex selection acting on male-male and female-female sexual selection.

---

ing to "mate-guard." An alternative explaining male's close attendance and following of females was that males stayed close to females when they were fertile in order to be close when females were ready to copulate. Both hypotheses had some support, and it is not unreasonable to think that close following by males of females might have more than one adaptive function. Once it was clear that, in many passerine species, females foray off of territories seeking copulations with extra-pair males, investigators looked for and found examples of male behavior seemingly directed at keeping "their" females at home. Some males attempted to "punish" their social mates that left territories during their fertile periods (Gowaty and Buschhaus 1998, Valera et al. 2003, Adler 2010, Kempenaers and Schlicht 2010). Punishment might be through hurting females in some way or by withholding parental care. That males withhold parental care from their social mates that copulate with extra-pair males was an early, keenly studied hypothesis; however, the prediction has rarely been met. Perhaps males are unable to recognize offspring not theirs, so that withholding food from some offspring is hard to do, or there are costs to all offspring when food is withheld from some offspring.

Empirical studies in socially monogamous species suggest that "mate-guarding" is common. However, alternative hy-

potheses for close following are seldom tested. Some studies indicate that mate-guarding may be energetically expensive, and in Eastern Bluebirds, experimental and comparative data across populations indicate that males only mate-guard after females have "gone missing" from territories the first time: in other words, mate-guarding by male Eastern Bluebirds is facultative, a male behavior that may be conditioned because of previous female behavior (Gowaty and Bridges 1991a). Whether females gain fitness benefits or fitness costs from male mate-guarding are other hypotheses seldom explored. Might fertile females gain benefits such as enhanced predator detection when her mate is closely following, or does mate-guarding increase the risk of predation? Studies of so-called mate-guarding would benefit from an observational or experimental approach in which investigators use strong inference to test simultaneously alternative hypotheses for close following of fertile females by males.

COPULATION

In most internally fertilizing vertebrate animals, males possess an intromittent organ, often called a penis, but only about 3 percent of bird species have them. Most vertebrates have internal fertilization, but birds are among the very few

Table 1. **Taxonomic distribution of intromittent organs (IO; + and − indicate presence or absence, respectively) and modes of fertilisation in the tetrapod vertebrates**

Class	Order	No. of species	IO	Fertilization
Amphibia[a]	Gymnophiona	150	+	internal
	Caudata	310	−	mainly internal
	Meantes	3	−	unknown
	Salienta	2510	−/+	mainly external
Reptilia[a]	Testudines	230	+	internal
	Rhynchocephalia	1	−	internal
	Squamata	5840	+	internal
	Crocodylia	21	+	internal
Aves[b]	Struthioniformes	10	+	internal
	Tinamiformes	46	+	internal
	Craciformes[c]	55	+/−	internal
	Anseriformes[d]	147	+	internal
	Psittaciformes[e]	358	+/−	internal
	Passeriformes[f]	5700	+/−	internal
	All other orders	2440	−	internal
Mammalia[g]	All orders	3500	+	internal

[a] Data from Goin et al. (1978).
[b] Number of species from Howard and Moore (1991); IO and mode of fertilisation from King (1981).
[c] Males do not have an IO in the Megapodiidae (Darryl Jones, personal communication).
[d] Males do have an IO in the Anhimidae (personal observation; *contra* Campbell and Lack 1985).
[e] Males have an IO in the two species *Coracopsis* (Wilkinson and Birkhead 1995); IOs are absent in all other Psittaciformes.
[f] Males have an IO in the two species of *Bubalornis* (Sushkin 1927); IOs are absent in all other Passeriformes.
[g] Data from Eckstein and Zuckerman (1956).

Figure 16.13. The reproduced table shows the taxonomic distribution of intromittent organs (Briskie and Montgomerie 1997). Most vertebrates are internally-fertilized, meaning that sperm must be transferred from inside males to inside females.
Table reproduced with permission of James Briskie from J. V. Briskie and R. Montgomerie 1997.

internally fertilizing vertebrates in which most species have no intromittent organ (fig. 16.13).

In the 97 percent of bird species lacking an intromittent organ, copulation is via a "cloacal kiss" (fig. 16.14), ornithological shorthand for the operational description of cloacal contact in which sperm is transferred from male to female tissues. During a cloacal kiss, females are as morphologically and physiologically active as males. To transfer sperm, the second compartment of their cloacae (fig. 16.15) must touch. The middle cloacal compartment is where the sex ducts of both females and males empty and where sperm must land to enter fallopian tubes of females. To achieve contact, both females and males must evert the second compartment of the cloacae through their third cloacal compartments and out their vents: what touches in the cloacal kisses are the tissues of cloacal compartments or, rarely, the elaborations of the proctodeum (a compartment of the cloacae) of males that can act as funnels, directing sperm onto the cloacal tissue of the females. During the cloacal kiss, female tissues take up the sperm via osmotic exchange. The process typically takes seconds—sometimes < one or two seconds—so that the actual copulation is hard to see. What field investigators often are able to see is a male briefly mounted on the back of a female, with the female's tail perhaps askew, but seldom are there signs similar to the dramatic, observable signs of ejaculation or orgasm that occur in many mammal species. One upshot of the way birds "do it" is that field observers can never quite be sure that the copulation posture actually results in the passage of sperm. Furthermore, the implications of cloacal kisses include that, because females are as active as males in copulation, forced copulation is difficult—if not impossible—in species without intromittent organs.

In bird species with intromittent organs, several hypotheses potentially explain adaptive function: (1) When copulation occurs on water, an intromittent organ may reduce the likelihood of water damage to sperm. (2) The intromittent organ may act as a balancing rod for males who might otherwise lose their footing while standing on females' backs. (3) Some flightless birds have intromittent organs, suggesting that evolutionary loss of intromittent phallus reduces flight costs. (4) The intromittent organ evolved to manipulate and control females during copulation (Gowaty 1997), a hypothesis that has recently gained attention through the research of ornithologist Patricia Brennan (box on page 513)

Figure 16.14. Northern Cardinals in a "cloacal kiss," copulating in an aviary during June 2010. *Photo courtesy of Geoff Hill.*

who studies the details of anatomy and physiology of forced copulation in ducks, all of which have intromittent organs.

But, the real puzzle in birds is not how intromittent organs evolved—because they are ancestral, but how they were lost in most of the avian lineages since the evolution of the first birds. Ornithologists James V. Briskie and Robert Montgom-

erie (1997) considered this problem. They described and evaluated several alternative hypotheses: (1) In birds, the cloacae for the urogenital and gastrointestinal systems are shared (fig. 16.15). Shared compartments of a cloaca have muscles for moving excretion products back and forth for storage and later ejection. Thus, birds may be particularly susceptible to sexually transmitted pathogens and parasites. If that is so, the loss of an intromittent organ could reduce transmission of disease-causing organisms. (2) When paternal investment is important to females and the success of her offspring, an intromittent organ may favor males in sperm competition, increasing "paternity confidence" and thus willingness of males to contribute care to offspring. (3) Females may have preferred males without intromittent organs. For some of these hypotheses, data are absent or weak; however, there was interesting correlative support for the female choice hypothesis. Under the assumptions that females only produce one egg a day (true of all birds except for those that lay an egg only every other day), and that a unique copulation is required to fertilize each egg a female produces (true in many birds, because sperm storage organs are rudimentary), if a male force-copulated a female, she could abandon an egg without catastrophic loss of her current reproductive success. The female choice hypothesis depended on the mechanisms by which females can resist male coercion or forced copulation. Conveniently, one of the constraints of avian life history is the one-egg-a-day rule. Add to that the assumption that, in species without intromittent organs, the physical contact of a cloacal kiss must involve cooperation between females and males for physical sperm transfer (Fitch and Shugart 1984). Add the assumption that birds fly, so that in many avian species females can often escape male attempts at coercive copulation just by flying away, and the female choice hypothesis takes on new power. With these assumptions one can predict that female birds—compared with less

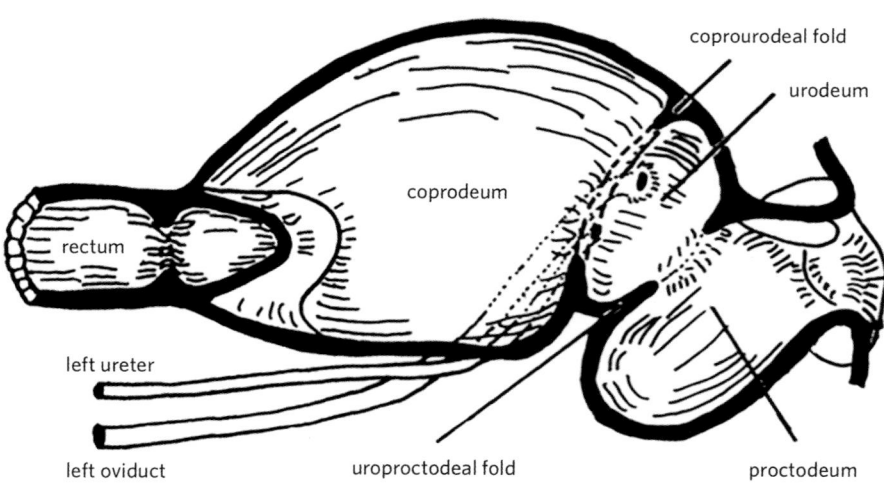

Figure 16.15. A stylized drawing of the cloacae of birds. The drawing captures the generalities of the organization of the avian cloaca. Both sexes have three compartments to their cloacae. *Figure reproduced with permission from the artist, S. P. Hubbell, from Gowaty and Buschhaus 1998.*

PATTY BRENNAN

Patty Brennan.

I had never thought about bird penises until I watched a Great Tinamou mating in the wild, and noticed the penis being drawn back into the male's cloaca afterward. I was completing my PhD at Cornell University at the time, and was intrigued by how little was known about avian pe-

nises, especially, why most birds didn't even have one. As a postdoctoral fellow, I studied genital evolution in ducks and discovered that penis length in males was associated with levels of forced copulations, and that female ducks have evolved convoluted vaginas that keep the male penis from fully everting during unwanted copulations. I also discovered that duck penises evert explosively, fully achieving erection and fertilization in less than a third of a second, allowing males to inseminate females even when they resist copulation. This research has become an iconic example of the evolutionary arms race that can result from sexual conflict over forced copulations that prevent females from exercising their choice of mate. My research also highlights the importance of examining genital coevolution, rather than the more common approach of explaining patterns of variation of male genitalia without considering the females. I continue to be puzzled by the loss of the penis in birds, and I have plans for further research in this area, even as I have expanded my studies of genital coevolution to other vertebrates.

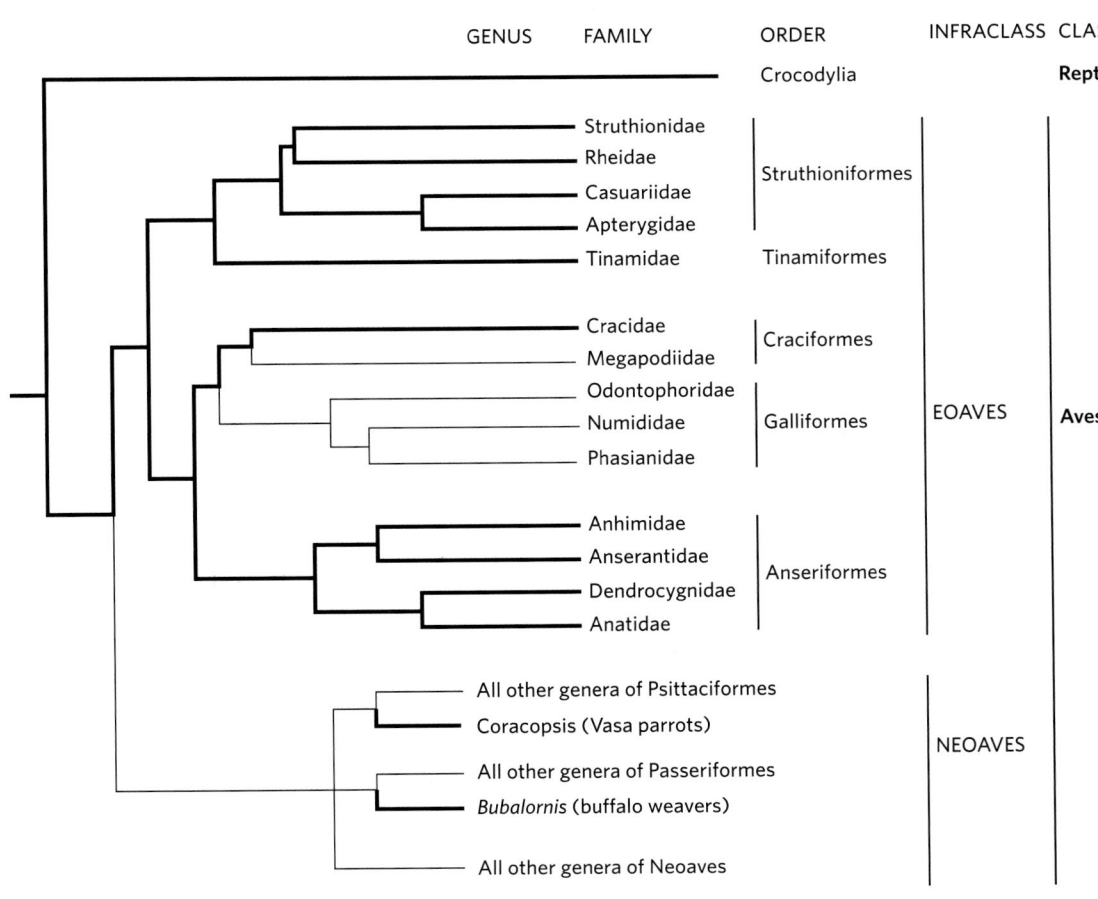

Figure 16.16. A phylogenetic tree that shows the occurrence of intromittent organs (thick lines) in the Crocodylia (class Reptilia) and in the families (infraclass Eoaves) and genera (infraclass Neoaves) of birds (class Aves). From Briskie and Montgomerie 1997. Courtesy of Jim Briskie.

Figure 16.17. Male-group attack on a female Mallard (*Anas platyrhyn-chos*). April 14, 2011, 17:43 in the UK. *Photo courtesy of Africa Gomez via CC-by license, http://therattlingcrow.blogspot.com/2015/03/forced -copulation.*

mobile organisms like mammals or weakly flying birds, e.g., turkeys—are often extremely difficult to force copulate. Thus, it is no surprise that in many avian species lacking intromittent organs, forced copulation is extremely unlikely, although male Hihi (*Notiomystis cincta*), a passerine species of New Zealand, have managed forced sex (Ewen et al. 2004) in the missionary position of face-to-face copulation.

Forced copulation is most well-studied in mallards—species with intromittent organs (box on page 499) that are notorious for what appears to be aggressive copulation—even on the wintering grounds, when males and females form pair bonds, often months before females are fertile (fig. 16.17). Given that females are unlikely to successfully produce eggs, incubate them, and have them hatch in winter, the functions of winter-time "forced copulations" often by a collaborating group of males, are better characterized as aggressive behavior, which raises the question of why these males are aggressive to females. The CODE hypothesis (box on page 515) predicts adaptive advantages for males that coerce females in winter, yielding fitness benefits for individual males during the breeding season. But, the adaptive arguments favoring collaborative male aggression against females in winter do not adequately explain why breeding season aggression and coercion are adaptive for males. In fact it is difficult to imagine what advantage there would be for a male to hurt his mate or prospective mate.

Even though forced copulation occurs in ducks and Hihi, in most passerines what looks like forced copulation might be better interpreted as intense aggressive interactions between females and males—the reason being that for sperm

transfer to occur in passerines requires that females be as active as males (Fitch and Shugart 1984). "Forced copulation" may not be so much copulation as male aggression against females, and it raises again the question of the adaptive advantages to males that hurt females (see the box on page 510 and the box on page 515). Male access to healthy females must be crucially important to successful reproduction of males. Thus, there is a paradox: How can hurting a female be a fitness advantage for males? Some theorists think aggression is a type of "punishment" that conditions female behavior for the advantage of the aggressor, which does not really explain why male aggression to females is more adaptive than say, male kindness and help to females, which might also condition female behavior without the risk of injury to them.

An alternative to the male-aggression-as-punishment hypothesis is "the killing time hypothesis," which is deduced from the switch point theorem (box on page 517). The switch point theorem (SPT) is a theory of mating based on a mathematically derived, algebraically expressed, "analytical" model of adaptive reproductive decision-making against a backdrop of stochastic demography. The SPT predicts that individuals (of both sexes) are flexible in their reproductive decisions, sometimes mating on encounter (as if indiscriminate) and sometimes waiting to mate for a better mate (as if choosy). The idea says that when individuals have a lower survival probability, they often will be more likely to enhance their fitness by indiscriminate mating than by holding out for a better mate. Thus, the killing time prediction is that males hurt potential mates in order to, at least momentarily, decrease female survival likelihood—to increase the likelihood that females will share gametes with them. This prediction, unpleasant as it is, is a quantitative explanation for why some males hurt their mates or potential mates. It is further interesting because females might kill a potential mate's time to manipulate or coerce him. In other words, the SPT says that mate coercion can occur by killing time or making time. "Killing time" might be mean, but "making time" might be kind. In both cases, whether males are killing females' time or females killing males' time, vulnerable potential mates are manipulated to trade time for fitness. The SPT's killing time hypothesis says that rejected individuals coerce potential mates into reproducing with them by introducing unavoidable opportunity costs into the lives of those they desire, but who do not desire them. Killing time is an extraordinary means of exploiting the sensory biases of potential mates.

Mating theory (Gowaty and Hubbell 2009) based on probabilistic demography is still relatively new, so some of the concepts will be unfamiliar, and none of the ideas have yet been tested in wild-living birds. However, another really interesting prediction—interesting because it is counterintuitive,

THE CODE HYPOTHESIS: A HYPOTHESIS LINKING FORCED COPULATION AND MONOGAMY

Aggressive or forced copulation is rare in most bird groups, but male aggression against females happens in many species. That forced copulation is seldom observed in birds, particularly passerines, is theoretically and empirically because of control of behavioral, morphological, and physiological mechanisms by which females can resist forced insemination and subsequent fertilization (Fitch and Shugart 1984). Such female traits include digestive epithelium lining the middle compartments of the cloaca, where sperm must get to in order to enter the female fallopian tubes, and powerful cloacal musculature that females use to eject fecal and urinary wastes as well as sperm. The resistance traits of females argue against the most common explanation for why males force-copulate females, namely that the forced copulation increases the chances that males will be successful at inseminating females. However, if female resistance mechanisms operate, then the "immediate fertilization enhancement" hypothesis for why males attempt forced copulation in birds is weakened. A German ornithologist (Heinroth 1911) and an American feminist (Brownmiller 1975) independently argued for an alternative hypothesis of advantage for males of forced or aggressive copulation. They argued that aggressive and forced copulation *creates a dangerous environment* for females, which induces females (ducks or humans) to trade sexual and social access for male protection against aggressive or rapacious males. The Heinroth/Brownmiller idea is now called the CODE hypothesis (Gowaty and Buschhaus 1998), which predicts that *males' creation of a dangerous environment* for females—the CODE—fosters male mating advantage via induction of monogamy in females. Theoretically these trades of protection for copulation promote social monogamy even in species with little or no paternal care, in which females are able to raise offspring without the help of males. The CODE hypothesis further predicts that aggressive and rapacious males could accrue selective advantages through "perverse reciprocal altruism" (a type of male col-

laboration to hurt females), through kin-selected benefits (when males harassed females that were later protected by his brothers or other male kin), or for direct benefits in which males gain a monogamously mated female partner in exchange for "protective services." The CODE hypothesis thus explains otherwise difficult-to-explain group male attacks on female Mallards in winter, when females are not fertile but when pair bonds form. The CODE hypothesis explaining social monogamy makes many predictions, all as a function of variation in females' abilities to resist coercion. For example, CODE predicts that aggressive copulation in birds is more common (1) in species with rather than without intromittent organs; (2) when male morphological structures can restrain females; (3) when females' are unable to fly or are confined; (4) when males manipulate females' reproductive decisions by controlling resources essential to females' reproduction and then "brokering" those resources to females, in effect making females trade copulations for access to food or nesting sites that males control; (5) when males are larger or stronger than females; (6) in colonial, rather than solitary species; and (7) when local kin groups consist primarily of male relatives, which happens when sons are more philopatric than daughters, which increases the likelihood that brothers but not sisters are surrounded by helpful kin. But the most interesting of the predictions from the idea that females can resist CODE is that extra-pair paternity is higher for females with the least vulnerability—not the most—to aggressive copulation. The last prediction emphasizes the fact that not all females in a population are equally vulnerable to male manipulation, male sexual coercion, or even to CODE. That not all females are equally vulnerable will suggest new study directions to those of you who realize that undiscovered and still poorly characterized variation among females within populations provides an opportunity for selection to operate among females as a function of their abilities to resist males' coercive attempts.

given the usual characterization of males as indiscriminate—is that the fanciest, most attractive males or those males defending the most attractive territorial resources will have the highest encounter rates with potentially mating females and thus will sometimes *reject* some females as mates and wait for a better mate (making such fancy males or those

defending superior territorial resources fit an operational definition of "choosy"). Thus, it is possible that young ornithologists studying this chapter may be the first investigators to observe and describe male mate choice in species with fancy, highly ornamented males or males defending superior resources.

THE FEMALE CONSTRAINT HYPOTHESIS

The female constraint hypothesis is a modification of the polygyny threshold hypothesis (PTH). Rather than representing environmental quality as in the PTH, here the x-axis is a combined function of females' intrinsic competence (say, perhaps, their eye-beak coordination affecting their foraging skill) along with the quality of their breeding environments (availability of food, refugia from predators, resources for nesting sites). As in the PTH, the y-axis represents a female's fitness (her survival and reproductive success relative to

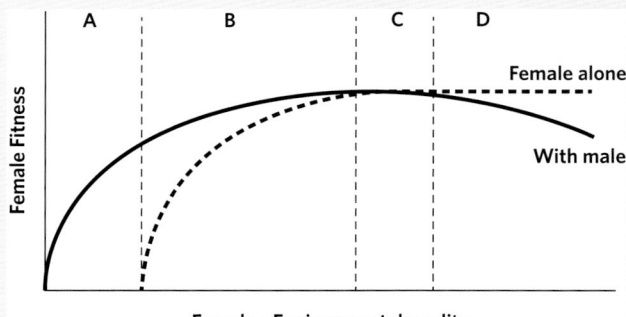

Female constraint hypothesis. *From Gowaty 1996a with permission of the author.*

other females in the population). The two lines on the graph represent the variation in female fitness when they are alone and when they are with males. The four sections—A, B, C, and D of the graph—indicate the importance to female fitness of male help. In A, females are dependent on male help. In B, females' dependence on males is determined by the combined variation in females and their environments: when female-environmental quality increases, female dependence on males decreases. In C, males neither help nor hinder female fitness; but in D, male "help" decreases female fitness. Some predictions of these ideas include that C and D females are more likely than B females to seek or accept extra-pair copulations and extra-pair paternity of their offspring. The female constraint hypothesis (Gowaty 1996a) makes additional predictions about the relative quality of extra-pair offspring and within-pair offspring as a function of mothers' requirement for male help. What is most interesting about this idea is that it emphasizes the importance of within-population variation among females and among males in organizing opportunities and requirements for both cooperation and conflict between females and males. Can you deduce other predictions from the simple assumptions of this graph?

COLLABORATION AND COOPERATION IN PAIRS AND THE ROLE OF MALES IN PARENTAL CARE

Parental care is something parents do to increase the survival or health of their offspring; other definitions include behavior of presumed parents that is *likely* to benefit the fitness of the parents (Clutton-Brock 1991). Parental care occurs in most bird species, except for the fewer than 90 species of interspecific brood parasites, species in which breeding females lay their eggs in the nests of heterospecifics, thus parasitizing the parenting behavior of other species.

When ornithologists study parental care, they describe male provisioning of females during egg-laying and incubation, sex differences and similarities in post-pairing territorial defense, sex differences in nest-building, incubation of eggs, brooding of chicks, feeding and defense of nestlings and tending to fledglings, and guiding fledglings to refugia and good feeding sites. Parent birds have a whole lot to do.

Biparental care is a hallmark of the most common avian mating system, social monogamy (Verner and Willson 1969).

Yet, uniparental care by either sex of parent is typical of some species. Uniparental female care occurs in species in which males display on leks, including cotingas, manakins, a few hummingbirds, grouse, bowerbirds, lyrebirds, and birds-of-paradise. Uniparental male care occurs in several species of shorebirds, and in kiwis, tinamous, cassowaries, and brush-turkey (the last four also being in the rare list of bird species with intromittent organs).

Authors even before David Lack, who founded the study of socio-ecology of birds, assumed that male care was essential to reproductive success of both sexes. Many ornithologists also assumed that male parental care is derived and female parental care the basic bird pattern (Silver et al. 1985, McKitrick 1992). But it turns out that phylogenetic analyses (Vanrhijn 1984, Vanrhijn 1991, Wesolowski 1994, 2004) are more consistent with male parental care being ancestral in birds, an interesting idea that ought to challenge the way most of us ask questions about the evolution and maintenance of parental care in birds. During the last 50 years, investigators have focused on the socio-ecological dynamics of paired males and females and often assumed that male

THE SPT AND THE KILLING TIME HYPOTHESIS

The switch point theorem (SPT) (Gowaty and Hubbell 2009) is an analytical, mathematical solution to the problem of the number of potential mates in a population that are acceptable or not to a focal individual. The answer depends on life history and demography: the likelihood of the focal individual's continued survival and the likelihood that the focal individual will encounter potential mates, the number of potential mates in the population, and the distribution of fitness of all potential mates. At any given time, an individual's probability of survival and probability of encountering potential mates indicate the time an individual has left for mating and reproduction, while the population's distribution of fitness indicates the likely fitness consequence of mating or not for an individual. The SPT's solution to the expected number of mates that are acceptable to a focal individual is a function that says individuals trade off time with fitness to make an adaptive decision about who is acceptable as a mate. The switch point is the point along an axis of all ranked potential mates from 1 (best) to worst (n, where n is the total number of potential mates in the population). The SPT proved that when a focal individual has a low probability of encountering potential

mates, its fitness is enhanced if it finds acceptable more of the potential mates than when its encounter probability is higher. More important is that when an individual's survival probability is lower, it will accept as a mate many more potential mates than when its survival probability is higher. These two predictions are intuitive, and experimental studies are often consistent with them (Ah-King and Gowaty 2016).

What the SPT says about mating coercion of one individual by another is one of its most arresting predictions: sexual coercion can efficiently occur when a coercive individual "kills the time" of a potential mate that resists mating. Lowering the resistive individual's encounter probability with other potential mates has the effect of "killing their time," making it more likely that they will accept a mating with a previously rejected suitor. Coercive behavior that lowers a resistive individual's survival probability also will efficiently "kill their time" (Gowaty and Hubbell 2010), making her or him much more likely to accept a mating with one they previously rejected. It's not a pleasant idea, but it does explain the conundrum of why an individual would hurt its mate or a prospective mate.

parental care evolved because females were unable to do all the work of caring for eggs and nestlings and fledglings alone (think about the PTH and its assumptions about male care and female's dependence on males). Clearly, however, a newer generation of ornithological scientists might ask questions about how females evolved to be the primary caregivers in some birds, and whether they wrestled for the role of major player in parental care or were left "holding the bag" when males deserted, if males did desert (see below).

What determines the dynamics of collaborative or unitary parental care is still not understood. Usually we ask questions like this: Is male help essential to their female partner's fitness? The answer we currently have is "sometimes, yes; sometimes, no" (fig 16.18), because there is within species and between species variation in the fitness payout to females of male help. Given that uniparental care by males is ancestral in birds, we might turn the usual question on its head and ask whether biparental care evolved because female help was/is essential for the male partner's fitness. A related question is about the social dynamics of desertion of a parent when biparental care is the norm. John Maynard Smith (1977) made a prospective analysis of parental desertion, taking into

account opportunity costs to the potentially deserting partner. His short, carefully reasoned paper is one I recommend students read and discuss together to get some good ideas about many of the remaining empirical questions about parenting in birds: When is it advantageous and when not for one parent to leave the other parent holding the bag of caring for the kids? More good field studies of this question would be a welcome addition to ornithology. What's curious in relation to the idea of "who is left holding the bag" is that, even when females engage in extra-pair mating, social mates continue to help raise offspring, even those not theirs. Intuitively, many humans think there must be a disadvantage of caring for others' offspring. It's as if we humans have failed to imagine constraints and opportunities from the perspective of birds, because we still lack good empirical support for why males care for offspring not theirs.

In some birds, biparental care is a "true partnership" of extraordinary collaboration (Black 2001). The "true partnership" species (box on page 519) include Sphenisciformes (penguins), Pelecaniformes (pelicans), and Procellariiformes (petrels and storm petrels). In passerines, the consensus is that females do most of the parenting (Verner and Willson 1969).

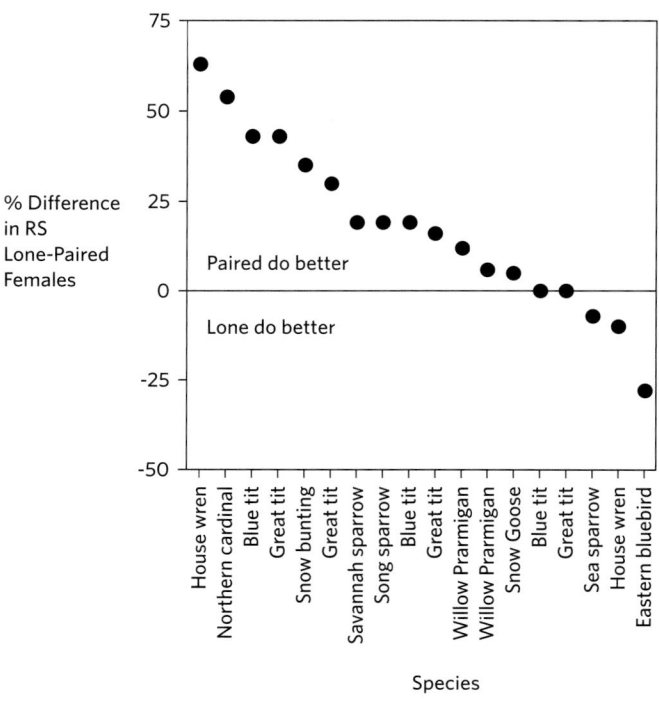

Figure 16.18. Difference scores in reproductive success (number of fledglings) of females in experimental studies of pair-male removal. *Data were from Bart and Tornes 1989; illustration from Gowaty 1996b.* The y-axis is the percent differences in RS for females who remained paired minus those females whose male partner was experimentally removed. *Courtesy of the author.*

Notably, however, in a great many species, particularly in temperate zones, males will feed females: males in some species feed females during nest-building, during the egg-laying period, and during incubation. What's most interesting is not the species-specificity of these patterns, but the within-species, between individual variations. Understanding why some males feed their mates and others do not is a question that is ideal for study from the perspective of socio-ecology.

Male parental care is absent in most of nature (Clutton-Brock 1991) but common in humans and in many bird species. Perhaps because of the similarities between human and avian monogamy, and the presence of male parental care in both, ornithologists, including the current author, have worked to understand the benefit of male parental care to females and their young from a perspective that asked about the adaptive significance of male parental care both to males and to their social mates. Birds are model organisms for the study of the most common form of human pairing (as distinct from "household formation strategies," which indicate that sometimes households have little to nothing to do with who mates with whom), so that one often finds anthropolo-

gists using "bird-centric" theories like the polygyny threshold hypothesis to study human mating patterns.

A note about the conservation of birds practicing "strict monogamy": under assumptions that mating is random, strict monogamy is hard to come by, and thus the lessons to be learned from those species that do live in strict monogamy is that the collaboration of the female and male is essential. It is easy to see that the powerful environmental constraints acting on Magellanic penguins provides the powerful selection that results in powerful collaborations of mates. Those same challenging environments are more and more under threat from climate warming, from habitat destruction, and even from tourism, as the work of P. Dee Boersma (box on page 504) emphasizes. Dee Boersma is among the greatest living conservationists: her highly successful ways of communicating about the lives of the penguins she studies are a model for all of us. If you are interested in conservation practice and how to be successful in conserving the living birds of the world, you will do well to get to know the work of Dr. Boersma (box on page 520).

NESTS, NEST-BUILDING, INCUBATION, FEEDING NESTLINGS, AND TENDING FLEDGLINGS

Nests play a crucial, central role in the lives of most bird species, none better described and researched than by the remarkable team of Nick and Elsie Collias (see the box on page 505). Birds' nests are reasonably characterized as an "extended phenotype" (Dawkins 1999) of birds, and the habit of nest-building is a key characteristic of many species (Collias and Collias 1964a, Collias 1964, 1997, Collias and Collias 1964b, Collias and Collias 2014). Bird nests, almost without exception, readily identify the bird species that built the nests. In many species of bird, the act of constructing nests is instinctual, an inborn trait, requiring little or no practice to produce a species-typical nest. Avian parents-to-be build nests, or otherwise prepare places in which to lay eggs and protect and contain nestlings (in altricial, nidicolous species). Nest design probably enhances the survival of adults as they incubate eggs. The types of nests are associated with modes of incubation. Megapodes—the mound-builders of Australasia—bury their eggs in large mounds of decaying vegetation, where the eggs are warmed passively. Parental care is often limited to male attendance at the mound, where he may adjust litter depth to modify mound temperatures. However, in the vast majority of birds, parents (either the mother or the father, or both parents) incubate eggs that they deposit in nests that they build in open cups, or in holes that

SOCIAL MONOGAMY IN MAGELLANIC PENGUINS (*SPHENISCUS MAGELLANICUS*)

By P. Dee Boersma

For over 30 years, I have studied the lives of Magellanic Penguins at Punta Tombo, Argentina. My helpers and I observed and recorded pairing and reproductive success of 6,090 pairs. Most socially monogamous pairs (55 percent) stay together for one season, but remarkably, a third of all pairs stay together for longer: 20 percent stay together for two seasons, 10 percent for three seasons. Pairs split up when they fail to hatch eggs. Some pairs, however, stayed together until their partner died. One pair was together for 17 years and three pairs for 15 years. But fewer than 1 percent of pairs were faithful to a single mate for 10 or more years, while the maximum lifespan for a Magellanic Penguin is about 30 years. Most males begin to breed when they are six or seven years old, while females begin at four years old. The remarkable durability of monogamous penguins is correlated with their success at raising chicks under some of the most arduous environmental conditions faced by breeding birds. Each year that a penguin pair breeds, they must care for eggs and nestlings for almost seven months, and because the foraging areas for Magellanics are miles offshore, one parent must continue incubating or brooding chicks for two weeks or more while the other parent is away foraging. On return, the foraging parent relieves the hungry parent from incubation and regurgitates food for their chicks. Successful reproduction is no easy task for Megellanics, requiring extraordinary collaboration and "trust" between partners, which is what penguin monogamy is all about. Males and females, unlike many passerines and shorebirds, really need each other and usually split their parenting duties evenly. When females are laying eggs, males defend the nest site. For Magellanic Penguins, both mates are persistent, and both sexes resist intrusions by interlopers and stick with their partner until their nest fails.

Extra-pair copulations and extra-pair offspring are extremely rare, even though the adult sex ratio is highly male-biased, giving females lots of opportunities to extra-pair copulate, which also means that female Magellanic Penguins always find a mate, but many males, year after year, go unmated. Penguin copulation is a balancing act. Forced insemination is difficult, because females can eject sperm by standing up and using the jet power stream of guano to also scour out sperm. As in passerines, females are active participants in successful insemination because the sperm cannot get far unless a female pulsates her cloaca, moving sperm into her reproductive tract. After a female accepts a copulation, she then waits quietly, not moving for a minute or longer as the sperm migrate farther up the reproductive track. Even though males cannot force-inseminate a female, they can make trouble, smashing eggs, crushing chicks, or otherwise bullying females. However, male troublemakers do not increase their chances of getting a female in the following year, and females do not copulate with these bullies, leaving curious naturalists with the question: Why do males smash eggs, crush chicks, and harass nesting females? Other unpaired male Magellanic Penguins take a different tack, sitting with a female after her pair-mate has gone to forage. Even friendly guys fail to get breeding bonuses in later seasons. Our data showed that 11 of 45 intruder males got a mate the following season, whereas 10 of 58 non-intruder males remained unmated in the following season. In 30 years my team has never seen a female copulate with an intruder. Our genetic parentage data indicate that females are faithful to their pair mate.

When there are fights in Magellanic Penguins, fights are intrasexual: males take on males and females fight females. When two females are fighting, the male does not help either female. When we experimentally removed winning females and allowed the loser female to return to the nest, the male readily accepted her. The winner female was heavier and in better condition than the loser (N = 10). Selection seems to have favored males that let the females decide the victor, because heavier females more often lay heavier eggs, which hatch more often than lighter eggs. Furthermore, females take the first long incubation stint (total incubation time is 40–42 days) so heavier females and their eggs are more likely to survive the long fast (two or more weeks) until her mate returns from foraging and relieves her so that she can forage. To recover body condition at Punta Tombo, Argentina, penguins forage hundreds of kilometers from their colony. When the male returns from his first long foraging trip, the female leaves and the male then incubates. Around hatching time, the partners alternate incubation every other day, probably an evolved adaptation allowing immediate feeding of newly hatched chicks. Mutual display calls of parents seem to induce chicks to hatch. Even with the amazing parenting choreography of Magellanic Penguins, 40 percent of chicks starve. Monogamy is not always successful for these penguins, but it is surely more successful than more complex breeding systems would be: ecological constraints guarantee social and genetic monogamy in Magellanic Penguins.

P. DEE BOERSMA
By Dee Boersma

P. Dee Boersma.

As a kid, I collected butterflies and moths. I enjoyed the chase, marveled at colors, textures, and the bizarre changes from plant-eating caterpillar to flying nectar-sucking adult. At ten, I wanted to be an entomologist, but neither my elementary school or high school taught insects. When I did take entomology in college, the course was largely about how to kill pests and not about their marvelous natural history. I quickly learned I was not interested in how to kill insects. Conversely, my ornithology course was about natural history and how to identify birds by sight and sound. I became a biology and conservation major because that was what connected me to the natural world.

I also learned that birds could tell us about human impacts on the environment. In my first field studies, I discovered that Galapagos Penguins were ocean sentinels of climate change, telling us about consequences of the extreme environmental variation associated with El Niño. Next I studied Fork-Tailed Storm Petrels and learned that their eggs and young were able to survive for days when, owing to storms or moonlight nights, parents did not return to their nest. Storm petrels feed at the ocean's surface and proved to be good reflectors of petroleum contamination. Penguins remain my focal species. My questions are, how do penguins do what they do, and what do they need to thrive? By studying penguins I've learned to appreciate the resiliency of nature, the variation among individuals, and how the environment drives behavioral and evolutionary change. Long-term studies are crucial to understanding how ecological systems work and what species need to survive and thrive. Humans need marine protected areas and terrestrial protected areas as much as do penguins and other creatures. We will better manage our planet if we understand and foster what other species need to survive and thrive. The challenge is to manage our own consumption and our numbers while protecting the natural world that penguins, other creatures, and we depend upon.

one or both parents dig, or in nesting cavities that prospective parents excavate or usurp from other cavity-nesters.

That avian eggs develop outside of mothers' bodies in nests is perhaps the key constraint, almost unarguably the most costly of any reproductive behavior in birds. Compared with pregnant mammal mothers that can simultaneously gestate and forage, incubating birds are constrained during a critical period of offspring development: not only does it take up time, making other activities impossible, it uses up individual energy stores, increases the basal metabolic rate of incubators, reduces parental immunity, increases parental risk of disease, sometimes lowers their instantaneous survival, and is dangerous, increasing the likelihood of predation (as in, being "a sitting duck"). It is also a crucial period during which the behavior of incubating adults may exert profound influences on the future phenotypes of their offspring, an emerging area of increasing interest to ornithologists (Rubenstein et al. 2015).

There is little doubt that, in mammals, maternal postzygotic influences via gestation and lactation on the phenotypes of offspring are greater than paternal influences. However, in birds, after an egg is laid, further control or influence over offspring phenotypes can be by fathers only, by mothers only, or both parents may collaborate, alternating incubation bouts, perhaps contending with each other over the control of offspring characteristics (a so-far unexplored hypothesis). In some reptiles, for instance, variation in incubation temperature determines the sex of offspring. While temperature-dependent sex ratio variation has failed to be discovered in birds, there are many other ways that variation in incubation temperature or even the tempo of parent's incubation behavior may affect the phenotypes of offspring. The mechanisms

NICK AND ELSIE COLLIAS
By Malcom Gordon UCLA

Nick and Elsie Collias. *Photograph courtesy of their daughter, Karen Whilden.*

Nicholas E. Collias (1914–2010) and Elsie C. Collias (1920–2006) were two of the world's leading students of bird behavior for almost the entire 57 years of their close and happy marriage. They were equal partners both personally and professionally. They were creative and original laboratory (aviary) experimentalists and field observers who asked and answered many important questions relating to social hierarchies in birds (especially pecking orders), vocal communications among birds (also some mammals), and external constructions (especially nests) of birds. Deep understanding both of the environmental conditions in which their subjects lived and of evolutionary contexts framed their questions. Clear thinking, careful and thorough data-gathering and analyses, and excellent writing characterized their work. They labored together on almost all projects, whether or not both of their names appeared on the research papers that resulted. Their published bibliographies include over 100 research papers, three substantial books, five research-based scientific documentary films, and an assortment of smaller projects.

They studied many different groups of birds, with greatest emphasis on the passerine weaverbirds (studied in aviaries and in the field in several parts of Africa) and the red junglefowl (which they studied both in the field in Asia and in feral populations in North America). They also carried out a variety of field studies on different groups in Europe, central and South America, Australia, and some islands in the Pacific.

They were both actively involved in multiple professional scientific societies, most notably the American Ornithologists' Union and the Animal Behavior Society (ABS). Nick was one of the founders of ABS. Both separately and together, they received an array of professional awards recognizing the quality and significance of their research.

of such "parental effects" are likely through induced epigenetic changes (i.e., changes in gene regulation induced by environmental inputs) producing developmental variation in offspring. What the phenotypic effects on offspring might be is today predicted from correlational studies, and is in most cases speculative (meaning a good idea that needs testing), but brimming with opportunities for theoretical and empirical attention from passionate naturalists (DuRant et al. 2013). For example, if one assumes that evolutionary success for any parent is "having offspring that have offspring," it is easy to predict that among the effects of variation in incubation temperatures and rhythms of incubation would be enhanced offspring survival (lifespan) probability (a complex parameter that depends on immunity genes and environmental and developmental conditions). Temperature variation during incubation of an egg may result in delays or accelerations in time to first breeding, or to differences in the likelihood of dispersal of offspring by sex, or to variation in energy metab-

olism or to posthatching growth. In other words, variation in incubation temperatures may exert effects on endocrine functions that could be organizational (setting cascades of developmental states into action) or regulatory (associated with adult cycles of oogenesis, spermatogenesis, molt, migration, etc.) when basal temperatures vary. Savvy students might also note the opportunity for strong inference testing of alternative hypotheses as well as the opportunity to cast efficient hypothetico-deductive hypotheses of within-species variation in incubation behavior based on the first principles of temperature dependence and metabolic theory and life histories.

Whenever one parent can raise the young without help from a collaborating partner, investigators assume that it is because of habitat quality, which releases one parent from the necessity of feeding nestlings, so that the other parent can or will desert. However, sometimes both pair members stay in the vicinity of a nest, but one parent may do all or most of the feeding, as in Indigo Buntings (Ritchison and

Little 2014). What is interesting is how few descriptive studies there are of the relative apportionment of caring duties to females versus males. The vast majority of modern studies focus on variation in male feeding of offspring as a function of observed extra-pair offspring in nests. The evidence that pair males "withdraw help" from feeding young is relatively often tested, but strong data in support of the idea are still sparse. It is as if males whose mates produced extra-pair young (EPO) stay around anyway to help feed offspring for more complex reasons than typically hypothesized. Perhaps males gain a reproductive success benefit guaranteeing that no nestlings die in a nest, including the within-pair offspring (WPO). Perhaps males that stay around a nest containing EPO gain other benefits more opaque to observers, such as the pair male's opportunity to mate with other females (Gowaty 1996b, 1996c). Perhaps you can think of other hypotheses as well.

Care of fledglings is understudied, requiring focus on moving targets; thus it is no surprise that there are relatively few studies of post-fledging care of offspring, so few that it is difficult to come to any strong conclusions about patterns within birds, or even just in passerines. Students wishing to gain insights into how to do an excellent study of post-fledging care should read Wheelwright and colleagues (2003) on savannah sparrows (*Passerculus sandwichensis*).

The curious perseveration of observers to the possible effects of males caring for extra-pair young dominates the current ornithological literature on parental care in birds. What is left partially studied is just exactly who does what when, and why they do it. Years ago, I was brought up short when I discovered that male Eastern Bluebirds living in different places in the southeastern US adjusted their behavior to nestlings depending on variation in mothers' abilities to capture insects. In one place insects were abundant; in the other the ground arthropods that bluebirds prefer were scarce because introduced red imported fire ants (*Solenopsis invicta*) ate the same ground arthropods bluebirds usually ate. In the place where food was limited, males adjusted their care of nestlings depending on the killing and capturing skill of their female mate. In that food-limited environment, males who were paired with less lucky or less skilled females increased their care of nestlings, while those males mated to more competent females did not increase their care. In the place without fire ants that still had rich arthropod resources, males did not adjust their care of nestlings at all. In other words, when males needed to help they did; when they did not need to help, they did not. Many questions then followed, most of which remain unanswered. These observations suggested that the behavior of males to females and females to males and both parents to offspring is likely more dynamic and more interesting than theorists have imagined. What is now needed are more quan-

titative and descriptive studies of how males and females interact with nestlings and fledglings under different natural or experimental conditions. There is more to paternal care than the cost of caring for an EPO, just as there seems to be more to study about the dynamics of collaboration over a shared benefit, namely the production of viable progeny.

KEY POINTS

- Socio-ecology is the study of how social and ecological environments shape—via selection—the reproductive behavior of individuals.
- Socio-ecologists use mostly the evolutionary logic of natural and sexual selection to organize deductive hypotheses with novel testable predictions about adaptive reproductive behavior of individuals in populations.
- The deductive character of hypotheses of natural and sexual selection provide a useful understanding of evolutionary principles that students can apply in many disciplines.
- Social monogamy and biparental care of young are emblematic of birds—and of humans—but extremely uncommon in other taxa, making birds "model organisms" for the study of behavior of great interest to humans. During the last half century, much research attention has focused on the relative and sometimes conflicted role of females and males in parental care. However, opportunities to further understand the collaborative and competitive roles of parents offer many untapped opportunities for students interested in the selection pressures associated with the power of parents, particularly, mothers, over the lives of offspring.
- Outstanding new and fascinating questions include those about the effect of between-individual, within-population variation in incubation and its epigenetic effects on the phenotypes of hatchlings, nestlings, and adult offspring.

KEY MANAGEMENT AND CONSERVATION IMPLICATIONS

- Individuals need diverse social experiences during development to "learn who they are" relative to other individuals, which allows them to make flexible and adaptive reproductive decisions. This means that management of captives might be aided if young animals had diverse social exposures to conspecifics.
- It is likely that all individuals need opportunities to make assessments of potential mates, thus managers of captive birds should allow both sexes the opportunity to evaluate alternative potential mates before assigning birds to reproductive partners.

- Global climate change, particularly global warming, will exert selection pressures on many aspects of avian reproductive behavior.
- Global warming is likely to exert profound epigenetic effects, changing the physiological, morphological, and behavioral phenotypes of birds and thus also their characteristic survival and reproductive success.

DISCUSSION QUESTIONS

1. Discuss the assumptions of the Orians (1969) polygyny threshold hypothesis and its descendants. Why is it important each time the PTH is tested to evaluate the assumptions before the predictions?

2. Very few bird species have intromittent organs, yet included among the bird species that do are some of the rare species with male-only parental care. Use what you may know or find out about these bird species to cast hypotheses for the adaptive significance of male-only care.

3. Among the most common characteristics of birds reported in virtually all natural histories, field guides to species, and even field guides to families of birds is the relative role of females and males in parental care. A reasonable homework assignment would be for students to collaborate making a table of the role of males and females in the jobs that make up reproductive behavior (Verner and Willson 1969). A new review, updating and expanding Verner and Willson (1969) will yield surprises. Different groups of students could review the literature on variation between species in different orders and families of birds to produce a group project.

4. Learn the format of scientific papers by reading and understanding papers in the ornithological literature. Pick out five or ten scientific papers about avian reproduction and read the papers to understand the methods thoroughly. Perhaps with collaborators in a group project, evaluate the methods in the paper relative to the questions the investigators set out to answer.

5. Some of you are hunters. There are no doubt serious effects of hunting on the later reproduction of some populations of game birds. Students might think about the kill rates/takes of hunters relative to the health and breeding decisions of remaining birds in a population. What might be the effects of "bag limits" for the reproductive decisions of survivors, and the possible effects on the health of survivors' offspring? In other words, short of wiping out populations, what are the possible downstream effects of hunting on, say, the reproductive success or reproductive decisions of survivors?

6. Parental effects via epigenetics is emerging as a hot topic. Consider some of the ways that parents can produce adaptive, fitness enhancing changes in offspring phenotypes via epigenetics (parental effects). There are hints in aid of discussion in the compensation hypothesis (Gowaty 2008).

References

Adler, M. (2010). Sexual conflict in waterfowl: Why do females resist extrapair copulations? Behavioral Ecology 21:182–192.

Ah-King, M., and P. A. Gowaty (2017). A conceptual review of mate choice: Stochastic demography, within-sex phenotypic plasticity, and individual flexibility. Ecology and Evolution 6:4607–4642.

Ar, A., and Y. Yomtov (1978). Evolution of parental care in birds. Evolution 32:655–669.

Bart, J., and A. Tornes (1989). Importance of monogamous male birds in determining reproductive success. Behavioral Ecology and Sociobiology 24 (2): 109–116.

Black, J. M. (1996). Introduction: Pair bonds and partnerships. In Partnerships in birds: The study of monogamy, vol. 6 (Oxford Ornithology Series), J. M. Black, Editor. Oxford University Press, Oxford, pp. 3–20.

Black, J. M. (2001). Fitness consequences of long-term pair bonds in barnacle geese: Monogamy in the extreme. Behavioral Ecology 12:640–645.

Bluhm, C. K., and P. A. Gowaty (2004a). Reproductive compensation for offspring viability deficits by female Mallards, Anas platyrhynchos. Animal Behaviour 68:985–992.

Bluhm, C. K., and P. A. Gowaty (2004b). Social constraints on female mate preferences in Mallards, Anas platyrhynchos, decrease offspring viability and mother productivity. Animal Behaviour 68:977–983.

Borgerhoff Mulder, M., and R. Schacht (2012). Human behavioural ecology: eLS. John Wiley & Sons, Ltd: Chichester. doi:10.1002/9780470015902.a0003671.pub2.

Brennan, P. L. R., R. O. Prum, K. G. McCracken, M. D. Sorenson, R. E. Wilson, and T. R. Birkhead (2007). Coevolution of male and female genital morphology in waterfowl. PLoS ONE 2 (5): e418.

Brennan, P. L. R., C. J. Clark, and R. O. Prum (2010). Explosive eversion and functional morphology of the duck penis supports sexual conflict in waterfowl genitalia. Proceedings of the Royal Society B: Biological Sciences 277:1309–1314.

Brennan, P. L. R., and R. O. Prum (2015). Mechanisms and evidence of genital coevolution: The roles of natural selection, mate choice, and sexual conflict. Cold Spring Harbor Perspectives in Biology 7 (7): a017749.

Briskie, J. V., and R. Montgomerie (1997). Sexual selection and the intromittent organ of birds. Journal of Avian Biology 28:73–86.

Brown, J. L. (1964). The evolution of diversity in avian territorial systems. Wilson Bulletin 76:160–169.

Brownmiller, S. (1975). Against our will: Men, women, and rape. Simon and Schuster, New York.

Clutton-Brock, T. (1991). The evolution of parental care. Princeton University Press, Princeton, NJ.

Collias, N. E. (1964). The evolution of nests and nest-building in birds. American Zoologist 1:175–190.

Collias, N. E. (1997). On the origin and evolution of nest building by passerine birds. The Condor 1:253–270.

Collias, E. C., and N. E. Collias (1964a). The development of nest-building behavior in a weaverbird. The Auk 81:42–52.

Collias, N. E., and E. C. Collias (1964b). Evolution of nest-building in the weaverbirds (Ploceidae). University of California Press, Berkeley.

Collias, N. E., and E. C. Collias (2014). Nest building and bird behavior. Princeton University Press, Princeton, NJ.

Darwin, C. (1859). The origin of species by means of natural selection. John Murray, London.

Darwin, C. (1871). Sexual selection and the descent of man. John Murray, London.

Darwin, C., and A. Wallace (1858). On the tendency of species to form varieties; and on the perpetuation of varieties and species by natural means of selection. Journal of the Proceedings of the Linnean Society of London. Zoology 3:45–62.

Davies, N. B. (1992). Dunnock behaviour and social evolution. Oxford University Press, Oxford.

Dawkins, R. (1999). The extended phenotype: The long reach of the gene. Oxford Paperbacks, Oxford.

Dennett, D. C. (1995). Darwin's dangerous idea: Evolution and the meanings of life. Simon and Schuster, New York.

DuRant, S. E., W. A. Hopkins, G. R. Hepp, and J. Walters (2013). Ecological, evolutionary, and conservation implications of incubation temperature-dependent phenotypes in birds. Biological Reviews 88:499–509.

Emlen, S. T., and L. W. Oring (1977). Ecology, sexual selection, and the evolution of mating systems. Science 197:215–223.

Ewen, J. G., D. P. Armstrong, B. Ebert, and L. H. Hansen (2004). Extra-pair copulation and paternity defense in the Hihi (or stitchbird) Notiomystis cincta. New Zealand Journal of Ecology 28:233–240.

Faircloth, B. C. (2008). An integrative study of social and reproductive systems in Northern Bobwhite (Colinus virginianus): A non-migratory, avian species bearing precocial young, PhD diss., University of Georgia, Athens.

Faircloth, B. C., W. E. Palmer, and J. P. Carroll (2005). Post-hatching brood amalgamation in Northern Bobwhites. Journal of Field Ornithology 76:175–182.

Fisher, R. A. (1930). The genetical theory of natural selection: A complete variorum edition. Oxford University Press, Oxford.

Fitch, M. A., and G. W. Shugart (1984). Requirements for a mixed reproductive strategy in avian species. American Naturalist 124 (1): 116–126.

Gowaty, P. A. (1981a). Aggression of breeding Eastern Bluebirds (Sialia sialis) toward their mates and models of intraspecific and interspecific intruders. Animal Behaviour 29:1013–1027.

Gowaty, P. A. (1981b). An extension of the Orians-Verner-Willson model to account for mating systems besides polygyny. American Naturalist 118:851–859.

Gowaty, P. A. (1985). Multiple parentage and apparent monogamy in birds. In Avian monogamy, P. A. Gowaty and D. W. Mock, Editors. Allen Press, Lawrence, KS, pp. 11–17.

Gowaty, P. A. (1996a). Battles of the sexes and origins of monogamy. In Partnerships in birds: The study of monogamy, J. M. Black, Editor. Oxford University Press, Oxford.

Gowaty, P. A. (1996b). Field studies of parental care in birds: New data focus questions on variation among females. In Parental care: Evolution, mechanisms, and adaptive significance, vol. 25, J. S. Rosenblatt and C. T. Snowdon, Editors (Series: Advances in the Study of Behavior, P. J. B. Slater and M. Milinski, Editors). Academic Press, San Diego, CA, pp. 25478–25530.

Gowaty, P. A. (1996c). Multiple mating by females selects for males that stay: Another hypothesis for social monogamy in passerine birds. Animal Behaviour 51:482–484.

Gowaty, P. A. (1997). Sexual dialectics, sexual selection, and variation in mating behavior. In Feminism and evolutionary biology, P. A. Gowaty, Editor. Chapman Hall, New York, pp. 351–613.

Gowaty, P. A. (2008). Reproductive compensation. Journal of Evolutionary Biology 21:1189–1200.

Gowaty, P. A., and W. C. Bridges (1991a). Behavioral, demographic, and environmental correlates of extrapair fertilizations in Eastern Bluebirds, Sialia sialis. Behavioral Ecology 2:339–350.

Gowaty, P. A., and W. C. Bridges (1991b). Nestbox availability affects extra-pair fertilizations and conspecific nest parasitism in Eastern Bluebirds, Sialia sialis. Animal Behaviour 41:661–675.

Gowaty, P. A., and N. Buschhaus (1998). Ultimate causation of aggressive and forced copulation in birds: Female resistance, the CODE hypothesis, and social monogamy. American Zoologist 38:207–225.

Gowaty, P. A., W. W. Anderson, C. K. Bluhm, L. C. Drickamer, Y.- K. Kim, and A. J. Moore (2007). The hypothesis of reproductive compensation and its assumptions about mate preferences and offspring viability. Proceedings of the National Academy of Sciences of the United States of America 104 (38): 15023–15027.

Gowaty, P. A., and S. P. Hubbell (2009). Reproductive decisions under ecological constraints: It's about time. Proceedings of the National Academy of Sciences USA 106:10017–10024.

Gowaty, P. A., and S. P. Hubbell (2010). Killing time: A mechanism of sexual selection and sexual conflict. In Primary sexual characters in animals, A. Cordoba-Aguilar and J. Leonard, Editors. Oxford University Press, Oxford, pp. 79–96.

Gowaty, P. A., and J. H. Plissner (2015). Eastern Bluebird (Sialia sialis). In Birds of North America online, A. Poole, Editor. Cornell Lab of Ornithology, Ithaca, NY.

Hamilton, W. D., and M. Zuk (1982). Heritable true fitness and bright birds: A role for parasites? Science 218:384–387.

Heinroth, O. (1911). Beitrage zur Biologie, namentlich Ethologie und Psychologie der Anatiden. In Fifth International Ornithological Congress, Ornithological Congress, Berlin, pp. 332–340.

Hinde, R. A. (1956). The biological significance of the territories of birds. Ibis 98:340–369.

Howard, H. E. (1920). Territory in bird life. J. Murray, London.

Hubbell, S. P., and L. K. Johnson (1987). Environmental variance in lifetime mating success, mate choice, and sexual selection. American Naturalist 130:91–112.

Karvonen, E., P. T. Rintamaki, and R. V. Alatalo (2000). Female-female aggression and female mate choice on black grouse leks. Animal Behaviour 59:981–987.

Kempenaers, B., and E. Schlicht (2010). Extra-pair behaviour. Springer-Verlag, Berlin.

Lack, D. (1968). Ecological adaptations for breeding in birds. Chapman and Hall, London.

Lawler, A. (2014). Why did the chicken cross the world?: The epic saga of the bird that powers civilization. Simon and Schuster, New York.

McKitrick, M. C. (1992). Phylogenetic analysis of avian parental care. The Auk 109 (4): 828–846.

Moreno, J. (2012). Parental infanticide in birds through early eviction from the nest: Rare or under-reported? Journal of Avian Biology 43 (1): 43–49.

Nice, M. M. (1941). The role of territory in bird life. American Midland Naturalist 26:441–487.

Noble, G. (1936). Courtship and sexual selection of the flicker (*Colaptes auratus luteus*). The Auk 53:269–282.

Orians, G. H. (1969). On the evolution of mating systems in birds and mammals. American Naturalist 103 (934): 589–603.

Orians, G. H. (2014). Snakes, sunrises, and Shakespeare: How evolution shapes our loves and fears. University of Chicago Press, Chicago.

Parker, G. A., R. Baker, and V. Smith (1972). The origin and evolution of gamete dimorphism and the male-female phenomenon. Journal of Theoretical Biology 36:529–553.

Platt, J. R. (1964). Strong inference. Science 146 (3642): 347–353.

Ritchison, G., and K. P. Little (2014). Provisioning behavior of male and female Indigo Buntings. Wilson Journal of Ornithology 126:370–373.

Rohwer, S. (1986). Selection for adoption versus infanticide by replacement "mates" in birds. *In* Current ornithology, R. Johnston, Editor. Springer, New York, pp. 353–395.

Rubenstein, D. R., H. Skolnik, A. Berrio, F. A. Champagne, S. Phelps, and J. Solomon (2015). Sex-specific fitness effects of unpredictable early life conditions are associated with DNA methylation in the avian glucocorticoid receptor. Molecular Ecology 25:1714–1725.

Sandell, M. I. (1998). Female aggression and the maintenance of monogamy: Female behaviour predicts male mating status in European Starlings. Proceedings of the Royal Society B: Biological Sciences 265:1307–1311.

Silver, R., H. Andrews, and G. F. Ball (1985). Parental care in an ecological perspective: A quantitative analysis of avian sub-families. American Zoologist 25:823–840.

Slagsvold, T. (1993). Female-female aggression and monogamy in Great Tits *Parus major*. Ornis Scandinavica 24:155–158.

Slagsvold, T., and J. T. Lifjeld (1994). Polygyny in birds: The role of competition between females for male parental care. American Naturalist 143:59–94.

Smith, J. M. (1977). Parental investment: A prospective analysis. Animal Behaviour 25:1–9.

Trivers, R. (1972). Parental investment and sexual selection. Biological Laboratories, Harvard University, Cambridge, MA.

Valera, F., H. Hoi, and A. Kristin (2003). Male shrikes punish unfaithful females. Behavioral Ecology 14:403–408.

Vanrhijn, J. (1984). Phylogenetical constraints in the evolution of parental care strategies in birds. Netherlands Journal of Zoology 34:103–122.

Vanrhijn, J. G. (1991). Mate guarding as a key factor in the evolution of parental care in birds. Animal Behaviour 41:963–970.

Veiga, J. P. (1990). Infanticide by male and female house sparrows. Animal Behaviour 39:496–502.

Verner, J., and M. F. Willson (1969). Mating systems, sexual dimorphism, and the role of male North American passerine birds in the nesting cycle. Ornithological Monographs 9:1–76.

Wallace, A. R. (1889). Darwinism: An exposition of the theory of natural selection with some of its applications. MacMillan, London.

Wesolowski, T. (1994). On the origin of parental care and the early evolution of male and female parental-roles in birds. American Naturalist 143:39–58.

Wesolowski, T. (2004). The origin of parental care in birds: A reassessment. Behavioral Ecology 15:520–523.

Wheelwright, N. T., K. A. Tice, and C. R. Freeman-Gallant (2003). Postfledging parental care in Savannah Sparrows: Sex, size and survival. Animal Behaviour 65:435–443.

Wittenberger, J. F., and R. L. Tilson (1980). The evolution of monogamy: Hypotheses and evidence. Annual Review of Ecology and Systematics 11:197–232.

CHAPTER 17

Social Systems

Susan B. McRae

Birds are highly social animals with complex behavioral repertoires, sophisticated communication systems, and extended parental care. Yet, the extent and degree of social interaction observed varies greatly among species according to a variety of ecological and genetic factors. Spatial and temporal distribution of food, seasonality in breeding, migratory status, type and abundance of predators, and intrinsic life history and hereditary factors contribute to shaping bird societies. Most birds aggregate with conspecifics for at least part of the year, often during migration and/or the non-breeding season. Some species form dense breeding aggregations (colonial nesting), where pairs benefit from the dilution of predation pressure.

In some tropical species, relatively solitary adult males occupy disparate display territories, "exploded leks," and after fledging, adults only ever interact with conspecifics during brief interludes of courtship and mating. At the other end of the extreme, many cooperative breeders live their entire lives in the company of conspecifics, often in extended family groups. In these cases, members of the more philopatric sex (typically males) never leave the natal range or territory. In many cooperatively breeding species, reproductively mature individuals forgo breeding for one or many years and help conspecifics to breed. In others, breeders cooperate in their nesting attempts; two or more reproductive females may engage in communal nesting. All of these represent variations of sociality observed in different taxa.

The study of sociality is inherently interesting because avian social evolution parallels processes that led to our own complex societies. Like us, birds are homeotherms with high metabolic rates, perception mainly in the visual and auditory realms, and complex forms of communication and behavior including sophisticated altruistic and parasitic interactions. Much of this chapter focuses on species in which individu-

als spend a substantial portion of their daily lives interacting with conspecifics, just as we do. We explore the causes and consequences of group living, with examples that depict the diversity of avian social systems. Spatial and temporal variation in group-living behavior is often determined by ecological constraints (Emlen 1982). However, variation in group structure, dominance, and altruism are more often related to age structure, sex ratio variation, and genetic relatedness among group members. Cooperatively breeding species are commonly found in ecologically sensitive areas (rare habitats, harsh and unpredictable climates). An extraordinary amount of research in this field over the last ~40 years has revealed a rich diversity of behavioral variation underlying social evolution and produced a general understanding of some of the main factors underlying sociality in birds (fig. 17.1). Social experience can have profound effects on individual fitness. Sociality also feeds back on population-level processes. Many questions still remain, including how to manage species with specific habitat requirements and how to maintain minimum group size and population stability in the face of low reproductive rates and declining population numbers.

Studying social systems in birds is challenging because of the variety and complexity of social structures found in avian taxa. Variability in group sizes, sex ratios, dominance structures, relatedness, etc. within systems makes it difficult to do meaningful experimental manipulations with adequate replication. Removals and other experimental manipulations risk upsetting group stability and create undesired disturbance. Highly social species are therefore challenging to study using traditional methods. Variation among social systems complicates comparison among species, but illustrates how evolution has shaped different solutions to similar environmental challenges. Structural complexity is part of what fascinates us about avian social systems.

Figure 17.1. White-fronted Bee-eaters live in family groups year-round in East Africa. Nonbreeding mature helpers cooperate in feeding young. This chapter explores why group living is advantageous, how individuals make decisions about whether or not to join a group, and how they make sacrifices while gaining benefits of sociality. *Photo by Warwick Tarboton.*

GROUP LIVING AND SOCIAL EVOLUTION

What are societies? Societies have been defined variously as groups of animals that defend a common resource, live together in breeding groups with more than one member of at least one sex, have significant brood care, display overlap of generations, and demonstrate reproductive inequalities based on social interactions and kinship. Widespread tendencies among birds to flock ephemerally, roost communally, and nest colonially reveal a predisposition for sociality that precedes the evolution of complex societies. Sociality can be considered to be more broadly inclusive of aggregations of conspecifics, or indeed heterospecifics, that flock for a common purpose for varying periods of time.

Sociality in birds can be viewed as a spectrum. Social interactions in the lives of individual birds vary by species, population, and individual circumstance (box on page 528). At one extreme are largely solitary species that spend the majority of their time foraging alone. Others only associate with conspecifics other than their mates in the nonbreeding season, such as in large flocks during migration. Yet others might gather in aggregations to **lek** in the breeding season (~eight months of the year in the case of tropical species, e.g., Andean Cock-of-the-rock, *Rupicola peruvianus*). Aggregation in breeding colonies is common in certain lineages of birds whose food sources preclude territoriality and that benefit through protection from predators. Some species have mul-

tiple successful breeding attempts in a season, and young of the year remain and associate with family members before dispersing. Cooperative breeders commonly have overlapping generations of long-lived individuals and queues for limited breeding positions. No cases of eusociality (characterized by permanent nonreproductive castes) have been described in birds.

FROM AGGREGATION TO SOCIAL GROUPS
Mostly Solitary Existence

Few species of birds are solitary. However, some species have only brief periods of association with parents, mates, and/or offspring. For example, some tropical bowerbirds live a largely solitary existence. Male Vogelkop Bowerbirds (*Amblyornis inornata*) defend territories in Western New Guinean rainforests. Food is abundant and regularly distributed, and individuals are able to forage in a relatively small area. They maintain and decorate an elaborate bower used to attract females for the purpose of mating. Interactions between mating partners are brief, and once completed, females depart to build a nest and raise young alone. Thus, apart from the period of parental care, adult females also live solitary lives (Frith and Frith 2004).

Australian Brushturkeys (*Alectura lathami*; order Galliformes, family Megapodiidae) represent a striking example of how the level of social interaction can vary throughout the life of an individual. Adults live solitary lives, but social interaction at the juvenile stage facilitates mate recognition. Solitary adult males invest effort in building, maintaining, and defending incubation mounds in which females lay their eggs. Mating occurs during these brief female visits. Chicks hatch asynchronously and, after about a 40-hour rest period, dig themselves out of the mound and run away, receiving no further parental care (Jones et al. 1995). Between two days and several weeks post-hatching, juvenile Brushturkeys aggregate with conspecifics. Aviary experiments using realistic robotic models of Brushturkey chicks revealed that both behavior and morphology were important in social recognition. Chicks innately responded to stereotypic pecking and scratching, approaching robots made to perform these movements (Göth and Evans 2004; fig. 17.2). Chicks also demonstrated perceptual biases for body regions reflecting at short wavelengths, such as found mostly in the bill and feet. Thus, social aggregation relies on the perception of specific visual cues. Moreover, chicks initiate social feeding and interactions without prior social experience demonstrating its innate origins (Göth and Jones 2003). Remarkably, the importance of aggregation and social interaction among immatures has also recently been hypothesized to be critical

THE SOCIAL LIFE OF A BIRD: THREE BIRDS, THREE LIFESTYLES

A male Andean Cock-of-the-rock is raised by a single mom in a brood of two in a solitary nest in the Peruvian rainforest. Upon reaching independence, he will forage mostly on fruit in the forest, occasionally joining mixed flocks of birds following army ants that scare up arthropod prey. Daily at dawn, and often also close to dusk, he will join a lek of ~12 males at a traditional clearing in the forest. For a little less than an hour, he will display by calling and jumping on a small court that he defends within the lek. Females visit these arenas to mate. It will be a few years before a female chooses to mate with him. After engaging in this energetic display, all of the birds depart the lek to forage for the remainder of the day. This diurnal pattern is repeated. His sister spends her days foraging. She visits a lek for a few days before building her mud nest close by, and raises her brood of two alone.

A Cliff Swallow in Nebraska is raised in a brood of five chicks, within a colony of 1,500 nests. After fledging, it will forage as an aerial insectivore by day in a loose flock, and rejoin flock-mates at night in a communal roost. Come fall in the northern hemisphere, the flock migrates to South America, where the swallows will find dipterans to feed on through the austral summer. It will return as a breeder to a nearby nesting colony in Nebraska the following spring. Gregarious year-round, the swallow will continue to travel back and forth with the flock for the rest of its life. If it survives its second and later migrations, it will breed in the same colony in multiple years, each time with a different mate.

A male Florida Scrub Jay is raised in a brood of four chicks, cared for by both parents as well as two older brothers (full sibs from last year's brood), and an unrelated male that joined the group the previous year. It will be fed by all members of his extended family for several weeks after leaving the nest. The Scrub Jay stays and helps defend the family territory through the winter, sharing the territorial resources. His two sisters disperse before spring. In his first

Male Andean Cock-of-the rock. *Photo by Nick Athanas.*

Nesting Cliff Swallow pairs. *Photo by C. R. Brown, used with permission.*

Florida Scrub Jay family group. *Photo by Reed Bowman.*

and second year posthatching, he and his brother stay and assist their parents in rearing younger siblings. The following year, he pairs with a two-year-old immigrant female, and they begin defending a territory budded-off from that of his parents'. At two years of age, he breeds for the first time next to his natal territory. He will remain here and breed with the same female for the rest of his life (~six years), receiving additional help as his family grows.

---

for mate recognition in the absence of **sexual imprinting** during parental interactions in the brood parasitic Brown-headed Cowbird (*Molothrus ater*) (Louder et al. 2015).

Nonbreeding Social Aggregations

The principal benefit of aggregating is safety. The risk of predation decreases in proportion to the number of other potential prey in the vicinity, a principle formally known as

the **dilution effect**. By associating with others, each member of a group reduces its individual risk of predation or zone of danger (Hamilton 1971). Sometimes increased prey density can attract the attention of additional predators (Vine 1973), but so long as the attack rate does not increase as quickly as flock size increases, the general principle that individuals are safer in larger flocks, "safety in numbers," holds true even in the absence of other mechanisms.

Figure 17.2. Robotic models made from taxidermic mounts of juvenile conspecifics were used to test movement preferences in juvenile Australian Brushturkeys. Representative frames depict two types of motor patterns: pecking (photos 1–3) and scanning (photos 4–6). Choice stimuli were presented simultaneously from different arms of a choice chamber. Box plots show the percent time (median, first and third quartiles, whiskers largest and smallest values) spent by chicks in each choice arm when stimuli presented were *A*, a pecking robot versus a static robot, and *B*, a pecking robot versus a scanning robot. Juvenile Brushturkeys showed a clear preference for pecking movements. *Adapted with permission from Göth and Evans 2004. C*, An adult male Australian Brushturkey on his incubation mound. *Photo by Ann Göth.*

The benefits of dilution are enhanced in birds by having "more eyes" watching for predators (Pulliam 1973, Lima 1995), resulting in a **vigilance effect**. Vigilant individuals might raise an alarm in the event a predator is detected. "Alarm calls" (see chapter 14) can evolve through reciprocal altruism or kin selection (box on page 533). They can benefit the caller directly by mobilizing an antipredator response by the entire flock. Game theoretical models have been developed that fit empirical data from individual scanning rates in winter feeding flocks of juncos and sparrows at risk of attack by a hawk (Caraco 1979, Pulliam et al. 1982, McNamara and Houston 1992). Scanning comes at the cost of feeding, and individuals are expected to optimize the time allocated to either activity. Empirically, individuals showed

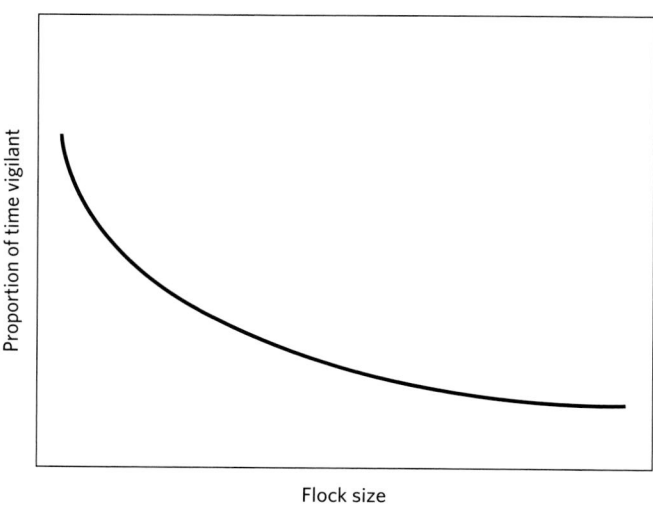

Figure 17.3. Individual vigilance rate is expected to be negatively related to group size. *After figure 1 in McNamara and Houston 1992.*

similar vigilance rates for given flock sizes. The models revealed that overall level of vigilance should increase with higher predation risk, but as was also observed in the field study (Pulliam et al. 1982), theory predicted that individual vigilance rate should be negatively related to group size (McNamara and Houston 1992; fig. 17.3).

Raising alarm in a flock can result in individuals taking flight in a disorderly array, leading to predator confusion. The **confusion effect** is an additional benefit of safety in numbers arising from the synchronized but chaotic response of individual members of a flock (McNamara and Houston 1992, Davies et al. 2012). Movement of many of individuals in haphazard directions makes it difficult or impossible for the predator to fix on any one, foiling its ability to make a capture. Susceptibility to predators using different strategies of hunting modifies predation risk, and may also impose selection on prey group sizes (Cresswell and Quinn 2010).

Seasonal Aggregations, Flocking

Many birds flock in the nonbreeding season, whether or not they are migratory, for safety and for locating and moving among patchily distributed sources of food. Seasonal flocks merge as birds migrate along traditional corridors. In some instances, this leads to the formation of mixed species flocks, where individuals benefit from enhanced safety through dilution in large flock sizes. A spectacular example can be seen during the fall migration of icterid blackbirds traversing the Midwest and overwintering in the southern United States. Icterids form a guild of seed-eaters that exploit large, rich patches of food in agricultural landscapes during the autumn harvest season. Mixed flocks can consist of hundreds of thousands of birds (see also chapter 22). Arctic breeding waterfowl also overwinter in flocks in the tens of thousands, grazing in shallow waterways and fields by day and roosting in wetlands at night (fig. 17.4).

Communal Roosts

Each evening at dusk, hundreds of Red-lored Amazons (*Amazona autumnalis*) descend on the branches of about two dozen large trees in the center of the village of Gamboa, Panama. Accompanied by a cacophony of contact calls, they arrive two by two, in mated pairs or small flocks of a few pairs together. Further chatter ensues as they arrange themselves among the branches. At daybreak, they depart in all directions, foraging on fruit they find dozens of miles away all over Panama's canal zone. Pairs will return to this traditional roost site every evening, 365 days per year, save for the breeding period, when more singletons arrive, reflecting overnight parental incubation duties (Berg and Angel 2006; fig. 17.5). In summer, pairs are often accompanied by their single surviving young of the year. Centrally located along the Panama Canal at the southern end of Lago Gatún, Gamboa's quiet neighborhood offers roost sites in tall trees. This parkland landscape is lit by street lamps overnight. Communal roosts like this one are commonly found in urban landscapes because of the predator deterrent effects of artificial lighting and human presence.

In addition to the safety of aggregating at traditional roost sites, benefits specific to communal roosts (box on page 532) include sharing information about food. Ward and Zahavi (1973) proposed the **information center hypothesis**, which formalized the notion that communal roosts should serve as a venue for communicating information about the location of quality food sources. In essence, poor or inexperienced foragers can follow successful foragers departing the roost when they return to food sources. Studies have been equivocal in finding support for this hypothesis as a driver of communal roosting or sociality, but that may be more a function of the difficulty of demonstrating information transfer scientifically (Brown 1998). As with any hypothesis, it is important to select a tractable system where it can be rigorously tested. Observational and experimental support for the information center hypothesis has been accumulated for carrion-feeders, particularly vultures and large corvids (Buckley 1997, Sonerud et al. 2001). These are large-bodied birds that tend to travel over long distances to find ephemeral, patchily distributed food (animal carcasses that persist over only a few days).

Ravens (*Corvus corax*) are highly social and use traditional roost sites. Ravens in Maine synchronized their departures from the roost, with individuals leaving in the same direc-

Figure 17.4. Migratory flocks of birds, some so dense they nearly block out the sun, are among the wonders of the world. Why do they gather in such great numbers? More than 100,000 waterfowl overwinter in the Pungo-Pocosin Lakes region of eastern North Carolina each year, where they congregate in feeding flocks in the food-rich wetlands. Shown are Snow Geese (*Chen caerulescens*), for which benefits to aggregating include opportunities to find mates and form pair-bonds on the wintering grounds before returning to their arctic breeding sites. *Photo by Keith Ramos.*

Figure 17.5. Mated pair of Red-lored Amazons returning to a communal roost in Panama. *Photo by Nick Athanas/antpitta.com.*

tion (Marzluff et al. 1996). Individuals captured and held for a time, then released at the roost, followed roost-mates to foraging areas. Others released at a food source joined communal roosts, and in 3/20 instances, led roost-mates to food. In a separate study in Wales, UK, sheep carcasses each baited with differently colored beads were fed-upon by Ravens. The colored beads could then be found in regurgitated pellets left by the successful foragers at the roost (Wright et al. 2003). The pattern of accumulation of color-bead-laden pellets at the roost site suggested the concentric spread of information from particular foragers in specific positions in the roost. Olfactory cues or observation of matted plumage could be used to identify successful foragers, but they may also advertise their discovery of a food source to naïve birds (Marzluff et al. 1996). Precisely how Ravens communicate information about food remains to be determined.

Communal roosting can entail thermal benefits. In Southern Africa, the temperature commonly dips below freezing at night during the dry season. Small-bodied resident species conserve body heat by huddling together with conspecifics while roosting to survive these cold nights. Some small-bodied species huddle with their social group in the open on tree branches (e.g., Swallow-tailed Bee-eaters, *Merops hirundineus*; White-backed Mousebirds, *Colius colius*; McKechnie and Lovegrove 2001). Others huddle while roosting in tree cavities that are defended year-round and represent an important and contested group resource (e.g., Green Wood Hoopoes, *Phoeniculus purpureus*; duPlessis and Williams 1994).

OVERVIEW OF SOCIAL STRUCTURES

Communal roosts: traditional shared roost sites, typically among conspecifics.

Colonial nesting: breeders in the same population build their nests in densely packed aggregations.

Cooperative breeding: in addition to two or more breeding birds, social units include individuals providing alloparental care that may or may not be breeders themselves.

- Helpers are often sexually mature offspring that remain with and assist their parents (with or without reproductive sharing) to rear subsequent broods.

- Cooperative breeding is typified by sustained group membership of subordinates, creating multigenerational families.

- Cooperative breeding can include help by unrelated group members with or without reproductive sharing.

- Helpers may alternatively be failed breeders that redirect care to nests of kin.

Plural breeding: breeding by multiple pairs within a social group, usually in separate nests, sometimes in a colony.

Joint nesting (also, communal nesting): two or more females lay in single nest; co-breeding females may share a male mate or nest jointly as plural breeders.

The ability of groups to maintain and defend roost cavities is recognized as an important ecological constraint for group-living woodpeckers in North America (Walters et al. 1992, Koenig et al. 2016). A different strategy for keeping warm at night is to build roost nests that insulate from the cold. In Africa, group-living, resident weaverbirds such as White-browed Sparrow-Weavers (*Plocepasser mahali*) build roost nests that are occupied singly, whereas Sociable Weavers (*Philetairus socius*) huddle in small groups in a large colonial nest (discussed below). The survival advantage of building roosting structures collectively is a factor driving sociality in these species (box on page 532).

There can be costs to aggregating in roosts. Individuals compete for the best positions in the center of the flock, since predation risk is higher at the edges ("selfish herd," Hamilton 1971). Higher positions should be preferred to avoid being defecated on by flock-mates. The risk of bird-to-bird transmission of parasites and pathogens is higher in dense flocks.

The persistence of West Nile virus (WNV) among migratory American Crows (*Corvus brachyrhynchos*), in the absence of mosquito vectors in winter, was proposed to be caused by fecal transmission in overwintering roosts (Hinton et al. 2015). More than 50 percent of crows at a large overwintering roost on the campus of the University of California at Davis were stained with feces. Crow carcasses salvaged in the region tested positive for WNV in the kidney and cloaca, demonstrating the capability to shed virus and potential for fecal-oral transmission, though feces collected in summer but not in winter tested positive for WNV (Hinton et al. 2015). Infected birds that are weakened by illness may be unable to compete for higher and more central positions in the roost.

Interactions within Groups

Social Dominance and Signaling

Another cost to aggregating is competition. By virtue of having more foragers searching, flocks may more efficiently locate profitable food sources, but group members must then share the resource. Dominance hierarchy formation in foraging flocks can lead to differential access to food (pecking order). To determine social rankings among members of small feeding flocks, behavioral researchers quantify the numbers of "wins" and "losses" (retreats) between individuals during altercations and produce a score matrix with which to rank individuals. This can be done in the field, or using captive birds in aviaries.

Plumage ornaments that honestly convey individual condition and fighting ability have been identified in some species. Harris's Sparrows (*Zonotrichia querula*) have varying amounts of black plumage on the ventral side of the neck and chest. The size and color penetration of this black "bib" correlates with social rank. Rohwer and Rohwer (1978) demonstrated that this trait in wintering Harris's Sparrows was correlated with linear dominance hierarchies that determine access to food and roost sites. However, a recent study of sexually dimorphic House Sparrows (*Passer domesticus*) in Europe, in which only males have black bibs, showed that testosterone level and social ranking were correlated with bill color but not bib size (Laucht et al. 2010). In general, plumage traits might be poor indicators, because an individual's physical condition and health can change rapidly, and feather growth adheres to circannual molt patterns (chapter 9).

Signals of quality might be expected to evolve in species that spend a large part of their lives interacting in flocks. Yet, this may not extend to conspecifics in densely packed social aggregations with scramble competition over resources and strong selection on individual recognition for purposes of breeding. In sexually dimorphic Red-billed Quelea (*Quelea quelea*), red plumage coloration in males shows extreme

KEY TERMS AND DEFINITIONS

Alloparental care: care provided to offspring by individuals other than the parents.

Altruism: an action that benefits the welfare of the recipient while incurring a cost to the actor, such as the provision of resources by one individual to another.

Kin selection: the process through which a behavioral adaptation is favored by enhanced survival and/or reproductive success of close relatives.

Reciprocity: the recipient of an altruistic act returns a favor to the individual that initiated it (in the case of delayed reciprocity, this can involve a time lag).

Reproductive sharing: shared parentage by group members of one sex (genetic polyandry or polygyny) or both (genetic polygynandry) in a single breeding attempt.

Reproductive skew: the degree to which reproductive sharing favors one dominant individual over others of the same sex.

Social dominance rank: status of an individual in relation to others in its group.

- Dominant individuals may display more, may be in better health or body condition, tend to be more aggressive, and are able to win more fights; dominance is often correlated with age.
- Subordinate individuals may show more submissive behavior, may be in poorer body condition or less healthy, or may simply be younger.

Social hierarchy: relative ranks of individuals within a social group, measured based on the sum of the outcomes of social interactions with other group members.

variability in brightness, prompting the suggestion that this might be an honest signal of condition used in mate choice (see chapter 11). Yet, field observations of Red-billed Quelea breeding in Zimbabwe indicated that males with brighter breast plumage and darker mask shade did not have higher breeding success (Dale 2000). A study of captive male Quelea raised in outdoor aviaries and observed as they expressed breeding plumage over the course of three years under a uniform *ad libitum* millet diet confirmed that plumage coloration was not related to individual condition. Furthermore, detailed analysis of photographic images of males revealed that the principal ornamental traits of breast plumage and face mask varied independently of one another (fig. 17.6). Collectively, the evidence suggests that selection on male plumage traits in this highly gregarious, irruptive migrant favors individuality and is independent of physical condition or breeding potential.

Coloration in soft parts, including areas of bare skin around the face such as wattles and eye rings, bills, and legs, may respond more rapidly to changes in condition. Color vividness of these ornaments is under sexual selection. Male Red Junglefowl (*Gallus gallus*) housed individually with a female or in flocks with a rival male and three females showed differential responses in comb size. Large combed males were dominant over smaller combed males, and comb size was positively associated with measures of immune function (Zuk and Johnsen 2000). Furthermore, dominants' combs grew and subordinates' combs shrank over the course of social housing, illustrating how these developmentally plastic ornaments can respond quickly to social cues.

The causal relationship between social interactions and expression of a sexually selected trait was demonstrated experimentally in the field in the Pukeko (New Zealand Purple Swamphen, *Porphyrio porphyrio*). On the North Island of New Zealand, Pukekos reside year-round in social groups of 4 to 13 individuals. Both sexes have a fleshy red shield (culmen) on the forehead that brightens and enlarges at the start of the breeding season and regresses in winter. The size and color of the shield signals condition, and shield size was correlated to social dominance (Dey et al. 2014). Behavioral observations were performed on 16 groups of resident Pukekos, 68 percent of which had been captured, measured, and banded with unique color-band combinations for individual identification. Perception of shield size was experimentally manipulated in one male from each of eight groups by blackening the outer 6 mm of the culmen with non-toxic paint (Dey et al. 2014). A control male from each of the other eight groups had red paint applied, matched to their natural shield color. Males with experimentally reduced shields incurred relatively more attacks and retreated significantly more often than control males. These behavioral interactions in turn affected the treated birds' actual shield size: in birds manipulated to appear to have reduced shields, but not in control birds, the actual shield size shrank significantly over the course of a week posttreatment (fig. 17.7; Dey et al. 2014). This experiment demonstrates that social interactions directly affect the size and appearance of badges of status. The proximate mechanism for this feedback remains to be determined in Pukeko, but shield color and growth is mediated via testosterone in the confamilial Common Moorhen (*Gallinula chloropus*) (Eens et al. 2000).

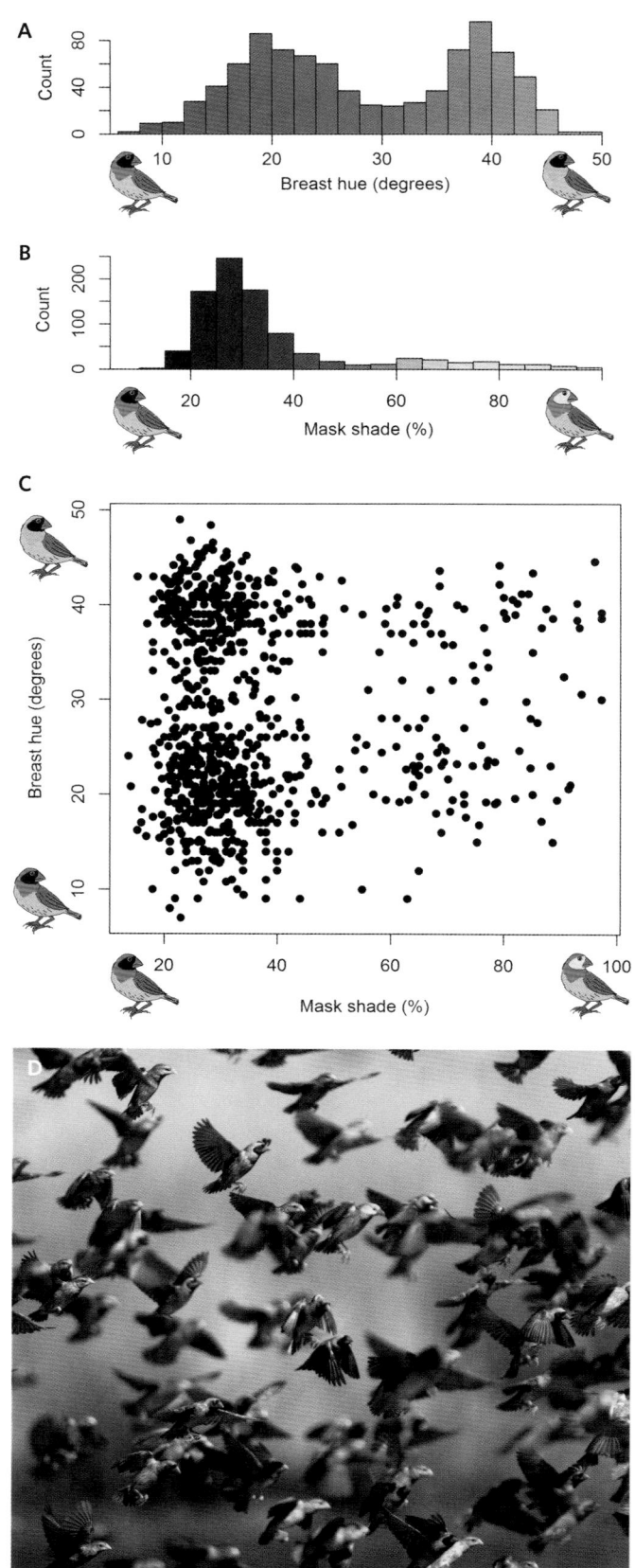

Figure 17.6. Red-billed Quelea show continuous variation in *A*, breast hue and *B*, mask shade; *C*, each parameter has a bimodal distribution, but these are not significantly correlated with one another, suggesting plumage pattern variation is selected for individuality. *Redrawn from Dale 2000 by James Dale.* (*n* = 897 males.) *D*, Red-billed Quelea flock in Zimbabwe. *Photo by Kim Wolhuter.*

Nonbreeding Sexual Segregation

In the nonbreeding season, unconstrained by the need to find mates and nest sites, individuals are distributed largely as a function of resource availability. In resident temperate species, competition over reduced food supplies in winter can be intense. Dominance hierarchies in social groups can cause marginalization of the weaker sex and lead to sexual segregation by microhabitat. Black-capped Chickadees (*Poecile atricapillus*, Paridae) are remarkable for their tolerance of the harsh winter climate in Canada. Resident Chickadees remain in tight social groups throughout the winter. Sexual segregation in feeding sites occurs based on social dominance, resulting in males foraging in preferred tiers of trees on low branches and close to trunks, and females probing high outer branches (fig. 17.8; Desrochers 1989). Temporary experimental removals of males resulted in females moving to lower and innermost branches previously occupied by dominant males, but they returned to the uppermost branches after the males were returned (Desrochers 1989).

Though migrating birds typically depart their breeding sites in mixed-sex flocks, when returning to their breeding grounds, the sexes are often segregated, with males arriving earlier (protandry). There is no doubt strong selection on males to arrive early on the breeding grounds to acquire and set up a good territory. However, proximally, variation in arrival time may arise because of differential access to food during the nonbreeding period, mediated through dominance interactions. Male birds are typically larger and tend to outcompete females for access to preferred feeding areas. This has been proposed as a possible driver of adult sex ratio, which tends to be male biased, mediated through sex-biased mortality (Marra and Holmes 2001).

Dominance interactions lead to habitat segregation at wintering sites in migrant American Redstarts (*Setophaga ruticilla*). At the birds' overwintering site in Jamaica, mangroves are preferred over scrub habitats for their abundance of dipteran insect food. That male American Redstarts exclude females from these preferred habitats through social dominance was demonstrated using removal experiments (Marra et al. 1993). Following the removal of territorial males, recolonizing individuals were significantly female-

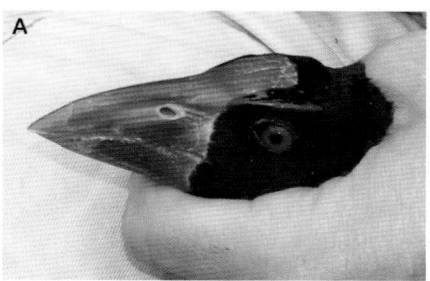

A

Figure 17.7. *Figure 17.7.* Experimental reduction in the appearance of red frontal shield size in Pukekos by *A*, blackening the outer edge to make it appear narrower (*photo by Cody Dey*), resulted in *B*, more aggressive behaviors being directed toward them by comparison with controls whose shields were marked similarly using a color identical to the natural shield color, and *C*, caused the actual shield width to decrease in size in response to the treatment, whereas shield width of controls did not change significantly. *Reprinted with permission from Dey et al. 2014.*

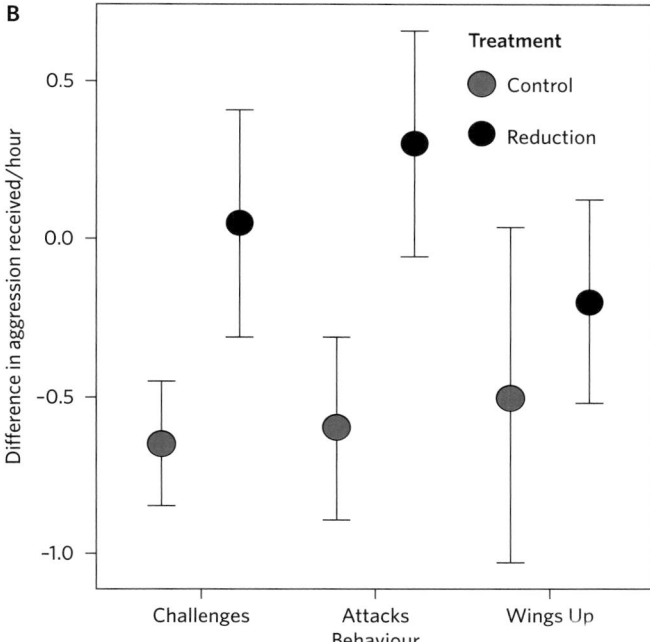

B

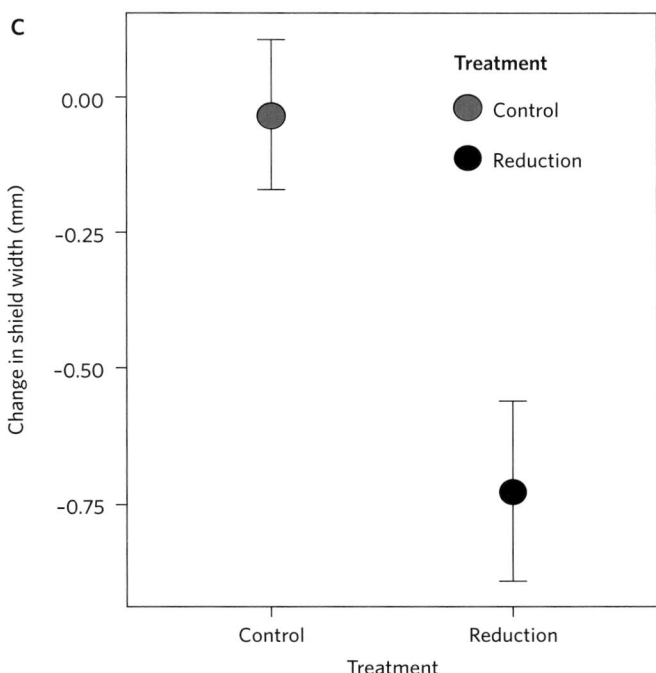

C

biased (fig. 17.9; Marra et al. 1993). Dominant male American Restarts initiate their northward migration significantly earlier than females, and have better survival rates (Marra and Holmes 2001). Significantly better reproductive success enjoyed by early-settling males was inferred to be caused by territoriality and differential access to resources in the non-breeding season. Difficulty in following individuals from breeding to nonbreeding areas has made this impractical to test, but new long distance tracking methods might make this possible in the future.

Group Mobilization in Response to Predators

Alarm calls (see chapter 14) are uttered to warn others of the presence of a predator in the vicinity. Some species have different calls to signify terrestrial versus aerial predators. Often comprising short, high-pitched **syllables**, difficult to locate due to pauses between, these calls are relatively conserved in structure among species. Consequently, they may be recognized by "eavesdropping" heterospecifics. This is not a disadvantage to the signaler. While alarm calling likely evolved as a mechanism of alerting close relatives to danger (evolved through kin selection), the benefits to individuals of coordinated antipredator responses of prey species extends to multispecies flocks.

Communities of certain Neotropical passerines regularly interact, allowing for complex behavioral interactions between individuals of different species over prolonged time scales. These tightly linked resident communities comprise species with large stable home ranges, as well as some species defending smaller territories within those home ranges, all of which feed on arthropods in the understory of Neotropical rainforests. They are united by their responses to members of a "sentinel" species whose distinctive alarm calls they are all drawn to (Munn and Terborgh 1979). In return for alerting others, the sentinels (typically flycatching members of the genera *Lanio*, shrike-tanagers, or *Thamnomanes*, antshrikes) benefit from gleaning arthropod prey scared up from the leaf litter by other flock members, "beater" species. Members of the sentinel species may also use "false" alarm calls to facilitate kleptoparasitism of food items displaced by

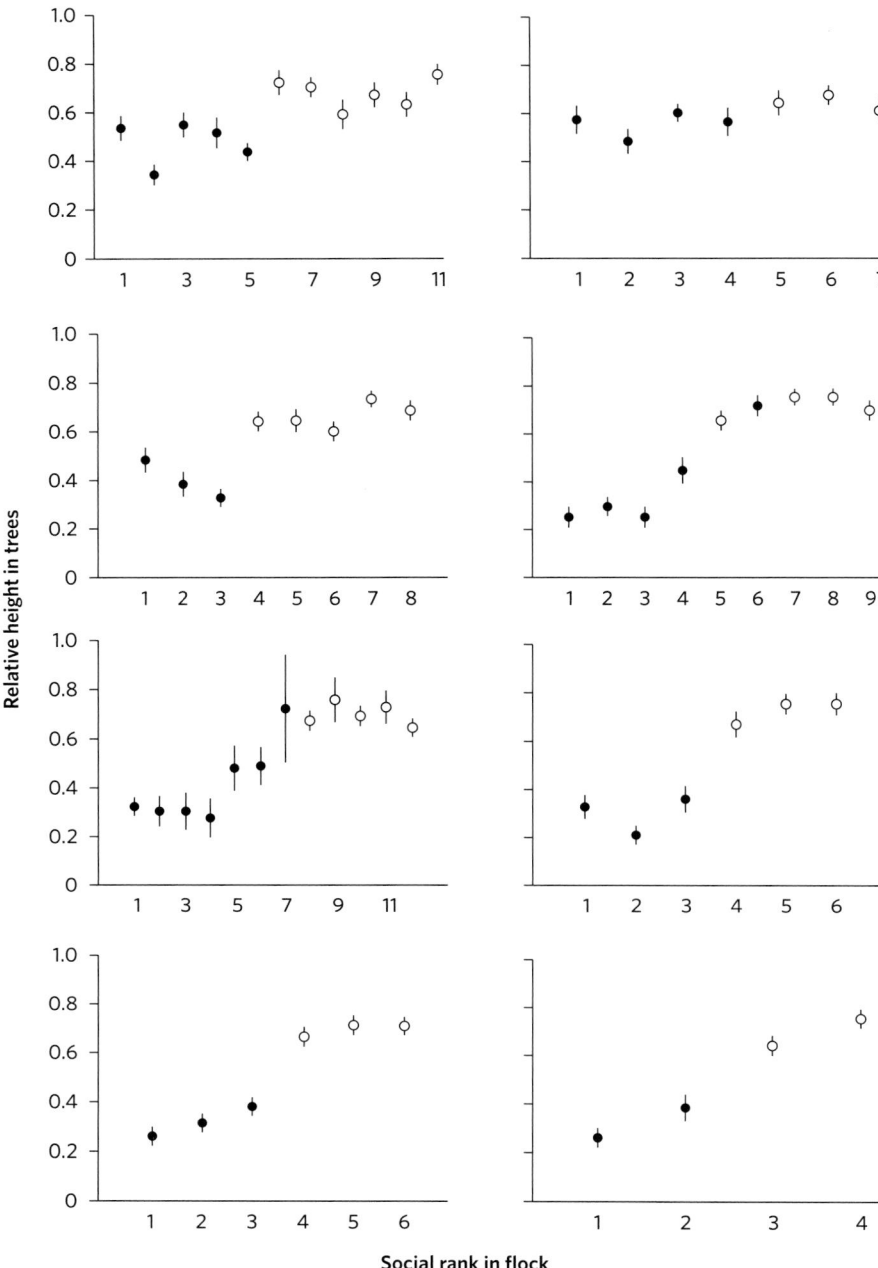

Figure 17.8. The relationship between social rank (1 most dominant) and average relative foraging height in eight flocks of Black-capped Chickadees overwintering in poplar woodlands. Height ranges from ground (0) to treetops (1.0). Males (●) were dominant to females (○) and foraged at preferred levels. Vertical bars show ±1 standard error. *Reprinted with permission from Desrochers 1989.*

beaters (Munn 1986). The interdependence of these species is akin to a symbiotic relationship, in that beater species mostly benefit from the alarms of sentinels, but sentinels can (more rarely) parasitize the beaters' foraging efforts. Group membership in these multispecies flocks is stable over years, and the community structure and species assemblages persist beyond the lifetimes of individuals (Martínez and Gomez 2013). Species composition, stability in territorial configurations, and continuity among generations are contingent on the environmental stability of the rainforest over the extent of the collective home range of the mixed species flock (~10 hect-

ares on average; Martínez and Gomez 2013). Community-level structuring is thus an outcome of individuals engaging in reciprocal interactions with heterospecifics over relatively long time scales.

BREEDING SOCIAL AGGREGATIONS

Benefits and Costs of Colonial Nesting

Many avian species nest in colonies. Nests of cliff-breeding seabirds like the Common Murre (*Uria aalge*) are often spaced just centimeters apart from one another. In addition to most

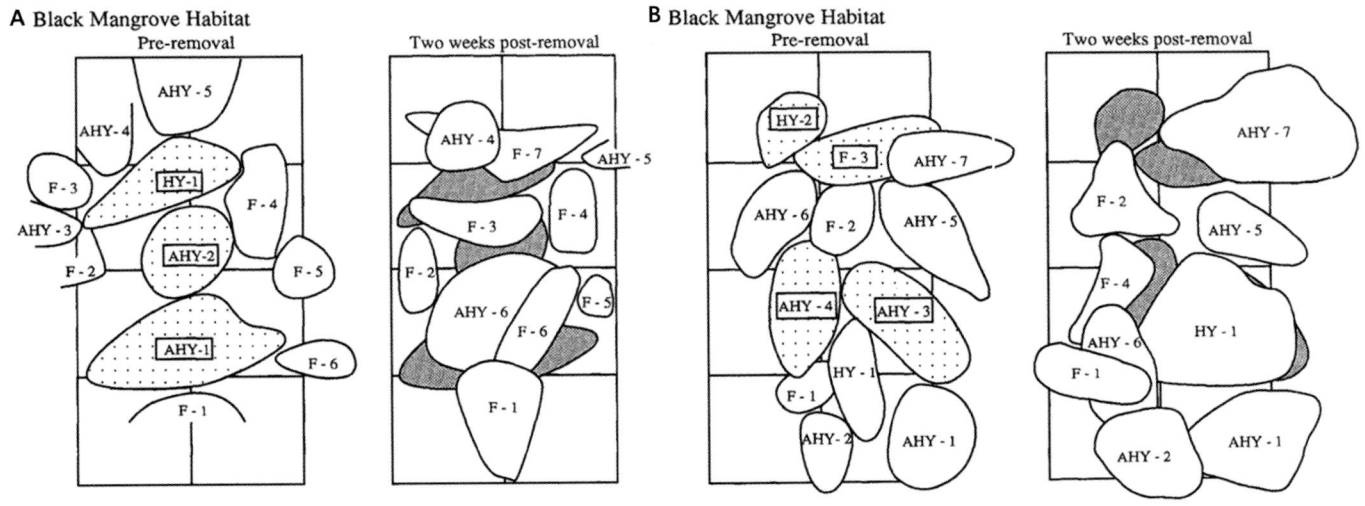

Figure 17.9. Older males (AHY—after hatch year) of migratory American Redstarts acquire superior feeding territories compared with females (F) and hatch-year males (HY) on their wintering grounds in Jamaica. Removal of dominant males from their territories at two locations in black mangrove (*A* and *B*) resulted in subdominant individuals shifting their territories into these preferred areas. Males were aged based on plumage differences; numbers signify individuals. *Reprinted with permission from Marra et al. 1993.*

auks and murres (Alcidae), penguins (Spheniscidae), pelicans (Pelecaniformes), most terns and gulls (Laridae), gannets and boobies (Sulidae), flamingos (Phoenicopteridae), hoatzins (Opisthocomidae), herons and egrets (Ardeidae), most swallows and martins (Hirundinidae), bee-eaters (Meropidae), and genera in a plethora of other families all typically nest in colonies.

Variation among taxa in colonial nesting tendency is based on ecological differences including migratory tendency, distribution and abundance of food during the breeding season, and nest predation rates. Colonies are often established at sites that are difficult to access by predators. Seabirds nest on cliff faces and small, uninhabited islands. Herons and egrets place their nests at the tops of snags in wooded swamps. Bee-eaters excavate deep burrows built into sandy riverbanks (fig. 17.10). Weaverbirds in arid zones of sub-Saharan Africa select thorny acacias that are defended with spines and backward-pointing barbs (evolved to deter herbivores but that are also aversive to nest predators). They have evolved to select the tips of the top branches as nest anchors, and often strip the ends of the branches of leaves, making the nests difficult to climb to (Crook 1963). While colonial nesting may have originated through selection of the same safe nest sites by conspecifics, the principles of safety in numbers outlined before also apply.

Eggs and nestlings are vulnerable, and the possibility of a collective response by parents toward nest predators likely selects for colonial nesting. **Mobbing** is a coordinated defense against or counterattack on a predator by multiple individuals. Mobbing involves diving, striking, pecking, vocal cries, and sometimes defecating on the predator, all with the intent of driving it away. The strategy is also used when a lone predator, such as a roosting owl or hawk, is detected by members of a foraging flock. Mobbing increases in effectiveness with more participants, and can involve heterospecifics attacking a common predator (Quinn and Ueta 2008). The advantage of collective vigilance enables colonial nesters to use mobbing effectively.

Cliff Swallows (*Petrochelidon pyrrhonota*) live in large flocks and nest colonially. The principal benefit is increased success of multiple foragers at finding spatiotemporally patchy swarms of insects. Chicks fledged from larger colonies tend to have higher survival rates. But there is also a cost: in larger, denser colonies, chicks suffer higher ectoparasite loads (Brown and Brown 2004). Hematophagous arthropods (Hemipteran "swallow bugs" and fleas) feed on chicks and cause them to fledge at significantly lower average weights reducing their probability of survival. Experimental fumigation of nest colonies was used to compare success rates of nests with and without swallow bugs. Mark-recapture has revealed that average within-year survival rates for juvenile Cliff Swallows were significantly related to colony size (fig. 17.11; Brown and Brown 2004). Chicks hatched from parasite-free nest colonies had higher average within-year survival rates than those hatched from ectoparasite-infested nests. Thus, colonial nesting in Cliff Swallows has both clear benefits and costs.

Figure 17.10. Colonial nesting dilutes predation risk and affords opportunities for coordinated nest defense. *A*, Carmine Bee-eaters (*Merops nubicoides*) flock in Katima Mulilo, Namibia; *B*, with excavated nest chambers in a riverbank in Shakawe, Botswana; *C*, Lesser flamingoes (*Phoenicopterus minor*) are colonial nesters and build; *D*, mud nests at Kamfers Dam near Kimberley, South Africa. *Photos by Warwick Tarboton.*

Nest clustering can also create conflict in the form of reproductive interference. **Conspecific brood parasitism** (CBP), where females lay one or more of their eggs in neighboring nests, is common in colonial nesting swallows (Brown and Brown 1998). Brood parasites benefit from leaving their young in the care of unsuspecting host parents, while hosts left to raise unrelated young pay a cost. This female reproductive strategy is not restricted to colonial nesters. CBP is found in diverse taxa, from Ostriches (*Struthio struthio*) (Kimwele and Graves 2003) to House Sparrows (*Passer domesticus*), and is common in waterfowl and galliforms (Yom-Tov 2001). CBP is favored by nesting synchrony and spatial proximity of nests, affording females opportunistic access to the nests of neighbors.

Female colonially breeding swallows have evolved an additional unique mode of parasitism: they can transfer by beak an egg from their own nest to a host nest (Cliff Swallows Brown and Brown 1988; and Cave Swallows *Petrochelidon fulva*, Weaver and Brown 2004). These 'sneaky' transfers occurred when hosts were away from the nest, and were known to have affected at least 6% of nests in colonies in Nebraska. Unlike parasitic laying that must occur during the host's own laying period to be successful, beak transfers were carried out at various times. In five cases observed directly the female had not yet begun incubation (Brown and Brown 1988). Parasitic female Cliff Swallows select host nests at a similar stage to their own, but with lower ectoparasite loads, where their chicks have better odds of survival. The long-term costs to hosts appear minimal; annual survival records for both males and females attending host nests were not significantly greater for enlarged clutch sizes (Brown and Brown 1998).

Reproductive interference by neighbors can also result in extra-pair fertilization (EPF). This represents a major cost of coloniality for some male Cliff Swallows. Males guard nests in preference to mate-guarding, possibly due to the high risk of brood parasitism that is also detrimental to their reproductive success (Brown 1998). Curiously, males that achieve

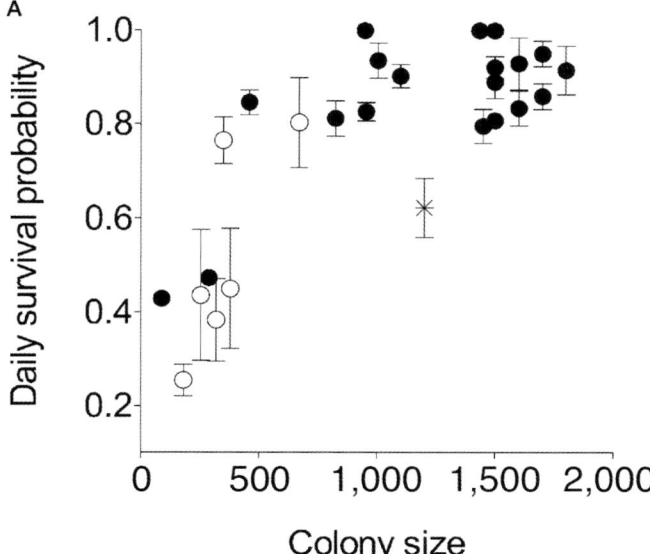

Figure 17.11. A, Daily survival probability (±SE) per colony for recently fledged Cliff Swallows in relation to colony size for nonfumigated colonies (o), fumigated colonies (•), and a fumigated colony in 1996 where extensive adult mortality occurred (*). Daily mean survival probability for all sites increased significantly with colony size (P<0.0001), and with colony size within nonfumigated sites (p = 0.02). *After Brown and Brown 2004; figure courtesy of C. R. Brown. B,* Cliff Swallow nest colony. *Photograph by C. R. Brown, used with permission.*

EPFs actually have lower survival (Brown and Brown 1998). While it was unclear in this study exactly why, it may be explained by those males investing in sexually selected traits (Safran et al. 2014). Within species, EPF rate tends to correlate with nest density (Westneat and Sherman 1997).

COOPERATIVE BREEDING

Cooperative breeding in birds encompasses a wide variety of social arrangements involving diverse mating systems and group structures. Often, it involves "nuclear families," with

breeding restricted to a dominant pair (high-skew societies) assisted by nonbreeding subordinate helpers. Alternatively, helpers may arrive from the ranks of closely related failed breeders from the vicinity. Plural breeders involve complex group structures, usually of patrilineal origin, where subordinate males remain with their parents but mate and breed in proximity, sometimes also helping at nearby nests. At the other extreme are cooperative groups composed of multiple breeders that nest jointly, often with nonrelatives (low-skew societies). In the following sections, we look at examples spanning this diversity of social organizations. The unifying characteristic is care of offspring by individuals other than their parents. Here, I broadly define cooperative breeders as species having alloparental care, including, but not exclusively, by nonbreeding adults.

Cooperation Arising from Social Conflict

Social groups can arise through sexual conflict among breeders. Dunnocks (also known as Hedge Sparrows, *Prunella modularis*) are small, gray-brown Eurasian garden passerines. They have secretive dispositions but extraordinary mating habits. Cooperation is an outcome of the conflict of interests between males and females. A female Dunnock paired to one male (**alpha**) can maximize her reproductive success by enlisting the assistance of another male (**beta**) by also copulating with him (Davies 1992). Satisfied of their opportunity for paternity, beta males will also provision chicks in the nest. The female benefits from **polyandry** through this additional paternal care. However, in spite of higher fledging success from both males provisioning, having to share paternity is a major cost to the alpha male. Males compete for mating access during the female's receptive period (Davies 1992; see also chapter 16). Intersexual conflict intensifies as males also attempt to maximize their reproductive success by mating polygynously. This can result in the coalescence of neighboring males' territories as they pursue additional mating opportunities (fig. 17.12; Davies 1989). When neither male is able to evict the other, and females, whose territories are exclusive, are unable to prevent their mates from sharing a second female, this results in the formation of a polygynandrous social group. Sexual conflict thus leads to a variable mating system, with females and males each trying to pursue their individual reproductive interests, and monogamy often emerging as an unstable compromise between the preferred options of polygyny for males and polyandry for females (Davies 1992).

Polyandrous male Dunnocks each contribute to feeding a common brood, in spite of their rivalry over paternity. This was one of the first ornithological field studies to confirm parentage by genetic analysis. As predicted from behavioral

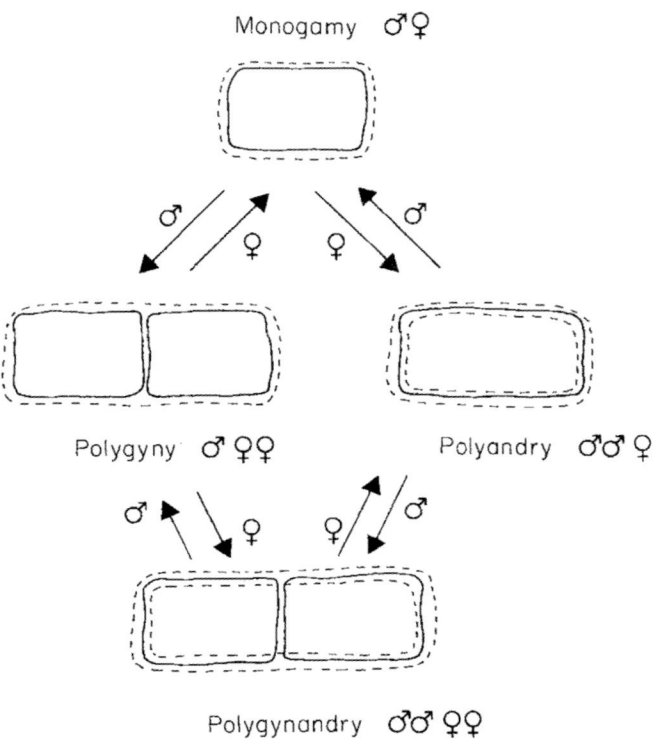

Monogamy ♂♀

Polygyny ♂ ♀♀   Polyandry ♂♂ ♀

Polygynandry ♂♂ ♀♀

Figure 17.12. Sexual conflict produces variation in social group structure in Dunnocks. Female territories shown as abstract shapes (___) are exclusive and may be defended by one or more (usually two) males (----). Arrows indicate the directions of change encouraged by alpha male and female behavior as they try to influence the social dynamic in their own interests. *Reprinted with permission from Davies 1989.*

observations, paternity of the brood was usually shared between alpha and beta males (Burke et al. 1989). Male Dunnocks use a simple rule: "feed chicks in the nest if I gained mating access during the laying period" (Davies 1992). Experiments in which a male was removed from polyandrous groups for three days before, during, or after the female's egg-laying period confirmed they could predict their paternity relatively well. Removed males fed chicks only if they gained mating access during the female's egg-laying period (Davies et al. 1992). Moreover, males paid attention to their relative access to females: a male's share of mating access during egg-laying was a better predictor of paternal effort than his total mating access, and this was dependent on the ability of alpha males to guard the female. By contrast, monogamous male Dunnocks that usually guard their females closely did not adjust their parental effort in response to short-term removal, in spite of losing paternity to extra-pair males (Davies et al. 1992). These experiments showed that individuals adjust their behavior according to the social dynamics of the group, and the behavior of other members,

in making decisions affecting their own reproductive success (box on page 541).

Cooperative **polyandry** also occurs in Galapagos Hawks (*Buteo galapagoensis*) (Faaborg et al. 1995). Here, groups of up to eight unrelated males (typically two to three) cooperate with one female to defend the breeding territory. Copulations and paternity are shared relatively equitably among group males. Similar associations of two or more males paired to one female are not uncommon among accipiters, the principal benefits being territory and nest defense. Dunnocks and other cooperatively polyandrous species thus represent an important link in social evolution, where sociality is rooted in individual reproductive strategies and there is limited cooperation among males but not among females. Cooperative **polygynandry** is a variable mating system where breeding units with varying numbers of males and females coexist within the same population. Here, all individuals cooperate in defending a group territory, and when there is more than one female, they typically lay in a communal nest (see Joint Nesting, below). However, as we will see, joint nesting can lead to conflict among females over partitioning of reproduction.

Helpers-at-the-Nest

"**Helpers-at-the-nest**" (hereafter, "helpers") are individuals other than parents that help provide care for nestling or juvenile birds. Alexander Skutch (1935) coined the term in describing his detailed observations of mature birds in Central America that did not breed but assisted in rearing the young of others. He also described earlier observations of "juvenile helpers" in British Common Moorhens that assisted in feeding their tiny siblings from a subsequent brood, but Skutch appeared to have been unaware of historical descriptions of Australian cooperative breeders dating back to the mid-1800s (see Boland and Cockburn 2002 for a review). The proportion of passerines living year-round in obvious social groups is greater in Australia. Consequently, most of the sophisticated experimental work and hypothesis testing to understand helper behavior prior to the 1970s was conducted down under.

Do Helpers Enhance Breeder Fitness?

Helpers participate in many parental care activities including territory defense, nest maintenance, feeding and care of young (chapter 16), and helpers often have positive effects on reproductive success of the breeders they assist. An early experimental study aimed at determining the magnitude of this effect was conducted on Grey-crowned Babblers (*Pomatostomus temporalis*) by Brown et al. (1982). These birds have four to six male helpers assisting the breeding pair. Experimental removal

DUNNOCK MATING SYSTEMS AND SEXUAL CONFLICT

By Nick Davies

Nick Davies. *Photo by Seb Chandler.*

"Dun" means dull brown, "ock" signifies little; true to its name, the Dunnock (*Prunella modularis*) seems the archetypal little brown bird, as it shuffles about in the undergrowth in search of small insects and seeds. Indeed, the Reverend F. O. Morris was so impressed with its modest appearance that in his *History of British Birds*, written in 1856, he recommended the Dunnock's "humble and homely deportment" and "sober and unpretending dress" as an ideal model for his parishioners to imitate "with advantage to themselves and benefit to others through an improved example." Our studies reveal that this recommendation was unfortunate; the Dunnock has an extraordinarily variable mating system, which arises from intense conflicts as individuals compete to maximize their reproductive success.

Prior to our work, it had already been shown that Dunnocks had variable mating arrangements, including monogamy, polyandry, and polygyny, but it was not known why. The stimulus for my study in the Cambridge University Botanic Garden was a new idea and a new technique. The new idea, from Robert Trivers and Geoff Parker in the 1970s, was that, in theory, there would often be sexual conflict over mating decisions and parental care. This revolutionized our view of mating systems; it led field workers to focus more on individual behavior and to expect variability in populations. The new technique of DNA profiles, developed by Alec Jeffreys in the 1980s, gave us the first opportunity for precise measures of parentage. An inspiring new idea and a powerful new technique made an intoxicating mix, and I felt that there was a whole new world to explore. Could the Dunnock's variable mating system be the outcome of sexual conflict? All we needed to do was to color-band a population, watch individuals in detail, and measure how their behavioral decisions influenced their reproductive success.

The most exciting years of the study were from 1988 to 1990, when Ben Hatchwell and I did the field observations and experiments, and Mike Bruford and Terry Burke analyzed the blood samples of parents and offspring to determine paternity and maternity. I can still remember how thrilled we all were to find that when an alpha and beta male had shared matings, this often led to mixed paternity in a brood. Usually field workers quickly discover that their birds know much more about what's going on than they do. But the DNA profiles gave us more detail than was available to the male Dunnocks themselves. For example, provided that both the alpha and the beta male gained matings, both would help to feed the brood, even in cases where, by chance, only one of them had success in the sperm lottery. Therefore, a male did not know his paternity for certain but used mating share as an indirect cue to guide his parental effort. Furthermore, a male's share of the matings correlated with his work share in chick feeding. We confirmed this was a causal link by temporary removals of males during the female's fertile period; experimental variation in the two males' mating share influenced their share in providing food for the brood (Davies et al. 1992).

Studies during the past forty years have shown that sexual conflict is often a powerful selective force in bird mating systems. We now need a better understanding of the ecological stage on which individuals play their behavior to discover why the outcomes of this conflict vary between different species.

of all but one helper per group left three caring adults in these groups compared with six to eight caring adults at unmanipulated control groups. Experimental groups had reduced reproductive success. Five out of nine experimental groups failed to produce any surviving young (mean fledglings per group = 0.8).

Eleven control groups produced on average three times as many fledglings (mean = 2.4; Brown et al. 1982). The complete failure of the majority of experimental nests highlights the importance of helper contributions to nest defense (see also Rabenold 1984).

In White-fronted Bee-eaters (*Merops bullockoides*), effects of group augmentation are linear: more helpers (up to five) enabled breeders to produce more fledglings (Emlen and Wrege 1988). However, research on plural breeding Chestnut-crowned Babblers (*Pomatostomus ruficeps*) in Australia has shown experimentally that the effect of helper number on productivity is not linear when group size is very large (Browning et al. 2012). Babblers live in groups of 3 to 23 (mean = 10) related and unrelated adults. Some of the social groups split into two to four breeding units that may nest hundreds of meters apart, reuniting again at the end of the breeding season. Nonbreeding individuals based their choice of which breeding unit to join on their kinship to breeders, not on the presence of other subordinates. Adult removal experiments can have undesired effects of altering group dynamics and stability. So, to determine effects of helper number while avoiding these pitfalls, one to three offspring were transferred between nests to create specific helper:brood size ratios, and standardized statistics were applied. Nest success at manipulated broods matched success at unmanipulated broods with the same helper:brood size ratios (Browning et al. 2012). Annual fledgling production was asymptotically related to breeding unit size (fig. 17.13; Browning et al. 2012). Breeding units of intermediate size produce chicks of greatest average fledging weights. Thus, helpers have positive effects on breeding success, but high numbers of helpers produce diminishing returns (in this case, helper:brood size ratios greater than 3).

Apart from increasing productivity, helpers may help defray the costs to breeders of reproducing. Breeders may benefit from help by being able to reduce their effort accordingly (compensatory effect), or by decreasing the interval between breeding attempts. Breeders may also benefit from help through increased survival and future fitness. A cross-species comparison of 27 cooperative breeders revealed that compensatory effects of helpers (load-lightening) tend to be seen in systems where nestling starvation rarely occurs, whereas additive benefits of help are common in systems where rates of nestling starvation are high (Hatchwell 1999). Both sexes showed compensatory reductions in care when nestling starvation was infrequent, but the effect was only significant for females (Hatchwell 1999; see also Female versus Male Helpers, below).

Considerable effort has been invested in determining correlates of cooperative breeding on a global scale. These can be divided into phylogenetic predispositions, ecological and environmental (climatic) factors, and life history traits. Broad patterns are summarized in the next sections.

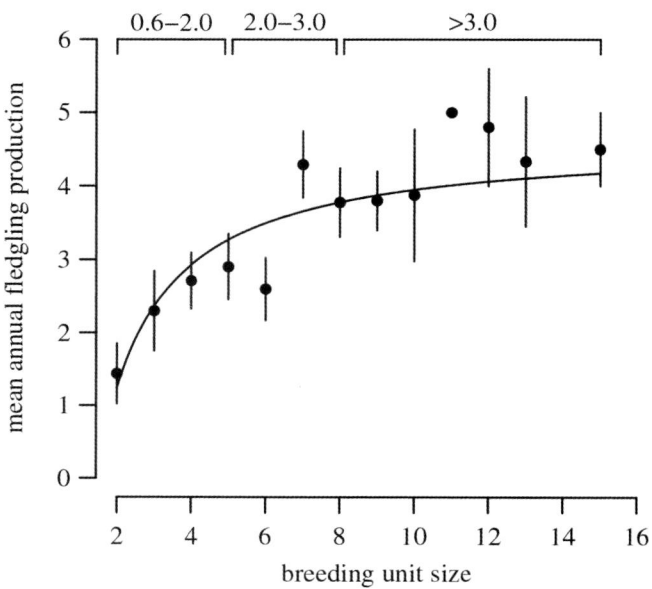

Figure 17.13. Helpers have positive effects on breeding success, but greater helper numbers give diminishing returns. An asymptotic increase in mean (±SE) annual fledgling productivity was related to the breeding unit size. Brackets at top show naturally occurring ranges of breeding unit to brood size ratios for the corresponding breeding unit sizes. *Reprinted with permission from Browning et al. 2012.*

Phylogenetic Predispositions to Sociality

An estimated 852 species of bird are cooperative breeders, approximately 9 percent of the species diversity representing 45 percent of bird families (Cockburn 2006). In some poorly studied species, cooperative breeding is inferred from phylogenetic affinity to cooperative species, but this estimate excludes 150 species known only rarely to exhibit cooperative breeding. Certain families of birds appear to be predisposed to sociality in general and cooperative breeding in particular. These include the Rallidae (moorhens and gallinules), Accipitridae (accipiters), Psittacidae (parrots and their allies), Coraciiformes (bee-eaters), Cuculiformes (specifically, South American lineages), Picidae (8 of the 21 species of woodpeckers in America), Ploceidae (weaverbirds), Leiothrichidae (laughingthrushes, Old World), Corvidae (crows and jays), and a large number of Australasian endemic oscine passerines related to corvids, including Australasian babblers, fairywrens, miners, apostlebirds, and others. In spite of an apparent degree of "phylogenetic inertia" (Edwards and Naeem 1993), ecological constraints and certain life history traits seem to also be important drivers of sociality.

Ecological Correlates of Sociality in Birds

Ecological constraints are thought to be implicated in the evolution of family living because of limitations on breed-

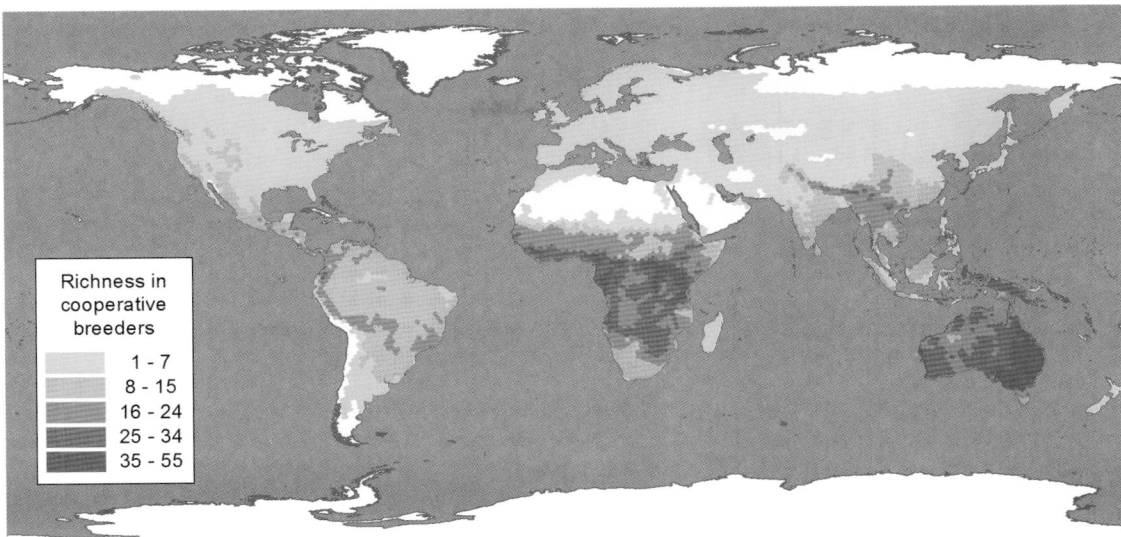

Figure 17.14. Global pattern of species richness of avian cooperative breeders. Colors show ranges in number of cooperatively breeding species. *Reprinted with permission from Feeney et al. 2013.*

ing vacancies (Emlen 1982). Cooperative breeders tend to be found in habitats referred to as savannah, shrubland, scrub, open woodland, and arid or semiarid. These are harsh environments, where resources are spatially or temporally unpredictable. Geographical information system technology enables us to view the relationship between habitat variables and species grouped by taxon or behavioral type on a global scale. For example, heat maps can be generated to plot the frequency of cooperative breeding onto global positions on a grid at high pixel densities. The global distribution of cooperative breeders among 9,310 non-marine species (95 percent of the world's birds) shows that cooperative breeders are more common in geographic locations with rich avifaunas, high mean annual temperatures, and low annual rainfall. Thus emerges a geographic trend for highly social species to be overrepresented in particular biomes that are more commonly found in the Southern Hemisphere (fig. 17.14; Jetz and Rubenstein 2011, Feeney et al. 2013).

Cooperative breeding is particularly common in the arid and semiarid regions of sub-Saharan Africa and Australia (Jetz and Rubenstein 2011, Feeney et al. 2013). A remarkable 277 (15 percent) species in the Afrotropical region, and 174 (12 percent) of species in the Australo-Pacific region are designated as cooperative breeders, with three or more adults regularly contributing parental care to young (Cockburn 2006). Cooperative breeders tend to be year-round residents, remaining in marginal habitat in the dry season. In these species, breeding behavior can be triggered within a day or two of the first major storm signaling the start of the rainy season. This contrasts with the strategy of some other species in arid subtropical regions that lead a nomadic lifestyle, subsisting in large flocks that travel by tracking precipitation among regions. These mobile flocks exhibit irruptive breeding in large

colonies only when the quantity of rain is sufficient to support a surplus of vegetative growth and food, e.g., granivores such as Zebra Finches (*Taeniopygia guttata*) in Australia (Zann 1996) and Red-billed Quelea in sub-Saharan Africa (Crook 1960). Rails and other waterbirds in sub-Saharan Africa also have nomadic populations that follow the rains and breed in ephemeral wetlands (Jamieson et al. 2000).

Comparatively fewer species inhabiting temperate areas are cooperative breeders. In North America, cooperative breeders tend to be found in rare or marginal habitats. For example, the Florida Scrub Jay (*Aphelocoma coerulescens*) is endemic to stunted oak scrubland that is restricted to fragments within the Florida peninsula. The species is so bound to this habitat that birds avoid any woodland with canopy. Likewise, the Red-cockaded Woodpecker (*Picoides borealis*) (see the box on pages 550–551) is a longleaf pine savannah specialist. Both of these habitats are fire-dependent; periodic natural wildfires reduce encroachment of woody undergrowth, maintaining the openness of the habitat and encouraging growth of fire-resistant tree species on which they each depend. Species obligately linked to fragile habitats of limited distribution are restricted in their ability to disperse. This in turn feeds back on their propensity for social living. It is no coincidence that some of the species of greatest conservation concern in North America are cooperative breeders. Both oak scrubland and longleaf pine savannah are now severely restricted. Both are found only in fragments in the eastern coastal plain. The low elevation of these rare habitats contributes additional risk from greater frequency and severity of storms affecting coastal areas and from forecasted sea level rise.

Fragile habitats clearly harbor disproportionate numbers of cooperative breeders, but recent research has explicitly addressed the effects of environmental unpredictability. Analysis

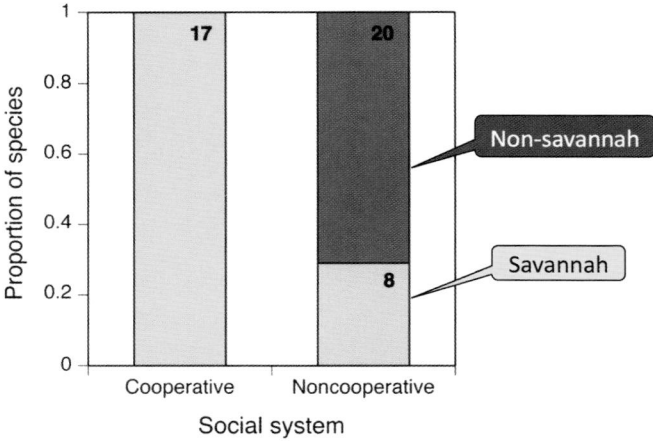

Figure 17.15. Proportions of cooperative breeding African Starlings in savannah (yellow bars) versus non-savannah habitats (green bar). *Reprinted with permission from Dustin R. Rubenstein, from Rubenstein and Lovette 2007.*

of inter- and intra-annual patterns of rainfall revealed that in Africa, climatic conditions suitable for breeding are relatively more unpredictable in savannah habitat than in deserts or forests (Rubenstein and Lovette 2007, Jetz and Rubenstein 2011). African starlings and weaverbird species show variability in their degree of sociality and group structure across the continent. Among 45 species of African starlings, 17 out of 25 species with helpers breed in savannah (fig. 17.15, Rubenstein and Lovette 2007). This relationship remained significant after controlling for phylogeny.

Ecological conditions that vary both seasonally and annually as a result of climatic stochasticity are proposed to select for a form of bet-hedging (Rubenstein 2011). Having extra caregivers at their disposal allows cooperative breeding groups to increase their reproductive success in seasons when prime conditions occur (Koenig and Walters 2015). Nevertheless, the relationship with environmental unpredictability appears inconsistent across taxa (Jetz and Rubenstein 2011). Hornbills (Bucerotidae) also inhabit a variety of dry and humid habitat types across Africa and Asia, and they nest in tree cavities. Yet, the occurrence of helpers among hornbill species was not related to unpredictability in rainfall (Gonzalez et al. 2013). In fact, a higher proportion of hornbill species were cooperative breeders in humid forests (21/42 species) than in savannahs or deciduous forests (4/19 species), but the relationship was not significant after controlling for phylogeny. Instead, only territoriality significantly predicted the occurrence of helpers in hornbills (Gonzalez et al. 2013). When seasonal and inter-year variation in precipitation and temperature were considered, a negative

relationship emerged between cooperative breeding and environmental variability (fig. 17.16; Gonzalez et al. 2013). In particular, among-year stability in rainfall emerged as a significant predictor of cooperative breeding in hornbills. Notably, hornbills have very much larger body size than the average passerine and may be less susceptible to environmental fluctuations. Hornbills are also renowned for a peculiar nesting habit: breeding females use mud and excrement to seal themselves into the nest chamber for the duration of the nesting cycle (Kemp 1995). This conceals the nest and likely evolved to reduce predation. The necessity for the female to be fed in addition to the brood has clear implications for the benefits of having provisioning helpers, at least in territories with sufficient resources to support them.

The discordance between studies reporting that cooperative breeding is related to environmental unpredictability on the one hand and stability on the other likely arises from other differences between taxa, including differences in feeding ecology related to body size, minimum home range size, and availability of suitable habitat (abundance and distribution of seasonal food as well as habitat saturation). Yet, this example highlights the importance of testing hypotheses in different lineages and of making comparisons among studies. To resolve whether or not there is a relationship between cooperative breeding and environmental stability warrants further study in other taxa exhibiting variation in social system. Differences in this regard are likely to be most striking between passerines and non-passerines. Collectively, non-passerines appear less sensitive to environmental variables (Jetz and Rubenstein 2011). Understanding how factors such as variation in annual precipitation among years will impact social evolution has important implications for species conservation in view of climate change.

Life History Correlates of Sociality

Social evolution theory predicts that cooperation can evolve through **kin selection**, and that helping is most likely to evolve in societies where genetic monogamy predominates. Evidence suggests that cooperative breeding has most commonly evolved in lineages with high mate fidelity. Comparative analyses of trait evolution are greatly facilitated by the availability of more complete and accurate avian phylogenies. Comparative analysis based on 267 species distributed over the entire bird phylogeny has shown that helping is more common in species with relatively lower frequencies of promiscuous mating. Overall, promiscuity rates were three times greater in noncooperative than in cooperative species (Cornwallis et al. 2010). Phylogenetically constrained analysis of the origins of the trait revealed that cooperative breeding had evolved 33 times, and been lost 20 times, in birds

Figure 17.16. Relation between cooperative breeding in hornbills and environmental uncertainty based on variation in A, temperature, and B, precipitation. Individual points in the scatterplots represent the climatic niche position of individual species, (●) cooperative and (○) noncooperative breeders. Tukey boxplots summarize data for each axis for cooperative (Coop) and noncooperative (Non) breeders. *Reprinted with permission from Gonzalez et al. 2013. C,* The Yellow-billed Hornbill (*Tockus leucomelas*) is a noncooperative species, inhabiting the semiarid zone of Southern Africa. *D,* A Southern Ground Hornbill (*Bucorvus leadbeateri*) social group in Kruger Park, South Africa, is led by a dominant breeding pair; individuals can live up to fifty years but there can be up to nine years between successful fledglings (Kemp 1995). *Photo by Warwick Tarboton.*

(fig. 17.17A; Cornwallis et al. 2010). Promiscuity rates were lower in noncooperative ancestors of cooperative breeders than in noncooperative ancestors of noncooperative breeders (fig. 17.17B; Cornwallis et al. 2010). Moreover, rates of promiscuity decreased during transitions to cooperative breeding, but reversions from cooperative breeding back to noncooperative breeding were associated with decreases in female promiscuity (fig. 17.17C; Cornwallis et al. 2010).

Variation in rates of promiscuity among avian lineages is also related to differences in mortality rate and life span (Arnold and Owens 2002, Dickinson et al. 2015). In cooperative breeders, individuals tend to be long-lived and commonly delay breeding for one or more years. Selection for "slow" life histories (long life, low rate of reproduction) is a characteristic of resident species with low dispersal rates, and is in part what determines the geographic bias for cooperative breeders in relatively warm areas of the world with year-round, albeit variable, food supplies. A more complete picture of the suite of characters predicting cooperative breeders emerges when we use comparative analysis combining independently derived observations. Cooperative breeders tend to be sedentary species living in warm climates, coming from lineages predisposed to longevity, having lower modal clutch sizes, low rates of promiscuity (Arnold and Owens 1998, 2002), and higher age at first reproduction (fig. 17.18A; Downing et al. 2015). Determining direction of causality between these correlated factors is challenging, but social living is clearly associated with high survival rates. Controlling for body mass, latitude, promiscuity and phylogenetic history, cooperative breeders have longer average annual survival rates than noncooperative breeders (fig. 17.18B; Downing et al. 2015).

Food is scarcer in the nonbreeding season, and supply maintenance of food resources is most critical at this time. The relationship between winter food supply, juvenile survivorship, and time to independence may be important in promoting delayed dispersal. In an attempt to understand the greater preponderance of cooperative breeders in the Southern Hemisphere, Russell (2000) found that average time to independence differed significantly between northern temperate passerines, which tend to remain dependent on their parents postfledging for less than a month, and tropical and southern temperate passerines, which commonly remain for one to three months. The underlying cause of this generality is not fully understood and warrants further investigation. However, it is clear that an extended period of dependence is a prerequisite for delaying dispersal beyond sexual maturation.

Dispersal and Philopatry

The evolution of cooperative breeding is perceived to be driven by two factors: the benefits of **delaying dispersal** and the benefits of **helping**. The diversity in social systems observed is attributable to differences in the degree and consequence of these two factors. In most cooperative breeders, at least a proportion of adults delay dispersal and behave as helpers, but these traits are not interdependent. We will see that delayed dispersal can occur without individuals investing help at their parents' nest (Ekman and Greisser 2002), and that failed breeders can help at the nests of relatives (Hatchwell et al. 2013).

Delayed Dispersal and Habitat Saturation

When suitable breeding habitat is limited, there are fewer opportunities for young adults to disperse and breed independently. Habitat saturation has been considered a major driver of the origin and evolution of cooperative care of young because it leads to retention of offspring on the natal territory (Emlen 1982). An experimental test of the importance of habitat saturation was conducted in a tiny (~15 g) tropical island endemic, the Seychelles Warbler (*Acrocephalus sechellensis*). Devastated by cats and other introduced predators, the entire world's population of 26 individuals was moved from the main island to the 29-hectare uninhabited island of Cousin in the Seychelles archipelago by conservation biologists with the International Council for Bird Preservation (now Birdlife International) in 1959. In 1968, intervention by the agency began, and after the removal of predators by 1973, the population had recovered to over 200 individuals. By this time, the number of breeding territories (±120) completely saturated the tiny island. It was in this year that the first instances of nonbreeding helpers were observed on some territories (fig. 17.19; Komdeur 1992).

Further increase in population size motivated the next stage in the population management plan: translocating 29 individuals to a second 68-hectare uninhabited island, Aride, in 1988, and two years later, the same number to a third island, 29-hectare Cousine Island, both situated nearby in the archipelago. Vacancies created on the first island, Cousin, by these removals were immediately filled. In most cases, the recruits were helpers from high-quality territories, demonstrating they would breed given the chance. Medium-quality territory vacancies were colonized only by birds originating from equivalent or lower-quality territories.

Translocated birds readily bred as monogamous pairs on high-quality territories in their new homes. As these islands in turn became saturated with territories, the Seychelles Warblers again bred cooperatively. However, importantly, the quality of available habitat factored heavily in individuals' decisions to disperse and breed versus stay and help on their natal territory (Komdeur 1992). Birds hatched in high-quality territories, based on monthly quantitative density

Figure 17.17. (opposite) Phylogenetic distribution of cooperative breeding and female promiscuity. *A*, Species labels and branches represent cooperatively breeding species (red) and noncooperatively breeding species (black). Blue circles indicate reconstructed ancestral values of promiscuity with circle size proportional to promiscuity rates; *B*, Promiscuity in noncooperative and cooperative ancestral lineages that gave rise to only noncooperative descendants (black bars), cooperative descendants (dark gray bars), or both (light-gray bars). Non-cooperative ancestors that led to cooperative descendants had lower promiscuity than those that produced noncooperative descendants; *C*, Percent changes (Mean ± SE of ancestral values) in promiscuity rates associated with transitions to and from cooperative breeding. Promiscuity rates decreased during transitions to cooperation, and decreased when cooperation broke down. *Reprinted with permission from Cornwallis et al. 2010.*

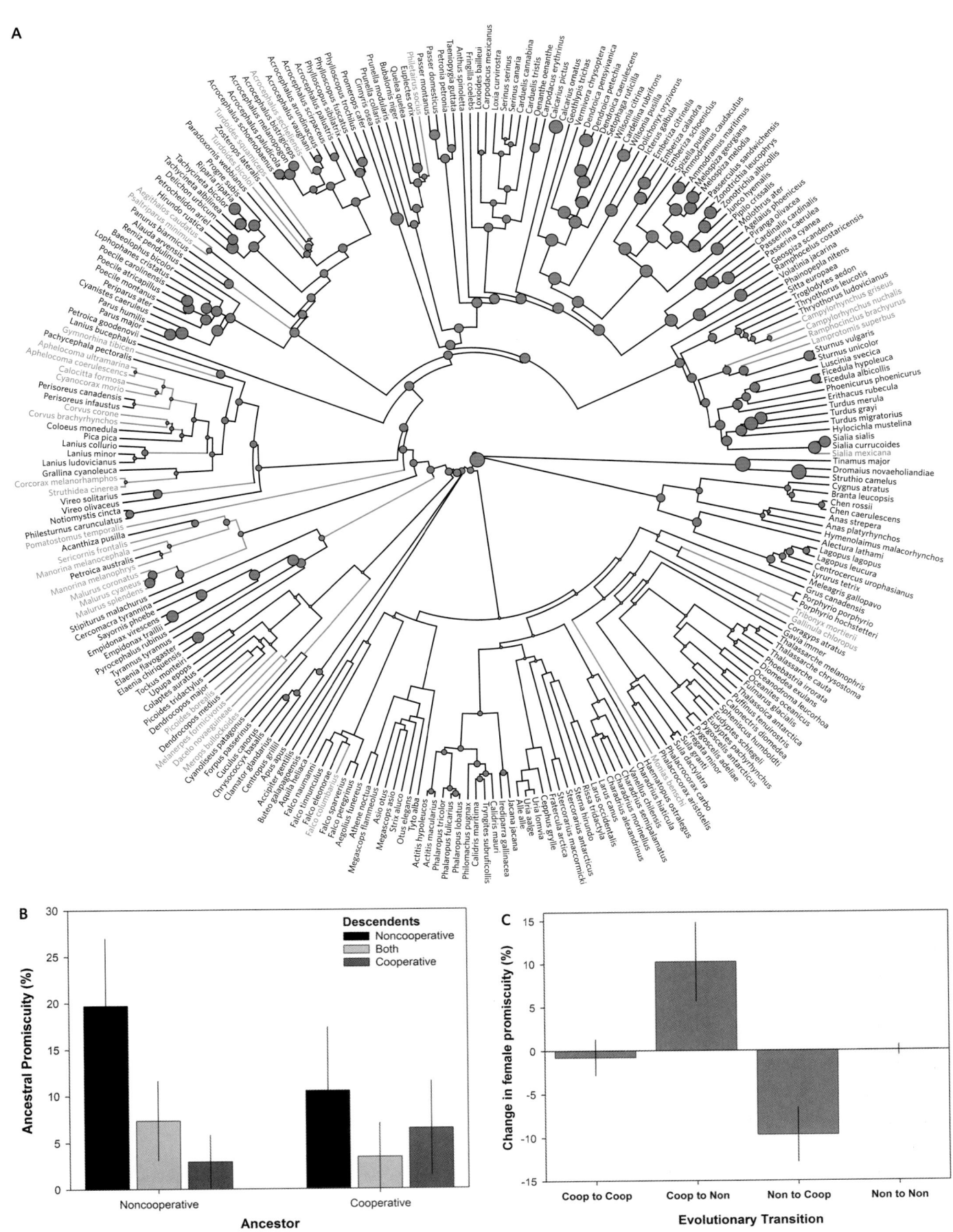

A

B

Ancestral Promiscuity (%)

Descendents
- Noncooperative
- Both
- Cooperative

Noncooperative | Cooperative

Ancestor

C

Change in female promiscuity (%)

Coop to Coop | Coop to Non | Non to Coop | Non to Non

Evolutionary Transition

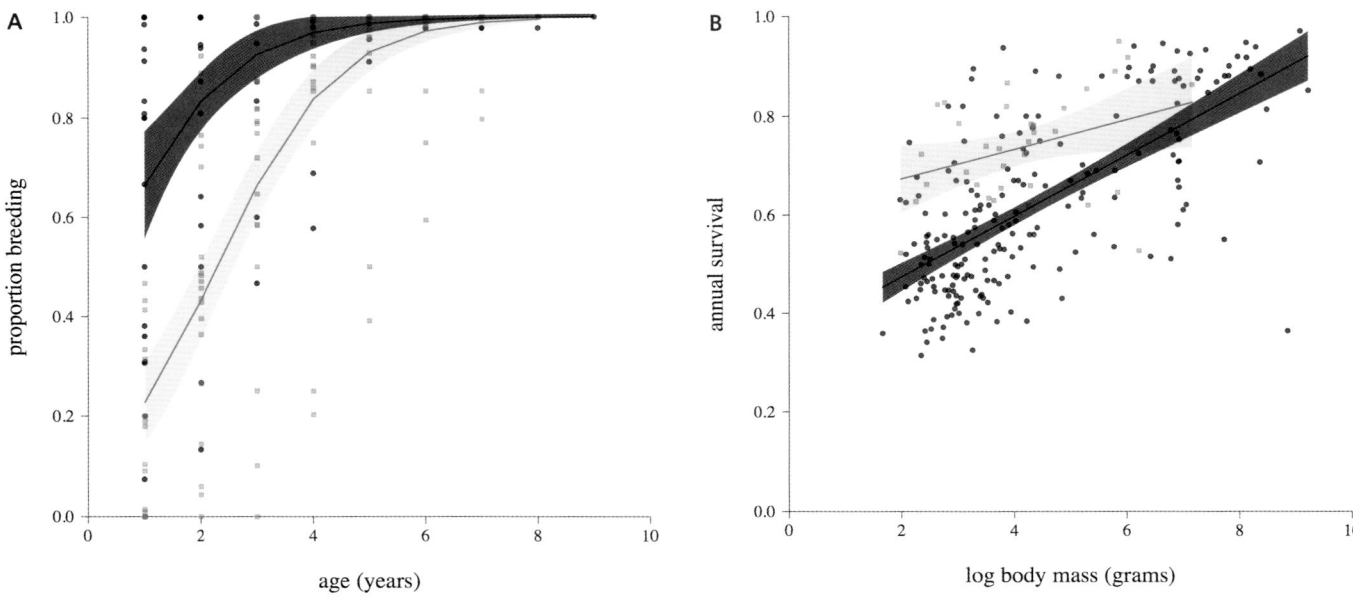

Figure 17.18. A, Mean age at first reproduction is later in cooperative breeders (blue/gray squares) than in noncooperative breeders (black circles). For each age in years, the proportion of individuals in each species is plotted. Regression lines are presented with 95 percent confidence intervals (*n* = 39 species); *B,* Cooperative breeders (blue/ gray squares) have higher survival rates than noncooperative breeders (black, circles) (*n* = 238 species). Regression lines are presented with 95 percent confidence intervals. *Reprinted with permission from Downing et al. 2015.*

estimates of available small foliage dipterans that the warblers glean (fig. 17.19B), delayed dispersal and became nonbreeding helpers on their natal territory when only low-quality habitat was available. In general, birds hatched in high-quality territories, and some in territories of intermediate quality, tended to delay dispersal, whereas those hatched on low-quality territories dispersed and bred in low-quality territories. Thus, while habitat saturation is clearly an important factor in the origin of cooperative breeding, individual dispersal decisions take account of habitat quality among other factors (Koenig et al. 1992).

Population densities have stabilized on the islands, and the Seychelles Warbler continues to be a facultative cooperative breeder. Helped by conservation efforts and translocations to two further islands in the archipelago, the world population is now ~2,750 adult birds distributed over five islands (Komdeur et al. 2016), and Birdlife International recently downlisted the species from globally endangered to vulnerable.

Resource-Based Philopatry

Variation in food quality or abundance can result in limited dispersal, as seen in the example above. Species that are food specialists are limited by the distribution of that resource and are more likely to live in social groups if the preferred food type is temporally or spatially unpredictable. This is especially true if food production is eruptive, such as in the fruit of masting trees. In California, Acorn Woodpecker (*Melanerpes formicivorus*) social groups accumulate seasonal stores of acorns in granaries. Families establish granaries in oak trees on their territories. All members of the group contribute to this communal resource, drilling holes in the bark and wedging an acorn tightly into each hole. Granaries at traditional sites contain from 2,000 to 10,000 holes, a valuable trove that group members collectively provision, maintain, and defend. Acorns are gathered and stored in the autumn. They provide essential food for the group during the lean winter months (fig. 17.20B; Koenig and Stacey 1990). Access to a granary is essential to winter survival, so there is strong selection for delaying dispersal. Moreover, average annual group reproductive success is strongly related to the relative abundance of the acorn crop in the previous year (fig. 17.20A; Koenig et al. 2016).

Staying without Helping

Benefits to future direct fitness can be sufficient to select for delayed dispersal without help in Siberian Jays (*Perisoreus infaustus*) that are year-round residents of the taiga biome in the northern extremes of Eurasia (sister species to the Grey Jay, *P. canadensis*, of the Canadian boreal forest). In passerines, males are generally more philopatric. Male Siberian Jays commonly delay dispersal from the natal territory, but

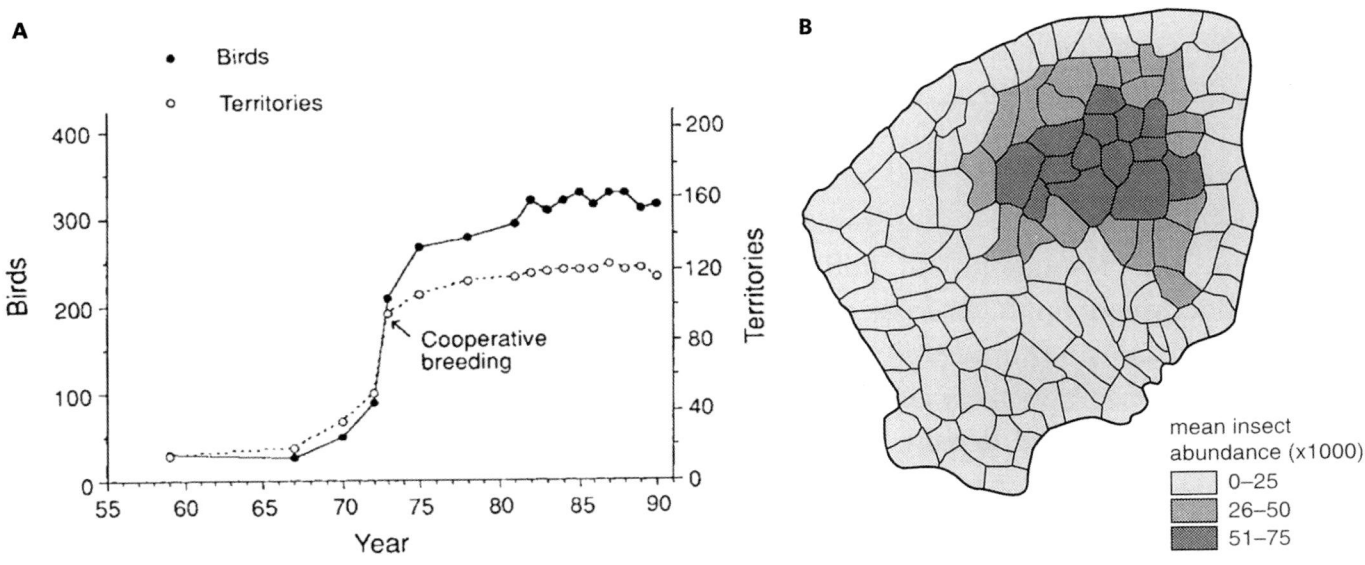

Figure 17.19. Seychelles Warblers began breeding cooperatively on Cousin Island once their population size reached the threshold limit of territory availability. *A*, The number of birds, and territories based on census data on Cousin Island 1959–1990. *Adapted with permission from Komdeur 1992. B*, Territory map showing the outline of Cousin Island and saturation of Seychelles Warbler territories (1986–1990). Territory quality is coded based on calculations of insect prey densities incorporating mean annual territory size, mean foliage cover of each plant species, and mean insect abundance per plant per unit leaf area. *Adapted with permission from Komdeur et al. 2016.*

they never assist in their parents' breeding attempts. Individuals remaining at home gain benefits of "nepotism," particularly in access to food during the harsh winter months (Ekman and Griesser 2016). Individuals that initially disperse at ~eight weeks of age and ones that delay dispersal for a year or more can both gain breeding positions, but delaying dispersal has lifetime fitness benefits. Radiotelemetry studies show higher survival rates from fledging to first year in delayed dispersers relative to early dispersers (Griesser et al. 2006). Group cohesion and the presence of both parents are both significant: removal experiments showed that subordinates whose father was replaced by a stepfather lost their advantage and most dispersed soon thereafter (Ekman and Griesser 2002). Though they delay breeding, late-dispersing young adult Siberian Jays have significantly higher lifetime breeding success compared with age-matched early-dispersers because of the benefits of group safety, familiarity, and access to group food sources (Ekman and Griesser 2016).

Patterns of Sex-Biased Dispersal and Incest Avoidance

When dispersal is constrained by the lack of suitable habitat, so that mature birds continue to live on their natal territory, a risk of inbreeding arises. Delayed dispersal is thought to have selected for incest avoidance in many cooperative breeders, but its expression varies among systems. In woodpeckers, incest avoidance is observed when dominant breeder vacancies are filled. Red-cockaded Woodpecker territory inheritance is patrilineal; a female can only inherit her natal territory upon the death of all of her male relatives (Walters and Garcia 2016). Adult members of Acorn Woodpecker social groups consist of a coalition of one to eight related males (brothers, fathers and sons) and a coalition of one to four related females (rarely more than two, sisters or a mother and her daughter). Thus, each subordinate member has kinship with same-sex members of the group, but the breeding male and female are not related to one another (Koenig et al. 2015). Breeding vacancies are usually filled from outside the group following power struggles. The death of a breeder triggers intense conflicts between same-sex coalitions (Koenig et al. 2016) that can persist over days in the case of territories with exceptional wealth in granary stores (Hannon et al. 1985). Power struggles over female breeding vacancies are also observed in White-browed Sparrow-Weavers that have matrilineal territorial inheritance, and in Southern Pied Babblers (*Turdoides bicolor*) that have patrilineal territory inheritance (Ridley 2016), providing strong evidence for ecological constraints as a root cause of sociality. Acorn Woodpeckers are remarkably faithful to their mates: no incidences of extra-pair paternity have been recorded (Haydock and Koenig 2003; box on pages 554–555).

In stark contrast to Acorn Woodpeckers, dominant breeder vacancies are always filled from within groups in Australian Superb Fairywrens (Mulder et al. 1994). Avoidance of

ECOLOGY AND CONSERVATION OF A HIGHLY SOCIAL SPECIES, THE RED-COCKADED WOODPECKER

Federally listed as endangered since 1968, the Red-cockaded Woodpecker, named for the faintly visible small red streak of plumage behind the temples of males, is resident only in longleaf pine savannah of the southeastern United States. It is reliant on 80- to 120-year-old pine trees in which it excavates roosting cavities that also serve as nests. The minimum diameter of the tree is age-dependent, and excavations can take a group of woodpeckers on the order of ten years to complete. Longleaf pine (*Pinus palustris*) is the preferred species, as the release of sticky resin from the sapwood at the entrance hole affords protection from rat snakes and other nest predators. Red-cockaded Woodpecker groups of two to seven individuals defend territories of 50–150 hectares. Families travel together as a foraging group and use clusters of roost chambers in different parts of the territory from night to night.

Logging, land conversion, and habitat fragmentation has severely limited the amount of suitable habitat left. For the woodpecker, the conundrum of requiring large territories and being habitat-saturated is compounded by the fact that distances between fragments of suitable habitat in some cases exceed normal dispersal distances. Habitat restoration is complicated by the need for mature trees, and an apparent aversion of the birds to midstory hardwoods (Conner et al. 2001).

Jeffrey Walters and his students at Virginia Tech have studied Red-cockaded Woodpeckers since 1980 in the Sandhills region, at Marine Corps Base Camp LeJeune in

A, David Allen of the North Carolina Wildlife Resources Commission installs an artificial nest cavity. *B*, Restoration of essential longleaf pine savannah habitat. *Photos A and B © NCWRC / Melissa McGaw.*

C, A female red-cockaded woodpecker at a natural nest cavity. *Photo by Michael McCloy.*

North Carolina, and at Eglin Air Force Base in the Florida Panhandle. Their long-term research has uncovered details of their family-group structure and formation, lifestyle as fire-dependent habitat specialists, and unusual life histories (Conner et al. 2001). Territories are extremely stable. The same family groups have occupied territories for generations, many since the beginning of the study (Walters and Garcia 2016).

An important practical contribution of their research was the recognition that Red-cockaded Woodpeckers would colonize artificial nest cavities. These additions were originally installed to test the hypothesis of habitat saturation versus the alternative of benefits of philopatry for cooperative breeding. The successful colonization of 18 out of 20 territories in which boxes were experimentally added led to the formation of 18 new breeding units, an important breakthrough to species conservation (Walters et al. 1992).

The US Forest Service, Department of Defense, and state partners participate in a US Fish and Wildlife Service–led species recovery plan monitoring Red-cockaded Woodpecker populations on publicly owned and private lands. Recognition of the importance of fire in maintaining this rare habitat has led to restoration of critical parcels through prescribed burns and midstory hardwood removal. The recovery plan also includes the installation of artificial nest cavities, "inserts" (purpose-built, with squirrel-proof metal front plates and short entrance tunnels), as a principal management strategy, as well as the translocation of juveniles between populations to increase genetic mixing and help mitigate inbreeding in small fragments. The increases seen in the number of woodpecker breeding units over the last 20 years has resulted from partnership between academic researchers, federal and state agencies, and private stakeholders (Walters and Garcia 2016). Their collaborative efforts are exemplary of sound conservation science. Using a metapopulation management approach is essential: ensuring sustainable regional persistence of populations must take account of movements between habitat fragments under different jurisdictions.

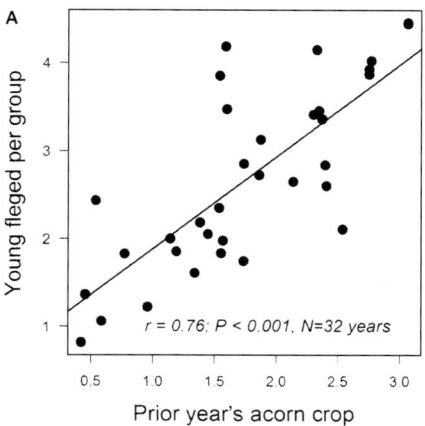

Figure 17.20. *A*, Mean number of Acorn Woodpecker young fledged per group in relation to the relative yield of acorns during the prior autumn. *Reprinted with permission from Koenig et al. 2016. B*, Acorn Woodpecker family group and part of an oak acorn granary at Hastings Reservation, in the Carmel Valley of California. *Photo by Bruce Lyon.*

Figure 17.21. Male Superb Fairywrens display to females on neighboring territories to try to gain extra-pair fertilizations (EPFs). *Photo by David Hollie reproduced with permission.*

incest in this species is achieved by females seeking matings elsewhere, resulting in 76 percent of offspring being sired by extra-group males (Mulder et al. 1994), and earning this species the distinction of the "world's most unfaithful bird" (fig. 17.21). Here, male kin (subordinate "auxiliaries"), sons of previous broods, queue for the dominant position. Deceased female breeders are commonly replaced by daughters. Extra-group males court females on the female's territory during their fertile period by performing a specialized "yellow-petal display" akin to presenting a bouquet (Mulder 1997). Stealthy mating behavior enables female Superb Fairywrens to control paternity of her offspring without disturbing group stability. In the ensuing days, females fly under cover of predawn darkness as males advertise with song during the dawn chorus (chapter 14) to copulate with the extra-group males advertising on their own territories by singing (Cockburn et al. 2009). Within populations, a few males tend to gain most of the extra-pair paternity (Mulder et al. 1994); timing of molt is correlated with male age, and an early pre-nuptial molt is a strong predictor of which extra-pair males are preferred by females (Mulder and Magrath 1993). Other fairywrens also have strong male philopatry and high extra-pair paternity (e.g., 24–52 percent of offspring in Splendid Fairywrens, *Malurus splendens melanotus*, Webster et al. 2004; 57 percent of offspring in Red-winged Fairywrens *M. elegans*, Brouwer et al. 2011). The degree to which paternity is sought within or outside the group in these species varies with female philopatry, consistent with incest avoidance.

Why Do Helpers Help?

Direct Benefits

In some species, maternity and/or paternity is shared among group members. Cooperative care in avian societies can in-

volve the full gamut of genetic breeding systems, from monogamous pairs with their offspring as helpers (nuclear families) to polygyny, polyandry, and polygynandry. That subordinate male helpers could gain paternity in the broods they helped rear was first recognized in tropical Stripe-backed Wrens (*Campylorhynchus nuchalis*) (Rabenold et al. 1990). Shared paternity varies among cooperative breeders with predominantly male helpers (see also Reproductive Partitioning and Skew, below).

Hypotheses for direct benefits of helping (box on page 555) have been tested extensively in cooperatively breeding Acorn Woodpeckers. Both sexes have relatively low levels of reproductive skew. Subordinates related to the opposite-sex breeder are nonbreeding helpers. In contrast, subordinates that are not related to the opposite-sex dominant will attempt to breed. Co-breeding males share paternity relatively equitably over multiple nests, although there is commonly high skew within a brood. Surprisingly, though, there is no evidence that males can assess accurately their paternity: relative effort in care is not related to paternity (Koenig et al. 2016). Joint-nesting females that rear young successfully lay relatively equal numbers of eggs and have similar success rates (Haydock and Koenig 2003). The **skills hypothesis** was not supported: the extent of helping by subordinates did not correlate with higher reproductive success later in life, and helping actually increased with age (Koenig and Walters 2011). Neither was the **pay-to-stay** hypothesis: helping subordinates were not more likely to inherit the breeding position than ones that provided little help, though subordinate males that remained longer on their natal territory were more likely to inherit. Instead, in Acorn Woodpeckers, the evidence is most consistent with the hypothesis of **indirect benefits** from helping to rear related young (Koenig and Walters 2011).

To test the pay-to-stay hypothesis in Superb Fairywrens, subordinate males were experimentally removed and held captive for 24 hours before release (Mulder and Langmore 1993). Subordinates are expected to help during breeding attempts, and their absence was noticed. Released helpers uttered alarm calls on return to their territory, and were joined immediately by family members. They were chased relentlessly by the dominant male, and punished with taunts and pecks (Mulder and Langmore 1993). Punishment may have been meted for abandonment of duty, or for return after perceived "dispersal." Nevertheless, this experiment contributes evidence in support of pay-to-stay and demonstrates that breeders pay attention to the activities of group members.

A remarkable example of exploitation of **delayed reciprocity** was observed in the White-winged Chough (*Corcorax melanorhamphos*), an Australian crow-like passerine that breeds in large social groups (Heinsohn 1991). White-winged Choughs are obligately cooperative: groups with fewer than three adults fail to breed successfully. Reproductive rates are low; groups with fewer than seven adults can produce only one surviving fledgling per year. Selection for **group augmentation** has led to instances of White-winged Choughs kidnapping the recently fledged young of neighbors. Kidnappings took the form of raids that resembled aggressive group contests. Usually the group perpetrating the kidnapping had a numerical advantage and lured the fledgling with food. Fledglings were reared to independence by the kidnappers, at which point they joined the ranks of the other adults in the dominance hierarchy. Neighboring groups competed over territory and were known to destroy the nesting attempts of rival groups. All of the fledglings that survived over a year in their new group helped at the nests of their kidnappers. Fledgling kidnappings have also been observed between groups of Southern Pied Babblers, specifically at the end of the breeding season by failed breeders (Ridley 2016).

Members of social groups can benefit directly from participating in group nest efforts as co-breeders, or by queuing for a breeding vacancy, inheriting a breeding position in either their group or a neighboring group. Male Pied Kingfishers (*Ceryle rudis*) can breed in their first year, but rarely have the opportunity. They can spend the first year as a nonbreeding "floater," or opt to assist at their parents' nest (primary helpers), or the nest of nonrelatives (secondary helpers). For indirect benefits to work, helpers have to enhance offspring fitness. Reyer (1984) showed that the aid provided by primary helpers does significantly increase fledging success. By contrast, secondary helpers provision nests at very low rates. In the absence of indirect benefits, secondary helpers have other incentives to help (Reyer 1984). One-year-old males that behaved as secondary helpers were significantly more likely than primary helpers to become breeders in their second year, sometimes evicting the male at the nest they helped and mating with the group female (Reyer 1990). Based on fitness calculations, being a secondary helper was only moderately less profitable than becoming a primary helper (Reyer 1984).

Indirect Benefits (Kin Selection)

Evolutionary thinking about social behavior was revolutionized by the development of **kin selection** theory. Hamilton's (1964) model showed that individuals could gain fitness benefits indirectly through help to related breeders. If an individual has no immediate prospects for independent reproduction, it can benefit indirectly by helping to rear young of close relatives, enhancing their survival and increasing the probability of their shared genes being passed on to the next generation. Even if offspring survival is not enhanced, the benefit of help can alternatively be gained through lightening the workload of the breeders, and their consequent enhanced survival and future reproductive success. By either mechanism, individuals can enhance their inclusive fitness, which over their lifetime is calculated as the sum of the reproductive success they achieve through breeding directly themselves (minus any benefit from help they receive in turn), and the reproductive success of relatives resulting from their assistance, with the latter multiplied by the coefficient of relatedness to the juveniles helped (see also **Hamilton's rule**, box on page 556).

White-fronted Bee-eaters (Coraciiformes) nest in colonies of several hundred birds. They excavate burrows in the sandy banks of East African rivers. In a study at Lake Nakuru National Park, Kenya, over half of the breeding pairs were assisted by one to five nonbreeding helpers. The social structure is complex: subunits of the large colonies, called "clans," consisted of several pairs and their helpers (3 to 17 members) (Emlen 1990). Clans also defended feeding territories up to seven kilometers away from the nest site. Each member found a home range within the group's territory where they hawked large flying insects. (They are named for the habit of catching hymenopteran prey, and are able to disarm the venomous stinger.)

White-fronted Bee-eater provisioning rates increased with helper number, and helpers had a positive effect on breeding success (Emlen 1990). Relatedness was a strong predictor of the likelihood of helping. Helpers were commonly males that fed younger full siblings, but some helped rear half sibs, a brother's offspring, grand-offspring, young siblings of their own parents, other more distant relatives, and occasionally unrelated young (Emlen and Wrege 1988). Related helpers gained significant indirect benefits through enhanced fledging rates of kin they provisioned (fig. 17.22; Emlen 1990). Relatedness to the offspring they provision is strongly correlated with helper provisioning rates in Red-cockaded Woodpeckers, Florida Scrub Jays, Chestnut-crowned Babblers, and a plethora of other cooperative breeders. The advantage to breeders of having helpers is sufficiently great in White-fronted Bee-eaters that older males have been observed to disrupt the breeding attempts of their grown sons in order to recruit them as "failed-breeder" helpers at their own nests. Remarkably, the indirect benefit to sons is so high that they tend not to resist (Emlen and Wrege 1992).

Localized dispersal based on territorial quality and structuring of "kin neighborhoods" underlies facultative cooperative breeding in Western Bluebirds (*Sialis mexicana*). Resident Western Bluebirds remain in extended kin groups in winter.

WALT KOENIG AND JANIS DICKINSON, PARTNERS IN LIFE AND IN EXPLORING SOCIAL EVOLUTION IN BIRDS

By Mark T. Stanback and Susan B. McRae

Walt Koenig and Acorn Woodpecker. *Photo by Bruce Lyon.*

For Walt Koenig and Janis Dickinson, studying cooperative breeding has been a family affair. Living and working together at the University of California, Berkeley's Hastings Reservation in California's Carmel Valley, they pursue a shared passion for uncovering the family secrets of birds.

For over 35 years, Walt and his collaborators have undertaken research into the behavior and ecology of Acorn Woodpeckers. In 1987, he and Ronald Mumme completed the monograph *Population Ecology of the Cooperatively Breeding Acorn Woodpecker*, describing their work on this system up to that point. In 1990, recognizing the value of comparing the findings of other studies of social species in order to understand general principles, Walt edited, with Peter Stacey, the ground-breaking *Cooperative Breeding in Birds: Long-Term Studies of Ecology and Behavior*, each chapter of which covered a different social bird selected from among long-term studies carried out around the world. This volume inspired many of us to pursue studies of social evolution in birds and guided us toward new systems and questions.

While the study of cooperative breeding has been his mainstay, Walt has also published on a wide range of topics related to social behavior and avian population ecology. He became particularly interested in the phenomenon of mast-fruiting, and the degree to which it varies spatially and temporally (known as "spatial synchrony"), after recognizing the extent to which the population of Acorn Woodpeckers at Hastings was dependent on the highly variable acorn crop.

Apart from Acorn Woodpeckers and acorn production of California oaks, he has also examined a variety of related questions in other species, including Northern Flickers, Red-billed and Yellow-billed Oxpeckers, Yellow-billed Magpies, European Starlings, Western Bluebirds, West Nile virus, dragonflies, periodical cicadas, monarch butterflies, and invasive gypsy moths and emerald ash borers. The beauty of this unusual range of study species and research interests is that every inquiry he makes and every paper he writes informs his ornithology.

Walt's insights into cooperative breeding are shaped by his ability to synthesize concepts into these many disparate avenues of inquiry. He is also a witty and entertaining speaker who can give a brilliant discourse on any of these topics peppered with hilariously funny anecdotes. Walt is known for tying ladders together to get to woodpecker nests on rotten limbs, so it should come as no surprise that John Alcock, in his textbook *Animal Behavior*, described him as an iconoclast.

Although some of Walt's inherent zaniness has undoubtedly rubbed off on Janis, she is quick to point out the sensibleness of studying a species that will nest in a box five feet off the ground. Janis has long been interested in the root causes of sociality. Having begun her career studying multiple mating and sperm competition in insects, she turned her sights to birds just as new genetic techniques were becoming available to measure relatedness. For over 25 years, Janis has led a study of Western Bluebirds, investigating ecological, genetic, and social factors that result in delayed dispersal and help in some but not all families. Indeed, her research has demonstrated that mistletoe berries constitute a form of family wealth that allows for delayed dispersal and family-group living. Her insights into sexual selection, sex ratios, and life history theory have, in turn, served to increase her effectiveness as professor of natural resources at Cornell University, Arthur A. Allen Director and Professor of Citizen Science at the Cornell Lab of Ornithology, positions she held between 2005 and 2018. In their focus on anthropogenic change and its consequences for biodiversity, she and her team must simultaneously consider landscape and climate change, spatial genetics, and individual variation. Janis also coordinated immensely popular Internet-based programs at the Lab: NestWatch, The YardMap Network, Project FeederWatch, and Celebrate Urban Birds. Janis has taken a leading role in communicating science to the public and in promoting the participation of bird enthusiasts in collecting ornithological data over broad geographic areas.

In the course of their respective long-term field studies, Walt and Janis have trained generations of students in field ornithology. They have each made unique contributions to our understanding of intra- and interfamilial interactions and how these affect population level dynamics. They have also striven to identify unifying principles in the field of avian social behavior, and to achieve this they have reached out to other scientists from around the world. Together, Walt and Janis have edited two books on cooperative breeding in birds and other vertebrates. The first, published in 2004, examined dimensions of avian sociality from its evolutionary origins to its proximate causes. The latest, in 2016, surveys long-term studies of social vertebrates, including revising and updating some of the original systems profiled in Stacey and Koenig (1990), as new data and analytical methods have made possible. These syntheses present the current state of the art, but also identify the gaps in our knowledge to move the field forward. Each stands as a classic, and will inspire future generations of students of ornithology.

Janis Dickinson watching Western Bluebirds at Hastings Reservation. *Photo by Walt Koenig.*

The presence of berry-rich mistletoe, clusters of which are found growing as epiphytes on the branches of native oaks, is an important resource sustaining group members on their territory through the winter months. Experiments in which mistletoe was removed from some territories revealed that mistletoe volume was a strong predictor of male philopatry (fig. 17.23A; Dickinson et al. 2014). Males that stayed during their first winter tended to disperse locally, and often budded-off a breeding territory from their parents'. Territorial budding is also observed in Florida Scrub Jays (Woolfenden and Fitzpatrick 1984). In addition to territory quality based on mistletoe wealth, male philopatry in Western Bluebirds

DIRECT BENEFITS OF HELPING

Subordinate breeding. Subordinates may breed while helping to rear the offspring of their parents or unrelated birds.

Pay-to-stay. Helping can be payment for being able to remain on the territory and share group resources. Helpers can be unrelated birds that help to gain acceptance by earning membership in the group.

Skills hypothesis. Helping enhances the development of parenting skills that will serve the helper in the future when it becomes a breeder (Brown 1987). Difficult to test experimentally, support for this hypothesis has been scarce (but see Komdeur (1996) for strong experimental support).

Delayed reciprocity. Reciprocal altruism (Trivers 1971) was first proposed as a mechanism promoting cooperative breeding by David Ligon (1983), who pointed out that by forming an association with offspring they help to rear, helpers could gain a workforce for their future breeding attempts. The recruitment of future helpers may be viewed as a form of delayed reciprocity because it may be a year or more before the favor is repaid. It is difficult to test predictions for this that are mutually exclusive of group augmentation or other hypotheses.

Group augmentation. Based on shared resources and defense, subordinates that remain on their natal territory tend to be in better physical condition than peers that disperse. This can increase their survival and enhance their future fitness. Their availability enables them to efficiently fill a breeding vacancy within the group or in an adjacent territory (queueing). Enhancing group size via helping increases these direct benefits.

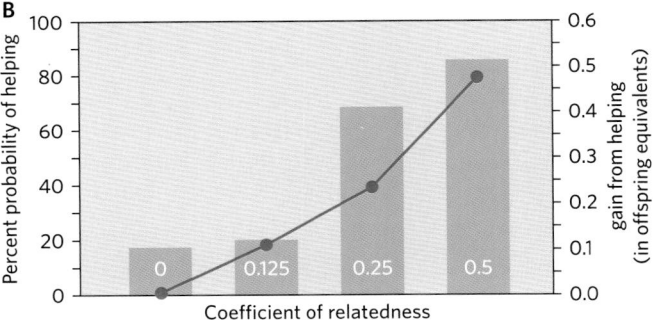

Figure 17.22. A, White-fronted Bee-eater helpers gain indirect fitness by provisioning related young at the nest. *Photo by Warwick Tarboton.* *B*, The indirect fitness benefit gained from helping is proportional to the coefficient of relatedness, *r*, between the helper and the offspring they help rear. The degree of relatedness predicts an individual's probability of helping. *Adapted with permission from Emlen et al. 1995.*

is contingent on the presence of at least one parent (Dickinson et al. 2014). The presence of both parents nearly doubled the likelihood that males would stay in the family group (fig. 17.23B; Dickinson et al. 2014). Sons commonly helped at the nests of their parents or brothers as first-year birds or as failed breeders, deriving indirect benefits (Dickinson 2004a; box on pages 554–555).

Help by Failed Breeders

Helping by breeders whose nests have been depredated is overwhelmingly kin-based, and can occur in species where all adults attempt breeding initially. Long-tailed Tits (*Aegithalos caudatus*) breed in woodland thickets and scrub across Eurasia. A study near Sheffield, England, revealed that helpers are recruited from local kin neighborhoods that arise through the tendency for male kin to settle in proximity to their natal territory (Hatchwell et al. 2001, Hatchwell 2009). All mated pairs initiate independent breeding, laying clutches of 8–11

eggs in their small domed nests. However, the risk of nest predation is high, and failed breeders join in feeding the nestlings of their kin. Such helping behavior by failed breeders is termed "redirected helping," and in the case of Long-tailed Tits the number of helpers at nests is significantly related to the level of nest predation and the length of the breeding season, which is subject to climatic variation. Limited by a short season, assisting in care for a relative's brood may be a better option than attempting a new nesting effort to replace one's own lost brood (Hatchwell et al. 2013). A high nest predation rate causes many individuals to achieve the majority or even all of their inclusive fitness through helping (box on page 558).

Though redirected care is costly to Long-tailed Tit helpers, indirect benefits of helping accrue primarily because of the increased survival of young at helped nests (Hatchwell et al. 2004). The presence of helpers also allows breeders, especially males, to reduce their own care ("load-lightening") and hence to have a higher chance of surviving to the following breeding season (Meade et al. 2010). Choice experiments

REPRODUCTIVE SUCCESS: A COOPERATIVE PERSPECTIVE

Direct fitness: an individual's lifetime reproductive success; the number of their own surviving offspring they contribute to the next generation.

Indirect fitness: the added contribution to offspring survival attributed to the helper, factored by the relatedness of the helper to the offspring.

Inclusive fitness: the sum of an individual's direct fitness and its indirect fitness.

Coefficient of relatedness, r: the estimated proportion of genes shared between two individuals; i.e., offspring share 50 percent of their genes with each parent, and full siblings share on average 50 percent of their genes with each other.

Hamilton's rule: help (caregiving behavior) will evolve when the following conditions are met:

$$r \times B > C$$

where r = coefficient of relatedness between care giver and recipient

B = benefit to the recipient of care

C = cost to the care giver

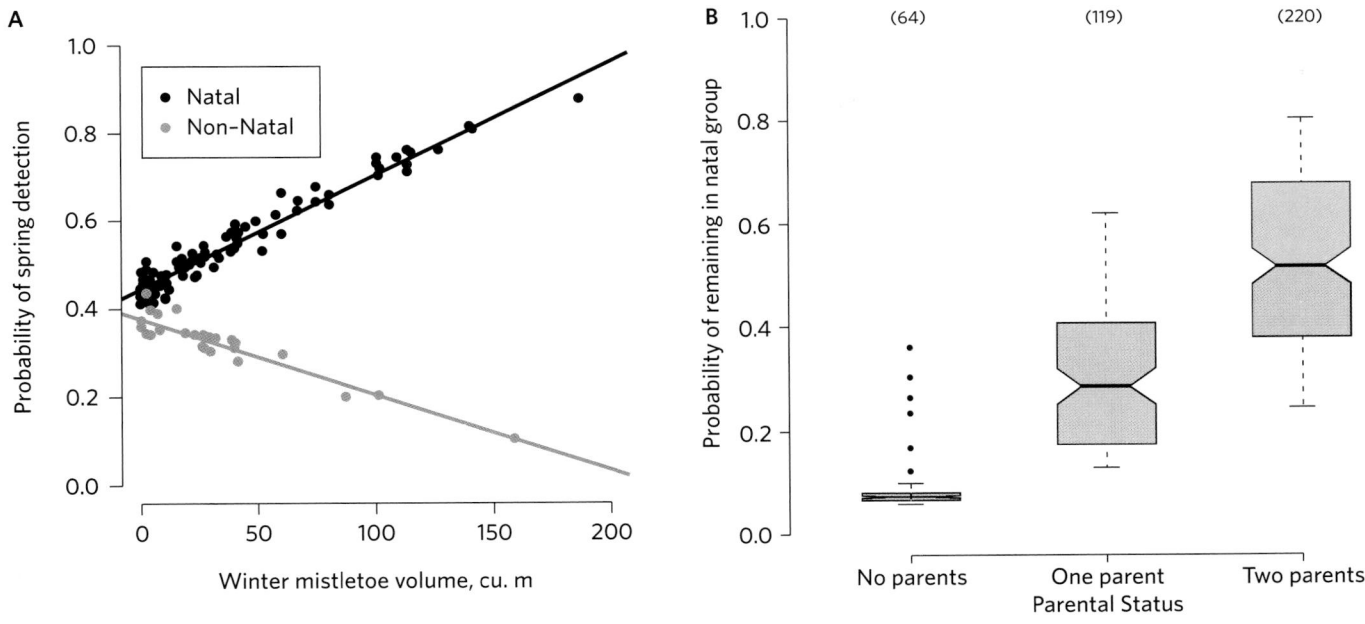

Figure 17.23. Male Western Bluebird philopatry is related to both territory quality and kinship to group breeders. *A,* The probability of detecting surviving male offspring in their natal territory the following spring was strongly positively related to mistletoe volume in the winter territory when they overwintered in their natal groups (black line, dots), but negatively related when they did not (gray line, dots); *B,* The probability that males remained in their natal group for the first winter was significantly related to the presence of parents. The presence of both parents doubled the average probability compared with only one. *Reprinted with permission from Dickinson et al. 2014.*

have shown that help by failed breeders is strongly kin-biased (Russell and Hatchwell 2001), but the unusual route to helping in Long-tailed Tits means that, although kin are preferred, helper relatedness to the brood is, on average, low (mean $r = 0.16$). However, the indirect benefits of increased productivity and survival of kin outweigh the survival costs that helpers incur through their helping behavior (Hatchwell et al. 2014), thereby satisfying Hamilton's rule for the evolution of costly cooperative behavior.

To selectively help kin requires kin recognition, and there is evidence that Long-tailed Tits have an auditory mechanism for doing so. Both sexes give a contact call for short-range communication. This *churr* call is more similar between close relatives, and in experiments, breeders responded differently to the playback of kin and non-kin. Cross-fostering experiments demonstrated that parameters of the *churr* calls were more similar between foster-sibs than between full siblings raised apart, providing the first evidence that these calls are primarily environmentally influenced (Sharp et al. 2005). Though Long-tailed Tits most often help related breeders with whom they have had previous direct association (e.g., as a sibling in the nest, an offspring, a parent or a recipient of helper care), the capacity to learn the *churr* call in the presence of kin at the nest provides a mechanism for learned kin recognition.

Vocal kin recognition could be important among cooperatively breeding passerines, generally. In plural-breeding Chestnut-crowned Babblers (*Pomatostomus ruficeps*), cooperative groups have distinctive contact calls (fig. 17.24A). Simultaneous playback to a foraging group of the calls of a group member and of an unfamiliar bird from a different group showed that members were significantly more likely to approach the familiar group member's call (fig. 17.24B, C; Crane et al. 2015). This adds to growing evidence from other studies that cooperative breeders commonly have group-distinctive call types, and that vocal signatures are important in kin recognition (Crane et al. 2015).

Cooperative Breeders in a Common Colonial Nest

Sociable Weavers benefit from collaboratively building large nests that have multiple nest chambers like an apartment complex, and resemble thatched roofs with individual entrances beneath (fig. 17.25). A single communal nest may house 5–300 individuals (van Dijk et al. 2014). The large multichambered communal nest provides shade on hot afternoons, but most important is its thermoregulatory function at night. Members roost within the chambers in small groups year-round; the huddled birds are insulated against the elements, particularly extreme cold at night during the dry season. Older, more dominant individuals occupy nest

LONG-TAILED TITS AND COOPERATIVE BREEDING
By Ben Hatchwell

Ben Hatchwell. *Photo by T. R. Birkhead.*

Long-tailed Tit and nestling. *Photo by B. J. Hatchwell.*

As I explored the countryside around home as a bird-obsessed kid, Long-tailed Tits were always a favorite. One memorable spring day, I found several of their exquisite nests and spent the rest of that Easter holiday taking copious but, in retrospect, rather useless notes on their nest-building behavior. At the time, I didn't know they were cooperative breeders; indeed, it wasn't until a year or two later, having read *The Selfish Gene*, that I even understood the evolutionary puzzle posed by cooperation. Scrolling forward a few years, for my PhD I studied Guillemot (*Uria aalge*, also known as Common Murre) population ecology with Tim Birkhead, but I was most enthused by their extra-pair copulation behavior. That interest in reproductive strategies was fueled further as a postdoc with Nick Davies, working on Dunnock mating systems. I then returned briefly to population biology, this time of Eurasian Blackbirds (*Turdus merula*) in a farming landscape around Wytham Wood outside Oxford, but when fishing around for a study system that would establish me as an independent researcher, Long-tailed Tits were the obvious choice. They were known to be cooperative breeders—David and Elizabeth Lack had described this in the 1950s—and subsequent studies in Wytham had revealed enough about their social system to convince me that more detailed study would be worthwhile.

Having moved to a lectureship at the University of Sheffield, I started watching Long-tailed Tits seriously, and remember vividly the thrill of seeing the first helper at a nest in 1994. Of course, one never knows how long a new project will last, but the possibility of it becoming a long-term study must have been at the back of my mind, because we have followed the monitoring protocols that I established in that first year ever since. The rationale for doing so came partly from my training in population biology, where consistent sampling is crucial, but also from the many inspirational studies in Peter Stacey and Walt Koenig's

chambers with the greatest thermoregulatory benefits (van Dijk et al. 2013). The effectiveness of the insulation depends on the depth of the thatch, which tends to be thicker toward the center of the colony. The contribution of members to building enhances the shared benefit of shelter, but some may benefit more than others, and this is related to colony structuring by kin group. Genetic analyses showed that average relatedness in Sociable Weaver colonies is low, but the nest chambers of kin within the communal structure are spa-

tially clustered, particularly for males (van Dijk et al. 2015). These kin clusters may facilitate helping between relatives as well as investment in the communal nest. In the rainy season, Sociable Weavers use the apartment-style chambers for breeding. In a long-term study at Benfontein, near Kimberley, South Africa, monogamous pairs have helpers that are usually close kin who typically assist their parents in raising later broods, gaining indirect but not direct fitness benefits from their cooperative behavior (Covas et al. 2006). Helpers

1990 book *Cooperative Breeding in Birds: Long-Term Studies of Ecology and Behavior.* Over the years, we have supplemented systematic monitoring with short-term behavioral experiments, but more disruptive experiments that disturb collection of long-term life history data we have conducted in a separate population.

So, apart from my teenage infatuation, why Long-tailed Tits? First, they have a simple cooperative system in which failed breeders redirect their care to help other breeders. This simplicity appealed to me, and "redirected" helping is an obvious transitional route to the more complex cooperative breeding systems of other species. Second, they are short-lived, so we could rapidly build up measures of lifetime reproductive success, from which the direct and indirect components of inclusive fitness can be determined. Ironically, I started with the expectation that Long-tailed Tit helpers were paying for the direct fitness benefit of living in winter flocks. This proved incorrect. Helping confers no direct fitness benefits, but rather is the product of kin selection. Helpers increase brood productivity and reduce breeders' reproductive costs, and crucially, long-term data and short-term experiments show that relatedness is the key driver of helping decisions. Understanding how the ecology and life history of Long-tailed Tits creates the opportunity for this kin-selected helping has been another major preoccupation, as has investigation of the mechanism of kin recognition underlying their kin-directed behavior.

As in any long-term study, there is a constant need to renew funding by asking new questions. Fortunately, the possibilities proliferate as the years go by, from quantitative genetics to the effects of climate change. The cast changes too. I have had the great fortune to work alongside many talented students and postdocs, each of whom has brought fresh insights and new vigor to the project, while also persuading me to keep following Long-tailed Tits for just a few years longer.

enhance breeding success, particularly in drier years under more food-limited conditions (Covas et al. 2008).

Large and bulky, Sociable Weaver nests are easily spotted and attract predators (Pale Chanting Goshawks, *Melierax canorus*, and especially snakes, such as the Boomslang *Dispholidus typus* and Cape Cobra *Naja nivea*). One snake can raid multiple nest chambers in quick succession. The benefits of dilution of nest predation may select for increased colony size, but this also necessitates cooperation among nonrela-

tives. Communal nests are liable to be subject to a "tragedy of the commons" scenario, since cheats might arise that use the resource but do not contribute to its maintenance. These enormous structures are built by inserting dry grasses stem by stem. Their construction and upkeep entails a considerable amount of work for the tiny 30-gram birds. Building is a highly male-biased activity, with 89 percent of the observations of marked birds building the thatch being males. However, individual males vary in their thatch-building effort: only ~50 percent of males were observed to contribute. Relatedness of males to other males using nest chambers in the vicinity was a significant predictor of individual nest-building activity (van Dijk et al. 2014). Therefore, in spite of low levels of relatedness in colonies, relatives are spatially and socially clustered by kin group, allowing for cooperation to function on a local level.

Conflict in Cooperative Societies
Reproductive Partitioning and Skew

Reproductive skew describes the manner in which reproduction is shared among adults of the same sex in social breeding groups. Most studies of kin-based cooperative breeders report strong reproductive skew in favor of the dominant pair, based on genetic determination of chick parentage. "High-skew societies" are common in cooperatively breeding birds, and include Red-cockaded Woodpeckers and Florida Scrub Jays. In spite of this, evidence suggests that at least some subordinates attempt reproduction in some systems.

The Southern Pied Babblers of South Africa have dominant monogamous breeding pairs, and fewer than 10 percent of chicks have a subordinate parent (Nelson-Flower et al. 2011). There is a strong incest taboo: subordinates of both sexes did not attempt to breed if the dominant of the opposite sex was their parent (Nelson-Flower et al. 2012). However, females fiercely contest the dominant breeding positions, attempting to evict one another (Raihani et al. 2010). Dominant females, identifiable by aggressive behavior toward subordinates and distinctive vocalizations during intergroup conflicts, destroyed eggs laid by subordinate females, successfully foiling most of their attempts to breed. Subordinates retaliated by destroying the eggs of the dominant female. If subordinate female reproduction is rarely successful, why do they attempt it? For females, reproductive options outside group are exceedingly rare, and living as a floater or competing for a dominant position are both risky strategies (Nelson-Flower et al. 2013). Gaining indirect fitness by feeding full siblings could be a better option. However, if the current male breeder is an immigrant, the offspring they help raise are not full siblings (related through mother only, so $r \sim 0.25$) and thus females might increase their fitness by trying to

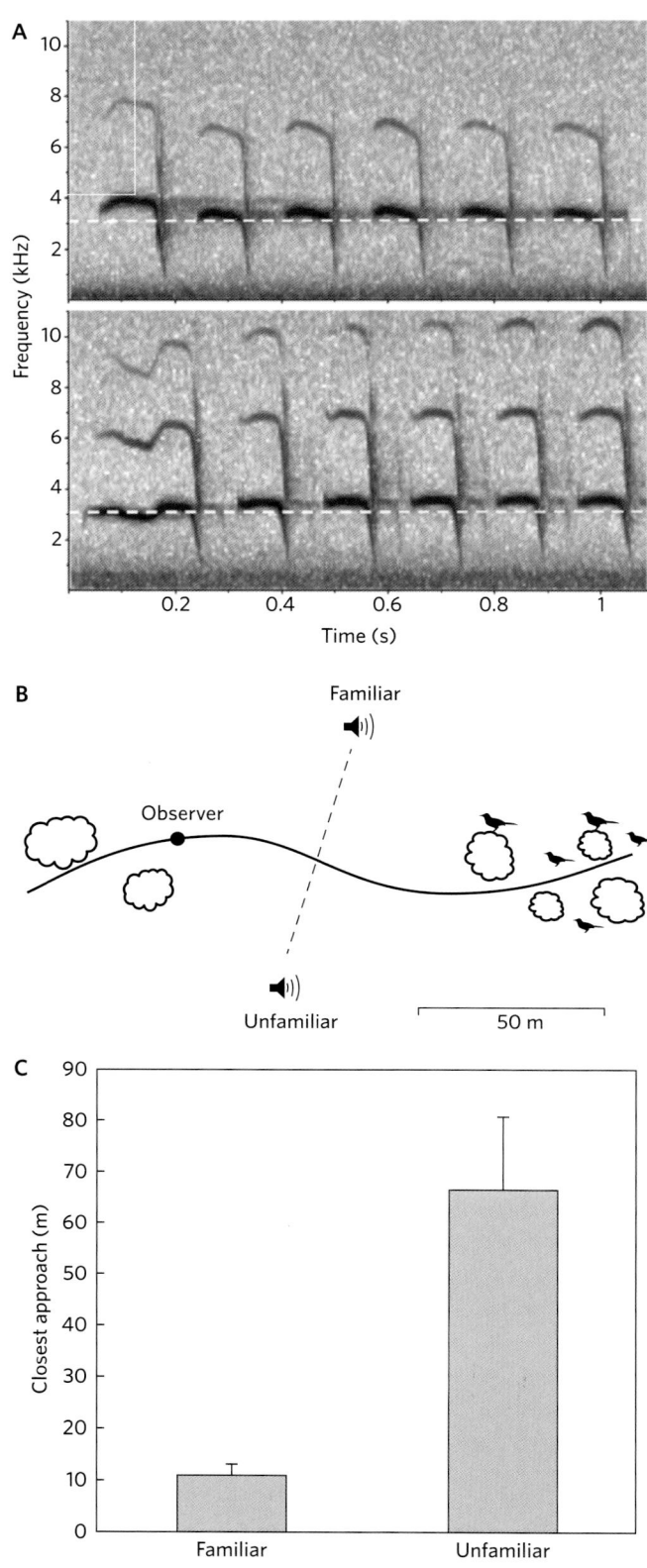

breed competitively (Nelson-Flowers et al. 2013). By contrast, dominant males of the Southern Pied Babbler rarely lost paternity to subordinates. Even with an immigrant dominant female, subordinate males' breeding attempts were rarely successful (Nelson-Flower et al. 2015). A low risk of paternity loss is likely offset by benefits of helping to subordinate males (fig. 17.26).

Joint Nesting

Joint nesting (also "communal nesting") occurs when two or more females lay in a single nest and participate in cooperative care of young. Though uncommon among cooperative breeders in general, joint nesting by two females occurs in Acorn Woodpeckers (Mumme et al. 1983), Seychelles Warblers (44 percent, Komdeur et al. 2016), and in some populations of Common Moorhens (McRae 1996a). Two or more females laying jointly in one nest is more common in a small but diverse array of species having high investment in pre-hatch paternal care and particularly male incubation. Same-sex group members may or may not be related, and reproduction is shared relatively equitably, especially among nonrelatives. Cooperative polygynandry occurs in a diverse but small number of taxa, including Ostriches (*Struthio camelus*, Kimwele and Graves 2003), some rails including Pukekos (Jamieson et al. 1994), Tasmanian Native Hens (*Gallinula mortierii*, Goldizen et al. 1998), and Common Moorhens (McRae and Burke 1996), and all four extant species of the cuckoo subfamily Crotophaginae of South and Central America—Groove-billed Anis (*Crotophaga sulcirostris*, Vehrencamp 1977), Smooth-billed Anis (*C. ani*, Schmaltz et al. 2008), Greater Anis (*C. major*, Riehl 2011), and Guira Cuckoos (*Guira guira*, Macedo et al. 2001, 2004)—as well as Yuhinas (*Yuhina brunneiceps*), small passerines endemic to Taiwan (Shen et al. 2012).

When related adult female Acorn Woodpeckers coexist in a group, they both lay in a joint nest. These are rarer than single female nests, and success rates are often compromised because of reproductive interference between co-breeding females. Before initiating her own laying bout, each female will toss the eggs of other females out of the nest, even if

Figure 17.24. Experimental tests of vocal recognition in Chestnut-crowned Babbler groups. *A*, Two spectrograms of long-distance contact calls of individuals from different groups illustrate differences in acoustical structuring. Differences in fundamental frequencies of repeat syllables can be seen with reference to the horizontal dotted line at 3 kHz. The call on the lower panel also has a modulated first syllable; *B*, Schematic of the dual playback experimental setup showing positions of speakers from which contact calls of group and non-group helper Chestnut-crowned Babblers were played simultaneously; *C*, Mean distance of closest approach to the speaker by focal groups in response to the playbacks. *Reprinted with permission from Crane et al. 2015.*

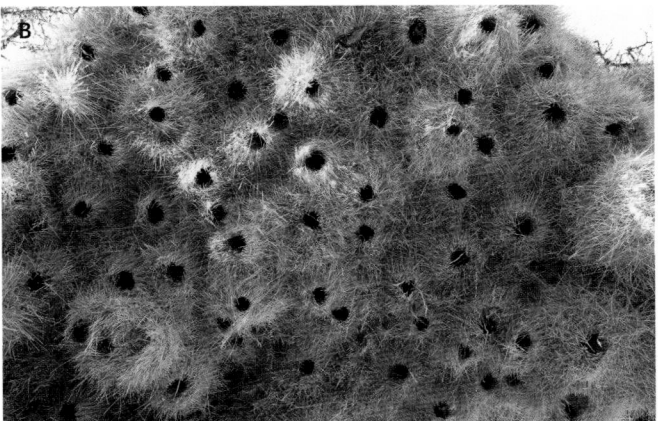

Figure 17.25. Sociable Weavers are cooperatively breeding colonial nesters. *A*, The large thatched communal nest is maintained year-round and insulates the roosting birds from extreme temperatures at night; *B*, The apartment-style chambers become breeding nests in the rainy season; *C*, Sociable Weavers build these enormous nest structures one grass stem at a time. *Photos by Warwick Tarboton.*

co-breeders are sisters or mother-daughter pairs (Mumme et al. 1983). This entails a substantial cost to group fitness, but per capita breeding success may be diminished by over-production if group resources are limited (Koenig et al. 1995). Egg tossing has also been reported in all four species of crotophagine cuckoos (Vehrencamp 1978, Macedo et al. 2001, 2004, Vehrencamp and Quinn 2004, Schmaltz et al. 2008, Riehl 2011), and Mexican Jays (*Aphelocoma ultramarina*) (Trail et al. 1981). Each female benefits by increasing the proportion of her own eggs raised; once a female begins laying, however, she cannot distinguish her own eggs from others. Thus, clutches of two or more females can coexist in the nest and be incubated communally.

Why do some species exhibit joint nesting but not others? Male nocturnal incubation is common (McRae 1996b, Vehrencamp 2000, Riehl 2012, Koenig et al. 2016), though not ubiquitous (e.g., Komdeur et al. 2016), among joint nesters. The provision of substantial paternal care, including male incubation, may be considered a limiting resource, and therefore a driver of the evolution of co-breeding by females. However, constraints on resources often limit group size in territorial cooperative breeders (Komdeur 1994, Koenig et al. 1995), and conflict over reproductive partitioning is expected in cases where enlarged broods produced by multiple females could decrease mean fitness of group members. That co-breeding females are first-order relatives in Seychelles Warblers, Acorn Woodpeckers and Common Moorhens, means they also gain indirect fitness from the success of the other female's young. Yet the preponderance of egg ejection suggests a cost, as well as limitations to egg recognition observed in many joint nesters. Thus, selection has favored co-breeding females that synchronize laying (McRae 1996a, Riehl 2010; fig. 17.27).

The social dynamics of joint-nesting Greater Anis contrasts sharply with those of traditional cooperative breeders in that group integrity relies exclusively on the direct benefits of cooperation (Riehl 2011). Greater Anis form breeding groups of two to five socially monogamous pairs that defend territories along the edges of major rivers and lakes in Panama's canal zone. As joint-nesting plural breeders, unrelated pairs contribute to a large elliptical nest (fig. 17.27B; Riehl and Jara 2009). Female offspring always disperse from the group. Breeders of both sexes are unrelated. Tests of genetic parentage reveal that reproductive partitioning is relatively equitable (Riehl 2011, 2012). All group females contribute to the communal nest. At least 75 percent offspring are produced though genetic monogamy, with up to 24 percent of offspring being produced through extra-pair paternity within or outside the group (fig. 17.28; Riehl 2012).

As many as 56 percent of Smooth-billed Ani eggs are tossed out of communal nests during the laying period, and egg losses

Figure 17.26. Southern Pied Babbler groups consist of 3 to 15 adults. Benefits of larger group sizes include enhanced predator detection and territory defense, higher reproductive success, and load-lightening benefits, while costs include competition over resources and reproduction (Ridley 2016). *Photo by Warwick Tarboton.*

Figure 17.27. Joint nest of *A*, Common Moorhens in Britain, where the co-breeding females are first-order relatives. *Photo by Susan McRae. B*, Greater Anis in Panama, where co-breeding females are unrelated. *Photo by Christina Riehl. C*, Three Greater Ani eggs at different stages of wear, as the vaterite erodes to reveal blue-green eggshells. *Photo by Christina Riehl.*

per capita increase with group, resulting in larger groups incubating smaller numbers of eggs (Schmaltz et al. 2008). Egg ejection by co-breeding female Greater Anis is also common: the first two to three eggs laid are at particularly high risk of ejection, because ejecting females that have not yet laid can be sure these are not their own. At the time of laying, Greater Ani eggs are completely covered with a chalky white coating of vaterite, a mineral deposit based on calcium carbonate (fig. 17.27B; Riehl and Jara 2009; seen also on freshly laid eggs of the related Guira Cuckoo, Macedo 2016). This soft coating is gradually abraded over the course of a few days to reveal a deep blue-green pigmented shell (fig. 17.27C). This transition clearly identifies newly laid eggs, but it may also help to conceal maternity in the competitive arena of the joint nest. Parasitic laying by females from outside the group is rarely successful: of 48 clutches monitored over four years, 40 percent contained eggs laid parasitically by non-group members (7 percent of 411 eggs; Riehl 2010), but only 1 percent of young sampled were products of CBP (Riehl 2012). Most parasitic eggs were laid outside of the brief window of the group females' laying period, and observations from natural and experimental parasitism showed that these eggs were usually recognized and ejected (Riehl 2010).

Cooperative societies are potentially vulnerable to exploitation by social parasites that obtain benefits without contributing. Species in which female reproductive strategies of joint nesting and conspecific brood parasitism are observed within the same population are particularly promising systems for studying the evolution of parasitism and cooperation (Gibbons 1986, McRae 1996a, Riehl 2012; box on page 564).

Of 213 species of cooperatively breeding birds for which data are available, 55 percent breed in nuclear family groups, and only 15 percent breed primarily with nonrelatives (Riehl 2013). Benefits to individuals of joining non-kin cooperative groups include enhanced survival, access to extra-pair copu-

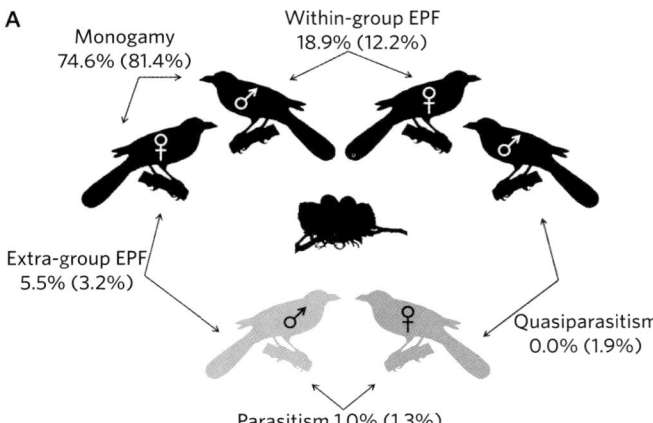

Figure 17.28. A, Percentages of nestlings in communal broods of Greater Anis produced monogamously, by extra-pair fertilizations (EPFs) within and outside of the group, by conspecific brood parasitism (CBP), and quasi-parasitism (CBP by a brood parasite that mated with a male group member). Group members are silhouetted in black, extra-group individuals in gray. Estimates inside and outside of brackets are from microsatellite genotyping analyses using separate sets of offspring. Parentage analyses with all adults and chicks genotyped (*n* = 201 nestlings), and with incomplete sampling of adults inside the brackets (*n* = 156 nestlings). *Reprinted with permission from Riehl 2012. B,* Greater Ani social group. *Photo by Nick Athanas.*

lations and queuing for future breeding opportunities (Riehl 2013). The remaining 30 percent breed in mixed groups of relatives and nonrelatives. For example, Central American White-throated Magpie-Jays (*Calocitta formosa*), breed in social groups with a dominant pair and helpers of both sexes, but females are more philopatric than males, have higher kinship, and are the predominant helpers. Female magpie-jays have the options of gaining direct fitness either within the group nest or as floaters through brood parasitism, or of gaining indirect fitness through help at the nests of relatives (Berg 2005). Social groups with mixed kin structures are most common in families with multiple origins of coop-

erative breeding, such as Rallidae, Psittacidae, Accipitridae, and Corvidae, and in species considered "obligate" cooperative breeders (no instances of successful lone pair breeding e.g., babblers, Timaliidae). By contrast, facultative cooperative breeders are most likely to cooperate with relatives. This arises through natal philopatry and dispersal patterns favoring kin neighborhoods (Hatchwell 2010).

As more systems are explored, the interplay of conflict and cooperation becomes more apparent. Just as joint nesting occurs occasionally in traditional cooperative breeders, unmated related helpers are rarely also found in predominantly non-kin based societies, such as in Guira Cuckoos, in which monogamy and within-group polygynandry are the most common mating arrangements (Quinn et al. 1994, Lima et al. 2011), and Taiwan Yuhinas (Shen et al. 2012). Interestingly, in Yuhinas, joint-nesting plural breeders with two to four monogamous cooperating pairs, the degree of breeding synchrony and cooperation were both found to increase, and the level of intragroup aggression decrease, with harsher environmental conditions (Shen et al. 2012). Long-term studies of these rarer forms of social organization promise to help shed light on less-well-understood aspects of social evolution.

Female versus Male Helpers

As already described, helpers are commonly close relatives of breeding birds. Passerines typically show female-biased dispersal, and sons are more likely to remain on their natal territory and help their parents. Though male helpers are more common, systems where both sexes help merit consideration. Females and males have different reproductive interests, so it might be expected that their helping contributions should differ. As more females in a group reproduce, more offspring are produced, so subordinate reproduction may be more constrained for females than for males. In kin-based systems, females may therefore be expected to contribute more help and have a larger impact on offspring fitness than males because, lacking direct benefits, indirect benefits gained through helping relatives are better than nothing.

Controlled experiments involving field manipulation have been the preferred method of testing hypotheses in behavioral ecology for the last four decades. However, experiments involving the manipulation of group membership have been problematic in studies of cooperative breeders because they tend to disrupt group stability. To avoid this pitfall, recent studies have used correlational data, but made use of novel statistical methods to account for the effects of multiple potential interacting variables.

The Red-winged Fairywren (*Malurus elegans*) is unusual among fairywrens, and Australian cooperative breeders in general, for having both male and female helpers. Brouwer

FOR CHRISTIE RIEHL, STUDYING THE SOCIAL LIVES OF BIRDS IS A LABOR OF LOVE

By Susan B. McRae

Christie Riehl climbing. *Photo by Bryson Voirin.*

Following the lives of birds was an occupation that started early for Christie Riehl, assistant professor at Princeton University. Christie grew up exploring the parks in New Orleans. It was a close-up and personal encounter with a Wood Thrush during these adventures that catalyzed a lifelong love of birds. Christie's parents nurtured her passion and encouraged her to take an ornithology course at Tulane University during her high school years. This led to a volunteer position as a field assistant on a project studying Swallow-tailed Kites in Louisiana. During this total-immersion experience, she decided to be a biologist. She learned firsthand the key to being a successful field ornithologist: focus and drive based on reward. "There's a special thrill about finding a bird's nest that never diminishes," she says.

The communal nests of Greater Anis are conspicuous in comparison to those of many other birds, and this was a factor in Christie's decision to study them. Arriving at Barro Colorado Island in Panama with her PhD supervisor, Martin Wikelski, she was immediately fascinated by a group of six iridescent blue Greater Anis building a nest together at the edge of Lago Gatún near the field station buildings. When she searched the Smithsonian Tropical Research Institute's library for information about them, she realized that, unlike the other anis, their behavior had not yet been described.

Previous researchers had been discouraged by the inaccessibility of their nests: whereas the other crotophagine cuckoos are terrestrial, Greater Anis always nest on islands or in branches overhanging water, requiring a boat to reach them. So, Christie decided to attempt to address this oversight. Her fate was sealed when she went out for the first time to try to catch the anis, and one flew into the mist net before she had finished setting it up. Catching anis is not always easy, though. Mist nets must be set up on poles in shallow water near the shoreline, and a team of assistants in boats must work together to remove birds from the nets. The population of crocodiles in Lago Gatún has grown over the ten years of Christie's study. She abandoned trapping at one site after a crocodile pulled a mist net with a just-netted ani down before the boats could reach it.

When watching Greater Anis, Christie feels like they are also observing her. Sociable and curious, they appear to monitor the actions of individuals around them. Given the extent of coordination that goes into building and laying in a group nest, and the cohesiveness of the group, sneaky egg-dumping by brood parasites was one of the results that Christie found more surprising. Parasitic laying by females outside the group had not been observed in the field before it was revealed through genetic analyses. "The parasites have to avoid being detected by the host group, so they're pretty good at sneaking by us too!"

Christie wants to know how breeding groups of nonrelatives get started in the first place. Perhaps the habit of conspecific brood parasitism has led to females habitually laying together in the same nest: an originally parasitic relationship might evolve into a cooperative one if, at some point, ecological conditions favored "parasites" that stayed at the nest to help care for the mixed clutch of offspring. Evolutionary origins of reproductive synchrony among co-breeders and cooperative care are unresolved. Christie will be pursuing these and other questions as she continues her studies of these fascinating birds.

et al. (2014) asked whether parents and other group members adjusted their care in relation to helper number and sex, and whether offspring benefited from higher provisioning rates in the case of more helpers. They measured contributions of each adult in terms of chick provisioning rate late in the nestling period when all group members are most likely to participate. The number and sex ratio of helpers varied greatly: groups had a dominant pair (putative breeders) and combinations of zero to four male and zero to five female helpers. Brouwer et al. (2014) found striking effects of helper

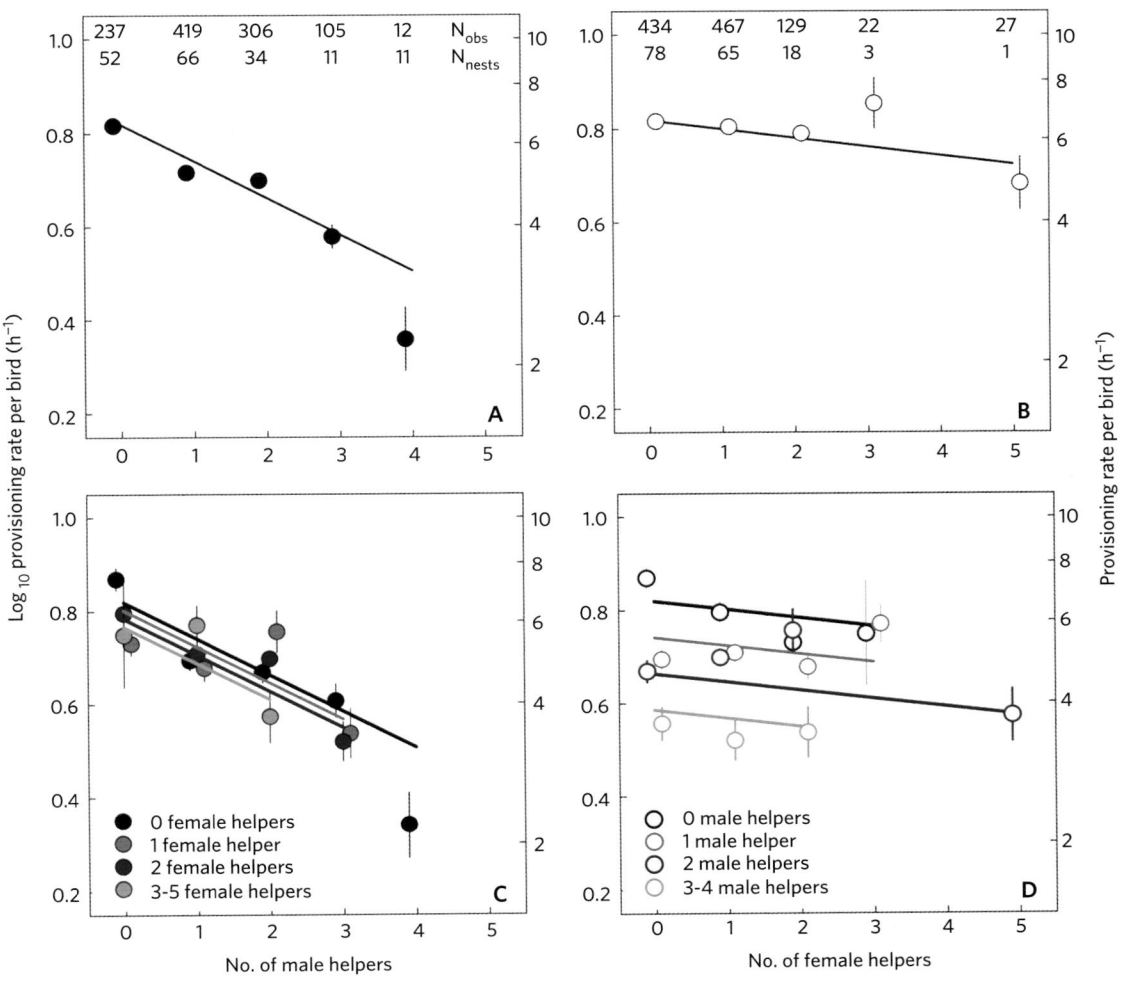

Figure 17.29. Provisioning rates by Red-winged Fairywrens decreased significantly in relation to the number of male helpers, but not in relation to the number of female helpers. The standardized mean (±SE) provisioning rate per bird per hour in relation to *A*, the number of male helpers in the group in the absence of female helpers; *B*, the number of female helpers in the group in the absence of male helpers; *C*, the number of male helpers in the group shown for varying number of female helpers; *D*, the number of female helpers in the group shown for varying number of male helpers. Provisioning rates were standardized for year, brood size, brood age, status, and number of opposite sex helpers. Numbers of observations (N obs) and number of nests (N nests) are indicated at the top. *Reprinted with permission from Brouwer et al. 2014.*

sex after controlling for brood size and age. Red-winged Fairy-wren breeders and helpers of both sexes significantly reduced their provisioning rate in relation to the number of male helpers in the group (fig. 17.29A and C), but there was no compensatory effect in provisioning by helpers in relation to female helper number (fig. 17.29B and D).

Possible effects of overall group size and territory quality were controlled-for statistically by taking into account the between- versus within-subject effects, so the load-lightening effect of male helpers appeared to represent plasticity in individual provisioning effort (Brouwer et al. 2014). The provisioning by female helpers was therefore additive, raising the total delivery rate of food. The number of female helpers had significant positive effects on fitness, measured in terms of offspring growth, nestling weight, fledging rate, and survival to one year of age (Brouwer et al. 2014). Why should individuals adjust their provisioning in relation to the number of male helpers but not the number of female helpers? Female helpers disperse on average earlier than males, and could do so before the offspring are independent. A small proportion (7 percent) of females in the study population also bred plurally, initiating a second nest on the group territory. Thus, female helpers may furnish additional short-term help, but pose a liability to the dominant breeders if they breed.

Sex Ratio Allocation

Trivers and Willard (1973) suggested that natural selection should favor parental ability to bias the sex of offspring at conception based on the offspring's expected future reproductive success. Given that helpers are usually strongly sex-biased, there has been considerable interest in whether birds might adaptively modify the sex of their offspring to favor the helping sex in populations in which helpers were advantageous. Therefore, females may be selected to bias the sex ratio of their broods according to social or ecological conditions. In cooperative breeders where helping is sex-dependent, females are predicted to be under selection to bias their broods in favor of the helping sex under conditions where helpers are beneficial. Empirical evidence for sex bias at the nestling stage was first reported in Red-cockaded Woodpeckers (Gowaty and Lennartz 1985). The possibility that helping might actually drive selection for production of more of the sex that helps (rather than assuming sex-bias among helpers was an outcome of differential mortality between the sexes) became known as the **"repayment hypothesis"** (Emlen et al. 1986). Thus, if sons "repay" through helping part of the cost to produce them, they become the 'cheaper' sex (Trivers and Willard 1973), resulting in selection for a brood sex ratio equilibrium that results in overproduction of males (Emlen et al. 1986).

One of the strongest tests of **sex allocation theory** in the context of cooperative breeding came from the Seychelles Warbler. This species is unusual among cooperative breeders in that helpers-at-the nest are more commonly female. The presence of one female helper substantially increases a pair's reproductive success if they are on a high-quality territory, but not if they are on a low-quality one. However, even on a good territory, adding a second female helper reduced a pair's reproductive success, possibly because the territory has insufficient food to support additional adults (Komdeur 1994). Tropical passerines lay small clutches, and this is exceptionally true in this species: the clutch size per female is one. Using a standard molecular technique for sexing chicks, Komdeur and colleagues found a strong skew in the primary sex ratio (clutch sex ratio, in this case determined at hatch). On high-quality territories, if a pair already had a female helper, the breeding female produced a male egg. But when a pair did not have a female helper, she produced a female. Thus, the helping sex was produced significantly more often than expected when the benefit of having a helper daughter was greatest. By contrast, pairs on poor territories produced significantly more male offspring (fig. 17.30). That the adjustment was conditional, depending heavily on territory quality, was demonstrated experimentally in the course of establishing a new population on Aride Island. Breeding pairs that were translocated from low- to high-quality

territories switched from producing sons (90 percent probability), to producing daughters (85 percent probability) (Komdeur et al. 1997). Biasing sex-allocation pays significant dividends in terms of fitness benefits to breeders fortunate to have high-quality territories (fig. 17.30; Komdeur 1998).

Other studies of cooperative breeders report significant effects of brood sex ratio bias (e.g., in Laughing Kookaburras, *Dacelo novaeguineae*, Legge et al. 2001; and Western Bluebirds, Dickinson 2004b). Failure to find predicted skews in the brood sex ratio of cooperatively breeding Acorn Woodpeckers, Superb Fairywrens, Sociable Weavers, and others may be because of lower intensity of selection on sex allocation (West et al. 2005). Even in Seychelles Warblers, the strong bias has eroded over time as ecological and social conditions have changed (Komdeur et al. 2016). Comparative analysis suggests that sex ratio adjustment is dependent on the magnitude of increase in breeder success caused by helpers (Griffin et al. 2005). While the ability of female birds to bias the sex ratio of their offspring is well-established, little is known about the mechanism by which females are able to so. In birds, females are the heterogametic sex; the female's gamete carries the sex determining W or Z chromosome. A possible mechanism is the production of a surplus of either Z- or W-chromosome-bearing eggs caused by adjustment by epigenetic factors (environmentally influenced heritable changes in genes) that modify development of functional gametes bearing one of the sex chromosomes (Rutkowska and Badyaev 2008).

Proximate Mechanisms in Social Evolution
Proximate Causes of Social Dominance

Dominance within social groups has long been thought to have a hormonal basis. Testosterone is associated with the development of male secondary sexual characteristics and aggressive interactions (see chapter 8) and is implicated in the development of social hierarchies in flocking species (Butterfield and Crook 1968, Laucht et al. 2010). However, in the White-browed Sparrow-Weaver, a resident ploceid weaver found in the semiarid zones of sub-Saharan Africa, luteinizing hormone (LH) rather than testosterone seems to be involved in dominance hierarchies. LH is normally involved in stimulating sexual maturation in both sexes. In White-browed Sparrow-Weavers, it appears to regulate aggression (Wingfield et al. 1992). Exhibiting a high-skew, nuclear family society, a dominant breeding pair and mature subordinates of both sexes, including mature offspring and sometimes unrelated males, cooperate year-round in defending the group territory. They maintain roost nests in a central tree or group of trees, and proclaim ownership of their territory in daily group choruses. Subordinates of both sexes assist in

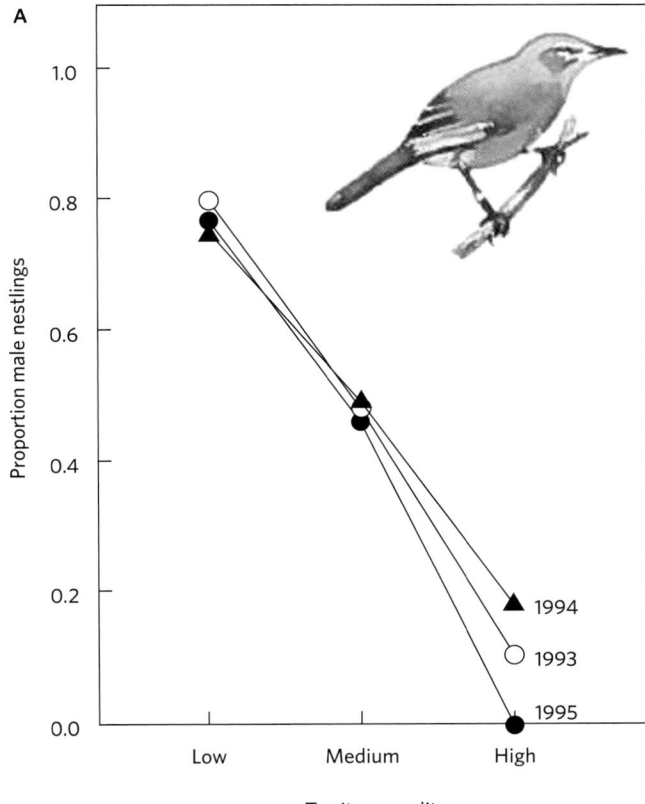

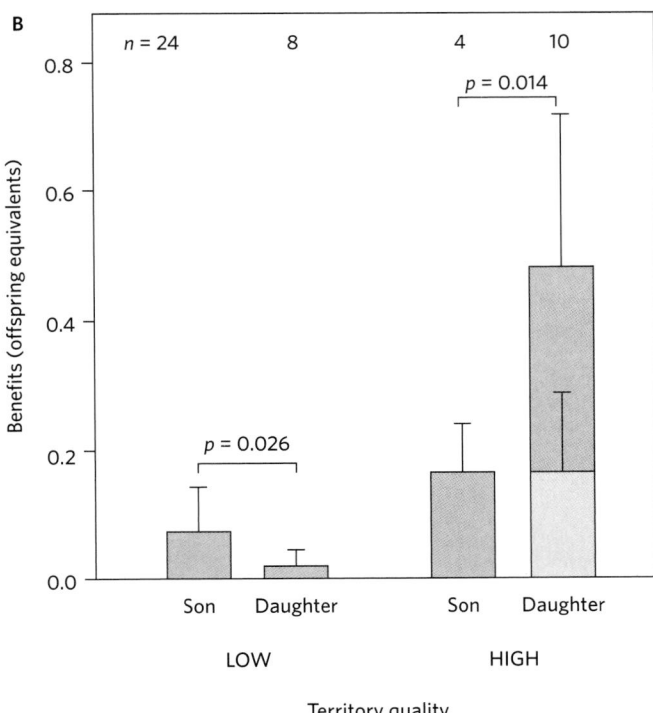

Figure 17.30. A, The probability of producing male nestlings in single egg clutches of Seychelles Warbler pairs in relation to the quality of their breeding territories (low, medium, high) in the absence of other young on the territory. *After Komdeur et al. 1997. B,* Inclusive fitness

rearing young, but females help more and tend to remain on their natal territory, while males are more likely to disperse and join a group of unrelated individuals. Unusually, this species is sex role–reversed by comparison with other cooperative breeders in having female territory inheritance and male dispersal.

To determine how aggression was mediated in White-browed Sparrow-Weavers, Wingfield et al. (1991) experimentally removed dominant males to destabilize group structure. This resulted in a power struggle, sometimes lasting days, as a new dominant male became established. Testosterone levels were unrelated to aggression. However, LH was transiently elevated in the plasma of new dominant males (Wingfield et al. 1992). To confirm that aggression is moderated by LH instead of testosterone, a second experiment was conducted. Territorial intrusions were simulated by introducing into a group's territory four or five conspecifics in a cage. Group members responded aggressively toward the intruders, displaying and vocalizing. They were then caught and blood sampled. Circulating LH became elevated in all group members, in particular the breeding females (Wingfield and Lewis 1993), suggesting incursions are a more severe threat to the status of the dominant female in this unusual matrilineal society. In spite of a great deal of interest in questions regarding the mechanisms governing antagonistic interactions and hierarchy formation, answers remain elusive.

Delayed Breeding by Subordinates: Constraint, Restraint, or Suppression?

Why don't all subordinates breed? The existence of mature helpers that do not attempt to breed has led researchers to ask whether, at a mechanistic level, subordinates experience breeding suppression ("psychological castration"; Brown 1987) or whether they lack sufficient social, physiological, or physical stimulation (such as by food availability or climatic or other environmental triggers) to breed (Reyer 1984). For example, the dominance structure may lead to differential access to resources necessary to allow individuals to achieve breeding condition.

In the Florida Scrub Jay, circulating levels of testosterone in nonbreeding male subordinates were lower than in breeding

benefits of focal breeding pairs that produce sons versus daughters on low- and high-quality territories. Indirect (dark bars) and direct (light bars) fitness benefits (±SD) expressed as the additional numbers of yearlings produced by the focal breeding pair through help provided by subordinate offspring helpers. *P*-values denote results of two-tailed Mann-Whitney *U* tests, based on *n* breeding pairs. *From Komdeur et al. 2016, based on analyses in Komdeur 1998.*

males, and their testes were smaller during relevant stages in the breeding cycle, but they were primed to an extent that they would be capable of breeding given appropriate stimulation (Schoech et al. 1996a). Levels of estrogen and LH were also similar between mature female breeders and nonbreeders (Schoech et al. 1996a). Gonadotropin-releasing hormone (GnRH) is naturally released by the hypothalamus and stimulates LH release from the pituitary gland. An experimental challenge was performed by injecting Florida Scrub Jays of different statuses within groups with GnRH; blood was sampled before and after the challenges. Breeders and nonbreeders responded with an increase in LH release, as measured from circulating levels in blood over time. The response in nonbreeders suggested that the hypothalamo-pituitary-gonadal endocrine axis is functional in nonbreeding subordinates (fig. 17.31; Schoech et al. 1996a). Therefore, it seems that subordinates have in place the physiological capability to trigger reproduction.

Hormones associated with stress (glucocorticoids) appear to be implicated in the social hierarchies of some cooperative breeders. The possibility that corticosterone might function in reproductive suppression was tested in both White-browed Sparrow-Weavers (Wingfield et al. 1991) and in Florida Scrub Jays (Schoech et al. 1997). Failure to find a significant difference in circulating levels of corticosterone in the blood of breeders and nonbreeders at a variety of stages in the reproductive cycle in each case led these researchers to reject this hypothesis.

Collectively, these results suggest that rather than suppression caused by intrinsic factors, delayed reproduction by subordinates is more likely caused by extrinsic factors such as the lack of access to resources or social cues. In Red-cockaded Woodpecker groups, sons of the female breeder have lower circulating testosterone than unrelated male subordinates (Khan et al. 2001). This supports the idea that stimulation through differential social interactions with the breeding pair might trigger physiological changes in breeding condition. Additional studies will be needed to verify this. The pattern that is emerging is that there is considerable variation among cooperative breeding systems in terms of endocrine regulation of dominance and breeding.

Proximate Causes of Helping Behavior

Proximate mechanisms governing helping behavior are not fully understood, but production of the pituitary hormone prolactin is strongly correlated with helping-at-the-nest. Prolactin (named for being secreted prior to lactation in female mammals) is associated with parental behavior in vertebrates. In birds, prolactin it is associated with development of the brood patch (defeathering and vascularization of ab-

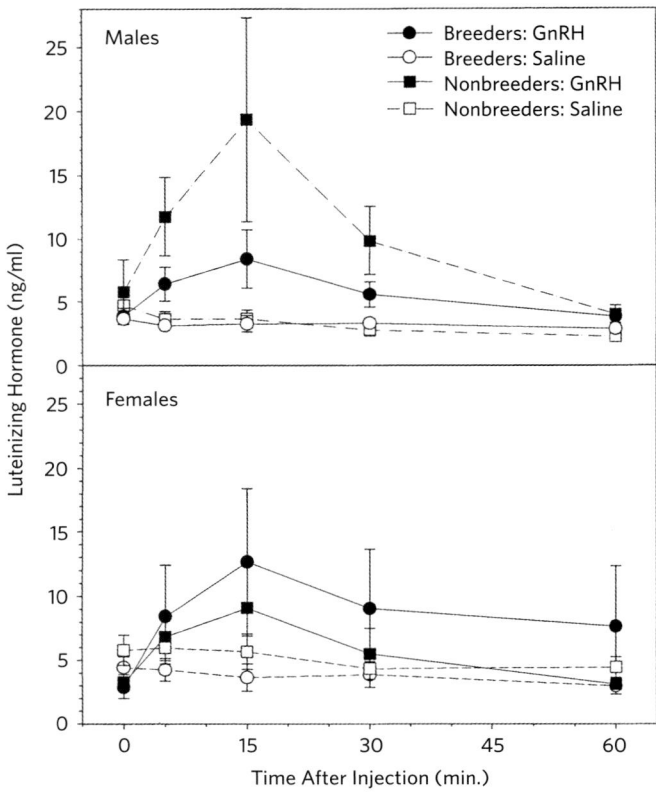

Figure 17.31. Results of gonadotrophic-releasing hormone (GnRH) challenge in breeders and nonbreeders in Florida Scrub Jay social groups. A rise in response to GnRH challenge of plasma levels of luteinizing hormone (LH) is indicative of readiness to breed. Both male and female subordinate nonbreeding members experienced a rise in circulating LH in response to GnRH challenge that was significantly greater than to saline challenge controls. *Reprinted with permission from Schoech et al. 1996a.*

dominal region in readiness for incubation) and brood provisioning, among other parental behaviors. In Florida Scrub Jays, breeders provision young at higher rates than helpers do. Among helpers and breeders, nestling feeding rates were correlated with circulating prolactin levels (Schoech et al. 1996b). In Red-cockaded Woodpeckers, prolactin increased in all group adults (breeding and nonbreeding males and breeding females) through incubation. However, levels declined slightly during the nestling phase in males as it continued to rise in females, suggesting a link to greater contact of females with nestlings through incubation and brooding (Khan et al. 2001).

The relationship between prolactin and parental behavior is taxonomically widespread. In Harris's Hawk (*Parabuteo unicinctus*), a cooperative breeder in the Sonoran Desert of Arizona—and the only bird known to hunt prey cooperatively (see chapter 15)—prolactin levels are also related to the degree of helping. During the nestling stage, the highest levels of circulating prolactin were measured in adult male helpers, and they also provisioned the most. As prolactin levels in breeders declined soon after hatching, helper levels increased, coinciding with their increase in provisioning effort (Vleck et al. 1991).

Sociality and Cognition

Sociality is associated with individual recognition and an aptitude for adaptive learning. Using techniques similar to those employed by primatologists, researchers studying Arabian Babblers in the Negev desert, Israel (Carlisle and Zahavi 1986), and Southern Pied Babblers in the Kalahari Desert of South Africa (Ridley 2016) have habituated these birds to the point of being able to observe them at distances of two to three meters without changing their behavior. This greatly enhances their ability to collect detailed observations and make these tractable systems for observation and manipulation. Observers remain still for hours during habituation. Southern Pied Babblers were habituated by enticing them with a small food reward. By placing the food on a top-loading balance, the investigators were able to weigh individuals daily at dawn. Relating these measures to behavioral interactions and controlling for variation in group size, sex ratio, and ecological variables, Amanda Ridley and her team were able to gain important insights into how condition and membership status affect an individual's behavior over long time frames. Specifically, they have shown that subordinate Southern Pied Babblers tend to delay dispersal for several years and benefit more by helping the dominant breeders (and gaining indirect fitness benefits) than by attempting to breed themselves.

Corvids and their relatives are renowned for their capacity for learning and memory. Social evolution has shaped a relatively large corvid brain skilled in problem solving (Taylor et al. 2012). American Crows have advanced cognitive skills and an extraordinary ability to recognize individuals, extending even to human facial recognition (Marzluff et al. 2010). Researchers wore distinctive face masks when capturing American Crows during a study on the campus of the University of Washington, Seattle. The process of capturing, measuring, and banding the birds does not harm them, but the perception of being captured by a predator is sufficiently traumatic to leave a lasting impression. Individuals wearing the same mask even up to two years later were avoided by the marked crows (Marzluff et al. 2010), demonstrating their superior long-term memory.

Imaging the brains of live anaesthetized crows has revealed specific neural pathways implicated in the visual discrimination task. Adult males were wild-caught in flocks or roosts by a person wearing a distinctive-looking "dangerous" mask and held in aviaries for four weeks while being cared for by persons in a different "caring" mask. (Mask types were switched such that some subjects learned a particular mask as "caring" while others learned the reverse.) On the day of the scan, the bird was injected with ^{18}F-fluorodeoxyglucose (^{18}F-FDG), a radiolabeled form of glucose that lights up areas of neural activity in the brain under positron emission tomography (PET scan). Subjects were placed in a cage and shown a person in either the "dangerous" or the "caring" mask every other minute for 15 minutes, then they were immediately anaesthetized and placed in the scanning chamber for imaging (fig. 17.32). A control group of crows subjected to the same procedures but shown no faces was also imaged.

Crows inspected the "dangerous" mask with a fixed gaze, rarely blinking. Brain regions illuminated in the PET scans have analogs in the mammalian brain that are known to be associated with emotional response, motivation, and conditioned fear learning. Crows shown the "dangerous" mask had a more consistent and overall greater area of activity involving areas in their thalamus and brainstem known to be associated with fear (nidopallium, arcopallium, amygdala; Marzluff et al. 2012). Birds exposed to the "caring" mask, had greater activity in areas of the vertebrate brain known to be involved in social interactions, associative learning, and hunger (preoptic area and medial striatum; Marzluff et al. 2012). Other brain areas activated suggested the crows related their visual experience in the context of memory (the visual cortex, nidopallium and mesopallium, and the associative striatum). Neural circuitry has thus developed in these highly social birds that is comparable to that of social mammals, including ourselves.

The brain is one of the last frontiers of scientific exploration. Bird brains are remarkably complex for their compact size. Sociality selects for sophisticated mechanisms of discrimination, complex communication systems, and problem-solving skills. Crows have been dubbed "avian Einsteins," and corvids are becoming important model systems for studies of the association between social experience and cognition (Kenward et al. 2011). Psittaciforms (parrots and their allies) are exceptional among birds in cognitive skills, and are also ubiquitously highly social (Toft and Wright 2015).

GROUP LIVING HAS BENEFITS AND COSTS

Sociality evolves for collective benefits of protection from predators, defense of resources or territory, and surviving in harsh environments. Group dwelling involves costs of having to share limited resources, higher rates of parasitism and

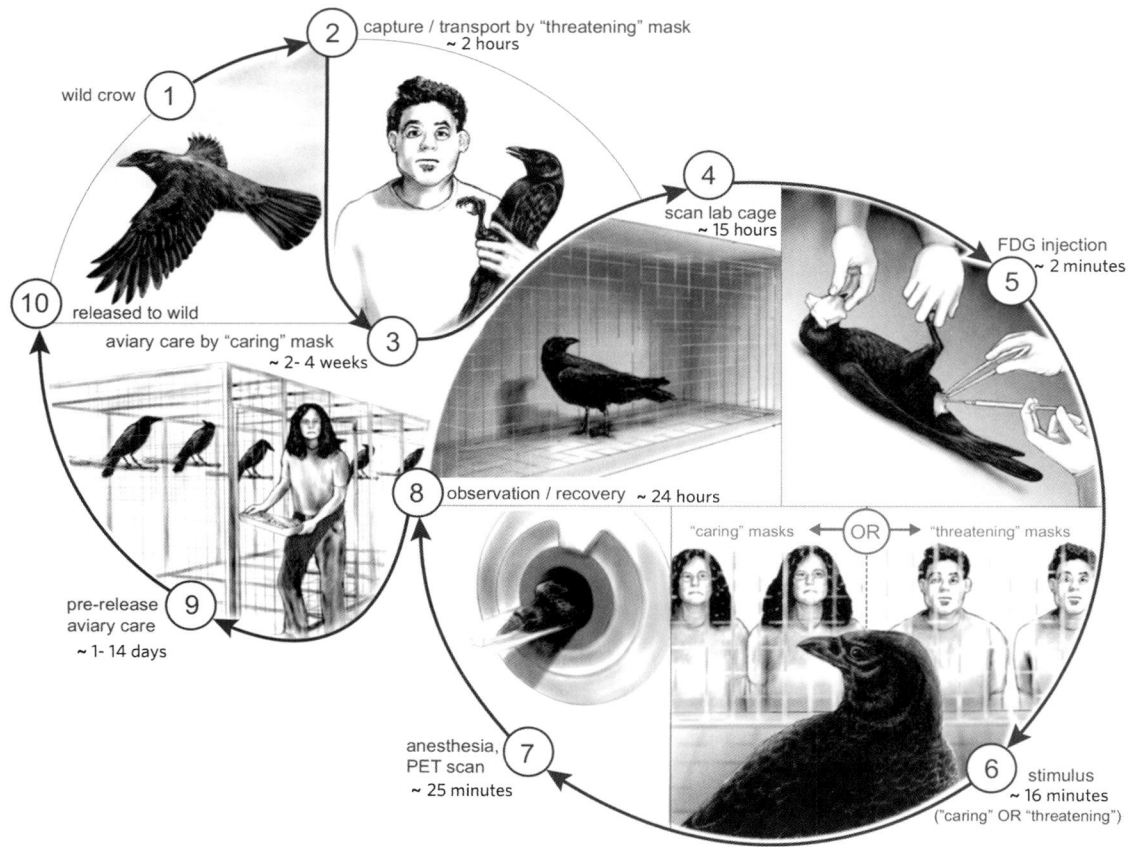

Figure 17.32. Experimental protocol used for brain imaging study. Adult male American Crows were (*1*) wild-caught in flocks or roosts by a person wearing a distinctive looking "dangerous" mask and (*2*) held in aviaries for four weeks, (*3*) cared for by persons in a different "caring" mask. After (*4*) habituation in a test cage, they were (*5*) injected with [^{18}F-]fluorodeoxyglucose (FDG). They were (*6*) shown a person in either the "dangerous" or the "caring" mask, then (*7*) they were immediately anaesthetized and placed in the scanning chamber for imaging. They were (*8, 9*) briefly held for recovery and observation in aviaries before (*10*) re-release. *Reprinted from Marzluff et al. 2012.*

disease, and reproductive interference, but also provides benefits to individuals, including higher reproductive success for themselves and their kin, leading to the establishment of long-lived resident family groups (dynasties), sometimes with accumulated territorial wealth. From social living emerges sophisticated cooperative behavior involving individual recognition and nepotism. Group living is subject to exploitation as well, where individuals that appear to be cooperating behave in their own interests at the expense of others, including members of their own social group. Strategies of deception include but are not limited to surreptitious activities such as extra-pair matings, brood parasitism, and even kidnapping. Yet, the benefits of living in groups typically outweigh the costs, as most birds show at least some degree of sociality during their adult lives, and many species spend their lives interacting with conspecifics on a daily basis. Few other animal taxa have as rich a diversity in social systems.

KEY POINTS

- Social evolution in birds has resulted in diverse social structures that vary from simple to complex, and from short-term seasonal aggregations to sustained multi-generational social units.
- Birds aggregate in flocks when the benefits (safety in numbers, collective vigilance) outweigh the costs (competition for resources, or an inability to defend resources).
- Within flocks, social dominance can determine access to resources (pecking order).
- Social interactions are both triggered by and can stimulate secretion of stress hormones and sex steroids.
- Cooperative breeders often live in harsh environments where independent breeding opportunities are limited, and helpers-at-the-nest are making the best of a bad situation.

- Additional helpers often increase offspring survival, and helping kin-related individuals increases the probability that shared genes will be passed on, enhancing inclusive fitness.
- Less commonly, non-kin-based societies evolve when the advantages of group defense of resources (or the nest) is essential to reproductive success.
- Highly social species in which individuals interact frequently have individual recognition and advanced cognitive abilities, including remarkable ability to learn and remember.

KEY MANAGEMENT AND CONSERVATION IMPLICATIONS

- Most species aggregate during at least part of the annual cycle, and in cases where the majority of the global species population aggregates in one place, these sites are designated important bird areas.
- Cooperative breeders are typically resident in fragile habitats and unpredictable environments to which they are obligately tied.
- Availability of suitable habitat limits global populations, and where their habitat type is rare or diminishing, cooperative breeders are more likely to be at risk of extinction.
- Long-term research into the behavior and ecology of populations is necessary to understand how ecological constraints, life history strategy, and genetics interact to shape population dynamics.

DISCUSSION QUESTIONS

1. Why do birds aggregate in flocks? What are the advantages for individuals?
2. Considering birds in parts of the world with varying abiotic factors, what seasonal patterns are seen in flock formation? What does this tell us about the annual cycles of these birds?
3. Are there specific traits that make certain bird taxa more likely to gather in flocks? Why do some birds roost and/or nest colonially while others do not?
4. What life history characteristics are typically associated with birds that breed cooperatively? Explain.
5. What are some of the principal ecological and environmental correlates of sustained group living and cooperative breeding? Do the same rules apply for all species? Why or why not?
6. Describe some of the routes to group living in birds. Why do they differ among species?
7. Are high levels of relatedness a requirement for sociality? How does relatedness affect the evolution of sociality?
8. Comparative analysis allows us to derive important relationships based on metadata from independent studies. In a sortable database worksheet, construct a table listing some of the different cooperatively breeding species in this chapter with the following headings: Species, habitat, food type (nonbreeding season), territorial/colonial, mean group size, helper sex (male, female, both), helper relatedness (related, unrelated, both), reproductive skew (separate for males and females, high/low), single/joint nest/plural nesting. Try sorting the sheet using each variable to identify ecological, social, and genetic factors that predict reproductive systems in cooperative breeders. Are there any obvious correlations? For a more in depth analysis, use the references below to include a larger number of species.

References

Alcock, J. (2005). Animal behavior: An evolutionary approach. 8th ed. Sinauer Associates, Inc. Publishers, Sunderland, MA.

Arnold, K. E., and I. P. F. Owens (1998). Cooperative breeding in birds: A comparative test of the life history hypothesis. Proceedings of the Royal Society of London B 265:739–745.

Arnold, K. E., and I. P. F. Owens (2002). Extra-pair paternity and egg dumping in birds: Life history, parental care and the risk of retaliation. Proceedings of the Royal Society of London B 269:1263–1269.

Berg, E. C. (2005). Parentage and reproductive success in the White-throated Magpie-Jay, *Calocitta formosa*, a cooperative breeder with female helpers. Animal Behaviour 70:375–385.

Berg, S. F., and R. R. Angel (2006). Seasonal roosts of Red-lored Amazons in Ecuador provide information about population size and structure. Journal of Field Ornithology 77:95–103.

Boland, C. R. J., and A. Cockburn (2002). Short sketches from the long history of cooperative breeding in Australian birds. Emu 102:9–17.

Brouwer, L., M. van de Pol, E. Atema, and A. Cockburn (2011). Strategic promiscuity helps avoid inbreeding at multiple levels in a cooperative breeder where both sexes are philopatric. Molecular Ecology 20:4796–4807.

Brouwer, L., M. van der Pol, and A. Cockburn (2014). The role of social environment on parental care: Offspring benefit more from the presence of female than male helpers. Journal of Animal Ecology 83:491–503.

Brown, C. R. (1998). Swallow summer. University of Nebraska Press, Lincoln, NE.

Brown, C. R., and M. B. Brown (1988). A new form of reproductive parasitism in Cliff Swallows. Nature 331:66–68.

Brown, C. R., and M. B. Brown (1996). Coloniality in the Cliff Swallow: The effect of group size on social behavior. University of Chicago Press, Chicago, IL.

Brown, C. R., and M. B. Brown (1998). Fitness components associated with alternative reproductive tactics in Cliff Swallows. Behavioral Ecology 9:158–171.

Brown, C. R., and M. B. Brown (2004). Group size and ectoparasitism affect daily survival probability in a colonial bird. Behavioral Ecology and Sociobiology 56:498–511.

Brown, J. L. (1987). Helping and communal breeding in birds: Ecology and evolution. Princeton University Press, Princeton, NJ.

Brown, J. L., E. R. Brown, S. D. Brown, and D. D. Dow (1982). Helpers: Effects of experimental removals on reproductive success. Science 215:421–422.

Browning, L. E., S. C. Patrick, L. A. Rollins, S. C. Griffiths, and A. F. Russell (2012). Kin selection, not group augmentation, predicts helping in an obligate cooperatively breeding bird. Proceedings of the Royal Society of London B 279:3861–3869.

Buckley, N. J. (1997). Experimental tests of the information-center hypothesis with Black Vultures (Coragyps atratus) and Turkey Vultures (Cathartes aura). Behavioral Ecology and Sociobiology 41:267–279.

Burke, T., N. B. Davies, M. W. Bruford, and B. J. Hatchwell (1989). Parental care and mating behavior of polyandrous Dunnocks Prunella modularis related to paternity by DNA fingerprinting. Nature 338:249–251.

Butterfield, P. A., and J. H. Crook (1968). The annual cycle of nest building and agonistic behavior in captive Quelea quelea with reference to endocrine factors. Animal Behaviour 16:308–317.

Caraco, T. H. (1979). Time budgeting and group size: A test of theory. Ecology 60:618–627.

Carlisle, T. R., and A. Zahavi (1986). Helping at the nest, allofeeding and social status in immature Arabian Babblers. Behavioral Ecology and Sociobiology 18:339–351.

Cockburn A. (2006). Prevalence of different modes of parental care in birds. Proceedings of the Royal Society of London B 273:1375–1383.

Cockburn, A., A. H. Dalziell, C. J. Blackmore, M. C. Double, H. Kokko, H. L. Osmond, N. R. Beck, M. L. Head, and K. Wells (2009). Superb Fairy-wren males aggregate into hidden leks to solicit extra-group fertilizations before dawn. Behavioral Ecology 20:501–510.

Conner, R. N., D. C. Rudolph, and J. R. Walters (2001). The Red-cockaded Woodpecker: Surviving in a fire-maintained ecosystem. University of Texas Press, Austin, TX.

Cornwallis, C., S. A. West, K. E. Davis, and A. S. Griffin (2010). Promiscuity and the evolutionary transition to complex societies. Nature 466:969–972.

Covas, R. A., A. Dalecky, A. Caizergues, and C. Doutrelant (2006). Kin associations and direct vs indirect fitness benefits in colonial cooperatively breeding Sociable Weavers Philetairus socius. Behavioral Ecology and Sociobiology 60:323–331.

Covas, R. A., M. A. duPlessis, and C. Doutrelant (2008). Helpers in colonial cooperatively breeding Sociable Weavers Philetairus socius contribute to buffer the effects of adverse breeding conditions. Behavioral Ecology and Sociobiology 63:103–112.

Crane, J. M. S., J. L. Pick, A. J. Tribe, E. Vincze, B. J. Hatchwell, and A. F. Russell (2015). Chestnut-crowned Babblers show

affinity for calls of removed group members: A dual playback without expectancy violation. Animal Behaviour 104:51–57.

Conner, R. N., D. C. Rudolph, and J. R. Walters (2001). The red-cockaded woodpecker: Surviving in a fire-maintained ecosystem. University of Texas Press, Austin, TX.

Cresswell, W., and J. L. Quinn (2010). Attack frequency, attack success and choice of prey group size for two predators with contrasting hunting strategies. Animal Behaviour 80:643–648.

Crook, J. H. (1960). Studies on the social behaviour of Quelea q. quelea (Linnaeus) in French West Africa. Behaviour 16:1–55.

Crook, J. H. (1963). A comparative analysis of nest structure in the weaver birds (Ploceinae). Ibis 105:238–262.

Dale, J. (2000). Ornamental plumage does not signal male quality in Red-billed Queleas. Proceedings of the Royal Society of London B 267:2143–2149.

Davies, N. B. (1989). Sexual conflict and the polygamy threshold. Animal Behaviour 38:226–234.

Davies, N. B. (1992). Dunnock behaviour and social evolution. Cambridge University Press, Cambridge, UK.

Davies, N. B., B. J. Hatchwell, T. Robson, and T. Burke (1992). Paternity and parental effort in Dunnocks Prunella modularis: How good are male chick-feeding rules? Animal Behaviour 43:729–745.

Davies, N. B., J. R. Krebs, and S. A. West (2012). An introduction to behavioral ecology. 4th ed. Wiley Press, Oxford.

Dawkins, R. (1976). The selfish gene. Paladin Granada Publishing, London.

Desrochers, A. (1989). Sex, dominance, and microhabitat use in wintering Black-capped Chickadees: A field experiment. Ecology 70:636–645.

Dey, C. J., J. Dale, and J. S. Quinn (2014). Manipulating the appearance of a badge of status causes changes in true badge expression. Proceedings of the Royal Society of London B 281:20132680.

Dickinson, J. L. (2004a). A test of the importance of direct and indirect fitness benefits for helping decisions in Western Bluebirds. Behavioral Ecology 15:233–238.

Dickinson, J. L. (2004b). Facultative sex ratio adjustment by Western Bluebird mothers with stay-at-home helpers-at-the-nest. Animal Behaviour 68:373–380.

Dickinson, J. L., E. D. Ferree, C. A. Stern, R. Swift, and B. Zuckerberg (2014). Delayed dispersal in Western Bluebirds: Teasing apart the importance of resources and parents. Behavioral Ecology 25: 843–885.

Downing, P. A., C. K. Cornwallis, and A. S. Griffin (2015). Sex, long life and the evolutionary transition to cooperative breeding in birds. Proceedings of the Royal Society of London, Series B 282: 20151663.

DuPlessis, M. A., and J. B. Williams (1994). Communal cavity roosting in Green Woodhoopes: Consequences for energy-expenditure and the seasonal pattern of mortality. The Auk 111:292–299.

Edwards, S. V., and S. Naeem (1993). The phylogenetic component of cooperative breeding in perching birds. American Naturalist 141:754–789.

Eens, M., E. Van Duyse, L. Berghman, and R. Pinxten (2000). Shield characteristics are testosterone-dependent in both male and female Moorhens. Hormones and Behavior 37:126–134.

Ekman, J., and M. Griesser (2002). Why offspring delay dispersal: Experimental evidence for a role of parental tolerance. Proceedings of the Royal Society of London B 269:1709–1713.

Ekman, J., and M. Griesser (2016). Siberian Jays: Delayed dispersal in the absence of cooperative breeding. In Cooperative breeding in vertebrates: Studies of ecology, evolution, and behavior, W. D. Koenig and J. L. Dickinson, Editors. Cambridge University Press, Cambridge, UK, pp. 6–18.

Emlen, S. T. (1982). The evolution of helping I: An ecological constraints model. American Naturalist 119:29–39.

Emlen, S. T. (1990). White-fronted Bee-eaters: Helping in a colonially nesting species. In Cooperative breeding in birds: Long-term studies of ecology and behavior, P. B. Stacey and W. D. Koenig, Editors. Cambridge University Press, Cambridge, UK, pp. 489–526.

Emlen, S. T., J. M. Emlen, and S. A. Levin (1986). Sex-ratio selection in species with helpers-at-the-nest. American Naturalist 127:1–8.

Emlen, S. T., and P. H. Wrege (1988). The role of kinship in helping decisions among White-fronted Bee-eaters. Behavioral Ecology and Sociobiology 23:305–315.

Emlen, S. T., and P. H. Wrege (1992). Parent-offspring conflict and the recruitment of helpers among bee-eaters. Nature 356:331–333.

Emlen, S. T., P. H. Wrege, and N. Demong (1995). Making decisions in the family: An evolutionary perspective. American Scientist 83:148–157.

Faaborg, J. P., G. Parker, L. DeLay, T. de Vries, J. C. Bednarz, S. Maria Paz, J. Naranjo, and T. A. Waite (1995). Confirmation of cooperative polyandry in the Galapagos Hawk Buteo galapagoensis. Behavioral Ecology and Sociobiology 36:83–90.

Feeney, W. E., I. Medina, M. Somveille, R. Heinsohn, M. L. Hall, R. A. Mulder, J. A. Stein, R. M. Kilner, and N. E. Langmore (2013). Brood parasitism and the evolution of cooperative breeding in birds. Science 342:1506–1508.

Frith, C. B., and D. W. Frith (2004). The bowerbirds. Oxford University Press, Oxford.

Gibbons, D. W. (1986). Brood parasitism and cooperative nesting in the Moorhen, Gallinula chloropus. Behavioral Ecology and Sociobiology 19:221–232.

Goldizen, A. W., D. A. Putland, and A. R. Goldizen (1998). Variable mating patterns in Tasmanian Native Hens Gallinula mortierii correlates of reproductive success. Journal of Animal Ecology 67:307–317.

Gonzalez, J.-C. T., B. C. Sheldon, and J. A. Tobias (2013). Environmental stability and the evolution of cooperative breeding in hornbills. Proceedings of the Royal Society of London B 280:1768–1776.

Gowaty, P. A., and M. R. Lennartz (1985). Sex ratios of nestling and fledgling Red-cockaded Woodpeckers (Picoides borealis) favor males. American Naturalist 126:347–353.

Göth, A., and D. N. Jones (2003). Ontogeny of social behaviour in the megapode Alectura lathami (Australian Brush-turkey). Journal of Comparative Psychology 117:36–43.

Göth, A., and C. S. Evans (2004). Social responses without early experience: Australian Brush-turkey chicks use specific visual cues to aggregate with conspecifics. Journal of Experimental Biology 207:2199–2208.

Greisser, M., M. Nystrand, and J. Ekman (2006). Reduced mortality selects for family cohesion in a social species. Proceedings of the Royal Society of London B 273:1881–1886.

Griffin, A. S., B. C. Sheldon, and S. A. West (2005). Cooperative breeders adjust offspring sex ratios top helpful helpers. American Naturalist 166:628–632.

Hamilton, W. D. (1964). The genetical evolution of social behaviour II. Journal of Theoretical Biology 7:17–52.

Hamilton, W. D. (1971). Geometry for the selfish herd. Journal of Theoretical Biology 31:295–311.

Hannon, S. J., R. L. Mumme, W. D. Koenig, and F. A. Pitelka (1985). Replacement of breeders and within-group conflict in the cooperatively breeding Acorn Woodpecker. Behavioral Ecology and Sociobiology 17:303–312.

Hatchwell, B. J. (1999). Investment strategies of breeders in avian cooperative breeding systems. American Naturalist 154:205–219.

Hatchwell, B. J. (2009). The evolution of cooperative breeding in birds: Kinship, dispersal and life history. Philosophical Transactions of the Royal Society of London B 364:3217–3227.

Hatchwell, B. J. (2010). Cryptic kin selection: Kin structure in vertebrate populations and opportunities for kin-directed cooperation. Ethology 116:203–216.

Hatchwell, B. J., C. Anderson, D. J. Ross, M. K. Fowlie, and P. G. Blackwell (2001). Social organization in cooperatively breeding Long-tailed Tits: Kinship and spatial dynamics. Journal of Animal Ecology 70:820–830.

Hatchwell, B. J., A. F. Russell, A. D. C. MacColl, D. J. Ross, M. K. Fowlie, and A. McGowan (2004). Helpers increase long-term but not short-term productivity in cooperatively breeding Long-tailed Tits. Behavioral Ecology 15:1–10.

Hatchwell, B. J., S. P. Sharp, A. P. Beckerman, and J. Meade (2013). Ecological and demographic correlate of helping behavior in a cooperatively breeding bird. Journal of Animal Ecology 82:486–494.

Hatchwell, B. J., P. R. Gullett, and M. J. Adams (2014). Helping in cooperatively breeding Long-tailed Tits: A test of Hamilton's rule. Philosophical Transactions of the Royal Society of London B 369:20130565.

Haydock, J., and W. D. Koenig (2003). Patterns of reproductive skew in the polygynandrous Acorn Woodpecker. American Naturalist 162:277–289.

Heinsohn, R. (1991). Kidnapping and reciprocity in cooperatively breeding White-winged Choughs. Animal Behaviour 41:1097–1100.

Hinton, M. G., W. K. Reisen, S. S. Wheeler, and A. K. Townsend (2015). West Nile virus activity in a winter roost of American Crows (Corvus brachyrhynchos): Is bird-to-bird transmission important in persistence and amplification? Journal of Medical Entomology 52:683–692.

Jamieson, I. G., J. S. Quinn, P. A. Rose, and B. N. White (1994). Shared paternity among non-relatives is a result of an egalitarian mating system in a communally breeding bird, the Pukeko. Proceedings of the Royal Society of London B 257:271–277.

Jamieson, I. G., S. B. McRae, M. Trewby, and R. E. Simmons (2000). High rates of conspecific brood parasitism and egg rejection in coots and moorhens in ephemeral wetlands in Namibia. The Auk 117:250–252.

Jetz, W., and D. R. Rubenstein (2011). Environmental uncertainty and the global biogeography of cooperative breeding in birds. Current Biology 21:72–78.

Jones, D. N., R. W. R. J. Dekker, and C. S. Roselaar (1995). The Megapodes. Oxford University Press, Oxford.

Kemp, A. L. (1995). The hornbills: Bucerotiformes. (Series: Bird Families of the World). Oxford University Press, Oxford.

Kenward, B., C. Schloegl, C. Rutz, A. A. S. Weir, T. Bugnar, and A. Kacelnik (2011). On the evolutionary and ontogenetic origins of tool-oriented behavior in New Caledonian Crows (Corvus moneduloides). Biological Journal of the Linnean Society 102:870–877.

Khan, M. Z., F. M. A. McNabb, J. R. Walters, and P. J. Sharp (2001). Patterns of testosterone and prolactin concentrations and reproductive behavior of helpers and breeders in the cooperatively breeding Red-cockaded Woodpecker (Picoides borealis). Hormones and Behavior 40:1–13.

Kimwele, C. N., and J. A. Graves (2003). A molecular genetic analysis of the communal nesting of the Ostrich (Struthio camelus). Molecular Ecology 12:229–236.

Koenig, W. D., and R. L. Mumme (1987). Population ecology of the cooperatively breeding Acorn Woodpeckers. Princeton University Press, Princeton, NJ.

Koenig, W. D., and P. B. Stacey (1990). Acorn Woodpeckers: Group-living and food shortage under contrasting ecological conditions. In Cooperative breeding in birds: Long-term studies of ecology and behavior, P. B. Stacey and W. D. Koenig, Editors. Cambridge University Press, Cambridge, UK, pp. 413–454.

Koenig, W. D., F. A. Pitelka, W. J. Carmen, R. L. Mumme, and M. T. Stanback (1992). The evolution of delayed dispersal in cooperative breeders. Quarterly Review of Biology 67:111–150.

Koenig, W. D., R. L. Mumme, M. T. Stanback, and F. A. Pitelka (1995). Patterns and consequences of egg destruction among joint-nesting Acorn Woodpeckers. Animal Behaviour 50:607–621.

Koenig, W. D., and J. L. Dickinson, Editors (2004). Ecology and evolution of cooperative breeding in birds. Cambridge University Press, Cambridge, UK.

Koenig, W. D., and E. L. Walters (2011). Age-related provisioning behavior in the cooperatively breeding Acorn Woodpecker: Testing the skills and pay-to-stay hypotheses. Animal Behaviour 82:437–444.

Koenig, W. D., and E. L. Walters (2015). Temporal variability and cooperative breeding: Testing the bet-hedging hypothesis in the Acorn Woodpecker. Proceedings of the Royal Society of London B 282:20151742.

Koenig, W. D., E. L. Walters, and J. Haydock (2015). Variable helper effects, ecological conditions, and the evolution of cooperative breeding in the Acorn Woodpecker. American Naturalist 178:145–158.

Koenig, W. D., and J. L. Dickinson, Editors (2016). Cooperative breeding in vertebrates: Studies of ecology, evolution, and behavior. Cambridge University Press, Cambridge, UK.

Koenig, W. D., E. L. Walters, and J. Haydock (2016). Acorn Woodpeckers: Helping at the nest, polygynandry, and dependence on a variable acorn crop. In Cooperative breeding in vertebrates: Studies of ecology, evolution and behavior, W. D. Koenig and J. L. Dickinson, Editors. Cambridge University Press, Cambridge, UK, pp. 217–236.

Komdeur, J. (1992). Importance of habitat saturation and territory quality for evolution of cooperative breeding in the Seychelles Warbler. Nature 358:493–495.

Komdeur, J. (1994). Experimental evidence for helping and hindering by previous offspring in the cooperative-breeding Seychelles Warbler Acrocephalus sechellensis. Behavioral Ecology and Sociobiology 34:175–184.

Komdeur, J. (1996). Influence of helping and breeding experience on reproductive performance in the Seychelles Warbler: A translocation experiment. Behavioral Ecology 7:326–333.

Komdeur, J. (1998). Long-term fitness benefits of egg sex modification by the Seychelles Warbler. Ecology Letters 1:56–62.

Komdeur, J., S. Daan, J. Tinbergen, and C. Mateman (1997). Extreme adaptive modification in sex ratio of the Seychelles Warbler's eggs. Nature 385:522–525.

Komdeur, J., T. Burke, H. Dugdale, and D. S. Richardson (2016). Seychelles Warblers: Complexities of the helping paradox. In Cooperative breeding in vertebrates: Studies of ecology, evolution and behavior, W. D. Koenig and J. L. Dickinson, Editors. Cambridge University Press, Cambridge, UK, pp. 197–216.

Laucht, S., B. Kempenaers, and J. Dale (2010). Bill color, not badge size, indicates testosterone-related information in House Sparrows. Behavioral Ecology and Sociobiology 64:1461–1471.

Legge, S., R. Heinsohn, M. C. Double, R. Griffiths, and A. Cockburn (2001). Complex sex allocation in the Laughing Kookaburra. Behavioral Ecology 12:524–533.

Ligon, J. D. (1983). Cooperation and reciprocity in avian social systems. American Naturalist 121:366–383.

Lima, M. R., R. H. Macedo, L. Muniz, A. Pacheco, and J. A. Graves (2011). Group composition, mating system, and relatedness in the communally breeding Guira Cuckoo (Guira guira) in central Brazil. The Auk 128:475–486.

Lima, S. L. (1995). Back to basics of anti-predatory vigilance: The group-size effect. Animal Behaviour 49:11–20.

Louder, M. I. M., M. P. Ward, W. M. Schelsky, M. E. Hauber, and J. P. Hoover (2015). Out on their own: A test of adult-assisted dispersal in fledgling brood parasites reveals solitary departures from hosts. Animal Behaviour 110:29–37.

Macedo, R. H. (2016). Guira Cuckoos: Cooperation, infanticide, and female reproductive investment in a joint-nesting species. In Cooperative breeding in vertebrates: Studies of ecology, evolution and behavior, W. D. Koenig and J. L. Dickinson,

Editors. Cambridge University Press, Cambridge, UK, pp. 257–271.

Macedo, R. H. F., M. O. Cariello, and L. Muniz (2001). Context and frequency of infanticide in communally breeding Guira Cuckoos. The Condor 103:170–175.

Macedo, R. H. F., M. O. Cariello, J. Graves, and H. Schwabl (2004). Reproductive partitioning in communally breeding Guira Cuckoos, Guira guira. Behavioral Ecology and Sociobiology 55:213–222.

Marra, P. P., R. T. Holmes, and T. W. Sherry (1993). Territorial exclusion by a temperate-tropical migrant warbler in Jamaica: A removal experiment with American Redstarts (Setophaga ruticilla). The Auk 110:565–572.

Marra, P. P., and R. T. Holmes (2001). Consequences of dominance-mediated habitat segregation in American Redstarts during the nonbreeding season. The Auk 118:92–104.

Martínez, A. E., and J. P. Gomez (2013). Are mixed-species bird flocks stable through two decades? American Naturalist 181:E53–E59

Marzluff, J. M., B. Heinrich, and C. S. Heinrich (1996). Raven roosts are mobile information centers. Animal Behaviour 51:89–103.

Marzluff, J. M., J. Walls, H. N. Cornel, J. Withey, and D. P. Craig (2010). Lasting recognition of threatening people by wild American Crows. Animal Behaviour 79:699–707.

Marzluff, J. M., R. Miyaoka, S. Minoshima, and D. J. Cross (2012). Brain imaging reveals neuronal circuitry underlying the crow's perception of human faces. Proceedings of the National Academy of Sciences USA 109:15912–15917.

McKechnie, A. E., and B. G. Lovegrove (2001). Thermoregulation and the energetic significance of clustering behavior in the White-backed Mousebird (Colius colius). Physiological and Biochemical Zoology 74:238–249.

McNamara, J. M., and A. I. Houston (1992). Evolutionarily stable levels of vigilance as a function of group size. Animal Behaviour 43:641–658.

McRae, S. B. (1996a). Family values: Costs and benefits of communal breeding in the Moorhen. Animal Behaviour 52:225–245.

McRae, S. B. (1996b). Brood parasitism in the Moorhen: Brief encounters between parasites and hosts and the significance of an evening laying hour. Journal of Avian Biology 27:311–320.

McRae, S. B., and T. Burke (1996). Intraspecific brood parasitism in the Moorhen: Parentage and parasite-host relationships determined by DNA fingerprinting. Behavioral Ecology and Sociobiology 38:115–129.

Meade, J., K.-B. Nam, A. P. Beckerman, and B. J. Hatchwell (2010). Consequences of "load-lightening" for future indirect fitness gains by helpers in a cooperatively breeding bird. Journal of Animal Ecology 79:529–537.

Morris, F. O. (1856). A history of British birds. Groombridge and Sons, London.

Mulder, R. A. (1997). Extra-group courtship displays and other reproductive tactics of Superb Fairy-wrens. Australian Journal of Zoology 45:131–143.

Mulder, R. A., and N. E. Langmore (1993). Dominant males punish helpers for temporary defection in Superb Fairy-wrens. Animal Behaviour 45:830–833.

Mulder, R. A., and M. J. L. Magrath (1993). Timing of prenuptial molt as a sexually selected indicator of male quality in Superb Fairy-wrens (Malurus cyaneus). Behavioral Ecology 5:393–400.

Mulder, R. A., P. O. Dunn, A. Cockburn, K. A. Lazenby-Cohen, and M. J. Howell (1994). Helpers liberate female Fairy-wrens from constraints on extra-pair mate choice. Proceedings of the Royal Society of London B 255:223–229.

Mumme, R. L., W. D. Koenig, and F. A. Pitelka (1983). Reproductive competition in the communal Acorn Woodpecker: Sisters destroy each other's eggs. Nature 306:583–584.

Munn, C. A. (1986). Birds that "cry wolf." Nature 319:143–145.

Munn, C. A., and J. W. Terborgh. (1979). Multi-species territoriality in Neotropical foraging flocks. The Condor 81:338–347.

Nelson-Flower, M. J., P. A. R. Hockey, C. O'Ryan, N. J. Raihani, M. A. du Plessis, and A. R. Ridley (2011). Monogamous dominant pairs monopolize reproduction in the cooperatively breeding Pied Babbler. Behavioral Ecology 22:559–565.

Nelson-Flower, M. J., P. A. R. Hockey, C. O'Ryan, and A. R. Ridley (2012). Inbreeding avoidance mechanisms: Dispersal dynamics in cooperatively breeding Southern Pied Babblers. Journal of Animal Ecology 81:876–883.

Nelson-Flower, M. J., P. A. R. Hockey, C. O'Ryan, S. English, A. M. Thompson, K. Bradley, R. Rose, and A. R. Ridley (2013). Costly reproductive competition between females in a monogamous cooperatively breeding bird. Proceedings of the Royal Society of London B 280:20130728.

Nelson-Flower, M. J., and A. R. Ridley (2015). Male-male competition is not costly to dominant males in a cooperatively breeding bird. Behavioral Ecology and Sociobiology 69:1997–2004.

Pulliam, H. R. (1973). On the advantages of flocking. Journal of Theoretical Biology 38:419–422.

Pulliam, H. R., G. H. Pyke, and T. Caraco (1982). The scanning behaviour of juncos: A game-theoretical approach. Journal of Theoretical Biology 95:89–103.

Quinn, J. L., and M. Ueta (2008). Protective nesting associations in birds. Ibis 150:146–167.

Quinn, J. S., R. Macedo, and B. N. White (1994). Genetic relatedness of communally breeding Guira Cuckoos. Animal Behaviour 47:515–529.

Rabenold, K. N. (1984). Cooperative enhancement of reproductive success in tropical Wren societies. Ecology 65:871–885.

Rabenold, P. P., K. N. Rabenold, W. H. Piper, J. Haydock, and S. W. Zack. (1990). Shared paternity revealed by genetic analysis in cooperatively breeding tropical Wrens. Nature 348:538–540.

Raihani, N. J., M. J. Nelson-Flower, K. A. Golabek, and A. R. Ridley (2010). Routes to breeding in cooperatively breeding Pied Babblers Turdoides bicolor. Journal of Avian Biology 41:681–686.

Reyer, U. (1984). Investment and relatedness: A cost-benefit analysis of breeding and helping in the Pied Kingfisher (Ceryle rudis). Animal Behaviour 32:1163–1178.

Reyer, U. (1990). Pied Kingfishers: Ecological causes and reproductive consequences of cooperative breeding. In Cooperative breeding in birds: Long-term studies of ecology and behavior, P. B. Stacey and W. D. Koenig, Editors. Cambridge University Press, Cambridge, UK, pp. 529–557.

Ridley, A. N. (2016). Southern Pied Babblers: The dynamics of conflict and cooperation in a group-living society. *In* Cooperative breeding in vertebrates: Studies of ecology, evolution and behavior, W. D. Koenig and J. L. Dickinson, Editors. Cambridge University Press, Cambridge, UK, pp. 115–132.

Riehl C. (2010). A simple rule reduces costs of extra-group parasitism in a communally breeding bird. Current Biology 20:1830–1833.

Riehl, C. (2011). Living with strangers: Direct benefits favor non-kin cooperation in a communally nesting bird. Proceedings of the Royal Society of London B 278:1728–1735.

Riehl C. (2012). Mating system and reproductive skew in a communally breeding cuckoo: Hard-working males do not sire more young. Animal Behaviour 84:707–714.

Riehl, C. (2013). Evolutionary routes to non-kin cooperative breeding in birds. Proceedings of the Royal Society of London B 280:20132245.

Riehl, C., and L. Jara (2009). Natural history and reproductive biology of the communally breeding Greater Ani (*Crotophaga major*) at Gatún Lake, Panama. Wilson Journal of Ornithology 121:679–687.

Rohwer, S., and F. C. Rohwer (1978). Status signaling in Harris Sparrows: Experimental deceptions achieved. Animal Behaviour 26:1012–1022.

Rubenstein, D. R. (2011). Spatiotemporal environmental variation, risk aversion, and the evolution of cooperative breeding as a bet-hedging strategy. Proceedings of the National Academy of Sciences USA 108:10816–10822.

Rubenstein, D. R., and I. J. Lovette (2007). Temporal environmental variability drives the evolution of cooperative breeding in birds. Current Biology 17:1414–1419.

Russell, A. J., and B. J. Hatchwell (2001). Experimental evidence for kin-biased helping in a cooperatively breeding vertebrate. Proceedings of the Royal Society of London B 268:2169–2174.

Russell, E. M. (2000). Avian life histories: Is extended parental care the southern secret? Emu 100:377–399.

Rutkowska, J., and A. V. Badyaev (2008). Meiotic drive and sex determination: Molecular and cytological mechanisms of sex ratio adjustment in birds. Philosophical Transactions of the Royal Society of London B 363:1675–1686.

Safran, R. J., J. S. Adelman, K. J. McGraw, and M. Hau (2014). Sexual signal exaggeration affects physiological state in male Barn Swallows. Current Biology 18:R461–R462.

Schmaltz, G., J. S. Quinn, and C. Lentz (2008). Competition and waste in the communally breeding Smooth-billed Ani: Effects of group size on egg-laying behavior. Animal Behaviour 76:153–162.

Schoech, S. J., R. L. Mumme, and J. C. Wingfield (1996a). Delayed breeding in the cooperatively breeding Florida Scrub-Jay (*Aphelocoma coerulescens*) inhibition or the absence of stimulation. Behavioral Ecology and Sociobiology 39:77–90.

Schoech, S. J., R. L. Mumme, and J. C. Wingfield (1996b). Prolactin and helping behavior in the cooperatively breeding Florida Scrub-Jay (*Aphelocoma coerulescens*). Animal Behaviour 52:445–456.

Schoech, S. J., R. L. Mumme, and J. C. Wingfield (1997). Corticosterone, reproductive status, and body mass in a cooperative breeder, the Florida Scrub-Jay (*Aphelocoma coerulescens*). Physiological Zoology 70:68–73.

Sharp, S. P., A. McGowan, M. J. Wood, and B. J. Hatchwell (2005). Learned kin recognition cues in a social bird. Nature 434:1127–1130.

Shen, S.-F., S. L. Vehrencamp, R. A. Johnstone, H.-C. Chen, S.-F. Chan, W.-Y. Liao, K.-Y. Lin, and H.-W. Yuan (2012). Unfavorable environment limits social conflict in *Yuhina brunneiceps*. Nature Communications 3:885. doi:10.1038/ncomms1894.

Skutch, A. F. (1935). Helpers at the nest. The Auk 52:257–273.

Sonerud, G. A., C. A. Smedshaug, and Ø. Bråthen (2001). Ignorant Hooded Crows follow knowledgeable roost-mates to food: Support for the information center hypothesis. Proceedings of the Royal Society of London B 268:827–831.

Stacey, P. B., and W. D. Koenig, Editors (1990). Cooperative breeding in birds: Long-term studies of ecology and behavior. Cambridge University Press, Cambridge, UK.

Taylor, A. H., R. Miller, and R. D. Gray (2012). New Caledonian Crows reason about hidden causal agents. Proceedings of the National Academy of Sciences USA 109:16389–16391.

Toft, C. A., and T. F. Wright (2015). Parrots of the wild. University of California Press, Oakland.

Trail, P. W., S. D. Strahl, and J. L. Brown (1981). Infanticide in relation to individual and flock histories in a communally breeding bird, the Mexican Jay (*Aphelocoma ultramarina*). American Naturalist 118:72–82.

Trivers, R. L. (1971). Evolution of reciprocal altruism. Quarterly Review of Biology 46:35–57.

Trivers, R. L., and D. E. Willard (1973). Natural selection of parental ability to vary the sex ratio of offspring. Science 179:90–92.

van Dijk, R. E., J. C. Kaden, A. Argüelles-Ticó, L. M. Beltran, M. Paquet, and R. Covas (2013). The thermoregulatory benefits of the communal nest of Sociable Weavers *Philetairus socius* are spatially structured within nests. Journal of Avian Biology 44:102–110.

van Dijk, R. E., J. C. Kaden, A. Argüelles-Ticó, D. A. Dawson, T. Burke, and B. J. Hatchwell (2014). Cooperative investment in public goods is kin directed in communal nests of social birds. Ecology Letters 17:1141–1148.

van Dijk, R. E., R. Covas, C. Doutrelant, C. N. Spottiswoode, and B. J. Hatchwell (2015). Fine-scale genetic structure reflects sex-specific dispersal strategies in a population of Sociable Weavers (*Philetairus socius*). Molecular Ecology 24:4296–4311.

Vehrencamp, S. L. (1977). Relative fecundity and parental effort in communally nesting Anis, *Crotophaga sulcirostris*. Science 197:403–405.

Vehrencamp, S. L. (1978). The adaptive significance of communal nesting in Groove-billed Anis (*Crotophaga sulcirostris*). Behavioral Ecology and Sociobiology 4:1–33.

Vehrencamp, S. L. (2000). Evolutionary routes to joint-female nesting in birds. Behavioral Ecology 11:334–344.

Vehrencamp, S. L., and J. S. Quinn (2004). Avian joint laying systems. *In* Ecology and evolution of cooperative breeding in birds, W. D. Koenig and J. Dickinson, Editors. Cambridge University Press, Cambridge, UK, pp. 177–196.

Vine, I. (1973). Detection of prey flocks by predators. Journal of Theoretical Biology 40:207–210.

Vleck, C. M., N. A. Mays, J. W. Dawson, and A. R. Goldsmith (1991). Hormonal correlates of parental and helping behavior in cooperatively breeding Harris' Hawks (*Parabuteo unicinctus*). The Auk 108:638–648.

Walters, J. R., C. K. Copeyon, and J. H. Carter III (1992). Test of the ecological basis of cooperative breeding in Red-cockaded Woodpeckers. The Auk 109:90–97.

Walters, J. R., and V. Garcia (2016). Red-cockaded Woodpeckers: Alternative pathways to breeding success. *In* Cooperative breeding in vertebrates: Studies of ecology, evolution and behavior, W. D. Koenig and J. L. Dickinson, Editors. Cambridge University Press, Cambridge, UK, pp. 58–76.

Ward, P., and A. Zahavi (1973). The importance of certain assemblages of birds as "information-centers" for food finding. Ibis 115:517–534.

Weaver, H. B., and C. R. Brown (2004). Brood parasitism and egg transfer in Cave Swallows (*Petrochelidon fulva*) and Cliff Swallows (*P. pyrrhonota*) in South Texas. The Auk 121:1122–1129.

Webster, M. S., K. A. Tarvin, E. M. Tuttle, and S. Pruett-Jones (2004). Reproductive promiscuity in the Splendid Fairy-wren: Effects of group size and reproduction by auxiliaries. Behavioral Ecology 15:907–915.

West, S. A., D. M. Shuker, and B. C. Sheldon (2005). Sex-ratio adjustment when relatives interact: A test of constraints on adaptation. Evolution 59:1211–1228.

Westneat, D. F., and P. W. Sherman (1997). Density and extra-pair fertilizations in birds: A comparative analysis. Behavioral Ecology and Sociobiology 41:205–215.

Wingfield, J. C., R. E. Hegner, and D. M. Lewis (1991). Circulating levels of luteinizing hormone and steroid hormones in relation to social status in the cooperatively breeding White-browed Sparrow-Weaver, *Plocepasser mahali*. Journal of Zoology 225:43–58.

Wingfield, J. C., R. E. Hegner, and D. M. Lewis (1992). Hormonal responses to removal of a breeding male in the cooperatively breeding White-browed Sparrow-Weaver, *Plocepasser mahali*. Hormones and Behavior 26:145–155.

Wingfield, J. C., and D. M. Lewis (1993). Hormonal and behavioral responses to simulated territorial intrusion in the cooperatively breeding White-browed Sparrow-Weaver, *Plocepasser mahali*. Animal Behaviour 45:1–11.

Woolfenden, G. E., and J. W. Fitzpatrick (1984). The Florida Scrub Jay: Demography of a cooperatively-breeding bird. Princeton University Press, Princeton, NJ.

Wright, J., R. E. Stone, and N. Brown (2003). Communal roosts as structured information centers in the Raven, *Corvus corax*. Journal of Animal Ecology 72:1003–1014.

Yom-Tov, Y. (2001). An updated list and some comments on the occurrence of intraspecific nest parasitism in birds. Ibis 143:133–143.

Zann, R. A. (1996). Zebra Finch: A synthesis of field and laboratory studies. Oxford University Press, Oxford.

Zuk, M., and T. Johnsen (2000). Social environment and immunity in male Red Jungle Fowl. Behavioral Ecology 11:146–153.

CHAPTER 18

Habitat Ecology

Matthew Johnson and Eric M. Wood

THE NORTHERN SPOTTED OWL: A PREVIEW OF LESSONS IN AVIAN HABITAT ECOLOGY

The Northern Spotted Owl (*Strix occidentalis caurina*) is one of the most well-studied bird species in the world (Gutiérrez et al. 1995, USFWS 2011), and also one of the most controversial. Conflict between conservationists and the timber industry erupted in the 1980s over the spotted owl's association with economically valuable old-growth forests of the Pacific Northwest in North America (Yaffee 1994), leading to debates, lawsuits, death threats, and eventually the large-scale ecosystem management plan called the Northwest Forest Plan (Davis et al. 2011). This owl has become a poster child for the tension between traditional environmentalism and economic development (Gup 1990). Habitat ecology is the centerpiece to bird conservation and is at the heart of the spotted owl saga; the owl's story provides a preview of seven major lessons of this chapter (fig. 18.1).

The Northern Spotted Owl is a denizen of old forests of the Pacific Northwest—places dominated by large coniferous trees, such as Douglas-fir (*Pseudotsuga menziesii*) and western hemlock (*Tsuga heterophylla*), and complex forest canopies with dead or damaged trees and downed logs (e.g., Forsman et al. 1984). But knowledge of simple habitat associations is insufficient. Spotted owls use forests with large (usually old) trees at several spatial scales (from nest trees to landscapes; Blakesley et al. 2005), but the availability of safe nesting sites and rodent prey underlie the selection of forest attributes (Sakai and Noon 1993, Zabel et al. 1995). These studies of spotted owl habitat associations highlight the first two lessons of this chapter: (1) We must be mindful that habitat is more than the vegetation around a bird; it is rooted in the resources and ecological conditions a bird needs for survival and reproduction. And (2), to balance species conserva-

tion and human needs, we must understand the nuances of "habitat selection"—the study of where birds live and why. Indeed, Franklin et al. (2000) found that in Northern California, where spotted owls eat woodrats (*Neotoma* spp.) associated with young forests and forest edges, owl reproduction and survival was highest where owl territories were composed of a mosaic of old and young forests that provide large trees for nesting and young patches with abundant prey. Such findings illustrate another lesson: (3) Estimating the capacity for a habitat to support survival and reproduction is arguably the best way to measure habitat quality (Johnson 2007). This work proved central to understanding spotted owl demography (USFWS 2011) and underscored a fourth major lesson: (4) The quality, quantity, and spatial distribution of habitats strongly influence population dynamics for

Figure 18.1. Nothern Spotted Owls (*Strix occidentalis caurina*). In this photo, you can see colored leg bands used to mark these birds for studies of their habitat and population dynamics, in this case on the Hoopa Valley Indian Reservation in Northern California. *Photo by J. Mark Higley, Hoopa Tribal Forestry.*

many bird species in the world (Newton 1998). The profound role of habitat in spotted owl conservation has prompted extensive research to map the distribution of the owls and their habitats. These maps began as simple hand-drawn range limits based on aerial imagery, but with the rapid improvement of statistical and spatial models (Elith et al. 2006, Phillips et al. 2006) and advances in remotely sensed data (e.g., high resolution satellite imagery; Ackers et al. 2015), avian ecologists are now producing high-resolution maps of bird distributions over broad spatial and temporal extents (e.g., Schumaker et al. 2014) including for the Northern Spotted Owl (Stralberg et al. 2009b). This provides another lesson for contemporary ornithologists: (5) New technologies can be harnessed to help measure habitat selection and habitat quality, and to map habitats in novel ways. However, with new technologies come discrepancies in their application, and controversy among owl researchers continues today over how to best link habitat selection, maps, and demography in order to further conservation goals (Loehle et al. 2015, Dunk et al. 2015, Bell et al. 2015). These are not just academic arguments, because conservation and land-use rules rely on accurate and meaningful maps of spotted owl habitat. This reliance exemplifies another lesson: (6) Habitat ecology strongly affects environmental policies, which carry enormous social and economic consequences. In fact, the term "critical habitat" is a keystone to the US Endangered Species Act (Camaclang et al. 2015), arguably the most powerful piece of wildlife conservation legislation in the world. In a landmark decision involving both Northern Spotted Owls and Red-cockaded Woodpeckers (*Leuconotopicus borealis*), the US Supreme Court confirmed that destroying critical habitat is just as serious as killing birds directly (*Babbitt v. Sweet Home* 1995). This decision not only represented a huge shift in environmental law and prompted modern wildlife conservation (Kareiva and Marvier 2015), it also exemplifies the relationship between science and law (Alagona 2013) evident in recent summaries of spotted owl conservation plans by the US Fish and Wildlife Service (USFWS 2011, 2012).

So where does this story end? There is both reason for concern and cause for celebration regarding the Northern Spotted Owl. Thankfully, a better understanding of habitat ecology has enabled modern forestry to, in some places, integrate timber extraction and owl habitat conservation (e.g., Forsman et al. 2011). However, the spotted owl needs more than just sufficient high-quality habitat. Demographic modeling has revealed the importance of local weather and climate, with cold wet winters diminishing owl reproduction (Franklin et al. 2000, Dugger et al. 2005, Glenn et al. 2011). Old forests may provide some protection from inclement weather, but these findings

raise the specter of how this species may fare under ongoing climate change (Carroll 2010). Added to the list of challenges facing this bird is the looming advance of a closely related species, the Barred Owl (*Strix varia*), into the spotted owl's range. Barred Owls appear to outcompete and depredate spotted owls in most, if not all, habitats (Wiens et al. 2014), and the Northern Spotted Owl continues to decline in most of its range (Forsman et al. 2011). Thus, despite the enormity of research on spotted owl habitat ecology, the latest findings reveal a final important lesson: (7) Other factors can interact with or in some cases overwhelm habitat in driving bird populations, and ornithologists must be aware of the range of processes affecting birds.

In this chapter, you will come to recognize the theory underlying these seven lessons illustrated by the story of the Northern Spotted Owl. First, we examine core definitions of habitat ecology and learn about the history of the discipline. Next, we introduce theories of habitat quality and selection before considering the consequences of habitat to bird populations. Then we explore some common methods for studying bird habitat, both in the field and with remote measurements. The chapter concludes with the implications of habitat ecology for modern avian conservation. Throughout, you will gain an introduction to the field of avian habitat ecology, becoming familiar with its development, importance, and opportunity, and its limitations.

HABITAT: ITS CORE DEFINITION AND RELEVANCE

As the story of the spotted owl illustrates, habitat is central to bird conservation because the resources and environmental conditions in a habitat affect birds' survival and reproduction (Bernstein et al. 1991, Pulliam 2000). It is no surprise, then, that ornithologists have long recognized the need to understand variation in habitat for birds (Block and Brennan 1993). Indeed, there is overwhelming evidence that the loss and degradation of habitat poses the greatest threat to bird species (fig. 18.2).

Despite the centrality of habitat to the discipline of avian ecology, confusion remains over how to best measure variation in habitats over space and time. Hall and her colleagues (Hall et al. 1997, Morrison et al. 2002) argued that some of this confusion stems from inconsistent and imprecise use of terms, which is unsurprising, given habitat's long history in ecology (Grinnell 1917, MacArthur et al. 1962, Whittaker et al. 1973, Johnson 2007). Hall et al. (1997) sought to provide standards, emphasized that habitat is species-specific, and established a **definition of habitat** as "the resources and conditions present in an area that produce occupancy—including

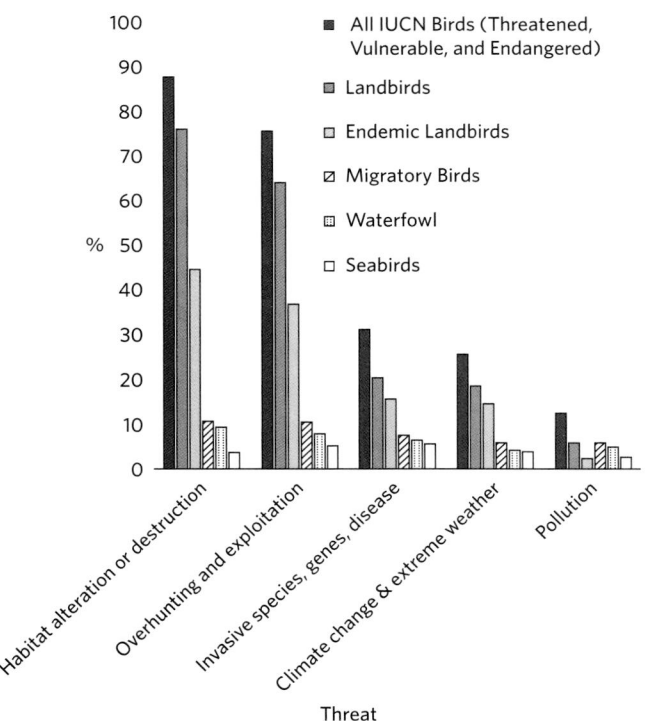

Figure 18.2. The top five threats facing birds species in the world, according to the International Union for the Conservation of Nature (IUCN) Red List of bird species (near threatened, vulnerable, and endangered, *n* = 2,364). Numbers sum to more than 100 percent because a species can be affected by more than one threat. Habitat alteration or destruction is the overwhelmingly most common threat faced and includes the following categories of human activity: agriculture and aquaculture, energy production and mining, human intrusions and disturbance, residential and commercial development, and transportation and service corridors.

survival and reproduction—by a given organism." This is the definition of habitat used in this chapter. Sometimes, habitat is crudely described as the vegetation where a bird species is typically found—e.g., bottomland hardwood habitat, tall grass prairie habitat, estuary habitat (Hutto 1985). This definition is of limited value because it depends on classification of often continuous resources or conditions into discrete categories, and it does not address the issue of habitat quality (Gaillard et al. 2010). Further, that definition does not recognize that habitat is species-specific, and it is insensitive to spatial and temporal scales. As the lesson of the spotted owl showed, we must be mindful that habitat is more than the vegetation around a bird; it is rooted in the **resources** and **ecological conditions** a bird needs for survival and reproduction. By studying the resources (e.g., food, cover, and nesting sites) and ecological conditions (e.g., climate, predation risk, and competition), we can better understand what

birds need. Critics of habitat ecology rightly emphasize that ornithologists too often rely on simple vegetation classifications, and that research should instead strive to focus on resources and constraints affecting bird fitness (Romesburg 1981, Morrison 2001).

Historical Perspective

Ornithologists have contributed substantively to the development of the habitat concept, perhaps in part because birds are generally conspicuous and easy to observe in their habitats, and they lay eggs in discrete nests that enable quantification of reproductive rates (Rotenberry 1981, Block and Brennan 1993). The primacy of habitat in ornithology has resulted in several eras of avian habitat ecology: cataloging habitats and natural history (box on page 581), quantitative ecology, wildlife–habitat relationships, and, most recently, spatial habitat modeling. We briefly summarize these eras here. For more comprehensive reviews of this history, see Karr (1980), Block and Brennan (1993) and Stauffer (2002).

The cataloging and natural history eras of habitat ecology began with Aristotle, with basic descriptions of the vegetation commonly associated with animals (Mayr 1982). It peaked with the insight and synthesis offered by Joseph Grinnell and his colleagues and students (Morrison et al. 2012). During this broad span of time, ornithologists made increasingly sophisticated qualitative descriptions of animals' habitats, with later work offering the lasting contribution of posing testable hypotheses about ecological factors that may, over evolutionary time, influence the distribution and adaptations in birds (Block and Brennan 1993).

The era of quantitative habitat ecology was prompted by the seminal work of Hutchinson and his student MacArthur. Hutchinson (1957) introduced the modern concept of a multidimensional ecological niche, meaning that several resources axes (such as gradients in forest canopy cover, elevation, and insect abundance) could describe a theoretical space within which a bird species can persist. MacArthur (1958) advanced these ideas with his classic study of eastern wood warblers (*Parulidae* spp.), showing that species use different habitats within the trees of eastern forests and bolstering the notion that species distributions are governed by the combined effects of biotic and abiotic factors.

The advent of more powerful computers propelled quantitative habitat ecology further, enabling sophisticated analyses of birds and their habitats. The work of James and her colleagues (e.g., James and Shugart 1970, James 1971, James and McCulloch 1990) is exemplary in that she used multivariate statistics to operationalize Hutchinson's niche ideas (Stauffer 2002). These advances triggered an era of wildlife habitat relationship studies that continued and evolved through the

MARGARET MORSE NICE

By Dr. Chris Tonra, The Ohio State University

Ornithologists and bird enthusiasts often are captivated by the exotic and rare, but it is more often in the common and ubiquitous that the complexity of nature is best examined. While in the early twentieth century ornithology was a field focused on lengthening the species list from far-flung locales, one woman saw the incredible discoveries that could be made in one's own garden. Margaret Morse Nice produced a sea change in ornithology by peering into the (not so) mundane daily lives of the birds we see every day.

In 1927, already an accomplished ornithologist and naturalist, Nice moved to Columbus, Ohio, after her husband took a faculty job at Ohio State University. Her family settled near the banks of the Olentangy River. They chose a home not for good construction, or immaculate grounds, but for the "great weed tangle that stretched between the yard and the river" that she called Interpont. The following spring, when she began banding her yard birds, she captured two birds that would change her life, and an entire scientific field: the male Song Sparrows (*Melospiza melodia*) she named Uno and 4M. Over the next nine years Nice spent countless hours watching these birds and their mates, neighbors, and decedents. She pulled back the curtain on a bird species in ways that had never been done before. As she put it: "Incredible as it may seem, almost complete ignorance reigned as to the life history of this abundant, friendly, and well-nigh universally distributed bird. I went to the books and read that this species has two notes beside the song, and that incubation lasted ten to fourteen days and was performed by both sexes—meager enough information and all of it wrong."

In her time watching and recording her observations of the Olentangy Song Sparrows she introduced ornithology to the complexity of avian territorial interactions and how despotic behaviors (often displayed by the "truculent and meddlesome" 4M) play a role in spatial arrangements. In the field of behavior she described individual variation in song repertoires, the fluidity of pair bonds, antipredator/parasite behavior, brood parasite behavior, and post-fledging care. She described studying the nest behavior and delineation of territories of songbirds, foreshadowing modern methods of monitoring breeding success and density (see example of one of her territory maps, right). In the words of the eminent German ornithologist Ernst Mayr, she "almost single-handedly initiated a new era in American ornithology and the only effective counter movement against the list-chasing movement."

Nice's gift of her words in communicating the wonder of these everyday birds to the rest of the world was in many ways as great as her contributions to science. She published hundreds of writings, many of them for public consumption, which engaged people in the appreciation of birds and bird-watching. It was her personal connection to the individual birds she followed that informed this passion. After the death of 4M she wrote: "For seven years I devoted myself to the study of the Song Sparrow. During the seasons of alternating hope and discouragement, fulfillment and bitter disappointment, there had been one great blessing—the dauntless cheer of this precious bird and the miracle of his long life." Translating these experiences into her writings, scientific and otherwise, made this 15-gram, drab little sparrow world famous. Nice often bucked convention with her writings, as she preferred the liberal use of language to paint a picture for the reader. She loathed the loss of colorful language in science, at one time noting, "Unfortunately, especially in the United States, it has become the fashion to write up researches so stiffly, matter-of-factly, and technically that all feeling and atmosphere have been banished from too many of them."

Perhaps most importantly, Nice became a high-profile influential woman in the male-dominated world of science in the first half of the twentieth century, in the process becoming an inspiration to young female science enthusiasts everywhere. Today the American Ornithologists Union annually gives the Margaret Morse Nice Award for research by female graduate students, and the Wilson Ornithological Society named its highest honor, for career achievements in ornithology, the Margaret Morse Nice Medal. Nice left an amazing and inspiring legacy when she left the world in 1974 to join 4M, Uno, and the many other birds whose lives she shared with the world. Not bad for spending time sitting on a camp stool under a maple tree with a notebook and "bird glasses," watching the sparrows in the garden.

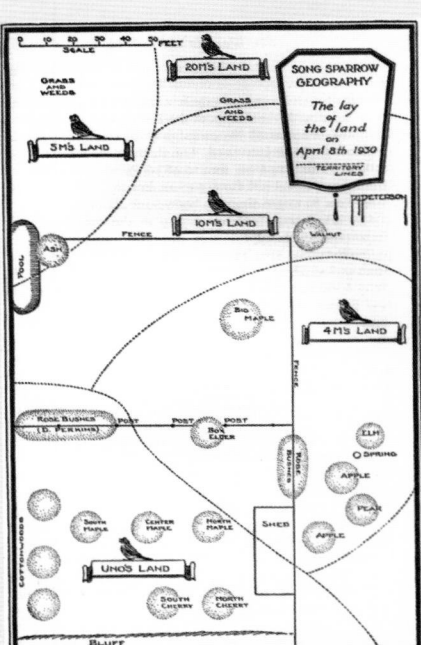

Figure from Nice 1967. Illustration by Roger Tory Peterson. Used by permission of the publisher.

1980s and 1990s (Verner et al. 1986, Scott et al. 2002) and paved the way for the emerging discipline of landscape ecology (Forman and Godron 1986) which enabled researchers to examine multiple scales of habitat associations—from the nest site to the entire range.

The current era of spatial habitat modeling combines concepts from the previous eras with large scale information from remotely sensed habitat data (e.g., from satellite or aerial imagery) and geographic information system (GIS) analyses. In some cases, ornithologists use habitat modeling to test hypotheses about how habitat attributes affect the selection of habitat by birds (e.g., Manly et al. 2002). In other cases, ornithologists conduct analyses to produce maps to predict how birds may respond to future land use changes or anticipated climate change (e.g., Benning et al. 2002, Warren and Seifert 2011). This remains an active field of research and conservation application for ornithologists, and we return to describe current methods later in this chapter.

Documenting where birds are and why they occupy those places is critical for understanding their biology, but ultimately we must link habitat distributions to bird populations to enable effective conservation (Boyce and McDonald 1999, Stauffer 2002). Therefore, alongside the rapid progress in the study of bird distribution, ornithologists have also advanced the study of habitat-specific demography (e.g., Holmes et al. 1996, Rodenhouse et al. 1997, Gaillard et al. 2010), and helped reveal the role of social cues and density-dependence in animals' selection of habitats (e.g., Stamps and Krishnan 2005, Seppänen et al. 2007, Schmidt et al. 2010), concepts we further develop in the following sections.

THEORY FOR HABITAT QUALITY AND SELECTION

Variation in habitat conditions over space and time affect reproduction and survival of individual birds (Brown 1969, Fretwell and Lucas 1970, Sutherland and Parker 1985), and this generates strong selective pressure for birds to select habitat accordingly (Cody 1985). For example, following an abrupt natural decline in the population of European Shags (a cormorant, *Phalacrocorax aristotelis*) on the Farne Islands northeast of England, remaining birds selected among the best available nesting sites, and high rates of reproduction enabled the population to grow until competition forced a greater percentage of birds to occupy poorer sites, pulling down average reproduction (Potts et al. 1980). Thus, the supply of habitat can regulate a bird's population (Newton 1998). It is no surprise, then, that ornithologists work to understand variation in **habitat quality** and **habitat selection** (Block and Brennan 1993, Johnson 2007).

Habitat Quality Theory

Generally, habitat quality is the capacity of the environment to provide conditions appropriate for individual and population persistence (for a more thorough review of habitat quality for birds, see Johnson 2007). While this simple description is intuitive, much is masked by considering habitat quality to relate to both individual- and population-level perspectives (Hobbs and Hanley 1990). For example, in forests of Tennessee that are managed in part for wood production, Boves et al. (2015) found that indicators of habitat quality for Cerulean Warblers (*Setophaga cerulea*) depended on whether they were calculated at the individual level (per capita) or population level (per unit area). In this system, forests that were moderately or heavily harvested were markedly lower in quality than unharvested forest from the perspective of individual birds (per capita rate of reproduction). This result might suggest that unharvested forests should be prioritized in areas targeted for Cerulean Warbler conservation. However, a high density of Cerulean Warblers in some of those same harvested plots more than compensated for the reduced per capita fecundity, so from a land manager's perspective, population-level density and reproductive rates (per unit area) could be maximized in harvested forest. This trade-off in quality and quantity of resources underscores the necessity to distinguish between habitat quality from the perspective of individual animals, which seek to maximize their own fitness, and habitat quality from the perspective of conservationists concerned with populations.

Birds occupying habitats that maximize their lifetime reproductive success will contribute the most to future generations; that is, habitat is a key contributor to an individual's fitness (Block and Brennan 1993, Franklin et al. 2000). Natural selection therefore favors individuals that are capable of distinguishing high- and low-quality habitats (see Habitat Selection Theory, below). Though fitness is an individual measure, Fretwell and Lucas (1970) combined the concepts of habitat and fitness into the notion that a habitat confers fitness on its occupants. Wiens (1992) considered this contribution to an organism's fitness the **habitat fitness potential**, which provides the theoretical basis for an individual-based definition of habitat quality. For example, Franklin et al. (2000) quantified habitat fitness potential for Northern Spotted Owls as the relative contribution to the overall population of individuals occupying a given habitat (fig. 18.3). Thus, habitat quality at the level of an individual bird is defined as the per capita contribution to population growth expected from a given habitat. This conceptualization of habitat quality places evolutionary fitness in a measurable, ecological context with variation that can be quantified over space and

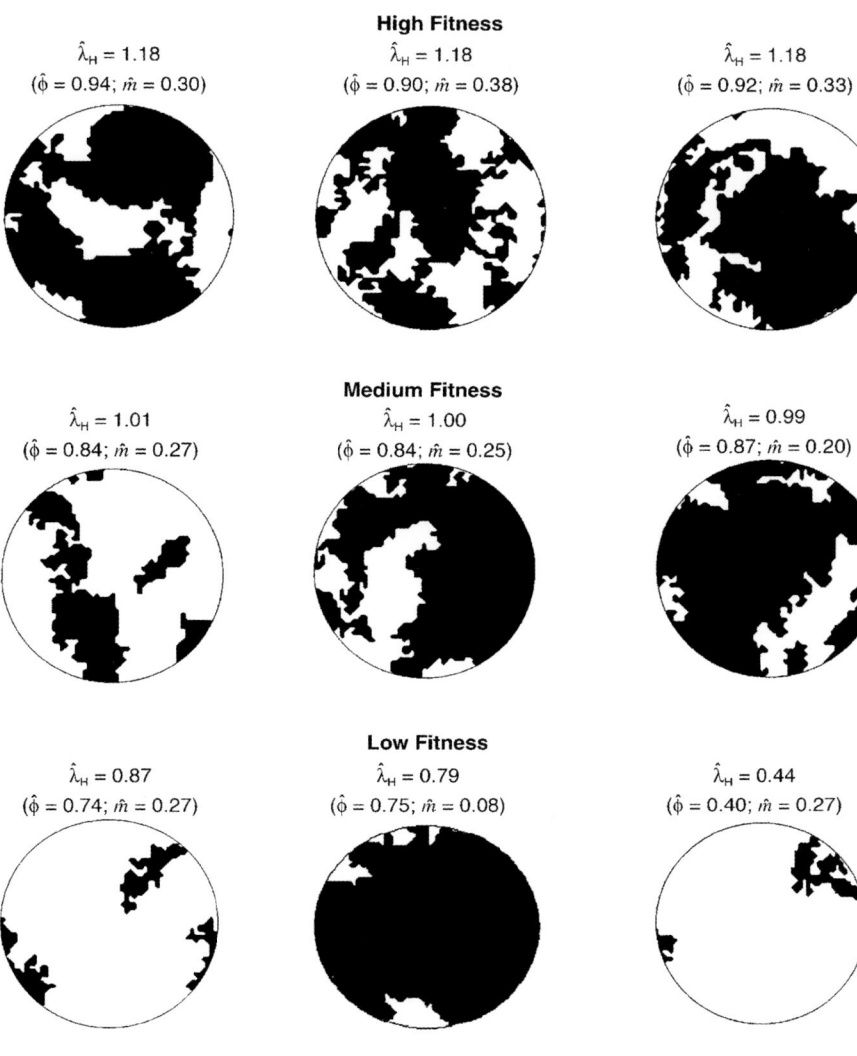

High Fitness

$\hat{\lambda}_H = 1.18$
($\hat{\phi} = 0.94$; $\hat{m} = 0.30$)

$\hat{\lambda}_H = 1.18$
($\hat{\phi} = 0.90$; $\hat{m} = 0.38$)

$\hat{\lambda}_H = 1.18$
($\hat{\phi} = 0.92$; $\hat{m} = 0.33$)

Medium Fitness

$\hat{\lambda}_H = 1.01$
($\hat{\phi} = 0.84$; $\hat{m} = 0.27$)

$\hat{\lambda}_H = 1.00$
($\hat{\phi} = 0.84$; $\hat{m} = 0.25$)

$\hat{\lambda}_H = 0.99$
($\hat{\phi} = 0.87$; $\hat{m} = 0.20$)

Low Fitness

$\hat{\lambda}_H = 0.87$
($\hat{\phi} = 0.74$; $\hat{m} = 0.27$)

$\hat{\lambda}_H = 0.79$
($\hat{\phi} = 0.75$; $\hat{m} = 0.08$)

$\hat{\lambda}_H = 0.44$
($\hat{\phi} = 0.40$; $\hat{m} = 0.27$)

Figure 18.3. Habitat quality for Northern Spotted Owls (*Strix occidentalis caurina*) is highest in territories that are a mixture of old forests for nesting sites and other forests that harbor abundant rodent prey. Circles depict habitat characteristics (within 0.71 km radius circles used to define Northern Spotted Owl territories) at three levels of habitat fitness potential in northwestern California. Dark areas are Northern Spotted Owl habitat; white areas are other vegetation types. Estimates of ϕ (apparent survival) and m (fecundity) are for owls ≥3 years old. *From Franklin et al. 2000.*

time (Coulson et al. 2006, Gaillard et al. 2010) and provides the conceptual underpinning for the definition of habitat quality used in this chapter.

As the density of birds in a habitat increases, competition intensifies and diminishes the average fitness conferred upon the occupants (Gaillard et al. 2010; see Consequences of Habitat to Bird Populations, below, for details on how this happens). Thus, theoreticians distinguish the quality of habitat in the absence of interactions with other organisms, called **fundamental habitat quality**, from the quality actually experienced by interacting occupants, called **realized habitat quality**. This distinction allows a more careful examination of how a bird should select among available habitats to maximize its fitness.

Habitat Selection Theory

Human activity fundamentally alters the Earth. Our needs for food, fiber, and energy drive global habitat loss, and yet,

we also have the capacity to engage in targeted conservation (Vitousek et al. 1997, Hoekstra et al. 2005, Steffen et al. 2011). Habitat is paramount to bird conservation; therefore, to balance bird conservation and human needs, we must understand the nuances of "habitat selection"—the study of where birds live and why. Habitat selection is a process that operates at the level of an individual bird (Krebs 2001). Two kinds of factors should be kept in mind when examining habitat selection in birds: (1) evolutionary (or ultimate) factors, which confer fitness consequences on habitat selection; and (2) behavioral (or proximate) factors, which describe the mechanism by which birds select some habitats over others (Krebs 2001).

Wiens (1985) provided a conceptual model to clarify the myriad variables that can affect avian habitat selection (fig. 18.4). A "habitat selection template" represents habitat preferences in a bird that result from both an individual's genetic makeup and learned preferences through experience (Sogge and Marshall 2000). For example, Klopfer (1963)

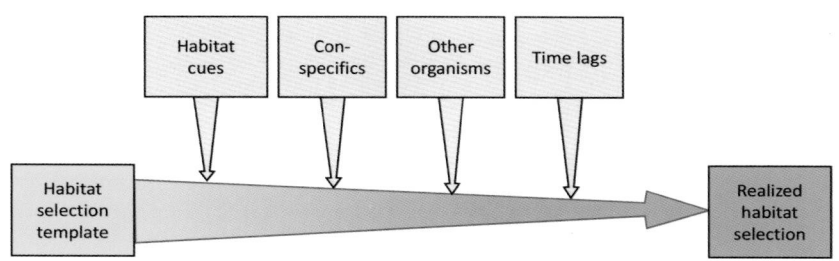

Figure 18.4. A conceptual model of the factors that may affect habitat selection at the local level. These contsraints operate to limit the range of habitats occupied by a species. *Modified from Wiens 1985.*

found that Chipping Sparrows (*Spizella passerina*) experimentally raised in the laboratory selected habitats containing pine branches rather than alternative sites containing oak branches, just as they do in the wild. However, birds raised in the laboratory with oak branches showed diminished selection of pine branches as adults, suggesting a genetic preference for pine could be modified by earlier experience (Krebs 2001). Realized habitat selection, describing what a bird actually selects, is modified from this habitat template by a series of other factors. The cues used by birds are habitat characteristics such as nest sites, cover, food supply, or predation risk that birds can use to make settlement decisions. Habitat selection results if the sum of stimuli from each of these factors exceeds a selection threshold (Cody 1985), which could be raised or lowered depending on the urgency for selection. For example, a migrating bird with depleted fat reserves may be far less picky in selecting habitats than a bird with large stores of fat. The lean bird's low selection threshold may prompt it to select a foraging patch with abundant food even if it poses risk of predation, whereas the fat bird may keep searching for something better (Hildén 1965).

Conspecifics

The presence of conspecifics (i.e., members of the same species) is one of the most important influences on habitat selection. In some cases, low population densities can decrease fitness, and the presence of conspecifics can actually favor habitat selection (see review of so-called Allee effects and other process operating at low density in Stephens and Sutherland 1999, Greene and Stamps 2001). For example, birds may benefit by joining a flock with shared vigilance and a reduced predation risk (Elgar 1989), or they may enjoy greater reproductive opportunities at high density because of improved mate choices (Stephens and Sutherland 1999). Also, a high density of conspecifics can be attractive to an individual if conspecifics are useful cues for resources insensitive to density-dependent competition (see reviews in Stamps 1991, Doligez et al. 2004, Valone 2007). Understanding these processes is important for conservationists, because habitat restoration often operates with a "if you build it, they will come" assumption. This assumption is invalid if a spe-

cies is reliant on conspecific attraction for habitat selection (Ahlering and Faaborg 2006). In these cases, simulating the presence of conspecifics can be an important conservation practice. For example, Black-capped Vireos (*Vireo atricapilla*), a bird endangered by loss of habitat and nest parasitism, have been successfully attracted to managed oak woodlands with playbacks of songs and calls (Ward and Schlossberg 2004). Likewise, Common Murres (*Uria aalge*) have been attracted to offshore breeding rocks with decoys of nesting birds, facilitating their return to historic breeding habitats after recovering from long-term population declines (Parker et al. 2007).

More commonly, conspecifics are competitors for limited resources, such as food or nest sites. Assuming that a bird should select the habitat that best increases its fitness, and that competition varies with density of conspecifics, we expect habitat selection to be also density-dependent (Rosenzweig 1981, Morris 1989, Gaillard et al. 2010). Under density-dependence, birds experience realized habitat quality affected by competition with conspecifics, and they choose habitats accordingly (box on page 585). If competitors are equal, this leads to an "ideal free distribution" (IFD; Fretwell and Lucas 1970), in which individuals are distributed among habitats that vary in fundamental habitat quality (also called intrinsic habitat quality or zero-density suitability, sensu Bernstein et al. 1991) such that all individuals experience the same realized habitat quality. In contrast, an "ideal despotic distribution" (IDD) emerges if individuals are unequal competitors, in which the preemption of resources or territories in the highest quality habitats ensures that the strongest competitors reap the greatest rewards (Parker and Sutherland 1986). In this case, at equilibrium the average fitness conferred by a habitat on its occupants—realized habitat quality—is lower in habitats with low fundamental habitat quality (box on page 585).

While the ideal free and ideal despotic models offer great heuristic value, they represent idealized abstractions that usually fail to account for observed distributions of animals (Tregenza 1995, Gaillard et al. 2010). Indeed, wild birds are not omniscient, not free to settle anywhere, and they generally face marked environmental stochasticity that can

IDEAL FREE AND IDEAL DESPOTIC DISTRIBUTIONS

Models of ideal free (*A*) and ideal despotic distributions (*B*) (*from Fretwell and Lucas 1970, Parker and Sutherland 1986, Bernstein et al. 1991*). Two habitats varying in quality are modeled, each showing a linear density-dependent decline in quality. Fundamental habitat quality is the "intrinsic" quality of a habitat in the absence of intraspecific competition; realized habitat quality accounts for negative effects of competition. In the ideal free distribution, the first six competitors select the rich habitat to maximize realized habitat quality conferred; the seventh chooses between the partially filled rich habitat and the empty poor habitat,

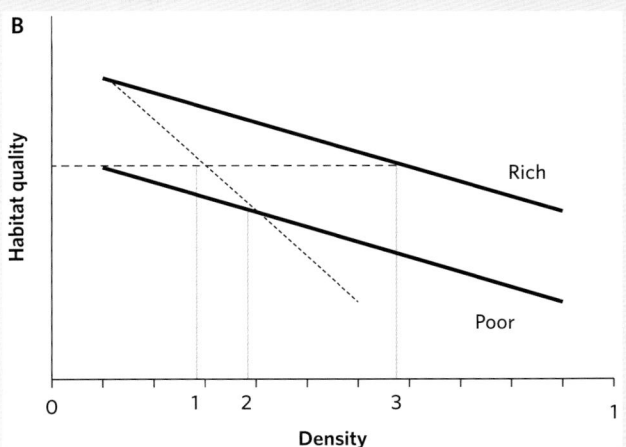

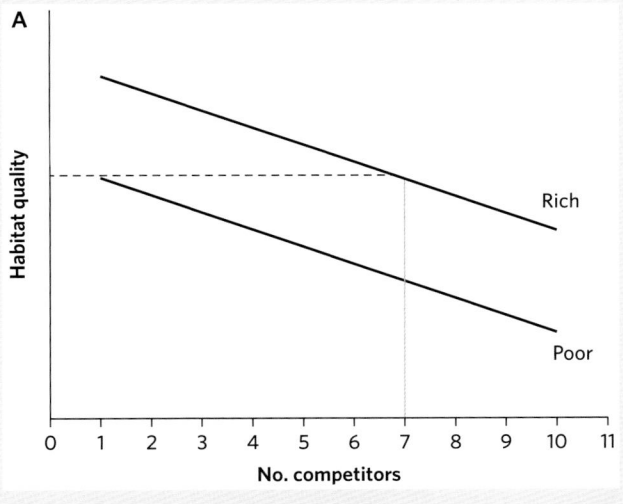

which offer the same realized habitat quality at densities of 7 and 1, respectively (depicted by horizontal dashed line). In the despotic distribution model, competitors are unequal. As density increases, weak competitors (heavy dashed line) suffer a steeper decline in realized habitat quality than do strong competitors. At density 1 in the rich habitat, strong competitors occupy the rich habitat only, but for weak competitors, realized habitat quality in the rich habitat has diminished to the level of fundamental habitat quality in the poor habitat. At density 2, it pays weak competitors to occupy the poor habitat exclusively. Strong competitors should not use the poor habitat until they reach density 3. *From Johnson 2007.*

change the relative value of habitat or prevent expected fitness from being realized (Gaillard et al. 2010). Therefore, there are numerous scenarios that can lead birds to select poor habitats and avoid rich ones (Rapport 1991, Railsback et al. 2003), including incomplete information (Shochat et al. 2002, Stamps et al. 2005), ecological traps (Battin 2004), time lags (Davis and Stamps 2004), a lack of high-quality habitat (Halpern et al. 2005), and others (Bernstein et al. 1991, Block and Brennan 1993, Kristan 2003), some of which we discuss in more detail later. Nonetheless, the distinction between the theoretical Fretwell-Lucas models is important because it reveals opposing prioritizations of habitats for managers. Under the ideal free model, fundamental habitat quality corresponds with density. Therefore, although all individuals receive the same reward at equilibrium, the habitats with the most birds are fundamentally highest in quality and should be prioritized for conservation. This convenience allows managers to simply count birds in different habitats and rank conservation priorities accordingly. Under a despotic distribution, the equilibrium density among fundamentally rich and poor habitats depends on the relative competitive abilities of strong and weak competitors. If weak competitors are much more influenced by competition than are strong competitors, the density of birds in poor habitats is likely to be higher than in rich habitats (Bernstein et al. 1991). In this case, density will be a misleading indicator of habitat quality as cautioned by Van Horne (1983), and prioritizing habitats should involve measuring individual birds' performance to assess variation in realized habitat quality. Thus, it behooves ornithologists to consider whether their study species are more likely to approximate an ideal free or despotic model by carefully considering the likelihood of strong variation in the competitive ability of individual birds owing to differences in age, sex, experience, or knowledge of the habitats.

Other Organisms

As with conspecifics, individuals of other species can contribute either positively or negatively to habitat selection. The presence and abundance of heterospecifics may introduce interspecific competition that could diminish available resources, reduce habitat quality, and discourage habitat selection. Experimental removals of heterospecifics have confirmed their negative influence on habitat selection by showing an expansion into previously unselected habitats (Sherry and Holmes 1988) or an increase in realized habitat quality (Connell 1983). For example when Martin and Martin (2001) experimentally removed Orange-crowned Warblers (*Oreothlypis celata*) from territories in Arizona, Virginia's Warblers (*Oreothlypis virginiae*) shifted their nest locations to sites indistinguishable from Orange-crowned Warbler nest sites, increased feeding rates during both the incubation and nestling periods, and enjoyed reduced nest predation rates, compared with control plots where Orange-crowned Warblers were present. The presence of predator species can also affect habitat selection. It has long been recognized that females select nest sites, in part, to minimize risk of predation (Forstmeier and Weiss 2004), but evidence also indicates that nest predators affect larger scale habitat selection (e.g., selection of territories or habitat patches). For example, Fontaine and Martin (2006) found that birds nested in higher densities in habitats from which rodent nest predators were experimentally removed. Predation also affects habitat selection of nonbreeding birds, as evidenced by studies of Bramblings (*Fringilla montifringilla*) and Western Sandpipers (*Calidris mauri*) that each show avoidance of areas with high risk of predation despite high food availability (Lindstrom 1990, Ydenberg et al. 2004). Fewer examples exist of habitat selection being influenced by parasitic species, but Forsman and Martin (2009) found evidence for selection of parasite-free space by hosts of the parasitic Brown-headed Cowbirds (*Molothrus ater*). There is also evidence that the geographic distribution of some shorebirds (a consequence of evolutionary habitat selection) is in part influenced by exposure to mosquitoes and the diseases they can transmit to birds (e.g., avian malaria; Mendes et al. 2005).

Negative effects of other species on a bird's selection of habitat may be the most obvious, but an accumulating number of empirical studies demonstrate that the presence of heterospecifics, like conspecifics, can provide information about habitat that is useful to birds in their selection of breeding sites (e.g., Doligez et al. 2004, Kivelä et al. 2014). For example, working on an island in the Baltic Sea, Forsman et al. (2009) found that the arriving migratory insectivorous birds showed selection for habitats with high densities of nonmigratory titmice (*Parus* spp.). In theory, the use of social information (from conspecifics and/or heterospecifics) should persist if the benefits outweigh the costs associated with selecting sites with competitors. More specifically, social information may pay if the environment is too variable for innate habitat preferences to reliably direct birds to high-quality habitats, but sufficiently predictable in time or space for observations of other birds to hold value later (Boulinier and Danchin 1997, Mönkkönen and Reunanen 1999, Doligez et al. 2003, Fletcher 2006, Seppanen et al. 2007).

Time Lags

Time lags in habitat selection result from a delay in a bird's response to changing environmental conditions (Wiens 1985). A common form of a time lag in habitat selection is site fidelity—the tendency to stay in or return to a previously used site or territory (Switzer 1993). The evolution of adaptive time lags is positively associated with homogeneity in territory quality, predictability of territory quality, short life spans, the cost of changing territories, and the probability of mortality in a habitat. However, many bird species show facultative or conditional site fidelity depending on previous nest success. For example, Dow and Fredga (1983) found that female Common Goldeneyes (*Bucephala clangula*) tend to return to sites in which they were previously successful, while they are likely to disperse after a failed nesting attempt. Changes to habitats introduced by human activity are evolutionarily novel, and some species' previously adaptive time lags may render site fidelity maladaptive under current conditions. For example, Walker et al. (2007) found that Greater Sage Grouse (*Centrocercus urophasianus*) leks disappeared from traditionally used sites following coal-bed natural gas development in Wyoming, but not until after an average of about four years. Likewise, Meyer et al. (2002) found that Marbled Murrelets (*Brachyramphus marmoratus*) took several years to abandon nesting sites in old-growth redwood forests after they were fragmented by ongoing harvesting of second-growth forests. A time lag can also delay a bird's selection of habitat that has improved because of human-caused habitat restoration, which can prompt the need for conservationists to overwhelm this time lag to encourage colonization (Ahlering et al. 2010), as described earlier for Black-capped Vireos and Common Murres.

For highly mobile organisms inhabiting dynamic environments, temporal and spatial scales can strongly affect patterns of habitat selection (Hildén 1965). For example, female Yellow-headed Blackbirds (*Xanthocephalus xanthocephalus*) select marshes based on the emergence of their aquatic insect prey, but they use vegetation density rather than food availability to select individual nesting locations (Orians and

Wittenberger 1991). In this system, prey availability can be forecasted accurately only at a coarse spatial scale, whereas vegetation density varies at a fine spatial scale and is more temporally stable. Thus, at the time of settling, the birds assess information about prey availability and vegetation at different spatial scales in correspondence to their temporal reliability. Indeed, gathering information about the environment is an important determinant of a bird's fitness (Dall et al. 2005), and there is growing empirical evidence that birds sample various habitats ("prospecting") before selecting (Reed et al. 1999, Kristan 2006). Birds selecting habitat in a dynamic landscape face a complex forecasting problem, in that they must choose a location with previous or current information that may not always predict later realized habitat quality (Gates and Gysel 1978, Best 1986, Misenhelter and Rotenberry 2000).

Hierarchical Habitat Selection

All organisms face restrictions in the information they can gather about their environment (Levin 1992), and the interaction between spatial scale and the ability to distinguish features at a distance create different habitat selection processes depending on whether a bird samples primarily at ground level or high above it (Kristan 2003). "Top-down" habitat choice is a hierarchical, sequential process, beginning with a bird flying over a landscape and deciding where to settle based on the habitat encountered (Hutto 1985). In this case, birds are best able to resolve course habitat variation, such as landscape features and general types of vegetation, before descending into a location, where they are better able to resolve finer variation, such as foraging sites in different species of trees (fig. 18.5). In contrast, a dispersing juvenile of

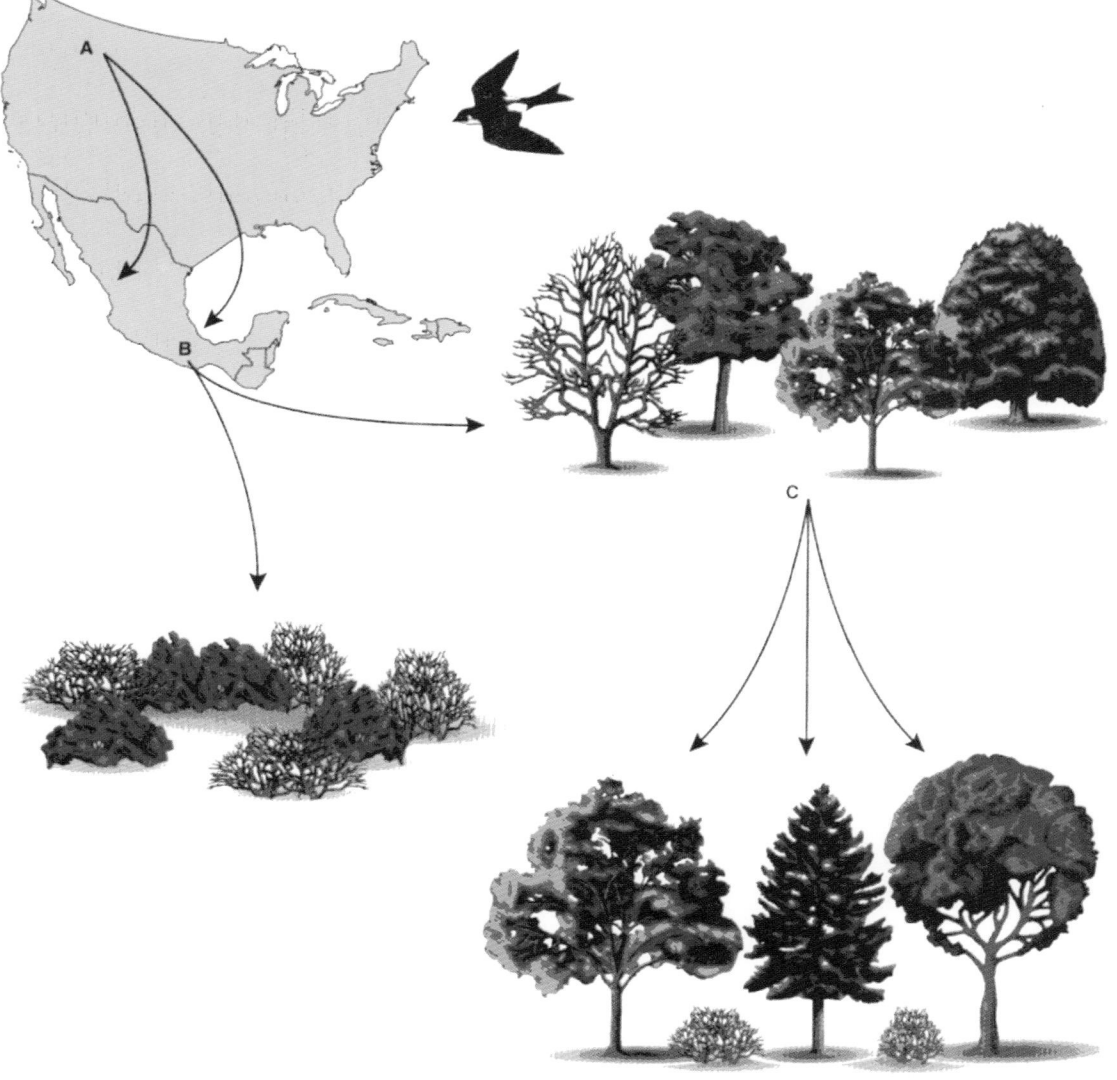

Figure 18.5. Hierarchical decision-making process for the choice of nonbreeding habitat by a migrating Violet-green Swallow (*Tachycineta thalassina*) in Mexico. At *A*, the bird selects whether to go to southern or western Mexico to overwinter. At *B*, it selects woodland or shrubland-dominated vegetation. And at *C*, it selects which of several treetops to occupy. *Modified from Hutto 1985; image from Krebs 2001.*

a nonmigratory species may move from tree to tree, habitat patch to habitat patch, and sample habitat at a variety of scales simultaneously ("bottom-up" habitat selection). This pattern of movement presents an "information barrier" (Forbes and Kaiser 1994), meaning that a bird lacks knowledge of habitat beyond what it has yet experienced. In these cases, a bird must balance the benefit of sampling distant habitats against the cost of movement, which may be particularly high for juvenile birds. Bottom-up selection may also be favored if local conditions are more consequential than those at larger spatial scales, such as woodpeckers selecting nesting trees within a forested landscape (Lawler and Edwards 2006). In either top-down or bottom-up habitat selection, the optimal habitat may go unused if it is not found (Kristan 2006). Thus, the constraints imposed by incomplete information available to birds at different spatial scales introduce another mechanism by which habitat selection may not follow the "ideal" models, leading birds to make maladaptive choices.

Consequences of Habitat to Bird Populations

It is widely recognized that spatial variation in habitat is central to the regulation of bird populations. This notion is based on two generally accepted tenets rooted in the theory of habitat quality and selection (Newton 1998). First, for any species, habitat varies in quality from place to place and, second, that as a bird's population size increases, individuals will select good habitat over poor ones. If these conditions hold, then it is inevitable that as population size increases, an increasing proportion of individuals will be pushed down the habitat gradient to poorer habitats where, by definition, their reproduction or survival is diminished (Rodenhouse et al. 1997, Newton 1998). Therefore, the average reproduction or survival will decline overall with increasing density, and rise again when population size is low. This habitat-mediated density-dependence is the essence of population regulation imposed by the quantity and quality of habitats available to birds. Empirical evidence for this pattern is widespread, especially for breeding birds. For example, Peregrine Falcons (*Falco peregrinus*, Mearns and Newton 1988) and Black-legged Kittiwakes (*Rissa tridactyla*, Coulson 1968), which both recovered from historically low population sizes, saw mean reproductive declines with increased population size. Similarly, Black-throated Blue Warblers (*Setophaga caerulescens*, Rodenhouse et al. 2003) and Tengmalm's Owls (*Aegolius funereus*, Korpimäki 1988) had increased breeding performance with lower population size. For more details of avian population ecology, see chapter 21.

Sources, Sinks, and Traps

Survival and reproduction in especially poor habitats may be insufficient to maintain local populations; thus poor habitats act as "sinks" in the landscape. The persistence of sinks on the landscape is thus reliant on emigration from other better habitats, acting as "sources" of surplus birds. This source-sink model is an enduring one in habitat ecology (Pulliam 1988, Pulliam et al. 1991), and has prompted other advances in our understanding of the roles of habitat selection and quality in animal population dynamics. For example, while birds appear to generally distinguish habitat quality well enough to show selection for sources over sinks, in some cases this is not so. These attractive sinks are called "ecological traps." Theory suggests that, under most circumstances, their presence in a landscape will drive a population to extinction, prompting an urgent need for conservationists to identify them and manage accordingly (Kristan 2003). Typical population modeling does not consider habitat selection explicitly and may mask the effects of ecological traps, leading to overly optimistic predictions about population persistence (Battin 2004). Ecological traps may arise if human-caused changes in the habitat decouple the habitat cues birds use to select habitat from the characteristics that afford realized habitat quality. Evolution may be too slow for birds to respond immediately to these human-caused changes, trapping birds into selecting bad habitats. Specifically, habitat alteration capable of creating an ecological trap must either alter the cues birds use to select habitats (increasing their attractiveness), sharply decrease the quality of a habitat associated with cues, or both (Robertson and Hutto 2006). For example, Weldon and Hadded (2005) found that Indigo Buntings (*Passerina cyanea*) actively selected edgy patches of habitat in South Carolina, where they suffered high rates of predation, presumably because the human-created edges created (evolutionarily recent) were highly attractive to nest predators (Weldon 2006, fig. 18.6).

Habitat Distribution

It is not just the quality and quantity of habitats that affect population dynamics of birds; it is also the spatial distribution of habitats in a landscape (Newton 1998). The distribution of some habitats, such as mountaintops and aquatic habitats are naturally patchy, but relentless human activity has fragmented other once-continuous habitats into small patches embedded in a matrix of land modified for people to grow crops, harvest other resources, and live, work, and play (Perfecto et al. 2009). Indeed, up to 40 percent of the Earth's ice-free terrestrial land surface is devoted to agriculture, 20 percent to managed forests, and almost 10 percent to human settlements, with only about 10 percent in protected areas (Foley et al. 2005). Thus, most bird species live in a mosaic of disturbed and (comparatively) undisturbed habitat patches. For many species, the individuals inhabiting

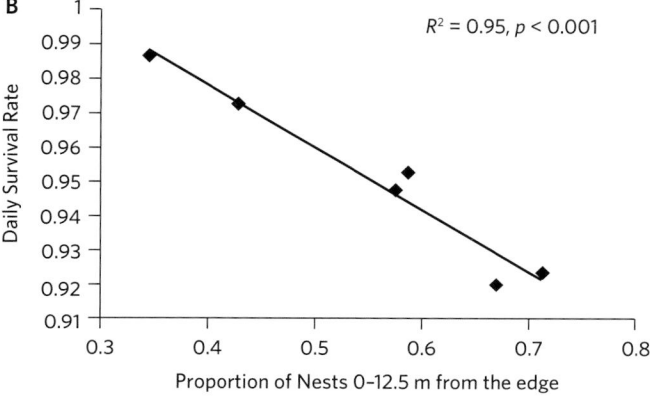

Figure 18.6. A, Indigo Bunting nest. *From Weldon and Haddad 2005; photo by A. Weldon. B,* Daily survival rates of Indigo Buntings declined as a function of the proportion of nests 0–12.5 m from the edge across nesting periods and treatments. *From Weldon 2006.*

long-term experiments in Amazonian forests show that bird extinction rates are surprisingly high in even the largest patches of forest, underscoring the importance of movement among patches to ensure metapopulation persistence (Ferraz et al. 2003). Metapopulation theory is therefore probably generally appropriate for many bird species, but it suffers from an implicit assumption that the matrix between patches is homogeneous and comparatively poor in quality. A growing body of research, sometimes called countryside biogeography, challenges this assumption (Mendenhall et al. 2014). This work places more focus on understanding how variable habitat conditions in the matrix contribute to metapopulation persistence by either supporting birds outside conventionally "intact" habitat patches or by improving the permeability of the matrix and enabling birds to move among habitat patches more successfully (Perfecto et al. 2009).

Habitat Complementarity

While metapopulation theory prompted ecologists to recognize the importance of habitat distribution in a landscape, the idea of habitat complementarity emphasizes how the distribution of different habitats sometimes matters to birds at much finer scales as well. This notion applies to bird species that require markedly different habitat attributes within a single phase of their annual cycle (Dunning et al. 1992). For example, breeding Ruffed Grouse (*Bonasa umbellus*) require unique habitats for their drumming mating displays (conifer forest with fallen logs), nesting sites (high shrub density to help conceal nests), and brooding habitats where chicks are reared (abundant forbs and insects as food supply). These must be arranged in close proximity to enable birds to complete their breeding cycle (Gullion 1988). A similar story applies to Northern Spotted Owls in California, which achieve higher per capita fitness potential in territories that are composed of a mixture of old forests for nesting sites and other forest types that harbor abundant rodent prey (Franklin et al. 2000; fig. 18.3).

Seasonal Shifts and Migratory Species

The consequences of using different habitats at different times are most pronounced for migratory birds. For example, many birds that breed in mature deciduous forests, such as the Wood Thrush (*Hylocichla mustelina*), shift to using shrubby habitats after young have fledged (Vitz and Rodewald 2006). These changes are likely caused by shifting resources and constraints over time. For example, the availability of protein-rich food such as insect larvae is especially important during the breeding season to foster nestling growth (Greenberg 1995), whereas dense cover and abundant fruit resources increase in importance after breeding so that birds can better

these patches may interact as "subpopulations," and the collection of them all, called a "metapopulation," may persist or fail following processes that transcend the fates of bird in individual patches (for further details of metapopulations, see chapters 20 and 21). Specifically, the extinction risk for a metapopulation depends on the balance between the loss of subpopulations because of local extinction and the gain (or regain) of subpopulations because of (re)colonization. For a given ensemble of subpopulations, an equilibrium exists in which loss and gain rates are equal, resulting in persistence of the metapopulation, even while individual subpopulations may blink in and out of existence. This equilibrium depends on the rate of movement among habitat patches, and the per-patch rate of local extinction (Kareiva and Marvier 2015). That local extinctions in patches (fragments) are a normal and inevitable part of ecosystem dynamics is now rarely questioned by ecologists (Perfecto et al. 2009). For example,

avoid predation and fuel their oncoming migrations (Vitz and Rodewald 2007, 2011). The complexities of habitat requirements over the annual cycle peak with long-distance migrants, such as those that breed in north temperate forests and migrate back and forth from tropical nonbreeding habitats. Although breeding habitats profoundly affect populations, nonbreeding habitat can be at least as important (Rodewald 2015). Sillett and Holmes (2002) showed that 85 percent of mortality of Black-throated Blue Warblers occurred during the migratory period, underscoring the imperative to protect habitat used by migrant birds en route between their breeding and nonbreeding ranges (Moore and Barrow 2000). Moreover, changes in population size in one season can affect density-dependent processes in another (Marra et al. 2015a), meaning the habitat that a bird selects in the breeding season can "carry over" and affect its nonbreeding survival, or vice versa (Norris et al. 2004, Marra et al. 2015b). For example, American Redstarts (*Setophaga ruticilla*) that overwinter in high quality mangrove forests arrive on the breeding grounds earlier, in better condition, and fledge more young than do birds that overwinter in lower quality scrubby habitats with a less reliable food supply (Marra et al. 1998, Norris et al. 2004, Cooper et al. 2015). These processes have been most studied in birds with conspicuous global migrations, but they likely occur for many bird species, and ornithologists must remain attentive to the role of habitat in bird populations throughout the annual cycle (Marra et al. 2015a; box on page 590).

MEASUREMENTS IN HABITAT ECOLOGY

The previous sections underscored the imperative for avian ecologists to understand habitat use, selection, and quality for birds throughout the annual cycle, which of course demands we accurately measure habitats and document how birds use and select them. In this section, we first review methods of measuring habitats, both remotely and on the ground, then provide an overview of current methods for quantifying habitat use, selection, and quality for birds in the field. A persistent theme throughout this section is that new technologies can be harnessed by ornithologists to help measure habitat in ways that advance our understanding of birds and how to conserve them.

Measuring Habitat Attributes

When it comes to measuring habitats, the critical challenge facing ornithologists is to determine which habitat attributes to measure and how to accurately quantify them. The traditional approach has been to measure habitat attributes at ground-based locations (e.g., breeding bird territory) and

DR. ENRIQUETA VELARDE
By Dr. Daniel Barton, Humboldt State University

Seabirds inhabit vast ocean basins, and their distributions and populations are influenced by the effects of basin-wide oceanographic processes, such as the El Niño/La Niña phenomena, on marine food availability. In contrast, local processes, such as the invasion of nesting islands by introduced predators or disturbance of nesting colonies, can also strongly affect seabirds, causing numerous extinctions and threatening many more. Thus, understanding seabirds and their conservation can require detailed knowledge of processes that occur at both of these levels.

Seabird ecologist Enriqueta Velarde, senior researcher at Universidad Veracruzana in Xalapa, Mexico, reveals how both region-wide and local processes affect seabirds, and how to use such information to conserve seabirds. Dr. Velarde has worked for almost 40 years on Isla Rasa in the Gulf of California, and during this time has essentially prevented the extinction of not one but two seabird species that principally nest there: Heermann's Gull (*Larus heermanni*) and Elegant Tern (*Thalasseus elegans*). Beginning in the early 1980s, Velarde and colleague Jesús Ramírez worked tirelessly to reduce human disturbance at Isla Rasa, followed by a campaign in the early 1990s to remove introduced rats and mice on the island. The success of these efforts not only secured the future of Heermann's Gull and Elegant Tern, but inspired a larger effort to conserve numerous island endemic taxa using the same techniques at a much larger scale, directly resulting in tangible conservation impacts.

Beyond preventing extinctions and inspiring a generation of conservation biologists, Velarde's work on the diet of seabirds nesting on Isla Rasa has shown how shifts in diet are a harbinger of oceanic changes to come. In El Niño years, the diets of Heermann's Gull, Elegant Tern, and California Brown Pelicans (*Pelecanus occidentalis californicus*) shift markedly from Pacific Sardine to Northern Anchovy, while in La Niña years, the converse occurs. These shifts impressively predict the success of the Gulf of California's sardine fisheries in the following year. Such information helps further understanding of large-scale oceanic processes and efforts to manage Gulf of California fisheries. The seabirds of Isla Rasa thus forecast change to come in a way that benefits humans—and Dr. Velarde first prevented their extinction and then showed us how we can learn something unique from them about their difficult-to-study marine habitats.

then statistically relate these findings to bird habitat use, selection, and quality (Morrison et al. 2012; see below for more details and additional references).

For decades, ornithologists have recognized that broad-scale habitat features are important drivers of bird distribution patterns and avian community organization. The distribution of many avian species mirrors the distribution of their habitats. Sometimes distributions are determined by the juxtaposition of multiple habitat types required during different life stages. For example, Golden-winged Warblers (*Vermivora chrysoptera*), a species of conservation concern, require early successional shrubby habitat for breeding throughout eastern and midwestern forests. However, recent work indicates that early successional habitats that are critical for nesting must be surrounded by dense forests necessary for foraging adults (Streby et al. 2012). If habitat attributes were not quantified at larger areas than the breeding territory, the importance of dense forest to breeding Golden-winged Warblers may have gone unrecognized. Therefore, ornithologists acknowledge that habitats should be quantified at multiple spatial extents (e.g., breeding locations and areas surrounding breeding sites). This idea requires considerable effort and, in many cases, is impossible to achieve using traditional on-the-ground measurements. To circumvent this challenge, ornithologists have increasingly adopted the use of data collected from remote sources (e.g., Palmeirim 1988, Gauthreaux-Jr. and Belser 2003, Gottschalk et al. 2005, Vierling et al. 2008). Linking habitat data on the ground with data from remotely

sensed measurements can enhance our ability to characterize avian habitat across large areas. This allows for a more complete understanding of habitat quality and its relation to bird demographics and distributional patterns than is possible with data from a single scale (Wiens and Rotenberry 1981, Saab 1999, Wood et al. 2016). It is critical to understand avian habitat quality and selection at multiple spatial scales. A spatial scale is a combination of extent (i.e., the area of habitat under study) and grain (i.e., the resolution at which a habitat is studied). Many species of birds select habitat hierarchically (see Hutto 1985 above), and therefore, understanding habitat quality and selection at multiple spatial scales underscores the importance of "landscape ecology" (i.e., study of ecological processes in the environment, Turner 1989) in ornithology.

Below, we provide a more in-depth overview of the ways in which ornithologists measure habitat on the ground and from remote data sources. We describe case studies and also detail a handful of advances in bird-habitat quantification that will likely aid our understanding of bird demographic and distribution patterns throughout the annual cycle. On the ground measurements (**ground-based measurements**) provide a fine-resolution overview of the immediate habitat that a bird experiences: the structure of vegetation, the diversity of plant species and food resources, and the presence of other species, such as predators, competitors, mutualists, and commensalists (Morrison et al. 2012), all of which influence habitat selection patterns by birds (Cody 1985). Once habitat attributes are measured during fieldwork, those data

Table 18.1. Examples of techniques that can be used by ornithologists for measuring territory-level habitat attributes that may influence breeding bird species habitat selection.

Habitat attributes	Reference	Technique
Vegetation structure		
Cover classes	Martin et al. 1997	Circular plot
Vegetation density	James 1971	Circular plot
Vertical vegetation structure	MacArthur and MacArthur 1961, Robel et al. 1970	Point sampling
Horizontal vegetation structure	Nudds 1977	Cover board
Vegetation composition		
Tree diversity	Mitchell 2001	Point center quarter
Species lists	Wood et al. 2011	Relevé
Food availability		
Insects	Cooper and Whitmore 1990, Johnson 2000	Branch clipping
Fruits and nuts	Koenig et al. 1994, Higgins et al. 2012	Seed traps and visual counts
Predators and Conspecifics	Ralph et al. 1995	Point counting

Table 18.2. **Examples of remote sensing data and techniques that can be used by ornithologists for measuring habitat attributes that may influence breeding bird species habitat selection.**

Remotely sensed habitat features	Reference	Data or Technique
Passive remote sensing		
Land cover	Homer et al. 2015	Land cover data
Habitat loss and fragmentation	Briant et al. 2010	MODIS
Habitat phenology	Toral et al. 2011	Landsat
Vegetation structure	Wood et al. 2012	Aerial photographs
Vegetation composition	Martin et al. 1998	Airborne hyperspectral data
Active remote sensing		
Vegetation Structure	Bergen et al. 2009	LIDAR and RADAR

are then commonly linked via statistical modeling to bird demographic and distributional patterns (Morrison et al. 2012). The traditional approach has been to define a sampling unit within some area of interest, be it a habitat patch, survey plot, or individual territory. The sampling unit is often centered on a bird survey location (e.g., a point count or a nest site), and information is collected regarding habitat attributes known or hypothesized to be relevant to the species (see table 18.1 for some common examples). Excellent reviews of techniques for measuring vegetation, prey availability, fruit and seed abundance, and other biotic and abiotic attributes of habitat relevant to birds are provided by Cooperrider et al. (1986) and Higgins et al. (2012). An overview of common approaches to surveying birds and their habitats is provided in chapter 31.

Remote Measurements

More recently, with the advent of technology and the emergence of the field of landscape ecology, ornithologists have adopted the use of data from remote sensing for characterizing habitat attributes at broad spatial and temporal extents. There are two typical approaches for using data from remote sensing: passive and active remote sensing (table 18.2). Here, we provide an overview of remote sensing approaches used by ornithologists. For further reading on the importance of remote sensing in biodiversity studies, see the excellent reviews by Nagendra (2001), Kerr and Ostrovsky (2003), Turner et al. (2003), Bradbury et al. (2005), Gillespie et al. (2008), and Pettorelli et al. (2014).

Passive   Passive remote sensing involves the use of sensors that measure the reflectance of natural radiation (e.g., reflected sunlight) by the Earth (Turner et al. 2003). The most common forms of these data are from aerial photography and satellite imagery (Turner et al. 2003, Pettorelli et al. 2014). Aerial photographs are generally acquired from cameras mounted on airplanes (Morgan et al. 2010) and more recently by unmanned aerial vehicles (i.e., drones, Anderson and Gaston 2013). Aerial photographs are typically very detailed images with resolutions as low as < 1m and are extremely useful for mapping fine-grained variability in vegetation across areas as large as the extent of the photographs (Fensham and Fairfax 2002). This is useful for describing habitat attributes important to birds over extents larger than are typically possible with ground-based surveys (Wood et al. 2012). In addition to the high tonal detail, aerial photographs are extremely useful for mapping and monitoring landscape change in regions where aerial photographs have been taken for decades (Morgan et al. 2010). For example, in Oklahoma, researchers were able to characterize woody-plant encroachment (by juniper, *Juniperus* spp.) in grassland habitats, which were then linked to declines in grassland birds (Coppedge et al. 2001). The other common form of passive remote sensing is from satellite imagery, which is used by ornithologists to map landcover and habitat. Typically, satellite images do not provide data at the resolution of aerial photographs (though, as an example, the resolution of the Quickbird Satellite is 0.65 m). Yet, the strength of satellite imagery is in the high temporal replicability of imagery over broad spatial extents. For example, a common satellite data source is derived from the Landsat thematic mapper, which provides multispectral imagery at a 30 m resolution over most of the Earth and has been used by ornithologists for characterizing habitat attributes that are then linked to bird distribution patterns (e.g., Culbert et al. 2012). The Landsat program launched its first earth-monitoring satellite in 1972, and since then, updated satellites have been put into Earth's orbit providing imagery at a given location every 16 days. The combination of relatively high spatial and temporal resolution of Landsat data has allowed ornithologists the opportunity to characterize both long- and short-term changes in habitat, which can be linked to bird distribution patterns (Knick and Rotenberry 2000).

Ecologists use data from passive remote sensing to describe land cover composition, fragmentation patterns, and habitat structure (Turner et al. 2003), and recent advances allow for fine resolution habitat mapping, characterization of vegetation composition, and phenology monitoring (Pettorelli et al. 2014). Traditionally, ecologists have used passive remote sensing to classify vegetation into categories ("vegetation types"), often converting raster-based imagery into polygons circumscribing similar vegetation to create so-called vegetation maps. Raster-based images are composed of pixels (i.e., grid dots), wherein each pixel is composed of a digital number (e.g., assigned color or black-and-white value). Polygons are vector-based data (i.e., geometric shapes, such as points, lines, or polygons), which ornithologists typically create using geospatial analysis programs to delineate boundaries of habitat patches or study areas. But as we learned earlier, habitat encompasses far more than the vegetation around a bird. So, ornithologists continue to expand the use of data from passive remote sensing sources to provide more nuanced habitat data, which has advanced our understanding of the role of habitat in structuring avian communities, populations, and individuals' habitat selection patterns. For example, "habitat structure"—generally defined as the physical arrangement of vegetation and other habitat attributes in space—is widely recognized to affect birds (Rotenberry 1985) but is time-consuming to measure, and thus is often only accomplished at small scales with on-the-ground field surveys. To address the issue, ornithologists have adapted the use of passive remote sensing approaches to measure habitat structure across broad spatial extents. One promising approach is image texture analysis. Raster-based images, whether satellite images or aerial photographs, are composed of tone, which is the reflectance value (i.e., digital number) of a particular cell. Texture refers to the spatial relationships of tonal values of neighboring pixels, and it is possible to quantify these relationships across an image (Haralick et al. 1973, Haralick 1979). Why texture analysis has proven important to ornithologists is because texture from remotely sensed images, such as air photos or satellite scenes, is correlated with heterogeneity in vegetation and habitat structure (Culbert et al. 2012, Wood et al. 2012, fig. 18.7), and in turn is an excellent predictor of bird species richness (St-Louis et al. 2006, 2009, 2014) and diversity (Wallis et al. 2016), habitat use patterns (Tuttle et al. 2006), habitat suitability (Bellis et al. 2008), and habitat quality (Wood et al. 2013).

Passive remote sensing using satellite imagery can also be useful for deriving measures of other habitat properties, such as net primary productivity and vegetation phenology (timing of plant growth). For example, imagery from several multispectral satellites has been used to calculate the normalized difference vegetation index (NDVI; Gottschalk et al. 2005, Higgins et al. 2012). NDVI is derived from the red to near-infrared reflectance ratio, which provides information about plant structure and growth because chlorophyll absorbs red light, and mesophyll scatters near-infrared light. Numerous multispectral satellites provide data used to calculate NDVI. Landsat data are widely used to calculate NDVI; another satellite platform that has proved valuable to ornithologists is the advanced very-high-resolution radiometer (AVHRR), which has a fairly coarse resolution (≥1 km), but a high temporal resolution (1 day). NDVI metrics calculated

using AVHRR data can thus be used to examine temporal changes in primary productivity, green-up, and the length of the growing season, as well as other relevant habitat attributes (Higgins et al. 2012). For example, Sanz et al. (2003) used NDVI metrics calculated using AVHRR data to show that global climate change is causing oak leaf-out to have occurred earlier in the Mediterranean region, and the peak date of caterpillar abundance is advancing accordingly. However, Pied Flycatchers (*Ficedula hypoleuca*) did not change their arrival time between 1980 and 2000, causing a mismatch in peak food supply and the birds' nesting cycle. Determining the broad-scale pattern of oak leaf emergence in relation to caterpillar abundance and Pied Flycatcher arrival at breeding locations would not have be possible without the use of satellite imagery for mapping vegetation change among years.

Active   In contrast to passive remote sensing, active remote sensing involves sensors that are mounted with devices that actively emit information such as laser (LiDAR) or motion pulses (RADAR), which are then bounced back to the sensor (Pettorelli et al. 2014). The use of data from active remote sensing platforms has exploded over the past decade, in particular because as remote sensing technology has advanced, active sensing techniques have been developed to gather structural features from below the canopy level (Martinuzzi et al. 2009). The advance of active remote sensing has proved incredibly important for bird-habitat studies, in particular for the ability to map fine-resolution habitat structural features across broad extents, especially during the breeding period, when forested systems typically form a dense canopy of tree cover. Two active remote sensing applications that we will highlight in this chapter are RADAR (Imhoff et al. 1997, Bergen et al. 2009) and LiDAR (Light Detection and Ranging; Bradbury et al. 2005, Vierling et al. 2008).

As described above, active remote sensing procedures have improved our ability to measure habitat attributes beneath the canopy over broad spatial extents. For example, in Wisconsin forests (USA), ornithologists were interested in describing patterns of bird species richness in relation to forest structural attributes. Many of the forest birds of the study area breed in the understory and respond strongly to differences in fine-resolution forest attributes, such as shrub and tree density. The ornithologists were interested in describing bird species richness over a broad spatial extent, and therefore it was necessary to use data from remote sensing sources. However, the forested system was dominated by tall, broad-canopy oak (*Quercus* spp.), hickory (*Carya* spp.), and maple (*Acer* spp.) trees, and passive remote sensing methodologies would not be effective at characterizing forest structure below the canopy. Therefore, the researchers used LiDAR to characterize fine-resolution differences in forest

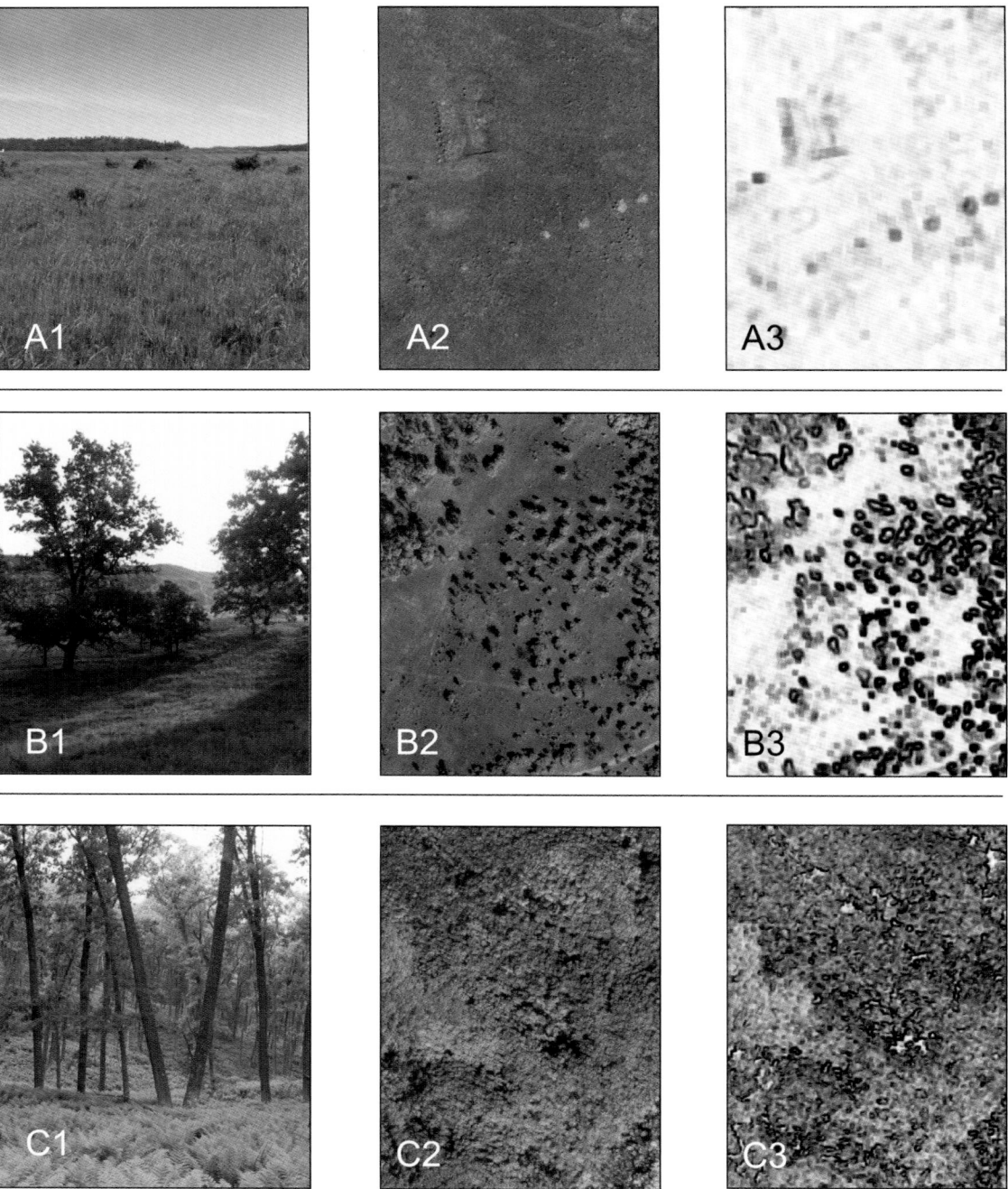

Figure 18.7. Three vegetation types across an open to closed tree canopy continuum: *A*, Grassland; *B*, Savanna; *C*, Woodland at Fort McCoy Installation in Wisconsin, USA. Each vegetation type depicted with (1) a ground photograph, (2) an infrared air-photograph, and (3) an infrared air-photograph processed for first-order variance (image texture).

structure, which were then related to patterns of breeding bird species richness (Lesak et al. 2010). Using a similar rationale to the Wisconsin system (i.e., to characterize understory vegetation structure, which are then linked to bird data), researchers have used LiDAR to describe density of birds in conifer forests of the Black Hills, South Dakota, USA (Clawges et al. 2008); foraging locations by woodpeckers (*Picidae* spp.) in Idaho conifer forests (Vierling et al. 2013); and territory quality for migratory birds in a mixed forest system in central New Hampshire (Goetz et al. 2010).

In addition to the LiDAR studies, ornithologists have used RADAR data to characterize bird habitat at broad spatial extents. For example, in northern Australian eucalyptus (*Myrtaceae* spp.) and melaleuca (*Melaleuceae* spp.) woodlands and forest, Imhoff et al. (1997) used Synthetic Aperture Radar (SAR) in combination with aerial photography to map bird habitat (i.e., vegetation structure). Building on these findings, Bergen et al. (2007) integrated SAR data with optical remote sensing data to describe bird habitat in forest habitats of northern Michigan. Bergen et al. (2007) found that including RADAR information, which characterized biomass, in their analyses improved habitat classification for forest bird species. LiDAR and RADAR data can capture redundant habitat information (Hyde et al. 2006). Nevertheless, the use of active remote sensing data has greatly improved our ability to characterize bird habitat across broad spatial extents (Bergen et al. 2009). What is next for remote sensing in ornithology habitat research? While this remains a "hot" topic of exploration, the next frontier of passive and active remote sensing will likely involve finer-resolution characterizations of habitat features that can be linked to bird distributions (e.g., discriminating plant species, Roth et al. 2015).

Measuring Habitat Use

Habitat use generally refers to a bird's use of physical and biological components in a habitat, such as the consumption of food or the occupation of a nest cavity. Habitat use therefore can be directly described from associating the presence of a species with habitat features, or from observations of how animals interact with habitat features (Gaillard et al. 2010). Use of a habitat does not necessarily indicate the conditions that are most preferred by a species, or those that are most strongly associated with fitness. Habitat use, rather, simply reflects what is being used by an individual. Often, to describe habitat use patterns, ecologists typically will measure habitat attributes in areas where birds are observed or where they are engaged in particular behaviors (e.g., foraging). Another approach is to map locations of individual birds marked with unique colored leg bands, which can show how birds use habitats (e.g., Gregory et al. 2004). Likewise, locations of birds obtained from radio (very high frequency, VHF) telemetry can reflect the frequency with which a bird uses different habitats within their home range or territory (Powell 2000).

Technological advances enable ornithologists to measure habitat use in new ways. Locations obtained from global position system (GPS) receivers small enough to be mounted on birds allow more accurate, fine-scaled habitat associations, and thus more realistic estimates of habitat use than some previous techniques (Cagnacci et al. 2010). Other advances in the remote detection of bird locations include the use of increasingly sophisticated remote or bird-borne cameras (O'Connell 2010, Gómez-Laich et al. 2015), telemetry transmitters that communicate with satellites or cell phone networks (Millspaugh et al. 2012), and other animal-borne sensors that can reveal a bird's location, such as "geolocators" (or global light location sensors) useful for tracking migratory birds too small to carry heavier, more precise technologies (e.g., Hallworth et al. 2015). For additional details on avian movement ecology and remote sensing, see chapters 19 and 31.

Measuring Habitat Selection

Habitat selection is an evolutionary response to a species' environment, resulting from a complex, hierarchical process of behavioral choices. Johnson (1980) defined habitat use to be selective if components of habitat are used disproportionately to their availability. Availability refers to a component being present and ready for immediate use; it must be accessible or obtainable by a bird (Hall et al. 1997, Gaillard et al. 2010).

Used, Unused, and Available

The methods ornithologists use to measure and infer habitat selection vary in their precision and applicability (Alldredge and Ratti 1986, Thomas and Taylor 1990, Jones 2001). Habitat selection is tested for wild birds in two main ways: comparing used habitats with unused habitats and comparing used habitats with available habitats (i.e., randomly-located plots) (Jones 2001). Used habitat is measured as described above. Available habitat refers to all habitat biologically available to a bird in the area of interest. Unused habitat is a subset of available habitat; it is available but reliably documented to not be used by a given species at a given time. For example, consider a cavity-nesting species such as a woodpecker: a tree with an active nest cavity is "used," trees within the species' range and large enough for a cavity are "available," and an "unused" tree is an available tree demonstrably lacking an active nest cavity. By statistically comparing either used vs. unused or used vs. available habitats, researchers can draw inference about habitat selection and avoidance (Manly et al. 2002). Thomas and Taylor (1990) provide a review of several study designs and common statistical techniques for these sorts of approaches.

There are pros and cons for used vs. unused and used vs. available approaches (Manly et al. 2002), and disagreements persist over which is more informative and less subject to violations of statistical assumptions or logical limitation. The reconciliation is far beyond the scope of this chapter, but we highlight a few points to consider when designing studies or interpreting their findings. With used vs. unused approaches, it is often difficult to confirm that a location is truly unused, since the detection probability of birds is rarely 100 percent (MacKenzie et al. 2009). In addition, absence from a particular habitat does not mean that the habitat is being avoided (Wiens 1989), since various constraints, including low population density and dispersal limits, may have a major effect on which habitats are used or not (Pulliam 2000, Jones 2001). Further, a bird's use of a particular habitat affects its use of other habitats (Thomas and Taylor 1990, Aebischer et al. 1993), and this introduces a lack of independence in data that can negatively affects the power of some statistical techniques (Jones 2001). Nonetheless, only by comparing used and unused locations can statistical models yield estimates that reflect the true probability a bird uses a given site, a metric enormously useful for managers making land use decisions (Manly et al. 2002). In contrast, used vs. available approaches have advantages but can only yield relative probabilities and indices of selection. Another disadvantage of used vs. available comparisons is that articulation of relevant habitat availability is problematic. For example, researchers routinely remind us that availability refers to both the accessibility and obtainability of resources, not just their abundance (Hall et al. 1997, Jones 2001). Blindly distributing random plots on a landscape (to provide sample units of available habitat) may not reflect true availability, and may result in uninformative results (e.g., a landbird selects terrestrial and avoids aquatic habitats). Johnson (1980) also emphasized the importance of scale and research question specificity in delineating available habitats. He identified four "orders" of habitat selection corresponding roughly to Hutto's (1985) depiction of top-down hierarchical habitat selection (fig. 18.5). Of course, these levels of selection are only landmarks belonging to a continuum from very fine to very coarse spatial scales (Mayor et al. 2009, Gaillard et al. 2010). With this approach, a researcher might wish to examine how birds select where to place their territory within a larger landscape (Johnson's "second order" of habitat selection) or how they select specific nest site locations within their home range (Johnson's "fourth order"), and these questions demand different sets of available habitats for comparison.

Statistical Procedures

In recent years, a growing number of statistical procedures for analyzing habitat selection have been proposed. Two main philosophies can be recognized (Gaillard et al. 2010). The first approach includes methods rooted in ecological niche theory. It typically aims to yield maps of spatial distribution, and it involves analysis of remotely sensed biophysical and habitat variables at used locations in correspondence with random available locations obtained in a geographic information system (GIS). Although this design has been referred to as "presence-only," analyses still involve comparisons between used and available locations. This approach yields models often labeled species distribution models (SDMs). One of the earliest SDMs involved generalized linear models to predict the distribution of the Rufous Scrub-bird (*Atrichornis rufescens*) using locality records of the species and remotely mapped environmental variables obtained from a GIS (Elith and Leathwick 2009). Much work involving SDMs continues with birds today. Armed with a sample of bird locations and environmental data from a GIS, ornithologists can use specialized modeling procedures (such as MaxEnt; Phillips et al.

2006) to develop predictive maps of bird distribution useful for conservation practitioners and as tools for making predictions of birds' responses to land use and climate change scenarios (Benning et al. 2002). An almost synonymous approach, usually called ecological niche modeling (ENM, Peterson et al. 1999, Warren and Seifert 2011), tends to place less emphasis on implications for population processes and more on the biophysical variables associated with a species distribution—the fundamental niche of a species (Peterson et al. 2001, Peterson and Soberón 2012). This philosophical approach provides useful and reliable description of the multivariate niche and is often used to map the habitat for a population or species at broader spatial and temporal scales in a manner consistent with the niche-based definition of habitat (Gaillard et al. 2010).

The second philosophical approach emphasizes drawing reliable inference on habitat attributes actively selected by the species studied: that is, testing for habitat selection (Gaillard et al. 2010). Typically, researchers use environmental data to distinguish locations where birds are observed (used locations) from the pool of available locations to reveal habitat selection (see review by Jones 2001), often by using a statistical regression technique (logistic regression or more advanced general linear models) to yield a resource selection function (RSF; Manly et al. 2002). Assuming that the null model of no selection corresponds with a proportional relationship between use and availability, one can statistically test whether a given habitat component is selected for, selected against (aka avoided), or not selected. This philosophical approach emphasizes hypothesis testing and enables quantifying the contribution of individual habitat attributes to a habitat selection (Gaillard et al. 2010). An important variant on this approach compares used and unused locations while explicitly acknowledging imperfect detection probability. Although relatively conspicuous, birds are mobile and can easily be overlooked (or not heard). Specialized statistical models can be adjusted for imperfect sampling or detection probabilities, yielding more rigorous estimates of occupancy (Royle et al. 2004, MacKenzie et al. 2009). While most analyses have emphasized habitat selection in space, birds also exhibit selection in time, and recent work has offered new analytical approaches that better incorporate temporal aspects of habitat selection (Porzig et al. 2014). However, all approaches for measuring habitat selection are insufficient for examining the fitness consequences of that selection. Thus, ornithologists also work to rigorously measure habitat quality.

Measuring Habitat Quality

There are two basic approaches to conceptualizing how to measure habitat quality. We can either assess habitat quality directly, by measuring attributes of the habitats themselves, or we can measure variables for individual birds and populations in different habitats to reveal variation in habitat quality. Measuring habitats directly is far less common than measuring indirectly, because measuring directly requires that we know which resources (e.g., specific food items, nest sites) are essential for the species of interest. This is difficult, because we often do not know exactly which resources and ecological conditions are most relevant for many species, and even when we do, they may be difficult (or impossible) to measure. Nonetheless, for some well-studied species, direct measures of habitat quality are possible. For example, Barnes et al. (1995) measured habitat quality for Northern Bobwhite (*Colinus virginianus*) by quantifying grass forage quality, food (insect) abundance, and shrub cover availability. This approach has also been used effectively by Piersma (2012) and his colleagues, who have investigated habitat quality for migratory Red Knots (*Calidris canutus*) by quantifying the availability of prey (mollusks and crustaceans) and competition with other birds. So, while it is possible to use direct measures to measure habitat quality, the far more common approach is to use indirect measures, which we highlight below.

Indirect Measurements—Demography

Most studies take the second conceptual approach and measure avian habitat quality indirectly by quantifying bird abundance, distribution, or performance among different habitats to assess variation in their quality (Johnson 2007). As explained earlier, habitat quality is best defined from an individual bird's perspective as the per capita rate of population increase expected from a given habitat. Thus, the roots of the concept are demographic, and habitat-specific measures of density, reproduction, and survival offer some of the best measures of habitat quality (e.g., Virkkala 1990, Holmes et al. 1996, Franklin et al. 2000, Murphy 2001, Persson 2003). Using the abundance (or density) of birds in habitat is the most common measure of habitat quality, since birds are relatively easy to survey (via point counts or mist nets for example; see chapter 31). Because birds reproduce in discrete nests, reproduction can be ascribed to individual birds and their habitats, rendering nest success, the number of young fledged, and other measures of reproduction common metrics of habitat quality; up to a third of bird habitat studies employ this approach (Johnson 2007). Fewer studies use adult survival as a measure of habitat quality, probably because of the large and lengthy data sets required to assess it rigorously. That said, an increasing number of studies now assess survival, perhaps because of the increased availability and power of survival analysis software (White and Burnham 1999, Murray 2006).

Indirect Measurements—Distribution

Numerous measures of bird distribution can be used to indicate habitat quality based on the theoretical Fretwell-Lucas models suggesting birds should select habitats to maximize their realized habitat quality (box on page 585). Thus, studies that confirm that some habitats are used disproportionate to their availability (ipso facto habitat selection) can reveal high-quality habitat. For example, Hall and Mannan (1999) examined habitat selection to determine what constituted high-quality habitat for Elegant Trogons (*Trogon elegans*) in southeastern Arizona, which highlighted the importance of Arizona sycamore trees (*Platanus wrightii*). Habitat selection models predict that, relative to low-quality habitats, high-quality habitats should be occupied for longer periods within a season and more consistently over years. Consequently, some investigators have used timing, duration, and frequency of habitat occupancy as measures of habitat quality (reviewed by Sergio and Newton 2003). For example, Ferrer and Donázar (1996) found that habitat occupancy was related to both resource availability and reproduction for Imperial Eagles (*Aquila heliaca*) in Spain. Despotic distribution models predict that dominant individuals should settle disproportionately in the highest quality habitats. Therefore, the ratio of behavioral classes among habitats (e.g., adult vs. young, male vs. female) could reveal variation in their quality (Railsback et al. 2003). For example, Rohwer (2004) used age ratios to show that despotic territorial behavior forced yearling male Hermit (*Setophaga occidentalis*) and Townsend's (*Setophaga townsendi*) Warblers into marginal high-elevation habitats for their first potential breeding season, and Marra (2000) found that ratios of dominant to subordinate age and sex classes of wintering American Redstarts varied markedly between high-quality (mangrove) and low-quality (scrub forest) habitats in Jamaica.

Indirect Measurements—Condition

All metrics of habitat quality reviewed in this chapter require measuring populations of birds, which can be problematic for species that are difficult to observe or capture and for birds using habitats only briefly, such as migratory species. As an alternative, some ornithologists have used measures of individual birds' physical condition as indicators of habitat quality. These include external, visible, and measurable features, such as body mass or visible fat deposits, and variables that rely on analysis of sampled tissues (especially blood). For example, Strong and Sherry (2001) found that body mass was a reliable indicator of habitat quality for Ovenbirds (*Seiurus aurocapilla*) wintering in Jamaica, and Seaman et al. (2006) showed that blood plasma metabolites indicated which habitats afforded the best stopover refueling opportunities for migrating Western Sandpipers in British Colombia and Washington. For further details of the pros and cons of measuring habitat quality with birds' physical condition, see Johnson (2007) and Homyack (2010).

Assumptions

All of these measures of habitat quality carry the assumption that their variation is a consequence of, rather than a cause for, different habitat selection. That is, variation in habitat attributes such as food supply and predation risk must lead to variation in the demographic rates, distribution, or physiological condition of birds. This may often be at least partially true, but can be violated if, for example, strong differences in individual bird quality cause variation in survival or reproduction and also cause birds to use different habitats. For example, Carrete et al. (2006) found that variation in reproduction for eagles (*Aquila chrysaetos* and *A. fasciata*) in southern Spain arose because of changes in bird quality (age) and not because of suitability for breeding per se. Preexisting differences in the condition of birds may also cause them to use different habitats, rendering measures of body condition misleading. For example, lean individuals may choose food-rich but risky habitats, while fat individuals may choose safer but food-poor habitats (Moore and Aborn 2000). In this case, local food supply and body fat would be inversely related, and good body condition would be a poor indicator of food-rich habitats.

More generally, Van Horne (1983) cautioned that the density of animals in a habitat can, in some cases, be a misleading indicator of habitat quality because habitat conditions favoring density, survival, and reproduction may not be the same (Franklin et al. 2000), which could lead to misleading measures of habitat quality if only one parameter is used to rank habitats. Since the publication of her influential and oft-cited paper, biologists recognize that robust measures of habitat quality require a thorough unraveling of habitat-specific measures of demography—i.e., density, reproduction, *and* survival measures in each habitat considered (Bock and Jones 2004). Time and money constraints rarely allow all of these measures to be obtained, but when multiple measure are obtained, novel insights sometimes emerge. For example, Murphy (2001) learned that annual productivity for Eastern Kingbirds (*Tyrannus tyrannus*) was lower in floodplain than in creek and upland habitats in the Charlotte Valley of central New York, but estimates of survival suggested that all three habitats were population sinks, and numbers were supplemented substantially by emigration from other areas.

Although some empirical evidence indicates that measures of simple abundance can mislead, Bock and Jones (2004) demonstrated that density is usually roughly correlated with habitat quality for breeding birds. Furthermore, decoupling of density and reproduction was not associated with most environmental and life history attributes as predicted by theory, although their results do suggest that discrepancies emerged most frequently in human-disturbed landscapes. Future work should explore whether density and survival covary over habitats (Johnson et al. 2006).

In this light, the question "which habitat is best?" can be reexamined by asking, how do we measure habitat quality for the relevant management unit (populations), when habitat selection is a process operating at the individual level? To explicate individual habitat quality for population management purposes, we must consider how temporal and spatial scales influence habitat choices and their demographic consequences (Wiens 1989). A habitat's quality can change rapidly for a given species, and care must be taken to understand when resources are most limited and when consequences of habitat occupancy most influence a population (Sherry and Holmes 1995). Sutherland (1998) and Runge and Marra (2005) developed models that articulate the temporal (seasonal) interactions of local habitat quality, availability, and demographics in birds. These models extended previous work describing how individual birds' choices of habitats (based on local quality) impact populations over shorter temporal windows (Orians and Wittenberger 1991, Goss-Custard et al. 1995). These models all demonstrate the delay between birds' habitat choices and their demographic consequences, which should prompt researchers to track birds' fates as long as feasible.

CONSERVATION APPLICATIONS, FROM HABITATS TO LANDSCAPES TO ECOSYSTEMS

Throughout this chapter we have discussed the importance of habitat for bird ecology and conservation. Indeed, the greatest threat to birds, and to biodiversity in general, is habitat loss (Foley et al. 2005; fig. 18.2). In this section, we specify how understanding bird-habitat relationships can advance avian conservation by helping us (1) prioritize habitats for conservation, (2) restore places that have already been degraded, (3) anticipate the response of birds to climate change, (4) incentivize habitat conservation via the provisioning of ecosystem services, and (5) guide key policies to protect habitats for birds. For additional details on bird conservation, see chapters 24–29.

Prioritizing Places and Habitats

The quantification of bird-habitat relationships has enabled conservation planners to use empirical evidence to prioritize habitats for protection or restoration. Conservation is always limited by time and money, so effective conservation

SPECIES DISTRIBUTION MODELS
By Dr. Brooke Bateman, University of Wisconsin–Madison

Species distribution models (SDMs) are commonly used to characterize and predict species current and future distributions. These models help identify the relationship between where a species occurs and the environmental conditions found at those locations, and to use this information to map a species distribution in geographical space. Maxent (Phillips and Dudík 2004) is a commonly used SDM algorithm that uses presence records of a species and environmental data relating to the locations where that species occurs. Absence data are not needed, as Maxent uses "background" points that are representative of the range of environmental conditions of the study area. The Maxent modeling algorithm discriminates what is unique about the environmental conditions associated with species presence data in comparison to the environmental conditions of the entire study area. With the availability of large datasets that contain bird occurrence data (e.g., e-Bird), we can use modeling techniques such as Maxent to map the potential

distribution of many bird species over large extents. With this information we can map species richness for many species using environmental data such as climate data (see figure). Pairing bird occurrence data with climate data allowed the researchers in this study to explore how climate change in the recent past affected where the potential breeding distributions occurred in the landscape. Given that species distributions are influenced by more than climate alone, these predictions are of a species "potential" distribution or areas where a species could possibly occur based solely on climate conditions. It is also important to note that SDMs use a limited amount of information and that species occurrences are influenced by many factors (e.g., biotic interactions) that may not be included in the model. However, these spatially explicit models of bird distributions can be used to identify priority areas for conservation initiatives and to help understand the potential effects of climate change on species.

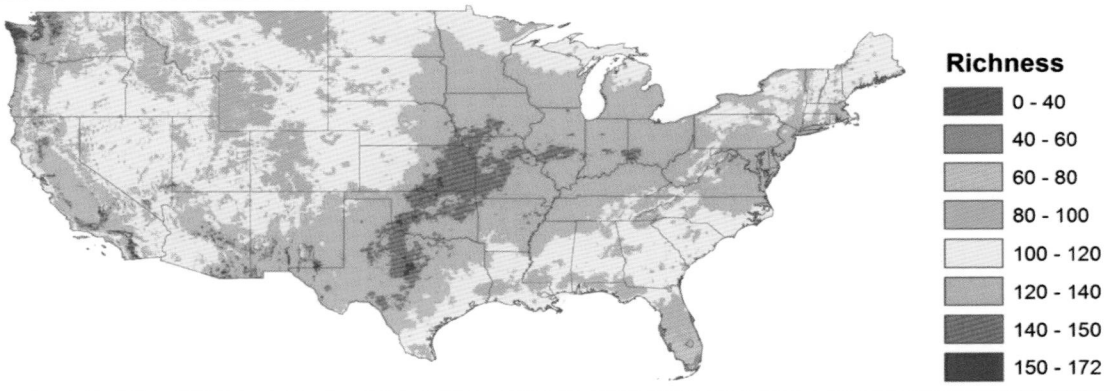

Adapted from Bateman et al. 2016. Potential species richness of 285 breeding bird species within the continental United States. Potential species richness is defined as the number of species that have suitable climate within a given grid cell.

action relies on planning that uses ecologically based methods for prioritizing actions (Pressey et al. 2007). Advances in the study of avian habitat selection, especially the increasing availability of empirically based, high-resolution species distribution models (box on page 599), are providing new opportunities to guide on-the-ground decision-making across a wide range of spatial scales (Bayliss et al. 2005, Seavy et al. 2012). For example, ornithologists integrated bird distribution models into conservation activities along the San Joaquin River in the Central Valley of California (Seavy et al.

2012). This work identified synergies in which conservation action for some bird species aligned with other ecological considerations such as flood control and fish habitat. However, they also found that optimal habitat conditions varied among groups of birds, demanding planners to balance trade-offs in predicted effects on target species. They concluded that one-on-one interactions between the ornithologists that develop the models and the decision-makers that use them are essential to best inform conservation and restoration (Seavy and Howell 2010). This work is clearly vital for bird conservation, but birds are often used as indicators of environmental condition in general (Hilty and Merenlender 2000), so bird-habitat relationships can typically be used to inform priorities for broader conservation agendas. For example, areas with a large proportion of unique bird species—Endemic Bird Areas—have been used to identify sites of global conservation priority (Wilson et al. 2006).

Ecological Restoration

Where habitat has already been degraded, conservationists can use understanding of bird-habitat relationships to optimize habitat restoration. For example, Kus (1998) used species-specific models of habitat suitability for endangered Least Bell's Vireo (*Vireo bellii pusillus*) to confirm that restored sites were developing vegetation characteristics of intact natural habitat, and found that occupation of restored sites for breeding took several years, but was accelerated by the presence of adjacent mature riparian habitat. Gardali and Holmes (2001) found that riparian bird species responded favorably to particular local restoration practices such as the number of tree species planted and the planting tree density of certain species. However, they also found that birds responded the amount of riparian forest in the surrounding landscape, again highlighting the importance of landscape-level conservation planning in habitat restoration.

Climate Change

In addition to habitat loss, climate change poses a major threat to bird conservation (fig. 18.2), and understanding bird-habitat relationships is vital for predicting how birds respond to climate change. For example, in New York, Zuckerberg et al. (2009) found that the population center for 129 bird species has shifted northward by over 4 km over the past 30 years, which provided strong evidence that birds are indeed shifting their ranges poleward in response to climate change. Further, as species shift in response to climate change, novel communities (i.e., those with no natural analog) are being formed (Stralberg et al. 2009a, Prince and Zuckerberg 2015). In the eastern portions of North America, the wintering distribution of 38 bird species responded strongly, with more southerly species shifting their ranges northward in response to milder winter temperatures (Prince and Zuckerberg 2015). While we now understand that species assemblages are responding strongly to climate change (Prince and Zuckerberg 2015), it remains unclear how climate change will affect avian biodiversity in the long run (Walther et al. 2002). Bird-habitat relationships are helping conservationists plan for climate change by enabling forecasts of bird distributions aimed at either protecting habitats likely to be needed in the future (e.g., Benitez-Lopez et al. 2014) or identifying climate refugia from anticipated climate change (e.g., Stralberg et al. 2015). While much work has focused on montane birds' responses to rising temperature (Elsen and Tingley 2015), modeling effects of estuarine habitat in the face of sea level rise is vital for coastal birds (Veloz et al. 2013). Other approaches have used species-habitat models to prioritize habitats that will maintain connectivity, which may be necessary for species to shift their distribution as the climate warms (Mazaris et al. 2013, Jones et al. 2016).

While global climate change is prompting gradual shifts in temperature and precipitation, more immediate and potentially more detrimental consequences to birds come from extreme weather events such as prolonged droughts, cold snaps, tornado outbreaks, or severe storms, which have greatly increased in frequency and intensity over the past few decades (Cai et al. 2014, Conrey et al. 2016), often with negative consequences on biodiversity (Parmesan et al. 2000, McCreedy and van Riper III 2014). For birds, extreme weather can lead to changes in distribution (Albright et al. 2009), disrupt avian community structure (Rittenhouse et al. 2010), alter migratory pathways (Streby et al. 2015), and lead to mass mortality events (McKechnie and Wolf 2009). Additionally, extreme weather can alter phenological relationships of migratory birds and their seasonal resources. There is strong evidence that migratory birds time their spring migrations to match the peak availability of caterpillars and other protein-rich prey (Graber and Graber 1983), and any disruptions in the delicate timing of these events could be detrimental for birds (Kellermann and van Riper 2015). For example, throughout Europe, there is evidence that migratory birds that do not arrive to their temperate breeding locations in coincidence with peaks in food availability have lower breeding success (Møller et al. 2008). Many insects in North America emerge coinciding with plant budburst, but Neotropical-Nearctic migrant birds are unable to adjust their migratory timing to match the early phenology of the insect food resources (Wood and Pidgeon 2015). Without the migratory birds present to consume herbivorous insects, vegetation damage is high, which highlights a potential negative consequence caused by extreme weather (Wood and Pidgeon 2015). In theory, understanding bird-habitat relationships could help conservationists seeking ways to mitigate effects of extreme weather, which remains an urgent research need. For highly mobile species like birds, movement away from extreme climatic events may be a viable response, even for species that normally show high site fidelity (Martin et al. 2007). Applying our understanding of bird-habitat relationships could help us prioritize refugia from weather extremes, which could occur, for example, at the edges of species' ranges or at key geographic locations along migratory pathways (Bateman et al. 2015).

Dynamic Conservation

While understanding avian habitat ecology can certainly help us respond to future conservation needs of birds, it can also be used to guide current, real-time conservation. For example, an innovative collaboration between The Nature Conservancy, Cornell Lab of Ornithology, and Point Blue Conservation Science uses data on bird distribution to predict where wetland management can best benefit shorebirds and waterfowl during their migrations. By crunching data from eBird, ecologists can overlay maps of bird distribution with aerial views of existing surface water, revealing where the birds' need for habitat is greatest (Robbins 2014). Funding from The Nature Conservancy then pays rice farmers in the birds' flight path to keep their fields flooded with irrigation water to provide habitat for migrating flocks. This work also underscores the power of economic incentives for the conservation of bird habitat, a topic that has received much recent attention by ornithologists and economists examining so-called ecosystem services (Whelan et al. 2008, Wenny et al. 2011).

Ecosystem Services

Conserving birds demands that we protect their habitats, so it is imperative that policies incentivize habitat conservation. Strong environmental policies can protect habitat on public lands and for species protected by government regulations, such the US Endangered Species Act, but incentives that apply for common birds and on private lands are also needed (Kareiva and Marvier 2007, Armsworth et al. 2012). Recent conservation research has emphasized that valuing ecosystem services can provide these incentives (Gómez-Baggethun et al. 2010). Ecosystem services are process that help sustain and fulfill human life (Daily 1997). Examples of ecosystem services provided by birds include pest control (a regulation service, Johnson et al. 2010), and seed dispersal (a supporting service, Hougner et al. 2006), and recreational and aesthetic value (a cultural service, Gürlük and Rehber 2008). By explicitly linking the provisioning of these services to bird habitat, an incentive for habitat conservation can materialize. For example, Kellermann at al. (2008) and Railsback and Johnson (2014) found that warblers in Jamaica help control economically damaging insect pests in coffee farms, and that the delivery of this pest control service was enhanced by the conservation of trees and forest patches both within and beyond the farms' boundaries. Cultural values, such as aesthetic appreciation, may be more difficult to quantify than other ecosystem services, but they can also provide incentives for habitat conservation. Neumann et al. (2009) conducted an analysis of real estate prices near the Great Meadows National Wildlife Refuge in Massachusetts, a popular place for bird-watching. They found that properties located close to the refuge had a demonstrable price premium, providing incentive for landowners and city planners to ensure the protection of habitats for birds and other wildlife.

Habitat and Policy

Many governments have recognized the central role of habitat protection in the conservation of biodiversity, passing legislation that provides protection of habitat considered "critical" (Camaclang et al. 2015). These laws, such as the US Endangered Species Act (ESA 1973), Australia's Environment Protection and Biodiversity Conservation Act (EPBC 1999), and Canada's Species at Risk Act (SARA 2002), necessarily combine with science to specify how these critical or essential habitats should be defined and designated for a given species. This law-science relationship and its effect on habitat policy was exemplified in a 1994 Supreme Court case concerning the Northern Spotted Owls mentioned at the start of this chapter, a case now regarded as one of the most important in American conservation history (Petersen 2002, Alagona 2013). The ESA makes it illegal to "harass, harm, pursue, hunt, shoot, wound, kill, trap, capture, collect or attempt to engage in any such conduct" toward a listed species, actions collectively called a "take" in the lexicon of ESA. In the 1994 case of *Babbitt v. Sweet Home Chapter of the Communities for a Great Oregon*, the Supreme Court ruled in a six to three majority that science indicates the word "harm" in this case must include habitat modification. This ruling firmly linked the legal protection of species with the protection of their habitats, and the ESA has since become a model for analogous legislation around the world (Camaclang et al. 2015). The concepts of endangered species and habitat, thus affected by this law-science relationship, are no longer possible to examine without referring to both science and the law (Ruhl 2007, Alagona 2013).

To advance the conservation of birds, our efforts must prioritize habitat conservation. Our view is that recognizing the reliance of birds on their habitats, incentivizing avian conservation by valuing ecosystem services, and aiming environmental policy toward the protection of species and their habitats are powerful tools for conservation and for improving human life. But these tools should be used alongside recognition of the intrinsic value of birds (Gavin et al. 2015), because regardless of how birds may benefit human livelihood and fulfillment, they are also simply our co-inhabitants on this planet.

KEY POINTS

- Habitat is an area with the resources and conditions that promote occupancy by individuals of a given species and allows those individuals to survive and reproduce.
- To balance species conservation and human needs, we must understand the nuances of **habitat selection**—the study of where birds live and why.
- Habitat quality is best considered from individual birds' perspectives as the per capita contribution to population growth expected from a given habitat. This is best measured as directly as possible; indirect measures are far more efficient, but their accuracy must be verified.
- The availability and quality of habitats often limit or regulate bird populations.
- Advances in remote measurements have significantly propelled the study of bird habitats, but they have also led researchers to focus on incomplete assessments of habitat, such as an overreliance on vegetation. Recent work is rightly focusing on the resources and conditions underlying spatial variation in the performance of birds.
- Habitat ecology strongly affects environmental policies, which carry enormous social and economic consequences.
- Other factors can interact with or in some cases overwhelm habitat in driving bird populations, and ornithologists must be aware of the range of processes affecting wild birds.

KEY MANAGEMENT AND CONSERVATION IMPLICATIONS

- Practitioners must understand the role of habitat in limiting and regulating bird populations.
- It is vital to recognize spatial and temporal variation in habitat quality to prioritize habitats for conservation effectively.
- Ornithologists should examine how global and landscape-scale dynamics can affect local habitat selection and quality in order to optimize habitat management, restoration, and acquisition.
- Traditional habitat conservation—identifying habitats to preserve, restore, and manage—remains an important tool for bird conservationists. But newer, innovative approaches are opening additional opportunities to integrate the needs of birds and people.
- The instrumental value of services that birds provide can be harnessed to incentivize conservation, while acknowledging the intrinsic value of birds.

DISCUSSION QUESTIONS

1. What is the difference between vegetation type and habitat, and why is the latter a better concept for understanding the distribution of birds?
2. How can the quality and quantity of habitat regulate a bird population—that is, cause it to rise when it is below a long-term average and decrease when it is above a long-term average?
3. Can you think of a scenario in which there is a discrepancy between habitat quality from the perspective of an individual bird and from a population perspective?
4. If conservationists are trying to identify the highest quality habitats for conserving a threatened bird species, why does it matter whether the species more closely follows an ideal free or an ideal despotic distribution?
5. How can the presence of conspecifics favor the selection of habitat by birds? How can the presence of conspecifics disfavor the selection of habitat by birds? What natural history attributes might make one of these outcomes more likely than the other for a species?
6. Birds are rarely 100 percent detectable. How does that reality affect the design of studies of avian habitat selection? Specifically, how does imperfect detection affect "used vs. unused" and "unused vs. available" study designs?
7. How can species distribution models prioritize habitats for avian conservation in the face of climate change?
8. What is the legal link between habitat and the Endangered Species Act?

References

Ackers, S. H., R. J. Davis, K. A. Olsen, and K. M. Dugger (2015). The evolution of mapping habitat for Northern Spotted Owls (*Strix occidentalis caurina*): A comparison of photo-interpreted, Landsat-based, and LiDAR-based habitat maps. Remote Sensing of Environment 156:361–373.

Aebischer, N. J., P. A. Robertson, and R. E. Kenward (1993). Compositional analysis of habitat use from animal radio-tracking data. Ecology 74:1313–1325.

Ahlering, M. A., and J. Faaborg (2006). Avian habitat management meets conspecific attraction: If you build it, will they come? The Auk 123:301–312.

Ahlering, M. A., D. Arlt, M. G. Betts, R. J. Fletcher Jr., J. J. Nocera, and M. P. Ward (2010). Research needs and recommendations for the use of conspecific-attraction methods in the conservation of migratory songbirds. The Condor 112:252–264.

Alagona, P. (2013). After the grizzly: Endangered species and the politics of place in California. University of California Press, Berkeley.

Albright, T. P., A. M. Pidgeon, C. D. Rittenhouse, M. K. Clayton, C. H. Flather, P. D. Culbert, B. D. Wardlow, and V. C. Radeloff (2009). Effects of drought on avian community structure. Global Change Biology 16:2158–2170.

Alldredge, J. R., and J. T. Ratti (1986). Comparison of some statistical techniques for analysis of resource selection. Journal of Wildlife Management 50:157–165.

Anderson, K., and K. J. Gaston (2013). Lightweight unmanned aerial vehicles will revolutionize spatial ecology. Frontiers in Ecology and the Environment 11:138–146.

Armsworth, P. R., S. Acs, M. Dallimer, K. J. Gaston, N. Hanley, and P. Wilson (2012). The cost of policy simplification in conservation incentive programs. Ecology Letters 15:406–414.

Babbitt, B. (1995) *Bruce Babbitt, Secretary of the Interior, et al., Petitioners v. Sweet Home Chapter of Communities for a Great Oregon et al.*, Supreme Court of the United States, no. 94–859, decided June 29, 1995.

Barnes, T. G., L. A. Madison, J. D. Sole, and M. J. Lacki (1995). An assessment of habitat quality for Northern Bobwhite in tall fescue-dominated fields. Wildlife Society Bulletin 23:231–237.

Bateman, B. L., A. M. Pidgeon, V. C. Radeloff, A. J. Allstadt, H. R. Akçakaya, W. E. Thogmartin, and P. J. Heglund (2015). The importance of range edges for an irruptive species during extreme weather events. Landscape Ecology 30:1095–1110.

Bateman, B. L., A. M. Pidgeon, V. C. Radeloff, J. VanDerWal, W. E. Thogmartin, S. J. Vavrus, and P. J. Heglund (2016). The pace of past climate change vs. potential bird distributions and land use in the United States. Global Change Biology 22:1130–1144.

Battin, J. (2004). When good animals love bad habitats: Ecological traps and the conservation of animal populations. Conservation Biology 18:1482–1491.

Bayliss, J. L., V. Simonite, and S. Thompson (2005). The use of probabilistic habitat suitability models for biodiversity action planning. Agriculture, Ecosystems and Environment 108:228–250.

Bell, D. M., M. J. Gregory, H. M. Roberts, R. J. Davis, and J. L. Ohmann (2015). How sampling and scale limit accuracy assessment of vegetation maps: A comment on Loehle et al. (2015). Forest Ecology and Management 358:361–364.

Bellis, L. M., A. M. Pidgeon, V. C. Radeloff, V. St-Louis, J. L. Navarro, and M. B. Martella (2008). Modeling habitat suitability for Greater Rheas based on satellite image texture. Ecological Applications 18:1956–1966.

Benitez-Lopez, A., J. Vinuela, I. Hervas, F. Suarez, and J. T. Garcia (2014). Modelling sandgrouse (*Pterocles* spp.) distributions and large-scale habitat requirements in Spain: Implications for conservation. Environmental Conservation 41:132–143.

Benning, T. L., D. LaPointe, C. T. Atkinson, and P. M. Vitousek (2002). Interactions of climate change with biological invasions and land use in the Hawaiian Islands: Modeling the fate of endemic birds using a geographic information system. Proceedings of the National Academy of Sciences 99:14246–14249.

Bergen, K. M., A. M. Gilboy, and D. G. Brown (2007). Multidimensional vegetation structure in modeling avian habitat. Ecological Informatics 2:9–22.

Bergen, K. M., S. J. Goetz, R. O. Dubayah, G. M. Henebry, C. T. Hunsaker, M. L. Imhoff, R. F. Nelson, G. G. Parker, and V. C. Radeloff (2009). Remote sensing of vegetation 3D structure for biodiversity and habitat: Review and implications for LiDAR

and radar spaceborne missions. Journal of Geophysical Research 114:G00E06.

Bernstein, C., J. R. Krebs, and A. Kacelnik (1991). Distribution of birds amongst habitats: Theory and relevance to conservation. *In* Bird population studies: Relevance to conservation and management. Oxford University Press, Oxford, pp. 317–345.

Best, L. B. (1986). Conservation tillage: Ecological traps for nesting birds? Wildlife Society Bulletin 14:308–317.

Blakesley, J. A., B. R. Noon, and D. R. Anderson (2005). Site occupancy, apparent survival, and reproduction of California spotted owls in relation to forest stand characteristics. Journal of Wildlife Management 69:1554–1564.

Block, W. M., and L. A. Brennan (1993). The habitat concept in ornithology. Current Ornithology 11:35–91.

Bock, C. E., and Z. F. Jones (2004). Avian habitat evaluation: Should counting birds count? Frontiers in Ecology and the Environment 2:403–410.

Boulinier, T., and E. Danchin (1997). The use of conspecific reproductive success for breeding patch selection in terrestrial migratory species. Evolutionary Ecology 11:505–517.

Boves, T. J., A. D. Rodewald, P. B. Wood, D. A. Buehler, J. L. Larkin, T. B. Wigley, and P. D. Keyser (2015). Habitat quality from individual-and population-level perspectives and implications for management. Wildlife Society Bulletin 39:443–447.

Boyce, M. S., and L. L. McDonald (1999). Relating populations to habitats using resource selection functions. Trends in Ecology and Evolution 14:268–272.

Bradbury, R. B., R. A. Hill, D. C. Mason, S. A. Hinsley, J. D. Wilson, H. Balzter, G. Q. A. Anderson, M. J. Whittingham, I. J. Davenport, and P. E. Bellamy (2005). Modelling relationships between birds and vegetation structure using airborne LiDAR data: A review with case studies from agricultural and woodland environments. Ibis 147:443–452.

Briant, G., V. Gond, and S. G. W. Laurance (2010). Habitat fragmentation and the desiccation of forest canopies: A case study from eastern Amazonia. Biological Conservation 143:2763–2769.

Brown, J. L. (1969). Territorial behavior and population regulation in birds. Wilson Bulletin 81:293–329.

Cagnacci, F., L. Boitani, R. A. Powell, and M. S. Boyce (2010). Animal ecology meets GPS-based radiotelemetry: A perfect storm of opportunities and challenges. Philosophical Transactions of the Royal Society of London B: Biological Sciences 365:2157–2162.

Cai, W., S. Borlace, M. Lengaigne, P. van Rensch, M. Collins, G. Vecchi, A. Timmermann, et al. (2014). Increasing frequency of extreme El Niño events due to greenhouse warming. Nature Climate Change 5:1–6.

Camaclang, A. E., M. Maron, T. G. Martin, and H. P. Possingham (2015). Current practices in the identification of critical habitat for threatened species. Conservation Biology 29:482–492.

Carrete, M., J. A. Sánchez-Zapata, J. L. Tella, J. M. Gil-Sánchez, and M. Moleón (2006). Components of breeding performance in two competing species: Habitat heterogeneity, individual quality and density-dependence. Oikos 112:680–690.

Carroll, C. (2010). Role of climatic niche models in focal-species-based conservation planning: Assessing potential effects of climate change on Northern Spotted Owl in the Pacific Northwest, USA. Biological Conservation 143:1432–1437.

Clawges, R., K. I. Vierling, L. Vierling, and E. Rowell (2008). The use of airborne LiDAR to assess avian species diversity, density, and occurrence in a pine/aspen forest. Remote Sensing of Environment 112:2064–2073.

Cody, M. L., Editor (1985). Habitat selection in birds. Academic Press, London.

Connell, J. H. (1983). On the prevalence and relative importance of interspecific competition: Evidence from field experiments. American Naturalist 122:661–696.

Conrey, R. Y., S. K. Skagen, A. A. Yackel Adams, and A. O Panjabi (2016). Extremes of heat, drought and precipitation depress reproductive performance in shortgrass prairie passerines. Ibis 158:614–629.

Cooper, R. J., and R. C. Whitmore (1990). Arthropod sampling methods in ornithology. Studies in Avian Biology 13:29–37.

Cooper, N. W., T. W. Sherry, and P. P. Marra (2015). Experimental reduction of winter food decreases body condition and delays migration in a long-distance migratory bird. Ecology 96:1933–1942.

Cooperrider, A. Y., R. J. Boyd, and H. R. Stuart, Editors (1986). Inventory and monitoring of wildlife habitat. US Department of the Interior, Bureau of Land Management, Denver, CO.

Coppedge, B. R., D. M. Engle, R. E. Masters, and M. S. Gregory (2001). Avian response to landscape change in fragmented southern Great Plains grasslands. Ecological Applications 11:47–59.

Coulson, J. C. (1968). Differences in the quality of birds nesting in the centre and on the edges of a colony. Nature 217:478–479.

Coulson, T., T. G. Benton, P. Lundberg, S. R. X. Dall, B. E. Kendall, and J.-M. Gaillard (2006). Estimating individual contributions to population growth: Evolutionary fitness in ecological time. Proceedings of the Royal Society of London Series B 273:547–555.

Culbert, P. D., V. C. Radeloff, V. St-Louis, C. H. Flather, C. D. Rittenhouse, T. P. Albright, and A. M. Pidgeon (2012). Modeling broad-scale patterns of avian species richness across the Midwestern United States with measures of satellite image texture. Remote Sensing of Environment 118:140–150.

Daily, G. C. (1997). Nature's services: Societal dependence on natural ecosystems. Island Press, Seattle.

Dall, S. R., L. A. Giraldeau, O. Olsson, J. M. McNamara, and D. W. Stephens (2005). Information and its use by animals in evolutionary ecology. Trends in Ecology and Evolution 20:187–193.

Davis, J. M., and J. A. Stamps (2004). The effect of natal experience on habitat preferences. Trends in Ecology and Evolution 19:411–416.

Davis, R. J., K. M. Dugger, S. Mohoric, L. Evers, and W. C. Aney (2011). Northwest Forest Plan—the first 15 years (1994 to 2008): Status and trend of Northern Spotted Owl populations and habitats. General Technical Report PNW-GTR-850. US Department of Agriculture, Forest Service, Pacific Northwest Research Station, Portland.

Doligez, B., C. Cadet, E. Danchin, and T. Boulinier (2003). When to use public information for breeding habitat selection? The role of environmental predictability and density dependence. Animal Behaviour 66:973–988.

Doligez, B., T. Pärt, E. Danchin, J. Clobert, and L. Gustafsson (2004). Availability and use of public information and conspecific density for settlement decisions in the Collared Flycatcher. Journal of Animal Ecology 73:75–87

Donovan, T. M., F. R. Thompson, J. Faaborg, and J. R. Probst (1995). Reproductive success of migratory birds in habitat sources and sinks. Conservation Biology 9:1380–1395.

Dow, H., and S. Fredga. 1983. Breeding and natal dispersal of the Goldeneye (*Bucephala clangula*). Journal of Animal Ecology 52:681–695.

Dugger, K. M., F. Wagner, R. G. Anthony, and G. S. Olson (2005). The relationship between habitat characteristics and demographic performance of Northern Spotted Owls in southern Oregon. The Condor 107:863–878.

Dugger, K. M., E. D. Forsman, A. B. Franklin, R. J. Davis, G. C. White, C. J. Schwarz, K. P. Burnham, et al. (2016). The effects of habitat, climate, and Barred Owls on long-term demography of Northern Spotted Owls. The Condor 118:57–116.

Dunk, J. R., B. Woodbridge, E. M. Glenn, R. J. Davis, K. Fitzgerald, P. Henson, D. W. LaPlante, et al. (2015). The scientific basis for modeling Northern Spotted Owl habitat: A response to Loehle, Irwin, Manly, and Merrill. Forest Ecology and Management 358:355–360.

Dunning, J. B., B. J. Danielson, and H. R. Pulliam (1992). Ecological processes that affect populations in complex landscapes. Oikos 65:169–175.

Elgar, M. A. (1989). Predator vigilance and group size in mammals and birds: A critical review of the empirical evidence. Biological Reviews 64:13–33.

Elith, J., C. H. Graham, R. P. Anderson, M. Dudik, S. Ferrier, A. Guisan, R. J. Hijmans, et al. (2006). Novel methods improve prediction of species' distributions from occurrence data. Ecography 2:129–151.

Elith, J., and J. R. Leathwick (2009). Species distribution models: Ecological explanation and prediction across space and time. Annual Review of Ecology, Evolution, and Systematics 40:677–697.

Elsen, P. R., and M. W. Tingley (2015). Global mountain topography and the fate of montane species under climate change. Nature Climate Change 5:772–776.

Fahrig, L. (2003). Effects of habitat fragmentation on biodiversity. Annual Review of Ecology, Evolution, and Systematics 34:487–515.

Fensham, R. J., and R. J. Fairfax (2002). Aerial photography for assessing vegetation change: A review of applications and the relevance of findings for Australian vegetation history. Australian Journal of Botany 50:415–429.

Ferraz, G., G. J. Russell, P. C. Stouffer, R. O. Bierregaard, S. L. Pimm, and T. E. Lovejoy (2003). Rates of species loss from Amazonian forest fragments. Proceedings of the National Academy of Sciences 100:14069–14073.

Ferrer, M., and J. A. Donazar (1996). Density-dependent fecundity by habitat heterogeneity in an increasing population of Spanish imperial eagles. Ecology 77:69–74.

Fletcher Jr., R. J. (2006). Emergent properties of conspecific attraction in fragmented landscapes. American Naturalist 168:207–219.

Foley, J. A., R. DeFries, G. P. Asner, C. Barford, G. Bonan, S. R. Carpenter, F. S. Chapin, et al. (2005). Global consequences of land use. Science 309:570–574.

Fontaine, J. J., and T. E. Martin (2006). Habitat selection responses of parents to offspring predation risk: An experimental test. American Naturalist 168:811–818.

Forbes, L. S., and G. W. Kaiser (1994). Habitat choice in breeding seabirds: When to cross the information barrier. Oikos 70:377–384.

Forman, R. T., and M. Godron (1986). Landscape ecology. John Wiley and Sons, New York.

Forsman, E. (2011). Population demography of Northern Spotted Owls: Published for the Cooper Ornithological Society. Vol. 40. University of California Press, Berkeley.

Forsman, E. D., E. C. Meslow, and H. M. Wight (1984). Distribution and biology of the Spotted Owl in Oregon. Wildlife Monographs 87:3–64.

Forsman, J. T., M. B. Hjernquist, and L. Gustafsson (2009). Experimental evidence for the use of density based interspecific social information in forest birds. Ecography 32:539–545.

Forsman, J. T., and T. E. Martin (2009). Habitat selection for parasite-free space by hosts of parasitic cowbirds. Oikos 118: 464–470.

Forsman, E. D., R. G. Anthony, K. M. Dugger, E. M. Glenn, A. B. Franklin, G. C. White, C. J. Schwarz, K. P. Burnham, D. R. Anderson, J. D. Nichols, J. E. Hines, J. B. Lint, R. J. Davis, S. H. Ackers, S. Andrews, B. L. Biswell, P. C. Carlson, L. V. Diller, S. A. Gremel, D. R. Herter, J. M. Higley, R. B. Horn, J. A. Reid, J. Rockweit, J. P. Schaberl, T. J. Snetisinger, and S. S. Sovern (2011). Population demography of northern spotted owls. Studies in Avian Biology No. 40. University of California Press, Berkeley, CA.

Forstmeier, W., and I. Weiss (2004). Adaptive plasticity in nest-site selection in response to changing predation risk. Oikos 104:487–499.

Franklin, A. B., D. R. Anderson, R. J. Gutiérrez, and K. P. Burnham (2000). Climate, habitat quality, and fitness in Northern Spotted Owl populations in northwestern California. Ecological Monographs 70:539–590.

Fretwell, S. D., and H. L. Lucas Jr. (1970). On territorial behavior and other factors influencing habitat distribution in birds. I: Theoretical development. Acta Biotheoretica 19:16–36.

Gaillard, J. M., M. Hebblewhite, A. Loison, M. Fuller, R. Powell, M. Basille, and B. Van Moorter (2010). Habitat-performance relationships: Finding the right metric at a given spatial scale. Philosophical Transactions of the Royal Society B: Biological Sciences 365:2255–2265.

Gardali, T., and A. L. Holmes (2011). Maximizing benefits from riparian revegetation efforts: Local- and landscape-level determinants of avian response. Environmental Management 48:28–37.

Gates, J. E., and L. W. Gysel (1978). Avian nest dispersion and fledging success in field-forest ecotones. Ecology 59:871–883.

Gauthreaux-Jr., S. A., and C. G. Belser (2003). Radar ornithology and biological conservation. The Auk 120:266–277.

Gavin, M. C., J. McCarter, A. Mead, F. Berkes, J. R. Stepp, D. Peterson, and R. Tang (2015). Defining biocultural approaches to conservation. Trends in Ecology and Evolution 30:140–145.

George, T. L., and D. S. Dobkin (2002). Effects of habitat fragmentation on birds in western landscapes: Contrasts with paradigms from the eastern United States. Studies in Avian Biology 25.

Gillespie, T. W., G. M. Foody, D. Rocchini, A. P. Giorgi, and S. Saatchi (2008). Measuring and modelling biodiversity from space. Progress in Physical Geography 32:203–221.

Glenn, E. M., R. G. Anthony, E. D. Forsman, and G. S. Olson (2011). Local weather, regional climate, and annual survival of the Northern Spotted Owl. The Condor 113:159–176.

Goetz, S. J., D. Steinberg, M. G. Betts, R. T. Holmes, P. J. Doran, R. Dubayah, and M. Hofton (2010). LiDAR remote sensing variables predict breeding habitat of a Neotropical migrant bird. Ecology 91:1569–1576.

Gómez-Baggethun, E., R. de Groot, P. Lomas, and C. Montes (2010). The history of ecosystem services in economic theory and practice: From early notions to markets and payment schemes. Ecological Economics 6:1209–1218.

Gómez-Laich, A., K. Yoda, C. Zavalaga, and F. Quintana (2015). Selfies of Imperial Cormorants (Phalacrocorax atriceps): What is happening underwater? PLoS ONE 10:e0136980.

Goss-Custard, J. D., R. W. G. Caldow, R. T. Clarke, and A. D. West (1995). Deriving population parameters from individual variations in foraging behaviour. II: Model tests and population parameters. Journal of Animal Ecology 64:277–289.

Gottschalk, T. K., F. Huettmann, and M. Ehlers (2005). Thirty years of analyzing and modelling avian habitat relationships using satellite imagery data: A review. International Journal of Remote Sensing 26:2631–2656.

Graber, J. W., and R. R. Graber (1983). Feeding rates of warblers in spring. The Condor 85:139–150.

Greenberg, R. (1995). Insectivorous migratory birds in tropical ecosystems: The breeding currency hypothesis. Journal of Avian Biology 26:260–264.

Greene, C. M., and J. A. Stamps (2001). Habitat selection at low population densities. Ecology 82:2091–2100.

Gregory, R. D., D. W. Gibbons, and P. F. Donald (2004). Bird census and survey techniques. In Bird ecology and conservation: A handbook of techniques, W. J. Sutherland, I. Newton, and R. E. Green, Editors. Cambridge University Press, Cambridge, UK, pp. 17–55.

Grinnell, J. (1917). The niche-relationships of the California Thrasher. The Auk 34:427–433.

Gullion, G. W. (1988). Aspen management for ruffed grouse. Economic and social development: A role for forests and forestry professionals. Proceedings of the Society of American Foresters National Convention (USA). October 18–21, 1987, Minneapolis, Minnesota, pp. 137–139.

Gup, T. (1990). Owl vs. man: Who gives a hoot? TIME magazine, accessed April 2016. http://content.time.com/time/magazine/article/0,9171,970447,00.html.

Gürlük, S., and E. Rehber (2008). A travel cost study to estimate recreational value for a bird refuge at Lake Manyas, Turkey. Journal of Environmental Management 88:1350–1360.

Gutiérrez, R. J., A. B. Franklin, and W. S. Lahaye (1995). Spotted Owl (*Strix occidentalis*). The Birds of North America Online, A. Poole, Editor. Cornell Lab of Ornithology, Ithaca, NY. Retrieved from the Birds of North America Online, http://bna .birds.cornell.edu/bna/species/179.

Hall, L. S., P. R. Krausman, and M. L. Morrison (1997). The habitat concept and a plea for standard terminology. Wildlife Society Bulletin 25:173–182.

Hall, L. S., and R. W. Mannan (1999). Multiscaled habitat selection by Elegant Trogons in southeastern Arizona. Journal of Wildlife Management 63:451–461.

Hallworth, M. T., T. S. Sillett, S. L. Van Wilgenburg, K. A. Hobson, and P. P. Marra (2015). Migratory connectivity of a Neotropical migratory songbird revealed by archival light-level geolocators. Ecological Applications 25:336–347.

Halpern, B. S., S. D. Gaines, and R. R. Warner (2005). Habitat size, recruitment, and longevity as factors limiting population size in stage-structured species. American Naturalist 165:82–94.

Haralick, R. M. (1979). Statistical and structural approaches to texture. Proceedings of the IEEE 67:786–804.

Haralick, R. M., K. Shanmugam, and I. H. Dinstein (1973). Textural features for image classification. IEEE Transactions on Systems, Man, and Cybernetics 3:610–621.

Higgins, K. F., K. J. Jenkins, G. K. Clambey, D. W. Uresk, D. E. Naugle, J. E. Norland, and W. T. Barker (2012). Techniques for wildlife investigations and management. In The Wildlife Techniques Manual, N. J. Silvy, Editor. Johns Hopkins University Press, Baltimore, MD, pp. 381–409.

Hildén, O. (1965). Habitat selection in birds: A review. Annales Zoologici Fennici 2:53–75. Finnish Zoological and Botanical Publishing Board.

Hilty, J., and A. Merenlender (2000). Faunal indicator taxa selection for monitoring ecosystem health. Biological Conservation 92:185–197.

Hobbs, N. T., and T. A. Hanley (1990). Habitat evaluation: Do use/availability data reflect carrying capacity? Journal of Wildlife Management 54:515–522.

Hoekstra, J. M., T. M. Boucher, T. H. Ricketts, and C. Roberts (2005). Confronting a biome crisis: Global disparities of habitat loss and protection. Ecology Letters 8:23–29.

Holmes, R. T., P. P. Marra, and T. W. Sherry (1996). Habitat-specific demography of breeding Black-throated Blue Warblers (*Dendroica caerulescens*): Implications for population dynamics. Journal of Animal Ecology 65:183–195.

Homer, C., J. Dewitz, L. Yang, S. Jin, P. Danielson, G. Xian, J. Coulston, N. Herold, J. Wickham, and K. Megown (2015). Completion of the 2011 National Land Cover database for the conterminous United States—representing a decade of land cover change information. Photogrammetric Engineering and Remote Sensing 81:345–354.

Homyack, J. A. (2010). Evaluating habitat quality of vertebrates using conservation physiology tools. Wildlife Research 37:332–342.

Hougner, C., J. Colding, and T. Söderqvist (2006). Economic valuation of a seed dispersal service in the Stockholm National Urban Park, Sweden. Ecological Economics 59:364–374.

Howell, C. A., S. C. Latta, T. M. Donovan, P. A. Porneluzi, G. R. Parks, and J. Faaborg (2000). Landscape effects mediate breeding bird abundance in Midwestern forests. Landscape Ecology 15:547–562.

Hutchinson, G. E. (1957). The multivariate niche. Cold Spring Harbor Symposia on Quantitative Biology 22:415–421.

Hutto, R. L. (1985). Habitat selection by nonbreeding, migratory land birds. In Habitat selection in birds, M. L. Cody, Editor. Academic Press, Orlando, FL, pp. 455–476.

Hyde, P., R. Dubayah, W. Walker, J. B. Blair, M. Hofton, and C. Hunsaker (2006). Mapping forest structure for wildlife habitat analysis using multi-sensor (LiDAR, SAR/InSAR, ETM+, Quickbird) synergy. Remote Sensing of Environment 102:63–73.

Imhoff, M. L., T. D. Sisk, A. Milne, G. Morgan, and T. Orr (1997). Remotely sensed indicators of habitat heterogeneity: Use of synthetic aperture radar in mapping vegetation structure and bird habitat. Remote Sensing of Environment 60:217–227.

James, F. C. (1971). Ordinations of habitat relationships among breeding birds. Wilson Bulletin 83:215–236.

James, F. C., and H. H. Shugart Jr. (1970). A quantitative method of habitat description. Audubon Field Notes 24:727–736.

James, F. C., and C. E. McCulloch (1990). Multivariate analysis in ecology and systematics: Panacea or Pandora's box? Annual Review of Ecology and Systematics 21:129–166.

Johnson, D. H. (1980). The comparison of usage and availability measurements for evaluating resource preference. Ecology 61: 65–71.

Johnson, M. D. (2000). Evaluation of an arthropod sampling technique for measuring food availability for forest insectivorous birds. Journal of Field Ornithology 71:88–109.

Johnson, M. D. (2007). Measuring habitat quality: A review. The Condor 109:489–504.

Johnson, M. D., T. W. Sherry, R. T. Holmes, and P. P. Marra (2006). Assessing habitat quality for a migratory songbird wintering in natural and agricultural habitats. Conservation Biology 20:1433–1444.

Johnson, M. D., J. L. Kellermann, and A. M. Stercho (2010). Pest control services by birds in shade and sun coffee in Jamaica. Animal Conservation 13:140–147.

Jones, J. (2001). Habitat selection studies in avian ecology: A critical review. The Auk 118:557–562.

Jones, K. R., J. E. Watson, H. P. Possingham, and C. J. Klein (2016). Incorporating climate change into spatial conservation prioritisation: A review. Biological Conservation 194:121–130.

Kareiva, P., and M. Marvier (2007). Conservation for the people. Scientific American 297:50–57.

Kareiva, P., and M. Marvier (2015). Conservation science: Balancing the needs of people and nature. 2nd ed. Roberts and Company, Englewood, CO.

Karr, J. R. (1980). History of the habitat concept in birds and the measurement of avian habitats. Acta XVII Congressus Internationalis Ornithologici, pp. 991–997.

Kellermann, J. L., M. D. Johnson, A. M. Stercho, and S. Hackett (2008). Ecological and economic services provided by birds on

Jamaican Blue Mountain coffee farms. Conservation Biology 22:1177–1185.

Kellermann, J. L., and C. van Riper III (2015). Detecting mismatches of bird migration stopover and tree phenology in response to changing climate. Oecologia 178:1227–1238.

Kerr, J. T., and M. Ostrovsky (2003). From space to species: Ecological applications for remote sensing. Trends in Ecology and Evolution 18:299–305.

Kivelä, S. M., J. T. Seppänen, O. Ovaskainen, B. Doligez, L. Gustafsson, M. Mönkkönen, and J. T. Forsman (2014). The past and the present in decision-making: The use of conspecific and heterospecific cues in nest site selection. Ecology 95:3428–3439.

Klopfer, P. (1963). Behavioral aspects of habitat selection: The role of early experience. Wilson Bulletin 75:15–22.

Knick, S. T., and J. T. Rotenberry (2000). Ghosts of habitats: Contributions of landscape change to current habitats used by shrubland birds. Ecology 81:220–227.

Koenig, W. D., J. M. H. Knops, W. J. Carmen, M. T. Stanback, and R. L. Mumme (1994). Estimating acorn crops using visual surveys. Canadian Journal of Forest Research 24:2105–2112.

Korpimaki, E. (1988). Effects of territory quality on occupancy, breeding performance and breeding dispersal in Tengmalm's owl. Journal of Animal Ecology 57:97–108.

Krebs, C. J. (2001). Ecology. 5th ed. Pearson/Benjamin-Cummings, San Francisco, CA.

Kristan III, W. B. (2003). The role of habitat selection behavior in population dynamics: Source–sink systems and ecological traps. Oikos 103:457–468.

Kristan III, W. B. (2006). Sources and expectations for hierarchical structure in bird-habitat associations. The Condor 108:5–12.

Kus, B. E. (1998). Use of restored riparian habitat by the endangered Least Bell's Vireo (*Vireo bellii pusillus*). Restoration Ecology 6:75–82.

La Sorte, F. A., W. M. Hochachka, A. Farnsworth, A. A. Dhondt, and D. Sheldon (2016). The implications of mid-latitude climate extremes for North American migratory bird populations. Ecosphere 7. doi:10.1002/ecs2.1261.

Lawler, J. J., and T. C. Edwards Jr. (2006). A variance-decomposition approach to investigating multiscale habitat associations. The Condor 108:47–58.

Lesak, A. A., V. C. Radeloff, T. J. Hawbaker, A. M. Pidgeon, T. Gobakken, and K. Contrucci (2010). Modeling forest songbird species richness using LiDAR-derived measures of forest structure. Remote Sensing of Environment 115:2823–2835.

Levin, S. A. (1992). The problem of pattern and scale in ecology: The Robert H. MacArthur award lecture. Ecology 73:1943–1967.

Lindstrom, Å. (1990). The role of predation risk in stopover habitat selection in migrating bramblings, *Fringilla montifringilla*. Behavioral Ecology 1:102–106.

Loehle, C., L. Irwin, B. F. Manly, and A. Merrill (2015). Range-wide analysis of northern spotted owl nesting habitat relations. Forest Ecology and Management 342:8–20.

MacArthur, R. H. (1958). Population ecology of some warblers of northeastern coniferous forests. Ecology 39:599–619.

MacArthur, R. H., and J. W. MacArthur (1961). On bird species diversity. Ecology 43:594–598.

MacArthur, R. H., J. W. MacArthur, and J. Preer (1962). On bird species diversity. II: Prediction of bird census from habitat measurements. American Naturalist 96:167–174.

MacKenzie, D. I., J. D. Nichols, M. E. Seamans, and R. J. Gutiérrez (2009). Modeling species occurrence dynamics with multiple states and imperfect detection. Ecology 90:823–835.

Manly, B. F., L. McDonald, D. Thomas, T. L. McDonald, and W. P. Erickson (2002). Resource selection by animals: Statistical design and analysis for field studies. 2nd ed. Springer Science and Business Media, New York.

Marra, P. P. (2000). The role of behavioral dominance in structuring patterns of habitat occupancy in a migrant bird during the nonbreeding season. Behavioral Ecology 11:299–308.

Marra, P. P., K. A. Hobson, and R. T. Holmes (1998). Linking winter and summer events in a migratory bird by using stable-carbon isotopes. Science 282:1884–1886.

Marra, P. P., E. B. Cohen, S. R. Loss, J. E. Rutter, and C. M. Tonra (2015a). A call for full annual cycle research in animal ecology. Biology Letters 11:2015.0552.

Marra, P. P., C. E. Studds, S. Wilson, T. S. Sillett, T. W. Sherry, and R. T. Holmes (2015b). Non-breeding season habitat quality mediates the strength of density-dependence for a migratory bird. Proceedings of the Royal Society B 282: 2015.0624.

Martin J., W. Kitchens, and J. Hines (2007). Natal location influences movement and survival of a spatially structured population of snail kites. Oecologia 153:291–301.

Martin, M., S. Newman, J. Aber, and R. Congalton (1998). Determining forest species composition using high spectral resolution remote sensing data. Remote Sensing of Environment 65:249–254.

Martin, P. R., and T. E. Martin (2001). Ecological and fitness consequences of species coexistence: A removal experiment with wood warblers. Ecology 82:189–206.

Martin, T. E., C. R. Paine, C. J. Conway, W. M. Hochachka, and W. Jenkins (1997). BBIRD Field Protocol. Montana Cooperative Wildlife Research Unit, University of Montana, Missoula. http://pica.wru.umt.edu/BBIRD/protocol/protocol.htm.

Martinuzzi, S. S., L. A. Vierling, W. A. Gould, M. J. Falkowski, J. S. Evans, A. T. Hudak, and K. T. Vierling (2009). Mapping snags and understory shrubs for a LiDAR-based assessment of wildlife habitat suitability. Remote Sensing of Environment 113:2533–2546.

Mayor, S. J., D. C. Schneider, J. A. Schaefer, and S. P. Mahoney (2009). Habitat selection at multiple scales. Ecoscience 16:238–247.

Mayr, E. (1982). Speciation and macroevolution. Evolution 36:1119–1132.

Mazaris, A. D., M. Papanikolaou, A. S. Barbet-Massin, F. Kallimanis, D. S. Jiguet, D. S. Schmeller, and J. D. Pantis (2013). Evaluating the connectivity of a protected areas' network under the prism of global change: The efficiency of the European Natura (2000) network for four birds of prey. PLoS ONE 8:e59640.

McCreedy, C., and C. van Riper III (2014). Drought-caused delay in nesting of Sonoran Desert birds and its facilitation of parasite- and predator-mediated variation in reproductive success. The Auk 132: 235–247.

McKechnie, A. E., and B. O. Wolf (2009). Climate change increases the likelihood of catastrophic avian mortality events during extreme heat waves. Biology Letters 6:253–256.

Mendenhall, C. D., D. S. Karp, C. F. Meyer, E. A. Hadly, and G. C. Daily (2014). Predicting biodiversity change and averting collapse in agricultural landscapes. Nature 509:213–217.

Mendes, L., T. Piersma, M. Lecoq, B. Spaans, and R. E Ricklefs (2005). Disease-limited distributions? Contrasts in the prevalence of avian malaria in shorebird species using marine and freshwater habitats. Oikos 109:396–404.

Meyer, C. B., S. L. Miller, and C. J. Ralph (2002). Multi-scale landscape and seascape patterns associated with Marbled Murrelet nesting areas on the US west coast. Landscape Ecology 17:95–115.

Millspaugh, J. J., D. C. Kesler, R. W. Kays, R. A. Gitzen, J. H. Schulz, C. T. Rota, C. M. B. Jachowski, J. L. Belant, and B. J. Keller (2012). Wildlife radiotelemetry and remote monitoring. Wildlife Techniques Manual 1:480–501.

Misenhelter, M. D., and J. T. Rotenberry (2000). Choices and consequences of habitat occupancy and nest site selection in Sage Sparrows. Ecology 81:2892–2901.

Mitchell, K. (2001). Quantitative analysis by the point-centered quarter method. Hobart and William Smith Colleges, people. hws.edu/mitchell/PCQM.pdf. Accessed October 15, 2011.

Møller, A. P., D. Rubolini, and E. Lehikoinen (2008). Populations of migratory bird species that did not show a phenological response to climate change are declining. Proceedings of the National Academy of Sciences USA 105:16195–200.

Mönkkönen, M., P. Helle, G. J. Niemi, and K. Montgomery (1997). Heterospecific attraction affects community structure and migrant abundances in northern breeding bird communities. Canadian Journal of Zoology 75:2077–2083.

Mönkkönen, M., and P. Reunanen (1999). On critical thresholds in landscape connectivity: a management perspective. Oikos 84:302–305.

Moore, F. R., and D. A. Aborn (2000). Mechanisms of en route habitat selection: How do migrants make habitat decisions during stopover? Studies in Avian Biology 20:34–42.

Moore, F. R., and W. Barrow (2000). Stopover ecology of Nearctic-Neotropical landbird migrants. Studies in Avian Biology 20. Cooper Ornithological Society.

Morgan, J. L., S. E. Gergel, and N. C. Coops (2010). Aerial photography: A rapidly evolving tool for ecological management. BioScience 60:47–59.

Morris, D. W. (1989). Density-dependent habitat selection: Testing the theory with fitness data. Evolutionary Ecology 3:80–94.

Morrison, M. L. (2001). A proposed research emphasis to overcome the limits of wildlife-habitat relationship studies. Journal of Wildlife Management 65:613–623.

Morrison, M. L., and L. S. Hall (2002). Standard terminology: Toward a common language to advance ecological understanding and application. In Predicting species occurrence: Issues of accuracy and scale, J. Scott, P. Heglund, M. Morrison, J. Haufler, M. Raphael, W. Wall, and F. Samson, Editors. Island Press, Covelo, CA, pp. 43–52.

Morrison, M. L., B. G. Marcot, and R. W Mannan (2002). Wildlife-habitat relationships: Concepts and applications. University of Wisconsin Press, Madison.

Morrison, M. L., B. Marcot, and W. Mannan (2012). Wildlife-habitat relationships: Concepts and applications. Island Press, Washington, DC.

Murphy, M. T. (2001). Habitat-specific demography of a long-distance, Neotropical migrant bird, the Eastern Kingbird. Ecology 82:1304–1318.

Murray, D. L. (2006). On improving telemetry-based survival estimation. Journal of Wildlife Management 70:1530–1543.

Nagendra, H. (2001). Using remote sensing to assess biodiversity. International Journal of Remote Sensing 22:2377–2400.

Neumann, B. C., K. J. Boyle, and K. P. Bell (2009). Property price effects of a national wildlife refuge: Great Meadows National Wildlife Refuge in Massachusetts. Land Use Policy 26:1011–1019.

Newton, I. (1998). Population limitation in birds. Academic Press, San Diego, CA.

Nice, Margaret Morse (1967). The watcher at the nest. Dover Publications, New York.

Norris, D. R., P. P. Marra, T. K. Kyser, T. W. Sherry, and L. M. Ratcliffe (2004). Tropical winter habitat limits reproductive success on the temperate breeding grounds in a migratory bird. Proceedings of the Royal Society of London B: Biological Sciences 271:59–64.

Nudds, T. D. (1977). Quantifying the vegetative structure of wildlife cover. Wildlife Society Bulletin 5:113–117.

O'Connell, A. F., J. D. Nichols, and K. U. Karanth, Editors (2010). Camera traps in animal ecology: Methods and analyses. Springer Science and Business Media, New York.

Orians, G. H., and J. F. Wittenberger (1991). Spatial and temporal scales in habitat selection. American Naturalist 137:S29–S49.

Palmeirim, J. M. (1988). Automatic mapping of avian species habitat using satellite imagery. Oikos 52:59–68.

Parker, G. A., and W. J. Sutherland (1986). Ideal free distributions when individuals differ in competitive ability: Phenotype-limited ideal free models. Animal Behaviour 34:1222–1242.

Parker, M. W., S. W. Kress, R. T. Golightly, H. R. Carter, E. B. Parsons, S. E. Schubel, J. A. Boyce, G. J. McChesney, and S. M. Wisely (2007). Assessment of social attraction techniques used to restore a Common Murre colony in central California. Waterbirds 30:17–28.

Parmesan, C., T. L. Root, and M. R. Willig (2000). Impacts of extreme weather and climate on terrestrial biota. Bulletin of the American Meteorological Society 81:443–450.

Perfecto, I., J. Vandermeer, and A. Wright (2009). Nature's matrix: Linking agriculture, conservation and food sovereignty. Earthscan Publishers, New York.

Persson, M. (2003). Habitat quality, breeding success and density in Tawny Owl Strix aluco. Ornis Svecica 13:137–143.

Petersen, S. (2002). Acting for endangered species: The statutory ark. University Press of Kansas, Lawrence.

Peterson, A. T. (2001). Predicting species' geographic distributions based on ecological niche modeling. The Condor 103:599–605.

Peterson, A. T., J. Soberón, and V. Sánchez-Cordero (1999). Conservatism of ecological niches in evolutionary time. Science 285:1265–1267.

Peterson, A. T., and J. Soberón (2012). Species distribution modeling and ecological niche modeling: Getting the concepts right. Natureza and Conservação 10:102–107.

Pettorelli, N., W. F. Laurance, T. G. O'Brien, M. Wegmann, H. Nagendra, and W. Turner (2014). Satellite remote sensing for applied ecologists: Opportunities and challenges. Journal of Applied Ecology 51:839–848.

Phillips, S. J., and M. Dudík (2004). A maximum entropy approach to species distribution modeling. In Proceedings of the Twenty-First International Conference on Machine Learning, Banff, Canada.

Phillips, S. J., R. P. Anderson, and R. E. Schapire (2006). Maximum entropy modeling of species geographic distributions. Ecological Modelling 190:231–259.

Piersma, T. (2012). What is habitat quality? Dissecting a research portfolio on shorebirds. In Birds and habitat: Relationships in changing landscapes, R. J. Fuller, Editor. Cambridge University Press, Cambridge, UK, pp. 383–407.

Porzig, E. L., N. E. Seavy, T. Gardali, G. R. Geupel, M. Holyoak, and J. M. Eadie (2014). Habitat suitability through time: Using time series and habitat models to understand changes in bird density. Ecosphere 5:1–16.

Potts, G. R., J. C. Coulson, and I. R. Deans (1980). Population dynamics and breeding success of the shag, Phalacrocorax aristotelis, on the Farne Islands, Northumberland. Journal of Animal Ecology 49:465–484.

Powell, R. A. (2000). Animal home ranges and territories and home range estimators. In Research techniques in animal ecology, L. Boitani and T. K. Fuller, Editors. Columbia University Press, New York, pp. 65–110.

Pressey, R. L., M. Cabeza, M. E. Watts, R. M. Cowling, and K. A. Wilson (2007). Conservation planning in a changing world. Trends in Ecology and Evolution 22:583–592.

Prince, K., and B. Zuckerberg (2015). Climate change in our backyards: The reshuffling of North America's winter bird communities. Global Change Biology 21:572–585.

Pulliam, H. R. (1988). Sources, sinks, and population regulation. American Naturalist 132:652–661.

Pulliam, H. R. (2000). On the relationship between niche and distribution. Ecology Letters 3:349–361.

Pulliam, H. R., and B. J. Danielson (1991). Sources, sinks, and habitat selection: A landscape perspective on population dynamics. American Naturalist S50–S66.

Railsback, S. F., H. B. Stauffer, and B. C. Harvey (2003). What can habitat preference models tell us? Tests using a virtual trout population. Ecological Applications 13:1580–1594.

Railsback, S. F., and M. D. Johnson (2014). Effects of land use on bird populations and pest control services on coffee farms. Proceedings of the National Academy of Sciences 111:6109–6114.

Ralph, C. J., S. Droege, and J. R. Sauer (1995). Managing and monitoring birds using point counts: standards and applications.

In Monitoring bird populations by point counts, C. J. Ralph, J. R. Sauer, and S. Droege, Editors. U.S. Forest Service General Technical Report PSW-GTR-149, pp. 161–168.

Rapport, D. J. (1991). Myths in the foundations of economics and ecology. Biological Journal of the Linnean Society 44:185–202.

Reed, J. M., T. Boulinier, E. Danchin, and L. W. Oring (1999). Informed dispersal. Current Ornithology 15:189–259.

Rittenhouse, C. D., A. M. Pidgeon, T. P. Albright, P. D. Culbert, M. K. Clayton, C. H. Flather, C. Huang, J. G. Masek, and V. C. Radeloff (2010). Avifauna response to hurricanes: Regional changes in community similarity. Global Change Biology 16:905–917.

Robbins, J. (2014) Paying farmers to welcome birds. New York Times, April 14.

Robel, R. J., J. N. Briggs, A. D. Dayton, and L. C. Hulbert (1970). Relationships between visual obstruction measurements and weight of grassland vegetation. Journal of Range Management 23:295–297.

Robertson, B. A., and R. L. Hutto (2006). A framework for understanding ecological traps and an evaluation of existing evidence. Ecology 87:1075–1085.

Rodenhouse, N. L., T. W. Sherry, and R. T. Holmes (1997). Site-dependent regulation of population size: A new synthesis. Ecology 78:2025–2042.

Rodenhouse, N. L., T. S. Sillett, P. J. Doran, and R. T. Holmes (2003). Multiple density–dependence mechanisms regulate a migratory bird population during the breeding season. Proceedings of the Royal Society of London B: Biological Sciences 270:2105–2110.

Rodewald, A. (2015). Demographic consequences of habitat. In Wildlife habitat conservation: Concepts, challenges, and solutions, M. L. Morrison and H. A. Mathewson, Editors. Johns Hopkins University Press, Baltimore, MD, pp. 19–23.

Rohwer, S. (2004). Using age ratios to infer survival and despotic breeding dispersal in hybridizing warblers. Ecology 85:423–431.

Romesburg, H. C. (1981). Wildlife science: Gaining reliable knowledge. Journal of Wildlife Management 45:293–313.

Rosenzweig, M. L. (1981). A theory of habitat selection. Ecology 62:327–335.

Rotenberry, J. T. (1981). Why measure bird habitat? The use of multivariate statistics in studies of wildlife habitat. US Forest Service, General Technical Report, RM-87, pp. 29–32.

Rotenberry, J. T. (1985). The role of habitat in avian community composition: Physiognomy or floristics? Oecologia 67:213–217.

Roth, K. L., D. A. Roberts, P. E. Dennison, S. H. Peterson, and M. Alonzo (2015). The impact of spatial resolution on the classification of plant species and functional types within imaging spectrometer data. Remote Sensing of Environment 171:45–57.

Royle, J. A., D. K. Dawson, and S. Bates (2004). Modeling abundance effects in distance sampling. Ecology 85:1591–1597.

Ruhl, J. B. (2007) Reconstructing the wall of virtue: Maxims for the co-evolution of environmental law and environmental science. Environmental Law 37:1063–1082.

Runge, J. P., M. C. Runge, and J. D. Nichols (2006). The role of local populations within a landscape context: Defining and classifying sources and sinks. American Naturalist 167:925–938.

Runge, M. C., and P. P. Marra. (2005). Modeling seasonal interactions in the population dynamics of migratory birds. *In* Birds of two worlds: The ecology and evolution of migration, R. Greenberg and P. Marra, Editors. Johns Hopkins University Press, Baltimore, MD, pp. 375–389.

Saab, V. (1999). Importance of spatial scale to habitat use by breeding birds in riparian forests: A hierarchical analysis. Ecological Applications 9:135–151.

Sakai, H. F., and B. R. Noon (1993). Dusky-footed woodrat abundance in different-aged forests in northwestern California. Journal Wildlife Management 57:373–382.

Sanz, Â., J. Potti, J. Moreno, S. Merino, J. Jose, Â. Abascal, and E.- Madrid (2003). Climate change and fitness components of a migratory bird breeding in the Mediterranean region. Global Change Biology 9:461–472.

Schmidt, K. A., S. R. X. Dall, and J. A. Van Gils (2010). The ecology of information: An overview on the ecological significance of making informed decisions. Oikos 119:304–316.

Schumaker, N. H., A. Brookes, J. R. Dunk, B. Woodbridge, J. A. Heinrichs, J. Lawler, C. Carroll, and D. LaPlante (2014). Mapping sources, sinks and connectivity using a simulation model of Northern Spotted Owls. Landscape Ecology 29:579–592.

Scott, J. M., P. J. Heglund, M. L. Morrison, J. B. Haufler, M. G. Raphael, W. A. Wall, and F. B. Samson (2002). Predicting species occurrences: Issues of scale and accuracy. Island Press, Washington, DC.

Seaman, D. A., C. G. Guglielmo, R. W. Elner, and T. D. Williams (2006). Landscape-scale physiology: Site differences in refueling rates indicated by plasma metabolite analysis in free-living migratory sandpipers. The Auk 123:563–574.

Seavy, N. E., and C.A. Howell (2010). How can we improve delivery of decision support tools for conservation and restoration? Biodiversity and Conservation 19:1261–1267.

Seavy, N. E., T. Gardali, G. H. Golet, D. Jongsomjit, R. Kelsey, S. Matsumoto, S. Paine, and D. Stralberg (2012). Integrating avian habitat distribution models into a conservation planning framework for the San Joaquin River, California, USA. Natural Areas Journal 32:420–426.

Seppänen, J.-T., J. T. Forsman, M. Mönkkönen, and R. L. Thomson (2007). Social information use is a process across time, space, and ecology, reaching heterospecifics. Ecology 88:1622–1633.

Sergio, F., and I. Newton (2003). Occupancy as a measure of habitat quality. Journal of Animal Ecology 72:857–865.

Sherry, T. W., and R. T. Holmes (1988). Habitat selection by breeding American Redstarts in response to a dominant competitor, the Least Flycatcher. The Auk 105:350–364.

Sherry, T. W., and R. T. Holmes (1995). Summer versus winter limitation of populations: What are the issues and what is the evidence. *In* Ecology and management of Neotropical migratory birds, T. E. Martin and D. M. Finch, Editors. Oxford University Press, New York, pp. 85–120.

Sherry, T. W., and R. T. Holmes (1996). Winter habitat quality, population limitation, and conservation of Neotropical-Nearctic migrant birds. Ecology 77:36–48.

Shochat, E., Z. Abramsky, B. Pinshow, and M. E. A. Whitehouse (2002). Density-dependent habitat selection in migratory passerines during stopover: What causes the deviation from IFD? Evolutionary Ecology 16:469–488.

Sillett, T. S., and R. T. Holmes (2002). Variation in survivorship of a migratory songbird throughout its annual cycle. Journal of Animal Ecology 71:296–308.

Sogge, M. K., and R. M. Marshall (2000). A survey of current breeding habitats. *In* Status, ecology, and conservation of the Southwestern Willow Flycatcher, D. M. Finch and S. H. Stoleson, Editors. General Technical Report RMRS-GTR-60. Ogden, UT: US Department of Agriculture, Forest Service, Rocky Mountain Research Station, pp. 43–56.

Stamps, J. A. (1991). The effect of conspecifics on habitat selection in territorial species. Behavioral Ecology and Sociobiology 28:29–36.

Stamps, J., and V. V. Krishnan (2005). Nonintuitive cue use in habitat selection. Ecology 86:2860–2867.

Stamps, J. A., V. V. Krishnan, and M. L. Reid (2005). Search costs and habitat selection by dispersers. Ecology 86:510–518.

Stauffer, D. F. (2002). Linking populations and habitats: Where have we been, where are we going? *In* Predicting species occurrence: Issues of accuracy and scale, J. Scott, P. Heglund, M. Morrison, J. Haufler, M. Raphael, W. Wall, and F. Samson, Editors. Island Press, Covelo, CA, pp. 53–62.

Steffen, W., Å. Persson, L. Deutsch, J. Zalasiewicz, M. Williams, K. Richardson, C. Crumley, et al. (2011). The Anthropocene: From global change to planetary stewardship. Ambio 40:739–761.

Stephens, P. A., and W. J. Sutherland (1999). Consequences of the Allee effect for behaviour, ecology and conservation. Trends in Ecology and Evolution 14:401–405.

St-Louis, V., A. M. Pidgeon, M. K. Clayton, B. A. Locke, D. Bash, and V. C. Radeloff (2009). Satellite image texture and a vegetation index predict avian biodiversity in the Chihuahuan desert of New Mexico. Ecography 32:468–480.

St-Louis, V., A. M. Pidgeon, T. Kuemmerle, R. Sonnenschein, V. C. Radeloff, M. K. Clayton, B. A. Locke, D. Bash, and P. Hostert (2014). Modelling avian biodiversity using raw, unclassified satellite imagery. Philosophical Transactions of the Royal Society B: Biological Sciences 369:2013.0197.

St-Louis, V., A. M. Pidgeon, V. C. Radeloff, T. J. Hawbaker, and M. K. Clayton (2006). High-resolution image texture as a predictor of bird species richness. Remote Sensing of Environment 105:299–312.

Stralberg, D., D. Jongsomjit, C. A. Howell, M. A. Snyder, J. D. Alexander, J. A. Wiens, and T. L. Root (2009a). Re-shuffling of species with climate disruption: A no-analog future for California birds? PLoS ONE 4:e6825.

Stralberg, D., K. E. Fehring, L. Y. Pomara, N. Nur, N. D. B. Adams, D. Hatch, G. R. Geupel, and S. Allen (2009b). Modeling nest-site occurrence for the Northern Spotted Owl at its southern range limit in central California. Landscape and Urban Planning 90:76–85.

Stralberg, D., E. M. Bayne, S. G. Cumming, P. Sólymos, S. J. Song, and F. K. Schmiegelow (2015). Conservation of future boreal forest bird communities considering lags in vegetation response to climate change: A modified refugia approach. Diversity and Distributions 21:1112–1128.

Streby, H. M., G. R. Kramer, S. M. Peterson, J. A. Lehman, D. A. Buehler, and D. E. Andersen (2015). Tornadic storm avoidance behavior in breeding songbirds. Current Biology 25:98–102.

Streby, H. M., J. P. Loegering, and D. E. Andersen (2012). Spot-mapping underestimates song-territory size and use of mature forest by breeding Golden-winged Warblers in Minnesota, USA. Wildlife Society Bulletin 36:40–46.

Strong, A. M., and T. W. Sherry (2001). Body condition of Swainson's Warblers wintering in Jamaica and the conservation value of Caribbean dry forests. Wilson Bulletin 113:410–418.

Sutherland, W. J. (1998). The effect of local change in habitat quality on populations of migratory species. Journal of Applied Ecology 35:418–421.

Sutherland, W. J., and G. A. Parker (1985). Distribution of unequal competitors. In Behavioural ecology: Ecological consequences of adaptive behavior, R. M Sibly and R. H. Smith, Editors. Blackwell Scientific, Oxford, UK, pp. 255–273.

Switzer, P. V. (1993). Site fidelity in predictable and unpredictable habitats. Evolutionary Ecology 7:533–555.

Tershy, B. R., C. J. Donlan, B. Keitt, D. Croll, J. A. Sanchez, B. Wood, M. A. Hermosillo, and G. Howald (2002). Island conservation in Northwest Mexico: A conservation model integrating research, education and exotic mammal eradication. In Turning the tide: The eradication of invasive species, C. R. Veitch and M. N. Clout, Editors. Auckland, New Zealand, Invasive Species Specialist Group of the World Conservation Union (IUCN), pp. 293–300.

Thomas, D. L., and E. J. Taylor (1990). Study designs and tests for comparing resource use and availability. Journal of Wildlife Management 54:322–330.

Toral, G. M., D. Aragonés, J. Bustamante, and J. Figuerola (2011). Using Landsat images to map habitat availability for waterbirds in rice fields. Ibis 153:684–694.

Travis, J. M. J. (2003). Climate change and habitat destruction: A deadly anthropogenic cocktail. Proceedings of the Royal Society of London B: Biological Sciences 270:467–473.

Tregenza, T. (1995). Building on the ideal free distribution. Advances in Ecological Research 26:253–307.

Turner, M. G. (1989). Landscape ecology: The effect of pattern on process. Annual Review of Ecology and Systematics 20:171–197.

Turner, W., S. Spector, N. Gardiner, M. Fladeland, E. Sterling, and M. Steininger (2003). Remote sensing for biodiversity science and conservation. Trends in Ecology and Evolution 18:306–314.

Tuttle, E. M., R. R. Jensen, V. A. Formica, and R. A. Gonser (2006). Using remote sensing image texture to study habitat use patterns: A case study using the polymorphic White-throated Sparrow (Zonotrichia albicollis). Global Ecology and Biogeography 15:349–357.

US Fish and Wildlife Service (2011). Revised recovery plan for the Northern Spotted Owl (Strix occidentalis caurina). US Fish and Wildlife Service, Portland, OR.

US Fish and Wildlife Service (2012). Designation of revised critical habitat for the Northern Spotted Owl, final rule. Federal Register 77:71876–72068.

Valone, T. J. (2007). From eavesdropping on performance to copying the behavior of others: A review of public information use. Behavioral Ecology and Sociobiology 62:1–14.

Van Horne, B. (1983). Density as a misleading indicator of habitat quality. Journal of Wildlife Management 47:893–901.

Velarde, E., and D. W. Anderson (1994). Conservation and management of seabird islands in the Gulf of California: Setbacks and successes. In Seabirds on islands: Threats, case studies and action plans, D. N. Nettleship, J. Burger, and M. Gochfeld, Editors. Birdlife International, Cambridge, UK, pp. 229–243.

Velarde, E., E. Ezcurra, and D. W. Anderson (2013). Seabird diets provide early warning of sardine fishery declines in the Gulf of California. Scientific Reports 3:1332.

Veloz, S. D., N. Nur, L. Salas, D. Jongsomjit, J. Wood, D. Stralberg, and G. Ballard (2013). Modeling climate change impacts on tidal marsh birds: Restoration and conservation planning in the face of uncertainty. Ecosphere 4:1–25.

Verner, J., M. L. Morrison, and C. J. Ralph (1986). Wildlife 2000: Modelling habitat relationships of terrestrial vertebrates. University of Wisconsin Press, Madison.

Verner, J., M. L. Morrison, and C. J. Ralph (2000). Modeling habitat relationships of terrestrial vertebrates. University of Wisconsin Press, Madison.

Vierling, K. T., L. A. Vierling, W. A. Gould, S. Martinuzzi, and R. M. Clawges (2008). LiDAR: Shedding new light on habitat characterization and modeling. Frontiers in Ecology and the Environment 6:90–98.

Vierling, L. A., K. T. Vierling, P. Adam, and A. T. Hudak (2013). Using satellite and airborne LiDAR to model woodpecker habitat occupancy at the landscape scale. PLoS ONE 8:e80988. doi:10.1371/journal.pone.0080988.

Virkkala, R. (1990). Ecology of the Siberian Tit Parus cinctus in relation to habitat quality: Effects of forest management. Ornis Scandinavica 21:139–146.

Vitousek, P. M., H. A. Mooney, J. Lubchenco, and J. M. Melillo (1997). Human domination of Earth's ecosystems. Science 277:494–499.

Vitz, A. C., and A. D. Rodewald (2006). Can regenerating clearcuts benefit mature-forest songbirds? An examination of post-breeding ecology. Biological Conservation 127:477–486.

Vitz, A. C., and A. D. Rodewald (2007). Vegetative and fruit resources as determinants of habitat use by mature-forest birds during the postbreeding period. The Auk 124:494–507.

Vitz, A. C., and A. D. Rodewald (2011). Influence of condition and habitat use on survival of post-fledging songbirds. The Condor 113:400–411.

Walker, B. L., D. E. Naugle, and K. E. Doherty (2007). Greater sage-grouse population response to energy development and habitat loss. Journal of Wildlife Management 71:2644–2654.

Wallis, C. I. B., D. Paulsch, J. Zeilinger, B. Silva, G. F. Curatola Fernández, R. Brandl, N. Farwig, and J. Bendix (2016). Contrasting performance of LiDAR and optical texture models in predicting avian diversity in a tropical mountain forest. Remote Sensing of Environment 174:223–232.

Walther, G. R., E. Post, P. Convey, A. Menzel, C. Parmesank, T. J. C. Beebee, J. M. Fromentin, O. H. I. F. Bairlein, C. Parmesan, and O. Hoegh-Guldberg (2002). Ecological responses to recent climate change. Nature 416:389–395.

Ward, M. P., and S. Schlossberg (2004). Conspecific attraction and the conservation of territorial songbirds. Conservation Biology 18:519–525.

Warren, D. L., and S. N. Seifert (2011). Ecological niche modeling in Maxent: The importance of model complexity and the performance of model selection criteria. Ecological Applications 21:335–342.

Weldon, A. J. (2006). How corridors reduce Indigo Bunting nest success. Conservation Biology 20:1300–1305.

Weldon, A. J., and N. M. Haddad (2005). The effects of patch shape on Indigo Buntings: Evidence for an ecological trap. Ecology 86:1422–1431.

Wenny, D., T. L. DeVault, M. D. Johnson, D. Kelly, Ç. H. Şekercioğlu, D. F. Tomback, and C. J. Whelan (2011). The need to quantify ecosystem services provided by birds. The Auk 128:1–14.

Whelan, C. J., D. G. Wenny, and R. J. Marquis (2008). Ecosystem services provided by birds. Annals of the New York Academy of Sciences 1134:25–60.

White, G. C., and K. P. Burnham (1999). Program MARK: Survival estimation from populations of marked animals. Bird Study 46:S120–S139.

Whittaker, R. H., S. A. Levin, and R. B. Root (1973). Niche, habitat, and ecotone. American Naturalist 107:321–338.

Wiens, J. A. (1985). Habitat selection in variable environments: Shrub-steppe birds. In Habitat selection in birds, M. L. Cody, Editors. Academic Press, New York, pp. 227–251.

Wiens, J. A. (1989). Spatial scaling in ecology. Functional Ecology 3:385–397.

Wiens, J. A. (1992). The ecology of bird communities. Vol. 1. Cambridge University Press, Cambridge, UK.

Wiens, J. A., and J. T. Rotenberry (1981). Habitat associations and community structure of birds in shrubsteppe environments. Ecological Monographs 51:21–42.

Wiens, J. D., R. G. Anthony, and E. D. Forsman (2014). Competitive interactions and resource partitioning between Northern Spotted Owls and Barred Owls in western Oregon. Wildlife Monographs 185:1–50.

Wilson, K. A., M. F. McBride, M. Bode, and H. P. Possingham (2006). Prioritizing global conservation efforts. Nature 440:337–340.

Wood, E. M., M. D. Johnson, and B. A. Garrison (2011). Quantifying bird-habitat using a relevé method. Transactions of the Western Section of the Wildlife Society 46:25–41.

Wood, E. M., A. M. Pidgeon, V. C. Radeloff, and N. S. Keuler (2012). Image texture as a remotely sensed measure of vegetation structure. Remote Sensing of Environment 121:516–526.

Wood, E. M., A. M. Pidgeon, V. C. Radeloff, and N. S. Keuler (2013). Image texture predicts avian density and species richness. PLoS ONE 8:e63211. doi:10.1371/journal.pone.0063211.

Wood, E. M., and A. M. Pidgeon (2015). Extreme variations in spring temperature affect ecosystem regulating services provided by birds during migration. Ecosphere 6. http://dx.doi.org /10.1890/ES15–00397.1.

Wood, E. M., S. E. B. Swarthout, W. M. Hochachka, J. L. Larkin, R. W. Rohrbaugh, K. V. Rosenberg, and A. D. Rodewald (2016). Intermediate habitat associations by hybrids may facilitate genetic introgression in a songbird. Journal of Avian Biology 47:508–520.

Yaffee, S. L. (1994). The wisdom of the spotted owl: Policy lessons for a new century. Island Press, Washington, DC.

Ydenberg, R. C., R. W. Butler, D. B. Lank, B. D. Smith, and J. Ireland (2004). Western Sandpipers have altered migration tactics as Peregrine Falcon populations have recovered. Proceedings of the Royal Society of London B: Biological Sciences 271:1263–1269.

Zabel, C. J., K. McKelvey, and J. P. Ward Jr. (1995). Influence of primary prey on home-range size and habitat-use patterns of Northern Spotted Owls (Strix occidentalis caurina). Canadian Journal of Zoology 73:433–439.

Zuckerberg, B., A. M. Woods, and W. F. Porter (2009). Poleward shifts in breeding bird distributions in New York State. Global Change Biology 15:1866–1883.

Birds on the Move
Ecology of Migration and Dispersal

Alice Boyle

Large-scale movements of birds are among the most impressive behaviors in the whole animal kingdom. Seasonal appearances and disappearances of birds have captured human imagination for at least as long as our written history. In Europe, observations of swallows foraging over lakes and ponds led to the belief that they burrowed into the mud under water during winter months (fig. 19.1). Other birds were believed to hibernate as many mammals do, and the replacement of breeding redstarts by overwintering robins in the same areas was thought to represent a "transmutation" or seasonal change in appearance of the same species (Aristotle). Ancient Mayans associated "rivers" of hawks passing over Central America as harbingers of the rainy season (Bassie-Sweet 2008). In the eighteenth century, naturalists proposed that birds went to the moon when they disappeared from Europe in winter. All of these observations instead represent large-scale movements of birds.

Movement is fundamental to interpreting most other aspects of avian biology because flight—the principal form of avian locomotion—is extremely costly (chapters 7 and 10). Thus, allocating energy and time to major movements can constrain other parts of a bird's life cycle, such as reproduction (chapter 16) and molt (chapter 9). The distance that birds fly influences the evolution of morphological traits such as wing and feather shape (chapter 6). The ability to fly also means that birds are among the most mobile animals in the world. Being able to fly means birds are better at crossing geographic barriers such as oceans and mountain ranges than many other taxa, permitting them to colonize every continent on Earth (chapter 4). Birds also take advantage of resources found at the highest latitudes and altitudes on earth, in part because of their ability to migrate seasonally away from those extreme environments.

Spectacular feats of long-distance movement are the most impressive of bird movements, but they represent only the most extreme examples of common behaviors. Movement types include foraging and other local movements within home ranges; one-way directional movements that we call **dispersal**; and cyclical, annual, return movements that we call **migration**. We refer to **nomadism** when birds engage in multiple, sequential dispersal events, and to **irruptions** when migratory tendency or distance varies a lot from year to year. Each of these movement types can occur over a wide span of distances, ranging from only a few tens of meters to thousands of kilometers.

While birds' mobility allows them to take advantage of most of the earth's surface, that same mobility imposes costs not experienced by more sedentary taxa. Some costs are a direct result of long movements—running out of energy while crossing inhospitable terrain, or colliding with human-made structures during nocturnal flights. Some costs are the indirect result of movement—risks associated with finding food and avoiding predators in unfamiliar sites en route, time spent in movement that precludes other activities, or lingering consequences of movement that affect future health and survival. Adaptations that enable a mobile lifestyle come at

Figure 19.1. Woodcut depicting fishermen hauling in "hibernating swallows" from a lake—an erroneous explanation proposed to explain the seasonal disappearances of birds, which is actually caused by their migration. From Olaus 1996.

evolutionary costs as well—investment in traits that enable long-distance movement constrains investment in clutch size, brain size, and behavioral flexibility. Determining the fitness costs and benefits of alternative strategies is essential to understanding why species exhibit different patterns of movement and why individuals sometimes engage in different movement strategies at different times of their lives. Studying bird migration and dispersal is difficult because the movements themselves take our study subjects away from earthbound researchers. However, technological advances are improving our understanding of movement patterns, making it an exciting time to study movement ecology.

In the pages ahead, we explore the incredible variety of patterns, distances, and gradients over which birds move. We then delve into migration and dispersal in greater detail, in each section highlighting studies investigating proximate and ultimate causes and ecological and evolutionary consequences of such movements. Final sections provide an overview of the major methods ornithologists have used to study migration, and the unique threats birds experience as a consequence of their mobility.

A TAXONOMY OF AVIAN MOVEMENTS

The diversity of animal movements (and indeed, even the dispersal of "sedentary" organisms like trees and fungi) can be understood as the outcome of differences in the animal's internal state that affect its motivation to move, navigational capacity, locomotion capacity, and the animal's external environmental circumstances (Nathan et al. 2008). These factors ultimately determine the multiple types, scales, modes, and directions of movement a bird makes over its lifetime. Even the most sedentary of birds make local movements to find food, commute to roosting sites, and find mates (fig. 19.2). Local movements have not been as well-studied as other avian movements, but as technology increases our ability to document movement at fine temporal and spatial scales, detailed movement tracks of foraging and "ranging" behavior now allow us to test hypotheses explaining variation in those movements (box on page 615). For small land birds, foraging movements may cover only a few hectares, but for others, notably pelagic seabirds such as Short-tailed Shearwaters (*Ardenna tenuirostris*), foraging trips routinely traverse more than 1,000 kilometers (Einoder et al. 2011). Even in species that do not travel long distances to find food, unpredictable excursions outside local areas probably occur more often than is generally appreciated. For instance, birds make long forays to avoid bad weather; in April 2013, Golden-winged Warblers apparently retraced 700 km of their migration journey only days after spring arrival just before severe storms hit their breeding sites (Streby et al. 2015). Birds can use long-distance forays to exploit ephemeral food resources or to gather information about the location of potential mates or high quality habitat. That information may later be used to guide disper-

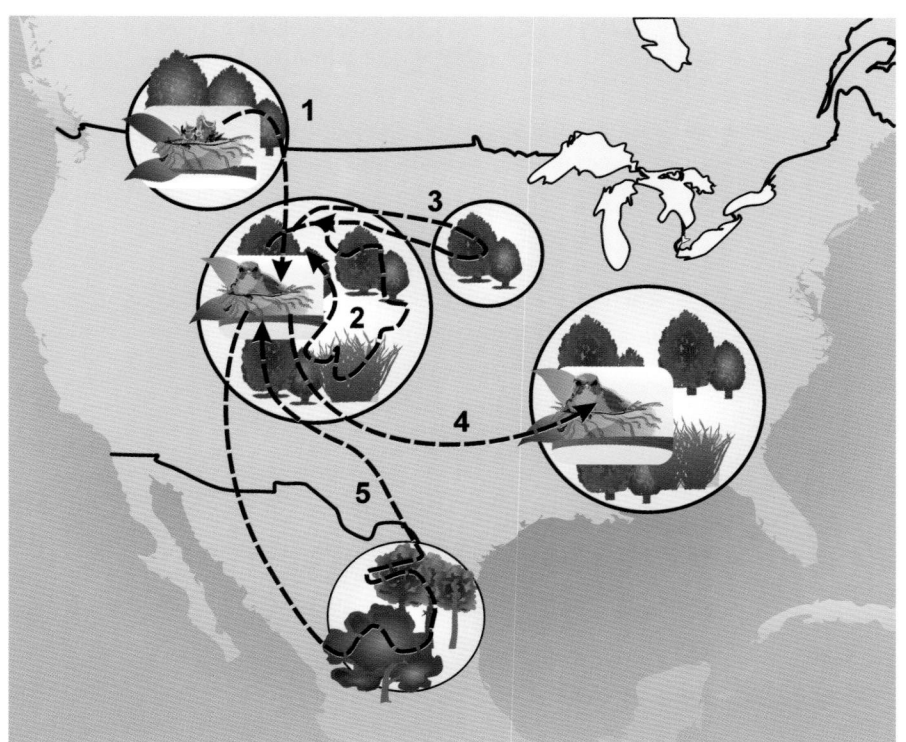

Figure 19.2. Most birds make many types of movement within their lifetimes. (*1*) Young birds disperse from their natal site to the site where they first reproduce (**natal dispersal**). (*2*) All birds make movements within their home ranges to forage, seek mates, thermoregulate, and avoid predation. (*3*) Sometimes birds make longer foraging excursions or **forays** well beyond the limits of their normal home range. (*4*) Although many species will breed in the same locations year after year, some species engage in **breeding dispersal**, which is a one-way movement to a new home range. (*5*) Seasonal, return movements between discrete breeding and nonbreeding areas are referred to as **migrations** in the ornithological literature.

TOOLS TO ANSWER THE AGE-OLD QUESTION: WHY DID THE CHICKEN CROSS THE ROAD?

By Virginia Winder

As the price and size of radio and satellite transmitters have decreased, ornithologists have been able to collect data on fine-scale movements of individual birds. Movement data are useful in linking habitat selection to demographic rates and generating vital information to inform conservation decisions. The outline of a home range provides information on *where* an animal is estimated to have been located, but space use within a home range is rarely uniform. Rather, some areas of the home range are heavily used, while others are infrequently used. Resource utilization functions (RUFs) allow researchers to link space use within an animal's home range to landscape features that can help determine *why* animals use certain areas more frequently than others (Marzluff et al. 2004). The analytical process includes four steps: (1) generate a home range polygon using movement data; (2) generate

a utilization distribution or "heat map," representing differential space use within the home range; (3) determine the landscape conditions at each cell within the home range for key resources hypothesized to predict habitat selection decisions; and (4) relate the intensity of space use to resource values on a cell-by-cell basis. We used RUFs to investigate how the installation of wind turbines affected space use decisions of female Greater Prairie Chickens (*Tympanuchus cupido*) in Kansas (Winder et al. 2014). Following construction of the wind energy facility, females avoided areas near wind turbines within their home ranges, even though they had shown no avoidance of eventual turbine sites prior to construction. Long-term persistence of avoidance behavior could result in extirpation of prairie-chickens from this site (see figure).

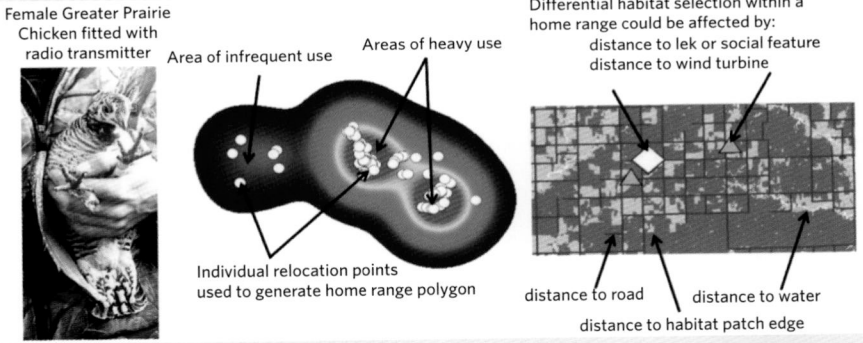

On the *left*, a Greater Prairie Chicken is fitted with a radio transmitter. This threatened species is dependent on tallgrass prairie of the Great Plains. The *center* image represents the modeled space use of birds based on points of known occurrence (white dots), determined using telemetry. The warmer the color, the more frequently birds are predicted to use those areas. By overlaying such "heat maps" on the map on the *right*, we can relate hot spots of space use to landscape features such as patches of native grassland (green) and nonhabitat (agricultural fields, beige), lek sites (orange triangles), roads (black lines), and wind turbines (white diamonds). *Images © Virginia Winder.*

sal decisions. Some longer-range forays blur the boundaries between migration and dispersal—two immensely important types of movement for birds—which are the topics of the rest of this chapter.

Dispersal is a one-way movement that takes a bird permanently away from its previous home range, and is the only movement that results in gene flow between populations. Virtually all species engage in **natal dispersal**, a one-way move-

ment from the location of the nests birds fledge from to the site where they first reproduce. Some birds also engage in **breeding dispersal**, a one-way movement that takes place between nesting bouts. Both types of dispersal can result in species-level range shifts (often, confusingly, called "migration" in other branches of biology; box on page 616). Key attributes of avian **migratory** behavior involve annual, cyclical, to-and-fro movements between different breeding and nonbreeding ranges, with some element of predictability in the direction of movements and the timing of departures and returns. When all individuals in a species engage in similar, often innately controlled movements, those migration or dispersal strategies are often treated as species-level traits. However, within species, different populations and individuals often vary greatly in their movements. **Partially migratory** species or populations comprise some individuals that migrate and others that do not. **Differential migrants** are species in which identifiable

groups of individuals differ in migratory distance or timing, or both (Terrill and Able 1988). Even though annual, cyclical migrations can span very large spatial scales, migrants often have low rates of breeding dispersal, resulting in high **site fidelity**, meaning they return to the same site (or even the same territory) in successive years. Hence, migration does not necessarily result in gene flow, unlike dispersal movements.

Understanding many aspects of migration and dispersal ecology requires that we study the behavioral choices made

WHEN DISPERSAL CAN'T KEEP PACE WITH CLIMATE CHANGE

Assisted "migration" (or colonization) is human-assisted establishment of populations outside their current range. It is a highly controversial conservation tool that is becoming increasingly relevant for species endangered by shrinking areas of suitable climatic conditions and those unable to disperse to the new areas where those same climates will occur in future decades. Land managers help populations disperse using the same methods as they do in reintroduction programs, except with added risks because the species being moved was never historically part of the community to which they are introduced. Because birds are relatively good dispersers, they have less need of assisted migration than do less mobile organisms. However, birds like the endan-

gered Hihi or Stitchbird (*Notiomystis cincta*) may require such dramatic intervention in order to survive (Chauvenet et al. 2013). The Hihi was extirpated on all but one tiny island adjacent to the North Island of New Zealand and has subsequently been reintroduced to a handful of other small islands and sanctuaries. Because climate drives population dynamics of this species, and climate change is pushing suitable habitat southward, the species will likely not persist without translocations to safe sites where climates are expected to be appropriate well into the future. In the case of the Hihi, survival might be dependent on translocation to New Zealand's South Island where they were not found historically (see figure).

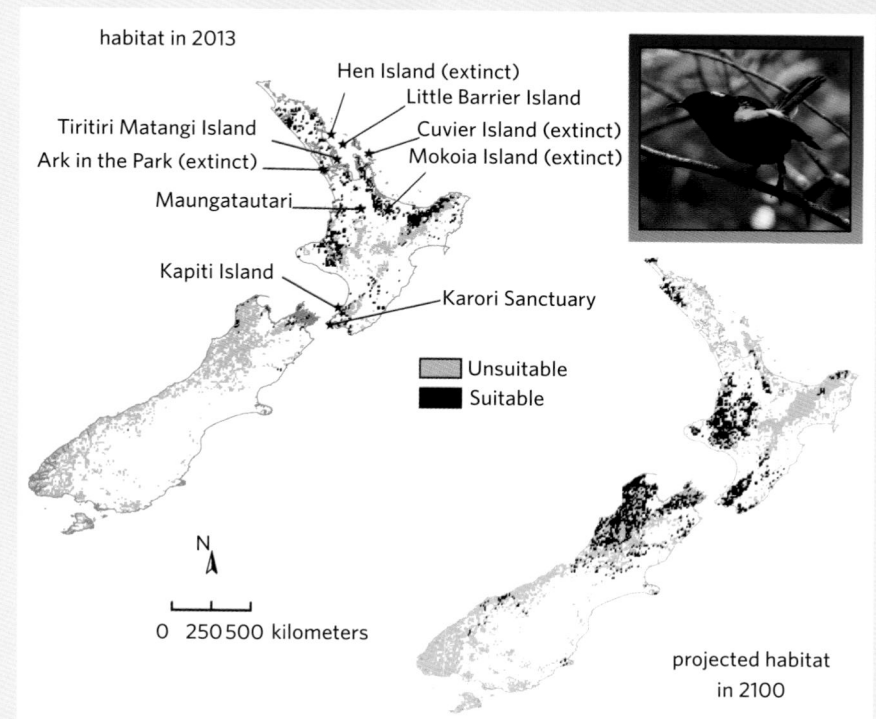

As with many native New Zealand birds, the Hihi (inset) was driven nearly to extinction by predators that accompanied European settlers. They now exist in the wild in only five of nine historic populations (starred locations). Black shading in upper left map depicts current distribution of suitable habitat, and gray shading depicts other native forests that are currently unsuitable because of climate. The lower right map depicts the projected distribution of suitable climatic conditions in the year 2100 under a "middle of the road" climate change scenario. Very little suitable habitat is projected to remain by 2100 in areas where the bird currently persists, and more suitable habitat may be found in the South Island, well beyond the dispersal capabilities of this species. *Hihi image © David A. Rintoul. Maps reproduced with permission from Alienor Chavenet and John Wiley & Sons, Inc.*

by individuals. Indeed, one definition of migratory movements is based on behavioral and physiological syndromes. Dingle (1996) defines migration as "persistent, straightened-out movement effected by the animal's own locomotory exertions . . . [that] depends on some temporary inhibition of station-keeping responses, but promotes their eventual disinhibition and recurrence." In other words, migrations differ from regular day-to-day movements by being more directional, and, while migrating, animals temporarily ignore cues (such as encountering a rich food source) that would normally cause them to stop moving. In birds, long-distance migrants almost certainly exemplify such behaviors in spades. Nevertheless, the application of Dingle's criteria to classify movement in wild animals is challenging because we lack detailed knowledge of the patterns and mechanistic control of most movements. Furthermore, by emphasizing attributes of the movement path and responsiveness to cues, this definition does not distinguish between one-way (i.e., dispersal) and round-trip movements (i.e., migration).

The differences between migration and dispersal are important to the study of ecology, evolution, and conservation. Scaling up from individuals to populations, a key feature of migration to avian population biology lies in the sometimes huge mortality costs; vast numbers of birds die while migrating. In contrast, dispersal is central to population biology because it is the mechanism by which individuals move in and out of populations and colonize new areas. Population biologists need to know how many birds that disappear from a population dispersed (i.e., emigrated) away, rather than died, and the extent to which the influx of dispersing (i.e., immigrating) individuals "rescue" that population demographically. Scaling up from populations to species, we encounter new definitions and reasons to differentiate between migration and dispersal. Selection for dispersal behavior can occur on different traits and for different reasons than selection for migratory behavior. Species can be highly **philopatric** (the propensity for young to return to breed for the first time at their natal site—the inverse of natal dispersal) or have high **site fidelity** (returning in successive years to the same breeding site—the inverse of breeding dispersal), yet migrate thousands of kilometers between breeding seasons. Conversely, species can forgo round-trip migrations but exhibit low site fidelity. It is impossible to infer anything about dispersal from looking at a species' range map, but range maps do tell us something about migration when breeding and nonbreeding ranges do not fully overlap—birds must be moving between areas seasonally for such geographic patterns to emerge. Classification of movement at broad geographic scales and at higher levels of biological organization becomes both more difficult and less accurate. Do we classify species as migrants or nonmigrants when migratory movements occur entirely within their year-round distribution? Such challenges are typical when applying categorical labels to traits that vary continuously in nature. Categories can be useful, but movements are flexible, individual-level behaviors, and researchers must apply criteria to distinguish between behaviors that make the most sense in the context of their scientific questions.

MIGRATION

Patterns

How common is migration? Estimates of the proportion of the world's bird species making migrations range from ~20 percent to over 50 percent (Fretwell 1972, Berthold 2001, Kirby et al. 2008). But those estimates represent minima for two reasons. First, the lower estimates are based on range-level characterizations (such as nonoverlapping breeding and wintering ranges), which eliminates all but the longest-distance, complete migrants. Second, the movements of many species are completely unknown. Thus, global estimates and most macroecological studies of migration do not include short-distance and partial migrants that constitute a large fraction of migratory bird diversity (Berthold 2001). The prevalence of migration varies geographically. In both the Old and New World, the proportion of migratory species increases steadily from subtropical latitudes toward the arctic (Newton and Dale 1996a, Newton and Dale 1996b, Kuo et al. 2013). Interestingly, the same patterns are not as strong in the Southern Hemisphere. Migration in southern continents (**austral migration**) tends to involve fewer complete, obligate, latitudinal migrant species, more partial migrants, and more complex spatial patterns (Chesser 1994; fig. 19.3). These hemispheric differences have been attributed to smaller landmasses of Southern Hemisphere continents, geographic idiosyncrasies of barriers and continental shape, and a stronger influence of precipitation on climatic seasonality in southern than in northern regions. Migrants are drawn from most major clades of extant birds, and most avian orders contain both migratory and nonmigratory representatives. However, certain clades, such as the New World Warblers (Parulidae) or shorebirds and plovers (Scolopacidae and Charadriidae) contain a preponderance of migrants, while others, such as the antbirds (Thamnophilidae) or the ratites are primarily nonmigratory.

Bird migrations traverse many types of gradients. Latitudinal (i.e., north–south) gradients are the most common, but birds also migrate along elevational gradients. **Altitudinal (or elevational) migrants** move up and down between breeding and nonbreeding ranges along the slopes of all major

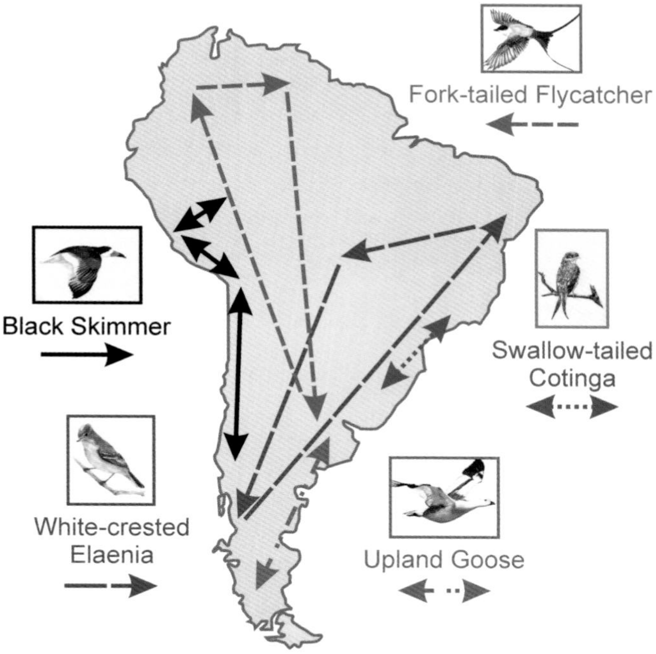

Figure 19.3. Examples of diverse patterns of seasonal migration within South America. References for each species are as follows: Fork-tailed Flycatcher (*Tyrannus savanna*, Jahn et al. 2013); Black Skimmer (*Rynchops niger*, Davenport et al. 2016); Swallow-tailed Cotinga (*Phibalura flavirostris*, Snow et al. 2016); White-crested Elaenia (*Elaenia albiceps*, Bravo and Cueto 2015); Upland Goose (*Chloephaga picta*, Carboneras and Kirwan 2016). *Map by Alex E. Jahn; drawings by Caitlin Wiechman.*

mountain ranges on every continent except Antarctica. In a few regions, ornithologists have estimated the percentage of species in a community that migrate altitudinally; those estimates are typically around 10–20 percent (Stiles and Skutch 1989, Johnson and Maclean 1994, Nocedal 1994, Burgess and Mlingwa 2000, Bildstein 2004, Boyle 2017). Rarely, birds migrate over longitudinal gradients, especially in regions with strong east–west precipitation gradients, such as in central North America (Prairie Falcons, *Falco mexicanus*), or in subtropical South America (Golden-rumped Euphonias, *Euphonia cyanocephala*; Areta and Bodrati 2010). The distance that birds migrate varies hugely, regardless of the gradients they traverse. In particular, some latitudinal migrants make migrations of as little as tens or hundreds of kilometers such as Lewis's Woodpecker (*Melanerpes lewis*). Others, such as Bar-tailed Godwits (*Limosa lapponica*), migrate tens of thousands of kilometers in nonstop oceanic flights of nine days. When breeding and wintering distributions overlap, determining what individuals do in the zone of overlap is not straightforward: Are some birds year-round residents, while the birds that breed at high latitudes migrate past them to lower lat-

itudes in what is known as **leapfrog migration?** Or do all populations shift synchronously toward lower latitudes or altitudes, known as **chain migrations?** Both patterns occur. For instance, tundra-breeding Peregrine Falcons (*Falco peregrinus*) migrate to southern South America, passing by shorter-distance migrants and residents of the same species along the way. In contrast, the most northerly-breeding Dunlin (*Calidris alpina*) spend their winters at more northerly wintering locations than do Dunlins breeding farther south, resulting in a chain migration pattern (fig. 19.4A). The extent to which breeding populations of migrants intermix with others during the nonbreeding period is termed **migratory connectivity**; when all individuals of a population overwinter together in the same region, and each population of the species occupies different wintering areas, the populations have very high connectivity, whereas when breeding populations intermix completed during winter, they have low connectivity (Boulet and Norris 2006) (fig. 19.4B). Knowledge of migratory connectivity is crucial for understanding how events that occur during one season influence migrants in later season (termed **carryover effects**), and for implementing **full-life-cycle conservation** (Marra et al. 2015).

Species in which all individuals migrate every year are considered **complete migrants**. In contrast, species of **partial migrants** comprise both migrants and year-round residents. Variation in migratory behavior can occur within populations (i.e., some individuals migrate, others remain resident), or at the population level (i.e., some populations are completely migratory, others are completely resident). Even within complete migrants, age or sex classes often vary in the distance they migrate; this is called **differential migration**. For instance, female Western Sandpipers travel farther south than do males, and within sexes, younger birds winter farther from the breeding grounds than do adults (Nebel et al. 2002). Differential migration is often associated with **protandry**— the tendency of males to return to breeding grounds earlier than do females. Protandry is common in male-territorial species, in which a male's reproductive success depends on the quality of the territory he can acquire and defend (chapter 16). Multiple variations on migratory behavior often occur within a single species. The common and widespread American Robin (which gets its scientific name *Turdus migratorius* for its migratory behavior) comprises long-distance latitudinal migrants that breed in Canada, altitudinal migrants that breed in mountains, and residents breeding in Mexico. Some robin populations are completely migratory, others are partially migratory, others do not migrate at all, and migration distances range from tens to thousands of kilometers.

Despite all the complexity and variation, latitudinal migrants often follow major "flyways" (fig. 19.5), which are

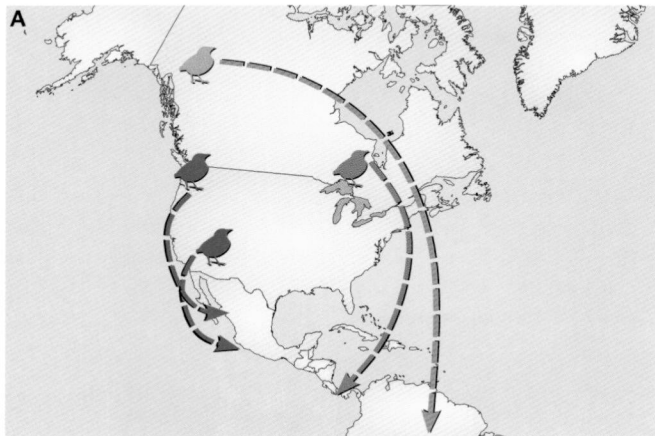

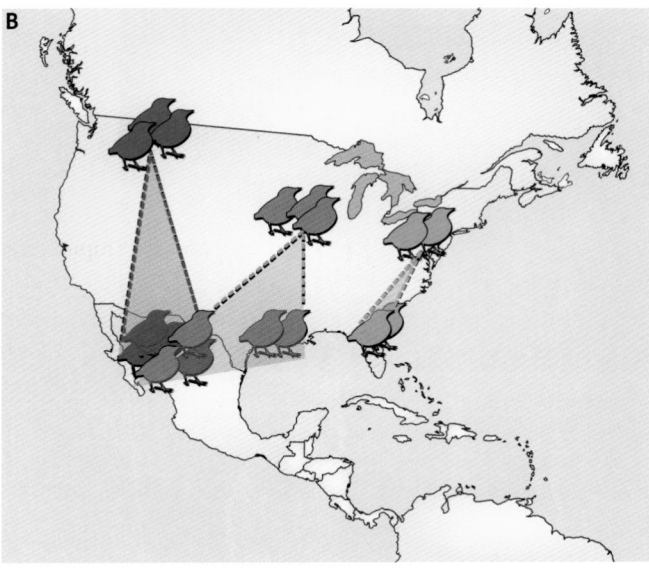

Figure 19.4. A, Different breeding populations of the same species (e.g., blue birds) sometimes all migrate similar distances, with northern-breeding birds spending the nonbreeding seasons at more northerly wintering locations than southern-breeding birds. Such patterns of migration are termed chain migrations. Alternatively, birds can exhibit "leapfrog" migration patterns in which a more northern-breeding species or population (e.g., orange birds) migrate over both the breeding and wintering locations of more southern-breeding species or populations (e.g., green birds). *B,* The degree of migratory connectivity refers to the extent to which different breeding populations mix during the winter in their nonbreeding range. The orange breeding populations in this figure display high migratory connectivity—all birds migrate to the same wintering areas. Severe habitat loss to that one wintering site could potentially wipe out the whole breeding population in the orange part of the range. By contrast, the green breeding populations display weak connectivity—the birds breeding in one area spread out over multiple wintering areas. The blue breeding populations are intermediate. They spread out a bit on their wintering grounds, and mix with the green birds in some parts of their nonbreeding range.

equivalent to superhighways used by large numbers of individuals and species when crossing continents. Flyways are most clearly defined where routes are constrained by geographic features. For example, the mountain ranges running down the spine of the Americas create uplift exploited by a river of migrating raptors each fall and spring that attracted the attention of the ancient Mayans. The location of mid-continental wetlands also concentrate waterfowl into crucial **stopover** areas where birds refuel mid-journey. Some long-distance migrants spend over 30 percent of their lives migrating, and of that time, they can spend 70 percent at stopover sites, where they must avoid dying from predation, weather, or starvation; refuel to successfully complete their migrations; and do all this on a tight time budget (Hedenstrom and Alerstam 1997, Sillett and Holmes 2002). Small land birds funnel through land bridges linking Europe and Africa, such as the Strait of Gibraltar and routes along the coasts of Israel and Egypt. By doing so, birds avoid the risks of open-water Mediterranean crossings. The direction of prevailing winds affects migration routes and can lead to different northward and southward patterns. Tiny, 12-gram Blackpoll Warblers (*Setophaga striata*) travel north in spring through forests of eastern North America to breed in Canada's boreal forests. Remarkably, their southward fall journeys trace a vast arc over the western Atlantic. Birds depart from the Maritime provinces and New England and do not make landfall until reaching the northeast coast of South America, exploiting northeasterly trade winds during the latter part of this journey (DeLuca et al. 2015). Such **loop migration** patterns are common, and can reflect seasonal and spatial differences in stopover habitat quality (such as food availability) as well as in-flight environmental conditions.

Proximate Causes and Control of Migration

Migration is controlled in different ways in different species. In some lineages, migration is culturally transmitted. Most waterbirds learn their migration routes from their elders—a fact that makes their migratory behavior very flexible in the face of changing environmental conditions. However, cultural transmission means that migratory behavior can be lost during severe population bottlenecks. One such loss occurred in the past century: all Whooping Cranes (*Grus americana*) today are descendants of only 15 migratory individuals alive in 1942. Through intensive captive-breeding, the species has recovered to a few hundred individuals. But when this species was first reintroduced to eastern North America, young cranes had to be trained to follow costumed caretakers in ultralight aircraft, eventually being led across the continent on their first migrations (Urbanek et al. 2010; fig. 19.6). Other types of **facultative migrations** do not involve cultural

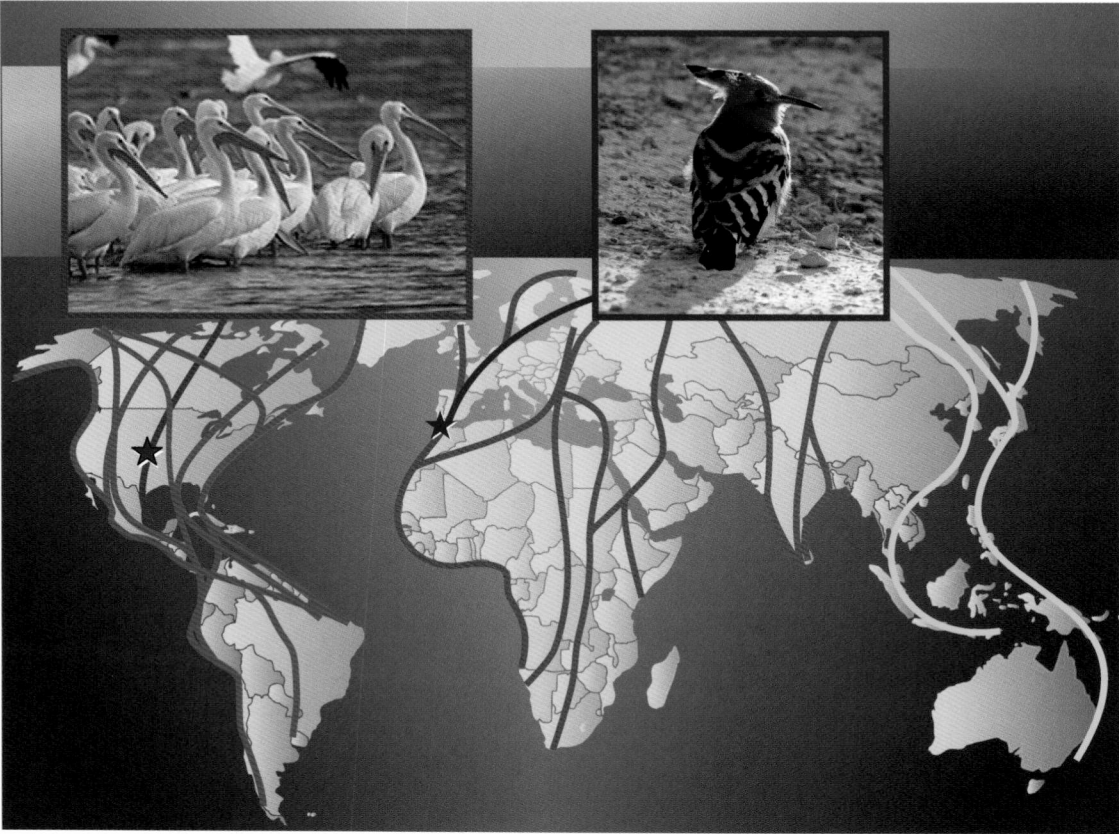

Figure 19.5. Examples of some of the major migratory flyways worldwide. Waterbirds such as American White Pelicans (*Pelecanus erythrorhynchos*, left) use isolated wetlands in the Great Plains as stopover sites in the central flyway (dark orange star). Hoopoes (*Upupa epops*, right), like many other European birds, converge on narrow land bridges or geographic features (dark purple star) that offer short overwater routes around the Mediterranean to reach wintering areas in Africa. *Photo of American White Pelicans by Mark Herse; Hoopoe image © 2005 David A. Rintoul.*

transmission, but are similar to the above example in that they involve flexible responses to interacting sets of external and internal cues.

Obligate migrations are controlled endogenously via genetic programs. This means that young birds raised from the egg in the lab can successfully complete their first migratory journey without ever interacting with conspecifics. Indeed, the adults of many long-distance migrant songbirds and shorebirds depart breeding areas well before their offspring, leaving young to complete development and make their first migrations alone, often during nocturnal flights. Given the appropriate environmental cues, obligate migrants are hard-wired to migrate every year. Apparently, the most efficient way to package the complex information required to time migration and navigate over unknown territory is in a hard-wired genetic program. Such programs control integrated sets of physiological and behavioral adaptations that are turned on and off twice each year. In a series of classic experiments on Eurasian Blackcaps (*Sylvia atricapilla*), Ber-

thold (box on page 622) demonstrated that migration can be a heritable behavior controlled by multiple genes. First, he brought individuals from completely migratory and completely resident populations into a "common garden." Birds from the migratory population exhibited migratory tendencies (measured via expression of pre-migratory restlessness or *Zugunruhe*; box on page 623) and those from the resident population did not. In doing so, he showed that environmental conditions are not the proximate cause of differences in migratory behavior in Blackcaps. Then, Berthold cross-fostered eggs from migratory individuals into the nests of residents and vice versa. Offspring of migratory parents still exhibited migratory tendencies, demonstrating that the causes of their behavioral differences were genetic. Berthold then interbred migrants and residents, creating a population that exhibited intermediate migratory tendencies. He performed a selection experiment, mating the most migratory individuals with each other and the least migratory together. In only six generations, he had fully migratory

Figure 19.6. Captive-reared Whooping Cranes learning their migration routes from human trainers in ultralight aircraft. *Photo by H. Ray, Operation Migration USA Inc.*

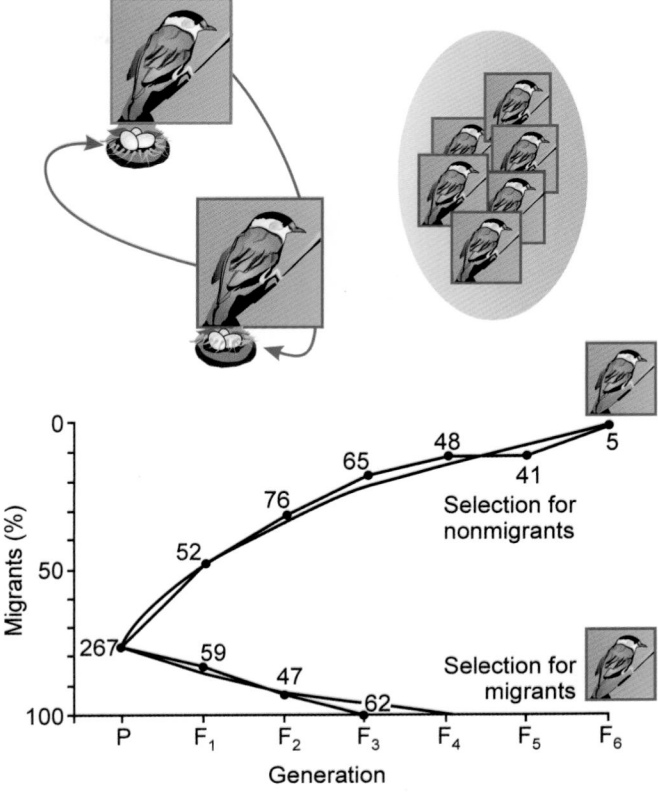

Figure 19.7. Some key experiments and results demonstrating the genetic basis for migratory behavior. Upper left, eggs from an obligate migrant population of Blackcaps (blue boxes) were reared by birds in a nonmigratory population (red boxes), and offspring exhibited the migratory tendencies of their biological rather than their adoptive parents. Upper right, birds from both migrant and nonmigrant populations were bred and raised in a common aviary, and offspring exhibited ancestral migrant tendencies. The lower panel depicts the outcome of selective breeding from a pool of hybrid individuals (crosses between migrant and resident populations) resulting quickly in fully migratory and resident lines. *Selection figure reproduced from Berthold et al. 1990 with permission from Springer.*

and fully resident populations once again (Berthold 1999; fig. 19.7).

What exactly does a genetically controlled migratory "program" consist of? It is the interrelated navigational components controlling timing, distance, and direction. The correct **migratory phenology**, or timing of migratory journeys, is critical to a migrant's fitness. Departing temperate breeding grounds too late puts birds at risk from possible death during early winter storms (Wellicome et al. 2014). Departing wintering areas too late can mean a late arrival to breeding grounds, compromising reproductive success (box on page 628). Correct timing also operates at finer time-scales, such as deciding when to depart stopover areas for risky legs of migratory journeys. Traveling the correct distance involves ending migrations at the right time. This does not merely involve stopping flying, because most birds must stop regularly to refuel along the way. Migration requires, by Dingle's definition, that birds stop ignoring the cues that would regularly induce them to resume "station-keeping" behavior. Ending migratory journeys involves responding to those cues once again. Migrating in the correct direction is of course key to going the correct distance, and requires that birds possess an internal compass. Navigational systems can involve serial sets of directions and distances can create an endogenous flight plan, but to integrate that plan, birds must also possess a refined map sense and a well-tuned internal clock. We know that birds use multiple sources of information to determine their locations, correctly sense direction, and regularly calibrate their navigation systems. These mechanisms include the ability to see and interpret information from polarized light and sun angles, sense geomagnetic sensitivities, and use celestial navigation, as well as possessing an accurate circannual clock and long-term

spatial memory (chapter 12). The extent to which different species use different navigational senses, and how birds integrate disparate sources of information, remains an area of active research (Alerstam 2006). The topic of avian navigation and orientation is a large and fascinating field of study (Wiltschko and Wiltschko 2003, Wiltschko and Wiltschko 2009, Deutschlander and Beason 2014, Holland 2014, Willemoes et al. 2015); unfortunately, a detailed discussion of this subject is beyond the scope of this chapter.

Beyond determining when and where to go, genetically controlled migratory programs also must include mechanisms that physiologically prepare birds for long flights and regulate and optimize moment-to-moment responses to internal and

DR. PETER BERTHOLD

Professor Dr. Peter Berthold. *Photo* © *Dr. Berthold.*

Dr. Peter Berthold is a German ornithologist who, as a boy, became fascinated with birds, observing them with binoculars left over from World War II. By the age of 16, he had joined the German Ornithological Society; he proceeded eventually to a PhD at the University of Tubingen, embarking formally on the study of bird migration. He spent his long and illustrious career at the Vogelwarte Radolfzell and is best known for his groundbreaking research on the genetic basis for migratory behavior. This work involved, at one time, immense aviaries in which Berthold kept up to 2,000 Blackcaps. Berthold has been incredibly prolific, writing dozens of books and ~450 scientific articles on many topics in avian biology. His *Bird Migration: A General Survey* (Berthold 2001) synthesizes much of his life's research and learning on migration. In recent years, he has devoted a good deal of his attention to conservation (see figure).

external information. Birds are sensitive to environmental cues that trigger neuroendocrine pathways, which, in turn, result in suites of physiological and behavioral pathways (Ramenofsky et al. 2012). In obligate migrants, the cues that trigger migratory behaviors are rarely the same as the factors that ultimately shape those migrations. Obligate migrants typically depart before experiencing the severe winter weather or food shortages that can act as ultimate selective factors shaping migratory behavior. Consequently, effective proximate cues convey reliable information about future en-

vironmental conditions. Changes in day length are the most important of such cues. The hypothalamic-pituitary-gonadal axis (HPG) responds to changing photoperiod (chapter 8), and captive migrants housed under naturally varying photoperiod increase their *Zugunruhe* and fat deposition during the seasons in which they would normally migrate (chapter 7). Because many events in the annual cycle are controlled by photoperiod, the nature and sensitivity of a bird's photoperiod response is dependent on both completion of previous events (e.g., breeding) and time of year, which birds monitor via an internal **circannual** clock. Although other exogenous cues tend to provide less-reliable information about future conditions, they nevertheless modulate responses of obligate migrants. Examples of alternate cues include changes in food, social conditions, or weather. Responses to these less-predictable cues are mediated via a different neuroendocrine pathway, the hypothalamic-pituitary-adrenal axis (HPA). The proximate causes and mechanisms controlling facultative migrations are not well understood, but likely, the proximate cues and ultimate processes are more tightly coupled in facultative migrants than in obligate migrants (Ramenofsky et al. 2012).

Facultative and obligate migration can be viewed as extremes along a continuum of varying dependence on innate and flexible programs, both employing a combination of cues and pathways for responding adaptively to environmental variation. Obligate migrants combine strongly directional, repeatable, consistently timed phases of their migration with less predictable, facultative phases. Young obligate migrant birds apparently learn during their first journeys and make more accurate directional decisions as they age. Even in Whooping Cranes, some elements of the migratory program are probably innate, because after their first ultralight-guided southbound trip, groups of juveniles departed on their northward spring migration at the correct time without human assistance. At the same time, facultative migrants respond to changing photoperiods via the HPG in many of the same ways that obligate migrants do. Thus, obligate and facultative migrants may differ primarily in how sensitive they are to environmental stimuli; possibly obligate migrants have lower response thresholds to internal or external cues that trigger migration. In wild birds, this threshold model would translate into a continuously varying ratio of obligate to facultative phases in a bird's migratory pathway (Newton 2012).

Although much remains to be learned about the control of irruptive migrations, we know that several different control mechanisms operate, and that the relative importance of those mechanisms varies among species. The consistent, predictable part of movements includes southward fall movements and northward spring movements each year. However,

HOW DO WE STUDY MIGRATION IN THE LAB? EMLEN FUNNELS AND BEYOND

Peter Berthold's classic experiments and many other important breakthroughs in migration ecology would not have been possible by only studying migration in the wild. Yet when we bring birds into the lab, we prevent them from migrating. So how do ornithologists measure migratory behavior in captive birds? In 1966, two brothers, Stephen and John Emlen, devised a simple device that allows us to quantify a bird's urge to migrate, and the direction of flight, were it not confined to a cage (Emlen and Emlen 1966). The device consists of a paper funnel covered with wire mesh, originally with an ink pad base but now consisting of thermal paper sensitive to abrasions (first figure). Nocturnal migrants get "antsy" when kept in cages overnight during migration seasons. When placed in an Emlen funnel, birds view the night sky through the mesh and they hop onto the paper funnel, leaving their tiny footprints behind. If birds can see the stars or a representation of the night sky, then their scratch marks cluster in a single direction. Researchers quantify the number and orientation of those marks to represent migratory urge and direction. New video-based techniques (rather than paper-based) are being developed to quantify the same behaviors (Delmore et al. 2016). In studies not aimed at understanding navigation and orientation, we can quantify the same nocturnal migratory restlessness, or *Zugunruhe*, by attaching electronic sensors to perches in normal bird cages, measuring nocturnal activity levels but not directions. In a few labs around the world, researchers can simulate real migratory flights by training birds to fly in wind tunnels. Huge fans create controlled airflow through a chamber roughly a meter wide and a few meters long. Some birds learn to fly for hours under these conditions (second figure). Researchers can then manipulate environmental conditions during flight, or diet, health, or other preflight treatments, measuring a response that

more accurately reflects the physical experience of migrating than proxies such as *Zugunruhe*.

Upper panel: Willow Warbler (*Phylloscopus trochilus*) in an Emlen funnel in Sweden. *Photo by David Toews.* Lower panel: Swainson's Thrush (*Catharus ustulatus*) flying in the wind tunnel at the Advanced Facility for Avian Research at Western University in London, Canada. *Photo by A. R. Gerson.*

in some years, large numbers of individuals spend the winter much farther south than usual. Instead of attempting to explain these irruptions as a unique behavior, we can focus on identifying the thresholds and specific combination of environmental cues that trigger birds to travel farther in some years than others.

The Origin and Evolution of Migration

Migration can involve feats of athleticism and navigation far beyond anything humans could ever achieve. How did such

a complex trait evolve? It would be natural to presume that migratory behavior would be strongly conserved, phylogenetically, arising a few times in avian evolutionary history and persisting in the descendants of those first migrants. The reality is just the opposite. Migration blinks on and off the avian tree of life like lights on a Christmas tree. Genera vary in migratory tendencies within families, species vary within genera, populations vary within species, and individuals vary within populations. We see sedentary lineages giving rise to migratory ones, and descendants of migrants becoming

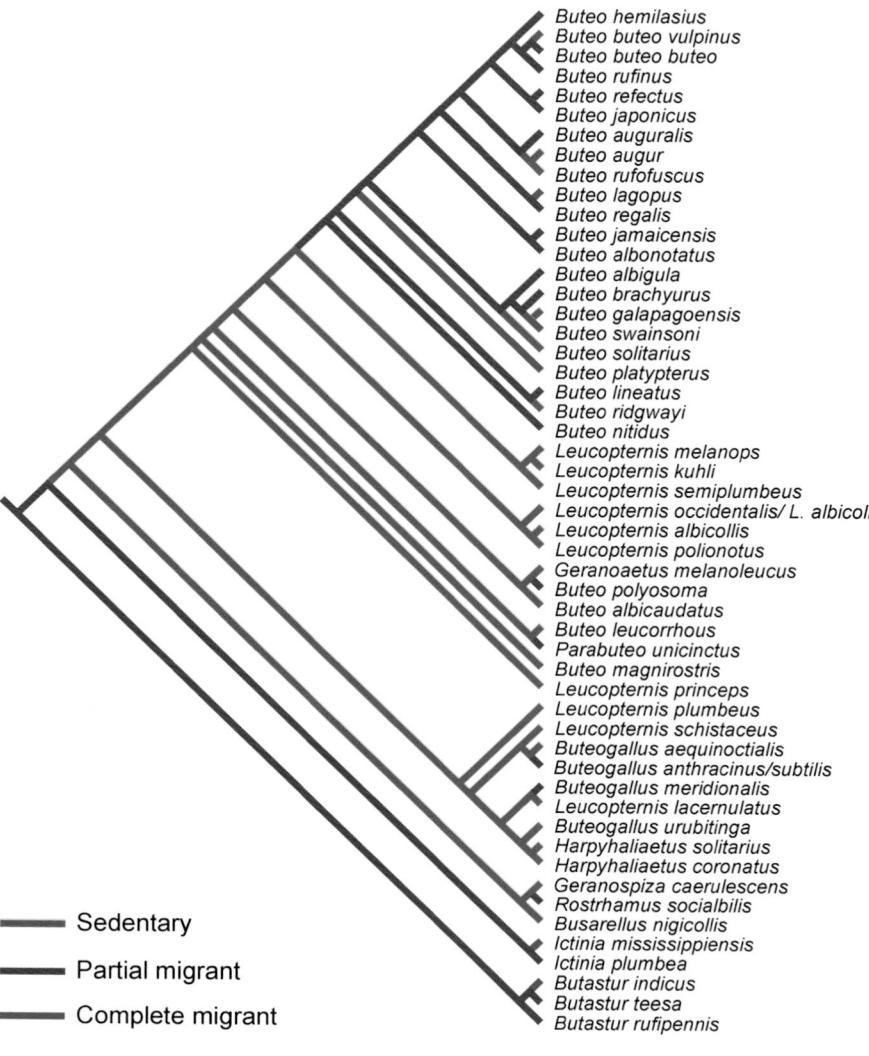

Buteo hemilasius
Buteo buteo vulpinus
Buteo buteo buteo
Buteo rufinus
Buteo refectus
Buteo japonicus
Buteo auguralis
Buteo augur
Buteo rufofuscus
Buteo lagopus
Buteo regalis
Buteo jamaicensis
Buteo albonotatus
Buteo albigula
Buteo brachyurus
Buteo galapagoensis
Buteo swainsoni
Buteo solitarius
Buteo platypterus
Buteo lineatus
Buteo ridgwayi
Buteo nitidus
Leucopternis melanops
Leucopternis kuhli
Leucopternis semiplumbeus
Leucopternis occidentalis/ L. albicollis
Leucopternis albicollis
Leucopternis polionotus
Geranoaetus melanoleucus
Buteo polyosoma
Buteo albicaudatus
Buteo leucorrhous
Parabuteo unicinctus
Buteo magnirostris
Leucopternis princeps
Leucopternis plumbeus
Leucopternis schistaceus
Buteogallus aequinoctialis
Buteogallus anthracinus/subtilis
Buteogallus meridionalis
Leucopternis lacernulatus
Buteogallus urubitinga
Harpyhaliaetus solitarius
Harpyhaliaetus coronatus
Geranospiza caerulescens
Rostrhamus socialbilis
Busarellus nigicollis
Ictinia mississippiensis
Ictinia plumbea
Butastur indicus
Butastur teesa
Butastur rufipennis

Sedentary

Partial migrant

Complete migrant

Figure 19.8. The migratory behavior of extant species varies between close relatives and changes rapidly over evolutionary time in the genus *Buteo* and its relatives (Amaral et al. 2009). In this phylogeny, many of the living species (names on right) and their ancestors (represented by the branches) are partially migratory (purple shading), and the most basal representative of this group was inferred to be a partial migrant. Partial migration gave rise to completely migratory species such as Rough-legged Hawks (*Buteo lagopus*), whose nearest relatives are all partial migrants. In other cases, the reverse happened and migratory behavior was lost entirely. For example, a partially migratory ancestor gave rise to the nonmigratory (red shading) Ridgway's Hawk (*Buteo ridgwayi*) endemic to Hispaniola. *Redrawn with permission from Fabio Raposo do Amaral.*

sedentary (fig. 19.8). For example, Hawaiian Hawks (*Buteo solitarius*) and Galapagos Hawks (*B. galapagoensis*) both likely descended from a migratory ancestor of Swainson's Hawks (*B. swainsoni*), which may have been blown off course during migration (Bildstein 2004). We have witnessed gains and losses of bird migration, even within the past few generations. Famously, a population of sedentary Californian House Finches (*Haemorhous mexicanus*) introduced onto Long Island, New York, became migratory, gradually increasing their migration distance over many generations (Able and Belthoff 1998). The urbanization of large parts of the world continues to affect migration behavior; urban populations of Eurasian Blackbirds (*Turdus merula*) are becoming less migratory, and this change in behavior is controlled genetically (Partecke and Gwinner 2007). Decreases in migratoriness are accompanied by physiological changes that allow earlier nesting by blackbirds living in cities than blackbirds living in forests.

Turning the migratory switch on or off is not the only way that birds respond evolutionarily in their migratory behavior. New migration routes can emerge rapidly as the result of changes in resource availability. Over only a few decades, about 10 percent of blackcaps breeding in Germany dramatically altered their migration direction, no longer departing in a SW direction to winter in Spain but departing in a west-northwesterly direction and overwintering in Great Britain, apparently in response to changing winter food availability caused by extensive bird feeding in England (Berthold et al. 1992; fig. 19.9). Selection can also act against evolutionary novelty in migratory behavior. Hybrid individuals at a zone of contact between genetically distinct western and eastern populations of Swainson's Thrush (*Catharus ustulatus*) exhibit intermediate migratory pathways over more mountainous terrain and different overwintering locations relative to the two parental populations—an outcome that may decrease fitness of hybrids (Delmore and Irwin 2014).

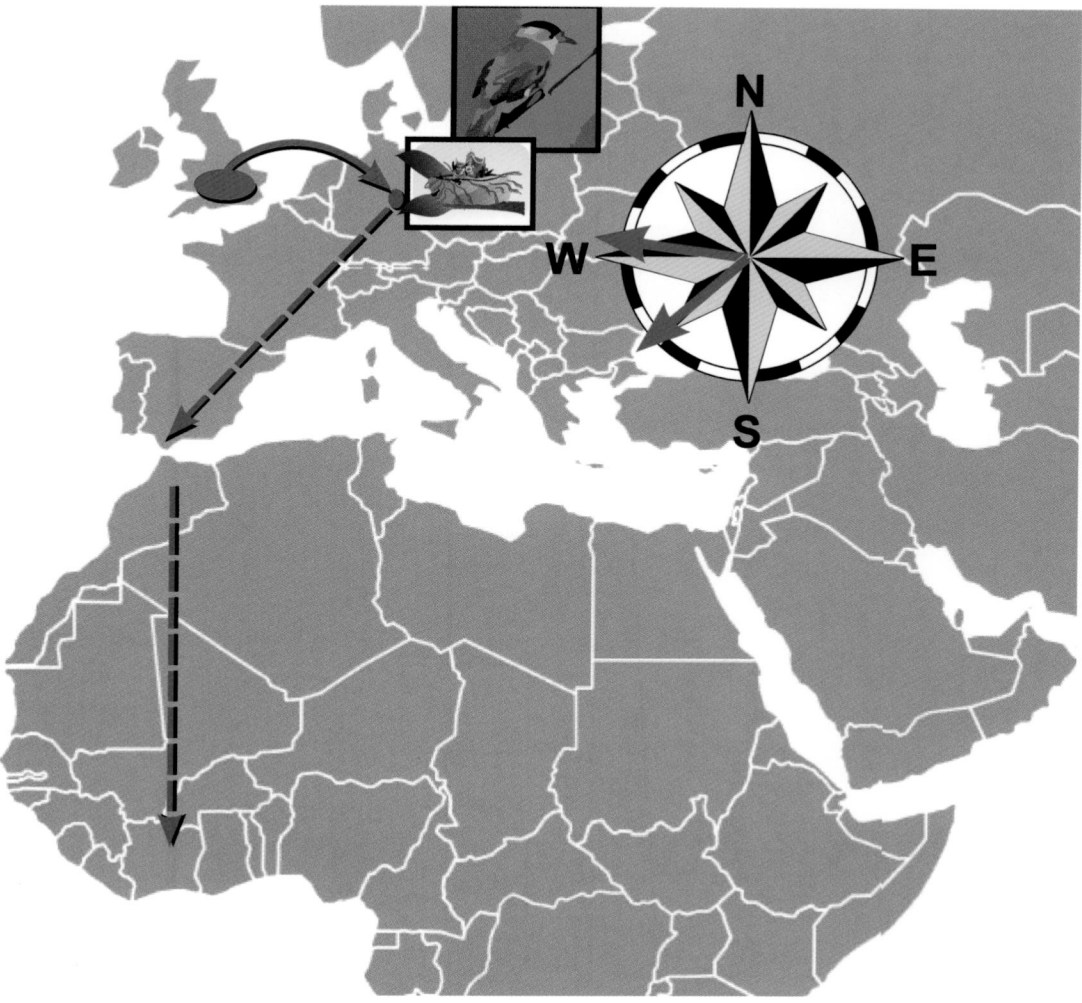

Figure 19.9. Experimental evidence that the orientation traits in Blackcaps exhibited rapid evolutionary change in recent decades. Birds breeding in Germany typically migrate southwest to Spain, and from there, south to Africa (green dashed arrow). Within recent decades, Blackcaps have begun overwintering in England. Berthold and colleagues captured birds during winter in England (red circle), bred them in captivity in Germany, and then conducted experiments with those captive-reared offspring. Young captive-bred birds from "English" parents oriented to the west toward England during the fall migration period (red arrow on compass), whereas young wild-caught birds from the same site oriented southwest toward Spain (green arrow).

Rapid changes in endogenously controlled migrations suggest that the whole suite of behaviors required to successfully migrate doesn't truly evolve independently over and over again. Rather, ornithologists are coming to the consensus that probably most, if not all, birds possess a migratory program that is easily switched on and off as ecological circumstances change (Zink 2011). Indeed, it is possible that some form of migration is ancestral for all of Aves. Avian fossil morphology suggests, however, that long-distance landbird migration was not prevalent until 15–25 million years ago, more or less concurrent with the great diversification of passerine birds (Steadman 2005). Over the evolutionary history of extant birds, the earth has experienced tremendous environmental fluctuations caused by glaciation events and longer-duration cycles of warmth and low seasonality alternating with cool, drier, more seasonal periods. With those climatic changes, birds have coped with major shifts in the distribution and extent of terrestrial habitats and extinction of other species (such as the Pleistocene megafauna or, more recently, Passenger Pigeons, *Ectopistes migratorius*) that profoundly influenced terrestrial communities. Many living genera including species of migrants and nonmigrants persisted through all these transitions with minimal morphological change, highlighting the extreme adaptability of birds and implying that migratory behavior must be flexible over longer time scales.

The extreme lability of migration has important implications for how we study the topic. For one thing, attempts to infer ancestral character states (required to map changes in migratory behavior onto a phylogeny; e.g., fig. 19.8) may not recover the full history of gains and losses of migratory behavior. If birds can change from being completely and obligately migratory, to being partially migratory, to resident in a handful of generations, then many changes in character state may have occurred since lineages diverged. In some ways, however, this possibility makes understanding the ecological drivers of migration easier. We can fairly safely assume that if migration persists today, then the behavior likely confers a fitness advantage under contemporary circumstances rather than being a product of past selection with few current fitness benefits.

For decades, ornithologists attempting to explain the ultimate causes of migration debated whether temperate-tropical latitudinal migrants were essentially temperate species that "vacationed" in the tropics, or essentially tropical species that made a quick jaunt up to more polar latitudes to take advantage of some temporarily abundant but seasonal resource. Theories of why and how migration evolves frequently rely on one or the other scenario. For instance, Levey and Stiles's (1992) "evolutionary precursor hypothesis" proposes that short-distance, facultative migrations within tropical regions were an intermediate step in the evolution of obligate, long-distance latitudinal migrations. Other theories assume northern ancestry; Bell (2000) invokes the fact that birds have less to gain by site fidelity during the nonbreeding season relative to breeding seasons, and that incipient migratory behavior is more likely to involve movements toward more climatically benign or less food-limited environments. We now know that both scenarios are true for different taxa. Hummingbirds, for example, originated and diversified in South America and reach astounding diversity within the Neotropics, particularly in the northern Andes. A few genera have evolved long-distance latitudinal migration and seasonally reach the high latitudes of North and South America (fig. 19.10). In contrast, North America's emberizine sparrows probably diversified in temperate regions. Many of the contemporary temperate-breeding species now make latitudinal migrations, and some reach tropical and subtropical regions (Barker et al. 2015). Whether a bird now winters or breeds in its ancestral home contributes little to understanding the

Figure 19.10. Hummingbirds diversified within tropical South America, where a dazzling array of resident species live today, including the Sword-billed Hummingbird (*Ensifera ensifera*; lower right) from the high Andes. Tropical resident ancestors gave rise to many species evolving altitudinal migration up and down tropical mountains, such as the Snowcap (*Microchera albocoronata*), pictured upper right. Additionally, longer-distance migration into temperate regions of both South and North America has evolved from tropical resident ancestors. Pictured lower left is the Green-backed Firecrown (*Sephanoides sephanoides*) that breeds in Chile and migrates east across the Andes, wintering in Argentina. Upper left is the Rufous Hummingbird, a ~3.3 gram bird whose migration—if measured in body lengths—is probably the longest in the world. Rufous Hummingbirds breed farther north than any other hummingbird, reaching 61°N in Alaska, and winter in central Mexico. *Images: Rufous Hummingbird © Ted Ardley Photography; Green-backed Firecrown courtesy of Wikipedia/Creative Commons Attribution-Share Alike 3.0 Unported license, https://commons.wikimedia.org/wiki/File: Picaflor_colibri_rubi.jpg; Swordbill and Snowcap photos by Richard C. Hoyer.*

ultimate ecological selection pressures that shape migration patterns, because we must simultaneously explain both why birds depart their breeding areas during the nonbreeding season and why birds then do not remain on wintering grounds to breed.

Ultimate Costs and Benefits of Migration

The benefits of migrating result from spatial and/or seasonal variation in needs or the availability of one or a combination of factors, including food, predation risk (especially on the most vulnerable of avian life stages—eggs and nestlings), climatic conditions suitable for reproduction and survival, and sometimes, the availability of potential mates. Each of these major ecological factors can also influence migration via a variety of mechanisms. For instance, food availability could drive migration via simple changes in abundance such as huge midsummer pulses of arthropod prey at high latitudes. In addition, the time available to forage is very seasonal at high latitudes because of changing day lengths, so birds can potentially feed more young over short time spans because they can forage for more hours each day (Alerstam et al. 2003). True "availability" of a food source can change seasonally based on changes in the abundance and behavior of other competing species. Competition can also change seasonally because of changes in the energetic needs of conspecifics; birds require more food per capita while engaging in energetically costly activities such as reproduction and molt. Finally, many birds undergo seasonal changes in nutritional needs that drive food preferences, so even if one resource remains relatively constant in time and space, food could drive migration if the different resources they depend on at different times of the year are produced in different places. Similar variations exist in explanations that depend on variation in predation risk or climate. Some predators are only a risk to adults, whereas others target eggs and nestlings. Predator communities, abundances, and food requirements can vary seasonally. Birds' vulnerability to predation changes seasonally based on molt and reproductive behavior. Birds' physiological tolerances change seasonally, influencing the range of temperature and precipitation conditions under which they can thrive (chapter 7).

For migration to increase fitness, the benefits must outweigh the costs. Some costs are obvious. The act of migrating is risky and many (particularly young) birds die along the way. Landbirds sometimes die in large numbers when they encounter unanticipated bad weather, especially when making long overwater crossings. Opportunistic reports of mass mortality indicate just how risky weather can be: 1.5 million Lapland Longspurs (*Calcarius lapponicus*) died during a March snowstorm in 1904; more than 10,000 Magnolia Warblers (*Setophaga magnolia*) caught in a storm off the Gulf Coast died in May 1951; 200,000 jays, thrushes, and warblers were killed over Lake Huron in May 1976; more than 20,000 corvids perished in dense fog off the Swedish coast in April 1985; and more than 40,000 birds of various species died in tornadoes off the Louisiana coast in April 1993 (Newton 2007b). Although it is difficult to document the magnitude of migration mortality during normal weather, an estimated half of all known deaths of satellite-tagged European-African migrant raptors occurred during migration, and mortality rates were six times higher during migration than during breeding or wintering (Klaassen et al. 2014). The ever-increasing human population is also increasing mortality risks for migrating birds. An estimated 6.8 million migrating birds die annually in the USA and Canada in collisions with communications towers (Longcore et al. 2012). Migrants also incur indirect mortality costs resulting from lack of familiarity with local conditions at stopover or upon arrival at their winter destination. Migrants may suffer disproportionately from predation risk or starvation than birds with more intimate knowledge of local resources and risks. Competition for food on tropical wintering grounds can **carry over** to influence migrants' condition and ability to return to breeding areas in a timely fashion (Cooper et al. 2015) (box on page 628). But migratory birds can also circumvent the costs of remaining resident. Extreme cold and food shortages kill many resident birds at temperate latitudes. To mitigate such risks, residents store fat as insurance against foraging uncertainty, which then makes them more susceptible to predation (Lima 1986, Rogers 2015).

Which of the many potential costs and benefits have been the most important in shaping bird migration behavior? Many combinations and variations have been proposed, particularly in the context of long-distance latitudinal migration. Predominant among those are the ideas of Cox (1968) and Von Haartman (1968), which postulate that resource competition—especially competition for food—underlies much of the interspecific variation in migratory behavior we see. Cox thought that migration evolved in species that were not as flexible in foraging behavior and used spatial segregation as a way to avoid competition during the times of year when food is scarcest. Alerstam and Enckell (1979) stressed that food limitation was likely to be most critical in species with relatively narrow foraging niches, because their specialization typically affords them a competitive advantage during the breeding season, and thus they can compensate for the costs of migration via high reproductive output. Fretwell (1980) argued that nest predation was the strongest driver of

BIRDS WITH BAGGAGE
By Henry Streby

Carryover effects are exactly what the name implies: conditions during one stage of a bird's life can carry over to affect the bird during a subsequent life stage. Studying these effects is a critical area of research because what happens to a bird during migration or winter helps us understand differences between individual breeding birds, or within the same individual over time. It is even possible that effort and energy expended during one breeding season can determine whether or how well a bird breeds in the subsequent year. Carryover effects are especially difficult to study in small migratory birds because of the challenges of observing individuals for more than one stage of their annual cycle. Therefore, ornithologists are often forced to piece things together from observations of different birds during different times of year or by inferring conditions from previous portions of the annual cycle using characteristics like current body condition or stable isotope analysis. Male American Redstarts (*Setophaga ruticilla*; see figure) that arrive later on the breeding grounds mate with females who lay smaller clutch sizes, and those pairs have lower nesting success generally (Norris et al. 2004a). However, those males then are more likely to interrupt fall migration to molt, and grow paler feathers—an important trait in mate attraction (Norris et al. 2004b). Males that control territories with less food on the wintering grounds end the winter in poorer condition and are late to leave wintering grounds on spring migration (Marra et al. 1998). Whether later migration departure leads to later spring arrival, or whether nesting success is a reliable measure of reproductive fitness, is yet to be determined in most songbirds, but if so, American Redstarts are an excel-

lent system in which to study carryover effects. There are plenty of pieces left to be placed in the puzzle of full-annual-cycle research in order to understand the frequency and importance of carryover effects in migratory songbirds. Lessons and models from nearly a century of full-annual-cycle research, conservation, and management in North American waterfowl provide an ideal starting point. Continued improvements in technology and the miniaturization of tracking devices are already making such work possible in some of the smallest songbirds. As such tools become ubiquitous in field data collection, the extraordinary work that has been completed to date, such as that with American Redstarts, will continue to be a fruitful area of research with important implications for the fields of evolution and conservation.

Male American Redstart. *Photo by Geoffrey Hill.*

variation in avian reproductive success, and competition for food was the strongest driver of mortality. Thus, high risks of nesting in the tropics would drive migration to temperate breeding areas. Subsequently, seasonal reductions in food that lead to intense competition drives the southward movements to less seasonal wintering areas (Fretwell 1980, 1985). In the "time allocation hypothesis," Greenberg (1980) incorporated many of the same ecological processes by proposing that migrants trade off juvenile mortality resulting from migration in favor of higher adult survival during the nonbreeding season. These ideas were derived from analysis of

interspecific variation in demographic rates consistent with the tropics being a good place to survive but a difficult place to reproduce.

Empirical Tests of Hypotheses Explaining Migration

Testing ultimate hypotheses for migration is particularly challenging in obligate, long-distance migration systems, both because of the huge spatial scales involved and because, when all individuals migrate, we cannot compare the fitness outcomes of migration and residency in the same species. However, in many species, individuals do vary in migratory distance. Researchers studying arctic-nesting shorebirds took advantage of that variation and found that across 29° of latitude, a migrant shorebird might expect to enjoy a reduction of 3.6 percent in nest predation risk for every additional 1° she traveled north in the spring (McKinnon et al. 2010; fig. 19.11). This result is consistent with ideas of Fretwell: that a primary benefit of northward spring migration is to exploit the relative safety of temperate latitude habitats during the most risky phase of a bird's life—before it can leave the nest.

A rich body of empirical studies of the Dark-eyed Junco (*Junco hyemalis*) also exploited differences in migratory distance to determine the relative importance of different costs and benefits shaping migration patterns. Unlike shorebirds that differ in breeding distribution, juncos of different ages and sexes travel different distances from their breeding range to spend the winter at different latitudes: females travel farther than males, and adults travel farther than young birds. In perhaps the most thorough set of empirical evaluations of

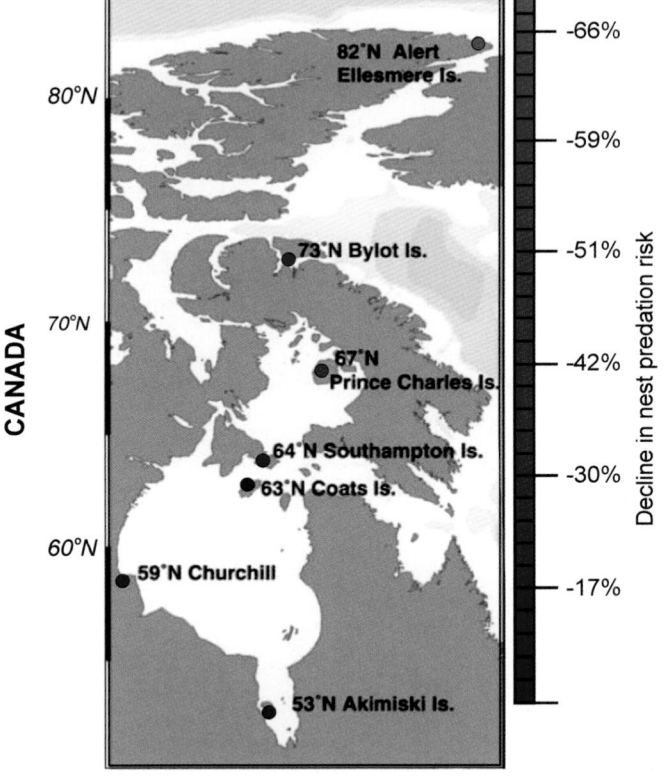

Figure 19.11. Shorebirds breeding at sites in the Canadian arctic may be trading off predation risk with the energetic costs and additional time constraints of flying and nesting farther north. Birds breeding at the northernmost site (Ellesmere Island, red dot) experience 66 percent lower nest predation risk than those breeding at the southernmost site at Akimiski Island (blue dot). *Redrawn with permission from the American Association for the Advancement of Science.*

alternative hypotheses explaining variation in migratory behavior, Ketterson and Nolan (1983) concluded that wintering distributions are most likely the outcome of groups of individuals optimizing migratory benefits and risks in ways that differ among age and sex classes. The mortality costs of longer migration are likely highest for young, inexperienced birds, which selects for shorter migrations. The benefits of remaining close to breeding areas and being able to return to them quickly and early in the spring are greater for males, as they are the sex that competes for territories (**arrival time hypothesis**; Myers 1981). Such benefits are likely strongest for young males because they have no prior territorial claims and must break into the competitive male hierarchy. Thus, the benefits of early arrival should select for shorter migrations in males and, particularly, in young males. Other factors, such as size-related differences in ability of birds to withstand cold and short-term fasting cannot explain the junco patterns because, although males are larger than females, smaller young birds would be expected to suffer greater costs of more northerly winter distribution and consequently migrate farther than adults (**body size hypothesis**; Ketterson and Nolan 1976). Likewise, if competition for winter food and dominance hierarchies dictates how far birds must travel to find wintering habitat, then, again, adults would be expected to winter farther north than the subordinate young (**dominance hypothesis**; Gauthreaux 1982). For groups of individuals whose costs of migration are relatively low, the benefits of traveling farther include reduced competition from conspecifics, or gradients in habitat suitability, with more southerly sites offering increased chances of survival because of food availability, climate, and/or reduced predation risk (Cristol et al. 1999). Such multifactor explanations are likely common to many, if not all, migrant species and highlight the importance of integrating the external factors that vary in space and time with internal factors (such as age and sex) that affect the fitness payoffs of different behaviors.

Similar types of individualistic trade-offs seem to underlie altitudinal migrations. Although the major hypotheses proposed to explain migration were formulated in the context of long-distance, latitudinal movements, elevational and latitudinal gradients parallel each in many important ways. For instance, although temperatures typically decline along both geographic gradients, and nest predation risk seems to be generally higher in the lowlands than at higher elevations, the patterns and magnitude of climatic seasonality typically are similar along elevational gradients, unlike latitudinal gradients. The altitudinal migration of American Dippers (*Cinclus mexicanus*) of western North American mountains appears to ultimately be driven by competition for high-quality territories that provide rich food resources, consistent with the hypotheses of Cox (1968), Von Haartman (1968), and Alerstam and Enckell (1979). In Dippers, some birds remain resident on low-elevation streams year round, overwintering together with migrants who breed on high-elevation streams. The residents raise more young each season and survive equally as well as the migrants, implying that migratory behavior may be a direct result of heightened competition and increased food demands during the breeding season, combined with variation in individual quality (Morrissey 2004, Mackas et al. 2010, Green et al. 2015).

In contrast to the mechanism of seasonal changes in needs, seasonal changes in the relative availability of food over elevational gradients has been proposed to explain tropical altitudinal migrations, many of which depend primarily on fruit and nectar year-round (Stiles 1980, Stiles 1988). Although seasonal peaks in the abundances of frugivores often occur at the same time as peaks in food availability (Loiselle and Blake 1991, Solórzano et al. 2000; fig. 19.12), the timing and locations of migrations often take birds away from elevations of high relative food abundance (Chaves-Campos 2004). Instead of simple food availability, diet-migration relationships can result from interactions between seasonal variation in rainfall and high food intake rates dictated by the nutritional content of fruit and nectar. Like many tropical montane species, some White-ruffed Manakins (*Corapipo altera*) breeding in wet montane forests of Central America migrate to low elevations during the nonbreeding season, while others remain resident year-round. In contrast to the juncos, males are more migratory than females, and birds in compromised condition are more likely to migrate (Boyle 2008). Downhill migration is synchronous with storms, and many species are more migratory in the wettest years (Boyle and Sigel 2015). Birds experience storm as stressors, undergoing short-term fasts, especially at high elevations (Boyle et al. 2010). The birds that spend the nonbreeding season in the lowlands must cope with less food but benefit from better weather, experiencing storms only half as severe as at higher elevations. Migrating altitudinally likely confers a survival advantage, but migrant males are less successful than residents at gaining access to display sites and females the following breeding season (Boyle et al. 2011).

Constraints and Challenges Imposed by Migration

Migratory behavior—particularly long-distance migration—is associated with a host of other adaptations ranging from physiological innovations (chapter 7) to shifts in life history strategies (chapter 16). Birds such as Bar-tailed Godwits (*Limosa lapponica*) are able to depart spring staging areas in the Wadden Sea with body masses 90 percent higher than during nonmigratory periods because of their remarkable ability

Figure 19.12. Male Resplendent Quetzal (*Pharomachrus mocinno*) perched near fruits of *Nectandra cufodontisii* (Lauraceae, avocado family). *Photo © Wes Hanson.*

to rapidly deposit fat stores, known as **hyperphagia**. Birds achieve remarkably high rates of fat deposition by increasing foraging rates, increasing digestive efficiency, and/or by shifts in diet (Bairlein and Gwinner 1994). For example, most North American insectivores shift to a carbohydrate-rich diet of fruit prior to fall migration that enables them to rapidly deposit fat (Parrish 1997). Long-distance migrants modulate the mass of other tissues as well. Pectoralis muscle increases, while organs not needed during long migratory flights (e.g., stomach and intestine) atrophy immediately prior to departure (Piersma et al. 1999). Additionally, burning lean mass for fuel during migration may be common, and function primarily to avoid dehydration, because lean mass catabolism releases water, whereas burning fat does not (Gerson and Guglielmo 2011). At the proximate level, fine-scale modulation of body composition must be integrated temporally into mechanisms controlling migratory behavior. At the ultimate level, the benefits of large fat stores trade off with reductions in flight efficiency and agility when birds are atypically heavy (Lima 1986, Rogers 2015). A fat bird is at greater risk from predators for the same reason that a "couch potato" runs more slowly than an Olympic athlete. Birds must integrate all benefits and costs when making a coordinated set of decisions, particularly during migratory stopover, where most birds accumulate the majority of their energy reserves and also spend much of the fuel they acquire (Cimprich et al. 2005, Cohen et al. 2014).

External and internal morphology differs between long-distance migrants and residents. Migrants have longer, more pointed wings, which increases the efficiency of long-distance flight at a cost to maneuverability (Mönkkönen 1995). A host of other less obvious morphological changes are also associated with migration; these include less graduated tails, shorter legs, lighter leg muscle mass, flatter skulls, and narrower bills (Winkler and Leisler 2005). Some adaptations can be interpreted as the outcome of trading off the mass of structures that support bipedal locomotion with those that support flight. Changes to skull morphology are related to overall reductions in brain size in migrants. Why might migrants, who have to find a way to thrive in multiple habitats over the year, have smaller brains than closely related residents, when big brains help birds cope with novelty (Sol et al. 2005)? Big brains are costly to grow and maintain, so the energetic demands of migration itself may constrain cognitive ability. Alternatively, or additionally, by moving between climatically benign locations the whole year, migrants may not require the same capacity for behavioral innovation as residents who must cope with whatever environmental challenges occur on their year-round home (Sol et al. 2010).

One major consequence of migration is the time constraints that movement imposes on annual cycles. Some migratory raptors spend as much or more time traveling to and from breeding grounds than they spend at either location (fig. 19.13). Because birds usually avoid multiple simultaneous, energetically costly activities, time spent in migration reduces time available for reproduction and molt. Even when the duration of migration is short, departing breeding grounds removes the possibility of more reproductive attempts for the season. Major consequences of temporal constraints on reproduction include fewer broods a female can raise each year and fewer renesting attempts following nest failure. A common pattern in landbirds is to molt following breeding but prior to southward migration. Attempting to both reproduce and molt during the short window of opportunity between spring and fall migration imposes further time constraints. In western North America, some species characterized as **molt migrants** reduce those constraints by breaking their fall migration and molting at an intermediate location (chapter 9). Still other species, such as Yellow-billed Cuckoos (*Coccyzus americanus*) sometimes

Figure 19.13. Swainson's Hawk (*Buteo swainsoni*). *Photo by Geoffrey Hill.*

circumvent the temporal constraints of migration on breeding by being **migratory double breeders**—these birds initiate fall migration after nesting in the north, then break migration to breed a second time in a more southerly location (Rohwer et al. 2009), a behavior that can also be characterized as long-distance, within-season breeding dispersal.

Time constraints are only one of several ways by which migration affects avian life histories. Conditions during previous seasons can influence when birds complete migratory journeys, the quality of breeding territories they are able to secure, the number of young they are able to raise, and even their chances of survival (e.g., Saino et al. 2004, Inger et al. 2010, Duriez et al. 2012, Catry et al. 2013). More types of these carryover effects are being documented as technology provides the means to gain detailed information on individuals over their complete annual cycles (box on page 628). Migration also allows birds to avoid some of the fitness costs incurred by residents. Tropical resident species typically lie at the slow end of the life history continuum, with low reproductive output each year (small clutch size, high rates of nest failure) but high adult survival (Martin 1996). North temperate residents tend to lie near the fast end, having larger clutch sizes, higher nest success, but low adult survival probably caused by harsh winter climates (Sherry and Holmes 1995). Latitudinal migrants would be predicted to enjoy the reproductive benefits of northerly breeding and the survival benefits of more southerly overwintering, minus the costs of reduced allocation to reproduction and the mortality costs of migration itself. Indeed, in five species of North American sparrows wintering together, clutch size did increase with breeding latitude, but temporal constraints also meant that medium- and long-distance migrants were confined to raising fewer broods and had fewer opportunities to renest. Annual survival was not lower in species with longer migratory journeys, indicating that the survival benefits of residency may only accrue when wintering ranges differ (Sandercock and Jaramillo 2002). While we know that migration is a major source of direct mortality, curiously, migrants exhibit slower rates of senescence after controlling for breeding latitude, age at first reproduction, and adult survival rate (Møller 2007). One potential explanation for this pattern is intriguing; the act of migrating may delay senescence as a by-product of physiological repairs following intensely athletic migratory journeys. We still have much to learn about the basic patterns of life history strategies for birds around the globe, migrants and residents alike. A better understanding of these patterns and the causes for regional and behavioral correlates in components of life history will undoubtedly reveal much about the rich diversity and multiple consequences of migratory behavior.

DISPERSAL

Patterns

Roughly half of the worlds' birds migrate, but all birds engage in some form of dispersal during their lifetimes. By far the most common type is natal dispersal—the permanent movement away from where a bird hatches to the site of first reproduction (Greenwood 1980). Indeed, natal dispersal is a "behavior" that birds share with most other organisms on earth, including those we think of as being sessile, such as plants or fungi. Less commonly, birds also engage in breeding dispersal—the one-way movement from one breeding site to another. In vertebrates, both natal and breeding dispersal is sex-biased. In birds, females usually are more likely to disperse, and to disperse farther than males (Clarke et al. 1997). For instance, young female Tree Swallows (*Tachycineta bicolor*) first breed an average of just over ~8 kilometers from the nest from which they fledged, whereas young males disperse on average fewer than 2.5 kilometers (Winkler et al. 2005). Subsequently, 14 percent of female Tree Swallows disperse between breeding sites, whereas only 4 percent of males make breeding dispersal movements (Winkler et al. 2004). Tree Swallows are typical for birds by displaying low **philopatry** while exhibiting high **site fidelity**—the inverse of natal dispersal and breeding dispersal, respectively (Greenwood and Harvey 1982). Despite making migrations of thousands of kilometers each year, many migrants return to within a few kilometers of their natal site to breed, and continue to do so, year after year. In some guilds and lineages, the sex bias in dispersal is reversed, most notably in many waterfowl (Clarke et al. 1997).

Dispersal tendency varies greatly among species. Great Reed Warblers (*Acrocephalus arundinaceus*) and Great Tits (*Parus major*) in England both exemplify high philopatry and site fidelity—67–69 percent of young males recruit as breeders at their natal site, and 92 percent of adults are then site faithful to previous breeding localities (Hansson et al. 2002, Andreu and Barba 2006). In contrast, many North American grassland birds appear to be highly dispersive, with few to no young birds returning to breed at natal sites, and adult site fidelity is often less than 10 percent (Jones et al. 2007). Dispersal distance also varies greatly among species. Mallards (*Anas platyrhynchos*) only disperse up to 1 kilometer between breeding attempts (Clark and Shutler 1999), while some Razorbills (*Alca torda*) move more than 1,000 kilometers from natal sites and between breeding attempts (Lavers et al. 2007). Often natal dispersal is longer than breeding dispersal; Bobolink (*Dolichonyx oryzivorus*) and Savannah Sparrow (*Passerculus sandwichensis*) natal dispersal is four to eight times farther than subsequent breeding dispersal distances (Fajardo et al. 2009). However, the reverse is sometimes true: breeding dispersal of Ortolan Buntings (*Emberiza hortulana*) is four times farther than their typical natal dispersal distances (Dale et al. 2005).

Over evolutionary time, dispersal behavior appears to be less labile than migration, meaning that researchers must take phylogenetic relationships into account when testing for associations between dispersal and other traits (Paradis et al. 1998). Dispersal is also strongly associated with body size; large birds are more dispersive than small birds. Abundant species with large geographic distributions are among the most philopatric and site faithful, and migrants tend to be more dispersive than residents (Paradis et al. 1998). Even fine-scale habitat differences can strongly influence dispersal behavior. Tropical understory birds are remarkably unwilling to cross barriers such as small roads and rivers, making them the least dispersive guild of terrestrial birds; over eight years and nearly 6,000 telemetry locations, none of 30 tagged Afro-tropical understory birds ever crossed a forest gap more than 15 meters wide (Newmark et al. 2010; fig. 19.14). Such "gap-shy" behavior can seriously limit gene flow and affect large-scale biogeographic patterns. For instance, in a comparison of understory and canopy-dwelling South American birds living on both the east and west slopes of the Andes, the understory species had considerably higher levels of genetic divergence, consistent with extremely low rates of dispersal across the major Andean mountain barrier (Burney and Brumfield 2009).

Breeding dispersal typically occurs between breeding seasons in different years, which means that in migrants, the traveling phases (and potentially some of the costs) of dispersal may be subsumed by seasonal migration to and from wintering quarters; on their return journeys, some birds simply settle in a different breeding location than the one they

Figure 19.14. Radio-tracking data from tropical birds such as this Orange Ground Thrush (*Zoothera gurneyi;* inset) reveal that forest fragmentation, such as in the East Usumbara Mountains of Tanzania, makes dispersal between forest patches extremely difficult. *Photos by William Newmark.*

occupied previously. Migratory movements may also provide information about the location of good breeding habitat useful in post-dispersal settlement decisions. However, breeding dispersal sometimes occurs within breeding seasons, between breeding attempts. In most cases, such within-year breeding dispersal movements are subject to more stringent time constraints because the climatic conditions suitable for breeding only last for a restricted period. Although we know relatively little about within-year dispersal, it appears that between- and within-year patterns of breeding dispersal are similar. For example, compared with between-year dispersal, the within-season breeding dispersal of Hoopoes (*Upupa epops*) is slightly less common and of slightly shorter distance (mean of ~0.5–0.8 versus ~1.5–2 kilometers), but shows the same patterns of sex bias (longer in females) and little relationship with previous reproductive success (Botsch et al. 2012).

Proximate Causes and Control of Dispersal

Like migration, the proximate factors that trigger dispersal involve both innate and facultative elements, and the flexibility in dispersal behavior varies among species. The ubiquity of natal dispersal, the highly conserved taxonomic patterns of sex-bias in dispersal tendency, and the heritability of dispersal and associated traits implies that the urge to disperse early in life is at least partially innate (Dingemanse et al. 2003). But to what extent are the timing, distance, and other elements of dispersal fixed rather than the outcomes of facultative decisions? When an individual's phenotype influences the costs and benefits of dispersal, and when those phenotypic

traits are heritable, then there is the potential for dispersal **syndromes** to emerge. Such syndromes integrate dispersal tendency with other behavioral and morphological traits and correlated genetic processes, resulting in individual-level differences in dispersal within species (Duckworth 2012) as described in a series of experiments with bluebirds (box on page 634). Aggression is heritable but is also modulated by maternal effects: under high cavity availability, males hatched from eggs laid late in the clutch, and those males were nonaggressive and philopatric. However, when cavities were scarce, males hatched from early laid eggs and those young males were both aggressive and dispersive (Duckworth 2009; box on page 635). Evidence for innate processes controlling dispersal decisions come from other birds as well. Snail Kites (*Rostrhamus sociabilis*) nest in wetlands either by lakeshores or in freshwater marshes. Young kites make natal dispersal movements of tens to hundreds of kilometers, but regardless of how far they moved, they usually selected the same kind of wetland in which to breed as that in which they themselves had been reared. Interestingly, such fixed preferences resulted in reproductive costs, not benefits; kites that dispersed to locations similar to their natal areas raised fewer young than kites that switched habitats (Fletcher et al. 2015).

Plenty of evidence also exists for a strong facultative component to dispersal tendency, timing, and direction, particularly in breeding dispersal. Breeding dispersal tendency varies with quality of breeding habitat (Lecomte et al. 2008), density of competitors, and condition of the bird (Kim et al. 2007). The most common correlate of breeding dispersal tendency

GO WHERE MOM TELLS YOU! MATERNAL EFFECTS AND BLUEBIRD DISPERSAL
By Henry Streby

Generally speaking, a maternal effect is the influence of female phenotype on the phenotype of her offspring, and such effects can be genetic, environmental, or a combination of both. More specifically, maternal effects are often discussed as when females influence morphological or behavioral phenotypes of their offspring through some physiological or behavioral response by the female to current environmental conditions. One can imagine the potential fitness benefits of being able to produce more male offspring when a population is female-biased or to produce more or less aggressive offspring in response to population density or perceived competition for limited resources. Dr. Renee Duckworth described the nature of maternal effects in Western Bluebirds (*Sialia mexicana*) in an elegant series of manipulations to the habitat of wild populations (Duckworth et al. 2015). Duckworth demonstrated that female Western Bluebirds can change the order in which they produce male and female offspring within a clutch in response to the quantity of nearby available nesting cavities. When cavities are abundant, females produced male offspring later in the clutch, which results in less-aggressive males that do not disperse far from their natal territory. When cavities are scarce, females produced male offspring earlier in the clutch, which results in more-aggressive males that disperse farther from natal territories (see figure). Western Bluebirds are secondary cavity nesters that colonize open forested landscapes, often a few years after a forest fire. The cavities they exploit are an ephemeral resource because cavities are often abundant when Western Bluebirds colonize an area but become less abundant as other animals occupy them, snags containing cavities decay or fall, or cavities become unusable for any other reason. Colonization of an area by Western Bluebirds often follows colonization by Mountain Bluebirds (*Sialia currucoides*), a closely related species that is typically displaced by the relatively aggressive Western Bluebirds. There is a fitness benefit to being an aggressive Western Bluebirds when colonizing a new area, and a strong relationship between long-distance dispersal and aggression assures colonizing Western Bluebirds can outcompete Mountain Bluebirds. However, high aggression is also associated with reduced parental care by male Western Bluebirds, resulting in relatively low fitness for aggressive males compared with less

aggressive males once an area has been colonized. All of this means that females benefit when they can respond to high local cavity availability by producing males that are less aggressive, shorter dispersers, and better at parental care, and when they can respond to low cavity availability by producing males that are more aggressive, longer dispersers, but could maybe benefit from a parenting class or two.

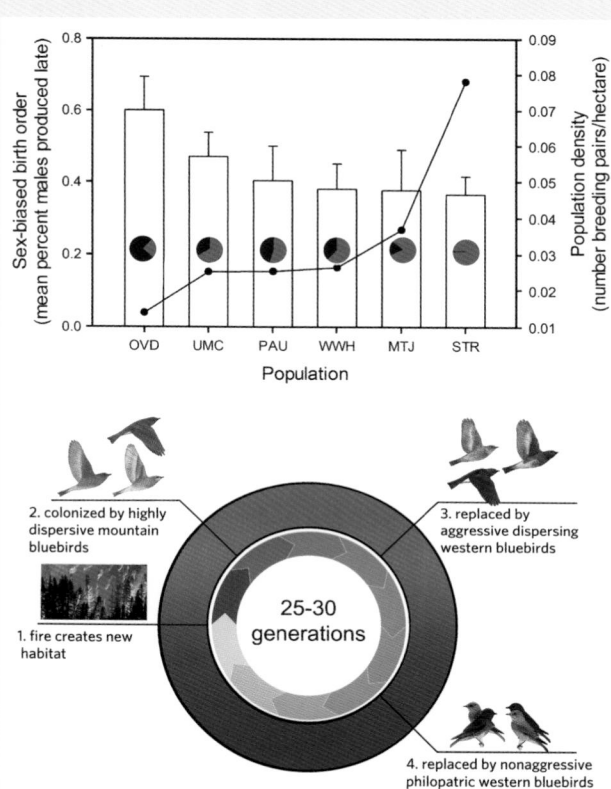

Bars in the upper figure depict the proportions of males produced later in clutches of Western Bluebirds in six different populations. When the density of conspecifics (dots connected by lines) was high, and fewer territories were occupied by Mountain Bluebirds (dark blue portions of pie charts), female Western Bluebirds laid more of their male offspring early in the clutch, giving rise to more aggressive and dispersive individuals. These maternal effects shape the cycle of colonization and replacement of Mountain Bluebirds with Western Bluebirds in fire-adapted habitats as depicted in the lower panel (Figure 1 and Figure 3 from Duckworth et al. 2015). *Images © Renee A. Duckworth.*

RENÉE DUCKWORTH

Dr. Renée Duck-
worth. *Image
© Dr. Renee A.
Duckworth.*

Dr. Renée Duckworth is an associate professor in ecology and evolutionary biology at the University of Arizona.
Dr. Duckworth investigates how behavior evolves and how
evolutionary changes in behavior influence population
dynamics and macroevolutionary processes. Her work
is exceptionally successful at linking developmental and
physiological processes to differences among individuals
in behavior, then using that information to explain ecological patterns and evolutionary processes at broad spatial
scales. Dr. Duckworth has received numerous early-career
awards for her work on the evolutionary ecology of birds—
particularly her bluebird research featured in this chapter.

is past reproductive success. Frequently, birds adopt a "win-stay, lose-switch" strategy in which adults that successfully fledge broods are site faithful, whereas those whose nests fail disperse to new breeding areas (Chalfoun and Schmidt 2012). Such a strategy implies that birds use past experience to predict future success and make movement decisions conditional on that likelihood. Facultative dispersal decisions are overlaid on fundamental differences among clades, guilds, and age and sex groups that likely reflect innate processes. Although few studies have explored how facultative and innate processes interact, species-level behavioral differences probably result from variability in cue-response thresholds, such as in migratory behavior.

Ultimate Causes of Dispersal

Because dispersal is a behavior common to all life on earth, a large body of theory exists to explain the selective advantages of such movements. The three general classes of dispersal

benefits are (1) inbreeding avoidance, (2) kin selection, and (3) variability in habitat quality or availability of mates (Clobert et al. 2009). Birds, as in other outcrossing, sexually reproducing organisms, have been selected to avoid mating with close relatives because inbreeding leads to the expression of rare, recessive alleles with adverse fitness consequences. Sex-biased dispersal can be an efficient way to maximize genetic benefits of inbreeding avoidance while minimizing costs, because only half the individuals disperse. Inbreeding avoidance is fundamental to explaining natal dispersal but is less applicable to breeding dispersal in most species; if a young female's natal dispersal is effective in moving her to new neighborhoods far away from her father, brothers, uncles, and cousins, then she would have little to gain in terms of inbreeding avoidance by dispersing again later in life.

Kin selection can affect dispersal tendencies in two opposing ways. As classically formulated, dispersing to breed far away from kin reduces competition among relatives for resources such as food, breeding sites, or mates. Reduced kin competition increases the fitness of relatives and, thus, the disperser's inclusive fitness (Hamilton and May 1977). In cooperatively breeding species, however, delayed dispersal may increase an individuals' inclusive fitness; by helping raise younger brothers and sisters, inexperienced birds can increase the odds of their genes being passed on to the next generation through their extended family (Koenig et al. 1992). Spatial and temporal variability in habitat quality (e.g., food or nest site availability, differences in predation risk, or microclimate) or social environment can also explain why animals disperse, either prior to or following first reproduction. Birds that require ephemeral resources for breeding sometimes make truly remarkable breeding dispersal movements. For example, the Australian Banded Stilt (*Cladorhynchus leucocephalus*) breeds along shallow inland salt lakes that form following unpredictable desert rains. Within a few days of heavy rains, birds arrive by the thousands from coastal areas hundreds of kilometers away (Pedler et al. 2014; fig. 19.15).

Sex-biased dispersal patterns may ultimately be shaped by mating system (chapter 17); the sex that has more to gain

Figure 19.15. Banded Stilts. *Photo by Stewart Monckton.*

from familiarity with breeding areas and local social hierarchies should be more philopatric. In roughly 70 percent of birds, dispersal is female-biased, and in more than 90 percent of bird species, males compete for resources and females select males on the basis of territory quality. Males would benefit greatly from detailed local knowledge of habitat quality and previous interactions with neighboring males, whereas females' fitness would be most strongly affected by their choice of male (Greenwood 1980). Consistent with these ideas is the dispersal behavior of species without male territorial mating systems. For instance, in the polygynandrous Smith's Longspur (*Calcarius pictus*), neither sex defends territories, multiple mating opportunities exist in close proximity, and males and females are equally likely to be site faithful (Craig et al. 2015). Likewise, no evidence of sex-biased dispersal exists in Brown-headed Cowbirds (*Molothrus ater*), a species that defends no territories and lays its eggs in other birds' nests (Anderson et al. 2012).

Female-biased dispersal in male territorial mating systems is also consistent with an alternative explanation. In territorial species, males can select any unoccupied site in which to establish themselves, whereas females are much more restricted. Females can only choose from sites that (a) males have already selected, and (b) have not already been selected by other females. This imbalance in the availability of breeding sites, rather than sex-related differences in benefits of site fidelity, may cause females to typically disperse more frequently, and to disperse farther, than do males (Arlt and Part 2008). African Houbara Bustards (*Chlamydotis undulata*) provided an opportunity to test this explanation. Young male bustards make longer dispersal movements than do females, and males of this large, terrestrial bird display together in leks. While both sexes incur mortality costs of dispersal, males incur higher costs than females, especially when dispersing long distances (Hardouin et al. 2012). Lekking behavior may explain the reverse-sex bias in this species because male reproductive success is not dependent on defending resources. Likely male bustard breeding site options are limited. Females, on the other hand, may benefit from intimate knowledge of nesting areas and would not be as constrained in breeding sites as would females with male resource-defense mating systems.

The costs of dispersal overlap broadly with the costs of migratory behavior, including risks caused by lack of local knowledge of food or predation risks, opportunity costs of time spent traveling and establishing in new locations, and energetic costs of moving itself. Likewise, selection for dispersal abilities involves trade-offs with a wide variety of morphological, cognitive, and behavioral traits (Danchin and Cam 2002, Nevoux et al. 2013).

Empirical Tests of Hypotheses Explaining Dispersal

Empirical tests of the causes of natal dispersal have focused primarily on inbreeding avoidance and kin selection. Indeed, the tendency of adults to be highly site faithful while natal dispersal is so ubiquitous implicates inbreeding avoidance—if parents breed repeatedly in the same locations, then their offspring have little option but to go elsewhere if they are to avoid mating with relatives. In migratory Savannah Sparrows (*Passerculus sandwichensis*) breeding on islands in the Bay of Fundy, young birds disperse fairly short distances, mostly returning to the same small archipelago from which they fledged (Wheelwright and Mauck 1998). The dispersal distances of parents and young were uncorrelated, indicating that, at least at small spatial scales, natal dispersal has low heritability. However, birds managed to avoid mating with their relatives; young birds avoided breeding near parents of the opposite sex, and siblings raised together dispersed to different areas, resulting in lower within-pair relatedness than expected by chance. A key prediction of the kin selection hypothesis is that when competition among kin is high, natal dispersal should be more frequent and/or longer. English Great Tits (*Parus major*) exemplify this pattern; males disperse a distance corresponding to a greater number of territories (although not necessarily a greater absolute distance) when local territory density is high (Greenwood et al. 1979). Competition generally influences fine-scale dispersal movements in tits; population density affects resource availability, which results in spatial and temporal variation in habitat quality, which, in turn, influences dispersal behavior (Matechou et al. 2015).

Empirical tests of breeding dispersal have provided support for all major explanations for dispersal. However, most studies have emphasized the importance of habitat variability. For instance, Black-throated Blue Warblers (*Setophaga caerulescens*) breed in both high-quality shrubby areas, and lower-quality areas with fewer shrubs. Both sexes make shorter breeding dispersal movements when they breed in high-quality habitat. Additionally, males disperse shorter distances as they age, presumably because they secure higher-quality territories (Cline et al. 2013). Additionally, many birds use public information to inform their assessments of relative habitat quality when making breeding dispersal decisions. Colonial-nesting Black-legged Kittiwakes (*Rissa tridactyla*) judge nesting habitat based not only on their own nest success, but also on the nest success of their neighbors; experimental manipulations of perceived habitat quality by removing eggs from some birds' nests resulted in neighboring birds being more likely to disperse if their own nests failed (Boulinier et al. 2008). After deciding to disperse, birds also use social information to choose post-dispersal habitat.

Grasshopper Sparrows (*Ammodramus savannarum*) breeding in North American midcontinental tallgrass prairies are highly dispersive between and within years (Williams and Boyle 2018). By broadcasting songs through loudspeakers, researchers experimentally manipulated the perceived density of territorial male sparrows. Relative to control plots, plots with loudspeakers attracted twice as many territorial males late in the breeding season. This result suggests that conspecifics may provide a signal of breeding habitat quality, especially in temporally constrained within-season dispersal (Andrews et al. 2015). Despite the importance of habitat variability, both inbreeding avoidance and kin selection can contribute to shaping breeding dispersal patterns. For instance, Red-cockaded Woodpeckers (*Leuconotopicus borealis*) are a communally breeding species that depend on old-growth longleaf pine in which to build nests that are used by cooperatively breeding groups of adults. In this system, both inbreeding avoidance and mate and habitat quality shape the dispersal decisions of females. Females disperse more frequently when their sons become breeders in their social group, consistent with females avoiding mating with close relatives. But females also disperse more frequently following nest failure, if they do not have helpers at their nests, and if their mates are of low quality. Thus, female Red-cockaded Woodpeckers may disperse to seek out better reproductive opportunities, independent of the presence of kin (Daniels and Walters 2000).

Consequences of Dispersal

Dispersal has many important consequences for other branches of avian biology. At the genetic level, the physical one-way movement of individual birds is a prerequisite to gene flow or **effective dispersal**. Dispersal does not ensure gene flow if post-dispersal breeding is unsuccessful, but gene flow cannot occur without dispersal of reproductive individuals. Only a few birds moving between populations each generation is sufficient to maintain high levels of genetic connectivity, and the rate of effective dispersal influences the potential of populations to adapt to local environmental conditions and, ultimately, to diverge into reproductively isolated species (Garant et al. 2005). Thus, differences in dispersal tendency and distance among lineages ultimately has a bearing on broad-scale patterns of species distribution and diversity patterns (Ghalambor et al. 2006).

The other main class of dispersal consequences is spatial. In the absence of dispersal, speciation could not happen for another, nongenetic reason: dispersal is the mechanism by which ranges expand to fill space and move into novel areas. This fact makes dispersal incredibly important for predicting distributional responses to ongoing climate change. For example, the latitudinal distribution of North American wintering bird species ranges is shifting north at roughly 1 km per year (La Sorte and Thompson 2007), the altitudinal ranges of Peruvian montane birds have shifted ~50 m uphill over 40 years (Forero-Medina et al. 2011), and New Guinean montane species have shifted 95–152 m uphill over a similar time period (Freeman and Class Freeman 2014; fig. 19.16). Dispersal is the mechanism by which distributional changes occur. Thus, while dispersal can swamp local adaptation when gene flow is high, thereby reducing speciation rates and diversification (Lenormand 2002, Claramunt et al. 2012), dispersal can also promote speciation via colonization of novel habitats.

Figure 19.16. Sclater's Whistler (*Pachycephala soror*, inset) is one of many tropical species whose distribution has been shifting upslope as climates change. In 1965 on Mount Karimui in New Guinea, this species' elevational range extended from 1,133 m up to 1,768 m (white shading). By 2012, both lower and upper distributional limits had shifted upward, and the species now occupies elevational between 1,243–1,903 m (red shading; Freeman and Class Freeman 2014). *Background photo by Benjamin Freeman; image of Sclater's Whistler by Jessica Nguyen.*

Even within a species' current distribution, the spatial consequences of dispersal are multifaceted. The rate of dispersal in and out of populations is a key parameter in metapopulation dynamic models. Within single populations, demographic studies either make assumptions about, or require independent estimates of, dispersal rates. Those estimates are rarely available, however, and, consequently, in populations where site fidelity is low, estimates of apparent survival are a poor estimate of true survival. Thus, dispersal severely inhibits the ability of wildlife managers to accurately estimate population growth rates and determine which parts of the annual cycle to target for conservation in order to achieve the most positive demographic response.

The dispersal of birds can have important implications for the species with which they interact. Dispersing birds can act as vectors for other taxa, such as in mutualist seed dispersal interactions. In the tropics, large, frugivorous birds are common, and most trees produce animal-dispersed seeds. While the plants themselves influence the movements of the birds through spatial variation in fruit production, the birds in turn shape the patterns of seed rain on the landscape (Levey et al. 2005). For example, highly mobile African Hornbills travel long distances, potentially moving the seeds of plants orders of magnitude farther than any other potential dispersers (Holbrook et al. 2002). Similarly, waterbirds routinely move propagules of aquatic plant species and macroinvertebrates between wetlands; a meta-analysis revealed that an astonishing one-third of duck and rail droppings contained at least one propagule, and one-third of those were viable (van Leeuwen et al. 2012). The resulting long-distance dispersal may be rare for plants, but may also be incredibly important for understanding population genetics and survival prospects in a changing world. Sometimes, however, bird movements result in the dispersal of taxa detrimental to humans or other bird populations. Avian dispersal and migration can move parasites and diseases between regions (Owen et al. 2006). Much remains to be learned about the importance of long-distance migrants in the spread of pathogens (chapter 23). While the physiological costs of long-distance movements may compromise or suppress birds' immune systems, long-distance movements may act to purge pathogens from populations when infection reduces survival (Altizer et al. 2011).

LINKS BETWEEN MIGRATION AND DISPERSAL BEHAVIOR

Migration and dispersal have been studied separately by ornithologists, likely because these two movement behaviors appear to vary independently among taxa. Both long-distance migrants and residents can be either incredibly site-faithful or highly dispersive, and sometimes, different populations of the same species differ in one but not the other type of movement. However, as knowledge of the patterns, proximate causes, and constraints on both types of behaviors increases, intriguing relationships between migration and dispersal are emerging. At very broad spatial scales, the predominant north–south direction of long-distance migrations worldwide appears to constrain the spread of avian lineages, ultimately contributing to broad-scale avian biogeographic patterns. Despite the fact that birds are so mobile, birds have rarely dispersed east–west between the New World and Old World (Bohning-Gaese et al. 1998). At finer spatial and taxonomic scales, we do see migration-imposed constraints on dispersal. Dispersal of eastern European White Storks (*Ciconia ciconia*) is oriented along a southeast–northwest axis more frequently than would be expected by chance. The southeast–northwest axis aligns with the direction of spring migratory journeys, again providing evidence that migration may constrain dispersal. Such patterns could arise if dispersal happens when some individuals simply stop their migrations a little too early, or overshoot previous breeding locations, continuing along the same course (Itonaga et al. 2010).

Scandinavian- and Siberian-breeding migrants have smaller breeding distributions than do residents (Bensch 1999). A potential explanation for those smaller distributions is that, in obligate migrants, long-distance dispersal and expansion of breeding ranges into new areas may require considerable modification to migratory programs. To successfully navigate to good wintering areas, birds colonizing new breeding ranges would have to adjust the direction and possibly the timing of migratory trips, or locate new overwintering locations. Newton (2012) noted differences in dispersal tendency between obligate and facultative migrant species and proposed that such differences result from differences in the ways migration is controlled. Species with more facultative components to their migrations exhibit far greater inter-individual variation in migration directions and distances, and those movements are shaped by contemporary local conditions. Thus, dispersal to novel areas for a facultative migrant would impose fewer navigational costs than in an obligate migrant whose migrations are less flexible. The **serial residency** hypothesis extends these ideas and reconciles differences in natal and breeding dispersal within obligate long-distance species (Cresswell 2014). In long-distance, obligate migrants, young birds typically navigate their first migrations based on intrinsic map and compass senses that are subject to considerable error. If young birds successfully make the return trip and breed the following summer, then those individuals have now much to gain from locating the same breeding area in subsequent years. Thus, the capacity to accurately retrace individual, migratory paths following a fairly haphazard journey made in their first year could be adaptive. Much evidence is consistent with predictions of the serial residency hypothesis. Many **vagrants**—birds that show up in unexpected places—are young birds making their first migratory flights (Newton 2007a). Additionally, long-distance migrant adults show remarkable individual-level consistency in their migration behavior (Stanley et al. 2012), yet population-level connectivity is often weak (Franks et al. 2012). The serial residency hypothesis can reconcile such individual predictability and population-level mixing if breeding populations are assembled from individuals ending up in the same place because of chance deviations in their innately controlled navigational systems. Finally, the serial residency hypothesis is consistent with there being no association between breeding dispersal and migration tendency within partial migrants, but longer natal dispersal in migrants than in residents (Alonso et al. 2000).

Nomadism

Nomadism has its own name and place in the avian movement literature, and the term is applied differently by different ornithologists. In some cases, species in which movements closely resemble migration are called nomadic. In other cases, the term refers to movements best considered as the most extreme expression of breeding dispersal. For example, individuals of some grassland-dependent species, such as Henslow's Sparrows (*Ammodramus henslowii*) and Cassin's Sparrows (*Peucaea cassinii*), breed in different locations each year, and even seemingly suitable habitat that was filled with song one year may be silent the following spring (Dornak et al. 2013). Likewise, many northern-breeding owls such as Boreal Owls (*Aegolius funereus*) routinely breed hundreds of kilometers from their previous territories. In such cases, individuals move independently and unpredictably from year to year. Birds may also sometimes be called nomadic when we lack sufficient information to understand and categorize patterns of individual- or population-level migration. For instance, Lawrence's Goldfinch (*Spinus lawrencei*) exhibits a generally east–west migration pattern at the species level, but the numbers of individuals and populations migrating varies from year to year, and even the direction of those movements is not consistent (Davis 1999). For many such "data-deficient" nomadic species, we may eventually understand the temporal and spatial patterns sufficiently to classify their movements more precisely.

Nomadism, as classically applied in the ornithological literature, refers to birds that engage in movements that are equivalent to spatially and temporally unpredictable migrations. Individuals move together in groups, resulting in negli-

Figure 19.17. White-winged Crossbill (*Loxia leucoptera*) on a Douglas-fir cone (*Pseudotsuga menziesii*). *Photo by Kurt Kirchmeier.*

gible genetic consequences, unlike dispersal. Birds of this type of nomadic species rarely breed in the same location twice, being incredibly flexible in their space use over time. Some of the best-studied avian nomads are the crossbill species inhabiting boreal forests of North America and Europe. Crossbills can breed at any time of year, and movements are triggered by changes in availability of conifer cones—the foods that have shaped crossbills' characteristic bill used in prying open cones to extract nutritious seeds (fig. 19.17). Crossbills can breed thousands of kilometers away from natal or previous breeding sites and exhibit seasonal irruptions driven by inter-annual variation in conifer seed production (Newton 2006). Birds reaching southern Europe during irruptions come from different breeding locations in different years but from similar breeding areas in the same year, meaning that breeding populations remain separate (Marquiss et al. 2012). Such cohesiveness is probably adaptive, because crossbill populations have differently shaped bills that match the traits of the conifer species on which they depend (Benkman et al. 2001). Thus, not only the availability of food but also the ability of crossbills to locate mates that share functionally important bill traits may be important factors shaping their movements (Smith et al. 2012).

STUDYING AVIAN MOVEMENTS

Studying the movement ecology of flying animals presents unique challenges because the behavior we want to understand takes our study organisms away from our study sites. Long-distance migrants often cross international boundaries and sometimes travel to different continents. Movement ecologists must cope with the logistical difficulties of determining where birds go when they leave our sites, the costs of our own travel, and sometimes language or cultural barriers that can impede collaborations between researchers at different ends of migratory journeys. Simply determining the patterns of movement remains a huge challenge for a majority of the earth's bird species, and determining the causes of differences in movements is even harder. Consequently, models of movement used in conservation are usually unreliable because of lack of detailed movement information (Beissinger and Westphal 1998). The need for better data on dispersal behavior, in particular, is frequently cited as a conservation priority (Donovan et al. 2002). The major methods we use to study movement differ in their resolution, cost, and effectiveness over different spatial scales, meaning that research questions and study species dictate the method most appropriate to any given study.

Humans have observed bird movements directly for as long as we have had written records. Ancient naturalists also noted the seasonal changes in local presence and absence of species that result from migratory behavior. Such observations are cheap and easily applied to most species, providing that observations are thorough and unbiased. Studies based on direct observations made by amateur ornithologists have made a resurgence because large-scale citizen science databases such as eBird (http://ebird.org) now capture unprecedented amounts of data (Sullivan et al. 2009). Direct counts of migrating raptors also continue to be an effective method for studying the spatial and temporal patterns of migration at the species level (Kim et al. 2015). Another form of direct observation that has been very useful in studies of migration for several decades is use of data collected by large-scale weather radars (fig. 19.18). Scientists have devised analytical tools to distinguish the signals of inanimate objects such as raindrops or hail from the radio signals reflected by animals, and even to be able to estimate body sizes, flight speeds, and wingbeat frequencies (Gauthreaux and Livingston 2006). Although species-level identification is rarely possible, radar has provided a wealth of valuable migration data, particularly on nocturnal movements, for which other forms of direct observation are extremely difficult (Schmaljohann et al. 2007).

A problem with observation of unmarked birds is that we cannot learn much about individual behavior. While observation can provide species-level or, in some cases, population-level data of mass movements, such methods are not useful for studying dispersal or any kind of partial or short-distance migration involving individualistic movements. The practice of marking birds with uniquely coded leg bands, flags, patagial tags, or other markers and recapturing or resighting them elsewhere has transformed our knowledge of avian movements. The earliest records of movements of marked

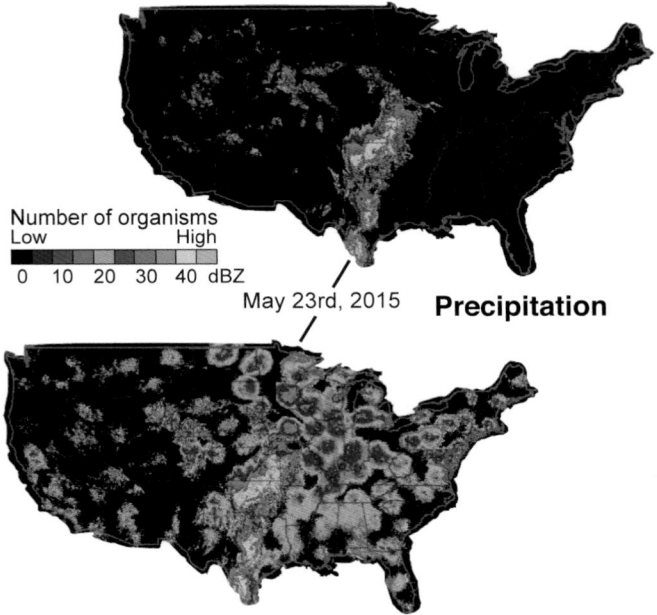

Number of organisms
Low   High

0   10   20   30   40 dBZ

May 23rd, 2015   **Precipitation**

Precipitation + Aerial Organisms

Figure 19.18. Weather radar images that you might find on your weather app have filtered data to only display precipitation (upper image). However, ornithologists can use the original, unfiltered data to visualize, quantify, and study the migration patterns of birds and other aerial organisms (such as bats and large insects). The blue circles visible in the lower image give an idea of the locations of the radar instruments, and show that on the evening of May 23, 2015, large numbers of migrant birds departed from stopover sites, especially in the Midwest and along the eastern seaboard as part of their spring migratory journeys. *Images by Jeff Kelly.*

individuals come from European falconers in the 1500s and 1600s whose birds turned up far from home. Bird banding began with Hans Mortensen in Denmark, who etched his return address on bands placed on the legs of waterfowl, starlings, and raptors. Mortensen's methods became the model for national or continent-wide banding schemes now in place throughout much of the world. Although only a small fraction of banded birds are ever recaptured, resighted, or killed, and reported, the sheer number of marked birds means that bird banding is still one of the most important sources of individual-based movement data available. For instance, band recoveries revealed how House Finches began to migrate latitudinally in the eastern United States (Able and Belthoff 1998) and the dispersal patterns of Smith's Longspurs (Craig et al. 2015), both described earlier in this chapter.

The primary problem with banding is that it is incredibly labor intensive. Consequently, indirect methods of studying individual movement behavior not requiring recaptures are often more cost-effective. Two primary indirect sources of movement information come from population genetics and analyses of stable isotopes and other chemical tracers (chapter 31). Comparisons of allele frequencies between populations provide indirect estimates of the degree of effective dispersal and population connectivity. However, because it takes relatively few dispersal events per generation to result in significant gene flow, genetic and ecological measures of movement tendency can yield very different results. For example, a study of 118 radio-tracked White-tailed Ptarmigan (*Lagopus leucura*) living in disjunct alpine areas of Vancouver Island revealed that birds typically never dispersed more than 1 km. However, populations separated by more than 100 kilometers were identified as genetically cohesive clusters (Fedy et al. 2008). Population-specific molecular markers can also provide direct evidence of the movement of specific individuals from one population to another, or can discriminate the breeding populations of migrants on their wintering or stopover grounds, required to establish patterns of connectivity. For instance, region-specific molecular markers for migrating Red-tailed Hawks (*Buteo jamaicensis*) revealed that two temporal peaks in the number of observed migrants resulted from differences in timing of two genetically distinct populations breeding in California and the Intermountain West (Hull et al. 2009).

The application of stable isotope methods to studying animal migration and dispersal has taken off in recent decades. As an animal feeds, it incorporates the isotopic "signature" (i.e., the ratio of heavy to more common forms of elements) of the feeding location. Because those ratios vary spatially, when we capture an animal and sample tissue, we can infer past locations by matching the ratios in tissues to known geographic variation on the landscape. Consequently, every capture becomes almost equivalent to a "recapture" because we gain geographic information for two times in a single bird's life. Bird migration and dispersal studies most commonly use the ratios of D:H (expressed as δ^2H; chapter 31) and $^{13}C:^{12}C$ ($\delta^{13}C$). Ornithologists most commonly measure those ratios from information in feather samples, providing geographic information from a narrow window of time when feathers were molted, because feathers are inert once grown (Hobson 2005). Given sufficient knowledge of the timing of molt, feather tissues can be useful in linking breeding and nonbreeding ranges. When we need information on a bird's location from a time of year other than during molt, then measuring isotopic ratios in continuously growing tissues that turn over at different rates is very useful. For instance, the claws of small birds turn over on the order of around three months, and red blood cells turn over every two weeks or so. Increasingly, ornithologists combine data from multiple isotopes and/or multiple tissues sampled from the same individuals to gain better temporal and spatial resolution. For example, researchers analyzed δ^2H, $\delta^{13}C$ and $\delta^{15}N$ in feathers, muscle, liver, and blood from three species of *Cinclodes* living in Chile. In combination, their data revealed both inter- and intraspecific variation in altitudinal migration patterns, as well as seasonal shifts in diet from marine to freshwater invertebrates (Newsome et al. 2015). Similarly, the combination of $\delta^{13}C$ data from blood and δ^2H from Black-and-white Warblers (*Mniotilta varia*) feathers sampled during stopover revealed not only the breeding destinations of those birds, but also that birds wintering in high-quality habitats migrated earlier than birds that wintered in arid, lower-quality habitats (Paxton and Moore 2015).

While genetic and isotopic data provide individual-level data without the need for recapture or resighting, both methods yield fairly imprecise geographic data. Isotope and genetic methods can allow us to assign birds to regions (e.g., at the scale of hundreds of kilometers) rather than specific locations (e.g., at scales of <10 kilometers). To gain fine-scale spatial data, ornithologists rely on attaching tags that transmit a signal and/or store geographic information. For decades, manual tracking using small radio telemetry tags has been a fundamental tool to obtain unbiased estimates of movement within small areas (e.g., box on page 615). Radio telemetry revolutionized our knowledge of bird movement, but has limits because tag life and, to some extent, detection range are determined by battery size. Tag mass should not exceed ~3–4 percent of bird mass to minimize birds' additional energetic costs and changes in behavior. Thus, both the duration and spatial scale of movement studies using standard radio telemetry is limited. Fortunately, tag technology is rapidly improving. Batteries are getting lighter, they last longer with micro-solar charging systems, and arrays of automated telemetry receivers placed on towers or other tall structures are greatly extending spatial scale. By strategically placing towers near migration corridors, we can now "see" small nocturnal birds departing on migratory flights and collect individual-level data on stopover duration (e.g., see discussion question 3 below) and other aspects of migration phenology (Bridge et al. 2011).

Technical innovations are expanding the scope of telemetry-type markers in other respects (chapter 31). Large birds can be fitted with GPS devices that send location data to researchers' email at preprogrammed intervals. Although costly, such devices provide very accurate spatial information practically anywhere in the world without requiring recapture. Thus, they are immensely useful for revealing the timing and patterns of migration and dispersal, and the inter- and intra-individual consistency of movements for larger birds (Therrien et al. 2014; fig. 19.19). However, such devices are too

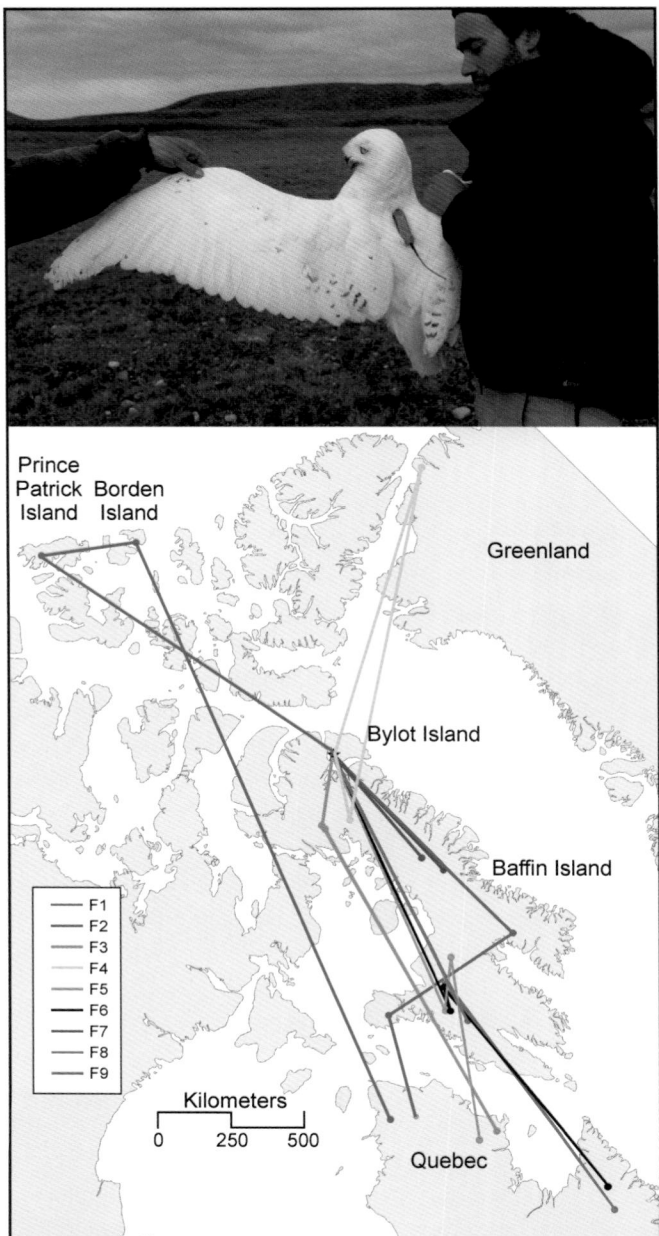

Figure 19.19. Satellite transmitter technology has transformed our ability to study the movements of larger birds, especially in remote locations. The photo shows a Snowy Owl (*Bubo scandiacus*) fitted with a satellite tag in the Canadian arctic. The map shows the resulting data. Each of the nine dots is a breeding location of a different Snowy Owl (each individual is a different colored line) over four successive years (Therrien et al. 2014). All the birds were captured breeding at the point marked with a star. Only one owl returned to breed at that same site in the following three years (individual F4, yellow line) after breeding as far away as northern Greenland in intervening years. In addition to documenting breeding dispersal of up to 2,224 km, satellite data provides much detailed information on routes and timing that were impossible to study until recently. Map reproduced with permission from Jean-François Therrien. *Snowy owl photo by Hilde Marie Johansen.*

heavy for most birds. Those that are too small to carry satellite transmitters can often carry tiny light-detecting geolocators that record the time of daylight hours. The combination of date and time of sunrise and sunset allows us to infer (±a couple hundred kilometers) the latitude and longitude of the bird from deployment to eventual recapture and recovery of the tag. This technology has recently exploded in popularity because it finally allows ornithologists to document patterns of migratory connectivity, migration routes and stopover locations, winter movements, and migration phenology, all at the individual level (McKinnon et al. 2013). Smaller birds can also wear tags that allow researchers to preprogram a limited number of GPS fixes at specified times and dates. The device then stores information onboard, meaning that bird must be recaptured to retrieve data. Thus, like geolocators, their utility is limited to birds that are site-faithful to breeding or wintering areas but provides finer-scale spatial resolution.

MOBILITY AS A LIABILITY

Birds pose unique conservation challenges and opportunities as a consequence of their mobility. To ensure full-life-cycle conservation of migrants, we must protect far more habitat than we do for a sedentary animal. Migrants require not only different breeding and nonbreeding habitat, but stopover areas that provide food and safety from predators. Unfortunately, for many migrants, we don't always know where to focus our conservation efforts (Runge et al. 2015). While we usually have a good idea of species' breeding ranges, we know less about wintering ranges, and we have woefully inadequate information on stopovers. For instance, the wintering location of Black Swifts (*Cypseloides niger*) was only recently discovered in northwestern Brazil (Beason et al. 2012). Effective conservation must also support dispersal capabilities. For habitat specialists, fragmentation can increase the mortality costs of dispersing or effectively eliminate dispersal in cases where birds lack the behavioral or morphological capabilities to cross human-dominated landscapes lacking corridors between habitat patches (Ford et al. 2001). Experimental evidence in tropical understory species demonstrates that the willingness or ability to cross stretches of open water is positively associated with likelihood of occupying isolated islands in the Panama Canal (Moore et al. 2008). The flip side is that highly dispersive species may avoid some of the negative consequences of fragmentation. Despite near elimination of their native grassland, birds such Henslow's and Grasshopper Sparrows have an uncanny ability to locate and breed in patches of newly restored habitat (Gill et al. 2006, Herkert 2007).

Even where suitable habitat allows for normal movements of migrants and dispersers throughout their life,

Figure 19.20. Building collisions are a major source of mortality for nocturnally migrating birds. The Fatal Light Awareness Program (FLAP) works to educate the public and policy makers on how to reduce bird-building collisions and rehabilitate injured birds. Volunteers collect dead and injured birds from near the bases of buildings during migration. This image shows a small fraction of the birds killed by building collisions in 2016, arranged with species experiencing strong population declines in the center. *Photo used by permission of FLAP Canada.*

human activities continue to make those movements more perilous (chapter 26). Nocturnally migrating long-distance migrants must cope with increasing light pollution. Cities, fishing boats, oil and gas rigs, communications towers, and fires now create nocturnal light environments very different from those in which avian navigational abilities evolved (Longcore and Rich 2004). Lights attract nocturnally migrating birds, especially on cloudy nights. Under certain weather conditions, migrating birds apparently become disoriented by lights and collide with lit obstacles in large numbers (fig. 19.20). Some of the species-level estimates of mortality are astounding—an estimated 9 percent of all Yellow Rails (*Coturnicops noveboracensis*), 8.9 percent of Swainson's Warblers (*Limnothlypis swainsonii*), and ~5 percent of Bay-breasted Warblers (*Setophaga castanea*) and Black-throated Blue Warblers (*Setophaga caerulescens*) are killed each year in collisions with communication towers alone (Longcore et al. 2013). An additional quarter of a million birds are estimated to die at monopole wind turbines each year in the United States, and those numbers are predicted to increase as turbines get taller (Loss et al. 2013). The number of birds killed each year by colliding with buildings and power lines in Canada is estimated at over 32 million, second only to the astounding 134 million birds killed by cats (Calvert et al. 2013). For a few species, researchers have been able to estimate the relative riskiness of different phases of the annual cycle, and migration is clearly a dangerous business; roughly 85 percent of the annual mortality of Black-throated Blue Warblers occurs during migration (Sillett and Holmes 2002).

For birds that successfully complete their migratory journeys and find suitable habitat at their destination, anthropogenic disruptions to global climate systems can result in additional problems. When migrations are endogenously controlled, migration phenology is remarkably predictable from year to year (Fraser et al. 2013). Arrival time ultimately determines the timing of peak offspring demand for food. For species that breed at high latitudes, the timing of migration likely evolved in the context of matching the timing of reproduction and food availability. Syncing those two events is critical because short growing seasons mean that prey are abundant only for brief periods. Thus, there is widespread concern that birds and their prey will experience **phenological mismatch** as the emergence of prey shifts earlier in response to local warming, but migrants continue to depart wintering grounds on schedules controlled by unchanging day-length cues. Consistent with phenological mismatch, European migrant arrival dates are not advancing as much as is the climate, and the species that are less flexible in

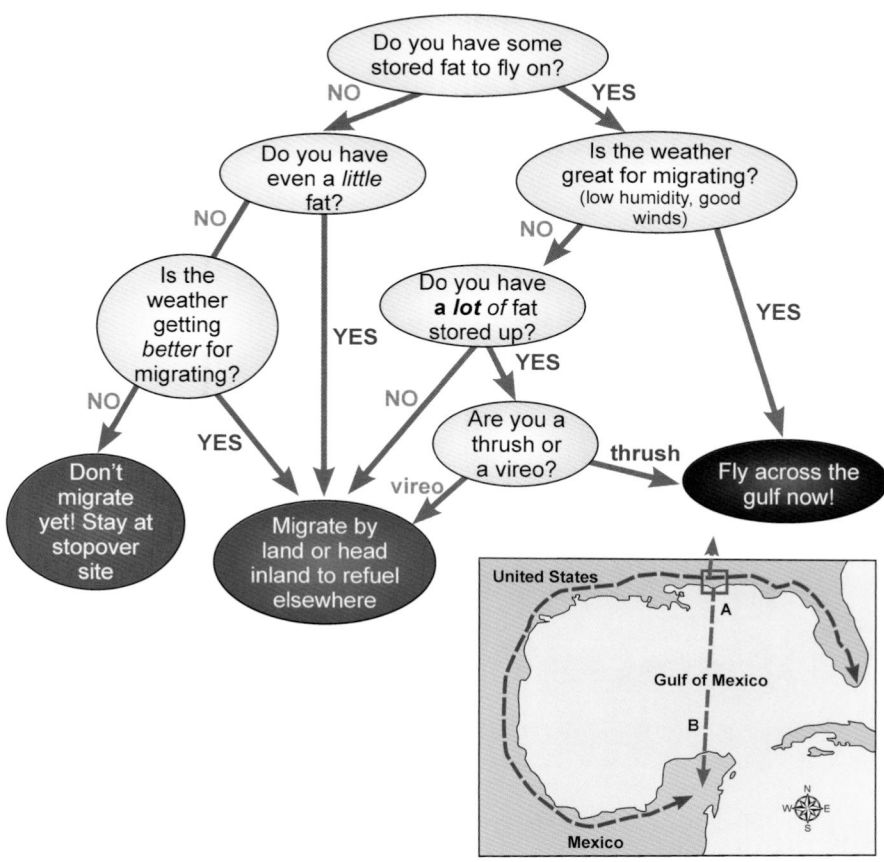

Figure 19.21. Successful migration requires that birds make good decisions. This flow chart represents the decisions made by migrating Swainson's Thrushes, Wood Thrushes, and Red-eyed Vireos. The map shows each of the possible outcomes (migrate over the Gulf of Mexico, blue; depart overland, green; or stay at stopover, red). *Redrawn from Deppe et al. 2015 with permission from Jill Deppe.*

migration timing are also those whose populations are declining (Saino et al. 2011).

Fortunately, variation in movement ecology within single species may buffer them from the myriad threats they experience over their annual cycle. In a study of 340 European-breeding species, partial migrants and species whose nonbreeding range is larger than their breeding range were less likely to be declining than either residents or migrants with small nonbreeding ranges (Gilroy et al. 2016). These results suggest that flexibility in movement behavior may allow birds to respond more quickly to novel challenges. Additionally or alternately, by spreading nonbreeding mortality risk over more areas, species may be more resilient to unpredictable risks affecting portions of their nonbreeding range. As both field and analytic methods improve, our understanding of the spatial scales and patterns of movement are increasing (Clark et al. 2004). As our knowledge of bird movements increases, ornithologists will be better equipped to manage bird populations and provide hope that movement strategies—the product of millennia of adaptation to ever-changing environmental contexts—will be flexible enough to cope with the current rapid pace of anthropogenic change.

KEY POINTS

- Birds are the most mobile creatures on earth; they exhibit an incredible diversity of movement patterns ranging from extremely long to very short distances, across gradients of latitude, elevation, and aridity, and those movements range from highly predictable, repeated movements, to one-time, unpredictable movements.
- Probably at least half of all bird species engage in annual, round-trip migrations; obligate long-distance migrants represent the most extreme of a continuum of behaviors, and most migrants are probably partial and/or facultative migrants.
- Migration has been "switched" on and off many times in the evolution history of birds, appearing when the ultimate benefits, such as availability of food, benign climatic conditions, or predator-free space, outweigh the considerable mortality risks, energetic costs, and temporal constraints.
- Dispersal differs from migration in being a one-way movement that can result in gene flow between populations; some members of all species engage in natal dispersal prior to their first breeding attempt, and in some species, birds make breeding dispersal movements between successive breeding attempts.

- Both migration and dispersal are controlled at proximate levels by both innate and facultative processes; knowing when to travel, where to go, and when to stop involves complex interactions between external cues, genetic predisposition, and physiological processes.
- While direct observation and bird banding have been extremely important sources of movement data, newer technology is rapidly increasing our ability to study movement at the individual level, making it possible to answer long-standing questions and revealing new and intriguing patterns.

KEY MANAGEMENT AND CONSERVATION IMPLICATIONS

- Because birds are so mobile, their migration and dispersal abilities mean that we must conserve far more land area than for resident animals, and we must maintain connectivity among fragmented habitats to maintain healthy populations.
- Full annual cycle conservation of migrants is particularly challenging, especially when birds cross geopolitical barriers requiring international partnerships—critical breeding, stopover, and wintering areas may be in different countries or continents.
- Climate change will likely affect movements in several ways, including promoting dispersal to new regions, affecting the match between of innately controlled migration phenology and food availability and, in facultative migrants, potentially leading to changes in the frequency, direction, or fitness payoffs of different strategies.
- Urbanization affects migratory behavior via selecting for less mobile strategies and impeding migrations by reducing habitat and increasing mortality through impacts with tall buildings and lighted structures.

DISCUSSION QUESTIONS

1. In contrast to birds, mammals exhibit male-biased dispersal—males typically disperse away from natal areas, whereas females remain near natal areas to breed. Discuss the potential causes for such a pattern, based on the ultimate causes of dispersal described in this chapter.
2. Some bird enthusiasts have been concerned that the practice of putting up hummingbird feeders in North America is leading to the loss of migratory behavior in some individuals. Use information in this chapter to argue both for and against the idea that human activities would affect migration in this way.

3. Successfully migrating long distances involves making a series of good decisions en route. How long to stay at stopover locations? When to leave? Which direction to travel? Decisions at key points can make a huge difference to a bird's immediate prospects of survival, how long it survives, and how many offspring it can produce in the future. Researchers placed radio transmitters on Red-eyed Vireos (*Vireo olivaceus*), Wood Thrushes (*Hylocichla mustelina*), and Swainson's Thrushes (*Catharus ustulatus*) at a stopover site in coastal Alabama (Deppe et al. 2015), detected departure times and directions using telemetry receivers along the Gulf Coast and detected arrival on the Yucatan peninsula. Birds chose to cross 1,000 km of open water to the Yucatan coast, or to depart overland, or to remain at stopover sites (fig. 19.21). The birds they studied made different decisions that depended on their fat stores, weather, and their species (vireos vs. two thrush species). Study the decision tree pictured here and discuss the following questions. What costs and benefits might birds be balancing when deciding whether to stay and increase fat stores to make the overwater crossing or depart, taking an overland route around the Gulf Coast? Low and dropping humidity is associated with favorable tail winds that may help explain the importance of weather variables in departure decisions. In what other ways might high humidity affect the costs of nocturnal flight? Vireos were less likely to cross the Gulf of Mexico than the thrushes. What might explain this pattern? (Hint: look up their range maps.)

4. Migration within the Southern Hemisphere is characterized by more facultative and shorter-distance migrations relative to birds breeding in the Northern Hemisphere. One potential driver of such differences is that precipitation patterns may be more important climatic drivers of Austral migrants because of generally milder and drier climates. No similar comparison of northern and southern dispersal tendencies yet exists for birds. If precipitation plays a greater role than temperature in shaping migrations in austral birds, then would you expect dispersal tendencies to also vary? Why or why not?

5. Migrant birds are among those in greatest decline in North America. Habitat destruction is the prime culprit in most species' declines, but migration confounds our ability to pinpoint problem locations and times during the annual cycle. Scientists have limited funds for research. Consider how you would decide how to spend funding if your goal was to conserve Dickcissels. Dickcissels breed in prairies of the midcontinental USA, wintering in Venezuela, where they eat grain crops and are persecuted by farmers. Dickcissels have moder-

ate breeding site fidelity and are parasitized heavily by Brown-headed Cowbirds. You can choose to learn only one of the following priorities for future research (Temple 2002): (a) whether or not birds rear second broods in more northerly parts of the breeding range after raising a brood further south earlier in the breeding season; (b) rates of survival during migration and on wintering grounds; (c) identity of nest predators; (d) the magnitude of financial incentives farmers in Venezuela might be accept in compensation for crop losses if they agree to avoid killing Dickcissels. Which research objective would you chose, and why?

References

Able, K. P., and J. R. Belthoff (1998). Rapid "evolution" of migratory behaviour in the introduced House Finch of eastern North America. Proceedings of the Royal Society of London B: Biological Sciences 265:2063–2071.

Alerstam, T. (2006). Conflicting evidence about long-distance animal navigation. Science 313:791–794.

Alerstam, T., and P. H. Enckell (1979). Unpredictable habitats and evolution of bird migration. Oikos 33:228–232.

Alerstam, T., A. Hedenström, and S. Åkesson (2003). Long-distance migration: Evolution and determinants. Oikos 103:247–260.

Alonso, J. C., M. B. Morales, and J. A. Alonso (2000). Partial migration, and lek and nesting area fidelity in female Great Bustards. The Condor 102:127–136.

Altizer, S., R. Bartel, and B. A. Han (2011). Animal migration and infectious disease risk. Science 331:296–302.

Amaral, F. R., F. H. Sheldon, A. Gamauf, E. Haring, M. Riesing, L. F. Silveira, and A. Wajntal (2009). Patterns and processes of diversification in a widespread and ecologically diverse avian group, the buteonine hawks (Aves, Accipitridae). Molecular Phylogenetics and Evolution 53:703–715.

Anderson, K. E., M. Fujiwara, and S. I. Rothstein (2012). Demography and dispersal of juvenile and adult Brown-headed Cowbirds (*Molothrus ater*) in the Eastern Sierra Nevada, California, estimated using multistate models. The Auk 129:307–318.

Andreu, J., and E. Barba (2006). Breeding dispersal of Great Tits *Parus major* in a homogeneous habitat: Effects of sex, age, and mating status. Ardea 94:45–58.

Andrews, J. E., J. D. Brawn, and M. P. Ward (2015). When to use social cues: Conspecific attraction at newly created grasslands. The Condor: Ornithological Applications 117:297–305.

Areta, J. I., and A. Bodrati (2010). A longitudinal migratory system within the Atlantic Forest: Seasonal movements and taxonomy of the Golden-rumped Euphonia (*Euphonia cyanocephala*) in Misiones (Argentina) and Paraguay. Ornitologia Neotropical 21:71–86.

Aristotle History of Animals VIII (IX) 49b.

Arlt, D., and T. Part (2008). Sex-biased dispersal: A result of a sex difference in breeding site availability. American Naturalist 171:844–850.

Bairlein, F., and E. Gwinner (1994). Nutritional mechanisms and temporal control of migratory energy accumulation in birds. Annual Review of Nutrition 14:187–215.

Barker, F. K., K. J. Burns, J. Klicka, S. M. Lanyon, and I. J. Lovette (2015). New insights into New World biogeography: An integrated view from the phylogeny of blackbirds, cardinals, sparrows, tanagers, warblers, and allies. The Auk 132:333–348.

Bassie-Sweet, K. (2008). Maya sacred geography and the creator deities. University of Oklahoma Press, Norman, OK.

Beason, J. P., C. Gunn, K. M. Potter, R. A. Sparks, and J. W. Fox (2012). The Northern Black Swift: Migration path and wintering area revealed. Wilson Journal of Ornithology 124:1–8.

Beissinger, S. R., and M. I. Westphal (1998). On the use of demographic models of population viability in endangered species management. Journal of Wildlife Management 62:821–841.

Bell, C. P. (2000). Process in the evolution of bird migration and pattern in avian ecogeography. Journal of Avian Biology 31:258–265.

Benkman, C. W., W. C. Holimon, and J. W. Smith (2001). The influence of a competitor on the geographic mosaic of co-evolution between crossbills and lodgepole pine. Evolution 55:282–294.

Bensch, S. (1999). Is the range size of migratory birds constrained by their migratory program? Journal of Biogeography 26:1225–1235.

Berthold, P. (1999). A comprehensive theory for the evolution, control and adaptability of avian migration. Ostrich 70:1–11.

Berthold, P. (2001). Bird migration: A general survey. 2nd ed. Oxford University Press, Oxford, UK.

Berthold, P., G. Mohr, and U. Querner (1990). Control and evolutionary potential of obligate partial migration: Results of a two-way selective breeding experiment with the Blackcap (Sylvia atricapilla). Journal Fur Ornithologie 131:33–45.

Berthold, P., A. J. Helbig, G. Mohr, and U. Querner (1992). Rapid microevolution of migratory behavior in a wild bird species. Nature 360:668–670.

Bildstein, K. L. (2004). Raptor migration in the Neotropics: Patterns, processes, and consequences. Ornitologia Neotropical 15:83–99.

Bohning-Gaese, K., L. I. Gonzalez-Guzman, and J. H. Brown (1998). Constraints on dispersal and the evolution of the avifauna of the Northern Hemisphere. Evolutionary Ecology 12:767–783.

Botsch, Y., R. Arlettaz, and M. Schaub (2012). Breeding dispersal of Eurasian Hoopoes (Upupa epops) within and between years in relation to reproductive success, sex, and age. The Auk 129:283–295.

Boulet, M., and D. R. Norris (2006). The past and present of migratory connectivity. Ornithological Monographs 61:1–13.

Boulinier, T., K. D. McCoy, N. G. Yoccoz, J. Gasparini, and T. Tveraa (2008). Public information affects breeding dispersal in a colonial bird: Kittiwakes cue on neighbours. Biology Letters 4:538–540.

Boyle, W. A. (2008). Partial migration in birds: Tests of three hypotheses in a tropical lekking frugivore. Journal of Animal Ecology 77:1122–1128.

Boyle, W. A. (2017). Altitudinal bird migration in North America. The Auk: Ornithological Advances 134:443–465.

Boyle, W. A., D. R. Norris, and C. G. Guglielmo (2010). Storms drive altitudinal migration in a tropical bird. Proceedings of the Royal Society B: Biological Sciences 277:2511–2519.

Boyle, W. A., C. G. Guglielmo, K. A. Hobson, and D. R. Norris (2011). Lekking birds in a tropical forest forgo sex for migration. Biology Letters 7:661–663.

Boyle, W. A., and B. J. Sigel (2015). Ongoing changes in the avifauna of La Selva Biological Station, Costa Rica: Twenty-three years of Christmas Bird Counts. Biological Conservation 188:11–21.

Bravo, S. P., and V. R. Cueto (2015). La migración vista como algo más que venir, reproducirse y partir: El rol funcional de Elaenia albiceps en los bosques andino patagónicos. In X Neotropical Ornithological Congress and XXII Congresso Brasileiro de Ornitologia, Manaus, Brazil.

Bridge, E. S., K. Thorup, M. S. Bowlin, P. B. Chilson, R. H. Diehl, R. W. Fléron, P. Hartl, et al. (2011). Technology on the move: Recent and forthcoming innovations for tracking migratory birds. Bioscience 61:689–698.

Burgess, N. D., and C. O. F. Mlingwa (2000). Evidence for altitudinal migration of forest birds between montane Eastern Arc and lowland forests in East Africa. Ostrich 71:184–190.

Burney, C. W., and R. T. Brumfield (2009). Ecology predicts levels of genetic differentiation in Neotropical birds. American Naturalist 174:358–368.

Calvert, A. M., C. A. Bishop, R. D. Elliot, E. A. Krebs, T. M. Kydd, C. S. Machtans, and G. J. Robertson (2013). A synthesis of human-related avian mortality in Canada. Avian Conservation and Ecology 8 (2): 11. http://dx.doi.org/10.5751/ACE-00581-080211.

Carboneras, C., and G. M. Kirwan (2016). Upland Goose (Chloephaga picta). In Handbook of the birds of the world alive, del Hoyo, J., A. Elliot, J. Sargatal, D. A. Christie, and E. de Juana, Editors. Lynx Edicions, Barcelona.

Catry, P., M. P. Dias, R. A. Phillips, and J. P. Granadeiro (2013). Carry-over effects from breeding modulate the annual cycle of a long-distance migrant: An experimental demonstration. Ecology 94:1230–1235.

Chalfoun, A. D., and K. A. Schmidt (2012). Adaptive breeding-habitat selection: Is it for the birds? The Auk 129:589–599.

Chauvenet, A. L. M., J. G. Ewen, D. Armstrong, and N. Pettorelli (2013). Saving the Hihi under climate change: A case for assisted colonization. Journal of Applied Ecology 50:1330–1340.

Chaves-Campos, J. (2004). Elevational movements of large frugivorous birds and temporal variation in abundance of fruits along an elevational gradient. Ornitologia Neotropical 15:433–445.

Chesser, R. T. (1994). Migration in South America: An overview of the austral system. Bird Conservation International 4:91–107.

Cimprich, D. A., M. S. Woodrey, and F. R. Moore (2005). Passerine migrants respond to variation in predation risk during stopover. Animal Behaviour 69:1173–1179.

Claramunt, S., E. P. Derryberry, J. V. Remsen Jr., and R. T. Brumfield (2012). High dispersal ability inhibits speciation in a continental radiation of passerine birds. Proceedings of the Royal Society B: Biological Sciences 279:1567–1574.

Clarke, A. L., B. E. Saether, and E. Roskaft (1997). Sex biases in avian dispersal: A reappraisal. Oikos 79:429–438.

Clark, R. G., and D. Shutler (1999). Avian habitat selection: Pattern from process in nest-site use by ducks? Ecology 80:272–287.

Clark, R. G., K. A. Hobson, J. D. Nichols, and S. Bearhop (2004). Dispersal and demography scaling up to the landscape and beyond. The Condor 106:717–719.

Cline, M. H., A. M. Strong, T. S. Sillett, N. L. Rodenhouse, and R. T. Holmes (2013). Correlates and consequences of breeding dispersal in a migratory songbird. The Auk 130:742–752.

Clobert, J., J. F. Le Galliard, J. Cote, S. Meylan, and M. Massot (2009). Informed dispersal, heterogeneity in animal dispersal syndromes and the dynamics of spatially structured populations. Ecology Letters 12:197–209.

Cohen, E. B., F. R. Moore, and R. A. Fischer (2014). Fuel stores, time of spring, and movement behavior influence stopover duration of Red-eyed Vireo *Vireo olivaceus*. Journal of Ornithology 155:785–792.

Cooper, N. W., T. W. Sherry, and P. P. Marra (2015). Experimental reduction of winter food decreases body condition and delays migration in a long-distance migratory bird. Ecology 96:1933–1942.

Cox, G. W. (1968). The role of competition in the evolution of migration. Evolution 22:180–192.

Craig, H. R., S. Kendall, T. Wild, and A. N. Powell (2015). Dispersal and survival of a polygynandrous passerine. The Auk 132:916–925.

Cresswell, W. (2014). Migratory connectivity of Palaearctic-African migratory birds and their responses to environmental change: The serial residency hypothesis. Ibis 156:493–510.

Cristol, D. A., M. C. Baker, and C. Corbone (1999). Differential migration revisited: Latitudinal segregation by age and sex class. Current Ornithology 15:33–88.

Dale, S., A. Lunde, and O. Steifetten (2005). Longer breeding dispersal than natal dispersal in the Ortolan Bunting. Behavioral Ecology 16:20–24.

Danchin, E., and E. Cam (2002). Can non-breeding be a cost of breeding dispersal? Behavioral Ecology and Sociobiology 51:153–163.

Daniels, S. J., and J. R. Walters (2000). Between-year breeding dispersal in Red-cockaded Woodpeckers: Multiple causes and estimated cost. Ecology 81:2473–2484.

Davenport, L. C., K. S. Goodenough, and T. Haugaasen (2016). Birds of two oceans? Trans-Andean and divergent migration of Black Skimmers (*Rynchops niger cinerascens*) from the Peruvian Amazon. PLoS ONE 11. https://doi.org/10.1371/journal.pone.0144994.

Davis, J. N. (1999). Lawrence's Goldfinch (*Spinus lawrencei*). *In* The birds of North America online, A. Poole, Editor. Cornell Lab of Ornithology, Ithaca, NY.

Delmore, K. E., and D. E. Irwin (2014). Hybrid songbirds employ intermediate routes in a migratory divide. Ecology Letters 17:1211–1218.

Delmore, Kira E., David P. L. Toews, Ryan R. Germain, Gregory L. Owens, and Darren E. Irwin (2016). The genetics of seasonal migration and plumage color. Current Biology. http://dx.doi.org/10.1016/j.cub.2016.06.015.

DeLuca, W. V., B. K. Woodworth, C. C. Rimmer, P. P. Marra, P. D. Taylor, K. P. McFarland, S. A. Mackenzie, and D. R. Norris (2015). Transoceanic migration by a 12 g songbird. Biology Letters 11 (4): 20141045. doi:10.1098/rsbl.2014.1045.

Deppe, J. L., M. P. Ward, R. T. Bolus, R. H. Diehl, A. Celis-Murillo, T. J. Zenzal, F. R. Moore, et al. (2015). Fat, weather, and date affect migratory songbirds' departure decisions, routes, and time it takes to cross the Gulf of Mexico. Proceedings of the National Academy of Sciences 112:E6331–E6338.

Deutschlander, M. E., and R. C. Beason (2014). Avian navigation and geographic positioning. Journal of Field Ornithology 85:111–133.

Dingemanse, N. J., C. Both, A. J. van Noordwijk, A. L. Rutten, and P. J. Drent (2003). Natal dispersal and personalities in Great Tits (*Parus major*). Proceedings of the Royal Society B: Biological Sciences 270:741–747.

Dingle, H. (1996). Migration: The biology of life on the move. Oxford University Press, New York.

Donovan, T. M., C. J. Beardmore, D. N. Bonter, J. D. Brawn, R. J. Cooper, J. A. Fitzgerald, R. Ford, et al. (2002). Priority research needs for the conservation of Neotropical migrant landbirds. Journal of Field Ornithology 73:329–339.

Dornak, L. L., N. Barve, and A. T. Peterson (2013). Spatial scaling of prevalence and population variation in three grassland sparrows. The Condor 115:186–197.

Duckworth, R. A. (2009). Maternal effects and range expansion: A key factor in a dynamic process? Philosophical Transactions of the Royal Society B: Biological Sciences 364:1075–1086.

Duckworth, R. A. (2012). Evolution of genetically integrated dispersal strategies. *In* Dispersal ecology and evolution, J. Clobert, M. Baguette, T. G. Benton, and D. E. Bowler, Editors. Oxford University Press, Oxford, UK, pp. 83–94.

Duckworth, R. A., V. Belloni, and S. R. Anderson (2015). Cycles of species replacement emerge from locally induced maternal effects on offspring behavior in a passerine bird. Science 347:875–877.

Duriez, O., B. J. Ens, R. Choquet, R. Pradel, and M. Klaassen (2012). Comparing the seasonal survival of resident and migratory oystercatchers: Carry-over effects of habitat quality and weather conditions. Oikos 121:862–873.

Einoder, L. D., B. Page, S. D. Goldsworthy, S. C. De Little, and C. J. A. Bradshaw (2011). Exploitation of distant Antarctic waters and close neritic waters by Short-tailed Shearwaters breeding in South Australia. Austral Ecology 36:461–475.

Emlen, S. T., and J. T. Emlen (1966). A technique for recording migratory orientation of captive birds. The Auk 83:361–367.

Fajardo, N., A. M. Strong, N. G. Perlut, and N. J. Buckley (2009). Natal and breeding dispersal of Bobolinks (*Dolichonyx oryzivorus*) and Savannah Sparrows (*Passerculus sandwichensis*) in an agricultural landscape. The Auk 126:310–318.

Fedy, B. C., K. Martin, C. Ritland, and J. Young (2008). Genetic and ecological data provide incongruent interpretations of population structure and dispersal in naturally subdivided populations of White-tailed Ptarmigan (*Lagopus leucura*). Molecular Ecology 17:1905–1917.

Fletcher, R. J., E. P. Robertson, R. C. Wilcox, B. E. Reichert, J. D. Austin, and W. M. Kitchens (2015). Affinity for natal environments by dispersers impacts reproduction and explains geographical structure of a highly mobile bird. Proceedings of the Royal Society of London B: Biological Sciences 282 (1814): 20151545. doi:10.1098/rspb.2015.1545.

Ford, H. A., G. W. Barrett, D. A. Saunders, and H. F. Recher (2001). Why have birds in the woodlands of Southern Australia declined? Biological Conservation 97:71–88.

Forero-Medina, G., J. Terborgh, S. J. Socolar, and S. L. Pimm (2011). Elevational ranges of birds on a tropical montane gradient lag behind warming temperatures PLoS ONE 6:e28535.

Franks, S. E., D. R. Norris, T. K. Kyser, G. Fernandez, B. Schwarz, R. Carmona, M. A. Colwell, et al. (2012). Range-wide patterns of migratory connectivity in the Western Sandpiper *Calidris mauri*. Journal of Avian Biology 43:155–167.

Fraser, K. C., C. Silverio, P. Kramer, N. Mickle, R. Aeppli, and B. J. M. Stutchbury (2013). A trans-hemispheric migratory songbird does not advance spring schedules or increase migration rate in response to record-setting temperatures at breeding sites. PLoS ONE 8:e64587.

Freeman, B. G., and A. M. Class Freeman (2014). Rapid upslope shifts in New Guinean birds illustrate strong distributional responses of tropical montane species to global warming. Proceedings of the National Academy of Sciences 111:4490–4494.

Fretwell, S. D. (1972). Populations in a seasonal environment. Princeton University Press, Princeton, NJ.

Fretwell, S. D. (1980). Evolution of migration in relation to factors regulating bird numbers. *In* Migrant birds in the Neotropics, A. Keast and E. S. Morton, Editors. Smithsonian Institution Press, Washington, DC, pp. 517–527.

Fretwell, S. D. (1985). Why do birds migrate? Inter and intraspecific competition in the evolution of bird migration: Contributions from population ecology. *In* Proceedings of the 18th International Ornithological Congress, pp. 630–637.

Garant, D., L. E. B. Kruuk, T. A. Wilkin, R. H. McCleery, and B. C. Sheldon (2005). Evolution driven by differential dispersal within a wild bird population. Nature 433:60–65.

Gauthreaux, S. A., Jr. (1982). The ecology and evolution of avian migration systems. *In* Avian biology, vol. 6, D. S. Farner, J. R. King, and K. C. Parkes, Editor. Academic Press, New York, pp. 93–168.

Gauthreaux, S. A., and J. W. Livingston (2006). Monitoring bird migration with a fixed-beam radar and a thermal-imaging camera. Journal of Field Ornithology 77:319–328.

Gerson, A. R., and C. G. Guglielmo (2011). Flight at low ambient humidity increases protein catabolism in migratory birds. Science 333:1434–1436.

Ghalambor, C. K., R. B. Huey, P. R. Martin, J. J. Tewksbury, and G. Wang (2006). Are mountain passes higher in the tropics? Janzen's hypothesis revisited. Integrative and Comparative Biology 46:5–17.

Gill, D. E., P. Blank, J. Parks, J. B. Guerard, B. Lohr, E. Schwartzman, J. G. Gruber, G. Dodge, C. A. Rewa, and H. F. Sears (2006). Plants and breeding bird response on a managed conservation reserve program grassland in Maryland. Wildlife Society Bulletin 34:944–956.

Gilroy, J. J., J. A. Gill, S. H. M. Butchart, V. R. Jones, and A. M. A. Franco (2016). Migratory diversity predicts population declines in birds. Ecology Letters 19 (3): 308–317. doi:10.1111/ele.12569.

Green, D. J., I. B. J. Whitehorne, H. A. Middleton, and C. A. Morrissey (2015). Do American Dippers obtain a survival benefit from altitudinal migration? PLoS ONE 10:e0125734.

Greenberg, R. (1980). Demographic aspects of long-distance migration. *In* Migrant birds in the Neotropics, A. Keast and E. S. Morton, Editors. Smithsonian Institution Press, Washington, DC, pp. 493–504.

Greenwood, P. J. (1980). Mating systems, philopatry and dispersal in birds and mammals. Animal Behaviour 28:1140–1162.

Greenwood, P. J., P. H. Harvey, and C. M. Perrins (1979). Role of dispersal in the Great Tit (*Parus major*): Causes, consequences, and heritability of natal dispersal. Journal of Animal Ecology 48:123–142.

Greenwood, P. J., and P. H. Harvey (1982). The natal and breeding dispersal of birds. Annual Review of Ecology and Systematics 13:1–21.

Hamilton, W. D., and R. M. May (1977). Dispersal in stable habitats. Nature 269:578–581.

Hansson, B., S. Bensch, D. Hasselquist, and B. Nielsen (2002). Restricted dispersal in a long-distance migrant bird with patchy distribution, the Great Reed Warbler. Oecologia 130:536–542.

Hardouin, L. A., M. Nevoux, A. Robert, O. Gimenez, F. Lacroix, and Y. Hingrat (2012). Determinants and costs of natal dispersal in a lekking species. Oikos 121:804–812.

Hedenstrom, A., and T. Alerstam (1997). Optimum fuel loads in migratory birds: Distinguishing between time and energy minimization. Journal of Theoretical Biology 189:227–234.

Herkert, J. R. (2007). Evidence for a recent Henslow's Sparrow population increase in Illinois. Journal of Wildlife Management 71:1229–1233.

Hobson, K. A. (2005). Stable isotopes and the determination of avian migratory connectivity and seasonal interactions. The Auk 122:1037–1048.

Holbrook, K. M., T. B. Smith, and B. D. Hardesty (2002). Implications of long-distance movements of frugivorous rain forest hornbills. Ecography 25:745–749.

Holland, R. A. (2014). True navigation in birds: From quantum physics to global migration. Journal of Zoology 293 (1): 1–15. doi:10.1111/jzo.12107.

Hull, J. M., H. B. Ernest, J. A. Harley, A. M. Fish, and A. C. Hull (2009). Differential migration between discrete populations of juvenile Red-tailed Hawks (*Buteo jamaicensis*). The Auk 126:389–396.

Inger, R., X. A. Harrison, G. D. Ruxton, J. Newton, K. Colhoun, G. A. Gudmundsson, G. McElwaine, M. Pickford, D. Hodgson, and S. Bearhop (2010). Carry-over effects reveal reproductive costs in a long-distance migrant. Journal of Animal Ecology 79:974–982.

Itonaga, N., U. Koppen, M. Plath, and D. Wallschlager (2010). Breeding dispersal directions in the White Stork (*Ciconia*

ciconia) are affected by spring migration routes. Journal of Ethology 28:393–397.

Jahn, A. E., D. J. Levey, V. R. Cueto, J. P. Ledezma, D. T. Tuero, J. W. Fox, and D. Masson (2013). Long-distance bird migration within South America revealed by light-level geolocators. The Auk 130:223–229.

Johnson, D. N., and G. L. Maclean (1994). Altitudinal migration in Natal. Ostrich 65:86–94.

Jones, S. L., J. S. Dieni, M. T. Green, and P. J. Gouse (2007). Annual return rates of breeding grassland songbirds. Wilson Journal of Ornithology 119:89–94.

Ketterson, E. D., and V. Nolan Jr. (1976). Geographic variation and its climatic correlates in sex-ratio of eastern wintering Dark-eyed Juncos (*Junco hyemalis hyemalis*). Ecology 57:679–693.

Ketterson, E. D., and V. Nolan Jr. (1983). The evolution of differential bird migration. Current Ornithology 1:357–402.

Kim, H.-K., M. Sendra Vega, M. Wahl, C. L. Puan, L. Goodrich, and K. L. Bildstein (2015). Relationship between the North Atlantic Oscillation and spring migration phenology of Broad-winged Hawks (*Buteo platypterus*) at Hawk Mountain Sanctuary, 1998–2013. Journal of Raptor Research 49:471–478.

Kim, S. Y., R. Torres, C. Rodriguez, and H. Drummond (2007). Effects of breeding success, mate fidelity and senescence on breeding dispersal of male and female Blue-footed Boobies. Journal of Animal Ecology 76:471–479.

Kirby, J. S., A. J. Stattersfield, S. H. M. Butchart, M. I. Evans, R. F. A. Grimmett, V. R. Jones, J. O'Sullivan, G. M. Tucker, and I. Newton (2008). Key conservation issues for migratory land- and waterbird species on the world's major flyways. Bird Conservation International 18:S49–S73.

Klaassen, R. H. G., M. Hake, R. Strandberg, B. J. Koks, C. Trierweiler, K.-M. Exo, F. Bairlein, and T. Alerstam (2014). When and where does mortality occur in migratory birds? Direct evidence from long-term satellite tracking of raptors. Journal of Animal Ecology 83:176–184.

Koenig, W. D., F. A. Pitelka, W. J. Carmen, R. L. Mumme, and M. T. Stanback (1992). The evolution of delated dispersal in cooperative breeders. Quarterly Review of Biology 67:111–150.

Kuo, Y. L., D. L. Lin, F. M. Chuang, P. F. Lee, and T. S. Ding (2013). Bird species migration ratio in East Asia, Australia, and surrounding islands. Naturwissenschaften 100:729–738.

La Sorte, F. A., and F. R. Thompson (2007). Poleward shifts in winter ranges of North American birds. Ecology 88:1803–1812.

Lavers, J. L., I. L. Jones, and A. W. Diamond (2007). Natal and breeding dispersal of Razorbills (*Alca torda*) in eastern North America. Waterbirds 30:588–594.

Lecomte, N., G. Gauthier, and J. F. Giroux (2008). Breeding dispersal in a heterogeneous landscape: The influence of habitat and nesting success in Greater Snow Geese. Oecologia 155:33–41.

Lenormand, T. (2002). Gene flow and the limits to natural selection. Trends in Ecology and Evolution 17:183–189.

Levey, D. J., and F. G. Stiles (1992). Evolutionary precursors of long-distance migration: Resource availability and movement patterns in Neotropical landbirds. American Naturalist 140:447–476.

Levey, D. J., B. M. Bolker, J. J. Tewksbury, S. Sargent, and N. M. Haddad (2005). Effects of landscape corridors on seed dispersal by birds. Science 309:146–148.

Lima, S. L. (1986). Predation risk and unpredictable feeding conditions: Determinants of body mass in birds. Ecology 67:377–385.

Loiselle, B. A., and J. G. Blake (1991). Temporal variation in birds and fruits along an elevational gradient in Costa Rica. Ecology 72:180–193.

Longcore, T., and C. Rich (2004). Ecological light pollution. Frontiers in Ecology and the Environment 2:191–198.

Longcore, T., C. Rich, P. Mineau, B. MacDonald, D. G. Bert, L. M. Sullivan, E. Mutrie, et al. (2012). An estimate of avian mortality at communication towers in the United States and Canada. PLoS ONE 7:e34025.

Longcore, T., C. Rich, P. Mineau, B. MacDonald, D. G. Bert, L. M. Sullivan, E. Mutrie, et al. (2013). Avian mortality at communication towers in the United States and Canada: Which species, how many, and where? Biological Conservation 158:410–419.

Loss, S. R., T. Will, and P. P. Marra (2013). Estimates of bird collision mortality at wind facilities in the contiguous United States. Biological Conservation 168:201–209.

Mackas, R. H., D. J. Green, I. B. J. Whitehorne, E. N. Fairhurst, H. A. Middleton, and C. A. Morrissey (2010). Altitudinal migration in American Dippers (*Cinclus mexicanus*): Do migrants produce higher quality offspring? Canadian Journal of Zoology-Revue Canadienne de Zoologie 88:369–377.

Marquiss, M. I. Newton, K. A. Hobson, and Y. Kolbeinsson (2012). Origins of irruptive migrations by Common Crossbills *Loxia curvirostra* into northwestern Europe revealed by stable isotope analysis. Ibis 154:400–409.

Marra, P. P., K. A. Hobson, and R. T. Holmes (1998). Linking winter and summer events in a migratory bird by using stable-carbon isotopes. Science 282:1884–1886.

Marra, P. P., E. B. Cohen, S. R. Loss, J. E. Rutter, and C. M. Tonra (2015). A call for full annual cycle research in animal ecology. Biology Letters 11:20150552. http://dx.doi.org/10.1098/rsbl.2015.0552.

Martin, T. E. (1996). Life history evolution in tropical and south temperate birds: What do we really know? Journal of Avian Biology 27:263–272.

Marzluff, J. M., J. J. Millspaugh, P. Hurvitz, and M. S. Handcock (2004). Relating resources to a probabilistic measure of space use: Forest fragments and Steller's Jays. Ecology 85:1411–1427.

Matechou, E., S. C. Cheng, L. R. Kidd, and C. J. Garroway (2015). Reproductive consequences of the timing of seasonal movements in a nonmigratory wild bird population. Ecology 96:1641–1649.

McKinnon, E. A., K. C. Fraser, and B. J. M. Stutchbury (2013). New discoveries in landbird migration using geolocators, and a flight plan for the future. The Auk 130:211–222.

McKinnon, L., P. A. Smith, E. Nol, J. L. Martin, F. I. Doyle, K. F. Abraham, H. G. Gilchrist, R. I. G. Morrison, and J. Bêty (2010). Lower predation risk for migratory birds at high latitudes. Science 327:327–328.

Møller, A. P. (2007). Senescence in relation to latitude and migration in birds. Journal of Evolutionary Biology 20:750–757.

Mönkkönen, M. (1995). Do migrant birds have more pointed wings? A comparative study. Evolutionary Ecology 9:520–528.

Moore, R. P., W. D. Robinson, I. J. Lovette, and T. R. Robinson (2008). Experimental evidence for extreme dispersal limitation in tropical forest birds. Ecology Letters 11:960–968.

Morrissey, C. A. (2004). Effect of altitudinal migration within a watershed on the reproductive success of American Dippers. Canadian Journal of Zoology-Revue Canadienne De Zoologie 82:800–807.

Morrissey, C. A. (2010). Altitudinal migration in American Dippers (*Cinclus mexicanus*): Do migrants produce higher quality offspring? Canadian Journal of Zoology-Revue Canadienne De Zoologie 88:369–377.

Myers, J. P. (1981). A test of three hypotheses for latitudinal segregation of the sexes in wintering birds. Canadian Journal of Zoology-Revue Canadienne De Zoologie 59:1527–1534.

Nathan, R., W. M. Getz, E. Revilla, M. Holyoak, R. Kadmon, D. Saltz, and P. E. Smouse (2008). A movement ecology paradigm for unifying organismal movement research. Proceedings of the National Academy of Sciences 105:19052–19059.

Nebel, S., D. B. Lank, P. D. O'Hara, G. Fernandez, B. Haase, F. Delgado, F. A. Estela, et al. (2002). Western Sandpipers (*Calidris mauri*) during the nonbreeding season: Spatial segregation on a hemispheric scale. The Auk 119:922–928.

Nevoux, M., D. Arlt, M. Nicoll, C. Jones, and K. Norris (2013). The short- and long-term fitness consequences of natal dispersal in a wild bird population. Ecology Letters 16 (4): 438–445. doi:10.1111/ele.12060.

Newmark, W. D., V. J. Mkongewa, and A. D. Sobek (2010). Ranging behavior and habitat selection of terrestrial insectivorous birds in north-east Tanzania: Implications for corridor design in the Eastern Arc Mountains. Animal Conservation 13:474–482.

Newsome, S. D., P. Sabat, N. Wolf, J. A. Rader, and C. M. del Rio (2015). Multi-tissue δ2H analysis reveals altitudinal migration and tissue-specific discrimination patterns in Cinclodes. Ecosphere 6:1–18.

Newton, I. (2006). Movement patterns of Common Crossbills *Loxia curvirostra* in Europe. Ibis 148:782–788.

Newton, I. (2007a). The migration ecology of birds. Academic Press, London.

Newton, I. (2007b). Weather-related mass-mortality events in migrants. Ibis 149:453–467.

Newton, I. (2012). Obligate and facultative migration in birds: Ecological aspects. Journal of Ornithology 153:S171–S180.

Newton, I., and L. Dale (1996a). Relationship between migration and latitude among west European birds. Journal of Animal Ecology 65:137–146.

Newton, I., and L. C. Dale (1996b). Bird migration at different latitudes in eastern North America. The Auk 113:626–635.

Nocedal, J. (1994). Local migrations of insectivorous birds in western Mexico: Implications for the protection and conservation of their habitats. Bird Conservation International 4:129–142.

Norris, D. R., P. P. Marra, T. K. Kyser, T. W. Sherry, and L. M. Ratcliffe (2004a). Tropical winter habitat limits reproductive success on the temperate breeding grounds in a migratory bird. Proceedings of the Royal Society of London B: Biological Sciences 271:59–64.

Norris, D. R., P. P. Marra, R. Montgomerie, T. K. Kyser, and L. M. Ratcliffe (2004b). Reproductive effort, molting latitude, and feather color in a migratory songbird. Science 306:2249–2250.

Olaus, M. (1996). Description of the northern peoples: Rome 1555. Hakluyt Society, London.

Owen, J., F. Moore, N. Panella, E. Edwards, R. Bru, M. Hughes, and N. Komar (2006). Migrating birds as dispersal vehicles for West Nile virus. Ecohealth 3:79–85.

Paradis, E., S. R. Baillie, W. J. Sutherland, and R. D. Gregory (1998). Patterns of natal and breeding dispersal in birds. Journal of Animal Ecology 67:518–536.

Parrish, J. D. (1997). Patterns of frugivory and energetic condition in Nearctic landbirds during autumn migration. The Condor 99:681–697.

Partecke, J., and E. Gwinner (2007). Increased sedentariness in European Blackbirds following urbanization: A consequence of local adaptation? Ecology 88:882–890.

Paxton, K. L., and F. R. Moore (2015). Carry-over effects of winter habitat quality on en route timing and condition of a migratory passerine during spring migration. Journal of Avian Biology 46:495–506.

Pedler, R. D., R. F. H. Ribot, and A. T. Bennett (2014). Extreme nomadism in desert waterbirds: Flights of the Banded Stilt. Biology Letters 10. doi:10.1098/rsbl.2014.0547.

Piersma, T., G. A. Gudmundsson, and K. Lilliendahl (1999). Rapid changes in the size of different functional organ and muscle groups during refueling in a long-distance migrating shorebird. Physiological and Biochemical Zoology 72:405–415.

Ramenofsky, M., J. Cornelius, and B. Helm (2012). Physiological and behavioral responses of migrants to environmental cues. Journal of Ornithology 153:181–191.

Rogers, C. M. (2015). Testing optimal body mass theory: Evidence for cost of fat in wintering birds. Ecosphere 6. doi:10.1890/ES14-00317.1.

Rohwer, S., K. A. Hobson, and V. G. Rohwer (2009). Migratory double breeding in Neotropical migrant birds. Proceedings of the National Academy of Sciences USA 106:19050–19055.

Runge, C. A., J. E. M. Watson, S. H. M. Butchart, J. O. Hanson, H. P. Possingham, and R. A. Fuller (2015). Protected areas and global conservation of migratory birds. Science 350:1255–1258.

Saino, N., T. Szep, R. Ambrosini, M. Romano, and A. P. Møller (2004). Ecological conditions during winter affect sexual selection and breeding in a migratory bird. Proceedings of the Royal Society of London B: Biological Sciences 271:681–686.

Saino, N., R. Ambrosini, D. Rubolini, J. von Hardenberg, A. Provenzale, K. Huppop, O. Huppop, et al. (2011). Climate warming, ecological mismatch at arrival and population decline in migratory birds. Proceedings of the Royal Society B: Biological Sciences 278:835–842.

Sandercock, B. K., and A. Jaramillo (2002). Annual survival rates of wintering sparrows: Assessing demographic consequences of migration. The Auk 119:149–165.

Schmaljohann, H., F. Liechti, and B. Bruderer (2007). Songbird migration across the Sahara: The non-stop hypothesis rejected!

Proceedings of the Royal Society B: Biological Sciences 274:735–739.

Sherry, T. W., and R. T. Holmes (1995). Summer versus winter limitation of populations: What are the issues and what is the evidence? *In* Ecology and management of Neotropical migratory birds: A synthesis and review of critical issues, T. E. Martin and D. M. Finch, Editors. Oxford University Press, New York, pp. 85–120.

Sillett, T. S., and R. T. Holmes (2002). Variation in survivorship of a migratory songbird throughout its annual cycle. Journal of Animal Ecology 71:296–308.

Smith, J. W., S. M. Sjoberg, M. C. Mueller, and C. W. Benkman (2012). Assortative flocking in crossbills and implications for ecological speciation. Proceedings of the Royal Society B: Biological Sciences 279:4223–4229.

Snow, D., E. de Juana, and C. J. Sharpe (2016). Swallow-tailed Cotinga (*Phibalura flavirostris*). *In* Handbook of the birds of the world alive, del Hoyo, J., A. Elliot, J. Sargatal, D. A. Christie, and E. de Juana, Editors. Lynx Edicions, Barcelona.

Sol, D., R. P. Duncan, T. M. Blackburn, P. Cassey, and L. Lefebvre (2005). Big brains, enhanced cognition, and response of birds to novel environments. Proceedings of the National Academy of Sciences 102:5460–5465.

Sol, D., N. Garcia, A. Iwaniuk, K. Davis, A. Meade, W. A. Boyle, and T. Székely (2010). Evolutionary divergence in brain size between migratory and resident birds. PLOS ONE 5:e9617.

Solórzano, S., S. Castillo, T. Valverde, and L. Ávila (2000). Quetzal abundance in relation to fruit availability in a cloud forest of southeastern Mexico. Biotropica 32:523–532.

Stanley, C. Q., M. MacPherson, K. C. Fraser, E. A. McKinnon, and B. J. M. Stutchbury (2012). Repeat tracking of individual songbirds reveals consistent migration timing but flexibility in route. PLoS ONE 7. https://doi.org/10.1371/journal.pone.0040688.

Steadman, D. W. (2005). The paleoecology and fossil history of migratory landbirds. *In* Birds of two worlds, C. H. Greenberg and P. P. Marra, Editors. Johns Hopkins University Press, Baltimore, MD.

Stiles, F. G. (1980). The annual cycle in a tropical wet forest hummingbird community. Ibis 122:322–343.

Stiles, F. G. (1988). Altitudinal movements of birds on the Caribbean slope of Costa Rica: Implications for conservation. *In* Tropical rainforests: Diversity and conservation, F. Alameda and C. M. Pringle, Editors. California Academy of Sciences, San Francisco, pp. 243–258.

Stiles, F. G., and A. F. Skutch (1989). A guide to the birds of Costa Rica. Cornell University Press, Ithaca, NY.

Streby, H. M., G. R. Kramer, S. M. Peterson, J. A. Lehman, D. A. Buehler, and D. E. Andersen (2015). Tornadic storm avoidance behavior in breeding songbirds. Current Biology 25:98–102.

Sullivan, B. L., C. L. Wood, M. J. Iliff, R. E. Bonney, D. Fink, and S. Kelling (2009). eBird: A citizen-based bird observation network in the biological sciences. Biological Conservation 142:2282–2292.

Temple, S. A. (2002). Dickcissel. *In* The birds of North America online, A. Poole, Editor. Retrieved from the birds of North America online. http://bna.birds.cornell.edu/bna/species/703. Cornell Lab of Ornithology, Ithaca, NY.

Terrill, S. B., and K. P. Able (1988). Bird migration terminology. The Auk 105:205–206.

Therrien, J. F., G. Gauthier, D. Pinaud, and J. Bety (2014). Irruptive movements and breeding dispersal of snowy owls: A specialized predator exploiting a pulsed resource. Journal of Avian Biology 45:536–544.

Urbanek, R. P., L. E. A. Fondow, S. E. Zimorski, M. A. Wellington, and M. A. Nipper (2010). Winter release and management of reintroduced migratory Whooping Cranes *Grus americana*. Bird Conservation International 20:43–54.

van Leeuwen, C. H. A., G. van der Velde, J. M. van Groenendael, and M. Klaassen (2012). Gut travellers: Internal dispersal of aquatic organisms by waterfowl. Journal of Biogeography 39:2031–2040.

Von Haartman, L. (1968). The evolution of resident versus migratory habit in birds: Some considerations. Ornis Fennica 45:1–7.

Wellicome, T. I., R. J. Fisher, R. G. Poulin, L. D. Todd, E. M. Bayne, D. T. T. Flockhart, J. K. Schmutz, K. De Smet, and P. C. James (2014). Apparent survival of adult Burrowing Owls that breed in Canada is influenced by weather during migration and on their wintering grounds. The Condor 116:446–458.

Wheelwright, N. T., and R. A. Mauck (1998). Philopatry, natal dispersal, and inbreeding avoidance in an island population of Savannah Sparrows. Ecology 79:755–767.

Willemoes, M., J. Blas, M. Wikelski, and K. Thorup (2015). Flexible navigation response in common cuckoos *Cuculus canorus* displaced experimentally during migration. Scientific Reports 5:16402.

Williams, E. J., and W. A. Boyle (2018). Patterns and correlates of within-season breeding dispersal in a declining grassland songbird. The Auk 135:1–14.

Wiltschko, R., and W. Wiltschko (2003). Avian navigation: From historical to modern concepts. Animal Behaviour 65:257–272.

Wiltschko, R., and W. Wiltschko (2009). Avian navigation. The Auk 126:717–743.

Winder, V. L., L. B. McNew, A. J. Gregory, L. M. Hunt, S. M. Wisely, and B. K. Sandercock (2014). Space use by female Greater Prairie-Chickens in response to wind energy development. Ecosphere 5. doi:10.1890/ES13-00206.1.

Winkler, D. W., P. H. Wrege, P. E. Allen, T. L. Kast, P. Senesac, M. F. Wasson, P. E. Llambias, V. Ferretti, and P. J. Sullivan (2004). Breeding dispersal and philopatry in the Tree Swallow. The Condor 106:768–776.

Winkler, D. W., P. H. Wrege, P. E. Allen, T. L. Kast, P. Senesac, M. F. Wasson, and P. J. Sullivan (2005). The natal dispersal of Tree Swallows in a continuous mainland environment. Journal of Animal Ecology 74:1080–1090.

Winkler, H., and B. Leisler (2005). To be a migrant: Ecomorphological burdens and chances. *In* Birds of two worlds: The ecology and evolution of migration, R. Greenberg and P. P. Marra, Editors. Johns Hopkins University Press, Baltimore, MD, pp. 79–86.

Zink, R. M. (2011). The evolution of avian migration. Biological Journal of the Linnean Society 104:237–250.

PART V

POPULATIONS AND ASSEMBLAGES

CHAPTER 20

Population Structure

Peter Arcese and Lukas F. Keller

Anyone fortunate enough to have observed a sample of individually identifiable birds from a single population in nature will note that despite one's first impressions—that most individuals of a species look alike and act similarly—detailed study will inevitably reveal that individual variation in behavioral, morphological, and other expressed traits is ubiquitous. To evolutionary ecologists, this variation represents the raw material on which natural and sexual selection act to shape individual life histories in all species, and it leads to many questions about the environmental, social, genetic, and developmental factors responsible for variation in traits such as beak shape, social status, and plumage coloration and about their roles in the differentiation of populations, subspecies, and species. To population ecologists, this variation also provides clues about the survival or reproductive probabilities of individuals and their potential contributions to population growth, because when the morphological, behavioral, or developmental traits of individuals are linked to fitness, the composition of populations with respect to those traits can affect how populations grow or decline in future or in response to environmental change. We hope to show in this chapter that an appreciation of both of these perspectives will often be necessary to pose, solve, and apply the

answers to many of the most intriguing questions asked by avian ecologists and evolutionary and conservation biologists and managers.

In birds, individual variation in phenotype, developmental state, and social status is often strikingly evident, or even advertised, by a bird's plumage and social behavior (fig. 20.1). In wintering populations of Harris's Sparrows (*Zonotrichia querula*), for example, Rohwer (1975) showed that the size of black "badges" displayed by individual birds predicted their priority access to food resources. By experimentally enhancing badge size in subordinate birds, Rohwer (1985) further showed that doing so enhanced their social status, raising many questions about how such signaling systems evolve and are maintained in nature. Understanding how this and other variable traits arise and are maintained in populations, and quantifying their effects on survival, reproduction, population growth, and the genetic composition of populations, represents a key puzzle that draws many researchers into the lives of individual birds, and it reinforces the importance of recognizing the potential significance of the remarkable but sometimes cryptic diversity of traits exhibited within species and populations.

With the advent of intensive studies of marked individuals by ornithological pioneers such as Margaret M. Nice (1937;

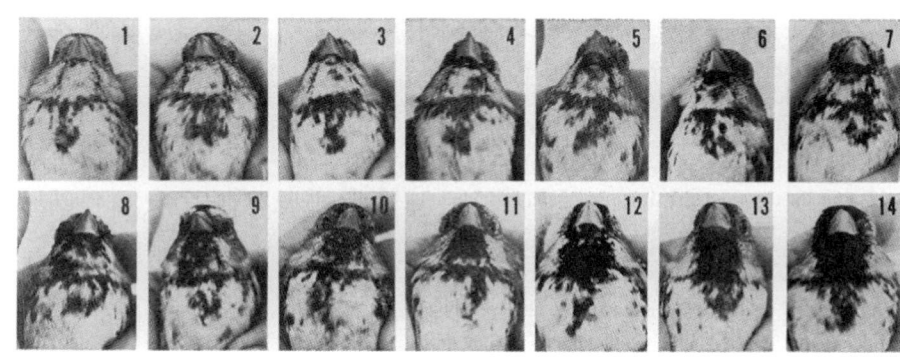

Figure 20.1. Harris's Sparrows vary in the number and arrangement of black feathers on their throat and breast. Larger "badges" signal high social status in the nonbreeding period and affect priority of access to food (Rohwer 1975). *Used with permission.*

Figure 20.2. Song Sparrows (*Melospiza melodia*) were immortalized by Nice (1937) as an archetypal, monogamous passerine in the first long-term study of individually marked birds. Using identical marking and newer genetic methods, Reid et al. (2011a,b) reported for the first time precise counts of the lifetime number of within- and extra-pair offspring produced by hundreds of marked birds in an island Song Sparrow population, revealing a complex polygynandrous mating system with profound consequences for local evolution. *Photo by Peter Arcese.*

see the boxes on pages 581 and 702), and the more recent application of geolocators and transmitters, we can now track birds over their entire lifetimes (chapters 19 and 31; fig. 20.2). These and other advances have allowed us to more precisely quantify trait variation in nature, and to estimate its consequences for individual variation in survival, reproduction, and lifetime fitness. Recognizing this variation as **population structure** is the first step in incorporating such differences among individuals into our thinking about the ecology, evolution, and management of birds.

Population structure refers to the makeup or composition of a population, including the ways in which individuals can be divided into groups based on their sex, age, life stage, phenotype, genotype, or any other trait by which individuals can be reliably assigned to a focal group of interest. Ornithologists assign individuals to such groups when theory or empirical results tell them that certain individuals within populations, by virtue of the traits they share, are likely to express similar

survival and reproductive rates. This means that if we can quantify links between the traits of individuals and the survival or reproductive rates those individual birds achieve in a given environment, we can also use theory to predict the ecological and evolutionary trends that we should expect to observe in the future, based on the prevalence and variety of traits represented and their effects on individual fitness.

Bird populations can also be structured in space, particularly when differences in distribution affect how individuals interact with their environment, including biotic interactions with other species such as predators, parasites, or competitors, and abiotic interactions with weather events or climatic regimes. The Song Sparrows studied by Nice (1937) and Reid et al. (2011a,b) represent just two of 24 diagnosable and 52 named Song Sparrow subspecies extant in North America, each of which includes individuals that vary in ways likely to affect their fate within populations (Arcese et al. 2002, table 1). These and other aspects of population structure have the potential to complicate predictions about the evolutionary and population dynamics of species at large spatial scales, because the distributions of traits within populations may affect individual and population fitness differently in different environments. Understanding the consequences of population structure therefore represents not only an outstanding puzzle for all curious naturalists, but also a substantial challenge for applied ornithologists and managers charged with conserving diverse and persistent populations in the future.

Overall, we hope this chapter helps you to recognize structure in populations and to speculate about how variation in morphology, distribution, social organization, and mating system can both arise from and drive demographic and evolutionary trends in nature. In our rapidly changing world, the ability to predict how populations will respond to human-induced land use and climate change and the assembly of novel biological communities will be critical to managing for the persistence of species and the services they provide to us.

Before we can delve further into population structure, however, we need to know what a population is. Ecologists, geneticists, managers, statisticians, and field ornithologists all define population differently. Because it is essential to understand these differences, we first describe how these approaches intersect and some of the mechanisms by which populations acquire structure. We conclude our chapter by describing how structure can develop in wild populations and why quantifying its effects on individuals can improve the predictions of quantitative and conceptual models in avian ecology, evolution, and management.

WHAT IS A POPULATION?

Populations are often defined as groups of conspecific individuals that share a common environment and have the capacity to interbreed (Krebs 2008). However, within this definition, population ecologists tend to emphasize the **demographic cohesion** of populations (Waples and Gaggiotti 2006): that is, the degree to which individuals respond similarly to variation in the environment and disperse among sites. In contrast, evolutionary ecologists emphasize genetic mixing within and between populations, which can be characterized as the **reproductive cohesion** of populations in space and time and measured as patterns of gene exchange among individuals at local scales and at the scale of a species range. Field and statistical ornithologists apply still different definitions that emphasize the suite of individuals from which their observations are drawn or inferences made, respectively. We elaborate these differences below because understanding each perspective is essential to integrate across subdisciplines in ornithology. Before doing so, however, we first explore some of the factors affecting reproductive and demographic cohesion to highlight the synergies between these paradigms and their relationship to the **eco-evolutionary dynamics** of populations overall.

Population ecologists often ask whether all individuals under study within a given focal area share a common environment, because empirical studies show that biotic and abiotic environmental factors are key drivers of variation in reproduction, survival, population growth, and species distribution (Lack 1954, Newton 1998; chapters 4, 7, 18, 21, 24). Where this assumption is supported, individuals can be treated as a single population for the purpose of predicting demographic trends. However, where spatial or temporal heterogeneity in the environment interacts with the traits of individuals sufficiently to affect their survival or reproductive rate, the composition of populations and traits of individuals may also affect demographic or evolutionary trends via their effects on individual fitness (e.g., box on page 454).

In contrast to population ecologists, evolutionary ecologists tend to focus on the capacity of individuals in populations to interbreed and the effects of natural and sexual selection on the representation of particular traits within and among them. For example, to the degree that Red Crossbills (*Loxia curvirostra*) are more or less likely to choose mates with similar bill shapes, the distribution of shapes inherited by their offspring may in turn affect the fitness of individuals and the populations they constitute by affecting feeding efficiency in different forest types (box on page 454). As a result, evolutionary ecologists often focus on the capacity of individuals in populations to interbreed, because any factor that reduces **random mating** between individuals, such as mate choice, represents a pathway by which populations can acquire genetic structure. In genetically structured populations, individuals may co-occur, experience a common environment, and have the capacity to interbreed, but *not* contribute equally to the genetic composition of populations or species in future. Moreover, in the case that the traits of individuals under selection are **heritable**, such as bill shape (Smith and Dhondt 1980, Grant 1983, Keller et al. 2001a), structured populations can be thought of as receiving continuous feedback from the environment, potentially affecting the distribution of traits in populations, facilitating local adaptation and the likelihood that they will occur in particular regions and persist over time (e.g., chapters 4, 7, 18). Overall, evolutionary and ecological paradigms employ different criteria to evaluate the isolation of populations within a species' range: the first focusing on the genetic cohesion of populations as evidenced by gene flow, and the second focusing on the demographic cohesion of populations as indicated by the dispersal of individuals and their effects on population size.

Deviations from random mating are particularly critical to this eco-evolutionary view of the world, because they affect the distributions of inherited traits in populations and their potential rates of change. For example, prior to widespread genotyping, ornithologists assumed that **socially monogamous** species were also **genetically monogamous**. More recently, genotyping has allowed us to quantify the contributions of individual birds to the genetic composition of populations and has revealed that genetic polygamy is common, including in many species thought previously to be monogamous (chapter 16; fig. 20.2). These discoveries have had major consequences for how we think about the evolutionary dynamics of populations because, under monogamy, we typically assume that all individuals have similar chances of passing genes to the next generation. In contrast, under polygamy, a relatively small number of males can sire a large fraction of the next generation (chapter 16), with potentially large effects on the diversity and range of variation in traits observed in populations.

Overall, we typically recognize the existence of structure in populations by characterizing individuals by one or more phenotypic, genetic, or developmental trait, which is both observable and known or assumed to affect fitness (table 20.1). Although delineating groups of individuals within populations has the potential to add substantial complexity to our conceptual model of population, if the traits used to delineate groups affect reproductive performance or survival in the individuals carrying them, accommodating this additional complexity may be necessary to predict

Table 20.1. **Known or assumed traits of individuals potentially contributing to population structure in the Song Sparrow.**

Trait	Source of Variation	Distribution of Variation	Effect on Fitness
Genotype			
Gender	I	D	K
Degree of Inbreeding	I	T, S, C	K
Phenotype			
Morphological			
Plumage	D, E, (I)	S, C, (D)	A
Wing length	D, E, I	T, S, C	A
Tarsus length	D, E, I	T, S, C	K
Bill length, depth, width	D, E, I	T, S, C	K
Behavioral			
Migratory behavior	(D), E, (I)	T, S, (D)	A
Natal dispersal	D, E, (I)	T, S, C	K
Extra-pair mating	D, E, I	(T), (S), C	K
Song type or number	D, (E)	T, S, D	K
Developmental			
Age	D	T, S, D	K
Social status	D, E	T, S, C	K

Sources of variation in listed traits can be developmental (D), environmental (E), or inherited (I) in origin, and observed as varying temporally (T) or spatially (S) in continuous (C) or discrete (D) distributions. In all cases, individual variation in survival or reproductive rate, and thus fitness, is known (K) or assumed (A) to be linked to the trait listed, which is also known to vary in the mean and distribution of values within populations temporally and spatially (e.g., Nice 1937, Arcese and Smith 1985, Germain and Arcese 2014, Keller et al. 1994, 2008, Marr et al. 2006, Reid et al. 2011a, 2011b, Smith et al. 2006, Wilson et al. 2007, 2011, and references in Arcese et al. 2002). Letters in parentheses denote hypothesized effect not yet demonstrated empirically.

the ecological and evolutionary dynamics of populations. As a consequence, while the term "population" is sufficient to describe interbreeding individuals that share a common environment, the delineation of demographically and reproductively cohesive **subpopulations** of individuals can help us quantify their effects on population structure.

Subpopulations and the Traits that Distinguish Them

In addition to genotype, external phenotype, developmental stage, and location, ecologists have also used **personality** to delineate subpopulations of individuals, reminding us that individuals in any given population may also be members of more or less cryptic subpopulations that differ in their potential contributions to population growth. A striking, though probably rare, case in point is the White-throated Sparrow (*Zonotrichia albicollis*), a species of north-central and eastern North American forest edge that occurs throughout its range as interbreeding subpopulations of white- and tan-striped individuals of each sex.

In 1961, Lowther (1961, 1962) suggested that white- and tan-striped plumage **morphs** of this species also varied predictably in aggressive and parental behavior, and that both traits appeared to be inherited in common. Theory tells

us that the co-expression of inherited traits will increase as the distance between the genes that code for them declines, leading to **genetic linkage** between the traits in question. Thorneycroft (1966) tested Lowther's idea using cytogenetics, a direct method of genotyping individuals by karyotype, to show that a chromosomal inversion included genes coding for variation in plumage and parental and aggressive behavior, effectively enforcing genetic linkage in these traits. But how are systems like this maintained in nature?

A host of studies have since elaborated this example to demonstrate that tan-striped females prefer to mate with white-striped males, leading to **negative assortative mating** among phenotypes and maintaining heterozygosity in populations overall (Tuttle et al. 2016; box on pages 661–662). These and other studies of individual birds have dramatically advanced understanding about the ways in which population genetic and demographic structure can interact to affect species distribution (box on pages 664–665), social organization (box on pages 668–669), and population dynamics (box on pages 670–671). All ornithologists must therefore consider that particular individuals observed in any given population may also be members of subpopulations that are more cryptic than white- and tan-striped sparrows, but nevertheless differ in demographic and evolutionary potential.

Linking Traits to Individuals, Subpopulations and Population Growth

When individual phenotype has the potential to affect fitness, it is also likely to influence population demography and evolutionary trajectory. This is because growth in a structured population will depend on the fraction of individuals constituting each delineated subpopulation and the **fitness value** of the traits they display. As a result, ornithologists often characterize differences in the reproductive and survival rates of delineated subpopulations and then estimate population growth by taking into account the relative abundance of each subpopulation in the overall population using matrix models, difference equations, or calculus implemented in computer models (Caswell 2001, Waples and Gaggiotti 2006, chapter 21).

An emphasis on **demographic cohesion** will often lead to the delineation of subpopulations based on age, experience, or developmental stage, or the role of the subpopulation with respect to immigration and emigration regionally (chapter 21). In contrast, an emphasis on **reproductive cohesion** is more likely to focus on behavioral or external phenotype, given known and assumed links to mating behavior, reproductive success, or survival. Although it is likely to be true that relatively few bird populations will ever be studied in sufficient detail to estimate demographic rates for all potentially influential subpopulations, existing examples from several well-studied species show us that quantifying the effects of population structure can influence predictions of species distribution (Pulliam 1988), the strength and direction of evolutionary change (boxes on pages 661–662, 664–665, 668–669), and the causes of long-term change in population size (box on pages 670–671; Norris et al. 2007, Blight et al. 2015). However, because populations in different parts of a species range may also experience different environments, we also need to consider that particular traits may have different fitness values in different environments and differ in frequency among populations as a consequence.

Population Structure and the Environment

Given individual heterogeneity in genotype, phenotype, and development in populations, we might also expect that local and landscape-level heterogeneity in the environment will interact with population structure to affect spatial variation in population growth and evolution (chapters 4 and 21). For example, Brown-headed Cowbirds (*Molothrus ater*) are a ubiquitous brood parasite that expanded its range in western North America after the introduction of domestic livestock

and forest clearing (Lowther 1993). Range expansion by this brood parasite has changed the biotic environment of many potential cowbird hosts, including the western subspecies of the Warbling Vireo (*Vireo gilvus swainsonii*; fig. 20.3). In contrast, the eastern subspecies (*V. g. gilvus*) has coexisted with cowbirds for generations, raising the possibility that these different evolutionary histories might also affect the ability of western populations to adapt to human-induced environmental change. In support of this possibility, Sealy et al. (2000) showed experimentally that eastern vireos punctured and ejected cowbird eggs from their nests as an adaptation to avoid parasitism. However, individuals in western populations uniformly accepted experimental and naturally laid cowbird eggs, and as a consequence have suffered dramatic declines in reproductive rate and population growth where cowbirds are common (Ward and Smith 2000). Where the ranges of these two subspecies meet, we might therefore expect the existence of cryptic subpopulations of individuals that either reject or accept cowbird eggs, based on lineage, and whose individual fitness and contributions to population growth in future vary with the intensity of parasitism locally. Overall, these findings show that populations with different evolutionary histories may differ in their abilities to adapt to environmental change, adding another layer of

Figure 20.3. Western and eastern subspecies of the Warbling Vireo (*Vireo gilvus swainsonii* and *V. g. gilvus*) are similar in appearance but differ in their responses to Brown-headed Cowbird (*Molothrus ater*) eggs in their nests. *V. g. gilvus* has coexisted with cowbirds for thousands of years and ejects cowbird eggs, whereas *V. g. swainsonii* accepts them and declines in abundance where cowbirds are common as a result. These evolved differences indicate the presence of marked spatial variation in population structure in vireos, with profound consequences for their demography, distribution, and persistence. *Photo by Dennis Paulson, Slater Museum.*

GENETIC POLYMORPHISM IN THE WHITE-THROATED SPARROW (*ZONOTRICHIA ALBICOLLIS*)

The White-throated Sparrow is a polymorphic species in which alternative phenotypes are genetically controlled and maintained in the population by disassortative mating and fitness trade-offs. In this species, males and females occur as one of two alternative white- or tan-crowned morphs. Morph is determined by a complex inversion (rearrangement) of chromosome 2, with white-crowned morphs being heterozygous for the inversion and tan-crowned morphs being homozygous and lacking the inversion. During the breeding season, the two morphs can be visually distinguished, with white-striped birds having more colorful crown plumage (*left*).

Dr. Elaina Tuttle and colleagues have studied White-throated Sparrows breeding in the Adirondack region of New York since 1988. Thousands of individuals have been genotyped, confirming that almost all white-striped birds are heterozygous for alternative forms of chromosome 2, as first described by Thorneycroft (1966, 1975). Of 1,989 genotyped birds, 3 (0.15 percent) "superwhite" birds were identified, which were homozygous for the rearranged form of chromosome 2. Genotyping also demonstrated nearly perfect disassortative mating among morphs, with only 17 of 1,116 pairs (1.5 percent) being assortative (Tuttle et al. 2016). These data suggest that the "superwhite" genotype is deleterious, promoting disassortative mating and stabilizing the genetic polymorphism.

Each year Tuttle and her team members also collected 80 or more hours of behavioral data per ~100 breeding pairs, located all nests, and sampled offspring for genetic paternity. Behavioral and genetic evidence indicate that white- and tan-striped morphs pursue distinct behavioral strategies, representing two extremes in reproductive trade-offs. White-striped males sing at high rates and are highly aggressive and promiscuous, engaging in high rates of extra-pair copulation. In contrast, tan-striped males are less aggressive, sing at

White- (*left*) and tan-striped morphs (*right*) of the White-throated Sparrow. White morphs have darker black lateral crown stripes (LCS), whiter median crown stripes (MCS) and throats, and brighter yellow superciliaries than tan-striped morphs. Both birds shown are males; in each morph, females tend to be slightly duller in coloration.

lower rates, and invest more in mate-guarding and paternal care (Tuttle 2003). White- and tan-striped males also prefer habitats that best suit their distinct behavioral strategies, with white-striped males establishing territories in areas of higher population density (Formica et al. 2004, Formica and Tuttle 2009). In parallel to males, white-striped females are aggressive and often sing, whereas tan-striped females do not display these male-like behaviors (Tuttle 2003). Along with disassortative mating, fitness trade-offs associated with behavioral differences likely contribute to maintaining the genetic polymorphism.

Most recently, genetic analyses have yielded exciting insights into the evolutionary history of the chromosomal rearrangement associated with morph identity. Genes within the inverted region are tightly linked, and function as a coadapted

complexity to the potential ways in which population structure, demography, and environment can interact (e.g., chapters 21 and 24).

Variation in the abiotic environment can also drive population demography via its interaction with the traits of individuals, especially in species with large geographic ranges. Consequently, ecologists, evolutionary and conservation biologists, and wildlife managers typically assume that most

species will occur as clusters of more or less distinct populations that vary in composition and, potentially, in their response to environmental change (chapters 4, 21, 24). However, because complex interactions like those described above for vireos and cowbirds depend on processes affecting the genetic structure of populations, understanding those processes can help us decide when we should include such effects in the way we manage populations.

Dr. Elaina M. Tuttle elaborated early work on the cytogenetics of the White-throated Sparrow, revealing remarkable links between individual genotype and phenotype, including differences in morphology and reproductive, parental and territorial behavior.

gene complex, or "supergene," to influence multiple aspects of morphology, physiology, and behavior. The inversion encompasses a large number of well-studied genes associated with physiology and behavior, notably including steroid hormone receptors (Tuttle et al. 2016). Tuttle et al. (2016) showed high levels of linkage disequilibrium and low rates of recombination within the inversion, which is promoting the maintenance of the supergene and leading to functional degradation of the rearranged form of chromosome 2, in a fashion analogous to sex chromosomes. Whole genome sequence data were also used to assess patterns of divergence between the White-throated Sparrow and its closest relatives. Those results suggest that the inversion was introduced to the White-throated Sparrow's genome through hybridization with an extinct species in the genus *Zonotrichia*, followed by introgression. These findings have shed new light on the evolutionary origins of supergenes and polymorphic species.

Populations as Units of Evolutionary Change

Demographic approaches to population structure are elaborated further in chapter 21 as stage-structured, age-structured, and individual-based models. As we consider those approaches, it is important to keep in mind that a complete understanding of population demography and evolution may require that we also evaluate the genetic processes potentially affecting population structure. In addition to their theoretical relevance, these genetic processes can affect how we manage populations and prioritize them for conservation.

Evolutionary and population geneticists typically assume that all species exhibit population structure at some spatial scale because of historic or present-day barriers to **genetically effective dispersal**, **non-random mating**, spatial or temporal heterogeneity in natural or sexual selection, **isolation by distance**, and the occurrence of **random genetic drift** or the existence of **founder effects** (defined below; Wright 1984, Avise 1994). For example, even in species with homogeneous distributions that are subject to a common environment, genetic structure can develop solely because of the finite nature of dispersal (Irwin et al. 2005). This is because the genetic relatedness of any two randomly selected individuals of a species will decline on average as the geographic distance between them increases, leading to isolation (Wright 1988, Wilson et al. 2011). By comparison, in species that occur in structured populations that are distributed heterogeneously in space and subject to different environments, a sophisticated and quantitative understanding of the genetic processes affecting populations may be needed to predict reliably how variation in the environment will affect population structure, demography, and evolution.

Population Isolation and Genetic Structure

The House Sparrow (*Passer domesticus*) is a species native to Europe that was widely introduced to North America and specializes on human-dominated landscapes (Lowther and Cink 2006). A reliance on humans means that House Sparrows live in "habitat islands" that, when sufficiently isolated from each other, become more likely to diverge genetically. Indeed, some of the earliest empirical observations of natural selection on morphology were made on this species (Bumpus 1899), and many studies have since elaborated the mechanisms by which House Sparrows have adapted to environmental variation across their novel range in North America (Lowther and Cink 2006), including to climate (fig. 20.4; Johnston and Selander 1964, Gould and Johnston 1972). Compared with habitat specialists, populations of habitat generalists tend to be less isolated because of their more regular or continuous occurrence across habitat types, which results in fewer barriers to dispersal, more regular genetic mixing among populations, and fewer opportunities for populations to diverge genetically overall (Avise 1994).

Overall, isolation provides a particularly useful framework in which to think about populations because of its effect on the rate at which genes are exchanged among them (Waples and Gaggiotti 2006). The term **genetically effec-**

A

average male sparrow size

smallest | 1 | 2 | 3 | 4 | 5 | 6 | 7 | 8 | largest

Figure 20.4. Geographic variation in the body size of the European House Sparrow (*Passer domesticus*) is broadly consistent with expectations of Bergman's and Allen's ecogeographic rules. *Modified from Gould and Johnston 1972; photo by Heyli Arcese.*

tive dispersal captures this idea by extending demographic definitions of natal and breeding dispersal (the movement of birds away from their natal sites or breeding sites, respectively; Greenwood 1980; chapter 19) to include only those dispersers that contribute genetic material to recipient populations. This refinement is critical because dispersal affects the genetic structure of populations only to the extent that it alters gene frequencies in the future (Williams 1966), leading to two population genetic **rules of thumb**. First, the likelihood that populations diverge genetically will increase as genetically effective dispersal is reduced, because high rates of genetic mixing tend to homogenize population structure.

Second, as genetically effective dispersal is reduced, gene frequencies within populations will increasingly become a product of natural selection, random genetic drift, inbreeding, and mutation (chapters 3 and 4; see below). Understanding something about the historic and present-day isolation of populations can therefore help us simplify some of the complicated ways in which population structure, demography, and environment influence the amazing diversity of bird species extant in our world and help predict how to maintain that diversity in future (chapters 24, 25, 26).

Having explored some of the factors affecting reproductive and demographic cohesion in populations, we can return to the question of how populations are defined in practice. Legal and quantitative definitions each take into account population structure.

Populations as Management Units

Many of the processes described above shape the ways in which managers define populations, including how they prioritize management investments, evaluate policy options, and designate legal protection. For example, **distinct population segments** are the smallest units of a species that can be protected under the US Endangered Species Act. They are distinguished by morphological, genetic, or other traits that allow us to assign individuals to a particular population or area. Similarly, populations identified as **evolutionarily significant units** tend to be geographically isolated from other populations and to differ from them genetically because of historical restrictions on gene flow or via natural or sexual selection and the evolution of locally adapted traits (Chan and Arcese 2002; chapters 4 and 25). Although these designations have seen limited application to birds, they offer clear links between the evolutionary origins of populations and decisions about their conservation (chapter 25).

Study and Sample Populations

Field ornithologists and statisticians often identify sample populations in similar ways; the key difference is the ways in which individuals are selected for observation or analysis (Morrison 2012). **Study populations**—individuals observed repeatedly in a defined area—rarely constitute demographically or genetically distinct units except in isolated habitat fragments or on islands. Study populations will almost always represent nonrandom samples of individuals with respect to the range of all trait variation in a species. In contrast, **sample populations** may represent random or non-random selections of individuals, potentially including subpopulations delineated by age, sex, genotype, or phenotype.

In general, we recommend defining "population" relative to the inferences being made and using "subpopulation" to refer to groups of individuals assignable to groups by traits such as age, gender, life stage, phenotype, genotype, or ecotype in the case that subpopulations are known to share traits in common that are linked to particular habitat or environmental conditions. We now provide detailed examples of the processes above to make clear that, in the "real world," almost all co-occur to varying degrees.

POPULATION DEMOGRAPHIC STRUCTURE
Gender, Age, and Individual Variation in Vital Rates

Bird populations invariably comprise subpopulations of individuals that are identifiable by traits such as gender, age, behavioral phenotype, or life stage (table 20.1, boxes on pages 661–662, 664–665, 668–669). Because these subpopulations often differ in ways that affect their distribution, survival, or reproductive rates, quantifying this variation in population structure and its links to demography represents a fundamental step in understanding avian life histories and in managing populations (chapters 4 and 25).

Gender is universally used by ornithologists to identify subpopulations in sexually dimorphic species, or with the aid of genetic tests in sexually cryptic species (Clutton-Brock and Isvaran 2007), and a long history of monitoring marked birds indicates that female birds survive less well than males on average (Lack 1966, Greenwood 1980; chapters 16 and 21). This finding is consistent with life history theory, given female-biased patterns of parental investment, and it can dramatically affect adult sex ratio, mating system, and life history as a result (Liker et al. 2014, Székely et al. 2014; chapter 16). Liker and Székely (2005) analyzed adult survival in 194 bird species to find that males typically survive ~6 percent better than females (0.672 vs. 0.636 ± 0.01 SE, respectively). However, because they also found that the magnitude of difference within species decreased as the intensity of male competition within species increased, they suggested that costs of male competition can offset the survival advantage males often experience by investing less in parental care.

Age-specific variation in survival and reproductive rate is also ubiquitous in birds, particularly when comparing yearlings with older animals. Subpopulations that comprise age-classes similar with respect to the demographic traits of interest are often identified as "stage-classes" in demographic models and typically represent birds that differ in experience, development, or residual reproductive value (e.g., Bouwhuis et al. 2012; chapters 16 and 21). Long-term marking studies of long-lived species have provided precise estimates of survival and allowed researchers to document increases with age, experience, and selective disappearance of poor-quality individuals from breeding populations, as well as declines in performance caused by senescence (fig. 20.5; Keller et al. 2008, Rebke et al. 2010). Even within age- or stage-classes, however, individual heterogeneity in performance can often be detected as fine-scale structure in populations.

For example, Pacific Black Brant (*Branta bernicla nigricans*) are long-lived migrant geese that have been studied in great detail for decades (box on pages 664–665). Most recently, Lindberg et al. (2013) examined 24 years of detailed data on individually marked brant to identify three classes of females that differed in age and "quality." Adult females represented a homogeneous subpopulation with respect to annual survival (0.85/yr). In contrast, goslings were assigned to subpopulations based on individual quality and survival (low- vs. high-quality goslings: 0.50 vs. 0.73/yr, respectively). The origins of these differences are yet to be resolved, but they may be linked to interactions of genetic and environmental influences on body size that affect the frac-

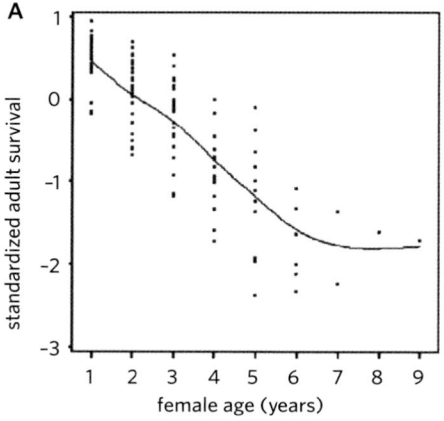
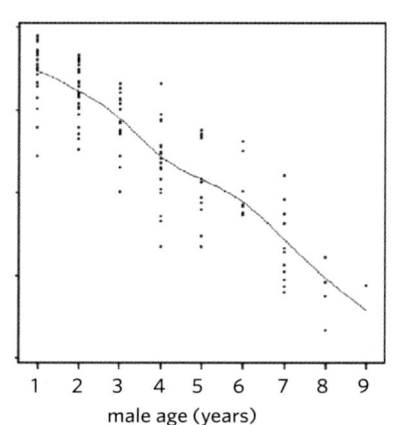

Figure 20.5. Age-specific variation in annual adult survival in *A*, female and *B*, male song sparrows, all standardized by "year" to account statistically for inter-annual variation in the environment. The estimates are the residuals of a null model adjusting for year of hatching, smoothed by cubic spline. *Graph from Keller et al. 2008.*

POPULATION STRUCTURE IN BLACK BRANT (*BRANTA BERNICLA NIGRICANS*)

Brant are highly social, colonially nesting geese that nest in coastal tundra, mainly in southwest Alaska, but extending from Canada to eastern Siberia (Lewis et al. 2013). As long as both mates are alive, pair bonds typically last for life. Most Black Brant migrate long distances to winter on the Pacific coast of Baja California, Mexico. Spring migration begins in late January, when birds begin to leave Mexico, and ends in May, when they arrive on the breeding grounds. Substantial mortality, especially of young of the year, occurs during migration (Ward et al. 2004, Sedinger and Chelgren 2007). Female brant are in part "capital breeders"; that is, females store fat and protein during migration stopovers in spring, which they use to meet their energy and nutritional needs during egg-laying and incubation.

Dr. Jim Sedinger has studied Black Brant at the Tutakoke River colony on the Bering Sea coast of the Yukon-Kuskokwim Delta, Alaska, since 1984, during which time 46,000 individuals have been marked with plastic bands that allow them to be re-sighted throughout the year. Along with those recorded by David Ward (US Geological Survey) and Sean Boyd (Environment Canada), more than 105,000 marked Black Brant have been observed or captured on the breeding colony, and 124,000 have been observed on winter and migration areas. These observations and the application of modern capture-mark-recapture methods (Lebreton et al. 1992, Williams et al. 2002; chapter 19) form the basis of Sedinger's ability to understand population structure and how fitness varies among individuals.

Female Black Brant are faithful to the breeding colony they first nest in, whereas males disperse among breeding colonies to accompany their mates (Lindberg et al. 1998, Sedinger et al. 2008). Site fidelity in females should lead to spatial variation in mitochondrial DNA among breeding areas (Shields 1990). In contrast, there is less geographic structure in nuclear DNA, which is transmitted by both males and females (e.g., Harrison et al. 2010). About 17 percent of young females that survive their first year fail to return to the Tutakoke colony (Sedinger et al. 2008) but Lemons et al. (2012) showed that 15 percent fewer

Dr. Jim Sedinger and colleagues have studied individually marked Black Brant for over 25 years to understand how long-term change in the environment and individual in phenotype influence age-specific reproduction, survival, and the aging process.

male goslings survive from hatching to fledging, so nearly all females that fail to return to nest were likely unable to find a mate because of the skewed sex ratio. In addition to females that may never breed, between 20 and 30 percent of females in the breeding population skip breeding each year (Sedinger et al. 2008). Females that nest in one year are more likely to do so the next, suggesting substantial heterogeneity in individual quality and fitness. The combination of permanent and temporary nonbreeders means that a substantial proportion of the population is absent from breeding areas each year. As a consequence, these individuals do not contribute to recruitment and are invisible to investigators that focus only on the breeding area.

Environmental and social factors are important determinants of fitness of Black Brant. Females rear goslings on specific areas within 30 km of the Tutakoke colony (Lindberg et al. 1998), which determines how rapidly goslings grow and their size at fledging (Nicolai and Sedinger 2012). The size of goslings at fledging governs their survival during their first year (Sedinger and Chelgren 2007) and whether they will breed if they do

---

tion of high- and low-quality individuals in populations, population growth, and potential harvest (box on pages 664–665).

Age-specific variation in reproduction is also apparent in some short-lived species. In a pedigreed population of Song Sparrows studied continuously since 1975 (table 20.1), reproductive success increased markedly in females from one to two years of age but declined in those four or more years old (Keller et al. 2008), similar to patterns observed in long-lived

species. But within age-classes, females that invested more in reproduction annually also survived less well (fig. 20.6), suggesting that some females maximized annual reproduction at the cost of a reduced lifespan (**fast lifestyle**), whereas others produced fewer young annually but lived longer as a consequence (**slow lifestyle**; fig. 20.6; Tarwater and Arcese 2017). Interestingly, variation in reproductive behavior is repeatable in several species including Song Sparrows (chap-

Individually marked Black Brant.

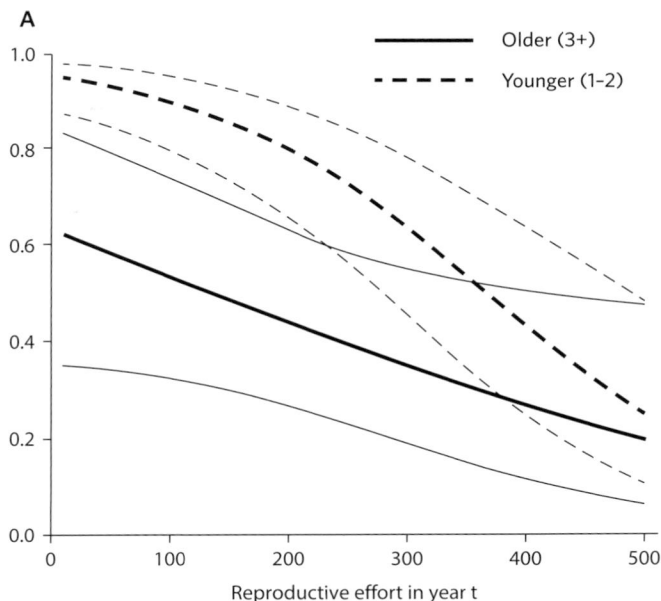

survive (Sedinger et al. 2004). The quality of wintering areas also governs whether Black Brant nest the next summer (inset 1; Sedinger et al. 2006, 2011). Thus, a decline in the overall quality of wintering areas, as in El Niño years (inset 1) or because of habitat alteration, reduces recruitment in Black Brant and population growth. Individuals that breed successfully are more likely to return to a higher quality wintering area in the next winter than those that did not breed (Sedinger et al. 2011), likely because family groups are socially dominant to individuals and pairs (Poisbleau et al. 2006). These links among social status, quality of wintering areas, and reproductive success increase the possibility that some lineages of Black Brant regularly experience high fitness while others experience lower fitness, consistent with heterogeneity in breeding noted above. Nicolai and Sedinger (2012) showed that females that lost their mates suffered lower survival than those that retained their mates (inset 2), and Leach (2015) found that mate loss also reduced the likelihood of breeding the next year. Both results demonstrate the importance of maintaining pair bonds in this monogamous species.

ter 16; Reid et al. 2011a, 2011b). How might such variation in individual performance arise in populations?

Individual Variation in Performance

Although the causes of repeatability in individual performance remain poorly understood, it has been linked in some species to variation in the environment experienced by individual birds during development (boxes on pages 661–662,

Figure 20.6. Female Song Sparrows that invested more in reproduction in a current year survived less often to the following year, particularly when young. Lines indicate predicted relationships and their 95 percent confidence intervals. *Photo by Peter Arcese. Graph by Peter Arcese and Corey Tarwater.*

664–665, 668–669, 670–671; Wilson et al. 2007). As in Black Brant, natal environment had lasting effects on survival, reproduction, and life span in Red-billed Choughs (*Pyrrhocorax pyrrhocorax*; Reid et al. 2003). In Common Guillemots (*Uria aalge*), repeatable differences in male and female performance were linked to foraging rates and length of the pair association (Lewis et al. 2006), similar to Black Brant (box on pages 670–671). In Song Sparrows, about half the repeatability in female reproductive rate was attributed to the quality of the nesting habitat that females occupied, but the remainder was attributed to **individual identity**, a catchall statistical term describing developmental and genetic effects unique to an individual (Germain and Arcese 2014).

Repeatable variation in aggressive, exploratory, and dispersal behavior, often measured as suites of personality traits or **behavioral syndromes**, have also been linked to variations in fitness, behavior, and demography in many species of birds (Sih et al. 2004, Pennisi 2016). Much evidence suggests that behavioral syndromes in birds can be inherited (Dingemanse et al. 2002), influenced by hormones during nestling development (box on pages 668–669) and indicative of population structure in individual life history via the fast-slow continuum (Ricklefs and Wikelski 2002). In the Great Tits (*Parus major*), Dingemanse et al. (2002) used artificial selection to create lines of fast, bold, proactive and slow, shy, reactive tits, indicating that variation in traits linked to these personality traits was inherited by the offspring produced by experimental pairs of male and female tits. These authors suggested that heritability in these traits was maintained in nature because the relative fitness value of the fast and slow lifestyles linked to these behavioral traits of individuals depended on annual variation in intraspecific competition, reminiscent of Duckworth's finding in bluebirds (box on pages 668–669). While it will be difficult or impossible to quantify the effects of individual heterogeneity in behavior and life history in some species, being aware of the potential for temporal and spatial variation in the structure of populations and the types of individuals that constitute them should nevertheless improve our ability to interpret and manage for the persistence of genetically diverse populations.

POPULATION STRUCTURE AND SOCIAL ORGANIZATION

Population structure and social organization are intimately linked because gender, age, experience, and behavioral phenotype can all influence the distribution and performance of individual birds and, cumulatively, drive temporal and spatial variation in demography as a consequence. Below we highlight a few of these links by scratching the surface of this particularly rich area of ornithology, elaborated in chapters throughout this book.

Density-Dependence, Dispersal, and the Population Surplus

The results of intraspecific competition for limiting resources are often observed in nature via the stratification of subpopulations across habitat types that vary in quality, where quality is defined as the expected fitness of an average individual in that site (Fretwell 1972, Germain and Arcese 2014). At fine spatial scales, such as within habitat patches or social groups, intraspecific competition often leads to dominance hierarchies that enforce priority of access to limiting resources, including mates (chapter 16). Hierarchies can have large effects on the fitness of dominant and subordinate individuals, particularly as resources become scarce and competition intensifies (chapters 17 and 21; fig. 20.1; Newton 1998). Moreover, because gender, age, and experience are ubiquitous predictors of social dominance in birds (Arcese and Smith 1985), intraspecific competition tends to accentuate population structure by contributing to gender, age, and phenotype-dependent biases in the distribution of subpopulations within habitat patches and across environmental gradients.

For example, dominance hierarchies formed during the nonbreeding period often result in adult males obtaining priority of access to habitats that offer the highest energetic return on foraging effort (chapters 15 and 17; Goss-Custard et al. 1995). In such cases, social factors can drive population structure by contributing to gender-biases in distribution, which may in turn affect survival in subpopulations of interest (Ketterson and Nolan 1983). Moreover, dominance hierarchies can restrict socially subordinate subpopulations of individuals to progressively poorer habitats as population density increases, regulating population growth as a consequence (Lomnicki 1988; chapters 17 and 21). In contrast, priority of access to high-quality habitat can enhance body condition and survival in socially dominant individuals and even confer advantages to socially dominant individuals via **carryover effects** on reproductive rate the following season (Norris et al. 2004; box on pages 664–665; chapter 17). However, high social status may also have its own associated costs, such as the increased investment in agonistic interactions necessary to maintain high social status in groups, which could explain the existence of more or less stable phenotype frequencies in populations over time (Rohwer and Ewald 1981).

Complex interactions among population density, development, and behavioral phenotype also contribute to spatial and temporal structure in bluebird populations by driving variation in dispersal and territorial behavior and influencing community membership (Duckworth et al. 2015). In a remarkable study system in the western United States, Duckworth and colleagues have shown that variation in food and nest site availability can affect aggressive competition for space by influencing development in nesting bluebirds via the androgens that females deposit in yolk. Yolk androgens are in turn linked to birth order and the aggressive and dispersal behavior of offspring in ways that enhance their ability to invade new breeding habitat. As a consequence, population growth rate and territorial and dispersal behavior vary systematically across the Western Bluebird (*Sialia mexicana*) populations as a function of habitat age (box on pages 668–669). But how do these advantages shift and strategies play out when nesting sites are abundant in certain habitats?

In territorial systems, preferred habitats often become saturated with breeding individuals that can preempt settlement by prospective immigrants (chapters 17 and 21; Lomnicki 1988). Preemption raises the potential for fitness trade-offs among the individuals relegated to nonbreeding subpopulations that develop as a consequence of habitat limitation, further contributing to population structure. Such trade-offs arise because as suitable breeding habitat becomes saturated with territorial breeders, attempts by preempted individuals to settle and breed in poor-quality habitat may be more detrimental to individual fitness than if those individuals were to pursue alternative tactics, such as dispersing farther in search of a favorable site, attempting to replace an existing owner forcibly, remaining nonterritorial and delaying breeding, or joining a cooperative group to aid related breeders or increase the chance of acquiring a breeding position in future (chapters 16 and 17). One demographic outcome of habitat limitation is therefore the creation of subpopulations of individuals that may contribute little or nothing to population growth, and thus represent a potential **population surplus** (chapter 21). A further evolutionary outcome of habitat limitation is to facilitate skew in the contributions of individuals to the genetic makeup of populations (chapter 16) and to give rise to alternative life history paths for individuals attempting to maximize their fitness given the social environment around them.

Floaters and Other Hangers-On

Limits on the availability of habitat, and in some species the time necessary to gain sufficient foraging skills to breed successfully, can lead to the formation of subpopulations of individuals that do not contribute to population growth via reproduction and may even disrupt breeding by territorial individuals (Piper et al. 2008; chapters 17 and 21). However, nonbreeding subpopulations can also contribute critically to population stability by providing a buffer against catastrophic mortality or reproductive failure among territorial

breeders (e.g., Brown 1969; chapter 21), particularly when socially subordinate sub-populations occupy habitats that experience different environmental limits on survival.

In the first case, socially subordinate nonbreeders excluded from breeding habitat, mating opportunities, or both, often persist at the interstitial areas of breeding territories as **floaters**, which are themselves ordered by social dominance into queues that regulate the likelihood of acquiring a breeding territory should an owner die (Smith 1978), or which actively intrude on the breeding territories to test the defensive capabilities of owners and attempt to depose those that are physically compromised by age, injury, or body condition (Arcese 1987, 1989a). First described in the Rufous-collared Sparrow (*Zonotrichia capensis*) by Smith (1978), floating populations are now known in a variety of bird species.

Although conflicts between floaters and territory holders are infrequently observed in nature, Piper et al. (2006) noted that 16–33 percent of evictions of territorial male Common Loons (*Gavia immer*) by nonterritorial floaters proved fatal for the displaced owner. In contrast, female loons seldom fought, perhaps because a male-bias in adult sex ratio meant that most females, but only a fraction of males, were able to secure breeding territories and mate annually. This difference in behavior suggests that the lifetime cost to individual fitness of failing to settle were higher for male than female loons. In contrast, in a saturated island population of Song Sparrows, deposed male territory owners often reentered the floating population and sometimes regained territories (Arcese 1989a). Although female floaters also occurred in this resident Song Sparrow population, all females eventually settled, often by inserting themselves as secondary females in polygamous groups after the dominant female had begun to incubate a first brood (Arcese 1989b). These examples emphasize that variation in the degree of habitat saturation, adult sex ratio, and the opportunity to engage in alternative mating tactics can contribute to temporal and spatial variation in population structure in nature.

Alternatively, nonterritorial individuals in many other species exist singly or in flocks outside the normal breeding range in habitats that offer the potential to survive and to develop the foraging or fighting skills necessary to breed in a subsequent year (chapters 16 and 21). In both of the cases above, individuals constituting nonbreeding subpopulations can often be distinguished by plumages that effectively signal their developmental stage to conspecific group members.

Social Structure and Mating System Evolution

A key evolutionary outcome of social and ecological limits on access to breeding habitat or status within groups is that it raises the possibility that new life history tactics arise as viable alternatives to fighting, perhaps to death, for advancement

BEHAVIORAL PHENOTYPE AS A DRIVER OF SPECIES DISTRIBUTION AND COMMUNITY STRUCTURE

Dr. Renée Duckworth has studied individually marked Western and Mountain Bluebirds to understand the evolution of individual variation in dispersal and roles of maternal effects, niche shifts and temporal variation in the environment on the coexistence and diversification of species.

Mountain and Western bluebirds (*Sialia currucoides* and *S. mexicana*, respectively) overlap in their breeding range in the northwestern United States. As secondary cavity nesters, bluebirds need a cavity to breed but, unlike primary cavity nesters such as woodpeckers, cannot make their own. Because natural cavities are rare in most habitats, bluebirds are in constant competition with each other and a diverse community of cavity-nesting birds and mammals for a potentially limiting resource.

Since 2001, Dr. Renée Duckworth has studied the competitive dynamics of Western and Mountain Bluebirds by monitoring multiple populations of banded birds in Montana. Duckworth and colleagues measured aggression in more than 1,000 bluebirds in the wild and banded over 10,000 individuals to maintain detailed records of social and familial relationships within and between populations. These data and parallel studies of physiology and development allowed Dr. Duckworth to explore the adaptive significance of species-specific and individual differences in aggression and their consequences for ecological and evolutionary dynamics.

In the US northwest prior to the widespread use of nest boxes, bluebird habitat was created mainly by fires that opened the understory and created snags. But in the absence of fire,

understory vegetation and trees recover and cause bluebirds to abandon patches as cavity density declines and forest regrowth eliminates meadows that bluebirds require to forage for insect prey. Thus, historically, habitats for bluebirds and their main limiting resource—nest cavities—were patchy and ephemeral (Duckworth 2014).

The ephemerality of bluebird habitat means that both species must colonize new patches continually to persist regionally, but each displays a distinct colonization strategy. Mountain Bluebirds are dispersive (Power and Lombardo 1996, Guinan et al. 2000) and frequently among the earliest colonists following fire (Hutto 1995, Schieck and Song 2006), whereas Western Bluebirds often delay colonization and are less dispersive and slower to find new habitat (inset 1; Kotliar et al. 2007, Saab et al. 2007). However, Western Bluebirds are more aggressive than Mountain Bluebirds and the former can rapidly displace the latter from habitat patches (Duckworth and Badyaev 2007). Moreover, within Western Bluebirds, two distinct dispersal strategies exist, such that more aggressive colonizing phenotypes are replaced by less aggressive and less dispersive phenotypes over time within patches (second figure; Duckworth et al. 2015).

This correlation between aggression and dispersal in Western Bluebirds is adaptive because aggressive birds have high fitness in new habitat patches where they outcompete earlier-arriving heterospecific competitors including Mountain Bluebirds for nest sites and also acquire larger territories (Duckworth 2008). In contrast, nonaggressive male Western Bluebirds are poor competitors but benefit from remaining in their natal population, where they gain territories by cooperating with relatives (Aguillon and Duckworth 2015). Nonaggressive males also do well in high-density populations during the later stages of the habitat cycle because they invest more in parental care than aggressive males and achieve higher fitness as a consequence (Duckworth 2006).

A switch in prevalence of these two dispersal strategies in Western Bluebird populations is driven by maternal effects that are induced by changes in the intensity of competi-

to breeding status in a current year. As suggested above, gender biases in social dominance, skews in the sex ratio of adults within populations, and the potential for females to raise young without the aid of male parental care can all influence the evolution of delayed maturation, group living, and

other alternative mating tactics (chapters 16 and 17). In particular, for species that habitually occur in saturated habitats, or where superior foraging skills, predator defense, or additional group members are required to raise offspring successfully, subpopulations of pre-reproductive individuals may exist as

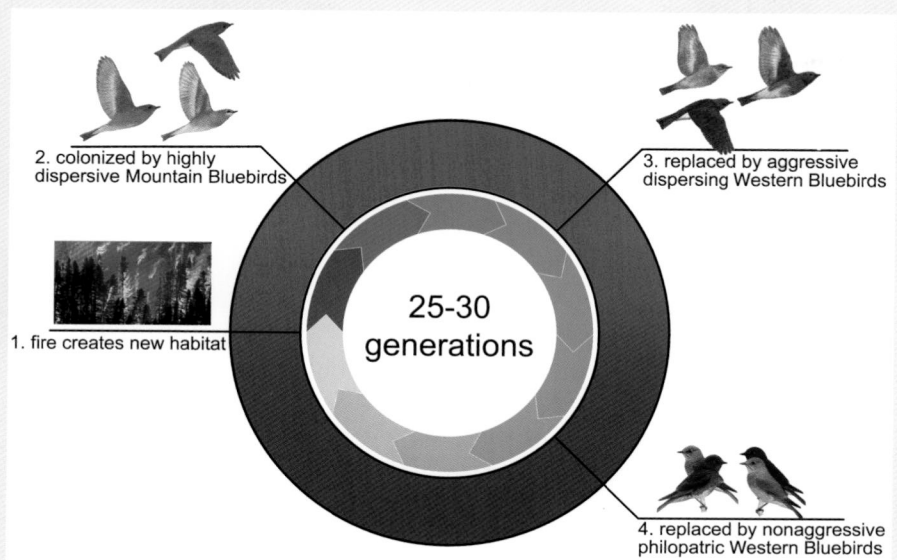

Cycles of bluebird species replacement in post-fire habitat. New habitat is produced by fire (1) and colonized first by Mountain Bluebirds (2). Mountain Bluebirds are replaced by aggressive dispersing Western Bluebirds (3), which in turn are replaced by nonaggressive philopatric individuals (4). Shading from red to purple indicates changes from high (early stages) to low (late stages) Western Bluebird aggression that are associated with increases in breeding density. The inner green ring indicates the predictable decrease in nest cavity availability as breeding density increases; darker arrows indicate greater resource availability. Post-fire forests provide habitat for up to 30 years.

1. fire creates new habitat

2. colonized by highly dispersive Mountain Bluebirds

3. replaced by aggressive dispersing Western Bluebirds

4. replaced by nonaggressive philopatric Western Bluebirds

25-30 generations

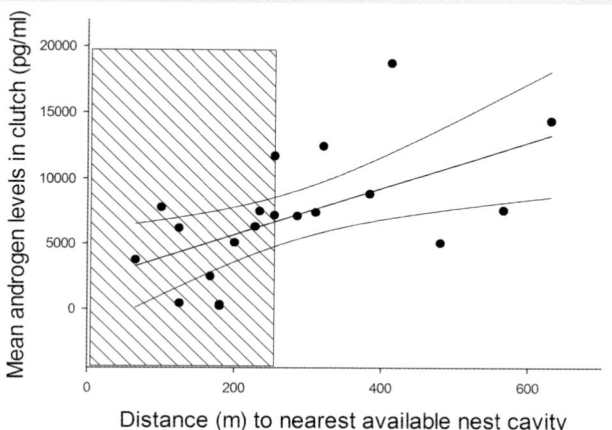

Androgen allocation to a clutch is a key link between sex-biased birth order and resource availability. Mean androgen levels were positively related to the distance to the nearest unoccupied nest cavity. The hatched area indicates distances within a typical bluebird territory. Shown are 95% confidence intervals.

tion (Duckworth et al. 2015). In older, high-density Western Bluebird populations, females produced sons that were more aggressive and more likely to disperse, whereas in newly colonized populations with a low density of Western Bluebirds, females tended to produce nonaggressive sons that remained philopatric. By experimentally manipulating cavity availability for some females but not others, Duckworth et al. (2015) showed that cavity availability, rather than population density itself, induced this maternal effect with remarkable consequences for the spatial and phenotypic structure of bluebird populations overall.

Although maternal effects are often thought to be induced by intraspecific competition for limiting resources, interactions with competitors other than bluebirds also triggers the maternal effect described above in Western Bluebirds. Conflicts with non-bluebird competitors at cavities often peak during egg production, whereas those with members of the same species were low at this time. Importantly, females on territories with extra nest cavities experienced less harassment from competitors, allocated less androgen to their eggs (third figure) and ultimately produced nonaggressive sons able to acquire a territory nearby. These patterns show that spatial and temporal variation of resource availability and biotic communities can have profound effects on the structure of bluebird populations and the makeup of avian communities via mechanistic links to the hormone levels in the eggs of female Western Bluebirds.

helpers in cooperative groups, or as **satellites** in lekking species that benefit from cooperative display (chapters 16 and 17), or even as **sneakers**, which appear to be maintained in populations as a consequence of frequency-dependent selection (box on pages 670–671). Although the evolution of alterna-

tive mating tactics has been a focus of behavioral ecologists for decades, the recent elaboration of examples of both cryptic (boxes on pages 664–665, 668–669) and apparent variation in behavioral and morphological phenotype in birds (boxes on pages 661–662, 670–671) make it clear that variation in social

BEHAVIORAL POLYMORPHISMS IN MALE MATING STRATEGIES IN RUFFS (*PHILOMACHUS PUGNAX*)

Dr. David Lank with a satellite male Ruff.
Photo by Clemens Küpper.

A central enigma of Ruffs is how three distinctly different kinds of males coexist in the same species. Answering that question has been a focus of David Lank's research since 1984. Lank started by trying to explain how two behavioral forms coexisted; then a third form, discovered by Friesian potato farmers (Jukema and Piersma 2006), made this question even more interesting.

Each spring, most male ruffs grow fancy display plumages and migrate from Africa to Scandinavia and northern Russia. In open marsh habitats, males with dark fancy plumages establish mating courts at display sites called leks (van Rhijn, 1991). Territorial males at leks fight with each other to gain dominance on adjacent courts about 1.5m². In contrast to these fancy males, female Ruffs resemble the drab sandpipers you often see on mudflats and beaches. Female Ruffs visit several leks, assess the displays of males, then mate with one or more of them (Lank et al. 2002) before leaving to nest and raise young alone. Since males offer no parental care, each male mates with as many females as possible to maximize its seasonal reproduc-

tive success. But in practice, while some males mate with many females, most mate with none, despite having flown all the way from Africa to display in the cold day after day.

Although leks exist in many species, Ruffs add a unique twist because two other male morphs also visit leks, often traveling in small flocks with mate-shopping females. Satellite males have fancy white-colored plumages. Instead of fighting to defend mating courts, white males are invited to join dark, territorial morphs on their display courts. On average, co-occupied courts attract more females than those with a single dark male (Hugie and Lank 1997, Widemo 1998), perhaps because the interactions between territorial and satellite males offer females information about male quality. However, females may mate with either male, or both. Territorial and satellite males thus maintain an uneasy balance by cooperating to attract females but competing to mate with them.

In contrast, the third type of male is a female mimic that lacks fancy plumage and does not engage in obvious displays (Jukema and Piersma 2006). These rare mimics hang out between lek courts but attempt to hop on the backs of females to mate with them as they crouch to solicit a mating by a displaying male. Mimics sometimes also crouch, perhaps to distract displaying males from real females.

In 1985, Lank began breeding sandpipers in captivity to establish inheritance patterns of morphs. The family trees Lank generated showed that dominant genes gave rise to male satellites or mimics. About 50 percent of the sons of each type had their male parent's morph; the rest were territorials (Lank et al. 1995, 2013). Recently, Lank's colleagues at Sheffield (UK), Graz (Austria), and in Sweden compared DNA sequences of the different male morphs to find consistent differences in a short autosomal inversion containing about 100 genes. This section of chromosome, with different inversion alleles for satellites and mimics, is the dominant Mendelian factor that controls develop-

structure represents a facet of population structure that can have profound consequences for the maintenance of biodiversity. Further studies of individual variation in behavior and mating tactics will be needed to understand how widespread such phenomena are in bird populations in general.

Metapopulations, Sources, and Sinks

Because the existence and relative abundance of surplus individuals in populations is expected to vary temporally with

population density and spatially with habitat quality and environment, social forces also affect population demographic and genetic structure at landscape scales (chapter 21). For example, to the degree that temporal and spatial variation in environment leads to variation in population growth and a population surplus, surplus individuals may represent immigrants to extinct or unsaturated populations elsewhere in a species range. Immigrant colonists to extirpated populations or unoccupied habitat fundamentally affect the evolution-

Territorial (left) and satellite ruff (middle). *Photo by Raymond Ng.* A female mimic (right). *Photo by Clemens Küpper.*

ment into each of three male morphs. Territorial males carry the uninverted ancestral sequence shared with other species of birds. As in White-throated Sparrows (box on pages 661–662), this region includes genes affecting differences in steroid hormones, which are likely to be intimately involved in the development of different morphs. This region also contains the MC1R, a gene, which codes for black vs. white plumages in several other bird species. Finally, one end of the inversion split a gene essential for mitosis. Since individuals cannot exist without a working copy of this gene, all satellites and mimics are heterozygotes, which explains the neat 50:50 Mendelian ratios Lank obtained by breeding them. But how do these three morphs coexist in nature?

Wild populations of Ruffs consist of stable equilibria of about 85 percent territorial, 14 percent satellite, and 1 percent female mimic males. This balance between territorials and satellites appears to be maintained by negative frequency-dependent selection, driven by interactions at the lek among territorial and satellite males, and females (Hugie and Lank 1997). What sets the equilibrium ratio remains unknown, but Lank believes it is related to competition between satellites at leks. Mimics appear to be limited to 1 percent of the population because of issues related to the maintenance of honest signals. Now that Lank has molecular markers for male type, he can address these questions empirically by comparing male and female genotypes to

the chicks they produce to assess paternity. These data will allow Lank to look for variation in the conditions affecting mating success of morphs to ask how male reproductive success varies within and between morphs. How did this system evolve?

Strong skews in mating success open up opportunities for alternative mating strategies to invade and spread within populations, because an alternative strategy does not have to do very well to exceed the expected success of an average male (Shuster and Wade 2003). Mutations, including inversions, may happen quite often, even though only a tiny fraction of them produce viable individuals. A combination of strong sexual selection and a rare chance event in history produced the unique situation we see in Ruffs today.

Future research may follow up on the gene expression and physiological pathways that control differences in morphology and behavior. It will be easier to isolate pathways by comparing differences in processes within a species than by comparing pathways between species. Such strong genetic determination of discrete differences in morphology and behavior is an exception rather than the rule in birds; many behaviors are plastic, and that flexibility can serve individuals well. But under rare conditions, selection disfavors intermediate or flexible forms, and fixed forms can persist, as we see in Ruffs. Footage and description of morphs, strategies, and lekking can be viewed at vimeo.com/144333522.

ary potential of the populations they found via the range of phenotypes and genotypes they represent (see below; chapters 3 and 4), and social organization can profoundly affect the probability of founding new populations and gene flow (chapters 17 and 19). Similarly, the influence of immigrants on extant populations will depend on their similarity to residents and their ability to produce offspring that are recruited in the breeding population. In either case, knowing how social organization affects the existence and mobility of a popu-

lation surplus helps us predict population growth and stability in nature and manage species for persistence at landscape scales. To do so, demographers and evolutionary ecologists have adopted conceptual models of populations distributed in space and linked by dispersal, referred to as metapopulations or population source and sink systems (chapter 21).

Source and sink systems include populations linked by dispersal but subject to increasingly different environments at progressively larger spatial scales (chapter 21). As a consequence, populations may show temporal variation in growth rate caused by peaks and valleys in environmental suitability (cf. Andrewartha and Birch 1954) or spatial variation in growth rate as a consequence of persistent spatial variation in the environment (cf. Pulliam 1988; fig. 20.7). In each case, **source** populations contribute a net surplus of emigrants to adjacent populations, supplementing them demographically and genetically, whereas **sink** populations rely on a net influx of immigrants to persist. Andrewartha and Birch (1954) described sink populations in poor years that became sources in good years, leading to the regional stabilization of population sizes overall. In contrast, where sink populations represent a net cost to regional population growth, regional declines become more likely as the fraction of sink to source populations increases (Pulliam 1988; chapter 21).

Although relatively few source-sink systems have been mapped at large spatial scales using empirical data on individual performance, Jewell and Arcese (2008) provided a method to do so using data on the distribution of Song Sparrows and

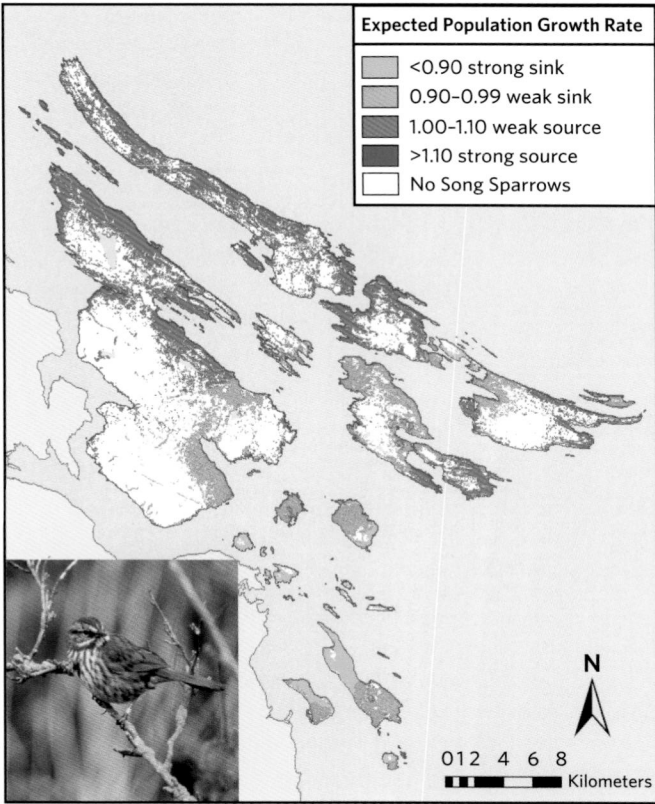

Figure 20.7. The predicted growth rate of island Song Sparrow populations in southeast British Columbia, Canada, given the documented impacts of Brown-headed Cowbirds on reproductive success (Jewell and Arcese 2008); no sparrows denotes areas where their predicted occurrence was < 0·70. *Song Sparrow photo by P. Arcese. Figure by P. Arcese, modified from Jewell and Arcese 2008.*

Brown-headed Cowbirds, demonstrating that cowbird distribution can drive sparrow dynamics by reducing the number of offspring they produce annually (Arcese et al. 1996). At regional scales, cowbirds may therefore drive source-sink dynamics by affecting the availability of a population surplus capable of augmenting adjacent populations via dispersal. Creating such maps gives land managers a spatial framework in which to consider human land use, cowbird distribution, and the persistence of host populations (fig. 20.7; Jewell et al. 2007, Jewell and Arcese 2008).

A more restrictive set of assumptions is applied to systems of metapopulations, described mathematically by Levins (1969). In this system, local populations experience periodic extinctions, but are recolonized by founders from adjacent extant populations. Levins's initial model had no explicit spatial structure, but more recent formulations allow colonists to come from adjacent populations at rates that depend on the distance between populations or existence of other barriers to dispersal. Levins's focus on extinction and recolonization is useful because it emphasizes the potential for founders to cause spatial variation in population genetic structure over a species range (see below; chapter 4). By comparison, dispersal in source and sink systems may enhance or dilute the genetic structure of populations depending on barriers to dispersal, gene flow, selection, and effective population size. We now explore further how genetic and demographic structure develop in populations, which affects our predictions about population demography and evolution, and therefore how we study and manage bird populations.

POPULATION GENETIC STRUCTURE

Just as evolution can be interpreted as both a pattern and a process (Stearns and Hoekstra 2005), geneticists view population structure as both a pattern and a process. The pattern-oriented view of population structure focuses on the amount of genetic variation and its distribution within and among subpopulations and individuals of a species (Templeton 2006), emphasizing in particular spatial patterns in genetic variation. Since most avian species do not occur in a single contiguous population but are instead structured into more or less isolated subpopulations that differ in size, most bird species exhibit some differences in the genetic variation at this level (Evans 1987; fig. 20.8). However, the degree to which subpopulations differ in genetic makeup varies widely among bird species.

For example, genetic differences among island populations of the Galapagos Sharp-beaked Ground Finches (*Geospiza difficilis*) and warbler finches (*Certhidea olivacea* and *C. fusca*) are pronounced (Petren et al. 2005; fig. 20.9), approaching in

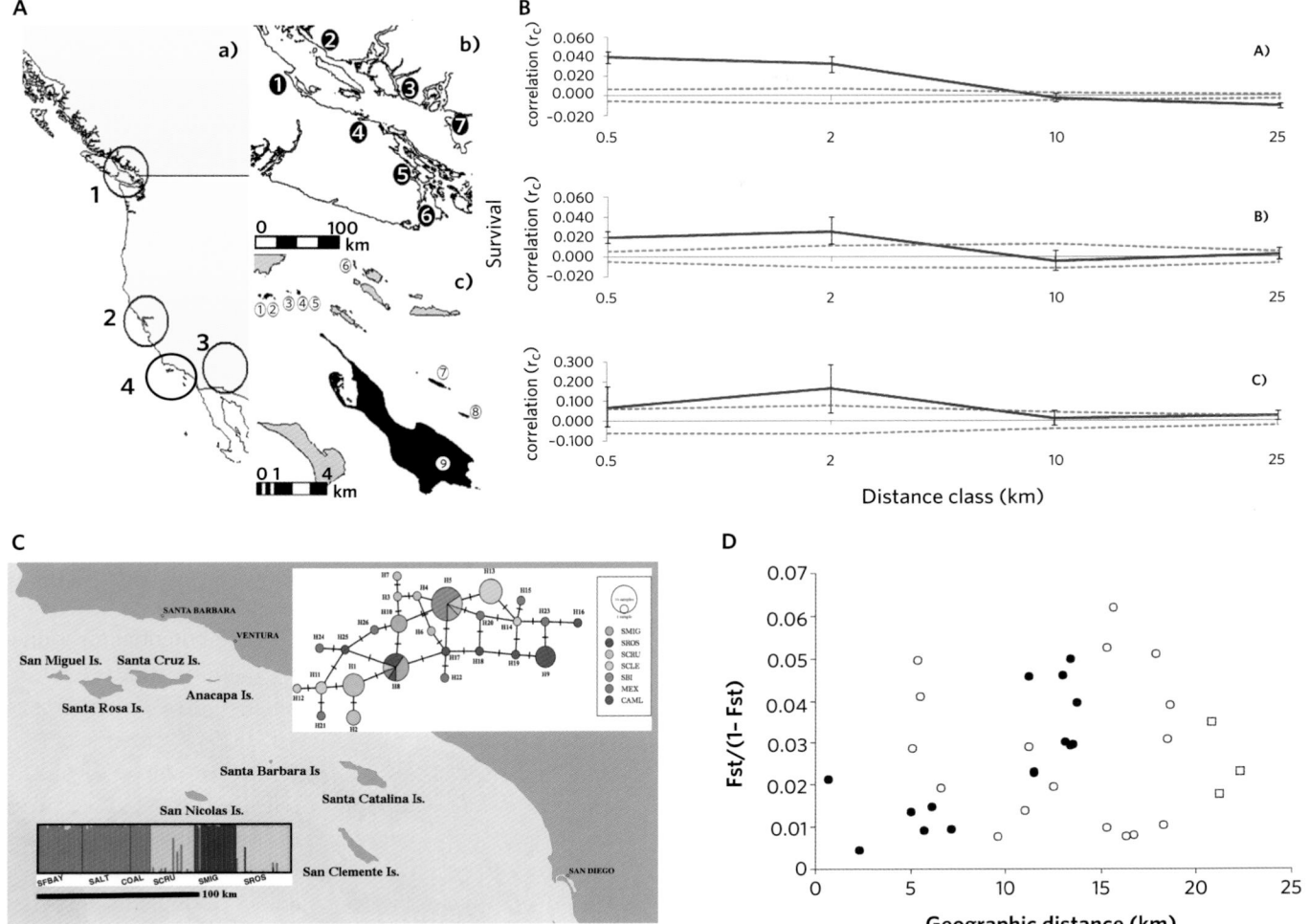

Figure 20.8. Sources of spatial variation in population genetic structure in Song Sparrows. *A* shows populations sampled for microsatellite or mtDNA variation in (*1*) the Georgia Basin of British Columbia and (*2*) San Francisco Bay, (*3*) Salton Sea and (*4*) Channel Islands of California (inset *a*). Insets (*b*) and (*c*) show sample populations on British Columbia mainland and Vancouver Island (1–7) and the Southern Gulf Islands (1–9), respectively. *B* shows the correlation (rc) between genetic and geographic distance calculated for individuals in the (A) San Francisco Bay, (B) Georgia Basin and (C) Salton Sea (Wilson 2008, Wilson et al. 2011; 95 percent CI shown as dashed lines, bootstrapped 95 percent CI around coefficient at each distance class). Positive genetic structure was detected at distances of <10 km, but were highest in California because of reduced gene flow across subspecific contact zones (Patten et al. 2004, Wilson et al. 2011). In British Columbia, island populations were more genetically diverged at similar spatial scales than were populations on mainland sites, suggesting water barriers reduced gene flow. *C* shows substantial genetic differentiation among three extant and two extinct Channel Island Song Sparrow populations (*M. m. graminea*; G"ST: 0.14–0.37; Wilson et al. 2015). Channel Island populations were also diverged from mainland populations of *M. m. heermanni* (G"ST: 0.30–0.64). Of 10 mtDNA haplotypes recovered in extant and extinct Channel Island populations, only two were shared between the northern and southern Channel Islands, and just one detected on the mainland. Wilson et al. (2015) used these and other results to suggest Song Sparrow populations on the northern Channel Islands are demographically independent. *D* shows genetic isolation by distance among 19 sample populations, representing three named subspecies endemic to tidal saltmarsh habitats of San Francisco Bay (*M. m. pusillula, M. m. maxillaris, M. m. samuelis*; black circles). In contrast, the absence of a similar pattern in two widespread subspecies outside the bay (*M. m. gouldii*, open circles; *M. m. heermanni*, squares) suggests a comparative absence of barriers to gene flow in upland areas (Chan and Arcese 2002, 2003). The existence of strong morphological differentiation among subspecies, which has often been linked to environmental factors likely to affect individual fitness, despite little evidence of differentiation at microsatellite loci, suggests (1) high gene flow and large effective sizes in most Song Sparrow populations, but also (2) that natural selection or phenotypic plasticity both contribute to marked morphological differentiation among populations and named subspecies (Chan and Arcese 2003, Patten et al. 2004, Wilson et al. 2011, 2015). *Material for figure by Amy G. Wilson.*

Figure 20.9. The genetic differentiation among populations varies widely in bird species, even among closely related species inhabiting the same regions. For example, both *A*, the Warbler Finch, and *B*, the Large Ground Finch, live on islands of the Galápagos archipelago, yet the Large Ground Finch exhibits very little genetic structure among islands, while the Warbler Finch exhibits pronounced population genetic structure, approaching levels typical of species. *Photos by Lukas Keller.*

2006). Because the process-oriented view of population structure focuses on these factors, we now review some of the processes that create that structure in bird populations.

Population Size and Genetic Drift

The size of subpopulations and how they vary in space and time are important determinants of population genetic structure for two reasons. First, all else being equal, large subpopulations contain more genetic variation than small subpopulations. This is a consequence of both the higher mutational input and the slower loss of existing genetic variation in large populations. With a constant mutation rate v per gene and generation, $2Nv$ new mutations occur each generation. Thus, the larger the population size N, the more new mutations are created each generation. Second, random genetic drift, a consequence of the Mendelian inheritance process creating variation in how often a particular allele is passed on to the next generation, leads to random changes in allele frequencies from one generation to the next (Templeton 2006). These random changes occur at a rate proportional to $1/2N$, thus they are more pronounced in smaller populations. Ultimately, random genetic drift leads to the loss of genetic variation in a population. A very small (<60 individuals) population of Floreana Mockingbirds (*Mimus trifasciatus*; fig. 20.10) in the Galápagos provides an example of the strength of genetic drift. Over a period of 100 years, genetic drift led to the loss of alleles with a frequency as high as 0.64 at the outset (Hoeck et al. 2010a). Because large populations lose less existing genetic variation to random genetic drift, and because they receive

Figure 20.10. The Floreana Mockingbird is a critically endangered bird species in the Galápagos Islands. Here, a juvenile is begging for food. The two small remaining populations of this species have experienced strong genetic drift over the past one and a half centuries, following the extinction of the biggest population (Hoeck et al. 2010a, 2010b). *Photo by Lukas Keller.*

magnitude differences reported among species (e.g., Hoeck et al. 2010a,b; see also fig. 20.8). In contrast, genetic variation across island populations of the Galapagos Small (*Geospiza fuliginosa*), Medium (*Geospiza fortis*), and Large Ground Finches (*Geospiza magnirostris*) display very small genetic differences, a pattern also shown in some species at large geographic scales (e.g., Milot et al. 2008; fig. 20.9).

These genetic patterns of population structure are the consequence of the passing of genetic material from one generation to the next. In sexually reproducing diploid organisms such as birds, genetic material is passed from one generation to the next in two steps. First, adults (zygotes) produce gametes, which then pair to form the next generation of diploid zygotes. Factors that influence which gametes are paired and which are not create population structure (Templeton

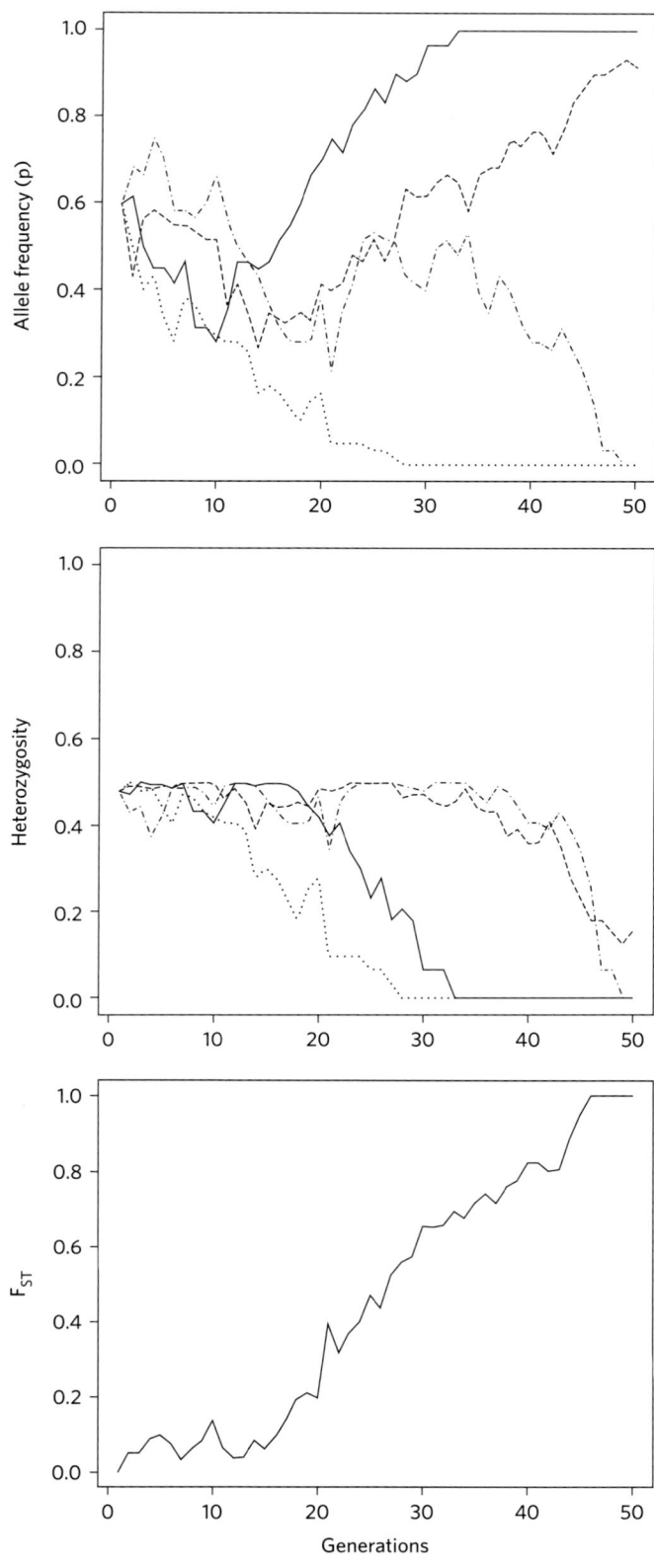

more mutational input, they contain more genetic variation than small populations. This theoretical prediction is generally borne out in data from bird populations (e.g., Hoeck et al. 2010b).

A second way in which population size contributes to population genetic structure is through the effects of random genetic drift on allele frequency divergence *among* populations. As noted above, the smaller the subpopulation, the more random genetic drift will lead to changes in allele frequency (Templeton 2006; fig. 20.11). The direction of change is random, so the frequency of a given allele will increase in some but decrease in other subpopulations. This results in genetic differentiation, or, equivalently, increased genetic variation among subpopulations. A study of House Sparrows provides an example of this effect. After a serious population decline that increased levels of random genetic drift across the whole of Finland, genetic differentiation among House Sparrow populations was about three times larger than before the decline (Kekkonen et al. 2011).

Inbreeding

The size of a subpopulation has an additional important effect: the smaller a subpopulation, the greater the likelihood that the gametes that are paired come from related individuals. In other words, everything else being equal, the smaller a subpopulation, the more inbreeding occurs (Keller and Waller 2002). Thus, inbreeding is particularly common in rare and captive-bred species. Inbreeding is an important aspect of population structure, because inbreeding reduces individual fitness on average, primarily by increasing homozygosity, including at loci with deleterious mutations. Deleterious mutations are constantly produced and can remain in populations for long periods, even when very detrimental. For example, a recessive lethal allele causing blindness in a small, insular Red-billed Chough population has existed for at least

Figure 20.11. Random genetic drift changes allele frequencies (*top*), thus reducing genetic variation within populations (*middle*), but increasing genetic differentiation among populations (*bottom*). Four isolated populations of size 30 each were simulated for 50 generations.

Results shown are for a single locus with an initial allele frequency of $p = 0.6$. Top panel shows the changes in allele frequencies over time caused by random genetic drift in each of the four populations. After 50 generations, all populations have allele frequencies substantially different from the original $p = 0.6$. After 33 generations, two of the populations have lost all their genetic variation and were fixed for one allele. These changes in allele frequencies led to a decrease in genetic variation within populations, as measured by expected heterozygosity (*middle*). At the same time, genetic differentiation among the four populations increased (*bottom*), as quantified using F_{ST} among the four populations. Note how F_{ST} approaches one as nearly all the genetic variation among individuals is converted to genetic variation among subpopulations (see text for more details on F_{ST}). *Graphic by the authors.*

15 years (Trask et al. 2016). Because inbreeding increases the phenotypic expression of deleterious alleles, resulting reductions in individual fitness are known as "inbreeding depression" and are measured as the decline in individual fitness for a given increase in inbreeding. For example, Keller et al. (1994) compared inbred and outbred Song Sparrows to show that severe winter weather eliminated inbred birds from the population much more often than it did outbred birds. Similarly, Marr et al. (2006) showed that inbred Song Sparrows abandoned incubation more often than did outbred birds in periods of heavy rain. These examples emphasize that inbreeding depression as a consequence of small population size can occur in a wide range of phenotypic traits.

Gene Flow

Through its effects both within and among populations, random genetic drift also causes a decrease of genetic variation within subpopulations, but an increase in genetic variation among sub-populations. An increase in variation among subpopulations occurs because genetic drift is random, and therefore uncorrelated in adjacent populations. In contrast, gene flow measures the exchange of genes among subpopulations and has exactly the opposite effects (Templeton 2006) as long as it also occurs randomly (Edelaar et al. 2008). The random exchange of genes counteracts genetic drift and decreases genetic differentiation among populations and increases genetic variation within subpopulations. Gene flow is therefore a powerful equalizer of genetic diversity. For example, gene flow quickly restored levels of genetic variation to a bottlenecked island population of Song Sparrows (Keller et al. 2001b), with the result that only three years after a severe population crash wiped out 95 percent of the population, within-population genetic diversity was back to levels measured prior to the crash. That finding corresponds to the relatively small genetic differences observed among Song Sparrow populations inhabiting islands of the Georgia Basin, Canada, and salt marshes of San Francisco Bay (fig. 20.8), where dispersal among adjacent sites occurs regularly (Wilson and Arcese 2008). In contrast, Song Sparrow populations of the Channel Islands, California, appear to only rarely share effective dispersers and are much more diverged as a consequence (fig. 20.8). Similarly, after the extinction of a large population of Floreana Mockingbirds, gene flow ceased its equalizing effects among populations, allowing the two remaining extant populations to diverge via random genetic drift (Hoeck et al. 2010a; fig. 20.10).

Because gene flow and random genetic drift typically have opposite effects, the spatial patterns of neutral genetic variation of a species are primarily determined by the balance between drift and gene flow (Templeton 2006). The

oldest and best known measure of this balance is F_{ST}, the proportion of genetic variation among individuals across an entire population that is to the consequence of genetic variation among subpopulations (Holsinger and Weir 2009). A high F_{ST} implies that much of the genetic variation across the entire population is caused by genetic variation among subpopulations. In other words, allele frequencies differ substantially among subpopulations, and individuals within a subpopulation are genetically similar. A high F_{ST} is expected when genetic drift, in the absence of much gene flow, has led to substantial allele frequency differences between populations and to little genetic variation within populations, as in some populations of Galapagos Mockingbirds (*Mimus parvulus*; Hoeck et al. 2010b). Conversely, low estimates of F_{ST} imply that gene flow has been high and/or that subpopulation sizes are big, and that little genetic drift has occurred. Such a pattern is typical of bird species that have very large population sizes and high dispersal abilities, such as the Little Auk (*Alle alle*; Wojczulanis-Jakubas et al. 2014). In such species, high genetically effective dispersal results in high rates of genetic mixing and thus homogeneous population structure. Overall, birds display a wide variety of population genetic structures (Avise 1994), ranging from substantial structure on very small spatial scales (e.g., Bertrand et al. 2014) to very little differentiation on a global scale (e.g., Kraus et al. 2012).

While F_{ST} and some of its derivatives (Meirmans and Hedrick 2011, Whitlock 2011) are informative measures of the **pattern** of population genetic structure, they provide less insight into the underlying **processes**. The main reason for this is that F_{ST} is neither a measure of gene flow nor a measure of random genetic drift alone, but a measure of the balance between drift and gene flow: $F_{ST} \approx 1/(4NM + 1)$. Because population genetic structure is a function of the product of the population size N and the fraction of migrants, m, it is often difficult to distinguish situations with high gene flow but low subpopulation sizes from the reverse. Moreover, many real populations are likely to violate the assumptions behind the equation given above, so that estimates of gene flow from patterns of F_{ST} will rarely be very reliable (Whitlock and McCauley 1999). Importantly, relating F_{ST} to Nm assumes that gene flow and genetic drift are in equilibrium, an assumption that is often violated because it can take dozens to thousands of generations to reach the equilibrium (Whitlock and McCauley 1999). For example, lower genetic differentiation among populations at high latitudes may, in part, still reflect a nonequilibrium situation caused by the ice ages (Wojczulanis-Jakubas et al. 2014). The equilibrium assumption is, however, not always unreasonable. A metapopulation of White-starred Robins (*Pogonocichla stellata*) in Kenya was in migration-drift equilibrium despite recent (~40 years) habitat fragmentation (Galbusera et al. 2004) and a comparison of museum and contemporary samples showed that mockingbirds on islands in Galápagos are in migration-drift equilibrium (Hoeck et al. 2010b).

Because of the difficulties of obtaining reliable estimates of gene flow from patterns of F_{ST} or its derivatives, alternative methods for estimating gene flow have been developed. One method, for example, looks for individuals with migrant ancestry in a population sample (Wilson and Rannala 2003). Although this method has its own limitations (Meirmans 2014), it does not assume migration-drift equilibrium and can also handle nonrandom gene flow between populations. That gene flow among populations may be nonrandom, for example because dispersing individuals settle in those environments that best match their phenotype, has received relatively little attention in the literature (Edelaar et al. 2008). However, increasingly more studies discover such nonrandom gene flow. For example, Galápagos warbler finches preferentially disperse to islands that feature habitat similar to their natal environment (Tonnis et al. 2005). Such findings are important because nonrandom gene flow can lead to unexpected results (e.g., a negative relationship between genetic and geographic distance in the warbler finches; Tonnis et al. 2005) and can promote rather than retard population differentiation and adaptation (Edelaar et al. 2008). Nonrandom gene flow thus has profound ecological and evolutionary consequences, with important implications for conservation and management. For example, habitat restoration may be less effective than expected if birds disperse to habitat similar to what they experienced at their natal sites.

Effective Population Size

Most equations that describe the effects of population size on population genetic structure assume that the population in question does not vary in size over time and exhibits an equal sex ratio and a distribution of offspring numbers that follows a Poisson distribution and so on. (Templeton 2006). Natural populations never fulfil all these assumptions. That said, as long as the effective population size (N_e) is used instead of the census population size (N), the equations can still be used. The effective population size N_e is estimated from the real populations under study to account for all the violations of the model assumptions. N_e of real populations can be estimated using demographic information (e.g., O'Connor et al. 2006), pedigree information (e.g., Ewing et al. 2008), or molecular markers (e.g., Hoeck et al. 2010b). In most avian populations, N_e is smaller than N. On average, the mean N_e/N ratio in birds is 0.65 (Waples et al. 2013). Note that there are different definitions of N_e that can lead to markedly different estimates in natural bird populations (e.g., Ewing et al. 2008). Hence, it is important to choose the appropriate N_e for the question at hand.

Natural and Sexual Selection

Another evolutionary process that can lead to population genetic structure is natural selection. When the magnitude or the direction of natural selection differs in space, population genetic structure emerges. For example, Garroway et al. (2013) found evidence that spatially varying selection associated with spatial variation in malaria infection risk led to fine-scale genetic structure in a Great Tit (*Parus major*) population. Spatially varying selection pressure is, however, not the only way in which selection can lead to population genetic structure.

Population genetic structure can also emerge when the amount of genetic variation for a trait under selection varies spatially. In such situations, the response to selection can also differ spatially, even in the absence of spatially heterogeneous selection pressure, leading to population genetic structure (Garant et al. 2005). Moreover, gene flow can interact with selection to create population genetic structure. Spatially homogeneous selection against large clutches, combined with differential gene flow from the mainland, resulted in substantial population genetic structure between two subpopulations of an island population of Great Tits (Postma and van Noordwijk 2005). These examples from well-studied Great Tit populations (Garant et al. 2005, Postma and van Noordwijk 2005, Garroway et al. 2013) illustrate that natural selection interacts with gene flow, genetic drift, and other evolutionary forces to create population genetic structure, which in turn affects how natural selection operates (Templeton 2006). All of these considerations also apply to sexual selection (e.g., Oh and Badyaev 2010, chapter 14). Interactions between selection, gene flow, genetic drift, and other evolutionary forces are evident not only at the population level but also at the genomic level: some parts of the avian genome show strong evidence of population genetic structure, while others do not (e.g., Zhan et al. 2015).

EVOLUTION IN STRUCTURED POPULATIONS

Both population demographic and population genetic structure affect the course of evolution (Charlesworth 1994, Lion et al. 2011). The literature on this topic is quite technical and beyond the scope of this chapter, yet many important questions about evolution depend on population structure. For example, whether a species adapts to the local environment in which it lives, or whether a single species will split into separate species (chapter 3) are questions that can be answered only with knowledge of population demographic and population genetic structure. The evolutionary outcome then depends on the way this structure interacts with range size, environmental variation, and natural and sexual selection to drive or maintain systematic differences between populations in the frequency of inherited traits affecting individual fitness, such as those linked to energy balance, breeding phenology, and morphology (chapters 7, 8, 16, 18).

KEY POINTS

- All populations comprise individuals that differ in gender, age, developmental stage, phenotype, and genotype. These differences allow us to identify subpopulations of individuals that represent demographic and genetic structure in populations.
- When the shared traits of individuals that constitute a given subpopulation also affect those individuals' probabilities of surviving, reproducing, or dispersing, quantifying those effects on populations and on groups of populations linked by dispersal can improve predictions related to population demography, evolution, and management.
- Population processes linked to social dominance, territoriality, dispersal, and mating behavior can profoundly affect the rates of divergence among populations, revealing intricate links between demographic and genetic structure of populations.
- Evolutionary and ecological paradigms employ different criteria to evaluate the isolation of populations within a species' range: the first focusing on the genetic cohesion of populations as evidenced by gene flow, and the second focusing on the demographic cohesion of populations by augmenting population size and reproductive rate.
- Although many measurable traits allow us to describe population structure in considerable detail for many species, cryptic variation in individual phenotype and genotype can be instrumental to understanding spatial and temporal variation in species distribution, behavior, and demography.
- Understanding the origins of population demographic and genetic structure is essential to contributing to biological diversity overall and to developing effective management plans that enhance population growth and persistence, maintain or promote the adaptive capacity of populations subject to environmental change, and conserve the adaptive and nonadaptive differences among populations distributed across a species' range.

KEY MANAGEMENT AND CONSERVATION IMPLICATIONS

- Spatial variation in the genetic structure of populations is widespread in nature but can arise via different mechanisms, often with profound implications for conservation and management.
- Many traits indicative of population genetic and demographic structure are easily quantified and well-known to be linked to individual variation in fitness. The advent of genetic tools and maturation of many long-term studies of individual birds also indicates that more cryptic differences among individuals also exist. When such differences among individuals also affect their probabilities of survival and reproduction, they will also have the potential to affect ecological and genetic processes with consequences for population differentiation, growth, and persistence.
- Adaptive differences among populations arise when spatial variation in natural or sexual selection prompt evolutionary change in traits linked to individual fitness. Management actions or landscape modifications that enhance gene flow between populations adapted to different micro- or macro-environments therefore have the potential to reduce individual fitness and population viability.
- Nonadaptive differences among populations can arise as a consequence of founder effects, random genetic drift, and inbreeding, which all become more influential as populations decline in size or become more isolated. Enhancing gene flow between such populations has the potential to homogenize genetic differences between them, erasing the effects of historical isolation, founder events, or population bottlenecks. Although the genetic homogenization of populations will not always affect fitness, it may eliminate genetic differences that nevertheless contribute to biological diversity.
- Population genetic structure can also arise rapidly in populations that are reduced in size or become isolated from other populations of the same species as a consequence of human actions in the environment. In such cases, management to reestablish historic levels of genetic variation has the potential to enhance individual fitness, population persistence, and the adaptive scope of populations subject to environmental change.
- Distinct population segments are the smallest unit of a species that can be protected under the US Endangered Species Act and are identified by morphological, genetic, or other traits of individuals that allow their reliable assignment to a particular population or region. Diag-

nosable differences among populations in traits linked to fitness may indicate the existence of locally adapted populations which, if lost, cannot be reestablished in ecological timeframes by reintroducing individuals from other population segments.

DISCUSSION QUESTIONS

1. Imagine that you are a biologist tasked with conserving a small, isolated population of birds in Idaho that was once part of a large, widely distributed population. Given concern about the potential for inbreeding and low genetic diversity to threaten the persistence of this population, you are considering translocating individuals from a larger, more genetically diverse population still extant in Oklahoma to augment genetic variation in the Idaho population. What are the potential positive and negative outcomes of the proposed translocation? What information would you need to make the best possible decision in the face of the trade-offs you have identified?

2. The US Endangered Species Act recognizes several levels of organization when identifying units that will receive legal protection. How might spatial variation in the traits of species help managers decide the appropriate level of organization at which to conserve species, subspecies, or distinct population segments? What would you want to know about the origins of variation in the traits under consideration before using them to make such decisions?

3. Population ecologists and wildlife managers often work together to predict how populations will respond to environmental changes of many different types. How might the spatial distribution and size of the populations under study determine whether or not they are considered to represent a single versus multiple entities?

4. The demographic structure of populations is often characterized by delineating subpopulations of individuals by gender and age. What factors might affect whether or not funds should be allocated to research designed to estimate the expected reproductive and survival rates for each of these delineated subpopulations versus considering them as a single population?

5. Develop and discuss arguments for why or why not managers should aim to maintain variation in behavioral and morphological phenotypes described within populations in nature. Assuming that you can't have all the necessary information about population structure, how would you make decisions about management? What is the minimum necessary information?

References

Aguillon, S. M., and R. A. Duckworth (2015). Kin aggression and resource availability influence phenotype-dependent dispersal in a passerine bird. Behavioral Ecology and Sociobiology 69:625–633.

Andrewartha, H. G., and L. C. Birch (1954). The distribution and abundance of animals. University of Chicago Press, Chicago, IL.

Arcese, P. (1987). Age, intrusion pressure and defence against floaters by territorial male Song Sparrows. Animal Behaviour 35:773–784.

Arcese, P. (1989a). Territory acquisition and loss in male Song Sparrows. Animal Behaviour 37:45–55.

Arcese, P. (1989b). Intrasexual competition and the mating system of primarily monogamous birds: The case of the Song Sparrow. Animal Behaviour 38:96–111.

Arcese, P., and J. N. M. Smith (1985). Phenotypic correlates and ecological consequences of dominance in Song Sparrows. Journal of Animal Ecology 54:817–830.

Arcese, P., J. N. M. Smith, and M. I. Hatch (1996). Nest predation by cowbirds and its consequences for passerine demography. Proceedings of the National Academy of Sciences USA 93:4608–4611.

Arcese, P., M. K. Sogge, A. B. Marr, and M. A. Patten (2002). Song Sparrow (Melospiza melodia). In The birds of North America online, A. Poole, Editor. Cornell Lab of Ornithology, Ithaca, NY.

Avise, J. C. (1994). Molecular markers, natural history, and evolution. Chapman and Hall, New York.

Badyaev, A. V. (2011). Origin of the fittest: Link between emergent variation and evolutionary change as a critical question in evolutionary biology. Proceedings of the Royal Society of London B: Biological Sciences 278:1921–1929.

Barrowclough, G. F. (1980). Gene flow, effective population sizes, and genetic variance components in birds. Evolution 34:789–798.

Bertrand, J. A. M., Y. X. C. Bourgeois, B. Delahaie, T. Duval, R. García-Jiménez, J. Cornuault, P. Heeb, B. Milá, B. Pujol, and C. Thébaud (2014). Extremely reduced dispersal and gene flow in an island bird. Heredity 112:190–196.

Blight, L. K., M. C. Drever, and P. Arcese (2015). A century of change in Glaucous-winged Gull (Larus glaucescens) populations in a dynamic coastal environment. The Condor 117:108–120.

Bouwhuis, S., R. Choquet, B. C. Sheldon, and S. Verhulst (2012). The forms and fitness cost of senescence: Age-specific recapture, survival, reproduction, and reproductive value in a wild bird population. American Naturalist 179: E15–E27.

Brown, J. L. (1969). The buffer effect and productivity in tit populations. American Naturalist 103:347–354.

Bumpus, H. C. (1899). The elimination of the unfit as illustrated by the introduced sparrow, Passer domesticus. Biological lectures delivered at the Marine Biological Laboratory of Wood's Hole 6:209–226.

Caswell, H. (2001). Matrix population models: Construction, analysis and interpretation. 2nd ed. Sinauer, Sunderland, MA.

Chan, Y. H.-L., and P. Arcese (2002). Subspecific differentiation and conservation of Song Sparrows (*Melospiza melodia*) in the San Francisco Bay region inferred by microsatellite loci analysis. The Auk 119:641–657.

Chan, Y. H.-L., and P. Arcese (2003). Morphological and microsatellite differentiation in *Melospiza melodia* (Aves) at a microgeographic scale. Journal of Evolutionary Biology 16:939–947.

Charlesworth, B. (1994). Evolution in age-structured populations. 2nd ed. Cambridge University Press, Cambridge, UK.

Clutton-Brock, T. H., and K. Isvaran (2007). Sex differences in ageing in natural populations of vertebrates. Proceedings of the Royal Society of London B 274:3097–3104. doi:10.1098/rspb.2007.1138.

Dingemanse, N., C. Both, P. J. Drent, K. van Oers, and A. J. van Noordwijk (2002). Repeatability and heritability of exploratory behaviour in great tits from the wild. Animal Behavior 64:929–938.

Duckworth, R. A. (2006). Behavioral correlations across breeding contexts provide a mechanism for a cost of aggression. Behavioral Ecology 17:1011–1019.

Duckworth, R. A. (2008). Adaptive dispersal strategies and the dynamics of a range expansion. American Naturalist 172:S4–S17.

Duckworth, R. A. (2014). Human-induced changes in the dynamics of species coexistence: An example with two sister species. *In* Avian urban ecology: Behavioural and physiological adaptations, D. Gil and H. Brumm, Editors. Oxford University Press, Oxford, pp. 181–191.

Duckworth, R. A., and A. V. Badyaev (2007). Coupling of dispersal and aggression facilitates the rapid range expansion of a passerine bird. Proceedings of the National Academy of Sciences USA 104:15017–15022.

Duckworth, R. A., V. Belloni, and S. R. Anderson (2015). Cycles of species replacement emerge from locally induced maternal effects in a passerine bird. Science 374:875–877.

Edelaar, P., A. M. Siepielski, and J. Clobert (2008). Matching habitat choice causes directed gene flow: A neglected dimension in evolution and ecology. Evolution 62:2462–2472.

Evans, P. G. H. (1987). Electrophoretic variability of gene products. *In* Avian genetics: A population and ecological approach, F. Cooke and P. A. Buckley, Editors. Academic Press, London.

Ewing, S. R., R. G. Nager, M. A. C. Nicoll, A. Aumjaud, C. G. Jones, and L. F. Keller (2008). Inbreeding and loss of genetic variation in a reintroduced population of Mauritius Kestrel. Conservation Biology 22:395–404.

Formica, V. A., R. A. Gonser, S. M. Ramsay, and E. M. Tuttle (2004). Spatial dynamics of alternative reproductive strategies: The role of neighbors. Ecology 85:1125–1136.

Formica, V. A., and E. M. Tuttle (2009). Examining the social landscapes of alternative reproductive strategies. Journal of Evolutionary Biology 12:2395–2408.

Fretwell, S. D. (1972). Populations in a seasonal environment. Princeton University Press, Princeton, NJ.

Galbusera, P., M. Githiru, L. Lens, and E. Matthysen (2004). Genetic equilibrium despite habitat fragmentation in an Afrotropical bird. Molecular Ecology 13:1409–1421.

Garant, D., L. E. B. Kruuk, T. A. Wilkin, R. H. McCleery, and B. C. Sheldon (2005). Evolution driven by differential dispersal within a wild bird population. Nature 433:60–65.

Garroway, C. J., R. Radersma, I. Sepil, A. W. Santure, I. De Cauwer, J. Slate, and B. C. Sheldon (2013). Fine-scale genetic structure in a wild bird population: The role of limited dispersal and environmentally based selection as causal factors. Evolution 67:3488–3500.

Germain, R. R., and P. Arcese (2014). Distinguishing individual quality from habitat preference and quality in a territorial passerine. Ecology 95:436–445.

Goss-Custard, J. D., R. W. G. Caldow, R. T. Clarke, S. E. A. V. Durell, and W. J. Sutherland (1995). Deriving population parameters from individual variations in foraging behaviour I: Empirical game theory distribution model of oystercatchers *Haematopus ostralegus* feeding on mussels *Mytilus edulis*. Journal of Animal Ecology 64:265–276.

Gould, S. J., and R. F. Johnston. (1972). Geographic variation. Annual Review of Ecology and Systematics 3:457–498.

Grant, P. R. (1983). Inheritance of size and shape in a population of Darwin's finches, *Geospiza conirostris*. Proceedings of the Royal Society of London B: Biological Sciences 220:219–236.

Greenwood, P. J. (1980). Mating systems, philopatry and dispersal in birds and mammals. Animal Behaviour 28:1140–1162.

Guinan, J. A., P. A. Gowaty, and E. K. Eltzroth (2000). Western Bluebird (*Sialia mexicana*). The Birds of North America 510:1–31.

Harrison, X. A., T. Tregenza, R. Inger, K. Colhoun, D. A. Dawson, G. A. Gudmundsson, D. J. Hodgson, G. J. Horsburgh, G. Mcelwaine, and S. Bearhop (2010). Cultural inheritance drives site fidelity and migratory connectivity in a long-distance migrant. Molecular Ecology 19:5484–5496.

Hoeck, P. E. A., M. A. Beaumont, K. E. James, B. R. Grant, P. R. Grant, and L. F. Keller (2010a). Saving Darwin's muse: Evolutionary genetics for the recovery of the Floreana Mockingbird. Biology Letters 6:212–215.

Hoeck, P. E. A., J. L. Bollmer, P. G. Parker, and L. F. Keller (2010b). Differentiation with drift: A spatio-temporal genetic analysis of Galápagos mockingbird populations (*Mimus* spp.). Philosophical Transactions of the Royal Society B 365:1127–1138.

Holsinger, K. E., and B. S. Weir (2009). Genetics in geographically structured populations: Defining, estimating and interpreting F_{ST}. Nature Reviews Genetics 10:639–650.

Hugie, D. M., and D. B. Lank (1997). The resident's dilemma: A female choice model for the evolution of alternative male reproductive strategies in lekking male Ruffs (*Philomachus pugnax*). Behavioral Ecology 8:218–225.

Hutto, R. L. (1995). Composition of bird communities following stand-replacement fires in northern rocky mountain (U.S.A.) conifer forests. Conservation Biology 9:1041–1058.

Irwin, D. E., S. Bensch, J. H. Irwin, and T. D. Price (2005). Speciation by distance in a ring species. Science 307:414–416.

Jewell, K. J., P. Arcese, and S. E. Gergel (2007). Robust prediction of species distribution: Spatial habitat models for a brood parasite. Biological Conservation 140:259–272.

Jewell, K. J., and P. Arcese (2008). Consequences of parasite invasion and land use on the spatial dynamics of host populations. Journal of Applied Ecology 45:1180–1188.

Johnston, R. F., and R. K. Selander (1964). House Sparrows: Rapid evolution of races in North America. Science 144:548–550.

Jukema, J., and T. Piersma (2006). Permanent female mimics in a lekking shorebird. Biology Letters 2:161–164.

Kekkonen, J., I. K. Hanski, H. Jensen, R. A. Väisänen, and J. E. Brommer (2011). Increased genetic differentiation in House Sparrows after a strong population decline: From panmixia towards structure in a common bird. Biological Conservation 144:2931–2940.

Keller, L. F., P. Arcese, J. N. M. Smith, W. Hochachka, and S. C. Stearns (1994). Selection against inbred Song Sparrows during a natural population bottleneck. Nature 372:356–357.

Keller, L. F., P. R. Grant, B. R. Grant, and K. Petren (2001a). Heritability of morphological traits in Darwin's finches: Misidentified paternity and maternal effects. Heredity 87:325–336.

Keller, L. F., K. J. Jeffery, P. Arcese, M. A. Beaumont, W. M. Hochachka, J. N. M. Smith, and M. W. Bruford (2001b). Immigration and the ephemerality of a natural population bottleneck: Evidence from molecular markers. Proceedings of the Royal Society of London B 268:1387–1394.

Keller, L. F., and D. M. Waller (2002). Inbreeding effects in wild populations. Trends in Ecology and Evolution 17:230–241.

Keller, L. F., J. M. Reid, and P. Arcese (2008). Testing evolutionary models of senescence in a natural population: Inbreeding and age effects on survival and reproduction in Song Sparrows (Melospiza melodia). Proceedings of the Royal Society of London B 275:597–604.

Ketterson, E. D., and V. Nolan Jr. (1983). The evolution of differential bird migration. In Current Ornithology, vol. 1, R. F. Johnston, Editor. Plenum Press, New York, pp. 357–402.

Kotliar, N. B., P. L. Kennedy, and K. Ferree (2007). Avifaunal responses to fire in southwestern montane forests along a burn severity gradient. Ecological Applications 17:491–507.

Kraus, R. H. S., P. van Hooft, H.-J. Megens, A. Tsvey, S. Y. Fokin, R. C. Ydenberg, and H. H. T. Prins (2012). Global lack of flyway structure in a cosmopolitan bird revealed by a genome wide survey of single nucleotide polymorphisms. Molecular Ecology 22:41–55.

Krebs, C. J. (2008). Ecology: The experimental analysis of distribution and abundance. 6th ed. Benjamin Cummings, San Francisco.

Lack, D. (1954). The natural regulation of animal numbers. Clarendon Press, Oxford, UK.

Lack, D. (1966). Population studies of birds. Clarendon Press, London.

Lank, D. B., C. M. Smith, O. Hanotte, T. A. Burke, and F. Cooke (1995). Genetic polymorphism for alternative mating behaviour in lekking male Ruff, Philomachus pugnax. Nature 378:59–62.

Lank, D. B., C. M. Smith, O. Hanotte, A. Ohtonen, S. Bailey, and T. Burke (2002). High frequency of polyandry in a lek mating system. Behavioral Ecology 13:209–215.

Lank, D. B., L. L. Farrell, T. Burke, T. Piersma, and S. B. McRae (2013). A dominant allele controls development into female mimic male and diminutive female Ruffs. Biology Letters 9:20130653. doi:10.1098/rsbl.2013.0653.

Leach, A. G. (2015). Fitness consequences of clutch size decisions and mate change in Black Brant. PhD diss., University of Nevada, Reno, NV.

Lebreton, J.-D., K. P. Burnham, J. Clobert, and D. R. Anderson (1992). Modeling survival and testing biological hypotheses using marked animals: A unified approach with case studies. Ecological Monographs 62:67–118.

Lemons, P. R., J. S. Sedinger, C. A. Nicolai, and L. W. Oring (2012). Sexual dimorphism, survival, and parental investment in relation to offspring sex in a precocial bird. Journal of Avian Biology 43:445–453.

Levins, R. (1969). Some demographic and genetic consequences of environmental heterogeneity for biological control. Bulletin of the Entomological Society of America 15:237–240.

Lewis, S., S. Wanless, D. A. Elston, M. D. Schultz, E. Mackley, M. Toit, J. G. Underhill, and M. Harris (2006). Determinants of quality in a long-lived colonial species. Journal of Animal Ecology 75:1304–1312.

Lewis, T. L., D. H. Ward, J. S. Sedinger, A. Reed, and D. V. Derksen (2013). Brant (Branta bernicla). In The birds of North America online, A. Poole, Editor. Cornell Lab of Ornithology, Ithaca, NY. http://bna.birds.cornell.edu/bna/species/337. doi:10.2173/bna.337.

Liker, A., and T. Székely (2005). Mortality costs of sexual selection and parental care in natural populations of birds. Evolution 59:890–897.

Liker, A., R. P. Freckleton, and T. Szekely (2014). Divorce and infidelity are associated with skewed adult sex ratios in birds. Current Biology 24:880–884.

Lindberg, M. S., J. S. Sedinger, R. F. Rockwell, and D. V. Derksen (1998). Population structure and natal and breeding movements in a Black Brant metapopulation. Ecology 79:1893–1904.

Lindberg, M. S., J. S. Sedinger, and J.-D. Lebreton (2013). Individual heterogeneity in Black Brant survival and recruitment with implications for harvest dynamics. Ecology and Evolution 3:4045–4056.

Lion, S., V. A. A. Jansen, and T. Day (2011). Evolution in structured populations: Beyond the kin versus group debate. Trends in Ecology and Evolution 26:193–201.

Lomnicki, A. (1988). Population ecology of individuals. Monographs in Population Biology 25. Princeton University Press, NJ.

Lowther, J. K. (1961). Polymorphism in the White-throated Sparrow, Zonotrichia albicollis (Gmelin). Canadian Journal of Zoology 39:281–292.

Lowther, J. K. (1962). Colour and behavioural polymorphism in the White-throated Sparrow, Zonotrichia albicollis (Gmelin). PhD diss., University of Toronto, Toronto.

Lowther, P. E. (1993). Brown-headed Cowbird (Molothrus ater). In The birds of North America online, A. Poole, Editor. Cornell Lab of Ornithology, Ithaca, NY. http://bna.birds.cornell.edu/bna/species/047.

Lowther, P. E., and C. L. Cink (2006). House Sparrow (*Passer domesticus*). *In* The birds of North America online, A. Poole, Editor. Cornell Lab of Ornithology, Ithaca, NY. http://bna.birds.cornell.edu.ezproxy.library.ubc.ca/bna/species/012.

Marr, A. B., P. Arcese, W. M. Hochchaka, J. M. Reid, and L. F. Keller (2006). Interactive effects of environmental effects and inbreeding in a wild bird population. Journal of Animal Ecology 75:1406–1415.

Meirmans, P. G. (2014). Nonconvergence in Bayesian estimation of migration rates. Molecular Ecology Resources 14:726–733.

Meirmans, P. G., and P. W. Hedrick (2011). Assessing population structure: F_{ST} and related measures. Molecular Ecology Resources 11:5–18.

Milot, E., H. Weimerskirch, and L. Bernatchez (2008). The seabird paradox: Dispersal, genetic structure and population dynamics in a highly mobile, but philopatric albatross species. Molecular Ecology 17:1658–1673.

Morrison, M. L. (2012). The habitat sampling and analysis paradigm has limited value in animal conservation: A prequel. Journal of Wildlife Management 76:438–450.

Newton, I. (1998). Population limitation in birds. Academic Press, London.

Nice, M. M. (1937). Studies in the life history of the Song Sparrow I: A population study of the Song Sparrow. Transactions of the Linnean Society of New York 4:1–247.

Nicolai, C. A., and J. S. Sedinger (2012). Are there trade-offs between pre- and post-fledging survival in black brent geese? Journal of Animal Ecology 81:788–797.

Norris, D. R., P. P. Marra, T. K. Kyser, T. W. Sherry, and L. M. Ratcliffe (2004). Tropical winter habitat limits reproductive success on the temperate breeding grounds in a migratory bird. Proceeding of the Royal Society B 271:59–64.

Norris, D. R., P. Arcese, D. Preikshot, D. F. Bertram, and T. K. Kyser (2007). Diet reconstruction and historic population dynamics in a threatened seabird. Journal of Applied Ecology 44:875–884.

O'Connor, K. D., A. B. Marr, P. Arcese, L. F. Keller, K. J. Jeffery, and M. W. Bruford (2006). Extra-pair fertilization and effective population size in the Song Sparrow (*Melospiza melodia*). Journal of Avian Biology 37:572–578.

Oh, K. P., and A. V. Badyaev (2010). Structure of social networks in a passerine bird: Consequences for sexual selection and the evolution of mating strategies. American Naturalist 176:E80–E89.

Patten, M. A., J. T. Rotenberry, and M. Zuk (2004). Habitat selection, acoustic adaptation, and the evolution of reproductive isolation. Evolution 58:2144–2155.

Pennisi, E. (2016). The power of personality. Science 352:644–647.

Petren, K., P. R. Grant, B. R. Grant, and L. F. Keller (2005). Comparative landscape genetics and the adaptive radiation of Darwin's finches: The role of peripheral isolation. Molecular Ecology 14:2943–2957.

Piper, W. H., C. Walcott, J. Mager, M. Perala, K. B. Tischler, Erin Harrington, A. J. Turcotte, M. Schwabenlander, and N. Banfield (2006). Prospecting in a solitary breeder: Chick production elicits territorial intrusions in common loons. Behavioral Ecology 17:881–888.

Piper, W. H., C. Walcott, J. N. Mager, and F. Spilker (2008). Fatal battles in common loons: A preliminary analysis. Animal Behaviour 75:1109–1115.

Poisbleau, M., H. Fritz, M. Valeix, P.-Y. Perroi, S. Dalloyau, and M. M. Lambrechts (2006). Social dominance correlates and family status in wintering Dark-bellied Brent geese, *Branta bernicla bernicla*. Animal Behaviour 71:1351–1358.

Postma, E., and A. J. van Noordwijk (2005). Gene flow maintains a large genetic difference in clutch size at a small spatial scale. Nature 433:65–68.

Power, H. W., and M. P. Lombardo (1996). Mountain Bluebird. *In* The Birds of North America, vol. 222, A. Poole and F. Gill, Editors. Birds of North America, Philadelphia, PA, pp. 1–21.

Pulliam, R. H. (1988). Sources, sinks, and population regulation. American Naturalist 132:652–661.

Rebke, M., T. Coulson, P. H. Becker, and J. W. Vaupela (2010). Reproductive improvement and senescence in a long-lived bird. Proceeding of the National Academy of Sciences USA. 107:7841–7846. doi:10.1073/pnas.1002645107.

Reid, J. M., E. M. Bignal, S. Bignal, D. I. McCracken, and P. Monaghan (2003). Environmental variability, life-history covariation and cohort effects in the Red-billed Chough *Pyrrhocorax pyrrhocorax*. Journal of Animal Ecology 72:36–46. doi:10.1046/j.1365 2656.2003.00673.x.

Reid, J. M., P. Arcese, R. J. Sardell, and L. F. Keller (2011a). Additive genetic variance, heritability, and inbreeding depression in male extra-pair reproductive success. American Naturalist 177:177–187.

Reid, J. M., P. Arcese, R. J. Sardell, and L. F. Keller (2011b). Heritability of female extra-pair paternity rate in Song Sparrows (*Melospiza melodia*). Proceedings of the Royal Society of London B 278:1114–1120.

Ricklefs, R., and M. Wikelski (2002). The physiology/life history nexus. Trends in Ecology and Evolution 17:462–468.

Rohwer, S. (1975). The social significance of avian winter plumage variability. Evolution 29:593–610.

Rohwer, S. (1985). Dyed birds achieve higher social status than controls in Harris' Sparrows. Animal Behaviour 33:1325–1331.

Rohwer, S., and P. W. Ewald (1981). The cost of dominance and advantage of subordination in a badge signaling system. Evolution 35:441–454.

Saab, V. A., R. E. Russell, and J. G. Dudley (2007). Nest densities of cavity-nesting birds in relation to postfire salvage logging and time since wildfire. The Condor 109:97–108.

Schieck, J., and S. J. Song (2006). Changes in bird communities throughout succession following fire and harvest in boreal forests of western North America: Literature review and meta-analysis. Canadian Journal Forest Research 36:1299–1318.

Sealy, S. G., A. J. Banks, and J. F. Chace (2000). Two subspecies of warbling vireo differ in their responses to cowbird eggs. Western Birds 31:190–194.

Sedinger, J. S., M. P. Herzog, and D. H. Ward (2004). Early environment and recruitment of Black Brant into the breeding population. The Auk 121:68–73.

Sedinger, J. S., D. H. Ward, J. L. Schamber, W. I. Butler, W. D. Eldridge, B. Conant, J. F. Voelzer, N. D. Chelgren, and M. P. Herzog (2006). Effects of El Niño on distribution and reproductive performance of Black Brant. Ecology 87:151–159.

Sedinger, J. S., and N. D. Chelgren (2007). Survival and breeding advantages of larger Black Brant goslings: Within and among cohort variation. The Auk 124:1281–1293.

Sedinger, J. S., N. D. Chelgren, M. S. Lindberg, and D. H. Ward (2008). Fidelity and breeding probability related to population density and individual quality in Black Brent geese (Branta bernicla nigricans). Journal of Animal Ecology 77:702–712.

Sedinger, J. S., J. L. Schamber, D. H. Ward, C. A. Nicolai, and B. Conant (2011). Carryover effects associated with winter location affect fitness, social status, and population dynamics in a long distance migrant. American Naturalist 178:E110–E123.

Shuster, S. M., and M. J. Wade (2003). Mating systems and strategies. Princeton University Press, Princeton, NJ.

Shields, G. F. (1990). Analysis of mitochondrial DNA of Pacific Black Brant (Branta bernicla nigricans). The Auk 107:620–623.

Sih, A., A. M. Bell, J. C. Johnson, and R. E. Ziemba (2004). Behavioral syndromes: An integrative overview. Quarterly Review of Biology 79:241–277.

Smith, J. N. M., and A. A. Dhondt (1980). Experimental confirmation of heritable morphological variation in a natural population of Song Sparrows. Evolution 34:1155–1158.

Smith, J. N. M., L. F. Keller, A. B. Marr, and P. Arcese, Editors (2006). Conservation and biology of small populations: The Song Sparrows of Mandarte Island. Oxford University Press, New York.

Smith, S. M. (1978). The "underworld" in a Territorial Sparrow: Adaptive strategy for floaters. American Naturalist 112:571–582.

Stearns, S. C., and R. F. Hoeckstra (2005). Evolution, an introduction. Oxford University Press, Oxford.

Székely, T., A. Liker, R. P. Freckleton, C. Fichtel, and P. M. Kappeler (2014). Sex-biased survival predicts adult sex ratio variation in wild birds. Proceedings of the Royal Society B 281:20140342.

Tarwater, C. E., and P. Arcese (2017). Age and years to death disparately influence reproductive allocation in a short-lived bird. Ecology, in press.

Templeton, A. R. (2006). Population genetics and microevolutionary theory. John Wiley and Sons, Hoboken, NJ.

Thorneycroft, H. B. (1966). Chromosomal polymorphism in the White-throated Sparrow, Zonotrichia albicollis (Gmelin). Science 154:1571–1572.

Thorneycroft, H. B. (1975). A cytogenetic study of the White-throated Sparrow, Zonotrichia albicollis (Gmelin). Evolution 29:611–621.

Tonnis, B., P. R. Grant, B. R. Grant, and K. Petren (2005). Habitat selection and ecological speciation in Galápagos warbler finches (Certhidea olivacea and Certhidea fusca). Proceedings of the Royal Society of London B 272:819–826.

Trask, A. E., E. M. Bignal, D. I. McCracken, P. Monaghan, S. B. Piertney, and J. M. Reid (2016). Evidence of the phenotypic expression of a lethal recessive allele under inbreeding in a wild population of conservation concern. Journal of Animal Ecology 85:879–891.

Tuttle, E. M. (2003). Alternative reproductive strategies in the polymorphic White-throated Sparrow: Behavioral and genetic evidence. Behavioral Ecology 14:425–432.

Tuttle, E. M., A. O. Bergland, M. L. Korody, M. S. Brewer, D. J. Newhouse, P. Minx, M. Stager, et al. (2016). Divergence and functional degradation of a sex chromosome-like supergene. Current Biology 26:344–350.

van Rhijn, J. G. (1991). The Ruff. Poyser, London.

Waples, R. S., and O. Gaggiotti (2006). What is a population? An empirical evaluation of some genetic methods for identifying the number of gene pools and their degree of connectivity. Molecular Ecology 15:1419–1431.

Waples, R. S., G. Luikart, J. R. Faulkner, and D. A. Tallmon (2013). Simple life-history traits explain key effective population size ratios across diverse taxa. Proceedings of the Royal Society B 280:20131339.

Ward, D., and J. N. M. Smith (2000). Brown-headed Cowbird parasitism results in a sink population in warbling vireos. The Auk 117:337–344.

Ward, D. H., J. A. Schmutz, J. S. Sedinger, K. S. Bollinger, P. P. Martin, and B. A. Anderson (2004). Temporal and geographic variation in survival of juvenile Black Brant. The Condor 106:263–274.

Whitlock, M. C. (2011). G'ST and D do not replace FST. Molecular Ecology 20:1083–1091.

Whitlock, M. C., and D. E. McCauley (1999). Indirect measures of gene flow and migration: FST ≠1/(4Nm+1). Heredity 82:117–125.

Widemo, F. (1998). Alternative reproductive strategies in the Ruff, Philomachus pugnax: A mixed ESS? Animal Behaviour 56:329–336.

Williams, B. K., J. D. Nichols, and M. J. Conroy (2002). Analysis and management of animal populations. Academic Press, London.

Williams, G. C. (1966). Adaptation and natural selection. Princeton University Press, Princeton, NJ.

Wilson, A. S. G. (2008). The role of insularity in promoting intraspecific differentiation in Song Sparrows. PhD diss., University of British Columbia, Vancouver, BC.

Wilson, A. G., and P. Arcese (2008). Factors influencing natal dispersal in an avian island metapopulation. Journal of Avian Biology 39:341–347.

Wilson, A. G., P. Arcese, Y. L. Chan, and M. A. Patten (2011). Micro-spatial genetic structure in Song Sparrows (Melospiza melodia). Conservation Genetics 12:213–222.

Wilson, A. G., Y. L. Chan, S. S. Taylor, and P. Arcese (2015). Genetic divergence of an avian endemic on the Californian Channel Islands. PLoS ONE 10:e0134471.

Wilson, G., and B. Rannala (2003). Bayesian inference of recent migration rates using multilocus genotypes. Genetics 163:1177–1191.

Wilson, S., D. R. Norris, A. G. Wilson, and P. Arcese (2007). Breeding experience and population density affect the ability of a songbird to respond to future climate variation Proceedings of the Royal Society of London B 274:2539–2545.

Wojczulanis-Jakubas, K., A. Kilikowska, A. M. A. Harding, D. Jakubas, N. J. Karnovsky, H. Steen, H. Strøm, et al. (2014). Weak population genetic differentiation in the most numerous Arctic seabird, the little auk. Polar Biology 37:621–630.

Wright, S. (1984). Evolution and the genetics of populations: Genetics and biometric foundations v. 4. Variability within and among natural populations. New ed. University of Chicago Press, Chicago, IL.

Wright, S. (1988). Surfaces of selective value revisited. The American Naturalist 131:115–123.

Zhan, X., A. Dixon, N. Batbayar, E. Bragin, Z. Ayas, L. Deutschova, J. Chavko, et al. (2015). Exonic versus intronic SNPs: Contrasting roles in revealing the population genetic differentiation of a widespread bird species. Heredity 114:1–9.

CHAPTER 21

Population Ecology

Leonard A. Brennan and William M. Block

Populations are more abstract conceptual entities than cells or organisms and are
somewhat more elusive, but they are nonetheless real.

—Eric Pianka (1978)

WHAT IS POPULATION ECOLOGY?

Understanding the factors that regulate the numbers and
distribution of species (i.e., a population) has long fascinated
ecologists, from early naturalists such as Joseph Grinnell
(box on page 686) and Alden Miller (box on page 687) to many
contemporary quantitative ecologists.

As defined in chapter 20, a population is a group of individuals of a species that occupy a particular place at a given
time. Typically, individuals within a group that is defined
as a population interbreed. Defining a population is scale-
dependent or determined by the extent (spatial area) or time
frame (temporal) that biologists or managers use.

Populations are structured hierarchically. The spatial
scale at which a population is defined can be somewhat flexible because it can be user-defined, within practical limits.
Often, ornithologists study a subset of a population of birds
that corresponds to a delineated study area. Whether or not
the population subset studied represents the greater population depends largely on the research question or management consideration and the sampling design (size, number,
and distribution of sampling units) (Thompson et al. 1998).

Ecology is the study of interrelationships of organisms
with their environment. Thus, population ecology is a branch
of scientific inquiry that strives to understand the mechanisms
that underlie and/or influence how populations change over
time. That is, what are the factors that predispose a population to increase, decrease, or remain stable? As simple as this
question seems at face value, answering it is quite a complex
undertaking, given the number of different biotic and abiotic
factors that influence populations.

DIFFERENT KINDS OF POPULATIONS

If one considers a widespread species with a disjunct distribution across its geographic range, inferences of population
status may vary within parts of its distribution. Disjunctions
in distributions can be related to demographic structure, genetic structure, or spatial structure. A demographic disjunction is the result of too few individuals immigrating into a
population to have an impact on vital rates such as growth
or death. A genetic disjunction is the result of individuals in
one population sharing a set of genetic attributes that are not
shared with individuals of a different group. Spatial disjunctions are usually the driver of demographic and genetic disjunctions (Wells and Richmond 1995).

Spatial disjunctions can be considered from the stand-
point of a **metapopulation** (Levins 1969, Hanski 1991). The
metapopulation concept has roots in island biogeography
(MacArthur and Wilson 1967; chapter 4) in that a population
is composed of a network of subpopulations that are linked
through dispersal and colonization events. Dispersal is a key
factor in that it provides demographic relief (e.g., immigration) and gene flow. Gene flow contributes to genetic variation, making a population more resilient to disease and other
mortality factors. Metapopulations meet one working definition of a population in that individuals interbreed, although
interbreeding *among* individuals of subpopulations is much
less than *within* a subpopulation. Metapopulations have four
underlying conditions: (1) each habitat patch must be able
to support a breeding population; (2) any subpopulation
must be vulnerable to extinction; (3) recolonization is possible; and (4) subpopulation dynamics are asynchronous

JOSEPH GRINNELL

Joseph Grinnell. *Used with the permission of The Museum of Vertebrate Zoology, University of California, Berkeley.*

Bird population, in kind and quantity, is controlled primarily by conditions of habitat.

—Grinnell (1922)

The foundation of our knowledge of avian population ecology can be traced to keen observations by early naturalists. Indeed, the works of Joseph Grinnell (February 27, 1877, to May 29, 1936) are foremost at providing the underpinnings of much of what we now know about bird populations in much of western North America. Grinnell was born at the Kiowa, Comanche, and Wichita Indian Agency on the Washita River in Indian Territory, where his father, Fordyce Grinnell, MD, was the agency physician. In 1880 the family moved to the Dakotas, where Joseph played with Sioux children and learned from them to observe and listen to nature. These lessons served him well throughout his career. Five years later, the family moved to Pasadena, California. This was far before southern California became a congested megalopolis, so Joseph had lots of room to roam and explore. Much of his time during his early years was spent collecting and cataloguing vertebrates. His collection started when he was 13, with a toad, and by the time he entered graduate school at Stanford University he had collected over 4,500 specimens—and over 20,000 over the course of his career. Grinnell was a vocal advocate for collecting, and as he wrote in his article, "Conserving the Collector" (Grinnell 1915), "I consider judicious collecting absolutely indispensable to serious ornithological research."

Grinnell's commitment and zeal as a collector served him well as the first director of the Museum of Vertebrate Zoology (MVZ) at the University of California, Berkeley. Annie Alexander, a philanthropist and naturalist in her own right, approached Grinnell with the idea of developing a museum that would rival the Smithsonian. They differed on where to house the museum, with Grinnell advocating for Stanford and Alexander insisting on UC Berkeley. Holding the purse strings, Alexander won, and the MVZ was established in 1908 in Berkeley, with Grinnell serving as director until his death.

Grinnell developed an intricate method for recording detailed field notes, a method still used today (Herman 1986). His old field notes are a gold mine of information that provides baselines for assessing changes in bird populations and distributions today. His philosophy was that you could never record too much information because you never knew when even the most casual observation might be relevant in the future. Many of his field notes are archived in the Bancroft Library, University of California, Berkeley.

Grinnell authored well over 500 scientific papers, many covering a wide variety of ornithological topics. Many papers described the fauna in various locations in western North America. Grinnell was certainly not reticent in sharing his philosophy on a number of topics, such as the value of collecting, the importance of museum collections for scientific research, the role of physical barriers influencing animal distributions, and the importance of natural resource management (Grinnell 1940). He was the first to apply the concept of the ecological niche when describing the environment of the California Thrasher (*Toxostoma redivivum*) (Grinnell 1917), a concept that holds true today. A bibliography of his works was published in the obituary written by his wife, Hilda Wood Grinnell (Grinnell 1940).

Joseph Grinnell was an active member in numerous professional societies. The professional societies spanned taxa including herpetofauna, birds, and mammals. He was president of the American Ornithologists' Union, Cooper Ornithological Society, and the American Society of Mammalogists, and vice president of the Ecological Society. He was on the editorial board for a number of journals, including serving as editor of the *Condor*.

Grinnell was compelled to become involved in natural resource policy issues, often to the point of alienating those who disagreed with him. After Congress reduced Yosemite National Park's area by 500 square miles to meet mining and ranching interests and Yosemite's Hetch Hetchy Valley was dammed to supply San Francisco's drinking water, he campaigned to park officials, arguing for more stringent conservation policies. Many of his ideas—from placing trained naturalists in national parks to prohibiting hunting and trapping park animals—are in place today.

The contributions of Joseph Grinnell to ornithology cannot be overstated. Beyond the research he conducted, his activity in professional societies, his service as director of the MVZ, and his influence on national park strategy, he left a lineage of ornithologists who are among the most influential in the world (http://mvz.berkeley.edu/Grinnell _Lineage.html). His legacy lives on!

ALDEN HOLMES MILLER

Alden Holmes Miller. *Used with the permission of The Museum of Vertebrate Zoology, University of California, Berkeley.*

Alden H. Miller (February 4, 1906, to October 9, 1965) was an ornithologist who conducted fundamental research on speciation, distribution, and reproductive biology of birds in the American West, the Neotropics, and Australia. Along with Joseph Grinnell, he coauthored *The Distribution of the Birds of California* (Grinnell and Miller 1944), a monograph and inventory of bird populations for that state. It is arguably one of the first and, up to that point, most comprehensive treatments of birdlife for a state in the United States.

Alden received his primary and high school education from the Los Angeles public schools. His father Loye Miller was a renowned professor of paleontology and zoology at UCLA who had a huge influence on Alden's life and education. Alden received his AB from UCLA in 1927, and then went on to Berkeley for his MA (1928) and PhD (1930). After finishing his PhD, he joined the University of California, Berkeley, zoology faculty and quickly climbed the academic ladder to become a full professor in 1945 at the age of 39. After the untimely death of Joseph Grinnell, Alden became the second director of the Museum of Vertebrate Zoology in 1940 and held this position until his death 25 years later.

Alden Miller published 258 scientific papers during his lifetime. His first paper, which was on Black-bellied Plovers (*Pluvialis squatarola*), was published when he was 18 years old (Mayr 1973). His last published paper (posthumous, in 1966) was on Northern Sage Sparrows (*Amphispiza belli nevadensis*). He is known primarily for major monographs on the analysis of speciation in two avian genera: shrikes (*Lanius*,

Miller 1931) and juncos (*Junco*, Miller 1941). Miller's monograph on juncos was based on more than 12,000 specimens of all species in this generic complex that is distributed across North America and south to Costa Rica and Panama.

A critically important—and largely unappreciated—aspect of Alden Miller's body of work was that he represents an important intellectual bridge between classical natural history and experimental biology of birds. As an ornithologist with feet firmly planted in both the natural history and experimental biology camps, Miller carried on with the Grinnellian tradition of personally collecting and preserving more than 12,000 museum specimens while also pioneering original research on reproductive physiology of birds. He was particularly interested in how the reproductive condition of birds was regulated in the tropics, where changes in day length were subtle to nonexistent across the seasons in comparison with reproductive condition of birds in temperate environments, where changes in day length across the seasons are profound (Miller 1959, Miller 1961).

In addition to his duties as professor of zoology and director of the Museum of Vertebrate Zoology at UC Berkeley, Alden held a variety of administrative positions, such as vice-chancellor for academic affairs and acting chair of the Department of Paleontology. His 26-year run as editor of the *Condor* is legendary by any standard. Alden was named a fellow of the American Ornithologists' Union (AOU) in 1939 and was president of the AOU from 1953 to 1956. He received the Brewster Medal, the most prestigious award of the AOU, in 1943. Miller also received similar recognition from the Cooper Ornithological Society; he was elected an honorary member in 1956. In 1957 he was elected to the National Academy of Sciences.

Two obituaries of Alden Miller (Davis 1967, Mayr 1973) noted that he strongly preferred to focus his intellectual efforts on original peer-reviewed publications that were based on analyses of his observations and data. He eschewed writing review papers and books, even when colleagues invited him to do so. Despite the lack of review papers and books, Alden Miller's body of work remains at the foundation of both classical and experimental ornithology and is still highly relevant today. Any serious student of modern ornithology would be wise to read and assimilate his publications, a complete list of which is available in Mayr (1973).

to minimize the chance that all will become extinct simultaneously. Metapopulation theory has become an incredibly popular concept in conservation biology. The hoped-for existence of metapopulations has been invoked to explain population dynamics of various taxa. Unfortunately, true avian metapopulations have rarely been confirmed with empirical data (Fronhofer et al. 2012). The Spotted Owl (*Strix occidentalis*) is a notable exception to a general lack of documented avian metapopulations (Noon and McKelvey 1992, LaHaye et al. 1994, Akçakaya and Raphael 1998). Of course, not all spatially divided or subdivided populations will function as metapopulations, because dispersal among the subpopulations is an essential component of a functional metapopulation.

Migrating birds may consist of groups of individuals of the same species that originated in different breeding populations. Individuals from these different breeding populations converge along migration corridors to become a migratory population (or populations, see chapter 19). Likewise, for birds on wintering grounds, where individuals from different breeding populations coalesce, these groups of birds are typically referred to as wintering populations. Hence, the prerequisite for individuals to interbreed does not hold for migrating populations or wintering populations of birds. Birds that stay in the same general area year-round and do not migrate are referred to as resident populations. Thus, it is extremely important to define the kind of population under consideration (breeding, migrating, wintering, or resident) to avoid confusion and ambiguity.

FACTORS THAT INFLUENCE POPULATIONS

Factors that influence populations can be either **endogenous** or **exogenous**. Endogenous factors are those that originate within the organism. Examples include hormonal changes that influence the onset of breeding (see chapter 8) and secretions from the pineal grand influencing circadian rhythms. Exogenous factors are those that originate outside of the organism. Food availability, photoperiod, and temperature might serve as external cues for birds to settle at a potential breeding spot and to initiate breeding (see chapter 16).

Intrinsic factors (those that originate within the population) and extrinsic factors (those that originate from outside the population) also influence populations. Intrinsic factors are built in to the genetic basis of each species, whereas extrinsic factors involve an interaction of a population with one or more species. Some population ecologists refer to these as deterministic (predictable from an internal mechanism) and stochastic (random from an external influence) factors. The main point to understand is that the factors that influence

the growth and decline of a population can originate from factors within the population as well as from factors outside the population.

Populations are dynamic over time in that they can grow as the result of births and immigration or shrink as the result of deaths and emigration. **Immigration** refers to the birds dispersing in to one population from another, whereas **emigration** refers to birds dispersing from one population to another. This process can be illustrated by the following equation:

$$N_{t+1} = N_t + b_t + i_t - d_t - e_t,$$

where N = population size, b = births, i = immigrations, d = deaths, e = emigrations, and t = time.

The magnitude of each of the four demographic parameters at a given time will define the population trajectory. That is, if the combination of births and immigration exceed the combination of deaths and emigration, the population will increase; if deaths and emigration are greater than the sum of births and immigration, the population will decrease; if births and immigration are roughly equal to deaths and emigration, the population is deemed stable.

Populations of birds have overlapping generations. Many species of insects are semelparous; that is, individuals reproduce only once in their lifetime and die and thus have nonoverlapping generations. Birds, on the other hand, are **iteroparous**, which means that they potentially live multiple years and attempt to breed multiple times over the course of their lifetimes. Some species (such as many cup-nesting passerines) will attempt to breed each year, whereas other, particularly long-lived species (such as many large raptors), may attempt to breed only in years when resources and conditions are favorable. Often the distinction between the two strategies relates to parental investment in rearing offspring, where birds and mammals invest considerable time feeding and caring for young to the point that the young can become self-sufficient. Iteroparous species produce a limited number of young; hence they invest considerable time and resources to assure their survival (see chapter 16).

Patterns of Population Growth

All populations are regulated by limiting factors. A limiting factor is defined as an environmental factor that is of critical importance in restricting the size of a population. For example, a lack of small mammal prey availability may be a limiting factor for a raptor population.

In the absence of limiting factors, a population can potentially exhibit an exponential growth rate over time. Various

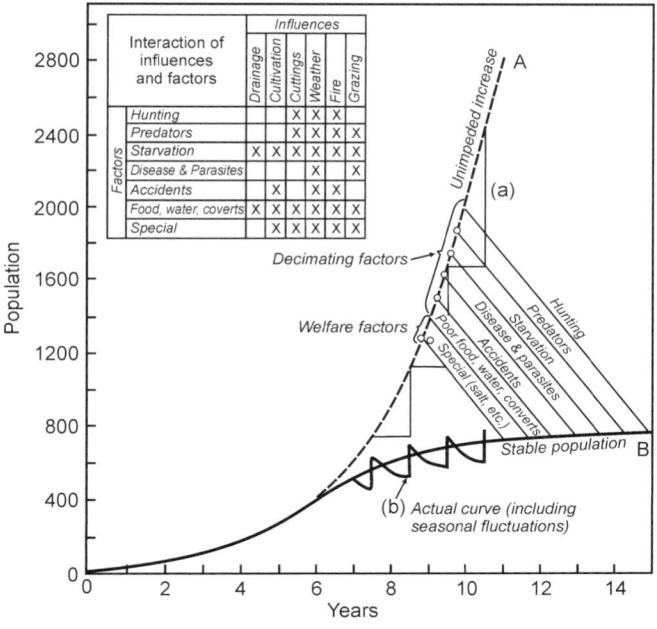

Figure 21.1. Hypothetical illustration of unlimited population growth and various factors that work to dampen population growth over time. *Adapted from Leopold 1933. © 1987 by the Board of Regents of the University of Wisconsin System. Reprinted courtesy of University of Wisconsin Press.*

limiting factors serve to dampen exponential growth over time and cause populations to fluctuate (fig. 21.1). A population with no constraints—or limiting factors—will grow exponentially, at least in theory. Consider, for example, the American Robin (*Turdus migratorius*), a species that produces on average four young in as many as two broods per year. Assuming that all of the offspring of a pair of American Robins survive over a decade, there will be 19,500,000 descendants in the population after ten years (Welty 1975). Of course, population numbers such as these do not become a biological reality because intrinsic (within-population, often also called endogenous) and extrinsic (external to the population, often also called exogenous) factors—as noted above—work to influence the number of individuals in a population by suppressing growth rates and dampening population numbers over time.

The growth rate of a population is typically expressed as *r*, a metric that quantifies the rate of change in population numbers over time. If a population has a per capita birth rate (*b*) that is greater than the per capita death rate (*d*), the population will increase (*b* > *d*); in contrast if *b* < *d*, the population will decrease.

The parameter *r* is somewhat difficult to calculate because it requires information about the number of individuals in a population, which is a challenging metric to determine because it must be based on a census or on estimate-based

sampling. Assuming that one has a reliable estimate of the number of individuals in a population, a constant growth rate can be calculated with the equation

$$\text{Growth rate or } r = N_t - N_0 \ / \ t - t_0 = \Delta N \ / \ \Delta t$$

where N = the number of individuals at time t, and N_0 = the initial number of individuals, t_0 is the initial time and Δ = the change in N and change in t over time. Instantaneous rate of increase or average rate of increase over time period t can also be expressed as

$$r = (\ln N_{t+1} - \ln N_t) \ / \ t$$

Building on the concept of the rate of population increase, the finite rate of increase, λ (lambda), can be used as an easily understandable metric that indicates whether a population is increasing (λ > 1) or decreasing (λ < 1). Lambda is typically calculated with the equation

$$r = \log_e \lambda \text{ or } \lambda = e^r$$

where *e* is the base of the natural logarithms, and *r* is the population growth rate (Pianka 1978). For iteroparous organisms, the exponential growth model is the following:

$$dN/dt = r \, N_t$$
$$N_{t+1} = N_t \, e^{r\,t}$$

Logistic Growth

Although populations have the potential to exhibit exponential growth, no population in the natural world can increase indefinitely over time (Malthus 1798). Therefore, understanding the concept of **exponential growth** is a useful background for understanding how **logistic growth** rates can dampen what would otherwise be unrealistic infinite population increases. The basic concept behind logistic population growth is that, after an initial rapid growth rate following an event such as dispersal or translocation into previously unoccupied but favorable conditions, the population reaches some kind of asymptote and more or less levels off (fig. 21.1).

The basic logistic growth model takes this form:

$$dN/dt = r \, N \, [(K-N)/K]$$

where *r* = rate of increase K − N = amount of "empty space" left in habitat and [(K−N)/K] = proportion of environment that is "empty."

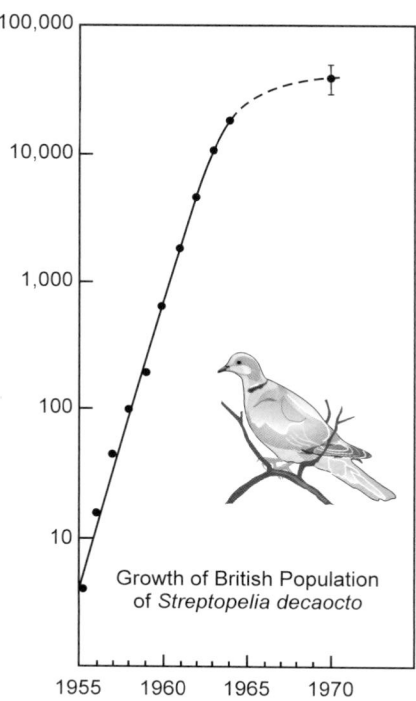

Growth of British Population
of *Streptopelia decaocto*

Figure 21.2. Logistic population growth of the Collared Turtledove in Great Britain. *Adapted from Hutchinson 1978. Used with permission of Yale University Press.*

There are classic examples of bird populations that have exhibited logistic population growth. In Great Britain, Hudson (1965, 1972) documented the rapid population increase of the Collared Turtledove (*Streptopelia decaocto*) for about a decade (1955–1963) followed by a rapid decline in this increase from 1963 to 1972; fig. 21.2). Another example of logistic population growth is from the introduction of the Ring-necked Pheasant (*Phasianus colchicus*), in which six females and two males were brought to Protection Island on the west coast of Washington State in 1937. In fewer than six years, there were at least 1,898 pheasants in the population, after which the growth rate of the population slowed dramatically (Einarson 1942).

The mechanism of carrying capacity is often used to explain why logistic population growth trajectories typically level off over time (fig. 21.1). The Malthusian concept that environmental factors will ultimately limit population growth, or even cause a population to decline, has been assumed to be true for centuries. More recently, the concept of carrying capacity has been used to define the key factor(s) that limit a population. While the origin of the term remains obscure (Sayre 2008), **carrying capacity** (typically abbreviated as *K*) has been used since the mid-twentieth century by population ecologists to define factors that limit the intrinsic population increases in free-ranging organisms, including birds. With the concept of carrying capacity as background, the idea of population regulation then leads to density-dependent and density-independent factors that limit populations (Lack 1954, 1966) (box on page 692).

Density-Independent and Density-Dependent Factors

Factors that influence population abundance such as weather or wildfire are considered **density-independent** because such factors influence the same proportion of organisms independent of their population density. Density-independent factors tend to be abiotic in nature, and are, as mentioned above, often related to disturbances related to weather (droughts, hurricanes, tornados), catastrophic wildfires (although drought can be a driving factor that sets the stage for such disturbances), and other abiotic factors such as excess heat (Guthery et al. 2005). Pollution from organochlorine pesticides such as DDT after World War II disrupted the metabolic pathway for depositing calcium on eggshells of high trophic-level predatory birds. This lack of calcium resulted in excessively thin eggshells and the death of chicks before they hatched. The result was decimated raptor populations such as those of the Peregrine Falcon (*Falco peregrinus*). This is a classic example of how a density-independent factor can operate independent of population abundance (Newton 1979).

Density-dependent factors, in contrast to density-independent factors, affect population numbers in proportion to population density. For example, reproductive rates, as well as mortality rates, of bird populations have been documented to vary in relation to population density. Typically, reproductive rates are higher when population density is lower, and conversely, mortality rates (especially in wintering populations) are lower when population density is lower.

During the course of his fifteen-year study of a Northern Bobwhite (*Colinus virginianus*) population in the midwestern United States, Errington (1945) observed that bobwhite production was greater during summers when the overwintering population that survived to the breeding season was low. Errington termed this dynamic **inversity**, even though the term **density-dependence** can be traced back to Verhulst (1838). The presence of density-dependence was later confirmed by a subsequent long-term study of a Northern Bobwhite population in Southern Illinois (Roseberry and Klimstra 1984; see fig. 21.3)

Despite its importance as a factor that plays a critical role in the dynamics and regulation of many populations, density-dependence is extremely difficult to detect in free-ranging populations of vertebrates. Birds are no exception to this difficulty, despite the fact that two classic instances of density-dependence were documented in wild Northern Bobwhite populations, as noted above. Detecting density-dependence is challenging because, first, a relatively long (at least 10–15 years minimum) time series of population estimates (or complete census counts, which are even more rare than long-term estimates) is required. The second challenge

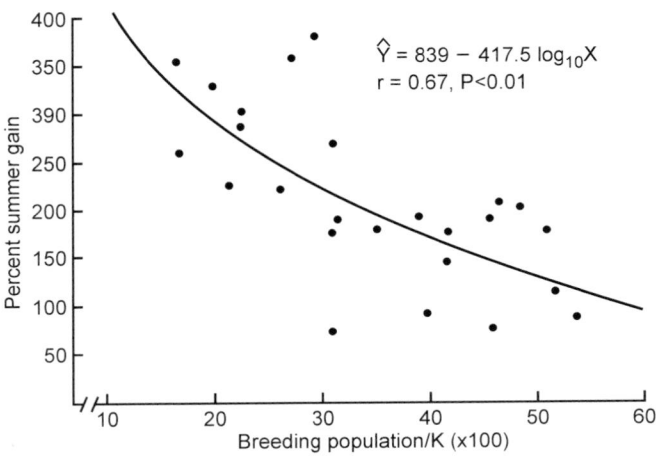

$$\hat{Y} = 839 - 417.5 \log_{10}X$$
$$r = 0.67, P<0.01$$

Figure 21.3. Graphical illustration of Northern Bobwhite density dependence. Note how summer gain (number of new birds produced during a breeding season) decreases as overall population size increases. *Adapted from Roseberry and Klimstra 1984, copyright © 1984 by the Board of Trustees, Southern Illinois University.*

to detecting density dependence is obtaining population census estimates that are both accurate (reflect actual abundance) and precise (minimal variance from sampling error). The third challenge is that, while a simple graphical analysis (fig. 21.3) can suggest the presence of density-dependence in a population, population simulations and other complex mathematical analyses are often required to understand the exact nature of the shape and form of the density-dependent relationship that influences a population (Lebreton and Giminez 2012). Even with long-term data that showed a density-

dependent relationship, Roseberry and Klimstra (1984:95) used four different approaches to population simulations in order to understand the dynamics in their data. Further simulation analyses using Northern Bobwhite population data from South Texas indicated that a very weak but biologically significant form of density-dependence evidently operates during the breeding season as well as on overwintering populations of Northern Bobwhites in this part of their range (DeMaso et al. 2012).

Life History Strategies and Population Dynamics

As part of their classic monograph on the theory of island biogeography, MacArthur and Wilson (1967) developed the terms **r-selection** and **K-selection** to explain the relative differences in life history strategies of free-ranging organisms. Despite being relative terms, *r*-selection versus *K*-selection are useful concepts that help us understand and explain how populations of birds make a living in the wild (table 21.1).

Population dynamics can vary widely among species, in part because of different life history strategies that govern patterns of growth and development, reproduction, parental investment, senescence, and death. Species that are considered *r*-selected are those that have life history strategies that maximize *r*, the population growth rate, such as early sexual maturity and large numbers of young. Whereas weak density-dependent relationships often operate in the background to regulate populations, *r*-selected species tend to be short-lived, opportunistic, and have high rates of population turnover. Species of birds that are considered *r*-selected have populations in which the rate of increase (or rate of decline) occurs rapidly. They increase their numbers rapidly

Table 21.1. Concepts and factors related to *r*-selection and *K*-selection. *Adapted from Pianka 1978. Used with permission.*

	r Selection	K Selection
Climate	Variable, uncertain	Fairly constant, predictable
Mortality	Often catastrophic Density independent	More directed Density dependent
Population size	Highly variable in time	Fairly constant in time
Intra- and interspecific competition	Variable, often lax	Usually keen
Selection factors	1. Rapid development 2. High rate of increase 3. Early reproduction 4. Small body size 5. Single reproduction 6. Many offspring	1. Slower development 2. Greater competitive ability 3. Delayed reproduction 4. Large body size 5. Multiple reproduction 6. Fewer offspring
Length of life	Short, usually <1 year	Longer, usually >1 year

DAVID LACK

David Lack. *Photo used with permission from Lack family.*

David Lack (July 16, 1910, to March 12, 1973) was a preeminent ornithologist focusing on evolutionary ecology, population biology, and density-dependent population regulation. He had a passion for birds, and by the age of 15, he had observed and recorded more than 100 species. He authored his first scientific paper before entering college. At Cambridge University in the early 1930s, Lack was disappointed to find that his zoology professors taught little about birds. In fact, only two professional ornithologists existed in all of Britain at that time. Thus, David Lack took it upon himself to create his own learning opportunities. As an undergraduate, he became the president of the Cambridge Ornithological Club and established a professional relationship with Sir Julian Huxley, an eminent evolutionary biologist. Huxley was a strong influence on Lack, inspiring him to study tropical birds.

Despite limited professional opportunities, Lack became the leading British ornithologist of his time. He spent most of the 1930s teaching high school, but used his free time to travel, conduct field work, and make professional contacts. During World War II he worked on the radar development program, skills that he later applied to the study of bird migration patterns. Following the war, he became director of the Edward Grey Institute of Field Ornithology.

Lack's work in ornithology was almost entirely based on field studies. He was one of the first to apply quantitative approaches to life history studies. Lack became increasingly interested in the relation of natural selection to population regulation, concluding that density-dependence was a primary force limiting bird populations. His work suggested that natural selection favored clutch sizes that ensured the greatest number of surviving young. In 1947, he published his classic study *Darwin's Finches*. His ideas on speciation, ecological isolation, group selection, migration, and the evolution of reproductive strategies are best summarized in his two most influential books, *The Natural Regulation of Animal Numbers* in 1954, and *Ecological Adaptations for Breeding in Birds* in 1968.

when environmental conditions are favorable, and their populations decline rapidly when environmental conditions deteriorate. Examples of classic *r*-selected birds are short-lived species such as the Blue Tit (*Parus caeruleus*; Dhondt 1989) and House Martin (*Delichon urbicum* Bryant 1989).

Species that are considered *K*-selected are those that have traits associated with maintaining populations close to *K*, carrying capacity. Population dynamics of *r*-selected species also are affected by environmental resources, but their reproductive capacity and short life span provoke different population responses to booms and busts in resource levels. The idea of a population being at some sort of equilibrium (Pianka 1978) is considered an important aspect of *K*-selection. Relatively strong density-dependent relationships often operate in the background to regulate *K*-selected populations. The individuals in a population of a *K*-selected species will at first increase over time and then fluctuate around environmental conditions. Populations of *K*-selected species are buffered by fluctuations in environmental conditions as a consequence of their long life spans and limited capacity for rapid growth. Examples of classic *K*-selected birds are long-lived species such as the Barnacle Goose (*Branta leucopsis*; Owen and Black 1989) and the Short-tailed Shearwater (*Puffinus tenuirostris*; Wooler et al. 1989).

MODELING AND ESTIMATING POPULATIONS

Why Model Bird Populations?

Models are simplified descriptions of complex systems. Models can be verbal, graphical, or numerical. Models are used to describe real-world situations based on either concepts or data. As such, they may be considered hypotheses that can be tested with empirical information. Some models are used to describe population trajectories, others are used to evaluate the factors influencing a population, and some are used to understand effects of specific human activities (e.g., management).

Models are commonly used to analyze avian population dynamics (how populations change over time). Models are used by population ecologists for a number of different reasons. For example, a population ecologist may want to examine how populations change as a result of human-induced environmental change (e.g., climate change impacts) by making simulated projections into the future (see chapter 31) and by defining boundaries of populations in the most realistic manner available. Alternatively, a population ecologist with a long-term (20–30 years or longer) data set may want to develop models that can explain or predict past dynamics, or predict population trajectories into the future. Finally, the idea of population viability analysis has received a great deal of attention from ornithologists who are interested in examining the probability that a population of a threatened or endangered species will persist through time or become extinct.

There are various population modeling approaches. Our purpose in this section is to provide an overview of the widely used approaches to modeling populations. Of course, birds figure prominently in the world of population modeling because they are widespread, conspicuous, and (sometimes) relatively easy to count and monitor. The classic population equations introduced earlier in this chapter are models. In this section we describe the applications of several contemporary examples of avian population models that have implications for conservation and management.

Types of Models

Three types of models are generally used in population ecology: theoretical, empirical, and decision-theoretical (Nichols 2001). Various approaches to model populations are applied to avian studies. Models range from the traditional Verhulst model describing logistic growth (described earlier in this chapter) in a population to age-structured Leslie matrix models (described below), and they can take many other forms. Complexity increases as one incorporates space and time and moves from a continuous to a disjunct population distribution (e.g., space-time models).

Theoretical Models

Theoretical models are used to develop constructs to explain flows and linkages between or among components of an ecological system. The Lotka-Volterra model was one of the first modern attempts to use a theoretical model to explain linkages between populations in a system. In the early 1920s, A. J. Lotka and V. Volterra independently deduced that populations of predators and prey (or parasites and hosts) could, at least in theory, oscillate and therefore persist through time (Hutchinson 1978:221). The model has also been applied to competitive species as well. It takes the form of two differential equations (Williams et al. 2002:161):

$$dN_1/dt = [r_1 - d_1N_2(t)]N_1(t)$$
$$dN_2/dt = [b_2N_1(t) - d_2]N_2(t)$$

where $N_1(t)$ is the number of prey (e.g., voles); $N_2(t)$ is the number of predators (e.g., owls); r_1 is the per capita prey rate of growth without predation; d_1N_2 is the mortality rate of predators; and b_2N_1 is the birth rate of prey. Both prey and predator population change depends on interactions between predator and prey. The model assumes that (1) the prey population has ample food; (2) food for the predator population depends on the size of the prey population; (3) population rate of change is proportional to its size; (4) the environment does not change to benefit one species; (5) genetic adaptation is inconsequential; and (6) predator appetite is not satiated. The Lotka-Volterra population model represents the starting point for modern population models of wild animals, including birds. It is based on the exponential growth equation developed by Verhulst (Hutchinson 1978:1–41) and was further developed by Nicholson and Bailey (1935) for modeling agricultural crop pests. One can easily argue that the roots of modern population modeling began with Lotka and Volterra, and they are therefore worth understanding for this point alone.

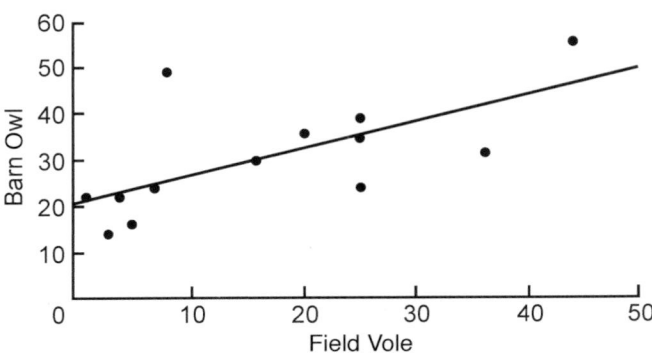

Figure 21.4. Numerical response of Barn Owl to field voles. Note how Barn Owl population size increases in relation to field vole population size. *Adapted from Hone and Sibly 2002. Used with permission.*

Figure 21.5. Trends in abundance of Barn Owl pairs and field vole abundance in southern Scotland. *Adapted from Taylor 1994. Used with permission.*

Hone and Sibly (2002) reanalyzed data presented by Taylor (1994) in his monograph of the Barn Owl (*Tyto alba*). Their analysis demonstrated a strong numerical response by the owl to its prey, the field vole (*Microtus agrestis*; fig. 21.4). This numerical response was further evidenced when tracking populations of owls and voles over time (fig. 21.5). Thus, the Barn Owl population did not follow a typical logistic growth pattern, but the owl population was modified by the population size of its primary prey.

Other Types of Theoretical Models

Reynolds et al. (2006) used a theoretical model to depict how various biotic and abiotic factors interacted to influence

Northern Goshawk (*Accipiter gentilis*) populations (fig. 21.6). As trophic relationships are complex and interacting, teasing apart the strength of these relationships is no easy task. This type of model provides a basis for a series of research studies to test predictions from the model, which can iteratively identify factors that can explain the most variation in population numbers.

In agricultural lands of the United Kingdom, Potts (1986) developed a theoretical simulation model (fig. 21.7) to predict Grey Partridge (*Perdix perdix*) density from 1968 to 1985 based on factors related to chick survival in relation to nest predator management and application of herbicides that indirectly affected arthropod foods needed by Grey Partridge chicks. Overall, the model did a good job of predicting actual Grey Partridge population density (fig. 21.8). When projected forward through time, the model predicted that Grey Partridge density would be approximately four times greater in areas where nest predators were controlled by gamekeepers (fig. 21.9).

There are two other types of theoretical approaches to modeling populations that have recently become popular: **state-space models** and **individual-based models**.

State-space models are designed to project population trends through time in relation to biotic and abiotic factors in the environment along with other factors intrinsic to a population. These models allow for an investigator to account for both random variation in the environment or demography and variation related to observer error in counts

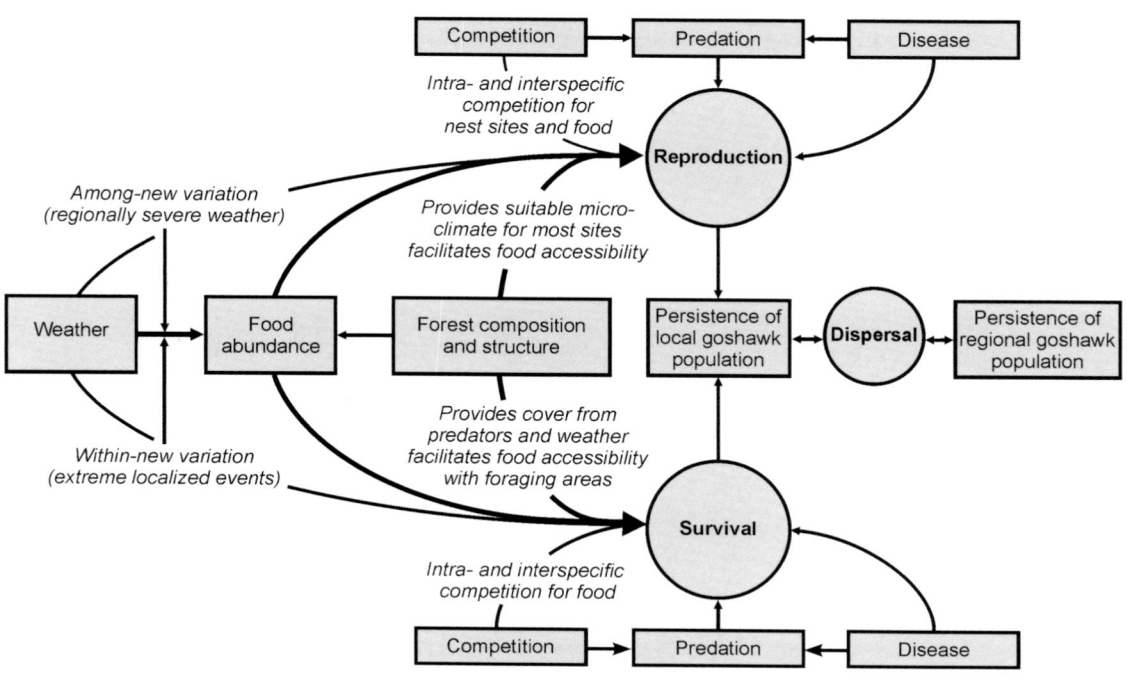

Figure 21.6. Logic and conceptual structure of a theoretical model to depict how various biotic and abiotic factors interact to influence Northern Goshawk populations. *Adapted from Reynolds et al. 2006.*

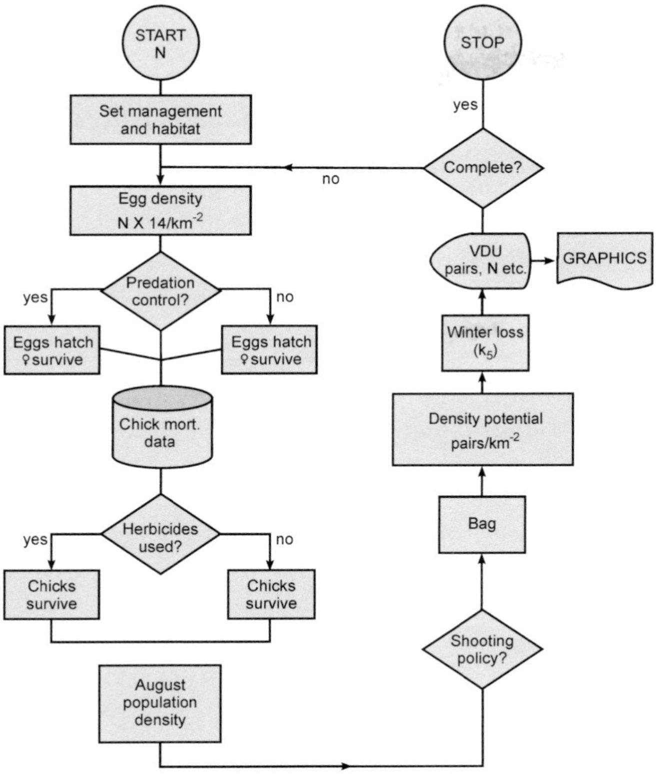

Figure 21.7. Sequence of steps and options in a theoretical simulation model illustrating factors that influence Grey Partridge populations in Sussex, United Kingdom. *Adapted from Potts 1986. Reprinted by permission of HarperCollins Publishers Ltd., © Potts 1986.*

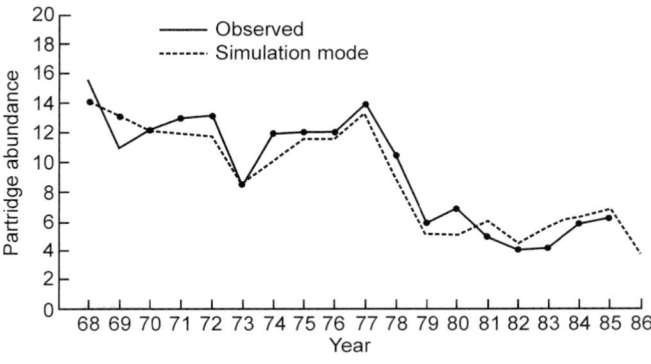

Figure 21.8. Observed (solid line) and predicted (dashed line) densities of Grey Partridge pairs based on output from the simulation model shown in Figure 21.7. *Adapted from Potts 1986. Reprinted by permission of HarperCollins Publishers Ltd., © Potts 1986.*

(Newman et al. 2014). State-space models are considered to be constructed from a "top-down" manner or perspective. State-space models take into account theories and known relationships that are hypothesized to drive population numbers and/or dynamics through time. Clark and Mangel (2000:140–160) provide a synthesis of how energetics, nutritional fuel loads and various other state variables can be combined in a set of models to predict timing, routes, and stopover events for migrating Western Sandpipers (*Calidris mauri*).

Individual-based models are relatively new to ecology and ornithology (Grimm 1999). In contrast to state-space models, individual-based models are built from a "bottom-up" approach, starting with the parts or individuals in a system or, in this case, a population. A primary goal when using individual-based models is to understand how a population's properties emerge from interactions among the individuals within that population. Railsback and Grimm (2012) described an individual-based model of Red-billed Wood Hoopoe (*Phoeniculus purpureus*) behaviors to identify three parameters for building a theory to explain factors that drive

the reproduction of this species: (1) group size of individuals within a territory; (2) extraterritorial scouting forays are by younger birds; and (3) the number of extraterritorial scouting forays is lowest in the months just before and just after breeding in December. Goss-Custard et al. (2006) used an individual-based model to examine the trade-offs between mudflat habitat loss and overwinter shorebird mortality and identified that restoring 10 percent of recent mudflat losses could mitigate the resulting shorebird mortality.

Empirical Models

Empirical or statistical models use data to make inferences from a data set. The process of fitting data to a statistical model is termed **parameter estimation**. The parameters estimated vary widely but may include population size or density, finite rate of population increase, emigration, immigration, and age-specific reproduction and survival. A wide variety of approaches are used, each with its merits and limitations.

Potts et al. (1980) developed and tested an empirical population model to predict the trend of the Shag (*Phalacrocorax aristotelis*) on the Farne Islands in the United Kingdom from 1930 to 1995 (fig. 21.10). It took nearly 30 years of data collection to fully test the veracity of the model, which pointed out that nest site availability was a major factor for predicting actual population numbers (Potts 2000).

Matrix population models are composed of cohorts with different fecundity and survival rates. Reproduction rates and survival probabilities can vary depending on the ages of the individuals. Understanding how age structure affects population trajectories requires information on age-specific reproduction and survival. The traditional approach to tracking age-specific populations is by using a life table. Two primary parameters included are survival and reproduction

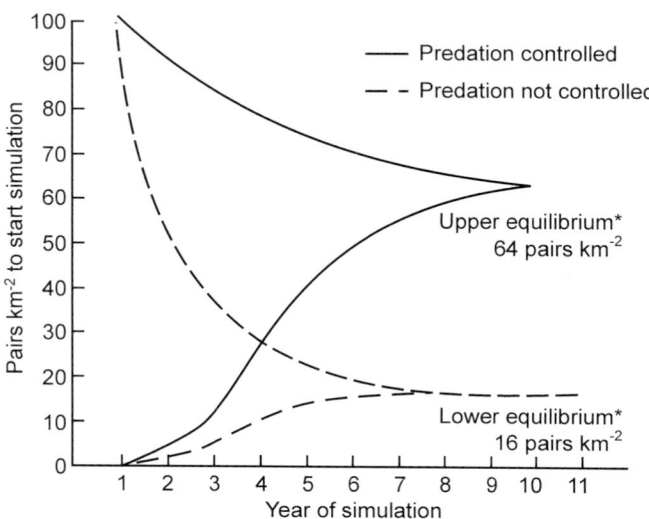

Figure 21.9. Computer simulations from the Grey Partridge population model (fig. 21.7) using different starting densities and then projecting through time with and without predator management. Note how Grey Partridge numbers converge at upper and lower equilibriums in relation to predator management, regardless of starting densities. *Adapted from Potts 1986. Reprinted by permission of HarperCollins Publishers Ltd., © Potts 1986.*

(table 21.2). Life tables are essentially the foundation of matrix models. Matrix models use a mathematical framework to estimate demographic parameters, with an ultimate goal of estimating the finite rate of population change or lambda (λ) (Caswell 2001). Leslie (1945) was the first to develop matrix population models. The basic matrix population model takes the form:

$$
\begin{bmatrix} n_1 \\ n_2 \\ n_3 \\ n_4 \end{bmatrix}_{t+1} = \begin{bmatrix} f_1 & f_2 & f_3 & f_4 \\ s_1 & 0 & 0 & 0 \\ 0 & s_2 & 0 & 0 \\ 0 & 0 & s_3 & 0 \end{bmatrix} \cdot \begin{bmatrix} n_1 \\ n_2 \\ n_3 \\ n_4 \end{bmatrix}_t
$$

where n_i is the population size for each age class i at times t and $t+1$, f_i is the fecundity rate for each age class (at four life stages; 0–1 year, 1–2 years, 2–3 years, and 3–4 years), and s_i is the survival probability. In this example there are four categories of fecundity and three categories of survival. There is always one less survival category than fecundity category because the third survival category ends in death. Matrix algebra is then used to move model projections of population abundance forward through time based on the number of individuals starting at time N_t (Caswell 2001). From this, we can then calculate lambda (λ), or the rate of population change as:

$$\lambda = \Sigma\ n_i(t+1)/\Sigma\ n_i(t)$$

Burnham et al. (1996) incorporated survival and recruitment rates into a Leslie matrix meta-analysis for the threatened Northern Spotted Owl (*S. o. caurina*). They used data from 11 study areas ranging in duration from four to nine years and encompassing much of the geographic range of the owl. They estimated that $\lambda = 0.9548$ (SE = 0.017). They rejected the null hypothesis of no change in population trajectory in favor of an alternative hypothesis that $\lambda < 1.0$, hence documenting a declining population. While the 0.95 value of λ is only a 5 percent difference from 1.0, this indicates a potential decline of about 5 percent per year, which is biologically significant. This serves as a powerful example of how the application of matrix models can inform the status of species of conservation concern and lead, one hopes, to conservation measures that can arrest and hopefully reverse the decline of such populations.

Decision-Theoretical Models

Decision-theoretical models are used to understand how anthropogenic perturbations may affect population parameter(s) of a species of interest. These model approaches can be considered from a **single "best" model approach** or from a **multiple-model approach** (Nichols 2001). Adaptive management models (sensu Walters 1986) are examples of decision-theoretical models. Here a model or set of models is developed with a hypothesis about how a population will respond to certain management actions. The actions are undertaken and population responses are monitored. Pending the outcome of the management activities on the population, management is continued, modified in an itera-

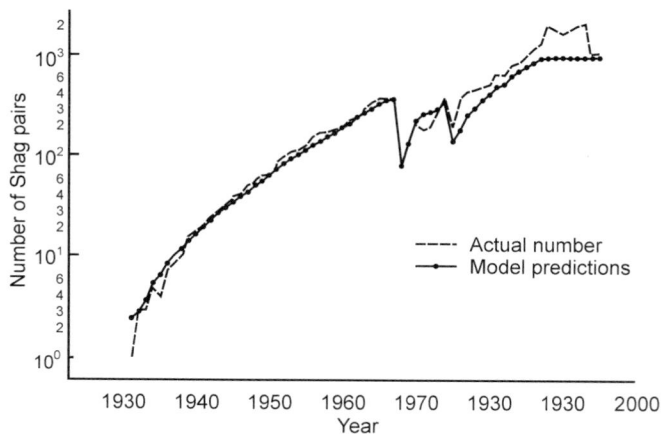

Figure 21.10. Output from an empirical population model to predict the trend of the Shag population on the Farne Islands in the United Kingdom from 1930 to 1995. Note how closely model predictions track actual population numbers. *Adapted from Potts 2000. Used with permission.*

Table 21.2. Structure of a life table used to track age-specific populations. *Adapted from Williams et al. 2002. Used with permission.*

i	l_i	S_i	N_i	b_i	$F_i = S_i b_{i+1}$	$B_i = N_i F_i$
0	1.000	0.250	1000	—	0.125	125
1	0.250	0.650	250	0.50	0.650	163
2	0.162	0.700	163	1.00	1.400	228
3	0.114	0.700	114	2.00	1.400	160
4	0.080	0.500	80	2.00	1.000	80
5	0.040	0.600	40	2.00	1.200	48
6.	0.024	0.000	24	2.00	—	0
6+	—	—	0	—	—	—

* Followed from birth (age class $i = 0$) until all have died (age class $i = 6$).

tive fashion, or terminated. The Grey Partridge model by Potts that was described earlier in this chapter could easily be transitioned into a decision-theoretical model if the model predictions were tested by experimental perturbations in reduction of nest predators that were predicted by the model to increase annual productivity and therefore partridge density.

METHODS FOR ESTIMATION OF POPULATION SIZE AND TRENDS

Ornithologists, conservation biologists, and wildlife managers often assess how many birds are present in a given area because they are interested in the status and trends of these populations. Two general approaches—censuses and surveys—are available to estimate populations. A **census** is a count that is a complete enumeration of all birds found within a particular area, whereas a **survey** is a sample count of individuals from which parameter estimates are derived. Censuses are rarely possible in natural settings because birds are mobile, secretive, and not detected with certainty. As a result, surveys are used much more widely than are censuses in field settings. With surveys, investigators can report simple counts (i.e., the number, frequency, or occupancy of birds detected) or adjusted counts, in which the simple count data are refined by accounting for the probability of detecting a bird when present. For example, consider a situation in which an investigator surveys an area and detects 30 White-headed Woodpeckers (*Picoides albolarvatus*), which constitutes a simple count. If the probability of detecting a woodpecker is 0.80, the adjusted count is $30/0.80 = 37.5$. As you can see, failure to account for detection probability may significantly underestimate population size.

Given the diversity of birds and variations in their ecologies, no one counting method is optimal for all species. Meth-

ods also vary with the scope of inference. That is, an ornithologist might use one technique to estimate a population within a small study area, but will use a different method across a larger region. Ornithologists have a number of tools at their disposal, and the exact method used depends on resources available and the rigor and robustness of the estimates they require. In the section below, we review and discuss some of these tools.

Indexing Trends over Broad Geographic Areas
Breeding Bird Survey

The North American Breeding Bird Survey (BBS; see chapter 30) was begun in 1966 by Chandler Robbins. The survey is basically a roadside survey conducted by volunteer observers and is a primary source of information on spatial and temporal patterns of population change for North American birds. Over 2,000 BBS routes are distributed throughout North America. Routes are organized into one of 62 strata corresponding to different physiographic areas (table 21.3). Each route consists of 50 stations. Breeding Bird Surveys typically occur in June, during the breeding season. Observers drive the route, stopping at each station and then recording all birds seen or heard during a three-minute sampling period.

The BBS is a monumental effort and the data collected can be used to evaluate trends in breeding bird abundance (Sauer et al. 2013). As with any effort this broad, coordination and standardization are real obstacles. Its usefulness lies in providing a broad index of population trends across both space and time. BBS data have also been used to show both range expansions and range contractions of various species of birds in the United States and parts of Canada.

Christmas Bird Count

Whereas the BBS is used to index breeding bird populations, the Christmas Bird Count (CBC) is used to index populations of wintering birds in North America and elsewhere. Begun in 1900, the CBC is the world's largest and oldest database on bird populations and is administered by the National Audubon Society (Bock and Root 1981). The early CBC included just 27 observers and 25 locations in the United States and Canada in 1900, but has grown to nearly 60,000 observers and 2,200 locations from the Arctic Circle to the waters off Tierra del Fuego in Patagonia. Each count is a daylong survey conducted within a fixed circle, 24 km in diameter. Surveys are conducted within a two-week window centered on Christmas Day. Observers record the species and numbers of all birds encountered during the daylong survey. As with the BBS, the CBC is a monumental effort and is fraught with numerous sources of variation and error. Whereas the intent of this broad-based survey is to strive for standardization,

Table 21.3. Physiographic stratification used in Breeding Bird Survey analyses. *Adapted from Bykstra 1981. Used with permission of the Cooper Ornithological Society.*

I. Northern Boreal Forest
    25 Open Boreal Forest
    28 Northern Spruce-Hardwoods
    29 Closed Boreal Forest

II. Eastern Deciduous Forest
    A. Appalachians
        8 Glaciated Coastal Plain
        10 Northern Piedmont
        12 Southern New England
        13 Ridge and Valley
        21 Cumberland Plateau
        22 Ohio Hills
        23 Blue Ridge Mountains
        24 Allegheny Plateau
        26 Adirondack Mountains
        27 Northern New England
    B. Interior Plains
        14 Highland Rim
        15 Lexington Plains
        16 Great Lakes Plains
        17 Driftless Area
        18 St. Lawrence River Plain
        19 Ozark-Ouachita Plateau
        20 Great Lakes Transition
        31 Till Plains

III. Southeastern Forest
    A. Coastal Plain
        1 Subtropical
        2 Floridian
        3 Coastal Flatwoods
        4 Upper Coastal Plain
        5 Mississippi Alluvial Plain
        6 Eastern Texas Prairies
        7 South Texas Brushlands
    B. Foothills
        11 Southern Piedmont

IV. Great Plains
    A. Northern Plains
        30 Aspen Parklands
        32 Dissected Till Plains

        37 Drift Prairie
        38 Missouri Coteau
        39 Great Plains Roughlands
        40 Black Prairie
    B. Southern Plains
        33 Osage Plain-Cross Timbers
        34 High Plains Border
        35 Staked and Pecos Plains
        36 High Plains
        53 Edwards Plateau

V. Rocky Mountains
    A. Basin and Deserts
        54 Colorado, Uinta Basins
        84 Pinyon-Juniper Woodlands
        85 Pitt-Klamath Plateau
        86 Wyoming Basin
        88 Great Basin
        89 Columbia Plateau
    B. Forested Mountains
        61 Black Hills
        62 Colorado Rockies
        63 High Plateaus of Utah
        64 Northern Rockies
        65 Dissected Rockies
        68 Canadian Rockies

VI. Pacific Mountains
    A. Cascade-Sierra Axis
        8 Sierra Nevada
        10 Cascade Mountains
    B. Pacific Ranges
        91 Central Valley
        92 California Foothills
        93 S. Pacific Rainforests
        94 N. Pacific Rainforests
        95 Los Angeles Ranges
VII. Southwestern Arid
        81 Mexican Highlands
        82 Sonoran Desert
        83 Mojave Desert

differences among observers, count effort, weather conditions, unusual habitats, access, sample size for rare species, difficulties with counting flocking species and data collection contribute considerable noise to the data set. The value of CBC data is to explore large-scale patterns in species distributions and trends over time. They should not be used to explore local trends. A value not so easily quantified is the social aspect, in that it allows participation of a large number of people in providing data on the status and trend of birds.

In this respect, the CBC represents one of the first endeavors of citizen science (see chapter 30).

European Breeding Bird Atlas

The European Breeding Bird Council initiated a project in the 1980s to produce a grid-based atlas depicting the distribution and abundance of breeding birds in Europe (Hagenmeijer and Blair 1997). Data to create the atlas were collected from 1985 to 1988. Birds were surveyed within a 50 km^2 grid cell and results mapped. The geographical extent of these efforts included all of Europe extending to Madeira, the Azores, Iceland, Svalbard, Novaya Zemlya, Franz Josef Land, and Transcaucasia, as well as European Russia east to the Ural Mountains. The atlas provides information in species accounts that include maps of the breeding distribution, histograms showing countries with the largest breeding populations, and species text. Many states in the United States have compiled breeding bird atlases over the years.

Estimating Populations within Restricted Geographical Areas

Mark-Resight

Mark-resight studies occur when a bird is identified with a unique marking and then resighted at a subsequent time. A number of methods are used to mark birds, including metal and colored bands (rings), patagial tags, neck collars, back tags, radio transmitters, dyes, and feather patterns (Calvo and Furness 1992). Typically, however, birds are marked with numbered or colored bands placed on their tarsi. Within the United States, the USGS Bird Banding Laboratory Office of Migratory Birds provides uniquely numbered bands to place on birds. Because band numbers are so small, ornithologists need to recapture a bird to read the numbers. To avoid having to recapture a bird, ornithologists will also use colored bands in different combinations to uniquely mark a bird. They can then view the color bands on the bird with binoculars to identify it.

Marking birds individually allows ornithologists to estimate a number of population parameters such as abundance, reproduction, survival, emigration, and immigration. The simplest way to estimate abundance is with the Lincoln-Petersen index:

$$N = T_t \times C_{t+1} / M_{t+1}$$

where N is the population estimate, T is the number of birds captured and tagged at time t, C is the number of birds captured at time t+1, and M is the number of birds tagged at time t recaptured at t+1. For example, if 200 Ring-necked Pheasants were captured, tagged, and released into an area, and a few days later 10 of 40 birds shot were tagged, the population, pre-shooting, would be

$$N = 200 \times 40 / 10 = 800$$

The Lincoln-Petersen index rests on a number of assumptions: (1) marks are not lost, (2) population is closed (no immigration, emigration, births or deaths), and (3) marks do not affect capture probabilities. These assumptions are often violated, and this has led to the development of more sophisticated mark-recapture models that are not subject to these assumptions.

Lincoln-Petersen methods do not require that individuals be given unique marks; the same mark can be used on all individuals. Not using unique marks, however, limits opportunities for gathering information. By marking individuals uniquely, one can also estimate a number of other population parameters and apply more rigorous analyses such as the matrix models described above.

Distance Sampling

Traditionally, ornithologists recorded the number of birds encountered within a fixed area, whether along a line transect, within a quadrat, or within a set radius from a fixed point. An underlying assumption here is that the probability of detecting a bird is constant within the sampling area. Emlen (1971) observed that the number of birds detected while walking a transect declined with increasing distance from the observer (fig. 21.11). This led him to propose a correction factor based on the problem of declining detectability as birds were detected farther from the observer. Note that this issue of detectability is somewhat different from detectability in occupancy estimation, as noted below. A number of factors can influence detectability, including vegetation, mode of detection, weather, and frequency and strength of cues (e.g., singing, calling, movements). Further, detectability varies among species (as shown in fig. 21.11), thereby requiring different correction factors.

With the advent of computers and the ability to use complex mathematical models, the science of distance sampling has made great strides. Initially, distance sampling was applied to line transects (Emlen 1971, Burnham et al. 1981). An observer walks along a transect while looking and listening for the species being sampled. Once a bird is detected, the observer records the distance and angle deviance off the transect line to the bird to calculate the horizontal distance to the bird from the transect line. Distance sampling rests on four key assumptions: (1) birds on the transect line are not missed; (2) birds are detected before moving; (3) distance and angles are measured accurately; and (4) sightings are independent events. Once sampling is completed, observers plot the probability of detecting a bird with increasing distance from the line transects (fig. 21.12). Depending on the shape of the distribution, one can fit the best mathematical model to estimate bird density. As the theory of distance sampling developed, application has been extended to point counts, strip transects, trapping webs, and cue counts (Buckland et al. 2001).

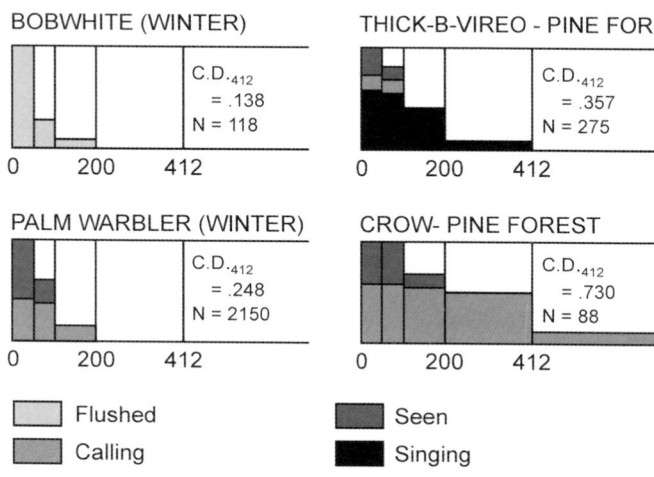

Figure 21.11. Types and distributions of detections laterally from the transect route for four species of birds. The value for the basal strip is plotted as 100 (full column height on the left of each graph) and values for the distal strips are proportions. The bobwhite sample is from a non-singing wintering population where the birds were flushed at close range; few detections were made beyond 100 feet (30 meters) laterally, and coefficient of detection (C.D.) values were low. The Palm Warbler sample of is a non-singing wintering population; tallies beyond 100 feet (30 meters) were auditory detections of call notes. The Thick-billed Vireo population represents a resident breeding population with moderately loud song that can be heard through pine forest vegetation for 200 to 300 feet (60 to ca. 100 meters). The crow sample represents a situation where visual detection falls off in the first 100 to 200 feet (30 to 60 meters), while auditory detections can be heard from 1,000 feet (300 meters) or more. *Adapted from Emlen 1971. Used with permission of the American Ornithologists' Union.*

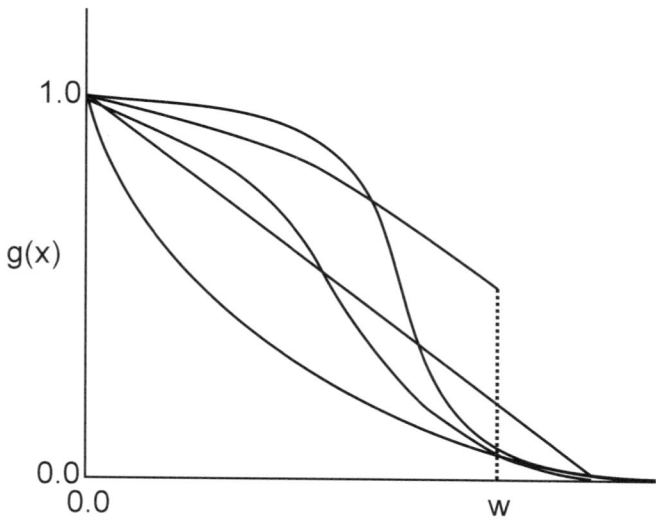

Figure 21.12. Several possible detection function shapes that can be derived from distance-sampling data. *Adapted from Burnham et al. 1981. Used with permission.*

Spot Mapping

Spot mapping is a technique used with territorial breeding birds. Standards for spot mapping have been established by the International Bird Census Committee (1969). Spot mapping is perhaps the most time-consuming bird count method (Bibby et al. 1992). The method occurs within fixed plots ranging in size from 10–20 hectares for woodland birds and 50–100 hectares in more open habitats. A gridded study plot map needs to be developed at the scale of approximately 1:2500. Ideally, prominent plot features such as vegetation and changes in topography are also mapped to aid in orientation. An observer then walks the plot slowly, noting detection of species under study and their activities. Up to 10 repeat visits, spaced perhaps a week apart, are needed to identify territories and estimate the number of pairs of birds on the plot. A set of standard symbols are used to denote bird activities and to identify bird territories (see examples, box 3.3 in Bibby et al. 1992). The ambiguity of identifying territories is reduced when conspecifics can be seen or heard singing simultaneously. The result is a map that depicts the location and number of territories on the plot such as the one developed by Kendeigh (1944) for the House Wren (*Troglodytes aedon*; fig. 21.13).

Occupancy Estimation

There are many situations where ornithologists are interested in tracking the abundance of uncommon, rare, or elusive species. With such species, methods such as mark-resight or distance sampling are often not suitable because it is difficult to obtain an adequate number of detections to obtain a reliable estimate of population abundance or density. The most important and useful thing about occupancy modeling is that it explicitly estimates detection probability and can correct for variable and/or imperfect detection. Because it relies on repeated visits, the approach allows for detection probability to be modeled as a function of time of day, observer, site/habitat characteristics, or other factors that could otherwise bias estimates.

Distance sampling requires a minimum of 40 detections to estimate density (Burnham et al. 1981). This works well for common species that are easily detected, but becomes problematic for rare species or those not easily detected. As a result, ornithologists can estimate populations or densities for common species but, often, not for rare species that are of conservation concern. In these cases, one option is to record the presence of a bird when encountered during sampling, whether at a point, along a transect, or within a fixed plot. The simplest way to report presence in this case is by frequency, or the number of times at least one individual of a species, irrespective of its abundance, is detected during each

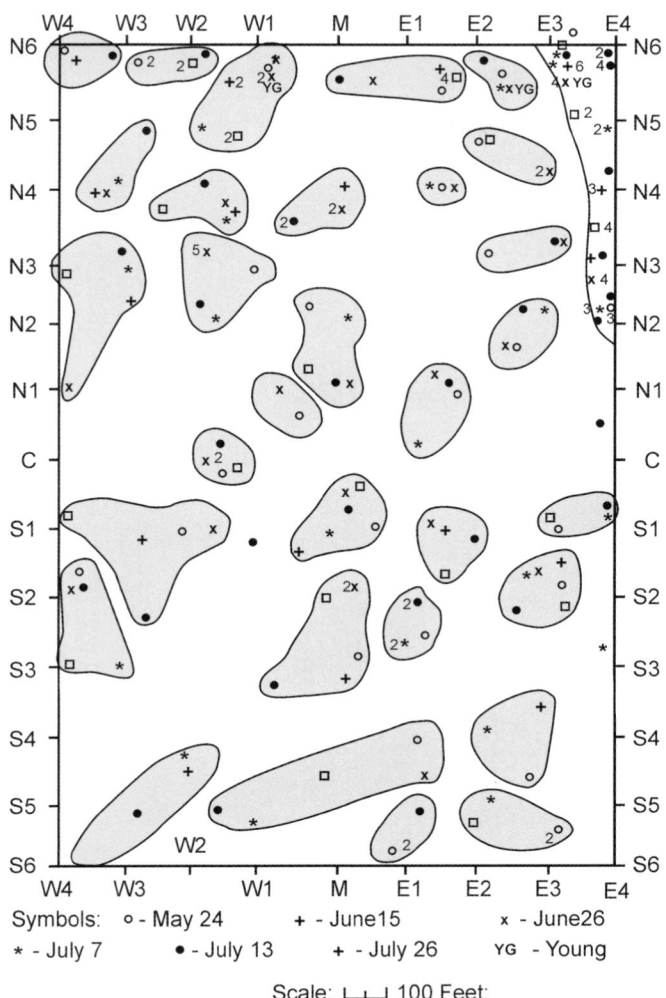

Figure 21.13. Composite map of House Wren territories and behavior based on spot-mapping observations of birds in the field. *Adapted from Kendeigh et al. 1944. Used with permission.*

sample (Verner 1984). For example, if a Nuttall's Woodpecker (*Picoides nuttallii*) is detected at 8 of 100 sampling points, its frequency would be reported as 8 percent because it was recorded at 8 percent of the points. This simple frequency metric, however, fails to account for the detection probability and the possibility that the observer failed to detect the woodpecker at points where it was present. MacKenzie and colleagues (2002) recognized this shortcoming and developed a method to incorporate detection probabilities to refine occupancy estimates. This method, which is commonly referred to as **occupancy modeling**, allows an investigator to collect presence and absence data from repeated surveys and then use these data to estimate two useful parameters: (1) occupancy rate (the proportion of sites surveyed where

the species of interest is present), and (2) the probability of detecting the species of interest, which is almost always less than 1. The models assume (1) a closed population, in that sites are occupied with no colonization or extinction from sample sites during survey period; (2) species are not falsely detected when not present; and (3) detecting a species at one sample site is independent of detecting it at another. To estimate detection probabilities, each sampling location must be sampled at least twice. Parameters estimated are the occupancy probability for the sampling unit (ψ) and probability of detecting the species given that it is present (p) for the sample. Assuming two visits to each sampling unit, one of four outcomes is possible. The probability of detecting the species on both occasions is

$$\psi_j\, p_j\, p_j.$$

If the species is detected only on the first visit, then the probability is

$$\psi_j\, p_j (1 - p_j).$$

If only detected on the last visit, then the probability is

$$p_j.$$

If birds are never detected, then the probability is

$$\psi_j\, (1 - p_j)\, (1 - p_j) + (1 - \psi_j).$$

These four probabilities sum to one because they are the only possible observations. The parameters, ψ and p, can be estimated in a maximum likelihood framework based on empirical observations from the sampling units.

A key assumption of using occupancy estimation is that it tracks the behavior of the population. That is, if occupancy rates increase, then the population is increasing; if occupancy rates decline, the population is declining. This assumption is rarely tested, but should be, to validate the use of occupancy as a surrogate for measuring a population directly.

Thus, if the assumptions noted above can be met, occupancy data can be used to track trends in rare or elusive species over time. Instead of some measure of abundance, or density, however, the metrics of interest are how occupancy rates and detection probabilities change over time. Additionally, covariates such as vegetation, temperature, and precipitation can be included in the occupancy analyses as factors that may help explain or interpret the variation in occupancy and detectability.

MARGARET MORSE NICE

Margaret Morse Nice. *Photo used with permission.*

Margaret Morse Nice (December 6, 1883, to June 26, 1974) was one of the premier ornithologists in the history of the profession. Her research on the life history of the Song Sparrow (*Melospiza melodia*) ranks among the landmark achievements in modern ornithology.

Margaret was born in Amherst, Massachusetts, the youngest of four siblings. She began taking notes on birds when she was 13 years old, and continued recording her observations of birds for more than six decades. She received her BA from Mount Holyoke University in 1906 and married Leonard B. Nice in 1908, after which they moved to Oklahoma where Leonard worked as a professor. The research for her MA in biology from Clark University in 1915 was the first comprehensive assessment of the foods eaten by the Northern Bobwhite (*Colinus virginianus*).

Margaret Nice pioneered the study of avian life histories at time in the early twentieth century when the foci of ornithological research was on collecting, description, and making lists. Along with Herbert Stoddard's comprehensive study of Northern Bobwhite life history, Nice's life history research on the Song Sparrow ushered in a new era of ornithological research that focused on populations of birds in relation to each other—and in relation to their environment—that continues to this day. It is an interesting coincidence that Stoddard's bobwhite research and Nice's sparrow research began around the same time, in the late 1920s.

During 1927, Margaret's husband was hired as a professor at Ohio State University. After settling in Columbus, Margaret started banding Song Sparrows and recording their behavior. Colored bands allowed her to delineate territories of individual birds, record their interactions with other Song Sparrows, and study them as individuals within a population. Her first major research papers on Song Sparrow life history were published in 1933 and 1934 in *Journal für Ornitholigie*, in German, because editors of American ornithological journals thought they were too long. With nearly 250 published papers, her lifetime research productivity was tremendous by any standard. Her book *Watcher at the Nest* (Nice 1939) made her early scientific research available to a broad audience.

Margaret's Song Sparrow research caught the attention of Aldo Leopold, who, in *A Sand County Almanac*, unfortunately referred to her life history research as "a study of the song sparrow conducted by an Ohio housewife." Leopold somewhat redeemed himself later in the same paragraph when he wrote, "In ten years she knew more about sparrow society, sparrow politics, sparrow economics, and sparrow psychology than anyone had ever learned about any bird. Science beat a path to her door. Ornithologists of all nations seek her counsel" (Leopold 1949:190).

Margaret was the first woman president of the Wilson Ornithological Club, and she was a fellow in the American Ornithologists' Union. The Wilson Ornithological Society established the Margaret Morse Nice Medal, their premier ornithological award, in 1997. The medal honors a lifetime of contributions to ornithology.

Details of Margaret Morse Nice's professional accomplishments and personal life are documented in her autobiography (Nice 1979), which was published posthumously. It is a must-read for anyone interested in the development of modern research on avian life history and population ecology.

CASE HISTORIES IN AVIAN POPULATION ECOLOGY

Much of this chapter has been devoted to the basic concepts and ideas that have evolved to form the foundation of avian population ecology. Illustrating how the basic concepts of pop-ulation ecology operate in nature provides a link between theory and application. In this section, we focus on case histories of four iconic species that illustrate many of these concepts in a real-world context. The first case history covers how population modeling was used to identify factors that influenced a threatened owl. The second case history addresses how ornithologists identified a set of limiting factors and are working to recover populations of an endangered woodpecker. The third case history tells the story of what happened when a species was introduced to new areas and how many unintended consequences ensued. The fourth and final case history illustrates the dynamics of a rapidly expanding and increasing species of dove in relation to several apparently interrelated factors.

Spotted Owl: Population Modeling for a Threatened Species

The Spotted Owl is an iconic species closely tied to old-growth coniferous forests of the western United States and parts of Canada and Mexico. Three subspecies are recognized: Northern, California (*S. o. occidentalis*), and Mexican (*S. o. lucida*). Forestry practices such as clear-cutting, as well as regional fragmentation and effects of competition from the Barred Owl (*Strix varia*) throughout much of the northern range of the species have altered much of its habitat to the point that it is no longer suitable for occupancy. As a result, two of the subspecies (Northern and Mexican) have been listed as threatened species under the Endangered Species Act, and the other (California) has been proposed to be listed. Given that all three subspecies are of conservation concern, managers need to know more about their population status. Initially, extensive surveys were conducted throughout the ranges of the subspecies to inventory owls and acquire basic information on population numbers and their geographic distributions. Whereas these early surveys were instructive, they provided little information that managers could use to understand population trends and propose corrective measures if warranted. Dr. R. J. Gutiérrez and his students initiated a landmark population demography study on the Northern Spotted Owl in Northern California. Essentially, this entailed a mark-resight study in which they captured owls and banded them with unique color bands. This enabled them to apply matrix modeling techniques to estimate fecundity, survival, and ultimately the finite rate of population change (λ). Soon, other researchers within the range of the Northern Spotted Owls initiated similar studies in Oregon and Washington. They quickly realized that results from each study were site-specific and provided only a glimpse of population trends across the range of the bird. To address this, the researchers convened a series of workshops that also included biometricians from Colorado State University to conduct meta-analyses to better understand range-wide trends (Burnham et al. 1996, Forsman et al. 2011). Many of the initial analyses were based on matrix modeling, and more recently (Forsman et al. 2011) a model-selection framework has been applied to the data (Burnham and Anderson 2002). Although the earlier analyses indicated a declining population, it was not until the Forsman et al. (2011) study appeared that analysts could incorporate covariates to potentially explain reasons underlying population declines. These analyses suggested that increasing numbers of Barred Owls and loss of habitat contributed to demographic declines. The presence of Barred Owls appeared to be the strongest and most consistent negative factor related to Spotted Owl survival. Although these results are not from an experiment and may not represent cause-effect relationships, they certainly suggest that Barred Owl invasion into the range of the Spotted Owl has influenced population declines.

Meta-analyses have also been completed to assess population trends for the California Spotted Owl (Franklin et al. 2004). Although analyses suggested that $\lambda < 1.0$ on four of the five study areas, confidence intervals for trend estimates overlapped 1.0, thus making the results equivocal. To date, no range-wide population assessment has been undertaken for the Mexican Spotted Owl, and the status of its population remains unknown.

Management of threatened and endangered species requires reliable information on their population status. Field and analytical methods are improving greatly, which enables ornithologists and managers to make informed decisions. Ornithologists and biometricians working on the Spotted Owl are in many ways at the forefront of developing new analytical approaches that can be applied to other species in the future. The Spotted Owl case history also shows the importance of incorporating spatial and temporal scale and demographic structure for understanding population trends across larger scales.

Red-cockaded Woodpecker: Recovery of an Endangered Species?

The Red-cockaded Woodpecker (*Picoides borealis*) was listed as an endangered species in 1970, three years before the Endangered Species Act was passed by the US Congress and signed into law by then-president Nixon. The current population abundance of Red-cockaded Woodpeckers is thought to be about 3 percent of what it was before Europeans settled the southeastern Atlantic coastal plain of the United States, which is where this species is found (see chapter 27).

Red-cockaded Woodpeckers have an unusual life history strategy in that they are one of the few, and perhaps the only, species of woodpecker that is an obligate cavity nester in live trees. This unusual life history strategy probably evolved as a predator-defense tactic. Nesting in live pine (*Pinus* spp.) trees allows the birds to maintain a set of resin wells around the cavity entrance by frequently pecking small wounds in the tree bark that release pine sap. The patches of pine sap around the cavity entrance are a deterrent to their primary nest predator, rat snakes (*Elaphe* spp.). Red-cockaded Woodpeckers have an uncanny ability to find living pine trees with red heart fungus (*Phellinus pini*) in the heartwood section of the tree. However, even with trees that have red heart fungus, it takes a pair of woodpeckers nearly a year to excavate a nest cavity.

Red-cockaded Woodpecker populations have declined by 97 percent because the longleaf pine (*P. palustris*) forest that provided their nesting and wintering habitat has declined by 99 percent (Simberloff 1993) since European settlement. Biologists thought that the result of this widespread decline in Red-cockaded Woodpecker population in relation to the corresponding loss of nesting and wintering habitat would only result in extinction. Fortunately, that is not how things have developed over the past five decades.

Three critical factors have converged to put the Red-cockaded Woodpecker on a path that looks like it might be a road to population recovery. These three factors (artificial nest cavities, translocation of individuals into unoccupied habitat, and habitat conservation and restoration) often come into play with respect to recovery of other endangered or declining bird species.

Constructing artificial nest cavities for Red-cockaded Woodpeckers had been discussed, and dismissed, for many years, until Hurricane Hugo devastated the Francis Marion National Forest in South Carolina in 1989. The loss of hundreds of Red-cockaded Woodpecker nest cavities motivated biologists to experiment with artificial nests (Kuvlesky et al. 2013). The widespread success of this project inspired biologists to implement construction of nest cavities elsewhere. The result was that Red-cockaded Woodpeckers readily used them.

The Red-Cockaded Woodpecker Recovery Plan (US Fish and Wildlife Service 2003) lists four biological rationales for translocating birds: (1) augmentation of a population in immediate danger of extinction, (2) reduction of isolation of groups, (3) restoration of birds to their historical range, and (4) management of genetics. Translocations are often done in conjunction with placement of artificial cavities in longleaf pine forest that otherwise has the structure provided by mature trees, with an open canopy and understory maintained by frequent prescribed fire. Translocation has become a successful tool for population restoration, especially for Red-cockaded Woodpecker populations that were small, isolated, and declining.

The past conservation and current restoration efforts to keep and expand the remnant patches of longleaf pine forest on the landscape has been the third key factor in Red-cockaded Woodpecker management. Fortunately, most of the remaining longleaf pine forests that were sustaining the Red-cockaded Woodpecker were on public lands such as national forests, wildlife refuges, and military bases. Private lands also are playing a key role in the conservation of this species, especially on properties that are managed for Northern Bobwhite hunting in the Southeastern Coastal Plain (Crawford and Brueckheimer 2012). Additionally, nonprofit conservation groups such as the Longleaf Alliance (http://www.longleafalliance.org) are playing a critical role in longleaf pine conservation and restoration.

In 1993 there were 4,700 known clusters (groups of nesting pairs ranging in size from 2 to > 50 nesting pairs) of Red-cockaded Woodpeckers. By 2006, this number has increased to at least 6,100, and this increase continues today. It will probably take until about 2050 to achieve Red-cockaded Woodpecker population recovery goals and remove it from the endangered species list. While at the time of this writing (2016) 2050 is about 35 years from now, it is important to also realize that this species has been listed as endangered for at least 45 years.

The broader take-home lesson about Red-cockaded Woodpecker conservation and management is that, when implemented correctly with a biologically based recovery plan, the Endangered Species Act can work to save species of birds and other animals from extinction. The Peregrine Falcon is an excellent case in point. The combination of translocating birds in the wild via hack sites and reducing the use of organochlorine chemicals has worked to recover this species. But it did not happen overnight. Rather it took more than half a century to identify the problem and then implement the conservation strategies needed to recover Peregrine Falcon populations. Even prior to the Endangered Species Act, the widespread use of nesting boxes had a huge and positive impact on Wood Duck (*Aix sponsa*) populations. The basic idea is to identify what the Endangered Species Act calls "critical habitat" and then work to overcome the other limiting factors that are driving populations of the species downward and eventually to extinction.

Common Myna: When a Good Idea Goes Bad

The Common Myna (*Acridotheres tristis*) is one of three bird species on the IUCN Species Survival Commission's list of the world's 100 worst invasive species. Native to Asia and, more specifically, Iran, Pakistan, India, Nepal, Bhutan, Bangladesh, Sri Lanka, and other Asian nations, the Common Myna has been introduced to Canada, Australia, Israel, New Zealand, New Caledonia, South Africa, United States (Hawai'i and Florida) and numerous islands in the Indian, Pacific, and Atlantic Oceans. Some introductions (e.g., Australia) have been purposeful, to control agricultural and garden insects. In other regions (e.g., Israel, South Africa), mynas are caged pets, prized for their melodic song, colorful plumage, and ability to mimic human speech (Peacock et al. 2007). Occasionally, they escape or are released, colonize urban and natural surroundings, and expand in range and numbers (Holzapfel et al. 2006, Peacock et al. 2007).

A member of the starling family (Sturnidae) the Common Myna is an omnivorous open-woodland bird that does well in urban environments. Mynas are ground-foraging birds, feeding primarily on insects and secondarily on seed and fruits (Senguta 1976). They nest in tree cavities and often are found in large communal roosts (Peacock et al. 2007).

The Common Myna has been implicated as a threat to native biodiversity through competition for nest and roost sites, a vector in the spread noxious weeds, and a threat to fruit crops by foraging (Peacock et al. 2007). They may also contribute to the spread of parasites (*Ornithonyssus* spp., which may cause dermatitis) and disease (avian malaria). Communal roosts can include several thousand birds, which may result in damage to trees, and the accumulation of fecal droppings results in unpleasant odors to humans. They have also been implicated in the spread of the noxious weed *Lantana camara* into native grasslands in Hawai'i.

There is some debate as to whether or not the Common Myna competes with native birds. Lowe et al. (2011) concluded that the Common Mynas had little competitive impact on resource use by native bird species in Sydney, Australia. They based this conclusion on the fact that Common Mynas tended to use more modified urban environments than native species, nest sites were plentiful, and they overlapped little in foraging locations with native species. In a broader nationwide analysis, Grarock et al. (2012) found a negative relationship between the establishment of the Common Myna and the long-term abundance of three cavity-nesting species (Sulphur-crested Cockatoo, *Cacatua galerita*; Crimson Rosella, *Platycercus elegans*; Laughing Kookaburra, *Dacelo novaeguineae*) and eight small bird species (Striated Pardalote, *Pardalotus striatus*; Rufous Whistler, *Pachycephala rufiventris*; Willie Wagtail, *Rhipidura leucophrys*; Grey Fantail, *Rhipidura albiscapa*; Magpie-Lark, *Grallina cyanoleuca*; House Sparrow, *Passer domesticus*; Silvereye, *Zosterops lateralis*; Common Blackbird, *Turdus merula*). Indeed, efforts are under way in island situations to eradicate the myna to favor native species. Canning (2011) reported a successful campaign to remove mynas from Frégate Island, Seychelles, to favor the endangered Seychelles Magpie-Robin (*Copsychus sechellarum*). On Moturoa Island within the Bay of Islands, New Zealand, Tindall et al. (2007) noted the increase of numerous bird species—most notably the Tui (*Prosthemadera novaeseelandiae*), Grey Warbler (*Gerygone igata*), and Blackbird—following myna removal.

Clearly, introduction of an exotic species may result in unintended consequences to native species, in the spread of disease and parasites and in nuisance to humans. The myna serves as but one example of a good idea that went awry. Many of the methods described in this chapter provide approaches to assess and track myna populations, and then to model trajectories into the future. This information can be used to understand the situation and to explore control measures should they become necessary. Humans should be more mindful when considering future introductions and try to better understand ramifications prior to initiating actions.

White-winged Dove: Why Are Their Populations Expanding?

The White-winged Dove (*Zenaida asiatica*) is a relatively large (150–170 grams) columbid that was historically distributed across parts of southern Arizona, New Mexico, and Texas in the United States, with a disjunct population in southeastern Florida (Brown et al. 1977). They are the second-most hunted dove in North America, after the Mourning Dove (*Z. macroura*). White-wings are typically more gregarious than Mourning Doves because they forage and roost in large flocks and have a predilection to roost and nest in citrus groves and other patches of relatively large trees in semitropical, semiarid environments.

From 1968 to 1993, the White-winged Dove population in the Lower Rio Grande Valley of South Texas was declining. The primary cause of the population decline was thought to be clearing native brush for agricultural crops other than citrus (George et al. 1994). During the same time (the 1980s) that the White-winged Dove population was declining in the Lower Rio Grande Valley, their numbers were increasing near San Antonio and in other parts of central Texas. Along with these increases was a northward range expansion where White-winged Doves had expanded their geographic range as far north as Amarillo and as far east as Houston by 1990 (George et al. 1994). The range expansion of the White-winged Dove now includes Oklahoma and is still taking place (Schwertner et al. 2002, Brennan et al. 2017). Individual White-winged Doves have even been documented at bird feeders in the maritime provinces of Canada, and in the Pacific Northwest (Schwertner et al. 2002).

In Texas, not only have White-winged Doves expanded their range to encompass most of the state (except the eastern Piney Woods), but their numbers have also increased dramatically. There are now more than half a million White-winged

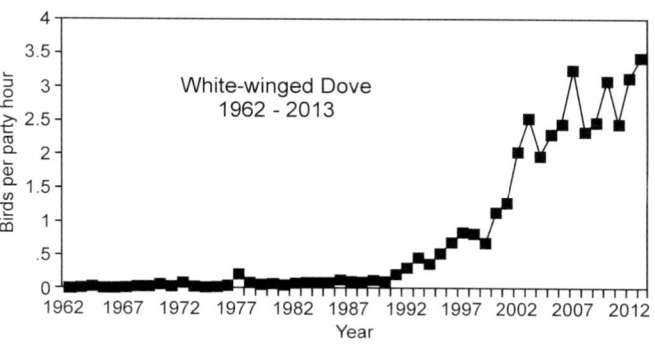

Figure 21.14. Trajectory of White-winged Dove abundance in Texas, based on Christmas Bird Count data. *From L.A. Brennan collection.*

Doves breeding in the San Antonio area, when they were rare there just four decades ago (Lockwood and Freeman 2004). Christmas Bird Count data from Texas show an exponential-like population increase that may or may not have reached an asymptote (fig. 21.14). This is a relatively rare phenomenon for a bird species that has not been introduced by humans to a new geographic region. For some unknown reason or reasons, White-winged Doves in Texas began increasing during the late 1980s and early 1990s, with Christmas Bird Count numbers more than quadrupling in little more than a decade (fig. 21.14).

There is no single, definitive factor that is universally considered responsible for the White-winged Dove range expansion and population increase. The most plausible explanation is a combination of warming climate and gradual shifts in vegetation composition and structure. For a semitropical species, a warming climate clearly has the potential to cause geographic range shifts, as we are seeing with other semitropical species such as the Green Jay (*Cyanocorax yncas*; Rappole et al. 2007). However, during the past 40 years, vegetation across much of Texas has also changed dramatically both in rural and in urban areas. In rangelands especially, vast regions have seen large increases in woody cover, especially from species such as mesquite (*Prosopis* spp.) as a result of cattle grazing, lack of prescribed fire, and other potential causes. Thus, it is not surprising that populations of birds with affinities for semitropical brushy rangeland habitat have responded positively to increases in woody brush cover at the landscape scale. In urban and suburban areas, the maturation of large trees, especially southern live oaks (*Quercus virginicus*) seems to have made these areas attractive to White-winged Doves, especially for nesting. White-winged Doves are commonly found in older or historic-district types of neighborhoods with mature oaks, hackberries, and similar broad-leaved deciduous tree species, and are absent from

new housing developments with little or no mature landscaping or woody vegetation (Brennan et al. 2017).

Although the causes of the White-winged Dove population expansion and increase may never be known from an absolute perspective, this case history is a classic example that shows how the distribution and abundance of bird populations is dramatic, and not static, in space and time. Changes in bird population dynamics may not always be as dramatic as this case history for the White-winged Dove demonstrates. However, as climate and vegetation change, so too will the distribution and abundance of wild bird populations.

WHAT IS THE FUTURE OF AVIAN POPULATION ECOLOGY?

Understanding how populations increase, decrease, and are regulated is basic to understanding the structure and dynamics of a species' ecology and the foundation for community ecology (Ricklefs 1973). Species do not live in a vacuum. Their populations are influenced by numerous intrinsic and extrinsic factors. Populations are not static over time; they can increase or decrease, or even go extinct. Unraveling the complexities that underlie population ecology is an endeavor that ornithologists have been engaged in for more than a century. The science of avian population ecology continues to evolve.

The importance of avian population ecology cannot be overstated. The myriad factors that affect and influence births, deaths, immigration, and emigration of bird populations are interrelated, synergistic, and, in some cases, intractable. Recent advances in mathematical modeling facilitated by enhanced computing capabilities have provided ornithologists with new sets of tools to test predictions of conceptual models with empirical information. However, sophisticated analytical tools and elaborate models cannot substitute for poor study design and lack of data. As we learn more about how populations behave, our understanding of complex ecological relationships will improve and our ability to manage and conserve populations of wild birds will be better informed.

KEY POINTS

- Population ecology is a branch of scientific inquiry that strives to understand the mechanisms that underlie and/or influence how populations change over time.
- The basic form for understanding population ecology is $N_{t+1} = N_t + b_t + i_t - d_t - e_t$, where N = population size, b = births, i = immigrations, d = deaths, e = emigrations, and t = time.

- Populations rarely grow exponentially for long periods of time and are regulated by density-dependent and density-independent factors that limit their size.
- Population models are critical tools for estimating population parameters and evaluating how populations behave.
- Ornithologists have developed a number of field techniques to estimate population sizes and trajectories.

KEY MANAGEMENT AND CONSERVATION IMPLICATIONS

- Understanding status and trends of species' populations allows managers to identify species at risk.
- Numerous factors influence populations, and recent advances in modeling and computing capabilities allow managers to focus on those with the greatest influence.
- Through various case studies, managers can learn what works to conserve populations.
- Our knowledge of population ecology continues to progress as research generates new information.

DISCUSSION QUESTIONS

1. What is population ecology and why is it important to study?
2. How do populations grow and what are some of the factors that might accelerate or limit growth?
3. Why is modeling an important tool for ornithologists, and what types of insights would they hope to gain from it?
4. What are some of the techniques used to sample bird populations, and what are some advantages and disadvantages of various approaches?
5. Are species introductions a good idea? If so, why? If not, why not?
6. How can managers or conservationists address population declines of species of concern?

References

Akçakaya, H. R., and M. G. Raphael (1998). Assessing human impact despite uncertainty: Viability of the Northern Spotted Owl metapopulation in the Pacific Northwest. Biodiversity and Conservation 7:875–984.

Bibby, C. J., N. D. Burgess, and D. A. Hill (1992). Bird census techniques. Academic Press, San Diego, CA.

Bock, C. E., and T. L. Root (1981). The Christmas Bird Count and avian ecology. Studies in Avian Biology 6:17–23.

Brennan, L. A., D. L. Williford, B. M. Ballard, W. P. Kuvlesky Jr., E. D. Grahmann, and S. J. DeMaso (2017). The resident and webless migratory game birds of Texas. Texas A&M University Press, College Station.

Brown, D. E., D. R. Blakenship, P. K. Evans, W. H. Keil Jr., G. L. Waggerman, and C. K. Winkler (1977). White-winged Dove. In Management of migratory shore and upland game birds in North America, G. C. Sanderson, Editor. International Association of Fish and Wildlife Agencies, Washington, DC, pp. 247–272.

Bryant, D. M. (1989). House Martin. In Lifetime reproduction in birds, I. Newton, Editor. Academic Press, London, pp. 75–88.

Buckland, S. T., D. R. Anderson, K. P. Burnham, J. L. Laake, D. L. Bourchers, and L. Thomas (2001). Introduction to distance sampling. Oxford University Press, Oxford.

Burnham, K. P., D. R. Anderson, and J. L. Laake (1981). Estimation of density from line transect sampling of biological populations. Wildlife Monographs 72:1–201.

Burnham, K. P., D. R. Anderson, and G. C. White (1996). Meta-analysis of vital rates for the Northern Spotted Owl. Studies in Avian Biology 17:92–101.

Burnham, K. P., and D. R. Anderson (2002). Model selection and multimodel inference. 2nd ed. Springer-Verlag, New York.

Bykstra, D. (1981). The North American Breeding Bird Survey. Studies in Avian Biology 6:34–41.

Calvo, B., and R. W. Furness (1992). A review of the use and the effects of marks and devices on birds. Ringing and Migration 13:129–151.

Canning, G. (2011). Eradication of the invasive Common Myna, Acridotheres tristis, from Fregate Island, Seychelles. Phelsuma 19:43–53.

Caswell, H. (2001). Matrix population models: Construction, analysis, and interpretation. 2nd ed. Sinauer, Sunderland, MA.

Clark, C. W., and M. Mangel (2000). Dynamic state variable models in ecology: Methods and applications. Oxford University Press, London.

Crawford, R. L., and W. R. Brueckheimer (2012). The legacy of a Red Hills hunting plantation: Tall Timbers Research Station and Land Conservancy. University Press of Florida, Gainesville.

Davis, J. (1967). In memoriam: Alden Holmes Miller. The Auk 84:192–202.

DeMaso, S. J., J. P. Sands, L. A. Brennan, F. Hernandez, and R. DeYoung (2012). Simulating density-dependent relationships in South Texas Northern Bobwhite populations. Journal of Wildlife Management 77:24–32.

Dhondt, A. A. (1989). Blue Tit. In Lifetime reproduction in birds, I. Newton, Editor. Academic Press, London, pp. 15–34.

Einarson, A. S. (1942). Specific results from Ring-necked Pheasant studies in the Pacific Northwest. Transactions of the North American Wildlife Conference 7:130–145.

Emlen, J. T. (1971). Population densities of birds derived from transect counts. The Auk 88:323–342.

Errington, P. L. (1945). Some contributions of a fifteen-year study of the Northern Bobwhite to a knowledge of population phenomena. Ecological Monographs 15:1–34.

Forsman, E. D., R. G. Anthony, K. M. Dugger, E. M. Glenn, A. B. Franklin, G. C. White, C. J. Schwarz, et al. (2011). Population demography of Northern Spotted Owls. Studies in Avian Biology 40.

Franklin, A. B., R. J. Gutiérrez, J. D. Nichols, M. E. Seamans, G. C. White, G. S. Zimmerman, J. E. Hines, et al. (2004). Population dynamics of the California Spotted Owl (*Strix occidentalis occidentalis*): A meta-analysis. Ornithological Monographs 54.

Fronhofer, E. A., A. Kubisch, F. M. Hilker, T. Hovestat, and H. J. Poethke (2012). Why are metapopulations so rare? Ecology 98:1967–1978.

George, R. R., R. E. Tomlinson, R. W. Engel-Wilson, G. L. Waggerman, and A. G. Sprat (1994). White-winged Dove. *In* Migratory shore and upland game bird management in North America, T. C. Tacha and C. E. Braun, Editors. International Association of Fish and Wildlife Agencies, Washington, DC, pp. 29–59.

Goss-Custard, J. D., N. K. Burton, N. A. Clark, P. N. Ferns, S. McGorty, C. J. Reading, M. M. Rehfisch, et al. (2006). Test of a behavior-based individual-based model: Response of shorebird mortality to habitat loss. Ecological Applications 16:2215–2222.

Grarock, K., C. R. Tidemann, J. Wood, and D. B. Lindenmayer (2012). Is it benign or is it a pariah? Empirical evidence for the impact of the Common Myna (*Acridotheres tristis*) on Australian birds. PLoS ONE 7 (7): e40622. doi:10.1371/journal .pone.0040622.

Grimm, V. (1999). Ten years of individual-based modeling in ecology: What have we learned and what could we learn in the future? Ecological Modelling 115:129–148.

Grinnell, H. W. (1940). Joseph Grinnell: 1877–1939. The Condor 42:3–34.

Grinnell, J. (1915). Conserve the collector. Science 41:229–232.

Grinnell, J. (1917). The niche relationships of the California Thrasher. The Auk 34:427–433.

Grinnell, J. (1922). The trend of avian populations in California. Science 56:671–676.

Grinnell, J., and A. H. Miller (1944). The distribution of the birds of California. Cooper Ornithological Club, Berkeley, CA.

Guthery, F. S., A. R. Rybak, S. D. Fuhlendorf, T. L. Hiller, S. G. Smith, W. H. Puckett Jr., and R. A. Baker (2005). Aspects of the thermal ecology of Bobwhites in north Texas. Wildlife Monographs 159.

Hagenmeijer, W. J. M., and M. J. Blair (1997). The EBCC Atlas of European breeding birds: Their distribution and abundance. T. and A. D. Poyser, London.

Hanski, I. (1991). Single-species metapopulation dynamics: Concepts, models and observations. Biological Journal of the Linnean Society 42:17–38.

Herman, S. G. (1986). The naturalist's field journal: A manual of instruction based on a system established by Joseph Grinnell. Buteo Books, Vermillion, SD.

Holzapfel, C., N. Levin, O. Hatzofe, and S. Kark (2006). Colonisation of the Middle East by the invasive Common Myna *Acri-*

dotheres tristis L., with special reference to Israel. Sandgrouse 28:44–51.

Hone, J., and R. M. Sibly (2002). Demographic, mechanistic and density-dependent determinants of avian population growth: A case study in an avian predator. Philosophical Transactions of the Royal Society of London B 357:1171–1177.

Hudson, R. (1965). The spread of the Collared Dove in Britain and Ireland. British Birds 58:105–139.

Hudson, R. (1972). Collared Doves in Britain and Ireland during 1965–1970. British Birds 65:139–155.

Hutchinson, G. E. (1978). An introduction to population ecology. Yale University Press, New Haven, CT.

International Bird Census Committee (IBCC) (1969). Recommendations for an international standard for a mapping method in bird census work. Bird Study 16:248–255.

Kendeigh, S. C. (1944). Measurement of bird populations. Ecological Monographs 14:67–106.

Kuvlesky Jr., W. P., L. A. Brennan, B. M. Ballard, T. A. Campbell, D. G. Hewitt, C. A. DeYoung, S. E. Henke, F. Hernandez, and F. C. Bryant (2013). Managing populations. *In* Wildlife management and conservation: Contemporary principles and practices, P. R. Krausmann and J. W. Cain III, Editors. Johns Hopkins University Press, Baltimore, MD, pp. 299–322.

Lack, D. (1947). Darwin's finches. Cambridge University Press, Cambridge, UK.

Lack, D. (1954). The natural regulation of animal numbers. Oxford University Press, London.

Lack, D. (1966). Population studies of birds. Oxford University Press, London.

Lack, D. (1968). Ecological adaptations for breeding in birds. Methuen, London.

LaHaye, W. S., R. J. Gutiérrez, and H. R. Akçakaya (1994). Spotted owl metapopulation dynamics in Southern California. Journal of Animal Ecology 63:775–785.

Lebreton, J. D., and O. Giminez (2012). Detecting and estimating density dependence in wildlife populations. Journal of Wildlife Management 77:12–23.

Leopold, A. (1933). Game management. Charles Scribner's Sons, New York.

Leopold, A. (1949). A Sand County almanac and sketches here and there. Oxford University Press, New York.

Leslie, P. H. (1945). The use of matrices in certain population mathematics. Biometrika 33:183–212.

Levins, R. (1969). Some demographic and genetic consequences of environmental heterogeneity for biological control. Bulletin of the Entomological Society of America 62:237–240.

Lockwood, M. W., and B. Freeman (2004). The TOS handbook of Texas birds. Texas A&M University Press, College Station.

Lowe, K. A., C. E. Taylor, and R. E. Major (2011). Do Common Mynas significantly compete with native birds in urban environments? Journal of Ornithology 152:909–921.

MacArthur, R. H., and E. O. Wilson (1967). The theory of island biogeography. Princeton University Press, Princeton, NJ.

MacKenzie, D. I., J. D. Nichols, G. B. Lachmann, S. Droege, J. A. Royle, and C. A. Langtimm (2002). Estimating site occupancy rates when detection probabilities are less than one. Ecology 83:2248–2255.

Malthus, T. R. (1798). An essay on the principles of population. *In* Populations, evolution and birth control: A collage of controversial ideas, G. Hardin, Editor (1964). Freeman, San Francisco, pp. 4–16.

Mayr, E. (1973). Alden Holmes Miller: 1906–1965. Biographical memoir. National Academy of Sciences, Washington, DC.

Miller, A. H. (1931). Systematic revision and natural history of the American shrikes (*Lanius*). University of California Publications in Zoology 38:11–242.

Miller, A. H. (1941). Speciation in the avian genus *Junco*. University of California Publications in Zoology 44:173–434.

Miller, A. H. (1959). Reproductive cycles in an equatorial sparrow. Proceedings of the National Academy of Sciences 45:1095–1100.

Miller, A. H. (1961). Molt cycles in Andean sparrows. The Condor 63:143–161.

Newman, K. B., S. T. Buckland, B. J. T. Morgan, R. King, D. L. Borchers, D. J. Cole, P. Besbeas, O. Gimenez, and L. Thomas (2014). Modelling population dynamics: Model formulation, fitting, and assessment using state-space models. Springer, New York.

Newton, I. (1979). Population ecology of raptors. Buteo Books, Vermillion, SD. Consolidated Amethyst Communications, Toronto.

Nice, M. M. (1939). Watcher at the nest. Dover Publications, New York.

Nice, M. M. (1979). Research is a passion with me: The autobiography of a bird lover. Consolidated Amethyst Communications, Toronto.

Nichols, J. D. (2001). Using models in the conduct of science and management of natural resources. *In* Modeling in natural resource management, T. M. Shenk and A. B. Franklin, Editors. Island Press, Covelo, CA, pp. 11–34.

Nicholson, A. J., and V. A. Bailey (1935). The balance of animal populations. Proceedings of the Zoological Society of London 3:551–598.

Noon, B. R., and K. S. McKelvey (1992). Stability properties of the Spotted Owl metapopulation in southern California. *In* The California spotted owl: A technical assessment of its current status. USDA Forest Service General Technical Report PSW-GTR-133. Pacific Southwest Research Station, Albany, CA, pp. 187–206.

Owen, M., and J. M. Black (1989). Barnacle Goose. *In* Lifetime reproduction in birds, I. Newton, Editor. Academic Press, London, pp. 349–362.

Peacock, D. S., B. J. Van Rensburg, and M. P. Robertson (2007). The distribution and spread of the invasive alien Common Myna, *Acridotheres tristis* L. (Aves: Sturnidae), in southern Africa. South African Journal of Science 103:465–473.

Pianka, E. R. (1978). Evolutionary ecology. 2nd ed. Harper and Row, New York.

Potts, G. R. (1986). The partridge: Pesticides, predation and conservation. Collins, London.

Potts, G. R. (2000). Using the scientific method to improve game bird management and research: Time. National Quail Symposium Proceedings 4:2–6.

Potts, G. R., J. C. Coulson, and I. R. Deans (1980). Population dynamics of the Shag (*Phalacrocorax aristotelis*) on the Farne Islands, Northumberland. Journal of Animal Ecology 49:465–484.

Railsback, S. F., and V. Grimm (2012). Agent-based and individual-based modeling: A practical introduction. Princeton University Press, Princeton, NJ.

Rappole, G. H., G. W. Blacklock, and J. Norwine (2007). Apparent rapid change in South Texas birds: Rapid response to climate change? *In* The changing climate of South Texas: Problems and prospects, impacts and implications, J. Norwine and K. John, Editors. CREST-RESSACA. Texas A&M University–Kingsville, pp. 133–145.

Reynolds, R. T., J. D. Wiens, and S. R. Salafsky (2006). A review and evaluation of factors limiting Northern Goshawk populations. Studies in Avian Biology 31:260–273.

Ricklefs, R. E. (1973). Ecology. Chiron Press, Newton, MA.

Roseberry, J. L., and W. D. Klimstra (1984). Population ecology of the Bobwhite. Southern Illinois University Press, Carbondale.

Sauer, J. R., W. A. Link, J. E. Fallon, K. L. Pardieck, and D. J. Ziolkowski Jr. (2013). The North American Breeding Bird Survey 1966–2011: Summary analysis and species accounts. North American Fauna 79:1–32.

Sayre, N. F. (2008). The genesis, history and limits of carrying capacity. Annals of the Association of American Geographers 98:120–134.

Schwertner, T. W., H. A. Mathewson, J. A. Roberson, M. Small, and G. L. Waggerman (2002). White-winged Dove (*Zenaida asiatica*). Species Account Number 170. *In* The Birds of North America Online, A. Poole, Editor. Cornell Laboratory of Ornithology, Ithaca, NY.

Senguta, S. (1976). Food and feeding ecology of the Common Myna, *Acridotheres tristis* (Linn.). Proceedings of the Indian National Science Academy 42:338–345.

Simberloff, D. (1993). Species-area and fragmentation effects on old-growth forests: Prospects for longleaf pine communities. Tall Timbers Fire Ecology Conference Proceedings 18:1–13.

Taylor, I. (1994). Barn Owls: Predator-prey relationships and conservation. Cambridge University Press, Cambridge, UK.

Thompson, W. L., G. C. White, and C. Gowan (1998). Monitoring vertebrate populations. Academic Press, San Diego, CA.

Tindall, D., C. J. Ralph, and M. N. Clout (2007). Changes in bird abundance following Common Myna control on a New Zealand island. Pacific Conservation Biology 13:202–212.

US Fish and Wildlife Service (2003). Recovery plan for the Red-cockaded Woodpecker (*Picoides borealis*). Second revision. US Fish and Wildlife Service, Atlanta, GA.

Verhulst, P. F. (1838). Notic sur loi que la population suit dans son accroissement. Correspondence in Mathematics and Physics 10:113–121.

Verner, J. (1984). Assessment of counting techniques. Current Ornithology 2:247–302.

Walters, C. J. (1986). Adaptive management of renewable resources. McGraw-Hill, New York.

Wells, J. V., and M. E. Richmond (1995). Populations, metapopulations, and species populations: What are they and who should care. Wildlife Society Bulletin 23:458–462.

Welty, J. C. (1975). The life of birds. 2nd ed. W. B. Saunders Co., Philadelphia, PA.

Williams, B. K., J. D. Nichols, and M. J. Conroy (2002). Analysis and management of animal populations. Academic Press, San Diego, CA.

Wooler, R. D., J. S. Bradley, I. J. Skira, and D. L. Serventy (1989). Short-tailed Shearwater. *In* Lifetime reproduction in birds, I. Newton, Editor. Academic Press, London, pp. 405–418.

CHAPTER 22

Assemblages and Communities

Shannon Farrell and Robert J. Cooper

Community ecology involves the study of the organization and functioning of communities—assemblages of interacting species. As with many areas of science in general and ecology in particular, birds have played a disproportionately large role in the development of community ecology, especially since the 1960s. Birds are conspicuous in their appearance and behaviors, and are therefore relatively easy to observe and count compared with many other organisms. Most birds are diurnally active, vocalize readily, and can even be closely observed when engaged in behaviors such as foraging, feeding, and nesting. Their life histories and distributions are generally well known. Therefore it is not surprising that birds have often been the focus of studies of animal communities. In particular, the work of Robert MacArthur and colleagues that was so influential in ecology frequently focused on avian community ecology (e.g., MacArthur 1958, MacArthur and MacArthur 1961, MacArthur 1965, MacArthur and Wilson 1967), as did the work of John Wiens and colleagues (Wiens 1989 and references therein; box on page 712). In this chapter, we explore this rich history by covering some basic concepts of avian community ecology, community assembly rules, types of species interactions, patterns of species diversity, and current topics in community ecology.

The distribution of a single species is the result of a variety of interacting factors including its evolutionary history, ecological tolerances, the geographical distribution of locations satisfying these tolerances, and how these factors interact with other species, spatial and temporal dynamics, and chance. Learning about avian assemblages and communities includes understanding the diversity of different species living together in an area, how they interact with each other over time and space, how these interactions limit or enable the coexistence of species, and how those interactions may influence the distributions, behavior, survival, and fecundity of participating species and consequently influence populations over time and space.

Bird species occurring in an area can interact in a variety of ways, from occasional encounters to frequent interactions that can influence each species. Interactions may take the form of competition, commensalism, mutualism, parasitism, predation, and neutralism. Species can also co-occur without interacting. Understanding which species co-occur and coexist, whether and how they interact, and how these interactions shape their populations and evolutionary trajectories is an essential piece of the puzzle for understanding the ecology, evolution, and diversity of birds.

There are several ways of subdividing and thinking about groups of species that occur together in an area. For our purposes, an **assemblage** is the group of individuals of all species that occur in the same specified place and time. For example, we might conduct a survey to identify all of the species that constitute the assemblage in an Adirondack bog in northern New York State. We may use this term when we are interested in thinking about, describing, or studying a group of species that occur in an area of interest, without knowing whether or how these species may interact with one another. A **community** is a set of individuals of multiple species **interacting ecologically** in a specified place and time (Ricklefs and Relyea 2014). A community may comprise species interacting in a variety of ways (some of which we will discuss in this chapter), from competing with one another to eating one another to cooperating with one another. We may also describe other collections of species in an area based on specific characteristics they share. A **guild** is a group of species that exploit the same class of resources in a similar way. For example, we might be interested in how the guild of seed-eating species in an area will respond to a drought and the resulting reduction in seed availability.

JOHN A. WIENS

John A. Wiens.
*Photo courtesy of
John Wiens.*

John A. Wiens (born August 29, 1939) is a leading ecologist in the areas of avian ecology, landscape ecology, and conservation. Although a pioneer in the area of landscape ecology, he is perhaps still best known for his important research on arid grassland and shrub-steppe bird communities of the western United States. He grew up in Oklahoma as an avid birder and obtained advanced degrees from the University of Oklahoma (MS) and the University of Wisconsin–Madison (PhD). He joined the faculty of Oregon State University, then the University of New Mexico, and finally Colorado State University, where he is still a University Distinguished Professor Emeritus.

Wiens's shrub-steppe bird work in the 1970s brought into question the role of interspecific competition in structuring biotic communities, a paradigm that pervaded the ecological literature at the time. He and his postdoctoral researcher, John Rotenberry, examined the avian communities of arid grassland and shrub-steppe ecosystems from every conceivable angle: diet, morphology, and especially habitat selection, which was studied at several spatial scales. Wiens followed up this research with his two-volume *Avian Community Ecology*, published in 1989, which is still the definitive work on this subject.

Increasingly, Wiens's work has taken a more applied, conservation focus. He worked (and continues to work) on marine bird response to the *Exxon Valdez* oil spill, and on other problem-based research on land use, habitat loss, and climate change, among other things. In 2002 he left academia to become the lead scientist for the Nature Conservancy, then their chief scientist, then the chief conservation science officer for Point Reyes Bird Observatory (now Point Blue) in California, a position he held until 2012. While with the Nature Conservancy, he noted that "much as we might like to think that conservation and science are joined at the hip, in reality they are not. . . . There remains a gap, or perhaps even a chasm, between the worlds of science and that of conservation practice and action. The challenge is to bridge that gap." He still works and writes, principally in the area of conservation, and at the time of this writing splits his time between his home in Corvallis, Oregon, and the University of Western Australia, where he is a Winthrop Research Professor.

Thinking about avian assemblages and communities requires careful consideration of the unique behaviors and life history characteristics of birds and how these can vary among species. Some bird species remain resident in an area throughout all parts of the year and across multiple years. Others make the same migration twice yearly, shifting locations between a breeding-season location and a winter location, while others move nomadically in response to resources or other environmental conditions (types of bird movements are discussed in more detail in chapter 17). Thus, when we think about assemblages, communities, or guilds, we need to consider where and how bird species may overlap in time and space dynamically. We must ask not only what species are occurring in an area of interest, but also what groups of species may occur at different times of year and how those groups may change seasonally or annually. In addition to birds changing their locations and habitats over time and among seasons, the habitats used by birds are also dynamic, changing over time. Conditions in a given location, from the structure of the vegetation to the types of predators present, can change gradually over time through succession or change dramatically and quickly because of an acute disturbance. Such changes to a location can fundamentally change what bird species occur there and how they interact. This chapter discusses the basic principles that set the foundation for understanding assemblages and communities, addresses the range of species interactions we can find within a community, considers how communities vary over space and time and how they change, and discusses some of the ways that scientists study assemblages and communities.

NICHE, HABITAT, AND RESOURCES

The basic principles of community ecology state that each species has its own unique niche, or set of conditions within which it exists (summarized by Schoener 2009). The term **niche** was first used extensively by Joseph Grinnell in 1917 in his paper "The niche-relationships of the California Thrasher." Grinnell used the term niche to describe the habitat in which an organism lives and the suite of behaviors and adaptations that allows an organism to survive and persist in that habitat (Grinnell 1917). Charles Elton defined a niche as an animal's place in the biotic environment in relation to its food and enemies, with a stronger emphasis on the organism's place in the trophic interactions of an ecosystem, largely based on the foraging behavior of organisms. For example, birds that are gleaning insectivores would be considered to occupy that foraging niche (Elton 1927). While Elton's focus was on trophic interactions, he brought the key idea of **interactions** into the niche concept, considering how one species affected, or was affected by, others. G. Evelyn Hutchinson updated this niche concept in 1957 to capture a wide array of elements that define the conditions and requirements that allow an organism to occur and persist. Hutchinson defined a niche as an n-dimensional hypervolume—a shape or space with many dimensions—where the dimensions represent the many biotic and abiotic conditions, and the levels of those conditions, that enable occurrence of an organism. For example, a species may have a specific range of temperatures within which it can live, as well as a specific range of rainfall conditions, food availability conditions, and so on. (Hutchinson 1957). Hutchinson's niche definition provided a conceptual framework that scientists expanded to explore **niche breadth**, the range of resources or conditions used by a given species; **niche overlap**, the overlap of resource and conditions used by different species; and **niche partitioning**, the separation or differentiation of resources and conditions used by different species (fig. 22.1). These concepts provide the groundwork for exploring the processes that influence community composition and species interactions.

The development of the niche concept has further developed into the key terms we use today. A species' **fundamental niche** is defined by the range of biotic and abiotic conditions that individuals of that species can tolerate and in which they can exist. All species have a range of abiotic conditions within which they can exist, such as temperature highs and lows, humidity, rainfall, or elevation. The range of conditions an organism can tolerate constrains the set of possible locations where it can potentially exist. However, a species may not occupy all possible locations where it could potentially exist based on the range of conditions it can tol-

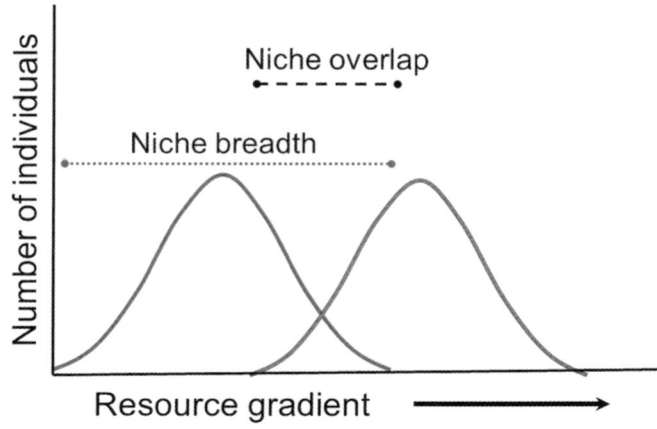

Figure 22.1. Schematic illustration of the niche for two different species along a gradient for a resource or environmental condition. Curves represent the number of individuals of each species that exist at that level of resources. The breadth of the niche represents the full range of conditions within which the species exists. The niche overlap reflects the region in which the two species use the same resource or environmental condition.

erate. Some locations with tolerable conditions may not be occupied by a species for other reasons, constraining a species to what we call their **realized niche**, or the range of conditions within which a species does exist and persist. The presence and behavior of other species, such as predators or competitors, can limit the ability of a species to access its entire fundamental niche, which we discuss further in the next section.

The physical location a species occupies within its realized niche is considered its **habitat**. Within a habitat, the availability and abundance of resources may be important for not only the existence but also the survival, reproduction, and ultimately persistence of a species in a location. **Resources** are elements of the environment that an organism consumes or uses that can influence individual survival or reproduction and thus can influence population growth and species persistence. We can think of resources as potential **limiting factors**, because their availability and accessibility can limit the ability of a species to survive, persist, or reproduce. Some resources may be highly abundant and thus not act to limit the ability of a species' population to persist, whereas others may put important limits on what species may occur in an area, how many different species can co-occur in an area, or how many individuals of any species may be supported. The availability of suitable nesting sites, for example, may limit the ability of a species to exist and persist in an area with otherwise acceptable conditions. In locations where multiple species of secondary cavity-nesters—species that require the presence of preexisting cavities for nesting—may occur, competition for this limited resource may limit the number of individuals and species that can co-occur in the area.

ASSEMBLY RULES AND COMMUNITY COMPOSITION

Imagine a forest that has recently undergone a large, intense wildfire. This disturbance has changed a large portion of the area from mature forest to the early stages of meadow. A range of biotic and abiotic factors can influence or constrain the set of species that can or will assemble in this location. We can think about the various conditions and circumstances that lead to the set of species we might expect to occupy that newly created meadow in a hierarchical fashion through the concept of **assembly rules**.

Evolutionary processes and historical events shape the fundamental niche of a species, its current traits, and current locations. For example, we know that the current distributions of some *Setophaga* warbler species have been partly shaped by Pleistocene glacial cycles. Physiological constraints shape the fundamental niche of a species, restricting species to regions where temperatures, elevations, or other abiotic conditions are physiologically tolerable and allow the

species to exist. Biological barriers may limit the distribution of a species. For example, hummingbird species may be limited to areas where their primary flowering food sources are located, thereby limiting the geographic distribution of the species. These factors shape a species' current distribution and, ultimately, result in the regional **species pool**: the finite set of species that are present in the region that can plausibly make an area, such as our newly created meadow, home (fig. 22.2).

A series of other filters will further influence what species in this available pool ultimately settle in the new meadow habitat. A species' ability to disperse will influence whether it can access this new location (bird movements, including dispersal, are discussed in detail in chapter 17). For our purposes, dispersal distances and barriers to dispersal are important for shaping community composition. Natal dispersal is the movement of birds from their natal area to a new area to occupy and reproduce. Breeding dispersal is the movement of adult birds from one breeding area to another between breeding seasons, irrespective of whether or where to they may migrate during the intervening nonbreeding season. Natal and breeding dispersal allow a species to occupy new areas. Dispersal distances vary among species, and dispersal ability

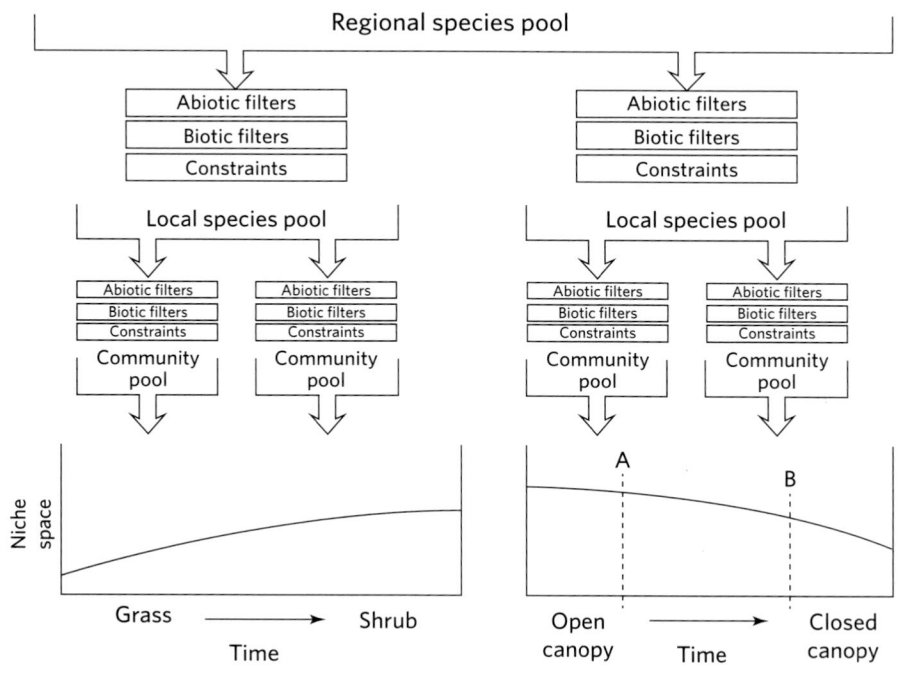

Figure 22.2. The species present in an area are those that occupy the area as a result of the filtering process at the local and regional levels. The type and number of species present across seral stages will also be a reflection of the size of the target area and the niche space available. The squares *A* and *B* indicate how species may change in type and total number over time as the system changes. *From Morrison 2009, used with permission.*

constrains the ability of each species to colonize an otherwise suitable area. Studies suggest that migrant species may disperse farther on average than resident species (Paradis et al. 1998) and that wing-tip length, bill depth, and tail graduation, all possible indicators of aerodynamics, are predictors of dispersal distance among species (Dawideit et al. 2009). However, dispersal distances can vary widely among migrants and residents. Spatial or temporal barriers may also prohibit or limit access to an area by some species. Physical features such as mountain ranges, deserts, large rivers, oceans, or other features may influence the ability of some species to access otherwise suitable locations, and thus may also act as filters controlling what species may settle in a given location. Some species may readily be able to traverse these features, or even use them as movement corridors or navigational tools to access other locations, while other species may be constrained in their ability to traverse these features.

The specific conditions a species selects, in part influenced by the set of resources a species requires, will influence the range of locations it may choose to occupy. For example, a species that requires mature trees for nesting or insects found on tall conifers for foraging will not find the necessary resources at a newly created meadow. We know from chapter 18 that factors that influence bird habitat selection can include a wide array of factors, from landscape conditions such as the amount of human development in the area to local microsite conditions such as suitable forks in tree branches for nest placement to the presence and density of conspecifics or heterospecifics. Some species may not occupy an area even when it is accessible, conditions appear to be within a tolerable range, and the right types of resources are present. For example, species that use forests for nesting habitat may fail to settle in forest patches with seemingly suitable conditions and available nesting sites and food, if the forest patch is too small. Area sensitivity may prevent some species from using a location, and thus factors such as patch size may constrain the set of species that can be part of an assemblage or community in a given location.

Species that can and do choose to settle in a location such as our newly created meadow will encounter other species. Interactions between these species, including predation and competition, will govern whether the species can exist and persist in this new location.

SPECIES INTERACTIONS

Colonization of an area by a species and continued occurrence of the species is shaped not only by the presence of the biotic and abiotic filters that influence a species' ability to access and settle in an area, or the presence of the right kind of

Interaction Type	Effect on Species 1	Effect on Species 2
Neutralism	0	0
Commensalism	+	0
Mutualism	+	+
Competition	-	-
Amensalism	0	-
Predation and parasitism	+	-

Figure 22.3. We can think about each of these categories of interactions based on which parties in the interaction experience a positive effect or negative effect.

resources that result in habitat selection by an organism. Occurrence and persistence of a species in an area is also mediated by species interactions. Species occurring in an area can interact in a variety of ways that shape their ability to persist and coexist in an area.

Species interactions can result in positive, negative, or neutral outcomes for each species involved in the interaction. There are six main types of interactions between species: **competition** and **amensalism**; **antagonism** (including **predation and parasitism**); and symbiotic relationships, including **mutualism**, **commensalism**, and **neutralism**. We can think about each of these categories of interactions based on which parties in the interaction experience a positive effect or negative effect (fig. 22.3). The consequences of these interactions shape the composition of assemblages and communities, influence the range or distribution of a species and affect populations, and can affect a species' genotypes, phenotypes, and evolutionary trajectory.

Competition

Resources in a habitat are typically not infinite, and thus individuals may compete for limited resources. When competition occurs, both individuals involved in the interaction may experience a reduction in fitness, such as a reduction in fecundity or survival, or the interaction may be experienced as an amensalism, in which one species may be negatively affected while the other experiences no effect. Competition may occur between individuals of the same species, called **intraspecific competition,** or between individuals of different species, called **interspecific competition.** Individuals that are more successful competitors will have increased success in passing their genes on within their population of conspecifics.

The process of interspecific competition can drive adaptation, separating the niches of would-be competitors and thus shaping the evolution of species and the composition of assemblages and communities.

The effects of competition may result from **exploitative competition**, the exploitation of available resources by one individual to an extent that it reduces the availability or quality of those resources to other individuals, or from direct interference in an individual's ability to access resources through **interference competition**. Exploitative competition can occur over any resource, such as food or suitable nesting sites, in which use by one species can reduce availability of that resource for another species. For cavity-nesting species, suitable nest cavities can be a limited resource subject to competition (see below). Territoriality in birds can be an example of both exploitative and interference competition. When a bird establishes and defends a territory, it uses resources such as space in a way that reduces available resources for others. But even if the territory holder does not use all of the resources, by preventing access to the space, it interferes with another individual's ability to access available food or nest-site resources within the territory, a form of interference competition. As an example of the two types of competition, Edworthy (2016) studied the endangered Forty-spotted Pardalote (*Pardalotus quadragintus*) and the more common Striated Pardalote (*Pardalotus striatus*), Tasmanian songbirds that compete for limited nest cavities. The resident Forty-spotted Pardalote nests earlier, leaving fewer available cavities for the partially migrant Striated Pardalote to use, an example of exploitative competition. However, Striated Pardalotes may be stronger competitors, securing 17 percent of available cavities and taking over about 10 percent of nest cavities already in use by Forty-spotted Pardalotes, an ex-

ample of interference competition. Some scientists are concerned that the dominance of Striated Pardalotes may play a part in the declines of the Forty-spotted Pardalote in many parts of Tasmania (fig. 22.4).

How does one species cope with another competing species? The **competitive exclusion principle**, also known as **Gause's principle**, asserts that two or more species with similar niches cannot coexist in a stable equilibrium (Gause 1934). The principle suggests that when multiple species compete for the same resources, one will be a better competitor and will experience greater survival and reproduction. The poorer competitor will follow one of two trajectories: (1) The poorer competitor may experience **competitive exclusion**, in which it may be eliminated or excluded from the area or potentially driven to local extinction; (2) the poorer competitor may adapt over time to minimize overlap in resource use with competitors, including the evolution of new traits that enable coexistence, called **character displacement** (fig. 22.5). The evolutionary process of character displacement is thus said to result in the ecological phenomenon of niche partitioning, or the distinction between species in their niche characteristics. Studies of several dabbling duck species suggest that breeding ducks that coexist on a coarse scale (e.g., co-occur on the same lake or pond), segregate along a gradient of body size. Researchers suggest some ducks have evolved to be larger to gain access to prey items in deeper waters, rather than compete for prey in locations where prey is accessible only to smaller ducks (Nudds and Bowlby 1984, Pöysä et al. 1994, Nudds et al. 2000). Alternatively, in flat lakes with limited variation in depth where small and large ducks may all be able to reach the sediment for food, how do they segregate to avoid competition for food resources? Research suggests that character displacement has resulted in differing lamellar

Figure 22.4. Examples and consequences of competition for nest cavities between Forty-spotted and Striated Pardalotes in southeastern Tasmania, Australia, 2013–2015. *A,* Forty-spotted Pardalotes in aerial combat on Bruny Island, Tasmania, Australia. *Photo by Alfred Schulte. B,* Striated Pardalote defending a recently usurped Forty-spotted Pardalote nest cavity. *Photo by Mick Brown. C,* Forty-spotted Pardalote eggs punctured and removed from a cavity by Striated Pardalotes (Edworthy 2016). *Photo by Amanda Edworthy.*

Character Displacement

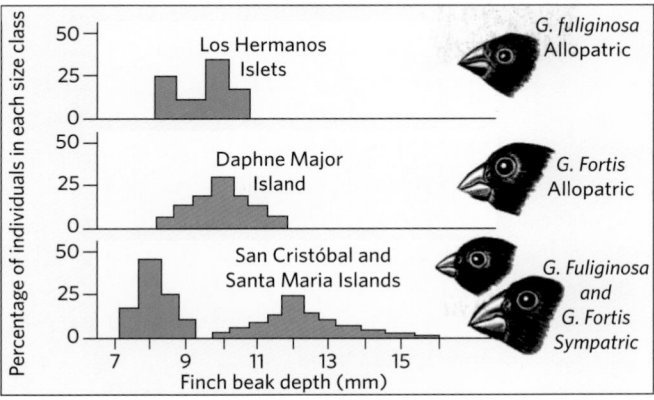

Figure 22.5. The evolution of new traits that enable coexistence, called character displacement. The two species of Darwin's finches (genus *Geospiza*) in the figure have bills of similar size where the finches are allopatric, each living on an island where the other does not occur. On islands where they are sympatric, the two species have evolved beaks of different sizes, one adapted to larger seeds, the other to smaller ones. *From Raven and Johnson 2002. Used with permission from McGraw-Hill Education.*

densities, meaning that differing filter sizes allow different species to obtain and process different-sized prey. Alternatively, in some systems we can see two or more competing species coexist in non-stable, or nonequilibrium conditions, which may be facilitated by dynamics in the system that alternately favor one competitor or another and change with time. This can include disturbance-mediated or predator-mediated coexistence, which we discuss later in the chapter.

Robert MacArthur (box on page 718) was among the first scientists to describe an example of apparent niche partitioning among five species wood warblers (MacArthur 1957, 1958). All five species occupied mature coniferous forests of northeastern North America during the breeding season. These species are congeneric (i.e., in the same genus), similar in size and shape, insectivorous, and appear to be highly similar in general habitat associations. It appeared that these species might be an exception to Gause's principle. MacArthur was interested in determining whether these five species did indeed occupy identical niches and coexist. He predicted that these species were likely to compete with each other for resources such as insects and thus would show evidence of niche partitioning. MacArthur devised an observational study, recording the feeding positions of each warbler species in the trees, including the distance down from the top of the tree and outward from the trunk, to investigate whether the warblers partitioned resources by foraging on different parts of the trees or using different foraging behaviors. MacArthur observed that each warbler species differed in its use of regions of the tree and in foraging style, concluding that "the

birds behave in such a way as to be exposed to different kinds of food," suggesting they were exploiting slightly different insect resources and, thus, occupying different niches (fig. 22.6).

In 1980, Bradford Lister observed foraging in four species of tits in the family *Paridae* coexisting in the pine forests at Thetford Chase, United Kingdom (Lister 1980). He found that during periods of scarce food, the tit species had more spatially distinct foraging locations. However, the extent of spatial overlap between the species increased as food availability increased. Lister's results, like MacArthur's, could be interpreted in multiple ways. We could conclude that the multiple species co-occurring in the same habitat were able to do so as a consequence of intense competition having driven them to partition resources to reduce overlap and allow them to feed without excessive competition. Alternatively, we could conclude that these species used different resources as a result of their evolutionary history, and their co-occurrence was a function of each species occurring in an area where its foraging resources were available, which happened to overlap with the foraging areas of interest for the other species as well, resulting in a large amount of spatial overlap among the species' foraging locations. It can be difficult to determine whether competition, character displacement, and niche partitioning drive co-occurrence in communities.

In fact, simply determining that two species do in fact compete for a resource is challenging. First, we must consider whether the resource of interest is available at levels that result in potential competition for that resource. In some cases, a resource may be sufficiently abundant that populations of multiple species within a community can exploit the resource without impeding the ability of other species to access or exploit the resource. Where a resource is or may be

Niche partitioning among five species of coexisting warblers

Figure 22.6. This diagram illustrates the niche partitioning among coexisting warbler species studied by MacArthur. MacArthur observed that each warbler species differed in its use of regions of the tree and in foraging style, suggesting they were exploiting slightly different insect resources and thus occupying different niches. *From Raven and Johnson 2002. Used with permission from McGraw-Hill Education.*

ROBERT H. MACARTHUR

Robert MacArthur. *Photo courtesy of Marlboro College.*

Robert H. MacArthur (April 7, 1930, to November 1, 1972) was a preeminent evolutionary ecologist of his time, best known for blending population and community ecology, biogeography, and genetics into a common fundamental theory that we now know as population biology. Much of his work featured birds as model organisms. He began his college education in mathematics, obtaining an undergraduate degree from Marlboro College in Vermont and a master's degree from Brown University. For his 1957 PhD, however, he blended his mathematical background with ecology, working under the direction of famed theoretical ecologist G. Evelyn Hutchinson at Yale University. His dissertation involved a field study of the foraging ecology of five warbler species, now commonly known as "MacArthur's warblers," which

exemplified the concepts of niche partitioning among a group of organisms with seemingly identical niches, thus allowing them to coexist. Although this famous article and the many others it influenced were to become controversial, there was no denying the elegant and powerful blend of mathematics and field ecology it featured. Notably, after obtaining his PhD, MacArthur spent the 1957–1958 academic year at Oxford University, furthering his field ornithological skills with prominent ornithologist David Lack.

From 1958 to 1965, MacArthur quickly advanced through the ranks to full professor at the University of Pennsylvania, then moved to Princeton University, where he remained until the end of his career. The 1960s were a period of intense activity for MacArthur, during which he authored and coauthored many articles in the general area of population biology. His publications often sought to describe and explain local to global patterns of species diversity, nearly all of them reflecting his unique skills as mathematician-naturalist. Many involved birds. None of these contributions, however, were more influential than the book *The Theory of Island Biogeography*, which he coauthored with his student E. O. Wilson. One of the first major contributions to the field of landscape ecology, the book also served as a cornerstone of the new field of conservation biology because of its relevance to understanding extinction and the planning of natural reserves.

Science suffered a great loss in 1972 when Robert MacArthur was taken from us at the young age of 42. But his influence lives on through his highly influential publications and through the work of his students, many of whom are among the most prominent ecologists of our day.

---

limited enough that use by one species may limit its availability to another, we would predict competition for that resource. In the case of the Forty-spotted Pardalote and the Striated Pardalote, nest cavities do indeed appear to be limited enough to result in competition. We can conduct studies to determine the resource utilization of one or more species and to compare differences in resource utilization to look for evidence of niche partitioning.

However, it is often difficult to determine whether differences in resources utilization are the result of competition-driven niche partitioning or character displacement, or simply the result of multiple species in an area that evolved

unique resource preferences as a result of other selection pressures and evolutionary trajectories. Connell (1980, 1983) noted that patterns of resource partitioning and coexistence that we observe today may be the result of the "ghost of competition past," or previous competitive pressure. But we cannot say with certainty that past competition was the driver of observed patterns of coexistence today. Species may evolve to exploit their unique niches because of a variety of forces that may not include competition with their coexisting neighbor throughout their evolutionary history. In the case of *Sylvia* warblers, researchers who conducted another study investigating the co-occurrence of several *Sylvia* warbler species in Northeast Catalonia observed that the density of the other *Sylvia* species did not appear to be an important factor associated with habitat use and that *Sylvia* coexistence is likely the result of ecological segregation—separation of their preferred niche—because of factors other than interspecific competition (Pons et al. 2008). How would we test the hypothesis that competition and resulting niche partitioning drives the ability of similar species to coexist? Testing such a hypothesis is likely to require manipulative experimentation to remove one or the other species on multiple sites and observe patterns of resource use.

Antagonistic Interactions: Predation and Parasitism

Antagonistic interactions are interactions that result in a positive outcome for one of the interacting species and a negative outcome for the other participant. In the case of predation, predators benefit by obtaining food resources by consuming a prey item, and the prey item experiences the ultimate negative effect, mortality. Parasitism results in a positive outcome for the parasite who obtains resources or other benefits from the host. But unlike predation, the host species may experience a range of negative effects including sublethal effects such as reduction in body condition or fecundity or, eventually, mortality. Parasitism and predation can act to shape the composition and dynamics of communities in several ways.

Predation

Predators can play a range of roles that influence what species occur in an area. The presence of a predator species may prevent a prey species from selecting an otherwise suitable

A REMOVAL EXPERIMENT DEMONSTRATES ECOLOGICAL AND FITNESS CONSEQUENCES OF COEXISTENCE FOR TWO WARBLER SPECIES

Species in an assemblage may or may not interact in ways that affect each other. Determining whether interactions occur and quantifying the fitness consequences of interactions based on observational studies alone can be challenging. Removal experiments allow researchers to investigate the consequences of coexistence by observing changes in resource use, behavior, survival, or reproduction of one species when the other is removed. Orange-crowned Warblers (*Vermivora celata*; OCWA) and Virginia's Warblers (*V. virginiae*; VIWA) in central Arizona have overlapping breeding territories and similar nesting, foraging, and predator ecology (Martin 1988, 1993, 1998). Martin and Martin (2001) removed unpaired territorial male OCWA or VIWA in study plots from 1996 to 1998 to examine whether coexistence of these two warblers affects access to nest sites or food resources and depredation of nests and adult females, and whether any ecological effects resulted in fitness consequences (i.e., effects on clutch size, number of young fledged, or adult survival). Where VIWA were removed, OCWA showed lower nest predation rates compared with control plots where VIWA were present. On plots where OCWA were removed, VIWA shifted their nest sites to sites much more similar to OCWA nest sites (see figure), increased feeding rates of hatched young, and showed reduced nest predation, compared with control plots where OCWA were present. Both species fledged 78–129 percent more young per nest on plots where the other species was removed, indicating that both species experience substantial fitness costs of coexistence.

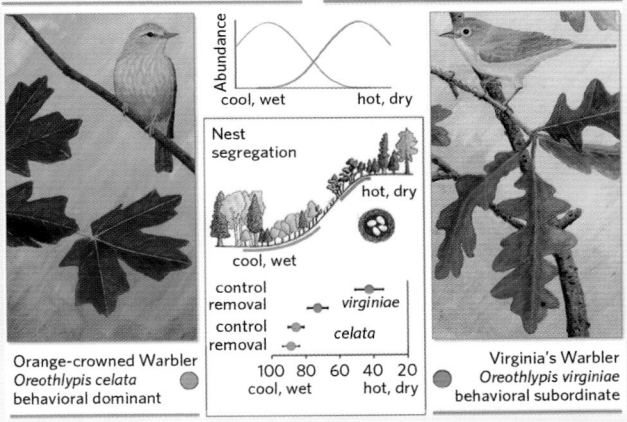

Orange-crowned Warbler
Oreothlypis celata ●
behavioral dominant

Virginia's Warbler
Oreothlypis virginiae ●
behavioral subordinate

The diagram depicts segregation of Orange-crowned (*Oreothlypis celata*, blue) and Virginia's (*O. virginiae*, red) warblers (*Parulidae*) along gradients of temperature and moisture during the breeding season. The larger Orange-crowned Warbler is dominant in aggressive interactions and excludes the smaller Virginia's Warbler from preferred nest sites in cool, moist regions of their breeding territories. The smaller Virginia's Warbler can tolerate hot, dry nesting conditions that the Orange-crowned cannot, leading to spatial partitioning of nest sites where their breeding territories overlap. Experimental removal of the opposite species resulted in the Virginia's Warblers shifting their nest sites into the cool, wet regions of the gradient in the absence of Orange-crowned Warblers. But no shift in Orange-crowned Warbler nest sites in the absence of Virginia's was observed (Martin 1998, Martin and Martin 2001). *Martin, Ghalambor, and Woods 2015. Copyright © 2015 by John Wiley & Sons, Inc. Reproduced with permission of John Wiley & Sons, Inc.*

area, or may prevent a species from persisting in an area it occupies by reducing survival and reproduction, resulting in local extirpation. However, predators can have more subtle, complex effects on an assemblage or community. For example, a generalist predator that depredates individuals of all or most prey species in a community may cause overall reductions in the abundance or density of all species present. Specialist predators that prefer a particular prey species can thus influence the abundance, density, or behavior of one species in the community while not affecting others in the same way. By influencing the abundance, density, or behavior of particular species in a community, the presence of predators can fundamentally influence the other interactions among species (box on page 721).

In some cases, predation may reduce the population of a dominant competitor enough to allow the coexistence of a poorer competitor in an area. This phenomenon is called predator-mediated coexistence. Predator-mediated coexistence is hypothesized to be a factor that can enhance the number of species that can coexist in an area by influencing the dynamics among the species. For example, predation may result in changes to the population of one or more species in the system, resulting in cascading differences in how the coexisting species interact. A study conducted by Kullberg and Ekman (2003) on nine Scandinavian islands suggested that predation by Eurasian Pygmy Owls (*Glaucidium passerinum*) influenced the potential for coexistence of multiple species in the family *Paridae*. The five islands without Pygmy Owls were occupied by only one of the species, the Coal Tit (*Periparus ater*). The four islands occupied by the Pygmy Owls were also occupied by the Coal Tit, but also the Willow Tit (*Poecile montanus*) and the Crested Tit (*Lophophanes cristatus*). The researchers suggest that where the Pygmy Owls are present, predation by the owls may reduce the density of Coal Tits, the dominant competitor, enabling the other *Parid* species to persist. The presence of predators can also result in changes to prey behavior that may influence the interactions among prey species. For example, predation pressure may lead individuals of a prey species to spend increased time hiding, being vigilant and watching for predators, and potentially decreased time actively foraging, seeking mates or conducting courtship activities, nest building, or feeding young, resulting in changes in their interactions with other species in the community.

Parasitism

Parasitism is an interaction that results in benefits to one participating species and deleterious effects for the other that can range from minor reductions in fitness to reduced survival or mortality. Parasitic interactions can take several forms in birds, from brood parasitism to kleptoparasitism.

Like predation, parasitism can influence the composition and dynamics of a community in several ways.

Brood parasitism, the behavior in which one bird lays an egg in the nest of another bird so that their offspring is raised by the host species, is one common form of parasitic interaction observed in birds. In chapter 16, you learned about reproductive behavior that includes intraspecific and interspecific brood parasitism. Because we are focused on interactions among different species in this chapter, we focus here on interspecific brood parasitism. Brood parasites may be considered obligate or non-obligate brood parasites. Obligate brood parasites are those that have evolved such that they must act as parasites and cannot raise their own young. Most interspecific brood parasites are obligate brood parasites, so parasitism is required for the survival and persistence of their species. This interspecific form of brood parasitism has evolved independently at least seven times in bird lineages. Some species of brood parasites are specialists, parasitizing only one or few host species, while others are generalists that can parasitize numerous other bird species. So how do brood parasitism interactions affect bird assemblages and communities? First, the species composition of an assemblage or community may influence the likelihood that a parasite will choose to join the community and use an area for breeding. For example, some research suggests that cowbirds may choose what areas to occupy based, at least in part, on what potential hosts may be present or how many hosts may be available to parasitize. But how do parasites then affect the community in which they choose to live and parasitize?

Many songbird populations are deleteriously affected by parasitism by Brown-headed Cowbirds (*Molothrus ater*) through a variety of mechanisms that cause negative effects to hosts, from minor reductions in fecundity to complete nest failure. Effects vary by host species; some hosts suffer relatively minor effects while others experience significant fitness costs that can affect populations and species persistence over time. Brown-headed Cowbird adults may damage or remove host eggs from a nest before laying an egg in the nest, leading to a direct reduction in host fecundity. Brown-headed Cowbird eggs typically have a short incubation period relative to hosts, meaning that cowbird eggs typically hatch before host eggs and are able to acquire food resources from parents and grow large before host eggs hatch. Once host eggs hatch, host young may be smaller, younger, and at a competitive disadvantage, and in many cases host young may be outcompeted for parental food deliveries or even trampled and killed by larger, more active cowbird young.

Another form of parasitism we observe in birds that benefits one species and negatively affects the other is called **kleptoparasitism**. The word kleptoparasitism literally means theft

A BEHAVIORALLY DRIVEN TROPHIC CASCADE BETWEEN OWLS AND SONGBIRDS

Predators can affect avian prey through direct predation or more indirectly, by influencing their behavior, movement, or activity levels. In systems with multiple tiers of predators, top predators can affect prey by altering the behavior of mesopredators.

Schmidt (2006) explored the possibility of a behaviorally driven trophic cascade from owls to the known nest predator, the white-footed mouse, to the Veery (*Catharus fuscescens*) in eastern temperate deciduous forests of southeastern New York. Previous studies have shown a close correlation between spatial activity in mice and nest predation rates on ground-nesting birds. Moonlight is a known deterrent of activity in nocturnal rodents. Schmidt investigated long-term nesting records for the Veery to test whether nest predation rates were correlated negatively with moonlight (the first figure below). For the first half of the lunar cycle (full moon to new moon) predation rates decreased with moonlight as predicted. To test whether these patterns in predation on Veery nests, associated with moonlight, were caused by differences in mouse activity as mice sought to avoid predation during high moonlight periods, Schmidt conducted an experiment. Schmidt presented owl vocalizations, a direct cue of predation risk, to white-footed mice (*Peromyscus leucopus*) during either a new or full moon. Mice reduced their activity in space by nearly two-thirds in response to playbacks of owl vocalizations during a full moon (the second figure below). However, neither moonlight nor the presence or

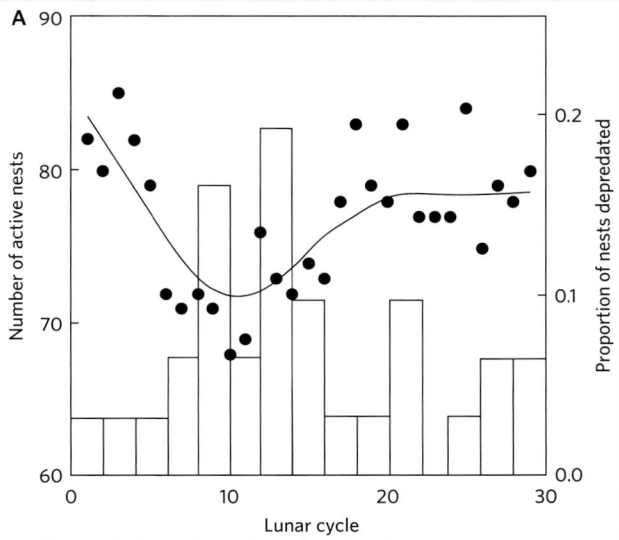

Pattern of Veery nest predation (1998–2004) over the lunar cycle. Each circle represents the nest daily mortality rate for a three-day interval whose mid-date is given next to its respective data point (day 1 = full moon). Note that the left-most data point begins just prior to the full moon (Schmidt 2006).

absence of owl calls had an effect on mouse space use when each cue varied alone. Mouse space-use behavior showed a change only when owl calls were present in conjunction with moonlight.

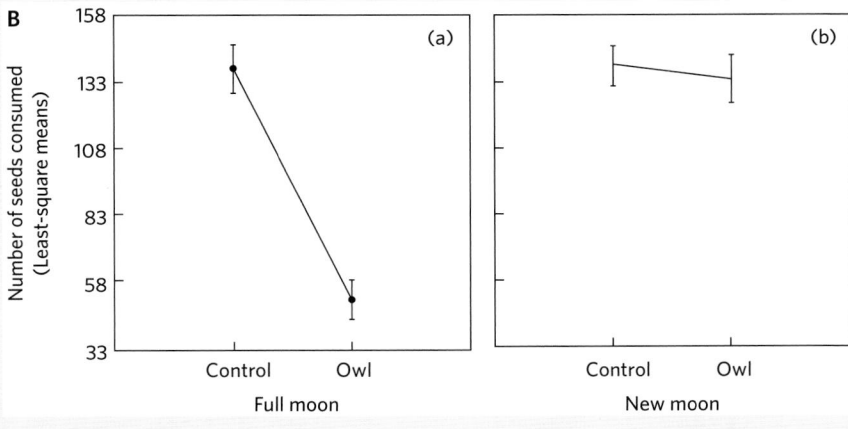

Number of oat seeds (least square mean ± SE) consumed by mice as a function of the two experimental treatments: moonlight (full vs. new) and owl playback (present vs. absence or control; Schmidt 2006).

parasitism. Kleptoparasitism is a form of feeding in which one individual takes food from another individual. Some species of birds feed in part or almost entirely by stealing food items from another species. For example, skuas, seabirds in the genus *Stercorarius*, obtain close to 95 percent of their food through kleptoparasitism. They commonly steal food from gulls, terns, and other seabirds. As such, it benefits skuas to live in communities where sufficient numbers of prospective victims as present. Frigatebirds, seabirds in the genus *Fregata,* use kleptoparasitism to steal an estimated 5 percent of their total food needs from other seabirds, obtaining their remaining food by foraging and catching food items themselves. The American Wigeon (*Mareca americana*) is nicknamed the Robber Duck or Poacher Duck because it steals vegetation it cannot reach from other waterfowl including Redheads (*Aythya americana*), Canvasbacks (*Aythya valisineria*), and American Coots (*Fulica americana*). This behavior benefits the thief, providing food items for which they did not need to search or hunt, and hinders the victim, who loses a food item they use energy and time to obtain. We can imagine how this interaction can shape communities. Species that rely heavily on stealing food from other species must find suitable communities in which to live that have a sufficient number of prospective species and individuals from which they can steal food.

Positive Interspecific Interactions: Mutualism and Commensalism

Competition between heterospecifics (i.e., individuals of different species) has long been the focus of research investigating coexistence of species. Research is increasingly finding examples that demonstrate that positive interactions among heterospecifics, including mutualisms and commensalisms, can also shape assemblages and communities and facilitate coexistence. The ways species may interact in positive, noncompetitive ways are numerous.

Mutualism

A mutualism is an interaction between two species in which both species obtain a positive effect from the interaction. When we think about mutualistic interactions, we often think of examples that include a bird species and another plant or animal species. For example, Oxpeckers, including two species in the genus *Buphagus*, provide cleaning symbiosis with large mammals hosts such as cattle, zebras, impalas, or rhinoceroses. Oxpeckers are phagophiles, feeding on ectoparasites of other species and consuming a blood meal from any wounds they may open, while the mammal hosts benefit from the removal of harmful ectoparasites.

When it comes to interactions between bird species, species may interact in several positive, noncompetitive ways.

Cooperation among individuals of the same species, particularly among highly related individuals, is observed in several avian behaviors including cooperative breeding and cooperative nest defense. However, cooperation among heterospecifics can be an important part of the puzzle in understanding the coexistence of many species in a community. We can observe cooperative, mutually beneficial interactions between species most commonly where we see aggregations or groupings of birds. One of the most common forms of aggregation we see in birds is the formation of flocks comprising conspecifics or heterospecifics. Being part of a flock brings a range of potential costs and benefits. You can read more about flock formation, behavior, and ecology in chapter 15 and chapter 17. For our purposes, mixed-species foraging flocks provide a clear example in which interactions among multiple species can provide benefits for multiple species participating, including improved foraging efficiency and enhanced protection from predation; with many individuals and many eyes looking for food and looking out for predators, all flock members can benefit. Colonial nesting is another example of an aggregation of multiple individuals and, in some cases, of multiple species. Nesting in an aggregation of multiple species can provide benefits and costs similar to those described for flocking (you can read more about colonial nesting in chapter 16). When considering communities and assemblages, we see that community composition may be influenced by the dynamics driving and resulting from colonial nesting and flocking. In cases where the benefits outweigh the costs, co-occurring with an aggregation of con- or heterospecifics may be beneficial.

Mobbing is a cooperative bird behavior employed by groups of conspecific or heterospecific individuals in which multiple individuals perform a joint antipredator assault in order to drive off, confuse, frighten, or otherwise disable a predator using calls, movements, chasing, or other behaviors. Cooperation between closely related conspecifics in a population is consistent with our understanding of kin selection. However, mobbing cooperation among highly unrelated heterospecifics is yet to be fully understood. We would predict that interactions among species in groups, like mobbing, must have positive or neutral fitness consequences for participants; otherwise we would not expect these behaviors to exist and persist. Living in close proximity to heterospecifics, and joining a mobbing group, may be beneficial if the mob is more effective at reducing predation risk than a single individual or species, thus providing a mutual benefit to participants. However, it is difficult to determine the ultimate fitness consequences for individuals engaging in these behaviors, and thus more study is needed to understand the proximate and ultimate drivers of these behaviors and how the costs and benefits of mutualistic interactions between species can act to shape communities.

Commensalism

Commensalism is a form of interaction in which one species obtains a positive benefit from the interaction while the other receives a neutral—neither a positive nor a negative—effect. We can find a variety of interactions among bird species that fall into this category.

We can observe a commensal interaction among bird species in cases where one or more species chooses to nest near or among aggressive heterospecifics, serving as a sound antipredator strategy. For example, in a colony of breeding seabirds at Triangle Island, off British Columbia, Canada, breeding success of Common Murres (*Uria aalge*) and Pelagic Cormorants (*Phalacrocorax pelagicus*) was highest in years when they shared the nesting area with several pairs of Peregrine Falcons (*Falco peregrinus*). The falcons aggressively defended their nests from larger predators such as Bald Eagles (*Haliaeetus leucocephalus*), and the nesting seabirds experienced the benefit of this defense in the form of fewer visits and attacks by eagles and other predators (Hipfner et al. 2011). In this case, the murres and cormorants obtain a benefit from living in a community that includes Peregrine Falcons, by deriving antipredator benefits. The falcons do not obtain any direct benefit or costs of this interaction. As with competition, it can be difficult to determine whether a commensal interaction is an important factor explaining the coexistence of two or more species. How might you design an experiment that would test the hypothesis that this antipredator commensalism was the primary factor explaining why these species occur in the same area?

Another class of interactions among bird species that is increasingly being investigated involves social information use. Individuals need to make many decisions about where to forage, find mates, set up territories or next sites, or access other important resources. Birds can use a variety of information to make choices about where to nest, where to forage, how to respond to predators, and other behaviors (you can read more about the process of habitat selection in chapter 18). It is often difficult or impossible to collect complete, perfect information to make optimal decisions about a prospective foraging or nesting location. The presence, behavior, and vocal communication of other birds, including other species, can be a useful and efficient source of information. Individuals of one species can benefit by observing the choices or behaviors of other species, or by eavesdropping and using information shared by other species, generally at no cost to the information provider. Consider the use of chickadee alarm calls by other species. Black-capped Chicka-

dees (*Poecile atricapilla*) produce informative alarm calls that provide information about the type and size of predators they detect (Templeton et al. 2005). Other species, such as Red-breasted Nuthatches (*Sitta canadensis*) use the informative chickadee alarm calls, responding appropriately based on the encoded information about the type and size of predator. Red-breasted Nuthatches benefit from living in a community with Black-capped Chickadees, in part by benefiting from eavesdropping on their alarm calls, at no apparent cost to the chickadees.

Another example in which we see information use forming a commensal relationship occurs in the process of nest site selection. Often, we expect that a species may prefer to avoid locations where they observe heterospecifics that share similar resource needs, because of the threat of competition. However, research has observed that many species respond positively to the presence of heterospecifics with similar resource needs, in some cases preferentially choosing locations where these heterospecifics are present, often referred to as habitat copying. Researchers Seppänen and Forsman (2007) conducted a fascinating experiment investigating the use of heterospecific social information in nest-site choices of migrant Collared Flycatchers (*Ficedula albicollis*) and Pied Flycatchers (*F. hypoleuca*) that observed resident Great Tits (*P. major*) and Blue Tits (*P. caeruleus*) on study sites in Sweden and Finland. They found that, particularly for later-arriving migrants, the migrant flycatchers frequently copied the nest-site choices of the resident tits, even when there were no detectable ecological differences between nest-site options (box on page 724).

Facilitation is another example in which one bird species may benefit from interacting with another, at no cost to the second participant. For example, primary cavity-nesters, such as many species of woodpeckers, actively excavate cavities in trees for foraging or nesting purposes. Secondary cavity-nesters, such a several species of chickadees (*Poecile* spp.), do not themselves excavate cavities but require existing cavities for nest sites. Thus they benefit from the presence of woodpecker species that facilitate their use of an area by creating the cavities that can later be used by secondary cavity-nesters. The presence, abundance, and activity of woodpeckers can facilitate the occurrence of secondary cavity-nesting species. The woodpeckers, meanwhile, neither benefit nor experience a negative effect from this service.

PATTERNS OF SPECIES DIVERSITY AND RICHNESS

The composition of assemblages and communities can be influenced by a range of spatial and temporal dynamics, from

INTERSPECIFIC SOCIAL LEARNING: NOVEL PREFERENCE CAN BE ACQUIRED FROM A COMPETING SPECIES

Studies by Seppänen and Forsman (2007) and Seppänen et al. (2010) demonstrated that one way co-occurring species interact is by sharing information. In their studies, they experimentally demonstrated that two migrant flycatcher species observed and copied the nest site preferences of resident tits (family *Paridae*).

The researchers set up pairs of nest boxes for these cavity-nesters on multiple study plots. The resident tits were the first to select nest sites and begin nesting. When a mating pair of resident Great Tits (*Parus major*) or Blue Tits (*Cyanistes caeruleus*) chose one of the two nest boxes in a pair to nest, the researchers gave that nest box a geometric symbol (randomly assigned on each study plot), to provide a neutral cue representing the chosen box (see first figure below). New pairs of empty boxes were presented to the later-arriving migrant Collared Flycatchers (*Ficedula albicollis*) and Pied Flycatchers (*F. hypoleuca*), with each box in the pair having one of the two symbols. By pairing the boxes, the researchers ensured that the boxes were otherwise in ecologically similar conditions, and the difference between them was only the neutral geometric symbol. In the 2010 experiment, researchers also documented the number of tit offspring, to see whether flycatchers chose the boxes with symbols that matched with those where the tit species produced the most offspring

Female flycatcher choices matching the tit preference became increasingly frequent as the season progressed, and females tended to choose boxes with symbols that matched those of tits who produced the greatest number of offspring

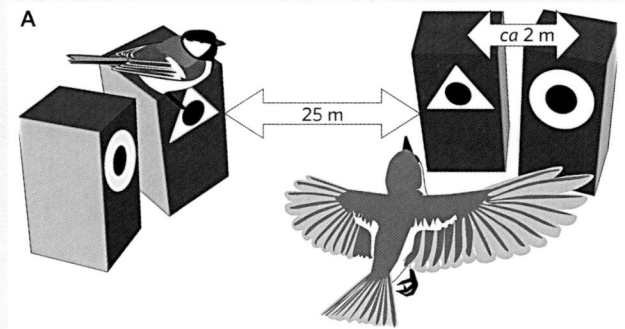

Schematic representation of the experimental design setup. A symbol was randomly assigned to the tit nest, and the alternative symbol was attached to the adjacent empty box (depicted on left of diagram). Two empty boxes with the alternative symbols were offered for arriving flycatchers approximately 25 m away, adjacent to each other (depicted on the right of diagram). The two boxes fall within the small territory defended by male flycatchers, forcing the female to choose in which to build its nest. The tit nest and its accompanying empty box were always visible from the flycatcher boxes (Seppänen et al. 2010).

(see second figure below). The researchers suggest that later-arriving females may find it more efficient to rely on information from tit preferences to select nest sites if they have limited time to assess conditions, survey options, and make a nest-site selection. Additionally, the female flycatchers may use the number of tit offspring as an indicator of the quality of the tit choices and thus decide to copy the tit nesting preferences accordingly.

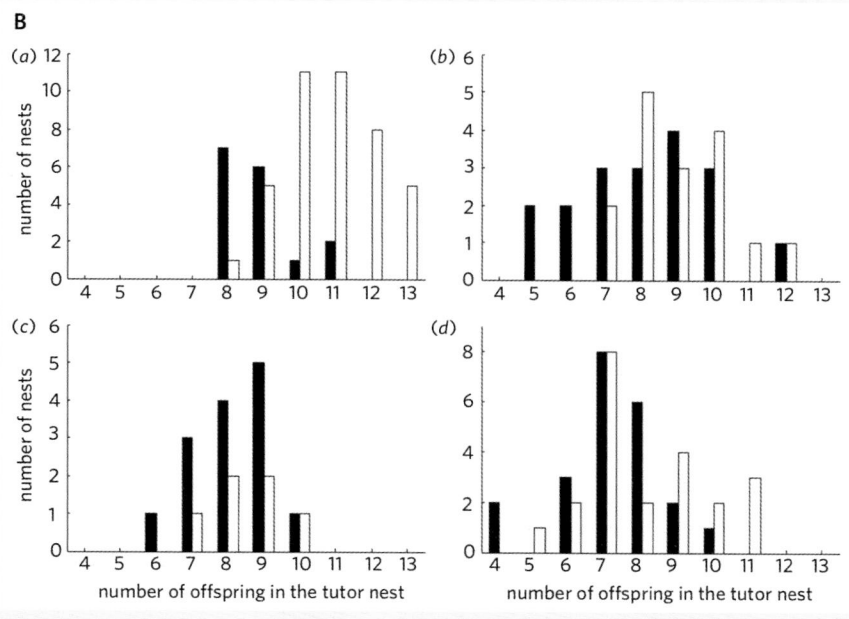

The choices of Pied Flycatcher females in the four study areas. White bars denote number of flycatcher nests in the nest boxes with a symbol matching the symbol on the tit tutor's nest box. Black bars denote nests in the nest boxes with an opposite symbol. X-axis shows the number of offspring in the tit tutor nest on the day when presence of nest material indicated flycatcher choice (Seppänen et al. 2010).

the size of a habitat patch to the change in biotic and abiotic conditions and resources over time in both predictable and unpredictable ways (you can read more about the broad-scale patterns of species distributions in chapter 20). We have discussed the factors and phenomena shaping species assemblages and communities in an area. But we can also examine broad spatial and temporal patterns of species diversity and richness across space.

Species-Area Curves and Island Biogeography

Spatial patterns can have important effects on community structure and can help us understand and predict patterns and plan conservation strategies. Bird species diversity, in fact the diversity of all species in any community, is predicted to increase with increasing size of a habitat patch. This is called the species-area relationship (fig. 22.7). This concept was first described by H. G. Watson in 1835, who noted that as the area of a county in England increases, the number of plant species within that area increases. Several hypotheses have been proposed to explain why this phenomenon occurs, including island biogeography theory and the habitat diversity hypothesis.

Island biogeography theory was developed through hypotheses and experiments conducted on islands, and later the concept was applied to think about habitat patches as islands. MacArthur and Wilson (1963) proposed the equilibrium model of island biogeography, which suggested that the number of species on an island is determined by the immigration of new species to the island and the extinction rate of species already present on the island; when these two rates are in balance, the number of species on the island is at equilibrium. This model assumes that the population size of each species is proportional to island size, with larger islands providing for larger populations. The researchers thus predicted that extinction rates on large islands would be lower than on small islands because of the larger population sizes. Therefore larger islands, at equilibrium, would have a greater number of species. Another explanation for the larger number of species in larger areas that does not require thinking of areas as islands is the habitat diversity hypothesis. This hypothesis suggests that larger areas are more likely to have a more diverse set of conditions resulting in more unique niches. Thus larger areas have an increased ability to support numerous species, compared with smaller, more homogeneous areas (Williams 1964). Alternatively, some researchers argue that this apparent pattern of more species present in larger islands or regions is largely the result of the process of the sampling or data collection method used to described the species diversity in an area. Studies have shown that as we increase the area we collect samples from, we *detect* more species, even when we know the area does not actually contain more species (Connor and McCoy 1979). Several mechanisms may explain this phenomenon. For example, in a larger area, more individuals of each species may be present; thus, for species that are difficult to detect, surveying an area that contains more individuals of that species may result in an increased likelihood of detecting at least one individual of the species.

Latitudinal Diversity Patterns

Another interesting pattern observed in bird species diversity follows a latitudinal gradient. Bird species diversity increases with proximity to the equator and decreases approaching the planetary poles (fig. 22.8). The causes of this pattern remain the subject of continued debate. Some researchers have suggested this pattern may be caused by historical and evolutionary processes such as a high speciation rate in equatorial regions during the Pleistocene (Ricklefs 2006, Martin and Tewksbury 2008; also see the box on page 726). However, recent genetic studies investigating patterns of avian diversification suggest locations and rates of high speciation have varied over time, and recent speciation rates may not reflect this latitudinal pattern (Jetz et al. 2012, Mannion et al. 2014, Rabosky et al. 2015).

Pianka (1966) was one of the first of many investigators to summarize the different hypotheses that explain the large species diversity in the tropics (Kricher 2011). Some researchers have suggested that because tropical climates tend to have more stable and consistent temperature—the climactic stability hypothesis—species are able to specialize on a narrow range of consistently available resources, promoting a diversity of specialists rather than few generalists adapted to cope with seasonally varying resources. The

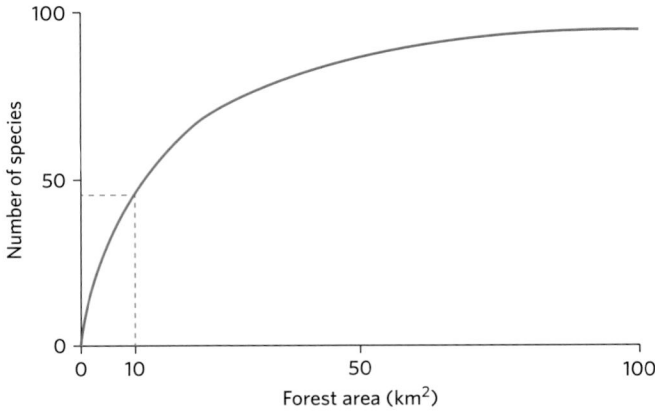

Figure 22.7. This diagram depicts the species-area relationship, in which more species are detected as the size of the sampled area increases.

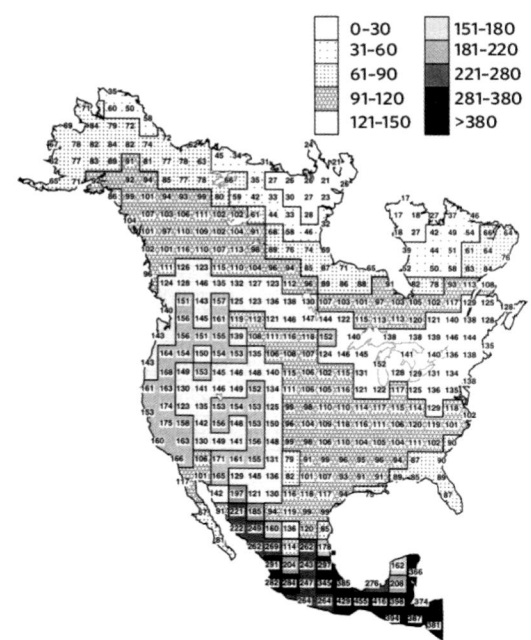

Figure 22.8. Geographic variation in the richness of breeding terrestrial bird species in the Nearctic and northern Neotropics. Numbers represent the number of bird species in each cell (Hawkins et al. 2016).

productivity hypothesis proposes that tropical latitudes tend to have greater primary productivity and insect biomass, due perhaps in part to their climatic stability, and thus may provide food resources sufficient to support more species. Tropical regions may also provide unique food resources such as large fruiting trees and large insects that provide unique niches less common in temperate regions for frugivorous birds (e.g., parrots, trogons, toucans, cotingas, and manakins) and species that rely in part on large insects (e.g., motmots, cuckoos, antbirds, woodcreepers, and some flycatchers). Similarly, tropical forests, with their many canopy layers, lianas, and epiphytes, may have greater structural habitat complexity than their temperate counterparts, providing a larger array of unique niches and thus greater opportunities for evolution of unique niche specializations and species diversification.

The competition hypothesis posits that resource availability in the tropics has resulted in high levels of competition, resulting in increased selection pressures for character displacement and niche partitioning, and thus driving diversification and speciation. The predation hypothesis states that predation pressure may be more intense in the tropics, and as we observed with coexisting species on Scandinavian islands, this predation pressure enabled coexistence of numerous species by reducing the abundance of competitors and allowing for more resource use overlap.

The time theory proposes that older communities will have more species than younger ones; because tropical regions did not undergo glaciation disturbance, they are therefore older ecosystems than temperate ecosystems and thus should have more species. Others have suggested that greater species diversity is a consequence of greater effective evolutionary time or evolutionary speed in the tropics, resulting from higher temperatures and thus higher energy in the ecosystem, yielding faster kinetics, shorter generation times, faster mutation rates, and faster selection at high temperatures (Rohde 1992, Brown 2014). Certainly, the latitudinal gradient in species diversity may be the result of multiple causes—a perfect storm of ecological and biogeographic conditions for speciation and diversity maintenance (Kricher 2011).

TROPICAL CONSERVATISM HYPOTHESIS

The tropical conservatism hypothesis proposed by Wiens and Donoghue (2004) to explain the latitudinal gradient in species diversity includes three main concepts:

1. Many groups of organisms originated in the tropics and have spread to temperate regions more recently or not at all. A clade that originated in the tropics would be expected to have more species because of the greater time available for speciation to occur in tropical regions.

2. One reason that many groups of organisms originated in the tropics is that tropical regions covered a greater geographical extent until relatively recently (approximately 30–40 mya), when temperate zones increased in size. If tropical regions covered a larger area for a long period of time, we would expect more groups of species to have originated in those regions.

3. Because of factors such as this combination of greater time and space for evolution of species in tropical regions, many species and clades are specialized for tropical climates, and the adaptations necessary to occupy and survive in temperate or cold conditions have evolved in fewer species.

Many groups of organisms do show the predicted pattern of biogeography, with a tropical origin and only more recent dispersal to temperate regions. Additionally, many distantly related groups show similar northern range limits, suggesting that the cold climate may act as a barrier to the invasion by clades with tropical origins.

Other, more complex patterns in diversity can be observed across smaller spatial extents. In fact, across a smaller spatial extent and grain, we see the opposite trend along the east coast of North America, where breeding forest bird species diversity increases as you move southward. However, once breeding species migrate to their wintering grounds and only winter residents remain in the eastern North American forests, this pattern disappears. Proposed explanations for this pattern posit that high levels of plant and insect productivity in northern forests during the summer season support high levels of diversity, but during the winter, this productivity is reduced by cold weather and snow (Rabenold 1993).

DYNAMICS

Spatial and temporal dynamics of ecosystems, that is, change in ecosystems over space and time, can have major effects on bird assemblages and communities. For example, a forest that experiences a wildfire will likely be occupied by a different community of birds after the fire than before, and this community will change over time as the burned forest transitions from bare ground to grassland, shrubland, young forest, and old forest.

Disturbance and Succession

Ecosystems can undergo slow gradual change over time or more rapid change. A **disturbance** is an event that causes sudden or significant changes in the organization of energy and resources in the system and may alter the trajectory of successional change, often causing conditions to shift toward a previous successional stage. Disturbances may be frequent or infrequent, predictable or unpredictable. For example, a rare severe hurricane can dramatically alter the vegetative structure and composition of a coastal forest or wetland. Alternatively, some ecosystems regularly experience annual or periodic wildfires. **Succession** is the gradual change of biotic and abiotic conditions, species composition, physical structure, and species interactions in an ecosystem over time. We may observe succession after a disturbance event such as a wildfire. Bird communities in an area may change dramatically as a result of a disturbance and then gradually over time as succession produces gradual changes to the resources and conditions in an area.

Some disturbances create conditions that can be used by numerous species of birds over time. In North America, grasslands or prairies, shrublands, savannas, early successional forests, and floodplains are ecosystems that experience regular or frequent disturbance that facilitates use by numerous avian species at varying times following the disturbance event. Suppression of agents of disturbance, such as suppression of fire or flooding, can act as a form of habitat loss by reducing the periodic availability of successional stages used by numerous bird species following disturbance events. For many bird species adapted to exploit disturbance-mediated habitats, abundances have declined, in some cases to a much greater extent than for species that primarily use areas with late successional conditions and little disturbance.

Consider an old-growth forest where wildfires rarely occur and weather events are rarely severe enough to significantly disturb the floristic or physical composition of the forest as a low-disturbance system. Alternatively, a site that experiences extremely frequent fire, flooding, or other catastrophic change can be considered a high-disturbance system. In both cases, we would expect that a small set of species have evolved to specialize on the limited number of unique niches available in these ecosystems. In the high-disturbance system, the set of available unique niches changes regularly over time and may tend to provide more total niches over time, though no niche will be present for long before it the system changes. The low-disturbance system provides a smaller set of niches that remain consistent over time. The **intermediate disturbance hypothesis** asserts that systems that experience a moderate level of disturbance, in frequency and/or intensity, exhibit a wide array of habitat and resources conditions as a result of disturbances, and thus can attain the highest level of species diversity compared with highly disturbed or undisturbed systems. Many researchers have demonstrated the positive effects of periodic wildfire on bird species diversity, supporting this hypothesis. However, for some species that are adapted to live in stable, late successional ecosystems, or in highly disturbed systems, and intermediate disturbance regime may not be optimal for their fitness and persistence.

STUDYING ASSEMBLAGES AND COMMUNITIES

To fully understand why even a single species occurs in an area, or what regulates its abundance, requires an understanding of physical and chemical conditions, resources available, the changing resource needs for the species at various life stages, the influence of species interactions, and how these factors affect birth, death, immigration, emigration, and migration. You have read about how many of these specific aspects of a species life history and ecology are studied by researchers in other chapters of this book. In this section, we focus on the ways that researchers describe and investigate assemblages and communities.

Researchers use a number of metrics to describe and quantify the composition of an assemblage or community. **Species**

diversity consists of two key component metrics, species richness and species evenness. **Species richness** describes the number of species in a community. Species richness is calculated by a simple count of how many different species are present in a study area of interest. However, we know that some species may be more abundant than others, so a simple count of the number of species present provides an incomplete picture of the composition of the community. A community with two Northern Cardinals (*Cardinalis cardinalis*) and 500 American Robins (*Turdus migratorius*) would have the same species richness count as a community with 230 Northern Cardinals and 250 American Robins. **Species evenness** helps address this important difference by including a measure of relative abundance for each species in the community (Hill 1973). A community with less variation in the relative abundance of individuals among the species is considered more even than one with more variation in relative abundance among species. So, using our previous example, the community with 230 Northern Cardinals and 250 American Robins would be considered more even than our community with 2 Northern Cardinals and 500 American Robins.

Comparing community diversity among different areas using these basic metrics is easy if richness or evenness is common across sites. However, communities can vary in both richness and evenness in complex ways (Rosenzweig 1995). Researchers have developed a number of mathematical metrics to describe community composition that bring these two components together into a single measure—a diversity index—that can be compared among areas of interest, including the Shannon-Wiener index (Shannon 1948, Chao and Shen 2003), the Simpson index (Simpson 1949), and the Gini-Simpson Index (Lande 1996, Magurran 2004). Each diversity index assigns different weightings to species richness and evenness, so a researcher may choose which index to use based on the goals and priorities of their comparison (Smith and Wilson 1996, Tuomisto 2012). Alternatively, in some cases it is more useful or appropriate to calculate a similarity index that compares similarity in community composition between or among sites (Leinster and Cobbold 2012).

It is important to note that when we talk about diversity, we most often think about species diversity, but ornithologists and avian ecologists may be interested in other kinds of diversity. The functioning of assemblages and communities may be influenced by other kinds of diversity including genetic diversity (within and among species), diversity of functional groups (such as foraging guilds like insectivores, seed-eaters, frugivores), or diversity of life forms at taxonomic levels other than species.

Researchers may investigate patterns of resources and habitat use to investigate similarities and differences in re-

sources use between species using tools such as resource utilization functions (RUFs), Resource Selection Probability Functions (RSPFs), occupancy models, or other habitat use models. These research and analysis approaches allow scientists to use data on bird locations and environmental and resource conditions at those locations to determine what resources a bird species is more or less likely to use. This allows scientists to identify resources and conditions associated with species use and how different levels of those resources may increase or decrease the likelihood of species use. While these approaches do not provide us with information about how or why a species uses a particular set of resources, it does provide us with essential information about patterns of resource use that can provide a basis for further investigation and the formation of testable hypotheses. Why does species *A* use areas with more canopy cover compared with species *B*? Is this a result of competition and niche partitioning between species *A* and *B*? Is it the result of physiological tolerances that differ between the two species? Contemporary techniques allow researchers to include the presence of a heterospecific as a variable of interest alongside vegetation or other environmental characteristics. These types of models provide tools to investigate co-occurrence by allowing researchers to assess whether the presence of one species is associated positively or negatively with the presence of another, providing the basis for further investigation into the nature of the interactions between the species.

Investigating the mechanisms or processes that shape species occurrence and species interactions usually requires experiments. Carefully designed experiments can test hypotheses about trophic interactions, predator-prey interactions, or even interspecific competition or commensalism interactions. For example, think back to the experiment described in the box on page 719. Researchers investigated the interactions between Orange-crowned Warblers and Virginia's Warblers, two species that coexist in some areas, by removing one or the other species on a set of study plots. They tested hypotheses including whether coexistence of these two similar species results in reduced access to nest sites and food resources for the species and whether such reductions result in fitness consequences for one or both species. This hypothesis would have been nearly impossible to test directly without conducting a manipulative experiment.

Food webs represent the flow and exchange of matter and energy among organisms in an ecosystem, from primary producers to top predators. Today, scientists have a range of tools that enable them to investigate trophic interactions, including where a species fits into a food web in an ecosystem, what prey it eats, and what predators may eat it. Stable isotope ratios, such as isotopes of nitrogen ($\delta15N$) and carbon

(δ13C) can be used to understand the structure and dynamics in communities, including estimating the trophic position of a species (Peterson et al. 1987, Post 2002). Stable carbon isotope ratios (δ13C) vary among plants because of differences in the inorganic carbon substrate and the photosynthetic pathway (C3 or C4) used in plant autotrophy (Smith and Epstein 1970). Carbon isotope ratios of the primary producers can be observed in the tissues of their consumers, a kind of you-are-what-you-eat effect. Thus, carbon isotope ratios provide information on the carbon/energy sources at the base of the food web. Unlike carbon, stable nitrogen isotope ratios (δ15N) of consumers are typically enriched by 2–4 percent relative to their diet, meaning that nitrogen isotope ratios can be used to estimate the trophic position of a consumer. Used together, these two isotopes allow a more complete picture of a food web and the place of an organism within it, which integrates all the different trophic pathways leading to an organism, representing the complexity of a food web rather than a simplified food chain.

Researchers are also using ecotoxicological studies to investigate trophic interactions. Bioaccumulation and biomagnification of persistent organic pollutants (POPs) that exist in the environment enables researchers to understand the flow and exchange of toxins through an ecosystem and food web, but also inadvertently provides a tool for researchers to understand the trophic position of an organism based on its toxicology. Biomagnification, a special case of bioaccumulation by which concentrations of chemicals in consumers exceed the concentrations in the consumer's prey, can occur at each step in a food chain. Thus, it is possible to determine the trophic position of a species based on accumulation of these chemicals (Hebert et al. 2000). Concerns over how global change will influence species distributions and interactions is one among many reasons driving the growing need for robust methods to quantify niches and niche differences among or within species. Many researchers are developing and proposing study designs and statistical frameworks to describe and compare niches using occurrence and spatial environmental data to detect niche overlap (Broennimann et al. 2012).

CONTEMPORARY CONSERVATION TOPICS

Avian assemblages and communities face a number of conservation challenges in our modern world. Anthropogenic activities—the activities of humans—can affect everything from ecosystem resources and conditions to species interactions within communities. While habitat loss is one of the most conspicuous effects humans have on birds, changing habitat conditions can also affect communities. For example, as we discussed above, human suppression of wildfire can change the disturbance regime of an ecosystem, changing conditions and thus fundamentally altering the bird species composition and interactions over time.

One major effect that humans can have on communities is changing the species composition by introducing new, nonnative species. Introduction of nonnative species can alter the species composition and lead to changes in interactions within the community. Researchers are now describing the development of **no-analog communities**, or communities for which we have no other naturally occurring example. When the composition of the community differs from any previously known or understood example, we have limited ability to predict the interactions that will occur and the trajectory of species within the community. In some cases, it is evident that introduced species can lead to significant negative effects on native species. For example, we know the introduced House Sparrow (*Passer domesticus*) exhibits antagonistic interactions with a number of native species. House Sparrows may aggressively attack and evict native Eastern Bluebirds (*Sialia sialis*), Purple Martins (*Progne subis*), and Tree Swallows (*Tachycineta bicolor*) from nest cavities, acting as a competitor for this limited resource and resulting in negative fitness consequences for these native species. Over time, the effects of nonnative species introductions can result in major changes to the species composition and species diversity of some bird communities, including a loss of overall diversity. Another significant conservation concern is a change in the abundance of one or more important species in a community. Some human activities may favor certain species over others. For example, attributes of suburban development, such as landscaped lawns of exotic grass, golf courses, and exotic shrub and tree species may provide additional forage for species such as Canada Geese (*Branta canadensis*); dramatic increases in species that successfully adapt to human development can cause increased competitive pressure or displace other species in the community.

In some communities, certain species may play a more important role in maintaining community structure and function than others. We often call these species **keystone species**. Think about the example we discussed earlier, in which the presence of Pygmy Owls played an important role in regulating the co-occurrence of numerous songbird species. Anthropogenic activities that cause reductions or local extirpations of keystone species can have significant effects on community structure and function. In some cases, loss of a single species can have a chain of effect on numerous other species, called a **trophic cascade**. A classic example of a trophic cascade was investigated by Hebblewhite et al. in 2005. In this study, researchers observed that areas with high human activity had lower wolf (*Canis lupus*) densities. Areas with lower wolf densities in turn exhibited greater elk (*Cervus elaphus*) density. Areas with greater elk densities showed lower willow (*Salix* spp.) production and consequently showed lower bird species diversity. Trophic cascades can mean significant, often unexpected, indirect effects across a community as a result of the loss or reduction in an important species within the community.

KEY POINTS

- Species occurring together may interact in a variety of ways, from competition resulting in mutual negative effects to mutualisms that result in positive benefits to both species. A community is the set of species in an area that interact in one of these many ways.
- Patterns of species occurrence and the resulting composition of communities are the result of an array of biotic and abiotic filters and species interactions.
- Interspecific interactions can significantly shape the realized niche of a species.
- Competition can play an important role in community composition, but can be difficult to study and demonstrate directly.
- Studies of species habitat and resource use provide important information on patterns of species occurrence. But, manipulative experiments may be necessary to test and detect the interaction processes influencing these patterns.
- Contemporary techniques such as use of stable isotopes and tracing of pollutants enable researchers to investigate and understand complex trophic interactions in communities

KEY MANAGEMENT AND CONSERVATION IMPLICATIONS

- Communities can be affected by a wide range of human activities.
- Though humans may view disturbances such as wildfire as undesirable phenomenon, disturbance can be important for maintaining species diversity.
- Introduction of nonnative species can alter not only species interactions within a community, but can also result in no-analog communities; these no-analog communities present management challenges, as researchers and managers do not yet have sufficient information on how species will interact in these systems.
- Changes in community composition through the addition or loss of a species, or significant increases or decreases in density of keystone species, can have

far-reaching effects on other species in the community, including trophic cascades.

DISCUSSION QUESTIONS

1. Consider the intermediate disturbance hypothesis. Should we aim to integrate regular, moderate disturbances into ecosystems to enhance bird diversity? Why or why not?
2. How might you design an experiment to test the hypothesis that two species are competing for food resources? How might you design an experiment to test for whether two species are competing for nest sites?
3. What are some roles of predation in bird communities?
4. What are some ways that human activities may affect communities, including effects on community composition and species interactions?
5. What attributes make a species a keystone species?
6. How do you think researchers and conservation and management planners should deal with no-analog communities?
7. Why do you think there so many species of birds in the tropics compared with temperate regions?

References

Broennimann, O., M. C. Fitzpatrick, P. B. Pearman, B. Petitpierre L. Pellissier, N. G. Yoccoz, W. Thuiller, et al. (2012). Measuring ecological niche overlap from occurrence and spatial environmental data. Global Ecology and Biogeography 21:481–497.

Brown, J. H. (2014). Why are there so many species in the tropics? Journal of Biogeography 41:8–22.

Chakra, M. A., C. Hilbe, and A. Traulsen (2014). Plastic behaviors in hosts promote the emergence of retaliatory parasites. Scientific Reports 4. doi:10.1038/srep04251.

Chao, A., and T.-J. Shen (2003). Nonparametric estimation of Shannon's index of diversity when there are unseen species in sample. Environmental and Ecological Statistics 10:429–443.

Connell, J. H. (1980). Diversity and the coevolution of competitors, or the ghost of competition past. Oikos 35:131–138.

Connell, J. H. (1983). On the prevalence and relative importance of interspecific competition: Evidence from field experiments. American Naturalist 122:661–696.

Connor, E. F., and E. D. McCoy (1979). The statistics and biology of the species-area relationship. American Naturalist 113:791–833.

Dawideit, B. A., A. B. Phillimore, I. Laube, B. Leisler, and K. Bohning-Gaese (2009). Ecomorphological predictors of natal dispersal distances in birds. Journal of Animal Ecology 78:388–395.

Edworthy, A. (2016). Competition and aggression for nest cavities between Striated Pardalotes and endangered Forty-spotted Pardalotes. The Condor 118:1–11.

Elton, C. (1927). Animal ecology. Sidgwick and Jackson, London.

Gause, G. F. (1934). The struggle for existence. Williams and Wilkins, Baltimore, MD.

Grinnell, J. (1917). The niche-relationships of the California Thrasher. The Auk 34:427–433.

Hawkins, B. A., E. E. Porter, and J. A. Felizola Diniz-Filho (2003). Productivity and history as predictors of the latitudinal diversity gradient of terrestrial birds. Ecology 84:1608–1623.

Hawkins, B. A., J. A. F. Diniz-Filho, C. A. Jaramillo, and S. A. Soeller (2016). Post-Eocene climate change, niche conservatism, and the latitudinal diversity gradient of New World birds. Journal of Biogeography 33:770–780.

Hebblewhite, M., C. A. White, C. G. Nietvelt, J. A. McKenzie, T. E. Hurd, J. M. Fryxell, S. E. Bayley, and P. C. Paquet (2005). Human activity mediates a trophic cascade caused by wolves. Ecology 86:2135–2144.

Hebert, C. E., K. A. Hobson, and J. L. Shutt (2000). Changes in food web structure affect rate of PCB decline in Herring Gull (Larus argentatus) eggs. Environmental Science and Technology 34:1609–1614.

Hill, M. O. (1973). Diversity and evenness: A unifying notation and its consequences. Ecology 54:427–432.

Hipfner, J. M., K. W. Morrison, and R. Darvill (2011). Peregrine Falcons enable two species of colonial seabirds to breed successfully by excluding other aerial predators. Waterbirds 34:82–88.

Hutchinson, G. E. (1957). Concluding remarks. Cold Spring Harbor Symposia on Quantitative Biology 22:415–427.

Jetz, W., G. H. Thomas, J. B. Joy, K. Hartmann, and A. O. Mooers (2012). The global diversity of birds in space and time. Nature 491:444–448.

Kricher, J. C. (2011). Tropical ecology. Princeton University Press, Princeton, NJ.

Kullberg, C., and J. Ekman (2000). Does predation maintain tit community diversity? Oikos 89:41–45.

Lande, R. (1996). Statistics and partitioning of species diversity, and similarity among multiple communities. Oikos 76:5–13.

Leinster, T., and C. A. Cobbold (2012). Measuring diversity: The importance of species similarity. Ecology 93:477–489.

Lister, B. C. (1980). Resource variation and the structure of British bird communities. Proceedings of the National Academy of Sciences 77:4185–4187.

MacArthur, R. H. (1957). Population ecology of some warblers of Northeastern coniferous forests. PhD diss., Yale University, New Haven, CT.

MacArthur, R. H. (1958). Population ecology of some warblers of northeastern coniferous forests. Ecology 39:599–619.

MacArthur, J. H. (1965). Patterns of species diversity. Biological Reviews 40:510–533.

MacArthur, R. H., and J. W. MacArthur (1961). On bird species diversity. Ecology 42:594–598.

MacArthur, R. H., and E. O. Wilson (1963). An equilibrium theory of insular zoogeography. Evolution 17:373–387.

MacArthur, J. H., and E. O. Wilson (1967). The theory of island biogeography. Princeton University Press, Princeton, NJ.

Magurran, A. E. (2004). Measuring biological diversity. Blackwell, Oxford, UK.

Mannion, P. D., P. Upchurch, R. B. Benson, and A. Goswami (2014). The latitudinal biodiversity gradient through deep time. Trends in Ecology and Evolution 29:42–50.

Martin, T. E. (1988). On the advantage of being different: Nest predation and the coexistence of bird species. Proceedings of the National Academy of Sciences USA 85:2196–2199.

Martin, T. E. (1993). Nest predation and nest sites: New perspectives on old patterns. BioScience 43:523–532.

Martin, T. E. (1998). Are microhabitat preferences of coexisting species under selection and adaptive? Ecology 79:656–670.

Martin, P. R., and T. E. Martin (2001). Ecological and fitness consequences of species coexistence: A removal experiment with wood warblers. Ecology 82:189–206.

Martin, P. R., and J. J. Tewksbury (2008). Latitudinal variation in subspecific diversification of birds. Evolution 62:2775–2788.

Martin, Lynn B., Cameron K. Ghalambor, and H. Arthur Woods, Editors (2015). Integrative organismal biology. John Wiley and Sons, Hoboken, NJ.

Morrison, M. L. (2009). Restoring wildlife: Ecological concepts and practical applications. 2nd ed. Island Press, Washington, DC.

Nudds, T. D., and J. N. Bowlby (1984). Predator-prey size relationships in North American dabbling ducks. Canadian Journal of Zoology 62:2002–2008.

Nudds, T. D., J. Elmberg, H. Pöysä, K. Sjöberg, and P. Nummi (2000). Ecomorphology in breeding Holarctic dabbling ducks: The importance of lamellar density and body length varies with habitat type. Oikos 91:583–588.

Paradis, E., S. R. Baillie, W. J. Sutherland, and R. D. Gregory (1998). Patterns of natal and breeding dispersal in birds. Journal of Animal Ecology 67:518–536.

Peterson, B. J., R. W. Howarth, and R. H. Garrett (1985). Multiple stable isotopes used to trace the flow of organic matter in estuarine food webs. Science 227:1361–1363.

Peterson, B. J., and B. Fry (1987). Stable isotopes in ecosystem studies. Annual Review of Ecology, Evolution, and Systematics 18:293–320.

Pianka, E. R. (1966). Latitudinal gradients in species diversity: A review of the concepts. American Naturalist 100:33–46.

Pons, P., J. M. Bas, R. Prodon, N. Roura-Pascual, and M. Clavero (2008). Territory characteristics and coexistence with heterospecifics in the Dartford Warbler Sylvia undata across a habitat gradient. Behavioral Ecology and Sociobiology 62:1217–1228.

Post, D. M. (2002). Using stable isotopes to estimate trophic position: Models, methods, and assumptions. Ecology 83:703–718.

Pöysä, H., J. Elmberg, P. Nummi, and K. Sjöberg (1994). Species composition of dabbling duck assemblages: Ecomorphological patterns compared with null models. Oecologia 98:193–200.

Rabenold, K. (1993). Latitudinal gradients in avian species diversity and the role of long-distance migration. Current Ornithology 10:247–274.

Rabosky, D. L., and H. Huang (2015). Minimal effects of latitude on present-day speciation rates in New World birds. Proceedings of the Royal Society B 282:2014–2889.

Raven, Peter H., and George B. Johnson (2002). Biology. 6th ed. McGraw-Hill, Boston, MA.

Ricklefs, R. E. (2006). Global variation in the diversification rate of passerine birds. Ecology 87:2468–2478.

Ricklefs, R. E., and R. Relyea (2014). Ecology: The economy of nature. 7th ed. W. H. Freeman, New York.

Rohde, K. (1992). Latitudinal gradients in species diversity: The search for the primary cause. Oikos 65:514–527.

Rosenzweig, M. L. (1995). Species diversity in space and time. Cambridge University Press, Cambridge, UK.

Schmidt, K. A. (2006). Non-additivity among multiple cues of predation risk: A behaviorally-driven trophic cascade between owls and songbirds. Oikos 113:82–90.

Schoener, T. W. (2009). Ecological niche. *In* The Princeton guide to ecology, S. A. Levin, Editor. Princeton University Press, Princeton, NJ, pp. 3–13.

Seppänen, J. T., and J. T. Forsman (2007). Interspecific social learning: Novel preference can be acquired from a competing species. Current Biology 17:1248–1252.

Seppänen, J. T., J. T. Forsman, M. Mönkkönen, I. Krams, and T. Salmi (2010). New behavioural trait adopted or rejected by observing heterospecific tutor fitness. Proceedings of the Royal Society B: Biological Sciences 278:1736–1741.

Shannon, C. E. (1948). A mathematical theory of communication. The Bell System Technical Journal 27:379–423 and 623–656.

Simpson, E. H. (1949). Measurement of diversity. Nature 163:688.

Smith, B. N., and S. Epstein (1970). Biogeochemistry of the stable isotopes of hydrogen and carbon in salt marsh biota. Plant Physiology 46:738–742.

Smith, B., and J. B. Wilson (1996). A consumer's guide to evenness indices. Oikos 76:70–82.

Templeton, C. N., E. Greene, and K. Davis (2005). Allometry of alarm calls: Black-capped chickadees encode information about predator size. Science 308:1934–1937.

Templeton, C. N., and E. Greene (2007). Nuthatches eavesdrop on variations in heterospecific chickadee mobbing alarm calls. Proceedings of the National Academy of Sciences 104:5479–5482.

Tuomisto, H. (2012). An updated consumer's guide to evenness and related indices. Oikos 121:1203–1218.

Watson, H. C. (1835). Remarks on the geographical distribution of British plants. Longman, Rees, Orme, Brown, Green and Longman, London.

Wiens, J. A. (1989). The ecology of bird communities. Cambridge University Press, Cambridge, UK.

Wiens, J. J., and M. J. Donoghue (2004). Historical biogeography, ecology and species richness. Trends in Ecology and Evolution 19:639–644.

Williams, C. B. (1964). Patterns in the balance of nature. Academic Press, London.

CHAPTER 23

Parasite and Disease Ecology

Jen C. Owen

Birds have evolved to exploit almost every environmental niche on earth and therefore are exposed to a vast community of disease-causing agents. However, until recently, the study of infectious diseases of wild birds, and the pathogens and parasites that cause them, has been a relatively small subdiscipline of ornithology. In the last several decades, we have seen emerging disease threats to wild birds, such as the introduction and rapid spread of West Nile virus (WNV) in North America from 1999 to 2004 and the subsequent population declines of several bird species (Naugle et al. 2005, LaDeau et al. 2007). In the 1990s, a unique strain of *Mycoplasma gallisepticum* (which causes conjunctivitis) emerged, leading to the dramatic decline of the eastern populations of House Finches (*Haemorhous mexicanus*) in North America (Dhondt et al. 2006). Populations of endemic species of birds in remote archipelagos such as the Galápagos and Hawaiian Islands have experienced declines from the simultaneous effects of disease and introduction of nonnative species (Warner 1968, Wikelski et al. 2004, Dvorak et al. 2012). There is also evidence that parasites can limit population growth, as demonstrated by a long-term study of the host-parasite interaction of the Red Grouse (*Lagopus lagopus scotica*) and the parasitizing trematode (*Trichostrongylus tenuis*) (Dobson and Hudson 1992, Hudson et al. 1992). Avian diseases deserve our attention—not only because of their impact on avian health and fitness, but also because of the risk they pose to human and ecosystem health.

USEFUL TERMINOLOGY AND CONCEPTS

Disease is the product of the interactions among the disease-causing agent, the host, and the environment. The word **host** refers to the organism/population that harbors the relevant agent of infection. The different types of hosts are defined in the box on page 734, along with other common terms used in this chapter. The two terms used to describe infectious, disease-causing agents are parasite and pathogen. **Parasites** are living organisms that live in or on a host from which the parasite obtains nourishment. The host typically bears some cost associated with the parasitism. **Pathogens** are any agents capable of causing disease and include viruses, bacteria, fungi, protozoa, and parasitic helminths. The terms are not synonymous; not all parasites are pathogenic, and not all pathogens live and feed on their host. For instance, ectoparasites such as ticks and hippoboscid flies are parasites, but not pathogens. However, the ectoparasites themselves may be a **vector,** which is a biological carrier of the pathogen.

Pathogens and parasites are typically grouped into two broad categories—micro- and macroparasites. As the names suggest, **microparasites** are microscopic or small-bodied and include viruses, fungi, bacteria, and protozoa. **Macroparasites** are larger-bodied, visible to the human eye, and include helminths and arthropods. Additionally, the two groups differ in their locations and modes of replication, which has implications for disease severity and host immunity. Microparasites have short generation times and replicate directly within one definitive host. They lead to acute (short and potentially severe) diseases that elicit a strong host immune response, which may provide lifelong protection from that same parasite. Macroparasites multiply outside the definitive host, either in an intermediate invertebrate or vertebrate host or in the environment. A disease is manifested only when the host accumulates or is infected with high numbers of parasites. Hence, macroparasite infections tend to present as chronic infections for which hosts typically do not develop long-lasting immunity.

Once a host has been exposed to and invaded by the pathogen, it is considered **infected** or having an **infection** with that particular pathogen. Organisms are constantly exposed to pathogens and parasites, but few succumb to infections.

COMMON TERMS USED IN THIS CHAPTER

Infection is when a host organism is successfully invaded by a pathogen or parasite.

Disease is when the host shows outward (clinical) signs of the infection.

Infectious is the state in which a host, vector, or object is capable of transmitting the pathogen or parasite to a susceptible individual.

Morbidity is another term that describes the host state when exhibiting clinical signs of infection (disease).

Susceptible describes the state of an individual in a population that has not been infected by the pathogen of interest.

Reservoir hosts are the species within a community that are the ultimate source of new infections for a particular pathogen or parasite and are required for its persistence.

Definitive hosts are the species within which the parasites reach maturity and reproduce.

Intermediate hosts are the obligatory species in which one or more the parasite's life stages are completed.

Reservoir competence is typically defined as an infected host's ability to successfully infect an arthropod during a blood meal.

Fomites are inanimate particles.

Pathogenicity is the infectious agent's ability to cause disease.

Virulence is the measure of disease severity in the host organism.

The term **disease** describes the host's state when it begins to show clinical signs and impairments from the infection. An organism can be infected with a pathogen and never develop a disease; for instance, many species of birds can be infected with West Nile virus but never become sick or show disease. **Infectious diseases**, the primary focus of this chapter, are those caused by agents that are transmissible from one organism to another. Noninfectious diseases are not transmissible, such as most cancers, autoimmune disorders, nutritional deficiencies, and heart disease in humans.

The transmission of a pathogen from an infectious individual to a susceptible host occurs either directly or indirectly. **Direct transmission** is via direct contact with the infectious individual and/or contact with infectious fluids/

particles, such as mucus and blood. **Indirect** transmission occurs when an individual is exposed to the infectious particle via an intermediate source, such as contaminated food or water, an arthropod vector, or an inanimate object (also called a fomite). Collectively these examples describe transmission between individuals in the same generation, or **horizontal** transmission. Some pathogens can be transmitted **vertically**; that is, an infected female will pass the pathogen directly to her offspring via the egg. Knowing the route of transmission for a host-pathogen system is essential for developing and implementing effective strategies for containing or preventing disease outbreaks.

Epidemiologists often characterize disease occurrence using the epidemiological triad (fig. 23.1A). An infectious disease is the product of the interactions among the infectious agent (pathogen), a susceptible host, and the environment that facilitates the survival and interactions of the host and the pathogen. If applicable, a vector is included as a separate contributing factor (fig. 23.1B).

Each factor has attributes that influence its role in the disease process. The pathogen's ability to persist and infect a host depends on its mode of transmission, dose (amount of pathogen that enters the host), pathogenicity (pathogen's ability to cause disease in the host), mode of replication, and host specificity. A host's exposure and susceptibility to a pathogen may be mediated by the host's behavior, immune system, genetics, age, or sex. Finally, environmental factors that may mediate the host-pathogen interaction include climate, geology, landscape, population density, and community structure. In vector-transmitted pathogens, the vector's receptivity to the pathogen, life span, dispersal distance, and host specificity can all influence its role in the disease process.

Visualizing the epidemiological triangle helps to identify the most cost-effective strategies for halting or preventing a disease outbreak. Breaking the link between any two factors in the triangle will eliminate disease. For instance, administering a vaccine to the susceptible host population or quarantining infectious individuals will break the connection between the host and pathogen. However, most diseases have fairly complex dynamics, and it is hard to implement any single measure to prevent outbreaks. This complexity is exemplified by **zoonotic** pathogens, which have an animal origin but are transmissible to and cause disease in humans. Some notable zoonoses (diseases caused by zoonotic pathogens) that originate in wild birds are West Nile virus and avian influenza viruses.

The presence of disease in a population can be characterized by the magnitude and extent of its occurrence. A disease that is an integral and constant part of the population or community, in which its annual or seasonal occurrence does not fluctuate, is considered **endemic** or **enzootic**, for a

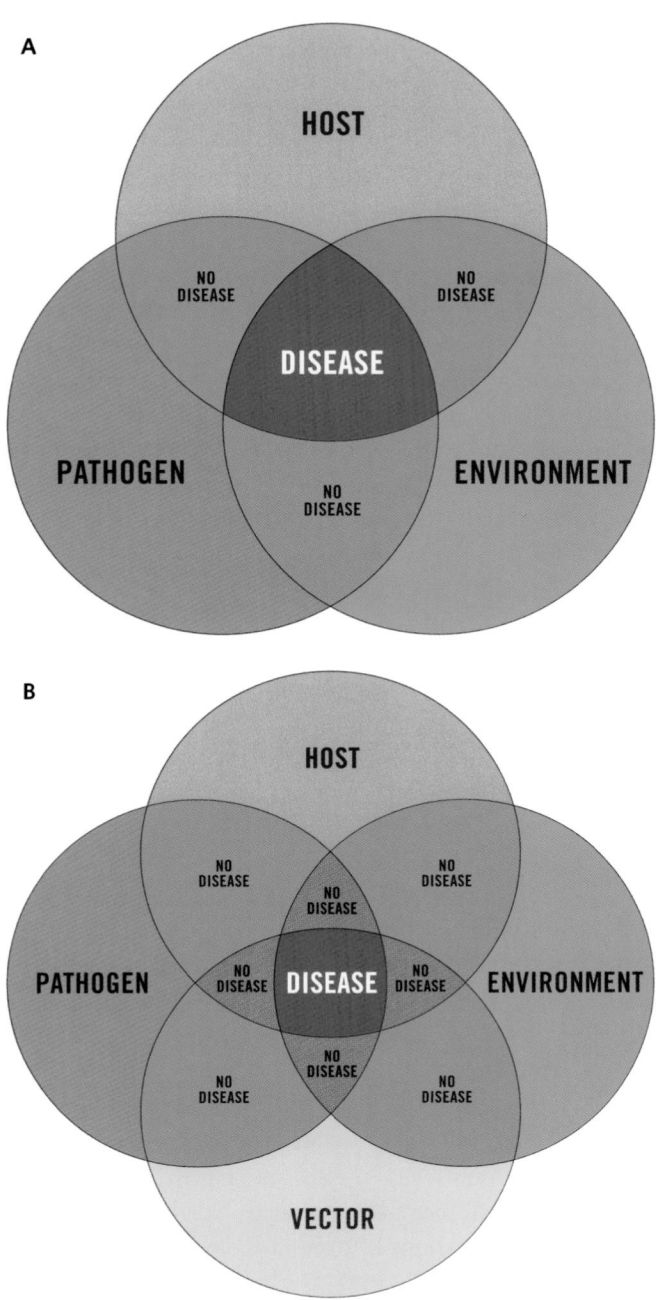

A

B

Figure 23.1. Factors underlying disease causation for *A*, non-vector, and *B*, vector transmitted pathogens. In the absence of any one factor, disease will not occur.

human or nonhuman population, respectively. If the number of cases is more than what is expected for a particular region or community, it is called a disease **epidemic** or **epizootic**. A term not commonly used is **epornitic**, an avian-specific term used to describe a disease epidemic in birds. A **pandemic** is when a disease has affected most regions of the world—a global-sized epidemic.

In the next section, I cover some specific pathogens and the diseases they cause in wild bird populations. The section is divided into microparasites and macroparasites, with each containing separate sections on specific classes of parasites and/or pathogens. For a more comprehensive coverage of avian diseases and parasites, see Atkinson et al. (2008) and Thomas et al. (2008).

MICROPARASITES

Viruses

Viruses are submicroscopic infectious agents that contain nucleic acid/genetic material in the form of a single-stranded RNA or double-stranded DNA molecule encased in a protein coat and, in the case of birds, almost always surrounded by a lipid bilayer membrane, or envelope. Viruses are unable to replicate on their own, and once they invade the susceptible host's cell, they will "hijack" the host's cell machinery to replicate and generate genetically identical copies. A broad range of viruses infects birds; some viruses are restricted to birds, and others can spill over into other taxonomic groups and pose a significant risk to domestic poultry and/or human populations.

Viruses are defined as DNA or RNA viruses; this nomenclature indicates the type of nucleic acid they contain and how the virus replicates within the host cell. When a DNA virus invades a host nucleus and inserts itself into the host DNA, it becomes the template for transcription of the mRNA. RNA viruses skip this step and are directly incorporated into the mRNA during the translation step. If you can recall your introductory biology course, when DNA is being translated into an mRNA, the enzyme DNA polymerase goes along the replicated strand and proofreads and corrects most mistakes, or mutations, detected. In contrast, during the translation process, RNA polymerase is not as effective at detecting mutations along the mRNA strand; hence, mutations are more frequent in RNA viral genomes.

Avian poxviruses are DNA viruses in the family Poxviridae that cause avian pox disease in birds. The virus has been isolated from wild birds on every continent except Antarctica. Regionally, prevalence is highest in warm and moist regions, which is why avian pox has become such a problem in places like Hawai'i. There are currently 13 recognized strains of poxvirus; each one varies in host specificity, and overall, they have been isolated from birds in 20 avian orders (van Riper and Forrester 2007). The transmission of poxvirus can occur through multiple routes, including from biting insects (primarily mosquitoes, but also mites, midges, and flies) that act as mechanical vectors or through direct contact with infectious individuals or fomites (i.e., bird feeders, dust) on which the virus can survive for months to years, even under extreme environmental conditions (van Riper and Forrester 2007). Note that transmission via a vector by mechanical means is not the same as biological transmission. The latter is when the virus is obligated to pass through the vector for development, while the former is when the vector simply transmits the pathogen; it does not reflect an obligatory relationship between vector and pathogen. This distinction also applies to nonviral pathogens.

Following infection, the poxvirus may incubate for one month to one year before clinical signs of infection are apparent. The disease manifests itself in two different forms. The most common is when wart-like lesions become localized in featherless areas of the body (i.e., legs, toes, the base of the bill, eyelid, keel; van Riper et al. 2002). Birds usually recover from these infections; but, in some cases, they may be permanently impaired through loss of an appendage or irreversible blindness. Less common is the diphtheritic form of a poxvirus infection, in which lesions occur throughout the mucosal lining of the respiratory tract and usually lead to rapid mortality. Birds surviving infection will develop strain-specific immunity, but may be susceptible to other strains of poxvirus (van Riper and Forrester 2007).

Avian pox outbreaks are more commonly documented in areas where the avian population does not have a long evolutionary history with the virus and where climatic conditions favor the vector population, such as the Galápagos (Deem et al. 2007), Canary (Smits et al. 2005), and Hawaiian Islands. Population impacts of the epizootics vary but are primarily mild and of short duration. Even during very severe outbreaks, when a large portion of the population is infected, few individuals exhibit severe lesions or the diphtheritic form that leads to death. Experimental infection studies demonstrate that the endemic avifauna of Hawai'i are highly susceptible to avian pox and more likely to die than introduced species that have evolved with the virus and developed resistance (van Riper et al. 2002, Atkinson et al. 2005). In the field, mortality of Hawaiian endemics was highest in the susceptible native birds that had the greatest geographic overlap with the mosquito vector population. Association with domestic poultry can also affect susceptibility to poxvirus. Domestic poultry are commonly infected with a variety of poxvirus strains; in the Canary Islands it was observed that species commonly found in proximity to poultry farms were less likely to develop pox lesions. Yet, species that did not currently or historically interact with domestic poultry were highly susceptible to poxvirus (Smits et al. 2005).

West Nile virus (WNV) is an RNA virus in the family Flaviviridae with other closely related and notable viruses such as St. Louis encephalitis virus (North America), Japanese encephalitis virus (Asia and Australia), and Murray Valley encephalitis virus (Australia). When WNV was first detected in the United States in 1999, it was thought to be St. Louis encephalitis virus, which has been circulating in avian populations since at least the 1930s. West Nile virus is the most widely distributed flavivirus in the world, occurring on every continent except Antarctica (Chancey et al. 2015). The enzootic cycle of WNV includes the mosquito vector (the biological carrier of the virus) and an avian **reservoir** host. WNV has a broad avian host range, with Passeriformes being important as amplification hosts needed to maintain the virus in nature. Likewise, while many species of mosquitoes can transmit the virus, *Culex* species are the principal vectors worldwide (van der Meulen et al. 2005).

Both the mosquito vector and avian host exhibit interspecific variation in ability to transmit the pathogen. In other words, they differ in vector capacity and reservoir competence. Given the same exposure dose of WNV, some mosquito species are more likely to become infected with the virus and transmit the virus when taking a blood meal on a naïve host, and we call this their vector capacity. Likewise, some species of birds, once infected, are more permissive for viral replication and produce a lot of virus without succumbing to the infection, making them more likely to infect a biting mosquito; we call this a competent reservoir. Experimental infection studies have identified American Robins (*Turdus migratorius*), Tufted Titmouse (*Baeolophus bicolor*), and Northern Cardinals (*Cardinalis cardinalis*), to name a few, as competent reservoirs for WNV (Owen et al. 2012, Kilpatrick et al. 2013, VanDalen et al. 2013).

WNV can cause high death rates in some species of passeriform birds, particularly species in the family Corvidae, although not exclusively (Owen and Garvin 2010). Some of the most notable mortality events in North America have occurred in populations of American Crows (*Corvus brachyrhynchos*) (Brault et al. 2004), Greater Sage-Grouse (*Centrocercus urophasianus*) (Naugle et al. 2005), American White Pelicans (*Pelecanus erythrorhynchos*) (Rocke et al. 2005), and raptors (Nemeth et al. 2006). The spread of WNV in the first five years following its introduction in the United States coincided with the decline of several avian populations throughout North America (LaDeau et al. 2007). The Greater Sage-Grouse is on the brink of being listed as an endangered species as a consequence of 80 percent loss of the population in the last decade, due in part to high mortality associated with a series of WNV outbreaks (Naugle et al. 2005). West Nile virus continues to impact avian populations in North America; a recent study by George et al. (2015) found that

23 of 49 species studied, particularly in the families Emberi-zidae, Fringillidae, and Vireonidae, experienced population declines in the years following the virus' emergence.

Avian influenza viruses (AIV) are RNA viruses in the family Orthomyxoviridae; they are technically called avian influenza A viruses, referring to their classification as a type A influenza virus, with the other types being B and C. Why is this classification important? Well, all type A viruses origi-nate from wild birds, but they can infect a variety of different animals, including humans (fig. 23.2). In fact, most human influenza infections (including the seasonal flu circulating each winter) are caused by type A influenza viruses, and these are also the cause of all human influenza pandemics. The most notable pandemics include the 1918 H1N1 Spanish flu, during which over 50 million people died around the world and, more recently, the 2009–2010 H1N1 (swine flu) that caused illness (< 0.02 percent mortality rate) in approxi-mately 50 to 90 million people worldwide. Aquatic birds be-longing to the orders Anseriformes and Charadriiformes are the primary source or reservoir for AIVs in the wild, but the virus has been detected in many other species belonging to at least 10 other avian orders (Stallknecht et al. 2007).

AIV is characterized by different subtypes, strains, and pathogenicities. For example, one subtype of avian influenza

virus is H5N1, which is denoted by the virus's unique surface proteins, hemagglutinin (H) and neuraminidase (N) that de-termine host specificity of the virus. Currently, 16 subtypes of H and 9 subtypes of N are recognized (Stallknecht et al. 2007). Because of the frequent mutation of RNA viruses, each subtype (e.g., H1N1, H5N1) can have multiple strains, in which small genetic differences in the H and N proteins influence the virus's ability to bind to host cells, thus affect-ing host specificity. Strains of influenza are further classified by their pathogenicity to domestic poultry—low-pathogenic avian influenza virus (LPAIV) results in mild to no observ-able signs of disease in poultry; the high-pathogenic strain (HPAIV) results in high morbidity and mortality (up to 100 percent) in domestic poultry and is the leading cause of death of domestic poultry worldwide (Stallknecht et al. 2007). Most HPAIVs are thought to evolve from the LPAIV of the same subtype, and all HPAIVs are either H5 or H7 sub-types (Webster et al. 1992).

AIV is transmitted fecal-orally, with infected birds capa-ble of shedding large amounts of virus into the environment (Stallknecht et al. 1990). The virus can persist and remain in-fectious in surface water for relatively long periods (up to six months) particularly in cold, low-salinity water (Brown et al. 2009). Naïve birds have the potential to become infected through direct contact with infectious birds or by ingesting contaminated water. Dabbling waterfowl (e.g., ducks in the genus *Anas*) are the most likely to get infected as a conse-quence of their foraging behavior and location in shallow water (Webster et al. 1992).

The pathogenicity of HPAIV to wild birds varies and is either host- or strain-specific. There are only a few documented major mortality events involving wild birds. In 1961, HPAIV H5N3 was isolated from Common Terns (*Sterna hirundo*) in South Africa and was linked to the mortality of over 1,300 terns in one colony (Becker 1966). From 2002 to 2005, HPAIV H5N1 was isolated from over 134 species of birds in 62 coun-tries. One mortality event occurred at Qinghai Lake in cen-tral China, where the death of over 6,000 wild birds within two months was linked to the virus (Chen et al. 2005, Liu et al. 2005). Mortality was first detected in, and had the great-est impact on, Bar-headed Geese (*Anser indicus*), resulting in an estimated 10 percent loss of the world's population. Many other species were affected but in smaller numbers, includ-ing the Great Cormorant (*Phalacrocorax carbo*), Ruddy Shel-duck (*Tadorna ferruginea*), Great Black-headed Gull (*Larus ichthyaetus*), and Brown-headed Gull (*Larus brunnicephalus*). While wild birds do not exhibit clinical signs of LPAIV in-fection, emerging data suggest that the infection may not be completely benign (van Gils et al. 2007, Latorre-Margalef et al. 2009).

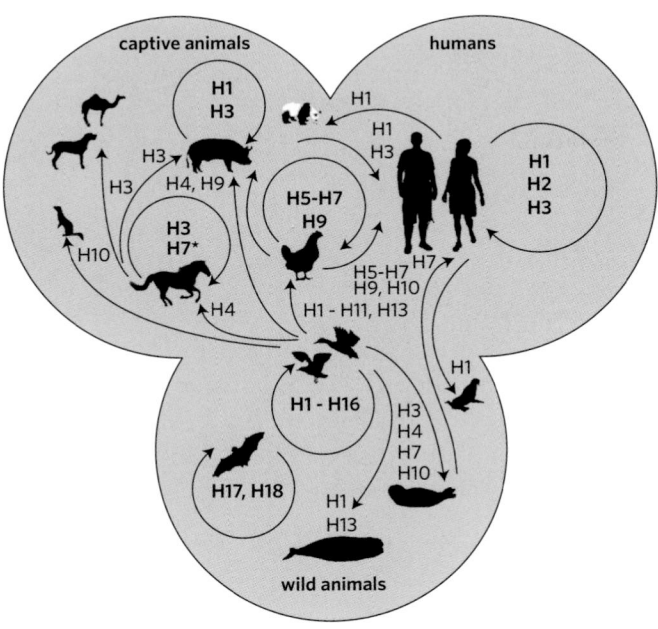

Figure 23.2. Reservoirs and interspecies transmission events of low-pathogenic avian influenza viruses (LPAIVs). Wild birds, particularly waterbirds, are the source of most circulating LPAIVs in nature. *From Short et al. 2015. Image available from http://www.sciencedirect.com /science/article/pii/S2352771415000026. Open access under CC BY 4.0.*

Bacteria

Bacteria are prokaryotes that, unlike viruses, are able to replicate on their own via binary fission. They contain genetic material in the form of DNA, but they do not contain a nucleus or membrane-bound organelles. Bacteria are ubiquitous on earth, and the majority are not pathogenic (i.e., do not cause disease in host). Those that are pathogenic are a common cause of diseases in wild birds and include enteropathogenic bacteria, which replicate in and cause disease of the intestinal tract (e.g., *Salmonella* spp.), and other non-enteropathogenic bacteria. Two of the latter that have had notable effects on avian populations include *Pasteurella multocida*, the causative agent of avian cholera (otherwise known as "fowl cholera") and *Mycoplasma gallisepticum*, the etiological agent of conjunctivitis in wild fringillid species.

Pasteurella multocida, with its various serotypes (different types within same species), has a global distribution. It has been isolated from bird species in most orders but affects primarily Anseriformes and Gruiformes (Samuel et al. 2007). Birds become infected through a variety of routes, including direct contact with infectious individuals or fomites, ingestion of contaminated water and food including infected carcasses, inhalation of aerosolized bacteria-laden water and respiratory droplets, and, albeit uncommon, even infected arthropod bites (Botzler 1991). The environment, particularly shallow water bodies that can host large numbers of migrating waterfowl, are a primary source of infection. The bacteria can survive for weeks to months in water, oil, and carcasses, which allows them to accumulate in the environment and facilitate cholera outbreaks (Friend 1999).

Once birds are exposed to the bacteria, they may or may not become infected, the likelihood of which depends on the dose concentration and frequency of exposure as well as the host's inherent susceptibility to the bacteria (Samuel et al. 2007). Once bacteria successfully invade and establish themselves in the host, they replicate rapidly, leading to massive bacteremia (bacteria in the bloodstream) that swiftly kills the bird. Given the acute nature of the disease, it is rare to find sick birds; detection of the bacteria in a population comes from incidental examinations of birds or large mortality events (see Samuel et al. 2007).

Known cholera outbreaks have been responsible for massive mortality of wintering waterfowl in western North America (Botzler 1991). One outbreak in 1970 and 1971 in California had mortality estimates of 7.5 million waterfowl over one winter (Rosen 1972). Despite these significant mortality events, the bacteria do not appear to cause population declines of waterfowl and coots, and the annual take by hunters still exceeds the losses from periodic cholera epizootics (Samuel et al. 2007).

Mycoplasma gallisepticum (MG) in wild birds is a unique lineage of the bacterial pathogen with the same name found in poultry. It was first identified in 1994 as the causative agent of conjunctivitis in House Finches in the eastern United States. Infected House Finches develop mild to severe conjunctivitis, symptoms of which include swollen eyelids, crusty lesions around the eye, and eye discharge (fig. 23.3). Systemic disease results in profound lethargy and weight loss; death is frequently caused by predation or starvation. MG is an extremely transmissible pathogen, but it does not survive well outside the host; conditions that are optimal for transmission are direct contact with infectious individuals and/or infectious fomites.

After the pathogen's emergence, it rapidly became an epizootic, spreading throughout the eastern population of House Finches, causing high morbidity and mortality (Fischer et al. 1997). Scientists were able to document the population decline of the finches before MG emergence through well-established citizen science bird monitoring programs (chapter 30) including Audubon's Christmas Bird Count (https://www.audubon.org/conservation/science/christmas-bird-count), the Cornell Lab's Project Feeder Watch (http://feederwatch.org/), and then through the quick initiation of the House Finch Disease Survey immediately following the bacteria's emergence (Hochachka and Dhondt 2000). With these data, scientists were able to document density-dependent population declines of House Finches (fig. 23.4; Hochachka and Dhondt 2000), demonstrating the important contributions of citizen scientists.

Figure 23.3. Mild (*left*) and severe (*right*) mycoplasmal conjunctivitis in female House Finches (*Haemorhous mexicanus*). *Photos courtesy of G. Hill (left) and Hilton Pond Center (right).*

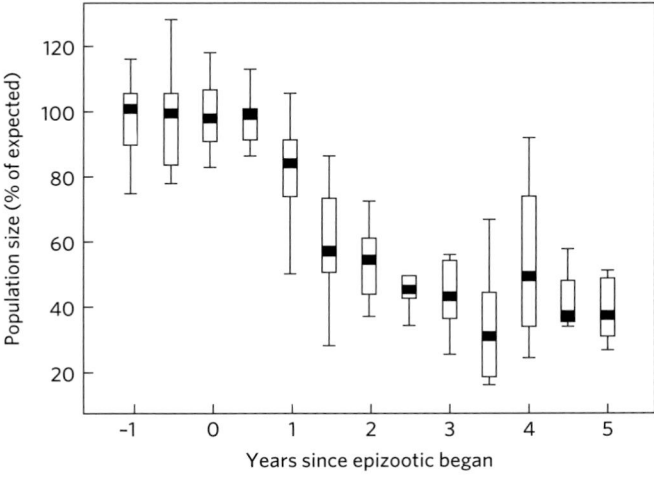

Figure 23.4. Population impacts of mycoplasmal conjunctivitis in House Finches (*Haemorhous mexicanus*). Changes in abundance of House Finches after the arrival of conjunctivitis in a region. The change in abundance is expressed as a percentage of the expected abundance if conjunctivitis did not affect House Finch numbers (see Hochachka and Dhondt 2000). Thick horizontal lines indicate medians and boxes span the interquartile range. Year 0 represents House Finch abundance in the last year before the epizootic reached the 20 percent threshold value in that specific area. *From Hochachka and Dhondt 2000. Copyright (2000) National Academy of Sciences, U.S.A.*

House Finches are not native to the eastern part of North America. Populations there were the result of an illegal release in Long Island, New York, in 1940, and their population expanded over the next five decades (Veit and Lewis 1996). After the emergence of MG and its rapid spread throughout the eastern population, the bacteria slowly spread westward and were first detected in the native western House Finches in the Great Plains in 2002 (Duckworth et al. 2003). Interestingly, the bacteria did not cause significant harm to the host (i.e., low virulence), and overall morbidity and mortality was much lower than observed in the far eastern population. Several hypotheses were put forward to explain this observation, but the one that is finding empirical support is the evolution of parasite virulence and the trade-off hypothesis, first proposed by Anderson and May (1982) and experimentally tested by Hawley et al. (2013). The evolutionary fitness of the parasite relies on its ability to replicate or reproduce and be transmitted from one host to the next, and transmission requires contact between an infected host and an uninfected, susceptible host. A highly virulent pathogen produces a lot of progeny in the infected host, which increases the probability that, upon contact, the infected host will transmit the pathogen to an uninfected bird. However, virulent pathogens also cause significant harm to the host, which may reduce their

contact rate with uninfected hosts—they are too sick to move or they die before they make contact. So, how does this relate to MG virulence in western House Finch populations? Well, it relates to spatial structure of the population (Messinger and Ostling 2009). The population density of the Great Plains House Finches was fairly low; hence, low virulence increases the pathogen's fitness by permitting infected birds to survive long enough and move far enough to contact a susceptible host. In dense populations, such as those in the eastern United States, high virulence would be selected for, since contact rate is not a limiting factor for transmission and high virulence ensures transmission. This exact scenario, high virulence and rapid spread of MG in eastern populations, was then observed when *Mycoplasma gallisepticum* independently emerged in the western United States, where House Finch numbers and densities were much higher (Hawley et al. 2013).

The bacteria have a wide host range; conjunctivitis has been observed in over 30 species, and the bacteria have been detected in far more species of passerines and near passerines (i.e., Columbiformes, Piciformes, and Passeriformes) (Dhondt et al. 2014). In addition to House Finches, other Fringillid species that are frequent visitors to bird feeders, such as American Goldfinch (*Spinus tristis*), Northern Cardinal, Purple Finch (*Haemorhous purpureus*), Evening Grosbeak (*Coccothraustes vespertinus*), and Pine Grosbeak (*Pinicola enucleator*) (Dhondt et al. 2014), are susceptible and exhibit conjunctivitis when experimentally infected with the bacteria, particularly American Goldfinches (Farmer et al. 2005).

Yet, despite this high susceptibility observed in controlled experimental infections, the prevalence of infected, free-living American Goldfinches is fairly low relative to House Finches. This disparity is not yet understood but may be because of longer-lasting, chronic infections in House Finches or more efficient transmission between House Finches relative to other species (Dhondt et al. 2014).

Bird-feeding stations have the potential to serve as hotspots for transmission of disease-causing agents; uninfected, susceptible individuals may be exposed to the bacteria through direct contact with infectious individuals or with infectious fomites, such as metal bird feeder parts where the bacteria can survive up to a couple of days (Hawley et al. 2007). In addition to MG, pathogens that are circulating and likely being transmitted between birds at feeding stations include *Salmonella typhimurium* and *Escherichia coli* bacteria, *Coccidia* and *Trichomonas* protozoa, *Aspergillus* fungi, and avian poxvirus (Brittingham and Temple 1988, Brittingham et al. 1988, Robinson et al. 2010). While there are correlative data that suggest infection prevalence is higher in birds that frequent feeding stations (Dhondt et al. 2007, Lawson et al.

DANA HAWLEY, ASSOCIATE PROFESSOR, VIRGINIA TECH UNIVERSITY

Dr. Dana Hawley, Associate Professor at Virginia Tech University in Richmond, Virginia, shares how she became a leading scientist investigating how individual behavior impacts disease dynamics in a unique host-pathogen system.

My interest in birds was first sparked as an undergraduate when I began to observe and study the fascinating social behaviors that many birds exhibit. I wondered why some individuals within a species were more social than others, and what consequences these behaviors might have in terms of disease susceptibility and spread. Although that was almost 20 years ago, my lab members and I continue to seek answers to these questions today. We use House Finches, a highly social and human-adapted bird species, as our model for understanding how individual behavior contributes to disease dynamics. House Finches congregate in large numbers at bird feeders during the nonbreeding season, creating the "perfect storm" for the spread of a debilitating eye disease, *Mycoplasma gallisepticum*. Because *M. gallisepticum* is highly contagious, we hypothesized that the most social House Finches—those individuals who consistently inter-

acted with many other members of the population—would have the highest risk of acquiring it. We also hypothesized that individual House Finches that relied more heavily on bird feeders, where the pathogen can thrive for short periods of time outside of a host, might also be at the highest risk for acquiring disease. We tested these hypotheses by using PIT-tag technology to track the behavior of free-living House Finches. We fitted the leg bands of each individual with a small chip (about the size of a grain of a rice) containing a unique bar code that automatically recorded each time a bird visited one of the monitored feeders. We then constructed social networks using the temporal patterns of feeder visitation: if two different birds were often seen feeding together, they were assumed to be closely linked in the social network. We monitored individuals throughout the fall and winter, when *M. gallisepticum* infections are most common, in order to track who acquired disease. In contrast to our predictions, House Finches that were most central in the social network did not have higher risk of acquiring disease. Instead, the extent of dependence on bird feeders was the strongest predictor of risk of infection: House Finches that spent the most time eating at our monitored feeders were at highest risk of disease. Our results suggest that bird feeders are more important for transmission of this pathogen than social interactions, which were previously thought to be the main culprit of disease spread. Bird feeders can have many positive benefits for bird populations, but our results suggest that, at least in the case of *M. gallisepticum* in House Finches, human feeding of birds can increase the risk of disease spread. More broadly, understanding how the diverse changes associated with urbanization (habitat change, species composition, bird feeding, etc.) influence disease spread is critically important as habitats across the globe continue to become increasingly urbanized.

2012), no studies directly link avian epizootics to feeding stations (Jones 2011).

Protozoa

Protozoa are single-celled microorganisms but are considerably larger than bacteria and viruses. The majority of protists are not parasites; however, a few do have the potential to cause disease in wild birds. A common group of parasitic protists are haemosporidian or blood parasites that infect the red blood cells of their hosts. In birds, blood parasites belong to three main families—Plasmodiidae, Haemopro-

teidae, and Leucocytozoidae. They are transmitted between birds via blood-sucking arthropods, including mosquitoes (*Plasmodium* spp., or avian malaria parasites), biting midges and hippoboscid flies (haemoproteids), and black flies (leucocytozoids). Of the bird species sampled to date (~half of the world's species; Valkiunas 2004), haemosporidian parasites have been detected in at least half. The haemoproteids are the most common followed by plasmodids and leucocytozoids. Passerines are the most commonly infected group, but most of the world's avian orders are known to be infected.

Three main stages characterize infection with a blood parasite—the prepatent period, during which the host is infected but the parasite is not yet detectable; parasitemia, during which parasite numbers increase and appear in the circulating red blood cells; and (presuming the individual survives acute infection) a latent period, during which parasite numbers decrease and become undetectable in the circulating blood but persist in the bird's tissues. These latent infections can reactivate, which typically occurs when the host's immune system is compromised and the parasites begin to replicate, leave the tissue, and reenter the circulating blood. These relapses frequently coincide with the breeding period, when the bird may be forced to make physiological trade-offs at the expense of its immune system (Sheldon and Verhulst 1996), and this immunosuppression promotes transmission of the parasite to naïve, susceptible offspring (Valkiunas 2004).

In most cases, infected birds do not develop outward signs of infection, which may be a consequence of the bird having a low-level infection or an inherent tolerance to the parasite (i.e., regardless of the intensity of the parasitemia, host fitness is not altered). If a bird is going to exhibit clinical signs of malaria, it will occur during the parasitemia period, and the primary pathogenic impact of the parasite is the destruction of the infected red blood cells by causing anemia, which can range from mild to severe. The reduced red blood cell volume also decreases hemoglobin levels and the oxygen-carrying capacity of the blood. High parasitemia may also cause a grossly enlarged liver and spleen that, in severe cases, may cause the organs to rupture.

Of the malarial parasites, *Plasmodium* spp. have had the largest impact on bird populations, with *Plasmodium relictum* being the most pathogenic protist to birds. *P. relictum* has a worldwide distribution (except Antarctica), a broad avian host range, and has been responsible for dramatic population declines and extinctions of birds around the world. Most notably, *P. relictum,* coupled with avian poxvirus, led to the disastrous population declines and extinctions of endemic avifauna in Hawai'i (Warner 1968). However, these two pathogens were not solely to blame for these outbreaks. It was actually the accidental introduction of the parasite's primary vector, *Culex quinquefasciatus,* in the early 1800s and then, many years later, the deliberate release and establishment of nonnative birds that were naturally resistant to the parasite but served as a source of the parasite, that led to the epizootic (van Riper et al. 1986, van Riper et al. 2002). These introductions have dramatically changed the ecology, distribution, and behavior of native and nonnative birds in the Hawaiian archipelago (box on page 742; fig. 23.5).

Even in the absence of morbidity and mortality, malarial parasites may have life history consequences for infected

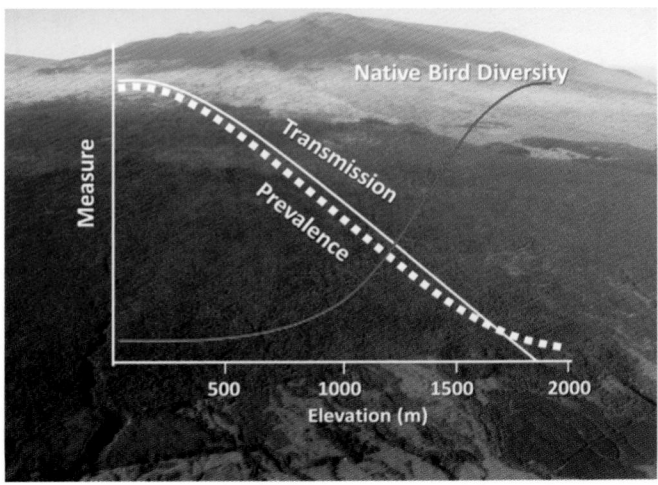

Figure 23.5. Windward, eastern slope of Mauna Kea volcano on the Island of Hawai'i. Native forest below 600 m (*bottom*) has mostly been lost to intensive agricultural development, while forest above approximately 1,800 m (light green band, *top*) has been severely degraded or lost to cattle ranching. Transmission and prevalence of both avian poxvirus and malaria are greatest at elevations below 1,000 m and decline as cooler temperatures limit mosquito populations and extrinsic development of malarial parasites. Refugia at elevations above approximately 1,500 m provide a safe haven from disease transmission for threatened and endangered species such as the 'Akiapōlā'au (*Hemignathus wilsoni*), Hawai'i Creeper (*Loxops mana*), and Hawai'i 'Ākepa (*Loxops coccineus*). This area is limited to a narrow band of forest just below the tree line, where remaining populations of native forest birds reach their highest diversity. *Courtesy of C. Atkinson.*

birds. Breeding female Great Reed Warblers (*Acrocephalus arundinaceus*) with high parasitemia of malarial blood parasites exhibited delayed arrival to breeding grounds, fewer fledged offspring, and fewer offspring that survived to reproductive age than uninfected or mildly parasitemic females (Asghar et al. 2011). Likewise, Western Bluebird (*Sialia mexicana*) males without active malarial infections sired more extra-pair offspring than actively infected males (Jacobs et al. 2015). And, in a controlled study, Marzal et al. (2005) compared reproductive success of Common House Martins (*Delichon urbicum*) infected with malarial blood parasites (*Haemoproteus prognei*) with conspecifics given an antimalarial in the same population. The treated birds had larger clutches and higher hatching and fledging success than heavily parasitized individuals. These observed fitness costs are certainly not universal; many studies have found no fitness cost of a malarial infection (Szöllősi et al. 2009, Knutie et al. 2013).

Fungi

Fungi are single-celled or multicellular organisms that are ubiquitous in nature, serving an important ecosystem

CARTER ATKINSON, ECOLOGIST, USGS, HAWAI'I VOLCANOES NATIONAL PARK

Dr. Carter Atkinson, Ecologist with the United States Geological Survey Hawai'i Volcanoes National Park in Hilo, Hawai'i shares how he became a leading scientist investigating the impacts of vector-borne diseases on the endemic avifauna of Hawai'i.

I've always had an interest in the natural world, but it wasn't until high school that I became enthralled with trying to identify resident birds by sight and sound in my rural Maryland neighborhood. Avian disease never crossed my mind, however, until the summer after finishing my undergraduate degree, when I was lucky to be hired as a field technician on the first comprehensive forest bird survey of the Hawai'an Islands. Seeing distributional anomalies among native species firsthand led to a budding interest and appreciation of the complex interactions among hosts, vectors, and pathogens, and led to my graduate work on vector-borne wildlife disease and eventual work in the Hawai'an Islands with the US Geological Survey.

Epizootiology of Avian Malaria in Hawai'i

The geographic isolation that fostered spectacular adaptive radiation of native Hawai'an forest birds eventually contributed to the demise of this remarkable avifauna as invasive plant and animal species followed the tide of human migration to the islands. Hawai'i was once one of few subtropical places on the planet free of mosquitoes and the wide range of avian pathogens that they transmit. Avian poxvirus, the first of two major vector-borne avian pathogens in Hawai'i, reached the islands sometime before the late nineteenth century, with avian malaria (*Plasmodium relictum*) following

in the early twentieth century. Both introductions probably came through intentional release of almost 40 species of nonnative passerines between 1870 and 1950—a period when native forest birds were rapidly disappearing throughout the islands from the effects of introduced predators and habitat alterations associated with agricultural and urban development. Both pathogens probably spread rapidly at lower elevations where *Culex quinquefasciatus* is active throughout the year, but spread more slowly at elevations above 1,000 m, where cooler temperatures begin to limit mosquito populations. While avian poxvirus is mechanically transmitted on the mouthparts of mosquitoes, *P. relictum* requires a temperature-dependent period of extrinsic development within the mosquito vector in order to become infectious. This temperature dependence also contributed to rapid drops in transmission at elevations above 1,000 m, particularly during the cooler winter months. The result was an altitudinal gradient of disease transmission that peaked in the warm lowlands and dropped precipitously at elevations above 1,500 m (see figure 23.5). While native species all but disappeared on the lower Hawai'an Islands, high elevation refugia on the higher islands of Kaua'i, Maui, and Hawai'i provided some protection from mosquitoes and disease transmission and allowed many highly susceptible native species to persist into the twenty-first century. These refugia are currently threatened by warming temperatures and changing rainfall patterns associated with global climate change. While some high-elevation populations on the island of Kaua'i are beginning to collapse in response to increased disease transmission, at least one common species (Hawai'i 'Amakihi, *Chlorodrepanis virens*) on Hawai'i Island is responding to intense selection pressure at lower elevations and evolving tolerance to infection with malaria. It remains to be seen whether remaining high-elevation populations of the most endangered species have the genetic diversity to adapt to increased disease transmission without direct intervention to control or eliminate mosquito vectors.

function as decomposers. Some fungi are pathogenic to plants and even fewer are pathogenic to animals. In animals they are commonly opportunistic pathogens; they can invade and proliferate in the host when there is an injury to tissue and when host immune defenses are compromised. Hence, fungal disease outbreaks are usually associated with unrelated

extrinsic or intrinsic stressors. Aspergillosis is one of the most common fungal diseases and is caused by a multicellular, saprophytic *Aspergillus* species, of which *A. fumigatus* has had the biggest impact on birds (Converse 2007). This fungus is zoonotic and poses a risk to humans handling infected birds.

Birds become infected when they inhale the fungal spores. In the aerobic environment of the host, the spores produce plaques of hyphae (i.e., vegetative growth structures) that colonize the tissues and lead to acute, localized infections. Systemic infections occur if the host inhales a high dose of the spores that overwhelm the host immune system or if the host already has a weakened immune system. Bird mortality from aspergillosis is common in captive wild birds—a consequence of the stressful conditions, such as inadequate nutrition, crowding, and human handling (Carrasco et al. 2001).

More often than not, systemic aspergillosis and other fungal infections are detected in birds with concurrent infections, making it difficult to attribute morbidity or mortality to a particular pathogen (Penrith et al. 1994). Wild-caught Magellanic Penguins (*Spheniscus magellanicus*) from southern Chile were brought into captivity, and within five months, 36 of the 46 birds died. While *P. relictum* was considered the leading cause of death, over 60 percent of the birds were positive for *Aspergillus*, as well as bacterial and helminth parasites (Fix et al. 1988).

Biotoxins

Biotoxins are nonliving substances produced and excreted by living organisms and are causative agents of disease. Botulinum toxin, a neurotoxin that causes botulism in birds, is produced by the spores of the bacteria *Clostridium botulinum*; it is the most lethal naturally occurring toxin on earth. Seven serotypes (i.e., distinct variations/strains of a microorganism or cell) of *C. botulinum*–derived neurotoxins exist, named as types A through G. Types C and E are responsible for large-scale epizootics in wild birds. Type C has a worldwide distribution, but type E is restricted to the Great Lakes region of North America.

While all birds are likely susceptible to botulinum toxins, some species are more likely to be exposed to the toxin than others, largely because of their foraging behavior and preferred prey. The *C. botulinum* bacteria lives in the soil and thrives under anaerobic and warm conditions, which favor the germination of the bacterial endospores (a structure within the bacterial cell that allows the cell to become dormant and withstand extreme conditions) and subsequent excretion of the botulinum toxin. The species most likely to be exposed to type C botulinum toxin are those that feed in shallow water or mudflats, such as the filter-feeding and dabbling ducks in the order Anseriformes and the shallow probing shorebirds in the order Charadriiformes. In contrast, bacteria that produce type E botulinum toxin are in the sediment on the lake bed and therefore quite deep. Fish-eating diving birds, such as grebes, loons, and cormorants, are the most likely to be exposed to type E because of their diet of bottom-dwelling fish and macroinvertebrates. Clinical signs of botulism include neurological disorders, inability to lift the neck (called limberneck), and partial paralysis. Once a bird becomes sick, they rarely recover; hence, it is one of the leading causes of death of waterbirds.

Type C botulism epizootics occur on a much broader scale and have been linked to the annual mortality of hundreds of thousands of waterbirds annually, particularly in western North America (Rocke and Friend 1999). However, because other diseases such as Newcastle disease (caused by Newcastle disease virus, or NDV) and avian cholera also affect waterbirds, the prevalence of botulism is likely underreported (Rocke and Bollinger 2007). Type E botulism outbreaks in the Great Lakes became fairly regular starting in the early 2000s. The invasion of nonnative benthic-foraging invertebrates (i.e., dreissenid mussels) and bottom-dwelling fish (i.e., Round Goby, *Neogobius melanostomus*) may play a role in the outbreaks; however, their involvement is not confirmed. Environmental factors are a more likely culprit, with the timing of outbreaks corresponding closely with drops in water levels and higher surface water temperature (Lafrancois et al. 2011).

MACROPARASITES

Parasitic Helminths

Parasitic helminths encompass all the wormlike, multicellular, eukaryotic macroparasites and include nematodes, flatworms, tapeworms, and trematodes. The helminth's development from egg to adult typically occurs in different host organisms (fig. 23.6). In general, sexual reproduction occurs in a definitive host, which then excretes the eggs into the environment, where an intermediate host for the parasite may ingest them. The egg and larval development occur in the intermediate hosts, potentially requiring multiple hosts to complete development. Birds are a common definitive host, particularly piscivorous wading birds (e.g., Ciconiiformes, Gaviiformes, Podicipediformes, and Pelecaniformes), which forage on invertebrate and vertebrate intermediate hosts.

While helminth infections are common in birds, they rarely make birds sick and frequently go undetected. Helminths, like other macroparasites, do not replicate within the definitive host, so parasite burdens depend on the number of parasites ingested over the host's life span; morbidity and mortality are commonly observed in older individuals within a population. However, juvenile birds with low parasite burdens may also exhibit significant morbidity and mortality because of their less-developed immune system.

In some instances helminths have led to epizootics that cause significant mortality in avian populations. Eustrongylidosis, a disease caused by several different species of

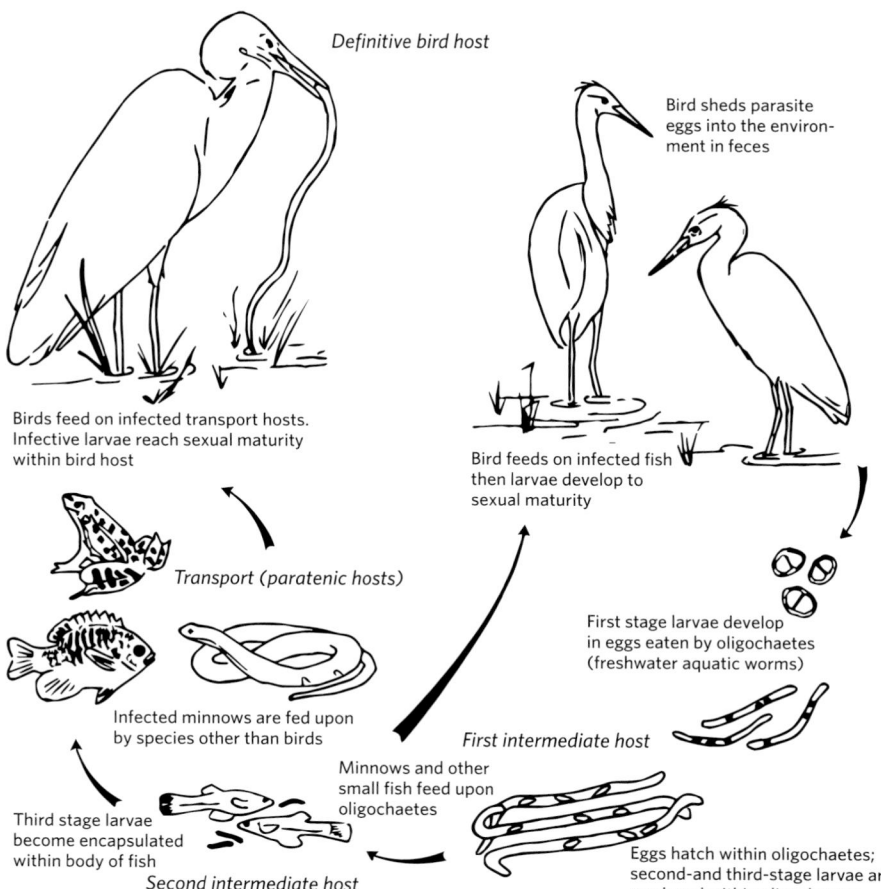

Definitive bird host

Bird sheds parasite eggs into the environment in feces

Figure 23.6. The life cycle of *Eustrongylides* spp., a helminth parasite in which wading birds (Ciconiiformes) are the definitive host. *From Cole 1999. By permission of U.S. Geological Survey Department of the Interior/USGS U.S. Geological Survey.*

Birds feed on infected transport hosts. Infective larvae reach sexual maturity within bird host

Bird feeds on infected fish then larvae develop to sexual maturity

Transport (paratenic hosts)

First stage larvae develop in eggs eaten by oligochaetes (freshwater aquatic worms)

Infected minnows are fed upon by species other than birds

First intermediate host

Minnows and other small fish feed upon oligochaetes

Third stage larvae become encapsulated within body of fish

Eggs hatch within oligochaetes; second-and third-stage larvae are produced within oligochaetes

Second intermediate host

Eustrongylides nematodes or roundworm parasites, is the most common cause of death of nestling wading birds (Spalding and Forrester 2009). In a breeding colony of herons and egrets in central Florida, *Eustrongylides ignotus* was the causative agent of disease that killed 80 percent of the nestlings (Spalding and Forrester 1993). Similar events have occurred throughout the United States and in Europe, and are often associated with anthropogenic eutrophication of our aquatic ecosystems (Caudill et al. 2014).

Helminths also have an impact on terrestrial ecosystems. A notable example is the host-parasite interaction of the Red Grouse and the nematode *Trichostrongylus tenuis* in England and Scotland. Unlike other macroparasites, *T. tenuis* does not have intermediate hosts; the grouse sheds eggs into the environment and, under appropriate climatic conditions (i.e., above −15°C and humid) (Connan and Wise 1994), the offspring develop through the four larval stages. The third infective stage larvae dwell in the low-lying vegetation and are ingested by the grouse while foraging, after which they reside in the bird's caeca, where they can impact digestive processes when parasite burden is high. The formative experimental work by Peter Hudson and colleagues (Dobson and

Hudson 1992) linked the annual fluctuations in the grouse population to the presence of the parasite, the first such example for population regulation by a parasite (fig. 23.7).

Ectoparasites

Anyone that has handled many birds is familiar with a variety of ectoparasites that are found on wild birds—from the creepy hippoboscid flies that crawl out onto your hand to the engorged ticks that may be found around a bird's head. If you look more closely, you may notice smaller ectoparasites on the bird's feathers—such as the feather-degrading bacteria and feather-chewing lice (see chapter 9). Ectoparasites are parasites that live on the external surface of the host, either attaching onto or burrowing into the skin or feather, and include flies, lice, mites, fleas, mosquitoes, and ticks.

Birds are commonly infested with ectoparasites, but these parasites rarely cause significant host morbidity and mortality, except in cases of high infestations. A bird's largest protection from ectoparasites is through behavioral defenses, such as abandoning infested nests and removing parasites through preening and allopreening. To prevent the attachment of parasites, birds may engage in activities such as hop-

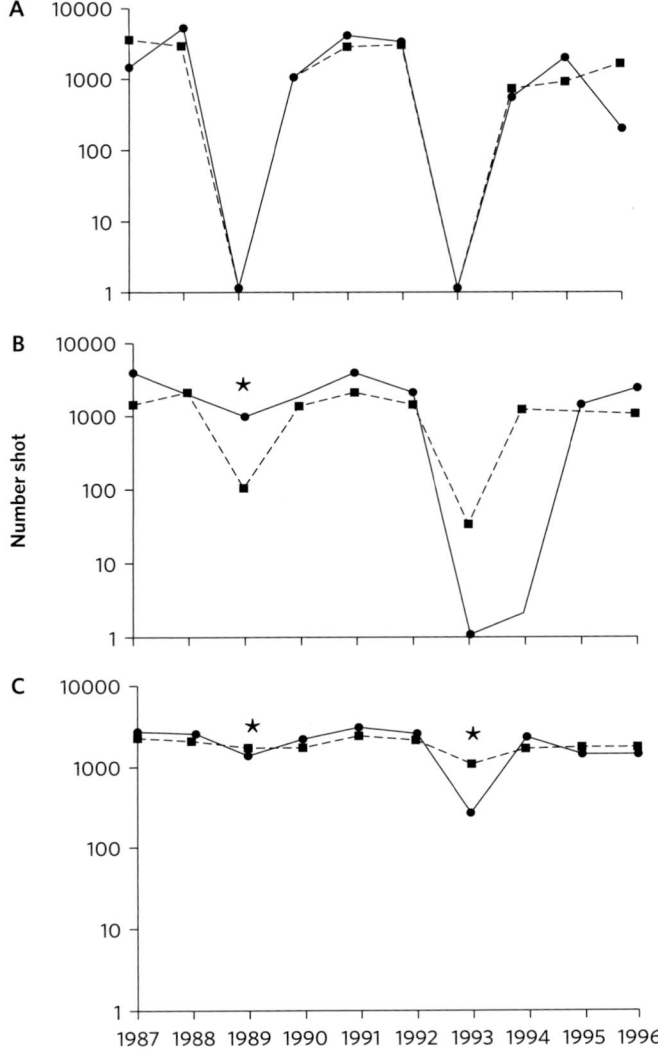

Figure 23.7. Population changes of Red Grouse (*Lagopus lagopus scoticus*), as represented through bag records in *A*, the two control sites; *B*, the two populations, each with a single anti-helminthic treatment; and *C*, the two populations with two treatments each. Asterisks represent the years of treatment when worm burdens in adult grouse were reduced. *From Hudson et al. 1998. Used with permission.*

ping, flapping wings, bill snapping, defensive postures (i.e., tucking bill and leg) (Hart 1997), and protecting feathers with secretions from the uropygial gland, which have insecticidal and antimicrobial properties (Magallanes et al. 2016). High infestations of ectoparasites are typically seen in birds that are not efficient at preening or have more exposure to the parasite, such as nestlings. Adults are vulnerable to infestations when they are weak or lethargic for other reasons and are unable to defend themselves. These high infestations lead to anemia from blood loss and a suppressed immune system and can make the individual more susceptible to

other infections. Of the many ectoparasites, ticks and parasitic botfly larvae have the most profound impact on nestling birds (Dudaniec and Kleindorfer 2006).

Dipteran (botfly) larvae (genus *Philornis*) feed on the bird's skin and can cause a condition called myiasis, where the larvae penetrate the subcutaneous layer of tissue and produce deep lesions (Little 2008). Nestlings, particularly altricial species, are especially susceptible. Bird nests provide an optimal environment by providing the adult flies with suitable substrate on which to lay eggs and then a source of nourishment via blood meals on nestlings for larval development. Once the larvae progress through all the larval stages, they migrate back into nest material to pupate. High infestations of the larvae can lead to high blood loss and deformation of the nares and bill, which directly affects foraging and survival of young birds (Galligan and Kleindorfer 2009, O'Connor et al. 2010). The extent of infestations varies among nests and bird species, but they can have devastating impacts on nestling survival (Dudaniec and Kleindorfer 2006).

One of the most notable impacts of *Philornis* parasitism is in the Galápagos Islands where *Philornis downsi*, a nonnative species, was first detected in bird nests in 1997. This parasite is considered to be one of the largest threats to landbird nestlings in the archipelago, with nestling mortality reaching 90 percent in affected Darwin finch populations (O'Connor et al. 2010, Kleindorfer et al. 2014). The species with the highest infestation rates, the Woodpecker Finch (*Camarhynchus pallidus*) and Warbler Finch (*Certhidea olivacea*), are suffering the greatest population declines (Dvorak et al. 2012). The parasite is thriving in the region, likely because of its ability to tolerate the climatic extremes characteristic of the Galápagos Islands (Koop et al. 2013).

Ticks (phylum Arthropoda, class Arachnida) are another common ectoparasite on birds. Exposure to ticks depends on the tick's foraging behavior and habitat preferences. Tick parasitism is most commonly seen in forest-dwelling and ground-feeding birds, which are most likely to come in contact with questing (host-seeking) ticks on the low-lying vegetation. Once a tick jumps onto the host, they crawl to featherless areas before attaching for a blood meal. Birds can rid themselves of many ticks through preening; hence, ticks on birds are most likely found in places that are difficult for the bird to reach, such as the inner ear, eyelids, and under the lower mandible (fig. 23.8). The greatest impact of tick infestations has been documented in nestlings of colonial, ground-nesting seabirds, particularly Sphenisciformes and Procellariiformes (Dietrich et al. 2011); high chick mortality and colony desertion have been linked to years of high tick infestations both in nests and on chicks (Haemig et al. 1998, Bergström et al. 1999).

Figure 23.8. Replete, larval hard ticks in the ear of a Lincoln's Sparrow (*Melospiza lincolnii*) captured during fall migration in mid-Michigan. *Photo courtesy of Z. Pohlen.*

Another important consequence of tick infestations on birds is the bird's ability to move pathogen-infected ticks over large geographic distances during migration (Hasle 2013). In many cases the birds act as mechanical carriers of the infected ticks without harboring the pathogen themselves. It is not uncommon to find ticks on birds during migration—particularly ground-dwelling birds, such as thrushes, have much higher infestation rates than other species. However, when considering all species captured, the prevalence of ticks on birds is fairly low, rarely exceeding 5 percent of birds examined. Moreover, very few of those are infected with a tick-borne pathogen, with infection rates varying by tick species, region, and season. The tick-borne pathogens detected in ticks attached to birds include tick-borne viruses and bacteria species in the genera *Borrelia, Babesia, Anaplasma,* and *Rickettsia.* Most notable is *Borrelia burgdorferi,* the causative agent of Lyme disease in humans, which has been isolated from *Ixodes* spp. ticks attached to migrating birds captured at stopover sites from around the world, including North America, Europe, and Asia (Olsén et al. 1995, Bjöersdorff et al. 2001, Ogden et al. 2008, Hamer et al. 2012).

For most tick-borne pathogens, the enzootic cycle is maintained by a mammalian host and a tick vector. However, birds are known to be infected with these pathogens and in some cases serve as the source of infection to blood-feeding ticks, although their role in the enzootic cycle of the pathogen is not well understood. The little we do know is that, much like West Nile virus, the reservoir competence of birds varies with the species of tick, pathogen, and avian host. For instance, when a naïve *Ixodes scapularis* tick takes a blood meal from a *B. burgdorferi*–infected American Robin, the tick has a moderate chance of becoming infected, whereas when feeding on an infected Gray Catbird (*Dumetella carolin-*

ensis) the tick is unlikely to become infected (Ginsberg et al. 2005). The same result was found when testing competence for *Anaplasma* bacteria (Johnston et al. 2013). An exception to this limited role of birds is found in seabird colonies, where the enzootic cycle of the bacteria *Borrelia garinii,* also a cause of Lyme disease, appears to be able to persist in the absence of mammals and is commonly isolated from both seabirds and *Ixodes uriae* ticks (Olsén et al. 1995). Most, if not all, tick-borne pathogens are nonpathogenic to wild birds.

BIRD MOVEMENT OF PATHOGENS

As reservoir hosts, birds can act as biological carriers of the pathogens discussed above as well as others and, if migratory, have the potential to disperse the pathogens long distances and introduce them into new geographic areas. Migrating birds have repeatedly been implicated in the spread, introduction, and establishment of pathogens around the world, such as West Nile virus, HPAIV, and *Plasmodium relictum,* to name a few (Peterson et al. 2003, Weber and Stilianakis 2007, Levin et al. 2013).

But before we implicate migrating birds in the spread of these and other pathogens, it is important to consider the biology of the bird, pathogen, and vector. For a bird to successfully move a pathogen and introduce or reintroduce the pathogen to a new geographic area, several criteria need to be met. First, once a bird is infected it must fly to the new area and, upon arrival, still be infectious (capable of infecting others through indirect or direct transmission). In addition, susceptible hosts and appropriate conditions must be present (environment, vector) for transmission to occur. While this scenario is clearly plausible, it is actually quite difficult to document its occurrence in wild populations. Given the transient nature of most infections, particularly with microparasites, attempting to detect an infectious bird during migration is much like looking for a needle in a haystack. Further, being able to track the movement of the infectious individual to another geographic area has both logistical and ethical challenges. The ethical dilemma is that if a bird is captured and infectious with a zoonotic pathogen, which poses a potential risk to humans, it cannot knowingly be released back into the wild where it may contribute to future infections. Tracking the intercontinental movements of an infectious bird for the entire duration of their migration requires sophisticated tracking technologies, which are not currently available for small birds. Recent studies have started to use these techniques: van Gils et al. (2007) isolated LPAIV from two Bewick's Swans (*Cygnus columbianus*), placed GPS collars on them as well as on three uninfected swans, and then tracked their movements and foraging behavior during fall migration. In short, they found that infected birds experienced a monthlong delay in their migration, as a consequence of lower refueling and moving shorter distances, compared with uninfected birds. This and other studies are beginning to advance our knowledge about the movement of infectious birds, but their influence is limited by small sample sizes (van Gils et al. 2007, Bridge et al. 2014).

Another approach to understanding a bird's motivation and capacity to continue to migrate while infectious is to take advantage of the well-established techniques for studying migratory behavior of wild-caught captive birds (see chapter 19). We conducted an experiment in which we induced wild-caught, hatch-year Gray Catbirds and Swainson's Thrushes (*Catharus ustulatus*) to enter migratory disposition by altering photoperiod (Owen et al. 2006). Once the birds were exhibiting migratory restlessness, or *Zugunruhe*, we inoculated them with WNV and then looked for changes in restlessness, or the proportion of the nighttime they were active during the three-day period when they were infectious (i.e., virus circulating in blood, or viremia). We found that all the catbirds' and 6 of the 10 Swainson's Thrushes' migratory restlessness did not change, and they continued to exhibit migratory behavior during this infectious period, which suggests that in the wild they could disperse the virus through migration. The remaining four thrushes stopped activity altogether during the viremic period, which tells us that there is a lot of intraspecific variation in response to infection.

The extent to which migrating birds are involved in the geographic spread of pathogens is far from understood (Altizer et al. 2011, Hall et al. 2014); however, it is clear that bird migration has broad implications for the health of birds, humans, and other wild and domestic animals. Bear in mind, migration is not the only mechanism by which birds can move pathogens. Postnatal dispersal, when large numbers of susceptible juvenile birds move away from their natal territory, may lead to localized amplification and movement of pathogens.

ENVIRONMENTAL CONDITIONS UNDERLYING PATHOGEN EMERGENCE AND DISEASE OUTBREAKS

Disturbances, both environmental and ecological, influence the occurrence, distribution, and severity of disease outbreaks. Consider the epidemiological triad/tetrad (fig. 23.1): disease occurs only when the host, pathogen, vector (if applicable), and suitable environment are present and interacting. Hence, changes in the environment have the potential to profoundly influence the dynamics of host-pathogen-vector interactions, largely by leading to circumstances in which a naïve host is

exposed to a novel pathogen for which it has no resistance, or by affecting the frequency of transmission. The drivers of environmental change include increasing globalization, changing land use practices (i.e., urbanization, agricultural, deforestation), agricultural intensification and expansion, climate change, introduction of nonnative species, and expanding human populations (Sehgal 2010).

Increasing globalization of human trade and travel is a leading factor responsible for the introduction of pathogens into areas with a naïve and susceptible host population. Intercontinental movements of humans and products by air was a likely pathway for the introduction of WNV into the Western Hemisphere in 1999; given the high volume of air traffic into the New York City area, it is quite probable that WNV-infected mosquitoes or larvae were inadvertently transported on or in a plane (i.e., container with water, tire wheels) that originated from the Middle East where the specific strain of WNV detected was endemic (Kilpatrick 2011). Even the accidental introduction of uninfected, nonnative arthropod vectors into new areas can lead to devastating epizootics of vector-borne pathogens. As described above, the accidental introduction of *Culex quinquefasciatus*, the southern house mosquito and principal vector for *Plasmodium relictum*, occurred in the early nineteenth century and by all historical accounts arrived on sailing cargo ships, although the exact year and origin of the ships is debated (Hughes and Porter 1956). Likewise, *C. quinquefasciatus* was introduced into the Galápagos in 1989, and while *P. relictum* has yet to be detected on the islands, it poses a significant threat and could be potentially devastating to the endemic and naïve Galápagos avifauna (Bataille et al. 2009).

Pathogens are also introduced through the intentional transport of birds, either for the exotic pet trade, translocation of free-living species, or trade of domestic poultry. As described by Carter Atkinson (box on page 742), *P. relictum* was likely introduced into the Hawaiian Islands multiple times through the deliberate release of nonnative species carrying this malarial parasite. Avian influenza virus outbreaks can be spread through the trade of infected poultry before the virus has been detected in the flock, an event more likely to occur in areas with less rigorous biosecurity practices.

Subtle changes to the environment can also have far-reaching consequences for disease emergence; the expansion or shift in the geographic and altitudinal distributions of the host, pathogen, or vector can expose a naïve host to a pathogen for the first time. Climate change and the global warming trends can lead to shifts in the geographic ranges of both birds and arthropod vectors. In Hawai'i, warmer temperatures are facilitating the principal vector's ability to survive and become established at higher altitudes, which exposes naïve endemic bird populations to the *Plasmodium relictum* for the first time (Benning et al. 2002). Just a few degrees Celsius increase in annual or seasonal temperatures can have profound effects on the development and reproduction of both the mosquito and parasite; it can speed up the mosquitoes' development from egg to adult, increase biting rate, extend their breeding season, and lead to higher reproduction and faster development of the parasite within the mosquito vector. Together, these increase the probability of an infected mosquito coming into contact with and infecting the susceptible host. Global warming also impacts the environment in ways that promote the persistence and/or survival of non-vector-borne pathogens, particularly for pathogens that can remain viable in the soil or water, such as those that cause aspergillosis, avian botulism, avian cholera, avian influenza, and Newcastle disease (Fuller et al. 2012).

In the absence of a novel exposure, epizootics can also occur when environmental conditions change leading to an increase in the frequency of transmission events. Changes in land use, such as agricultural intensification and expansion, can alter the contact rate between susceptible individuals and the pathogen by altering habitat availability, population densities, and movement rates. In multi-host disease systems, the increase in population of intermediate hosts can lead to disease outbreaks, even when all else is equal. For instance, anthropogenic eutrophication, caused by the excessive runoff of nutrients generated from agricultural practices, sewage, and erosion, has significantly changed the ecology of aquatic ecosystems. Freshwater snails, which serve as intermediate hosts for some trematode parasites that infect birds, are tolerant of and benefit from the buildup of high nutrients; hence, eutrophic conditions are frequently associated with disease outbreaks (Johnson and Carpenter 2008). Restoration of eutrophic ecosystems to reduce nutrient pollution is linked to the lower prevalence of helminth-infected birds (Coyner et al. 2003). Irrigation practices also influence disease dynamics by altering the availability of water that provides breeding areas for arthropod vectors. Japanese encephalitis virus, closely related to WNV, is spreading across Southeast Asia as a consequence of the increase in irrigation associated with rice production. The virus is maintained in nature by the principal vector, *Culex tritaeniorhynchus,* a species that commonly breeds in irrigated fields. The principal hosts include both pigs and wading birds such as herons and egrets (Erlanger et al. 2009).

WILD BIRDS AND ONE HEALTH

Disease dynamics within avian populations cannot be adequately understood or mitigated without a comprehensive understanding of the complex interactions occurring at the scale of the organism, population, community, landscape, and ecosystem. Further, it should also be apparent that the health of birds is inextricably linked to the health of humans, domestic animals, and the ecosystem. The formal recognition of this interconnectedness is the basis for the relatively new field—One Health—that takes a multidisciplinary approach to understanding the events and drivers of disease emergence and outbreaks as well as developing strategies to mitigate their impact.

Many bird-borne pathogens are zoonotic and pose a risk to human and domestic animal health. The expansion of human settlements and their subsequent encroachment into wildlife habitat has led to the repeated spillover of novel pathogens between human and wildlife populations. Agricultural practices have led to higher contact rates between wildlife and domestic animals, and this is exemplified by the outbreaks of avian influenza and Newcastle disease in poultry operations worldwide, which together pose the highest economic cost to poultry operations globally. The causative agent of Newcastle disease is avian paramyxovirus (APMV), which, similar to HPAIV, comes in highly virulent forms that can cause up to 100 percent mortality in domestic poultry. The source of APMV is wild birds, particularly cormorants; both Double-crested Cormorants (*Phalacrocorax auritus*) and Great Cormorants (*Phalacrocorax carbo*) are linked to Newcastle disease outbreaks. Newcastle disease does have health and population impacts on cormorants; juvenile Double-crested Cormorants (< 18 weeks) are highly susceptible to the virus, with mortality rates as high as 100 percent in infected colonies (Kuiken et al. 1999). This chapter provides just a primer of the pathogens and parasites infecting birds and the diseases they cause. There is much more to learn from existing literature, and even more research yet to be conducted, to fully understand how pathogens and parasites affect wild bird populations.

KEY POINTS

- Infectious disease of wild birds is the outcome of the interaction among the host, pathogen, environment, and sometimes an arthropod vector.
- Birds are commonly infected with one or more pathogens, including viruses, bacteria, fungi, protozoa, biotoxins, and helminths. Some of these, but not all, cause disease in birds and may lead to epizootics in which a large proportion of a population is affected.
- Even in the absence of outward signs of disease, being infected with a pathogen can have fitness costs and life history consequences.

- Pathogens can have profound impacts on birds; they can alter population structure and dynamics, community interactions, and ecosystem functioning. Likewise, larger-scale processes can feed back and influence occurrence of disease.
- Pathogen emergence and disease outbreaks are frequently driven by environmental and ecological disturbances. The invasion or introduction of nonnative species is one of the largest drivers of disease outbreaks in isolated bird populations.
- Some bird-borne pathogens are zoonotic, such as West Nile virus and highly pathogenic avian influenza virus, and pose a risk to humans and domestic animals.

KEY MANAGEMENT AND CONSERVATION IMPLICATIONS

- To manage or prevent a disease outbreak, you have to "break" only one connection of the epidemiological triad/tetrad. For instance, administering vaccines breaks the link between the pathogen and the host by preventing the host from becoming sick and/or infectious with the pathogen.
- Small, isolated island populations of endemic avifauna are the most vulnerable to the introduction of novel pathogens and/or vectors and have experienced the greatest population declines and extinctions compared with mainland species around the world.
- Environmental and ecological changes can exacerbate the frequency and severity of avian epizootics. The increasing globalization of trade and travel, changes in land use, and climate change have played significant roles in the emergence and spread of bird-borne pathogens.
- Given the complexity of host-parasite-environment-vector interactions, a multidisciplinary One Health approach is needed to fully understand the factors underlying disease emergence and to develop strategies to minimize risk to avian populations.
- Citizen scientists can make vital contributions to our understanding about disease in wild bird populations. By observing birds and collecting data, citizen scientists can document disease outbreaks and both the extent and spread of disease, as well as the magnitude of its impact on avian populations.

DISCUSSION QUESTIONS

1. You are tasked to identify the causative agent of a large mortality event of birds in a localized area. How would you approach the investigation? What information, data,

and samples would you collect? What are the limitations of your investigation and possible confounding factors that may affect your conclusion?

2. Based on what you learned in this chapter and the knowledge you already possess about climate change, generate a list of predictions about how climate change will affect disease dynamics in wild bird populations (i.e., change in disease severity, frequency, population impact, distribution).

3. Backyard bird feeding is a favorite pastime; yet, there is some evidence (box on page 740) that feeders may be potential hotspots for pathogen transmission. Few studies have been able to confirm the role that feeding stations play in pathogen transmission and associated disease outbreaks. How would you go about designing an experiment to test the direct effect of bird feeders on local disease dynamics for a particular host-parasite system?

4. Migratory birds have been implicated in the intercontinental spread of pathogens. Given what you have learned in previous chapters about migration, what characteristics of a migrating bird (i.e., behavior, physiology, flight, stopover/staging) would support or not support this claim/hypothesis?

5. Small, isolated populations of birds are at the greatest risk of extinction from a disease outbreak. What can be done to protect these vulnerable populations from the devastating impacts of disease? How would this be accomplished and what are the barriers to success?

6. The ultimate goal for emerging infectious diseases is to eradicate the pathogen altogether; however, this has been done successfully only twice in our history—smallpox in humans and rinderpest virus in ungulates. If we cannot eradicate a pathogen, what strategies do you think would be most effective at preventing a disease outbreak or minimizing its effects on a population, and why?

References

Altizer, S., R. Bartel, and B. A. Han (2011). Animal migration and infectious disease risk. Science 331:296–302.

Anderson, R. M., and R. M. May (1982). Coevolution of hosts and parasites. Parasitology 85:411–426.

Asghar, M., D. Hasselquist, and S. Bensch (2011). Are chronic avian haemosporidian infections costly in wild birds? Journal of Avian Biology 42:530–537.

Atkinson, C. T., J. K. Lease, R. J. Dusek, and M. D. Samuel (2005). Prevalence of pox-like lesions and malaria in forest bird communities on leeward Mauna Loa volcano, Hawaii. The Condor 107:537–546.

Atkinson, C. T., N. J. Thomas, and D. B. Hunter (2008). Parasitic diseases of wild birds. Wiley-Blackwell, Ames, IA.

Bataille, A., A. A. Cunningham, V. Cedeño, L. Patiño, A. Constantinou, L. D. Kramer, and S. J. Goodman (2009). Natural colonization and adaptation of a mosquito species in Galápagos and its implications for disease threats to endemic wildlife. Proceedings of the National Academy of Sciences 106:10230–10235.

Becker, W. B. (1966). The isolation and classification of Tern virus: Influenza A-Tern South Africa—1961. Journal of Hygiene 64:309–320.

Benning, T. L., D. LaPointe, C. T. Atkinson, and P. M. Vitousek (2002). Interactions of climate change with biological invasions and land use in the Hawaiian Islands: Modeling the fate of endemic birds using a geographic information system. Proceedings of the National Academy of Sciences 99:14246–14249.

Bergström, S., P. D Haemig, and B. Olsen (1999). Increased mortality of Black-browed Albatross chicks at a colony heavily-infested with the tick Ixodes uriae. International Journal for Parasitology 29:1359–1361.

Björsdorff, A., S. Bergström, R. F. Massung, P. D. Haemig, and B. Olsen (2001). Ehrlichia-infected ticks on migrating birds. Emerging Infectious Diseases 7:877–879.

Botzler, R. G. (1991). Epizootiology of avian cholera in wildfowl. Journal of Wildlife Diseases 27:367–395.

Brault, A. C., S. A. Langevin, R. A. Bowen, N. A. Panella, B. J. Biggerstaff, B. R. Miller, and N. Komar (2004). Differential virulence of West Nile strains for American crows. Emerging Infectious Diseases 10:2161–2168.

Bridge, E. S., J. F. Kelly, X. Xiao, J. Y. Takekawa, N. J. Hill, M. Yamage, E. U. Haque, M. A. Islam, T. Mundkur, and K. E. Yavuz (2014). Bird migration and avian influenza: A comparison of hydrogen stable isotopes and satellite tracking methods. Ecological Indicators 45:266–273.

Brittingham, M. C., and S. A. Temple (1988). Avian disease and winter bird feeding. Passenger Pigeon 50:195–203.

Brittingham, M. C., S. A. Temple, and R. M. Duncan (1988). A survey of the prevalence of selected bacteria in wild birds. Journal of Wildlife Diseases 24:299–307.

Brown, J. D., G. Goekjian, R. Poulson, S. Valeika, and D. E. Stallknecht (2009). Avian influenza virus in water: Infectivity is dependent on pH, salinity and temperature. Veterinary Microbiology 136:20–26.

Carrasco, L., J. Lima, D. Halfen, F. Salguero, P. Sánchez-Cordón, and G. Becker (2001). Systemic aspergillosis in an oiled Magallanic Penguin (Spheniscus magellanicus). Journal of Veterinary Medicine B 48:551–554.

Caudill, G., D. Wolf, D. Caudill, J. Brown, and V. Shearn-Bochsler (2014). A juvenile wading-bird mortality event in urban Jacksonville, Florida, associated with the parasite Eustrongylides. Florida Field Naturalist 42:108–113.

Chancey, C., A. Grinev, E. Volkova, and M. Rios (2015). The global ecology and epidemiology of West Nile virus. BioMed Research International 376230.

Chen, H., G. J. Smith, S. Y. Zhang, K. Qin, J. Wang, K. S. Li, R. G. Webster, J. S. Peiris, and Y. Guan (2005). Avian flu: H5N1 virus outbreak in migratory waterfowl. Nature 436:191–192.

Cole, R. A. (1999). Eustrongylidosis. In Field manual of wildlife diseases—general field procedures and diseases of birds, M. Friend, J. C. Franson, and E. A. Ciganovich, Editors. US Geological Survey, Washington, DC, pp. 223–228.

Connan, R., and D. Wise (1994). Further studies on the development and survival at low temperatures of the free living stages of Trichostrongylus tenuis. Research in Veterinary Science 57:215–219.

Converse, K. A. (2007). Aspergillosis. In Infectious diseases of wild birds, N. J. Thomas, D. B. Hunter, and C. T. Atkinson, Editors. Blackwell Publishing, Ames, IA, pp. 360–374.

Coyner, D. F., M. G. Spalding, and D. J. Forrester (2003). Epizootiology of Eustrongylides ignotus in Florida: Transmission and development of larvae in intermediate hosts. Journal of Parasitology 89:290–298.

Deem, S., M. Cruz, G. Jiménez-Uzcátegui, B. Fessl, R. Miller, and P. Parker (2007). Pathogens and parasites: An increasing threat to the conservation of Galapagos avifauna. Galápagos report 2007–2008:125–130.

Dhondt, A. A., A. V. Badyaev, A. P. Dobson, D. M. Hawley, M. J. Driscoll, W. M. Hochachka, and D. H. Ley (2006). Dynamics of mycoplasmal conjunctivitis in the native and introduced range of the host. EcoHealth 3:95–102.

Dhondt, A. A., K. V. Dhondt, D. M. Hawley, and C. S. Jennelle (2007). Experimental evidence for transmission of Mycoplasma gallisepticum in house finches by fomites. Avian Pathology 36:205–208.

Dhondt, A. A., J. C. DeCoste, D. H. Ley, and W. M. Hochachka (2014). Diverse wild bird host range of Mycoplasma gallisepticum in eastern North America. PLoS ONE 9:e103553.

Dietrich, M., E. Gómez-Díaz, and K. D. McCoy (2011). Worldwide distribution and diversity of seabird ticks: Implications for the ecology and epidemiology of tick-borne pathogens. Vector-Borne and Zoonotic Diseases 11:453–470.

Dobson, A. P., and P. J. Hudson (1992). Regulation and stability of a free-living host-parasite system: Trichostrongylus tenuis in Red Grouse II: Population models. Journal of Animal Ecology 61:487–498.

Duckworth, R. A., A. V. Badyaev, K. L. Farmer, G. E. Hill, S. R. Roberts, and K. Smith (2003). First case of Mycoplasma gallisepticum infection in the western range of the House Finch (Carpodacus mexicanus). The Auk: Ornithological Advances 120:528–530.

Dudaniec, R. Y., and S. Kleindorfer (2006). Effects of the parasitic flies of the genus Philornis (Diptera: Muscidae) on birds. Emu 106:13–20.

Dvorak, M., B. Fessl, E. Nemeth, S. Kleindorfer, and S. Tebbich (2012). Distribution and abundance of Darwin's finches and other land birds on Santa Cruz Island, Galápagos: Evidence for declining populations. Oryx 46:78–86.

Erlanger, T. E., S. Weiss, J. Keiser, J. Utzinger, and K. Wiedenmayer (2009). Past, present, and future of Japanese encephalitis. Emerging Infectious Diseases 15:1–7.

Farmer, K., G. Hill, and S. Roberts (2005). Susceptibility of wild songbirds to the House Finch strain of Mycoplasma gallisepticum. Journal of Wildlife Diseases 41:317–325.

Fischer, J. R., D. E. Stallknecht, P. Luttrell, A. A. Dhondt, and K. A. Converse (1997). Mycoplasmal conjunctivitis in wild songbirds: The spread of a new contagious disease in a mobile host population. Emerging Infectious Diseases 3:69–72.

Fix, A. S., C. Waterhouse, E. C. Greiner, and M. K. Stoskopf (1988). *Plasmodium relictum* as a cause of avian malaria in wild-caught Magellanic Penguins (*Spheniscus magellanicus*). Journal of Wildlife Diseases 24:610–619.

Friend, M. (1999). Avian cholera. *In* Field manual of wildlife diseases—general field procedures and diseases of birds, M. Friend, J. C. Franson, and E. A. Ciganovich, Editors. US Geological Survey, Washington, DC, pp. 75–92.

Fuller, T., S. Bensch, I. Müller, J. Novembre, J. Pérez-Tris, R. E. Ricklefs, T. B. Smith, and J. Waldenström (2012). The ecology of emerging infectious diseases in migratory birds: An assessment of the role of climate change and priorities for future research. EcoHealth 9:80–88.

Galligan, T. H., and S. Kleindorfer (2009). Naris and beak malformation caused by the parasitic fly, *Philornis downsi* (Diptera: Muscidae), in Darwin's Small Ground Finch, *Geospiza fuliginosa* (Passeriformes: Emberizidae). Biological Journal of the Linnean Society 98:577–585.

George, T. L., R. J. Harrigan, J. A. LaManna, D. F. DeSante, J. F. Saracco, and T. B. Smith (2015). Persistent impacts of West Nile virus on North American bird populations. Proceedings of the National Academy of Sciences 112:14290–14294.

Ginsberg, H. S., P. A. Buckley, M. G. Balmforth, E. Zhioua, S. Mitra, and F. G. Buckley (2005). Reservoir competence of native North American birds for the Lyme disease spirochete, *Borrelia burgdorferi*. Journal of Medical Entomology 42:445–449.

Haemig, P. D., S. Bergström, and B. Olsen (1998). Survival and mortality of Grey-headed Albatross chicks in relation to infestation by the tick *Ixodes uriae*. Colonial Waterbirds 21:452–453.

Hall, R. J., S. Altizer, and R. A. Bartel (2014). Greater migratory propensity in hosts lowers pathogen transmission and impacts. Journal of Animal Ecology 83:1068–1077.

Hamer, S. A., T. L. Goldberg, U. D. Kitron, J. D. Brawn, T. K. Anderson, S. R. Loss, E. D. Walker, and G. L. Hamer (2012). Wild birds and urban ecology of ticks and tick-borne pathogens, Chicago, Illinois, USA, 2005–2010. Emerging Infectious Diseases 18:1589–1595.

Hart, L. (1997). Behavioural defence. *In* Host-parasite evolution: General principals and avian models, D. H. Clayton and J. Moore, Editors. Oxford University Press, Oxford, UK, pp. 59–77.

Hasle, G. (2013). Transport of ixodid ticks and tick-borne pathogens by migratory birds. Frontiers in Cellular and Infection Microbiology 3:48.

Hawley, D., C. Jennelle, K. Sydenstricker, and A. Dhondt (2007). Pathogen resistance and immunocompetence covary with social status in House Finches (*Carpodacus mexicanus*). Functional Ecology 21:520–527.

Hawley, D. M., E. E. Osnas, A. P. Dobson, W. M. Hochachka, D. H. Ley, and A. A. Dhondt (2013). Parallel patterns of increased virulence in a recently emerged wildlife pathogen. PLoS Biology 11:e1001570.

Hochachka, W. M., and A. A. Dhondt (2000). Density-dependent decline of host abundance resulting from a new infectious disease. Proceedings of the National Academy of Sciences 97:5303–5306.

Hudson, P. J., D. Newborn, and A. P. Dobson (1992). Regulation and stability of a free-living host-parasite system: *Trichostrongylus tenuis* in Red Grouse I: Monitoring and parasite reduction experiments. Journal of Animal Ecology 61:477–486.

Hudson, P. J., A. P. Dobson, and D. Newborn (1998). Prevention of population cycles by parasite removal. Science 282:2256–2258.

Hughes, J. H., and J. E. Porter (1956). Dispersal of mosquitoes through transportation, with particular reference to immature stages. Mosquito News 16:106–111.

Jacobs, A. C., J. M. Fair, and M. Zuk (2015). Parasite infection, but not immune response, influences paternity in Western Bluebirds. Behavioral Ecology and Sociobiology 69:193–203.

Johnson, P. T. J., and S. R. Carpenter (2008). Influence of eutrophication on disease in aquatic ecosystems: Patterns, processes and predictions. *In* Infectious disease ecology: Effects of ecosystems on disease and of disease on ecosystems, R. S. Ostfeld, F. Keesing, and V. T. Eviner, Editors. Princeton University Press, Princeton, NJ, pp. 71–99.

Johnston, E., J. I. Tsao, J. D. Munoz, and J. Owen (2013). *Anaplasma phagocytophilum* infection in American Robins and Gray Catbirds: An assessment of reservoir competence and disease in captive wildlife. Journal of Medical Entomology 50:163–170.

Jones, D. (2011). An appetite for connection: Why we need to understand the effect and value of feeding wild birds. Emu 111:i–vii.

Kilpatrick, A. M. (2011). Globalization, land use, and the invasion of West Nile virus. Science 334:323–327.

Kilpatrick, A. M., R. J. Peters, A. P. Dupuis, M. J. Jones, P. Daszak, P. P. Marra, and L. D. Kramer (2013). Predicted and observed mortality from vector-borne disease in wildlife: West Nile virus and small songbirds. Biological Conservation 165:79–85.

Kleindorfer, S., K. J. Peters, G. Custance, R. Y. Dudaniec, and J. A. O'Connor (2014). Changes in *Philornis* infestation behavior threaten Darwin's finch survival. Current Zoology 60:542–550.

Knutie, S. A., J. L. Waite, and D. H. Clayton (2013). Does avian malaria reduce fledging success: An experimental test of the selection hypothesis. Evolutionary Ecology 27:185–191.

Koop, J. A., C. Le Bohec, and D. H. Clayton (2013). Dry year does not reduce invasive parasitic fly prevalence or abundance in Darwin's finch nests. Reports in Parasitology 3:11–17.

Kuiken, T., G. Wobeser, F. A. Leighton, D. M. Haines, B. Chelack, J. Bogdan, L. Hassard, R. A. Heckert, and J. Riva (1999). Pathology of Newcastle disease in Double-crested Cormorants from Saskatchewan, with comparison of diagnostic methods. Journal of Wildlife Diseases 35:8–23.

LaDeau, S. L., A. M. Kilpatrick, and P. P. Marra (2007). West Nile virus emergence and large-scale declines of North American bird populations. Nature 447:710–713.

Lafrancois, B. M., S. C. Riley, D. S. Blehert, and A. E. Ballmann (2011). Links between type E botulism outbreaks, lake levels, and surface water temperatures in Lake Michigan, 1963–2008. Journal of Great Lakes Research 37:86–91.

Latorre-Margalef, N., G. Gunnarsson, V. J. Munster, R. A. M. Fouchier, A. D. M. E. Osterhaus, J. Elmberg, B. Olsen, A. Wallensten, P. D. Haemig, and T. Fransson (2009). Effects of influenza A virus infection on migrating Mallard Ducks. Proceedings of the Royal Society B: Biological Sciences 276:1029–1036.

Lawson, B., R. A. Robinson, K. M. Colvile, K. M. Peck, J. Chantrey, T. W. Pennycott, V. R. Simpson, M. P. Toms, and A. A. Cunningham (2012). The emergence and spread of finch trichomonosis in the British Isles. Philosophical Transactions of the Royal Society B: Biological Sciences 367:2852–2863.

Levin, I., P. Zwiers, S. Deem, E. Geest, J. Higashiguchi, T. Iezhova, G. Jiménez-Uzcátegui, D. Kim, J. Morton, and N. Perlut (2013). Multiple lineages of avian malaria parasites (Plasmodium) in the Galapagos Islands and evidence for arrival via migratory birds. Conservation Biology 27:1366–1377.

Little, S. E. (2008). Myiasis in wild birds. In Parasitic diseases of wild birds, C. T. Atkinson, N. J. Thomas, and D. B. Hunter, Editors. Wiley-Blackwell, Oxford, UK, pp. 546–556.

Liu, J., H. Xiao, F. Lei, Q. Zhu, K. Qin, X. Zhang, X. Zhang, D. Zhao, G. Wang, and Y. Feng (2005). Highly pathogenic H5N1 influenza virus infection in migratory birds. Science 309:1206.

Magallanes, S., A. P. Møller, L. García-Longoria, F. de Lope, and A. Marzal (2016). Volume and antimicrobial activity of secretions of the uropygial gland are correlated with malaria infection in house sparrows. Parasites and Vectors 9:232.

Marzal, A., F. De Lope, C. Navarro, and A. P. Møller (2005). Malarial parasites decrease reproductive success: An experimental study in a passerine bird. Oecologia 142:541–545.

Messinger, S. M., and A. Ostling (2009). The consequences of spatial structure for the evolution of pathogen transmission rate and virulence. American Naturalist 174:441–454.

Naugle, D. E., C. L. Aldridge, B. L. Walker, K. E. Doherty, M. R. Matchett, J. McIntosh, T. E. Cornish, and M. S. Boyce (2005). West Nile virus and sage-grouse: What more have we learned? Wildlife Society Bulletin 33:616–623.

Nemeth, N., D. Gould, R. Bowen, and N. Komar (2006). Natural and experimental West Nile virus infection in five raptor species. Journal of Wildlife Diseases 42:1–13.

O'Connor, J. A., F. J. Sulloway, J. Robertson, and S. Kleindorfer (2010). Philornis downsi parasitism is the primary cause of nestling mortality in the critically endangered Darwin's Medium Tree Finch (Camarhynchus pauper). Biodiversity and Conservation 19:853–866.

Ogden, N. H., L. R. Lindsay, K. Hanincova, I. K. Barker, M. Bigras-Poulin, D. F. Charron, A. Heagy, et al. (2008). Role of migratory birds in introduction and range expansion of Ixodes scapularis ticks and of Borrelia burgdorferi and Anaplasma phagocytophilum in Canada. Applied and Environmental Microbiology 74:1780–1790.

Olsén, B., T. Jaenson, and S. Bergström (1995). Prevalence of Borrelia burgdorferi sensu lato-infected ticks on migrating birds. Applied and Environmental Microbiology 61:3082–3087.

Owen, J. C., F. Moore, N. Panella, E. Edwards, R. Bru, M. Hughes, and N. Komar (2006). Migrating birds as dispersal vehicles for West Nile virus. EcoHealth 3:79–85.

Owen, J. C., and M. C. Garvin (2010). Epizootology of West Nile virus in birds. In Avian ecology and conservation: A Pennsylvania focus with national implications, S. K. Majumdar, T. L. Master, M. Brittingham, R. M. Ross, R. Mulvihill, and J. Huffman, Editors. Pennsylvania Academy of Science, Easton, PA, pp. 304–314.

Owen, J. C., A. Nakamura, C. A. C. Coon, and L. B. Martin (2012). The effect of exogenous corticosterone on West Nile virus infection in Northern Cardinals (Cardinalis cardinalis). Veterinary Research 43:34–42.

Penrith, M. L., F. Huchzermeyer, S. C. De Wet, and M. Penrith (1994). Concurrent infection with Clostridium and Plasmodium in a captive King Penguin Aptenodytes patagonicus. Avian Pathology 23:373–380.

Peterson, A. T., D. A. Vieglais, and J. K. Andreasen (2003). Migratory birds modeled as critical transport agents for West Nile virus in North America. Vector Borne and Zoonotic Diseases 3:27–37.

Robinson, R. A., B. Lawson, M. P. Toms, K. M. Peck, J. K. Kirkwood, J. Chantrey, I. R. Clatworthy, A. D. Evans, L. A. Hughes, and O. C. Hutchinson (2010). Emerging infectious disease leads to rapid population declines of common British birds. PLoS ONE 5:e12215.

Rocke, T., and M. Friend (1999). Avian botulism. In Field manual of wildlife diseases—general procedures and diseases of birds, M. Friend, J. C. Franson, and E. A. Ciganovich, Editors. US Geological Survey, Washington, DC, pp. 271–282.

Rocke, T., K. Converse, C. Meteyer, and B. McLean (2005). The impact of disease in the American white pelican in North America. Waterbirds 28:87–94.

Rocke, T. E., and T. K. Bollinger (2007). Avian botulism. In Infectious diseases of wild birds, N. J. Thomas, D. B. Hunter, and C. T. Atkinson, Editors. Blackwell Publishing, Ames, IA, pp. 377–416.

Rosen, M. N. (1972). The 1970–71 avian cholera epornitic's impact on certain species. Journal of Wildlife Disease 8:75–78.

Samuel, M. D., R. G. Botzler, and G. A. Wobeser (2007). Avian cholera. In Infectious diseases of wild birds, N. J. Thomas, D. B. Hunter, and C. T. Atkinson, Editors. Blackwell Publishing, Ames, IA, pp. 239–269.

Sehgal, R. N. M. (2010). Deforestation and avian infectious diseases. Journal of Experimental Biology 213:955–960.

Sheldon, B. C., and S. Verhulst (1996). Ecological immunology: Costly parasite defences and trade-offs in evolutionary ecology. Trends in Ecology and Evolution 11:317–321.

Short, K. R., M. Richard, J. H. Verhagen, D. van Riel, E. J. A. Schrauwen, J. M. A. van den Brand, B. Mänz, R. Bodewes, and S. Herfst (2015). One health, multiple challenges: The interspecies transmission of avian influenza A virus. One Health 1:1–13.

Smits, J., J. L. Tella, M. Carrete, D. Serrano, and G. López (2005). An epizootic of avian pox in endemic Short-toed Larks (Calan-

drella rufescens) and Berthelot's Pipits (*Anthus berthelotti*) in the Canary Islands, Spain. Veterinary Pathology Online 42:59–65.

Spalding, M. G., and D. J. Forrester (1993). Pathogenesis of *Eustrongylides ignotus* (Nematoda: Dioctophymatoidea) in Ciconiiformes. Journal of Wildlife Diseases 29:250–260.

Spalding, M. G., and D. J. Forrester (2009). Eustrongylidosis. *In* Parasitic diseases of wild birds, C. T. Atkinson, N. J. Thomas, and D. B. Hunter, Editors. Wiley-Blackwell, Ames, IA, pp. 289–315.

Stallknecht, D. E., S. M. Shane, M. T. Kearney, and P. J. Zwank (1990). Persistence of avian influenza viruses in water. Avian Diseases 34:406–411.

Stallknecht, D. E., E. Nagy, D. B. Hunter, and R. D. Slemons (2007). Avian influenza. *In* Infectious diseases of wild birds, N. J. Thomas, D. B. Hunter, and C. T. Atkinson, Editors. Blackwell Publishing, Ames, IA, pp. 108–130.

Szöllősi, E., B. Rosivall, D. Hasselquist, and J. Török (2009). The effect of parental quality and malaria infection on nestling performance in the Collared Flycatcher (*Ficedula albicollis*). Journal of Ornithology 150:519–527.

Thomas, N. J., D. B. Hunter, and C. T. Atkinson (2008). Infectious diseases of wild birds. John Wiley and Sons, Hoboken, NJ.

Valkiunas, G. (2004). Avian malaria parasites and other haemosporidia. CRC Press, Boca Raton, FL.

VanDalen, K. K., J. S. Hall, L. Clark, R. G. McLean, and C. Smeraski (2013). West Nile virus infection in American Robins: New insights on dose response. PLoS ONE 8:e68537.

van der Meulen, K., M. Pensaert, and H. Nauwynck (2005). West Nile virus in the vertebrate world. Archives of Virology 150:637–657.

van Gils, J. A., V. J. Munster, R. Radersma, D. Liefhebber, R. A. Fouchier, and M. Klaassen (2007). Hampered foraging and migratory performance in swans infected with low-pathogenic avian influenza a virus. PLoS ONE 2:e184.

van Riper III, C., S. G. van Riper, M. L. Goff, and M. Laird (1986). The epizootiology and ecological significance of malaria in Hawaiian land birds. Ecological Monographs 56:327–344.

van Riper III, C., S. G. van Riper, W. R. Hansen, and S. Hackett (2002). Epizootiology and effect of avian pox on Hawaiian forest birds. The Auk: Ornithological Advances 119:929–942.

van Riper, C., and D. J. Forrester (2007). Avian pox. *In* Infectious diseases of wild birds, N. J. Thomas, D. B. Hunter, and C. T. Atkinson, Editors. Blackwell Publishing, Ames, IA, pp. 131–176.

Veit, R. R., and M. A. Lewis (1996). Dispersal, population growth, and the Allee effect: Dynamics of the House Finch invasion of eastern North America. American Naturalist 148:255–274.

Warner, R. E. (1968). The role of introduced diseases in the extinction of the endemic Hawaiian avifauna. The Condor 70:101–120.

Weber, T. P., and N. I. Stilianakis (2007). Ecologic immunology of avian influenza (H5N1) in migratory birds. Emerging Infectious Diseases 13:1139–1143.

Webster, R. G., W. J. Bean, O. T. Gorman, T. M. Chambers, and Y. Kawaoka (1992). Evolution and ecology of influenza A viruses. Microbiology and Molecular Biology Reviews 56:152–179.

Wikelski, M., J. Foufopoulos, H. Vargas, and H. Snell (2004). Galápagos birds and diseases: Invasive pathogens as threats for island species. Ecology and Society 9:5.

CHAPTER 24

Modern Climate Change and Birds

Benjamin Zuckerberg and Lars Pomara

Across the globe, climate influences the distribution and abundance of bird species and their habitats (chapters 18, 22). Although birds have numerous adaptations that allow them to accommodate changes in weather and climate, concern is mounting that human-induced climate change represents a unique and growing conservation threat. The most recent assessment of the Intergovernmental Panel on Climate Change (IPCC)—a global collaboration of scientists tasked with evaluating the scientific information on human-induced climate change—warns that the warming of the Earth's atmosphere and oceans is now unequivocal, because of continued increases in concentrations of greenhouse gases. These global changes in the climate system have resulted in altered water cycles, rising sea levels, diminishing snow and ice cover, and a higher frequency of extreme weather events (IPCC 2013). Most worrisome is that the rates of many of these observed changes are unprecedented over decades to millennia. In this chapter, we present an overview of some of the major changes in the Earth's climate and discuss the implications of these changes for bird behavior, phenology, population dynamics, distributions, and conservation.

A WARMING WORLD

Each of the last three decades (1980s, 1990s, and 2000s) has been successively warmer at the Earth's surface than any preceding decade since 1850. From 1880 to 2012, land and ocean surface temperatures have shown a warming trend of $0.85°C$ (Hartmann et al. 2013) (fig. 24.1). This global trend has resulted in the Earth being, on average, $0.78°C$ warmer in the early twenty-first century compared with the late nineteenth century. For many bird populations, this shift in the global climate reflects even more alarming changes in temperature at regional scales. For example, birds inhabiting and breeding in the more northerly latitudes are considered at greater

risk since many of these regions have been experiencing some of the more rapid increases in temperature (fig. 24.1). Across the Northern Hemisphere, the period of 1983 to 2012 was likely the warmest 30-year period of the last 1,400 years (IPCC 2013), and arctic temperatures have risen almost twice as quickly over the past 100 years compared with global temperature trends (Christensen et al. 2013). The rate of temperature change in relation to geographic gradients (referred to as **climate velocity**, $°C/km$) has been faster across land compared with oceans and fastest in low elevation, open habitats such as grasslands, heathlands, and wetlands (Loarie et al. 2009, Burrows et al. 2011). Although changes in the global climate systems occur as a result of natural phenomena (e.g., sun cycles, changes in the earth's orbit), it is increasingly clear that recent warming is driven by an increase in fossil fuel emissions and rising levels of CO^2 in the atmosphere. Over the next 100 years, the IPCC predicts that global surface temperature increase is likely to exceed $1.5°C$ under most future models and Representative Concentration Pathways (see the box on page 756). Under RCP 8.5, the Earth will experience substantial warming over all terrestrial regions by the end of the twenty-first century. The greatest warming will occur in northerly regions, including more than $4°C$ above current climate conditions. By the end of the twenty-first century, warming is predicted to exceed $4°C$ over most land areas, with much of northern North America and northern Eurasia exceeding $6°C$ (Diffenbaugh and Field 2013) (fig. 24.2). The results of these projections depict a sobering reality: the observed warming over the past 100 years pales in comparison with future changes.

COMPLEX CHANGES IN PRECIPITATION

Global temperatures have not been the only aspect of the Earth's climate system to change. Changes in the global

A

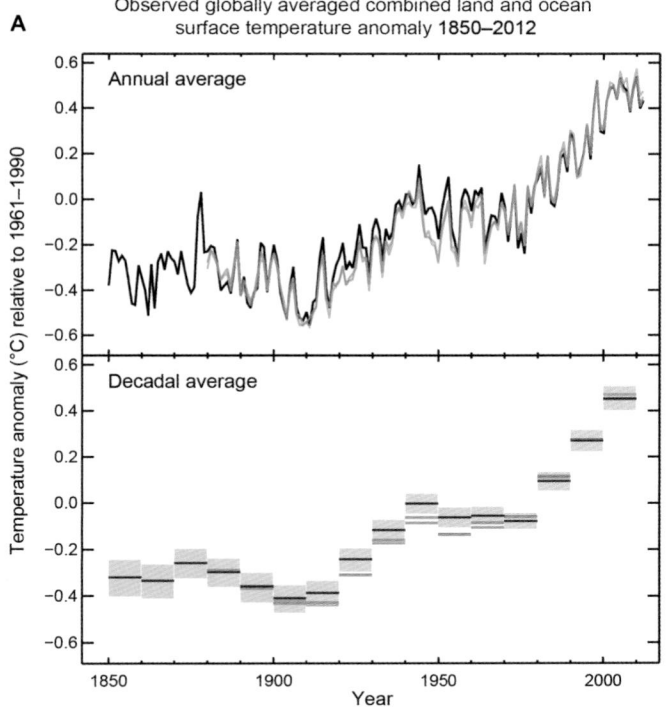

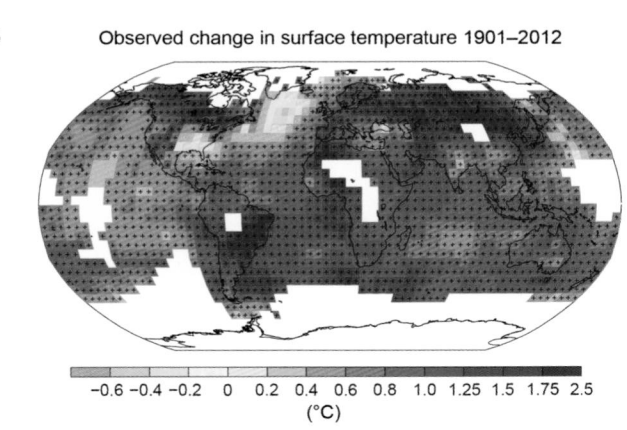

Figure 24.1. Rates of global warming showing *A*, observed global mean combined land and ocean surface temperature anomalies, from 1850 to 2012, and *B*, map of the observed surface temperature change from 1901 to 2012 derived from temperature trends determined by linear regression. Grid boxes where the trend is significant at the 10 percent level are indicated by a + sign. *Adapted from IPCC 2013.*

water cycle have resulted in significant, yet complicated, trends in precipitation. Since the mid-twentieth century, a general pattern has emerged of increasing annual precipitation in many parts of the world (IPCC 2013). Up until the early 2000s, tropical regions were the exception to this rule, and parts of South America and Africa showed increased drying and drought conditions in many areas. However, in-

creasing precipitation in the 2000s appeared to reverse the drying trend, especially in South America. Drying in other regions, such as Western Africa and East Asia, have continued unabated and exacerbated the contrast between wet and arid regions of the world. This poses a significant problem for birds occupying arid regions that have life histories that are critically in line with pulses in precipitation and resources. In the future, substantial changes in annual precipitation are expected for many regions of the world (Diffenbaugh and Field 2013), including increases over the high northern latitudes and decreases over the Mediterranean region and regions of southwestern South America, Africa, and Australia (fig. 24.2). Under most scenarios, it appears that future climate change will result in overall wetter conditions, but these trends are complex and many regions are expected to demonstrate significant decreases in annual precipitation.

RISING SEA LEVELS

Similar to rising global temperatures, the rate of sea level rise since the mid-nineteenth century has been greater than for any period during the previous two millennia (IPCC 2013). Between 1901 and 2010, global mean sea level rose by 0.19 m, and the rate of increase was almost twice as high in recent decades. Global mean sea level rise is considered a result of multiple drivers including ocean thermal expansion (the physical increase in volume of water as it warms) and changes in the Greenland and Antarctic ice sheets, but since the early 1970s, the combined effects of glacier mass loss and ocean thermal expansion explains about 75 percent of the observed mean sea level rise (Church et al. 2013). Global mean sea level will continue to rise during the next century. Under all RCP scenarios, the rate of sea level rise will exceed that observed during the pasty forty years as a consequence of increased ocean warming and continuing loss of mass from glaciers and ice sheets. For RCP8.5, the relative rise in sea level by the end of the century is predicted to be between 0.52 and 0.98 m, with a late-century rate of 8 to 16 mm per year. Under future projections, thermal expansion accounts for 30 to 55 percent global mean sea level rise, and glaciers 15 to 35 percent. One of the largest drivers of future sea level change is an increase in surface melting of the Greenland ice sheet that will eventually exceed any increase in snowfall. It is clear that these changes are most relevant for coastal breeding birds, and an estimated 70 percent of the coastlines worldwide are projected to experience significant sea level change (Church et al. 2013). Sea level rise of such magnitude will cause inundation of coastal areas, flooding, erosion, saltwater intrusion into estuaries and other freshwater coastal habitats, rising water tables, and habitat loss (Nicholls et al. 2007).

REPRESENTATIVE CONCENTRATION PATHWAYS

To predict future changes in climate, scientists use a variety of climate models based on a set of scenarios called Representative Concentration Pathways (RCPs) (Moss et al. 2008). In all RCPs, atmospheric CO_2 concentrations are higher in the year 2100 relative to present day. RCPs represent four greenhouse gas concentration trajectories and possible climate futures, all of which depend on the amount of greenhouse gases emitted in the years to come. The four RCPs are named after a range of radiative forcings that capture the change in energy in the atmosphere caused by GHG emissions in the year 2100 relative to pre-industrial values (+2.6, +4.5, +6.0, and +8.5 W/m², respectively).

REDUCED SNOW COVER AND RETREATING ICE

In many regions of the world occupied by birds, snow and ice cover the landscape during different parts of the year; these regions are considered part of the **cryosphere**. The cryo-

sphere represents those portions of Earth's surface where water is in solid form, including ice sheets, snow cover, glaciers, and permafrost. The components of the cryosphere play a crucial role in the Earth's climate and are some of the most sensitive to climate shifts. Over the last two decades, glaciers have diminished almost worldwide, the Greenland and Antarctic ice sheets have been losing mass, and Arctic sea ice and Northern Hemisphere spring snow cover have decreased in extent (IPCC 2013). Around the world, the rate of ice loss from glaciers has been increasing in recent years (fig. 24.3). Glaciers in the tropics and midlatitudes are particularly vulnerable and are retreating; striking examples of tropical glacier retreat have been documented in the Andes Mountains and in Indonesia and Kenya (Klein and Kincaid 2006, Rostom and Hastenrath 2007). In other regions, the glaciers of the Alps have lost nearly 50 percent of their surface area and mass since 1850 (Zemp et al. 2006), and 80 percent of glaciers in Montana's Glacier National Park have diminished significantly (Hall and Fagre 2003). At the same time that many glaciers have retreated, a similar pattern is playing out with sea ice. Between 1970 and 2012, Arctic sea ice extent decreased at a rate of 3.5 to 4.1 percent per decade, resulting in the loss of almost half a million km² of sea ice per decade (Vaughan et al. 2013). This decrease in Arctic sea ice has been most rapid in summer. In Antarctica, the story is more com-

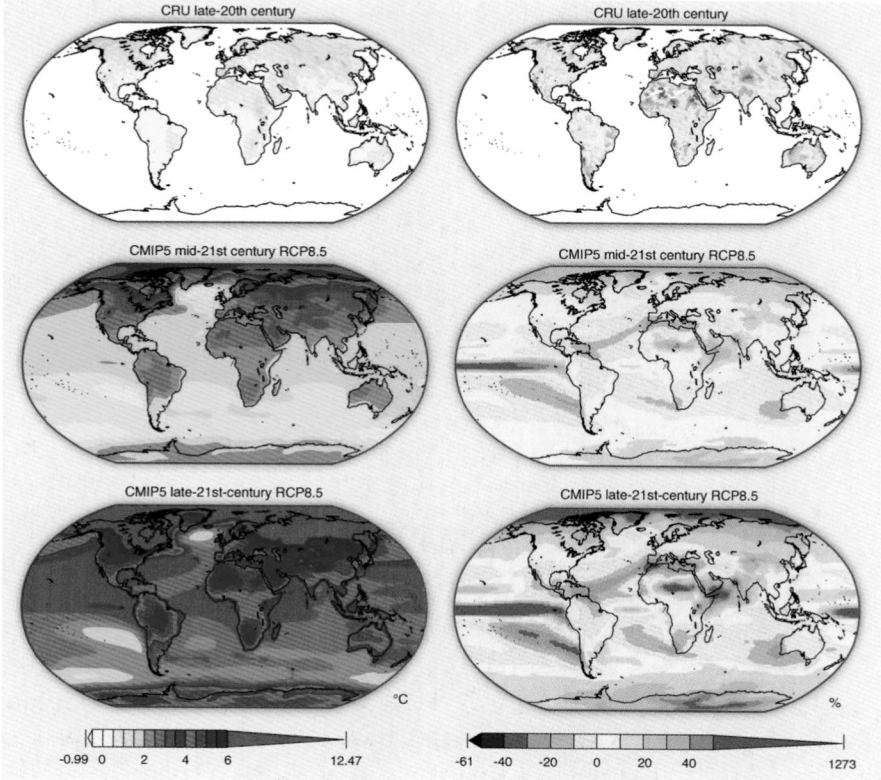

Figure 24.2. Observed and projected changes in annual temperature and precipitation for the late twentieth century calculated as 1986–2005 minus 1956–1975 (*top panel*). Differences in the mid-twenty-first-century period of the CMIP5 RCP8.5 ensemble, calculated as 2046–2065 minus 1986–2005 (*middle panel*). Differences in the late-twenty-first-century period of the CMIP5 RCP8.5 ensemble, calculated as 2081–2100 minus 1986–2005 (*bottom panel*). Values at the left and right extremes of the color bars give the minimum and maximum values (respectively) that occur across all of the periods. *Diffenbaugh and Field 2013, used with permission.*

Figure 24.3. Glacier recession has advanced globally, including in high-elevation tropical mountain ranges such as the Andes of South America. *A*, Ice loss between 1987 and 2011 is shown in red around the Pastoruri glacier in Peru's Huascaran National Park, in the Cordillera Blanca of the Andes. Glacier recession was determined by measuring changes in the Normalized Difference Snow Index between 1987 and 2011 Landsat TM imagery. *B*, Stream flows are strongly altered by increasing glacial recession rates, with negative ecological consequences for downstream habitat. *C*, Peatland patches near glaciers, an important high-elevation Andean habitat, are fragmenting and undergoing attrition. In the image, researchers collect peat samples. *D*, Observers visit the receding Pastoruri glacier on a day trip with a climate change theme. *Image analysis and photographs by Molly H. Polk and Kenneth R. Young.*

plicated, with some regions increasing and some decreasing in sea ice extent.

Both persistent and seasonal snow cover are important features for many cold-adapted birds, but snow cover regimes have changed dramatically in many parts of the Northern Hemisphere. Since the 1970s, snow cover extent has decreased 1.6 percent per decade for March and April, and 11.7 percent per decade for June (Derksen and Brown 2012). During this period, snow cover extent in the Northern Hemisphere did not show a statistically significant increase in any month. At the same time, permafrost temperatures have increased in most regions since the early 1980s. As an example, there has been a considerable reduction in permafrost thickness and areal extent in many parts of northern Russia.

As is the case for the Earth's changes in precipitation, it is difficult to make robust predictions regarding changes in the cryosphere. It is nonetheless considered very likely that Arctic sea ice cover will continue to shrink, Northern Hemisphere spring snow cover will decrease, and global glacier volumes will diminish during the twenty-first century as global temperature rises. Year-round reductions in Arctic sea ice extent are projected to range from 43 percent (RCP2.6) to 94 percent (RCP8.5) (IPCC 2013). By the end of the twenty-first century, the global glacier volume is projected to decrease by 15 to 55 percent for RCP2.6 and by 35 to 85 percent for RCP8.5 (with medium confidence in these predictions). The area of Northern Hemisphere spring snow cover is projected to decrease by 7 percent for RCP2.6 and by 25 percent in RCP8.5 over the next 100 years. It is virtually certain that permafrost extent at high northern latitudes will be reduced as global mean surface temperature increases, with projected decreases between 37 percent (RCP2.6) and 81 percent (RCP8.5).

EXTREME WEATHER: HEAT WAVES, DROUGHT, AND FLOODS

Changes in extreme weather and climate events, such as heat waves and droughts, are the most direct way bird populations can be affected by climate change. Climate change has already increased the number and strength of some of these extreme events (IPCC 2013). Although changes in the frequency and magnitude of extreme events have been observed, it is difficult to attribute a single extreme event (drought, storm, or flood) to climate change (Trenberth et al. 2015). As the characteristics of Earth's climate shift, however, we can expect a higher likelihood of certain anomalies. As expected with warming temperatures, unusually cold days and nights and frosts have become less frequent, while hot days and nights have become more frequent at a global scale (Christidis et al. 2011, IPCC 2013). This increase in the frequency of hot days has raised the concern of more frequent and intense heat waves (Duffy and Tebaldi 2012). For example, the summer of 2003 in Europe was the warmest summer on record since 1540, the record broken only by another heat wave seven years later—the 2010 heat wave in Russia (Barriopedro et al. 2011). The likelihood of such "mega" heat waves is predicted to increase by a factor of 5 to 10 within the next 40 years (Barriopedro et al. 2011). Higher temperatures lead to increased rates of evaporation, and even in regions

where precipitation has not declined, increases in temperatures and in surface evaporation can promote drought conditions. Regions with increased droughts include the Sahel of Central Africa, the Mediterranean, southern Africa and parts of southern Asia (IPCC 2013). In North America, there has been an increase in the concurrence of droughts and heatwaves across the United States (Mazdiyasni and AghaKouchak 2015), and there is growing concern that the future American Southwest and Central Plains will experience "megadrought" conditions far beyond the contemporary experience of bird populations inhabiting Western North America (Cook et al. 2015).

In addition to the growing conservation concern associated with drought, there is a strong possibility that many land regions are experiencing an increasing frequency or intensity of heavy precipitation events, which also pose conservation risks. Increased amounts of water vapor and atmospheric moisture have caused higher numbers of precipitation events and increased precipitation intensity leading to increases in flooding risk in many regions. However, the risk of flooding is not increasing homogeneously around the globe. Regions experiencing increases in the number and intensity of precipitation events are eastern parts of North and South America, northern Europe, and north-central Asia (IPCC 2013). Extreme precipitation events over most of the midlatitude and tropical regions will very likely become more intense and more frequent by the end of this century (IPCC 2013).

CHANGES IN SEASONALITY: EARLIER SPRINGS, DELAYED AUTUMNS

Climate-mediated seasonal transitions play a crucial role in avian life cycle events. For example, the onset of spring in temperate ecosystems is strongly associated with birds arriving to their breeding grounds, defending their territories, finding a mate, and building a nest. In many regions, however, climate change has altered the timing of these seasonal transitions, resulting in earlier springs caused by rising global temperatures (McCabe et al. 2012, Ault et al. 2015). Across the coterminous United States, regional trends in spring leaf emergence have shifted earlier from 0.8 to 1.6 days per decade, while flower emergence has shifted from 0.4 to 1.2 days earlier per decade (Ault et al. 2015). Many of these trends, however, are influenced by variations in climate variability (e.g., the El Niño-Southern Oscillation) and vary strongly across regions where spring onset has even been delayed in some areas (Schwartz et al. 2013). Future climate projections suggest that earlier springs will con-

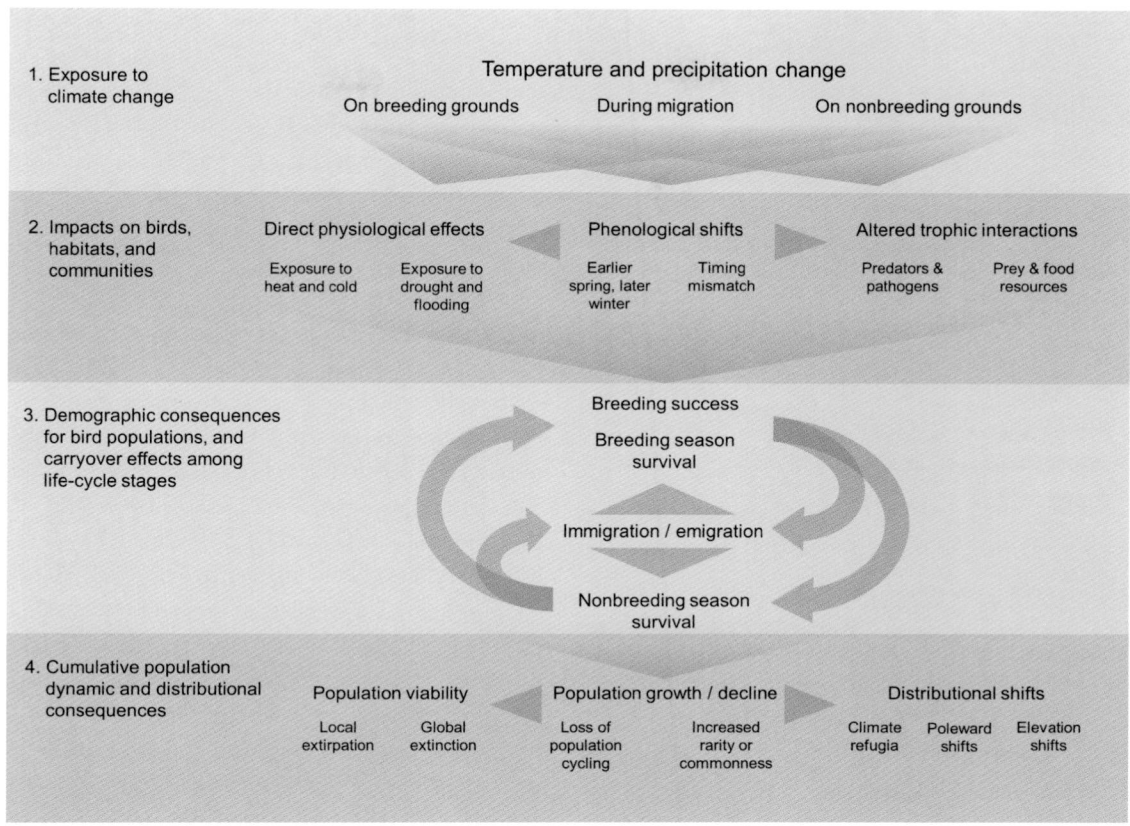

Figure 24.4. (*1*) Exposure to climate change can affect birds directly and indirectly through (*2*) physiological impact on the bird, changes in its habitat, and changes in its ecological community and relationships with other species. These effects in turn can (*3*) alter the demography of bird populations by increasing or decreasing rates of birth, death, immigration, and emigration; and effects during one stage of the life cycle may carry over into later stages. Ultimately, (*4*) these cumula-tive demographic consequences shape overall population growth and decline, which can cause shifts in a species' geographic distribution through local colonization and extirpation processes, as well as changes in long-term viability including the possibility of extinction. The impacts of climate change at stage 4 can also feed back into stages 2 and 3, for example by influencing community composition and interactions with other species.

tinue into the future, with estimates of a 23-day shift across the United States under the RCP8.5 scenario (Allstadt et al. 2015). Earlier spring onset can increase the vulnerability of plants and animals to subfreezing temperatures; these "false spring" events can damage new plant growth and potentially impact early arriving birds. The risk of false spring events is expected to increase in regions such as the Great Plains and Upper Midwest (Allstadt et al. 2015). Although there is strong consensus that a warming world promotes earlier springs, less is known about autumn phenology (Gallinat et al. 2015). In general, there are indications that leaf senescence has been delayed by rising global temperatures (Menzel et al. 2006, Ibanez et al. 2010), and combined with earlier springs, this represents a general extension in the growing season for many ecosystems.

A BIRD'S EYE VIEW OF CLIMATE CHANGE

As we have seen, modern climate change has resulted in significant changes in global temperatures, precipitation patterns, sea levels, extreme weather events, and seasonality. Bird populations in all regions of the globe, and in all terrestrial and ocean habitats, are likely to be affected by these ongoing changes. This is because the ability of birds to survive, reproduce, migrate, and persist over the long term are all influenced by weather and climate (fig. 24.4). Weather represents atmospheric conditions that birds must adjust to over short periods of time, whereas climate comprises characteristics averaged over relatively long periods of time (e.g., over several decades). Weather and climate both shape the habitats and ecological communities in which birds flourish and to which

TERRY L. ROOT

Terry L. Root is a senior fellow emerita at the Woods Institute for the Environment at Stanford University and a professor, by courtesy, in the Biology Department. She researches how wild animals and plants are changing with climate change. *Photo with permission from T. Root.*

The work of Dr. Terry L. Root focuses on how wild animals and plants are changing with climate change. Her early research focused on the role of climate on bird metabolic constraints and winter ranges, and was an early vindication of the use of citizen science in documenting bird ranges. Her current focus is on assessing the mass extinction of species with warming and the broader attribution of human impacts on shifting ranges and phenology for birds, mammals, and plants. She has been a lead author for the Third (2001) and Fourth (2007) Assessment Reports of the IPCC. Root is widely credited for early theories on how bird ranges were constrained by climate and has received numerous recognitions including the Spirit of Defenders Award for Science, Aldo Leopold Leadership Fellow, Pew Scholar in Conservation and the Environment, and a Presidential Young Investigator Award from the National Science Foundation.

mental demographic processes: birth, death, immigration, and emigration (Jenouvrier 2013). It is through these four processes that populations are structured, grow, decline, colonize new areas, or become extirpated (chapter 21). Because the geographic distribution of a species is nothing more than the cumulative result of many local extirpation and colonization events, climate variability influences species distributions through these demographic processes (chapter 4). Therefore, in asking how bird species might be affected by modern climate change, it is useful to consider different ways in which climate variability can influence demography and to consider how changes in demographic rates may shape populations as a whole.

How well birds survive and reproduce is strongly determined by access to critical resources such as food and shelter. These resources can be influenced by weather and increasing variability in climate. The most direct effects of climate variability may come during extreme events, when unusually cold, hot, wet, or dry spells expose birds to conditions that cause physiological stress. Indirect climatic effects include changes in food availability caused by, for example, influences on plant and invertebrate prey. As climate changes and extreme events become more common, these stresses may increase. Climate change may also cause changes in populations of predators and diseases, or change the exposure of bird populations to these natural enemies by modifying habitats or driving distributional changes that increase range overlap.

Ultimately, the effect of climate change on birds may be expressed in a variety of ways at different times of year, different stages of an organism's life history, in different geographic regions, and cumulatively over longer periods of time (fig. 24.4). For example, for the Black-throated Blue Warbler (*Setophaga caerulescens*), climate influences both adult survival on the Caribbean wintering grounds and fecundity on the temperate breeding grounds more than a thousand miles distant. Both of these effects contribute to overall population dynamics, which are consequently linked to long-term climate fluctuations in the Northern Hemisphere (Sillett et al. 2000). Cases like this one illustrate how climate change may influence a given bird species at multiple stages along the life trajectory. There are also ways in which these influences may carry over into subsequent life stages and potentially interact with one another in complex ways, sometimes with unexpected outcomes in terms of overall population growth or decline.

EARLIER ARRIVAL AND CHANGES IN MIGRATION PHENOLOGY

Phenology is the timing of regular life history events such as arrival to breeding grounds and nest-building in birds or annual fruiting and leaf-out events in plants. Climate influ-

they are adapted (chapters 18 and 22). Ultimately, climate change has the potential to affect birds directly through extreme events or long-term changes in climatic conditions, and indirectly by reshaping their habitats.

Climate change can impact a bird population by having some direct or indirect effect on one or more of the funda-

ences phenology in complex ways, and for birds, the most visible aspect of phenology is migration (Berthold et al. 2003, Newton 2010). The advancement of spring phenology has been one of the strongest signals of modern climate change and has important implications for bird migration (Walther et al. 2002, Cox 2010). Although day length is considered the ultimate driver of when birds begin to migrate, weather and climate can influence departure from wintering grounds, migration speed, and, ultimately, arrival to breeding grounds. In both Europe and North America, the timing of arrival to the breeding grounds has been linked to spring temperature and precipitation, large-scale climate variability (e.g., North Atlantic Oscillation), and wind speed and direction (Knudsen et al. 2007). Using data from banding/ringing stations, migration studies focus on long-term trends in **first arrival** or **median arrival** dates. First arrival dates are calculated to capture the front of the migration and the earliest arriving individuals, whereas median arrival dates reflect the arrival of the bulk of the population. Across multiple bird species and in many locations across the world, the earliest arriving individuals are arriving to their breeding grounds some six to eight days earlier over the past thirty years, while the rest of the population are arriving four to five days earlier (Pearce-Higgins and Green 2014; see the box on page 762). Not all species, or even populations within species, are showing similar advancements, but the systematic earlier arrival to the breeding grounds is strong evidence that birds are responding to earlier springs throughout the Northern Hemisphere.

The influence that climate has on migratory phenology is complicated by the distance a species must travel from overwintering to breeding grounds. Short-distance, temperate migrants are thought to overwinter in regions that have the best mixture of tolerable local winter conditions while still providing close access to resources on their breeding grounds (Somveille et al. 2015). As a result, the advancement in spring arrival for short-distance migrants is a direct consequence of warming temperatures and the earlier onset of spring in mid- and high latitudes (Swanson and Palmer 2009). For long-distance migrants, however, it is more likely that climatic conditions on overwintering grounds and migratory stopovers are influencing rates of migration. Studies on Pied Flycatchers (*Ficedula hypoleuca*) and Barn Swallows (*Hirundo rustica*) have shown that arrival to the breeding grounds in Europe is associated with weather conditions on the wintering grounds in Africa (Both 2010). Similarly, the timing of spring departure for American Redstarts (*Setophaga ruticilla*) overwintering in Jamaica is a product of rainfall, with wet springs promoting increased arthropod abundance and increased body condition, and leading to earlier departure for breeding grounds in North America (Studds and

Figure 24.5. The timing of spring departure for American Redstarts (*Setophaga ruticilla*) overwintering in Jamaica is strongly influenced by rainfall, with wet springs promoting increased arthropod abundance, increased body condition, and leading to earlier departure for breeding grounds in North America. Even though long-distance migrants have migratory tendencies that are driven by non-climatic cues like photoperiod, climate conditions on their nonbreeding grounds and migratory stopovers can influence their arrival to breeding grounds. *Photo taken by Dan Pancamo, Quintana, Texas, November 30, 2010, licensed under the Creative Commons Attribution-Share Alike 2.0 Generic license, https://commons.wikimedia.org/wiki/File:American_Redstart_of _Quintana_Texas1.jpg.*

Marra 2007, 2011) (fig. 24.5). Following departure from overwintering grounds, long-distance migrants are more susceptible to variation in vegetative productivity and extreme weather events, such as drought, that may contribute to prolonged stopover duration (Tottrup et al. 2012). Weather conditions may indeed help facilitate earlier departures, but for species that have to travel only short distances for migration, the conditions on overwintering grounds may be serving as a proxy for the conditions of their nearby breeding grounds.

As many birds are shifting their spring arrival, so too are nesting dates and breeding cycles shifting earlier. An earlier timing of breeding activity has led to the growing conservation concern of a decoupling between bird arrival and nesting with the availability of crucially important prey such as insects. This decoupling between the phenology of birds and the resources they depend on is referred to as **phenological mismatch** (Both et al. 2004), and these mismatches have important effects on both individual birds and populations (Pearce-Higgins and Green 2014). Studies on long-distance migrants have found that a failure to advance spring phenology could contribute to population declines (Both et al. 2006). Recently, however, studies have shown that the impact of phenological mismatches can be lessened by high local food availability (Dunn et al. 2011) and differences across habitats (Burger et al. 2012). In addition, population-level consequences may not be realized because of reduced competition among surviving juveniles during years when the mismatch is particularly strong (Reed et al. 2013). Altogether, there is strong potential for important consequences of phenological mismatch for birds and their prey, but the full extent of these consequences is likely to be influenced by factors such as habitat quality, seasonality of food resources, and competition.

MOVING POLEWARD AND SEEKING HIGHER GROUND IN A WARMING WORLD

In the face of environmental change, species can evolve new physiological tolerances to cope with altered climatic conditions, or they can move to maintain existing associations with the particular climates that define each species' **climatic niche**. As climate change occurs, we might expect vagile taxa, such as birds, to shift their distributions as they track their climatic niches through time. As such, the past forty years of rapid warming represents a large-scale experiment on how climate change influences bird ranges from regional to continental scales. If bird ranges are more or less limited by climate, then as climate conditions change, one would predict that bird distributions would follow (Pearson and Dawson 2003). An aspect of species ranges that are particularly sensitive to climate are **range boundaries** (Thomas 2010), and northerly range boundaries for many birds in the northern hemisphere tend to be correlated with long-term gradients in temperature and precipitation (Root 1988) (fig. 24.6; see the box on page 760). A fundamental prediction of climate change is that as global temperatures rise, bird range boundaries should respond by shifting to higher latitudes, i.e., toward the poles. These poleward shifts have been documented for birds in various regions of the world (Chen et al. 2011), including wintering and breeding birds in North America and Europe (Hitch and Leberg 2006, La Sorte and Thompson 2007, Mason et al. 2015). Based on multiple studies across the Northern Hemisphere, there has been a consistent poleward shift in bird ranges of 0.76 km per year, resulting in a mean shift of 15 km over the past 20 years (Pearce-Higgins and Green 2014). This does not mean that all birds have shifted northward, but it does suggest a trend that the majority of bird ranges have shifted in a direction that is concordant with a response to a warming world.

Past and future shifts in bird species ranges are likely to be more complex than poleward shifts. Across large geographic scales, birds respond differently to climate change because of species-specific sensitivities as well as the role of climatic factors other than temperature that might show more com-

JAMES PEARCE-HIGGINS

James Pierce-Higgins is the Director of Science at the British Trust for Ornithology. He researches the effects of climate change on bird populations and informs programs in bird conservation and climate change adaptation. *Photo with permission from the British Trust for Ornithology. Photo by Stephen H. Scheider.*

Dr. James Pearce-Higgins serves as the science director for the British Trust for Ornithology (BTO). Dr. Pearce-Higgins provides strategic oversight of BTO science and climate change research. His research has focused on changes in the abundance, distribution, and demography of bird populations and understanding the causes of population change to inform what is required to manage species and habitats sustainably. BTO's climate change research involves documenting the impacts of climate change on bird biodiversity, evaluating future impacts, and informing climate change adaptation efforts. Dr. Pierce-Higgins leads several multi-organizational efforts in climate change research throughout Europe. Before joining the BTO, Dr. Pierce-Higgins led a wide range of research projects on the impacts of weather and seasonality on upland bird demography. His current research in both applied and basic ornithology involves documenting the effects of climate change on avian biodiversity, undertaking projections of the future impact of climate change on bird distributions and abundance, and informing the development of climate change adaptation. He leads several research consortia on the ecological effects of climate change and coauthored *Birds and Climate Change: Impacts and Conservation Responses.*

Figure 24.6. Many wintering birds of North America have shifted northward over the past three decades. The Northern Cardinal (*Cardinalis cardinalis*) is a common denizen of people's backyards and has demonstrated a general shift northward in its range over time, potentially taking advantage of milder winters. *Photo taken by John Capella in Fishers, IN.*

plex variability over space and time (precipitation in particular). In support of the idea that range shifts can show complex directionality, recent studies have found that range changes do not always follow a poleward trajectory, and multidirectional range shifts in response to climate change are common for birds across the United States, Britain, and Australia (VanDerWal et al. 2013, Gillings et al. 2015, Bateman et al. 2016). In addition, it may take birds many years to "track" their suitable climate space. For example, a continental assessment of range shifts in wintering birds in North America found that individual species displayed highly variable responses to changes in winter temperature, and that many bird ranges took over 30 years to catch up to their changing climatic niche (La Sorte and Jetz 2012).

The influence of climate change on bird ranges is perhaps nowhere more evident than in montane ecosystems. Similar to poleward shifts, upslope shifts in elevation in response to warming temperatures are considered one of the strongest lines of evidence that species are responding to modern climate change (Parmesan 2006), but unlike range shifts, which are measured in kilometers, elevational shifts are much smaller and more variable. Upslope shifts in bird ranges average roughly 0.33 m per year, equivalent to a 6.6 m shift over 20 years (Pearce-Higgins and Green 2014). This may not seem like much, but changes over longer time periods can be significant. Over nearly a century of climate change, birds in the Sierra Nevada of California shifted their elevational ranges as they tracked long-term changes in precipitation and temperature (Tingley et al. 2009). In this case,

rising temperatures pushed bird populations upslope while increased precipitation pulled them downslope, resulting in a complex mixture of elevational shifts for individual species and regions (Tingley et al. 2012) (fig. 24.7). Similar patterns of upslope and downslope shifts over time have been found for birds in other montane ecosystems such as the Swiss Alps (Maggini et al. 2011). Small or insignificant shifts in elevation suggest that birds are not moving fast enough to track climate up in elevation, and in the case of montane environments, birds may be more limited by habitat availability or constrained in their dispersal abilities by the topographic

Figure 24.7. Bird species of the Sierra Nevada Mountains of California have responded to climate change by changing their distributions across elevational gradients. Changes in precipitation and temperature are complex and have resulted in both upslope and downslope changes in bird distributions. Over 100 years of change, birds like *A,* Western Scrub Jay (*Aphelocoma californica*) have shifted downslope, while other birds like *B,* Red-winged Blackbird shifted upslope (*Agelaius phoeniceus*). *Western Scrub Jay photo taken by Morgan Tingley. Red-winged Blackbird photo taken by Alan Wilson, September 26, 2006, licensed under the Creative Commons Attribution-Share Alike 2.0 Generic license, https://commons.wikimedia.org/wiki/File:Red_winged _blackbird_-_natures_pics.jpg.*

complexity (Elsen and Tingley 2015). In addition, differences exist between temperate and tropical locations. There is evidence that tropical montane birds are more responsive to changes in mean temperature than temperate zone montane birds, moving upslope at faster rates and showing less lag with temperature changes (Forero-Medina et al. 2011, Freeman and Freeman 2014).

EXTREME WEATHER AND ITS INFLUENCE ON BIRD POPULATIONS

Increasing frequency of extreme weather events such as drought, floods, and heat waves poses the most direct threat of climate change for bird populations. As these weather events are expected to become more severe and frequent in the coming years, understanding their impacts on bird demography and population viability is increasingly important. The probability of an individual bird successfully reproducing (or failing to reproduce) can be influenced by extreme weather events, with critical implications for populations as a whole.

Perhaps the most obvious effect of extreme weather on birds is drought. Although birds occupying warm and arid regions are adapted to variability in precipitation, the increasing frequency and magnitude of these extreme weather events can have far-reaching demographic consequences. For example, populations of the Florida Snail Kite (*Rostrhamus sociabilis plumbeus*), an endangered bird, are highly sensitive to droughts; during drought years, older adults are less likely to breed, while younger kites are more likely. This shift has had lasting effects on the structure of the population, as more experienced birds failed to breed during harsh environmental conditions (Reichert et al. 2012). The southwestern United States has been experiencing severe drought conditions, and in one case a Burrowing Owl (*Athene cunicularia*) population declined by 98 percent over a 16-year period as a result of the negative effects of decreased precipitation and increased air temperatures on physiological condition, reproductive output, and breeding phenology (McDonnell-Cruz and Wolf 2015). Across entire regions and countries, widespread drought and heat waves can impact entire bird communities. In the Great Plains of the United States, drought negatively affected bird communities by reducing the abundance and richness of Neotropical migrants by 13 percent and 6 percent, respectively (Albright et al. 2010). Similarly, a six-month heat wave in 2003 halted or reduced population growth for many bird species in France (Jiguet et al. 2006).

Just as birds are affected by droughts in arid regions, birds occupying the cryosphere can be affected by harsh winters and snowstorms. Extremely harsh winters can reduce juvenile and adult bird survival rates and compromise population persistence through mass mortality (Altwegg et al. 2006, Link and Sauer 2007). In arctic and antarctic ecosystems, snowstorms can have lasting effects on nest survival and overall reproductive success. For example, the frequency of snowstorms explained almost 30 percent of daily nest and colony productivity for Antarctic Petrels (*Thalassoica antarctica*) over a 20-year period (Descamps et al. 2015). Notably, chicks in poor condition were more likely to die during a snowstorm than chicks in good condition. By their very nature, however, extreme weather events such as storms normally occur over short time periods and may not always have lasting effects on bird survival. A detailed study of Black-tailed Godwits (*Limosa limosa limosa*) found that when a spring snowstorm coincided with their spring migration, many individual birds showed reverse migration, delayed breeding, and elevated metabolic costs. However, overall breeding success for the population was unaffected, and there was little carryover effect as a result of these late snowstorms (Senner et al. 2015). Ice-dependent seabirds, such as Southern Fulmars (*Fulmarus glacialoides*), that forage near the ice edge must deal with increasingly extreme years when sea ice area is reduced and the distance between foraging areas and the breeding colony is high (Jenouvrier et al. 2015) (fig. 24.8). During these extreme years, foraging trips were greater in distance and duration, resulting in less food delivered to chicks, reduced body condition, low breeding success, and reduced population growth rate. More experienced and successful breeders were better able to deal with these extremes, and as a result, years characterized by extreme loss of sea ice tended to exacerbate differences in individual behavior and flexibility (Jenouvrier et al. 2015). Similarly, projected future changes in ice conditions of the antarctic ecosystem are dire for Emperor Penguins (*Aptenodytes forsteri*), with models suggesting that by 2100 penguin populations will be suffering severe population declines resulting in a 19 percent decline of the global population (Jenouvrier et al. 2014) (fig. 24.8). Clearly, the ability of bird populations to adapt to extreme events is crucial to their long-term survival.

CLIMATE CHANGE AND BIRD COMMUNITIES

Climate exerts a strong influence on which species are present in a given place, and how bird communities characterized in this way change over time and across regions. That is, bird **species composition** is shaped in part by climate conditions and climate variability, along with other important factors (chapter 22). As the phenology and ranges of birds

Figure 24.8. Ice-dependent seabirds, such as A, Southern Fulmar (*Fulmarus glacialoides*) and B, Emperor Penguin (*Aptenodytes forsteri*) that forage near the ice edge must deal with increasingly extreme years when sea ice area is reduced and the distance between foraging areas and the breeding colony is high. Expected future changes in ice conditions of the Antarctic ecosystem suggest that C, large colonies of Emperor Penguins are likely to face catastrophic population declines. *Penguin photos by Erica Fitzpatrick. Southern Fulmar photo by JJ Harrison, May 28, 2012, licensed under the Creative Commons Attribution-Share Alike 2.0 Generic license, https://commons.wikimedia.org/wiki /File:Fulmarus_glacialoides_in_flight_-_SE _Tasmania.jpg.*

change over time, climate strongly influences how species interact with one another and with their environments and, ultimately, helps shape bird communities. Modern climate change is therefore likely to induce important changes in bird communities. Changes in species composition can be characterized in a number of ways. Some of the most familiar attributes of species composition in ecology are the related concepts of **species richness** and **diversity** (chapter 22). Simply put, some bird communities contain more species than others. Even given similar numbers of species, different bird communities may be composed of entirely different species, and thus function very differently. These patterns can be characterized using measures of **species turnover** and **beta-diversity**, which quantify changes in species composition over time or space. Ecological differences among communities can also be measured by examining the characteristics of their constituent species, such as diet or dispersal abilities. For example, high-latitude bird communities tend to have higher proportions of migratory species than do communities closer to the equator.

In this section we review how some of these community characteristics are expected to respond to climate change, what changes have already been observed in recent decades, and what the future might hold for bird communities. The detection of change in community characteristics over time and across large areas requires extensive, long-term data that are difficult or impossible for individual research teams to collect. The advent of **citizen science** programs, in which large numbers of people, often knowledgeable nonscientists, can contribute observations of birds, has aided such efforts immeasurably.

Changes in Species Richness, Functional Diversity, and Beta-Diversity

It is generally understood that bird species richness tends to be higher in warmer and wetter places than in cooler and drier places (chapter 4). The productivity of plants, which is usually higher under warm, humid conditions, has much to do with these patterns. For example, tropical rain forests have both very high plant productivity and very high bird species richness. These patterns produce expectations about the effects of climate change. For example, as climates warm in temperate regions, we might predict that species richness will increase, given no or little decline in precipitation. Similarly, reduced precipitation such as that predicted for some tropical regions may induce declines in species richness. Long-term data sets capable of detecting changes in richness over time are rare for tropical systems. In temperate regions such as Europe, long-term monitoring efforts have shown evidence of increasing species richness associated with increasing mean temperatures in recent decades (Davey et al. 2012).

Individual species respond in distinctive ways to climate change largely because relationships with the environment and behavioral repertoires vary from one species to another. That is, species differ from one another in ecological characteristics or **functional traits**. For example, range shifts in response to changing climates may take much longer for sedentary species with little ability to disperse long distances than for highly migratory birds, and poor dispersers may be especially disadvantaged if their habitat is highly fragmented. Across whole communities, this may result in a filtering of species over time according to functional traits and habitat relationships. Community composition may thus shift toward species with traits that allow adaptation to new climate conditions and away from those with disadvantageous traits. As this happens, there is potential for the loss of **functional diversity**—that is, a reduction in the variety of different ecological roles that are played by the collection of species in a community. Research in Europe has shown that habitat generalists have responded to a warming climate more favorably than have habitat specialists, leading to bird communities increasingly composed of generalists, as specialists have declined (Le Viol et al. 2012).

Another way of using species traits to describe communities, which has particular relevance for assessing climate change responses, involves measuring mean climate conditions across a species' range. The **species temperature index** (STI), for example, represents mean temperature at all locations where a species is known to occur within a study region. The **community temperature index** (CTI) in turn represents the mean STI for all bird species found in a given community or location. Communities may be composed mainly of warm-adapted species (high CTI) or cool-adapted species (low CTI). Long-term increases in CTI have been observed in North America and Europe, reflecting a combination of northward colonization by southerly species, losses of northerly species, and changes in the abundances of both of these groups (Devictor et al. 2008, Princé and Zuckerberg 2014).

Species composition is not fixed through time or across landscapes, but reflects cumulative patterns of individual species distributions and distributional changes in response to species' environments and to one another. Because of this, ecological communities tend not to move across landscapes as units in response to environmental change. Instead, landscapes show complex shifting mosaics of species composition, and particular communities may be unique to their time and place. While not the only factor, climatic variation plays an important role in these dynamics. One important expectation is that if climate change favors particular kinds of birds—such as warm-adapted species, habitat generalists, and good dispersers—then landscape-level, place-to-place diversity (**beta-diversity**) will decline, making different communities more similar and landscapes more homogeneous over time, even while species richness may increase locally. This observation has in fact been borne out in studies of Eu-

ropean birds over the past two decades (Le Viol et al. 2012, Davey et al. 2013).

Future Bird Communities under Climate Change

Novel climate conditions that have no present-day analog are expected to occur over large areas before the end of this century, particularly in the tropics, and we might expect that landscapes with novel climate conditions will support novel aggregations of species (Williams and Jackson 2007). One way to test this is to forecast the range shifts of species expected under climate change scenarios, given their particular climatic tolerances. Geographic shifts in climate suitability can be forecast in this way with **species distribution models** (Franklin 2009) (fig. 24.9). The results of many such models for different species can be used to estimate potential changes in species richness and rates of species turnover at individual locations, and across large regions (Peterson et al. 2002, Stralberg et al. 2009, Lawler et al. 2009). Such studies have suggested that most present-day communities will change as individual species distributions shift in response to climate change, which is consistent with what has been observed so far (fig. 24.10). While some species gains are predicted in northerly regions and at high elevations, species losses are also predicted in tropical regions and at low elevations (Lawler et al. 2009). Some community compositions found today may no longer occur at any location after sufficient climate change has occurred. Conversely, many locations in the future will support novel, or so-called **no-analog** bird communities that are not today found anywhere—that is, they have no present-day analog (fig. 24.10).

Given that species composition in many places will be novel in the near future, we can also expect that the dominance or importance of different kinds of interactions among bird species will also shift, resulting in novel community dynamics. New community compositions may result primarily from within-range shifts in abundance and local occurrence, in such a way that species with already overlapping distributions will come into contact more or less frequently. In other cases, species with previously nonoverlapping distributions may undergo range shifts that bring them into contact, contributing to the creation of no-analog communities. Some of the more negative effects that might occur in such cases could result from contact between closely related species or subspecies, which are more likely to compete for similar resources or, if sufficiently related, to hybridize. However, at least one study has estimated that the number of closely related species pairs expected to undergo climate-induced range overlap is relatively small (Krosby et al. 2015).

Climate change is likely to push ecosystems into new states, and under these conditions, the collections of species found in future landscapes may bear little resemblance to the bird communities of the past. New species combinations and interactions that are not yet within the experience of ecological science will be introduced, but these are difficult to predict, as are their ecosystem-level consequences. Of particular concern is the observation that there have been, to date, climate-change "winners" and "losers" among bird species, and the losers are expected to be increasingly threatened as climate change proceeds.

CLIMATE-SMART CONSERVATION FOR BIRDS

Birds face multiple threats as a result of human activities, including the negative effects of habitat loss and fragmentation, invasive species, pollution, and overexploitation of birds (chapters 25 and 26). Climate change is now added to this list of environmental stressors with which bird populations must contend (Thomas et al. 2004). While variations in climate are nothing new, it is the pace of modern, human-caused changes

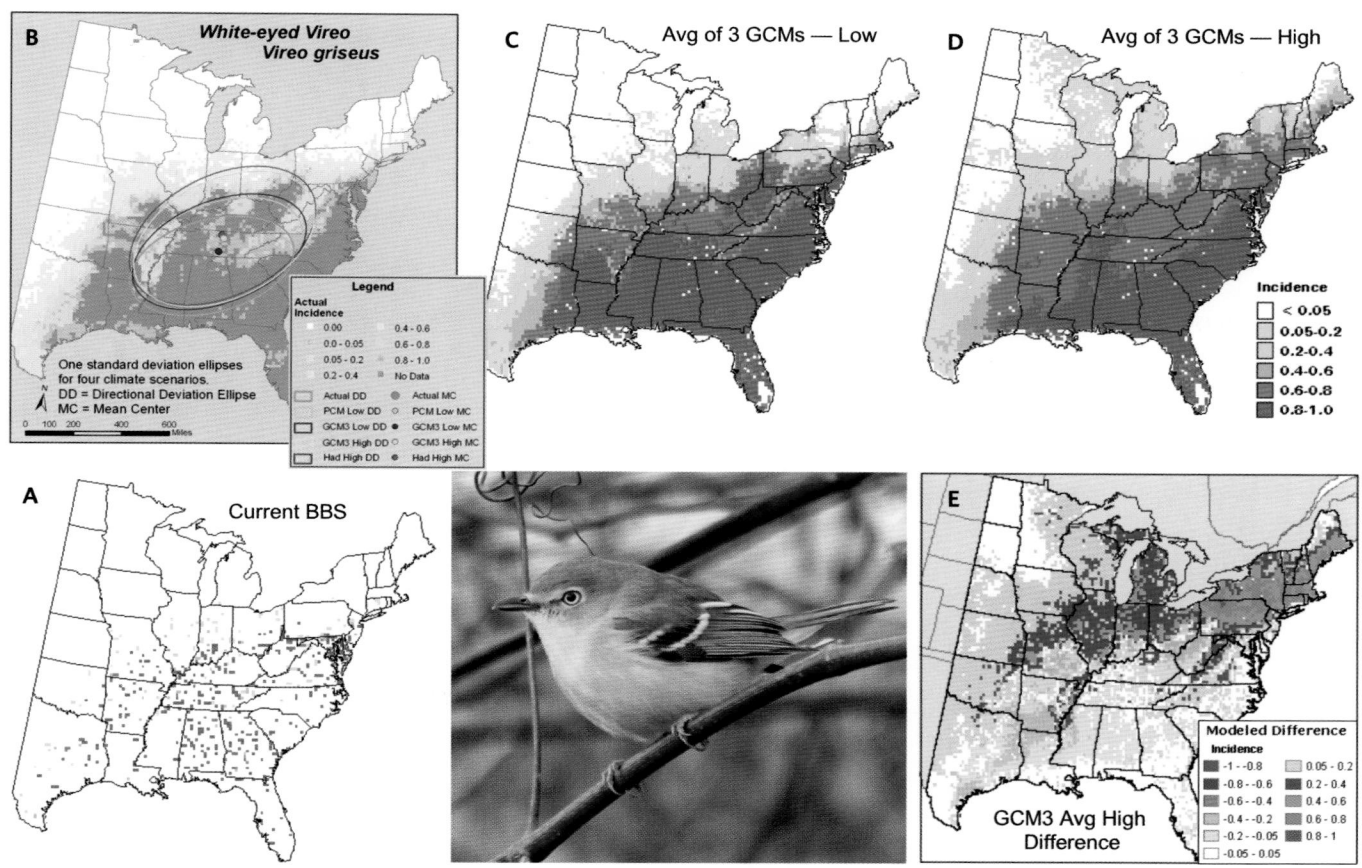

Figure 24.9. Species distribution modeling (SDM) can be a powerful tool for understanding how the distribution of suitable habitat and, potentially, the occurrence of a species is likely to shift in response to long-term changes in climate and other environmental factors. Here we illustrate this approach for White-eyed Vireo (*Vireo griseus*), adapted from the US Forest Service's Climate Change Bird Atlas, an online tool that allows users to explore SDMs for 147 bird species in the eastern United States (Matthews et al. 2011, Landscape Change Research Group 2014). These models do not give predictions for species occurrence in specific locations or landscapes, but do provide a generalized expectation of broad distributional shifts in habitat suitability, given expected climate change by the end of the twenty-first century. *A,* documented occurrences of White-eyed Vireo collected during the annual US Breeding Bird Survey (BBS) were used to relate species presence to local temperature, precipitation, forest vegetation, topography, and other conditions. *B,* Locations with conditions similar to those where the species was observed were mapped across the eastern United States. Points and ellipses show the expected shift in the centroid of the distribution under different climate change models and scenarios (see *C* and *D*). *C* and *D,* Using these species-environment relationships, the distribution of suitable habitat under predicted future (i.e., end of the twenty-first century) climate conditions was projected using three different general circulation models and two different greenhouse gas emissions scenarios (a high- and a low-emissions scenario). *E,* Expected difference in habitat suitability between the present day and the future prediction periods, under the high-emissions scenario (orange-brown areas show expected decline, and blue-green areas show expected increase). *Photo of White-eyed Vireo by Larry Clarfeld, in Vermont, where the species is very rare, in November of 2014.*

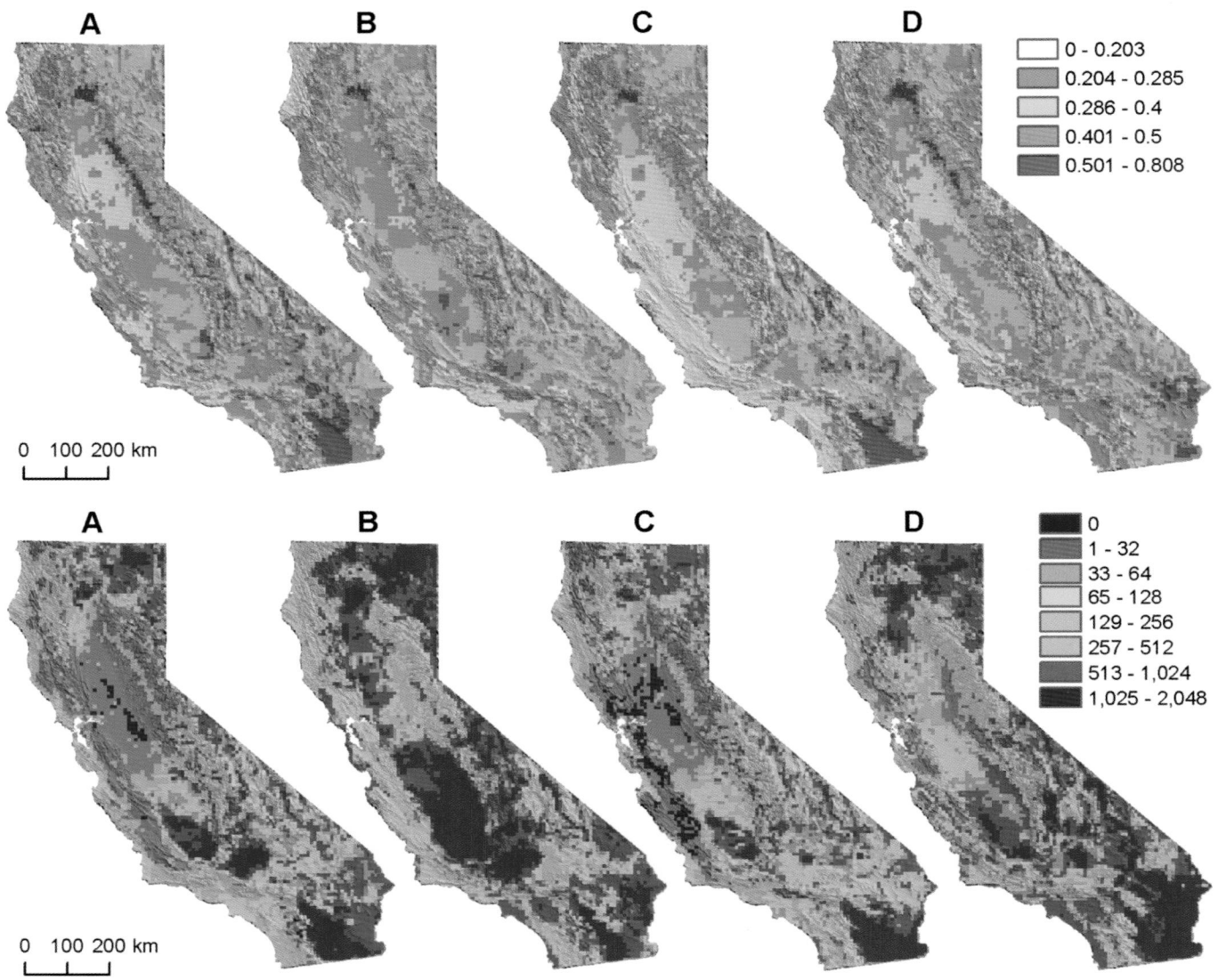

Figure 24.10. Change in bird species composition, and no-analog bird communities, predicted to result from future climate change in California (Stralberg et al. 2009). Differences in species composition between the present day and the middle of the twenty-first century (2038–2070) were predicted on the basis of species distribution models. The models used vegetation and climate conditions, together with presence and absence information for 60 different bird species, to characterize species-specific habitat quality across the state. Then, climate models were used to project future distributions of high and low habitat suitability, given expected climate change. In the upper panels, higher values indicate more predicted difference between present-day and future species composition at a given location (in this case, measured with the Bray-Curtis index). In the lower panels, each location was compared with every other location in California to quantify how unique the predicted future bird community in that location was in relation to present-day communities across the state. Lower values indicate fewer present-day analogues, and therefore represent "no-analog" future communities. The different maps from left to right result from the use of different combinations of particular species distribution models and climate general circulation models. While different models produced slightly different results, overall geographic patterns of predicted change were similar (species distribution models: maps *A* and *C* used generalized additive models; *B* and *D* used maximum entropy models. Climate models: maps *A* and *B* used the National Center for Atmospheric Research (NCAR) Community Climate System Model 3.0; *C* and *D* used the Geophysical Fluid Dynamics Laboratory (GFDL) General Circulation Model CM2.1). Both climate models were based on the Intergovernmental Panel on Climate Change (IPCC) SRES A2 scenario, which is a medium-high emissions scenario.

in climate that conservation practitioners find alarming (figs. 24.1, 24.2). In this century, these changes are expected to be more rapid and extreme than the natural climatic variation that bird populations have faced in their recent evolutionary history (Williams et al. 2007). If species for which the new conditions are unfavorable cannot adapt on this time frame, they are likely to decline in abundance and perhaps become threatened with local extirpations or even extinction (fig. 24.4). Unfortunately, many species that are already affected by stressors such as habitat loss may also be vulnerable to climate change, and these effects may combine in ways that are more harmful than the individual effects in isolation.

A large tool kit of bird conservation strategies has been developed in response to the threats posed by human activities (chapters 27 and 28). The conservation community is currently in the early stages of expanding this tool kit to address climate change. This response falls largely in two areas. First, the underlying causes of modern climate change must be addressed through efforts to reduce the amount of greenhouse gases being emitted and held in the Earth's atmosphere. These efforts are referred to as **climate change mitigation**. Ironically, some of these efforts, such as the conversion to greener energy sources like biofuels and wind power, may themselves have negative impacts on some birds. These trade-offs must be carefully considered in overall conservation strategies. Second, historical and ongoing greenhouse gas emissions have already committed us to a certain degree of unavoidable climate change. If threatened bird populations are going to survive, they must do so under these new conditions of rapid global change. Efforts to create conditions in which birds can effectively adapt to ongoing and future climate change are referred to as **climate change adaptation**.

Which Species Are Most Vulnerable?

One key approach to the conservation of birds has been to focus on the needs of those particular species that are being most negatively affected by human activities, usually considered threatened and endangered species. In the context of climate change, the first step in this process is to identify the species most likely to be negatively affected by the kinds of climate change they will experience in coming decades (Foden et al. 2013, Pacifici et al. 2015). Gauging vulnerability to climate change involves estimating a species' **sensitivity**, **exposure**, and **adaptive capacity**. Sensitivity refers to how negatively a species may respond to changes in temperature and precipitation, such as unusual drought conditions or heat waves. Sensitivity is gauged through a careful examination of the ecology of a species in relation to climate and weather data. Exposure refers to the future climate changes that the

species will actually experience and is typically gauged using climate models. Thus, some very sensitive species may not be greatly exposed within their limited range, and other species may be relatively insensitive to the kinds of climate variability to which they will likely be exposed. It is the combination of strong sensitivity and extensive exposure that poses the greatest risk. Adaptive capacity refers to the ability of species to overcome this risk through adaptive behaviors, such as dispersal to more favorable places, or through rapid evolutionary changes. Adaptive capacity can be strongly influenced by habitat conditions—for example, species that cannot disperse through highly fragmented habitat may have little capacity to colonize sites with more suitable climate conditions. A primary goal of vulnerability assessment is to prioritize the most vulnerable species for conservation attention and identify specific conservation and management efforts that stand the best chance of maximizing the adaptive capacity of those species (Nicotra et al. 2015).

Bird Conservation Strategies in a Changing Climate

Most strategies for conserving species in the face of climate change rely on approaches that have already been developed to address other major threats to biodiversity, but with some important modifications. Continuing to manage protected areas such as national parks, reserves, and wildlife refuges for biodiversity conservation using an ecosystem-level approach (chapters 27 and 28) should help to ensure that vulnerable species have suitable habitat where chances for adaptation to climate change are best, because other threats are minimized. These areas may provide suitable areas for colonization by species of conservation concern, allowing the distributions of those species to shift in response to climate change. However, as land use has intensified in the broader landscape, protected areas have in effect become more isolated and island-like, making movement among them more difficult (DeFries et al. 2005, Radeloff et al. 2010). Thus, climate change adds emphasis to the need for creating greater **connectivity** among protected areas. Connectivity can be provided by maintaining and enhancing habitat suitability outside of existing protected areas, particularly in key locations that can act as **corridors** or stepping-stones for wildlife movement among protected areas. This can be achieved through a range of large-scale planning and management approaches that may include establishing new protected areas in key locations and working with landowners in places that are not exclusively managed for biodiversity protection, to encourage land uses that are nonetheless consistent with this goal (see the box on page 770). Most of the Earth's surface will not be included within designated protected areas, and will experience varying intensities of resource use, even as climate change adaptation requires that species have large areas for movement and colonization in order to persist. Thus, enhancing the conservation value of "working lands and seas" should be a high priority, even while more strictly protected areas remain a backbone of the conservation landscape.

In order to maximize the benefit of conservation activities in the limited areas where they can be undertaken, these areas should be carefully selected. As climate changes, protecting threatened and vulnerable species requires that conservation actions take place not only where these species currently occur, but also where they are likely to have suitable climate conditions in the future. **Conservation planning**, which is concerned with prioritizing landscapes to efficiently achieve conservation goals, has only recently begun to address this complicating factor of potentially rapid change over time (Fuller et al. 2011). For example, expected geographic shifts in habitat suitability can be forecast with species distribution models (Franklin 2009). Combinations of such models for multiple species can be used to identify suites of locations that collectively can be expected to protect current and future habitat for the maximum number of species (Hannah et al. 2002).

Within areas that have been identified for some kind of conservation activity, a variety of ecosystem management and restoration activities may be undertaken, depending on the condition of existing habitat and the needs of particular species. Creating optimal habitat conditions for bird species of concern is likely to maximize their resilience and adaptive capacity to climate change in those landscapes. However, because multiple bird species with different specific habitat affinities may be of concern, an ecosystem approach to management is often called for (Christensen et al. 1996, West et al. 2009). Management activities to achieve this typically include establishing and maintaining diverse plant communities, controlling invasive predators, competitors, or nest parasites, and establishing ecological disturbance regimes (e.g., prescribed fire) within historical ranges of variability. However, climate change is also likely to push many ecosystems into new states, and maintaining formerly "normal" conditions may become untenable. Under these circumstances, effective bird species conservation will entail recognizing these inevitable changes and working with them. This means continuing to implement active management techniques that can provide suitable habitat for a wide range of species, even if the collection of species found in a given conservation landscape bears little resemblance to the bird communities of the past (West et al. 2009, Fuller et al. 2011).

LANDSCAPE CONSERVATION COOPERATIVES

The Landscape Conservation Cooperative (LCC) Network is an applied science partnership with the shared goal of conserving landscapes to sustain natural and cultural resources for current and future generations. The LCCs represent an effort to address large-scale conservation challenges, climate change being the premier example, by developing science-based conservation strategies that are coordinated across large landscapes and across administrative boundaries. Housed within the US Department of the Interior, the network's 22 self-directed cooperatives are associated with multistate regions that collectively include most of North America, and partners within each cooperative include federal, state, and tribal natural resource agencies and nongovernmental conservation organizations. At the time of this writing, the LCC network is a young institution that promises to become a significant platform for researchers and conservation practitioners to bring science addressing the consequences of climate change together with applied ecosystem management, helping to meet the challenges of biodiversity conservation in a rapidly changing world (LCC Network, 2014).

It is possible that climate change will be too rapid, and habitat too fragmented, for some species to colonize newly suitable areas quickly enough to avoid extinction. This has not yet been observed, but species at risk of this eventuality are likely to be habitat specialists already reduced in numbers by other human-caused threats, with strong climate change vulnerability and poor dispersal ability. **Assisted colonization**, or assisted migration, is a proposed approach that entails introducing such species to new areas, possibly outside their existing range, as climate change makes those areas suitable (Hoegh-Guldberg et al. 2008). While a promising idea, this method should be considered a last resort after other ecosystem management efforts have failed. The risk that species introduced to new environments may become invasive, and have negative impacts on other species, is well documented, though the likelihood of this happening will depend on the ecology of the species and environments in question. Carefully measuring extinction risk against possible negative effects to other at-risk species and broader ecosystem functions will be crucial in the ongoing consideration of this approach.

Effects of Climate Change Mitigation on Birds

Efforts to reduce greenhouse gas emissions and their atmospheric concentrations are a first priority in combating climate change. The net effect of these mitigation efforts will be beneficial for birds if they are successful in reducing underlying causes of climate change and rendering more extreme adaptation efforts unnecessary. Nonetheless, existing approaches to achieving this, particularly renewable energy production methods, can be detrimental to bird populations in other ways.

Carbon sequestration refers to the storage of carbon to prevent it from contributing to atmospheric greenhouse gas concentrations. The standing biomass of terrestrial and ocean ecosystems, such as that in forest trees and soils, is a primary form of carbon storage. The need for carbon sequestration can therefore help to create and maintain habitat for birds, particularly in forest habitats. Incentives for carbon sequestration efforts, however, do not necessarily result in land use decisions that improve conservation outcomes (Nelson et al. 2008). For example, some vegetation types, such as single-species tree plantations, can provide carbon storage but have little bird conservation value. Delivering on the potential for win-win solutions for both carbon sequestration and biodiversity protection depends on ensuring that adequate attention is given to the biodiversity impacts of particular sequestration approaches.

Wind power, hydroelectricity, solar power, and agricultural bioenergy production all promise to reduce carbon emissions by reducing reliance on fossil fuels, but they also involve land use changes and infrastructure development that can have negative effects on bird populations. These effects accrue through habitat loss, disruption of ecosystem processes, or direct effects of infrastructure such as bird collisions with wind turbines. Collisions with turbines may be the most apparent negative effect of large wind energy farms, particularly because raptors and other large birds such as geese and seabirds tend to be the most at risk, but some species can also suffer a less visible form of habitat loss when they naturally avoid areas with large towering structures, even if otherwise suitable habitat is present (Drewitt and Langston 2006, Pearce-Higgins and Green 2014). Solar energy production may pose the lowest risks among renewable energy options, whereas some of the most negative impacts on wildlife to date have resulted from the large-scale conversion of tropical forests for agricultural biofuel production. In addition to direct biodiversity losses associated with forest loss, the conversion of forests for biofuel production also releases sequestered carbon, negating the climate benefits of biofuel production under such circumstances (Fargione et al. 2008).

Minimizing the negative effects of renewable energy projects will involve careful siting of infrastructure away from areas of heavy wildlife use such as important migration corridors and large wetlands, as well as adopting improved technologies and methods. Bioenergy production may be made more benign, for example, by siting on existing agricultural lands and utilizing diverse mixes of native plant species that supply usable habitat for grassland and open-country birds. Even efforts to minimize negative impacts can be complicated by unexpected trade-offs—for example, newer wind turbines that are larger in size and therefore can replace the energy production capacity of several smaller turbines appear to result in a reduction of overall bird collision rates, but they may actually increase bat collision rates (Barclay et al. 2007). Ultimately, all methods of energy production have some effect on wildlife populations, and probably the best hope for minimizing these negative effects is to minimize energy consumption. While beyond the scope of this book, reducing per capita energy consumption particularly in wealthy nations is both a great societal challenge and a necessary centerpiece of efforts to address the hazards of climate change.

KEY POINTS

- In recent decades, timing of seasonal events (spring arrival and egg-laying dates) has advanced consistent with earlier spring phenology.
- There is evidence that a phenological mismatch between birds and their food resources has important implications for population dynamics (although some species and guilds are likely more vulnerable than others).
- Species are shifting their distributions (primarily poleward and upslope) over recent decades, leading to unanticipated species interactions.
- Extreme events (flooding, heat waves, ice storms) have been cautiously attributed to modern climate change and can have wide-ranging impact on avian demographics in tropical, temperate, and polar environments.
- Range shifts can lead to novel bird communities, and increases in species richness have been observed in northerly latitudes experiencing strong warming.
- Certain species or groups (seabirds, upland birds, woodland birds) are likely to be more sensitive to climate change because of their particular life history characteristics (although this could result from a sampling bias).

KEY MANAGEMENT AND CONSERVATION IMPLICATIONS

- Human-caused climate change is already affecting bird populations, and these effects will only strengthen over time; climate change adaptation efforts are needed to help ensure the survival of species vulnerable to the negative effects of climate change to which the world is already committed.
- Climate change vulnerability assessments can help to focus conservation attention on the most at-risk species and populations.
- Negative effects of climate change will compound the effects of other stressors such as habitat loss and fragmentation.
- Climate change places greater importance than ever on an all-lands, ecosystem management approach to bird conservation, with protected areas as well as working lands playing crucial roles.
- Climate change mitigation efforts such as the shift to low-carbon energy sources (wind, solar) are essential, but these efforts may themselves have negative impacts on bird populations (and other taxa), which should be addressed.

DISCUSSION QUESTIONS

1. When studying poleward shifts in bird distributions, researchers often focus on quantifying the range boundaries. Why might range boundaries be sensitive to climate change? What specific aspects of population dynamics would be responsible for expanding or retracting range boundaries?

2. In studies of phenological mismatch, researchers have found differential rates of advancement for trees, caterpillars, resident birds, migrant birds, and predators on birds. If all these groups are experiencing the same shifting climate conditions, what is the possible cause of these different rates of advancement?

3. How might behavioral plasticity be important for mitigating the population-level consequences of extreme weather (e.g., drought, storms, and heat waves)?

4. Why might birds in tropical montane ecosystems be more affected by rising temperatures than birds occupying temperate montane systems?

5. How might habitat fragmentation and climate change interact to cause more stress on bird populations than either of these two factors would cause in isolation from each other?

References

Albright, T. P., A. M. Pidgeon, C. D. Rittenhouse, M. K. Clayton, C. H. Flather, P. D. Culbert, B. D. Wardlow, and V. C. Radeloff (2010). Effects of drought on avian community structure. Global Change Biology 16:2158–2170.

Allstadt, A. J., A. M. Liebhold, D. M. Johnson, R. E. Davis, and K. J. Haynes (2015). Temporal variation in the synchrony of weather and its consequences for spatiotemporal population dynamics. Ecology 96:2935–2946.

Altwegg, R., A. Roulin, M. Kestenholz, and L. Jenni (2006). Demographic effects of extreme winter weather in the Barn Owl. Oecologia 149:44–51.

Ault, T. R., M. D. Schwartz, R. Zurita-Milla, J. F. Weltzin, and J. L. Betancourt (2015). Trends and natural variability of spring onset in the coterminous United States as evaluated by a new gridded dataset of spring indices. Journal of Climate 28:8363–8378.

Barclay, R. M., E. Baerwald, and J. Gruver (2007). Variation in bat and bird fatalities at wind energy facilities: Assessing the effects of rotor size and tower height. Canadian Journal of Zoology 85:381–387.

Barriopedro, D., E. M. Fischer, J. Luterbacher, R. Trigo, and R. Garcia-Herrera (2011). The hot summer of 2010: Redrawing the temperature record map of Europe. Science 332:220–224.

Bateman, B. L., A. M. Pidgeon, V. C. Radeloff, J. VanDerWal, W. E. Thogmartin, S. J. Vavrus, and P. J. Heglund (2016). The pace of past climate change vs. potential bird distributions and land use in the United States. Global Change Biology. doi:10.1111/gcb.13154.

Berthold, P., E. Gwinner, and E. Sonnenschein (2003). Avian migration. Springer, New York.

Both, C., A. V. Artemyev, B. Blaauw, R. J. Cowie, A. J. Dekhuijzen, T. Eeva, A. Enemar, et al. (2004). Large-scale geographical variation confirms that climate change causes birds to lay earlier. Proceedings of the Royal Society B: Biological Sciences 271:1657–1662.

Both, C., S. Bouwhuis, C. M. Lessells, and M. E. Visser (2006). Climate change and population declines in a long-distance migratory bird. Nature 441:81–83.

Both, C. (2010). Flexibility of timing of avian migration to climate change masked by environmental constraints en route. Current Biology 20:243–248.

Burger, C., E. Belskii, T. Eeva, T. Laaksonen, M. Magi, R. Mand, A. Qvarnstrom, et al. (2012). Climate change, breeding date and nestling diet: How temperature differentially affects seasonal changes in Pied Flycatcher diet depending on habitat variation. Journal of Animal Ecology 81:926–936.

Burrows, M. T., D. S. Schoeman, L. B. Buckley, P. Moore, E. S. Poloczanska, K. M. Brander, C. Brown, et al. (2011). The pace of shifting climate in marine and terrestrial ecosystems. Science 334:652–655.

Chen, I. C., J. K. Hill, R. Ohlemuller, D. B. Roy, and C. D. Thomas (2011). Rapid range shifts of species associated with high levels of climate warming. Science 333:1024–1026.

Christensen, N. L., A. M. Bartuska, J. H. Brown, S. Carpenter, C. D'Antonio, R. Francis, J. F. Franklin, J. A. MacMahon, R. F. Noss, and D. J. Parsons (1996). The report of the Ecological Society of America committee on the scientific basis for ecosystem management. Ecological Applications 6:665–691.

Christensen, J. H., K. Krishna Kumar, E. Aldrian, S.-I. An, I. F. A. Cavalcanti, M. de Castro, W. Dong, et al. (2013). Climate phenomena and their relevance for future regional climate change. *In* Climate change 2013: The physical science basis. Contribution of Working Group I to the Fifth Assessment Report of the Intergovernmental Panel on Climate Change, Stocker, T. F., D. Qin, G.-K. Plattner, M. Tignor, S. K. Allen, J. Boschung, A. Nauels, et al., Editors. Cambridge University Press, Cambridge, UK, pp. 1217–1308.

Christidis, N., P. A. Stott, and S. J. Brown (2011). The role of human activity in the recent warming of extremely warm daytime temperatures. Journal of Climate 24:1922–1930.

Church, J. A., P. U. Clark, A. Cazenave, J. M. Gregory, S. Jevrejeva, A. Levermann, M. A. Merrifield, et al. (2013). Sea level change. *In* Climate change 2013: The physical science basis. Contribution of Working Group I to the Fifth Assessment Report of the Intergovernmental Panel on Climate Change, Stocker, T. F., D. Qin, G.-K. Plattner, M. Tignor, S. K. Allen, J. Boschung, A. Nauels, et al., Editors. Cambridge University Press, Cambridge, UK, pp. 1137–1216.

Cook, B. I., T. R. Ault, and J. E. Smerdon (2015). Unprecedented 21st century drought risk in the American Southwest and Central Plains. Sciences Advances 1:1–7.

Cox, G. W. (2010). Bird migration and global change. 1st ed. Island Press, Washington, DC.

Davey, C. M., D. E. Chamberlain, S. E. Newson, D. G. Noble, and A. Johnston (2012). Rise of the generalists: Evidence for climate driven homogenization in avian communities. Global Ecology and Biogeography 21:568–578.

Davey, C. M., V. Devictor, N. Jonzen, Å. Lindström, and H. G. Smith (2013). Impact of climate change on communities: Revealing species' contribution. Journal of Animal Ecology 82:551–561.

DeFries, R., A. Hansen, A. C. Newton, and M. C. Hansen (2005). Increasing isolation of protected areas in tropical forests over the past twenty years. Ecological Applications 15:19–26.

Derksen, C., and R. Brown (2012). Spring snow cover extent reductions in the 2008–2012 period exceeding climate model projections. Geophysical Research Letters 39.

Descamps, S., A. Tarroux, O. Varpe, N. G. Yoccoz, T. Tveraa, and S. H. Lorentsen (2015). Demographic effects of extreme weather events: Snow storms, breeding success, and population growth rate in a long-lived Antarctic seabird. Ecology and Evolution 5:314–325.

Devictor, V., R. Julliard, D. Couvet, and F. Jiguet (2008). Birds are tracking climate warming, but not fast enough. Proceedings of the Royal Society B: Biological Sciences 275:2743–2748.

Diffenbaugh, N. S., and C. B. Field (2013). Changes in ecologically critical terrestrial climate conditions. Science 341:486–492.

Drewitt, A. L., and R. H. Langston (2006). Assessing the impacts of wind farms on birds. Ibis 148:29–42.

Duffy, P. B., and C. Tebaldi (2012). Increasing prevalence of extreme summer temperatures in the U.S. Climatic Change 111:487–495.

Dunn, P. O., D. W. Winkler, L. A. Whittingham, S. J. Hannon, and R. J. Robertson (2011). A test of the mismatch hypothesis: How is timing of reproduction related to food abundance in an aerial insectivore? Ecology 92:450–461.

Elsen, P. R., and M. W. Tingley (2015). Global mountain topography and the fate of montane species under climate change. Nature Climate Change 5:772–777.

Fargione, J., J. Hill, D. Tilman, S. Polasky, and P. Hawthorne (2008). Land clearing and the biofuel carbon debt. Science 319:1235–1238.

Foden, W. B., S. H. Butchart, S. N. Stuart, J.-C. Vié, H. R. Akçakaya, A. Angulo, L. M. DeVantier, A. Gutsche, E. Turak, and L. Cao (2013). Identifying the world's most climate change vulnerable species: A systematic trait-based assessment of all birds, amphibians and corals. PLoS ONE 8:e65427. doi:10.1371/journal.pone.0065427.

Forero-Medina, G., J. Terborgh, S. J. Socolar, and S. L. Pimm (2011). Elevational ranges of birds on a tropical montane gradient lag behind warming temperatures. PLoS ONE 6:e28535. https://doi.org/10.1371/journal.pone.0028535.

Franklin, J. (2009). Mapping species distributions: Spatial inference and prediction. Cambridge University Press, Cambridge, UK.

Freeman, B. G., and A. M. C. Freeman (2014). Rapid upslope shifts in New Guinean birds illustrate strong distributional responses of tropical montane species to global warming. Proceedings of the National Academy of Sciences USA 111:4490–4494.

Fuller, R. A., R. J. Ladle, R. J. Whittaker, and H. P. Possingham (2011). Planning for persistence in a changing world. *In* Conservation biogeography. John Wiley and Sons, Hoboken, NJ, pp. 161–189.

Gallinat, A. S., R. B. Primack, and D. L. Wagner (2015). Autumn, the neglected season in climate change research. Trends in Ecology and Evolution 30:169–176.

Gillings, S., D. E. Balmer, and R. J. Fuller (2015). Directionality of recent bird distribution shifts and climate change in Great Britain. Global Change Biology 21:2155–2168.

Hall, M. H. P., and D. B. Fagre (2003). Modeled climate-induced glacier change in Glacier National Park, 1850–2100. Bioscience 53:131–140.

Hannah, L., G. F. Midgley, and D. Millar (2002). Climate change-integrated conservation strategies. Global Ecology and Biogeography 11:485–495.

Hartmann, D. L., A. M. G. Klein Tank, M. Rusticucci, L. V. Alexander, S. Brönnimann, Y. Charabi, F. J. Dentener, et al. (2013). Observations: Atmosphere and surface. *In* Climate change 2013: The physical science basis. Contribution of Working Group I to the Fifth Assessment Report of the Intergovernmental Panel on Climate Change, Stocker, T. F., D. Qin, G.-K. Plattner, M. Tignor, S. K. Allen, J. Boschung, A. Nauels, et al., Editors. Cambridge University Press, Cambridge, UK, pp. 159–254.

Hitch, A. T., and P. L. Leberg (2006). Breeding distributions of North American bird species moving north as a result of climate change. Conservation Biology 21:534–539.

Hoegh-Guldberg, O., L. Hughes, S. McIntyre, D. Lindenmayer, C. Parmesan, H. Possingham, and C. Thomas (2008). Assisted colonization and rapid climate change. Science (Washington) 321:345–346.

Ibanez, I., R. B. Primack, A. J. Miller-Rushing, E. Ellwood, H. Higuchi, S. D. Lee, H. Kobori, and J. A. Silander (2010). Forecasting phenology under global warming. Philosophical Transactions of the Royal Society B: Biological Sciences 365:3247–3260.

IPCC (2013). Climate change 2013: The physical science basis. Contribution of Working Group I to the Fifth Assessment Report of the Intergovernmental Panel on Climate Change. Cambridge University Press, Cambridge, UK.

Jenouvrier, S. (2013). Impacts of climate change on avian populations. Global Change Biology 19:2036–2057.

Jenouvrier, S., M. Holland, J. Stroeve, M. Serreze, C. Barbraud, H. Weimerskirch, and H. Caswell (2014). Projected continent-wide declines of the Emperor Penguin under climate change. Nature Climate Change 4:715–718.

Jenouvrier, S., C. Peron, and H. Weimerskirch (2015). Extreme climate events and individual heterogeneity shape life-history traits and population dynamics. Ecological Monographs 85:605–624.

Jiguet, F., R. Julliard, C. D. Thomas, O. Dehorter, S. E. Newson, and D. Couvet (2006). Thermal range predicts bird population resilience to extreme high temperatures. Ecology Letters 9:1321–1330.

Klein, A. G., and J. L. Kincaid (2006). Retreat of glaciers on Puncak Jaya, Irian Jaya, determined from 2000 and 2002 IKONOS satellite images. Journal of Glaciology 52:65–79.

Knudsen, E., A. Linden, T. Ergon, N. Jonzen, J. O. Vik, J. Knape, J. E. Roer, and N. C. Stenseth (2007). Characterizing bird migration phenology using data from standardized monitoring at bird observatories. Climate Research 35:59–77.

Krosby, M., C. B. Wilsey, J. L. McGuire, J. M. Duggan, T. M. Nogeire, J. A. Heinrichs, J. J. Tewksbury, and J. J. Lawler (2015). Climate-induced range overlap among closely related species. Nature Climate Change 5:883–886.

Landscape Change Research Group (2014). Climate change atlas. Northern Research Station, US Forest Service, Delaware, OH. http://www.nrs.fs.fed.us/atlas.

Landscape Conservation Cooperative (LCC) Network (2014). Network strategic plan. Available at http://lccnetwork.org/strategic-plan.

La Sorte, F. A., and W. Jetz (2012). Tracking of climatic niche boundaries under recent climate change. Journal of Animal Ecology 81:914–925.

La Sorte, F. A., and F. R. Thompson (2007). Poleward shifts in winter ranges of North American birds. Ecology 88:1803–1812.

Lawler, J. J., S. L. Shafer, D. White, P. Kareiva, E. P. Maurer, A. R. Blaustein, and P. J. Bartlein (2009). Projected climate-induced faunal change in the Western Hemisphere. Ecology 90:588–597.

Le Viol, I., F. Jiguet, L. Brotons, S. Herrando, Å. Lindström, J. W. Pearce-Higgins, J. Reif, C. Van Turnhout, and V. Devictor (2012). More and more generalists: Two decades of changes in the European avifauna. Biology Letters 8:780–782.

Link, W. A., and J. R. Sauer (2007). Seasonal components of avian population change: Joint analysis of two large-scale monitoring programs. Ecology 88:49–55.

Loarie, S. R., P. B. Duffy, H. Hamilton, G. P. Asner, C. B. Field, and D. D. Ackerly (2009). The velocity of climate change. Nature 462:1052–1055. doi:10.1038/nature08649.

Maggini, R., A. Lehmann, M. Kery, H. Schmid, M. Beniston, L. Jenni, and N. Zbinden (2011). Are Swiss birds tracking climate change? Detecting elevational shifts using response curve shapes. Ecological Modelling 222:21–32.

Mason, S. C., G. Palmer, R. Fox, S. Gillings, J. K. Hill, C. D. Thomas, and T. H. Oliver (2015). Geographical range margins of many taxonomic groups continue to shift polewards. Biological Journal of the Linnean Society 115:586–597.

Matthews, S. N., L. R. Iverson, A. M. Prasad, and M. P. Peters (2011). Changes in potential habitat of 147 North American breeding bird species in response to redistribution of trees and climate following predicted climate change. Ecography 34:933–945.

Mazdiyasni, O., and A. AghaKouchak (2015). Substantial increase in concurrent droughts and heatwaves in the United States. Proceedings of the National Academy of Sciences USA 112:11484–11489.

McCabe, G. J., T. R. Ault, B. I. Cook, J. L. Betancourt, and M. D. Schwartz (2012). Influences of the El Nino Southern Oscillation and the Pacific Decadal Oscillation on the timing of the North American spring. International Journal of Climatology 32:2301–2310.

McDonnell-Cruz, K., and B. O. Wolf (2015). Rapid warming and drought negatively impact population size and reproductive dynamics of an avian predator in the arid Southwest. Global Change Biology 2016:237–253.

Menzel, A., T. H. Sparks, N. Estrella, E. Koch, A. Aasa, R. Ahas, K. Alm-Kubler, et al. (2006). European phenological response to climate change matches the warming pattern. Global Change Biology 12:1969–1976.

Moss, R., M. Babiker, S. Brinkman, E. Calvo, T. Carter, J. Edmonds, I. Elgizouli, et al. (2008). Towards new scenarios for analysis of emissions, climate change, impacts, and response strategies: IPCC Expert Meeting report: 19–21 September, 2007, Noordwijkerhout, The Netherlands. Intergovernmental Panel on Climate Change, Geneva, Switzerland.

Nelson, E., S. Polasky, D. J. Lewis, A. J. Plantinga, E. Lonsdorf, D. White, D. Bael, and J. J. Lawler (2008). Efficiency of incentives to jointly increase carbon sequestration and species conservation on a landscape. Proceedings of the National Academy of Sciences 105:9471–9476.

Newton, I. (2010). Bird migration. Collins, London.

Nicholls, R. J., R. S. J. Tol, and J. W. Hall (2007). Assessing impacts and responses to global-mean sea-level rise. In Human-induced climate change: An interdisciplinary assessment, M. E.

Schlesinger, H. Kheshgi, J. B. Smith, F. C. de la Chesnaye, J. M. Reilly, T. Wilson, C. D. Kolstad, Editors. Cambridge University Press, Cambridge, UK, pp. 119–134.

Nicotra, A. B., E. A. Beever, A. L. Robertson, G. E. Hofmann, and J. O'Leary (2015). Assessing the components of adaptive capacity to improve conservation and management efforts under global change. Conservation Biology 29:1268–1278.

Pacifici, M., W. B. Foden, P. Visconti, J. E. Watson, S. H. Butchart, K. M. Kovacs, B. R. Scheffers, D. G. Hole, T. G. Martin, and H. R. Akçakaya (2015). Assessing species vulnerability to climate change. Nature Climate Change 5:215–224.

Parmesan, C. (2006). Ecological and evolutionary responses to recent climate change. Annual Review of Ecology Evolution and Systematics 37:637–669.

Pearce-Higgins, J. W., and R. E. Green (2014). Birds and climate change: Impacts and conservation responses. Cambridge University Press, Cambridge, UK.

Pearson, R. G., and T. P. Dawson (2003). Predicting the impacts of climate change on the distribution of species: Are bioclimate envelope models useful? Global Ecology and Biogeography 12:361–371.

Peterson, A. T., M. A. Ortega-Huerta, J. Bartley, V. Sanchez-Cordero, J. Soberon, R. H. Buddemeier, and D. R. B. Stockwell (2002). Future projections for Mexican faunas under global climate change scenarios. Nature 416:626–629.

Princé, K., and B. Zuckerberg (2014). Climate change in our backyards: The reshuffling of North America's winter bird communities. Global Change Biology. doi:10.1111/gcb.12740.

Radeloff, V. C., S. I. Stewart, T. J. Hawbaker, U. Gimmi, A. M. Pidgeon, C. H. Flather, R. B. Hammer, and D. P. Helmers (2010). Housing growth in and near United States protected areas limits their conservation value. Proceedings of the National Academy of Sciences USA 107:940–945.

Reed, T. E., V. Grotan, S. Jenouvrier, B. E. Saether, and M. E. Visser (2013). Population growth in a wild bird is buffered against phenological mismatch. Science 340:488–491.

Reichert, B. E., C. E. Cattau, R. J. Fletcher, W. L. Kendall, and W. M. Kitchens (2012). Extreme weather and experience influence reproduction in an endangered bird. Ecology 93:2580–2589.

Root, T. (1988). Environmental-factors associated with avian distributional boundaries. Journal of Biogeography 15:489–505.

Rostom, R., and S. Hastenrath (2007). Variations of Mount Kenya's glaciers 1993–2004. Erdkunde 61:277–283.

Schwartz, M. D., T. R. Ault, and J. L. Betancourt (2013). Spring onset variations and trends in the continental United States: Past and regional assessment using temperature-based indices. International Journal of Climatology 33:2917–2922.

Senner, N. R., M. A. Verhoeven, J. M. Abad-Gomez, J. S. Gutierrez, J. C. E. W. Hooijmeijer, R. Kentie, J. A. Masero, T. L. Tibbitts, and T. Piersma (2015). When Siberia came to the Netherlands: The response of continental Black-tailed Godwits to a rare spring weather event. Journal of Animal Ecology 84:1164–1176.

Sillett, T., R. T. Holmes, and T. W. Sherry (2000). Impacts of a global climate cycle on population dynamics of a migratory songbird. Science 288:2040–2042.

Somveille, M., A. S. L. Rodrigues, and A. Manica (2015). Why do birds migrate? A macroecological perspective. Global Ecology and Biogeography 24:664–674.

Stralberg, D., D. Jongsomjit, C. A. Howell, M. A. Snyder, J. D. Alexander, J. A. Wiens, and T. L. Root (2009). Re-shuffling of species with climate disruption: A no-analog future for California birds? PLoS ONE 4:e6825. doi:6810.1371/journal.pone.0006825.

Studds, C. E., and P. P. Marra (2007). Linking fluctuations in rainfall to nonbreeding season performance in a long-distance migratory bird, Setophaga ruticilla. Climate Research 35:115–122.

Studds, C. E., and P. P. Marra (2011). Rainfall-induced changes in food availability modify the spring departure programme of a migratory bird. Proceedings of the Royal Society B: Biological Sciences 278:3437–3443.

Swanson, J. L., and J. S. Palmer (2009). Spring migration phenology of birds in the Northern Prairie region is correlated with local climate change. Journal of Field Ornithology 80:351–363.

Thomas, C. D. (2010). Climate, climate change and range boundaries. Diversity and Distributions 16:488–495.

Thomas, C. D., A. Cameron, R. E. Green, M. Bakkenes, L. J. Beaumont, Y. C. Collingham, B. F. Erasmus, M. F. De Siqueira, A. Grainger, and L. Hannah (2004). Extinction risk from climate change. Nature 427:145–148.

Tingley, M. W., W. B. Monahan, S. R. Beissinger, and C. Moritz (2009). Birds track their Grinnellian niche through a century of climate change. Proceedings of the National Academy of Sciences USA 106:19637–19643.

Tingley, M. W., M. S. Koo, C. Moritz, A. C. Rush, and S. R. Beissinger (2012). The push and pull of climate change causes heterogeneous shifts in avian elevational ranges. Global Change Biology 18:3279–3290.

Tottrup, A. P., R. H. G. Klaassen, M. W. Kristensen, R. Strandberg, Y. Vardanis, A. Lindstrom, C. Rahbek, T. Alerstam, and K. Thorup (2012). Drought in Africa caused delayed arrival of European songbirds. Science 338:1307.

Trenberth, K. E., J. T. Fasullo, and T. G. Shepherd (2015). Attribution of climate extreme events. Nature Climate Change 5:725–730.

VanDerWal, J., H. T. Murphy, A. S. Kutt, G. C. Perkins, B. L. Bateman, J. J. Perry, and A. E. Reside (2013). Focus on poleward shifts in species' distribution underestimates the fingerprint of climate change. Nature Climate Change 3:239–243.

Vaughan, D. G., J. C. Comiso, I. Allison, J. Carrasco, G. Kaser, R. Kwok, P. Mote, et al. (2013). Observations: Cryosphere. In Climate change 2013: The physical science basis. Contribution of Working Group I to the Fifth Assessment Report of the Intergovernmental Panel on Climate Change, Stocker, T. F., D. Qin, G.-K. Plattner, M. Tignor, S. K. Allen, J. Boschung, A. Nauels, et al., Editors. Cambridge University Press, Cambridge, UK, pp. 317–382.

Walther, G. R., E. Post, P. Convey, A. Menzel, C. Parmesan, T. J. C. Beebee, J. M. Fromentin, O. Hoegh-Guldberg, and F. Bairlein (2002). Ecological responses to recent climate change. Nature 416:389–395.

West, J. M., S. H. Julius, P. Kareiva, C. Enquist, J. J. Lawler, B. Petersen, A. E. Johnson, and M. R. Shaw (2009). US natural resources and climate change: Concepts and approaches for management adaptation. Environmental Management 44:1001–1021.

Williams, J., and S. Jackson (2007). Novel climates, no-analog communities, and ecological surprises. Frontiers in Ecology and the Environment 5:475–482.

Williams, J., S. Jackson, and J. Kutzbach (2007). Projected distributions of novel and disappearing climates by 2100 A.D. Proceedings of the National Academy of Sciences 104:5738–5742.

Zemp, M., W. Haeberli, M. Hoelzle, and F. Paul (2006). Alpine glaciers to disappear within decades? Geophysical Research Letters 33. doi:10.1029/2006GL026319.

PART VI

MANAGEMENT AND CONSERVATION

CHAPTER 25

Extinction and Endangerment

John M. Marzluff and Marco Restani

THE DARK BEFORE THE LIGHT

The diversity of birds was on the rise over 140 million years ago (mya) during the Mesozoic era (Chiappe and Dyke 2002, chapter 2). Toothed marine divers, such as *Hesperornis*, fished the seas of the Cretaceous period. Small and agile alvarezsaurids ran about the landmasses of North and South America. A great many Enantiornithes commanded inland, marine, and littoral environments worldwide (Feduccia 1996). These sparrow- to vulture-sized toothed birds were accomplished fliers with alulas and a jury-rigged triosseal canal formed in part by a broad scapula rather than the coracoid, a feature *"opposite"* that of modern birds (see chapter 6). Enantiornithes radiated into forms any birder would recognize today: creepers, long-legged waders, and fowl. A few ancestral forms of today's species also were evolving in the Late Cretaceous. Tinamous and ostriches had diverged from a basal Palaeognathae 84 mya and basal Neoaves and Galloanseraes diverged from an ancestral Neognathae 88 mya (Jarvis et al. 2014), setting the stage for the rapid diversification of all modern birds. But before adaptive radiation there was mass extinction.

The Cretaceous ended with a bang, literally. Approximately 65.5 mya, an asteroid some 10 km in diameter crashed into Chicxulub, on the Yucatán Peninsula of Mexico (fig. 25.1). The impact crater is 180–200 km in diameter. Its creation spawned tsunamis in the Gulf of Mexico, earthquakes of magnitude greater than 11, a massive cloud of superheated air, shock-melted rock that caused a global pulse of radiation, and sulfur aerosols that absorbed the sun's warming rays and acidified the rain (Schulte et al. 2010). The Earth was dark and some 10°C colder for years to decades. In short, it sucked to be alive. It was so bad that 60 percent of known species went extinct in what we now know as the latest, and one of the three largest, mass extinction events to devastate Earth's biota. Dinosaurs, marine and flying reptiles, ammonites, rudists, many land plants, and calcareous phytoplankton, as well as the early radiation of toothed birds, were extinguished.

As was the case following earlier mass extinctions, biodiversity rebounded rapidly and spectacularly after the Cretaceous (Nee and May 1997). The diversity of the group of animals that includes all modern birds and is presumed to have evolved from a common ancestor—a clade known as the Neoaves—exploded as 36 modern lineages diverged between 10–15 mya after the asteroid impact (Jarvis et al. 2014). Rapid speciation following a mass extinction is thought to occur because the surviving species are afforded entry into niches similar to those vacated by the exterminated. Surely this was the case for ancestral Neoaves, who quickly exploited the many niches previously occupied by the Enantiornithes.

Birds were not the only survivors of the asteroid. Some mammals also survived, and as birds diversified to fill new niches, so too did the mammals. As terrestrial and aquatic vertebrates diversified, one species in particular came to dominate Earth: humans. In only a few million years of evolution, humans have transformed nearly every corner of the planet, co-opted the nitrogen and carbon cycle, harnessed half of the world's supply of fresh water, and are widely seen as initiating the sixth mass extinction (Vitousek et al. 1997). Our actions are prompting geologists to consider defining a new epoch in the geological time scale, the Anthropocene (Corlett 2015, Monastersky 2015, Waters et al. 2016). The Anthropocene is characterized by an increase of atmospheric methane, an increase of atmospheric CO_2, the appearance of manufactured materials in the stratigraphic record, sea-level rise, and an increased rate of species extinction (Corcoran et al. 2014, Waters et al. 2016).

In this chapter, we explore the ongoing extinction crisis that characterizes humanity's domination of Earth. We will

Figure 25.1. Flying Pteranodons witness the asteroid impacting Earth at the end of the Cretaceous. *Painting by Don Davis. Public domain, http://www.donaldedavis.com/PARTS/K-TNASA.jpg.*

learn about some of the avian casualties; why, when, and where they went extinct; and the implications of their loss to ecosystems.

WELCOME TO THE ANTHROPOCENE

Imagine yourself an ancient, seafaring Indonesian or Polynesian. As you visited the islands of the Indian or Pacific Oceans 1000–2000 years ago (Wilmhurst et al. 2011), you would have been greeted by a tremendous diversity of birds. Their brilliant plumes and delightful songs would enthrall you and shape your culture (Amante-Helweg and Conant 2009). Those larger than you, such as the moa, might seem hazardous, while the slow or plump would fill your pot. As you and your ancestors lived among the island birds, your needs and the actions of the rats, dogs, and pigs that accompanied you reduced the abundance and caused the extinction of many. For example, on the Hawaiian Islands from approximately 400 AD to 1778 AD, when Europeans first visited the archipelago, 71 species or subspecies of birds went extinct (Banko and Banko 2009). A similar fate—extinction of half or more of all species—was typical of other island avifaunas during their first encounters with humans (Steadman 1995). In the tropical Pacific, 154 species, mostly birds of large size, those that nested on the ground, and those that were flight-less, were the first casualties (Boyer 2008, 2009). In total, over 1,000 species—10 percent of all birds—are thought to have gone extinct at this time (Duncan et al. 2013).

Now imagine you are a naturalist aboard a European ship during the Age of Discovery (early 1400s–early 1600s). The diversity of birds you would catalog would bring honor to your sponsors. Your name might be forever attached to a favorite species, such as Steller's Eider (*Polysticta stelleri*), Humboldt Penguin (*Spheniscus humboldti*), or Forster's Tern (*Sterna forsteri*). But the actions of your ship's crew and those that followed your path would set off a second wave of avian extinctions, as tasty and easily harvested birds satisfied your hunger and as new organisms such as goats, cats, mosquitos, and malaria were unleashed onto ecologically naïve island birds. In the last two centuries, Hawai'i again lost nearly 50 percent of its native bird life (24–35 species or subspecies out of 71 known from the time of Captain James Cook's encounter with the islands in 1778; Banko and Banko 2009). During this time, 28 species, mostly medium-sized birds that drank nectar or ate insects, were extinguished on all tropical Pacific islands (Boyer 2008). During the last 400–500 years, between 80 and 150 bird species are known to have gone extinct worldwide, nearly all from islands in the Pacific and Indian Oceans (Fuller 2001, Szabo et al. 2012). The range in the estimated number of extinctions reflects the uncertainty associated with when the last individual of a species had been seen (Diamond 1989), as well as the considerable variation in whether the authors considered only full species or species and subspecies (Fuller 2001, Boyer, 2009, Szabo et al. 2012). Regardless of the absolute count, the names of some birds, such as the large flightless pigeon relative from Mauritius known as the Dodo (*Raphus cucullatus*), are synonymous with extinction. Others are less familiar, such as the Black Mamo (*Drepanis funerea*), Mauritius Blue Pigeon (*Alectroenas nitidissima*), and Delalande's Coucal (*Coua delalandei*).

Given the magnitude of bird extinctions during the last two millennia, you may think the worst is over. Not true. While the pace of avian extinctions has fluctuated over time—for example dropping in the mid-1900s—it has recently increased. In the last 50 years—during these authors' lifetimes—at least 21 bird species, including nine from the Hawaiian Islands alone, have vanished from Earth (table 25.1). These losses represent approximately 2.1 extinctions every five years and a doubling of the rate that produced 80 species extinctions in the last 400 years (Fuller 2001).

The Intergovernmental Science-Policy Platform on Biodiversity and Ecosystem Services regularly assesses biological diversity. Their most recent assessment used the International Union for Conservation of Nature's (IUCN) Red List (www.iucnredlist.org) to quantify the pace of extinctions before and after 1900 (Pimm et al. 2014). For bird species described prior to 1900, 89 of 8,922 species went extinct, yielding an extinction rate of 49 species/species known/one million years. From 1900 to the present, this rate jumped to 132 (13 of 1,230 species described during this period went extinct). These rates dwarf the background extinction rate suffered by birds since the end of the Cretaceous and before humans evolved, which is estimated to be as low as 0.1 extinctions/species/one million years (Pimm et al. 2014). Whereas the total loss of species to date is well below the 60 percent lost at the end of the Cretaceous, the rate of species loss is now 1,000 times greater than the background rate since then (Nott et al. 1995, Pimm et al. 2014). More importantly, the recent rate of extinction greatly outpaces the rate at which new species of birds evolve, which is roughly 0.15 new species/species/one million years (Phillimore and Price 2008). To learn more about Stuart Pimm, one of the scientists involved in these grim calculations, see the box on page 782.

The magnitude of modern bird extinctions is large relative to what is known about most other forms of life. In 2015, the IUCN Red List assessed 79,837 species. Of the 10,424 bird species, 140 (1.3 percent) were classified as extinct. Mammals were similarly extinction prone (1.4 percent), but amphibians and reptiles appeared more resilient (<1 percent extinct), though perhaps also much less well known. Certainly many more amphibians and

Table 25.1. Bird species that have gone extinct during the last 55 years (1960–2015). List does not include those currently extinct in wild but held in captivity. Dates last seen from Fuller (2001), Banko and Banko (2009), and Szabo et al. (2012).

Species	Last Location	Last Seen
Alagoas Curassow (*Mitu mitu*)	NE Brazil	1988
Alaotra Grebe (*Tachybaptus rufolavatus*)	Madagascar	1985
Atitlán Grebe (*Podilymbus gigas*)	Guatemala	1986
Colombian Grebe (*Podiceps andinus*)	Colombia	1977
Eskimo Curlew (*Numenius borealis*)	Americas	1983
Imperial Woodpecker (*Campephilus imperialis*)	Mexico	1995
Ivory-billed Woodpecker (*Campephilus principalis*)	Cuba	1992
Guam Flycatcher (*Myiagra freycineti*)	Guam	1983
ʻŌʻāʻā (*Moho braccatus*)	Hawaiʻi	1987
Bushwren (*Xenicus longipes*)	New Zealand	1972
Bishop's ʻŌʻō (*Moho bishopi*)	Hawaiʻi	1981
White-chested White-eye (*Zosterops albogularis*)	Norfolk Island	1979
Kāmaʻo (*Myadestes myadestinus*)	Hawaiʻi	1985
Olomaʻo (*Myadestes obscurus*)	Hawaiʻi	1980
ʻŌʻū (*Psittirostra psittacea*)	Hawaiʻi	1989
Greater ʻAmakihi (*Hemignathus sagittirostris*)	Hawaiʻi	1969
Nukupuʻu (*Hemignathus lucidus*)	Hawaiʻi	1990s
Kākāwahie (*Paroremyza flammae*)	Hawaiʻi	1963
Poʻo-uli (*Melamprosops phaeosoma*)	Hawaiʻi	2004
Aldabra Bush Warbler (*Nesillas aldabrana*)	Aldabra atoll	1983
Bachman's Warbler (*Vermivora bachmanii*)	SE USA	1975

reptiles are today threatened with extinction than are birds and mammals (Pimm et al. 2014). Molluscs are the most extinction-prone taxon investigated (4.3 percent), and their demise signals a general degradation of Earth's aquatic systems. In Britain, declines and extinctions of birds for the last two to four decades of the twentieth century were fewer than those of butterflies and plants, which may portend an even more ominous future for birds because insects typically respond rapidly to changes in the environment (Thomas et al. 2004).

THE GEOGRAPHY OF EXTINCTION

Some of the most notable adaptive radiations of birds have occurred on island archipelagos. Individual islands typically vary in climate and topography, and they are characterized by abundant plant life, reduced competition from non-avian taxa, and limited predation, parasitism, and disease. Under

STUART PIMM

Stuart Pimm. *Photo by Rudi van Aarde.*

Stuart Pimm is the Doris Duke Chair of Conservation at Duke University. He is a world leader in the study of present-day extinctions and what can be done to prevent them. Pimm received his BSc from Oxford University (1971) and his PhD from New Mexico State University (1974). He has authored 300 scientific papers and four books. Professor Pimm has documented the rate of extinction in the Anthropocene, focusing especially on the loss of birds on Pacific Islands. His concern for saving what is left led him to direct SavingSpecies, a 501(c)(3) nonprofit organization that uses funds for carbon emissions offsets to fund local conservation groups to restore degraded lands in areas of exceptional tropical biodiversity. His international honors include the Tyler Prize for Environmental Achievement (2010) and the Dr. A. H. Heineken Prize for Environmental Sciences from the Royal Netherlands Academy of Arts and Sciences (2006). Below he describes a seminal event that has shaped his passion to help slow the sixth mass extinction event. In his words,

> I knew it had to go as soon as I saw it. Off the busy coastal road east of Rio de Janeiro in Brazil a dirt road took me northwards. I drove past dozens of black vultures waiting for the thin cattle to die. Only sparse grass covered the poor laterite soil. On either side, further from the road was lush tropical forest that soon closed the cattle pasture to a narrow gap.

> These humid coastal forests of Brazil once covered a million square kilometres. An exceptional number of endemic birds live here. Sadly, extinction threatens more of them than anywhere else in the Americas. That's because only about 7 percent of the forest remains, most of it in tiny fragments. To my right was one of the largest: the Reserva Biológica Unão. The narrow strip of cattle pasture through which I traveled made Unão an "island." That narrow strip was what had to go.

> Forest "islands" are remnant patches surrounded by a "sea" of former forest cleared for cattle grazing. We'd analysed data from the project established by Tom Lovejoy in the Amazon. Following forest clearing, substantial fractions of the bird species initially present in the forest went extinct within a decade or two in fragments smaller than 10 km^2.

> All this told me that this damnable cattle pasture might eventually be responsible for more bird extinctions than any other scar in the Americas. So, I founded a nonprofit, SavingSpecies (http://www.savingspecies.org). The mission was to raise the $300,000 to help our Brazilian friends, the Associação Mico-Leão Dourado (http://www.micoleao.org.br) buy the cattle pasture and restore it. Local school children helped plant native trees, which grew quickly in this warm, humid climate. A decade later, the forest restoration is obvious from space—one can see the forest coming back on Google Earth, by comparing the historical images.

> That model is one we're now implementing in Colombia, Ecuador, elsewhere in Brazil, and soon in Africa. It's CPR for Earth—connect, protect, and restore—and it affords a cost-effective solution to stop bird extinctions.

those conditions, colonizing birds radiated quickly and profusely on the Hawaiian Islands, the Galapagos, Tahiti and the Society Islands, the West Indies, New Zealand, Madagascar, and the nearby Seychelles and Mascarene Islands (Grant and Grant 2008, Pratt 2009). These former hotbeds of evolution are today's killing fields. The vast majority (78 percent) of extinctions since 1500 have occurred on oceanic and continental islands (fig. 25.2), usually within a century after being settled by technologically advanced people (fig. 25.3; Blackburn et al. 2004, Boyer 2008, Szabo et al. 2012). According to the IUCN Red List, nearly half of all avian extinctions have been recorded from Hawaii (22 species), Mauritius (20), and New Zealand (19). Since the start of the twentieth century, the rate of island extinctions has declined, while the rate of mainland extinctions has increased (Szabo et al. 2012). Overall, however, mainland extinctions remain the exception rather than the rule for birds. For example, we know of only six species extinctions on North America, five on South America, five on Africa, and one each on mainland Europe and Asia (Szabo et al. 2012).

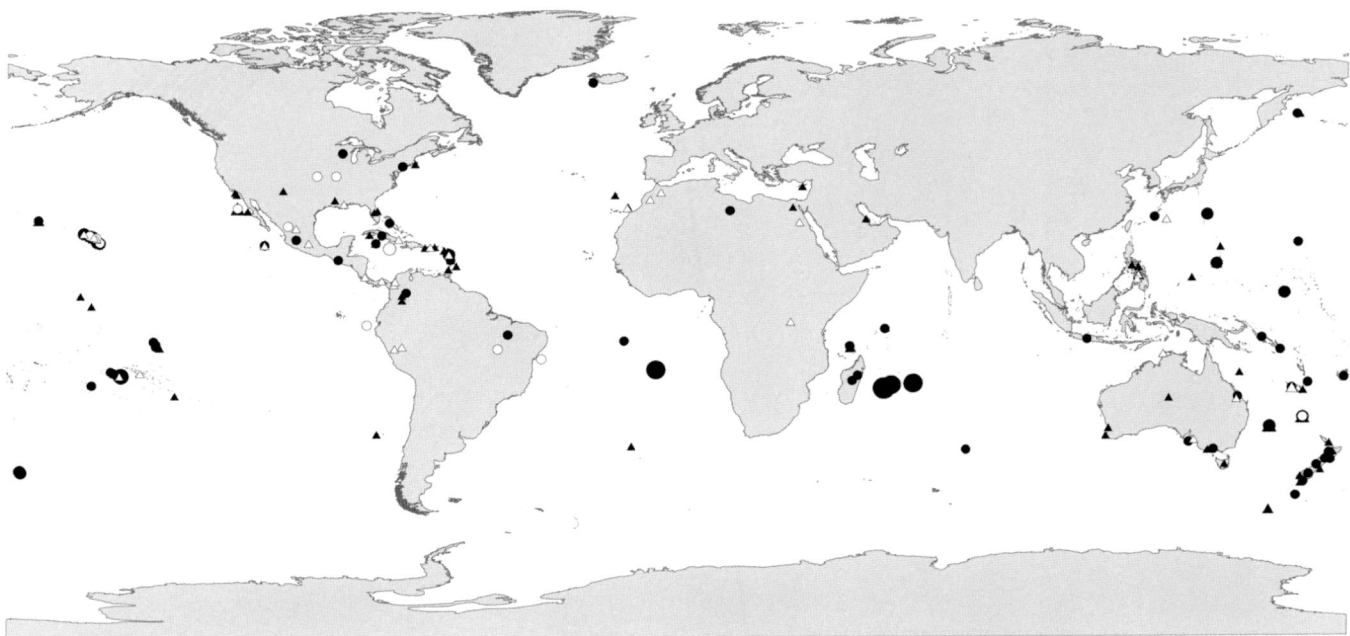

Figure 25.2. Geographic location of extinctions since 1500, demonstrating the propensity for island birds to go extinct in modern times. Circles indicate species, triangles indicate subspecies, open symbols indicate "extinct" taxa, and solid symbols indicate "possibly extinct" taxa. Larger symbols indicate larger numbers of taxa, as illustrated for species. *Reprinted with permission from Szabo et al. 2012, http://journals.plos.org/plosone/article?id=10.1371/journal.pone.0047080. Open access under CC BY license.*

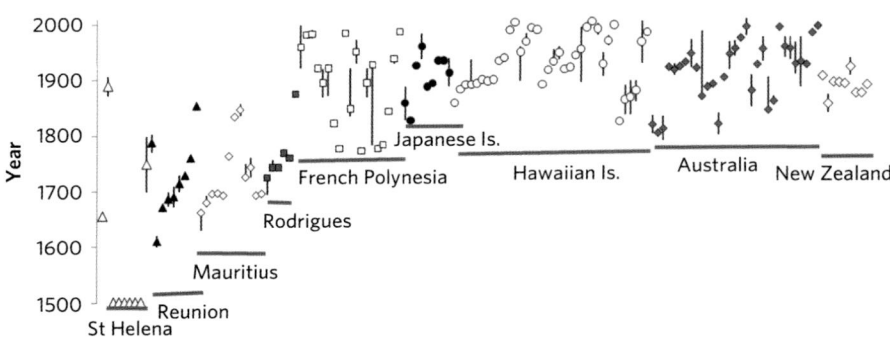

Figure 25.3. When people first established lasting settlements on islands, birds rapidly went extinct. This is demonstrated by the close correspondence between the date of species and subspecies extinctions (with vertical lines showing minimum and maximum estimate of date where available) for selected locations and the date of first human settlement lasting at least 0.5 years by a continental or continental-island based nation (gray line). *Reprinted with permission from Szabo et al. 2012, http://journals.plos.org/plosone/article?id=10.1371/journal.pone.0047080. Open access under CC BY license.*

Most avian extinctions have occurred in forests, especially moist, lowland, tropical, or subtropical forests. The IUCN Red List documents the habitats once occupied by 134 extinct birds. Over half of the extinctions (83 species) involved forest birds, and all but 20 of these occurred in tropical or subtropical locales. Shrublands (20 extinctions) and inland wetlands (20), which include rivers, lakes, marshes, bogs, and swamps, were the next most extinction-prone habitats. The birds of savannas (2 extinctions), grasslands (9),

cliffs and mountain peaks (2), and marine locales (14) suffered the least.

THE PHYLOGENY OF EXTINCTION

Ancient avian taxa that occur principally on tropical islands have been especially thinned by modern extinctions. Over one-third of all extinctions have involved parrots, rails, finches, and pigeons (fig. 25.4; Szabo et al. 2012). Evolutionarily distinct avian families and genera (those with few species) and taxa with long evolutionary histories have been disproportionately decimated (Gaston and Blackburn 1997, Russell et al. 1998, Frishkoff et al. 2014). The families including emus (Dromaiidae), dodo and solitaires (Raphidae), New Zealand wrens (Acanthisittidae), and Hawaiian honeycreepers (Mohoidae) have each lost more than half of their members in the last 500 years (Szabo et al. 2012). The selective extinction of species in unique and species-poor higher taxa likely reflects their small geographic ranges on individual islands or archipelagos (Gaston and Blackburn 1997, Russell et al. 1998). The relative selectivity of extinction is expected to weaken, however, because factors driving population declines now include widespread destruction of mainland and island habitats and global climate change, which are expected to affect many species regardless of phylogeny (see below; Russell et al. 1998).

A few families have largely escaped extinction. These include the babblers and parrotbills (Timaliidae; Szabo et al. 2012); crows, jays, and white-eyes (Corvidae and Zos-

Figure 25.4. Two extinct island parrots, the Black-fronted Parakeet from Tahiti (*above*) and the forest-dwelling Society Parakeet from the Society Islands (*below*). *Oil painting on panel by Errol Fuller.*

teropidae; Blackburn and Gaston 2005); and woodpeckers, flycatchers, and titmice (Picidae, Tyrannidae, and Paridae; Lockwood et al. 2000). The most recently evolved and most speciose order of birds, the Passeriformes, which includes two-thirds of all avian taxa, has suffered a relatively modest 44 percent of extinctions (Szabo et al. 2012).

THE ECOLOGY OF EXTINCTION

Species prone to extinction share a number of interrelated demographic and ecological traits. Extinction risk is highest in species that exist in small populations across a narrow geographic or elevational range (Rabinowitz 1981, Purvis et al. 2000, White and Bennett 2015). Most at-risk bird species live within geographic ranges smaller than 11,000 km^2 (Harris and Pimm 2008), which predisposes them to the detrimental effects of climate change, habitat loss, and catastrophe (White and Bennet 2015). Moreover, small populations possess less genetic variability and thus have limited ability to adapt to threats as they confront genetic drift more often than natural selection (Lande 1988, 1999).

Extinction-prone species typically have "slow" life history strategies characterized by long lifespans and slow reproductive rates. The slow-growing populations of these species cannot quickly rebound after being reduced by catastrophic, stochastic, or deterministic perturbations (Lande 1999, Lande et al. 2003). Species with slow life histories are frequently large-bodied habitat specialists that exist high on an ecosystem's trophic level (Gaston and Blackburn 1997, Bennett and Owens 1997, Purvis et al. 2000), and therefore are less able to adjust to changing resources (Shultz et al. 2005, Santisteban et al. 2012). Predators, scavengers, and other species at mid- to upper trophic levels are susceptible to fluctuations and contamination of prey populations (Diamond 1989, Walters et al. 2010, Wurster 2015) and are frequently persecuted by humans (Purvis et al. 2000). In addition to risks associated directly with small and slow-growing populations, large body size alone increases the risk of extinction. On islands, for example, large birds are also often flightless and therefore easy and profitable quarry for hunters and introduced predators (Boyer 2008, 2009, Duncan et al. 2013).

Endemism and ground nesting are also associated with extinction (Boyer 2009). Endemic avian species typically have small ranges, specialized habits, and are naïve to non-avian predators, competitors, and diseases. Ground-nesting birds may have been easily hunted by the early human colonists of islands (Boyer 2008, 2009), and today ground nesting may predispose continental birds to population declines, especially where agricultural intensification alters large tracts of habitat (Van Turnhout et al. 2010).

REASONS FOR THE CURRENT EXTINCTION CRISIS

Human population growth—there are now over seven billion of us—and increasing per capita consumption of finite resources ultimately extinguish other forms of life from Earth (Pimm et al. 2014). We are the leading cause for the modern extinction of birds (Szabo et al. 2012). The specific actions of our species that cause avian extinctions can be obtained by querying the IUCN's Red List of 140 extinct birds. This database suggests that our propensity to purposefully or accidentally introduce nonnative species to ecosystems, especially oceanic islands, led to the extinction of 84 bird species. Hunting, fishing, and trapping were implicated in the demise of 67 bird species. Predation by humans is the second leading cause of extinction on islands and the third leading cause on continents (Szabo et al. 2012). As we spread across the land, we also cleared forests, causing the extinction of 37 bird species, while the settlements and farms we carved from native vegetation precipitated the extinction of an additional 25 species. Agriculture, especially the transformation of natural land to cropland, is the most important driver of bird extinctions on continental landmasses (Szabo et al. 2012). Less important human actions that led to the extinction of birds in modern time include disturbance from recreation or work, suppression of fire, and pollution (table 25.2).

Table 25.2. **Factors associated with the extinction of bird species as identified on the IUCN Red List (www.redlist.org).**

Threat	Number of Extinctions
Invasive Species	84
Natural Resource Use	
Hunting/trapping	63
Logging and wood harvest	37
Gathering plants	2
Fishing/harvesting aquatics	4
Agriculture	
Crops	14
Ranching	11
Natural System Modification	
Fire suppression	9
Other	2
Climate Change / Weather	7
Human Disturbance	
Recreation	2
Work	2
Development (Housing and Urbanization)	3
Pollution	2
Geological Event	1

CASE HISTORIES ILLUSTRATING THE PROCESS OF EXTINCTION

As species become rare, be it from inherently slow life history strategies, reductions in their geographic range, or interactions with new predators or diseases, they typically begin a death spiral toward extinction. This spiral, known as an "extinction vortex" (Gilpen and Soule 1986), is a synergistic breakdown in genetic, demographic, and social function that creates a strong positive feedback that drives declining populations ever smaller (box on page 786).

No Room for Error

Framed by sharp volcanic peaks, Guatamala's mile-high Lake Atitlan fills a 380-meter-deep and ancient caldera with aquamarine waters. This stunning, if violent, place has attracted sightseers and anglers for decades. It was also the only place in the world where a large pied-billed grebe, known locally as the Poc or Atitlan Grebe (*Podilymbus gigas*), existed. In the 1960s, a small population of 200–300 grebes appeared stable, though vulnerable to extinction because they were flightless and their restricted, geothermally active range was attractive to people (LaBastille 1974, 1983). In 1965, the population was reduced to 80 birds as native people harvested the reeds for the country's famous floor mats. Grebes nest in reeds, and this activity reduced their already limited breeding habitat. The airline company Pan American also established a resort on the lake and introduced largemouth bass (*Micropterus salmoides*) to the waters to boost the locale's appeal to fishermen. The voracious bass ate many of the lake's native crabs and small fish, which the grebes fed on. Bass also preyed directly on grebe chicks, further depressing the population. A refuge was established for the grebes in 1966, and with habitat restoration and a diligent game warden, the population rebounded to about 200 in 1973. Unfortunately, the susceptibility of a small population to stochastic events was demonstrated in 1975 when the effects of an earthquake drained the lake, dropping the water level to 6 meters and leaving the refuge high and dry. This type of random event, known as "environmental stochasticity," is a fundamental aspect of the extinction vortex that can even extinguish a growing, but rare, population (Lande 1999). A nonnative pied-billed grebe species (*Podilymbus podiceps*), capable of flight, was introduced to the lake, possibly to bolster the population or to confuse the conservation efforts under way. In 1982 Edgar Bauer Jr., who was the game warden and leading advocate for the grebe, was assassinated. The grebe population declined to 32 in 1983 and showed signs of hybridizing with the introduced grebes, as many offspring could fly (LaBastille 1983, Hunter 1988). In 1987, the pure genetic stock of

THE EXTINCTION VORTEX

Gilpin and Soule (1986) suggested that small population size reduces population growth through four types of positive feedback loops. Not unlike the water swirling down a flushed toilet, as rare species decline in abundance they race faster and faster toward the end.

R Vortex. The R refers to population growth rate (r or R are the growth parameters in equations of population growth). As populations decline their growth rates become more variable, leading to occasionally strong decreases and increases in the population, which translates into more volatility in population size. Even in favorable conditions, volatile populations can go extinct by random events (Shaffer 1981, Lande 1999). In the R vortex, population volatility stems mostly from demographic stochasticity, the random and independent fluctuation in the survival and reproduction of individuals within a population (Lande et al. 2003). For example, a small population might experience a year in which many males die. This leads to a skewed sex ratio making it difficult to find a mate, which reduces the next year's reproductive success, lowering population size even further.

D Vortex. The D refers to dispersal, which is reduced as small populations become more and more isolated. In many species, populations are arranged across the geographic range into subpopulations, or demes. Breeding is regular among members within a deme, but dependent on dispersal across atypical habitat if it is to occur between members of different demes. When habitats suitable for a declining species are reduced they also typically become more isolated via habitat fragmentation (Fahrig 1997). This vortex occurs when the demes of a declining population become more isolated and dispersal among them is reduced. Reduced dispersal leads to smaller populations, which are more isolated and therefore decline, thereby increasing isolation and the decline toward extinction (Thomas 2000).

F Vortex. The F refers to the inbreeding coefficient from population genetics. As populations decline the number of individuals that successfully and equally contribute their genotypes to the next generation, which is referred to as the effective population size (Hartl 1981), also declines. As effective population size declines the probability of breeding among individuals that share substantial portions of their genotype increases. This inbreeding eventually reduces population performance, which is called "inbreeding depression" (Keller and Waller 2002). Inbreeding depression lowers effective population size still further, which increases inbreeding, so depression again nudges the population lower toward extinction.

A Vortex. The A refers to the ability of a population to adapt, which is reduced in small populations. As populations decline the genetic variability of the population is reduced because mutations are rare and genetic drift is common (Lynch 1996). Genetic variability is created by mutation and reduced by drift or natural selection. In small populations, drift is more important that natural selection at pruning the creativity of mutation (Lynch et al. 1995). Drift is random and therefore does not necessarily act to fit a population to its environment like natural selection does; therefore, as populations decline they are increasingly unable to track environmental change, which leads to further reduction in population size and greater drift.

Combination of Vortices. The four vortices described above do not necessarily act alone. More typically, declining populations suffer from some combination of several or all together. For example, declining populations that suffer from inbreeding depression also become less able to track environmental change as genetic drift increases in prominence. These species are said to suffer "mutational meltdown" (Lynch et al. 1993) in which deleterious genetic combinations drift into prominence within a population, driving it to smaller size and increasing its exposure to lethal or maladaptive forms of genes.

Podilymbus gigas hit the bottom of the R, and likely F, vortex and was declared extinct.

Two other range-restricted, rare grebes met similar fates. The Alaotra Grebe (*Tachybaptus rufolavatus*) was a poor flier restricted to shallow and reed-fringed Lake Alaotra, Madagascar. Its population declined as nonnative predatory fish and the competitive food-fish, tilapia (family Cichlidae), were introduced to the lake. The lake's water quality was simulta-neously reduced by runoff from nearby agriculture. When another grebe colonized the lake from Africa, it hybridized with the native grebe, further polluting the unique gene pool. With increasing human use of the lake and the advent of nylon fishing nets, the last blow to the species was likely a combination of poaching and entanglement in gill nets (BirdLife International 2011a). The species was declared extinct in 1985. The Columbian Grebe (*Podiceps andinus*) inhabited the Bogota wetlands high in the eastern Andes Mountains. It was abundant in 1945 on Lake Tota, but declined throughout the 1950s as wetlands were drained, reeds were harvested, water quality declined from agricultural siltation and pesticide pollution, and rainbow trout (*Oncorhynchus mykiss*) were introduced for sportfishing. It was last seen in 1977 (BirdLife International 2011b).

Into the Black Hole

Seafarers and early colonists of distant lands must have had immense appetites for fresh meat. Most large, flightless, or otherwise easily captured birds were overharvested, pushing them toward extinction. For example, the Dune Shearwater (*Puffinus holeae*), a ground-nesting colonial seabird native to the Canary Islands, appeared to vanish as soon as people colonized the islands (Rando and Alcover 2010). Tim Flannery (2001) has described this entrance to the extinction vortex as the "black hole between our ears." Our mouths, the hole Flannery laments, have indeed consumed a great many species. Early in our prehistory, native people killed off the largest birds ever known: the several dozen species of elephant birds from Madagascar and moa from New Zealand (Fuller 2001). The elephant birds' eggs, which were over 0.3 meters long and as large as seven ostrich eggs, were enough to feed a family. The largest moa stood over 4 meters tall. Island peoples butchered tens of thousands of moas, and the species were thought extinct by the early 1400s, after only 200 years of contact with ancient humans (Perry et al. 2014).

It seems anything that could profitably be caught was eaten. Most famously was the Dodo of Mauritius. This fat, apparently clumsy, flightless bird stood 1 meter tall. Although its taste was described as "disgusting" and their flesh was said to get tougher and less appealing the longer it was cooked, Portuguese and Dutch sailors killed and ate enough Dodos during the 1500s and 1600s to extinguish the species by 1690 (Fuller 2001). The Great Auk (*Alca impennis*), a large seabird of the north Atlantic, lasted a bit longer (fig. 25.5). Although distributed widely from Newfoundland to Norway, Great Auks bred on only a few islands off Newfoundland, Iceland, and St. Kilda. It was on these breeding grounds that the flightless birds were slaughtered for food and their feathers. European collectors of natural history also prized their eggs and skins. The last Great Auk was seen on Eldey Island, off the coast of Iceland in 1844. Fuller (2001) eulogized soon thereafter that "somewhere in the deep waters of the North Atlantic or in a hollow upon a wind-lashed reef, perhaps sheltering from the fury of the waves, the last great auk died."

It is plausible that large birds confined to narrow ranges, such as the moas, dodo, and auk, might be driven to extinction by overharvest. Amazingly, overharvest, in combination

Figure 25.5. The now-extinct Great Auk bred in colonies on islands in the cold, misty north Atlantic. *Oil painting on panel by Errol Fuller.*

with the action of the D vortex and the likely meltdown of social behavior, can also contribute to the extinction of some of the world's most common species. Martha, the last Passenger Pigeon (*Ectopistes migratorius*), died just before 1 p.m. on September 1, 1914 (fig. 25.6). She once belonged to a species that was among the world's most abundant. Her lineage was known for its immense flocks that, according to John James Audubon, would darken the skies in the 1830s and take days to pass (Jackson and Jackson 2007, Schultz et al. 2014). A hunter needed to harvest a mind-boggling 30,000 or more birds to win one of the many pigeon shoots organized in the 1800s. For decades market hunters supplied hundreds of thousands of carcasses annually to feed the growing human population of the fledgling United States. As flocks were decimated, hunting subsided, but by then the pigeon had entered the D vortex. Once-immense breeding colonies that had stretched for 65 or more kilometers (Fuller 2001) were fragmented, and dispersal among flocks presumably became more difficult. Small flocks may have winked out from stochastic events, or may have been unable to track the vagaries of mast, such as acorns, that the species fed on. When populations of social species decline, some critical behaviors, such as avoiding predators or finding mates and food, may become dysfunctional or unexpressed. This "Allee effect" (where a substantial minimum number of individuals is necessary for behaviors critical to any reproductive activity) and reliance on a large society predisposes species such as the Passenger Pigeon to a rapid spiral down the D vortex. In North America, hunters also pushed the once super-abundant Eskimo Curlew (*Numenius borealis*) into the D vortex, while on the Solomon Islands the easy capture of roosting Choiseul Pigeon (*Microgoura meeki*) precipitated their extinction.

Figure 25.6. Passenger Pigeons roamed the eastern United States in large flocks searching for mast nut crops produced by trees such as hickory, beech, and oak. Their social behavior led them to roost and breed colonially. *Painting by John James Audubon, public domain.*

Too Pretty

Our love of natural beauty and a burning desire to possess it doomed some of the most brilliantly colored birds to extinction. Ten species of parrot no longer grace this world with their raucous calls and dazzling plumes. Nearly all were reduced in part by capture for the pet trade. The Paradise Parrot (*Psephotus pulcherrimus*), for example, was in high demand as a cage bird by Londoners in the early twentieth century (Fuller 2001). The population of this locally common, exquisite, east Australian bird declined initially as its preferred food, native grass seed, was reduced as agriculture intensified. Introduced mammals—cats, rats, foxes, and other predators—also helped usher the long-tailed, blue, red, brown, and black beauty into the D vortex as its once large flocks were splintered. Several years of drought and bush fires further reduced its food resources. It is easy to imagine how Allee effects manifested an erosion of predator vigilance and reduced the parrot's ability to track ephemeral seed crops, thereby dragging the species to the bottom of the vortex. The last was seen in 1927.

The ancient Hawaiians coveted feathers. The dress of kings, princes, and warriors included helmets, capes, robes, and standards festooned with the feathers of thousands of birds captured by royal trappers (Amante-Helweg and Conant 2009). Red, yellow, and black feathers were especially prized. The three species of ʻŌʻō (*Moho* spp.; one each from the islands of Molokaʻi/Maui, Hawaiʻi, and Oʻahu; Banko and Banko 2009) sported bright yellow axillary plumes that made them frequent targets of trappers. Another honeycreeper, the Hawaiʻi Mamo (*Drepanis pacifica*), was also sought for its black and yellow feathers. The cloak of King Kamehameha alone included the feathers from an estimated 80,000 of these small, colorful songbirds (Conant 2005). Although trappers were instructed to only pluck feathers, the flesh of the ʻŌʻō was said to be delicious when fried in its own fat. The ancient harvest may have reduced the abundance of ʻŌʻō and Mamo, as only Bishop's ʻŌʻō (*Moho bishopi*) has survived past 1902 (Banko and Banko 2009). However, all species restricted to single islands flutter along the brink of extinction, and it seems that diseases, predators, and habitat changes brought to Hawaiʻi by European explorers and settlers also helped push these island beauties over the edge.

Island Crows

It seems impossible that crows, among the most widespread and adaptable of bird groups, could go extinct. However, at least four species that evolved on the Hawaiian Islands did not survive the Polynesian colonization (Fleischer and McIntosh 2001, Banko 2009) and two other island specialists are dangerously close behind.

ʻAlalā (Hawaiian crow, *Corvus hawaiiensis*) were revered by the ancient Hawaiians as *akua ʻaumāku*, animal manifestations of lesser gods or family spirits (Amante-Helweg and Conant 2009). The appreciation of the last species of crow as a sacred being, ancestral to particular families of Hawaiians, perhaps enabled it to survive in large numbers into the early 1900s. However, by the 1960s, their range was reduced to a small portion of the Big Island, and by 1970 fewer than 100 birds were estimated to be alive (Banko 2009). Those that remained were mostly isolated breeding pairs, deep within the D vortex. Allee effects likely contributed to the infrequent establishment of new pairs. Habitat loss and degradation from ranching, logging, and the actions of nonnative ungulates, as well as shooting by humans, pushed the population to 11 individuals in 1992 (Banko 2009), all of which existed on a private ranch in central Kona. At that time, in response to a lawsuit stipulating that the US government was doing too little to restore the crow (National Research Council 1992), it was decided that releasing captive-born crows should be attempted to augment the wild population. From 1993–1999,

27 juvenile ʻAlalā were released on the Kona ranch near the remaining wild adults. These juveniles came from breeding a small captive population of birds salvaged during the 1970s (typically <10 birds) and from the removal, incubation, and rearing of eggs from the last wild breeders. Survival of released juveniles was low due to a combination of predation by Hawaiian Hawks (*Buteo solitarius*) and feral cats, and disease (especially toxoplasmosis, which is caused by a protozoan, *Toxoplasma gondii*, spread by cat feces). The six juveniles that remained alive at the end of 1999 were recaptured and placed in captivity. The last wild ʻAlalā died in 2002. With improved captive-rearing facilities, today the captive population of ʻAlalā exceeds 60 birds and a new round of reintroductions has begun. It is hoped that newly secured refuges where predators are controlled will enable the intriguing voice and calm intellect of the ʻAlalā to once again thrill humans. Biologist Paul Banko experienced the loss of ʻAlalā in his lifetime (see the box on page 790).

Aga (Mariana Crow, *Corvus kubaryi*), are similar to ʻAlalā in being tropical forest crows that, while once relatively abundant, always occupied a range restricted to a few islands of the South Pacific. In recent times Aga lived only on the islands of Guam and Rota in the Mariana archipelago (National Research Council 1997). Aga were common throughout the 1940s and were considered agricultural pests and thus targets for suppression. However, they declined rapidly on Guam beginning in the 1960s, and the Guam population was estimated at only 357 birds in 1981 (Engbring and Ramsey 1984). The pattern of decline mirrored the decline of other birds on Guam, starting in the south and expanding to mid-island, leaving the birds only in the north, principally on Andersen Air Force Base (Savidge 1987). In 1996 only 20 wild Aga existed on Guam, deep within the R vortex as the sex ratio appeared strongly male-biased (National Research Council 1997). The last wild crow on Guam disappeared in 2011 (Amar and Esselstyn 2014). The principal reason for the Aga's decline was predation by the nonnative brown tree snake (*Boiga irregularis*), which was accidentally introduced to southern Guam in the 1940s as stowaways on military equipment passing through and to Guam after World War II (National Research Council 1997). The population of Aga on the nearby island of Rota, which fortunately is snake-free, is also rapidly declining. Clearing of forest habitat, illegal shooting of birds that island residents perceive are limiting their birthright to own land, and predation by cats appear responsible for endangering the species (USFWS 2005).

Novel Predators

The Aga was the last in a long list of bird species driven extinct on Guam because of brown tree snakes (fig. 25.7;

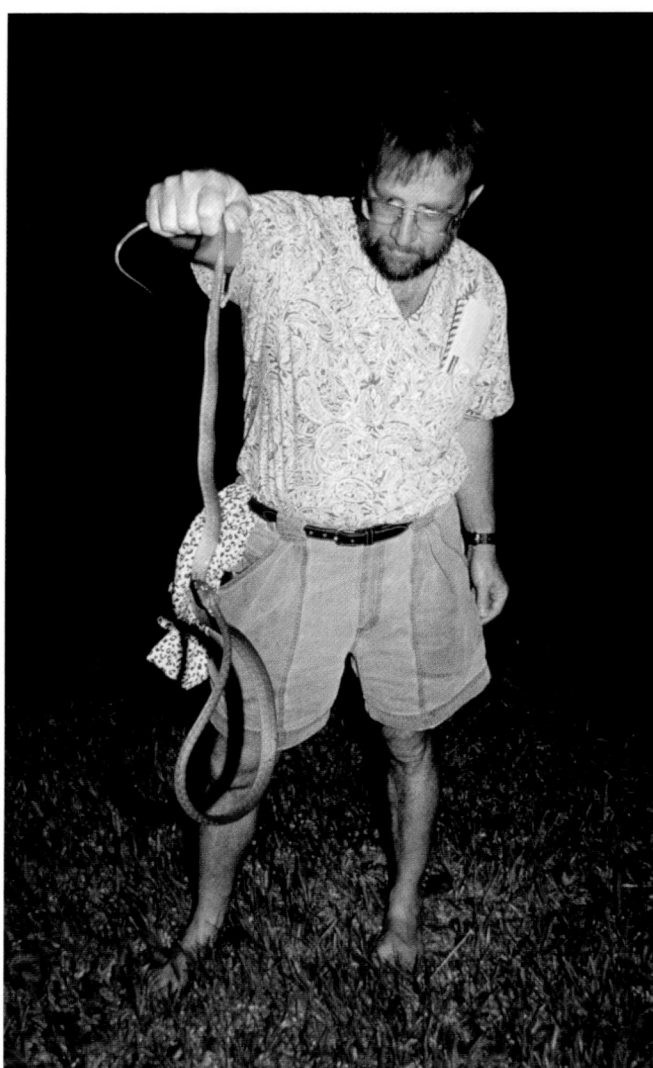

Figure 25.7. Dr. Tom Fritts holds a brown tree snake in Guam, where this reptile has nearly extinguished an entire island avifauna. *Photo by John Marzluff.*

Savidge 1987, National Research Council 1997). This arboreal, nocturnal snake reached densities 20 times greater than in their natural range: more than 100 snakes per hectare were estimated to occur on Guam when native bird, bat, and lizard faunas collapsed (National Research Council 1997). Their abundance and propensity to climb on electrical wires also led to widespread power outages on the island. By the mid-1980s, the White-throated Ground Dove (*Gallicolumba xanthonura*), Mariana Fruit Dove (*Ptilinopus roseicapilla*), Nightingale Reed Warbler (*Acrocephalus luscinius*), Rufous Fantail (*Rhipidura rufifrons*), Guam Flycatcher (*Myiagra freycineti*), Bridled White-eye (*Zosterops conspicillatus*), and Cardinal Myzomela (*Myzomela cardinalis*) were extirpated from Guam (Savidge 1987). Soon thereafter, the Guam Rail (*Gallirallus owstoni*) and Micronesian Kingfisher (*Todiramphus cinnamominus*) were

PAUL BANKO

Paul Banko. *Photo by Suzanna Valerie.*

Dr. Paul Banko, a research wildlife biologist with the United States Geological Survey, grew up in Hawai'i, where he observed and studied the extinction of its birds. Dr. Banko received both a BS and PhD from the University of Washington. His extensive studies of endangered Hawaiian birds suggest paths for their conservation. His father, Win Banko, was also a student of bird extinction, and together this father and son team helped make the world aware of the islands' losses. Here Paul describes growing up with extinction.

Experiencing extinction in real time is an unavoidable downside to a career in conservation biology in Hawai'i. Where else have 12 bird species disappeared from the wild in less than half a century? Although species not observed for decades will likely never be seen again, the 'Alalā, or Hawaiian Crow, will have a second chance when it is returned to the wild from captive breeding stock in 2016. I began my career in wildlife biology in the early 1970s under the mentorship of my dad, Win Banko (1920–2016), helping him with surveys of the 'Alalā. As the first biologist assigned by the US Fish and Wildlife Service to the Hawaiian bird extinction crisis, Win decided that the 'Alalā was a good species to start with because they were relatively conspicuous and everyone recognized a crow. Few people knew the other native forest birds. 'Alalā surveys led him to interview many cowboys, hunters, and others who worked or traveled in the forests of the Big Island of Hawai'i. These conversations made him aware of the steady retreat of crows from areas where they had once been common. He also gained the trust, if not always the appreciation, of the public in his efforts to solve the extinction problem. Finding only 57 'Alalā scattered on the flanks of Hualālai and Mauna Loa

volcanoes, together we also learned how challenging it can be to convince skeptics that on islands even a species of crow can be driven to the brink of extinction, despite membership in a family of birds renowned worldwide for adaptability and tenacious survival. The peril faced by the charismatic and engaging 'Alalā prompted Win to delve into the history of bird extinction in the Hawaiian Islands to better understand why some species persisted, even if at reduced numbers and range, while others succumbed quickly and completely to the onslaught of invasive threats and habitat loss.

I was very fortunate to have Win as my mentor because he was a natural historian who valued the long view of events and, like any good detective, reveled in the details. His painstaking compilation and analysis of museum specimen labels and the published and often obscure, unpublished literature of the 1800s and early 1900s, when many more species and local populations were observed and collected, produced a very complex and puzzling picture of Hawaiian bird extinctions since Western contact in 1778. Understanding these historical patterns was challenging enough without the added complexity of considering even earlier extinctions evident from ancient bird bones found in collapsed lava tubes, coastal sinkholes, and petrified sand dunes. Seeking clarity in the face of complexity is surely a common human endeavor, but the historical record when viewed carefully does not support simple explanations for the wholesale loss of bird communities. Thus, we began to consider not just extinction at the level of species but at the level of populations, which afforded us greater inferential power. We also reexamined the obvious: the dazzling diversity of beak morphology of the Hawaiian honeycreepers, the largest group of Hawaiian birds and one of the most dramatic examples of adaptive radiation and resource specialization. Specialization, we realized after analyzing the feeding ecology and other life history traits of each species, was key to understanding Hawaiian bird extinction because it was linked to reproductive capacity (lower for specialists, higher for generalists) and other important ecological characteristics. By thinking about the basic ecology of the birds, therefore, we were able to understand extinction in the context of vulnerabilities not only to novel predators and diseases but also to food competitors, food web disruption, and habitat change. I could never have reached these insights without the guidance and collaboration of Win. In my view, therefore, mentorship and teamwork contribute greatly to understanding and slowing extinction.

extinct in the wild. Today, only two native forest birds remain in the wild on Guam: the Micronesian Starling (*Aplonis opaca*) and the Mariana Swiftlet (*Aerodramus bartschi*).

The extinction of nearly all of Guam's avifauna by a single species of novel predator is a chilling testament to the ability of a nonnative species to harm naïve native birds. The more typical circumstances surrounding extinction involve the action of multiple population-limiting factors acting in concert. For example, development and predation by domestic cats often work together to extinguish many island species, such as the flightless Stephens Island Wren (*Traversia lyalli*) of New Zealand, which disappeared about 1900 (BirdLife International 2011c). Moreover, it is often a series of predator introductions that spell doom for island avifaunas. For example, extinction rates on islands usually peak when about 50 percent of the avifauna has been extinguished after six or more predator species have been introduced (Blackburn et al. 2005).

Predators have been especially harmful to the many species of rails (family Rallidae) that once existed across the Pacific Islands. These inconspicuous, but meaty, birds live in the rank undergrowth of wetlands, grasslands, shrublands, and forests. Having speciated marvelously across the Pacific, many became secondarily flightless. No other family of birds has suffered more extinction than have the rails (Steadman 1997). The rails of Chatham Island were extinguished by the combined actions of cats and habitat destruction by islanders creating pastures. Japanese soldiers occupying Wake Island and early mariners landing on Ascension Island are said to have hunted native rails to oblivion. The combined appetites of cats and rats spelled doom for the Tahitian Red-billed Rail (*Gallirallus pacificus*) and the Samoan Wood Rail (*Gallinula pacifica*). Rats extinguished the Kusaie Island Crake (*Porzana monasa*). The Hawaiian Rail (*Porzana sandwichensis*) also could not survive the many rats, dogs, and cats introduced to the islands. Laysan Rails (*Porzana palmeri*) lost most of their habitat to overgrazing by rabbits and guinea pigs introduced as human food in the 1890s, but did not vanish until the US Navy accidentally introduced rats to the island in 1943. The Lord Howe Swamphen (*Porphyrio albus*) fed too many whalers, early settlers, and visitors, who were able to kill them using only a stick. The Mauritius Red Hen (*Aphanapteryx bonasia*) put up a bit more of a fight. It too was easily killed with a stick, but the hunter first had to lure the bird within striking range, something apparently done by simply flapping a red flag (Fuller 2001).

In the Name of Progress

Many of the most famous bird extinctions in North America were the simple result of human greed and intolerance. Clearing, degrading, and converting native forests and grasslands have been important drivers of extinction and continue to be important agents endangering today's surviving birds (Czech and Krausman 1997, Diamond 1989, Wilcove et al. 1998).

When the pilgrims first settled Plymouth Colony in 1620 in what is today Massachusetts, United States, they began to clear forests for buildings and farm fields. The native prairie grouse, a subspecies of Greater Prairie Chicken known as the Heath Hen (*Tympanuchus cupido cupido*), likely benefited from these initial subtle transformations. This species occupied open barrens and quickly adapted to the edges between farm and forest that multiplied with the growth of the New England colonies. It was so abundant in the area soon to be transformed into Boston that laborers and servants were instructed by their employers not to serve them the bird more than a few times a week (Edey 1998). As farmland gave way to cities and towns, however, the bird's habitat declined and overharvest precipitated the demise of the Heath Hen. There were no regulations to manage hunting, and by the early 1800s the species was rare around Boston and generally extirpated from the rest of its range (Hough 1933). Hunting was suspended in 1837, and by 1839 the hen was only known from Martha's Vineyard, a small island off the Massachusetts coast. Residents of the island were permitted to hunt Heath Hens intermittently as their population fluctuated over the next several decades. The bird's lekking behavior, where males tooted and strutted on communal grounds to attract mates, made them especially vulnerable to harvest even when rare. By 1896 fewer than 100 birds existed. After a brush fire in 1906 reduced the cover used by the birds, an estimated 80 Heath Hens remained, which prompted local conservationists to establish a 650-hectare reserve for the birds. Stochastic processes then threatened to push the Heath Hen deep into the R Vortex. Although the reserve and a respite from fire and hunting enabled a rebound in hen numbers to an estimated 2,000 in 1916, a catastrophic brush fire and "an unprecedented flight of Goshawks (*Accipiter gentilis*)" knocked the population down to about 100 birds (Edey 1998). Numbers continued to decline, and by 1923 only 28 birds remained. The effective population size was certainly much lower because of the species' polygamous mating system (see chapter 20), in which not all males contribute equally to the next generation's gene pool. Highly skewed male reproductive success also negatively affected the related Greater Prairie Chicken in the Midwestern United States (Westemeier et al. 1998). In addition, Allee effects in Heath Hens appeared, as normal separation of males from females during incubation and brood-rearing stopped occurring (Edey 1998). The next stochastic event to hit the wild hens was disease; plague from domestic chickens spread to the reserve, and by 1928 only three birds, all males, survived.

Figure 25.8. The largest woodpecker that inhabited North America, the Ivory-billed Woodpecker, survived into the twentieth century. *Painting by John James Audubon, public domain.*

Only one male was alive by December 8, 1928. This bird, named Boomer, was captured and banded by Dr. Alfred Gross in 1931. Boomer was last seen in March 1931. His bands, one copper (A634024) and one aluminum (407880), were never retrieved (Edey 1998).

The largest woodpeckers in North America, the Ivory-billed (fig. 25.8) and Imperial (*Campephilus principalis, C. imperialis,* respectively), were unable to tolerate the rapid logging of virgin pine forests of the southeastern United States, Cuba, and the Mexican highlands. Both woodpeckers nested and roosted in large dead trees, the legacies of large living trees that were overharvested and soon to disappear across the landscape. Logging and an extreme sensitivity to human activities likely doomed the Ivory-billed Woodpecker in the United States, which probably went extinct in the 1950s. High profile but unconfirmed sightings (Jackson 2006, Sibley et al. 2006) of the species are occasionally made (e.g., Fitz-

patrick et al. 2005). The Cuban subspecies of the Ivory-billed Woodpecker was last seen in 1987 (www.iucnredlist.org). The Imperial Woodpecker was last seen in 1956 and also suffered from habitat loss and fragmentation; its nestlings were considered a delicacy of local Indians, thus hunting probably also contributed to its demise (Tanner 1964).

Flocks of Carolina Parakeets (*Conuropsis carolinensis*) enlivened the forests of the southeastern United States with their brilliant green, yellow, and orange plumage and chittering flight calls (fig. 25.9). Forest clearing reduced their nesting habitat, and shooting of the birds as they flocked to agricultural crops, such as corn planted by colonial Americans, reduced their once-great numbers. A few birds may have existed in South Carolina in the late 1930s, but the last official record of the species is from 1918. A pair of parakeets, named Lady Jane and Incas, was the last of their breed. They finished their long lives—over 30 years—in the Cincinnati Zoo. Incas was the last to go, dying a few months after his lifelong mate. His keepers, noting the pair's strong bond, cited grief, not old age, as the cause of death (Fuller 2001).

Figure 25.9. Flocks of brightly colored Carolina Parakeets once graced the farms and fields of the southeastern United States. *Painting by John James Audubon, public domain.*

The development of Florida's coastal cordgrass marshes for homes and the US Space Program caused the extinction of the Dusky Seaside Sparrow (*Ammodramus maritimus nigrescens*) (Post and Greenlaw 2009). Flooding marshes around Kennedy Space Center to control mosquitos was particularly destructive. This distinctive subspecies lived in a very restricted range and began declining in the 1950s; only 1,800 were estimated to survive in 1968, a population insufficient to keep the species out of the combined actions of D and R vortices. As the population was further fragmented, numbers crashed until, in the late 1980s, it was feared that only five birds remained. All were males. Though functionally extinct, the males were brought into captivity and crossbred with another subspecies of seaside sparrow. Humans launched a spacecraft from the former habitat of this species to walk on the moon in 1969. The last pure dusky sparrow died, blind in one eye and infertile, in 1987 (Post and Greenlaw 2009).

Poison and Persecution

Beginning with the Age of Discovery, seafarers, whalers, and sealers spent months to years at sea cramped in tight quarters working and surviving on a diet of brackish water and food devoid of variety and nourishment. Lacking refrigeration, ships' crews often introduced livestock to islands during their outgoing voyages so they had a source of fresh meat on return passages. Their goats, pigs, and cattle wreaked havoc on plant communities that evolved without vertebrate herbivory. The relatively cold, foggy, and windswept Guadalupe Island off the northwest coast of Mexico provided ample vegetation for goats, which were dropped off at the height of whaling and sealing during the early 1800s. In fewer than 50 years the goats were believed to number 100,000 (Moran 1996). Tragically, it was the fertile fishing grounds supporting an abundance of cetaceans and seals along the California and Mexico Coasts that doomed the Guadalupe Caracara (*Caracara lutosa*).

Caracaras are renowned for their fearless and bold demeanor, traits we seem to tolerate in only our species and not others. After the collapse of oceanic fish and whale stocks, herders who tended the goat population on Guadalupe Island shot and poisoned caracaras thought to kill young goats. By the mid-1870s, Guadalupe Island had attracted the attention of ornithologists, and many a collecting expedition outbound from California made its way to the island. Early records suggest that caracaras were still common during the earliest collecting trips, but in only 25 years, the species was rendered extinct. A two-month expedition in 1906 that placed freshly killed goat carcasses on open tablelands of the island failed to attract any caracaras (Thayer and Bangs 1908).

The demise of the Guadalupe Caracara serves as a stark reminder that overkill of an isolated and relatively small population of fearless birds can lead to extinction in a matter of generations. Moreover, the unrestrained zeal displayed by some museum collectors may have put the finishing touches on its extinction. Although only one caracara was observed during a collecting trip in 1897 (Kaeding 1905), a collector in late 1900 succeeded in killing 9 of 11 birds in a flock (Abbott 1933), which was the last confirmed sighting. The tragic irony is that many museum specimens of the Guadalupe Caracara were lost during the 1906 earthquake and fires in San Francisco. Only about 35 specimens remain, hardly enough to keep the lesson alive.

BACK FROM THE BRINK

The increased rate of bird extinction during the Anthropocene is shocking and alarming, but not all has been bad the past 50 years. There are some success stories. For example, Rachel Carson's stark warning in her classic book, *Silent Spring*, drew national attention to the indiscriminate use of pesticides, and the Environmental Defense Fund's victories in court over pesticide producers and dispensers catalyzed today's environmental movement (Wurster 2015). Biomagnification of pesticides in food chains and their subsequent negative effects on reproduction of upper trophic level birds, particularly Bald Eagles (*Haliaeetus leucocephalus*) and Peregrine Falcons (*Falco peregrinus*), prompted a slew of environmental legislation—the Clean Water Act, Clean Air Act, Endangered Species Act (ESA)—and the creation of the Environmental Protection Agency. DDT, in particular its derivative DDE, caused eggshell thinning such that adult females inadvertently broke their eggs during incubation (Hickey 1969, Grier 1982). Decades of poor reproduction of eagles and peregrines spiraled their populations into the R Vortex and garnered protection under the ESA. With restrictions on use and the banning of some pesticides, populations rebounded and both species were removed from the Endangered Species list in the 1990s. What saved these raptors was that their habitat, although polluted and needing time to cleanse, still existed to support recruits.

Herculean—and well-funded—efforts by government agencies and private organizations during the 1980s and 1990s saved the California Condor (*Gymnogyps californianus*) from certain extinction (fig. 25.10). This ancient species, with nearly a 3-meter wingspan, once again arcs across the skies over parts of southwestern North America. Condors are a textbook slow life history species: large, with high survival and low reproductive potential, traits that did not serve them well as humans began to dominate Earth. Although once having a range that extended along the entire west coast of North America, condors were restricted to southern California

Figure 25.10. The California Condor is threatened with extinction because, as it scavenges the remains of deer, pigs, and other hunted species, it ingests fragments of lead bullets. *Photo by Susan Haig.*

by the 1950s. Population declines brought about from the loss of carrion and an increase in persecution quickly brought condors to the brink of extinction—in 1982 the population numbered only 22 individuals. A focused captive breeding and release program has achieved great success: by 2014 the population of condors numbered 425 (219 in the wild and 206 in captivity (USFWS 2014). Reintroduced condors had begun breeding in the wild by 2003. Although lead poisoning, paucity of carrion, necessary supplemental feeding, and ingestion of microtrash by nestlings remain significant challenges, the species may continue its road to recovery if human attitudes and financial resources remain committed to the task (Walters et al. 2010).

Finally, a continuing theme throughout this chapter has been the immense toll taken by introduced predators on naïve island avifauna. Well, some countries have aggressively taken the fight to these invaders in attempts to return island ecosystems to their former state. For example, in 2001 New Zealand launched a rat poisoning campaign on Campbell Island, which was once used as a base for sealers and whalers in the 1800s and was occupied by the usual suspects—sheep, cattle, rats, and cats. Livestock removals were begun in the 1970s, and by 1999 the mysterious extirpation of cats was confirmed, possibly from regrowth of the island's dense vegetation hindering their foraging ability. Baiting trials for rats in 1999 were deemed a success, and the poisoning program became fully operational in 2001. Helicopters delivered baits containing brodifacoum, an anticoagulant highly effective against rodents. Post-poisoning surveys failed to detect rats, and the program was declared a success in 2006. Campbell Island was soon recolonized naturally by species that had been absent for decades: Grey-backed Storm Petrels (*Oceanites nereis*), White-chinned Petrels (*Procellaria aequinoctialis*), Campbell

Island Snipe (*Coenocorypha aucklandica*), and Campbell Island Pipits (*Anthus novaeseelandiae*) (McClelland 2011). The biggest winner on Campbell Island was a bird that would have had a harder time getting back on its own—the flightless Campbell Island Teal (*Anas nesiotis*). Teal had become extirpated from the Campbell Island and existed only on a nearby islet. In fact, it barely existed, with a global population estimated at only 25 pairs in 1998 (Gummer and Williams 1999). Two years later the IUCN categorized the species critically endangered. A captive breeding and release program delivered teal to Campbell Island in 2004 and 2005 (McClelland and Gummer 2006). Breeding on the island was confirmed in 2006 and the population grew quickly to over 200, prompting the IUCN to downgrade its listing to endangered. The successful rat eradication program on Campbell Island has served as a model for South Georgia (United Kingdom) and the Aleutians (United States).

THE FUTURE

Many of the next casualties of the sixth mass extinction, which is occurring during the Anthropocene, are likely to come from today's birds that are extremely rare. Of the approximately 10,000 bird species that grace today's skies, roughly 10–13 percent are considered so rare that they are threatened with extinction (Pimm et al. 1995, Chapin et al. 2000). In the next 100 years, Bird Life International (www.birdlife.org) estimates that one in eight bird species may go extinct. In their analysis of the IUCN Red List, they conclude that 217 bird species face "an extremely high risk of extinction in the immediate future." This "immediate future" is somewhere less than a century, and if this prognosis is realized, it would represent an unprecedented uptick in the rate of avian extinction to at least two species per year. That pace would double the number of extinctions recorded from 1900 to present, which, as Pimm et al. (2014) noted, was already a thousand times greater than the historical background rate.

As of this writing, there are five bird species that exist only in captivity. Each is under constant care in breeding facilities in the hopes that their native habitats will be improved and reintroduction can follow. However, there are fewer than 200 individuals of each species, so the next extinct species could very well be among them. These "walking dead" include the Socorro Dove (*Zenaida graysoni*) from Mexico, the Alagoas Curassow (*Mitu mitu*) from Brazil, the Hawaiian Crow, and two species from Guam, the Guam Rail and the Guam Kingfisher.

The birds prone to extinction in the future will likely share many attributes with those that have already perished. Island endemics and other species with limited geographic range will always be at risk, as will larger species high on the trophic system that typically have low reproductive capacity.

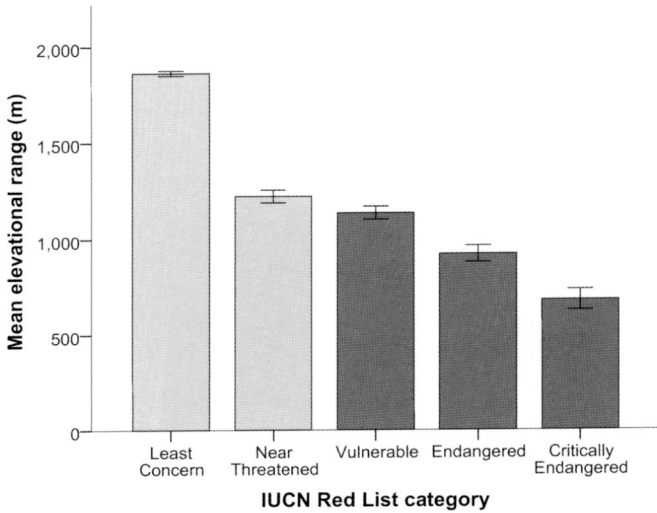

Figure 25.11. Birds with narrow elevational ranges are often at risk of future extinction (classified as vulnerable to critically endangered by the IUCN), in part because they have few options for adjusting to the challenge of climate change. Reprinted with permission from White and Bennett 2015, http://journals.plos.org/plosone/article?id=10.1371/journal .pone.0121849. Open access under CC BY license.

Species vulnerable to overexploitation remain at risk, especially in the developing world (Yiming and Wilcove 2005). However, habitat destruction threatens birds with a wide range of life history strategies and ecological traits, suggesting that future extinctions will also include birds not historically at risk (Wilcove et al. 1998, Manne et al. 1999).

Our modification of Earth's climate (chapter 24) is sure to join the list of leading causes for bird extinctions in the next century (Pimm et al. 2014). Birds that inhabit low elevations and a narrow range of elevations have few options to adjust to rapid habitat loss or climate-induced change in habitat suitability and are thus most threatened with extinction (fig. 25.11; White and Bennet 2015). If landbirds, for example, using montane habitats today, simply shift to higher elevations as Earth's climate warms, then 400–550 species may run out of habitat and go extinct by 2100 (Sekercioglu et al. 2008). Although these predictions are imprecise, dependent on complex climate and vegetation models and ignorant of possible adaptations birds may evolve in a warming world, they suggest that for every degree Earth's climate warms, 100–500 species of landbirds will go extinct. Already there is evidence that species unable to rapidly adjust their annual cycle to track changes in the phenology of critical resources, such as insects, fruits, and nuts, are in decline (Both et al. 2009, Møller et al. 2008). It is disconcerting that most projected future casualties are not seriously threatened today. For example, many seabirds that nest on low coral islands will have their colonies inundated by rising sea levels, driving species locally extinct.

The first casualty of climate change is a species just introduced to science (see the box on page 796). In 1996, Craig Benkman, a professor at the University of Wyoming, discovered a new type of bird that was recognized as a distinct species in 2017. The "Cassia Crossbill" that Dr. Benkman discovered in the South Hills of Idaho is called *Loxia sinesciurus*, the crossbill without squirrels (Benkman et al. 2009). This finch, whose mandibles cross to better extract seeds from pine cones, is a specialist on the seeds of a unique lodgepole pine (*Pinus contorta latifolia* Engelm.) that occurs over only 180 km² in Idaho. The crossbill is a bit larger than others and lives in nonmigratory flocks that utter distinct call notes and songs, which enables it to maintain flocks separately from other more vagrant crossbills. It is from flocks that mates are selected, so the gene pool of the South Hills birds remains distinct from other forms of red crossbill (Smith et al. 2012). As with other forms, this crossbill has coevolved with the pine tree it exploits (Benkman 2003). In the South Hills, cones of the lodgepole pine remain intact on the tree until the heat of a forest fire springs them open to release the seeds. Serotiny is typical of lodgepole pine, but elsewhere, closed cones are quickly removed by American red squirrels (*Tamiasciurus hudsonicus*). There are no marauding squirrels in the South Hills, so cones age on the branch, which allows crossbills to forage efficiently by seeking only older cones that have been weakened by the elements. A consistent supply of old (>10 years on the branch) cones are available every year, affording the crossbill the luxury of remaining resident. This has been a good arrangement for bird and tree, but in a warming world it portends trouble. When summer temperatures exceed 32°C, which they now regularly do in the South Hills, pine cones open and seeds fall to the ground, where they are lost to the crossbill. In response to declining food, the crossbill population is crashing; down 60 percent from 2003–2008 (Santisteban et al. 2012). Moving to other locales where pines exist is not an option because other crossbill forms and squirrels would be formidable competitors. As summer temperatures increase, more and more seeds will be shed and the South Hills Crossbill will decline deeper and deeper into the extinction vortex that has already claimed many recognized full species that specialize on narrowly distributed resources.

As the factors driving extinction change, so too will the characteristics of the future's extinct species. On islands, for example, if future introductions of predators are curtailed, ground-dwelling birds may no longer be especially prone to extinction because those susceptible to predation by nonnative mammals have already gone extinct (Blackburn et al. 2004). As humans conserve rather than overharvest island birds, we would also expect that future extinctions will not be biased

CRAIG BENKMAN

Craig Benkman. *Photo by Ashley Smith, courtesy of* Times-News *and* Magicvalley.com, *Twin Falls, ID.*

Professor Craig Benkman is an ornithologist and evolutionary ecologist at the University of Wyoming, where he holds the Robert Berry Distinguished Chair in Ecology. Dr. Benkman completed his bachelor's degree in biology at the University of California, Berkeley, in 1978; a Master of Science in biology at Northern Arizona University in 1982; and a PhD at the State University of New York, Albany, in 1985. He has explored interactions between crossbills and conifers in conifer forests from the Rocky Mountains to Hispaniola, the Mediterranean, and Southeast Asia.

Here, Dr. Benkman describes his discovery of a new species of bird that now is threatened with extinction.

On my drive to the ornithological conference in Boise in 1996 I made a discovery that still thrills me. Along the drive, I stopped to sample lodgepole pine cones in two isolated mountain ranges in southern Idaho, the South Hills, and Albion Mountains. Maps indicated there would be a small area of lodgepole pine and no red squirrels. I expected to find few crossbills and small poorly defended pine cones. However, the pine forests were extensive and crossbills abundant. The cones were distinctive, with enlarged scales at their distal ends where crossbills feed. With subsequent analyses I could detect the signature of a coevolutionary arms race between crossbills and lodgepole pine, which was driving their divergence. As I drove to the meeting in Boise, I knew I had found something special. I thought the crossbill would prove to represent a new species, and I told colleagues so. I think they ignored my claims. Yes, new species are being discovered, but not by field ornithologists working in the continental United States! Our behavioral and ongoing genetic studies indicate nearly complete reproductive isolation for this population and support its designation as a species: what I call *Loxia sinesciurus*, or the crossbill without tree squirrels. This discovery is bittersweet though. Its population declined by about 80 percent between 2003 and 2011. It has rebounded since, but its decline to perhaps fewer than 500 birds was likely caused by an exceptional number of hot summer days. Although the South Hills has experienced few hot days during the last several summers, allowing the crossbill to rebound along with its seed supply, soon hot summer days will not be exceptional. It will be heartbreaking.

toward ground-nesting, large, and flightless species as in the past (Boyer 2008). On continents, agricultural monocultures may revive the bias in extinction toward evolutionarily ancient lineages. In Costa Rica, for example, the birds able to thrive in agricultural systems were of more recent origin than were those species found in forests (Frishkoff et al. 2014). Agricultural intensification is also expected to favor arboreal-nesting resident or short-distance migrants (Van Turnhout et al. 2010). Interestingly, future extinction may be less biased toward large-bodied birds, as many fare well in the face of expanding agriculture (Van Turnhout et al. 2010).

The extinction of birds ripples through entire ecosystems in a process known as **coextinction** (Dunn et al. 2009). Parasites intimately coevolved with hosts die out as the host perishes. In this way, the mites, lice, and other microbiota unique to the Passenger Pigeon, Dodo, and Guadalupe Caracara are also extinct. Plants, such as the Hawai'i flowering vine 'ie'ie (*Freycinetia arborea*), that were pollinated by now-extinct Hawaiian honeycreepers such as the 'Ō'ū (*Psittirostra psittacea*), are today reliant on new pollinators or threatened with extinction (Cox 1983). Similarly, loulu palms (*Pritchardia* spp.) and hō'awa (*Pittosporum hosmeri*) that produce large nuts that can be dispersed only by large birds, such as the Hawaiian Crow, are also threatened by the lack of their coevolved partners (Culliney et al. 2012). As extinctions occur, the functions of the ecosystem from which they are lost shift and shrink (Boyer and Jetz 2014) as secondary ripple effects become apparent (Diamond 1989). Impoverished, the world's ecosystems become more vulnerable to future disturbances. Fragile ecosystems provide fewer and fewer of the services that benefit humanity, such as water purification, crop pollination, and soil fertility. In this way the cumulative effects of species extinction will come home to roost and directly endanger our own species. Future humans may find sustaining life more difficult, and as they disassociate from nature and the other species that shaped their cultural identities, they may, as Paul Shepard (1996) suggests, no longer know who they are.

Slowing species extinctions that currently characterizes our tenure on Earth will require unprecedented selfless co-operation among humans. The world will need to unite to reduce dependence on fossil fuels that, when burned, exac-erbate climate change, restrict the movement of invasive species, stem the loss of our remaining wild places, restore ecosystem integrity, and enhance the understanding and tol-erance of humanity for wildlife. These efforts may succeed only when we also agree to stem our own exponential popu-lation growth. Providing a suitable space for other life forms is necessary, but not sufficient to slow extinction. Large, inter-national investments will also be needed to recover species we already know to be at risk (Garnett et al. 2003). Increasingly these investments must target island ecosystems, which are at the tip of the modern extinction spear (Restani and Marzluff 2002, Leonard 2008). Organizations such as the IUCN, Bird-Life International, Partners in Flight, and the Recovery Office of the US Fish and Wildlife Service have prioritized actions necessary to slow extinction. It is up to conservationists to convince the world's leaders that it is time to act.

KEY POINTS

- Human activities are the main cause of extinction and endangerment in the Anthropocene, which harbors the sixth mass extinction event to rock Earth. A combi-nation of actions including overharvest and persecution, introduction of predators, habitat loss, secondary ripple effects, and stochastic environmental events drive species into an extinction vortex from which few escape.
- The previous five mass extinctions were natural and cat-astrophic. The overall loss of species during these events was much greater than current losses, but the pace was much slower than recorded in the Anthropocene.
- Island species—initially those that were large, colorful, or tasty and nesting on the ground, and later nectar and insect eaters of moderate size that were susceptible to disease—were lost disproportionately compared with mainland species during the Anthropocene.
- Species with slow life histories, narrow geographic ranges, and those that are predators or secondarily flight-less are especially prone to extinction.
- The very diverse family of rails that colonized many Pa-cific Islands has been especially decimated by extinction.
- There is some hope, but it comes with a huge price tag. Species have been extracted from the extinction vortex by poisoning campaigns directed at removing nonnative

predators, such as rats, from small islands. Once such habitats are restored, and rare species are increased by captive breeding, they can be reintroduced.

KEY MANAGEMENT AND CONSERVATION IMPLICATIONS

- Knowledge of limiting factors and species life history must guide conservation efforts. Overharvest and pes-ticide use are often regulated today. Other factors, such as the lead in rifle bullets, continue to poison rare birds including the California Condor.
- The challenge of climate change will require international cooperation and sacrifice on behalf on biological diversity.
- Rare species often accelerate to extinction because of the combined, synergistic, and positively reinforcing factors that constitute the extinction vortex. Some factors are deterministic and amenable to management, such as habitat loss, nonnative predators, and overharvest. Others are stochastic and beyond human control, such as earthquakes and strong weather events.
- When species go extinct, the functioning of their eco-systems is changed. Loss of birds, for example reduces pollination and dispersal of seeds, which leads to lower functional diversity and eventually will reduce the ser-vices humans derive from nature.
- Long-term funds and commitment are necessary to achieve restoration of critically endangered species.

DISCUSSION QUESTIONS

1. What, if anything, do you think should be done to stem human population growth and overconsumption?
2. De-extinction, or the return of lost species through gene-tic engineering, may be possible in the future. Should we bring back lost species? Would it be OK to restore pretty, tasty, or benign species such as Passenger Pigeons, but not those that are more challenging to humans, such as moas, or the large carnivorous terror birds that once hunted South America?
3. Peter Grant (1995) suggested that we should commemo-rate extinction, for example, with a national holiday to reflect on lost species. Do you agree? If so, what date would you select?
4. If we are unable to save all the rare species, how might we triage those that remain?
5. How much longer do you think humans will be extant?
6. Are humans prepared to share? How do industrial countries justify their holier-than-thou attitude toward undeveloped countries to conserve more than they have?

References

Abbott, C. G. (1933). Closing history of the Guadalupe Caracara. The Condor 35:10–14.

Amante-Helweg, V. L. U., and S. Conant (2009). Hawaiian culture and forest birds. In Conservation biology of Hawaiian forest birds, T. K. Pratt, C. T. Atkinson, P. C. Banko, J. D. Jacobi, and B. L. Woodworth, Editors. Yale University Press, New Haven, CT, pp. 59–79.

Amar, A., and J. A. Esselstyn (2014). Positive association between rat abundance and breeding success of the critically endangered Mariana Crow Corvus kubaryi. Bird Conservation International 24:192–200.

Banko, P. C. (2009). ʻAlalā. In Conservation biology of Hawaiian forest birds, T. K. Pratt, C. T. Atkinson, P. C. Banko, J. D. Jacobi, and B. L. Woodworth, Editors. Yale University Press, New Haven, CT, pp. 473–486.

Banko, W. E., and P. C. Banko (2009). Historic decline and extinction. In Conservation biology of Hawaiian forest birds, T. K. Pratt, C. T. Atkinson, P. C. Banko, J. D. Jacobi, and B. L. Woodworth, Editors. Yale University Press, New Haven, CT, pp. 25–58.

Benkman, C. W. (2003). Divergent selection drives the adaptive radiation of crossbills. Evolution 57:1176–1181.

Benkman, C. W., J. W. Smith, P. C. Keenan, and T. L. Parchman (2009). A new species of the Red Crossbill (Fringillidae: Loxia) from Idaho. The Condor 111:169–176.

Bennett, P. M., and I. P. Owens (1997). Variation in extinction risk among birds: Chance or evolutionary predisposition? Proceedings of the Royal Society of London B 264:401–408.

BirdLife International (2011a). http://www.birdlife.org/datazone/speciesfactsheet.php?id=3630.

BirdLife International (2011b). http://www.birdlife.org/datazone/speciesfactsheet.php?id=3642.

BirdLife International (2011c). http://www.birdlife.org/datazone/speciesfactsheet.php?id=3993.

Blackburn, T. M., P. Cassey, R. P. Duncan, K. L. Evans, and K. J. Gaston (2004). Avian extinction and mammalian introductions on oceanic islands. Science 305:1955–1958.

Blackburn, T. M., and K. J. Gaston (2005). Biological invasions and the loss of birds on islands: Insights into the idiosyncrasies of extinction. In Species invasions: Insights into ecology, evolution, and biogeography, D. F. Sax, J. J. Stachowicz, and S. D. Gaines, Editors. Sinauer, Sunderland, MA, pp. 85–110.

Blackburn, T. M., O. L. Pletchey, P. Cassey, and K. J. Gaston (2005). Functional diversity of mammalian predators and extinction in island birds. Ecology 86:2916–2923.

Both, C., M. van Asch, R. G. Bijlsma, A. B. van dem Burg, and M. E. Visser (2009). Climate change and unequal phenological changes across four trophic levels: Constraints or adaptations? Journal of Animal Ecology 78:73–83.

Boyer, A. G. (2008). Extinction patterns in the avifauna of the Hawaiian Islands. Diversity and Distributions 14:509–517.

Boyer, A. G. (2009). Consistent ecological selectivity through time in Pacific island avian extinctions. Conservation Biology 24:511–519.

Boyer, A. G., and W. Jetz (2014). Extinctions and the loss of ecological function in island bird communities. Global Ecology and Biogeography 23:679–688.

Chapin III, F. S., E. S. Zavaleta, V. T. Eviner, R. L. Naylor, P. M. Vitousek, H. L. Reynolds, D. U. Hooper, et al. (2000). Consequences of changing biodiversity. Nature 305:234–242.

Chiappe, L. M., and G. J. Dyke (2002). The Mesozoic radiation of birds. Annual Review of Ecology and Systematics 33:91–124.

Conant, S. (2005). Honeycreepers in Hawaiian material culture. In The Hawaiian honeycreepers: Drepanidinae, H. D. Pratt, Editor. Oxford University Press, Oxford, UK, pp. 278–284.

Corcoran, P. L., C. J. Moore, and K. Jazvac (2014). An anthropogenic marker horizon in the future rock record. GSA Today 24:4–8.

Corlett, R. T. (2015). The Anthropocene concept in ecology and conservation. Trends in Ecology and Evolution 30:36–41.

Cox, P. A. (1983). Extinction of the Hawaiian avifauna resulted in a change of pollinators for the ieie, Freycinetia arborea. Oikos 41:195–199.

Culliney, S., L. Pejchar, R. Switzer, and V. Ruiz-Gutierrez (2012). Seed dispersal by a captive corvid: The role of the ʻAlalā (Corvus hawaiiensis) in shaping Hawaiʻi's plant communities. Ecological Applications 22:1718–1732.

Czech, B., and P. R. Krausman (1997). Distribution and causation of species endangerment in the United States. Science 277:1116.

Diamond, J. M. (1989). The present, past and future of human-caused extinction. Philosophical Transactions of the Royal Society of London: B 325:469–478.

Duncan, R. P., A. G. Boyer, and T. M. Blackburn (2013). Magnitude and variation of prehistoric bird extinctions in the Pacific. Proceedings of the National Academy of Sciences USA 110:6436–6441.

Dunn, R. R., N. C. Harris, R. K. Colwell, L. P. Koh, and N. S. Sodhi (2009). The sixth mass coextinction: Are most endangered species parasites and mutualists? Proceedings of the Royal Society of London B 276:3037–3045.

Edey, M. A. (1998). The last stand of the Heath Hen. Intelligencer 39:155–174.

Engbring, J., and F. L. Ramsey (1984). Distribution and abundance of the forest birds of Guam: Results of a 1981 survey. US Fish and Wildlife Service, Washington, DC.

Fahrig, L. (1997). Relative effects of habitat loss and fragmentation on population extinction. Journal of Wildlife Management 61:603–610.

Feduccia, A. (1996). The origin and evolution of birds. Yale University Press, New Haven, CT.

Fitzpatrick, J. W., M. Lammertink, M. D. Luneau Jr., T. W. Gallagher, B. R. Harrison, G. M. Sparling, K. V. Rosenberg, et al. (2005). Ivory-billed Woodpecker (Campephilus principalis) persists in continental North America. Science 308:1460–1462.

Flannery, T. (2001). The eternal frontier. Atlantic Monthly Press, New York.

Fleischer, R. C., and C. E. Mcintosh (2001). Molecular systematics and biogeography of the Hawaiian avifauna. Studies in Avian Biology 22:51–60.

Frishkoff, L. O., D. S. Karp, L. K. M'Gonigle, C. D. Mendenhall, J. Zook, C. Kremen, E. A. Hadley, and G. C. Daily (2014). Loss of avian phylogenetic diversity in neotropical agricultural systems. Science 345:1343–1346.

Fuller, E. (2001). Extinct birds, rev. ed. Cornell University Press, Ithaca, NY.

Garnett, S., G. Crowley, and A. Balmford (2003). The costs and effectiveness of funding the conservation of Australian threatened birds. BioScience 53:658–665.

Gaston, K. J., and T. M. Blackburn (1997). Evolutionary age and risk of extinction in the global avifauna. Evolutionary Ecology 11:557–565.

Gilpin, M. E., and M. L. Soule (1986). Minimum viable populations: Processes of species extinction. In Conservation biology, M. L. Soule, Editor. Sinauer, Sunderland, MA, pp. 19–34.

Grant, P. R. (1995). Commemorating extinctions. American Scientist 83:420–422.

Grant, P. R., and B. R. Grant (2008). How and why species multiply. Princeton University Press, Princeton, NJ.

Grier, J. W. (1982). Ban of DDT and subsequent recovery of reproduction in Bald Eagles. Science 218:1232–1235.

Gummer, H., and M. Williams (1999). Campbell Island Teal: Conservation update. Wildfowl 50:133–138.

Harris, G., and S. L. Pimm (2008). Range size and extinction risk in forest birds. Conservation Biology 22:163–171.

Hickey, J. J., Editor (1969). Peregrine Falcon populations: Their biology and decline. University of Wisconsin Press, Madison, WI.

Hough, H. B. (1933). The Heath Hen's journey to extinction, 1792–1933. Dukes County Historical Society. Edgartown, MA.

Hunter, L. A. (1988). Status of the endemic Atitilan Grebe of Guatemala: Is it extinct? The Condor 90:906–912.

Jackson, J. A. (2006). Ivory-billed Woodpecker (Campephilus principalis): Hope, and the interfaces of science, conservation, and politics. The Auk 123:1–15.

Jackson, J. A., and B. J. S. Jackson (2007). Once upon a time in American ornithology. Wilson Journal of Ornithology 119:767–772.

Jarvis, E. D., S. Mirarab, A. J. Aberer, B. Li, P. Houde, C. Li, S. Y. Ho, et al. (2014). Whole-genome analyses resolve early branches in the tree of life of modern birds. Science 346:1320–1331.

Kaeding, H. B. (1905). Birds from the west coast of lower California and adjacent islands (concluded). The Condor 7:134–138.

Keller, L. F., and D. M. Waller (2002). Inbreeding effects in wild populations. Trends in Ecology and Evolution 17:230–241.

LaBastille, A. (1974). Ecology and management of the Atitlan Grebe, Lake Atitlan, Guatemala. Wildlife Monographs 37:1–66.

LaBastille, A. (1983). Drastic decline in Guatemala's Giant Pied-billed Grebe population. Environmental Management 11:346–348.

Lande, R. (1988). Genetics and demography in biological conservation. Science 241:1455–1460.

Lande, R. (1999). Extinction risks from anthropogenic, ecological, and genetic factors. In Genetics and the extinction of species, L. F. Landweber and A. P. Dobson, Editors. Princeton University Press, Princeton, NJ, pp. 1–22.

Lande, R., S. Engen, and B.-E. Saether (2003). Stochastic population dynamics in ecology and conservation. Oxford University Press, Oxford, UK.

Leonard Jr., D. L. (2008). Recovery expenditures for birds listed under the US Endangered Species Act: The disparity between mainland and Hawaiian taxa. Biological Conservation 141:2054–2061.

Lockwood, J. L., T. M. Brooks, and M. L. McKinney (2000). Taxonomic homogenization of the global avifauna. Animal Conservation 3:27–35.

Lynch, M. (1996). A quantitative-genetic perspective on conservation issues. In Conservation genetics, J. C. Avise and J. L. Hamrick, Editors. Chapman & Hall, New York, pp. 471–501.

Lynch, M., R. Burger, D. Butcher, and W. Gabriel (1993). The mutational meltdown in asexual populations. Journal of Heredity 84:339–344.

Lynch, M., J. Conery, and R. Burger (1995). Mutation accumulation and the extinction of small populations. The American Naturalist 146:489–518.

Manne, L. L., T. M. Brooks, and S. L. Pimm (1999). Relative risk of extinction of passerine birds on continents and islands. Nature 399:258–261.

McClelland, P. (2011) Campbell Island—pushing the boundaries of rat eradications. In Island invasives: Eradication and management, C. R. Veitch, M. N. Clout, and D. R. Towns, Editors. IUCN, Gland, Switzerland.

McClelland, P., and H. Gummer (2006). Reintroduction of the critically endangered Campbell Island Teal Anas nesiotis to Campbell Island, New Zealand. Conservation Evidence 3:61–63.

Møller, A. P., D. Rubolini, and E. Lehikoinen (2008). Populations of migratory bird species that did not show a phonological response to climate change are declining. Proceedings of the National Academy of Sciences USA 105:16195–16200.

Monastersky, R. (2015). Anthropocene: The human age. Nature 519:144–147.

Moran, R. V. (1996). The flora of Guadalupe Island, Mexico. Memoirs of the California Academy of Sciences 19:190.

National Research Council (1992). Scientific bases for preservation of the Hawaiian Crow. National Academy Press, Washington, DC.

National Research Council (1997). Scientific bases for preservation of the Mariana Crow. National Academy Press, Washington, DC.

Nee, S., and R. M. May (1997). Extinction and the loss of evolutionary history. Science 278:692–694.

Nott, M. P., E. Rogers, and S. Pimm (1995). Modern extinctions in the kilo-death range. Current Biology 5:14–17.

Perry, G. L. W., A. B. Wheeler, J. R. Wood, and J. M. Wilmshurst (2014). A high-precision chronology for the rapid extinction of New Zealand moa (Aves, Dinornithiformes). Quaternary Science Reviews 105:126–135.

Phillimore, A. B., and T. D. Price (2008). Density-dependent cladogenesis in birds. PLoS Biology 6: e71. doi:10.1371/journal.pbio.0060071.

Pimm, S. L., G. J. Russell, J. L. Gittleman, and T. M. Brooks (1995). The future of biodiversity. Science 269:347–350.

Pimm, S. L., C. N. Jenkins, R. Abell, T. M. Brooks, J. L. Gittleman, L. N. Joppa, P. H. Raven, C. M. Roberts, and J. O. Sexton (2014). The biodiversity of species and their rates of extinction, distribution, and protection. Science 334:1–10.

Post, W., and J. S. Greenlaw (2009). Seaside Sparrow (Ammodramus maritimus). In Birds of North America online, A. Poole, Editor. Cornell Lab of Ornithology, Ithaca, NY. http://bna.birds.cornell.edu/bna/species/127.

Pratt, T. K. (2009). Origins and evolution. *In* Conservation biology of Hawaiian forest birds, T. K. Pratt, C. T. Atkinson, P. C. Banko, J. D. Jacobi, and B. L. Woodworth, Editors. Yale University Press, New Haven, CT, pp. 3–24.

Purvis, A., J. L. Gittleman, G. Cowlishaw, and G. M. Mace (2000). Predicting extinction risk in declining species. Proceedings of the Royal Society of London B 267:1947–1952.

Rabinowitz, D. (1981). Seven forms of rarity. *In* The biological aspects of rare plant conservation, H. Synge, Editor. Wiley and Sons, Chichester, UK, pp. 205–217.

Rando, J. C., and J. A. Alcover (2010). On the extinction of the Dune Shearwater (*Puffinus holeae*) from the Canary Islands. Journal of Ornithology 151:365–369.

Restani, M., and J. M. Marzluff (2002). Funding extinction? Biological needs and political realities in the allocation of resources to endangered species recovery. BioScience 52:169–177.

Russell, G. J., T. M. Brooks, M. M., McKinney, and C. G. Anderson (1998). Present and future taxonomic selectivity in bird and mammal extinctions. Conservation Biology 12:1365–1376.

Santisteban, L., C. W. Benkman, T. Fetz, and J. W. Smith (2012). Survival and population size of a resident bird species are declining as temperature increases. Journal of Animal Ecology 81:352–363.

Savidge, J. A. (1987). Extinction of an island forest avifauna by an introduced snake. Ecology 68:660–668.

Shaffer, M. L. (1981). Minimum population sizes for species conservation. BioScience 31:131–134.

Schulte, P., L. Alegret, I. Arenillas, J. A. Arz, P. J. Barton, P. R. Bown, T. J. Bralower, et al. (2010). The Chicxulub asteroid impact and mass extinction at the Cretaceous-Paleogene boundary. Science 327:1214–1218.

Schultz, J. H., D. L. Otis, and S. A. Temple (2014). 100th anniversary of the Passenger Pigeon extinction: Lessons for a complex and uncertain future. Wildlife Society Bulletin 38:445–450.

Sekercioglu, C. H., S. H. Schneider, J. P. Fay, and S. R. Loarie (2008). Climate change, elevational range shifts, and bird extinctions. Conservation Biology 22:140–150.

Shepard, P. (1996). The Others: How animals made us human. Island Press, Covelo, CA.

Shultz, S., R. B. Bradbury, K. L. Evans, R. D. Gregory, and T. M. Blackburn (2005). Brain size and resource specialization predict long-term population trends in British birds. Proceedings of the Royal Society of London B 272:2305–2311.

Sibley, D. A., L. R. Bevier, M. A. Patten, and C. S. Elphick (2006). Comment on "Ivory-billed Woodpecker (*Campephilus principalis*) persists in continental North America." Science 311:1555a.

Smith, J. W., S. M. Sjoberg, M. C. Mueller, and C. W. Benkman (2012). Assortative flocking in crossbills and implications for ecological speciation. Proceedings of the Royal Society of London B 279:4223–4229.

Steadman, D. W. (1995). Prehistoric extinctions of Pacific Island birds: Biodiversity meets zooarchaeology. Science 267:1123–1131.

Steadman, D. W. (1997). Human-caused extinction of birds. *In* Biodiversity II: Understanding and protecting our biological resources, M. L. Reaka-Kudla, D. E. Wilson, and E. O. Wilson, Editors. National Academy Press, Washington, DC, pp. 139–161.

Szabo, J. K., N. Khwaja, S. T. Garnett, and S. H. M. Butchart (2012). Global patterns and drivers of avian extinctions at the species and subspecies level. PLoS ONE 7: e47080.

Tanner, J. T. (1964). The decline and present status of the Imperial Woodpecker of Mexico. The Auk 81:74–81.

Thayer, J. E., and O. Bangs (1908). The present state of the Ornis of Guadaloupe Island. The Condor 10:101–106.

Thomas, C. D. (2000). Dispersal and extinction in fragmented landscapes. Proceedings of the Royal Society B: Biological Sciences 267:139–145.

Thomas, J. A., M. G. Telfer, D. B. Roy, C. D. Preston, J. J. D. Greenwood, J. Asher, R. Fox, R. T. Clarke, and J. H. Lawton (2004). Comparative losses of British butterflies, birds, and plants and the global extinction crisis. Science 303:1879–1881.

USFWS (US Fish and Wildlife Service) (2005). Draft revised recovery plan for the Aga or Mariana Crow, *Corvus kubaryi*. USFWS, Portland, OR.

USFWS (US Fish and Wildlife Service) (2014). California Condor Monthly Status Report, October 31. USFWS. Pacific Southwest Region, Albuquerque, NM. https://www.fws.gov/cno/es/calcondor/CondorCount.cfm.

Van Turnhout, C. A. M., R. P. B. Foppen, R. S. E. W. Leuven, A. van Strien, and H. Siepel (2010). Life-history and ecological correlates of population change in Dutch breeding birds. Biological Conservation 143:173–181.

Vitousek, P. M., H. A. Mooney, J. Lubchenco, and J. M. Melillo (1997). Human domination of Earth's ecosystems. Science 277:494–499.

Walters, J. R., S. R. Derrickson, D. M. Fry, S. M. Haig, J. M. Marzluff, and J. M. Wunderlie Jr. (2010). Status of the California Condor (*Gymnogyps californianus*) and efforts to achieve its recovery. The Auk 127:969–1001.

Waters, C. N., J. Zalasiewicz, C. Summerhayes, A. D. Barnosky, C. Poirier, A. Gałuszka, A. Cearreta, et al. (2016). The Anthropocene is functionally and stratigraphically distinct from the Holocene. Science 351:137. doi:10.1126/science.aad2622.

Westemeier, R. L., J. D. Brawn, S. A. Simpson, T. L. Esker, R. W. Jansen, J. W. Walk, E. L. Kershner, J. L. Bouzat, and K. N. Paige (1998). Tracking the long-term decline and recovery of an isolated population. Science 282:1695–1698.

White, R. L., and P. M. Bennett (2015). Elevational distribution and extinction risk in birds. PLoS ONE 10:e0121849.

Wilcove, D. S., D. Rothstein, J. Dubow, A. Phillips, and E. Losos (1998). Quantifying threats to imperiled species in the United States. BioScience 48:607–615.

Wilmshurst, J. M., T. L. Hunt, C. P. Lipo, and A. J. Anderson (2011). High-precision radiocarbon dating shows recent and rapid initial human colonization of East Polynesia. Proceedings of the National Academy of Sciences USA 108:1815–1820.

Wurster, C. F. (2015). DDT wars. Oxford University Press, Oxford, UK.

Yiming, L., and D. S. Wilcove (2005). Threats to vertebrate species in China and the United States. BioScience 55:147–153.

CHAPTER 26

In Harm's Way

Scott R. Loss and Timothy J. O'Connell

A large and growing variety of **anthropogenic** (i.e., human-related) activities threaten birds from the most common to the most critically endangered. Unlike natural threats, such as depredation by native predators and death from storms, the vast majority of anthropogenic mortality factors for birds are recent, emerging since the nineteenth century or more recently. Many bird species are affected by several anthropogenic threats. For example, many Neotropical migrant songbirds face habitat loss on both breeding and wintering grounds, as well as along migratory pathways (i.e., stopover habitat), and they can also die by colliding with many types of man-made structures. In this chapter, we refer to **mortality** as the general phenomenon or overall amount of bird death from one or more causes, **mortality rate** as the number of deaths in an area or time period, and **fatality** as individual occurrences of bird death.

In general, anthropogenic threats are novel enough that bird populations have not evolved mechanisms to absorb their effects and are unlikely to in the short term. Furthermore, as the human population continues to grow, many of these threats are becoming even more pervasive. Therefore, many anthropogenic threats likely cause at least some **additive mortality** to bird populations, mortality that would not have otherwise occurred and cannot be compensated for by density-dependent population processes (i.e., mortality that is not **compensatory**). Identifying, understanding, and managing anthropogenic threats is crucial for preventing extinctions of endangered or otherwise rare species (chapter 25). Equally important is preventing the decline of common species to the level of endangerment; keeping common birds common requires studying and managing the anthropogenic harms they face. In this chapter, we review anthropogenic threats to birds from the conservation perspective of keeping common species common.

Several additional terms are used in this chapter. **Direct mortality sources** are threats that can immediately kill birds without intervening mechanisms (e.g., birds dying from colliding with structures). **Indirect mortality sources** typically kill birds after a time lag and via one or more intermediate mechanisms (Loss et al. 2015b). The most important examples of indirect mortality sources—climate change and habitat loss—profoundly influence bird distributions and abundance by determining the amount of habitable space and thus setting the carrying capacity for bird populations. The amount of death caused by direct mortality sources contributes to determining how close to this carrying capacity bird populations stay or recover to. Although mortality is the most severe effect of anthropogenic activities, government agencies usually regulate **take** of birds under policies such as the US Endangered Species Act and Migratory Bird Treaty Act, the Canadian Migratory Birds Convention Act, and the United Kingdom Wildlife and Countryside Act. Take includes mortality, as well as harassment, pursuit, injury, or otherwise harming birds (fig. 26.1). Finally, mortality and take can be **purposeful** (intentional) or **incidental** (accidental).

In this chapter, we focus on purposeful and incidental direct mortality sources. Major indirect threats are covered elsewhere, including climate change (chapter 24) and habitat loss (chapter 25). Space constraints prevent us from exhaustively covering every direct mortality source. We therefore focus on sources that have received substantial research or are increasingly relevant because of substantial public or conservation attention. The relatively nascent state of the field means that most information about anthropogenic mortality is descriptive, and there is tremendous opportunity for future research to test mortality mechanisms and effects using a hypothetico-deductive approach. Finally, interpreting the overall magnitude of impact for each mortality source and comparing different mortality sources would ideally include consideration of the total number of birds. Yet there are no such scientifically derived estimates for North America, and

Figure 26.1. A sign marking the entrance to an area of closed public lands containing a nesting population of US federally endangered Snowy Plovers (*Charadrius nivosus*) in the North Spit Area of Critical Environmental Concern near Coos Bay, Oregon, USA; signs such as these are meant to minimize all types of take of endangered species, including purposeful and incidental mortality, and in this case indirect harassment caused by beachgoer traffic. *Photo from Bureau of Land Management Oregon and Washington, https://commons.wikimedia.org /wiki/File:North_Spit_ACEC_(15596181489).jpg. Wikimedia Commons, open access.*

devising such an estimate is beyond the scope of this chapter. Nonetheless, to provide rough context for the severity of the mortality sources we describe, it is worth considering an often-cited (albeit speculative) estimate of 10–20 billion total birds in North America (USFWS 2002).

ANTHROPOGENIC MORTALITY SOURCES

Direct mortality sources fall into several categories, characterized by similar types of activities, human behaviors, and landscape types. In the following sections, we briefly summarize what is known about the mechanisms, drivers, and amount of mortality from different threats and, when they exist, solutions for reducing mortality. We cover several types of purposeful human-caused mortality as well as incidental mortality from energy development, production, and usage; poisoning by chemicals; other commercial and industrial activities; and human activities in our neighborhoods. Infectious diseases are another major source of anthropogenic mortality covered in detail in chapter 23.

Purposeful Mortality

Humans have hunted birds for tens of thousands of years (Peresani et al. 2011). Although there are significant exceptions (e.g., moas in New Zealand; Holdaway and Jacomb 2000), for much of human history and most species, mortality from

hunting was probably compensatory (Péron 2013), with population effects similar to those imposed by other predators. As human technology advanced, however, so did our ability to rapidly harvest animals, and exploitation of birds has often crossed over from compensatory to additive mortality. Birds have been hunted for food and feathers, to reduce depredation on crops (Saikku 1990), and to protect poultry from "chicken hawks" (Bildstein 2001) (fig. 26.2). As more people moved from rural areas to cities in North America, hunters supplied wild birds for markets in large eastern cities such as Boston, New York, and Philadelphia. During much of the nineteenth century, the only limits on the number of birds harvested were demand and ease of shipment. This market hunting was unsustainable, however, and populations of many North American species dwindled, often in concert with habitat loss. Perhaps the best studied among these examples is the now-extinct Passenger Pigeon (*Ectopistes migratorius*) (Halliday 1980), but we also lost the Great Auk (*Pinguinus impennis*) (Montevecchi and Kirk 1996), Carolina Parakeet (*Conuropsis carolinensis*) (Saikku 1990), and Eskimo Curlew (*Numenius borealis*) (Weidensaul 1999), largely to similar pressure (fig. 26.3). Populations of egrets, terns, shorebirds, and waterfowl were also reduced to extremely low levels in the late nineteenth and early twentieth centuries as these species were relentlessly hunted for food and feathers. In an attempt to reduce their depredations on livestock, poultry, and desirable game species, raptors were persecuted with bounties until the early twentieth century (Leopold 1933, Bildstein 2001). The large number of raptors recovering from gunshot wounds in wildlife rehabilitation centers today indicates that "chicken hawks" continue to be persecuted.

Regulated Hunting

Concern for declining species led to the first restrictions on hunting in the United States (Halliday 1980) and, ultimately, to the permitted system of legal hunting (Leopold 1933) that today is central to the conservation of both game and nongame birds in North America (box on page 806). Wildlife biologists monitor bird populations to help recover rare species, keep common species common, and ensure that game species are abundant enough to support demand from licensed hunters. Monitoring efforts (e.g., from the Waterfowl Breeding Population and Habitat Survey; USFWS 2015a) are regularly summarized and published so participation and harvest rates can be tracked through time. For example, Raftovich et al. (2012) reported total harvest numbers and hunters for all migratory game species in the United States and Canada during 2010 and 2011. Their summary included data on approximately 50 hunted species of ducks, geese, shorebirds, doves, and rails. Harvests ranged from 10,000 to 20,000 rails

Figure 26.2. A, Georg Flegel (1566–1638) captured the diversity of bird species that once constituted German table fare in *Früchte und Gemüse Stilleben mit Vögeln. Photo accessed from https://commons. wikimedia.org/wiki/File:Georg_Flegel_Fr%C3%BCchte-_und_Gem% C3%BCsestilleben_mit_V%C3%B6geln.jpg. Wikimedia Commons,* *public domain. B,* A young man poses with Red-tailed Hawks (*Buteo jamaicensis*) he shot on his farm. *From Bird Lore 1909, photo accessed from https://commons.wikimedia.org/wiki/File:Bird-lore_(1909) _(14563052639).jpg. Wikimedia Commons, public domain.*

to 14.8 million to 15.9 million ducks and 16.6 million to 17.2 million Mourning Doves (*Zenaida macroura*). Estimates like these depend on self-reporting by hunters and cooperation between state and federal agencies. Modern assessments of overall mortality from regulated hunting increasingly include birds that are crippled or killed but never recovered. Schulz et al. (2013) suggested that self-reported **crippling rates** can greatly underestimate total mortality from hunting.

Under unusual circumstances, nongame species protected under the Migratory Bird Treaty Act (1918) may be taken to support scientific research or reduce damage to personal property or other managed resources. Management of Double-crested Cormorant (*Phalacrocorax auritus*) in the Columbia River Estuary in the US Pacific Northwest provides an example of the latter. Charged under the Endangered Species Act to manage recovering populations of salmonid fish, the US Army Corps of Engineers found that the mission was untenable amid predation on juvenile fish by cormorants from a massive nesting colony nearby. The Army Corps developed a management plan for cormorants that aimed to cull 11,000 adults and destroy 26,000 nests in an attempt to reduce the breeding population to about 5,600 pairs (USACE 2015).

In Europe, at least 28 nations are signatories to the European Commission's Birds Directive (1979), the oldest environmental legislation in the European Union. Much like the widespread protections for birds that the Migratory Bird Treaty Act established in the western hemisphere, the Birds Directive outlines protections for populations and habitats of approximately 500 species in the European Union. Eighty-two of those species can be hunted, subject to restricted seasons and other limitations (EU 2009). The European Commission has worked with the nongovernmental organizations BirdLife International and the European Federation of Associations for Hunting and Conservation (FACE) to develop a sustainable hunting initiative that recognizes and seeks to maintain the important conservation role of regulated hunting within the confines of the Birds Directive. BirdLife International and FACE work with the EU to eradicate illegal killing, trapping, and trade in wild birds and encourage the enforcement of other laws protecting birds and bird habitats.

Despite cooperation at the highest levels, conflicts frequently arise regarding management of hunted species. In the United Kingdom, for example, hunting upland game birds is popular sport but the woodland and heather that supports game species is limited by fragmentation, urbanization, afforestation of nonnative pines, and overgrazing (Thompson et al. 1995). Private landowners who can derive revenue from providing hunting opportunities have incentive to manage their land for game species (Oldfield et al. 2003), and this management can provide ancillary benefits to nongame species. For example, managing for pen-reared Ring-necked Pheasants (*Phasianus colchicus*) produced a more open-canopy and densely vegetated understory in English woodlands that was associated with higher abundance of warblers (sum of six species) (Draycott et al. 2008). Game management has been

Figure 26.3. Four species of extinct North American birds for which unregulated hunting is largely to blame for their demise: *A*, Great Auk (*Pinguinus impennis*); *B*, Passenger Pigeon (*Ectopistes migratorius*); *C*, Carolina Parakeet (*Conuropsis carolinensis*); *D*, Eskimo Curlew (*Numenius borealis*). *Paintings by John James Audubon, public domain.*

highly controversial, however, because it typically also relies on the local eradication of Hen Harriers (*Circus cyaneus*) and other predators to promote reproductive success of the highly prized Red Grouse (*Lagopus lagopus*) (fig. 26.4). In addition to legal hunting of Hen Harriers for this purpose, illegal hunting has rendered the species rare in Great Britain. Some game managers have developed complex management sys-

tems in which Hen Harriers are tolerated on heather moorlands used for Red Grouse hunting, but only within specific population targets. Small populations of harriers on grouse hunting moorlands are supplemented with food to reduce depredation on young grouse. Through careful monitoring and translocation, excess harriers are captive-reared and released in other areas of the species' former range. This inten-

Figure 26.4. In Great Britain and Ireland *A*, Red Grouse (*Lagopus lagopus*) is a popular game bird, and management of moorlands to support grouse provides vital habitat for multiple nongame species. *B*, Although grouse management often involves predator control, including the legal eradication of Hen Harrier (*Circus cyaneus*), many harriers are killed illegally each year, and conservationists are working to develop management plans that benefit both species. *A, Photo by MPF, https://commons.wikimedia.org/wiki/File:2014-04-21_Lagopus _lagopus_scotica,_Hawsen_Burn_3.jpg, B, R. Zweers https://www .flickr.com/photos/65033273@N05/15714647009. Creative Commons CC BY 2.0 license.*

sive management suggests that Hen Harriers can be restored with minimal loss (< 10 percent in autumn abundance) in Red Grouse numbers (Elston et al. 2014).

Unregulated Hunting

The unregulated hunting of birds for food is a serious global problem, and the capture of birds for the pet trade can also be considered a special case of unregulated hunting because it effectively removes individuals from populations. Especially in Mediterranean countries, where they are concentrated during autumnal and vernal migration, birds can be vulnerable to unregulated hunting in large numbers, i.e., 11–33 million

annually (Brochet et al. 2016). Birds may be shot, captured in nets or other traps, or captured with lime glue spread on perches. Birds are lured with recordings of their calls and in many cases shot with automatic shotguns (BirdLife International 2008). Birds are killed indiscriminately, but it is often the highly migratory species (with accumulated fat) that are sought after for sale in local markets. Species such as Ortolan Bunting (*Emberiza hortulana*) and Eurasian Blackcap (*Sylvia atricapilla*) are heavily persecuted but remain abundant; European Turtle Dove (*Streptopelia turtur*) was downgraded in 2015 from Near Threatened to the more critical Vulnerable category on the European Red List (BirdLife International 2015; IUCN Red List categories in fig. 26.5). Illegal hunting of Afro-Palearctic migrants in autumn could be compensatory mortality, but the hunting of returning migrants in spring is more likely to be additive for some species (Vickery et al. 2014), and international conservation groups are working to better understand the scale of the problem and develop solutions. Brochet et al. (2016) emphasized greater coordination among countries to better track hunting pressure, strengthening penalties for illegal killing of migratory birds, and in some cases increasing enforcement of existing legislation.

In tropical regions, illegal hunting of wild birds can be widespread, and hunting pressure may be contributing to declines in critically rare species. In the illegal bushmeat trade of West Africa, birds are often uncommon in the markets. Whytock et al. (2014) suspected, however, that many birds

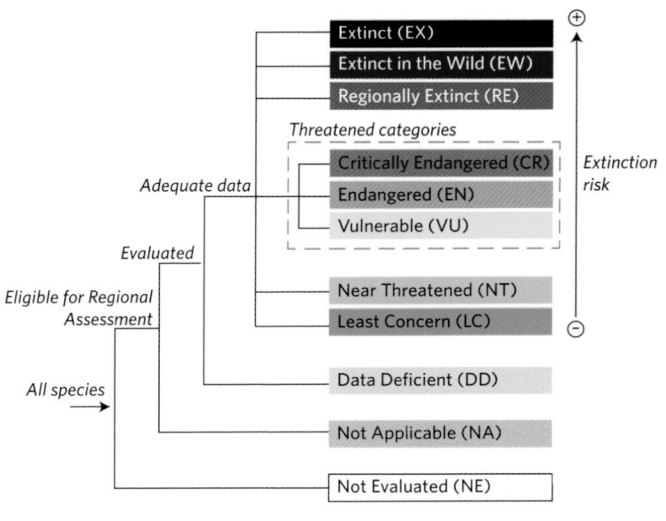

Figure 26.5. Structure of the International Union for the Conservation of Nature (IUCN) Red List categories for national and regional assessment of species extinction risk. The Global Red List category structure is similar except there is no category for Regionally Extinct (RE). *From IUCN 2000.*

THE ROLE OF MANAGED HUNTING IN THE CONSERVATION OF NORTH AMERICAN BIRDS

By Dr. R. Dwayne Elmore, Department of Natural Resource Ecology and Management, Oklahoma State University

Managed hunting (hereafter "hunting")—as opposed to illegal (e.g., poaching) and/or unregulated hunting for sport or subsistence—has several benefits to wildlife conservation including increased awareness and appreciation of wildlife, stakeholder involvement in political processes affecting wildlife, and landowner incentives to manage land for wildlife. One of the most important but frequently misunderstood benefits of hunting is the generated revenue directed toward wildlife conservation.

State agencies in the United States have authority to manage and issue licenses and permits for hunting of nonmigratory birds such as quail, grouse, and Wild Turkey (*Meleagris gallopavo*). These license sales directly contribute to wildlife management, land acquisition, and wildlife law enforcement. Many hunted species, however, are considered migratory and therefore regulated by the Migratory Bird Treaty Act of 1918. This international act between the United States, Mexico, Canada, Japan, and Russia was an important milestone in ending the unregulated harvests that were unsustainable for many species. The act has provisions allowing some reasonable and legitimate harvest provided the hunter has a valid permit pursuant to federal regulations. In the United States, the US Fish and Wildlife Service (USFWS) has regulatory authority to administer and enforce the act. Under the overarching USFWS statutes, individual states can set regulations that fall within federal limits. Hunting of migratory birds, particularly waterfowl and doves, is tightly regulated under an adaptive management framework that relies on reliable estimates of both harvest and species population numbers, ensuring that science has a prominent place in hunting regulations and that individual species harvest is sustainable.

In addition to required state licenses and permits, waterfowl hunters must possess a Federal Duck Stamp. The Federal Duck Stamp, instituted in 1934, generates over $22 million annually, of which 98 percent is used to purchase or protect lands for migratory birds (USFWS 2015b). Approximately 90 percent of all Federal Duck Stamps are purchased by waterfowl hunters (USFWS 2015c), but there is increasingly a move to encourage nonconsumptive users (e.g., birders and hikers) to purchase stamps.

Another important act for wildlife conservation is the Pittman-Robertson Act of 1937. This act established an 11 percent excise tax on ammunition, archery equipment, and

In the United States, hunting license sales and an excise tax on firearms and ammunition provide much of the financial support for state wildlife agencies to acquire and manage land. In this example, the desire to hunt Northern Bobwhite (*Colinus virginianus*) quail encourages land management for quail, including prescribed fire resulting in grassland, open shrubland, and savanna that provide vital habitat for multiple nongame species. *Photo from Florida Fish and Wildlife Conservation Commission, https://commons.wikimedia.org/wiki/File:Quail_Hunting_with_Dogs.jpg. Wikimedia Commons, open access.*

sporting firearms, and a 10 percent excise tax on handguns, to provide a stable funding mechanism for state wildlife agencies that match 25 percent of the federal appropriations (based on the land area of each state and the number of state licensed hunters). These funds are used for wildlife research projects, to acquire and manage land for wildlife, and to provide hunter education. In recent years, the Pittman-Robertson excise tax collections have exceeded $450 million per year (National Shooting Sports Foundation 2016). Notably, most state wildlife agencies receive no general state tax support and therefore license sales generate most of their funding for wildlife conservation, either directly through license revenue or indirectly through Pittman-Robertson allocations. Therefore, hunters and anglers largely provide the funding for state wildlife agencies in most states. While some state expenditures are directed toward hunted wildlife species, the acquisition and management of land and various research projects clearly benefits many wildlife species and stakeholder groups.

were being consumed by hunters while encamped in the forest, while mammals and reptiles were reserved for sale in the markets. In addition to hunter surveys and interviews, the researchers confirmed the presence of at least nine bird species in middens left in hunter camps in Cameroon (Whytock et al. 2014). The most numerous carcasses were large-bodied birds such as hornbills and raptors. In neighboring Nigeria, Atuo et al. (2015) examined the illegal hunting and trade of birds primarily for their perceived spiritual properties and use in traditional healing. Of 27 species from 13 families represented in the local trade, at least three (each among the top five most sought) were IUCN conservation priorities: Grey Parrot (*Psittacus erithacus*; Vulnerable), Hooded Vulture (*Necrosyrtes monachus*; Endangered), and Martial Eagle (*Polemaetus bellicosus*; Vulnerable).

The decline in Grey Parrot populations caused by the pet trade reveals another world of exploitation by humans, but one for which mortality is unintentional and population loss occurs even if birds survive. Annorbah et al. (2016) concluded that the population of Grey Parrots in Ghana had declined 90–99 percent since 1992. The authors considered capture for the pet trade to be nearly as serious a problem for Grey Parrots as habitat loss, adding that export mortality of captured birds averaged 50 percent (fig. 26.6). In Brazil, Alves et al. (2013) suggested that as many as 400 species of native birds could be sold in the illegal pet trade; 36 of the species confirmed as objects of the illegal pet trade were IUCN Red List species ranging from Near Threatened to Critically Endangered. In Peru, 17 Red List species were among 130 species available for purchase in local markets (Daut et al. 2015). In addition to the loss of individuals that are captured but do not survive, the pet trade can also supply birds likely to establish as invasive species in a new area. In Taiwan, Su et al. (2015) determined that 23 of the 25 naturalized nonnative species on the island can be found for sale in markets.

Incidental Mortality: Energy Sources

Energy is necessary for day-to-day human activities and the long-term maintenance and growth of human societies and economies. Exploration for and development, extraction, transport, and use of energy also have a wide range of unintended environmental consequences. Here, we focus primarily on energy sectors with well-documented and/or publicized impacts to birds. We also cover emerging energy sectors that, although not currently well-studied, have potential to affect large numbers of birds now or in the future. For context to this section, it is helpful to remember that the use of some energy sources may also benefit birds and the environment in some ways. For example, the expansion of solar, wind, and other renewable energy sources reduces the use and harmful effects of fossil fuels, which include habitat loss and degradation caused by extraction and transport activities and global climate change from burning of carbon dioxide and other greenhouse gases. Our current understanding of the positives and negatives of all energy sources—especially for newly emerging sources, such as large-scale solar and unconventional oil/gas—is insufficient to direct strategies that maximize benefits to humans while minimizing harmful effects on the environment.

Wind Energy

The global development of wind energy has increased dramatically in recent decades (fig. 26.7). In the United States, wind energy production in May 2017 was 84 gigawats, representing 8 percent of total U.S. energy generating capacity (EIA 2017). A further fivefold production increase is projected by 2030 (USDOE 2008). Because of this energy growth, the impact of wind energy on wildlife has become a major focus of research, policy, and management attention. Birds experience direct and indirect effects of wind energy, with the primary indirect effect being displacement as a consequence of habitat loss or alteration (Stevens et al. 2013, Shaffer and Buhl 2015). Direct mortality occurs when birds collide with wind turbines: between 140,000 and 644,000 birds are estimated to die this way in the United States annually (Johnson et al. 2016).

Factors affecting bird collision rates at wind facilities remain somewhat unclear. However, turbine placement undoubtedly influences mortality. Broad regional differences in

Figure 26.6. The African Grey Parrot (*Psittacus erithacus*) is a familiar sight in the pet trade, and this has threatened the continued existence of this species in the wild. *Photo by Dick Daniels. Image available from https://commons.wikimedia.org/wiki/User:DickDaniels/Birds_P_-_S# /media/File:African_Grey_Parrot_RWD2.jpg.*

Figure 26.7. Examples of wind farms in *A*, Ireland; *B*, The US Great Plains (Oklahoma) with Scissor-tailed Flycatcher (*Tyrannus forficatus*); *C*, The North Sea of Europe (Denmark). *Photos A and B by T. O'Connell; C, Wiki Media Commons, https://commons.wikimedia.org/wiki/File:Middelgrunden_wind_farm_2009-07-01_edit_filtered.jpg, original artist: K. Hansen.*

Figure 26.8. The Ivanpah Solar Power Facility in California is an example of a large-scale solar energy development in the US desert Southwest. Birds may perceive large arrays of reflecting mirrors as water bodies and collide when attempting to land or when foraging for insects near their surface; birds are also burned/singed when flying near light concentration towers at facilities such as Ivanpah that have a power tower in the center of a mirror array. *Photo credit: Wiki Media Commons, https://commons.wikimedia.org/wiki/File:Ivanpah_Solar_Power_Facility_from_the_air_2015.jpg, original artist: Jllm06.*

US mortality rates exist, with the greatest per-turbine mortality in California and mountains in the eastern United States (Loss et al. 2013a). Local features within wind facilities and at the scale of individual wind turbines are also likely to influence collision rates, but well-replicated studies of this topic are currently lacking for birds. Kiesecker et al. (2011) suggested that, in order to reduce impacts to birds, wind energy development should occur primarily in agricultural areas that are already human-modified.

Bird-turbine collision rates vary across species. However, at most wind facilities, monitoring is either not conducted at all or the duration and/or frequency of monitoring is insufficient to document the full range of species affected (Beston et al. 2015). The existing body of data suggests that raptors are disproportionately affected in the western United States, whereas few raptors, but far more Neotropical migrant songbirds, collide with turbines from the Great Plains eastward (Erickson et al. 2014, Beston et al. 2016). Population effects of turbines have not been documented except for a few well-studied raptor species in Europe (Carrete et al. 2009, Dahl et al. 2012). Several raptors (e.g., Ferruginous Hawk, *Buteo re-*

galis, and Golden Eagle, *Aquila chrysaetos*) appear most likely to experience declines in North America as a result of turbine collisions (Beston et al. 2016).

Solar Energy

Among renewable energy sources in the United States, solar energy is expected to have the greatest percentage growth by 2040 (USEIA 2014). Much of this development will occur in the US desert Southwest (Lovich and Ennen 2011). In addition to displacement caused by habitat alteration, birds are directly killed at solar plants by (1) collision with solar panels and reflecting mirrors and (2) singeing or burning of flying birds at light concentration towers (at plants with a power tower in the center of a mirror array) (fig. 26.8). Very few mortality studies at solar plants have been conducted, largely because of the relatively nascent state of large-scale solar development and because widespread recognition of the issue of bird mortality at solar plants is recent. However, a large variety of bird species appear to be affected (Kagan et al. 2014). Research on solar energy impacts will likely increase in the future because of the rapid rate of development and the location of some current and planned facilities in biodiverse ecosystems with protected species (Lovich and Ennen 2011).

Oil and Natural Gas Energy

By 2040, oil and natural gas are expected to eclipse coal as the leading global energy sources (IEA 2014; fig. 26.9). Notably, both the gas and oil industries are shifting from **conventional extraction** approaches (i.e., extraction of crude oil and associated natural gas liquids using vertical wells) to a variety of **unconventional extraction** approaches, including extraction of oil from oil sands (e.g., in Canada and Argentina) and extraction of both oil and gas from shale deposits with horizontal drilling and/or hydraulic fracturing (i.e., "fracking") in much of the United States. Few studies of impacts to birds have been conducted for any type of unconventional extraction activity. In particular, the much-publicized fracking approach remains virtually unstudied with respect to any of its environmental impacts (Butt et al. 2013, Souther et al. 2014), probably because the exceptionally rapid rate of expansion of these technologies has been recent.

For birds, the most-studied effect of oil and gas extraction is displacement caused by habitat alteration. Several grassland songbirds (e.g., Sprague's Pipit, *Anthus spragueii*; Baird's Sparrow, *Ammodramus bairdii*; and Chestnut-collared Longspur, *Calcarius ornatus*) avoid unconventional oil sites in the massive Bakken oil fields of North Dakota (Thompson et al. 2015, fig. 26.10), and Greater Sage Grouse (*Centrocercus urophasianus*) avoid conventional oil sites during winter in Wyoming (Holloran et al. 2015). Entire communities of breeding forest birds have been shown to be altered by conventional oil and gas development in Pennsylvania (Thomas et al. 2014). Specifically, this study showed that the guild of forest interior species decreased in abundance, while guilds of early successional and **synanthropic** (i.e., human-adapted) species increased in abundance near well sites; these responses contributed to homogenization of bird communities surrounding well sites in different regions.

Direct mortality of birds also occurs at oil and gas sites, but these impacts are even less studied than habitat-related effects. Between 500,000 and 1 million birds are estimated to drown each year at open pits and tanks used for disposing waste byproducts from oil production (Trail 2006). Greater than 1,000 birds at a time have died from attraction and burning at flares at oil and gas sites (Bjorge 1987, CBC News 2013). Birds can also become entrapped and die in oil and gas site heater treaters (Ramirez 2013). Finally, an unknown number of nocturnally migrating birds collide with lighted structures and drilling rigs at oil and gas sites.

Power Lines

Power lines connect energy production points to energy users. Large, high-voltage **transmission lines** carry electricity across long distances between energy generation facilities and power substations. Small low-voltage **distribution lines** carry electricity across shorter distances from substations to commercial, industrial, and residential users (fig. 26.11). Millions of kilometers of power lines exist, and the global power line network is growing at a rate of 5 percent per year (Jenkins et al. 2010) to feed energy sectors that are growing to supply the burgeoning human population.

In addition to causing habitat fragmentation, direct bird mortality occurs at power lines from both collisions and electrocution. Collisions occur primarily at transmission lines, while electrocution is most common at distribution lines. In the United States, 8–57 million birds die annually from power line collisions and 1–11 million birds die annually from electrocution (Loss et al. 2014c). Population-level effects of bird mortality at power lines are uncertain. For several large-bodied and/or weak-flying species—such as storks in Switzerland and grouse in Norway—power line collisions

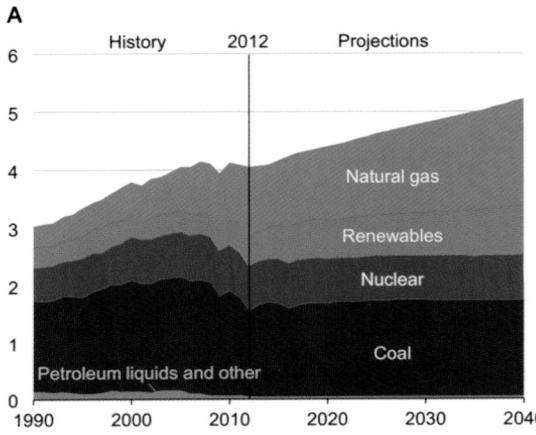
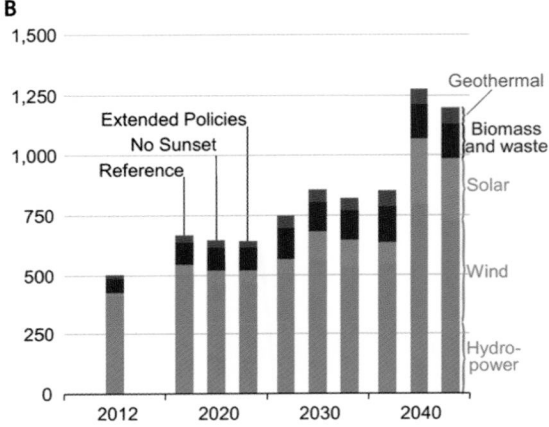

Figure 26.9. Projected use of *A,* all major energy sources in a single reference case (in trillions of kilowatt-hours) and *B,* major renewable energy sources in three references cases (in billions of kilowatt-hours). *From USEIA 2014.*

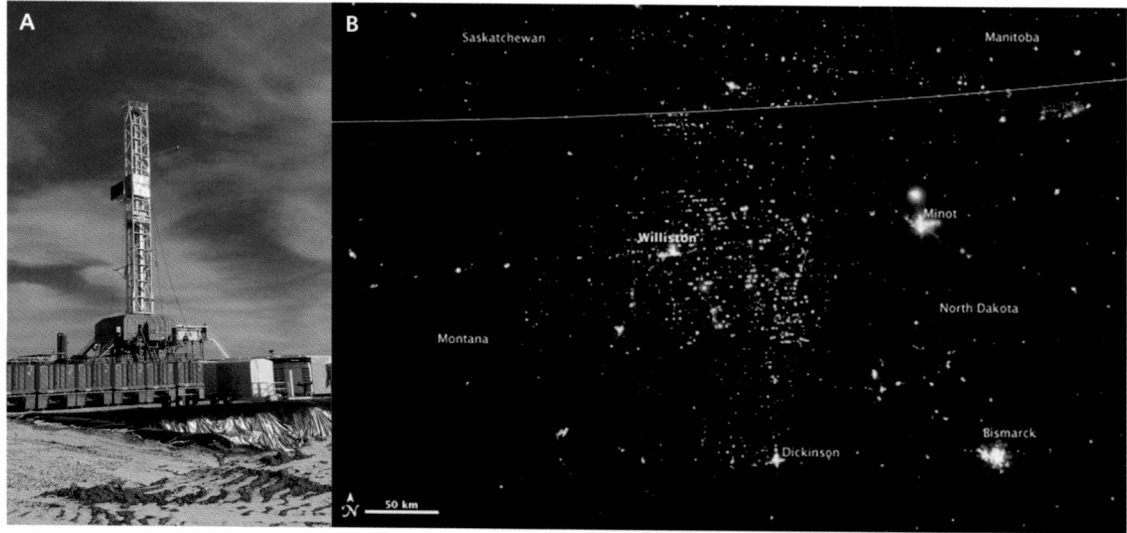

Figure 26.10. *A*, Oil drilling rig in the Bakken oil field of North Dakota, one of the largest producing oil fields on earth; *B*, Solar flares and lighting associated with infrastructure on the Bakken oil field is bright enough to be detected by orbiting satellites (the diffuse lights near Williston, North Dakota). *Photo credit: A, Wikimedia Commons,* *https://commons.wikimedia.org/wiki/File:Drilling_the_Bakken_forma tion_in_the_Williston_Basin.jpg, original artist: J. Doubek; B, Wikimedia Commons, https://commons.wikimedia.org/wiki/File:NASA.EOS.79810 .Suomi_NPP_night_bakken.vir_2012317.jpg, NASA Earth Observatory; J. Allen and R. Simmon.*

Figure 26.11. Typical transmission *A* and distribution *B* power lines; although both electrocution and collision of birds can occur at both line types, collisions occur primarily at transmission lines, while electrocutions typically occur at distribution lines. *Photo credit: A, Wiki Media Commons, https://commons .wikimedia.org/wiki/File:Electric_transmission _lines.jpg, original artist: Nixdorf; B, Wikimedia Commons, original artist: U.Y.B.A. Mikail.*

are estimated to cause a high proportion of all mortality (Bevanger 1995, Schaub and Pradel 2004). This indirectly suggests the potential for bird population declines as a result of collisions with power lines.

Substantial efforts have been made to identify and implement approaches to reduce bird mortality at power lines. In particular, the collaboration that exists among researchers, conservation agencies, and several power utilities—in the form of the Avian Power Line Interaction Committee (APLIC)—could provide a model for efforts to reduce bird mortality from other industrial activities. This collaboration has resulted in identification and publication of best practices for reducing collisions and electrocutions (APLIC 2006, 2012). For collisions with transmission lines, marking wires with flight diverters can substantially reduce mortality rates (Barrientos et al. 2011). For electrocution at distribution lines,

mortality can largely be averted by using nonmetal materials whenever possible and ensuring that the minimum spacing between energized parts exceeds the wingspan and body length of the largest birds potentially affected (APLIC 2006).

Other Energy-Related Effects

Other energy sources (e.g., coal and nuclear) also cause direct and indirect bird mortality. Birds collide with smokestacks at coal-fired power plants and cooling towers at coal and nuclear plants (Avery 1979, Erickson et al. 2005). Resource extraction for many energy types also causes indirect and direct effects. Coal mining—and especially the highly destructive mountaintop removal approach used in the Appalachian Mountains of the eastern United States (fig. 26.12)—degrades or completely removes large amounts of forest (Palmer et al. 2010), and thus threatens forest-inhabiting bird species,

Figure 26.12. Mountaintop removal is a highly destructive mining approach that results in degradation or complete removal of large amounts of bird habitat. *Photo credit: Wikimedia Commons, https:// commons.wikimedia.org/wiki/File:Euniceblast4.JPG, original artist: Roston.*

such as the Cerulean Warbler (*Setophaga cerulea*). Bird poisoning death occurs at coal and uranium mining sites (US-FWS 2008). Avian injury and death from both onshore and offshore oil spills during extraction and transport is a highly publicized form of direct bird mortality. Major offshore spills, such as the 1989 *Exxon Valdez* spill in Alaska and the 2010 Deepwater Horizon spill in the Gulf of Mexico, likely caused hundreds of thousands and greater than one million bird deaths, respectively (Piatt et al. 1990, National Audubon Society 2014; fig. 26.13). Though less publicized, small-scale spills may cause thousands of additional fatalities (Erickson et al. 2005). Most of these energy-related impacts have received

little study. Therefore, attempts to directly compare effects to other energy sources are unreliable.

Incidental Mortality: Poisoning by Industrial and Commercial Chemicals

Humans release a variety of poisons into the environment in association with industrial and commercial activities, and many of these have lethal or sublethal effects to birds. **Lethal** effects cause birds to die immediately or shortly after toxin exposure, whereas **sublethal** effects cause illness or alteration of physiological processes that influence behavior and/ or decrease long-term survival probability. In this section, we cover agricultural chemicals and lead poisoning, the two most studied groups of poisons that affect birds.

Agricultural Chemicals

More than 100 herbicides, insecticides, and fungicides used for agricultural purposes (hereafter **agricultural chemicals**) have lethal or sublethal effects when birds directly consume them or indirectly ingest them from contaminated food (Mineau et al. 2001). These agricultural chemicals kill 1 million to 4.4 million birds each year in Canada (Calvert et al. 2013). No comparable mortality estimates have been made for other countries. Lethal effects of agricultural chemicals, including occasional mass die-offs of multiple birds of numerous species, are the most obvious side effect of agricultural chemical use. Sublethal effects are less obvious and more difficult to study. Sublethal behavioral and physiological effects from pesticides can substantially affect bird populations in two ways: (1) directly (for individuals exposed) by altering reproduction, foraging, and predator avoidance behaviors; and (2) indirectly (for avian predators consuming exposed prey)

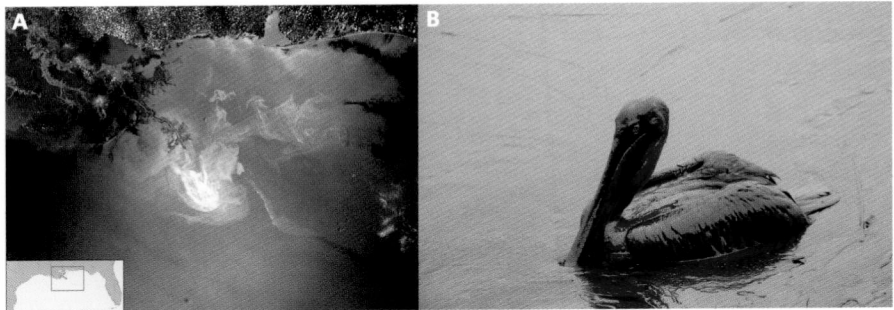

Figure 26.13. *A*, A satellite image of the 2010 Deepwater Horizon oil spill, the largest oil spill ever, along the US Gulf Coast off the states of Louisiana, Mississippi, and Alabama; *B*, A Brown Pelican (*Pelecanus occidentalis*) oiled as a result of the spill. *Photo credit: A, Wikimedia Commons, https://commons.wikimedia.org/wiki/File:Deepwater_Horizon_oil_spill_-_May_24,_2010.jpg, NASA, Goddard Space Flight Center, and MODIS Rapid Response Team; B, Wiki Media Commons, https://commons.wikimedia.org/wiki/File:Oiled_Pelicans.jpg, Louisiana Governor's Office of Homeland Security and Emergency Preparedness.*

by altering prey availability or causing predators to feed on poisoned prey items, thus increasing their own exposure (Walker 2003).

Documenting causal links between agricultural chemical use and avian population declines is difficult, largely because pesticide-use data in many countries is incomplete or nonexistent. Correlative studies indicate that patterns of broad-scale chemical use are related to large-scale population declines of several bird species. This pattern of decline has been shown for granivorous and insectivorous bird species exposed to granular insecticides in Canada (Mineau 2005a) and grassland bird species exposed to a variety of insecticides in the United States (Mineau and Whiteside 2013). Overall community diversity of farmland birds declined significantly in relation to fungicide application rates across nine European countries (Geiger et al. 2010). In the Netherlands, **neonicotinoids**—a family of chemicals similar to nicotine that have strong physiological effects on invertebrates, but not vertebrates—have likely caused the decline of many insectivorous bird species (Hallman et al. 2014). Although neonicotinoids have increasingly been used because they have minimal direct effect on vertebrates, this study suggested they indirectly affect insectivorous bird populations by reducing insect prey density.

There are many examples of countries banning the use of agricultural chemicals that are highly toxic to birds and other vertebrates. DDT—a chemical that is lipophilic (i.e., dissolves in fats, oils, and lipids) and bioaccumulates as DDE in bird tissues, causing thinned eggshells and reduced avian reproductive success—has been banned for agricultural uses in most developed countries, leading to associated increases in affected bird populations (e.g., Bald Eagles, *Haliaeetus leucocephalus*; Grier 1982). Carbofuran—a cholinesterase inhibitor that blocks the breakdown of acetylcholine—is one of the most toxic chemicals to birds, with as little as a single pellet being lethal. Many bird species have been found to be affected by carbofuran, both on North American breeding grounds and in wintering areas of Latin America (Richards 2011), including Bobolink (*Dolichonyx oryzivorus*) and Dickcissel (*Spiza americana*). Between 17 million and 91 million songbirds were estimated to be killed each year in the US Midwest at the peak of this chemical's use (Mineau 2005b). Carbofuran has been banned for all uses in Canada and Europe and for agricultural uses in the United States. Despite these success stories, the broad-scale application of many highly toxic chemicals continues today even in developed countries, and chemical regulation often lags behind in undeveloped countries.

Lead Poisoning

Lead is a highly toxic substance to both humans and wildlife that affects multiple physiological processes. Lead ingestion by birds can have lethal and sublethal effects, including behavioral changes and impaired body condition and reproduction (De Francisco et al. 2003). Lead enters the environment through a variety of point and nonpoint sources. **Point sources** are discrete locations of entry that include mining (of lead and other metals containing lead), industry (lead processing and smelting plants, refineries, power plants), manufacturing (auto parts, tires, electronics, etc.), and dumps. **Nonpoint sources** are characterized by a large number of diffuse locations of entry; these include paints in older buildings, car exhaust, gasoline, water pipes, gunshot pellets, and fishing weights.

The toxic effect of lead shot and fishing tackle ingestion by birds is perhaps the most studied mechanism of avian lead poisoning (reviewed by Haig et al. 2014). Between 1.5 million and 3 million waterfowl may have died annually from ingesting lead shot prior to its ban for waterfowl hunting in 1991 (Thomas et al. 1985). Birds of prey, such as the endangered California Condor (*Gymnogyps californianus*) (fig. 26.14), experience lethal and sublethal effects of feeding on animals killed by lead ammunition (Haig et al. 2014). Waterbirds, such as loons and swans, are susceptible to lead poisoning from ingesting fishing weights (Pokras et al. 2009). Seed-eating birds, such as doves and other upland game species, experience lead poisoning when ingesting lead pellets as grit to help grind

Figure 26.14. A California Condor (*Gymnogyps californianus*) released after recovering from lead poisoning; condors typically become exposed to lead when feeding on carcasses killed with lead ammunition. *Photo credit: Wikimedia Commons, https://commons.wikimedia.org/wiki/File:Condor_Released_at_Hopper_Mountain_NWR_(14521489764).jpg, US Fish and Wildlife Service Pacific Southwest Region.*

seeds in their gizzard (Fisher et al. 2006). In addition to the well-documented effects of ammunition and fishing tackle, birds can also experience lethal and sublethal effects from elevated environmental lead levels near mines (Beyer et al. 2004). Decades-long legacies of nonpoint lead pollution in urban areas (e.g., from paint and gasoline) can also elevate blood-lead levels and reduce body condition in birds (Roux and Marra 2007).

The use of lead has gradually been reduced in the United States. Leaded gasoline was banned for automobiles in 1995 but remains in use for small aircraft (USEPA 2015). Lead ammunition has been banned in the United States for waterfowl hunting since 1991. Bans of lead shot and fishing weights for all hunting and fishing uses have occurred on some federal and state lands (Scheuhammer and Norris 1995). However, lead ammunition remains widely used for hunting animals other than waterfowl. Haig et al. (2014) outlined a multifaceted approach to reducing avian mortality and morbidity from this route of lead exposure, including (1) federal and state/provincial policies banning lead ammunition, (2) voluntary reductions in use of lead ammunition and tackle, (3) giveaways or exchanges of non-lead ammunition/tackle to hunters and anglers, and (4) outreach and communication about the impacts of lead ammunition and tackle.

Incidental Mortality: Other Commercial and Industrial Activities

In addition to avian mortality associated with energy sources and chemicals, humans cause bird mortality through a broad variety of commercial and industrial activities. A conservation success story about one of these commercial sources of bird mortality—bird collisions with communications towers (fig. 26.15)—is covered in the box on page 814.

Fisheries Bycatch

As evidenced by many of the above examples, birds often die incidentally as a result of natural resource extraction and use. Seabirds provide an additional example of how human demand for seafood exposes birds to novel risks. One global estimate suggests that at least 400,000 birds drown in gill nets each year (Zydelis et al. 2013). In longline fishing, thousands of baited hooks are towed behind vessels. For species such as albatrosses and shearwaters that pluck prey at or just beneath the surface, such baits can be irresistible, and birds are hooked and often drown during the many hours that baits are deployed. In eastern Canada, for example, longline and gill net operations were implicated in the conservative estimate that at least 5,000 birds were killed from 1998 to 2011. The affected species varied regionally, with murres (*Uria* spp.) frequently killed in gill nets off Newfoundland, and Northern Fulmar (*Fulmarus glacialis*) caught both in gill nets and on longlines in Baffin Bay and Davis Strait (Hedd et al. 2015). On the Patagonian Shelf off the coast of Argentina, at least 10,000 fatalities were estimated to be caused from longline fishing from 1999 to 2001 (Favero et al. 2003). An estimated 4–6 percent of the local breeding population of Cory's Shearwater (*Calonectris diomedea*) was killed annually in the Spanish Mediterranean (Cooper et al. 2003). Robertson et al. (2014) reported that modified fishing gear that allowed hooks to sink deeper into the water posed less risk to albatrosses (*Thalassarche* spp.): From an annual estimate of 1,555 Black-browed Albatrosses (*T. melanophrys*) killed in Chilean waters in 2002, mortality was eliminated completely with the new baiting system.

Nest Losses to Hayfield Management

Birds can also be vulnerable to incidental mortality from human activities on land, such as the harvest of hayfields. These grasslands are specifically managed to be cut multiple times during the growing season. In temperate zones of the Northern Hemisphere, the first cutting of cool-season grasses often takes place in May with subsequent harvests from June through August. At least one and often two harvests overlap with nesting of grassland birds, with machinery present in

Figure 26.15. *A*, Birds collide with communications towers and their guy wires; guyed towers kill far more birds than unguyed towers, and replacing steady burning aviation warning lights with flashing lights significantly reduces mortality; *B*, A biologist monitors for dead birds under a guyed communications tower in Michigan, USA. *Photos by Joelle Gehring.*

BIRD COLLISIONS WITH COMMUNICATIONS TOWERS: PROGRESSING FROM RESEARCH TO POLICY CHANGE AND BIRD CONSERVATION

By Dr. Joelle Gehring, Federal Communications Commission

Bird collisions with communications towers were documented as early as 1949 and occasionally resulted in thousands of fatalities in a single night (Aronoff 1949). Recent meta-analysis of more than 30 studies estimated 6.8 million bird collision fatalities each year in the United States and Canada, including 350 species, with the majority long-distance Neotropical migrants (e.g., warblers, vireos, and thrushes), as well as some short to medium-distance migrants (sparrows and kinglets) (Longcore et al. 2012, 2013). Motivated by concerns about the potential adverse effects of 180 new public safety communications towers in Michigan, a multi-collaborator group (including the State of Michigan, federal agencies, universities, and a private consultant) embarked in 2003 on the first comprehensive, experimental, multiyear study of bird fatalities at communications towers.

Gehring et al. (2009) compared numbers of bird fatalities at towers with different heights, tower support systems, and lighting systems. This study found 16 times more bird fatalities at guyed towers than unguyed towers and more fatalities at tall towers than shorter towers. Critically important to reducing bird collisions with existing towers, the study found that towers lit at night only with flashing lights caused fewer bird fatalities than towers lit with non-flashing lights. The authors found no differences in bird fatality rates between towers lit with red or white flashing lights; however, avian fatalities were reduced by 50–71 percent when non-flashing red lights were extinguished.

Given that non-flashing lights and flashing lights were both required in red tower lighting systems, the industry, federal and state agencies, and bird conservation groups encouraged the Federal Aviation Administration (FAA) to evaluate safety for aircraft if non-flashing lights were extinguished (FAA 2007). In 2012, the FAA published a study documenting that extinguishing non-flashing lights on towers taller than 350 ft (107 m) was safe for aircraft (Patterson 2012). In 2015, the FAA issued standards that eliminated non-flashing lights from tower light systems approved after December 4, 2015 (FAA 2015). In March 2016, Transport Canada was also expected to eliminate the use of non-flashing lights on new towers taller than 350 ft. To reduce the 6.8 million bird collisions with towers built before the FAA policy change, tower owners will need to extinguish non-flashing lights or replace them with flashing lights, with the former option saving energy and maintenance costs as well as reducing carbon emissions. The Federal Communications Commission provides educational materials on the process of extinguishing non-flashing lights and encourages tower owners to pursue lighting changes to reduce migratory bird collisions.

More than a decade of collaboration among scientists, the communications tower industry, federal and state agencies, and conservation groups has resulted in new tower lighting options that reduce avian collisions by as much as 71 percent, while also reducing maintenance and energy costs for the industry and carbon emissions for the life of the tower. This case study emphasizes the value of rigorous, applied, scientific research conducted in collaboration with diverse stakeholders in order to resolve long-term conservation issues.

Migratory birds found under a communication tower in a single-night kill event during migration in the midwestern United States, including numerous warbler and vireo species such as Red-eyed Vireo (*Vireo olivaceus*), Ovenbird (*Seiurus aurocapilla*), Magnolia Warbler (*Setophaga magnolia*), and American Redstart (*Setophaga ruticilla*). *Photo by Andy Paulios, Wisconsin Department of Natural Resources.*

Figure 26.16. Hayfields can provide important habitat for grassland birds but the timing of harvest *A* during the peak of nesting season can have detrimental effects on species such as *B* Bobolink (*Dolichonyx oryzivorus*). Photo credit: A, Wikimedia Commons, https:// commons.wikimedia.org/wiki/File:Baling_hay _in_Vermont.jpg, original artist: Putneypics; B, Wikimedia Commons, https://commons .wikimedia.org/wiki/File:Bobolink-Male _(9052883003).jpg, original artist: CheepShot.

fields first for cutting and then for raking and baling. A recent attempt to model mortality in Canada suggested that approximately 2.2 million eggs and nestlings of five grassland bird species are lost each year to haying and other mechanical operations in fields (Tews et al. 2013). Bollinger et al. (1990) estimated that 51 percent of Bobolink eggs and nestlings were destroyed by the first cut in Illinois (fig. 26.16). Additional nest losses occurred as a consequence of abandonment by adults following the disturbance, increased nest predation caused by reduced concealment, and subsequent raking and baling. In sum, Bollinger et al. (1990) estimated that haying induced at least 94 percent nest loss. In addition, at least 50 percent of previously fledged juvenile Bobolinks were killed during haying. The overall influence of hayfield management on grassland birds is complex and variable, but with specific information on nesting phenology, management plans can be developed that are less destructive to grassland birds.

In Europe, correspondingly greater effort has been invested in developing haying regimes that reduce effects to nesting birds. Two common approaches involve (1) intensive efforts to find nests and mark the locations for farmers to avoid, and (2) delaying the onset of first haying to at least July 1. Grüebler et al. (2012) examined both of these approaches for conservation of nesting Whinchats (*Saxicola rubetra*) in Switzerland. For nests subjected to early-season haying, modeled daily survivorship plunged to 10 percent (i.e., 90 percent failure). A combined approach of protecting individual nests and delaying the onset of haying improved nest survivorship to at least 70 percent. Although the authors described individual nest protection as labor intensive and requiring highly skilled technicians, they recognized that small hayfields that support critically endangered species sometimes make the outcome worth the effort.

Fence Collisions

In working agricultural landscapes, even static structures such as fences can cause bird fatalities. The galliformes seem to be especially vulnerable to collision with wire fencing. In Scotland, Baines and Andrew (2003) confirmed 437 bird collisions with monitored wire mesh fence lines in a two-year study. Red Grouse (*Lagopus lagopus*) accounted for 42 percent of all collisions, followed by Black Grouse (*Tetrao tetrix*) at 29 percent and Capercaillie (*Tetrao urogallus*) at 20 percent. Marking fences with orange plastic netting made the fences more visible and reduced collisions by 68 percent. In western North America, the application of small, white, plastic clips to barbed wire fencing evidently made those fences more obvious to Lesser Prairie Chickens (*Tympanuchus pallidicinctus*). In that study, prior to fence marking, Wolfe et al. (2009) confirmed one fence collision fatality for every mile of fencing. Following fence marking, the researchers conducted 30 months of similar searches and had not confirmed a single fatality.

Entrapment in Open Pipes

Another recently recognized direct mortality source is open metal or polyvinyl chloride (PVC) pipes. Vertical pipes with an open top mimic a small snag that might be attractive to cavity-nesting or roosting birds. Often, however, birds that enter such pipes become trapped and die. Hathcock and Fair (2014) surveyed open pipes at the Los Alamos National Laboratory in New Mexico and found dead birds in 27 percent of 100 pipes searched. Most (61 percent) of the identifiable carcasses were Western Bluebird (*Sialis mexicana*), a secondary cavity-nester, but at least five other species were also confirmed.

Open pipes that can be attractive to birds are widespread in rural areas. In the Western United States, many access roads are gated, and gates are often flanked by open-topped pipes of ~10 cm in diameter. In addition to supports for gates, open pipes have been used for survey markers, signposts, marking mine locations, and ventilation for mines and irrigation systems. Extrapolating their findings to the number of potentially deadly open pipes in New Mexico, Hathcock and Fair (2014) estimated that 13,580 birds are killed in in the state each year. Conservation organizations are beginning to

Figure 26.17. Support pipes for fences and gates can be dangerous to birds if open-topped as in *A* instead of capped as in *B* because birds may mistake open pipes for roosting locations and become entrapped. *Photos by T. O'Connell.*

Figure 26.18. Free-ranging domestic cat populations are abundant in most areas, including *A* urban areas (as illustrated by a night-time trail camera image from Oklahoma, USA) and *B* protected natural areas (including this state park in Ohio, USA). *Photos by S. Loss.*

mount information campaigns to educate about the danger to birds from open pipes (fig. 26.17). For example, the National Audubon Society has produced an information sheet for landowners that introduces the problem and illustrates a wide variety of features that can entrap birds in pipes (Audubon California 2011). The solution to this problem, capping the pipes, is straightforward and inexpensive, but the ubiquity of the pipes, especially in the Western United States, continues to make this mortality a challenge for conservation.

Incidental Mortality in Our Neighborhoods

Many human behaviors and activities associated with our day-to-day activities can be lethal to birds. Virtually everyone—including the authors—has contributed to bird mortality from these activities, such as driving a car to the grocery store or not being able to prevent every last bird collision with the windows on their home. We hope that covering these threats to birds provides increased awareness of the many ways we accidentally harm birds on a day-to-day basis. We further hope to spur readers to, whenever possible, alter activities and behaviors in a way that minimizes unintended harm to birds.

Introduced Predators

The human introduction of nonnative species has profound effects on biological diversity and ecosystem functioning. Predation is the most direct and catastrophic effect of invasive nonnative species on birds. Examples abound of invasive predators decimating bird populations, with the most frequent and dramatic examples coming from oceanic islands. For example, the brown tree snake (*Boiga irregularis*) has caused the extinction of 12 endemic bird species on the island of Guam in the latter half of the twentieth century (Wiles et al. 2003). Introduction of the small Indian mongoose (*Herpestes auropunctatus*) to numerous islands around the world has caused the extinction or decline of several bird species (Lowe et al. 2000), presumably including the Jamaican Poorwill (*Siphonorhis americana*) (BirdLife International 2016). Depredation risk posed by the nonnative black rat (*Rattus rattus*) on the Hawaiian island of Oʻahu is so great as to have contributed to the evolution of higher nesting height by the endangered Oʻahu ʻElepaio (*Chasiempis ibidis*) (VanderWerf 2012). In a promising example of the amelioration of invasive predator effects, removal of black rats from Hawaiian seabird colonies has been shown to improve population productivity for species such as

Figure 26.19. A sample of the many extinct island bird species and subspecies whose decline to extinction was at least partly driven by predation by free-ranging domestic cats (Medina et al. 2011): *A*, Hawaiian Rail (*Porzana sandwichensis*) endemic to the Big Island of Hawai'i, USA; *B*, Choiseul Pigeon (*Microgoura meeki*) endemic to Choiseul Island, Solomon Islands; *C*, Hawaiian Crow (or 'Alalā) (*Corvus hawaiiensis*) endemic to the Big Island of Hawai'i, USA; *D*, Northern (Guadalupe) Flicker (*Colaptes auratus rufipileus*) endemic to Guadalupe Island, Mexico; *E*, Chatham Fernbird (*Megalurus rufescens*) endemic to Pitt and Mangere Islands, New Zealand. *Photo credit: A, https://commons .wikimedia.org/wiki/File:Hawaiian_Spotted_Rail.png; B, https://commons .wikimedia.org/wiki/File:Choiseul_Crested_Pigeon.jpg; C, Wikimedia Commons, https://commons.wikimedia.org/wiki/File:Corvus_hawaiiensis _FWS.jpg, US Fish and Wildlife Service; D, Wikimedia Commons, https:// commons.wikimedia.org/wiki/File:Colaptes_rufipileus_(Guadalupe _flicker).jpg, J. St. John); E, Wikimedia Commons, https://commons .wikimedia.org/wiki/File:Chatham_Fernbird.png; original artist: J. G. Keulemans).*

Wedge-tailed Shearwater (*Ardenna pacificus*) and Laysan Albatross (*Phoebastria immutabilis*) (Young et al. 2013).

Among the most widespread invasive predators on earth—and arguably, the invasive mammalian predator with the greatest collective effect on earth's wildlife (Doherty et al. 2016)—is one that we release into our own backyards and purposefully abandon into the wild to live on its own. Over the entire western hemisphere, there has never been a predator like the domestic cat (*Felis catus*). Even in regions with ecological analogs, such as the wildcat (*Felis silvestris*) in Europe, Asia, and Africa, domestic cat populations usually exist at unnaturally high levels (fig. 26.18). As with other invasive predators, cats have had devastating effects on oceanic islands, where they have caused the extinction of dozens of species of birds, small mammals, and reptiles (Medina et al. 2011; Doherty et al. 2016) (fig. 26.19). Population impacts of cats in continental areas are less clear because of uncertainty about the number of free-ranging cats in most areas and the difficulty of parsing out the effects of cats versus other human effects. Nonetheless, cats are likely the top source of direct anthropogenic mortality in North America, with billions and hundreds of millions of birds killed in the United States and Canada, respectively (Blancher 2013, Loss et al. 2013b). Furthermore, numerous studies illustrate that cats can reduce bird populations, even in mainland areas (van Heezik et al. 2010, Balogh et al. 2011, Loss and Marra 2017).

In most cases, the control and removal of invasive predators following introduction is logistically challenging because

Figure 26.20. Trap-neuter-return (TNR) colonies are organized efforts to subsidize cats with food and shelter following capture, steriliza-tion, and, sometimes, immunization; as illustrated in *A*, TNR cats often have one or both ears "tipped" to allow future identification; in addition to TNR colonies, countless unregulated and often hidden free-ranging cat feeding stations are maintained in *B*, backyards and *C*, public parks. *Photo credit: A, Wikimedia Commons, https://commons.wi-kimedia.org/wiki/File:Feral_cat_with_clipped_ear.jpg, FAL; B, Wikimedia Commons, https://commons.wikimedia.org/wiki/File:Female_feral_cat _with_its_kitten_feeding.JPG; C, Wikimedia Commons, https://commons .wikimedia.org/wiki/File:PikiWiki_Israel_28384_Cities_in_Israel.JPG, U. Steinwell.*

individuals are spread across large areas and are difficult to detect and/or trap because of their cryptic behavior. For cats, even the decision about whether to attempt to control and reduce populations is highly controversial. The management of free-ranging pet cat and feral cat populations is one of the most contentious issues facing policy makers and conserva-tion biologists today. This controversy arises because cats are popular pets and therefore not considered by many to be eco-logically harmful despite the fact that they kill native wildlife and are reservoirs for multiple diseases that infect humans and wildlife (Gerhold and Jessup 2013; VanWormer et al. 2013). A thorough coverage of cat management approaches is be-yond the scope of this chapter and covered elsewhere (Lohr and Lepczyk 2014, Loss and Marra 2017). Nevertheless, widely accepted and commonsense methods to reduce cat predation, such as keeping pet cats indoors and creating and enforcing laws that prevent pet abandonment, are crucial to minimiz-ing impacts to birds and improving cat welfare. Outlawing ac-tivities that support feral cat populations, both formally (e.g., trap-neuter-return colonies) and informally (e.g., impromptu feeding/sheltering stations) (fig. 26.20), should be strongly con-sidered, especially in proximity to important wildlife popula-tions and habitats, such as nesting colonies of threatened and endangered species like Piping Plovers (*Charadrius melodus*) and Least Terns (*Sternula antillarum*) and crucial stopover loca-tions that draw large numbers of fatigued migrants.

Buildings and Windows

Nearly everyone has heard the "thud" of a bird hitting a win-dow. Bird-building collisions—most of which are collisions with windows—are estimated as the second leading direct anthropogenic mortality source in North America, with 365–988 million birds killed in the United States (Loss et al. 2014a) and 16–42 million birds killed in Canada each year (Machtans

et al. 2013). Birds collide with many types of buildings, and the issue of birds colliding with small residential buildings is fundamentally different than that of collisions with large industrial or commercial buildings. Recent quantitative re-views (Machtans et al. 2013, Loss et al. 2014a) illustrate that, although the lowest collision rates occur at residences, resi-dences collectively cause significant amounts of mortality (159–378 million and 22–31 million birds killed per year in the United States and Canada, respectively). Skyscrapers, de-spite causing much greater per-building mortality, cause less than 1 percent of total collision mortality because there are far fewer skyscrapers. Several Neotropical migratory song-birds, including 12 wood warbler species, are most vulnerable to collisions at skyscrapers and low-rise buildings; the most vulnerable species at residences include a mix of Neotropical migrants and backyard "feeder birds" (Loss et al. 2014a).

As a result of uncertainty about how birds see their sur-roundings, finding effective solutions to building collisions remains difficult. For skyscrapers and large low-rise buildings, birds are attracted and disoriented by external lighting. There-fore, turning off building lights at night, especially during peak migration periods, can reduce collision rates (Field Museum 2008). The National Audubon Society's Lights Out Program—a national effort to convince building owners and managers to turn off excess lighting during key migratory periods—has ac-tive branches in nearly two dozen North American cities (Na-tional Audubon Society 2016). For newer large buildings, espe-cially on corporate, medical, and university campuses, there is a noticeable trend for increased use of large expanses of glass (fig. 26.21). These buildings are likely to be "super-killers," caus-ing dozens to hundreds of bird deaths each year and increasing the overall avian death toll from window collisions.

To reduce window collisions in the future, building de-signers could increasingly incorporate bird-friendly design

Figure 26.21. Many new buildings, especially on office, university, and medical campuses, use large unbroken expanses of glass that cause large numbers of bird collision fatalities; *A*, A window-killed Clay-colored Sparrow (*Spizella pallida*) in Stillwater, Oklahoma, USA; *B*, A Swainson's Thrush (*Catharus ustulatus*) in Cleveland, Ohio, USA. *Photos by Scott Loss.*

elements by giving greater weight to bird collision deterrence in "green building" certification systems. For the widely used Leadership in Energy and Environmental Design (LEED) certification standards—a system designed to have a positive impact on building occupants while promoting clean energy—there is currently a single credit for bird deterrence among ~100 other credits, including at least two credits ("quality view" and "daylight") that favor large windows and therefore potentially offset the deterrence credit (US Green Building Council 2015). Detailed collision-prevention steps for residential buildings and major outstanding research needs for this mortality source are covered in the box on page 820; a biosketch for Dr. Daniel Klem Jr., one of the pioneers of bird-building collision research, is found in the box on page 822.

Automobiles and Other Vehicles

Another familiar source of bird mortality that we experience regularly is the collision of birds with automobiles. This mortality occurs because road corridors provide favorable habitat, food sources, and/or movement corridors for birds (Bishop and Brogan 2013). Bird-auto collisions kill 89–340 million birds in the United States each year (Loss et al. 2014b) and 9–19 million birds per year in Canada during the breeding season alone (Bishop and Brogan 2013). Meta-analyses illustrate that reductions in the abundance of bird populations can occur near roadways (Benitez-Lopez et al. 2010). However, whether these declines are caused by collisions or are a collective result of the many indirect effects of roads (e.g., habitat alteration and noise, light, and air pollution) remains uncertain. Automobile collision mortality can contribute to roadside habitats being **population sinks** (i.e., subpopulations with net negative growth rates that require immigration from other populations to persist) (Mumme et al. 2000). Bird-automobile collisions remove many healthy and reproductively mature individuals from populations (Bujoczek et al. 2011) and can

cause high mortality rates and/or constitute a large proportion of all mortality for some species (Van der Zande et al. 1980). Automobile collisions therefore appear to have potential to cause bird population declines in some cases. Barn Owls (*Tyto alba*) in particular are known to experience high rates of mortality from automobile collisions throughout their near-global geographic range, and one study in Idaho estimated a mortality rate of six owls/km/year for this species (Boves and Belthoff 2012) (fig. 26.22).

Reducing bird collisions with automobiles is more difficult than managing many other anthropogenic mortality sources because the global road network is massive (>6.5 million kilometers in the United States alone; USDOT 2012), identifying

Figure 26.22. Barn Owls (*Tyto alba*) are killed in large numbers by colliding with vehicles throughout much of their nearly global geographic range; here, an automobile-killed Barn Owl is shown along a major interstate highway in Idaho, USA, where one of the most intensive studies of the issue has been conducted (Boves and Belthoff 2012). *Photo courtesy of J. Belthoff.*

PREVENTION OF AND FUTURE DIRECTIONS IN THE STUDY OF BIRD-BUILDING COLLISIONS
By Dr. Daniel J. Klem Jr., Biology Department, Muhlenberg College

Why do birds fly into windows? Birds behave as if clear and reflective windows are invisible to them. They stun, injure, or kill themselves trying to fly to habitat and sky visible through clear panes and reflected from mirrored panes. Most windows installed in human dwellings are reflective; even a perfectly clear pane will act as a mirror if the level of light behind the window is lower than the brighter outdoors.

How can bird-window collisions be prevented? Managing the human-built environment to protect birds from windows is both biologically and ethically reasonable. A bird the size of a sparrow need leave a perch only a meter away from a window to kill itself outright in a collision. The short-term solution to make existing windows safe is to retrofit them with external coverings or adhesives that make windows visible to birds. The long-term solution is manufacturing novel bird-safe sheet glass for use in remodeling and new construction. Remarkably few architects are aware of this issue and need to be encouraged to use their creative talents to design bird-friendly structures. To eliminate strikes it is crucial to apply visual patterns that birds see well to the outside surface of windows such that the elements uniformly cover the entire pane and are separated by 5 cm if oriented in horizontal rows and 10 cm if oriented in vertical columns (Klem 1990, Klem and Saenger 2013). Greater spacing between pattern elements increases the risk of a strike. Yet-to-be-manufactured are proven patterns using UV reflecting and absorbing

elements that birds see and humans do not. The use of UV signals to prevent collisions is likely the most elegant solution because it transforms windows into barriers that birds avoid while retaining the unobstructed view humans enjoy. However, researchers are still seeking to determine what specific combinations of UV and other wavelengths are most visible to different species.

What additional research is needed? Document which species are killed and in what numbers under what conditions at specific locations using standard data collection methods and analyses. Improve measurement of population size and mortality attributable to windows to determine how bird-window collisions influence the population health of specific species and birds in general. Determine the species-specific probability of surviving and recovering from a window strike. Decide on a standard method to measure the effectiveness of existing and yet-to-be developed new solutions to transform windows into barriers that birds will see and avoid. Define what level of protection is bird safe. Educate leaders and architects to effectively educate citizens about these unintended and unwanted deaths and, in so doing, stimulate manufacturers to offer bird-safe products for existing and new construction. Encourage separate and cooperative studies in the life sciences, architecture, economics, law, psychology, sociology, and recreation to enhance the promise of saving more bird lives from windows.

mortality problem areas is difficult, and determining what causes elevated mortality rates is challenging. Perhaps most challenging is implementing management regulations and approaches that are widely acceptable to (and observed by) motorists. Potential mortality reduction approaches are largely tied to previously identified correlates of mortality rates. For example, bird mortality can likely be reduced by lowering speed limits, erecting signage that alerts drivers, identifying and removing the food or cover that attracts birds to roadsides, and using deflectors that cause birds to fly above traffic (Boves and Belthoff 2012, Bishop and Brogan 2013).

Notably, birds also collide with vehicles other than automobiles, primarily trains (Dorsey et al. 2015) and aircraft (Sodhi 2002); however, in general these other mortality sources are less well-studied and overall mortality amounts

are not understood. Because of the public safety risks posed by bird strikes with aircraft, researchers have investigated and implemented numerous approaches to reducing collisions, including monitoring for flocks of birds near airstrips (Sodhi 2002), managing habitat (Barras and Seamans 2002), reducing food sources such as invasive earthworms (Seamans et al. 2015), broadcasting loud sounds (Swaddle et al. 2016), and in some cases, killing birds that pose the greatest collision risk (e.g., gulls; Dolbeer et al. 1993).

Plastics and Seabirds

Increasingly, human refuse affects wildlife in detrimental ways. The plastics that have added tremendously to modern society in terms of convenience and technology have a short useful lifespan before being discarded. Only a tiny fraction of

Figure 26.23. Necropsy of a Northern Fulmar (*Fulmarus glacialis*) at the National Wildlife Health Lab in the United States revealed these numerous bits of plastic from its stomach. *Photo by C. Meteyer, US Geological Survey.*

all plastics are recycled globally. The rest are released to the environment or transported to landfills. Through inefficiencies in waste disposal and intentional dumping, oceans accumulate plastics worldwide. We also ship plastic pellets across oceans from industrial sources to manufacturers of plastic products, and these pellets often escape from their shipping containers (Provencher et al. 2014). Much of this plastic floats at the surface as brightly colored flecks that are highly attractive to foraging seabirds. Presumably, the plastics resemble natural foods in shape, color, or both.

Of multiple problems stemming from plastics ingestion, a likely contributor to avian mortality is obstruction of the gastrointestinal tract (fig. 26.23). Pierce et al. (2004) describe this effect from necropsies of a Northern Gannet (*Morus bassanus*) and Great Shearwater (*Puffinus gravis*) that died in a rehabilitation center in Massachusetts. Confirmed in the gannet and likely for the shearwater, a bright red, plastic bottle cap prevented the passage of food through the gizzard or proventriculus, and both birds ultimately died of starvation.

Estimating the number of birds sickened, weakened, or killed by the ingestion of marine plastics is difficult. Birds found dead on beaches or delivered to rehabilitation centers likely represent only a small fraction of all victims. In pe-

lagic zones, most victims likely become weak and fall prey to predators and scavengers (Pierce et al. 2004). A recent review (Provencher et al. 2014) attempted to quantify plastics ingestion among Canadian seabirds. They reported that 39 of Canada's 91 seabird species had been confirmed to ingest plastics or incorporate plastics into a nest. They cautioned, however, that small sample sizes suggest underreporting of this issue.

Disease Transmission Caused by Feeding Birds and Other Wildlife

Even with the best intentions, humans can harm birds, and the most common interaction we have with birds is through the provisioning of food. Davies et al. (2012) estimated that 64 percent of British households engaged in some form of wild bird feeding. In the United States, annual expenditure on bird seed and feeders exceeds $9 billion (USDI 2014). Robb et al. (2008) outlined multiple complex and variable effects of bird feeding. For some species, supplemental feeding increases overwinter survivorship and increases bird body condition in spring. These positive outcomes have likely helped increase population size and promote northward range expansion of species such as Northern Cardinal (*Cardinal cardinalis*) (Robb et al. 2008). However, bird feeders can also negatively affect

DR. DANIEL J. KLEM JR.
By Peter G. Saenger, Muhlenberg College

Dr. Daniel J. Klem Jr., the Sarkis Acopian Professor for Ornithology and Conservation Biology at Muhlenberg College in Allentown, Pennsylvania, pioneered the study of bird-building collisions and has been studying the issue since the 1970s. *Photo by P. Saenger.*

Dr. Daniel Klem Jr. is the Sarkis Acopian Professor for Ornithology and Conservation Biology at Muhlenberg College in Allentown, Pennsylvania. Dan grew up in the Wyoming Valley of northeastern Pennsylvania, where he developed a keen interest in wildlife and natural history from an early age. Upon graduation with a bachelor of arts degree at Wilkes University, he continued his education at Hofstra University, earning a master of arts degree for his work on the digestive tract of Yellow-rumped Warblers. In 1979, Dan earned a PhD in zoology/ornithology, ethology, and biostatistics from Southern Illinois University at Carbondale. Among many accolades bestowed on Dr. Klem, he earned a doctor of science from his undergraduate alma mater in 2000.

Dan became fascinated with human influences on mortality in birds and began the fieldwork on his doctoral thesis,

"Biology of Collisions between Birds and Windows," in 1974. He invested an arduous five years in gathering data on window collisions from museum collections, literature searches, private homes, college campuses, and other buildings. He designed novel experiments to tease out the data needed to evaluate the danger that glass posed to birds. When his work led him to the conclusion that, in the United States, up to one billion birds died from window collisions each year, he began a lifelong campaign to bring this to the attention of the world. Thanks in large part to Dan Klem's tireless attention to this issue, there is today a groundswell of research and conservation work that aims to reduce the number of birds killed by collisions with glass.

Today, multiple scientists and conservation organizations—and innumerable private citizens working to reduce collisions at their own homes—are the legacy of Dan's early work and lifelong commitment to find solutions to the problem. Much work is still needed to develop solutions that are both economically viable and esthetically acceptable. Dan's field studies have shown that ultraviolet signals can be used to warn birds of the dangers of glass, and he holds two patents on this promising avenue of prevention. Building from his years of research and experience, glass and window film manufacturers are developing materials to help birds recognize glass as the solid barrier that it is. Dan continues to be a champion for the birds, and with his energetic and intense interest in solving this problem, further advancements should be expected.

---

some species by facilitating population growth of competitors and predators. For example, corvids such as Blue Jay (*Cyanocitta cristata*) benefit from feeding stations, but Blue Jays also depredate eggs and nestlings of other birds.

A more direct negative consequence of feeding birds is that artificially inflated densities of birds around feeding stations exposes birds to disease transmission. The best studied of these diseases is *Mycoplasma gallisepticum* in North American finches, primarily the House Finch (*Haemorhous mexicanus*) (Altizer et al. 2004). This disease results in swelling and ulcerations around the eyes, causing blindness and increasing susceptibility to predation (fig. 26.24). In addition to diseases transmitted among individuals, birds can also become sickened and die from toxins in the food itself. The ubiquitous *Aspergillus*

fungi, for example, can colonize grain and bird seed to produce lethal aflatoxins as metabolic byproducts (Dale et al. 2015).

KEY POINTS

- Almost all anthropogenic threats to birds are recent, emerging since the nineteenth century or more recently (e.g., most energy sources, large structures, and poisons). These threats are novel enough that natural selection has not yet resulted in individual and population level adaptations to deal with them. Therefore, many anthropogenic threats are likely causing at least some additive mortality (i.e., mortality that cannot be compensated for by population processes such as density dependence).

Figure 26.24. Disease transmission can be facilitated at feeding stations as a consequence of high bird densities, and occasionally, unsanitary conditions. This House Finch (*Carpodacus mexicanus*) is nearly blinded from *Mycoplasma gallisepticum* infection and trapped in a feeder. *Photo by T. O'Connell.*

- The combined effect of these novel threats is likely contributing—along with other indirect effects like climate change and habitat loss—to the widespread avian population declines observed throughout the world.
- We still know relatively little about most anthropogenic threats to birds, including the amount of direct mortality caused, the types and magnitude of indirect and sub-lethal effects, the factors influencing spatial and temporal variation in effects, and the best conservation management approaches to reduce harmful impacts.
- Additional research of understudied and emerging impacts (e.g., neonicotinoid poisons, solar energy generation facilities, and fracking to extract gas/oil) is needed to fully understand how humans are negatively affecting birds and how we can act to reduce that effect.
- Aside from purposeful hunting, anthropogenic threats to birds are unintentional. Humans do not erect buildings and energy infrastructure or allow their cats to roam free with the explicit intent of killing birds. Therefore, social, economic, and political goodwill should exist toward taking steps that reduce negative anthropogenic impacts to birds.

KEY MANAGEMENT AND CONSERVATION IMPLICATIONS

- Anthropogenic threats affect virtually all bird species, and most bird species are affected by multiple anthropogenic threats. Many threats in this chapter primarily af-fect still-common species in danger of becoming uncommon without further conservation actions.
- Long-distance migratory species are disproportionately affected by many anthropogenic threats (especially lighted structures), and because population-level consequences are likely to vary widely depending on seasonal timing of mortality, a major conservation challenge will be to understand how mortality sources collectively affect migratory populations throughout the annual/migratory cycle.
- Anthropogenic threats to endangered, threatened, or otherwise rare species are the most likely to trigger conservation concern in most jurisdictions. However, conserving the vast majority of bird species requires continuing the paradigm shift toward keeping common species common. This shift should encompass the numerous anthropogenic threats we place in birds' way and follow a precautionary approach to management (see next point).
- Given challenges associated with linking anthropogenic cause to biological impact—as well as the scientific impossibility of "proving" impacts—a precautionary principle should prevail, whereby proof of population-level effects is not required to implement steps to ameliorate anthropogenic threats.
- Keeping common species common (1) saves monetary resources that would otherwise be spent on expensive recovery programs for rare species, (2) preserves spectacles of abundance (e.g., mass migrations, feeding frenzies, and major breeding concentrations), and (3) maintains ecosystem services, such as carcass removal, population control of rodent and insect pests, and aesthetic enjoyment of biological abundance and diversity (chapter 29).
- Government legislation, such as the US Migratory Bird Treaty Act, provides policy makers with an important trigger for preventing and managing the incidental take of a wide variety of bird species, including still-common species (chapter 27). This legislation will be a crucial component of future efforts to minimize anthropogenic impacts to birds, to prevent common species from declining to rarity or becoming extinct, and to ensure that avian populations are sustained at or above current levels.
- Cooperation between scientists and citizens will be crucial to successfully reducing harm to birds in the future (Loss et al. 2015a). Because of the dramatic, relatable, and highly visible nature of many anthropogenic impacts, especially those that occur in our neighborhoods, integrating public participation in science (i.e., **citizen science**) shows great promise for improving understanding

and successful management of these threats. Likewise, scientific outreach to the public is necessary to provide education about types of negative anthropogenic effects and steps that can be taken to reduce these effects.

DISCUSSION QUESTIONS

1. Are there other harms placed in birds' way by humans that are not covered in this chapter? What are the likely correlates influencing variation in the magnitude of these threats?

2. What do you think should be some key components of an outreach/education effort to inform the public about harms to birds?

3. What are commonalities that arise across multiple anthropogenic threats (for example, in terms of the bird group[s] affected, and potential management steps to reduce impacts).

4. Assuming you are tasked with reducing overall anthropogenic impacts to birds, what criteria would you use to decide which threats to prioritize resources and effort toward?

5. What factors would you consider when deciding whether additional development of a renewable energy source is worth the added negative effect to birds?

6. Why do you think the management of free-ranging domestic cats is so controversial? What are potential ways to overcome this controversy and implement steps that benefit both birds and cats?

7. How would you design a study to determine the amount of mortality caused by an understudied anthropogenic threat like bird entrapment in open pipes? What about for bird collisions with solar panels? What biases would you try to control for and not be able to control for in each case?

References

Altizer, S., W. M. Hochachka, and A. A. Dhondt (2004). Seasonal dynamics of mycoplasmal conjunctivitis in eastern North American house finches. Journal of Animal Ecology 73:309–322.

Alves, R. R. N., J. R. D. F. Lima, and H. D. F. P. Araujo (2013). The live bird trade in Brazil and its conservation implications: An overview. Bird Conservation International 23:53–65.

Annorbah, N. N. D., N. J. Collar, and S. J. Marsden (2016). Trade and habitat change virtually eliminate the Grey Parrot *Psittacus erithacus* from Ghana. Ibis 158:82–91.

Aronoff, A. (1949). The September migration tragedy. Linnaean News-Letter 3:2.

Atuo, F. A., T. J. O'Connell, and P. U. Abanyam (2015). An assessment of socio-economic drivers of avian body parts trade in West African rainforests. Biological Conservation 191:614–622.

Audubon California (2011). Nests to die for: Open pipes. Audubon California, Kern River Preserve Landowner Stewardship Program. http://kern.audubon.org/death_pipes.htm.

Avery, M. L. (1979). Review of avian mortality due to collisions with manmade structures. Bird Control Seminars Proceedings Paper 2. US Fish and Wildlife Service, Ann Arbor, Michigan.

APLIC (Avian Power Line Interaction Committee) (2006). Suggested practices for avian protection on power lines: The State of the art in 2006. APLIC and the California Energy Commission, Washington, DC.

APLIC (Avian Power Line Interaction Committee) (2012). Reducing avian collisions with power lines: The state of the art in 2012. Edison Electrical Institute/APLIC, Washington, DC.

Baines, D., and M. Andrew (2003). Marking of deer fences to reduce frequency of collisions by woodland grouse. Biological Conservation 110:169–176.

Balogh A. L., T. B. Ryder, and P. P. Marra (2011). Population demography of Gray Catbirds in the suburban matrix: Sources, sinks, and domestic cats. Journal of Ornithology 152:717–726.

Barras, S. C., and T. W. Seamans (2002). Habitat management approaches for reducing wildlife use of airfields. USDA National Wildlife Research Center, Staff Publications, Paper 463. http://digitalcommons.unl.edu/cgi/viewcontent.cgi?article=1458&context=icwdm_usdanwrc.

Barrientos, R., J. C. Alonso, C. Ponce, and C. Palacin (2011). Meta-analysis of the effectiveness of marked wire in reducing avian collisions with power lines. Conservation Biology 25:893–903.

Benitez-Lopez, A., R. Alkemade, and P. A. Verweij (2010). The impacts of roads and other infrastructure on mammal and bird populations: A meta-analysis. Biological Conservation 143:1307–1316.

Beston, J. A., J. E. Diffendorfer, and S. R. Loss (2015). Insufficient sampling to identify species affected by turbine collisions. Journal of Wildlife Management 79:513–517.

Beston, J. A., J. E. Diffendorfer, S. R. Loss, and D. H. Johnson (2016). Prioritizing avian species for their risk of population level consequences from wind energy development. PLoS ONE 11:e0150813.

Bevanger, K. (1995). Estimates and population consequences of tetraonid mortality caused by collisions with high tension power lines in Norway. Journal of Applied Ecology 32:745–753.

Beyer, W. N., J. Dalgarn, S. Dudding, J. B. French, R. Mateo, J. Miesner, L. Sileo, and J. Spann (2004). Zinc and lead poisoning in wild birds in the tri-state mining district (Oklahoma, Kansas, and Missouri). Archives of Environmental Contamination and Toxicology 48:108–117.

Bildstein, K. L. (2001). Raptors as vermin: A history of human attitudes towards Pennsylvania's birds of prey. Endangered Species Update 18:124–128.

BirdLife International (2008). Illegal trade in European songbirds for food. Presented as part of the BirdLife State of the World's Birds website. http://www.birdlife.org/datazone/sowb/casestudy/294.

BirdLife International (2015). European Red List of birds. Office for Official Publications of the European Communities, Luxembourg.

BirdLife International (2016). Species factsheet: *Siphonorhis americana*. http://www.birdlife.org/datazone/speciesfactsheet.php?id=2373.

Bishop, C. A., and J. M. Brogan (2013). Estimates of avian mortality attributed to vehicle collisions in Canada. Avian Conservation and Ecology 8:2.

Bjorge, R. R. (1987). Bird kill at an oil industry flare stack in northwest Alberta. Canadian Field-Naturalist 101:346–350.

Blancher, P. J. (2013). Estimated number of birds killed by house cats (*Felis catus*) in Canada. Avian Conservation and Ecology 8:3.

Bollinger, E. K., P. B. Bollinger, and T. A. Gavin (1990). Effects of hay-cropping on eastern populations of the Bobolink. Wildlife Society Bulletin 18:142–150.

Boves, T. J., and J. R. Belthoff (2012). Roadway mortality of Barn Owls in Idaho, USA. Journal of Wildlife Management 76:1381–1392.

Brochet, A.-L., W. Van Den Bossche, S. Jbour, P. K. Ndang'ang'a, V. R. Jones, W. A. L. I. Abdou, A. R. Al-Hmoud, et al. (2016). Preliminary assessment of the scope and scale of illegal killing and taking of birds in the Mediterranean. Bird Conservation International 26:1–28.

Bujoczek, M., M. Ciach, and R. Yosef (2011). Road-kills affect avian population quality. Biological Conservation 144:1036–1039.

Butt, N., H. L. Beyer, J. R. Bennett, D. Biggs, R. Maggini, M. Mills, A. R. Renwick, and H. P. Possingham (2013). Biodiversity risks from fossil fuel extraction. Science 342:425–426.

Calvert, A. M., C. A. Bishop, R. D. Elliot, E. A. Krebs, T. M. Kydd, C. S. Machtans, and G. J. Robertson (2013). A synthesis of human-related avian mortality in Canada. Avian Conservation and Ecology 8:11.

Carrete, M., J. A. Sanchez-Zapata, J. R. Benitez, M. Lobon, and J. A. Donazar (2009). Large scale risk-assessment of wind-farms on population viability of a globally endangered long-lived raptor. Biological Conservation 142:2954–2961.

CBC News (2013). 7,500 songbirds killed at Canaport gas plant in Saint John. http://www.cbc.ca/news/canada/new-brunswick/7-500-songbirds-killed-at-canaport-gas-plant-in-saint-john-1.1857615.

Cooper, J., N. Baccetti, E. J. Belda, J. J. Borg, D. Oro, C. Papaconstantinou, and A. Sanchez (2003). Seabird mortality from longline fishing in the Mediterranean Sea and Macaronesian waters: A review and a way forward 67:57–64.

Dahl, E. L., K. Bevanger, T. Nygård, E. Røskaft, and B. G. Stokke (2012). Reduced breeding success in White-tailed Eagles at Smøla windfarm, western Norway, is caused by mortality and displacement. Biological Conservation 145:79–85.

Dale, L., T. O'Connell, and R. D. Elmore (2015). Aflatoxins in wildlife feed: Know how to protect wildlife. Oklahoma Cooperative Extension Service, fact sheet NREM 9021. http://pods.dasnr.okstate.edu/docushare/dsweb/Get/Document-9926/NREM-9021.pdf.

Daut, E. F., D. J. Brightsmith, A. P. Mendoza, L. Puhakka, and M. J. Peterson (2015). Illegal domestic bird trade and the role of export quotas in Peru. Journal for Nature Conservation 27:44–53.

Davies, Z. G., R. A. Fuller, M. Dallimer, A. Loram, and K. J. Gaston (2012). Household factors influencing participation in bird feeding activity: A national scale analysis. PLoS ONE 7: e39692.

De Francisco, N., J. D. Ruiz Troya, and E. I. Agüera (2003). Lead and lead toxicity in domestic and free living birds. Avian Pathology 32:3–13.

Doherty, T. S., A. S. Glen, D. G. Nimmo, E. G. Ritchie, and C. R. Dickman (2016). Invasive predators and global biodiversity loss. Proceedings of the National Academy of Sciences 113:11261–11265.

Dolbeer, R. A., J. L. Belant, and J. L. Sillings (1993). Shooting gulls reduces strikes with aircraft at John F. Kennedy International Airport. Wildlife Society Bulletin 21:442–450.

Dorsey, B., M. Olsson, and L. J. Rew (2015). Ecological effects of railways on wildlife. *In* Handbook of road ecology, R. van der Ree, D. J. Smith, and C. Grilo, Editors. Hoboken, NJ: John Wiley.

Draycott, R. A. H., A. N. Hoodless, and R. B. Sage (2008). Effects of pheasant management on vegetation and birds in lowland woodlands. Journal of Applied Ecology 45:334–341.

EIA (U.S. Energy Information Administration) (2017). Wind turbines provide 8% of U.S. generating capacity, more than any other renewable source. https://www.eia.gov/todayinenergy/detail.php?id=31032.

Elston, D. A., L. Spezia, D. Baines, and S. M. Redpath (2014). Working with stakeholders to reduce conflict—modeling the impact of varying Hen Harrier *Circus cyaneus* densities on Red Grouse *Lagopus lagopus* populations. Journal of Applied Ecology 51:1236–1245.

Erickson, W. P., G. D. Johnson, and D. P. Young Jr. (2005). A summary and comparison of bird mortality from anthropogenic causes with an emphasis on collisions. USDA Forest Service General Technical Report PSW-GTR-191, pp. 1029–1042. http://www.wingpowerenergy.com/wp-content/uploads/2012/07/birdmortality.pdf.

Erickson, W. P., M. M. Wolfe, K. J. Bay, D. H. Johnson, and J. L. Gehring (2014). A comprehensive analysis of small-passerine fatalities from collision with turbines at wind energy facilities. PLoS ONE 9:e107491.

EU (European Union) (2009). Directive 2009/147/EC of the European Parliament and of the Council of 30 November 2009 on the conservation of wild birds.

Favero, M., C. E. Khatchikian, A. Arias, M. P. S. Rodriguez, G. Canete, and R. Mariano-Jelicich (2003). Estimates of seabird by-catch along the Patagonian Shelf by Argentine longline fishing vessels, 1999–2001. Bird Conservation International 13:273–281.

FAA (Federal Aviation Administration) (2007). Obstruction Marking and Lighting. AC 70/7460-1K. https://www.faa.gov/documentLibrary/media/Advisory_Circular/AC%2070%207460-1K.pdf.

FAA (Federal Aviation Administration) (2015). Obstruction Marking and Lighting. AC 70/7460-1L. https://www.faa.gov/documentLibrary/media/Advisory_Circular/AC_70_7460-1L_.pdf.

Field Museum (2008). Turning off building lights reduces bird window-kill by 83 percent: Field Museum scientists release data from two-year study. The Field Museum, Chicago, IL. http://www.eurekalert.org/pub_releases/200205/fm-tob050802.php.

Fisher, I. J., D. J. Pain, and V. G. Thomas (2006). A review of lead poisoning from ammunition sources in terrestrial birds. Biological Conservation 131:421–432.

Gehring, J., P. Kerlinger, and A. M. Manville II (2009). Communication towers, lights and birds: Successful methods of reducing the frequency of avian collisions. Ecological Applications 19:505–514.

Geiger F., J. F. Bengtsson, W. W. Berendse, M. Weisser, M. B. Emmerson, P. Morales, J. Ceryngier, et al. (2010) Persistent negative effects of pesticides on biodiversity and biological control potential on European farmland. Basic and Applied Ecology 11:97–105.

Gerhold, R. W., and D. A. Jessup (2013). Zoonotic diseases associated with free-roaming cats. Zoonoses and Public Health 60:189–195.

Grier, J. W. (1982). Ban of DDT and subsequent recovery of reproduction in bald eagles. Science 218:1232–1235.

Grüebler, M. U., H. Schuler, P. Horch, and R. Spaar (2012). The effectiveness of conservation measures to enhance nest survival in a meadow bird suffering from anthropogenic nest loss. Biological Conservation 146:197–203.

Haig, S. M., J. D'Elia, C. Eagles-Smith, J. M. Fair, J. Gervais, G. Herring, J. W. Rivers, and J. H. Schulz (2014). The persistent problem of lead poisoning in birds from ammunition and fishing tackle. The Condor 116:408–428.

Halliday, T. R. (1980). The extinction of the Passenger Pigeon *Ectopistes migratorius* and its relevance to contemporary conservation. Biological Conservation 17:157–162.

Hallmann, C. A., R. P. Foppen, C. A. van Turnhout, H. de Kroon, and E. Jongejans (2014). Declines in insectivorous birds are associated with high neonicotinoid concentrations. Nature 511:341–343.

Hathcock, C. D., and J. M. Fair (2014). Hazards to birds from open metal pipes. Western North American Naturalist 74:228–230.

Hedd, A., P. M. Regular, S. I. Wilhelm, J-F. Rail, B. Drolet, M. Fowler, C. Pekarik, and G. J. Robertson (2015). Characterization of seabird bycatch in eastern Canadian waters, 1998–2011, assessed from onboard fisheries observer data. Aquatic Conservation: Marine and Freshwater Ecosystems. doi:10.1002/aqc.2551.

Holdaway, R. N., and C. Jacomb (2000). Rapid extinction of the moas (Aves: Dinornithiformes): Model, test, and implications. Science 287:2250–2254.

Holloran, M. J., B. C. Fedy, and J. Dahlke (2015). Winter habitat use of Greater Sage-grouse relative to activity levels at natural gas well pads. Journal of Wildlife Management 79:630–640.

IEA (International Energy Agency) (2014). World energy outlook 2014. International Energy Agency, Paris.

IUCN (International Union for Conservation of Nature) (2000). IUCN Red List categories and criteria. IUCN, Gland Switzerland. http://jr.iucnredlist.org/documents/redlist_cats_crit_en.pdf.

Jenkins, A. R., J. J. Smallie, and M. Diamond (2010). Avian collisions with power lines: A global review of causes and mitigation with a South African perspective. Bird Conservation International 20:263–278.

Johnson, D. H., S. R. Loss, K. S. Smallwood, and W. P. Erickson (2016). Avian fatalities at wind energy facilities in North America: A comparison of recent approaches. Human-Wildlife Interactions 10:7–18.

Kagan, R. A., T. C. Viner, P. W. Trail, and E. O. Espinoza (2014). Avian mortality at solar energy facilities in southern California: A preliminary analysis. National Fish and Wildlife Forensics Laboratory. http://www.ourenergypolicy.org/avian-mortality-at-solar-energy-facilities-in-southern-california-a-preliminary-analysis/.

Kiesecker, J. M., J. S. Evans, J. Fargione, K. Doherty, K. R. Foresman, T. H. Kunz, D. Naugle, N. P. Nibbelink, and N. D. Niemuth (2011). Win-win for wind and wildlife: A vision to facilitate sustainable development. PLoS ONE 6:e17566.

Klem Jr., D. (1990). Collisions between birds and windows: Mortality and prevention. Journal of Field Ornithology 61:120–128.

Klem Jr., D., and P. G. Saenger (2013). Evaluating the effectiveness of select visual signals to prevent bird-window collisions. Wilson Journal of Ornithology 125:406–411.

Leopold, A. (1933). Game management. Charles Scribner's Sons, New York.

Lohr, C. A., and C. A. Lepczyk (2014). Desires and management preferences of stakeholders regarding feral cats in the Hawaiian Islands. Conservation Biology 28:392–403.

Longcore, T., C. Rich, P. Mineau, B. MacDonald, D. G. Bert, L. M. Sullivan, E. Mutrie, et al. (2012). An estimate of avian mortality at communication towers in the United States and Canada. PLoS ONE 7:1–17.

Longcore, T., C. Rich, P. Mineau, B. MacDonald, D. G. Bert, L. M. Sullivan, E. Mutrie, et al. (2013). Avian mortality at communication towers in the United States and Canada: Which species, how many, and where? Biological Conservation 158:410–419.

Loss, S. R., T. Will, and P. P. Marra (2013a). Estimates of bird collision mortality at wind facilities in the contiguous United States. Biological Conservation 168:201–209.

Loss, S. R., T. Will, and P. P. Marra (2013b). The impact of free-ranging domestic cats on wildlife of the United States. Nature Communications 4:1396.

Loss S. R., T. Will, S. S. Loss, and P. P. Marra (2014a). Bird-building collisions in the United States: Estimates of annual mortality and species vulnerability. The Condor 16:8–23.

Loss, S. R., T. Will, and P. P. Marra (2014b). Estimation of annual bird mortality from vehicle collisions on roads in the United States. Journal of Wildlife Management 78:763–771.

Loss, S. R., T. Will, and P. P. Marra (2014c). Refining estimates of bird collision and electrocution mortality at power lines in the United States. PLoS ONE 9:e101565

Loss, S. R., S. S. Loss, T. Will, and P. P. Marra (2015a). Linking place-based citizen science with large-scale conservation research: A case study of bird-building collisions and the role of professional scientists. Biological Conservation 184:439–445.

Loss, S. R., T. Will, and P. P. Marra (2015b). Direct mortality of birds from anthropogenic causes. Annual Review of Ecology, Evolution, and Systematics 46:99–120.

Loss, S. R., and P. P. Marra (2017). Population impacts of free-ranging domestic cats on mainland vertebrates. Frontiers in Ecology and the Environment 15:502–509.

Lovich, J. E., and J. R. Ennen (2011). Wildlife conservation and solar energy development in the desert Southwest, United States. BioScience 61:982–992.

Lowe, S., M. Browne, and S. Boudjelas (2000). 100 of the world's worst invasive alien species: A selection from the global invasive species database. Invasive Species Specialist Group, International Union for Conservation of Nature. http://rewilding.org/rewildit/images/IUCN-GISP.pdf.

Machtans, C. S., C. H. R. Wedeles, and E. M. Bayne (2013). A first estimate for Canada of the number of birds killed by colliding with buildings. Avian Conservation and Ecology 8:6.

Medina, F. M., E. Bonnaud, E. Vidal, B. R. Tershy, E. S. Zavaleta, C. J. Donlan, B. S. Keitt, M. Le Corre, S. V. Horwath, and M. Nogales (2011). A global review of the impacts of invasive cats on island endangered vertebrates. Global Change Biology 17:3503–3510.

Mineau, P. (2005a). Patterns of bird species abundance in relation to granular insecticide use in the Canadian prairies. Ecoscience 12:267–278.

Mineau, P. (2005b). Direct losses of birds to pesticides—Beginnings of a quantification. In Bird conservation implementation and integration in the Americas: Proceedings of the Third International Partners in Flight Conference 2002, C. J. Ralph and T. D. Rich, Editors. USDA Forest Service, GTR-PSW-191, Albany, CA.

Mineau, P., A. Baril, B. T. Collins, J. Duffe, G. Joerman, and R. Luttick (2001). Pesticide acute toxicity reference values for birds. Reviews of Environmental Contamination and Toxicology 170:13–74.

Mineau, P., and M. Whiteside (2013). Pesticide acute toxicity is a better correlate of US grassland bird declines than agricultural intensification. PLoS ONE 8:e57457.

Montevecchi, W. A., and D. A. Kirk (1996). Great Auk (Pinguinus impennis). In The birds of North America, no. 260, A. Poole, Editor. Academy of Natural Sciences, Philadelphia, PA/American Ornithologists' Union, Washington, DC.

Mumme, R. I., S. J. Schoech, G. E. Woolfenden, and J. W. Fitzpatrick (2000). Life and death in the fast lane: Demographic consequences of road mortality in the Florida Scrub-Jay. Conservation Biology 14:501–512.

National Audubon Society (2014). More than one million birds died during Deepwater Horizon disaster. https://www.audubon.org/news/more-one-million-birds-died-during-deepwater-horizon-disaster.

National Audubon Society (2016). Lights Out. https://www.audubon.org/conservation/project/lights-out.

National Shooting Sports Foundation (2016). Pittman-Robertson Excise Tax—Fast Facts. http://nssf.org/factsheets/PDF/PittmanRobertsonFacts.pdf.

Oldfield, T. E. E., R. J. Smith, S. R. Harrop, and N. Leader-Williams (2003). Field sports and conservation in the United Kingdom. Nature 423:531–533.

Palmer, M. A., E. S. Bernhardt, W. H. Schlesinger, K. N. Eshleman, E. Foufoula-Georgiou, M. S. Hendryx, A. D. Lemly, et al. (2010). Mountaintop mining consequences. Science 327:148–149.

Patterson Jr., J. (2012). Evaluation of new obstruction lighting techniques to reduce avian fatalities. Technical Note DOT/FAA/TC-TN12/9. https://trid.trb.org/view.aspx?id=1222903.

Peresani, M., I. Fiore, M. Gala, M. Romandini, and A. Tagliacozzo (2011). Late Neanderthals and the intentional removal of feathers as evidenced from bird bone taphonomy at Fumane Cave 44 ky B.P., Italy. Proceedings of the National Academy of Sciences 108: 3888–3893.

Péron, G. (2013). Compensation and additivity of anthropogenic mortality: Life-history effects and review of methods. Journal of Animal Ecology 82:408–417.

Piatt, J. F., C. J. Lensink, W. Butler, M. Kendziorek, and D. R. Nysewander (1990). Immediate impact of the Exxon Valdez oil spill on marine birds. The Auk 107:387–397.

Pierce, K. E., R. J. Harris, L. S. Larned, and M. A. Pokras (2004). Obstruction and starvation associated with plastic ingestion in a Northern Gannet Morus bassanus and a Greater Shearwater Puffinus gravis. Marine Ornithology 32:187–189.

Pokras, M., M. Kneeland, A. Ludi, E. Golden, A. Major, R. Miconi, and R. H. Poppenga (2009). Lead objects ingested by Common Loons in New England. Northeastern Naturalist 16:177–182.

Provencher, J. F., A. L. Bond, and M. L. Mallory (2014). Marine birds and plastic debris in Canada: A national synthesis and a way forward. Environmental Reviews 23:1–13.

Raftovich, R. V., K. Wilkins, S. S. Williams, and H. L. Spriggs (2012). Migratory bird hunting activity and harvest for the 2010 and 2011 hunting seasons. US Fish and Wildlife Service Publications, Paper 360. http://digitalcommons.unl.edu/cgi/viewcontent.cgi?article=1359&context=usfwspubs.

Ramirez Jr., P. (2013). Migratory bird mortality in oil and gas facilities in Colorado, Kansas, Montana, Nebraska, North Dakota, South Dakota, Utah, and Wyoming. Department of the Interior, US Fish and Wildlife Service, Environmental Contaminants Program Report no. R6/726C/13. https://www.fws.gov/mountain-prairie/contaminants/papers/R6726C13.pdf.

Richards, N. (2011). Carbofuran and wildlife poisoning: Global perspectives and forensic approaches. Wiley, Hoboken, NJ.

Robb, G. N., R. A. McDonald, D. E. Chamberlain, and S. Bearhop (2008). Food for thought: Supplementary feeding as a driver of ecological change in avian populations. Frontiers in Ecology and the Environment 6:476–484.

Robertson, G., C. Moreno, J. A. Arata, S. G. Candy, K. Lawton, J. Valencia, B. Wienecke, R. Kirkwood, P. Taylor, and C. G. Suazo (2014). Black-browed Albatross numbers in Chile increase in response to reduced mortality in fisheries. Biological Conservation 169:319–333.

Roux, K. E., and P. P. Marra (2007). The presence and impact of environmental lead in passerine birds along an urban to rural land use gradient. Archives of Environmental Contamination and Toxicology 53:261–268.

Saikku, M. (1990). The extinction of the Carolina Parakeet. Environmental History Review 14:1–18.

Schaub, M., and R. Pradel (2004). Assessing the relative importance of different sources of mortality from recoveries of marked animals. Ecology 85:930–938.

Scheuhammer, A. M., and S. L. Norris (1995). A review of the environmental impacts of lead shotshell ammunition and lead fishing weights in Canada. Occasional Paper 88. Ottawa, Ontario: Canadian Wildlife Service.

Schulz, J. H., T. W. Bonnot, J. J. Millspaugh, and T. W. Mong (2013). Harvest and crippling rates of mourning doves in Missouri. Wildlife Society Bulletin 37:287–292.

Seamans, T. W., B. F. Blackwell, G. E. Bernhardt, and D. A. Potter (2015). Assessing chemical control of earthworms at airports. Wildlife Society Bulletin 39:434–442.

Shaffer, J. A., and D. A. Buhl (2015). Effects of wind-energy facilities on breeding grassland bird distributions. Conservation Biology 30:59–71.

Sodhi, N. S. (2002). Competition in the air: Birds versus aircraft. The Auk 119:587–595.

Souther, S., M. W. Tingley, V. D. Popescu, D. T. Hayman, M. E. Ryan, T. Graves, B. Hartl, and K. Terrell (2014). Biotic impacts of energy development from shale: Research priorities and knowledge gaps. Frontiers in Ecology and the Environment 12:330–338.

Stevens, T. K., A. M. Hale, K. B. Karsten, and V. J. Bennett (2013). An analysis of displacement from wind turbines in a wintering grassland bird community. Biodiversity and Conservation 22:1755–1767.

Su, S., P. Cassey, and T. M. Blackburn (2015). The wildlife pet trade as a driver of introduction and establishment in alien birds in Taiwan. Biological Invasions 18:215–229.

Swaddle, J., D. Moseley, M. Hinders, and E. P. Smith (2016). A sonic net excludes birds from an airfield: Implications for reducing bird strike and crop losses. Ecological Applications 26:339–345.

Tews, J. D., G. Bert, and P. Mineau (2013). Estimated mortality of selected migratory bird species from mowing and other mechanical operations in Canadian agriculture. Avian Conservation and Ecology 8:8.

Thomas, E. H., M. C. Brittingham, and S. H. Stoleson (2014). Conventional oil and gas development alters forest songbird communities. Journal of Wildlife Management 78:293–306.

Thomas, G. J., C. M. Perrins, and J. Sears (1985). Lead poisoning and waterfowl. Angling and Wildlife in Fresh Waters, ITE symposium no. 19. https://core.ac.uk/download/pdf/62149.pdf#page=7.

Thompson, D. B. A., A. J. MacDonald, J. H. Marsden, and C. A. Galbraith (1995). Upland heather moorland in Great Britain: A review of international importance, vegetation change and some objectives for nature conservation. Biological Conservation 71:163–178.

Thompson, S. J., D. H. Johnson, N. D. Niemuth, and C. A. Ribic (2015). Avoidance of unconventional oil wells and roads exacerbates habitat loss for grassland birds in the North American Great Plains. Biological Conservation 192:82–90.

Trail, P. W. (2006). Avian mortality at oil pits in the United States: A review of the problem and efforts for its solution. Environmental Management 38:532–544.

USACE (US Army Corps of Engineers) (2015). Double-crested Cormorant management plan to reduce predation of juvenile salmonids in the Columbia River Estuary. Final Environmental Impact Statement. US Army Corps of Engineers, Portland District, Portland, OR.

USDOE (US Department of Energy) (2008). 20 percent wind energy by 2030: Increasing wind energy's contribution to U.S. electricity supply. US Department of Energy, Office of Scientific and Technical Information, Oak Ridge, Tennessee.

USDI (US Department of the Interior), US Fish and Wildlife Service, and US Department of Commerce, US Census Bureau (2014). 2011 National survey of fishing, hunting, and wildlife-associated recreation. https://www.census.gov/prod/2012pubs/fhw11-nat.pdf.

USDOT (US Department of Transportation) (2012). Federal Highway Administration, Office of Highway Policy Information. Highway statistics 2011. http://www.fhwa.dot.gov/policyinformation/statistics/2011/hm10.cfm.

USEIA (US Energy Information Administration) (2014). Annual energy outlook 2014 with projections to 2040. US Energy Information Administration, Office of Integrated and International Energy Analysis, US Department of Energy, Washington, DC.

USEPA (US Environmental Protection Agency) (2015). Nonroad engines, equipment, and vehicles—aircraft. http://www3.epa.gov/otaq/aviation.htm.

USFWS (US Fish and Wildlife Service) (2002). Migratory bird mortality fact sheet. US Department of the Interior, Fish and Wildlife Service, Arlington, VA.

USFWS (US Fish and Wildlife Service) (2008). Region six environmental contaminants at pit lakes. https://www.fws.gov/mountain-prairie/contaminants/.

USFWS (US Fish and Wildlife Service) (2015a). Waterfowl population status, 2015. US Department of the Interior, Washington, DC.

USFWS (US Fish and Wildlife Service) (2015b). 2015–2016 Migratory bird hunting and conservation stamp. http://www.fws.gov/birds/get-involved/duck-stamp.php.

USFWS (US Fish and Wildlife Service) (2015c). National Wildlife Refuge System: How long has the federal government been setting aside lands for wildlife? http://www.fws.gov/refuges/about/acquisition.html.

US Green Building Council (2015). LEED credit library. http://www.usgbc.org/credit.

VanderWerf, E. A. (2012). Evolution of nesting height in an endangered Hawaiian forest bird in response to a non-native predator. Conservation Biology 26:905–911.

Van der Zande, A. N., W. J. Ter Keurs, and W. J. Van der Weijden (1980). The impact of roads on the densities of four bird species in an open field habitat—evidence of a long-distance effect. Biological Conservation 18:299–321.

van Heezik, Y., A. Smyth, A. Adams, and J. Gordon (2010). Do domestic cats impose an unsustainable harvest on urban bird populations? Biological Conservation 143:121–130.

VanWormer, E., P. A. Conrad, M. A. Miller, A. C. Melli, T. E. Carpenter, and J. A. Mazet (2013). *Toxoplasma gondii*, source to sea: Higher contribution of domestic felids to terrestrial parasite loading despite lower infection prevalence. EcoHealth 10:277–289.

Vickery, J. A., S. R. Ewing, K. W. Smith, D. J. Pain, F. Barlein, J. Skorpilová, and R. D. Gregory (2014). The decline of Afro-Palearctic migrants and an assessment of potential causes. Ibis 156:1–22.

Walker, C. H. (2003). Neurotoxic pesticides and behavioural effects upon birds. Ecotoxicology 12:307–316.

Weidensaul, S. (1999). Living on the wind: Across the hemisphere with migratory birds. North Point Press, New York.

Whytock, R. C., R. Buij, M. Z. Virani, and B. J. Morgan (2014). Do large birds experience previously undetected levels of hunting pressure in the forests of Central and West Africa? Oryx: 1–8. doi:10.1017/S0030605314000064.

Wiles, G. J., J. Bart, R. E. Beck, and C. F. Aguon (2003). Impacts of the brown tree snake: Patterns of decline and species persistence in Guam's avifauna. Conservation Biology 17:1350–1360.

Wolfe, D. H., M. A. Patten, and S. K. Sherrod (2009). Reducing grouse collision mortality by marking fences (Oklahoma). Ecological Restoration 27:141–143.

Young, L. C., E. A. VanderWerf, M. T. Lohr, C. J. Miller, A. J. Titmus, D. Peters, and L. Wilson (2013). Multi-species predator eradication within a predator-proof fence at Ka'ena Point, Hawai'i. Biological Invasions 15:2627–2638.

Zydelis, A., C. Small, and G. French (2013). The incidental catch of seabirds in gillnet fisheries: A global review. Biological Conservation 162:76–88.

CHAPTER 27

Conservation Tools and Strategies

Dylan C. Kesler, Elisabeth B. Webb, and Jeff R. Walters

Biological diversity, or **biodiversity**, encompasses both the range of species found on earth and variations among individuals (United Nations 1992), and in previous chapters we encountered the tremendous biodiversity present in class Aves. Chapter 25 reviewed losses of avian diversity occurring through extinctions, and chapter 26 described impending threats to birds and the potential for additional losses if swift conservation actions are not taken. Indeed, the stakes are high, because the biodiversity we encounter on Earth today is the result of billions of years of evolution, and once it is lost, it is gone forever.

Unfortunately, threats to biodiversity are multifaceted and complicated, and there is no single conservation solution. Avian conservation has taken many forms over the years, ranging from the enactment of broad protective legislation to restoring thousands of hectares of habitats to establishing captive breeding programs for some of the very last individuals of a species. Although the methods and approaches differ among situations, the end goal of these efforts is singular: to reduce the loss of biodiversity that occurs as a result of human impacts to the environment.

Birds and bird populations have been central to the development and implementation of numerous conservation tools and strategies. Strong focus on the conservation of avifauna is fortunate for birds, and it is likely the product of substantial public interest. As readers of this book have surmised, birds are phenomenally captivating; they are charismatic, colorful, noticeable, and ubiquitous throughout the world. These same characteristics have driven interest, and they underpinned the conservation concerns of the general public, scientists, and practitioners that will be further discussed in chapter 29. Indeed, declines in bird populations, and a few notable extinctions, garnered public attention at the beginning of the twentieth century—a time that many consider to be the dawn of the modern conservation era. Subsequently,

those situations were among the first addressed by foundational conservation legislation, and they motivated development of sophisticated and effective conservation tools and strategies that we explore in this chapter.

Birds benefit from the public and private attention that they generate; however, the same characteristics that make people marvel may also make birds challenging to manage and thereby render them vulnerable to extinction. In preceding chapters we reviewed the range of threats faced by birds, and it is evident that the gregarious, highly mobile, and wide-ranging characteristics exhibited by birds often put them in harm's way. In particular, migratory birds sometimes require access to habitats and resources that literally span the globe. For example, designing conservation strategies that address the needs of a Bar-tailed Godwit (*Limosa lapponica*) that migrates annually from the southern islands of New Zealand and Australia to the northern slopes of Siberia and Alaska and then back again requires conservationists to consider many factors encountered during the godwits' international transit (Battley et al. 2012). Like other migratory birds, the Bar-tailed Godwit requires secure nesting habitats, accessible and abundant foraging resources to fuel its migration across the Pacific Ocean, stopover habitats that are safe from predators, and additional foraging resources to support it on the wintering grounds. If any one of these resources is disturbed or unavailable, godwits may not survive through to the next season of the annual journey (fig. 27.1).

Similarly, we have learned that birds often are perched at the top of food chains, where they obtain the energy-dense nutrition needed to power their mobile, endothermic life histories. But there are challenges associated with being so high on food chains, because factors affecting species at lower levels can cause ripples that affect each successive consumer group and ultimately the top-ranking birds. For example, the Short-tailed Albatross (*Phoebastria albatrus*) is a large seabird

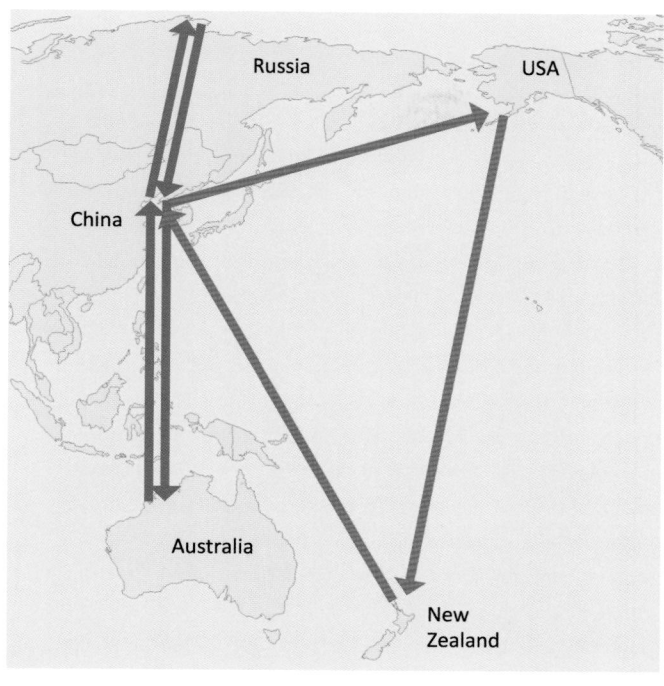

Figure 27.1. Map of generalized migratory pathways used by the Bar-tailed Godwit (*Limosa lapponica*). *Adapted from Battley et al. 2012.* The birds transit the entirety of the Pacific Ocean annually, when migrating from the southern islands of Australia and New Zealand to the most northern slopes of Siberia and Alaska. As such, conservation for the birds requires international coordination.

that forages on fish, squid, and other sea creatures, which are also elevated in the food chain because they consume other fish and zooplankton (Conners et al. 2015). Further, zooplankton consume phytoplankton. Although albatross benefit from the elevated energy density of the higher-order marine organisms they consume, the birds also are vulnerable to any number of factors that affect phytoplankton, zooplankton, squid, and fish. Survival of Short-tailed Albatross can be affected by something as seemingly remote to their existence as changes in the amount of sun entering the ocean. Changes in sunlight can affect the phytoplankton that act as primary producers, which in turn affects zooplankton, and eventually also the squid and fish that the birds eat. And for the same reason, developing tools and strategies to maintain healthy populations of the many birds that are so high on the food chain requires broadly considering the environments and other taxa on which they depend.

IDENTIFYING AND LISTING IMPERILED SPECIES

Detecting when a population is imperiled is an obvious initial step taken by conservation biologists intent on pre-

venting losses of avian diversity. Imperiled populations include those that may currently be threatened with extinction and those likely to be threatened in the future. Many state, national, and international organizations invest in what might seem to be a mundane task of developing lists of imperiled populations. However, the listing process is essential for identifying which species should receive limited conservation resources. Listing is challenging because it requires information about many organisms, something that is especially difficult to obtain for rare and elusive species. The International Union for Conservation of Nature (IUCN) attempts to evaluate the conservation status of every plant and animal species in the world every few years, with the goal of providing an international perspective on imperilment and risks of extinction (IUCN 2015). This monumental task involves conducting surveys, amassing and evaluating published literature, and communicating directly with biologists, conservation practitioners, and members of the general public about every species. As stated above, however, birds have a worldwide fan club of citizenry, conservationists, and scientists, and thus the database on birds is among the best available for any taxonomic group.

After data are gathered, the IUCN, and partner organizations like Birdlife International, evaluate the status of every species using the same set of criteria, which identify (1) whether a population is extremely small; (2) whether a population is in decline; (3) the size of the range in which a species can be found; and (4) whether population models indicate a high risk of extinction in coming years (IUCN 2001). Depending on the characteristics of each population, the organization assigns a level of imperilment that ranges from **Least Concern** to **Critically Endangered** and **Extinct in the Wild**. After the IUCN evaluates each species, it posts results of the listings in the Red List of Threatened and Endangered Species, which is publicly available on the Internet, with hopes that it will be a useful tool for conservation planning throughout the world (see the box on page 832; IUCN 2015).

National and local governments can establish policies and invest resources in on-the-ground conservation, which is something that IUCN cannot do under its international mission. National and state conservation policies, too, often begin with lists of imperiled species. For example, Canada develops a list of imperiled species under the Species and Recovery Act, which incorporates imperiled organisms that are assessed on a case-by-case basis. Similarly, the United States evaluates the status of species and lists those that are threatened or endangered with extinction under the Endangered Species Act of 1973 (fig. 27.2). New Zealand and Australia also maintain lists of imperiled species under the New Zealand Threat Classification System and the Environment Protec-

CRITICALLY ENDANGERED SPECIES: TUAMOTU KINGFISHER

The Tuamotu Kingfisher (*Todiramphus gambierii*) is an example of a bird that has been listed by IUCN as being "critically endangered," which means there is a high chance that the bird may go extinct without substantial conservation management. The very small population of 125–250 blue and white kingfishers persists only on the small atoll island of Niau, in the Tuamotu Archipelago of the South Pacific Ocean, and is likely threatened by introduced predatory mammals such as rats (*Rattus* spp.) and cats (*Felis cattus*), which did

not exist on the island until recent centuries when people arrived. After a series of studies that helped to identify where the birds persist and which factors potentially threaten them, conservationists from Niau Island, French Polynesia, the Ornithological Society of French Polynesian, and UNESCO Biosphere Reserve developed plans to help prevent the bird's extinction. Much of Niau Island is used for coconut production, and many of the remaining pairs of Tuamotu Kingfishers nest in trees within the coconut plantations. Thus, the conservation work involved outreach programs and cooperative work with the coconut farmers. Local conservation authorities also have considered the possibility of capturing a number of Tuamotu Kingfisher pairs and moving them to another island where a second population might be established (Kesler et al. 2012). The likelihood of extinction from localized catastrophic events like typhoons or tsunamis can be greatly reduced when there are multiple populations and when those populations are in disparate locations. However, such a "conservation introduction" is also controversial because there is the possibility that an introduced population of kingfishers might cause harm to another species at the introduction location. In order to avoid a situation wherein the kingfishers are released onto an island and cause harm to native populations, a number of scientific investigations are planned for the birds.

Tuamotu Kingfisher (*Todiramphus gambierii*). *Photo courtesy of Ryan Sarsfield.*

tion and Biodiversity Conservation Act, respectively. These governmental lists differ substantially from the IUCN Red List of Threatened and Endangered Species, because they are formulated by local authorities focused on the status of species in their jurisdiction. National and state governments can legislate legal protections for endangered populations and prescribe punishment for those who cause harm to imperiled species, as well as implement recovery planning and designate protections for habitats necessary for species persistence.

The Declining Population Paradigm

A substantial amount of effort is invested in ensuring that bird populations are never put at risk of extinction in the first place. Hunting regulations are established to reduce chances of overharvest, laws are enacted to regulate greater impacts to the environment from human activities, and lands and habitats are set aside and managed as national parks, reserves,

and refuges. Nonetheless, many bird populations have slipped into the declining trajectories that indicate that extinction may loom. Once a species is recognized as being in need of conservation attention, the next step often is to identify factors threatening the population, and to then use conservation tools and strategies appropriate to those factors to prevent further population declines and subsequent extinctions.

For many imperiled species, downward population trajectories can be reversed if the factors that created population declines in the first place are identified and addressed. In a noteworthy paper published in the *Journal of Animal Ecology* (1994), Graeme Caughley characterized these species as being in a **declining population paradigm** and identified a number of conservation tools appropriate to that condition. In general, species in the declining population paradigm are not on the very brink of extinction—they often number in the thousands of individuals and have a reasonable chance of recovering if the external factors causing population de-

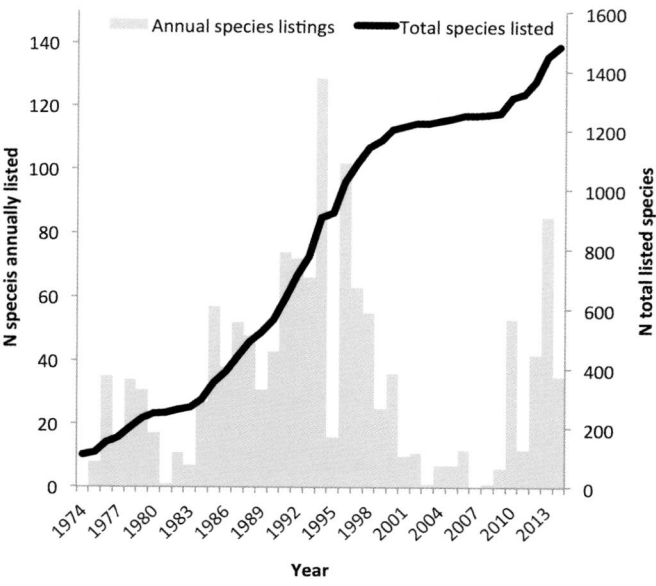

Figure 27.2. The history of species listings under the Endangered Species Act (United States) since 1973. The total number of species receiving conservation protection has increased since the inception of conservation legislation in the United States (see Puckett et al. 2016).

clines can be identified and addressed. Populations of species fitting the declining population paradigm may be shrinking because the environments and habitats in which they live are being reduced or degraded (carrying capacity issues), or because changes in their environments and habitats affect components of their life history such as survival, reproduction, and dispersal (vital rates issues). Graeme Caughley (see the box on page 834) also identified a **small population paradigm,** characterized by species on the very brink of extinction that are affected by a suite of additional factors, which we discuss later in this chapter.

Noted ecologists and conservation biologists have lumped declining population paradigm threats into a few useful categories. For example, Jared Diamond described what was called an "evil quartet," which included **habitat degradation and destruction, overharvest or overkill, invasive species,** and **synergies of extinctions** (Diamond 1984). Edward O. Wilson similarly presented a version of factors impinging on endangered species that was dubbed HIPPO because it identified **habitat destruction, invasive species, pollution, human overpopulation,** and **overharvest** as the primary causes of extinction (Wilson 2002). These categories become the underpinnings of the conservation tools used to help species recover from impending extinctions. Once the causes of imperilment are identified for a particular species, conservation practitioners can employ targeted tools that address the conservation issues at hand. In the following section, we dis-

cuss conservation tools that can be used to address species in the declining population paradigm.

Conservation Tools Used to Address Habitats

In recent years, the discipline of "landscape ecology" has emerged to provide conceptual representations of how patterns of habitats, landscapes, and regions change across space and time. Landscape ecologists have helped conservationists recognize that habitats can be affected by **loss, degradation,** and **fragmentation,** and all three have the potential to severely impact bird populations.

Habitat loss occurs when the resources and environmental conditions necessary for species persistence within a region are converted to other forms that provide fewer benefits to wild populations therein. Habitat loss results in reduction of **carrying capacity,** or the total achievable population size that can be sustained within a given area. For example, many of the grassland, savanna, and sage-steppe communities of North America have been converted to agricultural croplands or grazing lands for stock animals. These altered landscapes provide an abundance of food for humans, but they do not closely represent the grassland communities that historically characterized the region. Not surprisingly, a host of birds that rely on native grassland systems are now threatened with local disappearances and extinctions, such as Lesser Prairie Chickens (*Tympanuchus pallidicinctus*), Greater Sage Grouse (*Centrocercus urophasianus*), Sprague's Pipit (*Anthus spragueii*), and Chestnut-collared Longspurs (*Calcarius ornatus*) (Jones 2010, Sandercock et al. 2011).

The losses of key habitat types and the associated natural communities are among the most common threats to imperiled birds. Fortunately, habitat losses also can be some of the most easily remedied threats, at least in principle, through **habitat conservation** and management. Habitat conservation, the most obvious of conservation tools, can be applied to prevent further habitat loss when high-quality existing habitats are protected as refuges, critical habitats, and national parks.

In many cases, habitats are not lost altogether but merely degraded. Habitat loss and habitat degradation represent a continuum along a gradient of potential habitat use for a particular species, with the potential for lost or degraded habitat to transition to a form more likely to promote species persistence. **Degraded habitats** can sometimes be restored, and restoration takes a number of forms that reflect the many ways in which habitats may be degraded. In many cases, habitats are degraded through losses of key resources. For example, old-growth trees have been removed from forests around the world, and they are vital for nesting by species such as Red-cockaded Woodpeckers (*Leuconotopicus borealis*)

DR. GRAEME CAUGHLEY

Graeme Caughley. *Image courtesy of the Australian Academy of Science.*

Dr. Graeme Caughley (1937–1994), a native of New Zealand, worked in Australia as well as his native country. Dr. Caughley considered the tools and strategies used by conservation biologists and identified where and how conservationists might work to develop more effective tools and techniques

for preventing extinction. He summarized his concepts in a 1994 paper titled "Directions in Conservation Biology," which was published in the journal *Animal Ecology.* In his paper, Dr. Caughley succinctly ordered and described the range and types of conservation strategies and tools used by conservation biologists. He identified that, at the time, conservation biologist were working in two distinct subdisciplines, which he titled the "small population paradigm" and the "declining population paradigm." In general, these two paradigms addressed species that numbered in the tens or hundreds and were very near extinction and those in larger populations that were declining because of a list of external threats respectively. Importantly, the publication also identified areas wherein additional attention was needed in order to improve the effectiveness of conservation biology, and Dr. Caughley described ways that conservation biologists could connect and integrate their works. Graeme Caughley's opinions sparked additional debate and publications that subsequently led to broader considerations about conservation tools, strategies, and approaches. Dr. Caughley died within days of publication of his influential paper in 1994, although other conservation works, including a conservation biology textbook, were released posthumously.

and Marbled Murrelets (*Brachyramphus marmoratus*) in North America, and large parrot species in Australia (Blakesley et al. 1992, Bennett et al. 1993, Ripple et al. 2003).

Habitat degradation often also takes the form of reduced biodiversity that can restrict food availability for some species, or altered habitat structure that can affect foraging niches available to birds. For example, altered hydrology has degraded habitats in the Florida Everglades, negatively affecting the life cycle of the apple snail (*Pomacea* spp.) on which Snail Kites (*Rostrhamus sociabilis*) feed almost exclusively (fig. 27.3; Takekawa and Beissinger 1989). The resulting reduction in the distribution and abundance of the snail now imperils the kite population. However, restoring hydrology and maintaining historic water levels in Everglades wetlands has the potential to reverse habitat degradation by restoring water quality and the vegetative communities required to sustain apple snails and increase Snail Kite populations (Darby et al. 2015).

Conservationists may be able to employ any number of habitat management tools to restore degraded habitats. Often, habitats can be restored by reestablishing natural pro-

cesses that were eliminated or restricted by anthropogenic development. Over the course of the last century, people have prevented or reduced the occurrence of wildfire and have built dams, levees, and other structures to prevent rivers from flooding. Without these **disturbance processes**, the associated natural communities transition to later successional stages. Woody plants that are otherwise prevented from growing by fires can overtake grasslands, and sandy shores and water-scoured floodplains can become overgrown with vegetation when flooding is stopped. Birds that forage or nest in grasslands and savannas, or on sandy riverbanks, may therefore be negatively affected by changes to the vegetative community that occur when disturbance processes are disrupted.

Conservation practitioners have worked to restore natural disturbance processes in recent decades. They have used prescribed fire and controlled burning, and they have opened dams and restored natural flow regimes on streams and floodplains. In some cases, as with the endangered Yuma Clapper Rail (*Rallus longirostris yumanensis*), prescribed burning has been used as a surrogate disturbance process when

Figure 27.3. The Snail Kite (*Rostrhamus sociabilis*) is an extreme food specialist, with the habit of foraging almost exclusively on native apple snails (*Pomacea paludosa*) and a related introduced snail (*P. maculata*) in the wetland habitats of Florida, Cuba, and Mexico. Changes to hydrological flow regimes and vegetation in Florida wetlands caused population declines that resulted in greatly reduced kite populations and a subsequent endangered species act listing of Endangered. *Photo courtesy of Will Randall.*

controlled flooding was not possible (Conway et al. 2010). Similarly, for example, the Cape Sable Seaside Sparrow (*Ammodramus maritimus mirabilis*) inhabits seasonally flooded grassland habitats in the Florida Everglades. The habitats have been heavily altered by water control projects constructed to redirect surface waters and reduce flooding. Recently, a multibillion dollar restoration effort was undertaken to restore the historic quality, quantity, timing, and distribution of water over a vast area of south Florida, recreating the sheet flow characterized as the "river of grass" (National Research Council 2014). The intention in recreating historic hydrology is to restore habitats in order to provide for the needs of the sparrow, the Snail Kite, and the many other species from the region that have been affected by habitat degradation. Similarly, controlled flooding and simulated sandbar creation along the banks of rivers has created habitats for Least Terns (*Sternula antillarum*), Piping Plovers (*Charadrius melodus*), and other birds that nest on sandy substrates in the midcontinent of North America (Sherfy et al. 2009, 2011).

Another example of the effects of habitat loss and degradation on imperiled birds is found in the northwestern portions of North America, where the coniferous forests have been used for timber production and harvest over the course of the last century, resulting in the reduction of the old-growth stands of trees that are a dominant component of the ecosystem on which many birds depend. In the 1990s, the Northern Spotted Owl (*Strix occidentalis caurina*) and the Marbled Murrelet became central to the debate about timber industry impacts to the region when environmentalists concerned about biodiversity preservation clashed with others who were invested in timber production (fig. 27.4). The conflict brought about one of the greatest and most comprehensive environmental management plans ever enacted. The Northwest Forest Plan was a broad plan that protected habitats by addressing timber management and ecological restoration across nearly 10 million hectares. The enactment of the plan engaged a number of government agencies, indigenous tribes, and private organizations, which have since cooperatively managed the northwestern forests for combined ecological health and economic purposes (Tuchmann et al. 1996).

For some imperiled birds, conservation practitioners employ a series of tools to address factors restricting population growth. The Red-cockaded Woodpecker exemplifies just such a species, in this case one that suffered from habitat loss and multiple forms of habitat degradation (fig. 27.5). The Red-cockaded Woodpecker is endemic to the pine savannas that once covered much of the southeastern United States. They are cooperative breeders (see chapter 17) that live in family groups, and they have the unique natural history of excavating cavities for roosting and nesting in living pine trees, especially longleaf pine (*Pinus palustris*). Like the Spotted Owl, they are dependent on old-growth forests, as only old trees have sufficient heartwood to house a cavity. The once vast longleaf pine forests have been reduced to less than 3 percent of their original extent as a consequence of logging, clearing for agriculture, and other development—97 percent of the original habitat has been lost and the remaining remnants are fragmented across the southeastern portion of the United States. Most of what remains was degraded from the perspective of the Red-cockaded Woodpecker because of the absence of old trees. As a result, this once-abundant, continuously distributed species has been reduced to a small number of isolated populations in the fragments of remaining habitat (Conner et al. 2001).

The Red-cockaded Woodpecker also has been greatly affected by another form of habitat degradation resulting from changes to the disturbance regime of the longleaf pine ecosystem on which they depend. Historically, fires that were ignited primarily by lightning strikes maintained an ecosystem condition characterized by a rich, highly diverse ground cover dominated by grasses and herbaceous plants, a sparse

Figure 27.4. As forests in Canada have been fragmented in recent decades by timber harvest, interior forest-dwelling Northern Spotted Owls *A* have been put in more frequent contact with Barred Owls *C*, which more frequently use forest edges. As a result, the two species are hybridizing and rearing "Sparred" Owls *B*, which carry the genes of both species. Conservation practitioners are concerned that the evolutionary trajectory of the Spotted Owl may be disrupted as its genes are combined with those of the Barred Owl. *Photos courtesy of Jared Hobbs.*

Figure 27.5. The Red-cockaded Woodpecker (*Leuconotopicus borealis*) is listed as Endangered in the southeastern United States. Over the course of the last several decades, remnant populations of the birds have been stabilized, and some have even been declared as recovered through a series of focused conservation strategies that included habitat protections, restoring key nesting and roosting cavity resources, prescribed burning, and translocation. *Photo courtesy of Kevin R. Rose.*

mid-story of fire-adapted oaks and small pines, and a pine canopy. Suppression of fire for several decades resulted in a dramatic change in habitat structure. Without fire, a dense hardwood mid-story formed and shaded a sparse, depauperate ground cover. Encroachment of hardwood mid-story causes Red-cockaded Woodpeckers to abandon their cavities, and they avoid foraging in areas with tall, dense mid-story. As a result, the birds are unable to persist in **fire-suppressed habitat**.

Fire suppression and harvesting of old trees persisted into the 1980s, and the Red-cockaded Woodpecker appeared headed for extinction, as most populations were declining and none were increasing. Since that time, however, the species has recovered remarkably thanks to the use of management tools designed to address habitat degradation. Rather than suppress fire, managers now vigorously apply **prescribed burning** to mimic the natural historical fire regime, restoring habitat structure back to its original state

(Conner et al. 2001). Research also indicated that populations were limited by availability of nesting and roosting cavities (Walters et al. 1992), and techniques to construct artificial cavities in living pine trees were developed to provide cavities in areas where none were available. This technique proved highly effective in increasing populations, essentially by increasing key resources that restricted carrying capacity. Studies subsequently found that interpopulation dispersal and colonization of unoccupied habitat by woodpeckers was limited, and **translocation techniques** were developed to augment small populations and to reintroduce Red-cockaded Woodpeckers to suitable habitat from which they had been extirpated. Because of these efforts to address habitat loss and degradation, many populations have increased greatly, four to the point that they have exceeded recovery goals and been declared recovered from the threat of extinction.

Habitat Fragmentation, Edge Effects, and Reserve Design

In addition to the losses and degradation of habitats described above, habitats also can be affected by **habitat fragmentation**. Fragmentation occurs when smaller and smaller portions of native habitats are left while surrounding regions are converted to other less suitable habitat types. The dissection of large habitat blocks into smaller sections can result in patches of native communities that are disjunct and disconnected from other similar habitat types. Fragmentation typically accompanies habitat loss and changes the shape, or **configuration**, of habitats at a landscape or regional level. At the same time, the populations of birds inhabiting the fragmented habitats also are subdivided, and often isolated on the remaining small habitat patches. Once isolated, the subpopulations lose connections with other subpopulations, and they can blink out of existence without the inflow of new genes and mates from other areas. For example, above we describe the conversion of the native North American grasslands to agricultural habitats. The agricultural conversions did not occur as one contiguous sweep from east to west, however. Rather, as farms and ranches were established, small remaining patches of native grasslands were distributed across the entire region, disconnected from other native grassland blocks and surrounded by a "hostile matrix" of production agriculture.

The Greater Prairie Chicken (*Tympanuchus cupido*) was historically distributed across the entire midcontinent of North America. But as the grasslands were divided and converted into agricultural lands, populations of Greater Prairie Chickens also became fragmented and reduced, and primarily persisted in small sections of remnant grasslands distributed throughout the region. The associated separated and small subpopulations of Greater Prairie Chickens then began to disappear, leaving only a few populations with substantial numbers. Below, we discuss how conservationists are using translocation to help ameliorate the effects of habitat fragmentation—and Greater Prairie Chickens are among the species benefited by capturing birds and moving them among the subpopulations to simulate the **demographic and genetic connectivity** that historically characterized a region with vast and connected prairies.

A similar set of concepts about large and small habitat patches emerged from **island biogeography theory**, which was broadly presented by Robert MacArthur and E. O. Wilson in their 1967 book, The Theory of Island Biogeography. The authors described that larger islands had greater species richness than smaller islands, and that islands farther from mainlands also had fewer species. They hypothesized that the pattern was caused by lower chances of extinction for larger populations on larger islands than for smaller populations on smaller islands, and that dispersers from the mainland are more likely to survive transits to islands closer to mainlands than to distant islands. The theory was subsequently tested in several studies, including a seminal set by Wilson's graduate student Dan Simberloff (Simberloff and Wilson 1968).

The island biogeography theory led some to suggest that when designing reserves for imperiled species and communities, conservation practitioners should attempt to create larger reserves, rather than small ones. In this case, they likened reserves to the islands studied by Wilson and others. To the contrary, however, biologists Dan Simberloff and Lawrence Able pointed out that sometimes two different smaller reserves might encompass different communities and thus smaller reserves might also be beneficial. The disagreement about reserve design continued with the **single large or several small** debate (termed the SLOSS debate), wherein some advocated for fewer larger reserves, and others advocated for higher numbers of smaller reserves. As an addition, others suggested that separate smaller reserves could be connected with habitat corridors to help them function more like single larger reserves. Similarly, when multiple smaller reserves are being selected, conservation planners have the opportunity to select reserves that complement, rather than duplicate, each other.

Several decades later, most practitioners would likely agree that a range of forces and factors require consideration when designing reserve systems and protected areas (Williams et al. 2005). Habitat loss and fragmentation has the potential to reduce habitat areas and to change configurations of large contiguous blocks of habitat. Edge effects (see below) also may present challenges to species that evolved in the absence of edge-associated predators, competitors, and para-

sites. And finally, larger reserves are better than small ones, but a number of representative reserves from across a range of conditions might also be required in order to maximize biodiversity.

A second aspect of habitat fragmentation relates to the **edge-to-area ratio** of remaining patches of native habitats. The geometric relationship between the outside perimeter, or edge, and the internal area for any polygon is directly related to scale. Larger patches have a lower edge-to-area ratio, whereas smaller ones have higher ratios. For example, the perimeter of a square habitat that is one hectare in size is 400 meters, whereas two smaller one-half hectares squares with the same internal area have a total perimeter of 566 meters. Because of the different configurations, the two smaller squares of habitat have vastly more edge for every meter of internal habitat.

These differences in habitat configurations affect birds because, in some cases, habitat edges are preferred by nest predators and brood parasites. For example, the Brown-headed Cowbird (*Molothrus ater*) posed just such a threat in North America, where large contiguous forests historically characterized the landscape, and many birds evolved in the interiors of those large forests. As agriculture moved into the region and forests were fragmented for croplands and pastures, more and more edge habitats were created. In chapter 16 we discussed brood parasites, including the Brown-headed Cowbird, which lay eggs in the nests of other birds and do not care for their own young. Cowbird parasitism reduces reproduction in the host species, as the parasite outcompetes the host nestlings for food items brought to the nest, reducing the chances of host nestling survival. Brown-headed Cowbirds are associated with open habitats, and they therefore can affect other birds nesting on the edges of forests. As forested habitats have been fragmented in recent decades, fewer large blocks of forest remain, and the many remaining smaller forest patches have increased edge-to-area ratios, which leads to increased nest parasitism rates (Robinson et al. 1995). In response to the effects of brood parasitism, conservationists have implemented control programs that trapped and removed Brown-headed Cowbirds from areas with sensitive species—a conservation strategy that has been demonstrated to reduce parasitism rates in some situations (Stutchbury 1997, Ortega et al. 2005, Smith et al. 2013). Notably, however, cowbird removal programs are not always successful, and they do not address the underlying causal issues associated with habitat fragmentation (Rothstein and Peer 2005).

Recent interactions between Barred Owls (*Strix varia*) and Northern Spotted Owls in the Northwestern regions of North America present another example of habitat fragmentation effects. Historically, Barred Owls were restricted to eastern portions of the continent, and they persisted in naturally fragmented habitats, whereas the slightly smaller, congeneric Spotted Owls inhabited larger and more contiguous forests of the Pacific Northwest. Recently, Barred Owls have arrived in the west, however, and their movement from their eastern range is thought to have been facilitated by forest openings and fragmentation caused by logging and timber harvests (fig. 27.4; Dugger et al. 2011).

As the Barred Owls began to interact with Northern Spotted Owls, the two species started to hybridize and create viable offspring with genes from both species, which have been labeled "Sparred Owls." Now, conservationists are concerned that the interbreeding of the two species may threaten the continued persistence of the Northern Spotted Owl as a species because the gene pool is becoming diluted with Barred Owl genes (fig. 27.4). They are seeking ways to manage the situation. One solution would be to restore the contiguous forests that are most suitable for Northern Spotted Owls, which may result in less Spotted Owl exposure to the Barred Owls that inhabit edge habitats. But restoring large sections of contiguous forests at the continent scale is challenging because it would require substantial changes in land use, and the benefits of such a strategy would accrue only over long time periods. Others are experimenting with programs that directly target and remove Barred Owls where they have invaded to test whether the approach benefits remaining pairs of Spotted Owls, and early results indicate that Barred Owl removal combined with habitat conservation may slow or even reverse population declines of the endangered Spotted Owls (Weins et al. 2016). Removal programs can, however, create situations in which species become consistently reliant on continued conservation action (box on page 840).

STRATEGIES AND TOOLS FOR POPULATIONS IMPERILED BY FACTORS AFFECTING VITAL RATES

Some populations of birds continue to decline despite having access to what appear to be extensive suitable habitats and abundant life-sustaining resources. Over the years, however, conservation biologists have identified that mysterious declines in populations can be caused by factors that affect **population vital rates**, such as reproduction and survival. Avian reproduction is complex, being characterized by incubation and brood-rearing stages, a fledgling-rearing stage, and then subsequent survival of juveniles and adults. Birds sit on nests for extended periods, which can increase the vulnerability of incubating adults to predators, and eggs, nestlings, and fledglings also are vulnerable to a number of factors. Below, we discuss some of the tools used by conservation practitioners to address predators and introduced competitors, nest

parasites, diseases, overharvest, and pollution, which are identified in Wilson's HIPPO and Diamond's "evil quartet," and are the primary factors responsible for adverse effects on vital rates in birds.

Tools for Controlling Predators and Parasites

Predators can consume enough eggs, chicks, or adults to cause imperilment in some bird populations, necessitating conservation responses. Although predators represent a major source of annual mortality among birds, particularly for incubating females and juveniles, many species have evolved with predators and thus have natural defenses that offset predator effects and keep populations relatively stable. Exotic predators are widely recognized as being catastrophic for some species, but the impacts of native predators also can be exaggerated or magnified as a consequence of other human-associated changes in the system. Changes in predator density or predator community, native or exotic, can reduce rates of reproduction or survival sufficiently that avian populations are not sustained. In these cases, **predator control** is sometimes implemented as a conservation tool to reduce predation rates and allow avian populations to recover.

For example, in 1997 the US Fish and Wildlife Service estimated the Alaska-breeding population of Steller's Eider (*Polysticta stelleri*) included fewer than 2,500 birds, causing them to list the population as Threatened under the Endangered Species Act. The apparent population decline and range contraction of Steller's Eider was linked to a number of factors, including changes in predation pressure on nests (fig. 27.6). The primary nesting area for Steller's Eider in North America is located on the North Slope of Alaska (Quackenbush et al. 1995, Obritschkewitsch et al. 2001), where nest failure rates exceeded 85 percent in some years, primarily as a result of predation. In an effort to increase nest success, those responsible for the bird's recovery identified a number of approaches to protect the eider population. Efforts were made to reduce the availability of anthropogenic food sources, such as landfills, that attracted and maintained artificially high levels of predators near nesting areas. Further, a program was implemented to control arctic foxes (*Vulpes lagopus*), which were considered a major egg predator, near Barrow, Alaska, from 2005 to 2014 (Safine 2015). The direct removal program appears to have increased Steller's Eider nest success in general; however, other factors including annual variation in avian predator abundance and food availability for artic fox continue to confound this relationship (Safine 2015).

Especially dramatic examples derive from Oceania, where many islands were completely devoid of mammalian predators prior to the arrival of humans. The inhabitant birds lost many of the antipredator traits characteristic of continental populations while in isolation and without predators. Many

CONSERVATION-RELIANT SPECIES: KIRTLAND'S WARBLER

Although population increases of Red-cockaded Woodpeckers represent a conservation success story, the bird remains dependent on human assistance, even in recovered populations. Managers continue to provide birds with artificial cavities and to burn hundreds of thousands of hectares annually to maintain habitat. Michael Scott and colleagues have termed species that will continue to require some form

Kirtland's Warbler (*Setophaga kirtlandii*). *Photo courtesy of Joel A. Trick, USFWS.*

of management in the foreseeable future until the threats that led to their imperilment can be eliminated **conservation-reliant** (Scott et al. 2005). In the case of the Red-cockaded Woodpecker, the ultimate recovery goal is to establish self-sustaining populations that are no longer conservation-reliant (USFWS 2003), for example, to provide a sufficient old-growth component within the forests they inhabit that the birds can excavate their own cavities, eliminating the need for artificial cavities. In other cases, however, eliminating the causes of imperilment is not realistic, and thus some species likely will remain conservation-reliant in perpetuity. For example, the Kirtland's Warbler (*Setophaga kirtlandii*) is an endangered species restricted to early successional jack pine (*Pinus banksiana*) forests in Michigan. Anthropogenic changes to the landscape in which these forests occur preclude the fires that formerly maintained this habitat, and brood parasitism by the Brown-headed Cowbirds that have invaded in response to these landscape changes reduce reproduction far below the level necessary to sustain warbler populations (Bocetti et al. 2012). The species' reproduction is dependent on annual removal of cowbirds, and its habitat is dependent on harvesting and replanting of jack pine. Although there is hope that cowbird removal programs may be reduced in the future, there is little hope that the species can be recovered in the sense of establishing self-sustaining populations without remaining reliant on substantial jack pine management. A large fraction of endangered species in the United States are similarly conservation-reliant (Scott et al. 2010).

species were flightless and others lacked fear, for example, and, as such insular populations were particularly vulnerable when nonnative predators did arrive. Upon the eventual arrival of humans, island birds went extinct at a rate 12 times greater than their continental counterparts (Halliday 1978).

Control of invasive island predators thus is an important conservation strategy to protect or restore remaining native avifauna on islands. For example, the Tuamotu Sandpiper (*Prosobonia cancellata*) once was distributed throughout the Tuamotu Archipelago of eastern Polynesia, but it is now restricted to only four small islands because of introduced rats (*Rattus* sp.), which were brought on boats sailed by humans arriving from the continents. The Tuamotu Sandpiper population is estimated at fewer than 1,500 individuals, and conservation strategies to protect this endangered bird include

preventing rats from colonizing the remaining rat-free islands and eradicating predators (including rats, feral cats, and hogs) from additional islands where the birds can be reintroduced (Pierce and Blanvillain 2004).

The flightless Campbell Island Teal (*Anas nesiotis*) provides a more extreme example. The bird is endemic to the small Campbell Island, and it was considered extinct as a consequence of the impact of Norway rats (*Rattus norvegicus*) introduced to the island by whalers. However, more than 100 years after being declared extinct, a remnant population of approximately 25 breeding pairs was discovered on nearby Dent Island, which had not been colonized by rats (Gummer and Williams 1999). A **captive breeding program** was established to increase the overall number of Campbell Island Teal, and the New Zealand Department of Conservation implemented the world's

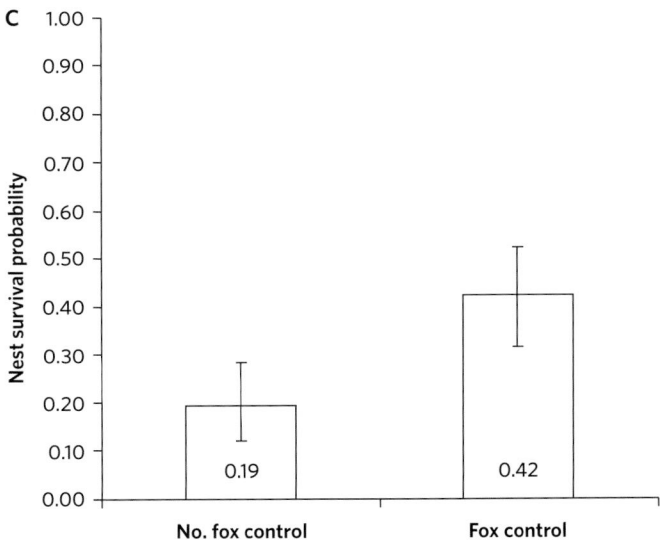

Figure 27.6. The range of the Steller's Eider (*Polysticta stelleri*) *A* extends from Siberia into the Arctic Coastal Plain of Northern Alaska. The birds are listed as Threatened in the United States, and a number of management approaches have been implemented, including controlling predatory arctic fox (*Vulpes lagopus*) *B* populations. Over the course of six years, foxes have been controlled on the breeding grounds, and the efforts appear to be improving nest success and reproduction *C*. *Photos courtesy of Ted Swem; graph data courtesy of D. Safine, USFWS.*

largest rat eradication program. They dropped more than 120 tons of rat poison from helicopters, such that the entire island was treated (McClelland 2014). After this tremendous effort, Campbell Island was designated rat-free, and approximately 100 Campbell Island Teal were reintroduced back to their native island. The birds have since established a self-sustaining population estimated at over 200 individuals, causing the IUCN to downgrade the species Red List status from Critically Endangered to Endangered (McClelland 2011).

The Campbell Island Teal program is an especially dramatic example of the recipe for success in addressing vital rates issues: identify the vital rate whose alteration is responsible for population decline, identify the factor responsible for that alteration, and eliminate that factor. Once that is achieved, the species generally will recover through the natural process of density-dependent population growth. This example also illustrates the effort that may be required to eliminate the factor responsible for the vital rate problem.

Addressing Human Impacts on Bird Populations

Humans can have direct impacts on avian vital rates and populations by exploiting them as a resource, or indirect impacts by contaminating the environment. Direct exploitation by humans is in fact a leading cause of the extinctions of avian species recorded to date. In principle, exploitation is an easy problem to fix: reduce exploitation to the point where overharvest no longer occurs. Indirect impacts of contamination of the environment are similarly easily remedied in theory: eliminate the pollution that is reducing reproduction or survival to the point that vital rates return to levels necessary for populations to sustain themselves. However, **conservation legislation** is the appropriate tool to address human exploitation, and it can be challenging to develop and is not easily applied in practice.

Early in the twentieth century, legislation became one of the first conservation tools available for addressing the decline in bird populations, which at the time was caused by overhunting and commercial sale of migratory birds for both food and fashion. Upland game birds, waterfowl, and indeed some shorebird species were harvested and sold as food in rapidly expanding urban areas in the early 1900s. Further, over 64 bird species, including the Snowy Egret (*Egretta thula*) were harvested for feathers and plumes that were sold as decorative ladies' millinery supplies (McCally 1999). The extensive harvests began to affect bird populations and alarm the public, which responded by insisting that government act to protect the birds. **The Lacey Act**, which was passed in 1900, was the first federal law designed to protect wildlife in North America, and it regulated interstate or foreign commerce of wildlife. In large part it prevented hunters from illegally harvesting game in one state and then escaping prosecution by crossing state lines and selling those animals in other states. The act was immediately effective, and early prosecutions were directed at large-scale interstate trafficking of illegally harvested wildlife, including an Illinois trafficking ring that shipped 22,000 illegally taken ducks, grouse, and quail. As a follow-up, the **Weeks McLean Act** was passed in 1913. In addition to establishing additional market hunting regulations, it granted the United States secretary of agriculture the power to designate nationwide hunting seasons.

The example of the effects of the chemical dichlorodiphenyltrichloroethane, or DDT, on birds in North America illustrates use of legislation to address an environmental pollution issue affecting avian populations. In the mid-twentieth century, DDT was widely used as a pesticide. Because this chemical and its breakdown products persist in the environment and accumulate in fatty tissues of animals, the contaminants tend to increasingly concentrate in species higher on the food chain, through a process known as **bioaccumulation**. Thus species like Peregrine Falcon (*Falco peregrinus*), Bald Eagle (*Haliaeetus leucocephalus*), Osprey (*Pandion haliaetus*), and Brown Pelican (*Pelecanus occidentalis*) at the top of contaminated food chains accumulated high levels of DDT. The chemical affected eggshell formation, such that the birds laid eggs whose shells were too thin and cracked when incubating adults sat on them. The resulting reduction in productivity caused populations of these species to decline rapidly.

Half the battle in such cases is identifying which vital rate has been altered to cause population declines and what factor has altered that vital rate. In the case of DDT, substantial effort and research were required to identify that the effects of DDT on reproduction was the cause of declining predatory bird populations. In this case, once the science definitively identified the problem, the required legislative solution—a ban on the use of DDT—was quickly adopted. As with the Campbell Island Teal, once the factor adversely affecting the key vital rate was eliminated, populations of all of the affected species grew rapidly, resulting in another success story in avian conservation.

Of course, the legislative tool required to address human impacts is not always as readily enacted as in the DDT case. The California Condor (*Gymnogyps californianus*) provides an illustrative example. Condors were extirpated from the wild in 1987, at which point the population consisted of just 27 cap-

tive birds. The primary cause of the population decline was increased mortality caused by **lead poisoning**, the main source of which was lead ammunition in carcasses on which these scavengers fed. As of this writing, there are now more than 400 condors, and more than 200 in the wild, so the condor recovery program is seemingly a success. However, lead has not been removed from the environment, and as a result the wild population persists only because program managers regularly capture the wild birds, test them for lead poisoning, and treat those that are ill with chelation therapy, which involves the intravenous application of chemicals that bind with lead, so it can be excreted by the bird (Walters et al. 2010). For example, 437 birds were tested and 66 treated for lead poisoning over a five-year period in just one of the four wild populations (Parish et al. 2007). Without this effort, the birds would quickly be extirpated from the wild once again (Walters et al. 2010).

Several policy changes have been enacted to reduce exposure of condors to lead. In Arizona the state agency provides recreational game hunters with copper bullets, which pose no threat to condors. California has banned the use of lead ammunition for big game and upland game bird hunting within the range of the condor. However, the condor has a very slow life history, characterized by a long lifespan and low annual productivity (two chicks in three years per pair, at best). As a result, annual mortality of adults needs to be under 5 percent for populations to be self-sustaining. One carcass from an animal shot with lead ammunition contains enough lead to kill a dozen condors, thus a single contaminated carcass can potentially kill more than 5 percent of the current wild California Condor population. It thus is not surprising that the policies enacted to date have not been able to reduce mortality sufficiently. Eliminating lead from the environment requires a more extensive legislative solution, which in the past has been difficult to achieve because of powerful political lobbies opposing restrictions on use of lead ammunition. However, California passed a total ban on use of lead ammunition, to take effect in 2018. Effects of lead on scavenging birds extend beyond condors: in a recent study of two vulture species in eastern North America, all 108 birds tested exhibited evidence of chronic exposure to lead (Behmke et al. 2015).

THE SMALL POPULATION PARADIGM

In his landmark 1994 publication about conservation paradigms, Graeme Caughley also identified a situation encountered by species that are quite near to extinction, which he described as the **small population paradigm**. Species in the small population paradigm often number in the tens or hundreds, and populations are so very small that they are vulnerable to factors quite different from what caused initial population declines. For example, when populations are very small, inbreeding, or mating among relatives, is more likely to occur. Inbreeding can result in loss of **genetic diversity** (a form of biodiversity itself) within populations. In addition, a high frequency of mating between very close relatives may result in reduced reproduction and survival, a phenomenon known as **inbreeding depression**. Also, the normal population fluctuations caused by annual variation in weather and other aspects of the environment, known as **environmental stochasticity**, which reduce numbers in bad years and increase them in good years, may have disastrous consequences for very small populations: a few bad years in a row could reduce a very small population to zero. In populations numbering just tens of individuals, chance events, that is, bad luck experienced by a number of individuals regardless of genotype or phenotype, are termed **demographic stochasticity**. These events can result in extinction—when all remaining offspring are of a single sex for example. Finally, in species that number in the hundreds, if all of the remaining individuals are spread thinly across the landscape, they may never encounter another individual with whom to mate—a situation described as an **Allee effect** (Allee 1931). Other forms of Allee effects in which some aspect of life history becomes dysfunctional at very small population size, exacerbating population decline, also have been observed.

In chapter 25 we learned that bird populations in the small population paradigm can be the most critically endangered, because the factors affecting them can be so severe that extinction is imminent without direct and serious intervention by conservationists. Unlike many species in the declining population paradigm, those in the small population paradigm will not recover if enough habitat is protected or restored, if hunting is stopped, or if nests are protected from predators. That is, the factors that caused decline in the first place will still have to be addressed, but so too will the issues specific to small populations. The situation has been termed the **extinction vortex**, because the smaller the population gets, the more difficult it can be to recover, and population trajectories accelerate toward complete extinction. Accordingly, a different set of quite specific conservation tools has been developed for the smallest populations, including captive breeding programs, population viability analyses, and, in the most dire of situations, crossbreeding.

Strategies for Conserving Extremely Small Populations

In recent decades, zoos and other dedicated breeding facilities have oriented their missions toward conservation, and by doing so they have provided a new set of tools that can be used to help prevent species' extinction. The numbers of birds in wild populations are often much larger than those that can be held in captivity, and so the use of **captive breeding** is reserved for only the direst circumstances. Nonetheless, captive breeding facilities and knowledgeable caretakers can remove birds from the threats of the outside world and skilled caretakers can strategically pair and mate key individuals to maintain genetic diversity in subsequent generations. Captive breeding programs have the goals of creating safety populations and also breeding individuals for release back into the wild.

Above we discuss the conservation cases of the California Condor, Peregrine Falcon, and Campbell Island Teal, which all benefited from captive breeding programs in zoos and other institutions. The birds were reintroduced into the wild and now persist with substantial conservation investment. Other species too are precariously perched near extinction, and captive breeding is one tool used to address their plight. The Guam Kingfisher (*Todiramphus cinnamominus*) and the flightless Guam Rail (*Gallirallus owstoni*) are examples of species that might no longer exist if not for investments by captive breeders (fig. 27.7). Both species originally inhabited the central Pacific Island of Guam. Like all oceanic islands, the plants and animals that inhabit Guam arrived there as dispersers from distant continents and from other islands. The 16 native bird species on Guam then evolved without rats, cats, or predatory snakes, which were all subsequently introduced to the island, first when Micronesians arrived in canoes, and later when other people landed in ships and airplanes from Europe, the Americas, and Asia.

The brown tree snake (*Boiga irregularis*) was one of the invasive species introduced to Guam shortly after the end of the World War II. The snake likely arrived as a stowaway in cargo that was transported from Indonesia. Then, between the 1950s and the 1980s, observers noted declines in bird numbers on Guam, although they did not realize that the introduced snake was the cause. Indeed, many doubted that a single invasive snake could cause what was later termed the "extinction of an island forest avifauna" by biologist Julie Savidge when she realized that the introduced snakes were causing bird populations to crash (Savidge 1987). In response, the United States added seven birds to its list of threatened and endangered in 1984, and conservation practitioners initiated measures to prevent the extinction of the island's remaining bird species. As part of that effort, 29 Guam Kingfishers and 22 Guam Rails were captured and brought into breeding programs just before they were extirpated from the wild (Haig et al. 1995). Since then, the extinction of these species has been prevented only by the **species survival plan** that was implemented by captive breeders.

Some of the techniques developed for husbandry of domesticated animals can be used as conservation tools for

A

Figure 27.7. A, The Guam Kingfisher and *B*, the Guam Rail were placed in a captive breeding program when the invasive brown tree snake caused populations to decline and eventually disappear entirely from their native range on the Pacific Island of Guam. Without the captive breeding programs, the kingfisher likely would be completely extinct. *Photos courtesy of James D. Jenkins (kingfisher) and Susan M. Haig (rail).*

wild birds in captive breeding programs. For example, breeding specialists have used pedigree information, or the information about an individual's family and relatives, to manage the genetics of captive animals. For centuries, breeders of domesticated birds, such as chickens (*Gallus gallus*), selected

and bred individuals with certain desirable traits, and they used pedigree information to identify relatives and to avoid inbreeding. Those same pedigree-management tools have become highly important for managing captive populations of wild birds, which often originate from a very few individuals when they are first brought in to captivity. To preserve as much genetic diversity as possible, breeders have developed sophisticated pedigrees and genetic management plans for entire populations of captive birds. Nonetheless, the longer the species is restricted to a small captive population, the more genetic diversity is lost and the less likely the species will be able to thrive upon release back into the wild (Frankham et al. 2002). So there is a constant incentive to quickly reestablish wild populations that can expand and better protect genetic representation.

Conservation practitioners often are interested in evaluating the risks of extinction, either for small populations that remain in the wild, or for populations of captive individuals that are released, and **population viability analysis** (PVA) is a conservation tool that has been developed to address just such questions. In short, PVAs are most often implemented as simulations, in which fictitious computer-generated populations are allowed to live for a certain amount of time (see the box on page 846). The simulated populations in PVAs encounter simulations of real-world forces, including **deterministic forces** and **stochastic forces** that can cause them to grow or shrink. The deterministic forces are obvious and their effects are well known. Alterations in habitats and resources, reproductive success, or mortality rates cause simulated populations to grow or shrink. On the other hand, stochastic forces include unpredictable and random events such as catastrophic typhoons or other weather events that may eliminate large numbers of individuals or cause nesting and reproduction to fail for an entire year. Stochastic events may also increase populations, for example if conditions for reproduction are favorable for several consecutive years. They also are not well captured by single deterministic equations, and so PVAs are used to expose simulated populations to stochastic forces and to identify what happens, on average, or to construct probabilities of extinction or persistence.

A PVA, for example, was used to simulate the establishment of a wild population of Guam Kingfishers and to evaluate risks of extinction that might be faced under different conditions and different management scenarios (fig. 27.8). Conservation practitioners were interested in identifying the best way to establish a wild population—by releasing a large cohort from captivity all at once or by releasing smaller cohorts of birds spread across many successive years. They also wanted to know whether the establishment of multiple wild populations of birds would further reduce chances of extinc-

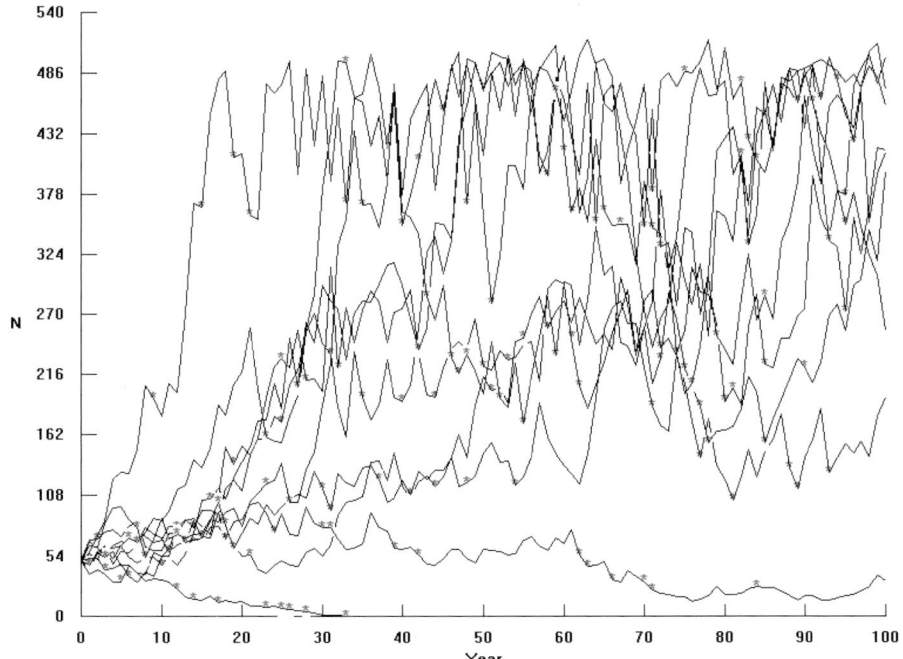

Figure 27.8. Plots of population viability analysis simulations of Guam Kingfishers. Each line represents the trajectory of a simulated population that progressed from the start year (*left*) to the final year (*right*) of the simulation. The line that intersected with the y-axis (bottom of the frame) represents a simulated population that went extinct after a number of stochastic events (red dots). Those progressing to the right of the frame were predicted to have remained extant. *Graph courtesy of Dylan C. Kesler.*

tion. Obviously, answering such questions with traditional experiments was not possible, since the entire extant population of the kingfisher remained in captivity, and so they turned to PVA. In the PVA, computer-generated populations of kingfishers were released onto a fictitious island as releases of 10 birds each year for four years, whereas others simulations included many more birds each year, and included releases lasting up to nine years. Some simulated populations also included releases on a second island, and occasional movements of birds between the two computer-generated populations. The fictitious computer-generated populations were allowed to persist for 200 years, during which time they encountered a number of simulated typhoons and other stochastic events and environmental challenges. In the end, results led researchers to determine that the way in which birds are released into the wild—as many birds over just a few years or fewer birds over extended periods—had little to do with the probability of extinction. However, the establishment of a second population and movements among islands did substantially increase chances of population persistence and the retention of genetic diversity (Laws and Kesler 2012).

Above, we discuss how **hybridization** between an endangered species and another species, or subspecies, can create a situation in which the genes of the endangered species might be diluted or overwhelmed. In some cases those hybridizations can be cause for concern because the genes of the endangered population may begin to shift, such as with the "Sparred Owls" of northwestern North America. In other cases—especially when a species is on the brink of

extinction—hybridization can be used as a conservation tool. **Extinction** is a demographic event that occurs when the last individual of a species dies. But in some cases, conservationists have argued that when there are just a few individuals of a species (or subspecies) left, perhaps they should be interbred with other species (or subspecies) to preserve some semblance of their genome.

The Dusky Seaside Sparrow (*Ammodramus maritimus nigrescens*) provides an example of how hybridization might be used to prevent the absolute extinction of a subspecies. The sparrow historically inhabited the salt marshes of Florida and the St. Johns River. By the 1970s, the population of birds had declined to perilously low numbers because of habitat losses and other factors. By 1979, only six individuals were known to exist, and all were males. Thus, in order to preserve some semblance of the Dusky Seaside Sparrow, five of those males were captured and put into a captive breeding program where they were crossbred with another subspecies of Seaside Sparrows (*A. m. peninsulae*). Unfortunately, the captive breeding program for the Dusky Seaside Sparrow encountered challenges and did not succeed. But the experience with the birds, and with other endangered mammals, indicates that hybridization may be a viable tool for conservationists addressing taxa at the very brink of extinction (Zink and Kale 1995).

Conservation Strategies for Migratory Birds

Migrant birds that undertake movements that transit continents and hemispheres pose special challenges to conservation

DR. ROBERT C. LACY

Robert C. Lacy. *Photo courtesy of Robert C. Lacy.*

Dr. Robert C. Lacy is a senior conservation scientist for the Chicago Zoological Society and is on the faculty of the Committee on Evolutionary Biology at the University of Chicago. He has served as chair of the International Union for Conservation of Nature (IUCN) Conservation Breeding Specialist Group (2003–2011), and he was chair of an IUCN Species Conservation Planning Task Force that developed a new framework for strategic planning for species conservation. Dr. Lacy helped develop the pedigree analysis methods used for the management of captive populations, including some of the most endangered birds. He then implemented his methods in freely available software that is now used to guide the management of genetics in captive breeding conservation programs. Similarly, his freely available population viability analysis software (Vortex) is also used by conservationists, wildlife managers, researchers, and students throughout the world to help guide risk assessments and conservation planning. Dr. Lacy continues to work on research and conservation tools to facilitate management of captive breeding programs, reintroductions, and translocation efforts.

practitioners. Threats to avian populations can differ among wintering areas, migration stopover sites, and breeding locations, making identification of these threats and the conservation solutions to address them challenging. Oftentimes, international cooperation is needed to monitor populations

and implement conservation actions for long-distance migrants. Hence, for migratory species, legislation again is an important conservation tool (see the box on page 847).

Early conservation efforts designed to address dwindling avian populations often focused on threats at breeding areas and to nest success. However, in recent years conservation practitioners have attempted to address the full spectrum of habitats needed by migrants, and increasing attention has been aimed at **migration stopover sites**, where migrant birds rest and refuel during their journey. Particular attention has focused on migration stopover sites used by entire populations of birds, because even though the sites may be used for only a few days each year, they have the potential to affect the vast majority of individuals in a population.

For example, every spring Red Knots (*Calidris canutus*) migrate 15,000 km from their wintering grounds in Tierra del Fuego in southern South America to northern Arctic breeding areas. One of the primary migration stopover locations during spring migration occurs in the Delaware Bay, where Red Knots forage almost exclusively on horseshoe crab (*Limulus polyphemus*) eggs. Over 90 percent of the Red Knot population is estimated to use Delaware Bay as a migration stopover site, and stopover timing usually coincides with peak abundance of horseshoe crab eggs that Red Knots rely on for the lipid reserves necessary to complete migration (Niles et al. 2008). Recent declines in the availability of horseshoe crab eggs, caused by overharvest of horseshoe crabs for bait by the fishing industry, has been linked to reduced mass gain of Red Knots while stopping at Delaware Bay and subsequent lower adult survival: this is considered to be one of the main factors contributing to the declining Red Knot population (McGowan et al. 2011). Within a five-year time period (ending in 2003), researchers documented an over 50 percent decrease in Red Knot populations, leading the USFWS to list one Red Knot subpopulation as Threatened under the Endangered Species Act and several states to impose limits on horseshoe crab harvest in an attempt to increase populations (Baker et al. 2004).

Given that migratory birds often depend on habitats in multiple countries and continents to meet their annual life cycle needs, comprehensive conservation efforts require international coordination and cooperation. One of the first actions to promote international avian conservation in the Nearctic was the **Migratory Bird Treaty Act of 1918**, which represented an agreement between the United States and Canada to protect migratory game bird species from overhunting. The act makes it illegal to "pursue, hunt, take, capture, kill, possess, sell, purchase, barter, import, export, or transport any migratory bird, or any part, nest, or egg, or any such

DR. SUSAN M. HAIG

Susan M. Haig. *Photo courtesy of Tom Haig.*

Dr. Susan Haig began her scientific career at a time when conservation tools for the small-population paradigm were being developed. She dedicated the next several decades to guiding research that integrated science, conservation theory, and policy. Some of Dr. Haig's earliest projects, carried out at the Smithsonian Institution, developed conservation genetic methods and population viability analyses to guide captive breeding programs for critically endangered Pacific island birds. Subsequently, she expanded her focus to include wild populations and established the Conservation Genetics Laboratory at the US Geological Survey's Forest and Rangeland Ecosystem Science Center. There, she continued her use of molecular genetic tools to identify issues of concern to imperiled populations, evaluate population connectivity, and assess animal movement on a global scale.

Dr. Haig has directed or served on many recovery efforts that advocated for using strong scientific approaches to endangered species recovery. She has written about genetic implications of species listings under the US Endangered Species Act and wrote the initial petition for the Piping Plover

(*Charadrius melodus*) to be considered endangered in Canada. Dr. Haig directed the US Fish and Wildlife Service Piping Plover Recovery Team for 12 years and has overseen implementation of the International Piping Plover Breeding and Winter Census every five years since 1991. She is an invited member of the IUCN Conservation Breeding Specialist Group and the IUCN Reintroduction Specialist Group. Dr. Haig recently completed a term as president of the American Ornithologists' Union.

Dr. Haig's current projects continue the integration of theoretical and applied aspects of small population recovery. She has worked on the genetic aspects of Spotted Owl (*Strix occidentalis*) conservation for almost 25 years and continues to study various genetic, demographic, and ecological issues related to bringing California Condors (*Gymnogyps californianus*) back to the Pacific Northwest. Her ongoing work on Red-cockaded Woodpeckers (*Leuconotopicus borealis*) provides a template that illustrates the value in bringing together long-term data from various aspects of population biology to more precisely diagnose and resolve issues in recovery.

Throughout her career, Dr. Haig has invested substantially in mentorship to ensure that the next generation of conservation scientist is well equipped. As a professor of wildlife ecology at Oregon State University, and in other roles, she has taught courses in the United States, Mexico, Brazil, Kenya, and Namibia to graduate students and conservation professionals on conservation theory, genetics, and demography of small populations and on the application of conservation tools to facilitate recovery. Along with a circle of colleagues, Dr. Haig also established the Migratory Connectivity Project (www.migratoryconnectivityproject.org), which, among other accomplishments, sponsors hands-on and intensive workshops that have provided training to more than 100 graduate students and professionals interested in animal movements and conservation.

---

bird, unless authorized under permit." Other countries later joined the Migratory Bird Treaty, including Mexico (1936), Japan (1972), and Russia (1978), promoting the conservation of birds across international boundaries; the act now broadly protects nearly every migrant species in North America. More recent efforts to promote international conservation were enacted as part of the **Convention on International Trade in Endangered Species of Wild Fauna and Flora** (CITES), which was agreed upon by 80 countries, in 1973,

and is designed to regulate the international trade of imperiled species. As of 2015, CITES protections have extended to more than 5,600 animals globally, including approximately 1,400 bird species.

Another example of a conservation strategy that requires international cooperation is the Partners in Flight program, which is a collaborative effort among federal, state, and local government agencies, conservation groups, industry, the scientific community, and private individuals in Canada, Mexico,

DR. IAN JAMIESON
By Rebecca Laws

Ian Jamieson.
Photo courtesy of Frances Anderson.

Dr. Ian Jamieson, a New Zealander, began his career as a behavioral ecologist and subsequently moved to conservation biology. Located at the University of Otago in Dunedin, his work drastically changed the way New Zealand managed its many endangered birds. Dr. Jamieson advocated for a "two-pronged" approach to conservation, whereby threatened species are first **secured** from extinction, by mitigating threats, and then managed for **recovery**, by creating opportunities for population growth (Jamieson 2015). In New Zealand, translocations are central to both aspects of this approach. Beginning in the 1990s, the New Zealand Department of Conservation undertook a series of projects to eradicate rats and other introduced mammals from small islands, and followed by repopulating those islands with birds that had been extirpated by the introduced mammals. The island refuges sheltered a number of species while larger populations rapidly declined on the mainland. At the time, it was hypothesized that because many of New Zealand's bird species had already gone through tight **population bottlenecks**, removing the impediments to population growth (threatening processes) should enable populations to recover on their own (Craig et al. 2000). These predictions hinged on the suggestion that such populations were unlikely to suffer from the adverse effects of inbreeding depression and loss of genetic variation brought about through translocation, although these hypotheses were largely untested (Jamieson et al. 2006).

To evaluate the effects of inbreeding and genetic diversity on bottlenecked New Zealand species, and therefore the capacity for these effects to impede recovery of translocated populations, Dr. Jamieson established a research program collecting detailed survival and pedigree data, along with taking blood samples to compare the DNA of translocated individuals. These data showed that translocated populations still suffered the effects of inbreeding depression even after rapid population expansion post-translocation (Jamieson 2011). Some species also suffered loss of genetic diversity through the course of translocation (Jamieson 2009). Based on this work, Dr. Jamieson built a series of tools and strategies for improving long-term translocation success and managing translocated populations. Some examples are provided below (see also Jamieson 2015).

To improve survival of translocated populations, the right birds need to be selected to translocate. Long-term radio-tracking was used to evaluate which factors led to the dispersal or death of endangered Saddlebacks (*Philesturnus carunculatus*) following translocation (Masuda and Jamieson 2012). This information could then be used to select the most appropriate age classes for future translocations. Dr. Jamieson also showed that the source population used for translocation was important. He investigated how birds from populations on rat-free sanctuaries reacted to model rats compared with the reactions of birds from the mainland, where rats were present. Naïve birds were less likely to recognize and respond to rats as predators, making them far less viable candidates for future translocation to mainland sanctuaries (Jamieson and Ludwig 2012).

Once translocated populations are established, maintaining genetic diversity is essential to ensuring a healthy genetic population post-translocation. Dr. Jamieson's work included the development of a computer model that simulated the loss of allelic diversity over several overlapping generations, assessing the benefit of alternate translocation regimes going forward (Weiser et al. 2012). He also developed Bayesian PVA models to evaluate a metapopulation strategy for managing translocated and remnant populations of Takahe (*Porphyrio hochstetteri*) (Hegg et al. 2003). Together these approaches have informed the metapopulation strategy of translocated and remnant populations of threatened species in New Zealand and around the world.

and the United States. The Partners in Flight program was initiated in 1990 in response to concerns over declines in many landbird populations, particularly Neotropical migrants, and seeks to promote collaboration between public and private organizations in the western hemisphere. Within Partners in Flight, regional working groups develop region-specific bird conservation plans that identify problems, synthesize information, and generate conservation solutions for use by biologists, as well as the general public. Since the inception of the program, the Partners in Flight mission has expanded to include all landbird species and is focused on three primary goals: helping species at risk, keeping common species common, and developing conservation partnerships to achieve these objectives. The Partners in Flight program also developed a species-assessment process, which involves evaluating conservation vulnerability of landbird species and then prioritizing species based on their vulnerability status and conservation importance (Carter et al. 2000)

Reintroductions, Translocations, and Assisted Colonization

In this chapter, we have presented examples of imperiled bird species that were challenged by habitat fragmentation, such as the Red-cockaded Woodpecker and Greater Prairie-chicken, and others, like the California Condor, Campbell Island Teal, and Peregrine Falcon, that were benefited by captive breeding programs. In these and other cases, conservation practitioners have used **translocation** techniques to facilitate demographic and genetic connections among disjunct population fragments and **reintroductions** to reestablish wild populations (see the box on page 848). Reintroductions are a subset of translocations that occur when a population has been extirpated from a portion of its range and birds from elsewhere, or from captive breeding facilities, are used to reestablish a wild population. Translocations include the purposeful movement of captured birds from one wild population to another, for the purposes of augmenting an existing population or introducing genetic diversity (Ewen et al. 2012).

Reintroduction programs can be complicated by a number of factors, including the suitability of habitats at release sites, the availability of individuals for release, and the ability to capture, hold, and move birds safely. Further, when release cohorts comprise captive-reared birds, additional investments may be required to train individuals to survive in the wild. Nonetheless, the effort-intensive technique of holding and releasing individual birds may be the only viable option for reestablishing wild populations and for facilitating genetic and demographic connections, and successful reintroductions may have intercepted the extinction of a number of species.

The Laysan Duck (*Anas laysanensis*) is an example of a species that has been translocated from one location to another, with the intent of reintroducing it to historically occupied habitats where additional populations can be established. Although the birds once ranged across most of the Hawaiian archipelago, they only persisted on the small atoll island of Laysan by the 1990s. In 2002, 42 individuals were captured and translocated via ship to the distant island of Midway, where a second population was established (Reynolds and Klavitter 2006). By creating two populations of the birds, conservationists reduced the possibility of extinction that might occur if a single catastrophic event, such as a typhoon or tsunami, struck the island of Laysan. Then in 2014, 28 Laysan Ducks were captured on Midway Atoll and translocated to a third location, Kure Atoll, where a tertiary population was established. Chances of extinction have been greatly reduced for the Laysan Duck because of these translocation programs.

Prospects of global change have caused biologists to consider newer and broader conservation approaches that were not previously embraced. In years past, for example, many conservationists were strongly opposed to the release of any species into any location outside of its historical range for fear that it might become invasive and thereby devastate native populations at the release site. However, when habitats within a native range become entirely unsuitable, or when climate changes cause natural communities to shift from one region to another, conservationists have considered the use of **conservation introductions** and **assisted migration**, whereby species are introduced into areas in which they have not occurred historically. In the years to come, and as conservation programs become more aggressive, we may see more projects based on these and other new conservation baselines (Jachowski et al. 2014).

KEY POINTS

- Threatened and endangered birds can present particular challenges to conservation practitioners because of their mobility, extensive distributions, specialized life history requirements, and high-energy life history. Threats to avian biodiversity are numerous, complex, and often require more than one conservation strategy coordinated among multiple partners across vast landscapes to address declining avian populations.

- National and state lists of imperiled species are developed with combinations of listing criteria and case-specific information, and they can form the basis of legal protections intended to prevent extinction. At the international level, the International Union for Conservation of Nature uses a criteria-based approach to evaluate and identify which bird species are threatened with endangerment and extinction, and it creates the Red List of Threatened and Endangered Species.

- Species characterized by the **declining population paradigm** often number in the thousands, and conservation strategies are generally aimed at restoring key resources, including habitats, foraging resources, and nesting resources, and at removing or preventing threats to survival and reproduction.

- Species in the **small population paradigm** often number in the tens or hundreds, and they are dually affected by the factors that caused population declines in the first place and a special set of factors impinging on the smallest of populations. Conservation strategies for species characterized by the small population paradigm aim at protecting the last representatives of a species, augmenting the smallest of populations, managing genetics to retain genetic diversity and prevent inbreeding, and identifying methods to avoid near-term extinction.

- Habitat conservation and restoration are frequently used strategies to address threats to avian populations caused by habitat loss, degradation, and fragmentation. Effective habitat conservation requires consideration of habitat area, configuration (including edge-to-area ratio), and restoration of ecological processes, including periodic disturbance, to mitigate threats to avian populations.

- Development and implementation of legislation can be an effective tool for reversing direct impacts from human activities such as overharvest and pollution, or for coordinating international conservation activities.

- Active intervention, in the form of predator or invasive species control, captive breeding programs, or reintroduction/translocation, is sometimes necessary to address factors contributing to low avian population vital rates, limited distribution, hybridization, or low genetic diversity.

- Extinction risk, as well as potential effectiveness of proposed conservation strategies in allowing population persistence, is often assessed using **population viability analysis (PVA)**. These population simulations predict population persistence over time under a variety of potential scenarios and incorporate deterministic and stochastic factors known to influence avian populations.

DISCUSSION QUESTIONS

1. Conservation practitioners develop lists of species that are believed to be threatened with extinction. Those lists can be based on quantitative criteria, such as popula-

tion size, or on case-by-case analyses and expert opinions. What are the potential costs and benefits of each approach?
2. Which institutions, organizations, or government entities are responsible for developing and implementing conservation strategies in your country? What about in other countries? Who performs these duties at the international level?
3. How do conservation tools addressing populations of birds differ from those used for plants, mammals, or fish?
4. With regard to conservation-reliant species, what are bounds of the ethical, scientific, and practical considerations for indefinite efforts needed to prevent extinction?
5. For the region in which you live, are there natural processes that have been disrupted by anthropogenic activities? Do those disruptions affect birds? If so, how might those disturbance processes be restored or replicated?
6. What are the laws that protect birds from overharvest? What are the resources and habitats needed by birds in your region?
7. How do conservation strategies intended to address resident bird species differ from those aimed at protecting migratory birds?

References

Allee, W. C. (1931). Animal aggregations, a study in general sociology. University of Chicago Press, Chicago, IL.

Baker, A. J., P. M. González, T. Piersma, L. J. Niles, I. L. S. do Nascimento, P. W. Atkinson, N. A. Clark, C. D. T. Minton, M. K. Peck, and G. Aarts (2004). Rapid population decline in Red Knots: Fitness consequences of decreased refueling rates and late arrival in Delaware Bay. Proceedings of the Royal Society B 271:875–882.

Battley, P. F., N. Warnock, T. L. Tibbitts, and R. E. Gill (2012). Contrasting extreme long-distance migration patterns in Bar-tailed Godwits Limosa lapponica. Journal of Avian Biology 43:21–32.

Behmke, S., J. Fallon, A. E. Duerr, A. Lehner, J. Buchweitz, and T. Katzner (2015). Chronic lead exposure is epidemic in obligate scavenger populations in eastern North America. Environment International 79:51–55.

Bennett, A. F., L. F. Lumsden, and A. O. Nicholls (1993). Tree hollows as a resource for wildlife in remnant woodlands: Spatial and temporal patterns across the northern plains of Victoria, Australia. Pacific Conservation Biology 1:222–235.

Bocetti, C. I., D. D. Goble, and J. M. Scott (2012). Using conservation management agreements to secure postrecovery perpetuation of conservation-reliant species: The Kirtland's Warbler as a case study. Bioscience 62:874–879.

Carter, M. F., W. C. Hunter, D. N. Pashley, and K. V. Rosenberg (2000). Setting conservation priorities for landbirds in the United States: The Partners in Flight approach. The Auk 117:541–548.

Caughley, G. (1994). Directions in conservation biology. Journal of Animal Ecology 63:215–244.

Conner, R. N., D. C. Rudloph, and J. R. Walters (2001). The Red-Cockaded Woodpecker: Surviving in a fire-maintained ecosystem. University of Texas Press, Austin.

Conners, M. G., E. L. Hazen, D. P. Costa, and S. A. Shaffer (2015). Shadowed by scale: Subtle behavioral niche partitioning in two sympatric, tropical breeding albatross species. Movement Ecology 3:38.

Conway, C. J., C. P. Nadeau, and L. Piest (2010). Fire helps restore natural disturbance regime to benefit rare and endangered marsh birds endemic to the Colorado River. Ecological Applications 20:2024–2035.

Craig, J., S. Anderson, M. Clout, B. Creese, N. Mitchell, J. Ogden, M. Roberts, and G. Usher (2000). Conservation issues in New Zealand. Annual Review of Ecology and Systematics 31:61–78.

Darby, P. C., D. L. DeAngelis, S. S. Romanach, K. Suir, and J. Bridevaux (2015). Modeling apple snail population dynamics on the Everglades landscape. Landscape Ecology 30:1497–1510.

Diamond, J. (1984). "Normal" extinctions of isolated populations. In Extinctions, M. Nitecki, Editor. Proceedings of the Sixth Annual Spring Systematics Symposium. University of Chicago Press, Chicago, IL.

Dugger, K. M., R. G. Anthony, and L. S. Andrews (2011). Transient dynamics of invasive competition: Barred Owls, Spotted Owls, habitat, and the demons of competition present. Ecological Applications 21:2459–2468.

Ewen, J. G., D. P. Armstrong, K. A. Parker, and P. J. Seddon (2012). Reintroduction biology: Integrating science and management. Wiley-Blackwell, Oxford, UK.

Frankham, R., D. Briscoe, and J. Ballou (2002). Introduction to conservation genetics. Cambridge University Press, Cambridge, UK.

Gummer, H., and M. Williams (1999). Campbell Island Teal: Conservation update. Wildfowl 50:133–138.

Haig, S., J. Ballou, and N. Casna (1995). Genetic identification of kin in Micronesian Kingfishers. Journal of Heredity 86:423–431.

Halliday, T. (1978). Vanishing birds. Holt, Rinehart and Winston, New York.

Hegg, D., D. MacKenzie, and I. G. Jamieson (2013). Use of Bayesian population viability analysis to assess multiple management decisions in the recovery program for the endangered Takahe Porphyrio hochstetteri. Oryx 47:144–152.

IUCN (2001). 2001 IUCN Red List Categories and Criteria, version 3.1. http://www.iucnredlist.org.

IUCN (2015). The IUCN Red List of Threatened Species. Version 2015-4. http://www.iucnredlist.org.

Jachowski, D. S., D. C. Kesler, D. A. Steen, and J. R. Walters (2014). Redefining baselines in endangered species recovery. The Journal of Wildlife Management 79:3–9.

Jamieson, I. (2009). Loss of genetic diversity and inbreeding depression in New Zealand's threatened bird species. Science for Conservation 293:1–59.

Jamieson, I. (2011). Founder effects, inbreeding, and loss of genetic diversity in four avian reintroduction programs. Conservation Biology 25:115–123.

Jamieson, I. (2015). Significance of population genetics for managing small natural and reintroduced populations in New Zealand. New Zealand Journal of Ecology 39:1–18.

Jamieson, I., G. Wallis, and J. Briskie (2006). Inbreeding and endangered species management: Is New Zealand out of step with the rest of the world? Conservation Biology 20:38–47.

Jamieson, I., and K. Ludwig (2012). Rat-wise robins quickly lose fear of rats when introduced to a rat-free island. Animal Behaviour 82:225–229.

Jones, S. L. (2010). Sprague's Pipit (Anthus spragueii) Conservation Plan. US Fish and Wildlife Service, Denver, CO.

Kesler, D. C., A. S. Cox, G. Albar, A. Gouni, J. Mejeur, and C. Plassé (2012). Translocation of Tuamotu Kingfishers, postrelease exploratory behavior, and harvest effects on the donor population. Pacific Science 66:467–480.

Laws, R. J., and D. C. Kesler (2014). An evaluation of Guam Micronesian Kingfisher population persistence under multiple translocation scenarios. University of Missouri Research Report, Columbia, MO.

MacArthur, R., and E. Wilson (1967). The theory of island biogeography. Princeton University Press, Princeton, NJ.

Masuda, B., and I. Jamieson (2012). Age-specific differences in settlement rates of saddlebacks (Philesturnus carunculatus) reintroduced to a fenced mainland sanctuary. New Zealand Journal of Ecology 36:123–130.

McCally, D. (1999). The Everglades: An environmental history. University Press of Florida, Gainesville.

McClelland, P. (2011). Campbell Island—pushing the boundaries of rat eradications. In Island invasives: Eradication and management, C. R. Veitch, M. N. Clout, and D. R. Towns, Editors. IUCN, Gland, Switzerland.

McClelland, P. (2014). Campbell Island Teal—the restoration of an island and the saving of a species. In Conservation through aviculture: ISBBC 2007: Proceedings of the IV International Symposium on Breeding Birds in Captivity, M. M. Lamont, Editor. TerraFauna, Canada.

McGowan, C. P., J. E. Hines, J. D. Nichols, J. E. Lyons, D. R. Smith, K. S. Kalasz, L. J. Niles, et al. (2011). Demographic consequences of migratory stopover: Linking red knot survival to horseshoe crab spawning abundance. Ecosphere 2:1–22.

National Research Council of the National Academies (2014). Progress toward restoring the everglades: The fifth biennial review—2014. National Academies Press, Washington, DC.

Niles, L. J., H. P. Sitters, A. D. Dey, P. W. Atkinson, A. J. Baker, K. A. Bennett, R. Carmona, et al. (2008). Status of the Red Knot in the western hemisphere. Studies in Avian Biology 36.

Obritschkewitsch, T., P. D. Martin, and R. S. Suydam (2001). Breeding biology of Steller's Eiders nesting near Barrow, Alaska, 1999, 2000. US Fish and Wildlife Service Technical Report NAES-TR-01 04, Fairbanks, AK.

Ortega, C. P., J. F. Chace, and B. D. Peers (2005). Management of cowbirds and their hosts: Balancing science, ethics, and mandates. Ornithological Monographs 57.

Parish, C. N., W. R. Heinrich, and W. G. Hunt (2007). Lead exposure, diagnosis and treatment in California Condors released in Arizona. In California Condors in the 21st Century, A. Mee and L. S. Hall, Editors. Series in Ornithology, no. 2. American Ornithologists' Union and Nuttall Ornithological Club, Washington, DC, pp. 97–108.

Pierce, R. J., and C. Blanvillain (2004). Current status of the endangered Tuamotu Sandpiper or Titi *Prosobonia cancellata* and recommended actions for its recovery. Wader Study Group Bulletin 105:93–100.

Puckett, E. E., D. C. Kesler, and D. N Greenwald (2016). Taxa, petitioning agency, and lawsuits affect time spent awaiting listing under the US Endangered Species Act. Biological Conservation 201:220–229.

Quakenbush, L. T., R. S. Suydam, K. M. Fluetsch, and C. L. Donaldson (1995). Breeding biology of Steller's Eiders nesting near Barrow, Alaska, 1991–1994. US Fish and Wildlife Service Technical Report NAES-TR 95 03, Fairbanks, AK.

Reynolds, M., and J. Klavitter (2006). Translocation of wild Laysan duck *Anas laysanensis* to establish a population at Midway Atoll National Wildlife Refuge, United States and US Pacific Possession. Conservation Evidence 3:6–8.

Ripple, W. J., S. K. Nelson, and E. M. Glenn (2003). Forest landscape patterns around Marbled Murrelet nest sites in the Oregon Coast Range. Northwestern Naturalist 84:80–89.

Robinson, S. K., F. R. Thompson III, T. M. Donovan, D. R. Whitehead, and J. Faaborg (1995). Regional forest fragmentation and the nesting success of migratory birds. Science 267:1987–1990.

Rothstein, S. I., and B. D. Peer (2005). Conservation solutions for threatened and endangered cowbird (*Molothrus* spp.) hosts: Separating fact from fiction. Ornithological Monographs 57:98–114.

Safine, D. E. (2015). Breeding ecology of Steller's and spectacled eiders nesting near Barrow, Alaska, 2013–2014. US Fish and Wildlife Service, Fairbanks Fish and Wildlife Office, Fairbanks, AK. Technical report.

Sandercock, B. K., K. Martin, and G. Segelbacher (2011). Ecology, conservation, and management of grouse. University of California Press, Oakland, CA.

Savidge, J. (1987). Extinction of an island forest avifauna by an introduced snake. Ecology 68:660–668.

Scott, J. M., D. D. Goble, J. A. Wiens, D. S. Wilcove, M. Bean, and T. Male (2005). Recovery of imperiled species under the Endangered Species Act: The need for a new approach. Frontiers in Ecology and the Environment 3:383–389.

Scott, J. M., D. D. Goble, A. M. Haines, J. A. Wiens, and M. C. Neel (2010). Conservation-reliant species and the future of conservation. Conservation Letters 3:91–97.

Sherfy, M. H., J. H. Stucker, and M. J. Anteau (2009). Missouri River emergent sandbar habitat monitoring plan—a conceptual framework for adaptive management. US Geological Survey, Northern Prairie Wildlife Research Center, Jamestown, ND.

Sherfy, M. H., J. H. Stucker, and D. A. Buhl (2011). Selection of nest-site habitat by interior least terns in relation to sandbar construction. The Journal of Wildlife Management 76:363–371.

Simberloff, D., and E. Wilson (1968). Experimental zoogeography of islands: The colonization of empty islands. Ecology 50:861–879.

Smith, K. N., A. J. Campomizzi, M. L. Morrison, and R. N. Wilkins (2013). Managing Brown-headed Cowbirds to sustain abundance of Black-capped Vireos. Wildlife Society Bulletin 37:281–286.

Stutchbury, B. (1997). Effects of female cowbird removal on reproductive success of Hooded Warblers. Wilson Bulletin 109:74–81.

Takekawa, J. E., and S. R. Beissinger (1989). Cyclic drought, dispersal and the conservation of the Snail Kite in Florida: Lessons in critical habitat. Conservation Biology 3:302–311.

Tuchmann, E. T., K. D. Commaughton, L. E. Freedman, and C. B. Moriwaki (1996). The Northwest Forest Plan, a report to the president and congress. USDA. Office of Forest and Economic Assistance, Washington, DC.

United Nations (1992). Convention on Biological Diversity. *In* Treaty Collection, Rio de Janeiro, Brazil. https://treaties.un.org/pages/ViewDetails.aspx?src=TREATY&mtdsg_no=XXVII-8&chapter=27&lang=en.

USFWS (US Fish and Wildlife Service) (2003). Recovery Plan for the Red-cockaded Woodpecker (*Picoides borealis*). US Fish and Wildlife Service, Atlanta, GA.

Villard, M. A., M. K. Trzcinski, and G. Merriam (1999). Fragmentation effects on forest birds: Relative influence of woodland cover and configuration on landscape occupancy. Conservation Biology 13:774–783.

Walters, J., C. Copeyon, and J. Carter III (1992). Test of the ecological basis of cooperative breeding in Red-cockaded Woodpeckers. The Auk 109:90–97.

Walters, J. R., S. R. Derrickson, D. Michael Fry, S. M. Haig, J. M. Marzluff, and J. M. Wunderle Jr. (2010). Status of the California Condor (*Gymnogyps californianus*) and efforts to achieve its recovery. The Auk 127:969–1001.

Weins, J. D., K. M. Dugger, K. E. Lewicki, and D. C. Simon (2016). Effects of experimental removal of Barred Owls on population demography of Northern Spotted Owls in Washington and Oregon—2015 Progress Report, 1–24. USGS. https://pubs.usgs.gov/of/2016/1041/ofr20161041.pdf.

Weiser, E., C. Grueber, and I. Jamieson (2012). AlleleRetain: A program to assess management options for conserving allelic diversity in small isolated populations. Molecular Ecology Resources 12:1161–1167.

Williams, J., C. ReVelle, and S. Levin (2005). Spatial attributes and reserve design models: A review. Environmental Modeling and Assessment 10:163–181.

Wilson, E. O. (2002). The future of life. Knopf, New York.

Zink, R. M., and H. W. Kale (1995). Conservation genetics of the extinct Dusky Seaside Sparrow *Ammodramus maritimus nigrescens*. Biological Conservation 74:69–71.

CHAPTER 28

Ecosystem and Landscape Management and Planning

Kerri T. Vierling and S. Mažeika P. Sullivan

Many bird species have different habitat needs across their life cycles, and it is therefore important to consider management at broad spatial extents ranging from ecosystems to landscapes. For instance, a Neotropical migrant that breeds in North America might select a reproductive territory that provides nesting habitat with suitable cover and forage. However, these breeding habitat requirements likely differ from suitable stopover habitat used during the course of migration, which also may differ from overwintering habitat. Thus, management for even a single species across its entire life cycle requires that we understand (1) which habitats are necessary for the different stages of a bird's life history, (2) how those habitats are distributed across ecosystems and landscapes, and (3) how to effectively manage ecosystems and landscapes to benefit bird populations and communities in light of natural and human disturbances and in the context of diverse social and cultural values.

The objective of this chapter is to provide a foundation on which to examine current challenges in bird management at ecosystem and landscape scales. The chapter is organized according to the following sections:

Management Frameworks
Ecosystem and Landscape Principles
Ecosystem- and Landscape-Based Case Studies: Freshwater
    Aquatic Ecosystems, Coniferous Forests, and Grasslands
Additional Management Considerations
Conclusions

Management objectives range from efforts focused on single species to efforts aimed at multispecies assemblages. To illustrate how ecological principles are considered in management contexts, we present a suite of case studies that illustrate approaches to bird management in specific ecosystem and landscape types. Our case studies are complemented by contributions from experts who present bird conservation challenges in Canada and South Africa in light of different ecological, social, and cultural contexts (see the box on pages 862–863 and the box on pages 870–871). At the end of the chapter, we discuss additional management considerations, including cultural perspectives to management, management across avian life cycles, management within multiple-use public lands, and alternative management strategies.

MANAGEMENT FRAMEWORKS

Avian management is an exercise in decision-making that involves choosing actions that are most likely to best achieve objectives at the appropriate spatial scale. Structured decision-making is a widely used approach that informs conservation and management strategies of bird species and communities. Structured decision-making (fig. 28.1) involves multiple steps that include (1) identifying the problem and its context (which may include a combination of social, legal, and ecological perspectives), (2) determining the objectives, (3) developing a set of alternative actions, (4) evaluating the consequences of each of those alternative actions, and (5) incorporating values to determine trade-offs associated with different options before decisions are made as to the action that best meets the objectives (Runge 2011). The social or legal context of a question (e.g., a question focused on a federally endangered species) may frame the nature of the problem and objectives, and there may be multiple objectives (Runge 2011). The preferred action that best addresses the objective(s) relative to the values incorporated into the decision analysis is identified; there can be numerous feedbacks that occur within the decision-making process.

Another commonly utilized framework for management of wildlife populations is an adaptive-management approach (Garton et al. 2012), which is a special case of struc-

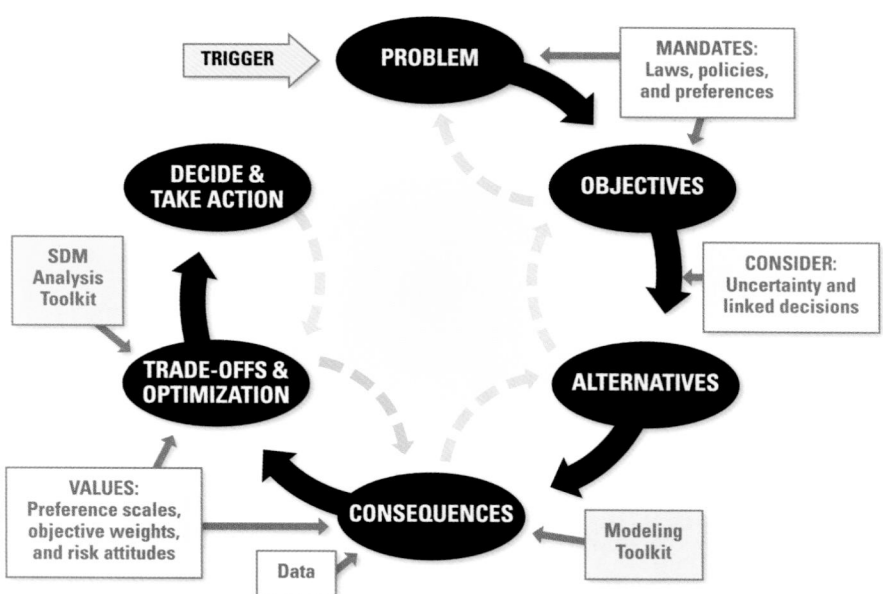

Figure 28.1. Conceptual diagram of the structured decision-making process. Social values and policies may frame objectives, and multiple alternative actions are developed to address the objectives. Alternative actions have different consequences, and trade-offs for each alternative action are identified prior to decision-making. Toolkits are a collection of resources (e.g., modeling software, GIS, other analytical products) that assist decision-makers in assessing multiple scenarios and their outcomes. Benefits of structural decision-making include a transparent process leading to decisions that are clear, documented, and repeatable. *Figure from Runge et al. 2016.*

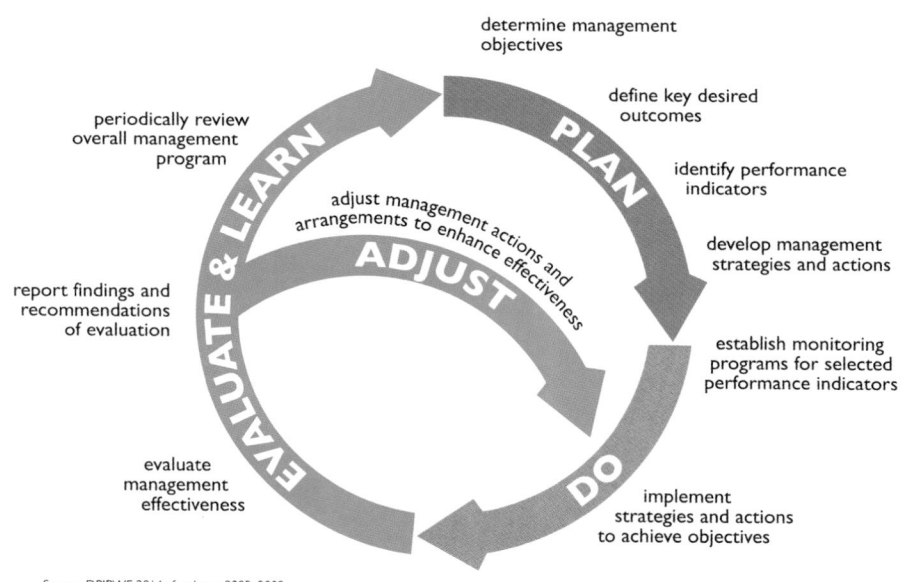

Figure 28.2. Conceptual diagram of the adaptive-management cycle. This cycle is an iterative process that places management in a research framework. *From DPIPWE 2014, after Jones 2005, 2009.*

tured decision-making. Adaptive management is an iterative process whereby (1) a management objective is identified, (2) management strategies and actions are determined with explicit attention to the desired outcomes and measures of effectiveness, (3) strategies and actions are implemented, (4) monitoring is conducted to determine the effectiveness of the strategies and actions, and (5) feedback from monitoring is continually integrated into the management plan (fig. 28.2). Adaptive management can be applied in situations in which there are competing hypotheses about how the system func-

tions (Lyons et al. 2008); adaptive management is often used in cases where the uncertainty about management outcomes is high. Adaptive management seeks to improve the effectiveness of management actions over time by continuously monitoring outcomes and modifying actions, and can be applied across spatial extents from small, discrete habitat areas (i.e., patches) to ecosystems to landscapes.

Adaptive management and decision-analysis tools such as structured decision-making can be used together to address conservation and management issues associated with birds.

For example, structured decision-making is widely used by the US Fish and Wildlife Service to assess extinction risks for listing decisions and identify optimal recovery actions. In the Migratory Bird Program, structured decision-making is involved in many of the regulatory decisions (e.g., waterfowl harvest) (Williams et al. 2009, Williams and Brown 2012). Adaptive management and decision-analysis tools can be used together to address multiple management goals related to multispecies management. For instance, these frameworks have been used for the management of the endangered Whooping Crane (*Grus americana*), the threatened Piping Plover (*Charadrius melodus*), and the Interior Least Tern (*Sternula antillarum athalassos*) in the Platte River in Nebraska (Smith 2011), where both social (water use) and legal (Endangered Species Act recovery goals) issues are important aspects influencing management decisions (Smith 2011).

Structured decision-making and adaptive resource management are but two approaches of many used to inform conservation and management strategies of bird species and communities. Environmental risk assessments, such as multicriteria decision-analysis and comparative risk assessment, represent other decision-making strategies used to balance scientific findings with input from stakeholders and other factors. In all cases, management is driven by a combination of ecological and human factors (e.g., social perceptions, legislation, cultural values, valuation of ecosystem services) that drive management objectives (Slocombe 1998).

As human populations increase, changes in ecosystem and landscape structure and function are inevitable, and the role of effective management frameworks will become increasingly important to avian conservation. There are currently over 900 species of breeding birds in the United States alone, and bird populations in many habitats within the United States are declining (North American Bird Conservation Initiative 2014; figs. 28.3A, B). Human activity can alter ecosystems and landscapes, and these changes will influence birds across levels of biological organization (i.e., individuals to population to communities). Changes in avian population viability, community structure over time, or functional roles can be strongly influenced by human-mediated changes to ecosystems and landscapes. Thus, it is important to understand how alterations to ecosystems and landscapes affect birds and, conversely, how changes in bird communities might influence ecosystems and landscapes via losses of important avian functional roles such as seed dispersal, scavenging, and pollination among others (see below and chapter 29 on ecosystem services). To understand the intricate relationships between birds and ecosystems and landscapes, we subsequently provide a brief overview of ecosystem principles and landscape fundamentals.

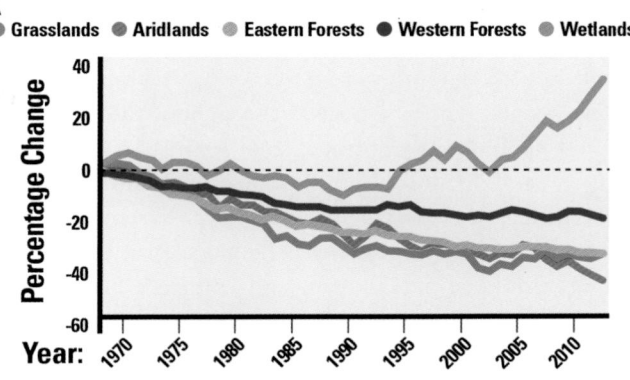

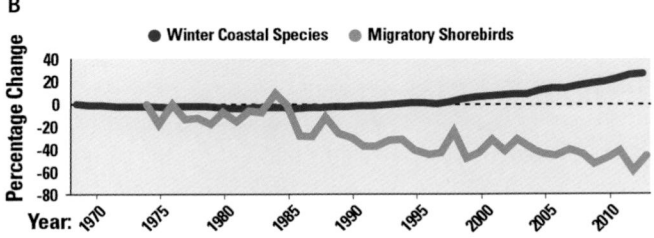

Figure 28.3. A, The percentage change of breeding birds in five inland ecosystem types in the United States. These trends are based on bird species that are breeding-habitat obligates in grasslands, aridlands, eastern and western forests, and wetlands in the United States. *B,* The percentage change of breeding birds in coastal environments in the United States. These trends are based on birds that are breeding-habitat obligates in coastal environments. *From the North American Bird Conservation Initiative 2014, State of the Birds 2014 Report.*

ECOSYSTEM AND LANDSCAPE PRINCIPLES

Ecosystem Basics

Birds are integral parts of functional ecosystems, and thus effective ecosystem management should protect both bird species and the important ecosystem functions they provide. Ecosystems are maintained by a suite of processes related to the cycling of energy and nutrients. Both living (e.g., producers, consumers, decomposers) and nonliving (e.g., organic, inorganic, climatic characteristics) elements form ecosystems. Food webs (i.e., consumer-resource systems) are central properties of ecosystems, depicting complex energetic networks. Birds span multiple feeding relationships—also known as foraging guilds or feeding groups—that represent critical food-web linkages. In a simplified food web, trophic levels are clearly distinguished. At the first trophic level, primary producers (plants, algae, and some bacteria) produce organic plant material via photosynthesis. Herbivores are primary consumers that feed solely on plants and represent the second trophic level. The third trophic level is represented by secondary consumers such as carnivores and scavengers (i.e., predators) that feed on primary consumers. Apex predators are those that reside at the top of the food chain. For example, species of raptors are often considered top predators, whose trophic impacts cascade throughout the ecosystem (Nislow et al. 1999, Parrish et al. 2001).

In general, food webs are typically much more complex, whereby a gradient of trophic positions exists between complete autotrophs (i.e., those that obtain their sole source of carbon from the atmosphere) and complete heterotrophs (e.g., birds) that must feed to assimilate organic matter. Food webs provide specific information on consumer-resource relationships but can also provide insight into other types of interspecific relationships. For instance, Ospreys (*Pandion haliaetus*) are apex predators (fig. 28.4), but in addition to direct predation, Ospreys and other raptors can exert indirect effects on prey populations by establishing a "landscape of fear" (Brown and Kotler 2004) that can influence prey behavior, predator-prey dynamics, and community structure.

Understanding these types of trophic and interspecific relationships can therefore improve our understanding of how ecosystem and landscape changes might influence the relationships important for individual species' conservation and management as well as community dynamics.

The wide array of food-web positions occupied by birds allows them to play an important role in maintaining ecosystem function (Lundberg and Moberg 2003). Birds are considered "mobile links"—organisms that actively move and connect habitats and ecosystems in space and time. The ecological functions of birds include three major mobile linkages: genetic, resource, and process (Şekercioğlu 2006). Genetic linkers include seed dispersers and pollinators, such as frugivores and nectarivores that transport genetic material among plants and habitats. Resource linkers primarily refer to birds as nutrient depositors through the transport of minerals, nutrients, and even contaminants from one ecosystem to another (Blais et al. 2007). Process linkers include those related

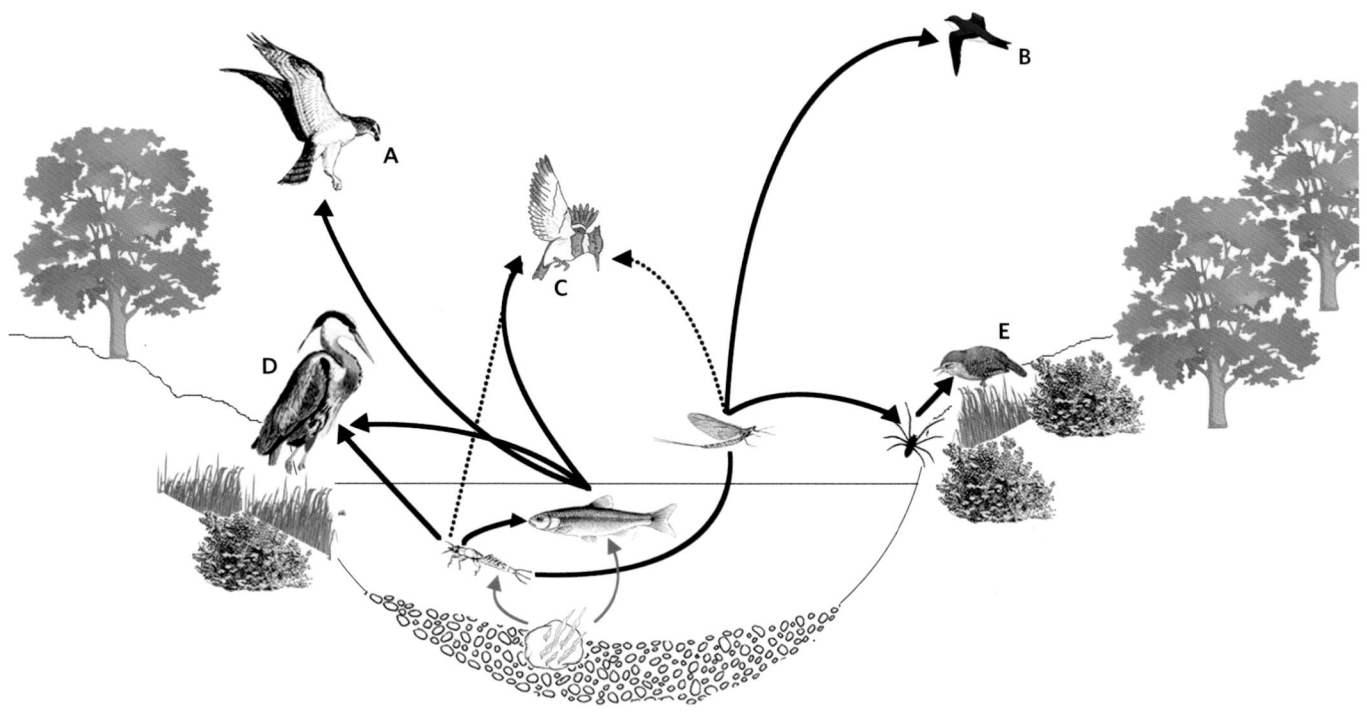

Figure 28.4. Schematic representation of ecosystem functions relative to example trophic processes that birds can encompass in a stream ecosystem. Solid arrows represent primary/obligate feeding relationships; dashed arrows represent facultative feeding relationships. *A*, Ospreys (*Pandion haliaetus*) belong to the **piscivores:water ambushers** foraging guild and are apex predators in river ecosystems; *B*, Tree Swallows (*Tachycineta bicolor*) are **insectivores:air screeners** and feed on a mixture of both emergent aquatic insects and terrestrial aerial insects (Alberts et al. 2013); *C*, Belted Kingfishers (*Ceryle alcyon*) belong to the **piscivore:water plunger** foraging guild, although they also feed on crayfish and aquatic insects (see Sullivan et al. 2006); *D*, Great Blue Herons (*Ardea herodias*) are classified as **piscivores:water ambushers**; *E*, Various species of wrens (e.g., House Wrens, *Troglodytes aedon*) are **insectivores:ground and low-canopy gleaners**. Although they are not known to feed directly from the water, they consume spiders that are known to be highly reliant on aquatic resources (Tagwireyi and Sullivan 2015). Through their foraging and movements, these and other avian species integrate aquatic and terrestrial environments.

Figure 28.5. Examples of the main types of avian ecosystem-service providers. *A*, Seed disperser: Black-mandibled Toucan (*Ramphastos ambiguus*) (Las Cruces, Costa Rica); *B*, Pollinator: Snowy-bellied Hummingbird (*Amazilia edward*) (Las Cruces, Costa Rica); *C*, Nutrient depositor: Gentoo Penguin (*Pygoscelis papua*) (Port Lockroy, Antarctica); *D*, Grazer: Cackling Goose (*Branta hutchinsii*) (California, USA); *E*, Insectivore: Golden-crowned Warbler (*Basileuterus culicivorus*) (Las Cruces, Costa Rica); *F*, Raptor: Bald Eagle (*Haliaeetus leucocephalus*) (Alaska, USA); *G*, Scavenger: Andean Condor (*Vultur gryphus*) (Patagonia, Chile); *H*, Ecosystem engineer: Slaty-tailed Trogon (*Trogon massena*) (Pipeline Road, Panama). *Reprinted from Trends in Ecology and Evolution 21 (8); C. H. Şekercioğlu, Increasing awareness of avian ecological function, p. 467, © 2016, with permission from Elsevier.*

to both trophic and non-trophic processes. Insectivores, scavengers, and raptors are all important trophic-process linkers because of their role in connecting habitats and ecosystems as primary or secondary consumers (fig. 28.5). Non-trophic-process linkers include ecosystem engineers that modify the environment, such as through the construction of cavities (e.g., woodpeckers) and burrows (e.g., some owls and bee-eaters) for nesting, which in turn can be used be a variety of other species (Daily et al. 1993).

Multiple abiotic factors interact with food webs and communities within an ecosystem. Disturbances (e.g., fire, insect outbreaks, storms, disease, changes in vegetation via human activity) are important processes because they can influence how different abiotic and biotic components within ecosystems interact (e.g., DellaSalla and Hanson 2015). Disturbance effects on ecosystems may vary because of the intensity and frequency of the disturbance, the time since the disturbance and its predictability (i.e., are species "adapted" to a disturbance, such as waterbirds are to seasonal floods in rivers), the geographic area and extent affected by the disturbance, and the land-use activities both before and after the disturbance (e.g., Morrison et al. 2006). The disturbance history of an ecosystem, in turn, will influence ecosystem function (e.g., Bengtsson et al. 2000, Knick and Rotenberry 2000).

The interaction of different disturbance regimes and their effects on ecosystems is well illustrated by perturbations that have influenced the endangered Red-cockaded Woodpecker

(*Picoides borealis*; fig. 28.6A) in the southeastern United States (fig. 28.6B). Red-cockaded Woodpeckers occur in mature longleaf pine (*Pinus palustris*) forests (fig. 28.6C) that were historically maintained by frequent low-severity fires (Jackson 1994). In recent decades, a suite of disturbances has altered Red-cockaded Woodpecker habitat, including fire, logging, and Hurricane Hugo (1989). Management actions have been evaluated via adaptive-management frameworks (e.g., Moore and Conroy 2006) to assist with Red-cockaded Woodpecker recovery goals. These actions range from efforts to increase longleaf pine regeneration to creating artificial cavities in trees, and have varied depending on the extent and type of disturbance across Red-cockaded Woodpecker breeding habitat (Williams and Lipscomb 2002, Bainbridge et al. 2011).

Landscape Ecology Basics and the Concept of Spatial Scale

Landscapes are not homogeneous environments but rather consist of patches of habitats arranged within a dominant land cover (i.e., the matrix), and birds are affected by both patch and matrix dynamics. Patches are the habitats that are associated with a specific species or set of species, and are relatively homogeneous areas that are distinguishable from their surroundings. Landscape composition includes patch richness and diversity (table 28.1) and generally reflects the number and proportions of different patch types in a landscape

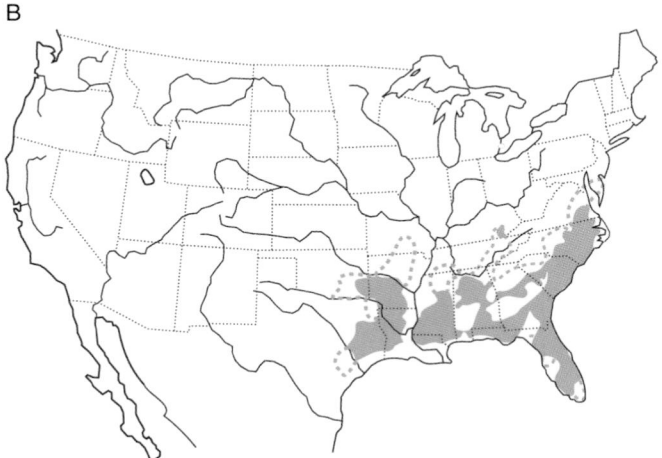

Figure 28.6. A, Red-cockaded Woodpecker; *B*, Year-round range in green, with dotted lines representing historic range; *C*, Pine breeding habitat. Variability in disturbance regimes (fire, logging, hurricanes) can alter ecosystem dynamics. These disturbances can vary both spatially and temporally, from short-term (years) to longer term (decades), and suitable habitats are patchily distributed throughout this range. *Photos courtesy of Geoffrey Hill; map from Birds of North America Online http://bna.bird.cornell.edu/bna, maintained by the Cornell Lab of Ornithology.*

(Fahrig et al. 2011). In contrast, landscape configuration describes the spatial characteristics and arrangements of these patches (Fahrig et al. 2011; table 28.1). The study of landscape ecology—which considers both landscape composition and configuration and examines the spatial relationships among interacting patches or ecosystems (Wu and Hobbs 2007)—has become central to avian ecology and conservation (table 28.1).

Because landscape patterns affect animal movement and demographic parameters (e.g., survival rate, fecundity), landscape ecology has been widely used as a theoretical basis for nature conservation (e.g., Baker 1989). For instance, patch size and shape are important characteristics that might influence nature reserve design (Diamond 1975; fig. 28.7). The design principles first proposed by Diamond (1975) incorporated patch characteristics and spatial arrangement. Using species extinction rates as the response metric, Diamond (1975) notes that larger patches are "better" for birds (i.e., extinction rates are low in such patches) compared with smaller patches (example A, fig. 28.7). Similarly, Diamond (1975) proposes that (1) a single large patch is better than several smaller patches of the same area (B, fig. 28.7); (2) patches close together are more beneficial to birds than patches that are farther away from each other (C, D, fig. 28.7); (3) patches connected by a corridor are better than those that are disconnected (E, fig. 28.7); and (4) patches that are circular are of greater benefit to birds than elongated patches (F, fig. 28.7).

While patch size and shape are undoubtedly important in the management of birds, the matrix surrounding patches must also be considered. Patches with similar characteristics may support markedly different bird communities depending on the matrix. For example, Kennedy et al. (2010) examined bird communities in Jamaican forests where forest fragment characteristics were similar but were embedded in one of three different matrices: an agricultural matrix, a peri-urban matrix (low-density residential housing), and a matrix dominated by bauxite mining. Kennedy et al. (2010) found that matrix type strongly influenced bird diversity with a lower percentage of resident species in forest fragments surrounded by bauxite and peri-urban matrices compared with forest fragments surrounded by agriculture (fig. 28.8). This study suggests that the matrix is important, influencing resource availability both inside and outside forest fragments. Additionally, results from this study indicate that species-specific responses to matrices influenced different foraging guilds differently within forest fragments surrounded by distinct matrix types.

Processes like fragmentation may result in smaller patches, the creation of edges, or the creation of corridors. Some species, such as the ovenbird (*Seiurus aurocapilla*) are area-sensitive; that is, they require large expanses of contigu-

Table 28.1. Characteristics of landscapes that influence avian populations and communities. Modified from Morrison et al. 2006.

Term	Definition	Example	Citation
Patch richness and diversity	Number of different patch types and patch heterogeneity within a landscape	Higher diversity of patch types influences bird diversity in riverine landscapes	Sullivan et al. (2007)
Patch dynamics	Changes in patches over space and time as a result of (1) types of disturbance, (2) frequency and severity of disturbances, and (3) post-disturbance vegetation succession	Grassland patches may be diverse, due to interactions between grazing intensities, fire histories, and post-disturbance vegetation succession	Fuhlendorf and Engle (2001)
Fragmentation	Division of contiguous patches of habitat into smaller patches that may be isolated from one another	Fragments of tropical forest decline in bird species richness over time as they become isolated from continuous forest	Turner and Corlett (1996)
Corridors	Environments or habitats in the landscape that connect isolated patches to each other; these are often linear	Riparian corridors support populations of Neotropical migrants	Gentry et al. (2006)
Edge effects	Opening (or edge) that influences microclimate or biotic communities of a patch	Clearcuts in forests; adjacent forested environments at the interface of clearcuts are warmer, drier, and support different vegetation and bird communities	Murcia (1995), Chalfoun et al. (2002)

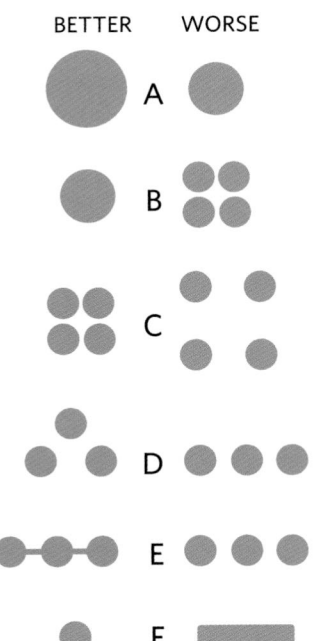

BETTER   WORSE

A
B
C
D
E
F

Figure 28.7. Patch sizes and arrangements for suggested nature reserve designs (Diamond 1975). In all cases, the designs in the left column would result in lower species extinction rates than the designs on the right. *Reprinted from Biological Conservation, vol. 7, J. M. Diamond, The island dilemma: Lessons of modern biogeographic studies for the design of natural reserves, p. 143, copyright (1975), with permission from Elsevier.*

ous forest in which to breed (Van Horn and Donovan 2011). Disturbances that create gaps or decrease the amount of contiguous forest may have negative consequences for these species (e.g., Donovan et al. 1995), in part because edges can be associated with an increase in predators and brood parasites (e.g., Brown-headed Cowbirds, *Molothrus ater*; Donovan et al. 1997). Corridors are typically narrow, linear connections between patches, and narrow corridors inherently have a high amount of edge. It is important to note that the effectiveness of corridors may be species-specific and parameters may differ in width and environmental setting (e.g., Askins et al. 2012).

Birds move within and among habitat patches, ecosystems, and landscapes during their foraging, reproductive, and migratory activities, and the ability of birds to move might be a function of individual patch characteristics as well as ecosystem or landscape context. For instance, permeability is the degree to which organisms can pass through patches, and this can be facilitated through habitat modifications. Tremblay and St. Clair (2011), for example, suggest that improving the permeability of urban landscapes for songbirds may be achieved by increasing connectivity and minimizing gaps in vegetation.

Concepts of spatial scale are inherent in studies of bird movement and ecology because individual birds, populations, and communities operate across multiple spatial scales (fig. 28.9). Spatial scale is a combination of grain and extent.

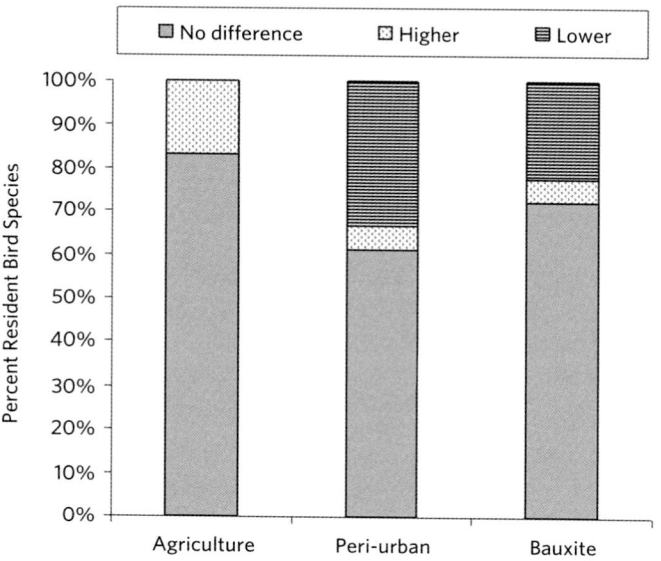

Figure 28.8. Percentages of resident bird species in agricultural, peri-urban, or bauxite mining landscapes that exhibited significantly lower abundance, higher abundance, or no abundance difference relative to forested landscapes over the study period (*n* = 36 species). *Reprinted from Ecological Monographs, vol. 80, C. M. Kennedy et al.,* Landscape matrix and species traits mediate responses of Neotropical resident birds to forest fragmentation in Jamaica, *p. 661, copyright (2010), with permission from John Wiley and Sons, Inc.*

Grain size refers to the resolution of environmental features, and extent refers to the area of land under consideration. Spatial scale is a function of the scale at which the organism perceives its environment, and its perception of the environment can be different than our human perception.

Effective ecosystem and landscape management requires consideration of a number of additional perspectives. First, there is no "typical" size for an ecosystem or landscape; the perception of an ecosystem or landscape by a raptor may differ fundamentally from the perception of a hummingbird. The concept of scale is emphasized, because different spatial scales are likely to be important for different bird species and because species forage and select habitat at various spatial scales (i.e., hierarchy theory; Wiens 1985, Kotliar and Wiens 1990). Additionally, the spatial extent at which management activities are likely to be implemented is important (Morrison et al. 2006). Finally, social, cultural, and economic factors may also vary with spatial scale and can influence management.

ECOSYSTEM AND LANDSCAPE-BASED CASE STUDIES
Overview of Case Studies

In this section, we focus on ecosystem and landscape management for three different systems: aquatic ecosystems, coniferous forests, and grasslands. We chose these ecosystem and landscape types as representative of some of the management challenges faced where various disturbance regimes

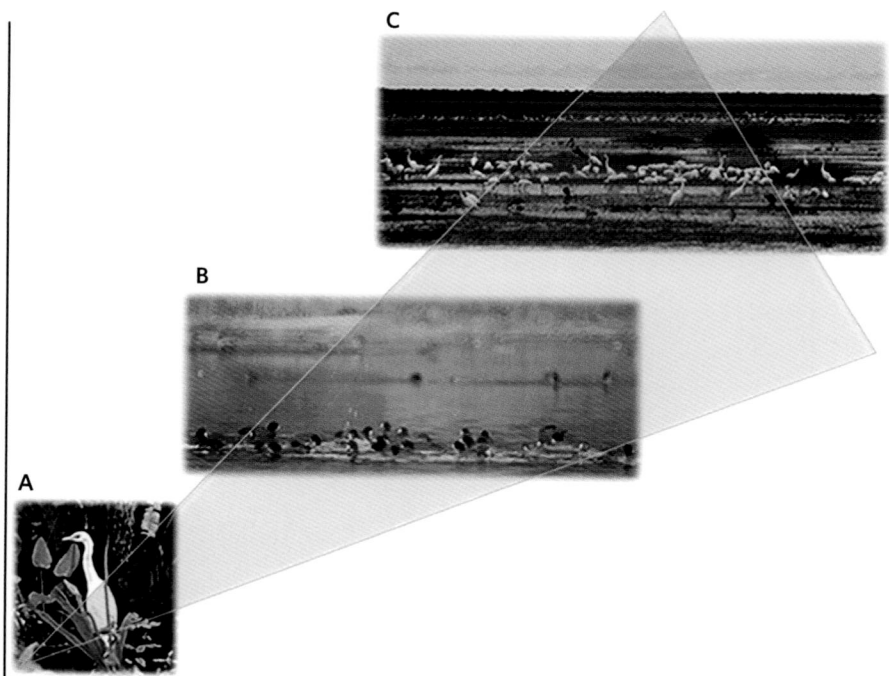

Figure 28.9. Conceptual relationship of the gradient among spatial scales (from individual habitat to landscape) and biological levels of organization of birds (from individuals to communities). The block arrow represents increasing spatial extents from habitat/patch to ecosystem to landscape. *A,* Great Egret (*Ardea alba*) in the Florida Everglades (USA); *B,* A flock of American Coots (*Fulica americana*) in an Idaho lake (USA); *C,* A mixed flock of waterbirds in a Botswana wetland complex. Understanding the levels of biological organization at which birds most commonly interact with their environment is critical in managing birds across spatial extents. *Photos and graphics courtesy of S. M. P. Sullivan.*

interact with anthropogenic stressors. Implicit in our case studies is the idea that various management frameworks (e.g., adaptive management, structured decision-making) contribute to the management actions discussed below. We start each case study with a broad context relating to the importance of the ecosystem or landscape to birds, bird population trends, and disturbance processes critical to ecosystem function. We follow up with management activities focused on either a single bird species or communities that reflect current management and conservation activities at the ecosystem to landscape scales.

Freshwater Aquatic Ecosystems

History and Context of Importance to Birds   Wetland and other freshwater habitats (e.g., streams, rivers, lakes) are critically important to avian populations because of their diversity of food and habitat resources (Buckton and Ormerod 2002, Mitsch and Gosselink 2007, Sullivan et al. 2007). Birds, because of their daily and seasonal movement patterns, might depend on multiple aquatic ecosystems across their life cycle. Conservation of important breeding and wintering areas is critical, but it is also important to manage aquatic stopover habitats that are used by migratory waterbirds as fueling and resting locations (e.g., Newton 2006, such as many river and wetlands throughout the American Midwest.

Multiple factors affect bird use of wetlands, including the quality, quantity, extent, temporal availability, and temperature of the water; the availability of food and refuge habitat (e.g., patchiness or openness of vegetation), and the presence or absence of predators, among others (Baschuk et al. 2012, Glisson et al. 2015, Holopainen et al. 2015). Likewise, the quality of stream and river habitat, including chemical water quality and physical condition, is an important determinant of waterbird use of fluvial (i.e., flowing water) systems (Ormerod and Tyler 1993, Sullivan et al. 2006). Because of the strong relationships between waterbirds and water quality and aquatic habitat, some species have been used as "indicators" of ecosystem condition. For example, the American Dipper (*Cinclus mexicanus*; fig. 28.10A) is a year-round resident of mountainous watersheds across much of western North America (fig. 28.10B) that feeds on aquatic macroinvertebrate larvae and small fish (Ealey 1977, Ormerod 1985). American Dippers have been shown to be associated with environmental characteristics from patch to landscape scales (Sullivan and Vierling 2012). This species is also sensitive to changes in water quality and aquatic habitat and can be absent in areas of poor water quality (Ormerod et al. 1991, Morrissey et al. 2004). Thus, this species is considered to be an indicator of unpolluted waters.

Management of wetlands may benefit specific waterfowl species, but also benefits ecosystem function because water-

Figure 28.10. A, The American Dipper (*Cinclus mexicanus*) is a year-round resident of *B,* streams and rivers of western North America and is often regarded as an indicator of water quality. *Photo courtesy of timjhopwood.com. Map from Birds of North America Online http://bna .bird.cornell.edu/bna, maintained by the Cornell Lab of Ornithology.*

birds can be key movers of nutrients, plants (seeds), and invertebrates among waterbodies and over larger geographic areas (Clausen et al. 2002, Green et al. 2002). For instance, waterfowl play an important role in the dispersal of aquatic organisms via their movements across landscapes (Green et al. 2002, Figuerola et al. 2003, van Leeuwen et al. 2012) (fig. 28.11). Green et al. (2008) suggest that birds may be particularly important in facilitating the recolonization of inverte-

GRAEME CUMMING: LANDSCAPE ECOLOGY OF WATERBIRDS AND CONSERVATION CHALLENGES FOR SOUTHERN AFRICAN DUCKS

Graeme Cumming. *Photo by Katharina Lauterbach.*

I am a landscape ecologist (see figure on the left) who works on problems relating to scale and spatial variation in landscape ecology and conservation. One of my recent projects has focused on waterbirds and wetlands, which are one of the world's most endangered habitats. Wetland conservation and management are huge challenges, particularly for species that use many wetland sites across the landscape. Movements of migratory species from Europe to Africa are supposedly facilitated by governmental responsibilities outlined in the Convention on Biodiversity, the Convention on Migratory Species, the RAMSAR convention, and AEWA (the Africa-Eurasia Waterbird Agreement). In water-scarce southern Africa, however, there is little coordinated policy or management for far-ranging wetland-dependent species.

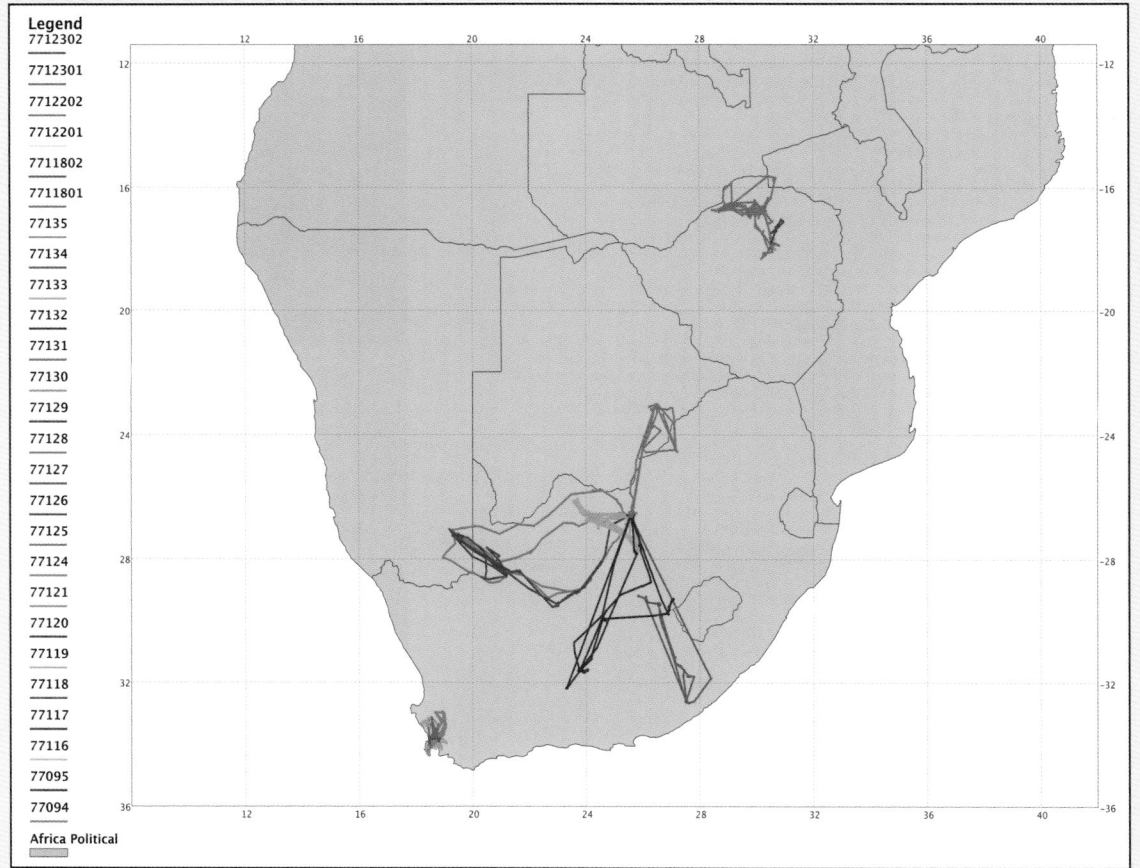

Movement tracks of Egyptian Geese tagged at Strandfontein (south), Barberspan, and Lake Manyame (north), as recorded using satellite GPS telemetry. The Barberspan birds move differently from those at the other two sites. *Reprinted with permission from Cumming et al. 2012a.*

We have been studying the causes and consequences of the movements of ducks around southern Africa using a combination of bird counts, atlas data, ringing data, and satellite telemetry. Apart from the Knob-billed Duck, the 16 species of southern African ducks are not long-distance migrants. Many do, however, undertake migrations to undergo wing-feather molt. Ducks lose all of their wing feathers at the same time, and the three to five weeks a year when they cannot fly is a high-vulnerability period. Ducks are also constrained, when breeding, to remain at their nest site, and the ducklings must be guarded by the parents until they are old enough to fly for themselves. Breeding—from making a nest to full fledging of the offspring—takes roughly three months. In between breeding and wing-feather molt, most species move across the landscape in flocks in a seminomadic fashion.

Our results show that ducks move differently in different locations, with some populations remaining in a relatively small area and others ranging much farther (see figure on previous page). Using telemetry and satellite data, one of my PhD students, Dominic Henry, has found that Egyptian Geese (see figure below; actually a shelduck) and Red-billed Teal (*Anas erythrorhyncha*) appear to predict improvements in habitat quality, moving toward areas as they green up; another student, Chevonne Reynolds, has demonstrated through field sampling and feeding trials that ducks are capable of moving relatively large quantities of seeds and other plant and animal propagules, including many potentially invasive species, around the landscape. Our work also shows that ducks are potentially important vectors of avian influenza. Influenza viruses are transmitted through the water, and ducks are often asymptomatic carriers. They can pass influenza viruses through feces and mucus via the water column to other wetland species that may be more susceptible, or infect domestic poultry and domestic ostriches by sharing their water and food supplies.

From a landscape management perspective, habitat must be maintained across sufficiently large areas and in sufficient quality to retain the many ecological roles that ducks play: as a food source for other species (including people), as herbivores that keep wetlands more open, as dispersers that both facilitate the recolonization of seasonal wetlands by plants and invertebrates and spread invasive species, and as vectors of parasites and pathogens. Management actions must consider different life history stages and demands. For example, key stepping-stone seasonal wetlands that help connect permanent waterbodies across arid areas must be maintained; and habitats that are used for only part of the year may nonetheless be vital for the long-term persistence of waterbird populations. Molting and breeding sites are of particular conservation importance, and maintaining habitat for specialists, such as the declining Pygmy Goose (*Nettapus auritus*, which depends on *Nymphaea* water lilies) and (diving) Maccoa Duck (*Oxyura maccoa*), is also of high priority. At the same time, potentially negative impacts and conflict with humans must be mitigated across the landscape by managing proximity problems: feeding by Egyptian and Spur-wing Geese (*Plectropterus gambensis*) on crops and golf courses, infection of domestic ostriches with avian influenza, and the introduction of invasive plant and animal species to farm dams and other waterways in which they may not be effectively managed. Wetlands and waterbirds must be conserved together, but achieving the necessary coordination and habitat protection across wide areas of a developing country is extremely difficult. For further information on this topic, see Cumming et al. 2011, Cumming et al. 2012a,b, Cumming et al. 2013, and Reynolds et al. 2015.

An Egyptian Goose (*Alopochen aegyptiacus*) carrying one of Microwave Telemetry's 30g solar-powered satellite GPS transmitters. The transmitter is attached using a backpack harness of Teflon ribbon. The device is set to record accurate coordinates every two hours and to upload these to the Argos data system, via a satellite link, every four days. The duck also has a metal ring (left foot) and a color ring (right foot) to facilitate reidentification. *Photo courtesy of Graeme Cumming.*

Macroinvertebrate propagules in droppings

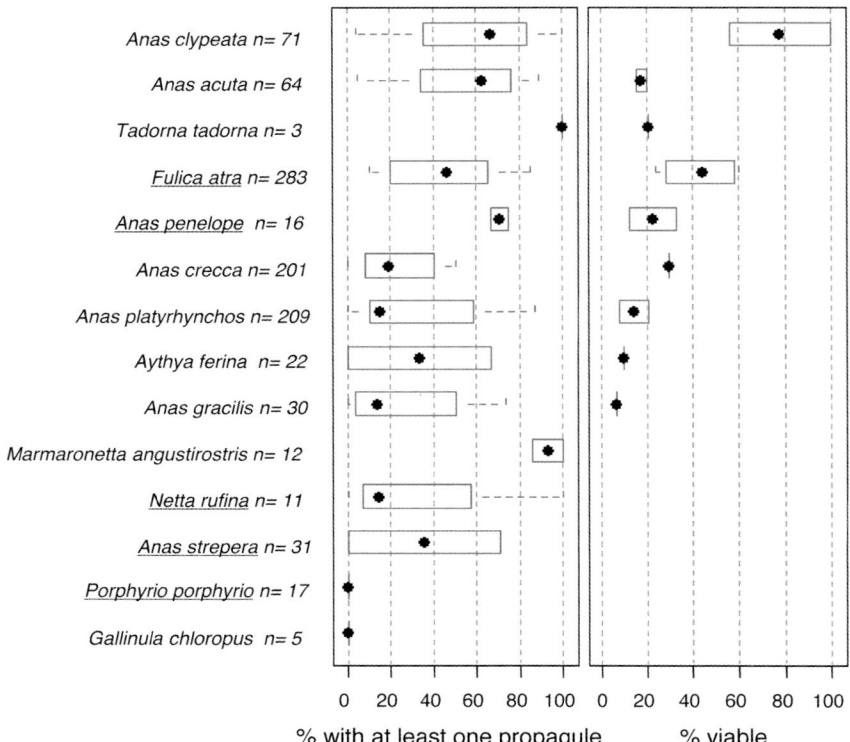

Anas clypeata n= 71
Anas acuta n= 64
Tadorna tadorna n= 3
Fulica atra n= 283
Anas penelope n= 16
Anas crecca n= 201
Anas platyrhynchos n= 209
Aythya ferina n= 22
Anas gracilis n= 30
Marmaronetta angustirostris n= 12
Netta rufina n= 11
Anas strepera n= 31
Porphyrio porphyrio n= 17
Gallinula chloropus n= 5

0 20 40 60 80 100   0 20 40 60 80 100

% with at least one propagule   % viable

Figure 28.11. Percentage of droppings with at least one intact macroinvertebrate propagule, and the percentage of these propagules that was viable (based on collections of droppings in the field only). On the vertical axes, bird species are ranked according to decreasing quantitative dispersal capacity, calculated from both the prevalence and viability of propagules in their droppings. *n* denotes the number of droppings collected. Underlined species are considered predominantly herbivores. *Reprinted from Journal of Biogeography 39 (11), C. H. A. van Leeuwen et al., Gut travellers: internal dispersal of aquatic organisms by waterfowl, p. 2034, © 2012 Blackwell Publishing, Ltd., with permission from John Wiley and Sons.*

brates and plants with limited drought resistance in arid climates. In this way, management activities targeting bird conservation can also protect and maintain critical ecosystem functions.

Anthropogenic disturbances to wetlands, rivers, and lakes include (1) water extraction for irrigation, livestock, household use, and drinking; (2) conversion to farmland; (3) channelization and flow alterations to prevent flooding; (4) impoundments for water storage and power generation; (5) climate change; and (6) pollution with nutrients, plastics, and heavy metals. These and other factors have led wetland and other aquatic systems to be among the world's most endangered habitats (Dudgeon et al. 2006, Vörösmarty et al. 2010). Fifty percent (29 of 58) of all US bird species (excluding Hawai'i and territories) that were listed either as federally threatened or endangered or on the US Fish and Wildlife Service 1995 List of Migratory Nongame Birds of Management Concern are wetland or aquatic dependent (Erwin et al. 2000).

However, there are signs of recovery within the continental United States (see fig. 28.3A); 87 obligate freshwater breeding bird species currently show strong population growth, with more than a 40 percent gain since 1968 (North American Bird Conservation Initiative 2014). These and other improvements are largely a legacy of important conservation actions and legislations including the Clean Water Act and

the Farm Bill's conservation provisions. Joint public-private cooperative conservation and management efforts that have protected or restored wetlands have contributed to reversing declines and provide important evidence of the potential for management success. Below we provide an example of how management activities at ecosystem and landscape scales can help promote management goals for Wood Ducks (*Aix sponsa*), a species of conservation and management interest.

Wood Duck   The entire annual cycle of the Wood Duck (fig. 28.12A)—which ranges broadly across North America and into the Caribbean (fig. 28.12B)—is dependent on forested wetland complexes that include live forest, rivers, oxbows and backwaters, riparian corridors, beaver ponds, and emergent vegetation. Bottomland hardwood wetlands of the American southeast provide critical habitat (fig. 28.12C), where large populations of Wood Ducks overwinter and where some populations breed as well (NRCS 2012). Forested wetlands characterized by seasonal or semipermanent water and a high density and structural diversity of trees provide critical nesting sites and abundant food resources (acorns and other forest mast, aquatic invertebrates, seeds) (Kaminski et al. 1993, Foth et al. 2014).

Although Wood Duck populations have recovered since the early twentieth century, habitat loss and fragmentation and reductions in habitat quality are ongoing issues of concern.

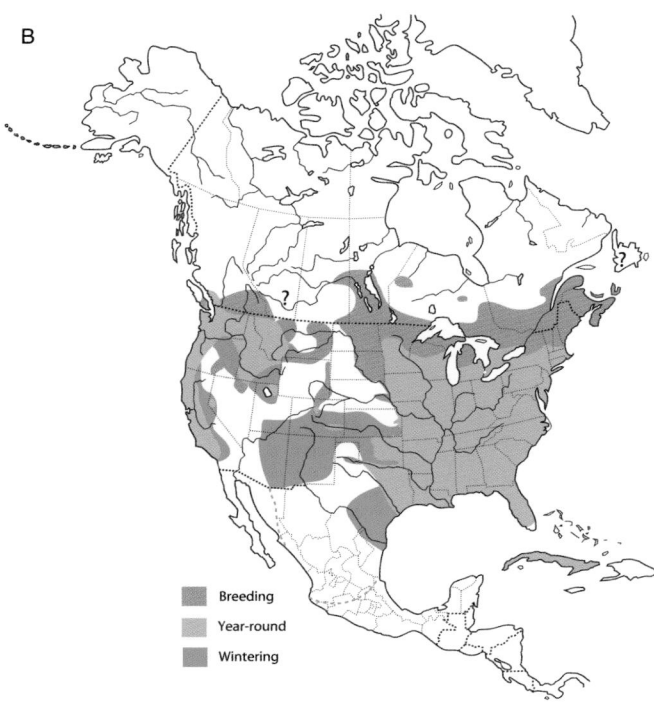

Figure 28.12. A, Male Wood Duck. *Photo courtesy of Mike Sweet / United States Fish and Wildlife Service. B*, Wood Duck range map. *From Birds of North America Online http://bna.bird.cornell.edu/bna, maintained by the Cornell Lab of Ornithology. C*, In addition to bayous, swamps, and rivers, the historic floodplain of the Mississippi River contained 24 million acres of hardwood forests. The remaining 4.4 million forested acres provide vital habitat for wildlife. *Photo © Byron Jorjorian.*

For example, the lower Mississippi Alluvial Valley (MAV) is particularly important to Wood Ducks, yet by the turn of the twenty-first century, only 25 percent of the original forest remained in the MAV (Twedt and Loesch 1999). Reductions in habitat quality have been caused in part by altered flooding regimes, which have changed tree-species composition from desirable oak species that produce small acorns (i.e., bottomland red oaks, *Quercus* spp.) to the more water-tolerant overcup oak (*Quercus lyrata*) (Dugger and Fredrickson 1992). This change in oak species composition can limit Wood Duck food availability, because overcup oak produces large acorns that are unsuitable food for waterfowl (Dugger and Fredrickson 1992). Because Wood Ducks also require specific water depths for foraging (≤ 30 cm is often considered ideal), altered flood regimes stemming from levees, dykes, ditches, greentree reservoirs (bottomland hardwood forest artificially flooded in the fall and winter), and other artificial structures may also affect the ability of Wood Ducks to effectively forage (Foth et al. 2014). Thus, important management options include restoring the natural hydrology of previously drained forested wetlands, reestablishing bottomland hardwood or emergent herbaceous wetlands, and increasing the size of habitat blocks (i.e., reducing habitat fragmentation and increasing permeability) (Twedt and Loesch 1999, Kaminski et al. 2003, NRCS 2012, Foth et al. 2014). Conservation and restoration efforts of bottomland hardwood forests that are advantageous to Wood Ducks also benefit other birds, as well as numerous wildlife species that are dependent on these ecosystems (Fredrickson 1978, Twedt and Loesch 1999).

Coniferous Forests

History and Context of Importance to Birds

Forest ecosystems are important for bird conservation and management, and multiple factors influence the size, shape, and vegetation composition of forested habitats used by birds. In the United States, coniferous forests generally dominate western landscapes, whereas deciduous forests dominate eastern landscapes (Yahner et al. 2012). Obligate breeding bird species within deciduous forests have declined approximately 30 percent since 1968, and obligate breeding bird species within western coniferous forests have declined approximately 20 percent in that same time period (North American Bird Conservation Initiative 2014; fig. 28.3A).

Forest structure and composition are generally affected by natural processes (e.g., disease, insect and fungal infestations, fire), anthropogenic activities (harvest, reforestation, and the associated secondary successional processes), climate, and interactions among these factors (Finch et al. 1997). For instance, wildfire strongly influences forest ecosystem structure and function. The spatial extent of fires, time since fire, and severity of fire can have myriad impacts on bird populations and communities (DellaSala and Hanson 2015). Fires burn heterogeneously across landscapes and can lead to a patchwork of burn severities, ranging from unburned to low-moderate severity to high severity; these patches of burn severity are likely to support distinct bird communities. High-severity fires might create high numbers of snags (i.e., standing dead or dying trees), which are utilized by woodpeckers as nesting habitat; however, these same fires initially decrease nesting habitat for ground and shrub-nesting birds after the fire (Saab and Powell 2005). Edges created by the fire between unburned and burned patches might also influence nest success within a burned patch (e.g., Vierling et al. 2008).

Birds might also be affected by silvicultural (harvest) treatments, which can vary in size, treatment type (e.g., clear-cuts, shelterwood cuts, etc.), and time since treatment. For instance, Kendrick et al. (2015) examined how different silvicultural treatments and time since treatment influenced bird communities in the Missouri Ozarks and found that bird communities were strongly influenced by interactions between treatment types and the time since treatment. Kendrick et al. (2015) suggest that managers could use their data either to guide management decisions for specific species of interest or to address multiple-species management goals.

Climate change also affects forest structure and composition. Mountain pine beetles (Dendroctonus ponderosae) are responsible for creating large swaths of dead trees via outbreaks in the western United States and British Columbia (Kurz et al. 2008; fig. 28.13). The increase in infestation is partially caused by higher winter temperatures, which have allowed mountain pine beetle overwinter survivorship to increase (Kurz et al. 2008). Collectively, the distribution, size, and shape of the beetle-killed forest patches (fig. 28.13) influence bird communities within this region. For instance, Drever et al. (2009) note differences in foraging-guild population responses based on different stages of mountain pine beetle infestations (e.g., red and gray attacks shown in fig. 28.13). Bark insectivores responded most strongly to mountain pine

beetle infestations compared with other insectivore guilds (Drever et al. 2009).

Forests are often actively managed via fire and harvest, and below we briefly discuss two examples of management for avian conservation. First, we describe how active management has influenced Kirtland's Warbler (Setophaga kirtlandii) habitat. Second, we examine how management activities have been used to address the needs of multiple species in western coniferous forests.

Kirtland's Warbler

Kirtland's Warbler is a federally endangered species, and multiple causal factors have contributed to its decline. This small-bodied insectivore (fig. 28.14A) relies on early successional jack pine (Pinus banksiana) forests during the breeding season in the central United States (fig. 28.14B). Historically, jack pine forests burned frequently, and Kirtland's Warblers generally occupy young (between 5 and 23 years old) regenerating jack pine forests (Bocetti et al. 2014). The restriction to this successional stage is related to both their nesting and foraging habitat requirements. Kirtland's Warblers nest in dense herbaceous cover and forage on the ground and near trees with lower live branches (fig. 28.14C). Lower live branches are important for this species because they provide foraging habitat as well as cover for nests and fledglings (Donner et al. 2008).

Kirtland's Warblers are commonly found on jack pine plantations and in burned sites. Because suitable habitat is ephemeral, the spatial and temporal aspects of Kirtland's Warbler habitat must be considered. For instance, Donner et al. (2008) examined the factors influencing population increases and the distribution of male Kirtland's Warblers in a matrix of patches that had undergone logging and wildfire disturbances. They found that male Kirtland's Warblers shifted to plantations as burned patches declined in quality. The spatial arrangement and amount of habitat was influential in explaining population increases, and Donner et al. (2008) emphasized the importance of considering both temporal aspects of management (i.e., when habitats might decline in quality) and the spatial arrangement of suitable habitat in a landscape. These approaches may be partnered with removal of Brown-headed Cowbirds (Molothrus ater), a brood parasite that negatively influences Kirtland's Warbler nest success. Adaptive-management strategies have been used to assess whether different approaches to habitat management for Kirtland's Warbler have been effective, and habitat management over time has broadened from a single-species approach to one that considers the ecosystem as a whole (Bocetti et al. 2014).

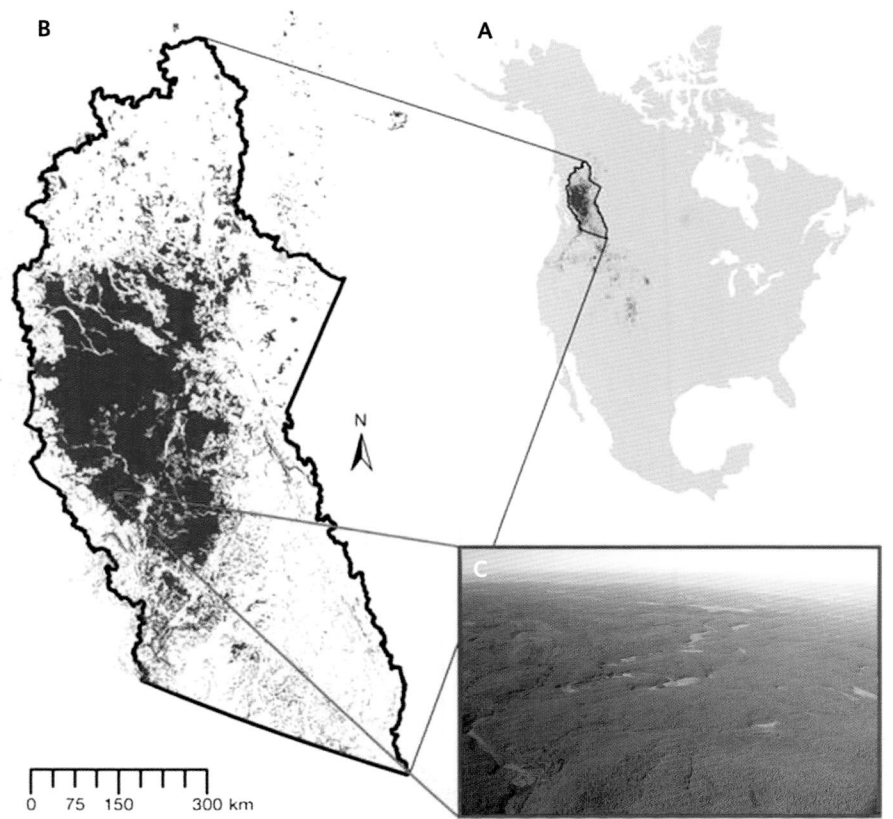

Figure 28.13. A, Distribution of the mountain pine beetle outbreak in North America in 2006; *B,* The extent of the outbreak within western Canada; *C,* An aerial photograph depicting patterns of beetle infestation; the red represents the first year after beetle kill, and the gray represents older areas of beetle-caused mortality. *Photo by Joan Westfall, Entopath Management Ltd. Reprinted by permission from Macmillan Publishers Ltd: Nature, vol. 452, W. A. Kurz et al., Mountain pine beetle and forest carbon feedback to climate change, p. 988. copyright (2008).*

Cavity Users in Western Coniferous Forests

Tree cavities provide important breeding and roosting habitat for multiple bird species, and cavity-nesters are a group of birds that rely on these tree hollows for breeding sites. In western North American forests, primary cavity excavators (e.g., woodpeckers) create tree cavities that are later used by a suite of species known as secondary cavity users (Martin et al. 2004). Secondary cavity users require tree cavities for either nesting or roosting (e.g., Blanc and Walters 2008, Gentry and Vierling 2008), and include species such as Mountain Bluebirds (*Sialia currucoides*), Buffleheads (*Bucephala albeola*), and Flammulated Owls (*Otus flammeolus*; Aitken and Martin 2007). The relative importance of cavity excavators differs across forest types because tree hollows formed by natural processes (i.e., limb breakage) may be more common in some forest types (e.g., Robles and Martin 2014). Nevertheless, woodpeckers are important for creating habitat for multiple species of vertebrates via their role as ecosystem engineers (Jones et al. 1994).

The majority of woodpeckers require snags in which to nest, because snags often have pockets of decayed wood that is soft enough for drilling (e.g., Lorenz et al. 2015). Snag management is, therefore, an important component to the management of cavity-nesters, and the size and distribution of snags is a critical consideration. Because woodpeckers are the main creator of tree cavities in many forested landscapes (Martin et al. 2004), management of forests to maximize the diversity of habitats needed by various cavity-excavating species can be an important management goal (Drever et al. 2008, Cooke and Hannon 2011). Management to benefit diverse woodpecker communities might incorporate a variety of harvest practices (pre- and post-fire), prescribed fire, snag management, and even artificial snag creation.

Grasslands

History and Context of Importance to Birds

Grasslands (ecosystems dominated by grasses and forbs) are widely distributed throughout North America as well as in other parts of the world. Samson and Knopf (1994) were among the first to summarize the threats faced by grassland birds, noting that ~80–99 percent of native grasslands had been lost in North America. Since then, multiple studies have documented significant declines in North American grassland bird populations since European settlement (e.g., Vickery and Herkert 1999, Brennan and Kuvlesky 2005). Threats to grassland birds include not only loss of habitat but also degradation of habitat caused by multiple factors including grazing and

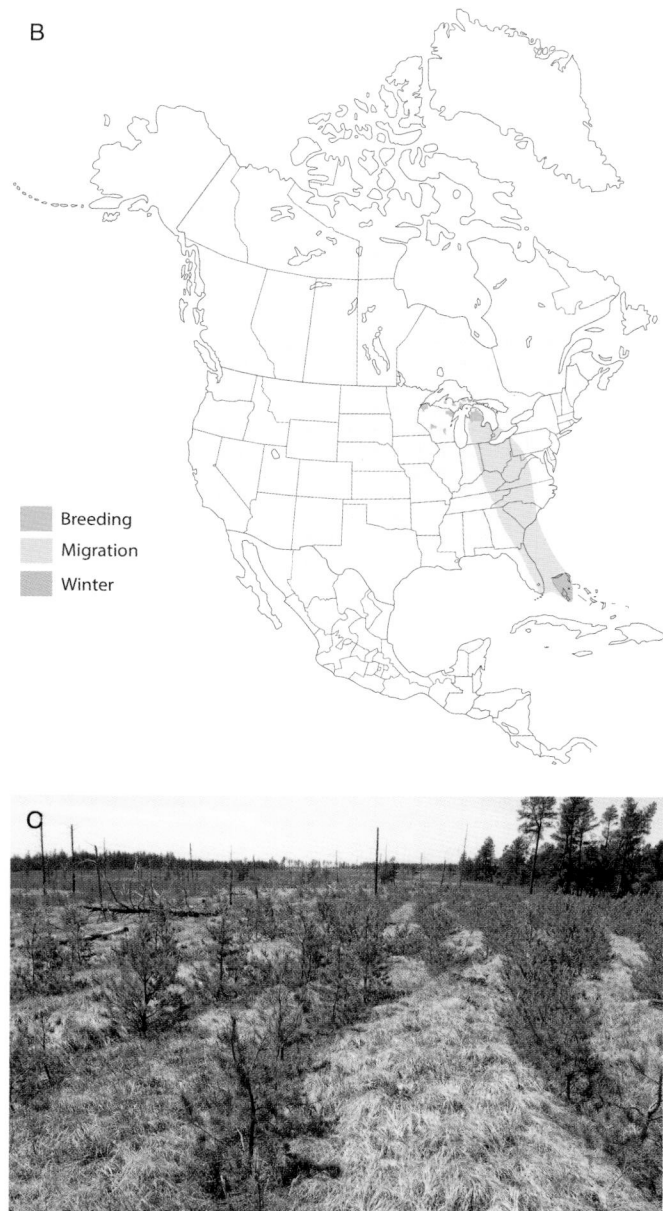

Figure 28.14. A, Male Kirtland's Warbler. *Photo courtesy of Ron Austing.*
B, Range map. *Map from Birds of North America Online http://bna.bird*
.cornell.edu/bna, maintained by the Cornell Lab of Ornithology. C, Young
jack pine plantation approximately three to four years after planting.
Photo courtesy of US Forest Service, Huron-Manistee National Forests.
Kirtland's Warblers typically begin occupying breeding habitat five
years after planting and the stand may be occupied for up to 10
breeding seasons before the site becomes unsuitable. Managers try
to mimic natural processes by leaving snags and clumps of standing
trees.

fire management, climate change, and factors associated with certain farmland management practices (North American Bird Conservation Initiative 2009). Grassland conversion was common during European settlement, a time when native grasslands were generally converted to large monocultures. The homogenization of these habitats generally has resulted in a loss of bird diversity (Best et al. 1995). Additionally, reproductive success has been noted to be low in these types of habitats (Rodenhouse and Best 1983), presumably because nests are easier to access by predators in these structurally simple row-crop environments. More recent studies have emphasized the benefits of spatial and temporal heterogeneity in

fire and grazing activities for grassland bird species (Fuhlendorf and Engle 2001, Hovick et al. 2015).

Although widespread declines were observed after initial European settlement, recent reports indicate that population trends have somewhat stabilized for obligate grassland breeding birds (North American Bird Conservation Initiative 2009, 2014). This is likely thanks to a variety of farm conservation programs. For instance, farmlands in the Conservation Reserve Program (CRP) are left fallow for 10–15 years, which can benefit birds through improved water quality or wildlife habitat (Warner et al. 2012). For example, Herkert (2007) noted that CRP enrollment has likely had a positive effect on

Henslow's Sparrows (*Ammodramus henslowii*; fig. 28.15); however, Osborne and Sparling (2013) found that Henslow's Sparrows had variable responses to different CRP management practices. Osborne and Spalding (2013) also observed that Henslow's Sparrows positively responded to both patch (e.g., litter depth) and landscape (the amount of cropland within 250 m of survey points) characteristics, which emphasizes the role of spatial scale and matrices in such assessments.

The landscape context surrounding both farmlands and grasslands is particularly important to consider because conservation buffers (i.e., the noncrop vegetation at the field margin), roads, and houses are all factors that can influence the distribution of nesting birds and breeding quality of these grasslands (fig. 28.16). The types of conservation buffers that immediately surround an area might increase nest density and reproductive success of some grassland species (Conover et al. 2011); conversely, conservation buffers located at the field margin might support higher numbers of nest predators and negatively influence reproduction (Ellison et al. 2013). Multiple grassland birds are area-sensitive (Ribic et al. 2009) and are thus sensitive to habitat fragmentation that results from roads. The proximity of grasslands to houses or suburban edges is also an important factor influencing birds because the densities of potential nest predators can be higher around human structures, and these grasslands might function as population sinks where mortality exceeds reproduction (Vierling 2000).

For these reasons, it is important that grassland characteristics and the surrounding matrix be considered in management of grassland birds. Below, we discuss a case study that emphasizes why perspectives from ecosystem and landscape management are important for conservation of grassland birds.

Bobolinks in Grasslands and Hayfields

Bobolinks (*Dolichonyx oryzivorus*; fig. 28.17) are declining across their range in North America (Sauer et al. 2014) and are particularly sensitive to activities within their breeding habitat as well as at the broader landscape scale (Renfrew et al. 2015). Bobolinks historically nested in vast expanses of mixed-grass and tallgrass prairies, but loss of these ecosystems have prompted them to nest in hayfields and meadows. Although hayfields can support relatively high densities of grassland birds such as Bobolinks, they often function as population sinks. If mowing occurs while nests are active, nest failure typically approaches 100 percent; the only successful nests are either in small, unmowed patches of grass, or where the young fledge prior to mowing activities (Bollinger et al. 1990, Vierling 2000). Thus, hayfields likely function as ecological traps, which result from a mismatch between the process of habitat selection and

the quality of habitat (Battin 2004). Because birds are unlikely to evolve the ability to recognize that hayfields are likely to be mowed and are poor-quality nesting habitats, management of hayfield-nesting birds has focused on manipulations of mowing dates. In Boulder, Colorado (USA), losses to hayfield-nesting bird communities were minimal (~5 percent) when haying was delayed into late June to early July (Vierling 1997). In Vermont (USA), Perlut et al. (2011) describe a program by which farmers were financially compensated to delay their second-cut harvest; this allowed productivity of Bobolinks to increase from zero to 2.8 fledglings per female per year and illustrated that hayfield mowing management was the primary determinant of reproductive success. Perlut et al. (2011) also note that a critical component contributing to the success of the program was open communication between the managers and the landowners.

In addition to managing breeding habitat itself, the surrounding matrix may have an effect on the quality of breeding habitat for this species. Bobolinks often prefer to nest in sites that are farther away from forest or wooded edges (Bollinger and Gavin 2004). Tree rows have been shown to provide habitat for nest predators; thus, tree-row removal might benefit a suite of grassland birds. Ellison et al. (2013) removed rows of trees located at the edges of grasslands and found that nesting densities of Bobolinks increased. The effect of tree removal on nest success varied temporally and spatially, but Ellison et al. (2013) suggest that tree-row removal might be an effective management action in some situations to conserve grassland birds.

ADDITIONAL MANAGEMENT CONSIDERATIONS

Cultural Perspectives in Management

Ecosystems and landscapes will continue to be modified as the human population grows, and the current 2015 human population estimate of over 7 billion people will likely increase to almost 10 billion people by 2050 (United Nations 2015). Intensification of agriculture, conversion of forests for fuel or building, and other human activities will have long-term effects on avian communities via habitat loss and habitat degradation. Additionally, it is increasingly likely that multiple uses of the remaining lands for recreation and resource extraction will add to current management challenges. Therefore, a greater understanding of social values and cultural contexts is important for natural resource management into the future (Ressurreicao et al. 2012).

Culture is a fundamental and broad-reaching concept within the social sciences, traditionally thought to encompass three domains: action, perception/ideological, and material

KATHY MARTIN: LANDSCAPE ECOLOGY OF ALPINE BIRDS AND CONSERVATION CHALLENGES FOR MOUNTAIN BIODIVERSITY

Kathy Martin.
Photo courtesy of University of British Columbia.

I am an alpine ecosystem and population ecologist (see figure above) who conducts research on how birds solve the problems of living and breeding successfully in mountain ecosystems. Mountains occur on all the continents, constituting 24 percent of the land base, globally and also in the Americas. One-quarter of mountain ecosystems worldwide occur in temperate zones, located predominantly in Europe, Asia, and the Americas. Temperate mountains are snowy habitats where climate and topographic factors such as elevation, aspect, and slope determine the extent of the alpine zone. Living on mountains, often called

"sky islands," has critical implications for the fitness of individuals, population demography, and various aspects of seasonal connectivity for wildlife. These sky islands differ in size, shape, and distribution across landscapes, and changes to the availability and distribution of these sky islands may have wide-ranging effects on bird communities.

At high elevations, birds must develop coping mechanisms as they encounter extreme and variable conditions such as delayed snowmelt or storm events throughout the breeding and postbreeding seasons, which makes it difficult to adapt to a common set of environmental conditions. Given the limited knowledge of wildlife communities and habitat associations for alpine landscapes, it is easy to discover new ecological relationships for mountain wildlife, even in North America! For example, from our field-based surveys and an extensive literature review, I was surprised to learn that 35 percent of the birds that breed in the continental portions of Canada and the United States use mountain ecosystems for at least one part of their annual cycle (breeding, migration, or winter), and that all major high-elevation habitats are important for the full life cycle conservation of our avifauna (see figure below).

In North America, birds use mountains year-round, with the period of highest biodiversity occurring during fall migration. One of our recent research projects focused on the extent of use of mountain environments by birds in northwestern North America. We conducted bird surveys in mountain habitats over four

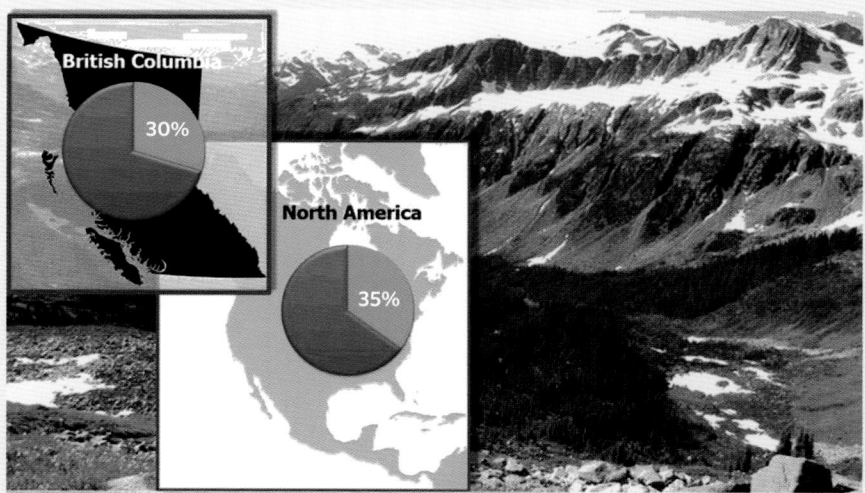

High elevation habitats are used seasonally by > 1/3 of the avifauna of British Columbia and of all North America

Extent of avian use of mountain areas. *Reprinted from Biological Conservation, vol. 192, W. A. Boyle and K. Martin. The conservation value of high elevation habitats to North American migrant birds. doi:10.1016/j.biocon.2015.10.008, copyright 2015 with permission from Elsevier.*

years in British Columbia, Canada, during August to October. We detected a remarkable diversity of birds (95 species in 30 families, 18,965 individuals) using alpine and subalpine habitats and upper montane forest (continuous forest close to tree line) for migration stopovers. We recorded the highest number of species in subalpine habitats, while montane forests supported the highest abundance of birds. However, alpine habitats supported 48 of the 95 species observed in fall, including birds that breed at high latitudes and alpine habitats, but also grassland and open country birds. We found that the long-distance migrants (birds that generally spend their winters in Central or South America) were more abundant earlier in the season (peaking in mid-August) and short-distance migrants (spend their winters in southern Canada or the United States) reached their peak numbers in mid-September. One-quarter of these species are on lists of continental or national conservation concern because they are red-listed in British Columbia (e.g., Northern Goshawk laingi subspecies [Accipiter gentilis laingi]) or they occur on the Partners in Flight Species of Special Concern list because they are common species now in steep decline in abundance, or species of high Tri-National concern (e.g., Prairie Falcon [Falco mexicanus], Band-tailed Pigeon [Patagioenas fasciata], Rufous Hummingbird [Selasphorus rufus], Pine Siskin [Spinus pinus]).

Our alpine ecosystem research highlights the importance of mountain habitats to migrating birds for at least three months of the year, a period equivalent to the length of the breeding season for most temperate habitat species. As low-elevation migration habitats such as riparian zones are lost or degraded, the importance of mountain habitats for migration stopovers will likely increase. This is especially important for resident species and short-distance migrants that spend all of their life cycle in North America.

Alpine habitats are experiencing globally significant increases in temperature and extreme weather, and tree- and shrub-lines are rising, which will reduce the amount, quality, and configuration of alpine habitats. The limited information about current avian use of alpine areas makes it difficult to predict the future for birds using mountains. Increasing climatic variability in alpine ecosystems will result in more challenging environmental conditions for many birds breeding and migrating in alpine landscapes. Our alpine ecosystem studies emphasize the need for effective conservation of networks of fragile mountain habitats. For additional information on the conservation of mountain birds, see Martin 2001, Martin and Wiebe 2004, Martin 2013, Martin 2014, Boyle and Martin 2015, Boyle et al. 2015, and Jackson et al. 2015.

(Milton 1996). Human values toward wildlife can vary across cultural contexts, and differences in value orientations (i.e., sets of basic beliefs about wildlife) can form the basis for conflict related to wildlife issues (Teel and Manfredo 2010). Cross-cultural understandings are critical for successful management of avian populations, particularly when management occurs across borders and restoration or conservation priorities must be established (Manfredo et al. 2009, Tella and Hiraldo 2014, Allan et al. 2015).

Management across the Life Cycle and across Political Boundaries

Swainson's Hawk (Buteo swainsoni, fig. 28.18A) is an ideal species with which to examine concepts associated with management across the life cycle and over large spatial extents. Swainson's Hawks are Neotropical migrants that breed in North America and cross multiple political boundaries before arriving on their wintering grounds in the Pampas of Argentina (fig. 28.18B). Kochert et al. (2011) fitted over 40 Swainson's Hawks with satellite transmitters and recorded their locations and stopover sites as they migrated to Argentina. Habitat requirements for the annual life cycle of this species span numerous countries and jurisdictional boundaries, necessitating international cooperation for effective and sustainable management.

Swainson's Hawks use grassland habitats throughout their annual cycle; grassland management can therefore have significant influences on their distribution, density, reproduction, and survival. Whereas some types of cultivation and farmland practices might benefit Swainson's Hawks through increased nesting substrates or food accessibility, other agricultural activities might negatively influence Swainson's Hawk populations (Bechard et al. 2010). For instance, Swainson's Hawks consume small mammals as a primary food source during the breeding season but shift to feeding almost exclusively on grasshoppers during the nonbreeding season (Bechard et al. 2010). Sarasola and Negro (2005) note that grasshoppers constitute ~98 percent of the diet of wintering Swainson's Hawks in Argentina. However, pesticide use in Argentina at overwintering sites caused massive mortalities of Swainson's Hawks in the 1990s. Although policy measures and other agreements have reduced the application of pesticides in some areas (Goldstein et al. 1999), habitat loss, alteration, and impairment continue to be conservation challenges (Bechard et al. 2010).

Management within Multiple-Use Public Lands

In addition to managing ecosystems and landscapes across international boundaries, intranational land ownership (e.g.,

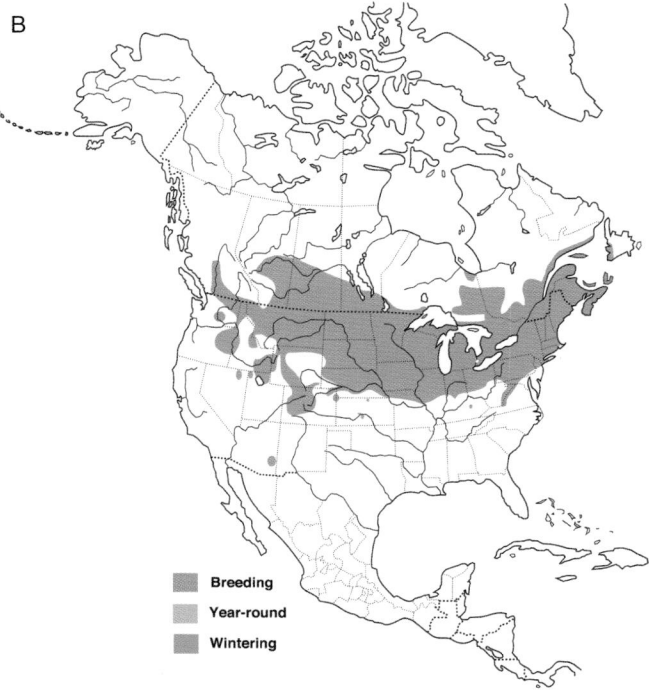

Figure 28.15. Relationship between the percentage of a counties' land area that is enrolled in the Conservation Reserve Program and county-level population trend of Henslow's Sparrows as indicated by data from Illinois' Spring Bird Count (USA). Population trend is expressed as the change in the number of birds per 100 party hours per year (1987–2004) for counties with relatively low, moderate, and high levels of Conservation Reserve Program enrollment. For some surveys, groups of observers form a party, and the number of party hours reflects the amount of time the party spent doing the survey. *Reprinted from* Journal of Wildlife Management, *vol. 71, J. Herkert, Evidence for a recent Henslow's Sparrow population increase in Illinois 2007. Open access under CC BY license.*

Figure 28.16. An aerial view of an agricultural landscape. Note the different sizes and shapes of the fields as well as the landscape matrix, which includes forest, roads, and houses. *From http://www.nrcs.usda.gov/wps/portal/nrcs/main/national/landuse/#.*

Figure 28.17. A, Male Bobolink in breeding plumage. *Photo courtesy of Geoffrey Hill. B, Bobolinks are declining across their breeding range. Map from Birds of North America Online http://bna.bird.cornell.edu/bna, maintained by the Cornell Lab of Ornithology.*

within the United States) might present similar challenges. For instance, the distribution of public and private lands is distributed unevenly across the United States (fig. 28.19). In contrast to private lands, animal populations and communities on public lands must be managed in light of both ecological processes and societal uses of those lands, which may include grazing, recreation, timber harvest, or mining. Since many species occur on both private and public lands, ecosystem and landscape management needs to incorporate the

demands of multiple, diverse stakeholders that may have different values and cultural perspectives.

Management of species that occur on both private and public lands must therefore involve partnerships between citizens and land-management agencies. These partnerships occur at the regional, national, and international levels and have become increasingly important for bird management. Landscape Conservation Cooperatives (LCCs), Migratory Bird Joint Ventures (JVs), Partners in Flight (PIF), and the

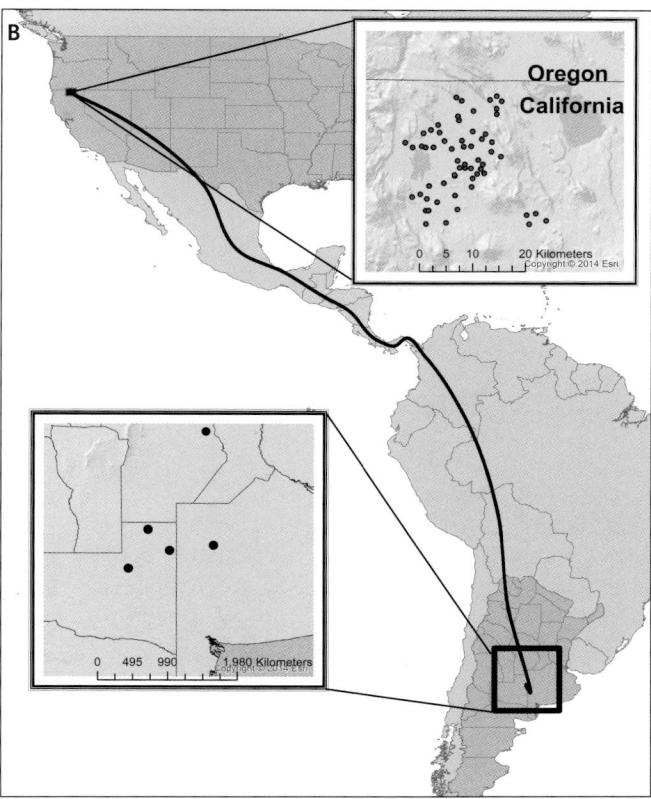

Figure 28.18. A, Swainson's Hawk. *Photo courtesy of the United States Fish and Wildlife Service. B,* Annual migration routes used by Swainson's Hawks annually. *Figure courtesy of Chris Briggs.*

North American Bird Conservation Initiative are but a few examples of multiple-partner initiatives that benefit birds and their habitats. For example, Migratory Bird Joint Ventures are regional partnerships focused on bird habitat conservation (www.mbjv.org). These partnerships may include federal and state agencies, tribes, nongovernmental organizations (NGOs), individuals, or corporations. To date, JVs have resulted in collaborations with over 5,700 partners, and have resulted in more than 23 million acres of habitat protection and restoration across North America (www.mbjv.org).

Alternative Management Approaches and Strategies

The formulation of a management goal is often the result of a federal decision (e.g., listing under the ESA) or via plans generated by the relevant stakeholders or states (i.e., State Wildlife Action Plans). However, time or logistical constraints often limit the ability to manage the needs of multiple individual species at any given time (Wiens et al. 2008). Whereas researchers continually contribute information that improves understanding of the interrelationships between biotic and abiotic processes and bird populations and communities, it is often difficult to assess the effect of management actions on multispecies assemblages.

Surrogate species approaches have been proposed as a way to address multispecies management challenges. Surrogate species are subsets of species used as representatives of a broader community or of specific environmental conditions (Caro and O'Doherty 1999). Surrogate species approaches include but are not limited to indicator species, flagship species, and umbrella species. Indicator species can be a useful management tool by enabling managers to delineate a specific ecosystem or landscape type, identify the relative environmental condition of a habitat, or monitor pollution or climate change. Indicator species can be used as an "early warning system" by scientists and conservation managers (reviewed in Carignan and Villard 2002). Flagship species are often chosen for their charismatic nature and public appeal, as well as the protection conferred on other species via their protection (Caro and O'Doherty 1999). Umbrella species are thought to protect co-occurring species; ideally, the habitat

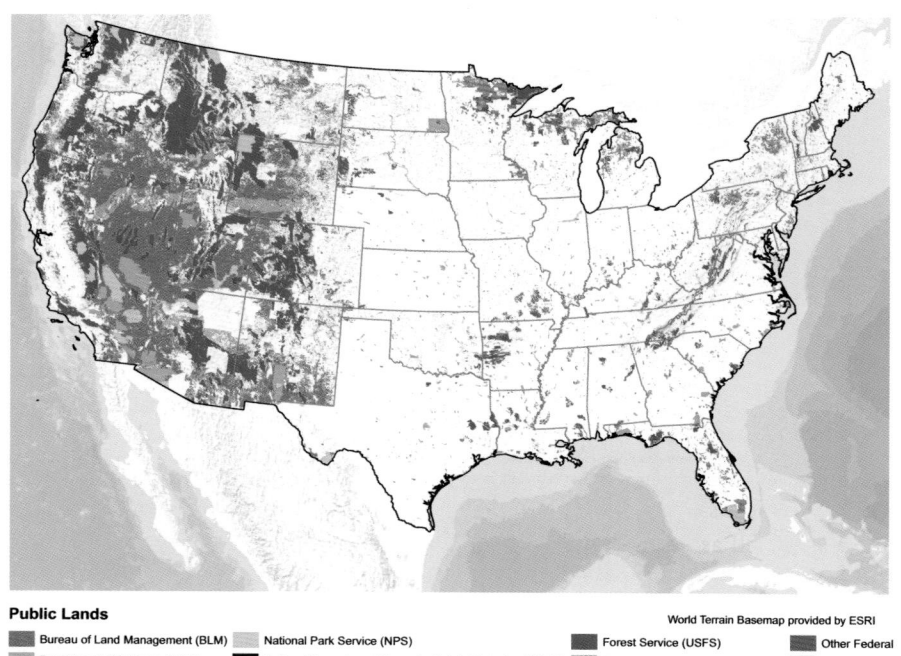

Figure 28.19. The distribution of public lands within the United States. Management activities differ geographically across regions of the United States based on the various activities that are allowed on public versus private lands. *From North American Bird Conservation Initiative 2011, State of the Birds 2011 Report.*

Public Lands

Bureau of Land Management (BLM)	National Park Service (NPS)	Forest Service (USFS)	Other Federal
Department of Defense (DOD)	National Oceanic and Atmospheric Administration (NOAA)	Fish and Wildlife Service (USFWS)	State Land

World Terrain Basemap provided by ESRI

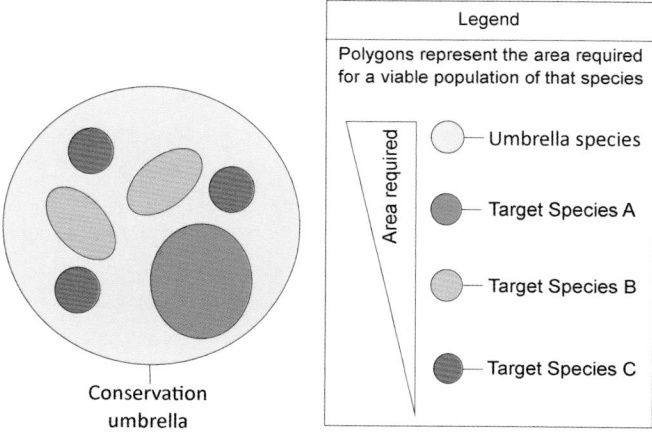

Figure 28.20. Umbrella species concept; the habitat protected within the spatial extent needed by an umbrella species should encompass the habitat needs of other target species. *Graphics courtesy of Andrés Mejia from Mejia 2015.*

protected within the spatial extent needed by an umbrella species should encompass the habitat needs of other species (Suter et al. 2002; fig. 28.20).

Assessments are mixed on the efficacy of surrogate species approaches. These approaches will not be applicable or useful in all situations because of species-specific differences in habitat requirements and the spatial scales at which

surrogate approaches might be applied (Wiens et al. 2008). Challenges with the concept of indicator species, for example, might include the explicit definition of what the presence, absence, or abundance of the species is "indicating." Additionally, the species must be linked to an environmental condition (or set of conditions) in a scientifically robust manner that justifies its use as a proxy.

Similarly, there are mixed assessments as to the utility of the umbrella species concept. For instance, American Woodcocks (*Scolopax minor*) are generally found in early successional forests, and Masse et al. (2015) suggested that the American Woodcock might be an appropriate umbrella species for early-successional forest birds because American Woodcock singing grounds exhibited 1.5 times greater bird community diversity than non-singing grounds. In contrast, White et al. (2013) note that forest bird diversity could be maximized by management that created within- and between-stand-level heterogeneity instead of using an umbrella species approach (using primarily the California Spotted Owl, *Strix occidentalis occidentalis*, as the umbrella species). The use of umbrella species is likely to be limited in effectiveness in many situations because different species typically respond to habitat variables and processes in distinct ways and at a variety of spatial scales that may not be captured by a single species or even a handful of species (Lindenmayer et al. 2002).

CONCLUSIONS

Worldwide, over 21 percent of bird species are currently extinction-prone and 6.5 percent are functionally extinct (i.e., only marginally contributing to ecosystem services); some projections suggest that by 2100, 6–14 percent of all bird species will be extinct, and 7–25 percent will be functionally extinct (Şekercioğlu et al. 2004). Moreover, whereas the global number of individual birds has experienced a 20–25 percent reduction since 1500 (Gaston et al. 2003), only 1.3 percent of bird species have gone extinct during this same time period (BLI 2000), indicating that avian populations and the ecosystem services they provision are declining faster than species extinctions suggest (Luck et al. 2003). Effective ecosystem and landscape management strategies are thus critical for sustaining both avian populations and communities as well as the ecosystem functions they provide.

Although we have discussed examples of distinct ecosystems and landscapes and their importance to bird management, habitat loss is only one of multiple contributors to declines in bird populations and the ecosystem services they provision. Exploitation (e.g., overhunting), disease, habitat fragmentation, environmental contamination, and climate change are among the many other factors that can lead to bird species' declines (see chapters 25 and 26). Şekercioğlu et al. (2004) report that half of threatened species are in danger from these and other factors that operate independent of habitat loss, and that scavengers, piscivores, herbivores, omnivores, granivores, and frugivores are most at risk.

Effective and long-term management and planning strategies, therefore, must consider a wide range of stressors that extend beyond habitat loss and that incorporate ecological processes that link habitats across spatial extents, from patches to ecosystems to landscapes and beyond. This is particularly important for birds, which are highly mobile organisms that commonly cross ecosystem and landscape boundaries during their daily, seasonal, and annual movements. To develop effective management strategies, multiple stakeholders must be involved. For many migratory species, this will involve multinational teams where cross-cultural perspectives will be critical.

Although managing ecosystems and landscapes to benefit birds is challenging, it is an exciting and dynamic field. We have introduced some basic concepts and types of management approaches, yet it is important to recognize that many ecosystem management approaches are in their developmental stages. Of additional note, other management approaches, such as ecosystem restoration, have also experienced significant success. For instance, the Kissimmee River Restoration Project (Florida, USA) is an example of a large-scale ecosys-

tem restoration project to reconnect a channelized river with its floodplain by reestablishing floodplain hydrology that more closely mimics historical conditions (Toth et al. 1998). The restoration of naturally fluctuating water levels, seasonal hydroperiods, and native vegetation communities is expected to attract waterbirds. In fact, the abundance and species richness of both wading birds and waterfowl have shown a positive response since the beginning of the project in 2001 (Cheek et al. 2014).

In conclusion, multiple management frameworks can help guide conservation and management decisions, and ecosystem processes critical for bird species and communities must be considered across a hierarchy of spatial scales. Ecosystem and landscape management approaches hold significant potential for effective conservation, and the incorporation of management actions with explicit consideration of cultural and social contexts is essential as we manage birds across both private and public lands, as well as across geopolitical boundaries.

KEY POINTS

- Structured decision-making and adaptive management are common approaches used to inform conservation and management strategies of bird species and communities.
- Approaches for conservation and management of birds are dependent on spatial scale: (1) the scale at which birds perceive their environment, (2) the scale at which disturbances operate (habitat patch, ecosystem, landscape), and (3) the scale at which management objectives are addressed. The temporal scale of management is also an important consideration.
- Management objectives are diverse and include management for (1) a single species of birds, (2) bird community diversity, and (3) the functional roles of birds and the ecosystem services they provision.
- The incorporation of social and cultural contexts will be increasingly important in avian conservation and management at all spatial scales.
- Surrogate species approaches have been proposed to aid conservation efforts, and those approaches are still being evaluated for efficacy.

KEY MANAGEMENT AND CONSERVATION IMPLICATIONS

- Ecological and anthropogenic disturbances that influence ecosystems and landscapes have a range of effects on different bird species and communities, their ecological roles, and the ecosystem services they provision; what benefits one species or group may not benefit others.

- Landscape composition and configuration can profoundly affect management outcomes.
- Knowledge of how ecosystem and landscape characteristics influence all aspects of the life cycles of birds (breeding, migration, wintering) is an essential step in developing management actions and effective, long-term conservation plans.
- Increasing human populations, associated land use changes, and large-scale factors such as climate change are likely to pose continuing challenges for management of avian populations and communities.
- Around the globe, there have been significant declines in breeding birds that are habitat obligates. Although habitat loss is one major factor influencing declines, multiple other factors are also responsible. Thus, conservation actions increasingly need to consider the multitude of social and ecological factors that interact to contribute to bird declines.

DISCUSSION QUESTIONS

1. What are the key elements of structured decision-making? What are the key elements of an adaptive management approach? How are these approaches similar and how are they different?
2. How are ecological processes at different spatial scales and extents related to bird diversity and management?
3. To what degree are single-species management strategies compatible with multispecies management strategies?
4. In what ways are the social and cultural contexts of ecosystem and landscape management important to bird conservation?
5. How is identifying important avian ecosystem-service providers critical to ecosystem and landscape management and planning?
6. Birds are highly mobile organisms that commonly cross ecosystem and landscape boundaries. What are the implications of this mobility for management and planning at ecosystem and landscape scales?

References

Aitken, K. E. H., and K. Martin (2007). The importance of excavators in hole-nesting communities: Availability and use of natural tree holes in old mixed forests of western Canada. Journal of Ornithology 148:S425–S434.

Alberts, J. M., S. M. P. Sullivan, and A. Kautza (2013). Riparian swallows as integrators of landscape change in a multiuse river system: Implications for aquatic-to-terrestrial transfers of contaminants. Science of the Total Environment 463:42–50.

Allan, J. D., S. D. P. Smith, P. B. McIntyre, C. A. Joseph, C. E. Dickinson, A. L. Marino, R. G. Biel, et al. (2015). Using cultural ecosystem services to inform restoration priorities in the

Laurentian Great Lakes. Frontiers in Ecology and the Environment 13:418–424.

Askins, R. A., C. M. Folsom-O'Keefe, and M. C. Hardy (2012). Effects of vegetation, corridor width and regional land use on early successional birds on powerline corridors. PLoS ONE 7 (2): e31520. doi:10.1371/journal.pone.0031520.

Bainbridge, B. K., A. Baum, D. Saenz, and C. K. Cory (2011). Red-cockaded Woodpecker cavity-tree damage by Hurricane Rita: An evaluation of contributing factors. Southeastern Naturalist 10:11–24.

Baker, W. L. (1989). Landscape ecology and the nature reserve design in the Boundary Waters canoe area, Minnesota. Ecology 70:23–35. doi:10.2307/1938409.

Baschuk, M. S., N. Koper, D. A. Wrubleski, and G. Goldsborough (2012). Effects of water depth, cover and food resources on habitat use of marsh birds and waterfowl in boreal wetlands of Manitoba, Canada. Waterbirds 35:44–55.

Battin, J. (2004). When good animals love bad habitats: Ecological traps and the conservation of animal populations. Conservation Biology 18:1482–1491.

Bechard, M. J., C. S. Houston, J. H. Sarasola, and A. S. England (2010). Swainson's Hawk (Buteo swainsoni). In The birds of North America, no. 265, A. Poole, Editor. Academy of Natural Sciences, Philadelphia, PA/American Ornithologists' Union, Washington, DC. doi:10.2173/bna.265.

Bengtsson, J., S. G. Nilsson, A. Franc, and P. Menozzi (2000). Biodiversity, disturbances, ecosystem function and management of European forests. Forest Ecology and Management 132:39–50.

Best, L. B., K. E. Freemark, J. J. Dinsmore, and M. Camp (1995). A review and synthesis of habitat use by breeding birds in agricultural landscapes of Iowa. American Midland Naturalist 134:1–29.

Blais, J. M., R. W. Macdonald, D. Mackey, E. Webster, C. Harvey, and J. P. Smol (2007). Biologically mediated transport of contaminants to aquatic systems. Environmental Science and Technology 41:1075–1084.

Blanc, L. A., and J. R. Walters (2008). Cavity-nest webs in a long-leaf pine ecosystem. The Condor 110:80–92.

BLI (2000). Threatened birds of the world. Lynx Edicions and BirdLife International, Barcelona, Spain.

Bocetti, C. I., D. M. Donner, and H. F. Mayfield (2014). Kirtland's Warbler (Setophaga kirtlandii). In The birds of North America, no. 19, A. Poole, Editor. Academy of Natural Sciences, Philadelphia, PA/American Ornithologists' Union, Washington, DC. doi:10.2173/bna.19.

Bollinger, E. K., P. B. Bollinger, and T. A. Gavin (1990). Effects of hay-cropping on eastern populations of the Bobolink. Wildlife Society Bulletin 18:142–150.

Bollinger, E. K., and T. A. Gavin (2004). Responses of nesting bobolinks (Dolichonyx oryzivorus) to habitat edges. The Auk 121:767–776.

Boyle, W. A., B. K. Sandercock, and K. Martin (2015). Patterns and drivers of intraspecific variation in avian life history along elevational gradients: A meta-analysis. Biological Reviews 91 (2): 469–482. doi:10.1111/brv.12180.

Boyle, W. A., and K. Martin (2015). The conservation value of high elevation habitats to North American migrant birds. Biological

Conservation 192:461–476. http://dx.doi.org/10.1016/j.biocon .2015.10.008.

Brennan, L. A., and W. P. Kuvlesky Jr. (2005). North American grassland birds: An unfolding conservation crisis? Journal of Wildlife Management 69:1–13.

Brown, J. S., and B. P. Kotler (2004). Hazardous duty pay and the foraging cost of predation. Ecology Letters 7:999–1014.

Buckton, S. T., and S. J. Ormerod (2002). Global patterns of diversity among the specialist birds of riverine landscapes. Freshwater Biology 47:695–709.

Carignan, V., and M.-A. Villard (2002). Selecting indicator species to monitor ecological integrity: A review. Environmental Monitoring and Assessment 78:45–61.

Caro, T. M., and G. O'Doherty (1999). On the use of surrogate species in conservation biology. Conservation Biology 13:805–814.

Chalfoun, A. D., F. R. Thompson, and M. J. Ratnaswamy (2002). Nest predators and fragmentation: A review and meta-analysis. Conservation Biology 16:306–318.

Cheek, M. D., G. E. Williams, S. G. Bousquin, J. Colee, and S. L. Melvin (2014). Interim response of wading birds (Pelecaniformes and Ciconiiformes) and waterfowl (Anseriformes) to the Kissimmee River Restoration Project, Florida, USA. Restoration Ecology 22:426–434.

Clausen, P., B. A., Nolet, A. D. Fox, and M. Klaassen (2002). Long-distance endozoochorous dispersal of submerged macrophyte seeds by migratory waterbirds in northern Europe: A critical review of possibilities and limitations. Acta Oecologica 23:191–203.

Conover, R. R., S. J. Dinsmore, and L. W. Burger Jr. (2011). Effects of conservation practices on bird nest density and survival in intensive agriculture. Agriculture, Ecosystems, and Environment 141:126–132.

Cooke, H. A., and S. J. Hannon (2011). Do aggregated harvests with structural retention conserve the cavity web of old upland forest in the boreal plains? Forest Ecology and Management 261:662–674.

Cumming, G. S., A. Caron, C. Abolnik, G. Catolli, L. Bruinzeel, C. E. Burger, K. Cecchettin, et al. (2011). The ecology of influenza A viruses in wild birds in southern Africa. Ecohealth 8:4–13.

Cumming, G. S., N. Gaidet, and M. Ndlovu (2012a). Towards a unification of movement ecology and biogeography: Conceptual framework and a case study on Afrotropical ducks. Journal of Biogeography 39:1401–1411.

Cumming, G. S., M. Paxton, J. King, and H. Beuster (2012b). Foraging guild membership explains variation in waterbird responses to the hydrological regime of an arid-region flood-pulse river in Namibia. Freshwater Biology 57:1202–1213.

Cumming, G. S., M. Ndlovu, G. L. Mutumi, and P. A. Hockey (2013). Responses of an African wading bird community to resource pulses are related to foraging guild and food-web position. Freshwater Biology 58:79–87.

Daily, G. C., P. R. Ehrlich, and N. M. Haddad (1993). Double keystone bird in a keystone species complex. Proceedings of the National Academy of Sciences USA 90:592–594.

DellaSala, D. A., and C. T. Hanson, Editors (2015). The ecological importance of mixed-severity fires: Nature's phoenix. Elsevier, Amsterdam.

Diamond, J. M. (1975). The island dilemma: Lessons of modern biogeographic studies for the design of natural reserves. Biological Conservation 7:129–146.

Donner, D. M., J. R. Probst, and C. A. Ribic (2008). Influence of habitat amount, arrangement, and use on population trend estimates of male Kirtland's Warblers. Landscape Ecology 23:467–480.

Donovan, T. M., F. R. Thompson III, J. Faaborg, and J. Probst (1995). Reproductive success of migratory birds in habitat sources and sinks. Conservation Biology 9:1380–1395.

Donovan, T. M., P. W. Jones, E. M. Annand, and F. R. Thompson III (1997). Variation in local-scale edge effects: Mechanisms and landscape context. Ecology 78:2064–2075.

DPIPWE (2014). Draft Tasmanian Wilderness World Heritage Area management plan 2014. www.dpipwe.tas.gov.au.

Drever, M. C., K. E. H. Aitken, A. R. Norris, and K. Martin (2008). Woodpeckers as reliable indicators of bird richness, forest health, and harvest. Biological Conservation 141:624–634.

Drever, M. C., J. R. Goheen, and K. Martin (2009). Species-energy theory, pulsed resources, and regulation of avian richness during a mountain pine beetle outbreak. Ecology 90:1095–1105.

Dudgeon, D., A. H. Arthington, M. O. Gessner, Z. I. Kawabata, D. J. Knowler, C. Leveque, R. J. Naiman, et al. (2006). Freshwater biodiversity: Importance, threats, status and conservation challenges. Biological Reviews 81:163–182.

Dugger, K. M., and L. H. Fredrickson (1992). Life history and habitat needs of the wood duck. US Department of the Interior Fish and Wildlife Service, Fish and Wildlife Leaflet 13.1.6. Waterfowl Management Handbook, Washington, DC.

Ealey, D. M. (1977). Aspects of the ecology and behaviour of a breeding population of Dippers (Cinclus mexicanus: Passeriformes) in southern Alberta. Master's thesis, University of Alberta, Edmonton, Canada.

Ellison, K. S., C. A. Ribic, D. W. Sample, M. J. Fawcett, and J. D. Dadisman (2013). Impacts of tree rows on grassland birds and potential nest predators: A removal experiment. PLoS ONE 8:e59151.

Erwin, R. M., M. K. Laubhan, J. E. Cornely, and D. M. Bradshaw (2000). Managing wetlands for waterbirds: How managers can make a difference in improving habitat to support a North American bird conservation plan. In Strategies for bird conservation, The Partners in Flight planning process: Proceedings of the 3rd Partners in Flight Workshop, October 1–5, 1995, Cape May, NJ, R. P. Bonney, D. Cooper, R. J. Niles, Editors. Department of Agriculture, Forest Service, Rocky Mountain Research Station, Ogden, UT.

Fahrig, L., J. Baudry, L. Brotons, F. G. Burel, T. O. Crist, R. J. Fuller, C. Sirami, G. M. Siriwardena, and J. Martin (2011). Functional landscape heterogenetiy and animal biodiversity in agricultural landscapes. Ecology Letters 14:101–112.

Figuerola, J., A. J. Green, and L. Santamaria (2003). Passive internal transport of aquatic organisms by waterfowl in Donana, southwest Spain. Global Ecology and Biogeography 12:427–436.

Finch, D. M., J. L. Ganey, W. Yong, R. T. Kimball, and R. Sallabanks (1997). Effects and interactions of fire, logging, and grazing. In USDA Forest Service General Technical Report

RM-GTR-292. Rocky Mountain Forest and Range Experiment Station, Fort Collins, CO.

Foth, J. R., J. N. Straub, R. M. Kaminski, J. B. Davis, and T. D. Leininger (2014). Aquatic invertebrate abundance and biomass in Arkansas, Mississippi, and Missouri bottomland hardwood forests during winter. Journal of Fish and Wildlife Management 5:243–251.

Fredrickson, H. (1978). Lowland hardwood wetlands: Current status and habitat values for wildlife. In Wetland functions and values: The state of our understanding, P. E. Greeson, J. R. Clark, and J. E. Clark, Editors. Proceedings of the National Symposium on Wetlands Lake Buena Vista, Florida November 7–10, pp. 296–306. American Water Resources Association National Wetlands Technical Council.

Fuhlendorf, S. D., and D. M. Engle (2001). Restoring heterogeneity on rangelands: Ecosystem management based on evolutionary grazing patterns. Bioscience 51:625–631.

Garton, E. O., J. S. Horne, J. L. Aycrigg, and J. T. Ratti (2012). Research and experimental design. In The wildlife techniques manual: Management, N. J. Silvy, Editor. Johns Hopkins University Press, Baltimore, MD.

Gaston, K. J., T. M. Blackburn, and K. K. Goldewijk (2003). Habitat conversion and global avian biodiversity loss. Proceedings of the Royal Society B: Biological Sciences 270:1293–1300.

Gentry, D. J., D. L. Swanson, and J. D. Carlisle (2006). Species richness and nesting success of migrant forest birds in natural river corridors and anthropogenic woodlands in Southeastern South Dakota. The Condor 108:140–153.

Gentry, D. J., and K. T. Vierling (2008). Reuse of cavities during the breeding and nonbreeding season in old burns in the Black Hills, South Dakota. American Midland Naturalist 160:413–429.

Glisson, W. J., R. S. Brady, A. T. Paulios, S. K. Jacobi, and D. J. Larkin (2015). Sensitivity of secretive marsh birds to vegetation condition in natural and restored wetlands in Wisconsin. Journal of Wildlife Management 79:1101–1116.

Goldstein, M. I., T. E. Lacher, M. E. Zaccagnini, M. L. Parker, and M. J. Hopper (1999). Monitoring and assessment of Swainson's Hawks in Argentina following restrictions on monocrotophos use, 1996–1997. Ecotoxicology 8:215–224.

Green, A. J., J. Figuerola, and M. I. Sanchez (2002). Implications of waterbird ecology for the dispersal of aquatic organisms. Acta Oecologica International Journal of Ecology 23:177–189.

Green, A. J., K. M. Jenkins, D. Bell, P. J. Morris, and R. T. Kingsford (2008). The potential role of waterbirds in dispersing invertebrates and plants in arid Australia. Freshwater Biology 53:380–392.

Herkert, J. R. (2007). Evidence for a recent Henslow's Sparrow population increase in Illinois. Journal of Wildlife Management 71:1229–1233.

Holopainen, S., C. Arzel, L. Dessborn, J. Elmberg, G. Gunnarsson, P. Nummi, H. Poysa, and K. Sjoberg (2015). Habitat use in ducks breeding in boreal freshwater wetlands: A review. European Journal of Wildlife Research 61:339–363.

Hovick, T. J., R. D. Elmore, S. D. Fuhlendorf, D. M. Engle, and R. G. Hamilton (2015). Spatial heterogeneity increases diversity and stability in grassland bird communities. Ecological Applications 25:662–672.

Jackson, J. A. (1994). Red-cockaded Woodpecker (Picoides borealis). In The birds of North America, no. 85, A. Poole, Editor. Academy of Natural Sciences, Philadelphia, PA/American Ornithologists' Union, Washington, DC. doi:10.2173/bna.85.

Jackson, M. M., S. E. Gergel, and K. Martin (2015). Effects of climate change on habitat availability and configuration for an endemic coastal bird. PLoS ONE 10 (11): e0142110. doi:10.1371/journal.pone.0142110.

Jones, C. G., J. H. Lawton, and M. Shachak (1994). Organisms as ecosystem engineers. Oikos 69:373–386.

Jones, G. (2005). Is the management plan achieving its objectives? In Protected area management: Principles and practice, G. Worboys, M. Lockwood, and T. De Lacy, Editors. Oxford University Press, Melbourne.

Jones, G. (2009). The adaptive management system for the Tasmanian Wilderness World Heritage Area—linking management planning with effectiveness evaluation. In Adaptive Environmental Management. A Practitioner's Guide, C. Allan and G. Stankey, Editors. Co-published by Springer and CSIRO Publishing, Dordrecht, The Netherlands, and Collingswood, Australia.

Kaminski, R. M., R. W. Alexander, and B. D. Leopold (1993). Wood Duck and Mallard winter microhabitats in Mississippi hardwood bottomlands. Journal of Wildlife Management 57:562–570.

Kaminski, R. M., J. B. Davis, H. W. Essig, P. D. Gerard, and K. J. Reinecke (2003). True metabolizable energy for Wood Ducks from acorns compared to other waterfowl foods. Journal of Wildlife Management 67:542–550.

Kendrick, S. W., P. A. Porneluzi, F. R. Thompson III, D. L. Morris, J. M. Haslerig, and J. Faaborg (2015). Stand-level bird response to experimental forest management in the Missouri Ozarks. Journal of Wildlife Management 79:50–59.

Kennedy, C. M., P. P. Marra, W. F. Fagan, and M. C. Neel (2010). Landscape matrix and species traits mediate responses of Neotropical resident birds to forest fragmentation in Jamaica. Ecological Monographs 80:651–669.

Knick, S. T., and J. T. Rotenberry (2000). Ghosts of habitats past: Contribution of landscape change to current habitats used by shrubland birds. Ecology 81:220–227.

Kochert, M. N., M. R. Fuller, L. S. Schueck, L. Bond, M. J. Bechard, B. Woodbridge, G. L. Holroyd, M. S. Martell, and U. Banasch (2011). Migration patterns, use of stopover areas, and austral summer movements of Swainson's Hawks. The Condor 113:89–106.

Kotliar, N. B., and J. A. Wiens (1990). Multiple scales of patchiness and patch structure: A hierarchical framework for the study of heterogeneity. Oikos 59:253–260.

Kurz, W. A., C. C. Dymond, G. Stinson, G. J. Rampley, E. T. Neilson, A. L. Carroll, T. Ebata, and L. Safranyik (2008). Mountain pine beetle and forest carbon feedback to climate change. Nature 452:987–990.

Lindenmayer, D. B., A. D. Manning, P. L. Smith, H. P. Possingham, J. Fischer, I. Oliver, and M. A. McCarthy (2002). The focal-species approach and landscape restoration: A critique. Conservation Biology 16:338–345.

Lorenz, T. J., K. T. Vierling, T. R. Johnson, and P. C. Fischer (2015). Choice or constraint? The role of wood hardness in limiting nest site selection in North American woodpeckers. Ecological Applications 25:1016–1033.

Luck, G. W., G. C. Daily, and P. R. Ehrlich (2003). Population diversity and ecosystem services. Trends in Ecology and Evolution 18:331–336.

Lundberg, J., and F. Moberg (2003). Mobile link organisms and ecosystem functioning: Implications for ecosystem resilience and management. Ecosystems 6:87–98.

Lyons, J. E., M. C. Runge, H. P. Laskowski, and W. L. Kendall (2008). Monitoring in the context of structured-decision making and adaptive management. Journal of Wildlife Management 72:1683–1692.

Manfredo, M. J., J. J. Vaske, P. J. Brown, D. J. Decker, and E. A. Duke, Editors (2009). Wildlife and society: The science of human dimensions. Island Press, Washington, DC.

Martin, K. (2001). Wildlife communities in alpine and subalpine habitats. In Wildlife-habitat relationships in Oregon and Washington, D. H. Johnson and T. A. O'Neil, Editors. Oregon University Press, Corvallis, pp. 239–260.

Martin, K. (2013). The ecological values of mountain environments and wildlife. In The impacts of skiing and related winter recreational activities on mountain environments, Bentham Publishers, Sharjah, UAE, pp. 3–29. doi:10.2174/9781608054886113010004.

Martin, K. (2014). Avian strategies for living at high elevation: Life-history variation and coping mechanisms in mountain habitats. Ibis—Conference Proceeding. http://www.bou.org.uk/bouproc-net/uplands/martin.pdf.

Martin, K., and K. L. Wiebe (2004). Coping mechanisms of alpine and arctic breeding birds: Extreme weather and limitations to reproductive resilience. Integrative and Comparative Biology 44:177–185.

Martin, K., K. E. H. Aitken, and K. L. Wiebe (2004). Nest sites and nest webs for cavity-nesting communities in interior British Columbia, Canada: Nest characteristics and nest partitioning. The Condor 106:5–19.

Masse, R. J., B. C. Tefft, and S. R. Mcwilliams (2015). Higher bird abundance and diversity where American Woodcock sing: Fringe benefits of managing forests for woodcock. Journal of Wildlife Management 79:1378–1384.

Mejia, A. (2015). Umbrella animals citizen engagement. http://www.slideshare.net/amejiapizano/umbrella-animals-citizen-engagement-project-by-andres-mejia-pizano.

Milton, K. (1996). Environmentalism and cultural theory: Exploring the role of anthropology in environmental discourse. Routledge, New York.

Mitsch, W. J., and J. G. Gosselink (2007). Wetlands. 5th ed. John Wiley and Sons, New York.

Moore, C. T., and M. J. Conroy (2006). Optimal regeneration planning for old-growth forest: Addressing scientific uncertainty in endangered species recovery through adaptive management. Forest Science 52:155–172.

Morrison, M. L., B. G. Marcot, and R. W. Mannan (2006). Wildlife-habitat relationships: Concepts and applications. 3rd ed. Island Press, Washington, DC.

Morrissey, C. A., L. I. Bendell-Young, and J. E. Elliott (2004). Linking contaminant profiles to the diet and breeding location of American dippers using stable isotopes. Journal of Applied Ecology 41:502–512.

Murcia, C. (1995). Edge effects in fragmented forests—implications for conservation. Trends in Ecology and Evolution 10:58–62.

Newton, I. (2006). Can conditions experienced during migration limit the population levels of birds? Journal of Ornithology 147:146–166.

Nislow, K. H., C. L. Folt, and D. L. Parrish (1999). Favorable foraging locations for young Atlantic salmon: Application to habitat and population restoration. Ecological Applications 9:1085–1099.

North American Bird Conservation Initiative, US Committee (2009). The state of the birds 2009 report. US Department of Interior, Washington, DC.

North American Bird Conservation Initiative, US Committee (2011). The state of the birds 2011 report. US Department of Interior, Washington, DC.

North American Bird Conservation Initiative, US Committee (2014). The state of the birds 2014 report. US Department of Interior, Washington, DC.

NRCS (2012). Fish and wildlife habitat management publication on Wood Duck (Aix sponsa). N. R. C. Service, Editor. USDA Natural Resources Conservation Service, Wildlife Habitat Management Institute, and the Wildlife Habitat Council.

Ormerod, S. J. (1985). The diet of breeding Dippers (Cinclus mexicanus) and their nestlings in the catchment of the River Wye, mid-Wales: A preliminary study by faecal analysis. Ibis 127:316–331.

Ormerod, S. J., J. Ohalloran, S. D. Gribbin, and S. J. Tyler (1991). The ecology of dippers Cinclus cinclus in relation to stream acidity in upland Wales: Breeding performance, calcium physiology and nestling growth. Journal of Applied Ecology 28:419–433.

Ormerod, S. J., and S. J. Tyler (1993). Birds as indicators of changes in water quality. In Birds as monitors of environmental change, R. W. Furness and J. J. D. Greenwood, Editors. Chapman and Hall, London, pp. 179–216.

Osborne, D. C., and D. W. Sparling (2013). Multi-scale associations of grassland birds in response to cost-share management of conservation reserve program fields in Illinois. Journal of Wildlife Management 77:920–930.

Parrish, J. K., M. Marvier, and R. T. Paine (2001). Direct and indirect effects: Interactions between Bald Eagles and Common Murres. Ecological Applications 11:1858–1869.

Perlut, N. G., A. M. Strong, and T. J. Alexander (2011). A model for integrating wildlife science and agri-environmental policy in the conservation of declining species. Journal of Wildlife Management 75:1657–1663.

Renfrew, R., A. M. Strong, N. G. Perlut, S. G. Martin, and T. A. Gavin (2015). Bobolink (Dolichonyx oryzivorus). In The birds of North America, no. 176, A. Poole, Editor. Academy of Natural Sciences, Philadelphia, PA/American Ornithologists' Union, Washington DC. doi:10.2173/bna.176.

Ressurreicao, A., J. Gibbons, M. Kaiser, T. P. Dentinho, T. Zarzycki, C. Bentley, M. Austen, et al. (2012). Different cultures, different values: The role of cultural variation in public's WTP for marine species conservation. Biological Conservation 145:148–159.

Reynolds, C., N. A. Miranda, and G. S. Cumming (2015). The role of waterbirds in the dispersal of aquatic alien and invasive species. Diversity and Distributions 21 (7): 744–754.

Ribic, C. A., R. R. Koford, J. R. Herkert, D. H. Johnson, N. D. Niemuth, D. E. Naugle, K. K. Bakker, D. W. Sample, and R. B. Renfrew (2009). Area sensitivity in North American grassland birds: Patterns and processes. The Auk 126: 233–244.

Robles, H., and K. Martin (2014). Habitat-mediated variation in the importance of ecosystem engineers for secondary cavity nesters in a nest web. PLoS ONE 9 (2): e90071. doi:10.1371/journal.pone.0090071.

Rodenhouse, N. L., and L. B. Best (1983). Breeding ecology of vesper sparrows in corn and soybean fields. American Midland Naturalist 110: 265–275.

Runge, M. C. (2011). An introduction to adaptive management for threatened and endangered species. Journal of Fish and Wildlife Management 2:220–233.

Runge M. C., A. M. Romito, G. Breese, J. F. Cochrane, S. J. Converse, M. J. Eaton, M. A. Larson, et al., Editors (2016). Introduction to structured decision making. US Fish and Wildlife Service, National Conservation Training Center, Shepherdstown, WV.

Saab, V. A., and H. D. W. Powell (2005). Fire and avian ecology in North America: Process influencing pattern. Studies in Avian Biology 30:1–13.

Samson, F., and F. Knopf (1994). Prairie conservation in North America. Bioscience 44:418–421.

Sarasola, J. H., and J. J. Negro (2005). Hunting success of wintering Swainson's Hawks: Environmental effects on timing and choice of foraging method. Canadian Journal of Zoology/Revue Canadienne De Zoologie 83 (10): 1353–1359.

Sauer, J. R., J. E. Hines, J. E. Fallon, K. L. Pardieck, D. J. Ziolkowski Jr., and W. A. Link (2014). The North American breeding bird survey, results and analysis 1966–2013. Version 01.30.2015. USGS Patuxent Wildlife Research Center, Laurel, MD.

Şekercioğlu, C. H. (2006). Increasing awareness of avian ecological function. Trends in Ecology and Evolution 21:464–471.

Şekercioğlu, C. H., G. C. Daily, and P. R. Ehrlich (2004). Ecosystem consequences of bird declines. Proceedings of the National Academy of Sciences USA 101:18042–18047.

Slocombe, D. S. (1998). Defining goals and criteria for ecosystem-based management. Environmental Management 22:483–493.

Smith, C. B. (2011). Adaptive management on the central Platte River—science, engineering, and decision analysis to assist in the recovery of four species. Journal of Environmental Management 92:1414–1419.

Sullivan, S. M. P., M. C. Watzin, and W. C. Hession (2006). Effects of geomorphic condition and habitat quality on measures of Belted Kingfisher (Ceryle alcyon) reproductive success. Waterbirds: The International Journal of Waterbird Biology 29:258–270.

Sullivan, S. M. P., M. C. Watzin, and W. S. Keeton (2007). A riverscape perspective on habitat associations among riverine bird assemblages in the Lake Champlain Basin, USA. Landscape Ecology 22:1169–1186.

Sullivan, S. M. P., and K. T. Vierling (2012). Exploring the influences of multiscale environmental factors on the American dipper Cinclus mexicanus. Ecography 35:624–636.

Suter, W., R. F. Graf, and R. Hess (2002). Capercaillie (Tetrao urogallus) and avian biodiversity: Testing the umbrella-species concept. Conservation Biology 16:778–788.

Tagwireyi, P., and S. M. P. Sullivan (2015). Distribution and trophic dynamics of riparian tetragnathid spiders in a large river system. Marine and Freshwater Research 67:309–318.

Teel, T. L., and M. J. Manfredo (2010). Understanding the diversity of public interests in wildlife conservation. Conservation Biology 24:128–139.

Tella, J. L., and F. Hiraldo (2014). Illegal and legal parrot trade shows a long-term, cross-cultural preference for the most attractive species increasing their risk of extinction. PLoS ONE 9:10.

Toth, L. A., S. L. Melvin, D. A. Arrington, and J. Chamberlain (1998). Hydrologic manipulations of the channelized Kissimmee river—implications for restoration. Bioscience 48:757–764.

Tremblay, M. A., and C. C. St. Clair (2011). Permeability of a heterogenous urban landscape to the movement of forest songbirds. Journal of Applied Ecology 48:679–688.

Turner, I. M., and R. T. Corlett (1996). The conservation value of small, isolated fragments of lowland tropical rain forest. TREE 11:330–333.

Twedt, D. J., and C. R. Loesch (1999). Forest area and distribution in the Mississippi alluvial valley: Implications for breeding bird conservation. Journal of Biogeography 26:1215–1224.

United Nations (2015). World population prospects: The 2015 revision, key findings and advance tables. Department of Economic and Social Affairs, Population Division. Working Paper ESA/P/WP.241, New York.

Van Horn, M. A., and T. M. Donovan (2011). Ovenbird (Seiurus aurocapilla). In The birds of North America, no. 088, A. Poole, Editor. Academy of Natural Sciences, Philadelphia, PA/American Ornithologists' Union, Washington, DC. doi:10.2173/bna.88.

van Leeuwen, C. H. A., G. van der Velde, J. M. van Groenendael, and M. Klaassen (2012). Gut travellers: Internal dispersal of aquatic organisms by waterfowl. Journal of Biogeography 39:2031–2040.

Vickery, P. C., and J. R. Herkert (1999). Ecology and conservation of grassland birds of the western hemisphere. Studies in Avian Biology no. 19. Cooper Ornithological Society, Camarillo, CA.

Vierling, K. T. (1997). The effects of suburbanization and haying on the reproductive success of grassland birds breeding in hayfields in Boulder, Colorado. Final Report for Boulder Open Space Department, Boulder, CO.

Vierling, K. T. (2000). Source and sink population dynamics of Red-winged Blackbirds in a rural/suburban landscape. Ecological Applications 10:1211–1218.

Vierling, K. T., L. B. Lentile, and N. Nielsen-Pincus (2008). Preburn characteristics and woodpecker use of burned coniferous forests. Journal of Wildlife Management 72:422–427.

Vörösmarty, C. J., P. B. McIntyre, M. O. Gessner, D. Dudgeon, A. Prusevich, P. Green, S. Glidden, et al. (2010). Global threats to human water security and river biodiversity. Nature 467 (7315): 555–561. doi:10.1038/nature09440.

Warner, R. E., J. W. Walk, and J. R. Herkert (2012). Managing farmlands for wildlife. *In* The wildlife techniques manual: Management, N. J. Silvy, Editor. Johns Hopkins University Press. Baltimore, MD.

White, A. M., E. F. Zipkin, P. N. Manley, and M. D. Schlesinger (2013). Conservation of avian diversity in the Sierra Nevada: Moving beyond a single-species management focus. PLoS ONE 8:e63088.

Wiens, J. H. (1985). Habitat selection in variable environments: Shrub-steppe birds. *In* Habitat selection in birds, M. L. Cody, Editor. Academic Press, Orlando, FL, pp. 227–251.

Wiens, J. H., G. D. Hayward, R. S. Holthausen, and M. J. Wisdom (2008). Using surrogate species and groups for conservation planning and management. Bioscience 58:241–252.

Williams, B. K., R. C. Szaro, and C. D. Shapiro (2009). Adaptive management: The US Department of the Interior technical guide. Adaptive Management Working Group, US Department of the Interior, Washington, DC.

Williams, B. K., and E. D. Brown (2012). Adaptive management: The US Department of the Interior applications guide. Adaptive Management Working Group, US Department of the Interior, Washington, DC.

Williams, T. M., and D. J. Lipscomb (2002). Natural recovery of Red-cockaded Woodpecker cavity trees after Hurricane Hugo. Southern Journal of Applied Forestry 26:197–206.

Wu, J., and R. Hobbs, Editors (2007). Key topics in landscape ecology. Cambridge University Press, Cambridge, UK.

Yahner, R. H., C. G. Mahan, and A. D. Rodewald (2012). Managing forests for wildlife. *In* The wildlife techniques manual: Management, N. J. Silvy, Editor. Johns Hopkins University Press, Baltimore, MD.

CHAPTER 29

The Social and Economic Worth of Birds

Christopher A. Lepczyk, Paige S. Warren, and Michael W. Strohbach

BIRDS IN HUMAN CULTURE

Among the wild organisms of the world, few taxonomic groups hold as strong a place in human life and society as birds. Birds are common symbols in many cultures and can be used to demonstrate such attributes as strength (eagles) and wisdom (owls). Likewise, birds serve central purposes in fairy tales and myths within many cultures. For instance, many children learn about storks delivering babies, and the word "nevermore" is quickly attributable to Edgar Allan Poe's raven. Birds and bird names find their way into popular songs across nearly every genre of music (e.g., "Mockingbird" by Eminem; "El Condor Pasa" by Simon and Garfunkel; "When Doves Cry" by Prince; "Strange Bird" by Jimmy Buffett; "Ornithology" by Charlie Parker) and from well-known proverbs (e.g., "a bird in the hand is worth two in the bush"). Even professional sports teams abound with bird names (e.g., St. Louis Cardinals, Toronto Blue Jays), and roads are commonly named after birds (e.g., Bluebird Way) or bird-related aspects (e.g., *halemanu*—Hawaiian for birdhouse). In fact, we would be hard-pressed to find many places or cultures that do not have birds as a basic element of their society. However, birds' ubiquity in appellations is only part of their importance.

Why do birds have such a strong social and cultural recognition? Surely part of the reason includes that many birds are brightly colored or conspicuous, fly, and are of a body size easily detected and recognized by people. Moreover, a number of bird species readily live in areas where people do, and they frequent bird feeders and birdhouses. In fact, millions of people around the world feed and watch birds in the vicinity of their own homes or places of work. But more generally, people experience birds in a greater number of ways in their everyday lives compared with other animal species.

For instance, many bird species provide basic resources for humans. Chief among those present-day resources is food.

A wide variety of birds have been domesticated for food, and selective breeding has been conducted throughout the world to improve growth rates, meat quality, the efficiency with which feed is converted into lean muscle, and other attributes designed to produce food for people. For instance, the 2012 United States Census of Agriculture (a complete census of all farms and farmers conducted every five years by the United States Department of Agriculture) reported that more than 8 billion chickens and nearly 300 million turkeys were sold for food in the United States alone (National Agricultural Statistics Service 2014). These estimates count neither the variety of other poultry raised on farms (e.g., 23 million ducks, 27 million quail, and ~8 million pheasants) nor the eggs of many species. In fact, the United States produces approximately 50 billion eggs per year. When considered together, the meat and eggs produced by domestic birds are a key component of human diets around the world, with many by-products also being used for such purposes as pet food and insulation (e.g., feathers).

Although we typically think of domestic birds as the ones that we consume, the reality is that humans also harvest a number of wild birds throughout the world for food. In the United States, these wild birds are typically what we call game birds and include species such as migratory waterfowl (ducks, geese, and swans), wild turkeys, quail, pheasant, doves, and grouse. Every five years the United States Fish and Wildlife Service, under the auspices of the United States Census Bureau, conducts the National Survey of Fishing, Hunting, and Wildlife-Associated Recreation to determine basic statistics about how many people hunt, who these people are, and how much they spend to hunt. The most recent survey in 2011 found that nearly 3.1 million hunters pursued wild turkeys, 1.5 million hunted pheasant, 800,000 hunted quail, 800,000 hunted grouse/prairie chickens, 1.4 million hunted ducks, 1.3 million hunted doves, and 800,000 hunted geese (US Department of the Interior et al. 2014). Migratory bird hunters alone spent 23 million days hunting birds, totaling $1.8 billion on bird-related trips and equipment (US Department of the Interior et al. 2014). Thus, hunting provides a source of food and recreation for people while also supporting local economies.

Beyond just food, birds provide a number of other products that people use in their lives (see, e.g., the box on page 899), but the product of widest use to humans is feathers. Historically, bird feathers, such as those from geese, were used as quills for writing (fig. 29.1). In fact, the word "pen" is derived from the Latin word *penna*, meaning feather. Feathers have been used regularly to adorn haute couture fashion, such as women's hats, and even today feathers can be found in clothing and jewelry. Feathers were also used in different cultures as signs of royalty, such as the *kāhili* in Native Hawaiian culture (box on page 884). But one of the main uses of feathers, both today and in the past, is to keep people warm. Because of their excellent heat-trapping capabilities (particularly down feathers), feathers are used in bedding material such as down comforters, winter coats, and sleeping bags. Today the feathers are predominantly from domestic

species, but the function remains the same. Beyond clothing, feathers have several other common uses in society. For instance, feathers are used as fletching on arrows and regularly used to make flies for fly-fishing (fig. 29.1).

People do not only use birds, but also keep them as pets. Depending on the source of information, somewhere between 3.7 million (American Veterinary Medical Association 2012) and 6.1 million (APPA 2015) households in the United States alone have pet birds, totaling over 14 million individual birds (APPA 2015). While not as popular as dogs and cats, birds do make up a substantial component of pets and the pet trade. Just as we have done for cats and dogs, people have selectively bred certain birds for their own interests, most notably the domestic pigeon, which is descended from Rock Doves (*Columba livia*). In fact, in many parts of the world the hobby of breeding pigeons is a common pastime, particularly Europe. Such pigeon fanciers have developed an extremely wide array of phenotypes for a variety of purposes (fig. 29.2). Even Charles Darwin was a pigeon fancier, using pigeons to explain many parts of the theory of natural selection (Darwin 1859).

Although people have derived great benefit from birds, their use as a resource has resulted in many species becoming imperiled and others going extinct (see chapter 25). Many of the world's flightless birds, particularly those on islands,

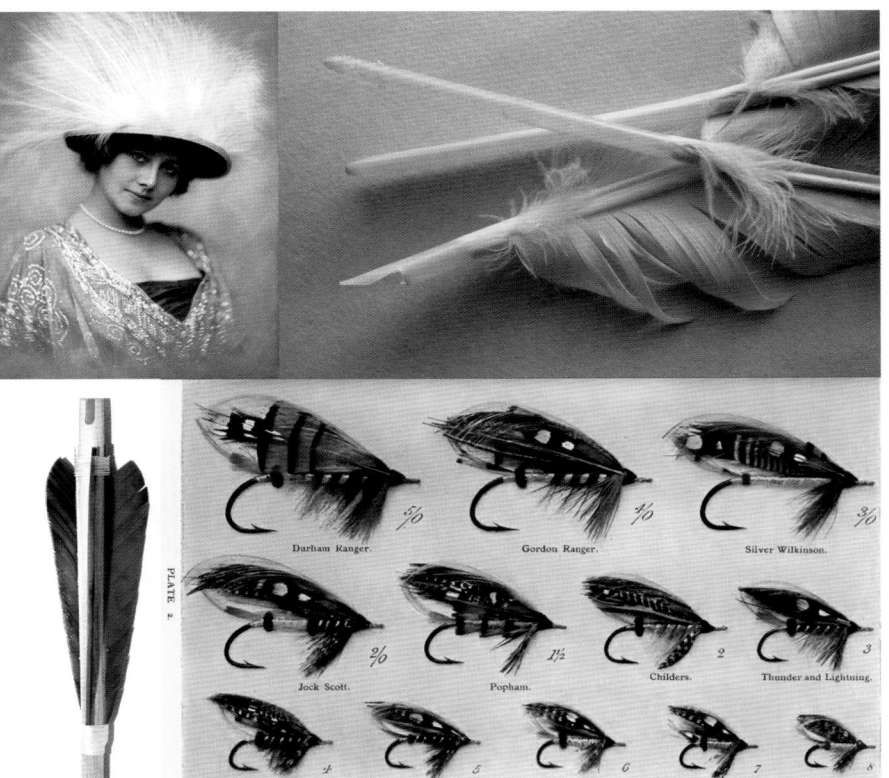

Figure 29.1. Some of the many ways feathers are used by people. Clockwise from top left is a woman's hat by Jan Langhans (July 9, 1851–March 22, 1928), *public domain Wikimedia Commons*; a set of three quills from rough to finished product *by Jonathunder, Wikimedia Commons, CC BY-SA 3.0, https:// upload.wikimedia.org/wikipedia/commons/2 /22/3quills.jpg*; fly fishing flies; and fletching on an arrow *by Simon A. Eugster Wikimedia Commons, CC BY 3.0.*

BIRDS IN HAWAIIAN MATERIAL CULTURE: FEATHER ARTIFACTS

By Sheila Conant

"The feathers of birds were the most valued possessions of the ancient Hawaiians."

David Malo (1951: 76)

Many cultures have used bird feathers to create beautiful ceremonial clothing and other types of artifacts. These objects were often symbols of social rank and wealth, and some were used in a variety of ceremonies, many of them sacred. This was certainly the case in Hawaiian culture when Europeans first came into contact with it in 1778 (Brigham 1899, 1903, Malo 1951, Buck 1957, Force and Force 1968). The astonishing array of precontact feather artifacts from Hawai'i collected on the voyages of Captain James Cook (Kaeppler 1978) provide us with an idea of the nature of Hawaiian feather work before it was influenced by outside cultures (Rose 1978, 1979, Rose et al. 1993, Amante-Helweg and Conant 2009).

During the annual Makahiki festival, when war was prohibited, feathers were given to members of the chiefly class (*ali'i*) of Hawaiian society. The most powerful chiefs could command a great many feathers. Bundles of feathers

Example of a *kāhili* from Queen Emma's Summer Palace on O'ahu. *Photo courtesy of Sheila Conant.*

collected during this festival were then made into a variety of objects, including large cloaks called *'ahu'ula*, shorter capes, and feathered helmets called *mahiole*. Anthropologists have speculated that *'ahu'ula* and *mahiole* were worn in battle to protect warriors from weapons such as spears and clubs. An important class of artifacts is *kāhili*, or feathered standards (see figure), which have been compared to coats of arms. These *kāhili* accompanied individuals of the chiefly class wherever they went, and ranged in size from about one to seven meters in height. Malo (1951) wrote, "Where the king went, there went his *kāhili* bearer (pa'a kahili or lawe kāhili) . . . and where he stopped there stopped also the kāhili bearer."

The red, yellow, and black feathers used in the most valuable and significant objects were taken from small forest birds, including the 'I'iwi (*Vestiaria coccinea*), Hawai'i 'Ō'ō (*Moho nobilis*), and the Mamo (*Drepanis pacifica*). *Kia manu* (bird catchers) used several techniques to capture these small birds that spend most of their time in the upper canopy (often well above 20 meters). Feathers of roosters (*Gallus gallus*), Red-tailed Tropicbird (*Phaethon rubricauda*), and Great Frigatebirds (*Fregata minor*) were also used.

How did the collection of feathers during the precontact era affect bird populations? We will never know. The seabird species whose feathers were used still exist in good numbers, but of the forest birds whose red, yellow, and black feathers were used in large numbers, only the 'I'iwi is extant today. Even though their numbers were orders of magnitude greater in the precontact era than they are today, capturing forests birds was extremely challenging. In addition, *'ahu'ula*, *mahiole*, and *kāhili* were made for and used by only the chiefly class, whose numbers were very small in comparison with the number of common people. In comparison with habitat destruction, disease, and introduced predators, which have decimated forest bird populations during the last 240 or so years, collecting feathers may not have had enough impact to cause extinctions.

have gone extinct because of human exploitation. We can look at the Dodo (*Raphus cucullatus*; Roberts and Solow 2003) and the variety of New Zealand moa and Hawaiian *moa nalo* (or "lost fowl"; family Anatidae; James and Olson 1983, Olson and James 1982) as prime examples of such flightless birds

that no longer grace the Earth. Even birds that were once thought too numerous to extinguish were overharvested during the **market hunting era** of the United States. During that period, which ended in the beginning of the twentieth century, animals were hunted commercially, leading to pre-

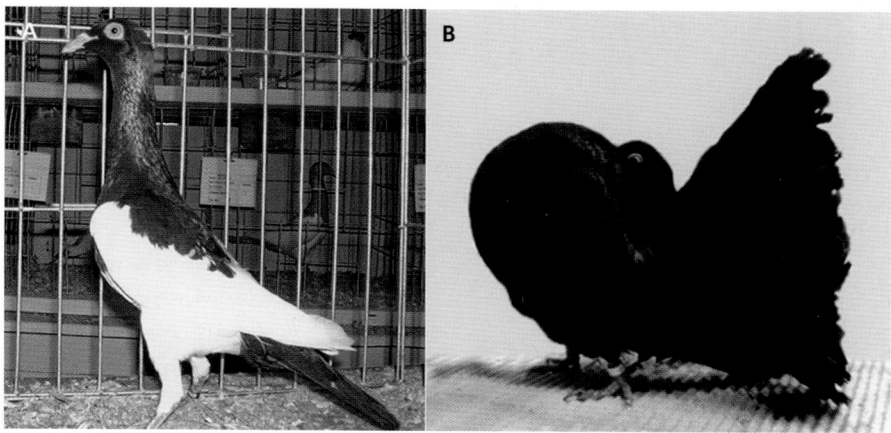

Figure 29.2. The wide variety of domestic pigeons produced through selective breeding. *A*, An English Magpie. *Photo by Graham Manning (using Kodak Z700 digital camera). CC BY-SA 3.0, http://creativecommons.org /licenses/by-sa/3.0) or GFDL, http://www .gnu.org/copyleft/fdl.html), via Wikimedia Commons. B*, A Black Fantail. *Photo by Jim Gifford (originally posted to Flickr as fantail (black self)), CC BY-SA 2.0 http://creativecommons .org/licenses/by-sa/2.0, via Wikimedia Commons.*

cipitous declines in many species and to some going extinct. Perhaps the most illustrative example of the effects of market hunting was the loss of the Passenger Pigeon (*Ectopistes migratorius*), which was once estimated to number over a billion individuals. In a mere matter of decades, it was reduced to a small population in northern Michigan, with the final individual, named Martha, dying in the Cincinnati Zoo in 1914 (chapter 25). The plume trade was another source of population decline for many shorebirds, seabirds, and wetland-complex birds in the late 1800s (chapter 25). During this time, feathers were in great demand by the millenary industry for use in women's hats, which led to millions of birds being harvested each year. Not only were the birds being harvested unsustainably, but many were hunted during the breeding season, as birds like egrets had nuptial plumages that were highly sought after. By 1915, plumes sold for $32 an ounce—the same price of gold at that time. Only through the enactment of a number of national laws did the market hunting era ultimately come to an end (Redekop 2012). While many bird populations have recovered in the intervening century, many others, such as Whooping Cranes (*Grus americana*), have faced a slower recovery. Today, one growing concern is the use of wild birds in the pet trade, both in terms of humane treatment and the impact on wild populations (Baker et al. 2013). One particularly vulnerable group is parrots. Of the world's nearly 400 extant parrot species, 40 are considered endangered and all but two parrot species are regulated by the Convention on International Trade in Endangered Species (CITES) of Wild Fauna and Flora.

While history is replete with examples of bird-human interactions having tragic outcomes, people do have a strong affinity for birds and spend a considerable amount of money observing and interacting with them. For instance, many people engage in what the British call **twitching**, whereby they pursue sightings of rare birds, often driving long distances for a chance to observe a vagrant bird or a new oc-

currence in a state or country. Many people also engage in ecotourist bird trips, on which they may spend thousands of dollars, in order to see rare or difficult to find birds. Perhaps the most extreme case is a group of birders that seek to record all the birds in a country (or other given location) within one year. This endeavor, called a **Big Year** (read the book of the same name by Mark Obmascik—it is funnier than the 2011 movie) drives birders literally to the ends of the world to visit remote islands like Attu in Alaska for a chance to record all of the birds in North America.

Beyond birding, though, people view birds positively and enjoy opportunities to observe them, as evident in the large numbers of people that engage in bird-watching, feeding birds in backyard feeders, and managing their property to create attractive habitats for birds. As is the case with many things in society, however, not all birds are enjoyed or seen in a positive light. Some species of birds are viewed as pests because of their effects on agriculture, such as House Sparrows (*Passer domesticus*) (see also the box on page 900) and American Crows (*Corvus brachyrhynchos*). Others are viewed as unpleasant because of their appearance or feeding habits, such as vultures. And yet others are disliked simply for overabundance and the damage their feces cause to architecture, such as Rock Doves in cities. But those negatively viewed species are far outweighed by the many species that people enjoy seeing and interacting with on a regular basis. Given the significant role that birds play in ecosystems (box on page 898 and box on page 900, and chapter 22), such a positive view is important.

BIRDS IN COUPLED HUMAN-NATURAL SYSTEMS

As we have seen, birds can be a source of inspiration, awe, and basic sustenance, and birds can be viewed as a nuisance or as a source of materials that humans find valuable. From the birds' point of view, people are creators or destroyers of habitat and can deplete or increase food and nesting resources. Moreover, people can introduce new species (e.g., novel predators such as rats and cats) or diseases to ecosystems, either intentionally or unintentionally, that have negative repercussions for birds. People even devote their time to bird hunting or bird conservation, or both. Looking at these examples, it becomes obvious that interactions between people and birds are complex and diverse.

Ecologists have long worked to understand how people and nature interact. Early ideas of ecosystems and systems modeling grappled with people as modifiers of energy flow through ecosystems (Odum 1959, Goudie 2000). Over the past 50 years, the idea of an ecosystem has expanded and shifted somewhat in terminology, but still embraces the central idea that people and nature are inherently interconnected and influence (as well as are influenced by) each other. Today, ecologists often describe these relationships in terms of a **coupled human-natural system** (CHANS) framework (Liu et al. 2007) or **socio-ecological system** (Ostrom 2009), the latter being essentially a synonym of CHANS. For example, understanding processes such as the success of Rock Doves in colonizing cities across the globe (Aronson et al. 2014), farmland birds disappearing from Europe's changing agricultural landscapes (Shrubb 2003), or the range expansion of the Northern Cardinal (*Cardinalis cardinalis*; Robb et al. 2008) is not possible without including humans in the equation. In light of such broadscale human influences as climate change, pollution, or the introduction of new species (Kareiva and Marvier 2015), ecologists now recognize that there are virtually no places on the planet where the effects of people on nature are absent, even in ecosystems that appear to be pristine (Vitousek 1997, Goudie 2000).

The CHANS framework (fig. 29.3) provides a useful way to integrate both the natural system (i.e., the abiotic and biotic components of ecosystems and their interactions) and the human system (e.g., culture, human behavior, institutions, economy) for understanding how people and birds are interrelated. For instance, the CHANS framework incorporates all the goods and benefits humans receive from ecosystems, which we call **ecosystem services** (Daily 1997), as well as the negative impacts of ecosystems on humans, which we call **ecosystem disservices** (Dunn 2010). Human influences on the natural system are often characterized as **long-term press disturbances**, like urban growth (Lepczyk et al. 2007, 2008), or **short-term pulse disturbances**, like tilling a field. Some human influences on the natural system are intentional and aim to maintain the production of ecosystem services, such as agriculture and forest management. Other press and pulse disturbances, such as habitat fragmentation or pollution, harm natural systems. Finally, humans can cause an ecosystem to change in a fundamental way, called a **state change**, such that ecosystem services and disservices change, which in turn feed back to influence the human system (including individuals, social and political groups, institutions, etc.). We discuss several of these in greater detail in the sections ahead.

THE SOCIAL CONTEXT OF BIRD-ORIENTED ACTIONS

The deep affinities that many people have for birds can lead them to take many different kinds of actions, ranging from small-scale individual decisions to feed birds at home (Lepczyk et al. 2004a) to organizing larger-scale habitat restoration or policy actions (see Participation in Conservation Organizations below and the box on page 887 and the box on page 888). Some human actions that affect birds may be unintentional (e.g., gardening practices), whereas others are intentional but can have unintended consequences (e.g.,

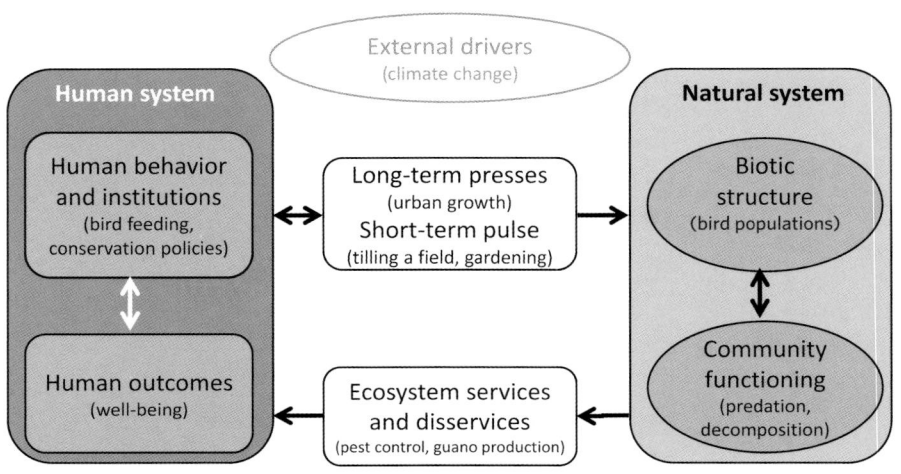

Figure 29.3. A coupled human-natural systems (CHANS) framework, useful for understanding the relationships and interactions between humans and birds (examples in parentheses). *Modified from Taylor 2007.*

ROSALIE BARROW EDGE: A PASSION FOR BIRDS WITH FAR-REACHING CONSEQUENCES

Rosalie Barrow Edge, founder of Hawk Mountain Sanctuary, standing by the entrance. *Photo courtesy of Hawk Mountain Sanctuary Archives, http://hawkmountain.org.*

Rosalie Edge (1877–1962, born Mable Rosalie Barrow) was raised in New York, a scion of privilege and wealth. She discovered birding as an adult, "as she puttered in the soil with the children" at the summer home she shared with Charles Edge on the Long Island Sound (Furmansky 2009). She subsequently spent many hours birding in Central Park, just as many birders continue to do today. As her interest in birds grew, she also began to learn about the threats to their continued existence. She had developed and honed the tools of activism as a local leader in the women's suffrage movement in the early 1910s. Thus, she was ready to leap into action in 1929, when she received a pamphlet critical of the National Association of Audubon Societies (NAAS, a forerunner of the National Audubon Society) for its failure to advocate measures to save species like the Whooping Crane, Trumpeter Swan, and the Bald Eagle (Furmansky 2009). She eventually founded the Emergency Conservation Committee, a small organization that goaded and pushed groups like NAAS to take a more aggressive stance on conservation of unpopular species such as raptors. At the time, hawks and other birds of prey were widely seen as vicious predators of domestic chickens and competitors for game birds like quail and grouse. Edge is most often remembered for her role in the founding of Hawk Mountain Sanctuary in 1934, the first known refuge for birds of prey (see hawkmountain.org). The sanctuary protects a key site in the annual hawk migrations through Pennsylvania. Prior to its protection, hunters regularly visited the ridge during migration, blasting hundreds of hawks out of the sky at a time. In the 1940s, Edge went on to successfully push for the formation of several large national parks, including Yosemite and Kings Canyon. Her tactics were considered overly aggressive by many at the time. However, shortly before her death in 1962, Edge had reconciled with the National Audubon Society and was honored with a standing ovation at the annual meeting of the society (Furmansky 2009).

when bird feeding elevates densities of nest predators or increases risk of disease; Dhondt et al. 2014). While human affinity toward birds influences many of our actions, three widespread ones that are perhaps of greatest note for the contemporary ornithologist are bird-watching, habitat management for birds in residential yards, and participation in conservation organizations.

Bird-watching

One of the largest and fastest growing recreational activities in the United States is bird-watching (fig. 29.4). In 2011, nearly 47 million people in the United States actively watched birds, taking a special interest in birds around their homes or traveling for the expressed purpose of wildlife watching (US Department of the Interior et al. 2014). These active bird-watchers represented about 15 percent of the nation's population that year. While roughly 41 million people watched birds on their property or within a mile of their household, nearly 18 million people traveled to watch birds (US Department of the Interior et al. 2014). Some of these travelers are twitchers, but many are simply those interested in observing common birds in their state or region. Not only does a large portion

THEODORE ROOSEVELT JR.: A POLITICIAN PASSIONATE ABOUT PEOPLE AND THE ENVIRONMENT

President Theodore Roosevelt during a visit to Yosemite National Park in 1904. *Underwood & Underwood, Copyright Claimant, 1904. Retrieved from the Library of Congress, https://www.loc.gov/item /2013650922, March 10, 2016.*

Theodore Roosevelt (1858–1919) was one of the most important environmental leaders in the history of the United States. During his presidency (1901–1909), 150 million acres of forest reserves, 51 federal bird reserves, and 5 national parks were created and the United States Forest Service was established. For Roosevelt, flora and fauna represented public goods, and he was willing to fight for environmental issues even against the resistance of his own party (Brinkley 2009, Redekop 2012).

His passion for the "out-of-doors natural history," and especially birds, started as a child, and he remained a keen naturalist, outdoors person, and hunter throughout his life. As the founding president (1888–1893) of the Boone and Crockett Club of hunters and anglers, he combined politics with his passion for nature and strongly lobbied for the Forest Reserve Act of 1891, which allowed the president to declare forest reserves (Brinkley 2009, Redekop 2012).

Roosevelt made very clear that conservation has benefits for humans. In his State of the Union address of 1901 he said, "Forest protection is not an end of itself; it is a means to increase and sustain the resources of our country and the industries which depend on them." Forest preserves should protect "native flora and fauna" and allow humans to find "rest, health and recreation" (Roosevelt 1901). In the words of today, forest protection is a means to protect biodiversity and to increase and sustain the ecosystem services forests provide for people.

of the population watch birds, but as mentioned earlier, they spend a large amount of money on it. Recent estimates found that in the United States people spent $54.9 billion per year on wildlife-watching (including bird-watching) activities and equipment (US Department of the Interior et al. 2014). These monies are not simply paying for binoculars and gas (or airfare in some instances) to travel to bird-watching sites, but also include monies spent on food and lodging. In fact, many states have wildlife-watching programs and bird-watching trails designed specifically to attract the public and bird-watchers to a given location. The result is that bird-watching is responsible for contributing a significant amount of money to local economies. A number of cities in North America see an economic bump during spring migration, when millions of birds fly north for breeding and birders flock to migration hotspots like Point Pelee on Lake Erie and Dauphin Island on

the Gulf Coast. Year-round, activities like bird festivals draw thousands of people, such as the annual Whooping Crane festival at Wheeler National Wildlife Refuge. (For a complete list of the many bird festivals throughout the United States, the American Birding Association maintains an active list at https://www.aba.org/festivals/.)

The stereotype of a bird-watcher, particularly in the United States, is of a white, retired, and relatively wealthy person, a stereotype somewhat supported by the most recent Survey of Fishing, Hunting, and Wildlife-Associated Recreation (US Department of the Interior et al. 2014). As a result, birding organizations are increasingly trying to engage a broader range of people in bird-watching. For instance, the Audubon Society has located nature centers in urban centers and established programs to reach out to urban youth. Similarly, nature centers like the Urban Ecology Center in Mil-

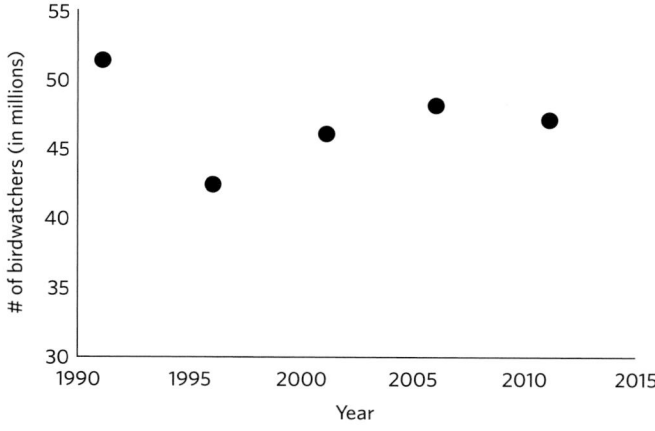

Figure 29.4. Historical trends in bird-watching, 1991–2011, based on USFWS Hunting, Fishing, and Wildlife Rec surveys. Note that surveys for 1991 and 1996 did not ask people if they specifically watched birds on trips away from home. As a result, data before and after 2001 are not strictly comparable.

waukee, Wisconsin, have intentionally situated themselves in urban areas to showcase and encourage activities like bird-watching to an audience that is predominantly inner city and nonwhite. Cornell's Laboratory of Ornithology has developed a multilingual program called Celebrate Urban Birds that targets urban communities, aiming especially to appeal to Latinos in cities. Beyond these educational programs are such bird-watching programs as the Great Backyard Bird Count, sponsored jointly by Audubon and the Cornell Lab of Ornithology, in which people provide valuable data on long-term population trends (Guralnik and VanCleve 2005, Silvertown 2009, chapter 30). Similar programs exist in other countries, for example in the United Kingdom (Garden Bird-Watch by the British Trust for Ornithology) or in Germany (Stunde der Gartenvögel by the NABU). In some places, artists are being employed to increase public awareness about the value of birds. A case in point is in Baltimore, Maryland, where local artist Iandry Randriamondroso created a series of murals to increase awareness of hidden bird habitats in the city and to engage more people in watching local birds even in a highly built up urban area (fig. 29.5). While many programs have been developed to increase bird-watching (and it is becoming more popular in the United States), it enjoys widespread appeal in many other countries around the world. For instance, in the United Kingdom bird-watching is a national pastime and a very common activity for families of all walks of life (Davies et al. 2009).

Habitat Management for Birds in Residential Yards

People regularly modify their land or property in ways to enhance their home based on aesthetics, lifestyle, and per-

sonal interests, as well as social norms (our perceptions of the expectations of our neighbors and peers). Whether planting a fruit tree, putting in a small pond, or even just placing a container garden on a porch, these activities all can result in habitat modification. Residential yards and other privately owned green spaces constitute both a vast portion of urban green spaces (Loram et al. 2007) and the places where many people have their daily experiences of nature. In 2005, lawns alone covered 16.4 million hectares in the United States, constituting more acreage than the top eight irrigated crops combined (Milesi et al. 2003), and nearly half as much land as is currently protected in United States national parks (34 million hectares, National Park Service n.d.). Yards can provide habitats used by a wide variety of birds, perhaps an unexpectedly large portion of the birds in a given region (Daniels and Kirkpatrick 2006, Lerman and Warren 2011, Aronson et al. 2014). In fact, the National Wildlife Federation has an entire program focused on certifying backyards as wildlife habitat and provides property owners with guides on how to manage their property to benefit wildlife (Widows and Drake 2014). Similarly, many Cooperative Extension programs of land grant universities have extensive information on how to manage private lands, including backyards, for the benefit of wildlife.

Many people make intentional choices to attract birds to their yards, including providing feeders, birdhouses, and birdbaths, and intentionally gardening or planting bird-friendly plants (Lepczyk et al. 2004a, Davies et al. 2009). These small-scale actions in individual yards may seem too small to have significant effects on bird populations. However,

Figure 29.5. "B'MORE Birds," a city-funded art project by Iandry Randriamondroso in Baltimore, Maryland, draws the attention of urban residents to the birds with whom they share the city (http://www.baltimorearts.org/1-for-public-art-bmore-birds-project-complete/). *Photo used with permission of the artist.*

if we consider the collective effect of many individuals living close by one another, and sometimes modifying their property in similar ways (e.g., providing bird food), it becomes apparent how many seemingly small-scale, individual actions could affect the types of birds present or absent as well as their population dynamics (Lepczyk et al. 2004a, Warren et al. 2008, Goddard et al. 2010). In other words, if all land-owners that live nearby one another manage their yards in a similar manner, the collective effect of the neighborhood can be quite large and thus have important ramifications for avian populations. While many of these actions aim to have positive impacts on birds, residential yards can also be hazardous places for birds as a result of cat predation, window collisions, and pesticide use (Lepczyk et al. 2004a, 2004b, Loss et al. 2012; see chapter 26).

The kinds of yards, and therefore the kinds of bird habitat, found in a neighborhood are the result of a complex interplay of individual or household-level decisions, social norms, economic constraints, geographic location, and institutional forces (box on pages 890–891). In other words, the yards we see are not solely a result of the inhabitants' personal preferences, but instead may be shaped by legacies from previous residents or pressure to live up to societal or neighborhood norms (Larsen and Harlan 2006, Larson et al. 2009, Clarke et al. 2013, Polsky et al. 2014, Strohbach et al. 2014). In addition, people with access to greater financial means may modify their property in a markedly different manner than those with limited financial resources (Kinzig et al. 2005, Strohbach et al. 2009, Lubbe et al. 2010, Belaire et al. 2014). Given the importance of landowners' decisions on the ecol-

PHOENIX: A CASE STUDY OF YARDS AND BIRDS

Phoenix, Arizona, provides a nice example of the factors shaping yards and the birds that occupy them. Most yards in Phoenix and many other arid cities in the United States can be classified as either **mesic**, with lawns and water-loving plants, or **xeric**, with drought-tolerant plants and gravel in place of lawns (see figure). Plants in xeric yards are not necessarily native, but they are more similar to the plants in the Sonoran Desert, which surrounds the city (Walker et al. 2009, Lerman and Warren 2011). For example, xeric yards generally contain more shrubs and cacti. Birds in xeric yards are also more likely to be Sonoran Desert specialist species unique to the region, while those in mesic yards tend to be broadly distributed generalist species, found in many other parts of the United States or even the world (see figure; Lerman and Warren 2011). It is not just the birds that are different. Xeric yards harbor spiders, beetles, bees, and other insects that are more similar to those found in the desert (McIntyre and Hostetler 2001, Bang et al. 2012). However, temperatures are cooler in mesic yards, in part as a function of increased water use (Stabler et al. 2005).

If xeric yards are so apparently good for birds, why doesn't everyone have one? It's important to recognize that the type of yard is not always up to the individual homeowner. Social scientists with the Central Arizona Phoenix Long-term Ecological Research project (https://sustainability.asu.edu/caplter/) divide the processes affecting yard vegetation into **top-down** and **bottom-up** processes (Larsen and Harlan 2006). Top-down effects on yards include citywide water conservation policies dictating the proportion of drought-resistant plantings in new developments, or developers' decisions about landscaping when marketing new homes to prospective buyers (Kirby 2000). In existing neighborhoods, homeowner's associations may dictate yard maintenance standards in great detail (e.g., the maximum height of plants) and can exact penalties for not adhering to the guidelines (Lerman et al. 2012).

Figure 29.6. Common example of a backyard bird feeder with American Goldfinches and House Finches. *Photo courtesy of Geoffrey E. Hill.*

ogy of their properties, this is an area of increasingly fruitful research. Ultimately, however, while the arrangement of vegetation in yards sets an important stage for which species may be found there, other management activities like bird feeding and providing nest boxes can have important ramifications for bird populations.

Bottom-up processes affecting yard vegetation include individual decisions about what to plant and how to maintain it, but even these decisions are generally made within the context of social norms, cultural traditions, and the demands of lifestyle (Warren et al. 2008, Larson et al. 2009). In Phoenix, wealthier neighborhoods have greater plant diversity and plant cover as well as cooler neighborhoods in the summer heat (Hope et al. 2003, Kinzig et al. 2005, Jenerette et al. 2007, 2011). All of these factors likely contribute to a tendency to see more Sonoran Desert birds and higher bird diversity (Kinzig et al. 2005, Lerman and Warren 2011) in wealthy neighborhoods. Greater native bird diversity is increasingly found in wealthy neighborhoods in other cities as well, though the processes generating this pattern may be different in different cities (Warren et al. 2010, Davis et al. 2012). Thus, many of the processes affecting birds in Phoenix yards may have analogous effects in other regions.

Two different types of yards in Phoenix, Arizona. Xeric yards *A*, with drought-tolerant desert vegetation, which is not always native to the region, appear to support more Sonoran Desert specialist birds like this *C*, Cactus Wren (*Campylorhynchus brunneicapillus*) and *D*, Verdin (*Auriparus flaviceps*). By contrast, *B*, mesic yards, dominated by grass and water-loving plants, are associated with broadly distributed generalist and invasive species like *E*, the Great-tailed Grackle (*Quiscalus mexicanus*) and *F*, European Starling (*Sturnus vulgaris*). Yard A photo courtesy of Christofer Bang and B courtesy of Kelli Larson; bird photos are courtesy Eyal Shochat, used by permission of the photographers.

Bird Feeding

Bird feeding is likely one of the most frequent ways that people interact with birds and other wildlife (fig. 29.6). In the United States the most recent estimate found that nearly 53 million people feed birds (US Department of the Interior et al. 2014) and in the United Kingdom about 12 million households (48 percent; Davies et al. 2009) feed wild birds. Drilling down to more local levels indicates that bird feeding may be even more common, as 76 percent of people surveyed in Phoenix, Arizona, and 82 percent of people survey in Michigan out of the general public engaged in feeding

(Lepczyk et al. 2012). While the propensity of people to feed birds varies across the United States, perhaps as a function of the harshness of the winter (Lepczyk et al. 2012), there are no places where it is absent. Interestingly, it has been suggested that people are more likely to feed birds in areas with higher bird diversity (Fuller et al. 2012). This relationship with bird diversity suggests an intriguing hypothesis that people experience positive feedbacks from seeing a variety of birds at their feeders, leading to a greater interest in feeding birds. Evidence does suggest that people engaged in more positive activities related to birds do have a greater positive view of birds (Lepczyk et al. 2002).

Unlike the relationships found between peoples' socioeconomic context and the vegetation on their property, bird feeding does not appear to demonstrate any similar strong associations. For instance, when bird feeding was investigated along an urban to rural gradient in Michigan, there was only a slight correspondence between bird-related activities on private lands and sociological variables such as age, gender, and occupation (Lepczyk et al. 2004a). Similarly, in a study comparing residents in Phoenix, Arizona, with participants from the Michigan study, age was a consistent correlate of bird feeding across the regions, with older respondents being more likely to feed birds in all locations. This finding is unsurprising, since bird-watching, particularly bird-watching close to the home, is a more common activity among people in their middle years or older (US Department of the Interior et al. 2014). Moreover, lower-income residents in the Arizona-Michigan comparison were just as likely to feed birds as high-income residents, but they were more likely to report feeding inexpensive foods (e.g., leftover bread or tortillas) (Lepczyk et al. 2012). Foods like sugar water or commercial seed, which are costlier, were more commonly reported by higher income residents (Lepczyk et al. 2012).

While sociodemographic factors do not explain as much about bird feeding as one might imagine, geography does. For instance, in Michigan, whether one lived in a rural, suburban, or urban location (i.e., a rural-urban gradient) was a stronger predictor of one's participation in bird-related activities than socioeconomic background. Specifically, urban landowners had a greater density of bird feeders and birdhouses and owned more cats, but planted or maintained vegetation at a lower frequency compared to rural landowners (Lepczyk et al. 2004a, 2004b). This greater density of bird feeders and birdhouses in the urban landscape is consistent with the broadly reported observation that bird densities are greater in urban than in rural or wildlands areas (Marzluff 2001, McKinney 2002, Shochat et al. 2006).

Bird feeding concentrates energy-dense food resources in the environment in a relatively easy-access location. Most bird food is already partially prepared for consumption (e.g., seeds removed from the stem of a plant, sugar water concentrated and in large quantities), requiring less handling time. While there can be risk to individual birds associated with foraging at a bird feeder, there is often a trade-off between the high reward of food relative to other risks. Given that bird feeding provides this concentration of energy in the environment, it is important to ask, what impact does all this bird feeding have on the birds? Most studies to date indicate that additional food leads to increased populations for those species that visit feeders (Fuller et al. 2012, Robb et al. 2008). For example, when scientists attempted to provide food the way individual landowners feed birds (Jansson et al. 1981), the winter survival probabilities of Willow Tits (*Parus montanus*) and Crested Tits (*Lophophanes cristatus*) increased. Likewise, the northern expansion of both the Northern Cardinal (*Cardinalis cardinalis*) and American Goldfinch (*Carduelis tristis*) in North America has been attributed in part to bird feeding (Morneau et al. 1999). While an extensive body of research indicates that providing supplemental food to birds does affect their survival and population, relatively little work has examined the specific mechanisms that drive these changes.

Some have argued that bird feeding, which at first glance seems likely to benefit birds, needs to be critically evaluated as a potential source of indirect negative impacts on birds. The high densities of birds clustering around feeders leads to greater contact, with potential for increased transmission of disease (chapter 23). Bird feeding and other human-provided food resources may also contribute to increasing the dominance of nonnative species (Shochat et al. 2010). For instance, in an experimental test of bird feeding in New Zealand, the addition of bird feeders markedly increased the number nonnative and invasive birds, serving as an instrument of community change (Galbraith et al. 2015). Another potential problem is that most people who feed birds provide commercial seed mixtures, a food that appeals primarily to generalist bird species (fig. 29.6; Lepczyk et al. 2004a, 2004b, 2012), with fewer respondents providing food for specialists (e.g., sugar water for nectar feeders or thistle seed for finches). However, it remains unclear whether predation risk is affected by the presence of bird feeders. The risk per individual for predation may actually decline around feeders as a function of dilution effects and increased vigilance (Dunn and Tessaglia 1994), and one study found that risk of cat predation declined in the presence of bird feeders (Woods et al. 2003).

Supplemental Nesting Resources

Aside from bird feeding, one of the other main ways people commonly interact with birds on their property is by providing nesting resources, usually in the form of a birdhouse or next box. While nest boxes are a resource that fewer birds use than supplemental bird food, they do provide a great benefit to those species that are cavity nesters or use nest boxes. For instance, Eastern Bluebirds (*Sialia sialis*) and Carolina Wrens (*Thryothorus ludovicianus*) commonly use nest boxes. Aside from just placing birdhouses on one's property, landowners can also keep dead and dying trees, for use by cavity-nesting birds (e.g., woodpeckers). Primary cavity-nesters like woodpeckers depend on having dead and decaying wood in yards. But there are legitimate risks of maintaining dead and decaying wood near homes, play areas, and streets. Trees that woodpeckers use for nesting are generally rated by certified arborists as having a higher probability of failure (Kane et al. 2015). Large dead or decaying limbs, which provide prime nesting habitat for a wide variety of species, can be more likely to fall and to cause greater damage when they do fall. Reducing the size of a dead limb through partial cutting can reduce these risks while maintaining habitat (Kane et al. 2015).

Unlike bird feeding, less work has evaluated the socioeconomic breakdown of who provides birdhouses. In the same Michigan study as illustrated before, nearly half of the populace surveyed had at least one birdhouse on their property (Lepczyk et al. 2004a). While people living in rural locations had a greater number of birdhouses on their property, urban dwellers had a much greater density. Hence, we again see that urban locations are likely concentrating supplemental resources, which could have an effect on bird population and community dynamics. Just as age and geographic location mattered with bird feeding, so too does it matter with regard to providing supplemental nesting, as people engaged in this activity were a bit older and tended to live in either rural or urban settings (Lepczyk et al. 2004a).

Effects of Other Yard-Management Practices

People interact with and manage their property in a variety of ways that can simply be thought of as habitat modification.

These modifications can ultimately influence the types and numbers of birds present (fig. 29.7). As we have illustrated, the most common of these activities include managing the plants or vegetation and providing supplemental food and shelter. While we have a broad understanding of these activities, in terms of gross levels of how much and where, we are only beginning to understand how they may affect avian population and community dynamics (e.g., Galbraith et al. 2015). In fact, one of the next challenges in avian ecology is to understand how collective choices people make on their property can translate into population and community outcomes.

Aside from the three main activities discussed above, people engage in many more activities that directly or indirectly affect birds. For instance, owning outdoor pets, particularly cats, negatively affects birds (discussed further in chapter 26; Lepczyk et al. 2004b, Loyd et al. 2013). Similarly, the use of pesticides and herbicides can directly affect the growth and development of birds (Bishop et al. 2000) as well as reducing food availability through affecting invertebrate prey resources (Pinowski et al. 1994, Newton 1995). Notably, the use of pesticides and herbicides is often decided by lawn care service providers, some of whom are hired not by private homeowners but by homeowner's associations. Even when the decisions about chemical applications are made by residents, these decisions can appear perverse. A case in point is that residents with xeric yards in Phoenix were more likely to apply pesticides than those with mesic yards (Larson et al. 2009), even though arthropods tend to be more abundant in mesic yards (Bang et al. 2012). The effects of these and many other yard management practices on birds have yet to be quantified.

Landscape Scale

In contrast to the local scale of yards, people also interact with birds through the ways that they use and alter land at large spatial scales, which we often term the **landscape scale**. (Note that within ecological theory, the concept of a **landscape** is dependent on the processes being studied and,

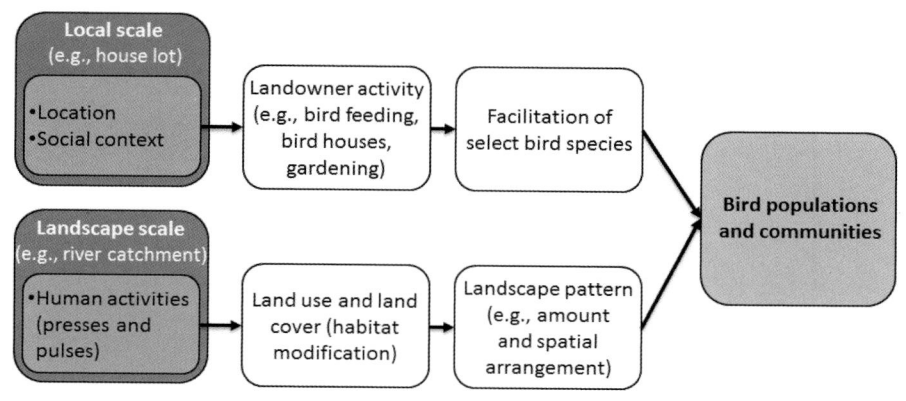

Figure 29.7. The relationship between people and birds at the local and landscape scales.

as a result, does not necessarily indicate a large spatial extent, whereas in practice, landscape scale often refers to large spatial extent; typically a collection of residential yards and other land uses constitute a landscape.) In particular, people have extensively modified the types of ecosystems or habitats (often described by land cover) that birds live in for our own benefits (fig. 29.7).

Broad changes in land cover over the past 50 years have resulted in marked changes in bird species across the United States. For instance, as landscapes become more degraded (i.e., move away from pristine conditions), bird diversity generally decreases, and many species' abundances declines (Lepczyk et al. 2008, Pidgeon et al. 2007, Strohbach et al. 2014). In other parts of the world, intensification of agriculture has also led to a loss in species richness and abundance (e.g., Shrubb 2003). Likewise, as people expand where they live on the landscape, that expansion has complex repercussions for many birds. Recent work viewing the development of houses over many decades has suggested that, initially, the diversity of birds increases until it reaches some apex (Pidgeon et al. 2014). After reaching maximum diversity, the number of unique species of birds begins to decline (fig. 29.8). In other words, through a multi-decadal lens we can see that the expansion of human housing is leading to the rise and then fall of bird diversity.

Aside from the effects of how broad spatial patterns in land cover and land use translate into outcomes for many bird species, a great deal of ecologists' work has focused on urban ecosystems (or what we outside of ecology typically call cities; see the box on page 895). Even though people have greatly altered or changed many landscapes, cities still house a remarkable amount of bird diversity. In fact, nearly 20 percent of the world's bird diversity has been found in just 54 cities from around the world (Aronson et al. 2014). Thus, while cities are often viewed as concrete jungles, they do in fact house great numbers of birds, including some that are threatened or rare. Notably, however, the threatened and rare species that do occur in cities are usually found in remnant patches of native habitats that fall within the city and are not generally found in city centers or industrial locations.

Participation in Conservation Organizations

Should we protect birds and their habitats because they are useful or because it is our moral duty to preserve them for future generations? Conservationists have been debating this question since the advent of nature conservation movements in the nineteenth century (Hays 1959, Schmoll 2005). In essence, the heated debate on using ecosystem services as a basis for conservation (see below) and the idea of accepting novel ecosystems are in many ways the evolution of

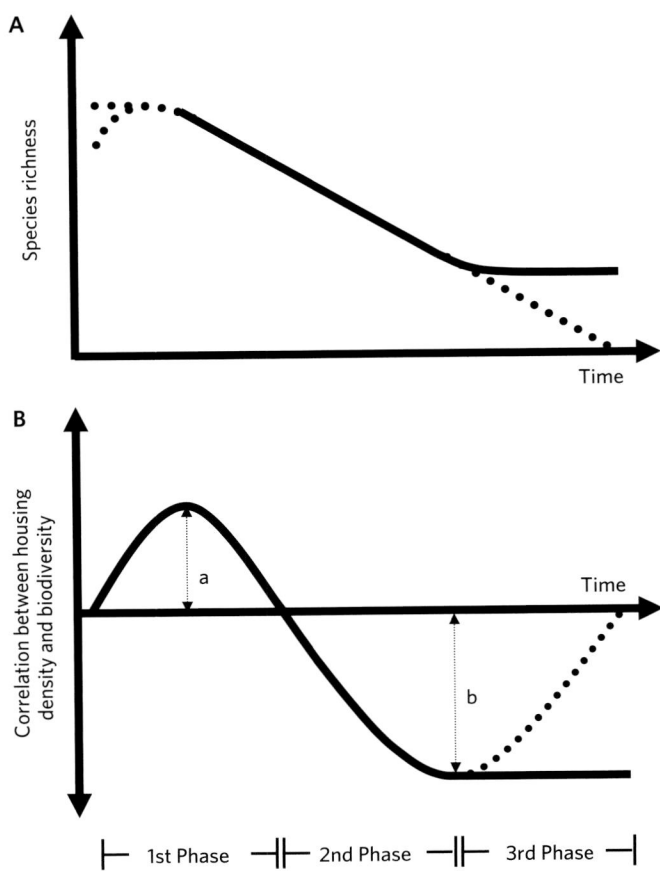

Figure 29.8. The effect of housing density on birds through time and the corresponding trajectory of species richness showing A, species richness trajectory of a given site over time and B, the nature of the relationship between housing density and bird biodiversity over time (a, maximum positive association between housing density and birds; b, maximum negative effect of housing density on birds). Under this model, houses are located initially in fertile, highly productive areas, and settlement exerts a neutral or slightly positive effect on bird diversity primarily because of the correlation of housing density and biodiversity with productivity (1st Phase). As development increases over time, the relationship changes such that negative effects associated with houses dominate the relationship (2nd Phase). In the 3rd Phase, the effect depends on the final settlement density. If all areas are completely developed, then the correlation will approach zero. *Graphic from Pidgeon et al. 2014.*

this same argument (Hunter et al. 2014, Kareiva and Marvier 2015). Whatever the motives were, bird enthusiasts had started to organize in societies and interest groups by the end of the nineteenth century (table 29.1) and have lobbied for animal rights and conservation in national and international politics ever since. Because birds live and move across national borders, their conservation quickly became a matter of international politics. In 1916 the **Migratory Bird**

CHARLES NILON: A PIONEER IN URBAN ECOLOGY

Charles Nilon. *Photo courtesy of C. Nilon.*

Ecologists first began focusing on cities and urban areas in the 1960s and 1970s, a time when there was great concern about the environment and people were moving to and from cities. However, after this period, cities became less of a focus for ecological research and began to fade away as a central focus of research. It was during this interim in urban ecological research that Charles (Charlie) Nilon began pursuing his doctorate in urban wildlife at the State University of New York's (SUNY) College of Environmental Science and Forestry. Charlie arrived at SUNY, where the faculty was one with wide interest in urban systems, after receiving degrees from both Morehouse and Yale. Under the mentorship of his doctoral adviser Larry Van Druff, as well as Rowan Rowntree and Ralph Sanders of the USDA Forest Service, he was able to develop a project that integrated social, economic, and wildlife information within an urban ecosystem. These ideas helped shape Charlie into an urban wildlife ecologist, with a large focus on urban birds and urban mammals.

Charlie moved on to the University of Missouri Columbia, where he has continued not only researching urban wildlife, but has also written a number of influential books and research papers. For instance, Charlie, along with Alan Berkowitz and Karen Hollweg, edited *Understanding Urban Ecosystems: A New Frontier for Science and Education*, one of the first "new" urban ecology books, in 2003 (Springer-Verlag). The uniqueness of this book was not only its focus on the resurgence of urban ecology; it also intentionally addressed environmental justice and some other topics within the broader context of urban ecosystems. Charlie helped broaden the concept of urban ecosystems to include a larger component of people and has worked tirelessly to bring diversity to ecology, both through working with undergrads and graduate students of color and working in and with neighborhoods where people of low income and people of color live. He has been an active member of many programs within the Ecological Society of America to broaden participation by underrepresented groups and was the recent recipient of Commitment to Human Diversity in Ecology Award. Charlie's passion for urban birds and mammals and commitment to increase diversity in the sciences have been an inspiration to several generations of scientists and continues to be for all of us working in urban systems.

Treaty between Canada and the United States was enacted and became the first international conservation legislation (see also chapter 27). Subsequently, a number of other treaties followed that sought to protect birds (and nature), involving almost every country on the globe (table 29.2).

While there are many conservation organizations, treaties, and people involved in bird conservation, there still remains more to do. In particular, while bird-watchers and bird enthusiasts spend billions of dollars on bird-related activities, trips, and products, they contribute relatively little money directly toward conservation, the way that hunters in the United States do. For instance, Pittman-Robertson funds (officially the result of the Federal Aid in Wildlife Restoration Act of 1937) are generated from an 11 percent tax added to the cost of firearms, ammunition, and archery equipment that covers a wide range of conservation and management activities, such as hunter and wildlife education, research, and habitat acquisition and management. Likewise, hunting waterfowl requires the purchase of a federal duck stamp, the funds of which are used directly to manage habitat for waterfowl. Given that the funds are used directly for wildlife, many ornithological clubs and birding groups recommend the purchase of duck stamps, even by nonhunters. Although previous attempts have been made to tax outdoor recreational equipment (e.g., Teaming with Wildlife Act) in a manner similar to that used in hunting, to date they have failed. However, outdoor recreation equipment taxes do exist at the state level, such as in Missouri, where by law one-eighth of a percent (0.125 percent) of sales tax is dedicated to conservation.

Table 29.1. Some of the oldest organizations founded by conservationists and bird enthusiasts, sorted by the year they were founded. Data from http://www.birdlife.org and, in the case of Massachusetts Audubon, from http://www.massaudubon.org.

Name	Country	Year founded	Members	Website
Royal Society for the Protection of Birds (founding name, Society for the Protection of Birds)	UK	1889	1,090,000	www.rspb.org.uk
Massachusetts Audubon Society	USA	1896	100,000	http://www.massaudubon.org
Naturschutzbund Deutschland (founding name, Bund für Vogelschutz)	Germany	1899	450,000	www.nabu.de
Vogelbescherming	Netherlands	1899	153,000	http://www.vogelbescherming.nl
National Audubon Society	USA	1905	400,000	http://www.audubon.org
Ligue pour la Protection des Oiseaux	France	1912	44,000	https://www.lpo.fr
Forest & Bird	New Zealand	1923	40,000	www.forestandbird.org.nz
Česká společnost ornitologická	Czech Republic	1926	2,100	http://www.cso.cz
BirdLife South Africa (founding name, South African Ornithological Society)	South Africa	1930	2,000	http://www.birdlife.org.za
Wild Bird Society of Japan	Japan	1934	47,000	http://www.wbsj.org

ECOSYSTEM SERVICES AND DISSERVICES

As touched on earlier, ecosystem services are the goods and benefits that humans receive from ecosystems and that constitute human well-being (Daily 1997). The ecosystem services concept was formalized toward the end of the twentieth century, in order to increase public awareness for the dependence of humans on ecosystems and biodiversity (Gomez-Baggethun et al. 2010) and allow for further approaches at economic valuation (Daily 1997), but the idea is by no means new (see also the box on page 899 and the box on page 888). However, by the late twentieth century, environmental depletion and destruction had reached unprecedented levels and created a sense of urgency. Proponents of the ecosystem services concept argued that if we can ascribe values to supportive functions of ecosystems, then we will have stronger justification and incentives for conservation (e.g., Kareiva and Marvier 2015). This idea became highly influential, especially after the United Nations–led Millennium Ecosystem Assessment assessed the state of the world's ecosystems and put ecosystem services on the policy agenda (Gomez-Baggethun et al. 2010). The Millennium Ecosystem Assessment also introduced the now-common classification of ecosystem services into four groups: **provisioning** (e.g., food and timber production), **regulating** (e.g., flood control by riparian forests), **cultural** (e.g., providing a sense of place), and **supporting** (e.g., nutrient cycling; Hassan et al. 2005).

Considering the success of the ecosystem services concept, it should come as no surprise that ornithologists have recently called for quantifying and evaluating the ecosystem services provided by birds in order to support their conservation (Wenny et al. 2011, Whelan et al. 2015). The most important services provided by birds, other than their direct use (e.g., **provisioning** of feathers or meat), are pest control (of insects, small mammals, and weeds), seed dispersal (fig. 29.9), pollination, scavenging (box on page 898), and nutrient cycling (Whelan et al. 2015, box on page 899). All of these services can be classified as **regulating** services, with the exception of nutrient cycling, which is considered a **supporting** service (Sekercioglu 2006, Green and Elmberg 2013). For their roles in recreational hunting, bird-watching, ecotourism, or inspiration for art, birds can also be considered to provide **cultural** ecosystem services (Green and Elmberg 2013).

The Millennium Ecosystem Assessment classification may not be the most useful for birds. A more simple classification is one that simply separates ecosystem services provided by birds into **behavior-driven services** (pest control, pollination, seed dispersal, scavenging, and guiding humans to resources) and **product-driven services** (nests and nutrient cycling; services stemming from the direct use

Table 29.2. Important international agreements for the conservation of birds.

Name	Enacted	Purpose	Member States	Source
Migratory Bird Treaty	1916	Protecting birds "useful to man or [. . .] harmless" through a uniform system of conservation.	6 (Sept. 2015)	http://www.treaty-accord.gc.ca/text-texte.aspx?id=101587
Convention on Wetlands of International Importance (Ramsar Convention)	1975	Conservation and wise use of wetlands with an emphasis on waterfowl in particular through the establishment of protected areas.	169 (Sept. 2015)	http://www.ramsar.org
Convention on International Trade in Endangered Species of Wild Fauna and Flora (Washington Convention)	1975	End or limit trade of wild, endangered species, including birds.	181 (July 2015)	https://cites.org/
Birds Directive	1979	Protection of endangered birds and establishment of protected areas throughout the European Union.	28 (all EU member states)	http://ec.europa.eu/environment/nature/legislation/birdsdirective/index_en.htm
Convention on the Conservation of European Wildlife and Natural Habitats (Bern Convention)	1982	Conservation of wild flora and fauna and their natural habitats, within the members of the Council of Europe, Belarus, and some African countries.	51 (July 2015)	http://www.coe.int/en/web/bern-convention/home
Convention on the Conservation of Migratory Species of Wild Animals	1983	Conservation of migratory species across their ranges and life cycles.	122 (Oct. 2015)	http://www.cms.int/
Convention on Biological Diversity	1993	Conservation and sustainable use of biological diversity.	196 (Feb. 2016)	https://www.cbd.int
Agreement on the Conservation of Albatrosses and Petrels	2004	Coordinating activity for the conservation of albatrosses and petrels.	13 (Jan. 2009)	http://www.acap.aq

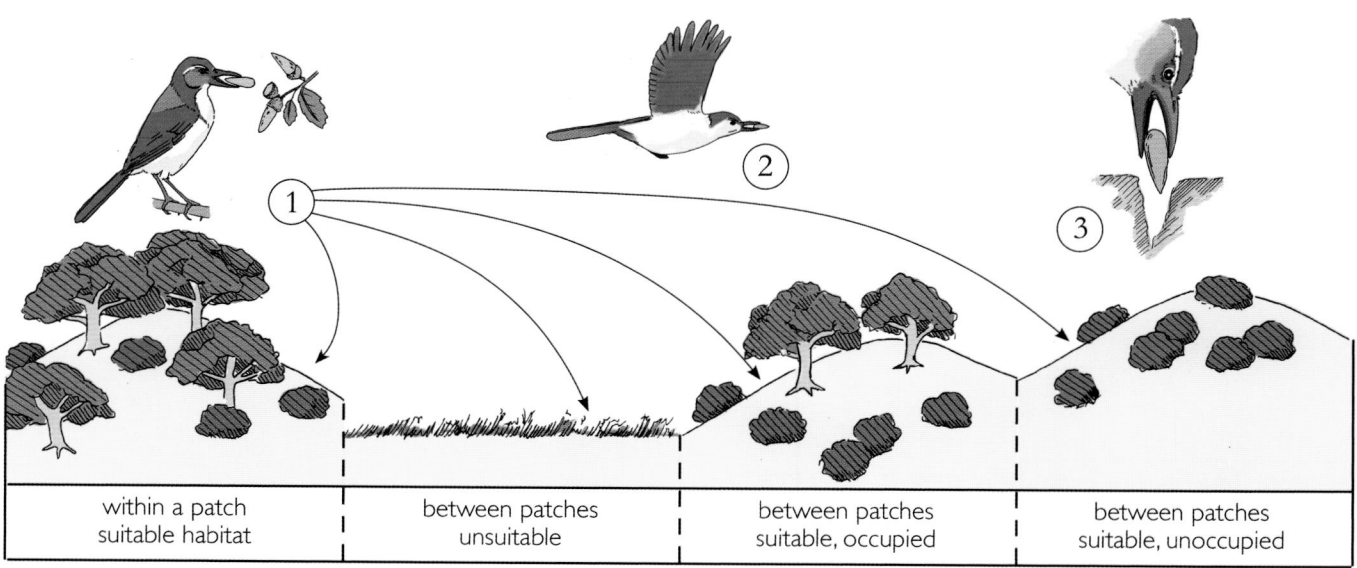

| within a patch suitable habitat | between patches unsuitable | between patches suitable, occupied | between patches suitable, unoccupied |

Figure 29.9. Seed dispersal is an important ecosystem service provided by birds. Common among corvids like jays and nutcrackers, and especially important for oaks and pines, is the scatter-hoarding process. As illustrated for oaks and jays, (1) birds gather seeds, (2) transport them in their bill, and (3) deposit them in spatially distributed caches for later consumption. Because not all seeds are recovered and consumed, some can germinate and grow into new oaks. *From Pesendorfer et al. 2016, illustration by Emily Underwood.*

VULTURES

Vultures may not be pretty, but they provide the essential ecosystem service of contributing to decomposition of animal corpses. In India, this service was not fully valued until it was nearly lost. All three populations of the resident vulture species in India, Oriental White-backed Vulture (*Gyps bengalensis*), Long-billed Vulture (*G. indicus*), and Slender-billed Vulture (*G. tenuirostris*) declined dramatically in the 1990s (98 percent loss between 1992 and 2007; Cuthbert et al. 2011). The use of an anti-inflammatory drug, diclofenac, has been identified as a major culprit in the vultures' decline. The drug has been used safely in humans, but once it began to be broadly applied in livestock (Green et al. 2004, Oaks et al. 2004, Shultz et al. 2004), it became a disaster for the vultures. The drug rapidly (within days) kills vultures that consume carcasses of animals treated with diclofenac (Oaks et al. 2004, Cuthbert et al. 2011). What was even more surprising was that the loss of the vultures led to an accumulation of carcasses at dumps (Markandya et al. 2008, Subramanian 2011). In other words, other scavenging birds were not as effective as vultures at removing carrion. As a result there was an increase in the feral dog population, which also feed on the carcasses. This rise in feral dog numbers led to fears about potential increases in outbreaks of rabies and other diseases, in a country that already had the highest rate of rabies infections in the world (Markandya et al. 2008).

The ramifications of the loss of vultures go beyond these critical public health concerns. There are economic effects on bone collectors who collect bones of cattle after they have been cleaned by scavengers to supply the fertilizer industry (Markandya et al. 2008). And there have been effects on religious practices. Members of the Parsi community have a centuries-old practice of laying their dead on Towers of

Parsi Tower of Silence in Mumbai, India. Vulture numbers have declined dramatically in recent decades, affecting also religious practices. *From http://www.gutenberg.org/files/27260/27260 -h/27260-h.htm#Page_128, East of Suez: Ceylon, India, China and Japan, published in February 1907.*

Silence, where the corpses are devoured by carrion-eating birds (see figure). In Mumbai, it has become necessary for Parsis to replace the vultures with solar concentrators (Markandya et al. 2008).

The loss of these services provided by vultures has generated several potential feedback mechanisms, increasing support for vulture conservation and restoration efforts. In 2006, diclofenac was banned across the region (Ogada et al. 2012), but evidence is mixed as to the effectiveness of the ban and potential for recovery of the species (Cuthbert et al. 2011, Ogada et al. 2012). The international response to the loss of the vultures has led to a captive breeding program (Subramanian 2011, Ogada et al. 2012), calling attention to declines in other vulture species in Asia and Africa, where declines are caused by factors aside from the use of diclofenac (Ogada et al. 2012). Still, the vultures remain functionally extinct in India (Ogada et al. 2012), and the ecosystem services they once provided are no longer available.

---

of the birds were not considered; Whelan et al. 2008). Furthermore, because of the special role birds have as mobile links in ecosystems, the services they provide can also be classified as **genetic linkers** (seed dispersers, pollinators), **resource linkers** (nutrient depositors), **trophic process linkers** (insectivores, raptors, scavengers), and **non-trophic process linkers** (cavity-nesters, burrow-nesters; Sekercioglu 2006).

While it is relatively easy to name ecosystem services provided by birds and to classify them, the quantification and economic valuation of these services remains challenging. An increasing variety of tool sets for valuation are available

(Kareiva and Marvier 2015), however, that can also be applied to birds. Some birds, and their products that can be traded, have an exchange value that can be used for their valuation, such as guano (box on page 899). Other services, however, are more challenging to value. For instance, using an experimental setting, the pest control service of Great Tits (*Parus major*) was evaluated for an apple orchard (Mols and Visser 2007) in which the birds reduced caterpillar damage by up to 50 percent (but economic evaluation was not estimated). Likewise, on coffee plantations, birds were estimated to provide pest control services between $44 and $105 per hectare (Kellermann et al. 2008).

GUANO

Guano is an organic fertilizer that is rich in nitrogen and phosphate. The word stems from the Quechua *huanu,* referring to any kind of manure used to fertilize crops but especially to seabird feces on rocky islands off the coast of Peru. Along this coast, in the nutrient-rich upwelling currents of the Pacific and under dry climatic conditions, feces of Guanay Cormorant (*Phalacrocorax bougainvillii*), Peruvian Booby (*Sula variegata*), Peruvian Pelican (*Pelecanus thagus;* see figure) and other seabirds had accumulated into guano deposits of great depth. Andean cultures have used guano to fertilize their crops for centuries. In the nineteenth century, after Alexander von Humboldt helped make guano known in Europe, it became a globally traded commodity (Cushman 2013). Guano from Peru boosted agricultural productivity in Europe and the United States in the nineteenth century, but was also essential for the chemical industry and the production of explosives (Whelan et al. 2008). Although it was quite obvious that birds were directly responsible for depositing guano, the ecology of the system, and in particular the needs of the guano-producing birds, were largely being ignored. Therefore, by the 1880s, the deposits were almost entirely exhausted.

In the beginning of the twentieth century, guano production was revived with help from biologists and conservationists who studied the coastal ecosystems in order to use them more sustainably. In what today would be called ecosystem service evaluation, Coker (1908) emphasized the economic importance of bird conservation. He estimated the number of pairs of birds needed to produce one ton of guano (28 pairs), took the value of a ton of guano (US$40) and estimated the annual value of a pair of birds to be US$1.43, which today would be US$37.66 (estimated for 2015 based on the Consumer Price Index; Minneapolis Fed n.d.).

Pesruvian Pelicans (*Pelecanus thagus*) on the Islas Ballestas, Peru. The feces of these and other piscivorous seabirds are called guano and make excellent fertilizer. *Photo by Alex E. Proimos.*

The expert-facilitated revival of guano extraction in the beginning of the twentieth century was short-lived, however. After production of artificial nitrogen fertilizers took off and other phosphorous deposits were discovered, the prices for guano dropped. In addition, strong El Niño events (weather anomalies leading to shifts of the nutrient-rich ocean currents) caused massive disruptions in the coastal marine ecosystems. Most importantly, an increase in fishing since the 1950s depleted the food source of the seabirds, in particular the stock of anchoveta (*Engraulis ringens*) (Whelan et al. 2008, Cushman 2013).

Since 2009, the guano islands of Peru are part of a conservation area, but guano is still extracted in a sustainable way (IUCN 2013). Although production is far from the amounts extracted in the nineteenth and mid-twentieth centuries, it has actually increased recently, driven by demands of the globally expanding organic farming industry (Cushman 2013).

Oftentimes the material value of birds is low. For example, the value of a Bluethroat (*Luscinia svecica*) was estimated to be just 3 cents when taking only the material value into account (Vester 1983; conversion from 1983 DM to 2015 US$ with Marcuse n.d. and Minneapolis Fed n.d.). However, estimating how much it would cost to replace the entire ecosystem services provided by a Bluethroat in its expected lifetime of five years (e.g., pest control, recreation) resulted in the total worth increasing to $1,264. Similarly, estimating how much it would cost to replace the ecosystem service (called

replacement cost) of seed dispersal services by a pair of Eurasian Jays (*Garrulus glandarius*) yielded an estimated between $4,900 and $22,000 (Hougner et al. 2006).

Other valuation techniques have also been applied to birds. For example, by asking people how much they spend on bird food (called **revealed preference**) and how much they would be willing to spend on conservation and reduction of birds (called **willingness to pay**) a group of scientists estimated the economic importance of birds to citizens of Berlin, Germany, and Seattle, Washington, United States (Clucas et al.

2014). The scientists found that the economic value of enjoying birds was around $120 million in Seattle and $70 million in Berlin. People's cultural background, however, strongly influenced the values they ascribe to certain bird. Hence, on average, the willingness to pay to support finches was positive in Berlin and Seattle, whereas the willingness to pay to support corvids (jays and crows) was lower in Berlin and actually negative in Seattle, where people supported measures to reduce the corvid population (Clucas et al. 2014). Yet another estimation approach is the **travel cost method**, in which real expenses are determined in order to calculate the overall economic worth of an activity, like attending a bird festival. Us-

ing the travel cost method, the value of traveling to see White Storks (*Ciconia ciconia*) at their colony was estimated at about $60 per person (Czajkowski et al. 2014). Notably, while many of these economic valuation approaches have been utilized, they vary greatly in their assumptions and approximations of reality (Kareiva and Marvier 2015).

In contrast to ecosystem services, **ecosystem disservices** refers to the negative effects of nature on humans (Dunn 2010). At the local scale, birds can cause great damage to crops, fruit plantations, or aquaculture, but general damage is rather low compared with that caused by rodents and insects (Price and Nickum 1995, Whelan et al. 2015, Bom-

A WAR ON SPARROWS

During the so-called Great Leap Forward in the 1950s and early 1960s, the Communist Party in China started a campaign against the "four pests." With the intention of increasing hygiene but also reducing grain losses, the campaign called for the eradication of rats, flies, mosquitos, and sparrows (Shapiro 2001). Sparrows of the genus *Passer* are mostly granivorous, making them likely to cause harvest losses. House Sparrows (*Passer domesticus*), for example, have been estimated to cause grain losses between 5 and 30 percent (Southern 1945, Dawson 1970). Therefore it seemed common sense that if sparrows were wiped out, grain losses would be reduced. In China, the Eurasian Tree Sparrow (*Passer montanus*) was the main target of the four pests campaign (Summers-Smith 1988). The persecution of the birds even involved children (see figure) and was extremely successful in exterminating almost the whole population (Summers-Smith 1988, Shapiro 2001). What the initiators of the campaign had not considered, however, is that grain is not the sole food of sparrows, as their diet also includes agricultural weeds and insect pests (Southern 1945, Whelan et al. 2008). The consequences of the eradication were therefore not gains but serious losses in crop production, tragically adding to the existing food shortages (Summers-Smith 1988, Shapiro 2001). By 1960, the sparrow had been taken off the list of "four pests" and replaced by bedbugs (Shapiro 2001).

Chinese propaganda poster "Everybody comes to beat sparrows" by Bi Cheng, 1956. During the 1950s, the Communist Party campaigned to wipe out the "four pests," one of which was sparrows. Even children were enlisted in the campaign to kill the animals. *Poster from the IISH / Stefan R. Landsberger Collections http:// chineseposters.net/.*

ford and Sinclair 2002). For fruit plantations and aquaculture, measures such as exclusion netting have proven to be an effective and simple countermeasure (Price and Nickum 1995, Bomford and Sinclair 2002). However, today our understanding of birds (and nature in general) is more nuanced, such that we consider birds entangled in complex interactions, with even "pest" bird species recognized to be of importance. In fact, management decisions based on incomplete knowledge of these ecological connections can have disastrous effects (box on page 900).

Birds are also potentially dangerous for aviation. The emergency landing of US Airways Flight 1549 in the Hudson River in New York City in 2009, for example, was caused by a collision with Canada Geese (*Branta canadensis*; Marra et al. 2009). Global costs of so-called **bird strikes** (i.e., the collision between birds and airplanes) to airlines are estimated to be at least $1.2 billion a year (Allan 2000). Given this cost, potential habitats in and around airports are strongly managed in order reduce the chance of bird strikes (Blackwell et al. 2009), but such measures are often not carried out to their full potential (Allan 2000, Marra et al. 2009).

Finally, birds can transmit pathogens to humans (i.e., zoonotic diseases). These pathogens include enteropathogens (e.g., *Salmonella*) and pathogens causing West Nile fever, Lyme disease, and influenza A (Reed et al. 2003). Because birds are mobile links (Sekercioglu 2006), they can also effectively and quickly spread pathogens across large distances, with undesired and unintended effects for humans.

The concept of evaluating services and disservices provided to humans is by no means new. The now almost forgotten discipline of **economic ornithology**, which emerged in the nineteenth century and disappeared after the 1950s, had at its core the quantification and evaluation of the harmful and beneficial effects of birds on agriculture, fishing, and forestry (Kronenberg 2014, Whelan et al. 2015; box on page 899). But why did economic ornithology disappear in the 1950s? The main reasons were that the focus on the direct positive or negative effects was often too narrow and the benefits from sustainable use and conservation of birds were outweighed by more profitable land use practices. In particular, the boost in agricultural production since the 1950s (the **green revolution**) rendered insignificant whatever impacts birds had on crop yield (Kronenberg 2014, Whelan et al. 2015). The lesson to be learned from the failure of economic ornithology is that nonsentimental, human-centered approaches to conservation and land management are not always successful. Birds have an intrinsic value that cannot be captured by assessing their economic worth. In other words, successful bird conservation should appeal to peoples' hearts, minds, and wallets (Johnson and Hackett 2016).

KEY POINTS

- Birds are one of the most important groups of species to people, providing cultural value, economic livelihood, and personal joy.
- While each person differs in how and why they interact with birds, as a whole, birds are positively perceived by people.
- People engage in a variety of activities in their yards and gardens that affect birds, including activities that intentionally attract birds to yards, such as bird feeding and installing birdhouses.
- Decisions about yard management are shaped by many factors beyond individual preferences and do not always yield an inhabitant's ideal yard.
- Bird-oriented activities, like bird feeding, have a mix of effects, some of which are positive for bird populations but some of which may negatively affect bird populations.
- The collective effect of small-scale yard-level management actions can be large, given the large and growing total area covered by residential yards.
- People's affinity for birds and the ecosystem services birds provide has motivated many people to take conservation actions that have benefited birds and other wildlife species.
- The social and ecological factors that influence bird populations can be captured in a coupled human and natural systems framework.
- Though birds are well-loved animals, people have caused many species to decline or go extinct, and even in the face of numerous policies, many species still face challenges.

KEY MANAGEMENT AND CONSERVATION IMPLICATIONS

- People play a fundamentally important role in the way they interact with and manage land, including the property they live on.
- The choices people make in terms of how, when, and what they do on their property can directly and indirectly affect the bird species present on their land.
- Understanding how human actions lead to changes in bird species types and numbers is of critical importance. In particular, disentangling the mechanistic reasons for how our actions affect birds at different scales is important for refining conservation and management activities and policies.
- People depend on birds for many parts of their livelihood and have strong affinities for seeing and interacting with birds.

- Birds provide an important taxonomic group to use for conservation, management, and public outreach.

DISCUSSION QUESTIONS

1. How do you think past uses of birds have affected the way humans regard them today? How do your own views on birds differ from those of your parents and grandparents?

2. What are ecosystem services, how are they classified, and what are the main services that birds provide? Can you think of some ecosystem services provided by birds that add to your own quality of life?

3. In what ways has human society negatively affected bird species?

4. Do you think that, on balance, it is a good idea for people to feed birds? Why or why not? In what ways might bird feeding be beneficial or detrimental to bird species?

5. What are ways that people's activities can positively and negatively affect bird populations on their property? How can people's activities at the landscape scale affect birds?

6. What are approaches that can be used to estimate the value of birds? Can you think of a way to use one of these approaches to estimate your own value of birds?

7. What are the moral and economic justifications for protecting birds? Do you think it is appropriate to justify bird conservation by calculating the economic value of birds to people?

8. Much of our current understanding of the effects on birds of people's land management practices is based on correlation. What might be gained by doing experimental manipulations? Can you design an experiment to test one of these effects?

References

Allan, J. R. (2000). The costs of bird strikes and bird strike prevention. *In* Human conflicts with wildlife: Economic considerations, L. Clark, Editor. USDA National Wildlife Research Center Symposia, paper 18. http://digitalcommons.unl.edu/nwrchumanconflicts/18.

Amante-Helweg, V. L. U., and S. Conant (2009). Hawaiian culture and forest birds. *In* Conservation biology of Hawaiian forest birds: Implications for island avifauna, T. K. Pratt, C. T. Atkinson, P. C. Banko, J. D. Jacobi, and B. L. Woodworth, Editors. Yale University Press, New Haven, CT.

American Veterinary Medical Association (2012). U.S. pet ownership and demographic sourcebook. AVMA, Schaumburg, IL.

APPA (2015). 2015–2016 APPA national pet owners survey. American Pet Products Association, Greenwich, CT.

Aronson, M. F. J., F. A. La Sorte, C. H. Nilon, M. Katti, M. A. Goddard, C. A. Lepczyk, P. S. Warren, et al. (2014). A global analysis of the impacts of urbanization on bird and plant diversity reveals key anthropogenic drivers. Proceedings of the Royal Society of London B 281:20133330.

Baker, S. E., R. Cain, F. Van Kesteren, Z. A. Zommers, N. D'Cruze, and D. W. Macdonald (2013). Rough trade: Animal welfare in the global wildlife trade. BioScience 63:928–938.

Bang, C., S. H. Faeth, and J. L. Sabo (2012). Control of arthropod abundance, richness, and composition in a heterogeneous desert city. Ecological Monographs 82:85–100.

Belaire, J. A., C. J. Whelan, and E. S. Minor (2014). Having our yards and sharing them too: The collective effects of yards on native bird species in an urban landscape. Ecological Applications 24:2132–2143.

Bishop, C. A., B. Collins, P. Mineau, N. M. Burgess, W. F. Read, and C. Risley (2000). Reproduction of cavity-nesting birds in pesticide-sprayed apple orchards in Southern Ontario, Canada, 1988–1994. Environmental Toxicology and Chemistry 19:588–599.

Blackwell, B. F., T. L. DeVault, E. Fernández-Juricic, and R. A. Dolbeer (2009). Wildlife collisions with aircraft: A missing component of land-use planning for airports. Landscape and Urban Planning 93:1–9.

Bomford, M., and R. Sinclair (2002). Australian research on bird pests: Impact, management and future directions. Emu 102:29–45.

Brigham, W. T. (1899). Hawaiian feather work. Memoirs of the Bernice Pauahi Bishop Museum of Polynesian Ethnology and Natural History 1 (1). Bishop Museum Press, Honolulu, HI.

Brigham, W. R. (1903). Additional notes on Hawaiian feather work. Memoirs of the Bernice Pauahi Bishop Museum of Polynesian Ethnology and Natural History 1 (5). Bishop Museum Press, Honolulu, HI.

Brinkley, D. (2009). The wilderness warrior: Theodore Roosevelt and the crusade for America. Harper Collins, New York.

Buck, P. H. (1957). Arts and crafts of Hawaii, Section XI: Religion. Bernice P. Bishop Museum Special Publication 45, Bishop Museum Press, Honolulu, HI.

Clarke, L. W., G. D. Jenerette, and A. Davila (2013). The luxury of vegetation and the legacy of tree biodiversity in Los Angeles, CA. Landscape and Urban Planning 116:48–59.

Clucas, B., S. Rabotyagov, and J. M. Marzluff (2014). How much is that birdie in my backyard? A cross-continental economic valuation of native urban songbirds. Urban Ecosystems 18:251–266.

Coker, R. E. (1908). The fisheries and the guano industry of Peru. Proceedings of the Fourth International Fishery Congress, Washington, DC.

Cushman, G. T. (2013). Guano and the opening of the Pacific World: A global ecological history. Cambridge University Press, New York.

Cuthbert, R. J., V. Prakash, M. Saini, S. Upreti, D. Swarup, A. Das, R. E. Green, and M. Taggart (2011). Are conservation actions reducing the threat to India's vulture populations? Current Science 101:1480–1484.

Czajkowski, M., M. Giergiczny, J. Kronenberg, and P. Tryjanowski (2014). The economic recreational value of a white stork nesting colony: A case of "stork village" in Poland. Tourism Management 40:352–360.

Daily, G. C. (1997). Nature's services: Societal dependence on natural ecosystems. Island Press, Washington, DC.

Daniels, G. D., and J. B. Kirkpatrick (2006). Does variation in garden characteristics influence the conservation of birds in suburbia? Biological Conservation 133:326–335.

Darwin, C. (1859). On the origin of species by means of natural selection, or the preservation of favoured races in the struggle for life. W. Clowes and Sons, London.

Davies, Z. G., R. A. Fuller, A. Loram, K. N. Irvine, V. Simsa, and K. J. Gaston (2009). A national scale inventory of resource provision for biodiversity within domestic gardens. Biological Conservation 142:761–771.

Davis, A. Y., J. A. Belaire, M. A. Farfan, D. Milz, E. R. Sweeney, S. R. Loss, and E. S. Minor (2012). Green infrastructure and bird diversity across an urban socioeconomic gradient. Ecosphere 3 (11): 105. doi:10.1890/ES1200126.1.

Dawson, D. G. (1970). Estimation of grain loss due to sparrows (Passer domesticus) in New Zealand. New Zealand Journal of Agricultural Research 13:681.

Dhondt, A. A., J. C. DeCoste, D. H. Ley, and W. M. Hochachka (2014). Diverse wild bird host range of Mycoplasma gallisepticum in Eastern North America. PLoS ONE 9 (7): e103553. doi:10.1371/journal.pone.0103553.

Dunn, E. H., and D. L. Tessaglia (1994). Predation of birds at feeders in winter (Depredación de aves en comederos durante el invierno). Journal of Field Ornithology 65:8–16.

Dunn, R. R. (2010). Global mapping of ecosystem disservices: The unspoken reality that nature sometimes kills us. Biotropica 42:555–557.

Force, R. W., and M. Force (1968). Art and artifacts of the 18th century. Bishop Museum Press, Honolulu, HI.

Fuller, R. A., K. N. Irvine, Z. G. Davies, P. R. Armsworth, and K. J. Gaston (2012). Interactions between people and birds in urban landscapes. Studies in Avian Biology 45:249–266.

Furmansky, D. Z. (2009). Rosalie Edge, hawk of mercy. University of Georgia Press, Athens.

Galbraith, J. A., J. R. Beggs, D. N. Jones, and M. C. Stanley (2015). Supplementary feeding restructures urban bird communities. Proceedings of the National Academy of Sciences 112:E2648–E2657.

Goddard, M. A., A. J. Dougill, and T. G. Benton (2010). Scaling up from gardens: Biodiversity conservation in urban environments. Trends in Ecology and Evolution 25:90–98.

Gomez-Baggethun, E., R. de Groot, P. L. Lomas, and C. Montes (2010). The history of ecosystem services in economic theory and practice: From early notions to markets and payment schemes. Ecological Economics 69:1209–1218.

Goudie, A. (2000). The human impact on the natural environment. MIT Press, Cambridge, MA.

Green, A. J., and J. Elmberg (2013). Ecosystem services provided by waterbirds. Biological Reviews 89:105–122.

Green, R. E., I. Newton, S. Shultz, A. A. Cunningham, M. Gilbert, D. J. Pain, and V. Prakash (2004). Diclofenac poisoning as a cause of vulture population declines across the Indian subcontinent. Journal of Applied Ecology 41:793–800.

Guralnick, R., and J. Van Cleve (2005). Strengths and weaknesses of museum and national survey data sets for predicting regional species richness: Comparative and combined approaches. Diversity and Distributions 11:349–359.

Hassan, R., R. Scholes, and N. Ash (2005). Ecosystems and human well-being: Current state and trends. Island Press, Washington, DC.

Hays, S. P. (1959). Conservation and the gospel of efficiency. Harvard University Press, Cambridge, MA.

Hope, D., C. Gries, W. Zhu, W. F. Fagan, C. L. Redman, N. B. Grimm, A. Nelson, C. Martin, and A. Kinzig (2003). Socioeconomics drive urban plant diversity. Proceedings of the National Academy of Sciences USA 100:8788–8792.

Hougner, C., J. Colding, and T. Söderqvist (2006). Economic valuation of a seed dispersal service in the Stockholm National Urban Park, Sweden. Ecological Economics 59:364–374.

Hunter, M. L., K. H. Redford, and D. B. Lindenmayer (2014). The complementary niches of anthropocentric and biocentric conservationists. Conservation Biology 28:641.

IUCN (2013). A bird droppings biodiversity paradise—the Guano Islands and Capes National Reserve System, Peru. IUCN Fact Sheet. http://www.iucn.org/about/work/programmes/gpap_home/pas_gpap/paoftheweek/gpap_paamerica/?13669/A-bird-droppings-biodiversity-paradise—the-Guano-Islands-and-Capes-National-Reserve-System-Peru.

James, H. F., and S. L. Olson (1983). Flightless birds. Natural History 92:30–40.

Jansson, C., J. Ekman, and A. von Brömssen (1981). Winter mortality and food supply in tits Parus spp. Oikos 37:313–322.

Jenerette, G. D., S. L. Harlan, A. Brazel, N. Jones, L. Larsen, and W. L. Stefanov (2007). Regional relationships between surface temperature, vegetation, and human settlement in a rapidly urbanizing ecosystem. Landscape Ecology 22:353–365.

Jenerette, G. D., S. L. Harlan, W. L. Stefanov, and C. A. Martin (2011). Ecosystem services and urban heat riskscape moderation: Water, green spaces, and social inequality in Phoenix, USA. Ecological Applications 21:2637–2651.

Johnson, M. D., and S. C. Hackett (2016). Why birds matter economically: Values, markets, and policies. In Why birds matter, C. H. Sekercioglu, D. G. Wenny, and C. J. Whelan, Editors. University of Chicago Press, Chicago, IL, pp. 27–48.

Kaeppler, A. L. (1978). Artificial curiosities. Bernice Pauahi Bishop Museum Special Publication 65. Bernice Pauahi Bishop Museum Press, Honolulu, HI.

Kane, B., P. S. Warren, and S. B. Lerman (2015). A broad scale analysis of tree risk, mitigation and potential habitat for cavity-nesting birds. Urban Forestry and Urban Greening 14:1137–1146.

Kareiva, P. M., and M. Marvier (2015). Conservation science. 2nd ed. Roberts and Company, Greenwood Village, CO.

Kellermann, J. L., M. D. Johnson, A. M. Stercho, and S. C. Hackett (2008). Ecological and economic services provided by birds on Jamaican Blue Mountain coffee farms. Conservation Biology 22:1177–1185.

Kinzig, A. P., P. Warren, C. Martin, D. Hope, and M. Katti (2005). The effects of human socioeconomic status and cultural characteristics on urban patterns of biodiversity. Ecology and Society 10.

Kirby, A. (2000). All new, improved! Cities 17 (1): 1–5.

Kronenberg, J. (2014). What can the current debate on ecosystem services learn from the past? Lessons from economic ornithology. Geoforum 55:164.

Larsen, L., and S. L. Harlan (2006). Desert dreamscapes: Residential landscape preference and behavior. Landscape and Urban Planning 78:85–100.

Larson, K. L., D. Casagrande, S. L. Harlan, and S. T. Yabiku (2009). Residents' yard choices and rationales in a desert city: Social priorities, ecological impacts, and decision tradeoffs. Environmental Management 44:921–937.

Lepczyk, C. A., A. G. Mertig, and J. Liu (2002). Landowner perceptions and activities related to birds across rural-to-urban landscapes. In Avian landscape ecology. Proceedings of the 2002 Annual UK-IALE Conference, D. Chamberlain and A. Wilson, Editors. http://iale.uk/node/29.

Lepczyk, C. A., A. G. Mertig, and J. Liu (2004a). Assessing landowner activities that influence birds across rural-to-urban landscapes. Environmental Management 33:110–125.

Lepczyk, C. A., A. G. Mertig, and J. Liu (2004b). Landowners and cat predation across rural-to-urban landscapes. Biological Conservation 115:191–201.

Lepczyk, C. A., R. B. Hammer, V. C. Radeloff, and S. I. Stewart (2007). Spatiotemporal dynamics of housing growth hotspots in the North Central U.S. from 1940 to 2000. Landscape Ecology 22:939–953.

Lepczyk, C. A., C. H. Flather, V. C. Radeloff, A. M. Pidgeon, R. B. Hammer, and J. Liu (2008). Human impacts on regional avian diversity and abundance. Conservation Biology 22:405–446.

Lepczyk, C. A., P. S. Warren, L. Machabée, A. P. Kinzig, and A. Mertig (2012). Who feeds the birds? A comparison between Phoenix, Arizona and Southeastern Michigan. In Urban bird ecology and conservation. Studies in Avian Biology 45, C. Lepczyk and P. Warren, Editors. University of California Press, Berkeley.

Lerman, S. B., and P. S. Warren (2011). The conservation value of residential yards: Linking birds and people. Ecological Applications 21:1327–1339.

Lerman, S. B., V. K. Turner, and C. Bang (2012). Homeowner associations as a vehicle for promoting native urban biodiversity. Ecology and Society 17 (4): 45. doi:10.5751/ES-05175 170445.

Liu, J., T. Dietz, S. R. Carpenter, C. Folke, M. Alberti, C. L. Redman, S. H. Schneider, et al. (2007). Coupled human and natural systems. Ambio 36:639–649.

Loram, A., J. Tratalos, P. H. Warren, and K. J. Gaston (2007). Urban domestic gardens (X): The extent and structure of the resource in five major cities. Landscape Ecology 22:601–615.

Loss, S. R., T. Will, and P. P. Marra (2012). Direct human-caused mortality of birds: Improving quantification of magnitude and assessment of population impact. Frontiers in Ecology and the Environment 10:357–364.

Loyd, K. A. T., S. M. Hernandez, J. P. Carroll, K. J. Abernathy, and G. J. Marshall (2013). Quantifying free-roaming domestic cat predation using animal-borne video cameras. Biological Conservation 160:183–189.

Lubbe, C. S., S. J. Siebert, and S. S. Cilliers (2010). Political legacy of South Africa affects the plant diversity patterns of urban domestic gardens along a socio-economic gradient. Scientific Research and Essays 5:2900–2910.

Malo (1951). Hawaiian antiquities (Moolelo Hawaii). Bernice P. Bishop Museum Special Publication, Honolulu, HI.

Marcuse, H. (n.d.). Historical dollar-to-marks currency conversion page. http://www.history.ucsb.edu/faculty/marcuse/projects/currency.htm.

Markandya, A., T. Taylor, A. Longo, M. N. Murty, S. Murty, and K. Dhavala (2008). Counting the cost of vulture decline: An appraisal of the human health and other benefits of vultures in India. Ecological Economics 67:194–204.

Marra, P. P., C. J. Dove, R. Dolbeer, N. F. Dahlan, M. Heacker, J. F. Whatton, N. E. Diggs, C. France, and G. A. Henkes (2009). Migratory Canada Geese cause crash of US Airways Flight 1549. Frontiers in Ecology and the Environment 7:297–301.

Marzluff, J. M., R. Bowman, and R. Donnelly (2001). A historical perspective on urban bird research: Trends, terms, and approaches. In Avian ecology in an urbanizing world, J. M. Marzluff, R. Bowman and R. Donnelly, Editors. Kluwer Academic Publishers, Boston, MA, pp. 1–18.

McIntyre, N. E., and M. E. Hostetler (2001). Effects of urban land use on pollinator (Hymenoptera: Apoidea) communities in a desert metropolis. Basic and Applied Ecology 2:209–218.

McKinney, M. L (2002). Urbanization, biodiversity, and conservation. BioScience 52:883–890.

Milesi, C., C. D. Elvidge, R. R. Nemani, and S. W. Running (2003). Assessing the impact of urban land development on net primary productivity in the southeastern United States. Remote Sensing of Environment 86:401–410.

Minneapolis Fed (n.d.). Consumer price index (estimate) 1800–. https://www.minneapolisfed.org/community/teaching-aids/cpi-calculator-information/consumer-price-index-1800.

Mols, C. M. M., and M. E. Visser (2007). Great Tits (Parus major) reduce caterpillar damage in commercial apple orchards. PLoS ONE 2:e202.

Morneau, F., R. Decarie, R. Pelletier, D. Lambert, J. L. DesGranges, and J. P. Savard (1999). Changes in breeding bird richness and abundance in Montreal parks over a period of 15 years. Landscape and Urban Planning 44:111–121.

National Agricultural Statistics Service (2014). Census of agriculture 2012: United States summary and state data. Vol. 1. Geographic Area Series, Part 51. US Department of Agriculture, Washington, DC.

National Park Service (n.d.). Frequently asked questions. https://www.nps.gov/aboutus/faqs.htm.

Newton, I. (1995). The contribution of some recent research on birds to ecological understanding. Journal of Animal Ecology 64:675–696.

Oaks, J. L., M. Gilbert, M. Z. Virani, R. T. Watson, C. U. Meteyer, B. A. Rideout, H. L. Shivaprasad, et al. (2004). Diclofenac residues as the cause of vulture population decline in Pakistan. Nature 427:630–633.

Odum, E. P. (1959). Fundamentals of ecology. W. B. Saunders, Philadelphia, PA.

Ogada, D. L., F. Keesing, and M. Z. Virani (2012). Dropping dead: Causes and consequences of vulture population declines worldwide. Annals of the New York Academy of Sciences 1249:57–71.

Olson, S. L., and H. F. James (1982). Prodomus of the fossil avifauna of the Hawaiian Islands. Smithsonian Contributions to Zoology 365:1–59.

Ostrom, E. (2009). A general framework for analyzing sustainability of social-ecological systems. Science 325:419–422.

Pesendorfer, M. B., T. S. Sillett, W. D. Koenig, and S. A. Morrison (2016). Scatter-hoarding corvids as seed dispersers for oaks and pines: A review of a widely distributed mutualism and its utility to habitat restoration. The Condor 118:215–237.

Pidgeon, A. M., V. C. Radeloff, C. H. Flather, C. A. Lepczyk, M. K. Clayton, T. J. Hawbaker, and R. B. Hammer (2007). Associations of forest bird species richness pattern with housing and landscape patterns across the USA. Ecological Applications 17:1989–2010.

Pidgeon, A. M., C. H. Flather, V. C. Radeloff, C. A. Lepczyk, N. S. Keuler, E. Wood, S. I. Stewart, and R. B. Hammer (2014). Systematic temporal patterns in the relationship between housing development and forest bird biodiversity. Conservation Biology 28:1291–1301.

Pinowski, J., M. Barkowska, A. H. Kruszewicz, and A. G. Kruszewicz (1994). The causes of mortality of eggs and nestlings of Passer spp. Journal of Biosciences 19:441–451.

Polsky, C., J. M. Grove, C. Knudson, P. M. Groffman, N. Bettez, J. Cavender-Bares, S. J. Hall, et al. (2014). Assessing the homogenization of urban land management with an application to US residential lawn care. Proceedings of the National Academy of Sciences 112:4432–4437.

Price, I. M., and J. G. Nickum (1995). Aquaculture and birds: The context for controversy. Colonial Waterbirds 18:33.

Redekop, B. W. (2012). The environmental leadership of Theodore Roosevelt. In Environmental leadership: A reference handbook, D. R. Gallagher, Editor. SAGE, Thousand Oaks, CA. doi:10.4135/9781452218601.n11.

Reed, K. D., J. K. Meece, J. S. Henkel, and S. K. Shukla (2003). Birds, migration and emerging zoonoses: West Nile virus, Lyme disease, influenza A and enteropathogens. Clinical Medicine and Research 1:5–12.

Robb, G. N., R. A. McDonald, D. E. Chamberlain, and S. Bearhop (2008). Food for thought: Supplementary feeding as a driver of ecological change in avian populations. Frontiers in Ecology and the Environment 6:476–484.

Roberts, D. L., and A. R. Solow (2003). Flightless birds—When did the dodo become extinct? Nature 426:245.

Roosevelt, T. (1901). First Annual Message, December 3, 1901. Online by Gerhard Peters and John T. Woolley, The American Presidency Project. http://www.presidency.ucsb.edu/ws/?pid=29542.

Rose, R. G. (1978). Symbols of sovereignty: Feather girdles of Tahiti and Hawaii. Pacific Anthropological Records no. 28. Department of Anthropology, Bernice P. Bishop Museum, Honolulu, HI.

Rose, R. G. (1979). On the origin and diversity of "Tahitian fly whisks." In Exploring the visual art of Oceania: Australia, Melanesia, Micronesia, and Polynesia, S. M. Mead, Editor. University Press of Hawaii, Honolulu.

Rose, R. G., S. Conant, and E. P. Kjellgren (1993). Hawaiian standing kāhili in Bishop Museum: An ethnological and biological analysis. Journal of the Polynesian Society 102:273–304.

Schmoll, F. (2005). Indication and identification. In Germany's nature: Cultural landscapes and environmental history, T. M. Lekan and T. Zeller, Editors. Rutgers University Press, New Brunswick, NJ.

Sekercioglu, C. H. (2006). Increasing awareness of avian ecological function. Trends in Ecology and Evolution 21:464–471.

Shapiro, J. (2001). Mao's war against nature: Politics and the environment in revolutionary China. Cambridge University Press, Cambridge, UK.

Shochat, E., P. S. Warren, S. H. Faeth, N. E. McIntyre, and D. Hope (2006). From patterns to emerging processes in mechanistic urban ecology. Trends in Ecology and Evolution 21:186–191.

Shochat, E., S. B. Lerman, J. M. Anderies, P. S. Warren, S. H. Faeth, and C. H. Nilon (2010). Invasion, competition, and biodiversity loss in urban ecosystems. Bioscience 60:199–208.

Shrubb, M. (2003). Birds, scythes, and combines: A history of birds and agricultural change. Cambridge University Press, Cambridge, UK.

Shultz, S., H. S. Baral, S. Charman, A. A. Cunningham, D. Das, G. R. Ghalsasi, M. S. Goudar, et al. (2004). Diclofenac poisoning is widespread in declining vulture populations across the Indian subcontinent. Proceedings of the Royal Society of London Series B: Biological Sciences 271:S458–S460.

Silvertown, J. (2009). A new dawn for citizen science. Trends in Ecology and Evolution 24:467–471.

Southern, H. N. (1945). The economic importance of the House Sparrow, Passer domesticus L.: A review. Annals of Applied Biology 32:57–67.

Stabler, L. B., C. A. Martin, and A. J. Brazel (2005). Microclimates in a desert city were related to land use and vegetation index. Urban Forestry and Urban Greening 3:137–147.

Strohbach, M. W., D. Haase, and N. Kabisch (2009). Birds and the city: Urban biodiversity, land use, and socioeconomics. Ecology and Society 14.

Strohbach, M. W., A. Hrycyna, and P. S. Warren (2014). 150 years of changes in bird life in Cambridge, Massachusetts from 1860 to 2012. Wilson Journal of Ornithology 126:192–206.

Subramian, M. (2011). India's vanishing vultures. Virginia Quarterly Review 87:28–47.

Summers-Smith, D. (1988). The sparrows: A study of the genus Passer. T. and A. D. Poyser, Calton, UK.

Taylor, P., Editor (2007). Integrative science for society and environment: A strategic research initiative. Report developed by Research Initiatives Subcommittee of the LTER Planning Process Conference Committee and the Cyberinfrastructure Core Team.

US Department of the Interior, US Fish and Wildlife Service, and US Department of Commerce, US Census Bureau (2014). 2011

National survey of fishing, hunting, and wildlife-associated recreation. https://www.census.gov/prod/2012pubs/fhw11-nat.pdf.

Vester, F. (1983). Der Wert eines Vogels. Kösel-Verlag, München.

Vitousek, P. M., H. A. Mooney, J. Lubchenco, and J. M. Melillo (1997). Human domination of Earth's ecosystems. Science 277:494–499.

Walker, J. S., N. B. Grimm, J. M. Briggs, C. Gries, and L. Dugan (2009). Effects of urbanization on plant species diversity in central Arizona. Frontiers in Ecology and the Environment 7:465–470.

Warren, P. S., S. B. Lerman, and N. D. Charney (2008). Plants of a feather: Spatial autocorrelation of gardening practices in suburban neighborhoods. Biological Conservation 141:3–4.

Warren, P. S., S. Harlan, C. Boone, S. B. Lerman, E. Shochat, and A. P. Kinzig (2010). Urban ecology and human social organization. In Urban ecology, K. Gaston, Editor. Cambridge University Press, Cambridge, UK.

Wenny, D. G., T. L. DeVault, M. D. Johnson, D. Kelly, C. H. Sekercioglu, D. F. Tomback, and C. J. Whelan (2011). The need to quantify ecosystem services provided by birds. The Auk 128:1.

Whelan, C. J., D. G. Wenny, and R. J. Marquis (2008). Ecosystem services provided by birds. Annals of the New York Academy of Sciences 1134:25–60.

Whelan, C. J., C. H. Sekercioglu, and D. G. Wenny (2015). Why birds matter: From economic ornithology to ecosystem services. Journal of Ornithology 156:227–238.

Widows, S. A., and D. Drake (2014). Evaluating the National Wildlife Federation's Certified Wildlife Habitat™ program. Landscape and Urban Planning 129:32–43.

Woods, M., R. A. Mcdonald, and S. Harris (2003). Predation of wildlife by domestic cats Felis catus in Great Britain. Mammal Review 33:174–188.

PART VII

THE SCIENCE AND PRACTICE OF ORNITHOLOGY

CHAPTER 30

Pathways in Ornithology

Melanie R. Colón, Ashley M. Long, Lori A. Blanc, and Caren B. Cooper

The first contributors to our understanding of birds were amateur scientists and natural history enthusiasts with training and incomes derived from nonscientific pursuits. Professionally employed scientists from other fields (e.g., botany, medicine) also added to our early ornithological knowledge. The study of museum collections helped stimulate global interest in avian biology and behavior, and over time, ornithology evolved into an independent field of study. Today, a variety of career opportunities are available to those interested in working with birds professionally. In this chapter, we discuss the development of the field, the diversity of career opportunities, and the skills needed to succeed as a professional ornithologist. In addition, we describe the continued role of amateurs in increasing our understanding of avifauna. Throughout the chapter, professional and amateur ornithologists describe their own experiences and offer advice to students interested in studying birds.

ORNITHOLOGY AS A PROFESSION

History of the Profession

People have studied birds throughout human history, but it wasn't until the eighteenth and nineteenth centuries that ornithology emerged as a professional discipline (Farber 1982). Before the professionalization of the field, early ornithologists were amateurs—individuals who derived their incomes from other pursuits but engaged in bird study as an avocation. While many early amateur ornithologists were men who lacked formal scientific training (e.g., traders, colonists, explorers, soldiers; but see the box on page 910), others were scientists who earned their livings in different disciplines. For example, William Turner was a physician and botanist who wrote the first book devoted entirely to birds (Turner 1544). Early amateur ornithologists collected, preserved, and described the bird species, nests, and eggs they encountered

at home and while traveling. They also kept birds in captivity and documented their behaviors. In fact, much of what we currently know about birds' song, learning, and migratory behavior came from observations made by early aviculturists, and many of the techniques we use to capture birds for research today (see chapter 31) evolved from those used by early bird trappers (Bub 1991, Birkhead and van Balen 2008).

In the eighteenth and nineteenth centuries, most professional ornithologists worked in museums, and their interests in classification, distribution, and natural history determined the initial scientific trajectory of ornithology as it developed into an independent profession. However, amateur ornithologists concurrently contributed to the development of the field. For instance, Alexander Wilson defined the study of birds in North America with his publication *American Ornithology* (1808–1814), a nine-volume series in which he described and classified 268 American species. Prior to writing the series, Wilson worked as a weaver, a peddler, a land surveyor, and a schoolteacher. Unlike most professional ornithologists of his day, Wilson did not rely entirely on museum specimens, but rather spent considerable effort observing wild birds, which enabled him to also record notes about their habitats and behaviors. Today, he is widely considered the Father of American Ornithology (Burtt and Davis 2013).

In the late 1800s, government agencies began encouraging ornithological study and employing professional ornithologists. For example, the Section of Economic Ornithology (later called the Division of Biological Survey) tasked amateurs and professionals alike to observe the feeding and migratory habits of bird species that might act as agricultural pests. By the early 1900s, a growing environmental movement prompted the founding of advocacy groups focused on bird protection (e.g., Massachusetts Audubon) and the preservation of natural areas by government agencies (e.g., National Park Service), which eventually increased the diversity of paid positions for ornithologists. Meanwhile, universities began offering formal courses in bird study, and Arthur A. Allen became the first recognized professor of ornithology (box on page 911). After World War II, the size and number of universities increased, as did the number of undergraduate and graduate degrees available in the life sciences (Geiger 1986, Snyder 1993). The number of full-time, paid positions in ornithology also grew, including a variety of fields wherein academics used birds as key taxa in research (e.g., animal behavior, ecology, biogeography; Barrow 1995, Coulson 2003). Conservation-based nonprofit organizations increasingly employed ornithologists, as did government institutions,

PIONEERING WOMEN IN ORNITHOLOGY

In the nineteenth century, men dominated both the scientific and hobbyist aspects of ornithology, and educational and professional opportunities for women in science were rare. Nevertheless, a select group of women broke the traditional gender stereotypes of the period and made important contributions to the field (Weidensaul 2007). Here are just a few:

Graceanna Lewis was a widely recognized systematist who lectured in ornithology and zoology. In addition to publishing several journal articles and authoring *Natural History of Birds* (1868), she also first identified the Unicolored Blackbird (*Agelaius cyanopus*) as a new species using study skins. In 1870, she became one of the first women elected to the Academy of Natural Sciences. Lewis's mentor, John Cassin, named the White-edged Oriole (*Icterus graceannae*) in her honor.

Martha A. Maxwell was a pioneer naturalist who documented the fauna of Colorado through taxidermy. Maxwell also recorded details about the habitats and behaviors of her subjects, which she incorporated into her taxidermic scenes. Most of her collection was lost to fire or decay, but the techniques and style she developed are still in use today. Maxwell also discovered the Rocky Mountain Screech Owl (*Megascops asio maxwelliae*).

Althea R. Sherman was a self-taught ornithologist who spent over 30 years studying birds. She published more than 70 articles on 38 species, but she was best known for her work with Chimney Swifts (*Chaetura pelagica*). In 1932, Sherman erected a tower on her property to attract the birds and was among the first to observe and describe helpers-at-the-nest or cooperative breeding (Sherman 1924). In 1912, she became the fourth woman elected as a member of the American Ornithologists' Union (AOU).

Mabel O. Wright was an early pioneer in avian conservation who helped popularize ornithology through her writings, including *Birdcraft: A Field Book of Two Hundred Song, Game, and Water Birds* (1895) and *Citizen Bird* (1897; coauthored with Elliot Coues). She also founded and was president of the Connecticut Audubon Society and served as a writer

Martha Maxwell. In the Workroom, ca. 1876. *Image from Library of Congress, Prints and Photographs Division.*

and editor for *Bird-Lore*, a bimonthly magazine published by the Audubon Society until 1940. Wright was among the first women elected as a member of the AOU.

Florence M. Bailey was the first woman elected as a fellow of the AOU (in 1929), and she won the organization's highest award (the Brewster Medal) in 1931 for her work on the birds of the US Southwest. She surveyed birds throughout the west with her husband (a naturalist for the Biological Survey) and published *Birds through an Opera Glass* (1890), *Birds of the Western United States* (1902), and *Birds of New Mexico* (1928), among other works.

For more information on these and other pioneering women in ornithology see Bonta (1985), Benson (1986), Bailey (1994), Barrow (2000), and Weidensaul (2007). Also read about Margaret Morse Nice in chapters 18 and 21. Today, women represent ~20 percent of professionally employed ornithologists, and numbers are increasing each year. References to their work can found throughout this textbook.

ARTHUR A. ALLEN
By Michael S. Webster

This simple fact says a lot: when Arthur "Doc" Allen was hired at Cornell University in 1912, after completing his dissertation on the breeding ecology of Red-winged Blackbirds (*Agelaius phoeniceus*), he became a faculty member in the Department of Entomology and Limnology. Clearly ornithology was not widely recognized as a science in the early twentieth century. Allen would make it his life's work to change that. He started in 1915 by creating the nation's first program to offer advanced degrees in ornithology. Allen stayed at Cornell until his retirement in 1953, and over the years several thousand students enrolled in his popular ornithology courses, many of them going on to careers studying birds. Yet Allen's greatest impact stretched far beyond his research and classroom teaching; he was a tireless promoter of bird study, using public lectures, radio broadcasts, films, and books to bring birds to the public. A seminal point came in 1929 when Allen was asked to help test a relatively new technology—sound-synched film (i.e., "talkies")—by filming birds singing in a local park. The experiment was a success, but just as importantly it gave Allen the brilliant idea that this technology could be used to record the songs and calls of birds. This epiphany grew into the Library of Natural Sounds (now the Macaulay Library), which in 1932 began producing

Arthur A. Allen, Tanner Expedition, 1935. *Photo from Arthur A. Allen papers, 21-18-1255. Division of Rare and Manuscript Collections, Cornell University Library.*

a series of popular phonograph recordings of bird sounds. Today Allen's legacy lives on in the Cornell Lab of Ornithology, furthering its mission to interpret and conserve the earth's biological diversity through research, education, and citizen science focused on birds.

and by the end of the twentieth century, the number of professionally employed ornithologists was nearly 100 times greater than it had been midcentury (Coulson 2003).

Current Opportunities in the Field

Today, there are many career paths for professional ornithologists. While some positions afford long-term dedication to bird-specific topics (e.g., avian taxonomy), others provide opportunities to pursue the biological and wildlife sciences with ornithology serving as a component of the broader work (e.g., animal behavior). Professional ornithologists have a wide range of expertise, and educational requirements for ornithological positions exist along a continuum from self-trained to terminal graduate degrees. However, employers typically require a BS for entry-level positions, an MS for higher-level research, and a PhD to lead independent research. Most institutions do not offer formal degrees in ornithology, but rather provide avenues for students to pursue their interests in birds via programs in biology, zoology, ecology, wildlife manage-

ment, geography, statistics, and education, among others. In the following section, we provide an introduction to various career tracks available to aspiring ornithologists.

Museums

Museum collections have always played an important role in the advancement of ornithological knowledge, but the way we use collections for research and education has changed substantially over time. Early professionals focused on specimen collections to document bird diversity and understand evolutionary relationships among species. These topics continue to be of interest, however, today's ornithologists also use collections to identify population declines (e.g., Winker 1996), examine the effects of contaminants on eggshell quality (e.g., Anderson and Duzan 1978), study ectoparasites and zoonotic diseases (e.g., Foster 1969), and understand the role of ultraviolet reflectance in avian reproductive biology (e.g., Moksnes and Roskaft 1995), among other topics. Professional ornithologists working at museums use molecular techniques, stable isotopes, advanced microscopy, gas chromatography, and sound analyses to answer questions that span large spatial and temporal scales, and museum collections are linked by online databases that amplify the possibilities for collaboration and scope of study (Suarez and Tsutsui 2004, Winker 2004, Joseph 2011; chapter 31).

Museums generally employ administrators, curators (e.g., box on page 912), researchers (e.g., box on page 913), educators, designers, and public relations specialists. However, the number of staff and their job duties depend on the museum size and type. For example, employment with a large federally funded or privately operated museum may afford a professional ornithologist the opportunity to perform curatorial duties (i.e., acquisition and maintenance of museum specimens) and research that is bird-specific, while those employed by a small, university-sponsored museum may oversee collections of multiple taxonomic groups. Employment for museum-based ornithologists usually requires a BS for entry-level positions and an advanced degree for independent research, education, and design positions. Participation in paid or volunteer internships is highly encouraged and can provide experience necessary to be competitive for full-time job placement.

Academia

Professional ornithologists in academia work as professors and instructors, research scientists, and extension specialists. Through a traditional faculty position, professors and researchers are often required to establish externally funded research programs and publish their lab's work in peer-reviewed journals and technical reports. Professors and instructors may also teach a variety of classes (e.g., ornithology, study design, wildlife management, statistics) and participate in student advising and mentorship of colleagues. Faculty positions are highly competitive and typically require a PhD, though some two-year colleges may hire faculty with MS degrees. Previous research and teaching experience and an established publication record are also required. Academic institutions also hire professional ornithologists with BS, MS, or PhD degrees to serve in research, education, and outreach capacities (e.g., teaching assistants, laboratory supervisors, field technicians).

Several types of public and private academic institutions hire professional ornithologists, including two- and four-year colleges and universities that award AS and BS degrees.

SCOTT V. EDWARDS, MUSEUM CURATOR

Scott V. Edwards

I am a professor of evolutionary biology at Harvard University, where I have been a faculty member for 12 years. I am also the curator of a large ornithology collection in the Museum of Comparative Zoology, where one can view the breadth of biodiversity up close. I received my BA in biology, my PhD in zoology, and did a postdoctoral fellowship in immunogenetics. I enjoy the mix of research, teaching,

outreach, and community service that my job entails. My research focuses on the evolutionary biology of birds, using genomics approaches to understand their population genetics and evolutionary history. I particularly enjoy conducting fieldwork, working with organisms in their natural habitats and bringing samples back to the laboratory for analysis. Increasingly, professors like me are bringing their research into their classrooms and providing research opportunities for undergraduates—something that I thoroughly enjoy. I chose my career path because it offered a work routine in which every day was different. Being a professor also allows me to take students into the field and lab to give them a better sense of how biologists collect data. My advice to students at all stages and backgrounds is to get involved with a research project at your school or university. Most students don't realize that the real work of a museum goes on behind the scenes, out of public view, in the vast storerooms where the research collections reside. It's easy to volunteer in your local museum and become familiar with the diversity of birds, their sizes, shapes, skeletons, and plumage. For me, it was much easier to understand biological principles while doing research than in the classroom.

CARLA J. DOVE, FORENSIC ORNITHOLOGIST

Carla J. Dove

I am the program director at the Smithsonian Institution's Feather Identification Lab, a specialized lab that develops and applies forensic techniques to determine the origin of fragmented bird feathers. I received an AAS in natural resource management, a BS in wildlife biology, an MS in systematics, evolution, and population biology, and a PhD in environmental science and public policy. The most important aspect of our lab's research is the identification of birds struck by aircraft, which can lead to improved safety through the design of safer engines and windscreens. The data we provide on avian species identification is used to create databases that predict bird movements, provide bird hazard warnings, and develop management plans that discourage birds' use of airfields. I find it rewarding to help field biologists and colleagues in the military identify bird remains, though it can be quite challenging when there is little evidence available to make definitive determinations. In addition to research, I advise students, publish and present research findings, and participate in outreach and educational events. I followed this career path because I have always had a love of nature, and I had several teachers who served as role models and showed me that I could make a career out of studying birds. I would recommend that anyone interested in this career path study math, value quality of life over financial wealth, and take advantage of career opportunities like volunteering, working entry-level jobs, and listening to the advice of mentors.

The latter may also award advanced MS and PhD degrees, depending on the extent of their available coursework and faculty-directed research programs. Colleges and universities classified as research institutions give high priority to research and graduate education (e.g., box on page 914). Some research institutions are also land grant schools, which maintain agricultural field stations and cooperative extension services that employ faculty and staff to bridge the gap between research and private citizens. Liberal arts universities place greater emphasis on teaching than research institutions do, and they expect faculty members to devote significant time to undergraduate education, experience, and mentorship (e.g., box on page 915). Those interested in academic positions should research the missions of the different institution types to ensure their career choice aligns with their interests and expectations.

Government

Jobs are also available for professional ornithologists in the government sector. Government positions exist at federal (e.g., US Fish and Wildlife Service, US Forest Service, military installations; e.g., the first box on page 916), state (e.g., fish and game departments, transit services; e.g., the second box on page 916), and local levels (e.g., city level management). Employees on this career track may focus exclusively on birds, but many more are hired to work across taxa, fill interdisciplinary roles (e.g., wildlife management, conservation, policy, invasive species control), and interact with the public. Professional ornithologists employed by government agencies have many titles, such as research or wildlife biologist, game or nongame biologist, interpreter, naturalist, game warden, and field technician. Employees typically have a BS or MS degree, and students may benefit from gaining work experience or completing internships as part of government programs (e.g., Pathways, Student Career Experience Program, Research Experience for Undergraduates, Career Internship Program) while finishing their coursework. Professional ornithologists who fill leadership roles within government agencies have similar duties (e.g., research, administration, supervision) and qualifications as professional ornithologists in academia (i.e., PhD, previous research experience, established publication record). Government employees may receive more generous

KAREN L. WIEBE, PROFESSOR

Karen L. Wiebe

years of research experience as a postdoc working on a variety of bird species from raptors to grouse, I was hired to teach courses at the undergraduate and graduate levels and to do research in ornithology and behavioral ecology. My job consists of about 40 percent teaching, 40 percent research, and 20 percent administration. I encourage undergraduate students to work or volunteer on summer research projects to see if such things as living in a field camp and collecting data are attractive to them and whether they want to pursue graduate studies. I enjoy the variability in my job, which also involves sharing research findings with colleagues at international conferences and spending time in a remote field camp handling birds. I have the freedom to study the topics I find interesting, but it is sometimes challenging to obtain grant money to support my research costs and graduate students. Publishing in high-quality scientific journals is, therefore, needed to demonstrate that my research is innovative and novel so that I can be competitive for grants. To be a strong candidate for an academic research job, it is important for students to pursue a doctorate in graduate school, to get publications from their dissertation work, and to have some teaching experience.

I am a professor at the University of Saskatchewan, a medium-sized research institution. I did a four-year undergraduate degree (BSc) and then a PhD in biology that took a further four and a half years to complete. After three more

---

benefits (e.g., fringe, vacation days, paid holidays, severance, pension, regular raises), have greater job security, and retire earlier compared with ornithologists working in other tracks, but they may also have less flexible work hours, less academic freedom, or more bureaucratic constraints.

Government employees working with birds may explore a variety of topics. For example, professionals at the United States Geological Survey's (USGS) Patuxent Wildlife Research Center monitor the status and trends of birds by coordinating survey efforts and analyzing data collected during the annual North American Breeding Bird Survey (discussed in more detail later in the chapter). The United States Department of Agriculture's National Wildlife Disease Program examines avian influenza in waterfowl and shorebird populations and collaborates with agency and university personnel to provide biosafety guidelines for researchers, poultry farmers, pet owners, hunters, and private citizens who interact with wild birds. Professional ornithologists at the state level are required to identify research priorities, serve on technical committees, draft management plans, promote bird conservation through outreach, compile state bird lists, and address time-sensitive and region-specific issues. In the case of the

federally endangered Kirtland's Warbler (*Setophaga kirtlandii*), for example, state biologists with the Michigan Department of Natural Resources work with the US Fish and Wildlife Service (USFWS), US Forest Service (USFS), and other stakeholders to develop and implement conservation plans that reduce immediate threats to the species (e.g., habitat loss, brood parasitism) and educate the public about biodiversity. The contributions of professional ornithologists employed by the government span many facets of avian research and management, and employment in this track may be especially rewarding for those interested in the applied aspect of ornithology.

Nonprofit Organizations

Several types of nonprofit organizations (NPOs) employ professional ornithologists (e.g., the first box on page 917 and the second box on page 917). The goals and objectives of NPOs vary widely and span conservation, management, policy, and education. Ornithological positions with NPOs are as varied as the NPOs themselves. NPOs hire people with a wide range of educational backgrounds as administrators, researchers, field and geographical information systems (GIS) technicians,

MICHAEL W. BUTLER, PROFESSOR

Michael W. Butler

I am an assistant professor at Lafayette College, a small liberal arts college in Pennsylvania. I received a BA in biology and physics, an MS in raptor biology, and a PhD in biology. I typically teach two mid-level courses and an upper-level seminar each year and conduct field-based research year-round with undergraduate students interested in ecological physiology. I chose this career path because I genuinely enjoy teaching, and I didn't want my career to depend solely on securing major grants within my first five years as a professor. Given my employment at an undergraduate-focused institution, I have high turnover in research assistants and must continually train new researchers. This can be extremely time-consuming, but my undergraduates constantly surprise me and exceed my expectations, making even this minor frustration a fulfilling part of my job. It's exciting to watch my students' development and realize that I may be, in part, responsible for their decisions to pursue careers in ornithology for themselves! If you are considering a career similar to mine, you will need to attend graduate school and make teaching experience a priority. You will also want to take advantage of university-sponsored courses and workshops designed to help graduate students develop strong teaching skills. During graduate school, focus on developing your own teaching style, and be prepared to write a well-constructed philosophy of teaching statement. During the interview, clearly establish that your main focus is teaching, but that you can also lead a successful undergraduate research program that will result in multiple publications, some of which have students as coauthors.

land managers, outreach biologists, editors, web developers, and fund-raisers. NPOs also offer volunteer opportunities for amateurs to contribute to research and management and provide volunteer and paid internships for aspiring ornithologists to gain experience prior to employment or pursuance of an advanced degree. Those interested in working or volunteering for an NPO could search for opportunities with the Audubon Society, American Bird Conservancy, British Trust for Ornithology, BirdLife International, Ducks Unlimited, The Nature Conservancy, The Peregrine Fund, Pheasants Forever, Pronatura, the World Wildlife Fund, and various scientific societies (e.g., American Ornithological Society, The Wildlife Society) and bird observatories (e.g., Long Point Bird Observatory, Bird Conservancy of the Rockies), to name a few.

NPO employees have contributed greatly to our knowledge of avian systematics, ecology, and management, and they often work independently and in collaboration with other organizations and private landowners to protect bird populations and their habitats. Many NPOs provide resources for species identification, breeding biology, and habitat preservation and restoration. Additionally, NPOs often manage large-scale volunteer efforts to examine population changes. For example, scientists with the National Audubon Society use long-term data collected by amateurs across the county to predict the influence of climate change on the distributions of North American bird species. Professional ornithologists employed by NPOs also operate banding stations throughout the world that provide information on avian morphology, migration, dispersal, survival, reproductive success, disease, contamination, and behavior. Given the breadth of research and conservation-driven work done by NPOs, professional ornithologists may find the nonprofit track a rewarding path to pursue.

Private-sector Opportunities

Many professional ornithologists seek employment in private industry as environmental consultants (e.g., box on page 918) or zoo staff. Others work as tour guides, photographers, artists, writers, and in other positions wherein they engage in avian conservation or environmental outreach. These positions

SUSAN K. SKAGEN, FEDERAL RESEARCH WILDLIFE BIOLOGIST

Susan K. Skagen

I have been in my current position at the United States Geological Survey as a research wildlife biologist for 27 years. I received my BS in zoology, an MS in biology, a PhD joint major in zoology and wildlife ecology, and completed my postdoc- toral experience in environmental studies. I chose my career path by following my interests in ecology and ornithology. I now conduct research on the effects of climate change on avian populations and communities, including birds dependent on wetlands and grasslands across the North American interior region. My fieldwork has focused on migration stopover ecology in dynamic wetland and riparian ecosystems and reproductive performance of declining prairie bird populations. I do much of my work in collaboration with highly talented graduate students, undergraduates, and postdoctoral research associates. My career is rewarding, because there are always interesting new questions to pursue, skills to learn, and collaborative opportunities. If you are interested in a career like mine, I recommend developing a combination of field-oriented, theoretical, modern, and classical quantitative skills. I also encourage you to engage in the profession in a variety of ways that enable you to network and let people get to know you. This could include going to annual meetings, giving presentations, and participating in scientific and conservation society committees and activities.

RYAN S. JONES, STATE WILDLIFE MANAGEMENT BIOLOGIST

Ryan S. Jones

I am a wildlife management biologist for the Missouri Department of Conservation. I received a BS and an MS in wildlife biology, and I have been in my current position for five years. Throughout my career, I have been fortunate to work with numerous bird species, including several species of grouse, and I have worked in unique and rare habitats. My current job duties include managing state-owned lands within a four-county district with a focus on natural community restoration. This includes primarily the use of prescribed fire and tree thinning to manipulate the composition of the vegetation to that of pre-European settlement. These habitats support a wide range of native species of all taxa. The most rewarding part of my job is selecting a tract of land in need of restoration or enhancement, envisioning what it could look like with management, administering the steps of the restoration, and then watching the increase in the diversity of flora and fauna of the natural community. As with most wildlife jobs, the most challenging aspect of my position is prioritizing efforts within the limited hours available, but knowing I have made a positive contribution to conservation is very rewarding. The best advice I could give someone wanting to follow this career path would be to gain as much practical and varied experience as you can through internships, education, and job opportunities.

FELIPE CHÁVEZ-RAMÍREZ, DIRECTOR OF CONSERVATION PROGRAMS

Felipe Chávez-Ramírez

I am an avian ecologist and director of conservation programs with the Gulf Coast Bird Observatory. I obtained a BS in biology and an MSc and PhD in wildlife ecology. I first decided to work for an NPO because I wanted to work on real-world research and conservation problems. Working for an NPO allows the flexibility to participate in conservation activities that may not fit well within other job descriptions and requires that you wear many "hats." Conservation NPOs require participation in not just biological work but also in social and policy arenas. The types of day-to-day activities can be interesting and challenging; some days you may trap birds for a research project, other days you can sit in a board room doing strategic planning for a species or area that needs conservation. It is rewarding when you can see that your efforts have generated information that will allow for better management and protection of birds and habitats. There are many challenges in working with an NPO, as we must work hard to procure funding to support our research, conservation, and education efforts. If you are interested in the NPO career path, I recommend participating on a variety of projects and activities outside the classroom to gain a better understanding of the varied day-to-day work while getting valuable experience.

ILIANA A. PEÑA, DIRECTOR OF CONSERVATION

Iliana A. Peña

I am the director of conservation for Audubon Texas, a state office of the National Audubon Society. I received a BS in wildlife and fisheries sciences and an MS in range and wildlife management. I have been in my current position for seven years, inheriting a small but long-lived coastal conservation program. In my position, I use applied science to help inform coastal colonial waterbird management. I also use creative market-based strategies to help reverse grassland bird declines while supporting ranchers and contracting with city leaders to implement conservation strategies necessary for community planning and maintenance. A professor during my undergraduate work once said that if you cannot effectively communicate the conservation work with people and communities, then you have already undermined your effort. That advice is the truest statement that I can share, especially in any region where the majority of land is privately owned. Every day I have the pleasure of working with landowners, city officials, NPOs, state and federal agency partners, donors, corporate and foundation leaders, and the general public. I strive to communicate conservation needs, develop partner synergies and collaborations, and raise funds in a sustainable manner to support and grow the work. The hours can be long, but I have the pleasure of working alongside many dedicated people who, like me, love what they are doing.

require a wide range of education and experience levels, and they can provide opportunities similar to those previously described. They may also offer greater flexibility in hours, variety of tasks, or financial motivation not afforded to those pursuing other career tracks. For example, a field technician working for an environmental consulting company may be more selective about the type of position and number of months worked per year, given a higher pay rate than a similar position on an NPO- or university-sponsored project. For private-sector employees pursuing self-employment or freelance work, success may depend on natural talent, personal investment, motivation, professional networking, reputation, or regular travel. This can be daunting but may ultimately provide opportunities for growth not available elsewhere.

Regardless of the career path, most professional ornithologists in the private sector promote avian conservation and management through their work. For example, consultants with Western EcoSystems Technology, Inc., and Wyoming Wildlife Consultants collaborated with researchers at the University of Wyoming to examine the short-term impacts of

wind energy development on Greater Sage Grouse (*Centrocercus urophasianus*) fitness (Lebeau et al. 2014). Photographers, painters, writers, and sculptors who depict birds in their work educate and inspire others to engage in conservation (e.g., box on page 919) and often support conservation directly through donation of their works to fund-raising programs (e.g., Artists for Conservation). Bird tour guide companies (e.g., Victor Emanuel Nature Tours, Field Guides, WINGS) provide bird enthusiasts with field experiences in destinations that would be difficult to navigate otherwise. As with all other tracks, the private sector provides professionals with many and varied opportunities that span all imaginable and necessary components the field of ornithology has to offer.

PREPARING FOR A PROFESSIONAL CAREER IN ORNITHOLOGY

Career preparation and hands-on experience are necessary for anyone pursuing a professional career in ornithology. This may include formal coursework, internships, workshops, tem-

J. H. CARTER III, ENVIRONMENTAL CONSULTANT

J. H. Carter III

I have been an environmental consultant for more than 39 years, since before there was an "environmental consulting" industry. My path into environmental consulting was largely a function of being in the right place at the right time. At an early age, I lived near a large Red-cockaded Woodpecker (RCW; *Picoides borealis*) colony and developed an interest in ornithology. Events in the 1970s (i.e., the Endangered Species Act, listing of the RCW as federally endangered, and

the first RCW Symposium) led to my MS and PhD research on RCW biology. My first consulting job in 1976 was on an RCW highway survey project. I later expanded my consulting resume to include wetlands and additional rare species. Today, students can pursue careers in environmental consulting with four-year or advanced degrees and internships that position them for jobs. If you are hoping to enter and succeed in this field, I offer four recommendations: (1) learn to identify dominant plant species and vegetative communities so you can understand what vegetation types and conditions say about soils, hydrology, species associations, and disturbance regimes; (2) be prepared to continually learn as new research is published, techniques improve, and regulations change; and (3) develop strong written and verbal communication skills, and the ability to work well with people. Attention to detail in communications will be important, because your proposals, reports, and emails will be reviewed by supervisors, clients, and regulators, and can be evidence in legal proceedings; so (4) develop strong personal ethics and a professional communication style. Environmental consultants must help balance the needs of clients, regulators, and the environment. This means sometimes working in situations that are contentious, so you must have self-confidence and experience in addition to a strong ethical foundation.

JANE KIM, ARTIST AND SCIENTIFIC ILLUSTRATOR

Jane Kim

I am an artist, a scientific illustrator, and the creator of Ink Dwell, an art studio with a mission to inspire people to love and protect the Earth, one work of art at a time. I earned a BFA in design, followed by a professional certificate in scientific illustration. Science illustration is an essential part of science, and I love that through images we can easily understand complex systems, learn about the microscopic, bring the extinct back to life, and imagine the future. By combining classical techniques of science illustration with the impact of fine art, I create iconic images that carry a powerful message of education and conservation. Birds add color and sound to our environments, and bird imagery is particularly evocative because it reminds us about our connection to nature. If you are interested in a career combining science and art, I recommend pursuing internship opportunities (e.g., Bartels Science Illustration Internship, Cornell Lab of Ornithology), science illustration programs (e.g., California State University–Monterey Bay, University of Washington), and organizations like the Guild of Natural Science Illustrators, all of which can be found online.

porary and volunteer positions, and self-directed learning. In this section, we discuss general and discipline-specific proficiencies sought by employers of professional ornithologists and identify pathways for obtaining these skills. Though our focus is ornithology, the information provided herein is applicable to other jobs in the life sciences and may benefit amateur researchers who are interested in birds or other wildlife.

Foundational Skills

Learning to identify bird species by sight and sound can be personally rewarding, and proficient bird identification is necessary for some positions in ornithology. However, ornithology encompasses more than just bird-watching (or "birding"; box on page 920). Professional ornithologists must also have well-developed skills in observation, scientific literacy, study design, data collection, analyses and interpretation, and communication of results.

Observation is an instinctive behavior, but the ability to practice focused attention on the natural world is at the core of being an ornithologist and a scientist (Sagarin and Pauchard 2012). While both ornithologists and bird-watchers use observational clues (e.g., plumage, vocalizations) to determine the species and sex of a given bird, advanced observational skills guided by careful study design and data collection protocols are necessary to produce information that leads to a deeper understanding of complex natural systems and evolutionary processes. In addition, ornithologists must read and interpret peer-reviewed scientific literature (i.e., the published literature) to build off one another's knowledge and develop new questions and methodologies that further our understanding of birds and their environments.

Understanding effective study design and statistical methods is also an important component of the study of birds. Recent graduates are often deficient in the quantitative and analytical skills necessary to interpret ecological processes and patterns and apply that understanding to natural resource management (Burger and Leopold 2001, Kendall and Gould 2002). Students who excel in this area have an advantage when seeking employment or admission to graduate school. Additionally, those seeking jobs in ornithology and other science fields need to communicate science effectively both orally and in writing. Ornithologists regularly describe their research to other scientists at professional meetings and by publishing in peer-reviewed journals. However, as natural resource issues become more central to society, and as the number and diversity of stakeholders who interact with scientists grows, there is an increasing need for ornithologists to communicate with policy makers and the public through different avenues (e.g., social media, popular press articles, public forums).

In addition to the skills described above, employers seek applicants with demonstrated experience in project management, collaboration, professionalism, and self-motivation. These are often referred to as "soft skills," but aspiring scientists should not underestimate the importance of these attributes. Soft skills are integral to the practice of science, and the Association of American Colleges and Universities (AACU 2013) noted that employers are concerned about whether the United States is producing college graduates who have the soft skills, knowledge, and personal responsibility to contribute to a constantly changing workplace—particularly in the areas of science, technology, engineering,

JED BURTT

Jed Burtt

You become an ornithologist by watching birds, learning to recognize them by sight and sound, and discovering what they do and where they hang out. Take notes. Listing species is a good start, and you can submit your lists to eBird, but soon you will want to know more information (e.g., how do birds catch prey or pluck seeds from plants). Learn by watching. Read your notes. Create a hypothesis about why birds might do what you saw. Test your hypothesis the next time you are in the field by making focused observations. Discuss your observations and discoveries with classmates, teachers, and friends who watch birds. As a young man, I participated in Christmas Bird Counts and completed and published two Breeding Bird Surveys, both programs that still welcome young birders. I chose a college with an ornithology course, a small museum, an ornithology library, and a field station. I took the necessary science courses and went to graduate school on a teaching fellowship. I earned a master's degree by extending my undergraduate summer research and a PhD examining coloration of warblers. At Ohio Wesleyan University, I taught undergraduates about birds. My former students now advocate for birds from government and environmental organizations. Some are museum curators, and some have become teachers in middle schools, high schools, and even colleges and universities.

and math. Kubik (2009) surveyed federal wildlife biology managers, researchers, educators, and field specialists and found that the most important skill for wildlife biologists in the future is the ability to continually learn and rapidly and effectively apply new knowledge to new situations. Closely ranked in importance were the effective application of digital technology, professional networking, collaboration, the ability to think and act strategically, and the flexibility and creativity to effectively deal with change and solve problems. Students can develop these skills by seeking out experiences beyond the classroom throughout their college years.

Gaining Practical Experience

Students can gain valuable experience, in part, through coursework, but exposure and practice obtained through undergraduate research, internships, temporary positions, volunteerism, and citizen science provide increased opportunities for growth. Hands-on experience allows students to develop skills and knowledge to make themselves more employable, but it also allows individuals to discover the types of work they find the most—and the least—rewarding. Extracurricular experience also provides students with opportunities to see how science is used in real-world settings and exposes them to a variety stakeholder needs and interests.

Undergraduate Research

Undergraduate research can provide an excellent opportunity to gain technical, hands-on skills in the laboratory or the field and, more importantly, to learn about the scientific process. Students can identify available opportunities by contacting their professors, departments, or undergraduate research offices. Additionally, some universities have databases where faculty advertise undergraduate research positions. Students conducting undergraduate research should actively take steps to ensure that they don't just provide labor in the lab or field (e.g., simple micro-pipetting or data entry) without having an academic dimension to the experience. Students should read grant proposals and publications that provide background for the lab's research and engage in lab discussions of related literature. Making these extra efforts will help students learn the language and process of scientific research, making them more competitive for future opportunities. Finally, students conducting research should present their findings at scientific conferences with organizations that encourage undergraduate research (e.g., Wilson Ornithological Society).

Internships and Temporary Positions

Seasonal laboratory or fieldwork generally consists of helping researchers and field biologists collect data or conduct wildlife and habitat management activities. It is common for undergraduate students to work in seasonal jobs during the summer and for several years after graduation to gain a broad base of experience. Seasonal field jobs don't necessarily pay generously, but they do provide excellent opportunities to experience different ecosystems, research environments, species, study types (e.g., applied versus basic science), and laboratory and field techniques. Though some positions may require a BS degree, we recommend that undergraduate students pursue internships and temporary positions with birds and other wildlife as early as possible. A variety of widely used job boards advertise seasonal positions online through universities and professional societies (e.g., Ecological Society of America, Ornithology Exchange, Society of Conservation Biology, Texas A&M University Wildlife and Fisheries Sciences, The Wildlife Society).

Volunteer Work

One of the easiest ways to gain experience is to volunteer. Paid positions may be competitive, and students may be limited in their time and ability to travel to distant job locations. However, many graduate students, professors, and professionals working with NPOs and government agencies are happy to provide volunteer opportunities to undergraduates who are interested in learning. Volunteering—particularly for students who are reliable, eager to learn, and willing to work hard—can lead to paid seasonal positions and may increase one's competitiveness when applying for graduate programs or full-time employment. To find volunteer opportunities, students should ask within their academic departments, contact local agencies, or visit online job boards.

Course-Related Research

Academic coursework and real-world, collaborative studies are not necessarily mutually exclusive experiences. Indeed, research and service learning programs that are integrated with formal coursework provide students with experiential learning opportunities that synthesize scientific observational skills with content knowledge (e.g., Leiser and Reilley 2015). Such experiences also provide hands-on opportunities to develop important practical skills (Millenbah and Millspaugh 2003), enabling students to become more aware of their own learning process and increase their academic achievement and career motivation (Hiller and Kitsantas 2015, Leiser and Reilley 2015). For example, students at Texas A&M and Sul Ross State universities have opportunities to engage in active research projects involving fox squirrels (*Sciurus niger*) and Scaled Quail (*Callipepla squamata*) through the academic lesson plans of wildlife ecology, habitat management, wildlife management techniques, and population dynamics courses. These projects allow students to gain valuable technological experience with radiotelemetry, GIS, and global positioning system (GPS) technology, as well as other field-based practices (e.g., habitat measurements, capture and handling techniques, population estimation; McCleery et al. 2005). The Bird-Window Collisions Project (box on page 922) and Project Owlnet (box on page 923) also offer opportunities for students to engage in research or service learning as part of formal coursework, but these projects are not limited to a single university. Instead, they bring together a network of students across campuses, allowing for data collection across a wider spatial scale. Students should consult with professors when choosing courses to see whether opportunities for curriculum-based research programs or service learning exist. Those without access to such opportunities can take initiative and talk with professors about ways to integrate research into their current curriculum or independent projects, or they can take advantage of hundreds of publicly available citizen science opportunities (discussed later in the chapter).

THE CONTINUED ROLE OF AMATEURS IN ORNITHOLOGY

No branch of biology has felt the hand of the amateur like ornithology.
—Harold F. Mayfield (1979)

Modern Amateur Scientists

With the wide array of career options now available to ornithologists, one might conclude there is no longer a role for amateurs in ornithology. However, today's amateurs continue to make valuable contributions to ornithological research through study design, data collection, dissemination of results, and participation in scientific societies. Amateurs use their birding and observational skills to conduct scientific research in addition to—rather than as part of—their paid positions. They are motivated by a genuine interest in birds and a passion for scientific discovery. Most are educated in other disciplines, and some lack formal scientific

CONDUCTING REAL-WORLD SCIENCE IN THE CLASSROOM: THE BIRD-WINDOW COLLISIONS PROJECT
By Steve B. Hager and Brad J. Cosentino

Building characteristics and the land cover surrounding homes and large office complexes affect variation in the number of bird-window collisions (BWCs). However, little is known about whether BWCs are affected at a regional scale by patterns of urbanization. We assessed how BWCs depend on building characteristics and land use patterns at local and regional scales. In fall 2014, professional researchers and university students at 40 sites and nearly 300 buildings throughout North America used standardized research protocols to document collision mortality, evaluate building characteristics, and measure land cover features at local and landscape scales. Overall, we found more than 300 bird carcasses (range = 0–34 per building) representing over 70 species. Building size was the most important driver of BWCs, but the strength of the effect of building size on BWCs depended on levels of urbanization. Bird mortality was highest at large commercial buildings containing high levels of window glass, but the positive relationship between BWCs and building size was stronger in rural than urban areas. We created the Bird-Window Collisions Project to assess our research goal and to provide valuable field and classroom-based research opportunities for students. In the field, students learned how

Photo by Stephen B. Hager.

to conduct standardized surveys for bird carcasses resulting from window strikes and to measure environmental variables hypothesized to influence BWCs. In the classroom, students and professors studied BWCs by completing a case study on assessing the drivers of BWCs, and they studied the conceptual background of BWCs in formal journal clubs.

C. SCOTT WEIDENSAUL, NATURALIST, WRITER, AMATEUR ORNITHOLOGIST

C. Scott
Weidensaul

Ornithology has long embraced serious amateurs, and I'm part of that tradition. Despite my lack of an academic degree in science, I've been deeply involved in ornithology for years, as both a communicator and a field researcher. I was an avid birder as a kid, and a college ornithology course cemented

my interest in the science of birds. As a natural history writer for the past 30 years, I've made ornithology and conservation (especially migration) a specialty—but over the decades I've also been drawn ever more deeply into doing science myself. In the 1980s, I began assisting Hawk Mountain Sanctuary's raptor research program, doing banding and radio-telemetry. I soon had my own banding permit, and in addition to working with diurnal raptors, by the late 1990s I was overseeing what grew into one of the largest owl-migration projects in the country, focusing on Northern Saw-whet Owls (*Aegolius acadicus*). That project, in turn, is part of a continental network of about 125 banding stations, called Project Owlnet, which I help manage. I am a cofounder of Project SNOW-storm, a collaboration among dozens of researchers studying the winter ecology of Snowy Owls (*Bubo scandiacus*). I am also one of the few hummingbird banders in North America, with a particular focus on the rapid evolution of new eastern migratory routes and wintering areas of several species of western hummingbirds. And recently, several colleagues and I launched an ambitious project to track the migration of many of the birds that nest in Alaska's 54 million acres of national park land.

training altogether, but there is no denying their important contributions to the field. For example, Harold Mayfield, perhaps the most well-known amateur ornithologist, served as the president of the three largest ornithological societies in North America, wrote over 300 scholarly papers, and received several awards for his work in ornithology—all while employed as the personnel director at a glass company. Though many amateurs continue to conduct independent research (e.g., boxes on pages 923 and 927), more are now involved in networked research programs that bring together citizens with varying skill levels to answer questions across multiple scales (i.e., citizen science). In the remainder of this chapter, we focus on the types of citizen science programs available and the questions they can answer.

Amateur Ornithologists in Citizen Science

Citizen science programs allow participants to learn more about birds and engage in different stages of the scientific process (Bonney et al. 2009). Participation in most of these programs is contributory, wherein volunteers collect and analyze data. However, there are also collaborative and co-created programs in which amateurs have greater control in

the development and direction of projects (e.g., identification of problems, development of hypotheses, interpretation of results, dissemination of findings; Shirk et al. 2012).

The Christmas Bird Count (CBC) is the longest-running citizen science bird survey in the world. The Audubon Society established the CBC as an alternative to a bird hunt that was popular in the nineteenth century. On Christmas Day in 1900, the Audubon Society organized and supervised 25 bird counts in cities across North America. Today, tens of thousands of people participate in the CBC each year in cities throughout the Americas. During the CBC, volunteers follow assigned routes within established 15-mile-diameter circles and count all birds seen or heard throughout the day. Following the count, the Audubon Society compiles the data and makes all information available online for use in individual and collaborative research. These data have been used by hundreds of researchers and are especially useful for examining population trends of wintering birds.

The Breeding Bird Survey (BBS) is a research and monitoring program similar to the CBC that is conducted during the peak bird breeding season. The USGS initiated the program in 1966 in response to concerns over the effects of pesticides

J. DREW LANHAM, PROFESSOR OF WILDLIFE ECOLOGY

J. Drew Lanham

I am a rare bird: a black birder. Unfortunately there aren't very many people of color who do what I do. I don't think it's because young black and brown people aren't captivated by birds or nature or because white people have a greater appreciation for those things. Like E. O. Wilson says, I think it's born in all of us. We simply need to nurture what's inside. Nurture for me was growing up in the middle of a national forest on a family farm. That environment was the cauldron that nurtured my nature. My mother was a biology teacher. My father taught earth science. I spent my days playing outdoors and wishing I could be like the feathered ones that sang from the trees. I marveled at the avian ability to sail from place to place. To me, flight was magic. As I prepared for college, I envisioned wildlife biology as a career. Others didn't understand why a young black man would pursue such a "white" profession. So for the first half of my college career, I put my passion aside to pursue a degree in something sensible and more appropriately "black." I studied mechanical engineering, but there was one problem. I was miserable. In a moment of clarity, I changed my major to zoology and let birds and other wild things become the driving force in my life. Studying ornithology in graduate school tattooed the love deeper on my heart. Now I'm a professor of wildlife ecology at Clemson University, an ornithologist, and, forevermore, a birder. As a black American I am intrigued with how ethnicity influences how we see nature and conserve natural resources. One of my missions is to broaden the audience of those who see and care for nature, by using birds and their conservation issues as inspirational vehicles to connect others to the outdoors and advocate for their protection. I call this "coloring the conservation conversation."

on bird populations. Each year, at the peak of avian breeding, thousands of amateurs conduct three-minute point counts (see chapter 31) along more than 4,000 preestablished survey routes. The data generated from these surveys provide information on the population trends of over 400 species, which researchers use to examine the impacts of land cover and land use changes as well as the effects of chemical contaminants and climate change on populations. Similar motivation sparked the formation of the Nest Record Card Program at the Cornell Lab of Ornithology (CLO) in 1966. Volunteers of this program monitored nests and recorded the details of each visit on nest cards. The Nest Record Card Program merged with the Birdhouse Network (to which volunteers contributed data specific to cavity-nesting birds) in 1997, and together these became NestWatch in 2008 (fig. 30.1). Today, NestWatch has over 350,000 nest record cards, with at least one for every species that breeds in North America. These cards have been used in over 130 publications (Phillips and Dickinson 2009), many highlighting natural history and nesting activities of species for which little else is known, and other studies examining a range of topics including brood parasitism, nesting phenology, patterns in clutch size, and changes in breeding distribution (reviewed by Cooper et al. 2015).

Since the establishment of the CBC, BBS, and the Nest Record Card program, citizen science programs in ornithology have increased in number, size, and scope. Advances in technology (e.g., phone apps, Internet) have enabled broader participation in citizen science programs, including several ways to engage online (table 30.1). Over the past few decades, hundreds of thousands of people have participated in local, regional, and national programs (e.g., the second box on page 927). Several websites catalog projects and help interested participants find programs that fit their individual needs (e.g.,

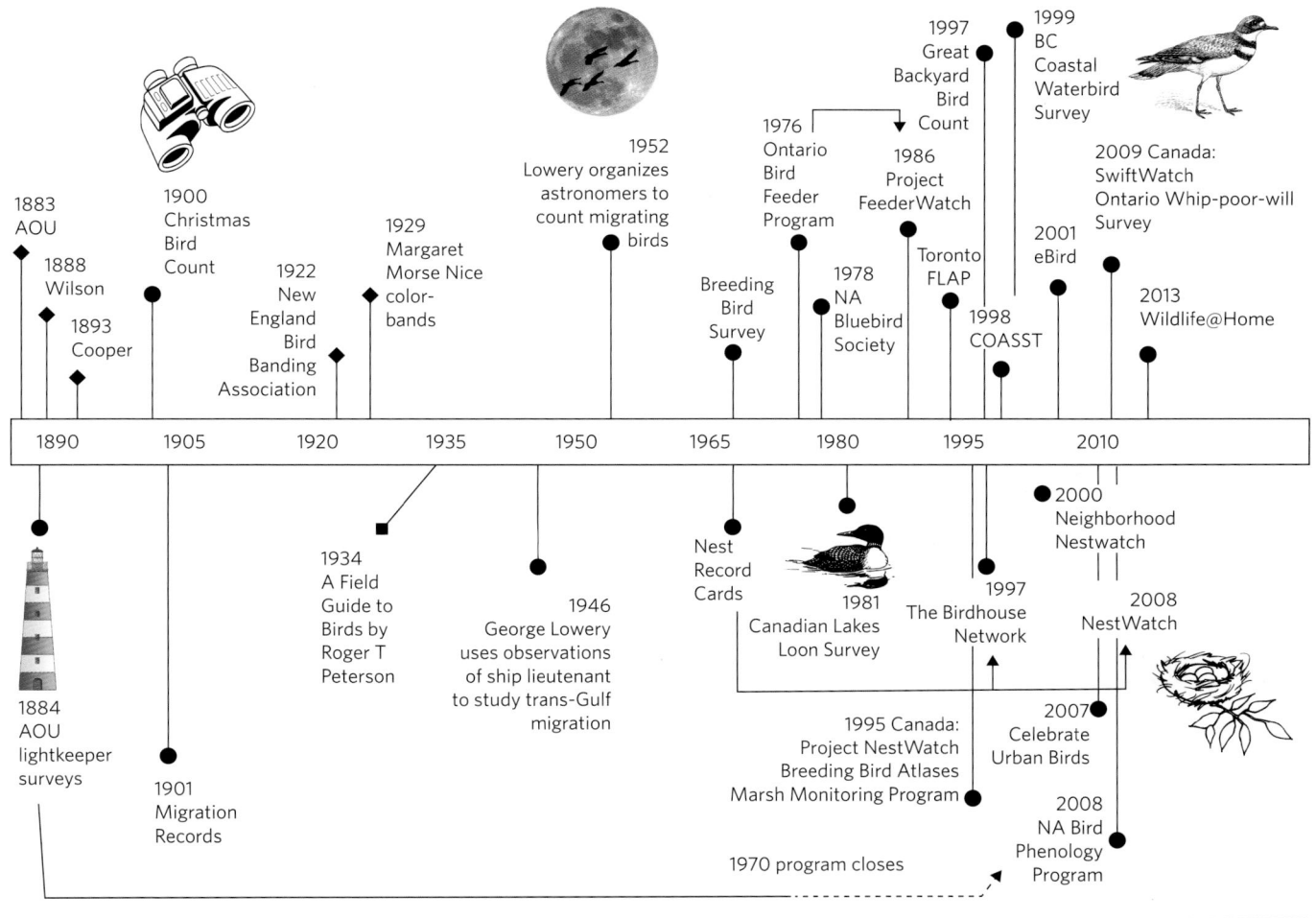

Figure 30.1. Timeline of key organizations, activities, and contributions of ornithologists that blurred the lines of professional and amateur science (diamonds), as well as start dates of citizen science projects focused on birds (circles). For example, the Christmas Bird Count be-gan in 1900 and continues to the present day. The Nest Record Cards began in the mid-1960s and later merged with Birdhouse Network, before both programs merged to form NestWatch in 2008. *Creative Commons CC0.*

Zooniverse, SciStarter, CLO). In addition, CLO, in collaboration with others, developed and hosts an online toolkit to guide individuals and organizations in developing and maintaining their own citizen science projects.

While bird-watchers vary in skill and expertise, technologies have helped address issues of data quality in citizen science programs. New technologies allow for easy submission of photographic documentation (Bonter and Cooper 2012) and geo-referenced information. For exclusively online projects, cyberinfrastructure includes consensus tools in which each classification, tagging, or transcription activity is performed by multiple participants until a minimum consensus is achieved (Matsunaga et al. 2014). In addition, web-based smart forms can proof data upon entry, and spatiotemporal profiles with checklists for specific regions and time frames (Kelling et al. 2011) allow experts to flag submissions outside of these profiles for review (Sullivan et al. 2009, Wiggins et al. 2014).

A global citizen science project that began in 2002, eBird, has engaged well over three million bird-watchers, who have cumulatively contributed over 280 million observations from almost two million locations. eBird data have been used in over 100 publications, including those reporting discoveries related to migratory pathways and speed (La Sorte et al. 2013, 2014a), avian responses to urbanization (La Sorte et al. 2014b), and shifts in species distributions with climate change (Hurlbert and Liang 2012). To verify the quality of submissions, eBird has more than 800 regional experts who contact participants with flagged records and judge unusual reports.

The Value of Citizen Science in Ornithology and Natural Resource Management

A common misconception is that citizen science is primarily a tool for outreach and education. Though citizen science certainly does fill that role, the scientific impact of citizen science

Table 30.1. **Examples of citizen science projects.**

Volunteer Activity	Projects
Field projects	
Banding and resighting	Monitoring Avian Productivity and Survival (MAPS), field stations, Neighborhood NestWatch
Nest monitoring	NestWatch, Neighborhood NestWatch, OspreyWatch
Checklisting	eBird, state breeding bird atlas programs, Breeding Bird Survey (BBS), Christmas Bird Count (CBC), Project Feeder Watch, Celebrate Urban Birds, Great Backyard Bird Count
Collecting specimens	Fatal Light Awareness Program (FLAP), Sparrow Swap, Seabird Ecological Assessment Network (SEANET)
Sound recording	Macaulay Library
Measurements, photo-documentation	Coastal Observation and Seabird Survey Team (COASST), Birds and Berries
Report deformities	Big Garden Beak Watch (UK), Rouge River Bird Observatory
Online projects	
Aligning sequences of genetic data	Phylo
Tagging behavior	PenguinWatch, Wildlife@Home, CondorWatch, Great Horned Owl Vocal Study
Transcribing data	North American Bird Phenology Program, Smithsonian Transcription Projects

in ornithological research is quite substantial. Professional ornithologists benefit from the contributions of amateurs to citizen science programs, because, collectively, amateurs can obtain data over larger spatial and temporal scales that would be otherwise impossible to acquire. The misconception persists, in part, because many scientific papers do not explicitly acknowledge when they contain contributions from citizen science projects. For instance, none of the papers in Knudsen et al.'s (2011) review about migratory birds and climate change used the term "citizen science," even though approximately half included data from citizen science activities in their analyses. In fact, about half of what we currently know about migratory birds and climate change comes from citizen science data (Cooper et al. 2014). For example, Dunn and Winkler (1999) examined more than 3,000 nest records from citizen science projects in Canada and the United States to investigate Tree Swallow (*Tachycineta bicolor*) responses to climate change. They found that from 1951 to 1991, Tree Swallows incrementally advanced the timing of their breeding across North America. Dunn and Winkler associated the shift to increased surface air temperatures during the same decades, which likely affected the abundance of food resources available to the birds.

Researchers also use data from citizen science programs to examine species distributions, migration, breeding biology, urban ecology, emerging diseases, management efforts, and invasive species, among other topics (Cooper et al. 2012). In addition to the many research questions that can be addressed with citizen science data, natural resource managers can use these data to determine management priorities and inform management decisions. At local scales, decisions about natural resource management are made and actions implemented more quickly when citizen scientists are involved in the collection and analysis of monitoring data (Danielsen et al. 2010). However, citizen science data can also be used for effective natural resource management at larger scales. For instance, the Nature Conservancy (TNC) pays rice farmers in California's Central Valley to flood their fields during the migration of shorebirds and waterfowl as part of their market-based program BirdReturns (fig. 30.2). TNC determines which fields to flood based on remotely sensed images of surface water and forecasts of probable bird locations derived from eBird records (Johnston et al. 2015). The flooded fields then act as "pop-up" wetlands that provide stopover habitat for over 220,000 birds during migration. Projects like BirdReturns demonstrate the value of broad-scale citizen science data to inform habitat management for birds.

KEY POINTS

- Amateur ornithologists pursue the study of birds as a pastime or hobby rather than a paid profession. Historically, amateurs were important to the evolution of

SCOTT R. AND ANN B. SWENGEL, AMATEUR ORNITHOLOGISTS

Scott R. and Ann B. Swengel

Ann and I are amateur ornithologists who have published more than 50 papers about survey techniques, long-term population trends, community ecology, and conservation of birds and insects. I have a BS in zoology and a BA in geology. Ann studied anthropology and earned an undergraduate degree in the classics. In the 1980s and 1990s, I worked as an aviculturist and later as the curator of birds at the International Crane Foundation. During that time, Ann worked various jobs with flexible hours so we could carry out and publish research "on the side." Ann now works a full-time office job, and I work from home, which allows us to dedicate our free time to fieldwork, data management, and writing papers. For example, in Swengel and Swengel (1999) we examined the abundance of grassland songbirds and prairie butterflies at over 100 study sites across six states, and in Swengel (2015), Ann described how managing for grassland birds benefits monarchs. According to Ann, "Science is determined by the nature of the activity or product, not the person who did it."

KAYCEE LICHLITER, CITIZEN SCIENTIST

Kaycee Lichliter

My love for nature, combined with an inquisitive mind, led me to obtain an associate degree in natural resource management in 1999. Working through my local Audubon Society, I learned about bluebirds and nest box monitoring, and by 2004, I was managing a trail of the society's nest boxes at University of Virginia's Blandy Experimental Farm. Intense work and dedication has grown the trail to currently include 132 nest boxes and 35 volunteers. Our data is shared with various organizations, including Cornell Lab of Ornithology's NestWatch citizen science project. To further enrich my life with birds, I volunteered at various bird banding stations. This encouraged me to attend a specialized program sponsored by Point Reyes Bird Observatory, which involved concentrated classroom and field study. After completing the course, I established, operated, and co-managed a constant-effort mist-netting station with the Monitoring Avian Productivity and Survivorship (MAPS) program for five years along with seven other volunteers. Each year, I start my Christmas celebration by participating with my "bird-friends" in the local Christmas Bird Count. Searching for and following birds has taken me to some distant and beautiful places, like Alaska, Costa Rica, Canada, and even my own backyard. Wherever I go, birds grace the environment. My involvement in citizen science projects has provided me the privilege of saying I have volunteered in order that future generations might have the same opportunities to enjoy birds and nature that I have had.

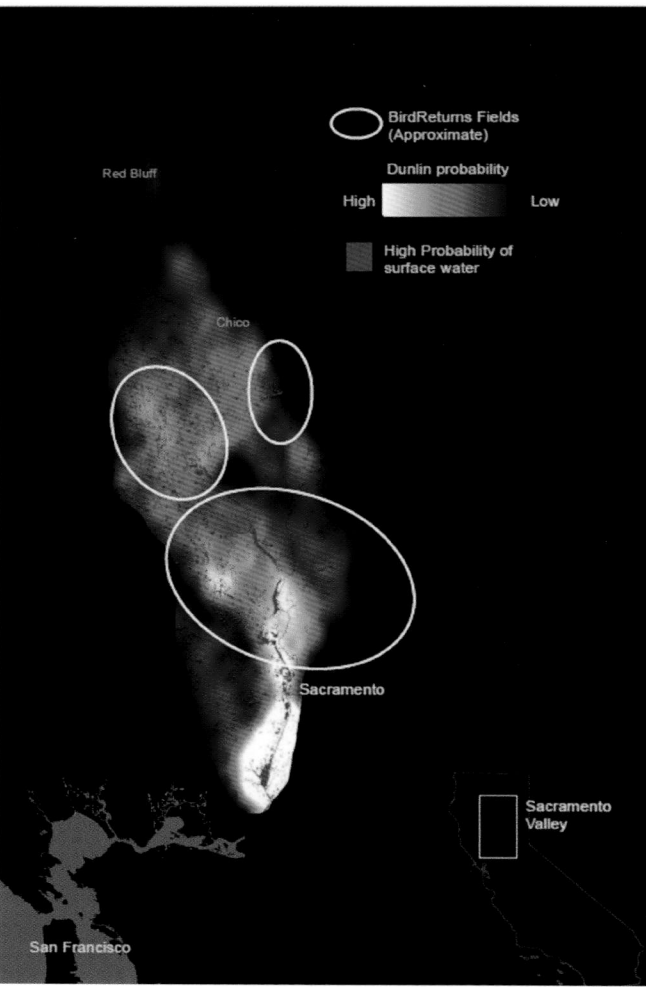

Figure 30.2. Dunlin (*Calidris alpine*) probabilities during spring migration on the Pacific Flyway as estimated with data from eBird, overlaid with the probability of surface water, as estimated using NASA Land-Sat data (Reiter et al. 2015). White ovals show the approximate areas where "pop-up" wetlands were created via the BirdReturns program of The Nature Conservancy. *The Nature Conservancy, Point Blue Conservation Science, Cornell Lab of Ornithology.*

ornithology as a scientific field, and their early works laid a foundation for much of what we know today. Amateurs continue to add to our collective understanding of birds through independent research and as citizen scientists.
- Today's professional ornithologists work in museums, universities, government agencies, nonprofit organizations, and the private sector. The positions they hold in these different sectors vary, require differing levels of experience and education, and offer a diversity of benefits.
- Positions in ornithology require keen observational skills, scientific literacy, an understanding of study design, data analysis and interpretation, and the ability to communi-

cate results effectively. They also require a suite of soft skills that can be gained through extracurricular activities.
- Students can gain experience by working with graduate students, professors, and other ornithological professionals through seasonal internships, paid and unpaid research opportunities, and volunteering. They can also get experience through classroom research and citizen science projects.
- Citizen science projects involve amateurs and hobbyists in the scientific process. They are a key resource for the study, planning, and implementation of avian research and natural resource management. Citizen science projects exist at many scales, some of which would be unavailable to researchers otherwise.

KEY MANAGEMENT AND CONSERVATION IMPLICATIONS

- Professional ornithologists are often involved in the management and conservation of bird populations. They collect and analyze data that can be used to inform management and policy decisions; they engage in on-the-ground management activities; and they educate and inform others about birds and bird conservation.
- Amateur ornithologists can also engage in management and conservation as independent researchers, citizen scientists, and volunteers. When they do, management decisions and activities at local scales can come together quickly with positive results for natural resources. Additionally, their contributions provide data at spatial and temporal scales not typically available to independent researchers.

DISCUSSION QUESTIONS

- Identify three aspects of your career goals that are important to you. How do these align with the variety of career tracks described in this chapter?
- Think of different ways in which you might acquire or improve "soft skills." How might these skills be beneficial to you in your career as a scientist or otherwise?
- Based on what you've seen in your classes, on campus, and in your community, imagine a citizen science project that could be carried out as part of an ornithology course. Discuss possible methods that might be employed to collect the data.

References

AACU (Association of American Colleges and Universities) (2013). It takes more than a major: Employer priorities for college learning and student success. http://www.aacu.org/liberaleducation/le-sp13/hartresearchassociates.cfm.

Anderson, W. L., and R. E. Duzan (1978). DDE residues and eggshell thinning in Loggerhead Shrikes. Wilson Bulletin 90:215–220.

Bailey, M. J. (1994). American women in science: A biographical dictionary. ABC-CLIO, Santa Barbara, CA.

Barrow Jr., M. V. (1995). Gentlemanly specialists in the age of professionalism: The first century of ornithology at Harvard's Museum of Comparative Zoology. In Contributions to the history of North American ornithology, W. E. Davis Jr. and J. A. Jackson, Editors. Nuttall Ornithological Club, Cambridge, MA.

Barrow Jr., M. V. (2000). A passion for birds: American ornithology after Audubon. Princeton University Press, Princeton, NJ.

Benson, M. (1986). Martha Maxwell: Rocky Mountain naturalist. University of Nebraska Press, Lincoln, NE.

Birkhead, T. R., and S. van Balen (2008). Bird keeping and the development of ornithological science. Archives of Natural History 35:281–305.

Bonney, R., C. B. Cooper, J. Dickinson, S. Kelling, T. Phillips, K. V. Rosenberg, and J. Shirk. (2009). Citizen science: A developing tool for expanding science knowledge and scientific literacy. BioScience 59:977–984.

Bonta, M. M. (1985). Graceanna Lewis: Portrait of a Quaker. Quaker History 74:27–40.

Bonter, D. N., and C. B. Cooper (2012). A process for improving data quality and a strategy for ensuring sustainability in a citizen science project. Frontiers in Ecology and Environment 10:305–307.

Bub, H. (1991). Bird trapping and bird banding. Cornell University Press, Ithaca, NY.

Burger, L., and B. D. Leopold (2001). Integrating mathematics and statistics into undergraduate wildlife programs. Wildlife Society Bulletin 29:1024–1030.

Burtt Jr., E. H., and W. E. Davis Jr. (2013). Alexander Wilson: The Scot who founded American ornithology. Harvard University Press, Cambridge, MA.

Cooper, C. B. (2014). Is there weekend bias in clutch-initiation dates from citizen science: Implications for studies of avian breeding phenology. International Journal of Biometeorology 58:1415–1419.

Cooper, C. B., W. M. Hochachka, and A. A. Dhondt (2012). The opportunities and challenges of citizen science as a tool for ecological research. In Citizen science: Public participation in environmental research, J. L. Dickinson and R. Bonney, Editors. Cornell University Press, Ithaca, NY.

Cooper, C. B., J. S. Shirk, and B. Zuckerberg (2014). The invisible prevalence of citizen science in global research: Migratory birds and climate change. PLoS ONE 9:e106508.

Cooper, C. B., R. Bailey, and D. Leech (2015). The role of citizen science in studies of avian reproduction. In Nests, eggs, and incubation, C. Deeming and J. Reynolds, Editors. Oxford University Press, Oxford, UK.

Coulson, J. (2003). Ornithology and ornithologists in the twentieth century. In A concise history of ornithology, M. Walters, Editor. Yale University Press, New Haven, CT.

Danielsen, F., N. D. Burgess, P. M. Jensen, and K. Pirhofer-Walzl (2010). Environmental monitoring: The scale and speed of implementation varies according to the degree of people's involvement. Journal of Applied Ecology 47:1166–1168.

Dunn, P. O., and D. W. Winkler (1999). Climate change has affected the breeding date of tree swallows throughout North America. Proceedings of the Royal Society of London B: Biological Sciences 266:2487–2490.

Farber, P. L. (1982). The emergence of ornithology as a scientific discipline: 1760–1850. D. Reidel, Dordrecht, Netherlands.

Foster, M. S. (1969). Synchronized life cycles in the Orange-crowned Warbler and its mallophagan parasites. Ecology 50:315–323.

Geiger, R. L. (1986). To advance knowledge: The growth of American research universities. Oxford University Press, Oxford, UK.

Hiller, S. E., and A. Kitsantas (2015). Fostering student metacognition and motivation in STEM through citizen science programs. In Metacognition: Fundaments, applications, and trends, A. Peña-Ayala, Editor. Springer International, New York.

Hurlbert, A. H., and Z. Liang (2012). Spatiotemporal variation in avian migration phenology: Citizen science reveals effects of climate change. PLoS ONE 7:e31662.

Johnston, A., D. Fink, M. D. Reynolds, W. M. Hochachka, B. L. Sullivan, N. E. Bruns, E. Hallstein, M. S. Merrifield, S. Matsumoto, and S. Kelling (2015). Abundance models improve spatial and temporal prioritization of conservation resources. Ecological Applications 25:1749–1756.

Joseph, L. (2011). Museum collections in ornithology: Today's record of avian biodiversity for tomorrow's world. Emu 111:i–xii.

Kelling, S., J. Yu, J. Gerbracht, and W. K. Wong (2011). Emergent filters: Automated data verification in a large-scale citizen science project. In Proceedings of the IEEE eScience 2011 Computing for Citizen Science Workshop, Ithaca, NY.

Kendall, W. L., and W. R. Gould (2002). An appeal to undergraduate wildlife programs: Send scientists to learn statistics. Wildlife Society Bulletin 30:623–627.

Knudsen E., A. Lindén, C. Both, N. Jonzén, F. Pulido, N. Saino, W. J. Sutherland, et al. (2011). Challenging claims in the study of migratory birds and climate change. Biological Reviews 86:928–946.

Kubik, G. H. (2009). Projected futures in competency development and applications: A Delphi study of the future of the wildlife biology profession. PhD diss., University of Minnesota, Minneapolis, MN.

La Sorte, F. A., D. Fink, W. M. Hochachka, J. P. DeLong, and S. Kelling (2013). Population-level scaling of avian migration speed with body size and migration distance for powered fliers. Ecology 94:1839–1847.

La Sorte, F. A., D. Fink, W. M. Hochachka, A. Farnsworth, A. D. Rodewald, K. V. Rosenberg, B. L. Sullivan, et al. (2014a). The role of atmospheric conditions in the seasonal dynamics of North American migration flyways. Journal of Biogeography 41:1685–1696.

La Sorte, F. A., M. W. Tingley, and A. H. Hurlbert (2014b). The role of urban and agricultural areas during avian migration: An assessment of within-year temporal turnover. Global Ecology and Biogeography 23:1225–1234.

Lebeau, C. W., J. L. Beck, G. D. Johnson, and M. J. Holloran (2014). Short-term impacts of wind energy development on greater sage-grouse fitness. Journal of Wildlife Management 78:522–530.

Leiser, J. K., and E. Reilley (2015). Service learning enhances classroom experiences but does not inflate grades: A lesson from courses in ecology and biology. Journal for Civic Commitment 23:1–16.

Matsunaga, A., A. Mast, and J. A. B. Fortes (2014). Reaching consensus in crowdsourced transcription of biocollections information. *In* 10th IEEE International Conference on e-Science, Guarujá, SP, Brazil.

Mayfield, H. F. (1979). The amateur in ornithology. The Auk 96:168–171.

McCleery, R. A., R. R. Lopez, L. A. Harveson, N. J. Silvy, and R. D. Slack (2005). Integrating on-campus wildlife research projects into the wildlife curriculum. Wildlife Society Bulletin 33:802–809.

Millenbah, K. F., and J. J. Millspaugh (2003). Using experiential learning in wildlife courses to improve retention, problem solving, and decision-making. Wildlife Society Bulletin 31:127–137.

Moksnes, A., and E. Roskaft (1995). Egg-morphs and host preference in the Common Cuckoo (*Cuculus canorus*): An analysis of cuckoo and host eggs from European museum collections. Journal of Zoology (London) 236:625–648.

Phillips, T., and J. Dickinson (2009). Tracking the nesting success of North America's breeding birds through public participation in NestWatch. Proceedings of the 4th International Partners in Flight Conference: Tundra to Tropics, pp. 633–640.

Reiter, M. E., N. Elliott, S. Veloz, D. Jongsomjit, C. M. Hickey, M. Merrifield, and M. D. Reynolds (2015). Spatio-temporal patters of open surface water in the Central Valley of California 2000–2011: Drought, landcover and waterbirds. Journal of the American Water Resources Association 51:1722–1738.

Sagarin, R., and A. Pauchard (2012). Observation and ecology: Broadening the scope of science to understand a complex world. Island Press, Washington, DC.

Sherman, A. R. (1924). Animal aggregations: A reply. The Condor 26:85–88.

Shirk, J. L., H. L. Ballard, C. C. Wilderman, T. Phillips, A. Wiggins, R. Jordan, E. McCallie, et al. (2012). Public participation in scientific research: A framework for deliberate design. Ecology and Society 17:29.

Snyder, T. D., Editor (1993). 120 years of American education: A statistical portrait. National Center for Education. US Department of Education. Office of Educational Research and Improvement, Darby, PA.

Suarez, A. V., and N. D. Tsutsui (2004). The value of museum collections for research and society. BioScience 54:66–74.

Sullivan, B., C. Wood, M. Iliff, R. Bonney, D. Fink, and S. Kelling (2009). eBird: A citizen-based bird observation network in the biological sciences. Biological Conservation 142:2282–2292.

Swengel, A. B. (2015). Massive monarchs: The prairie chicken effect. Wisconsin Entomological Society Newsletter 42:6–7.

Swengel, S. R., and A. B. Swengel (1999). Correlations in abundance of grassland songbirds and prairie butterflies. Biological Conservation 90:1–11.

Turner, W. (1544). Avium praecipuarum quarum apud Plinium et Aristolelem mentio est, brevis et succincta historia. Gymnicus, Cologne, Germany.

Weidensaul, S. (2007). Of a feather: A brief history of American birding. Harcourt, Orlando, FL.

Wiggins, A., and K. Crowston (2011). From conservation to crowdsourcing: A typology of citizen science. *In* Proceedings of the 44th annual Hawaii International Conference on Systems Sciences, Koloa, HI.

Wiggins, A., C. Lagoze, W. K. Wong, and S. Kelling (2014). A sensor network approach to managing data quality in citizen science. *In* Second AAAI Conference on Human Computation and Crowdsourcing, Pittsburg, PA.

Winker, K. (1996). The crumbling infrastructure of biodiversity: The avian example. Conservation Biology 10:703–707.

Winker, K. (2004). Natural history museums in a post biodiversity era. BioScience 54:455–459.

CHAPTER 31

Fundamental Methods

Geoffrey Geupel, John Dumbacher, and Maureen Flannery

Because birds are conspicuous, exceptionally vocal, and found most everywhere, they have long been used to teach the principles of ecology, evolution, and animal behavior, and they continue to be model organisms for advancing and understanding conservation and natural resource management (Martin 1992, 1995, Bock 1997, Piatt et al. 2007). Methods for the study of birds have received the most attention of any vertebrate in both scientific and popular literature and continue to evolve at a rapid rate. In this chapter, we describe some of the fundamental techniques used widely by researchers, students, conservation practitioners, and citizen scientists. We begin with standardized methodologies commonly used in the field for both research and monitoring. Following field methods we discuss common laboratory techniques, which focus on current advances in molecular genetics. Museum collections are a permanent historical record of individual birds at a specific location during a specific time and provide the foundation for the systematics of birds. We explore various modern uses of museum specimens and their relevant data. Finally, we present some examples of how data from these techniques are used to address conservation and management issues.

FIELD METHODOLOGIES

Many coordinated efforts to develop standardized field methods have been described for landbirds (Ralph et al. 1993, 1995, 1996, Geupel and Warkentin 1995, Bibby et al. 2000, Ralph and Dunn 2004), nonbreeding waterbirds (Loges et al. 2014), shorebirds (Bart et al. 2005), seabirds (Ainley and Boekelheide 1990, Slater 1995, CCAMLR 2014), marsh birds (Conway 2011), and many individual species of conservation concern (e.g., Halterman et al. 2015). Standardized approaches are critical, as they provide data that are comparable among studies, both across regions and through time.

Journal Keeping

Understanding the natural history of birds, "where they are, where they live, what they eat, why they behave the way they do," has intrigued students of natural sciences and amateurs alike for centuries (Herman 2002). Careful observational studies of birds has long been a cornerstone of human understanding of ecological systems and of conservation science, and a guiding force in resource management. Joseph Grinnell, the founder and director of the Museum of Vertebrate Zoology at the University of California, Berkeley, from 1908 to 1939, made extraordinary contributions to ornithology and conservation biology (Bock 1997). His legendary, thorough technique for keeping field notes and species accounts in a field journal, known as the "Grinnell method," is still taught and practiced today thanks to the efforts of Steve Herman (1986) (fig. 31.1). The practice of recording field observations in a journal not only substantially increases a student's ability to observe and interpret behavior but also increases our knowledge of a species' autecology. Furthermore, the art of sketching in the field is resurging, and strengthens the knowledge and appreciation of natural history and journaling (Laws 2016). The natural history of species and individuals can vary depending on habitat, ecological conditions, location, and various other factors (Nur et al. 2008). Unfortunately, recent emphasis on statistics, modeling, and other factors at many institutions of higher learning is limiting field opportunities for students. Consequently, important life history information is often neglected in current literature, and conservation actions and species management recommendations suffer accordingly (Tewksbury et al. 2014).

Visual Aids

The most basic tools of field ornithology are binoculars and spotting scopes. Binoculars are used for just about every

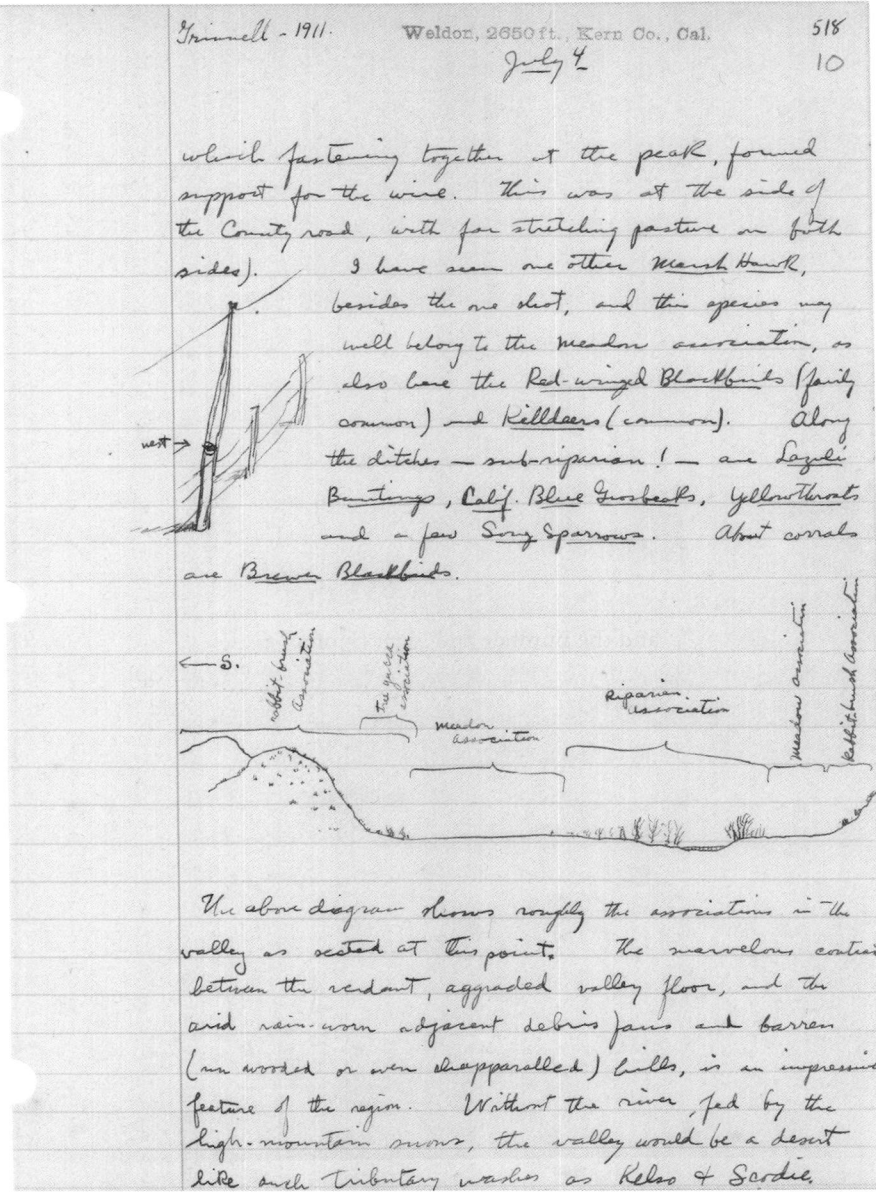

Figure 31.1. The Grinnell Method. Each journal page must include the writer's name, date including year, title, locality, and page number. Other information recorded in the journal can include species lists, habitat descriptions, behavioral observations, and information about collections made. *Photo used with the permission of the Museum of Vertebrate Zoology, University of California, Berkeley.*

type of observation, while spotting scopes are preferred for larger birds (e.g., raptors, shorebirds, waterfowl, and seabird colonies) that are more stationary and/or at greater distance from the observer. The types and cost of binoculars and scopes are highly variable, and unbiased information comparing models is regularly published in popular bird magazines (e.g., *Living Bird*) and online (Cooper 2016). In general, most ornithologists prefer medium-quality or better binoculars (which cost over $1,000) that are waterproof. Magnification should be at least seven times normal and no more than 10. High magnification can be a problem, as the field of view is more limited, which can make it difficult to find and track a moving bird. High-powered binoculars also

reduce the amount of light, are more difficult to hold steady, and subject the user to eye fatigue and headaches after extended use. The diameter of the objective lens should be at least 35 mm, as it impacts brightness and color detectability of the object in view.

Checklists

Many people become interested in the study of birds and natural history from watching and keeping track of the species they observe. Keeping checklists of birds is a simple and effective way to learn and study birds. A standardized checklist, limited by time and location, can become a useful technique for increasing our understanding of bird ecology and conservation. Checklists of birds and other species are often used in a **bioblitz**, an intense period of biological surveying that attempts to record all the living species within a designated area. The presence or absence of certain species can provide insight into the ecological integrity of an area and is useful for conservation planning (Chase and Geupel 2005). A recent development is the web- and app-based checklist tracking system called eBird, which allows birders to keep track of their own observations and at the same time share millions of observations across the globe (Sullivan et al. 2009, Wood et al. 2011). Exponential increases in data quantity and quality (e.g., number of individuals of each species reported) and the evolution of eBird into a cooperative partnership of citizen scientists and experts has allowed the production of remarkable high-resolution visualizations and novel insights into the movement of birds across the western hemisphere (Sullivan et al. 2014).

Standardized Bird Count Surveys

Many field techniques for use by professionals, students, volunteers, and citizen scientists have been developed to survey birds. These methods provide metrics such as population size and density, population trends, relative abundance, range, occupancy, habitat relationship, species richness, and species diversity (see reviews in Ralph and Scott 1981, Verner 1995).

Breeding bird atlases (atlasing) are typically one-time survey efforts engaging citizen scientists to inventory the breeding bird population of a particular geographic area (normally selected using political boundaries such as a county or state, divided into blocks). If the effort is repeated, changes in distribution can be documented. Other atlas projects have also included surveys during the nonbreeding season (e.g., Unitt 2004). With the aid of the Internet, novel approaches are being devised to make atlasing a more cost-effective program that will be capable of fulfilling multiple objectives (Tulloch et al. 2013).

Area search surveys are similar to atlases in that they are well-suited for groups of volunteers (Hewish and Loyn 1989). They differ in that time and area are constrained by habitat instead of political boundaries to limit observer bias (Ralph et al. 1993). Many waterbird and eBird observations/checklists may be considered area search surveys if the observer notes the area covered, the amount of effort (time), and the number and identity of all bird species encountered (Sullivan et al. 2009). Surveys that detect multiple focal species (fig. 31.2) in a targeted habitat type can aid land managers in planning within an adaptive management framework (Chase and Geupel 2005, Alexander et al. 2007).

Territorial mapping (also called "spot" or "flush" mapping) has been used by hundreds of volunteers annually in North

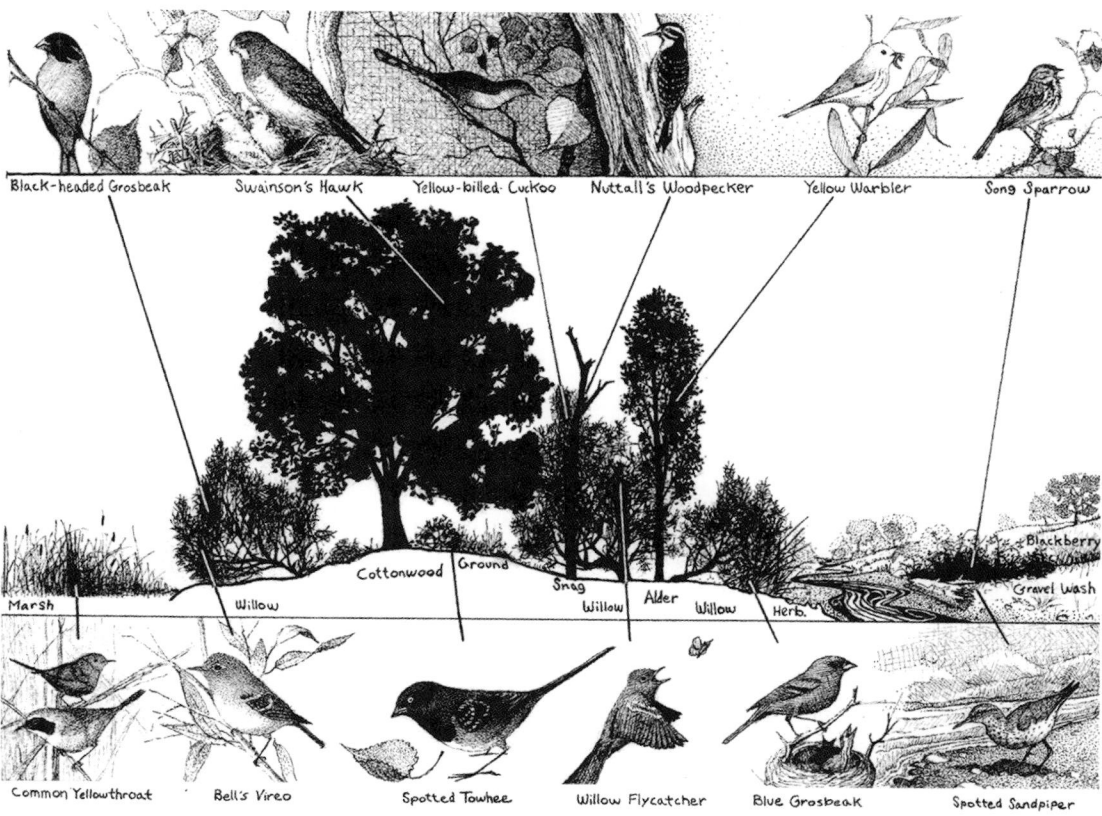

Figure 31.2. California riparian focal species that represent a range of habitat seral stages and elements. *From RHJV 2004, Chase and Geupel 2005. Illustration © Zac Denning.*

America (Resident Bird Counts) and Britain (Common Bird Census, Bibby et al. 2000) for decades. The technique involves repeated visits by observers to specific plots to record the locations of singing and counter-singing individuals of the same species on a map. These data allow observers to delineate individual bird territories. Other variations include flushing singing males, uniquely color-banding individuals and resighting, and locating nests. Results include breeding status and density, habitat use, and phenology. This technique is not applicable for quiet, nonterritorial, or wide-ranging species or species whose territories are large compared with the study area (Verner 1995). It is recommended that each plot be visited a minimum of eight visits per breeding season. While subject to considerable analyst and observer variability (Verner and Milne 1990), better observer training and GIS mapping (Witham and Kimball 1996) may reduce these biases. When used in conjunction with color-banding of individuals, absolute densities, adult survivorship, site fidelity, fecundity, and recruitment information may be determined (Geupel and DeSante 1990, Porzig et al. 2012, 2014).

Raptor counts are standardized counts of migrating raptors and other soaring species. Often performed by trained volunteers at geographic locations where migratory routes form bottlenecks, these counts have provided long-term population trend estimates in eastern North America (Bednarz et al. 1990, Allen et al. 1996) and Veracruz, Mexico (Ruelas Inzunza et al. 2000), as well as insight into the effects of weather on raptor movement in western North America (Hall et al. 1992). The Peregrine Fund recently published pioneering research that involved the development of new "low-tech and relatively inexpensive methods" for detecting and studying little-known tropical forest raptors (Whitacre 2010). Motion-sensitive trail cameras placed at bait piles can document eagle and other wildlife use, and they are being used at 150 remote sites across the Appalachians, from Maine to Alabama (http://www.appalachianeagles.org).

Line transects involve a skilled observer moving along a predetermined route of a known distance. Transects are normally used in open habitats, such as grasslands, irrigated pasture, marine or lake habitats (for seabirds and waterbirds), or occasionally on roads (for raptors), where an observer has excellent visibility and can move at a consistent rate (Emlen 1971, Spear et al. 1992). Recording distance of sighted birds in relationship to the line using two observers can create more robust data and analytical techniques.

Point count surveys, such as the Breeding Bird Survey (BBS), a volunteer-based continent-wide program (Sauer et al. 2013), are by far the most widely used and well-described method for estimating indices and trends in bird abundance, density, richness, and diversity, and for determining habitat associations (see Ralph and Scott 1981, Ralph et al. 1995). Most observers and monitoring programs now estimate exact distance to all bird detections (visual and audible) to measure observer bias, determine effective detection distances, and estimate abundance (Buckland et al. 2001, Rosenstock et al. 2002) (fig. 31.3). However identification of all songs and calls while estimating distance consistently requires extensive in-the-field training; fixed-radius counts have been found to be more user-friendly and better constitute a satisfactory index to bird abundance (Hutto 2016). For greater inference, locations of survey stations should be distributed randomly using a spatially balanced grid system, such as the generalized random tessellation stratified survey design (Stevens and Olsen 2004). Many analytical approaches now exist that aid in the analysis of distance sampling (Royle and Nichols 2003, Thomas et al. 2010) and occupancy modeling (Welsh et al. 2013, Hayes and Monfils 2015). Using these data models, current and future bird distribution, occurrence, and occupancy in relation to habitat, climate, and other multiple factors have been pursued to guide land management and conservation efforts (Verner et al. 1986, MacKenzie et al. 2006, Wiens et al. 2009, Stralberg et al. 2009, Veloz et al. 2015).

Autonomous recording units (ARUs) are sound-recording devices programmed to record sounds of birds and other vocal animals. They are growing in popularity for situations in which trained observers are limited, and they may have advantages over human-conducted surveys in that the records are permanent, allow 24-hour data collection and automated species identification, and may minimize observer bias (Acevedo and Villanueva-Rivera 2006, Celis-Murillo et al. 2009). ARUs, when compared with human observers, performed equally or better depending on species and habitat type (see summary in Alquezar and Machado 2015) although their cost-effectiveness remains unclear (Hutto and Stutzman 2009). For seabirds breeding on remote islands, ARUs were effective at monitoring breeding presence and reproductive success after predator removal (Buxton and Jones 2012).

Determining Vital Rates

To interpret population changes and make informed conservation decisions, data on vital rates—productivity (birth rate), survivorship (death rate, recruitment rate), and movement (dispersal, immigration and emigration), as well as where birds are limited during their life cycle—are required. Through the study of these primary population parameters at different times of the year, investigators may be forewarned of reductions in abundance before the actual declines are observed (Pienkowski 1991, Temple and Wiens 1989). These vital rates may also be used to calculate bird population growth rates (Lloyd et al. 2005) and population viability analysis

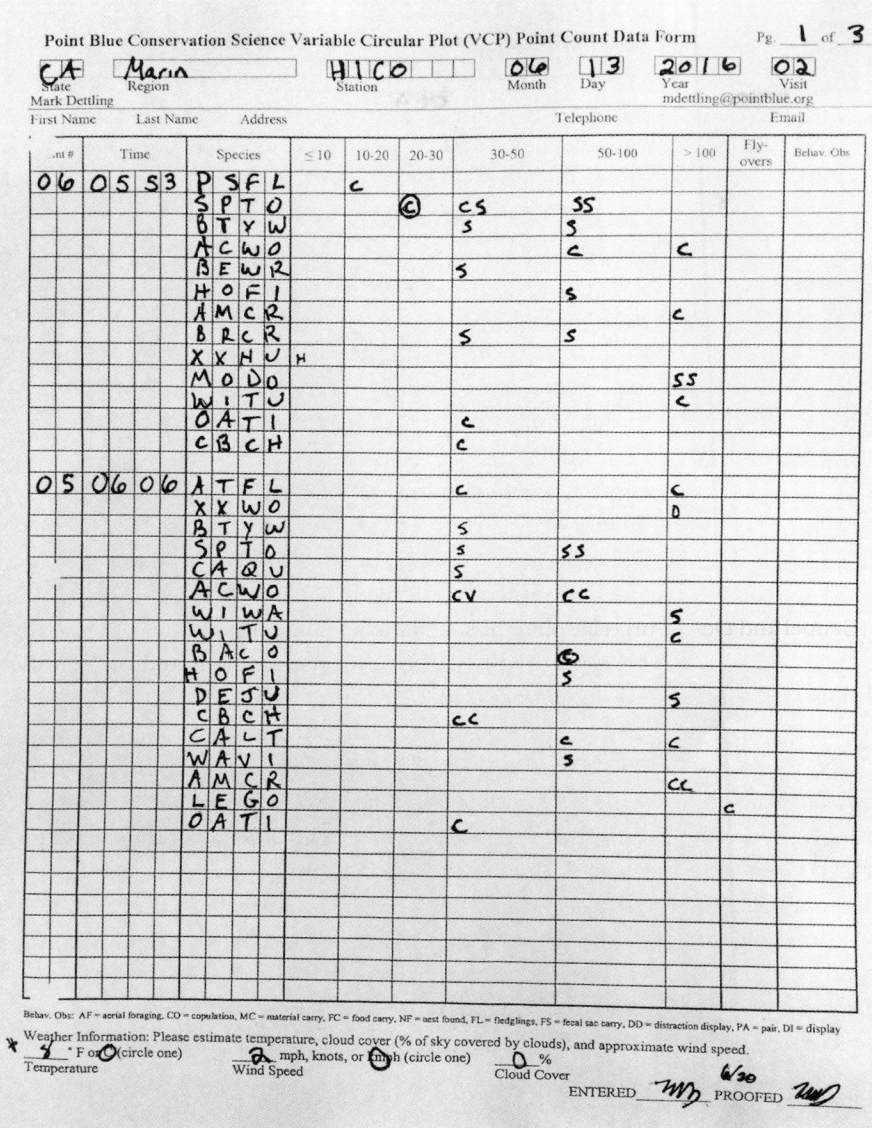

Figure 31.3. An example of a point count data form that records species (four-letter code), type of detection (call, song, visual), and distance. *Courtesy of Point Blue Conservation Science.*

(Shaffer 1990). Determining vital rates normally requires the capture and marking of wild birds.

Capturing and marking wild birds requires training and a permit (see permits at the end of the chapter). However it can provide a wealth of life history information, including longevity, breeding condition, molt cycles, plumage, body condition, energetics, range (e.g., stable isotopes), and trophic interactions (Ralph et al. 1993, Pyle 1997). There are many methods for capturing wild birds, including using mist nets (fig. 31.4), bow nets, cannon nets, funnel traps, nooses, decoys, and others (for review see Bub 1991). Capturing, banding, and aging land birds in an array of mist nets provides indices of productivity (DeSante and Geupel 1987, DeSante et al. 2015), within-season and among-year survivorship rates (Clobert et al. 1987, Peach and Baillie 1991, Chase et al. 1997),

return rates (Johnson and Geupel 1996), migration (Dunn and Hussell 1995), overwinter survivorship, and habitat preferences (Latta and Faaborg 2009; box on page 938). These demographic indices have been the focus of large continental programs in both Europe and North America (Baillie 1990, DeSante 1992). Furthermore, mist nets can be implemented in a way that minimizes injury and risk to captured birds (Spotswood et al. 2012). Targeted captures of individuals may also be useful for other purposes, such as color-banding (to identify age classes or individuals without recapture), feather and blood sampling, and applying electronic tags that can track movement (GLS, GPS, and PIT tags described below).

Nest locating and monitoring can provide important life history information that is lacking for many bird species, especially those with cryptic nest locations that are subject

Figure 31.4. Sofia Trujillo, a field biologist intern from El Salvador, removes an Allen's Hummingbird (*Selasphorus sasin*) from a mist net in coastal California. *Photo © Mark Dettling, Point Blue Conservation Science.*

to high predation rates, such as open-cup nesting passerines and secretive marsh species. Productivity (typically number of fledglings per pair, but see Mayfield 1975 for sources of bias in nests found at different stages) is measured directly by locating nests and monitoring their outcome. Nests are often monitored in specific plots, and the associated vegetation around the nest is assessed, enabling correlation of breeding productivity with habitat conditions and/or management practices (Martin 1992). Methods for locating the nests of open-cup nesting passerines using behavioral cues and determining the outcome are described in Martin and Geupel (1993). Grassland species can be found by careful observation of flushing females using rope-dragging (Davis 2003). Locating and monitoring nests may cause nest failure (Gotmark 1992), but this can be minimized or eliminated by following specific guidelines (Martin and Geupel 1993). Nests of a given species become easier to locate after learning female calls and behaviors and after developing a search image, a pattern that is learned through experience.

Cavity-nesting bird species, including primary (most woodpeckers), secondary (flycatchers, bluebirds, and wood ducks), and pendulant nesters (wrens, orioles, bushtits), are often easier to locate but more difficult to monitor. Burrow-nesting species such as owls, seabirds, and others can also be difficult to monitor without disturbing the integrity of the nest structure. New technologies using fiber optics have increased the accuracy and ease of monitoring cavity and burrow nests (Boland and Phillips 2005, Geleynse et al. 2016).

Most shorebird nests are relatively easy to find using behavioral cues in known habitats (Bart and Johnston 2012).

Waterfowl are typically found by flushing incubating hens using chains or heavy ropes dragged slowly behind all-terrain vehicles (Ball et al. 1995). Many historic and current researchers monitor seabird colonies using blinds and maps of colonies created before birds begin breeding (Ainley and Boekelheide 1990, Slater 1995). More recently, unmanned aircraft (drones) have been used for monitoring colonial seabirds (Sardà-Palomera et al. 2012) and waterbirds (Vas et al. 2015). The use of video and remotely operated cameras has revolutionized nest monitoring of remote seabird colonies (Southwell and Emmerson 2015). When used to study landbirds, this method allows the identification of the nest predators (Ribic et al. 2012) but may also influence nesting success rates (Richardson et al. 2009).

Tracking the movement and dispersal of birds, an extremely mobile species, has intrigued and confounded researchers for centuries. Recent technological advances have greatly increased our understanding of when and where birds go. Adult breeding and winter philopatry for most species is thought to be relatively high and is often used to estimate adult survival (Blakesley et al. 2006, DeSante et al. 2015). Poor breeding success, habitat quality, and food availability can influence an individual's philopatry (Johnson and Sherry 2001, Blakesley et al. 2006). Estimates of natal dispersal distances and juvenile recruitment rates in open populations are strongly influenced by the extent and shape of the areas sampled (Cooper et al. 2008). New tracking technologies (discussed below) may help in the determination of site fidelity, dispersal patterns, migration routes, and use of feeding stations.

In well-studied nonmigratory (resident) populations using marked individuals (using a unique combinations of colored leg bands), natal dispersal distances, estimated using known dispersal events (Baker et al. 1995), may have been influenced by inbreeding avoidance (Daniels and Walters 2000). Woltmann et al. (2012) characterized bird natal dispersal for the first time using molecular genetic parentage analyses. Using stable-hydrogen isotope ratios in feathers, Studds et al. (2008) suggest that habitat occupancy during the first nonbreeding season helps determine the latitude at which a species of Neotropical-Nearctic migratory bird may breed throughout its life. The use of feather samples collected at banding stations distributed across North America have provided novel insight as to where migrant songbirds breed and winter and the routes in between (Ruegg et al. 2014; box on page 939). Passive integrated transponders, or PIT tags, are lightweight (0.1 grams) tags that are attached to leg bands of birds. Each tag transmits a unique identification number to a reading device that contains a circuit board and antenna (radio frequency identification, RFID) built into a "wired" bird feeder or other site of regular visitation. When a bird with a tag vis-

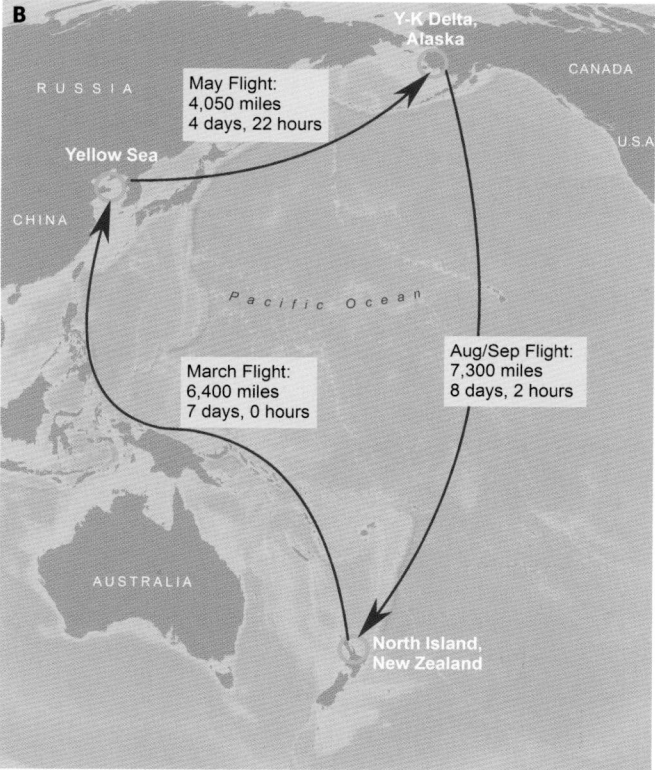

Figure 31.5. A, Female Bar-tailed Godwit (*Limosa lapponica*) in flight. *Photo © Steve N. G. Howell. B,* The remarkable transcontinental journey of a Bar-tailed Godwit, tracked by satellite transmitter crossing the Pacific Ocean in fall and spring migration. *Figure by Ben Sullender/ Audubon Alaska.*

The migration of birds and how they navigate and energetically survive flights of thousands of kilometers are some of "the most extraordinary migratory feats on the planet" (Able 1999, Bonter et al. 2009, Deluca et al. 2015). However, understanding of migratory birds' year-round ecology, stopover use, and movement remains patchy despite recent advances (Carlisle et al. 2009, Faaborg et al. 2010). For decades, radio devices have been used to track migrating birds, typically on larger species that could handle radio packs and for which their general location is accessible (less than 5 km away). More recently, the miniaturization of batteries has allowed smaller species, such as small shorebirds and passerines, to be tracked (Warnock and Takekawa 2003). The use of satellite transmitters that upload data to satellites at regular intervals was instrumental in the reintroduction of California Condors (Walters et al. 2010), linking curlews from breeding to wintering grounds (Page et al. 2014), and monitoring extreme long-distance (transcontinental) movements of godwits (Battley et al. 2012) (fig. 31.5). More recently, transmitters small enough to be placed on thrushes (White et al. 2005) are now miniaturized and can be used on new world warblers (Hallworth and Marra 2015). Automated radiotelemetry arrays can provide insights into passerine stopover ecology at large scales (Woodworth et al. 2015). Novel use of GPS data loggers on free-ranging birds and environmental information recorded by unmanned aerial systems are being used to investigate behavior and space use by raptors (Rodríguez et al. 2012).

The use of extremely lightweight geolocation devices, including global location sensors (GLS tags), that record sunlight at regular intervals and allow the location of a bird to be calculated (day length varies with latitude and solar noon with longitude) has revolutionized the ability to track long-distance migrant song birds. However the method requires the recapture of the individual to download data from the tag (Stutchbury et al. 2009). While exact location is not known, general breeding and winter provenance can be determined from relatively small sample sizes and has been used to track sparrows migrating from California to Alaska (Seavy et al. 2012), and six Blackpoll Warblers (averaging 12 grams) that were estimated to have traveled, on average, over 2,500 kilometers in 62 hours (Deluca et al. 2015).

Correlating rates of birds captured during migration with trends in breeding populations has been a focus of Bird Studies Canada for decades (Dunn and Hussell 1995) and of other select sites (islands, peninsulas, and habitat bottlenecks) where the number of birds arriving daily can be estimated (Pyle et al. 1994, Dunn et al. 2004, Carlisle et al. 2005). Effects of weather at any particular site make interpretation of changes in numbers difficult. Coordination among banding

its such a site, the bird's identity is recorded, along with the date and time of the visit (Bridge and Bonter 2011). Innovative researchers in Antarctica have implanted PIT tags and placed time-depth recorders on penguins. The penguins are then directed by fencing over automated weighbridges near breeding colonies to measure foraging trip duration, foraging strategies, meal size, adult body mass, and nesting success (Ballard et al. 2001, Ainley et al. 2004, Ballard et al. 2010).

sites, data sharing, and the use of weather and marine radar may help in the future.

The use of weather surveillance radar to detect large mass movements of migrating birds has been well documented (Gauthreaux and Belser 2003) and explains the impacts of wind patterns (Van Doren et al. 2015) and stopover habitat (Packett and Dunning 2009). More recently, the Cornell Laboratory of Ornithology has made efforts to direct volunteers to migration stopover sites during these mass movements to identify the species at these stopover sites (http://birdcast.info /research/).

Other Considerations

Study design is crucial for implementing a successful field investigation of birds that will be publishable and/or applicable to management in the future. It is highly recommend that a field ornithologist consult with a biostatistician/quantitative ecologist on such subjects as observer bias (Bart et al. 2004), inference (randomization), number of sample units, sample size, and study duration in relation to the hypotheses and objectives of the study before beginning any fieldwork. Study design, and the selection of methods to be used, varies depending on the goals that have been set by the investigator(s) and/or the program sponsor. It is important to understand the types of objectives that can be addressed from the implementation of each technique and the number of years required to obtain results. Nur et al. (1999) link field methods described above to variables measured and years needed to meet a specific objective. Latta et al. (2005) link basic questions with monitoring objectives and field techniques for North American landbirds. Lambert et al. (2009) offer 10 steps to increase the conservation value of an avian monitoring program.

Many have debated the difference between research and monitoring. Typically the difference is that research is used to identify mechanisms or options (hypothesis testing), while monitoring typically focuses on making better management decisions in an adaptive management context. Knowing the appropriate spatial and temporal scales at which to collect data and understanding how the information will be used (objectives) are the most important considerations. Hutto and Belote (2013) recommend classifying monitoring into four categories—surveillance, implementation, effectiveness, and ecological—which are dependent on the objectives. Adaptive frameworks have been proposed that allow monitoring programs to evolve iteratively as new information emerges and research questions change (Lindenmayer and Likens 2009).

Setting up a successful bird monitoring program requires detailed protocols, data collection, and management (Oakley et al. 2003). Objectives of avian monitoring programs have

INTERNATIONAL AVIAN MONITORING THROUGH A COLLABORATIVE NETWORK OF CONSTANT-EFFORT MIST NETTING STATIONS

The Institute for Bird Populations (http://www.birdpop .org/index.php) coordinates two innovative continent-wide demographic monitoring programs in North America. The Monitoring Avian Productivity and Survivorship (MAPS) program is a collaborative effort among public agencies, nongovernmental groups, and individuals that consistently run mist netting stations in a coordinated and standardized manner. Since 1989, more than 1,200 MAPS stations spread across nearly every state and Canadian province have collected more than two million bird capture records. The Monitoreo de Sobrevivencia Invernal (MoSI) program is a similar collaborative effort that focuses on the mark and recapture of Neotropical migrant birds on their wintering grounds. Since 2002, the MoSI program has operated more than 200 stations in 15 countries. These programs provide insights into important conservation questions such as

* What factors drive avian population declines?
* Where are problems most acute, on the breeding or nonbreeding grounds?
* What drives differences in trends among particular regions or habitats?
* What are the relationships among population change and weather, climate, and habitat loss?
* What can we do to reverse declines?

been well described (Nur et al. 1999). It is recommended that existing and new programs closely align with current bird conservation and management priorities (NABCI 2007). Monitoring is also fundamental to the adaptive management that many public and private land management agencies are mandated to pursue (Salafsky et al. 2001). While most would agree there is great value in setting explicit hypotheses for research or objectives for monitoring, many long-term avian programs have provided benefits that were not explicitly stated at their origins. An example is the use of Breeding Bird Survey data: originally designed to detect continental trends in bird populations (Sauer et al. 2013), it is currently used to set continental population objectives (Rich et al. 2004) and

WHERE BIRDS GO AND HOW WE KNOW

Combining field methods (capture and release of wild birds) with laboratory techniques (molecular genetics) can provide important insights into a long-distance migrant songbird's life history. Nonlethal feather samples collected from birds captured at MAPS (breeding) and MoSI (wintering) stations were used to obtain high-resolution genetic markers. These markers provided novel information about where species are wintering and breeding and the routes they use in passage (Ruegg et al. 2014; see figure).

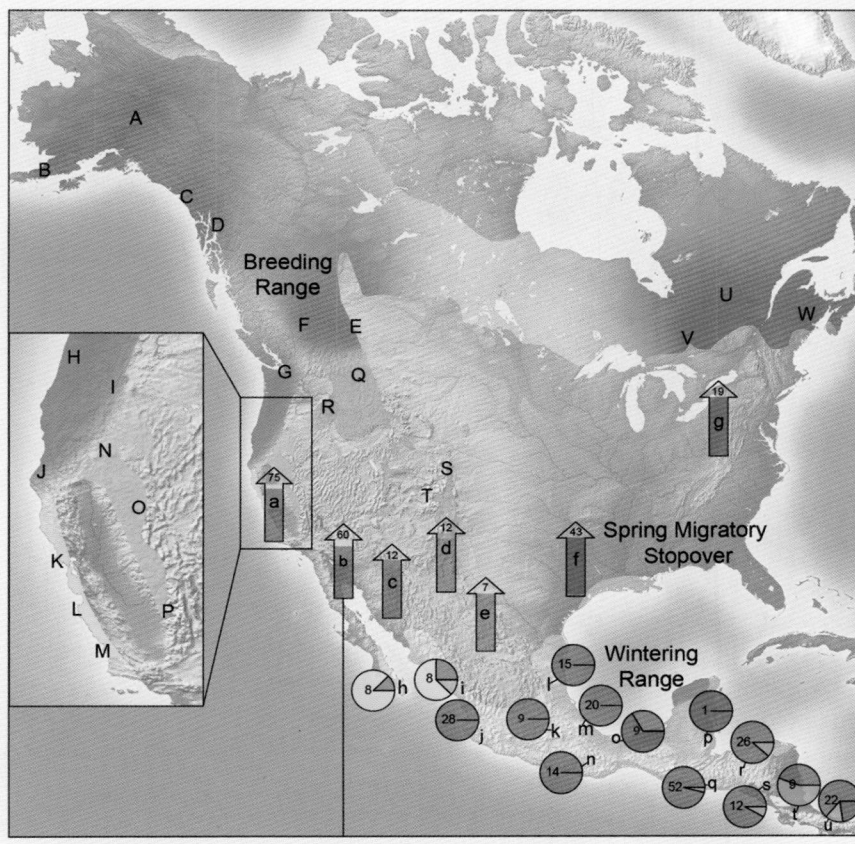

Distinct subpopulations of Wilson's Warbler (different colors), with migratory routes based, in part, on feather samples collected at MAPS and MoSI stations. *Figure from Ruegg et al. 2014. This figure was created by Eric Anderson and Kristen Ruegg using genetic data attained from bird feathers collected by hundreds of volunteer bird banders.*

predict bird species vulnerability to future climate change (Langham et al. 2015, Gorzo et al. 2016). Another example—banding data designed to discover where birds go—is now being used to determine vital rates of 158 North American landbird species (DeSante et al. 2015). A robust long-term avian monitoring program should direct future research and conservation efforts and provide insight into future unknown questions (e.g., Porzig et al. 2012). The use of standard methodologies and best data management practices (see below) that allow multiple investigators and amateurs to combine and share data have allowed new data-intensive approaches (Kelling et al. 2009, Sullivan et al. 2009, 2014), including mod-eling changes in bird density into the future in relation to climate change (Veloz et al. 2015).

Data Storage and Management

Best data management practices for field data include a plan for daily data entry, proofing, curation, and description of the data (metadata). The Avian Knowledge Network (AKN, http://www.avianknowledge.net/) is a collaborative partnership-based network that pursues the conservation of birds and their habitats through the curation, distribution, and sharing of data based on best data management practices (Martin and Ballard 2010). Using data centers (nodes) that

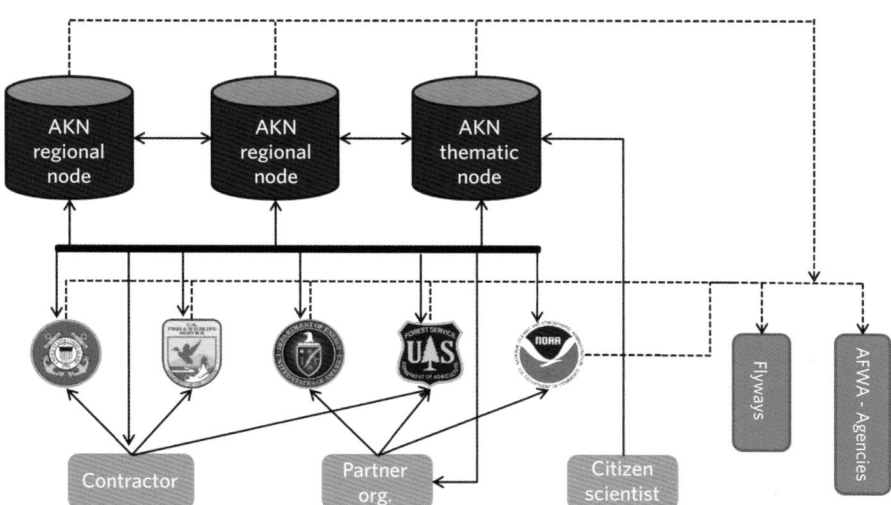

Figure 31.6. A schematic diagram depicting how data flow from collaborators to data centers, or "nodes," of the Avian Knowledge Network, and from nodes to data consumers: partner organizations, management agencies, contractors, researchers, etc. (http://www.avianknowledge.net/). *Courtesy of Point Blue Conservation Science.*

represent regional geographies (e.g., the Midwest Avian Data Center) or types of data (e.g., banding), the AKN hosts quantitative web-based decision support tools (Alexander et al. 2009) (fig. 31.6). It uses a data schema originally developed to facilitate the discovery, retrieval, and integration of information on museum specimens. For security and protection of publication rights, every record is assigned a data access level set by the researcher/data provider and governed by a data sharing policy. Many AKN nodes now offer free online data entry and curation for commonly used field methods (e.g., point counts surveys).

The use of large scale avian data sets is allowing investigators and bird conservation initiatives to ask questions at scales that were not possible before. Web-based tools and publications that predict the present and future occurrence under different climate models can help agencies plan for future conditions at regional and continental scales (Stralberg et al. 2009, Veloz et al. 2015, Langham et al. 2015)—scales that include the entire life history of a migratory species.

LAB TECHNIQUES

In all scientific fields, researchers are restricted to the types of data they can collect and analyze. Breakthroughs in multiple fields—especially genetics, stable isotope analysis, physiology, development, and computation—have allowed amazing advances in all fields of biology. Rather than try to review these vast and growing fields, we focus on a selection of recent advances that have disproportionate impacts and that are quickly becoming standard practice in ornithology.

Molecular Genetics

Molecular genetics is among the most rapidly growing fields and has been used in ornithology for studies of systematics,

taxonomy, population genetics, behavior, conservation, disease discovery and surveillance, and basic evolutionary biology. The work is now aided by lab methods that can sample hundreds or even thousands of genetic loci or sequence and assemble entire avian genomes (Toews et al. 2016).

For the last two decades, traditional Sanger DNA sequencing has been used in combination with the polymerase chain reaction (PCR) primers to sequence single genes or DNA markers (e.g., mitochondrial primers found in Kocher et al. 1989), and has been the foundation of genetic work in birds. Sanger sequencing can very accurately sequence a single targeted gene or genome region up to about 1,000 bases. Genetic sequence data can be elongated by sequencing multiple adjacent regions.

Recent developments in massively parallel sequencing platforms allow researchers to effectively sequence thousands to hundreds of millions of sequences in just a couple of days using platforms such as PacBio's SMRT technology and Illumina HiSeq and MiSeq machines. Each of these platforms has different strengths, but the data can be used to cost-effectively sequence sets of genome-wide markers for a variety of target loci studies or for assembling whole avian genomes (Glenn 2011, Quail et al. 2012). As data set size increases, more and more computing power and time are required to handle, assemble, analyze, and interpret these data.

Several important avian genomes have been sequenced and are relatively well annotated, including the chicken (*Gallus gallus*) (Hillier et al. 2004), Zebra Finch (*Taeniopygia guttata*) (Warren et al. 2010), domestic Turkey (*Meleagris gallopavo*) (Dalloul et al. 2010), Budgerigar (*Melopsittacus undulates*) (Ganapathy et al. 2014), and Medium Ground Finch (*Geospiza fortis*) (Lamichhaney et al. 2015), although the list of good-quality avian genomes is growing fast (Zhang et al. 2014a). The average bird genome size is about 1.36 pg of DNA, or about 1.33 gigabases

(Gb) of sequence (Gregory 2015). This is smaller than most other terrestrial vertebrate genomes, in part because of a reduction of highly repetitive elements, but also because of the deletion of large genome segments and some genes (Zhang et al. 2014b), which provides advantages to the study of avian genomes. The overall genomic structure and gene synteny are relatively conserved across birds, and this too allows for easier work and comparison at the genome level than with most other tetrapod taxa.

The resulting field of **genomics** seeks to understand the structure and function of entire genomes. Likewise, **transcriptomics** focuses on sequencing the expressed RNAs in a particular individual or tissue type, allowing for very precise measurement of differences in expression or gene sequence that relate to physiological, behavioral, developmental, or various other functional traits. Acquiring sequence data for genomics studies has progressed relatively rapidly, and analytical tools and approaches are being developed to make use of the data (Toews et al. 2016).

Uses for Genetic Data in Ornithology: Taxonomy and Systematics

The fields of taxonomy and systematics focus on identifying and naming bird species, quantifying biodiversity for conservation and other purposes, and elucidating the relationships among species, including the reconstruction of evolutionary trees, or phylogenies. Genetic data can be extremely useful for systematics because the genome contains vast information. We understand much about how DNA evolves, and it is usually possible to obtain data from the same gene regions from birds of different populations or species. This work requires gene regions that are variable enough to contain information, but not so variable that multiple changes at a single site can obscure its history (sometimes called saturation) or so variable that it is difficult to align homologous regions from multiple populations or species. Thus, researchers must choose gene regions that are evolving at the appropriate speed for the particular study. Many modern systematic studies are analyzing tens to hundreds of thousands of mutant sites, often called single nucleotide polymorphisms or SNPs (Toews et al. 2016). The field is currently in flux, as bio-informaticists and computer programmers are developing new tools to analyze these large datasets and construct phylogenetic trees (Stamatakis 2014, Kozlov et al. 2015; box on page 942).

Uses for Genetic Data in Ornithology: Population Genetics

The field of population genetics typically asks questions about the relatedness, connectedness, and history of different populations within a single species. Using aligned genetic data sets that are similar to those used for phylogenetics, researchers can estimate population parameters such as effective population size (N_e), genetic interchange among populations (called migration or m), coefficients of natural selection (s), understand population structure using analysis of variance and derivatives, and understand the history of population bottlenecks. Many questions in population genetics are geographically explicit and can incorporate biogeographic analyses about physical or ecological barriers to gene flow, or how genotypes are distributed across landscapes. Such genetic approaches have been critical for the field of conservation, which seeks to identify distinct population segments (i.e., conservation units), understand the genetic effects of inbreeding and bottlenecks, and offer viable management solutions (e.g., Haig et al. 2011, Manel and Holderegger 2013). The field of conservation genetics uses the tools of population genetics largely to understand genetic challenges within declining species and achieve meaningful recoveries (Shafer et al. 2015).

Uses for Genetic Data in Ornithology: Pathogen Discovery and Surveillance

DNA and RNA from parasites and pathogens can often be recovered alongside avian genetic material, allowing novel methods for detecting and monitoring pathogens. Pathogens can often be detected by the presence of their DNA or RNA in host tissues using PCR that is targeted to regions of the parasite genome. Carefully chosen PCR primers can show whether a host is infected and identify the pathogen strain involved. For example, primers have been designed that can amplify strains of *Plasmodium*, *Haemoproteus*, and *Leucocytozoon* avian malaria parasites (Hellgren et al. 2004, Richard et al. 2002), and these methods have been effective for detecting malaria from live birds as well as museum skins and 30-year-old blood smears. For infections with unknown causes, investigators can extract genomic DNA from infected avian tissues, and with ultra-high-throughput sequencing, enough pathogen sequences can be recovered to assemble its entire genome. For example, researchers who studied parrots with fatal proventricular disease were able to isolate and assemble the entire genome of an avian bornavirus that was then shown to cause the disease (Kistler et al. 2008; box on page 943). These methods have led to powerful surveys of avian disease and discovery of many new pathogens.

Other Uses of Genetic Data

Avian genetics has a variety of other important uses. Proponents of DNA barcoding use a single rapidly evolving locus (cytochrome oxidase I of the mitochondrial genome) as a quick and efficient way to identify the species of birds or bird

LARGE GENOME DATASETS REVEAL RELATIONSHIPS AMONG AVIAN ORDERS

One of the most vexing problems in avian systematics has been resolving the relationships among bird orders in the Neoaves. This large group of extant avian orders (all but the Paleognathae and the Galloanseres) appears to have radiated rapidly around the time of the Cretaceous-Paleogene extinction event. Thus, we would expect the available genetic signal of relationships would be small, and largely swamped by the number of mutations that have occurred during the subsequent 60–65 million years of evolution. To solve the problem, a huge number of researchers collaborated to sequence and analyze the whole genomes of 48 species representing all orders of Neoaves and multiple passerine families and outgroups (Jarvis et al. 2014; see figure). The resulting data set required new phylogenetic tools and huge amounts of computing time, but found excellent support for many of the early branches of modern birds. Because the data and tools are new and the number of taxa is still relatively small, this should still be considered a work in progress, and yet it has great promise for resolving these short branches on the avian tree. Additional large datasets with more taxa continue to be published (e.g., Prum et al. 2015) and continue to refine our understanding of phylogenetics at the level of avian orders.

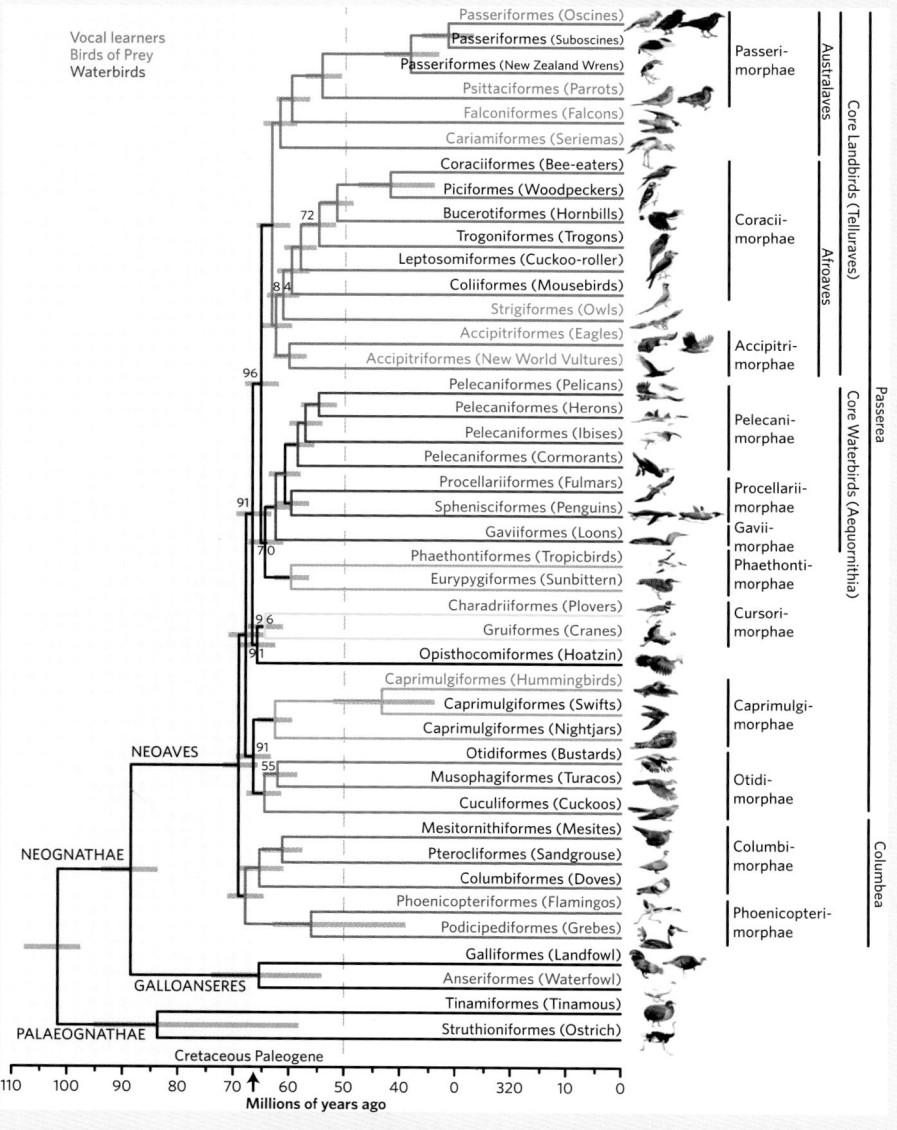

The genome-scale phylogeny of birds. *From Jarvis et al. 2014.* This is the total-evidence nucleotide tree inferred from maximum-likelihood analyses, showing time scale on the x-axis and the diversity of avian orders and proposed names for higher-order taxa on the right.

USING MASSIVELY PARALLEL GENETIC SEQUENCING TO RECOVER NOVEL AVIAN PATHOGENS

Proventricular dilatation disease (PDD) is a fatal disorder affecting over 50 species of parrots and their relatives as well as species in five other avian orders. The disease was first described in the 1970s as Macaw Wasting Disease, and it typically causes autonomic nerve damage in the upper and middle digestive tract, which leads to difficulty swallowing and processing food. Similar nerve damage can be seen in the brain, spinal cord, heart, peripheral nerves, and smooth muscle. Although a viral pathogen was suspected, none could be found in earlier work. Two groups extracted RNA from birds suffering from PDD, and sequenced thousands of RNA fragments from tissues that are heavily affected by the disease and should be rich in viral RNA (Kistler et al. 2008,

Honkavuori et al. 2008; see figure). Computer algorithms helped removed sequences that originated from host birds, and remaining fragments were assembled and BLASTed to databases. Both groups recovered fragments that matched Bornaviridae—a negative sense single-stranded RNA virus family that was previously not reported from birds. The strains from parrots were only 64 percent similar to known bornaviruses, although the gene content and order clearly placed it within the Bornaviridae. The virus was named avian bornavirus (ABV). Since the initial discovery, many more bornaviruses have been recovered from diverse avian groups (Payne et al. 2011), and some of these cause very different pathologies.

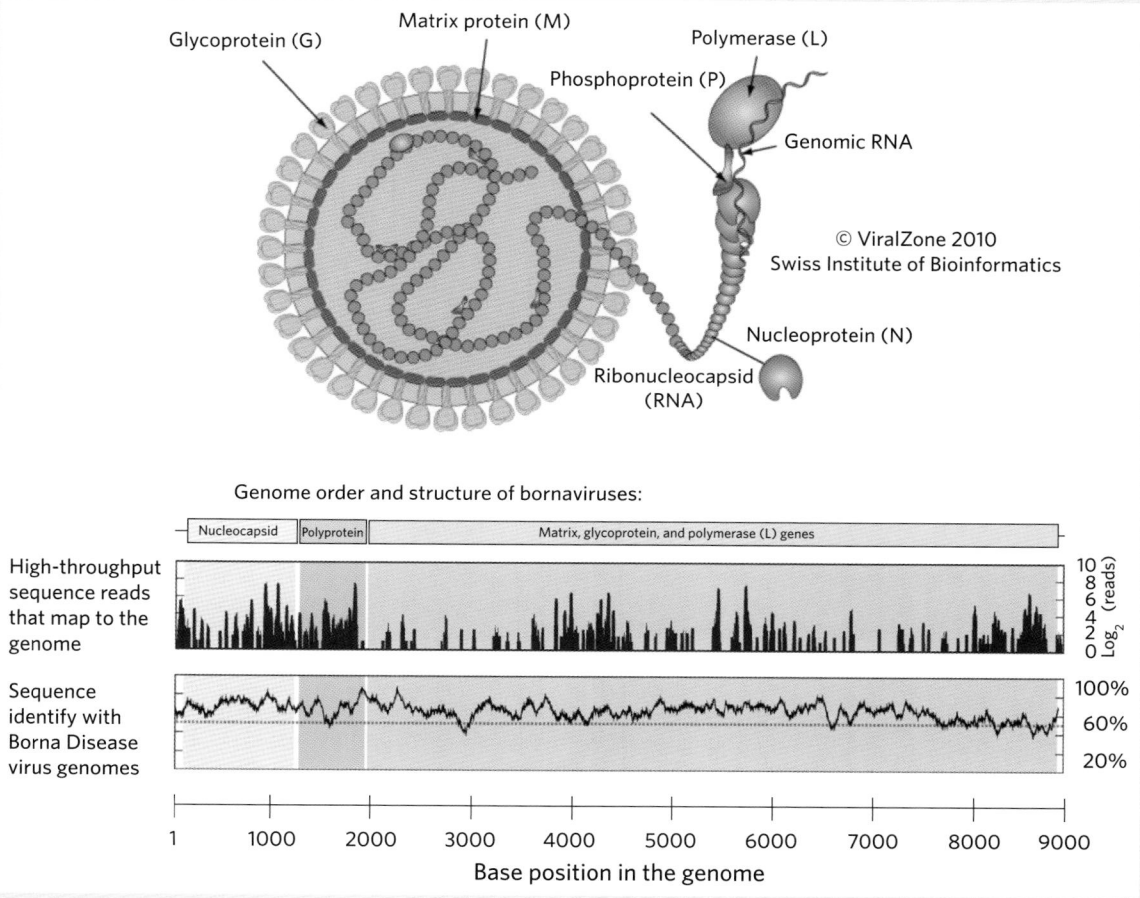

Avian bornavirus genome, as assembled from next-generation sequence reads and additional PCR sequences. Above is a representation of the virus and its parts. Below the diagram of the virus, the top bar shows the gene order and structure of ABV as determined from sequencing; the second bar shows coverage in the Illumina sequence reads, and the bottom bar shows the ABV sequence identity with borna disease virus. *Figure based on material in Kistler et al. 2008, and the virus representation is taken from ViralZone online (http://viralzone.expasy.org/), by permission of ViralZone, SIB Swiss Institute of Bioinformatics.*

parts and provide simple preliminary phylogenetic analyses (Lijtmaer et al. 2011, Lijtmaer et al. 2012). Genetics has been used to accurately assign parentage in avian clutches and has demonstrated that the social mating system (most commonly monogamy in birds) is not always the genetic mating system (i.e., that there are significant extra-pair fertilizations or conspecific egg dumping) (Hughes 1998). Genetics can be used to determine the sex of birds belonging to monomorphic species, "ancient" DNA can be extracted from specimens belonging to extinct species or populations (Fleischer et al. 2006), and there are many emerging uses for monitoring climate change and adaptation (Sheldon 2010).

Stable Isotopes

Common chemical elements often have more than one stable form or isotope that differs in molecular weight because of different numbers of neutrons in the atom, and isotopes are found at predictable frequencies in the environment. Biological and physical processes may alter the ratio of stable isotopes, for example the ratio of ^{15}N to ^{14}N increases with increasing trophic level, and the ratio of ^{13}C to ^{12}C increases as you move from the sea to onshore environments. Birds are constantly sequestering chemicals from their diet and using them to produce feathers, claws, bills, bones, and other tissues. Thus, the ratio of common isotopes of carbon, nitrogen, hydrogen, oxygen, and sulfur can be estimated from bird tissues and used to gather a variety of information about (1) avian diet and trophic level (Bond and Jones 2009, Boecklen et al. 2011); (2) the movements and origin of migratory birds through different habitats (Hobson et al. 2014a, Hobson et al. 2014b); and (3) the relative contribution of endogenous versus exogenous nutrients to reproduction, growth, and other processes (Hobson et al. 1997). This work requires careful design to sample the most appropriate tissues (e.g., feathers that were produced in the wintering grounds but can be sampled in the breeding grounds) and requires multiple controls. Lab work requires specialized equipment for extracting and accurately measuring quantities of rare isotopes.

Toxicology

Humans increasingly use and discharge significant volumes of unnatural chemicals into the environment (see chapter 26, In Harm's Way). Wildlife accumulate many of these chemicals and store them in their tissues, and some are toxic and have clear toxic effects. Like pathogens, each may accumulate in or affect distinct tissues, and each requires a different method for detecting and quantifying the presence of toxins. Few environmental toxins have been thoroughly studied, and the field is ripe for competent chemists who wish to study the accumulation and effects of various chemicals on

bird and other wildlife populations. Some of the best-studied toxic chemicals in birds include lead—a chemical that has accumulated primarily from sport hunting and fishing (Scheuhammer and Templeton 1998). Lead and other heavy metals are persistent and many bioaccumulate in the food chain, so predators and scavengers are especially susceptible to poisoning (Behmke et al. 2015). The California Condor is a classic example of a species pushed to the brink of extinction by lead poisoning (Finkelstein et al. 2012). Other sources of avian toxicity include agricultural pesticides, organophosphates, dioxins (Schecter et al. 2006), long-acting anticoagulant rodenticides (Thomas et al. 2011), oil spills (Leighton 1993), and even the veterinary drug diclofenac (Ogada et al. 2012), just to name a few.

Most of these environmental chemicals can be detected in the lab using mass spectrometry. Mass spectrometry (MS) typically bombards chemical samples with ionizing radiation that breaks the molecules into fragments or individual elements, ionizes the fragments, and measures the deflection of gas-phase ions in an electric field, yielding measurements of mass-to-charge ratio for each ion or fragment. From the mass spectra, chemists can typically identify the chemicals in the sample and can often measure total or relative amounts of each. Depending on the chemical's properties, it can either be purified in advance of the MS analyses or, for some chemicals, can be separated using front-end machines. For example, gas chromatography MS (GC-MS) uses a front-end machine that separates chemicals by their volatility and performs MS on the eluting gasses. Liquid chromatography (LC-MS) separates chemicals by their adhesion to a column in the presence of different solvents, and performs MS on eluted liquid that is sprayed into the ionization chamber. Direct-probe MS simply dries a purified substance on a probe that is placed into the ionization chamber. There are many options for fine-tuning these and other techniques for specific chemicals and studies. Most of these techniques require collaboration and training with highly competent chemists and laboratories with specialized equipment.

Lab Protocols and Methodologies

For any of these lab methods, the first step is collecting and properly preserving tissues from birds. The best and most versatile method is flash freezing tissue in clean vials in liquid nitrogen. This prevents any contaminants from buffers or preservatives from altering the chemical structure or composition of the sample (i.e., for DNA, RNA, toxicology, stable isotopes, etc.), and cryogenic freezing in liquid nitrogen effectively halts chemical and biological processes that breakdown the sample. RNA, DNA, and just about every other chemical is preserved intact in a sample frozen at $-80°C$ or in liquid nitrogen. When field freezing is not possible, or when researchers cannot maintain the chain of freezing, then buffers or preservatives can be used to protect the most critical elements in samples. The downstream analyses will determine which buffers might work best, but 95 percent or 100 percent ethanol does a fair job of preserving DNA in tissues for at least a few years to decades, and probably much longer if they are kept frozen once they return to the lab. For RNA, multiple buffers are available, including RNA*later*® and several other proprietary buffers that allow short-term storage at room temperature or refrigeration, but all require freezing for long-term preservation.

Standard protocols are also available for DNA and RNA extraction, and most labs utilize prepackaged kits that digest the tissue and lyse the cells, precipitate DNA using ethanol and a resin that binds to DNA, wash away other proteins and chemicals, and elute the DNA from the resin in an aqueous solution. DNA is typically stored in buffered water (usually pH 7.6–8.0) that may contain a weak salt solution (often the preservative EDTA), and frozen. Standard kits are available for RNA extraction, but extra cleanliness and precautions are required for RNA, as enzymes that degrade RNA are ubiquitous and act rapidly to destroy RNA. Studies of other chemicals or environmental contaminants will depend largely on the type of chemical studied, but usually standard field samples can be subsampled or used, and often even standard preserved museum specimens can be sampled (e.g., for stable isotopes, heavy metals, and many other stable chemicals). All of these laboratory techniques speak to the value of museum collections and the importance of properly lodging vouchers and samples in museums where samples are well documented, organized, curated, and shared with others.

THE MUSEUM COLLECTION

The scientific museum collection constitutes the most complete, tangible, and permanent record of life on earth. Once collected, specimens are preserved in various ways that allow them to be stored, studied, and sampled for hundreds or even thousands of years, and the original collector need not even imagine their future uses. Because of the high cost of fieldwork, preparation, and maintenance, as well as various permitting and ethical hurdles, modern collections seek to preserve the maximum amount of material and data per individual bird collected (Winker 2000).

There are a variety of specimen types in ornithological collections, and ideally a single individual bird can provide multiple different specimens. These include traditional stuffed "round" skins, spread wings, skeletons, alcohol or fluid-preserved specimens, eggs and nests, tissues (usually optimized for DNA, RNA, or protein), media (including recordings, video, photos, field notes), and other special collections or digital material (e.g., genetic sequence data, or micro-CT scans) (fig. 31.7).

A typical scientific research collection houses a large series of individuals of each species or subspecies from a variety of different locations. Each specimen represents an individual of that species at a particular place at a specific time. Large numbers of individual specimens and their associated data offer researchers an adequate sample size to examine variation among individuals, within a population, or among species. Documenting morphological variation in age, sex, and plumage including molt sequences, seasonal changes, and feather wear would not be possible without the opportunity to examine and take standard measurements from a large series of individuals. These series are particularly important in highlighting geographic differences, including changes in distribution over time and genetic variation. Large series of specimens within museum collections also provide valuable data to document biodiversity, habitat loss, climate change, and invasive species over an expansive period of time (Suarez and Tsutsui 2004). This wealth of information from specimens has proven invaluable in directing conservation management decisions, thus emphasizing the importance of maintaining these collections for future generations (Carmi et al. 2016, Rocha et al. 2014).

Building museum collections is a long process that sometimes attracts criticism (Rocha et al. 2014, Winker et al. 2010). Today, most modern museum collecting is focused on specific research questions or takes advantage of salvaged specimens killed by other anthropogenic causes, such as roadkill, wind farms, outdoor cats, and window collisions. Museum collections also grow through the acquisition of specimens from rehabilitation centers, zoological parks, and those confiscated from prosecuted cases involving poaching or the illegal trade of animal parts. Any focused collecting is conducted under permits, including federal, state, and local, and is often completed in conjunction with area conservation and management agencies with a specific question in mind. Despite these important sources of specimens, there are still situations that require general specimen collecting—including surveys of poorly known regions or taxa. Museum scientists must always make sure that their collecting practices have no lasting impact on birds at both the population and the species level. In fact, the number of specimens collected for scientific purposes is quite small; all the bird specimens of all species, collected over the past 250 years and currently held in museums throughout the world, number approximately six to seven million (VertNet 2015, GBIF 2015). This number pales in comparison with the estimate of the number of birds

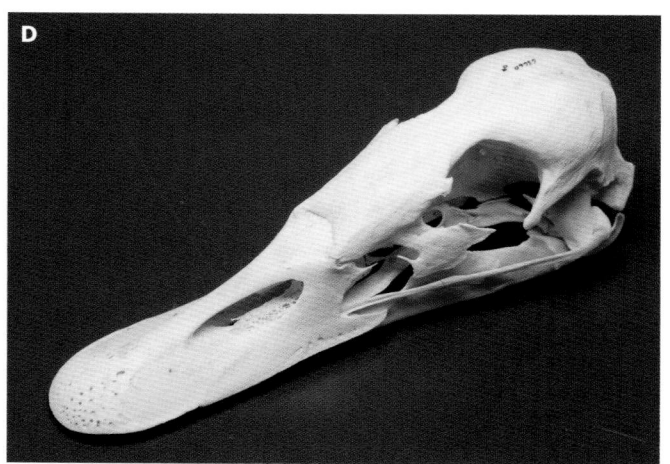

SOLITARY VIREO
Vireo solitarius

Nests in oaks,
manzanita,
buck-brush, etc.

Figure 31.7. Various types of museum specimens: A, study skins; B, eggs; C, fluid-preserved specimens; D, skulls and skeletons; E, nests; and eggs; F, spread wings; G, live mounts. *Photos © Jack Dumbacher, California Academy of Sciences.*

killed each year from window collisions in the United States alone, estimated at 365 to 988 million (Loss et al. 2014; see chapter 26 for further details). Specimens retained in museums provide continued scientific value for hundreds of years to come.

Uses of Collections

Museum collections are available for researchers from around the world to utilize both the data and the specimens themselves through database searches, specimen loans, and on-site visits. Museum collections are especially useful for undergraduate and graduate students by offering them ac-

cess to a wealth of information that would be impossible to gather during the short time that students are in school.

Most ornithology collections have digitized their specimen data and made them accessible to researchers through online search engines. Individual collections can usually be searched through an institutional website. The most efficient method of gathering data from many museum collections at once is to search one of the available collaborative databases that provide biodiversity data free on the web. Two projects that serve scientific ornithological data, including specimens and observations, are VertNet (www.VertNet.org) and the Global Biodiversity Information Facility (www.GBIF.org). Data users should be sure to read and comply with the data-use policies of both the institution providing the data and the collaborative platform from which it is served.

Traditional use of bird study skins focused on morphological examination and measurement to determine taxonomic position. Although avian taxonomy is relatively well resolved, study skins still offer researchers a variety of opportunities to investigate novel aspects of bird evolution and life history. For example, a recent publication by Danner et al. (2014) used bill, tarsus, and wing measurements from 1,488 bird study skins from eight museum collections in the United States to investigate morphological divergence between island and mainland populations of Song Sparrows (*Melospiza melodia*), a species for which taxonomy is well-known. The availability of vouchered museum specimens provides researchers with physical objects for use in studies that improve the overall scientific knowledge of birds (box on pages 948–949). Current studies also combine morphological analysis with genetic data to tease out the finer details of avian biology.

Museum collections also serve as a valuable resource for personnel from government agencies and biological consulting companies. Skeletal specimens housed in an ornithology collection can be used by archaeologists to more accurately identify specimens that they have encountered during an archeological dig or at a construction site (O'Conner and O'Conner 2008). Wildlife forensic specialists often reference museum research collections to identify specimens that have been involved in aviation strikes (Dove and Coddington 2015), were confiscated, or are part of an ongoing investigation. Researchers that study the effects of wind turbines also utilize museum collections to identify species from feathers that they have found during surveys. Without a large series of samples from various species to use as reference material, these different types of identifications are difficult to make.

Modern museum collections typically archive tissues for future genetic analysis. These tissue samples can include muscle, heart, liver, and other internal organs, depending

PETER PYLE AND THE IDENTIFICATION GUIDE TO NORTH AMERICAN BIRDS

Every bander knows how important it is to collect as much data as possible from each bird caught in a net. Each museum specimen also provides a wealth of information about the individual bird based on measurements, molt patterns, age, and sex. Peter Pyle, who began banding birds in the backyard with his father at the age of four, knew that all the information previously published on species identification, aging, and sexing needed to be accessible to researchers working in the field. The *Identification Guide to North American Birds*, Parts 1 and 2, represent the body of work that Peter produced to provide banders throughout North America with a comprehensive reference for use while examining a bird in the hand (Pyle 1997, 2008). The guides offer field ornithologists in-depth information on species and subspecies identification, methods for aging based on molt limits, and information on sexing all species (see figure). Over the years, Peter examined more than 50,000 museum specimens at over 20 collections, recording measurements, feather wear, and molt strategies and limits, among other things. Combining his field observations with previously published literature and the specimen collection data, including location, sex, and age, he produced the most comprehensive reference guide to aging and sexing birds in the hand. Peter likes to emphasize examining each feather for molt and plumage generation in order to collect the most accurate data possible on each bird examined. Parts 1 and 2 of the *Identification Guide* should be available in each ornithology classroom, at every banding station, and in all museum collections.

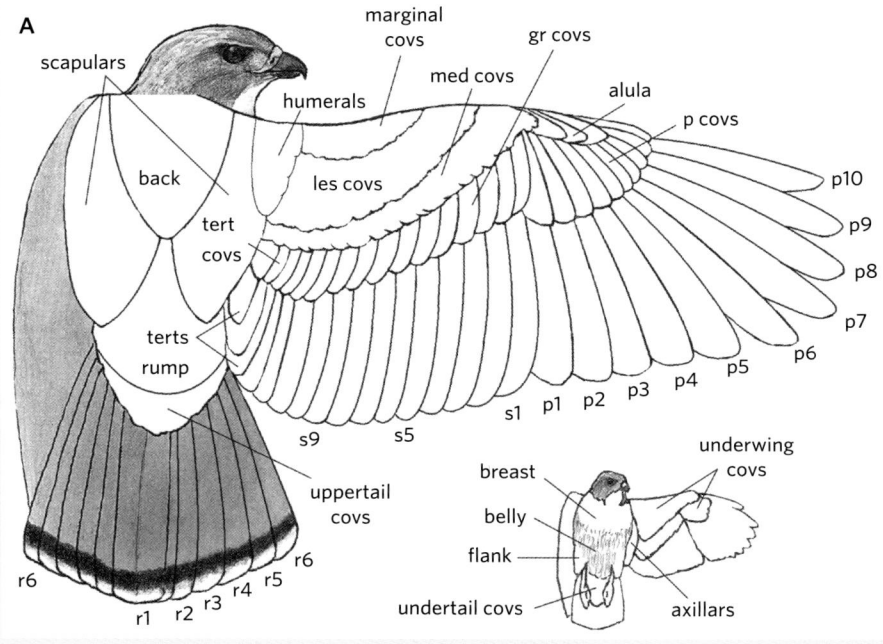

on the protocols of the individual collections. Tissue samples can be stored in various buffers such as ethyl alcohol, RNA*later*®, etc. They can also be stored without buffers and simply frozen. Most samples are stored in either ultracold freezers (−80°C) or liquid nitrogen tanks for long-term use. Researchers can request a small subsample of archived tissues for use in their projects with a simple loan request.

Although fresh tissue samples are preferred for genetic work, DNA can also be isolated from preserved specimens, even those that have been in the museum collection for hundreds of years (Wandeler et al. 2007, McCormack et al. 2015; box on page 950). This is often referred to as "ancient" DNA and is considered a destructive sampling process. Collection management policies usually have strict requirements for loan requests that involve destructively sampling study skins, bones, or eggs. Sampling, processing, and analysis protocols should be well developed in order to acquire approval for these types of requests. All DNA samples from study

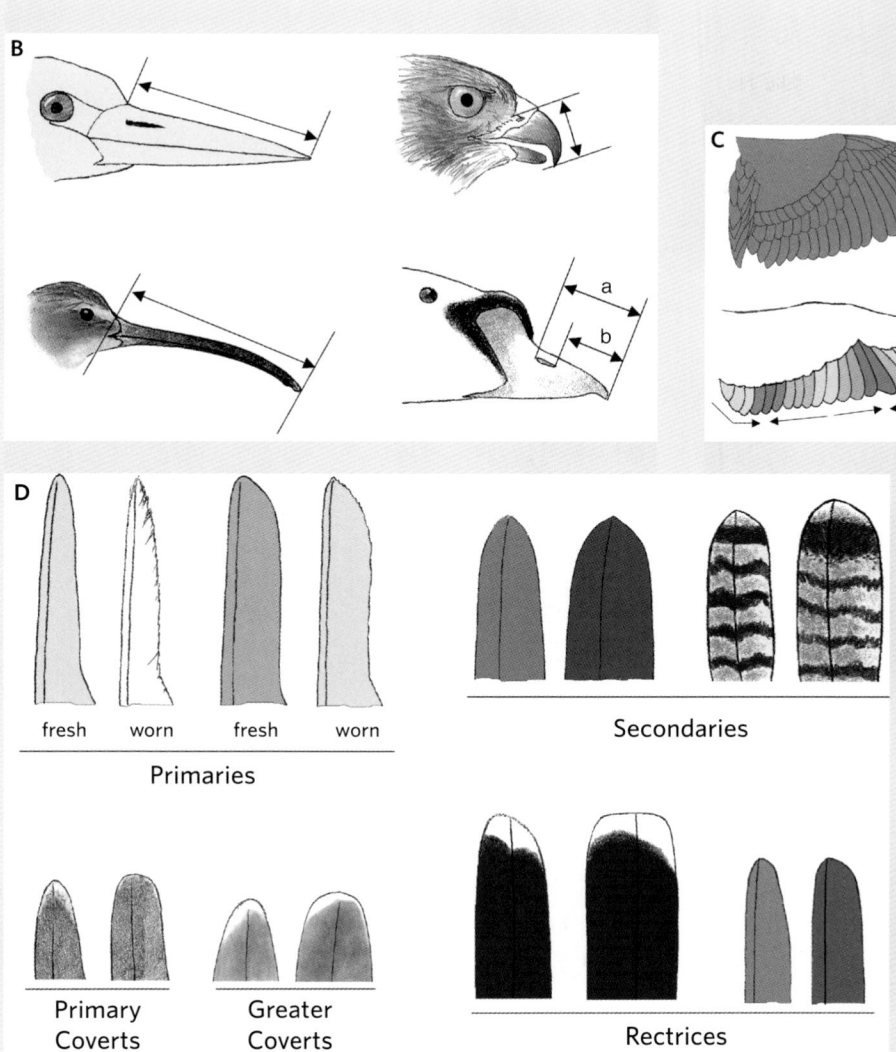

In parts 1 and 2 of the *Identification Guide to North American Birds* (Pyle 1997, 2008; www.slatecreekpress.com), Peter Pyle standardized the measurements and terminology used in identifying, aging, and sexing birds in the hand. Here are a few examples of illustrations from the guides: *A*, Anatomy of feathers for studying plumage and molt limits; *B*, examples of different culmen measurements for various species; *C*, wing molt strategies; *D*, feather shapes and patterns for aging and sexing. *Illustrations © Siobhan Ruck.*

skins should be processed in a specific laboratory designated for ancient DNA. Certain tissues have low overall DNA yields because the DNA is usually degraded and needs to be amplified in very short fragment sizes. Thus, protocols need to match ancient DNA stringency, including adequate controls and dedicated laboratories, in order for the results to be accepted by the scientific community (Wandeler et al. 2007).

Feather, tissue, and bone samples can also provide valuable information in the form of stable isotopes (Hobson

2011). Since stable isotope analysis requires the destruction of samples in order obtain the stable isotope ratios, most museums will require a destructive sampling request. Museums are more likely to allow sampling of bird contour or body feathers since they are abundant and can easily be sampled without affecting any future scientific use of the specimen. Flight feathers (both primaries and secondaries) offer additional information about migratory patterns, since the timing of flight feather molt often differs from the timing of

NOVEL USES OF MUSEUM SPECIMENS: THE ORIGIN OF AVIAN POX IN GALÁPAGOS

While studying wild birds on the Galápagos Islands, Dr. Patricia Parker was interested in the origin of the avian pox lesions that she was seeing on the feet of birds in the field. Since *Avipoxvirus*, a genus of DNA virus, is a serious pathogen when found within isolated island populations, she was particularly concerned about how the virus arrived on the Galápagos and whether it might drive endemic Galápagos birds to extinction. To determine the history of *Avipoxvirus* on the Galápagos, she and her colleagues examined 3,973 museum specimens of finches and mockingbirds collected in the Galápagos between 1891 and 1906 (Parker et al. 2011). Of the specimens examined, 226 displayed nodules that were consistent with *Avipoxvirus* (see figure). Nodules were found only on specimens collected during or after 1899. With permission from the museum, she took small clippings of the lesions from 59 specimens and was able to do histopathology and extract and sequences of *Avipoxvirus* DNA. Her findings suggested that the *Avipoxvirus* sequences were most closely related to forms found in canaries, and that these were found primarily on islands inhabited by human colonists, primarily on San Cristobal Island and others with human settlements. Based on the collection dates and the DNA results, she was able to conclude that, shortly before 1899, early settlers may have brought caged pet birds that were infected with the canarypox form of the virus to the islands.

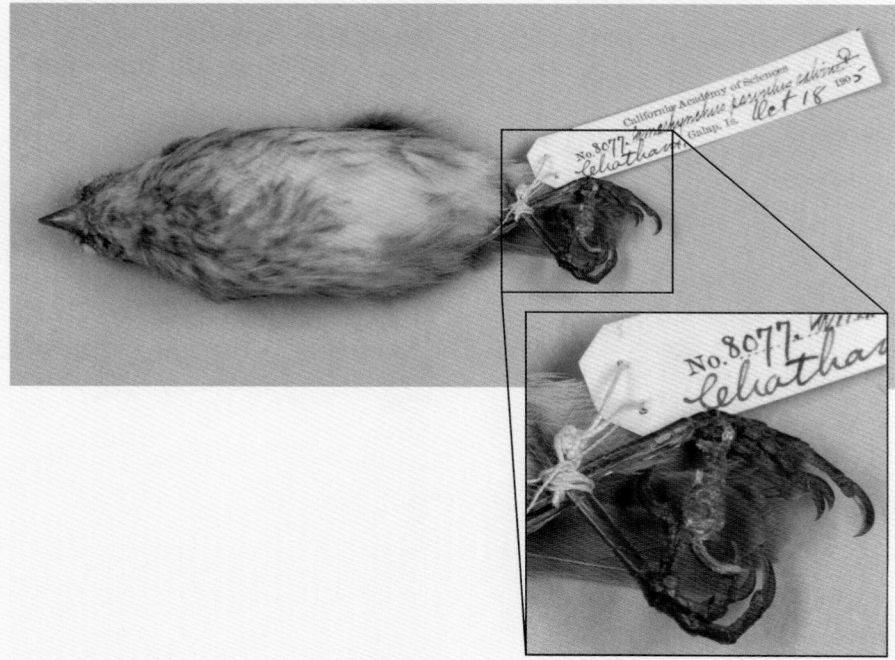

Image of a Galápagos finch specimen showing a large avian pox lesion on the left foot and no lesions on the right foot. *From the collections at the California Academy of Sciences. Photo © Jack Dumbacher, California Academy of Sciences.*

body feather molt (see chapter 9 for more details). However, it is more difficult to obtain permission to dissect sections of flight feathers from museum specimens for isotope analysis. The need to sample flight feathers must be well justified in any sampling request made to a museum collection.

Current technologies offer researchers new opportunities in ornithology. With the development of high-resolution X-ray computed tomography (CT), researchers are now able to examine new aspects of avian anatomy (Quayle et al. 2014,

Soons et al. 2015; box on page 951). High-resolution CT is a powerful method for visualizing, measuring, and disseminating morphological information from a wide range of museum specimens, including study skins and fluid-preserved specimens. CT scanning allows the internal anatomy of rare or delicate specimens to be investigated without causing any damage.

Museum specimens are also valuable resources for studying the effects of environmental contaminants on birds.

NEW TECHNOLOGIES: CT-SCANS FOR STUDYING FUNCTIONAL MORPHOLOGY

New technologies in the field, laboratory, and museum offer innovative methods for answering complicated scientific research questions. The work of Soons et al. (2015) provides just one example of the use of modern technology in addressing questions of adaptation in evolution. The team of researchers investigated the mechanical relationships between beak shape, stress dissipation, and fracture avoidance in the well-studied Galápagos finches. These species, often referred to as "Darwin's finches" are well known for their diversity of beak form and function and provide excellent examples of evolution through natural selection. In their study, the team used micro-computed tomography scans (µCT scans) and muscle dissections of fluid-preserved specimens (both historical and recently collected specimens) to show that bill shape and stress correlates with the risk of fracture (see figure). More specifically, they recorded different thicknesses in bill keratin, directly correlated to feeding strategies. These thicker areas of the bill play an important role in preventing fractures during the crushing and biting movements required for seed eating. This new information about

beak evolution may help explain natural patterns of selection in wild birds, especially populations of Galápagos finches.

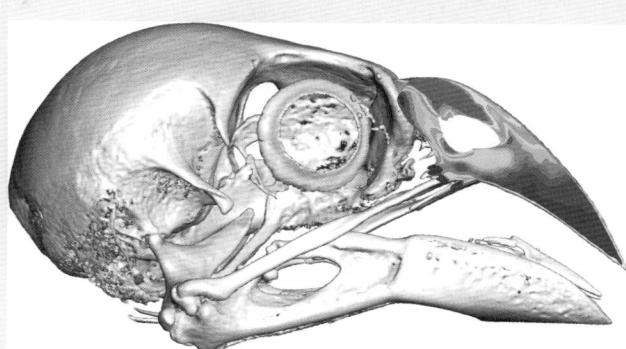

Schematic representation of a µCT scan of the head of a Medium Ground Finch (*Geospiza fortis*) with the results of a finite element model superimposed on the beak. Warmer colors indicate high von Mises stresses and cold colors indicate low stresses. The head of the finch was scanned by Dominique Adriaens (Ghent University). *Image published with permission from Anthony Herrel, Muséum National D'Histoire Naturelle, Paris, France.*

The most important study utilizing museum specimens to study toxins was completed in 1968. Hickey and Anderson (1968) examined more than 1,700 egg specimens from 39 museums from around the world. They were investigating the possible cause of decline in reproductive success of the then-endangered Peregrine Falcon (*Falco peregrinus*). Their study, measuring eggshell thickness, suggested that the eggs of several raptor species started to decrease in thickness in 1947 with the use of dichlorodiphenyltrichloroethane (DDT). As a result of this work, DDT was banned for use in the United States in 1971, and the Peregrine Falcon was delisted from the Endangered Species List in 1999.

Scientific researchers are not the only users of museum collections. Many collections allow artists to access the specimens held in their care. These artists use the collections as reference material for fine-scale drawings in field guides and other popular publications (Howell and Webb 1995). Bird study skins and mounts provide artists with a close-up view of plumage and soft color (bills and legs), parts that are not readily available from photographs or viewing birds in the field. Accuracy in illustrations of birds included in field guides provides birders a valuable resource for identifying birds in the field. Some museum collections also allow the

commercial use of museum specimens, with some artists using specimens as reference material or inspirational material for paintings or sculptures (fig. 31.8).

Building and Maintaining Collections

Although much growth in modern museum collections comes from the salvage of specimens from various sources, there still exists the need to actively collect specimens in the field (box on page 953). Methods for field collecting of specimens for museums are similar to those outlined in the beginning of the chapter for general fieldwork; however, there are some additional methods used to collect specimens. Museum collectors generally use mist nets and other traps to first capture the birds alive. Following the terms outlined in each permit—generally a maximum number of individuals from each location—collectors then decide which individuals to add to the museum collection through euthanasia, following the American Ornithologists' Union guidelines to the use of wild birds in research (Fair et al. 2010).

Given the difficulty of placing nets in some habitat types, and complicated bird behavior, mist nets are not always the most effective or efficient method of capturing birds. For example, forest canopy and grassland species are particularly

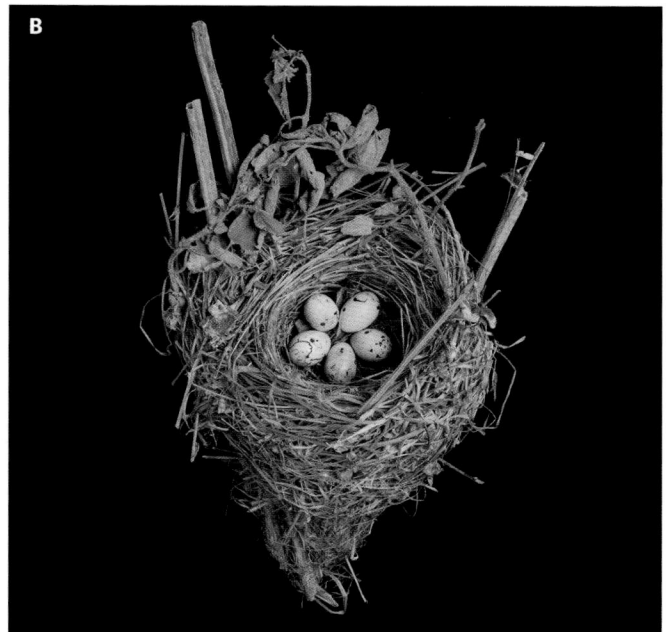

Figure 31.8. Specimens in museum collections are often used by artists as reference material or for inspiration. Here are a few examples: *A, © Tiffany Bozic,* Triangle of Love, *2014; B,* photograph of Tricolored Blackbird (*Agelaius tricolor*) nest and eggs, *© Sharon Beals 2015; C, ©Isabella Kirkland,* GONE, *2004; D,* plate of manakin species by Sophie Webb from *A Field Guide to the Birds of Peru. From Schulenberg et al. 2007.*

ROLLO HOWARD BECK (1870–1950), MUSEUM COLLECTOR

Rollo Beck (front) skinning birds in a makeshift laboratory aboard the schooner, Academy, around 1905. *Image MSS429, Joseph Slevin papers, © California Academy of Sciences.*

The world's museum collections owe a debt of gratitude to the early collectors. Many risked their lives traveling to dangerous places to collect the specimens and data that provide the foundation for our understanding of bird diversity and the material that supports many ongoing studies. Foremost among these collectors was Rollo Howard Beck (1870–1950). Beck grew up in agricultural lands near Los Gatos and Santa Clara, California, learned to skin birds, joined the local Cooper Ornithological Club, and corresponded and traded birds with notable ornithologists including Robert Ridgway and Charles Bendire. He eventually joined the Webster-Harris

expedition to the Galápagos Islands in 1897, funded by Sir Lionel Walter Rothschild, and that experience resulted in the California Academy of Sciences hiring him to lead their expedition to the Galápagos in 1905 (see figure). Afterward, Beck continued to collect for the California Academy of Sciences, as well as for Joseph Grinnell at Berkeley's Museum of Vertebrate Zoology. Frank Chapman and Leonard Sanford from the American Museum took note of his work, and they offered him more money to collect for them, eventually sending him to Alaska with Arthur Cleveland Bent and Alexander Wetmore. Sanford and Chapman then hired Beck to lead the Brewster-Sanford expeditions around South America by ship, finishing in the islands of the Caribbean. His work was so prolific, and his specimens so extraordinarily well-prepared, that he was chosen to organize and lead the Whitney South Seas expeditions. For over eight years he traveled throughout the South Pacific from Tahiti to Pitcairn Island to New Zealand, Solomon Islands, and New Guinea and just about everywhere in between. All told, about 58,000 specimens in museums bear his name as collector, and more were collected by his colleagues who traveled with him on some of his expeditions. His specimens have provided the foundation for several classic works, including David Lack's book, *Darwin's Finches*, on community ecology; Ernst Mayr's many works on birds of Melanesia and the South Pacific (which helped form many of his ideas regarding the nature of species and speciation); and Robert Cushman Murphy's *Oceanic Birds of South America* (Dumbacher and West 2010).

difficult to capture in mist nets. Where authorized, firearms can be a useful method for the lethal take of individual birds. Firearms are particularly useful for collecting larger birds that are not generally caught in mist nets.

Museum professionals generally decide what species to collect based on specific research questions or to fill holes in the existing research collection. When in the field, it is common practice to preserve as much of each individual specimen as possible (skin, post-cranial skeleton, tissues, stomach contents, parasites, etc.) (Winker 2000). Collectors cannot anticipate future uses for museum specimens, so they generally preserve as much material as possible in the hopes that it will provide valuable information to future biologists (see the box on page 955). Collecting can also be focused on specific ques-

tions, thus preserving specimens strictly to answer those questions (e.g., blood samples to study avian malaria). In most cases, specimens are prepared in the field, and the types of specimens prepared will depend on the specific questions being asked and the protocols of each individual institution.

Data collected with each specimen increases the scientific value. When collecting specimens, researchers should record as much information as possible, including species, age, sex, breeding condition, fat, stomach contents, molt limits, etc. The two most important pieces of data associated with each specimen are a date and a location. These two pieces of information tell a researcher that a species was recorded from a certain place at a certain time. In the field, researchers should always record latitude and longitude coordinates for each specimen. A specimen collection can be retroactively georeferenced; however, the data are more accurate if the geographic locations are recorded at the time that a specimen is collected.

Proper care and maintenance of a museum collection ensures that the specimens will exist for hundreds if not thousands of years and remain available for scientific research. There are several key aspects to proper collections care, including appropriate storage, environmental conditions, pest management, and data management.

Museum storage facilities differ in size and scope but are generally designed with proper long-term storage of specimens in mind. Specimens should be stored in climate-controlled facilities within rooms and cases that provide adequate protection from light, pests, dust, dirt, water, and natural disasters. Storage facilities must provide proper environmental conditions to ensure the long-term preservation of specimens. Temperature, relative humidity, and light should be well controlled so that specimens are not damaged. It is most important that these environmental conditions remain stable and not fluctuate. Large fluctuations in temperature, humidity, and exposure to light cause damage to museum specimens such as mold growth, fading, cracking, and pest infestations.

Pests were historically managed using poisons that protected the specimens but posed potential hazards to humans. Modern museum pest management includes a comprehensive protocol called **integrated pest management** or IPM. Most IPM plans restrict the consumption and storage of food near museum collections. These plans also include a large component of pest monitoring to quickly identify pests before damage can be done. Most pests are controlled through regular monitoring and freezing specimens if pests are found.

Since there is a long history of chemical use in museum collections, it is important for researchers to take protective measures when working with historical museum collections. The most common chemicals found in ornithology collections are inorganic arsenic and formalin. These chemicals are known carcinogens and must be handled with care and proper personal protective equipment (PPE) including latex gloves. Food or drink should not be consumed where arsenic or formalin may be present.

Data Management

Museum collection data and specimens are most useful if they are accessible to a large group of users. A collection has no use if no one knows that the specimens exist. Building and maintaining a database is one of the most important aspects of keeping a museum collection.

Several databases have been developed to help museum professionals organize and provide data to their users. Two

examples of databases used by professionals are Specify (www.specifysoftware.org) and Arctos (www.arctosdb.org). These platforms were developed to provide the most consistent data across taxa. These databases do require some knowledge of programming. While Specify and Arctos may be useful for large collections, smaller collections often do not need a complicated database. In these cases, simpler programs such as Microsoft Access or Filemaker (www.filemaker.com) may work well for the smaller museum collection.

No matter which database a museum uses to store their data, unless the data are available to the public and to the researchers that need them, that database is useless. There are several online portals that serve museum records so that they are accessible for all. In addition to VertNet and GBIF, another online portal that provides data to users is iDigBio (www.idigbio.org), which serves not only specimen data but also images and corresponding media.

PERMITTING

It is the responsibility of each researcher to possess the proper permits, whether they are working with live birds in the field or laboratory, museum specimens, or tissues in the laboratory. There are several levels of permitting, and all must be carefully complied with and followed, including the regular submission of annual reports.

IACUC

Most universities and museums require the approval of their Institutional Animal Care and Use Committee (IACUC) before researchers can conduct studies with any birds or bird parts. IACUC approval is necessary to ensure that animals are being treated ethically and that the researcher understands the ethical issues involved in working with live or dead birds in scientific research. Most institutions today require IACUC approval even for observational studies in which no birds are captured or handled, to ensure that research activities minimally impact wild birds. IACUC approval is generally required when applying for additional federal and local permits to work with birds.

Federal Permits

The most important federal permit for working with birds in the United States is a United States Fish and Wildlife Service (USFWS) Migratory Bird Treaty Act (MBTA) permit. It is illegal for any person to possess any bird parts, including loose feathers, bones, and eggs, without an MBTA permit. Even if the birds are found dead and parts are salvaged, the person must have an MBTA permit, given that one cannot prove that they did not kill the bird to collect those parts. USFWS MBTA permits generally renew every few years and cover activities such as scientific collecting, salvage, import, and export. Other federal permits are required to work on federal lands. If research is to be conducted in a designated national park, national seashore, national monument, national recreation area, national forest, or national wildlife refuge, researchers need to apply for specific permits from each federal entity that protects land.

If a species is regulated by the United States Department of Agriculture (USDA) a researcher does need to possess specific USDA permits to import, export, or transport specimens within the United States. These permits are particularly important if specimens originated in an area with exotic Newcastle disease or avian influenza.

State Permits

In addition to any federal permits, researchers must also possess permits specific to each state within which they plan to work. Most state fish and wildlife agencies require scientific collecting permits for work within their state. These permits can be very specific but generally allow collection and salvage of specimens within the state.

Within a state there are also government agencies that may or may not have strict restrictions on permits within their managed property. In most states, the state park system requires a specific application to conduct scientific research within a park. These permits are often the only means that smaller land managers, such as state parks, use to keep track of the research and the results of work conducted in their area.

Local Permits

A few local landowners or land managers require permits to work on their land. These can include county or regional parks as well as private companies serving as land managers. Ethical best practices require that researchers contact and obtain permission from local landowners before working on their land, for both documentation and safety.

Collecting permits issued by the USFWS, states, and other federal land managers generally specify the species and the number of individuals allowed under the permit. If a species

THE VALUE OF LARGE SERIES IN COLLECTIONS: WESTERN BARRED OWLS

In 2004, ornithologists assembled to assess the status of the federally listed Northern Spotted Owl (*Strix occidentalis caurina*). Data suggested that the abundance of Barred Owls (*Strix varia*), which invaded the Spotted Owl's range from the eastern United States, was negatively correlated with Northern Spotted Owls and that they might be negatively impacting the Northern Spotted Owl populations. Museum researchers were given permission to collect a small number of Barred Owl specimens in Northern California to document the invasion, and they collaborated with wildlife managers who could study the response of Northern Spotted Owls to Barred Owl removals. Initial results suggested that Spotted Owls quickly reoccupied territories from which Barred Owls were removed (Diller et al. 2016), so additional removals were permitted for experimental purposes. With the larger series of Barred Owl specimens available to museums (see figure) additional studies were made possible—including studies of molt, diet (stomach contents), toxicology and exposure to rodenticides, genetics of their invasion, and studies of gene flow and interbreeding between Barred Owls and Spotted Owls. The data generated from these removals have been critical in guiding US Fish and Wildlife Service plans for Northern Spotted Owl recovery and has led to a much larger

A series of Barred Owl specimens collected and prepared from field studies of Barred Owls in northern California. *Photo © Jack Dumbacher, California Academy of Sciences.*

strategic removal study that involves three states, multiple museums, and various stakeholders. Many of these studies are ongoing, and as more specimens are stored in museums, researchers continue to find novel and fascinating uses for such large series of natural bird populations.

is specially protected it will be restricted within a general collecting permit. Most scientific collecting permits allow for unlimited salvage of any species found dead along beaches, roads, or buildings.

Permits for Special Uses

There are many special permits that are required for specific research activities. When conducting any research with birds, please be sure to acquire all the necessary permits, including specialized permits, prior to beginning the work.

Banding Permits

The capture, handling, and banding of wild birds requires a federal banding permit and, in most states, a collecting or similar permit. Federal permits and bands are issued by USGS Bird Banding Laboratory at the Patuxent Wildlife Research Center. Master and subpermits are offered only to well-qualified individuals with specific research or monitoring objectives. References from permitted banders are also required (for more information see https://www.pwrc.usgs.gov/bbl/). Handling, banding, or using lures or song playbacks for federally or state listed species normally require additional permits, training, and experience with that species. Federal banding permits also allow for the temporary possession and transportation of bird carcasses until they can be deposited with an authorized museum or institution of higher learning. The North American Banding Council (http://www.nabanding.net/) was formed in 1998 "to promote sound and ethical bird-banding practices and techniques." They offer training materials, including an exceptionally thorough study guide on all aspects of banding (NABCI 2001) as well as methods for specific taxonomic groups. A three-level certification process (assistant, bander, and trainer) is also offered and may help in obtaining a banding permit.

Eagle Permits

It is illegal for anyone to possess parts from eagles without a specific USFWS eagle permit. Any eagle specimen found dead cannot be salvaged but must be sent to the National Eagle Repository, where it will be distributed to Native Americans for use in their traditional customs. When working with other endangered or protected species, there are often specific permits required by the USFWS and the individual states (see the box on page 955).

Importing and Exporting

A specific endangered species permit is required to import and export any specimens of endangered species. For species listed under the Convention on International Trade in Endangered Species (CITES), special permits are required to move specimens across borders.

KEY POINTS

- There are many diverse methods and approaches for studying birds in the field and lab, depending on the scientific questions being asked and the avian groups being studied. Investigators are encouraged to explicitly state their objectives and have them reviewed, and to consult with biostatisticians on such topics as sample size, observer bias, and inference.
- Searchable online databases are incredibly important for sharing data. Thus, best data management practices need to be followed.
- Technology is growing rapidly and has had huge effects on the fields of avian genetics, stable isotopes, and dealing with large data sets such as those from museum collections, bird monitoring stations, and citizen science repositories.
- Museum specimens are a permanent record of bird life. It is important to maintain data-rich voucher specimens, store them in public museums, and make them available for scientific research. It is impossible to predict all of the studies that they can aid in the future.

KEY MANAGEMENT AND CONSERVATION IMPLICATIONS

The complexity of management and conservation issues that we face today requires each of us to use multiple approaches, tools, and techniques in ornithology. Throughout this chapter we have introduced many different field, lab, and museum techniques, each of which has its own unique applications to conservation and management. In summary, here are a few examples:

- Many of the techniques described in this chapter are used for surveying and monitoring bird population size and trends, which is critical for detecting avian declines or range shifts. Relating changes in population to management and normal fluctuations over time will be critical for providing insight into causes and our ability to reverse declines. Relating population change to climatic variables and the mechanisms behind them (e.g., changes in phenology) will help considerably in our attempt to conserve bird populations in the future.
- In a time with increasingly limited funding and more amateur birders than ever, many citizen science data-

bases can be mined for data on bird population health and current trends.

- Genetics are often used for identifying legally protectable units, such as distinct species, subspecies, or discrete population segments.
- Researchers can identify critical habitats by studying bird movements through direct observation and stable isotope studies, as well as other approaches.
- Toxin and disease studies can help identify particular threats to avian health.
- The museum collection provides valuable historic data on birds and bird populations, which are often consulted for conservation purposes to identify historical genetic losses and bottlenecks, document populations that might no longer exist, and document some of the factors that affected birds at the place and time that the specimens were collected.
- Museum collections also house specimens of extinct species. These are currently the only source of genetic information for programs under way to "de-extinct" avian genes, bird populations or even species.

DISCUSSION QUESTIONS

1. What are some of the most common field methods used for studying birds, and what parameters can they measure? Which parameters do they fail to measure? How should field methods be adapted to season, size of bird, and hypothesis being studied?

2. In designing a monitoring program, what are some of the most important factors to consider? What could be done to best ensure its ability to inform management decisions?

3. Your favorite local bird species may be in trouble. Two populations from nearby parks may have disappeared, and a third population appears to be in decline. Officials are considering listing it as endangered. Describe how you might design studies that address the following questions: What are local population sizes, and are they really in decline? Are you losing genetic variation if only one population persists? Did they migrate to the same regions, or have similar diet? What might be affecting their viability? Consider all of the potential resources available to help you answer your questions (museum specimens, banding records, citizen science data sets) and propose field or laboratory studies that you could spearhead.

4. A bird struck your glass window and lays dead on the ground. Its plumage appears different than anything you see in your field guide. What resources could you use to identify it? How could you determine its age and sex?

Describe the permits that are required for you to keep it. If you give it to your local museum, what material and data should the museum preserve? If it still doesn't match any specimens in museums, what studies could you do to determine what it is, or to describe a new species?

References

Able, K. P. (1999). How birds migrate: Flight behavior, energetics, and navigation. *In* Gatherings of angels: Migrating birds and their ecology, K. P. Able, Editor. Cornell University Press, New York.

Acevedo, M. A., and L. J. Villanueva-Rivera (2006). Using automated digital recording systems as effective tools for the monitoring of birds and amphibians. Wildlife Society Bulletin 34:211–214.

Ainley, D. G., and R. J. Boekelheide (1990). Seabirds of the Farallon Islands: Ecology, structure, and dynamics of an upwelling system community. Stanford University Press, Redwood City, CA.

Ainley, D. G., C. A. Ribic, G. Ballard, S. Heath, I. Gaffney, B. J. Karl, K. J. Barton, P. R. Wilson, and S. Webb (2004). Geographic structure of Adélie Penguin populations: Overlap in colony-specific foraging areas. Ecological Monographs 74:159–178.

Alexander, J. D., N. E. Seavy, and P. E. Hosten (2007). Using conservation plans and bird monitoring to evaluate ecological effects of management: An example with fuels reduction activities in southwest Oregon. Forest Ecology and Management 238:375–383.

Alexander, J. D., J. L. Stevens, G. R. Geupel, and T. C. Will (2009). Decision support tools: Bridging the gap between science and management. Proceedings of the Fourth International Partners in Flight Conference: Tundra to Tropics 13:283–291.

Allen, P. E., L. J. Goodrich, and K. L. Bildstein (1996). Within- and among-year effects of cold fronts on migrating raptors at Hawk Mountain, Pennsylvania, 1934–1991. The Auk 113:329–338.

Alquezar, R. D., and R. B. Machado (2015). Comparisons between autonomous acoustic recordings and avian point counts in open woodland savanna. Wilson Journal of Ornithology 127:712–772.

Baillie, S. R. (1990). Integrated populations monitoring of breeding birds in Britain and Ireland. Ibis 132:151–166.

Baker, M., N. Nur, and G. R. Geupel (1995). Correcting biased estimates of dispersal and survival due to limited study area: Theory and an application using Wrentits. The Condor 97:663–674.

Ball, I. J., R. L. Eng, and S. K. T. Ball (1995). Population density and productivity of ducks on large grassland tracts in northcentral Montana. Wildlife Society Bulletin 23:767–773.

Ballard, G., D. G. Ainley, C. A. Ribic, and K. R. Barton (2001). Effect of instrument attachment on foraging trip duration and nesting success of Adélie Penguins. The Condor 103:481–490.

Ballard, G., K. M. Dugger, N. Nur, and D. G. Ainley (2010). Foraging strategies of Adélie Penguins: Adjusting body condition to cope with environmental variability. Marine Ecology Progress Series 405:287–302.

Bart, J., S. Droege, P. Geissler, B. Peterjohn, and C. J. Ralph (2004). Density estimation in wildlife surveys. Wildlife Society Bulletin 32:1242–1247.

Bart, J., B. Andres, S. Brown, G. Donaldson, B. Harrington, V. Johnston, S. Jones, R. I. G. Morrison, and S. Skagen (2005). The Program for regional and international shorebird monitoring (PRISM). *In* Bird conservation implementation and integration in the Americas: Proceedings of the Third International Partners in Flight conference, C. J. Ralph and T. D. Rich, Editors. USDA Forest Service General Technical Report PSW GTR 191. https://www.treesearch.fs.fed.us/pubs/31463.

Bart, J. R., and V. H. Johnston, Editors (2012). Arctic shorebirds in North America: A decade of monitoring. Studies in Avian Biology 44. University of California Press, Oakland.

Battley, P. F., N. Warnock, T. L. Tibbitts, R. E. Gill, T. Piersma, C. J. Hassell, and A. C. Riegen (2012). Contrasting extreme long-distance migration patterns in Bar-tailed Godwits (*Limosa lapponica*). Journal of Avian Biology 43:21–32.

Bednarz, J. C., D. Klem Jr., L. J. Goodrich, and S. E. Senner (1990). Migration counts of raptors at Hawk Mountain, Pennsylvania, as indicators of population trends, 1934–1986. The Auk 107:96–109.

Behmke, S., J. Fallon, A. E. Duerr, A. Lehner, J. Buchweitz, and T. Katzner (2015). Chronic lead exposure is epidemic in obligate scavenger populations in eastern North America. Environment International 79:51–55.

Bibby, C. J., N. D. Burgess, D. H. Hill, and S. H. Mustoe (2000). Bird census techniques. 2nd ed. Academic Press, London.

Blakesley, J. A., D. R. Anderson, and B. R. Noon (2006). Breeding dispersal in the California Spotted Owl. The Condor 10:71–81.

Bock, C. E. (1997). The role of ornithology in conservation of the American West. The Condor 99:1–6.

Boecklen, W. J., C. T. Yarnes, B. A. Cook, and A. C. James (2011). On the use of stable isotopes in trophic ecology. Annual Review Ecology and Systematics 42:411–440.

Boland, C. R. J., and R. M. Phillips (2005). A small, lightweight, and inexpensive "burrowscope" for viewing nest contents of tunnel nesting birds. Journal of Field Ornithology 76:21–26.

Bond, A. L., and I. L. Jones (2009). A practical introduction to stable-isotope analysis for seabird biologists: Approaches, cautions and caveats. Marine Ornithology 37:183–188.

Bonter, D. N., T. M. Donovan, and E. W. Brooks (2009). Daily mass changes in landbirds during migration stopover on the south shore of Lake Ontario. The Auk 124:122–133.

Bridge, E. S., and D. N. Bonter (2011). A low-cost radio frequency identification device for ornithological research. Journal of Field Ornithology 82:52–59.

Bub, H. (1991). Bird trapping and bird banding: A handbook for trapping methods all over the world. Cornell University Press, New York.

Buckland, S. T., D. R. Anderson, K. P. Burnham, J. L. Laake, D. L. Borchers, and L. Thomas (2001). Introduction to distance sampling: Estimating abundance of biological populations. Oxford University Press, Oxford, UK.

Buxton, R. T., and I. L. Jones (2012). Measuring nocturnal seabird activity and status using acoustic recording devices: Applications for island restoration. Journal of Field Ornithology 83:47–60.

Carlisle, J. D., G. S. Kaltenecker, and D. L. Swanson (2005). Stopover ecology of autumn landbird migrants in the Boise foothills of southwestern Idaho. The Condor 107:244–258.

Carlisle, J. D., S. K. Skagen, B. E. Kus, C. van Riper III, K. L. Paxtons, and J. F. Kelly (2009). Landbird migration in the American West: Recent progress and future research directions. The Condor 111:211–225.

Carmi, O., C. C. Witt, A. Jaramillo, and J. P. Dumbacher (2016). Phylogeography of the Vermilion Flycatcher species complex: Multiple speciation events, shifts in migratory behavior, and an apparent extinction of a Galápagos-endemic bird species. Molecular Phylogenetics and Evolution 102:150–173.

CCAMLR (2014). Commission for the Conservation of Antarctic Marine Living Resources ecosystem monitoring program—standard methods. Hobart, Australia. https://www.ccamlr.org/en/system/files/CEMP%20Standard%20Methods%20Jun%202014.pdf.

Celis-Murillo, A., J. L. Deppe, and M. F. Allen (2009). Using soundscape recordings to estimate bird species abundance, richness, and composition. Journal of Field Ornithology 80:64–78.

Chase, M. K., N. Nur, and G. R. Geupel (1997). Survival, productivity, and abundance in a Wilson's Warbler population. The Auk 114:354–366.

Chase, M. K., and G. R. Geupel (2005). The use of avian focal species for conservation planning in California. *In* Proceedings of the Third International Partners in Flight conference, C. J. Ralph and T. D. Rich, Editors. USDA Forest Service General Technical Report PSW-GTR 191. https://www.treesearch.fs.fed.us/pubs/31668.

Clobert, J., J. D. Lebertron, and D. Alliane (1987). A general approach to survival estimates by recaptures or resightings of marked birds. Ardea 75:133–142.

Conway, C. J. (2011). Standardized North American marsh bird monitoring protocol. Waterbirds 34:319–346.

Cooper, D. (2016). The best binoculars for birds, nature, and the outdoors. The Wirecutter. http://thewirecutter.com/reviews/the-best-binoculars/.

Cooper, C. B., S. J. Daniels, and J. R. Walters (2008). Can we improve estimates of juvenile distance and survival? Ecology 89:3349–3361.

Dalloul, R. A., J. A. Long, A. V. Zimin, L. Aslam, K. Beal, L. Bloomberg, P. Bouffard, D. W. Burt, O. Crasta, and R. Crooijmans (2010). Multi-platform next-generation sequencing of the Domestic Turkey (*Meleagris gallopavo*): Genome assembly and analysis. PLoS Biology 8:e1000475.

Daniels, S. J., and J. R. Walters (2000). Inbreeding depression and its effects on natal dispersal in Red-cockaded Woodpeckers. The Condor 102:482–491.

Danner, R. M., R. Greenberg, and T. S. Sillett (2014). The implications of increased body size in the Song Sparrows of the California Islands. Monographs of the Western North American Naturalist 7:348–356.

Davis, S. K. (2003). Nesting ecology of mixed-grass prairie songbirds in southern Saskatchewan. Wilson Bulletin 115:119–130.

Deluca, W., B. K. Woodworth, C. C. Rimmer, P. P. Marra, P. D. Taylor, K. P. McFarland, S. A. Mackenzie, and D. R. Norris (2015). Transoceanic migration by a 12 g songbird. Biology Letters 11:20141045. http://dx.doi.org/10.1098/rsbl.2014.1045.

DeSante, David F. (1992). Monitoring avian productivity and survivorship (MAPS): A sharp, rather than blunt, tool for monitoring and assessing landbird populations. *In* Wildlife 2001: Populations, D. R. McCullough and R. H. Barrett, Editors. Elsevier Applied Science, London.

DeSante, D. F., and G. R. Geupel (1987). Landbird productivity in central coastal California: The relationship to annual rainfall and a reproductive failure in 1986. The Condor 89:636–653.

DeSante, D. F., D. R. Kaschube, and J. F. Saracco (2015). Vital rates of North American landbirds. The Institute for Bird Populations. www.VitalRatesOfNorthAmericanLandbirds.org.

Diller, V. L., K. A. Hamm, D. A. Early, D. W. Lamphear, D. Katie, C. B. Yackulic, C. J. Schwarz, P. C. Carlson, and T. L. McDonald (2016). Demographic response of Northern Spotted Owls to Barred Owl removal. Journal of Wildlife Management 80:691–707. doi:10.1002/jwmg.1046.

Dove, C. J., and C. P. J. Coddington (2015). Forensic techniques identify the first record of Snowy Owl (*Bubo scandiacus*) feeding on a Razorbill (*Alca torda*). Wilson Journal of Ornithology 127:503–506.

Dumbacher, J. P., and B. West (2010). Collecting Galápagos and the Pacific: How Rollo Howard Beck shaped our understanding of evolution. Proceedings of the California Academy of Sciences 61:211–243.

Dunn, E. H., and D. J. Hussell (1995). Using migration counts to monitor landbird populations: Review and evaluation of current status. *In* Current ornithology, D. M. Power, Editor. Plenum Press, New York.

Dunn, E. H., J. D. Hussell, C. M. Francis, and J. D. McCracken (2004). A comparison of three count methods for monitoring songbird abundance during spring migration: Capture, census, and estimated totals. *In* Monitoring bird populations using mist nets, C. J. Ralph and E. H. Dunn, Editors. Studies in Avian Biology 29. Cooper Ornithological Society, Lawrence, KS.

Emlen, J. T. (1971). Population densities of birds derived from transect counts. The Auk 88:323–342.

Faaborg, J., R. T. Holmes, A. D. Anders, K. L. Bildstein, K. M. Dugger, S. A. Gauthreaux Jr., P. Heglund, et al. (2010). Conserving migratory land birds in the New World: Do we know enough? Ecological Applications 20:398–418.

Fair, J. M., E. Paul, J. Jones, A. B. Clark, C. Davie, and G. Kaiser (2010). Guidelines to the use of wild birds in research. Ornithological Council, Washington, DC.

Finkelstein, M. E., D. F. Doak, D. George, J. Burnett, J. Brandt, M. Church, J. Grantham, and D. R. Smith (2012). Lead poisoning and the deceptive recovery of the critically endangered California Condor. Proceedings of the National Academy of Sciences 109:11449–11454.

Fleischer, R. C., J. J. Kirchman, J. P. Dumbacher, L. Bevier, C. Dove, N. C. Rotzel, S. V. Edwards, M. Lammertink, K. J. Miglia, and W. S. Moore (2006). Mid-Pleistocene divergence of Cuban and North American Ivory-billed Woodpeckers. Biology Letters 2:466–469.

Ganapathy, G., J. T. Howard, J. M. Ward, J. Li, B. Li, Y. Li, Y. Xiong, et al. (2014). High-coverage sequencing and annotated assemblies of the Budgerigar genome. GigaScience 3:11.

Gauthreaux Jr., S. A., and C. G. Belser (2003). Radar ornithology and biological conservation. The Auk 120:266–277.

GBIF (2015). Secretariat: GBIF Backbone Taxonomy. http://www.gbif.org/species/1.

Geleynse, D. M., E. Nol, D. M. Burke, and K. A. Elliott (2016). Brown Creeper (*Certhia americana*) demographic response to hardwood forests managed under the selection system. Canadian Journal of Forest Research 46:499–507.

Geupel, G. R., and D. F. DeSante (1990). Incidence and determinants of double brooding in Wrentits. The Condor 92:67–75.

Geupel, G. R., and I. Warkentin (1995). Field methods for monitoring population parameters of landbirds in Mexico. *In* Conservation of Neotropical migratory birds in Mexico. Proceedings of a Symposium in Veracruz, Mexico, November 1993, M. Wilson and S. Sader, Editors. Maine Agricultural and Forest Experiment Station Miscellaneous Publication 727. Orono, Maine.

Glenn, T. C. (2011). Field guide to next-generation DNA sequencers. Molecular Ecology Resources 11:759–769.

Gorzo, J. M., A. M. Pidgeon, W. E. Thogmartin, A. J. Allstadt, V. C. Radeloff, P. J. Heglund, and S. J. Vavrus (2016). Using the North American Breeding Bird Survey to assess broad-scale response of the continent's most imperiled avian community, grassland birds, to weather variability. The Condor 118:502–512.

Gotmark, F. (1992). The effect of investigator disturbance on nesting birds. Current Ornithology 9:63–104.

Gregory, T. R. (2015). Animal genome size database. http://www.genomesize.com.

Haig, S. M., W. M. Bronaugh, R. S. Crowhurst, J. D'Elia, C. A. Eagles-Smith, C. W. Epps, B. Knaus, et al. (2011). Genetic applications in avian conservation. The Auk 128:205–229.

Hall, L. S., A. M. Fish, and M. L. Morrison (1992). The influence of weather on hawk movements in coastal Northern California. Wilson Bulletin 104:447–461.

Hallworth, M. T., and P. P. Marra (2015). Miniaturized GPS tags identify non-breeding territories of a small breeding migratory songbird. Scientific Reports 5:11069. doi:10.1038/srep11069.

Halterman, M., M. J. Johnson, J. A. Holmes, and S. A. Laymon (2015). A natural history summary and survey protocol for the western distinct population segment of the Yellow-billed Cuckoo: US Fish and Wildlife techniques and methods. https://www.fws.gov/southwest/es/Documents/R2ES/YBCU_SurveyProtocol_FINAL_DRAFT_22Apr2015.pdf.

Hayes, D. B., and M. J. Monfils (2015). Occupancy modeling of bird point counts: Implications of mobile animals. Journal of Wildlife Management 79:1361–1368.

Hellgren, O., J. Waldenström, and S. Bensch (2004). A new PCR assay for simultaneous studies of *Leucocytozoon*, *Plasmodium*, and *Haemoproteus* from avian blood. Journal of Parasitology 90:797–802.

Herman, S. G. (1986). The naturalist field journal, based on the method by J. Grinnell. Buteo Books, Vermillion, SD.

Herman, S. G. (2002). Wildlife biology and natural history: Time for a reunion. Journal of Wildlife Management 66:933–946.

Hewish, M. J., and R. H. Loyn (1989). Popularity and effectiveness of four survey methods for monitoring populations of Australian land birds. R.A.O.U. Report no. 55. Royal Australasian Ornithologists Union, Moonee Ponds, Australia.

Hickey, J. J., and D. W. Anderson (1968). Chlorinated hydrocarbons and eggshell changes in raptorial and fish-eating birds. Science 162:271–273.

Hillier, L. D. W., W. Miller, E. Birney, W. Warren, R. C. Hardison, C. P. Ponting, P. Bork, D. W. Burt, M. A. M. Groenen, and M. E. Delany (2004). Sequence and comparative analysis of the chicken genome provide unique perspectives on vertebrate evolution. Nature 432:695–716.

Hobson, K. A. (2011). Isotopic ornithology: A perspective. Journal of Ornithology 152:49–66.

Hobson, K. A., K. D. Hughes, and P. J. Ewins (1997). Using stable-isotope analysis to identify endogenous and exogenous sources of nutrients in eggs of migratory birds: Applications to Great Lakes contaminants research contaminants research. The Auk 114:467–478.

Hobson, K. A., S. L. Van Wilgenburg, J. Faaborg, J. D. Toms, C. Rengifo, A. L. Sosa, Y. Aubry, and R. Brito Aguilar (2014a). Connecting breeding and wintering grounds of Neotropical migrant songbirds using stable hydrogen isotopes: A call for an isotopic atlas of migratory connectivity. Journal of Field Ornithology 85:237–257.

Hobson, K. A., S. L. V. Wilgenburg, T. Wesołowski, M. Maziarz, R. G. Bijlsma, A. Grendelmeier, and J. W. Mallord (2014b). A multi-isotope (δ^2 H, δ^{13} C, δ^{15} N) approach to establishing migratory connectivity in Palearctic-Afrotropical migrants: An example using Wood Warblers Phylloscopus sibilatrix. Acta Ornithologica 49:57–69.

Honkavuori, K. S., H. L. Shivaprasad, B. L. Williams, P.-L. Quan, M. Hornig, C. Street, G. Palacios, et al. (2008). Novel borna virus in psittacine birds with proventricular dilatation disease. Emerging Infectious Diseases 14:1883–1886.

Howell, S. N. G., and S. Webb (1995). A guide to the birds of Mexico and Northern Central America. Oxford University Press, Oxford, UK.

Hughes, C. (1998). Integrating molecular techniques with field methods in studies of social behavior: A revolution results. Ecology 79:383–399.

Hutto, R. L. (2016). Should scientists be required to use a model-based solution to adjust for possible distance-based detectability bias? Ecological Applications 26:1287–1294.

Hutto, R. L., and R. J. Stutzman (2009). Humans versus autonomous recording units: A comparison of point-count results. Journal of Field Ornithology 80:387–398.

Hutto, R. L., and R. T. Belote (2013). Distinguishing four types of monitoring based on the questions they address. Forest Ecology and Management 289:183–189.

Jarvis, E. D., S. Mirarab, A. J. Aberer, B. Li, P. Houde, C. Li, S. Y. W. Ho, et al. (2014). Whole-genome analyses resolve early branches in the tree of life of modern birds. Science 346:1320–1331.

Johnson, M., and G. R. Geupel (1996). The importance of productivity to a Swainson's Thrush population. The Condor 98:133–141.

Johnson, M. D., and T. W. Sherry (2001). Effects of food availability on the distribution of migratory warblers among habitats in Jamaica. Journal of Animal Ecology 70:546–560.

Kelling, S., W. M. Hochachka, D. Fink, M. Riedewald, R. Caruana, G. Ballard, and G. Hooker (2009). Data-intensive science: A new paradigm for biodiversity studies. BioScience 59:613–620.

Kistler, A. L., A. Gancz, S. Clubb, P. Skewes-Cox, K. Fischer, K. Sorber, C. Y. Chiu, et al. (2008). Recovery of divergent avian bornaviruses from cases of proventricular dilatation disease: Identification of a candidate etiologic agent. Virology Journal 5:88.

Kocher, T. D., W. K. Thomas, A. Meyer, S. V. Edwards, S. Pääbo, F. X. Villablanca, and A. C. Wilson (1989). Dynamics of mitochondrial DNA evolution in animals: Amplification and sequencing with conserved primers. Proceedings of the National Academy of Sciences 86:6196–6200.

Kozlov, A. M., A. J. Aberer, and A. Stamatakis (2015). ExaML version 3: A tool for phylogenomic analyses on supercomputers. BioInformatics 31:2577–2579.

Lambert, J. D., T. P. Hodgman, E. J. Laurent, G. L. Brewer, M. J. Iliff, and R. Dettmers (2009). The Northeast bird monitoring handbook. American Bird Conservancy, The Plains, VA. http://www.nebirdmonitor.org.

Lamichhaney, S., J. Berglund, M. S. Almén, K. Maqbool, M. Grabherr, A. Martinez-Barrio, M. Promerová, et al. (2015). Evolution of Darwin's finches and their beaks revealed by genome sequencing. Nature 518:371–375.

Langham, G. M., J. G. Schuetz, T. Distler, C. U. Soykan, and C. Wilsey (2015). Conservation status of North American birds in the face of future climate change. PLoS ONE 10:e0135350.

Latta, S., C. J. Ralph, and G. R. Geupel (2005). Strategies for the conservation monitoring of resident landbirds and wintering Neotropical migrants in the Americas. Ornitologia Neotropica 16:163–174.

Latta, S. C., and J. Faaborg (2009). Benefits of studies of overwintering birds for understanding resident bird ecology and promoting development of conservation capacity. Conservation Biology 23:286–293.

Laws, J. M. (2016). The Laws Guide to nature drawing and journaling. Heyday Books, CA.

Leighton, F. A. (1993). The toxicity of petroleum oils to birds. Environmental Reviews 1:92–103.

Lijtmaer, D. A., K. C. R. Kerr, A. S. Barreira, P. D. N. Hebert, and P. L. Tubaro (2011). DNA barcode libraries provide insight into continental patterns of avian diversification. PLoS ONE 6:e20744.

Lijtmaer, D. A., K. C. R. Kerr, M. Y. Stoeckle, and P. L. Tubaro (2012). DNA barcoding birds: From field collection to data analysis. Methods in Molecular Biology 858:127–152.

Lindenmayer, D. B., and G. E. Likens (2009). Adaptive monitoring: A new paradigm for long-term research and monitoring. Trends in Ecology and Evolution 24:482–486.

Lloyd, P., T. E. Martin, R. L. Redmond, U. Langner, and M. M. Hart (2005). Linking demographic effects of habitat fragmentation across landscapes to continental source-sink dynamics. Ecological Applications 15:1504–1514.

Loges, B. W., B. G. Tavernia, A. M. Wilson, J. D. Stanton, J. H. Herner-Thogmartin, J. Casey, J. M. Coluccy, et al. (2014). National protocol framework for the inventory and monitoring of nonbreeding waterbirds and their habitats, an Integrated Waterbird Management and Monitoring Initiative (IWMM) approach. Natural Resources Program Center, Fort Collins, CO.

Loss, S. R., T. Will, S. S. Loss, and P. P. Marra (2014). Bird-building collisions in the United States: Estimates of annual mortality and species vulnerability. The Condor 116:8–23.

MacKenzie, D. I., J. D. Nichols, J. A. Royle, K. H. Pollock, L. L. Bailey, and J. E. Hines (2006). Occupancy estimation and modeling: Inferring patterns and dynamics of species occurrence. Elsevier, Burlington, MA.

Manel, S., and R. Holderegger (2013). Ten years of landscape genetics. Trends in Ecology and Evolution 28:614–621.

Martin, T. E. (1992). Breeding productivity considerations: What are the appropriate features for management? In Ecology and conservation of Neotropical migrant landbirds, J. M. Haga and D. W. Johnston, Editors. Smithsonian Institution Press, Washington, DC.

Martin, T. E. (1995). Summary: Model organisms for advancing and understanding of ecology and land management. In Ecology and management of Neotropical migratory birds: A synthesis and review of critical issues, T. E. Martin and D. M. Finch, Editors. Oxford University Press, New York.

Martin, T. E., and G. R. Geupel (1993). Nest monitoring plots: Methods for locating nests and monitoring success. Journal of Field Ornithology 64:507–519.

Martin, E., and G. Ballard (2010). Data management best practices and standards for biodiversity data applicable to bird monitoring data. US North American Bird Conservation Initiative Monitoring Subcommittee. http://www.nabci-us.org/.

Mayfield, H. F. (1975). Suggestions for calculating nest success. Wilson Bulletin 87:456–466.

McCormack, J. E., W. L. E. Tsai, and B. C. Faircloth (2015). Sequence capture of ultraconserved elements from bird museum specimens. Molecular Ecology Resources 16. doi:10.1111/1755-0998.12466.

NABCI (2001). North American bander's study guide. North American Banding Council Publication. http://www.nabanding.net/pubs.html.

NABCI (US North American Bird Conservation Initiative Monitoring Subcommittee) (2007). Opportunities for Improving Avian Monitoring. US North American Bird Conservation Initiative Report. Division of Migratory Bird Management, US Fish and Wildlife Service, Arlington, VA. http://www.nabci-us.org/.

Nur, N., S. Jones, and G. R. Geupel (1999). A statistical guide to data analysis of avian monitoring programs. US Department of the Interior, Fish and Wildlife Service, Biological Technical Publication BTP-R6001 1999, Washington, DC.

Nur, N., G. Ballard, and G. R. Geupel (2008). Regional analysis of riparian bird species response to vegetation and local habitat features. Wilson Journal of Ornithology 120:840–855.

Oakley, K. L., L. P. Thomas, and S. G. Fancy (2003). Guidelines for long-term monitoring protocols. Wildlife Society Bulletin 31:1000–1003.

O'Conner, T. P., and T. O'Conner (2008). The archeology of animal bones, no. 4. Texas A&M University Press, College Station.

Ogada, D. L., F. Keesing, and M. Z. Virani (2012). Dropping dead: Causes and consequences of vulture population declines worldwide. Annals of the New York Academy of Sciences 1249:57–71.

Packett, D. L., and J. B. Dunning (2009). Stopover habitat selection by migrant landbirds in a fragmented forest-agricultural landscape. The Auk 126:579–589.

Page, G. W., N. Warnock, T. L. Tibbitts, D. Jorgensen, C. A. Hartman, and L. E. Stenzel (2014). Annual migratory patterns of Long-billed Curlews in the American West. The Condor 116:50–61.

Parker, P. G., E. L. Buckles, H. Farrington, K. Petren, N. K. Whiteman, R. E. Ricklefs, J. L. Bollmer, and G. Jiménez-Uzcátegui (2011). 110 years of Avipoxvirus in the Galápagos Islands. PLoS ONE 6:e15989.

Payne, S., L. Covaleda, G. Jianhua, S. Swafford, J. Baroch, P. J. Ferro, B. Lupiani, J. Heatley, and I. Tizard (2011). Detection and characterization of a distinct Bornavirus lineage from healthy Canada Geese (Branta canadensis). Journal of Virology 85:12053–12056.

Peach, W. J., and S. Baillie (1991). Population changes on constant effort sites 1989–1990. British Trust for Ornithology News 173:12–14.

Piatt, J. F., W. J. Sydeman, and F. Wiese (2007). Introduction: A modern role of seabirds as indicators. Marine Ecology Progress Series 352:199–204.

Pienkowski, M. W. (1991). Using long-term ornithological studies in setting targets for conservation in Britain. Ibis 133:62–75.

Porzig, E. L., K. E. Dybala, T. Gardali, G. Ballard, G. R. Geupel, and J. A. Wiens (2012). Forty-five years and counting: Reflections from the Palomarin Field Station on the contribution of long-term monitoring and recommendations for the future. The Condor 113:713–723.

Porzig, E. L., N. E. Seavy, T. Gardali, G. R. Geupel, M. Holyoak, and J. M. Eadie (2014). Habitat suitability through time: Using time series and habitat models to understand changes in bird density. Ecosphere 5:12. http://dx.doi.org/10.1890/ES13-00166.1.

Prum, R. O., J. S. Berv, A. Dornburg, D. J. Field, J. P. Townsend, E. M. Lemmon, and A. R. Lemmon (2015). A comprehensive phylogeny of birds (Aves) using targeted next-generation DNA sequencing. Nature 526:569–573.

Pyle, P. (1997). Identification guide to North American birds. Part 1: Columbidae to Ploceidae. Slate Creek Press, Bolinas, CA. www.slatecreekpress.com.

Pyle, P. (2008). Identification guide to North American birds. Part 2: Anatidae to Alcidae. Slate Creek Press, Point Reyes Station, CA. www.slatecreekpress.com.

Pyle, P., N. Nur, and D. F. DeSante (1994). Trends in nocturnal migrant landbird populations at Southeast Farallon Island, California, 1968–1992. In A century of avifaunal change in western North America, J. R. Jehl Jr. and N. K. Johnson, Editors. Studies in Avian Biology 15. Cooper Ornithological Society, Lawrence, KS.

Quail, M. A., M. Smith, P. Coupland, T. D. Otto, S. R. Harris, T. R. Connor, A. Bertoni, H. P. Swerdlow, and Y. Gu (2012). A tale of three next generation sequencing platforms: Comparison of

Ion Torrent, Pacific Biosciences and Illumina MiSeq sequencers. BioMed Central Genomics 13:341.

Quayle, M. R., D. G. Barnes, O. L. Kaluza, and C. R. McHenry (2014). An interactive three dimensional approach to anatomical description—the jaw musculature of the Australian Laughing Kookaburra (*Dacelo novaeguineae*). PeerJ 2:e355–e355. doi:10.7717/peerj.355.

Ralph, C. J., and M. J. Scott, Editors (1981). Estimating numbers of terrestrial birds. Studies in Avian Biology 6. Cooper Ornithological Society, Lawrence, KS.

Ralph, C. J., G. R. Geupel, P Pyle, T. E. Martin, and D. F. DeSante (1993). Handbook of field methods for monitoring landbirds. USDA Forest Service General Technical Report PSW-GTR 144. https://www.fs.fed.us/psw/publications/documents/psw_gtr144/psw_gtr144.pdf.

Ralph, C. J., S. Droege, and J. R. Sauer (1995). Managing and monitoring birds using point counts: Standards and applications. *In* Monitoring bird populations by point counts. USDA Forest Service General Technical Report PSW-GTR 149, Albany, CA.

Ralph, C. J., G. R. Geupel, P. Pyle, T. E. Martin, D. F. DeSante, and B. Mila (1996). Manual de metodos de campo para el monitoreo de aves terrestres. USDA Forest Service General Technical Report PSW-GTR 159, Albany, CA.

Ralph, C. J., and E. H. Dunn, Editors (2004). Monitoring bird populations using mist nets. Studies in Avian Biology 29. Cooper Ornithological Society, Lawrence, KS.

RHJV (2004). The riparian bird conservation plan: A strategy for reversing the decline of riparian associated birds in California, California Partners in Flight. http://www.prbo.org/calpif/pdfs/riparian_v-2.pdf.

Ribic, C., F. R. Thompson, F. Richard, and P. J. Pietz, Editors (2012). Video surveillance of nesting birds. Studies in Avian Biology 43. University of California Press, Oakland, CA.

Rich, T. D., C. J. Beardmore, H. Berlanga, P. J. Blancher, M. S. W. Bradstreet, G. S. Butcher, D. W. Demarest, et al. (2004). Partners in Flight North American landbird conservation plan. Cornell Lab of Ornithology, Ithaca, New York. Partners in Flight. http://www.partnersinflight.org/cont_plan/.

Richard, F. A., R. N. M. Sehgal, H. I. Jones, and T. B. Smith (2002). A comparative analysis of PCR-based detection methods for avian malaria. Journal of Parasitology 88:819–822.

Richardson, T. W., T. Gardali, and S. Jenkin (2009). Review and meta-analysis of camera effects on avian nest success. Journal of Wildlife Management 73:287–293.

Rocha, L. A., A. Aleixo, G. Allen, F. Almeda, C. C. Baldwin, M. V. L. Barclay, J. M. Bates, et al. (2014). Specimen collection: An essential tool. Science 344:814–815.

Rodríguez, A., J. J. Negro, M. Mulero, C. Rodríguez, J. Hernández-Pliego, and J. Bustamante (2012). The eye in the sky: Combined use of unmanned aerial systems and GPS data loggers for ecological research and conservation of small birds. PLoS ONE 7:e50336.

Rosenstock, S. S., D. R. Anderson, K. M. Giesen, T. Leukering, and M. F. Carter (2002). Landbird counting techniques: Current practices and an alternative. The Auk 119:46–53.

Royle, J. A., and J. D. Nichols (2003). Estimating abundance from repeated presence-absence data or point counts. Ecology 84:777–790.

Ruegg, K., E. Anderson, K. Paxton, V. Apkenas, S. Lao, R. B. Siegel, D. F. DeSante, F. Moore, and T. Smith (2014). Mapping migration in a songbird using high-resolution genetic markers. Molecular Ecology 23:5726–5739.

Ruelas Inzunza, E., S. W. Hoffman, L. J. Goodrich, and R. Tingay (2000). Conservation strategies for the world's largest known raptor migration flyway: Veracruz, the river of raptors. *In* Raptors at risk: Proceedings of the 5th World Conference on Birds of Prey and Owl, R. D. Chancellor and B. U. Meyburg, Editors. World Working Group on Birds of Prey and Owl/Hancock House, Towcester, UK.

Salafsky, N., R. Margoluis, and K. Redford (2001). Adaptive management: A tool for conservation practitioners. Biodiversity Support Program, a Consortium of World Wildlife Fund, The Nature Conservancy, and World Resources Institute, Washington, DC.

Sardà-Palomera, F., G. Bota, C. Viñolo, O. Pallarés, V. Sazatornil, L. Brotons, S. Gomáriz, and F. Sardà (2012). Fine-scale bird monitoring from light unmanned aircraft systems. Ibis 154:177–183.

Sauer, J. R., W. A. Link, J. E. Fallon, K. L. Pardieck, and D. J. Ziolkowski (2013). The North American breeding bird survey 1966–2011: Summary analysis and species accounts. North American Fauna 79:1–32.

Schecter, A., L. Birnbaum, J. J. Ryan, and J. D. Constable (2006). Dioxins: An overview. Environmental Research 101:419–428.

Scheuhammer, A. M., and D. M. Templeton (1998). Use of stable isotope ratios to distinguish sources of lead exposure in wild birds. Ecotoxicology 7:37–42.

Schulenberg, T. S., D. F. Stotz, D. F. Lane, J. P. O'Niell, and T. A. Parker III, Editors (2007). Birds of Peru. Princeton University Press, NJ.

Seavy, N. E., D. L. Humple, R. L. Cormier, and T. Gardali (2012). Establishing the breeding provenance of a temperate-wintering North American passerine, the Golden-crowned Sparrow, using light-level geolocation. PLoS ONE 7:e34886.

Shafer, A. B. A., J. B. W. Wolf, P. C. Alves, L. Bergström, M. W. Bruford, I. Brännström, G. Colling, et al. (2015). Genomics and the challenging translation into conservation practice. Trends in Ecology and Evolution 30:78–87.

Shaffer M. (1990). Population viability analysis. Conservation Biology 4:39–40.

Sheldon, B. C. (2010). Genetic perspectives on the evolutionary consequences of climate change in birds. *In* Effects of climate change on birds, A. P. Møller, W. Fiedler, and P. Berthold, Editors. Oxford University Press, Oxford, UK.

Slater, L. (1995). Monitoring populations and productivity of seabirds at colonies in lower Cook Inlet, Alaska, in 1993 and 1994. Unpublished Report. US Fish and Wildlife Service, Homer, AK.

Soons, J., A. Genbrugge, J. Podos, D. Adriaens, P. Aerts, J. Dirckx, and A. Herrel (2015). Is beak morphology in Darwin's finches tuned to loading demands? PLoS ONE 10:e0129479–e0129479.

Southwell, C., and L. Emmerson (2015). Remotely-operating camera network expands Antarctic seabird observations of key breeding parameters for ecosystem monitoring and management. Journal for Nature Conservation 23:1–8.

Spear, L., N. Nur, and D. G. Ainley (1992). Estimating absolute densities of flying seabirds using analyses of relative movement. The Auk 109:385–389.

Spotswood, E. N., K. R. Goodman, J. Carlisle, R. L. Cormier, D. L. Humple, J. Rousseau, S. L Guers, and G. G. Barton (2012). How safe is mist netting? Evaluating the risk of injury and mortality to birds. Methods in Ecology and Evolution 3:29–38.

Stamatakis, A. (2014). RAxML Version 8: A tool for phylogenetic analysis and post-analysis of large phylogenies. BioInformatics 30:1312–1313.

Stevens, D. L., and A. R. Olsen (2004). Spatially balanced sampling of natural resources. Journal of the American Statistical Association 99:262–278.

Stralberg, D., D. Jongsomjit, C. A. Howell, M. A. Snyder, J. D. Alexander, J. A. Wiens, and T. L. Root (2009). Re-shuffling of species with climate disruption: A no-analog future for California birds. PLoS ONE 4:e6825.

Studds, C. E., T. K. Kyser, and P. P. Marra (2008). Natal dispersal driven by environmental conditions interacting across the annual cycle of a migratory songbird. Proceedings of the National Academy of Sciences 105:2929–2933.

Stutchbury, B. J. M., S. A. Tarof, T. Done, E. Gow, P. M. Kramer, J. Tautin, J. W. Fox, and V. Afanasyev (2009). Tracking long-distance songbird migration by using geolocators. Science 323:896–896.

Suarez, A. V., and N. D. Tsutsui (2004). The value of museum collections for research and society. BioScience 54:66–74.

Sullivan, B. L., C. L. Wood, J. I. Marshall, R. E. Bonney, D. Fink, and S. Kelling (2009). eBird: A citizen-based bird observation network in the biological sciences. Biological Conservation 142:2282–2292.

Sullivan, B. L., J. L. Aycrigg, J. H. Barry, R. E. Bonney, N. Bruns, C. B. Cooper, T. Damoulas, et al. (2014). The eBird enterprise: An integrated approach to development and application of citizen science. Biological Conservation 169:31–40.

Temple, S. A., and J. A. Wiens (1989). Bird populations and environmental changes: Can birds be bio-indicators? American Birds 43:260–270.

Tewksbury, J. J., J. G. T. Anderson, J. D. Bakker, T. J. Billo, P. W. Dunwiddie, M. J. Groom, S. E. Hampton, et al. (2014). Natural history's place in science and society. BioScience 64:300–310.

Thomas, L., S. T. Buckland, E. A. Rexstad, J. L. Laake, S. Strindberg, S. L. Hedley, J. R. B. Bishop, T. A. Marques, and K. P. Burnham (2010). Distance software: Design and analysis of distance sampling surveys for estimating population size. Journal of Applied Ecology 47:5–14.

Thomas, P. J., P. Mineau, R. F. Shore, L. Champoux, P. A. Martin, L. K. Wilson, G. Fitzgerald, and J. E. Elliott (2011). Second generation anticoagulant rodenticides in predatory birds: Probabilistic characterisation of toxic liver concentrations and implications for predatory bird populations in Canada. Environment International 37:914–920.

Toews, D. P. L., L. Campagna, S. A. Taylor, C. N. Balakrishnan, D. T. Baldassarre, P. E. Deane-Coe, M. G. Harvey, et al. (2016). Genomic approaches to understanding population divergence and speciation in birds. The Auk: Ornithological Advances 133:13–30.

Tulloch, A. I. T., H. P. Possingham, L. N. Joseph, J. Szabo, and T. G. Martin (2013). Realising the full potential of citizen science monitoring programs. Biological Conservation 165:128–138.

Unitt, P. (2004). San Diego County bird atlas. Proceedings of the San Diego Society of Natural History 39, Ibis Publishing, Temecula, CA.

Van Doren, B. M., D. Sheldon, J. Geevarghese, W. M. Hochachka, and A. Farnsworth (2015). Autumn morning flights of migrant songbirds in the northeastern United States are linked to nocturnal migration and winds aloft. The Auk 132:105–118.

Vas, E., A. Lescroël, O. Duriez, G. Boguszewski, and D. Grémillet (2015). Approaching birds with drones: First experiments and ethical guidelines. Biology Letters 11:20140754. http://dx.doi.org/10.1098/rsbl.2014.0754.

Veloz, S., L. Salas, B. Altman, J. Alexander, D. Jongsomjit, N. Elliott, and G. Ballard (2015). Improving effectiveness of systematic conservation planning with density data. Conservation Biology 29:1217–1227.

Verner, J. (1995). Assessment of counting techniques. Current Ornithology 2:247–301.

Verner, J., M. L. Morrison, and C. J. Ralph, Editors (1986). Wildlife 2000: Modeling habitat relationships of terrestrial vertebrates. University of Wisconsin Press, Madison, WI.

Verner, J., and K. A. Milne (1990). Analyst and observer variability in density estimates from spot mapping. The Condor 92:313–325.

VertNet (2015). Aves specimens from all data providers. http://portal.vertnet.org/search?q=Aves.

Walters, J. R., S. R. Derrickson, D. M. Fry, S. M. Haig, J. M. Marzluff, and J. M. Wunderle (2010). Status of the California Condor (Gymnogyps californianus) and efforts to achieve its recovery. The Auk 127:969–1001.

Wandeler, P., P. E. A. Hoeck, and L. F. Keller (2007). Back to the future: Museum specimens in population genetics. Trends in Ecology and Evolution 22:634–642.

Warnock, N., and J. Y. Takekawa (2003). Use of radio telemetry in studies of shorebirds: Past contributions and future directions. Wader Study Group Bulletin 100:138–150.

Warren, W. C., D. F. Clayton, H. Ellegren, A. P. Arnold, L. W. Hillier, A. Künstner, S. Searle, et al. (2010). The genome of a songbird. Nature 464:757–762.

Welsh, A. H., D. B. Lindenmayer, and C. F. Donnelly (2013). Fitting and interpreting occupancy models. PLoS ONE 8:e5201.

Whitacre, D. F., Editor (2010). Neotropical birds of prey biology and ecology of a forest raptor community. Cornell University Press, New York.

White, J. D., T. Gardali, F. R. Thompson, and J. Faaborg (2005). Resource selection by juvenile Swainson's Thrushes during the postfledging period. The Condor 107:388–401.

Wiens, J. A., D. Stralberg, D. Jongsomjit, C. A. Howell, and M. A. Snyder (2009). Niches, models, and climate change: Assessing the assumptions and uncertainties. Proceedings of the National Academy of Sciences 106:19729–19736.

Winker, K. (2000). Obtaining, preserving, and preparing bird specimens. Journal of Field Ornithology 71:250–297.

Winker, K., J. Michael Reed, P. Escalante, R. A. Askins, C. Cicero, G. E. Hough, and J. Bates (2010). The importance, effects, and ethics of bird collecting. The Auk 127:690–695.

Witham, J. W., and A. J. Kimball (1996). Use of a geographic information system to facilitate analysis of spot-mapping data (La utilización de sistemas de información geográfica para facilitar el análisis de datos en mapas de puntos). Journal of Field Ornithology 67:367–375.

Woltmann, S., T. W. Sherry, and B. R. Kreiser (2012). A genetic approach to estimating natal dispersal distances and self-recruitment in resident rainforest birds. Journal of Avian Biology 43:33–42.

Wood, C., B. Sullivan, M. Iliff, D. Fink, and S. Kelling (2011). eBird: Engaging birders in science and conservation. PLoS Biology 9:e1001220.

Woodworth, B. K., G. W. Mitchell, R. Norris, C. M. Francis, and P. D. Taylor (2015). Patterns and correlates of songbird movements at an ecological barrier during autumn migration assessed using landscape—and regional-scale automated radio-telemetry. Ibis 156:326–339.

Zhang, G., E. D. Jarvis, and M. T. P. Gilbert (2014a). A flock of genomes. Science 346:1308–1309.

Zhang, G., C. Li, Q. Li, B. Li, D. M. Larkin, C. Lee, J. F. Storz, et al. (2014b). Comparative genomics reveals insights into avian genome evolution and adaptation. Science 346:1311–1321.

Index